MORE STUDENTS EARN A's AND B's WHEN THEY USE THE LEARNSMART ADVANTAGE TOOLS.

SMARTBOOK®

The first—and only—adaptive reading experience designed to transform the way students read.

> Engages students with a personalized reading experience

> Ensures students retain knowledge

LEARNSMART®

The market-leading **adaptive study tool** proven to strengthen memory recall, increase class retention, and boost grades.

> Moves students beyond memorizing

> Allows instructors to align content with their goals

> Allows instructors to spend more time teaching higher-level concepts

LEARNSMART PREP®

An adaptive course preparation tool that quickly and efficiently helps students prepare for college-level work.

> Levels out student knowledge

> Keeps students on track

LEARNSMART ACHIEVE®

A learning system that continually adapts and provides learning tools to teach students the concepts they don't know.

> Adaptively provides learning resources

> A time management feature ensures students master course material to complete their assignments by the due date

LEARNSMART LABS®

LearnSmart Labs® is a super-adaptive simulated lab experience that brings meaningful scientific exploration to students. Through a series of adaptive questions, LearnSmart Labs identifies a student's knowledge gaps and provides resources to quickly and efficiently close those gaps. Once the student has mastered the necessary basic skills and concepts, they engage in a highly realistic simulated lab experience that allows for mistakes and the execution of the scientific method.

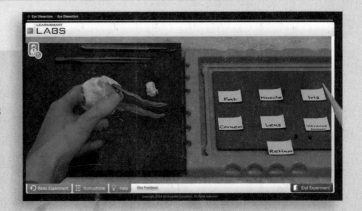

HOLE'S HUMAN ANATOMY& PHYSIOLOGY

FOURTEENTH EDITION

DAVID SHIER

WASHTENAW COMMUNITY COLLEGE

JACKIE BUTLER

GRAYSON COLLEGE

RICKI LEWIS

ALDEN MARCH BIOETHICS INSTITUTE

DIGITAL AUTHORS

LESLIE DAY

NORTHEASTERN UNIVERSITY

JULIE PILCHER

UNIVERSITY OF SOUTHERN INDIANA

HOLE'S HUMAN ANATOMY & PHYSIOLOGY, FOURTEENTH EDITION

Published by McGraw-Hill Education, 2 Penn Plaza, New York, NY 10121. Copyright © 2016 by McGraw-Hill Education. All rights reserved. Printed in the United States of America. Previous editions © 2013, 2010 and 2007. No part of this publication may be reproduced or distributed in any form or by any means, or stored in a database or retrieval system, without the prior written consent of McGraw-Hill Education, including, but not limited to, in any network or other electronic storage or transmission, or broadcast for distance learning.

Some ancillaries, including electronic and print components, may not be available to customers outside the United States.

This book is printed on acid-free paper.

2 3 4 5 6 7 8 9 0 DOW/DOW 1 0 9 8 7 6 5

ISBN: 978-0-07-802429-0
MHID: 0-07-802429-3

Senior Vice President, Products & Markets: *Kurt L. Strand*
Vice President, General Manager, Products & Markets: *Marty Lange*
Vice President, Content Production & Technology Services: *Kimberly Meriwether David*
Managing Director: *Michael S. Hackett*
Director of Marketing: *James F. Connely*
Brand Manager: *Amy L. Reed*
Senior Product Developer: *Fran Simon*
Director, Product Development: *Rose Koos*
Marketing Manager: *Jessica Cannavo*
Program Manager: *Angela R. FitzPatrick*
Content Project Managers: *Jayne Klein/Sherry Kane*
Buyer: *Sandy Ludovissy*
Design: *Tara McDermott*
Content Licensing Specialists: *John Leland/Lenny Behnke*
Cover Image: *©Hero Images/Getty Images (student and computer); ©micron/Getty Images (paramedic); ©Comstock Images/Jupiter Images (ER team); ©The McGraw-Hill Companies, Inc./Dennis Strete, photographer (background)*
Compositor: *ArtPlus Ltd.*
Typeface: *10/12 Times LT Std*
Printer: *R. R. Donnelley*

All credits appearing on page or at the end of the book are considered to be an extension of the copyright page.

Library of Congress Cataloging-in-Publication Data

Shier, David.
    Hole's human anatomy & physiology / David Shier, Washtenaw Community College, Jackie Butler, Grayson College, Ricki Lewis, Alden March Bioethics Institute.—Fourteenth edition.
        pages cm
    Includes index.
    ISBN 978-0-07-802429-0 (alk. paper)
    1. Human physiology. 2. Human anatomy. I. Butler, Jackie. II. Lewis, Ricki. III. Title.
IV. Title: Hole's human anatomy and physiology. V. Title: Human anatomy & physiology.
    QP34.5.S49 2016
    612–dc23   2014021876

The Internet addresses listed in the text were accurate at the time of publication. The inclusion of a website does not indicate an endorsement by the authors or McGraw-Hill Education, and McGraw-Hill Education does not guarantee the accuracy of the information presented at these sites.

www.mhhe.com

BRIEF CONTENTS

About the Authors iv | Acknowledgments vi | Updates and Additions vii | Dynamic New Art Program xii | Learn, Practice, Assess xiv | McGraw-Hill Connect Anatomy & Physiology xvi | Teaching and Learning xviii | Contents xix | Clinical Connections xxiv

PREVIEW
FOUNDATIONS FOR SUCCESS 1

P.1 Overview 2
P.2 Strategies for Success 2

UNIT 1
LEVELS OF ORGANIZATION 9

1 Introduction to Human Anatomy and Physiology 9
2 Chemical Basis of Life 57
3 Cells 82
4 Cellular Metabolism 120
5 Tissues 148

UNIT 2
SUPPORT AND MOVEMENT 177

6 Integumentary System 177
7 Skeletal System 199
8 Joints of the Skeletal System 267
9 Muscular System 291

UNIT 3
INTEGRATION AND COORDINATION 359

10 Nervous System I: Basic Structure and Function 359
11 Nervous System II: Divisions of the Nervous System 389
12 Nervous System III: Senses 443
13 Endocrine System 487

UNIT 4
TRANSPORT 527

14 Blood 527
15 Cardiovascular System 556
16 Lymphatic System and Immunity 616

UNIT 5
ABSORPTION AND EXCRETION 649

17 Digestive System 649
18 Nutrition and Metabolism 694
19 Respiratory System 730
20 Urinary System 768
21 Water, Electrolyte, and Acid-Base Balance 803

UNIT 6
THE HUMAN LIFE CYCLE 823

22 Reproductive Systems 823
23 Pregnancy, Growth, and Development 867
24 Genetics and Genomics 905

Appendices 925
Glossary 938
Credits 961
Index 965

DAVID SHIER
Washtenaw Community College

David Shier has more than thirty years of experience teaching anatomy and physiology, primarily to premedical, nursing, dental, and allied health students. He has effectively incorporated his extensive teaching experience into another student-friendly revision of *Hole's Essentials of Human Anatomy and Physiology* and *Hole's Human Anatomy and Physiology*. His interest in physiology and teaching began with a job as a research assistant at Harvard Medical School from 1976–1979. He completed his Ph.D. at the University of Michigan in 1984, and served on the faculty of the Medical College of Ohio from 1985–1989. He began teaching at Washtenaw Community College in 1990. David has recent experience in online course delivery, including recording lectures for so-called "flipped" classrooms. He has also been interested in the relationship between pedagogy and assessment, and the use of tools traditionally associated with assessment (e.g., lab quizzes) as pedagogical tools, often associated with group activities.

JACKIE BUTLER
Grayson College

Jackie Butler's professional background includes work at the University of Texas Health Science Center conducting research about the genetics of bilateral retinoblastoma. She later worked at Houston's M. D. Anderson Hospital investigating remission in leukemia patients. A popular educator for more than thirty years at Grayson College, Jackie has taught microbiology and human anatomy and physiology for health science majors. Her experience and work with students of various educational backgrounds have contributed significantly to another revision of *Hole's Essentials of Human Anatomy and Physiology* and *Hole's Human Anatomy and Physiology*. Jackie Butler received her B.S. and M.S. degrees from Texas A&M University, focusing on microbiology, including courses in immunology and epidemiology.

RICKI LEWIS
Alden March Bioethics Institute

Ricki Lewis's career communicating science began with earning a Ph.D. in Genetics from Indiana University in 1980. It quickly blossomed into writing for newspapers and magazines, and writing the introductory textbook *Life*. Since then she has taught a variety of life science courses and has authored the textbook *Human Genetics: Concepts and Applications* and books about gene therapy, stem cells, and scientific discovery. She is a genetic counselor for a large medical practice, teaches a graduate online course in "Genethics" at Albany Medical College, and writes for Medscape, the Multiple Sclerosis Discovery Forum, and Scientific American. Ricki writes the popular DNA Science blog at Public Library of Science and is a frequent public speaker.

Digital Authors

Leslie Day
Northeastern University

Leslie Day earned her B.S. in Exercise Physiology from UMass Lowell, an M.S. in Applied Anatomy & Physiology from Boston University, and a Ph.D. in Biology from Northeastern University with her research on the kinematics of locomotion. She currently works as an Assistant Clinical Professor in the Physical Therapy Department of Northeastern University with her main teaching role in Gross Anatomy and Neuro-anatomy courses. Students enjoy her clinical teaching style and use of technology. She has received the teaching with technology award three times and in 2009 was awarded the Excellence in Teaching Award. She has been asked to speak about teaching with technology at national conferences and to give workshops on gross anatomy to a variety of professionals. She has also worked as a personal trainer both in local fitness facilities and at clients' homes, a strength and conditioning coach for collegiate athletic teams, an Assistant Groups Exercise Director for Healthworks and Group Exercise, and Fitness Director of three sites for Gold's Gym.

Julie C. Pilcher
University of Southern Indiana

Julie Pilcher began teaching during her graduate training in Biomedical Sciences at Wright State University, Dayton, Ohio. She found, to her surprise, that working as a teaching assistant held her interest more than her research. Upon completion of her Ph.D. in 1986, she embarked on her teaching career, working for many years as an adjunct in a variety of schools as she raised her four children. In 1998, she began full-time at the University of Southern Indiana, Evansville. Her work with McGraw-Hill began several years ago, doing reviews of textbook chapters and lab manuals. More recently, she has been involved in content development for LearnSmart. In her A&P course at USI, she has also used Connect and has enjoyed the challenge of writing some of her own assignments. When the opportunity arose to become more involved in the authoring of digital content for McGraw-Hill, she could not pass it up. Based on her own experience, students are using more and more online resources, and she is pleased to be part of that aspect of A&P education.

ACKNOWLEDGMENTS

Any textbook is the result of hard work by a large team. Although we directed the revision, many "behind-the-scenes" people at McGraw-Hill were indispensable to the project. We would like to thank our editorial team of Marty Lange, Michael Hackett, Jim Connely, and Fran Simon; Jessica Cannavo, Marketing Manager, and our production team, which included Jayne Klein, Sandy Ludovissy, Tara McDermott, and John Leland; and most of all, John Hole, for giving us the opportunity and freedom to continue his classic work. We also thank our wonderfully patient families for their support.

David Shier, Jackie Butler, Ricki Lewis

REVIEWERS

We would like to acknowledge the valuable contributions of all professors and their students who have provided detailed recommendations for improving chapter content and illustrations throughout the revision process for each edition. Hundreds of professors have played a vital role in building a solid foundation for *Hole's Human Anatomy & Physiology*.

Patricia Adumanu Ahanotu, *Georgia Perimeter College*
Jeffrey Bell, *Northland Community and Technical College*
Richard A. Bennett, *University of Southern Indiana*
Gladys Delancey-Bolding, *Georgia Perimeter College*
Anita Brownstein, *Career Institute of Health and Technology*
William M. Clark, *Lone Star College Kingwood*
Amy Warenda Czura, *Suffolk County Community College*
Joseph D'Silva, *Norfolk State University, Norfolk, VA*
Betsy L. Diegel, *Davenport University*
Beverly Wilson Dunham, *WorWic Community College*
Georgia Everett, *Ivy Tech Community College*
Theresa Felten, *Polk State College*
Mary Catherine Flath, *Ashland Community and Technical College*
Larry M. Frolich, *Miami Dade College*
Kristin L. Gosselink, *The University of Texas at El Paso*

Melissa L. Greene, *Northwest Mississippi Community College*
Melissa M. Haswell, *Davenport University*
Kelli Hayes, *Pasco Hernando Community College*
Jean Jackson, *Bluegrass Community and Technical College*
Mark Jaffe, *Nova Southeastern University*
Suzanne Kempke, *St. Johns River State College*
Beth Ann Kersten, *State College of Florida Sarasota–Manatee*
Elizabeth Kozak, *Lewis University*
Mary LaCasse, *Hesser College/Kaplan*
Daudi K. Langat, *Labette Community College*
Craighton S. Mauk, *Gateway Community and Technical College*
Patricia Mote, *Georgia Perimeter College*
Ali Mustafa, *Hesser College/Kaplan*
Necia Morgan Nicholas, *Calhoun Community College*

DEDICATION

*This book is dedicated with much affection and appreciation
to Sherrie and Terry Martin, colleagues and friends whose
passion for improving the learning experiences of their students
has enriched our perspectives as teachers and as authors.*

UPDATES AND ADDITIONS

Global Changes

- WHOLE PICTURE overview replaces chapter vignettes.
- Career Corner describes a specific career opportunity for each chapter.
- Updated small boxes.
- New art program throughout.

- We have consistently avoided the names of specific individuals in boxes and clinical application pieces. We feel that the interest gained by including names is outweighed by the need to instill in our students the importance of patient confidentiality.

SELECTED SPECIFIC CHANGES AT-A-GLANCE

Chapter	Topic	Change	Rationale
1	Fluid compartments	Introduced in more detail in text and with new figure 1.5	Clarity, detail
1	Organ systems	Reorganized and rewritten, old figure 1.5 relocated	Clarity, detail
1	Body regions	Terminology updated, new photos accompany figure 1.24	Clarity, consistency with usage
2	Atoms and elements	Text rewritten	Clarity
2	Polar bonds	Text rewritten	Clarity
2	Acidosis and alkalosis	Rewritten introduction of terms	Clarity
2	Protein structure	Figure 2.19 redesigned	Clarity
3	Membrane proteins	Integral, transmembrane, and peripheral proteins better described in text and in figure 3.7	Clarity, detail
3	Organelles	Endoplasmic reticulum, Golgi apparatus, and polysome sections rewritten	Clarity, accuracy
3	Organelles	Phospholipid bilayer membrane structure emphasized where appropriate	Accuracy, detail
3	Overview of secretion processes	Figure 3.11 redrawn	Accuracy, clarity
3	Nonmotile cilia	Description expanded and added to text	Update
3	Osmosis	Section rewritten	Clarity, accuracy
4	Catalysts	Enzymes described as organic catalysts	Clarity, accuracy, detail
4	Metabolic cycle	New figure 4.9 shows how a metabolic pathway can form a cycle, prior to introducing the citric acid cycle	Clarity
4	Protein synthesis	Series of figures redone	Clarity
5	Thin sections	New figure 5.2 to help students understand orientation of micrographs	Clarity
5	Connective tissue	Table 5.6 reorganized to clarify cellular versus matrix components	Clarity
5	Adipose tissue	New discussion of brown fat	New information
6	Squamous cell carcinoma	Figure added to Clinical Application 6.1	Expanded discussion
6	Nails	Figure 6.6 expanded to include a longitudinal section and a dorsal view	Clarity
6	Hair follicles	Hair bulge included in discussion and in art	Update
6	Sweat glands	Merocrine (eccrine) terminology explained	Clarity

—Continued

UPDATES AND ADDITIONS

Chapter	Topic	Change	Rationale
6	Apocrine sweat glands	Expanded discussion	New information
7	Types of bones	Figure 7.1 color coded to help students identify bone locations	Clarity
7	Intramembranous bones	Expanded discussion, table 7.1 steps rewritten, and new figure 7.8	Clarity
7	Vertebral column	Figure 7.37 color coded to help students identify bone locations	Clarity
7	Sulcus	Term added to table 7.4	Clarity
8	Joint movements	Nonaxial, uniaxial, and multiaxial added to discussion and to table 8.1	Terminology update
8	Joint movements	Lateral flexion added	Terminology update
9	Organization of muscle	New table 9.1 replaces part of figure 9.2	Clarity
9	Muscle locations	Descriptions do not use the term "extend" because of potential confusion with muscle action	Clarity
9	Thick and thin muscle filaments	Figure 9.4 redone	Accuracy, clarity
9	Stimulus for contraction	Reorganized and rewritten to include sodium and potassium ion gradients and the concept of an action potential	Clarity
9	Myoglobin	Description rewritten	Clarity, accuracy
9	Lactic acid	Relationship to lactate clarified and role in muscle fatigue rewritten	Accuracy, update
9	Interaction of skeletal muscles	Rewritten section on agonist, antagonist, prime mover, and synergist	Clarity, clinical relevance
9	Agonist versus prime mover	Rewritten	Clarity
9	Major skeletal muscles	New section on popular versus anatomical terminology	Clarity
9	Muscle actions	Terms for movements from chapter 8 are used throughout (e.g., elevates instead of raises)	Clarity, consistency
9	Scalene muscles	Added to list of muscles and actions, with reference to role in breathing	Detail
9	Ligamentum nuchae	Added to text and table 9.3	Detail
9	Movements at shoulder and hip	Shoulder and hip flexion and extension are clarified in their respective sections	Clarity
9	Pelvic floor	Central tendon now included in text, table 9.12, and figure 9.36	Accuracy, consistency
10	Ganglia and nuclei	Introduced with classification of neurons	Clarity
10	Electrical synapses	Introduced in new boxed material	Detail
10	Action potential	Section rewritten with redesigned figures showing relationship to graded potentials and threshold	Detail, clarity
10	Refractory period	Section rewritten	Clarity
10	Neurotransmitters	Rewritten section on excitatory and inhibitory effects of neurotransmitters	Clarity, detail

SELECTED SPECIFIC CHANGES AT-A-GLANCE —*Continued*

Chapter	Topic	Change	Rationale
11	Brain and spinal cord	Sections are reordered with brain first	Clarity
11	Brain	Improved lateral brain figure 11.8a and related pieces	Clarity
11	CSF	Reworked discussion	Clarity
11	Memory	Section rewritten	Clarity
11	Brainstem	Figure 11.11 revised with new location icon	Clarity
11	Types of sleep	Rewritten	Clarity
11	Cranial nerves	Section introduction rewritten	Clarity
11	Cranial nerves	New figure 11.25 with enlargement of olfactory nerve	Clarity, accuracy
11	Spinal nerves	Section on cauda equina rewritten	Accuracy
11	Segmental innervation	New figures 11.18 and 11.19 showing relationship among levels	Clinical relevance
11	Autonomic nervous system	New description of reciprocal innervation and functional relationships	Clarity
11	Autonomic nervous system	Figure 11.40 redesigned	Clarity, consistency
12	Visual pathways	New figure 12.40	Clarity
12	Vibration transfer in middle ear	Figure 12.12 redesigned	Clarity
12	Light	Rewritten without reference to particles or waves	Clarity, level
12	Refraction	Figure 12.35 redesigned	Accuracy
13	Chemical structures of hormones (throughout chapter)	Figures color matched with figures from chapter 2 (Chemistry)	Clarity
13	Mechanism of T3 and T4	Possible specific membrane transport mechanism added to discussion	Update
13	Phosphorylation	Text discussion expanded	Clarity, accuracy
13	Insulin and glucagon	Section partially rewritten	Clarity
13	Diabetes mellitus	New Clinical Application 13.4	Clarity, consistency
13	Stress response	Section partially rewritten	Clarity
14	Normal range of values	Blue box added	Clinical significance
14	Red blood cell formation	Nuclear extrusion clarified in text and in figure 14.4	Accuracy
14	Platelets	Section rewritten	Clarity, accuracy
14	Plasma	Section on plasma lipids expanded	Clinical significance
14	Coagulation	Terminology includes "tissue factor pathway" and "contact activation pathway"	Update
14	Antigens and antibodies	Section rewritten	Clarity
15	Pericardial membranes	Revised figure 15.3b	Clarity
15	Electrocardiogram	Updated figures for both normal and abnormal ECGs	Accuracy
15	SA node and depolarization pathway	Figure 15.18 redrawn	Accuracy

—Continued

UPDATES AND ADDITIONS

Chapter	Topic	Change	Rationale
15	Clinical Application 15.2	Rewritten sections on aneurysm and on edema	Clarity
15	Aortic bodies	Sensing blood pH added as a function	Accuracy, clinical relevance
16	Lymphatic vessels	Cisterna chyli added to text	Accuracy
16	Lymphatic capillaries	Figure 16.8 redesigned	Clarity
16	Primary and secondary immune response	"Anamnestic" and "antibody titer" added to discussion.	Accuracy
16	Hypersensitivity responses	New table 16.10	Clarity, clinical relevance
16	HIV/AIDS	Updated Clinical Application 16.1	Clinical relevance
17	Gut microbiome	Examples now appear throughout chapter	Clinical significance
17	Pharynx	Figure 17.13 redesign shows relationship of pharyngeal muscles and lumen more clearly	Clarity, accuracy
17	Swallowing	Figure 17.14 muscles are now shown in section in context of mucous membranes to better illustrate process	Clarity
17	Mesentery	Figure 17.31 redrawn with sectional plane and enlargement	Clarity
18	Obesity	Updated Clinical Application 18.1	Clinical relevance
18	Vitamin D requirements	Updated table 18.6	Clinical relevance
18	Water-soluble vitamins	Table 18.8 RDA values updated	Clinical relevance
18	Major minerals	Table 18.9 RDA values updated	Clinical relevance
18	Trace elements	Table 18.10 RDA values updated	Clinical relevance
18	Dietary supplements	New material and table added to Clinical Application 18.2	Clinical relevance
19	Pharynx	Section rewritten to include subdivisions	Clarity, accuracy
19	Air Pollution	New section added to Clinical Application 19.2	Clinical relevance
19	Inspiration	Scalene muscles added	Detail, accuracy
19	Respiratory volumes and capacities	Figure 19.26 color-coded to highlight relationships	Clarity
19	Alveolar ventilation	Section expanded	Clarity
19	Factors affecting breathing	Section rewritten	Clarity
19	Carbon monoxide	Clinical Application 19.3 rewritten	Update
19	Gas transport	Section on carbon dioxide transport rewritten, including its importance in acid-base balance	Update, clarity
20	Vasa recta	Introduced earlier under renal blood vessels	Accuracy, clarity
20	Cortical versus juxtamedullary nephrons	Introduced earlier in text and in new figure 20.6	Clarity
20	Nephron structure	New diagrammatic figure 20.10	Clarity
20	Overview of urinary system structures	New figure 20.12	Clarity

SELECTED SPECIFIC CHANGES AT-A-GLANCE —*Continued*

Chapter	Topic	Change	Rationale
20	Glomerular filtration	Section reorganized	Clarity
20	Basic renal processes	Redesigned summary figure 20.19	Update
20	Potassium secretion	Section rewritten and figure 20.21 simplified	Clarity, update
20	Urine concentration	Countercurrent exchanger added to expanded discussion of vasa recta	Clarity, accuracy
21	Thirst	New material in section on osmoreceptors regarding osmolarity, osmolality, and their relationship to osmotic pressure	Clarity
22	Sperm cell structure	Enzymes on membrane as well as in acrosome described in text	Accuracy
22	Follicle maturation	Section rewritten to describe process taking almost 300 days	Update
22	Follicle maturation	New figure 22.22 portrays events in ovary during follicle maturation	Accuracy, clarity
22	Female reproductive cycle	Section rewritten to describe role of various hormones on follicle development and ovulation, micrographs added to figure 22.31	Update, clarity
22	Breast cancer	Updated Clinical Application 22.4	Accuracy, update
22	Birth control	Section rewritten	Update
23	Fertilization	Redrawn figure 23.3 and changes to text better depict role of sperm enzymes	Clarity
23	Embryonic development	Section reordered for better flow	Clarity
24	Prenatal testing	Text discussion rewritten	Update
24	Genes and genomes	Section reworked	Clarity

DYNAMIC NEW ART PROGRAM

Every piece of art has been updated to make it more vibrant, three-dimensional, and instructional. The authors examined every figure to ensure it was engaging and accurate. The fourteenth edition's art program will help with understanding the key concepts of anatomy and physiology.

Realistic, three-dimensional figures provide depth and orientation.

Colors highlighting atomic nuclei complement the atom colors in molecular models.

Process portrayed more accurately.

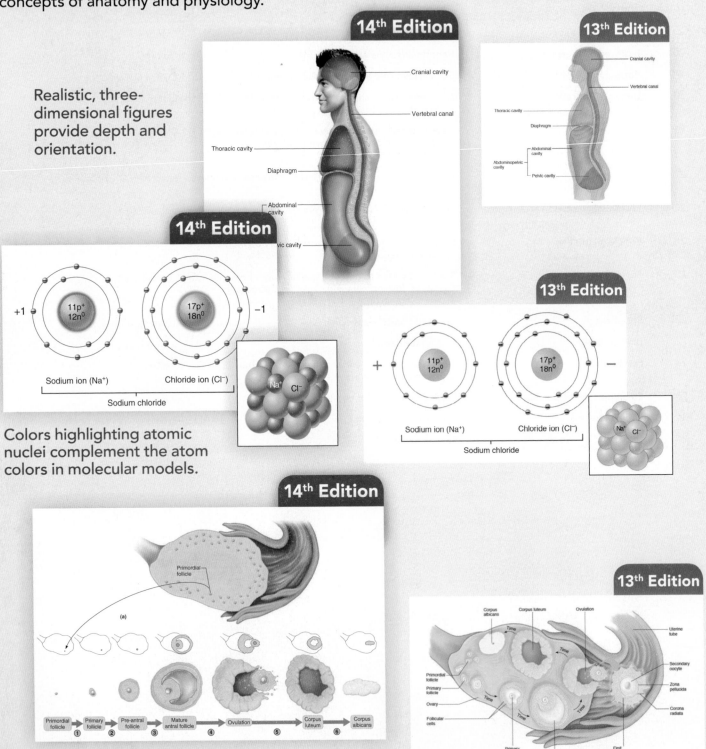

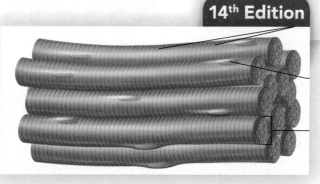

14th Edition

13th Edition

Line art for micrographs is three-dimensional to help visualize more than just the flat microscopic sample.

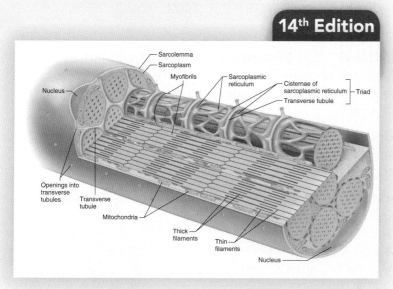

14th Edition

Sarcolemma
Sarcoplasm
Myofibrils
Sarcoplasmic reticulum
Cisternae of sarcoplasmic reticulum
Transverse tubule
Triad
Nucleus
Openings into transverse tubules
Transverse tubule
Mitochondria
Thick filaments
Thin filaments
Nucleus

13th Edition

Myofibrils
Cisternae of sarcoplasmic reticulum
Transverse tubule
Nucleus
Sarcoplasmic reticulum
Openings into transverse tubules
Mitochondria
Thick and thin filaments
Sarcoplasm
Nucleus
Sarcolemma

This longitudinal section shows the interior structures of a muscle fiber revealing more detail of the myofibrils, and thick and thin filaments.

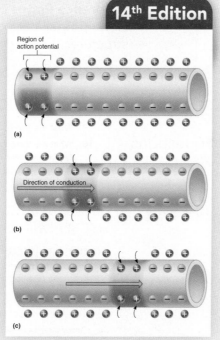

14th Edition

Region of action potential
(a)
Direction of conduction
(b)
(c)

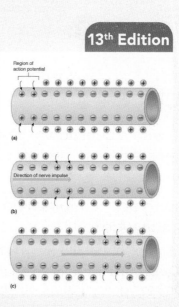

13th Edition

Region of action potential
(a)
Direction of nerve impulse
(b)
(c)

Color follows the movement of the action potential.

LEARN, PRACTICE, ASSESS

Learn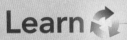

Learning tools to help the student succeed. . .

Check out the *Chapter Preview, Foundations for Success*, on page 1. The Chapter Preview was specifically designed to help the student **LEARN** how to study. It provides helpful study tips.

Learning Outcomes open chapters, and are closely linked to Chapter Assessments and Integrative Assessments/Critical Thinking questions found at the end of each chapter. Learning Outcomes are also tied to Connect content.

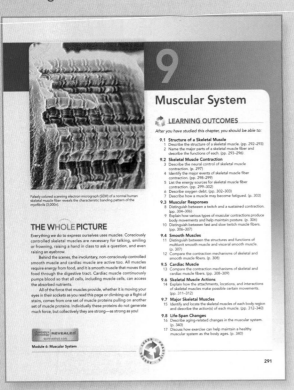

Falsely colored scanning electron micrograph (SEM) of a normal human skeletal muscle fiber reveals the characteristic banding pattern of the myofibrils (3,000×).

THE WHOLE PICTURE

Module 6: Muscular System

291

NEW! The WHOLE Picture gives an introduction to the chapter. It answers the question: "What is the big picture of how this chapter relates to Human Anatomy and Physiology"?

Anatomy and Physiology Revealed (APR) icon at the beginning of each chapter tells which system in APR applies to this chapter.

Understanding Words helps the student remember scientific word meanings. Examine root words, stems, prefixes, suffices, pronunciations, and build a solid anatomy and physiology vocabulary.

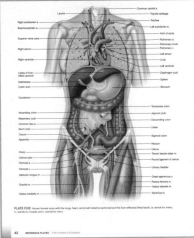

PLATE FIVE Human female torso with the lungs, heart, and small intestine sectioned and the liver reflected (lifted back). (*a.* stands for artery, *m.* stands for muscle, and *v.* stands for vein.)

42 REFERENCE PLATES | THE HUMAN ORGANISM

Reference Plates offer vibrant detail of body structures.

Practice

Practice with a question or series of questions after major sections. They will test understanding of the material.

Interesting applications help practice and apply knowledge. . .

Figure Questions allow an additional assessment. Found on key figures throughout the chapter.

FIGURE 9.8 **AP|R** Neuromuscular junction. (*a*) A neuromuscular junction includes the end of a motor neuron and the motor end plate of a muscle fiber. (*b*) Micrograph of a neuromuscular junction (500×).

How does neurotransmitter released into the synaptic cleft reach the muscle fiber membrane?

Answers can be found in Appendix G.

A few hours after death, the skeletal muscles partially contract, fixing the joints. This condition, called *rigor mortis*, may continue for seventy-two hours or more. It results from an increase in membrane permeability to calcium ions, which promotes cross-bridge formation, and a decrease in availability of ATP in the muscle fibers, which prevents myosin release from actin. Thus, the actin and myosin filaments of the muscle fibers remain linked until the proteins begin to decompose.

Boxed information applies ideas and facts in the narrative to clinical situations.

Career Corners, new to this edition, introduce interesting career opportunities.

CAREER CORNER
Massage Therapist

The middle-aged woman feels something give way in her left knee as she comes down from a jump in her dance class. She limps away between her much younger classmates, in great pain.

The frequent jumping followed by lateral movements caused patella tendinitis, or "jumper's knee." The woman injured in dance class went home and used "RICE" treatment—rest, ice, compression, and elevation—and three days later, at her weekly appointment with a massage therapist for stress relief, mentioned the injury.

Clinical Applications present disorders, physiological responses to environmental factors, and other topics of general interest.

Genetics Connections explore the molecular underpinnings of familiar as well as not so familiar illnesses. Read about such topics as ion channel disorders, muscular dystrophy, and cystic fibrosis.

CLINICAL APPLICATION 9.1
Myasthenia Gravis

In an autoimmune disorder, the immune system attacks part of the body. In myasthenia gravis (MG), the part attacked is the muscular system. The name means "grave muscular weakness." The body produces antibodies that target receptors for the neurotransmitter acetylcholine on skeletal muscle cells at neuromuscular junctions. People with MG have one-third the normal number of acetylcholine receptors on their skeletal muscle cells. On a whole-body level, this causes weak and easily fatigued muscles.

MG affects hundreds of thousands of people worldwide, mostly women beginning in their twenties or thirties, and men in their sixties and seventies. The specific symptoms depend upon the site of attack. For 85% of patients, the disease causes generalized muscle weakness. Many people develop a characteristic flat smile and nasal voice and have difficulty chewing and swallowing due to affected facial and neck muscles. Limb weakness is common. About 15% of patients experience the illness only in the

muscles surrounding their eyes. The disease reaches crisis level when respiratory muscles are affected, requiring a ventilator to support breathing. MG does not affect sensation or reflexes.

Most people with MG live a normal life span, with symptoms that are controlled with a combination of treatments that include the following:

• Drugs that inhibit acetylcholinesterase, the enzyme that normally breaks down acetylcholine, thus increasing levels of the neurotransmitter.

• Immunosuppressant drugs such as corticosteroids, cyclosporine, and rituximab, that decrease production of antibodies.

• Intravenous antibodies that bind and inactivate the antibodies causing the damage.

• Plasma exchange that rapidly removes the damaging antibodies from the bloodstream, helping people in crisis.

FROM SCIENCE TO TECHNOLOGY 5.2
Tissue Engineering: Building a Replacement Bladder

If an appliance part is damaged or fails, replacing it is simple. Not so for the human body. Donor organs and tissues for transplant are in short supply, so in the future spare parts may come from tissue engineering. In this technology, a patient's cells, extracellular matrix, and other biochemicals are grown with a synthetic scaffold to form an implant. The cells come from the patient, so the immune system does not reject them. Tissue engineering has provided skin, cartilage, bone, and blood vessels. Combining engineered tissues into structures that can replace organs is where the creativity comes in. Consider the replacement bladder.

Each year in the United States, about 10,000 people need their urinary bladders repaired or replaced. Typically a urologic surgeon replaces part of the bladder with part of the large intestine. However, the function of the intestine is to absorb, and the function of the bladder is to hold waste. Tissue engineering is providing a better replacement bladder. The natural organ is balloon-

like, with layers of smooth muscle, connective tissue, and a lining called urothelium.

Researchers pioneered replacement bladders in children who have birth defects in which the malfunctioning bladder can harm the kidneys. Each patient donated a postage-stamp-size sample of bladder tissue that consisted of about a million cells, from which the researchers separated two types of progenitor cells—for smooth muscle and urothelium—and let them divide in culture in a specific "cocktail" of growth factors. Within seven weeks the million cells had divided to yield 1.5 billion cells, which were seeded onto domes made of a synthetic material. The confluent layers of cells that formed were attached to the lower portions of the patients' bladders, after removing the upper portions. The scaffolds degenerated over time, leaving new bladders built from the patients' own cells. Today tissue-engineered bladders are also used in adults whose bladders have been removed to treat cancer.

Assess

Tools to help make the connection and master anatomy & physiology!

Chapter Assessments check understanding of the chapter's learning outcomes.

Integrative Assessments/Critical Thinking questions connect and apply information from previous chapters as well as information within the current chapter.

Chapter Summary Outlines help review the chapter's main ideas.

CHAPTER ASSESSMENTS

9.1 Structure of a Skeletal Muscle
1 Describe the difference between a tendon and an aponeurosis. (p. 292)
2 Describe how connective tissue is part of the structure of a skeletal muscle. (p. 293)
3 Distinguish among deep fascia, subcutaneous fascia, and subserous fascia. (p. 293)
4 Identify the major parts of a skeletal muscle fiber and describe the functions of each. (pp. 293–296)

9.2 Skeletal Muscle Contraction
5 A neurotransmitter _____. (p. 297)
a. binds actin filaments, causing them to slide
b. diffuses across a synaptic cleft from a neuron to a

9.3 Muscular Responses
13 Define *threshold stimulus*. (p. 304)
14 Distinguish between a twitch and a sustained contraction. (p. 304)
15 Define *motor unit* and explain how the number of fibers in a unit affects muscular contractions. (p. 305)
16 Which of the following describes addition of muscle fibers to take part in a contraction? (p. 305)
a. summation
b. recruitment
c. tetany
d. twitch
17 Explain how a skeletal muscle can be stimulated to produce a sustained contraction. (p. 306)

INTEGRATIVE ASSESSMENTS/CRITICAL THINKING

Outcomes 7.10, 7.11, 9.7
1. Several important nerves and blood vessels course through the muscles of the gluteal region. To avoid the possibility of damaging such parts, intramuscular injections are usually made into the lateral, superior portion of the gluteus medius. What landmarks would help you locate this muscle in a patient?

Outcomes 9.1, 9.2
2. Millions of people take drugs called statins to lower serum cholesterol levels. In a small percentage of people taking these drugs, muscle pain, termed myopathy, is an adverse effect. In a small percentage of these individuals, the condition progresses to rhabdomyolysis, in which the sarcolemma breaks down.
a. Describe the structure and state the function of the sarcolemma.
b. Physicians can measure a patient's levels of creatine

Outcome 9.2
4. As lactic acid and other substances accumulate in an active muscle, they stimulate pain receptors, and the muscle may feel sore. How might the application of heat or substances that dilate blood vessels help relieve such soreness?

Outcomes 9.2, 9.3, 9.5
5. Why do you think athletes generally perform better if they warm up by exercising lightly before a competitive event?

Outcomes 9.2, 9.3, 9.6
6. Following an injury to a nerve, the muscles it supplies with motor nerve fibers may become paralyzed. How would you explain to a patient the importance of moving the disabled muscles passively or contracting them with electrical stimulation?

Chapter Summary

9.1 Structure of a Skeletal Muscle (page 292)

Skeletal muscles are composed of nervous, vascular and various other connective tissues, as well as skeletal muscle tissue.

1. Connective tissue coverings
   a. **Fascia** covers each skeletal muscle.
   b. Other connective tissues surround skeletal muscle cells and groups of cells within the muscle's structure (epimysium, perimysium, endomysium).
   c. Fascia is part of a complex network of connective tissue that extends throughout the body.
2. Skeletal muscle fibers
   a. Each skeletal muscle fiber is a single muscle cell, the unit of contraction.
   b. Muscle fibers are cylindrical cells with many nuclei.
   c. The cytoplasm contains mitochondria, sarcoplasmic reticulum, and **myofibrils** of **actin** and **myosin**.

d. The arrangement of the actin and myosin filaments causes striations forming repeating patterns of **sarcomeres.** (I bands, Z lines, A bands, H zone and M line)
e. **Troponin** and **tropomyosin** molecules associate with actin filaments.
g. **Transverse tubules** extend from the cell membrane into the cytoplasm and are associated with the **cisternae** of the **sarcoplasmic reticulum.**

9.2 Skeletal Muscle Contraction (page 297)

Muscle fiber contraction results from a sliding movement of actin and myosin filaments overlapping that shortens the muscle fiber.

1. **Neuromuscular junction**
   a. The functional connection between a neuron and another cell is a **synapse.** The neuromuscular junction is a synapse.

McGraw-Hill Connect® Anatomy & Physiology integrated learning platform provides auto-graded assessments, a customizable, assignable eBook, an adaptive diagnostic tool, and powerful reporting against learning outcomes and level of difficulty—all in an easy-to-use interface. **Connect Anatomy & Physiology** is specific to this book and can be completely customized to a course and specific learning outcomes, connecting to just the material the students need to know.

Save time with auto-graded assessments and tutorials

Fully editable, customizable, auto-graded interactive assignments using high quality art from the textbook, animations, and videos from a variety of sources go way beyond multiple choice. Assignable content is available for every Learning Outcome in the book.

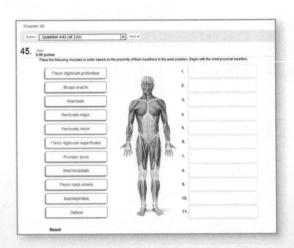

New! Unique Interactive Question Types for Each Chapter!

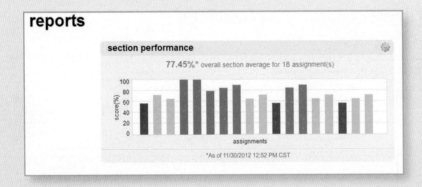

Gather assessment information

Generate powerful data related to student performance against learning outcomes, specific topics, level of difficulty, and more.

All **Connect** content is pre-tagged to Learning Outcomes for each chapter as well as topic, section, Bloom's Level, and Human Anatomy and Physiology Society (HAPS) Learning Outcomes to assist in both filtering out unneeded questions for ease of creating assignments and in reporting on students' performance. This will enhance effective assessment of student learning by allowing alignment of learning activities to peer-reviewed standards from an international organization.

Hi. We are Leslie Day and Julie Pilcher. We've joined the author team as digital authors. We know it's frustrating when online homework questions don't match the textbook. That's where digital authors come in! We ensure that our online resources correlate directly with the information and learning tools found in the textbook. Here are some of the question types we've created for the 14th edition!

Integrated activities use a series of interactive questions to apply new knowledge and/or see how different processes are related. Some integrated questions also include animations to help the visual learner.

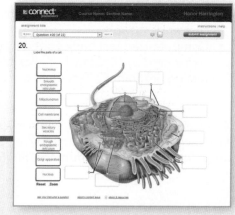

Labeling allows identification of structures using figures from the text. Also available are questions in which functions or descriptions are used as drag-and-drop labels.

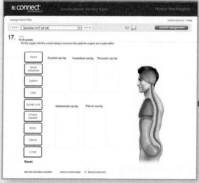

Classification questions ask for placement of terms into appropriate categories, to recognize the difference and similarities in structures.

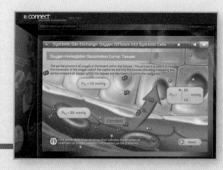

Concept Overview Interactives combine multiple concepts into one big-picture summary. These striking, visually dynamic presentations offer a review of previously covered material in a creatively designed environment to emphasize how individual parts fit together in the understanding of a larger mechanism or concept.

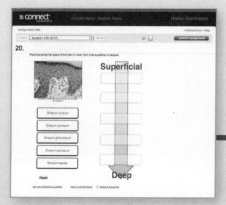

Sequence questions have been designed around ordering physiological processes or some anatomical topics, such as listing structures' locations from proximal to distal.

TEACHING AND LEARNING

McGraw-Hill's Presentation Tools

Presentation Materials for Lecture and Lab—incorporate customized lectures, visually enhanced tests and quizzes, compelling course websites, or attractive printed support materials.

- A complete set of pre-made PowerPoints linking **Anatomy & Physiology Revealed** to text material is now available.
- A complete set of animation embedded PowerPoint slides is now available.
- Along with our online digital library containing photos, artwork, and animations, we now also offer **FlexArt**. FlexArt allows the instructor to customize artwork.
- Computerized test bank edited by the **Author Team** is powered by McGraw-Hill's flexible electronic testing program EZ Test Online.

To obtain any of these tools, contact your McGraw-Hill Learning Technology rep at http://catalogs.mhhe.com/mhhe/findRep.do.

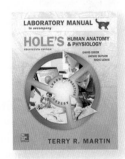

Laboratory Manual for Hole's Human Anatomy & Physiology, **Fourteenth Edition,** by Terry R. Martin, Kishwaukee College, is designed to accompany the Fourteenth edition of *Hole's Human Anatomy & Physiology*. It is available in a cat and fetal pig version

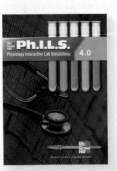

Ph.I.L.S. 4.0 has been updated! Users have requested and we are providing five new exercises (Respiratory Quotient, Weight & Contraction, Insulin and Glucose Tolerance, Blood Typing, and Anti-Diuretic Hormone). Ph.I.L.S. 4.0 is the perfect way to reinforce key physiology concepts with powerful lab experiments. Created by Dr. Phil Stephens at Villanova University, this program offers 42 laboratory simulations that may be used to supplement or substitute for wet labs. All 42 labs are self-contained experiments—no lengthy instruction manual required. Users can adjust variables, view outcomes, make predictions, draw conclusions, and print lab reports. This easy-to-use software offers the flexibility to change the parameters of the lab experiment. There is no limit!

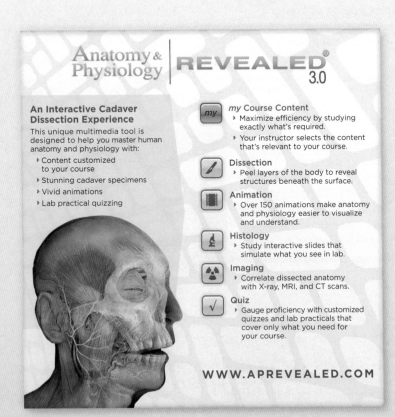

Anatomy & Physiology Revealed® is now available in Cat and Fetal Pig versions!

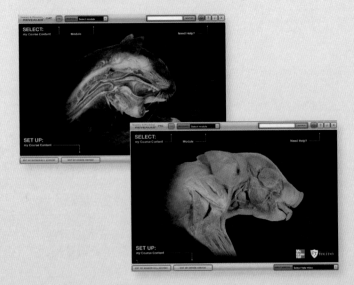

CONTENTS

About the Authors   iv | Acknowledgments   vi | Updates and Additions   vii | Dynamic New Art Program   xii |
Learn, Practice, Assess   xiv | McGraw-Hill Connect Anatomy & Physiology   xvi | Teaching and Learning   xviii |
Contents   xix | Clinical Connections   xxiv

PREVIEW
FOUNDATIONS FOR SUCCESS   1

P.1 Overview   2
P.2 Strategies for Success   2
CHAPTER SUMMARY   7
CHAPTER ASSESSMENTS   8
INTEGRATIVE ASSESSMENTS/CRITICAL THINKING   8

UNIT 1
LEVELS OF ORGANIZATION   9

CHAPTER **1**
Introduction to Human Anatomy and Physiology   9

1.1   Origins of Medical Science   10
1.2   Anatomy and Physiology   11
1.3   Levels of Organization   12
1.4   Characteristics of Life   14
1.5   Maintenance of Life   14
1.6   Organization of the Human Body   18
1.7   Life-Span Changes   27
1.8   Anatomical Terminology   27
        Some Medical and Applied Sciences   30
CHAPTER SUMMARY   33
CHAPTER ASSESSMENTS   35
INTEGRATIVE ASSESSMENTS/CRITICAL THINKING   36

REFERENCE PLATES 1–25
The Human Organism   37

CHAPTER **2**
Chemical Basis of Life   57

2.1   The Importance of Chemistry in Anatomy and Physiology   58
2.2   Structure of Matter   58
2.3   Chemical Constituents of Cells   68

CHAPTER SUMMARY   79
CHAPTER ASSESSMENTS   80
INTEGRATIVE ASSESSMENTS/CRITICAL THINKING   81

CHAPTER **3**
Cells   82

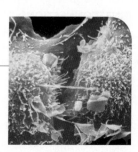

3.1   Cells Are the Basic Units of the Body   83
3.2   A Composite Cell   83
3.3   Movements Into and Out of the Cell   99
3.4   The Cell Cycle   107
3.5   Control of Cell Division   110
3.6   Stem and Progenitor Cells   112
3.7   Cell Death   113
CHAPTER SUMMARY   116
CHAPTER ASSESSMENTS   117
INTEGRATIVE ASSESSMENTS/CRITICAL THINKING   119

CHAPTER **4**
Cellular Metabolism   120

4.1   Metabolic Processes   121
4.2   Control of Metabolic Reactions   123
4.3   Energy for Metabolic Reactions   125
4.4   Cellular Respiration   127
4.5   Nucleic Acids and Protein Synthesis   132

4.6 Changes in Genetic Information 142

CHAPTER SUMMARY 144
CHAPTER ASSESSMENTS 146
INTEGRATIVE ASSESSMENTS/
CRITICAL THINKING 147

CHAPTER **5**

Tissues 148

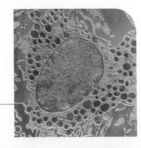

5.1 Cells Are Organized Into
Tissues 149

5.2 Epithelial Tissues 150
5.3 Connective Tissues 159
5.4 Types of Membranes 168
5.5 Muscle Tissues 170
5.6 Nervous Tissues 170

CHAPTER SUMMARY 173
CHAPTER ASSESSMENTS 175
INTEGRATIVE ASSESSMENTS/CRITICAL THINKING 176

UNIT 2
SUPPORT AND MOVEMENT 177

CHAPTER **6**

Integumentary System 177

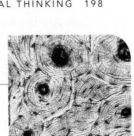

6.1 Skin and Its Tissues 178
6.2 Accessory Structures of
the Skin 184
6.3 Skin Functions 189
6.4 Healing of Wounds and
Burns 190
6.5 Life-Span Changes 192

INNERCONNECTIONS: INTEGUMENTARY SYSTEM 194

CHAPTER SUMMARY 195
CHAPTER ASSESSMENTS 197
INTEGRATIVE ASSESSMENTS/CRITICAL THINKING 198

CHAPTER **7**

Skeletal System 199

7.1 Bone Shape and Structure
200
7.2 Bone Development and
Growth 202
7.3 Bone Function 210
7.4 Skeletal Organization 212
7.5 Skull 215
7.6 Vertebral Column 226
7.7 Thoracic Cage 231
7.8 Pectoral Girdle 231
7.9 Upper Limb 233
7.10 Pelvic Girdle 238
7.11 Lower Limb 241
7.12 Life-Span Changes 245

INNERCONNECTIONS: SKELETAL SYSTEM 247

CHAPTER SUMMARY 246
CHAPTER ASSESSMENTS 250
INTEGRATIVE ASSESSMENTS/CRITICAL THINKING 251

REFERENCE PLATES 26–54
Human Skull 252

CHAPTER **8**

Joints of the Skeletal System
267

8.1 Types of Joints 268
8.2 Types of Joint Movements
274
8.3 Examples of Synovial Joints
277
8.4 Life-Span Changes 285

CHAPTER SUMMARY 288
CHAPTER ASSESSMENTS 289
INTEGRATIVE ASSESSMENTS/CRITICAL THINKING 290

CHAPTER **9**

Muscular System 291

9.1 Structure of a Skeletal
Muscle 292
9.2 Skeletal Muscle Contraction
297
9.3 Muscular Responses 303
9.4 Smooth Muscle 307
9.5 Cardiac Muscle 308
9.6 Skeletal Muscle Actions 310
9.7 Major Skeletal Muscles 312
9.8 Life-Span Changes 340

INNERCONNECTIONS: MUSCULAR SYSTEM 341

CHAPTER SUMMARY 340
CHAPTER ASSESSMENTS 344
INTEGRATIVE ASSESSMENTS/CRITICAL THINKING 346

REFERENCE PLATES 55–75
Surface Anatomy and Cadaver Dissection 347

UNIT 3
INTEGRATION AND COORDINATION 359

CHAPTER **10**

Nervous System I: Basic Structure and Function 359

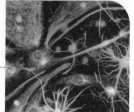

10.1	Overview of the Nervous System 360	
10.2	General Functions of the Nervous System 361	
10.3	Description of Cells of the Nervous System 363	
10.4	Classification of Cells of the Nervous System 363	
10.5	The Synapse 371	
10.6	Cell Membrane Potential 372	
10.7	Synaptic Transmission 378	
10.8	Impulse Processing 382	

INNERCONNECTIONS: NERVOUS SYSTEM 385

CHAPTER SUMMARY 383
CHAPTER ASSESSMENTS 387
INTEGRATIVE ASSESSMENTS/CRITICAL THINKING 388

CHAPTER **11**

Nervous System II: Divisions of the Nervous System 389

11.1	Overview of Divisions of the Nervous System 390
11.2	Meninges 390
11.3	Ventricles and Cerebrospinal Fluid 391
11.4	Brain 395
11.5	Spinal Cord 409
11.6	Peripheral Nervous System 418
11.7	Autonomic Nervous System 431
11.8	Life-Span Changes 438

CHAPTER SUMMARY 438
CHAPTER ASSESSMENTS 441
INTEGRATIVE ASSESSMENTS/CRITICAL THINKING 442

CHAPTER **12**

Nervous System III: Senses 443

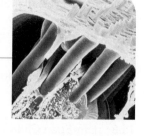

12.1	Introduction to Sensory Function 444
12.2	Receptors, Sensation, and Perception 444
12.3	General Senses 446
12.4	Special Senses 452
12.5	Life-Span Changes 482

CHAPTER SUMMARY 482
CHAPTER ASSESSMENTS 485
INTEGRATIVE ASSESSMENTS/CRITICAL THINKING 486

CHAPTER **13**

Endocrine System 487

13.1	General Characteristics of the Endocrine System 488
13.2	Hormone Action 489
13.3	Control of Hormonal Secretions 496
13.4	Pituitary Gland 498
13.5	Thyroid Gland 503
13.6	Parathyroid Glands 506
13.7	Adrenal Glands 508
13.8	Pancreas 514
13.9	Other Endocrine Glands 517
13.10	Stress and Its Effects 517
13.11	Life-Span Changes 519

INNERCONNECTIONS: ENDOCRINE SYSTEM 522

CHAPTER SUMMARY 521
CHAPTER ASSESSMENTS 525
INTEGRATIVE ASSESSMENTS/CRITICAL THINKING 526

UNIT 4
TRANSPORT 527

CHAPTER **14**

Blood 527

14.1	Characteristics of Blood 528
14.2	Blood Cells 529
14.3	Plasma 541
14.4	Hemostasis 542
14.5	Blood Groups and Transfusions 548

CHAPTER SUMMARY 552
CHAPTER ASSESSMENTS 554
INTEGRATIVE ASSESSMENTS/CRITICAL THINKING 555

CHAPTER **15**

Cardiovascular System   556

15.1   Overview of the Cardiovascular System   557
15.2   The Heart   557
15.3   Blood Vessels   576
15.4   Blood Pressure   582
15.5   Paths of Circulation   590
15.6   Arterial System   591
15.7   Venous System   602
15.8   Life-Span Changes   608

INNERCONNECTIONS: CARDIOVASCULAR SYSTEM   611

CHAPTER SUMMARY   610
CHAPTER ASSESSMENTS   614
INTEGRATIVE ASSESSMENTS/CRITICAL THINKING   615

CHAPTER **16**

Lymphatic System and Immunity   616

16.1   Lymphatic Pathways   617
16.2   Tissue Fluid and Lymph   620
16.3   Lymphatic Tissues and Lymphatic Organs   621
16.4   Body Defenses Against Infection   625
16.5   Life-Span Changes   640

INNERCONNECTIONS: LYMPHATIC SYSTEM   643

CHAPTER SUMMARY   644
CHAPTER ASSESSMENTS   647
INTEGRATIVE ASSESSMENTS/CRITICAL THINKING   648

UNIT 5
ABSORPTION AND EXCRETION   649

CHAPTER **17**

Digestive System   649

17.1   Overview of the Digestive System   650
17.2   Mouth   653
17.3   Salivary Glands   658
17.4   Pharynx and Esophagus   660
17.5   Stomach   661
17.6   Pancreas   668
17.7   Liver   669
17.8   Small Intestine   676
17.9   Large Intestine   683
17.10 Life-Span Changes   688

INNERCONNECTIONS: DIGESTIVE SYSTEM   689
CHAPTER SUMMARY   690
CHAPTER ASSESSMENTS   692
INTEGRATIVE ASSESSMENTS/CRITICAL THINKING   693

CHAPTER **18**

Nutrition and Metabolism   694

18.1   Carbohydrates   695
18.2   Lipids   696
18.3   Proteins   698
18.4   Energy Expenditures   701
18.5   Appetite Control   704
18.6   Vitamins   705
18.7   Minerals   711
18.8   Healthy Eating   718

18.9   Life-Span Changes   723
CHAPTER SUMMARY   724
CHAPTER ASSESSMENTS   727
INTEGRATIVE ASSESSMENTS/CRITICAL THINKING   729

CHAPTER **19**

Respiratory System   730

19.1   Overview of the Respiratory System   731
19.2   Organs of the Respiratory System   731
19.3   Breathing Mechanism   741
19.4   Control of Breathing   750
19.5   Alveolar Gas Exchanges   753
19.6   Gas Transport   758
19.7   Life-Span Changes   762

INNERCONNECTIONS: RESPIRATORY SYSTEM   763

CHAPTER SUMMARY   762
CHAPTER ASSESSMENTS   766
INTEGRATIVE ASSESSMENTS/CRITICAL THINKING   767

CHAPTER **20**

Urinary System   768

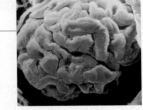

20.1   Overview of the Urinary System   769
20.2   Kidneys   769
20.3   Urine Formation   775
20.4   Storage and Elimination of Urine   791
20.5   Life-Span Changes   797

INNERCONNECTIONS: URINARY SYSTEM 799
CHAPTER SUMMARY 798
CHAPTER ASSESSMENTS 801
INTEGRATIVE ASSESSMENTS/
CRITICAL THINKING 802

CHAPTER **21**

Water, Electrolyte, and Acid-Base Balance 803

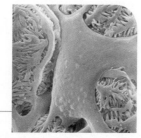

21.1   The Balance Concept 804

21.2   Distribution of Body Fluids 804
21.3   Water Balance 806
21.4   Electrolyte Balance 809
21.5   Acid-Base Balance 812
21.6   Acid-Base Imbalances 817
21.7   Compensation 819
CHAPTER SUMMARY 819
CHAPTER ASSESSMENTS 821
INTEGRATIVE ASSESSMENTS/CRITICAL THINKING 822

UNIT 6
THE HUMAN LIFE CYCLE 823

CHAPTER **22**

Reproductive Systems 823

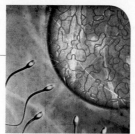

22.1   Meiosis and Sex Cell Production 824
22.2   Organs of the Male Reproductive System 826
22.3   Hormonal Control of Male Reproductive Functions 837
22.4   Organs of the Female Reproductive System 839
22.5   Hormonal Control of Female Reproductive Functions 849
22.6   Mammary Glands 852
22.7   Birth Control 854
22.8   Sexually Transmitted Infections 859
INNERCONNECTIONS: REPRODUCTIVE SYSTEMS 861
CHAPTER SUMMARY 860
CHAPTER ASSESSMENTS 865
INTEGRATIVE ASSESSMENTS/CRITICAL THINKING 866

CHAPTER **23**

Pregnancy, Growth, and Development 867

23.1   Fertilization 868
23.2   Pregnancy and the Prenatal Period 870
23.3   Postnatal Period 894
23.4   Aging 898
CHAPTER SUMMARY 900
CHAPTER ASSESSMENTS 903
INTEGRATIVE ASSESSMENTS/CRITICAL THINKING 904

CHAPTER **24**

Genetics and Genomics 905

24.1   Genes and Genomes 906
24.2   Modes of Inheritance 908
24.3   Factors That Affect Expression of Single Genes 912
24.4   Multifactorial Traits 912
24.5   Matters of Sex 914
24.6   Chromosome Disorders 916
24.7   Genomics and Health Care 920
CHAPTER SUMMARY 922
CHAPTER ASSESSMENTS 923
INTEGRATIVE ASSESSMENTS/CRITICAL THINKING 924

APPENDIX **A** 925
APPENDIX **B** 926
APPENDIX **C** 927
APPENDIX **D** 930
APPENDIX **E** 931
APPENDIX **F** 935
APPENDIX **G** 937

Glossary 938
Credits 961
Index 965

CLINICAL CONNECTIONS

Clinical Applications and From Science to Technology

CHAPTER 1
1.1: Ultrasonography and Magnetic Resonance Imaging: A Tale of Two Patients 13

CHAPTER 2
2.1: Radioactive Isotopes Reveal Physiology 61
2.2: Ionizing Radiation: From the Cold War to Yucca Mountain 62
2.3: CT Scanning and PET Imaging 78

CHAPTER 3
3.1: Faulty Ion Channels Cause Disease 89
3.2: Disease at the Organelle Level 94
3.1: Stem Cells to Study and Treat Disease 114

CHAPTER 4
4.1: Inborn Errors of Metabolism 126
4.1: The Human Metabolome 124
4.2: DNA Profiling Frees a Prisoner 136

CHAPTER 5
5.1: The Body's Glue: The Extracellular Matrix 161
5.2: Abnormalities of Collagen 163
5.1: Nanotechnology Meets the Blood-Brain Barrier 150
5.2: Tissue Engineering: Building a Replacement Bladder 172

CHAPTER 6
6.1: Indoor Tanning and Skin Cancer 182
6.2: Hair Loss 186
6.3: Acne 188
6.4: Elevated Body Temperature 191

CHAPTER 7
7.1: Fractures 208
7.2: Preventing Fragility Fractures 212
7.3: Disorders of the Vertebral Column 230

CHAPTER 8
8.1: Replacing Joints 282
8.2: Joint Disorders 286

CHAPTER 9
9.1: Myasthenia Gravis 298
9.2: Use and Disuse of Skeletal Muscles 308
9.3: TMJ Syndrome 315

CHAPTER 10
10.1: Migraine 362
10.2: Multiple Sclerosis 367
10.3: Factors Affecting Impulse Conduction 379

10.4: Opiates in the Human Body 382
10.5: Drug Addiction 384

CHAPTER 11
11.1: Traumatic Brain Injury 392
11.2: Cerebrospinal Fluid Pressure 396
11.3: Parkinson Disease 404
11.4: Brain Waves 410
11.5: Uses of Reflexes 414
11.6: Spinal Cord Injuries 419
11.7: Spinal Nerve Injuries 430

CHAPTER 12
12.1: Treating Pain 450
12.2: Mixed-Up Senses—Synesthesia 453
12.3: Smell and Taste Disorders 456
12.4: Getting a Cochlear Implant 463
12.5: Hearing Loss 465
12.6: Refraction Disorders 476

CHAPTER 13
13.1: Abusing Hormones to Improve Athletic Performance 494
13.2: Growth Hormone Ups and Downs 501
13.3: Disorders of the Adrenal Cortex 513
13.4: Diabetes Mellitus 516
13.1: Treating Diabetes 518

CHAPTER 14
14.1: Universal Precautions 530
14.2: Leukemia 540
14.3: Deep Vein Thrombosis 547
14.1: Blood Typing and Matching: From Serology to DNA Chips 549

CHAPTER 15
15.1: Arrhythmias 574
15.2: Blood Vessel Disorders 584
15.3: Hypertension 589
15.4: Exercise and the Cardiovascular System 591
15.5: Molecular Causes of Cardiovascular Disease 609
15.6: Coronary Artery Disease 610
15.1: Replacing the Heart—From Transplants to Stem Cell Implants 567
15.2: Altering Angiogenesis 577

CHAPTER 16
16.1: Immunity Breakdown: HIV/AIDS 641
16.1: Immunotherapy 633

CHAPTER 17
17.1: Dental Caries 658
17.2: A Common Problem: Heartburn 667
17.3: Hepatitis 674
17.4: Gallbladder Disease 675
17.5: Disorders of the Large Intestine 687

17.1: Replacing the Liver 671

CHAPTER 18
18.1: Obesity 703
18.2: Dietary Supplements—Proceed with Caution 718
18.3: Nutrition and the Athlete 720

CHAPTER 19
19.1: The Effects of Cigarette Smoking on the Respiratory System 735
19.2: Lung Irritants 743
19.3: Respiratory Disorders That Decrease Ventilation: Bronchial Asthma and Emphysema 750
19.4: Exercise and Breathing 754
19.5: Effects of High Altitude 756
19.6: Disorders that Impair Gas Exchange 757

CHAPTER 20
20.1: Glomerulonephritis 777
20.2: The Nephrotic Syndrome 785
20.3: Chronic Kidney Disease 792
20.4: Kidney Stones 793
20.5: Urinalysis: Clues to Health 797

CHAPTER 21
21.1: Water Balance Disorders 810
21.2: Sodium and Potassium Imbalances 813

CHAPTER 22
22.1: Prostate Cancer 834
22.2: Male Infertility 836
22.3: Female Infertility 853
22.4: Treating Breast Cancer 856

CHAPTER 23
23.1: Some Causes of Birth Defects 882
23.2: Human Milk—The Perfect Food for Human Babies 894
23.3: Living to 100—And Beyond 900
23.1: Assisted Reproductive Technologies 870
23.2: Preimplantation Genetic Diagnosis 875

CHAPTER 24
24.1: Down Syndrome 917
24.2: Exome Sequencing 921

A photo on the opening page for each chapter generates interest.

The Whole Picture presents a snapshot of the chapter content and explains how it relates to what new knowledge you will be adding to your knowledge base.

THE WHOLE PICTURE

Students often wonder why they are required to take anatomy and physiology in preparation for a career as a health-care professional. An understanding of homeostasis and normal anatomy and physiology helps the health-care professional recognize disease as it occurs in their patients.

Students should remember that among the reasons they are taking this course is to build a solid foundation for caring for their future patients.

This digital tool, as indicated below and with the APR icons within the chapters, allows you to explore the human body in depth through simulated dissection of cadavers and histology preparations. It also offers animations on chapter concepts.

Anatomy & Physiology **REVEALED**®
aprevealed.com

PREVIEW

Foundations for Success

The Preview Chapter not only provides great study tips to offer a foundation for success, but it also offers tips on how to utilize this particular text. Those tips will be found in boxes just like this.

 ## LEARNING OUTCOMES

After you have studied this chapter, you should be able to:

P.1 Overview
1 Explain the importance of an individualized approach to learning (p. 2)

P.2 Strategies for Success
2 Summarize what you should do before attending class. (p. 2)
3 Identify student activities that enhance the classroom experience. (p. 6)
4 List and describe several study techniques that can facilitate learning new material. (p. 6)

Each chapter has a list of learning outcomes indicating the knowledge you should gain as you work through the chapter. (Note the blue learning arrow.) These outcomes are intended to help you master the similar outcomes set by your instructor. The outcomes will be tied directly to assessments of knowledge gained.

UNDERSTANDING WORDS

ana-, up: *anatomy*—the study of breaking up the body into its parts.

multi-, many: *multi*tasking—performing several tasks simultaneously.

physio-, relationship to nature: *physiology*—the study of how body parts function.

This section introduces building blocks of words that your instructor may assign. Learning them is a good investment of your time, because they can be used over and over and apply to many of the terms you will use in your career. Inside the back cover and on the facing page is a comprehensive list of these prefixes, suffixes, and root words.

P.1 | Overview

Studying the human body can be overwhelming at times. The new terminology, used to describe body parts and how they work, can make it seem as if you are studying a foreign language. Learning all the parts of the body, along with the composition of each part, and how each part fits with the other parts to make the whole requires memorization. Understanding the way each body part works individually, as well as body parts working together, requires a higher level of knowledge, comprehension, and application. Identifying underlying structural similarities, from the macroscopic to the microscopic levels of body organization, taps more subtle critical thinking skills. This chapter will catalyze success in this active process of learning. (Remember that while the skills and tips discussed in this chapter relate to learning anatomy and physiology, they can be applied to other subjects.)

Learning occurs in different ways or modes. Most students use several modes (multimodal), but are more comfortable and use more effectively one or two, often referred to as learning styles. Some students prefer to read the written word to remember it and the concept it describes or to actually write the words; others learn best by looking at visual representations, such as photographs and drawings. Still others learn most effectively by hearing the information or explaining it to someone else. For some learners, true understanding remains elusive until a principle is revealed in a laboratory or clinical setting that provides a memorable context and engages all of the senses. This text accommodates the range of learning styles. Read-write learners will appreciate the lists, definitions (glossary), and tables. Visual learners will discover many diagrams, flow charts, and figures, all with consistent and purposeful use of color. For example, a particular bone is always the same color in figures where bones are color coded. Auditory learners will find pronunciations for new scientific terms to help sound them out, and kinesthetic learners can relate real-life examples and applications to their own acitvities.

The first section of each chapter is an overview that tells you what to expect and why the subject is important.

After each major section, a question or series of questions tests your understanding of the material and enables you to practice using the new information. (Note the green practice arrow.) If you cannot answer the question(s) you should reread that section, being on the lookout for the answer(s).

 PRACTICE

1 List some difficulties a student may experience when studying the human body.

2 Describe the ways that people learn.

Major divisions within a chapter are called "A-heads." They are numbered sequentially, and are titled in very large turquoise type. A-heads identify major content areas.

P.2 | Strategies for Your Success

Many of the strategies for academic success are common sense, but it might help to review them. You may encounter new and helpful methods of learning.

The major divisions are subdivided into "B-heads," which are identified by large, purple type. These will help you organize the concepts upon which the major divisions are built.

Before Class

Before attending class, prepare by reading and outlining or taking notes on the assigned pages of the text. If outlining, leave adequate space between entries to allow room for note-taking during lectures. Or, fold each page of notes taken before class in half so that class notes can be written on the blank side of the paper across from the reading notes on the same topic. This strategy introduces the topics of the next class discussion, as well as new terms. Some students team a vocabulary list with each chapter's notes. Take the notes from the reading to class and expand them. At a minimum, the student should at least skim through the text, reading A-heads, B-heads, and the chapter summary to become acquainted with the topics and vocabulary before class.

Sometimes in your reading you will be directed back to a related concept, discussed in an earlier chapter, to help you better understand the new concept that is being explained.

RECONNECT
To Chapter 1, Homeostasis, pages 15–18.

Students using this book and taking various courses are often preparing for careers in health care. In some cases students may be undecided as to a specific area or specialty. The Career Corner presents a description of a particular career choice with each chapter. If it doesn't describe a career that you seek, perhaps it will give you a better sense of what some of your coworkers and colleagues do! You may even discover something exciting that you didn't know about!

In a "stroke," or *cerebrovascular accident* (CVA), a sudden interruption in blood flow in a vessel supplying brain tissues damages the cerebrum. The affected blood vessel may rupture, bleeding into the brain, or be blocked by a clot. In either case, brain tissues downstream from the vascular accident die and some loss of function may occur. Temporary interruption in cerebral blood flow, perhaps by a clot that quickly breaks apart, produces a much less serious *transient ischemic attack* (TIA), sometimes called a ministroke.

As you read, you may feel the need for a "study break." Sometimes you may just need to "chill out." Other times, you may just need to shift gears. Try the following! Throughout the book are shaded boxes within the flow of the text, Clinical Application boxes, and From Science to Technology boxes that present sidelights to the main focus of the text. Indeed, some of these may cover topics that your instructor chooses to highlight. Read them! They are interesting, informative, and a change of pace.

The opposite of looking back and reconnecting is looking ahead. A Glimpse Ahead applies concepts being discussed in the particular section of the text to future learning. This feature tells how the information learned here will carry over and be incorporated into understanding the functioning of other body systems.

A GLIMPSE AHEAD | To Chapters 9 and 10

The energy we must expend just to stay alive is called the basal metabolic energy. The body uses close to 40% of the basal metabolic energy to actively transport sodium and potassium ions across cell membranes. Imagine learning that 40% of your household budget went for one item—it had better be important! In this case it is. The concentration gradients for sodium and potassium ions that the sodium/potassium pumps establish throughout the body are essential for muscle and nerve cells to function. Chapters 9 and 10 further discuss the functioning of these important cell types.

CAREER CORNER
Radiologic Technologist

At age fifty-two the woman is younger than most of the others having their bone mineral density measured. The gynecologist advised a baseline test to assess the health of the patient's skeleton because her parents had osteoporosis.

A radiology technologist conducts the test. She explains the procedure to the patient, then positions her on her back on a padded table, fully clothed. The scanner passes painlessly over the patient's hip and lower spine, emitting low-dose X rays that form images of the bones. Spaces on the scan indicate osteopenia, the low bone mineral density that may be a prelude to osteoporosis.

Radiology technologists administer medical imaging tests, such as ultrasound and magnetic resonance imaging (MRI), as well as mammography and the three-dimensional X-ray cross sections of computerized tomography (CT). The technologists protect patients from radiation with drapes and positioning, and operate scanning devices to produce the highest-quality images from which a radiologist can accurately diagnose an illness or injury.

A registered radiologic technologist completes two years of training at a hospital or a two- or four-year program at a college or university, and must pass a national certification exam.

CLINICAL APPLICATION 9.3

TMJ Syndrome

Temporomandibular joint (TMJ) syndrome causes facial pain, headache, ringing in the ears, a clicking jaw, insomnia, teeth sensitive to heat or cold, backache, dizziness, and pain in front of the ears. A misaligned jaw or grinding or clenching the teeth can cause TMJ by stressing the temporomandibular joint, which is the articulation between the mandibular condyle of the mandible and the mandibular fossa of the temporal bone. Loss of coordination of these structures affects the nerves that pass through the neck and jaw region, causing the symptoms.

Getting enough sleep and drinking enough water can help prevent symptoms of TMJ, and eating soft foods, applying ice packs, using relaxation techniques, and massaging affected muscles can alleviate symptoms. A physical therapist can recommend exercises that stretch and relax the jaw, which may help some people. Sitting for long hours in one position can cause or worsen TMJ.

Doctors diagnose TMJ syndrome using an electromyograph, in which electrodes record muscle activity in four pairs of head and neck muscle groups. Several treatments are available. The National Institute of Dental and Craniofacial Research recommends that treatments not permanently alter the teeth or jaw. Low doses of certain antidepressants, or injections of botulinum toxin or corticosteroids, may help. Using a procedure called arthrocentesis, a physician might remove fluid accumulating in the affected joint. Another treatment is an oral appliance fitted by a dentist that fine-tunes the action of jaw muscles to form a more comfortable bite. An oral appliance, also known as a bite guard or stabilization splint, is a piece of plastic that fits over the top or bottom teeth. Very rarely, surgery may be required to repair or replace a joint.

FROM SCIENCE TO TECHNOLOGY 4.1

The Human Metabolome

A generation ago, prehealth profession students had to memorize a complex chart of biochemical pathways that represent all of the energy reactions in a cell. The cellular respiration pathways ran down the center, with branches radiating outward and in some places interconnecting into a giant web. Today, several technologies as well as the ability to store massive amounts of data have made possible the Human Metabolome Database.

"Metabolome" refers to all of the small molecules that are part of metabolism in a cell, tissue, organ, or an entire organism. The database is a vast, annotated catalog of those molecules, "metabolites." The government of Canada is supporting the effort to search all published papers and books that describe metabolites and link that information with experimental data. The techniques of electrophoresis and chromatography are used to separate metabolites, and mass spectrometry (MS) and

nuclear magnetic resonance (NMR) spectroscopy describe their chemical characteristics.

Biochemists estimate that human cells have at least 2,500 different metabolites, but fewer than half have been identified. Far fewer have been analyzed for their concentrations in different cell types under different conditions. In the Human Metabolome Database, each entry has an electronic "MetaboCard" that includes 90 data fields, half with clinical data (such as associated diseases and drug interactions) and half with biochemical data (such as pathways and enzymes that interact with the metabolite). Each entry is also hyperlinked to other databases, interfacing with 1,500 drugs and 3,600 foods and food additives. The information in the Human Metabolome Database is being used in drug discovery, toxicology, transplant monitoring, clinical chemistry, disease diagnosis, and screening of newborns for metabolic disorders.

Remember when you were very young and presented with a substantial book for the first time? You were likely intimidated by its length, but were reassured that there were "a lot of pictures." There are a lot of illustrations in this book as well, all designed to help you master the material.

Photographs and Line Art

Sometimes subdivisions have so many parts that the book goes to a third level, the "C-head." This division is identified in a slightly smaller, bold, black font.

Photographs provide a realistic view of anatomy.

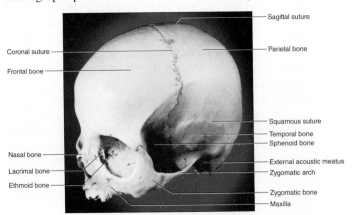

Line art can present different positions, layers, or perspectives.

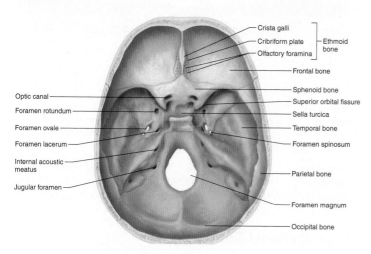

Flow Charts

Flow charts depict sequences of related events, steps of pathways, and complex concepts, easing comprehension. Other figures may show physiological processes.

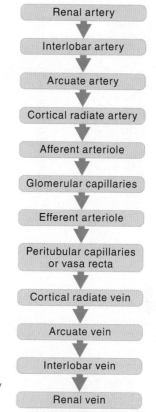

FIGURE 20.8
Pathway of blood through the blood vessels of the kidney and nephron.

Macroscopic to Microscopic

Many figures show anatomical structures in a manner that is macroscopic to microscopic (or vice versa).

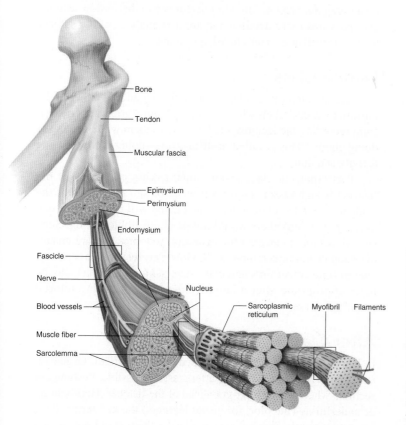

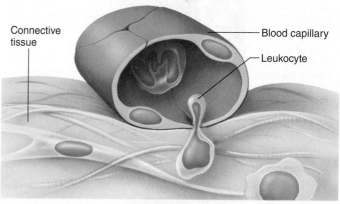

FIGURE 14.15 In a type of movement called diapedesis, leukocytes squeeze between the endothelial cells of a capillary wall and enter the tissue space outside the blood vessel.

 What is a monocyte called once it has left the bloodstream and entered the tissues?

Answer can be found in Appendix G.

Figure questions encourage you to think about what you are seeing and make connections between the visual representation and the words in the text.

Anatomical Structures

Some figures illustrate the locations of anatomical structures.

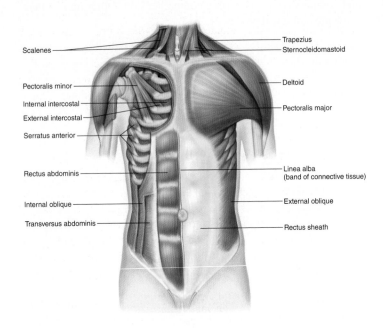

Other figures illustrate the functional relationships of anatomical structures.

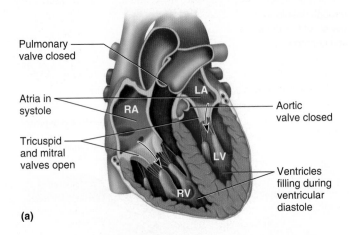

(a)

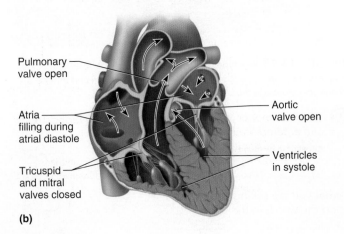

(b)

TABLE 5.4	Exocrine Glandular Secretions	
Type	Description of Secretion	Example
Merocrine glands	A fluid product released through the cell membrane by exocytosis	Salivary glands, pancreatic glands, sweat glands of the skin
Apocrine glands	Cellular product and portions of the free ends of glandular cells pinch off during secretion	Mammary glands, ceruminous glands lining the external acoustic meatus
Holocrine glands	Disintegrated entire cells filled with secretory products	Sebaceous glands of the skin

Organizational Tables

Organizational tables can help "put it all together," but are not a substitute for reading the text or having good notes.

It is critical that you attend class regularly, and be on time—even if the instructor's notes are posted online, and the information is in the textbook. For many learners, hearing and writing new information is a better way to retain facts than just scanning notes on a computer screen. Attending lectures and discussion sections also provides more detailed and applied analysis of the subject matter, as well as a chance to ask questions.

During Class

Be alert and attentive in class. Take notes by adding either to the outline or notes taken while reading. Auditory learners benefit from recording the lectures and listening to them while driving or doing chores. This is called **multitasking**—doing more than one activity at a time.

Participate in class discussions, asking questions of the instructor and answering questions he or she poses. All of the students are in the class to learn, and many will be glad someone asked a question others would not be comfortable asking. Such student response can alert the instructor to topics that are misunderstood or not understood at all. However, respect class policy. Due to time constraints and class size, asking questions may be more appropriate after a large lecture class or during tutorial (small group) sessions.

After Class

In learning complex material, expediency is critical. Organize, edit, and review notes as soon after class as possible, fleshing out sections where the lecturer got ahead of the listener. Highlighting or underlining (in color, for visual learners) the key terms, lists, important points and major topics make them stand out, which eases both daily reviews and studying for exams.

Lists

Organizing information into lists or categories can minimize information overload, breaking it into manageable chunks. For example, when studying the muscles of the thigh it is easier to learn the insertion, origin, action, and nerve supply of the four muscles making up the *quadriceps femoris* as a group, because they all have the same insertion, action at the knee, and nerve supply . . . they differ only in their origins.

Mnemonic Devices

Another method for remembering information is the **mnemonic device.** One type of mnemonic device is a list of words, forming a phrase, in which the first letter of each word corresponds to the first letter of each word that must be remembered. For example, *Frequent parade often tests soldiers' endurance* stands for the skull bones *f*rontal, *p*arietal, *o*ccipital, *t*emporal, *s*phenoid, and *e*thmoid. Another type of mnemonic device is a word formed by the first letters of the items to be remembered. For example, *ipmat* represents the stages in the cell cycle: *i*nterphase, *p*rophase, *m*etaphase, *a*naphase, and *t*elophase.

Study Groups

Forming small study groups helps some students. Together the students review course material and compare notes. Working as a team and alternating leaders allows students to verbalize the information. Individual students can study and master one part of the assigned material, and then explain it to the others in the group, which incorporates the information into the memory of the speaker. Hearing the material spoken aloud also helps the auditory learner. Be sure to use anatomical and physiological terms, in explanations and everyday conversation, until they become part of your working vocabulary, rather than intimidating jargon. Most important of all—the group must stay on task, and not become a vehicle for social interaction. Your instructor may have suggestions or guidelines for setting up study groups.

Flash Cards

Flash cards may seem archaic in this computer age, but they are still a great way to organize and master complex and abundant information. The act of writing or drawing on a note card helps the

tactile learner. Master a few new cards each day, and review cards from previous days, and use them all again at the end of the semester to prepare for the comprehensive final exam. They may even come in handy later, such as in studying for exams for admission to medical school or graduate school. Divide your deck in half and flip half of the cards so that the answer rather than the question is showing. Mix them together and shuffle them. Get used to identifying a structure or process from a description as well as giving a description when provided with a process or structure. This is more like what will be expected of you in the real world of the health-care professional.

Manage Your Time

Many of you have important obligations outside of class, such as jobs and family responsibilities. As important as these are, you still need to master this material on your path to becoming a health-care professional. Good time management skills are therefore essential in your study of human anatomy and physiology. In addition to class, lab, and study time, multitask. Spend time waiting for a ride, or waiting in a doctor's office, reviewing notes or reading the text.

Daily repetition is helpful, so scheduling several short study periods each day can replace a last-minute crunch to cram for an exam. This does not take the place of time to prepare for the next class. Thinking about these suggestions for learning now can maximize study time throughout the semester, and, hopefully, lead to academic success. A working knowledge of the structure and function of the human body provides the foundation for all careers in the health sciences.

 PRACTICE

3 Why is it important to prepare before attending class?

4 Name two ways to participate in class discussions.

5 List several aids for remembering information.

Chapter Summary

A summary of the chapter provides an outline to review major ideas and is a tool for organizing thoughts.

P.1 Overview (page 2)

Try a variety of methods to study the human body.

P.2 Strategies for Success (page 2)

While strategies for academic success seem to be common sense, you might benefit from reminders of study methods.

1. Before class
   Read the assigned text material prior to the corresponding class meeting.
   a. Reconnects refer back to helpful, previously discussed concepts.
   b. A Glimpse Ahead applies current learning to future topics.
   c. Shaded boxes present sidelights to the main focus of the text.
   d. Photographs, line art, flow charts, and organizational tables help in mastery of the materials.

2. During class
   Take notes and participate in class discussions.
3. After class
   a. Organize, edit, and review class notes.
   b. Mnemonic devices aid learning.
      (1) The first letters of the words to remember begin words of an easily recalled phrase.
      (2) The first letters of the items to be remembered form a word.
   c. Small study groups reviewing and vocalizing material can divide and conquer the learning task.
   d. Flash cards help the tactile learner.
   e. Time management skills encourage scheduled studying, including daily repetition instead of cramming for exams.

CHAPTER ASSESSMENTS

Chapter assessments that are tied directly to the learning outcomes allow you to self assess your mastery of the material. (Note the purple assess arrow.)

P.1 Overview

1 Explain how students learn in different ways. (p. 2)

P.2 Strategies for Success

2 Methods to prepare for class include _____. (p. 2)
   a. reading the chapter
   b. outlining the chapter
   c. taking notes on the assigned reading
   d. making a vocabulary list
   e. all of the above

3 Describe how you can participate in class discussions. (p. 6)
4 Forming the phrase "*I passed my anatomy test.*" To remember the cell cycle (interphase, prophase, metaphase, anaphase, telophase) is an example of a _____. (p. 7)
5 Name a benefit and a drawback of small study groups. (p. 7)
6 Explain the value of repetition in learning and preparation for exams. (p. 7)

A textbook is inherently linear. This text begins with chapter 1 and ends with chapter 24. Understanding physiology and the significance of anatomy, however, requires you to be able to recall previous concepts. Critical thinking is all about linking previous concepts with current concepts under novel circumstances, in new ways. Toward this end, we have included in the Integrative Assessment/Critical Thinking section references to sections from earlier chapters. Making connections is what it is all about!

INTEGRATIVE ASSESSMENTS/CRITICAL THINKING

Outcomes P.1, P.2

1. Which study methods are most successful for you?

Outcome P.2

2. Design a personalized study schedule.

Check out McGraw-Hill online resources that can help you practice and assess your learning.

McGraw-Hill Connect® Interactive Questions Reinforce your knowledge using assigned interactive questions.

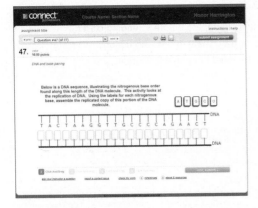

Connect Integrated Activity Practice your understanding.

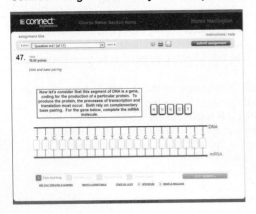

McGraw-Hill LearnSmart® Discover which concepts you have mastered and which require more attention with this personalized, adaptive learning tool.

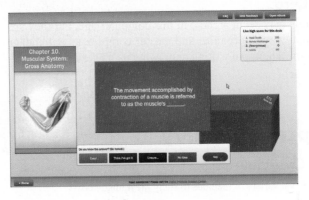

McGraw-Hill Anatomy & Physiology | REVEALED®
Go more in depth using virtual dissection of a cadaver.

1

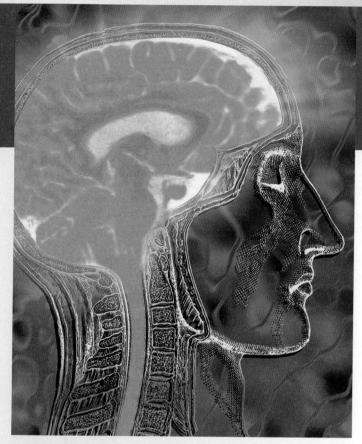

Your brain enables you to learn, to practice, and to assess your understanding—whether of a textbook, or how to handle a medical emergency.

Introduction to Human Anatomy and Physiology

 LEARNING OUTCOMES

After you have studied this chapter, you should be able to:

1.1 Origins of Medical Science
1 Identify some of the early discoveries that lead to our current understanding of the human body. (p. 10)

1.2 Anatomy and Physiology
2 Explain how anatomy and physiology are related. (p. 11)

1.3 Levels of Organization
3 List the levels of organization in the human body and the characteristics of each. (p. 12)

1.4 Characteristics of Life
4 List and describe the major characteristics of life. (p. 14)
5 Give examples of *metabolism*. (p. 14)

1.5 Maintenance of Life
6 List and describe the major requirements of organisms. (p. 14)
7 Explain the importance of homeostasis to survival. (p. 15)
8 Describe the parts of a homeostatic mechanism and explain how they function together. (p. 16)

1.6 Organization of the Human Body
9 Identify the locations of the major body cavities. (p. 18)
10 List the organs located in each major body cavity. (p. 18)
11 Name and identify the locations of the membranes associated with the thoracic and abdominopelvic cavities. (p. 20)
12 Name the major organ systems, and list the organs associated with each. (p. 22)
13 Describe the general function of each organ system. (p. 22)

1.7 Life-Span Changes
14 Identify changes related to aging, from the microscopic to the whole-body level. (p. 27)

1.8 Anatomical Terminology
15 Properly use the terms that describe relative positions, body sections, and body regions. (p. 27)

THE WHOLE PICTURE

Human anatomy and physiology are the studies of the human body and how it works. Our bodies are communities of cells, which are the microscopic units of living organisms. Cells are specialized to take on specific and necessary responsibilities, and together they maintain an environment within the body in which they can all live.

Learning anatomy and physiology requires familiarity with the language used to describe structures and functions. Cells aggregate and interact to form tissues, which in turn layer and fold and intertwine to form organs, which in turn connect into organ systems.

Mastering the principles of anatomy and physiology not only will give you a new appreciation for your day-to-day activities, talents, strengths, and health, but will provide a foundation for you to help your future patients, for those of you going into health care.

aprevealed.com

Module 1: Body Orientation

append-, to hang something: *appendicular*—pertaining to the upper limbs and lower limbs.

cardi-, heart: peri*cardium*—membrane that surrounds the heart.

cerebr-, brain: *cerebrum*—largest part of the brain.

cran-, helmet: *cranial*—pertaining to the part of the skull that surrounds the brain.

dors-, back: *dorsal*—position toward the back of the body.

homeo-, same: *homeo*stasis—maintenance of a stable internal environment.

-logy, the study of: physio*logy*—study of body functions.

meta-, change: *metabolism*—chemical changes in the body.

nas-, nose: *nasal*—pertaining to the nose.

orb-, circle: *orbital*—pertaining to the portion of the skull that encircles an eye.

pariet-, wall: *parietal* membrane—membrane that lines the wall of a cavity.

pelv-, basin: *pelvic* cavity—basin-shaped cavity enclosed by the pelvic bones.

peri-, around: *peri*cardial membrane—membrane that surrounds the heart.

pleur-, rib: *pleural* membrane—membrane that encloses the lungs within the rib cage.

-stasis, standing still: homeo*stasis*—maintenance of a stable internal environment.

super-, above: *superior*—referring to a body part located above another.

-tomy, cutting: ana*tomy*—study of structure, which often involves cutting or removing body parts.

1.1 | Origins of Medical Science

Our understanding of the human body has a long and interesting history (fig. 1.1). Our earliest ancestors must have been curious about how their bodies worked. At first they probably thought mostly about injuries and illnesses, because healthy bodies demand little attention from their owners. Just as we do today, primitive people suffered aches and pains, injured themselves, bled, broke bones, developed diseases, and contracted infections.

At first, healers relied heavily on superstitions and notions about magic. However, as they tried to help the sick, these early medical workers began to discover useful ways of examining and treating the human body. They observed the effects of injuries, noticed how wounds healed, and examined dead bodies to determine the causes of death. They also found that certain herbs and potions could relieve coughs, headaches, and other common problems. These long-ago physicians began to wonder how these substances, the forerunners of modern drugs, affected body functions.

People began asking more questions and seeking answers, setting the stage for the development of modern medical science. Techniques for making accurate observations and performing careful experiments evolved, allowing knowledge of the human body to expand rapidly.

This new knowledge of the structure and function of the human body required a new, specialized language. Early medical providers devised many terms to name body parts, describe their locations, and explain their functions. These terms, most of which originated from Greek and Latin, formed the basis for the language of anatomy and physiology. (A list of some of the modern medical and applied sciences appears on pp. 30–32.)

Study of corpses was forbidden in Europe during the Middle Ages, but dissection of dead bodies became a key part of medical education in the twentieth century. Today, cadaver dissection remains an important method to learn how the body functions and malfunctions, and autopsies are commonly depicted on television crime dramas. However, the traditional gross anatomy course in medical schools is sometimes supplemented with learning from body parts already dissected by instructors (in contrast to students doing this) as well as with computerized scans of cadavers, such as the Visible Human Project from the National Library of Medicine and Anatomy and Physiology Revealed available with this textbook.

Much of what is known now about the human body is based on *scientific method*, an approach to investigating the natural world. It is part of a general process called scientifc inquiry. Scientific

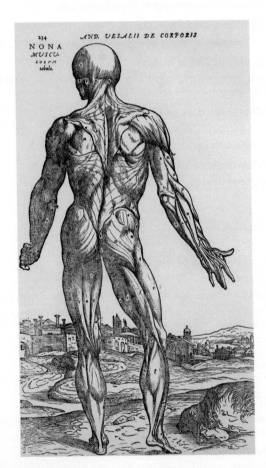

FIGURE 1.1 The study of the human body has a long history, as this illustration from the second book of *De Humani Corporis Fabrica* by Andreas Vesalius, issued in 1543, indicates. Note the similarity to the anatomical position (described on p. 27).

method consists of testing a hypothesis and then rejecting or accepting it, based on the results of experiments or observations. This method is described in greater detail in Appendix A, Scientific Method, but it is likely that aspects of its application are already familiar.

Imagine buying a used car. The dealer insists it is in fine shape, but the customer discovers that the engine doesn't start. That's an experiment! It tests the hypothesis: If this car is in good shape, then it will start. When the car doesn't start, the wary consumer rejects the hypothesis and doesn't buy the car.

Rather than giving us all the answers, science eliminates wrong explanations. Our knowledge of the workings of the human body reflects centuries of asking questions, testing, rejecting, and sometimes accepting hypotheses. New technologies provide new views of anatomy and physiology, so that knowledge is always growing. One day you may discover something previously unknown about the human body!

 PRACTICE

1 What factors probably stimulated an early interest in the human body?

2 What types of activities helped promote the development of modern medical science?

3 What is the role of a hypothesis in the scientific method?

1.2 | Anatomy and Physiology

Two major areas of medical science, **anatomy** (ah-nat'o-me) and **physiology** (fiz"e-ol'o-je), address how the body maintains life. Anatomy, from the Greek for "a cutting up," examines the **structures,** or morphology, of body parts—their forms and organization. Physiology, from the Greek for "relationship to nature," considers the **functions** of body parts—what they do and how they do it. Although anatomists rely more on examination of the body and physiologists more on experimentation, together their efforts have provided a solid foundation for understanding how our bodies work.

It is difficult to separate the topics of anatomy and physiology because anatomical structures make possible their functions. Body parts form a well-organized unit—the **human organism.** Each part contributes to the operation of the unit as a whole. This functional role arises from the way the part is constructed. For example, the arrangement of bones and muscles in the human hand, with its long, jointed fingers, makes grasping possible. The heart's powerful muscular walls contract and propel blood out of the chambers and into blood vessels, and heart valves keep blood moving in the proper direction. The shape of the mouth enables it to receive food; tooth shapes enable teeth to break solid foods into pieces; and the muscular tongue and cheeks are constructed in a way that helps mix food particles with saliva and prepare them for swallowing (fig. 1.2).

As ancient as the fields of anatomy and physiology are, we are always learning more. For example, researchers recently used imaging technology to identify a previously unrecognized part of the brain, the planum temporale, which enables people to locate sounds in space. Many discoveries today begin with investigations

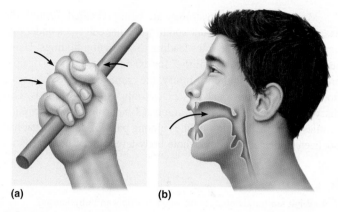

(a)   (b)

FIGURE 1.2 The structures of body parts make possible their functions: (a) The hand is adapted for grasping and (b) the mouth for receiving food. (Arrows indicate movements associated with these functions.)

at the molecular or cellular level. In this way, researchers have discovered that certain cells in the small intestine bear the same taste receptor proteins found on the tongue. At both locations, the receptors detect molecules of sugar. The cells in the tongue provide taste sensations, whereas the cells in the intestines help regulate the digestion of sugar. The discovery of the planum temporale is anatomical; the discovery of sweet receptors in the intestine is physiological.

 CAREER CORNER
Emergency Medical Technician

The driver turns a corner and suddenly swerves as a cat dashes into the road. She slams on the brakes but hits a parked car, banging her head against the steering wheel. Onlookers call 911, and within minutes an ambulance arrives.

The driver of the ambulance and another emergency medical technician (EMT) leap out and run over to the accident scene. They open the driver side door and quickly assess the woman's condition by taking her vital signs. She is bleeding from a laceration on her forehead, and is conscious but confused.

The EMTs carefully place a restraint at the back of the woman's neck and move her onto a board, then slide her into the ambulance. While one EMT drives, the other rides in the back with the patient and applies pressure to the cut. At the hospital the EMTs document the care provided and clean and restock the ambulance.

EMTs care for ill or injured people in emergency situations, and transport patients, such as from a hospital to a nursing home. The work is outdoors and indoors and requires quick thinking as well as strength. Requirements vary by state, but all EMTs must be licensed. Basic EMTs take 120 to 150 hours of training; paramedic EMTs take 1,200 to 1,800 hours of training. Paramedics may give injections, set up intravenous lines, and give more medications than can basic EMTs.

Many nuances of physiology are being revealed through the examination of genes that function in particular cell types under particular conditions, sometimes leading to surprising findings. Using such "gene expression profiling," for example, researchers discovered that after a spinal cord injury, the damaged tissue releases a flood of proteins previously associated only with skin wounds. This discovery suggests new drug targets. Comparing gene expression profiles can reveal commonalities among pairs of diseases that had not been suspected based on whole-body-level observations.

 PRACTICE

4 What are the differences between anatomy and physiology?

5 Why is it difficult to separate the topics of anatomy and physiology?

6 List several examples that illustrate how the structure of a body part makes possible its function.

7 How are anatomy and physiology both old and new fields?

1.3 | Levels of Organization

Early investigators, limited in their ability to observe small structures such as cells, focused their attention on larger body parts. Studies of small structures had to await invention of magnifying lenses and microscopes, about 400 years ago. These tools revealed that larger body structures were made up of smaller parts, which, in turn, were composed of even smaller ones.

All materials, including those that comprise the human body, are composed of chemicals. Chemicals consist of tiny particles called **atoms,** which are composed of even smaller **subatomic particles.** Atoms can join to form **molecules,** and small molecules may combine to form larger **macromolecules.**

In humans and other organisms, the basic unit of structure and function is a **cell.** Although individual cells vary in size and shape, all share certain characteristics. Cells of complex organisms, including those of humans, contain structures called **organelles** (or"gan-elz') that carry on specific activities. Organelles are composed of assemblies of large molecules, including proteins, carbohydrates, lipids, and nucleic acids. Most human cells contain a complete set of genetic instructions, yet use only a subset of them, allowing cells to specialize. All cells share the same characteristics of life and must meet certain requirements to stay alive.

Specialized cells assemble into layers or masses that have specific functions. Such a group of cells is called a **tissue.** Groups of different tissues form **organs**—complex structures with specialized functions—and groups of organs that function closely together comprise **organ systems.** Interacting organ systems make up an **organism.**

A body part can be described at different levels. The heart, for example, consists of muscle, fat, and nervous tissue. These tissues, in turn, are constructed of cells, which contain organelles. All of the structures of life are, ultimately, composed of chemicals (fig. 1.3). Clinical Application 1.1 describes two technologies used to visualize body parts based on body chemistry.

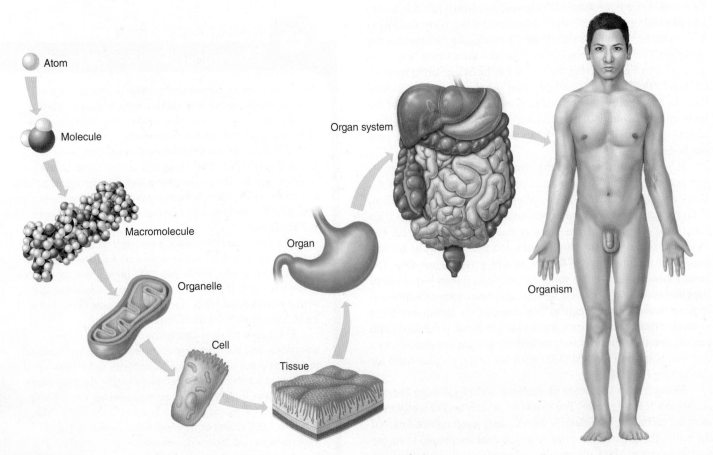

FIGURE 1.3 The human body is composed of parts within parts, with increasing complexity.

The two patients enter the hospital medical scanning unit hoping for opposite outcomes. Vanessa Q., who has suffered several pregnancy losses, hopes that an ultrasound exam will reveal that her current pregnancy is progressing normally. Michael P., a sixteen-year-old who has excruciating headaches, is to undergo a magnetic resonance (MR) scan to assure his physician (and himself!) that the cause of the headaches is not a brain tumor.

Ultrasound and magnetic resonance scans are noninvasive procedures that provide images of soft internal structures. Ultrasonography uses high-frequency sound waves beyond the range of human hearing. A technician gently presses a device called a transducer, which emits sound waves, against the skin and moves it slowly over the surface of the area being examined, which in this case is Vanessa's abdomen (fig. 1A).

Prior to the exam, Vanessa drank several glasses of water. Her filled bladder will intensify the contrast between her uterus (and its contents) and nearby organs because as the sound waves from the transducer travel into the body, some of the waves reflect back to the transducer when they reach a border between structures of slightly different densities. Other sound waves continue into deeper tissues, and some of them are reflected back by still other interfaces. As the reflected sound waves reach the transducer, they are converted into electrical impulses that are amplified and used to create a sectional image of the body's internal structure on a viewing screen. This image is a sonogram (fig. 1B).

Glancing at the screen, Vanessa smiles. The image reveals the fetus in her uterus, heart beating and already showing budlike structures that will develop into arms and legs. She happily heads home with a video of the fetus.

Vanessa's ultrasound exam takes only a few minutes, whereas Michael's MR scan takes an hour. First, Michael receives an injection of a dye that provides contrast so that a radiologist examining the scan can distinguish certain brain structures. Then, the motorized platform on which Michael lies moves into a chamber surrounded by a powerful magnet and a special radio antenna. The chamber, which looks like a metal doughnut, is the MR imaging instrument. As Michael settles back, closes his eyes, and listens to the music through earphones, a technician activates the device.

The magnet generates a magnetic field that alters the alignment and spin of certain types of atoms within Michael's brain. At the same time, a second rotating magnetic field causes particular types of atoms (such as the hydrogen atoms in body fluids and organic compounds) to release weak radio waves with characteristic frequencies. The nearby antenna receives and amplifies the radio waves, which are then processed by a computer. Within a few minutes, the computer generates a sectional image based on the locations and concentrations of the atoms being studied (fig. 1C). The device continues to produce data, painting portraits of Michael's brain from different angles.

Michael and his parents nervously wait two days for the expert eyes of a radiologist to interpret the MR scan. Happily, the scan shows normal brain structure. Whatever is causing Michael's headaches, it is not a brain tumor—at least not one large enough to be imaged.

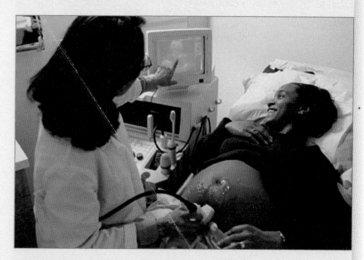

FIGURE 1A Ultrasonography uses reflected sound waves to visualize internal body structures.

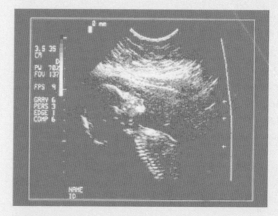

FIGURE 1B This image resulting from an ultrasonographic procedure reveals a fetus in the uterus.

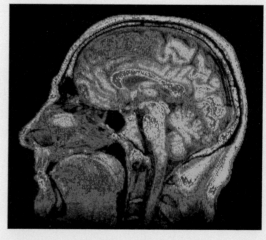

FIGURE 1C This falsely colored MR image of a human head shows the brain (sagittal section, see fig. 1.23).

TABLE 1.1	Levels of Organization	
Level	Example(s)	Representative Illustration(s)
Subatomic particles	Electrons, protons, neutrons	Figure 2.1
Atom	Hydrogen atom, lithium atom	Figure 2.3
Molecule	Water molecule, glucose molecule	Figures 2.7, 2.11
Macromolecule	Protein molecule, DNA molecule	Figures 2.19, 2.21
Organelle	Mitochondrion, Golgi apparatus, nucleus	Figure 3.3
Cell	Muscle cell, nerve cell	Figures 5.29, 5.32
Tissue	Simple squamous epithelium, bone	Figures 5.3, 5.27
Organ	Skin, femur, heart, kidney	Figures 6.2, 7.52, 15.2, 20.1
Organ system	Integumentary system, skeletal system, digestive system	Figure 1.20
Organism	Human	Figure 1.20

Chapters 2–6 discuss the levels of organization of the human body in greater detail. Chapter 2 describes the atomic and molecular levels; chapter 3 presents organelles and cellular structures and functions; chapter 4 explores cellular metabolism; chapter 5 describes tissues; and chapter 6 presents the skin and its accessory organs as an example of an organ system. The remaining chapters describe the structures and functions of the other organ systems in detail. Table 1.1 lists the levels of organization and some corresponding illustrations in this textbook.

 PRACTICE

8 How does the human body illustrate levels of organization?

9 What is an organism?

10 How do body parts at different levels of organization vary in complexity?

1.4 | Characteristics of Life

We think of the qualities that constitute the state of being alive at moments like the birth of a baby, as an injury happens, or at the time of death following a long illness. Although this textbook addresses the human body, all types of organisms share the most fundamental characteristics of life.

As living organisms, we can respond to our surroundings. Our bodies grow, eventually becoming able to reproduce. We gain energy by ingesting (taking in), digesting (breaking down), absorbing, and assimilating the nutrients in food. The absorbed substances circulate throughout the internal environment of our bodies. We can then, by the process of respiration, use the energy in these nutrients for such vital functions as growth and repair. Finally, we excrete wastes. Taken together, these physiological events that obtain, release, and use energy are a major part of **metabolism** (mĕ-tab′o-liz-m), the collection of chemical reactions in cells that support life. Table 1.2 summarizes the characteristics of life.

 PRACTICE

11 What are the characteristics of life?

12 Which physiological events constitute metabolism?

1.5 | Maintenance of Life

Nearly all body structures and functions work in ways that maintain life. The exception is an organism's reproductive system, which perpetuates the species.

Requirements of Organisms

Human life depends upon the following environmental factors:

1. **Water** is the most abundant substance in the body. It is required for a variety of metabolic processes, and it provides the environment in which most of them take place. Water also carries substances in organisms and is important in regulating body temperature.
2. **Food** refers to substances that provide organisms with necessary chemicals (nutrients) in addition to water. Nutrients supply energy and raw materials for building new living matter.

TABLE 1.2	Characteristics of Life			
Process	Examples	Process	Examples
Movement	Change in position of the body or of a body part; motion of an internal organ	Digestion	Breakdown of food substances into simpler forms that can be absorbed and used
Responsiveness	Reaction to a change inside or outside the body	Absorption	Passage of substances through membranes and into body fluids
Growth	Increase in body size without change in shape	Circulation	Movement of substances in body fluids
Reproduction	Production of new organisms and new cells	Assimilation	Changing of absorbed substances into different chemical forms
Respiration	Obtaining oxygen, removing carbon dioxide, and releasing energy from foods (some forms of life do not use oxygen in respiration)	Excretion	Removal of wastes produced by metabolic reactions

TABLE 1.3	Requirements of Organisms				
Factor	Characteristic	Use	Factor	Characteristic	Use
Water	A chemical substance	For metabolic processes, as a medium for metabolic reactions, to transport substances, and to regulate body temperature	Heat	A form of energy	To help regulate the rates of metabolic reactions
Food	Various chemical substances	To supply energy and raw materials for the production of necessary substances and for the regulation of vital reactions	Pressure	A force	Atmospheric pressure for breathing; hydrostatic pressure to help circulate blood
Oxygen	A chemical substance	To help release energy from food substances			

3. **Oxygen** is a gas that makes up about one-fifth of the air. It is used to release energy from nutrients. The energy, in turn, is used to drive metabolic processes.

4. **Heat** is a form of energy present in our environment. It is also a product of metabolic reactions and our body temperature depends in part on heat from the chemical reactions taking place in the body. Furthermore, the amount of heat present in the body partly controls the rate at which these reactions occur. *Temperature* is a measure of the amount of heat.

5. **Pressure** is an application of force on an object or substance. For example, the force acting on the outside of a land organism due to the weight of air above it is called *atmospheric pressure*. In humans, this pressure plays an important role in breathing. Similarly, organisms living under water are subjected to *hydrostatic pressure*—a pressure a liquid exerts—due to the weight of water above them. In complex animals, such as humans, heart action produces blood pressure (another form of hydrostatic pressure), which keeps blood flowing through blood vessels.

The human organism requires water, food, oxygen, heat, and pressure, but these factors alone are not enough to ensure survival. Both the quantities and the qualities of such factors are also important. Table 1.3 summarizes the major requirements of organisms.

Homeostasis

Homeostasis (ho″me-ō-sta′sis) refers to the body's ability to keep its internal conditions stable, such that its cells can survive. To this end, all cells, whether as part of a tissue, an organ, or an organ system, make some specific contribution.

Most of the earth's residents are unicellular, or single-celled. The most ancient and abundant unicellular organisms are the bacteria. Their cells do not have membrane-bound organelles. Some unicellular organisms have organelles that are as complex as our own. This is the case for the amoeba (fig. 1.4). It survives and reproduces as long as its lake or pond environment is of a tolerable temperature and composition, and the amoeba can obtain food. With a limited ability to move, the amoeba depends upon the conditions in its lake or pond environment to stay alive.

In contrast to the amoeba, adult humans are composed of about 37 trillion cells that maintain their own environment—our bodies.

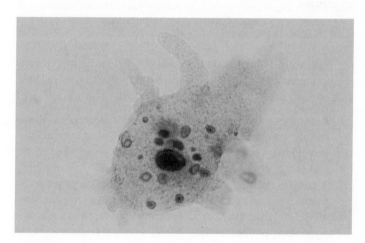

FIGURE 1.4 The amoeba is an organism consisting of a single, complex cell (100×).

Our cells, as parts of organs and organ systems, interact in ways that keep this **internal environment** relatively constant, despite an ever-changing outside environment.

The internal environment is so-named because it is found *inside* our bodies. It consists of the water and dissolved substances *outside* of our cells, called the **extracellular fluid.** Extracellular fluid includes the liquid portion of the blood, the **plasma,** and the **interstitial fluid,** or tissue fluid, which is found outside of the blood vessels. Interstitial fluid is in direct contact with cells throughout the body. All living cells contain water, along with substances dissolved in it, called **intracellular fluid.** Structures called *cell membranes* separate intracellular and extracellular fluids. They are discussed in chapter 3 (pp. 86–89). The internal environment protects our cells (and us!) from external changes (fig. 1.5).

Homeostasis is so important that it requires most of our metabolic energy. The interstitial fluid, which bathes cells in the body, is the environment to which those cells are most directly exposed, but the composition of the interstitial fluid is in equilibrium with the composition of the blood plasma, so both contribute to the internal environment. This relationship also explains why a simple blood test can provide important information about what is going on in the body's internal environment.

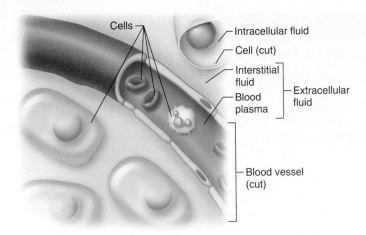

FIGURE 1.5 Intracellular and extracellular fluids. The extracellular fluid constitutes the internal environment of the body.

The body maintains homeostasis through a number of self-regulating control systems, or **homeostatic mechanisms.** These mechanisms share three components (fig. 1.6):

1. **Receptors** are on the lookout. They provide information about specific conditions (stimuli) in the internal environment. A receptor may be as small as a cell or even a protein that is part of a cell.
2. A **control center,** or decision-maker, that includes a **set point,** which is a particular value, such as body temperature at 37°C (Celsius) or 98.6°F (Fahrenheit). Note: More metric equivalents can be found in Appendix B, Metric Measurement System and Conversions. Metric units are used throughout this text.
3. **Effectors,** such as muscles or glands, take action. They cause appropriate responses.

Most homeostatic mechanisms operate by a process called **negative feedback.** In a negative feedback mechanism, effectors are activated that can return conditions toward normal. As this happens, the deviation from the set point progressively lessens, and the effectors gradually shut down. Such a response is called a negative feedback mechanism because the deviation from the set point is corrected (moves in the opposite or negative direction)

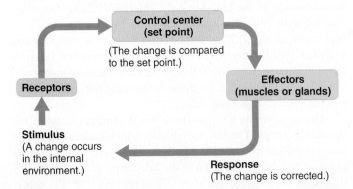

FIGURE 1.6 A homeostatic mechanism monitors a particular aspect of the internal environment and corrects any changes back to the value indicated by the set point.

and because the correction reduces the action of the effectors. This latter aspect is important because it prevents a correction from going too far.

To better understand the idea of maintaining a stable internal environment, imagine a room equipped with a furnace and an air conditioner. If the room temperature is to remain near 20°C (68°F) despite changes in the outside temperature, the thermostat must be adjusted to a set point of 20°C. A thermostat is sensitive to temperature changes, so it will signal the furnace to start and the air conditioner to stop whenever the room temperature drops below the set point. If the temperature rises above the set point, the thermostat will stop the furnace and start the air conditioner. These actions maintain a relatively constant temperature in the room (fig. 1.7).

A homeostatic mechanism similar to a thermostat regulates body temperature in humans (fig. 1.8). The body's "thermostat" is a temperature-sensitive region in a control center of the brain called the hypothalamus. In healthy persons, the set point of this body thermostat is at or near 37°C (98.6°F).

If a person becomes overheated, thermoreceptors (temperature receptors) throughout the body detect the change, and in response the hypothalamus initiates a series of actions that dissipate body heat. Sweat glands in the skin secrete watery perspiration that evaporates from the surface, carrying away heat and cooling the skin. At the same time, blood vessels in the skin dilate. This allows more blood that carries heat from deeper tissues to reach the surface, where the heat is lost to the outside.

If a person is exposed to a cold environment and the body temperature begins to drop, thermoreceptors detect the change and the hypothalamus initiates heat-conserving and heat-generating activities. Blood vessels in the skin constrict, reducing blood flow and enabling deeper tissues to retain heat. At the same time, small groups of muscle cells may be stimulated to contract involuntarily, an action called shivering that produces heat, which helps warm the body. Chapter 6 discusses body temperature regulation in more detail (p. 189).

Another homeostatic mechanism regulates the concentration of the sugar glucose in the blood. In this case, cells of an organ called the pancreas determine the set point. If the concentration of blood glucose increases following a meal, the pancreas detects this change and releases a chemical (insulin) into the blood. Insulin allows glucose to move from the blood into various body cells and to be stored in the liver and muscles. As this occurs, the concentration of blood glucose decreases, and as it reaches the normal set point, the pancreas decreases its release of insulin. If, on the other hand, blood glucose concentration falls too low, the pancreas detects this change and secretes a different chemical (glucagon) that releases stored glucose into the blood. Chapter 13 (pp. 514–515) discusses regulation of blood glucose concentration in more detail (see fig. 13.36, p. 515).

Human physiology offers many other examples of negative feedback mechanisms, which all work in the same basic way just described. Just as many anatomical terms are used in all areas of anatomy, so the basic principles of physiology apply in all organ systems.

In some cases, homeostatic mechanisms operate by **positive feedback,** in which a change is not reversed but intensified, and

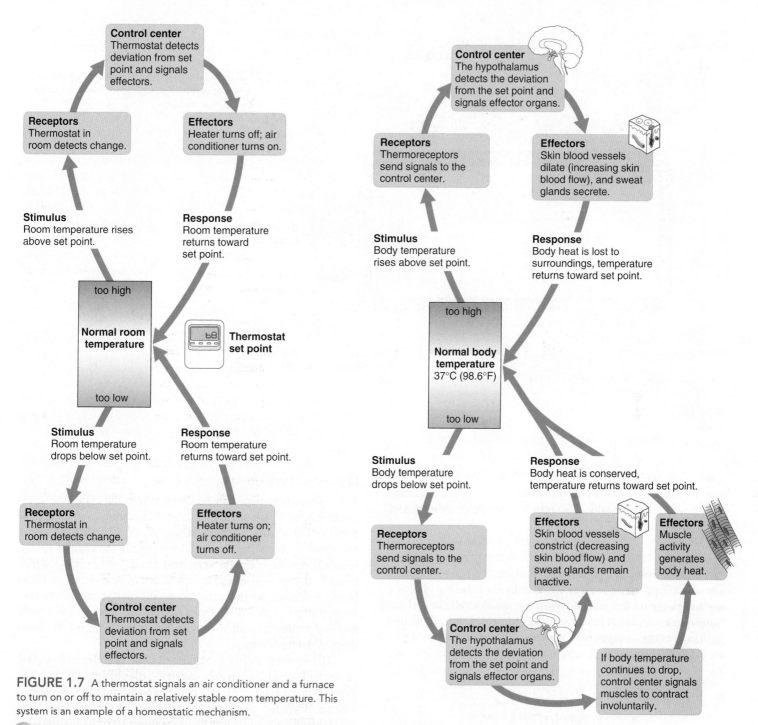

Control center
Thermostat detects deviation from set point and signals effectors.

Receptors
Thermostat in room detects change.

Effectors
Heater turns off; air conditioner turns on.

Stimulus
Room temperature rises above set point.

Response
Room temperature returns toward set point.

too high

Normal room temperature

too low

Thermostat set point

Stimulus
Room temperature drops below set point.

Response
Room temperature returns toward set point.

Receptors
Thermostat in room detects change.

Effectors
Heater turns on; air conditioner turns off.

Control center
Thermostat detects deviation from set point and signals effectors.

FIGURE 1.7 A thermostat signals an air conditioner and a furnace to turn on or off to maintain a relatively stable room temperature. This system is an example of a homeostatic mechanism.

Q *What would happen to room temperature if the set point were turned up?*

Answer can be found in Appendix G.

Control center
The hypothalamus detects the deviation from the set point and signals effector organs.

Receptors
Thermoreceptors send signals to the control center.

Effectors
Skin blood vessels dilate (increasing skin blood flow), and sweat glands secrete.

Stimulus
Body temperature rises above set point.

Response
Body heat is lost to surroundings, temperature returns toward set point.

too high

Normal body temperature
37°C (98.6°F)

too low

Stimulus
Body temperature drops below set point.

Response
Body heat is conserved, temperature returns toward set point.

Receptors
Thermoreceptors send signals to the control center.

Effectors
Skin blood vessels constrict (decreasing skin blood flow) and sweat glands remain inactive.

Effectors
Muscle activity generates body heat.

Control center
The hypothalamus detects the deviation from the set point and signals effector organs.

If body temperature continues to drop, control center signals muscles to contract involuntarily.

FIGURE 1.8 The homeostatic mechanism that regulates body temperature.

the effector activity is initially increased rather than turned off. In blood clotting, for example, certain chemicals stimulate more clotting, which minimizes bleeding (see chapter 14, p. 544). Preventing blood loss following an injury is critical to sustaining life. Similarly, a positive feedback mechanism increases the strength of uterine contractions during childbirth (see chapter 23, p. 890).

Positive feedback mechanisms usually produce unstable conditions, which might not seem compatible with homeostasis. However, the few examples of positive feedback associated with health have very specific functions and are short-lived.

Organ systems contribute to homeostasis in different ways. For example, resources brought in by the digestive and respiratory

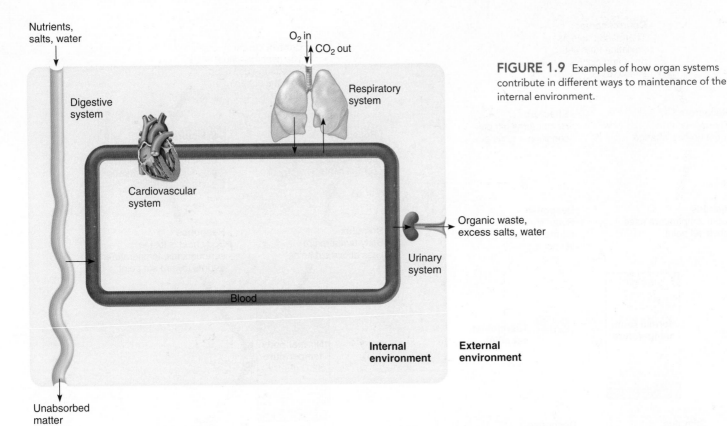

Nutrients,
salts, water

Digestive
system

O_2 in
CO_2 out

Respiratory
system

Cardiovascular
system

Organic waste,
excess salts, water

Urinary
system

Blood

Internal
environment

External
environment

Unabsorbed
matter

FIGURE 1.9 Examples of how organ systems contribute in different ways to maintenance of the internal environment.

systems are delivered to all body cells by the cardiovascular system. The same blood that brings needed nutrients to cells carries away waste products, which are removed by the respiratory and urinary systems (fig. 1.9).

Homeostatic mechanisms maintain a relatively constant internal environment, yet physiological values may vary slightly in a person from time to time or from one person to another. Therefore, both normal values for an individual and the idea of a **normal range** for the general population are clinically important. Numerous examples of homeostasis are presented throughout this book, and normal ranges for a number of physiological variables are listed in Appendix C, Laboratory Tests of Clinical Importance.

 PRACTICE

13 Which requirements of organisms does the external environment provide?

14 What is the function of pressure in the body?

15 Why is homeostasis so important to survival?

16 Describe three homeostatic mechanisms.

1.6 | Organization of the Human Body

The human organism is a complex structure composed of many parts working together to maintain homeostasis. The human body's major features include cavities, various types of membranes, and organ systems.

Body Cavities

The human organism can be divided into an **axial** (ak′se-al) **portion,** which includes the head, neck, and trunk, and an **appendicular** (ap″en-dik′u-lar) **portion,** which includes the upper and lower limbs. Within the axial portion are the **cranial cavity,** which houses the brain; the **vertebral canal** (spinal cavity), which contains the spinal cord and is surrounded by sections of the backbone (vertebrae); the **thoracic** (tho-ras′ik) **cavity;** and the **abdominopelvic** (ab-dom′ĭ-no-pel′vik) **cavity.** The organs within these last two cavities are called **viscera** (vis′er-ah). **Figure 1.10** shows these major body cavities.

The thoracic cavity is separated from the abdominopelvic cavity by a broad, thin muscle called the **diaphragm** (di′ah-fram). When it is at rest, this muscle curves upward into the thorax like a dome. When it contracts during inhalation, it presses down upon the abdominal viscera. The wall of the thoracic cavity is composed of skin, skeletal muscles, and bones. Within the thoracic cavity are the lungs and a region between the lungs, called the **mediastinum** (me″de-as-ti′num). The remaining thoracic viscera—heart, esophagus, trachea, and thymus—are within the mediastinum.

The abdominopelvic cavity, which includes an upper abdominal portion and a lower pelvic portion, extends from the diaphragm to the floor of the pelvis. Its wall primarily consists of skin, skeletal muscles, and bones. The viscera within the **abdominal cavity** include the stomach, liver, spleen, gallbladder, kidneys, and the small and large intestines.

The **pelvic cavity** is the portion of the abdominopelvic cavity enclosed by the pelvic bones. It contains the terminal end of the large intestine, the urinary bladder, and the internal reproductive organs.

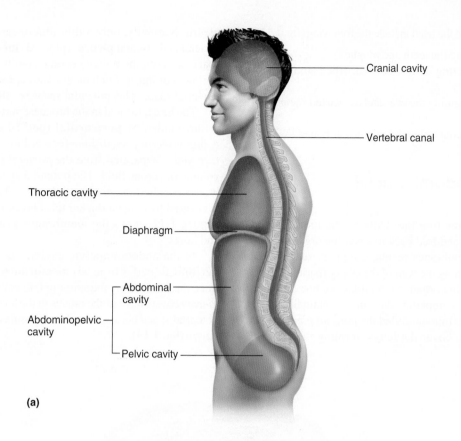

Cranial cavity

Vertebral canal

Thoracic cavity

Diaphragm

Abdominopelvic cavity
— Abdominal cavity

— Pelvic cavity

(a)

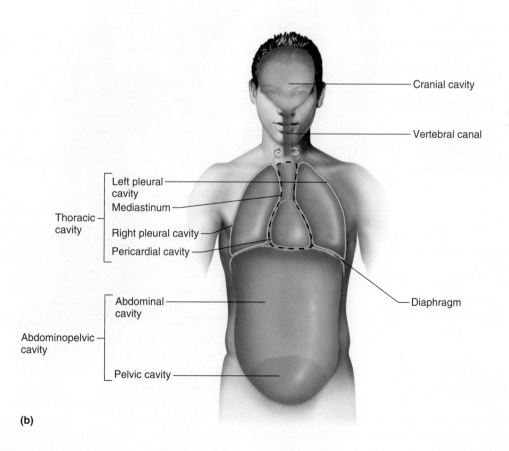

Cranial cavity

Vertebral canal

Thoracic cavity
— Left pleural cavity
— Mediastinum
— Right pleural cavity
— Pericardial cavity

Diaphragm

Abdominopelvic cavity
— Abdominal cavity

— Pelvic cavity

(b)

FIGURE 1.10 **AP|R** Major body cavities. (*a*) Lateral view. (*b*) Anterior view.

Smaller cavities within the head include the following (fig. 1.11):

1. *Oral cavity,* containing the teeth and tongue
2. *Nasal cavity,* connecting with several air-filled sinuses (see fig. 7.21, p. 217)
3. *Orbital cavities,* containing the eyes and associated skeletal muscles and nerves
4. *Middle ear cavities,* containing the middle ear bones

Thoracic and Abdominopelvic Membranes

Thin **serous membranes** line the walls of the thoracic and abdominopelvic cavities and fold back to cover the organs within these cavities. These membranes secrete a slippery serous fluid that separates the layer lining the wall of the cavity (parietal layer) from the layer covering the organ (visceral layer). For example, the right and left thoracic compartments, which contain the lungs, are lined with a serous membrane called the *parietal pleura.* This membrane folds back to cover the lungs, forming the *visceral pleura.* Normally, only a thin film of serous fluid separates the parietal and visceral **pleural** (ploo′ral) **membranes.** However, the space between them may become significantly larger as a result of illness or injury. Such membranes are said to be separated by a potential space. This potential space is called the *pleural cavity.*

The heart, located in the broadest portion of the mediastinum, is surrounded by **pericardial** (per″ĭ-kar′de-al) **membranes.** A thin *visceral pericardium* (epicardium) covers the heart's surface and is separated from the *parietal pericardium* by a small volume of serous fluid. The potential space between these membranes is called the *pericardial cavity.* The parietal pericardium is covered by a much thicker third layer, the *fibrous pericardium.* Figure 1.12 shows the membranes associated with the heart and lungs.

In the abdominopelvic cavity, the membranes are called **peritoneal** (per″-ĭ-to-ne′al) **membranes.** A *parietal peritoneum* lines the wall of the abdominopelvic cavity, and a *visceral peritoneum* covers most of the organs in the abdominopelvic cavity. The potential space between these membranes is called the *peritoneal cavity* (fig. 1.13).

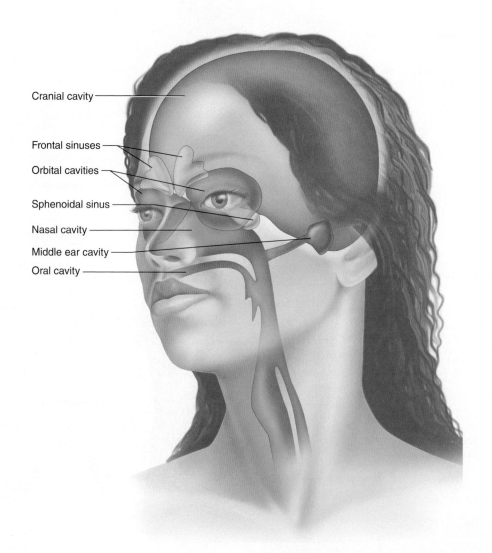

Cranial cavity

Frontal sinuses

Orbital cavities

Sphenoidal sinus

Nasal cavity

Middle ear cavity

Oral cavity

FIGURE 1.11 The cavities in the head include the cranial, oral, nasal, orbital, and middle ear cavities, as well as several sinuses.

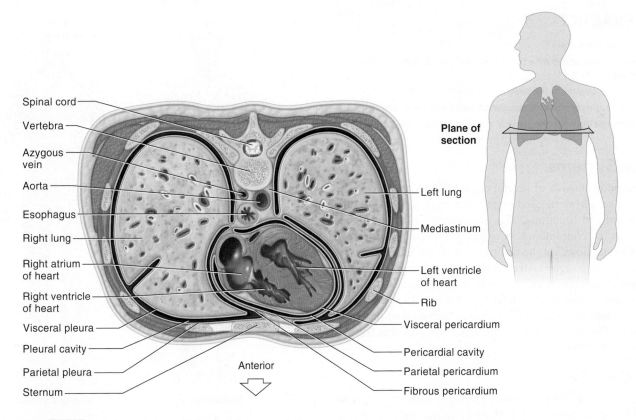

Spinal cord
Vertebra
Azygous vein
Aorta
Esophagus
Right lung
Right atrium of heart
Right ventricle of heart
Visceral pleura
Pleural cavity
Parietal pleura
Sternum

Plane of section

Left lung
Mediastinum
Left ventricle of heart
Rib
Visceral pericardium
Pericardial cavity
Parietal pericardium
Fibrous pericardium

Anterior

FIGURE 1.12 AP|R A transverse section through the thorax reveals the serous membranes associated with the heart and lungs (superior view).

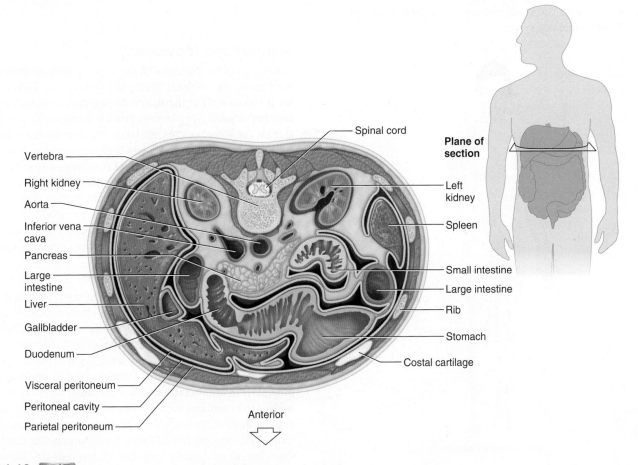

Spinal cord

Plane of section

Vertebra
Right kidney
Aorta
Inferior vena cava
Pancreas
Large intestine
Liver
Gallbladder
Duodenum
Visceral peritoneum
Peritoneal cavity
Parietal peritoneum

Left kidney
Spleen
Small intestine
Large intestine
Rib
Stomach
Costal cartilage

Anterior

FIGURE 1.13 AP|R Transverse section through the abdomen (superior view).

17 What are the viscera?

18 Which organs occupy the thoracic cavity? The abdominal cavity? The pelvic cavity?

19 Name the cavities of the head.

20 Describe the membranes associated with the thoracic cavity.

21 Distinguish between the parietal and visceral peritoneum.

Organ Systems

A human body consists of several organ systems, each of which includes a set of interrelated organs that work together to provide specialized functions. The maintenance of homeostasis depends on the coordination of organ systems. A figure called **"InnerConnections"** at the end of some chapters ties together the ways in which organ systems interact. As you read about each organ system, you may want to consult the illustrations and cadaver photos of the human torso in reference plates 1–25 at the end of this chapter (pp. 38–56) and locate some of the structures described in the text. The introduction to the organ systems that follows describes overall functions.

Body Covering

The organs of the **integumentary** (in-teg-u-men′tar-e) **system** (fig. 1.14) include the skin and accessory organs such as the hair, nails, sweat glands, and sebaceous glands. These parts protect underlying tissues, help regulate body temperature, house a variety of sensory receptors, and synthesize certain products. Chapter 6 discusses the integumentary system.

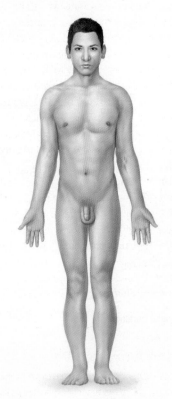

Integumentary system

FIGURE 1.14 AP|R The integumentary system covers the body.

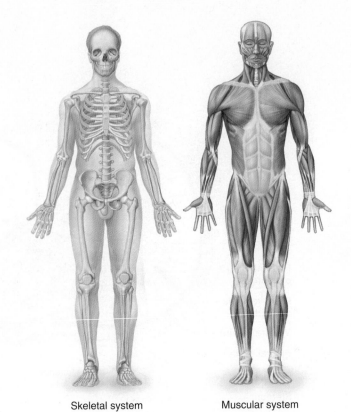

Skeletal system   Muscular system

FIGURE 1.15 AP|R The skeletal and muscular systems provide support and movement.

Support and Movement

The organs of the skeletal and muscular systems support and move body parts. The **skeletal** (skel′ĕ-tal) **system** (fig. 1.15) consists of the bones as well as the ligaments and cartilages that bind bones together at joints. These parts provide frameworks and protective shields for softer tissues, serve as attachments for muscles, and act together with muscles when body parts move. Tissues within bones also produce blood cells and store inorganic salts.

The muscles are the organs of the **muscular** (mus′ku-lar) **system** (fig. 1.15). By contracting and pulling their ends closer together, muscles provide the forces that move body parts. Muscles also help maintain posture and are the primary source of body heat. Chapters 7, 8, and 9 discuss the skeletal and muscular systems.

Integration and Coordination

For the body to act as a unit, its parts must be integrated and coordinated. The nervous and endocrine systems control and adjust various organ functions from time to time, maintaining homeostasis.

The **nervous** (ner′vus) **system** (fig. 1.16) consists of the brain, spinal cord, nerves, and sense organs. Nerve cells within these organs use a bioelectrical signal called an impulse (an action potential) in combination with a chemical signal (a neurotransmitter) to communicate with one another and with muscles and glands. Each neurotransmitter produces a rapid, relatively short-term effect, making it well suited for situations that require immediate, but not necessarily long-lasting, responses. Some nerve cells act in concert with specialized sensory receptors that can detect

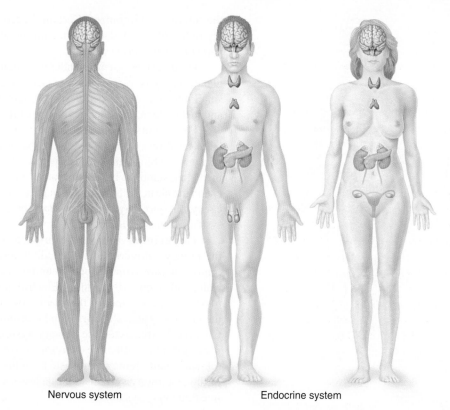

Nervous system   Endocrine system

FIGURE 1.16 **AP|R** The nervous and endocrine systems integrate and coordinate body functions.

changes inside and outside the body. Other nerve cells receive the signals from these sensory units and interpret and act on the information. Still other nerve cells carry signals from the brain or spinal cord to muscles or glands, causing them to contract or to secrete products, respectively, in response. Chapters 10 and 11 discuss the nervous system, and chapter 12 discusses sense organs.

The **endocrine** (en′do-krin) **system** (fig. 1.16) includes all the glands that secrete chemical messengers, called *hormones.* Hormones, in turn, travel away from the glands in body fluids such as interstitial fluid and blood. A particular hormone affects only a particular group of cells, called its *target cells.* A hormone alters the metabolism of its target cells. Hormonal effects last longer than those of neurotransmitters, making them well suited for responses that need to be maintained.

Organs of the endocrine system include the pituitary, thyroid, parathyroid, and adrenal glands, as well as the pancreas, ovaries, testes, pineal gland, and thymus. These are discussed further in chapter 13.

A GLIMPSE AHEAD | To Chapters 10 and 13

For both the nervous system and the endocrine system, it is essential that the cells being controlled are able to respond to the chemical stimulation (either by the neurotransmitter or by the hormone). This response requires special chemical "receptors" on the cells, without which the cells cannot respond. Many drugs act by binding receptors to stimulate a response. This is the case for beta agonists, which asthma patients use in inhalants. Many other drugs block receptors to prevent an action. Such drugs include the beta blockers many heart disease patients use and certain cancer drugs.

Transport

Two organ systems transport substances throughout the internal environment. The **cardiovascular** (kahr″de-o-vas′ku-lur) **system** (fig. 1.17) includes the heart, arteries, capillaries, veins, and blood. The heart is a muscular pump that helps force blood through the blood vessels. Blood carries gases, nutrients, hormones, and wastes. It carries oxygen from the lungs and nutrients from the digestive organs to all body cells, where these substances are used in metabolic processes. Blood also carries hormones from endocrine glands to their target cells and carries wastes from body cells to the excretory organs, where the wastes are removed from the blood and released to the outside. Blood and the cardiovascular system are discussed in chapters 14 and 15.

The **lymphatic** (lim-fat′ik) **system** (fig. 1.17) is the other transport system and is closely associated with the cardiovascular system. It is composed of the lymphatic vessels, lymph fluid, lymph nodes, thymus, and spleen. This system transports some of the fluid from the spaces in tissues (tissue fluid) back to the bloodstream and carries certain fatty substances away from the digestive organs. Cells of the lymphatic system, called lymphocytes, defend the body against infections by removing pathogens (disease-causing microorganisms and viruses) from tissue fluid. The lymphatic system is discussed in chapter 16.

Absorption and Excretion

Organs in several systems absorb nutrients and oxygen and excrete wastes. The organs of the **digestive** (di-jest′tiv) **system** (fig. 1.18), discussed in detail in chapter 17, receive foods and then break down food molecules into simpler forms that can be absorbed into the internal environment. Certain digestive organs (chapter 17, pp. 665, 666–669) also produce hormones and thus function as parts of the endocrine system.

The digestive system includes the mouth, tongue, teeth, salivary glands, pharynx, esophagus, stomach, liver, gallbladder, pancreas, small intestine, and large intestine. Chapter 18 discusses nutrition and metabolism, considering the fate of foods in the body.

The organs of the **respiratory** (re-spi′rah-to″re) **system** (fig. 1.18) move air in and out of the body and exchange gases between the blood and the air. Specifically, oxygen passes from air in the lungs into the blood, and carbon dioxide leaves the blood and enters the air in the lungs and then moves out of the body. The nasal cavity, pharynx, larynx, trachea, bronchi, and lungs are parts of this system, discussed in chapter 19.

The **urinary** (u′rĭ-ner″e) **system** (fig. 1.18) consists of the kidneys, ureters, urinary bladder, and urethra. The kidneys remove wastes from blood and assist in maintaining the body's water and electrolyte balance. (Electrolytes are chemicals, related to salts.) The product of these activities is urine. Other parts of the urinary system store urine and transport it to outside the body. Chapter 20 discusses the urinary system. Sometimes the urinary system is called the *excretory system*. However, **excretion** (ek-skre′shun), removal of waste from the body, is also a function of the respiratory system and, to a lesser extent, the digestive and integumentary systems.

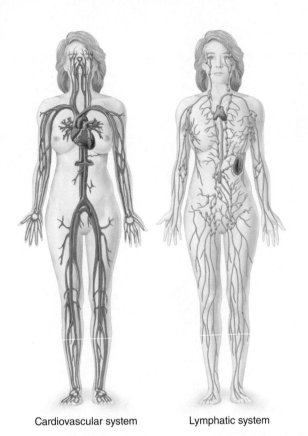

Cardiovascular system   Lymphatic system

FIGURE 1.17 AP|R The cardiovascular and lymphatic systems transport fluids.

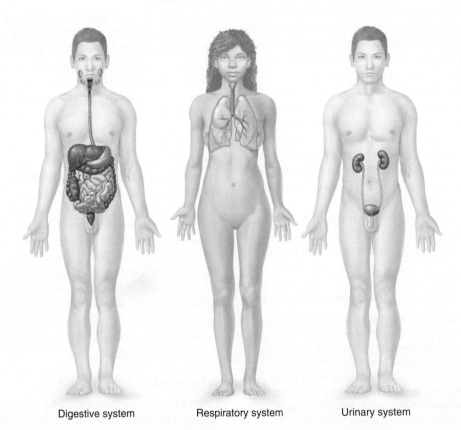

Digestive system   Respiratory system   Urinary system

FIGURE 1.18 AP|R The digestive, respiratory, and urinary systems absorb nutrients, take in oxygen and release carbon dioxide, and excrete wastes.

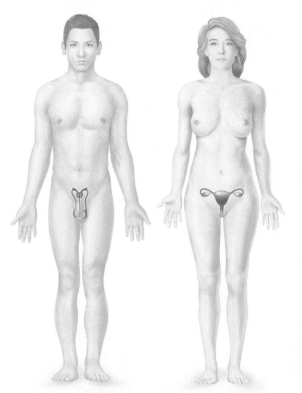

Male reproductive system   Female reproductive system

FIGURE 1.19 **AP|R** The reproductive systems manufacture and transport sex cells. The female reproductive system provides for prenatal development and childbirth.

Reproduction

Reproduction (re″pro-duk′shun) is the process of producing off-spring (progeny). Cells reproduce when they divide and give rise to new cells. The **reproductive** (re″pro-duk′tiv) **system** (fig. 1.19) of an organism, however, produces whole new organisms like itself (see chapters 22 and 23).

The male reproductive system includes the scrotum, testes, epididymides, ductus deferentia, seminal vesicles, prostate gland, bulbourethral glands, urethra, and penis. These structures produce and maintain the male sex cells, or sperm cells (spermatozoa). The male reproductive system also transfers these cells into the female reproductive tract and produces male sex hormones.

The female reproductive system consists of the ovaries, uterine tubes, uterus, vagina, clitoris, and vulva. These organs produce and maintain the female sex cell (egg cells or ova), transport the female's egg cell in the female reproductive system, and receive the male's sperm cells, which may fertilize an egg. The female reproductive system also supports development of an embryo, carries a fetus to term, functions in the birth process, and produces female sex hormones.

Table 1.4 summarizes the organ systems, the major organs that comprise them, and their major functions, in the order in which they are presented in this book. **Figure 1.20** illustrates the organ systems in humans.

 PRACTICE

22 Name the major organ systems and list the organs of each system.

23 Describe the general functions of each organ system.

TABLE 1.4	Organ Systems	
Organ System	**Major Organs**	**Major Functions**
Integumentary	Skin, hair, nails, sweat glands, sebaceous glands	Protect tissues, regulate body temperature, support sensory receptors
Skeletal	Bones, ligaments, cartilages	Provide framework, protect soft tissues, provide attachments for muscles, produce blood cells, store inorganic salts
Muscular	Muscles	Cause movements, maintain posture, produce body heat
Nervous	Brain, spinal cord, nerves, sense organs	Detect changes, receive and interpret sensory information, stimulate muscles and glands
Endocrine	Glands that secrete hormones (pituitary gland, thyroid gland, parathyroid glands, adrenal glands, pancreas, ovaries, testes, pineal gland, and thymus)	Control metabolic activities of body structures
Cardiovascular	Heart, arteries, capillaries, veins	Move blood through blood vessels and transport substances throughout body
Lymphatic	Lymphatic vessels, lymph nodes, thymus, spleen	Return tissue fluid to the blood, carry certain absorbed food molecules, defend the body against infection
Digestive	Mouth, tongue, teeth, salivary glands, pharynx, esophagus, stomach, liver, gallbladder, pancreas, small and large intestines	Receive, break down, and absorb food; eliminate unabsorbed material
Respiratory	Nasal cavity, pharynx, larynx, trachea, bronchi, lungs	Intake and output of air, exchange of gases between air and blood
Urinary	Kidneys, ureters, urinary bladder, urethra	Remove wastes from blood, maintain water and electrolyte balance, store and eliminate urine
Reproductive	Male: scrotum, testes, epididymides, ductus deferentia, seminal vesicles, prostate gland, bulbourethral glands, urethra, penis	Produce and maintain sperm cells, transfer sperm cells into female reproductive tract
	Female: ovaries, uterine tubes, uterus, vagina, clitoris, vulva	Produce and maintain egg cells, receive sperm cells, support development of an embryo, and function in birth process

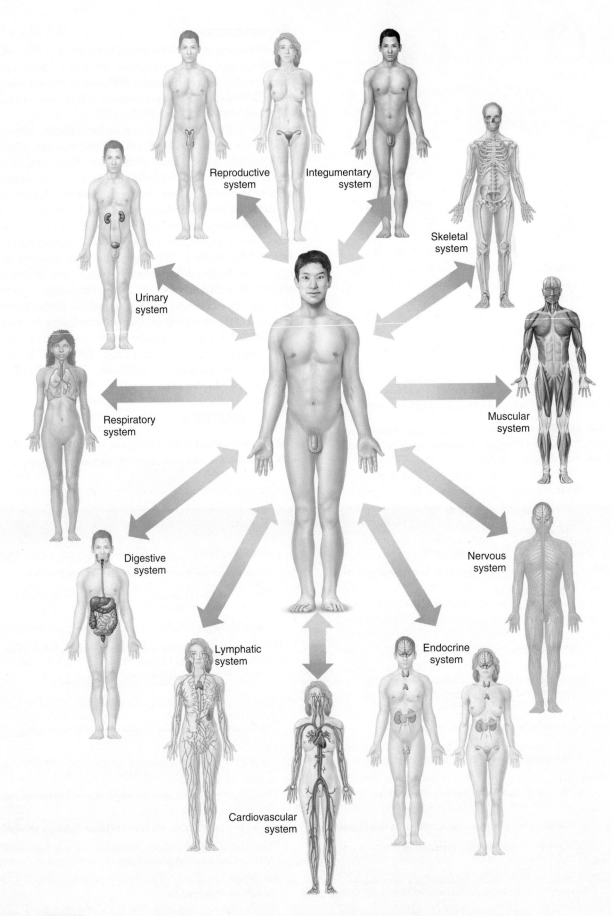

FIGURE 1.20 AP|R The organ systems in humans interact in ways that maintain homeostasis.

Reproductive system

Integumentary system

Skeletal system

Urinary system

Muscular system

Respiratory system

Digestive system

Nervous system

Lymphatic system

Endocrine system

Cardiovascular system

1.7 | Life-Span Changes

Aging refers to the changes in the body that occur with the passage of time. It is the process of becoming mature or old, and is a part of life. Because the passage of time is inevitable, so, too, is aging, despite common claims for the anti-aging properties of various diets, cosmetics, pills, and skin-care products.

Aging happens everywhere in the body, from the microscopic to the whole-body level. Although programmed cell death begins in the fetus, as structures form, we are usually not very aware of aging until the third decade of life, when a few gray hairs, faint lines etched into facial skin, and minor joint stiffness in the morning remind us that time marches on. A woman over the age of thirty-five attempting to conceive a child might be shocked to learn that she is of "advanced maternal age," because the chances of conceiving an offspring with an abnormal chromosome number increase with the age of the egg. In both sexes, by the fourth or fifth decade, as hair color fades and skin wrinkles, the first signs of adult-onset disorders may appear, such as elevated blood pressure that one day may be considered hypertension, and slightly high blood glucose that could become type 2 diabetes mellitus. A person with a strong family history of heart disease, coupled with unhealthy diet and exercise habits, may be advised to change his or her lifestyle, and perhaps begin taking a drug to lower serum cholesterol levels. The sixth decade sees grayer or whiter hair, more and deeper skin wrinkles, and a waning immunity that makes vaccinations against influenza and other infectious diseases important. Yet many, if not most, people in their sixties and older have sharp minds and are capable of many types of physical activities.

Changes at the tissue, cell, and molecular levels explain the familiar signs of aging. Decreased production of the connective tissue proteins collagen and elastin account for the stiffening of skin, and diminished levels of subcutaneous fat are responsible for wrinkling. Proportions of fat to water in the tissues change, with the percentage of fats increasing steadily in women, and increasing until about age sixty in men. These alterations explain why the elderly metabolize certain drugs at different rates than do younger people. As a person ages, tissues atrophy, and as a result, organs shrink.

Cells mark time too, many approaching the end of a limited number of predetermined cell divisions as their chromosome tips whittle down. Such cells reaching the end of their division days may enlarge or die. Some cells may be unable to build the apparatus that pulls apart replicated chromosomes in a cell on the verge of division. Impaired cell division slows wound healing, yet at the same time, the inappropriate cell division that underlies cancer becomes more likely. Certain subcellular functions lose efficiency, including repair of DNA damage and transport of substances into and out of cells. Aging cells are less efficient at extracting energy from nutrients and breaking down aged or damaged cell parts.

As changes at the tissue level cause organ-level signs of aging, certain biochemical changes fuel cellular aging. Lipofuscin and ceroid pigments accumulate when a cell can no longer prevent the formation of oxygen free radicals. A protein called beta amyloid may build up in the brain, contributing, in some individuals, to the development of Alzheimer disease. A generalized metabolic slow-down results from a dampening of thyroid gland function, impairing glucose use, the rate of protein synthesis, and production of digestive enzymes. At the whole-body level, we notice slowed metabolism as diminished tolerance to cold, weight gain, and fatigue.

Several investigations are identifying key characteristics, particularly gene variants, which people who live more than 100 years share. These fortunate individuals, called centenarians, fall into three broad groups: about 20% of them never get the diseases that kill most people; 40% get these diseases but at much older ages than average; and the other 40% live with and survive the more common disorders of aging. Environmental factors are important, too—another trait centenarians share is never having smoked.

Our organs and organ systems are interrelated, so aging-related changes in one influence the functioning of others. Several chapters in this book conclude with a "Life-Span Changes" section that discusses changes specific to particular organ systems. These changes reflect the natural breakdown of structure and function that accompanies the passage of time, as well as events in our genes ("nature") and symptoms or characteristics that might arise as a consequence of lifestyle choices and circumstances ("nurture").

 PRACTICE

24 Define aging.

25 List some aging-related changes at the microscopic and whole-body levels.

1.8 | Anatomical Terminology

To communicate effectively with one another, investigators over the ages have developed a set of terms with precise meanings for discussing the human body and its parts. These terms concern the relative positions of body parts, refer to imaginary planes along which cuts may be made, or describe body regions. Using such terms assumes that the body is in the **anatomical position**—standing erect; face forward; upper limbs at the sides, palms forward.

Relative Position

Terms of relative position are used to describe the location of one body part with respect to another. They include the following (many of these terms are illustrated in figure 1.21):

1. **Superior** means a part is above another part. (The thoracic cavity is superior to the abdominopelvic cavity.)
2. **Inferior** means a part is below another part. (The neck is inferior to the head.)
3. **Anterior** (ventral) means toward the front. (The eyes are anterior to the brain.)
4. **Posterior** (dorsal) means toward the back. (The pharynx is posterior to the oral cavity.)
5. **Medial** refers to an imaginary midline dividing the body into equal right and left halves. A part is medial if it is closer to midline than another part. (The nose is medial to the eyes.)
6. **Lateral** means toward the side, away from midline. (The ears are lateral to the eyes.)
7. **Bilateral** refers to paired structures, one on each side. (The lungs are bilateral.)

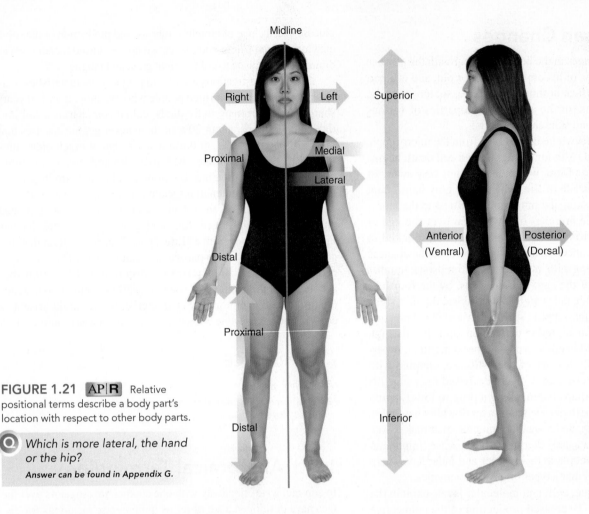

FIGURE 1.21 AP|R Relative positional terms describe a body part's location with respect to other body parts.

Q *Which is more lateral, the hand or the hip?*
Answer can be found in Appendix G.

8. **Ipsilateral** refers to structures on the same side. (The right lung and the right kidney are ipsilateral.)
9. **Contralateral** refers to structures on the opposite side. (A patient with a fractured right leg would have to bear weight on the contralateral—in this case, left—lower limb.)
10. **Proximal** describes a part closer to a point of attachment to the trunk than another body part. (The elbow is proximal to the wrist.) Proximal may also refer to another reference point such as the proximal tubules, which are closer to the filtering structures in the kidney.
11. **Distal** is the opposite of proximal. It means a particular body part is farther from a point of attachment to the trunk. (The fingers are distal to the wrist.) *Distal* may also denote another reference point, such as decreased blood flow distal to blockage of a coronary artery.
12. **Superficial** means near the surface. (The epidermis is the superficial layer of the skin.)
13. **Deep** describes more internal parts. (The dermis is the deep layer of the skin.)

Body Sections

To observe the relative locations and arrangements of internal parts, it is necessary to cut, or section, the body along various planes (figs. 1.22 and 1.23). The following terms describe such planes and sections:

1. **Sagittal** refers to a lengthwise cut that divides the body into right and left portions. If a sagittal section passes along the midline and divides the body into equal parts, it is called median (midsagittal). A sagittal section lateral to midline is called parasagittal.
2. **Transverse** (horizontal) refers to a cut that divides the body into superior and inferior portions.
3. **Frontal** (coronal) refers to a section that divides the body into anterior and posterior portions.

Sometimes a cylindrical organ such as a blood vessel is sectioned. In this case, a cut across the structure is called a *cross section,* an angular cut is called an *oblique section,* and a lengthwise cut is called a *longitudinal section* (fig. 1.24).

Body Parts and Regions

Parts of the axial portion of the body can be described using terms similar to the names of the cavities within them. The trunk includes the thorax (chest), the pelvis (area associated with the hips), and the abdomen (area below the chest and above the pelvis).

The appendicular portion can be described in more detail as well. The upper limbs include the arms (from shoulder to elbow) and the forearms (from elbow to hand). Similarly, the lower limbs include the thigh (from hip to knee) and the leg (from knee to foot).

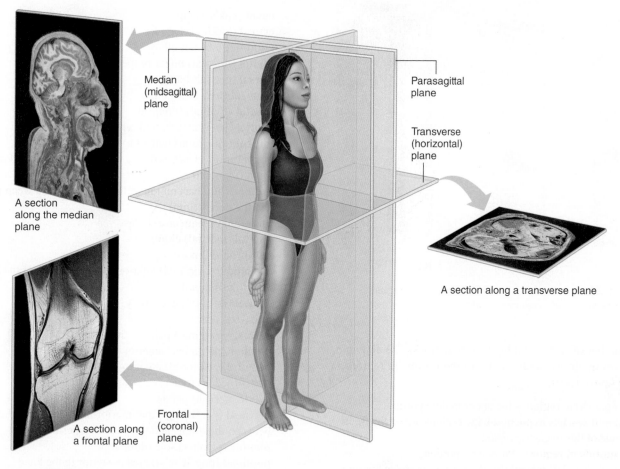

Median
(midsagittal)
plane

Parasagittal
plane

Transverse
(horizontal)
plane

A section
along the median
plane

A section along a transverse plane

A section along
a frontal plane

Frontal
(coronal)
plane

FIGURE 1.22 AP|R Observation of internal parts requires sectioning the body along various planes.

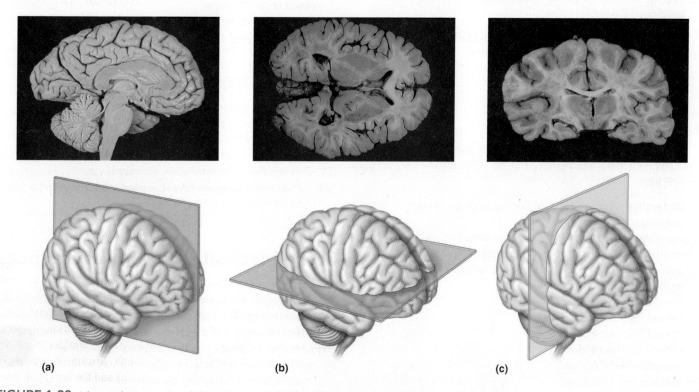

(a)

(b)

(c)

FIGURE 1.23 A human brain sectioned along (a) a sagittal plane, (b) a transverse plane, and (c) a frontal plane.

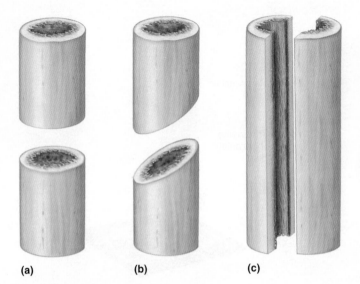

FIGURE 1.24 Cylindrical parts may be cut in (a) cross section, (b) oblique section, or (c) longitudinal section.

(a)   (b)   (c)

A number of terms designate body regions. The abdominal area, for example, is subdivided into the following regions, as shown in figure 1.25a:

1. The **epigastric region** is the upper middle portion.
2. The **right** and **left hypochondriac regions** are on the right/left side of the epigastric region.
3. The **umbilical region** is the central portion.
4. The **right** and **left lateral (lumbar) regions** are on the right/left side of the umbilical region.
5. The **pubic (hypogastric) region** is the lower middle portion.
6. The **right** and **left inguinal (iliac) regions** are on the right/left side of the pubic region.

The abdominal area may also be subdivided into the following four quadrants, as figure 1.25b illustrates:

1. Right upper quadrant (RUQ)
2. Right lower quadrant (RLQ)
3. Left upper quadrant (LUQ)
4. Left lower quadrant (LLQ)

The following adjectives are commonly used when referring to various body parts and regions. **Figure 1.26** illustrates some of these regions.

abdominal (ab-dom′ĭ-nal) region between the thorax and pelvis
acromial (ah-kro′me-al) point of the shoulder
antebrachial (an″te-bra′ke-al) forearm
antecubital (an″te-ku′bĭ-tal) anterior surface of the elbow
axillary (ak′sĭ-ler″e) armpit
brachial (bra′ke-al) arm
buccal (buk′al) cheek
carpal (kar′pal) wrist
celiac (se′le-ak) abdomen
cephalic (sĕ-fal′ik) head
cervical (ser′vĭ-kal) neck
costal (kos′tal) ribs

coxal (kok′sal) hip
crural (krōōr′al) leg
cubital (ku′bĭ-tal) elbow
digital (dij′ĭ-tal) finger or toe
dorsal (dor′sal) back
femoral (fem′or-al) thigh
frontal (frun′tal) forehead
genital (jen′i-tal) external reproductive organs
gluteal (gloo′te-al) buttocks
groin (groin) depressed area of the abdominal wall near the thigh (inguinal)
lumbar (lum′bar) region of the lower back between the ribs and the pelvis (loin)
mammary (mam′er-e) breast
mental (men′tal) chin
nasal (na′zal) nose
occipital (ok-sip′ĭ-tal) inferior posterior region of the head
oral (o′ral) mouth
orbital (or′bi-tal) eye cavity
otic (o′tik) ear
palmar (pahl′mar) palm of the hand
patellar (pah-tel′ar) anterior part of the knee
pectoral (pek′tor-al) anterior chest
pedal (ped′al) foot
pelvic (pel′vik) pelvis
perineal (per″ĭ-ne′al) the inferior-most region of the trunk between the thighs and the buttocks (perineum)
plantar (plan′tar) sole of the foot
popliteal (pop″lĭ-te′al) area posterior to the knee
sacral (sa′kral) posterior region between the hip bones
sternal (ster′nal) middle of the thorax, anteriorly
sural (su′ral) calf of the leg
tarsal (tahr′sal) ankle
umbilical (um-bil′ĭ-kal) navel
vertebral (ver′te-bral) spinal column

 PRACTICE

26 Describe the anatomical position.
27 Using the appropriate terms, describe the relative positions of several body parts.
28 Describe three types of body sections.
29 Describe the nine regions of the abdomen.
30 Explain how the names of the abdominal quadrants describe their locations.

Some Medical and Applied Sciences

cardiology (kar″de-ol′o-je) Branch of medical science dealing with the heart and heart diseases.
dermatology (der″mah-tol′o-je) Study of skin and its diseases.
endocrinology (en″do-krĭ-nol′o-je) Study of hormones, hormone-secreting glands, and associated diseases.
epidemiology (ep″ĭ-de″me-ol′o-je) Study of the factors that contribute to determining the distribution and frequency of health-related conditions within a defined human population.

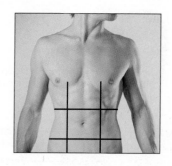

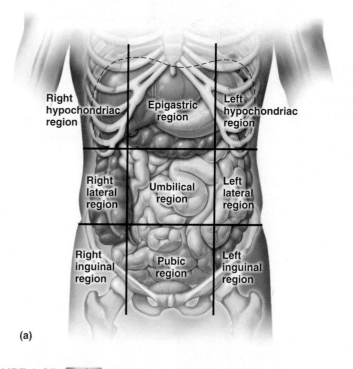

Right
hypochondriac
region

Epigastric
region

Left
hypochondriac
region

Right
lateral
region

Umbilical
region

Left
lateral
region

Right
inguinal
region

Pubic
region

Left
inguinal
region

(a)

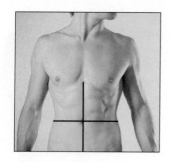

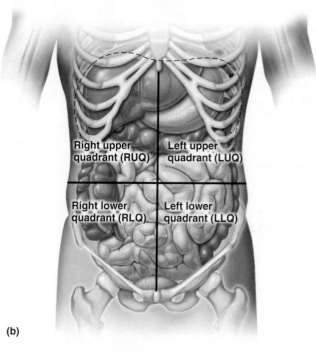

Right upper
quadrant (RUQ)

Left upper
quadrant (LUQ)

Right lower
quadrant (RLQ)

Left lower
quadrant (LLQ)

(b)

FIGURE 1.25 AP|R The abdominal area is commonly subdivided in two ways: (a) into nine regions and (b) into quadrants

gastroenterology (gas″tro-en″ter-ol′o-je) Study of the stomach and intestines, as well as their diseases.

geriatrics (jer″e-at′riks) Branch of medicine dealing with older individuals and their medical problems.

gerontology (jer″on-tol′o-je) Study of the process of aging and the various problems of older individuals.

gynecology (gi″nĕ-kol-o-je) Study of the female reproductive system and its diseases.

hematology (hem″ah-tol′o-je) Study of the blood and blood diseases.

histology (his-tol′o-je) Study of the structure and function of tissues (microscopic anatomy).

immunology (im″u-nol′o-je) Study of the body's resistance to disease.

neonatology (ne″o-na-tol′o-je) Study of newborns and the treatment of their disorders.

nephrology (nĕ-frol′o-je) Study of the structure, function, and diseases of the kidneys.

neurology (nu-rol′o-je) Study of the nervous system in health and disease.

obstetrics (ob-stet′riks) Branch of medicine dealing with pregnancy and childbirth.

oncology (ong-kol′o-je) Study of cancers.

ophthalmology (of″thal-mol′o-je) Study of the eyes and eye diseases.

orthopedics (or″tho-pe′diks) Branch of medicine dealing with the muscular and skeletal systems and their problems.

otolaryngology (o″to-lar″in-gol′o-je) Study of the ears, throat, larynx, and their diseases.

pathology (pah-thol′o-je) Study of structural and functional changes within the body associated with disease.

pediatrics (pe″de-at′riks) Branch of medicine dealing with children and their diseases.

pharmacology (fahr″mah-kol′o-je) Study of drugs and their uses in the treatment of diseases.

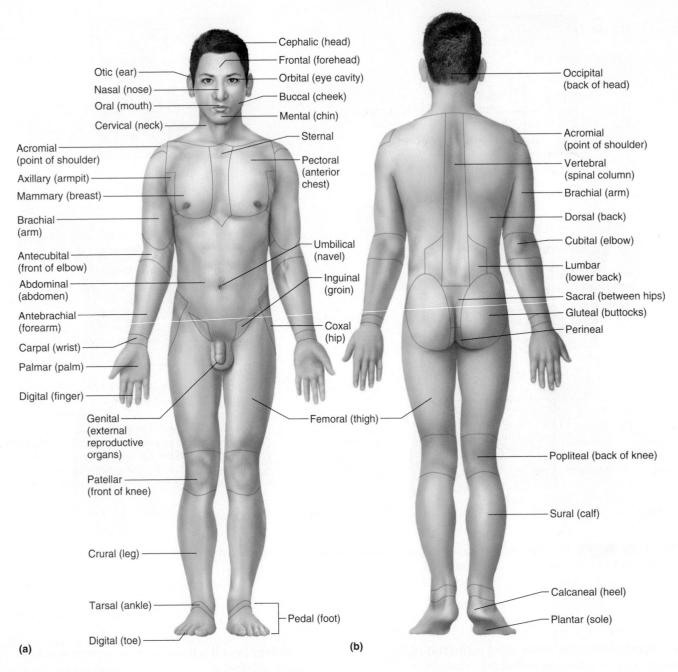

Labels on figure (a) Anterior regions:

Cephalic (head)
Frontal (forehead)
Otic (ear)
Orbital (eye cavity)
Nasal (nose)
Buccal (cheek)
Oral (mouth)
Mental (chin)
Cervical (neck)
Sternal
Acromial (point of shoulder)
Pectoral (anterior chest)
Axillary (armpit)
Mammary (breast)
Brachial (arm)
Umbilical (navel)
Antecubital (front of elbow)
Inguinal (groin)
Abdominal (abdomen)
Antebrachial (forearm)
Coxal (hip)
Carpal (wrist)
Palmar (palm)
Digital (finger)
Genital (external reproductive organs)
Femoral (thigh)
Patellar (front of knee)
Crural (leg)
Tarsal (ankle)
Pedal (foot)
Digital (toe)

Labels on figure (b) Posterior regions:

Occipital (back of head)
Acromial (point of shoulder)
Vertebral (spinal column)
Brachial (arm)
Dorsal (back)
Cubital (elbow)
Lumbar (lower back)
Sacral (between hips)
Gluteal (buttocks)
Perineal
Popliteal (back of knee)
Sural (calf)
Calcaneal (heel)
Plantar (sole)

(a)   (b)

FIGURE 1.26 AP|R Some terms used to describe body regions. (a) Anterior regions. (b) Posterior regions.

podiatry (po-di′ah-tre) Study of the care and treatment of the feet.

psychiatry (si-ki′ah-tre) Branch of medicine dealing with the mind and its disorders.

radiology (ra″de-ol′o-je) Study of X rays and radioactive substances, as well as their uses in diagnosing and treating diseases.

toxicology (tok″sĭ-kol′o-je) Study of poisonous substances and their effects on physiology.

urology (u-rol′o-je) Branch of medicine dealing with the urinary and male reproductive systems and their diseases. ■

Chapter Summary

1.1 Origins of Medical Science (page 10)

1. Early interest in the human body probably developed as people became concerned about injuries and illnesses.
2. Early doctors began to learn how certain herbs and potions affected body functions.
3. The idea that humans could understand forces that caused natural events led to the development of modern science.
4. A set of terms originating from Greek and Latin formed the basis for the language of anatomy and physiology.
5. Much of what we know about the human body was discovered using the scientific method.

1.2 Anatomy and Physiology (page 11)

1. **Anatomy** deals with the form and organization of body parts.
2. **Physiology** deals with the functions of these parts.
3. The function of a part depends upon the way it is constructed.

1.3 Levels of Organization (page 12)

The body is composed of parts that can be considered at different levels of organization.

1. Matter is composed of **atoms,** which are composed of **subatomic particles.**
2. Atoms join to form **molecules.**
3. **Organelles** consist of aggregates of interacting large molecules (**macromolecules).**
4. **Cells,** composed of organelles, are the basic units of structure and function of the body.
5. Cells are organized into layers or masses called **tissues.**
6. Tissues are organized into **organs.**
7. Organs form **organ systems.**
8. Organ systems constitute the **organism.**
9. These parts vary in complexity progressively from one level to the next.

1.4 Characteristics of Life (page 14)

Characteristics of life are traits all organisms share.

1. These characteristics include
   a. Movement—changing body position or moving internal parts.
   b. Responsiveness—sensing and reacting to internal or external changes.
   c. Growth—increasing in size without changing in shape.
   d. Reproduction—producing offspring.
   e. Respiration—obtaining oxygen, using oxygen to release energy from foods, and removing gaseous wastes.
   f. Digestion—breaking down food substances into forms that can be absorbed.
   g. Absorption—moving substances through membranes and into body fluids.
   h. Circulation—moving substances through the body in body fluids.
   i. Assimilation—changing substances into chemically different forms.
   j. Excretion—removing body wastes.

2. **Metabolism** is the acquisition and use of energy by an organism.

1.5 Maintenance of Life (page 14)

The structures and functions of body parts maintain the life of the organism.

1. Requirements of organisms
   a. **Water** is used in many metabolic processes, provides the environment for metabolic reactions, and carries substances.
   b. Nutrients from **food** supply energy, raw materials for building substances, and chemicals necessary in vital reactions.
   c. **Oxygen** is used in releasing energy from nutrients; this energy drives metabolic reactions.
   d. **Heat** is part of our environment and is a product of metabolic reactions; heat helps control rates of these reactions.
   e. **Pressure** is an application of force; in humans, atmospheric and hydrostatic pressures help breathing and blood movements, respectively.
2. Homeostasis
   a. If an organism is to survive, the conditions within its body fluids must remain relatively stable.
   b. The tendency to maintain a stable **internal environment** is called **homeostasis.**
   c. **Homeostatic mechanisms** involve sensory **receptors,** a **control center** with a **set point,** and **effectors.**
   d. Homeostatic mechanisms include those that regulate body temperature, blood pressure, and blood glucose concentration.
   e. Homeostatic mechanisms employ **negative feedback.**

1.6 Organization of the Human Body (page 18)

1. Body cavities
   a. The **axial portion** of the body contains the cranial cavity and vertebral canal, as well as the **thoracic** and **abdominopelvic cavities,** which are separated by the **diaphragm.**
   b. The organs within thoracic and abdominopelvic cavities are called **viscera.**
   c. Other body cavities include the oral, nasal, orbital, and middle ear cavities.
2. Thoracic and abdominopelvic membranes
   Parietal **serous membranes** line the walls of these cavities; visceral serous membranes cover organs within them. They secrete serous fluid.
   a. Thoracic membranes
      (1) **Pleural membranes** line the thoracic cavity and cover the lungs.
      (2) **Pericardial membranes** surround the heart and cover its surface.
      (3) The pleural and pericardial cavities are potential spaces between these membranes.
   b. Abdominopelvic membranes
      (1) **Peritoneal membranes** line the abdominopelvic cavity and cover the organs inside.
      (2) The peritoneal cavity is a potential space between these membranes.

3. Organ systems

The human organism consists of several organ systems. Each system includes interrelated organs.

a. Integumentary system
   (1) The **integumentary system** covers the body.
   (2) It includes the skin, hair, nails, sweat glands, and sebaceous glands.
   (3) It protects underlying tissues, regulates body temperature, houses sensory receptors, and synthesizes substances.

b. Skeletal system
   (1) The **skeletal system** is composed of bones and the ligaments and cartilages that bind bones together.
   (2) It provides framework, protective shields, and attachments for muscles; it also produces blood cells and stores inorganic salts.

c. Muscular system
   (1) The **muscular system** includes the muscles of the body.
   (2) It moves body parts, maintains posture, and produces body heat.

d. Nervous system
   (1) The **nervous system** consists of the brain, spinal cord, nerves, and sense organs.
   (2) It receives signals from sensory receptors, interprets this information, and acts by causing muscles or glands to respond.

e. Endocrine system
   (1) The **endocrine system** consists of glands that secrete hormones.
   (2) Hormones help regulate metabolism by stimulating target tissues.
   (3) It includes the pituitary gland, thyroid gland, parathyroid glands, adrenal glands, pancreas, ovaries, testes, pineal gland, and thymus.

f. Cardiovascular system
   (1) The **cardiovascular system** includes the heart, which pumps blood, and the blood vessels, which carry blood to and from body parts.
   (2) Blood carries oxygen, nutrients, hormones, and wastes.

g. Lymphatic system
   (1) The **lymphatic system** is composed of lymphatic vessels, lymph nodes, the thymus, and the spleen.
   (2) It transports lymph from tissue spaces to the bloodstream and carries certain fatty substances away from the digestive organs. Lymphocytes defend the body against disease-causing agents.

h. Digestive system
   (1) The **digestive system** receives foods, breaks down nutrients into forms that can pass through cell membranes, and eliminates unabsorbed materials.
   (2) Some digestive organs produce hormones.
   (3) The digestive system includes the mouth, tongue, teeth, salivary glands, pharynx, esophagus, stomach, liver, gallbladder, pancreas, small intestine, and large intestine.

i. Respiratory system
   (1) The **respiratory system** takes in and releases air and exchanges gases between the blood and the air.
   (2) It includes the nasal cavity, pharynx, larynx, trachea, bronchi, and lungs.

j. Urinary system
   (1) The **urinary system** includes the kidneys, ureters, urinary bladder, and urethra.
   (2) It filters wastes from the blood and helps maintain fluid and electrolyte balance.

k. Reproductive systems
   (1) The **reproductive system** enables an organism to produce progeny.
   (2) The male reproductive system produces, maintains, and transports male sex cells. It includes the scrotum, testes, epididymides, ductus deferentia, seminal vesicles, prostate gland, bulbourethral glands, urethra, and penis.
   (3) The female reproductive system produces, maintains, and transports female sex cells; it also supports the development and birth of offspring. It includes the ovaries, uterine tubes, uterus, vagina, clitoris, and vulva.

1.7 Life-Span Changes (page 27)

Aging occurs from conception on and has effects at the cell, tissue, organ, and organ system levels.

1. The first signs of aging are noticeable in one's thirties. Female fertility begins to decline during this time.
2. In the forties and fifties, adult-onset disorders may begin.
3. Skin changes reflect less elastin, collagen, and subcutaneous fat.
4. Older people may metabolize certain drugs at different rates than younger people.
5. Cells divide a limited number of times. As DNA repair falters, mutations may accumulate.
6. Oxygen free radical damage produces certain pigments. Metabolism slows, and beta amyloid protein may build up in the brain.

1.8 Anatomical Terminology (page 27)

Investigators use terms with precise meanings to effectively communicate with one another.

1. Relative position
   These terms describe the location of one part with respect to another part.
2. Body sections
   Body sections are planes along which the body may be cut to observe the relative locations and arrangements of internal parts.
3. Body regions
   Special terms designate various body regions.

CHAPTER ASSESSMENTS

1.1 Origins of Medical Science

1 Describe how an early interest in the human body eventually led to the development of modern medical science. (p. 10)

1.2 Anatomy and Physiology

2 Distinguish between anatomy and physiology. (p. 11)
3 Explain the relationship between the form and function of body parts and give three examples. (p. 11)

1.3 Levels of Organization

4 Describe the relationship between each of the following pairs: molecules and cells, tissues and organs, organs and organ systems. (p. 12)

1.4 Characteristics of Life

5 Which characteristics of life can you identify in yourself? (p. 14)
6 Identify those characteristics of living organisms that relate to metabolism. (p. 14)

1.5 Maintenance of Life

7 Compare your own needs for survival with the requirements of organisms described in the chapter. (pp. 14–15)
8 Explain the relationship between homeostasis and the internal environment. (p. 15)
9 Describe a general physiological control system, including the role of negative feedback. (p. 16)
10 Explain the control of body temperature. (p. 16)
11 Describe the homeostatic mechanisms that help regulate blood pressure and blood glucose—what do they have in common and how are they different? (p. 16)

1.6 Organization of the Human Body

12 Explain the difference between the axial and appendicular portions of the body. (p. 18)
13 Identify the cavities within the axial portion of the body. (p. 18)
14 Define *viscera*. (p. 18)
15 Describe the mediastinum and its contents. (p. 18)
16 Name the body cavity that houses each of the following organs: (p. 18)
   a. Stomach   f. Rectum
   b. Heart   g. Spinal cord
   c. Brain   h. Esophagus
   d. Liver   i. Spleen
   e. Trachea   j. Urinary bladder
17 List the cavities of the head and the contents of each cavity. (p. 20)
18 Distinguish between a parietal and a visceral membrane. (p. 20)
19 Describe the general contribution of each of the organ systems to maintaining homeostasis. (pp. 22–25)

20 List the major organs that compose each organ system and identify their functions. (pp. 22–25)

1.7 Life-Span Changes

21 Describe physical changes associated with aging that occur during each decade past the age of thirty. (p. 27)
22 List age-associated changes that occur at the molecular, cellular, tissue and/or organ levels. (p. 27)

1.8 Anatomical Terminology

23 Write complete sentences using each of the following terms to correctly describe the relative locations of specific body parts: (pp. 27–28)
   a. Superior   h. Ipsilateral
   b. Inferior   i. Contralateral
   c. Anterior   j. Proximal
   d. Posterior   k. Distal
   e. Medial   l. Superficial
   f. Lateral   m. Peripheral
   g. Bilateral   n. Deep
24 Sketch the outline of a human body, and use lines to indicate each of the following sections: (p. 28)
   a. Sagittal
   b. Transverse
   c. Frontal
25 Sketch the abdominal area, and indicate the locations of the following regions: (p. 30)
   a. Epigastric   d. Lateral
   b. Hypochondriac   e. Pubic
   c. Umbilical   f. Inguinal
26 Sketch the abdominal area, and indicate the location of the following regions: (p. 30)
   a. Right upper quadrant
   b. Right lower quadrant
   c. Left upper quadrant
   d. Left lower quadrant
27 Provide the common name for the region to which each of the following terms refers: (p. 30)
   a. Acromial   n. Orbital
   b. Antebrachial   o. Otic
   c. Axillary   p. Palmar
   d. Buccal   q. Pectoral
   e. Celiac   r. Pedal
   f. Coxal   s. Perineal
   g. Crural   t. Plantar
   h. Femoral   u. Popliteal
   i. Genital   v. Sacral
   j. Gluteal   w. Sternal
   k. Inguinal   x. Tarsal
   l. Mental   y. Umbilical
   m. Occipital   z. Vertebral

INTEGRATIVE ASSESSMENTS/CRITICAL THINKING

Outcomes 1.2, 1.3, 1.4, 1.5

1. Which characteristics of life does a computer have? Why is a computer not alive?

Outcomes 1.2, 1.3, 1.4, 1.5, 1.6

2. Put the following in order, from smallest and simplest, to largest and most complex, and describe their individual roles in homeostasis: organ, molecule, organelle, atom, organ system, tissue, organism, cell, macromolecule.

Outcomes 1.4, 1.5

3. What environmental conditions would be necessary for a human to survive on another planet?

Outcomes 1.5, 1.6, 1.7

4. In health, body parts interact to maintain homeostasis. Illness can threaten the maintenance of homeostasis, requiring treatment. What treatments might be used to help control a patient's (a) body temperature, (b) blood oxygen level, and (c) blood glucose level?

5. How might health-care professionals provide the basic requirements of life to an unconscious patient? Describe the body parts involved in the treatment, using correct directional and regional terms.

Outcome 1.6

6. Suppose two individuals develop benign (noncancerous) tumors that produce symptoms because they occupy space and crowd adjacent organs. If one of these persons has the tumor in the thoracic cavity and the other has the tumor in the abdominopelvic cavity, which person would be likely to develop symptoms first? Why? Which might be more immediately serious? Why?

7. If a patient complained of a "stomachache" and pointed to the umbilical region as the site of discomfort, which organs located in this region might be the source of the pain?

 # ONLINE STUDY TOOLS

Connect Interactive Questions Reinforce your knowledge using assigned interactive questions covering levels of organization, anatomical terminology, homeostasis, body organization, and more.

Connect Integrated Activity Can you help a surgeon localize different sites for incisions?

LearnSmart Discover which chapter concepts you have mastered and which require more attention. This adaptive learning tool is personalized, proven, and preferred.

Anatomy & Physiology Revealed Go more in depth into the human body by exploring body regions, planes of sections, and organ systems.

The Human Organism

The following series of plates includes illustrations of the major organs of the human torso and human cadaver photos. The first plate shows the anterior surface of the human torso and reveals the muscles on one side. Then, plates 2–7 expose deeper organs, including those in the thoracic, abdominal, and pelvic cavities. Plates 8–25 are photographs of sagittal sections, transverse sections, and anterior views of the torso of a human cadaver. These plates will help you visualize the proportional relationships between the major anatomical structures of actual specimens.

Variations exist in anatomical structures among humans. The illustrations in this textbook represent normal (normal means the most common variation) anatomy.

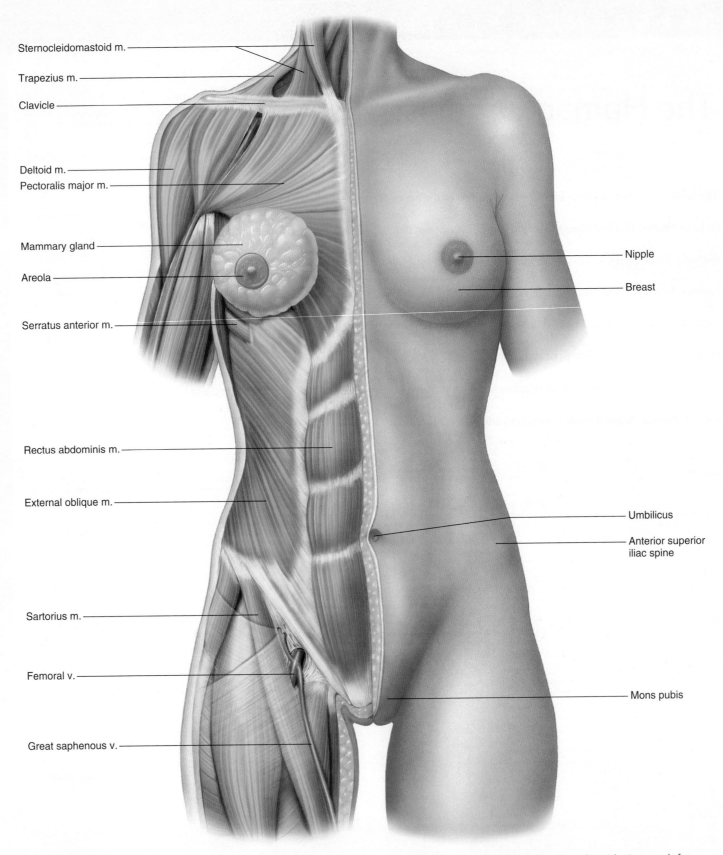

Sternocleidomastoid m.

Trapezius m.

Clavicle

Deltoid m.

Pectoralis major m.

Mammary gland

Areola

Serratus anterior m.

Rectus abdominis m.

External oblique m.

Sartorius m.

Femoral v.

Great saphenous v.

Nipple

Breast

Umbilicus

Anterior superior
iliac spine

Mons pubis

PLATE ONE Human female torso showing the anterior surface on one side and the superficial muscles exposed on the other side. (*m.* stands for *muscle*, and *v.* stands for *vein*)

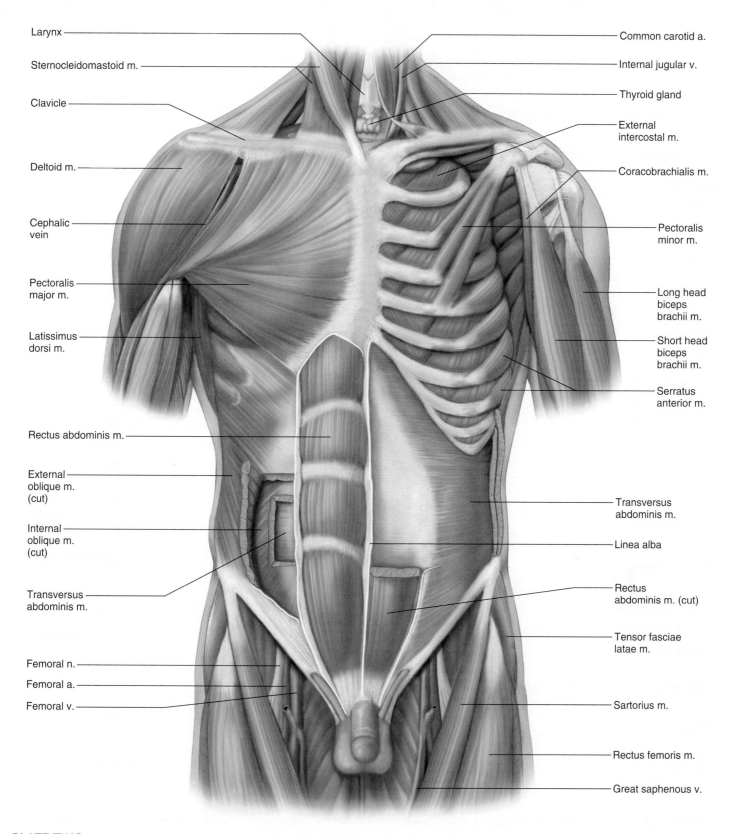

Larynx

Sternocleidomastoid m.

Clavicle

Deltoid m.

Cephalic vein

Pectoralis major m.

Latissimus dorsi m.

Rectus abdominis m.

External oblique m. (cut)

Internal oblique m. (cut)

Transversus abdominis m.

Femoral n.

Femoral a.

Femoral v.

Common carotid a.

Internal jugular v.

Thyroid gland

External intercostal m.

Coracobrachialis m.

Pectoralis minor m.

Long head biceps brachii m.

Short head biceps brachii m.

Serratus anterior m.

Transversus abdominis m.

Linea alba

Rectus abdominis m. (cut)

Tensor fasciae latae m.

Sartorius m.

Rectus femoris m.

Great saphenous v.

PLATE TWO Human male torso with the deeper muscle layers exposed. (*a.* stands for *artery*, *m.* stands for *muscle*, *n.* stands for *nerve*, and *v.* stands for *vein*)

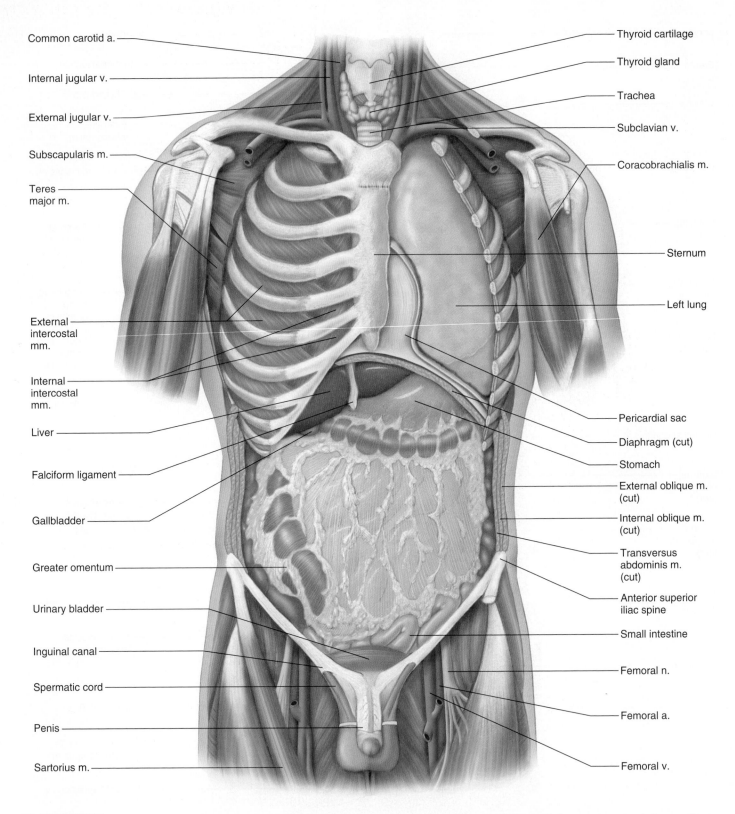

Common carotid a.

Internal jugular v.

External jugular v.

Subscapularis m.

Teres major m.

External intercostal mm.

Internal intercostal mm.

Liver

Falciform ligament

Gallbladder

Greater omentum

Urinary bladder

Inguinal canal

Spermatic cord

Penis

Sartorius m.

Thyroid cartilage

Thyroid gland

Trachea

Subclavian v.

Coracobrachialis m.

Sternum

Left lung

Pericardial sac

Diaphragm (cut)

Stomach

External oblique m. (cut)

Internal oblique m. (cut)

Transversus abdominis m. (cut)

Anterior superior iliac spine

Small intestine

Femoral n.

Femoral a.

Femoral v.

PLATE THREE Human male torso with the deep muscles removed and the abdominal viscera exposed. (*a.* stands for *artery*, *m.* stands for *muscle*, *mm.* stands for *muscles*, *n.* stands for *nerve*, and *v.* stands for *vein*)

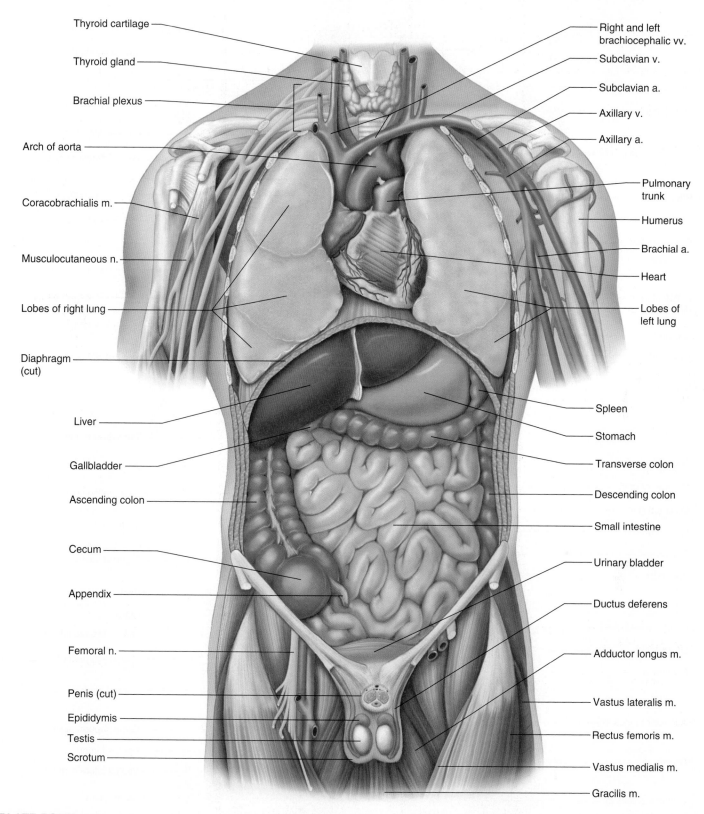

Thyroid cartilage

Thyroid gland

Brachial plexus

Arch of aorta

Coracobrachialis m.

Musculocutaneous n.

Lobes of right lung

Diaphragm (cut)

Liver

Gallbladder

Ascending colon

Cecum

Appendix

Femoral n.

Penis (cut)

Epididymis

Testis

Scrotum

Right and left brachiocephalic vv.

Subclavian v.

Subclavian a.

Axillary v.

Axillary a.

Pulmonary trunk

Humerus

Brachial a.

Heart

Lobes of left lung

Spleen

Stomach

Transverse colon

Descending colon

Small intestine

Urinary bladder

Ductus deferens

Adductor longus m.

Vastus lateralis m.

Rectus femoris m.

Vastus medialis m.

Gracilis m.

PLATE FOUR Human male torso with the thoracic and abdominal viscera exposed. (*a.* stands for *artery*, *m.* stands for *muscle*, *n.* stands for *nerve*, *v.* stands for *vein*, and *vv.* stands for *veins*)

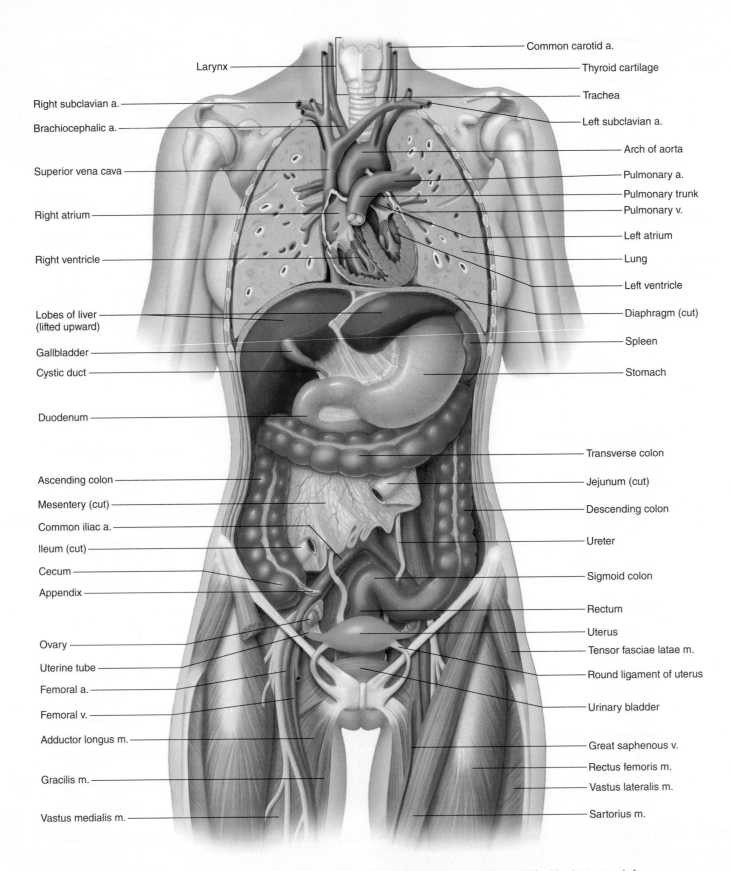

Larynx

Right subclavian a.

Brachiocephalic a.

Superior vena cava

Right atrium

Right ventricle

Lobes of liver
(lifted upward)

Gallbladder

Cystic duct

Duodenum

Ascending colon

Mesentery (cut)

Common iliac a.

Ileum (cut)

Cecum

Appendix

Ovary

Uterine tube

Femoral a.

Femoral v.

Adductor longus m.

Gracilis m.

Vastus medialis m.

Common carotid a.

Thyroid cartilage

Trachea

Left subclavian a.

Arch of aorta

Pulmonary a.

Pulmonary trunk

Pulmonary v.

Left atrium

Lung

Left ventricle

Diaphragm (cut)

Spleen

Stomach

Transverse colon

Jejunum (cut)

Descending colon

Ureter

Sigmoid colon

Rectum

Uterus

Tensor fasciae latae m.

Round ligament of uterus

Urinary bladder

Great saphenous v.

Rectus femoris m.

Vastus lateralis m.

Sartorius m.

PLATE FIVE Human female torso with the lungs, heart, and small intestine sectioned and the liver reflected (lifted back). (*a.* stands for *artery,* *m.* stands for *muscle,* and *v.* stands for *vein*)

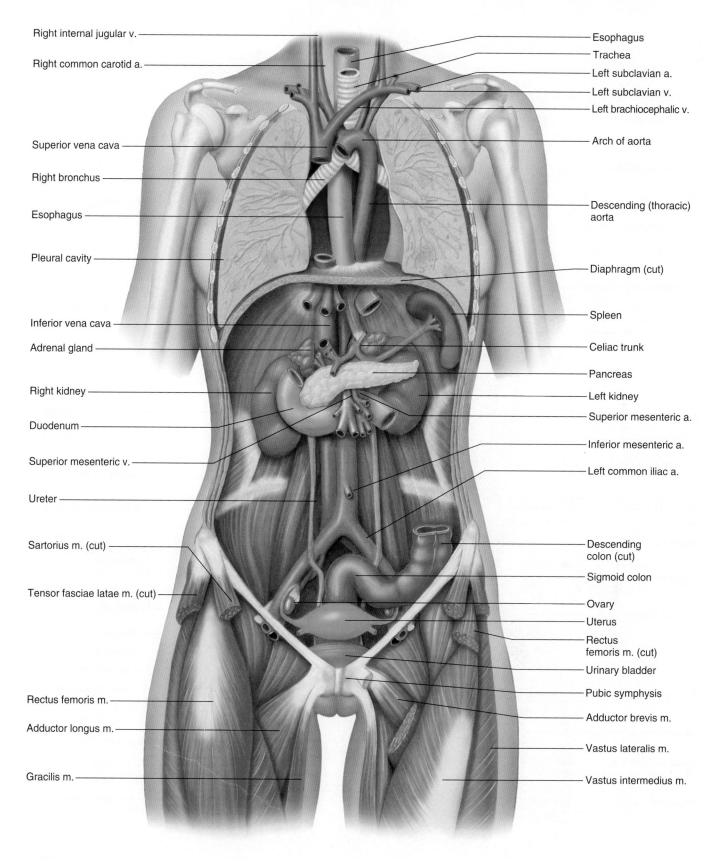

Right internal jugular v.

Right common carotid a.

Superior vena cava

Right bronchus

Esophagus

Pleural cavity

Inferior vena cava

Adrenal gland

Right kidney

Duodenum

Superior mesenteric v.

Ureter

Sartorius m. (cut)

Tensor fasciae latae m. (cut)

Rectus femoris m.

Adductor longus m.

Gracilis m.

Esophagus

Trachea

Left subclavian a.

Left subclavian v.

Left brachiocephalic v.

Arch of aorta

Descending (thoracic) aorta

Diaphragm (cut)

Spleen

Celiac trunk

Pancreas

Left kidney

Superior mesenteric a.

Inferior mesenteric a.

Left common iliac a.

Descending colon (cut)

Sigmoid colon

Ovary

Uterus

Rectus femoris m. (cut)

Urinary bladder

Pubic symphysis

Adductor brevis m.

Vastus lateralis m.

Vastus intermedius m.

PLATE SIX Human female torso with the heart, stomach, liver, and parts of the intestine and lungs removed. (*a.* stands for *artery, m.* stands for *muscle,* and *v.* stands for *vein*)

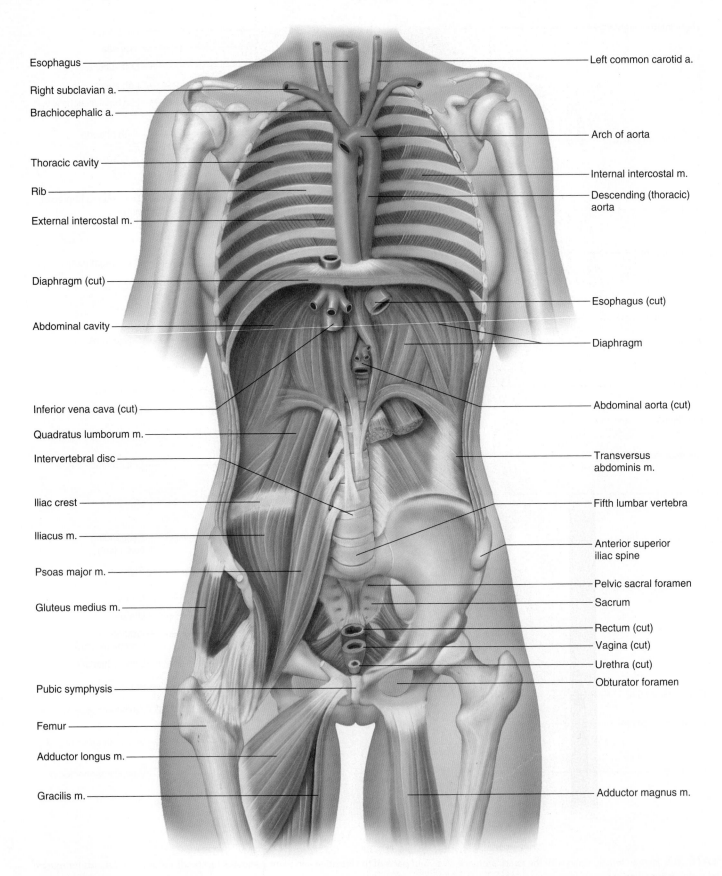

Esophagus

Right subclavian a.

Brachiocephalic a.

Thoracic cavity

Rib

External intercostal m.

Diaphragm (cut)

Abdominal cavity

Inferior vena cava (cut)

Quadratus lumborum m.

Intervertebral disc

Iliac crest

Iliacus m.

Psoas major m.

Gluteus medius m.

Pubic symphysis

Femur

Adductor longus m.

Gracilis m.

Left common carotid a.

Arch of aorta

Internal intercostal m.

Descending (thoracic) aorta

Esophagus (cut)

Diaphragm

Abdominal aorta (cut)

Transversus abdominis m.

Fifth lumbar vertebra

Anterior superior iliac spine

Pelvic sacral foramen

Sacrum

Rectum (cut)

Vagina (cut)

Urethra (cut)

Obturator foramen

Adductor magnus m.

PLATE SEVEN Human female torso with the thoracic, abdominal, and pelvic viscera removed. (*a.* stands for *artery* and *m.* stands for *muscle*)

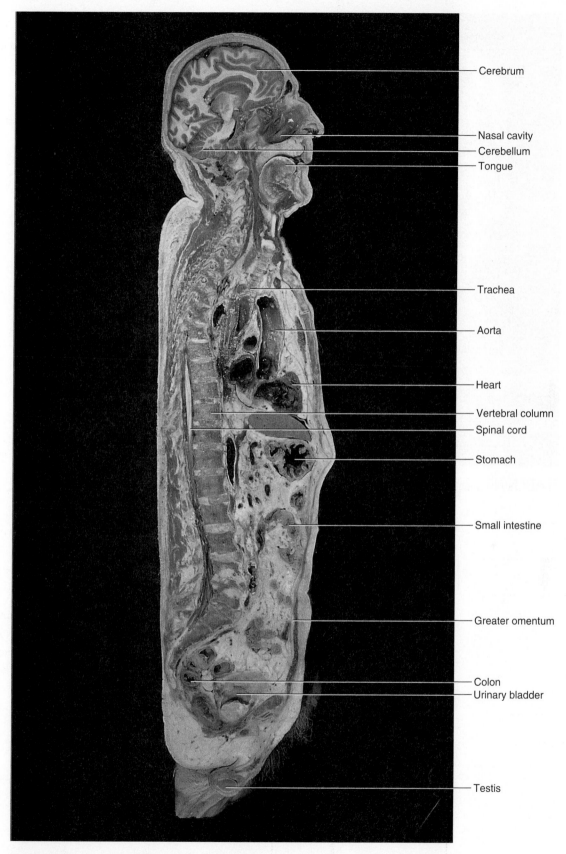

Cerebrum

Nasal cavity

Cerebellum

Tongue

Trachea

Aorta

Heart

Vertebral column

Spinal cord

Stomach

Small intestine

Greater omentum

Colon

Urinary bladder

Testis

PLATE EIGHT Sagittal section of the head, neck, and trunk.

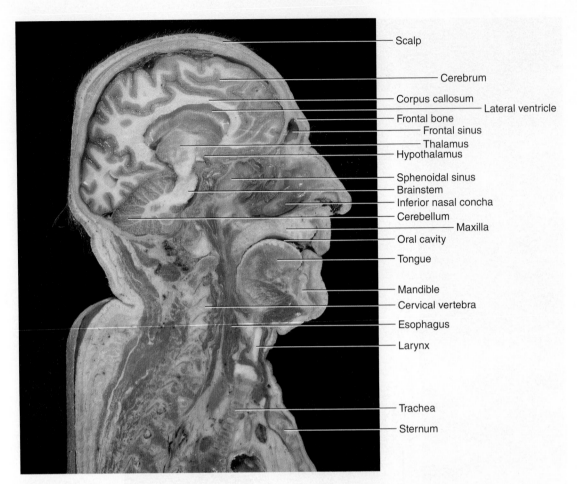

Scalp

Cerebrum

Corpus callosum

Lateral ventricle

Frontal bone

Frontal sinus

Thalamus

Hypothalamus

Sphenoidal sinus

Brainstem

Inferior nasal concha

Cerebellum

Maxilla

Oral cavity

Tongue

Mandible

Cervical vertebra

Esophagus

Larynx

Trachea

Sternum

PLATE NINE Sagittal section of the head and neck.

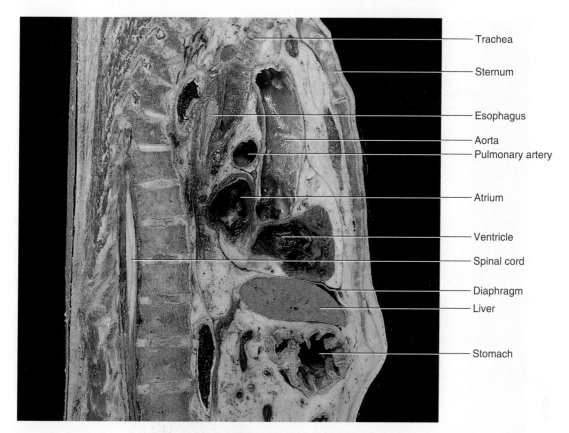

Trachea

Sternum

Esophagus

Aorta
Pulmonary artery

Atrium

Ventricle

Spinal cord

Diaphragm
Liver

Stomach

PLATE TEN Viscera of the thoracic cavity, sagittal section.

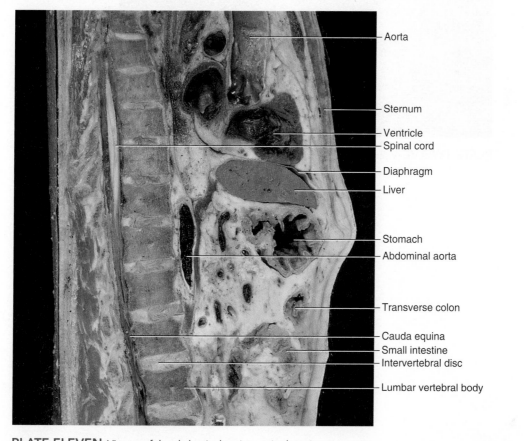

Aorta

Sternum

Ventricle
Spinal cord

Diaphragm
Liver

Stomach
Abdominal aorta

Transverse colon

Cauda equina
Small intestine
Intervertebral disc

Lumbar vertebral body

PLATE ELEVEN Viscera of the abdominal cavity, sagittal section.

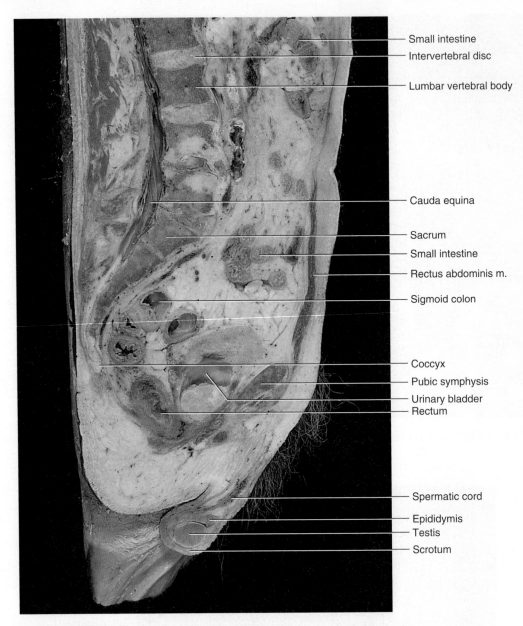

Small intestine
Intervertebral disc
Lumbar vertebral body
Cauda equina
Sacrum
Small intestine
Rectus abdominis m.
Sigmoid colon
Coccyx
Pubic symphysis
Urinary bladder
Rectum
Spermatic cord
Epididymis
Testis
Scrotum

PLATE TWELVE Viscera of the pelvic cavity, sagittal section. (*m.* stands for muscle)

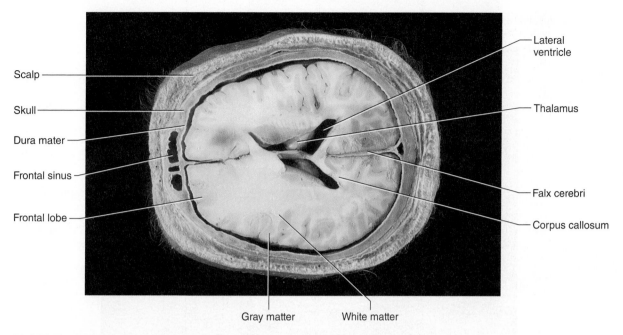

Scalp

Skull

Dura mater

Frontal sinus

Frontal lobe

Lateral ventricle

Thalamus

Falx cerebri

Corpus callosum

Gray matter   White matter

PLATE THIRTEEN Transverse section of the head above the eyes, superior view.

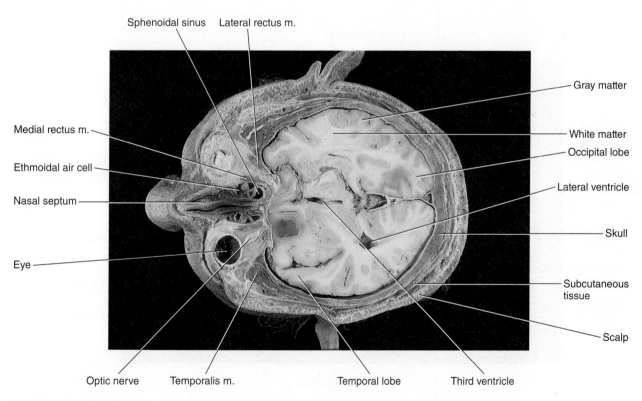

Sphenoidal sinus   Lateral rectus m.

Medial rectus m.

Ethmoidal air cell

Nasal septum

Eye

Gray matter

White matter

Occipital lobe

Lateral ventricle

Skull

Subcutaneous tissue

Scalp

Optic nerve   Temporalis m.   Temporal lobe   Third ventricle

PLATE FOURTEEN Transverse section of the head at the level of the eyes, superior view. (*m.* stands for muscle)

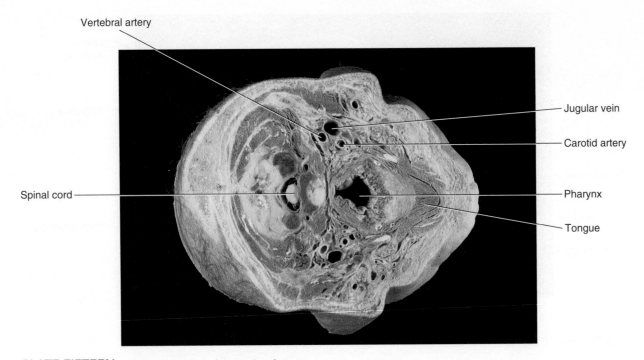

Vertebral artery

Jugular vein

Carotid artery

Spinal cord

Pharynx

Tongue

PLATE FIFTEEN Transverse section of the neck, inferior view.

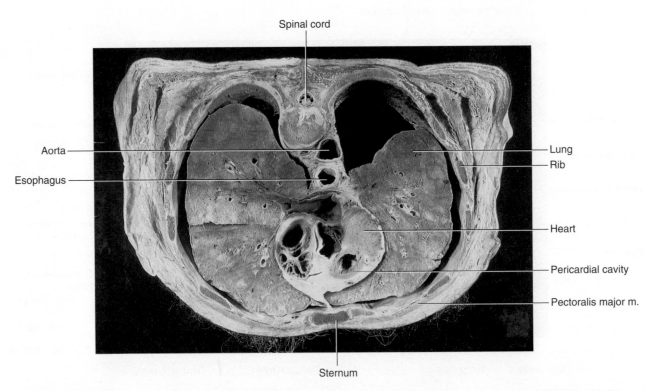

Spinal cord

Aorta

Esophagus

Lung

Rib

Heart

Pericardial cavity

Pectoralis major m.

Sternum

PLATE SIXTEEN Transverse section of the thorax through the base of the heart, superior view. (*m.* stands for muscle)

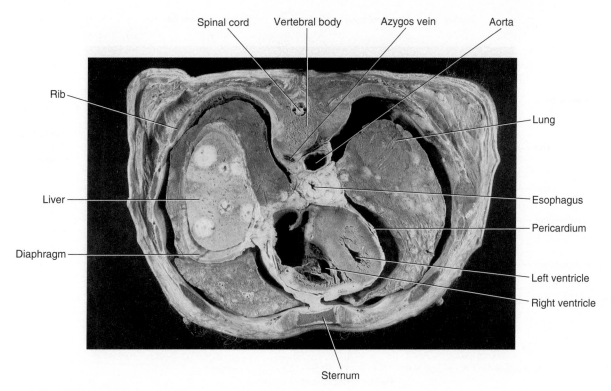

Spinal cord   Vertebral body   Azygos vein   Aorta

Rib

Lung

Liver

Esophagus

Pericardium

Diaphragm

Left ventricle

Right ventricle

Sternum

PLATE SEVENTEEN Transverse section of the thorax through the heart, superior view.

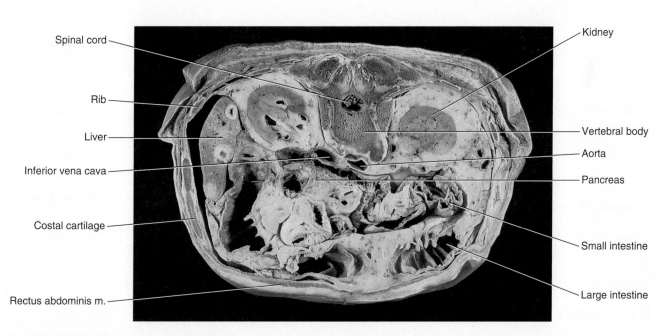

Spinal cord

Kidney

Rib

Liver

Vertebral body

Inferior vena cava

Aorta

Pancreas

Costal cartilage

Small intestine

Rectus abdominis m.

Large intestine

PLATE EIGHTEEN Transverse section of the abdomen through the kidneys, superior view. (*m.* stands for muscle)

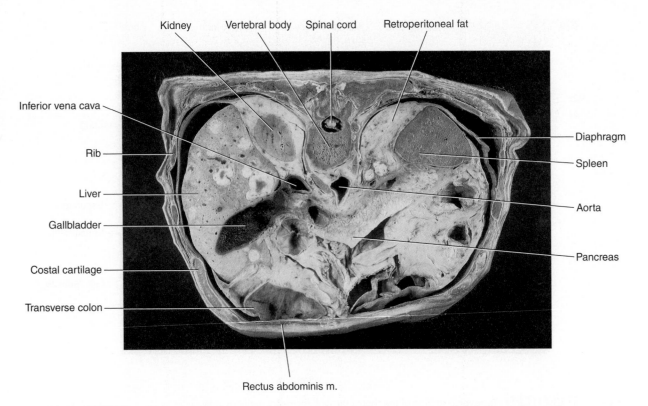

Kidney   Vertebral body   Spinal cord   Retroperitoneal fat

Inferior vena cava

Rib

Liver

Gallbladder

Costal cartilage

Transverse colon

Diaphragm

Spleen

Aorta

Pancreas

Rectus abdominis m.

PLATE NINETEEN   Transverse section of the abdomen through the pancreas, superior view. (*m.* stands for muscle)

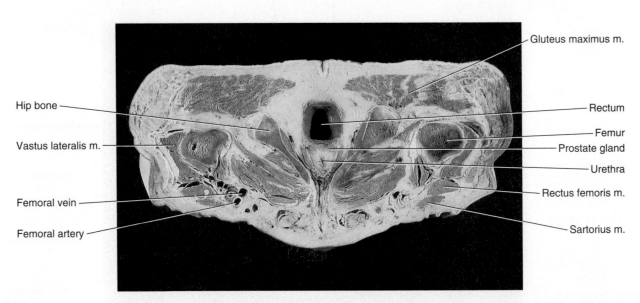

Gluteus maximus m.

Hip bone

Vastus lateralis m.

Femoral vein

Femoral artery

Rectum

Femur

Prostate gland

Urethra

Rectus femoris m.

Sartorius m.

PLATE TWENTY   Transverse section of the male pelvic cavity, superior view. (*m.* stands for muscle)

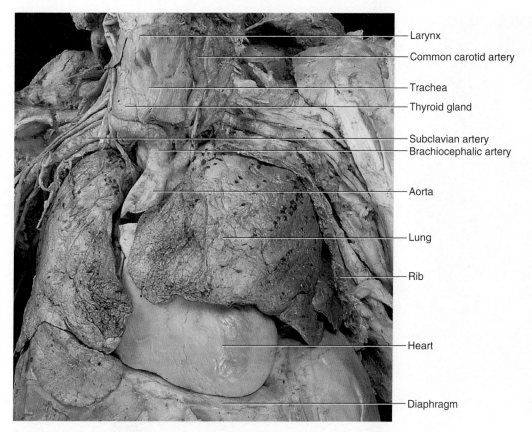

Larynx
Common carotid artery
Trachea
Thyroid gland
Subclavian artery
Brachiocephalic artery
Aorta
Lung
Rib
Heart
Diaphragm

PLATE TWENTY-ONE Thoracic viscera, anterior view. (Brachiocephalic veins have been removed to better expose the brachiocephalic artery and the aorta.)

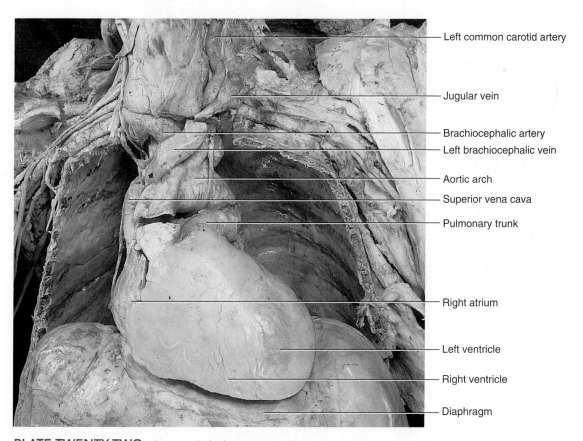

Left common carotid artery
Jugular vein
Brachiocephalic artery
Left brachiocephalic vein
Aortic arch
Superior vena cava
Pulmonary trunk
Right atrium
Left ventricle
Right ventricle
Diaphragm

PLATE TWENTY-TWO Thorax with the lungs removed, anterior view.

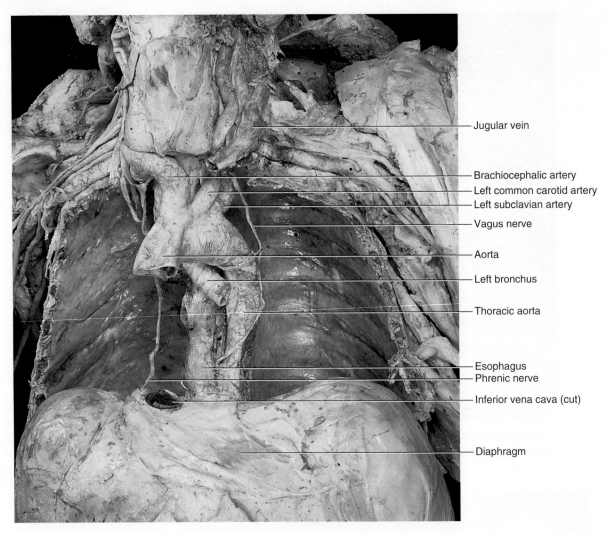

— Jugular vein

— Brachiocephalic artery
— Left common carotid artery
— Left subclavian artery
— Vagus nerve

— Aorta

— Left bronchus

— Thoracic aorta

— Esophagus
— Phrenic nerve
— Inferior vena cava (cut)

— Diaphragm

PLATE TWENTY-THREE Thorax with the heart and lungs removed, anterior view.

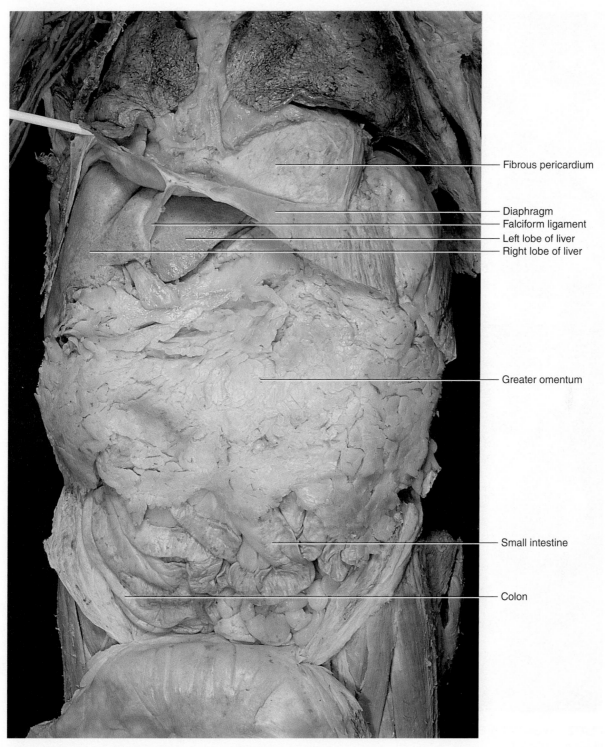

Fibrous pericardium

Diaphragm
Falciform ligament
Left lobe of liver
Right lobe of liver

Greater omentum

Small intestine

Colon

PLATE TWENTY-FOUR Abdominal viscera, anterior view.

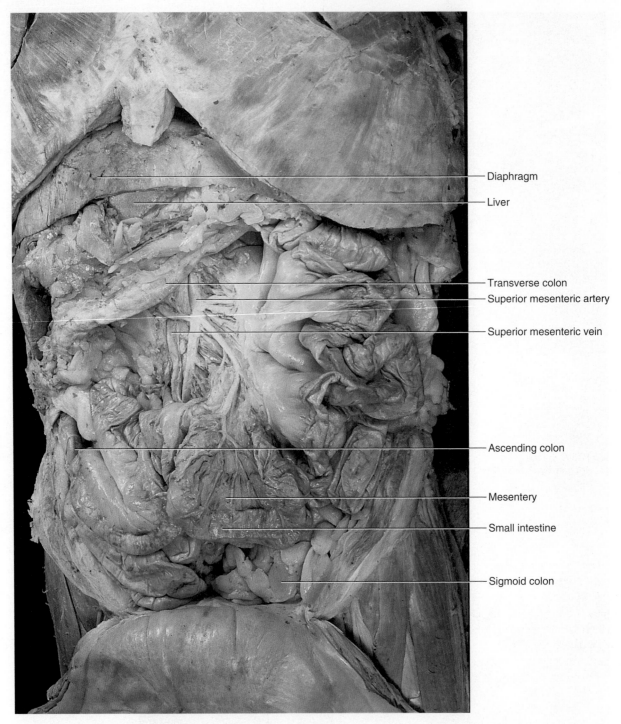

Diaphragm

Liver

Transverse colon

Superior mesenteric artery

Superior mesenteric vein

Ascending colon

Mesentery

Small intestine

Sigmoid colon

PLATE TWENTY-FIVE Abdominal viscera with the greater omentum removed, anterior view. (Small intestine has been displaced to the left.)

This partial model of DNA, the blueprint for an organism, shows carbon atoms black, oxygen red, nitrogen blue, phosphorus yellow, and hydrogen white.

Chemical Basis of Life

LEARNING OUTCOMES

After you have studied this chapter, you should be able to:

2.1 The Importance of Chemistry in Anatomy and Physiology

1 Give examples of how the study of living materials requires an understanding of chemistry. (p. 58)

2.2 Structure of Matter

2 Describe the relationships among matter, atoms, and compounds. (pp. 58–59)

3 Describe how atomic structure determines how atoms interact. (pp. 60 and 63–65)

4 Explain how molecular and structural formulas symbolize the composition of compounds. (pp. 62 and 64)

5 Describe three types of chemical reactions. (p. 66)

6 Describe the differences among acids, bases, and salts. (p. 66)

7 Explain the pH scale. (p. 67)

8 Explain the function of buffers. (p. 68)

2.3 Chemical Constituents of Cells

9 List the major inorganic chemicals common in cells and explain the function(s) of each. (p. 68)

10 Describe the general functions of the main classes of organic molecules in cells. (pp. 70–76)

THE WHOLE PICTURE

At the cellular level of organization, biology, in a sense, becomes chemistry. A cell's working parts—its organelles—are intricate assemblies of molecules. Because the molecules that build the cells that build tissues and organs are themselves composed of atoms, the study of anatomy and physiology begins with chemistry.

Chemistry in the body can show up in surprising ways. The ability of your nervous system to send signals at speeds of over 100 meters per second depends on the movement of chemical substances called ions. The force that your muscles generate to lift your textbook, or to walk up a flight of stairs, comes entirely from chemicals called proteins in your muscles pulling at each other at a microscopic level.

Module 2: Cells & Chemistry

UNDERSTANDING WORDS

bio-, life: *bio*chemistry—branch of science dealing with the chemistry of life forms.

di-, two: *di*saccharide—a molecule composed of two bonded simple sugar units.

glyc-, sweet: *glyc*ogen—complex carbohydrate composed of glucose molecules bonded in a particular way.

iso-, equal: *iso*tope—atom that has the same atomic number as another atom but a different atomic weight.

lip-, fat: *lip*ids—group of organic compounds that includes fats.

-lyt, break down: electro*lyte*—substance that breaks down and releases ions in water.

mono-, one: *mono*saccharide—a molecule consisting of a single simple sugar unit.

poly-, many: *poly*unsaturated—molecule with two or more double bonds between its carbon atoms.

sacchar-, sugar: mono*sacchar*ide—a molecule consisting of a single simple sugar unit.

syn-, together: *syn*thesis—process by which chemicals join to form new types of chemicals.

2.1 | The Importance of Chemistry in Anatomy and Physiology

Chemistry considers the composition of substances and how they change. It is possible to study anatomy without much reference to chemistry. However understanding the basics of chemistry is essential for understanding physiology, because body functions result from cellular functions that, in turn, result from chemical changes. The human body consists of chemicals, including salts, water, carbohydrates, lipids, proteins, and nucleic acids. The food that we eat, liquids that we drink, and medications that we take are chemicals.

Not only are physiological processes based on chemistry, but all anatomical structures are as well. Bones are in large part a composition of calcium, phosphorus, and proteins. Muscles, mostly formed of proteins, are held together and attached to bones by yet other types of proteins. Even water, the main component of the intracellular and extracellular fluids, is a chemical.

As interest in the chemistry of living organisms grew and knowledge of the subject expanded, a field of life science called biological chemistry, or **biochemistry,** emerged. Biochemistry has been important not only in helping to explain physiological processes but also in developing ways to detect, diagnose, and treat disease.

 PRACTICE

1 Why is a knowledge of chemistry essential to understanding physiology?

2 What is biochemistry?

2.2 | Structure of Matter

Matter is anything that has weight and takes up space. This includes all the solids, liquids, and gases in our surroundings as well as in our bodies. All matter consists of particles that are organized in specific ways. Table 2.1 lists some particles of matter and their characteristics.

Elements and Atoms

The simplest examples of matter with specific chemical properties are the **elements** (el'ĕ-mentz). At present, 98 naturally occurring elements are known and at least 20 more have been created in the laboratory. Among these elements are such common materials as iron, copper, silver, gold, aluminum, carbon, hydrogen, and oxygen. Some elements exist in a pure form, but these and other elements

---

Strictly speaking, matter has mass (rather than weight) and takes up space. Mass refers to the amount of a substance, whereas weight refers to how heavy it is. If your weight on earth is 150 pounds, on the moon it would be only 25 pounds, but your mass (in kilograms) would be the same in both places. That is, your composition is the same on the moon as it is on earth and takes up the same volume of space, but you weigh less on the moon, because the force of gravity is lower there. Because we are dealing with life on earth and constant gravity, we can consider mass and weight as roughly equivalent. Many of our students find it easier to think in terms of weight rather than mass.

TABLE 2.1	Some Particles of Matter		
Name	**Characteristic**	**Name**	**Characteristic**
Atom	Smallest particle of an element that has the properties of that element	Neutron (n^0)	Relatively large particle within an atom; about the same weight as a proton; uncharged and thus electrically neutral; found within an atomic nucleus
Electron (e^-)	Extremely small particle within an atom; almost no weight; carries a negative electrical charge and is in constant motion around an atomic nucleus	Ion	Particle, formed from an atom, that is electrically charged because it has gained or lost one or more electrons
Proton (p^+)	Relatively large particle within an atom; carries a positive electrical charge and is found within an atomic nucleus	Molecule	Particle formed by the chemical union of two or more atoms

are more commonly parts of chemical combinations called **compounds** (kom'powndz).

Elements the body requires in large amounts—such as carbon, hydrogen, oxygen, nitrogen, sulfur, and phosphorus—are termed **bulk elements.** These elements make up more than 95% (by weight) of the human body (table 2.2). Elements required in small amounts are called **trace elements.** Many trace elements are important parts of enzymes, which are proteins that regulate the rates of chemical reactions in living organisms. Some elements that are toxic in large amounts, such as arsenic, may be vital in very small amounts, and these are called **ultratrace elements.**

An **atom** (at'om) is the smallest unit of an element that has the chemical properties of that element. The atoms that make up each element are chemically identical but they differ from the atoms that make up other elements. Atoms vary in size, weight, and the ways they interact with other atoms. Some atoms can combine with atoms like themselves or with other atoms by forming attractions called **chemical bonds,** while other atoms cannot form such bonds.

Atomic Structure

An atom consists of a central portion called the **nucleus** (nu'kle-us) and one or more **electrons** (e-lek'tronz) that constantly move around the nucleus (fig. 2.1). The nucleus contains one or more relatively large particles, **protons** (pro'tonz) and usually **neutrons**

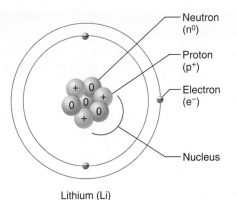

Lithium (Li)

FIGURE 2.1 AP|R An atom consists of subatomic particles. In an atom of the element lithium, three electrons encircle a nucleus that consists of three protons and four neutrons.

TABLE 2.2	Elements in the Human Body (By Weight)	
Major Elements	**Symbol**	**Approximate Percentage of the Human Body**
Oxygen	O	65.0
Carbon	C	18.5
Hydrogen	H	9.5
Nitrogen	N	3.2
Calcium	Ca	1.5
Phosphorus	P	1.0 — 99.9%
Potassium	K	0.4
Sulfur	S	0.3
Chlorine	Cl	0.2
Sodium	Na	0.2
Magnesium	Mg	0.1
Trace Elements		
Cobalt	Co	
Copper	Cu	
Fluorine	F	
Iodine	I	— less than 0.1%
Iron	Fe	
Manganese	Mn	
Zinc	Zn	

TABLE 2.3	Atomic Structure of Elements 1 Through 12							
Element	Symbol	Atomic Number	Approximate Atomic Weight	Protons	Neutrons	First Shell Electrons	Second Shell Electrons	Third Shell Electrons
Hydrogen	H	1	1	1	0	1		
Helium	He	2	4	2	2	2 (inert)		
Lithium	Li	3	7	3	4	2	1	
Beryllium	Be	4	9	4	5	2	2	
Boron	B	5	11	5	6	2	3	
Carbon	C	6	12	6	6	2	4	
Nitrogen	N	7	14	7	7	2	5	
Oxygen	O	8	16	8	8	2	6	
Fluorine	F	9	19	9	10	2	7	
Neon	Ne	10	20	10	10	2	8 (inert)	
Sodium	Na	11	23	11	12	2	8	1
Magnesium	Mg	12	24	12	12	2	8	2

For more detail, see Appendix D, Periodic Table of the Elements

(nu′tronz). Protons and neutrons are about equal in weight, but they have different electrical properties. Each proton carries a single, positive electrical charge (p$^+$). Neutrons are uncharged and thus are electrically neutral (n^0). Electrons, so small that they have almost no weight, carry a single, negative charge (e$^-$) (fig. 2.1).

The nucleus contains protons, so this part of an atom is always positively charged. However, the number of electrons outside the nucleus equals the number of protons. Therefore, a complete atom has no net charge and is thus electrically neutral.

The atoms of different elements have different numbers of protons. The number of protons in the atoms of a particular element is called its **atomic number.** Hydrogen, for example, whose atoms have one proton, has atomic number 1; carbon, whose atoms have six protons, has atomic number 6.

The weight of an atom of an element is primarily due to the protons and neutrons in its nucleus, because the electrons are so light. For this reason, the number of protons plus the number of neutrons in each of an element's atoms essentially equals the **atomic weight** of that atom. The atomic weight of a hydrogen atom, which has only one proton and no neutrons, is approximately 1. The atomic weight of a carbon atom, with six protons and six neutrons, is approximately 12 (table 2.3).

Isotopes

All the atoms of a particular element have the same atomic number because they have the same number of protons and electrons. However, the atoms of an element vary in the number of neutrons in their nuclei; thus, they vary in atomic weight. For example, all oxygen atoms have eight protons in their nuclei. Some, however, have eight neutrons (atomic weight 16), others have nine neutrons (atomic weight 17), and still others have ten neutrons (atomic weight 18). Atoms that have the same atomic number but different atomic weights are called **isotopes** (i′so-tōpz) of an element. A sample of an element is likely to include more than one isotope,

so the atomic weight of the element is often considered to be the average weight of the isotopes present. (See Appendix D, Periodic Table of the Elements.)

The ways atoms interact reflect their numbers of electrons. An atom has the same number of electrons and protons, so all the isotopes of a particular element have the same number of electrons and chemically react in the same manner. For example, any of the isotopes of oxygen can have the same function in metabolic reactions.

Isotopes of an element may be stable, or they may have unstable atomic nuclei that decompose, releasing energy or pieces of themselves until they reach a stable form. Such unstable isotopes are called *radioactive,* and the energy or atomic fragments they emit are called *atomic radiation.* Elements that have radioactive isotopes include oxygen, iodine, iron, phosphorus, and cobalt. Some radioactive isotopes are used to detect and treat disease (From Science to Technology 2.1).

Atomic radiation includes three common forms called alpha (α), beta (β), and gamma (γ). Each type of radioactive isotope produces one or more of these forms of radiation. Alpha radiation consists of particles from atomic nuclei, each of which includes two protons and two neutrons, that move slowly and cannot easily penetrate matter. Beta radiation consists of much smaller particles (electrons) that travel faster and more deeply penetrate matter. Gamma radiation is a form of energy similar to X-radiation and is the most penetrating form of atomic radiation. From Science to Technology 2.2 examines how radiation that moves electrons can affect human health.

 PRACTICE

3 What is the relationship between matter and elements?

4 Which elements are most common in the human body?

5 Where are electrons, protons, and neutrons located within an atom?

6 What is an isotope?

7 What is atomic radiation?

Vicki L. arrived early at the nuclear medicine department of the health center. As she sat in an isolated cubicle, a doctor in full sterile dress approached with a small metal canister marked with warnings. The doctor carefully unscrewed the top, inserted a straw, and watched as the young woman sipped the fluid within. It tasted like stale water but was a solution containing a radioactive isotope, iodine-131.

Vicki's thyroid gland had been removed three months earlier, and this test was to determine whether any active thyroid tissue remained. The thyroid is the only part of the body to metabolize iodine, so if Vicki's body retained any of the radioactive drink, it would mean that some of her cancerous thyroid gland remained. By using a radioactive isotope, her physicians could detect iodide uptake using a scanning device called a scintillation counter (fig. 2A). Figure 2B illustrates iodine-131 uptake in a complete thyroid gland.

The next day, Vicki returned for the scan, which showed that a small amount of thyroid tissue was left and functioning. Another treatment would be necessary. Vicki would drink enough radioactive iodide to destroy the remaining tissue. This time, she drank the solution in an isolation room lined with paper to keep her from contaminating the floor, walls, and furniture. The same physician administered the radioactive iodide. Vicki's physician had this job because his thyroid had been removed many years earlier, because he, too, had cancer, and therefore the radiation couldn't harm him.

After two days in isolation, Vicki went home with a list of odd instructions. She was to stay away from her children and pets, wash her clothing separately, use disposable utensils and plates, and flush the toilet three times each time she used it. These precautions would minimize her contaminating her family—mom was radioactive!

Iodine-131 is a medically useful radioactive isotope because it has a short half-life, a measurement of the time it takes for half of an amount of an isotope to decay to a nonradioactive form. The half-life of iodine-131 is 8.1 days. With the amount of radiation in Vicki's body dissipating by half every 8.1 days, after three months hardly any would be left. If all went well, any remaining cancer cells would leave her body along with the radioactive iodine.

Isotopes of other elements have different half-lives. The half-life of iron-59 is 45.1 days; that of phosphorus-32 is 14.3 days; that of cobalt-60 is 5.26 years; and that of radium-226 is 1,620 years.

A form of thallium-201 with a half-life of 73.5 hours is commonly used to detect disorders in the blood vessels supplying the heart muscle or to locate regions of damaged heart tissue after a heart attack.

Gallium-67, with a half-life of 78 hours, is used to detect and monitor the progress of certain cancers and inflammatory illnesses. These medical procedures inject the isotope into the blood and follow its path using detectors that record images on paper or film.

Radioactive isotopes are also used to assess kidney function, estimate the concentrations of hormones in body fluids, measure blood volume, and study changes in bone density. Cobalt-60 is a radioactive isotope used to treat some cancers. The cobalt emits radiation that damages cancer cells more readily than it does healthy cells.

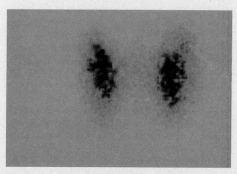

(a)

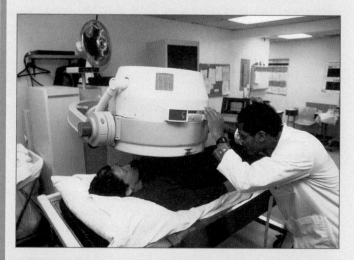

FIGURE 2A Scintillation counters detect radioactive isotopes.

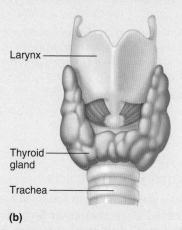

Larynx

Thyroid gland

Trachea

(b)

FIGURE 2B (*a*) A scan of the thyroid gland twenty-four hours after the patient receives radioactive iodine. Note how closely the scan in (*a*) resembles the shape of the thyroid gland shown in (*b*).

Alpha, beta, and gamma radiation are called ionizing radiation because their energy removes electrons from atoms (fig. 2C). Electrons dislodged by ionizing radiation can affect nearby atoms, disrupting physiology at the chemical level in a variety of ways—causing cancer, clouding the lens of the eye, and interfering with normal growth and development.

In the United States, some people are exposed to very low levels of ionizing radiation, mostly from background radiation, which originates from natural environmental sources (table 2A). For people who live near sites of atomic weapons manufacture, exposure is greater. Epidemiologists are investigating medical records that document illnesses linked to long-term exposure to ionizing radiation in a 1,200-square-kilometer area in Germany.

The lake near Oberrothenback, Germany, which appears inviting, harbors enough toxins to kill thousands of people. It is polluted with heavy metals, low-level radioactive chemical waste, and 22,500 tons of arsenic. Radon, a radioactive by-product of uranium, permeates the soil. Many farm animals and pets that have drunk from the lake have died. Cancer rates and respiratory disorders among the human residents nearby are well above normal.

The lake in Oberrothenback was once a dump for a factory that produced "yellow cake," a term for processed uranium ore,

BOX TABLE 2A	Sources of Ionizing Radiation
Background (Natural environmental)	Cosmic rays from space
	Radioactive elements in earth's crust
	Rocks and clay in building materials
	Radioactive elements naturally in the body (potassium-40, carbon-14)
Medical and dental	X rays
	Radioactive substances
Other	Atomic and nuclear weapons
	Mining and processing radioactive minerals
	Radioactive fuels in nuclear power plants
	Radioactive elements in consumer products (luminescent dials, smoke detectors, color TV components)

used to build atomic bombs for the former Soviet Union. In the early 1950s, nearly half a million workers labored here and in surrounding areas in factories and mines. Records released in 1989, after the reunification of Germany, reveal that workers were given perks, such as alcoholic beverages and better wages, to work in the more dangerous areas. The workers paid a heavy price: many died of lung ailments.

Until recently, concern over the health effects of exposure to ionizing radiation centered on the U.S. government's plan to transport tens of thousands of metric tons of high-level nuclear waste from 109 reactors around the country for burial beneath Yucca Mountain, in Nevada, which is about 100 miles from Las Vegas, by 2021. The waste, stored near the reactors, was to be buried in impenetrable containers under the mountain by robots. In the reactors, nuclear fuel rods contain uranium oxide, which produces electricity as it decays to plutonium, which gives off gamma rays. Periodically the fuel rods must be replaced, and the spent ones buried. In 2010 the federal government ended funding for the controversial Yucca Mountain plan. The search continues for an isolated place to store the nuclear waste.

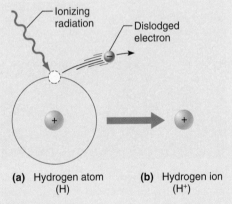

(a) Hydrogen atom (H) (b) Hydrogen ion (H⁺)

FIGURE 2C Ionizing radiation removes electrons from atoms. (a) Ionizing radiation may dislodge an electron from an electrically neutral hydrogen atom. (b) Without its electron, the hydrogen atom becomes a positively charged hydrogen ion (H⁺).

Molecules and Compounds

Two or more atoms may combine to form a distinctive type of particle called a **molecule. A molecular formula** is shorthand used to depict the numbers and types of atoms in a molecule. It consists of the symbols of the elements in the molecule with numerical subscripts that indicate how many atoms of each element are present. For example, the molecular formula for water is H_2O, which indicates two atoms of hydrogen and one atom of

oxygen in each molecule. The molecular formula for the sugar glucose, $C_6H_{12}O_6$, indicates six atoms of carbon, twelve atoms of hydrogen, and six atoms of oxygen.

If atoms of the same element combine, they produce molecules of that element. Gases of hydrogen (H_2), oxygen (O_2), and nitrogen (N_2) consist of such molecules. If atoms of different elements combine, molecules of compounds form. Two atoms of hydrogen, for example, can combine with one atom of oxygen to produce a

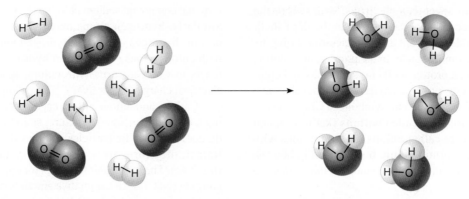

FIGURE 2.2 Under certain conditions, hydrogen molecules can combine with oxygen molecules, forming water molecules.

molecule of the compound water (H_2O), as **figure 2.2** shows. Table sugar, baking soda, natural gas, beverage alcohol, and most drugs are compounds.

A molecule of a compound always consists of definite types and numbers of atoms. A molecule of water (H_2O), for instance, always has two hydrogen atoms and one oxygen atom. If two hydrogen atoms combine with two oxygen atoms, the compound formed is not water, but hydrogen peroxide (H_2O_2).

Bonding of Atoms

Atoms combine with other atoms by forming links called **bonds.** Chemical bonds result from interactions of electrons.

The electrons of an atom occupy one or more regions of space called **electron shells** that encircle the nucleus. Because electrons have a level of energy characteristic of the particular shell they are in, the shells are also called *energy shells.* Each electron shell can hold a limited number of electrons. The maximum number of electrons that each of the first three shells can hold for elements of atomic number 18 and under is:

First shell (closest to the nucleus)	2 electrons
Second shell	8 electrons
Third shell	8 electrons

More complex atoms may have as many as eighteen electrons in the third shell.

Simplified diagrams such as those in **figure 2.3** are used to show electron configuration in atoms. The single electron of a hydrogen atom is in the first shell; the two electrons of a helium atom fill its first shell; and of the three electrons of a lithium atom, two are in the first shell and one is in the second shell (fig. 2.3*a*). Lower energy shells, closer to the nucleus, must be filled first. Figure 2.3*b* depicts a more complex atom, of the element sodium, and how the atomic number derives from the number of protons and the atomic weight from the number of protons and neutrons.

The number of electrons in the outermost shell of an atom determines whether it will react with another atom. Atoms react in a way that leaves the outermost shell completely filled with electrons, achieving a more stable structure. This is called the **octet rule,** because, except for the first shell, eight electrons are required to fill the shells in most of the atoms important in living organisms.

Atoms such as helium, whose outermost electron shells are filled, already have stable structures and are chemically inactive

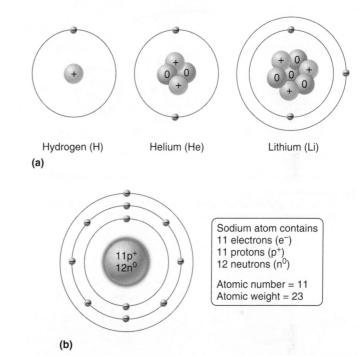

Hydrogen (H)   Helium (He)   Lithium (Li)

(a)

Sodium atom contains
11 electrons (e^-)
11 protons (p^+)
12 neutrons (n^0)

Atomic number = 11
Atomic weight = 23

(b)

FIGURE 2.3 Electrons orbit the atomic nucleus. (*a*) The single electron of a hydrogen atom moves within its first shell. The two electrons of a helium atom fill its first shell. Two of the three electrons of a lithium atom are in the first shell, and one is in the second shell. (*b*) A sodium atom has 11 protons and 12 neutrons in its nucleus, and 11 electrons that move within three shells.

or **inert** (they cannot form chemical bonds). Atoms with incompletely filled outer shells, such as those of hydrogen or lithium, tend to gain, lose, or share electrons in ways that empty or fill their outer shells. In this way, they achieve stable structures.

Atoms that gain or lose electrons become electrically charged and are called **ions** (i'onz). An atom of sodium, for example, has eleven electrons: two in the first shell, eight in the second shell, and one in the third shell. This atom tends to lose the electron from its outer (third) shell, which leaves the second (now the outermost) shell filled and the new form stable (**fig. 2.4*a***). In the process, sodium is left with eleven protons (11^+) in its nucleus and only ten electrons (10^-). As a result, the atom develops a net electrical charge of $^+1$ and is called a sodium ion, symbolized Na^+.

A chlorine atom has seventeen electrons, with two in the first shell, eight in the second shell, and seven in the third shell. An atom of this type tends to accept a single electron, filling its outer (third) shell and becoming stable. In the process, the chlorine atom is left with seventeen protons (17^+) in its nucleus and eighteen electrons (18^-). As a result, the atom develops a net electrical charge of -1 and is called a chloride ion, symbolized Cl^-.

Positively charged ions are called **cations** (kat'i-onz), and negatively charged ions are called **anions** (an'i-onz). Ions with opposite charges attract, forming **ionic bonds** (i-on'ik bondz). Sodium ions (Na^+) and chloride ions (Cl^-) unite in this manner to form the compound sodium chloride (NaCl), or table salt (fig. 2.4*b*). Similarly, hydrogen atoms may lose their single electrons and become hydrogen ions (H^+). Hydrogen ions can form ionic bonds with chloride ions (Cl^-) to form hydrogen chloride (HCl, which reacts in water to form hydrochloric acid). Some ions have an electrical charge greater than 1—for example, Ca^{+2} (or Ca^{++}).

Cations and anions attract each other in all directions, forming a three-dimensional structure, so ionically bound compounds do not form specific particles and they do not exist as molecules. Rather, they form arrays, such as crystals of sodium chloride (fig. 2.4*c*). The molecular formulas for compounds such as sodium chloride (NaCl) give the relative amounts of each element.

Atoms may also bond by sharing electrons rather than by gaining or losing them. A hydrogen atom, for example, has one electron in its first shell but requires two electrons to achieve a stable structure. It may fill this shell by combining with another hydrogen atom in such a way that the two atoms share a pair of electrons. As **figure 2.5*a*** shows, the two electrons then encircle the nuclei of both atoms, filling the outermost shell, and each atom becomes stable. A chemical bond between atoms that share electrons is called a **covalent bond** (ko'va-lent bond). Figure 2.5*b* shows the covalent bonding that joins the atoms of a very familiar molecule—water.

Usually atoms of each element form a specific number of covalent bonds. Hydrogen atoms form one bond, oxygen atoms form two bonds, nitrogen atoms form three bonds, and carbon atoms form four bonds. Atomic symbols and lines can be used to represent the bonding capacities of these atoms, as follows:

Atomic symbols and lines show how atoms bond and are arranged in various molecules. One pair of shared electrons, a single covalent bond, is depicted with a single line. Sometimes atoms may share two pairs of electrons (a double covalent bond), or even three pairs (a triple covalent bond), represented by two and three lines, respectively. Illustrations of this type, called **structural formulas** (fig. 2.6), are useful, but they cannot adequately capture the three-dimensional forms of molecules. In contrast, figure 2.7 shows a three-dimensional (space-filling) representation of a water molecule.

In ionic bonds, one or more electrons of one atom are pulled entirely toward another atom. In covalent bonds, atoms share electrons equally. In between these two extremes lies the covalent bond in which electrons are shared, but are not shared equally, such that the shared electrons move more toward one of the bonded atoms. This results in a molecule with an uneven distribution of charges. Such a molecule is called **polar.** Unlike an ion, a polar molecule has an equal number of protons and electrons, but more of the electrons are at one end of the molecule, making that end slightly negative, while the other end of the molecule is slightly positive. Typically, polar covalent bonds form where hydrogen atoms bond to oxygen or nitrogen atoms. Water is a polar molecule (fig. 2.8*a*).

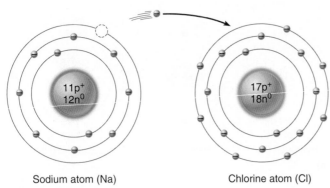

Sodium atom (Na)   Chlorine atom (Cl)

(a) Separate atoms
If a sodium atom loses an electron to a chlorine atom, the sodium atom becomes a sodium ion (Na^+), and the chlorine atom becomes a chloride ion (Cl^-).

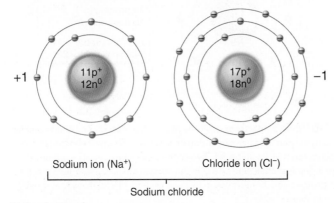

+1   11p$^+$ 12n^0 17p$^+$ 18n^0 -1

Sodium ion (Na$^+$)   Chloride ion (Cl$^-$)

Sodium chloride

(b) Bonded ions
These oppositely charged particles attract electrically and join by an ionic bond.

(c) Salt crystal
Ionically bonded substances form arrays such as a crystal of NaCl.

FIGURE 2.4 **AP|R** An ionic bond forms when (*a*) one atom loses and another atom gains one or more electrons and then (*b*) oppositely charged ions attract. (*c*) Ionically bonded substances may form crystals.

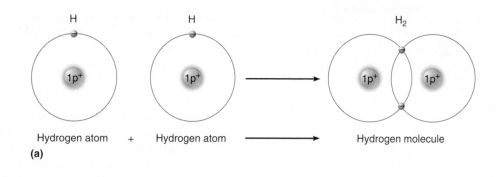

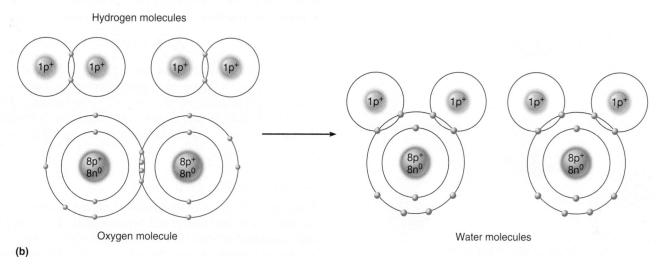

Hydrogen atom   +   Hydrogen atom   →   Hydrogen molecule

(a)

Hydrogen molecules

Oxygen molecule

Water molecules

(b)

FIGURE 2.5 A covalent bond forms when (*a*) two hydrogen atoms share a pair of electrons to form a hydrogen molecule. (*b*) Hydrogen molecules can combine with oxygen molecules, forming water molecules. The overlapping shells represent the shared electrons of covalent bonds.

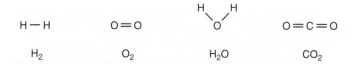

H—H   O=O $\overset{\displaystyle H \qquad H}{\underset{\displaystyle O}{\diagdown \diagup}}$ O=C=O

H_2 O_2 H_2O CO_2

FIGURE 2.6 Structural and molecular formulas depict molecules of hydrogen, oxygen, water, and carbon dioxide. Note the double covalent bonds. (Triple covalent bonds are also possible between some atoms.)

FIGURE 2.7 A three-dimensional model represents this water molecule (H_2O). The white parts represent the hydrogen atoms, and the red part represents oxygen.

The attraction of the positive hydrogen end of a polar molecule to the negative nitrogen or oxygen end of another polar molecule is called a **hydrogen bond.** These bonds are relatively weak. For example, below 0°C, the hydrogen bonds between water molecules shown in figure 2.8*b* are strong enough to form ice. As the temperature rises, increased molecular movement breaks the hydrogen bonds, and water becomes liquid. At body temperature, hydrogen bonds are important in protein and nucleic acid structure. In these cases, many hydrogen bonds form between polar regions of a single, very large molecule. Together, these individually weak bonds provide strength.

PRACTICE

8 Distinguish between a molecule and a compound.

9 What is an ion?

10 Describe two ways that atoms may combine with other atoms.

11 What is a molecular formula? A structural formula?

12 Distinguish between an ion and a polar molecule.

Chemical Reactions

Chemical reactions form or break bonds between atoms, ions, or molecules. The starting materials changed by the chemical reaction are called **reactants** (re-ak′tantz). The atoms, ions, or molecules formed at the reaction's conclusion are called **products.** When

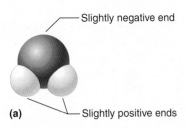

Slightly negative end

Slightly positive ends

(a)

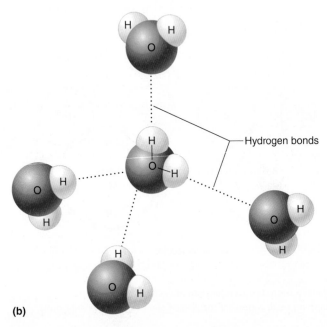

Hydrogen bonds

(b)

FIGURE 2.8 Water is a polar molecule. (*a*) Water molecules have equal numbers of electrons and protons but are polar because the electrons are shared unequally, creating slightly negative ends and slightly positive ends. (*b*) Water molecules form hydrogen bonds with each other.

How does the number of protons in a water molecule compare with the number of electrons?

Answer can be found in Appendix G.

two or more atoms, ions, or molecules bond to form a more complex structure, as when hydrogen and oxygen atoms bond to form molecules of water, the reaction is called **synthesis** (sin′thĕ-sis). Such a reaction can be symbolized as:

$$A + B \rightarrow AB$$

If the bonds of a reactant molecule break to form simpler molecules, atoms, or ions, the reaction is called **decomposition** (de-kom″po-zish′un). For example, molecules of water can decompose to yield the products hydrogen and oxygen. Decomposition is symbolized as:

$$AB \rightarrow A + B$$

Synthetic reactions, which build larger molecules from smaller ones, are particularly important in growth of body parts and repair of worn or damaged tissues. Decomposition reactions digest nutrient molecules into molecules small enough to be absorbed into the bloodstream in the small intestine.

In a third type of chemical reaction, an **exchange reaction** (replacement reaction), parts of two different types of molecules trade positions as bonds are broken and new bonds are formed. The reaction is symbolized as:

$$AB + CD \rightarrow AD + CB$$

An example of an exchange reaction is an acid reacting with a base, producing water and a salt. The following section discusses this type of reaction.

Many chemical reactions are reversible. This means the product or products can change back to the reactant or reactants. A **reversible reaction** is symbolized using a double arrow:

$$A + B \rightleftarrows AB$$

Whether a reversible reaction proceeds in one direction or another depends on the relative proportions of reactant (or reactants) and product (or products) as well as the amount of energy available. **Catalysts** (kat′ah-listz) are molecules that influence the rates (not the direction) of chemical reactions but are not consumed in the process.

Acids, Bases, and Salts

When ionically bound substances are placed in water, the ions are attracted to the positive and negative ends of the water molecules and tend to leave each other, or dissociate. In this way, the polarity of water dissociates the salts in the internal environment. Sodium chloride (NaCl), for example, ionizes into sodium ions (Na^+) and chloride ions (Cl^-) in water (fig. 2.9). This reaction is represented as:

$$NaCl \rightarrow Na^+ + Cl^-$$

The resulting solution has electrically charged particles (ions), so it conducts an electric current. Substances that release ions in water are, therefore, called **electrolytes** (e-lek′tro-lītz).

Electrolytes that dissociate to release hydrogen ions (H^+) in water are called **acids** (as′idz). For example, in water, the compound hydrochloric acid (HCl) releases hydrogen ions (H^+) and chloride ions (Cl^-):

$$HCl \rightarrow H^+ + Cl^-$$

Substances that release ions that combine with hydrogen ions are called **bases.** The compound sodium hydroxide (NaOH) in water releases hydroxide ions (OH^-), which can combine with hydrogen ions to form water. Thus, sodium hydroxide is a base:

$$NaOH \rightarrow Na^+ + OH^-$$

(*Note:* Some ions, such as OH^-, consist of two or more atoms. However, such a group behaves as a unit and usually remains unchanged during a chemical reaction.)

Bases can react with acids to neutralize them, forming water and electrolytes called **salts.** For example, hydrochloric acid and sodium hydroxide react to form water and sodium chloride:

$$HCl + NaOH \rightarrow H_2O + NaCl$$

Table 2.4 summarizes the three types of electrolytes.

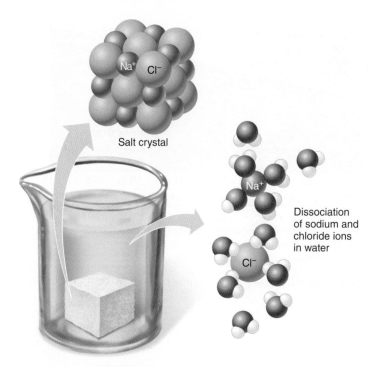

FIGURE 2.9 AP|R The polar nature of water molecules dissociates sodium chloride (NaCl) in water, releasing sodium ions (Na⁺) and chloride ions (Cl⁻).

Acid and Base Concentrations

Concentrations of acids and bases affect the chemical reactions that constitute many life processes, such as those controlling breathing rate. Therefore, the concentrations of these substances in body fluids are of special importance.

Hydrogen ion concentration can be measured in grams of ions per liter of solution. However, because hydrogen ion concentration can vary over a wide range (gastric juice has 0.01 grams H⁺/liter; household ammonia has 0.00000000001 grams H⁺/liter), shorthand called the **pH scale** is used. This system tracks the number of decimal places in a hydrogen ion concentration without writing them out. For example, a solution with a hydrogen ion concentration of 0.1 grams per liter has a pH of 1.0; a concentration of 0.01 g H⁺/L has pH 2.0; 0.001 g H⁺/L is pH 3.0; and so forth. Each whole number on the pH scale, which extends from 0 to 14, represents a tenfold difference in hydrogen ion concentration. As the hydrogen ion concentration increases, the pH number decreases. For example,

a solution of pH 6 has ten times the hydrogen ion concentration as a solution with pH 7. Therefore, small changes in pH reflect large changes in hydrogen ion concentration.

In pure water, which ionizes only slightly, the hydrogen ion concentration is 0.0000001 g/L, and the pH is 7.0. Water ionizes to release equal numbers of acidic hydrogen ions and basic hydroxide ions, so it is *neutral.*

$$H_2O \rightarrow H^+ + OH^-$$

Many bases are present in body fluids, but because of the way bases react in water, the concentration of hydroxide ions is a good estimate of the total base concentration. The concentrations of hydrogen ions and hydroxide ions in body fluids are always in balance such that if one increases, the other decreases, and vice versa. Solutions with more hydrogen ions than hydroxide ions are *acidic.* That is, acidic solutions have pH values less than 7.0 (fig. 2.10). Solutions with fewer hydrogen ions than hydroxide ions are *basic* (alkaline); they have pH values greater than 7.0.

Table 2.5 summarizes the relationship between hydrogen ion concentration and pH. Chapter 21 (pp. 812–813) discusses the regulation of hydrogen ion concentrations in the internal environment.

Negative feedback mechanisms such as those described in chapter 1 (p. 16) regulate the pH of the internal environment within a narrow pH range. Illness results when pH changes. The normal pH of blood, for example, is 7.35 to 7.45 (pH values can be decimals, in this case between 7 and 8). Because the pH of blood is normally slightly alkaline, we need terms other than *acidic* and *alkaline* to refer to abnormal blood pH. Blood pH of 7.5 to 7.8 is called **alkalosis** (al″kah-lo′sis), and makes one feel agitated and dizzy. This can be caused by breathing rapidly at high altitudes, taking too many antacids, high fever, anxiety, or mild to moderate vomiting that rids the body of stomach acid. **Acidosis** (as′ĭ-do′sis),

TABLE 2.4	Types of Electrolytes	
	Characteristic	**Examples**
Acid	Substance that releases hydrogen ions (H⁺)	Carbonic acid, hydrochloric acid, acetic acid, phosphoric acid
Base	Substance that releases ions that can combine with hydrogen ions	Sodium hydroxide, potassium hydroxide, magnesium hydroxide, sodium bicarbonate
Salt	Substance formed by the reaction between an acid and a base	Sodium chloride, aluminum chloride, magnesium sulfate

TABLE 2.5	Hydrogen Ion Concentrations and pH	
Grams of H⁺ per Liter	**pH**	
0.00000000000001	14	
0.0000000000001	13	
0.000000000001	12	
0.00000000001	11	
0.0000000001	10	
0.000000001	9	
0.00000001	8	Increasingly basic
0.0000001	7	Neutral—neither acidic nor basic
0.000001	6	Increasingly acidic
0.00001	5	
0.0001	4	
0.001	3	
0.01	2	
0.1	1	
1.00	0	

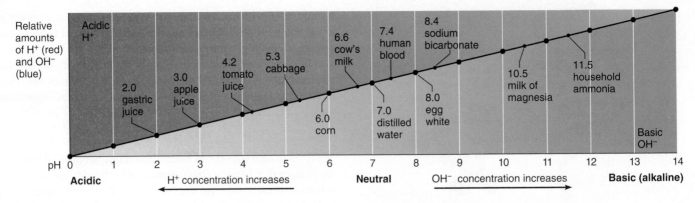

FIGURE 2.10 The pH scale reflects the hydrogen ion (H⁺) concentration. As the concentration of H⁺ increases, a solution becomes more acidic and the pH value decreases. As the concentration of ions that bond with H⁺ (such as hydroxide ions) increases, a solution becomes more basic (alkaline) and the pH value increases. The pH values of some common substances are shown.

 How does the hydrogen ion concentration compare between a solution of pH 6.4 and a solution of pH 8.4?

Answer can be found in Appendix G.

in which blood pH falls to 7.0 to 7.3, makes one feel disoriented and fatigued, and breathing may become difficult. This condition can result from severe vomiting that empties the alkaline small intestinal contents, diabetes, brain damage, impaired breathing, and lung and kidney disease.

Buffers are chemicals that resist pH change. They combine with hydrogen ions when these ions are in excess, or they donate hydrogen ions when these ions are depleted. Chapter 21 discusses buffers (pp. 814–815).

 PRACTICE

13 Describe three types of chemical reactions.

14 Compare the characteristics of an acid, a base, and a salt.

15 What does the pH scale measure?

16 What is a buffer?

2.3 | Chemical Constituents of Cells

Chemicals, including those that take part in metabolism (the cell's energy reactions), are of two general types. **Organic** (or-gan′ik) compounds have carbon and hydrogen. All other chemicals are **inorganic** (in′or-gan′ik). Many organic molecules have long chains or ring structures that can form because of a carbon atom's ability to form four covalent bonds.

The word *organic* has several meanings. Technically, it refers to carbon-containing chemical compounds (except carbon dioxide), as opposed to inorganic compounds that do not contain carbon. "Organic" may also refer to a substance obtained from an organism, or, even more generally, indicate a fundamental characteristic. In agriculture, "organic" refers to growing crops using only naturally occurring chemicals.

Inorganic substances usually dissolve in water and dissociate, forming ions. Therefore, they are *electrolytes*. Many organic compounds dissolve in water, but most dissolve in organic liquids such as ether or alcohol. Most organic compounds that dissolve in water do not release ions and are therefore called *nonelectrolytes*.

Inorganic Substances

Common inorganic substances in cells include water, oxygen, carbon dioxide, and inorganic salts.

Water

Water (H_2O) is the most abundant compound in living material and accounts for about two-thirds of the weight of an adult human. It is the major component of blood and other body fluids, including fluids in cells.

When substances dissolve in water, the polar water molecules separate molecules of the substance, or even dissociate them into ions. These liberated particles are much more likely to react. Consequently, most metabolic reactions occur in water.

Water is also important in transporting chemicals in the body. Blood, which is mostly water, carries oxygen, sugars, salts, vitamins, and other vital substances from organs of the digestive and respiratory systems to cells. Blood also carries waste materials, such as carbon dioxide and urea, from cells to the lungs and kidneys, respectively, which remove them from the blood and release them outside the body.

Water absorbs and carries heat. Blood brings heat released from muscle cells during exercise from deeper parts of the body to the surface. At the body surface, skin cells secrete water as part of sweat that can release heat by evaporation.

The body regularly gains and loses water, but it must be present in the correct concentration in the extracellular fluid, to maintain homeostasis. Such a condition, where gains and losses are equal, is called **water balance.**

Oxygen

Molecules of oxygen gas (O_2) enter the internal environment through the respiratory organs and are carried throughout the body by the blood, especially by red blood cells. In cells, organelles use oxygen to release energy from nutrient molecules. The energy then drives the cell's metabolic activities. A continuing supply of oxygen is necessary for cell survival and, ultimately, for the survival of the organism.

Carbon Dioxide

Carbon dioxide (CO_2) is a simple, carbon-containing inorganic compound. It is a waste product of the release of energy during certain metabolic reactions. As carbon dioxide moves from cells into the interstitial fluid and blood, most of it reacts with water to form a weak acid (carbonic acid, H_2CO_3). This acid ionizes, releasing hydrogen ions (H^+) and bicarbonate ions (HCO_3^-), which the blood carries to the respiratory organs. There, the chemical reactions reverse, and carbon dioxide gas is produced and is then exhaled.

Inorganic Salts

Inorganic salts are abundant in body fluids. They are the sources of many necessary ions, including ions of sodium (Na^+), chloride (Cl^-), potassium (K^+), calcium (Ca^{+2}), magnesium (Mg^{+2}), phosphate (PO_4^{-2}), carbonate (CO_3^{-2}), bicarbonate (HCO_3^-), and sulfate (SO_4^{-2}). These ions play important roles in metabolism, helping to maintain proper water concentrations in body fluids, controlling pH, blood clotting, bone development, energy transfer

NO (nitric oxide) and CO (carbon monoxide) are two small chemicals that can harm health, yet are also important to normal physiology. NO is found in smog, cigarettes, and acid rain. CO is a colorless, odorless, gas that is deadly when it leaks from home heating systems or exhaust pipes in closed garages. However, NO and CO are important biological messenger molecules. NO is involved in digestion, memory, immunity, respiration, and circulation. CO functions in the spleen, which recycles old red blood cells, and in the parts of the brain that control memory, smell, and vital functions.

in cells, and muscle and nerve functions. The body regularly gains and loses these electrolytes, but they must be present in certain concentrations, in the extracellular fluid, to maintain homeostasis. Such a condition, where gains and losses are equal, is called **electrolyte balance.** Table 2.6 summarizes the functions of some of the inorganic substances in the body.

 PRACTICE

17 What are the general differences between an organic molecule and an inorganic molecule?

18 What is the difference between an electrolyte and a nonelectrolyte?

19 Define electrolyte balance.

TABLE 2.6	Inorganic Substances Common in the Body	
Substance	**Symbol or Formula**	**Functions**
I. Inorganic Molecules		
Water	H_2O	Major component of body fluids (chapter 21, p. 805); medium in which most biochemical reactions occur; carries various chemical substances (chapter 14, p. 541); helps regulate body temperature (chapter 6, p. 189)
Oxygen	O_2	Used in release of energy from glucose molecules (chapter 4, p. 131)
Carbon dioxide	CO_2	Waste product that results from metabolism (chapter 4, p. 129); reacts with water to form carbonic acid (chapter 19, p. 752)
II. Inorganic Ions		
Bicarbonate ions	HCO_3^-	Help maintain acid-base balance (chapter 21, p. 814)
Calcium ions	Ca^{+2}	Necessary for bone development (chapter 7, p. 207); muscle contraction (chapter 9, p. 298), and blood clotting (chapter 14, fig. 14.19, p. 545)
Carbonate ions	CO_3^{-2}	Component of bone tissue (chapter 7, p. 211)
Chloride ions	Cl^-	Help maintain water balance (chapter 21, p. 805)
Hydrogen ions	H^+	pH of the internal environment (chapters 19 and 20, pp. 752 and 787)
Magnesium ions	Mg^{+2}	Component of bone tissue (chapter 7, p. 211); required for certain metabolic processes (chapter 18, p. 714)
Phosphate ions	PO_4^{-3}	Required for synthesis of ATP and nucleic acids (chapter 4, pp. 126 and 133); component of bone tissue (chapter 7, p. 211); help maintain polarization of cell membranes (chapter 10, p. 374)
Potassium ions	K^+	Required for polarization of cell membranes (chapter 10, p. 372)
Sodium ions	Na^+	Required for polarization of cell membranes (chapter 10, p. 372); help maintain water balance (chapter 21, p. 809)
Sulfate ions	SO_4^{-2}	Help maintain polarization of cell membranes (chapter 10, p. 374) and acid-base balance (chapter 21, p. 805)

Organic Substances

Important groups of organic chemicals in cells include carbohydrates, lipids, proteins, and nucleic acids.

Carbohydrates

Carbohydrates (kar′bo-hi′drātz) provide much of the energy that cells require. They also supply materials to build certain cell structures and they often are stored as reserve energy supplies.

Carbohydrates are water-soluble molecules that include atoms of carbon, hydrogen, and oxygen. Most of these molecules have twice as many hydrogen as oxygen atoms, which is the ratio of hydrogen to oxygen in water molecules (H_2O). This ratio is easy to see in the molecular formulas of the carbohydrates glucose ($C_6H_{12}O_6$) and sucrose ($C_{12}H_{22}O_{11}$).

Carbohydrates are classified by size. Simple carbohydrates, or **sugars,** include the **monosaccharides** (mon″o-sak′ah-rīdz) (single sugars) and **disaccharides** (di-sak′ah-rīdz) (double sugars). A monosaccharide may include from three to seven carbon atoms, in a straight chain or a ring (fig. 2.11). Monosaccharides include the five-carbon sugars ribose and deoxyribose, as well as the six-carbon sugars glucose, dextrose (a form of glucose), fructose, and galactose (fig. 2.12a). Disaccharides consist of two 6-carbon units (fig. 2.12b). Sucrose (table sugar) and lactose (milk sugar) are disaccharides.

Complex carbohydrates, also called **polysaccharides** (pol″e-sak′ah-rīdz), are built of simple carbohydrates linked to form larger molecules of different sizes (fig. 2.12c). Cellulose is a polysaccharide abundant in plants. It is made of many bonded glucose molecules. Humans cannot digest cellulose. It is considered to be dietary fiber, passing through the gastrointestinal tract without being broken down and absorbed into the bloodstream. Plant starch is another type of polysaccharide. Starch molecules consist of highly branched chains of glucose molecules connected differently than in cellulose. Humans easily digest starch.

Humans synthesize a polysaccharide similar to starch called *glycogen,* which is stored in the liver and skeletal muscles. Its molecules also are branched chains of sugar units; each branch consists of up to a dozen glucose units.

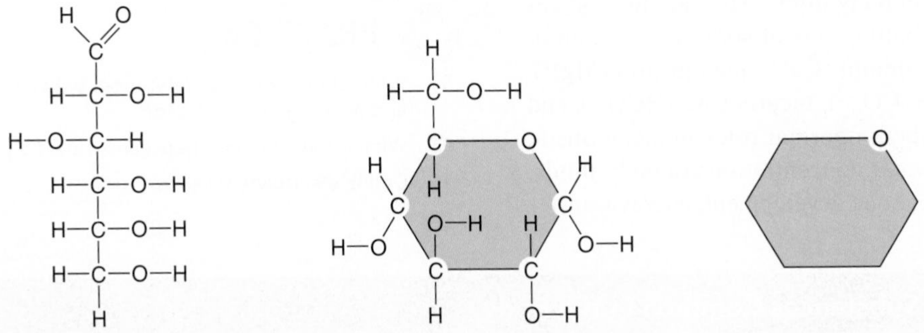

(a) Some glucose molecules ($C_6H_{12}O_6$) have a straight chain of carbon atoms.

(b) More commonly, glucose molecules form a ring structure.

(c) This shape symbolizes the ring structure of a glucose molecule.

FIGURE 2.11 Structural formulas depict a molecule of glucose ($C_6H_{12}O_6$).

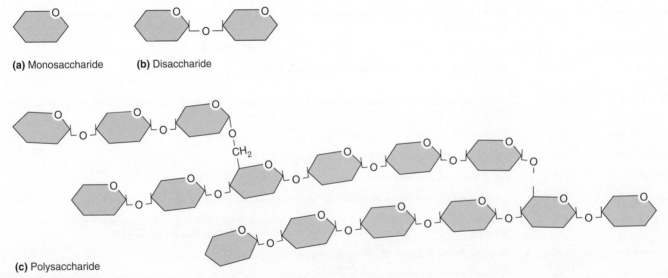

(a) Monosaccharide

(b) Disaccharide

(c) Polysaccharide

FIGURE 2.12 Carbohydrate molecules vary in size. (*a*) A monosaccharide molecule consists of one 6-carbon atom building block. (*b*) A disaccharide molecule consists of two of these building blocks. (*c*) A polysaccharide molecule consists of many of these building blocks.

Lipids

Lipids (lip'idz) are a group of organic chemicals that are insoluble in water but soluble in organic solvents, such as ether and chloroform. Lipids include a number of compounds, such as fats, phospholipids, and steroids, that have vital functions in cells and are important constituents of cell membranes (see chapter 3, p. 86). The most abundant lipids are the *fats,* primarily used to supply energy for cellular activities. Fat molecules can supply more energy gram for gram than can carbohydrate molecules.

Like carbohydrates, fat molecules are composed of carbon, hydrogen, and oxygen atoms. However, fats have a much smaller proportion of oxygen than do carbohydrates. The formula for the fat *tristearin*, $C_{57}H_{110}O_6$, illustrates these characteristic proportions.

The building blocks of fat molecules are **fatty acids** (fat'e as'idz) and **glycerol** (glis'er-ol). Although the glycerol portion of every fat molecule is the same, there are many types of fatty acids and, therefore, many types of fats. All fatty acid molecules include a carboxyl group (—COOH) at the end of a chain of carbon atoms.

Fatty acids differ in the lengths of their carbon atom chains, which usually have an even number of carbon atoms. The fatty acid chains also may vary in the ways the carbon atoms join. In some molecules, single carbon-carbon bonds link all the carbon atoms. This type of fatty acid is called a **saturated fatty acid;** that is, each carbon atom binds as many hydrogen atoms as possible and is thus saturated with them. In contrast, the chains of **unsaturated fatty acids** have one or more double bonds between carbon atoms. Fatty acids with one double bond are called *mono-unsaturated fatty acids,* and those with two or more double bonds are *polyunsaturated fatty acids* (fig. 2.13).

A glycerol molecule combines with three fatty acid molecules to form a single fat molecule, or *triglyceride* (figs. 2.14 and 2.15*a*). The fatty acids of a triglyceride may have different lengths and degrees of saturation, making the fats very diverse. Fat molecules that have only saturated fatty acids are called **saturated fats** (sat'u-rāt"ed fatz), and those that have unsaturated fatty acids are called **unsaturated fats** (unsat'u-rāted fatz). Each type of fat molecule has distinct properties.

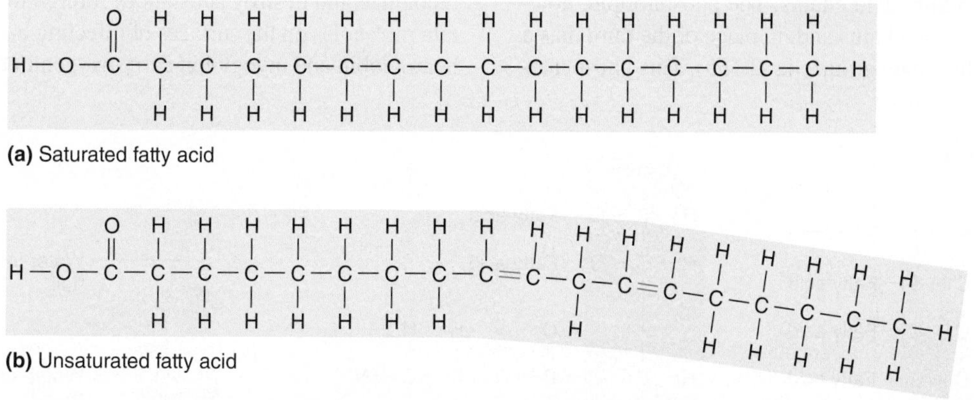

FIGURE 2.13 Fatty acids. (*a*) A molecule of saturated fatty acid and (*b*) a molecule of unsaturated fatty acid. Double bonds between carbon atoms are shown in red. They cause a "kink" in the shape of the molecule.

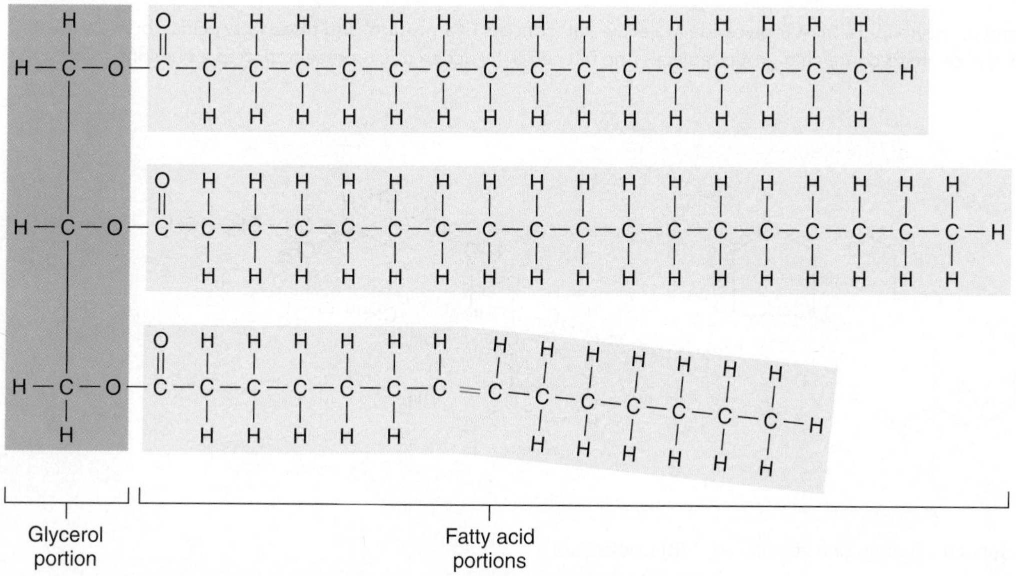

FIGURE 2.14 A triglyceride molecule (fat) consists of a glycerol portion and three fatty acid portions. This is an example of an unsaturated fat. The double bond between carbon atoms in the unsaturated fatty acid is shown in red.

Saturated fats are more abundant in fatty foods that are solids at room temperature, such as butter, lard, and most animal fats. Unsaturated fats are in foods that are liquid at room temperature, such as soft margarine and seed oils (corn, grape, sesame, soybean, sunflower, and peanut). Coconut and palm kernel oils are unusual in that they are high in saturated fats but are liquids at room temperature. The most heart-healthy fats are olive and canola (rapeseed) oils, which are monounsaturated—that is, they have one carbon-carbon double bond.

Manufacturers add hydrogen atoms to certain vegetable oils to make them harder and easier to use. This process, called hydrogenation, produces fats that are partially unsaturated and also "trans." ("Trans" refers to atoms in a molecule on opposite sides of a backbone-like structure, like stores on opposite sides of a street. Atoms on the same side—like stores on the same side of a street—are called "cis.")

A *phospholipid* molecule is similar to a fat molecule in that it includes a glycerol and fatty acid chains. The phospholipid, however, has only two fatty acid chains and, in place of the third, has a portion containing a phosphate group (fig. 2.15b). This phosphate-containing part is soluble in water (hydrophilic) and forms the "head" of the molecule, whereas the fatty acid portion is insoluble in water (hydrophobic) and forms a "tail" (fig. 2.15c). The fact that phospholipids are both attracted to and repelled by water allows them to form biological membranes, discussed in chapter 3 (pp. 86–87).

Steroid molecules are complex structures that include connected rings of carbon atoms (fig. 2.16). Among the more important steroids are: cholesterol, found in all body cells and used to synthesize other steroids; sex hormones, such as estrogen, progesterone, and testosterone; and several hormones from the adrenal glands. Chapters 13, 18, 20, 21, and 22 discuss these steroids. Table 2.7 summarizes the molecular structures and characteristics of lipids.

Proteins

Proteins (pro'tenz) have a great variety of functions. The human body has more than 200,000 types of proteins. Some are structural materials, energy sources, and chemical messengers (hormones). Other proteins combine with carbohydrates (forming glycoproteins) and function as receptors on cell surfaces, allowing cells to respond to specific types of molecules that bind to them. Antibody proteins recognize and destroy substances foreign to the body, such as certain molecules on the surfaces of infecting bacteria. Proteins such as hemoglobin and myoglobin carry oxygen in the blood and muscles,

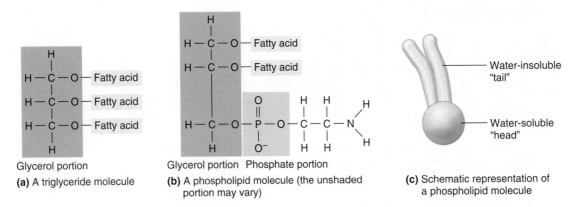

(a) A triglyceride molecule

(b) A phospholipid molecule (the unshaded portion may vary)

(c) Schematic representation of a phospholipid molecule

FIGURE 2.15 Fats and phospholipids. (a) A triglyceride molecule (fat) consists of a glycerol and three fatty acids. (b) In a phospholipid molecule, a phosphate-containing group replaces one fatty acid. (c) Schematic representation of a phospholipid.

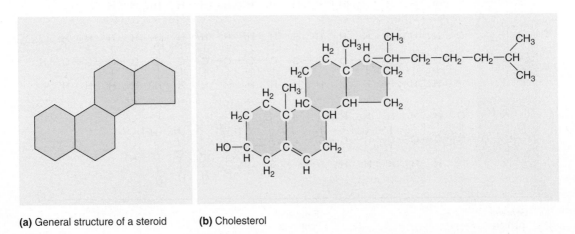

(a) General structure of a steroid

(b) Cholesterol

FIGURE 2.16 Steroid structure. (a) The general structure of a steroid. (b) The structural formula for cholesterol, a steroid widely distributed in the body and a component of cell membranes.

TABLE 2.7	Important Groups of Lipids	
Group	Basic Molecular Structure	Characteristics
Triglycerides	Three fatty acid molecules bound to a glycerol molecule	Most common lipid in the body; stored in fat tissue as an energy supply; fat tissue also provides insulation beneath the skin
Phospholipids	Two fatty acid molecules and a phosphate group bound to a glycerol molecule (may also include a nitrogen-containing molecule attached to the phosphate group)	Used as structural components in cell membranes; large amounts are in the liver and parts of the nervous system
Steroids	Four connected rings of carbon atoms	Widely distributed in the body with a variety of functions; includes cholesterol, sex hormones, and certain hormones of the adrenal glands

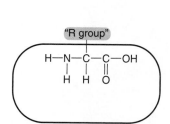

(a) General structure of an amino acid. The portion common to all amino acids is within the oval. It includes the amino group (—NH₂) and the carboxyl group (—COOH). The "R group", or the "rest of the molecule," varies and is what makes each type of amino acid unique.

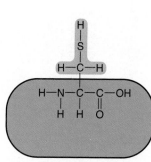

(b) Cysteine. Cysteine has an R group that contains sulfur.

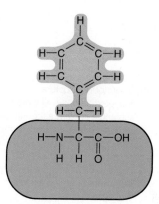

(c) Phenylalanine. Phenylalanine has a complex R group.

FIGURE 2.17 Amino acid structure. (*a*) An amino acid has an amino group, a carboxyl group, and a hydrogen atom that are common to all amino acid molecules, and a specific R group, all bonded to a central carbon. (*b*) and (*c*) Some representative amino acids and their structural formulas. Each type of amino acid molecule has a particular shape due to its different R group.

respectively, and actin and myosin are contractile proteins that provide muscle action. Many proteins play vital roles in metabolism as **enzymes** (en′zīmz), which are catalysts in living systems. That is, they speed specific chemical reactions without being consumed. (Enzymes are discussed in chapter 4, pp. 123–124.)

Proteins are similar to carbohydrates and lipids in that they consist of atoms of carbon, hydrogen, and oxygen. In addition, all proteins contain nitrogen atoms, and some contain sulfur atoms. The building blocks of proteins are **amino acids** (ah-me′no as′idz).

Twenty types of amino acids comprise proteins in organisms. Amino acid molecules have an amino group (—NH₂) at one end and a carboxyl group (—COOH) at the other end. Between these groups is a single carbon atom known as the *central carbon*. This central carbon is bonded to a hydrogen atom and to another group of atoms called a *side chain* or *R group* ("R" may be thought of as the "Rest of the molecule"). The composition of the R group distinguishes one type of amino acid from another (fig. 2.17).

Proteins have complex three-dimensional shapes, called **conformations,** yet they are assembled from simple chains of amino acids connected by peptide bonds. These are covalent bonds that link the amino end of one amino acid with the carboxyl end of another. **Figure 2.18** shows two amino acids connected by a peptide bond. The resulting molecule is a dipeptide. Adding a third amino acid creates a tripeptide. Many amino acids connected in this way constitute a polypeptide.

Proteins have four levels of structure: primary, secondary, tertiary, and quaternary. The *primary structure* is the amino acid sequence of the polypeptide chain (fig. 2.19*a*). The primary structure may range from fewer than 100 to more than 5,000 amino acids. The amino acid sequence is characteristic of a particular protein. Hemoglobin, actin, and an antibody protein have very different amino acid sequences.

FIGURE 2.18 A peptide bond (red) joins two amino acids.

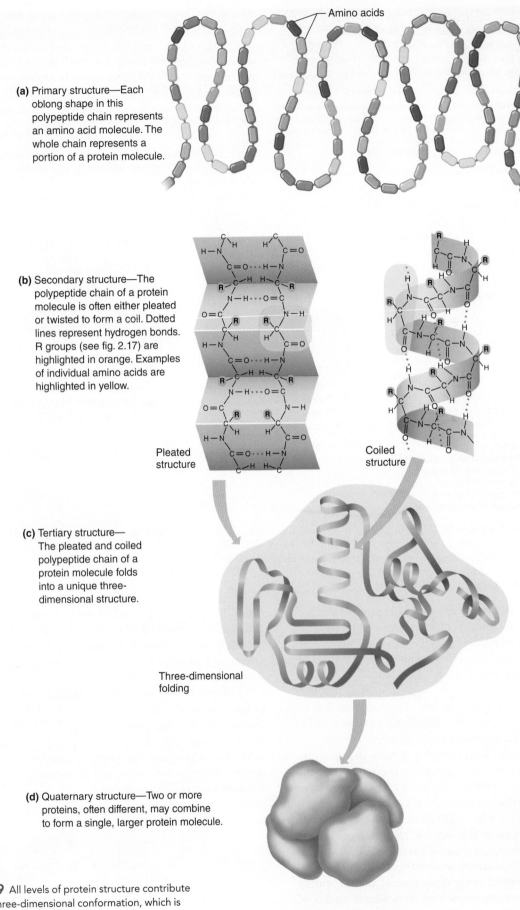

Amino acids

(a) Primary structure—Each oblong shape in this polypeptide chain represents an amino acid molecule. The whole chain represents a portion of a protein molecule.

(b) Secondary structure—The polypeptide chain of a protein molecule is often either pleated or twisted to form a coil. Dotted lines represent hydrogen bonds. R groups (see fig. 2.17) are highlighted in orange. Examples of individual amino acids are highlighted in yellow.

Pleated structure

Coiled structure

(c) Tertiary structure— The pleated and coiled polypeptide chain of a protein molecule folds into a unique three-dimensional structure.

Three-dimensional folding

(d) Quaternary structure—Two or more proteins, often different, may combine to form a single, larger protein molecule.

FIGURE 2.19 All levels of protein structure contribute to the overall three-dimensional conformation, which is vital to the protein's function.

In the *secondary structure*, the polypeptide chain either forms a springlike coil (alpha helix) or folds back and forth on itself (beta-pleated sheet) or into other shapes (fig. 2.19*b*). Secondary structure arises from hydrogen bonding. Recall that polar molecules result when electrons are not shared equally in certain covalent bonds. In amino acids, this unequal sharing results in slightly negative oxygen and nitrogen atoms and slightly positive hydrogen atoms. Hydrogen bonding between oxygen and hydrogen atoms in different parts of the polypeptide chain determines the secondary structure. Depending on the sequence of amino acids, a single very long polypeptide may include helices, sheets, and less common shapes with such colorful names as *zippers, hairpins, loops,* and *fingers*. These general shapes that arise from secondary structure are called *motifs*.

Hydrogen bonding and even covalent bonding between atoms in different parts of a polypeptide can impart another, larger level of folding, the *tertiary structure* (fig. 2.19*c*). The points of attachment in the tertiary structure are farther apart along the amino acid sequence than are the attachments that form the secondary structure. Altogether, the primary, secondary, and tertiary structures contribute to a protein's distinct conformation, which determines its function. Some proteins are long and fibrous, such as the keratins that form hair and the threads of fibrin that knit a blood clot. Myoglobin and hemoglobin are globular proteins, as are many enzymes.

For some proteins, slight reversible changes in conformation are part of their normal functions. For example, some of the proteins that interact to contract muscle exert a pulling force as a result of such a shape change, leading to movement. The reversibility of these changes enables the protein to function repeatedly.

Various treatments can more dramatically change or *denature* the secondary and tertiary structures of a protein's conformation. Because the primary structure (amino acid sequence) remains, sometimes the protein can regain its shape when normal conditions return. High temperature, radiation, pH changes, and certain chemicals (such as urea) can denature proteins.

A familiar example of irreversible protein denaturation is the response of the protein albumin to heat. This happens when cooking an egg white, which consists mostly of albumin. Some chemical treatments that curl hair also cause protein denaturation. The chemicals first break apart the tertiary structure formed when sulfur-containing amino acids attract each other within keratin molecules. The denaturation relaxes the hair. When the chemicals are washed out of the set hair, the sulfur bonds reform, but in different places. The appearance of the hair changes.

A GLIMPSE AHEAD | To Chapter 21

Crack an egg and place the egg white into some vinegar (a strong acid) for a few days, and it will turn hard and white just as if it were cooked, because the albumin protein denatures. Similarly, body fluids with too high or too low pH can denature the proteins that carry out cellular metabolism, disrupting their functions. Chapter 21 describes the homeostatic mechanisms for maintaining the pH of the internal environment.

Not all proteins are single polypeptide chains. In some proteins, several polypeptide chains are connected in a fourth level, or *quaternary structure*, to form a very large molecule (fig. 2.19*d*). Hemoglobin is a quaternary protein made up of four separate polypeptide chains.

A protein's conformation determines its function. The amino acid sequence and interactions among the amino acids in a protein determine the conformation. Thus, it is the amino acid sequence of a protein that determines its function in the body. **Nucleic acids** (nu-kle′ik as′idz) carry the instructions, in the form of *genes* (see chapter 4, p. 132), that control a cell's activities by encoding the amino acid sequences of proteins.

Nucleic Acids

The very large and complex nucleic acids include atoms of carbon, hydrogen, oxygen, nitrogen, and phosphorus, which form building blocks called **nucleotides** (nu′kle-o-tīdz). Each nucleotide consists of a 5-carbon sugar (ribose or deoxyribose), a phosphate group, and one of several nitrogen-containing organic bases, called **nitrogenous bases** (**fig. 2.20**). Such nucleotides, in a chain, form a polynucleotide (**fig. 2.21**).

Protein misfolding can cause disease. In some cases of cystic fibrosis, a protein cannot fold into its final form, which prevents it from anchoring in the cell membrane, where it would normally control the movement of chloride ions. Certain body fluids thicken, which impairs respiration and digestion. Illnesses called transmissible spongiform encephalopathies, which includes "mad cow disease," result when a type of protein called a prion folds into an abnormal, infectious form—that is, it converts normal prion protein into the pathological form, which riddles the brain with holes. Alzheimer disease results from the cutting of a protein called beta amyloid into pieces of a certain size. The proteins misfold, attach, and accumulate, forming structures called plaques in parts of the brain controlling memory and cognition. Spool-like structures in cells, called proteasomes, normally dismantle misfolded proteins before they can accumulate and cause illness.

A human body has more than 200,000 types of proteins, but only about 20,325 genes. The numbers are not the same because parts of some genes encode sequences of amino acids found in more than one type of protein. The fact that proteins outnumber genes is a little like assembling a large and diverse wardrobe from a few basic pieces of clothing.

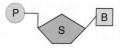

FIGURE 2.20 A nucleotide consists of a 5-carbon sugar (S = sugar), a phosphate group (P = phosphate), and a nitrogenous base (B = base).

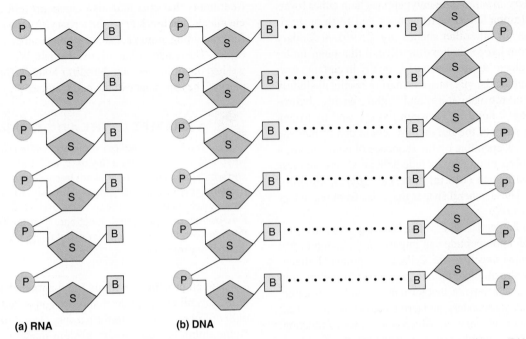

(a) RNA **(b) DNA**

FIGURE 2.21 A schematic representation of nucleic acid structure. A nucleic acid molecule consists of (a) one (RNA) or (b) two (DNA) polynucleotide chains. DNA chains are held together by hydrogen bonds (dotted lines) and they twist, forming a double helix. That the sugars of each chain point in opposite directions affects the way that the information in genes is "read." Chapter 4 (pp. 132–134, 136) discusses gene structure and function, and chapter 24 (p. 906) covers heredity.

There are two major types of nucleic acids. **RNA** (ribonucleic acid) is composed of nucleotides that have ribose sugar. Most RNA molecules are single-stranded polynucleotide chains, but they can fold into shapes that enable them to interact with DNA. The second type of nucleic acid, **DNA** (deoxyribonucleic acid), has deoxyribose sugar. DNA is a double polynucleotide chain wound into a double helix. Hydrogen bonds hold the two chains together. **Figure 2.22** compares the structures of ribose and deoxyribose, which differ by one oxygen atom. DNA and RNA also differ in that DNA molecules store the information for protein synthesis and RNA molecules use this information to construct specific protein molecules. RNA molecules are of several types. These will be discussed further in chapter 4 (pp. 138–139).

DNA molecules have a unique ability to make copies of, or replicate, themselves. They replicate prior to cell division, and each newly formed cell receives an exact copy of the original cell's

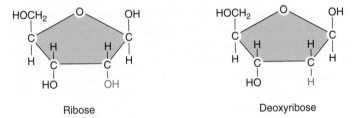

Ribose   Deoxyribose

FIGURE 2.22 The molecules of ribose and deoxyribose differ by a single oxygen atom.

DNA molecules. Chapter 4 (pp. 138–142) discusses the storage of information in nucleic acid molecules, use of the information to manufacture protein molecules, and how these proteins control metabolic reactions.

Table 2.8 summarizes the four groups of organic compounds. **Figure 2.23** shows three-dimensional (space-filling) models of some important molecules, illustrating their shapes. From Science to Technology 2.3 describes two techniques used to view human anatomy and physiology.

Recall that water molecules are polar. Many larger molecules have polar regions where nitrogen or oxygen bond with hydrogen. Such molecules, including carbohydrates, proteins, and nucleic acids, dissolve easily in water. They are water soluble, or hydrophilic ("liking" water). Molecules that do not have polar regions, such as triglycerides and steroids, do not dissolve in water ("oil and water don't mix"). Such molecules do dissolve in lipid and are said to be lipophilic ("liking" lipid). Water solubility and lipid solubility are important factors in drug delivery and in movements of substances throughout the body.

 PRACTICE

20 Compare the chemical composition of carbohydrates, lipids, proteins, and nucleic acids. How does an enzyme affect a chemical reaction?

21 What is likely to happen to a protein molecule exposed to intense heat or radiation?

22 What are the functions of DNA and RNA? ■

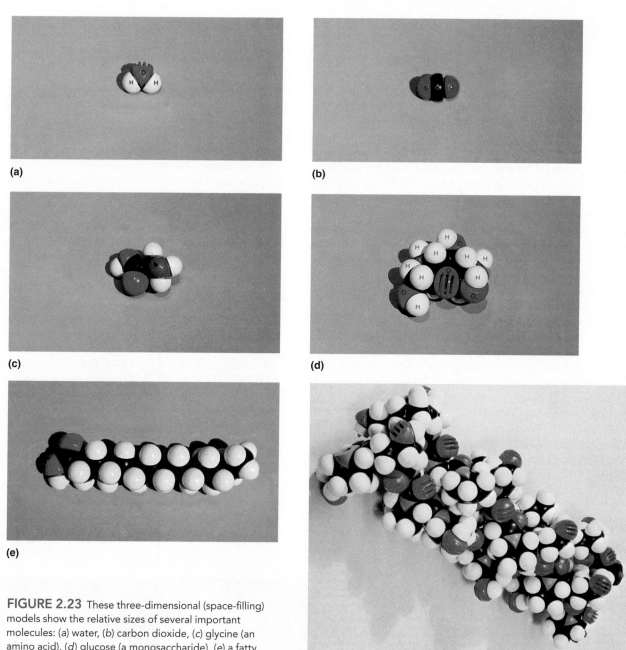

(a)

(b)

(c)

(d)

(e)

FIGURE 2.23 These three-dimensional (space-filling) models show the relative sizes of several important molecules: (*a*) water, (*b*) carbon dioxide, (*c*) glycine (an amino acid), (*d*) glucose (a monosaccharide), (*e*) a fatty acid, and (*f*) collagen (a protein). White = hydrogen, red = oxygen, blue = nitrogen, black = carbon.

(f)

TABLE 2.8	Organic Compounds in Cells			
Compound	**Elements Present**	**General Form**	**Functions**	**Examples**
Carbohydrates	C,H,O	Monosaccharide Disaccharide Polysaccharide	Provide energy, cell structure	Glucose Sucrose Glycogen
Lipids	C,H,O (often P)	Triglycerides Phospholipids Steroids	Provide energy, cell structure	Fat Cholesterol
Proteins	C,H,O,N (often S)	Polypeptide chain	Provide cell structure, enzymes, energy	Albumins, hemoglobin
Nucleic acids	C,H,O,N,P	Polynucleotide chain	Store information for the synthesis of proteins, control cell activities	RNA, DNA

Physicians use two techniques—computerized tomography (CT) scanning and positron emission tomography (PET) imaging—to paint portraits of anatomy and physiology, respectively.

In CT scanning, an X-ray-emitting device is positioned around the region of the body being examined. At the same time, an X-ray detector is moved in the opposite direction on the other side of the body. As these parts move, an X-ray beam passes through the body from hundreds of different angles. Because tissues and organs of varying composition absorb X rays differently, the intensity of X rays reaching the detector varies from position to position. A computer records the measurements from the X-ray detector and combines them mathematically, creating a three-dimensional sectional image of internal body parts (fig. 2D). In contrast, ordinary X-ray techniques produce two-dimensional images known as radiographs, X rays, or films.

A CT scan clearly differentiates soft tissues of slightly different densities, such as the liver and kidneys, which a conventional X-ray image cannot distinguish. In this way, a CT scan can detect abnormal tissue, such as a tumor. For example, a CT scan can tell whether a sinus headache that does not respond to antibiotic therapy is caused by a drug-resistant infection or by a tumor.

PET imaging uses radioactive isotopes that naturally emit positrons, which are atypical positively charged electrons, to detect biochemical activity in a specific body part. Useful isotopes in PET imaging include carbon-11, nitrogen-13, oxygen-15, and fluorine-18. When one of these isotopes releases a positron, it interacts with a nearby negatively charged electron. The two particles destroy each other in an event called annihilation. At the moment of destruction, two gamma rays form and move apart. Special equipment detects the gamma radiation.

To produce a PET image of biochemically active tissue, a person is injected with a metabolically active compound that includes a bound positron-emitting isotope. To study the brain, for example, a person is injected with glucose containing fluorine-18. After the brain takes up the isotope-tagged compound, the person rests the head within a circular array of radiation detectors. A device records each time two gamma rays are emitted simultaneously and travel in opposite directions (the result of annihilation). A computer collects and combines the data and generates a cross-sectional image. The image indicates the location and relative concentration of the radioactive isotope in different regions of the brain and can be used to study those parts metabolizing glucose.

PET images reveal the parts of the brain affected in such disorders as Huntington disease, Parkinson disease, epilepsy, and Alzheimer disease, and they are used to study blood flow in vessels supplying the brain and heart. The technology is invaluable for detecting the physiological bases of poorly understood behavioral disorders, such as obsessive-compulsive disorder. In this condition, a person repeatedly performs a certain behavior, such as washing hands, showering, locking doors, or checking to see that the stove is turned off. PET images of people with this disorder reveal intense activity in two parts of the brain that are quiet in the brains of unaffected individuals. Knowing the site of altered brain activity can help researchers develop more directed drug therapy.

In addition to highlighting biochemical activities behind illness, PET scans allow biologists to track normal brain physiology. Figure 2E shows different patterns of brain activity associated with different experiences.

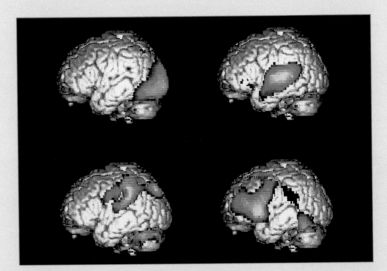

FIGURE 2E These PET images demonstrate brain changes that accompany different kinds of brain function. The upper left pattern represents sight, the upper right represents hearing, the lower left shows a tactile response such as feeling the bumps of braille print, and the lower right shows thoughts expressed as words.

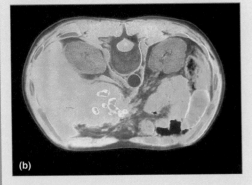

FIGURE 2D CT scans of (a) the head and (b) the abdomen.

Chapter Summary

2.1 The Importance of Chemistry in Anatomy and Physiology (page 58)

Chemistry deals with the composition of substances and how it changes. The human body is composed of chemicals. **Biochemistry** is the chemistry of living organisms.

2.2 Structure of Matter (page 58)

Matter is anything that has weight and takes up space.

1. Elements and atoms
   a. Naturally occurring matter on earth is composed of ninety-two **elements.**
   b. Elements usually combine to form **compounds.**
   c. Elements are composed of **atoms.**
   d. Atoms of different elements vary in size, weight, and ways of interacting.
2. Atomic structure
   a. An atom consists of **electrons** surrounding a **nucleus,** which has **protons** and **neutrons.** The exception is hydrogen, which has only a proton in its nucleus.
   b. Electrons are negatively charged, protons positively charged, and neutrons uncharged.
   c. A complete atom is electrically neutral.
   d. The **atomic number** of an element is equal to the number of protons in each atom; the **atomic weight** is equal to the number of protons plus the number of neutrons in each atom.
3. Isotopes
   a. **Isotopes** are atoms with the same atomic number but different atomic weights (due to differing numbers of neutrons).
   b. All the isotopes of an element react chemically in the same manner.
   c. Some isotopes are radioactive and release atomic radiation.
4. Molecules and compounds
   a. Two or more atoms may combine to form a **molecule.**
   b. A **molecular formula** represents the numbers and types of atoms in a molecule.
   c. If atoms of the same element combine, they produce molecules of that element.
   d. If atoms of different elements combine, they form molecules called compounds.
5. Bonding of atoms
   a. When atoms form links called **bonds,** they gain, lose, or share electrons.
   b. Electrons occupy space in areas called **electron shells** that encircle an atomic nucleus.
   c. Atoms with completely filled outer shells are **inert,** whereas atoms with incompletely filled outer shells gain, lose, or share electrons and thus become stable.
   d. Atoms that lose electrons become positively charged **(cations);** atoms that gain electrons become negatively charged **(anions).**
   e. Ions with opposite charges attract and join by **ionic bonds;** atoms that share electrons join by **covalent bonds.**
   f. A **structural formula** represents the arrangement of atoms in a molecule.
   g. **Polar** molecules result from an unequal sharing of electrons.
   h. **Hydrogen bonds** may form between polar molecules.

6. Chemical reactions
   a. In a chemical reaction, bonds between atoms, ions, or molecules break or form. Starting materials are called **reactants;** the resulting atoms or molecules are called **products.**
   b. Three types of chemical reactions are **synthesis,** in which large molecules build up from smaller ones; **decomposition,** in which molecules break down; and **exchange** reactions, in which parts of two different molecules trade positions.
   c. Many reactions are **reversible.** The direction of a reaction depends upon the proportion of reactants and products and the energy available.
   d. **Catalysts** (enzymes) influence the rate (not the direction) of the reaction.
7. Acids, bases, and salts
   a. Compounds that ionize in water are **electrolytes.**
   b. Electrolytes that release hydrogen ions are **acids,** and those that release hydroxide or other ions that react with hydrogen ions are **bases.**
   c. Acids and bases react to form water and electrolytes called **salts.**
8. Acid and base concentrations
   a. **pH** represents the concentration of hydrogen ions (H^+) and hydroxide ions (OH^-) in a solution.
   b. A solution with equal numbers of H^+ and OH^- is neutral and has a pH of 7.0; a solution with more H^+ than OH^- is acidic (pH less than 7.0); a solution with fewer H^+ than OH^- is basic (pH greater than 7.0).
   c. A tenfold difference in hydrogen ion concentration separates each whole number in the pH scale.
   d. **Buffers** are chemicals that resist pH change.

2.3 Chemical Constituents of Cells (page 68)

Molecules containing carbon and hydrogen atoms are **organic** and are usually nonelectrolytes; other molecules are **inorganic** and are usually **electrolytes.**

1. Inorganic substances
   a. Water is the most abundant compound in the body. Many chemical reactions take place in water. Water transports chemicals and heat and helps release excess body heat.
   b. Oxygen releases energy for metabolic activities from glucose and other molecules.
   c. Carbon dioxide is produced when certain metabolic processes release energy.
   d. Inorganic salts provide ions needed in a variety of metabolic processes.
   e. Electrolytes must be present in certain concentrations inside and outside of cells.
2. Organic substances
   a. **Carbohydrates** provide much of the energy cells require and are built of simple **sugar** molecules.
   b. **Lipids,** such as fats, phospholipids, and steroids, supply energy and are used to build cell parts; their building blocks are molecules of **glycerol** and **fatty acids.**
   c. **Proteins** serve as structural materials, energy sources, hormones, cell surface receptors, antibodies, and **enzymes** that speed chemical reactions without being consumed.
      (1) The building blocks of proteins are **amino acids.**

(2) Proteins vary in the numbers and types of their constituent amino acids; the sequences of these amino acids; and their three-dimensional structures, or **conformations.**

(3) Primary structure is the amino acid sequence. Secondary structure comes from attractions between amino acids that are close together in the primary structure. Tertiary structure reflects attractions of far-apart amino acids and folds the molecule.

(4) The amino acid sequence determines the protein's conformation.

(5) The protein's conformation determines its function.

(6) Exposure to excessive heat, radiation, electricity, or certain chemicals can denature proteins.

d. **Nucleic acids** constitute genes, the instructions that control cell activities, and direct protein synthesis.

(1) The two types are **RNA** and **DNA.**

(2) Nucleic acid building blocks are **nucleotides.**

(3) DNA molecules store information that cell parts use to construct specific proteins.

(4) RNA molecules help synthesize proteins.

(5) DNA molecules are replicated, and an exact copy of the original cell's DNA is passed to each of the newly formed cells resulting from cell division.

 CHAPTER ASSESSMENTS

2.1 The Importance of Chemistry in Anatomy and Physiology

1 Define *chemistry*. (p. 58)

2 Explain the difference between chemistry and biochemistry. (p. 58)

2.2 Structure of Matter

3 Define *matter*. (p. 58)

4 Define *compound*. (pp. 58–59)

5 List the four most abundant elements in the human body. (p. 59)

6 Explain the relationship between elements and atoms. (p. 59)

7 Identify the major parts of an atom and indicate where they are found within an atom. (pp. 59–60)

8 Distinguish between protons and neutrons. (p. 60)

9 Explain why a complete atom is electrically neutral. (p. 60)

10 Distinguish between atomic number and atomic weight. (p. 60)

11 Define *isotope*. (p. 60)

12 Define *atomic radiation*. (p. 60)

13 Explain the relationship between molecules and compounds. (pp. 62–63)

14 Explain how electrons are distributed within the electron shells of atoms. (p. 63)

15 Explain why some atoms are chemically inert. (p. 63)

16 An ionic bond forms when _____. (p. 64)
   a. atoms share electrons
   b. positively charged and negatively charged parts of covalent molecules attract
   c. ions with opposite electrical charges attract
   d. two atoms exchange protons

17 A covalent bond forms when _____. (p. 64)
   a. atoms share electrons
   b. positively charged and negatively charged parts of covalent molecules attract
   c. ions with opposite electrical charges attract
   d. two atoms exchange protons

18 Distinguish between a single covalent bond and a double covalent bond. (p. 64)

19 Show the difference between the molecular formula and the structural formula of a specific compound. (pp. 62 and 64)

20 Explain how a hydrogen bond forms. (p. 65)

21 Identify three major types of chemical reactions. (p. 66)

22 Define *reversible reaction*. (p. 66)

23 Define *catalyst*. (p. 66)

24 Define *electrolyte, acid, base,* and *salt*. (pp. 66–67)

25 Explain pH and how to use the pH scale. (p. 67)

26 Define *buffer*. (p. 68)

2.3 Chemical Constituents of Cells

27 Distinguish between inorganic and organic substances. (p. 68)

28 Distinguish between electrolytes and nonelectrolytes. (p. 68)

29 Describe the functions of water and oxygen in the human body. (pp. 68–69)

30 List several ions that cells require and identify their functions. (p. 69)

31 Define *electrolyte balance*. (p. 69)

32 Describe the general characteristics of carbohydrates. (p. 70)

33 Distinguish between simple and complex carbohydrates. (p. 70)

34 Describe the general characteristics of lipids. (p. 71)

35 List the three main types of lipids found in cells. (pp. 71–72)

36 Explain the difference between saturated and unsaturated fats. (p. 71)

37 A hydrophilic molecule dissolves in _____. (p. 72)
   a. lipid but not water
   b. water but not lipid
   c. neither lipid nor water
   d. both lipid and water

38 List at least three functions of proteins. (p. 72)

39 Describe the function of an enzyme. (p. 73)

40 Identify the four levels of protein structure. (pp. 73–75)

41 Identify two types of macromolecules in which hydrogen bonds are important parts of the structure. (pp. 75–76)

42 Describe how the change in shape of a protein may be either abnormal or associated with normal function. (p. 75)

43 Describe the general characteristics of nucleic acids. (p. 75)

44 Explain the general functions of nucleic acids. (p. 76)

INTEGRATIVE ASSESSMENTS/CRITICAL THINKING

Outcome 2.2

1. The thyroid gland metabolizes iodine, the most common form of which has a molecular weight of 127 (iodine-127). A physician wants to use a radioactive isotope of iodine (iodine-131) to test whether a patient's thyroid gland is metabolizing normally. Based on what you know about how atoms react, do you think this physician's plan makes sense?

2. How would you reassure a patient about to undergo CT scanning for evaluation of a tumor and fears becoming a radiation hazard to family members?

Outcomes 2.2, 2.3

3. What acidic and basic substances do you encounter in your everyday activities? What acidic foods do you eat regularly? What basic foods do you eat?

Outcome 2.3

4. A man on a very low-fat diet proclaims to his friend, "I'm going to get my cholesterol down to zero!" Is this desirable? Why or why not?

5. How would you explain the importance of amino acids and proteins in a diet to a person following a diet composed primarily of carbohydrates?

6. Explain why the symptoms of many inherited diseases result from abnormal protein function.

7. A friend, while frying eggs, points to the change in the egg white (which contains a protein called albumin) and explains that if the conformation of a protein changes, it will no longer have the same properties and will lose its ability to function. Do you agree or disagree with this statement?

 # ONLINE STUDY TOOLS

Connect Interactive Questions Reinforce your knowledge using assigned interactive questions covering atomic structure, and inorganic and organic substances.

Connect Integrated Activity Maintaining pH is important to homeostasis. How well do you understand acids, bases, and buffers?

LearnSmart Discover which chapter concepts you have mastered and which require more attention. This adaptive learning tool is personalized, proven, and preferred.

Anatomy & Physiology Revealed View animations on atomic structure, chemical bonds, and electrolytes.

3

Cells

This falsely colored scanning electron micrograph depicts two cells emerging from one after cell division (5,100×).

LEARNING OUTCOMES

After you have studied this chapter, you should be able to:

3.1 Cells Are the Basic Units of the Body
1 Explain how cells differ from one another. (p. 83)

3.2 A Composite Cell
2 Describe the general characteristics of a composite cell. (pp. 83 and 85)
3 Explain how the components of a cell's membrane provide its functions. (pp. 86–87)
4 Describe each kind of cytoplasmic organelle and explain its function. (pp. 90–93)
5 Describe the various cellular structures that are parts of the cytoskeleton and explain their functions. (pp. 94–96)
6 Describe the cell nucleus and its parts. (p. 97)

3.3 Movements Into and Out of the Cell
7 Explain how substances move into and out of cells. (pp. 99–106)

3.4 The Cell Cycle
8 Describe the cell cycle. (p. 107)
9 Explain how a cell divides. (pp. 108–110)

3.5 Control of Cell Division
10 Describe several controls of cell division. (pp. 110–111)

3.6 Stem and Progenitor Cells
11 Explain how stem cells and progenitor cells make possible growth and repair of tissues. (pp. 112–113)
12 Explain how two differentiated cell types can have the same genetic information, but different appearances and functions. (p. 113)

3.7 Cell Death
13 Discuss apoptosis. (p. 113)
14 Distinguish between apoptosis and necrosis. (p. 113)
15 Describe the relationship between apoptosis and mitosis. (p. 113)

THE WHOLE PICTURE

The cell is the smallest unit of life, and an average human consists of about 37 trillion of them. Inside your body, in what we call the *internal environment,* conditions are ideal for all cells. This doesn't happen by accident. The 290 different varieties of cells making up a human body work with each other, as tissues, organs (such as the heart), and organ systems (such as the cardiovascular system), to maintain the internal environment that keeps them and you alive, despite ever-changing conditions on the outside.

Most cells are part of a larger structure, such as a tissue or an organ, whereas others, such as certain white blood cells, move about. Each cell, like an organ or organ system, is a wonder of organization, but on a much smaller scale. An outer layer protects each cell. Inside a cell, a genetic headquarters oversees biochemical activities, and a production line that winds through a cell packages and routes various substances for use inside the cell or for export. Tiny parts that act as power plants continually extract energy from digested nutrients to keep the cell functioning. A cell even has its own garbage removal apparatus. Cells are, simply, the basis of life.

UNDERSTANDING WORDS

apo-, away, off, apart: *apo*ptosis—a form of cell death in which cells are shed from a developing structure.

cyt-, cell: *cyt*oplasm—fluid (cytosol) and organelles between the cell membrane and nuclear envelope.

endo-, within: *endo*plasmic reticulum—membranous complex within the cytoplasm.

hyper-, above: *hyper*tonic—solution that has a greater osmotic pressure than the cytosol.

hypo-, below: *hypo*tonic—solution that has a lesser osmotic pressure than the cytosol.

inter-, between: *inter*phase—stage between mitotic divisions of a cell.

iso-, equal: *iso*tonic—solution that has an osmotic pressure equal to that of the cytosol.

lys-, to break up: *lys*osome—organelle containing enzymes that break down proteins, carbohydrates, and nucleic acids.

mit-, thread: *mit*osis—stage of cell division when chromosomes condense.

phag-, to eat: *phag*ocytosis—process by which a cell takes in solid particles.

pino-, to drink: *pino*cytosis—process by which a cell takes in tiny droplets of liquid.

pro-, before: *pro*phase—first stage of mitosis.

-som, body: ribo*som*e—tiny, spherical organelle composed of protein and RNA that supports protein synthesis.

vesic-, bladder: *vesic*le—small, saclike organelle that contains substances to be transported within the cell or secreted.

3.1 | Cells Are the Basic Units of the Body

Cells vary considerably in size, which we measure in units called *micrometers* (mi'kro-me"terz). A micrometer equals one thousandth of a millimeter and is symbolized μm. A human egg cell is about 140 μm in diameter and is just barely visible to an unaided eye. An egg is large when compared to a red blood cell, which is about 7.5 μm in diameter, or the most common types of white blood cells, which are 10 to 12 μm in diameter. Smooth muscle cells are 20 to 500 μm long. Figure 3.1 compares four specialized cell types.

Cells generally have distinctive three-dimensional forms that make possible their functions (fig. 3.2). For instance, nerve cells that have threadlike extensions many centimeters long conduct bioelectric impulses from one part of the body to another. Epithelial cells that are part of the skin are thin, flattened, and tightly packed, somewhat like floor tiles. They form a barrier that shields underlying tissue. Muscle cells, slender and rodlike, contract and pull structures to which they are attached closer together.

Cells with specialized characteristics are termed **differentiated.** Such specialized cells form from less specialized cells that divide and express specific genes. A cell is like the Internet, harboring a vast store of information in its genome. However, like a person accessing only a small part of the Internet, a cell uses only some of the information in its genome to build structures and to carry out both the basic functions of life as well as its specialized functions.

 PRACTICE

1. What unit of measurement is used in describing cell size?
2. What is a differentiated cell?

3.2 | A Composite Cell

It is not possible to describe a typical cell, because cells vary greatly in size, shape, content, and function. We can, however, consider a hypothetical composite cell that includes many well-studied cell structures (fig. 3.3).

The three major parts of a cell—the **nucleus** (nu'kle-us), the **cytoplasm** (si'to-plazm), and the **cell membrane**—if appropriately stained, are easily seen under the light microscope. In many cell types the nucleus is innermost and is enclosed by a thin membrane called the *nuclear envelope.* The nucleus contains the genetic material (DNA), which directs the cell's functions.

The cytoplasm includes specialized structures called cytoplasmic **organelles** (or-gan-elz) that are suspended in a liquid called **cytosol** (si'to-sol). Organelles divide the labor in a cell by

 CAREER CORNER

Cytotechnologist

The woman has been tired for months, but blames it on frequent respiratory infections. When she notices several bruises, she visits her primary care physician, who sends a sample of her blood to a laboratory for analysis.

The cytotechnologist, who prepares a microscope slide from the blood sample and examines it, finds that there are too many white blood cells and too few red blood cells, and suspects that the patient has leukemia—a blood cancer.

A cytotechnologist recognizes signs of illness in cells. The job requires a bachelor's degree in biology, chemistry, or a related field that includes one to two years of an accredited program in cytotechnology. Further certification is available for Specialist in Cytotechnology and Technologist in Molecular Pathology, which may open up jobs in education and management. Cytotechnologists work independently but may consult with pathologists.

Cytotechnologists work in hospitals, doctors' offices, outpatient clinics, and diagnostic laboratories. Some tests performed on cells use automated equipment to identify cancerous or infected cells, but others require the cytotechnologist's judgment, based on experience with and knowledge of cell biology.

A career in cytotechnology can be very rewarding, because detecting disease on the cellular level is critical to accurate diagnosis of many conditions. Cytotechnologists can also work in research settings or sell laboratory equipment.

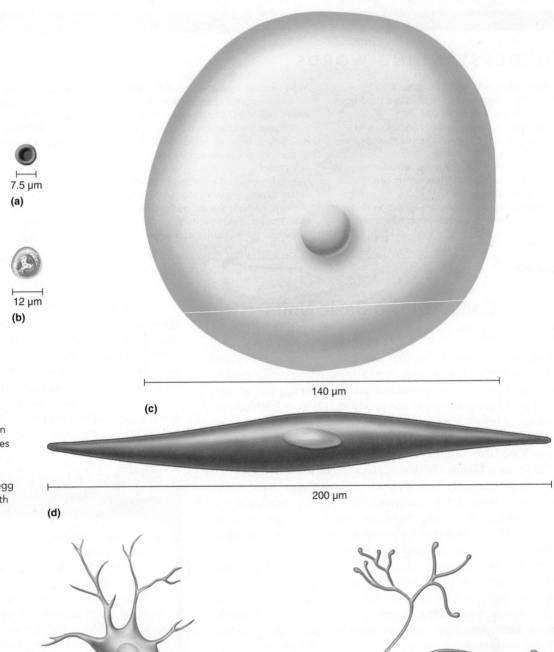

FIGURE 3.1 Cells vary considerably in size. This illustration shows the relative sizes of four types of cells. (*a*) Red blood cell, 7.5 μm in diameter; (*b*) white blood cell, 10–12 μm in diameter; (*c*) human egg cell, 140 μm in diameter; (*d*) smooth muscle cell, 20–500 μm in length.

7.5 μm
(a)

12 μm
(b)

140 μm
(c)

200 μm
(d)

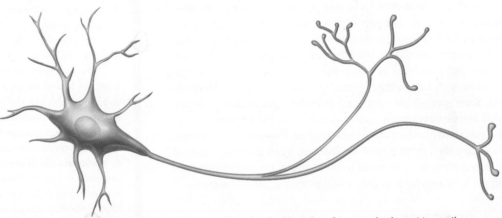

(a) A nerve cell's long extensions enable it to conduct electrical impulses from one body part to another.

(b) The sheetlike organization of epithelial cells enables them to protect underlying cells.

(c) The alignment of contractile proteins within muscle cells enables them to contract, pulling closer together the structures to which they attach.

FIGURE 3.2 Cells vary in shape and function.

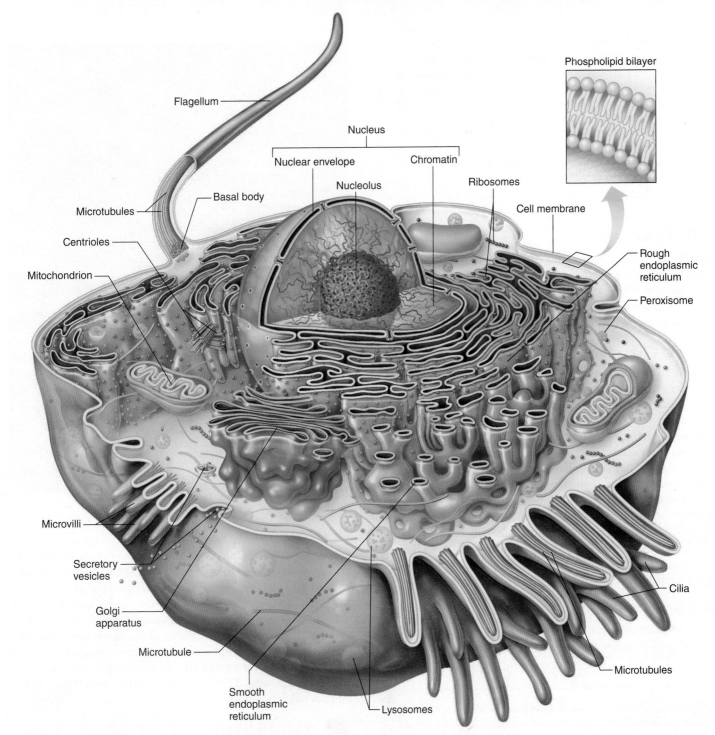

Phospholipid bilayer

Flagellum

Nucleus

Nuclear envelope

Chromatin

Nucleolus

Ribosomes

Cell membrane

Basal body

Microtubules

Centrioles

Rough endoplasmic reticulum

Mitochondrion

Peroxisome

Microvilli

Cilia

Secretory vesicles

Golgi apparatus

Microtubule

Microtubules

Smooth endoplasmic reticulum

Lysosomes

FIGURE 3.3 AP|R A composite cell illustrates the organelles and other structures found in cells. Specialized cells differ in the numbers and types of organelles, reflecting their functions. Organelles are not drawn to scale.

 Name three cellular structures composed of or including membranes.

Answer can be found in Appendix G.

partitioning off certain areas or providing specific functions, such as dismantling debris, extracting energy from nutrients, or packaging secretions. Cells with different specializations have different numbers of particular organelle types. Organelles are not static and still, as figure 3.3 might suggest. Some organelles move within the cell as part of their function, and even those that stay in one place are the sites of ongoing biochemical activity. A cell is a busy place! The cell membrane (also called a plasma membrane) contains the cytoplasm, which surrounds the nucleus.

PRACTICE

3 Name the major parts of a cell.

4 State the general functions of the cytoplasm and nucleus.

Cells with nuclei, such as those of the human body (except red blood cells), are termed *eukaryotic,* meaning "true nucleus." In contrast are the *prokaryotic* ("before nucleus") cells of bacteria. Although bacterial cells lack nuclei and other membrane-bound organelles and thus have fewer components than eukaryotic cells, the bacteria are widespread and have existed much longer than eukaryotic cells. Viruses are simpler than cells. They consist of genetic material in a protein coat and cannot reproduce outside of a host cell (the cell they infect).

Cell Membrane

The cell membrane is the outermost limit of a cell. Not just a simple boundary, the cell membrane is an actively functioning part of the living material. Many important metabolic reactions take place on its inner and outer surfaces, and it includes molecules that enable cells to communicate with each other and to interact.

General Characteristics

The cell membrane is extremely thin. It is visible only with the aid of an electron microscope (fig. 3.4). The cell membrane is flexible and somewhat elastic. The intricate features of a cell membrane, with its many outpouchings and infoldings, greatly increase the surface area of the cell. The cell membrane quickly seals tiny breaks; but if it is extensively damaged, the cell dies.

In addition to maintaining the integrity of the cell, the cell membrane controls the entrance and exit of substances, allowing some in while excluding others. A membrane that functions in this manner is *selectively permeable* (per′me-ah-bl). The cell membrane is crucial because it is a conduit between the cell and the extracellular fluids in the body's internal environment. It allows the cell to receive and respond to incoming messages, in a process called **signal transduction.**

Membrane Structure

The cell membrane is mainly composed of roughly equal amounts of lipids and proteins, with some carbohydrate. The lipid com-

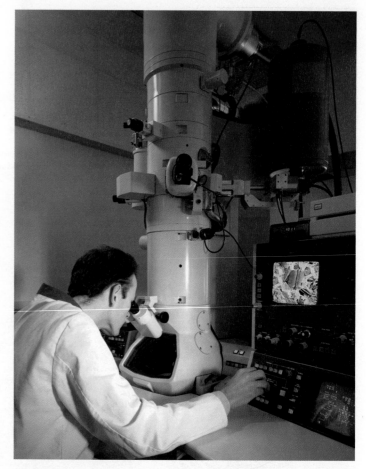

FIGURE 3.4 A transmission electron microscope.

ponent provides the physical properties, whereas the proteins provide specific functions. The basic framework of the cell membrane is a double layer, or **bilayer,** of phospholipid molecules (see fig. 2.15) that self-assemble so that their water-soluble (hydrophilic) "heads," containing phosphate groups, form the intracellular and extracellular surfaces of the membrane, and their water-insoluble (hydrophobic) "tails," consisting of fatty acid chains, make up the interior (the middle) of the membrane (see fig. 3.3 inset and fig. 3.6). The lipid molecules collectively form a thin but stable fluid film.

 RECONNECT
To Chapter 2, Lipids, page 72.

The middle portion of the cell membrane consists largely of the fatty acid portions of the phospholipid molecules, which make it oily. Molecules soluble in lipids, such as oxygen, carbon dioxide, and steroid hormones, can pass through the phospholipid bilayer easily. However, the bilayer is impermeable to water-soluble molecules, such as amino acids, sugars, proteins, nucleic acids, and various ions. Many cholesterol molecules embedded in the interior of the membrane also help make it impermeable to water-soluble substances. Additionally, the relatively rigid structure of the cholesterol molecules helps stabilize the cell membrane. However, some lipids and the proteins can move, so the cell membrane is described as a "fluid mosaic," to recognize its dynamic nature.

The maximum effective magnification possible using a light microscope is about 5,000×. A confocal microscope is a type of light microscope that passes white or laser light through a pinhole and lens to impinge on the object, which greatly enhances resolution (ability to distinguish fine detail). A transmission electron microscope (TEM) provides an effective magnification of nearly 1,000,000×, whereas a scanning electron microscope (SEM) can provide about 50,000×. Photographs of microscopic objects (micrographs) produced using the light microscope and the transmission electron microscope are typically two-dimensional, but those obtained with the scanning electron microscope have a three-dimensional quality (fig. 3.5). Scanning probe microscopes work differently from light or electron microscopes. They move a sharp tipped probe over a surface. Selective sensors detect up and down movements of the probe, and translate the movements into an image. These microscopes can image clusters of individual atoms and molecules.

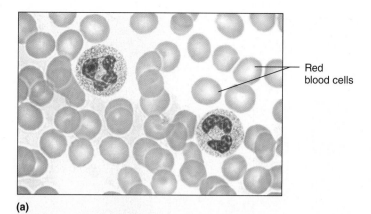

(a)

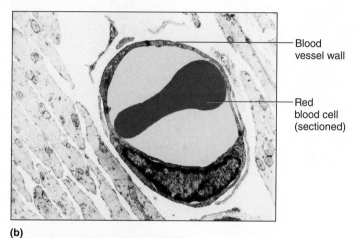

(b)

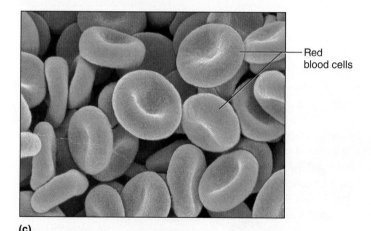

(c)

FIGURE 3.5 Human red blood cells as viewed using (*a*) a light microscope (1,200×), (*b*) a transmission electron microscope (12,000×), and (*c*) a scanning electron microscope (3,900×).

Membrane Proteins

A cell membrane includes only a few types of lipid molecules but many types of proteins (fig. 3.7), which provide specialized functions. Membrane proteins are classified by shape, their positions within the phospholipid bilayer, and function (table 3.1).

Proteins may be shaped like globs, rods, or fibers. Based on position, a membrane protein may be integral or peripheral. An integral protein extends into the lipid bilayer and may protrude from one or both sides. If it extends through both sides, it is called

a transmembrane protein. In contrast, peripheral membrane proteins associate with one side of the bilayer, from an outside attachment rather than being embedded within it.

A GLIMPSE AHEAD | **To Chapter 13**

Many transmembrane proteins are tightly coiled rods that function as *receptors*. They bind to specific incoming molecules, such as hormones, triggering responses from within the cell.

Certain compact and globular integral proteins provide routes for small molecules and ions to cross the otherwise impermeable phospholipid bilayer. Some of these proteins form "pores" that admit water. Other membrane proteins are highly selective and form channels that admit only particular types of ions. In nerve cells, for example, selective channels control the movements of sodium and potassium ions across the cell membrane (see chapter 10, pp. 375–376). Clinical Application 3.1 discusses how abnormal ion channels cause disease.

Peripheral proteins have a variety of functions. Many act as receptors responding to extracellular signals. Others are enzymes (see chapter 4, pp. 123–124), and many of these are parts of signal transduction pathways. Other peripheral proteins enable certain cells to touch or bind. Peripheral proteins also indicate a cell's identity. Carbohydrate groups attached to some peripheral proteins form glycoproteins that branch from a cell's surface, providing distinctive features that enable cells to recognize and interact with each other. These cell-cell interactions are important as cells aggregate and attach to form tissues during development, and throughout life.

Cell surface glycoproteins also mark the cells of an individual as "self," and distinguish particular differentiated cell types. The immune system can distinguish "self" cell surfaces from "nonself" cell surfaces that may indicate a potential threat, such as infection. Blood and tissue typing for transfusions or transplants consider the patterns of proteins and glycoproteins on cell surfaces.

Often cells must interact dynamically and transiently, rather than form permanent attachments. Proteins called **cellular adhesion molecules,** or CAMs for short, guide cells on the move. Consider a white blood cell moving in the bloodstream to the site of an injury, where it is required to fight infection. Imagine that such a cell must reach a woody splinter embedded in a person's palm (fig. 3.8). Once near the splinter, the white blood cell must slow down in the turbulence of the bloodstream. A type of CAM

TABLE 3.1	Types of Membrane Proteins
Protein Type	**Function**
Integral proteins	Form pores, channels, and carriers in the cell membrane; transduce signals
Peripheral proteins	
Receptor proteins	Respond to extracellular signals
Enzymes	Catalyze chemical reactions
Cell surface proteins	Establish self
Cellular adhesion molecules	Enable cells to stick to each other

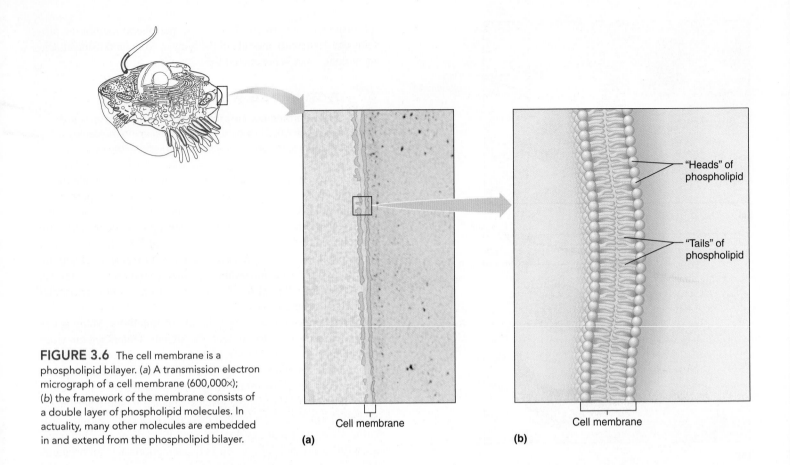

FIGURE 3.6 The cell membrane is a phospholipid bilayer. (*a*) A transmission electron micrograph of a cell membrane (600,000×); (*b*) the framework of the membrane consists of a double layer of phospholipid molecules. In actuality, many other molecules are embedded in and extend from the phospholipid bilayer.

(a)

(b)

Cell membrane

"Heads" of phospholipid

"Tails" of phospholipid

Cell membrane

Extracellular side of membrane

Glycolipid

Glycoprotein

Double layer (bilayer) of phospholipid molecules

Transmembrane protein

Integral proteins

Peripheral proteins

Cholesterol molecules

Hydrophobic fatty acid "tail"

Hydrophilic phosphate "head"

Cytoplasmic side of membrane

Phospholipid bilayer

FIGURE 3.7 AP|R The cell membrane is composed primarily of phospholipids (and some cholesterol), with proteins embedded throughout the lipid bilayer. Parts of the membrane-associated proteins that extend from the outer surface help to establish the identity of the cells as parts of a specific tissue, organ, and person.

called a *selectin* does this by coating the white blood cell and providing traction. The white blood cell slows to a roll and binds to carbohydrates on the inner capillary surface. Clotting blood, bacteria, and decaying tissue at the injury site release biochemicals (chemoattractants) that attract the white blood cell. Finally, a type of CAM called an *integrin* contacts an adhesion receptor protein protruding into the capillary space near the splinter. The integrin then pushes up through the capillary cell membrane, grabbing the passing slowed white blood cell and directing it between the tilelike cells of the capillary wall. White blood cells collecting at

What do abnormal pain intensity, irregular heartbeats, and cystic fibrosis have in common? All result from abnormal ion channels in cell membranes.

Ion channels are protein-lined tunnels in the phospholipid bilayer of a biological membrane. These passageways permit electrical signals to pass in and out of membranes as ions (charged particles). Many ion channels open or close like a gate in response to specific ions under specific conditions. These situations include a change in electrical forces across the membrane, binding of a molecule, or receiving biochemical messages from inside or outside the cell. In order for our cells to function normally, membranes must allow certain types of ions in and out at specific rates. Ion channels control these movements. Ion channels are specific for calcium (Ca^{+2}), sodium (Na^+), potassium (K^+), or chloride (Cl^-). A cell membrane may have a few thousand ion channels specific for each of these ions. Ten million or more ions can pass through an ion channel in one second! Drugs may act by affecting ion channels (table 3A), and abnormal ion channels cause certain disorders, including the following:

Unusual Pain and Sodium Channels

Because ion channels are built of proteins, genetic mutations (changes) can affect their functioning. At least ten genes encode proteins that form sodium channels, and mutations in them affect the sensation of pain. Mutations that hamper the functioning of the sodium channels cause inability to feel pain, whereas mutations that enable channels to function too efficiently cause extreme pain conditions.

A boy living in Pakistan was completely unable to feel pain. He was a performer, stabbing knives through his arms and walking on hot coals to entertain astonished crowds. Several of his relatives had the condition, too. He died at age 13 from jumping off a roof! In contrast, in erythermalgia, or "burning man syndrome," sodium channels become hypersensitive, causing extreme pain in the hands and feet, so that just putting on socks becomes unbearable. In another condition, "paroxysmal extreme pain disorder," the sodium channels stay open too long, causing excruciating pain in the rectum, jaw, and eyes. Researchers are using the information from these genetic studies to develop new painkillers.

Long-QT Syndrome and Potassium Channels

Four children in a Norwegian family were born deaf, and three of them died at ages four, five, and nine. All of the children had inherited from unaffected carrier parents "long-QT syndrome associated with deafness." They had abnormal potassium channels in the heart muscle and in the inner ear. In the heart, the malfunctioning channels disrupted electrical activity, causing a fatal disturbance to the heart rhythm. In the inner ear, the abnormal channels caused an increase in the extracellular concentration of potassium ions, impairing hearing.

Cystic Fibrosis and Chloride Channels

A seventeenth-century English saying, "A child that is salty to taste will die shortly after birth," described the consequence of abnormal chloride channels in cystic fibrosis (CF), inherited from carrier parents. The major symptoms—difficulty breathing, frequent severe respiratory infections, and a clogged pancreas that disrupts digestion—result from buildup of extremely thick mucous secretions.

Abnormal chloride channels in cells lining the lung passageways and ducts of the pancreas cause the symptoms of CF. The primary defect in the chloride channels also causes sodium channels to malfunction. The result: very salty sweat and abnormally thick mucus. Experimental gene therapy supplies the cells lining a patient's lungs with instructions to produce normal chloride channels.

TABLE 3A	Ion Channels and Drug Action
Target	**Action**
Calcium channels	Antihypertensives Antiangina (chest pain)
Sodium channels	Antiarrhythmias, diuretics Local anesthetics
Chloride channels	Anticonvulsants Muscle relaxants
Potassium channels	Antihypertensives, antidiabetics (non-insulin-dependent)

an injury site produce inflammation and, with the dying bacteria, form pus. (The role of white blood cells in body defense is discussed further in chapter 14 on page 538.)

Cellular adhesion is critical to many functions. CAMs guide cells surrounding an embryo to grow toward maternal cells and form the placenta, the supportive organ linking a pregnant woman to the fetus (see fig. 23.10 p. 878). Sequences of CAMs help establish the connections between nerve cells that underlie learning and memory.

Abnormal cellular adhesion affects health. Lack of cellular adhesion, for example, eases the journey of cancer cells as they spread from one part of the body to another. Arthritis may occur when white blood cells are captured by the wrong adhesion molecules and inflame a joint where there isn't an injury.

 PRACTICE

5 What is a selectively permeable membrane?

6 Describe the chemical structure of a cell membrane.

7 What are some functions of cell membrane proteins?

8 What happens in cellular adhesion?

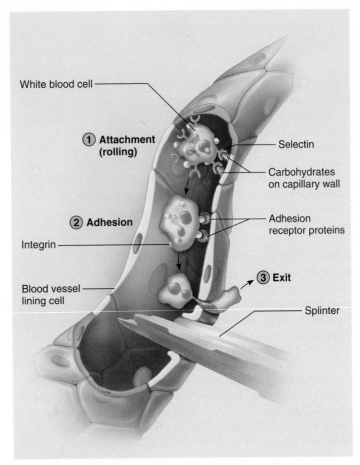

White blood cell

① **Attachment (rolling)**

Selectin

Carbohydrates on capillary wall

② **Adhesion**

Integrin

Adhesion receptor proteins

Blood vessel lining cell

③ **Exit**

Splinter

FIGURE 3.8 Cellular adhesion molecules (CAMs) direct white blood cells to injury sites, such as this splinter. (*1*) CAMs called selectins coat the white blood cell, slowing it. (*2*) CAMs called adhesion receptor proteins on the inside face of the blood vessel lining cell bind CAMs called integrins on the white blood cell, directing it between cells of the wall (*3*).

Cytoplasm

Cytoplasm usually appears clear with scattered specks when viewed through a light microscope. The higher magnification of a transmission electron microscope (see fig. 3.4) reveals vast and complex networks of membranes and organelles suspended in the cytosol. In addition to organelles, cytoplasm contains chemicals called *inclusions*. These usually are in a cell temporarily. Inclusions include stored nutrients, such as glycogen and lipids, and pigments, such as melanin in the skin. Cytoplasm also contains abundant protein rods and tubules that form a supportive framework called the **cytoskeleton** (si′to-skel-ĕ-ton).

Cytoplasmic Organelles

The activities of a cell occur largely in its cytoplasm, where nutrient molecules are received, processed, and used in metabolic reactions. The following are cytoplasmic organelles and their specific functions:

1. **Ribosomes** (ri′bo-sōmz) are tiny, spherical structures composed of protein and RNA. They provide a structural support and enzymatic activity to link amino acids to synthesize proteins (see chapter 4, p. 140). Unlike many of the other organelles, ribosomes are not composed of or contained in membranes. They are scattered in the cytoplasm and also bound to the endoplasmic reticulum, another organelle.

2. The **endoplasmic reticulum** (en-do-plaz′mik re-tik′u-lum) **(ER)** is a complex organelle composed of membranous flattened sacs, cylinders, and fluid-filled bubblelike sacs called vesicles (**fig. 3.9**). These parts are interconnected, and they interact with the cell membrane, the nuclear envelope, and certain other organelles. ER is widespread in the cytoplasm, providing a tubular transport system for molecules throughout the cell. It winds from the nucleus out toward the cell membrane, like a rail system leading from a large city out to the suburbs. The ER participates in the synthesis of protein and lipid molecules. These molecules may leave the cell as secretions or be used in the cell for such functions as producing new ER or cell membrane as the cell grows.

   The outer surface of some ER is studded with many ribosomes that give the ER a textured appearance when viewed with an electron microscope. Such endoplasmic reticulum is termed *rough ER*. Proteins synthesized on ribosomes associated with rough ER then move through the tubules of the endoplasmic reticulum, where they fold into their characteristic three-dimensional shapes, and then go to another organelle, the Golgi apparatus, for further processing.

   Some ER lacks ribosomes, and is called *smooth ER*. It appears more cylindrical than the rough ER (fig. 3.9). Along the smooth ER are enzymes that are important in synthesizing lipids, absorbing fats from the digestive tract, and breaking down drugs. Lipids are synthesized in the smooth ER and are added to proteins arriving from the rough ER. Smooth ER is especially abundant in liver cells that break down alcohol and drugs.

3. **Vesicles** (ves′ĭ-kelz) are membranous sacs that store or transport substances within a cell. They vary in size and contents. Larger vesicles that contain mostly water form when part of the cell membrane folds inward and pinches off, bringing material from outside the cell into the cytoplasm. Smaller vesicles shuttle material from the rough ER to the Golgi apparatus as part of secretion. Overall, the transport of substances into and out of cells by fleets of vesicles is called *vesicle trafficking*.

4. A **Golgi apparatus** (gol′je ap″ah-ra′tus) is a stack of five to eight flattened, membranous sacs called *cisternae* that resemble a stack of pancakes. This organelle refines, packages, and delivers proteins synthesized on the rough ER (**fig. 3.10**). A cell may have several Golgi apparatuses.

   Proteins arrive at a Golgi apparatus enclosed in vesicles composed of membrane from the ER. These sacs fuse to the membrane at the innermost end of the Golgi apparatus, which is an area specialized to receive proteins. Previously, in the ER, sugar molecules were attached to these protein molecules, forming glycoproteins.

   As the glycoproteins pass from layer to layer through the Golgi stacks, they are modified chemically. For example, sugar molecules may be added or removed, or proteins shortened. When the altered glycoproteins reach the outermost layer, they are packaged in bits of Golgi apparatus membrane that bud off and form transport vesicles. Such a vesicle may then move to the cell membrane, where it

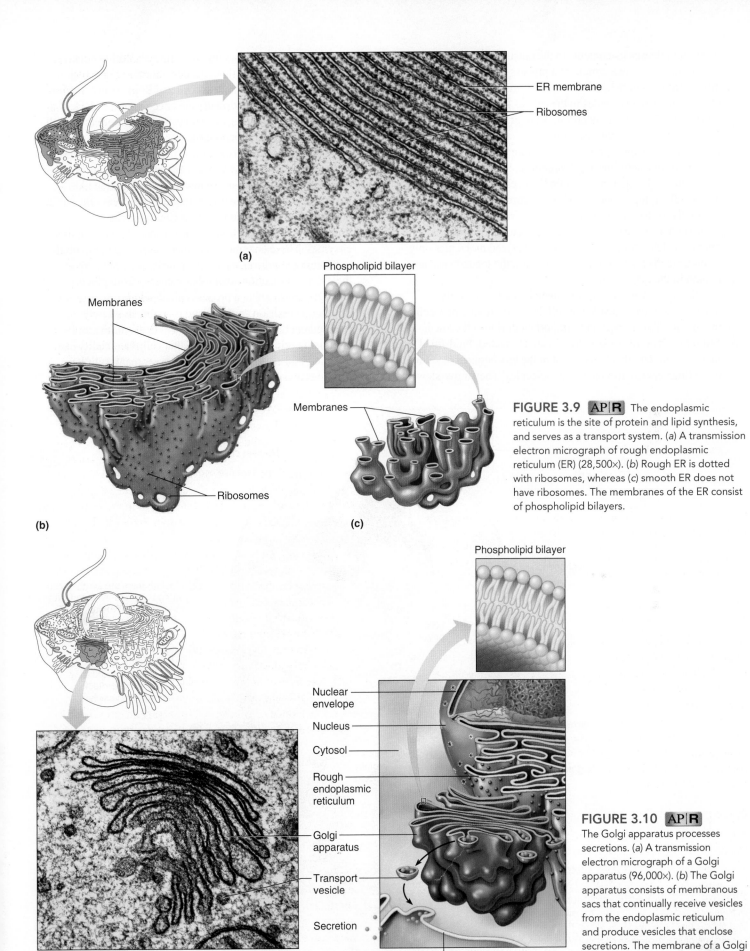

(a)

ER membrane

Ribosomes

Phospholipid bilayer

Membranes

Membranes

Ribosomes

(b)

(c)

FIGURE 3.9 AP|R The endoplasmic reticulum is the site of protein and lipid synthesis, and serves as a transport system. (*a*) A transmission electron micrograph of rough endoplasmic reticulum (ER) (28,500×). (*b*) Rough ER is dotted with ribosomes, whereas (*c*) smooth ER does not have ribosomes. The membranes of the ER consist of phospholipid bilayers.

Phospholipid bilayer

Nuclear envelope

Nucleus

Cytosol

Rough endoplasmic reticulum

Golgi apparatus

Transport vesicle

Secretion

(a)

(b) Cell membrane

FIGURE 3.10 AP|R The Golgi apparatus processes secretions. (*a*) A transmission electron micrograph of a Golgi apparatus (96,000×). (*b*) The Golgi apparatus consists of membranous sacs that continually receive vesicles from the endoplasmic reticulum and produce vesicles that enclose secretions. The membrane of a Golgi apparatus is a phospholipid bilayer.

fuses and releases its contents to the outside of the cell as a secretion. This is an example of a process called *exocytosis,* discussed in section 3.3. Other vesicles may transport glycoproteins to organelles in the cell, as **figure 3.11** shows for the process of milk secretion. A Golgi apparatus tags and sorts the molecules that travel through it so that they can be delivered to the appropriate places within the cell or marked for export, like activities in a warehouse.

Some cells, including certain liver cells and white blood cells (lymphocytes), secrete glycoprotein molecules as rapidly as they are synthesized. However, certain other cells, such as those that manufacture protein hormones, release vesicles containing newly synthesized molecules only when the cells are stimulated. Otherwise, the loaded vesicles remain in the cytoplasm.

Secretory vesicles that originate in the ER not only release substances outside the cell, but also provide new cell membrane. This is especially important during cell growth.

5. **Mitochondria** (mi″to-kon′dre-ah) are elongated, fluid-filled sacs 2–5 μm long that house most of the biochemical reactions that extract energy from nutrient molecules. They move slowly in the cytoplasm and can divide. A mitochondrion contains a small amount of DNA that encodes information for making a few types of proteins and specialized RNA. However, most proteins that mitochondria use are encoded in the DNA of the nucleus. These proteins are synthesized elsewhere in the cell and then enter the mitochondria.

A mitochondrion (mi″to-kon′dre-on) has two layers—an outer membrane and an inner membrane. The inner membrane folds extensively inward, forming shelf like partitions called *cristae* that dramatically increase the surface area on which chemical reactions can occur (fig. 3.12). From the cristae arise small stalked particles that contain enzymes. These enzymes and others dissolved in the fluid in the mitochondrion, called the matrix, control many of the chemical reactions that release energy from glucose and other nutrients in a process called *cellular respiration.* The mitochondrion captures and transfers this newly released energy into special chemical bonds of the molecule **adenosine triphosphate** (ATP), that cells can readily use (chapter 4, p. 126). ATP is used to power many cellular activities and can therefore be thought of as "cellular energy."

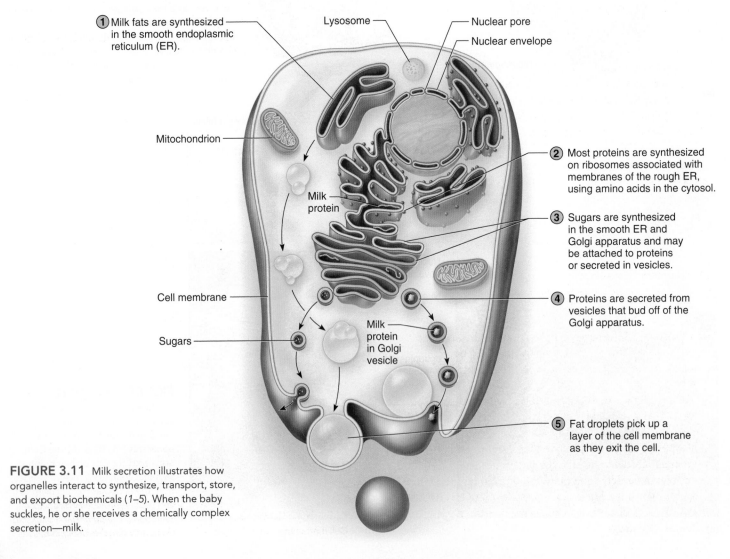

1. Milk fats are synthesized in the smooth endoplasmic reticulum (ER).

Lysosome

Nuclear pore

Nuclear envelope

Mitochondrion

Milk protein

2. Most proteins are synthesized on ribosomes associated with membranes of the rough ER, using amino acids in the cytosol.

3. Sugars are synthesized in the smooth ER and Golgi apparatus and may be attached to proteins or secreted in vesicles.

Cell membrane

Sugars

Milk protein in Golgi vesicle

4. Proteins are secreted from vesicles that bud off of the Golgi apparatus.

5. Fat droplets pick up a layer of the cell membrane as they exit the cell.

FIGURE 3.11 Milk secretion illustrates how organelles interact to synthesize, transport, store, and export biochemicals (*1–5*). When the baby suckles, he or she receives a chemically complex secretion—milk.

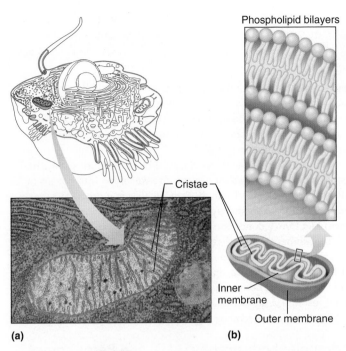

Phospholipid bilayers

Cristae

Inner membrane

Outer membrane

(a)

(b)

FIGURE 3.12 AP|R A mitochondrion is a major site of energy reactions. (*a*) A transmission electron micrograph of a mitochondrion (28,000×). (*b*) Cristae partition this saclike organelle. The mitochondrial membranes consist of phospholipid bilayers.

Because the mitochondrion supplies ATP, it is called the "powerhouse" of the cell (Clinical Application 3.2).

A typical cell has about 1,700 mitochondria, but cells with very high energy requirements, such as skeletal muscle cells, have many thousands of mitochondria. This abundance is why common symptoms of illnesses affecting mitochondria are exercise intolerance and weak, flaccid muscles. Some cells, such as red blood cells, lack mitochondria.

 PRACTICE

9 Distinguish between organelles and inclusions.

10 What are the functions of the endoplasmic reticulum?

11 Describe how the Golgi apparatus functions.

12 Why are mitochondria called the "powerhouses" of cells?

6. **Lysosomes** (li′so-sōmz) are the "garbage disposals" of the cell, where enzymes dismantle debris. The process of a cell's disposing of its own trash is called "autophagy," which means "eating self."

Lysosomes can be difficult to identify because their shapes vary, but many are small, membranous sacs (fig. 3.13). Lysosomes bud off of sections of Golgi membrane. They contain powerful enzymes that break down proteins, carbohydrates, and nucleic acids, including those of foreign particles that the cell has taken in. For example, lysosomes digest bacteria that certain white blood cells engulf. In liver cells, lysosomes break down cholesterol, toxins, and drugs.

Lysosomes destroy worn cellular parts. In certain scavenger cells they may engulf and digest entire smaller body cells that have been damaged. How the lysosomal membrane is able to withstand being digested is not well understood, but this organelle sequesters enzymes that can function only under very acidic conditions, preventing them from destroying the cellular contents around them. Human lysosomes contain 43 types of enzymes. An abnormality in just one type of lysosomal enzyme can be devastating to health (see Clinical Application 3.2).

7. **Peroxisomes** (pĕ-roks′ĭ-sōmz) are membranous sacs that resemble lysosomes in size and shape. They are in all human cells but are most abundant in cells of the liver and kidneys. Peroxisomes contain enzymes, called *peroxidases,* that catalyze metabolic reactions that release hydrogen peroxide (H_2O_2), which is toxic to cells. Peroxisomes also contain an enzyme called *catalase,* which decomposes hydrogen peroxide.

The outer surface of a peroxisome membrane contains some forty types of enzymes, which catalyze a variety of biochemical reactions, including:

- synthesis of bile acids used in fat digestion
- breakdown of lipids called very-long-chain fatty acids
- degradation of rare biochemicals
- detoxification of alcohol

Abnormal peroxisomal enzymes can greatly affect health (Clinical Application 3.2).

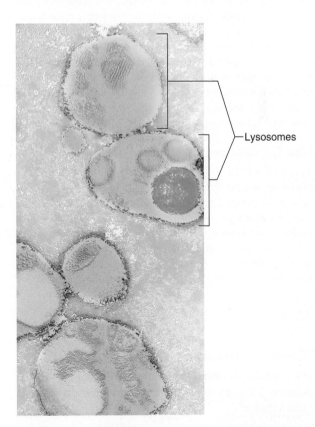

Lysosomes

FIGURE 3.13 AP|R Falsely colored transmission electron micrograph of lysosomes (27,000×). Lysosomes are membranous sacs that contain enzymes that dismantle debris in a process called autophagy.

German physiologist Rudolph Virchow envisioned disease at the cellular level in the 1850s. Today, treatments for many disorders are a direct result of understanding a disease process at the cellular level. Following are three examples of how abnormalities in organelles cause whole-body symptoms.

MELAS and Mitochondria

Sharon was small, easily fatigued, slightly developmentally delayed, and had difficulty with schoolwork. She also had seizures, and at age eleven she suffered a stroke. An astute physician who observed Sharon's mother, Lillian, suspected that the girl's symptoms were related and resulted from abnormal mitochondria, the organelles that house the biochemical reactions that extract energy from nutrients.

The doctor noticed that Lillian was uncoordinated and had numb hands. Another clue was that Lillian had frequent migraine headaches, as did her two sisters and one brother. All four siblings also had diabetes mellitus and muscle weakness. Blood tests revealed that mother and daughter had elevated levels of biochemicals (pyruvic acid and lactic acid) that indicated they were unable to extract maximal energy from nutrients. Their muscle cells had abnormal mitochondria. Accumulation of these mitochondria in smooth muscle cells in blood vessel walls in the brain caused Sharon's stroke, migraines, and seizures.

The affected family members had MELAS, which stands for the major symptoms—mitochondrial encephalomyopathy, *lactic acidosis*, and strokelike episodes. Their mitochondria cannot synthesize some of the proteins required to carry out the energy reactions. The mutant gene is part of the DNA in mitochondria, and Lillian's mother transmitted it to all of her children. Mitochondria are inherited only from the mother because the mitochondria are excluded from the part of a sperm that enters an egg cell. For this reason, Lillian's brother will not pass MELAS to his children.

Krabbe Disease and Lysosomes

Hunter was born in 1997. At first he cried frequently and had difficulty feeding, and his limbs were stiff. He became less alert, and his motor skills stopped progressing. When he was nine months old, Hunter was diagnosed with Krabbe disease, which he had inherited from his parents, who are carriers. Hunter's body could not make a lysosomal enzyme that is necessary to break down galactolipids into chemicals nerve cells use to produce myelin, a mixture of fatty molecules and proteins that electrically insulates the nerve cells. Without the enzyme certain galactolipids cannot be broken down, and they accumulate in the cells that make myelin. Accumulation of the toxic biochemicals damaged Hunter's nervous system. He ceased moving and responding, lost hearing and vision, and was tube fed.

Hunter lived for eight years. Had he been born today, he would have been tested for Krabbe disease, along with dozens of other "inborn errors of metabolism," with a few drops of blood from his heel taken shortly after birth. A cord blood stem cell transplant from a donor could have prevented his symptoms.

Lysosomal storage diseases, like Krabbe, result from a deficiency of one of the organelle's enzymes. There are several types of these rare conditions that affect about 10,000 people worldwide. Treatments use several strategies: replacing the enzyme or gene, using a drug to reduce the biochemical buildup, or using a drug that can unfold and correctly refold a misfolded enzyme.

Adrenoleukodystrophy (ALD) and Peroxisomes

For young Lorenzo, the first sign of adrenoleukodystrophy was disruptive behavior in school. When he became lethargic, weak, and dizzy, his teachers and parents realized that his problem was not just temper tantrums. His skin darkened, blood sugar levels plummeted, heart rhythm altered, and the levels of electrolytes in his body fluids changed. He lost control over his limbs as his nervous system continued to deteriorate. Lorenzo's parents took him to many doctors. Finally, one of them tested the child's blood for an enzyme normally manufactured in peroxisomes.

Lorenzo's peroxisomes lacked the second most abundant protein in the outer membrane, which normally transports an enzyme into the organelle. The enzyme controls breakdown of a type of very-long-chain fatty acid. Without the enzyme, the fatty acid builds up in cells in the brain and spinal cord, stripping these cells of their myelin. Without the myelin sheaths, the nerve cells cannot transmit messages fast enough. Death comes in a few years. Boys inherit ALD from carrier mothers.

A 1992 film, *Lorenzo's Oil*, told the story of Lorenzo's parents' efforts to develop a mixture of oils to slow the buildup of the very-long-chain fatty acids. Lorenzo lived much longer than his doctors expected, to age thirty, either due to the oil or the excellent supportive care. At the time of Lorenzo's death, he could not talk or see and communicated with finger movements and eye blinks. Today, ALD is treated with stem cell transplants or experimental gene therapy that delivers a functional version of the mutant gene.

Other Cellular Structures

Several other cellular structures are not organelles, but are parts of the cytoskeleton. They are vitally important for the ability of cells to divide and to move.

1. A **centrosome** (sen′tro-sōm) (central body) is a structure located in the cytoplasm near the nucleus. It is nonmembranous and consists of two cylinders, called *centrioles*, built of tube-like structures called microtubules organized as nine groups of three. The centrioles usually lie at right angles to each other. During cell division, the centrioles migrate to either side of the nucleus, where they produce spindle fibers that pull on and distribute *chromosomes* (kro′mo-sōmz), which carry DNA information to the newly forming cells (**fig. 3.14**). Centrioles also produce the internal parts of cell membrane projections called *cilia* and *flagella*.

2. **Cilia** (sing., *cilium*) and **flagella** (sing., *flagellum*) are motile extensions of the cell membranes of certain cell types. They are structurally similar and differ mainly in their length and abundance. Internally, both cilia and flagella consist of nine groups of three microtubules with two additional microtubules in the center, forming a distinct cylindrical pattern.

"Cilia" means "eyelashes" in Latin. They fringe the free surfaces of some epithelial cells. Each cilium is a hairlike structure about 10 µm long, which attaches just beneath the cell membrane to a modified centriole called a *basal body*. Cilia extend from cells in distinct patterns. They move in a coordinated "to-and-fro" manner so that rows of cilia beat one after the other, generating a wave that sweeps across the surface. Ciliary action moves an egg toward the uterus, and early in development controls movement of cells as they form organs. Waving cilia propel mucus over the lining of the respiratory tract (fig. 3.15). Chemicals in cigarette smoke destroy cilia, which impairs the respiratory tract's ability to expel bacteria. Infection may result.

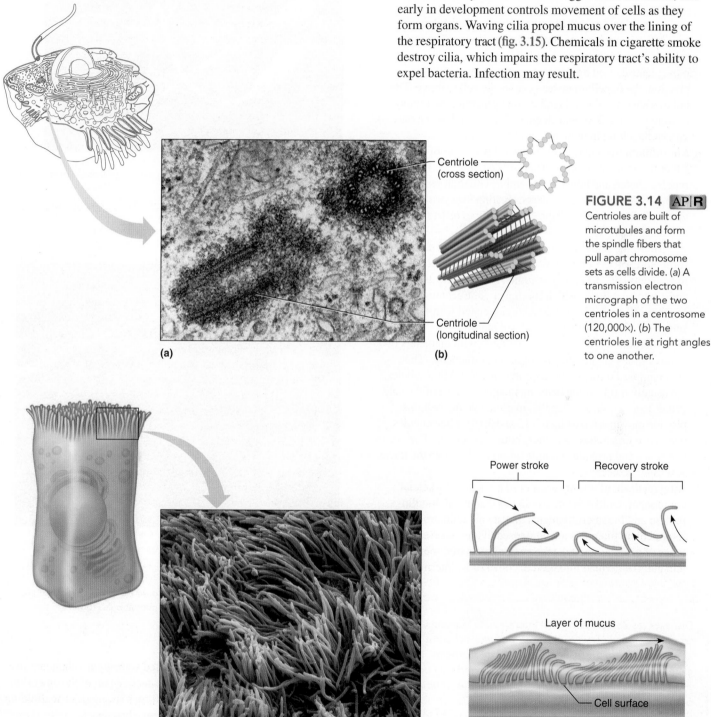

Centriole (cross section)

FIGURE 3.14 AP|R Centrioles are built of microtubules and form the spindle fibers that pull apart chromosome sets as cells divide. (*a*) A transmission electron micrograph of the two centrioles in a centrosome (120,000×). (*b*) The centrioles lie at right angles to one another.

Centriole (longitudinal section)

(a)

(b)

Power stroke   Recovery stroke

Layer of mucus

Cell surface

(a)

(b)

FIGURE 3.15 AP|R Cilia are sweeping hairlike extensions. (*a*) They fringe certain cells, such as those forming the inner lining of the respiratory tubes (5,400×). (*b*) Cilia that can move have a power stroke and a recovery stroke that create a "to-and-fro" movement that sweeps fluids across the tissue surface. Nonmotile cilia receive signals.

The cilia that wave secretions out of the respiratory system or move an egg toward the uterus are called motile cilia. Another type of this organelle, called a primary or nonmotile cilium, functions as a "cellular antenna," sensing signals and sending them into cells to control growth and maintain tissues. Compared to the many motile cilia that fringe cells, primary cilia are shorter and may be one per cell. Nearly all human cell types have them and diseases in which they are abnormal typically affect several organ systems. Such conditions are called "ciliopathies"—sick cilia.

A flagellum is much longer than a cilium. Cilia and flagella work in the same way, but in contrast to the sweeping movements caused by cilia, where the cells are fixed in position, the flagellum instead causes the cell to move. The tail of a sperm cell is a flagellum that generates swimming movements (fig. 3.16 and chapter 22, p. 829). It is the only known flagellum in humans.

3. **Microfilaments** and **microtubules** are two types of threadlike structures in the cytoplasm. They are distinguished by protein type, diameter, and how they assemble. Other proteins connect microtubules and microfilaments, creating the cytoskeleton that provides strength and the cell's ability to resist force and maintain shape.

Microfilaments are tiny rods of the protein *actin* that form meshworks or bundles and provide certain cellular movements (fig. 3.17). For example, microfilaments constitute *myofibrils*, which shorten or contract muscle cells. In other cell types, microfilaments associated with the inner surface of the cell membrane aid cell motility.

Microtubules are long, slender tubes with diameters two or three times greater than those of microfilaments. They are composed of the globular protein *tubulin*. Microtubules are usually rigid, which helps maintain the shape of the cell (fig. 3.18). In cilia and flagella, microtubule interactions provide movement (see figs. 3.15 and 3.16). Microtubules also move organelles and other cellular structures. They form centrioles and provide conduits for organelles, like the tracks of a roller coaster.

4. **Intermediate filaments** are a third class of cytoskeletal component. Unlike the microtubules that are all tubulin protein and the microfilaments that are all actin protein, intermediate filaments are composed of any of several types of proteins and take the general form of dimers (protein pairs) entwined into nested, coiled rods. Intermediate

Diseases result from abnormal intermediate filaments. In dystrophic epidermolysis bullosa, skin layers separate due to abnormal keratin proteins in intermediate filaments. In giant axonal neuropathy, intermediate filaments normally built of a protein called gigaxonin overfill motor neurons, robbing children of mobility. Most striking is progeria, in which abnormal lamin intermediate filaments cause extremely rapid aging. An affected child resembles a very elderly person by adolescence, and typically dies soon after from diseases that usually strike the very aged.

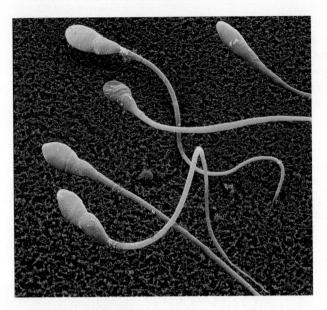

FIGURE 3.16 AP|R Flagella form the tails of these human sperm cells.

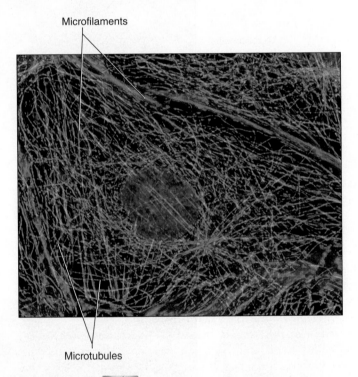

Microfilaments

Microtubules

FIGURE 3.17 AP|R A falsely colored transmission electron micrograph of microfilaments (red) and microtubules (green) in the cytoplasm (500×).

filaments made of proteins called *keratins* are abundant in the actively dividing cells in the deepest part of the outer skin layer, the epidermis. Here they form a strong inner scaffolding that helps the cells attach to form a barrier.

In all cells, intermediate filaments composed of proteins called *lamins* support the inner surface of the membranous envelope that defines the nucleus. Lamins interact with DNA, influencing which genes a cell uses to manufacture proteins.

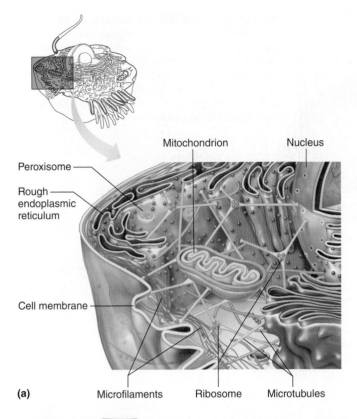

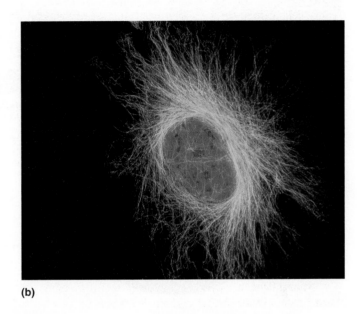

(a)   Microfilaments   Ribosome   Microtubules   (b)

Labels on figure (a): Mitochondrion, Nucleus, Peroxisome, Rough endoplasmic reticulum, Cell membrane

FIGURE 3.18 AP|R The cytoskeleton provides an inner scaffolding for a cell. (*a*) Microfilaments built of actin and microtubules built of tubulin help maintain the shape of a cell by forming an internal framework beneath the cell membrane and in the cytoplasm. (*b*) In this falsely colored micrograph, the cytoskeleton is green, yellow, and red (3,000×).

 PRACTICE

13 How do lysosomes function?

14 Describe the functions of microfilaments and microtubules.

Cell Nucleus

A nucleus is a relatively large, usually spherical, structure that contains the genetic material (DNA) that directs the activities of the cell. The extremely long molecules of DNA are complexed with proteins to form fibers called **chromatin.**

The nucleus is enclosed in a double-layered **nuclear envelope,** which consists of an inner and an outer lipid bilayer membrane. These two membranes have a narrow space between them, but are joined at places that surround openings called **nuclear pores.** These pores are not bare holes, but channels whose walls consist of multiple copies of thirty types of proteins. Nuclear pores allow certain dissolved substances to move between the nucleus and the cytoplasm (fig. 3.19). A nuclear pore lets out molecules of messenger RNA that carry genetic information, but does not let out the DNA, which must stay in the nucleus to maintain the genetic information.

The nucleus contains a fluid (nucleoplasm) in which other structures are suspended. These structures include the following:

1. Chromatin consists of the cell's 46 **chromosomes,** each of which contains DNA wound around many proteins, like a very long thread wound around multiple spools (see fig. 4.20,

p. 135). When cell division begins, chromatin fibers coil so tightly that the individual chromatin fibers become visible, when stained properly, under the light microscope.

The DNA molecules include genes, the information for synthesis of proteins. The tightness with which chromatin locally folds varies along the chromosomes, depending upon which genes are being accessed for their information at a particular time. The position of chromatin in the nucleus is not random, but reflects which genes are active.

2. A **nucleolus** (nu-kle′o-lus) ("little nucleus") is a small, dense body largely composed of RNA and protein. It has no surrounding membrane and is formed in specialized regions of certain chromosomes. The nucleolus is the site of ribosome production. Once ribosomes form, they migrate through the nuclear pores to the cytoplasm. A cell may have more than one nucleolus. The nuclei of cells that synthesize abundant protein, such as those of glands, may have especially large nucleoli.

Table 3.2 summarizes the structures and functions of cell parts.

 PRACTICE

15 How are the nuclear contents separated from the cytoplasm?

16 What is the function of the nucleolus?

17 What is chromatin?

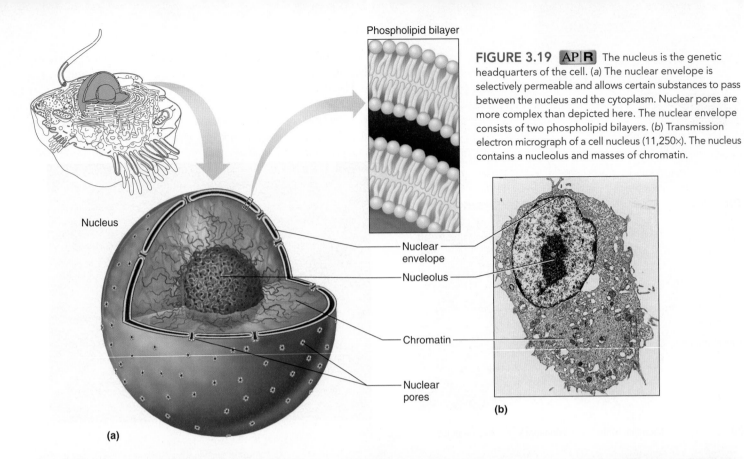

Phospholipid bilayer

FIGURE 3.19 AP|R The nucleus is the genetic headquarters of the cell. (a) The nuclear envelope is selectively permeable and allows certain substances to pass between the nucleus and the cytoplasm. Nuclear pores are more complex than depicted here. The nuclear envelope consists of two phospholipid bilayers. (b) Transmission electron micrograph of a cell nucleus (11,250×). The nucleus contains a nucleolus and masses of chromatin.

Nucleus

Nuclear envelope

Nucleolus

Chromatin

Nuclear pores

(a)

(b)

TABLE 3.2	Structures and Functions of Cell Parts	
Cell Parts	**Structure**	**Function**
Cell membrane	Membrane mainly composed of protein and lipid molecules	Maintains integrity of the cell, controls the passage of materials into and out of the cell, and provides for signal transduction
Ribosomes	Particles composed of protein and RNA molecules	Synthesize proteins
Endoplasmic reticulum	Complex of connected, membrane-bound sacs, canals, and vesicles	Transports materials within the cell, provides attachment for ribosomes, and synthesizes lipids
Vesicles	Membranous sacs	Contain substances that recently entered the cell, store and transport newly synthesized molecules
Golgi apparatus	Group of flattened, membranous sacs	Packages and modifies protein molecules for transport and secretion
Mitochondria	Membranous sacs with inner partitions	Release energy from food molecules and convert the energy into a usable form (ATP)
Lysosomes	Membranous sacs	Contain enzymes capable of digesting worn cellular parts or substances that enter cells
Peroxisomes	Membranous sacs	Contain enzymes called peroxidases, important in the breakdown of many organic molecules
Centrosome	Nonmembranous structure composed of two rodlike centrioles	Helps distribute chromosomes to new cells during cell division, initiates formation of cilia
Cilia	Motile projections attached to basal bodies beneath the cell membrane	Some cilia propel fluids over cell surface; others are sensory
Flagellum	Projection attached to a basal body beneath the cell membrane	Enables sperm cell to move
Microfilaments and microtubules	Thin rods and tubules	Support cytoplasm, help move substances and organelles within the cytoplasm
Nuclear envelope	Selectively permeable double membrane that separates the nuclear contents from the cytoplasm	Maintains the integrity of the nucleus, controls the passage of materials between the nucleus and cytoplasm
Chromatin	Fibers composed of protein and DNA molecules	Carries information for synthesizing proteins
Nucleolus	Dense, nonmembranous body composed of protein and RNA molecules	Site of ribosome formation

3.3 | Movements Into and Out of the Cell

The cell membrane is a selective barrier that controls which substances enter and leave the cell. Oxygen and nutrient molecules enter through this membrane, whereas carbon dioxide and other wastes leave through it. Movements of substances into and out of cells use *physical* (or passive) processes, such as diffusion, osmosis, facilitated diffusion, and filtration, and *physiological* (or active) processes, such as active transport, endocytosis, and exocytosis, which require energy from ATP. Understanding the mechanisms that transport substances across the cell membrane is important for understanding many aspects of physiology.

A GLIMPSE AHEAD | **To Chapter 20**

The details of membrane transport are important beyond the cellular level. The functional units of the kidneys—the nephrons—use filtration, diffusion, facilitated diffusion, osmosis, and active transport in forming urine. Activities across nephron cell membranes are important because substances that do not exit in the urine contribute to the extracellular fluid. In this way, the kidneys play a major role in maintaining the internal environment.

Diffusion

Diffusion (dǐ-fu′zhun) (also called simple diffusion) is the tendency of atoms, molecules, and ions in a liquid or air solution to move from areas of higher concentration to areas of lower concentration, thus becoming more evenly distributed, or more *diffuse*. Diffusion occurs because atoms, molecules, and ions are in constant motion. Each particle travels in a separate path along a straight line until it collides with another particle and bounces off in a new direction until it collides again and changes direction once more. As a result, there is a net movement of particles from an area of higher concentration to an area of lower concentration. This difference in concentrations is called a *concentration gradient,* and atoms, molecules, and ions are said to diffuse down a concentra-tion gradient. With time, the concentration of a given substance becomes uniform throughout a solution. In this situation, called *diffusional equilibrium* (dǐ-fu′zhun-ul e″kwi-lib′re-um), random movements continue but there is no further net movement, and the concentration of a substance remains uniform throughout the solution. Diffusional equilibrium is a little like people at a party moving back and forth between two rooms, and the number of people in each room remaining constant, even though they may be different individuals in each room at different times.

> Random motion mixes molecules. At body temperature, small molecules such as water move at more than a thousand miles per hour. However, the internal environment is crowded from a molecule's point of view. A single molecule may collide with other molecules a million times each second.

Sugar (a solute) put into a glass of water (a solvent) illustrates diffusion (fig. 3.20). The sugar at first remains in high concentration at the bottom of the glass. As the sugar molecules move, they may collide or miss each other. They are less likely to collide where there are fewer of them, so sugar molecules gradually diffuse from areas of higher concentration and eventually become uniformly distributed in the water.

Diffusion of a substance across a cell membrane can occur only if (1) the cell membrane is permeable to that substance and (2) a concentration gradient exists such that the substance is at a higher concentration on one side of the membrane or the other (fig. 3.21). Consider oxygen and carbon dioxide. Cell membranes are permeable to both. In the body, oxygen diffuses into cells and carbon dioxide diffuses out of cells, but equilibrium is never reached. Intracellular oxygen is always low because oxygen is constantly used up in metabolic reactions. Extracellular oxygen is maintained at a high level by homeostatic mechanisms in the respiratory and cardiovascular systems. In this way, a concentration gradient always allows oxygen to diffuse into cells (fig. 3.22).

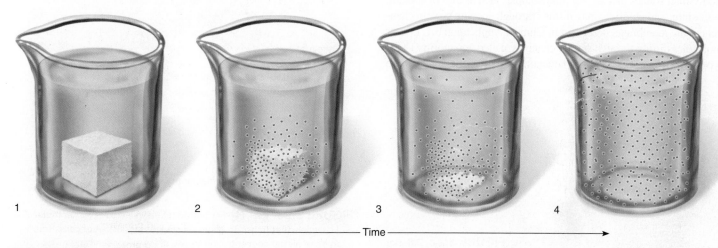

Time

FIGURE 3.20 A dissolving sugar cube illustrates diffusion. (*1–3*) A sugar cube placed in water slowly disappears as the sugar molecules dissolve and then diffuse from regions where they are more concentrated toward regions where they are less concentrated. (4) Eventually, the sugar molecules are distributed evenly throughout the water.

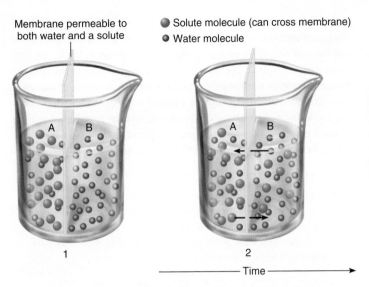

Membrane permeable to both water and a solute

● Solute molecule (can cross membrane)
● Water molecule

— Time —

1   2   3

FIGURE 3.21 AP|R Diffusion is a passive movement of molecules. (*1*) A membrane permeable to water and solute molecules separates a container into two compartments. Compartment *A* contains both types of molecules, while compartment *B* contains only water molecules. (*2*) As a result of molecular motions, solute molecules diffuse from compartment *A* into compartment *B*; water molecules diffuse from compartment *B* into compartment *A*. (*3*) Eventually, equilibrium is reached.

The level of carbon dioxide, produced as a waste product of metabolism, is always higher inside cells. Homeostasis maintains a lower extracellular carbon dioxide level, so a concentration gradient always favors carbon dioxide diffusing out of cells (fig. 3.22).

Diffusional equilibrium does not normally occur in organisms. Instead, organisms tend to reach a *physiological steady state,* where concentrations of diffusing substances are unequal but stable.

A number of factors influence the diffusion rate, but those most important in the body are distance, the concentration gradient, and temperature. In general, diffusion is more rapid over shorter distances, larger concentration gradients, and at higher temperatures. Homeostasis maintains all three of these factors at optimum levels.

Facilitated Diffusion

Some of the previous examples considered hypothetical membranes with specific permeabilities. For the cell membrane, permeability is more complex because of its selective nature. Lipid-soluble substances, such as oxygen, carbon dioxide, steroids, urea, ethanol, and general anesthetics, freely cross the cell membrane by simple diffusion. Small solutes that are not lipid-soluble, such as ions of sodium, potassium, and chloride, may diffuse through specific protein channels in the membrane, described in Clinical Application 3.1. (Water molecules diffuse through similar channels, called aquaporins.) This type of movement follows the concentration gradient, and because

it requires membrane proteins it is considered a form of **facilitated diffusion** (fah-sil″ĭ-tāt′ed dĭ-fu′zhun) meaning that it is "helped" (fig. 3.23). (Facilitated diffusion is also called facilitated transport.)

Facilitated diffusion includes not only ion and water channels, but also proteins that function as carriers (also known as transporters) to bring larger water-soluble molecules, such as glucose and amino acids, across the cell membrane down a concentration gradient. (Most sugars and amino acids are insoluble in lipids, and

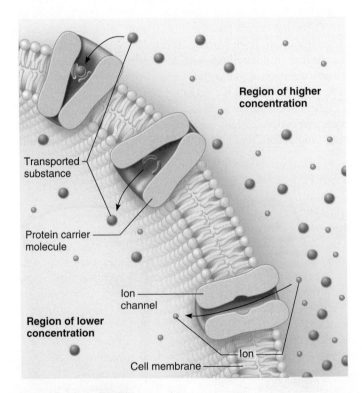

Region of higher concentration

Transported substance

Protein carrier molecule

Ion channel

Region of lower concentration

Ion

Cell membrane

FIGURE 3.23 AP|R Facilitated diffusion includes carrier molecules and channels that transport some substances into or out of cells, from a region of higher concentration to one of lower concentration. The larger spheres depict transported substances that cross through protein carrier molecules, and the small spheres represent ions, which cross the membrane through ion channels.

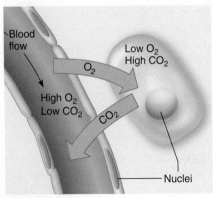

Blood flow

O_2

High O_2
Low CO_2

Low O_2
High CO_2

CO_2

Nuclei

FIGURE 3.22 Diffusion enables oxygen to enter cells and carbon dioxide to leave.

they are too large to pass through cell membrane channels.) In the facilitated diffusion of glucose, for example, a glucose molecule combines with a specific carrier molecule at the surface of the cell membrane. This union of glucose and its carrier molecule changes the shape of the carrier in a way that moves glucose to the inner surface of the membrane. Here, the glucose is released, and the carrier molecule returns to its original shape and may pick up another glucose molecule. The hormone *insulin,* discussed in chapter 13 (pp. 514–515), promotes facilitated diffusion of glucose through the membranes of certain cells.

Facilitated diffusion can move molecules only down a concentration gradient and is specific for a particular solute (a glucose carrier can only carry glucose, for example). The number of carrier molecules or channels limits the rate at which solutes can cross the membrane.

Osmosis

Osmosis (oz-mo′sis) is the movement of water across a selectively permeable membrane into a compartment containing solute that cannot cross that membrane (impermeant solute). One can think of impermeant solute as solute that is "trapped" on one side of the membrane. If impermeant solute is present on both sides of the membrane, water will move by osmosis to the side with the greater impermeant solute concentration. Osmosis is sometimes summarized as "water follows salt," but water will move by osmosis toward any impermeant solute.

A number of mechanisms explaining osmosis have been proposed, but they are beyond the scope of this discussion. Just remember that if there is a difference in solute concentration across a cell membrane or a even a layer of cells, water will flow by osmosis into the compartment with the higher concentration of impermeant (trapped) solute.

In figure 3.24, the selectively permeable membrane is permeable to water (the solvent) and impermeable to protein (the solute in this example). The protein concentration is greater in compartment A. Therefore, water moves from compartment B across the selectively permeable membrane and into compartment A by osmosis. Protein, on the other hand, cannot move out of compartment A because the selectively permeable membrane is impermeable to it.

Note in figure 3.24 that as osmosis occurs, the level of water on side A rises. This ability of osmosis to generate enough pressure to lift a volume of water is called osmotic pressure. When the height of the column of water in A reaches a certain point, the downward force (*hydrostatic pressure*) will prevent any further net osmotic movement of water. This is a condition known as osmotic equilibrium. Because the solute cannot cross the membrane, osmotic equilibrium must be achieved by the osmotic movement of water alone.

The greater the concentration of impermeant solute particles (protein in this example) in a solution, the greater the osmotic pressure of that solution. Water always tends to move by osmosis toward solutions of greater osmotic pressure. That is, water moves toward regions of trapped solute—whether in a lab exercise or in the body.

Cell membranes are generally permeable to water, so water equilibrates by osmosis throughout the body, and the concentration of water and solutes everywhere in the intracellular and extracellular fluids is essentially the same. Therefore, the osmotic

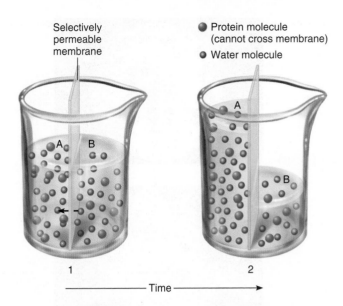

FIGURE 3.24 AP|R Osmosis. (*1*) A selectively permeable membrane separates the container into two compartments. At first, compartment A contains a higher concentration of protein than compartment B. Water moves by osmosis from compartment B into compartment A. (*2*) The membrane is impermeable to proteins, so equilibrium can only be reached by movement of water. As water accumulates in compartment A, the water level on that side of the membrane rises.

pressure of the intracellular and extracellular fluids is the same. Any solution, such as a 0.9% NaCl solution (normal saline), that has the same osmotic pressure as body fluids is called **isotonic.** Cells will not change size in this solution.

Solutions that have a higher osmotic pressure than body fluids are called **hypertonic.** If cells are put into a hypertonic solution, there will be a net movement of water by osmosis out of the cells into the surrounding solution, and the cells shrink. Conversely, cells put into a **hypotonic** solution, which has a lower osmotic pressure than body fluids, gain water by osmosis and swell. Although cell membranes are somewhat elastic, the cells may swell so much that they burst. In the case of red blood cells, this is called *hemolysis*. Figure 3.25 illustrates the effects of the three types of solutions on red blood cells.

Filtration

Filtration (fil-tra′shun) is another process that forces molecules through membranes by exerting pressure. Filtration is commonly used to separate solids from water. One method is to pour a mixture of solids and water onto filter paper in a funnel (fig. 3.26). The paper is a porous membrane through which the small water molecules pass, leaving the larger solid particles behind. Hydrostatic pressure forces the water molecules through to the other side. An example of filtration is making coffee by the drip method, using a filter paper cone or basket.

In the body, tissue fluid forms by filtration when water and dissolved substances are forced out through the thin, porous walls of blood capillaries, but larger particles such as plasma protein molecules remain inside (fig. 3.27). The force for this movement that forms tissue fluid comes from blood pressure, generated largely

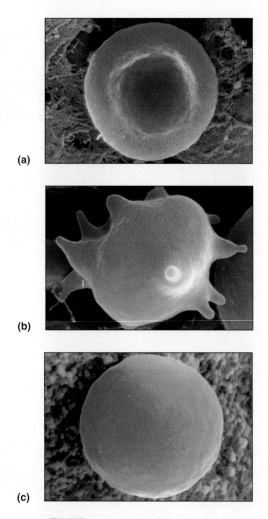

(a)

(b)

(c)

FIGURE 3.25 AP|R (a) In an isotonic solution, equal volumes of water enter and leave the red blood cells, so they maintain their characteristic sizes and shapes. (b) In a hypertonic solution, more water leaves than enters, so cells shrink. (c) In a hypotonic solution, more water enters than leaves, so cells swell and may burst (15,000×).

FIGURE 3.26 In filtration of water and solids, gravity forces water through filter paper, while tiny openings in the paper do not allow the solids to pass through. This process is similar to the drip method of preparing coffee.

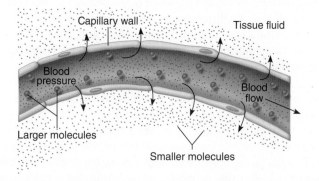

FIGURE 3.27 AP|R In filtration in the body, blood pressure forces smaller molecules through tiny openings in the capillary wall. The larger molecules remain inside.

by heart action, which is greater inside the vessel than outside it. However, the large proteins left inside capillaries oppose filtration by drawing water into blood vessels by osmosis, preventing the formation of excess tissue fluid, which causes a condition called *edema.* Although a heart beating is an active body process, filtration is considered passive because it can occur from the pressure of gravity alone. It is a little like the difference between preparing coffee by forcing water through coffee grounds using a French press (active) or by the drip method (passive). Filtration is discussed further in chapters 15 (p. 581) and 20 (p. 779).

 PRACTICE

18 List the types of substances that most readily diffuse through a cell membrane.

19 Explain the differences among diffusion, facilitated diffusion, and osmosis.

20 Distinguish among isotonic, hypertonic, and hypotonic solutions.

21 Explain how filtration occurs in the body.

Active Transport

When molecules or ions pass through cell membranes by diffusion or facilitated diffusion, their net movement is from regions of higher concentration to regions of lower concentration. In some situations, however, the net movement of particles passing through membranes is in the opposite direction, from a region of lower concentration to one of higher concentration. Sodium ions, for example, can diffuse slowly through cell membranes. Yet the concentration of these ions is many times greater outside cells (in the extracellular fluid) than inside cells (in the intracellular fluid). This is because sodium ions are continually moved through the cell membrane from regions of lower concentration (inside) to regions of higher concentration (outside). Movement against a concentration gradient is called **active transport** (ak'tiv trans'port). It requires energy, which comes from cellular metabolism, that is released as ATP molecules split.

Active transport is similar to facilitated diffusion in that it uses carrier molecules within cell membranes. As **figure 3.28** shows, these carrier molecules are proteins that have binding sites that combine with the specific particles being transported. Their union triggers release of energy from ATP, which alters the shape of the carrier protein, driving the "passenger" molecules through the membrane. Once on the other side, the transported particles are released, and the carrier molecules can return to their original shapes and accept new passenger molecules at their binding sites. Because these carrier proteins transport substances from regions of lower concentration to regions of higher concentration, they are called "pumps." Many pumps can transport more than one type of ion. A sodium/potassium pump, for example, transports sodium ions out of cells and potassium ions into cells.

Particles moved across cell membranes by active transport include sugars, amino acids, and ions of sodium, potassium, calcium, and hydrogen. Some of these substances are actively transported into cells, and others are actively transported out. Movements of this type are important to cell survival, particularly maintenance of homeostasis. Subsequent chapters describe some of these movements as they apply to specific organ systems.

A GLIMPSE AHEAD | To Chapters 9 and 10

The energy we must expend just to stay alive is called the basal metabolic energy. The body uses close to 40% of the basal metabolic energy to actively transport sodium and potassium ions across cell membranes. Imagine learning that 40% of your household budget went for one item—it had better be important! In this case it is. The concentration gradients for sodium and potassium ions that the sodium/potassium pumps establish throughout the body are essential for muscle and nerve cells to function. Chapters 9 and 10 further discuss the functioning of these important cell types.

Endocytosis

Energy from ATP may be used to move substances into or out of a cell without actually crossing the cell membrane. Different processes transport substances in each direction across the cell membrane. **Endocytosis** (en"do-si-to'sis) conveys molecules or other particles, that are too large to enter a cell by other means, to the inside of the cell in a vesicle that forms from a section of the cell membrane budding inward.

The three forms of endocytosis are pinocytosis, phagocytosis, and receptor-mediated endocytosis. In **pinocytosis** (pi"-no-si-to'sis), cells take in tiny droplets of liquid from their surroundings, as a small portion of cell membrane indents (invaginates) **(fig. 3.29)**. The open end of the tubelike part that forms then seals off and detaches from the cell membrane, producing a small vesicle about 0.1 μm in diameter in the cytoplasm. For a time, the vesicular membrane separates its contents from the rest of the cell. Eventually, the vesicular membrane breaks down, releasing the liquid inside into the cytoplasm. In this way, a cell can take in water and the molecules dissolved in it, such as proteins, that otherwise might be too large to enter.

Phagocytosis (fag"o-si-to'sis) is similar to pinocytosis, but the cell takes in solids rather than liquids. Certain types of cells, including some white blood cells, are called **phagocytes** because they can take in solid particles such as bacteria and cellular debris. When a phagocyte first encounters such a particle, the particle attaches to the cell membrane. This stimulates a portion of the membrane to project outward, surround the particle, and slowly draw it inside the cell. The part of the membrane surrounding the solid detaches from the cell's surface, forming a vesicle, called a *phagosome,* that contains the particle **(fig. 3.30)**. A phagosome may be several micrometers in diameter.

Usually, a lysosome joins a phagosome, forming a *phagolysosome.* The lysosomal digestive enzymes decompose the contents **(fig. 3.31)**. The products of this decomposition may then diffuse out of the phagolysosome and into the cytoplasm, where they may be used as raw materials in metabolic processes. Exocytosis, discussed in the next section, may expel any remaining residue. In this way, phagocytic cells dispose of foreign objects, such as dust particles;

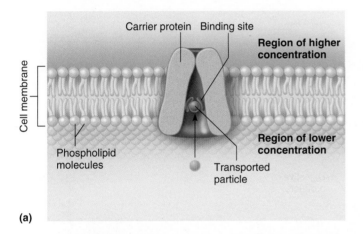

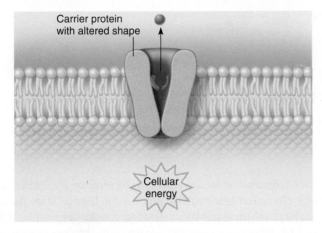

FIGURE 3.28 AP|R Active transport moves molecules against their concentration gradient. (*a*) During active transport, a molecule or an ion combines with a carrier protein, whose shape changes as a result. (*b*) This process, which requires cellular energy, transports the particle across the cell membrane.

Identify the form of cellular energy that powers active transport.
Answer can be found in Appendix G.

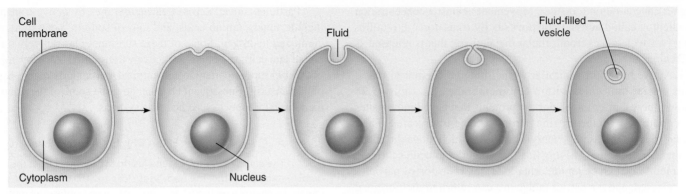

FIGURE 3.29 A cell may use pinocytosis to take in a tiny droplet of fluid from its surroundings. The fluid is encapsulated in a vesicle that forms from indentation of the cell membrane.

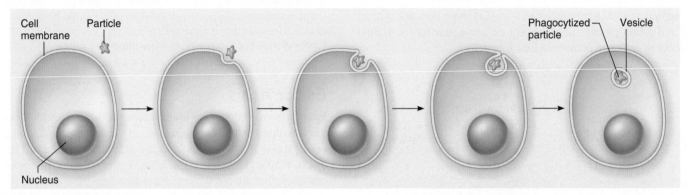

FIGURE 3.30 AP|R A cell may use phagocytosis to take in a solid particle from its surroundings. The particle is encapsulated in a portion of the cell membrane that indents.

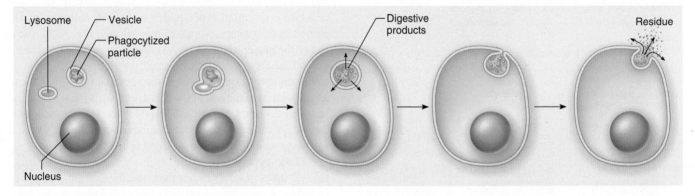

FIGURE 3.31 AP|R When a lysosome envelops a vesicle that contains a phagocytized particle, its digestive enzymes may destroy the particle. The products of this intracellular digestion diffuse into the cytoplasm. Exocytosis may expel any residue.

remove damaged cells or cell parts that are no longer functional; or destroy disease-causing microorganisms. Phagocytosis is an important line of defense against infection.

Pinocytosis and phagocytosis engulf nonspecifically. In contrast is the more discriminating **receptor-mediated endocytosis,** which moves very specific types of particles into the cell. This process uses protein molecules that extend through the cell membrane and are exposed on its outer surface. These proteins are receptors to which specific molecules from the fluids surrounding the cell can bind and selectively enter the cell, while other types of molecules are left outside (fig. 3.32). Molecules that bind specifically to receptors are called *ligands*.

Cholesterol molecules enter cells by receptor-mediated endocytosis. Cholesterol molecules synthesized in liver cells are packaged into large spherical particles called *low-density lipoproteins* (LDL). An LDL particle has a coating that contains a binding protein called *apolipoprotein-B*. The membranes of various body cells (including liver cells) have receptors for apolipoprotein-B. When the liver releases LDL particles into the blood, cells with apolipoprotein-B receptors recognize and bind the LDL particles. Formation of the receptor-ligand combination stimulates the cell membrane to indent and form a vesicle around the LDL particle. The vesicle then transports the LDL particle to a lysosome, where enzymes digest it and release the cholesterol molecules for cellular use.

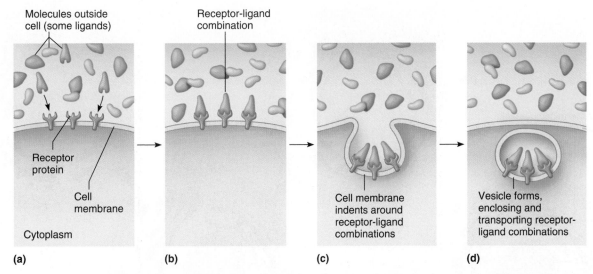

(a)   **(b)**   **(c)**   **(d)**

FIGURE 3.32 Receptor-mediated endocytosis brings specific molecules into a cell. (*a, b*) A specific molecule (ligand) binds to a receptor protein, forming a receptor-ligand combination. (*c*) The binding of the ligand to the receptor protein stimulates the cell membrane to indent. (*d*) Continued indentation forms a vesicle, which encloses and then transports the molecule into the cytoplasm.

Receptor-mediated endocytosis is particularly important because it allows cells with the appropriate receptors to remove and process specific types of substances from their surroundings, even when these substances are present in very low concentrations. In short, receptor-mediated endocytosis provides specificity (fig. 3.32).

More than 25 million people in the United States take cholesterol-lowering drugs called statins. The drugs inhibit an enzyme, HMG-CoA reductase, which cells use to produce cholesterol—in addition to the cholesterol that we eat. Feedback is at play. When levels of the enzyme drop with taking the drug, liver cells are stimulated to make more LDL receptors. With statin use not only does the body make less cholesterol, but the more abundant LDL receptors remove cholesterol from the bloodstream more efficiently.

Exocytosis

Exocytosis (ex-o-si-to′sis) is not a simple reversal of endocytosis, because the details of the steps differ. However, the overall result is the opposite: packaging substances made in the cell into vesicles that fuse with the cell membrane and are released outside the cell. Nerve cells use exocytosis to release the neurotransmitter chemicals that signal other nerve cells, muscle cells, or glands (fig. 3.33). Many cell types use exocytosis to secrete proteins.

Transcytosis

Endocytosis brings a substance into a cell, and exocytosis transports a substance out of a cell. Another process, **transcytosis** (tranz-si-to′sis), combines endocytosis and exocytosis to selectively and rapidly transport a substance or particle from one end of a cell to the other (fig. 3.34). Transcytosis moves substances across barriers formed by tightly connected cells. The process occurs in normal physiology and in disease.

Transcytosis enables the healthy immune system to monitor pathogens in the small intestine, protecting against some forms of food poisoning. It takes place in rare M cells that are among the epithelial cells lining the small intestine. M cells are so-named because the sides that face into the intestine have microfolds that maximize their surface area. The other sides of the M cells appear punched in, forming pockets where immune system cells gather. An M cell can bind to and take in a bacterium from the intestinal side by endocytosis, then transport it through the cell to the side that faces the immune system cells, where exocytosis releases it. The immune system cells bind parts of the bacterium, and, if they recognize surface features of a pathogen, they signal other cells to mature into antibody-producing cells. The antibodies are then secreted into the bloodstream and travel back to the small intestine, where they destroy the infecting bacteria before the person becomes ill.

HIV, the virus that causes AIDS, uses transcytosis to cross epithelial cells in the anus, mouth, and female reproductive tract (fig. 3.34). The virus enters certain cells in mucous secretions, and the secretions then carry the infected cells to an epithelial barrier. Near these lining cells, viruses rapidly exit the infected cells and are quickly enveloped by the lining cell membranes in receptor-mediated endocytosis. HIV particles are ferried, in vesicles, through the lining cell, without infecting (taking over) the cell, to exit from the cell membrane on the other side of the cell. After transcytosis, the HIV particles enter cells beyond the epithelial barrier and infection begins.

Another type of transport between cells uses vesicles called exosomes that bud from one cell and then travel to, merge with, and enter other cells. Exosomes are only 30 to 100 nanometers (billionths of a meter) in diameter. They may carry proteins, lipids, and RNA, and have been identified in many cell types. They remove debris, transport immune system molecules from cell to cell, and provide a vast communication network among cells. Table 3.3 summarizes the types of movements into and out of the cell.

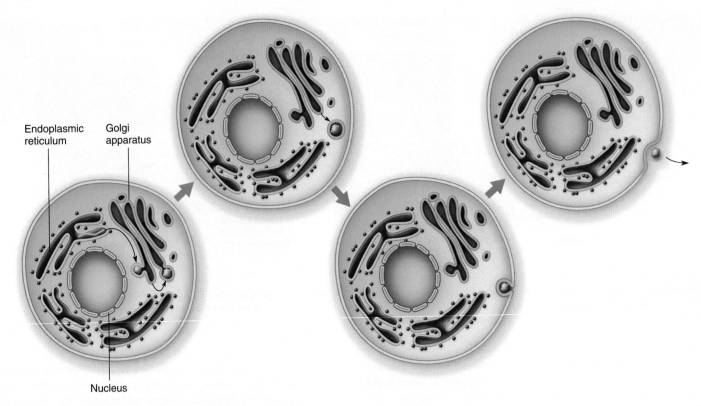

Endoplasmic reticulum

Golgi apparatus

Nucleus

FIGURE 3.33 Exocytosis releases particles, such as newly synthesized proteins, from cells.

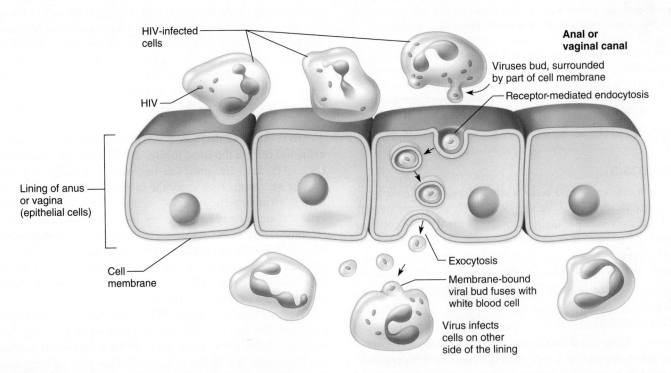

HIV-infected cells

HIV

Lining of anus or vagina (epithelial cells)

Cell membrane

Anal or vaginal canal

Viruses bud, surrounded by part of cell membrane

Receptor-mediated endocytosis

Exocytosis

Membrane-bound viral bud fuses with white blood cell

Virus infects cells on other side of the lining

FIGURE 3.34 Transcytosis transports HIV across the lining of the anus or vagina.

TABLE 3.3 Movements Into and Out of the Cell

Process	Characteristics	Source of Energy	Example
I. Passive (Physical) Processes			
A. Diffusion	Molecules move through the phospholipid bilayer from regions of higher concentration toward regions of lower concentration.	Molecular motion	Exchange of oxygen and carbon dioxide in the lungs
B. Facilitated diffusion	Ions move through channels, or molecules move by carrier proteins, across the membrane from a region of higher concentration to one of lower concentration.	Molecular motion	Movement of glucose through a cell membrane
C. Osmosis	Water molecules move through a selectively permeable membrane toward the solution with more impermeant solute (greater osmotic pressure).	Molecular motion	Distilled water entering a cell
D. Filtration	Smaller molecules are forced through porous membranes from regions of higher pressure to regions of lower pressure.	Hydrostatic pressure	Molecules leaving blood capillaries
II. Active (Physiological) Processes			
A. Active transport	Carrier molecules transport molecules or ions through membranes from regions of lower concentration toward regions of higher concentration.	ATP	Movement of various ions and amino acids through membranes
B. Endocytosis			
1. Pinocytosis	Membrane engulfs droplets containing dissolved molecules from surroundings.	ATP	Uptake of water and solutes by all body cells
2. Phagocytosis	Membrane engulfs solid particles from surroundings.	ATP	White blood cell membrane engulfing bacterial cell
3. Receptor-mediated endocytosis	Membrane engulfs selected molecules combined with receptor proteins.	ATP	Cell removing cholesterol-containing LDL particles from its surroundings
C. Exocytosis	Vesicles fuse with membrane and release contents outside of the cell.	ATP	Protein secretion, neurotransmitter release
D. Transcytosis	Receptor-mediated endocytosis and exocytosis combine to ferry particles through a cell.	ATP	HIV crossing a cell layer

 PRACTICE

22 How does a cell maintain unequal concentrations of ions on opposite sides of a cell membrane?

23 How are facilitated diffusion and active transport similar? How are they different?

24 What is the difference between pinocytosis and phagocytosis?

25 Describe receptor-mediated endocytosis.

26 What does transcytosis accomplish?

3.4 | The Cell Cycle

The series of changes that a cell undergoes, from the time it forms until it divides, is called the **cell cycle** (fig. 3.35). This cycle may seem straightforward—a newly formed cell grows for a time, and then divides in half to form two new cells, called daughter cells, which, in turn, may grow and divide. However, the specific events of the cell cycle are quite complex. For ease of study, the cell cycle is considered in distinct stages: interphase, **mitosis** (mi-to′sis; division of the nucleus), and **cytokinesis** (si″to-ki-ne′sis; division of the cytoplasm). Then the resulting daughter cells may change further, becoming specialized.

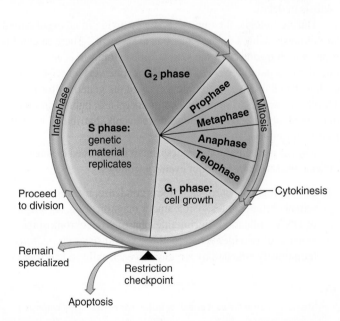

FIGURE 3.35 The cell cycle is divided into interphase, when cellular components duplicate, and cell division (mitosis and cytokinesis), when the cell splits in two, distributing its contents into two daughter cells. Interphase is divided into two gap phases (G₁ and G₂), when specific molecules and structures duplicate, and a synthesis phase (S), when DNA replicates. Mitosis can be considered in stages—prophase, metaphase, anaphase, and telophase.

The actions of several types of proteins form "checkpoints" that control the cell cycle. One particularly important checkpoint determines a cell's fate, whether it will: continue in the cell cycle and divide; stay specialized and alive, yet not divide; or die.

Interphase

Once thought to be a time of rest, **interphase** is actually a very active period. During interphase, the cell grows and maintains its routine functions as well as its contributions to the internal environment.

If the cell is developmentally programmed to divide, it must amass important biochemicals and duplicate much of its contents so that two cells can form from one. The cell must replicate all of its DNA (discussed in chapter 4, pp. 134 and 136) and synthesize and assemble the parts of membranes, ribosomes, lysosomes, peroxisomes, and mitochondria.

Interphase is divided into phases based on the sequence of activities. DNA is replicated during S phase (S stands for synthesis) and is bracketed by two G phases, G_1 and G_2 (G stands for gap or growth). Structures other than DNA are synthesized during the G phases. Cellular growth occurs then, too (fig. 3.35).

Mitosis

Mitosis and cytokinesis constitute a form of cell division that occurs in somatic (nonsex) cells and produces two daughter cells from an original cell (fig. 3.36). These new cells are genetically identical, each with the full complement of 46 chromosomes that is typical of humans. In contrast to mitosis is *meiosis,* a second form of cell division that occurs only in the cells that give rise to sex cells (sperm and eggs). Meiosis halves the chromosome number. In this way, when a sperm fertilizes an egg, the total number of 46 chromosomes is restored. Chapter 22 (pp. 824–825) considers meiosis in detail.

During mitosis, the nuclear contents divide in an event called karyokinesis, which means "nucleus movement." Then, in cytokinesis, which means "cell movement," the cytoplasm is apportioned into the two daughter cells. Mitosis must be very precise so that each new cell receives a complete copy of the genetic information. The chromosomes were duplicated in interphase, but it is in mitosis that the identical chromosome sets are evenly distributed between the two forming cells.

Mitosis is a continuous process, but it is described in stages that indicate the sequence of major events as follows:

1. In **prophase** the chromatin fibers condense, making the individual chromosomes visible under magnification. As a result of DNA replication (during interphase), each chromosome consists of two identical structures, called *sister chromatids,* temporarily attached by a region on each called a *centromere.*

> Mitosis is sometimes called cellular reproduction, because it results in two cells from one—the cell reproduces. This may be confusing, because meiosis is the prelude to human sexual reproduction. Both mitosis and meiosis are forms of *cell division,* with similar steps but different outcomes, and occurring in different types of cells.

The centrioles of the centrosome replicate just before the onset of mitosis, and during prophase, the two newly formed pairs of centrioles move to opposite sides of the cell (fig. 3.36a). Soon the nuclear envelope and the nucleolus disperse and are no longer visible. Microtubules are assembled from tubulin proteins in the cytoplasm, and these structures associate with the centrioles and chromosomes. A spindle-shaped array of microtubules (spindle fibers) forms between the centrioles as they move apart (fig. 3.36b).

2. In **metaphase** spindle fibers attach to the centromeres so that a fiber accompanying one chromatid attaches to one centromere and a fiber accompanying the other sister chromatid attaches to its centromere (fig. 3.36c). The chromosomes move along the spindle fibers, and microtubules help align them about midway between the centrioles.

3. In **anaphase** the centromeres of the chromatids separate, and the sister chromatids are now considered individual chromosomes. The separated chromosomes move in opposite directions, again as the result of microtubule activity. The spindle fibers shorten and pull their attached chromosomes toward the centrioles at opposite sides of the cell (fig. 3.36d).

4. In **telophase,** the final stage of mitosis, the chromosomes complete their migration toward the centrioles. It is much like the reverse of prophase. As the identical sets of chromosomes approach their respective centrioles, they begin to elongate and unwind from rodlike structures to threadlike structures. A nuclear envelope forms around each chromosome set, and nucleoli become visible within the newly formed nuclei. Finally, the microtubules disassemble into free tubulin molecules (fig. 3.36e).

Table 3.4 summarizes these stages of mitosis.

Cytoplasmic Division

Cytoplasmic division (cytokinesis) begins during anaphase when the cell membrane starts to constrict around the middle, which it continues to do through telophase. The musclelike contraction of a ring of actin microfilaments pinches off two cells from one (fig. 3.37).

TABLE 3.4	Major Events in Mitosis
Stage	**Major Events**
Prophase	Chromatin condenses into chromosomes; centrioles move to opposite sides of cytoplasm; nuclear membrane and nucleolus disperse; microtubules assemble and associate with centrioles and the two sister chromatids making up each chromosome.
Metaphase	Spindle fibers from the centrioles attach to the centromeres of the sister chromatids of each chromosome; chromosomes align midway between the centrioles.
Anaphase	Centromeres separate, and sister chromatids move apart, with each chromatid now an individual chromosome; spindle fibers shorten and pull these new individual chromosomes toward the centrioles.
Telophase	Chromosomes elongate and form chromatin threads; nuclear membranes form around each chromosome set; nucleoli form; microtubules break down.

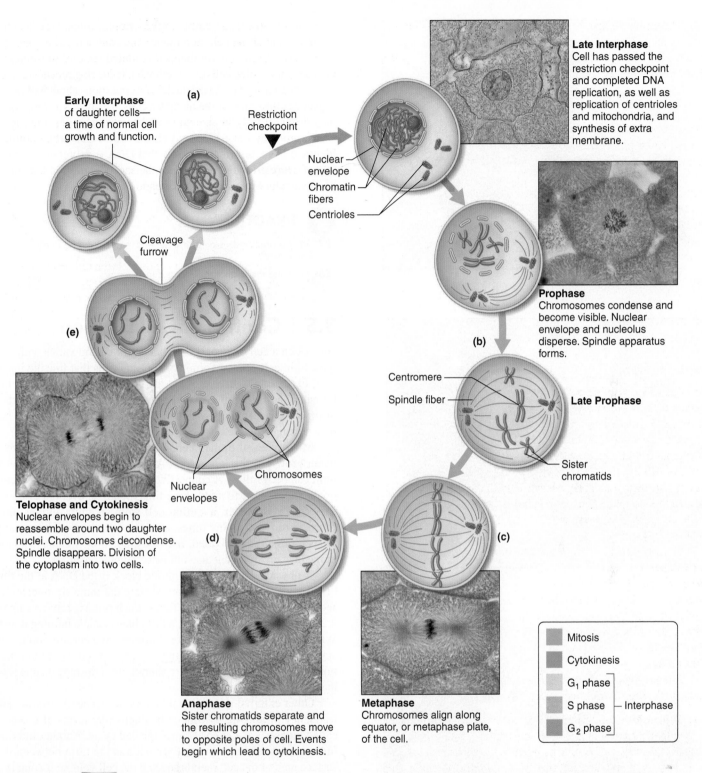

Early Interphase of daughter cells— a time of normal cell growth and function.

(a)

Restriction checkpoint ▼

Nuclear envelope
Chromatin fibers
Centrioles

Late Interphase
Cell has passed the restriction checkpoint and completed DNA replication, as well as replication of centrioles and mitochondria, and synthesis of extra membrane.

Prophase
Chromosomes condense and become visible. Nuclear envelope and nucleolus disperse. Spindle apparatus forms.

(b)

Centromere
Spindle fiber

Late Prophase

Sister chromatids

Cleavage furrow

(e)

Nuclear envelopes

Chromosomes

Telophase and Cytokinesis
Nuclear envelopes begin to reassemble around two daughter nuclei. Chromosomes decondense. Spindle disappears. Division of the cytoplasm into two cells.

(d)

(c)

Anaphase
Sister chromatids separate and the resulting chromosomes move to opposite poles of cell. Events begin which lead to cytokinesis.

Metaphase
Chromosomes align along equator, or metaphase plate, of the cell.

	Mitosis
	Cytokinesis
	G_1 phase
	S phase — Interphase
	G_2 phase

FIGURE 3.36 AP|R Mitosis and cytokinesis produce two cells from one. (*a*) During interphase, before mitosis, chromosomes are visible only as chromatin fibers. A single pair of centrioles is present, but not visible at this magnification. (*b*) In prophase, as mitosis begins, chromosomes have condensed and are easily visible when stained. The centrioles have replicated, and each pair moves to an opposite end of the cell. The nuclear envelope and nucleolus disappear, and spindle fibers associate with the centrioles and the chromosomes. (*c*) In metaphase, the chromosomes line up midway between the centrioles. (*d*) In anaphase, the centromeres are pulled apart by the spindle fibers, and the chromatids, now individual chromosomes, move in opposite directions. (*e*) In telophase, chromosomes complete their migration and become chromatin, the nuclear envelope re-forms, and microtubules disassemble. Cytokinesis, which began during anaphase, continues during telophase. Not all chromosomes are shown in these drawings. (Micrographs approximately 360×)

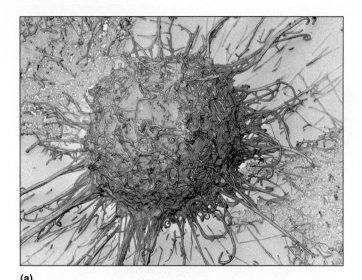
(a)

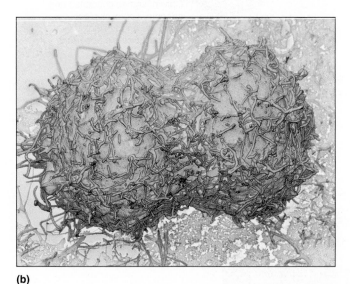
(b)

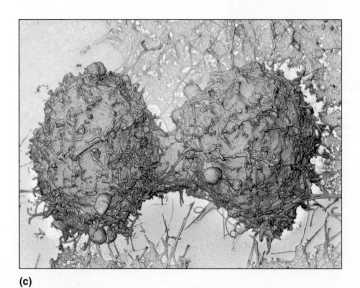
(c)

FIGURE 3.37 Cytoplasmic division (cytokinesis) is seen in these scanning electron micrographs (*a–c.* 2,200×).

The microfilaments assemble in the cytoplasm and attach to the inner surface of the cell membrane. The contractile ring forms at right angles to the microtubules that pulled the chromosomes to opposite ends of the cell during mitosis. As the ring pinches, it separates the two newly formed nuclei and apportions about half of the organelles into each of the daughter cells. The newly formed cells may differ slightly in size and number of organelles and inclusions, but they have identical chromosomes and thus contain identical DNA information. How that DNA is expressed (used to manufacture proteins) determines the specialization of the cell, a point we return to at the chapter's end.

 PRACTICE

27 Why is precise division of the genetic material during mitosis important?

28 Describe the events that occur during mitosis.

3.5 | Control of Cell Division

How often a cell divides is strictly controlled and varies with cell type. Skin cells, blood-forming cells, and cells that line the intestine, for example, divide often and continually. In contrast, the immature cells that give rise to neurons divide a specific number of times, and then cease—neurons are specialized and alive, but they no longer divide.

Most types of human cells divide from forty to sixty times when grown in the laboratory. Adherence to this cell division limit can be startling. A connective tissue cell from a human fetus divides thirty-five to sixty-three times, with an average of about fifty times. However, a similar cell from an adult divides only fourteen to twenty-nine times, as if the cell "knows" how many times it has already divided. In a body, however, signals from the immediate environment also influence mitotic potential.

A physical basis for this mitotic clock is the DNA at the tips of chromosomes, called *telomeres*. Here the same six-nucleotide sequence repeats hundreds of times. Each mitosis removes up to 1,200 nucleotides from the tip, a little like a candle burning down. When the chromosome tips are shortened to a certain point, this signals the cell to cease dividing. Severe psychological or emotional stress can hasten telomere shortening. This may be one way that stress can harm health.

Other external and internal factors influence the timing and frequency of mitosis. Within cells, fluctuating levels of proteins called kinases and cyclins control the cell cycle. Another internal influence on readiness for cell division is the ratio between the surface area of the cell membrane and the cell volume it contains; that is, cell size. The larger the cell, the more nutrients it requires to maintain the activities of life. However, the ability of a cell to take up nutrients (by diffusion and various transport processes) is limited by the surface area of the cell membrane. Because volume increases faster than does surface area, a cell can grow too large to efficiently obtain nutrients. Cell division solves this growth problem. The resulting daughter cells are smaller than the original cell and thus have a more favorable surface area-to-volume relationship. The smaller cells require less energy and fewer nutrients, and diffusion into and out of them is faster.

External controls of cell division include hormones and growth factors. Hormones are biochemicals manufactured in a gland and transported in the bloodstream to a site where they exert an effect. Hormones signal mitosis in the lining of a woman's uterus each month, building up the tissue to nurture a possible pregnancy. Similarly, a pregnant woman's hormones stimulate mitosis in her breasts when their function as milk-producing glands will soon be required.

Growth factors function like hormones but act closer to their sites of synthesis. Epidermal growth factor, for example, stimulates growth of new skin beneath a scab. Salivary glands also produce this growth factor, which is why an animal's licking a wound may speed healing.

Some people who have cancer may benefit from drugs that affect growth factors. Granulocyte colony stimulating factor (G-CSF, sold under several brand names) is given as a drug to boost white blood cell counts, which plummet during chemotherapy. In contrast, anti-angiogenesis drugs work oppositely on vascular endothelial growth factor (VEGF), cutting off a tumor's blood supply.

Space availability is another external factor that influences the timing and rate of cell division. Healthy cells do not divide if they are surrounded by other cells. This response to crowding is called contact (density-dependent) inhibition.

Control of cell division is crucial to health. With too infrequent mitoses, an embryo could not develop, a child could not grow, and wounds would not heal. However, too frequent mitoses, or those that continue unabated, produce an abnormal growth, or neoplasm, which may form a disorganized mass called a **tumor.**

Tumors are of two types. A *benign* tumor remains in place like a lump, and if it enlarges will eventually interfere with the function of healthy tissue. A *malignant,* or cancerous, tumor looks different from a benign tumor. A malignant tumor is invasive, extending into surrounding tissue. A growing malignant tumor may roughly resemble a crab with outreaching limbs. (The word "cancer" comes from the Latin for "the crab.") Cancer cells, if not stopped, eventually reach the circulation and spread, or metastasize, to other sites in the body. **Table 3.5** lists characteristics of cancer cells, and **figure 3.38** illustrates how cancer cells infiltrate healthy tissue.

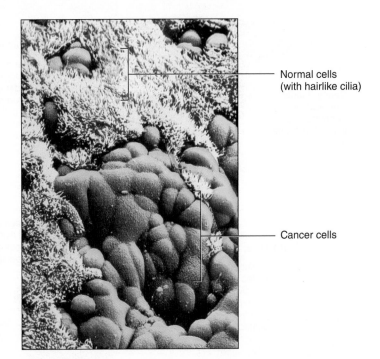

FIGURE 3.38 A cancer cell is rounder and less specialized than surrounding healthy cells. It secretes biochemicals that cut through nearby tissue (invasiveness) and other biochemicals that stimulate extension of blood vessels that nurture the tumor's growth (angiogenesis) (2,200×).

Normal cells (with hairlike cilia)

Cancer cells

Cancer is a collection of disorders distinguished by their site of origin, the affected cell type, mutations (genetic variants) present in the affected cells, and differences in gene expression. Many cancers are treatable with surgery, radiation, chemicals (chemotherapy), or immune system substances used as drugs. A newer, more targeted approach to treating cancer is to use molecules that bind to receptors unique to, or unusually abundant on, cancer cells, blocking the signals to divide.

Two major types of genes cause cancer. **Oncogenes** are abnormal variants of genes that normally control the cell cycle, but are overexpressed, increasing cell division rate. **Tumor suppressor genes** normally hold mitosis in check. When tumor suppressor genes are removed or inactivated, this lifts control of the cell cycle, and uncontrolled cell division leading to cancer results (**fig. 3.39**). Cancer cells are said to be "immortal" because they do not cease dividing at the 40 to 60 division limit that other cells follow.

TABLE 3.5	Characteristics of Cancer Cells
Loss of cell cycle control	
Heritability (a cancer cell divides to form more cancer cells)	
Transplantability (a cancer cell implanted into another individual will cause cancer to develop)	
Dedifferentiation (loss of specialized characteristics)	
Loss of contact inhibition	
Ability to induce local blood vessel formation (angiogenesis)	
Invasiveness	
Ability to metastasize (spread)	

As the number of mutations (genetic changes) increases a cancer progresses. Such mutations are described as being either "drivers" or "passengers." Just as a driver of a vehicle takes it to the destination, a driver mutation provides the selective growth advantage to a cell that makes it cancerous. Just as a passenger goes along for the ride, a passenger mutation does not cause or propel the cancer's growth or spread. A cancer generally has two to eight driver mutations, but more than 99% of the mutations in cancer cells are passengers. Tumors vary greatly in the numbers of each type of mutation.

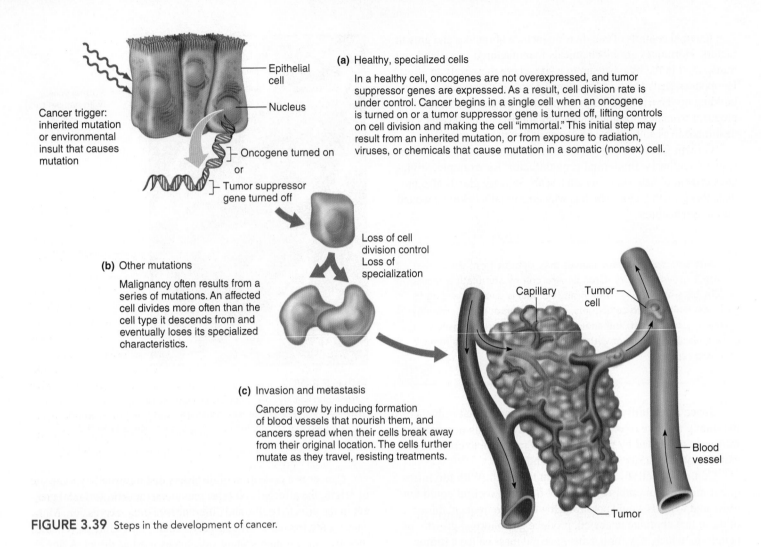

(a) Healthy, specialized cells

In a healthy cell, oncogenes are not overexpressed, and tumor suppressor genes are expressed. As a result, cell division rate is under control. Cancer begins in a single cell when an oncogene is turned on or a tumor suppressor gene is turned off, lifting controls on cell division and making the cell "immortal." This initial step may result from an inherited mutation, or from exposure to radiation, viruses, or chemicals that cause mutation in a somatic (nonsex) cell.

Epithelial cell

Nucleus

Cancer trigger: inherited mutation or environmental insult that causes mutation

Oncogene turned on
or
Tumor suppressor gene turned off

Loss of cell division control
Loss of specialization

(b) Other mutations

Malignancy often results from a series of mutations. An affected cell divides more often than the cell type it descends from and eventually loses its specialized characteristics.

Capillary

Tumor cell

(c) Invasion and metastasis

Cancers grow by inducing formation of blood vessels that nourish them, and cancers spread when their cells break away from their original location. The cells further mutate as they travel, resisting treatments.

Blood vessel

Tumor

FIGURE 3.39 Steps in the development of cancer.

Environmental factors, such as exposure to toxic chemicals or radiation, may induce cancer by altering (mutating) oncogenes and tumor suppressor genes in body (somatic) cells. Cancer may also be the consequence of a failure of normal programmed cell death (apoptosis), resulting in overgrowth.

 PRACTICE

29 How do cells vary in their rates of division?

30 Which factors control the number of times and the rate at which cells divide?

31 How can too infrequent or too frequent cell division affect health?

32 What is the difference between a benign and a malignant tumor?

33 What are two ways that genes cause cancer?

3.6 | Stem and Progenitor Cells

Cells come from preexisting cells, by the processes of mitosis and cytokinesis. Cell division explains how a fertilized egg develops into an individual consisting of trillions of cells, but not how all of these cells become specialized. There are at least 290 specialized types of cells, including 14 types that are unique to the embryo or fetus. The process of specialization is called *differentiation.*

A body must generate new cells in order to grow and to repair injured tissues. A **stem cell** satisfies this requirement by dividing mitotically to yield either two daughter cells like itself that can continue dividing without specializing, or one daughter cell that is a stem cell and one that is partially specialized. One defining characteristic of a stem cell is its ability, called *self-renewal,* to divide to give rise to other stem cells. A stem cell can also differentiate as any of many cell types, given appropriate biochemical signals.

A partly specialized cell that is the daughter of a stem cell but is intermediate between a stem cell and a fully differentiated cell is termed a **progenitor cell.** It is said to be "committed" because its daughter cells can become any of a restricted number of cell types (fig. 3.40). For example, a neural stem cell divides to give rise to progenitor cells that in turn give rise to the specialized neurons and neuroglia that make up neural tissue. Such neural progenitor cells, however, would not give rise to cell types that are part of muscle or bone tissue. All of the differentiated cell types in a human body can be traced back through lineages of progenitor and stem cells.

Stem cells and progenitor cells are described in terms of their potential, according to the possible fates of their daughter cells. A fertilized egg and cells of the very early embryo, when it is a small ball of cells, are **totipotent,** which means that their daughter cells can specialize as any cell type. In contrast, stem cells present

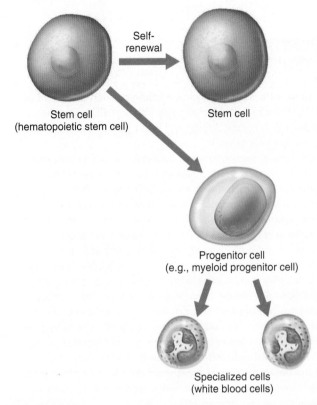

FIGURE 3.40 Stem cells and progenitor cells. A true stem cell divides mitotically to yield two stem cell daughters, or a stem cell and a progenitor cell, which may show the beginnings of differentiation. Progenitor cells give rise to progenitors or more differentiated cells of a restricted lineage.

PRACTICE

34 Distinguish between a stem cell and a progenitor cell.

35 Distinguish between totipotent and pluripotent.

36 Explain how cells differentiate.

3.7 | Cell Death

A cell that does not divide or differentiate has another option—death. **Apoptosis** (ap″o-to′sis) is a form of cell death also called "programmed cell death" because it is a normal part of development. Apoptosis sculpts organs from tissues that naturally overgrow. In the fetus, apoptosis carves away webbing between developing fingers and toes, prunes extra brain cells, and preserves only those immune system cells that recognize the body's cell surfaces. If it weren't for apoptosis, a child's lung or liver couldn't grow to adult size and maintain its characteristic form. Apoptosis is also protective. After a sunburn, this form of cell death peels away damaged skin cells that might otherwise turn cancerous.

In contrast to apoptosis is *necrosis*, which is cell death resulting from damage. The steps of the two processes are very different. Apoptosis is a normal event, whereas necrosis is not.

Apoptosis is a fast, orderly, contained destruction that packages cellular remnants into membrane-enclosed pieces that are then removed. It is a little like packaging up the contents of a messy room into plastic bags.

Like mitosis, apoptosis is a continuous, stepwise process. It begins when a "death receptor" on the doomed cell's cell membrane receives a signal to die. Within seconds, enzymes called *caspases* are activated inside the cell, where they cut up various cell components. These enzymes:

- destroy enzymes that replicate and repair DNA
- activate enzymes that cut DNA into similar-size pieces
- dismantle the cytoskeletal threads that support the nucleus, which collapses, condensing the DNA within
- fracture mitochondria, which release molecules that trigger further caspase activity, cut off the cell's energy supply, and destroy other organelles
- abolish the cell's ability to adhere to other cells
- transport certain phospholipids from the inner surface of the cell membrane to the outside, where they attract phagocytes that break down debris

A cell dying from apoptosis has a characteristic appearance (fig. 3.42). It rounds up as contacts with other cells are cut off, and the cell membrane undulates, forming bulges called *blebs*. The nucleus bursts under the multiple strains, releasing same-sized DNA pieces. Mitochondria decompose. Finally, the cell shatters. Almost instantly, pieces of membrane encapsulate the fragments, which prevents the signaling that triggers inflammation. Within an hour of the first release of caspases, the cell that underwent apoptosis is gone.

Mitosis and apoptosis are synchronized throughout development, maturation, and aging, and as a result, tissues and organs neither overgrow nor shrink. Disruptions in either process can cause cancer.

later in development, as well as progenitor cells, are **pluripotent,** which means that their daughter cells can follow any of several pathways, but not all of them. **Figure 3.41** depicts the lineages of skin and nervous tissue, which share a progenitor.

Some stem cells in the adult body were set aside during embryonic or fetal development as repositories for future healing. Many organs have very small populations of stem or progenitor cells, which are stimulated to divide when injury or illness occurs. For example, one in 10,000 to 15,000 bone marrow cells is a hematopoietic stem cell, which can give rise to blood and several other cell types. Some stem cells or progenitor cells may travel from bone marrow to replace damaged or dead cells in response to signals sent from injured or diseased tissues elsewhere.

All cells in the human body (except red blood cells, which expel their nuclei), have the same set of genetic instructions, but as cells specialize, they use some genes and ignore others. An immature bone cell (osteoblast) forms from a progenitor cell by manufacturing proteins necessary to bind bone mineral, as well as alkaline phosphatase, an enzyme required for bone formation. An immature muscle cell (myoblast), in contrast, forms from a muscle progenitor cell and accumulates the contractile proteins that define a muscle cell. The term *blast* is used for immature differentiated cells, such as osteoblast and myoblast. The osteoblast does not produce contractile proteins, just as the myoblast does not produce mineral-binding proteins and alkaline phosphatase. The process of differentiation is like using only some of the information in a database. From Science to Technology 3.1 looks at how stem cells are being investigated for use in health care.

In the human body, lineages of dividing stem cells and progenitor cells produce the specialized (differentiated) cell types that assemble and interact to form tissues and organs. Stem and progenitor cells are essential for growth and healing (fig. 3A). Stem cell technology is part of an emerging field, called regenerative medicine, that harnesses the body's ability to generate new cells to treat certain diseases and injuries.

Stem cells to treat disease come from donors or from the patient. Donor stem cells include umbilical cord stem cells saved from newborns that are used to treat a variety of blood disorders and certain metabolic conditions. Stem cells derived from a patient have two sources: their natural sites or cultured from "reprogrammed" differentiated cells.

A use of stem cells from their natural site is an autologous bone marrow transplant, in which a person's immune system is destroyed with drugs or radiation after stem cells are set aside. The stem cells are then infused to repopulate the bone marrow.

Reprogramming differentiated cells is a promising approach to producing therapeutic stem and progenitor cells. A fibroblast taken from a skin sample, for example, can be given genetic instructions to produce key proteins that return the cell to a state that resembles a stem cell from an embryo. Then a cocktail of specific biochemicals is added to guide differentiation. The altered cell divides in culture, specializing and passing on its new characteristics to its daughter cells.

In the future, reprogrammed cells may be used therapeutically, but for now they are a research tool. Rett syndrome, an inherited condition that affects females, illustrates what researchers can learn from experiments using reprogrammed cells. Girls with Rett syndrome move their hands in a characteristic and uncontrollable way, and gradually lose muscle tone and the ability to speak. Head growth slows, and many girls become disabled by the time they reach their teens. The disease is difficult to study at the cellular level in neurons taken from a patient, because this cell type does not divide. However, neurons can be continually cultured from reprogrammed cells. Reprogrammed neurons cultured from skin samples taken from four girls with the syndrome were too small, with too few connections and abnormal signaling. Drugs that helped mice with a form of Rett syndrome also corrected the defect in the reprogrammed cells from the four girls. The next step: developing a treatment based on what researchers observe in the reprogrammed cells.

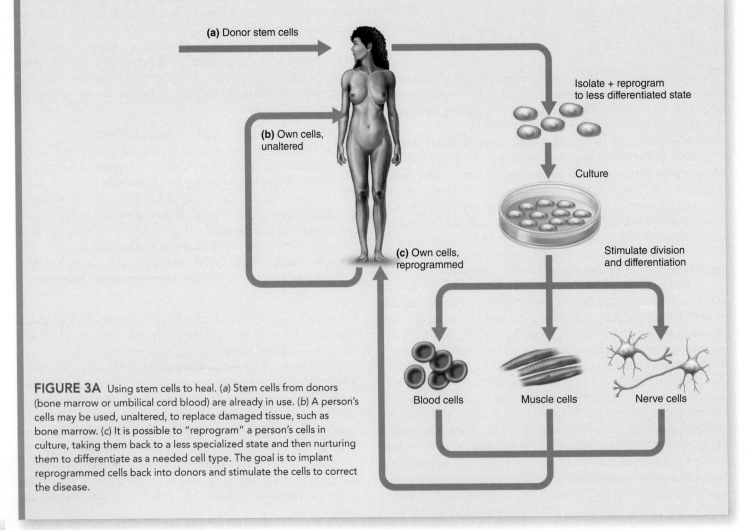

FIGURE 3A Using stem cells to heal. (*a*) Stem cells from donors (bone marrow or umbilical cord blood) are already in use. (*b*) A person's cells may be used, unaltered, to replace damaged tissue, such as bone marrow. (*c*) It is possible to "reprogram" a person's cells in culture, taking them back to a less specialized state and then nurturing them to differentiate as a needed cell type. The goal is to implant reprogrammed cells back into donors and stimulate the cells to correct the disease.

(a) Donor stem cells

(b) Own cells, unaltered

Isolate + reprogram to less differentiated state

Culture

(c) Own cells, reprogrammed

Stimulate division and differentiation

Blood cells

Muscle cells

Nerve cells

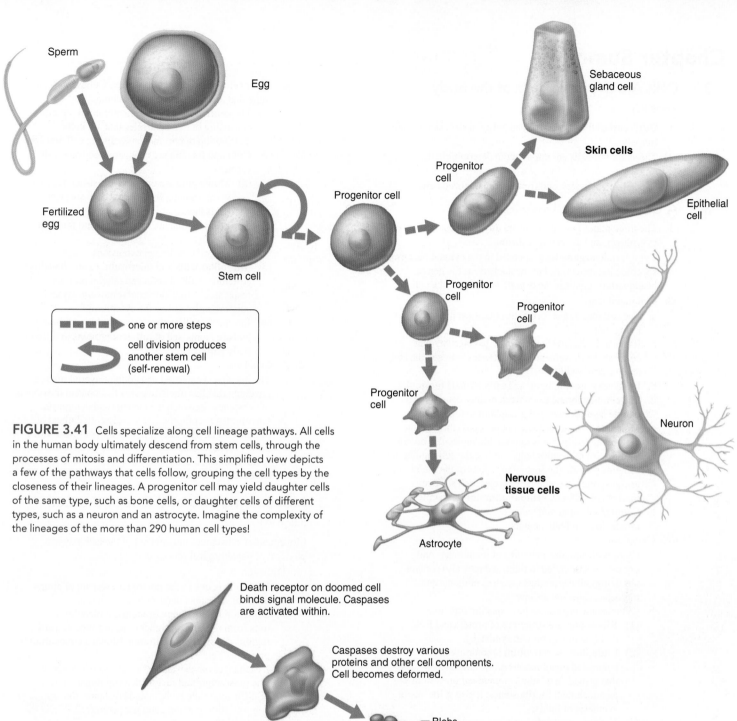

FIGURE 3.41 Cells specialize along cell lineage pathways. All cells in the human body ultimately descend from stem cells, through the processes of mitosis and differentiation. This simplified view depicts a few of the pathways that cells follow, grouping the cell types by the closeness of their lineages. A progenitor cell may yield daughter cells of the same type, such as bone cells, or daughter cells of different types, such as a neuron and an astrocyte. Imagine the complexity of the lineages of the more than 290 human cell types!

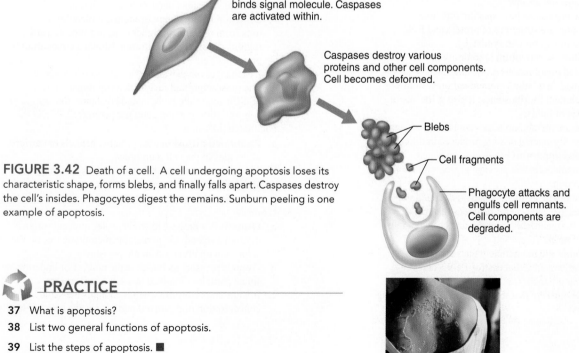

FIGURE 3.42 Death of a cell. A cell undergoing apoptosis loses its characteristic shape, forms blebs, and finally falls apart. Caspases destroy the cell's insides. Phagocytes digest the remains. Sunburn peeling is one example of apoptosis.

PRACTICE

37 What is apoptosis?

38 List two general functions of apoptosis.

39 List the steps of apoptosis. ∎

Chapter Summary

3.1 Cells Are the Basic Units of the Body

(page 83)

1. **Differentiated** cells vary considerably in size, shape, and function.
2. The shapes of cells are important in determining their functions.
3. Specialized cells descend from less specialized cells.

3.2 A Composite Cell (page 83)

1. The three major parts of a cell are the **nucleus,** the **cytoplasm,** and the **cell membrane.**
2. Cytoplasmic **organelles** suspended in the **cytosol** perform specific vital functions, but the nucleus, as the genetic headquarters, controls the overall activities of the cell.
3. Cell membrane
   a. The cell membrane forms the outermost limit of the living material.
   b. It acts as a selectively permeable passageway that controls the movements of substances between the cell and its surroundings.
   c. It includes protein, lipid, and carbohydrate molecules.
   d. The cell membrane framework mainly consists of a double layer (bilayer) of phospholipid molecules.
   e. Molecules that are soluble in lipids pass through the membrane easily, but water-soluble molecules do not.
   f. Cholesterol molecules help stabilize the cell membrane.
   g. Proteins provide the special functions of the membrane, as transporters, receptors, enzymes, cell surface markers of self, and **cellular adhesion molecules.**
   h. Cellular adhesion molecules oversee some cell interactions and movements.
4. Cytoplasm
   a. Cytoplasm contains networks of membranes and organelles suspended in fluid, and may also contain nonliving cellular products, such as nutrients and pigments, called inclusions.
   b. Cytoplasmic organelles have specific functions.
      (1) **Ribosomes** are structures of protein and RNA that function in protein synthesis.
      (2) **Endoplasmic reticulum** is composed of connected membranous sacs, canals, and vesicles that provide a tubular communication system and an attachment for ribosomes; it also is the site of synthesis of lipids.
      (3) **Vesicles** are membranous sacs containing substances that recently entered or were produced in the cell.
      (4) The **Golgi apparatus** is a stack of flattened, membranous sacs that package glycoproteins for secretion.
      (5) **Mitochondria** are membranous sacs containing enzymes that catalyze the reactions that release energy from nutrient molecules and change it into a usable form.
      (6) **Lysosomes** are membranous sacs containing digestive enzymes that destroy debris and worn-out organelles.
      (7) **Peroxisomes** are membranous, enzyme-containing vesicles.

   c. Other cellular structures are not organelles, including some that are parts of the cytoskeleton.
      (1) The **centrosome** is a nonmembranous structure consisting of two centrioles that aid in the distribution of chromosomes during cell division.
      (2) **Cilia** and **flagella** are extensions on some cell surfaces.
         (a) Motile cilia are tiny, hairlike structures that wave, moving fluids across cell surfaces. Cells have many motile cilia.
         (b) Primary cilia are one per cell and transmit signals.
         (c) Flagella are longer extensions.
      (3) **Microfilaments** and **microtubules** are threadlike structures built of actin and tubulin proteins, respectively, that aid cellular movements and support and stabilize the cytoplasm.
      (4) Intermediate filaments serve a variety of functions and are composed of different proteins in different body parts.
5. Cell nucleus
   a. The nucleus is enclosed in a double-layered **nuclear envelope** that has **nuclear pores** that control movement of substances between the nucleus and cytoplasm.
   b. **Chromatin** is composed of loosely coiled fibers of protein and DNA that coil tightly into **chromosomes** during cell division.
   c. A **nucleolus** is a dense body of protein and RNA where ribosome synthesis occurs.

3.3 Movements Into and Out of the Cell

(page 99)

Movement of substances into and out of the cell may use physical or physiological processes.

1. Diffusion
   a. **Diffusion** is due to the random movement of atoms, molecules, or ions in air or liquid solution.
   b. Diffusion is movement of atoms, molecules, or ions from regions of higher concentration toward regions of lower concentration (down a concentration gradient).
   c. It exchanges oxygen and carbon dioxide in the body.
   d. The most important factors determining the rate of diffusion in the body include distance, the concentration gradient, and temperature.
2. Facilitated diffusion
   a. **Facilitated diffusion** uses protein channels or carrier molecules in the cell membrane.
   b. This process moves substances such as ions, sugars, and amino acids from regions of higher concentration to regions of lower concentration.
3. Osmosis
   a. **Osmosis** is a process in which water molecules move through a selectively permeable membrane toward the solution with greater osmotic pressure.
   b. Osmotic pressure increases as the number of impermeant solute particles dissolved in a solution increases.
   c. A solution is **isotonic** when it contains the same concentration of dissolved particles as the cell contents.

d. Cells lose water when placed in **hypertonic solutions** and gain water when placed in **hypotonic solutions.**
4. Filtration
  a. In **filtration,** molecules move through a membrane from regions of higher hydrostatic pressure toward regions of lower hydrostatic pressure.
  b. Blood pressure filters water and dissolved substances through porous capillary walls.
5. Active transport
  a. **Active transport** moves molecules or ions from regions of lower concentration to regions of higher concentration.
  b. It requires ATP and carrier molecules in the cell membrane.
6. **Endocytosis**
  a. In **pinocytosis,** a cell membrane engulfs tiny droplets of liquid.
  b. In **phagocytosis,** a cell membrane engulfs solid particles.
  c. In **receptor-mediated endocytosis,** receptor proteins combine with specific molecules in the cell surroundings. The membrane engulfs the combinations.
7. Exocytosis
  a. **Exocytosis** has the reverse effect of endocytosis but the steps differ.
  b. In exocytosis, vesicles containing secretions fuse with the cell membrane, releasing the substances to the outside.
8. Transcytosis
  a. **Transcytosis** combines endocytosis and exocytosis.
  b. In transcytosis, a substance or particle crosses a cell.
  c. Transcytosis is specific.

3.4 The Cell Cycle (page 107)

1. The **cell cycle** includes interphase, **mitosis,** and **cytokinesis.**
2. Interphase
  a. **Interphase** is the stage when a cell grows, DNA replicates, and new organelles form.
  b. It terminates when the cell begins mitosis.
3. Mitosis
  a. Mitosis is the division and distribution of DNA to daughter cells.
  b. The stages of mitosis are **prophase, metaphase, anaphase,** and **telophase.**
4. In cytokinesis the cytoplasm divides, distributing organelles to the two daughter cells.

3.5 Control of Cell Division (page 110)

1. Cell division capacities vary greatly among cell types.
2. Chromosome tips that shorten with each mitosis provide a mitotic clock, usually limiting the number of divisions to about fifty.
3. Internal and external factors control cell division.
4. As a cell grows, its surface area increases to a lesser degree than its volume, and eventually the surface area limits the cell's ability to take in nutrients. Cell division restores the more favorable surface area–volume relationship.
5. Growth factors and hormones stimulate cell division.
6. Cancer is the consequence of a loss of cell cycle control.

3.6 Stem and Progenitor Cells (page 112)

1. A **stem cell** divides to yield another stem cell and a partially differentiated **progenitor cell.**
2. Cells that may give rise to any differentiated cell type are **totipotent.** Cells with more restricted fates are **pluripotent.**
3. Stem cells are present in adult organs and migrate from the bone marrow to replace damaged cells.
4. As cells specialize, they express different sets of genes that provide their distinct characteristics.

3.7 Cell Death (page 113)

1. **Apoptosis** is a form of cell death that is part of normal development and growth.
2. It is a fast, orderly multistep process that begins when a cell surface receptor receives a signal to die. Caspases start a chain reaction that cuts up the cell into membrane-bounded pieces. A phagocyte destroys the remains.
3. Apoptosis and mitosis are synchronized throughout development, maturation, and aging.

 CHAPTER ASSESSMENTS

3.1 Cells Are the Basic Units of the Body

1 What are the units used to measure cells? (p. 83)
  a. millimeters
  b. centimeters
  c. micrometers
  d. decimeters
2 Describe three types of differentiated cells. (p. 83)

3.2 A Composite Cell

3 The three major parts of a cell are _____. (p. 83)
  a. the nucleus, the nucleolus, and the nuclear envelope
  b. the nucleus, the cytoplasm, and the cell membrane
  c. the lysosomes, ribosomes, and vesicles

  d. the endoplasmic reticulum, the Golgi apparatus, and ribosomes
4 Distinguish between the cytoplasm and the cytosol of a cell. (p. 83)
5 Explain the general function of organelles. (pp. 83 and 85)
6 Define *selectively permeable.* (p. 86)
7 Describe the structure of a cell membrane and explain how this structural organization provides the membrane's function. (pp. 86–87)
8 List three functions of membrane proteins. (pp. 87–89)
9 State a way that cellular adhesion is essential to health and a way that abnormal cellular adhesion harms health. (p. 89)
10 Distinguish between organelles and inclusions. (p. 90)

11 Match the following structures with their descriptions: (pp. 90–96)

(1) Golgi apparatus
(2) mitochondria
(3) peroxisomes
(4) cilia
(5) smooth endoplasmic reticulum
(6) cytoskeleton
(7) vesicles
(8) ribosomes

A. sacs that contain enzymes that catalyze a variety of specific biochemical reactions
B. structures on which protein synthesis occurs
C. structures that house the reactions that release energy from nutrients
D. a network of microfilaments and microtubules that supports and shapes a cell
E. a structure that modifies, packages, and exports glycoproteins
F. membrane-bound sacs
G. a network of membranous channels and sacs where lipids are synthesized
H. hairlike structures that extend from certain cell surfaces and wave about

12 List the parts of the nucleus and explain why each is important. (p. 97)

3.3 Movements Into and Out of the Cell

13 Distinguish between active and passive mechanisms of movement across cell membranes. (p. 99)
14 Match the movements into and out of the cell on the left with their descriptions on the right. (pp. 99–105)

(1) simple diffusion
(2) facilitated diffusion
(3) filtration
(4) osmosis
(5) active transport
(6) endocytosis
(7) exocytosis

A. the cell membrane engulfs a particle or substance, drawing it into the cell in a vesicle
B. movement down a concentration gradient through an ion channel or with a carrier protein, without energy from ATP
C. movement down a concentration gradient through the phospholipid bilayer
D. a particle or substance leaves a cell when the vesicle containing it merges with the cell membrane
E. movement against a concentration gradient with a carrier protein and energy from ATP
F. hydrostatic pressure forces small substances through a membrane
G. water moves through a selectively permeable membrane into a region of greater concentration of impermeant solute

15 Define *osmosis*. (p. 101)
16 Distinguish among hypertonic, hypotonic, and isotonic solutions. (pp. 101–102)

17 Explain how phagocytosis differs from receptor-mediated endocytosis. (pp. 103–104)
18 Explain how transcytosis combines endocytosis and exocytosis. (p. 105)

3.4 The Cell Cycle

19 The period of the cell cycle when DNA replicates is_____. (p. 108)
   a. G_1 phase
   b. G_2 phase
   c. S phase
   d. prophase
20 Explain why interphase is not a period of rest for a cell. (p. 108)
21 Explain how meiosis differs from mitosis. (p. 108)
22 Describe the events of mitosis in sequence. (p. 108)
23 _____occur simultaneously. (pp. 108–110)
   a. G_1 phase and G_2 phase
   b. Interphase and mitosis
   c. Cytokinesis and telophase
   d. Prophase and metaphase

3.5 Control of Cell Division

24 List five factors that control when and if a cell divides. (pp. 110–111)
25 Explain why it is important for the cell cycle to be highly regulated. (p. 111)
26 Discuss the consequences of too little cell division and too much cell division. (p. 111)
27 Distinguish between the ways that mutations in oncogenes and tumor suppressor genes cause cancer. (p. 111)

3.6 Stem and Progenitor Cells

28 Define *differentiation*. (p. 112)
29 A stem cell _____. (p. 112)
   a. self-renews
   b. dies after fifty divisions
   c. is differentiated
   d. gives rise only to fully differentiated daughter cells
30 Which of the following is true? (p. 112)
   a. Progenitor cells are totipotent and stem cells are differentiated.
   b. Stem cells are totipotent and progenitor cells are differentiated.
   c. Differentiated cells are pluripotent until they specialize.
   d. Stem cells in the early embryo are totipotent and progenitor cells are pluripotent.
31 Describe a general function of stem cells in the body. (p. 112)

3.7 Cell Death

32 Explain how apoptosis (cell death) can be a normal part of development. (p. 113)
33 Provide an example of apoptosis. (p. 113)
34 Distinguish between necrosis and apoptosis. (p. 113)
35 List the steps of apoptosis. (p. 113)
36 Describe the relationship between apoptosis and mitosis. (p. 113)

INTEGRATIVE ASSESSMENTS/CRITICAL THINKING

Outcomes 2.3, 3.3

1. Liver cells are packed with glucose. What mechanism could transport more glucose into a liver cell? Why would only this mode of transport work?

2. What characteristic of cell membranes may explain why fat-soluble substances such as chloroform and ether rapidly affect cells?

Outcomes 3.1, 3.2, 3.6

3. For experimental stem cell therapy, state the part of a cell reprogrammed to function like that of a stem cell and stimulated to differentiate in a particular way.

Outcome 3.2

4. Organelles compartmentalize a cell. What advantage does this offer a large cell? Cite two examples of organelles and the activities they compartmentalize.

5. Exposure to tobacco smoke immobilizes and destroys motile cilia. How might this effect explain why smokers have an increased incidence of coughing and respiratory infections?

Outcome 3.3

6. Which process—diffusion, osmosis, or filtration—is used in the following situations?
   a. Injection of a drug hypertonic to the tissues stimulates pain.
   b. The urea concentration in the dialyzing fluid of an artificial kidney is decreased.
   c. A person with extremely low blood pressure stops producing urine.

Outcome 3.6

7. Reports in the media about stem cells usually state that they "turn into any kind of cell in the body." Explain why this statement is only partially correct, including a description of how a stem cell maintains the population of stem cells.

 # ONLINE STUDY TOOLS

Connect Interactive Questions Reinforce your knowledge using assigned interactive questions covering the cell membrane, organelles, movement through the cell membrane, and stem cells.

Connect Integrated Activity Can you predict the effect diseases or drugs might have on the cell membrane or organelles?

LearnSmart Discover which chapter concepts you have mastered and which require more attention. This adaptive learning tool is personalized, proven, and preferred.

Anatomy & Physiology Revealed Go more in depth into the human body by viewing parts of a cell under a microscope and viewing animations of the cell cycle.

Cellular Metabolism

LEARNING OUTCOMES

After you have studied this chapter, you should be able to:

4.1 Metabolic Processes
1 Compare and contrast anabolism and catabolism. (pp. 121–123)

4.2 Control of Metabolic Reactions
2 Describe the role of enzymes in metabolic reactions. (p. 123)
3 Explain how metabolic pathways are regulated. (pp. 124–125)

4.3 Energy for Metabolic Reactions
4 Explain how ATP stores chemical energy and makes it available to a cell. (p. 126)

4.4 Cellular Respiration
5 Explain how the reactions of cellular respiration release chemical energy. (p. 127)
6 Describe the general metabolic pathways of carbohydrate metabolism. (pp. 127–131)

4.5 Nucleic Acids and Protein Synthesis
7 Describe how DNA molecules store genetic information. (p. 132)
8 Describe how DNA molecules are replicated. (p. 134)
9 Explain how protein synthesis relies on genetic information. (p. 136)
10 Compare and contrast DNA and RNA. (p. 136)
11 Describe the steps of protein synthesis. (pp. 139–142)

4.6 Changes in Genetic Information
12 Describe how genetic information can be altered. (p. 142)
13 Explain how a mutation may or may not affect an organism. (p. 143)

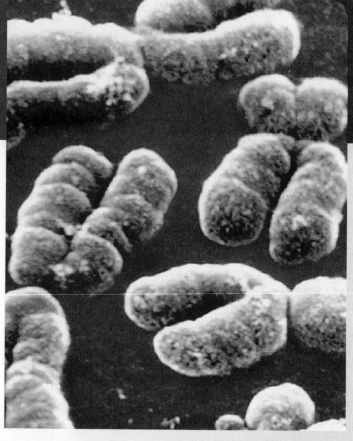

Chromosomes consist of long molecules of DNA complexed with proteins. The DNA sequences instruct cells to build specific proteins—including enzymes essential to metabolism (36,000×).

THE WHOLE PICTURE

In every human cell, even in the most sedentary individual, thousands of chemical reactions essential to life take place every second. Many metabolic reactions occur one after the other in a sequence, with the products of one reaction serving as starting materials for the next. These reactions form pathways and cycles that intersect where they share intermediate compounds. Special types of proteins called enzymes control the rate of each reaction, enabling them to proceed fast enough to sustain life. The sum total of chemical reactions in the cell constitutes cellular metabolism.

Anatomy & Physiology | REVEALED®
aprevealed.com

Module 2: Cells & Chemistry

UNDERSTANDING WORDS

aer-, air: *aerobic* respiration—respiratory process that requires oxygen.

an-, without: *anaerobic* respiration—respiratory process that does not require oxygen.

ana-, up: *anabolism*—cellular processes in which smaller molecules are built up into larger ones.

cata-, down: *catabolism*—cellular processes that break down larger molecules into smaller ones.

co-, with: *coenzyme*—substance that unites with a protein to complete the structure of an active enzyme molecule.

de-, undoing: *deamination*—process that removes nitrogen-containing portions of amino acid molecules.

mut-, change: *mutation*—change in genetic information.

-strat, spread out: sub*strate*—substance upon which an enzyme acts.

sub-, under: *substrate*—substance upon which an enzyme acts.

-zym, causing to ferment: *enzyme*—protein that speeds up a chemical reaction without itself being consumed.

4.1 | Metabolic Processes

Metabolic reactions and pathways are of two types: anabolism and catabolism. In **anabolism** (ăh-nab′o-liz″-ĕm), small molecules are built up into larger ones, requiring energy. In **catabolism** (kă-tab′o-liz″-ĕm), larger molecules are broken down into smaller ones, releasing energy. About 60% of the energy released as large molecules are dismantled escapes as heat. The rest of the energy is used to build molecules and to drive various activities of the cell.

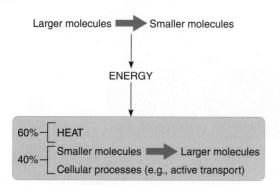

Anabolism

Anabolism provides all the materials a cell requires for maintenance, growth, and repair. For example, a type of anabolic process called **dehydration synthesis** (de″hi-dra′shun sin′the-sis) joins many simple sugar molecules (monosaccharides) to form larger molecules of glycogen, which store energy in their chemical bonds. When a runner eats pasta the night before a race, digestion breaks down the plant-based complex carbohydrates in the food to monosaccharides. These smaller molecules are absorbed into the bloodstream, which carries the energy-rich molecules to body cells. Here, dehydration synthesis joins the monosaccharides to form glycogen, which stores energy that the runner may need later, as the finish line nears. First, two monosaccharides join to form a disaccharide. When monosaccharide units join, an —OH (hydroxyl group) from one monosaccharide molecule and an —H (hydrogen atom) from an —OH group of another are removed. As the —H and —OH react to produce a water molecule, the mono-

saccharides are joined by a shared oxygen atom, as **figure 4.1** shows (read from left to right). As the process repeats, the molecular chain extends, forming a polysaccharide.

Dehydration synthesis also links glycerol and fatty acid molecules in fat cells (adipose tissue) to form triglyceride molecules. **Figure 4.2** shows (read from left to right) how a triglyceride forms as three hydrogen atoms are removed from a glycerol molecule

CAREER CORNER

Personal Trainer

The 45-year-old man's physician advised him to start an exercise program to lose weight, so the man joined a gym. But he hadn't been to one in many years, and the rows of machines looked daunting. So he hired a personal trainer to develop an exercise routine that would be just what the doctor ordered.

A personal trainer assesses a client's fitness level and guides the client in using specific machines and doing mat exercises. The trainer offers advice on which weight-lifting machines to use, how many repetitions to begin with, and how often to increase repetitions. The trainer might advise the client to lift weights only every other day, so that muscles will have time to recover from the microscopic tears caused by weight lifting. Using a mat, a trainer might lead a client through a series of exercises to strengthen the core abdominal muscles.

The personal trainer is part coach, part cheerleader, encouraging exercisers to push their limits. Trainers work in athletic clubs, at corporate fitness centers, in senior centers, and at other types of facilities that have exercise equipment. Some personal trainers even train clients at their homes.

Minimal requirements to become a personal trainer include a high school diploma, cardiopulmonary resuscitation (CPR) training, and completion of a personal trainer course, which generally takes from a few months to a year. Passing a certification exam is required. Personal trainers tend to be outgoing, friendly people who enjoy helping others become physically fit.

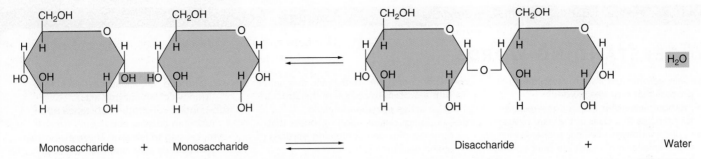

FIGURE 4.1 Building up and breaking down molecules. A disaccharide is formed from two monosaccharides in a dehydration synthesis reaction (arrows pointing to the right). In the reverse reaction, hydrolysis, a disaccharide is broken down into two monosaccharides (arrows pointing to the left).

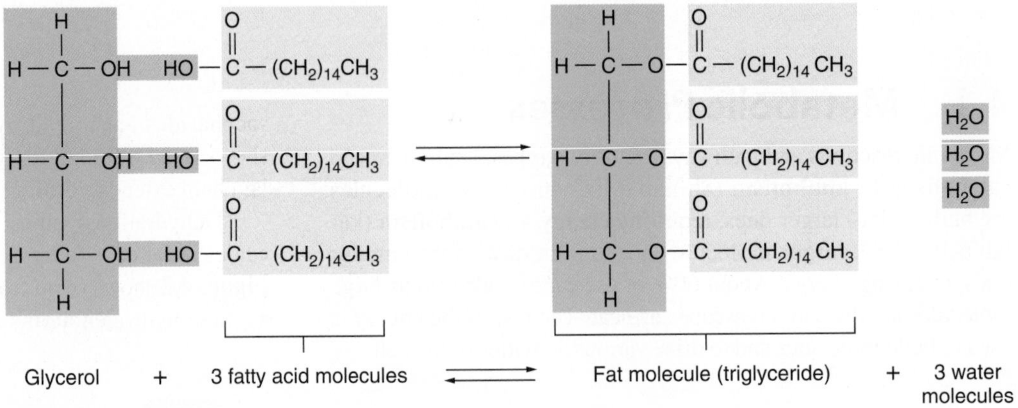

FIGURE 4.2 Forming a fat. A glycerol molecule and three fatty acid molecules react, yielding a fat molecule (triglyceride) in a dehydration synthesis reaction (arrows pointing to the right). In the reverse reaction, hydrolysis, a triglyceride is broken down into three fatty acids and a glycerol (arrows pointing to the left).

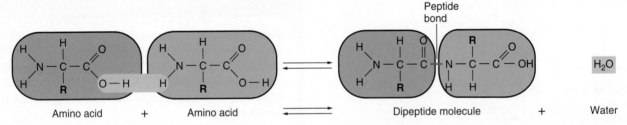

FIGURE 4.3 Peptide bonds link amino acids. When dehydration synthesis unites two amino acid molecules, a peptide bond forms between a carbon atom and a nitrogen atom, resulting in a dipeptide molecule (arrows pointing to the right). In the reverse reaction, hydrolysis, a dipeptide molecule is broken down into two amino acids (arrows pointing to the left).

and an —OH group is removed from each of three fatty acid molecules. The result of this reaction is three water molecules and a single fat molecule whose glycerol and fatty acid portions are bound by shared oxygen atoms.

In cells, dehydration synthesis also builds protein molecules by joining amino acid molecules. When two amino acid molecules are united, an —OH from the —COOH group of one and an —H from the —NH$_2$ group of another are removed. A water molecule forms, and the amino acid molecules join by a bond between a carbon atom and a nitrogen atom (fig. 4.3; read from left to right). This type of bond, called a *peptide bond*, holds the amino acids together. Two such bound amino acids form a *dipeptide*, and many joined in a chain form a *polypeptide*. Generally, a polypeptide consisting of 50 or more amino acid molecules is called a *protein*, although the boundary between polypeptides and proteins is not

precisely defined. Some proteins consist of more than one polypeptide chain. The large protein that carries oxygen in the blood (hemoglobin), for example, consists of four polypeptide chains, each wound into a globular shape.

Nucleic acids are also formed by dehydration synthesis. This process is described later in the chapter.

Catabolism

Catabolic reactions break down larger molecules into smaller ones. An example of catabolism is **hydrolysis** (hi-drol′ĭ-sis), which can decompose carbohydrates, lipids, and proteins. A water molecule is used for each bond that breaks. Hydrolysis of a disaccharide, for instance, yields two monosaccharide molecules (see fig. 4.1; read from right to left). The bond between the simple sugars breaks, and the water molecule supplies a hydrogen atom to one sugar molecule and a hydroxyl group to the other. Hydrolysis is the reverse of dehydration synthesis.

Hydrolysis breaks down carbohydrates into monosaccharides; fats into glycerol and fatty acids (see fig. 4.2; read from right to left); proteins into amino acids (see fig. 4.3; read from right to left); and nucleic acids into nucleotides. It does not occur automatically, even though in the body water molecules are readily available to provide the necessary —H and —OH. For example, water-soluble substances, such as the disaccharide sucrose (table sugar), *dissolve* in a glass of water but do not undergo hydrolysis. Like dehydration synthesis, hydrolysis requires specific enzymes, discussed in the next section, Control of Metabolic Reactions.

Many of the reactions of metabolism are reversible. However, the enzyme that speeds, or catalyzes, an anabolic reaction may be different from the enzyme that catalyzes the corresponding catabolic reaction.

Both catabolism and anabolism must be carefully controlled so that the breakdown or energy-releasing reactions occur at rates adjusted to the requirements of the building up or energy-utilizing reactions. Any disturbance in this balance is likely to damage or kill cells. From Science to Technology 4.1 discusses a systematic approach to identifying all the molecules that are part of metabolism in a human body.

 PRACTICE

1 What are the general functions of anabolism and catabolism?

2 What type of molecule is formed by the anabolism of monosaccharides? Of glycerol and fatty acids? Of amino acids?

3 Distinguish between dehydration synthesis and hydrolysis.

Metabolic reactions and pathways can be subgrouped. *Intermediary metabolism* refers to the processes that obtain, release, and use energy. Another way to classify metabolic reactions is by their necessity. *Primary metabolites* are products of metabolism essential to survival. *Secondary metabolites* are not essential to survival, but may provide an advantage or enhancement. Secondary metabolites are best studied in plants, where they usually help to defend against predators because they are toxins. Some of our most successful drugs are plant secondary metabolites. The vinca alkaloids, for example, protect the rosy Madagascar periwinkle that produces them by sickening animals that eat the vegetation, but we use these biochemicals to treat cancer. Their effect is to destabilize microtubule formation.

4.2 | Control of Metabolic Reactions

Different types of cells carry out specialized metabolic processes, but all cells perform certain basic reactions, such as the buildup and breakdown of carbohydrates, lipids, proteins, and nucleic acids. These common reactions include hundreds of very specific chemical changes that must occur in particular sequences. **Enzymes** control the rates of these metabolic reactions.

Enzyme Action

Metabolic reactions require energy (*activation energy*) before they proceed, just like any chemical reaction. This is why in laboratory experiments heat is used to increase the rates of chemical reactions. Heat energy increases the rate at which molecules move and the frequency of their collisions. These collisions increase the likelihood of interactions among the electrons of the molecules that can form new chemical bonds. The temperature conditions in cells are usually too mild to adequately promote the reactions of life. Enzymes make these reactions possible.

Most enzymes are globular proteins that catalyze specific chemical reactions in cells by lowering the activation energy required to start these reactions. Enzymes can speed metabolic reactions by a factor of a million or more. This acceleration of chemical reaction rate is called *catalysis,* and an enzyme is an organic catalyst.

Enzymes are required in small amounts, because as they work, they are not consumed and can, therefore, function repeatedly. Because of this, a few enzyme molecules can have a powerful effect. Each enzyme type is specific, acting only on a particular molecule, called its **substrate** (sub'strāt). For example, the substrate of an enzyme called *catalase* (found in the peroxisomes of liver and kidney cells) is hydrogen peroxide, a toxic by-product of certain metabolic reactions. This enzyme's only function is to decompose hydrogen peroxide into water and oxygen, an action that helps prevent accumulation of hydrogen peroxide, which damages cells.

The action of the enzyme catalase is obvious when using hydrogen peroxide to cleanse a wound. Injured cells release catalase, and when hydrogen peroxide contacts them, bubbles of oxygen are set free. The resulting foam removes debris from inaccessible parts of the wound.

Many enzyme names are derived from the names of their substrates, with the suffix *-ase*. A lipid-splitting enzyme is a *lipase,* a protein-splitting enzyme a *protease,* and a starch (amylum)-splitting enzyme an *amylase.* Enzyme names may be specific. For example, *s*ucrase splits the sugar sucrose, *maltase* splits the sugar maltose, and *lactase* splits the sugar lactose.

Each enzyme must be able to "recognize" its specific substrate. This ability to identify a substrate depends upon the shape of an enzyme molecule. That is, each enzyme's polypeptide chain twists and coils into a unique three-dimensional conformation that fits the particular shape of its substrate molecule.

 RECONNECT
To Chapter 2, Proteins, pages 72–75.

During an enzyme-catalyzed reaction, a region of the enzyme molecule called the **active site** temporarily combines with a specific part of the substrate, forming an enzyme–substrate complex. (Most enzymes have only one active site.) This interaction strains chemical bonds in the substrate in a way that makes a particular chemical reaction require less energy to occur. When it does, the enzyme is

A generation ago, prehealth profession students had to memorize a complex chart of biochemical pathways that represent all of the energy reactions in a cell. The cellular respiration pathways ran down the center, with branches radiating outward and in some places interconnecting into a giant web. Today, several technologies as well as the ability to store massive amounts of data have made possible the Human Metabolome Database.

"Metabolome" refers to all of the small molecules that are part of metabolism in a cell, tissue, organ, or an entire organism. The database is a vast, annotated catalog of those molecules, "metabolites." The government of Canada is supporting the effort to search all published papers and books that describe metabolites and link that information with experimental data. The techniques of electrophoresis and chromatography are used to separate metabolites, and mass spectrometry (MS) and nuclear magnetic resonance (NMR) spectroscopy describe their chemical characteristics.

Biochemists estimate that human cells have at least 2,500 different metabolites, but fewer than half have been identified. Far fewer have been analyzed for their concentrations in different cell types under different conditions. In the Human Metabolome Database, each entry has an electronic "MetaboCard" that includes 90 data fields, half with clinical data (such as associated diseases and drug interactions) and half with biochemical data (such as pathways and enzymes that interact with the metabolite). Each entry is also hyperlinked to other databases, interfacing with 1,500 drugs and 3,600 foods and food additives. The information in the Human Metabolome Database is being used in drug discovery, toxicology, transplant monitoring, clinical chemistry, disease diagnosis, and screening of newborns for metabolic disorders.

released in its original form, able to bind another substrate molecule (fig. 4.4). Many enzyme-catalyzed reactions are reversible and in some cases the same enzyme catalyzes both directions. An enzyme-catalyzed reaction can be summarized as follows:

$$\text{Substrate} + \text{Enzyme} \rightarrow \begin{array}{c}\text{Enzyme–}\\\text{substrate}\\\text{complex}\end{array} \rightarrow \text{Product} + \begin{array}{c}\text{Enzyme}\\\text{(unchanged)}\end{array}$$

The rate of an enzyme-catalyzed reaction depends partly on the number of enzyme and substrate molecules in the cell. The reaction is faster if the concentration of the enzyme or the concentration of the substrate increases. The efficiency of different types of enzymes varies greatly. Some enzymes can catalyze only a few reactions per second, whereas others can catalyze hundreds of thousands.

The shapes (conformations) of enzymes are critical to their functions. Exposure to excessive heat, radiation, electricity, certain chemicals, or fluids with extreme pH values can denature, or alter, the conformation of an enzyme. Many enzymes become inactive at 45°C, and nearly all are denatured at 55°C. Some poisons denature enzymes.

Metabolic Pathways

Cellular metabolism includes hundreds of different chemical reactions, each controlled by a specific type of enzyme. Enzyme-catalyzed reactions form pathways when the product of one reaction is the substrate of another reaction. **Metabolic pathways** lead to the synthesis or breakdown of particular biochemicals (fig. 4.5).

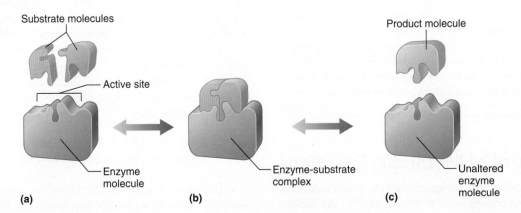

Substrate molecules

Product molecule

Active site

Enzyme molecule

Enzyme-substrate complex

Unaltered enzyme molecule

(a)   (b)   (c)

FIGURE 4.4 AP R An enzyme-catalyzed reaction. (Many enzyme-catalyzed reactions, as depicted here, are reversible.) In the forward reaction (dark-shaded arrows), (a) the shapes of the substrate molecules fit the shape of the enzyme's active site. (b) When the substrate molecules temporarily combine with the enzyme, a chemical reaction proceeds. (c) The result is a product molecule and an unaltered enzyme. The active site changes shape as the substrate binds, such that formation of the enzyme–substrate complex is more like a hand fitting into a glove, which has some flexibility, than a key fitting into a lock.

FIGURE 4.5 A metabolic pathway consists of a series of enzyme-controlled reactions leading to formation of a product.

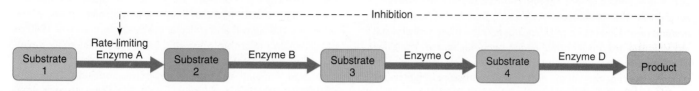

FIGURE 4.6 A negative feedback mechanism may control a rate-limiting enzyme in a metabolic pathway. The product of the pathway inhibits the enzyme.

In many metabolic pathways, a regulatory enzyme that catalyzes one step sets the rate for the entire sequence of reactions. The number of molecules of such a regulatory enzyme is limited. Consequently, these enzymes can become saturated when the substrate concentration exceeds a certain level. Once this happens, increasing the substrate concentration no longer affects the reaction rate. Because the enzyme does not work at high substrate concentrations, it is termed a **rate-limiting enzyme.** Such an enzyme is often the first in the series of reactions that comprises the metabolic pathway (fig. 4.6). This position is important because an intermediate product of the pathway might accumulate if an enzyme occupying another position in the sequence were rate-limiting.

For some metabolic pathways, the product inhibits the rate-limiting regulatory enzyme. This type of control is an example of negative feedback. Accumulating product inhibits the pathway, and synthesis of the product falls. When the concentration of product decreases, the inhibition lifts and more product is synthesized. In this way, a single enzyme can control a whole pathway, stabilizing the rate of product formation (fig. 4.6). Clinical Application 4.1 discusses "inborn errors of metabolism," which result from deficit or lack of an enzyme.

RECONNECT
To Chapter 1, Homeostasis, page 15.

Cofactors and Coenzymes

An enzyme may be inactive until it combines with a nonprotein component called a **cofactor,** which helps the active site fold into its appropriate conformation or helps bind the enzyme to its substrate. A cofactor may be an ion of an element, such as copper, iron, or zinc, or a small organic molecule, called a **coenzyme** (ko-en′zīm). Many coenzymes are composed of vitamin molecules or incorporate altered forms of vitamin molecules.

Vitamins are essential organic molecules that human cells cannot synthesize (or may not synthesize in sufficient amounts) and therefore must come from the diet. Vitamins provide coenzymes that can, like enzymes, function repeatedly. This is why cells require small amounts of vitamins. Coenzyme A, for example, is derived from the vitamin pantothenic acid and is necessary for one of the reactions of cellular respiration, discussed in the next section. Chapter 18 (pp. 708–711) discusses vitamins further.

 PRACTICE

4 Explain enzyme action in a metabolic reaction.

5 Explain how an enzyme "recognizes" its substrate.

6 List factors that can denature enzymes.

7 How does a rate-limiting enzyme illustrate negative feedback?

8 What is the role of a cofactor?

4.3 | Energy for Metabolic Reactions

Energy is the capacity to change something; it is the ability to do work. Therefore, we recognize energy by what it can do. Common forms of energy are heat, light, sound, electrical energy, mechanical energy, and chemical energy.

Energy cannot be created or destroyed, but it can change from one form to another. For example, compact fluorescent and light-emitting diode bulbs change electrical energy into heat and light. An automobile engine changes the chemical energy in gasoline to heat and mechanical energy.

Cellular respiration is the process that transfers energy from molecules such as glucose and makes it available for cellular use. The chemical reactions of cellular respiration must occur in a particular sequence, each one controlled by a different enzyme. Some of these enzymes are in the cell's cytosol, whereas others are in the mitochondria. The precision of activity and location within the cell suggests that the enzymes are physically positioned in the exact sequence as that of the reactions they control. The enzymes responsible for some of the reactions of cellular respiration are located in tiny, stalked particles on the membranes (cristae) in the mitochondria (see chapter 3, pp. 92–93).

Energy transfer powers the activities of life at the cellular level. Therefore, all metabolic reactions involve energy in some form. Most metabolic reactions use chemical energy.

ATP Molecules

Adenosine triphosphate (ATP) is a molecule that carries energy in a form that the cell can use. It is the primary energy-carrying molecule in a cell. Even though a cell has other energy carriers, cells quickly die without enough ATP. Each ATP molecule consists of three main parts—an adenine, a ribose, and three phosphates in a chain (fig. 4.7). The second and third phosphates of ATP are attached by high-energy bonds, and the chemical energy stored

In a type of inherited disease called an inborn error of metabolism, a deficient or absent enzyme causes a block in the biochemical pathway that it catalyzes. As a result, the biochemical that the enzyme normally acts upon builds up, and the biochemical resulting from the enzyme's normal action becomes scarce. The effect of an impaired enzyme in a metabolic pathway is like a twist in a garden hose that cuts off the flow of water: pressure builds behind the block, but no water comes out after the block.

Inborn errors of metabolism were initially described with the help of a mother of two young physically and intellectually disabled children in Norway, in 1931. The mother noted a musty odor to the children's urine, and her husband then mentioned it to an acquaintance who was a physician interested in biochemistry. Intrigued, the doctor analyzed the foul urine in a lab at the University of Oslo, with help from the mother, who brought him samples. The physician identified the problem in the children's metabolism—later named phenylketonuria (PKU)—and then found it among people in mental institutions.

In PKU, a missing or nonworking enzyme blocks the metabolic pathway that converts the amino acid phenylalanine into another amino acid, tyrosine. Excess phenylalanine spills into the urine and the blood, and poisons the brain. In the early 1960s a physician and microbiologist who had PKU in his family developed a test to detect the disease using drops of blood taken from a newborn's heel. It was the first newborn screening test to identify biochemicals in blood (using an analytical chemistry technique called mass spectrometry) to detect inborn errors of metabolism. Understanding PKU enabled researchers to develop a "medical food" that severely restricts phenylalanine, countering the biochemical abnormality. If begun very soon after birth, this special diet, followed for many years and sometimes for life, can completely prevent the symptoms of intellectual disability.

Today newborns are screened for dozens of inborn errors of metabolism. Some of these disorders, such as PKU, have dietary treatments, and others may respond better to different treatments if they are begun early.

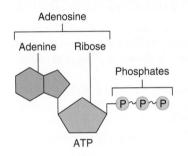

FIGURE 4.7 ATP is the currency for cellular energy. An ATP (adenosine triphosphate) molecule consists of an adenine, a ribose, and three phosphates. The wavy lines connecting the last two phosphates represent high-energy chemical bonds.

in one or both high-energy bonds may be quickly transferred to another molecule in a metabolic reaction. Energy from the breakdown of ATP powers cellular work such as skeletal muscle contraction, active transport across cell membranes, secretion, and many other functions.

An ATP molecule that loses its terminal (end) phosphate becomes an **adenosine diphosphate (ADP)** molecule, which has only two phosphates. ADP can be converted back into ATP using energy released from cellular respiration to attach a third phosphate, in a process called **phosphorylation** (fos″fōr-ĭ-la′shun). As figure 4.8 shows, ATP and ADP molecules shuttle back and forth between the energy-releasing reactions of cellular respiration and the energy-utilizing reactions of the cell.

Release of Chemical Energy

Most metabolic processes require chemical energy stored in ATP. This form of energy is initially held in the chemical bonds that link atoms into molecules and is released when these bonds break. Burning a marshmallow over a campfire releases the chemical energy held in the bonds of the molecules that make up the marshmallow as heat and light. Similarly, when we eat a marshmallow, the digestive system first breaks the sweet-tasting sugar sucrose into its constituent molecules of glucose and fructose. Inside cells, fructose ultimately reacts to form glucose. A process called **oxidation** (ok″sĭ-da′shun) releases the energy from the glucose, which is harnessed to power cellular metabolism.

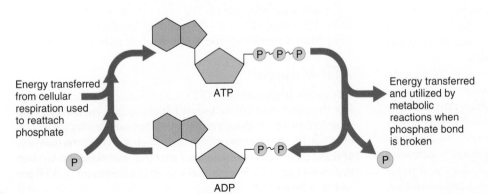

FIGURE 4.8 ATP provides energy for metabolic reactions. Cellular respiration generates ATP.

Oxidation of substances inside cells and the burning of substances outside them have important differences. Burning in nonliving systems (such as starting a fire in a fireplace) usually requires a great deal of energy to begin, and most of the energy released escapes as heat or light. In cells, enzymes initiate oxidation by lowering the activation energy. Also, by transferring energy to ATP, cells are able to capture almost half of the energy that is released in the form of chemical energy. About 60% of the energy escapes as heat, which helps maintain body temperature.

 PRACTICE

9   What is energy?

10 Define cellular respiration.

11 How does cellular oxidation differ from burning?

4.4 | Cellular Respiration

Cellular respiration occurs in three distinct, yet interconnected, series of reactions: **glycolysis** (gli-kol′ĭ-sis), the **citric acid cycle,** and the **electron transport chain** (oxidative phosphorylation). Glycolysis and the electron transport chain are pathways similar to those depicted in figures 4.5 and 4.6. **Figure 4.9** shows how a metabolic pathway can form a cycle if the final product reacts to replenish the original substrate.

Cellular respiration requires glucose and oxygen. The products of these reactions include carbon dioxide (CO_2), water, and energy. Although most of the energy is lost as heat, almost half is captured as ATP. **Figure 4.10** shows how the three components of cellular respiration connect.

Cellular respiration includes **aerobic** (a″er-ōb′ik) reactions, which require oxygen, and **anaerobic** (an-a″er-ōb′ik) reactions, which do not. For each glucose molecule decomposed completely

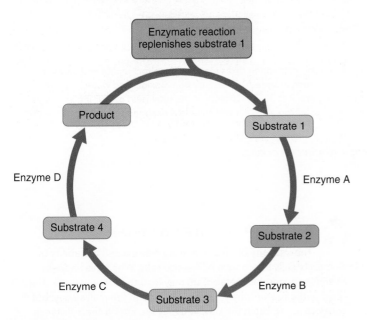

FIGURE 4.9 A metabolic cycle. A metabolic pathway forms a metabolic cycle when the product undergoes a reaction that replenishes the initial substrate.

by cellular respiration, up to thirty-eight molecules of ATP can be produced. All but two of the ATP molecules form in the aerobic reactions.

Glycolysis

Both aerobic and anaerobic pathways begin with glycolysis. "Glycolysis" means "the breaking of glucose." It is a series of ten enzyme-catalyzed reactions that break down the 6-carbon glucose molecule into two 3-carbon pyruvic acid molecules. Glycolysis occurs in the cytosol, and because it does not require oxygen, it is sometimes referred to as the *anaerobic phase of cellular respiration.*

Three phases occur during glycolysis (fig. 4.11):

1. First, two phosphate groups are added to a glucose molecule, one at each end, in a step called phosphorylation. This step requires energy from two ATPs, which are used to "prime" the glucose so that it is activated for some of the energy-releasing reactions that will happen.
2. Second, the 6-carbon glucose molecule is cleaved into two 3-carbon molecules.
3. Third, the electron carrier NADH is produced, ATP is synthesized, and two 3-carbon pyruvic acid molecules result.

Some of the reactions of glycolysis release hydrogen atoms. The electrons of these hydrogen atoms contain much of the energy from the chemical bonds of the original glucose molecule. To keep this energy from glucose in a form the cell can use, these hydrogen atoms are passed in pairs to molecules of the hydrogen carrier NAD^+ (nicotinamide adenine dinucleotide). In this reaction, two of the electrons and one hydrogen nucleus bind to NAD^+ to form NADH. The remaining hydrogen nucleus (a hydrogen ion) is released as follows:

$$NAD^+ + 2H \rightarrow NADH + H^+$$

NADH delivers these high-energy electrons to the electron transport chain elsewhere in the mitochondria, where most of the ATP will be synthesized. Therefore, NADH is called an electron carrier. Four ATPs are also synthesized directly. Subtracting the two ATPs used in the priming step gives a net yield of two ATPs per molecule of glucose from glycolysis. The reactions of glycolysis are described in more detail in Appendix E, Cellular Respiration (see fig. E.1).

 PRACTICE

12 What are the final products of cellular respiration?

13 Distinguish the aerobic from the anaerobic reactions.

14 What are the three phases of glycolysis?

Anaerobic Reactions

For glycolysis to continue, NADH + H^+ must be able to deliver electrons to the electron transport chain, replenishing the cellular supply of NAD^+. In the presence of oxygen, this is exactly what happens. Oxygen acts as the final electron acceptor at the end of the electron transport chain, enabling the chain to continue processing electrons and recycling NAD^+.

Glycolysis

(1) The 6-carbon sugar glucose is broken down in the cytosol into two 3-carbon pyruvic acid molecules with a net gain of 2 ATP and the release of high-energy electrons.

Citric Acid Cycle

(2) The 3-carbon pyruvic acids generated by glycolysis enter the mitochondria separately. Each loses a carbon (generating CO_2) and is combined with a coenzyme to form a 2-carbon acetyl coenzyme A (acetyl CoA). More high-energy electrons are released.

(3) Each acetyl CoA combines with a 4-carbon oxaloacetic acid to form the 6-carbon citric acid, for which the cycle is named. For each citric acid, a series of reactions removes 2 carbons (generating 2 CO_2's), synthesizes 1 ATP, and releases more high-energy electrons. The figure shows 2 ATP resulting directly from 2 turns of the cycle per glucose molecule that enters glycolysis.

Electron Transport Chain

(4) The high-energy electrons still contain most of the chemical energy of the original glucose molecule. Special carrier molecules bring the high-energy electrons to a series of enzymes that transfer much of the remaining energy to more ATP molecules. The electrons eventually combine with hydrogen ions and an oxygen atom to form water. The function of oxygen as the final electron acceptor in this last step is why the overall process is called aerobic respiration.

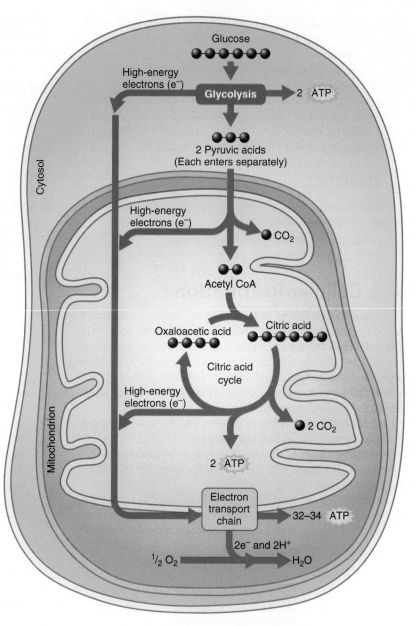

FIGURE 4.10 Glycolysis takes place in the cytosol and does not require oxygen. Aerobic respiration takes place in the mitochondria and only in the presence of oxygen. The products of glycolysis and the aerobic pathway of cellular respiration include ATP, heat, carbon dioxide, and water. Glycolysis generates 2 ATP, the citric acid cycle generates 2 ATP, and the electron transport chain generates 32–34. Therefore, the total yield of ATP molecules per glucose molecule is 36–38, depending on the cell type. (The ratio of the mitochondrion to surrounding cytosol volume is exaggerated to emphasize the metabolic pathways.)

Q *Identify the part of the cell where glycolysis occurs and where aerobic respiration occurs.*
Answer can be found in Appendix G.

Under anaerobic conditions the electron transport chain has no oxygen, and therefore nowhere to unload its electrons, and as a result it can no longer accept new electrons from NADH. As an alternative, $NADH + H^+$ can give its electrons and hydrogens back to pyruvic acid in a reaction that forms **lactic acid.** Although this reaction regenerates NAD^+, the buildup of lactic acid eventually inhibits glycolysis, and ATP production declines. The lactic acid diffuses into the blood, and when oxygen levels return to normal the liver converts the lactic acid back into pyruvic acid, which can finally enter the aerobic pathway.

A GLIMPSE AHEAD | To Chapter 9

The intensity of muscle activity may exceed the ability of the muscle cells to produce ATP aerobically. When this happens, the shift to anaerobic metabolism may lead to muscle fatigue. This is part of the reason why you can walk around the mall all day, but if you try to run to catch the bus, you may tire after a short distance. Chapter 9 further discusses muscle fatigue and its relationship to exercise patterns.

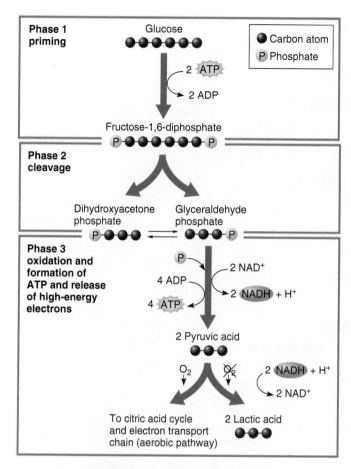

FIGURE 4.11 AP|R Glycolysis breaks down glucose in three phases: (1) phosphorylation, (2) splitting, and (3) production of NADH and ATP. Each glucose molecule broken down by glycolysis yields a net gain of 2 ATP.

PRACTICE

15 What is the role of oxygen in cellular respiration?

16 Under what conditions does a cell produce lactic acid?

Aerobic Reactions

If enough oxygen is available, the pyruvic acid generated by glycolysis can continue through the aerobic pathways (see fig. 4.10). These reactions include the synthesis of **acetyl coenzyme A** (as′ĕ-til ko-en′zīm A) or acetyl CoA, the citric acid cycle, and the electron transport chain. In addition to carbon dioxide and water, the aerobic reactions yield up to thirty-six ATP molecules per glucose. The theoretical yield of the aerobic reactions is up to thirty-six ATP per glucose molecule. In fact, more energy may be required to complete these reactions than once thought. Estimates taking this into account indicate a yield of ATP less than the theoretical maximum.

The aerobic reactions begin with pyruvic acid produced in glycolysis moving from the cytosol into the mitochondria (fig. 4.12). From each pyruvic acid molecule, enzymes inside the

mitochondria remove two hydrogen atoms, a carbon atom, and two oxygen atoms, generating NADH and a CO_2 and leaving a 2-carbon acetic acid. The acetic acid then combines with a molecule of coenzyme A to form acetyl CoA. CoA "carries" the acetic acid into the citric acid cycle.

> Ingestion of the element arsenic kills cells by blocking the action of the enzyme (pyruvate dehydrogenase) that catalyzes the reaction that produces acetyl CoA from pyruvic acid. Given in a single large dose, arsenic causes chest pain, vomiting, diarrhea, shock, coma, and death. Many small doses cause dark skin lesions that feel as if they are burning, numbness in hands and feet, and eventually skin cancer. Such gradual poisoning, called arsenicosis, may progress to paralysis and organ failure. Arsenicosis can result from contact with pesticides or other environmental pollutants. The world's largest outbreak, however, is due to natural exposure to arsenic in groundwater in India and Bangladesh.

Citric Acid Cycle

The citric acid cycle begins when a 2-carbon acetyl CoA molecule combines with a 4-carbon oxaloacetic acid molecule to form the 6-carbon citric acid and CoA (fig. 4.12). The citric acid is changed through a series of reactions back into oxaloacetic acid. The CoA can be used again to combine with acetic acid to form acetyl CoA. The cycle repeats as long as the mitochondrion receives oxygen and pyruvic acid.

The citric acid cycle has three important consequences:

1. One ATP is produced directly for each citric acid molecule that goes through the cycle.
2. For each citric acid molecule, eight hydrogen atoms with high-energy electrons are transferred to the hydrogen carriers NAD$^+$ and the related molecule FAD (flavine adenine dinucleotide):

$$NAD^+ + 2H \rightarrow NADH + H^+$$

$$FAD + 2H \rightarrow FADH_2$$

3. As the 6-carbon citric acid reacts to form the 4-carbon oxaloacetic acid, two carbon dioxide molecules are produced.

The carbon dioxide produced by the formation of acetyl CoA and in the citric acid cycle dissolves in the cytoplasm, diffuses from the cell, and enters the bloodstream. Eventually, the respiratory system excretes the carbon dioxide. The details of the citric acid cycle are covered in Appendix E, Cellular Respiration (see fig. E.2).

Electron Transport Chain

The hydrogen and high-energy electron carriers (NADH and FADH$_2$) generated by glycolysis and the citric acid cycle now hold most of the energy contained in the original glucose molecule. To couple this energy to ATP synthesis, the high-energy electrons are handed off to the electron transport chain, which is a series of

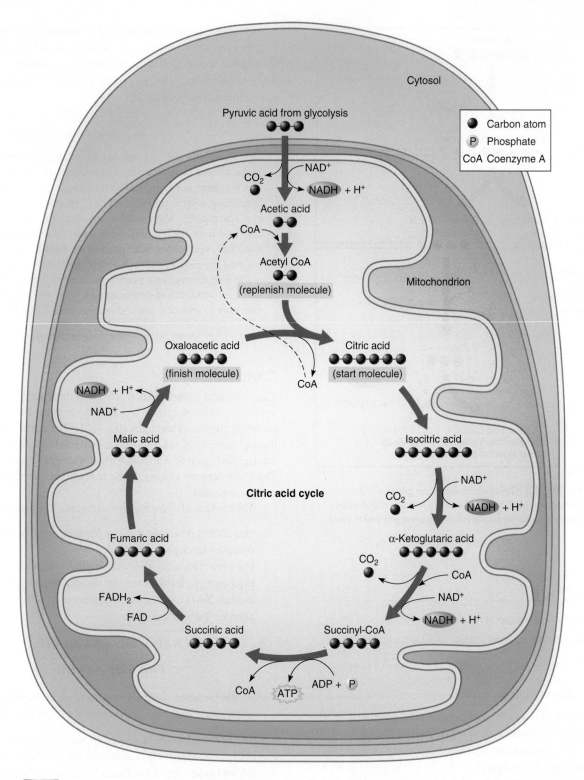

FIGURE 4.12 AP|R Each turn of the citric acid cycle (two "turns" or citric acids per glucose) produces one ATP directly and two CO_2 molecules. Eight hydrogens with high-energy electrons are released. (The ratio of the mitochondrial to cytosolic volumes is exaggerated to emphasize the metabolic pathways.)

enzyme complexes that pass electrons. Complexes of the electron transport chain dot the folds of the inner mitochondrial membranes (see chapter 3, pp. 92–93), which, if stretched out, may be forty-five times as long as the cell membrane in some cells. As an electron passes along the electron transport chain, it gradually loses energy.

That energy is transferred to ATP synthase, an enzyme complex that uses the energy to add a phosphate to ADP to form ATP (**fig. 4.13**). These reactions are known as oxidation/reduction reactions.

Neither glycolysis nor the citric acid cycle uses oxygen directly, although they are part of the aerobic metabolism of

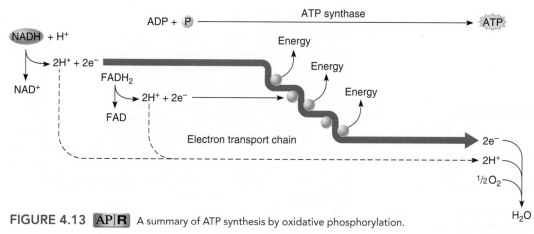

FIGURE 4.13 $\boxed{\text{AP}|\text{R}}$ A summary of ATP synthesis by oxidative phosphorylation.

glucose. Instead, the final enzyme of the electron transport chain gives up a pair of electrons that combine with two hydrogen ions (provided by the hydrogen carriers) and an atom of oxygen to form a water molecule:

$$2e^- + 2H^+ + 1/2\, O_2 \rightarrow H_2O$$

Oxygen is therefore the final electron "carrier." In the absence of oxygen, electrons cannot continue to pass through the electron transport chain, and the aerobic reactions of cellular respiration stop. Appendix E, Cellular Respiration discusses the details of the electron transport chain (see fig. E.3). **Figure 4.14** summarizes the steps in glucose metabolism.

Cyanide is a deadly poison because it halts ATP production in cells. It binds to an iron atom that is part of the enzyme that enables NADH from the citric acid cycle to transfer electrons to oxygen. Cyanide is absorbed through the skin, gastrointestinal tract, and respiratory tract, and exposure to it can kill in minutes. One source of cyanide is bitter almonds (not the sweet type that people prefer), which produce a compound called amygdalin that an enzyme in the human small intestine breaks down, releasing the poison. Cyanide is encountered in certain industrial processes, including metal plating, gold extraction, and in the raw materials for plastics. Rat poison and fumigants also contain cyanide.

Carbohydrate Storage

Metabolic pathways are interconnected in ways that enable certain molecules to enter more than one pathway. For example, carbohydrate molecules from foods may enter catabolic pathways and be used to supply energy, or they may enter anabolic pathways and be stored or react to form some of the twenty different amino acids (fig. 4.15).

Excess glucose in cells may enter anabolic carbohydrate pathways and be linked into storage forms such as glycogen. Most cells can produce glycogen; liver and muscle cells store the greatest amounts. Following a meal, when blood glucose concentration is relatively high, liver cells obtain glucose from the blood and

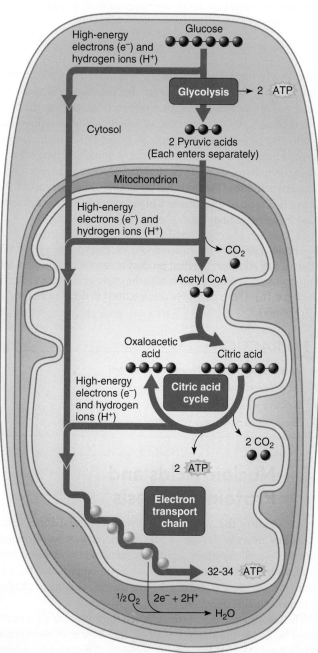

FIGURE 4.14 An overview of cellular respiration, including the net yield of ATP at each step per molecule of glucose.

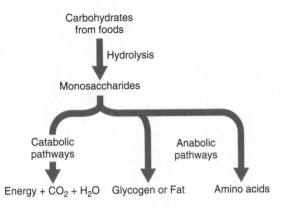

Carbohydrates
from foods

↓

Hydrolysis

Monosaccharides

Catabolic pathways / Anabolic pathways

Energy + CO_2 + H_2O — Glycogen or Fat — Amino acids

FIGURE 4.15 Hydrolysis breaks down carbohydrates from foods into monosaccharides. The resulting molecules may enter catabolic pathways and be used as energy sources, or they may enter anabolic pathways and be stored as glycogen or fat, or react to yield amino acids.

synthesize glycogen. Between meals, when blood glucose concentration is lower, the reaction reverses, and glucose is released into the blood. This mechanism ensures that cells throughout the body have a continual supply of glucose to support cellular respiration.

Glucose can also react to form fat molecules, which are later deposited in adipose tissue. This happens when a person takes in more carbohydrates than can be stored as glycogen or are required for normal activities. The body has an almost unlimited capacity to perform this type of anabolism. This is why overeating carbohydrates can increase body fat.

This section has considered the metabolism of glucose. Lipids and proteins can also be broken down to release energy for ATP synthesis. For all three, the final process is aerobic respiration, and the most common entry point is into the citric acid cycle as acetyl CoA (fig. 4.16). These pathways are described in detail in chapter 18 (pp. 697–699).

 PRACTICE

17 State the products of the aerobic reactions.

18 List the products of the citric acid cycle.

19 Explain the function of the electron transport chain.

20 Discuss fates of glucose other than cellular respiration.

4.5 | Nucleic Acids and Protein Synthesis

Enzymes control the metabolic pathways essential for cell survival. They oversee the use of all four major groups of organic molecules—carbohydrates, lipids, nucleic acids, and many other proteins important in physiology, such as blood proteins, the proteins that form muscle and connective tissues, and the antibodies that protect against infection.

The information that instructs a cell to synthesize a particular protein is held in the sequence of building blocks of **deoxyribonucleic acid (DNA),** the genetic material. The correspondence between a unit of DNA information and a particular amino acid constitutes the **genetic code** (jĕ-net′ik kōd).

Genetic Information

Children resemble their parents because of inherited traits, but what passes from parents to a child is genetic information, in the form of DNA molecules from the parents' sex cells. Long molecules of DNA and associated proteins are organized into chromosomes. The opening image for this chapter (p. 120) shows human chromosomes. As an offspring develops, mitosis passes the information in the DNA sequences of the chromosomes to new cells. Genetic information "tells" cells how to construct a great variety of protein molecules, each with a specific function. A DNA sequence that contains the information for making a particular polypeptide is called a **gene** (jēn).

 RECONNECT
To Chapter 3, Mitosis, page 108.

The complete set of genetic instructions in a cell, including the genes as well as other sequences, constitutes the **genome** (jēnōm′). Only a small part of the human genome encodes protein and it is called the *exome* (x-ōm). Much of the rest of the genome controls which proteins are produced in a particular cell under particular circumstances and the amounts produced, which is called *gene expression*. Studies of gene expression can reveal molecules and pathways that are part of physiological processes. Chapter 24 (pp. 906–907) discusses the human genome.

The human genome includes about 20,325 protein-encoding genes among 3.2 billion DNA nucleotides in each set of 23 chromosomes. Most cells have two chromosome sets. Sequencing the first human genomes took fifteen years and billions of dollars. Tens of thousands of people have had their genomes sequenced. It takes less than a day, but interpretation of the findings takes longer. Health-care professionals are learning how to incorporate information from the human genome into clinical practice.

Nucleotides are the building blocks of nucleic acids, as discussed in chapter 2 (pp. 75–76). A nucleotide consists of a 5-carbon sugar (ribose or deoxyribose), a phosphate group, and one of several nitrogenous bases (fig. 4.17). DNA and RNA nucleotides form long strands (polynucleotide chains) as dehydration synthesis alternately joins their sugars and phosphates, forming a "backbone" structure (fig. 4.18).

A DNA molecule is double-stranded, consisting of two polynucleotide chains. The nitrogenous bases project from the sugarphosphate backbone of one strand and bind, or pair, by hydrogen bonding to the nitrogenous bases of the second strand (fig. 4.19). The resulting molecular structure is like a ladder in which the rails represent the alternating sugar and phosphate backbones of the two strands and the rungs represent the paired nitrogenous bases. The sugars forming the two backbones point in opposite directions. For this reason, the two strands are called *antiparallel*.

A DNA molecule has a highly regular structure because the bases pair in only two combinations. A DNA base is one of four types: adenine (A), thymine (T), cytosine (C), or guanine (G).

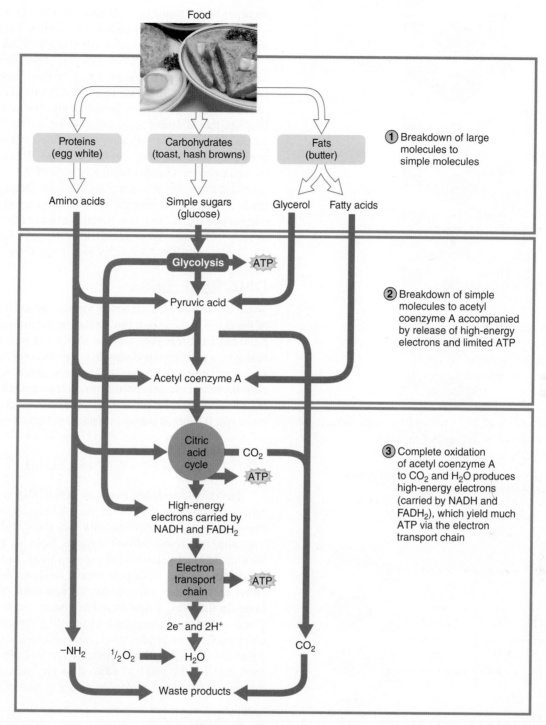

FIGURE 4.16 A summary of the breakdown (catabolism) of proteins, carbohydrates, and fats.

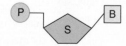

FIGURE 4.17 Each nucleotide of a nucleic acid consists of a 5-carbon sugar (S); a phosphate group (P); and an organic, nitrogenous base (B).

A and G are **purines** (pu′rēnz), and they consist of two organic ring structures. T and C are **pyrimidines** (pe-rimi-denz), and they have a single organic ring structure. A binds to T, and G binds to C. That is, a purine always binds to a pyrimidine, and this is what establishes the constant width of the DNA molecule. These pairs—A with T, and G with C—are called **complementary base pairs** (fig. 4.20a). The sequence of one DNA strand can always be derived from the other by following the "base-pairing rules." If the

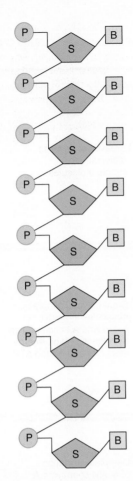

FIGURE 4.18 A polynucleotide chain consists of nucleotides connected by a sugar-phosphate backbone.

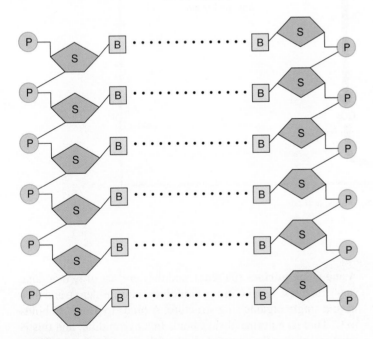

FIGURE 4.19 DNA is double-stranded, consisting of two polynucleotide chains. Hydrogen bonds (dotted lines) hold the nitrogenous bases of one strand to their partners on the other strand. The sugars point in opposite directions—that is, the strands are antiparallel.

sequence of one strand of the DNA molecule is G, A, C, T, then the complementary strand's sequence is C, T, G, A. It is the nucleotide sequence that provides the information in a molecule of DNA.

The double-stranded DNA molecule twists, forming a double helix (fig. 4.20b). A single gene may be thousands or even millions of bases long. In the nucleus, DNA is wound around proteins called *histones* clustered in groups of eight to form chromatin, a little like thread wound around spools (fig. 4.20b). The position of the chromatin plays a still poorly understood role in which genes are being accessed to synthesize proteins. Histones and other molecules come on and off different parts of the genome as some genes are expressed and others are silenced. During mitosis chromatin condenses, forming chromosomes that are visible under the microscope (fig. 4.20c). Investigators can use DNA sequences to identify individuals (From Science to Technology 4.2). Appendix F, A Closer Look at DNA and RNA Structures, shows more detail.

DNA Replication

When a cell divides, each newly formed cell must receive a copy of the entire genome so it can synthesize the proteins necessary to build basic cellular parts and metabolize, as well as the proteins that give a cell type its distinctive features. All cells except the sperm and egg receive two copies of the genome, in two sets of chromosomes, one inherited from each parent. DNA replication (re″pli-ka′shun) is the process that creates an exact copy of a DNA molecule. It happens during interphase of the cell cycle.

 RECONNECT
To Chapter 3, The Cell Cycle, page 108.

The double-stranded structure of the DNA molecule makes replication possible. As DNA replication begins, hydrogen bonds break between the complementary base pairs of the two strands. Then the strands unwind and separate, exposing unpaired bases. New nucleotides pair with the exposed bases, forming hydrogen bonds. An enzyme, DNA polymerase, catalyzes this base pairing. Other enzymes then knit together the new sugar-phosphate backbone. In this way, a new strand of complementary nucleotides forms and extends along each of the old (original) strands. Two complete DNA molecules result, each with one new and one original strand (fig. 4.21). During mitosis, the two DNA molecules that form the two chromatids of each of the chromosomes separate so that one of these DNA molecules passes to each of the new cells.

 PRACTICE

21 Define *genetic code*.
22 Define *gene*.
23 What is the structure of DNA?
24 Explain why DNA must replicate.
25 List the steps of DNA replication.

Genetic Code

Genetic information specifies the correct sequence of amino acids in a polypeptide chain. Each of the twenty different types of amino acids is represented in a DNA molecule by a triplet code, consisting

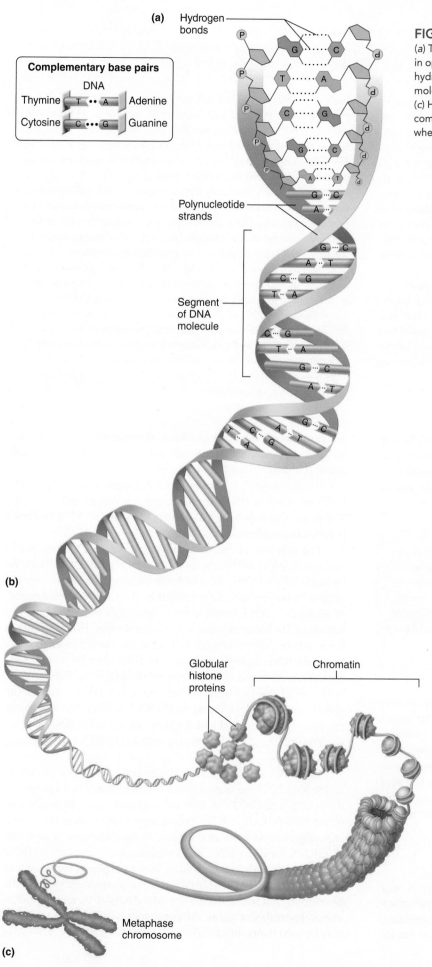

(a) Hydrogen bonds

Complementary base pairs

DNA

Thymine T •• A Adenine

Cytosine C ••• G Guanine

Polynucleotide strands

Segment of DNA molecule

(b)

Globular histone proteins

Chromatin

Metaphase chromosome

(c)

FIGURE 4.20 AP|R DNA and chromosome structure. (*a*) The two polynucleotide strands of a DNA molecule point in opposite directions (antiparallel) and are held together by hydrogen bonds between complementary base pairs. (*b*) The molecular "ladder" of a DNA molecule twists into a double helix. (*c*) Histone proteins enable the long double helix to assume more compact forms (chromosome and chromatin). Histones move when sections of the DNA are accessed for gene expression.

The human genome sequence differs from person to person because it includes 3.2 billion bits of information. Techniques called DNA profiling (or fingerprinting) compare the most variable parts of the genome among individuals for several purposes—to identify remains at crime scenes or after natural disasters; to confirm or rule out "blood" relationships; and, increasingly, to establish innocence when other types of evidence are questionable. The Innocence Project is a national litigation and public policy organization that provides DNA testing to people who claim that they have been wrongfully convicted. The Innocence Project has exonerated hundreds of people since it began in 1992.

One exonerated man had served four and a half years of a twenty-five-year sentence for rape when DNA profiling established his innocence. He and a friend had become suspects after a woman in Houston identified them as the men who had raped and threatened her with a gun, leaving her in a field. The two young men supplied saliva and blood samples, from which DNA profiles

were done and compared to DNA profiles from semen found in the victim and in her car. At the trial, an employee of the crime lab doing the DNA analysis testified that the probability that the suspect's DNA matched that of the evidence by chance was 1 in 694,000—a number that led jurors to convict him, even though he did not fit the victim's description of her assailant.

A DNA profile analyzes thirteen parts of the genome that vary in most populations. Usually this is sufficient information to rule out a suspect. Using these criteria, the suspect's DNA at first seemed to match the evidence. The problem, though, wasn't in the DNA, but in the population to which it was compared.

Proclaiming his innocence all along, the man had asked right away for an independent DNA test, but was told he couldn't afford one. Two journalists began investigating the Houston crime laboratory. They sent information on a few cases to a professor of criminology, who immediately saw the errors made in the suspect's DNA analysis. Retesting his DNA, and comparing it to a relevant population, proved his innocence.

---

of sequences of three nucleotides. That is, the sequence C, G, T in a DNA strand represents one type of amino acid; the sequence G, C, A represents another type. Other sequences encode instructions for beginning or ending the synthesis of a protein molecule, and for determining which genes are accessed for their information.

The genetic code is said to be universal because all species use the same DNA base triplets to specify the same amino acids. Researchers deciphered the code in the 1960s. When the media mentions an individual's genetic code, they really are referring to the sequence of DNA bases comprising a certain gene or genome—not the genetic code (the correspondence between DNA triplet and amino acid).

DNA is in the nucleus, and protein synthesis occurs in the cytoplasm. Because the cell must keep a permanent copy of the genetic instructions, a copy of the genetic information must move from the nucleus into the cytoplasm for the cell to use it. This copying and transfer of genetic information is a little like downloading a document: the original remains on the server. **RNA (ribonucleic acid)** molecules accomplish the transfer of information. They can exit the nucleus because they are single-stranded and much shorter than the DNA molecules that comprise chromosomes.

RNA Molecules

RNA molecules differ from DNA molecules in several ways (table 4.1). RNA nucleotides have ribose rather than deoxyribose sugar. Like DNA, RNA nucleotides each have one of four nitrogenous bases, but whereas adenine, cytosine, and guanine nucleo-

tides are part of both DNA and RNA, thymine nucleotides are only in DNA. In place of thymine nucleotides, RNA molecules have uracil (U) nucleotides (fig. 4.22 and Appendix F, A Closer Look at DNA and RNA Structures). In RNA U pairs with A (fig. 4.23). Different types of RNA have different size ranges and functions. The process of copying DNA information into an RNA sequence is called **transcription** (trans-krip′-shun).

The first step in delivering information from the nucleus to the cytoplasm is the synthesis of **messenger RNA (mRNA).** As its name suggests, this form of RNA carries a gene's information on how to build a specific polypeptide. As mRNA is synthesized, its nucleotides form complementary base pairs with one of the two strands of DNA that encodes a particular protein. However, just as the words in a sentence must be read in the correct order to make sense, the base sequence of a strand of DNA must be "read" in the correct direction and from the correct starting point. Furthermore, only one of the two antiparallel strands of DNA contains the genetic message. An enzyme called RNA polymerase recognizes the correct DNA strand and the right direction for RNA synthesis. The "sentence" always begins with the mRNA base sequence AUG (fig. 4.24).

In mRNA synthesis, RNA polymerase binds to a promoter, which is a DNA base sequence that begins a gene. Then a section of the double-stranded DNA unwinds and pulls apart, exposing part of the base sequence. RNA polymerase moves along the strand, exposing other sections of the gene. At the same time, a molecule of mRNA forms as RNA nucleotides complementary to those along the DNA strand are joined. For example, if the sequence of DNA bases is TACCCGAGG, the complementary bases in the developing mRNA molecule are AUGGGCUCC, as figure 4.24 shows. For different genes, different strands of the DNA molecule may be used to manufacture RNA.

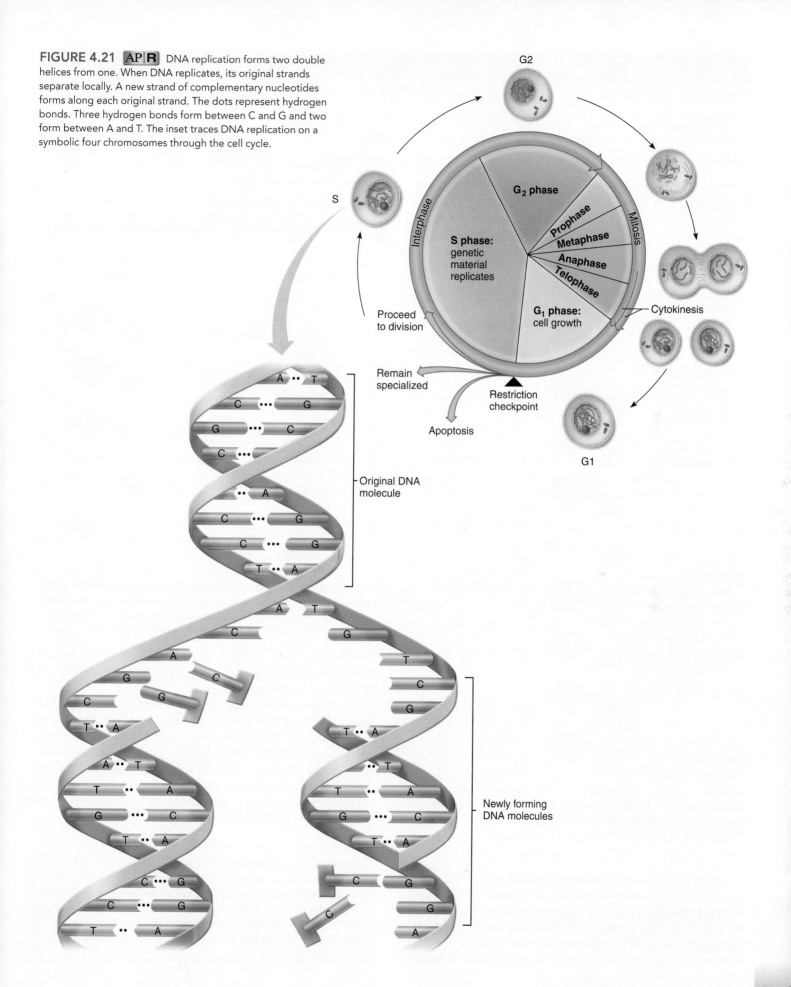

FIGURE 4.21 AP|R DNA replication forms two double helices from one. When DNA replicates, its original strands separate locally. A new strand of complementary nucleotides forms along each original strand. The dots represent hydrogen bonds. Three hydrogen bonds form between C and G and two form between A and T. The inset traces DNA replication on a symbolic four chromosomes through the cell cycle.

G2

S

Interphase

G₂ phase

Prophase

Metaphase

Mitosis

Anaphase

Telophase

S phase: genetic material replicates

G₁ phase: cell growth

Cytokinesis

Proceed to division

Remain specialized

Restriction checkpoint

Apoptosis

G1

Original DNA molecule

Newly forming DNA molecules

TABLE 4.1 A Comparison of DNA and RNA Molecules

	DNA	RNA
Main location	Part of chromosomes, in nucleus	Cytoplasm
5-carbon sugar	Deoxyribose	Ribose
Basic molecular structure	Double-stranded	Single-stranded
Nitrogenous bases included	Cytosine, guanine, adenine, thymine	Cytosine, guanine, adenine, uracil
Major functions	Contains genetic code for protein synthesis, replicates prior to mitosis	Messenger RNA carries transcribed DNA information to cytoplasm and acts as template for synthesis of protein molecules; transfer RNA carries amino acids to messenger RNA; ribosomal RNA provides structure for ribosomes

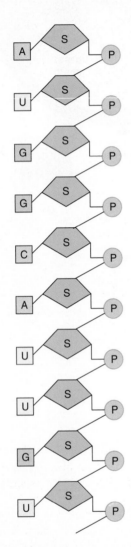

FIGURE 4.22 RNA differs from DNA in that it is single-stranded, contains ribose rather than deoxyribose, and has uracil (U) rather than thymine (T) as one of its four bases.

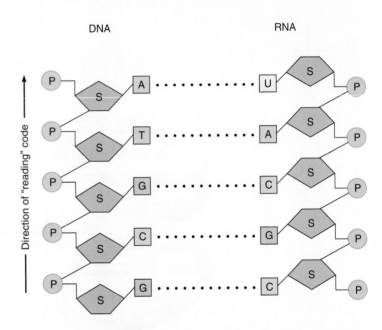

FIGURE 4.23 **AP|R** Transcription of RNA from DNA. When an RNA molecule is synthesized beside a strand of DNA, complementary nucleotides bond as in a double-stranded DNA molecule, with one exception: RNA contains uracil nucleotides (U) in place of thymine nucleotides (T).

Transcription of this gene is complete for now, although genes may be transcribed over and over.

Each amino acid in a protein is specified by a series of three bases in DNA and then by a series of three bases in mRNA, called **codons** (ko′donz). To complete protein synthesis, mRNA must leave the nucleus and associate with a ribosome. There, the mRNA is translated from the "language" of nucleic acids to the "language" of amino acids. This process is fittingly called **translation** (fig. 4.24). A cell has many ribosomes, and many mRNAs can be transcribed from a gene, so protein synthesis is a very fast and efficient process.

Most genes are transcribed in "pieces" called *exons*. Sections called *introns* are spliced out of messenger RNA molecules before the molecules are used to synthesize proteins. Cells can transcribe and assemble different combinations of exons from a particular gene, creating slightly different forms of a protein that can function in different cell types under specific conditions.

RNA polymerase continues to move along the DNA strand, exposing successive sections of the base sequence, until it reaches a special DNA base sequence (termination signal) that indicates the end of the gene. At this point, the RNA polymerase releases the newly formed mRNA molecule and leaves the DNA. The DNA then rewinds and assumes its previous double helix structure.

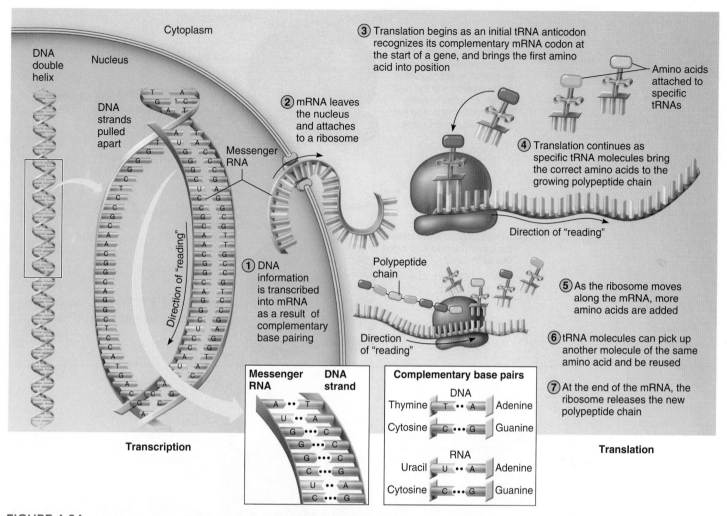

Transcription

Cytoplasm

DNA double helix

Nucleus

DNA strands pulled apart

Messenger RNA

Direction of "reading"

① DNA information is transcribed into mRNA as a result of complementary base pairing

② mRNA leaves the nucleus and attaches to a ribosome

③ Translation begins as an initial tRNA anticodon recognizes its complementary mRNA codon at the start of a gene, and brings the first amino acid into position

Amino acids attached to specific tRNAs

④ Translation continues as specific tRNA molecules bring the correct amino acids to the growing polypeptide chain

Direction of "reading"

Polypeptide chain

Direction of "reading"

⑤ As the ribosome moves along the mRNA, more amino acids are added

⑥ tRNA molecules can pick up another molecule of the same amino acid and be reused

⑦ At the end of the mRNA, the ribosome releases the new polypeptide chain

Messenger RNA   DNA strand

Complementary base pairs

DNA

Thymine T • • A Adenine

Cytosine C • • • G Guanine

RNA

Uracil U • • A Adenine

Cytosine C • • • G Guanine

Translation

FIGURE 4.24 Protein synthesis. DNA information is transcribed into mRNA, which in turn is translated into a sequence of amino acids. The insets show complementary base pairs in DNA and RNA.

Q *How are DNA, RNA, and proteins alike, and how are they different?*
Answer can be found in Appendix G.

Protein Synthesis

Synthesizing a protein molecule requires that the specified amino acid building blocks in the cytoplasm align in the proper sequence along an mRNA, and then attach. A second type of RNA molecule, called **transfer RNA (tRNA),** aligns amino acids in a way that enables enzymes to bond them to each other. A tRNA molecule consists of seventy to eighty nucleotides and has a complex three-dimensional shape, somewhat like a cloverleaf. Like mRNA, tRNA is transcribed in the nucleus and then sent into the cytoplasm.

The two ends of the tRNA molecule are "connectors," bringing together mRNA and the growing chain of amino acids (fig. 4.24). At one end, each tRNA molecule forms hydrogen bonds with a particular amino acid. At least one type of tRNA specifies each of the twenty amino acids. An amino acid must be activated for a tRNA to pick it up. Special enzymes catalyze this step. ATP provides the energy for an amino acid and its tRNA to bond (fig. 4.25). The other end of each tRNA molecule includes a specific three-nucleotide

sequence, called the **anticodon,** unique to that type of tRNA. An anticodon bonds only to the complementary mRNA codon. In this way, the appropriate tRNA carries its amino acid to the correct place in the mRNA sequence (fig. 4.25).

The genetic code specifies more than enough information to encode the proteins that a human body requires. Although only twenty types of amino acids need be encoded, the four types of bases can form sixty-four different mRNA codons. Therefore, some amino acids correspond to more than one codon (table 4.2). Three of the codons do not have a corresponding tRNA. They provide a "stop" signal, indicating the end of protein synthesis, much like the period at the end of this sentence. The mRNA sequence AUG is the "initiation codon," specifying the amino acid methionine and the start of the encoded protein. Sixty-one different tRNAs are specific for the remaining sixty-one codons, which means that more than one type of tRNA can correspond to the same amino acid type.

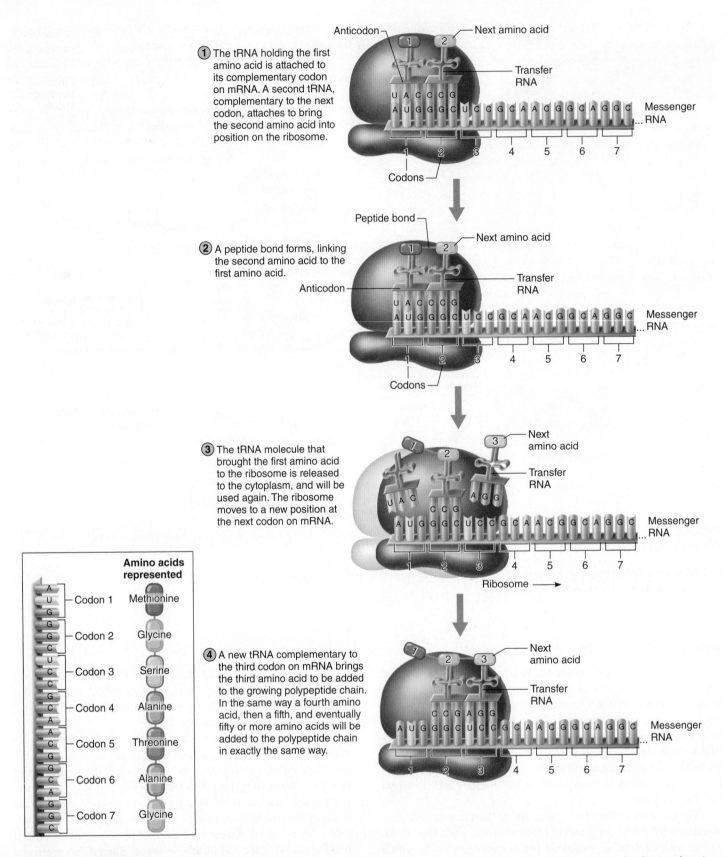

① The tRNA holding the first amino acid is attached to its complementary codon on mRNA. A second tRNA, complementary to the next codon, attaches to bring the second amino acid into position on the ribosome.

② A peptide bond forms, linking the second amino acid to the first amino acid.

③ The tRNA molecule that brought the first amino acid to the ribosome is released to the cytoplasm, and will be used again. The ribosome moves to a new position at the next codon on mRNA.

④ A new tRNA complementary to the third codon on mRNA brings the third amino acid to be added to the growing polypeptide chain. In the same way a fourth amino acid, then a fifth, and eventually fifty or more amino acids will be added to the polypeptide chain in exactly the same way.

Amino acids represented

Codon 1 — Methionine
Codon 2 — Glycine
Codon 3 — Serine
Codon 4 — Alanine
Codon 5 — Threonine
Codon 6 — Alanine
Codon 7 — Glycine

FIGURE 4.25 **AP|R** A closer look at protein synthesis. Molecules of transfer RNA (tRNA) attach to and carry specific amino acids, aligning them in the sequence determined by the codons of mRNA. These amino acids, connected by peptide bonds, form a polypeptide chain of a protein molecule. Protein synthesis occurs on ribosomes. The inset shows some examples of the correspondence between mRNA codons and the specific amino acids they encode.

The binding of tRNA and mRNA occurs in close association with a ribosome. A ribosome is an organelle that is a tiny particle of two unequal-sized subunits composed of **ribosomal RNA (rRNA)** and protein molecules. The smaller subunit of a ribosome binds to a molecule of mRNA near the first codon. A tRNA with the complementary anticodon brings its attached amino acid into position, temporarily joining to the ribosome. A second tRNA, complementary to the second mRNA codon, then binds (with its activated amino acid) to an adjacent site on the ribosome. The first tRNA molecule releases its amino acid, providing the energy for a peptide bond to form between the two amino acids (fig. 4.25). This process repeats as the ribosome moves along the mRNA, adding amino acids one at a time to the extending polypeptide chain. The enzymatic activity necessary for bonding of the amino acids comes from ribosomal proteins and some RNA molecules (ribozymes) in the larger subunit of the ribosome. This subunit also holds the growing chain of amino acids.

Protein synthesis is economical. A molecule of mRNA usually associates with several ribosomes at the same time. Multiple ribosomes can associate with mRNA at different points along its length, simultaneously synthesizing the same protein. Such a group of ribosomes is called a *polyribosome*. Several copies of the protein, each in a different stage of formation, may be present at any given moment. As the polypeptide forms, proteins called *chaperones* fold it into its unique shape, and when the process is completed, the polypeptide is released as a separate functional molecule. The tRNA molecules, ribosomes, mRNA, and the enzymes can function repeatedly in protein synthesis. Cells have a protein misfolding response that ensures that proteins folded incorrectly are destroyed or refolded in the proper conformation.

ATP molecules provide the energy for protein synthesis. A protein may consist of many hundreds of amino acids and the energy from three ATP molecules is required to link each amino acid to the growing chain. Much of a cell's energy supply supports protein synthesis. Table 4.3 summarizes protein synthesis.

The number of molecules of a particular protein that a cell synthesizes is generally proportional to the number of corresponding mRNA molecules. The rate at which mRNA is transcribed from DNA in the nucleus and the rate at which enzymes (ribonucleases) destroy the mRNA in the cytoplasm therefore control protein synthesis.

Certain types of proteins called transcription factors activate certain genes, moving aside the surrounding histone proteins to expose the promoter DNA sequences that start genes. Extracellular signals such as hormones and growth factors activate transcription factors. These actions are called "chromatin remodeling," and they control which proteins a cell produces and how many copies form under particular conditions. A connective tissue cell might have many mRNAs representing genes that encode the proteins collagen and elastin; a muscle cell would have abundant mRNAs encoding contractile proteins, such as actin and myosin.

 PRACTICE

26 Explain how genetic information is carried from the nucleus to the cytoplasm.

27 List the steps of protein synthesis.

28 Explain how gene expression is controlled.

TABLE 4.2	Codons (mRNA Three-Base Sequences)						

		SECOND LETTER						
		U	**C**	**A**	**G**		
U	UUU	phenylalanine (phe)	UCU	serine (ser)	UAU	tyrosine (tyr)	UGU	cysteine (cys)	**U**
	UUC		UCC		UAC		UGC		**C**
	UUA	leucine (leu)	UCA		UAA	STOP	UGA	STOP	**A**
	UUG		UCG		UAG	STOP	UGG	tryptophan (trp)	**G**
C	CUU	leucine (leu)	CCU	proline (pro)	CAU	histidine (his)	CGU	arginine (arg)	**U**
	CUC		CCC		CAC		CGC		**C**
	CUA		CCA		CAA	glutamine (gln)	CGA		**A**
	CUG		CCG		CAG		CGG		**G**
A	AUU	isoleucine (ile)	ACU	threonine (thr)	AAU	asparagine (asn)	AGU	serine (ser)	**U**
	AUC		ACC		AAC		AGC		**C**
	AUA		ACA		AAA	lysine (lys)	AGA	arginine (arg)	**A**
	AUG	START methionine (met)	ACG		AAG		AGG		**G**
G	GUU	valine (val)	GCU	alanine (ala)	GAU	aspartic acid (asp)	GGU	glycine (gly)	**U**
	GUC		GCC		GAC		GGC		**C**
	GUA		GCA		GAA	glutamic acid (glu)	GGA		**A**
	GUG		GCG		GAG		GGG		**G**

First Letter (left margin) / *Third Letter* (right margin)

TABLE 4.3 — Protein Synthesis

Transcription (In the Nucleus)

1. RNA polymerase binds to the DNA base sequence of a gene.

2. This enzyme unwinds and exposes part of the DNA molecule.

3. RNA polymerase moves along one strand of the exposed gene and catalyzes synthesis of an mRNA, whose nucleotides are complementary to those of the strand of the gene.

4. When RNA polymerase reaches the end of the gene, the newly formed mRNA is released.

5. The DNA rewinds and closes the double helix.

6. The mRNA passes through a pore in the nuclear envelope and enters the cytoplasm.

Translation (In the Cytoplasm)

1. A ribosome binds to the mRNA near the codon at the beginning of the messenger strand.

2. A tRNA molecule that has the complementary anticodon brings its amino acid to the ribosome.

3. A second tRNA brings the next amino acid to the ribosome.

4. A peptide bond forms between the two amino acids, and the first tRNA is released.

5. This process repeats for each codon in the mRNA sequence as the ribosome moves along its length, forming a chain of amino acids.

6. The growing amino acid chain folds into the unique conformation of a functional protein.

7. The completed protein molecule is released. The mRNA, ribosome, and tRNA are recycled.

4.6 | Changes in Genetic Information

Remarkably, we are more alike than different—human genome sequences are 99.9% the same among individuals. The tenth of a percent of the human genome that can vary from person to person includes rare DNA sequences that affect health or appearance, as well as common DNA base variations that do not exert any observable effects.

Nature of Mutations

Like typing a long text message, the long process of DNA replication can result in errors, adding a different nucleotide base than the one specified to a new DNA molecule. These errors produce changes, called **mutations** (mu-ta'shunz), in the DNA sequence.

To visualize the concept of mutation, imagine a simplified DNA sequence of twelve adenines:

AAAAAAAAAAAA

The sequence corresponds to an mRNA sequence of twelve uracils:

UUUUUUUUUUUU

which specifies four linked molecules of the amino acid phenylalanine.

Now imagine that the process of mutation changes the fourth DNA base in the sequence of AAA . . . to a C:

AAACAAAAAAAA

The corresponding mRNA—UUUGUUUUUUUU—now encodes the amino acid sequence phenylalanine-valine-phenylalanine-phenylalanine. (Use table 4.2 to assign amino acids to mRNA codon sequences.) The genetic information has changed.

A mutation that replaces one type of amino acid with a different kind is called a missense mutation. A mutation that changes an amino acid–encoding codon into a stop codon prematurely shortens the encoded protein, and is called a nonsense mutation. An example of a nonsense mutation is a change of CGA (arginine) to UGA (stop).

Mutations occur in two general ways—they may be spontaneous or induced. A spontaneous mutation arises from the chemical tendency of free nitrogenous bases to exist in two slightly different structures. For extremely short times, a base can be in an unstable form. If, by chance, such an unstable base is inserted into newly forming DNA, an error in sequence will be generated and perpetuated when the strand replicates. Another replication error that can cause mutation is when the existing (parental) DNA strand slips, adding nucleotides to or deleting nucleotides from the sequence.

Induced mutations are a response to exposure to certain chemicals or radiation, called **mutagens** (mu'tah-jenz). A familiar mutagen is ultraviolet radiation, which is part of sunlight. Prolonged exposure to ultraviolet radiation can cause an extra bond to form between two adjacent thymine DNA bases that are part of the same DNA strand in a skin cell. This extra bond kinks the double helix, causing an incorrect base to be inserted during replication. The cell harboring such a mutation may not be affected, may be so damaged that it dies and peels off, or it may become cancerous. This is one way that too much sun exposure can cause skin cancer (see Clinical Application 6.1, p. 182). Mutagens are also found in hair dye, food additives, smoked meats, flame retardants, and cigarette smoke.

Both spontaneous and induced mutations can harm health if they alter the amino acid sequence of the encoded protein so that it malfunctions or isn't produced at all, and there is not enough of the normal form of the protein to compensate. For example, the muscle weakness of Duchenne muscular dystrophy results from a mutation in the gene encoding the protein dystrophin. This protein normally enables muscle cell membranes to withstand the force of contraction. The mutation may be a missing or changed nucleotide base or absence of part or all of the dystrophin gene. Without the normal protein, muscle cells collapse and muscles throughout the body weaken and break down. **Figure 4.26** shows how the change

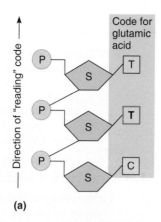
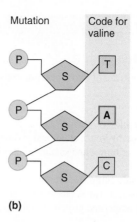

FIGURE 4.26 An example of mutation. (*a*) The DNA code for the amino acid glutamic acid is CTT. (*b*) If the first T is changed to A, the DNA code changes to CAT, which specifies the amino acid valine. If this happens to both copies of the DNA that encodes the sixth amino acid in a subunit of the protein hemoglobin, sickle cell disease results. The abnormal hemoglobin makes red blood cells sticky and bendable, and they lodge in narrow blood vessels, blocking circulation and causing great pain.

of one base causes another inherited illness, sickle cell disease. Rarely, a mutation confers an advantage. One such helpful mutation protects against HIV infection. In these individuals, the receptors (cell surface molecules) to which HIV must bind in order to enter cells are incomplete. HIV cannot enter.

Once DNA changes, the new sequence is transmitted every time the cell in which it originated divides. If that cell is an egg or sperm, then the change is passed to the next generation. We return to this point in the next section.

Protection Against Mutation

Cells can detect mutations and correct them. Special "repair enzymes" recognize and remove mismatched nucleotides and fill the resulting gap with the accurate, complementary nucleotides. This mechanism, called the DNA damage response or **DNA repair,** restores the original DNA sequence.

Disorders of DNA repair can make life difficult. Xeroderma pigmentosum, for example, causes extreme sun sensitivity. A child with the condition must completely cover up, applying sunblock on any exposed skin to prevent freckles, sores, and cancer. Special camps and programs allow these children to play outdoors at night, away from the danger of the sun.

The nature of the genetic code protects against mutation, to a degree. Sixty-one codons specify the twenty types of amino acids, and therefore, some amino acids correspond to more than one codon type. Usually, two or three codons specifying the same amino acid differ only in the third base of the codon. A mutation that changes the third codon base can encode the same amino acid. For example, the DNA triplets GGA and GGG each specify the amino acid

proline. If a mutation changes the third position of GGA to a G, the amino acid for that position in the encoded protein does not change—it is still proline.

If a mutation alters a base in the second position, the substituted amino acid is often similar in overall shape to the correct one, and the protein is not changed enough to affect its function. (An important exception is the mutation shown in fig. 4.26.) Yet another protection against mutation is that a person has two copies of each chromosome, and therefore of each gene. If one copy is mutated, the other may provide enough of the gene's normal function to maintain health. (This is more complicated for the sex chromosomes, X and Y, discussed in chapter 24, pp. 914–916.)

Timing of a mutation influences effects on health. A mutation in a sperm cell, egg cell, or fertilized ovum is repeated in every cell of the individual, causing an inherited disease or variant of a trait. A mutation in one cell of an embryo might be devastating because much of the body is still to develop, and many cells will inherit the mutation. In contrast, a mutation in a body cell of an adult would most likely have no effect because it would be only one among trillions of cells that do not have the mutation. However, if such a somatic (body cell) mutation confers a faster cell cycle and therefore cells bearing the mutation have a division advantage, cancer can result.

 PRACTICE

29 What is a mutation?

30 How do mutations arise?

31 How do mutations affect health or appearance?

32 Describe protections against mutation. ■

Chapter Summary

4.1 Metabolic Processes (page 121)

Metabolic reactions and pathways are of two types: **anabolism** and **catabolism.**

1. Anabolism
   a. Anabolism builds molecules.
   b. In **dehydration synthesis,** hydrogen atoms and hydroxyl groups are removed, water forms, and smaller molecules bind by sharing atoms.
   c. Anabolic reactions include complex carbohydrates synthesized from monosaccharides, fats synthesized from glycerol and fatty acids, and proteins synthesized from amino acids.
2. Catabolism
   a. Catabolism breaks down molecules.
   b. In **hydrolysis,** a water molecule adds a hydrogen atom to one portion of a molecule and a hydroxyl group to a second portion, breaking the bond between these parts.
   c. Catabolic reactions include complex carbohydrates decomposed into monosaccharides, fats decomposed into glycerol and fatty acids, and proteins broken down into amino acids.

4.2 Control of Metabolic Reactions (page 123)

Metabolic processes have many steps that occur in a specific sequence and are interconnected.

1. Enzyme action
   a. Metabolic reactions require energy to start.
   b. Most enzymes are proteins that increase the rate of (catalyze) specific metabolic reactions.
   c. Enzymes are usually named according to their substrates, with *-ase* at the end.
   d. An enzyme acts when its **active site** temporarily combines with the **substrate,** altering its chemical structure. This enables the substrate to react, forming a product. The enzyme is released in its original form.
   e. The rate of enzyme-controlled reactions depends upon the numbers of enzyme and substrate molecules and the efficiency of the enzyme.
   f. Enzymes can be denatured by heat, radiation, electricity, chemicals, and extreme pH values.
2. Metabolic pathways
   a. A sequence of enzyme-controlled reactions is a **metabolic pathway.**
   b. Enzyme-catalyzed reactions form pathways when a reaction's product is another's substrate.
   c. A **rate-limiting enzyme** may regulate a metabolic pathway.
   d. A negative feedback mechanism in which the product of a pathway inhibits the regulatory enzyme may control the regulatory enzyme.
   e. The rate of product formation usually remains stable in a pathway because of feedback mechanisms.

3. Cofactors and coenzymes
   a. **Cofactors** are additions to some enzymes that are necessary for their function.
   b. A cofactor may be an ion or a small organic molecule called a **coenzyme.**
   c. Some vitamins, the sources of coenzymes, cannot be synthesized by human cells in adequate amounts.

4.3 Energy for Metabolic Reactions (page 125)

Energy is a capacity to produce change or to do work. Forms of energy include heat, light, sound, and electrical, mechanical, and chemical energies. The reactions of **cellular respiration** use and release chemical energy.

1. ATP molecules
   a. Energy is captured in the bond of the terminal phosphate of **ATP.**
   b. Captured energy is released when the terminal phosphate bond of an ATP molecule breaks.
   c. ATP that loses its terminal phosphate becomes **ADP.**
   d. ADP can become ATP by capturing energy and a phosphate.
   e. ATP is the primary energy-carrying molecule in a cell.
2. Release of chemical energy
   a. Most metabolic processes use chemical energy released when molecular bonds break.
   b. The energy liberated from glucose during cellular respiration is used for metabolism.
   c. Enzymes in the cytoplasm and mitochondria control cellular respiration.

4.4 Cellular Respiration (page 127)

Cellular respiration transfers energy from molecules such as glucose and makes it available for cellular use. This process occurs in three distinct, interconnected series of reactions: **glycolysis,** the **citric acid cycle,** and the **electron transport chain.** Some of the steps require oxygen (**aerobic**) and some do not (**anaerobic**). Up to 38 ATPs form from breakdown of one glucose molecule.

1. Glycolysis
   a. Glycolysis, the first step of glucose catabolism, occurs in the cytosol and does not require oxygen.
   b. The three stages of glycolysis release and transfer some energy to ATP.
   c. Some of the energy is in the form of high-energy electrons attached to hydrogen carriers.
2. Anaerobic reactions (absence of oxygen)
   a. In the anaerobic reactions, NADH and H^+ donate electrons and hydrogens to pyruvic acid, generating **lactic acid.**
   b. Lactic acid builds up, eventually inhibiting glycolysis and ATP formation.
   c. When oxygen returns, in liver cells lactic acid reacts to form pyruvic acid.

3. Aerobic reactions (presence of oxygen)
   a. The second phase of glucose catabolism occurs in the mitochondria and requires oxygen.
   b. These reactions include the citric acid cycle and the electron transport chain.
   c. Considerably more energy is transferred to ATP during the aerobic reactions than during glycolysis.
   d. The products of the aerobic reactions of cellular respiration are heat, carbon dioxide, water, and energy.
   e. The citric acid cycle decomposes molecules, releases carbon dioxide, releases hydrogen atoms that have high-energy electrons, and forms ATP.
   f. High-energy electrons from hydrogen atoms enter an electron transport chain. Energy released from the chain is used to form ATP.
   g. Each metabolized glucose molecule yields up to thirty-eight ATP molecules.
   h. Excess carbohydrates may enter anabolic pathways and be linked into and stored as glycogen or react to produce fat.

4.5 Nucleic Acids and Protein Synthesis

(page 132)

DNA molecules contain and maintain information that tells a cell how to synthesize polypeptides, which comprise or associate to form proteins. The correspondence between a sequence of DNA nucleotides and a particular amino acid is the genetic code.

1. Genetic information
   a. DNA information specifies inherited traits.
   b. A **gene** is a portion of a DNA molecule that includes, in its nucleotide base sequence, the genetic information for making a polypeptide.
   c. The DNA nucleotides from both strands pair in a complementary fashion, joining the two strands. A binds T, and G binds C.
   d. All of the DNA in a cell is called the genome. The protein-encoding part is called the exome.
2. DNA replication
   a. Each new cell requires a copy of the original cell's genetic information.
   b. DNA molecules are replicated during interphase of the cell cycle.
   c. Each new DNA molecule consists of one old strand and one new strand.

3. Genetic code
   a. Three nucleotides in a DNA sequence code for one type of amino acid.
   b. RNA molecules transfer genetic information from the nucleus to the cytoplasm.
4. RNA molecules
   a. RNA molecules are usually single-stranded, have ribose instead of deoxyribose, and uracil in place of thymine.
   b. RNA synthesis is **transcription.**
   c. Messenger RNA (mRNA) molecules, synthesized in the nucleus, have a nucleotide sequence complementary to that of an exposed strand of DNA.
   d. Messenger RNA molecules move into the cytoplasm, associate with ribosomes, which consist of proteins and ribosomal RNAs, and are templates for the synthesis of polypeptide molecules in the process of **translation.**
5. Protein synthesis
   a. Molecules of transfer RNA (tRNA) position amino acids along a strand of mRNA.
   b. A ribosome binds to an mRNA and allows a tRNA to recognize its correct position on the mRNA.
   c. The ribosome has enzymes required for the synthesis of the polypeptide and holds the polypeptide until it is completed.
   d. As the polypeptide forms, it folds into a unique shape.
   e. ATP provides the energy for protein synthesis.

4.6 Changes in Genetic Information (page 142)

Mutation is a change in a DNA sequence.

1. Nature of mutations
   a. Mutations are rare and alter health or appearance.
   b. Mutations may be spontaneous or induced.
   c. A protein synthesized from an altered DNA sequence may or may not function normally.
   d. DNA changes are transmitted when the cell divides.
2. Protection against mutation
   a. DNA repair enzymes can correct some forms of DNA damage.
   b. The genetic code protects against effects of some mutations.
   c. A mutation in a sex cell, fertilized egg, or embryo may have more effects than a later mutation because a greater proportion of cells bear the mutation.

CHAPTER ASSESSMENTS

4.1 Metabolic Processes

1 Distinguish between anabolism and catabolism. (p. 121)
2 Distinguish between dehydration synthesis and hydrolysis. (pp. 121–122)
3 Give examples of a dehydration synthesis reaction and a hydrolysis reaction. (pp. 121–123)

4.2 Control of Metabolic Reactions

4 Describe how an enzyme interacts with its substrate. (pp. 123–124)
5 Define *active site*. (p. 123)
6 State two factors that influence the rate of an enzyme-catalyzed reaction. (p. 124)
7 A cell has _____ types of enzymes. (p. 124)
  a. 3
  b. hundreds of
  c. thousands of
  d. millions of
8 Explain the importance of a rate-limiting enzyme. (p. 125)
9 Describe how negative feedback involving a rate-limiting enzyme controls a metabolic pathway. (p. 125)
10 Define *cofactor*. (p. 125)
11 Discuss the relationship between a coenzyme and a vitamin. (p. 125)

4.3 Energy for Metabolic Reactions

12 Define *energy*. (p. 125)
13 Explain the importance of ATP and its relationship to ADP. (pp. 125–126)
14 Explain how the oxidation of molecules inside cells differs from the burning of substances outside cells. (pp. 126–127)

4.4 Cellular Respiration

15 State the overall function of cellular respiration. (p. 127)
16 Distinguish between anaerobic and aerobic phases of cellular respiration. (p. 127)
17 Match the part of cellular respiration to the associated activities. (pp. 127–131)

  (1) electron transport chain
  (2) glycolysis
  (3) citric acid cycle

  A. glucose molecules are broken down into pyruvic acid
  B. carrier molecules and enzymes extract energy and store it as ATP, releasing water and heat
  C. pyruvic acid molecules enter mitochondria, and become acetyl CoA that enter this series of reactions

18 Identify the final acceptor of the electrons released in the reactions of cellular respiration. (p. 131)
19 Excess glucose in cells may be linked and be stored as _____. (pp. 131–132)

4.5 Nucleic Acids and Protein Synthesis

20 The genetic code is _____. (p. 132)
  a. the bonding of purine to pyrimidine
  b. the correspondence between DNA triplet and amino acid
  c. the correspondence between DNA triplet and RNA triplet
  d. the controls that determine where the instructions for a polypeptide chain start and stop
21 DNA information provides instructions for the cell to _____. (p. 132)
  a. manufacture carbohydrate molecules
  b. extract energy
  c. manufacture RNA from amino acids
  d. synthesize protein molecules
22 Distinguish among a gene, an exome, and a genome. (p. 132)
23 Define *gene expression*. (p. 132)
24 If a DNA strand has the sequence ATGCGATCCGC then the sequence on the complementary DNA strand is _____. (pp. 133–134)
  a. ATGCGATCCGC
  b. TACGCTAGGCG
  c. UACGCUAGGCG
  d. AUGCGAUCCGC
25 Explain why DNA replication is essential. (p. 134)
26 Describe the events of DNA replication. (p. 134)
27 Identify the part of a DNA molecule that encodes information. (pp. 134–136)
28 Calculate the number of amino acids that a DNA sequence of twenty-seven nucleotides encodes. (p. 136)
29 List three ways that RNA differs from DNA. (p. 136)
30 If one strand of a DNA molecule has the sequence of ATTCTCGACTAT, the complementary mRNA has the sequence _____. (p. 136)
  a. ATTCTCGACTAT
  b. AUUCUCGACUAU
  c. TAAGAGCTGATA
  d. UAAGAGCUGAUA
31 Distinguish between transcription and translation. (pp. 136–138)
32 Distinguish the functions of mRNA, rRNA, and tRNA. (pp. 136–141)
33 Describe the function of a ribosome in protein synthesis. (p. 141)
34 List the steps of protein synthesis. (pp. 136–141)

4.6 Changes in Genetic Information

35 Explain how a mutation can alter the sequence of amino acids in a polypeptide. (p. 142)
36 Discuss two major ways that mutation occurs. (p. 142)
37 Explain why DNA repair is necessary. (p. 143)
38 Discuss three ways that the genetic code protects against the effects of a mutation. (p. 143)

INTEGRATIVE ASSESSMENTS/CRITICAL THINKING

Outcomes 2.2, 4.1

1. How can the same molecule be both a reactant (starting material) and a product of a biochemical pathway?

Outcomes 2.3, 3.2, 4.2

2. What effect might changes in the pH of body fluids or body temperature that accompany illness have on cells?

Outcomes 2.3, 4.1, 4.4

3. Michael P. was very weak from birth, with poor muscle tone, difficulty breathing, and great fatigue. By his third month, he began having seizures. Michael's medical tests were normal except for one: his cerebrospinal fluid (the fluid that bathes the brain and spinal cord) was unusually high in glucose. Hypothesizing that the boy could not produce enough ATP, doctors gave him a diet rich in certain fatty acids that caused the cellular respiration pathway to resume at the point of acetyl CoA formation. Michael rapidly improved. Explain what caused his symptoms.

Outcomes 4.3, 4.4

4. In cyanide poisoning, levels of ATP in the brain plummet, but levels of lactic acid increase markedly. Explain how both effects occur.

5. A student is accustomed to running 3 miles at a leisurely jogging pace. In a fitness class, she has to run a mile as fast as she can. Afterward, she is winded and has sharp pains in her chest and leg muscles. What has she experienced, in terms of energy metabolism?

Outcome 4.5

6. Consider the following DNA sequence:

   CATGTGTAGTCTAAA

   a. Write the sequence of the DNA strand that would be replicated from this one.
   b. Write the sequence of the RNA molecule that would be transcribed from the DNA strand.
   c. State how many codons the sequence specifies.
   d. State how many amino acids the sequence specifies.
   e. Use table 4.2 to write the sequence of amino acids that this DNA sequence encodes.

7. Some antibiotic drugs fight infection by interfering with DNA replication, transcription, or translation in bacteria. Indicate whether each of the following effects is on replication, transcription, or translation:
   a. Rifampin binds to bacterial RNA polymerase.
   b. Streptomycin binds bacterial ribosomes, disabling them.
   c. Quinolone blocks an enzyme that prevents bacterial DNA from unwinding.

 ONLINE STUDY TOOLS

 |ANATOMY & PHYSIOLOGY

Connect Interactive Questions Reinforce your knowledge using assigned interactive questions covering metabolism, cellular respiration, and protein synthesis.

Connect Integrated Activity Practice your understanding by building a protein using a provided DNA base sequence.

LearnSmart Discover which chapter concepts you have mastered and which require more attention. This adaptive learning tool is personalized, proven, and preferred.

Anatomy & Physiology Revealed Go more in depth, and view animations on cellular respiration, DNA structure and replication, and protein synthesis.

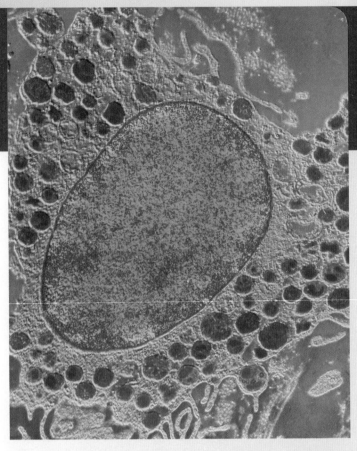

5

Tissues

LEARNING OUTCOMES

After you have studied this chapter, you should be able to:

5.1 Cells Are Organized Into Tissues
1 Describe how cells are organized into tissues. (p. 149)
2 Identify the intercellular junctions in tissues. (p. 149)
3 List the four major tissue types in the body. (p. 149)

5.2 Epithelial Tissues
4 Describe the general characteristics and functions of epithelial tissue. (pp. 150–152)
5 Name the types of epithelium and identify an organ in which each is found. (pp. 152–156)
6 Explain how glands are classified. (pp. 156 and 158)

5.3 Connective Tissues
7 Describe the general characteristics of connective tissue. (pp. 159–162)
8 Compare and contrast the components, cells, fibers, and extracellular matrix (where applicable) in each type of connective tissue. (pp. 163–168)
9 Describe the major functions of each type of connective tissue. (pp. 163–168)

5.4 Types of Membranes
10 Describe and locate each of the four types of membranes. (pp. 168–169)

5.5 Muscle Tissues
11 Distinguish among the three types of muscle tissue. (p. 170)

5.6 Nervous Tissues
12 Describe the general characteristics and functions of nervous tissue. (pp. 170–171)

Falsely colored transmission electron micrograph (TEM) of a portion of a mast cell. The large, orange oval is the cell's nucleus, and the red granules contain histamine and heparin (4,800×).

THE WHOLE PICTURE

In all complex organisms, cells are organized into tissues, which are layers or groups of similar cells. Tissues associate, assemble, and interact to form organs that have specialized functions. The study of tissues (histology) will assist understanding in later discussions of the physiology of organs and organ systems and their conbtributions to homeostasis.

Anatomy & Physiology | REVEALED
aprevealed.com

Module 3: Tissues

UNDERSTANDING WORDS

adip-, fat: *adip*ose tissue—tissue that stores fat.

chondr-, cartilage: *chondr*ocyte—cartilage cell.

-cyt, cell: osteo*cyt*e—bone cell.

epi-, upon, after, in addition: *epi*thelial tissue—tissue that covers all free body surfaces.

-glia, glue: neuro*glia*—cells that support neurons; part of nervous tissue.

hist-, web, tissue: *hist*ology—study of composition and function of tissues.

hyal-, resemblance to glass: *hyal*ine cartilage—flexible tissue containing chondrocytes.

inter-, among, between: *inter*calated disc—band between adjacent cardiac muscle cells.

macr-, large: *macr*ophage—large phago-cytic cell.

neur-, nerve: *neur*on—nerve cell.

os-, bone: *os*seous tissue—bone tissue.

phag-, to eat: *phag*ocyte—cell that engulfs and destroys foreign particles.

pseud-, false: *pseud*ostratified epithelium—tissue with cells that appear to be in layers, but are not.

squam-, scale: *squam*ous epithelium—tissue with flattened or scalelike cells.

strat-, layer: *strat*ified epithelium—tissue with cells in layers.

stria-, groove: *stria*tions—alternating light and dark cross-markings in skeletal and cardiac muscle cells.

5.1 | Cells Are Organized Into Tissues

Cells are organized into **tissues** (tish′uz). Some cells in the body, such as blood cells, are separated from each other in fluid-filled spaces or intercellular (in″ter-sell′u-lar) spaces. Many other cell types, however, are tightly packed, with structures called **intercellular junctions** that connect their cell membranes.

In one type of intercellular junction, called a *tight junction*, the membranes of adjacent cells converge and fuse. The area of fusion surrounds the cell like a belt, and the junction closes the space between the cells. Tight junctions typically join cells that form sheetlike layers, such as those that line the inside of the digestive tract. The linings of tiny blood vessels in the brain consist of cells held tightly together (From Science to Technology 5.1).

Another type of intercellular junction, called a *desmosome*, rivets or "spot welds" skin cells, enabling them to form a reinforced structural unit. The cell membranes of certain other cells, such as those in heart muscle and muscle of the digestive tract, are interconnected by tubular channels called *gap junctions*. These channels link the cytoplasm of adjacent cells and allow ions, nutrients (such as sugars, amino acids, and nucleotides), and other small molecules to move between them (fig. 5.1). Table 5.1 summarizes intercellular junctions.

Tissues can be distinguished from each other by variations in cell size, shape, organization, and function. The tissues of the human body include four major types: *epithelial, connective, muscle,* and *nervous*. These tissues associate, assemble, and interact to form organs that have even more specialized functions. Table 5.2 compares the four major tissue types. This chapter examines epithelial and connective tissues in detail, and introduces muscle and nervous tissues. Chapter 9 discusses muscle tissue in more detail, and chapters 10, 11, and 12 examine nervous tissue.

Tissues are three-dimensional structures in the body. Micrographs are photos of extremely thin slices (sections) of prepared tissue specimens. The advantage of the thin slices is that light passes through them more readily, making structures easier

CAREER CORNER

Tissue Recovery Technician

It is 5 AM and the tissue recovery technician reaches for her cell phone. A young man has died in a car crash from head trauma, but his skin is intact and healthy. The technician drives to a funeral home just a few miles out of the city, relieved that she doesn't have to travel as far as she did the last time she was on call. She is available to recover tissue for five 24-hour periods each month.

At the funeral home, the technician puts on mask and gloves, sterilizes the area of the skin of the corpse that she will excise, and then removes the tissue, using surgical instruments. She stores and packages the tissue, then ships it to a tissue bank, carefully recording details of the procedure. The technician is comfortable working with a corpse. She knows that even though the accident had a tragic end, several burn victims may receive grafts from the donor, after the tissue bank determines which patients will be most compatible with the grafts.

Recovering tissue, which is stored, is less stressful than recovering organs, which must be used quickly. In addition to skin, tissue banks store bone, corneas, tendons, ligaments, cartilage, blood vessels, and heart valves. A tissue recovery technician therefore must have excellent fine motor skills and knowledge of anatomy. The job also requires standing for long periods and possibly lifting bodies.

Getting a job as a tissue recovery technician requires completing a certification program from the American Association of Tissue Banks as a "certified tissue bank specialist." On-the-job training is typically provided. Tissue recovery technicians may be medical students or emergency medical technicians, or they might hold associates degrees in surgical technology.

A tissue recovery technician must feel comfortable working with corpses. In addition to funeral homes, the job may take the technician to hospital operating rooms or coroners' offices.

to see and photograph, but the relationship of the structures in three dimensions may not be obvious. **Figure 5.2** shows how a tubular structure would appear (a) in an oblique section and (b) in a longitudinal section (see also fig. 1.24, page 30). Throughout this chapter, three-dimensional line drawings (for example, fig. 5.3a) are included with each micrograph (for example, fig. 5.3b, c) to emphasize the orientation and distinguishing characteristics of the specific tissues, as well as locator icons (as in fig. 5.3) indicating where in the body the particular tissue may be found.

PRACTICE

1 What is a tissue?

2 What are the different types of intercellular junctions?

3 List the four major types of tissue.

5.2 | Epithelial Tissues AP|R

General Characteristics

Epithelial (ep″ĭ-the′le-al) **tissues** are found throughout the body. Epithelium covers the body surface and organs, forms the inner lining of body cavities, lines hollow organs, and composes glands. It always has a *free (apical) surface* exposed to the outside or internally to an open space. A thin, extracellular layer called the **basement membrane** anchors epithelium to underlying connective tissue.

Cancer cells secrete a substance that dissolves basement membranes, enabling the cells to invade tissue layers. Cancer cells also produce fewer adhesion proteins, or none at all, which allows them to spread into surrounding tissue.

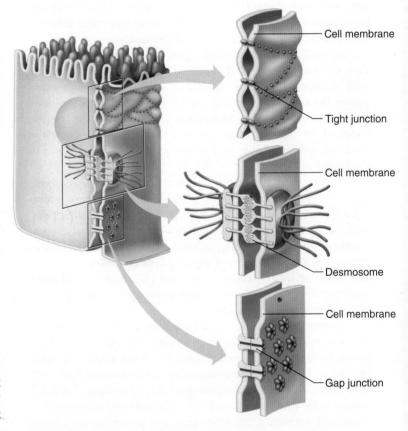

Cell membrane

Tight junction

Cell membrane

Desmosome

Cell membrane

Gap junction

FIGURE 5.1 Intercellular junctions. Tight junctions fuse cell membranes, desmosomes are "spot welds," and gap junctions form channels linking the cytoplasm of adjacent cells.

Which intercellular junction is the most likely to allow substances to move from one cell to another?

Answer can be found in Appendix G.

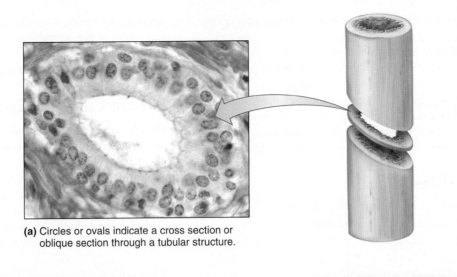

(a) Circles or ovals indicate a cross section or oblique section through a tubular structure.

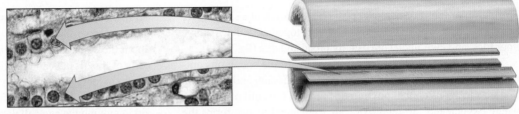

(b) Rows of cells with a space in between indicate a longitudinal section through a tube.

FIGURE 5.2 Tissue sections.
(a) Oblique section (600×).
(b) Longitudinal section (165×).

TABLE 5.1	Types of Intercellular Junctions	
Type	**Characteristics**	**Example**
Tight junctions	Close space between cells by fusing cell membranes	Cells that line the small intestine
Desmosomes	Bind cells by forming "spot welds" between cell membranes	Cells of the outer skin layer
Gap junctions	Form tubular channels between cells that allow exchange of substances	Muscle cells of the heart and digestive tract

TABLE 5.2	Tissues		
Type	**Function**	**Location**	**Distinguishing Characteristics**
Epithelial	Protection, secretion, absorption, excretion	Cover body surface, cover and line internal organs, compose glands	Lack blood vessels, cells readily divide, cells are tightly packed
Connective	Bind, support, protect, fill spaces, store fat, produce blood cells	Widely distributed throughout the body	Mostly have good blood supply, cells are farther apart than epithelial cells, with extracellular matrix in between
Muscle	Movement	Attached to bones, in the walls of hollow internal organs, heart	Able to contract in response to specific stimuli
Nervous	Conduct impulses for coordination, regulation, integration, and sensory reception	Brain, spinal cord, nerves	Cells communicate with each other and other body parts

As a rule, epithelial tissues lack blood vessels. However, nutrients diffuse to epithelium from underlying connective tissues, which have abundant blood vessels.

Epithelial cells readily divide, so injuries heal rapidly as new cells replace lost or damaged ones. For example, skin cells and the cells that line the stomach and intestines are continually damaged and replaced.

Epithelial cells are tightly packed. In many places, desmosomes attach epithelial cells, forming protective barriers in such structures as the outer layer of the skin and the lining of the mouth. Other epithelial functions include secretion, absorption, and excretion.

Epithelial tissues are classified according to cell shape and the number of cell layers. Epithelial tissues composed of thin, flattened cells are *squamous;* those with cubelike cells are *cuboidal;* and those with elongated cells are *columnar.* Epithelium composed of a single layer of cells is *simple* and with two or more layers of cells, *stratified.* In the following descriptions, modifications of the free surfaces of epithelial cells reflect their specialized functions.

 PRACTICE

4 List the general characteristics of epithelial tissue.

5 Explain how epithelial tissues are classified.

Simple Squamous Epithelium

Simple squamous (skwa'mus) **epithelium** consists of a single layer of thin, flattened cells. These cells fit tightly together, somewhat like floor tiles, and their nuclei are usually broad and thin (fig. 5.3).

Substances pass rather easily through simple squamous epithelium. This tissue is common at sites of diffusion and filtration. Simple squamous epithelium lines the air sacs (alveoli) of the lungs where oxygen and carbon dioxide are exchanged. It also forms the walls of capillaries, lines the insides of blood and lymph vessels, and is part of the membranes that line body cavities and cover the viscera. However, because it is so thin and delicate, simple squamous epithelium is easily damaged.

Simple Cuboidal Epithelium

Simple cuboidal epithelium consists of a single layer of cube-shaped cells. Most of these cells have centrally located, spherical nuclei (fig. 5.4).

Simple cuboidal epithelium lines the follicles of the thyroid gland, covers the ovaries, and lines the kidney tubules and ducts of certain glands, where the free surface faces the hollow channel or **lumen.** In the kidneys, it functions in tubular secretion and tubular reabsorption; in glands, it secretes glandular products.

Simple Columnar Epithelium

Simple columnar epithelium is composed of a single layer of cells that are longer than they are wide and whose nuclei are typically at about the same level, near the basement membrane (fig. 5.5). The cells of this tissue can be ciliated or nonciliated. Motile *cilia,* 7 to 10 μm in length, extend from the free surfaces of the cells, and they move constantly (see figure 3.15, p. 95). In the female, motile cilia aid in moving the egg cell through the uterine tube to the uterus.

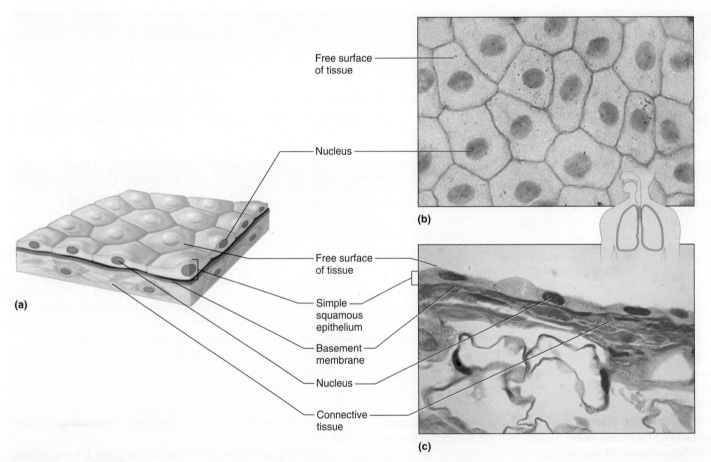

Free surface of tissue

Nucleus

(b)

Free surface of tissue

Simple squamous epithelium

Basement membrane

Nucleus

Connective tissue

(a)

(c)

FIGURE 5.3 AP|R Simple squamous epithelium consists of a single layer of tightly packed, flattened cells. (a) Idealized representation of simple squamous epithelium. Micrographs of (b) a surface view (250×) and (c) a side view of a section through simple squamous epithelium (250×).

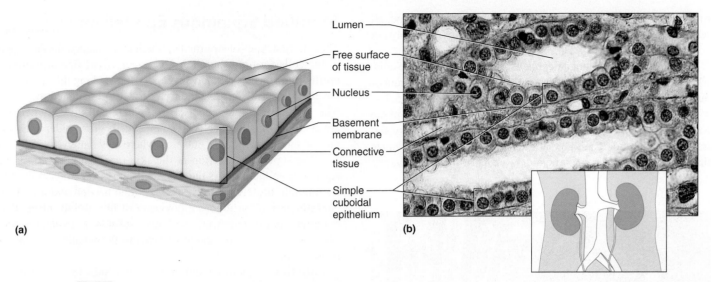

FIGURE 5.4 AP|R Simple cuboidal epithelium consists of a single layer of tightly packed, cube-shaped cells. (*a*) Idealized representation of simple cuboidal epithelium. (*b*) Micrograph of a section through simple cuboidal epithelium (165×).

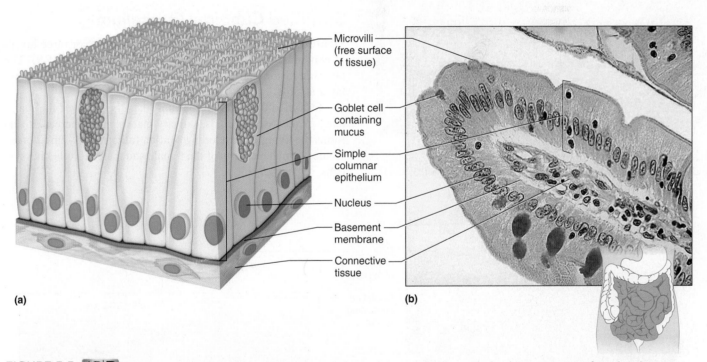

FIGURE 5.5 AP|R Simple columnar epithelium consists of a single layer of elongated cells. (*a*) Idealized representation of simple columnar epithelium. (*b*) Micrograph of a section through simple columnar epithelium (400×).

Nonciliated simple columnar epithelium lines the uterus and portions of the digestive tract, including the stomach and small and large intestines. Because its cells are elongated, this tissue is thick, which enables it to protect underlying tissues. Simple columnar epithelium also secretes digestive fluids and absorbs nutrients from digested food. Cells specialized for absorption typically have many tiny, cylindrical processes, called *microvilli,* extending from their free surfaces. Microvilli are 0.5 to 1.0 μm long, and they increase the surface area of the cell membrane where it is exposed to substances being absorbed (**fig. 5.6**).

Specialized flask-shaped glandular cells are scattered among the cells of simple columnar epithelium. These cells, called *goblet cells,* secrete a protective fluid called *mucus* onto the free surface of the tissue (see fig. 5.5).

Pseudostratified Columnar Epithelium

The cells of **pseudostratified** (soo″do-strat′ĭ-fīd) **columnar epithelium** appear stratified or layered, but they are not. A layered effect occurs because the nuclei are at two or more levels in the row of aligned cells. However, the cells, which vary in shape, all reach the basement membrane, even though some of them may not contact the free surface.

Pseudostratified columnar epithelial cells commonly have cilia, which extend from their free surfaces. Goblet cells scattered throughout this tissue secrete mucus, which the cilia sweep away (fig. 5.7).

Pseudostratified columnar epithelium lines the passages of the respiratory system. Here, the mucous-covered linings are sticky and trap dust and microorganisms that enter with the air. The cilia move the mucus and its captured particles upward and out of the airways.

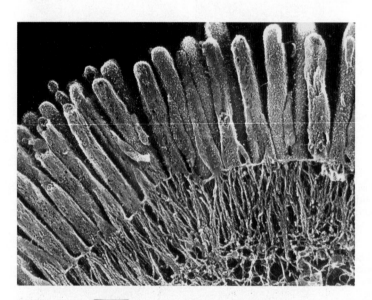

FIGURE 5.6 **AP|R** A scanning electron micrograph of microvilli, which fringe the free surfaces of some columnar epithelial cells (33,000×).

Stratified Squamous Epithelium

The different types of stratified epithelium are named for the shape of the cells forming the outermost layer. **Stratified squamous epithelium** consists of many layers of cells, making this tissue relatively thick. Cells nearest the free surface are flattened the most, whereas those in the deeper layers, where cell division occurs, are cuboidal or columnar. As the newer cells grow, older ones are pushed farther and farther outward, where they flatten (fig. 5.8).

The superficial portion of the skin (epidermis) is stratified squamous epithelium. As the older cells are pushed outward, they accumulate proteins called *keratins,* then harden and die. This "keratinization" produces a covering of dry, tough, protective material that prevents water and other substances from escaping from underlying tissues and blocks chemicals and microorganisms from entering.

Stratified squamous epithelium also lines the oral cavity, esophagus, vagina, and anal canal. In these parts, the tissue is not keratinized; it stays soft and moist, and the cells on its free surfaces remain alive.

Stratified Cuboidal Epithelium

Stratified cuboidal epithelium consists of two or three layers of cuboidal cells that form the lining of a lumen (fig. 5.9). The layering of the cells provides more protection than the single layer affords.

Stratified cuboidal epithelium lines the ducts of the mammary glands, sweat glands, salivary glands, and pancreas. It also forms the lining of developing ovarian follicles and seminiferous tubules, which are parts of the female and male reproductive systems, respectively.

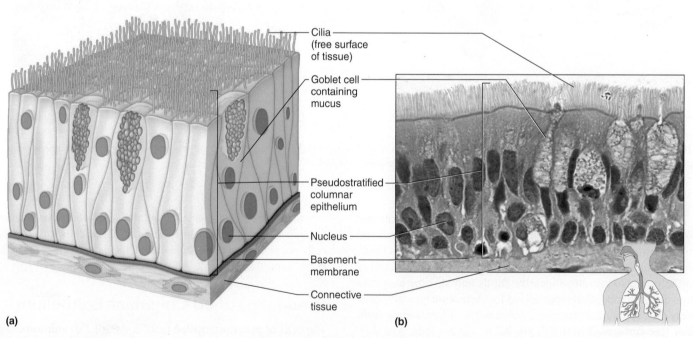

Cilia (free surface of tissue)

Goblet cell containing mucus

Pseudostratified columnar epithelium

Nucleus

Basement membrane

Connective tissue

(a)

(b)

FIGURE 5.7 **AP|R** Pseudostratified columnar epithelium appears stratified because the cell nuclei are located at different levels. (*a*) Idealized representation of pseudostratified columnar epithelium. (*b*) Micrograph of a section through pseudostratified columnar epithelium (1,000×).

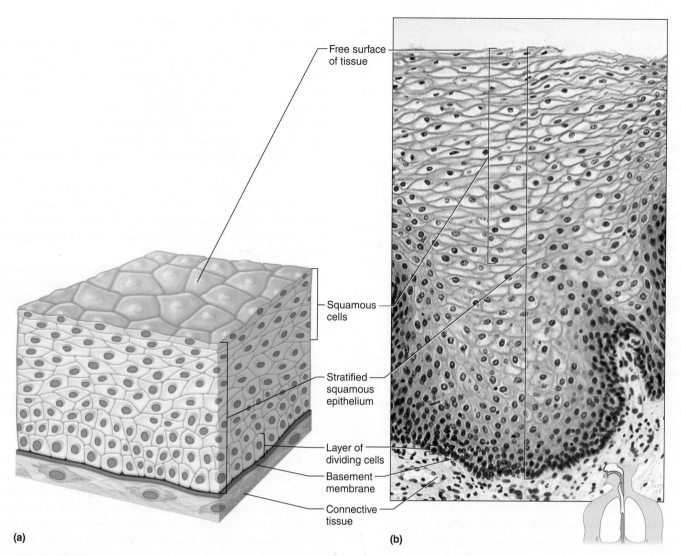

Free surface
of tissue

Squamous
cells

Stratified
squamous
epithelium

Layer of
dividing cells

Basement
membrane

Connective
tissue

(a)

(b)

FIGURE 5.8 AP|R Stratified squamous epithelium consists of many layers of cells. (*a*) Idealized representation of stratified squamous epithelium. (*b*) Micrograph of a section through stratified squamous epithelium (65×).

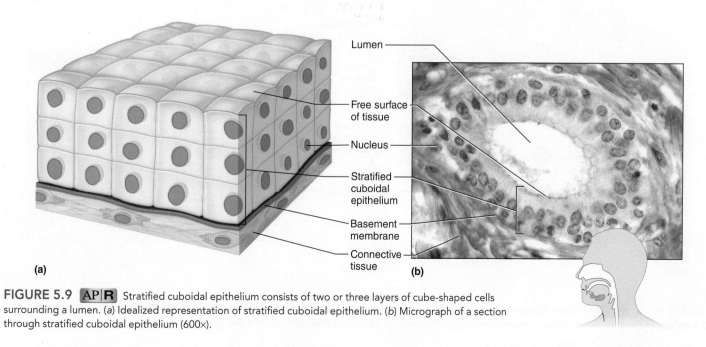

Lumen

Free surface
of tissue

Nucleus

Stratified
cuboidal
epithelium

Basement
membrane

Connective
tissue

(a)

(b)

FIGURE 5.9 AP|R Stratified cuboidal epithelium consists of two or three layers of cube-shaped cells surrounding a lumen. (*a*) Idealized representation of stratified cuboidal epithelium. (*b*) Micrograph of a section through stratified cuboidal epithelium (600×).

Stratified Columnar Epithelium

Stratified columnar epithelium consists of several layers of cells (fig. 5.10). The superficial cells are elongated, whereas the basal layers consist of cuboidal cells. Stratified columnar epithelium is found in part of the male urethra and lining the larger ducts of exocrine glands.

Transitional Epithelium

Transitional epithelium (uroepithelium) is specialized to change in response to increased tension. It forms the inner lining of the urinary bladder and lines the ureters and the superior urethra. When the wall of one of these organs contracts, the tissue consists of several layers of irregular-shaped cells. When the organ is distended, however, the tissue stretches, and the cells elongate. While distended, the tissue appears to contain only a few layers of cells (fig. 5.11). In addition to providing an expandable lining, transitional epithelium forms a barrier that helps prevent the contents of the urinary tract from diffusing back into the internal environment.

Up to 90% of human cancers are *carcinomas*, growths that originate in epithelium. Most carcinomas begin on surfaces that contact the external environment, such as skin, linings of the airways in the respiratory tract, or linings of the stomach or intestines in the digestive tract. Carcinomas may also arise internally, such as in a duct in a breast or in the prostate gland.

 PRACTICE

6 Describe the structure of each type of epithelium.
7 Describe the special functions of each type of epithelium.

Glandular Epithelium

Glandular epithelium is composed of cells specialized to produce and secrete substances into ducts or into body fluids. Such cells are usually found within columnar or cuboidal epithelium, and one or more of these cells constitute a *gland*. Glands that secrete their products into ducts that open onto surfaces, such as the skin or the lining of the digestive tract, are called **exocrine glands.** Glands that secrete their products into tissue fluid or blood are called **endocrine glands.** (Chapter 13 discusses endocrine glands.)

An exocrine gland may consist of a single epithelial cell (unicellular gland), such as a mucous-secreting goblet cell, or it may be composed of many cells (multicellular gland). In turn, the multicellular glands can be structurally subdivided into two groups—simple and compound.

A *simple gland* communicates with the surface by means of a duct that does not branch before reaching the glandular cells or secretory portions. A *compound gland* has a duct that branches repeatedly before reaching the secretory portion. These two types of glands can be further classified according to the shapes of their secretory portion. Glands that consist of epithelial-lined tubes are called *tubular glands;* those whose terminal portions form saclike dilations are called *alveolar glands* (acinar glands). The secretory portions may also branch or coil. **Figure 5.12** illustrates several types of exocrine glands classified by structure. **Table 5.3** summarizes the types of exocrine glands, lists their characteristics, and provides an example of each type.

Exocrine glands are also classified according to the ways these glands secrete their products. Glands that release fluid products by exocytosis are called **merocrine** (mer′o-krin) **glands.** Glands that lose small portions of their glandular cell bodies during secretion are called **apocrine** (ap′o-krin) **glands.** Glands that release entire cells are called **holocrine** (ho′lo-krin) **glands.**

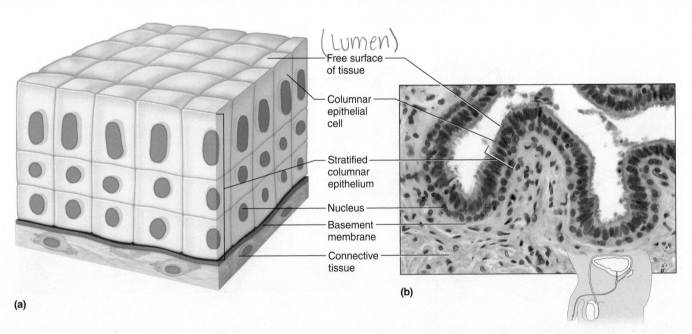

(Lumen)
Free surface of tissue
Columnar epithelial cell
Stratified columnar epithelium
Nucleus
Basement membrane
Connective tissue

(a)

(b)

FIGURE 5.10 **AP|R** Stratified columnar epithelium consists of a superficial layer of columnar cells overlying several layers of cuboidal cells. (a) Idealized representation of stratified columnar epithelium. (b) Micrograph of a section through stratified columnar epithelium (230×).

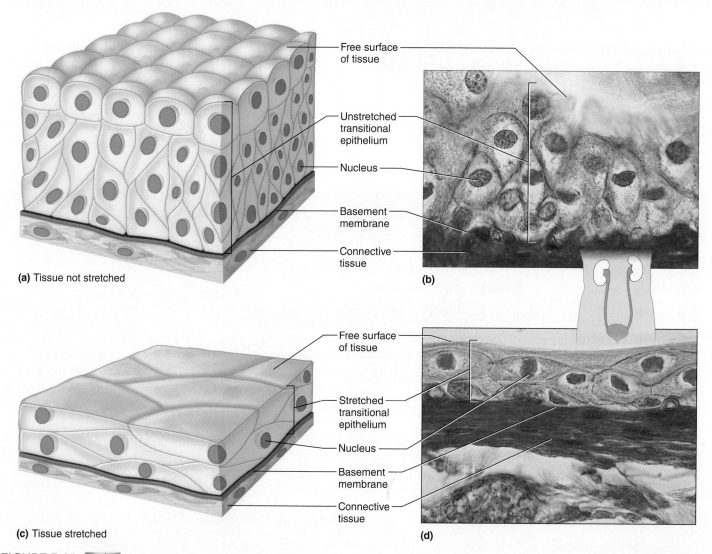

Free surface
of tissue

Unstretched
transitional
epithelium

Nucleus

Basement
membrane

Connective
tissue

(a) Tissue not stretched

(b)

Free surface
of tissue

Stretched
transitional
epithelium

Nucleus

Basement
membrane

Connective
tissue

(c) Tissue stretched

(d)

FIGURE 5.11 AP|R Transitional epithelium. (*a* and *c*) Idealized representations of transitional epithelium. (*b* and *d*) Micrographs of sections through transitional epithelium (675×). When the smooth muscle in the organ wall contracts, transitional epithelium is unstretched and consists of many layers. When the organ is distended, the tissue stretches and appears thinner.

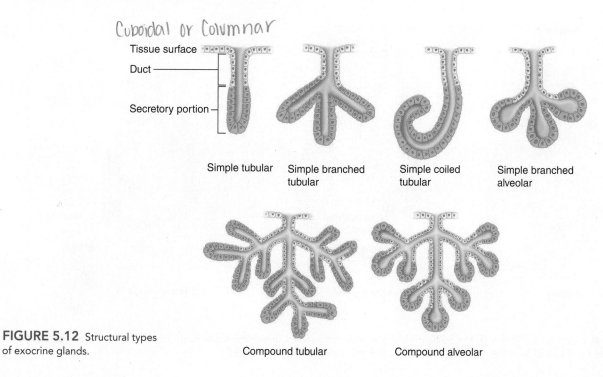

Cuboidal or Columnar

Tissue surface

Duct

Secretory portion

Simple tubular

Simple branched
tubular

Simple coiled
tubular

Simple branched
alveolar

Compound tubular

Compound alveolar

FIGURE 5.12 Structural types
of exocrine glands.

After release, the cells containing accumulated secretory products disintegrate, liberating their secretions (figs. 5.13 and 5.14). Table 5.4 summarizes these glands and their secretions.

RECONNECT

To Chapter 3, Movements Into and Out of the Cell, page 105.

Most exocrine secretory cells are merocrine, and they can be subclassified based on their secretion of serous fluid or mucus. **Serous fluid** is typically watery and slippery. Serous cells secreting this fluid, which lubricates, are commonly associated with the visceral and parietal membranes of the thoracic and abdomino-pelvic cavities. The thicker fluid, **mucus,** is rich in the glycoprotein *mucin*. Cells in the inner linings of the digestive, respiratory,

TABLE 5.3	Types of Exocrine Glands	
Type	**Characteristics**	**Example**
Unicellular glands	A single secretory cell	Mucous-secreting goblet cell (see fig. 5.5)
Multicellular glands	Glands that consist of many cells	
Simple glands	Glands that communicate with the surface by means of ducts that do not branch before reaching the secretory portion	
1. Simple tubular gland	Straight tubelike gland that opens directly onto surface	Intestinal glands of small intestine (see fig. 17.3)
2. Simple coiled tubular gland	Long, coiled, tubelike gland; long duct	Merocrine (sweat) glands of skin (see figs. 6.10 and 6.11)
3. Simple branched tubular gland	Branched, tubelike gland; duct short or absent	Gastric glands (see fig. 17.19)
4. Simple branched alveolar gland	Secretory portions of gland expand into saclike compartments along duct	Sebaceous gland of skin (see fig. 5.14)
Compound glands	Glands that communicate with surface by means of ducts that branch repeatedly before reaching the secretory portion	
1. Compound tubular gland	Secretory portions are coiled tubules, usually branched	Bulbourethral glands of male (see fig. 22.4)
2. Compound alveolar gland	Secretory portions are irregularly branched tubules with numerous saclike outgrowths	Mammary glands (see fig. 23.28)

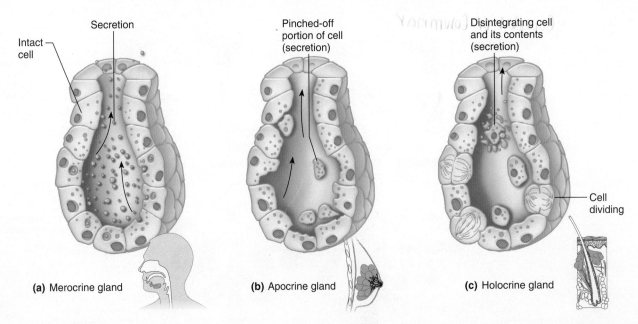

FIGURE 5.13 **AP|R** Glandular secretions. (*a*) Merocrine glands release secretions without losing cytoplasm. (*b*) Apocrine glands lose small portions of their cell bodies during secretion. (*c*) Holocrine glands release entire cells filled with secretory products.

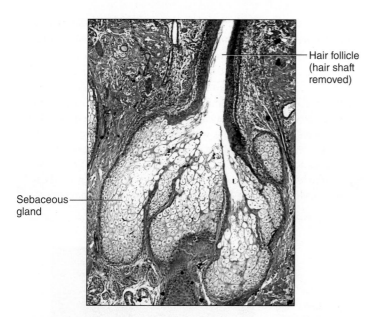

FIGURE 5.14 The sebaceous gland associated with a hair follicle is a simple-branched alveolar gland that secretes entire cells (40×).

Hair follicle (hair shaft removed)

Sebaceous gland

TABLE 5.4	Exocrine Glandular Secretions	
Type	Description of Secretion	Example
Merocrine glands	A fluid product released through the cell membrane by exocytosis	Salivary glands, pancreatic glands, sweat glands of the skin
Apocrine glands	Cellular product and portions of the free ends of glandular cells pinch off during secretion	Mammary glands, ceruminous glands lining the external acoustic meatus
Holocrine glands	Disintegrated entire cells filled with secretory products	Sebaceous glands of the skin

and reproductive systems secrete abundant mucus, which is protective. Mucous cells and goblet cells secrete mucus, but in different parts of the body. Table 5.5 summarizes the characteristics of the different types of epithelial tissues.

 PRACTICE

8 Distinguish between exocrine and endocrine glands.
9 Explain how exocrine glands are classified.
10 Distinguish between serous fluid and mucus.

5.3 | Connective Tissues AP|R

General Characteristics

Connective (kŏ-nek′tiv) **tissues** comprise much of the body and are the most abundant type of tissue by weight. They bind structures, provide support and protection, serve as frameworks, fill spaces, store fat, produce blood cells, protect against infections, and help repair tissue damage.

Connective tissue cells are farther apart than epithelial cells, and they have abundant **extracellular matrix** (eks″trah-sel′u-lar ma′triks) between them. This extracellular matrix is composed of *protein fibers* and a *ground substance* consisting of nonfibrous protein and other molecules, and fluid. The consistency of the extracellular matrix varies from fluid to semisolid to solid. The ground substance binds, supports, and provides a medium through which substances may be transferred between the blood and cells of the

TABLE 5.5	Epithelial Tissues		
Type	Description	Function	Location
Simple squamous epithelium	Single layer, flattened cells	Filtration, diffusion, osmosis, covers surface	Air sacs of lungs, walls of capillaries, linings of blood and lymph vessels, part of the membranes lining body cavities and covering viscera
Simple cuboidal epithelium	Single layer, cube-shaped cells	Protection, secretion, absorption	Surface of ovaries, linings of kidney tubules, and linings of ducts of certain glands
Simple columnar epithelium	Single layer, elongated cells	Protection, secretion, absorption	Linings of uterus, stomach, and intestines
Pseudostratified columnar epithelium	Single layer, elongated cells	Protection, secretion, movement of mucus and substances	Linings of respiratory passages
Stratified squamous epithelium	Many layers, top cells flattened	Protection	Superficial portion of skin and linings of oral cavity, vagina, and anal canal
Stratified cuboidal epithelium	2 or 3 layers, cube-shaped cells	Protection	Linings of ducts of mammary glands, sweat glands, salivary glands, and pancreas
Stratified columnar epithelium	Top layer of elongated cells, lower layers of cube-shaped cells	Protection, secretion	Part of the male urethra and lining of larger ducts of excretory glands
Transitional epithelium	Many layers of cube-shaped and elongated cells	Distensibility, protection	Inner lining of urinary bladder and linings of ureters and part of urethra
Glandular epithelium	Unicellular or multicellular	Secretion	Salivary glands, sweat glands, endocrine glands

tissue. Some connective tissues, such as bone and cartilage, are rigid. Loose connective tissue and dense connective tissue are more flexible. Clinical Application 5.1 discusses the extracellular matrix and its relationship to disease.

Most connective tissue cells can divide. Connective tissues typically have good blood supplies and are well nourished, but the density of blood vessels (vascularity) varies.

Major Cell Types

Connective tissues include a variety of cell types. Some of them are called *fixed cells* because they reside in the specific connective tissue type for an extended period. These include fibroblasts and mast cells. Other cells, such as macrophages, are *wandering cells.* They move through and appear in tissues temporarily, usually in response to an injury or infection.

Fibroblasts (fi′bro-blastz) are the most common type of fixed cell in connective tissues. These large, star-shaped cells produce fibers by secreting proteins into the extracellular matrix of connective tissues (fig. 5.15).

Macrophages (mak′ro-fājez), or histiocytes, originate as white blood cells (see chapter 14, p. 537) and are almost as numerous as fibroblasts in some connective tissues. They are usually attached to fibers but can detach and actively move about. Macrophages are specialized to carry on phagocytosis. As scavenger cells, they can clear foreign particles from tissues, providing an important defense against infection (fig. 5.16). Macrophages play additional roles in immunity (see chapter 16, p. 629).

Mast cells are large and widely distributed in connective tissues, where they are usually near blood vessels (fig. 5.17). They release *heparin,* a compound that prevents blood clotting. Mast

cells also release *histamine,* which is a substance that promotes some of the reactions associated with inflammation and allergies, such as asthma and hay fever (see chapter 16, p. 636).

Release of histamine stimulates inflammation by dilating the small arterioles that feed capillaries, the tiniest blood vessels. The increased blood flow, with the resulting swelling and redness, is inhospitable to infectious bacteria and viruses and also dilutes toxins. Inappropriate histamine release as part of an allergic response can be most uncomfortable. Allergy medications called antihistamines counter this misplaced inflammation.

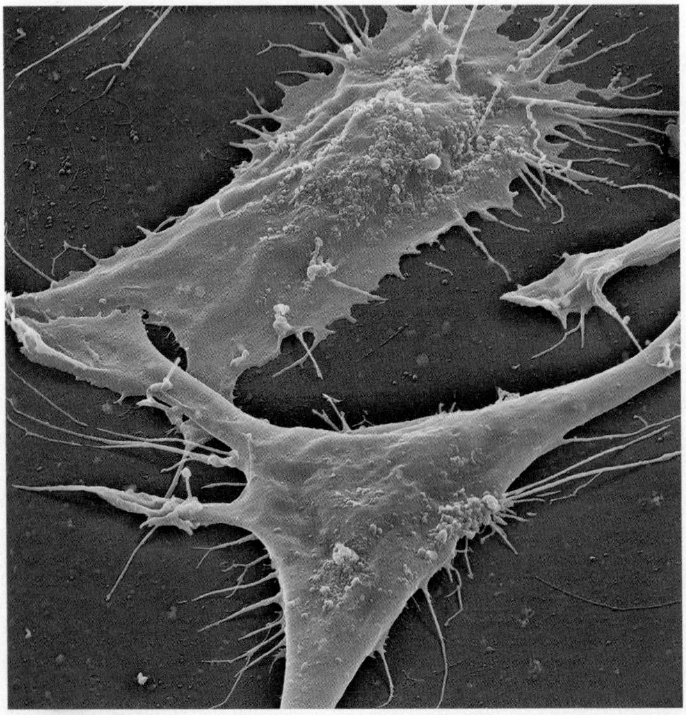

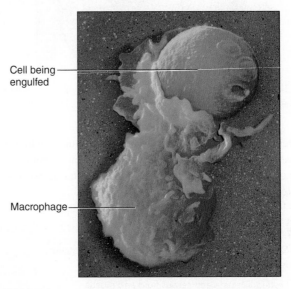

Cell being engulfed

Macrophage

FIGURE 5.16 Macrophages are scavenger cells common in connective tissues. This scanning electron micrograph shows a macrophage engulfing a bacterium (7,500×).

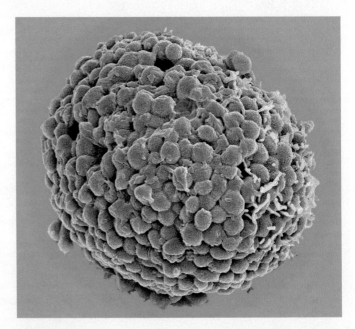

FIGURE 5.15 Scanning electron micrograph of fibroblasts, the most abundant cell type of connective tissue (1,500×).

FIGURE 5.17 Scanning electron micrograph of a mast cell, which releases heparin and histamine (1,750×).

The Body's Glue: The Extracellular Matrix

The extracellular matrix (ECM) is more than a "filler" between cells. It is a complex and changing mix of molecules that modifies the tissue to suit different organs and conditions. The ECM serves as a scaffolding to organize cells into tissues, and also relays the biochemical signals that control cell division, cell differentiation, tissue repair, and cell migration.

The ECM has two basic components: the basement membrane that anchors epithelium to underlying connective tissues, and the rest of the material between cells, called the interstitial matrix. The basement membrane is mostly composed of tightly packed collagen fibers from which large, cross-shaped glycoproteins called laminins extend. The laminins (and other glycoproteins such as fibronectin, the proteoglycans, and tenascin) traverse the interstitial matrix and contact receptors, called integrins, on other cells (fig. 5A). In this way, the ECM connects cells into tissues. At least twenty types of collagen and precursors of hormones, enzymes, growth factors, and immune system biochemicals (cytokines) comprise the various versions of the ECM. The precursor molecules are activated under certain conditions.

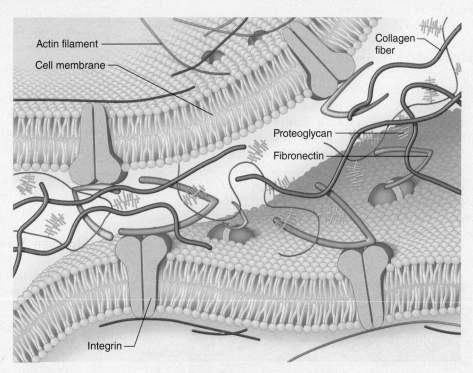

FIGURE 5A The extracellular matrix (ECM) is a complex and dynamic meshwork of various proteins and glycoproteins. Collagen is abundant. Other common components include integrins that anchor the ECM to cells, proteoglycans, and fibronectin. The ECM may also include precursors of growth factors, hormones, enzymes, and cytokines. It is vital to maintaining the specialized characteristics of tissues and organs.

The components of the ECM are always changing, as its cells synthesize proteins while enzymes called proteases break down specific proteins. The balance of components is important to maintaining and repairing organ structure. Disrupt the balance, and disease can result. Here are three common examples:

Cancer

The spread of a cancerous growth takes advantage of the normal ability of fibroblasts to contract as they close a wound, where they are replaced with normal epithelium. Chemical signals from cancer cells make fibroblasts (an abundant type of connective tissue cell) a more contractile cell type (myofibroblasts), and they take on the characteristics of cancer cells. At the same time, alterations in laminins loosen the connections of the fibroblasts to surrounding cells. This abnormal flexibility enables the changed fibroblasts to migrate, helping the cancer spread.

Liver Fibrosis

In fibrosis, which is a part of all chronic liver diseases, collagen deposition increases so that the ECM exceeds its normal 3% of the organ. Healthy liver ECM sculpts a framework that supports the epithelial and vascular tissues of the organ. In liver fibrosis, hepatic stellate cells secrete collagen fibers in the areas where the epithelium and blood vessels meet, in response to a dam-

aging agent such as a virus, alcohol, or a toxic drug. Such limited fibrosis seals off the affected area, preventing spread of the damage. But if the process continues—if an infection is not treated or the noxious stimulus not removed—the ECM grows and eventually blocks the interaction between liver cells and the bloodstream. Scarring of liver tissue occurs, with loss of normal function, resulting in a dangerous condition called *cirrhosis*.

Heart Failure and Atherosclerosis

The heart's ECM organizes cells into a three-dimensional network that coordinates their contractions into the rhythmic heartbeat necessary to pump blood. This ECM consists of collagen, fibronectin, laminin, and elastin surrounding cardiac muscle cells and myofibroblasts and is also in the walls of arteries. Some forms of heart failure reflect imbalances of collagen production and degradation. In heart muscle, as in the liver, the natural response of ECM buildup is to wall off an area where circulation is blocked, but if it continues, the extra scaffolding stiffens the heart, which can lead to heart failure. During a myocardial infarction (heart attack), collagen synthesis and deposition increase in affected and nonaffected heart parts, which is why damage can continue even after pain stops. In atherosclerosis, excess ECM accumulates on the interior linings of arteries, blocking blood flow.

Connective Tissue Fibers

Fibroblasts produce three types of connective tissue fibers: collagen fibers, elastic fibers, and reticular fibers. Collagen and elastic fibers are the most abundant.

Collagen (kolah-jen) **fibers** are thick threads of the protein *collagen*, the major structural protein of the body. Collagen fibers are grouped in long, parallel bundles, and they are flexible but only slightly elastic (fig. 5.18). More importantly, they have great tensile strength—that is, they can resist considerable pulling force. Thus, collagen fibers are important components of body parts that hold structures together, such as **ligaments** (which connect bones to bones) and **tendons** (which connect muscles to bones).

Tissue containing abundant collagen fibers is called *dense connective tissue.* It appears white, and for this reason collagen fibers of dense connective tissue are sometimes called white fibers. *Loose connective tissue,* on the other hand, has sparse collagen fibers. Clinical Application 5.2 describes disorders that result from abnormal collagen.

> When skin is exposed to prolonged and intense sunlight, connective tissue fibers lose elasticity, and the skin stiffens and becomes leathery. In time, the skin may sag and wrinkle. Collagen injections may temporarily smooth out wrinkles. However, collagen applied as a cream to the skin does not combat wrinkles because collagen molecules are far too large to penetrate the skin.

Elastic fibers are composed of a springlike protein called *elastin.* These fibers branch, forming complex networks in various tissues. Elastic fibers are weaker than collagen fibers, but they are easily stretched or deformed and will resume their original lengths and shapes when the force acting upon them is removed. Elastic fibers are abundant in body parts normally subjected to stretching, such as the vocal cords and air passages of the respiratory system. They are also called yellow fibers, because tissues amply supplied with them appear yellowish (fig. 5.18).

Reticular fibers are thin collagen fibers. They are highly branched and form delicate supporting networks in a variety of tissues, including those of the spleen. Table 5.6 summarizes the components of connective tissue.

PRACTICE

11. What are the general characteristics of connective tissue?
12. What are the major types of cells in connective tissue?
13. What is the primary function of fibroblasts?
14. What are the characteristics of collagen and elastin?

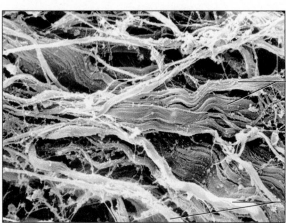

FIGURE 5.18 Scanning electron micrograph of collagenous fibers (shades of white to gray) and elastic fibers (yellow) (4,100×).

Collagen fibers

Elastic fibers

Categories of Connective Tissues

Connective tissue is divided into two major categories. *Connective tissue proper* includes loose connective tissue (areolar, adipose, reticular) and dense connective tissue (dense regular, dense irregular, elastic). The *specialized connective tissues* include cartilage, bone, and blood. The following sections describe each type of connective tissue.

TABLE 5.6	Components of Connective Tissue	
Component	**Characteristics**	**Function**
Cellular		
Fibroblasts	Widely distributed, large, star-shaped cells	Secrete proteins that become fibers
Macrophages	Motile cells sometimes attached to fibers	Clear foreign particles from tissues by phagocytosis
Mast cells	Large cells, usually located near blood vessels	Release substances that may help prevent blood clotting (heparin) and promote inflammation (histamine)
Extracellular matrix		
Collagen fibers (white fibers)	Thick, threadlike fibers of collagen with great tensile strength	Hold structures together
Elastic fibers (yellow fibers)	Bundles of microfibrils embedded in elastin	Provide elastic quality to parts that stretch
Reticular fibers	Thin fibers of collagen	Form delicate supportive networks within tissues
Ground substance	Nonfibrous protein and other molecules, and varying amounts of fluid	Fills in spaces around cells and fibers

Abnormalities of Collagen

Much of the human body consists of the protein collagen. It accounts for more than 60% of the protein in bone and cartilage and provides 50% to 90% of the dry weight of skin, ligaments, tendons, and the dentin of teeth. Collagen is in the eyes, blood vessel linings, basement membranes, and connective tissue. Because of collagen's abundance and wide distribution in the body, defects in this protein cause a variety of medical problems.

Collagen is especially vulnerable to disruption because it has an extremely precise structure. A slight alteration that might exert little noticeable effect on other proteins can destroy collagen's molecular structure. Collagen is produced from a precursor molecule called procollagen. Three polypeptide chains coil and entwine into a triple helix to form procollagen.

Once secreted from the cell, the helices are trimmed. The resulting pieces continue to associate outside the cell, building the networks of collagen fibers that hold the body together. Collagen is rapidly synthesized and assembled into its rigid architecture. Many types of mutations can disrupt the protein's structure, including missing procollagen chains, kinks in the triple helix, failure to cut procollagen, and defects in aggregation outside the cell.

Table 5A details some collagen disorders. Knowing which specific mutations cause disorders offers a way to identify the condition before symptoms arise. This can be helpful if early treatment can follow. A person who has a high risk of developing hereditary osteoporosis, for example, might take calcium supplements before symptoms appear. Aortic aneurysm is a more serious connective tissue disorder that can be presymptomatically detected if an underlying mutation is discovered. In aortic aneurysm, a weakened aorta (the largest blood vessel in the body, which emerges from the heart) swells and may burst. Knowing that the mutant gene has not been inherited can ease worries. Knowing that it has been inherited can warn affected individuals to have frequent ultrasound exams to detect aortic weakening early enough for corrective surgery.

TABLE 5A	Collagen Disorders	
Disorder	**Molecular Defect**	**Signs and Symptoms**
Chondrodysplasia	Collagen chains are too wide and asymmetric	Stunted growth; deformed joints
Dystrophic epidermolysis bullosa	Breakdown of collagen fibrils that attach skin layers to each other	Stretchy, easily scarred skin; lax joints
Hereditary osteoarthritis	Substituted amino acid in collagen chain alters shape	Painful joints
Marfan syndrome	Too little fibrillin, an elastic connective tissue protein	Long limbs; sunken chest; lens dislocation; spindly fingers; weakened aorta
Osteogenesis imperfecta type I	Too few collagen triple helices	Easily broken bones; deafness; blue sclera (whites of the eyes)
Stickler syndrome	Short collagen chains	Joint pain; degeneration of retina and fluid around it

Areolar Connective Tissue

Areolar (ah-re′o-lar) **connective tissue** forms delicate, thin membranes throughout the body. The cells of this tissue, mainly fibroblasts, are located some distance apart and are separated by a gel-like ground substance that contains collagen and elastic fibers that fibroblasts secrete (fig. 5.19).

Areolar connective tissue is found in the subcutaneous layer beneath the skin and surrounding organs. It underlies most layers of epithelium, where its many blood vessels nourish nearby epithelial cells.

Adipose (Fat) Tissue

Adipose (ad′ĭ-pōs) **tissue** develops when certain cells (adipocytes) store fat in droplets in their cytoplasm. At first, these cells resemble fibroblasts, but as they accumulate fat, they enlarge, and their nuclei are pushed to one side (fig. 5.20). When adipocytes become so abun-

dant that they crowd out other cell types, they form adipose tissue. This tissue lies beneath the skin, in spaces between muscles, around the kidneys, behind the eyeballs, in certain abdominal membranes, on the surface of the heart, and around certain joints. Adipose tissue cushions joints and some organs, such as the kidneys. It also insulates beneath the skin, and it stores energy in fat molecules.

Adipose tissue can be described as white adipose tissue (white fat) or brown adipose tissue (brown fat). White fat stores nutrients for nearby cells to use in the production of energy. Brown fat cells have many mitochondria that can break down nutrients to generate heat to warm the body. Infants have brown fat mostly on their backs to warm them. Adults have brown fat in the armpits, neck, and around the kidneys.

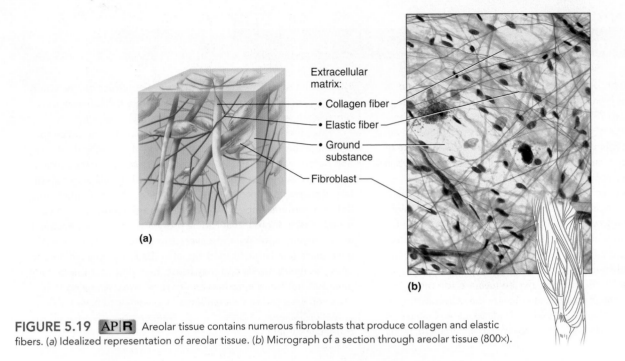

Extracellular matrix:
• Collagen fiber
• Elastic fiber
• Ground substance
Fibroblast

(a)

(b)

FIGURE 5.19 AP|R Areolar tissue contains numerous fibroblasts that produce collagen and elastic fibers. (*a*) Idealized representation of areolar tissue. (*b*) Micrograph of a section through areolar tissue (800×).

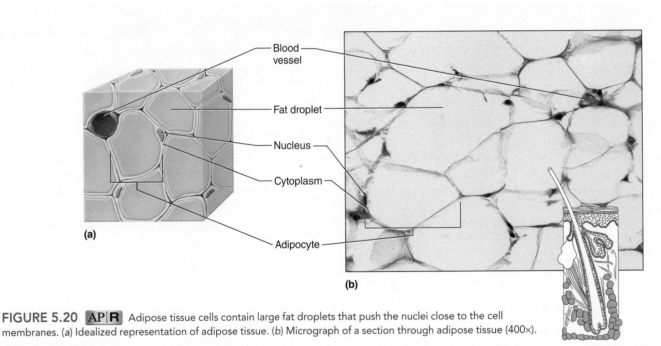

Blood vessel

Fat droplet

Nucleus

Cytoplasm

(a)

Adipocyte

(b)

FIGURE 5.20 AP|R Adipose tissue cells contain large fat droplets that push the nuclei close to the cell membranes. (*a*) Idealized representation of adipose tissue. (*b*) Micrograph of a section through adipose tissue (400×).

Reticular Connective Tissue

Reticular connective tissue is composed of thin, reticular fibers in a three-dimensional network. It helps provide the framework of certain internal organs, such as the liver and spleen (fig. 5.21).

Dense Regular Connective Tissue

Dense regular connective tissue consists of many closely packed, thick, collagen fibers; a fine network of elastic fibers; and a few cells, mostly fibroblasts. Collagen fibers of dense regular connective tissue are very strong, enabling the tissue to withstand pulling

forces (fig. 5.22). It often binds body structures as part of *tendons* and *ligaments*. The blood supply to dense regular connective tissue is poor, slowing tissue repair. This is why a sprain, which damages tissues surrounding a joint, may take considerable time to heal.

Dense Irregular Connective Tissue AP|R

Fibers of **dense irregular connective tissue** are thicker, interwoven, and more randomly distributed than fibers of dense regular connective tissue. The irregularly placed fibers allow the tissue to sustain tension exerted from many different directions. Dense irregular connective tissue is in the dermis, which is the deep skin layer.

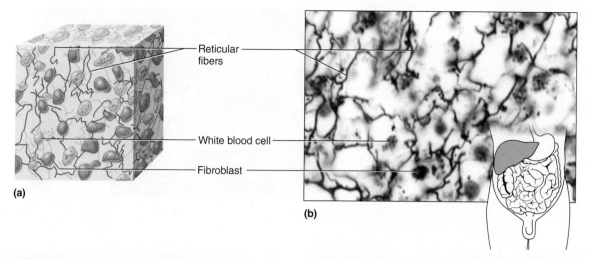

Reticular
fibers

White blood cell

Fibroblast

(a)

(b)

FIGURE 5.21 AP|R Reticular connective tissue is a network of thin reticular fibers, which contains numerous fibroblasts and white blood cells. (a) Idealized representation of reticular connective tissue. (b) Micrograph of a section through reticular connective tissue (1,000×).

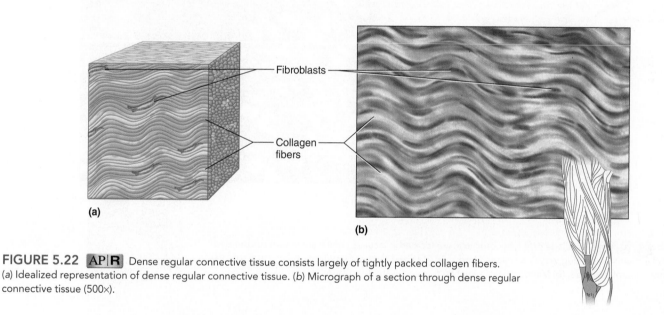

Fibroblasts

Collagen
fibers

(a)

(b)

FIGURE 5.22 AP|R Dense regular connective tissue consists largely of tightly packed collagen fibers. (a) Idealized representation of dense regular connective tissue. (b) Micrograph of a section through dense regular connective tissue (500×).

 A GLIMPSE AHEAD | **To Chapter 9**

Dense irregular connective tissue surrounds individual skeletal muscles *(fascia)*, and separates each muscle into bundles of skeletal muscle cells called *fascicles.* Each muscle cell is surrounded by areolar connective tissue.

Elastic Connective Tissue

Elastic connective tissue mainly consists of yellow, elastic fibers in parallel strands or in branching networks. Between these fibers are collagen fibers and fibroblasts. This tissue is found in the attachments between bones of the spinal column (ligamenta flava). It is also in the layers within the walls of certain hollow internal organs, including the larger arteries; some portions of the heart; and the larger airways, where it imparts an elastic quality (**fig. 5.23**).

PRACTICE

15 Differentiate between areolar connective tissue and dense connective tissue.

16 What are the functions of adipose connective tissue?

17 Distinguish between reticular and elastic connective tissues.

Cartilage

Cartilage (kar′ti-lij) is a rigid connective tissue. It provides support, frameworks, and attachments; protects underlying tissues; and forms structural models for many developing bones.

Cartilage extracellular matrix is abundant and is largely composed of collagen fibers embedded in a gel-like ground substance. This ground substance is rich in a protein-polysaccharide

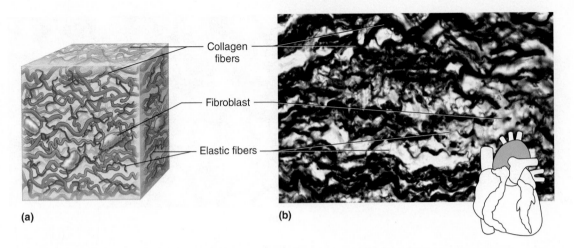

FIGURE 5.23 AP|R Elastic connective tissue contains many elastic fibers with collagen fibers between them. (*a*) Idealized representation of elastic connective tissue. (*b*) Micrograph of a section through elastic connective tissue (160×).

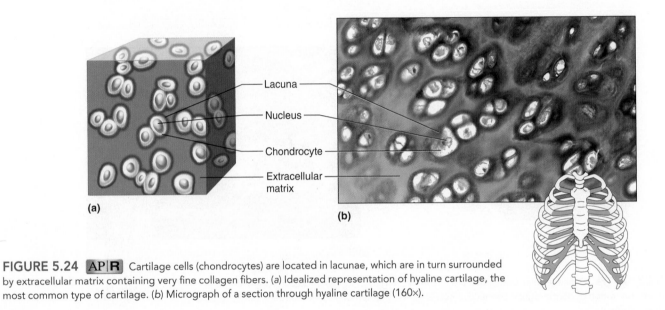

FIGURE 5.24 AP|R Cartilage cells (chondrocytes) are located in lacunae, which are in turn surrounded by extracellular matrix containing very fine collagen fibers. (*a*) Idealized representation of hyaline cartilage, the most common type of cartilage. (*b*) Micrograph of a section through hyaline cartilage (160×).

complex *(chondromucoprotein)* and contains a large volume of water. Cartilage cells, called **chondrocytes** (kon′dro-sītz), occupy small chambers called *lacunae* and lie completely within the extracellular matrix.

A cartilaginous structure is enclosed in a covering of connective tissue called *perichondrium.* Although cartilage tissue lacks a direct blood supply, blood vessels are in the surrounding perichondrium. Cartilage cells near the perichondrium obtain nutrients from these vessels by diffusion, aided by the water in the extracellular matrix. This lack of a direct blood supply is why torn cartilage heals slowly and why chondrocytes do not divide frequently. The three types of cartilage are distinguished by their types of extracellular matrix.

Hyaline cartilage (fig. 5.24), the most common type, has very fine collagen fibers in its extracellular matrix and looks somewhat like white glass. It is found on the ends of bones in many joints, in the soft part of the nose, and in the supporting rings of the respiratory passages. Parts of an embryo's skeleton begin as hyaline cartilage "models" that bone gradually replaces. Hyaline cartilage is

also important in the development and growth of most bones (see chapter 7, p. 204).

Elastic cartilage (fig. 5.25) is more flexible than hyaline cartilage because its extracellular matrix has a dense network of elastic fibers. It provides the framework for the external ears and parts of the larynx.

Fibrocartilage (fig. 5.26), a very tough tissue, has many collagen fibers. It is a shock absorber for structures subjected to pressure. For example, fibrocartilage forms pads (intervertebral discs) between the individual bones (vertebrae) of the spinal column. It also cushions bones in the knees and in the pelvic girdle.

Bone

Bone (osseous tissue) is the most rigid connective tissue. Its hardness is largely due to mineral salts, such as calcium phosphate and calcium carbonate, between cells. This extracellular matrix also contains abundant collagen fibers, which are flexible and reinforce the mineral components of bone.

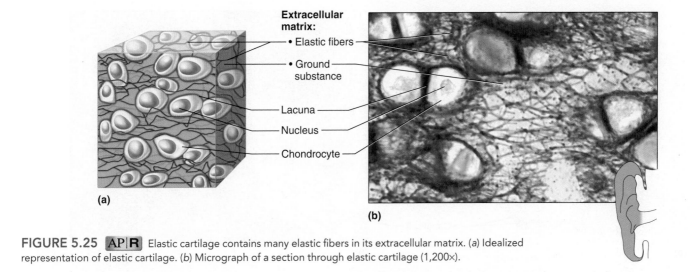

Extracellular matrix:
- Elastic fibers
- Ground substance

Lacuna

Nucleus

Chondrocyte

(a)

(b)

FIGURE 5.25 AP|R Elastic cartilage contains many elastic fibers in its extracellular matrix. (*a*) Idealized representation of elastic cartilage. (*b*) Micrograph of a section through elastic cartilage (1,200×).

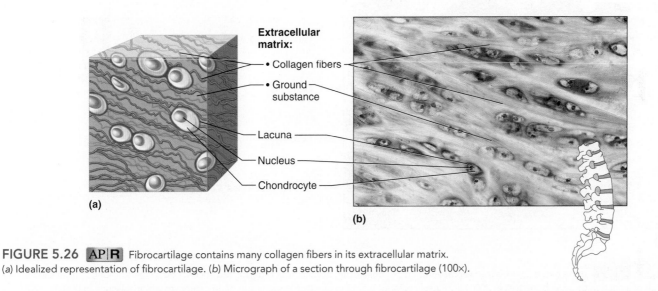

Extracellular matrix:
- Collagen fibers
- Ground substance

Lacuna

Nucleus

Chondrocyte

(a)

(b)

FIGURE 5.26 AP|R Fibrocartilage contains many collagen fibers in its extracellular matrix. (*a*) Idealized representation of fibrocartilage. (*b*) Micrograph of a section through fibrocartilage (100×).

Bone internally supports body structures. It protects vital structures in the cranial and thoracic cavities and is an attachment for muscles. Bone also contains red marrow, which forms blood cells. It stores and releases inorganic chemicals such as calcium and phosphorus.

In compact bone, cells called *osteoblasts* deposit bony matrix in thin layers called *lamellae,* which form concentric patterns around longitudinal tubes called *central,* or *Haversian, canals* which contain capillaries. Once osteoblasts are in lacunae surrounded by matrix, they are called *osteocytes* and are rather evenly spaced within the lamellae (fig. 5.27).

The osteocytes and layers of extracellular matrix, concentrically clustered around a central canal, form a cylinder-shaped unit called an *osteon,* also known as a Haversian system. Many of these units cemented together make up the more solid appearing compact bone that forms the outer portion of a bone (see chapter 7, p. 201).

Each central canal contains blood vessels, so every bone cell is fairly close to a nutrient supply. In addition, the bone cells have

many cytoplasmic processes that extend outward and pass through tiny tubes in the extracellular matrix called *canaliculi.* Gap junctions attach these cellular processes to the processes of nearby cells. As a result, materials can move rapidly between blood vessels and bone cells. Thus, despite its inert appearance, bone is an active tissue. Injured bone heals much more rapidly than does injured cartilage.

The interior portion of a bone is composed of spongy bone in which bone matrix is deposited around osteocytes, forming bony plates with spaces between them. These spaces lighten the weight of the bone and provide spaces for bone marrow. (The microscopic structure of bone is described in more detail in chapter 7, pp. 201–202.)

Blood

Blood, another type of connective tissue, is composed of cells suspended in a fluid extracellular matrix called *plasma.* These cells include *red blood cells, white blood cells,* and cellular fragments

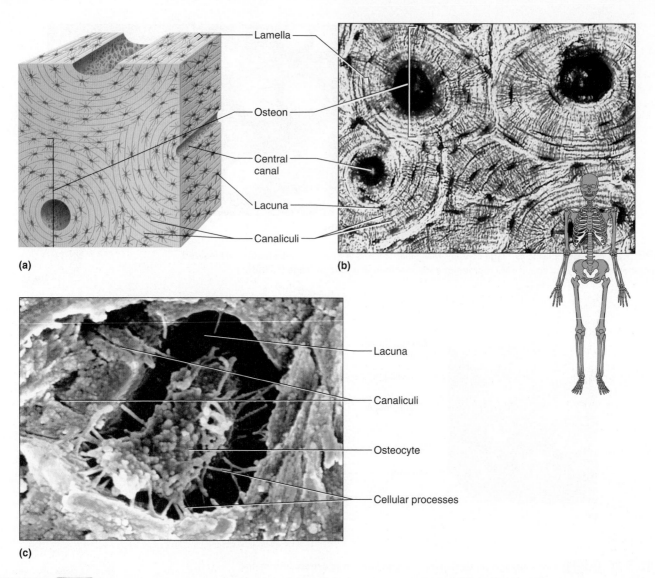

(a)

(b)

(c)

FIGURE 5.27 AP|R Bone tissue. (*a*) Idealized representation of bone tissue. Bone matrix is deposited in concentric layers around central canals. (*b*) Micrograph of a section through bone tissue (200×). (*c*) Falsely colored scanning electron micrograph of an osteocyte within a lacuna (6,000×).

called *platelets* (fig. 5.28). Red blood cells transport gases; white blood cells fight infection; and platelets are involved in blood clotting. Most blood cells form in special tissues (hematopoietic tissues) in red marrow within the hollow parts of certain bones. Chapter 14 describes blood.

Red blood cells are the only type of blood cells that function entirely in the blood vessels. In contrast, white blood cells typically migrate from the blood through capillary walls to connective tissues, where they carry on their major activities until they die. Table 5.7 lists the characteristics of the connective tissues.

 PRACTICE

18 Describe the general characteristics of cartilage.

19 Explain why injured bone heals more rapidly than does injured cartilage.

20 What are the major components of blood?

5.4 | Types of Membranes

After discussing epithelial and connective tissues, sheets of cells called membranes are better understood because they can include both of these tissue types. Many **epithelial membranes** are thin structures that are composed of epithelium and underlying connective tissue. They cover body surfaces and line body cavities. The three major types of epithelial membranes are *serous, mucous,* and *cutaneous.*

Serous (se′rus) **membranes** line the body cavities that do not open to the outside and reduce friction between the organs and cavity walls. They form the inner linings of the thorax and abdomen, and they cover the organs in these cavities (see figs. 1.12 and 1.13, p. 21). A serous membrane consists of a layer of simple squamous epithelium (mesothelium) and a thin layer of areolar connective tissue. Cells of a serous membrane secrete watery *serous fluid,* which helps lubricate membrane surfaces.

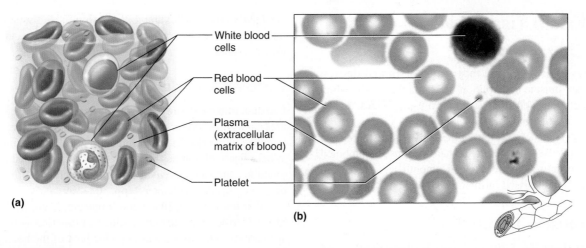

White blood cells

Red blood cells

Plasma (extracellular matrix of blood)

Platelet

(a)

(b)

FIGURE 5.28 **AP|R** Blood tissue consists of red blood cells, white blood cells, and platelets suspended in plasma. (*a*) Idealized representation of blood. (*b*) Micrograph of a sample of blood (1,000×).

Q *What is the consistency of the extracellular matrix of blood?*
Answer can be found in Appendix G.

TABLE 5.7	Connective Tissues		
Type	Description	Function	Location
Areolar connective tissue	Cells in fluid-gel matrix	Binds organs	Beneath the skin, surrounds organs
Adipose connective tissue	Cells in fluid-gel matrix	Protects, insulates, and stores fat	Beneath the skin, around the kidneys, behind the eyeballs, on the surface of the heart
Reticular connective tissue	Cells in fluid-gel matrix	Supports	Walls of liver and spleen
Dense regular connective tissue	Cells in fluid-gel matrix	Binds body parts	Tendons, ligaments
Dense irregular connective tissue	Cells in fluid-gel matrix	Sustains tissue tension	Dermis
Elastic connective tissue	Cells in fluid-gel matrix	Provides elastic quality	Connecting parts of the spinal column, in walls of arteries and airways
Hyaline cartilage	Cells in solid-gel matrix	Supports, protects, provides framework	Ends of bones, nose, and rings in walls of respiratory passages
Elastic cartilage	Cells in solid-gel matrix	Supports, protects, provides flexible framework	Framework of external ear and part of larynx
Fibrocartilage	Cells in solid-gel matrix	Supports, protects, absorbs shock	Between bony parts of spinal column, parts of pelvic girdle, and knee
Bone	Cells in solid matrix	Supports, protects, provides framework	Bones of skeleton, middle ear
Blood	Cells and platelets in fluid matrix	Transports gases, defends against disease, clotting	Throughout the body in a closed system of blood vessels and heart chambers

Mucous (mu'kus) **membranes** line the cavities and tubes that open to the outside of the body. These include the oral and nasal cavities and the tubes of the digestive, respiratory, urinary, and reproductive systems. A mucous membrane consists of epithelium overlying a layer of areolar connective tissue. The type of epithelium varies with the location of the membrane. For example, stratified squamous epithelium lines the oral cavity, pseudostratified columnar epithelium lines part of the nasal cavity, and simple columnar epithelium lines the small intestine. Goblet cells within a mucous membrane secrete *mucus*.

Another epithelial membrane is the **cutaneous** (ku-ta'ne-us) **membrane,** more commonly called *skin*. It is part of the integumentary system described in detail in chapter 6.

A type of membrane composed entirely of connective tissues is a **synovial** (sĭ-no've-al) **membrane.** It lines joints and is discussed further in chapter 8 (p. 271).

 PRACTICE

21 Name the four types of membranes, and explain how they differ.

5.5 | Muscle Tissues AP|R

General Characteristics

Muscle tissues are contractile; they can shorten and thicken. As they contract, muscle cells pull at their attached ends, which moves body parts. The cells that comprise muscle tissues are also called *muscle fibers* because they are elongated.

Approximately 40% by weight of the body is skeletal muscle, and almost another 10% is smooth and cardiac muscle combined. The three types of muscle tissue, skeletal, smooth, and cardiac, are introduced here and discussed further in chapter 9.

Skeletal Muscle Tissue

Skeletal muscle tissue (fig. 5.29) forms muscles that typically attach to bones and can be controlled by conscious effort. For this reason, it is also called *voluntary* muscle tissue. Skeletal muscle cells are long—up to or more than 40 mm in length—and narrow—less than 0.1 mm in width. These threadlike cells have alternating light and dark cross-markings called *striations*. Each cell has many nuclei (multinucleate). For a skeletal muscle cell to contract, it must be stimulated by a nerve cell. Then the muscle cell relaxes when stimulation stops. Skeletal muscles move the head, trunk, and limbs and enable us to make facial expressions, write, talk, and sing, as well as chew, swallow, and breathe.

Smooth Muscle Tissue

Smooth muscle tissue (fig. 5.30) is called smooth because its cells do not have striations. Smooth muscle cells are shorter than those of skeletal muscle and are spindle-shaped, each with a single, centrally located nucleus. This tissue comprises the walls of hollow internal organs, such as the stomach, intestines, urinary bladder, uterus, and blood vessels. Unlike skeletal muscle, smooth muscle usually cannot be stimulated to contract by conscious effort. Thus, its actions are *involuntary*. For example, smooth muscle tissue moves food through the digestive tract, constricts blood vessels, and empties the urinary bladder.

Cardiac Muscle Tissue

Cardiac muscle tissue is only in the heart (fig. 5.31). Its cells, striated and branched, are joined end-to-end, and interconnected in complex networks. Each cardiac muscle cell has a single nucleus. Where one cell touches another cell is a specialized intercellular junction called an *intercalated disc,* seen only in cardiac tissue and discussed further in chapter 9 (pp. 308–309).

Cardiac muscle, like smooth muscle, is controlled involuntarily. Cardiac muscle can continue to function without nervous stimulation. This tissue makes up the bulk of the heart and pumps blood through the heart chambers and into blood vessels.

 PRACTICE

22 List the general characteristics of muscle tissue.

23 Distinguish among skeletal, smooth, and cardiac muscle tissues.

5.6 | Nervous Tissues AP|R

Nervous (ner′vus) **tissues** are found in the brain, spinal cord, and peripheral nerves. The basic cells are called *neurons,* and they are highly specialized. Neurons sense certain types of changes in their surroundings. Incoming signals stimulate cellular processes called *dendrites,* which may cause neurons to respond by conducting electrical impulses along cellular processes called *axons* to other neurons or to muscles or glands (fig. 5.32). As a result of the patterns by which neurons communicate with each other and with muscle and gland cells, they can coordinate, regulate, and integrate many body functions.

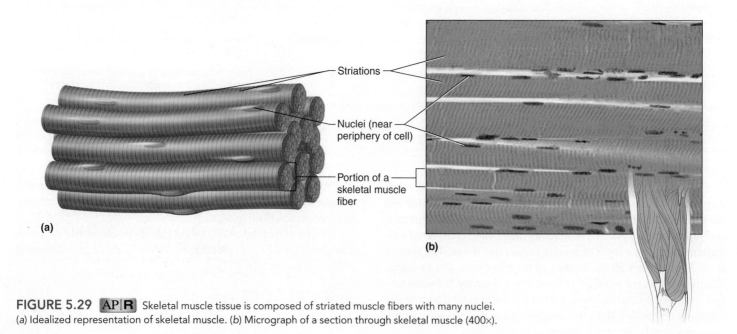

Striations

Nuclei (near periphery of cell)

Portion of a skeletal muscle fiber

(a)

(b)

FIGURE 5.29 AP|R Skeletal muscle tissue is composed of striated muscle fibers with many nuclei. (a) Idealized representation of skeletal muscle. (b) Micrograph of a section through skeletal muscle (400×).

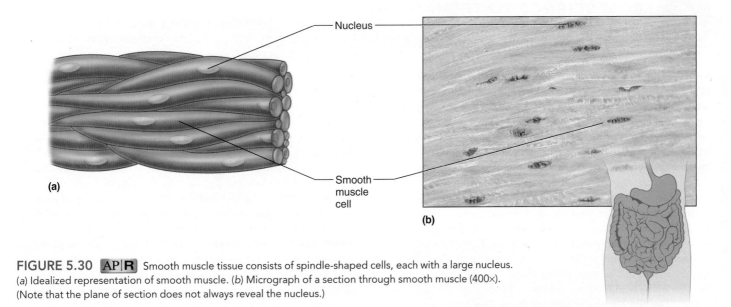

FIGURE 5.30 AP|R Smooth muscle tissue consists of spindle-shaped cells, each with a large nucleus. (a) Idealized representation of smooth muscle. (b) Micrograph of a section through smooth muscle (400×). (Note that the plane of section does not always reveal the nucleus.)

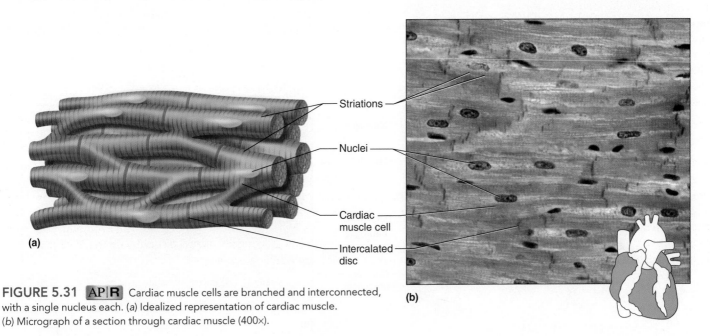

FIGURE 5.31 AP|R Cardiac muscle cells are branched and interconnected, with a single nucleus each. (a) Idealized representation of cardiac muscle. (b) Micrograph of a section through cardiac muscle (400×).

In addition to neurons, nervous tissue includes abundant *neuroglia,* shown in figure 5.32. Neuroglia divide and are crucial to the functioning of neurons. These cells support and bind the components of nervous tissue, carry on phagocytosis, and help supply growth factors and nutrients to neurons by connecting them to blood vessels. They also play a role in cell-to-cell communications. Chapter 10 discusses nervous tissue.

Table 5.8 summarizes the general characteristics of muscle and nervous tissues. From Science to Technology 5.2 discusses tissue engineering, part of a field called regenerative medicine.

 PRACTICE

24 Describe the general characteristics of nervous tissue.

25 Distinguish between neurons and neuroglia. ∎

The cells of different tissues vary greatly in their abilities to divide. Cells that divide continuously include the epithelial cells of the skin and the inner lining of the digestive tract, and the connective tissue progenitor cells that form blood cells in red bone marrow. However, skeletal and cardiac muscle cells and nerve cells do not usually divide at all after differentiating.

Fibroblasts respond rapidly to injuries by increasing in number and fiber production. They help to repair tissues that have limited abilities to regenerate. For instance, fibroblasts form scar tissue where a heart attack occurs. Many organs include pockets of stem or progenitor cells that can divide and replace damaged, differentiated cells, under certain conditions.

If an appliance part is damaged or fails, replacing it is simple. Not so for the human body. Donor organs and tissues for transplant are in short supply, so in the future spare parts may come from tissue engineering. In this technology, a patient's cells, extracellular matrix, and other biochemicals are grown with a synthetic scaffold to form an implant. The cells come from the patient, so the immune system does not reject them. Tissue engineering has provided skin, cartilage, bone, and blood vessels. Combining engineered tissues into structures that can replace organs is where the creativity comes in. Consider the replacement bladder.

Each year in the United States, about 10,000 people need their urinary bladders repaired or replaced. Typically a urologic surgeon replaces part of the bladder with part of the large intestine. However, the function of the intestine is to absorb, and the function of the bladder is to hold waste. Tissue engineering is providing a better replacement bladder. The natural organ is balloon-

like, with layers of smooth muscle, connective tissue, and a lining called urothelium.

Researchers pioneered replacement bladders in children who have birth defects in which the malfunctioning bladder can harm the kidneys. Each patient donated a postage-stamp-size sample of bladder tissue that consisted of about a million cells, from which the researchers separated two types of progenitor cells—for smooth muscle and urothelium—and let them divide in culture in a specific "cocktail" of growth factors. Within seven weeks the million cells had divided to yield 1.5 billion cells, which were seeded onto domes made of a synthetic material. The confluent layers of cells that formed were attached to the lower portions of the patients' bladders, after removing the upper portions. The scaffolds degenerated over time, leaving new bladders built from the patients' own cells. Today tissue-engineered bladders are also used in adults whose bladders have been removed to treat cancer.

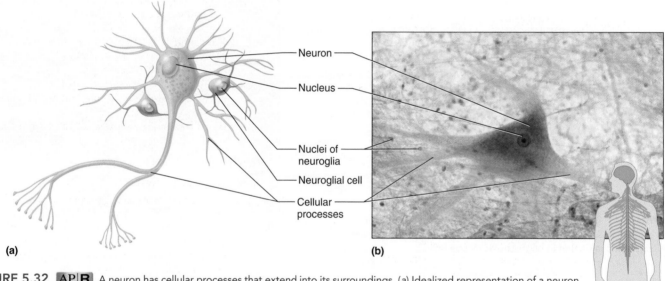

(a)   (b)

FIGURE 5.32 AP|R A neuron has cellular processes that extend into its surroundings. (*a*) Idealized representation of a neuron and neuroglia. (*b*) Micrograph of a section through nervous tissue (350×). Note that only the nuclei of the neuroglia are stained.

TABLE 5.8	Muscle and Nervous Tissues		
Type	**Description**	**Function**	**Location**
Skeletal muscle tissue	Long, threadlike cells, striated, many nuclei	Voluntary movements of skeletal parts	Muscles usually attached to bones
Smooth muscle tissue	Shorter cells, single, central nucleus	Involuntary movements of internal organs	Walls of hollow internal organs
Cardiac muscle tissue	Branched cells, striated, single nucleus	Heart movements	Heart muscle
Nervous tissue	Cell with cytoplasmic extensions	Sensory reception and conduction of electrical impulses	Brain, spinal cord, and peripheral nerves

Chapter Summary

5.1 Cells Are Organized Into Tissues (page 149)

1. Cells are organized in layers or groups to form **tissues.**
2. Specialized **intercellular junctions** (tight junctions, desmosomes, and gap junctions) connect cells.
3. The four major types of human tissue are epithelial, connective, muscle, and nervous.

5.2 Epithelial Tissues (page 150)

1. General characteristics
   a. Epithelial tissue covers all free body surfaces, forms the inner lining of body cavities, lines hollow organs, and is the major tissue of glands.
   b. A **basement membrane** anchors epithelium to connective tissue. Epithelial tissue lacks blood vessels, has cells that are tightly packed, and is continuously replaced.
   c. It protects, secretes, absorbs, and excretes.
   d. Epithelial tissues are classified according to cell shape and the number of cell layers.
2. **Simple squamous epithelium**
   a. This tissue consists of a single layer of thin, flattened cells through which substances pass easily.
   b. It functions in the gas exchange in the lungs and lines blood vessels, lymph vessels, and is part of the membranes lining body cavities and covering viscera.
3. **Simple cuboidal epithelium**
   a. This tissue consists of a single layer of cube-shaped cells.
   b. It carries on secretion and absorption in the kidneys and various glands.
4. **Simple columnar epithelium**
   a. This tissue is composed of elongated cells whose nuclei are near the basement membrane.
   b. It lines the uterus and digestive tract, where it functions in protection, secretion, and absorption.
   c. Absorbing cells often possess microvilli.
   d. This tissue has goblet cells that secrete mucus.
5. **Pseudostratified columnar epithelium**
   a. This tissue appears stratified because the nuclei are at two or more levels.
   b. Its cells may have cilia that move mucus over the surface of the tissue.
   c. It lines the respiratory passages.
6. **Stratified squamous epithelium**
   a. This tissue is composed of many layers of cells; the top layers are flattened.
   b. It protects underlying cells from harmful environmental effects.
   c. It is the superficial portion of the skin and lines the oral cavity, esophagus, vagina, and anal canal.
7. **Stratified cuboidal epithelium**
   a. This tissue is composed of two or three layers of cube-shaped cells.
   b. It lines the ducts of the mammary glands, sweat glands, salivary glands, and pancreas.
   c. It functions in protection.
8. **Stratified columnar epithelium**
   a. The cells in the top layer of this tissue are column-shaped. Cube-shaped cells make up the bottom layers.
   b. It is in part of the male urethra and lining of the larger ducts of exocrine glands.
   c. This tissue functions in protection and secretion.
9. **Transitional epithelium**
   a. This tissue is specialized to stretch.
   b. It lines the urinary bladder, ureters, and superior urethra.
   c. It helps prevent the contents of the urinary passageways from diffusing out.
10. **Glandular epithelium**
   a. Glandular epithelium is composed of cells specialized to secrete substances.
   b. A gland consists of one or more cells.
      (1) **Exocrine glands** secrete into ducts.
      (2) **Endocrine glands** secrete into tissue fluid or blood.
   c. Exocrine glands are classified according to the organization of their cells.
      (1) Simple glands have ducts that do not branch before reaching the secretory portion.
      (2) Compound glands have ducts that branch repeatedly before the secretory portion.
      (3) Tubular glands consist of epithelium-lined tubes.
      (4) Alveolar glands consist of saclike dilations connected to the surface by narrowed ducts.
   d. Exocrine glands are classified according to the composition of their secretions.
      (1) **Merocrine glands** secrete watery fluids without loss of cytoplasm. Most secretory cells are merocrine.
         (a) Serous cells secrete watery fluid.
         (b) Mucous cells secrete **mucus.**
      (2) **Apocrine glands** lose portions of their cells during secretion.
      (3) **Holocrine glands** release cells filled with secretions.

5.3 Connective Tissues (page 159)

1. General characteristics
   a. Connective tissue connects, supports, protects, provides frameworks, fills spaces, stores fat, produces blood cells, protects against infection, and helps repair damaged tissues.
   b. Connective tissue cells usually have considerable **extracellular matrix** between them.
   c. This extracellular matrix consists of fibers and a ground substance.
2. Major cell types
   a. **Fibroblasts** produce collagen and elastic fibers.
   b. **Macrophages** are phagocytes.
   c. **Mast cell**s may release heparin and histamine.
3. Connective tissue fibers
   a. **Collagen** fibers have great tensile strength.
   b. Elastic fibers are composed of **elastin** and are stretchy.
   c. Reticular fibers are fine collagen fibers.
4. Categories of connective tissues
   a. Connective tissue proper includes loose connective tissue (areolar, adipose, reticular) and dense connective tissue (dense regular, dense irregular, elastic).
   b. Specialized connective tissues include cartilage, bone, and blood.
5. **Areolar connective tissue**
   a. Areolar connective tissue forms thin membranes between organs and binds them.

b. It is beneath the skin and surrounds organs.

6. **Adipose tissue**

   a. Adipose tissue is a specialized form of connective tissue that stores fat, cushions, and insulates.

   b. It is found beneath the skin; in certain abdominal membranes; and around the kidneys, heart, and various joints.

7. **Reticular connective tissue**

   a. Reticular connective tissue largely consists of thin, branched reticular fibers.

   b. It supports the walls of the liver and spleen.

8. **Dense regular connective tissue**

   Dense regular connective tissue is largely composed of strong, collagen fibers that bind structures as parts of tendons and ligaments.

9. **Dense irregular connective tissue**

   Dense irregular connective tissue has thicker, randomly distributed collagen fibers and is found in the dermis.

10. **Elastic connective tissue**

    Elastic connective tissue is mainly composed of elastic fibers and imparts an elastic quality to the walls of certain hollow internal organs such as the lungs and blood vessels.

11. **Cartilage**

    a. Cartilage provides a supportive framework for various structures.

    b. Its extracellular matrix is composed of fibers and a gel-like ground substance.

    c. It lacks a direct blood supply and is slow to heal.

    d. Most cartilaginous structures are enclosed in a perichondrium, which contains blood vessels.

    e. Major types are **hyaline cartilage, elastic cartilage, and fibrocartilage.**

    f. Cartilage is at the ends of various bones; in the ear; in the larynx; and in the pads between the bones of the spinal column, pelvic girdle, and knees.

12. **Bone**

    a. The extracellular matrix of bone contains mineral salts and collagen.

    b. The cells of compact bone are arranged in a bony matrix of concentric circles around central canals, whereas the cells of spongy bone are embedded in bony plates with spaces between the plates. Canaliculi connect the cells.

    c. It is an active tissue that heals rapidly.

13. **Blood**

    a. Blood is composed of cells suspended in fluid.

    b. Blood cells are formed by special tissue in the hollow parts of certain bones.

5.4 Types of Membranes (page 168)

1. Epithelial membranes

   a. **Serous membranes**

      (1) Serous membranes line body cavities that do not open to the outside and cover the organs in these cavities.

      (2) They are composed of epithelium and areolar connective tissue.

      (3) Cells of serous membranes secrete watery serous fluid that lubricates membrane surfaces.

   b. **Mucous membranes**

      (1) Mucous membranes line cavities and tubes opening to the outside of the body.

      (2) They are composed of epithelium and areolar connective tissue.

      (3) Cells of mucous membranes secrete mucus.

   c. The **cutaneous membrane** is the external body covering commonly called skin.

2. **Synovial membranes** are composed of connective tissue only, and line joints.

5.5 Muscle Tissues (page 170)

1. General characteristics

   a. Muscle tissue contracts, moving structures attached to it.

   b. Muscle cells are also called muscle fibers.

   c. Three types are skeletal, smooth, and cardiac muscle tissues.

2. **Skeletal muscle tissue**

   a. Muscles containing this tissue usually attach to bones and can be controlled by conscious effort.

   b. Muscle cells are long and threadlike, containing several nuclei, with alternating light and dark cross-markings (striations).

   c. A muscle cell contracts when stimulated by a nerve cell, then relaxes when no longer stimulated.

3. **Smooth muscle tissue**

   a. This tissue of spindle-shaped cells, each with one nucleus, is in the walls of hollow internal organs.

   b. It is involuntarily controlled.

4. **Cardiac muscle tissue**

   a. This tissue is found only in the heart.

   b. Striated cells, each with a single nucleus, are joined by intercalated discs and form branched networks.

   c. Cardiac muscle tissue is involuntarily controlled.

5.6 Nervous Tissues (page 170)

1. **Nervous tissue** is in the brain, spinal cord, and peripheral nerves.

2. Neurons

   a. Neurons sense changes and respond by conducting electrical impulses to other neurons or to muscles or glands.

   b. They coordinate, regulate, and integrate body activities.

3. Neuroglia

   a. Some of these cells bind and support nervous tissue.

   b. Others carry on phagocytosis.

   c. Still others connect neurons to blood vessels.

   d. Some are involved in cell-to-cell communication.

CHAPTER ASSESSMENTS

5.1 Cells Are Organized Into Tissues

1 Define *tissue*. (p. 149)

2 Describe three types of intercellular junctions. (p. 149)

3 Which of the following is a major tissue type in the body? (p. 149)
   a. epithelial
   b. connective
   c. muscle
   d. all of the above

5.2 Epithelial Tissues

4 A general characteristic of epithelial tissues is that _____. (p. 151)
   a. numerous blood vessels are present
   b. cells are spaced apart
   c. cells divide rapidly
   d. there is much extracellular matrix between cells

5 Distinguish between simple epithelium and stratified epithelium. (p. 152)

6 Explain how the structure of simple squamous epithelium provides its function. (p. 152)

7 Match the epithelial tissue on the left to an organ in which the tissue is found. (pp. 152–156)

   (1) simple squamous epithelium   A. lining of intestines
   (2) simple cuboidal epithelium   B. lining of ducts of mammary glands
   (3) simple columnar epithelium   C. lining of urinary bladder
   (4) pseudostratified columnar epithelium   D. salivary glands
   (5) stratified squamous epithelium   E. air sacs of lungs
   (6) stratified cuboidal epithelium   F. respiratory passages
   (7) stratified columnar epithelium   G. part of male urethra
   (8) transitional epithelium   H. lining of kidney tubules
   (9) glandular epithelium   I. superficial portion of the skin

8 Distinguish between an exocrine gland and an endocrine gland. (p. 156)

9 Describe how glands are classified according to the structure of their ducts and the organization of their cells. (p. 156)

10 A gland that secretes substances by exocytosis is a(n) _____ gland. (p.156)
   a. merocrine
   b. apocrine
   c. holocrine
   d. mammary

5.3 Connective Tissues

11 Discuss the general characteristics of connective tissue. (pp. 159–160)

12 Define *extracellular matrix* and *ground substance*. (p. 159)

13 Describe three major types of connective tissue cells. (p. 160)

14 Thick fibers that have great tensile strength and are flexible, but only slightly elastic, are_____ fibers. (p. 162)
   a. reticular
   b. elastic
   c. collagen
   d. mucin

15 Describe areolar connective tissue, and indicate where it is found in the body. (p. 163)

16 Explain how the amount of adipose tissue in the body reflects diet. (p. 163)

17 Contrast dense regular and dense irregular connective tissues. (pp. 164–165)

18 Explain why injured dense regular connective tissue and cartilage are usually slow to heal. (pp. 164–166)

19 Distinguish between reticular and elastic connective tissues. (pp. 164–165)

20 Name the major types of cartilage, and describe their differences and similarities. (p. 166)

21 Describe how bone cells are organized in bone tissue. (p. 167)

22 Explain how bone cells receive nutrients. (p. 167)

23 The fluid extracellular matrix of blood is called _____. (p. 167)
   a. mucus
   b. serous fluid
   c. synovial fluid
   d. plasma

5.4 Types of Membranes

24 Contrast the structure of epithelial membranes to that of synovial membranes. (pp. 168–169)

25 Identify locations in the body of the four types of membranes. (pp. 168–169)

5.5 Muscle Tissues

26 Describe the general characteristics of muscle tissues. (p. 170)

27 Compare and contrast skeletal, smooth, and cardiac muscle tissues in terms of location, cell appearance, and control. (p. 170)

5.6 Nervous Tissues

28 Describe the general characteristics of nervous tissue. (p. 170)

29 Distinguish between the functions of neurons and neuroglia. (pp. 170–171)

INTEGRATIVE ASSESSMENTS/CRITICAL THINKING

Outcomes 3.2, 3.6, 5.1, 5.2, 5.3, 5.5, 5.6

1. Tissue engineering combines living cells with synthetic materials to create functional substitutes for human tissues. What components would you use to engineer replacement for (a) skin, (b) bone, (c) muscle, and (d) blood?

Outcomes 3.2, 3.5, 5.2

2. In the lungs of smokers, a process called metaplasia occurs where normal lining cells of the lung are replaced by squamous metaplastic cells (many layers of squamous epithelial cells). Functionally, why is this an undesirable body reaction to tobacco smoke?

Outcomes 3.4, 3.5, 5.2, 5.3, 5.5, 5.6

3. Cancer-causing agents (carcinogens) usually act on dividing cells. Which of the four tissues would carcinogens most influence? Least influence?

Outcomes 5.2, 5.4

4. Mucous cells may secrete excess mucus in response to irritants. What symptoms might this produce in the (a) digestive tract or (b) respiratory passageway?

Outcome 5.3

5. Disorders of collagen are characterized by deterioration of connective tissues. Why would such diseases produce widely varying symptoms?

6. Collagen and elastin are added to many beauty products. What type of tissues are they normally part of?

7. Joints such as the shoulder, elbow, and knee contain considerable amounts of cartilage and dense regular connective tissue. How does this composition explain that joint injuries are often slow to heal?

8. Answer the following questions with respect to the micrograph below (80×). (a) Identify the organ depicted. (b) What type of tissue is depicted (green arrow, yellow arrow)? (c) To what cell does the arrow point (red arrow, black arrow)?

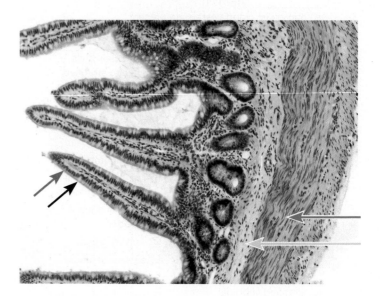

 ## ONLINE STUDY TOOLS

connect (plus+) |ANATOMY & PHYSIOLOGY | LEARNSMART | Anatomy & Physiology REVEALED aprevealed.com

Connect Interactive Questions Reinforce your knowledge using assigned interactive questions covering types and characteristics of tissues and glands.

Connect Integrated Activity Could you help a pathologist identify the anatomical source of a biopsied tissue sample?

LearnSmart Discover which chapter concepts you have mastered and which require more attention. This adaptive learning tool is personalized, proven, and preferred.

Anatomy & Physiology Revealed Go more in depth into the human body by using a virtual microscope to view a variety of tissues that are labeled.

Integumentary System

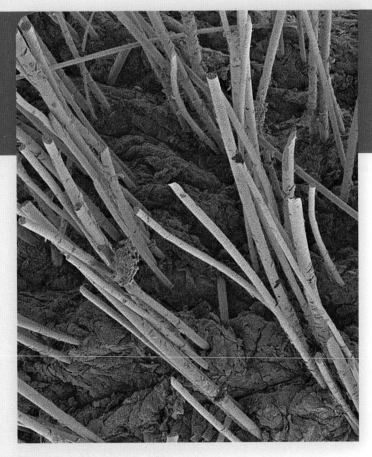

Scanning electron micrograph (SEM) of hair (22×). The light, bristlelike structures are hairs, and the pink at the base of the hairs is skin.

THE WHOLE PICTURE

Two or more types of tissues structurally connected and performing shared specialized functions constitute an organ. The skin, the largest organ in the body by weight, and its various accessory structures (hair, nails, sensory receptors, and glands) make up the integumentary system. Skin is a strong yet flexible covering of our bodies.

 LEARNING OUTCOMES

After you have studied this chapter, you should be able to:

6.1 Skin and Its Tissues

1 Describe the structure of the layers of the skin. (pp. 178–184)

2 Summarize the factors that determine skin color. (p. 183)

6.2 Accessory Structures of the Skin

3 Describe the accessory structures associated with the skin. (pp. 184–187)

4 Explain the functions of each accessory structure of the skin. (pp. 184–187)

6.3 Skin Functions

5 List various skin functions, and explain how the skin helps regulate body temperature. (p. 189)

6.4 Healing of Wounds and Burns

6 Describe wound healing. (p. 190)

7 Distinguish among the types of burns, including a description of healing with each type. (p. 192)

6.5 Life-Span Changes

8 Summarize life-span changes in the integumentary system. (pp. 192–195)

alb-, white: *albinism*—lack of pigment in skin, hair, and eyes.

cut-, skin: sub*cut*aneous—beneath the skin.

derm-, skin: *derm*is—inner layer of the skin.

epi-, upon, after, in addition: *epi*dermis—outer layer of the skin.

follic-, small bag: hair *follic*le—tubelike depression in which a hair develops.

hol-, entire, whole: *hol*ocrine gland—gland that discharges the entire cell containing the secretion.

kerat-, horn: *kerat*in—protein produced as epidermal cells die and harden.

melan-, black: *melan*in—dark pigment produced by certain cells.

por-, passage, channel: *por*e—opening by which a sweat gland communicates to the skin's surface.

seb-, grease: *seb*aceous gland—gland that secretes an oily substance.

sudor-, (sweat) *sudor*iferous glands: Exocrine glands that secrete sweat.

6.1 | Skin and Its Tissues

The skin is composed of several types of tissues (**fig. 6.1**). Also known as the cutaneous membrane, the skin includes two distinct layers: epithelial tissue overlying connective tissue. The outer layer, called the **epidermis** (ep″i-der′mis), is composed of stratified squamous epithelium. The inner layer, or **dermis** (der′mis), is thicker than the epidermis and is made up of connective tissue containing collagen and elastic fibers, smooth muscle tissue, nervous tissue, and blood. A *basement membrane* anchored to the dermis by short fibrils separates the two skin layers.

Beneath the dermis, masses of areolar and adipose tissues bind the skin to underlying organs. These tissues are not part of the skin. They form the **subcutaneous** (sub″ku-ta′ne-us) **layer,** or hypodermis (**fig. 6.2**). The collagen and elastic fibers of this layer are continuous with those of the dermis. Most of these fibers run parallel to the surface of the skin, extending in all directions. As a result, no sharp boundary separates the dermis and the subcutaneous layer. The adipose tissue of the subcutaneous layer insulates, helping to conserve body heat. The subcutaneous layer also contains the major blood vessels that supply the skin. Branches of these vessels form a network (rete cutaneum) between the dermis and the subcutaneous layer. They, in turn, give off smaller vessels that supply the dermis above and the underlying adipose tissue.

> *Intradermal injections* are administered into the skin. *Subcutaneous injections* are administered through a hollow needle into the subcutaneous layer beneath the skin. Subcutaneous injections and *intramuscular injections*, administered into muscles, are also called hypodermic injections.
>
> Some substances are introduced through the skin by means of an adhesive transdermal patch that includes a small reservoir containing the drug. The drug passes from the reservoir through a permeable membrane at a known rate. It then diffuses into the epidermis and enters the blood vessels of the dermis. Transdermal patches deliver drugs that protect against motion sickness, alleviate chest pain associated with heart disease, and lower blood pressure. A transdermal patch that delivers nicotine is used to help people stop smoking.

 PRACTICE

1. Name the tissues in the outer and inner layers of the skin.
2. Name the tissues in the subcutaneous layer beneath the skin.
3. What are the functions of the subcutaneous layer?

Epidermis

The epidermis is composed entirely of stratified squamous epithelium, and therefore it lacks blood vessels. Cells in the deepest layer of the epidermis, called the *stratum basale* (stratum germinativum), are near the dermis and are nourished by dermal blood vessels. As the cells (basal cells) of this layer divide and grow, the older epidermal cells (keratinocytes) are pushed away from the dermis toward the skin surface. The farther the cells move, the poorer their nutrient supply becomes, and in time, they die.

FIGURE 6.1 An organ, such as the skin, is composed of several types of tissues (30×).

— Stratified squamous epithelium

— Dense irregular connective tissue

— Adipose tissue

What other tissues, besides those labeled in this figure, are found in the skin?

Answer can be found in Appendix G.

The cell membranes of older skin cells thicken and develop many desmosomes that fasten them to each other (see chapter 5, p. 149). At the same time, these older cells begin to harden, in a process called **keratinization** (ker"ah-tin"i-za'shun). Strands of tough, fibrous, waterproof keratin proteins are synthesized and stored in the cell. As a result, many layers of tough, tightly packed dead cells accumulate in the epidermis, forming an outermost layer called the *stratum corneum.* These dead cells are eventually shed. Rubbing the skin briskly with a towel sheds dead cells.

Cells of the epidermis can die if they cannot receive nutrients from blood vessels in the dermis. This may happen when a bedridden person lies in one position for a prolonged period—the weight of the body pressing against the bed blocks the skin's blood supply. If cells die (necrosis), the tissues break down, and a pressure ulcer (also called a *decubitus ulcer* or *bedsore*) may appear. If the person moves, shear forces and friction from rubbing may damage the tissue further.

Pressure ulcers usually form in the skin overlying bony projections, such as on the hip, heel, elbow, or shoulder. Frequently changing body position or massaging the skin to stimulate blood flow in regions associated with bony prominences can prevent pressure ulcers. For a paralyzed person who cannot feel pressure or respond to it by shifting position, caregivers must turn the body often to prevent pressure ulcers. Motorized beds and other specialized equipment can periodically shift the patient, lowering the risk of developing pressure ulcers. Keeping affected areas of skin clean, eating a healthy diet, and not smoking can minimize the discomfort and damage from pressure ulcers.

The structural organization of the epidermis varies from region to region. It is thickest on the palms of the hands and the soles of the feet, where it may be 0.8–1.4 millimeters (mm) thick. In most areas, only four layers are distinguishable. They are the *stratum basale* (*stratum germinativum,* or basal cell layer), the deepest layer; the *stratum spinosum;* the *stratum granulosum;* and the *stratum corneum,* a fully keratinized outermost layer. An additional layer, the *stratum lucidum* (between the stratum granulosum and the stratum corneum) is in the thickened skin of the palms and soles. The cells of these layers change shape as they are pushed toward the surface (fig. 6.3).

In body regions other than the palms and soles, the epidermis is usually thin, averaging 0.07–0.12 mm. The stratum lucidum may be missing where the epidermis is thin. Table 6.1 describes the characteristics of each layer of the epidermis.

In healthy skin, production of epidermal cells in the stratum basale closely balances loss of dead cells from the stratum corneum. As a result, skin does not completely wear away. The rate of cell division increases where the skin is rubbed or pressed regularly, causing the growth of thickened areas called *calluses* on the palms and soles and keratinized conical masses on the toes called *corns.*

In psoriasis, a chronic skin disease, cells in the epidermis divide seven times more frequently than normal. Excess cells accumulate, forming bright red patches covered with silvery scales, which are keratinized cells. Medications used to treat cancer, such as methotrexate, are used to treat severe cases of psoriasis. Immune suppressing medications, such as topical corticosteroids, and drugs that block the actions of an inflammatory protein called tumor necrosis factor, are used for treatment of chronic psoriasis. About 7.5 million people in the United States (about 2% of the population) and 125 million people worldwide (about 2-3%) have psoriasis. People with dark skin are less likely to develop the condition.

CAREER CORNER
Burn Technician

The young woman was fleeing a burning house when she noticed a cat trapped inside, scratching frantically against a basement window. The woman grabbed a rock, smashed the window, and reached in for the cat, just as flame whooshed from the side, scorching her arms. First responders soon arrived, and within minutes they had transported the woman to a nearby medical center that, fortunately, had a burn unit.

At the emergency department, in a sterile environment, a physician quickly determined that the burns were full thickness, destroying both skin layers. A burn technician assisted the physician in removing dead skin and other tissue (debridement), and then gently cleaned the areas and applied sterile cloths. Over the following few days, the burn technician changed dressings, debrided and drained the affected areas, and helped the patient with hydrotherapy, which is a shower where the patient sits as warm water flows over the burns to aid healing and lower the risk of infection.

As an integral part of the burn treatment team, the burn technician assists with procedures, including grafting healthy skin from the patient to cover the burned areas and escharotomy, which is making a cut to release the pressure from accumulating fluid in surrounding areas where damaged skin has lost its elasticity. Without the cut, blood supply can be blocked.

The burn technician changes a patient's dressings several times a day, monitors the graft site and the areas of donor skin, and applies antimicrobial ointments to the affected and surrounding areas. Careful record-keeping and being alert to complications are important parts of the job.

Requirements include a high school diploma, basic cardiac life support certification, and certification as a nursing assistant or more advanced nursing training. Many burn technicians are nursing students or nurses. The job requires the ability to stand for long periods with frequent bending and lifting. A good rapport with people is helpful, because burn patients may be traumatized by their experience.

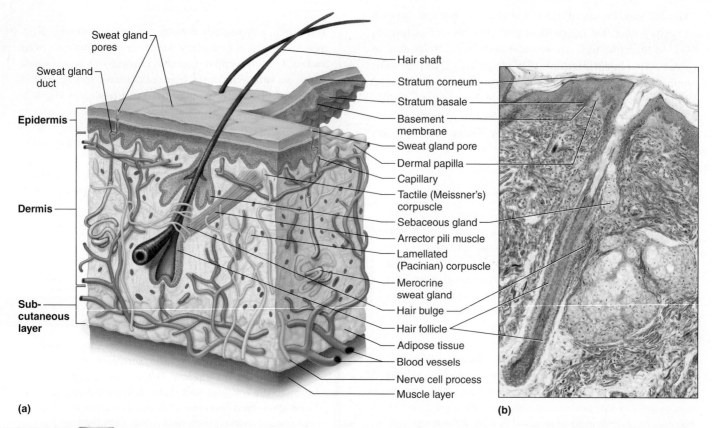

FIGURE 6.2 AP|R Skin. (*a*) The skin (epidermis and dermis) overlays the subcutaneous ("beneath the skin") layer and its underlying muscle. (*b*) A light micrograph depicting the layered structure of the skin (75×).

TABLE 6.1	Layers of the Epidermis	
Layer	**Location**	**Characteristics**
Stratum corneum	Outermost layer	Many layers of keratinized, dead epithelial cells that are flattened and non-nucleated
Stratum lucidum (only present in thick skin)	Between stratum corneum and stratum granulosum on soles and palms	Cells appear clear; nuclei, organelles, and cell membranes are no longer visible
Stratum granulosum	Beneath the stratum corneum	Three to five layers of flattened granular cells that contain shrunken fibers of keratin and shriveled nuclei
Stratum spinosum	Beneath the stratum granulosum	Many layers of cells with centrally located, large, oval nuclei and developing fibers of keratin; cells becoming flattened
Stratum basale (basal cell layer)	Deepest layer	A single row of cuboidal or columnar cells that divide and grow; this layer also includes melanocytes

The epidermis has important protective functions. It shields the moist underlying tissues against excess water loss, mechanical injury, and the effects of harmful chemicals. When intact, the epidermis also keeps out disease-causing microorganisms (pathogens).

 PRACTICE

4 Describe the composition of the epidermis.

5 Distinguish between the stratum basale and the stratum corneum.

6 List the protective functions of the epidermis.

Specialized cells in the epidermis called **melanocytes** produce the pigment **melanin** (mel′ah-nin) from the amino acid tyrosine in organelles called *melanosomes* (**fig. 6.4a**). Melanin provides skin color. Melanin also absorbs ultraviolet radiation in sunlight, which would otherwise cause mutations in the DNA of skin cells and other damaging effects. Clinical Application 6.1 discusses one consequence of excess sun exposure—skin cancer.

Melanocytes lie in the stratum basale of the epidermis. They are the only cells that can produce melanin, but the pigment may also appear in nearby epidermal cells. Melanocytes have long, pigment-containing cellular extensions that pass upward between

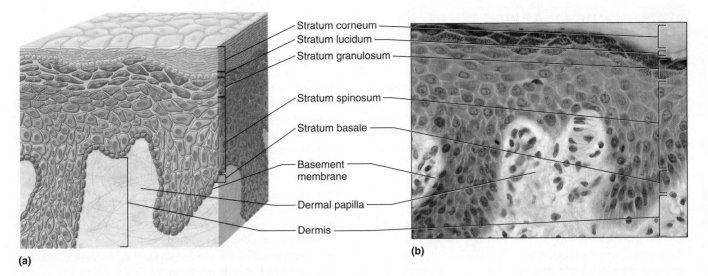

FIGURE 6.3 AP|R Epidermis of thick skin. (*a*) The layers of the epidermis are distinguished by changes in cells as they are pushed toward the surface of the skin. (*b*) Light micrograph of skin (500×).

Where is thick skin found on the body?

Answer can be found in Appendix G.

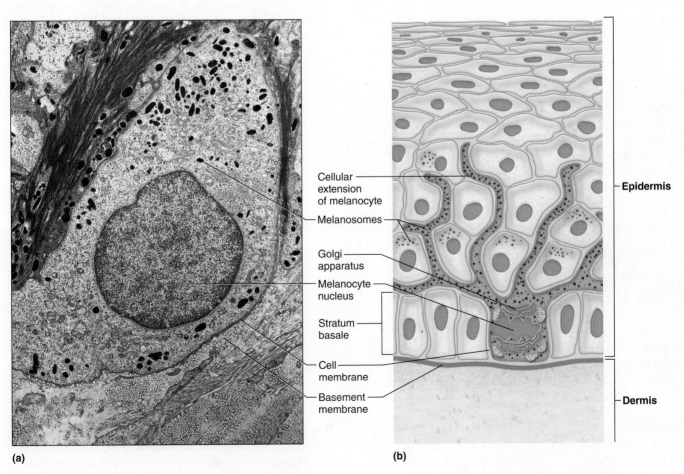

FIGURE 6.4 A melanocyte. (*a*) Transmission electron micrograph of a melanocyte with pigment-containing granules (melanosomes) (4,500×). (*b*) A melanocyte may have pigment-containing extensions that pass between epidermal cells and transfer pigment into them. Much of the melanin is deposited above the nucleus, where the pigment can absorb UV radiation from outside before the DNA is damaged.

A deep, dark tan was once considered desirable. In the 1960s a teenager might have spent hours on a beach, skin glistening with oil, maybe even using a reflecting device to concentrate sun exposure on the face. Today more than 40 million people in the United States use indoor tanning beds to temporarily darken their skin, but at a health cost. Rates of new skin cancer cases, particularly among young women, have skyrocketed since 1992, paralleling the increased use of tanning beds by this group.

Usually DNA repair, discussed in chapter 4 (p. 143), protects against sun overexposure. Solar radiation activates a gene that encodes a protein called p53 that normally mediates harmful effects of environmental insults in various tissues. In the skin, p53 stimulates a series of responses to sunning: keratinocytes produce signaling molecules that promote the redness (erythema) and swelling of inflammation. Meanwhile, melanocytes make more melanin, which enters keratinocytes. Tanning results. Use of tanning booths is dangerous because it bathes the skin in doses of ultraviolet radiation that can overwhelm the skin's natural protection against cancer.

Researchers hypothesize that the tanning response evolved about a million years ago, when our ancestors lived near the equator. Biology may also explain why we like to sunbathe—it stimulates keratinocytes to release beta-endorphin, a molecule related to opiates that promotes a sense of well-being.

Skin cancer begins when the sun (or tanning bed) exposure overwhelms the protection provided by p53. Most skin cancers arise in nonpigmented epithelial cells in the deep layer of the epidermis or from melanocytes. Skin cancers originating from epithelial cells are called *cutaneous carcinomas* (basal cell carcinoma or squamous cell carcinoma); those arising from melanocytes are *cutaneous melanomas* (melanocarcinomas or malignant melanomas) (fig. 6A).

Cutaneous carcinomas are the most common type of skin cancer, affecting mostly light-skinned people over forty years of age regularly exposed to sunlight. It typically develops from a hard, dry, scaly growth with a reddish base. The lesion may be flat or raised and usually firmly adheres to the skin, appearing most often on the neck, face, or scalp. Surgical removal or radiation treatment usually cures these slow-growing cancers.

A cutaneous melanoma is pigmented with melanin, often with a variety of colored areas—variegated brown, black, gray, or blue. A melanoma usually has irregular rather than smooth outlines and may feel bumpy. The "ABCDE" rule provides a checklist for melanoma: A for asymmetry; B for border (irregular); C for color (more than one); D for diameter (more than 6 millimeters); and E for elevation. Melanoma accounts for only 4% of skin cancers but for 80% of skin cancer deaths.

People of any age may develop a cutaneous melanoma. These cancers are caused by short, intermittent exposure to high-intensity sunlight, such as when a person who usually stays indoors occasionally sustains a blistering sunburn.

Light-skinned people who burn rather than tan are at higher risk of developing a cutaneous melanoma. The cancer usually appears in the skin of the trunk, especially the back, or the limbs, arising from normal-appearing skin or from a mole (nevus). The lesion spreads horizontally through the skin, but eventually may thicken and grow downward, invading deeper tissues. Surgical removal during the horizontal growth phase arrests the cancer in six of every seven cases. However, once the lesion thickens and deepens, it becomes more difficult to treat, and the survival rate is low. Melanomas in people who regularly use tanning beds tend to be thicker and more widely spread at the time of diagnosis than other early melanomas.

To reduce risk of developing skin cancer, avoid exposure to high-intensity sunlight in the environment and tanning beds, use sunscreens and sunblocks, and examine the skin regularly. Report any "ABCDE" lesions to a physician.

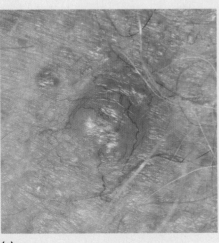
(a)

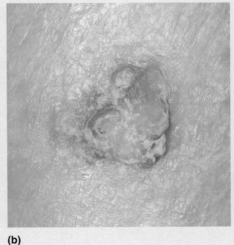
(b)

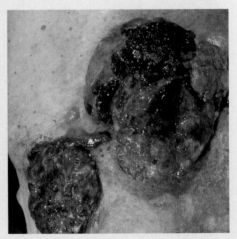
(c)

FIGURE 6A Skin cancer. (*a*) Basal cell carcinoma. (*b*) Squamous cell carcinoma. (*c*) Malignant melanoma.

neighboring epidermal cells (fig. 6.4*b*). These extensions transfer melanin granules in melanosomes into neighboring keratinocytes. The neighboring cells may accumulate more melanin than the melanocytes do.

Skin color is determined by heredity, as well as by environmental and physiological factors. Regardless of racial origin, all people have about the same number of melanocytes in their skin. Differences in skin color result from differences in the amount of melanin these cells produce, which is controlled by several genes. The more melanin, the darker the skin. The distribution and size of pigment granules in melanocytes also influence skin color. The granules in very dark skin are single and large; those in lighter skin aggregate in clusters of two to four smaller granules. People who inherit mutations in melanin genes have a condition called *albinism,* which includes nonpigmented skin. Albinism affects people of all races (fig. 6.5).

Environmental factors such as sunlight, ultraviolet light from sunlamps, and X rays affect skin color. These factors rapidly darken existing melanin, and they stimulate melanocytes to produce more pigment and transfer it to nearby epidermal cells within a few days. Unless exposure to sunlight continues, the tan fades as pigmented keratinocytes that have been pushed toward the skin surface wear away.

Blood in the dermal vessels adds color to the skin. When blood is well oxygenated, the blood pigment hemoglobin is bright red, making the skin of light-complexioned people appear pinkish. When the blood oxygen concentration is low, hemoglobin is dark red, and the skin appears bluish—a condition called *cyanosis.*

The state of the blood vessels also affects skin color. If the vessels are dilated, more blood enters the dermis, reddening the skin of a light-complexioned person. This may happen when a person is overheated, embarrassed, or under the influence of alcohol. Conversely, conditions that constrict blood vessels cause the skin to lose this reddish color. Thus, if body temperature drops abnormally or if a person is frightened, the skin may appear pale.

A yellow-orange plant pigment called *carotene,* found in yellow vegetables, can give skin a yellowish cast if a person consumes too much of it. The skin color results from accumulation of carotene in the adipose tissue of the dermis and subcutaneous layers. Diseases may also affect skin color. A yellowish skin tone can indicate *jaundice,* which is a consequence of liver malfunction.

A newborn who develops the yellowish skin of jaundice shortly after birth may have a blood incompatibility, an underdeveloped liver, or an inborn error of metabolism. An observant British hospital nurse discovered a treatment for newborn jaundice in 1958. When taking her tiny charges out in the sun, she noticed that a child whose skin had a yellow pallor developed normal pigmentation when he lay in sunlight. However, the part of the child's body covered by a diaper and therefore not exposed to the sun remained yellow. Further investigation showed that sunlight enables the body to break down bilirubin, the liver substance that accumulates in the skin. Today, newborns who develop persistently yellowish skin may have to lie under artificial "bili lights" for a few days, clad only in protective goggles.

PRACTICE

7 What is the function of melanin?

8 How do genetic factors influence skin color?

9 Which environmental factors influence skin color?

10 How do physiological factors influence skin color?

Dermis

The boundary between the epidermis and the dermis is typically uneven, because ridges from the epidermis project inward and cone-shaped *dermal papillae* extend from the dermis into the spaces between the ridges (see figs. 6.2 and 6.3). Dermal papillae increase the surface area where epidermal cells receive oxygen and nutrients from dermal capillaries.

Dermal papillae are found in the skin all over the body, but they are most abundant in the hands and feet. The friction ridges that the dermal papillae form leave a patterned impression—a fingerprint— when a finger presses against a surface. Fingerprints are used to identify individuals because they are unique. Genes determine general fingerprint patterns. Subtle, individual variations called papillary minutiae form as the fetus presses the developing ridges against structures within the uterus. No two fetuses move exactly alike, so even the fingerprints of identical twins are not exactly the same.

The dermis binds the epidermis to the underlying tissues. Although no distinct boundary is visible, the dermis has two layers. The upper or *papillary layer* is composed of areolar connective tissue. The lower or *reticular layer* is dense irregular connective tissue that includes tough collagen fibers and elastic fibers in a gel-like ground substance. Networks of these fibers give the skin toughness and elasticity. On average, the dermis is 1.0–2.0 mm thick; however, it may be as thin as 0.5 mm or less, such as on the eyelids, or as thick as 3.0 mm, such as on the soles of the feet.

The dermis also contains muscle fibers. Some regions, such as the skin that encloses the testes (scrotum), contain many smooth

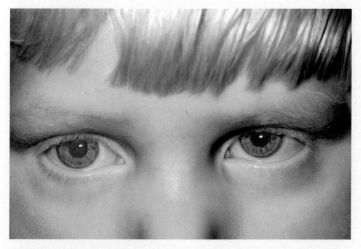

FIGURE 6.5 The pale irises, skin, and hair of a person with albinism reflect lack of melanin. Albinism is inherited.

muscle cells that can wrinkle the skin when they contract. Other smooth muscle in the dermis is associated with accessory structures such as hair follicles and glands. Many skeletal muscle fibers are anchored to the dermis in the skin of the face. They help produce the voluntary movements associated with facial expressions.

Dermal blood vessels supply nutrients to all skin cells. These vessels help regulate body temperature, as explained later in this chapter on page 189.

Nerve cell processes are scattered throughout the dermis. Motor nerve cell processes conduct impulses to dermal muscles and glands, and sensory nerve cell processes conduct impulses away from specialized sensory receptors (see fig. 6.2).

The dermis contains different types of sensory receptors. One type of sensory receptor in the deeper dermis, lamellated (Pacinian) corpuscles, responds to heavy pressure, whereas another type in the upper dermis, tactile (Meissner's) corpuscles, senses light touch and texture. Still other receptors (free nerve endings) extend into the epidermis and respond to temperature changes or to factors that can damage tissues. Sensory receptors are discussed in chapter 12 (p. 446). The dermis also contains accessory structures, including hair follicles, sebaceous (oil-producing) glands, and sweat glands.

> To make a tattoo, needles inject inks into the dermis. The color is permanent, because dermal cells are not shed, unlike cells of the epidermis. To remove a tattoo, a laser shatters the ink molecules, and the immune system removes the resulting debris. Before laser removal became available in the late 1980s, unwanted tattoos were scraped, frozen, or cut away—all painful procedures.

PRACTICE

11 What types of tissues make up the dermis?

12 What are the functions of these tissues?

6.2 | Accessory Structures of the Skin

Accessory structures of the skin originate from the epidermis and include nails, hair follicles, and skin glands. As long as accessory structures remain intact, severely burned or injured dermis can regenerate.

Nails

Nails are protective coverings on the ends of the fingers and toes. Each nail consists of a *nail plate* that overlies a surface of skin called the *nail bed.* Specialized epithelial cells continuous with the epithelium of the skin produce the nail bed. The whitish, thickened, half-moon-shaped region (lunula) at the base of a nail plate is the most active growing region. The epithelial cells here divide, and the newly formed cells become keratinized. This gives rise to tiny, keratinized scales that become part of the nail plate, pushing it forward over the nail bed. In time, the plate extends beyond the end of the nail bed and with normal use gradually wears away (fig. 6.6).

Hair Follicles

Hair is present on all skin surfaces except the palms, soles, lips, nipples, and parts of the external reproductive organs. However, in some places it is not well developed. For example, hair on the forehead is very fine in many people.

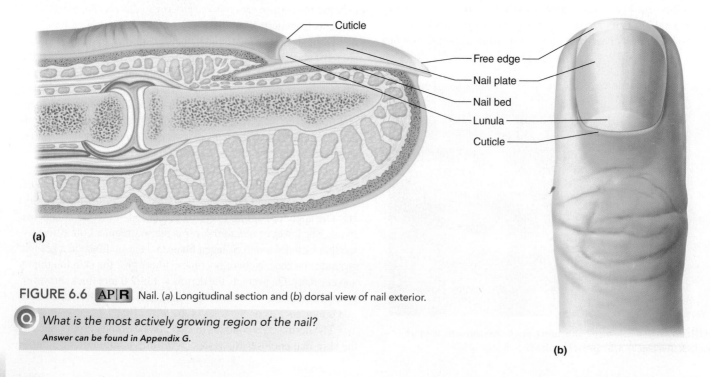

FIGURE 6.6 AP|R Nail. (*a*) Longitudinal section and (*b*) dorsal view of nail exterior.

What is the most actively growing region of the nail?
Answer can be found in Appendix G.

Hairs cycle through phases of active growth and being inactive. At any time, 90% of hair is in the growth phase. Each hair develops from a group of epidermal stem cells at the base of a tubelike depression called a **hair follicle** (hār fol′ĭ-kl). These stem cells originate from a region near the bottom of the hair follicle, known as the *hair bulge,* and migrate downward. The follicle, containing the *hair root,* extends from the surface into the dermis and sometimes into the subcutaneous layer. The epidermal cells in the *hair bulb* at its base are nourished from dermal blood vessels in a projection of connective tissue (hair papilla) at the deep end of the follicle. As these epidermal cells divide and grow, they push older cells toward the surface. The cells that move upward and away from the nutrient supply become keratinized and die. Their remains constitute the structure of a developing *hair shaft* that extends away from the skin surface. In other words, a hair is composed of dead epidermal cells (figs. 6.7 and 6.8). Both hair and epidermal cells develop from the same types of stem cells.

A healthy person loses from twenty to 100 hairs a day as part of the normal growth cycle of hair. A hair typically grows for two to six years, then does not grow for two to three months anchored in its follicle. Later, a new hair begins to grow from the base of the follicle, pushing the old hair outward until it drops off. If hairs shed from the scalp are not replaced, baldness results. Clinical Application 6.2 discusses several types of hair loss.

Genes determine hair color by directing the type and amount of pigment that epidermal melanocytes produce. Dark hair has more of the brownish-black **eumelanin** (u-mel′ah-nin), whereas blonde hair and red hair have more of the reddish-yellow **pheomelanin** (fe″o-mel′ah-nin). The white hair of a person with *albinism* lacks melanin altogether. A mixture of pigmented hairs and unpigmented hairs appears gray.

A bundle of smooth muscle cells, forming the *arrector pili muscle* (see figs. 6.2a and 6.7a), attaches to each hair follicle. When the muscle contracts, a short hair in the follicle stands on end. If a person is emotionally upset or very cold, nervous stimulation causes the arrector pili muscles to contract, producing gooseflesh, or goose bumps. Each hair follicle is also associated with one or more sebaceous (oil-producing) glands.

Skin Glands

Sebaceous glands (se-ba′shus glandz) (see fig. 6.2) contain groups of specialized epithelial cells and are usually associated with hair follicles. They are holocrine glands (see chapter 5, p. 156), and

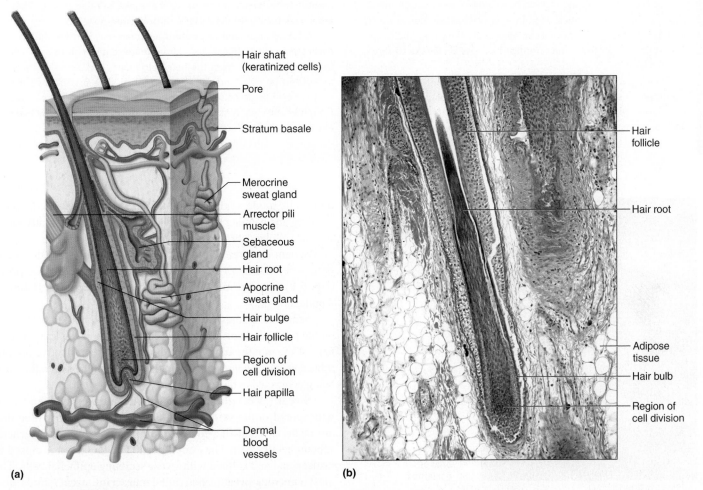

(a)

Hair shaft (keratinized cells)
Pore
Stratum basale
Merocrine sweat gland
Arrector pili muscle
Sebaceous gland
Hair root
Apocrine sweat gland
Hair bulge
Hair follicle
Region of cell division
Hair papilla
Dermal blood vessels

(b)

Hair follicle
Hair root
Adipose tissue
Hair bulb
Region of cell division

FIGURE 6.7 AP|R Hair follicle. (a) A hair grows from the base of a hair follicle when epidermal cells divide and older cells move outward and become keratinized. (b) Light micrograph of a hair follicle (175×).

About 1.4 billion people worldwide are bald. The most common type of baldness in adults is pattern baldness, in which the top of the head loses hair. Pattern baldness affects 40 million men and 21 million women in the United States and is also common elsewhere. The women tend to be past menopause, when lowered amounts of the hormone estrogen contribute to hair loss, which occurs more evenly on the scalp than it does in men. Pattern baldness is called *androgenic alopecia* because it is associated with testosterone, an androgenic (male) hormone. Analysis of scalp samples from the hairless and hairy parts of affected men's heads reveals that hair progenitor cells are lacking in the bald regions, although stem cells from which they descend are present. Abnormal hormone levels that mimic menopause may cause hair loss in young women.

Another type of baldness is *alopecia areata,* in which the body manufactures antibodies that attack the hair follicles. This action results in oval bald spots in mild cases but complete loss of scalp and body hair in severe cases. About 6.5 million people in the United States have alopecia areata.

Temporary hair loss has several causes. Lowered estrogen levels shortly before and after giving birth may cause a woman's hair to fall out in clumps. Taking birth control pills, cough medications, certain antibiotics, vitamin A derivatives, antidepressants, and many other medications can also cause temporary hair loss. A sustained high fever may prompt hair loss six weeks to three months later.

Many people losing their hair seek treatment (fig. 6B). One treatment is minoxidil (Rogaine), a drug originally used to lower high blood pressure. Rogaine causes new hair to grow in 10% to 14% of cases, and in 90% of people, it slows hair loss. However,

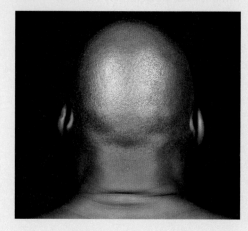

FIGURE 6B
Being bald can be beautiful, but many people with hair loss seek ways to grow hair.

when a person stops taking it, any new hair falls out. Hair transplants move hair follicles from a hairy body part to a bald part. They work. Several other approaches, however, can damage the scalp or lead to infection. These include suturing on hair pieces and implants of high-density artificial fibers. Products called "thinning hair supplements" are ordinary conditioners that make hair feel thicker. They are concoctions of herbs and the carbohydrate polysorbate.

A future approach to treating baldness may harness the ability of stem cells to divide and differentiate to give rise to new hair follicles. Stem cells from the hair bulge can produce hair as well as epidermal cells and sebaceous glands. The first clue to the existence of these cells was that new skin in burn patients arises from hair follicles. Manipulating stem cells could someday treat extreme hairiness (hirsutism) as well as baldness.

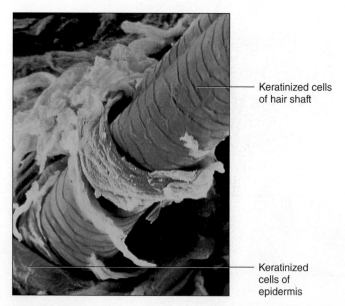

Keratinized cells of hair shaft

Keratinized cells of epidermis

FIGURE 6.8 This scanning electron micrograph shows a hair emerging from its follicle (875×).

their cells produce globules of a fatty material that accumulate, swelling and bursting the cells. The resulting mixture of fatty material and cellular debris is called *sebum.*

Sebum is secreted into hair follicles through short ducts and helps keep the hairs and the skin soft, pliable, and waterproof (fig. 6.9). Acne results from excess sebum secretion (Clinical Application 6.3).

Sebaceous glands are scattered throughout the skin but are not on the palms and soles. In some regions, such as the lips, the corners of the mouth, and parts of the external reproductive organs, sebaceous glands open directly to the surface of the skin rather than being connected to hair follicles.

Sweat (swet) **glands,** also called sudoriferous glands, are widespread in the skin. Each gland consists of a tiny tube that originates as a ball-shaped coil in the deeper dermis or superficial subcutaneous layer. The coiled portion of the gland is closed at its deep end and is lined with sweat-secreting epithelial cells. The most numerous sweat glands called **merocrine sweat glands,** also known as *eccrine sweat glands,* respond throughout life to body temperature elevated by environmental heat or physical exercise

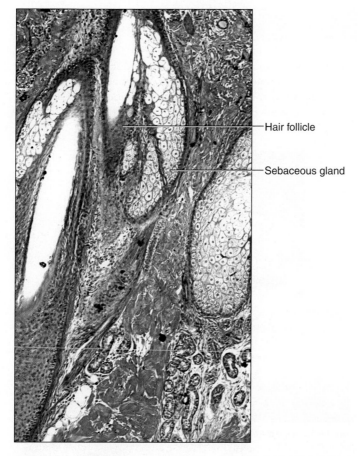

FIGURE 6.9 **AP|R** A sebaceous gland secretes sebum into a hair follicle, shown here in oblique section (200×).

(fig. 6.10). These sweat glands, although located all over the body, are abundant on the forehead, neck, and back, where they produce profuse sweat on hot days or during intense physical activity. They also release the moisture that appears on the palms and soles when a person is emotionally stressed.

A tube (duct) that opens at the surface as a *pore* carries the sweat that the merocrine sweat glands secrete (fig. 6.11). Sweat is mostly water, but it also contains small amounts of salts and wastes such as urea and uric acid. Thus, sweating is also an excretory function.

The secretions of certain sweat glands, called **apocrine** (ap′o-krin) **sweat glands,** develop a scent as skin bacteria metabolize them (see fig. 6.10). Although these glands are called apocrine, they secrete by exocytosis. Apocrine sweat glands become active at puberty and can wet certain areas of the skin when a person is emotionally upset, frightened, or in pain. They are also active during sexual arousal. Apocrine sweat glands are most numerous in axillary regions and the groin. Ducts of these sweat glands open into hair follicles. The secretions of these sweat glands include proteins and lipids, which, when metabolized by bacteria in the skin, produce body odor.

Other sweat glands are structurally and functionally modified to secrete specific fluids, such as the ceruminous glands of the external acoustic meatus that secrete ear wax (see chapter 12, p. 456) and the female mammary glands that secrete milk (see chapter 23, pp. 891–893). Table 6.2 summarizes the types of skin glands.

PRACTICE

13 Describe the structure of the nail bed.

14 Explain how a hair forms.

15 What produces goose bumps?

16 What is the function of the sebaceous glands?

17 Distinguish between merocrine sweat glands and apocrine sweat glands.

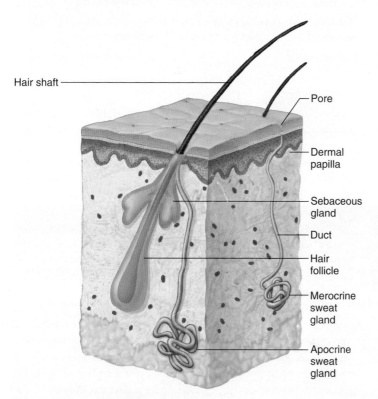

FIGURE 6.10 **AP|R** Note the difference in location of the ducts of the merocrine and apocrine sweat glands.

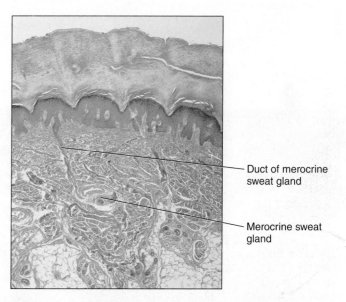

FIGURE 6.11 **AP|R** Light micrograph of the skin showing a merocrine sweat gland with its duct (25×).

Acne

Many young people are all too familiar with *acne vulgaris*, a disorder of the sebaceous glands (fig. 6C). Excess sebum and squamous epithelial cells clog the glands, producing blackheads and whiteheads (collectively comedones). The blackness is not dirt but results from the accumulated cells blocking light. In addition, the clogged sebaceous gland provides an attractive environment for anaerobic bacteria. Their presence signals the immune system to trigger inflammation. The inflamed, raised area is a pimple (pustule).

Acne is a common skin disease, affecting 80% of people at some time between the ages of eleven and thirty. It is usually hormonally induced. Just before puberty, the adrenal glands increase production of androgens, which stimulate increased secretion of sebum. At puberty, sebum production surges again. Acne usually develops because the sebaceous glands are extra responsive to androgens, but in some cases, androgens may be produced in excess.

Sometimes, acne can lead to scarring. Fortunately, several highly effective treatments are available, and dermatologists can tailor combinations of them to help individual patients.

What to Do—And Not Do

Acne is not caused by uncleanliness or eating too much chocolate or greasy food. Although cleansing products containing soaps, detergents, or astringents can remove surface sebum, they do not stop the flow of oil that contributes to acne. Abrasive products are harmful because they irritate the skin and increase inflammation.

Most acne treatments take weeks to months to work. Women with acne are sometimes prescribed certain types of birth control pills because the estrogens counter androgen excess. Isotretinoin is a very effective derivative of vitamin A but has side effects and causes birth defects. Systemic antibiotics can treat acne by clearing bacteria from sebaceous glands. Topical treatments include tretinoin (another vitamin A derivative), salicylic acid (an aspirin solution), and benzoyl peroxide. Exposure to blue light and chemical peels are other acne treatments.

Treatment for severe acne requires a doctor's care. Laser treatments, dermabrasion, or surgery may be used to remove scars caused by severe acne. Drug combinations are tailored to the severity of the condition (table 6A).

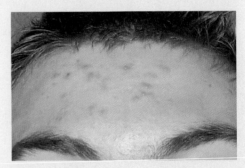

FIGURE 6C Acne is a common skin condition usually associated with a surge of androgen activity.

TABLE 6A	Acne Treatments (by Increasing Severity)
Condition	**Treatment**
Noninflammatory comedonal acne (blackheads and whiteheads)	Topical tretinoin or salicylic acid
Papular inflammatory acne	Topical antibiotic
Widespread blackheads and pustules	Topical tretinoin and systemic antibiotic
Severe cysts	Systemic isotretinoin
Explosive acne (ulcerated lesions, fever, joint pain)	Systemic corticosteroids

TABLE 6.2	Skin Glands		
Type	**Description**	**Function**	**Location**
Sebaceous glands	Groups of specialized epithelial cells	Keep hair soft, pliable, waterproof	Near or connected to hair follicles, everywhere but on palms and soles
Merocrine sweat glands	Abundant sweat glands with odorless secretion	Lower body temperature	Originate in deep dermis or subcutaneous layer and open to surface on forehead, neck, and back
Apocrine sweat glands	Less numerous sweat glands with secretions that develop odors	Wet skin during pain, fear, emotional upset, and sexual arousal	Near hair follicles in armpit and groin
Ceruminous glands	Modified sweat glands	Secrete ear wax	External acoustic meatus
Mammary glands	Modified sweat glands	Secrete milk	Breasts

6.3 | Skin Functions

The skin is one of the most versatile organs of the body and is vital in maintaining homeostasis. As a protective covering, the skin prevents many harmful substances, as well as microorganisms, from entering the body. Skin also retards water loss by diffusion from deeper tissues. It houses sensory receptors; contains *epidermal dendritic* (den-drit′ik) *cells*, also known as Langerhans cells, that play a role in initiating the immune response by phagocytizing harmful microorganisms; and excretes small quantities of wastes.

The skin plays a role in the production of vitamin D, which is necessary for normal bone and tooth development. Some skin cells produce vitamin D precursor (dehydrocholesterol), which when exposed to sunlight changes to an inactive form of vitamin D (cholecalciferol). In the liver and kidneys the inactive form is modified and becomes active vitamin D (calcitriol). The skin also helps regulate body temperature.

Regulation of Body Temperature

The regulation of body temperature is vitally important because even slight shifts can disrupt the rates of metabolic reactions. Normally, the temperature of deeper body parts remains close to a set point of 37°C (98.6°F). The maintenance of a stable temperature requires that the amount of heat the body loses be balanced by the amount it produces. The skin plays a key role in the homeostatic mechanism that regulates body temperature.

RECONNECT
To Chapter 1, Homeostasis, pages15–18.

Heat Production and Loss

Heat is a product of cellular metabolism. The major heat producers are the more active cells of the body, such as skeletal and cardiac muscle cells and cells of the liver.

When body temperature rises above the set point, the nervous system stimulates structures in the skin and other organs to release heat. For example, during physical exercise, active muscles release heat, which the blood carries away. The warmed blood reaches the part of the brain (the hypothalamus) that controls the body's temperature set point, which signals muscles in the walls of dermal blood vessels to relax. As these vessels dilate (vasodilation), more blood enters them, and some of the heat the blood carries escapes to the outside. At the same time, deeper blood vessels contract (vasoconstriction), diverting blood to the surface, and the skin reddens. The heart is stimulated to beat faster, moving more blood out of the deeper regions.

The primary means of body heat loss is **radiation** (ra-de-a′shun), by which infrared heat rays escape from warmer surfaces to cooler surroundings. These rays radiate in all directions, much like those from the bulb of a heat lamp.

Conduction and convection release less heat than does radiation. In **conduction** (kon-duk′shun), heat moves from the body directly into the molecules of cooler objects in contact with its surface. For example, heat is lost by conduction into the seat of a chair when a person sits down. The heat loss continues as long as the chair is cooler than the body surface touching it. Heat is also lost by conduction to the air molecules that contact the body. As air becomes heated, it moves away from the body, carrying heat with it, and is replaced by cooler air moving toward the body. This type of continuous circulation of air over a warm surface is **convection** (kon-vek′shun).

Still another means of body heat loss is **evaporation** (e-vap″o-ra′shun). When the body temperature rises above normal, the nervous system stimulates merocrine sweat glands to release sweat onto the surface of the skin. As this fluid evaporates (changes from a liquid to a gas), it carries heat away from the surface, cooling the skin.

When body temperature drops below the set point, as may occur in a very cold environment, the brain triggers different responses in the skin structures. Muscles in the walls of dermal blood vessels are stimulated to contract. This action decreases the flow of heat-carrying blood through the skin, which loses color, and helps reduce heat loss by radiation, conduction, and convection. At the same time, merocrine sweat glands remain inactive, decreasing heat loss by evaporation. If body temperature continues to drop, the nervous system may stimulate muscle cells in the skeletal muscles throughout the body to contract slightly. This action requires an increase in the rate of cellular respiration, which releases heat as a by-product. If this response does not raise the body temperature to normal, small groups of muscles may rhythmically contract with greater force, causing the person to shiver, generating more heat. Figure 6.12 summarizes the body's temperature-regulating mechanism, and Clinical Application 6.4 examines two causes of elevated body temperature.

Problems in Temperature Regulation

The body's temperature-regulating mechanism does not always operate satisfactorily, and the consequences may be dangerous. For example, air can hold only a limited volume of water vapor, so on a hot, humid day, the air may become nearly saturated with water. At such times, the sweat glands may be activated, but the sweat cannot quickly evaporate. The skin becomes wet, but the person remains hot and uncomfortable. In addition, if the air temperature is high, heat loss by radiation is less effective. If the air temperature exceeds body temperature, the person may gain heat from the surroundings, elevating body temperature even more. Body temperature may rise high enough to cause a condition called **hyperthermia.** The skin becomes dry, hot, and flushed. The person becomes weak, dizzy, and nauseous, with headache and a rapid, irregular pulse.

Hypothermia, or lowered body temperature, can result from prolonged exposure to cold or as part of an illness. It can be extremely dangerous. Hypothermia begins with shivering and a feeling of coldness. If not treated, it progresses to mental confusion, lethargy, loss of reflexes and consciousness, and, eventually, the shutdown of major organs. If the temperature in the body's core drops just a few degrees, fatal respiratory failure or heart arrhythmia may result. However, the extremities can withstand drops of 20°F to 30°F below normal.

Certain people are at higher risk for developing hypothermia due to less adipose tissue in the subcutaneous layer beneath the skin (less insulation). These include the very old, very thin individuals, and the homeless. The very young with undeveloped nervous systems have difficulty regulating their body temperature. Dressing appropriately and staying active in the cold can prevent hypothermia. A person suffering from hypothermia must be warmed gradually so that respiratory and cardiovascular functioning remain stable.

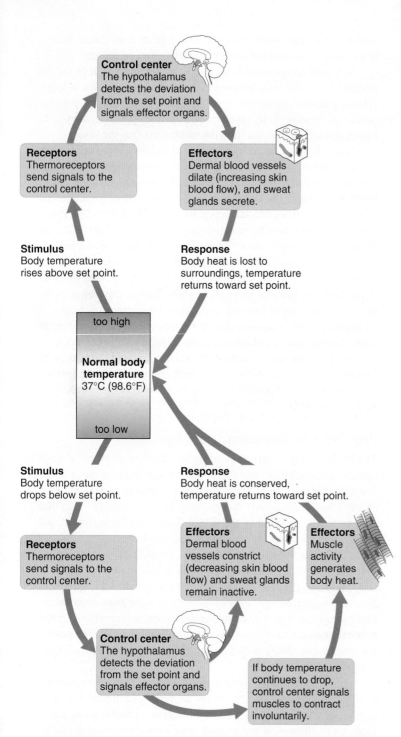

FIGURE 6.12 Body temperature regulation is an example of homeostasis.

Control center
The hypothalamus detects the deviation from the set point and signals effector organs.

Receptors
Thermoreceptors send signals to the control center.

Effectors
Dermal blood vessels dilate (increasing skin blood flow), and sweat glands secrete.

Stimulus
Body temperature rises above set point.

Response
Body heat is lost to surroundings, temperature returns toward set point.

too high

Normal body temperature
37°C (98.6°F)

too low

Stimulus
Body temperature drops below set point.

Response
Body heat is conserved, temperature returns toward set point.

Receptors
Thermoreceptors send signals to the control center.

Effectors
Dermal blood vessels constrict (decreasing skin blood flow) and sweat glands remain inactive.

Effectors
Muscle activity generates body heat.

Control center
The hypothalamus detects the deviation from the set point and signals effector organs.

If body temperature continues to drop, control center signals muscles to contract involuntarily.

Hypothermia is intentionally induced during certain surgical procedures to protect vital organs from decreased oxygen supply that may occur during the operations. For example, in heart surgery, body temperature may be lowered to between 78°F (26°C) and 89°F (32°C), which lowers the body's metabolic rate so that cells require less oxygen. Hypothermia for surgery is accomplished by packing the patient in ice or by removing blood, cooling it, and returning it to the patient.

 PRACTICE

18 What, besides body temperature regulation, are some functions of the skin?

19 How is body heat produced?

20 How does the body lose excess heat?

21 How does the skin help regulate body temperature?

6.4 | Healing of Wounds and Burns

Inflammation is a normal response to injury or stress. Blood vessels in affected tissues dilate and become more permeable, allowing fluids to leak into the damaged tissues. Inflamed skin may become reddened, swollen, warm, and painful to touch. However, the dilated blood vessels provide the tissues with more nutrients and oxygen, which aids healing. The specific events in the healing process depend on the nature and extent of the injury.

 A GLIMPSE AHEAD | **To Chapters 14, 15, and 16**

Vasodilation increases blood flow to an inflamed area. Cells in the blood help to eliminate dangerous microorganisms.

Cuts

Breaks in the skin must heal to prevent infection and loss of blood. If a break in the skin is shallow, epithelial cells of the epidermis along its margin are stimulated to divide more rapidly than usual. The newly formed cells fill the gap.

If an injury extends into the dermis or subcutaneous layer, blood vessels break and the released blood forms a clot in the wound. A clot consists mainly of a fibrous protein (fibrin) that forms from another protein in the plasma, blood cells, and platelets trapped in the protein fibers. Tissue fluids seep into the area and dry. The blood clot and the dried tissue fluids form a *scab* that covers and protects underlying tissues. Epithelial cells proliferate beneath the scab, bridging the wound. Before long, fibroblasts migrate into the injured region and begin secreting collagen fibers that bind the edges of the wound. Suturing (stitching) or otherwise closing a large break in the skin speeds this process. In addition, the connective tissue matrix releases *growth factors* that stimulate certain cells to divide and regenerate the damaged tissue.

As healing continues, blood vessels extend beneath the scab. Phagocytic cells remove dead cells and other debris. Eventually, the damaged tissues are replaced, and the scab sloughs off. In some cases, extensive production of collagen fibers may form an elevation above the normal epidermal surface, called a *scar*.

In large, open wounds, healing may be accompanied by formation of small, rounded masses called *granulations* that develop in the exposed tissues. A granulation consists of a new branch of a blood vessel and a cluster of collagen-secreting fibroblasts that the vessel nourishes. In time, some of the blood vessels are resorbed, and the fibroblasts move away, leaving a scar largely composed of collagen fibers. **Figure 6.13** shows the stages in the healing of a wound.

Elevated Body Temperature

It was a warm June morning when the harried and hurried father strapped his five-month-old son Bryan into the backseat of his car and headed for work. Tragically, the father forgot to drop his son off at the babysitter's. When his wife called him at work late that afternoon to inquire why the child was not at the sitter's, the shocked father realized his mistake and hurried down to his parked car. But it was too late—Bryan had died. Left for ten hours in the car in the sun, all windows shut, the baby's temperature had quickly soared. Two hours after he was discovered, the child's temperature still exceeded 41°C (106°F).

Sarah L.'s case of elevated body temperature was more typical. She awoke with a fever of 40°C (104°F) and a sore throat. At the doctor's office, a test revealed that Sarah had a *Streptococcus* infection. The fever was her body's attempt to fight the infection.

The true cases of Bryan and Sarah illustrate two reasons why body temperature may rise—inability of the **temperature homeostatic** mechanism to handle an extreme environment and an immune system response to infection. In Bryan's case, sustained exposure to very high heat overwhelmed his temperature-regulating mechanism, resulting in hyperthermia. Body heat built up faster than it could dissipate, and body temperature rose, even though the set point of the thermostat was normal. His blood vessels dilated so greatly in an attempt to dissipate the excess heat that after a few hours, his cardiovascular system collapsed.

Fever is a special case of hyperthermia in which temperature rises in response to an elevated set point. In fever, molecules on the surfaces of the infectious agents (usually bacteria or viruses) stimulate phagocytes to release a substance called interleukin-1 (also called endogenous pyrogen, meaning "fire maker from within"). The bloodstream carries interleukin-1 to the hypothalamus, where it raises the set point controlling temperature. In response, the brain signals skeletal muscles to contract (shivering) and increase heat production, blood flow to the skin to decrease, and sweat glands to decrease secretion. As a result, body temperature rises to the new set point, and fever develops. The increased body temperature helps the immune system kill the pathogens.

Rising body temperature requires different treatments, depending on the degree of elevation. Hyperthermia in response to exposure to intense, sustained heat should be rapidly treated by administering liquids to replace lost body fluids and electrolytes, sponging the skin with water to increase cooling by evaporation, and covering the person with a refrigerated blanket. Fever can be lowered with ibuprofen or acetaminophen, or aspirin in adults. Some health-care professionals believe that a slightly elevated temperature should not be reduced (with medication or cold baths) because it may be part of a normal immune response. A high or prolonged fever, however, requires medical attention.

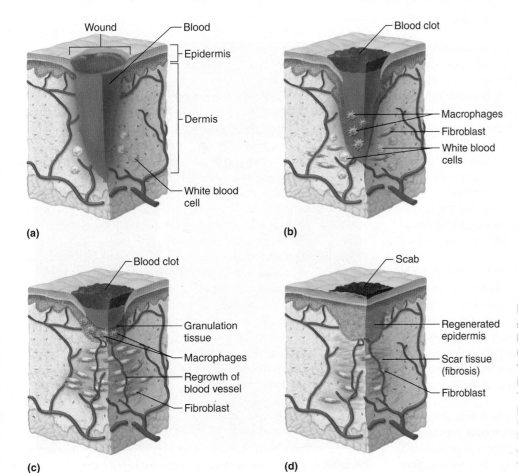

(a)

(b)

(c)

(d)

FIGURE 6.13 Healing of a wound. (*a*) When skin is injured, blood escapes from dermal blood vessels, and (*b*) a blood clot soon forms. (*c*) Blood vessels send out branches, and fibroblasts migrate into the area. The fibroblasts produce new collagen fibers. (*d*) The scab formed by the blood clot and dried tissue fluid protects the damaged region until the skin is mostly repaired. Then, the scab sloughs off. Scar tissue continues to form, elevating the epidermal surface. (Note: Cells are not drawn to scale.)

Burns

Burns are classified by the extent of tissue damage. Slightly burned skin, such as from a minor sunburn, becomes inflamed, warms and reddens (erythema) as dermal blood vessels dilate. Mild edema may swell the exposed tender skin, and a few days later the surface layer of skin may peel. Any burn injuring only the epidermis is a *superficial partial-thickness* (first-degree) *burn.* Healing usually takes a few days to two weeks, with no scarring.

A burn that destroys some epidermis as well as some underlying dermis is a *deep partial-thickness* (second-degree) *burn.* Fluid escapes from damaged dermal capillaries, accumulating beneath the outer layer of epidermal cells, forming blisters. The injured region becomes moist and firm and may vary in color from dark red to waxy white. Such a burn most commonly occurs as a result of exposure to hot objects, hot liquids, flames, or burning clothing.

Acute sunburn (solar erythema) is an inflammatory reaction of the skin to excessive exposure to ultraviolet radiation in sunlight. The skin becomes very red, swollen, and painful. Microscopic skin changes begin within a half hour of intense sun exposure, including damage to cells in the upper, epidermal layer of the skin, and swelling of blood vessels in the deeper, dermal layer. Discomfort peaks 6 to 48 hours after exposure. Within a few days the skin may peel, as surface cells die and are shed. Peeling, an example of apoptosis (programmed cell death), prevents cancer from developing by ridding the body of susceptible cells.

The extent of healing of a deep partial-thickness burn depends upon stem cells that are associated with accessory structures of the skin. These structures include hair follicles, sweat glands, and sebaceous glands. They survive the injury because although they are derived from the epidermis, they extend into the dermis. During healing, the stem cells divide, and their daughter cells grow out onto the surface of the exposed dermis, spread over it, and differentiate as new epidermis. In time, the skin usually completely recovers, and scar tissue does not develop unless an infection occurs.

A burn that destroys the epidermis, dermis, and the accessory structures of the skin is called a *full-thickness* (third-degree) *burn.* The injured skin becomes dry and leathery, and it may vary in color from red to black to white.

A full-thickness burn usually results from immersion in hot liquids or prolonged exposure to hot objects, flames, or corrosive chemicals. Because most of the epithelial cells in the affected region are destroyed, spontaneous healing occurs by growth of epithelial cells inward from the margin of the burn. If the injured area is extensive, treatment may require removing a thin layer of skin from an unburned region of the body and transplanting it to the injured area. This procedure is an example of an *autograft,* a transplant within the same individual.

If the injured area is too extensive to replace with skin from other parts of the body, cadaveric skin from a skin bank may be used to cover the injury. Such an *allograft* (from person to person) is a temporary covering that shrinks the wound, helps prevent infection,

and preserves deeper tissues. In time, after healing begins, an autograft may replace the temporary covering, as skin becomes available from healed areas. Skin grafts can leave extensive scars.

Various skin substitutes may also be used to cover the injured area. Artificial skin typically consists of a collagen framework seeded with a patient's own skin cells (fig. 6.14).

The treatment of a burn patient requires estimating the extent of the body's affected surface. Physicians use the "rule of nines," subdividing the skin's surface into regions, each accounting for 9% (or some multiple of 9%) of the total surface area (fig. 6.15). This estimate is important in planning to replace body fluids and electrolytes lost from injured tissues and for covering the burned area with skin or skin substitutes.

PRACTICE

22 What is the tissue response to inflammation?

23 How does a scab slough off?

24 Which type of burn is most likely to leave a scar? Why?

6.5 | Life-Span Changes

We are more aware of aging-related changes in skin than in other organ systems, because we can easily see them. Aging skin affects appearance, temperature regulation, and vitamin D formation.

The epidermis thins as the decades pass. As the cell cycle slows, epidermal cells grow larger and more irregular in shape,

FIGURE 6.14 Tissue-engineered (artificial) skin.

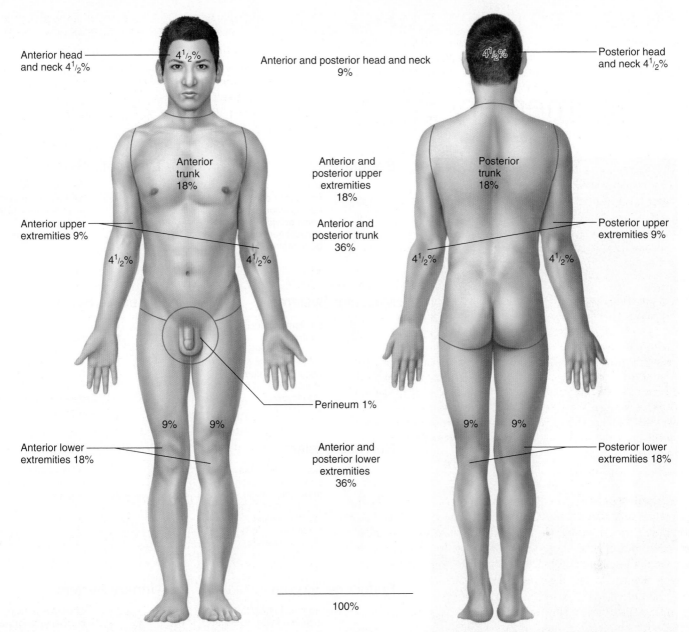

FIGURE 6.15 As an aid for estimating the extent of damage burns cause, the body is subdivided into regions, each representing 9% (or some multiple of 9%) of the total skin surface area.

but are fewer. Skin may appear scaly because, at the microscopic level, more sulfur–sulfur bonds form within keratin molecules. Patches of pigment commonly called "age spots" or "liver spots" appear and grow (fig. 6.16). These are sites of oxidation of fats in the secretory cells of apocrine and eccrine glands and reflect formation of oxygen free radicals.

The dermis becomes reduced as synthesis of the connective tissue proteins collagen and elastin slows. The combination of a shrinking dermis and loss of some fat from the subcutaneous layer results in wrinkling and sagging of the skin. Wound healing slows because the skin has fewer fibroblasts. Some of the changes in the skin's appearance result from specific deficits. The decrease in oil from sebaceous glands dries the skin.

Various treatments temporarily smooth facial wrinkles. "Botox" is an injection of a very dilute solution of botulinum toxin. Produced by the bacterium *Clostridium botulinum*, the toxin causes food poisoning. It also blocks nerve activation of certain muscle cells, including the facial muscles that control smiling, frowning, and squinting. After three months, though, the facial nerves contact the muscles at different points, and the wrinkles return. (Botox used at higher doses to treat neuromuscular conditions can cause adverse effects.) Other anti-wrinkle treatments include chemical peels and dermabrasion to reveal new skin surface; collagen injections; and transplants of subcutaneous fat from the buttocks to the face.

Integumentary System

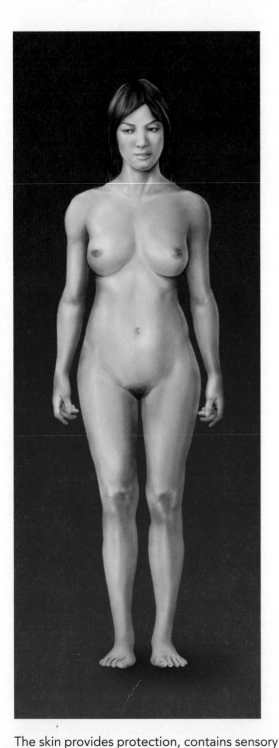

The skin provides protection, contains sensory organs, and helps control body temperature.

Skeletal System

Vitamin D, production of which begins in the skin, helps provide calcium needed for bone matrix.

Muscular System

Involuntary muscle contractions (shivering) work with the skin to control body temperature. Muscles act on facial skin to create expressions.

Nervous System

Sensory receptors provide information about the outside world to the nervous system. Nerves control the activity of sweat glands.

Endocrine System

Hormones help to increase skin blood flow during exercise. Other hormones stimulate either the synthesis or the decomposition of subcutaneous fat.

Cardiovascular System

Skin blood vessels play a role in regulating body temperature.

Lymphatic System

The skin, acting as a barrier, provides an important first line of defense for the immune system.

Digestive System

Excess calories may be stored as subcutaneous fat. Vitamin D, production of which begins in the skin, stimulates dietary calcium absorption.

Respiratory System

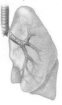

Stimulation of skin receptors may alter respiratory rate.

Urinary System

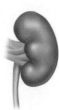

The kidneys help compensate for water and electrolytes lost in sweat.

Reproductive System

Sensory receptors play an important role in sexual activity and in the suckling reflex.

FIGURE 6.16 Aging-associated changes are obvious in the skin.

The skin's accessory structures also show signs of aging. Slowed melanin production whitens hair as the hair shafts become increasingly transparent. Hair growth slows, the hairs thin, and the number of follicles decreases. Males may develop pattern baldness. A diminished blood supply to the nail beds impairs their growth, dulling and hardening them. Sensitivity to pain and pressure diminishes with age as the number of sensory receptors falls. A ninety-year-old's skin has only one-third the number of such sensory receptors as the skin of a young adult.

The ability to control temperature falters as the number of sweat glands in the skin falls, as the capillary beds that surround sweat glands and hair follicles shrink, and as the ability to shiver declines. In addition, the number of blood vessels in the deeper layers decreases, as does the ability to shunt blood toward the body's interior to conserve heat. As a result, an older person is less able to tolerate the cold and cannot regulate heat. In the winter, an older person might set the thermostat ten to fifteen degrees higher than a younger person would. Fewer blood vessels in and underlying the skin account for the pale complexions of some older individuals. Changes in the distribution of blood vessels also contribute to development of pressure sores in a bedridden person whose skin does not receive adequate circulation.

Aging of the skin is also related to skeletal health. The skin is the site of formation of vitamin D from its precursor, which requires exposure to the sun. Vitamin D is necessary for absorption of calcium, needed for bone structure. Many older people do not get outdoors much, and the wavelengths of light that are important for vitamin D formation do not readily penetrate glass windows. In addition, older skin has a diminished ability to activate the vitamin. Therefore, homebound seniors can benefit from vitamin D supplements to help maintain bone structure.

 PRACTICE

25 What changes occur with age in the epidermis and dermis?

26 How do the skin's accessory structures change over time?

27 Why do older people have more difficulty controlling body temperature than do younger people? ∎

Chapter Summary

6.1 Skin and Its Tissues (page 178)

Skin is composed of an **epidermis** and a **dermis** separated by a basement membrane. A **subcutaneous layer,** not part of the skin, lies beneath the dermis. The subcutaneous layer is composed of areolar tissue and adipose tissue that helps conserve body heat. This layer contains blood vessels that supply the skin.

1. Epidermis
   a. The epidermis is stratified squamous epithelium that lacks blood vessels.
   b. The deepest layer, called the stratum basale, contains cells that divide and grow.
   c. Epidermal cells undergo **keratinization** as they are pushed toward the surface.
   d. The outermost layer, called the stratum corneum, is composed of dead epidermal cells.
   e. Production of epidermal cells in the stratum basale balances the rate at which they are lost at the surface.
   f. The epidermis protects underlying tissues against water loss, mechanical injury, and the effects of harmful chemicals.
   g. **Melanin,** a pigment produced from the amino acid tyrosine, provides skin color and protects underlying cells from the effects of ultraviolet light.
   h. **Melanocytes** transfer melanin to nearby epidermal cells.
   i. All humans have about the same number of melanocytes. Skin color is largely due to the amount of melanin in the epidermis.
      (1) Each person inherits genes for melanin production.
         (a) Dark skin is due to genes that cause large amounts of melanin to be produced; lighter skin is due to genes that cause lesser amounts of melanin to form.
         (b) Mutant genes may cause a lack of melanin in the skin.
      (2) Environmental factors that influence skin color include sunlight, ultraviolet light, and X rays. These factors darken existing melanin and stimulate additional melanin production.
      (3) Physiological factors influence skin color.
         (a) The oxygen content of the blood in dermal vessels may cause the skin of light-complexioned persons to appear pinkish or bluish.
         (b) Carotene in the subcutaneous layer may cause the skin to appear yellowish.
         (c) Disease may affect skin color.

2. Dermis
   a. Cone-shaped dermal papillae contain blood capillaries that provide epidermal cells with oxygen and nutrients.
   b. Dermal papillae create friction ridges that leave a fingerprint when a finger presses against a surface.

c. The dermis has two layers.
   (1) The upper papillary layer is areolar connective tissue.
   (2) The lower reticular layer is dense irregular connective tissue that binds the epidermis to underlying tissues.
d. The dermis contains smooth and skeletal muscle tissues.
e. Dermal blood vessels supply nutrients to skin cells and help regulate body temperature.
f. Nervous tissue is scattered throughout the dermis.
   (1) Some dermal nerve cell processes conduct impulses to muscles and glands of the skin.
   (2) Other dermal nerve cell processes are associated with sensory receptors in the skin.

6.2 Accessory Structures of the Skin (page 184)

1. **Nails**
   a. **Nails** are protective covers on the ends of fingers and toes.
   b. They consist of keratinized epidermal cells.
2. Hair follicles
   a. Hair covers nearly all regions of the skin.
   b. Each hair develops from epidermal cells at the base of a tubelike **hair follicle.**
   c. As newly formed cells develop and grow, older cells are pushed toward the surface and undergo keratinization.
   d. A hair usually grows for a while, rests, and then is replaced by a new hair.
   e. Hair color is determined by genes that direct the type and amount of pigment in hair cells.
   f. A bundle of smooth muscle cells and one or more sebaceous glands are attached to each hair follicle.
3. Skin glands
   a. **Sebaceous glands** secrete sebum, which softens and waterproofs both the skin and hair.
   b. Sebaceous glands are usually associated with hair follicles.
   c. **Sweat glands** are located in nearly all regions of the skin.
   d. Each sweat gland consists of a coiled tube.
   e. **Merocrine sweat glands,** located on the forehead, neck, back, palms, and soles, respond to elevated body temperature and produce sweat that is primarily water, but also contains salts and waste products.
   f. **Apocrine sweat glands,** located in the axillary regions and groin, moisten the skin when a person is emotionally upset, scared, in pain, or sexually aroused and produce sweat containing proteins and lipids.

6.3 Skin Functions (page 189)

The skin is vital in maintaining homeostasis. Regulation of body temperature is vital because heat affects the rates of metabolic reactions. Normal temperature of deeper body parts is close to a set point of 37°C (98.6°F).

1. The skin is a protective covering, retards water loss, and houses sensory receptors.
2. The skin produces vitamin D precursor.

3. The skin helps regulate body temperature.
   a. Heat production and loss
      (1) Heat is a by-product of cellular respiration.
      (2) When body temperature rises above normal, more blood enters dermal blood vessels and the skin reddens.
      (3) Heat is lost to the outside by **radiation, conduction, convection,** and **evaporation.**
      (4) Sweat gland activity increases heat loss by evaporation.
      (5) When body temperature drops below normal, dermal blood vessels constrict, causing the skin to lose color, and merocrine sweat glands to become inactive.
      (6) If body temperature continues to drop, skeletal muscles involuntarily contract; this increases cellular respiration, which produces additional heat.
   b. Problems in temperature regulation
      (1) Air can hold a limited volume of water vapor.
      (2) When the air is saturated with water, sweat may fail to evaporate and body temperature may remain elevated.
      (3) **Hyperthermia** is a body temperature above 101°F. It causes weakness, dizziness, nausea, headache, and a rapid, irregular pulse.
      (4) **Hypothermia** is lowered body temperature. It causes shivering, mental confusion, lethargy, loss of reflexes and consciousness, and eventually major organ failure.

6.4 Healing of Wounds and Burns (page 190)

Skin injuries trigger **inflammation.** The affected area becomes red, warm, swollen, and tender.

1. A cut in the epidermis is filled in by dividing epithelial cells. Clots close deeper cuts, sometimes leaving a scar where connective tissue produces collagen fibers, forming an elevation above the normal epidermal surface. Granulations form as part of the healing process in large, open wounds.
2. A superficial partial-thickness burn heals quickly with no scarring. The area is warm and red. A burn penetrating to the dermis is a deep partial-thickness burn. It blisters. Deeper skin structures help heal this more serious type of burn. A full-thickness burn is the most severe and may require a skin graft.
3. The "rule of nines" is used to estimate the extent of the body's affected surface in determining burn treatment.

6.5 Life-Span Changes (page 192)

1. Aging skin affects appearance as "age spots" or "liver spots" appear and grow, along with wrinkling and sagging.
2. Due to changes in the number of sweat glands and shrinking capillary beds in the skin, elderly people are less able to tolerate the cold and cannot regulate heat well.
3. Older skin has a diminished ability to produce vitamin D, which is necessary for skeletal health.

CHAPTER ASSESSMENTS

6.1 Skin and Its Tissues

1 The epidermis on the soles of the feet is composed of layers of _____ tissue. (p. 178)
2 Distinguish between the epidermis and the dermis. (p.178)
3 Explain the functions of the subcutaneous layer. (p. 178)
4 Explain what happens to epidermal cells as they undergo keratinization. (p. 179)
5 Place the layers of the epidermis in order (1–5) from the outermost layer to the layer attached to the dermis by the basement membrane. (p. 179)

_____ stratum spinosum

_____ stratum corneum

_____ stratum basale

_____ stratum lucidum

_____ stratum granulosum

6 Describe the function of melanocytes. (pp. 180 and 183)
7 Discuss the function of melanin, other than providing color to the skin. (p. 180)
8 Explain how environmental factors affect skin color. (p. 183)
9 Describe three physiological factors that affect skin color. (p. 183)
10 Name the tissue(s) in each of the two layers of the dermis and describe their functions. (p. 183)
11 Review the functions of dermal nervous tissue. (p. 184)

6.2 Accessory Structures of the Skin

12 Describe how nails are formed. (p. 184)
13 Distinguish between a hair and a hair follicle. (p. 185)
14 Review how hair color is determined. (p. 185)
15 Explain the function of sebaceous glands. (p. 186)
16 The sweat glands that respond to elevated body temperature and are commonly found on the forehead, neck, and back are _____ sweat glands. (pp. 185–187)
   a. sebaceous
   b. holocrine
   c. merocrine
   d. apocrine
17 Compare and contrast merocrine and apocrine sweat glands. (pp. 186–187)

6.3 Skin Functions

18 Functions of the skin include _____. (p. 189)
   a. retarding water loss
   b. body temperature regulation
   c. sensory reception
   d. all of the above
19 Describe the functions of skin not mentioned in question 18. (p. 189)
20 Explain the importance of body temperature regulation. (p. 189)
21 Describe the role of the skin in promoting the loss of excess body heat. (p. 189)
22 Match each means of losing body heat with its description. (p. 189)

   (1) radiation
   (2) conduction
   (3) convection
   (4) evaporation

   A. fluid changes from liquid to a gas
   B. heat moves from body directly into molecules of cooler objects in contact with its surface
   C. heat rays escape from warmer surfaces to cooler surroundings
   D. continuous circulation of air over a warm surface

23 Explain how air saturated with water vapor may interfere with body temperature regulation. (p. 189)
24 Describe the body's responses to decreasing body temperature. (p. 189)

6.4 Healing of Wounds and Burns

25 Distinguish between the healing of shallow and deeper breaks in the skin. (p. 190)
26 Distinguish among first-, second-, and third-degree burns. (p. 192)
27 Describe possible treatments for a third-degree burn. (p. 192)

6.5 Life-Span Changes

28 Discuss three effects of aging on skin. (pp. 192–195)

INTEGRATIVE ASSESSMENTS/CRITICAL THINKING

Outcomes 1.5, 6.2, 6.3

1. What methods might be used to cool the skin of a child experiencing a high fever? For each method you list, identify the means by which it promotes heat loss—radiation, conduction, convection, or evaporation.

Outcomes 5.3, 6.1

2. Why would collagen and elastin added to skin creams be unlikely to penetrate the skin—as some advertisements imply they do?

Outcomes 5.3, 6.1, 6.3

3. A premature infant typically lacks subcutaneous adipose tissue. Also, the surface area of an infant's body is relatively large compared to its volume. How do these factors affect the ability of an infant to regulate its body temperature?

Outcome 6.1

4. Which of the following would result in the more rapid absorption of a drug: a subcutaneous injection or an intradermal injection? Why?

5. Everyone's skin contains about the same number of melanocytes even though people come in many different colors. How is this possible?

Outcomes 6.1, 6.2, 6.3, 6.4

6. Using the rule of nines, estimate the extent of damage for an individual whose body, clad only in a short nightgown, was burned when the nightgown caught fire as she was escaping from a burning room. What special problems might result from the loss of the individual's functional skin surface? How might this individual's environment be modified to compensate partially for such a loss?

Outcomes 6.1, 6.3, 6.4

7. As a rule, a superficial partial-thickness burn is more painful than one involving deeper tissues. How would you explain this observation?

ONLINE STUDY TOOLS

 |

Connect Interactive Questions Reinforce your knowledge using assigned interactive questions covering the structure of skin and its accessory organs.

Connect Integrated Activity Practice your understanding of skin as you look at the process of wound healing.

LearnSmart Discover which chapter concepts you have mastered and which require more attention. This adaptive learning tool is personalized, proven, and preferred.

Anatomy & Physiology Revealed Go more in depth into the human body by exploring the skin and its histology.

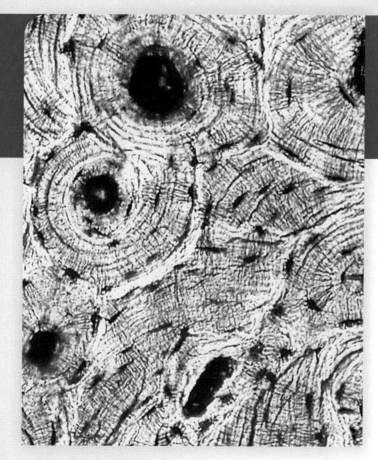

Micrograph of compact bone (200×).

THE WHOLE PICTURE

A bone may appear to be inert because of nonliving material in the extracellular matrix of bone tissue. However, bone also includes active, living tissues: bone tissue, cartilage, dense connective tissue, blood, and nervous tissue. Bones are not only alive, but also multifunctional. Bones, the organs of the skeletal system, support and protect softer tissues, provide points of attachment for muscles, house blood-producing cells, and store inorganic salts.

Skeletal System

 LEARNING OUTCOMES

After you have studied this chapter, you should be able to:

7.1 Bone Shape and Structure
1 Classify bones according to their shapes, and name an example from each group. (p. 200)
2 Describe the macroscopic and microscopic structure of a long bone, and list the functions of these parts. (pp. 200–202)

7.2 Bone Development and Growth
3 Distinguish between intramembranous and endochondral bones, and explain how such bones develop and grow. (pp. 202–207)
4 Describe the effects of sunlight, nutrition, hormonal secretions, and exercise on bone development and growth. (pp. 207–210)

7.3 Bone Function
5 Discuss the major functions of bones. (pp. 210–211)

7.4 Skeletal Organization
6 Distinguish between the axial and appendicular skeletons, and name the major parts of each. (pp. 212–215)

7.5 Skull–7.11 Lower Limb
7 Locate and identify the bones and the major features of the bones that comprise the skull, vertebral column, thoracic cage, pectoral girdle, upper limb, pelvic girdle, and lower limb. (pp. 215–245)
8 Describe the differences between male and female skeletons. (p. 241)

7.12 Life-Span Changes
9 Describe life-span changes in the skeletal system. (pp. 245–246)

aprevealed.com

UNDERSTANDING WORDS

acetabul-, vinegar cup: *acetabulum*—depression of the hip bone that articulates with the head of the femur.

ax-, axis: *axial* skeleton—upright portion of the skeleton that supports the head, neck, and trunk.

-blast, bud, a growing organism in early stages: osteo*blast*—cell that will form bone tissue.

canal-, channel: *canal*iculus—tubular passage.

carp-, wrist: *carp*als—wrist bones.

-clast, break: osteo*clast*—cell that breaks down bone tissue.

clav-, bar: *clav*icle—bone that articulates with the sternum and scapula.

condyl-, knob, knuckle: *condyl*e—rounded, bony process.

corac-, a crow's beak: *corac*oid process—beaklike process of the scapula.

cribr-, sieve: *cribr*iform plate—portion of the ethmoid bone with many small openings.

crist-, crest: *crist*a galli—bony ridge that projects upward into the cranial cavity.

fov-, pit: *fov*ea capitis—pit in the head of a femur.

glen-, joint socket: *glen*oid cavity—depression in the scapula that articulates with the head of a humerus.

inter-, among, between: *inter*vertebral disc—structure between vertebrae.

intra-, inside: *intra*membranous bone—bone that forms within sheetlike masses of connective tissue.

lamell-, thin plate: *lamell*a—thin, bony plate.

meat-, passage: external acoustic *meat*us—canal of the temporal bone that leads inward to parts of the ear.

odont-, tooth: *odont*oid process—toothlike process of the second cervical vertebra.

poie-, make, produce: hemato*poie*sis—process that forms blood cells.

7.1 | Bone Shape and Structure

The bones of the skeletal system vary greatly in size and shape. However, bones are similar in structure, development, and function.

Bone Shapes

Bones are classified according to their shapes—long, short, flat, or irregular (fig. 7.1).

- **Long bones** have long longitudinal axes and expanded ends. Examples of long bones are the forearm and thigh bones.
- **Short bones** are cubelike, with roughly equal lengths and widths. The bones of the wrists and ankles are this type. A special type of short bone is a **sesamoid bone,** or **round bone.** This type of bone is usually small and nodular and embedded in a tendon adjacent to a joint, where the tendon is compressed. The kneecap (patella) is a sesamoid bone.
- **Flat bones** are platelike structures with broad surfaces, such as the ribs, the scapulae, and some bones of the skull.
- **Irregular bones** have a variety of shapes, and most are connected to several other bones. Irregular bones include the vertebrae that compose the backbone, and many facial bones.

Parts of a Long Bone

The femur, the long bone in the thigh, illustrates the structure of a bone (fig. 7.2). At each end of a long bone is an expanded portion called an **epiphysis** (e-pif′ĭ-sis) (pl., *epiphyses*), which articulates (or forms a joint) with another bone. One epiphysis, called the proximal epiphysis, is nearest to the torso. The other, called the distal epiphysis, is farthest from the torso. On its outer surface, the articulating portion of the epiphysis is coated with a layer of hyaline cartilage called **articular cartilage** (ar-tik′u-lar kar′tĭ-lij). The shaft of the bone is called the **diaphysis** (di-af′ĭ-sis). The **metaphysis** (met″ah-fi′-sis) is the widening part of the bone between the diaphysis and the epiphysis.

A bone is enclosed by a tough, vascular covering of dense connective tissue called the **periosteum** (per″e-os′te-um), except for the articular cartilage on its ends. The periosteum is firmly attached to the bone, and the periosteal fibers are continuous with connected ligaments and tendons. The periosteum also helps form and repair bone tissue.

A bone's shape makes possible its functions. Bony projections called *processes,* for example, provide sites for attachment

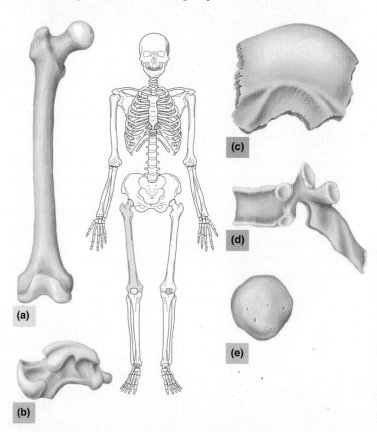

FIGURE 7.1 Bones are classified by shape. (*a*) The femur of the thigh is a long bone, (*b*) the talus of the ankle is a short bone, (*c*) a parietal bone of the skull is a flat bone, (*d*) a vertebra of the backbone is an irregular bone, and (*e*) the patella of the knee is a sesamoid bone. The location icon of the whole skeleton highlights the bones used as examples for classification.

of ligaments and tendons; grooves and openings are passageways for blood vessels and nerves; and a depression of one bone might articulate with a process of another.

The wall of the diaphysis is mainly composed of tightly packed tissue called **compact bone,** or cortical bone. This type of bone has a continuous extracellular matrix with no gaps (fig. 7.3*a*). The epiphyses, on the other hand, are largely composed of **spongy bone,** or cancellous bone, with thin layers of compact bone on their surfaces (fig. 7.3*b*). Spongy bone consists of many branching bony plates called **trabeculae** (trah-bek'u-le). Irregular connecting spaces between these plates help reduce the bone's weight. The bony plates are most highly developed in the regions of the epiphyses subjected to compressive forces. Both compact and spongy bone are strong and resist bending.

Most bones have compact bone overlying spongy bone, with the relative amounts of each varying in the differently shaped bones. Short, flat, and irregular bones typically consist of a mass of spongy bone either covered by a layer of compact bone or sandwiched between plates of compact bone (fig. 7.3*c*).

Compact bone in the diaphysis of a long bone forms a tube with a hollow chamber called the **medullary cavity** (med'u-lār"e kav'ĭ-te) that is continuous with the spaces of the spongy bone. A thin membrane containing bone-forming cells, called **endosteum** (en-dos'tē-um), lines these spaces and the medullary cavity. A specialized type of soft connective tissue called **marrow** (mar'o) fills the spongy bone spaces and the medullary cavity. The two forms of marrow, red and yellow, are described later in this chapter (see also fig. 7.2).

Microscopic Structure

Recall from chapter 5 (p. 167) that bone cells called **osteocytes** (os'te-o-sītz) are in tiny, bony chambers called *lacunae.* Osteocytes exchange substances with nearby cells by means of cellular processes passing through *canaliculi.* The extracellular matrix of bone tissue is largely collagen and inorganic salts. Collagen gives bone its strength and resilience, and inorganic salts make it hard and resistant to crushing.

Compact Bone

In compact bone, the osteocytes and layers of extracellular matrix called *lamellae* are concentrically clustered around a *central* canal (Haversian canal) forming a cylinder-shaped unit called an **osteon** (os'te-on), also called an Haversian system (figs. 7.4 and 7.5). Many of these units together form the substance of compact bone. The osteons run longitudinally with the axis of the bone, functioning as weight-bearing pillars, resisting compression.

Each central canal contains blood vessels and nerves surrounded by loose connective tissue. Blood in these vessels nourishes

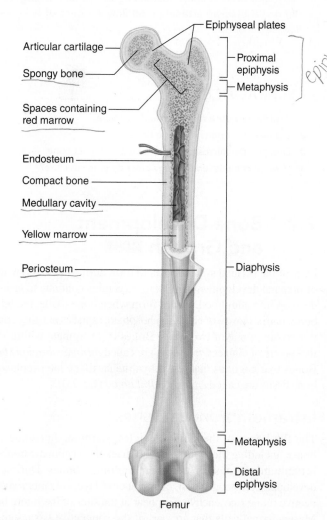

Articular cartilage
Spongy bone
Spaces containing red marrow
Endosteum
Compact bone
Medullary cavity
Yellow marrow
Periosteum

Epiphyseal plates
Proximal epiphysis
Metaphysis

Diaphysis

Metaphysis
Distal epiphysis

Femur

FIGURE 7.2 AP|R Major parts of a long bone.

Q *Name the bone.*
Answer can be found in Appendix G.

At age fifty-two the woman is younger than most of the others having their bone mineral density measured. The gynecologist advised a baseline test to assess the health of the patient's skeleton because her parents had osteoporosis.

A radiology technologist conducts the test. She explains the procedure to the patient, then positions her on her back on a padded table, fully clothed. The scanner passes painlessly over the patient's hip and lower spine, emitting low-dose X rays that form images of the bones. Spaces on the scan indicate osteopenia, the low bone mineral density that may be a prelude to osteoporosis.

Radiology technologists administer medical imaging tests, such as ultrasound and magnetic resonance imaging (MRI), as well as mammography and the three-dimensional X-ray cross sections of computerized tomography (CT). The technologists protect patients from radiation with drapes and positioning, and operate scanning devices to produce the highest-quality images from which a radiologist can accurately diagnose an illness or injury.

A registered radiologic technologist completes two years of training at a hospital or a two- or four-year program at a college or university, and must pass a national certification exam.

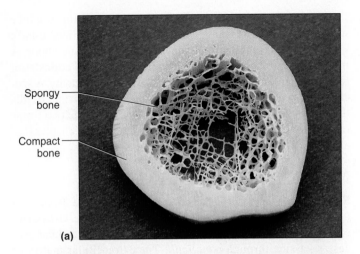

Spongy bone

Compact bone

(a)

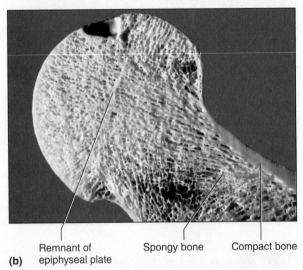

(b) Remnant of epiphyseal plate   Spongy bone   Compact bone

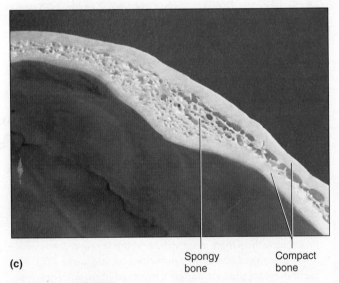

(c)   Spongy bone   Compact bone

FIGURE 7.3 **AP|R** Compact bone and spongy bone. (*a*) In a femur, the wall of the diaphysis consists mostly of compact bone. (*b*) The epiphyses of the femur contain spongy bone enclosed by a thin layer of compact bone. (*c*) This skull bone contains a layer of spongy bone sandwiched between plates of compact bone.

 What type of section has been cut from a long bone in (a)?
Answer can be found in Appendix G.

bone cells associated with the central canal via gap junctions between osteocytes.

Central canals extend longitudinally through bone tissue, and transverse *perforating canals* (Volkmann's canals) connect the central canals. Perforating canals contain larger blood vessels and nerves by which the smaller blood vessels and nerve fibers in central canals communicate with the surface of the bone and the medullary cavity (see fig. 7.4).

Spongy Bone

Spongy bone, like compact bone, is composed of osteocytes and extracellular matrix, but the bone cells do not aggregate around central canals. Instead, the cells lie in the trabeculae and get nutrients from substances diffusing into the canaliculi that lead to the surfaces of these thin, bony plates.

> Severe bone pain is one symptom of sickle cell disease, which is inherited. Under low oxygen conditions, abnormal hemoglobin (an oxygen-carrying protein) bends the red blood cells that contain it into sickle shapes, obstructing circulation. Radiographs can reveal blocked arterial blood flow in bones of sickle cell disease patients.

PRACTICE

1 Explain how bones are classified by shape.
2 List five major parts of a long bone.
3 Describe the microscopic structure of compact bone.
4 How do compact and spongy bone differ in structure?

7.2 | Bone Development and Growth AP|R

Parts of the skeletal system begin to form during the first few weeks of prenatal development, and bony structures continue to grow and develop into adulthood. Bones form when bone tissue, including a bony matrix mostly of calcium phosphate, replaces existing connective tissue in one of two ways. Bones that originate within sheetlike layers of connective tissues are called *intramembranous bones*. Bones that begin as masses of hyaline cartilage later replaced by bone tissue are called *endochondral bones* (fig. 7.6).

Intramembranous Bones

The flat bones of the skull, clavicles, sternum, and some facial bones, including the mandible, maxillae, and zygomatic bones, are **intramembranous** (in″trah-mem′brah-nus) **bones**. During their development (osteogenesis), membranelike layers of embryonic connective tissue (mesenchyme) appear at the sites of the future bones. Mesenchymal cells that are part of the connective tissues enlarge and further differentiate into bone-forming cells called **osteoblasts** (os′te-o-blasts), which, in turn, deposit bony matrix around themselves. Dense networks of blood vessels supply these connective tissues. Spongy bone forms in all directions along the blood vessels.

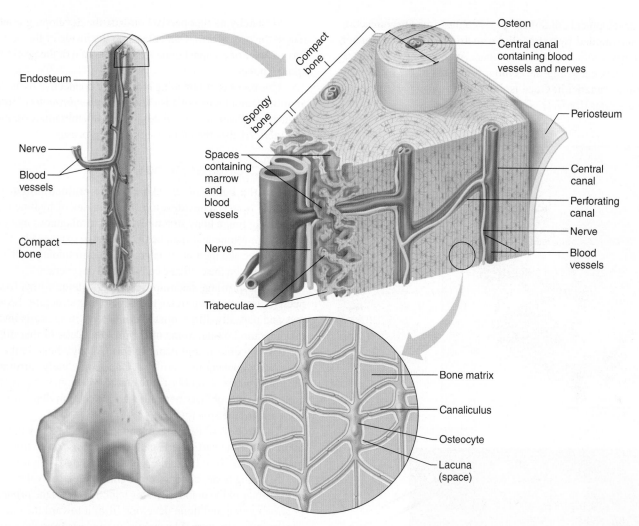

FIGURE 7.4 AP|R Compact bone is composed of osteons cemented together by bone matrix. Extensions from osteocytes communicate through tunnel-like canaliculi. (Note: Drawings are not to scale.)

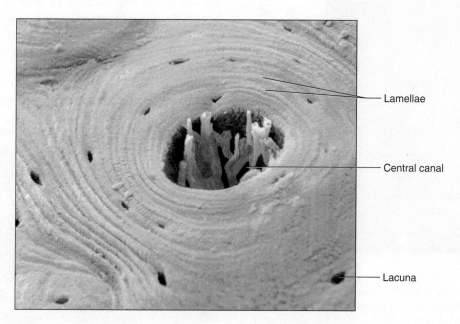

FIGURE 7.5 AP|R Scanning electron micrograph of a single osteon in compact bone (395×).

As development continues, the osteoblasts may become completely surrounded by extracellular matrix, and in this manner, they become secluded within lacunae. At the same time, extracellular matrix enclosing the cellular processes of the osteoblasts gives rise to canaliculi. Once isolated in lacunae, these cells are called *osteocytes* (fig. 7.7).

Mesenchyme that persists outside the developing bone gives rise to the periosteum. Osteoblasts on the inside of the periosteum form a layer of compact bone over the surface of the newly formed spongy bone.

This process of replacing embryonic connective tissue to form an intramembranous bone is called *intramembranous ossification*. Figure 7.8 illustrates the process of intramembranous ossification, and table 7.1 lists the major steps of the process.

Endochondral Bones

Most of the bones of the skeleton are **endochondral** (en′do-kon′dral) **bones.** They develop from masses of hyaline cartilage shaped like future bony structures. These cartilaginous models grow rapidly for a time and then begin to change extensively. Cartilage cells enlarge and their lacunae grow. The surrounding matrix breaks down, and soon the cartilage cells die and degenerate.

As the cartilage decomposes, a periosteum forms from connective tissue that encircles the developing structure. Blood vessels and partially differentiated connective tissue cells invade the disintegrating tissue. Some of the invading cells further differentiate into osteoblasts and begin to form spongy bone in the spaces previously housing the cartilage. Once completely surrounded by the bony matrix, osteoblasts are called osteocytes. As ossification continues, osteoblasts beneath the periosteum deposit compact bone around the spongy bone.

The process of forming an endochondral bone by the replacement of hyaline cartilage is called *endochondral ossification*. Its major steps are listed in table 7.1 and illustrated in figure 7.9.

In a long bone, bone tissue begins to replace hyaline cartilage in the center of the diaphysis. This region is called the *primary ossification center,* and bone develops from it toward the ends of the cartilaginous structure. Meanwhile, osteoblasts from the periosteum deposit compact bone around the primary ossification center. The epiphyses of the developing bone remain cartilaginous and continue to grow. Later, *secondary ossification centers* appear in the epiphyses, and spongy bone forms in all directions from them. As spongy bone is deposited in the diaphysis and in the epiphysis, a band of cartilage called the **epiphyseal plate** (ep″ĭ-fiz′e-al plāt) remains between the two ossification centers (see figs. 7.2, 7.3b, and 7.9).

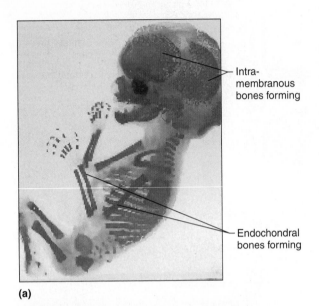

Intra-
membranous
bones forming

Endochondral
bones forming

(a)

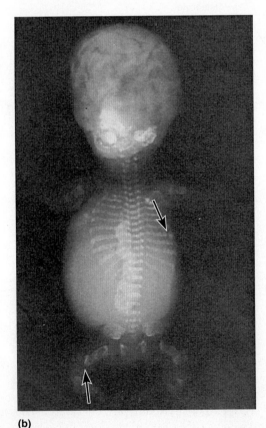

(b)

FIGURE 7.6 Fetal skeleton. (*a*) Note the stained developing bones of this fourteen-week fetus. (*b*) Bones can fracture even before birth. This fetus has numerous broken bones (arrows) because of an inherited defect in collagen called osteogenesis imperfecta, which causes brittle bones.

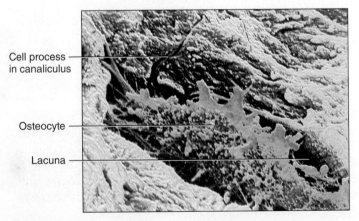

Cell process
in canaliculus

Osteocyte

Lacuna

FIGURE 7.7 Scanning electron micrograph (falsely colored) of an osteocyte isolated in a lacuna (4,700×).

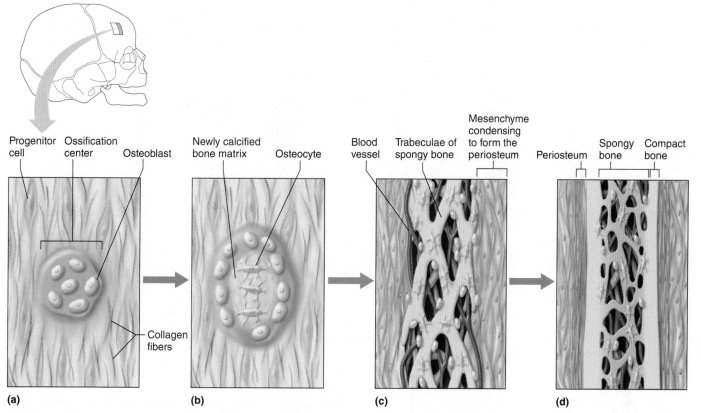

Progenitor cell | **Ossification center** | **Osteoblast**

Newly calcified bone matrix | **Osteocyte**

Blood vessel | **Trabeculae of spongy bone** | **Mesenchyme condensing to form the periosteum** | **Periosteum** | **Spongy bone** | **Compact bone**

Collagen fibers

(a)   (b)   (c)   (d)

FIGURE 7.8 Major stages in the embryonic and fetal development of intramembranous bone (a-d).

TABLE 7.1	Major Steps in Bone Development	
Intramembranous Ossification		**Endochondral Ossification**
1. Sheets of embryonic connective tissue (mesenchyme) appear at the sites of future bones.		1. Masses of hyaline cartilage form models of future bones.
2. Mesenchymal cells differentiate into osteoblasts, which deposit bone matrix.		2. Cartilage tissue breaks down. Periosteum develops.
3. Dense networks of blood vessels supply the developing spongy bone.		3. Blood vessels and differentiating osteoblasts from the periosteum invade the disintegrating tissue.
4. Osteoblasts become osteocytes when bony matrix completely surrounds them.		4. Osteoblasts form spongy bone in the space occupied by cartilage.
5. Mesenchyme on the surface of each developing structure condenses to form periosteum.		5. Osteoblasts beneath the periosteum deposit compact bone.
6. Osteoblasts on the inside of the periosteum deposit compact bone over the spongy bone.		6. Osteoblasts become osteocytes when bony matrix completely surrounds them.

Growth at the Epiphyseal Plate

In a long bone, the diaphysis is separated from the epiphysis by an epiphyseal plate. The cartilaginous cells of the epiphyseal plate form four layers, each of which may be several cells thick, as shown in figure 7.10. The first layer, or *zone of resting cartilage,* is closest to the end of the epiphysis. It is composed of resting cells that do not actively participate in growth. This layer anchors the epiphyseal plate to the bony tissue of the epiphysis.

The second layer of the epiphyseal plate, or *zone of proliferating cartilage,* includes rows of many young cells undergoing mito-sis. As new cells appear and as extracellular matrix forms around them, the cartilaginous plate thickens.

The rows of older cells, left behind when new cells appear, form the third layer, or *zone of hypertrophic cartilage,* enlarging and thickening the epiphyseal plate still more. Consequently, the entire bone lengthens. At the same time, invading osteoblasts, which secrete calcium salts, accumulate in the extracellular matrix adjacent to the oldest cartilage cells, and as the extracellular matrix calcifies, the cartilage cells begin to die.

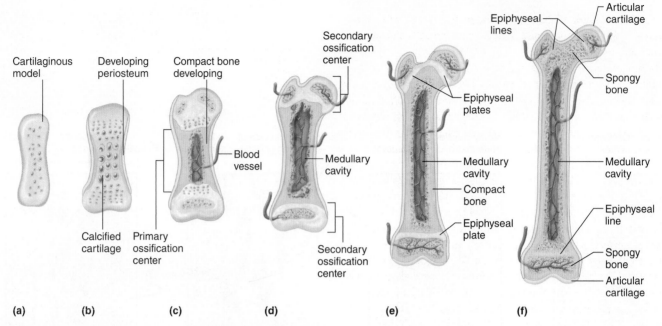

FIGURE 7.9 Major stages (*a–d* fetal, *e* child) in the development of an endochondral bone. In an (*f*) adult, when bone growth ceases, an epiphyseal line is what remains of the epiphyseal plate. (Relative bone sizes are not to scale.)

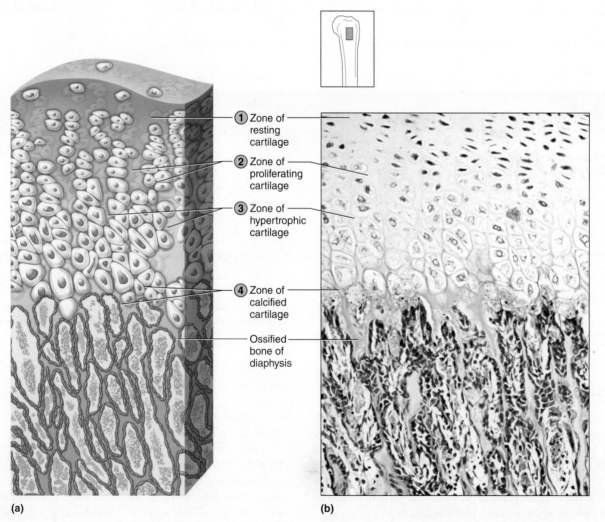

FIGURE 7.10 Epiphyseal plate. (*a*) The cartilaginous cells of an epiphyseal plate lie in four layers, each of which may be several cells thick. (*b*) A micrograph of an epiphyseal plate (100×).

The fourth layer of the epiphyseal plate, or *zone of calcified cartilage,* is thin. It is composed of dead cartilage cells and calcified extracellular matrix.

In time, large, multinucleated cells called **osteoclasts** (os′te-o-klasts) break down the calcified matrix. These large cells originate from the fusion of single-nucleated white blood cells called monocytes (see chapter 14, p. 537). Osteoclasts secrete an acid that dissolves the inorganic component of the calcified matrix, and their lysosomal enzymes digest the organic components. Osteoclasts also phagocytize components of the bony matrix. After osteoclasts remove the extracellular matrix, bone-building osteoblasts invade the region and deposit bone tissue in place of the calcified cartilage.

A long bone continues to lengthen while the cartilaginous cells of the epiphyseal plates are active. However, once the ossification centers of the diaphysis and epiphyses meet and the epiphyseal plates ossify, lengthening is no longer possible in that end of the bone.

A developing bone thickens as compact bone is deposited on the outside, just beneath the periosteum. As this compact bone forms on the surface, osteoclasts erode other bone tissue on the inside (fig. 7.11). The resulting space becomes the medullary cavity of the diaphysis, which later fills with marrow.

The bone in the central regions of the epiphyses and diaphysis remains spongy, and hyaline cartilage on the ends of the epiphyses persists throughout life as articular cartilage. Table 7.2 lists the ages at which various bones ossify.

If a child's long bones are still growing, a radiograph will reveal epiphyseal plates (fig. 7.12). If an epiphyseal plate is damaged as a result of a fracture before it ossifies, elongation of that long bone may prematurely cease, or if growth continues, it may be uneven. For this reason, injuries to the epiphyses of a young person's bones are of special concern. Surgery may be used on an epiphysis to equalize growth of bones developing at very different rates.

 PRACTICE

5 Describe the development of an intramembranous bone.

6 Explain how an endochondral bone develops.

7 Describe the four layers in an epiphyseal plate.

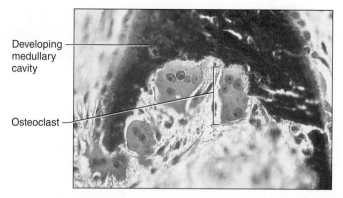

FIGURE 7.11 Micrograph of bone-resorbing osteoclasts (800×).

Labels: Developing medullary cavity; Osteoclast

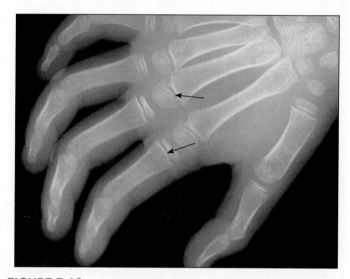

FIGURE 7.12 Radiograph of the right hand. Epiphyseal plates (arrows) in a child's bones indicate that the bones are still lengthening.

Homeostasis of Bone Tissue

After the intramembranous and endochondral bones form, the actions of osteoclasts and osteoblasts continually remodel them. **Bone remodeling** occurs throughout life as osteoclasts resorb bone tissue and osteoblasts replace the bone. These opposing processes of *resorption* and *deposition* occur on the surfaces of the endosteum and periosteum. The rate of bone remodeling is not uniform among the types of bones; remodeling of spongy bone is faster than that

TABLE 7.2	Ossification Timetable		
Age	**Occurrence**	**Age**	**Occurrence**
Third month of prenatal development	Ossification in long bones begins.	15 to 18 years (females) 17 to 20 years (males)	Bones of the upper limbs and scapulae completely ossified.
Fourth month of prenatal development	Most primary ossification centers have appeared in the diaphyses of long bones.	16 to 21 years (females) 18 to 23 years (males)	Bones of the lower limbs and hip bones completely ossified.
Birth to 5 years	Secondary ossification centers appear in the epiphyses of long bones.	21 to 23 years (females) 23 to 25 years (males)	Bones of the sternum, clavicles, and vertebrae completely ossified.
5 to 12 years (females) 5 to 14 years (males)	Ossification rapidly spreads from the ossification centers.	By 23 years (females) By 25 years (males)	Nearly all bones completely ossified.

CLINICAL APPLICATION 7.1

Fractures

When seven-year-old Jacob fell from the tree limb, he had been hanging about eight feet from the ground. He landed in a crumpled heap, crying, with his right leg at an abnormal angle. Emergency medical technicians immobilized the leg and took Jacob to the emergency department at the nearest hospital, where an X ray indicated a broken tibia. He spent the next six weeks with his leg in a cast, and the bone continued to heal over several months. By the next summer, Jacob was again climbing trees—but more carefully.

Many of us have had fractured, or broken, bones. A fracture is classified by its cause and the nature of the break. For example, a break due to injury is a traumatic fracture, whereas one resulting from disease is a *spontaneous*, or *pathologic*, *fracture*. A broken bone exposed to the outside by an opening in the skin is termed a *compound (open) fracture*. It has the added danger of infection, because microorganisms can enter through the broken skin. A break protected by uninjured skin is a *closed fracture*. Figure 7A shows several types of traumatic fractures.

Repair of a Fracture

When a bone breaks, blood vessels in it also break, and the periosteum is likely to tear. Blood from the broken vessels spreads through the damaged area and soon forms a *hematoma*, which is a localized collection of blood that usually is clotted. Vessels in surrounding tissues dilate, swelling and inflaming tissues.

Within days or weeks, developing blood vessels and large numbers of osteoblasts originating from the periosteum invade the hematoma. The osteoblasts rapidly divide in the regions close to the new blood vessels, building spongy bone nearby. Granulation tissue develops, and in regions farther from a blood supply, fibroblasts produce masses of fibrocartilage. Meanwhile, phagocytic cells begin to remove the blood clot as well as any dead or damaged cells in

 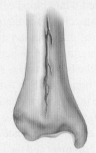 

A *greenstick* fracture is incomplete, and the break occurs on the convex surface of the bend in the bone.

A *fissured* fracture is an incomplete longitudinal break.

A *comminuted* fracture is complete and fragments the bone.

A *transverse* fracture is complete, and the break occurs at a right angle to the axis of the bone.

An *oblique* fracture occurs at an angle other than a right angle to the axis of the bone.

A *spiral* fracture is caused by excessive twisting of a bone.

FIGURE 7A Types of fractures.

the affected area. Osteoclasts also appear and resorb bone fragments, aiding in "cleaning up" debris.

In time, fibrocartilage fills the gap between the ends of the broken bone. This mass, a soft (cartilaginous) callus, is later replaced by bone tissue in much the same way that the hyaline cartilage of

of compact bone. The total mass of bone tissue in an adult skeleton normally remains nearly constant because of the tight regulation of bone remodeling. These ongoing processes replace 10% to 20% of the skeleton each year. Clinical Application 7.1 describes what happens when a bone breaks.

Factors Affecting Bone Development, Growth, and Repair

A number of factors influence bone development, growth, and repair. These include nutrition, exposure to sunlight, hormonal

secretions, and physical exercise. For example, vitamin D is necessary for proper absorption of dietary calcium in the small intestine. Without this vitamin, dietary calcium is poorly absorbed and the inorganic salt portion of bone matrix will lack calcium, softening and thereby deforming bones. In children, this condition is called *rickets*, and in adults, it is called *osteomalacia*.

Vitamin D is scarce in natural foods, except for eggs, but it is readily available in milk and other dairy products fortified with vitamin D. An inactive form of vitamin D also forms when dehydrocholesterol, produced by skin cells or obtained in the diet and carried by the blood to the skin, is exposed to ultraviolet light.

a developing endochondral bone is replaced. That is, the cartilaginous callus breaks down, blood vessels and osteoblasts invade the area, and a hard (bony) callus fills the space.

Typically, more bone is produced at the site of a healing fracture than is necessary to replace the damaged tissues. Osteoclasts remove the excess, and the result is a bone shaped much like the original. Figure 7B shows the steps in the healing of a fracture.

If the ends of a broken bone are close together, healing is faster than if they are far apart. Physicians can help the bone-healing process. The first casts to immobilize fractured bones were introduced in Philadelphia in 1876, and soon after, doctors began using screws and plates internally to align healing bone parts. Today, orthopedic surgeons also use rods, wires, and nails. These devices have become lighter and smaller; many are built of titanium. A device called a hybrid fixator treats a broken leg using metal pins internally to align bone pieces. The pins are anchored to a metal ring device worn outside the leg.

Some bones naturally heal more rapidly than others. The long bones of the upper limbs, for example, may heal in half the time required by the long bones of the lower limbs, as Jacob was unhappy to discover. However, his young age would favor quicker healing.

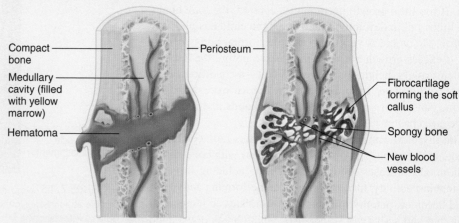

(a) Blood escapes from ruptured blood vessels and forms a hematoma.

(b) Spongy bone forms in regions close to developing blood vessels, and fibrocartilage forms in more distant regions.

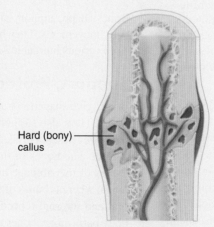

(c) A hard (bony) callus replaces the fibrocartilage.

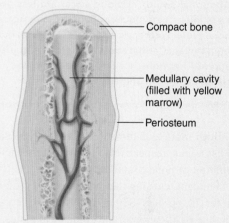

(d) Osteoclasts remove excess bony tissue, restoring new bone structure much like the original.

FIGURE 7B Major steps (a–d) in repair of a fracture.

This inactive form of vitamin D is then processed in the liver and kidneys to become the active form of vitamin D.

Vitamins A and C are also required for normal bone development and growth. Vitamin A is necessary for osteoblast and osteoclast activity during normal development. This is why deficiency of vitamin A may retard bone development. Vitamin C is required for collagen synthesis. Deficiency inhibits bone development. Osteoblasts cannot produce enough collagen in the extracellular matrix of the bone tissue, and as a result, bones are abnormally slender and fragile.

About 90% of the protein that is part of bone is collagen. Less abundant bone proteins are important too.

- Osteocalcin is activated by vitamin K to bind calcium, which in bone is part of the compound hydroxyapatite, the main component of bone matrix.
- Osteonectin binds hydroxyapatite and collagen and stimulates mineral crystal deposition in bone.
- Osteopontin speeds bone remodeling.
- Bone morphogenetic proteins include growth factors that induce bone and cartilage formation. They are used in spinal fusion procedures.

Hormones secreted by the pituitary gland, thyroid gland, parathyroid glands, and ovaries or testes affect bone growth and development. The pituitary gland secretes **growth hormone,** which stimulates division of cartilage cells in the epiphyseal plates. If too little growth hormone is secreted, the long bones of the limbs fail to develop normally, and the child has *pituitary dwarfism.* He or she is very short but has normal body proportions. If excess growth hormone is released before the epiphyseal plates ossify, height may exceed 8 feet—a condition called *pituitary gigantism.* In an adult, secretion of excess growth hormone causes *acromegaly,* in which the hands, feet, and jaw enlarge (see chapter 13, p. 500).

The thyroid hormone thyroxine stimulates replacement of cartilage in the epiphyseal plates of long bones with bone tissue. This hormone increases cellular metabolism, including stimulating osteoblast activity. In contrast to the bone-forming activity of thyroid hormone, parathyroid hormone stimulates an increase in the number and activity of osteoclasts, which break down bone (see chapter 13, pp. 505–507).

Abnormal swings in blood sugar are less likely in individuals who exercise regularly. Part of the reason may be that the bone protein osteocalcin appears to act as a hormone, released from bones after the stress of exercise. Osteocalcin works with another hormone, insulin, to help the body regulate blood sugar levels more precisely.

Both male and female sex hormones (called testosterone and estrogens, respectively) from the testes and ovaries promote formation of bone tissue. Beginning at puberty, these hormones are abundant, causing the long bones to grow considerably (see chapter 22, pp. 838–839 and 849). However, sex hormones also stimulate ossification of the epiphyseal plates, and consequently they stop bone lengthening at a relatively early age. The effect of estrogens on the epiphyseal plates is somewhat stronger than that of testosterone. For this reason, females typically reach their maximum heights earlier than males.

Physical stress also stimulates bone growth. For example, when skeletal muscles contract, they pull at their attachments on bones, and the resulting stress stimulates the bone tissue to thicken and strengthen (hypertrophy). Conversely, with lack of exercise, the same bone tissue becomes thinner and weaker. This condition, called atrophy, is why the bones of athletes are usually stronger and heavier than those of nonathletes (fig. 7.13). Atrophy is also why fractured bones immobilized in casts may shorten. Clinical Application 7.2 describes the importance of calcium, vitamin D, and exercise for maintaining healthy bones.

 PRACTICE

8 Explain how nutritional factors affect bone development.

9 What effects do hormones have on bone growth?

10 How does physical exercise affect bone structure?

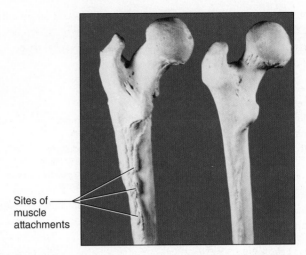

FIGURE 7.13 Note the increased amount of bone at the sites of muscle attachments in the femur on the left. The thickened bone is better able to withstand the forces resulting from muscle contraction.

7.3 | Bone Function

Bones shape, support, and protect body structures, as well as aid body movements. They house tissue that produces blood cells, and store various inorganic salts.

Support, Protection, and Movement

Bones give shape to structures such as the head, face, thorax, and limbs. They also support and protect. For example, the bones of the lower limbs, pelvis, and vertebral column support the body's weight. The bones of the skull protect the eyes, ears, and brain. Bones of the rib cage and shoulder girdle protect the heart and lungs, whereas bones of the pelvic girdle protect the lower abdominal and internal reproductive organs. Whenever limbs or other body parts move, bones and muscles interact.

Blood Cell Formation

The process of blood cell formation, called **hematopoiesis** (hem″ah-to-poi-e′sis), or hemopoiesis, begins in the yolk sac, which lies outside the embryo (see chapter 23, p. 874). Later in development, blood cells are manufactured in the liver and spleen, and still later, they form in bone marrow.

Marrow is a soft, netlike mass of connective tissue in the medullary cavities of long bones, in the irregular spaces of spongy bone, and in the larger central canals of compact bone tissue. The two types of marrow are red and yellow. In *red marrow,* red blood cells (erythrocytes), white blood cells (leukocytes), and blood platelets form. The color comes from the red, oxygen-carrying pigment **hemoglobin** in red blood cells.

The distribution of the two types of marrow changes with age. In an infant, the cavities of most bones house red marrow, but with time, yellow marrow replaces much of it. *Yellow marrow* stores fat and does not produce blood cells. In an adult, red marrow is primarily found in the spongy bone of the skull, ribs, sternum, clavicles, vertebrae, and hip bones. If the blood cell supply is deficient, some yellow marrow may become red marrow and produce blood cells. Then, the red marrow reverts to yellow marrow when the deficiency is corrected. Chapter 14 (pp. 531, 533, 536, and 539) discusses blood cell formation in more detail.

Bone marrow transplants have been used for more than half a century to enable people with certain cancers to tolerate high levels of chemotherapy drugs. Replacing bone marrow is also a way to treat an inherited disorder of blood cells (such as sickle cell disease) or a disease of the descendants of these cells that function outside the circulation (such as the brain disease adrenoleukodystrophy).

In a bone marrow transplant, a hollow needle and syringe remove normal red marrow cells from the spongy bone of a donor, or stem cells (which can give rise to specialized blood cells) are separated out from the donor's bloodstream. Stem cells from the umbilical cord of a newborn can be used in place of bone marrow and are less likely to stimulate an immune response in the recipient.

Donors are selected based on their cells having a pattern of molecules on their surfaces that closely matches that of the recipient's cells. In 30% of bone marrow transplants from donors, the donor is a blood relative. The cells are injected into the bloodstream of the recipient, whose own marrow has been intentionally destroyed with radiation or chemotherapy. If all goes well, the donor cells travel to the spaces within bones that red marrow normally occupies, where they replenish the blood supply with healthy cells. About 5% of patients die from infection because their immune systems reject the transplant, or because the transplanted tissue attacks the recipient, a condition called graft-versus-host disease.

Safer than a bone marrow transplant for some conditions is an autologous ("self") stem cell transplant. Stem cells are taken from a patient's bloodstream, and a mutation is corrected or healthy cells are separated and cultured. Or the bone marrow is stimulated to make stem cells, which are isolated. Then high doses of chemotherapy or radiation are used to destroy the rest of the bone marrow in the patient's body, and the patient's stem cells are infused. They migrate to the bone marrow and reconstitute a disease-free blood-forming system that the patient's immune system does not reject.

Inorganic Salt Storage

Recall that the extracellular matrix of bone tissue includes collagen and inorganic mineral salts. The salts account for about 70% of the extracellular matrix by weight and are mostly small crystals of a type of calcium phosphate called *hydroxyapatite*.

The human body requires calcium for a number of vital metabolic processes, including muscle cell contraction, nerve cell conduction, and blood clot formation. When the blood is low in calcium, parathyroid hormone stimulates osteoclasts to break down bone tissue, releasing calcium salts from the extracellular matrix into the blood. Very high blood calcium inhibits osteoclast activity, and calcitonin from the thyroid gland stimulates osteoblasts to form bone tissue, storing excess calcium in the extracellular matrix (fig. 7.14).

This response is particularly important in developing bone matrix in children. The details of the homeostatic mechanism that controls calcium levels in the blood are in chapter 13, pp. 505–508.

In addition to storing calcium and phosphorus (as calcium phosphate), bone tissue contains smaller amounts of magnesium, sodium, potassium, and carbonate ions. Bones also accumulate certain harmful metallic elements such as lead, radium, and strontium, which are not normally present in the body but are sometimes accidentally ingested.

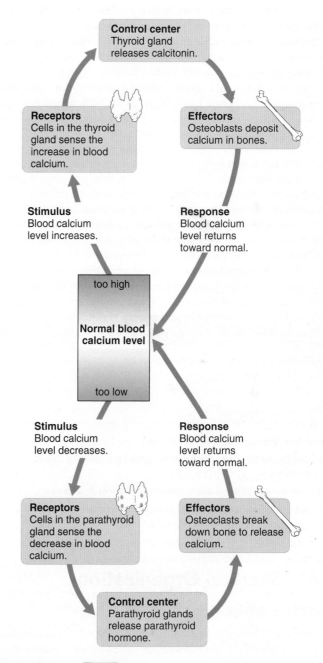

FIGURE 7.14 **AP|R** Hormonal regulation of bone calcium deposition and resorption.

What three components of a homeostatic mechanism (see fig. 1.6, p. 16) does this figure show?

Answer can be found in Appendix G.

Skeletal health is a matter of balance. Before age thirty, cells that form new bone counter cells that degrade it, so that living bone is in a constant state of remodeling. Then the balance shifts so that bone is lost, especially in women past menopause, due to hormonal changes. This imbalance may progress to osteopenia or the more severe osteoporosis (fig. 7C).

A "fragility fracture" is a telltale sign of dangerously low bone density. This is a fracture that happens after a fall from less than standing height, which a strong, healthy skeleton could resist. Fragility fractures occur in more than 2 million people in the United States each year, yet despite this warning sign, only one-fourth to one-third of them are followed up with bone scans and treatment to build new bone tissue. Since 1995, five new drugs have become available to treat osteoporosis. One class, the bisphosphonates, actually builds new bone.

Osteopenia and osteoporosis are common. The surgeon general estimates that half of all people over age fifty have one of these conditions, which amounts to 10 million with osteoporosis and another 35 million with osteopenia. Screening is advised for all individuals over age sixty-five, as well as for those with risk factors. The most telling predictor is a previous fragility fracture. Other risk factors include a family history of osteoporosis, recent height loss, and older age.

Osteopenia and osteoporosis are not just concerns of older people approaching retirement age. People of all ages can take steps to avoid developing these preventable conditions. Researchers think that what puts people at risk is failure to attain maximal possible bone density by age thirty. To keep bones as

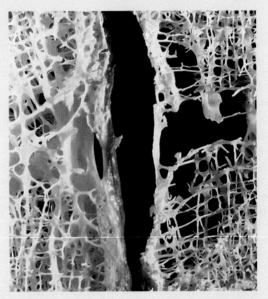

FIGURE 7C Spaces in bones enlarge when a person has osteoporosis. The portion of a vertebra on the left is normal; the one on the right has been weakened by osteoporosis.

strong as possible for as long as possible, it is essential to get at least 30 minutes of exercise daily (some of which should be weight-bearing), consume enough daily calcium (1,000–1,200 mg) and vitamin D (600-4,000 IU), and not smoke. There is much you can do to promote skeletal health—at any age.

PRACTICE

11 Name the major functions of bones.

12 Distinguish between the functions of red marrow and yellow marrow.

13 Explain regulation of the concentration of blood calcium.

14 List the substances normally stored in bone tissue.

7.4 | Skeletal Organization

Number of Bones

The number of bones in an adult human skeleton is often reported to be 206 (table 7.3), but the number varies because people may lack certain bones or have extra ones. For example, the flat bones of the skull usually grow together and tightly join along irregular lines called *sutures*. In some people, extra bones called sutural bones (wormian bones) develop in these sutures (fig. 7.15). Extra small, round sesamoid bones may develop in tendons, where they reduce friction in places where tendons pass over bony prominences.

Divisions of the Skeleton

For purposes of study, it is convenient to divide the skeleton into two major portions—an axial skeleton and an appendicular skeleton (fig. 7.16). The **axial skeleton** consists of the bony and cartilaginous parts that support and protect the organs of the head, neck, and trunk. These parts include the following:

1. The **skull** is composed of the *cranium* (brain case) and the *facial bones.*

2. The **middle ear bones** transfer sound vibrations to the hearing receptors.

3. The **hyoid** (hi'oid) **bone** is located in the neck between the lower jaw and the larynx (fig. 7.17). It does not articulate with any other bones but is fixed in position by muscles and ligaments. The hyoid bone supports the tongue and is an attachment for certain muscles that help move the tongue during swallowing. It can be felt approximately a finger's width above the anterior prominence of the larynx.

4. The **vertebral column**, or spinal column, consists of many **vertebrae** separated by cartilaginous *intervertebral discs.* This column forms the central axis of the skeleton. Near its distal end, five vertebrae fuse to form the **sacrum** (sa'krum),

TABLE 7.3 Bones of the Adult Skeleton

1. Axial Skeleton			2. Appendicular Skeleton		
a. Skull		22 bones	**a. Pectoral girdle**		4 bones
8 cranial bones	14 facial bones		scapula 2		
frontal 1	maxilla 2		clavicle 2		
parietal 2	palatine 2		**b. Upper limbs**		60 bones
occipital 1	zygomatic 2		humerus 2		
temporal 2	lacrimal 2		radius 2		
sphenoid 1	nasal 2		ulna 2		
ethmoid 1	vomer 1		carpal 16		
	inferior nasal concha 2		metacarpal 10		
	mandible 1		phalanx 28		
b. Middle ear bones		6 bones	**c. Pelvic girdle**		2 bones
malleus 2			hip bone 2		
incus 2			**d. Lower limbs**		60 bones
stapes 2			femur 2		
c. Hyoid		1 bone	tibia 2		
hyoid bone 1			fibula 2		
d. Vertebral column		26 bones	patella 2		
cervical vertebra 7			tarsal 14		
thoracic vertebra 12			metatarsal 10		
lumbar vertebra 5			phalanx 28		
sacrum 1			**Total**		206 bones
coccyx 1					
e. Thoracic cage		25 bones			
rib 24					
sternum 1					

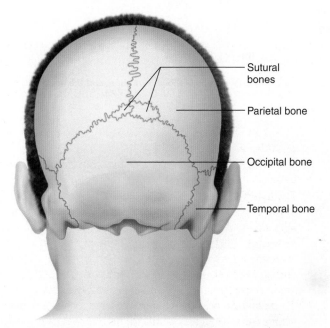

FIGURE 7.15 Sutural (wormian) bones are extra bones that sometimes develop in sutures between the flat bones of the skull.

- Sutural bones
- Parietal bone
- Occipital bone
- Temporal bone

part of the pelvis. A small tailbone formed by the fusion of four vertebrae and called the **coccyx** (kok′siks) is attached to the end of the sacrum.

5. The **thoracic cage** protects the organs of the thoracic cavity and the upper abdominal cavity. It is composed of twelve pairs of **ribs,** which articulate posteriorly with the thoracic vertebrae. It also includes the **sternum** (ster′num), or breastbone, to which most of the ribs are attached anteriorly.

The **appendicular skeleton** consists of the bones of the upper and lower limbs and the bones that anchor the limbs to the axial skeleton. It includes the following:

1. The **pectoral girdle** is formed by a **scapula** (scap′u-lah), or shoulder blade, and a **clavicle** (klav′ĭ-k′l), or collarbone, on both sides of the body. The pectoral girdle connects the bones of the upper limbs to the axial skeleton and aids in upper limb movements.

2. Each **upper limb** consists of a **humerus** (hu′mer-us), or arm bone; two forearm bones—a **radius** (ra′de-us) and an **ulna** (ul′nah)—and a hand. The humerus, radius, and ulna articulate with each other at the elbow joint. At the distal end of the radius and ulna is the hand. There are eight **carpals** (kar′palz), or wrist bones. The five bones of the palm are called **metacarpals** (met″ah-kar′palz), and the fourteen finger bones are called **phalanges** (fah-lan′jēz; sing., *phalanx,* fa′lanks).

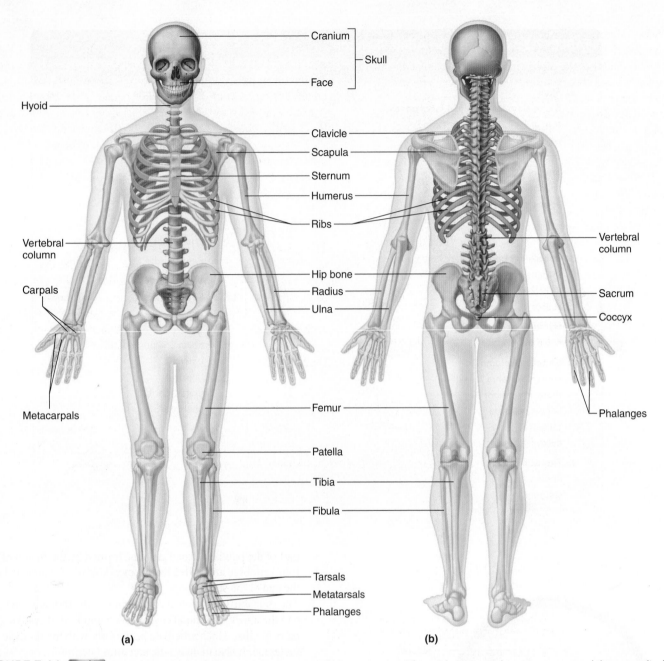

FIGURE 7.16 [AP|R] Major bones of the skeleton. (*a*) Anterior view. (*b*) Posterior view. The axial portion is shown in orange, and the appendicular portions are shown in yellow.

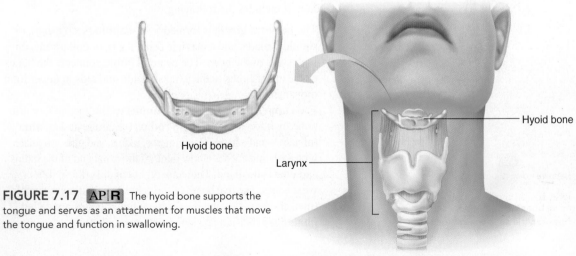

FIGURE 7.17 [AP|R] The hyoid bone supports the tongue and serves as an attachment for muscles that move the tongue and function in swallowing.

3. The **pelvic girdle** is formed by two hip bones attached to each other anteriorly and to the sacrum posteriorly. They connect the bones of the lower limbs to the axial skeleton and, with the sacrum and coccyx, form the **pelvis,** which protects the lower abdominal and internal reproductive organs.

4. Each **lower limb** consists of a **femur** (fe′mur), or thigh bone; two leg bones—a large **tibia** (tib′e-ah), or shin bone, and a slender **fibula** (fib′u-lah)—and a foot. The femur and tibia articulate with each other at the knee joint, where the **patella** (pah-tel′ah), or kneecap, covers the anterior surface. At the distal ends of the tibia and fibula is the foot. There are seven **tarsals** (tahr′salz), or ankle bones. The five bones of the instep are called **metatarsals** (met″ah-tar′salz), and the fourteen bones of the toes (like those of the fingers) are called **phalanges.** Table 7.4 defines some terms used to describe skeletal structures.

 PRACTICE

15 Distinguish between the axial and appendicular skeletons.

16 List the bones of the axial skeleton and of the appendicular skeleton.

7.5 | Skull AP|R

A human skull usually consists of twenty-two bones that, except for the lower jaw, are firmly interlocked along immovable joints called sutures. Eight of these interlocked bones make up the cranium and fourteen form the facial skeleton. The **mandible** (man′dĭ-b′l), or lower jawbone, is a movable bone held to the cranium by ligaments (figs. 7.18 and 7.19). Some facial and cranial bones together form the orbit of the eye (fig. 7.20). Plates 26–54 on pages 252–266 show a set of photographs of the human skull and its parts.

Cranium

The **cranium** (kra′ne-um) encloses and protects the brain, and its surface provides attachments for muscles that make chewing and head movements possible. Some of the cranial bones contain air-filled cavities called *paranasal sinuses,* lined with mucous membranes and connected by passageways to the nasal cavity (fig. 7.21). Sinuses reduce the weight of the skull and increase the intensity of the voice by serving as resonant sound chambers.

TABLE 7.4	Terms Used to Describe Skeletal Structures	
Term	**Definition**	**Example**
Condyle (kon′dīl)	Rounded process that usually articulates with another bone	Occipital condyle of the occipital bone (fig. 7.22)
Crest (krest)	Narrow, ridgelike projection	Iliac crest of the ilium (fig. 7.49)
Epicondyle (ep″ĭ-kon′dīl)	Projection situated above a condyle	Medial epicondyle of the humerus (fig. 7.44)
Facet (fas′et)	Small, nearly flat surface	Rib facet of a thoracic vertebra (fig. 7.37b)
Fissure (fish′ūr)	Cleft or groove	Inferior orbital fissure in the orbit of the eye (fig. 7.19)
Fontanel (fon″tah-nel′)	Soft spot in the skull where membranes cover the space between bones	Anterior fontanel between the frontal and parietal bones (fig. 7.32)
Foramen (fo-ra′men)	Opening through a bone that usually serves as a passageway for blood vessels, nerves, or ligaments	Foramen magnum of the occipital bone (fig. 7.22)
Fossa (fos′ah)	Relatively deep pit or depression	Olecranon fossa of the humerus (fig. 7.44b)
Fovea (fo′ve-ah)	Tiny pit or depression	Fovea capitis of the femur (fig. 7.52b)
Head (hed)	Enlargement on the end of a bone	Head of the humerus (fig. 7.44)
Linea (lin′e-ah)	Narrow ridge	Linea aspera of the femur (fig. 7.52b)
Meatus (me-a′tus)	Tubelike passageway within a bone	External acoustic meatus of the temporal bone (fig. 7.19)
Process (pros′es)	Prominent projection on a bone	Mastoid process of the temporal bone (fig. 7.19)
Ramus (ra′mus)	Branch or similar extension	Ramus of the mandible (fig. 7.30a)
Sinus (si′nus)	Cavity within a bone	Frontal sinus of the frontal bone (fig. 7.21)
Spine (spīn)	Thornlike projection	Spine of the scapula (fig. 7.42a, b)
Sulcus (sul′kus)	Furrow or groove	Intertubucular sulcus of the humerus (fig. 7.44)
Suture (soo′cher)	Interlocking line of union between bones	Lambdoid suture between the occipital and parietal bones (fig. 7.19)
Trochanter (tro-kan′ter)	Relatively large process	Greater trochanter of the femur (fig. 7.52a)
Tubercle (tu′ber-kl)	Knoblike process	Tubercle of a rib (fig. 7.40)
Tuberosity (tu″bĕ-ros′ĭ-te)	Knoblike process usually larger than a tubercle	Radial tuberosity of the radius (fig. 7.45a)

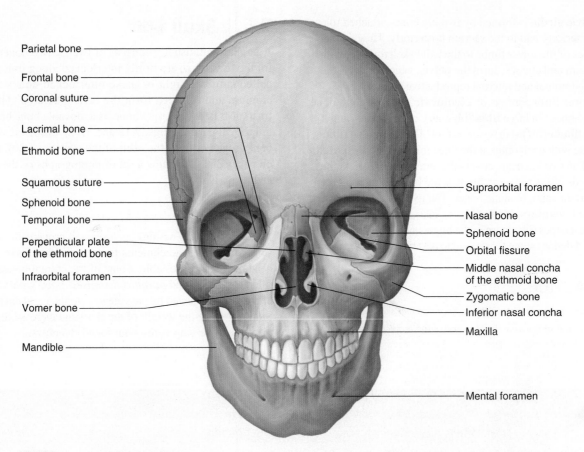

Parietal bone
Frontal bone
Coronal suture
Lacrimal bone
Ethmoid bone
Squamous suture
Sphenoid bone
Temporal bone
Perpendicular plate of the ethmoid bone
Infraorbital foramen
Vomer bone
Mandible

Supraorbital foramen
Nasal bone
Sphenoid bone
Orbital fissure
Middle nasal concha of the ethmoid bone
Zygomatic bone
Inferior nasal concha
Maxilla
Mental foramen

FIGURE 7.18 AP|R Anterior view of the skull.

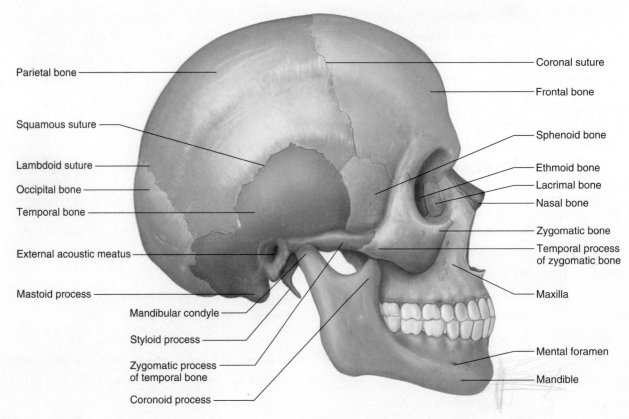

Parietal bone
Squamous suture
Lambdoid suture
Occipital bone
Temporal bone
External acoustic meatus
Mastoid process
Mandibular condyle
Styloid process
Zygomatic process of temporal bone
Coronoid process

Coronal suture
Frontal bone
Sphenoid bone
Ethmoid bone
Lacrimal bone
Nasal bone
Zygomatic bone
Temporal process of zygomatic bone
Maxilla
Mental foramen
Mandible

FIGURE 7.19 AP|R Right lateral view of the skull.

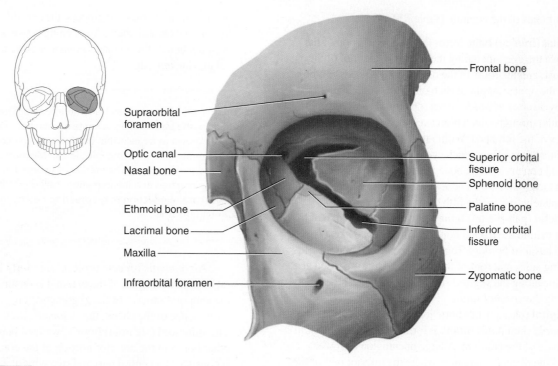

FIGURE 7.20 AP|R The orbit of the eye includes both cranial and facial bones.

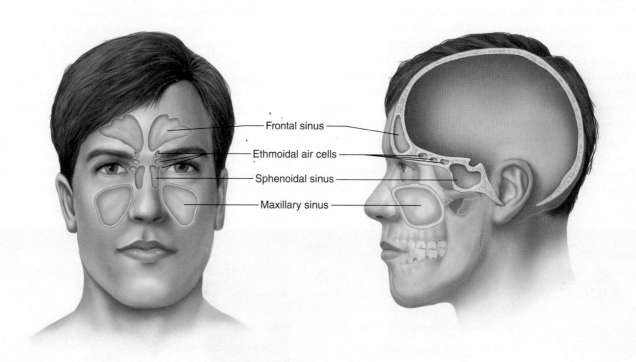

FIGURE 7.21 AP|R Locations of the paranasal sinuses.

The eight bones of the cranium (table 7.5) are:

1. The **frontal** (frun'tal) **bone** forms the anterior portion of the skull above the eyes, including the forehead, the roof of the nasal cavity, and the roofs of the orbits (bony sockets) of the eyes. On the upper margin of each orbit, the frontal bone is marked by a *supraorbital foramen* (or *supraorbital notch* in some skulls) through which blood vessels and nerves pass to the tissues of the forehead. Within the frontal bone are two *frontal sinuses*, one above each eye near the midline (fig. 7.21). The frontal bone is a single bone in adults, but it develops in two parts (see fig. 7.32b). These halves grow together and usually completely fuse by the fifth or sixth year of life.

2. One **parietal** (pah-ri'ĕ-tal) **bone** is located on each side of the skull just behind the frontal bone. Each is shaped like a curved plate and has four borders. Together, the parietal bones form the bulging sides and roof of the cranium. They are fused at the midline along the *sagittal suture*, and they meet the frontal bone along the *coronal suture*.

3. The **occipital** (ok-sip'i-tal) **bone** joins the parietal bones along the *lambdoid* (lam'doid) *suture*. It forms the back of the skull and the base of the cranium. A large opening on its lower surface is the *foramen magnum*, where the inferior part of the brainstem connects with the spinal cord. Rounded processes called *occipital condyles*, located on each side of the foramen magnum, articulate with the first vertebra (atlas) of the vertebral column.

4. A **temporal** (tem'por-al) **bone** on each side of the skull joins the parietal bone along a *squamous suture*. The temporal bones form parts of the sides and the base of the cranium. Located near the inferior margin is an opening, the *external acoustic* (auditory) *meatus*, which leads inward to parts of the ear. The temporal bones also house the internal ear structures and have depressions called the *mandibular fossae* (glenoid fossae) that articulate with condyles of the mandible. Below each external acoustic meatus are two projections—a rounded *mastoid process* and a long, pointed *styloid process* (see fig. 7.20). The mastoid process provides an attachment for certain muscles of the neck, whereas the styloid process anchors muscles associated with the tongue and pharynx. An

opening near the mastoid process, the *carotid canal*, transmits the internal carotid artery. An opening between the temporal and occipital bones, the *jugular foramen*, accommodates the internal jugular vein (fig. 7.22).

> The mastoid process may become infected. The tissues in this region of the temporal bone contain a number of interconnected spaces known as air cells that are lined with mucous membranes that communicate with the middle ear. Microorganisms from an infected middle ear *(otitis media)* can spread into the air cells, causing infection and inflammation, called *mastoiditis.* The danger is that the infection may spread to nearby membranes that surround the brain.

A *zygomatic process* projects anteriorly from the temporal bone in the region of the external acoustic meatus. It joins the temporal process of the zygomatic bone and helps form the prominence of the cheek, the *zygomatic arch* (fig. 7.22).

5. The **sphenoid** (sfe'noid) **bone** is wedged between several other bones in the anterior portion of the cranium (fig. 7.23). It consists of a central part and two winglike structures that extend laterally toward each side of the skull. This bone helps form the base of the cranium, the sides of the skull, and the floors and sides of the orbits. Along the midline within the cranial cavity, a portion of the sphenoid bone indents to form the saddle-shaped *sella turcica* (sel'ah tur'si-ka) (Turk's saddle). In this depression lies the pituitary gland, which hangs from the base of the brain by a stalk.

   The sphenoid bone also contains two *sphenoidal sinuses* (see fig. 7.21). These lie side by side and are separated by a bony septum that projects downward into the nasal cavity.

6. The **ethmoid** (eth'moid) **bone** is located in front of the sphenoid bone (fig. 7.24). It consists of two masses, one on each side of the nasal cavity, joined horizontally by thin *cribriform* (krib'rĭ-form) *plates*. These plates form part of the roof of the nasal cavity, and nerves associated with the sense of smell pass through tiny openings (*olfactory foramina*) in them. Portions of the ethmoid bone also form sections of the

TABLE 7.5	Cranial Bones	
Name and Number	**Description**	**Special Features**
Frontal (1)	Forms forehead, roof of nasal cavity, and roofs of orbits	Supraorbital foramen, frontal sinuses
Parietal (2)	Form side walls and roof of cranium	Fused at midline along sagittal suture
Occipital (1)	Forms back of skull and base of cranium	Foramen magnum, occipital condyles
Temporal (2)	Form side walls and floor of cranium	External acoustic meatus, mandibular fossa, mastoid process, styloid process, zygomatic process
Sphenoid (1)	Forms parts of base of cranium, sides of skull, and floors and sides of orbits	Sella turcica, sphenoidal sinuses
Ethmoid (1)	Forms parts of roof and walls of nasal cavity, floor of cranium, and walls of orbits	Cribriform plates, perpendicular plate, superior and middle nasal conchae, ethmoidal air cells, crista galli

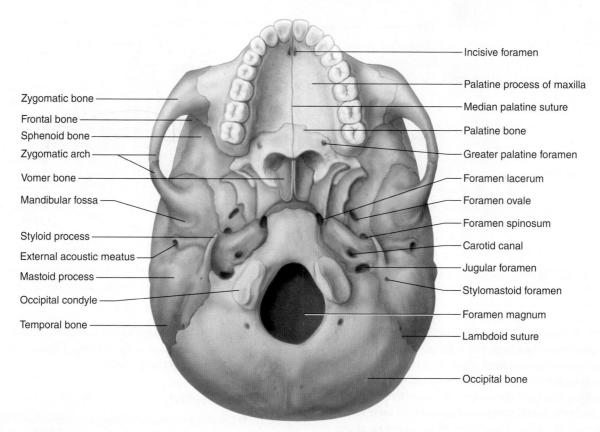

Zygomatic bone
Frontal bone
Sphenoid bone
Zygomatic arch
Vomer bone
Mandibular fossa
Styloid process
External acoustic meatus
Mastoid process
Occipital condyle
Temporal bone

Incisive foramen
Palatine process of maxilla
Median palatine suture
Palatine bone
Greater palatine foramen
Foramen lacerum
Foramen ovale
Foramen spinosum
Carotid canal
Jugular foramen
Stylomastoid foramen
Foramen magnum
Lambdoid suture
Occipital bone

FIGURE 7.22 AP|R Inferior view of the skull.

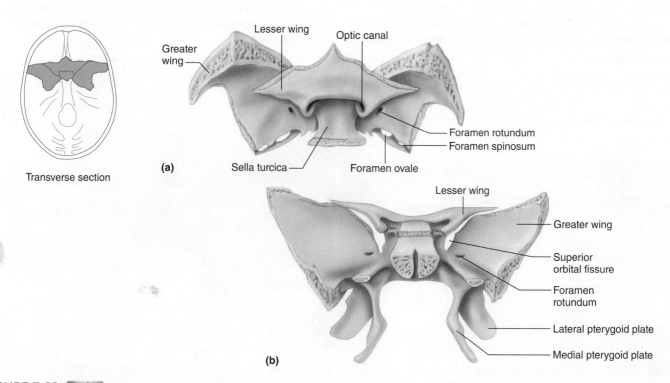

Transverse section

(a)

Greater wing
Lesser wing
Optic canal
Foramen rotundum
Foramen spinosum
Foramen ovale
Sella turcica

(b)

Lesser wing
Greater wing
Superior orbital fissure
Foramen rotundum
Lateral pterygoid plate
Medial pterygoid plate

FIGURE 7.23 AP|R The sphenoid bone. (*a*) Superior view. (*b*) Posterior view.

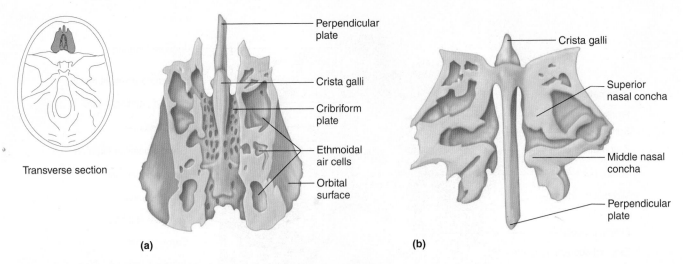

FIGURE 7.24 AP|R The ethmoid bone. (*a*) Superior view. (*b*) Posterior view.

cranial floor, orbital walls, and nasal cavity walls. A *perpendicular plate* projects downward in the midline from the cribriform plates to form most of the nasal septum.

Delicate, scroll-shaped plates called the *superior nasal concha* (kong′kah) and the *middle nasal concha* project inward from the lateral portions of the ethmoid bone toward the perpendicular plate. These bony plates support mucous membranes that line the nasal cavity. The mucous membranes, in turn, begin moistening, warming, and filtering air as it enters the respiratory tract. The lateral portions of the ethmoid bone contain many small spaces, the *ethmoidal air cells,* that together form the ethmoidal sinus (see fig. 7.21). Figure 7.25 shows various structures in the nasal cavity.

Projecting upward into the cranial cavity between the cribriform plates is a triangular process of the ethmoid bone called the *crista galli* (kris′tă gal′li) (cock's comb). Membranes that enclose the brain attach to this process. Figure 7.26 shows a view of the floor of the cranial cavity.

Facial Skeleton

The **facial skeleton** consists of thirteen immovable bones and a movable lower jawbone. In addition to forming the basic shape of the face, these bones provide attachments for muscles that move the jaw and control facial expressions.

The bones of the facial skeleton are as follows:

1. The **maxillae** (mak-sil′e; sing., *maxilla,* mak-sil′ah) form the upper jaw; together they form the keystone of the face, because the other immovable facial bones articulate with them.

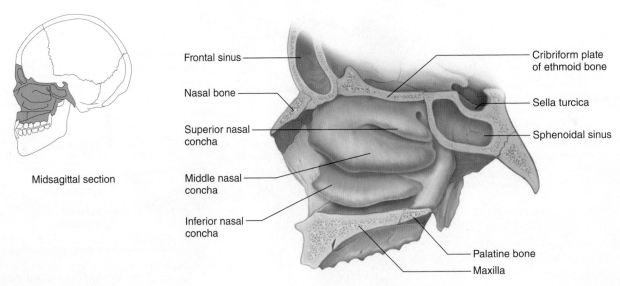

FIGURE 7.25 Lateral wall of the nasal cavity.

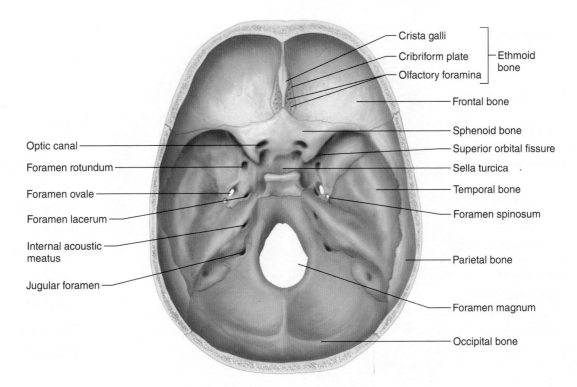

Crista galli
Cribriform plate
Olfactory foramina
} Ethmoid bone

Frontal bone

Sphenoid bone

Superior orbital fissure

Optic canal
Foramen rotundum
Foramen ovale
Foramen lacerum
Internal acoustic meatus
Jugular foramen

Sella turcica

Temporal bone

Foramen spinosum

Parietal bone

Foramen magnum

Occipital bone

FIGURE 7.26 **AP|R** Floor of the cranial cavity, viewed from above.

Portions of the maxillary bones compose the anterior roof of the mouth *(hard palate),* the floors of the orbits, and the sides and floor of the nasal cavity. They also contain the sockets of the upper teeth. Inside the maxillae, lateral to the nasal cavity, are *maxillary sinuses* (see fig. 7.21). These spaces are the largest of the sinuses, and they extend from the floor of the orbits to the roots of the upper teeth. During development, portions of the maxillary bones called *palatine processes* grow together and fuse along the midline, or median palatine suture. This structure forms the anterior section of the hard palate (see fig. 7.22).

A *cleft palate* results from incomplete fusion of the palatine processes of the maxillae by the time of birth. An infant with a cleft palate may have trouble sucking a bottle due to the opening between the oral and nasal cavities. A temporary prosthetic device (artificial palate) worn in the mouth, or a special type of nipple placed on a bottle, can help the child eat and drink until surgery corrects the cleft. This surgery is best performed between the ages of twelve and eighteen months.

The inferior border of each maxillary bone projects downward, forming an *alveolar* (al-ve′o-lar) *process.* Together these processes form a horseshoe-shaped *alveolar arch* (dental arch). Teeth occupy cavities (dental alveoli) in this arch (see chapter 17, p. 655). Dense connective tissue binds teeth to the bony sockets.

2. The L-shaped **palatine** (pal′ah-tīn) **bones** are located behind the maxillae (**fig. 7.27**). The horizontal portions of these bones form the posterior section of the hard palate and the floor of the nasal cavity. The perpendicular portions of the bones help form the lateral walls of the nasal cavity.
3. The **zygomatic** (zi″go-mat′ik) **bones** are responsible for the prominences of the cheeks below and to the sides of the eyes. These bones also help form the lateral walls and the floors of the orbits. Each bone has a *temporal process,* which extends posteriorly to join the zygomatic process of a temporal bone (see fig. 7.20).

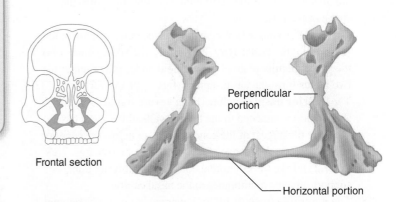

Frontal section

Perpendicular portion

Horizontal portion

FIGURE 7.27 The horizontal portions of the palatine bones form the posterior section of the hard palate, and the perpendicular portions help form the lateral walls of the nasal cavity.

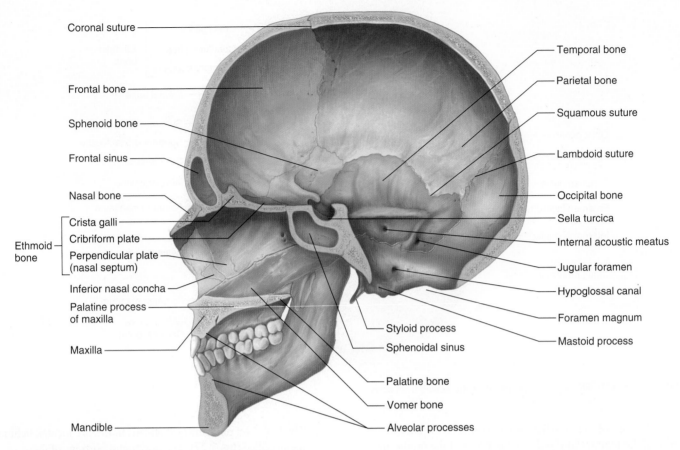

FIGURE 7.28 AP|R Sagittal section of the skull.

Labels on figure:
Coronal suture
Frontal bone
Sphenoid bone
Frontal sinus
Nasal bone
Crista galli
Cribriform plate — Ethmoid bone
Perpendicular plate (nasal septum)
Inferior nasal concha
Palatine process of maxilla
Maxilla
Mandible
Styloid process
Sphenoidal sinus
Palatine bone
Vomer bone
Alveolar processes
Temporal bone
Parietal bone
Squamous suture
Lambdoid suture
Occipital bone
Sella turcica
Internal acoustic meatus
Jugular foramen
Hypoglossal canal
Foramen magnum
Mastoid process

4. A **lacrimal** (lak′rĭ-mal) **bone** is a thin, scalelike structure located in the medial wall of each orbit between the ethmoid bone and the maxilla (see fig. 7.20). A groove in its anterior portion leads from the orbit to the nasal cavity, providing a pathway for a channel that carries tears from the eye to the nasal cavity.

5. The **nasal** (na′zal) **bones** are long, thin, and nearly rectangular (see fig. 7.18). They lie side by side and are fused at the midline, where they form the bridge of the nose. These bones are attachments for the cartilaginous tissues that form the shape of the nose.

6. The thin, flat **vomer** (vo′mer) **bone** is located along the midline within the nasal cavity. Posteriorly, the vomer bone joins the perpendicular plate of the ethmoid bone, and together the two bones form the nasal septum (figs. 7.28 and 7.29).

7. The **inferior nasal conchae** (kong′ke) are fragile, scroll-shaped bones attached to the lateral walls of the nasal cavity. They are the largest of the conchae and are below the superior and middle nasal conchae of the ethmoid bone (see figs. 7.18 and 7.25). Like the ethmoidal conchae, the inferior conchae support mucous membranes in the nasal cavity.

8. The **mandible** (man′dĭ-bl), or lower jawbone, is a horizontal, horseshoe-shaped body with a vertical, flat *ramus* projecting upward at each end. The rami are divided into a posterior

mandibular condyle and an anterior *coronoid* (kor′o-noid) *process* (fig. 7.30). The mandibular condyles articulate with the mandibular fossae of the temporal bones, whereas the coronoid processes provide attachments for muscles used in chewing. Other large chewing muscles are inserted on the lateral surfaces of the rami. On the superior border of the mandible, the *alveolar processes* contain the hollow sockets (dental alveoli) that bear the lower teeth.

On the medial side of the mandible, near the center of each ramus, is a *mandibular foramen*. This opening admits blood vessels and a nerve, which supply the roots of the lower teeth. Dentists inject anesthetic into the tissues near this foramen to temporarily block impulse conduction and desensitize teeth on that side of the jaw. Branches of the blood vessels and the nerve emerge from the mandible through the *mental foramen,* which opens on the outside near the point of the jaw. They supply the tissues of the chin and lower lip.

Table 7.6 describes the fourteen facial bones. **Figure 7.31** shows features of these bones on radiographs. **Table 7.7** lists the major openings *(foramina)* and passageways through bones of the skull, as well as their general locations and the structures that pass through them.

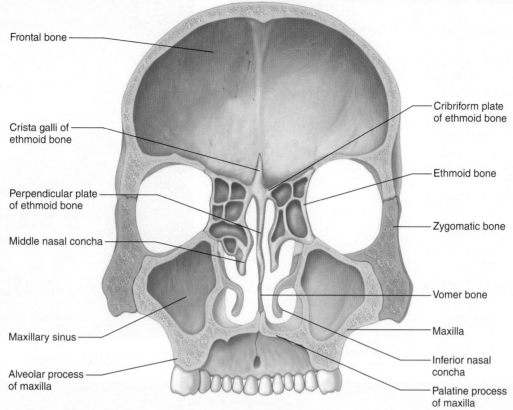

Frontal bone

Crista galli of ethmoid bone

Perpendicular plate of ethmoid bone

Middle nasal concha

Maxillary sinus

Alveolar process of maxilla

Cribriform plate of ethmoid bone

Ethmoid bone

Zygomatic bone

Vomer bone

Maxilla

Inferior nasal concha

Palatine process of maxilla

FIGURE 7.29 Frontal section of the skull (posterior view).

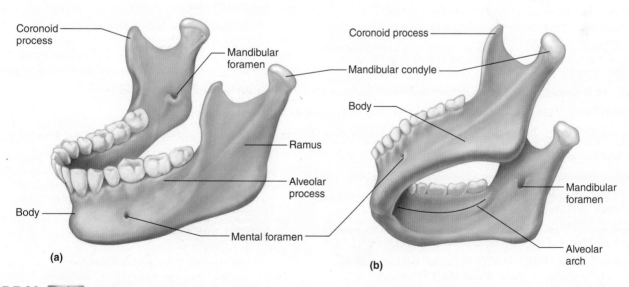

Coronoid process

Mandibular foramen

Ramus

Alveolar process

Body

Mental foramen

(a)

Coronoid process

Mandibular condyle

Body

Mandibular foramen

Alveolar arch

(b)

FIGURE 7.30 AP|R Mandible. (a) Left lateral view. (b) Inferior view.

TABLE 7.6 Bones of the Facial Skeleton

Name and Number	Description	Special Features
Maxilla (2)	Form upper jaw, anterior roof of mouth, floors of orbits, and sides and floor of nasal cavity	Alveolar processes, maxillary sinuses, palatine process
Palatine (2)	Form posterior roof of mouth and floor and lateral walls of nasal cavity	
Zygomatic (2)	Form prominences of cheeks and lateral walls and floors of orbits	Temporal process
Lacrimal (2)	Form part of medial walls of orbits	Groove that leads from orbit to nasal cavity
Nasal (2)	Form bridge of nose	
Vomer (1)	Forms inferior portion of nasal septum	
Inferior nasal concha (2)	Extend into nasal cavity from its lateral walls	
Mandible (1)	Forms lower jaw	Body, ramus, mandibular condyle, coronoid process, alveolar process, mandibular foramen, mental foramen

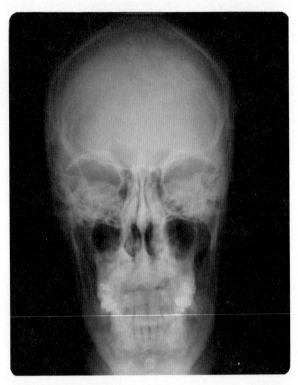

FIGURE 7.31 AP|R Radiograph of the skull. Anterior view.

TABLE 7.7 Passageways Through Bones of the Skull

Passageway	Location	Major Structures Passing Through
Carotid canal (fig. 7.22)	Inferior surface of the temporal bone	Internal carotid artery, veins, and nerves
Foramen lacerum (fig. 7.22)	Floor of cranial cavity between temporal and sphenoid bones	Branch of pharyngeal artery (in life, opening is largely covered by fibrocartilage)
Foramen magnum (fig. 7.26)	Base of skull in occipital bone	Inferior part of brainstem connecting to spinal cord, also certain arteries
Foramen ovale (fig. 7.22)	Floor of cranial cavity in sphenoid bone	Mandibular division of trigeminal nerve and veins
Foramen rotundum (fig. 7.26)	Floor of cranial cavity in sphenoid bone	Maxillary division of trigeminal nerve
Foramen spinosum (fig. 7.26)	Floor of cranial cavity in sphenoid bone	Middle meningeal blood vessels and branch of mandibular nerve
Greater palatine foramen (fig. 7.22)	Posterior portion of hard palate in palatine bone	Palatine blood vessels and nerves
Hypoglossal canal (fig. 7.28)	Near margin of foramen magnum in occipital bone	Hypoglossal nerve
Incisive foramen (fig. 7.22)	Incisive fossa in anterior portion of hard palate	Nasopalatine nerves, openings of vomeronasal organ
Inferior orbital fissure (fig. 7.20)	Floor of the orbit	Maxillary nerve and blood vessels
Infraorbital foramen (fig. 7.20)	Below the orbit in maxillary bone	Infraorbital blood vessels and nerves
Internal acoustic meatus (fig. 7.26)	Floor of cranial cavity in temporal bone	Branches of facial and vestibulocochlear nerves and blood vessels
Jugular foramen (fig. 7.26)	Base of the skull between temporal and occipital bones	Glossopharyngeal, vagus and accessory nerves, and blood vessels
Mandibular foramen (fig. 7.30)	Inner surface of ramus of mandible	Inferior alveolar blood vessels and nerves
Mental foramen (fig. 7.30)	Near point of jaw in mandible	Mental nerve and blood vessels
Optic canal (fig. 7.20)	Posterior portion of orbit in sphenoid bone	Optic nerve and ophthalmic artery
Stylomastoid foramen (fig. 7.22)	Between styloid and mastoid processes	Facial nerve and blood vessels
Superior orbital fissure (fig. 7.20)	Posterior wall of orbit	Oculomotor, trochlear, and abducens nerves and ophthalmic division of trigeminal nerve
Supraorbital foramen (fig. 7.18)	Upper margin or orbit in frontal bone	Supraorbital blood vessels and nerves

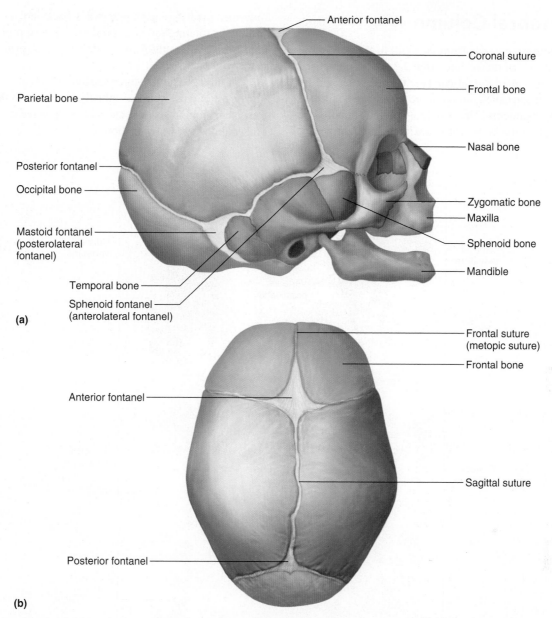

FIGURE 7.32 Fontanels. (a) Right lateral view and (b) superior view of the infantile skull.

Infantile Skull

At birth, the skull is incompletely developed, with fibrous membranes connecting the cranial bones. These membranous areas of incomplete intramembranous ossification are called **fontanels** (fon″tah-nel′z), or, more commonly, soft spots (fig. 7.32). They permit some movement between the bones so that the developing skull is partially compressible and can slightly change shape. This action, called *molding,* eases passage of an infant's skull through the birth canal. Eventually, the fontanels close as the cranial bones grow together. The posterior fontanel usually closes about two months after birth; the sphenoidal fontanel closes at about three months; the mastoid fontanel closes near the end of the first year; and the anterior fontanel may not close until the middle or end of the second year.

Other characteristics of an infantile skull include a small face with a prominent forehead and large orbits. The jaw and nasal cavity are small, the sinuses are incompletely formed, and the frontal bone is in two parts (reference plate 51, p. 265). The skull bones are thin, but they are also somewhat flexible and thus are less easily fractured than adult bones.

 PRACTICE

17 Locate and name each of the bones of the cranium.

18 Locate and name each of the facial bones.

19 Explain how an adult skull differs from that of an infant.

7.6 | Vertebral Column

The **vertebral column** extends from the skull to the pelvis and forms the vertical axis of the skeleton (fig. 7.33). It is composed of many bony parts called **vertebrae** (ver'tĕ-bre) separated by masses of fibrocartilage called *intervertebral discs* and connected to one another by ligaments. The vertebral column supports the head and the trunk of the body, yet is flexible enough to permit movements, such as bending forward, backward, or to the side and turning or rotating on the central axis. It also protects the spinal cord, which passes through a *vertebral canal* formed by openings in the vertebrae.

An infant has thirty-three separate bones in the vertebral column. Five of these bones eventually fuse to form the sacrum, and four others join to become the coccyx. As a result, an adult vertebral column has twenty-six bones.

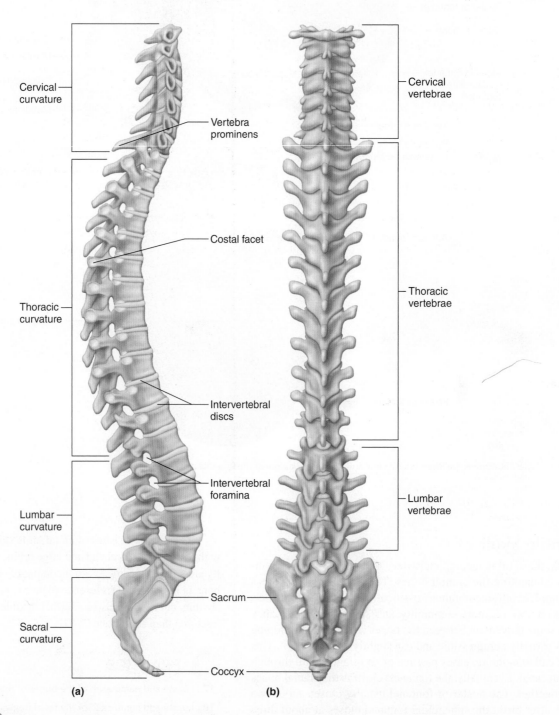

Cervical curvature

Vertebra prominens

Costal facet

Thoracic curvature

Intervertebral discs

Intervertebral foramina

Lumbar curvature

Sacral curvature

Cervical vertebrae

Thoracic vertebrae

Lumbar vertebrae

Sacrum

Coccyx

(a)

(b)

FIGURE 7.33 **AP|R** The curved vertebral column consists of many vertebrae separated by intervertebral discs. (*a*) Right lateral view. (*b*) Posterior view.

Normally, the vertebral column has four curvatures, which give it a degree of resiliency. The names of the curves correspond to the regions where they are located, as figure 7.33 shows. The *thoracic* and *sacral curvatures* are concave anteriorly and are called primary curves. The *cervical curvature* in the neck and the *lumbar curvature* in the lower back are convex anteriorly and are called secondary curves. The cervical curvature develops when a baby begins to hold up its head, and the lumbar curvature develops when the child begins to stand.

A Typical Vertebra

The vertebrae in different regions of the vertebral column have special characteristics, but they also have features in common. A typical vertebra has a drum-shaped *body,* which forms the thick, anterior portion of the bone (fig. 7.34). A longitudinal row of these vertebral bodies supports the weight of the head and trunk. The intervertebral discs, which separate adjacent vertebrae, are fastened to the roughened upper and lower surfaces of the vertebral bodies. These discs cushion and soften the forces generated by such movements as walking and jumping, which might otherwise fracture vertebrae or jar the brain. The bodies of adjacent vertebrae are joined on their anterior surfaces by *anterior longitudinal ligaments* and on their posterior surfaces by *posterior longitudinal ligaments.*

Projecting posteriorly from each vertebral body are two short stalks called *pedicles* (ped´ĭ-k´lz). They form the sides of the *vertebral foramen*. Two plates called *laminae* (lam´ĭ-ne) arise from the pedicles and fuse in the back to become a *spinous process*. The pedicles, laminae, and spinous process together complete a bony *vertebral arch* around the vertebral foramen, through which the spinal cord passes.

If the laminae of the vertebrae fail to unite during development, the vertebral arch remains incomplete, causing a condition call *spina bifida*. The contents of the vertebral canal protrude. This problem occurs most frequently in the lumbrosacral region. Spina bifida is associated with folic acid deficiency in certain genetically susceptible individuals.

Between the pedicles and laminae of a typical vertebra is a *transverse process,* which projects laterally and posteriorly. Various ligaments and muscles are attached to the dorsal spinous process and the transverse processes. Projecting upward and downward from each vertebral arch are *superior* and *inferior articular processes*. These processes bear cartilage-covered facets by which each vertebra is joined to the one above and the one below it.

On the lower surfaces of the vertebral pedicles are notches that align with adjacent vertebrae to help form openings called *intervertebral foramina* (in˝ter-ver´tĕ-bral fo-ram´ĭ-nah). These openings provide passageways for spinal nerves (see fig. 7.33).

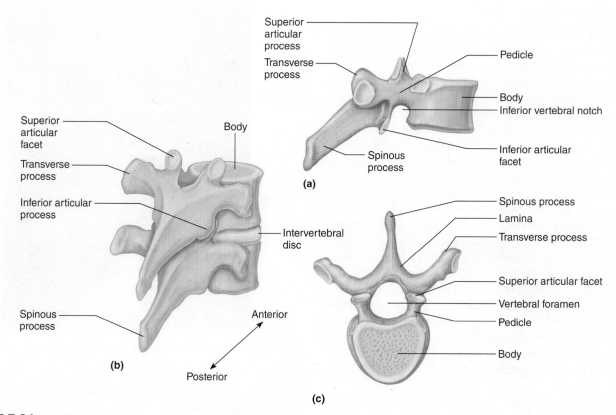

FIGURE 7.34 Typical thoracic vertebra. (*a*) Right lateral view. (*b*) Adjacent vertebrae join at their articular processes. (*c*) Superior view.

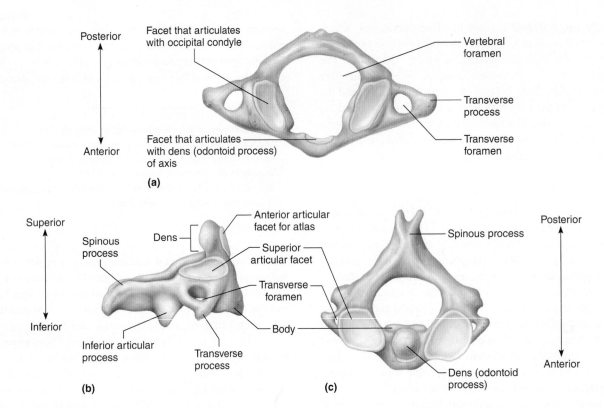

FIGURE 7.35 **AP|R** Atlas and axis. (*a*) Superior view of the atlas. (*b*) Right lateral view and (*c*) superior view of the axis.

Cervical Vertebrae

Seven **cervical vertebrae** comprise the bony axis of the neck. These are the smallest vertebrae, but their bone tissues are denser than those in any other region of the vertebral column.

The transverse processes of the cervical vertebrae are distinctive because they have *transverse foramina,* which are passageways for arteries leading to the brain. Also, the spinous processes of the second through the sixth cervical vertebrae are uniquely forked (bifid). These processes provide attachments for muscles.

The spinous process of the seventh vertebra is longer and protrudes beyond the other cervical spines. It is called the *vertebra prominens,* and because it can be felt through the skin, it is a useful landmark for locating other vertebral parts (see fig. 7.33).

Two of the cervical vertebrae, shown in figure 7.35, are of special interest. The first vertebra, or **atlas** (at′las), supports the head. It has practically no body or spine and appears as a bony ring with two transverse processes. On its superior surface, the atlas has two kidney-shaped *facets,* which articulate with the occipital condyles.

The second cervical vertebra, or **axis** (ak′sis), bears a toothlike *dens* (odontoid process) on its body. This process projects upward and lies in the ring of the atlas. As the head is turned from side to side, the atlas pivots around the dens (fig. 7.35). Figure 7.36 is a radiograph showing the cervical vertebrae.

Thoracic Vertebrae

The twelve **thoracic vertebrae** are larger than those in the cervical region. Their transverse processes project posteriorly at sharp angles. Each vertebra has a long, pointed spinous process, which

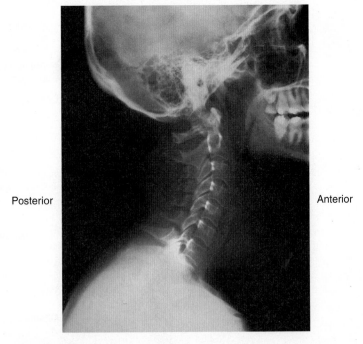

FIGURE 7.36 Radiograph of the cervical vertebrae.

slopes downward, and a facet on each side of its body which articulates with a rib head.

Beginning with the third thoracic vertebra and moving inferiorly, the bodies of these bones increase in size. This adapts them to bear increasing loads of body weight.

Lumbar Vertebrae

The five **lumbar vertebrae** in the small of the back (loin) support more weight than the superior vertebrae and have larger and stronger bodies. Compared to other types of vertebrae, the thinner transverse processes of these vertebrae project laterally, whereas their short, thick spinous processes are nearly horizontal where they project posteriorly. Figure 7.37 compares the structures of the cervical, thoracic, and lumbar vertebrae.

Sacrum

The **sacrum** (sa'krum) is a triangular structure at the base of the vertebral column. It is composed of five vertebrae that develop separately but gradually fuse between ages eighteen and thirty (fig. 7.38). In some individuals only four vertebrae fuse to form the sacrum, and the fifth vertebra becomes a sixth lumbar vertebra. The spinous processes of these fused bones form a ridge of tubercles, called the *median sacral crest.* Nerves and blood vessels pass through rows of openings, called the *posterior sacral foramina,* located to the sides of the tubercles.

The sacrum is wedged between the hip bones of the pelvis and joins them at its *auricular surfaces* by the fibrocartilage of the *sacroiliac* (sa"kro-il'e-ak) *joints.* The pelvic girdle transmits the body's weight to the legs at these joints (see fig. 7.16).

The sacrum forms the posterior wall of the pelvic cavity. The upper anterior margin of the sacrum, which represents the body of the first sacral vertebra, is called the *sacral promontory* (sa'kral prom'on-to"re). A physician performing a vaginal examination can feel this projection and use it as a guide in determining the size of the pelvis. This measurement is helpful in estimating how easily an infant may be able to pass through a woman's pelvic cavity.

The vertebral foramina of the sacral vertebrae form the *sacral canal,* which continues through the sacrum to an opening of variable size at the tip, called the *sacral hiatus* (hi-a'tus). This foramen exists because the laminae of the last sacral vertebra are not fused. On the ventral surface of the sacrum, four pairs of *anterior sacral foramina* provide passageways for nerves and blood vessels.

The painful condition spondylolisthesis results from a vertebra that slips out of place over the vertebra beneath it. Most commonly the fifth lumbar vertebra slides forward over the body of the sacrum. Spondylolisthesis may be present at birth, be a consequence of small stress fractures of the vertebra between the superior and inferior articular processes (spondylolysis), or, most commonly, be associated with the loss of moisture in intervertebral discs that accompanies aging. Gymnasts, football players, and other athletes and dancers who flex, extend, or rotate their vertebral columns excessively and forcefully are at elevated risk of developing spondylolisthesis.

Coccyx

The **coccyx** (kok'siks), or tailbone, is the lowest part of the vertebral column and is usually composed of four vertebrae that fuse between the ages of twenty-five and thirty (fig. 7.38).

Variations in individuals include three to five coccygeal vertebrae with typically the last three fused. In the elderly, the coccyx may fuse to the sacrum. Ligaments attach the coccyx to the margins

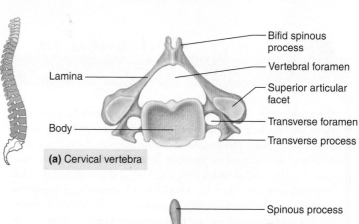

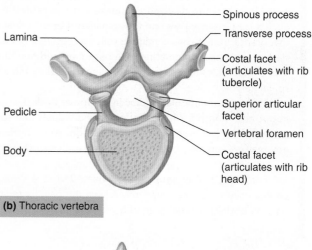

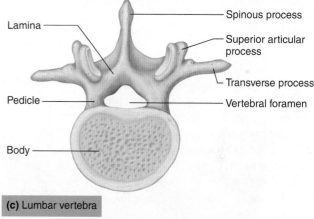

FIGURE 7.37 AP|R Superior view of (*a*) a cervical vertebra, (*b*) a thoracic vertebra, and (*c*) a lumbar vertebra.

of the sacral hiatus. Sitting presses on the coccyx, and it moves forward, acting like a shock absorber. Sitting down with great force, as when slipping and falling on ice, can fracture or dislocate the coccyx. The coccyx is also an attachment for the muscles of the pelvic floor. Table 7.8 summarizes the bones of the vertebral column, and Clinical Application 7.3 discusses disorders of the vertebral column.

 PRACTICE

20 Describe the structure of the vertebral column.
21 Explain the difference between the vertebral column of an adult and that of an infant.
22 Describe a typical vertebra.
23 Explain how the structures of cervical, thoracic, and lumbar vertebrae differ.

Changes in the intervertebral discs may cause painful problems. Each disc is composed of a tough, outer layer of fibrocartilage (annulus fibrosus) and an elastic central mass (nucleus pulposus). With age, these discs degenerate. The central masses lose firmness, and the outer layers thin, weaken, and crack. Taking a fall or lifting a heavy object can exert enough pressure to break the outer layers of the discs and squeeze out the central masses. Such a rupture may press on the spinal cord or on spinal nerves that branch from it. This condition, called a *ruptured,* or *herniated disc,* may cause inflammation with back pain and numbness, or loss of muscular function in the parts innervated by the affected spinal nerves.

The pain of a herniated disc in the lower back decreases in about one month for half of affected individuals and for most others within six months. Pain medications and physical therapy may be very helpful to strengthen muscles surrounding the injured area.

For about 10% of people with herniated discs, a surgical procedure called a laminectomy may relieve the pain by removing a portion of the posterior arch of a vertebra, which reduces pressure on nerve tissues. Then a surgeon may perform a microdiscectomy to remove the herniated disc material.

Poor posture, injury, or disease can affect the curvatures of the vertebral column. An exaggerated thoracic curvature causes rounded shoulders and a hunchback. This condition, called *kyphosis,* is seen in adolescents who undertake strenuous athletic activities. Unless corrected before bone growth completes, the condition can permanently deform the vertebral column.

The vertebral column may develop an abnormal lateral curvature, placing one hip or shoulder lower than the other, which may displace or compress the thoracic and abdominal organs. This condition, called *scoliosis,* is most common in adolescent females. It may accompany poliomyelitis, rickets, or tuberculosis, or have an unknown cause. An accentuated lumbar curvature is called *lordosis,* or swayback.

As a person ages, the intervertebral discs shrink and become more rigid, and compression is more likely to fracture the vertebral bodies. Consequently, height may decrease, and the thoracic curvature of the vertebral column may become accentuated, bowing the back.

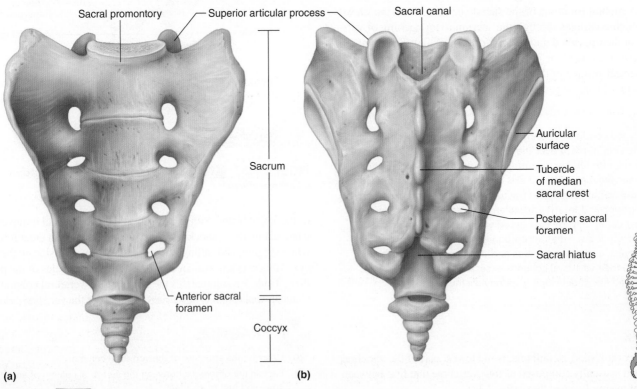

FIGURE 7.38 **AP|R** Sacrum and coccyx. (*a*) Anterior view. (*b*) Posterior view.

TABLE 7.8	Bones of the Vertebral Column					
Bones	Number	Special Features	Bones	Number	Special Features
Cervical vertebra	7	Transverse foramina; facets of atlas that articulate with occipital condyles of skull; dens of axis that articulates with atlas; spinous processes of second through sixth vertebrae are bifid	Lumbar vertebra	5	Large bodies; thinner transverse processes that project laterally; short, thick spinous processes that project posteriorly nearly horizontal
Thoracic vertebra	12	Transverse processes that project posteriorly at sharp angles; pointed spinous processes that slope downward; facets that articulate with ribs	Sacrum	5 vertebrae fused	Posterior sacral foramina, auricular surfaces, sacral promontory, sacral canal, sacral hiatus, anterior sacral foramina
			Coccyx	4 vertebrae fused	Attached by ligaments to the margins of the sacral hiatus

7.7 | Thoracic Cage

The **thoracic cage** includes the ribs, the thoracic vertebrae, the sternum, and the costal cartilages that attach the ribs to the sternum (fig. 7.39). These bones support the pectoral girdle and upper limbs, protect the viscera in the thoracic and upper abdominal cavities, and play a role in breathing.

Ribs

The usual number of **ribs** is twenty-four—one pair attached to each of the twelve thoracic vertebrae. Some individuals have extra ribs associated with their cervical or lumbar vertebrae.

The first seven rib pairs, called the *true ribs* (vertebrosternal ribs), join the sternum directly by their costal cartilages. The remaining five pairs are called *false ribs* because their cartilages do not reach the sternum directly. Instead, the cartilages of the upper three false ribs (vertebrochondral ribs) join the cartilages of the seventh rib, whereas the lower two rib pairs do not attach to the sternum. These lower two pairs (or in some individuals, the lower three pairs) are called *floating ribs* (vertebral ribs).

A typical rib has a long, slender shaft that curves around the chest and slopes downward (fig. 7.40). On the posterior end is an enlarged *head* by which the rib articulates with a facet on the body of its own vertebra and with the body of the next higher vertebra. The neck of the rib is flattened, lateral to the head, where ligaments attach. A *tubercle,* close to the head of the rib, articulates with the transverse process of the vertebra.

The costal cartilages are composed of hyaline cartilage. They are attached to the anterior ends of the ribs and continue in line with them toward the sternum.

Sternum

The **sternum** (ster'num), or breastbone, is located along the midline in the anterior portion of the thoracic cage. It is a flat, elongated bone that develops in three parts—an upper *manubrium* (mah-nu'bre-um), a middle *body,* and a lower *xiphoid* (zif'oid) *process* that projects downward (see fig. 7.39).

The sides of the manubrium and the body are notched where they articulate with costal cartilages. The manubrium also articulates with the clavicles by facets on its superior border. It usually remains as a separate bone until middle age or later, when it fuses to the body of the sternum.

> The manubrium and body of the sternum lie in different planes, so their line of union projects slightly forward. This projection, at the level of the second costal cartilage, is called the sternal angle (angle of Louis). It is commonly used as a clinical landmark (see fig. 7.39a).

The xiphoid process begins as a piece of cartilage. It slowly ossifies, and by middle age it usually fuses to the body of the sternum.

 PRACTICE

24 Which bones compose the thoracic cage?

25 Describe a typical rib.

26 What are the differences among true, false, and floating ribs?

27 Name the three parts of the sternum.

7.8 | Pectoral Girdle

The **pectoral** (pek'tor-al) **girdle,** or shoulder girdle, is composed of four parts—two clavicles (collarbones) and two scapulae (shoulder blades) (fig. 7.41). Although the word *girdle* suggests a ring-shaped structure, the pectoral girdle is an incomplete ring. It is open in the back between the scapulae, and the sternum separates its bones in front. The pectoral girdle supports the upper limbs and is an attachment for several muscles that move them.

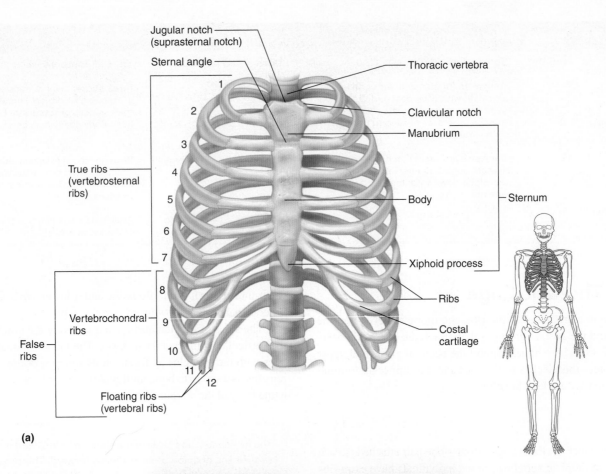

Jugular notch
(suprasternal notch)

Sternal angle

1

2

3

True ribs
(vertebrosternal
ribs)

4

5

6

7

8

Vertebrochondral
ribs

9

False
ribs

10

11

12

Floating ribs
(vertebral ribs)

Thoracic vertebra

Clavicular notch

Manubrium

Body

Sternum

Xiphoid process

Ribs

Costal
cartilage

(a)

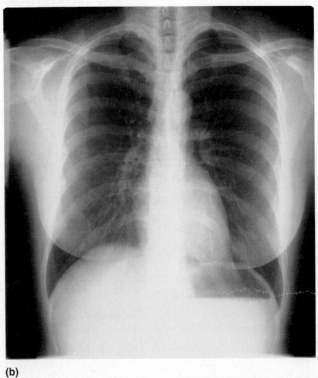

(b)

FIGURE 7.39 AP|R The thoracic cage includes (*a*) the thoracic vertebrae, the sternum, the ribs, and the costal cartilages that attach the ribs to the sternum. (*b*) Radiograph of the thoracic cage, anterior view. The light region behind the sternum and above the diaphragm is the heart.

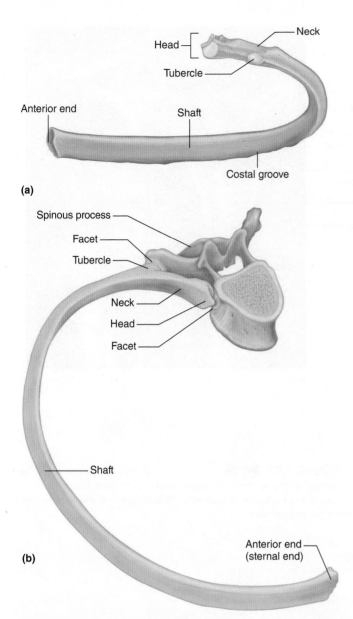

Head

Neck

Tubercle

Anterior end

Shaft

Costal groove

(a)

Spinous process

Facet

Tubercle

Neck

Head

Facet

Shaft

Anterior end
(sternal end)

(b)

FIGURE 7.40 A typical rib. (*a*) Posterior view. (*b*) Articulations of a rib with a thoracic vertebra (superior view).

Clavicles

The **clavicles** (klavˈĭ-k'lz) are slender, rodlike bones with elongated S-shapes (fig. 7.41). Located at the base of the neck, they run horizontally between the manubrium and the scapulae. The sternal (or medial) ends of the clavicles articulate with the manubrium, and the acromial (or lateral) ends join processes of the scapulae.

The clavicles brace the freely movable scapulae, helping to hold the shoulders in place. They also provide attachments for muscles of the upper limbs, chest, and back. The clavicle is structurally weak because of its elongated double curve. If compressed lengthwise due to abnormal pressure on the shoulder, it is likely to fracture.

Scapulae

The **scapulae** (skapˈu-le) are broad, somewhat triangular bones located on either side of the upper back. They have flat bodies with

concave anterior surfaces **(fig. 7.42)**. The posterior surface of each scapula is divided into unequal portions by a *spine.* Above the spine is the *supraspinous fossa,* and below the spine is the *infraspinous fossa.* This spine leads to an *acromion* (ah-kroˈme-on) *process* that forms the tip of the shoulder. The acromion process articulates with the clavicle and provides attachments for muscles of the upper limb and chest. A *coracoid* (korˈah-koid) *process* curves anteriorly and inferiorly to the acromion process. The coracoid process also provides attachments for upper limb and chest muscles. On the lateral surface of the scapula between these processes is a depression called the *glenoid cavity* (glenoid fossa of the scapula). It articulates with the head of the arm bone (humerus).

The scapula has three borders. The *superior border* is on the superior edge. The *axillary,* or *lateral border,* is directed toward the upper limb. The *vertebral,* or *medial border,* is closest to the vertebral column, about 5 cm away.

 PRACTICE

28 Which bones form the pectoral girdle?

29 What is the function of the pectoral girdle?

7.9 | Upper Limb

The bones of the upper limb form the framework of the arm, forearm, and hand. They also provide attachments for muscles and interact with muscles to move limb parts. These bones include a humerus, a radius, an ulna, carpals, metacarpals, and phalanges **(fig. 7.43)**.

Humerus

The **humerus** is a long bone that extends from the scapula to the elbow. At its upper end is a smooth, rounded *head* that fits into the glenoid cavity of the scapula **(fig. 7.44)**. Just below the head are two processes—a *greater tubercle* on the lateral side and a *lesser tubercle* on the anterior side. These tubercles provide attachments for muscles that move the upper limb at the shoulder. Between them is a narrow furrow, the *intertubercular sulcus* (intertubercular groove), through which a tendon passes from a muscle in the arm (biceps brachii) to the shoulder.

The narrow depression along the lower margin of the head that separates it from the tubercles is called the *anatomical neck.* Just below the head and the tubercles of the humerus is a tapering region called the *surgical neck,* so named because fractures commonly occur there. Near the middle of the bony shaft on the lateral side is a rough V-shaped area called the *deltoid tuberosity.* It provides an attachment for the muscle (deltoid) that raises the upper limb horizontally to the side.

At the lower end of the humerus are two smooth *condyles*—a knoblike *capitulum* (kah-pitˈu-lum) on the lateral side and a pulley-shaped *trochlea* (trokˈle-ah) on the medial side. The capitulum articulates with the radius at the elbow, whereas the trochlea joins the ulna.

Above the condyles on either side are *epicondyles,* which provide attachments for muscles and ligaments of the elbow.

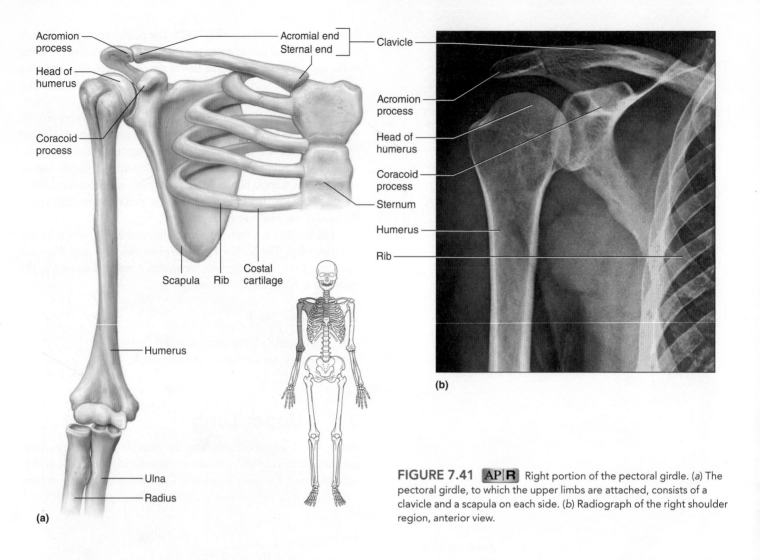

FIGURE 7.41 **AP|R** Right portion of the pectoral girdle. (*a*) The pectoral girdle, to which the upper limbs are attached, consists of a clavicle and a scapula on each side. (*b*) Radiograph of the right shoulder region, anterior view.

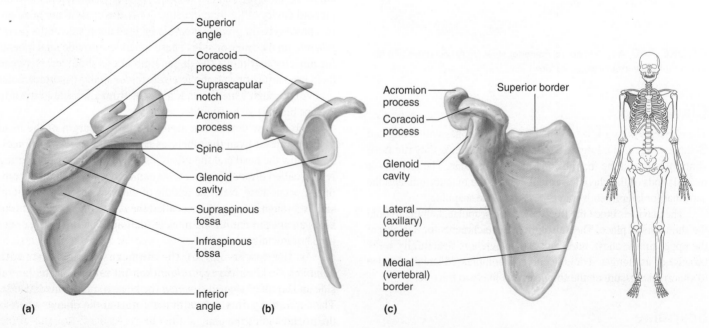

FIGURE 7.42 **AP|R** Right scapula. (*a*) Posterior surface. (*b*) Lateral view showing the glenoid cavity that articulates with the head of the humerus. (*c*) Anterior surface.

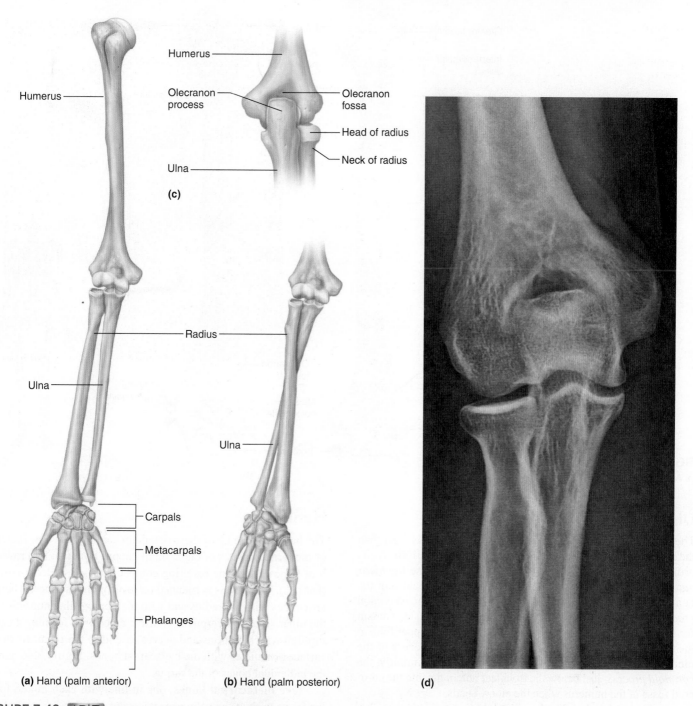

(a) Hand (palm anterior) **(b)** Hand (palm posterior) **(d)**

FIGURE 7.43 AP|R Right upper limb. (*a*) Anterior view with the hand, palm anterior and (*b*) with the hand, palm posterior. (*c*) Posterior view of the right elbow. (*d*) Radiograph of the right elbow and forearm, anterior view.

Between the epicondyles anteriorly is a depression, the *coronoid fossa,* that receives a process of the ulna (coronoid process) when the elbow bends. Another depression on the posterior surface, the *olecranon* (o″lek′ra-non) *fossa,* receives a different ulnar process (olecranon process) when the elbow straightens.

Radius

The **radius,** located on the thumb side of the forearm, is somewhat shorter than its companion, the ulna (fig. 7.45). The radius extends

from the elbow to the wrist and crosses over the ulna when the hand is turned so that the palm faces backward.

A thick, disclike *head* at the upper end of the radius articulates with the capitulum of the humerus and a notch of the ulna (radial notch). This arrangement allows the radius to rotate.

On the radial shaft just below the head is a process called the *radial tuberosity.* It is an attachment for a muscle (biceps brachii) that bends the upper limb at the elbow. At the distal end of the radius, a lateral *styloid* (sti′loid) *process* provides attachments for ligaments of the wrist.

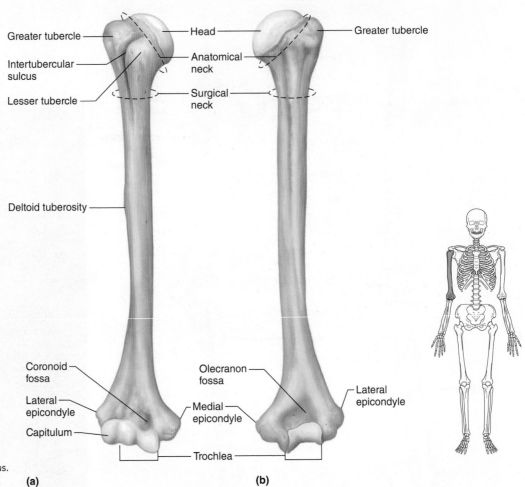

FIGURE 7.44 AP|R Right humerus.
(a) Anterior surface. (b) Posterior surface. **(a)** **(b)**

Ulna

The **ulna,** located on the medial side of the forearm, is longer than the radius and overlaps the end of the humerus posteriorly. At its proximal end, the ulna has a wrenchlike opening, the *trochlear notch* (semilunar notch), that articulates with the trochlea of the humerus. A process lies on either side of this notch. The *olecranon process,* located above the trochlear notch, provides an attachment for the muscle (triceps brachii) that straightens the upper limb at the elbow. During this movement, the olecranon process of the ulna fits into the olecranon fossa of the humerus. Similarly, the *coronoid process,* just below the trochlear notch, fits into the coronoid fossa of the humerus when the elbow bends.

At the distal end of the ulna, its knoblike *head* articulates laterally with a notch of the radius (ulnar notch) and with a disc of fibrocartilage inferiorly (fig. 7.45). This disc, in turn, joins a wrist bone (triquetrum). A medial *styloid process* at the distal end of the ulna provides attachments for ligaments of the wrist.

Many a thirtyish parent of a young little leaguer or softball player becomes tempted to join in. But if he or she has not pitched in many years, sudden activity may break the forearm. Forearm pain while pitching is a signal that a fracture could happen. Medical specialists advise returning to the pitching mound gradually. Start with twenty pitches, five days a week, for two to three months before regular games begin. By the season's start, 120 pitches per daily practice session should be painless.

Hand

The hand is made up of the wrist, palm, and fingers. The skeleton of the wrist consists of eight small **carpal bones** in two rows of four bones each. The resulting compact mass is called a *carpus* (kar'pus). The carpus is rounded on its proximal surface, where it articulates with the radius and with the fibrocartilaginous disc on the ulnar side. The carpus is concave anteriorly, forming a canal through which tendons and nerves extend to the palm. Its distal surface articulates with the metacarpal bones. **Figure 7.46** names the individual bones of the carpus.

Five **metacarpal bones,** one in line with each finger, form the framework of the palm or *metacarpus* (met″ah-kar'pus) of the hand. These bones are cylindrical, with rounded distal ends that form the knuckles of a clenched fist. The metacarpals articulate proximally with the carpals and distally with the phalanges. The metacarpal on the lateral side is the most freely movable; it permits the thumb to oppose the fingers when grasping something. These bones are numbered 1 to 5, beginning with the metacarpal of the thumb (fig. 7.46).

The **phalanges** are the finger bones. Three are in each finger—a proximal, a middle, and a distal phalanx—and two are in the thumb. (The thumb lacks a middle phalanx.) Thus, each hand has fourteen finger bones. **Figure 7.47** depicts a rare inherited condition in which extra toes and/or fingers develop. **Table 7.9** summarizes the bones of the pectoral girdle and upper limbs.

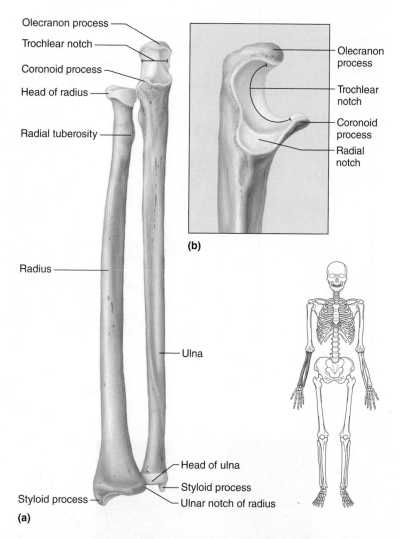

Olecranon process
Trochlear notch
Coronoid process
Head of radius

Radial tuberosity

Radius

Ulna

Olecranon process
Trochlear notch
Coronoid process
Radial notch

(b)

Head of ulna
Styloid process
Ulnar notch of radius
Styloid process

(a)

FIGURE 7.45 **AP|R** Right radius and ulna. (*a*) The head of the radius articulates with the radial notch of the ulna, and the head of the ulna articulates with the ulnar notch of the radius. (*b*) Lateral view of the proximal end of the ulna.

TABLE 7.9	Bones of the Pectoral Girdle and Upper Limbs	
Name and Number	**Location**	**Special Features**
Clavicle (2)	Base of neck between sternum and scapula	Sternal end, acromial end
Scapula (2)	Upper back, forming part of shoulder	Body, spine, acromion process, coracoid process, glenoid cavity
Humerus (2)	Arm, between scapula and elbow	Head, greater tubercle, lesser tubercle, intertubercular sulcus, anatomical neck, surgical neck, deltoid tuberosity, capitulum, trochlea, medial epicondyle, lateral epicondyle, coronoid fossa, olecranon fossa
Radius (2)	Lateral side of forearm, between elbow and wrist	Head, radial tuberosity, styloid process, ulnar notch
Ulna (2)	Medial side of forearm, between elbow and wrist	Trochlear notch, olecranon process, coronoid process, head, styloid process, radial notch
Carpal (16)	Wrist	Two rows of four bones each
Metacarpal (10)	Palm	One bone in line with each finger and thumb
Phalanx (28)	Finger	Three phalanges in each finger; two phalanges in each thumb

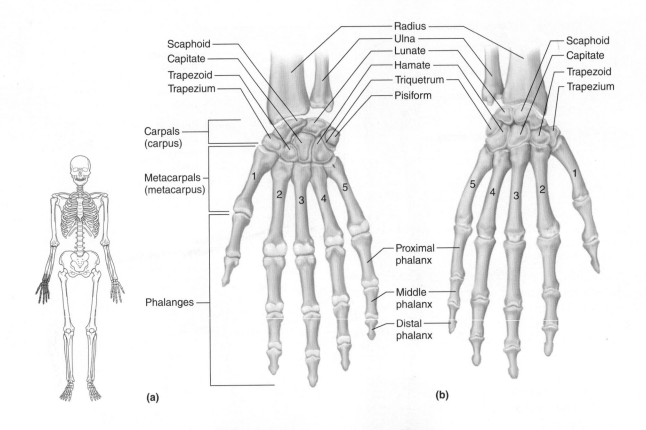

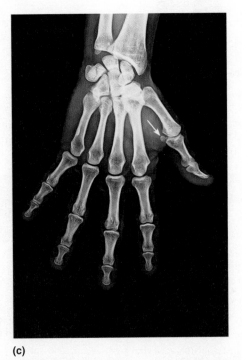

FIGURE 7.46 **AP|R** Right hand.
(a) Anterior view. (b) Posterior view.
(c) Radiograph, posterior view.
(Note: The arrow points to a sesamoid
bone associated with the first metacarpal
in the radiograph.)

PRACTICE

30 Locate and name each of the bones of the upper limb.

31 Explain how the bones of the upper limb articulate.

7.10 | Pelvic Girdle

The **pelvic girdle** consists of the two hip bones, also known as coxal bones (ossa coxae), pelvic bones or innominate bones, which articulate with each other anteriorly and with the sacrum posteriorly (fig. 7.48). The sacrum, coccyx, and pelvic girdle form

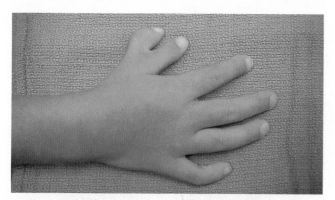

FIGURE 7.47 Polydactyly ("extra digits") is an inherited trait. Hands like these are rarely seen, because in many cases of polydactyly the extra fingers and/or toes are surgically removed shortly after birth.

the bowl-shaped *pelvis*. The pelvic girdle supports the trunk of the body; provides attachments for the lower limbs; and protects the urinary bladder, the distal end of the large intestine, and the internal reproductive organs. The body's weight is transmitted through the pelvic girdle to the lower limbs and then onto the ground.

Hip Bones

Each hip bone develops from three parts—an ilium, an ischium, and a pubis (fig. 7.49). These parts fuse in the region of a cup-shaped cavity called the *acetabulum* (as″ĕ-tab′u-lum). This depression, on the lateral surface of the hip bone, receives the rounded head of the femur or thigh bone.

The **ilium** (il′e-um), the largest and most superior portion of the hip bone, flares outward, forming the prominence of the hip. The margin of this prominence is called the *iliac crest*. The smooth, concave surface on the anterior aspect of the ilium is the *iliac fossa*.

Posteriorly, the ilium joins the sacrum at the *sacroiliac joint*. Anteriorly, a projection of the ilium, the *anterior superior iliac spine*, can be felt lateral to the groin. This spine provides attachments for ligaments and muscles and is an important surgical landmark.

> A common injury in contact sports such as football is bruising the soft tissues and bone associated with the anterior superior iliac spine. Wearing protective padding can prevent this painful injury, called a *hip pointer*.

On the posterior border of the ilium is a *posterior superior iliac spine*. Below this spine is a deep indentation, the *greater sciatic notch,* through which a number of nerves and blood vessels pass.

The **ischium** (is′ke-um), which forms the lowest portion of the hip bone, is L-shaped, with its angle, the *ischial tuberosity*, pointing posteriorly and downward. This tuberosity has a rough surface that provides attachments for ligaments and lower limb muscles. It also supports the weight of the body during sitting. Above the ischial tuberosity, near the junction of the ilium and ischium, is a sharp projection called the *ischial spine*. Like the sacral promontory, this spine, which can be felt during a vaginal examination, is used as a guide for determining pelvis size. The distance between the ischial spines is the shortest diameter of the pelvic outlet.

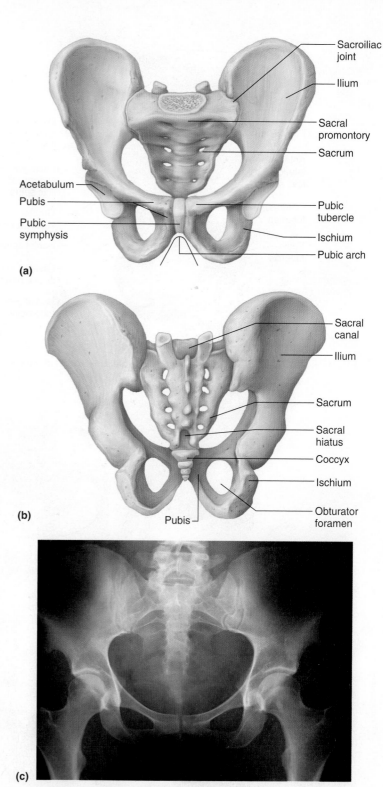

(a)

(b)

(c)

FIGURE 7.48 AP|R Pelvic girdle. (*a*) Anterior view. (*b*) Posterior view. The pelvic girdle provides an attachment for the lower limbs and together with the sacrum and coccyx forms the pelvis. (*c*) Radiograph of the pelvic girdle, anterior view.

The **pubis** (pu′bis) constitutes the anterior portion of the hip bone. The two pubic bones come together at the midline to form a joint called the *pubic symphysis* (pu′bik sim′fĭ-sis). The angle these bones form below the symphysis is the *pubic arch* (fig. 7.50).

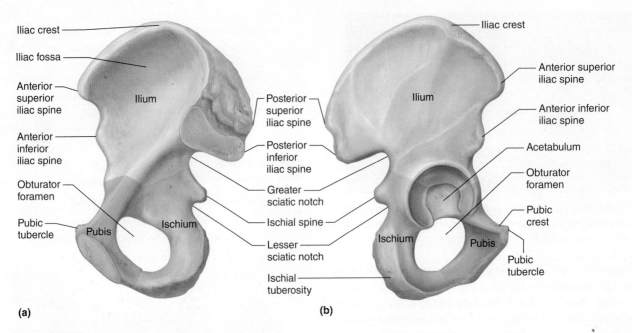

(a)

(b)

FIGURE 7.49 AP|R Right hip bone. (*a*) Medial surface. (*b*) Lateral view.

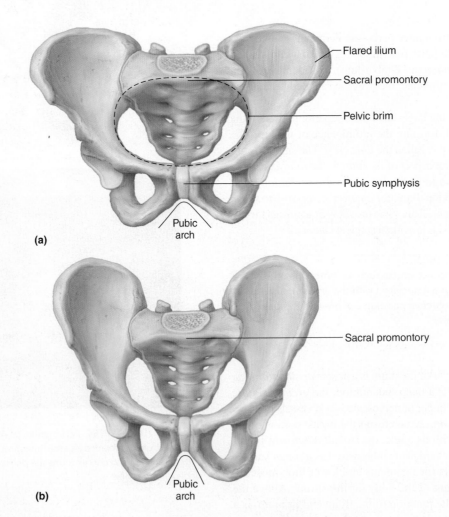

(a)

(b)

FIGURE 7.50 AP|R The female pelvis is usually wider in all diameters and roomier than that of the male. (*a*) Female pelvis. (*b*) Male pelvis.

Is the radiograph in fig. 7.48c taken of a male or female?
Answer can be found in Appendix G.

A portion of each pubis passes posteriorly and downward to join an ischium. Between the bodies of these bones on either side is a large opening, the *obturator foramen,* which is the largest foramen in the skeleton. An obturator membrane covers and nearly closes this foramen (see figs. 7.48 and 7.49).

True and False Pelves

If a line were drawn along each side of the pelvis from the sacral promontory downward and anteriorly to the upper margin of the symphysis pubis, it would mark the *pelvic brim* (linea terminalis). This margin separates the lower, or lesser (true), pelvis from the upper, or greater (false), pelvis (fig. 7.50).

Superior to the pelvic brim, the **false pelvis** is bounded posteriorly by the lumbar vertebrae, laterally by the flared parts of the iliac bones, and anteriorly by the abdominal wall. The false pelvis helps support the abdominal organs.

Inferior to the pelvic brim, the **true pelvis** is bounded posteriorly by the sacrum and coccyx and laterally and anteriorly by the lower ilium, ischium, and pubis bones. This portion of the pelvis surrounds a short, canal-like cavity. The superior opening of the cavity, the *pelvic inlet,* is at the boundary between the true and false pelves. The inferior opening of the cavity is the *pelvic outlet.* An infant passes through this cavity during childbirth.

Differences Between Male and Female Pelves

Some basic structural differences distinguish the male and the female pelves, even though it may be difficult to find all of the "typical" characteristics in any one individual. These differences arise from the function of the female pelvis as a birth canal. In most females the iliac bones are more flared than those in males. Consequently, the female hips are typically broader than the male hips. The angle of the female pubic arch is usually greater, the distance between the ischial spines and the ischial tuberosities is greater, and the sacral curvature is shorter and flatter. Thus, the female pelvic cavity is usually wider in all diameters than that of the male. Also, the bones of the female pelvis are typically lighter, more delicate, and show less evidence of muscle attachments although the attachments are the same in both sexes (fig. 7.50). Table 7.10 summarizes some of the differences between the male and female skeletons.

 PRACTICE

32 Locate and name each bone that forms the pelvis.

33 Name the bones that fuse to form a hip bone.

34 Distinguish between the false pelvis and the true pelvis.

35 How are male and female pelves different?

7.11 | Lower Limb

The bones of the lower limb form the frameworks of the thigh, leg, and foot. They include a femur, a tibia, a fibula, tarsals, metatarsals, and phalanges (fig. 7.51).

Femur

The **femur,** or thigh bone, is the longest bone in the body and extends from the hip to the knee. A large, rounded *head* at its proximal end projects medially into the acetabulum of the hip bone (fig. 7.52). On the head, a pit called the *fovea capitis* marks the attachment of a ligament. Just below the head are a constriction, or *neck,* and two large processes—a superior, lateral *greater*

TABLE 7.10	Differences Between the Male and Female Skeletons	
Part	**Male Differences**	**Female Differences**
Skull	Larger, heavier, more conspicuous muscle attachment	Smaller, more delicate, less evidence of muscle attachment
mastoid process	Larger	Smaller
supraorbital ridge	More prominent	Less prominent
chin	More squared	More pointed
jaw angle	Angle of ramus about 90 degrees	Angle of ramus greater than 125 degrees
forehead	Shorter	Taller
orbit	Superior border thicker, blunt edge	Superior border thinner, sharp edge
palate	Wider U-shape	Narrower U-shape
Pelvis	Hip bones heavier, thicker, more evidence of muscle attachment	Hip bones lighter, less evidence of muscle attachment
obturator foramen	More oval	More triangular
acetabulum	Larger	Smaller
pubic arch	Narrow, sharper angle	Broader, flatter angle
sacrum	Narrow, sacral promontory projects more forward, sacral curvature bends less sharply posteriorly	Wide, sacral curvature bends sharply posteriorly
coccyx	Less movable	More movable
cavity	Narrow and long, more funnel-shaped	Wide, distance between ischial spines and ischial tuberosities is greater

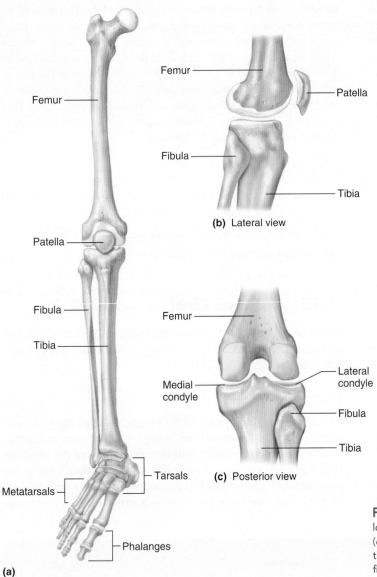

(b) Lateral view

(c) Posterior view

(a)

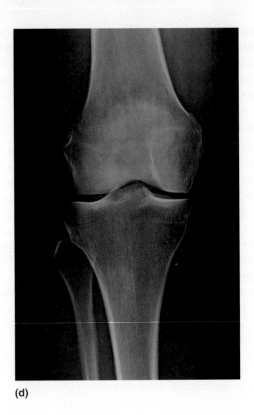

(d)

FIGURE 7.51 AP|R Right lower limb. (*a*) Anterior view of the lower limb. (*b*) Lateral view of the knee. (*c*) Posterior view of the knee. (*d*) Radiograph of the knee (anterior view), showing the ends of the femur, tibia, and fibula. Thinner areas of bone, such as part of the head of the fibula and the patella, barely show in this radiograph.

trochanter and an inferior, medial *lesser trochanter.* These processes provide attachments for muscles of the lower limbs and buttocks. On the posterior surface in the middle third of the shaft is a longitudinal crest called the *linea aspera.* This rough strip is an attachment for several muscles.

At the distal end of the femur, two rounded processes, the *lateral* and *medial condyles,* articulate with the tibia of the leg. A patella also articulates with the femur on its distal anterior surface.

On the medial surface at its distal end is a prominent *medial epicondyle,* and on the lateral surface is a *lateral epicondyle.* These projections provide attachments for muscles and ligaments.

Patella

The **patella,** or kneecap, is a flat sesamoid bone located in a tendon that passes anteriorly over the knee (see fig. 7.51). The patella, because of its position, controls the angle at which this tendon continues toward the tibia. Therefore it functions in lever actions associated with lower limb movements.

The patella can slip to one side as a result of a blow to the knee or a forceful unnatural movement of the leg. This painful condition is called a *patellar dislocation.* Doing exercises that strengthen muscles associated with the knee and wearing protective padding can prevent knee displacement. Unfortunately, once the soft tissues that hold the patella in place are stretched, patellar dislocation tends to recur.

Tibia

The **tibia,** or shin bone, is the larger of the two leg bones and is located on the medial side. Its proximal end is expanded into *medial* and *lateral condyles,* which have concave surfaces and articulate with the condyles of the femur (fig. 7.53). Below the condyles, on the anterior surface, is a process called the *tibial tuberosity,* which provides an attachment for the *patellar ligament* (a continuation of the patella-bearing tendon). A prominent *anterior crest* extends downward from the tuberosity and attaches connective tissues in the leg.

At its distal end, the tibia expands to form a prominence on the medial side of the ankle called the *medial malleolus* (mah-le'o-lus), which is an attachment for ligaments. On its lateral side is a depression that articulates with the fibula. The inferior surface of the tibia's distal end articulates with a large bone (the talus) in the ankle.

Fibula

The **fibula** is a long, slender bone located on the lateral side of the tibia. Its ends are slightly enlarged into a proximal *head* and a distal *lateral malleolus* (fig. 7.53). The head articulates with the tibia just below the lateral condyle, but it does not enter into the knee joint and does not bear any body weight. The lateral malleolus articulates with the ankle and protrudes on the lateral side.

Foot

The foot is made up of the ankle, the instep, and the toes. The ankle or *tarsus* (tahr'sus) is composed of seven **tarsal bones.** One of these bones, the **talus** (ta'lus), can move freely where it joins the tibia and fibula, forming the ankle. The largest of the tarsals, the **calcaneus** (kal-ka'ne-us), or heel bone, is below the talus where it projects backward to form the base of the heel. The calcaneus helps support body weight and provides an attachment, the *calcaneal tuberosity,* for muscles that move the foot. Figures 7.54 and 7.55 name the bones of the tarsus.

The instep or *metatarsus* (met"ah-tahr'sus) consists of five elongated **metatarsal bones,** which articulate with the tarsus. They are numbered 1 to 5, beginning on the medial side (fig. 7.55). The heads at the distal ends of these bones form the ball of the foot. The tarsals and metatarsals are bound by ligaments, forming the arches of the foot. A longitudinal arch extends from the heel to the toe, and a transverse arch stretches across the foot. These arches provide a stable, springy base for the body. Sometimes, however, the tissues that bind the metatarsals weaken, producing fallen arches, or flat feet.

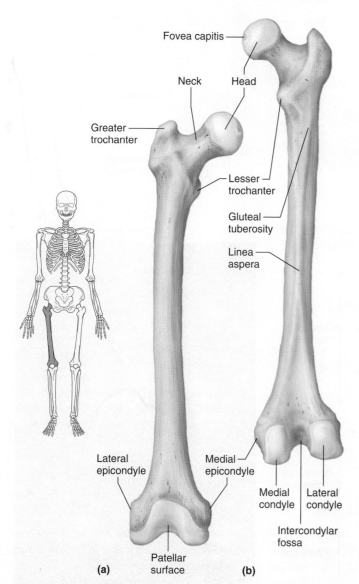

FIGURE 7.52 AP|R Right femur. (a) Anterior surface. (b) Posterior surface.

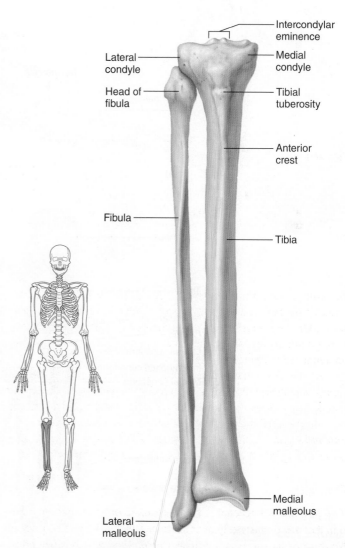

FIGURE 7.53 AP|R Bones of the right leg, anterior view.

FIGURE 7.54 AP|R Right foot. (a) Radiograph view from the medial side. (b) The talus moves freely where it articulates with the tibia and fibula.

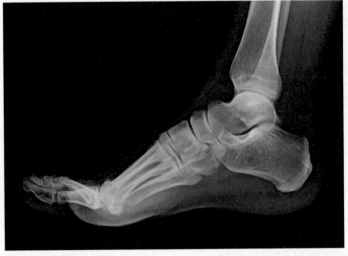
(a)

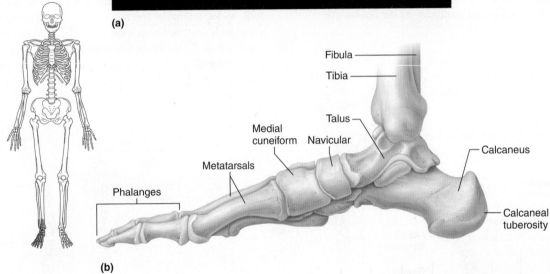
(b)

Fibula
Tibia
Talus
Calcaneus
Calcaneal tuberosity
Navicular
Medial cuneiform
Metatarsals
Phalanges

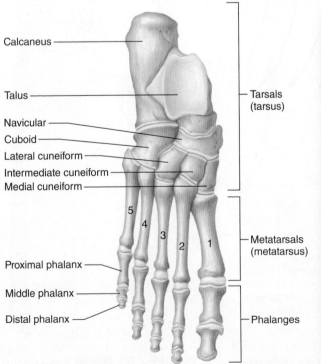
(a)

Calcaneus
Talus
Navicular
Cuboid
Lateral cuneiform
Intermediate cuneiform
Medial cuneiform
Tarsals (tarsus)
5 4 3 2 1
Metatarsals (metatarsus)
Proximal phalanx
Middle phalanx
Distal phalanx
Phalanges

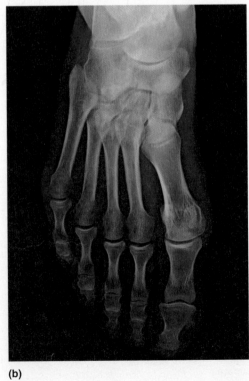
(b)

FIGURE 7.55 AP|R Right foot. (a) Viewed superiorly. (b) Radiograph of the foot viewed superiorly.

TABLE 7.11 Bones of the Pelvic Girdle and Lower Limbs

Name and Number	Location	Special Features
Hip bone (2)	Hip, articulating with the other hip bone anteriorly and with the sacrum posteriorly	Ilium, iliac crest, anterior superior iliac spine, ischium, ischial tuberosity, ischial spine, obturator foramen, acetabulum, pubis
Femur (2)	Thigh, between hip and knee	Head, fovea capitis, neck, greater trochanter, lesser trochanter, linea aspera, lateral condyle, medial condyle, gluteal tuberosity, intercondylar fossa
Patella (2)	Anterior surface of knee	A flat sesamoid bone located within a tendon
Tibia (2)	Medial side of leg, between knee and ankle	Medial condyle, lateral condyle, tibial tuberosity, anterior crest, medial malleolus, intercondylar eminence
Fibula (2)	Lateral side of leg, between knee and ankle	Head, lateral malleolus
Tarsal (14)	Ankle	Freely movable talus that articulates with leg bones; calcaneus that forms the base of the heel; five other tarsal bones bound firmly together
Metatarsal (10)	Instep	One bone in line with each toe, bound by ligaments to form arches
Phalanx (28)	Toe	Three phalanges in each toe, two phalanges in great toe

The **phalanges** of the toes are shorter but otherwise similar to those of the fingers and align and articulate with the metatarsals. Each toe has three phalanges—a proximal, a middle, and a distal phalanx—except the great toe, which has only two because it lacks the middle phalanx (fig. 7.55). Table 7.11 summarizes the bones of the pelvic girdle and lower limbs.

> An infant with two casts on her feet is probably being treated for clubfoot, a common birth defect in which the foot twists out of its normal position. Clubfoot probably results from arrested development during fetal existence, but the precise cause is not known.

 PRACTICE

36 Locate and name each of the bones of the lower limb.

37 Explain how the bones of the lower limb articulate.

38 Describe how the foot is adapted to support the body.

7.12 | Life-Span Changes

Aging-associated changes in the skeletal system are apparent at the cellular and whole-body levels. Most obvious is the incremental decrease in height that begins at about age thirty, with a loss of about 1/16 of an inch a year. In the later years, compression fractures in the vertebrae may contribute significantly to loss of height (fig. 7.56). Overall, as calcium levels fall and bone material gradually vanishes, the skeleton loses strength, and the bones become brittle and increasingly prone to fracture. However, the continued ability of fractures to heal reveals that the bone tissue is still alive and functional.

Components of the skeletal system and individual bones change to different degrees and at different rates over a lifetime. Gradually, osteoclasts come to outnumber osteoblasts, which means that bone is eaten away in the remodeling process at a faster rate than it is

FIGURE 7.56 The bones change to different degrees and at different rates over a lifetime.

replaced. The result is more spaces in bones. The bone thins, its strength waning. Bone matrix changes, with the ratio of mineral to protein increasing, making bones more brittle and prone to fracture. Beginning in the third decade of life, bone matrix is removed faster than it is laid down. By age thirty-five, we start to lose bone mass.

Trabecular bone, due to its spongy, less compact nature, shows the changes of aging first. It thins, increasing in porosity and weakening the overall structure. The vertebrae consist mostly of trabecular bone. It is also found in the upper part of the femur, whereas the shaft is more compact bone. The fact that trabecular bone weakens sooner than compact bone destabilizes the upper part of the femur, which is why it is a commonly broken bone among the elderly.

Compact bone loss begins at around age forty and continues at about half the rate of loss of trabecular bone. As remodeling continues throughout life, older osteons disappear as new ones are built next to them. With age, the osteons may coalesce, further weakening the overall structures as gaps form.

Bone loss is slow and steady in men, but in women it is clearly linked to changing hormone levels. In the first decade following menopause, women lose 15% to 20% of trabecular bone, which is two to three times the rate of loss in men and in premenopausal women. During the same time, compact bone loss is 10% to 15%, which is three to four times the rate of loss in men and premenopausal women. By about age seventy, both sexes are losing bone at about the same rate. By very old age, a woman may have only half the trabecular and compact bone mass as she did in her twenties, whereas a very elderly man may have one-third less bone mass.

Falls among the elderly are common and have many causes (table 7.12). The most common fractures, after vertebral compression and hip fracture, are of the wrist, leg, and pelvis. Aging-related increased risk of fracture usually begins at about age fifty. Healing slows, and pain from a broken bone may persist for months. Recommendations for preserving skeletal health include

TABLE 7.12	Possible Reasons for Falls Among the Elderly
Overall frailty
Decreased muscle strength
Decreased coordination
Side effects of medication
Slowed reaction time due to stiffening joints
Poor vision and/or hearing
Disease (cancer, infection, arthritis)

avoiding falls, taking calcium supplements, getting enough vitamin D, avoiding carbonated beverages (phosphates deplete bone), and getting regular exercise.

 PRACTICE

39 Why is bone lost faster with aging than it is replaced?

40 Which bones most commonly fracture in the elderly? ■

Chapter Summary

7.1 Bone Shape and Structure (page 200)

Bones are grouped according to their shapes—long, short, flat, or irregular.

1. Parts of a long bone
   a. **Epiphyses** at each end are covered with articular cartilage and articulate with other bones.
   b. The shaft of a bone is called the **diaphysis.**
   c. The region between the epiphysis and diaphysis is the **metaphysis.**
   d. Except for the **articular cartilage,** a bone is covered by a **periosteum.**
   e. **Compact bone** has a continuous extracellular matrix with no gaps.
   f. **Spongy bone** has irregular interconnecting spaces between bony plates called trabeculae.
   g. Both compact and spongy bone are strong and resist bending.
   h. The diaphysis contains a **medullary cavity** filled with **marrow.**
2. Microscopic structure
   a. **Osteocytes** are in bony chambers called lacunae.
   b. Compact bone contains **osteons** held together by bone matrix.
   c. Central canals in compact bone contain blood vessels that nourish the cells of osteons.
   d. Perforating canals in compact bone connect central canals transversely and communicate with the bone's surface and the medullary cavity.
   e. Diffusion from the surface of thin bony plates nourishes cells of spongy bones.

7.2 Bone Development and Growth (page 202)

1. **Intramembranous bones**
   a. Flat bones of the skull, the clavicles, the sternum, and some facial bones are intramembranous bones.
   b. They develop from sheetlike layers of embryonic connective tissues (mesenchyme).
   c. **Osteoblasts** within the membranous layers form bone tissue.
   d. Osteoblasts surrounded by extracellular matrix are called osteocytes.
   e. Mesenchyme outside the developing bone gives rise to the periosteum.
2. **Endochondral bones**
   a. Most of the bones of the skeleton are endochondral.
   b. They develop as hyaline cartilage that bone tissue later replaces.
   c. The primary ossification center appears in the diaphysis, whereas secondary ossification centers appear in the epiphyses.
   d. An **epiphyseal plate** remains between the primary and secondary ossification centers.
3. Growth at the epiphyseal plate
   a. An epiphyseal plate consists of layers of cells: zone of resting cartilage, zone of proliferating cartilage, zone of hypertrophic cartilage, and zone of calcified cartilage.
   b. The epiphyseal plates are responsible for bone lengthening.
   c. Long bones continue to lengthen until the epiphyseal plates are ossified.
   d. Growth in bone thickness is due to ossification beneath the periosteum.
   e. The action of **osteoclasts** forms the medullary cavity.
4. Homeostasis of bone tissue
   a. Osteoclasts and osteoblasts continually remodel bone.
   b. The total mass of bone remains nearly constant.
5. Factors affecting bone development, growth, and repair
   a. Deficiencies of vitamin A, C, or D result in abnormal bone development.
   b. Insufficient secretion of pituitary growth hormone may result in dwarfism; excessive secretion may result in gigantism.

Skeletal System

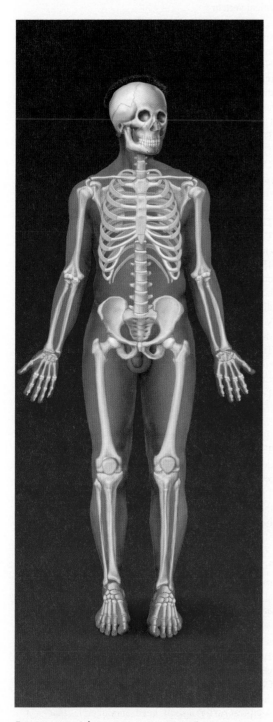

Bones provide support, protection, and movement and also play a role in calcium balance.

Integumentary System

Vitamin D, production of which begins in the skin, plays a role in calcium absorption and availability for bone matrix.

Muscular System

Muscles pull on bones to cause movement.

Nervous System

Proprioceptors sense the position of body parts. Pain receptors warn of trauma to bone. Bones protect the brain and spinal cord.

Endocrine System

Some hormones act on bone to help regulate blood calcium levels.

Cardiovascular System

Blood transports nutrients to bone cells. Bone helps regulate plasma calcium levels, important to heart function.

Lymphatic System

Cells of the immune system originate in the bone marrow.

Digestive System

Absorption of dietary calcium provides material for bone matrix.

Respiratory System

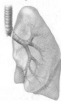

Ribs and muscles work together in breathing.

Urinary System

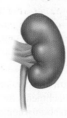

The kidneys and bones work together to help regulate blood calcium levels.

Reproductive System

The pelvis helps support the uterus during pregnancy. Bones provide a source of calcium during lactation.

c. Thyroxine stimulates replacement of cartilage in the epiphyseal plates with bone.

d. Male and female sex hormones promote bone formation and stimulate ossification of the epiphyseal plates.

e. Physical stress stimulates bone growth.

7.3 Bone Function (page 210)

1. Support, protection, and movement
   a. Bones shape and form body structures.
   b. Bones support and protect softer, underlying tissues.
   c. Bones and muscles interact, producing movement.
2. Blood cell formation
   a. At different stages in life, **hematopoiesis** occurs in the yolk sac, the liver, the spleen, and the red bone marrow.
   b. Red marrow houses developing red blood cells, white blood cells, and blood platelets.
3. Inorganic salt storage
   a. The extracellular matrix of bone tissue contains abundant calcium phosphate in the form of hydroxyapatite.
   b. When blood calcium ion concentration is low, osteoclasts resorb bone, releasing calcium salts.
   c. When blood calcium ion concentration is high, osteoblasts are stimulated to form bone tissue and store calcium salts.
   d. Bone stores small amounts of sodium, magnesium, potassium, and carbonate ions.
   e. Bone tissues may accumulate lead, radium, or strontium.

7.4 Skeletal Organization (page 212)

1. Number of bones
   a. Usually an adult human skeleton has 206 bones, but the number may vary.
   b. Extra bones include sutural bones and sesamoid bones.
2. Divisions of the skeleton
   a. The skeleton can be divided into axial and appendicular portions.
   b. The **axial skeleton** consists of the **skull, hyoid bone, vertebral column,** and **thoracic cage.**
   c. The **appendicular skeleton** consists of the **pectoral girdle, upper limbs, pelvic girdle,** and **lower limbs.**

7.5 Skull (page 215)

The skull consists of twenty-two bones, which include eight cranial bones and fourteen facial bones.

1. Cranium
   a. The **cranium** encloses and protects the brain and provides attachments for muscles.
   b. Some cranial bones contain air-filled paranasal sinuses that help reduce the weight of the skull.
   c. Cranial bones include the **frontal bone, parietal bones, occipital bone, temporal bones, sphenoid bone,** and **ethmoid bone.**
2. **Facial skeleton**
   a. Facial bones form the basic shape of the face and provide attachments for muscles.
   b. Facial bones include the **maxillae, palatine bones, zygomatic bones, lacrimal bones, nasal bones, vomer bone, inferior nasal conchae,** and **mandible.**
3. Infantile skull
   a. Incompletely developed bones, connected by **fontanels,** enable the infantile skull to change shape slightly during childbirth.

b. Infantile skull bones are thin, somewhat flexible, and less easily fractured.

7.6 Vertebral Column (page 226)

The **vertebral column** extends from the skull to the pelvis and protects the spinal cord. It is composed of **vertebrae** separated by intervertebral discs. An infant has thirty-three vertebral bones and an adult has twenty-six. The vertebral column has four curvatures—cervical, thoracic, lumbar, and sacral.

1. A typical vertebra
   a. A typical vertebra consists of a body, pedicles, laminae, spinous process, transverse processes, and superior and inferior articular processes.
   b. Notches on the upper and lower surfaces of the pedicles on adjacent vertebrae form intervertebral foramina through which spinal nerves pass.
2. Cervical vertebrae
   a. **Cervical vertebrae** comprise the bones of the neck.
   b. Transverse processes have transverse foramina.
   c. The **atlas** (first vertebra) supports the head.
   d. The dens of the **axis** (second vertebra) provides a pivot for the atlas that allows the head to turn from side to side.
3. Thoracic vertebrae
   a. **Thoracic vertebrae** are larger than cervical vertebrae.
   b. Their transverse processes project posteriorly at sharp angles.
   c. Their long spinous processes slope downward, and facets on the sides of the vertebral bodies articulate with the heads of ribs.
4. Lumbar vertebrae
   a. Vertebral bodies of **lumbar vertebrae** are large and strong.
   b. Their transverse processes project laterally, and their spinous processes project posteriorly nearly horizontally.
5. Sacrum
   a. The **sacrum,** formed of five fused vertebrae, is a triangular structure that has rows of dorsal sacral foramina.
   b. It is united with the hip bones at the sacroiliac joints.
   c. The sacral promontory provides a guide for determining the size of the pelvis.
6. Coccyx
   a. The **coccyx,** composed of four fused vertebrae, forms the lowest part of the vertebral column.
   b. It acts as a shock absorber when a person sits and is an attachment for muscles of the pelvic floor.

7.7 Thoracic Cage (page 231)

The **thoracic cage** includes the ribs, thoracic vertebrae, sternum, and costal cartilages that attach the ribs to the sternum. It supports the pectoral girdle and upper limbs, protects viscera, and functions in breathing.

1. Ribs
   a. Twelve pairs of **ribs** are attached to the twelve thoracic vertebrae.
   b. Costal cartilages of the true ribs join the sternum directly; those of the false ribs join indirectly or not at all.
   c. A typical rib has a shaft, head, and tubercles that articulate with the thoracic vertebrae.
2. Sternum
   a. The **sternum** consists of a manubrium, body, and xiphoid process.

b. It articulates with costal cartilages of the ribs and the clavicles.

7.8 Pectoral Girdle (page 231)

The **pectoral girdle** is composed of two clavicles and two scapulae. It forms an incomplete ring that supports the upper limbs and provides attachments for muscles that move the upper limbs.

1. Clavicles
   a. **Clavicles** are rodlike bones that run horizontally between the sternum and shoulders.
   b. They help hold the shoulders in place and provide attachments for muscles.
2. Scapulae
   a. The **scapulae** are broad, triangular bones each with a body, spine, acromion process, coracoid process, glenoid cavity, supraspinous and infraspinous fossae, superior border, axillary border, and vertebral border.
   b. Each articulates with the humerus of each upper limb and provides attachments for muscles of the upper limbs and chest.

7.9 Upper Limb (page 233)

Bones of the upper limb provide the framework for the limb and provide the attachments for muscles that move the limb.

1. Humerus
   a. The **humerus** extends from the scapula to the elbow.
   b. It has a head, greater tubercle, lesser tubercle, intertubercular sulcus, anatomical neck, surgical neck, deltoid tuberosity, capitulum, trochlea, epicondyles, coronoid fossa, and olecranon fossa.
2. Radius
   a. The **radius** is on the thumb side of the forearm between the elbow and wrist.
   b. It has a head, radial tuberosity, styloid process, and ulnar notch.
3. Ulna
   a. The **ulna,** on the medial side of the forearm, is longer than the radius and overlaps the humerus posteriorly.
   b. It has a trochlear notch, olecranon process, coronoid process, head, styloid process, and radial notch.
4. Hand
   a. The wrist has eight **carpals.**
   b. The palm has five **metacarpals.**
   c. The five fingers have fourteen **phalanges.**

7.10 Pelvic Girdle (page 238)

The **pelvic girdle** consists of two hip bones that articulate with each other anteriorly and with the sacrum posteriorly. The sacrum, coccyx, and pelvic girdle form the pelvis. The girdle provides support for body weight and attachments for muscles and protects visceral organs.

1. Hip bones
   Each hip bone consists of an ilium, ischium, and pubis, fused in the region of the acetabulum.
   a. Ilium
      (1) The **ilium,** the largest portion of the hip bone, joins the sacrum at the sacroiliac joint.
      (2) It has an iliac crest with anterior and posterior superior iliac spines and iliac fossae.
   b. Ischium
      (1) The **ischium** is the lowest portion of the hip bone.
      (2) It has an ischial tuberosity and ischial spine.
   c. Pubis
      (1) The **pubis** is the anterior portion of the hip bone.
      (2) Pubic bones are joined anteriorly at the pubic symphysis.
2. True and false pelves
   a. The **false pelvis** is superior to the pelvic brim; the **true pelvis** is inferior to the pelvic brim.
   b. The false pelvis helps support abdominal organs; the true pelvis functions as a birth canal.
3. Differences between male and female pelves
   a. Differences between male and female pelves reflect the function of the female pelvis as a birth canal.
   b. Usually the female pelvis is more flared; pubic arch is broader; distance between the ischial spines and the ischial tuberosities is greater; and sacral curvature is shorter.

7.11 Lower Limb (page 241)

Bones of the lower limb provide the framework for the limb and provide the attachments for muscles that move the limb.

1. Femur
   a. The **femur** extends from the hip to the knee.
   b. It has a head, fovea capitis, neck, greater trochanter, lesser trochanter, linea aspera, lateral condyle, and medial condyle.
2. Patella
   a. The **patella** is a sesamoid bone in the tendon that passes anteriorly over the knee.
   b. It controls the angle of this tendon and functions in lever actions associated with lower limb movements.
3. Tibia
   a. The **tibia** is located on the medial side of the leg.
   b. It has medial and lateral condyles, tibial tuberosity, anterior crest, and medial malleolus.
   c. It articulates with the talus of the ankle.
4. Fibula
   a. The **fibula** is located on the lateral side of the tibia.
   b. It has a head that proximally articulates with the tibia and distally the lateral malleolus articulates with the ankle, but the fibula does not bear body weight.
5. Foot
   a. The ankle includes the **talus,** the **calcaneus,** and five other **tarsals.**
   b. The instep has five **metatarsals.**
   c. The five toes have fourteen **phalanges.**

7.12 Life-Span Changes (page 245)

Aging-associated changes in the skeleton are apparent at the cellular and whole-body levels.

1. Incremental decrease in height begins at about age thirty.
2. Gradually, bone loss exceeds bone replacement.
   a. In the first decade following menopause, bone loss occurs more rapidly in women than in men or premenopausal women. By age seventy, both sexes are losing bone at about the same rate.
   b. Aging increases risk of bone fractures.

CHAPTER ASSESSMENTS

7.1 Bone Shape and Structure

1　List four groups of bones based on their shapes, and give an example from each group. (p. 200)

2　Sketch a typical long bone, and label its epiphyses, diaphysis, medullary cavity, periosteum, and articular cartilages. Designate the locations of compact and spongy bone. (pp. 200–201)

3　Discuss the functions of the parts labeled in the sketch you made for question 2. (pp. 200–201)

4　Distinguish between the microscopic structure of compact bone and spongy bone. (pp. 201–202)

5　Explain how central canals and perforating canals are related. (p. 202)

7.2 Bone Development and Growth

6　Explain how the development of intramembranous bone differs from that of endochondral bone. (pp. 202–204)

7　_____ are bone cells in lacunae, whereas _____ are bone-forming cells and _____ are bone-resorbing cells. (pp. 202–207)

8　Explain the function of an epiphyseal plate. (pp. 205–207)

9　Place the zones of cartilage in an epiphyseal plate in order (1–4), with the first zone attached to the epiphysis. (p. 205)
　_____ zone of hypertrophic cartilage
　_____ zone of calcified cartilage
　_____ zone of resting cartilage
　_____ zone of perforating cartilage

10　Explain how osteoblasts and osteoclasts regulate bone mass. (p. 207)

11　Describe the effects of vitamin deficiencies on bone development and growth. (pp. 207–209)

12　Explain the causes of pituitary dwarfism and gigantism. (p. 210)

13　Describe the effects of thyroid and sex hormones on bone development and growth. (p. 210)

14　Physical exercise pulling on muscular attachments to bone stimulates _____ . (p. 210)

7.3 Bone Function

15　Provide several examples to illustrate how bones support and protect body parts. (p. 210)

16　Describe the functions of red and yellow bone marrow. (p. 210)

17　Explain the mechanism that regulates the concentration of blood calcium ions. (p. 211)

18　List three metallic elements that may be abnormally stored in bone. (p. 211)

7.4 Skeletal Organization

19　Bones of the head, neck, and trunk compose the _____ skeleton; bones of the limbs and their attachments compose the _____ skeleton. (pp. 212–213)

7.5 Skull–7.11 Lower Limb

20　Name the bones of the cranium and the facial skeleton. (pp. 218–222)

21　Explain the importance of fontanels. (p. 225)

22　Describe a typical vertebra, and distinguish among the cervical, thoracic, and lumbar vertebrae. (pp. 227–229)

23　Describe the locations of the sacroiliac joint, the sacral promontory, and the sacral hiatus. (p. 229)

24　Name the bones that comprise the thoracic cage. (p. 231)

25　The clavicle and scapula form the _____ girdle, whereas the hip bones and sacrum form the _____ girdle. (pp. 231 and 238)

26　Name the bones of the upper limb, and describe their locations. (pp. 233–236)

27　Name the bones that comprise the hip bone. (p. 239)

28　Explain the major differences between the male and female skeletons. (p. 241)

29　Name the bones of the lower limb, and describe their locations. (pp. 241–245)

30　Match the parts listed on the left with the bones listed on the right. (pp. 218–243)

(1)	coronoid process	A.	ethmoid bone
(2)	cribriform plate	B.	frontal bone
(3)	foramen magnum	C.	mandible
(4)	mastoid process	D.	maxilla
(5)	palatine process	E.	occipital bone
(6)	sella turcica	F.	temporal bone
(7)	supraorbital notch	G.	sphenoid bone
(8)	temporal process	H.	zygomatic bone
(9)	acromion process	I.	femur
(10)	deltoid tuberosity	J.	fibula
(11)	greater trochanter	K.	humerus
(12)	lateral malleolus	L.	radius
(13)	medial malleolus	M.	scapula
(14)	olecranon process	N.	sternum
(15)	radial tuberosity	O.	tibia
(16)	xiphoid process	P.	ulna

7.12 Life-Span Changes

31　Describe the changes brought about by aging in trabecular bone. (p. 245)

INTEGRATIVE ASSESSMENTS/CRITICAL THINKING

Outcomes 1.6, 7.2, 7.3, 7.5

1. How might the condition of an infant's fontanels be used to evaluate skeletal development? How might the fontanels be used to estimate intracranial pressure (pressure in the cranial cavity)?

Outcomes 1.7, 7.1, 7.2, 7.12

2. Why are incomplete, longitudinal fractures of bone shafts (greenstick fractures) more common in children than in adults?

Outcomes 5.3, 7.1, 7.5

3. How does the structure of a bone make it strong yet lightweight?

Outcomes 5.3, 7.2

4. If a young patient's forearm and elbow are immobilized by a cast for several weeks, what changes would you expect to occur in the bones of the upper limb?

Outcomes 7.2, 7.3, 7.9, 7.11

5. When a child's bone is fractured, growth may be stimulated at the epiphyseal plate. What problems might this extra growth cause in an upper or lower limb before the growth of the other limb compensates for the difference in length?

Outcomes 7.2, 7.10, 7.12

6. Suppose archeologists discover human skeletal remains in Ethiopia. Examination of the bones suggests that the remains represent four types of individuals. Two of the skeletons have bone densities that are 30% less than those of the other two skeletons. The skeletons with the lower bone mass also have broader front pelvic bones. Within the two groups defined by bone mass, smaller skeletons have bones with evidence of epiphyseal plates, but larger bones have only a thin line where the epiphyseal plates should be. Give the age group and gender of the individuals in this find.

Outcomes 7.6, 7.12

7. Why do elderly persons often develop bowed backs and appear to lose height?

 ## ONLINE STUDY TOOLS

Connect Interactive Questions Reinforce your knowledge using assigned interactive questions covering skeletal tissue, ossification, and parts of the skeleton.

Connect Integrated Activity Can you help a forensic investigator identify and assemble a skeleton from scattered bones at a crime scene or identify bones on an X ray?

LearnSmart Discover which chapter concepts you have mastered and which require more attention. This adaptive learning tool is personalized, proven, and preferred.

Anatomy & Physiology Revealed Go more in depth into the human body by viewing X rays and CT scans of the skeleton.

Human Skull

The following set of reference plates will help you locate some of the more prominent features of the human skull. As you study these photographs, it is important to remember that individual human skulls vary in every characteristic. Also, the photographs in this set depict bones from several different skulls.

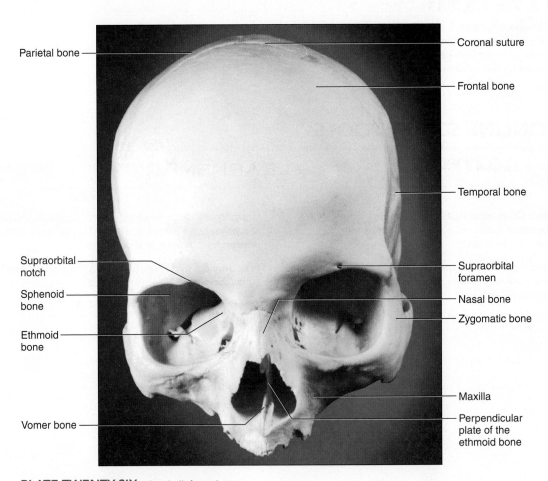

Parietal bone

Coronal suture

Frontal bone

Temporal bone

Supraorbital notch

Supraorbital foramen

Sphenoid bone

Nasal bone

Zygomatic bone

Ethmoid bone

Vomer bone

Maxilla

Perpendicular plate of the ethmoid bone

PLATE TWENTY-SIX The skull, frontal view.

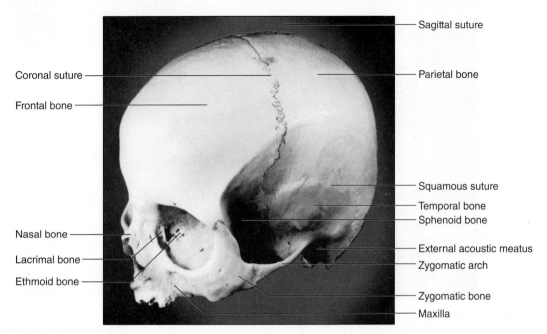

Labels for top image (left anterolateral view):
- Sagittal suture
- Coronal suture
- Frontal bone
- Parietal bone
- Squamous suture
- Temporal bone
- Sphenoid bone
- Nasal bone
- Lacrimal bone
- Ethmoid bone
- External acoustic meatus
- Zygomatic arch
- Zygomatic bone
- Maxilla

PLATE TWENTY-SEVEN The skull, left anterolateral view.

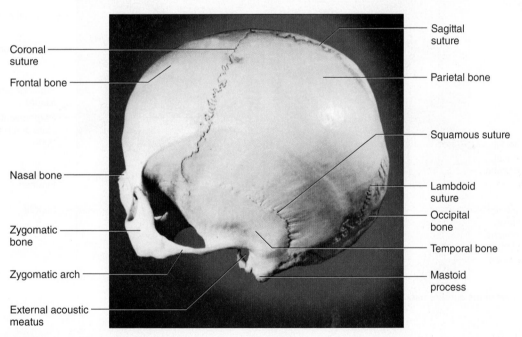

Labels for bottom image (left posterolateral view):
- Coronal suture
- Frontal bone
- Nasal bone
- Zygomatic bone
- Zygomatic arch
- External acoustic meatus
- Sagittal suture
- Parietal bone
- Squamous suture
- Lambdoid suture
- Occipital bone
- Temporal bone
- Mastoid process

PLATE TWENTY-EIGHT The skull, left posterolateral view.

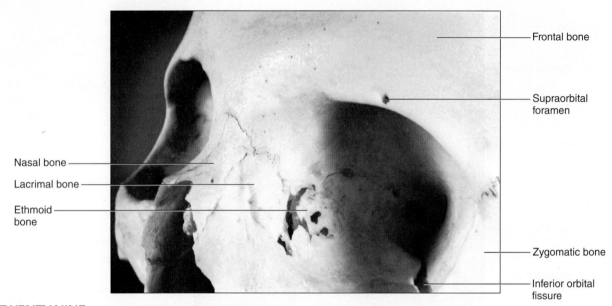

Frontal bone

Supraorbital
foramen

Nasal bone

Lacrimal bone

Ethmoid
bone

Zygomatic bone

Inferior orbital
fissure

PLATE TWENTY-NINE Bones of the left orbital region.

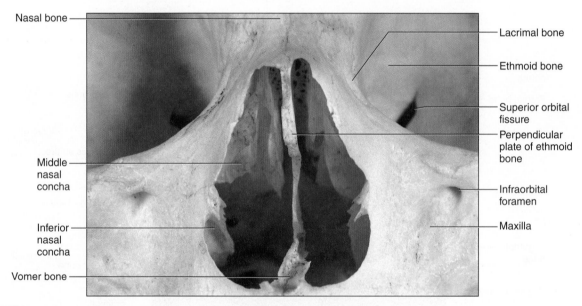

Nasal bone

Lacrimal bone

Ethmoid bone

Superior orbital
fissure

Perpendicular
plate of ethmoid
bone

Infraorbital
foramen

Maxilla

Middle
nasal
concha

Inferior
nasal
concha

Vomer bone

PLATE THIRTY Bones of the anterior nasal region.

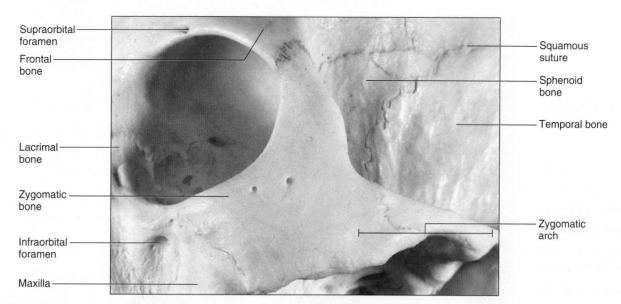

Supraorbital foramen

Frontal bone

Lacrimal bone

Zygomatic bone

Infraorbital foramen

Maxilla

Squamous suture

Sphenoid bone

Temporal bone

Zygomatic arch

PLATE THIRTY-ONE Bones of the left zygomatic region.

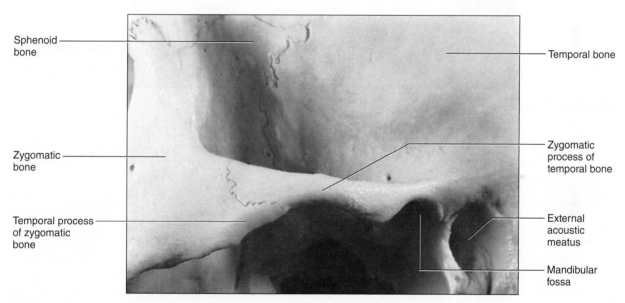

Sphenoid bone

Zygomatic bone

Temporal process of zygomatic bone

Temporal bone

Zygomatic process of temporal bone

External acoustic meatus

Mandibular fossa

PLATE THIRTY-TWO Bones of the left temporal region.

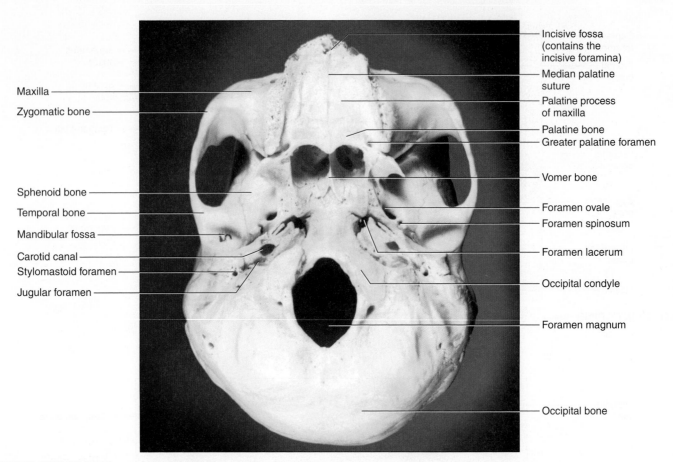

Incisive fossa
(contains the
incisive foramina)

Median palatine
suture

Palatine process
of maxilla

Palatine bone

Greater palatine foramen

Vomer bone

Foramen ovale

Foramen spinosum

Foramen lacerum

Occipital condyle

Foramen magnum

Occipital bone

Maxilla

Zygomatic bone

Sphenoid bone

Temporal bone

Mandibular fossa

Carotid canal

Stylomastoid foramen

Jugular foramen

PLATE THIRTY-THREE The skull, inferior view.

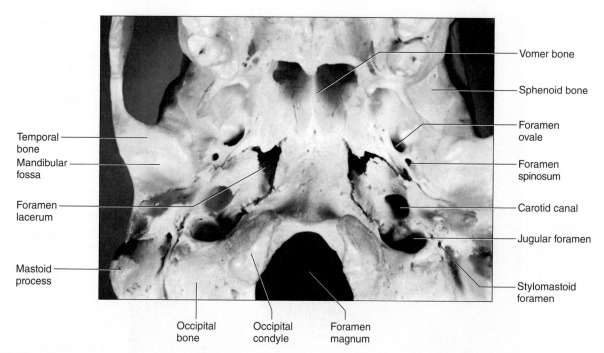

Vomer bone

Sphenoid bone

Foramen ovale

Foramen spinosum

Carotid canal

Jugular foramen

Stylomastoid foramen

Temporal bone

Mandibular fossa

Foramen lacerum

Mastoid process

Occipital bone

Occipital condyle

Foramen magnum

PLATE THIRTY-FOUR Base of the skull, sphenoid region.

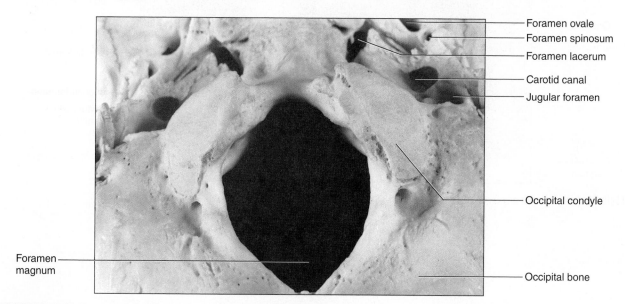

Foramen ovale
Foramen spinosum
Foramen lacerum

Carotid canal
Jugular foramen

Occipital condyle

Foramen magnum

Occipital bone

PLATE THIRTY-FIVE Base of the skull, occipital region.

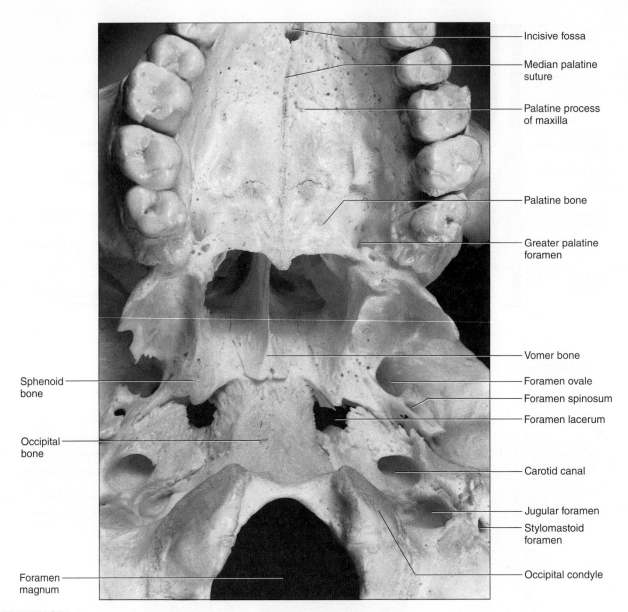

Incisive fossa

Median palatine suture

Palatine process of maxilla

Palatine bone

Greater palatine foramen

Vomer bone

Foramen ovale

Foramen spinosum

Foramen lacerum

Carotid canal

Jugular foramen

Stylomastoid foramen

Occipital condyle

Sphenoid bone

Occipital bone

Foramen magnum

PLATE THIRTY-SIX Base of the skull, maxillary region.

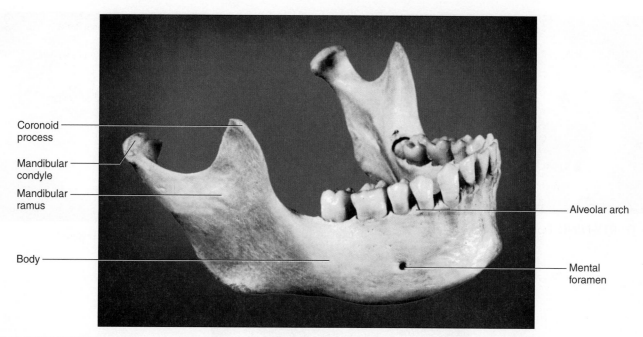

Coronoid
process

Mandibular
condyle

Mandibular
ramus

Body

Alveolar arch

Mental
foramen

PLATE THIRTY-SEVEN Mandible, right lateral view.

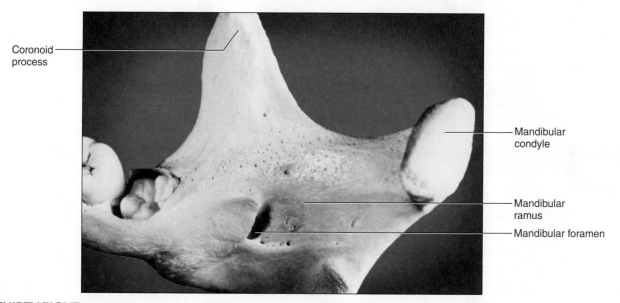

Coronoid
process

Mandibular
condyle

Mandibular
ramus

Mandibular foramen

PLATE THIRTY-EIGHT Mandible, medial surface of right ramus.

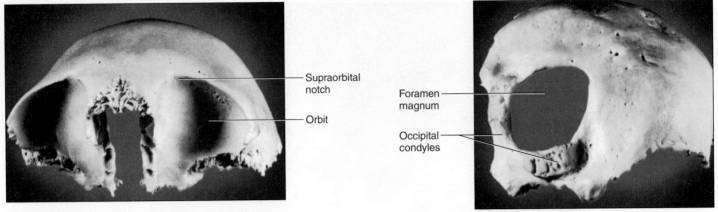

Supraorbital notch

Orbit

PLATE THIRTY-NINE Frontal bone, anterior view.

Foramen magnum

Occipital condyles

PLATE FORTY Occipital bone, inferior view.

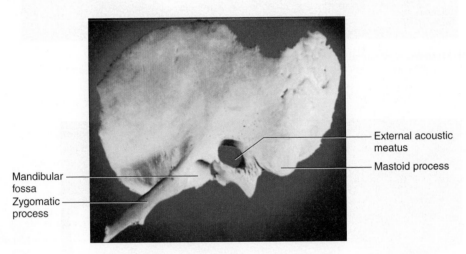

External acoustic meatus

Mastoid process

Mandibular fossa

Zygomatic process

PLATE FORTY-ONE Temporal bone, left lateral view.

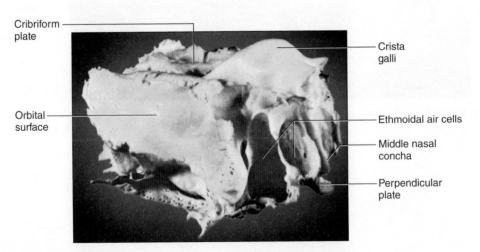

Cribriform plate

Crista galli

Orbital surface

Ethmoidal air cells

Middle nasal concha

Perpendicular plate

PLATE FORTY-TWO Ethmoid bone, right lateral view.

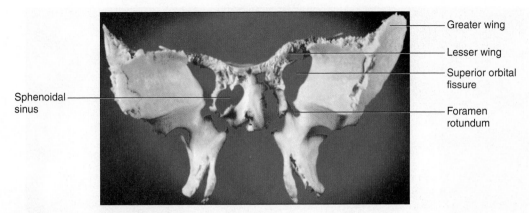

PLATE FORTY-THREE Sphenoid bone, anterior view.

— Greater wing

— Lesser wing

— Superior orbital fissure

Sphenoidal sinus —

— Foramen rotundum

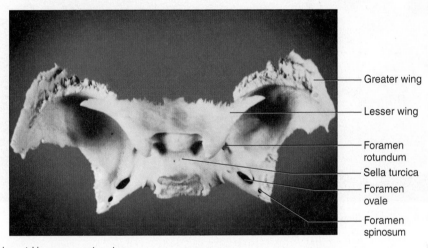

PLATE FORTY-FOUR Sphenoid bone, superior view.

— Greater wing

— Lesser wing

— Foramen rotundum

— Sella turcica

— Foramen ovale

— Foramen spinosum

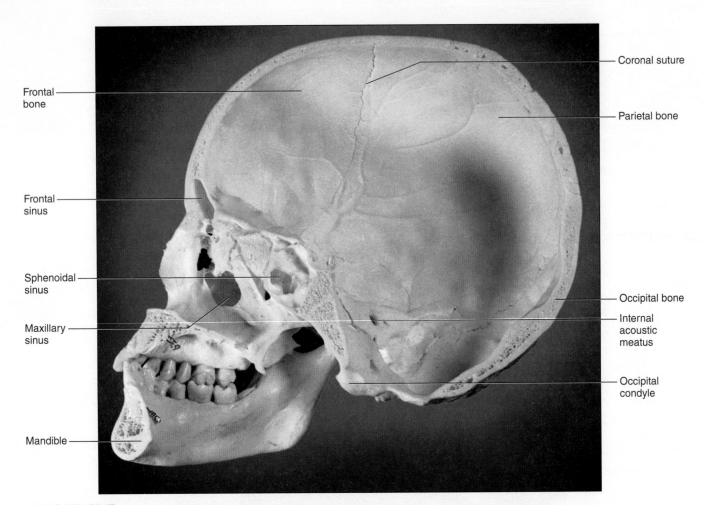

Frontal bone

Frontal sinus

Sphenoidal sinus

Maxillary sinus

Mandible

Coronal suture

Parietal bone

Occipital bone

Internal acoustic meatus

Occipital condyle

PLATE FORTY-FIVE The skull, sagittal section.

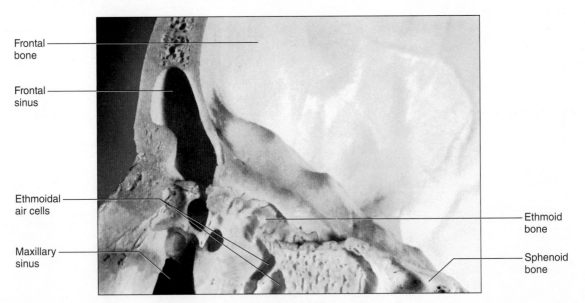

Frontal bone

Frontal sinus

Ethmoidal air cells

Maxillary sinus

Ethmoid bone

Sphenoid bone

PLATE FORTY-SIX Ethmoidal region, sagittal section.

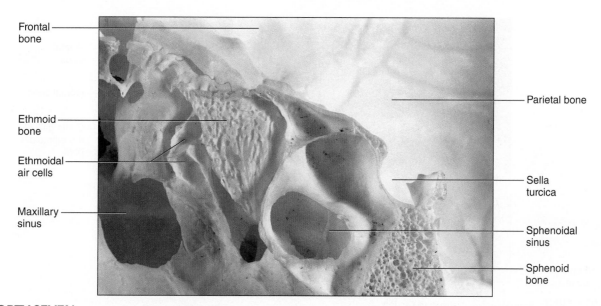

Frontal bone

Ethmoid bone

Ethmoidal air cells

Maxillary sinus

Parietal bone

Sella turcica

Sphenoidal sinus

Sphenoid bone

PLATE FORTY-SEVEN Sphenoidal region, sagittal section.

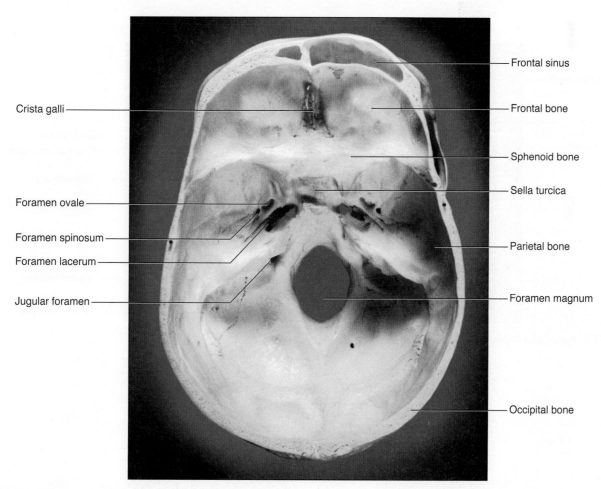

Crista galli

Foramen ovale

Foramen spinosum

Foramen lacerum

Jugular foramen

Frontal sinus

Frontal bone

Sphenoid bone

Sella turcica

Parietal bone

Foramen magnum

Occipital bone

PLATE FORTY-EIGHT The skull, floor of the cranial cavity.

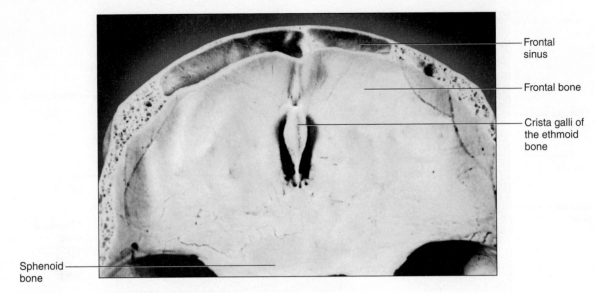

Frontal sinus

Frontal bone

Crista galli of the ethmoid bone

Sphenoid bone

PLATE FORTY-NINE Frontal region, transverse section.

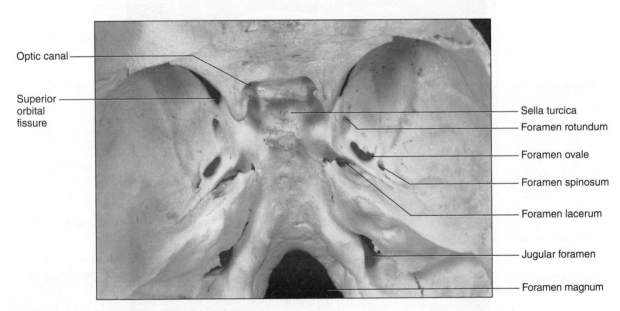

Optic canal

Superior orbital fissure

Sella turcica

Foramen rotundum

Foramen ovale

Foramen spinosum

Foramen lacerum

Jugular foramen

Foramen magnum

PLATE FIFTY Sphenoidal region, floor of the cranial cavity.

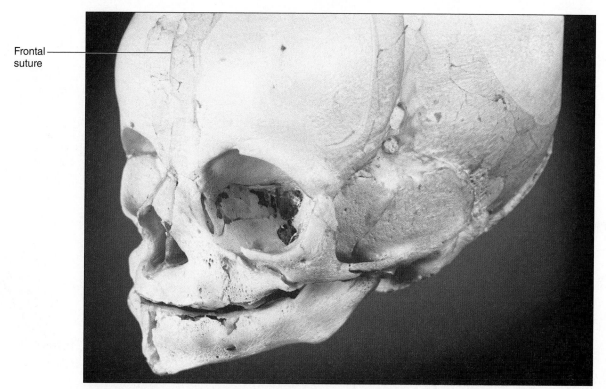

Frontal
suture

PLATE FIFTY-ONE Skull of a fetus, left anterolateral view.

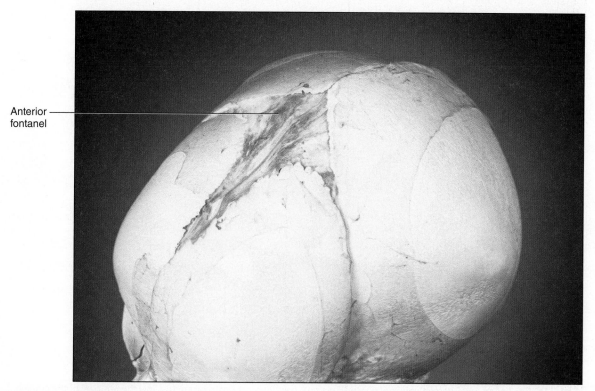

Anterior
fontanel

PLATE FIFTY-TWO Skull of a fetus, left superior view.

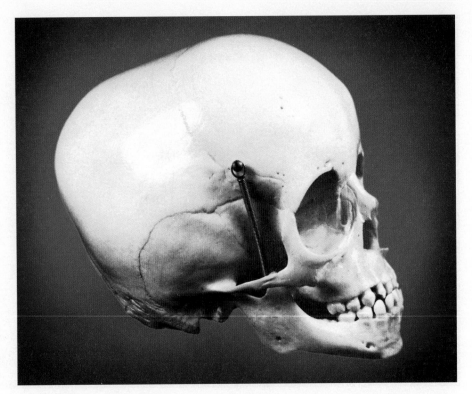

PLATE FIFTY-THREE Skull of a child, right lateral view.

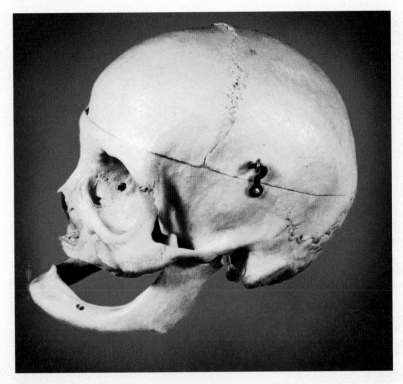

PLATE FIFTY-FOUR Skull of an aged person, left lateral view. (This skull has been cut postmortem to allow the removal of the cranium.)

Arthritis has inflamed the joints in these fingers. Drugs and replacement joints are used to treat this painful condition.

Joints of the Skeletal System

 LEARNING OUTCOMES

After you have studied this chapter, you should be able to:

8.1 Types of Joints

1 Explain how joints can be classified according to the type of tissue that binds the bones together and the degree of movement possible at the joint. (p. 268)

2 Describe how bones of fibrous joints are held together, and name an example of each type of fibrous joint. (p. 268)

3 Describe how bones of cartilaginous joints are held together, and name an example of each type of cartilaginous joint. (p. 269)

4 Describe the general structure of a synovial joint. (pp. 269–272)

5 Distinguish among the six types of synovial joints, and name an example of each type. (p. 272)

8.2 Types of Joint Movements

6 Explain how skeletal muscles produce movements at joints, and identify several types of joint movements. (pp. 274–277)

8.3 Examples of Synovial Joints

7 Describe the shoulder joint, and explain how its articulating parts are held together. (pp. 277 and 279)

8 Describe the elbow, hip, and knee joints, and explain how their articulating parts are held together. (pp. 279–284)

8.4 Life-Span Changes

9 Describe life-span changes in joints. (p. 285)

THE WHOLE PICTURE

Joints, or articulations, are functional junctions between bones. They bond parts of the skeletal system, make possible bone growth, permit parts of the skeleton to change shape during childbirth, and enable the body to move in response to skeletal muscle contractions.

Module 5: Skeletal System

UNDERSTANDING WORDS

anul-, ring: *anular* ligament—ring-shaped band of connective tissue below the elbow joint that encircles the head of the radius.

arth-, joint: *arthrology*—study of joints and ligaments.

burs-, bag, purse: prepatellar *bursa*—fluid-filled sac between the skin and the patella.

glen-, joint socket: *glenoid* cavity—depression in the scapula that articulates with the head of the humerus.

labr-, lip: glenoidal *labrum*—rim of fibrocartilage attached to the margin of the glenoid cavity.

ov-, egglike: synovial fluid—thick fluid in a joint cavity that resembles egg white.

sutur-, sewing: *suture*—type of joint in which flat bones are interlocked by a set of tiny bony processes.

syn-, with, together: *synchondrosis*—type of joint in which the bones are held together by cartilage.

syndesm-, band, ligament: *syndesmosis*—type of joint in which the bones are held together by long fibers of connective tissue.

8.1 | Types of Joints

Joints, which are structures that connect bones, vary considerably in structure and function. However, they can be classified structurally by the type of tissue that binds the bones at each junction. Three general groups are fibrous joints, cartilaginous joints, and synovial joints.

Joints can also be grouped functionally according to the degree of movement possible at the bony junctions. In this scheme, joints are classified as immovable (synarthrotic), slightly movable (amphiarthrotic), and freely movable (diarthrotic). At some diarthrotic joints, movement can occur over considerable distances, such as flexion and extension of the elbow. In contrast, certain other joints, such as the joint between the sacrum and the ilium, move freely, but only for short distances. The structural and functional classification schemes overlap somewhat. Currently, structural classification is the one most commonly used.

 PRACTICE

1 How are joints classified?

Fibrous Joints

Fibrous (fi′brus) **joints** are so named because the dense connective tissue holding them together includes many collagen fibers. These joints are between bones in close contact. The three types of fibrous joints are:

1. In a **syndesmosis** (sin″des-mo′sis), the bones are bound by a sheet (*interosseous membrane*) or bundle of dense connective tissue (*interosseous ligament*). This junction is flexible and may be twisted, so the joint may permit slight movement and thus is amphiarthrotic (am″fe-ar-thro′tik). A syndesmosis lies between the tibia and fibula (fig. 8.1).
2. **Sutures** (soo′cherz) are only between flat bones of the skull, where the broad margins of adjacent bones grow together and unite by a thin layer of dense connective tissue called a *sutural ligament*. Recall from chapter 7 (p. 225) that the infantile skull is incompletely developed, with several of the

bones connected by membranous areas called *fontanels* (see fig. 7.32, p. 225). These areas allow the skull to change shape slightly during childbirth, but as the bones continue to grow, the fontanels close, and sutures replace them. With time, some of the bones at sutures interlock by tiny bony processes. Such a suture is in the adult human skull where the parietal and occipital bones meet to form the lambdoid suture. Sutures are immovable, and therefore they are synarthrotic (sin′ar-thro′tik) joints (figs. 8.2 and 8.3).

3. A **gomphosis** (gom-fo′sis) is a joint formed by the union of a cone-shaped bony process in a bony socket. The peglike root of a tooth fastened to a maxilla or the mandible by a *periodontal ligament* is such a joint. This ligament surrounds the tooth root and firmly attaches it to the bone with bundles of thick collagen fibers. A gomphosis is a synarthrotic joint (fig. 8.4).

 PRACTICE

2 Describe three types of fibrous joints.
3 What is the function of the fontanels?

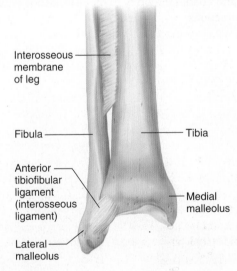

FIGURE 8.1 The articulation between the tibia and fibula is an example of a syndesmosis.

Labels: Interosseous membrane of leg; Fibula; Tibia; Anterior tibiofibular ligament (interosseous ligament); Medial malleolus; Lateral malleolus

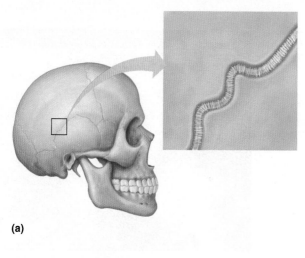

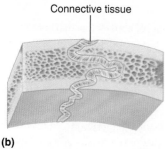

Connective tissue

FIGURE 8.2 Fibrous joints. (*a*) The fibrous joints between the bones of the skull are immovable and are called sutures. (*b*) A thin layer of connective tissue connects the bones at the suture.

Cartilaginous Joints

Hyaline cartilage or fibrocartilage connects the bones of **cartilaginous** (kar″tĭ-laj′ĭnus) **joints.** The two types are:

1. In a **synchondrosis** (sin″kon-dro′sis), bands of hyaline cartilage unite the bones. Many of these joints are temporary structures that disappear during growth. An example of a synchondrosis is that part of an immature long bone where a band of hyaline cartilage (the epiphyseal plate) connects an epiphysis to a diaphysis. This cartilage band participates in bone lengthening and, in time, is replaced with bone. When ossification completes, and the epiphyseal plate becomes an epiphyseal line, usually before the age of twenty-five years, the joint becomes a *synostosis,* a bony joint. The synostosis is synarthrotic (see fig. 7.9, p. 206).

   Another synchondrosis lies between the manubrium and the first rib, directly united by costal cartilage (fig. 8.5). This joint is also synarthrotic, but permanent. Most of the joints between the sternum and the costal cartilages of ribs 2 through 7 are synovial.

2. The articular surfaces of the bones at a **symphysis** (sim′fĭ-sis) are covered by a thin layer of hyaline cartilage, and the bones are connected by a pad of fibrocartilage. Limited movement occurs at such a joint whenever forces compress or deform the fibrocartilage pad. An example of this type of joint is the pubic symphysis between the pubic bones of the pelvis, which allows maternal pelvic bones to shift, as a result of hormone changes affecting the fibrocartilage pad during pregnancy, when an infant passes through the birth canal (fig. 8.6*a*).

The joint formed by the bodies of two adjacent vertebrae separated and connected by an intervertebral disc is also a symphysis (fig. 8.6*b* and reference plate 11, p. 47). Each intervertebral disc is composed of a band of fibrocartilage (annulus fibrosus) that surrounds a gelatinous core (nucleus pulposus). The disc absorbs shocks and helps equalize pressure between the vertebrae when the body moves. Because each disc is slightly flexible, the combined movement of many of the joints in the vertebral column allows the back to bend forward or to the side or to twist. The intervertebral discs are amphiarthrotic joints because they allow slight movements.

 PRACTICE

4   Describe two types of cartilaginous joints.

5   What is the function of an intervertebral disc?

Synovial Joints

Most joints of the skeletal system are **synovial** (sĭ-no′ve-al) **joints,** and because they allow free movement, they are diarthrotic (di″ar-thro′tik). These joints are more complex structurally than fibrous or cartilaginous joints. They consist of articular cartilage; a joint capsule; and a synovial membrane, which secretes synovial fluid.

General Structure of a Synovial Joint

The articular ends of the bones in a synovial joint are covered with a thin layer of hyaline cartilage. This layer, the **articular cartilage,** resists wear and minimizes friction when it is compressed as the joint moves (fig. 8.7).

 CAREER CORNER
Physical Therapy Assistant

The forty-eight-year-old man joined an over-thirty basketball league, and now he regrets his enthusiasm. He lies on the gym floor, in pain, thanks to an overambitious jump shot. He had felt a sudden twinge in his knee, and now it is red and swelling.

In the emergency department of the nearest hospital, an orthopedist sends the man for an MRI scan, which reveals a small tear in the anterior cruciate ligament (ACL). The patient is willing to give up basketball and doesn't want surgery, so he goes for physical therapy twice a week for a month. A physical therapy assistant leads him through a series of exercises that will restore full mobility.

The therapy, under the supervision of a physical therapist, begins with deep knee bends and progresses to lifting weights, stepping, squats, and using a single-leg bicycle that isolates and builds up the injured leg. The physical therapy assistant gives her patient exercises to do daily, at home. The therapy builds the muscles around the knee to compensate for the hurt ACL, restoring full range of motion.

A physical therapy assistant must complete a two-year college program and pass a certification exam. Physical therapy assistants work in hospitals, skilled nursing facilities, private homes, schools, fitness centers, and workplaces.

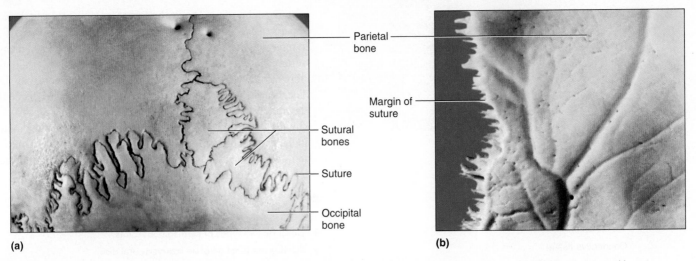

FIGURE 8.3 Cranial sutures. (*a*) Sutures between the parietal and occipital bones of the skull. (*b*) The inner margin of a suture (parietal bone). The grooves on the inside of this parietal bone mark the paths of blood vessels near the brain's surface.

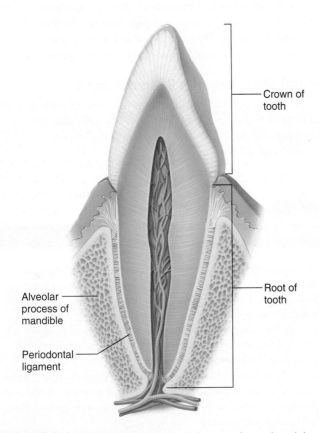

FIGURE 8.4 The articulation between the root of a tooth and the mandible is a gomphosis.

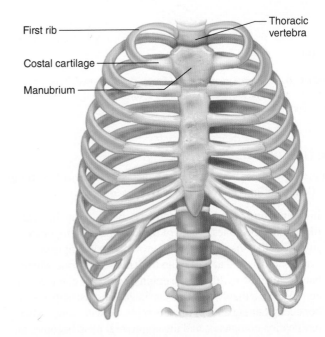

FIGURE 8.5 The articulation between the first rib and the manubrium is a synchondrosis.

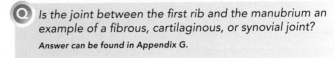

Is the joint between the first rib and the manubrium an example of a fibrous, cartilaginous, or synovial joint?

Answer can be found in Appendix G.

A tubular **joint capsule** (articular capsule) that has two distinct layers holds together the bones of a synovial joint. The outer layer largely consists of dense connective tissue, whose fibers attach to the periosteum around the circumference of each bone of the joint near its articular end. Thus, the outer fibrous layer of the capsule completely encloses the other parts of the joint. It is,

however, flexible enough to permit movement and strong enough to help prevent the bones from being pulled apart.

Bundles of strong, tough collagen fibers called **ligaments** (lig'ah-mentz) reinforce the joint capsule and help bind the articular ends of the bones. Some ligaments appear as thickenings in the fibrous layer of the capsule, whereas others are *accessory structures*

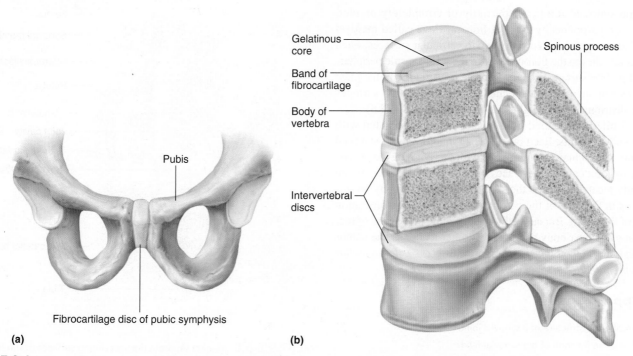

Gelatinous core

Band of fibrocartilage

Body of vertebra

Intervertebral discs

Spinous process

Pubis

Fibrocartilage disc of pubic symphysis

(a)

(b)

FIGURE 8.6 Fibrocartilage composes (a) the pubic symphysis and (b) the intervertebral discs of the vertebrae.

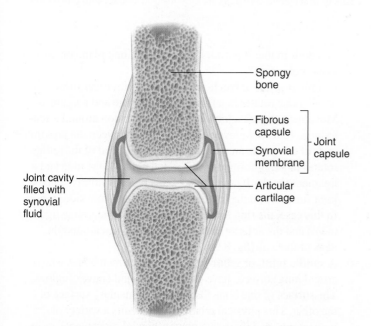

Spongy bone

Fibrous capsule

Synovial membrane

Joint capsule

Articular cartilage

Joint cavity filled with synovial fluid

FIGURE 8.7 The generalized structure of a synovial joint.

called the *synovial cavity,* into which the synovial membrane secretes a clear, viscous fluid called **synovial fluid.** In some regions, the surface of the synovial membrane has villi as well as larger folds and projections that extend into the cavity. Besides filling spaces and irregularities of the joint cavity, these extensions increase the surface area of the synovial membrane. The synovial membrane may also store adipose tissue and form movable fatty pads in the joint. This multifunctional membrane also reabsorbs fluid, which is important when a joint cavity is injured or infected. Synovial fluid contains stem cells, which may function in ligament regeneration following injury.

Synovial fluid has a consistency similar to uncooked egg white, and it moistens and lubricates the smooth cartilaginous surfaces of the joint. It also helps supply articular cartilage with nutrients obtained from blood vessels of the synovial membrane. The volume of synovial fluid in a joint cavity is usually just enough to cover the articulating surfaces with a thin film of fluid. The volume of synovial fluid in the cavity of the knee is 0.5 mL or less.

located outside the capsule. In either case, these structures help prevent excessive movement at the joint. That is, the ligament is relatively inelastic, and it tightens when the joint is stressed.

The inner layer of the joint capsule consists of a shiny, vascular lining of loose connective tissue called the **synovial membrane.** This membrane, only a few cells thick, covers all of the surfaces within the joint capsule, except the areas the articular cartilage covers. The synovial membrane surrounds a closed sac

A physician can determine the cause of joint inflammation or degeneration (arthritis) by aspirating a sample of synovial fluid from the affected joint using a procedure called arthrocentesis. Bloody fluid may indicate trauma to the joint. Cloudy, yellowish fluid may indicate rheumatoid arthritis, and crystals in the synovial fluid may signal gout. If the fluid is cloudy but red-tinged and containing pus, a bacterial infection may be present that requires prompt treatment. Normal synovial fluid has 180 or fewer leukocytes (white blood cells) per mL. If the leukocyte count exceeds 2,000, the fluid is infected.

Some synovial joints are partially or completely divided into two compartments by discs of fibrocartilage called **menisci** (me-nis′ke) (sing., *meniscus*) between the articular surfaces. Each meniscus attaches to the fibrous layer of the joint capsule peripherally, and its free surface projects into the joint cavity. In the knee joint, crescent-shaped menisci cushion the articulating surfaces and help distribute body weight onto these surfaces (fig. 8.8).

Fluid-filled sacs called **bursae** (ber′se) are associated with certain synovial joints. Each bursa has an inner lining of synovial membrane, which may be continuous with the synovial membrane of a nearby joint cavity. These sacs contain synovial fluid and are commonly located between the skin and underlying bony prominences, as in the case of the patella of the knee or the olecranon process of the elbow. Bursae cushion and aid the movement of tendons that glide over bony parts or over other tendons. The names of bursae indicate their locations. Figure 8.8 shows a *suprapatellar bursa,* a *prepatellar bursa,* and an *infrapatellar bursa.*

 PRACTICE

6 Describe the structure of a synovial joint.

7 What is the function of the synovial fluid?

Types of Synovial Joints

The articulating bones of synovial joints have a variety of shapes that allow different types of movement. Based upon their shapes and the movements they permit, these joints can be classified into six major types—ball-and-socket joints, condylar joints, plane joints, hinge joints, pivot joints, and saddle joints.

1. A **ball-and-socket joint,** or **spheroidal joint,** consists of a bone with a globular or slightly egg-shaped head that articulates with the cup-shaped cavity of another bone. Such a joint allows a wider range of motion than does any other type, permitting movements in all planes (multiaxial movement), as well as rotational movement around a central axis. The hip and shoulder have joints of this type (fig. 8.9*a*).

2. In a **condylar joint,** or **ellipsoidal joint,** the ovoid condyle of one bone fits into the elliptical cavity of another bone, as in the joints between the metacarpals and phalanges (fig. 8.9*b*). This type of joint permits back-and-forth and side-to-side movement in two planes (biaxial movement), but not rotation.

3. The articulating surfaces of **plane joints,** or **gliding joints,** are nearly flat or slightly curved. These joints allow sliding or back-and-forth motion and twisting movements (nonaxial movement). Most of the joints in the wrist and ankle, as well as those between the articular processes of vertebrae, belong to this group (fig. 8.9*c*). The sacroiliac joints and the joints formed by ribs 2 through 7 connecting with the sternum are also plane joints.

4. In a **hinge joint,** the convex surface of one bone fits into the concave surface of another, as in the elbow and the joints of the phalanges (fig. 8.9*d*). Such a joint resembles the hinge

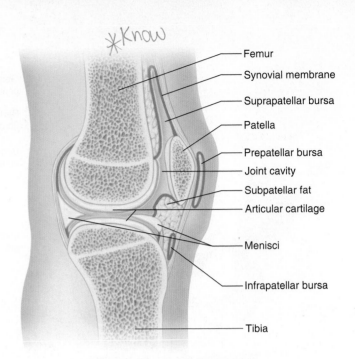

*Know

FIGURE 8.8 Menisci separate the articulating surfaces of the femur and tibia. Several bursae are associated with the knee joint. Synovial spaces in this and other line drawings in this chapter are exaggerated. In actuality, articulating cartilages are essentially in contact with one another.

of a door in that it permits movement in one plane only (uniaxial movement).

5. In a **pivot joint,** or **trochoid joint,** the cylindrical surface of one bone rotates in a ring formed of bone and a ligament. Movement at such a joint is limited to rotation around a central axis (uniaxial movement). The joint between the proximal ends of the radius and the ulna, where the head of the radius rotates in a ring formed by the radial notch of the ulna and a ligament (anular ligament), is of this type. Similarly, a pivot joint functions in the neck as the head turns from side to side. In this case, the ring formed by a ligament (transverse ligament) and the anterior arch of the atlas rotates around the dens of the axis (fig. 8.9*e*).

6. A **saddle joint,** or **sellar joint,** forms between bones whose articulating surfaces have both concave and convex regions. The surface of one bone fits the complementary surface of the other. This physical relationship permits a variety of movements, mainly in two planes (biaxial movement), as in the case of the joint between the carpal (trapezium) and the metacarpal of the thumb (fig. 8.9*f*).

Table 8.1 summarizes the types of joints.

 PRACTICE

8 Name six types of synovial joints.

9 Describe the structure of each type of synovial joint.

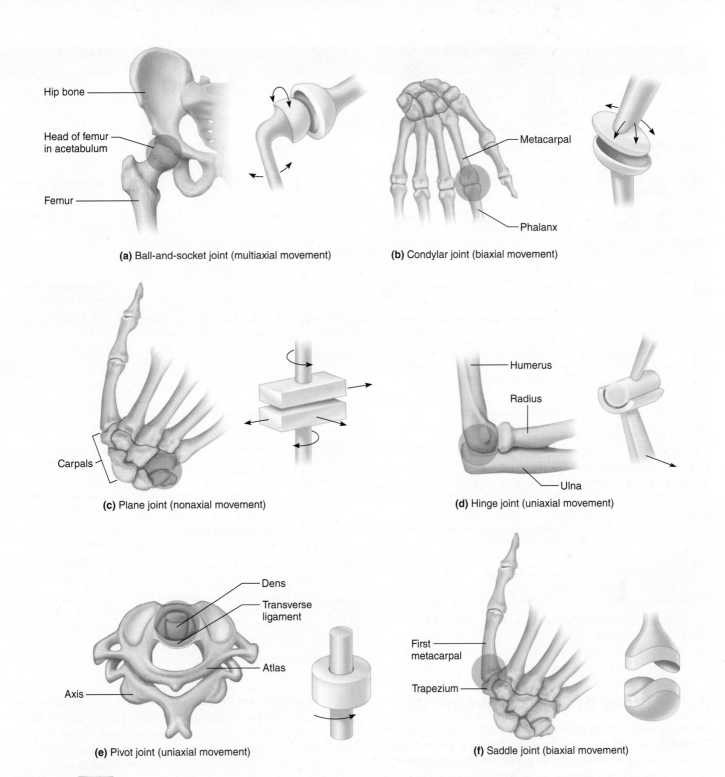

(a) Ball-and-socket joint (multiaxial movement)

Hip bone

Head of femur in acetabulum

Femur

(b) Condylar joint (biaxial movement)

Metacarpal

Phalanx

(c) Plane joint (nonaxial movement)

Carpals

(d) Hinge joint (uniaxial movement)

Humerus

Radius

Ulna

(e) Pivot joint (uniaxial movement)

Dens

Transverse ligament

Atlas

Axis

(f) Saddle joint (biaxial movement)

First metacarpal

Trapezium

FIGURE 8.9 AP|R Types and examples of synovial (freely movable) joints (*a-f*).

TABLE 8.1 | Types of Joints

Type of Joint	Description	Possible Movements	Example
Fibrous	Articulating bones fastened together by a thin layer of dense connective tissue containing many collagen fibers		
1. *Syndesmosis* (amphiarthrotic)	Bones bound by interosseous ligament	Joint flexible and may be twisted	Tibiofibular articulation
2. *Suture* (synarthrotic)	Flat bones united by sutural ligament	None	Parietal bones articulate at sagittal suture of skull
3. *Gomphosis* (synarthrotic)	Cone-shaped process fastened in bony socket by periodontal ligament	None	Root of tooth united with mandible
Cartilaginous	Articulating bones connected by hyaline cartilage or fibrocartilage		
1. *Synchondrosis* (synarthrotic)	Bones united by bands of hyaline cartilage	None	Joint between the first rib and the manubrium
2. *Symphysis* (amphiarthrotic)	Articular surfaces of bones are covered by hyaline cartilage and the bones are connected by a pad of fibrocartilage	Limited movement, as when the back is bent or twisted	Joints between bodies of vertebrae
Synovial (diarthrotic)	Articulating ends of bones surrounded by a joint capsule; articular bone ends covered by hyaline cartilage and separated by synovial fluid		
1. *Ball-and-socket*	Ball-shaped head of one bone articulates with cup-shaped socket of another	Movements in all planes (multiaxial), including rotation	Shoulder, hip
2. *Condylar*	Oval-shaped condyle of one bone articulates with elliptical cavity of another	Variety of movements in two planes (biaxial), but no rotation	Joints between metacarpals and phalanges
3. *Plane*	Articulating surfaces are nearly flat or slightly curved	Sliding or twisting (nonaxial movement)	Joints between various bones of wrist and ankle
4. *Hinge*	Convex surface of one bone articulates with concave surface of another	Flexion and extension (uniaxial)	Elbow and joints of phalanges
5. *Pivot*	Cylindrical surface of one bone articulates with ring of bone and ligament	Rotation (uniaxial)	Joint between proximal ends of radius and ulna
6. *Saddle*	Articulating surfaces have both concave and convex regions; surface of one bone fits the complementary surface of another	Variety of movements, mainly in two planes (biaxial)	Joint between carpal and metacarpal of thumb

8.2 | Types of Joint Movements

Skeletal muscle action produces movements at synovial joints. Typically, one end of a muscle is attached to a less movable or relatively fixed part on one side of a joint, and the other end of the muscle is fastened to a more movable part on the other side. When the muscle contracts, its fibers pull its movable end *(insertion)* toward its fixed end *(origin)*, and a movement occurs at the joint.

The following terms describe movements at joints that occur in different directions and in different planes (figs. 8.10, 8.11, and 8.12):

flexion (flek′shun) Bending parts at a joint so that the angle between them decreases and the parts come closer together (bending the knee).

extension (ek-sten′shun) Moving parts at a joint so that the angle between them increases and the parts move farther apart (straightening the knee).

hyperextension (hi″per-ek-sten′shun) A term sometimes used to describe the extension of the parts at a joint beyond the anatomical position (bending the head back beyond the upright position); often used to describe an abnormal extension beyond the normal range of motion, resulting in injury.

dorsiflexion (dor″si-flek′shun) Movement at the ankle that brings the foot closer to the shin (rocking back on one's heels).

plantar flexion (plan′tar flek′shun) Movement at the ankle that brings the foot farther from the shin (walking or standing on one's toes).

abduction (ab-duk′shun) Moving a part away from the midline (lifting the upper limb horizontally to form a right angle with the side of the body) or from the axial line of the limb (spreading the fingers or toes). Abduction of the head and neck and bending of the trunk to the side may be termed *lateral flexion.*

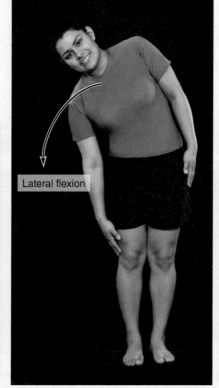

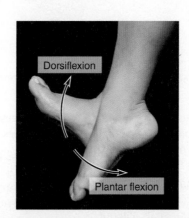

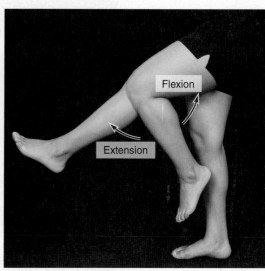

FIGURE 8.10 **AP|R** Joint movements illustrating abduction, adduction, lateral flexion, extension, and flexion.

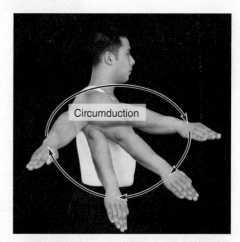

FIGURE 8.11 **AP|R** Joint movements illustrating dorsiflexion, plantar flexion, circumduction, rotation, supination, and pronation.

Name the bone that is turning on its longitudinal axis in this illustration of rotation.

Answer can be found in Appendix G.

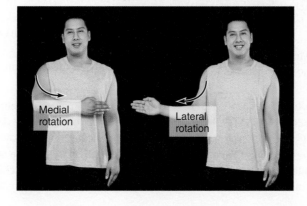

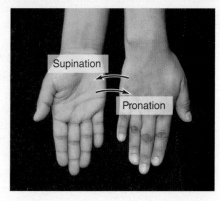

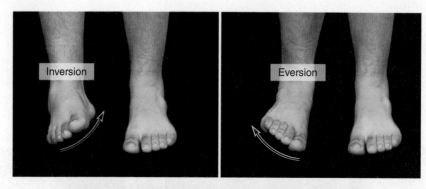

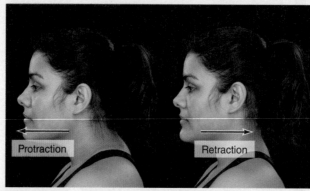

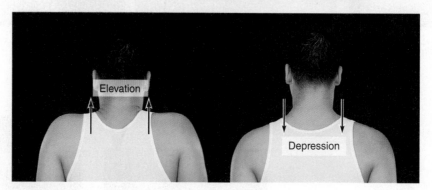

FIGURE 8.12 AP|R Joint movements illustrating inversion, eversion, protraction, retraction, elevation, and depression.

adduction (ah-duk'shun) Moving a part toward the midline (returning the upper limb from the horizontal position to the side of the body) or toward the axial line of the limb (moving the fingers or toes closer together).

rotation (ro-ta'shun) Moving a part around an axis (twisting the head from side to side). Medial (internal) rotation is the turning of a limb on its longitudinal axis so its anterior surface moves toward the midline, whereas lateral (external) rotation is the turning of a limb on its longitudinal axis in the opposite direction.

circumduction (ser″kum-duk'shun) Moving a part so that its end follows a circular path (moving the finger in a circular motion without moving the hand).

supination (soo″pĭ-na'shun) Rotation of the forearm so the palm is upward or facing anteriorly (in anatomical position). Supine refers to the body lying face up.

pronation (pro-na'shun) Rotation of the forearm so the palm is downward or facing posteriorly (in anatomical position). Prone refers to the body lying face down.

eversion (e-ver'zhun) Turning the foot so the plantar surface faces laterally.

inversion (in-ver'zhun) Turning the foot so the plantar surface faces medially.

protraction (pro-trak'shun) Moving a part forward (thrusting the head forward).

retraction (rĕ-trak'shun) Moving a part backward (pulling the head backward).

elevation (el-e-vā'shun) Raising a part (shrugging the shoulders).

depression (de-presh'un) Lowering a part (drooping the shoulders).

Description of movements of body parts is complex. At times, it will suffice to include the descriptive term of the movement followed by the part that is moving. For example, the deltoid muscle "abducts arm." Because the action of a muscle is at the insertion, the action of the biceps brachii muscle, for example, is sometimes described as "flexes forearm at the elbow." We have elected to use the more anatomically correct description of the change in geometry at the joint, "flexes elbow," to describe the action of the biceps brachii muscle. Table 8.2 lists information on several joints.

 PRACTICE

10 Describe how movement occurs at a joint when a muscle contracts.

11 What terms describe movements at synovial joints?

8.3 | Examples of Synovial Joints

The shoulder, elbow, hip, and knee are large, freely movable joints. Although these joints have much in common, each has a unique structure that makes possible its specific function.

Shoulder Joint

The **shoulder joint** is a ball-and-socket joint that consists of the rounded head of the humerus and the shallow glenoid cavity of the scapula. The coracoid and acromion processes of the scapula protect these parts, and dense connective tissue and muscle hold them together.

The joint capsule of the shoulder is attached along the circumference of the glenoid cavity and the anatomical neck of the humerus. Although it completely envelops the joint, the capsule is

very loose, and by itself is unable to keep the bones of the joint in close contact. However, muscles and tendons surround and reinforce the capsule, keeping together the articulating parts of the shoulder (fig. 8.13).

> The tendons of several muscles intimately blend with the fibrous layer of the shoulder joint capsule, forming the *rotator cuff*, which reinforces and supports the shoulder joint. Some sports-related movements can injure the rotator cuff. Injury to this area can make activities of daily living difficult. If rest, pain medication, and physical therapy do not help, surgery may be necessary.

The ligaments of the shoulder joint, some of which help prevent displacement of the articulating surfaces, include the following (fig. 8.14):

1. The **coracohumeral** (kor"ah-ko-hu'mer-al) **ligament** is composed of a broad band of connective tissue that connects the coracoid process of the scapula to the greater tubercle of the humerus. It strengthens the superior portion of the joint capsule.

2. The **glenohumeral** (gle"no-hu'mer-al) **ligaments** include three bands of fibers that appear as thickenings in the ventral wall of the joint capsule. They extend from the edge of the glenoid cavity to the lesser tubercle and the anatomical neck of the humerus.

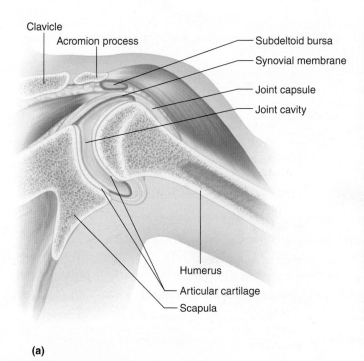

Clavicle
Acromion process
Subdeltoid bursa
Synovial membrane
Joint capsule
Joint cavity
Humerus
Articular cartilage
Scapula

(a)

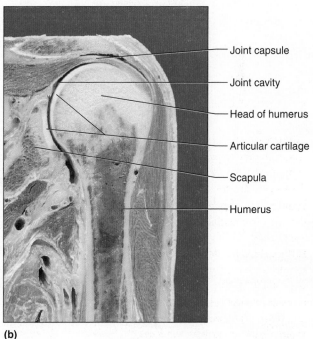

Joint capsule
Joint cavity
Head of humerus
Articular cartilage
Scapula
Humerus

(b)

FIGURE 8.13 AP|R Left shoulder joint. (*a*) The shoulder joint allows movements in all directions. A bursa is associated with this joint. (*b*) Photograph of the shoulder joint.

 The photograph (fig. 8.13b) is of what type of section (sagittal, transverse, or frontal)?
Answer can be found in Appendix G.

TABLE 8.2 | Joints of the Body

Joint	Location	Type of Joint	Type of Movement
Skull	Cranial and facial bones	Suture, fibrous	Immovable; synarthrotic
Temporomandibular	Temporal bone, mandible	Modified hinge, synovial	Elevation, depression, protraction, retraction; diarthrotic
Atlantooccipital	Atlas, occipital bone	Condylar, synovial	Flexion, extension; diarthrotic
Atlantoaxial	Atlas, axis	Pivot, synovial	Rotation; diarthrotic
Intervertebral	Between vertebral bodies	Symphysis, cartilaginous	Slight movement; amphiarthrotic
Intervertebral	Between articular processes	Plane, synovial	Flexion, extension, slight rotation; diarthrotic
Sacroiliac	Sacrum and ilium	Plane, synovial	Sliding movement; diarthrotic
Vertebrocostal	Vertebrae and ribs	Plane, synovial	Sliding movement during breathing; diarthrotic
Sternoclavicular	Sternum and clavicle	Plane, synovial	Sliding movement when shrugging shoulders; diarthrotic
Sternocostal	Sternum and rib 1	Synchondrosis, cartilaginous	Immovable; synarthrotic
Sternocostal	Sternum and ribs 2–7	Plane, synovial	Sliding movement during breathing; diarthrotic
Acromioclavicular	Scapula and clavicle	Plane, synovial	Protraction, retraction, elevation, depression, rotation; diarthrotic
Shoulder (glenohumeral)	Humerus and scapula	Ball-and-socket, synovial	Flexion, extension, adduction, abduction, rotation, circumduction; diarthrotic
Elbow (humeroulnar)	Humerus and ulna	Hinge, synovial	Flexion, extension; diarthrotic
Elbow (humeroradial)	Humerus and radius	Plane, synovial	Sliding movement; diarthrotic
Proximal radioulnar	Radius and ulna	Pivot, synovial	Rotation; diarthrotic
Distal radioulnar	Radius and ulna	Pivot, synovial	Pronation, supination; diarthrotic
Wrist (radiocarpal)	Radius and carpals	Condylar, synovial	Flexion, extension, adduction, abduction, circumduction; diarthrotic
Intercarpal	Adjacent carpals	Plane, synovial	Sliding movement, adduction, abduction, flexion, extension; diarthrotic
Carpometacarpal	Carpal and metacarpal 1	Saddle, synovial	Flexion, extension, adduction, abduction; diarthrotic
Carpometacarpal	Carpals and metacarpals 2–5	Condylar, synovial	Flexion, extension, adduction, abduction, circumduction; diarthrotic
Metacarpophalangeal	Metacarpal and proximal phalanx	Condylar, synovial	Flexion, extension, adduction, abduction, circumduction; diarthrotic
Interphalangeal	Adjacent phalanges	Hinge, synovial	Flexion, extension; diarthrotic
Pubic symphysis	Pubic bones	Symphysis, cartilaginous	Slight movement; amphiarthrotic
Hip	Hip bone and femur	Ball-and-socket, synovial	Flexion, extension, adduction, abduction, rotation, circumduction; diarthrotic
Knee (tibiofemoral)	Femur and tibia	Modified hinge, synovial	Flexion, extension, slight rotation when flexed; diarthrotic
Knee (femoropatellar)	Femur and patella	Plane, synovial	Sliding movement; diarthrotic
Proximal tibiofibular	Tibia and fibula	Plane, synovial	Sliding movement; diarthrotic
Distal tibiofibular	Tibia and fibula	Syndesmosis, fibrous	Slight rotation during dorsiflexion; amphiarthrotic
Ankle (talocrural)	Talus, tibia, and fibula	Hinge, synovial	Dorsiflexion, plantar flexion, slight circumduction; diarthrotic
Intertarsal	Adjacent tarsals	Plane, synovial	Inversion, eversion; diarthrotic
Tarsometatarsal	Tarsals and metatarsals	Plane, synovial	Sliding movement; diarthrotic
Metatarsophalangeal	Metatarsal and proximal phalanx	Condylar, synovial	Flexion, extension, adduction, abduction; diarthrotic

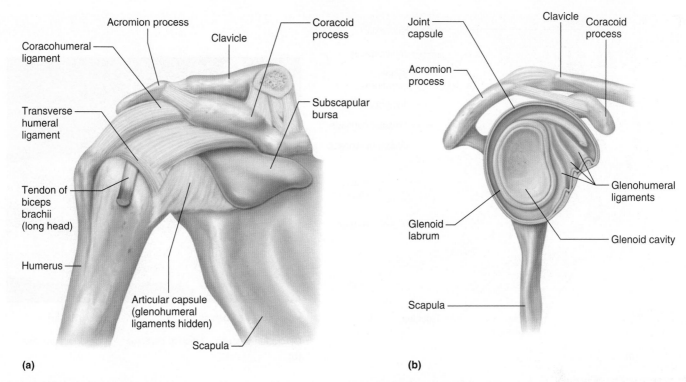

(a)

- Coracohumeral ligament
- Acromion process
- Clavicle
- Coracoid process
- Subscapular bursa
- Transverse humeral ligament
- Tendon of biceps brachii (long head)
- Humerus
- Articular capsule (glenohumeral ligaments hidden)
- Scapula

(b)

- Joint capsule
- Clavicle
- Coracoid process
- Acromion process
- Glenohumeral ligaments
- Glenoid labrum
- Glenoid cavity
- Scapula

FIGURE 8.14 Ligaments associated with the right shoulder joint. (a) Ligaments hold together the articulating surfaces of the shoulder (frontal view). (b) The glenoid labrum is composed of fibrocartilage (lateral view of the shoulder joint with the humerus removed).

3. The **transverse humeral ligament** consists of a narrow sheet of connective tissue fibers that runs between the lesser and the greater tubercles of the humerus. Together with the intertubercular sulcus of the humerus, the ligament forms a canal (retinaculum) through which the long head of the biceps brachii muscle passes.

The *glenoid labrum* (gle'noid la'brum) is composed of fibrocartilage. It is attached along the margin of the glenoid cavity and forms a rim with a thin, free edge that deepens the cavity.

Several bursae are associated with the shoulder joint. The major ones include the *subscapular bursa* between the joint capsule and the tendon of the subscapularis muscle, the *subdeltoid bursa* between the joint capsule and the deep surface of the deltoid muscle, the *subacromial bursa* between the joint capsule and the undersurface of the acromion process of the scapula, and the *subcoracoid bursa* between the joint capsule and the coracoid process of the scapula (see figs. 8.13 and 8.14). Of these, the subscapular bursa is usually continuous with the synovial cavity of the joint cavity, and although the others do not communicate with the joint cavity, they may be connected to each other.

The shoulder joint is capable of a wide range of movement, due to the looseness of its attachments and the large articular surface of the humerus compared to the shallow depth of the glenoid cavity. These movements include flexion, extension, adduction, abduction, rotation, and circumduction. Motion occurring simultaneously in the joint formed between the scapula and the clavicle may also aid such movements.

The shoulder joint is somewhat weak because the bones are mainly held together by supporting muscles rather than by bony structures and strong ligaments. Consequently, the articulating surfaces may become displaced or dislocated easily. Such a *dislocation* can occur with a forceful impact, as when a person falls on an outstretched arm. This movement may press the head of the humerus against the lower part of the joint capsule where its wall is thin and poorly supported by ligaments. Dislocations most commonly affect joints of the shoulders, knees, fingers, and jaw.

Elbow Joint

The **elbow joint** is a complex structure that includes two articulations—a hinge joint between the trochlea of the humerus and the trochlear notch of the ulna and a plane joint between the capitulum of the humerus and a shallow depression (fovea) on the head of the radius. A joint capsule completely encloses and holds together these unions (fig. 8.15). Ulnar and radial collateral ligaments thicken the two joints, and fibers from a muscle (brachialis) in the arm reinforce its anterior surface.

The **ulnar collateral ligament,** a thick band of dense connective tissue, is located in the medial wall of the capsule. The anterior portion of this ligament connects the medial epicondyle of the humerus to the medial margin of the coronoid process of the ulna. Its posterior part is attached to the medial epicondyle of the humerus and to the olecranon process of the ulna (fig. 8.16a).

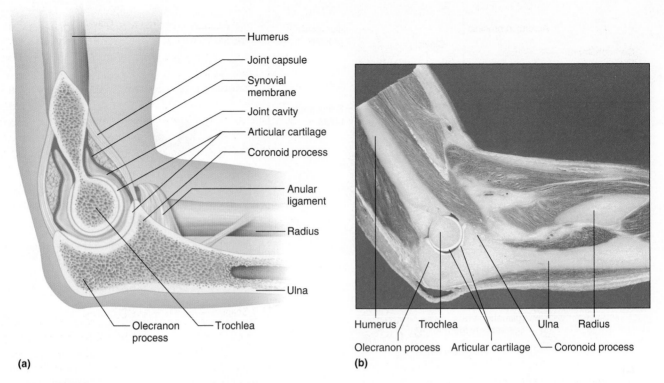

FIGURE 8.15 AP|R Left elbow joint. (*a*) The elbow joint allows hinge movements, as well as pronation and supination of the hand. (*b*) Photograph of the elbow joint (sagittal section).

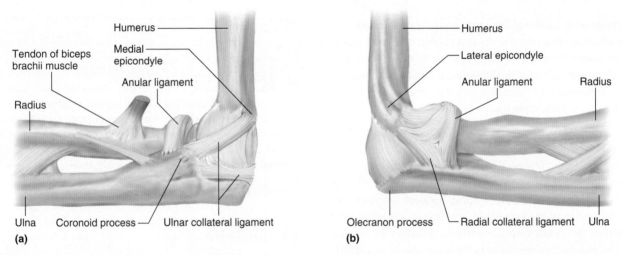

FIGURE 8.16 Ligaments associated with the right elbow joint. (*a*) The ulnar collateral ligament, medial view, and (*b*) the radial collateral ligament strengthen the capsular wall of the elbow joint, lateral view.

The **radial collateral ligament,** which strengthens the lateral wall of the joint capsule, is a fibrous band extending between the lateral epicondyle of the humerus and the *anular ligament of the radius.* The anular ligament, in turn, attaches to the margin of the trochlear notch of the ulna, and it encircles the head of the radius, keeping the head in contact with the radial notch of the ulna (fig. 8.16*b*). The elbow joint capsule encloses the resulting radio-ulnar joint so that its function is closely associated with the elbow.

The synovial membrane that forms the inner lining of the elbow capsule projects into the joint cavity between the radius and ulna and partially divides the joint into humerus–ulnar and humerus–radial portions. Also, varying amounts of adipose tissue form fatty pads between the synovial membrane and the fibrous layer of the joint capsule. These pads help protect nonarticular bony areas during joint movements.

The only movements that can occur at the elbow between the humerus and ulna are hinge-type movements—flexion and extension. The head of the radius, however, is free to rotate in the anular ligament. This movement allows pronation and supination of the forearm.

PRACTICE

12 Which parts help keep together the articulating surfaces of the shoulder joint?

13 What factors allow an especially wide range of motion in the shoulder?

14 Which structures form the hinge joint of the elbow?

15 Which parts of the elbow permit pronation and supination of the hand?

A procedure called *arthroscopy* is commonly used to diagnose and treat injuries to the shoulder, elbow, and knee. The technique enables a surgeon to visualize the interior of a joint and perform procedures, guided by the image on a video screen. An arthroscope is a thin, tubular instrument about 25 cm long containing optical fibers that transmit an image. The surgeon inserts the device through a small incision in the joint capsule that is first filled with saline to provide a good view. Arthroscopy is much less invasive than conventional surgery. Some runners have undergone uncomplicated arthroscopy and raced several weeks later.

Arthroscopy can also help to rapidly diagnose infection. Guided by an arthroscope, the surgeon samples a small piece of the synovial membrane and extracts and examines DNA for bacterial sequences, such as from the bacterium that causes Lyme disease (*Borrelia burgdorferi*). Sampling the synovial membrane may provide valuable information because a variety of bacteria can infect joints, and choosing the appropriate antibiotic, based on knowing the type of bacterium, is crucial for fast and complete recovery.

Hip Joint

The **hip joint** is a ball-and-socket joint that consists of the head of the femur and the cup-shaped acetabulum of the hip bone (fig. 8.17). A ligament (ligamentum capitis) attaches to a pit (fovea capitis) on the head of the femur and to connective tissue in the acetabulum. This attachment, however, seems to have little importance in holding the articulating bones together, but rather carries blood vessels to the head of the femur.

A horseshoe-shaped ring of fibrocartilage (acetabular labrum) at the rim of the acetabulum deepens the cavity of the acetabulum. It encloses the head of the femur and helps hold it securely in place. In addition, a heavy, cylindrical joint capsule reinforced with still other ligaments surrounds the articulating structures and connects the neck of the femur to the margin of the acetabulum (fig. 8.18).

The major ligaments of the hip joint include the following (fig. 8.19):

1. The **iliofemoral** (il"e-o-fem'o-ral) **ligament** consists of a Y-shaped band of strong fibers that connects the anterior inferior iliac spine of the hip bone to a bony line (intertrochanteric line) extending between the greater and lesser trochanters of the femur. The iliofemoral ligament is the strongest ligament in the body.
2. The **pubofemoral** (pu"bo-fem'o-ral) **ligament** extends between the superior portion of the pubis and the iliofemoral ligament. Its fibers also blend with the fibers of the joint capsule.
3. The **ischiofemoral** (is"ke-o-fem'o-ral) **ligament** consists of a band of strong fibers that originates on the ischium just posterior to the acetabulum and blends with the fibers of the joint capsule.

Muscles surround the joint capsule of the hip. The articulating parts of the hip are held more closely together than those of the shoulder, allowing considerably less freedom of movement. The structure of the hip joint, however, still permits a wide variety of movements, including flexion, extension, adduction, abduction, rotation, and circumduction. The hip is one of the joints most frequently replaced (Clinical Application 8.1).

Knee Joint

The **knee joint** is the largest and most complex of the synovial joints. It consists of the medial and lateral condyles at the distal end of the femur and the medial and lateral condyles at the proximal

(a)   (b)

FIGURE 8.17 Hip joint. (*a*) The ball-like head of the femur. (*b*) The acetabulum provides the socket for the head of the femur in the hip joint.

Surgeons use synthetic materials to replace joints severely damaged by arthritis or injury. Metals such as steel and titanium are used to replace larger joints, whereas flexible silicone polymers are often used to replace smaller joints. Such artificial joints must be durable yet not provoke immune system rejection. They must also allow normal healing to occur and not move surrounding structures out of their normal positions. Ceramic materials are used in about 5% of hip replacements. More than two dozen joint replacement models are in use by more than a million people, mostly in the hip.

A surgeon inserts a joint implant in a procedure called implant resection arthroplasty. The surgeon first removes the surface of the joint bones and excess cartilage. Next, the centers of the tips of abutting bones are hollowed out, and the stems of the implant inserted. The movable part of the implant lies between the bones, aligning them yet allowing them to move. Bone cement fixes the

implant in place. Finally, the surgeon repairs the tendons, muscles, and ligaments. A year of physical therapy may be necessary to fully benefit from replacement joints.

Newer joint replacements use materials that resemble natural body chemicals. Hip implants, for example, may bear a coat of hydroxyapatite, which interacts with natural bone. Instead of filling in spaces with bone cement, some investigators are testing a variety of porous coatings that allow bone tissue to grow into the implant area.

Three-dimensional (3D) printing technology is useful in creating a replacement joint that closely matches an individual's anatomy. A biomedical engineer applies the technology to CT or MRI scans to create and customize a 3D model to guide surgeons in performing the replacement. The precision that the technology offers can decrease surgical time, speed recovery, lengthen the lifetime of the part, and increase range of motion.

end of the tibia. In addition, the femur articulates anteriorly with the patella. Although the knee articulations between the condyles of the femur and tibia function largely as a modified hinge joint (allowing flexion and extension), they allow some rotation when the knee is flexed. The joint between the femur and patella is a plane joint.

The *joint capsule* of the knee is relatively thin, but ligaments and the tendons of several muscles greatly strengthen it. For example, the fused tendons of several muscles in the thigh cover the capsule anteriorly. Fibers from these tendons descend to the patella, partially enclose it, and continue downward to the tibia. The capsule attaches to the margins of the femoral and tibial condyles as well as between these condyles (fig. 8.20).

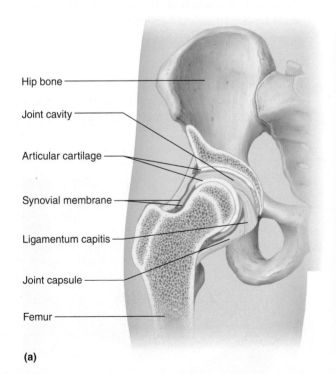

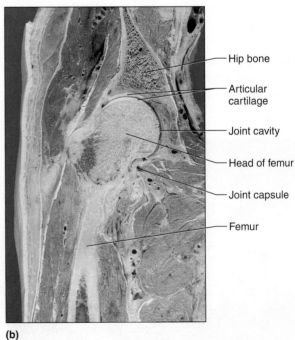

FIGURE 8.18 Right hip joint. (*a*) A ring of cartilage in the acetabulum and a ligament-reinforced joint capsule hold together the hip joint. (*b*) Photograph of the hip joint (frontal section).

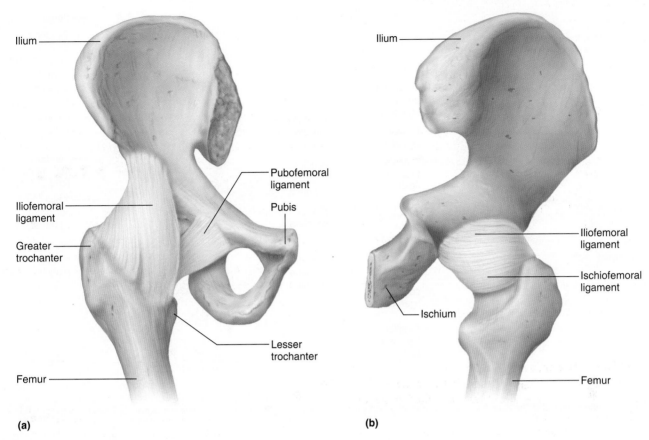

(a)

(b)

FIGURE 8.19 The major ligaments of the right hip joint. (*a*) Anterior view. (*b*) Posterior view.

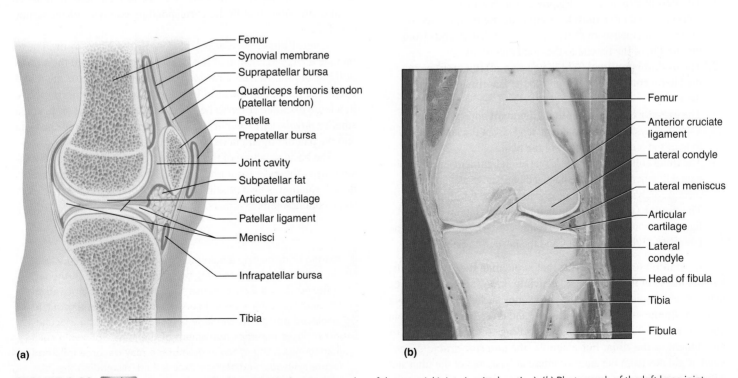

(a)

(b)

FIGURE 8.20 [AP|R] Knee joint. (*a*) The knee joint is the most complex of the synovial joints (sagittal section). (*b*) Photograph of the left knee joint (frontal section).

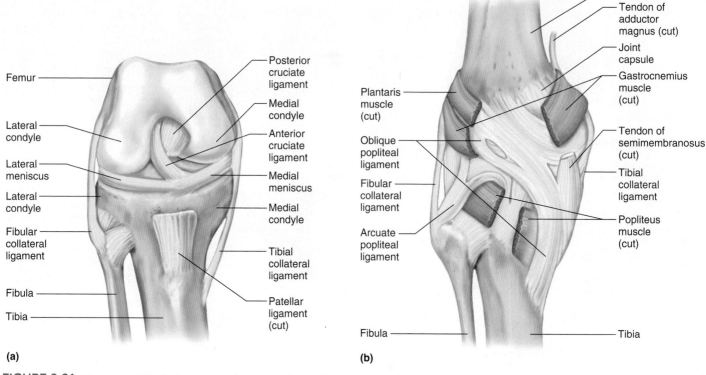

FIGURE 8.21 Ligaments within the knee joint help to strengthen it. (*a*) Anterior view of right bent knee (patella removed). (*b*) Posterior view of left knee.

The ligaments associated with the joint capsule that help keep the articulating surfaces of the knee joint in contact include the following (fig. 8.21):

1. The **patellar** (pah-tel′ar) **ligament** is a continuation of a tendon from a large muscle group in the thigh (quadriceps femoris). It consists of a strong, flat band that extends from the margin of the patella to the tibial tuberosity.
2. The **oblique popliteal** (ŏ′blēk pop-lit′e-al) **ligament** connects the lateral condyle of the femur to the margin of the head of the tibia.
3. The **arcuate** (ar′ku-āt) **popliteal ligament** appears as a Y-shaped system of fibers that extends from the lateral condyle of the femur to the head of the fibula.
4. The **tibial collateral** (tib′e-al kŏ-lat′er-al) **ligament** (medial collateral ligament) is a broad, flat band of tissue that connects the medial condyle of the femur to the medial condyle of the tibia.
5. The **fibular** (fib′u-lar) **collateral ligament** (lateral collateral ligament) consists of a strong, round cord located between the lateral condyle of the femur and the head of the fibula.

In addition to the ligaments that strengthen the joint capsule, two ligaments in the joint, called **cruciate** (kroo′she-āt) **ligaments,** help prevent displacement of the articulating surfaces. These strong bands of fibrous tissue stretch upward and cross between the tibia and the femur. They are named according to their positions of attachment to the tibia. For example, the *anterior cruciate ligament* originates from the anterior intercondylar area of the tibia and extends to the lateral condyle of the femur. The *posterior cruciate ligament* connects the posterior intercondylar area of the tibia to the medial condyle of the femur.

Two fibrocartilaginous *menisci* separate the articulating surfaces of the femur and tibia and help align them. Each meniscus is roughly C-shaped, with a thick rim and a thinner center, and attaches to the head of the tibia. The medial and lateral menisci form depressions that fit the corresponding condyles of the femur (fig. 8.21).

Several bursae are associated with the knee joint. These include a large extension of the knee joint cavity called the *suprapatellar bursa,* located between the anterior surface of the distal end of the femur and the muscle group (quadriceps femoris) above it; a large *prepatellar bursa* between the patella and the skin; and a smaller *infrapatellar bursa* between the proximal end of the tibia and the patellar ligament (see fig. 8.8).

The basic structure of the knee joint permits flexion and extension, as is the case for a hinge joint. However, when the knee is flexed, rotation is also possible. Clinical Application 8.2 discusses some common joint disorders.

Tearing or displacing a meniscus is a common knee injury, usually resulting from forcefully twisting the knee when the leg is flexed (fig. 8.22). Because the meniscus is composed of fibrocartilage, this type of injury heals slowly. Also, a torn and displaced portion of cartilage jammed between the articulating surfaces impedes movement of the joint. Following such a knee injury, the synovial membrane may become inflamed (acute synovitis) and secrete excess fluid, distending the joint capsule so that the knee swells above and on the sides of the patella.

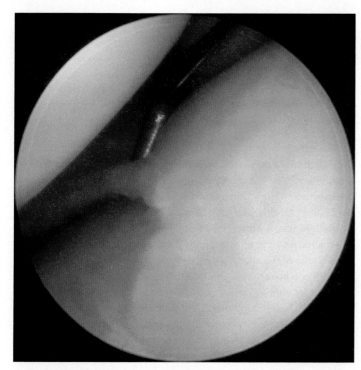

FIGURE 8.22 Arthroscopic view of a torn meniscus in the knee and arthroscopic scissors. Fibrocartilage does not heal well, so in many cases of torn meniscus the only treatment option is to cut out the damaged portion.

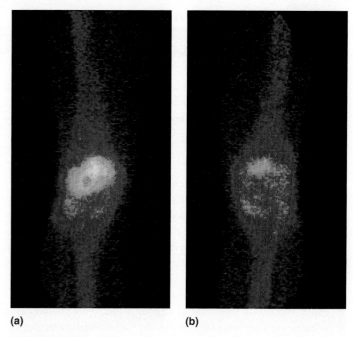

(a)   (b)

FIGURE 8.23 Nuclear scan of (a) a healthy knee and (b) an arthritic knee. The different colors in (b) indicate changes within the tissues associated with degeneration.

 PRACTICE

16 Which structures help keep the articulating surfaces of the hip together?

17 What types of movement does the hip joint permit?

18 What types of joints are in the knee?

19 Which parts help hold together the articulating surfaces of the knee?

8.4 | Life-Span Changes

Joint stiffness is an early sign of aging. By the fourth decade, a person may notice that the first steps each morning become difficult. Changes in collagen structure lie behind the increasing stiffness (fig. 8.23). Range of motion may diminish. However, joints age slowly, and exercise can lessen or forestall stiffness.

The fibrous joints are the first to change, as the four types of fontanels close the bony plates of the skull at two, three, twelve, and eighteen to twenty-four months of age. Other fibrous joints may accumulate bone matrix over time, bringing bones closer together, even fusing them. Fibrous joints strengthen over a lifetime.

Synchondroses that connect epiphyses to diaphyses in long bones disappear as the skeleton grows and develops. Another synchondrosis is the joint that links the first rib to the manubrium (sternum). As water content decreases and deposition of calcium salts increases, this cartilage stiffens. Ligaments lose their elas-

ticity as the collagen fibers become more tightly cross-linked. Breathing may become labored, and movement more restrained.

Aging also affects symphysis joints, which consist of a pad of fibrocartilage sandwiched between thin layers of hyaline cartilage. In the intervertebral discs, less water diminishes the flexibility of the vertebral column and impairs the ability of the soft centers of the discs to absorb shocks. The discs may even collapse on themselves slightly, contributing to the loss of height in the elderly. The stiffening spine gradually restricts the range of motion.

Loss of function in synovial joints begins in the third decade of life, but progresses slowly. Fewer capillaries serving the synovial membrane slows the circulation of synovial fluid, and the membrane may become infiltrated with fibrous material and cartilage. As a result, the joint may lose elasticity, stiffening. More collagen cross-links shorten and stiffen ligaments, affecting the range of motion. This may, in turn, upset balance and retard the ability to respond in a protective way to falling, which may explain why older people are more likely to be injured in falls than younger individuals.

Using joints, through activity and exercise, can keep them functional longer. Disuse hampers the nutrient supply to joints, which hastens stiffening. Paradoxically, this can keep people from exercising, when this is exactly what they should be doing.

 PRACTICE

20 Describe the loss of function in synovial joints as a progressive process. ■

Joints must support weight, provide a variety of body movements, and are used frequently. Trauma, overuse, infection, a misdirected immune system attack, and degeneration can injure joints. Here is a look at some common joint problems.

Sprains

Sprains result from overstretching or tearing the connective tissues, including cartilage, ligaments, and tendons associated with a joint, but they do not dislocate the articular bones. Usually forceful wrenching or twisting sprains a wrist or ankle. For example, inverting an ankle too far can sprain it by stretching the ligaments on its lateral side. Severe injuries may pull these structures loose from their attachments.

A sprained joint is painful and swollen, restricting movement. Immediate treatment of a sprain is rest; more serious cases require medical attention. However, immobilization of a joint, even for a brief period, causes bone resorption and weakens ligaments. Exercise may strengthen a joint.

Bursitis

Overuse of a joint or stress on a bursa may cause *bursitis,* an inflammation of a bursa. The bursa between the heel bone (calcaneus) and the Achilles tendon may become inflamed as a result of a sudden increase in physical activity using the feet. Bursitis is treated with rest. Medical attention may be necessary.

Arthritis

Arthritis causes inflamed, swollen, and painful joints. More than a hundred different types of arthritis affect 50 million people in the United States. Arthritis can also be part of other syndromes (table 8A). The most common types of arthritis are rheumatoid arthritis (RA), osteoarthritis, and Lyme arthritis.

Rheumatoid Arthritis (RA)

Rheumatoid arthritis, an autoimmune disorder (a condition in which the immune system attacks the body's healthy tissues), is painful and debilitating. The synovial membrane of a joint becomes inflamed and thickens, forming a mass called a pannus. Then, the articular cartilage is damaged, and fibrous tissue infiltrates, interfering with joint movements. In time, the joint may ossify, fusing the articulating bones (bony ankylosis). Joints severely damaged by RA may be surgically replaced.

RA may affect many joints or only a few. It is usually a systemic illness, accompanied by fatigue, muscular atrophy, anemia, and osteoporosis, as well as changes in the skin, eyes, lungs, blood vessels, and heart. RA usually affects adults, but there is a juvenile form.

Osteoarthritis

Osteoarthritis, a degenerative disorder, is the most common type of arthritis (fig. 8A). It usually occurs with aging, and its prevalence is expected to greatly increase by 2020 due to the aging of the population. An inherited form of osteoarthritis may appear as early as one's thirties. A person may first become aware of osteoarthritis when a blow to the affected joint produces intense pain. Gradually the area of the affected joint deforms. Arthritic fingers become gnarled, or a knee may bulge.

In osteoarthritis, articular cartilage softens and disintegrates gradually, roughening the articular surfaces. Joints become painful, with restricted movement. For example, arthritic fingers may lock into place while a person is playing the guitar or tying a shoelace. Osteoarthritis most often affects joints used the most, such as those of the fingers, hips, knees, and lower vertebral column.

If a person with osteoarthritis is overweight or obese, the first treatment is usually an exercise and dietary program to lose weight. Nonsteroidal anti-inflammatory drugs (NSAIDs) have been used for many years to control osteoarthritis symptoms. NSAIDs called COX-2 inhibitors relieve inflammation without the gastrointestinal side effects of older drugs, but they are prescribed only to people who do not have risk factors for cardiovascular disease, to which some of these drugs are linked.

Lyme Arthritis

Lyme disease, a bacterial infection passed in a tick bite, causes intermittent arthritis of several joints, usually weeks after the initial symptoms of rash, fatigue, and flulike aches and pains. Lyme arthritis was first observed by a resident of Lyme, Connecticut, who noticed that many of her young neighbors had what appeared to be the very rare juvenile form of rheumatoid arthritis. She contacted Yale University rheumatologist Allen Steere, who traced the illness to bacteria-bearing ticks. Antibiotic treatment that begins as soon as the early symptoms are recognized may prevent Lyme arthritis.

Other types of bacteria that cause arthritis include common *Staphylococcus* and *Streptococcus* species, *Neisseria gonorrhoeae* (which causes the sexually transmitted infection gonorrhea), and *Mycobacterium* (which causes tuberculosis). Arthritis may also be associated with AIDS, because the immune system breakdown raises the risk of infection by bacteria that can cause arthritis.

TABLE 8A Different Types of Arthritis

Some More-Common Forms of Arthritis

Type	Incidence in the United States
Osteoarthritis	20.7 million
Rheumatoid arthritis	2.1 million

Some Less-Common Forms of Arthritis

Type	Incidence in the United States	Age of Onset	Symptoms
Gout	1.6 million (85% male)	>40	Sudden onset of extreme pain and swelling of a large joint
Juvenile rheumatoid arthritis	100,000	<18	Joint stiffness, often in knee
Scleroderma	300,000	30–50	Skin hardens and thickens
Systemic lupus erythematosus	500,000 (>90% female)	teens–50s	Fever, weakness, upper body rash, joint pain
Kawasaki disease	Hundreds of cases in local outbreaks	6 months–11 years	Fever, joint pain, red rash on palms and soles, heart complications
Strep A infection	100,000	any age	Confusion, body aches, shock, low blood pressure, dizziness, arthritis, pneumonia
Lyme disease	15,000	any age	Arthritis, malaise, neurologic and cardiac manifestations

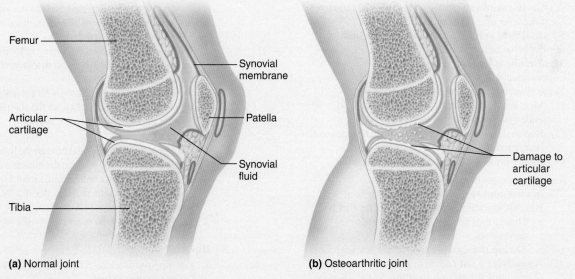

(a) Normal joint   (b) Osteoarthritic joint

FIGURE 8A   In osteoarthritis, an inherited defect in collagen, trauma, or prolonged wear and tear destroys joints. The (a) normal knee as compared to the (b) osteoarthritic knee, in which articular cartilage is breaking down.

Chapter Summary

8.1 Types of Joints (page 268)

Joints are classified according to structure (the type of tissue that binds the bones) and function (the degree of movement possible).

1. Fibrous joints
   a. Bones at **fibrous joints** are tightly fastened to each other by a layer of dense connective tissue with many collagen fibers.
   b. There are three types of fibrous joints.
      (1) A **syndesmosis** has bones bound by long connective tissue fibers.
      (2) A **suture** is where flat bones are united by a thin layer of connective tissue and are interlocked by a set of bony processes.
      (3) A **gomphosis** is formed by the union of a cone-shaped bony process with a bony socket.
2. Cartilaginous joints
   a. A layer of cartilage holds together bones of **cartilaginous joints.**
   b. There are two types of cartilaginous joints.
      (1) A **synchondrosis** occurs where bones are united by hyaline cartilage that may disappear as a result of bone growth.
      (2) A **symphysis** is found where articular surfaces of the bones are covered by hyaline cartilage and the bones are connected by a pad of fibrocartilage.
3. Synovial joints
   a. **Synovial joints** have a more complex structure than other types of joints.
   b. **Articular cartilage** covers articular ends of bones in a synovial joint.
   c. A **joint capsule** strengthened by **ligaments** holds bones together.
   d. A **synovial membrane** that secretes synovial fluid lines the inner layer of a joint capsule.
   e. **Synovial fluid** moistens, provides nutrients, and lubricates the articular surfaces.
   f. **Menisci** divide some synovial joints into compartments.
   g. Some synovial joints have fluid-filled **bursae.**
      (1) Most bursae are located between the skin and underlying bony prominences.
      (2) Bursae cushion and aid movements of tendons over bony parts.
      (3) Bursae are named according to their locations.
   h. There are six types of synovial joints.
      (1) In a **ball-and-socket joint,** the globular head of a bone fits into the cup-shaped cavity of another bone.
         a. These joints permit a wide variety of movements.
         b. The hip and shoulder are ball-and-socket joints.
      (2) A **condylar joint** consists of an ovoid condyle of one bone fitting into an elliptical cavity of another bone.
         a. This joint permits movement in two planes.
         b. The joints between the metacarpals and phalanges are condylar.
      (3) Articular surfaces of **plane joints** are nearly flat.
         a. These joints permit the articular surfaces to slide back and forth.
         b. Most of the joints of the wrist and ankle are plane joints.
      (4) In a **hinge joint,** the convex surface of one bone fits into the concave surface of another bone.
         a. This joint permits movement in one plane only.
         b. The elbow and the joints of the phalanges are the hinge type.
      (5) In a **pivot joint,** a cylindrical surface of one bone rotates within a ring of bone and ligament.
         a. This joint permits rotational movement.
         b. The articulation between the proximal ends of the radius and the ulna is a pivot joint.
      (6) A **saddle joint** forms between bones that have complementary surfaces with both concave and convex regions.
         a. This joint permits a variety of movements.
         b. The articulation between the carpal and metacarpal of the thumb is a saddle joint.

8.2 Types of Joint Movements (page 274)

1. Muscles acting at synovial joints produce movements in different directions and in different planes.
2. Joint movements include **flexion, extension, dorsiflexion, plantar flexion, abduction, adduction, rotation, circumduction, supination, pronation, eversion, inversion, protraction, retraction, elevation,** and **depression.**

8.3 Examples of Synovial Joints (page 277)

1. **Shoulder joint**
   a. The shoulder joint is a ball-and-socket joint that consists of the head of the humerus and the glenoid cavity of the scapula.
   b. A cylindrical joint capsule envelops the joint.
      (1) The capsule is loose and by itself cannot keep the articular surfaces together.
      (2) It is reinforced by surrounding muscles and tendons.
   c. Several ligaments help prevent displacement of the bones.
   d. Several bursae are associated with the shoulder joint.
   e. Its parts are loosely attached, so the shoulder joint permits a wide range of movements.
2. **Elbow joint**
   a. The elbow has a hinge joint between the humerus and the ulna and a plane joint between the humerus and the radius.
   b. Collateral ligaments reinforce the joint capsule.
   c. A synovial membrane partially divides the joint cavity into two portions.
   d. The joint between the humerus and the ulna permits flexion and extension only.
3. **Hip joint**
   a. The hip joint is a ball-and-socket joint between the femur and the hip bone.
   b. A ring of fibrocartilage deepens the cavity of the acetabulum.
   c. The articular surfaces are held together by a heavy joint capsule reinforced by ligaments.
   d. The hip joint permits a wide variety of movements.
4. **Knee joint**
   a. The knee joint includes a modified hinge joint between the femur and the tibia and a plane joint between the femur and the patella.

b. Ligaments and tendons strengthen the thin joint capsule.
c. Several ligaments, some in the joint capsule, bind the articular surfaces of the joint bones.
d. Two menisci separate the articulating surfaces of the femur and the tibia.
e. Several bursae are associated with the knee joint.
f. The knee joint permits flexion and extension; when the knee is flexed, some lower limb rotation is possible.

8.4 Life-Span Changes (page 285)

1. Joint stiffness is often the earliest sign of aging.
   a. Collagen changes cause the feeling of stiffness.
   b. Regular exercise can lessen the effects.

2. Fibrous joints are the first to begin to change and strengthen over a lifetime.
3. Synchondroses of the long bones disappear with growth and development.
4. Changes in symphysis joints of the vertebral column diminish flexibility and decrease height.
5. Over time, synovial joints lose elasticity.

CHAPTER ASSESSMENTS

8.1 Types of Joints

1 Describe how joints are classified. (p. 268)
2 A _____ is a fibrous joint with bones bound by long connective tissue fibers, whereas a _____ is a fibrous joint where flat bones are united by a thin layer of connective tissue. (p. 268)
3 Describe a gomphosis, and name an example. (p. 268)
4 Compare the structures of a synchondrosis and a symphysis. (p. 269)
5 Explain how the joints between vertebrae permit movement. (p. 269)
6 Draw the general structure of a synovial joint, labeling all the main parts. (pp. 269–272)
7 Describe how a joint capsule may be reinforced. (p. 270)
8 Explain the function of a synovial membrane. (p. 271)
9 Explain the function of synovial fluid. (p. 271)
10 Define *meniscus*. (p. 272)
11 Define *bursa*. (p. 272)
12 Describe the six types of synovial joints, and name an example of each type. (p. 272)
13 Describe the movements permitted by each type of synovial joint. (p. 272)

8.2 Types of Joint Movements

14 Joint movements occur when a muscle contracts and the muscle fibers pull the muscle's more movable end of attachment, the _____ , toward its less movable or relatively fixed end, the _____ . (p. 274)
15 Match the movements listed on the left with the descriptions listed on the right. (pp. 274–276)

(1)	rotation	A. turning the palm upward
(2)	supination	B. decreasing the angle between parts
(3)	extension	C. moving a part forward
(4)	eversion	D. moving a part around an axis
(5)	protraction	E. moving a part toward midline
(6)	flexion	F. turning the foot so the plantar surface faces laterally
(7)	pronation	
(8)	abduction	G. increasing the angle between parts
(9)	depression	H. lowering a part
(10)	adduction	I. turning the palm downward
		J. moving a part away from midline

8.3 Examples of Synovial Joints

16 Name the parts that compose the shoulder joint. (p. 277)
17 Name the major ligaments associated with the shoulder joint. (pp. 277 and 279)
18 Explain why the shoulder joint permits a wide range of movements. (p. 279)
19 Name the parts that comprise the elbow joint. (p. 279)
20 Name the major ligaments associated with the elbow joint. (pp. 279–280)
21 Describe the movements permitted by the elbow joint. (pp. 280–281)
22 Name the parts that comprise the hip joint. (p. 281)
23 Describe how the articular surfaces of the hip joint are held together. (p. 281)
24 Explain why there is less freedom of movement in the hip joint than in the shoulder joint. (p. 281)
25 Name the parts that comprise the knee joint. (pp. 281–282)
26 Describe the major ligaments associated with the knee joint. (p. 284)
27 Explain the function of the menisci of the knee. (p. 284)
28 Describe the locations of the bursae associated with the knee joint. (p. 284)

8.4 Life-Span Changes

29 Describe the process of aging as it contributes to the stiffening of fibrous, cartilaginous, and synovial joints. (p. 285)

INTEGRATIVE ASSESSMENTS/CRITICAL THINKING

Outcomes 5.3, 8.1, 8.3, 8.4

1. How would you explain to an athlete why damaged joint ligaments and cartilages are so slow to heal following an injury?

Outcomes 8.1, 8.3, 8.4

2. How would you explain to a person with a dislocated shoulder that the shoulder is likely to become more easily dislocated in the future?

Outcomes 8.1, 8.3

3. Based upon your knowledge of joint structures, which do you think could be more satisfactorily replaced by a prosthetic device—a hip joint or a knee joint? Why?

4. Compared to the shoulder and hip joints, in what ways is the knee joint especially vulnerable to injuries?

 # ONLINE STUDY TOOLS

Connect Interactive Questions Reinforce your knowledge using assigned interactive questions covering joint structure and function.

Connect Integrated Activity Can you help a physical therapist pick activities for a patient based on specific desired joint movements?

LearnSmart Discover which chapter concepts you have mastered and which require more attention. This adaptive learning tool is personalized, proven, and preferred.

Anatomy & Physiology Revealed Go more in depth into the human body as you view the anatomy of specific joints. Also view animations of the movement that occurs at various joints.

Falsely colored scanning electron micrograph (SEM) of a normal human skeletal muscle fiber reveals the characteristic banding pattern of the myofibrils (3,000×).

9

Muscular System

 LEARNING OUTCOMES

After you have studied this chapter, you should be able to:

9.1 Structure of a Skeletal Muscle
1. Describe the structure of a skeletal muscle. (pp. 292–293)
2. Name the major parts of a skeletal muscle fiber and describe the functions of each. (pp. 293–296)

9.2 Skeletal Muscle Contraction
3. Describe the neural control of skeletal muscle contraction. (p. 297)
4. Identify the major events of skeletal muscle fiber contraction. (pp. 298–299)
5. List the energy sources for skeletal muscle fiber contraction. (pp. 299–302)
6. Describe oxygen debt. (pp. 302–303)
7. Describe how a muscle may become fatigued. (p. 303)

9.3 Muscular Responses
8. Distinguish between a twitch and a sustained contraction. (pp. 304–306)
9. Explain how various types of muscular contractions produce body movements and help maintain posture. (p. 306)
10. Distinguish between fast and slow twitch muscle fibers. (pp. 306–307)

9.4 Smooth Muscle
11. Distinguish between the structures and functions of multiunit smooth muscle and visceral smooth muscle. (p. 307)
12. Compare the contraction mechanisms of skeletal and smooth muscle cells. (p. 308)

9.5 Cardiac Muscle
13. Compare the contraction mechanisms of skeletal and cardiac muscle cells. (pp. 308–309)

9.6 Skeletal Muscle Actions
14. Explain how the attachments, locations, and interactions of skeletal muscles make possible certain movements. (pp. 311–312)

9.7 Major Skeletal Muscles
15. Identify and locate the skeletal muscles of each body region and describe the action(s) of each muscle. (pp. 312–340)

9.8 Life-Span Changes
16. Describe aging-related changes in the muscular system. (p. 340)
17. Discuss how exercise can help maintain a healthy muscular system as the body ages. (p. 340)

THE WHOLE PICTURE

Everything we do to express ourselves uses muscles. Consciously controlled skeletal muscles are necessary for talking, smiling or frowning, raising a hand in class to ask a question, and even raising an eyebrow.

Behind the scenes, the involuntary, non-consciously-controlled smooth muscle and cardiac muscle are active too. All muscle requires energy from food, and it is smooth muscle that moves that food through the digestive tract. Cardiac muscle continuously pumps blood so that all cells, including muscle cells, can access the absorbed nutrients.

All of the force that muscles provide, whether it is moving your eyes in their sockets as you read this page or climbing up a flight of stairs, comes from one set of muscle proteins pulling on another set of muscle proteins. Individually these proteins do not generate much force, but collectively they are strong—as strong as you!

Anatomy & Physiology | **REVEALED**®
aprevealed.com

Module 6: Muscular System

calat-, something inserted: inter*cal*ated disc—membranous band that connects cardiac muscle cells.

erg-, work: syn*erg*ist—muscle that works with a prime mover, producing a movement.

fasc-, bundle: *fasc*iculus—bundle of muscle fibers.

-gram, something written: myo*gram*—recording of a muscular contraction.

hyper-, over, more: muscular *hyper*trophy—enlargement of muscle fibers.

inter-, between: *inter*calated disc—membranous band that connects cardiac muscle cells.

iso-, equal: *iso*tonic contraction—contraction during which the tension in a muscle remains unchanged.

laten-, hidden: *lat*ent period—period between a stimulus and the beginning of a muscle contraction.

myo-, muscle: *myo*fibril—contractile fiber of a muscle cell.

reticul-, a net: sarcoplasmic *reticul*um—network of membranous channels within a muscle fiber.

sarco-, flesh: *sarco*plasm—substance (cytoplasm) within a muscle fiber.

syn-, together: *syn*ergist—muscle that works with a prime mover, producing a movement.

tetan-, stiff: *tetan*ic contraction—sustained muscular contraction.

-tonic, stretched: iso*tonic* contraction—contraction during which the tension of a muscle remains unchanged.

-troph, well fed: muscular hyper*troph*y—enlargement of muscle fibers.

voluntar-, of one's free will: *voluntar*y muscle—muscle that can be controlled by conscious effort.

9.1 | Structure of a Skeletal Muscle

This chapter focuses mostly on skeletal muscles, which attach to bones and to the skin of the face and are under conscious control. A skeletal muscle is an organ of the muscular system. It is composed primarily of skeletal muscle tissue, nervous tissue, blood, and other connective tissues.

Connective Tissue Coverings

An individual skeletal muscle is separated from adjacent muscles and held in position by layers of dense connective tissue called **fascia** (fash'e-ah). This connective tissue surrounds each muscle and may project beyond the ends of its muscle fibers, forming a cordlike **tendon.** Fibers in a tendon may intertwine with those in the periosteum of a bone, attaching the muscle to the bone. Or, the connective tissues associated with a muscle form broad, fibrous sheets called **aponeuroses** (ap"o-nu-ro'sēz), which may attach to bone or the coverings of adjacent muscles (figs. 9.1 and 9.2).

A tendon or the connective tissue sheath of a tendon (tenosynovium) may become painfully inflamed and swollen following an injury or the repeated stress of athletic activity. *Tendinitis* affects the tendon and *tenosynovitis* affects the connective tissue sheath of the tendon. Most commonly affected are the tendons associated with the joint capsules of the shoulder, elbow, hip, and knee and those involved with moving the wrist, hand, thigh, and foot.

The layer of connective tissue that closely surrounds a skeletal muscle is called the *epimysium,* which in some areas of the body may merge with the surrounding deep fascia. Another layer of connective tissue, called the *perimysium,* extends inward from the epimysium and separates the muscle tissue into small sections. These sections contain bundles of skeletal muscle fibers called *fascicles* (fasciculi). Each muscle fiber within a fascicle (fasciculus)

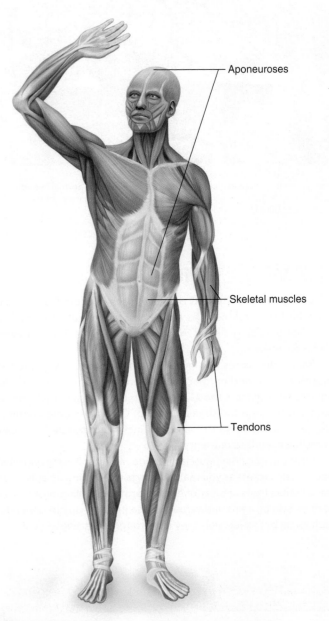

FIGURE 9.1 Tendons attach muscles to bones, whereas aponeuroses attach muscles to other muscles or to bone.

Aponeuroses

Skeletal muscles

Tendons

TABLE 9.1	Muscle Structures		
Structures in order of increasing complexity	**Description**	**Location**	
Thick and thin filaments	Structures containing the contractile proteins actin and myosin	Intracellular	
Myofibril	Overlapping parallel groups of thick and thin filaments in a repeating pattern; the underlying basis for the striation pattern	Intracellular	
Muscle fiber	A single muscle cell, multinucleated and may be many centimeters long	Within a muscle, surrounded by a layer of connective tissue called endomysium	
Fascicle	A bundle of muscle fibers within a muscle	Within a muscle, surrounded by a layer of connective tissue called perimysium	
Muscle	A bundle of fascicles	Surrounded by a layer of connective tissue called epimysium and in some cases an additional layer called muscle fascia	

lies within a layer of connective tissue in the form of a thin covering called *endomysium* (fig. 9.2, fig. 9.3, and table 9.1). Layers of connective tissue, therefore, enclose and separate all parts of a skeletal muscle. This organization allows the parts to move somewhat independently. Many blood vessels and nerves pass through these layers.

A *compartment* is the space that contains a particular group of muscles, blood vessels, and nerves, all tightly enclosed by fascia. The limbs have many such compartments. If an injury causes fluid, such as blood from an internal hemorrhage, to accumulate in a compartment, the pressure inside will rise. The increased pressure, in turn, may interfere with blood flow into the region, reducing the supply of oxygen and nutrients to the affected tissues. This condition, called compartment syndrome, often produces severe, unrelenting pain. Persistently elevated compartmental pressure may irreversibly damage the enclosed muscles and nerves. Treatment for compartment syndrome may require an immediate surgical incision through the fascia (fasciotomy) to relieve the pressure and restore circulation.

The fascia associated with each individual organ of the muscular system is part of a complex network of fasciae that extends throughout the body. The portion of the network that surrounds the muscles is called *deep fascia*. It is continuous with the *subcutaneous fascia* that lies just beneath the skin, forming the subcutaneous layer described in chapter 6 (p. 178). The network is also continuous with the *subserous fascia* that forms the connective tissue layer of the serous membranes covering organs in various body cavities and lining those cavities (see chapter 5, p. 168).

Skeletal Muscle Fibers

Recall from chapter 5 (p. 170) that a skeletal muscle fiber is a single muscle cell (see fig. 5.29, p. 170). Each fiber forms from many undifferentiated cells that fuse during development. The resulting multinucleated muscle fiber is a thin, elongated cylinder with

rounded ends that attach to the connective tissues associated with a muscle. Just beneath the muscle cell membrane *(sarcolemma),* the cytoplasm *(sarcoplasm)* of the fiber contains many small, oval nuclei and mitochondria. The sarcoplasm also has abundant long, parallel structures called **myofibrils** (mi″o-fi′-brilz) (fig. 9.4a).

The myofibrils play a fundamental role in the muscle contraction mechanism. They consist of two types of protein filaments: thick filaments composed of the protein **myosin** (mi′o-sin), and thin filaments composed primarily of the protein **actin** (ak′tin). (Two other thin filament proteins, troponin and tropomyosin, will

CAREER CORNER
Massage Therapist

The middle-aged woman feels something give way in her left knee as she comes down from a jump in her dance class. She limps away between her much younger classmates, in great pain.

The frequent jumping followed by lateral movements caused patella tendinitis, or "jumper's knee." The woman injured in dance class went home and used "RICE" treatment—rest, ice, compression, and elevation—and three days later, at her weekly appointment with a massage therapist for stress relief, mentioned the injury.

Over the next few weeks, the massage therapist helped by applying light pressure to the injured area, which stimulated circulation, and applied friction in a transverse pattern to break up scar tissue and relax the muscles. She also massaged the quadriceps to improve flexibility.

A massage therapist manipulates soft tissues, using combinations of pressing, stroking, kneading, compressing, and vibrating, to relieve pain and reduce stress. Many cultures have practiced massage therapy, dating back 5,000 years. Modern massage therapists complete programs that include 300 to 1,000 hours of class time, hands-on practice, and continuing education. A massage therapist may work in a pediatrics or orthopedics practice, a spa, a hospital, or a fitness center.

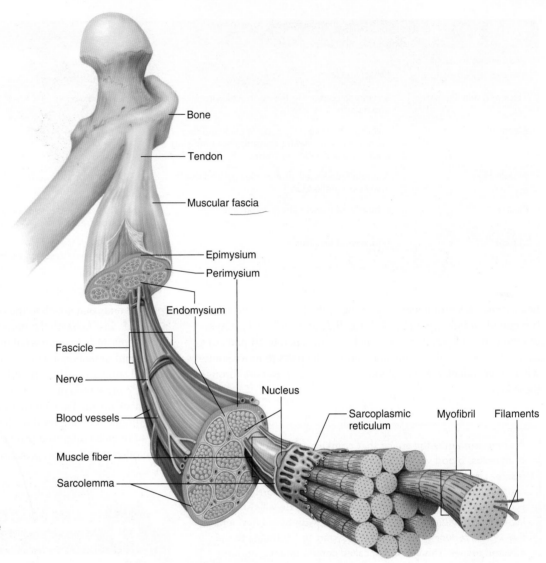

Bone

Tendon

Muscular fascia

Epimysium

Perimysium

Endomysium

Fascicle

Nerve

Nucleus

Blood vessels

Sarcoplasmic reticulum

Myofibril   Filaments

Muscle fiber

Sarcolemma

FIGURE 9.2 **AP|R** A skeletal muscle is composed of a variety of tissues, including layers of connective tissue. Fascia covers the surface of the muscle, epimysium lies beneath the fascia, and perimysium extends into the structure of the muscle where it separates fascicles. Endomysium separates individual muscle fibers.

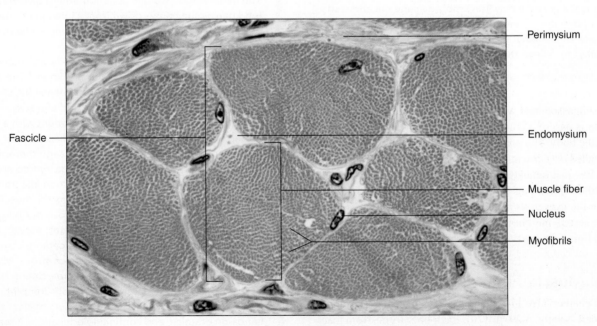

Perimysium

Fascicle

Endomysium

Muscle fiber

Nucleus

Myofibrils

FIGURE 9.3 Light micrograph of a fascicle (fasciculus) surrounded by its connective tissue sheath, the perimysium. Muscle fibers within the fascicle are surrounded by endomysium (320×).

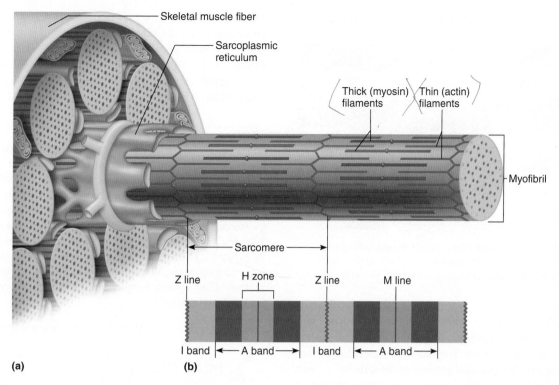

(a)　　　　　**(b)**

FIGURE 9.4 Skeletal muscle fiber. (*a*) A skeletal muscle fiber contains numerous myofibrils, each consisting of (*b*) repeating units called sarcomeres. The characteristic striations of a sarcomere reflect the organization of actin and myosin filaments.

be discussed later.) The organization of these filaments produces the alternating light and dark striations characteristic of skeletal muscle (and cardiac muscle) fibers. The striations form a repeating pattern of units called **sarcomeres** (sar′ko-mērz) along each muscle fiber. The myofibrils may be thought of as sarcomeres joined end to end (fig. 9.4*a*). Muscle fibers, and in a way muscles themselves, are basically collections of sarcomeres, discussed later in this chapter as the functional units of muscle contraction.

The striation pattern of skeletal muscle has two main parts. The first, the *I bands* (the light bands), are composed of thin actin filaments held by direct attachments to structures called *Z lines,* which appear in the center of the I bands. The second part of the striation pattern consists of the *A bands* (the dark bands), which are composed of thick myosin filaments overlapping thin actin filaments (fig. 9.4*b*).

The A band consists not only of a region where thick and thin filaments overlap, but also a slightly lighter central region (*H zone*) consisting only of thick filaments. The A band includes a thickening known as the *M line,* which consists of proteins that help hold the thick filaments in place (fig. 9.4*b*). The myosin filaments are also held in place by the Z lines and are attached to them by a large protein called **titin** (connectin) (fig. 9.5). A sarcomere extends from one Z line to the next.

Thick filaments are composed of many molecules of myosin. Each myosin molecule consists of two twisted protein strands with globular parts called heads that project outward along their lengths. Thin filaments consist of double strands of actin twisted into a helix. Actin molecules are globular, and each has a binding site to which the heads of a myosin molecule can attach (fig. 9.6).

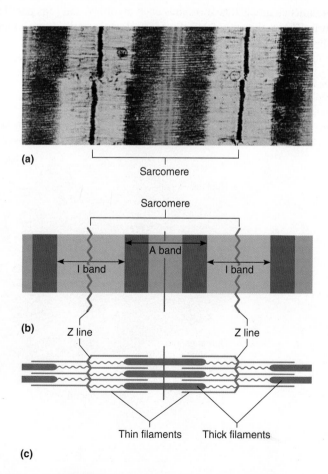

(a)

(b)

(c)

FIGURE 9.5 **AP|R** A sarcomere. (*a*) Micrograph (16,000×). (*b–c*) The relationship of thin and thick filaments in a sarcomere.

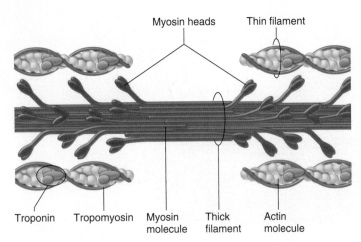

FIGURE 9.6 Thick filaments are composed of the protein myosin, and thin filaments are primarily composed of the protein actin. Myosin molecules have heads that extend toward nearby actin filaments.

Two other types of protein, **troponin** and **tropomyosin,** associate with actin filaments. Troponin molecules have three protein subunits and are attached to actin. Tropomyosin molecules are rod-shaped and occupy the longitudinal grooves of the actin helix. Each tropomyosin is held in place by a troponin molecule, forming a troponin-tropomyosin complex (fig. 9.6).

Within the sarcoplasm of a muscle fiber is a network of membranous channels that surrounds each myofibril and runs parallel to it. These channels form the **sarcoplasmic reticulum,** which corresponds to the endoplasmic reticulum of other cells (see figs. 9.2 and 9.4). A set of membranous channels, the **transverse tubules** (T tubules), extends into the sarcoplasm as invaginations continuous with the sarcolemma and contains extracellular fluid. Each transverse tubule lies between two enlarged portions of the sarcoplasmic reticulum called **cisternae.** These three structures form a **triad** near the region where the actin and myosin filaments overlap (fig. 9.7).

Muscle fibers and the connective tissues associated with them are flexible, but they can tear if overstretched. This type of injury, common in athletes, is called a *muscle strain*. The severity of the injury depends on the degree of damage the tissues sustain. In a mild strain, only a few muscle fibers are injured, the fascia remains intact, and little function is lost. In a severe strain, many muscle fibers as well as fascia tear, and muscle function may be lost completely. A severe strain is very painful and is accompanied by discoloration and swelling of tissues due to torn blood vessels. Surgery may be required to reconnect the separated tissues.

 PRACTICE

1 Describe how connective tissue is associated with a skeletal muscle.
2 Describe the general structure of a skeletal muscle fiber.
3 Explain why skeletal muscle fibers appear striated.
4 Explain the physical relationship between the sarcoplasmic reticulum and the transverse tubules.

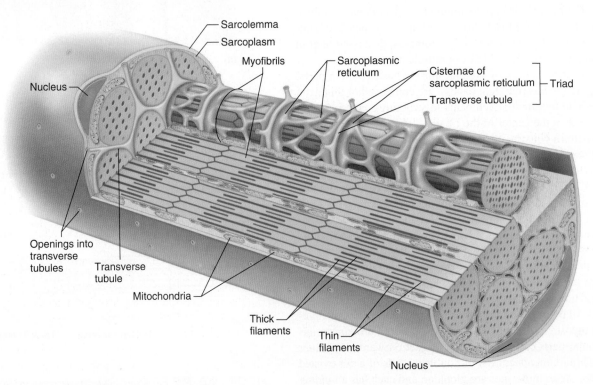

FIGURE 9.7 Within the sarcoplasm of a skeletal muscle fiber is a network of sarcoplasmic reticulum and a system of transverse tubules.

9.2 | Skeletal Muscle Contraction

A muscle fiber contraction is a complex interaction of several cellular and chemical constituents. The result is a movement within the myofibrils in which the filaments of actin and myosin slide past one another, shortening the sarcomeres. When this happens, the muscle fiber shortens and pulls on its attachments.

> Actin, myosin, troponin, and tropomyosin are abundant in muscle cells. Scarcer proteins are also vital to muscle function. A rod-shaped muscle protein called dystrophin, for example, accounts for only 0.002% of total muscle protein in skeletal muscle, but its absence causes the devastating inherited disorder Duchenne muscular dystrophy, which only affects males. Dystrophin binds to the inside face of muscle cell membranes, supporting them against the powerful force of contraction. Without dystrophin, muscle cells lose their normal structure and die. Abnormalities in several other types of proteins that bind to dystrophin cause other inherited muscle diseases.

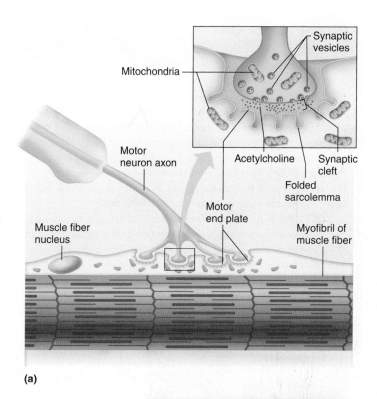

(a)

Neuromuscular Junction

Recall from chapter 5 (p. 170) that neurons establish communication networks throughout the body based on their specialized structures and functions. Each neuron has a process called an axon, which extends from the cell body and can conduct electrical impulses called *action potentials* (described in chapter 10, pp. 375–377).

Each cell that a neuron controls is connected functionally (but not physically) to the end of an axon, in much the same way that you can talk into a cell phone although your mouth is not in direct physical contact with it. The site of this functional connection is called a **synapse.** Neurons communicate with the cells that they control by releasing chemicals, called **neurotransmitters** (nu″ro-trans-mit-erz), at a synapse. The neurotransmitter molecules then diffuse a very short distance to the cell being controlled, where they have a specific effect.

Neurons that control effectors, including skeletal muscle fibers, are called **motor neurons**. Normally a skeletal muscle fiber contracts only upon stimulation by its motor neuron.

The synapse where a motor neuron axon and a skeletal muscle fiber meet is called a **neuromuscular junction** (myoneural junction). Here, the muscle fiber membrane is specialized to form a **motor end plate,** where nuclei and mitochondria are abundant and the sarcolemma is extensively folded (fig. 9.8).

A small gap called the **synaptic cleft** separates the membrane of the neuron and the membrane of the muscle fiber. The cytoplasm at the distal end of the axon is rich in mitochondria and contains many tiny vesicles (synaptic vesicles) that store neurotransmitters.

Stimulus for Contraction

Acetylcholine (ACh) is the neurotransmitter that motor neurons use to control skeletal muscle contraction. ACh is synthesized in the cytoplasm of the motor neuron and is stored in synaptic vesicles near the distal end of its axon. When an action potential

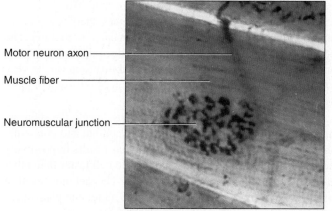

(b)

FIGURE 9.8 AP|R Neuromuscular junction. (*a*) A neuromuscular junction includes the end of a motor neuron and the motor end plate of a muscle fiber. (*b*) Micrograph of a neuromuscular junction (500×).

How does neurotransmitter released into the synaptic cleft reach the muscle fiber membrane?
Answers can be found in Appendix G.

reaches the end of the axon, some of these vesicles release acetylcholine into the synaptic cleft (fig. 9.8).

Acetylcholine diffuses rapidly across the synaptic cleft and binds to specific protein molecules (receptors) in the muscle fiber membrane, increasing the membrane permeability to sodium and potassium ions. Recall from chapter 3 (p. 103) that active transport of sodium and potassium ions across the cell membrane creates a condition in which the sodium ion concentration is higher outside the cell and the potassium ion concentration is higher

In an autoimmune disorder, the immune system attacks part of the body. In myasthenia gravis (MG), the part attacked is the muscular system. The name means "grave muscular weakness." The body produces antibodies that target receptors for the neurotransmitter acetylcholine on skeletal muscle cells at neuromuscular junctions. People with MG have one-third the normal number of acetylcholine receptors on their skeletal muscle cells. On a whole-body level, this causes weak and easily fatigued muscles.

MG affects hundreds of thousands of people worldwide, mostly women beginning in their twenties or thirties, and men in their sixties and seventies. The specific symptoms depend upon the site of attack. For 85% of patients, the disease causes generalized muscle weakness. Many people develop a characteristic flat smile and nasal voice and have difficulty chewing and swallowing due to affected facial and neck muscles. Limb weakness is common. About 15% of patients experience the illness only in the muscles surrounding their eyes. The disease reaches crisis level when respiratory muscles are affected, requiring a ventilator to support breathing. MG does not affect sensation or reflexes.

Most people with MG live a normal life span, with symptoms that are controlled with a combination of treatments that include the following:

- Drugs that inhibit acetylcholinesterase, the enzyme that normally breaks down acetylcholine, thus increasing levels of the neurotransmitter.
- Immunosuppressant drugs, such as corticosteroids, cyclosporine, and rituximab, that decrease production of antibodies.
- Intravenous antibodies that bind and inactivate the antibodies causing the damage.
- Plasma exchange that rapidly removes the damaging antibodies from the bloodstream, helping people in crisis.

inside the cell. Also remember that an ion will diffuse across a cell membrane only if two conditions are met: there is a concentration gradient and the membrane is permeable to that ion. The sodium/potassium pump maintains the concentration gradients for these two ions, but membrane permeability is normally very low, so in the non-stimulated state the extent of diffusion of these ions across the membrane is very low.

When a motor neuron stimulates a muscle cell, the situation changes. Membrane permeability to both sodium and potassium ions increases temporarily in a pattern that results in positively charged ions (sodium) entering the muscle cell faster than other positively charged ions (potassium) leave. This net movement of positive charges into the muscle cell at the motor end plate opens nearby sodium channels on the sarcolemma.

The result of this second set of channels opening is an impulse, an action potential. In much the same way that an impulse

When the bacterium *Clostridium botulinum* grows in an anaerobic (oxygen-poor) environment, such as in a can of unrefrigerated food, it produces a toxin. If a person ingests the toxin, the release of acetylcholine from axon terminals at neuromuscular junctions is prevented. Symptoms of such food poisoning include nausea, vomiting, and diarrhea; headache, dizziness, and blurred or double vision; and finally, weakness, hoarseness, and difficulty swallowing and, eventually, inability to breathe. Physicians can administer an antitoxin substance that binds to and inactivates botulinum toxin in the bloodstream, stemming further symptoms, although not correcting damage already done. Small amounts of botulinum toxin are used to treat migraine headaches and to temporarily paralyze selected facial muscles, smoothing wrinkles (Botox).

is conducted along the axon of a neuron, an impulse now spreads throughout the muscle cell. This electrical impulse is what triggers the release of calcium ions from the sarcoplasmic reticulum, leading to muscle contraction. Clinical Application 9.1 discusses myasthenia gravis, in which the immune system attacks certain neuromuscular junctions.

Excitation-Contraction Coupling AP|R

The connection between stimulation of a muscle fiber and contraction is called *excitation-contraction coupling*, and it involves an increase in calcium ions in the cytosol. At rest, the sarcoplasmic reticulum has a high concentration of calcium ions compared to the cytosol. This is due to active transport of calcium ions (calcium pump) in the membrane of the sarcoplasmic reticulum. In response to stimulation, the membranes of the cisternae become more permeable to these ions, and the calcium ions diffuse out of the cisternae into the cytosol of the muscle fiber (see fig. 9.7).

 RECONNECT
To Chapter 3, Active Transport, page 103.

When a muscle fiber is at rest, the troponin-tropomyosin complexes block the binding sites on the actin molecules and thus prevent the formation of linkages with myosin heads (fig. 9.9 ①). As the concentration of calcium ions in the cytosol rises, however, the calcium ions bind to the troponin, changing its shape (conformation) and altering the position of the tropomyosin. The movement of the tropomyosin molecules exposes the binding sites on the actin filaments, allowing linkages to form between myosin heads and actin, forming cross-bridges (fig. 9.9 ② and ③).

 RECONNECT
To Chapter 2, Proteins, pages 72–75.

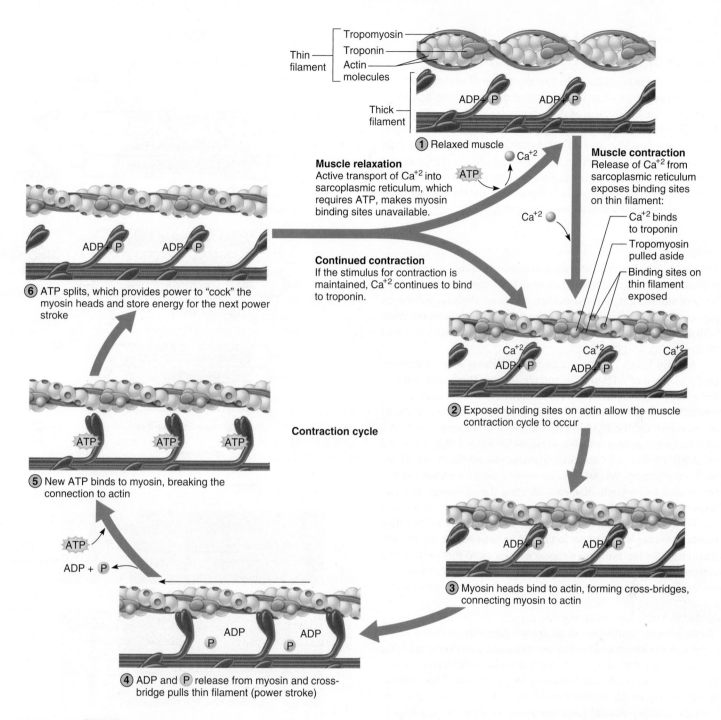

Thin filament
— Tropomyosin
— Troponin
— Actin molecules

Thick filament

ADP+ P   ADP+ P

① Relaxed muscle

Muscle relaxation
Active transport of Ca^{+2} into sarcoplasmic reticulum, which requires ATP, makes myosin binding sites unavailable.

Ca^{+2}

ATP

Ca^{+2}

Muscle contraction
Release of Ca^{+2} from sarcoplasmic reticulum exposes binding sites on thin filament:

— Ca^{+2} binds to troponin
— Tropomyosin pulled aside
— Binding sites on thin filament exposed

Continued contraction
If the stimulus for contraction is maintained, Ca^{+2} continues to bind to troponin.

Ca^{+2} Ca^{+2} Ca^{+2}
ADP+ P   ADP+ P

② Exposed binding sites on actin allow the muscle contraction cycle to occur

Contraction cycle

ADP+ P   ADP+ P

ADP+ P   ADP+ P

⑥ ATP splits, which provides power to "cock" the myosin heads and store energy for the next power stroke

ATP   ATP   ATP

⑤ New ATP binds to myosin, breaking the connection to actin

ATP

ADP + P

ADP   ADP
P   P

④ ADP and P release from myosin and cross-bridge pulls thin filament (power stroke)

③ Myosin heads bind to actin, forming cross-bridges, connecting myosin to actin

FIGURE 9.9 **AP|R** The sliding filament model. *(1)* Relaxed muscle. *(2)* and *(3)* When calcium ion concentration rises, binding sites on actin filaments open and myosin heads bind to the actin, forming cross-bridges. *(4)* Upon binding to actin, myosin heads spring from the cocked position and pull on actin filaments. *(5)* ATP binds (but is not yet broken down), causing the myosin heads to release from the actin filament. *(6)* ATP breakdown provides energy to "cock" the unattached myosin heads. As long as ATP and calcium ions are present, the cycle continues. When the calcium ion concentration in the cytosol is low, the muscle is relaxed.

The Sliding Filament Model of Muscle Contraction

The sarcomere is considered the functional unit of skeletal muscles because contraction of an entire skeletal muscle can be described in terms of the shortening of the sarcomeres of its muscle fibers. According to the **sliding filament model,** when sarcomeres shorten, the thick and thin filaments do not change length. Rather, they slide past one another, with the thin filaments moving toward the center of the sarcomere from both ends. As this occurs, the H zones and the I bands narrow; the regions of overlap widen; and the Z lines move closer together, shortening the sarcomere (fig. 9.10).

Cross-Bridge Cycling

The force that shortens the sarcomeres comes from cross-bridges pulling on the thin filaments. A myosin head can attach to an actin binding site forming a cross-bridge, and bend slightly, pulling on the actin filament. Then the head can release, straighten, combine with another binding site further down the actin filament, and pull again (see fig. 9.9 ②–⑥).

Myosin heads contain the enzyme **ATPase,** which catalyzes the breakdown of ATP to ADP and phosphate. This reaction transfers energy (see chapter 4, p. 126) that provides the force for muscle contraction. Breakdown of ATP puts the myosin head in a "cocked" position (see fig. 9.9 ⑥). When a muscle is stimulated to contract, a cocked myosin head attaches to actin (see fig. 9.9 ③), forming a cross-bridge that pulls the actin filament toward the center of the sarcomere (see fig. 9.9 ④). This movement causes a greater overlap of the actin and myosin filaments, shortens the sarcomere and thus shortens the muscle (fig. 9.10). When another ATP binds, the myosin head first detaches from the actin binding site (see fig. 9.9 ⑤), then breaks down the ATP to return to the cocked position (see fig. 9.9 ⑥). This cross-bridge cycle may repeat as long as ATP is present and action potentials release ACh at that neuromuscular junction.

Relaxation

When nervous stimulation of the muscle fiber ceases (when action potentials no longer reach the axon terminal), two events relax the muscle fiber. First, an enzyme called **acetylcholinesterase** rapidly decomposes acetylcholine remaining in the synapse. This enzyme, which is also on the membranes of the motor end plate, prevents a single action potential from continuously stimulating a muscle fiber. Second, when ACh breaks down, the stimulus to the sarcolemma and the membranes of the muscle fiber ceases. The calcium pump (which requires ATP) quickly moves calcium ions back into the sarcoplasmic reticulum, decreasing the calcium ion concentration of the cytosol. The cross-bridge linkages break (see fig. 9.9 ⑥—this also requires ATP, although it is not broken down in this step), and tropomyosin rolls back into its groove, preventing cross-bridge attachment (see fig. 9.9 ①). Consequently, the muscle fiber relaxes. Table 9.2 summarizes the major events leading to muscle contraction and relaxation.

It is important to remember that ATP is necessary for both muscle contraction and for muscle relaxation. The trigger for contraction is the increase in cytosolic calcium in response to stimulation by ACh from a motor neuron.

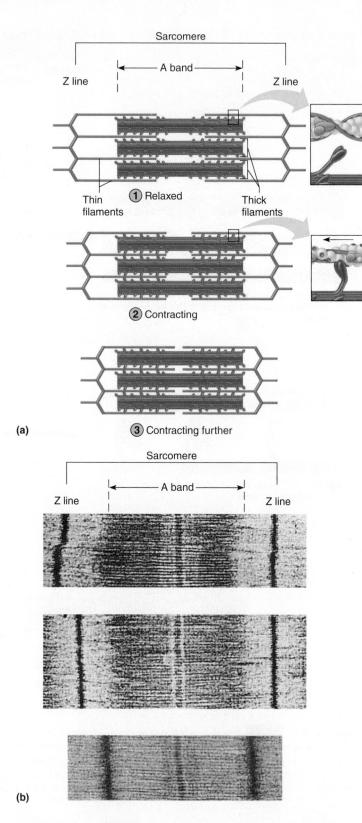

FIGURE 9.10 AP|R When a skeletal muscle contracts (a), individual sarcomeres shorten as thick and thin filaments slide past one another. (b) Transmission electron micrograph showing a sarcomere shortening during muscle contraction (23,000×).

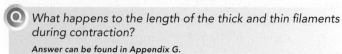

What happens to the length of the thick and thin filaments during contraction?

Answer can be found in Appendix G.

TABLE 9.2 Major Events of Muscle Contraction and Relaxation

Muscle Fiber Contraction	Muscle Fiber Relaxation
1. An action potential is conducted down a motor neuron axon.	1. Acetylcholinesterase decomposes acetylcholine, and the muscle fiber membrane is no longer stimulated.
2. The motor neuron terminal releases the neurotransmitter acetylcholine (ACh).	2. Calcium ions are actively transported into the sarcoplasmic reticulum.
3. ACh binds to ACh receptors on the muscle fiber.	3. ATP breaks cross-bridge linkages between actin and myosin filaments without breakdown of the ATP itself.
4. The sarcolemma is stimulated, an action potential is generated, and the impulse is conducted over the surface of the muscle fiber and deep into the fiber through the transverse tubules.	4. Breakdown of ATP "cocks" the myosin heads.
5. The impulse reaches the sarcoplasmic reticulum, and calcium channels open.	5. Troponin and tropomyosin molecules block the interaction between myosin and actin filaments.
6. Calcium ions diffuse from the sarcoplasmic reticulum into the cytosol and bind to troponin molecules.	6. Muscle fiber remains relaxed, yet ready until stimulated again.
7. Tropomyosin molecules move and expose specific sites on actin where myosin heads can bind.	
8. Actin and myosin link, forming cross-bridges.	
9. Thin (actin) filaments are pulled toward the center of the sarcomere by pulling of the cross-bridges, increasing the overlap of the thin and thick filaments.	
10. The muscle fiber shortens as contraction occurs.	

A few hours after death, the skeletal muscles partially contract, fixing the joints. This condition, called *rigor mortis*, may continue for seventy-two hours or more. It results from an increase in membrane permeability to calcium ions, which promotes cross-bridge formation, and a decrease in availability of ATP in the muscle fibers, which prevents myosin release from actin. Thus, the actin and myosin filaments of the muscle fibers remain linked until the proteins begin to decompose.

 PRACTICE

5   Describe a neuromuscular junction.

6   Explain how a motor neuron action potential can trigger a skeletal muscle fiber contraction.

7   List four proteins associated with myofibrils, and explain their structural and functional relationships.

8   Explain how the filaments of a myofibril interact during muscle contraction.

Energy Sources for Contraction

The energy that powers the interaction between actin and myosin filaments as muscle fibers contract comes from ATP molecules. However, a muscle fiber has only enough ATP to contract briefly, and must regenerate ATP.

The initial source of energy available to regenerate ATP from ADP and phosphate is **creatine phosphate.** Like ATP, creatine phosphate includes a high-energy phosphate bond. Whenever sufficient ATP is present, an enzyme in the mitochondria (creatine phosphokinase) promotes the synthesis of creatine phosphate, which stores excess energy in its phosphate bond (fig. 9.11).

Creatine phosphate is four to six times more abundant in muscle fibers than ATP, but it cannot directly supply energy to a cell's energy-utilizing reactions. Instead, as ATP is decomposed to ADP, the phosphate from creatine phosphate molecules is transferred to these ADP molecules, quickly converting them back into ATP. The amount of ATP and creatine phosphate in a skeletal muscle, however, is usually not sufficient to support maximal muscle activity for more than about ten seconds during an intense contraction. As a result, the muscle fibers in an active muscle soon use cellular respiration of glucose to synthesize ATP. Typically, a muscle stores glucose in the form of glycogen.

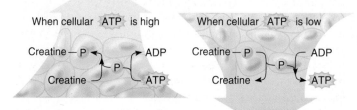

FIGURE 9.11 Creatine phosphate may be used to replenish ATP stores when ATP levels in a muscle cell are low.

Oxygen Supply and Cellular Respiration

Recall from chapter 4 (pp. 127–128) that glycolysis, the early phase of cellular respiration, occurs in the cytosol and is *anaerobic,* not requiring oxygen. This phase only partially breaks down energy-supplying glucose and yields only two ATP molecules for each molecule of glucose. The complete breakdown of glucose occurs in the mitochondria and is *aerobic,* requiring oxygen. This process, which includes the complex series of reactions of the *citric acid cycle* and *electron transport chain,* produces many ATP molecules (see chapter 4, pp.129–131).

Blood carries the oxygen necessary to support the aerobic reactions of cellular respiration from the lungs to body cells. Oxygen is carried in red blood cells, where it is loosely bound to molecules of hemoglobin, the pigment responsible for the red color of blood. In regions of the body where the oxygen concentration is low, oxygen is released from hemoglobin and becomes available for the aerobic reactions of cellular respiration.

Another pigment, **myoglobin,** is synthesized in muscle cells and imparts the reddish brown color of skeletal muscle tissue. Like hemoglobin, myoglobin can loosely bind oxygen and, in fact, has a greater attraction for oxygen than does hemoglobin. Myoglobin's ability to temporarily store oxygen increases the amount of oxygen available in the muscle cells to support aerobic respiration (**fig. 9.12**). Oxygen storage in myoglobin is important because blood flow may decrease when contracting muscle fibers compress blood vessels.

Oxygen Debt

When a person is resting or moderately active, the respiratory and cardiovascular systems can usually supply sufficient oxygen to the skeletal muscles to support the aerobic reactions of cellular respiration. However, when skeletal muscles are used more strenuously, these systems may not be able to supply enough oxygen to sustain the aerobic reactions of cellular respiration. Under these conditions muscles may be limited to the anaerobic reactions to produce ATP. This shift in metabolism is referred to as the anaerobic threshold, or the **lactic acid threshold.**

In the anaerobic reactions, glycolysis breaks down glucose molecules to yield *pyruvic acid,* which would normally enter the citric acid cycle. When the oxygen supply is low, however, the pyruvic acid reacts to produce *lactic acid,* which enters the bloodstream and eventually reaches the liver (see chapter 4, p. 129). In liver cells, reactions requiring ATP synthesize glucose from lactic acid (**fig. 9.13**). In the body, lactic acid dissociates rapidly to form lactate ion (lactate) and hydrogen ion, both of which leave muscle cells by facilitated diffusion. For simplicity we refer only to lactic acid in this discussion.

During strenuous exercise, available oxygen is primarily used to synthesize ATP for muscle contraction rather than to make ATP for reacting lactic acid to yield glucose. Consequently, as lactic acid accumulates, a person develops an **oxygen debt** or *excess postexercise oxygen consumption* (EPOC) that must be repaid at a later time. The degree of oxygen debt includes the amount of oxygen that

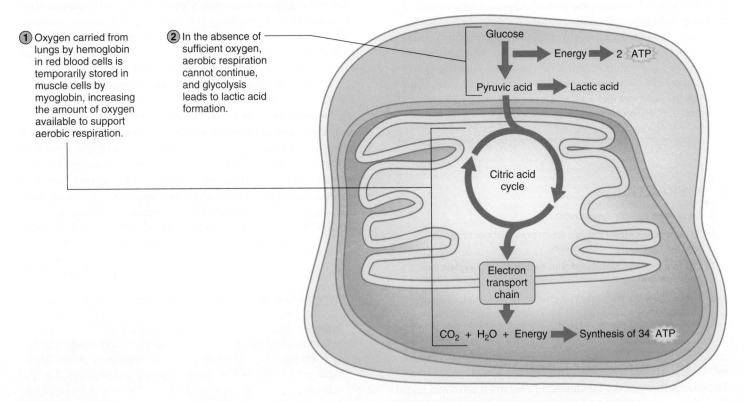

① Oxygen carried from lungs by hemoglobin in red blood cells is temporarily stored in muscle cells by myoglobin, increasing the amount of oxygen available to support aerobic respiration.

② In the absence of sufficient oxygen, aerobic respiration cannot continue, and glycolysis leads to lactic acid formation.

Glucose → Energy → 2 ATP

Pyruvic acid → Lactic acid

Citric acid cycle

Electron transport chain

CO_2 + H_2O + Energy → Synthesis of 34 ATP

FIGURE 9.12 The oxygen required to support the aerobic reactions of cellular respiration is carried in the blood and stored in myoglobin. The maximum number of ATPs generated per glucose molecule varies with cell type; in skeletal muscle, it is 36 (2 + 34). In the absence of sufficient oxygen, anaerobic reactions generate only 2 ATP per glucose molecule and use pyruvic acid to produce lactic acid.

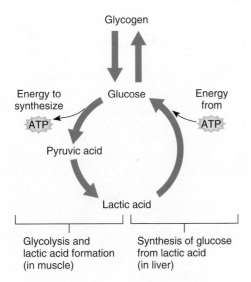

Glycogen

Glucose

Energy to synthesize
ATP

Energy from
ATP

Pyruvic acid

Lactic acid

Glycolysis and lactic acid formation (in muscle)

Synthesis of glucose from lactic acid (in liver)

FIGURE 9.13 Liver cells can synthesize glucose from lactic acid produced anaerobically by muscles.

liver cells require to convert the accumulated lactic acid to glucose, as well as the amount that the muscle cells require to resynthesize sufficient ATP and creatine phosphate to restore their original concentrations. The degree of oxygen debt also reflects the oxygen required to restore blood and tissue oxygen levels to preexercise levels.

The metabolic capacity of a muscle may change with athletic training. With high-intensity exercise, which depends more on glycolysis for ATP, a muscle will synthesize more glycolytic enzymes, and its capacity for glycolysis will increase. With aerobic exercise, the muscles' capacity for the aerobic reactions of cellular respiration increases.

The runners are on the starting line, their muscles primed for a sprint. The starting horn sounds. Energy comes first from residual ATP, but almost instantly, creatine phosphate begins donating high-energy phosphates to ADP, regenerating ATP. Meanwhile, oxidation of glucose potentially produces more ATP, but because the runner cannot take in enough oxygen to meet the high demand, most ATP is generated in glycolysis. Lactic acid forms, and as the runner crosses the finish line, her liver is actively converting lactic acid back to pyruvic acid. In her muscles, creatine phosphate levels begin to return to normal.

Muscle Fatigue

A muscle exercised persistently for a prolonged period may lose its ability to contract, which is a condition called *fatigue*. It has a number of potential causes, including decreased blood flow, ion imbalances across the sarcolemma from repeated stimulation, and psychological loss of the desire to continue the exercise.

For many years fatigue associated with anaerobic exercise has been attributed to increased lactic acid production. As muscle metabolism shifts from aerobic to anaerobic ATP production,

lactic acid begins to accumulate in muscles and to appear in the bloodstream (lactic acid threshold). In the body, lactic acid dissociates rapidly into lactate ion (lactate) and hydrogen ion. Recent studies on the effect of both hydrogen ion and lactate suggest that the relationship between lactic acid and muscle fatigue may not be as clear-cut as once thought. Research in this area is ongoing.

The strenuous exercise of aerobic training stimulates new capillaries to extend into muscles, supplying more oxygen and nutrients to the muscle fibers. Such physical training also adds mitochondria, which split and enlarge to increase their abundance. Both the adaptations of increased blood supply and more mitochondria boost the ability of muscle fibers to produce ATP aerobically.

Occasionally a muscle fatigues and cramps at the same time. A cramp is a sustained, painful, involuntary muscle contraction. The cause of muscle cramps is not fully understood. One hypothesis suggests that cramps may result from changes in electrolyte concentration in the extracellular fluid surrounding the muscle fibers and their motor neurons, triggering uncontrolled stimulation of the muscle.

Heat Production

All active cells generate heat, which is a by-product of cellular respiration. Muscle tissue constitutes such a large proportion of total body mass that it is a major source of heat.

Less than half of the energy released in cellular respiration is transferred to ATP; the rest becomes heat. Blood transports the heat from muscle contraction throughout the body, which helps to maintain body temperature. Homeostatic mechanisms promote heat loss when the temperature of the internal environment begins to rise (see chapters 1 and 6, p. 16 and 189, respectively).

 PRACTICE

9 What are the sources of energy used to regenerate ATP?

10 What are the sources of oxygen required for the aerobic reactions of cellular respiration?

11 How do lactic acid and oxygen debt relate to muscle fatigue?

12 What is the relationship between cellular respiration and heat production?

9.3 | Muscular Responses

One way to observe muscle contraction is to remove a single muscle fiber from a skeletal muscle of a small animal, such as a frog, and connect it to a device that senses and records changes in the fiber's length. An electrical stimulator is usually used to promote muscle contraction.

Threshold Stimulus

In the laboratory, when an isolated muscle fiber is exposed to a series of electrical stimuli of increasing strength, the fiber remains unresponsive until a certain strength of stimulation called the **threshold stimulus** (thresh'old stim'u-lus) is applied. Once threshold is reached, an action potential is generated, resulting in an impulse that spreads throughout the muscle fiber, releasing enough calcium ions from the sarcoplasmic reticulum to activate cross-bridge formation and contract the fiber. A single action potential conducted down a motor neuron normally releases enough ACh to bring the muscle fibers to threshold, generating an impulse.

Recording of a Muscle Contraction

The contractile response of a single muscle fiber to a single impulse is called a **twitch.** A twitch consists of a period of contraction, during which the fiber pulls at its attachments, followed by a period of relaxation, during which the pulling force declines. These events can be recorded in a pattern called a myogram (fig. 9.14). A twitch has a brief delay between the time of stimulation and the beginning of contraction. This is the **latent period,** which in human muscle may be less than 2 milliseconds.

The length to which a muscle fiber is stretched before stimulation affects the force it will develop. If a skeletal muscle fiber is stretched well beyond its normal resting length, the force will decrease. This is because sarcomeres of that fiber become so extended that some myosin heads cannot reach binding sites on the thin filaments and cannot contribute to contraction. Conversely, at very short fiber lengths, the sarcomeres become compressed, and further shortening is not possible (fig. 9.15). During normal activities, muscle fibers contract at their optimal lengths. Some activities, such as walking up stairs two at a time or lifting something from an awkward position, put fibers at a disadvantageous length and compromise muscle performance.

A GLIMPSE AHEAD | To Chapter 15

Cardiac muscle does not normally decrease its force of contraction when stretched. This ability allows the heart to respond to conditions of greater demand, such as exercise.

> Increasing the stimulus above threshold will not result in a stronger twitch in a muscle fiber. However, twitches from a given fiber may vary in strength for a number of reasons, including the length to which the fiber is stretched and the development of fatigue.

Understanding how individual muscle fibers contract is important for understanding how muscles work, but such contractions by themselves are of little significance in day-to-day activities. Rather, the actions we need to perform usually require the contribution of multiple muscle fibers simultaneously. To record how a whole muscle responds to stimulation, a skeletal muscle can be removed from a small animal, such as a frog, and mounted on a special device. The muscle is then electrically stimulated, and when it contracts, it pulls on a lever. The lever's movement is recorded as a myogram, which reflects the combined twitches of muscle fibers contracting, so it looks like the twitch contraction depicted in figure 9.14.

Sustained contractions of whole muscles enable us to perform everyday activities, but the force generated by those contractions must be controlled. For example, holding a styrofoam cup of coffee firmly enough to keep it from slipping through our fingers, but not so forcefully as to crush it, requires precise control of contractile

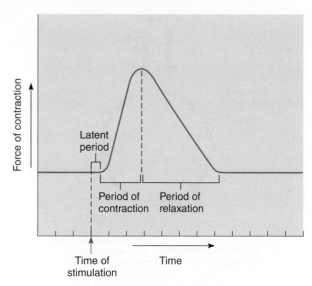

FIGURE 9.14 A myogram of a single muscle twitch.

force. In the whole muscle, the force developed reflects (1) the frequency at which individual muscle fibers are stimulated and (2) how many fibers take part in the overall contraction of the muscle.

Summation

The force that a muscle fiber can generate is not limited to the maximum force of a single twitch (fig. 9.16a). A muscle fiber exposed to a series of stimuli of increasing frequency reaches a point when it is unable to completely relax before the next stimulus in the series arrives. When this happens, the individual twitches begin to combine, and the contraction becomes sustained. In such a *sustained contraction,* the force of individual twitches combines by the process of **summation** (fig. 9.16b). At higher frequencies of stimulation, as the time spent in relaxation becomes very brief, a condition called *partial tetany* results. When the resulting forceful, sustained contraction lacks even partial relaxation, it is called a **complete (fused) tetanic contraction** (tetanus) (fig. 9.16c). Complete tetany does not occur in the body, but can be demonstrated in the laboratory.

Recruitment of Motor Units

While summation increases the force of contraction of a single muscle fiber, a whole muscle can generate more force if more muscle fibers are involved in the contraction. Most muscle fibers have only one motor end plate. Motor neuron axons, however, are densely branched, which enables one such axon to connect to many muscle fibers. Together, a motor neuron and the muscle fibers it controls constitute a *motor unit* (mo'tor u'nit) (fig. 9.17). The number of muscle fibers in a motor unit varies considerably. The fewer muscle fibers in the motor units, however, the more precise the movements that can be produced in a particular muscle. For example, the motor units of the muscles that move the eyes may include fewer than ten muscle fibers per motor unit and can produce very slight movements. Conversely, the motor units of the large muscles in the back may include a hundred or more muscle fibers. Movements of these motor units are larger-scale than those of the eye.

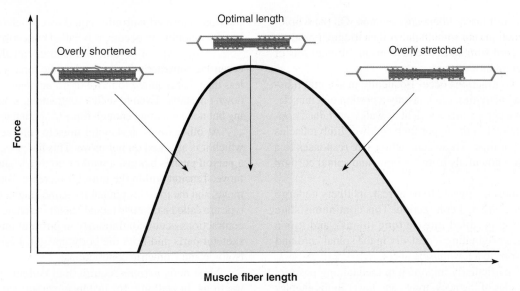

FIGURE 9.15 The force a muscle fiber can generate depends on the length to which it is stretched when stimulated.

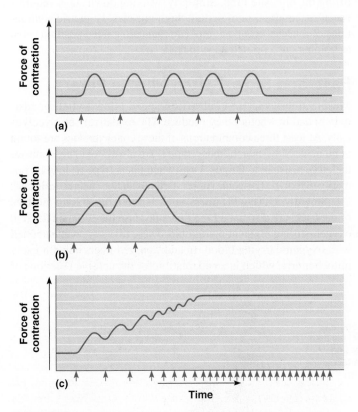

FIGURE 9.16 Myograms of (a) a series of twitches, (b) summation, and (c) a tetanic contraction. Stimulation frequency increases from one myogram to the next.

Each motor unit is a functional unit, because an impulse in its motor neuron will contract all of the fibers in that motor unit simultaneously. A whole muscle is composed of many such motor units controlled by different motor neurons. Like muscle fibers, motor neurons must be brought to threshold before an impulse is generated. Some motor neurons are more easily brought to threshold than others. If only the more easily stimulated motor neurons

are involved, few motor units contract. At higher intensities of stimulation, other motor neurons respond, and more motor units are activated. Such an increase in the number of activated motor units is called *multiple motor unit summation,* or **recruitment** (re-kroo-t'ment). As the intensity of stimulation increases, recruitment of motor units continues until finally all motor units are activated in that muscle.

Sustained Contractions

During sustained contractions smaller motor units, whose neurons have smaller-diameter axons, are more easily stimulated and are recruited first. Larger motor units, whose neurons have larger-diameter axons, respond later and with greater force. Summation and recruitment together can produce a sustained contraction of increasing strength.

Typically, many action potentials are triggered in a motor neuron, and so individual twitches do not normally occur. Partial

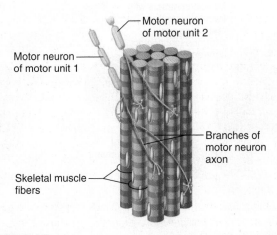

FIGURE 9.17 Two motor units. The muscle fibers of a motor unit are innervated by a single motor neuron and may be distributed throughout the muscle.

tetanic contractions of muscle fibers are common. On the whole-muscle level, contractions are smooth rather than irregular or jerky because the spinal cord stimulates contractions in different sets of motor units at different moments.

Partial tetanic contractions occur frequently in skeletal muscles during everyday activities, often in only a portion of a muscle. For example, when a person lifts a weight or walks, sustained contractions are maintained in the upper limb or lower limb muscles for varying lengths of time. These contractions are responses to a rapid series of action potentials from the brain and spinal cord on motor neurons.

Even when a muscle appears to be at rest, its fibers undergo a certain degree of sustained contraction. This continuous state of partial contraction is called **muscle tone** (tonus), and it is a response to impulses originating repeatedly in the spinal cord and conducted along axons to a few muscle fibers.

Muscle tone is particularly important in maintaining posture. Tautness in the muscles of the neck, trunk, and lower limbs enables a person to hold the head upright, stand, or sit. If tone is suddenly lost, such as when a person loses consciousness, the body collapses. Muscle tone is maintained in health but is lost if motor nerve axons are cut or if diseases interfere with impulse conduction.

> Skeletal muscles contracting forcefully may generate up to 50 pounds of pull for each square inch of muscle cross section. Consequently, large muscles such as those in the thigh can pull with several hundred pounds of force. Occasionally, this force is so great that the tendons of muscles pull away from their attachments to the bones.

Types of Contractions

Sometimes muscles shorten when they contract. For example, if a person lifts an object, the muscles remain taut, their attached ends pull closer together, and the object is moved. This type of contraction is termed **isotonic** (equal force—change in length), and because shortening occurs, it is called **concentric.**

Another type of isotonic contraction, called a lengthening or an **eccentric contraction,** occurs when the force a muscle generates is less than that required to move or lift an object, as in laying a book down on a table. Even in such a contraction, cross-bridges are working but not generating enough force to shorten the muscle.

At other times, a skeletal muscle contracts, but the parts to which it is attached do not move. This happens, for instance, when a person pushes against a wall or holds a yoga pose but does not move. Tension within the muscles increases, but the wall does not move, and the muscles remain the same length. Contractions of this type are called **isometric** (equal length—change in force). Isometric contractions occur continuously in postural muscles that stabilize skeletal parts and hold the body upright. **Figure 9.18** illustrates isotonic and isometric contractions.

Most body actions require both isotonic and isometric contractions. In walking, for instance, certain leg and thigh muscles contract isometrically and keep the limb stiff as it touches the ground, while other muscles contract isotonically, bending and lifting the opposite limb. Similarly, walking down stairs requires eccentric contraction of certain thigh muscles. Isometric contractions of muscles around a joint can maintain a fixed position, such as holding out a printed page to read.

Fast- and Slow-Twitch Muscle Fibers

Muscle fibers vary in contraction speed (slow-twitch or fast-twitch) and in whether they produce ATP oxidatively or glycolytically. At least three combinations of these characteristics are found in humans. Slow-twitch fibers (type I) are always oxidative and are therefore resistant to fatigue. Fast-twitch fibers (type II) may be primarily glycolytic (fatigable) or primarily oxidative (fatigue resistant).

Slow-twitch (type I) fibers, such as those in the long muscles of the back, are also called *red fibers* because they contain the red, oxygen-storing pigment myoglobin. These fibers are well supplied with oxygen-carrying blood. In addition, red fibers contain many mitochondria, which is an adaptation for the aerobic reactions of

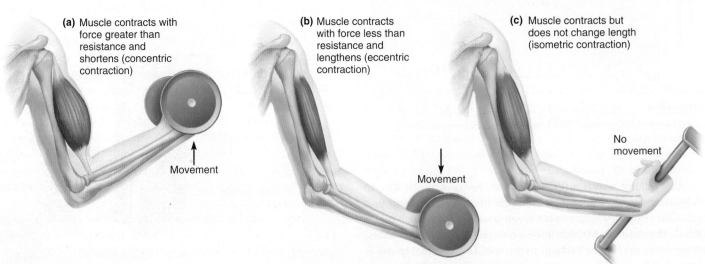

FIGURE 9.18 Types of muscle contractions. (*a* and *b*) Isotonic contractions include concentric and eccentric contractions. (*c*) Isometric contractions occur when a muscle contracts but does not shorten or lengthen.

cellular respiration. Red fibers have a high respiratory capacity and can generate ATP fast enough to keep up with the ATP breakdown that occurs when they contract. Because of the high ATP production, red fibers can function for long periods without fatiguing.

Fast-twitch glycolytic fibers (type IIb) are also called *white fibers* because they have less myoglobin and have a poorer blood supply than red fibers. White fibers are in certain hand muscles as well as in muscles that move the eye. These fibers have fewer mitochondria and thus have a reduced respiratory capacity. However, they have a more extensive sarcoplasmic reticulum to store and reabsorb calcium ions, and their ATPase is faster than that of red fibers. White muscle fibers can contract rapidly because of these factors, although they are more susceptible to fatigue.

A type of white fiber, the fast-twitch fatigue-resistant fibers (type IIa), are also called *intermediate fibers*. These fibers have the fast-twitch speed associated with white fibers with an intermediate oxidative capacity.

All muscles include a combination of fiber types, although some muscles may have mostly one fiber type. The speed of contraction and aerobic capacities of the fibers reflect the specialized functions of the muscle. For example, muscles that move the eyes contract about ten times faster than those that maintain posture, and the muscles that move the limbs contract at intermediate rates. Slowing of eye movements may be an early sign of certain neurological diseases. Clinical Application 9.2 discusses noticeable effects of muscle use and disuse.

 PRACTICE

13 Define *threshold stimulus.*

14 Distinguish between a twitch and a sustained contraction.

15 Define *muscle tone.*

16 Explain the differences between isometric and isotonic contractions.

17 Distinguish between fast-twitch and slow-twitch muscle fibers.

9.4 | Smooth Muscle

The contractile mechanisms of smooth and cardiac muscle are similar to those of skeletal muscles. However, the cells of these tissues have important structural and functional distinctions.

Smooth Muscle

Recall from chapter 5 (p. 170) that smooth muscle cells are shorter than the fibers of skeletal muscle, and they have single, centrally located nuclei. Smooth muscle cells are elongated with tapering ends and contain filaments of actin and myosin in myofibrils that extend throughout their lengths, but these filaments are thin and more randomly distributed than those in skeletal muscle fibers. Therefore, smooth muscle cells lack striations and appear "smooth" under the microscope. Smooth muscle cells do not have transverse tubules, and their sarcoplasmic reticula are not well developed.

The two major types of smooth muscle are multiunit and visceral. In **multiunit smooth muscle,** the muscle cells are less well organized and function as separate units, independent of neighboring cells. Smooth muscle of this type is found in the irises of the eyes and in the walls of large blood vessels. Typically, multiunit smooth muscle contracts only after stimulation by neurons or certain hormones.

Visceral smooth muscle (single-unit smooth muscle) is composed of sheets of spindle-shaped cells held in close contact by gap junctions. The thick portion of each cell lies next to the thin parts of adjacent cells. Cells of visceral smooth muscle respond as a single unit. When one cell is stimulated, the impulse conducted over its surface may excite adjacent cells that, in turn, stimulate others. Some visceral smooth muscle also displays *rhythmicity*—a pattern of spontaneous repeated contractions.

These two features of visceral smooth muscle—conduction of impulses from cell to cell and rhythmicity—are largely responsible for the wavelike motion called **peristalsis** of certain tubular organs (see chapter 17, p. 651). Peristalsis consists of alternate contractions and relaxations of the longitudinal and circular muscles. These movements help mix the contents of a tube and push them along its length. In the intestines, for example, peristaltic waves move masses of partially digested food and help to mix them with digestive fluids. Peristalsis in the ureters moves urine from the kidneys to the urinary bladder.

Visceral smooth muscle is the more common type of smooth muscle. It is found in the walls of hollow organs, such as the stomach, intestines, urinary bladder, and uterus. Most of the smooth muscle in the walls of these organs is of two thicknesses. The cells of the outer coats are oriented longitudinally, whereas those of the inner coats are circular. The muscular layers change the sizes and shapes of the organs as they contract and relax.

Vascular smooth muscle, which is a form of visceral smooth muscle, is found in the walls of many of the smaller blood vessels, where it plays a role in controlling blood flow and blood pressure. The function of vascular smooth muscle will be discussed further in chapter 15 (pp. 580–581 and 586).

Smooth Muscle Contraction

Smooth muscle contraction resembles skeletal muscle contraction in a number of ways. Both mechanisms reflect reactions of actin and myosin. Both are triggered by membrane impulses and release of calcium ions. Finally, both use energy from ATP molecules. However, smooth muscle and skeletal muscle also differ in their actions. Smooth muscle cells do not have troponin, the protein that binds to calcium ions in skeletal muscle. Instead, smooth muscle uses a protein called *calmodulin,* which binds to calcium ions released when its cells are stimulated, activating contraction. In addition, much of the calcium necessary for smooth muscle contraction diffuses into the cell from the extracellular fluid.

Acetylcholine, the neurotransmitter in skeletal muscle, as well as *norepinephrine,* affect smooth muscle. Each of these neurotransmitters stimulates contraction in some smooth muscle and inhibits contraction in other smooth muscle. The discussion of the autonomic nervous system in chapter 11 (pp. 431–438) describes these actions in greater detail.

Hormones affect smooth muscle by stimulating or inhibiting contraction in some cases and altering the degree of response to neurotransmitters in others. For example, during the later stages of childbirth, the hormone oxytocin stimulates smooth muscle in the wall of the uterus to contract (see chapter 23, p. 890).

Skeletal muscles respond to use and disuse. Forcefully exercised muscles enlarge, or *hypertrophy*. Unused muscles *atrophy*, decreasing in size and strength.

The way a muscle responds to use also depends on the type of exercise. Aerobic activity, such as in swimming and running, activates slow-twitch, fatigue-resistant red fibers. In response, these fibers develop more mitochondria and the muscles that they comprise develop more extensive capillary networks, both of which increase fatigue-resistance during prolonged exercise.

In forceful exercise, such as weightlifting, a muscle may exert more than 75% of its maximum tension, using predominantly the muscle's fast-twitch, fatigable white fibers. In response, existing muscle fibers synthesize new filaments of actin and myosin, and as muscle fiber diameters increase, the entire muscle enlarges.

The strength of a contraction is directly proportional to the diameter of the muscle fibers. Therefore, an enlarged muscle can contract more strongly than before. However, such a change in response to forceful exercise does not increase the muscle's ability to resist fatigue during aerobic activities such as running or swimming.

If regular muscle use stops, capillary networks shrink, and muscle fibers lose some mitochondria. Actin and myosin filaments diminish, and the entire muscle atrophies. Injured limbs immobilized in casts, or accidents or diseases that interfere with nervous stimulation, also cause muscle atrophy. An unexercised muscle may shrink to less than one-half its usual size in a few months.

In functional electrical stimulation, physical therapists apply electrodes to the skin over an injured muscle. The electrical current triggers action potentials in motor neurons controlling nearby muscle fibers, contracting those muscle fibers. This technique can prevent tissue atrophy, increase muscle strength, reduce pain, and promote healing by increasing blood supply.

Muscle fibers whose motor neurons are severed not only shrink but also may fragment, and in time fat or fibrous tissue replaces them. However, reinnervation of such a muscle within the first few months following an injury can restore function.

New technologies can compensate for some muscle loss. "Targeted muscle reinnervation," for example, can tap into the neuromuscular system to assist a person who has lost an upper limb. A surgeon reattaches muscles from a severed arm to the patient's chest wall, then uses electromyography to detect the electrical activity that still reaches those muscles. The information is sent to a microprocessor built into an attached prosthetic arm, where a "neural-machine interface" enables the patient to move the "myoelectric prosthetic" arm at will, just as he or she would consciously direct the movement of the missing part.

Stretching of smooth muscle can also trigger contractions. This response is particularly important to the function of visceral smooth muscle in the walls of certain hollow organs, such as the urinary bladder and the intestines. For example, when partially digested food stretches the wall of the intestine, contractions move the contents further along the intestine.

Smooth muscle is slower to contract and relax than skeletal muscle, yet smooth muscle can forcefully contract longer with the same amount of ATP. Unlike skeletal muscle, smooth muscle cells can change length without changing tautness. This ability enables smooth muscle in the stomach and intestinal walls to stretch as these organs fill, yet maintain a constant pressure inside the organs.

PRACTICE

18 Describe the two major types of smooth muscle.

19 What special characteristics of visceral smooth muscle make peristalsis possible?

20 How is smooth muscle contraction similar to skeletal muscle contraction?

21 How do the contraction mechanisms of smooth and skeletal muscles differ?

9.5 | Cardiac Muscle

Cardiac muscle is found only in the heart. It is composed of striated cells joined end to end, forming interconnected, branching, three-dimensional networks. Each cell contains a single nucleus and many filaments of actin and myosin similar to those in skeletal muscle. A cardiac muscle cell also has a well-developed sarcoplasmic reticulum, a system of transverse tubules, and many mitochondria. However, the cisternae of the sarcoplasmic reticulum of a cardiac muscle cell are less developed and store less calcium than those of a skeletal muscle cell. On the other hand, the transverse tubules of cardiac muscle cells are larger than those in skeletal muscle cells, and they release many calcium ions into the sarcoplasm in response to a single impulse.

The calcium ions in transverse tubules come from the fluid outside the cardiac muscle cell. In this way, extracellular calcium partially controls the strength of cardiac muscle contraction and enables cardiac muscle cells to contract longer than skeletal muscle fibers can.

The opposing ends of cardiac muscle cells are connected by cross-bands called *intercalated discs*. These bands are complex intercellular junctions that include components of desmosomes and gap junctions.

Drugs called calcium channel blockers are used to treat irregular heart rhythms. They do this by blocking ion channels that admit extracellular calcium into cardiac muscle cells.

FIGURE 9.19 AP|R The intercalated discs of cardiac muscle, shown in this transmission electron micrograph, bind cells and allow ions to move between the cells (12,500×).

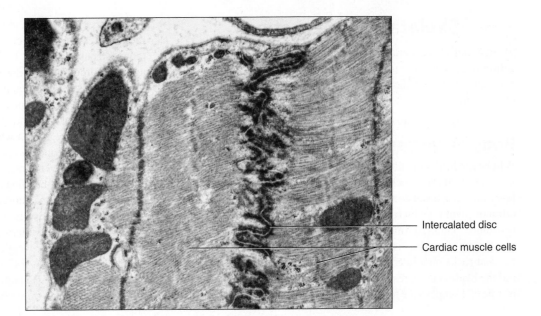

Intercalated disc

Cardiac muscle cells

TABLE 9.3	Characteristics of Muscle Tissues		
	Skeletal	**Smooth**	**Cardiac**
Dimensions			
Length	Up to 30 cm	30–200 μm	50–100 μm
Diameter	10–100 μm	3–6 μm	14 μm
Major location	Skeletal muscles	Walls of hollow organs	Wall of the heart
Major function	Movement of bones at joints; maintenance of posture	Movement of walls of hollow organs; peristalsis; vasoconstriction	Pumping action of the heart
Cellular characteristics			
Striations	Present	Absent	Present
Nucleus	Multiple nuclei	Single nucleus	Single nucleus
Special features	Transverse tubule system is well developed	Lacks transverse tubules	Transverse tubule system is well developed; intercalated discs connect cells
Mode of control	Voluntary	Involuntary	Involuntary
Contraction characteristics	Contracts and relaxes relatively rapidly	Contracts and relaxes relatively slowly; some types self-exciting; rhythmic	Network of cells contracts as a unit; self-exciting; rhythmic; remains refractory until contraction ends

Intercalated discs help join cells and transmit the force of contraction from cell to cell, and also allow ions to diffuse between the cells. This ion movement allows action potentials to travel from cell to cell (see fig. 5.31, p. 171, and fig. 9.19). In this way, when one portion of the cardiac muscle network is stimulated, the impulse passes rapidly to other cells of the network, and the whole structure contracts as a unit or *syncytium* (sin-sish′e-um); that is, the network responds to stimulation in an all-or-none manner.

Cardiac muscle is self-exciting and rhythmic. A pattern of contraction and relaxation repeats, generating the rhythmic contraction of the heart. Also, cardiac muscle becomes nonresponsive (refractory) after stimulation until the contraction ends. Therefore, sustained or tetanic contractions do not occur in the

heart muscle. Table 9.3 summarizes characteristics of the three types of muscles.

RECONNECT

To Chapter 5, Figure 5.1, Intercellular Junctions, page 150.

PRACTICE

22 How is cardiac muscle similar to skeletal muscle?

23 How does cardiac muscle differ from skeletal muscle?

24 What is the function of intercalated discs?

25 What characteristic of cardiac muscle causes the heart to contract as a unit?

9.6 | Skeletal Muscle Actions

Skeletal muscles generate a great variety of body movements. The action of each muscle mostly depends upon the type of joint it is associated with and the way the muscle is attached on either side of that joint.

Body Movement

Whenever limbs or other body parts move, bones and muscles interact as simple mechanical devices called **levers** (lev′erz). A lever has four basic components: (1) a rigid bar or rod, (2) a fulcrum or pivot on which the bar turns, (3) an object moved against resistance, and (4) a force that supplies energy to move the bar.

A pair of scissors is a lever. The handle and blade form a rigid bar that rocks on a fulcrum near the center (the screw). The material the blades cut represents the resistance, while the person on the handle end supplies the force to cut the material.

Figure 9.20 shows the three types of levers, which differ in their component arrangements. A first-class lever's parts are like those of a pair of scissors. Its fulcrum is located between the resistance and the force, making the sequence of components resistance–fulcrum–force. Other examples of first-class levers are seesaws and hemostats (devices used to clamp blood vessels).

The parts of a second-class lever are in the sequence fulcrum–resistance–force, as in a wheelbarrow. The parts of a third-class lever are in the sequence resistance–force–fulcrum. Eyebrow tweezers or forceps used to grasp an object illustrate this type of lever.

The actions of bending and straightening the upper limb at the elbow illustrate bones and muscles functioning as levers. When the upper limb bends, the forearm bones represent the rigid bar, the elbow joint is the fulcrum, the hand is moved against the resistance provided by the weight, and the force is supplied by muscles on the anterior side of the arm (fig. 9.21a). One of these muscles, the *biceps brachii,* is attached by a tendon to a projection (radial tuberosity) on the *radius* bone in the forearm, a short distance

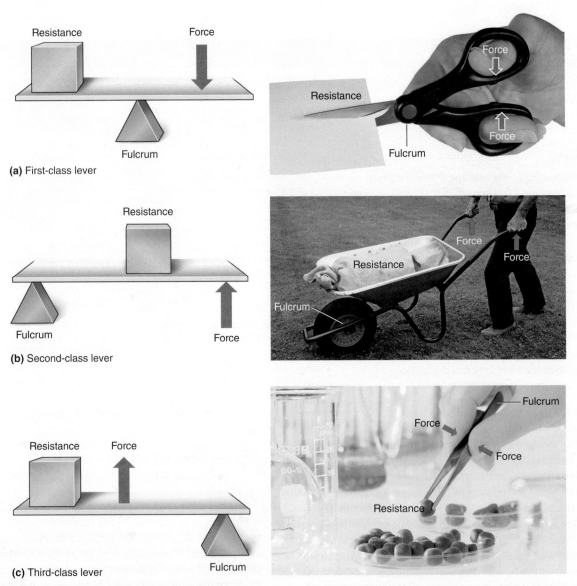

(a) First-class lever

(b) Second-class lever

(c) Third-class lever

FIGURE 9.20 Three types of levers. (a) A first-class lever is used in a pair of scissors, (b) a second-class lever is used in a wheelbarrow, and (c) a third-class lever is used in a pair of forceps.

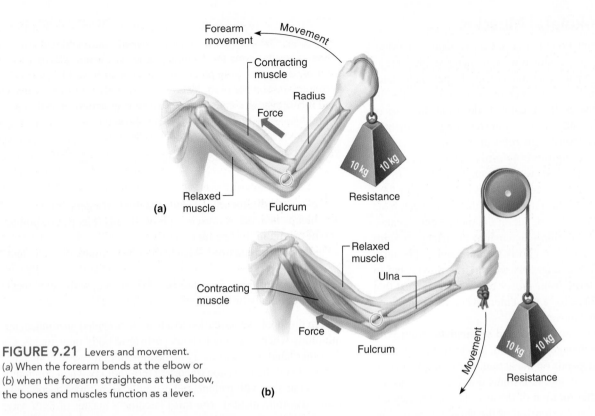

FIGURE 9.21 Levers and movement. (a) When the forearm bends at the elbow or (b) when the forearm straightens at the elbow, the bones and muscles function as a lever.

below the elbow. Because the parts of this lever are arranged in the sequence resistance–force–fulcrum, it is a third-class lever.

When the upper limb straightens at the elbow, the forearm bones again serve as the rigid bar, the hand moves against the resistance by pulling on the rope to raise the weight (fig. 9.21*b*), and the elbow joint serves as the fulcrum. However, in this case the *triceps brachii,* a muscle located on the posterior side of the arm, supplies the force. A tendon of this muscle attaches to a projection (olecranon process) of the ulna bone at the point of the elbow. Because the parts of the lever are arranged resistance–fulcrum–force, it is a first-class lever.

The human body also has a second-class lever (fulcrum–resistance–force). The fulcrum is the temporomandibular joint. Muscles supply the resistance, attaching to a projection (coronoid process) and body of the mandible. These muscles resist or oppose opening the mouth. The muscles attached to the chin area of the mandible provide the force that opens the mouth.

Levers provide a range of movements. Levers that move limbs, for example, produce rapid motions, whereas others, such as those that move the head, help maintain posture with minimal effort.

Origin and Insertion

Recall from chapter 8 (p. 274) that one end of a skeletal muscle is typically fastened to a relatively immovable or fixed part on one side of a joint, and the other end is connected to a movable part on the other side of that joint. The less movable end is called the **origin** of the muscle, and the more movable end is called its **insertion.** When a muscle contracts, its insertion is pulled toward its origin (fig. 9.22).

Some muscles have more than one origin or insertion. The *biceps brachii* in the arm, for example, has two origins. This is reflected in its name *biceps,* meaning "two heads." The *head* of a

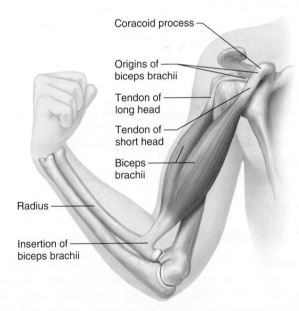

FIGURE 9.22 The biceps brachii has two heads that originate on the scapula. A tendon inserts this muscle on the radius.

muscle is the part nearest its origin. As figure 9.22 shows, one head of the muscle is attached to the coracoid process of the scapula, and the other head arises from a tubercle above the glenoid cavity of the scapula. The muscle runs along the anterior surface of the humerus and is inserted by a single tendon on the radial tuberosity of the radius. When the biceps brachii contracts, its insertion is pulled toward its origin, and the elbow flexes.

Interaction of Skeletal Muscles

Most skeletal muscles function in groups. Consequently, a particular body movement requires contracting more than a single muscle. After learning to make a particular movement, the person initiates the movement consciously, and the nervous system stimulates the appropriate group of muscles. It is possible to determine the roles of particular muscles by carefully observing the resulting movement.

A number of terms describe the roles of muscles in performing particular actions. An **agonist** (ag′o-nist) causes an action, and an **antagonist** (an-tag′o-nist) works against the action. In the example of elbow flexion, the biceps brachii is the agonist for flexion and the triceps brachii is the antagonist to that movement. Note that the role of a muscle is dependent on the movement, because in the opposite example of elbow extension, the triceps brachii is the agonist and the biceps brachii is the antagonist. The term *agonist* is often used interchangeably with **prime mover,** but in many applications the term *prime mover* refers to an agonist that provides most of the force for a movement when more than one muscle contributes to that movement.

In most cases other muscles, called **synergists,** contribute to an action by helping the agonist. Again, however, the relationship between muscles depends on the action. For example, pectoralis major, a chest muscle, and latissimus dorsi, a back muscle, are synergistic for medial rotation of the arm. However, they are antagonistic for flexion and extension of the shoulder. Similarly, two muscles on the lateral forearm, the flexor carpi radialis and the extensor carpi radialis longus, are synergistic for abduction of the hand, yet they are antagonistic for flexion and extension of the wrist. Thus, any role of a muscle must be learned in the context of a particular movement.

 PRACTICE

26 Explain how parts of the upper limb form a first-class lever and a third-class lever.

27 Distinguish between the origin and the insertion of a muscle.

28 Define *agonist*.

29 What is the function of an antagonist? a synergist?

9.7 | Major Skeletal Muscles

This section discusses the locations, actions, origins, and insertions of some of the major skeletal muscles. The tables that summarize the information concerning groups of these muscles also include the names of nerves that supply (innervate) the individual muscles. Chapter 11 (pp. 420–421) presents the origins and pathways of these nerves.

Figures 9.23 and 9.24 show the locations of superficial skeletal muscles—that is, those near the surface. The names of muscles often describe them. A name may indicate a muscle's size, shape, location, action, number of attachments, or the direction of its fibers, as in the following examples:

pectoralis major A muscle of large size (major) in the pectoral region (chest).
deltoid Shaped like a delta or triangle.

extensor digitorum Extends the digits (fingers or toes).
biceps brachii A muscle with two heads (biceps), or points of origin, in the brachium (arm).
sternocleidomastoid Attached to the sternum, clavicle, and mastoid process.
external oblique Located near the outside, with fibers that run obliquely or in a slanting direction.

In this book we have tried to adhere to accepted anatomical terminology when referring to body parts or to body movements. On the other hand, you will likely be dealing not only with colleagues (who rely on the precision of correct terminology when communicating) but also with patients (who simply, sometimes desperately, want to communicate). You must become a master of both ways of communicating. The bottom line is being able to communicate accurately with colleagues and effectively with patients.

Note that in the anatomical position some actions have already occurred, such as supination of the forearm and hand, extension of the elbow, and extension of the knee. The muscle actions described in the following section consider the entire range of movement at each joint, and do not presume that the starting point is the anatomical position.

Some muscles have more than one origin or more than one insertion. The wide range of attachments of some of the larger muscles has the effect of giving those muscles different, sometimes opposing, actions, depending on which portion of the muscle is active. These cases are described in the appropriate sections and tables.

Muscles of Facial Expression

A number of small muscles beneath the skin of the face and scalp enable us to communicate feelings through facial expression. Many of these muscles are located around the eyes and mouth, and they make possible such expressions as surprise, sadness, anger, fear, disgust, and pain. As a group, the muscles of facial expression connect the bones of the skull to connective tissue in regions of the overlying skin. Figure 9.25 and reference plate 66 (p. 352) show these muscles, and table 9.4 lists them. The muscles of facial expression include the following:

Epicranius   Zygomaticus major
Orbicularis oculi   Zygomaticus minor
Orbicularis oris   Platysma
Buccinator

The **epicranius** (ep″ĭ-kra′ne-us) covers the upper part of the cranium and includes two muscular parts—the *frontalis*

> At times, there are discrepancies between anatomical terminology and the terms the general public uses when referring to body parts. For example, someone with a bruised thigh may complain of a sore leg, instead of correctly referring to the thigh. Similarly, someone may describe an action as flexing a muscle, when really the muscle is contracting and the corresponding joint, such as the elbow, is flexing.

on test ✱

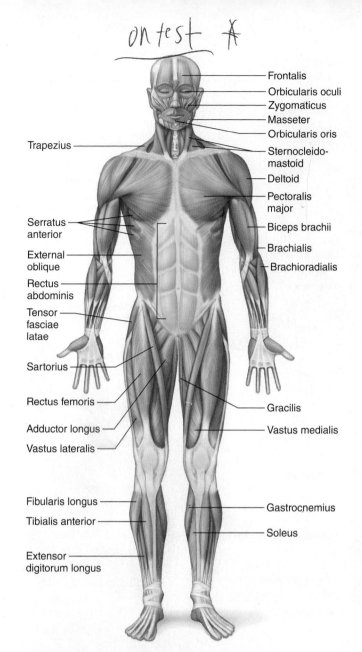

FIGURE 9.23 Anterior view of superficial skeletal muscles.

Frontalis
Orbicularis oculi
Zygomaticus
Masseter
Orbicularis oris
Trapezius
Sternocleido-mastoid
Deltoid
Pectoralis major
Serratus anterior
Biceps brachii
External oblique
Brachialis
Rectus abdominis
Brachioradialis
Tensor fasciae latae
Sartorius
Rectus femoris
Gracilis
Adductor longus
Vastus medialis
Vastus lateralis
Fibularis longus
Gastrocnemius
Tibialis anterior
Soleus
Extensor digitorum longus

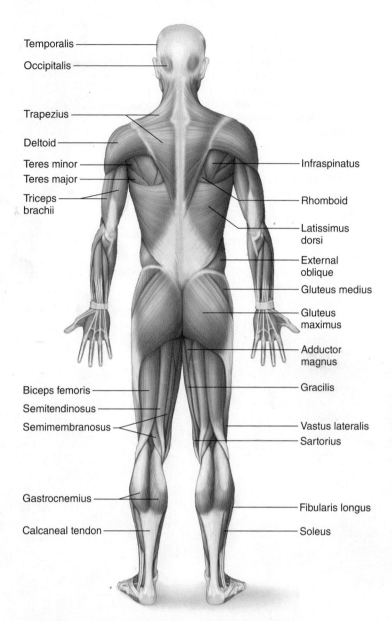

FIGURE 9.24 Posterior view of superficial skeletal muscles.

Temporalis
Occipitalis
Trapezius
Deltoid
Teres minor
Teres major
Triceps brachii
Infraspinatus
Rhomboid
Latissimus dorsi
External oblique
Gluteus medius
Gluteus maximus
Adductor magnus
Biceps femoris
Gracilis
Semitendinosus
Semimembranosus
Vastus lateralis
Sartorius
Gastrocnemius
Fibularis longus
Calcaneal tendon
Soleus

TABLE 9.4	Muscles of Facial Expression AP|R			
Muscle	**Origin**	**Insertion**	**Action**	**Nerve Supply**
Epicranius	Occipital bone	Skin around eye	Elevates eyebrow as when surprised	Facial nerve
Orbicularis oculi	Maxillary and frontal bones	Skin around eye	Closes eye as in blinking	Facial nerve
Orbicularis oris	Muscles near the mouth	Skin of central lip	Closes lips, protrudes lips as for kissing	Facial nerve
Buccinator	Alveolar processes of maxilla and mandible	Orbicularis oris	Compresses cheeks	Facial nerve
Zygomaticus major	Zygomatic bone	Skin and muscle at corner of mouth	Elevates corner of mouth as when smiling	Facial nerve
Zygomaticus minor	Zygomatic bone	Skin and muscle at corner of mouth	Elevates corner of mouth as when smiling	Facial nerve
Platysma	Fascia in upper chest	Skin and muscles below mouth; mandible	Depresses lower lip and angle of mouth as when pouting	Facial nerve

FIGURE 9.25 AP|R

Muscles of the head and face. (a) Muscles of facial expression and mastication; isolated views of (b) the temporalis and buccinator muscles and (c) the lateral and medial pterygoid muscles.

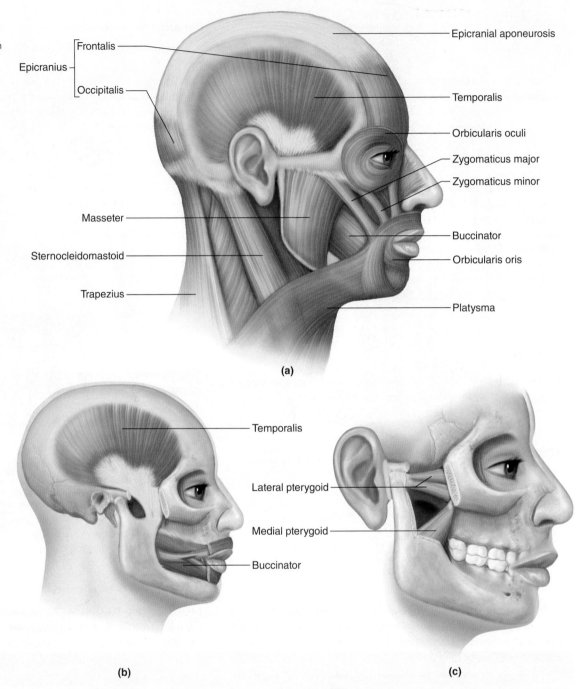

Epicranius
Frontalis
Occipitalis

Epicranial aponeurosis
Temporalis
Orbicularis oculi
Zygomaticus major
Zygomaticus minor
Buccinator
Orbicularis oris
Platysma

Masseter
Sternocleidomastoid
Trapezius

(a)

Temporalis
Lateral pterygoid
Medial pterygoid
Buccinator

(b)

(c)

(frun-ta′lis), which lies over the frontal bone, and the *occipitalis* (ok-sip″ĭ-ta′lis), which lies over the occipital bone. These muscles are united by a broad, tendinous membrane called the *epicranial aponeurosis,* which covers the cranium like a cap. Contraction of the epicranius raises the eyebrows and horizontally wrinkles the skin of the forehead, as when a person expresses surprise. Headaches often result from sustained contraction of this muscle.

The **orbicularis oculi** (or-bik′u-la-rus ok′u-li) is a ringlike band of muscle, called a *sphincter muscle,* that surrounds the eye. It lies in the subcutaneous tissue of the eyelid and closes or blinks the eye. At the same time, it compresses the nearby tear gland, or *lacrimal gland,* aiding the flow of tears over the surface of the eye. Contraction of the orbicularis oculi also causes the folds, or crow's

feet, that radiate laterally from the corner of the eye. Chapter 12 (pp. 469–470) describes the muscles that move the eye.

The **orbicularis oris** (or-bik′u-la-rus o′ris) encircles the mouth. It lies between the skin and the mucous membranes of the lips, extending upward to the nose and downward to the region between the lower lip and chin. The orbicularis oris is also called the kissing muscle because it closes and puckers the lips.

The **buccinator** (buk′sĭ-na″tor) is located in the wall of the cheek. Its fibers are directed forward from the bones of the jaws to the angle of the mouth, and when they contract, the cheek is pulled inward. This action helps hold food in contact with the teeth when a person is chewing. The buccinator also aids in blowing air out of the mouth, and for this reason, it is also called the trumpeter muscle.

TMJ Syndrome

Temporomandibular joint (TMJ) syndrome causes facial pain, headache, ringing in the ears, a clicking jaw, insomnia, teeth sensitive to heat or cold, backache, dizziness, and pain in front of the ears. A misaligned jaw or grinding or clenching the teeth can cause TMJ by stressing the temporomandibular joint, which is the articulation between the mandibular condyle of the mandible and the mandibular fossa of the temporal bone. Loss of coordination of these structures affects the nerves that pass through the neck and jaw region, causing the symptoms.

Getting enough sleep and drinking enough water can help prevent symptoms of TMJ, and eating soft foods, applying ice packs, using relaxation techniques, and massaging affected muscles can alleviate symptoms. A physical therapist can recommend exercises that stretch and relax the jaw, which may help some people. Sitting for long hours in one position can cause or worsen TMJ.

Doctors diagnose TMJ syndrome using an electromyograph, in which electrodes record muscle activity in four pairs of head and neck muscle groups. Several treatments are available. The National Institute of Dental and Craniofacial Research recommends that treatments not permanently alter the teeth or jaw. Low doses of certain antidepressants, or injections of botulinum toxin or corticosteroids, may help. Using a procedure called arthrocentesis, a physician might remove fluid accumulating in the affected joint. Another treatment is an oral appliance fitted by a dentist that fine-tunes the action of jaw muscles to form a more comfortable bite. An oral appliance, also known as a bite guard or stabilization splint, is a piece of plastic that fits over the top or bottom teeth. Very rarely, surgery may be required to repair or replace a joint.

TABLE 9.5	Muscles of Mastication AP|R			
Muscle	**Origin**	**Insertion**	**Action**	**Nerve Supply**
Masseter	Zygomatic arch	Lateral surface of ramus of mandible	Elevates and protracts mandible	Trigeminal nerve
Temporalis	Temporal bone	Coronoid process and anterior ramus of mandible	Elevates and retracts mandible	Trigeminal nerve
Medial pterygoid	Sphenoid, palatine, and maxillary bones	Medial surface of ramus of mandible	Elevates mandible and moves it from side to side	Trigeminal nerve
Lateral pterygoid	Sphenoid bone	Anterior surface of mandibular condyle	Depresses and protracts mandible and moves it from side to side	Trigeminal nerve

The **zygomaticus** (zi″go-mat′ik-us) **major** and **minor** run from the zygomatic arch downward to the corner of the mouth. When they contract, the corner of the mouth is drawn upward, such as in smiling or laughing.

The **platysma** (plah-tiz′mah) is a thin, sheetlike muscle whose fibers run from the chest upward over the neck to the face. It pulls the angle of the mouth downward, as in pouting. The platysma also helps open the mouth.

Muscles of Mastication

Four pairs of muscles attached to the mandible produce chewing movements. Three pairs of these muscles close the lower jaw, which happens in biting. The fourth pair of muscles of mastication can lower the jaw, cause side-to-side grinding motions of the mandible, and pull the mandible forward so that it protrudes. Figure 9.25 and reference plate 66 (p. 352) show the muscles of mastication, and table 9.5 lists them. They include the following:

Masseter   Medial pterygoid
Temporalis   Lateral pterygoid

The **masseter** (mas-se′ter) is a thick, flattened muscle that can be felt just in front of the ear when the teeth are clenched. Its fibers run downward from the zygomatic arch to the mandible. The masseter raises the jaw, but it can also control the rate at which the jaw falls open in response to gravity (fig. 9.25a).

The **temporalis** (tem-po-ra′lis) is a fan-shaped muscle located on the side of the skull above and in front of the ear. Its fibers, which also raise the jaw, pass downward beneath the zygomatic arch to the mandible (fig. 9.25a and b). Tensing this muscle is associated with temporomandibular joint syndrome, discussed in Clinical Application 9.3.

The **medial pterygoid** (ter′ĭ-goid) runs back and downward from the sphenoid, palatine, and maxillary bones to the ramus of the mandible. It closes the jaw (fig. 9.25c) and moves it from side to side.

The fibers of the **lateral pterygoid** run from the region just below the mandibular condyle forward to the sphenoid bone. This muscle can open the mouth, pull the mandible forward to make it protrude, and move the mandible from side to side (fig. 9.25c).

TABLE 9.6 Muscles That Move the Head and Vertebral Column AP|R

Muscle	Origin	Insertion	Action	Nerve Supply
Sternocleidomastoid	Anterior surface of sternum and upper surface of clavicle	Mastoid process of temporal bone	Each laterally flexes neck to the same side, rotates head to the opposite side. Together they flex neck forward; also aid in forceful inhalation.	Accessory, C2 and C3 cervical nerves
Splenius capitis	Ligamentum nuchae and spinous processes of seventh cervical and upper thoracic vertebrae	Occipital bone	Rotates head, laterally flexes and extends neck	Cervical nerves
Semispinalis capitis	Processes of lower cervical and upper thoracic vertebrae	Occipital bone	Extends neck, rotates head	Cervical and thoracic spinal nerves
Scalenes	Transverse processes of cervical vertebrae	Superior and lateral surfaces of first two ribs	Flex neck, elevate first two ribs during forceful inhalation	Cervical spinal nerves
Quadratus lumborum	Iliac crest	Upper lumbar vertebrae and twelfth rib	Extends lumbar region of vertebral column	Thoracic and lumbar spinal nerves
Erector spinae (divides into three groups)				
Iliocostalis (lateral) group				
Iliocostalis lumborum	Iliac crest	Lower six ribs	Extends lumbar region of vertebral column	Lumbar spinal nerves
Iliocostalis thoracis	Lower six ribs	Upper six ribs	Holds spine erect	Thoracic spinal nerves
Iliocostalis cervicis	Upper six ribs	Fourth through sixth cervical vertebrae	Extends cervical region of vertebral column	Cervical spinal nerves
Longissimus (intermediate) group				
Longissimus thoracis	Lumbar vertebrae	Thoracic and upper lumbar vertebrae and ribs 9 and 10	Extends thoracic region of vertebral column	Spinal nerves
Longissimus cervicis	Fourth and fifth thoracic vertebrae	Second through sixth cervical vertebrae	Extends cervical region of vertebral column	Spinal nerves
Longissimus capitis	Upper thoracic and lower cervical vertebrae	Mastoid process of temporal bone	Extends and rotates head	Cervical spinal nerves
Spinalis (medial) group				
Spinalis thoracis	Upper lumbar and lower thoracic vertebrae	Upper thoracic vertebrae	Extends vertebral column	Spinal nerves
Spinalis cervicis	Ligamentum nuchae and seventh cervical vertebra	Axis	Extends vertebral column	Spinal nerves

Muscles That Move the Head and Vertebral Column

Paired muscles in the neck and back flex and extend the neck, and rotate the head and hold the torso erect (figs. 9.26 and see fig. 9.28, table 9.6). They include the following:

Sternocleidomastoid   Quadratus lumborum
Splenius capitis   Erector spinae
Semispinalis capitis
Scalenes

The **sternocleidomastoid** (ster″no-kli″do-mas′toid) is a long muscle in the side of the neck that runs upward from the thorax to the base of the skull behind the ear. When the sternocleidomastoid on one side contracts, the face turns to the opposite side.

When both muscles contract, the head is pulled toward the chest. If other muscles fix the head in position, the sternocleidomastoids can raise the sternum, aiding forceful inhalation (see fig. 9.28 and table 9.6).

The **splenius capitis** (sple′ne-us kap′ĭ-tis) is a broad, straplike muscle in the back of the neck. It connects the base of the skull to the vertebrae in the neck and upper thorax. A splenius capitis acting singly rotates the head and laterally flexes the neck. Acting together, these muscles bring the head into an upright position (fig. 9.26 and table 9.6).

The **semispinalis capitis** (sem″e-spi-na′lis kap′ĭ-tis) is a broad, sheetlike muscle running upward from the vertebrae in the neck and thorax to the occipital bone and mastoid processes of the temporal bones. It extends and laterally flexes the neck or rotates the head (fig. 9.26 and table 9.6).

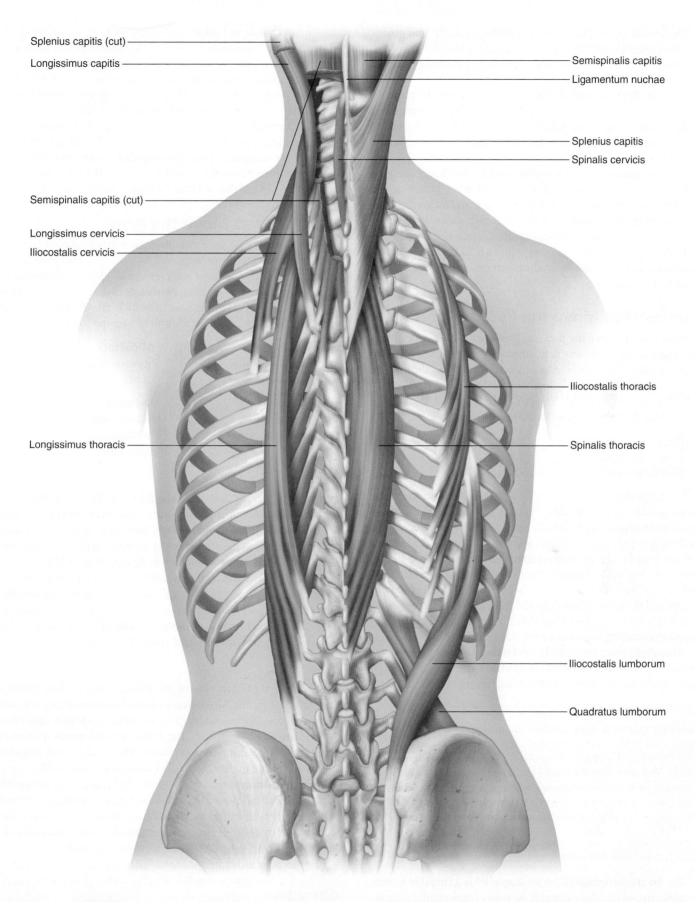

Splenius capitis (cut)

Longissimus capitis

Semispinalis capitis (cut)

Longissimus cervicis

Iliocostalis cervicis

Longissimus thoracis

Semispinalis capitis

Ligamentum nuchae

Splenius capitis

Spinalis cervicis

Iliocostalis thoracis

Spinalis thoracis

Iliocostalis lumborum

Quadratus lumborum

FIGURE 9.26 **AP|R** Deep muscles of the back and the neck help move the head (posterior view) and hold the torso erect. The splenius capitis and semispinalis capitis are removed on the left to show underlying muscles.

Three **scalene** (ska'len) muscles (anterior, middle, and posterior) are synergistic with the sternocleidomastoid for flexion of the neck. They also assist with forceful breathing by elevating the first two ribs (figs. 9.27 and **9.28**).

The **quadratus lumborum** (kwod-ra'tus lum-bo'rum) is located in the lumbar region. When the quadratus lumborum muscles on both sides contract, the vertebral column is extended. When the muscle on only one side contracts, the vertebral column is flexed laterally.

Erector spinae (ĕ-rek'tor spi'ne) muscles run longitudinally along the back, with origins and insertions at many places on the axial skeleton. These muscles extend the neck and rotate the head and maintain the erect position of the vertebral column. In the neck region they are aided by an elastic ligament (*ligamentum nuchae*), which limits flexion of the neck and helps to hold the head upright. Erector spinae can be subdivided into lateral, intermediate, and medial groups (table 9.6).

Muscles That Move the Pectoral Girdle

The muscles that move the pectoral girdle are closely associated with those that move the arm. A number of these chest and shoulder muscles connect the scapula to nearby bones and move the scapula upward, downward, forward, and backward (figs. 9.27, 9.28, and fig. 9.29; reference plates 68, 69, pp. 354–355; table 9.7). Muscles that move the pectoral girdle include the following:

Trapezius	Levator scapulae
Rhomboid major	Serratus anterior
Rhomboid minor	Pectoralis minor

The **trapezius** (trah-pe'ze-us) is a large, triangular muscle in the upper back that runs horizontally from the base of the skull and the cervical and thoracic vertebrae to the shoulder. Its fibers are organized into three groups—upper, middle, and lower. Together these fibers rotate the scapula. The upper fibers acting alone raise the scapula and shoulder, such as when shrugging the shoulders to express a feeling of indifference. The middle fibers pull the scapula toward the vertebral column, and the lower fibers draw the scapula and shoulder downward. When other muscles fix the shoulder in position, the trapezius can pull the head backward or to one side (see fig. 9.24).

Rhomboid (rom-boid') **major** and **minor** connect the vertebral column to the scapula. Both retract and elevate the scapula. Rhomboid major can also rotate the scapula downward (see fig. 9.27).

A small, triangular region, called the *triangle of auscultation*, is located in the back where the trapezius overlaps the superior border of the latissimus dorsi and the underlying rhomboid major. This area, near the medial border of the scapula, enlarges when a person bends forward with the arms folded across the chest. By placing the bell of a stethoscope in the triangle of auscultation, a physician can usually clearly hear the sounds of the respiratory organs.

The **levator scapulae** (le-va'tor scap'u-le) is a straplike muscle that runs almost vertically through the neck, connecting the cervical vertebrae to the scapula. It elevates the scapula (see figs. 9.27 and 9.29).

The **serratus anterior** (ser-ra'tus an-te're-or) is a broad, curved muscle located on the side of the chest. It arises as fleshy, narrow strips on the upper ribs and continues along the medial wall of the axilla to the ventral surface of the scapula. It pulls the scapula downward and anteriorly and is used to thrust the shoulder forward, as when pushing something (see fig. 9.28).

The **pectoralis** (pek'tor-a'lis) **minor** is a thin, flat muscle that lies beneath the larger pectoralis major. It runs laterally and upward from the ribs to the scapula and pulls the scapula forward and downward. When other muscles fix the scapula in position, the pectoralis minor can raise the ribs and thus aid forceful inhalation (see fig. 9.28).

Muscles That Move the Arm

The arm is one of the more freely movable parts of the body because muscles connect the humerus to regions of the pectoral girdle, ribs, and vertebral column. These muscles can be grouped according to their primary actions—flexion, extension, abduction, and rotation (figs. 9.29, **9.30, 9.31**; reference plates 67, 68, 69, pp. 353–355; **table 9.8**). Muscles that move the arm include the following:

Flexors	**Abductors**
Coracobrachialis	Supraspinatus
Pectoralis major	Deltoid
Extensors	**Rotators**
Teres major	Subscapularis
Latissimus dorsi	Infraspinatus
	Teres minor

Flexors

The movement of flexion of the shoulder may be less obvious than at other joints. Movements of the humerus forward and upward flex the shoulder. The **coracobrachialis** (kor"ah-ko-bra'ke-al-is) runs from the scapula to the middle of the humerus along its medial surface. It flexes and adducts the arm (see figs. 9.30 and 9.31).

The **pectoralis major** is a thick, fan-shaped muscle in the upper chest. Its fibers run from the center of the thorax through the armpit to the humerus. This muscle primarily pulls the arm forward and across the chest. It can also rotate the humerus medially and adduct the arm from a raised position (see fig. 9.28).

Extensors

The movement of extension of the shoulder may be less obvious than at other joints. Movements of the humerus backward and upward extend the shoulder. The shoulder is already extended in the anatomical position. The **teres** (te'rēz) **major** connects the scapula to the humerus. It extends the shoulder and can also adduct and rotate the arm medially (see figs. 9.27 and 9.29).

The **latissimus dorsi** (lah-tis'i-mus dor'si) is a wide, triangular muscle that curves upward from the lower back, around the side, and to the armpit. It can extend the shoulder and adduct the arm and rotate the humerus medially. It also pulls the shoulder downward and back. The actions of pulling the arm back in swimming, climbing, and rowing use this muscle (see figs. 9.27 and 9.30).

Abductors

The **supraspinatus** (su"prah-spi'na-tus) is located in the depression above the spine of the scapula on its posterior surface. It

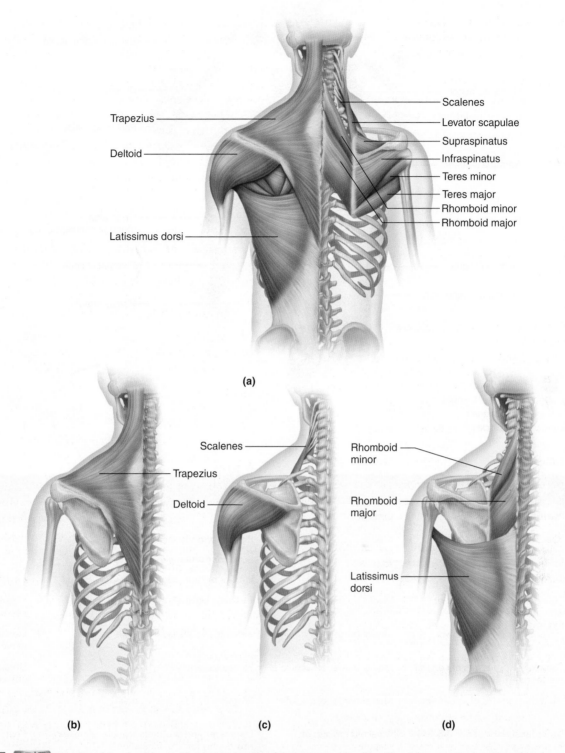

FIGURE 9.27 AP|R Muscles of the shoulder and back. (*a*) Muscles of the posterior shoulder. The right trapezius is removed to show underlying muscles. Isolated views of (*b*) trapezius, (*c*) deltoid, and (*d*) rhomboid and latissimus dorsi muscles.

connects the scapula to the greater tubercle of the humerus and abducts the arm (see figs. 9.27 and 9.29).

The **deltoid** (del′toid) is a thick, triangular muscle that covers the shoulder joint. It connects the clavicle and scapula to the lateral side of the humerus and abducts the arm. The deltoid's posterior fibers can extend the shoulder, and its anterior fibers can flex the shoulder (see fig. 9.27).

A humerus fractured at its surgical neck may damage the axillary nerve that supplies the deltoid muscle (see fig. 7.44, p. 236). If this occurs, the muscle is likely to shrink and weaken. To test the deltoid for such weakness, a physician may ask a patient to abduct the arm against some resistance and maintain that position for a time.

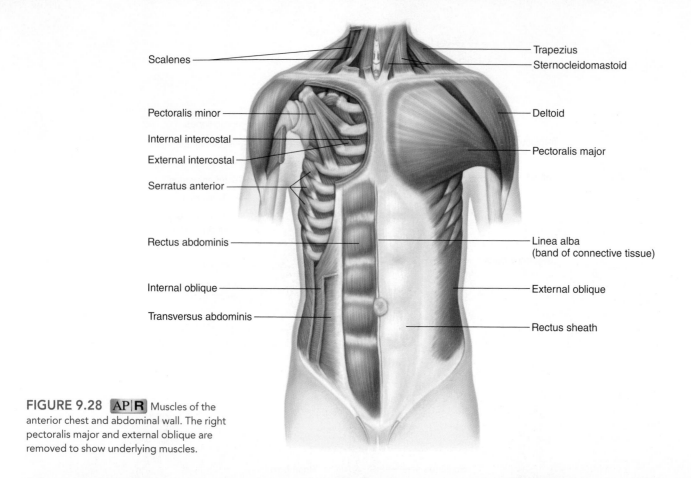

FIGURE 9.28 AP|R Muscles of the anterior chest and abdominal wall. The right pectoralis major and external oblique are removed to show underlying muscles.

Labels (clockwise from top left):
Scalenes — Trapezius — Sternocleidomastoid — Pectoralis minor — Deltoid — Internal intercostal — Pectoralis major — External intercostal — Serratus anterior — Rectus abdominis — Linea alba (band of connective tissue) — Internal oblique — External oblique — Transversus abdominis — Rectus sheath

TABLE 9.7	Muscles That Move the Pectoral Girdle AP|R			
Muscle	**Origin**	**Insertion**	**Action**	**Nerve Supply**
Trapezius	Occipital bone, ligamentum nuchae, and spinous processes of seventh cervical and all thoracic vertebrae	Clavicle, spine, and acromion process of scapula	Rotates and retracts scapula. Superior portion elevates scapula. Inferior portion depresses scapula	Accessory nerve
Rhomboid major	Spinous processes of upper thoracic vertebrae	Medial border of scapula	Retracts, elevates, and rotates scapula	Dorsal scapular nerve
Rhomboid minor	Spinous processes of lower cervical vertebrae	Medial border of scapula	Retracts and elevates scapula	Dorsal scapular nerve
Levator scapulae	Transverse processes of cervical vertebrae	Medial border of scapula	Elevates scapula	Dorsal scapular and cervical nerves
Serratus anterior	Anterior surfaces of upper ribs	Medial border of scapula	Protracts and rotates scapula	Long thoracic nerve
Pectoralis minor	Anterior surface of ribs 3-5	Coracoid process of scapula	Depresses and protracts scapula, elevates ribs during forceful inhalation	Pectoral nerve

Rotators

The **subscapularis** (sub-scap'u-lar-is) is a large, triangular muscle that covers the anterior surface of the scapula. It connects the scapula to the humerus and rotates the arm medially (fig. 9.31).

The **infraspinatus** (in"frah-spi'na-tus) occupies the depression below the spine of the scapula on its posterior surface. The fibers of this muscle attach the scapula to the humerus and rotate the arm laterally (see fig. 9.29).

The **teres minor** is a small muscle connecting the scapula to the humerus. It rotates the arm laterally (see figs. 9.27 and 9.29).

Muscles That Move the Forearm

Most forearm movements are produced by muscles that connect the radius or ulna to the humerus or pectoral girdle. A group of muscles located along the anterior surface of the humerus flexes the forearm at the elbow, whereas a single posterior muscle

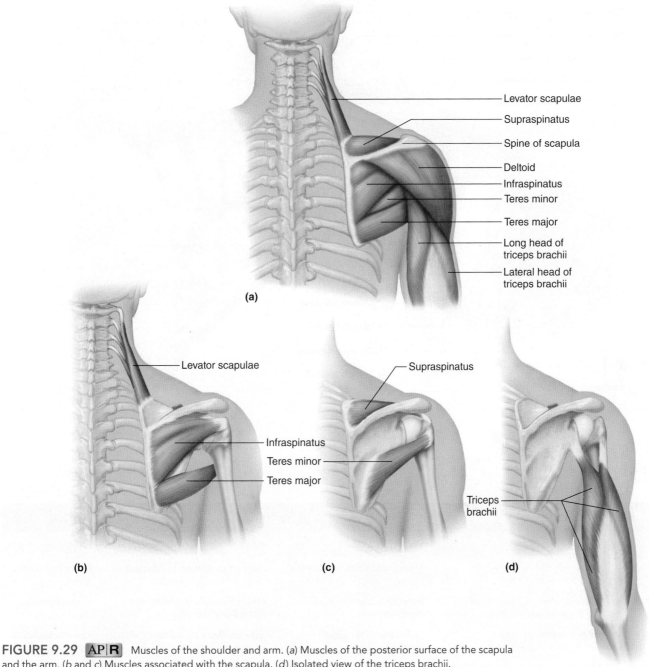

FIGURE 9.29 AP|R   Muscles of the shoulder and arm. (*a*) Muscles of the posterior surface of the scapula and the arm. (*b* and *c*) Muscles associated with the scapula. (*d*) Isolated view of the triceps brachii.

extends this joint. Other muscles cause movements at the radioulnar joint and rotate the forearm.

The muscles that move the forearm are shown in figures 9.31, 9.32, 9.33, and 9.34, in reference plates 68, 69, and 70 (pp. 354–356), and are listed in table 9.9, grouped according to their primary actions. They include the following:

Flexors	Extensor	Rotators
Biceps brachii	Triceps brachii	Supinator
Brachialis		Pronator teres
Brachioradialis		Pronator quadratus

Flexors

The **biceps brachii** (bi′seps bra′ke-i) is a fleshy muscle that forms a long, rounded mass on the anterior side of the arm. It connects the scapula to the radius and flexes the elbow and rotates the hand laterally (supination), as when a person turns a doorknob or screwdriver (see fig. 9.31).

The **brachialis** (bra′ke-al-is) is a large muscle beneath the biceps brachii. It connects the shaft of the humerus to the ulna and is the strongest flexor of the elbow (see fig. 9.31).

The **brachioradialis** (bra″ke-o-ra″de-a′lis) connects the humerus to the radius. It aids in flexing the elbow (see fig. 9.32).

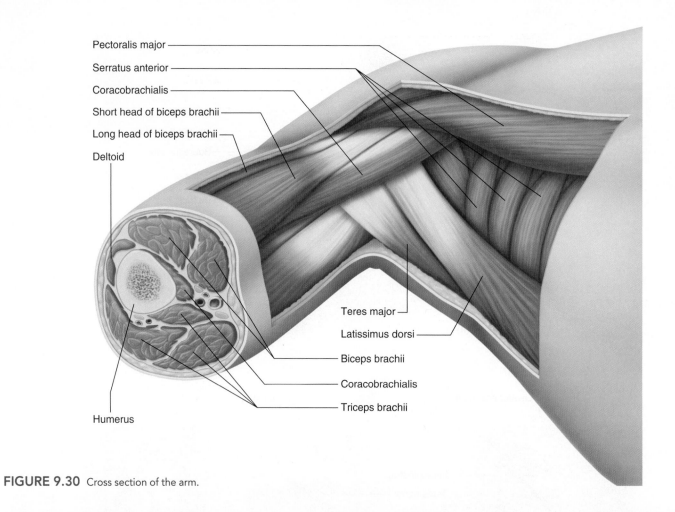

Pectoralis major

Serratus anterior

Coracobrachialis

Short head of biceps brachii

Long head of biceps brachii

Deltoid

Humerus

Teres major

Latissimus dorsi

Biceps brachii

Coracobrachialis

Triceps brachii

FIGURE 9.30 Cross section of the arm.

TABLE 9.8	Muscles That Move the Arm APᏒ			
Muscle	**Origin**	**Insertion**	**Action**	**Nerve Supply**
Coracobrachialis	Coracoid process of scapula	Medial mid-shaft of humerus	Flexes shoulder and adducts arm	Musculocutaneus nerve
Pectoralis major	Clavicle, sternum, and costal cartilages of upper ribs	Intertubercular sulcus of humerus	Flexes shoulder, adducts, and rotates arm medially	Pectoral nerve
Teres major	Lateral border of scapula	Intertubercular sulcus of humerus	Extends shoulder, adducts, and rotates arm medially	Lower subscapular nerve
Latissimus dorsi	Spines of sacral, lumbar, and lower thoracic vertebrae, iliac crest, and lower ribs	Intertubercular sulcus of humerus	Extends shoulder, adducts, and rotates arm medially, or depresses and retracts shoulder	Thoracodorsal nerve
Supraspinatus	Supraspinous fossa of scapula	Greater tubercle of humerus	Abducts arm	Suprascapular nerve
Deltoid	Acromion process, spine of the scapula, and the clavicle	Deltoid tuberosity of humerus	Lateral portion abducts arm Anterior portion flexes shoulder Posterior portion extends shoulder	Axillary nerve
Subscapularis	Anterior surface of scapula	Lesser tubercle of humerus	Rotates arm medially	Subscapular nerve
Infraspinatus	Infraspinous fossa of scapula	Greater tubercle of humerus	Rotates arm laterally	Upper and lower suprascapular nerves
Teres minor	Lateral border of scapula	Greater tubercle of humerus	Rotates arm laterally	Axillary nerve

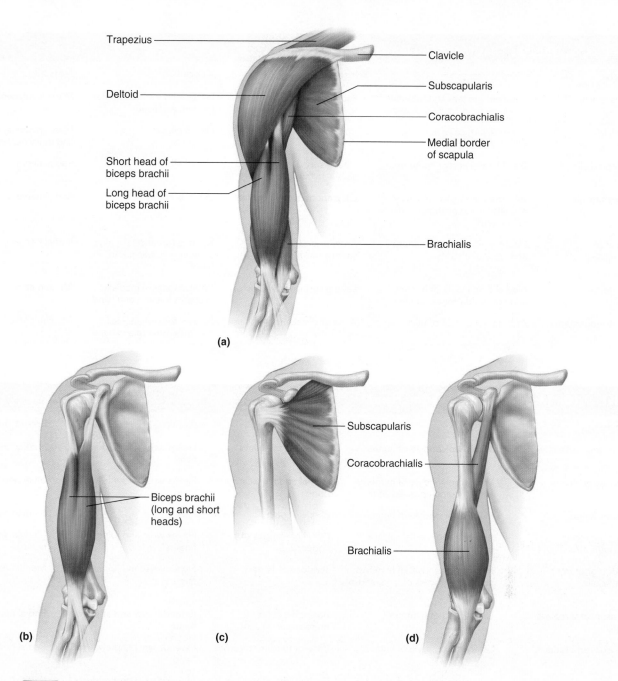

Trapezius

Clavicle

Deltoid

Subscapularis

Coracobrachialis

Medial border
of scapula

Short head of
biceps brachii

Long head of
biceps brachii

Brachialis

(a)

Biceps brachii
(long and short
heads)

Subscapularis

Coracobrachialis

Brachialis

(b)　　　　　　　　(c)　　　　　　　　(d)

FIGURE 9.31 AP|R   Muscles of the shoulder and arm. (*a*) Muscles of the anterior shoulder and the arm, with the rib cage removed. (*b, c,* and *d*) Isolated views of muscles associated with the arm.

Extensor

The **triceps brachii** (tri´seps bra´ke-i) has three heads and is the only muscle on the back of the arm. It connects the humerus and scapula to the ulna and is the primary extensor of the elbow (see figs. 9.29 and 9.30).

Rotators

The **supinator** (su´pĭ-na-tor) is a short muscle whose fibers run from the ulna and the lateral end of the humerus to the radius. It assists the biceps brachii in rotating the forearm laterally, such as when turning the hand so the palm faces upward (supination) (see fig. 9.32).

The **pronator teres** (pro-na´tor te´rēz) is a short muscle connecting the ends of the humerus and ulna to the radius. It rotates the arm medially, such as when turning the hand so the palm faces downward (pronation) (see fig. 9.32).

The **pronator quadratus** (pro-na´tor kwod-ra´tus) runs from the distal end of the ulna to the distal end of the radius. It assists the pronator teres in rotating the arm medially (see fig. 9.32).

Muscles That Move the Hand

Movements of the hand include movements of the wrist and fingers. Many of the implicated muscles originate from the distal end of the humerus and from the radius and ulna. The two major

TABLE 9.9 Muscles That Move the Forearm AP|R

Muscle	Origin	Insertion	Action	Nerve Supply
Biceps brachii	Coracoid process and tubercle above glenoid cavity of scapula	Radial tuberosity of radius	Flexes elbow and supinates forearm and hand	Musculocutaneous nerve
Brachialis	Anterior surface of humerus	Coronoid process of ulna	Flexes elbow	Musculocutaneous, median, and radial nerves
Brachioradialis	Distal lateral end of humerus	Lateral surface of radius above styloid process	Flexes elbow	Radial nerve
Triceps brachii	Tubercle below glenoid cavity and lateral and posterior surfaces of humerus	Olecranon process of ulna	Extends elbow	Radial nerve
Supinator	Lateral epicondyle of humerus and crest of ulna	Anterior and lateral surface of radius	Rotates forearm laterally; supinates forearm and hand	Radial nerve
Pronator teres	Medial epicondyle of humerus and coronoid process of ulna	Lateral surface of radius	Rotates forearm medially; pronates forearm and hand	Median nerve
Pronator quadratus	Anterior distal end of ulna	Anterior distal end of radius	Rotates forearm medially; pronates forearm and hand	Median nerve

TABLE 9.10 Muscles That Move the Hand

Muscle	Origin	Insertion	Action	Nerve Supply
Flexor carpi radialis	Medial epicondyle of humerus	Base of second and third metacarpals	Flexes wrist and abducts hand	Median nerve
Flexor carpi ulnaris	Medial epicondyle of humerus and olecranon process of ulna	Carpal and metacarpal bones	Flexes wrist and adducts hand	Ulnar nerve
Palmaris longus	Medial epicondyle of humerus	Fascia of palm	Flexes wrist	Median nerve
Flexor digitorum profundus	Anterior and medial surface of ulna	Distal phalanges of fingers 2–5	Flexes wrist and joints of fingers	Median and ulnar nerves
Flexor digitorum superficialis	Medial epicondyle of humerus, coronoid process of ulna, and radius	Tendons of fingers	Flexes wrist and joints of fingers	Median nerve
Extensor carpi radialis longus	Lateral end of humerus	Base of second metacarpal	Extends wrist and abducts hand	Radial nerve
Extensor carpi radialis brevis	Lateral epicondyle of humerus	Base of third metacarpal	Extends wrist and abducts hand	Radial nerve
Extensor carpi ulnaris	Lateral epicondyle of humerus and proximal, posterior ulna	Base of fifth metacarpal	Extends wrist and adducts hand	Radial nerve
Extensor digitorum	Lateral epicondyle of humerus	Posterior surface of phalanges in fingers 2–5	Extends joints of fingers	Radial nerve

groups of these muscles are flexors on the anterior side of the forearm and extensors on the posterior side. Figures 9.32, 9.33, 9.34, and reference plate 70 (p. 356) show these muscles, and table 9.10 lists them. The muscles that move the hand include the following:

Flexors
Flexor carpi radialis
Flexor carpi ulnaris
Palmaris longus
Flexor digitorum profundus
Flexor digitorum superficialis

Extensors
Extensor carpi radialis longus
Extensor carpi radialis brevis
Extensor carpi ulnaris
Extensor digitorum

Flexors

The **flexor carpi radialis** (flek'sor kar-pi' ra"de-a'lis) is a fleshy muscle located medially on the anterior side of the forearm. It runs from the distal end of the humerus into the hand, where it is attached to metacarpal bones. The flexor carpi radialis flexes the wrist and abducts the hand (see fig. 9.32).

The **flexor carpi ulnaris** (flek'sor kar-pi' ul-na'ris) is located along the medial border of the forearm. It connects the distal end of the humerus and the proximal end of the ulna to carpal and metacarpal bones. It flexes the wrist and adducts the hand (see fig. 9.32).

The **palmaris longus** (pal-ma′ris long′gus) is a slender muscle located on the medial side of the forearm between the flexor carpi radialis and the flexor carpi ulnaris. It connects the distal end of the humerus to fascia of the palm and flexes the wrist (see fig. 9.32).

The **flexor digitorum profundus** (flek′sor dij″ĭ-to′rum pro-fun′dus) is a large muscle that connects the ulna to the distal phalanges. It flexes the distal joints of the fingers, such as when making a fist (see fig. 9.34).

The **flexor digitorum superficialis** (flek′sor dij″ĭ-to′rum su″per-fish″e-a′lis) is a large muscle located beneath the flexor carpi ulnaris. It arises by three heads—one from the medial epicondyle of the humerus, one from the medial side of the ulna, and one from the radius. It is inserted in the tendons of the fingers and flexes the fingers and, by a combined action, flexes the wrist (see fig. 9.32).

More than thirty muscles provide fingertip dexterity and hand movements. To track the movements required to send a text message, researchers recorded the electrical activity and fingertip force in seven muscles of the index fingers of volunteers as they pushed their fingers against a surface. The researchers observed two patterns of muscle activation, indicating two types of movement—light tapping from an angle versus direct downward pressure on one key. The act of texting entails a key-locating "tap" followed by a more direct push. The switch from one type of movement to another is so fast and fluid that we are not aware of it.

Extensors

The **extensor carpi radialis longus** (eks-ten′sor kar-pi′ra″de-a′lis long′gus) runs along the lateral side of the forearm, connecting the humerus to the hand. It extends the wrist and assists in abducting the hand (see figs. 9.33 and 9.34).

The **extensor carpi radialis brevis** (eks-ten′sor kar-pi′ ra″de-a′lis brev′ĭs) is a companion of the extensor carpi radialis longus and is located medially to it. This muscle runs from the humerus to the metacarpal bones and extends the wrist. It also assists in abducting the hand (see figs. 9.33 and 9.34).

The **extensor carpi ulnaris** (eks-ten′sor kar-pi′ ul-na′ris) is located along the posterior surface of the ulna and connects the humerus to the hand. It extends the wrist and assists in adducting the hand (see figs. 9.33 and 9.34).

The **extensor digitorum** (eks-ten′sor dij″ĭ-to′rum) runs medially along the back of the forearm. It connects the humerus to the posterior surface of the phalanges and extends the fingers (see figs. 9.33 and 9.34).

A structure called the *extensor retinaculum* consists of a group of heavy connective tissue fibers in the fascia of the wrist (see fig. 9.33). It connects the lateral margin of the radius with the medial border of the styloid process of the ulna and certain bones of the wrist. The retinaculum gives off branches of connective tissue to the underlying wrist bones, creating a series of sheathlike compartments through which the tendons of the extensor muscles pass to the wrist and fingers.

Muscles of the Abdominal Wall

The walls of the chest and pelvic regions are supported directly by bone, but those of the abdomen are not. Instead, the anterior and lateral walls of the abdomen are composed of layers of broad, flattened muscles. These muscles connect the rib cage and vertebral column to the pelvic girdle. A band of tough connective tissue, called the **linea alba** (lin′e-ah al′bah), extends from the xiphoid process of the sternum to the pubic symphysis. It is an attachment for some of the abdominal wall muscles.

Contraction of the abdominal muscles decreases the volume of the abdominal cavity and increases the pressure inside, which is transferred upwards to the thoracic cavity. This action helps move air out of the lungs during forceful exhalation and also aids in defecation, urination, vomiting, and childbirth.

The abdominal wall muscles are shown in figure 9.35, reference plate 67 (p. 353), and are listed in table 9.11. They include the following:

External oblique	Transversus abdominis
Internal oblique	Rectus abdominis

The **external oblique** (eks-ter′nal ŏ-blēk), is a broad, thin sheet of muscle whose fibers slant downward from the lower ribs to the pelvic girdle and the linea alba. When this muscle contracts, it tenses the abdominal wall and compresses the contents of the abdominal cavity. Similar in structure and function is the **internal oblique** (in-ter′nal ŏ-blēk) which is a broad, thin sheet of muscle beneath the external oblique. Its fibers run up and forward from the pelvic girdle to the lower ribs.

The **transversus abdominis** (trans-ver′sus ab-dom′ĭ-nis) forms a third layer of muscle beneath the external and internal obliques. Its fibers run horizontally from the lower ribs, lumbar vertebrae, and ilium to the linea alba and pubic bones. It functions in the same manner as the external and internal obliques.

The **rectus abdominis** (rek′tus ab-dom′ĭ-nis) is a long, strap-like muscle that connects the pubic bones to the ribs and sternum. Three or more fibrous bands cross the muscle transversely, giving it a segmented appearance. The muscle functions with other abdominal wall muscles to compress the contents of the abdominal cavity, and it also helps to flex the vertebral column.

Muscles of the Pelvic Floor

Two muscular sheets form the floor of the pelvis—a deeper **pelvic diaphragm** and a more superficial **urogenital diaphragm.** The pelvic diaphragm forms the floor of the pelvic cavity, and the urogenital diaphragm fills the space within the pubic arch. Figure 9.36 and table 9.12 show the muscles of the male and female pelvic floors. They include the following:

Pelvic Diaphragm	**Urogenital Diaphragm**
Levator ani	Superficial transversus perinei
Coccygeus	Bulbospongiosus
	Ischiocavernosus
	Sphincter urethrae

Pelvic Diaphragm

The **levator ani** (le-va′tor ah-ni′) muscles form a thin sheet across the pelvic outlet. They are connected at the midline posteriorly by

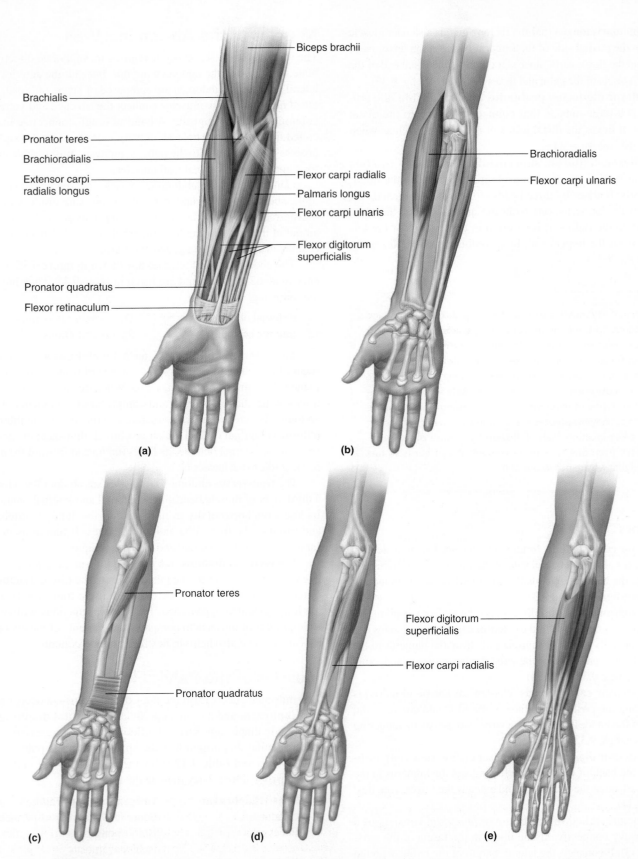

FIGURE 9.32 AP|R Muscles of the arm and forearm. (*a*) Muscles of the anterior forearm. (*b–e*) Isolated views of muscles associated with the anterior forearm.

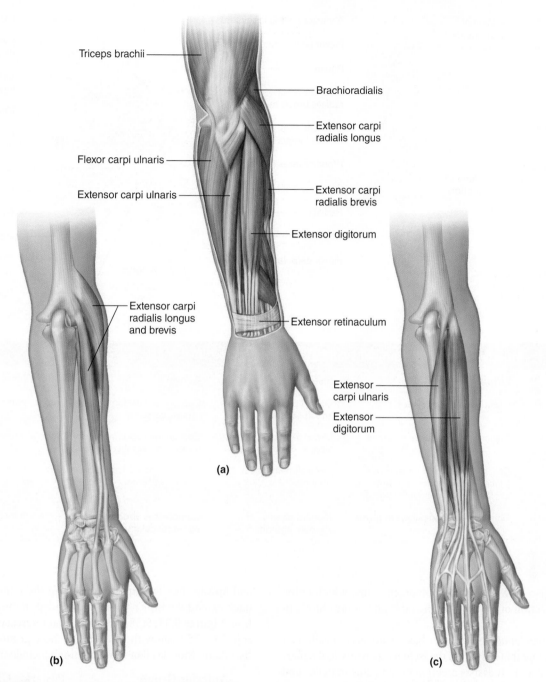

Triceps brachii

Flexor carpi ulnaris

Extensor carpi ulnaris

Extensor carpi radialis longus and brevis

Brachioradialis

Extensor carpi radialis longus

Extensor carpi radialis brevis

Extensor digitorum

Extensor retinaculum

Extensor carpi ulnaris

Extensor digitorum

(a)

(b)

(c)

FIGURE 9.33 **AP|R** Muscles of the arm and forearm. (*a*) Muscles of the posterior forearm. (*b* and *c*) Isolated views of muscles associated with the posterior forearm.

a *central tendon* that extends from the tip of the coccyx to the anal canal. Anteriorly, they are separated in the male by the urethra and the anal canal, and in the female by the urethra, vagina, and anal canal. These muscles help support the pelvic viscera and provide sphincterlike action in the anal canal and vagina.

An *external anal sphincter* under voluntary control and an *internal anal sphincter* formed of involuntary muscle fibers of the intestine encircle the anal canal and keep it closed.

The **coccygeus** (kok-sij′e-us) is a fan-shaped muscle that spans from the ischial spine to the coccyx and sacrum. It aids the levator ani.

Urogenital Diaphragm

The **superficial transversus perinei** (soo′per-fish′al trans-ver′sus per″ĭ-ne′i) consists of a small bundle of muscle fibers that passes medially from the ischial tuberosity along the posterior border of the urogenital diaphragm. It assists other muscles in supporting the pelvic viscera.

In the male, the **bulbospongiosus** (bul″bo-spon″je-o′sus) muscles are united surrounding the base of the penis and assist in emptying the urethra. In the female, these muscles are separated by the vagina medially, and constrict the vaginal opening. These

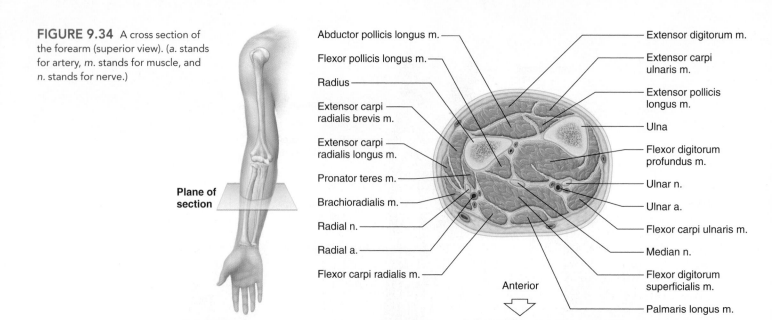

FIGURE 9.34 A cross section of the forearm (superior view). (*a.* stands for artery, *m.* stands for muscle, and *n.* stands for nerve.)

Abductor pollicis longus m.
Flexor pollicis longus m.
Radius
Extensor carpi radialis brevis m.
Extensor carpi radialis longus m.
Pronator teres m.
Brachioradialis m.
Radial n.
Radial a.
Flexor carpi radialis m.

Extensor digitorum m.
Extensor carpi ulnaris m.
Extensor pollicis longus m.
Ulna
Flexor digitorum profundus m.
Ulnar n.
Ulnar a.
Flexor carpi ulnaris m.
Median n.
Flexor digitorum superficialis m.
Palmaris longus m.

Anterior

TABLE 9.11	Muscles of the Abdominal Wall AP|R			
Muscle	Origin	Insertion	Action	Nerve Supply
External oblique	Outer surfaces of lower 8 ribs	Outer lip of iliac crest and linea alba	Compresses abdomen, flexes and rotates vertebral column	Intercostal nerves 7–12
Internal oblique	Iliac crest and inguinal ligament	Lower 3-4 ribs, linea alba, and crest of pubis	Compresses abdomen, flexes and rotates vertebral column	Intercostal nerves 7–12
Transversus abdominis	Costal cartilages of lower 6 ribs, processes of lumbar vertebrae, lip of iliac crest, and inguinal ligament	Linea alba and crest of pubis	Compresses abdomen, flexes and rotates vertebral column	Intercostal nerves 7–12
Rectus abdominis	Crest of pubis and pubic symphysis	Xiphoid process of sternum and ribs 5-7	Compresses abdomen, flexes vertebral column	Intercostal nerves 7–12

muscles can also slow the flow of blood in veins, which helps maintain an erection of the penis in the male and of the clitoris in the female.

The **ischiocavernosus** (is″ke-o-kav″er-no′sus) muscle is a tendinous structure that runs from the ischial tuberosity to the margin of the pubic arch. It assists erection of the penis in males and the clitoris in females.

The **sphincter urethrae** (sfingk′ter u-re′thrē) are muscles that arise from the margins of the pubic and ischial bones. Each arches around the urethra and unites with the one on the other side. Together they act as a sphincter that closes the urethra by compression and opens it by relaxation, helping control the flow of urine.

Muscles That Move the Thigh

The muscles that move the thigh are attached to the femur and to part of the pelvic girdle. (Important exceptions are the sartorius and rectus femoris, described later.) The muscles can be separated into an anterior group that primarily flexes the hip and a posterior group that extends the hip and abducts or rotates the thigh.

The movement of flexion and extension of the hip may be less obvious than at other joints. Movements of the femur forward and upward flex the hip. Movement of the femur down or to the back extend the hip. The hip is extended in the anatomical position. **Figures 9.37, 9.38, 9.39, 9.40** and reference plates 71 and 72 (pp. 356–357) show the muscles in these groups, and **table 9.13** lists them. Muscles that move the thigh include the following:

Anterior Group	Posterior Group
Psoas major	Gluteus maximus
Iliacus	Gluteus medius
	Gluteus minimus
	Piriformis
	Tensor fasciae latae

Another group of muscles, attached to the femur and pelvic girdle, adducts the thigh. This group includes the following:

Pectineus	Adductor magnus
Adductor brevis	Gracilis
Adductor longus	

Anterior Group

The **psoas** (so′as) **major** is a long, thick muscle that connects the lumbar vertebrae to the femur. It flexes the hip (see fig. 9.37).

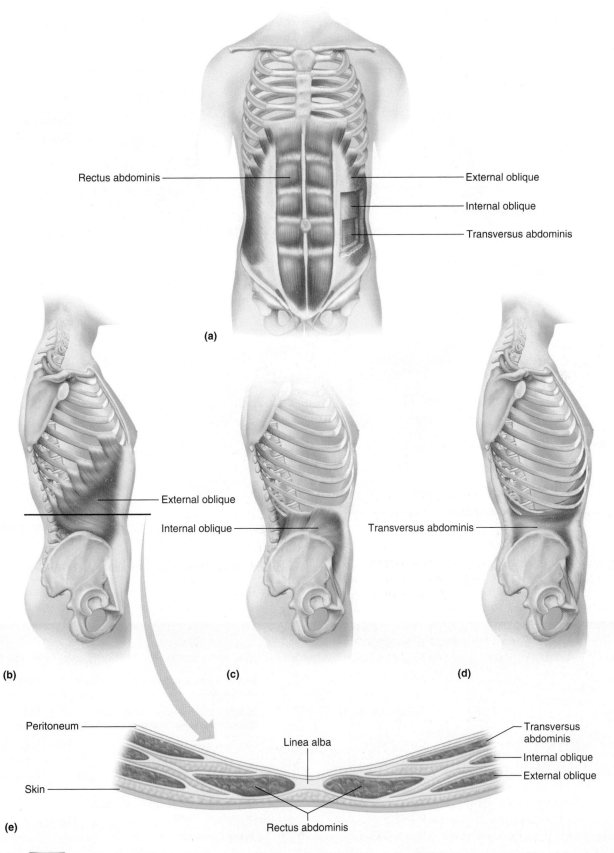

FIGURE 9.35 AP|R Muscles of the abdominal wall. (*a–d*) Isolated muscles of the abdominal wall. (*e*) Transverse section through the abdominal wall.

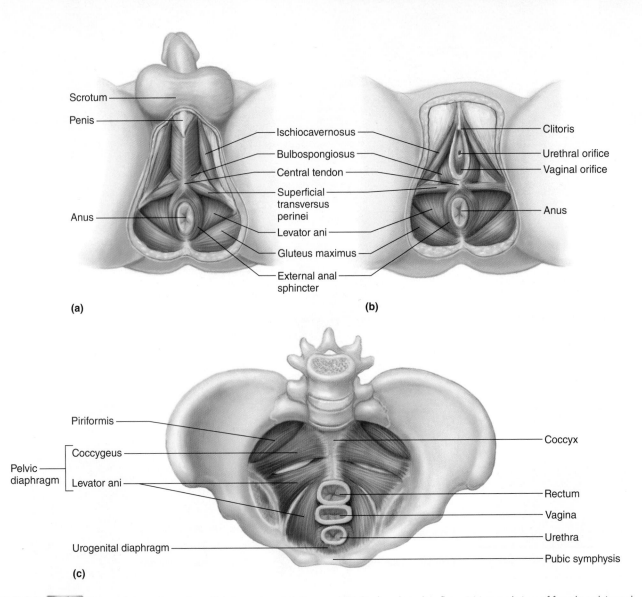

FIGURE 9.36 **AP|R** External view of muscles of (*a*) the male pelvic floor and (*b*) the female pelvic floor. (*c*) Internal view of female pelvic and urogenital diaphragms.

TABLE 9.12	Muscles of the Pelvic Floor				
Muscle	**Origin**	**Insertion**	**Action**	**Nerve Supply**
Levator ani	Pubis and ischial spine	Coccyx	Supports pelvic viscera and compresses anal canal	Pudendal nerve
Coccygeus	Ischial spine	Sacrum and coccyx	Supports pelvic viscera and provides sphincterlike action in anal canal and vagina	S4 and S5 nerves
Superficial transversus perinei	Ischial tuberosity	Central tendon	Supports pelvic viscera	Pudendal nerve
Bulbospongiosus	Central tendon	Males: Corpus cavernosa of penis	Males: Assists emptying of urethra and assists in erection of penis	Pudendal nerve
		Females: Corpus cavernosa of clitoris	Females: Constricts vagina and assists in erection of clitoris	
Ischiocavernosus	Ischial tuberosity	Males: Corpus cavernosa of penis	Males: Contributes to erection of the penis	Pudendal nerve
		Females: Corpus cavernosa of clitoris	Females: Contributes to erection of the clitoris	
Sphincter urethrae	Margins of pubis and ischium	Fibers of each unite with those from other side	Closes urethra	Pudendal nerve

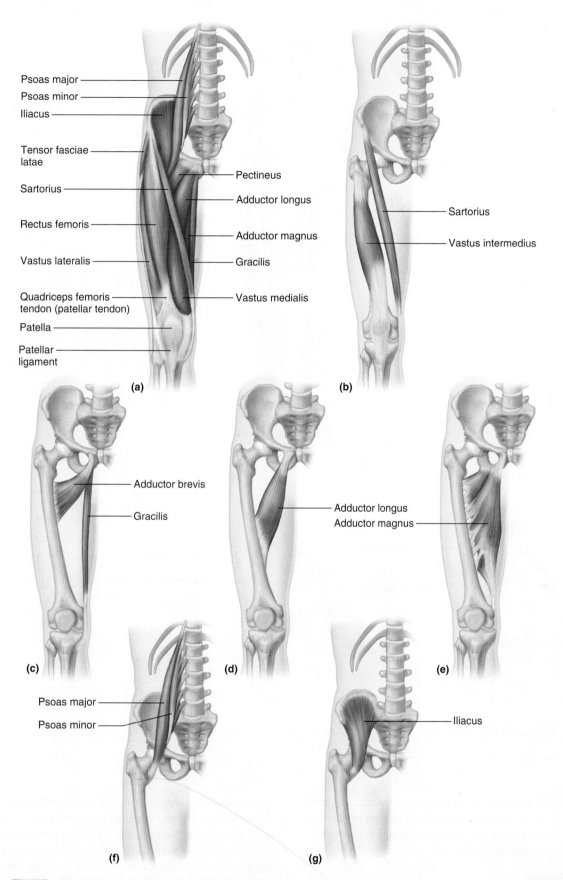

FIGURE 9.37 [AP|R] Muscles of the thigh and leg. (*a*) Muscles of the anterior right thigh. Isolated views of (*b*) the vastus intermedius, (*c–e*) adductors of the thigh, (*f–g*) flexors of the hip.

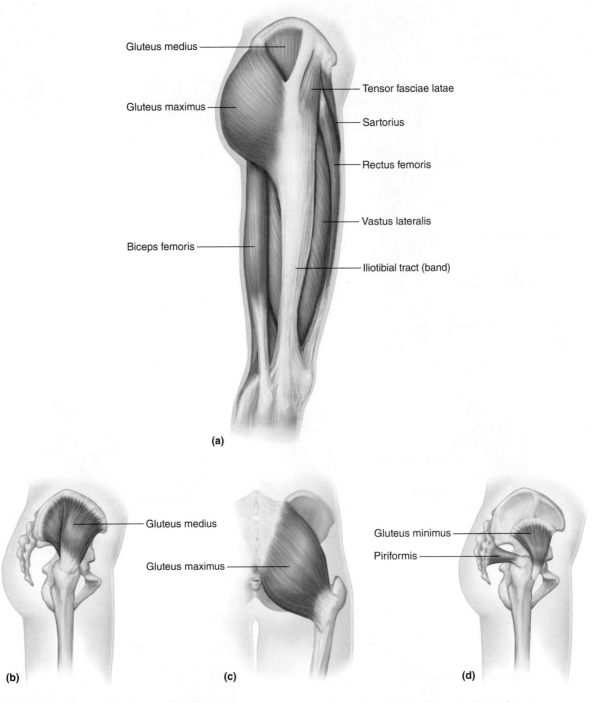

Gluteus medius

Gluteus maximus

Biceps femoris

Tensor fasciae latae

Sartorius

Rectus femoris

Vastus lateralis

Iliotibial tract (band)

(a)

Gluteus medius

Gluteus maximus

(b)

(c)

Gluteus minimus

Piriformis

(d)

FIGURE 9.38 Muscles of the thigh and leg. (*a*) Muscles of the lateral right thigh. (*b–d*) Isolated views of the gluteal muscles.

The **iliacus** (il′e-ak-us), a large, fan-shaped muscle, lies along the lateral side of the psoas major. The iliacus and the psoas major are the primary flexors of the hip, and they advance the lower limb in walking movements (see fig. 9.37).

Posterior Group

The **gluteus maximus** (gloo′te-us mak′si-mus) is the largest muscle in the body and covers a large part of each buttock. It connects the ilium, sacrum, and coccyx to the femur by fascia of the thigh

and extends the hip. The gluteus maximus helps to straighten the lower limb at the hip when a person walks, runs, or climbs. It is also used to raise the body from a sitting position (see fig. 9.38).

The **gluteus medius** (gloo′te-us me′de-us) is partly covered by the gluteus maximus. Its fibers run from the ilium to the femur, and they abduct the thigh and rotate it medially (see fig. 9.38).

The **gluteus minimus** (gloo′te-us min′ĭ-mus) lies beneath the gluteus medius and is its companion in attachments and functions (see fig. 9.38).

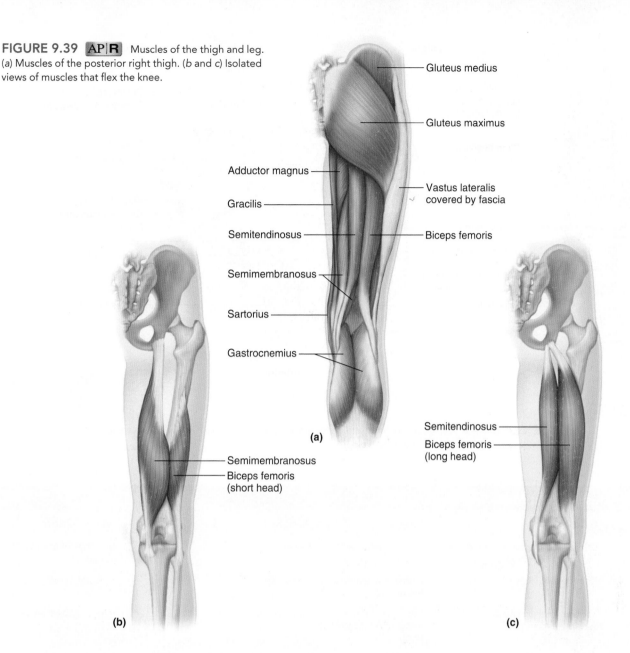

FIGURE 9.39 AP|R Muscles of the thigh and leg. (*a*) Muscles of the posterior right thigh. (*b* and *c*) Isolated views of muscles that flex the knee.

- Gluteus medius
- Gluteus maximus
- Adductor magnus
- Gracilis
- Semitendinosus
- Vastus lateralis covered by fascia
- Biceps femoris
- Semimembranosus
- Sartorius
- Gastrocnemius

(a)

- Semimembranosus
- Biceps femoris (short head)

(b)

- Semitendinosus
- Biceps femoris (long head)

(c)

The **piriformis** (pir-ĭ-for′mis) is shaped like a pyramid and located inferior to the gluteus minimus. It abducts and laterally rotates the thigh and is part of the posterior group of muscles that stabilizes the hip.

The **tensor fasciae latae** (ten′sor fash′e-e lah-tē) connects the ilium to the iliotibial tract (fascia of the thigh), which continues downward to the tibia. This muscle flexes the hip and abducts and rotates the thigh medially (see fig. 9.38).

> The gluteus medius and gluteus minimus help support and maintain the normal position of the pelvis. If these muscles are paralyzed on one side as a result of injury or disease, the pelvis tends to drop on the contralateral side when that foot is raised. Consequently, the person walks with a waddling movement called the *gluteal gait*.

Thigh Adductors

The **pectineus** (pek-tin′e-us) muscle runs from the spine of the pubis to the femur. It flexes the hip and adducts the thigh (see fig. 9.37).

The **adductor brevis** (ah-duk′tor brev′ĭs) is a short, triangular muscle that runs from the pubic bone to the femur. It adducts the thigh and assists in flexing the hip (see fig. 9.37).

The **adductor longus** (ah-duk′tor long′gus) is a long, triangular muscle that runs from the pubic bone to the femur. It adducts the thigh and assists in flexing the hip and rotating the thigh laterally (see fig. 9.37).

The **adductor magnus** (ah-duk′tor mag′nus) is the largest adductor of the thigh. It is a triangular muscle that connects the ischium to the femur. It adducts the thigh and portions assist in flexing or extending the hip (see fig. 9.37).

The **gracilis** (gras′il-is) is a long, straplike muscle that passes from the pubic bone to the tibia. It adducts the thigh and flexes the knee (see fig. 9.37).

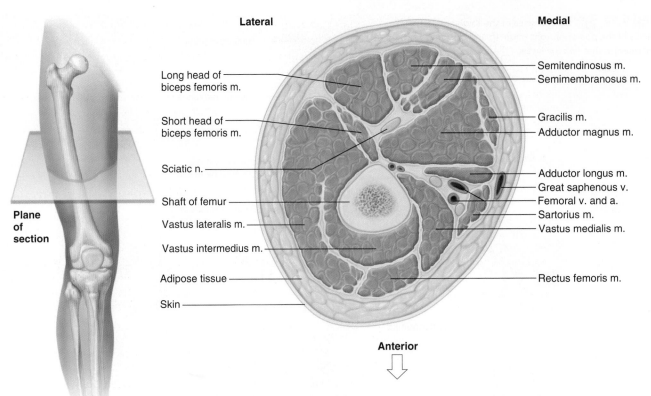

Lateral

Long head of biceps femoris m.

Short head of biceps femoris m.

Sciatic n.

Shaft of femur

Vastus lateralis m.

Vastus intermedius m.

Adipose tissue

Skin

Medial

Semitendinosus m.

Semimembranosus m.

Gracilis m.

Adductor magnus m.

Adductor longus m.

Great saphenous v.

Femoral v. and a.

Sartorius m.

Vastus medialis m.

Rectus femoris m.

Plane of section

Anterior

FIGURE 9.40 A cross section of the thigh (superior view). (*a.* stands for artery, *m.* stands for muscle, *n.* stands for nerve, and *v.* stands for vein.)

TABLE 9.13	Muscles That Move The Thigh AP|R			
Muscle	**Origin**	**Insertion**	**Action**	**Nerve Supply**
Psoas major	Bodies and transverse processes of lumbar vertebrae	Lesser trochanter of femur	Flexes hip	Branches of L1–3 nerves
Iliacus	Iliac fossa of ilium	Lesser trochanter of femur	Flexes hip	Femoral nerve
Gluteus maximus	Sacrum, coccyx, and posterior surface of ilium	Posterior surface of femur and fascia of thigh	Extends hip	Inferior gluteal nerve
Gluteus medius	Lateral surface of ilium	Greater trochanter of femur	Abducts and rotates thigh medially	Superior gluteal nerve
Gluteus minimus	Lateral surface of ilium	Greater trochanter of femur	Abducts and rotates thigh medially	Superior gluteal nerve
Piriformis	Anterior surface of sacrum	Greater trochanter of femur	Abducts and rotates thigh laterally	L5, S1, and S2 nerves
Tensor fasciae latae	Anterior iliac crest	Iliotibial tract (fascia of thigh)	Abducts thigh, flexes hip, and rotates thigh medially	Superior gluteal nerve
Pectineus	Spine of pubis	Femur distal to lesser trochanter	Flexes hip and adducts thigh	Obturator and femoral nerves
Adductor brevis	Pubic bone	Posterior surface of femur	Adducts thigh and flexes hip	Obturator nerve
Adductor longus	Pubic bone near pubic symphysis	Posterior surface of femur	Adducts thigh and flexes hip	Obturator nerve
Adductor magnus	Ischial tuberosity	Posterior surface of femur	Adducts thigh, posterior portion extends and anterior portion flexes hip	Obturator and branch of sciatic nerves
Gracilis	Lower edge of pubic symphysis	Proximal, medial surface of tibia	Adducts thigh and flexes knee	Obturator nerve

TABLE 9.14 | Muscles That Move the Leg AP|R

Muscle	Origin	Insertion	Action	Nerve Supply
Hamstring Group				
Biceps femoris	Ischial tuberosity and linea aspera of femur	Head of fibula	Flexes knee, rotates leg laterally, and extends hip	Tibial nerve
Semitendinosus	Ischial tuberosity	Proximal, medial surface of tibia	Flexes knee, rotates leg medially, and extends hip	Tibial nerve
Semimembranosus	Ischial tuberosity	Medial condyle of tibia	Flexes knee, rotates leg medially, and extends hip	Tibial neve
Sartorius	Anterior superior iliac spine	Medial surface of tibia	Flexes knee and hip, abducts and rotates thigh laterally, and rotates leg medially	Femoral nerve
Quadriceps Femoris Group				
Rectus femoris	Anterior superior iliac spine and margin of acetabulum	Patella by tendon, which continues as patellar ligament to tibial tuberosity	Extends knee, flexes hip	Femoral nerve
Vastus lateralis	Greater trochanter and posterior surface of femur	Patella by tendon, which continues as patellar ligament to tibial tuberosity	Extends knee	Femoral nerve
Vastus medialis	Medial surface of femur	Patella by tendon, which continues as patellar ligament to tibial tuberosity	Extends knee	Femoral nerve
Vastus intermedius	Anterior and lateral surfaces of femur	Patella by tendon, which continues as patellar ligament to tibial tuberosity	Extends knee	Femoral nerve

Muscles That Move the Leg

The muscles that move the leg connect the tibia or fibula to the femur or to the pelvic girdle. They fall into two major groups—those that flex the knee and those that extend it. The muscles of these groups are shown in figures 9.37, 9.38, 9.39, 9.40, in reference plates 71 and 72 (pp. 356–357), and are listed in table 9.14. Muscles that move the leg include the following:

Flexors
Biceps femoris
Semitendinosus
Semimembranosus
Sartorius

Extensors
Quadriceps femoris group

Flexors

As its name implies, the **biceps femoris** (bi′seps fem′or-is) has two heads, one attached to the ischium and the other attached to the femur. This muscle passes along the back of the thigh on the lateral side and connects to the proximal ends of the fibula and tibia. The biceps femoris is one of the hamstring muscles, and its tendon (hamstring) feels like a lateral ridge behind the knee. This muscle flexes the knee, rotates the leg laterally, and extends the hip (see figs. 9.38 and 9.39).

The **semitendinosus** (sem″e-ten′dĭ-no-sus) is another hamstring muscle. It is a long, bandlike muscle on the back of the thigh toward the medial side, connecting the ischium to the proximal end of the tibia. The semitendinosus is so named because it becomes tendinous in the middle of the thigh, continuing to its

> Strenuous running or kicking motions can tear the tendinous attachments of the hamstring muscles to the ischial tuberosity. Internal bleeding from damaged blood vessels that supply the muscles usually occurs with this painful injury, commonly called a *pulled hamstring.*

insertion as a long, cordlike tendon. It flexes the knee, rotates the leg medially, and extends the hip (see fig. 9.39).

The **semimembranosus** (sem″e-mem′brah-no-sus) is the third hamstring muscle and is the most medially located muscle in the back of the thigh. It connects the ischium to the tibia and flexes the knee, rotates the leg medially, and extends the hip (see fig. 9.39).

The **sartorius** (sar-to′re-us) is an elongated, straplike muscle that passes obliquely across the front of the thigh and then descends over the medial side of the knee. It connects the ilium to the tibia and flexes the knee and the hip. It can also abduct the thigh, rotate the thigh laterally, and rotate the leg medially (see figs. 9.37 and 9.38).

Extensor

The large muscle group called the **quadriceps femoris** (kwod′rĭ-seps fem′or-is) occupies the front and sides of the thigh and is the primary extensor of the knee. It is composed of four parts—*rectus femoris, vastus lateralis, vastus medialis,* and *vastus intermedius* (see figs. 9.38 and 9.40). These parts connect the ilium and femur to a common *patellar tendon,* which passes over the front of the knee and attaches to the patella. This tendon then continues as the *patellar ligament* to the tibia. Because the rectus femoris originates on the ilium, it can flex the hip.

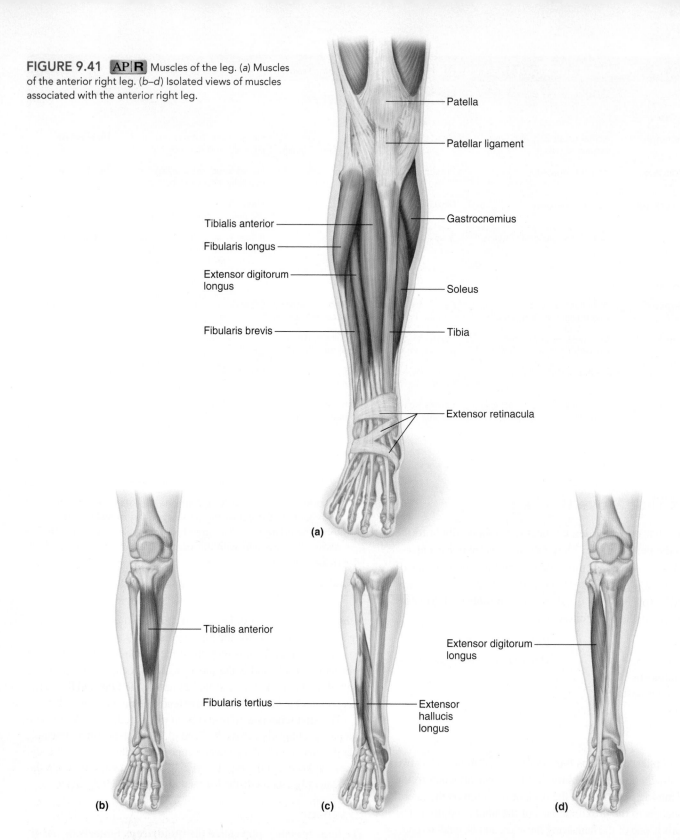

FIGURE 9.41 **AP|R** Muscles of the leg. (*a*) Muscles of the anterior right leg. (*b–d*) Isolated views of muscles associated with the anterior right leg.

Patella

Patellar ligament

Tibialis anterior

Fibularis longus

Extensor digitorum longus

Fibularis brevis

Gastrocnemius

Soleus

Tibia

Extensor retinacula

(a)

Tibialis anterior

Fibularis tertius

(b)

Extensor hallucis longus

(c)

Extensor digitorum longus

(d)

Muscles That Move the Foot

Movements of the foot include movements of the ankle and toes. A number of muscles that move the foot are in the leg. They attach the femur, tibia, and fibula to bones of the foot and move the foot upward (dorsiflexion) or downward (plantar flexion) and turn the foot so the plantar surface faces medially (inversion) or laterally (eversion). These muscles are shown in **figures 9.41, 9.42, 9.43,** and 9.44, in reference plates 73, 74, 75 (pp. 357–358), and are listed in **table 9.15.** Muscles that move the foot include the following:

Dorsal Flexors	**Invertor**
Tibialis anterior	Tibialis posterior
Fibularis tertius	
Extensor digitorum longus	
Extensor hallucis longus	

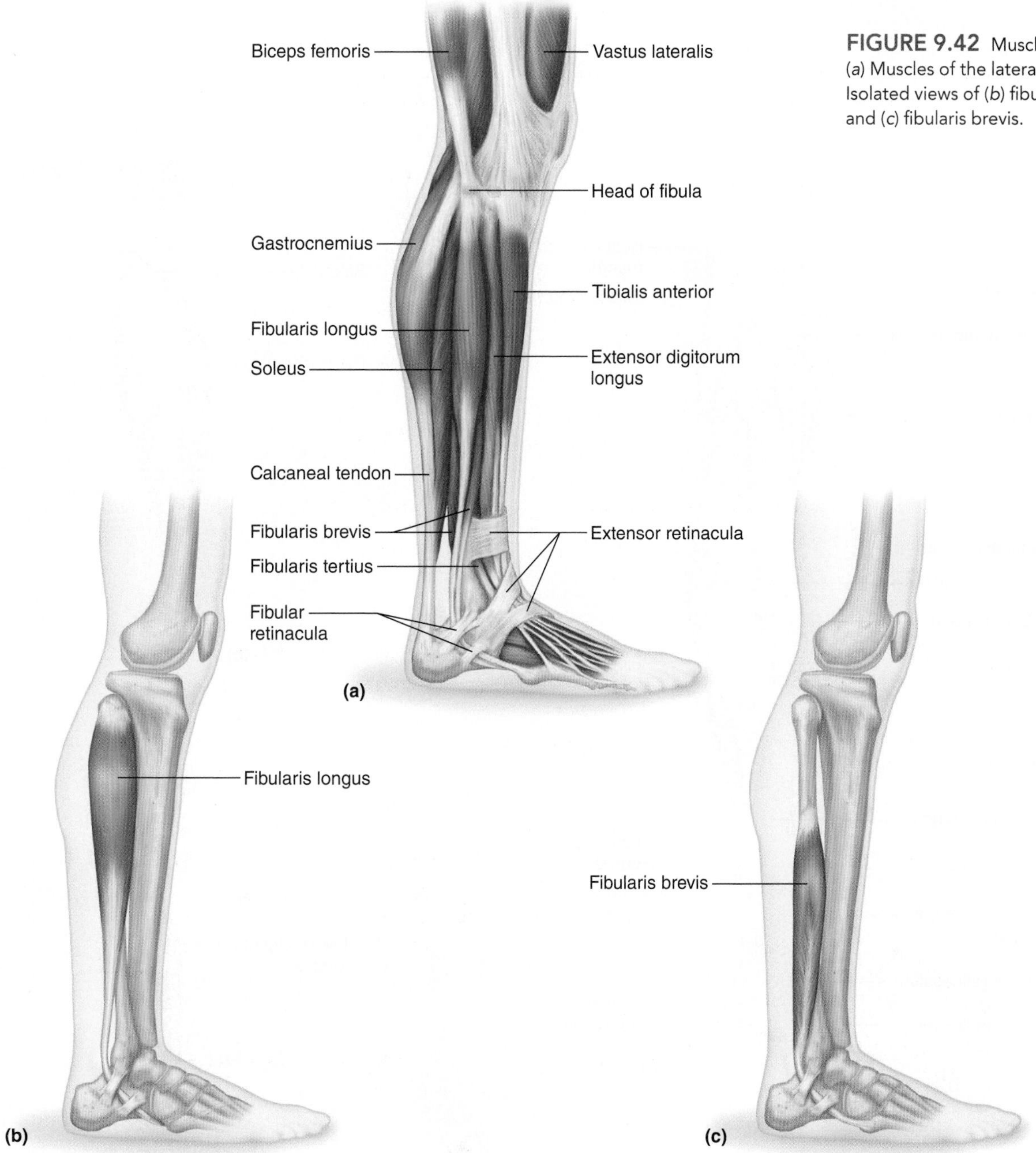

FIGURE 9.42 Muscles of the leg. (a) Muscles of the lateral right leg. Isolated views of (b) fibularis longus and (c) fibularis brevis.

Biceps femoris

Vastus lateralis

Head of fibula

Gastrocnemius

Tibialis anterior

Fibularis longus

Extensor digitorum longus

Soleus

Calcaneal tendon

Fibularis brevis

Extensor retinacula

Fibularis tertius

Fibular retinacula

(a)

Fibularis longus

(b)

Fibularis brevis

(c)

Plantar Flexors
Gastrocnemius
Soleus
Plantaris
Flexor digitorum longus

Evertor
Fibularis longus

Dorsal Flexors

The **tibialis anterior** (tib″e-a′lis ante′re-or) is an elongated, spindle-shaped muscle located on the front of the leg. It arises from the surface of the tibia, passes medially over the distal end of the tibia, and attaches to bones of the foot. Contraction of the tibialis anterior causes dorsiflexion and inversion of the foot (see fig. 9.41).

The **fibularis** (peroneus) **tertius** (fib″u-la′ris ter′shus) is a muscle of variable size that connects the fibula to the lateral side of the foot. It functions in dorsiflexion and eversion of the foot (see fig. 9.41).

The **extensor digitorum longus** (eks-ten′sor dij″ĭ-to′rum long′gus) is situated along the lateral side of the leg just behind the tibialis anterior. It arises from the proximal end of the tibia and the shaft of the fibula. Its tendon divides into four parts as it passes over the front of the ankle. These parts continue over the surface of the foot and attach to the four lateral toes. The actions of the extensor digitorum longus include dorsiflexion of the foot, eversion of the foot, and extension of the toes (see figs. 9.41 and 9.42).

The **extensor hallucis longus** (eks-ten′sor hal′lu-sis long′gus) connects the anterior fibula with the great toe. Actions include extension of the great toe, dorsiflexion and inversion of the foot (see fig. 9.41).

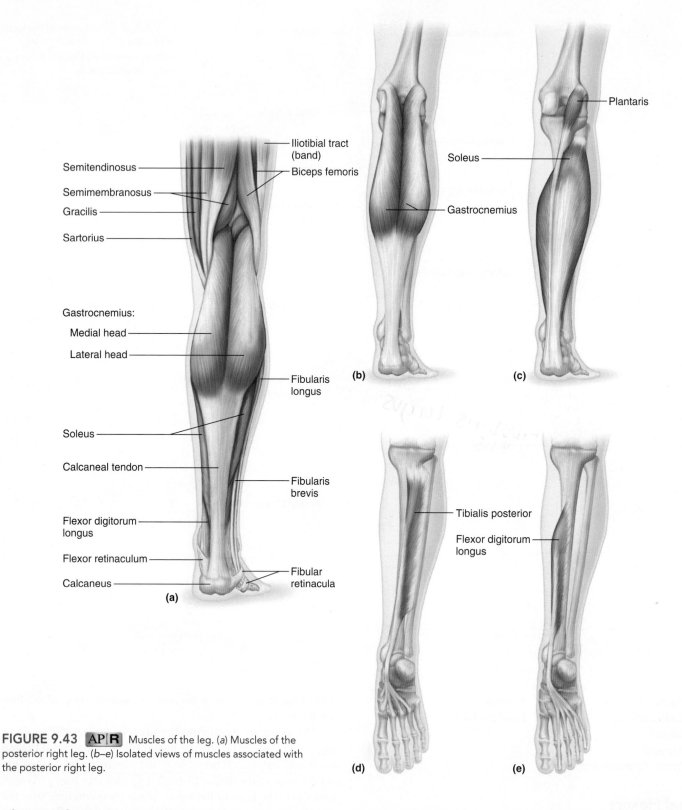

Semitendinosus

Semimembranosus

Gracilis

Sartorius

Gastrocnemius:

   Medial head

   Lateral head

Soleus

Calcaneal tendon

Flexor digitorum
longus

Flexor retinaculum

Calcaneus

Iliotibial tract
(band)

Biceps femoris

Fibularis
longus

Fibularis
brevis

Fibular
retinacula

(a)

Soleus

Gastrocnemius

(b)

Plantaris

Soleus

(c)

Tibialis posterior

Flexor digitorum
longus

(d)   **(e)**

FIGURE 9.43 **AP|R** Muscles of the leg. (a) Muscles of the posterior right leg. (b–e) Isolated views of muscles associated with the posterior right leg.

Plantar Flexors

The **gastrocnemius** (gas″trok-ne′me-us) on the back of the leg forms part of the calf. It arises by two heads from the femur. The distal end of this muscle joins the strong *calcaneal tendon* (Achilles tendon), which descends to the heel and attaches to the calcaneus. The gastrocnemius is a powerful plantar flexor of the foot that aids in pushing the body forward when a person walks or runs. It also flexes the knee (see figs. 9.42 and 9.43).

Strenuous athletic activity may partially or completely tear the calcaneal (Achilles) tendon. This injury occurs most frequently in middle-aged athletes who run or play sports that involve quick movements and directional changes. A torn calcaneal tendon usually requires surgical treatment.

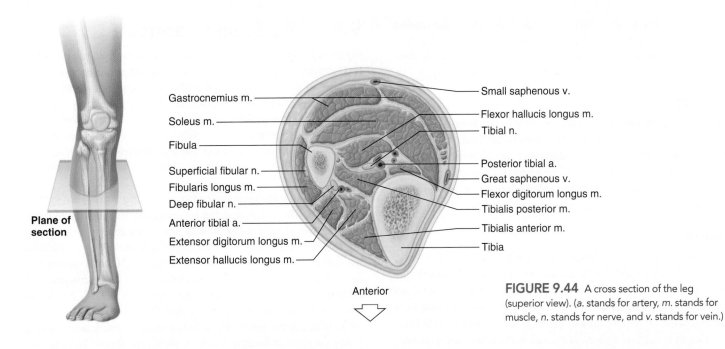

Gastrocnemius m.
Soleus m.
Fibula
Superficial fibular n.
Fibularis longus m.
Deep fibular n.
Anterior tibial a.
Extensor digitorum longus m.
Extensor hallucis longus m.

Small saphenous v.
Flexor hallucis longus m.
Tibial n.
Posterior tibial a.
Great saphenous v.
Flexor digitorum longus m.
Tibialis posterior m.
Tibialis anterior m.
Tibia

Anterior

FIGURE 9.44 A cross section of the leg (superior view). (*a.* stands for artery, *m.* stands for muscle, *n.* stands for nerve, and *v.* stands for vein.)

Plane of section

TABLE 9.15	Muscles That Move the Foot AP|R			
Muscle	**Origin**	**Insertion**	**Action**	**Nerve Supply**
Tibialis anterior	Lateral condyle and lateral surface of tibia	Tarsal bone (medial cuneiform) and first metatarsal	Dorsiflexion and inversion of foot	Deep fibular nerve
Fibularis tertius	Anterior surface of fibula	Dorsal surface of fifth metatarsal	Dorsiflexion and eversion of foot	Deep fibular nerve
Extensor digitorum longus	Lateral condyle of tibia and anterior surface of fibula	Dorsal surfaces of middle and distal phalanges of four lateral toes	Dorsiflexion of foot, extends toes	Deep fibular nerve
Extensor hallucis longus	Anterior surface of fibula	Distal phalanx of the great toe	Extends great toe, dorsiflexion and inversion of foot	Deep fibular nerve
Gastrocnemius	Lateral and medial condyles of femur	Posterior surface of calcaneus	Plantar flexion of foot, flexes knee	Tibial nerve
Soleus	Head and shaft of fibula and posterior surface of tibia	Posterior surface of calcaneus	Plantar flexion of foot	Tibial nerve
Plantaris	Femur	Calcaneus	Plantar flexion of foot, flexes knee	Tibial nerve
Flexor digitorum longus	Posterior surface of tibia	Distal phalanges of four lateral toes	Plantar flexion and inversion of foot, flexes four lateral toes	Tibial nerve
Tibialis posterior	Lateral condyle and posterior surface of tibia and posterior surface of fibula	Tarsal and first metatarsal	Plantar flexion and inversion of foot	Tibial nerve
Fibularis longus	Lateral condyle of tibia and head and shaft of fibula	Tarsal bone (medial cuneiform) and first metatarsal	Plantar flexion and eversion of foot, also supports arch	Superficial fibular nerve

The **soleus** (so'le-us) is a thick, flat muscle located beneath the gastrocnemius, and together these two muscles form the calf of the leg. The soleus arises from the tibia and fibula, and it inserts at the heel by way of the calcaneal tendon. It acts with the gastrocnemius to cause plantar flexion of the foot (see figs. 9.42 and 9.43).

The **plantaris** (plan-ta′ris) connects the femur to the heel, where it inserts with the gastrocnemius and soleus via the calcaneal tendon. When the plantaris contracts it plantar flexes the foot, and because it crosses the knee joint, it also flexes the knee.

The **flexor digitorum longus** (flek′sor dij″ĭ-to′rum long′gus) runs from the posterior surface of the tibia to the foot. Its tendon passes along the plantar surface of the foot. There the tendon divides into four parts that attach to the distal phalanges of the four lateral toes. This muscle assists in plantar flexion of the foot, flexion of the four lateral toes, and inversion of the foot (see fig. 9.43).

Invertor

The **tibialis posterior** (tib"e-a'lis pos-tēr'e-or) is the deepest of the muscles on the back of the leg. It connects the fibula and tibia to the ankle bones by means of a tendon that curves under the medial malleolus. This muscle assists in inversion and plantar flexion of the foot (see fig. 9.43). *Extensor hallucis longus* also inverts the foot, because it pulls up on the medial portion (see fig. 9.41).

Evertor

The **fibularis** (peroneus) **longus** (fib"u-la'ris long'gus) is a long, straplike muscle located on the lateral side of the leg. It connects the tibia and the fibula to the foot by means of a stout tendon that passes behind the lateral malleolus. It everts the foot, assists in plantar flexion, and helps support the arch of the foot (see figs. 9.42 and 9.44).

Fascia in various regions of the ankle thicken to form retinacula, as in the wrist. Anteriorly, for example, *extensor retinacula* connect the tibia and fibula as well as the calcaneus and fascia of the sole. These retinacula form sheaths for tendons crossing the front of the ankle (see figs. 9.41 and 9.42).

Posteriorly, on the inside, a *flexor retinaculum* runs between the medial malleolus and the calcaneus and forms sheaths for tendons passing beneath the foot (see fig. 9.43). *Fibular retinacula* connect the lateral malleolus and the calcaneus, providing sheaths for tendons on the lateral side of the ankle (see fig. 9.42).

 PRACTICE

30 Which muscles provide facial expression? ability to chew? head movements?

31 Which muscles move the pectoral girdle? the arm, forearm, and hand?

32 Which muscles move the thigh, legs, and foot?

9.8 | Life-Span Changes

Signs of aging in the muscular system begin to appear in one's forties, although a person can still be active. At a microscopic level, supplies of the molecules that enable muscles to function—myoglobin, ATP, and creatine phosphate—decline. Gradually, the muscles become smaller, drier, and capable of less forceful contraction. Connective tissue and adipose cells begin to replace some muscle tissue. By age eighty, nearly half the muscle mass has atrophied, due to a decline in motor neuron activity. Diminishing muscular strength slows reflexes.

Exercise can help maintain a healthy muscular system throughout life, countering the less effective oxygen delivery that results from the decreased muscle mass that accompanies aging. Exercise can even lead to formation of new muscle by stimulating skeletal muscle cells to release interleukin-6 (IL-6), a type of proinflammatory molecule called a cytokine. The IL-6 stimulates satellite cells, which function as muscle stem cells. They divide and migrate, becoming incorporated into the muscle fiber. Exercise also maintains the flexibility of blood vessels, which helps to keep blood pressure at healthy levels. A physician should be consulted before starting any exercise program.

According to the National Institute on Aging, exercise should include strength training and aerobics. Strength training consists of weight lifting or using a machine that works specific muscles against a resistance, performed so that the same muscle is not exercised on consecutive days. Strength training increases muscle mass, and the resulting stronger muscles can alleviate pressure on the joints, which may lessen arthritis pain. Aerobic exercise improves oxygen use by muscles and increases endurance.

 PRACTICE

33 What changes are associated with an aging muscular system?

34 Describe two types of recommended exercise. ■

Chapter Summary

9.1 Structure of a Skeletal Muscle (page 292)

Skeletal muscles are composed of nervous, vascular, and various other connective tissues, as well as skeletal muscle tissue.

1. Connective tissue coverings
    a. **Fascia** covers each skeletal muscle.
    b. Other connective tissues surround cells and groups of cells within the muscle's structure (epimysium, perimysium, endomysium).
    c. Fascia is part of a complex network of connective tissue that extends throughout the body.
2. Skeletal muscle fibers
    a. Each skeletal muscle fiber is a single muscle cell, the unit of contraction.
    b. Muscle fibers are cylindrical cells with many nuclei.
    c. The cytoplasm contains mitochondria, sarcoplasmic reticulum, and **myofibrils** of **actin** and **myosin.**

    d. The arrangement of the actin and myosin filaments causes striations forming repeating patterns of **sarcomeres.** (I bands, Z lines, A bands, H zone, and M line)
    e. **Troponin** and **tropomyosin** molecules associate with actin filaments.
    f. **Transverse tubules** extend from the cell membrane into the cytoplasm and are associated with the **cisternae** of the **sarcoplasmic reticulum.**

9.2 Skeletal Muscle Contraction (page 297)

Muscle fiber contraction results from a sliding movement of actin and myosin filaments overlapping that shortens the muscle fiber.

1. **Neuromuscular junction**
    a. The functional connection between a neuron and another cell is a **synapse.** The neuromuscular junction is a synapse.

Muscular System

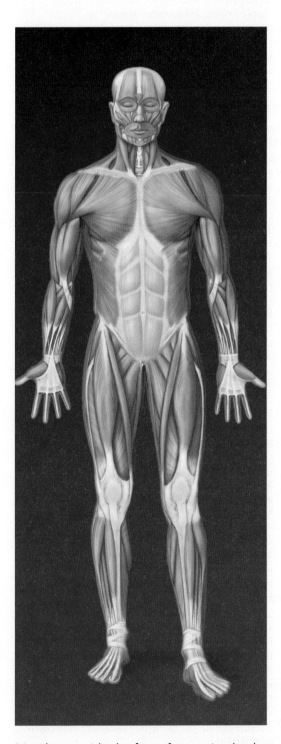

Muscles provide the force for moving body parts.

Integumentary System

The skin increases heat loss during skeletal muscle activity.

Skeletal System

Bones provide attachments that allow skeletal muscles to cause movement.

Nervous System

Neurons control muscle contractions.

Endocrine System

Hormones help increase blood flow to exercising skeletal muscles.

Cardiovascular System

The heart pumps as a result of cardiac muscle contraction. Blood flow delivers oxygen and nutrients and removes wastes.

Lymphatic System

Muscle action pumps lymph through lymphatic vessels.

Digestive System

Skeletal muscles are important in swallowing. The digestive system absorbs nutrients needed for muscle contraction.

Respiratory System

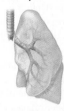

Breathing depends on skeletal muscles. The lungs provide oxygen for muscle cells and excrete carbon dioxide.

Urinary System

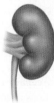

Skeletal muscles help control expulsion of urine from the urinary bladder.

Reproductive System

Skeletal muscles are important in sexual activity.

b. **Motor neurons** stimulate muscle fibers to contract.

c. The **motor end plate** of a muscle fiber lies on one side of a neuromuscular junction.

d. In response to an action potential, the end of a motor neuron axon releases a **neurotransmitter,** which diffuses across the junction and stimulates the muscle fiber.

2. Stimulus for contraction

a. **Acetylcholine** released from the end of a motor neuron axon stimulates a muscle fiber.

b. Stimulation causes a muscle fiber to conduct an impulse that travels over the surface of the sarcolemma and reaches the deep parts of the fiber by means of the transverse tubules.

3. Excitation-contraction coupling

a. In response to stimulation the sarcoplasmic reticulum releases calcium ions.

b. Calcium ions combine with troponin, causing the tropomyosin to shift and expose active sites on the actin for myosin binding.

c. Cross-bridges form between myosin and actin, and the actin filaments move inward, shortening the sarcomere.

4. The **sliding filament model** of muscle contraction

a. The sarcomere is the functional unit of skeletal muscle.

b. When the overlapping thick and thin myofilaments slide past one another, the sarcomeres shorten. The muscle contracts.

5. Cross-bridge cycling

a. A myosin head can attach to an actin binding site to form a cross-bridge which pulls on the actin filament. The myosin head can then release the actin and attach to another active binding site farther down the actin filament and pull again.

b. The breakdown of ATP releases energy that provides the repetition of the cross-bridge cycle.

6. Relaxation

a. **Acetylcholinesterase** rapidly decomposes acetylcholine remaining in the synapse, preventing continuous stimulation of a muscle fiber.

b. The muscle fiber relaxes when calcium ions are transported back into the sarcoplasmic reticulum.

c. Cross-bridge linkages break and do not re-form—the muscle fiber relaxes.

7. Energy sources for contraction

a. ATP supplies the energy for muscle fiber contraction.

b. **Creatine phosphate** stores energy that can be used to synthesize ATP as it is decomposed.

c. Active muscles require cellular respiration for energy.

8. Oxygen supply and cellular respiration

a. Anaerobic reactions of cellular respiration yield few ATP molecules, whereas aerobic reactions of cellular respiration provide many ATP molecules.

b. Hemoglobin in red blood cells carries oxygen from the lungs to body cells.

c. **Myoglobin** in muscle cells temporarily stores some oxygen.

9. **Oxygen debt**

a. During rest or moderate exercise, oxygen is sufficient to support the aerobic reactions of cellular respiration.

b. During strenuous exercise, oxygen deficiency may develop, and lactic acid may accumulate as a result of the anaerobic reactions of cellular respiration.

c. The oxygen debt includes the amount of oxygen required to react accumulated lactic acid to form glucose and to restore supplies of ATP and creatine phosphate.

10. Muscle fatigue

a. A fatigued muscle loses its ability to contract.

b. Muscle fatigue may be due in part to the effects of accumulation of lactic acid.

c. Athletes usually produce less lactic acid than nonathletes because of their increased ability to supply oxygen and nutrients to muscles.

11. Heat production

a. Muscular contraction generates body heat.

b. Most of the energy released by cellular respiration is lost as heat.

9.3 Muscular Responses (page 303)

1. **Threshold stimulus** is the minimal stimulus needed to elicit a muscular contraction.

2. Recording of a muscle contraction

a. A **twitch** is a single, short contraction of a muscle fiber.

b. A myogram is a recording of the contraction of an electrically stimulated isolated muscle or muscle fiber.

c. The **latent period** is the time between stimulus and responding contraction.

d. The length to which a muscle is stretched before stimulation affects the force it will develop.

   (1) Normal activities occur at optimal length.

   (2) Too long or too short decreases force.

e. Sustained contractions are more important than twitch contractions in everyday activities.

3. **Summation**

a. A rapid series of stimuli may produce summation of twitches and sustained contraction.

b. At higher frequencies of stimulation, contraction with little time for relaxation is called partial tetany.

c. Forceful, sustained contraction without any relaxation is a **complete** (fused) **tetanic contraction.**

4. **Recruitment** of motor units

a. One motor neuron and the muscle fibers associated with it constitute a motor unit.

b. Muscles whose motor units have few muscle fibers produce finer movements.

c. Motor units respond in an all-or-none manner.

d. At low intensity of stimulation, relatively few motor units contract.

e. At increasing intensities of stimulation, other motor units are recruited until the muscle contracts with maximal tension.

5. Sustained contractions

a. Tetanic contractions are common in everyday activities.

b. Even when a whole muscle appears at rest, some of its fibers undergo sustained contraction. This is called **muscle tone.**

6. Types of contractions

a. One type of contraction called **isotonic** occurs when a muscle contracts and its ends are pulled closer together. Because the muscle shortens, it is called a **concentric contraction.**

b. In another type of isotonic contraction, the force a muscle generates is less than that required to move or lift an object. This lengthening contraction is an **eccentric contraction.**

c. When a muscle contracts but its attachments do not move, the contraction is **isometric.**

d. Most body movements involve both isometric and isotonic contractions.

7. Fast- and slow-twitch muscle fibers

a. The speed of contraction is related to a muscle's specific function.

b. Slow-contracting, or red, muscles can generate ATP fast enough to keep up with ATP breakdown and can contract for long periods.

c. Fast-contracting, or white, muscles have reduced ability to carry on the aerobic reactions of cellular respiration and tend to fatigue rapidly.

9.4 Smooth Muscle (page 307)

The contractile mechanisms of smooth and cardiac muscle are similar to those of skeletal muscle.

1. Smooth muscle cells

a. Smooth muscle cells contain filaments of myosin and actin.

b. They lack transverse tubules, and the sarcoplasmic reticula are not well developed.

c. Types include **multiunit smooth muscle** and **visceral smooth muscle.**

d. Visceral smooth muscle displays rhythmicity.

e. **Peristalsis** aids movement of material through hollow organs.

2. Smooth muscle contraction

a. In smooth muscle, calmodulin binds to calcium ions and activates the contraction mechanism.

b. Both acetylcholine and norepinephrine are neurotransmitters for smooth muscles.

c. Hormones and stretching affect smooth muscle contractions.

d. With a given amount of energy, smooth muscle can maintain a contraction longer than skeletal muscle.

e. Smooth muscle can change length without changing tautness.

9.5 Cardiac Muscle (page 308)

1. Cardiac muscle contracts for a longer time than skeletal muscle because transverse tubules supply extra calcium ions.

2. Intercalated discs connect the ends of cardiac muscle cells and hold the cells together.

3. A network of cells contracts as a unit and responds to stimulation in an all-or-none manner.

4. Cardiac muscle is self-exciting, rhythmic, and remains refractory to further stimulation until a contraction is completed.

9.6 Skeletal Muscle Actions (page 310)

1. Body movement

a. Bones and muscles function together as **levers.**

b. A lever consists of a rod, a fulcrum (pivot), a resistance, and a force that supplies energy.

c. Parts of a first-class lever are arranged resistance–fulcrum–force; of a second-class lever, fulcrum–resistance–force; and of a third-class lever, resistance–force–fulcrum.

2. Origin and insertion

a. The less movable end of attachment of a skeletal muscle to a bone is its **origin,** and the more movable end is its **insertion.**

b. Some muscles have more than one origin or insertion.

3. Interaction of skeletal muscles

a. Skeletal muscles function in groups.

b. A muscle that causes a specific movement is an **agonist** for that movement; **antagonists** can resist a movement; **synergists** work together to perform a specific movement.

c. Smooth movements depend upon antagonists giving way to the actions of agonists.

9.7 Major Skeletal Muscles (page 312)

Muscle names often describe sizes, shapes, locations, actions, number of attachments, or direction of fibers.

1. Muscles of facial expression

a. These muscles lie beneath the skin of the face and scalp and are used to communicate feelings through facial expression.

b. They include the **epicranius, orbicularis oculi, orbicularis oris, buccinator, zygomaticus major, zygomaticus minor,** and **platysma.**

2. Muscles of mastication

a. These muscles are attached to the mandible and are used in chewing.

b. They include the **masseter, temporalis, medial pterygoid,** and **lateral pterygoid.**

3. Muscles that move the head and vertebral column

a. Muscles in the neck and back move the head.

b. They include the **sternocleidomastoid, splenius capitis, semispinalis capitis, scalenes, quadratus lumborum,** and **erector spinae.**

4. Muscles that move the pectoral girdle

a. Most of these muscles connect the scapula to nearby bones and are closely associated with muscles that move the arm.

b. They include the **trapezius, rhomboid major, rhomboid minor, levator scapulae, serratus anterior,** and **pectoralis minor.**

5. Muscles that move the arm

a. These muscles connect the humerus to various regions of the pectoral girdle, ribs, and vertebral column.

b. They include the **coracobrachialis, pectoralis major, teres major, latissimus dorsi, supraspinatus, deltoid, subscapularis, infraspinatus,** and **teres minor.**

6. Muscles that move the forearm

a. These muscles connect the radius and ulna to the humerus and pectoral girdle.

b. They include the **biceps brachii, brachialis, brachioradialis, triceps brachii, supinator, pronator teres,** and **pronator quadratus.**

7. Muscles that move the hand

a. These muscles arise from the distal end of the humerus and from the radius and ulna.

b. They include the **flexor carpi radialis, flexor carpi ulnaris, palmaris longus, flexor digitorum profundus, flexor digitorum superficialis, extensor carpi radialis longus, extensor carpi radialis brevis, extensor carpi ulnaris,** and **extensor digitorum.**
c. An extensor retinaculum forms sheaths for tendons of the extensor muscles.

8. Muscles of the abdominal wall
   a. These muscles connect the rib cage and vertebral column to the pelvic girdle.
   b. They include the **external oblique, internal oblique, transversus abdominis,** and **rectus abdominis.**

9. Muscles of the pelvic floor
   a. These muscles form the floor of the pelvic cavity and fill the space of the pubic arch.
   b. They include the **levator ani, coccygeus, superficial transversus perinei, bulbospongiosus, ischiocavernosus,** and **sphincter urethrae.**

10. Muscles that move the thigh
    a. These muscles are attached to the femur and to some part of the pelvic girdle.
    b. They include the **psoas major, iliacus, gluteus maximus, gluteus medius, gluteus minimus, piriformis, tensor fasciae latae, pectineus, adductor brevis, adductor longus, adductor magnus,** and **gracilis.**

11. Muscles that move the leg
    a. These muscles connect the tibia or fibula to the femur or pelvic girdle.
    b. They include the **biceps femoris, semitendinosus, semimembranosus, sartorius, rectus femoris, vastus lateralis, vastus medialis,** and **vastus intermedius.**

12. Muscles that move the foot
    a. These muscles attach the femur, tibia, and fibula to various bones of the foot.
    b. They include the **tibialis anterior, fibularis tertius, extensor digitorum longus, extensor hallucis longus, gastrocnemius, soleus, plantaris, flexor digitorum longus, tibialis posterior,** and **fibularis longus.**
    c. Retinacula form sheaths for tendons passing to the foot.

9.8 Life-Span Changes (page 340)

1. Beginning in one's forties, supplies of ATP, myoglobin, and creatine phosphate begin to decline.
2. By age eighty, muscle mass may be halved. Reflexes slow. Adipose cells and connective tissue replace some muscle tissue.
3. Exercise is beneficial in maintaining muscle function.

CHAPTER ASSESSMENTS

9.1 Structure of a Skeletal Muscle

1 Describe the difference between a tendon and an aponeurosis. (p. 292)
2 Describe how connective tissue is part of the structure of a skeletal muscle. (p. 293)
3 Distinguish among deep fascia, subcutaneous fascia, and subserous fascia. (p. 293)
4 Identify the major parts of a skeletal muscle fiber and describe the functions of each. (pp. 293–296)

9.2 Skeletal Muscle Contraction

5 A neurotransmitter _____. (p. 297)
   a. binds actin filaments, causing them to slide
   b. diffuses across a synaptic cleft from a neuron to a muscle cell
   c. transports ATP across the synaptic cleft
   d. breaks down acetylcholine at the synapse
6 Describe the neuromuscular junction. (p. 297)
7 Describe the neural control of skeletal muscle contraction. (p. 297)
8 Identify the major events that occur during skeletal muscle fiber contraction. (pp. 298–299)
9 Explain how ATP and creatine phosphate function in skeletal muscle fiber contraction. (pp. 299–302)
10 Describe how oxygen is supplied to skeletal muscle. (p. 302)
11 Describe how oxygen debt may develop. (pp. 302–303)
12 Explain how a muscle may become fatigued, and how a person's physical condition may affect tolerance to fatigue. (p. 303)

9.3 Muscular Responses

13 Define *threshold stimulus.* (p. 304)
14 Distinguish between a twitch and a sustained contraction. (p. 304)
15 Define *motor unit* and explain how the number of fibers in a unit affects muscular contractions. (p. 305)
16 Which of the following describes addition of muscle fibers to take part in a contraction? (p. 305)
   a. summation
   b. recruitment
   c. tetany
   d. twitch
17 Explain how a skeletal muscle can be stimulated to produce a sustained contraction. (p. 306)
18 Distinguish between a complete tetanic contraction and muscle tone. (pp. 304–306)
19 Distinguish between concentric and eccentric contractions, and explain how each is used in body movements. (p. 306)
20 Distinguish between fast- and slow-twitch muscle fibers. (pp. 306–307)

9.4 Smooth Muscle

21 Distinguish between multiunit smooth muscle and visceral smooth muscle. (p. 307)
22 Define *peristalsis* and explain its function. (p. 307)
23 Compare the characteristics of skeletal and smooth muscle cell contractions. (pp. 307–308)

9.5 Cardiac Muscle

24 Compare the characteristics of skeletal and cardiac muscle cell contractions. (pp. 308–309)

9.6 Skeletal Muscle Actions

25 Describe a lever, and explain how its parts may be arranged to form first-class, second-class, and third-class levers. (pp. 310–311)

26 Explain how limb movements function as levers. (pp. 310–311)

27 Distinguish between a muscle's origin and its insertion. (p. 311)

28 Define *agonist*, *antagonist*, and *synergist*. (p. 312)

9.7 Major Skeletal Muscles

29 Match the muscle with its description or action. (pp. 312–340)

(1) buccinator	A. inserted on the coronoid process of the mandible
(2) epicranius	B. elevates the corner of the mouth
(3) lateral pterygoid	C. can elevate and adduct the scapula
(4) platysma	D. can extend the neck
(5) rhomboid major	E. consists of two parts—the frontalis and the occipitalis
(6) splenius capitis	F. compresses the cheeks
(7) temporalis	G. runs over the neck from the chest to the face
(8) zygomaticus major	H. pulls the jaw from side to side
(9) biceps brachii	I. primary extensor of the elbow
(10) brachialis	J. depresses and retracts the shoulder
(11) deltoid	K. abducts the arm
(12) latissimus dorsi	L. rotates the arm laterally
(13) pectoralis major	M. flexes and adducts the arm
(14) pronator teres	N. rotates the arm medially
(15) teres minor	O. strongest flexor of the elbow
(16) triceps brachii	P. strongest supinator of the forearm and hand
(17) biceps femoris	Q. inverts the foot
(18) external oblique	R. a member of the quadriceps femoris group
(19) gastrocnemius	S. a plantar flexor of the foot
(20) gluteus maximus	T. compresses the contents of the abdominal cavity
(21) gluteus medius	U. largest muscle in the body
(22) gracilis	V. a hamstring muscle
(23) rectus femoris	W. adducts the thigh
(24) tibialis anterior	X. abducts the thigh

30 Label as many muscles as you can identify in these photos of a model whose muscles are enlarged by exercise. Describe the action of each muscle identified. (pp. 312–340)

9.8 Life-Span Changes

31 Describe three aging-related changes in the muscular system. (p. 340)

32 Explain the benefits of exercise for maintaining muscular health while aging. (p. 340)

INTEGRATIVE ASSESSMENTS/CRITICAL THINKING

Outcomes 7.10, 7.11, 9.7

1. Several important nerves and blood vessels course through the muscles of the gluteal region. To avoid the possibility of damaging such parts, intramuscular injections are usually made into the lateral, superior portion of the gluteus medius. What landmarks would help you locate this muscle in a patient?

Outcomes 9.1, 9.2

2. Millions of people take drugs called statins to lower serum cholesterol levels. In a small percentage of people taking these drugs, muscle pain, termed myopathy, is an adverse effect. In a small percentage of these individuals, the condition progresses to rhabdomyolysis, in which the sarcolemma breaks down.

   a. Describe the structure and state the function of the sarcolemma.
   b. Physicians can measure a patient's levels of creatine phosphokinase to track the safety of using a statin. Enzyme levels that exceed 10 times normal indicate possible rhabdomyolysis. Explain what this elevated enzyme level indicates about the physiology of the muscle cell.
   c. Explain why a dusky, dark color in the urine, resulting from the presence of myoglobin, also indicates muscle breakdown.

Outcomes 9.1, 9.2, 9.3

3. What steps might be taken to minimize atrophy of skeletal muscles in patients confined to bed for prolonged times?

Outcome 9.2

4. As lactic acid and other substances accumulate in an active muscle, they stimulate pain receptors, and the muscle may feel sore. How might the application of heat or substances that dilate blood vessels help relieve such soreness?

Outcomes 9.2, 9.3, 9.5

5. Why do you think athletes generally perform better if they warm up by exercising lightly before a competitive event?

Outcomes 9.2, 9.3, 9.6

6. Following an injury to a nerve, the muscles it supplies with motor nerve fibers may become paralyzed. How would you explain to a patient the importance of moving the disabled muscles passively or contracting them with electrical stimulation?

Outcomes 9.3, 9.7

7. Following childbirth, a woman may lose urinary control (incontinence) when sneezing or coughing. Which muscles of the pelvic floor should be strengthened by exercise to help control this problem?

ONLINE STUDY TOOLS

Connect Interactive Questions Reinforce your knowledge using assigned interactive questions covering muscle structure, the process of muscle contraction, and identification of skeletal muscles.

Connect Integrated Activity Can you predict the effects on muscle function of different drugs, toxins, and neuromuscular diseases?

LearnSmart Discover which chapter concepts you have mastered and which require more attention. This adaptive learning tool is personalized, proven, and preferred.

Anatomy & Physiology Revealed Go more in depth into the human body by exploring cadaver dissections of assigned skeletal muscles and viewing animations of their actions.

Surface Anatomy and Cadaver Dissection

The following set of reference plates, made up of surface anatomy photos and cadaver dissection photos, is presented to help you locate some of the more prominent surface features in various regions of the body. For the most part, the labeled structures on the surface anatomy photos are easily seen or palpated (felt) through the skin. As a review, you may want to locate as many of these features as possible on your own body. The cadaver dissection photos reveal the structures located beneath the skin.

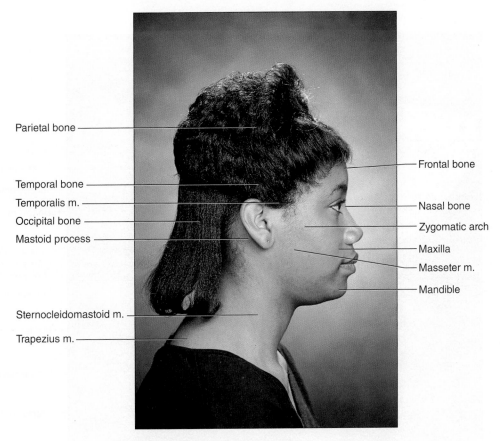

PLATE FIFTY-FIVE Surface anatomy of head and neck, lateral view. (*m.* stands for muscle.)

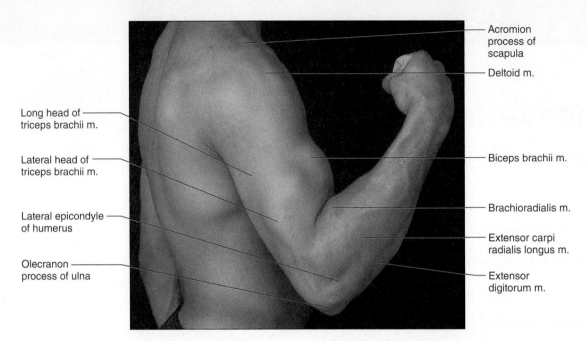

Acromion
process of
scapula

Deltoid m.

Long head of
triceps brachii m.

Biceps brachii m.

Lateral head of
triceps brachii m.

Brachioradialis m.

Lateral epicondyle
of humerus

Extensor carpi
radialis longus m.

Olecranon
process of ulna

Extensor
digitorum m.

PLATE FIFTY-SIX Surface anatomy of upper limb and thorax, lateral view. (*m.* stands for muscle.)

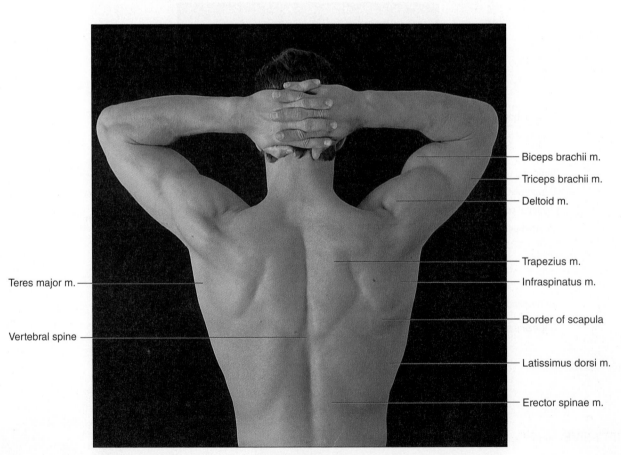

Biceps brachii m.

Triceps brachii m.

Deltoid m.

Teres major m.

Trapezius m.

Infraspinatus m.

Border of scapula

Vertebral spine

Latissimus dorsi m.

Erector spinae m.

PLATE FIFTY-SEVEN Surface anatomy of back and upper limbs, posterior view. (*m.* stands for muscle.)

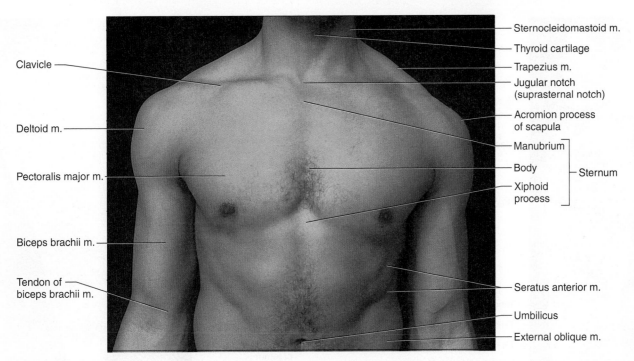

Clavicle

Deltoid m.

Pectoralis major m.

Biceps brachii m.

Tendon of
biceps brachii m.

Sternocleidomastoid m.

Thyroid cartilage

Trapezius m.

Jugular notch
(suprasternal notch)

Acromion process
of scapula

Manubrium

Body ⎤
⎥ Sternum
Xiphoid ⎦
process

Seratus anterior m.

Umbilicus

External oblique m.

PLATE FIFTY-EIGHT Surface anatomy of torso and arms, anterior view. (*m.* stands for muscle.)

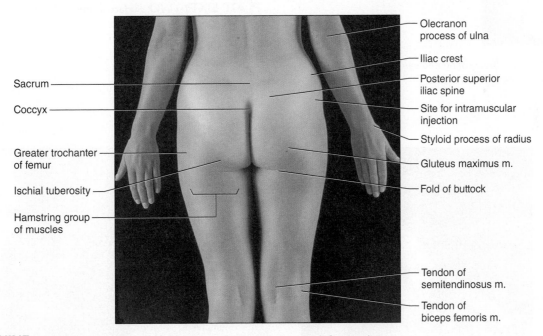

Sacrum

Coccyx

Greater trochanter
of femur

Ischial tuberosity

Hamstring group
of muscles

Olecranon
process of ulna

Iliac crest

Posterior superior
iliac spine

Site for intramuscular
injection

Styloid process of radius

Gluteus maximus m.

Fold of buttock

Tendon of
semitendinosus m.

Tendon of
biceps femoris m.

PLATE FIFTY-NINE Surface anatomy of torso and thighs, posterior view. (*m.* stands for muscle.)

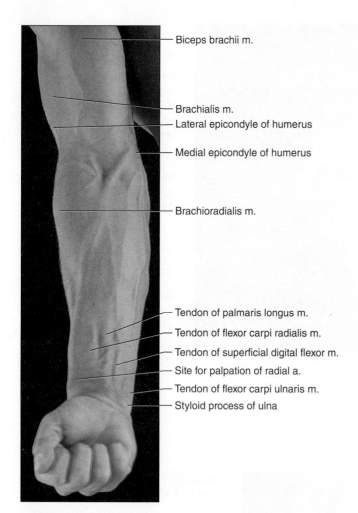

- Biceps brachii m.
- Brachialis m.
- Lateral epicondyle of humerus
- Medial epicondyle of humerus
- Brachioradialis m.
- Tendon of palmaris longus m.
- Tendon of flexor carpi radialis m.
- Tendon of superficial digital flexor m.
- Site for palpation of radial a.
- Tendon of flexor carpi ulnaris m.
- Styloid process of ulna

PLATE SIXTY Surface anatomy of right forearm, anterior view. (*m.* stands for muscle and *a.* stands for artery.)

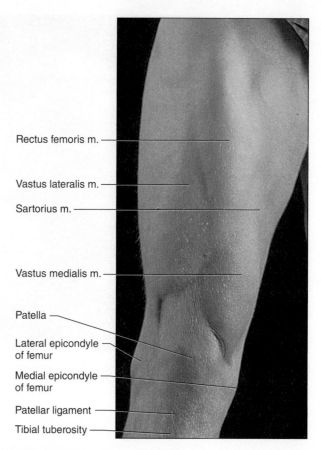

- Rectus femoris m.
- Vastus lateralis m.
- Sartorius m.
- Vastus medialis m.
- Patella
- Lateral epicondyle of femur
- Medial epicondyle of femur
- Patellar ligament
- Tibial tuberosity

PLATE SIXTY-TWO Surface anatomy of right knee and surrounding area, anterior view. (*m.* stands for muscle.)

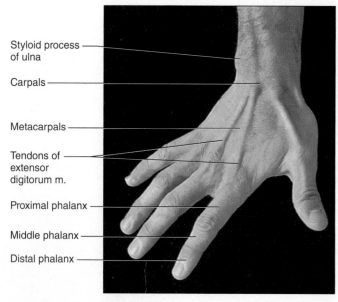

- Styloid process of ulna
- Carpals
- Metacarpals
- Tendons of extensor digitorum m.
- Proximal phalanx
- Middle phalanx
- Distal phalanx

PLATE SIXTY-ONE Surface anatomy of the right hand. (*m.* stands for muscle.)

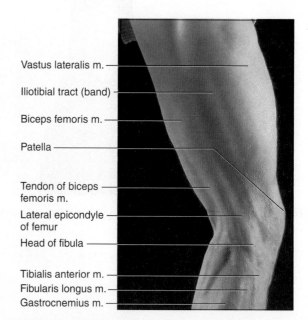

- Vastus lateralis m.
- Iliotibial tract (band)
- Biceps femoris m.
- Patella
- Tendon of biceps femoris m.
- Lateral epicondyle of femur
- Head of fibula
- Tibialis anterior m.
- Fibularis longus m.
- Gastrocnemius m.

PLATE SIXTY-THREE Surface anatomy of right knee and surrounding area, lateral view. (*m.* stands for muscle.)

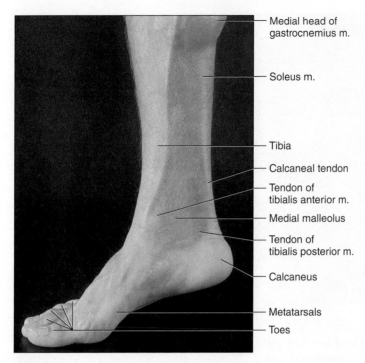

Medial head of
gastrocnemius m.

Soleus m.

Tibia

Calcaneal tendon

Tendon of
tibialis anterior m.

Medial malleolus

Tendon of
tibialis posterior m.

Calcaneus

Metatarsals

Toes

PLATE SIXTY-FOUR Surface anatomy of right foot and leg, medial view. (*m.* stands for muscle.)

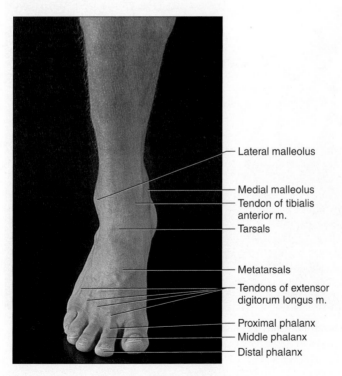

Lateral malleolus

Medial malleolus

Tendon of tibialis
anterior m.

Tarsals

Metatarsals

Tendons of extensor
digitorum longus m.

Proximal phalanx

Middle phalanx

Distal phalanx

PLATE SIXTY-FIVE Surface anatomy of right foot. (*m.* stands for muscle.)

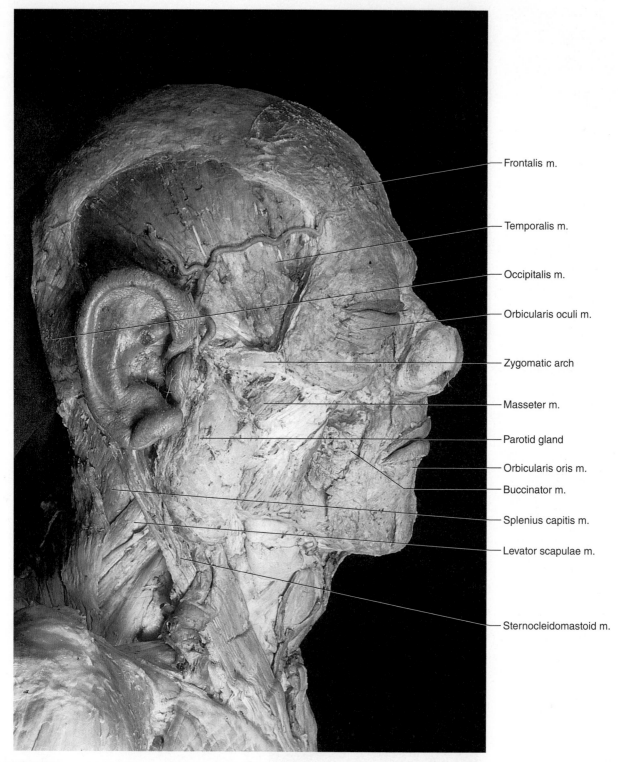

- Frontalis m.
- Temporalis m.
- Occipitalis m.
- Orbicularis oculi m.
- Zygomatic arch
- Masseter m.
- Parotid gland
- Orbicularis oris m.
- Buccinator m.
- Splenius capitis m.
- Levator scapulae m.
- Sternocleidomastoid m.

PLATE SIXTY-SIX Lateral view of the head. (*m.* stands for muscle.)

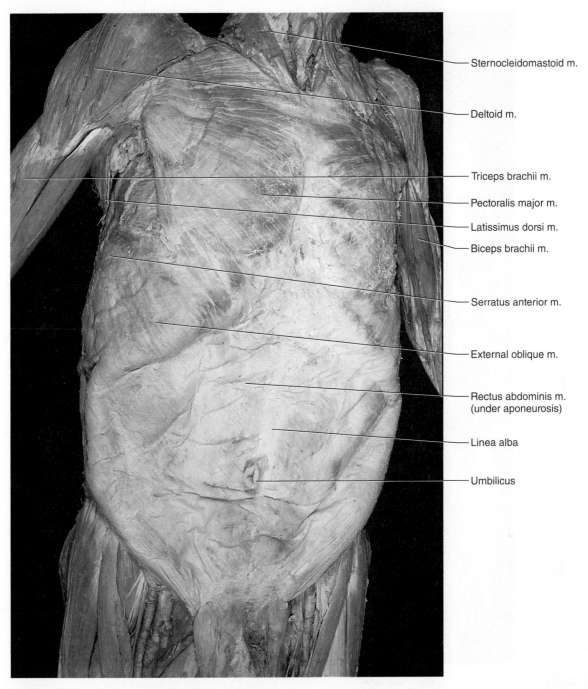

— Sternocleidomastoid m.

— Deltoid m.

— Triceps brachii m.

— Pectoralis major m.

— Latissimus dorsi m.

— Biceps brachii m.

— Serratus anterior m.

— External oblique m.

— Rectus abdominis m.
(under aponeurosis)

— Linea alba

— Umbilicus

PLATE SIXTY-SEVEN Anterior view of the trunk. (*m.* stands for muscle.)

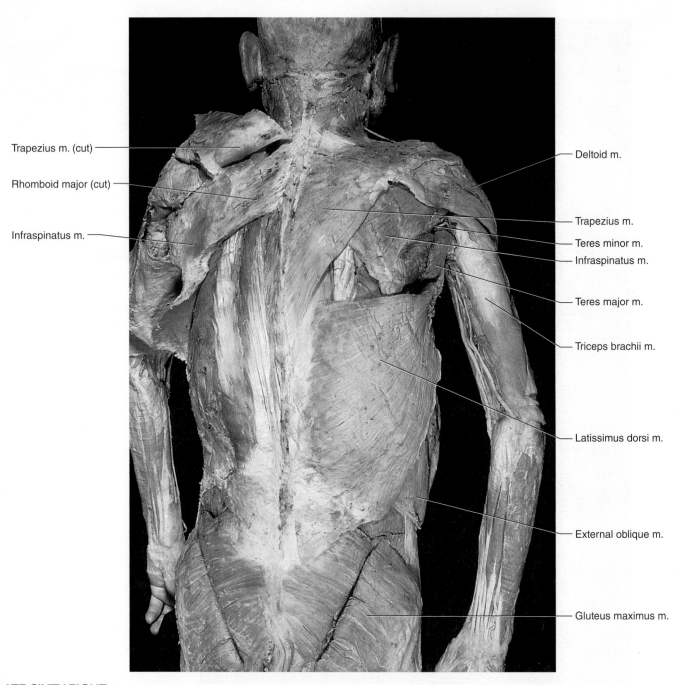

Trapezius m. (cut)

Rhomboid major (cut)

Infraspinatus m.

Deltoid m.

Trapezius m.

Teres minor m.

Infraspinatus m.

Teres major m.

Triceps brachii m.

Latissimus dorsi m.

External oblique m.

Gluteus maximus m.

PLATE SIXTY-EIGHT Posterior view of the trunk, with deep thoracic muscles exposed on the left. (*m.* stands for muscle.)

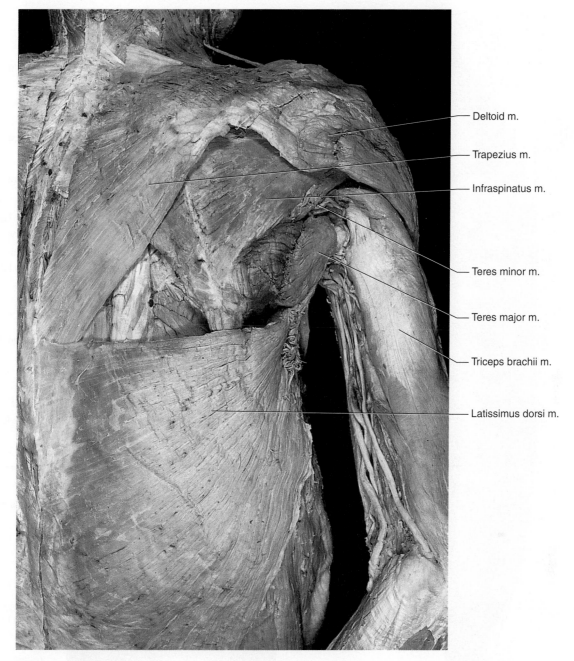

— Deltoid m.

— Trapezius m.

— Infraspinatus m.

— Teres minor m.

— Teres major m.

— Triceps brachii m.

— Latissimus dorsi m.

PLATE SIXTY-NINE Posterior view of the right thorax and ar*m*. (*m*. stands for muscle.)

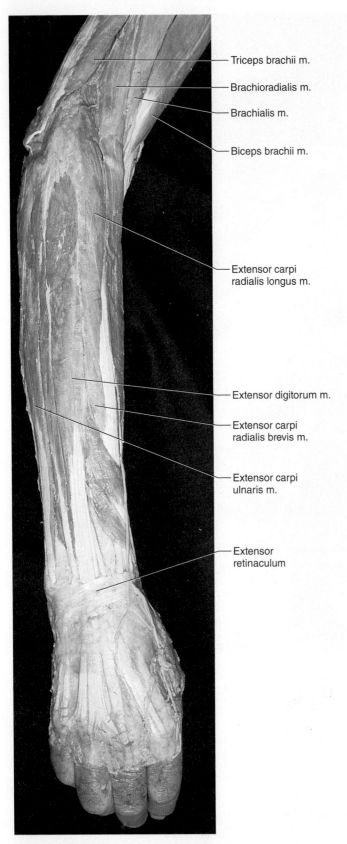

Triceps brachii m.

Brachioradialis m.

Brachialis m.

Biceps brachii m.

Extensor carpi radialis longus m.

Extensor digitorum m.

Extensor carpi radialis brevis m.

Extensor carpi ulnaris m.

Extensor retinaculum

PLATE SEVENTY Posterior view of the right forearm and hand. (*m.* stands for muscle.)

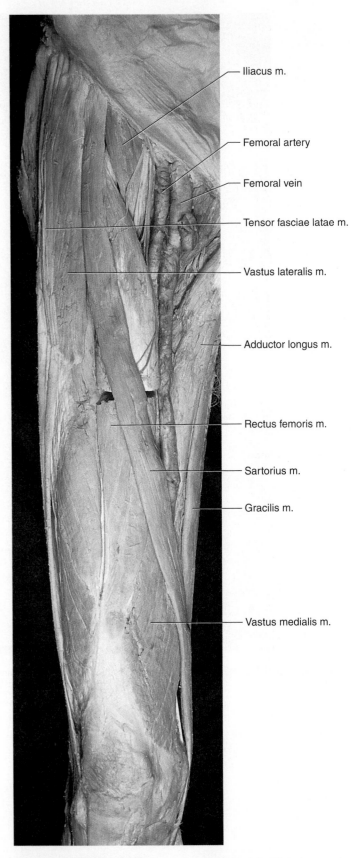

Iliacus m.

Femoral artery

Femoral vein

Tensor fasciae latae m.

Vastus lateralis m.

Adductor longus m.

Rectus femoris m.

Sartorius m.

Gracilis m.

Vastus medialis m.

PLATE SEVENTY-ONE Anterior view of the right thigh. (*m.* stands for muscle.)

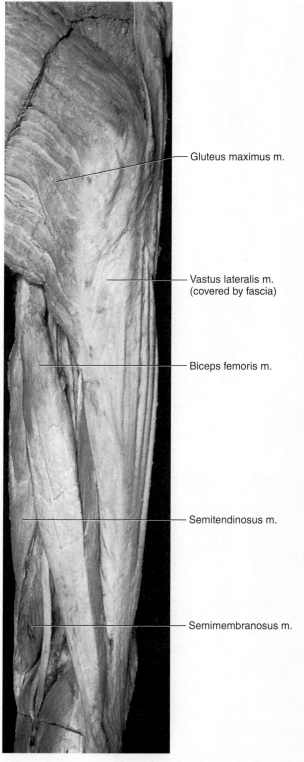

- Gluteus maximus m.

- Vastus lateralis m.
  (covered by fascia)

- Biceps femoris m.

- Semitendinosus m.

- Semimembranosus m.

PLATE SEVENTY-TWO Posterior view of the right thigh. (*m.* stands for muscle.)

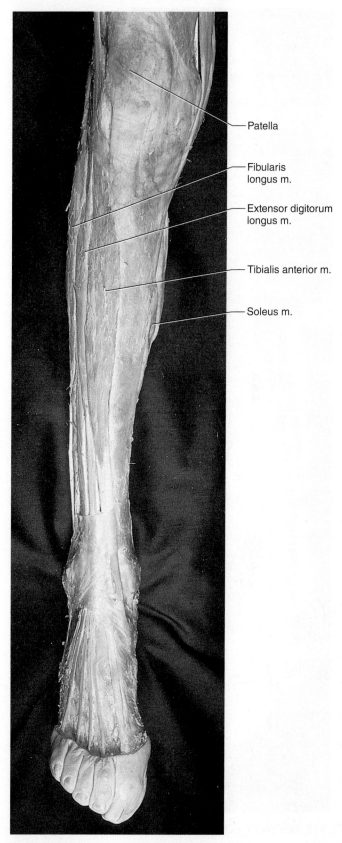

- Patella

- Fibularis
  longus m.

- Extensor digitorum
  longus m.

- Tibialis anterior m.

- Soleus m.

PLATE SEVENTY-THREE Anterior view of the right leg. (*m.* stands for muscle.)

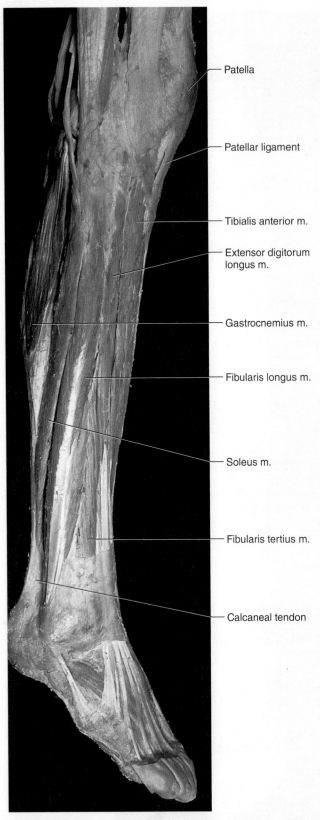

Patella

Patellar ligament

Tibialis anterior m.

Extensor digitorum
longus m.

Gastrocnemius m.

Fibularis longus m.

Soleus m.

Fibularis tertius m.

Calcaneal tendon

PLATE SEVENTY-FOUR Lateral view of the right leg. (*m.* stands for muscle.)

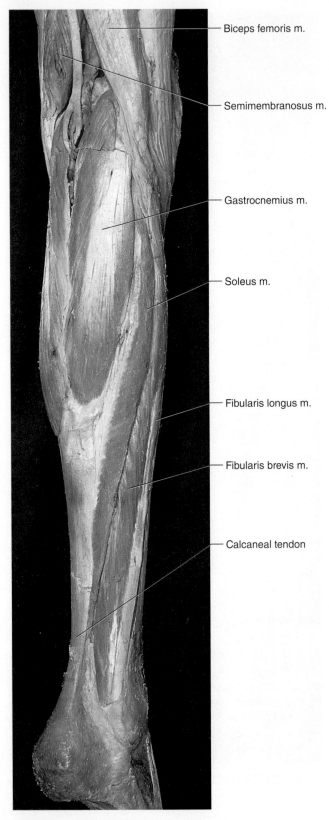

Biceps femoris m.

Semimembranosus m.

Gastrocnemius m.

Soleus m.

Fibularis longus m.

Fibularis brevis m.

Calcaneal tendon

PLATE SEVENTY-FIVE Posterior view of the right leg. (*m.* stands for muscle.)

10

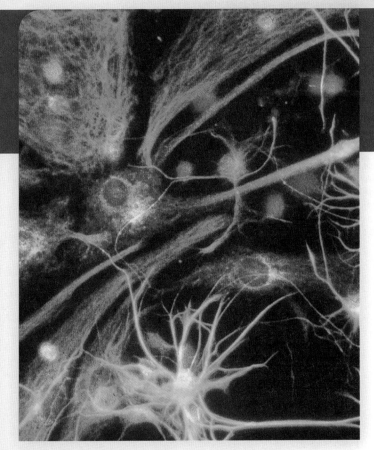

These progenitor cells will give rise to astrocytes (green) that supply neurons with nutrients. In this immunofluorescent light micrograph, cell nuclei are stained blue (1,150×).

Nervous System I
Basic Structure and Function

 LEARNING OUTCOMES

After you have studied this chapter, you should be able to:

10.1 Overview of the Nervous System
1 Describe the general functions of the nervous system. (p. 360)
2 Identify the two types of cells that comprise nervous tissue. (p. 360)
3 Identify the two major groups of nervous system organs. (p. 361)

10.2 General Functions of the Nervous System
4 List the functions of sensory receptors. (p. 361)
5 Describe how the nervous system responds to stimuli. (p. 361)

10.3 Description of Cells of the Nervous System
6 Describe the parts of a neuron. (p. 363)
7 Describe the relationships among myelin, the neurilemma, and nodes of Ranvier. (p. 363)
8 Distinguish between the sources of white matter and gray matter. (p. 363)

10.4 Classification of Cells of the Nervous System
9 Identify structural and functional differences among neurons. (pp. 363–368)
10 Identify the types of neuroglia in the central nervous system and their functions. (pp. 368–369)
11 Describe the role of Schwann cells in the peripheral nervous system. (p. 370)

10.5 The Synapse
12 Explain how information passes from a presynaptic neuron to a postsynaptic cell. (pp. 371–372)

10.6 Cell Membrane Potential
13 Explain how a cell membrane becomes polarized. (p. 372)
14 Describe the events leading to the generation of an action potential. (p. 375)
15 Explain how action potentials move down an axon. (pp. 375–377)
16 Compare impulse conduction in myelinated and unmyelinated neurons. (p. 378)

10.7 Synaptic Transmission
17 Identify the changes in membrane potential associated with excitatory and inhibitory neurotransmitters. (p. 379)
18 Explain what prevents a postsynaptic cell from being continuously stimulated. (p. 381)

10.8 Impulse Processing
19 Describe the basic ways in which the nervous system processes information. (pp. 382–383)

THE WHOLE PICTURE

Snap your fingers! In the time it took to do that, a decision made in a part of your brain that controls skeletal muscles resulted in impulses along motor neuron axons to the muscles in your hand, releasing acetylcholine (ACh) at neuromuscular junctions. As soon as the muscles contracted during the "snap," a decision in the brain stopped the action. Impulses ceased, enzymes broke down the ACh, active transport carried calcium back into storage in the muscle cells, and your hand relaxed.

Think about how quickly these events unfolded. Then focus on all of the activities going on in your body while reading this passage. Your nervous system exerts precise control over many of the body's functions, and is responsible for your awareness of some of what is happening.

apreyealed.com

Module 7: Nervous System

UNDERSTANDING WORDS

astr-, starlike: *astrocyte*—star-shaped neuroglia.

ax-, axle: *axon*—cylindrical nerve process that conducts impulses away from a neuron cell body.

bi-, two: *bipolar* neuron—neuron with two processes extending from the cell body.

dendr-, tree: *dendrite*—branched nerve process that serves as the receptor surface of a neuron.

ependym-, tunic: *ependyma*—neuroglia that line spaces in the brain and spinal cord.

-lemm, rind or peel: neuri*lemma*—sheath that surrounds the myelin of a nerve cell process.

moto-, moving: *motor* neuron—neuron that stimulates a muscle to contract or a gland to release a secretion.

multi-, many: *multi*polar neuron—neuron with many processes extending from the cell body.

oligo-, few: *oligo*dendrocyte—small type of neuroglia with few cellular processes.

peri-, all around: *peripheral* nervous system—portion of the nervous system that consists of the nerves branching from the brain and spinal cord.

saltator-, a dancer: *saltatory* conduction—impulse conduction in which the impulse seems to jump from node to node along the nerve fiber.

sens-, feeling: *sensory* neuron—neuron that can be stimulated by a sensory receptor and conducts impulses into the brain or spinal cord.

syn-, together: *synapse*—junction between two neurons.

uni-, one: *uni*polar—neuron with only one process extending from the cell body.

10.1 | Overview of the Nervous System

The nervous system oversees all that we do and determines who we are. Through a vast communicating network of cells and the information that they send and receive, the nervous system can detect changes affecting the body, make decisions, and stimulate muscles or glands to respond. Typically, these responses counteract the effects of the changes, and in this way, the nervous system helps maintain homeostasis.

The nervous system is composed predominantly of neural tissue, but also includes blood vessels and connective tissue. Neural tissue consists of two cell types: nerve cells, or **neurons** (nu′ronz), and **neuroglia** (nu-ro′gle-ah) (or neuroglial cells). Clinical Application 10.1 discusses how environmental changes may trigger migraine headaches, a common medical problem attributed to the nervous system that may involve its blood supply as well as neurons.

Neurons are specialized to react to physical and chemical changes in their surroundings. Small cellular processes called **dendrites** (den′drītz) receive the input. A longer process called an **axon** (ak′son), or nerve fiber, carries the information away from the cell in the form of bioelectric signals, called **impulses** (action potentials), which allow the neuron to communicate with other neurons and with cells outside the nervous system (fig. 10.1). Typically, axons within the nervous system are not isolated, but bundled in groups. In the peripheral nervous system, such bundles of axons are called *nerves*. In the central nervous system (brain and spinal cord) they are called *tracts*.

Neuroglia are found throughout the nervous system, and in the brain they greatly outnumber neurons. It was once thought that neuroglia only fill spaces and surround or support neurons. Today we know that they have many other functions, including nourishing neurons and sending and receiving chemical messages.

Neuroglia assemble in a special, protective way in the brain. The blood-brain barrier offers an example. Most capillaries (the

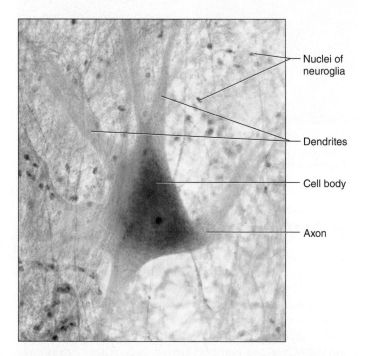

FIGURE 10.1 AP|R Neurons are the structural and functional units of the nervous system (600×). The dark spots in the area surrounding the neuron are the nuclei of neuroglia. Note the neuron processes (dendrites and a single axon).

🔘 *What structure forms the outer portion of the axon and dendrites of a neuron?*

Answer can be found in Appendix G.

smallest blood vessels) are "leaky," allowing small molecules to enter or leave the bloodstream. The cells that form capillaries in the brain, in contrast, are much more tightly connected, thanks partly to neuroglia called astrocytes. The barrier that this specialized architecture provides shields delicate brain tissue from chemical fluctuations, blocking entry to many substances (see From Science to Technology 5.1, p. 150). Drug developers must consider the barrier when formulating drugs that act in the brain, including chemicals that let the drug through.

An important part of the nervous system at the cellular level is not a cell at all, but the small space between a neuron and the cell(s) with which it communicates, called a **synapse** (sin'aps). Much of the work of the nervous system is to send and receive electrochemical messages across synapses. Biological messenger molecules called **neurotransmitters** (nu"ro-trans-mit'erz) convey this neural information.

The organs of the nervous system can be divided into two groups. One group, consisting of the brain and spinal cord, forms the **central nervous system (CNS).** The other group, composed of the nerves (cranial and spinal nerves) that connect the central nervous system to other body parts, is the **peripheral nervous system (PNS)** (fig. 10.2).

PRACTICE

1 Name two cell types in neural tissue.

2 Name two groups of nervous system organs.

10.2 | General Functions of the Nervous System

The three general functions of the nervous system—receiving information, deciding what to do, and acting on those decisions—are termed sensory, integrative, and motor. Structures called **sensory receptors** at the ends of neurons in the peripheral nervous system (peripheral neurons) provide the sensory function of the nervous system (see chapter 11, p. 411). These receptors gather information by detecting changes inside and outside the body. They monitor external environmental factors such as light and sound intensities as well as the temperature, oxygen concentration, and other conditions of the body's internal environment.

Sensory receptors convert (or transduce) their information into impulses, which are then conducted along peripheral nerves to the CNS. There the signals are integrated. That is, they are brought together, creating sensations, adding to memory, or helping produce thoughts. Following integration, conscious or subconscious decisions are made and then acted upon by means of motor functions.

Neurons that conduct impulses from the CNS to responsive structures called *effectors* carry out the motor functions of the nervous system. These effectors are outside the nervous system and include muscles and glands whose actions are either controlled or modified by nerve activity. The motor portion of the PNS can be subdivided into the somatic and the autonomic nervous systems. The **somatic nervous system** communicates voluntary (conscious) instructions originating in the CNS to skeletal muscles, causing contraction. The **autonomic nervous system** communicates instructions from the CNS that control viscera, such as the heart and various glands, and thus causes involuntary subconscious actions.

PRACTICE

3 List the general functions of the nervous system.

An action potential is often referred to as a "nerve impulse." This may be misleading, because in anatomy the term *nerve* refers to a bundle of axons, not an individual cell. A similar situation arises when referring to a neuron as a "nerve cell." We may sometimes use the terms *nerve impulse* and *nerve cell* because they are familiar, but we feel it is important to point out this departure from strict anatomical terminology.

CAREER CORNER
Pharmacist

Shortly before the man is discharged from the cardiac care unit of a hospital, a nurse emails prescriptions to the supermarket pharmacy near the patient's home. By the time the man's daughter arrives at the pharmacy a few hours later, not only are all five drugs packaged and ready to go, but the pharmacist explains what each drug does and how it should be taken. She also answers the daughter's questions about potential interactions of the prescription drugs with over-the-counter medications that her father sometimes takes.

A pharmacist is a health-care professional who measures and packages prescription drugs and explains to patients or their caregivers what each drug is for, how it should be taken, and possible adverse effects. In many instances a pharmacist can knowledgeably answer the same questions about a drug that a doctor might be asked.

In college, a future pharmacist should take a variety of science and math courses. The profession requires knowledge of chemistry (especially organic chemistry), biology, physics, and anatomy and physiology. Attention to detail is a critical skill—an error in dispensing a drug can have tragic consequences.

To become a pharmacist, one must earn a doctor of pharmacy, or "pharmD" degree, and pass state licensure exams. A pharmD is a graduate degree. Admission to pharmacy school requires at least two years of college and a passing score on the pharmacy college admission test. The four-year program includes 7 to 10 rotations of 4 to 6 weeks each, during which the student works in various clinical and pharmaceutical settings. Following graduate school a pharmacist might acquire training in a specialty, such as psychopharmaceuticals, chemotherapy, or working with radioactive drugs, in a one- to two-year residency program.

Pharmacists work in retail drug, grocery, and "big box" stores; at hospitals as part of a clinical team; and in long-term care facilities such as rehabilitation centers and nursing homes. They may work for the government, such as at the Food and Drug Administration, or as sales representatives for pharmaceutical companies.

The signs of a migraine are unmistakable—a pounding head, typically in one area, waves of nausea, shimmering images in the peripheral visual field called an "aura," and extreme sensitivity to light or sound. Inherited susceptibilities and environmental factors probably cause migraines. Environmental triggers include sudden exposure to bright light, eating a particular food (chocolate, red wine, nuts, and processed meats top the list), lack of sleep, stress, high altitude, stormy weather, and excessive caffeine or alcohol intake. Hormonal influences may also be involved, because about three-quarters of the 300 million people who suffer from migraines worldwide are women between the ages of 15 and 55. Many women get migraines just before menstruation.

A migraine attack may last from 4 to 72 hours. It is due to a phenomenon called "cortical spreading depression," in which an intense wave of excitation followed by a brief period of unresponsiveness in certain neurons stimulates areas at the base of the brain to produce pain sensations. The excitation and dampening of the activity level of these neurons also triggers changes in blood flow in the brain that were once thought to be the direct cause of migraine.

Drugs called triptans can very effectively halt a migraine attack, but must be taken as soon as symptoms begin. Triptans block the release of neurotransmitter from certain neurons. Because triptans constrict blood vessels throughout the body, making them dangerous for some people, newer migraine drugs have been developed that block the specific neurotransmitter that these neurons release (calcitonin gene-related peptide), better targeting the therapeutic effect.

A newer treatment option is "transcranial magnetic stimulation," which is used to treat migraine among the third of sufferers who experience aura. A device is held to the back of the head, and a button is pressed that sends magnetic energy into the brain.

Several drugs developed to treat other conditions are used on a long-term, daily basis to lessen the frequency of migraines. These drugs include certain antidepressants, anticonvulsants, and drugs used to treat high blood pressure (calcium channel blockers and beta blockers). A physician considers an individual's family and health history before prescribing these drugs to prevent migraine.

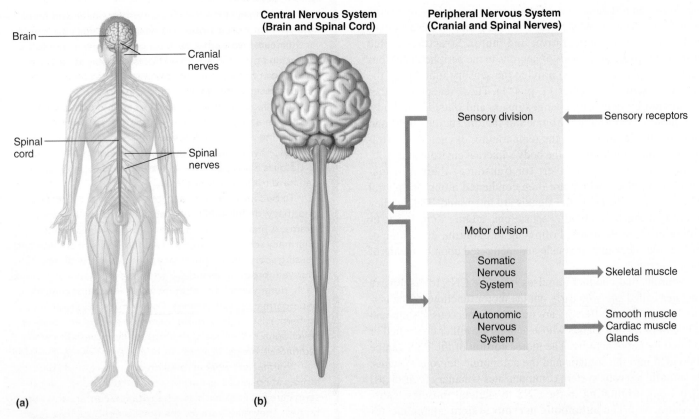

FIGURE 10.2 A diagrammatic representation of the nervous system. (*a*) The nervous system includes the central nervous system (brain and spinal cord) and the peripheral nervous system (cranial nerves and spinal nerves). (*b*) The nervous system receives information from sensory receptors and initiates responses through effector organs (muscles and glands).

10.3 | Description of Cells of the Nervous System

Neurons vary in size and shape. They may differ in the lengths and sizes of their axons and dendrites and in the number of processes. Despite this variability, neurons share certain features. Every neuron has a **cell body,** dendrites, and an axon. **Figure 10.3** shows some of the other structures common to neurons.

A neuron's cell body (soma or perikaryon) contains granular cytoplasm, mitochondria, lysosomes, a Golgi apparatus, and many microtubules. A network of fine threads called **neurofilaments** extends into the axon and supports it. Scattered throughout the cytoplasm are many membranous packets of **chromatophilic substance** (Nissl bodies), which consist mainly of rough endoplasmic reticulum. Cytoplasmic inclusions in neurons include glycogen, lipids, and pigments such as melanin. Near the center of the neuron cell body is a large, spherical nucleus with a conspicuous nucleolus.

Dendrites are typically highly branched, providing receptive surfaces with which processes from other neurons communicate. (In some types of neurons, the cell body provides such a receptive surface.) Some dendrites have tiny, thornlike spines (dendritic spines) on their surfaces, which are contact points for other neurons.

A neuron may have many dendrites, but no more than one axon. In most neurons the axon arises from the cell body as a cone-shaped thickening called the *axon hillock.* The cytoplasm of the axon includes many mitochondria, microtubules, and neurofibrils (ribosomes are found only in the cell body). The axon may give off branches, called *collaterals.* Near its end, an axon may have many fine extensions, each with a specialized ending called an *axon terminal.* The axon terminal ends as a *synaptic knob* close to the receptive surface of another cell, separated only by a space called the **synaptic cleft.** The general pattern is that neurons receive input through the dendrites and the cell body, and send output in the form of an impulse conducted away from the cell body, down the axon.

An axon, in addition to conducting impulses, conveys biochemicals and organelles, which can be quite a task in these long cells. In this activity, called *axonal transport,* movement occurs in both directions between the cell body and the ends of the axon. For example, enzymes required for neurotransmitter synthesis are produced in the cell body and transported to the axon terminals. Old organelles and other cellular components may be transported in the reverse direction to be recycled. It is a highly regulated process.

In the PNS, neuroglia called **Schwann cells** encase the large axons of peripheral neurons in lipid-rich sheaths. These tight coverings form as Schwann cell membranes wind and wrap around axons. The layers are composed of **myelin** (mi'ě-lin), which consists of several types of lipids and proteins. Myelin gives the cell membranes of Schwann cells a higher proportion of lipid than other cell membranes. This coating is called a *myelin sheath.* The parts of the Schwann cells that contain most of the cytoplasm and the nuclei remain outside the myelin sheath and comprise a **neurilemma** (nur″ĭlem′ah)**,** or *neurilemmal sheath,* which surrounds the myelin sheath. Narrow gaps in the myelin sheath between Schwann cells are called **nodes of Ranvier** (fig. 10.4).

Schwann cells also enclose, but do not wind around, the smallest axons of peripheral neurons. Consequently, these axons do not have myelin sheaths. Instead, the axon or a group of axons may lie partially or completely in a longitudinal groove of a Schwann cell.

Axons that have myelin sheaths are called *myelinated* (medullated) axons, and those that do not have these sheaths are *unmyelinated axons* (fig. 10.5). Myelinated axons conduct impulses rapidly compared to unmyelinated axons. Groups of myelinated axons appear white. The *white matter* in the brain and spinal cord gets its color from masses of myelinated axons. In the CNS, myelin is produced by a type of neuroglia called an **oligodendrocyte** rather than by a Schwann cell. In the brain and spinal cord, myelinated axons do not have neurilemmae.

Unmyelinated nerve tissue appears gray. Thus, the *gray matter* in the CNS contains many unmyelinated axons and neuron cell bodies. Clinical Application 10.2 discusses multiple sclerosis, a condition in which neurons in the brain and spinal cord lose their myelin.

 PRACTICE

4 Describe a neuron.

5 Explain how an axon in the peripheral nervous system becomes myelinated.

Myelin begins to form on axons during the fourteenth week of prenatal development. At the time of birth, many axons are not completely myelinated. All myelinated axons have begun to develop sheaths by the time a child starts to walk, and myelination continues into adolescence.

Excess myelin seriously impairs nervous system functioning. In Tay-Sachs disease, deficiency of a lysosomal enzyme causes myelin to accumulate, burying neurons in lipid. An affected child begins to show symptoms by six months of age, gradually losing sight, hearing, and muscle function until death occurs by age four. Thanks to genetic screening among people of eastern European descent and other groups who are most likely to carry this mutation, Tay-Sachs disease is extremely rare.

Too little myelin is devastating, too. Clinical Application 3.2 (p. 94) describes adrenoleukodystrophy, in which myelin vanishes in the brains and spinal cords of boys.

10.4 | Classification of Cells of the Nervous System

The cells of nervous tissue (neurons and neuroglia) are intimately related. They descend from the same neural stem cells and remain associated throughout their existence.

Classification of Neurons

Neurons can be classified into three major groups based on *structural differences*, as **figure 10.6** shows. Each type of neuron is specialized to conduct an impulse in one direction.

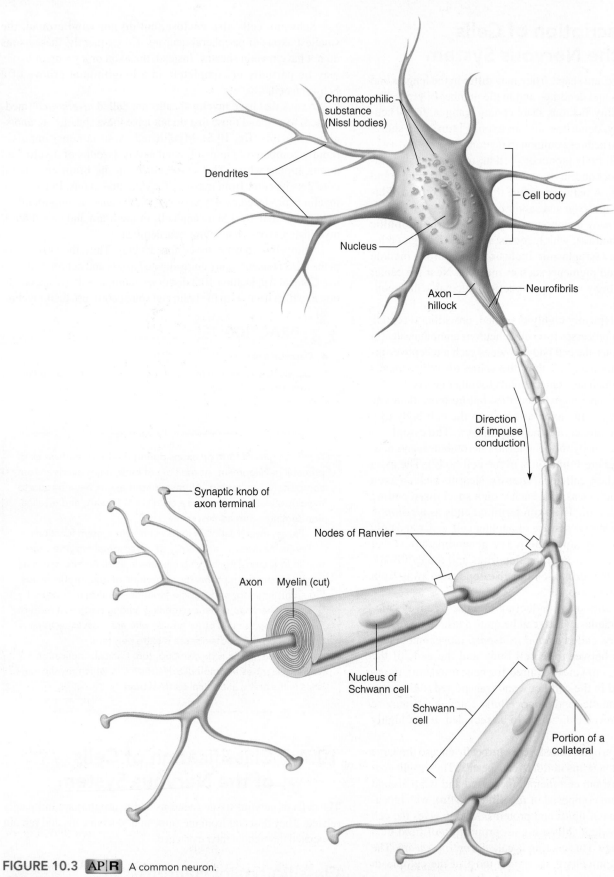

FIGURE 10.3 **AP|R** A common neuron.

> *The genes in the nucleus of a neuron are the same as the genes in the nuclei of its Schwann cells. What causes these cells to have such different form and function?*
>
> **Answer can be found in Appendix G.**

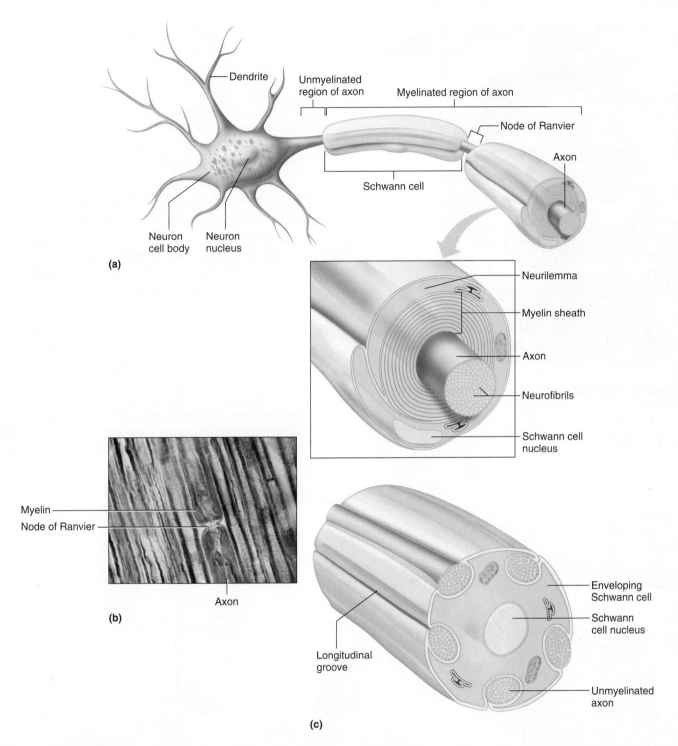

FIGURE 10.4 AP|R A myelinated axon. (*a*) The part of a Schwann cell that winds tightly around an axon forms the myelin sheath. The cytoplasm and nucleus of the Schwann cell, remaining on the outside, form the neurilemma. (*b*) Light micrograph of a myelinated axon (longitudinal section) (650×). (*c*) An axon lying in a longitudinal groove of a Schwann cell does not have a myelin sheath.

1. A **multipolar neuron** has many processes arising from its cell body. Only one is an axon; the rest are dendrites. Most neurons whose cell bodies lie within the brain or spinal cord are of this type. Some multipolar neurons associated with the autonomic nervous system neurons aggregate in specialized masses of nerve tissue called **ganglia,** which are located outside the brain and spinal cord. Others are found in specialized parts of the eyes. The neuron illustrated in figure 10.3 is multipolar.

2. The cell body of a **bipolar neuron** has only two processes, one arising from either end. Although these processes are similar in structure, one is an axon and the other is a dendrite. Bipolar neurons are found in specialized parts of the eyes, nose, and ears.

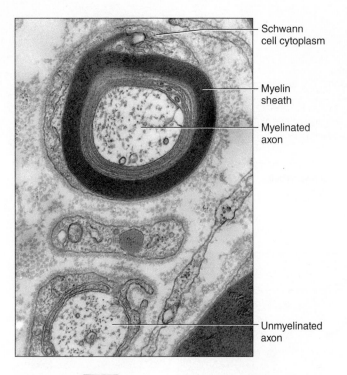

Schwann
cell cytoplasm

Myelin
sheath

Myelinated
axon

Unmyelinated
axon

FIGURE 10.5 AP|R A falsely colored transmission electron micrograph of myelinated and unmyelinated axons in cross section (30,000×).

3. Each **unipolar neuron** has a single process extending from its cell body. (These are also called pseudounipolar, because they start out with two processes that merge into one during development.) A short distance from the cell body, this process divides into two branches, which really function as a single axon: One branch (peripheral process) has dendrites near a peripheral body part. The other branch (central process) enters the brain or spinal cord. The cell bodies of most unipolar neurons are found in ganglia.

Neurons can also be classified by *functional differences* into the following groups, depending on whether they carry information into the CNS, completely within the CNS, or out of the CNS (fig. 10.7).

1. **Sensory neurons** (afferent neurons) conduct impulses from peripheral body parts into the brain or spinal cord. At their distal ends, the dendrites of these neurons or specialized structures associated with them act as sensory receptors, detecting changes in the outside world (for example, eyes, ears, or touch receptors in the skin) or in the body (for example, temperature or blood pressure receptors). When sufficiently stimulated, sensory receptors trigger impulses that travel on sensory neuron axons into the brain or spinal cord. Most sensory neurons are unipolar, as shown in figure 10.7, although some are bipolar and others are multipolar.

2. **Interneurons** (also called association or internuncial neurons) lie within the brain or spinal cord. They are multipolar and form links with other neurons. Interneurons relay information

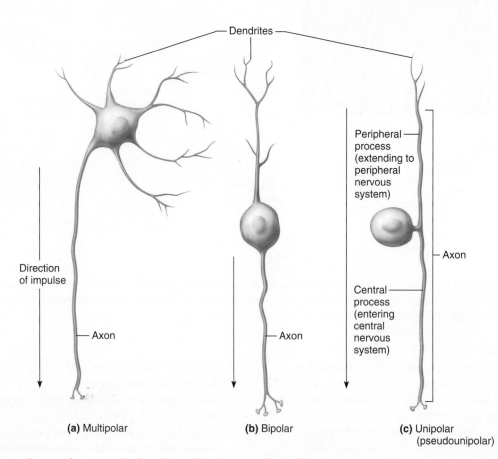

Dendrites

Peripheral process (extending to peripheral nervous system)

Axon

Central process (entering central nervous system)

Direction of impulse

Axon

Axon

(a) Multipolar

(b) Bipolar

(c) Unipolar (pseudounipolar)

FIGURE 10.6 Structural types of neurons include (*a*) the multipolar neuron, (*b*) the bipolar neuron, and (*c*) the unipolar neuron.

Multiple sclerosis (MS) is a disorder of the CNS that affects 2.5 million people worldwide, and 400,000 in North America. In addition to overt nervous system symptoms, affected individuals experience disability, mood problems such as depression, and great fatigue. Four subtypes of MS are recognized, based on the pattern of symptomatic periods over time. The two most common subtypes are relapsing-remitting (with episodes) and secondary progressive (when symptoms do not abate).

In MS, the myelin coating in various sites throughout the brain and spinal cord becomes inflamed due to an immune response and is eventually destroyed, leaving hard scars, called scleroses, that block the underlying neurons from conducting impulses. Muscles that no longer receive input from motor neurons stop contracting, and eventually, they atrophy. Symptoms reflect the specific neurons affected. Short-circuiting in one part of the brain affects fine coordination in one hand; in another part of the brain, malfunctioning neurons alter vision.

The first symptoms of MS are often blurred or double vision and weak, numb limbs that can make walking and standing difficult. Because in many cases these symptoms are intermittent, diagnosis may take a while. It is based on symptoms and repeated magnetic resonance (MR) scans that track the development of lesions. A diagnostic workup for MS might also include a lumbar puncture to rule out infection and an evoked potential test to measure electrical signals sent from the brain. About 70% of affected individuals first notice symptoms between the ages of twenty and forty; the earliest known age of onset is three years, and the latest, sixty-seven years. Some affected individuals eventually become permanently paralyzed. Women are twice as likely to develop MS as men, and Caucasians are more often affected than people of other races.

MS may develop when particular infections in certain individuals stimulate T cells (a type of white blood cell that takes part in immune responses) in the periphery, which then cross the blood-brain barrier. Here, these "myelin-reactive" T cells attack myelin-producing cells through a flood of inflammatory molecules and by stimulating other cells to produce antibodies against myelin.

A virus may lie behind the misplaced immune attack that is MS. Evidence includes the observations that viral infection can cause repeated bouts of symptoms and that MS is much more common in some geographical regions (the temperate zones of Europe, South America, and North America) than others, suggesting a pattern of infection.

Some people have such mild symptoms of MS that no treatment is required. In others, drugs can help to manage symptoms. Drugs to decrease bladder spasms can temper problems of urinary urgency and incontinence. Antidepressants are sometimes prescribed, and short-term steroid drugs are used to shorten the length of acute disabling relapses. Muscle relaxants ease stiffness and spasms.

Several drugs are used for long-term treatment of MS. Beta interferons are immune system biochemicals that diminish the intensity of flare-ups, but they may cause flulike adverse effects. Beta interferons are self-injected once to several times a week.

Glatiramer is an alternative to beta interferon that is prescribed if the course of the disease is "relapsing remitting," with periodic flare-ups. Glatiramer is self-injected daily and dampens the immune system's attack on myelin. It consists in part of myelin basic protein, the most abundant protein of myelin. In response to the drug, T cells decrease inflammation. Glatiramer also stimulates increased production of brain-derived neurotrophic factor, which protects axons.

Mitoxantrone is another drug that halts the immune response against CNS myelin, but because it can damage the heart, it is typically used in severe cases of MS and only for up to two years. Another drug, natalizumab, prevents T cells from binding blood vessels in the brain, also quelling the abnormal immune response against myelin. It too may have rare but serious adverse effects.

Experimental approaches attempt to temper the body's attack on its own myelin. "Personalized T cell immunotherapy" redirects a patient's immune response to attack the myelin-reactive T cells. Autologous bone marrow transplant, in which a person's immune system is destroyed and then reconstituted, apparently rids the body of myelin-reactive T cells. It works best for patients with an inflammatory relapsing-remitting form of the disease. Many patients treated with bone marrow transplants remain symptom-free more than a decade later.

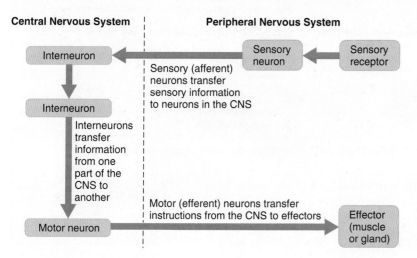

Central Nervous System | **Peripheral Nervous System**

Sensory (afferent) neurons transfer sensory information to neurons in the CNS

Interneurons transfer information from one part of the CNS to another

Motor (efferent) neurons transfer instructions from the CNS to effectors

FIGURE 10.7 Neurons are classified by function as well as structure. Sensory (afferent) neurons carry information into the central nervous system (CNS), interneurons are completely within the CNS, and motor (efferent) neurons carry instructions to effectors.

from one part of the brain or spinal cord to another. That is, they may conduct incoming sensory information to appropriate regions for processing and interpreting. Other instructions are transferred to motor neurons. The cell bodies of some interneurons aggregate in specialized masses of nervous tissue called **nuclei** (singular, *nucleus*). Nuclei are similar to ganglia, but are within the CNS.

3. **Motor neurons** (efferent neurons) are multipolar and conduct impulses out of the brain or spinal cord to effectors. For example, when motor neurons stimulate muscle cells, the muscle cells contract; when motor neurons stimulate glands, the glands release secretions.

   Motor neurons of the somatic nervous system (see fig. 10.2) that control skeletal muscle contraction are under voluntary (conscious) control. Those that control cardiac and smooth muscle contraction and the secretions of glands are part of the autonomic nervous system and are largely under involuntary control.

Table 10.1 summarizes the classification of neurons.

PRACTICE

6 Describe the types of neurons based on structural characteristics.

7 Indicate the function of sensory neurons, interneurons, and motor neurons.

Classification of Neuroglia

Neuroglia were once thought to be mere bystanders to neural function, providing scaffolding and controlling the sites at which neurons contact one another (figs. 10.8 and 10.9). These important cells have additional functions. In the embryo, neuroglia guide neurons to their positions and may stimulate them to specialize. Neuroglia also produce the growth factors that nourish neurons and remove excess

ions and neurotransmitters that accumulate between neurons. In cell culture experiments, certain types of neuroglia (astrocytes) signal neurons to form and maintain synapses.

Neuroglia of the CNS

The four types of CNS neuroglia are astrocytes, oligodendrocytes, microglia, and ependyma:

1. As their name implies, **astrocytes** are star-shaped cells. They are commonly found between neurons and blood vessels, where they provide support and hold structures together with abundant cellular processes. Astrocytes aid metabolism of certain substances, such as glucose, and they may help regulate the concentrations of important ions, such as potassium ions, in the interstitial space of nervous tissue. Astrocytes also respond to injury of brain tissue and form a special type of scar tissue, which fills spaces and closes gaps in the CNS. These multifunctional cells also have a nutritive function, regulating movement of substances from blood vessels to neurons and bathing nearby neurons in growth factors. Astrocytes play an important role in the blood-brain barrier, which restricts movement of substances between the blood and the CNS. Gap junctions link astrocytes to one another, forming protein-lined channels through which calcium ions travel, possibly stimulating neurons.

2. **Oligodendrocytes** resemble astrocytes but are smaller and have fewer processes. They form in rows along axons (nerve fibers), and myelinate these axons in the brain and spinal cord.

   Unlike the Schwann cells of the PNS, oligodendrocytes can send out a number of processes, each of which forms a myelin sheath around a nearby axon. In this way, a single oligodendrocyte may myelinate many axons. However, these cells do not form neurilemmae.

3. **Microglia** are small cells and have fewer processes than other types of neuroglia. These cells are scattered throughout the CNS, where they help support neurons and phagocytize

TABLE 10.1	Types of Neurons	
A. CLASSIFIED BY STRUCTURE		
Type	**Structural Characteristics**	**Location**
1. Multipolar neuron	Cell body with many processes, one of which is an axon, the rest dendrites	Most common type of neuron in the brain and spinal cord; also found in ganglia of the autonomic nervous system
2. Bipolar neuron	Cell body with a process arising from each end, one axon and one dendrite	In specialized parts of the eyes, nose, and ears
3. Unipolar neuron	Cell body with a single process that divides into two branches and functions as an axon	Found in ganglia
B. CLASSIFIED BY FUNCTION		
Type	**Functional Characteristics**	**Structural Characteristics**
1. Sensory neuron	Conducts impulses from receptors in peripheral body parts into the brain or spinal cord	Most unipolar, some bipolar, some multipolar
2. Interneuron	Relays information between neurons in the brain and spinal cord	Multipolar
3. Motor neuron	Conducts impulses from the brain or spinal cord out to effectors—muscles or glands	Multipolar

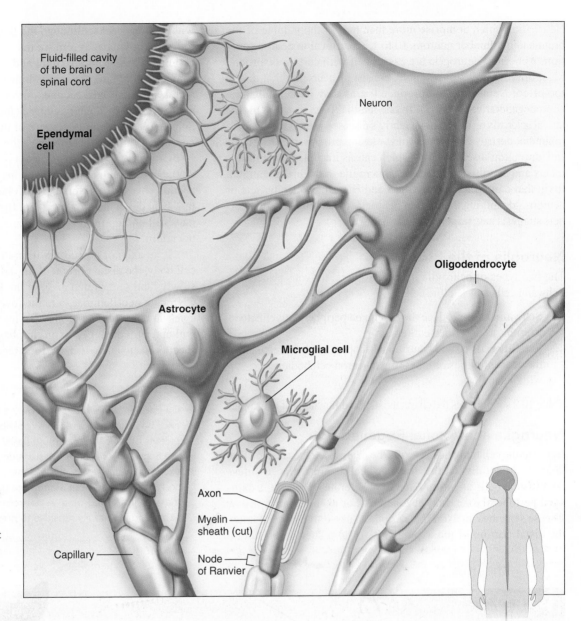

Fluid-filled cavity of the brain or spinal cord

Ependymal cell

Neuron

Oligodendrocyte

Astrocyte

Microglial cell

Axon

Myelin sheath (cut)

Capillary

Node of Ranvier

FIGURE 10.8 Types of neuroglia in the central nervous system include the astrocyte, oligodendrocyte, microglial cell, and ependymal cell. Motile cilia are on most ependymal cells during development and early childhood, but in the adult are mostly on ependymal cells in the ventricles of the brain (spaces where cerebrospinal fluid forms).

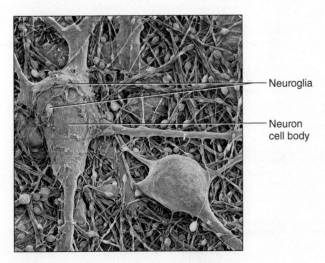

Neuroglia

Neuron cell body

FIGURE 10.9 A scanning electron micrograph of a neuron cell body and some of the neuroglia associated with it (5,000×).

bacterial cells and cellular debris. They usually proliferate whenever the brain or spinal cord is inflamed because of injury or disease.

4. **Ependyma** are cuboidal or columnar cells in shape and may have cilia. They form the inner lining of the *central canal* that extends downward through the spinal cord. Ependymal cells also form a one-cell-thick epithelial-like membrane that covers the inside of spaces in the brain called *ventricles* (see chapter 11, pp. 391–394). Here, gap junctions join ependymal cells, allowing free exchange between cells. The ependymal layer itself is porous, allowing substances to diffuse freely between the interstitial fluid of the brain tissues and the fluid (cerebrospinal fluid) in the ventricles.

Ependymal cells also cover the specialized capillaries called *choroid plexuses* associated with the ventricles of the brain. Here they help regulate the composition of the cerebrospinal fluid.

Neuroglia, which comprise more than half of the volume of the brain and outnumber neurons 10 to 1, are critical to neuron function. Abnormal neuroglia are associated with certain disorders. Most brain tumors, for example, consist of neuroglia that divide too often. Neuroglia that produce toxins may lie behind some neurodegenerative disorders. In one familial form of amyotrophic lateral sclerosis (ALS, or Lou Gehrig's disease), astrocytes release a toxin that destroys motor neurons, causing progressive weakness. In Huntington disease (HD), which causes uncontrollable movements and cognitive impairment, microglia in the brain release a toxin that damages neurons. ALS and HD affect specific sets of neurons. Identifying the roles of neuroglia in nervous system disorders suggests new targets for treatments.

Neuroglia of the PNS

The two types of neuroglia in the peripheral nervous system are Schwann cells and satellite cells:

1. **Schwann cells** produce the myelin on peripheral myelinated neurons, as described earlier.
2. **Satellite cells** provide nutritional support and help regulate the concentrations of ions around neuron cell bodies within ganglia.

Table 10.2 summarizes the characteristics and functions of neuroglia.

Neuroglia and Axonal Regeneration

Injury to the cell body usually kills the neuron, and because mature neurons do not divide, the destroyed cell is not replaced unless neural stem cells are stimulated to proliferate. However, a damaged peripheral axon may regenerate. For example, if injury or disease separates an axon in a peripheral nerve from its cell body, the distal portion of the axon and its myelin sheath deteriorate within a few weeks, although the Schwann cells and neurilemma remain. Macrophages remove the fragments of myelin and other cellular debris. The proximal end of the injured axon develops sprouts shortly after the injury. Influenced by nerve growth factors that nearby neuroglia secrete, one of these sprouts may grow into a tube formed by the remaining Schwann cells. At the same time, Schwann cells along the length of the regenerating portion form new myelin around the growing axon.

Growth of a regenerating axon is slow (up to 4 millimeters per day), but eventually the new axon may reestablish the former connection (fig. 10.10). Nerve growth factors, secreted by neuroglia, may help direct the growing axon. However, the regenerating axon may still end up in the wrong place, so full function often does not return.

If an axon of a neuron within the CNS is separated from its cell body, the distal portion of the axon will degenerate, but more slowly than a separated axon in the PNS. However, axons in the CNS lack neurilemmae, and the myelin-producing oligodendrocytes do not proliferate following injury. Consequently, the proximal end of a damaged axon that begins to grow has no tube of sheath cells to guide it. Therefore, regeneration is unlikely.

If a peripheral nerve is severed, it is important that the two cut ends be connected as soon as possible so that the regenerating sprouts of the axons can more easily reach the tubes formed by the Schwann cells on the distal side of the gap. When the gap exceeds 3 millimeters, the regenerating axons may not line up correctly with the distal side of the gap and may form a tangled mass called a *neuroma*. A neuroma is composed of sensory axons and is painfully sensitive to pressure.

TABLE 10.2	Types of Neuroglia	
Type	**Characteristics**	**Functions**
CNS		
Astrocytes	Star-shaped cells between neurons and blood vessels	Structural support, formation of scar tissue, transport of substances between blood vessels and neurons, communicate with one another and with neurons, mop up excess ions and neurotransmitters, induce synapse formation
Oligodendrocytes	Shaped like astrocytes, but with fewer cellular processes, in rows along axons	Form myelin sheaths in the brain and spinal cord, produce nerve growth factors
Microglia	Small cells with few cellular processes and found throughout the CNS	Structural support and phagocytosis (immune protection)
Ependyma	Cuboidal and columnar cells in the lining of the ventricles of the brain and the central canal of the spinal cord	Form a porous layer through which substances diffuse between the interstitial fluid of the brain and spinal cord and the cerebrospinal fluid
PNS		
Schwann cells	Cells with abundant, lipid-rich membranes that wrap tightly around the axons of peripheral neurons	Form myelin sheaths
Satellite cells	Small, cuboidal cells that surround cell bodies of neurons in ganglia	Support ganglia

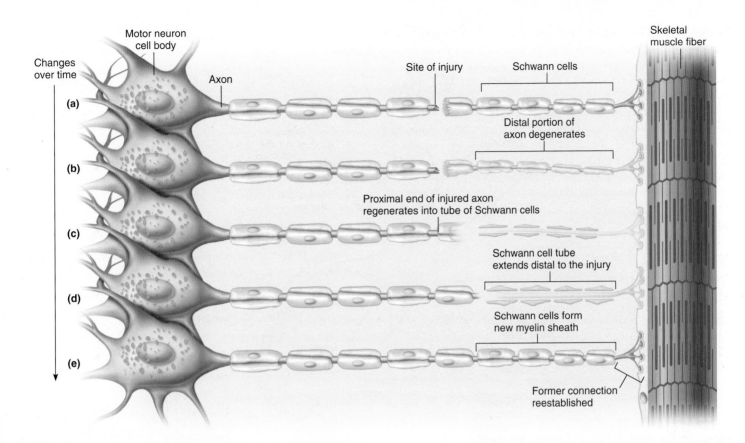

FIGURE 10.10 If a myelinated axon is injured, the following events may occur over several weeks to months: (*a*) The proximal portion of the axon may survive, but (*b*) the portion distal to the injury degenerates. (*c* and *d*) In time, the proximal portion may develop extensions that grow into the tube formed by the remaining Schwann cells and (*e*) possibly reestablish the former connection. Nerve growth factors that neuroglia secrete assist in the regeneration process.

Neurons do not divide. New neural tissue arises from neural stem cells, which give rise to neural progenitor cells that can give rise to neurons or neuroglia. In the adult brain, the rare neural stem cells are in a region called the dentate gyrus and near the fluid-filled ventricles.

Neural stem cells were discovered in the 1980s, in songbirds—the cells were inferred to exist because the numbers of neurons waxed and waned with the seasons, peaking when the birds learned songs. Moving songbirds far from food, forcing them to sing longer, resulted in more brain neurons, thanks to the stem cells. In the 1990s, researchers identified stem cells in brain slices from marmosets and tree shrews given a drug that marks dividing cells. Then stem cells were discovered in humans when a researcher learned of patients taking a drug to mark their tongue and larynx cancer cells. The drug also marked neural stem cells. When five of the patients donated their brains, researchers were able to see human neural stem cells for the first time. Today, human neural stem and progenitor cells are being used to screen drugs and are being delivered as implants to experimentally treat a variety of brain disorders. One day, a person's neural stem cells may be coaxed to help heal from within.

10.5 | The Synapse

Neurons communicate with one another (or with other cells) at synapses (fig. 10.11). When you get a text message, the person texting is the sender and you are the receiver. Similarly, the neuron conducting an impulse to the synapse is the sender, or **presynaptic** neuron. The neuron receiving input at the synapse is the **postsynaptic** neuron. (The postsynaptic cell could also be a muscle or glandular cell.) A *synaptic cleft*, or gap, separates the two cells, which are connected functionally, but not physically. The process by which the impulse in the presynaptic neuron signals the postsynaptic cell is called **synaptic transmission.** As a result of synaptic transmission, the presynaptic neuron stimulates or inhibits a postsynaptic cell (fig. 10.12).

Synaptic transmission is a one-way process carried out by *neurotransmitters*. An impulse travels along the axon of the presynaptic neuron to the axon terminal. Most axons have several rounded synaptic knobs at their terminals, which dendrites do not have. These knobs have arrays of membranous sacs, called synaptic vesicles, that contain neurotransmitter molecules. When an impulse reaches a synaptic knob, voltage-sensitive calcium channels open and calcium diffuses inward from the extracellular fluid. The increased calcium concentration inside the cell initiates a series of events that fuses the synaptic vesicles with the cell membrane, where they release their neurotransmitter by exocytosis.

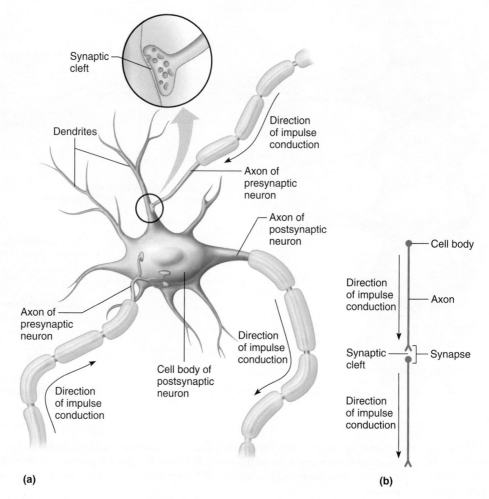

(a)

(b)

FIGURE 10.11 AP|R For an impulse to continue from one neuron to another, it must cross the synaptic cleft. A synapse usually separates an axon and a dendrite or an axon and a cell body.

Once the neurotransmitter binds to receptors on a postsynaptic cell, the action is either excitatory (turning a process on) or inhibitory (turning a process off). The net effect on the postsynaptic cell depends on the combined effect of the excitatory and inhibitory inputs from as few as 1 to 10,000 or more presynaptic neurons.

 PRACTICE

8 Name and describe four types of neuroglia.

9 What are some functions of neuroglia?

10 Explain how an injured peripheral axon might regenerate.

11 Describe synaptic transmission.

10.6 | Cell Membrane Potential

A cell membrane is usually electrically charged, or *polarized,* so that the inside is negatively charged with respect to the outside. This polarization is due to an unequal distribution of positive and negative ions across the membrane. It is important in the conduction of impulses in both muscle fibers and neurons.

Distribution of Ions

Potassium ions (K⁺) are the major intracellular positive ion (cation), and sodium ions (Na⁺) are the major extracellular cation. The distribution is created largely by the sodium/potassium pump (Na⁺/K⁺pump), which actively transports sodium ions out of the cell and potassium ions into the cell. It is also in part due to channels in the cell membrane that determine membrane permeability to these ions. These channels, formed by membrane proteins, can be selective; that is, a particular channel may allow only one type of ion to pass through and exclude all other ions of different size and charge. Thus, even though concentration gradients are present for sodium and potassium, the ability of these ions to diffuse across the cell membrane depends on the presence of channels.

 RECONNECT
To Chapter 3, Cell Membrane, pages 86–87.

Some channels are always open, whereas others may be either open or closed, somewhat like a gate. Both chemical and electrical factors can affect the opening and closing of these *gated channels* (fig. 10.13).

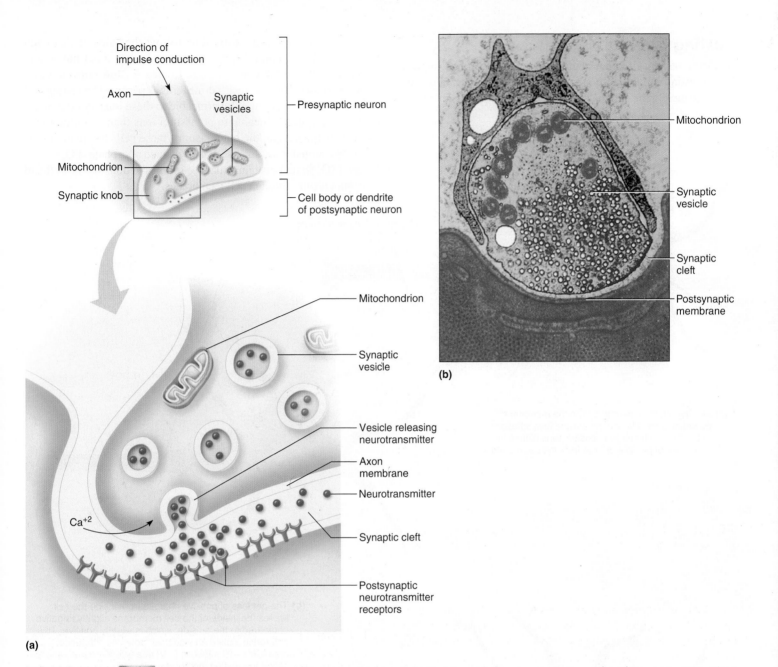

FIGURE 10.12 AP|R The synapse. (a) When an impulse reaches the synaptic knob at the end of an axon, synaptic vesicles release a neurotransmitter that diffuses across the synaptic cleft. (b) A transmission electron micrograph of a synaptic knob filled with synaptic vesicles (37,500×).

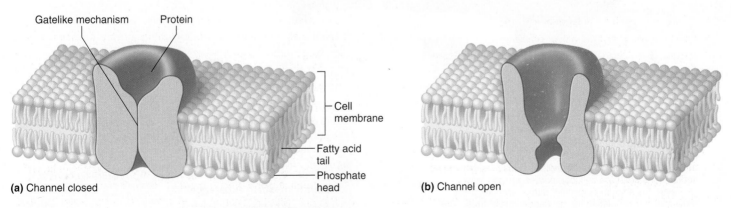

FIGURE 10.13 A gatelike mechanism can (a) close or (b) open some of the channels in cell membranes through which ions pass.

Resting Potential

A resting nerve cell is not being stimulated to send an impulse. Under resting conditions, non-gated (always open) channels determine the membrane permeability to sodium and potassium ions.

Sodium and potassium ions follow the rules of diffusion described in chapter 3 (pp. 98–100) and show a net movement from areas of high concentration to areas of low concentration across a membrane as their permeabilities permit. The resting cell membrane is only slightly permeable to these ions, but the membrane is more permeable to potassium ions than to sodium ions (fig. 10.14a). Also, the cytoplasm of these cells has many negatively charged ions (anions), which include phosphate (PO_4^{-2}), sulfate (SO_4^{-2}), and proteins, that are synthesized inside the cell and cannot diffuse through cell membranes.

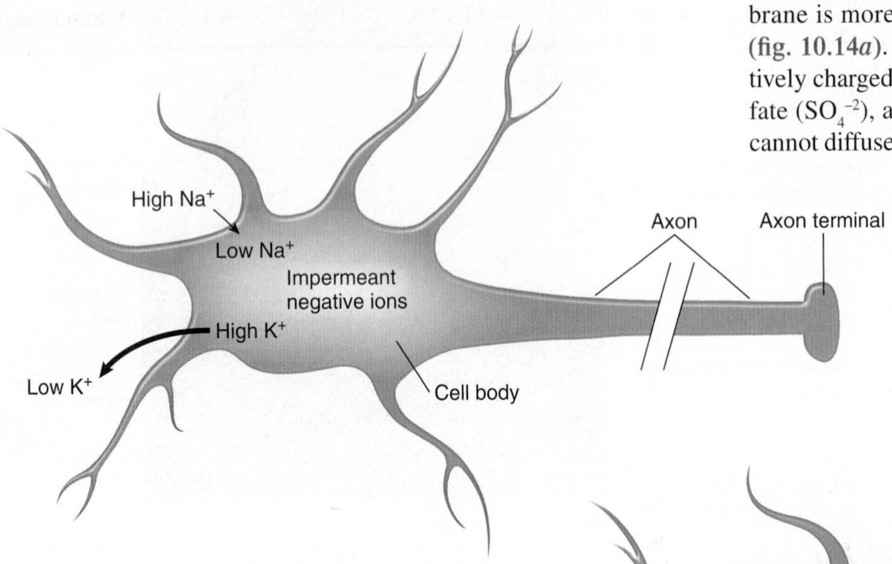

(a) In a hypothetical neuron before the membrane potential is established, potassium ions diffuse out of the cell faster than sodium ions diffuse in. A net loss of positive charge from the cell results.

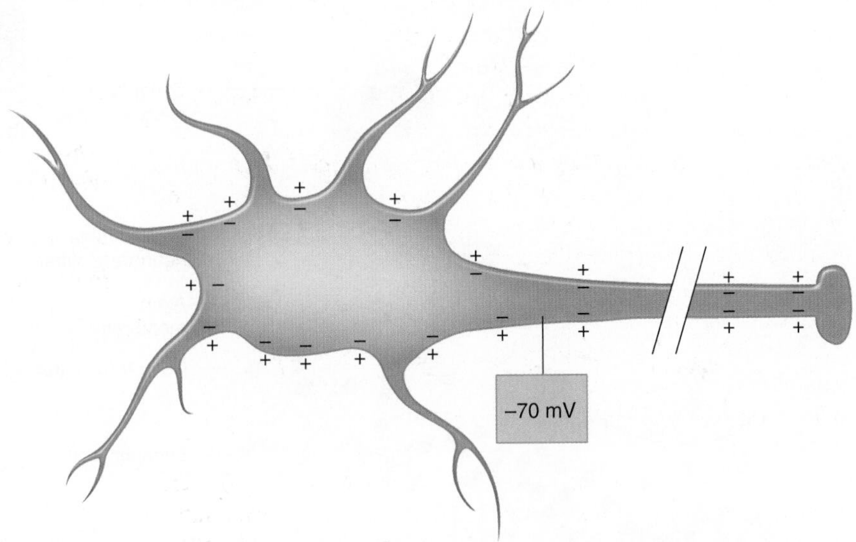

(b) The net loss of positive charges from inside the cell leaves the inside of the cell membrane slightly negative compared to the outside, which is slightly positive. This difference, called an electrical "potential difference," measures –70 millivolts (mV) in a typical neuron, and is called the resting membrane potential.

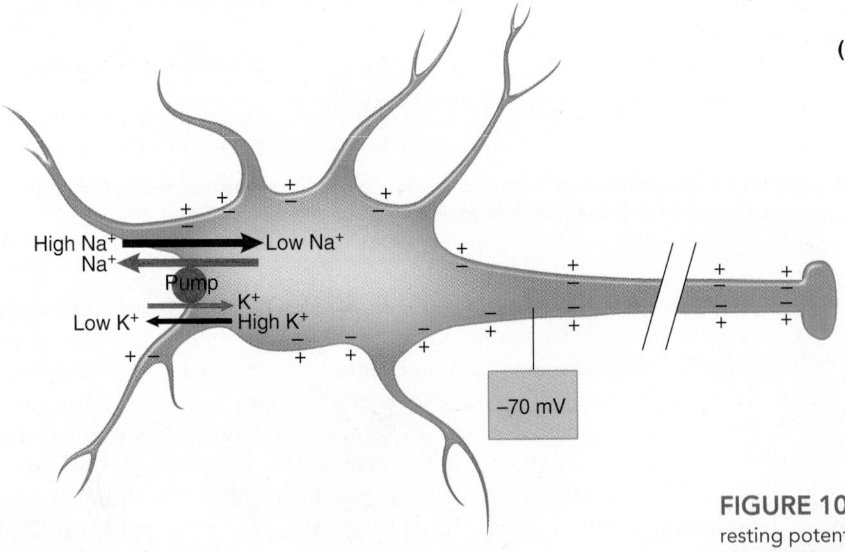

(c) The membrane potential, negative on the inside of the membrane, aids sodium diffusion into the cell, and opposes potassium diffusion out of the cell. As a result, slightly more sodium ions enter the cell than potassium ions leave. However, the sodium/potassium pump balances these movements, maintaining the concentrations of these ions and the resting membrane potential.

FIGURE 10.14 The resting potential. (a) Conditions that lead to the resting potential. (b) In the resting neuron, the inside of the membrane is negative relative to the outside. (c) The Na$^+$/K$^+$ pump maintains the concentration gradients for Na$^+$ and K$^+$ ions.

Q *Constant activity of the Na$^+$/K$^+$ pump requires a constant supply of which substance?*

Answer can be found in Appendix G.

If we consider a hypothetical neuron, before a membrane potential has been established, we would expect potassium to diffuse out of the cell more rapidly than sodium could diffuse in. This means that every millisecond (as the membrane potential is being established in our hypothetical cell), a few more positive ions leave the cell than enter it (fig. 10.14*a*). As a result, the outside of the membrane gains a slight surplus of positive charges, and the inside reflects a surplus of the impermeant negatively charged ions. This situation separates positive and negative electrical charges between the inside and outside surfaces of the cell membrane (fig. 10.14*b*). All this time, the cell continues to expend metabolic energy in the form of ATP to actively transport sodium and potassium ions in opposite directions, thus maintaining the concentration gradients for those ions responsible for their diffusion in the first place.

The difference in electrical charge between two points is measured in units called volts. It is called a potential difference because it represents stored electrical energy that can be used to do work at some future time. The potential difference across the cell membrane is called the **membrane potential** (transmembrane potential) and is measured in millivolts.

In the case of a resting neuron, one that is not sending impulses or responding to other neurons, the membrane potential is termed the **resting potential** (resting membrane potential) and has a value of –70 millivolts (fig. 10.14*b*). The negative sign is relative to the inside of the cell and is due to the excess negative charges on the inside of the cell membrane. To understand how the resting potential provides the energy for sending an impulse down the axon, we must first understand how neurons respond to signals called stimuli.

With the resting membrane potential established, a few sodium ions and potassium ions continue to diffuse across the cell membrane. The negative membrane potential helps sodium ions enter the cell despite sodium's low permeability, but it hinders potassium ions from leaving the cell despite potassium's higher permeability. The net effect is that three sodium ions "leak" into the cell for every two potassium ions that "leak" out. The Na^+/K^+ pump exactly balances these leaks by pumping three sodium ions out for every two potassium ions it pumps in (fig. 10.14*c*).

Recall from chapter 1 (p. 22) that neurons can conduct electrical signals. The key to understanding how this happens is the **action potential**, which is a rapid change in the membrane potential, first in a positive direction, then in a negative direction, returning to the resting potential (figure 10.15). When a neuron conducts an electrical current, that current is in the form of a series of action potentials occurring in sequence along the axon, from the cell body to the axon terminal. The next sections discuss how all of this happens.

Local Potential Changes

Neurons are excitable; that is, they can respond to changes in their surroundings. Some neurons, for example, detect changes in temperature, light, or pressure outside the body, whereas others respond to signals from inside the body, often from other neurons. In either case, such changes or stimuli usually affect the membrane potential in the region of the membrane exposed to the stimulus, causing a local potential change.

Typically, the environmental change affects the membrane potential by opening a gated ion channel. The effect will depend on the ion that can pass through the channel. If, as a result, the membrane potential becomes more negative than the resting potential, the membrane is *hyperpolarized*. If the membrane becomes less negative (more positive) than the resting potential, the membrane is *depolarized*.

Local potential changes are graded. This means that the degree of change in the resting potential is directly proportional to the intensity of the stimulation. For example, if the membrane is being depolarized, the greater the stimulus, the greater the depolarization. If and only if neurons are sufficiently depolarized, the membrane potential reaches a level called the **threshold potential,** which is approximately –55 millivolts in a neuron. If threshold is reached, an **action potential** results.

In many cases, a depolarizing stimulus is not sufficient to bring the postsynaptic cell to threshold (fig. 10.16*a*). Such a subthreshold depolarization will not result in an action potential.

Ion Movements During Action Potentials

If the presynaptic neurons release more neurotransmitter, or if other neurons that synapse with the same cell have an additive effect on depolarization, threshold may be reached, and an action potential result. The mechanism uses another type of ion channel, a voltage-gated sodium channel that opens when threshold is reached (fig. 10.16*b*).

In a multipolar neuron, the first part of the axon, the cone-shaped axon hillock or *initial segment,* is often referred to as the **trigger zone** because it contains many such voltage-gated sodium channels. At the resting membrane potential, these sodium channels remain closed, but when threshold is reached, they open for an instant, briefly increasing sodium permeability (fig. **10.17***a*). Sodium ions diffuse inward through the open sodium channels, down their concentration gradient, aided by the attraction of the sodium ions to the negative electrical state on the inside of the membrane (fig. 10.17*b*).

As the sodium ions diffuse inward, the membrane potential at the trigger zone changes from its resting value and momentarily becomes positive on the inside (still considered depolarization). At the peak of the action potential, the membrane potential may reach +30 mV or more (fig. 10.17*b*).

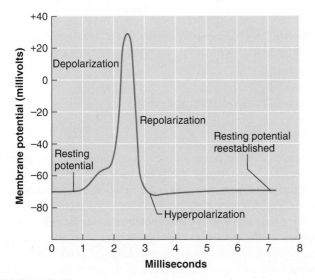

FIGURE 10.15 A recording of an action potential.

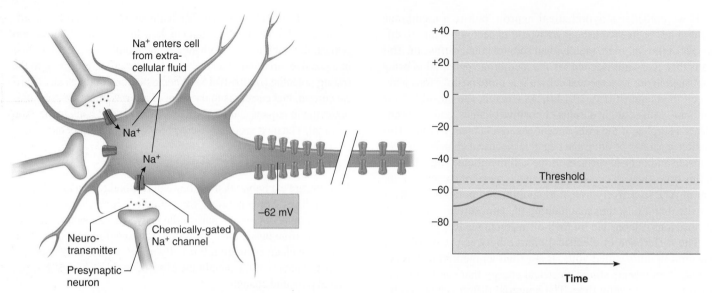

(a) If sodium or potassium channels open, more of that particular ion will cross the cell membrane, altering the resting membrane potential. This illustration depicts the effect of sodium channels opening in response to a neurotransmitter. As sodium ions enter the cell, the membrane potential becomes more positive (or less negative), changing from −70 millivolts to −62 millivolts in this example. This change in a positive direction is called depolarization. Here the depolarization is subthreshold, and does not generate an action potential.

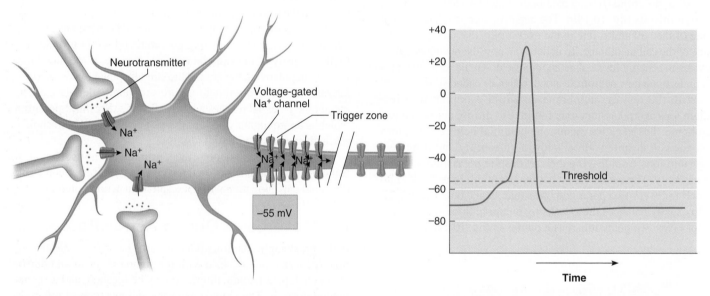

(b) If sufficient sodium ions enter the cell and the membrane potential depolarizes to threshold (here −55 millivolts), another type of sodium channel opens. These channels are found along the axon, especially near the origin in an area called the "trigger zone." Opening of these channels triggers the action potential.

FIGURE 10.16 Action potentials. (*a*) A subthreshold depolarization will not result in an action potential. (*b*) Multiple stimulation by presynaptic neurons may reach threshold, opening voltage-gated channels at the trigger zone.

The voltage-gated sodium channels quickly close, but at almost the same time, slower voltage-gated potassium channels open and briefly increase potassium permeability. As potassium ions diffuse outward through the open potassium channels, the inside becomes negatively charged once more. The membrane is thus repolarized (note in fig. 10.17*c* that it hyperpolarizes for an instant). The voltage-gated potassium channels then close as well. These actions quickly reestablish the resting potential, which remains in the resting state until it is stimulated again (see fig. 10.17*a*). The number of sodium and potassium ions crossing the membrane during an action potential is extremely small, although the bioelectric effect is quite significant. The active transport mechanism in the membrane restores and maintains the original concentrations of sodium and potassium ions.

Axons are capable of action potentials, but the cell body and dendrites are not. An action potential at the trigger zone causes an electric current to flow a short distance down the axon, which stimulates the adjacent membrane to reach its threshold level, triggering another action potential. The second action potential causes another

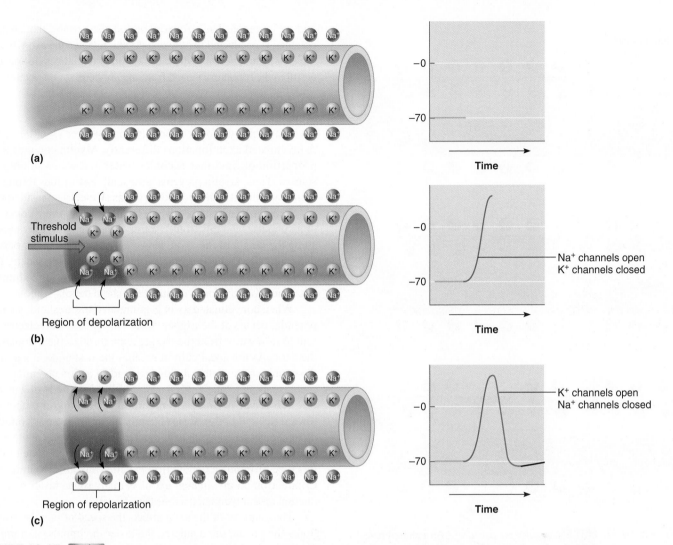

(a)

(b)

Threshold stimulus

Region of depolarization

Na⁺ channels open
K⁺ channels closed

(c)

Region of repolarization

K⁺ channels open
Na⁺ channels closed

Time

Time

Time

FIGURE 10.17 **AP|R** At rest (a), the axon membrane potential is about –70 millivolts. When the membrane reaches threshold (b), voltage-sensitive sodium channels open, some Na⁺ diffuses inward, and the membrane is depolarized. Soon afterward (c), voltage-sensitive potassium channels open, K⁺ diffuses out, and the membrane is repolarized. (Negative ions not shown.)

electric current to flow farther down the axon. This sequence of events results in a series of action potentials occurring sequentially all the way to the end of the axon without decreasing in amplitude, even if the axon branches. The propagation of action potentials (or impulse conduction) continues along the axon (fig. 10.18).

Conduction of an impulse along an axon is similar to conduction of an impulse in a muscle fiber mentioned in chapter 9, page 304. In the muscle fiber, stimulation at the motor end plate triggers an impulse to travel over the surface of the fiber and down into its transverse tubules. See table 10.3 for a summary of the events leading to the conduction of a series of action potentials down an axon.

All-or-None Response

An action potential is an all-or-none response. In other words, if a neuron responds at all, it responds completely. Thus, an impulse is conducted whenever a stimulus of threshold intensity or above is applied to an axon and all impulses conducted on that axon are the same strength. A greater intensity of stimulation produces more impulses per second, not a stronger impulse.

TABLE 10.3	Events Leading to Impulse Conduction

1. Nerve cell membrane maintains resting potential by diffusion of Na⁺ and K⁺ down their concentration gradients as the cell pumps them up the gradients.

2. Neurons receive stimulation, causing local potentials, which may sum to reach threshold.

3. Sodium channels in the trigger zone of the axon open.

4. Sodium ions diffuse inward, depolarizing the membrane.

5. Potassium channels in the membrane open.

6. Potassium ions diffuse outward, repolarizing the membrane.

7. The resulting action potential causes an electric current that stimulates adjacent portions of the membrane.

8. Action potentials occur sequentially along the length of the axon.

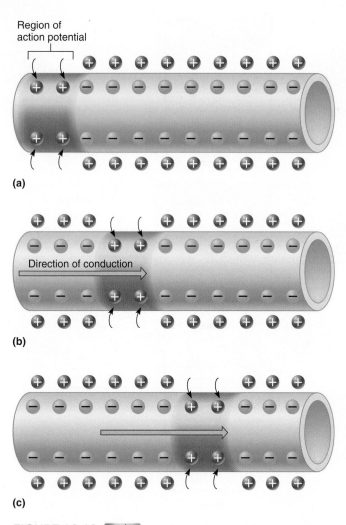

Region of
action potential

(a)

Direction of conduction

(b)

(c)

FIGURE 10.18 AP|R Impulse conduction. (*a*) An action potential in one region stimulates the adjacent region, and (*b* and *c*) a wave of action potentials moves along the axon.

Refractory Period

The number of impulses per second that an axon can generate is limited, because during an impulse, that part of the axon becomes unresponsive to another normal threshold stimulus. This brief period, called the **refractory period,** has two parts. During the *absolute refractory period,* which lasts about 1/1,000 of a second (1 millisecond), the axon's voltage-gated sodium channels are temporarily not responsive at all, and the axon cannot be stimulated. A *relative refractory period* follows, as the membrane reestablishes its resting potential. During this time, even though repolarization is incomplete, a stimulus of higher than usual intensity may trigger an impulse.

Rapidly, the intensity of stimulation required to trigger an impulse decreases until the axon's original excitability is restored. This return to the resting state usually takes from 1 to 3 milliseconds.

The refractory period limits how many action potentials may be generated in a neuron in a given period. Remembering that the action potential takes about a millisecond, and adding the time of the relative refractory period, the maximum theoretical frequency of impulses in a neuron is about 700 per second. In the body, this limit is rarely achieved—frequencies of about 100 impulses per sec-

ond are common. The refractory period also ensures that an action potential is conducted in only one direction, down the axon, because the area upstream from where the action potential has just occurred is still in the refractory period from the previous action potential.

Impulse Conduction

An unmyelinated axon conducts an impulse over its entire surface. A myelinated axon functions differently. Myelin contains a high proportion of lipid that excludes water and water-soluble substances. Thus, myelin prevents almost all flow of ions through the membrane that it encloses and serves as an electrical insulator.

It might seem that the myelin sheath would prevent conduction of an impulse, and this would be true if the sheath were continuous along the length of the axon. However, nodes of Ranvier between Schwann cells or oligodendrocytes interrupt the sheath (see fig. 10.3). At these nodes, the axon membrane has channels for sodium and potassium ions that open during a threshold depolarization.

When a myelinated axon is stimulated to threshold, an action potential occurs at the trigger zone. This causes a bioelectric current to flow away from the trigger zone through the cytoplasm of the axon. As this local current reaches the first node, it stimulates the membrane to its threshold level, and an action potential occurs there, sending a bioelectric current to the next node downstream (the refractory period prevents impulses from going backward toward the cell body). Consequently, as an impulse is conducted along a myelinated axon, action potentials occur only at the nodes. Because the action potentials appear to jump from node to node, this type of impulse conduction is called **saltatory conduction.** Conduction on myelinated axons is many times faster than conduction on unmyelinated axons (fig. 10.19).

The diameter of the axon affects the speed of impulse conduction—the greater the diameter, the faster the impulse. An impulse on a comparatively thick, myelinated axon, such as that of a motor neuron associated with a skeletal muscle, might travel 120 meters per second, whereas an impulse on a thin, unmyelinated axon, such as that of a sensory neuron associated with the skin, might move only 0.5 meter per second. Clinical Application 10.3 discusses factors that influence impulse conduction.

 PRACTICE

12 Summarize how a resting potential is achieved.

13 Explain how a polarized axon responds to stimulation.

14 List the major events of an action potential.

15 Define *refractory period.*

16 Explain how impulse conduction differs in myelinated and unmyelinated axons.

10.7 | Synaptic Transmission

Released neurotransmitter molecules diffuse across the synaptic cleft and bind to receptors on the postsynaptic cell membrane. When neurotransmitters bind these receptors, they cause ion channels in the postsynaptic cells to open. Ion channels that respond to neurotransmitter molecules are called *chemically gated,* in

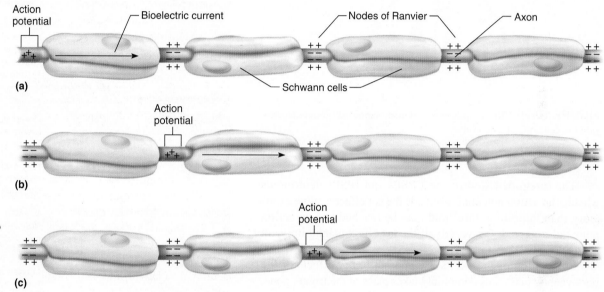

FIGURE 10.19 On a myelinated axon, an impulse appears to jump from node to node. (a)–(c) show the action potential progressing down the axon.

contrast to the voltage-gated ion channels that participate in action potentials. Changes in chemically gated ion channels create local potentials, called **synaptic potentials,** which enable one neuron to affect another.

Neurotransmitters that increase postsynaptic membrane permeability to sodium ions will bring the postsynaptic membrane closer to threshold and may trigger impulses. Such neurotransmitters are **excitatory.** Neurotransmitters that make reaching threshold less likely are called **inhibitory,** because they decrease the chance that an impulse will occur.

Some relatively uncommon synapses, called *electrical synapses,* are in certain parts of the brain and eyes. Electrical synapses involve direct exchange of ions between neurons through gap junctions (see chapter 5, p. 149). Synaptic transmission always occurs, because the postsynaptic cell does not need to reach threshold.

Synaptic Potentials

Synaptic potentials can depolarize or hyperpolarize the receiving cell membrane. For example, if a neurotransmitter binds to a postsynaptic receptor and opens sodium ion channels, the ions diffuse inward, depolarizing the membrane, possibly triggering an action potential. This type of membrane change is called an **excitatory postsynaptic potential** (EPSP), and it lasts for about 15 milliseconds.

If a different neurotransmitter binds other receptors and increases membrane permeability to potassium ions, these ions diffuse outward, hyperpolarizing the membrane. Because an action potential is now less likely to occur, this change is called an **inhibitory postsynaptic potential** (IPSP). Some inhibitory neurotransmitters open chloride ion channels. In this case, if sodium ions enter the cell, negative chloride ions are free to follow, opposing the depolarization.

In the brain and spinal cord, each neuron may receive the synaptic knobs of a thousand or more axons on its dendrites and cell body (fig. 10.20). Furthermore, at any moment, some of the postsynaptic potentials are excitatory on a particular neuron, while others are inhibitory.

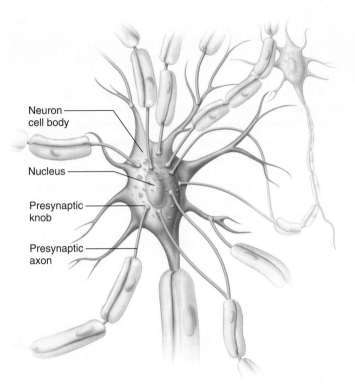

Neuron
cell body

Nucleus

Presynaptic
knob

Presynaptic
axon

FIGURE 10.20 The synaptic knobs of many axons may communicate with the cell body of a neuron.

The integrated sum of the EPSPs and IPSPs determines whether an action potential results. If the net effect is more excitatory than inhibitory, threshold may be reached and an action potential triggered. Conversely, if the net effect is inhibitory, an action potential does not occur.

Summation of the excitatory and inhibitory effects of the postsynaptic potentials commonly takes place at the trigger zone. This is usually in a proximal region of the axon, but in some sensory neurons it may be in the distal peripheral process. This region has an especially low threshold for triggering an action potential. In this way, the trigger zone, as its name implies, serves as a decision-making part of the neuron.

Neurotransmitters

The nervous system produces at least 100 different types of neurotransmitters in the brain alone. Some neurons release only one type, whereas others produce two or three. Neurotransmitters include *acetylcholine,* which stimulates skeletal muscle contractions (see chapter 9, p. 297); a group of compounds called *monoamines* (such as epinephrine, norepinephrine, dopamine, and serotonin), which are modified amino acids; a group of unmodified *amino acids* (such as glycine, glutamic acid, aspartic acid, and gamma-aminobutyric acid—GABA); and a large group of *peptides* (such as enkephalins and substance P), which are short chains of amino acids.

Peptide neurotransmitters are synthesized in the rough endoplasmic reticulum of a neuron cell body and transported in vesicles down the axon to the nerve cell terminal. Other neurotransmitters are synthesized in the cytoplasm of the nerve cell terminal and stored in vesicles. When an action potential passes along the membrane of a

synaptic knob, it increases the membrane's permeability to calcium ions by opening calcium ion channels. Calcium ions diffuse inward, and in response, some of the synaptic vesicles fuse with the presynaptic membrane and release their contents by exocytosis into the synaptic cleft (see fig. 10.12). If multiple action potentials reach the synaptic knob, more calcium will enter. The more calcium that enters the synaptic knob, the more vesicles release neurotransmitter. Table 10.4 lists the major neurotransmitters and their actions. Tables 10.5 and 10.6, respectively, list disorders and drugs that alter neurotransmitter levels.

TABLE 10.4	Some Neurotransmitters and Representative Actions	
Neurotransmitter	**Location**	**Major Actions**
Acetylcholine	CNS	Controls skeletal muscle actions
	PNS	Stimulates skeletal muscle contraction at neuromuscular junctions; may excite or inhibit at autonomic nervous system synapses
Biogenic amines		
Norepinephrine	CNS	Creates a sense of well-being; low levels may lead to depression
	PNS	May excite or inhibit autonomic nervous system actions, depending on receptors
Dopamine	CNS	Creates a sense of well-being; deficiency in some brain areas associated with Parkinson disease
	PNS	Limited actions in autonomic nervous system; may excite or inhibit, depending on receptors
Serotonin	CNS	Primarily inhibitory; leads to sleepiness; action is blocked by LSD, enhanced by selective serotonin reuptake inhibitor antidepressant drugs
Histamine	CNS	Release in hypothalamus promotes alertness
Amino acids		
GABA	CNS	Generally inhibitory
Glutamate	CNS	Generally excitatory
Neuropeptides		
Enkephalins, endorphins	CNS	Generally inhibitory; reduce pain by inhibiting substance P release
Substance P	PNS	Excitatory; pain perception
Gases		
Nitric oxide	CNS	May play a role in memory
	PNS	Vasodilation

TABLE 10.5	Disorders Associated with Neurotransmitter Imbalances	
Condition	**Symptoms**	**Imbalance of Neurotransmitter in Brain**
Clinical depression	Debilitating, inexplicable sadness	Deficient norepinephrine and/or serotonin
Epilepsy	Seizures, loss of consciousness	Excess GABA leads to excess norepinephrine and dopamine
Huntington disease	Cognitive and behavioral changes, loss of coordination, uncontrollable dancelike movements, death	Deficient GABA
Hypersomnia	Excessive sleeping	Excess serotonin
Insomnia	Inability to sleep	Deficient serotonin
Mania	Elation, irritability, overtalkativeness, increased movements	Excess norepinephrine
Parkinson disease	Tremors of hands, slowed movements, muscle rigidity	Deficient dopamine
Schizophrenia	Inappropriate emotional responses, hallucinations	Deficient GABA leads to excess dopamine
Tardive dyskinesia	Uncontrollable movements of facial muscles	Deficient dopamine

TABLE 10.6	Drugs That Alter Neurotransmitter Levels		
Drug	**Neurotransmitter Affected***	**Mechanism of Action**	**Effect**
Tryptophan	Serotonin	Stimulates neurotransmitter synthesis	Sleepiness
Reserpine	Norepinephrine	Decreases packaging of neurotransmitter into vesicles	Decreases blood pressure
Curare	Acetylcholine	Blocks receptor binding	Muscle paralysis
Valium	GABA	Enhances receptor binding	Decreases anxiety
Nicotine	Acetylcholine	Activates receptors	Increases alertness
	Dopamine	Elevates levels	Sense of pleasure
Cocaine	Dopamine	Blocks reuptake	Euphoria
Tricyclic antidepressants	Norepinephrine	Blocks reuptake	Antidepressant
	Serotonin	Blocks reuptake	Antidepressant
Monoamine oxidase inhibitors	Norepinephrine	Blocks enzymatic degradation of neurotransmitter in presynaptic cell	Antidepressant
Selective serotonin reuptake inhibitors	Serotonin	Blocks reuptake	Antidepressant, anti-anxiety agent
Dual reuptake inhibitors	Serotonin, norepinephrine	Blocks reuptake	Mood elevation

*Others may be affected as well.

RECONNECT
To Chapter 3, Exocytosis, page 105.

A vesicle becomes part of the cell membrane after it releases its neurotransmitter. Endocytosis eventually returns the membrane material to the cytoplasm, where it can provide material to form new secretory vesicles. Table 10.7 summarizes this process, which is called *vesicle trafficking.*

To keep signal duration short, enzymes in synaptic clefts and on postsynaptic membranes rapidly decompose some neurotransmitters. The enzyme **acetylcholinesterase,** for example, decomposes acetylcholine on postsynaptic membranes. Other neurotransmitters are transported back into the synaptic knob of the presynaptic neuron or into nearby neurons or neuroglia, in a process called *reuptake.* The enzyme **monoamine oxidase** inactivates the monoamine neurotransmitters epinephrine and norepinephrine after reuptake. This enzyme is found in mitochondria in the synaptic knob. Destruction or removal of neurotransmitter prevents continuous stimulation of the postsynaptic neuron.

Neuropeptides

Neurons in the brain or spinal cord synthesize **neuropeptides.** These peptides act as neurotransmitters or as *neuromodulators,* which are substances that alter a neuron's response to a neurotransmitter or block the release of a neurotransmitter.

Opiate drugs, such as morphine, heroin, codeine, and opium, are potent painkillers derived from the poppy plant. These drugs alter pain perception making it easier to tolerate, and they elevate mood.

The human body produces opiates, called endorphins (for "endogenous morphine"), that are peptides. Like the poppy-derived opiates that they structurally resemble, endorphins influence mood and perception of pain.

Endorphins were discovered in the 1960s at the University of California, San Francisco, in dried pituitary glands, but their functions were not described until 1971, when researchers at Stanford University and the Johns Hopkins School of Medicine exposed pieces of brain tissue to morphine. The morphine was radioactively labeled (some of the atoms were radioactive isotopes) so researchers could follow its destination in the brain.

The morphine bound receptors on neurons that transmit pain. Why, the investigators wondered, would an animal's brain cells have receptors for a plant chemical? One explanation was that a mammal's body could manufacture opiates. The opiate receptors, then, would normally bind the body's opiates (the endorphins) but would also bind the chemically similar compounds from poppies. Researchers have since identified several types of endorphins in the human brain and associated their release with situations involving pain relief, such as acupuncture, analgesia to mother and child during childbirth, and the feeling of well-being that accompanies exercise. PET scans reveal endorphins binding opiate receptors after conditioned athletes run for two hours.

Endorphins explain why some people addicted to opiate drugs, such as heroin, experience withdrawal pain when they stop taking the drug. Initially the body interprets the frequent binding of heroin to its endorphin receptors as an excess of endorphins. To lower the level, the body slows its own production of endorphins. Then, when the person stops taking the heroin, the body becomes short of opiates (heroin and endorphins). The result is pain.

Opiate drugs can be powerfully addicting when they are taken repeatedly by a person who is not in pain. These same drugs, however, are extremely useful in dulling severe pain, particularly in terminal illnesses and after surgery.

TABLE 10.7	Events Leading to Neurotransmitter Release

1. Action potential passes along an axon and over the surface of its synaptic knob.

2. Synaptic knob membrane becomes more permeable to calcium ions, and they diffuse inward.

3. In the presence of calcium ions, synaptic vesicles fuse to synaptic knob membrane.

4. Synaptic vesicles release their neurotransmitter by exocytosis into the synaptic cleft.

5. Synaptic vesicles become part of the membrane.

6. The added membrane provides material for endocytotic vesicles.

Among the neuropeptides are the *enkephalins,* which are present throughout the brain and spinal cord. Each enkephalin molecule is a chain of five amino acids. Synthesis of enkephalins increases during periods of painful stress, and they bind to the same receptors in the brain (opiate receptors) as the narcotic morphine. Enkephalins relieve pain sensations and probably have other functions. Another morphinelike peptide, *beta endorphin,* is found in the brain and cerebrospinal fluid. It acts longer than enkephalins and is a much more potent pain reliever. Clinical Application 10.4 discusses opiates in the human body.

Substance P is a neuropeptide that consists of eleven amino acids and is widely distributed. It functions as a neurotransmitter (or perhaps as a neuromodulator) in the neurons that conduct impulses associated with pain into the spinal cord and on to the brain. Enkephalins and endorphins may relieve pain by inhibiting the release of substance P from these neurons.

 PRACTICE

17 Distinguish between an EPSP and an IPSP.

18 Describe the net effects of EPSPs and IPSPs.

19 Explain the function of a neurotransmitter.

20 Define neuropeptide.

10.8 | Impulse Processing

The way the nervous system collects, processes, and responds to information reflects, in part, the organization of neurons and axons in the brain and spinal cord.

Neuronal Pools

Interneurons, which are the neurons completely within the CNS, are organized into **neuronal pools.** These are groups of neurons that synapse with each other and perform a common function, even though their cell bodies may be in different parts of the CNS. Each neuronal pool receives input from neurons (which may be part of other pools), and each pool generates output. Neuronal pools may have excitatory or inhibitory effects on other pools or on peripheral effectors.

As a result of incoming impulses and neurotransmitter release, a particular neuron of a neuronal pool may be excited by some presynaptic neurons and inhibited by others. If the net effect is excitatory, threshold may be reached, and an impulse triggered. If the net effect is excitatory, but subthreshold, an impulse will not be triggered.

Repeated impulses on excitatory presynaptic neurons may result in the increased release of neurotransmitter in response to an impulse, making it more likely to bring the postsynaptic cell to threshold. This phenomenon is called **facilitation** (fah-sil″ĭ-tă′shun).

Convergence

Any single neuron in a neuronal pool may receive input from two or more other neurons. Axons originating from different neurons leading to the same postsynaptic neuron exhibit **convergence** (kon-ver′jens).

Incoming impulses often represent information from various sensory receptors that detect changes. Convergence allows the nervous system to collect, process, and respond to information.

Convergence makes it possible for a neuron to sum impulses from different sources. For example, if a neuron receives what would be subthreshold stimulation from one input neuron, it may reach threshold if it receives additional stimulation from a second input neuron. Thus, impulses on this neuron may reflect summation of input from two sources (fig. 10.21a). Such output may reach a particular effector and evoke a response.

Divergence

A neuron has a single axon, but axons may branch at several points. Therefore, impulses conducted by a neuron of a neuronal pool may exhibit **divergence** (di-ver′jens) by reaching several other neurons. One neuron stimulates two others, each of which stimulates several others, and so forth. Such a pattern of diverging axons allows an impulse to reach increasing numbers of neurons within the pool (fig. 10.21b).

As a result of divergence, an impulse originating from a single neuron in the CNS may stimulate several motor units in a skeletal muscle to contract. Similarly, an impulse originating from a sensory

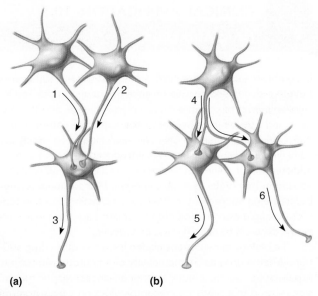

(a) (b)

FIGURE 10.21 Impulse processing in neuronal pools. (a) Axons of neurons 1 and 2 converge to the cell body of neuron 3. (b) The axon of neuron 4 diverges to the cell bodies of neurons 5 and 6.

receptor may diverge and reach several different regions of the CNS, where the information can be processed and evoke a response.

The nervous system enables us to experience the world and to think and feel emotion. This organ system is also sensitive to outside influences. Clinical Application 10.5 discusses one way that an outside influence can affect the nervous system—drug addiction.

PRACTICE

21 What is a neuronal pool?

22 Define *facilitation*.

23 What is convergence?

24 What is the relationship between divergence and amplification? ■

CHAPTER SUMMARY

10.1 Overview of the Nervous System (page 360)

1. The nervous system is a network of cells that sense and respond to stimuli in ways that maintain homeostasis.
2. The nervous system is composed of neural tissue (**neurons** and **neuroglia**), blood vessels, and connective tissue.
3. Neurons have processes that receive (**dendrites**) and send (**axons**) bioelectric signals (**impulses**).
4. Biological messenger molecules (**neurotransmitters**) convey this neural information across spaces (**synapses**) between neurons and the cells they affect.
5. Organs of the nervous system are divided into the **central** and **peripheral nervous systems.**

10.2 General Functions of the Nervous System (page 361)

1. **Sensory receptors** detect changes in internal and external body conditions.
2. Integrative functions gather sensory information and make decisions that affect motor functions.
3. Motor impulses stimulate effectors to respond.
    a. The motor portion of the PNS that carries out voluntary activities is the **somatic nervous system.**
    b. The motor portion of the PNS that carries out involuntary activities is the **autonomic nervous system.**

Drug abuse and addiction have a long history. A 3,500-year-old Egyptian document decries reliance on opium. In the 1600s, a smokable form of opium enslaved many Chinese, and the Japanese and Europeans discovered the addictive nature of nicotine. During the American Civil War, morphine was a widely used painkiller, and soon cocaine was introduced to relieve veterans addicted to morphine. LSD, originally used in psychotherapy, was abused in the 1960s as a hallucinogen. PCP was an anesthetic before being abused in the 1980s. An increasing problem today is that people who are addicted to opiate-based painkillers may switch to heroin because it is easier to obtain.

The biology of neurotransmission helps to explain drug addiction. When a drug alters the activity of a neurotransmitter on a postsynaptic neuron, it either halts or enhances synaptic transmission. A drug that binds to a receptor, blocking a neurotransmitter from binding, is called an *antagonist*. A drug that activates the receptor, triggering an action potential, or that helps a neurotransmitter to bind, is called an *agonist*. The effect of a drug depends upon whether it is an antagonist or an agonist; on the particular behaviors the affected neurotransmitter normally regulates; and in which parts of the brain the drug affects neurotransmitters and their binding to receptors. Many addictive substances bind to receptors for the neurotransmitter dopamine, in a brain region called the nucleus accumbens.

With repeated use of an addictive substance, the number of receptors it targets can decline. When this happens, the person must use more of the drug to feel the same effect. For example, neural pathways that use the neurotransmitter norepinephrine control arousal, dreaming, and mood. Amphetamine enhances norepinephrine activity, thereby heightening alertness and mood. Amphetamine's structure is so similar to that of norepinephrine that it binds to norepinephrine receptors and triggers the same changes in the postsynaptic membrane.

Cocaine has a complex mechanism of action, both blocking reuptake of norepinephrine and binding to molecules that transport dopamine to postsynaptic cells. The drug valium causes relaxation and inhibits seizures and anxiety by helping GABA, an inhibitory neurotransmitter used in a third of the brain's synapses, bind to receptors on postsynaptic neurons. Valium is therefore a GABA agonist.

Nicotine is highly addictive, and the chemicals in cigarette smoke can destroy health. An activated form of nicotine binds postsynaptic nicotinic receptors that normally receive acetylcholine. When sufficient nicotine binds, a receptor channel opens, allowing positive ions into the cell (fig. 10A). When a certain number of positive ions enter, the neuron releases dopamine from its other end, which provides the pleasurable feelings associated with smoking. A drug that helps some people stop smoking (varenicline, or Chantix) works by partially activating the nicotinic receptors, which decreases both the enjoyment of smoking and the craving.

When a smoker increases the number of cigarettes smoked, the number of nicotinic receptors increases. This happens because nicotine binding impairs the recycling of receptor proteins, so that receptors are produced faster than they are taken apart. After a period of steady nicotine exposure, many of the receptors malfunction and no longer admit the positive ions that trigger dopamine release. Receptor malfunction may be why as time goes on it takes more nicotine to produce the same effects—a hallmark of addiction.

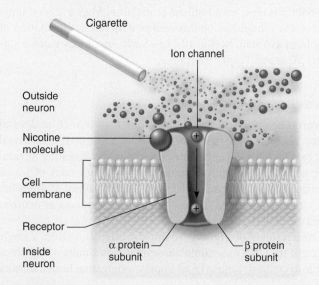

FIGURE 10A Nicotine binds and transiently alters postsynaptic receptors that normally bind the neurotransmitter acetylcholine. As a result, positive ions enter the cell, triggering dopamine release. With frequent smoking, receptors accumulate and soon become nonfunctional. Nicotine's effects on the nervous system are complex.

Nervous System

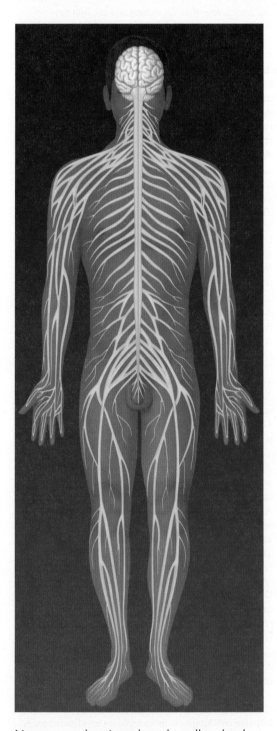

Nerves conduct impulses that allow body systems to communicate.

Integumentary System

Sensory receptors provide the nervous system with information about the outside world.

Skeletal System

Bones protect the brain and spinal cord and help maintain plasma calcium, which is important to neuron function.

Muscular System

The nervous system controls movement and processes information about the position of body parts.

Endocrine System

The nervous system controls secretion of many hormones.

Cardiovascular System

The nervous system helps control blood flow and blood pressure.

Lymphatic System

Stress may impair the immune response.

Digestive System

The nervous system can influence digestive function.

Respiratory System

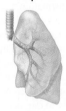

The nervous system alters respiratory activity to control oxygen levels and blood pH.

Urinary System

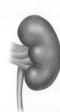

The nervous system plays a role in urine production and elimination.

Reproductive System

The nervous system plays a role in egg and sperm formation, sexual pleasure, childbirth, and nursing.

10.3 Description of Cells of the Nervous System (page 363)

1. Neurons vary in size, shape, sizes and lengths of axons and dendrites, and number of dendrites.
2. A neuron includes a **cell body,** cell processes, and the organelles usually found in cells.
3. **Neurofilaments** support axons.
4. **Chromatophilic substance** is mostly rough ER and is scattered throughout the cytoplasm of neurons.
5. Dendrites and the cell body provide receptive surfaces.
6. A single axon arises from the cell body and may be enclosed in a myelin sheath and a **neurilemma.**
7. White matter consists of myelinated axons, and gray matter consists of unmyelinated axons and cell bodies.

10.4 Classification of Cells of the Nervous System (page 363)

1. Classification of neurons
   a. Neurons are structurally classified as **multipolar, bipolar,** or **unipolar.**
   b. Neurons are functionally classified as **sensory neurons, interneurons,** or **motor neurons.**
2. Classification of neuroglia
   a. Neuroglia are abundant and have several functions.
   b. They fill spaces, support neurons, hold nervous tissue together, help metabolize glucose, help regulate potassium ion concentration, produce myelin, carry on phagocytosis, rid synapses of excess ions and neurotransmitters, nourish neurons, and stimulate synapse formation.
   c. They include **astrocytes, oligodendrocytes, microglia,** and **ependymal cells** in the CNS and **Schwann cells** and **satellite cells** in the PNS.
   d. Malfunctioning neuroglia can cause disease.
   e. Neuroglia are involved in axonal regeneration.
      (1) If a neuron cell body is injured, the neuron is likely to die; neural stem cells may proliferate and produce replacements.
      (2) If a peripheral axon is severed, its distal portion will die, but under the influence of nerve growth factors, the proximal portion may regenerate and reestablish connections, if a tube of remaining Schwann cells guides it.
      (3) Significant regeneration is not likely in the CNS.

10.5 The Synapse (page 371)

A synapse is a functional connection between two cells. A synaptic cleft is the gap between parts of two cells at a synapse. **Synaptic transmission** is the process by which the impulse in the **presynaptic** neuron signals the **postsynaptic** cell.

1. An impulse is conducted along the axon to a synapse.
2. Axons have synaptic knobs at their distal ends that release neurotransmitters.
3. The neurotransmitter is released when an impulse reaches the end of an axon, and the neurotransmitter diffuses across the synaptic cleft.
4. A neurotransmitter reaching a postsynaptic neuron or other cell may be excitatory or inhibitory.

10.6 Cell Membrane Potential (page 372)

A cell membrane is usually polarized as a result of an unequal distribution of ions on either side. Channels in membranes that allow passage of some ions but not others control ion distribution.

1. Distribution of ions
   a. Membrane ion channels, formed by proteins, may be always open or sometimes open and sometimes closed.
   b. Potassium ions pass more readily through resting neuron cell membranes than do sodium and calcium ions.
   c. A high concentration of sodium ions is on the outside of the membrane, and a high concentration of potassium ions is on the inside.
2. **Resting potential**
   a. Large numbers of negatively charged ions, which cannot diffuse through the cell membrane, are inside the cell.
   b. In a resting cell, more positive ions leave the cell than enter it, causing the inside of the cell membrane to develop a negative charge with respect to the outside.
3. Local potential changes
   a. Stimulation of a membrane affects its resting potential in a local region.
   b. The membrane is depolarized if it becomes less negative; it is hyperpolarized if it becomes more negative.
   c. Local potential changes are graded and subject to summation.
   d. Reaching **threshold potential** triggers an **action potential.**
4. Action potentials
   a. At threshold, sodium channels open and sodium ions diffuse inward, depolarizing the membrane.
   b. At almost the same time, potassium channels open and potassium ions diffuse outward, repolarizing the membrane.
   c. This rapid change in potential is an action potential.
   d. Active transport reestablishes and maintains the original concentrations of ions across the cell membrane.
   e. The propagation of action potentials along a nerve fiber is also called impulse conduction.
5. All-or-none response
   a. An action potential is an all-or-none response. An action potential is generated if a stimulus of threshold intensity is applied to an axon.
   b. All the impulses conducted on an axon are the same strength.
6. Refractory period
   a. The **refractory period** is a brief time during the passage of an impulse when the membrane is unresponsive to an ordinary stimulus.
   b. During the absolute refractory period, the membrane cannot respond to a stimulus; during the relative refractory period, the membrane can respond to a high-intensity stimulus.
7. Impulse conduction
   a. An unmyelinated axon conducts impulses that travel over its entire surface.
   b. A myelinated axon conducts impulses that travel from node to node (**saltatory conduction**).
   c. Impulse conduction is more rapid on myelinated axons with large diameters.

10.7 Synaptic Transmission (page 378)

Neurotransmitter molecules diffuse across the synaptic cleft and react with receptors in the postsynaptic neuron membrane.

1. Synaptic potentials
   a. Some neurotransmitters can depolarize the postsynaptic membrane, possibly triggering an action potential. This is an **excitatory postsynaptic potential** (EPSP).
   b. Others hyperpolarize the membrane, inhibiting an action potential. This is an **inhibitory postsynaptic potential** (IPSP).
   c. EPSPs and IPSPs are summed in a trigger zone of the neuron.
2. Neurotransmitters
   a. The nervous system produces at least 100 different types of neurotransmitters in the brain alone.
   b. Calcium ions diffuse into synaptic knobs in response to action potentials, releasing neurotransmitters.
   c. Neurotransmitters are quickly decomposed or removed from synaptic clefts.
3. Neuropeptides
   a. **Neuropeptides** are chains of amino acids.
   b. Some neuropeptides are neurotransmitters or neuromodulators.
   c. They include enkephalins, endorphins, and substance P.

10.8 Impulse Processing (page 382)

The way impulses are processed reflects the organization of neurons in the brain and spinal cord.

1. Neuronal pools
   a. Neurons are organized into **neuronal pools** in the CNS.
   b. Each pool receives input, processes the information, and may generate output.
   c. Each neuron in a pool may receive excitatory and inhibitory stimuli.
   d. A neuron is facilitated when it receives subthreshold stimuli and becomes more excitable.
2. **Convergence**
   a. Impulses from two or more axons may converge on a single postsynaptic neuron.
   b. Convergence enables a neuron to sum input from different sources.
3. **Divergence**
   a. Impulses from a presynaptic neuron may reach several postsynaptic neurons.
   b. Divergence allows a single neuron reaching threshold to have potentially widespread effects.

CHAPTER ASSESSMENTS

10.1 Overview of the Nervous System

1 Describe how the nervous system detects change associated with the body and reacts to that change to maintain homeostasis. (p. 360)
2 Distinguish between neurons and neuroglia. (p. 360)
3 Which of the following descriptions is accurate? (p. 360)
   a. A neuron has a single dendrite, which sends information.
   b. A neuron has a single axon, which sends information.
   c. A neuron has many axons, which receive information.
   d. A neuron has many dendrites, which send information.
4 Explain the difference between the central nervous system (CNS) and the peripheral nervous system (PNS). (p. 361)

10.2 General Functions of the Nervous System

5 List three general functions of the nervous system. (p. 361)
6 Distinguish a sensory receptor from an effector. (p. 361)
7 Distinguish between the types of activities that the somatic and autonomic nervous systems control. (p. 361)

10.3 Description of Cells of the Nervous System

8 Match the part of a neuron on the left with the description on the right (pp. 360–363):

   (1) dendrites
   (2) chromatophilic substance
   (3) axon
   (4) cell body
   (5) neurofilaments

   A. fine threads in an axon
   B. part of neuron from which axon and dendrites extend
   C. highly branched, multiple processes that may have spines
   D. conducts impulses
   E. rough endoplasmic reticulum

9 Explain how Schwann cells encase large axons, including the formation of myelin, the neurilemma, and the nodes of Ranvier. (p. 363)

10 What do Schwann cells and oligodendrocytes have in common, and how do they differ? (p. 363)
11 Distinguish between myelinated and unmyelinated axons. (p. 363)

10.4 Classification of Cells of the Nervous System

12 Describe the three types of neurons classified on the basis of structure. (pp. 363–366)
13 Describe the three types of neurons classified on the basis of function (pp. 366–368)
14 List six functions of neuroglia. (p. 368)
15 Describe the neuroglia of the CNS. (pp. 368–369)
16 Explain how malfunctioning neuroglia can harm health. (pp. 369–370)
17 Describe the neuroglia of the PNS. (p. 370)
18 Explain how an injured neuron may regenerate. (p. 370)

10.5 The Synapse

19 The _____ _____ brings the impulse to the synapse, whereas the _____ _____ on the other side of the synapse is stimulated or inhibited as a result of the synaptic transmission. (p. 371)
20 Explain how information is passed from a presynaptic neuron to a postsynaptic cell. (p. 372)
21 Diffusion of which of the following ions into the synaptic knob triggers the release of neurotransmitter? (p. 372)
   a. Na^+
   b. Ca^{+2}
   c. Cl^-
   d. K^+

10.6 Cell Membrane Potential

22 Define *resting potential*. (p. 375)

23 Distinguish among polarized, hyperpolarized, and depolarized. (p. 375)
24 Explain why the "trigger zone" of a neuron is named as such. (p. 375)
25 List in correct order the changes that occur during an action potential. (pp. 375–377)
26 Explain the relationship between an action potential and impulse conduction down an axon. (pp. 376–377)
27 Define *refractory period.* (p. 378)
28 Explain the importance of the nodes of Ranvier and conduction in myelinated fibers as opposed to conduction in unmyelinated fibers. (p. 378)

10.7 Synaptic Transmission
29 Distinguish between excitatory and inhibitory postsynaptic potentials. (p. 379)
30 Explain how enzymes within synaptic clefts and reuptake of neurotransmitter prevents continuous stimulation of the postsynaptic cell. (p. 381)

10.8 Impulse Processing
31 Explain what determines the output of a neuronal pool in terms of input neurons, excitation, and inhibition. (p. 382)
32 Define *facilitation.* (p. 383)
33 Distinguish between convergence and divergence. (p. 383)

INTEGRATIVE ASSESSMENTS/CRITICAL THINKING

Outcomes 10.3, 10.4

1. Why are rapidly growing cancers that originate in nervous tissue more likely to be composed of neuroglia than of neurons?

Outcomes 10.3, 10.4, 10.6

2. In Tay-Sachs disease, an infant rapidly loses nervous system functions as neurons in the brain become covered in too much myelin. In multiple sclerosis, cells in the CNS have too little myelin. Identify the type of neuroglia implicated in each of these conditions.

Outcomes 10.4, 10.5, 10.7

3. How would you explain the following observations?
   a. When motor nerve fibers in the leg are severed, the muscles they innervate become paralyzed; however, in time, control over the muscles often returns.
   b. When motor nerve fibers in the spinal cord are severed, the muscles they control become permanently paralyzed.

Outcomes 10.5, 10.6, 10.7

4. Drugs that improve early symptoms of Alzheimer disease do so by slowing the breakdown of acetylcholine in synaptic clefts in certain parts of the brain. From this information, suggest a neurotransmitter imbalance that lies behind Alzheimer disease.

Outcome 10.6

5. People who inherit familial periodic paralysis often develop very low blood potassium concentrations. How would you explain that the paralysis may disappear quickly when potassium ions are administered intravenously?

 # ONLINE STUDY TOOLS

Connect Interactive Questions Reinforce your knowledge using assigned interactive questions covering the structure of nervous tissue, the generation and conduction of impulses, and the basics of impulse integration.

Connect Integrated Activity Can you assist a neurology patient in understanding the actions of drugs that alter neuron function and impulse conduction?

LearnSmart Discover which chapter concepts you have mastered and which require more attention. This adaptive learning tool is personalized, proven, and preferred.

Anatomy & Physiology Revealed Go more in depth into the functioning of the human body by viewing animations on synapses and action potentials.

11

Nervous System II

*Divisions of the
Nervous System*

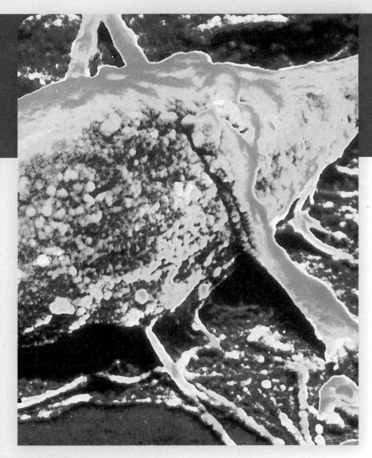

Falsely colored scanning electron micrograph (SEM) of the cell body of a single neuron of the human cerebral cortex—the outer gray matter of the brain (7,200×).

THE WHOLE PICTURE

All neurons conduct action potentials, and all of theses action potentials are the same. Yet, the nervous system can process a wide variety of information from the external environment, including sight, sounds, and touch on the surface of the skin. The nervous system can also interpret information from receptors that sense changes in the internal environment and can activate effectors to correct those changes. Among the most fascinating aspects of nervous system function are the abilities of the brain to store memories and to process conscious thought.

All of these things are accomplished by neurons working in much the same way, but serving different functions within the brain, the spinal cord, or the peripheral nerves. This is the chapter in which your brain gets to learn about itself and the other parts of the nervous system!

Module 7: Nervous System

LEARNING OUTCOMES

After you have studied this chapter, you should be able to:

11.1 Overview of Divisions of the Nervous System
1 Describe the relationship among the brain, brainstem, and spinal cord. (p. 390)

11.2 Meninges
2 Describe the coverings of the brain and spinal cord. (pp. 390–391)

11.3 Ventricles and Cerebrospinal Fluid
3 Discuss the formation and function of cerebrospinal fluid. (pp. 391–395)

11.4 Brain
4 Describe the development of the major parts of the brain and explain the functions of each part. (p. 395)
5 Distinguish among sensory, association, and motor areas of the cerebral cortex. (pp. 399–402)
6 Discuss hemisphere dominance. (p. 402)
7 Explain the stages in memory storage. (pp. 402–403)
8 Explain the functions of the limbic system and the reticular formation. (pp. 403, 405, and 407)

11.5 Spinal Cord
9 Describe the structure of the spinal cord and its major functions. (pp. 409–410)
10 Describe a reflex arc and reflex behavior. (pp. 411–413)

11.6 Peripheral Nervous System
11 Distinguish between the major parts of the peripheral nervous system. (p. 418)
12 Describe the structure of a peripheral nerve and how its fibers are classified. (pp. 418–421)
13 Identify the cranial nerves and list their major functions. (pp. 419–424)
14 Explain how spinal nerves are named and their functions. (pp. 424–425)

11.7 Autonomic Nervous System
15 Characterize the autonomic nervous system. (p. 431)
16 Distinguish between the sympathetic and the parasympathetic divisions of the autonomic nervous system. (pp. 431–432)
17 Compare sympathetic and parasympathetic nerve pathways. (pp. 431–432)
18 Explain how the different autonomic neurotransmitters affect visceral effectors. (p. 433)

11.8 Life-Span Changes
19 Describe aging-associated changes in the nervous system. (p. 438)

cephal-, head: en*cephal*itis—inflammation of the brain.

chiasm-, cross: optic *chiasma*—X-shaped structure produced by the crossing over of optic nerve fibers.

flacc-, flabby: *flaccid* paralysis—loss of tone in muscles innervated by damaged axons.

funi-, small cord or fiber: *funiculus*—major nerve tract or bundle of myelinated axons within the spinal cord.

gangli-, swelling: *gangli*on—mass of neuron cell bodies.

mening-, membrane: *meninges*—membranous coverings of the brain and spinal cord.

plex-, interweaving: choroid *plexus*—mass of specialized capillaries associated with spaces in the brain.

11.1 | Overview of Divisions of the Nervous System

The central nervous system (CNS) consists of the brain and the spinal cord. The brain is the largest and most complex part of the nervous system. It oversees many aspects of physiology, such as sensation and perception, movement, and thinking. The brain includes the two cerebral hemispheres, the diencephalon, the brainstem, and the cerebellum, all described in detail in section 11.4. The brain contains about one hundred billion (10^{11}) multipolar neurons as well as countless branches of the axons by which these neurons communicate with each other and with neurons elsewhere in the nervous system.

Recall from chapter 10 (p. 363) that areas of the nervous system containing mostly neuron cell bodies (and unmyelinated axons) appear gray and are called gray matter, whereas areas containing myelinated axons appear white and are called white matter. Both the brain and the spinal cord have gray matter and white matter. In the brain, the outer layers of the cerebral hemispheres and cerebellum are largely gray matter. White matter, representing interconnecting axons, is found deeper, with islands of gray matter

located throughout. In the spinal cord, in contrast, gray matter (the cell bodies of neurons) is found more centrally, with white matter more peripheral and consisting of axons extending up to the brain or down from the brain.

The brain connects to the spinal cord through the brainstem. Axons conducting impulses up to the brain are bundled in ascending tracts. Those conducting impulses down through the spinal cord are in descending tracts. Both the brain and the spinal cord connect to the peripheral nervous system (PNS) via peripheral nerves.

Bones, membranes, and fluid surround the organs of the CNS. The brain lies in the cranial cavity of the skull, and the spinal cord occupies the vertebral canal in the vertebral column. Beneath these bony coverings, membranes called **meninges,** located between the bone and the soft tissues of the nervous system, protect the brain and spinal cord (fig. 11.1*a*).

11.2 | Meninges

The meninges (sing., *meninx*) have three layers—dura mater, arachnoid mater, and pia mater (fig. 11.1*b*). The **dura mater** is the outermost layer. It is primarily composed of tough, white, dense

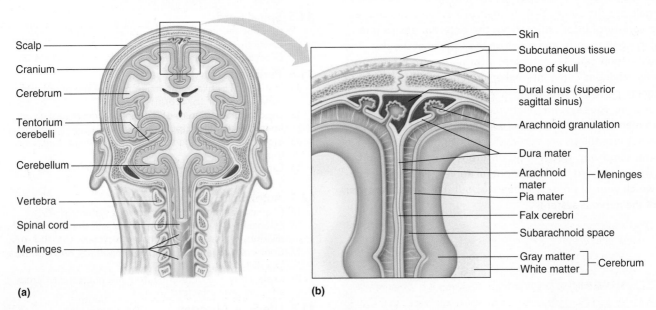

FIGURE 11.1 AP|R Meninges. (*a*) Membranes called meninges enclose the brain and spinal cord. (*b*) The meninges include three layers: dura mater, arachnoid mater, and pia mater.

connective tissue and contains many blood vessels and nerves. It attaches to the inside of the cranial cavity and forms the internal periosteum of the surrounding skull bones (see reference plate 13, p. 49).

In some regions, the dura mater extends inward between lobes of the brain and forms supportive and protective partitions (table 11.1). In other areas, the dura mater splits into two layers, forming channels called *dural sinuses,* shown in figure 11.1*b*. Venous blood flows through these channels as it returns from the brain to vessels leading to the heart.

The dura mater continues into the vertebral canal as a strong, tubular sheath that surrounds the spinal cord. It is attached to the cord at regular intervals by a band of pia mater (denticulate ligaments) that extends the length of the spinal cord on either side. The dural sheath terminates as a blind sac at the level of the second sacral vertebra, below the end of the spinal cord. The sheath around the spinal cord is not attached directly to the vertebrae but is separated by an *epidural space,* which lies between the dural sheath and the bony walls (fig. 11.2). This space contains blood vessels, loose connective tissue, and adipose tissue that pad the spinal cord.

> A blow to the head may rupture some blood vessels associated with the brain, and the escaping blood may collect beneath the dura mater. This condition, called *subdural hematoma,* can increase pressure between the rigid bones of the skull and the soft tissues of the brain. Unless the accumulating blood is promptly evacuated, compression of the brain may lead to functional losses or even death.

The **arachnoid mater** is a thin, weblike membrane that does not have blood vessels and is located between the dura and pia maters. It spreads over the brain and spinal cord but generally does not dip into the grooves and depressions on their surfaces. Many thin strands extend from its undersurface and attach to the pia mater. A *subarachnoid space* between the arachnoid and pia maters contains the clear, watery **cerebrospinal fluid** (ser″ĕ-bro-spi′nal floo′id), or **CSF.**

The subarachnoid space of the meninges completely surrounds the brain and spinal cord, so these organs in effect float in the CSF. The fluid protects the brain and spinal cord by absorbing forces that might otherwise jar and damage their delicate tissues. Clinical Application 11.1 discusses traumatic brain injury, which vividly illustrates the importance of the cushioning function of CSF.

The **pia mater** is thin and contains many nerves, as well as blood vessels that nourish the underlying cells of the brain and spinal cord. The pia mater is attached to the surfaces of these organs and follows their irregular contours, passing over the high areas and dipping into the depressions.

> *Meningitis,* an inflammation of the meninges usually caused by bacterial or viral infection of the CSF, affects the arachnoid and pia maters and sometimes the dura mater, mostly in children. Complications include visual loss, hearing loss, paralysis, and intellectual disability. It may be fatal. Children are vaccinated against *Haemophilus influenza* type b, which was once the most common bacterial cause of meningitis.

TABLE 11.1	Partitions Formed by the Dura Mater
Partition	**Location**
Falx cerebelli	Separates the right and left cerebellar hemispheres
Falx cerebri	Extends downward into the longitudinal fissure, and separates the right and left cerebral hemispheres (fig. 11.1*b*)
Tentorium cerebelli	Separates the occipital lobes of the cerebrum from the cerebellum (fig. 11.1*a*)

 PRACTICE

1 Describe the meninges.

2 Name the layers of the meninges.

3 Explain the location of cerebrospinal fluid.

11.3 | Ventricles and Cerebrospinal Fluid

CSF is formed in four interconnected cavities called **ventricles** (ven′trĭ-klz) that lie in the cerebral hemispheres and brainstem (fig. 11.3 and reference plates 13 and 14, p. 49). These spaces are filled with CSF and are continuous with the central canal of the spinal cord, which extends the full length of the cord (although in most adults it is at least partially closed).

 CAREER CORNER
Occupational Therapist

The man with amyotrophic lateral sclerosis (ALS, or Lou Gehrig's disease) had been growing frustrated with his increasing inability to carry out the activities of daily living. He couldn't use his hands, and his wrists were growing weaker. A visit from an occupational therapist greatly improved both his independence and his spirit.

The occupational therapist showed the man how to continue to use a bathroom sink by supporting his weight on his arms, and how to use mirrors to compensate for his neck stiffness. The therapist was comforting and practical as he showed the man how to repurpose metal salad tongs to hold toilet paper to care for his bathroom needs.

An occupational therapist helps a person maintain normal activities while struggling with a disease, injury, disability, or other limitation. The therapist evaluates the patient's situation and how it is likely to change, sets goals, researches and presents interventions and adaptive equipment that may help, and assesses results. The therapist may also instruct family members and caregivers on how to assist the patient.

Occupational therapists work in health-care facilities, schools, home health services, and nursing homes. They must have a master's degree in occupational therapy and state licensure.

A traumatic brain injury (TBI) results from mechanical force such as from a fall, accident, attack, or sports-related injury. According to the Brain Trauma Foundation, TBI in the United States is the leading cause of death and disability from ages one to forty-four years, and is responsible for 52,000 deaths per year. More than 5 million people have such injuries, which are classified as mild, mild repetitive, or severe.

Mild TBI, also known as a concussion, produces loss of consciousness or altered mental status. Its effects are more psychological than neurological, and it does not appear to cause lasting damage. Symptoms include disturbed sleep, ringing in the ears, memory lapse, balance problems, irritability, and sensitivity to light and sound. These physical symptoms are heightened if the person also suffers from depression or post-traumatic stress disorder (PTSD). Mild TBI may cause PTSD if, as the brain hits the skull, the injury generates a shearing force that impairs the prefrontal cortex's control of a region called the amygdala so that it becomes overactive. As a result, the person cannot let go of psychological trauma, which is the definition of PTSD.

A sports-related form of mild repetitive TBI is chronic traumatic encephalopathy (CTE). It results from many small injuries over time, rather than a single violent blow. The first report of the condition in a medical journal appeared in 2005, regarding a player for the National Football League, but in the 1980s CTE was recognized in boxers and wrestlers, in whom it was called "punch-drunk syndrome" and "dementia pugilistica." Symptoms typically begin years after the first of the repetitive head injuries associated with the sport. They include depression, impulsive and erratic behavior, headaches, dizziness, memory loss, dementia, and loss of executive function (ability to process information and make decisions). These difficulties stem from mechanical trauma to the cortex, hematomas (bleeding) in the subcortex, vasospasm, ischemia, and sudden movement of the skull, which tears axons. The brains of people who have died with CTE show changes characteristic of Alzheimer disease. Studies of CTE in football players who wear helmets with sensors to record the forces applied to their brains show that a college football player receives on average 950 hits to the head in a season (fig. 11A).

Severe TBI is seen in combat situations, where the cause and pattern of damage is called "blast-related brain injury." The damage results from a change in atmospheric pressure, a violent release of energy (sound, heat, pressure, or electromagnetic waves), and sometimes exposure to a neurotoxin released from the blast. Rocket-propelled grenades, improvised incendiary devices, and land mines are the primary causes of this most dangerous type of TBI. The brain is initially jolted forward at a force exceeding 1,600 feet per second, and then is hit again as air in the cranium rushes forward.

FIGURE 11A Chronic traumatic encephalopathy is a form of mild repetitive traumatic brain injury seen in football players.

A problem in treating blast-related brain injury is recognizing it swiftly, because symptoms may not appear until hours after the violent event. Whereas one soldier immediately after the blast might be blind or deaf or unable to move or speak, another soldier who has suffered similar injuries to the soft tissue of the brain might not show such effects until later. The effects of blast-related brain injury are lasting. Studies on veterans of the Vietnam War indicate injury-related cognitive decline years after the injury.

The two *lateral ventricles* are the largest. The first ventricle is in the left cerebral hemisphere and the second ventricle is in the right cerebral hemisphere. They extend anteriorly and posteriorly into the cerebral hemispheres.

A narrow space that constitutes the *third ventricle* is in the midline of the brain beneath the corpus callosum, which is a bridge of axons that links the two cerebral hemispheres. This ventricle communicates with the lateral ventricles through openings *(interventricular foramina)* in its anterior end.

The *fourth ventricle* is in the brainstem, just anterior to the cerebellum. A narrow canal, the *cerebral aqueduct* (aqueduct of Sylvius), connects the third ventricle to the fourth ventricle and

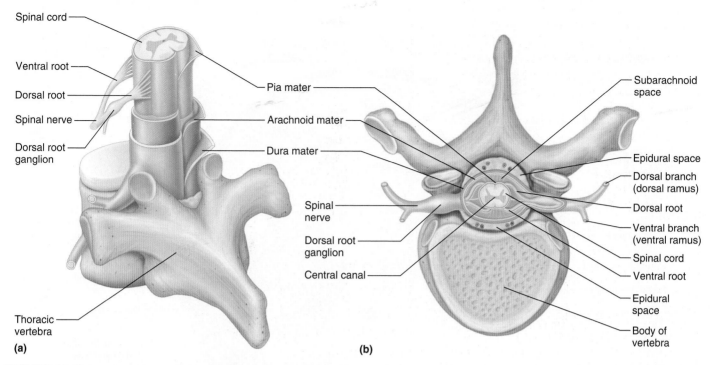

FIGURE 11.2 Meninges of the spinal cord. (*a*) The dura mater ensheathes the spinal cord. (*b*) Tissues forming a protective pad around the cord fill the epidural space between the dural sheath and the bone of the vertebra.

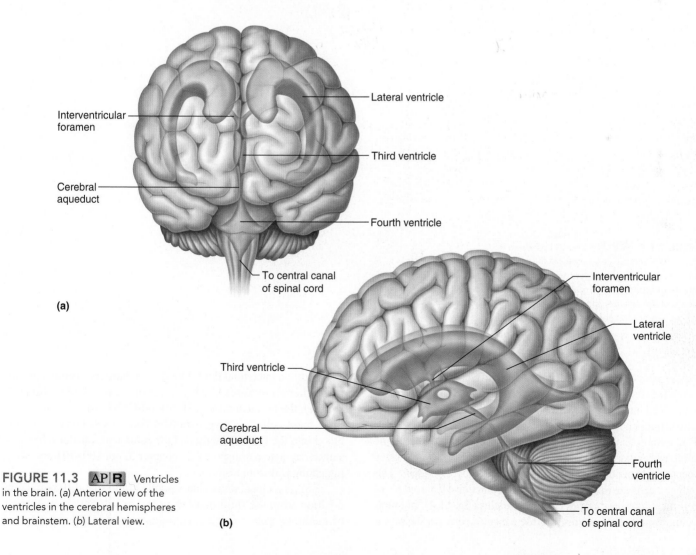

FIGURE 11.3 AP|R Ventricles in the brain. (*a*) Anterior view of the ventricles in the cerebral hemispheres and brainstem. (*b*) Lateral view.

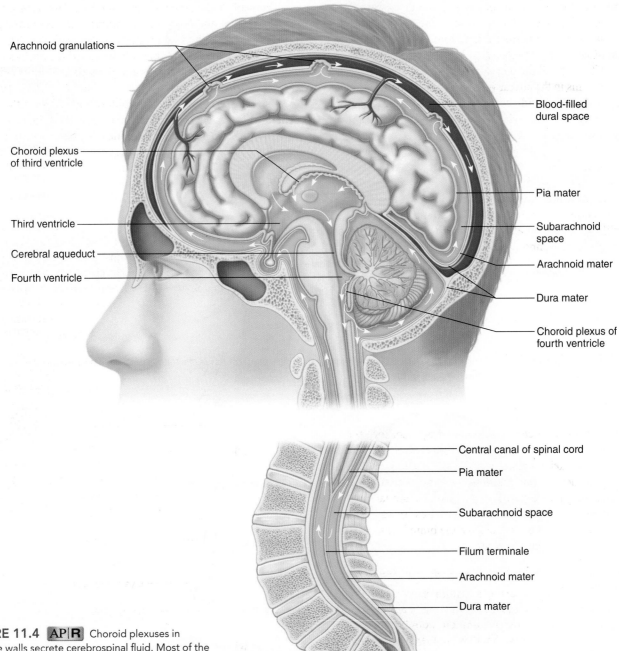

Arachnoid granulations

Choroid plexus
of third ventricle

Third ventricle

Cerebral aqueduct

Fourth ventricle

Blood-filled
dural space

Pia mater

Subarachnoid
space

Arachnoid mater

Dura mater

Choroid plexus of
fourth ventricle

Central canal of spinal cord

Pia mater

Subarachnoid space

Filum terminale

Arachnoid mater

Dura mater

FIGURE 11.4 AP|R Choroid plexuses in ventricle walls secrete cerebrospinal fluid. Most of the fluid circulates through the ventricles and enters the subarachnoid space, and is reabsorbed into the blood of the dural sinuses through arachnoid granulations. (Spinal nerves are not shown.)

passes lengthwise through the brainstem. This ventricle is continuous with the central canal of the spinal cord and has openings in its roof that lead into the subarachnoid space of the meninges.

Structures called **choroid plexuses** (ko′roid plek′sus-ez) secrete CSF. Choroid plexuses are tiny, reddish, cauliflower-like masses of specialized capillaries from the pia mater, covered by a single layer of specialized ependymal cells (see chapter 10, pp. 368–369). The ependymal cells are joined closely by tight junctions. Choroid plexuses project into the cavities of the ventricles (fig. 11.4). In much the same way that astrocytes provide a barrier between the blood

and the brain interstitial fluid (blood-brain barrier), ependymal cells in the choroid plexuses block passage of water-soluble substances between the blood and the CSF (blood-CSF barrier). At the same time, these cells selectively transfer certain substances from the blood into the CSF by facilitated diffusion and transfer other substances by active transport (see chapter 3, pp. 100–101 and 103), regulating CSF composition.

CSF is a clear, somewhat viscous liquid that differs in composition from the fluid that leaves the capillaries in other parts of the body. Specifically, it contains a greater concentration of sodium and lesser concentrations of glucose and potassium than do other extracellular fluids. Its function is nutritive as well as protective. CSF helps maintain a stable ionic concentration in the CNS and provides a pathway to the blood for waste. Clinical Application 11.2 discusses the pressure that CSF generates.

Most CSF forms in the lateral ventricles, from where it slowly circulates into the third and fourth ventricles. Small amounts enter the central canal of the spinal cord, but most CSF circulates through the subarachnoid space of both the brain and the spinal cord (fig. 11.4) by passing through the wall of the fourth ventricle near the cerebellum.

Humans secrete nearly 500 milliliters of CSF daily. However, only about 140 milliliters are in the nervous system at any time, because CSF is continuously reabsorbed into the blood through tiny, fingerlike structures called *arachnoid granulations* that project from the subarachnoid space into the blood-filled dural sinuses (fig. 11.4).

 PRACTICE

4 Where are the ventricles of the brain located?

5 How does CSF form?

6 Describe the pattern of CSF circulation.

11.4 | Brain

The **brain** contains neural centers associated with sensory functions and is responsible for sensations and perceptions. It issues motor commands to skeletal muscles and carries on higher mental functions, such as memory and reasoning. The brain also contains neural centers and pathways that coordinate muscular movements, and others that regulate visceral activities. In addition to overseeing the function of the entire body, the brain is responsible for characteristics such as personality.

> The term "muscle fiber" refers to an entire skeletal muscle cell, as discussed in chapter 9 (p. 293). The term "fiber" is also used in discussing the nervous system, but here it does not refer to an entire cell. A "nerve fiber" is another term for the axon of a neuron.

Brain Development

The basic structure of the brain reflects the way it forms during early (embryonic) development. It begins as the neural tube that gives rise to the central nervous system. The portion that becomes the brain has three major cavities, or vesicles, at one end—the *forebrain* (prosencephalon), *midbrain* (mesencephalon), and *hindbrain* (rhombencephalon) (fig. 11.5). Later, the forebrain divides into anterior and posterior portions (telencephalon and diencephalon, respectively), and the hindbrain partially divides into two parts (metencephalon and myelencephalon). The resulting five cavities persist in the mature brain as the fluid-filled *ventricles* and the tubes that connect them. Cells of the tissue surrounding the spaces differentiate into the structural and functional regions of the brain.

The wall of the anterior portion of the forebrain gives rise to the *cerebrum* and *basal nuclei,* whereas the posterior portion forms a section of the brain called the *diencephalon.* The region the midbrain produces continues to be called the *midbrain* in the adult structure, and the hindbrain gives rise to the *cerebellum, pons,* and *medulla oblongata* (fig. 11.6 and table 11.2). Together, the midbrain, pons, and medulla oblongata comprise the **brainstem** (brān'stem), which attaches the brain to the spinal cord.

On a cellular level, the brain develops as specific neurons attract others by secreting growth factors. In the embryo and fetus, the brain overgrows, and then apoptosis (programmed cell death) destroys excess cells.

Structure of the Cerebrum

The **cerebrum** (sĕr'ē-brum), which develops from the anterior portion of the forebrain, is the largest part of the mature brain. It consists of two large masses, or **cerebral hemispheres** (ser'ĕ-bral hem'i-sfērz), which are essentially mirror images of each

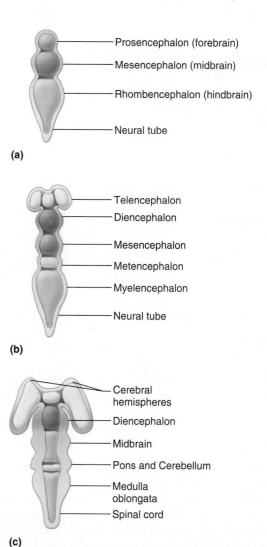

FIGURE 11.5 Brain development. (*a*) The brain develops from a tubular structure with three cavities. (*b*) The cavities persist as the ventricles and their interconnections. (*c*) The wall of the tube gives rise to various regions of the brain, brainstem, and spinal cord.

Cerebrospinal fluid (CSF) is secreted and reabsorbed continuously, which keeps the fluid pressure in the ventricles relatively constant. However, infection, a tumor, or a blood clot can interfere with the fluid's circulation, increasing pressure in the ventricles (intracranial pressure or ICP). The pressure can collapse cerebral blood vessels, slowing blood flow. Brain tissues forced against the skull may be injured.

A *lumbar puncture* (spinal tap) measures CSF pressure. A physician inserts a fine, hollow needle into the subarachnoid space between the third and fourth or between the fourth and fifth lumbar vertebrae—below the end of the spinal cord (fig. 11B). An instrument called a *manometer* measures the pressure of the fluid, which is usually about 130 millimeters of water (10 millimeters of mercury). At the same time, samples of CSF may be withdrawn and tested for abnormal constituents. Red blood cells in the CSF, for example, may indicate a hemorrhage in the central nervous system. A temporary drain inserted into the subarachnoid space between the fourth and fifth lumbar vertebrae can relieve pressure.

In a fetus or infant whose cranial sutures have not yet united, increasing ICP may enlarge the cranium, causing *hydrocephalus*, or "water on the brain" (fig. 11C). A shunt to relieve hydrocephalus drains fluid away from the cranial cavity and into the digestive tract, where it is either reabsorbed into the blood or excreted.

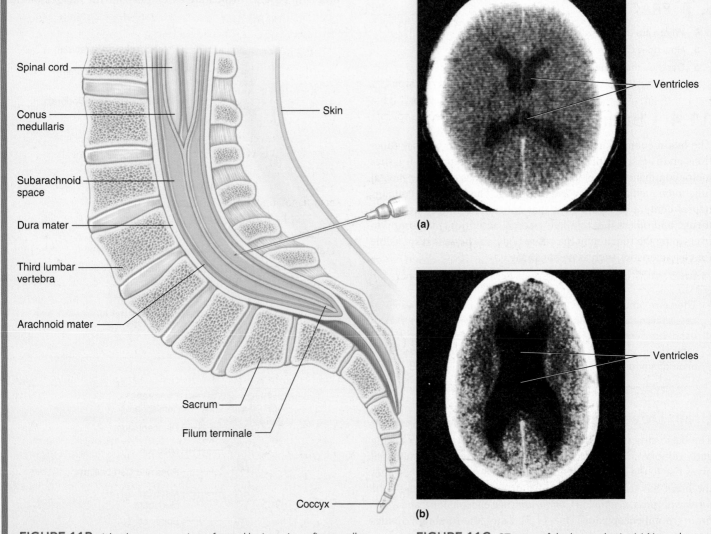

FIGURE 11B A lumbar puncture is performed by inserting a fine needle, usually below the fourth lumbar vertebra, and withdrawing a sample of CSF from the subarachnoid space. (For clarity, spinal nerves are not shown.)

FIGURE 11C CT scans of the human brain. (*a*) Normal ventricles. (*b*) Ventricles enlarged by accumulated fluid.

TABLE 11.2 Structural Development of the Brain

Embryonic Vesicle	Spaces Produced	Regions of the Brain Produced
Forebrain (prosencephalon)		
Anterior portion (telencephalon)	Lateral ventricles	Cerebrum Basal nuclei
Posterior portion (diencephalon)	Third ventricle	Thalamus Hypothalamus Posterior pituitary gland Pineal gland
Midbrain (mesencephalon)	Cerebral aqueduct	Midbrain
Hindbrain (rhombencephalon)		
Anterior portion (metencephalon)	Fourth ventricle	Cerebellum, pons
Posterior portion (myelencephalon)	Fourth ventricle	Medulla oblongata

other (fig. 11.7 and reference plate 9, p. 46). A broad, flat bundle of axons called the **corpus callosum** connects the cerebral hemispheres. A layer of dura mater called the *falx cerebri* separates them (see fig. 11.1*b*).

Many ridges or convolutions, called **gyri** (ji′ri) (sing., *gyrus*), separated by grooves, mark the cerebrum's surface. Generally, a shallow to somewhat deep groove is called a **sulcus** (sul′kus; pl. *sulci,* sul′si)**,** and a very deep groove is called a **fissure.** The pattern of these elevations and depressions is complex, but, despite individual variations, is similar in all normal brains. For example, a *longitudinal fissure* separates the right and left cerebral hemispheres; a *transverse fissure* separates the cerebrum from the cerebellum; and sulci divide each hemisphere into lobes (see figs. 11.6 and 11.7).

A fetus or newborn with *anencephaly* has a face and lower brain structures but lacks most higher brain structures. A newborn with this anomaly typically survives only a day or two.

Anencephaly is a type of neural tube defect (NTD). It occurs at about the twenty-eighth day of prenatal development, when a sheet of tissue that normally folds to form the neural tube, which develops into the CNS, remains open at the top. In *spina bifida,* an opening is farther down the neural tube, and in severe cases may cause paralysis from that point downward. In some cases surgery, before or shortly after birth, can partially correct spina bifida. Taking folic acid supplements just before and during pregnancy can lower the risk of an NTD.

In a disorder called *lissencephaly* ("smooth brain"), a newborn has a smooth cerebral cortex, completely lacking convolutions. Absence of a protein early in prenatal development prevents certain neurons from migrating in the brain, which blocks formation of convolutions. The child is profoundly disabled intellectually, with frequent seizures and other neurological problems.

Most of the five lobes of the cerebral hemispheres (fig. 11.7) are named after the skull bones that they underlie. The lobes include the following:

1. The **frontal lobe** forms the anterior portion of each cerebral hemisphere. It is bordered posteriorly by a *central sulcus* (fissure of Rolando), which passes out from the longitudinal fissure at a right angle, and inferiorly by a *lateral sulcus* (fissure of Sylvius), which exits the undersurface of the brain along its sides.
2. The **parietal lobe** is posterior to the frontal lobe and is separated from it by the central sulcus.
3. The **temporal lobe** lies inferior to the frontal and parietal lobes and is separated from them by the lateral sulcus.
4. The **occipital lobe** forms the posterior portion of each cerebral hemisphere and is separated from the cerebellum by a shelf-like extension of dura mater called the *tentorium cerebelli.* The occipital lobe and the parietal and temporal lobes have no distinct boundary.
5. The **insula** (island of Reil) is a lobe deep within the lateral sulcus and is so named because it is covered by parts of the frontal, parietal, and temporal lobes. A *circular sulcus* separates the insula from the other lobes.

(Some sources include a sixth lobe, the limbic lobe, which is deeper than the other lobes. This area is discussed as the "limbic system" later in this chapter.)

A thin layer of gray matter (2 to 5 millimeters thick) called the **cerebral cortex** (ser′ĕ-bral kor′teks) constitutes the outermost portion of the cerebrum. It covers the gyri, dipping into the sulci and fissures. The cerebral cortex contains nearly 75% of all the neuron cell bodies in the nervous system.

Just beneath the cerebral cortex is a mass of white matter that makes up the bulk of the cerebrum. This mass contains bundles of myelinated axons that connect neuron cell bodies of the cortex with other parts of the nervous system. Some of these fibers pass from one cerebral hemisphere to the other by way of the corpus callosum, and others carry sensory or motor impulses from the cortex to areas of gray matter deeper in the brain or to the spinal cord.

In a "stroke," or *cerebrovascular accident* (CVA), a sudden interruption in blood flow in a vessel supplying brain tissues damages the cerebrum. The affected blood vessel may rupture, bleeding into the brain, or be blocked by a clot. In either case, brain tissues downstream from the vascular accident die and some loss of function may occur. Temporary interruption in cerebral blood flow, perhaps by a clot that quickly breaks apart, produces a much less serious *transient ischemic attack* (TIA), sometimes called a ministroke.

Functions of the Cerebral Cortex

The cerebral cortex provides higher brain functions: interpreting impulses from sense organs, initiating voluntary muscular movements, storing information as memory, and retrieving this informa-

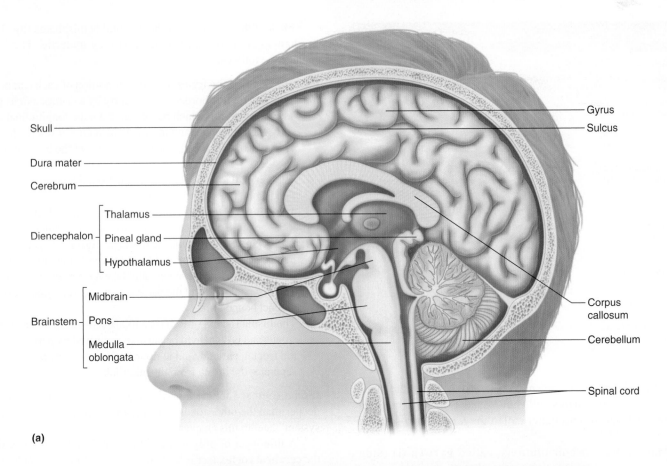

(a)

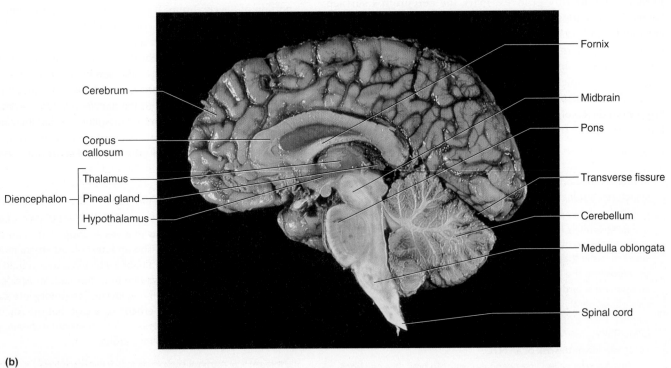

(b)

FIGURE 11.6 **AP|R** Sagittal section of the brain and spinal cord, including the uncut medial surface of the right cerebral hemisphere. (*a*) The major portions of the brain include the cerebrum, the diencephalon, the cerebellum, and the brainstem. (*b*) Photo of a midsagittal section through a human brain.

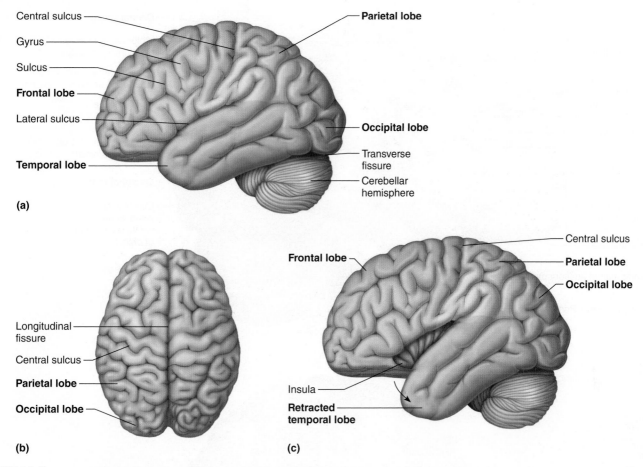

FIGURE 11.7 Colors in this figure distinguish the lobes of the cerebral hemispheres. (*a*) Lateral view of the left hemisphere. (*b*) Hemispheres viewed from above. (*c*) Lateral view of the left hemisphere with the insula exposed.

tion in reasoning. The cerebral cortex is also the part of the brain responsible for intelligence and personality.

Functional Regions of the Cortex

A variety of techniques are used to study the specific functions of regions of the cerebral cortex. For example, From Science to Technology 2.3, figure 2E (p. 78) shows how PET scans reveal the frontal cortex turning sensations of sight, hearing, and touch into concepts represented by words. Clues to cerebral functioning also come from investigating deficits in people who have suffered brain disease or injury.

In other studies, areas of cortices have been exposed surgically and stimulated mechanically or electrically. Researchers observe the responses in certain muscles or the specific sensations that result. Based on such investigations, researchers have divided the cerebral cortex into sensory, association, and motor areas that overlap somewhat. Each of these areas is a collection of neurons that work together to provide a particular brain function.

Sensory Areas of the Cortex

Sensory areas in several lobes of the cerebrum receive and interpret impulses from sensory receptors, producing feelings or sensations. For example, the sensations of temperature, touch, pressure, and pain in the skin arise in the postcentral gyri of the parietal lobes posterior to the central sulcus (**fig. 11.8**). The posterior parts of the

occipital lobes provide vision, whereas the superior posterior portions of the temporal lobes contain the centers for hearing. The sensory areas for taste are near the bases of the lateral sulci and include part of the insula. The sense of smell arises from centers deep in the temporal lobes.

Sensory fibers from the PNS cross over in the spinal cord or the brainstem. Thus, the centers in the right central hemisphere interpret impulses originating from the left side of the body, and vice versa. However, the sensory areas concerned with vision receive impulses from both eyes, and those concerned with hearing receive impulses from both ears.

Not all sensory areas are bilateral. The *sensory speech area,* also called *Wernicke's area,* is in the temporal lobe adjacent to the parietal lobe, near the posterior end of the lateral sulcus, usually in the left hemisphere (fig. 11.8). This area is important for understanding and formulating written and spoken language.

Association Areas of the Cortex

Association areas are neither primarily sensory nor motor. They connect with each other and with other brain structures. Association areas occupy the anterior portions of the frontal lobes and are widespread in the lateral portions of the parietal, temporal, and occipital lobes. Association areas analyze and interpret sensory experiences and help provide memory, reasoning, verbalizing, judgment, and emotions (fig. 11.8).

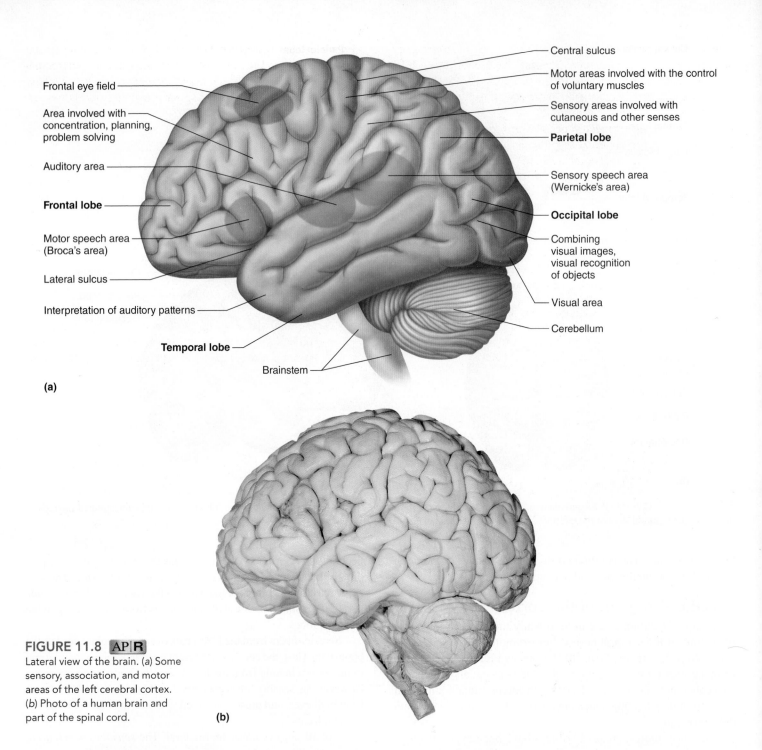

FIGURE 11.8 AP|R
Lateral view of the brain. (*a*) Some sensory, association, and motor areas of the left cerebral cortex. (*b*) Photo of a human brain and part of the spinal cord.

(a)

Central sulcus

Motor areas involved with the control of voluntary muscles

Sensory areas involved with cutaneous and other senses

Parietal lobe

Sensory speech area (Wernicke's area)

Occipital lobe

Combining visual images, visual recognition of objects

Visual area

Cerebellum

Frontal eye field

Area involved with concentration, planning, problem solving

Auditory area

Frontal lobe

Motor speech area (Broca's area)

Lateral sulcus

Interpretation of auditory patterns

Temporal lobe

Brainstem

(b)

The association areas of the frontal lobes provide higher intellectual processes, such as concentrating, planning, and complex problem solving. The anterior and inferior portions of these lobes (prefrontal areas) control emotional behavior and produce awareness of the possible consequences of behaviors. These abilities are also collectively called executive function.

The parietal lobes have association areas that help interpret sensory information and aid in understanding speech and choosing words to express thoughts and feelings. Awareness of the form of objects, including one's own body parts, stems from the posterior regions of these lobes.

The association areas of the temporal lobes and the regions at the posterior ends of the lateral sulci interpret complex sensory experiences, such as those needed to understand speech and to read. These regions also store memories of visual scenes, music, and other complex sensory patterns.

The occipital lobes have association areas adjacent to the visual centers. These are important in analyzing visual patterns and combining visual images with other sensory experiences, such as when one person recognizes another.

The functions of the insula are not as well known as those of the other lobes because of the insula's location deep within the cerebrum. As mentioned earlier, it is the location of some sensory areas processing taste information. Studies that use functional MRI scanning suggest that the insula serves as a crossroads for translating sensory information into appropriate emotional responses, such as feeling disgust at the sight of something unpleasant, or a feeling of joy when hearing a symphony or when biting into a slice of pizza. Some researchers hypothesize that the insula is, in some complex way, responsible for some of the qualities that make us human.

Association areas often interact. The area where the occipital, parietal, and temporal lobes meet processes sensory information from all three of these association areas. It plays a role in integrating visual, auditory, and other sensory information and then interpreting a situation. For example, you hear a familiar voice, look up from your notes, see a friend from class, and realize that it is time for your study group.

Motor Areas of the Cortex

The *primary motor areas* of the cerebral cortex lie in the precentral gyri of the frontal lobes just in front of the central sulcus and in the anterior wall of this sulcus (fig. 11.8). The nervous tissue in these regions contains many large *pyramidal cells,* named for their pyramid-shaped cell bodies.

Impulses from the pyramidal cells move downward through the brainstem and into the spinal cord on descending tracts. Most of the nerve fibers in these tracts cross over from one side of the brain to the other within the brainstem.

Impulses conducted on these pathways in special patterns and frequencies are responsible for movements in skeletal muscles. More specifically, as **figure 11.9** shows, cells in the upper portions of the motor areas send impulses to muscles in the thighs and

> A person with *dyslexia* sees letters separately and must learn to read differently than people whose nervous systems can group letters into words. Three to 10% of people have dyslexia. The condition probably has several causes, with inborn visual and perceptual skills interacting with the way the child learns to read. Dyslexia is not related to intelligence. Many brilliant thinkers with dyslexia were "slow" in school because educators did not know how to help them.

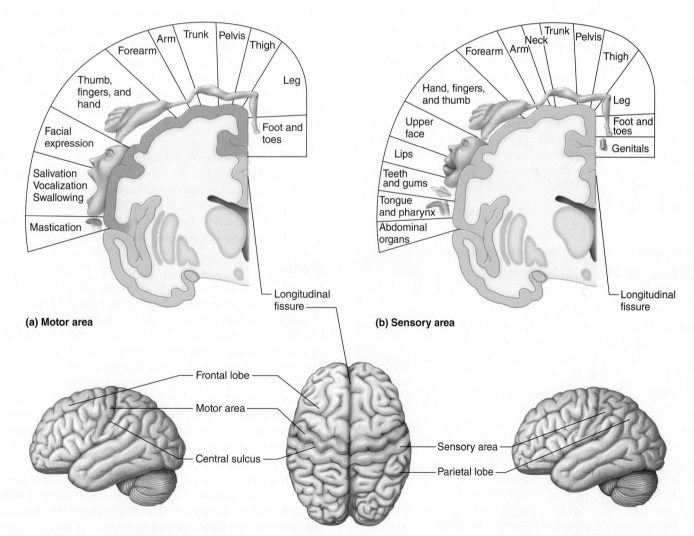

FIGURE 11.9 Functional regions of the cerebral cortex. (*a*) Motor areas that control voluntary muscles (only left hemisphere shown). (*b*) Sensory areas involved with cutaneous and other senses (only left hemisphere shown).

legs; those in the middle portions control muscles in the arms and forearms; and those in lower portions activate muscles of the head, face, and tongue. The relative (distorted) size of each area in the figure reflects the extent of the cortex devoted to it.

In addition to the primary motor areas, certain other regions of the frontal lobe control motor functions. For example, a region called the *motor speech area,* also known as *Broca's area,* is in the frontal lobe, usually in the left hemisphere, just anterior to the primary motor cortex and superior to the lateral sulcus. The motor speech area is important in generating the complex pattern of muscular actions of the mouth, tongue, and larynx, which make speech possible (see fig. 11.8). Bundles of axons directly and indirectly connect the motor speech area to the sensory speech area (Wernicke's area). Other regions of the cerebral cortex take part in the speech process, but basically Broca's area provides motor instructions for written or spoken communication, whereas Wernicke's area ensures that the communication makes sense. For example, a person with an injury to Broca's area may be able to understand spoken words but be unable to speak. A person with damage to Wernicke's area could initiate speech, but without control of its content.

In the superior part of the frontal lobe is a region called the *frontal eye field.* The motor cortex in this area controls voluntary movements of the eyes and eyelids. Nearby is the cortex responsible for movements of the head that direct the eyes. Another region just anterior to the primary motor area controls the muscular movements of the hands and fingers that make such skills as writing possible (see fig. 11.8). Table 11.3 summarizes the functions of the cerebral lobes.

An injury to a motor area of the cerebral cortex may impair the ability to produce purposeful muscular movements. Such a condition that affects use of the upper and lower limbs, head, or eyes is called *apraxia.* When apraxia affects the speech muscles, disrupting speaking ability, it is called *aphasia.*

TABLE 11.3 Functions of the Cerebral Lobes

Lobe	Functions
Frontal lobes	Association areas carry on higher intellectual processes for concentrating, planning, complex problem solving, and judging the consequences of behavior.
	Motor areas control movements of voluntary skeletal muscles.
Parietal lobes	Sensory areas provide sensations of temperature, touch, pressure, and pain involving the skin.
	Association areas function in understanding speech and in using words to express thoughts and feelings.
Temporal lobes	Sensory areas are responsible for hearing.
	Association areas interpret sensory experiences and remember visual scenes, music, and other complex sensory patterns.
Occipital lobes	Sensory areas are responsible for vision.
	Association areas combine visual images with other sensory experiences.

PRACTICE

7 How does the brain form during early development?

8 Describe the cerebrum.

9 List the general functions of the cerebrum.

10 Where in the brain are the sensory areas located?

11 Explain the functions of association areas.

12 Where in the brain are the motor areas located?

Hemisphere Dominance

Both cerebral hemispheres participate in basic functions, such as receiving and analyzing sensory impulses, controlling skeletal muscles on opposite sides of the body, and storing memory. However, one side usually acts as a *dominant hemisphere* for certain other functions.

In most people, the left hemisphere is dominant for the language-related activities of speech, writing, and reading. It is also dominant for complex intellectual functions requiring verbal, analytical, and computational skills. In other people, the right hemisphere is dominant, and in some, the hemispheres are equally dominant.

The nondominant hemisphere specializes in nonverbal functions, in addition to carrying on basic functions. Nonverbal functions include motor tasks that require orientation of the body in space, understanding and interpreting musical patterns, and visual experiences. The nondominant hemisphere also provides emotional and intuitive thought processes. For example, the region in the nondominant hemisphere that corresponds to the motor speech area does not control speech, but it influences the emotional aspects of spoken language.

Nerve fibers of the *corpus callosum,* which connect the cerebral hemispheres, enable the dominant hemisphere to control the motor cortex of the nondominant hemisphere. These fibers also transfer sensory information reaching the nondominant hemisphere to the general interpretative area of the dominant one, where the information can be used in decision making.

Memory

Memory, one of the most astonishing capabilities of the brain, is the consequence of learning. Whereas learning is the acquisition of new knowledge, memory is the persistence of that learning, with the ability to access it at a later time. Two stages of memory, short term and long term, have been recognized for many years. They differ in characteristics other than duration.

Short-term, or "working," memories are thought to involve neurons connected in a circuit and stimulated so rapidly that the likelihood of neurons in the circuit reaching threshold is temporarily increased. As long as the circuit is facilitated, the memory persists, but if the stimulus is removed, the circuit becomes inactive.

Long-term memory can hold much more information than short-term memory and lasts a lifetime. Some long-term memory establishes new synaptic connections through increased branching of axons and dendrites. In another mechanism, **long-term potentiation,** very rapid repeated stimulation of the same neurons increases the number of postsynaptic neurotransmitter receptors and causes physical changes at the synapse that make synaptic transmission more effective. Both mechanisms—new synaptic connections and repeated stimulation of the same neurons—fulfill two requirements of long-term memory. First, sufficient synapses form to encode an almost limitless number of memories. Each of the 10 billion neurons in the cortex can make tens of thousands of synaptic connections to other neurons, forming 60 trillion links. Second, a certain pattern of synapses can persist for years.

Eventually memories are stored in various parts of the cerebral cortex in a process called **memory consolidation.** The hippocampus plays an important role in directing memory information to the appropriate location in the cortex, although it does not actually store memories. Another area of the temporal lobe, the amygdala, assigns value to a memory, such as whether it was pleasant.

Unusual behaviors and skills of people who have damaged the hippocampus have taught researchers much about this intriguing part of the brain. In 1953, a surgeon removed parts of the hippocampus and the amygdala of a young man to relieve his severe epilepsy. His seizures became less frequent, but he suffered a profound loss in the ability to consolidate short-term memories into long-term ones. As a result, events in his life faded quickly from his memory. He was unable to recall any events that took place since surgery, living always as if it was the 1950s. He would read the same magazine article repeatedly with renewed interest each time.

Basal Nuclei

The **basal nuclei,** also called the basal ganglia, are masses of gray matter deep within the cerebral hemispheres. They include the *caudate nucleus,* the *putamen,* and the *globus pallidus,* and they develop from the anterior portion of the forebrain (**fig. 11.10**). The basal nuclei produce the inhibitory neurotransmitter *dopamine.* The neurons of the basal nuclei interact with other brain areas, including the motor cortex, thalamus, and cerebellum. These interactions, through a combination of stimulation and inhibition, facilitate voluntary movement. Clinical Application 11.3 discusses Parkinson disease, in which neurons in the basal nuclei degenerate.

 PRACTICE

13 What is hemisphere dominance?

14 What are the functions of the nondominant hemisphere?

15 Distinguish between short-term and long-term memory.

16 What is the function of the basal nuclei?

Diencephalon

The **diencephalon** (di″en-sef′ah-lon) develops from the posterior forebrain and is located between the cerebral hemispheres and superior to the brainstem (see figs. 11.6 and 11.10). It surrounds the third ventricle and is largely composed of gray matter. In the diencephalon, a dense mass called the **thalamus** (thal′ah-mus) bulges into the third ventricle from each side. Another region of

the diencephalon that includes many nuclei is the **hypothalamus** (hi″po-thal′ah-mus). It lies inferior to the thalamic nuclei and forms the lower walls and floor of the third ventricle (see reference plates 9 and 13, pp. 46 and 49).

Other parts of the diencephalon include (1) the **optic tracts,** which originate from the **optic chiasma** (formed by some of the optic nerve fibers crossing over); (2) the **infundibulum** (pituitary stalk), which is a conical process behind the optic chiasma to which the pituitary gland is attached; (3) the **posterior pituitary gland,** which hangs from the floor of the hypothalamus; (4) the **mammillary** (mam′ĭ-lar″e) **bodies,** which are two rounded structures behind the infundibulum; and (5) the **pineal gland,** which forms as a cone-shaped projection from the roof of the diencephalon (see chapter 13, p. 516).

The thalamus is a selective gateway for sensory impulses ascending from other parts of the nervous system to the cerebral cortex. It receives all sensory impulses (except those associated with the sense of smell) and channels them to appropriate regions of the cortex for interpretation. In addition, all regions of the cerebral cortex can communicate with the thalamus by means of descending fibers.

The thalamus relays sensory information by synchronizing action potentials. Consider vision. An image on the retina stimulates the *lateral geniculate nucleus* (LGN) region of the thalamus, which then sends action potentials to a part of the visual cortex. Those action potentials are synchronized—fired simultaneously—by the LGN's neurons only if the stimuli come from a single object, such as a bar. If the stimulus is two black dots, the resulting thalamic action potentials are not synchronized. The synchronicity of action potentials, therefore, may be a way that the thalamus selects which stimuli to relay to higher brain structures. Therefore, the thalamus is not only a messenger but also an editor.

Pathways connect the hypothalamus to the cerebral cortex, thalamus, and parts of the brainstem so that it can receive impulses from them and send impulses to them. The hypothalamus maintains homeostasis by regulating a variety of visceral activities and by linking the nervous and endocrine systems.

The hypothalamus regulates:

1. heart rate and arterial blood pressure
2. body temperature
3. water and electrolyte balance
4. control of hunger and body weight
5. control of movements and glandular secretions of the stomach and intestines
6. production of neurosecretory substances that stimulate the pituitary gland to release hormones that help regulate growth, control various glands, and influence reproductive physiology
7. sleep and wakefulness

Structures in the region of the diencephalon also are important in controlling emotional responses. Parts of the cerebral cortex in the medial parts of the frontal and temporal lobes connect with the hypothalamus, thalamus, basal nuclei, and other deep nuclei. These structures form a complex called the **limbic system.** It controls emotional experience and expression and can modify the way a person acts, producing such feelings as fear, anger, pleasure, and sorrow. The limbic system reacts to potentially

Actor Michael J. Fox was in his late twenties when his wife noticed the first sign of Parkinson disease (PD)—he leaned when walking. When one of his fingers began twitching, Fox consulted a physician, and so began the journey toward his diagnosis. Of the approximately 6 million people worldwide who have PD, only 10% develop symptoms before the age of forty, like Michael J. Fox did.

Fox kept his diagnosis private, but by the late 1990s his co-workers began to notice symptoms that emerged when medication wore off—rigidity, a shuffling and off-balance gait, and poor small-motor control. Fox's face became masklike, a characteristic called hypomimia. It took effort to speak, a symptom called hypophia. Even though his brain could string thoughts into sentences, the muscles of his jaw, lips, and tongue could not utter them. He also developed micrographia, or small handwriting. Fox founded the Michael J. Fox Foundation for Parkinson's Research, and he continues to act on television.

In PD, neurons degenerate in the substantia nigra area of the brainstem. *Substantia nigra* means "large black area," named for the dark pigment that the neurons release as a by-product of synthesizing the neurotransmitter dopamine. When these neurons degenerate, less dopamine reaches synapses with neurons in the striatum of the basal nuclei. The decrease in dopamine causes the motor symptoms of PD. Some patients develop other symptoms, including depression, dementia, constipation, incontinence, sleep problems, and orthostatic hypotension (dizziness upon standing).

Until recently PD was diagnosed based on neurological symptoms alone, because MRI and CT scans appear normal. However, an imaging technology called SPECT (single photon emission computed tomography) uses a chemical compound similar to cocaine that binds specifically to dopamine transporter molecules in the striatum, marking them. This compound used with SPECT is called a DaTscan. It distinguishes tremors that are specifically due to PD, which is useful both for diagnosis and in research to ensure that people in clinical trials to evaluate PD treatments actually have PD.

So far, no treatments can cure or slow the course of PD, but replacing or enhancing use of dopamine can temporarily alleviate symptoms. The standard treatment for many years has been levodopa, a precursor to dopamine that crosses the blood-brain barrier. In the brain, levodopa is converted to dopamine. Levodopa provides temporary relief from twitching and rigidity.

Drug treatment for PD becomes less effective over time. By a feedback mechanism the brain senses the external supply of dopamine and decreases its own production, so that eventually higher doses of levodopa are needed to achieve the effect. Taking too much levodopa leads to another condition, tardive dyskinesia, that produces uncontrollable facial tics and spastic extensions of the limbs. Tardive dyskinesia may result from effects of excess dopamine in brain areas other than those affected in PD.

Surgery can alleviate Parkinson's symptoms. Fox underwent thalamotomy, in which an electrode caused a lesion in his thalamus that calmed violent shaking in his left arm. Another surgical procedure, pallidotomy, causes lesions in the globus pallidus internus, a part of the basal nuclei, and the approach is also used on an area posterior to the thalamus. Deep brain stimulation from implants of electrodes may also control some symptoms.

Researchers are turning to cells in a patient's body as sources of dopamine. For example, cells in the eye's retinal pigment epithelium (chapter 12, pp. 478–479) can be cultured with biochemicals that stimulate them to produce dopamine or levodopa. Perhaps cells can be sampled from the patient without impairing vision and implanted in the substantia nigra. Neural stem and progenitor cells may also be useful. Implants of fetal dopamine-producing cells performed in the late 1990s alleviated symptoms in some patients for several years.

PD has been attributed to use of certain drugs, exposure to pesticides, and frequent violent blows to the head (fig. 11D), perhaps combined with inherited susceptibility. For example, people with a certain genetic makeup are two to six times more likely than other individuals to develop PD after prolonged exposure to any of a dozen types of pesticides. However, most cases of PD are not directly inherited. One gene that causes PD when mutant encodes a protein called alpha-synuclein. The abnormal protein folds improperly, forming deposits in the brain called Lewy bodies (fig. 11E).

FIGURE 11D Professional boxers are at higher risk of developing Parkinson disease (PD) from repeated blows to the head. Muhammad Ali has PD from many years of head injuries. Michael J. Fox, an actor, not a boxer, first experienced symptoms of PD at age 29, which is unusual.

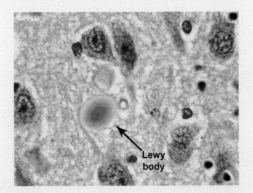

FIGURE 11E The chemical composition of Lewy bodies, characteristic of the brains of some people with Parkinson disease, may provide clues to the cause of the condition. Lewy bodies include alpha-synuclein and other components.

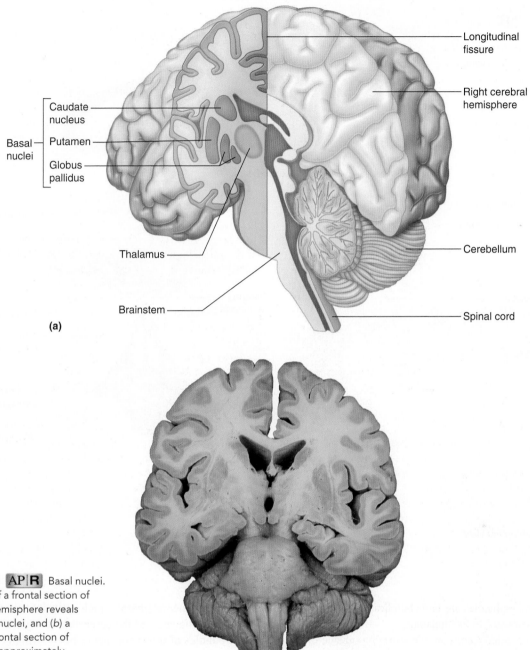

FIGURE 11.10 AP|R Basal nuclei.
(a) A dorsal view of a frontal section of
the left cerebral hemisphere reveals
some of the basal nuclei, and (b) a
ventral view of a frontal section of
a cadaver brain at approximately
the same position as in (a).

life-threatening upsets in a person's physical or psychological
condition. By causing pleasant or unpleasant feelings about experiences, the limbic system guides behavior that may increase the
chance of survival. In addition, parts of the limbic system interpret
sensory impulses from the receptors associated with the sense of
smell (olfactory receptors).

Brainstem

The **brainstem** connects the brain to the spinal cord. It consists
of the midbrain, pons, and medulla oblongata. These structures
include many tracts of nerve fibers and masses of gray matter
called *nuclei* (see figs. 11.6, 11.10, and 11.11).

Midbrain

The **midbrain** (mesencephalon) is a short section of the brainstem
between the diencephalon and the pons. It contains bundles of
myelinated nerve fibers that join lower parts of the brainstem and
spinal cord with higher parts of the brain. The midbrain includes
several masses of gray matter that serve as reflex centers. It also
contains the *cerebral aqueduct* that connects the third and fourth
ventricles (fig. 11.12).

Two prominent bundles of nerve fibers on the underside of the
midbrain comprise the *cerebral peduncles*. These fibers include
descending tracts and are the main motor pathways between the
cerebrum and lower parts of the nervous system (see fig. 11.11).

FIGURE 11.11 AP|R Brainstem.
(a) Ventral view of the brainstem. (b) Dorsal view of the brainstem with the cerebellum removed, exposing the fourth ventricle.

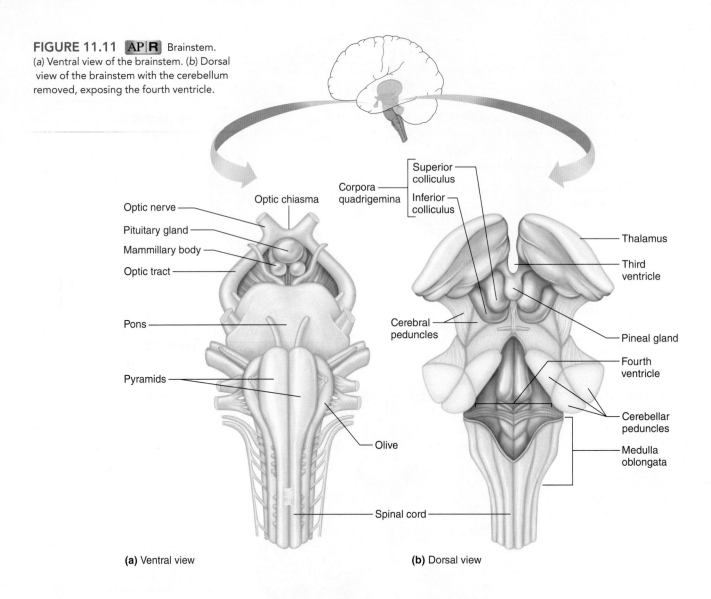

(a) Ventral view

(b) Dorsal view

Beneath the cerebral peduncles are large bundles of sensory fibers that carry impulses upward to the thalamus.

Two pairs of rounded knobs on the superior surface of the midbrain mark the location of four nuclei, known collectively as *corpora quadrigemina.* The upper masses (superior colliculi) contain the centers for certain visual reflexes, such as those responsible for moving the eyes to view something as the head turns. The lower ones (inferior colliculi) contain the auditory reflex centers that operate when it is necessary to move the head to hear sounds more distinctly (see fig. 11.11).

Near the center of the midbrain is a mass of gray matter called the *red nucleus.* This nucleus communicates with the cerebellum and with centers of the spinal cord, and it plays a role in reflexes that maintain posture. It appears red because it is richly supplied with blood vessels.

Pons

The **pons** appears as a rounded bulge on the underside of the brainstem where it separates the midbrain from the medulla oblongata (see fig. 11.11). The dorsal portion of the pons consists mostly of

longitudinal nerve fibers, which relay information between the medulla oblongata and the cerebrum. Its ventral portion contains large bundles of transverse nerve fibers, which conduct impulses from the cerebrum to centers within the cerebellum.

Several nuclei of the pons relay sensory information from peripheral nerves to higher brain centers. Other nuclei may function with centers of the medulla oblongata to control breathing.

Medulla Oblongata

The **medulla oblongata** (mĕ-dul′ah ob″long-ga′tah) is an enlarged continuation of the spinal cord, extending from the level of the foramen magnum to the pons (see fig. 11.11). Its dorsal surface flattens to form the floor of the fourth ventricle, and its ventral surface is marked by descending tracts, most of whose fibers cross over at this level. On each side of the medulla oblongata is an oval swelling called the *olive,* from which a large bundle of nerve fibers arises and passes to the cerebellum.

The ascending and descending nerve fibers connecting the brain and spinal cord must pass through the medulla oblongata because of its location. As in the spinal cord, the white matter of the medulla surrounds a central mass of gray matter. Here, however, the gray matter breaks up into nuclei separated by nerve fibers. Some of these nuclei relay ascending impulses to the other side of the brainstem and then on to higher brain centers. The *nucleus gracilis* and the *nucleus cuneatus,* for example, receive sensory impulses from fibers of ascending tracts and pass them on to the thalamus or the cerebellum.

Other nuclei in the medulla oblongata control vital visceral activities. These reflex centers include the following:

1. Peripheral nerves conduct impulses originating in the **cardiac center** to the heart, where they increase or decrease heart rate.
2. Certain cells of the **vasomotor center** initiate impulses that affect smooth muscle in the walls of blood vessels and stimulate them to contract, constricting the vessels (vasoconstriction) and thereby increasing blood pressure. A decrease in the activity of these cells can produce the opposite effect—dilation of the blood vessels (vasodilation) and a consequent drop in blood pressure.
3. The **respiratory center** maintains the basic rhythm of breathing and adjusts the rate and depth of breathing to meet changing needs.

Some nuclei in the medulla oblongata are centers for certain nonvital reflexes, such as those associated with coughing, sneezing, swallowing, and vomiting. However, because the medulla also contains vital reflex centers, injuries to this part of the brainstem are often fatal.

Reticular Formation

Scattered throughout the medulla oblongata, pons, and midbrain is a complex network of nerve fibers associated with tiny islands of gray matter. This network, the **reticular formation** (rĕ-tik′u-

lar fōr-ma′shun), or reticular activating system, extends from the superior portion of the spinal cord into the diencephalon (fig. 11.12). Its neurons connect centers of the hypothalamus, basal nuclei, cerebellum, and cerebrum with all of the major ascending and descending tracts.

When sensory impulses reach the reticular formation, it responds by activating the cerebral cortex into a state of wakefulness. Without this arousal, the cortex remains unaware of stimulation and cannot interpret sensory information or carry on thought processes. Decreased activity in the reticular formation results in sleep. If the reticular formation is injured and ceases to function, the person remains unconscious, even with strong stimulation. This is called a comatose state. The term *ascending reticular activating system* refers to this functional aspect of the reticular formation. Without it the cerebral cortex cannot function consciously.

The reticular formation filters incoming sensory impulses. Impulses judged to be important, such as those originating in pain receptors, are passed on to the cerebral cortex, while others are disregarded. This selective action of the reticular formation frees the cortex from what would otherwise be a continual bombardment of sensory stimulation and allows it to concentrate on more significant information. Because the cerebral cortex can also activate the reticular formation, intense cerebral activity keeps a person awake. In addition, the reticular formation, through connections with the basal nuclei and the motor cortex, influences motor activities to ensure that skeletal muscles move together evenly. The reticular formation also inhibits or enhances certain spinal reflexes.

> A person in a persistent vegetative state may occasionally be awake and responsive to stimuli, but is not aware; a person in a coma does not appear awake and is not aware. In some cases following a severe injury, a person will become comatose and then gradually enter a persistent vegetative state. Persistent vegetative state and coma are also seen in the end stage of neurodegenerative disorders such as Alzheimer disease; when there is an untreatable mass in the brain, such as a blood clot or tumor; and in anencephaly, when a newborn lacks higher brain structures.

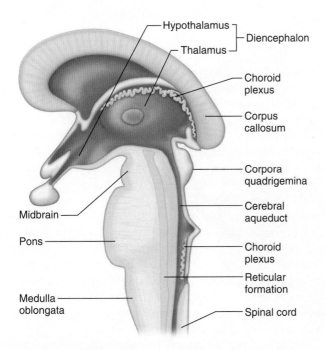

FIGURE 11.12 AP|R The reticular formation (shown in gold) extends from the superior portion of the spinal cord into the diencephalon.

Labels: Hypothalamus, Thalamus — Diencephalon; Choroid plexus; Corpus callosum; Corpora quadrigemina; Cerebral aqueduct; Choroid plexus; Reticular formation; Spinal cord; Midbrain; Pons; Medulla oblongata

Types of Sleep

The two types of normal sleep are *non-rapid eye movement* (non-REM) and *rapid eye movement* (REM). Non-REM sleep occurs when a person is very tired, and it reflects decreasing activity of the reticular formation. It is restful, dreamless, and accompanied by reduced blood pressure and respiratory rate. Non-REM sleep may range from light to heavy and is described in three stages. The third stage is also known as slow-wave sleep. It may last from seventy to ninety minutes. Non-REM and REM sleep alternate.

REM sleep is also called "paradoxical sleep" because some areas of the brain are active. As its name implies, the eyes can be seen rapidly moving beneath the eyelids. Cats and dogs in REM sleep sometimes twitch their limbs. In humans, REM sleep usually lasts from five to fifteen minutes. This "dream sleep" is important. If a person lacks REM sleep for just one night, sleep on the next night makes up for it. During REM sleep, heart and respiratory rates are irregular. Marijuana, alcohol, and certain other drugs, such as benzodiazepines, interfere with REM sleep. **Table 11.4** describes several disorders of sleep.

 PRACTICE

17 What are the major functions of the thalamus? of the hypothalamus?

18 How may the limbic system influence a person's behavior?

19 Which vital reflex centers are located in the brainstem?

20 What is the function of the reticular formation?

21 Describe two types of sleep.

Cerebellum

The **cerebellum** (ser″ĕ-bel′um) is a large mass of tissue inferior to the occipital lobes of the cerebrum and posterior to the pons and medulla oblongata (fig. 11.13). It consists of two lateral hemispheres partially separated by a layer of dura mater called the *falx cerebelli*. A structure called the *vermis* connects the cerebellar hemispheres at the midline.

Like the cerebrum, the cerebellum is primarily composed of white matter with a thin layer of gray matter, the **cerebellar cortex,** on its surface. This cortex doubles over on itself in a series of complex folds that have myelinated nerve fibers branching into them. A cut into the cerebellum reveals a treelike pattern of white matter, called the *arbor vitae,* surrounded by gray matter. A number

TABLE 11.4	Sleep Disorders	
Disorder	**Symptoms**	**Percent of Population**
Fatal familial insomnia	Inability to sleep; emotional instability; hallucinations; stupor; coma; death within thirteen months of onset around age fifty; both slow-wave and REM sleep abolished	Very rare
Insomnia	Inability to fall or remain asleep	10%
Narcolepsy	Abnormal REM sleep causes extreme daytime sleepiness, begins between ages of fifteen and twenty-five	0.02–0.06%
Obstructive sleep apnea syndrome	Upper airway collapses repeatedly during sleep, blocking breathing; snoring and daytime sleepiness	4–5%
Parasomnias	Sleepwalking; sleeptalking; and night terrors	<5% of children
REM-sleep behavior disorder	Excessive motor activity during REM sleep, which disturbs continuous sleep	Very rare
Restless legs syndrome	Brief, repetitive leg jerks during sleep; leg pain forces person to get up several times a night	5.5%
Sleep paralysis	Inability to move for up to a few minutes after awakening or when falling asleep	Very rare

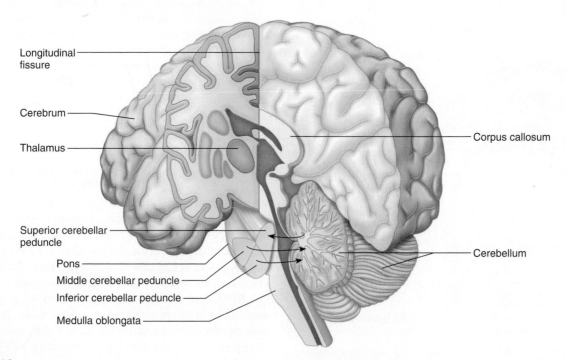

FIGURE 11.13 The cerebellum, located inferior to the occipital lobes of the cerebrum, communicates with other parts of the nervous system by means of the cerebellar peduncles.

of nuclei lie deep within each cerebellar hemisphere. The largest and most important is the *dentate nucleus.*

The cerebellum communicates with other parts of the CNS by means of three pairs of nerve tracts called **cerebellar peduncles** (ser″ĕ-bel′ar pe-dung′kls) (fig. 11.13). One pair, the *inferior peduncles,* brings sensory information concerning the position of body parts such as limbs and joints to the cerebellum via the spinal cord and medulla oblongata. The *middle peduncles* send information from the cerebral cortex about the desired position of these body parts. After integrating and analyzing the information from these two sources, the cerebellum sends correcting impulses from the dentate nucleus via the *superior peduncles* to the thalamus and eventually to the motor cortex (fig. 11.13). These corrections are incorporated into motor impulses that travel downward through the pons, medulla oblongata, and spinal cord to lower motor neurons in the appropriate patterns to move the body in the desired way.

Overall, the cerebellum integrates sensory information concerning the position of body parts and coordinates skeletal muscle activity and maintains posture. It receives sensory impulses from receptors in muscles, tendons, and joints (proprioceptors) and from special sense organs, such as the eyes and ears. For example, the cerebellum uses sensory information from the semicircular canals of the inner ears concerning the motion and position of the head to help maintain equilibrium (see chapter 12, p. 466). Damage to the cerebellum can cause tremors, inaccurate movements of voluntary muscles, loss of muscle tone, a reeling walk, and loss of equilibrium.

Table 11.5 summarizes the characteristics and functions of the major parts of the brain. Clinical Application 11.4 discusses how brain waves reflect brain activity.

PRACTICE

22 Where is the cerebellum located?

23 What are the major functions of the cerebellum?

24 What types of receptors provide information to the cerebellum?

11.5 | Spinal Cord

The **spinal cord** is a slender column of nervous tissue that is continuous with the brain and extends downward through the vertebral canal. The spinal cord originates where nervous tissue leaves the cranial cavity at the level of the foramen magnum (see reference plate 15, p. 50). The cord tapers to a point and terminates near the intervertebral disc that separates the first and second lumbar vertebrae (fig. 11.14a).

Structure of the Spinal Cord

The spinal cord consists of thirty-one segments, each of which gives rise to a pair of **spinal nerves.** These nerves branch to various body parts and connect them with the CNS.

In the neck region, a thickening in the spinal cord, called the *cervical enlargement,* supplies nerves to the upper limbs. A similar thickening in the lower back, the *lumbar enlargement,* gives off nerves to the lower limbs. Just inferior to the lumbar enlargement, the spinal cord tapers to a structure called the *conus medullaris.* From this tip, nervous tissue, including axons of both motor and sensory neurons, extends downward within the vertebral canal to become spinal nerves at the remaining lumbar and sacral levels. The resulting structure resembles a horse's tail, and is called the *cauda equina.* A thin cord of connective tissue originating from the pia mater and

TABLE 11.5	Major Parts of the Brain	
Part	**Characteristics**	**Functions**
1. Cerebrum	Largest part of the brain; two hemispheres connected by the corpus callosum	Controls higher brain functions, including interpreting sensory impulses, initiating muscular movements, storing memory, reasoning, and intelligence
2. Basal nuclei (basal ganglia)	Masses of gray matter deep within the cerebral hemispheres	Relay stations for motor impulses originating in the cerebral cortex and passing into the brainstem and spinal cord; facilitate and help coordinate voluntary movement
3. Diencephalon	Includes masses of gray matter (thalamus and hypothalamus)	The thalamus is a relay station for sensory impulses ascending from other parts of the nervous system to the cerebral cortex; the hypothalamus helps maintain homeostasis by regulating visceral activities and by linking the nervous and endocrine systems
4. Brainstem	Connects the cerebrum to the spinal cord	
a. Midbrain	Contains masses of gray matter and bundles of nerve fibers that join the spinal cord to higher regions of the brain	Contains reflex centers that move the eyes and head; maintains posture
b. Pons	A bulge on the underside of the brainstem that contains masses of gray matter and nerve fibers	Relays impulses between the medulla oblongata and cerebrum; helps regulate rate and depth of breathing
c. Medulla oblongata	An enlarged continuation of the spinal cord that extends from the foramen magnum to the pons and contains masses of gray matter and nerve fibers	Conducts ascending and descending impulses between the brain and spinal cord; contains cardiac, vasomotor, and respiratory control centers and various nonvital reflex control centers
5. Cerebellum	A large mass of tissue inferior to the cerebrum and posterior to the brainstem; includes two lateral hemispheres connected by the vermis	Communicates with other parts of the CNS by tracts; integrates sensory information concerning the position of body parts; coordinates muscle activities and maintains posture

Brain waves are recordings of fluctuating electrical changes in the brain. To obtain such a recording, electrodes are positioned on the surface of a surgically exposed brain (an electrocorticogram, ECoG) or on the outer surface of the head (an electroencephalogram, EEG). These electrodes detect electrical changes in the extracellular fluid of the brain in response to changes in potential among large groups of neurons. The resulting signals from the electrodes are amplified and recorded. Brain waves originate from the cerebral cortex but also reflect activities in other parts of the brain that influence the cortex, such as the reticular formation. Because the intensity of electrical changes is proportional to the degree of neuronal activity, brain waves vary markedly in amplitude and frequency between sleep and wakefulness.

Brain waves are classified as alpha, beta, theta, and delta waves (fig. 11F). *Alpha waves* are recorded most easily from the posterior regions of the head and have a frequency of 8–13 cycles per second. They occur when a person is awake but resting, with the eyes closed. These waves disappear during sleep, and if the wakeful person's eyes open, higher-frequency beta waves replace the alpha waves.

Beta waves have a frequency of more than 13 cycles per second. Most are recorded in the anterior region of the head. They occur when a person is actively engaged in mental activity or is under tension.

Theta waves have a frequency of 4–7 cycles per second and occur mainly in the parietal and temporal regions of the cerebrum. They are normal in children but do not usually occur in adults. However, some adults produce theta waves in early stages of sleep or at times of emotional stress.

Delta waves have a frequency below 4 cycles per second and happen during sleep. They originate from the cerebral cortex when the reticular formation is not activating it.

Brain wave patterns can be useful for diagnosing disease conditions, such as distinguishing types of seizure disorders (epilepsy) and locating brain tumors. Brain waves are also used to detect *brain death*, in which neuronal activity ceases. In some countries including the United States, an EEG that lacks waves (isoelectric EEG) verifies brain death. However, drugs that greatly depress brain functions must be excluded as the cause of the flat EEG pattern before confirming brain death.

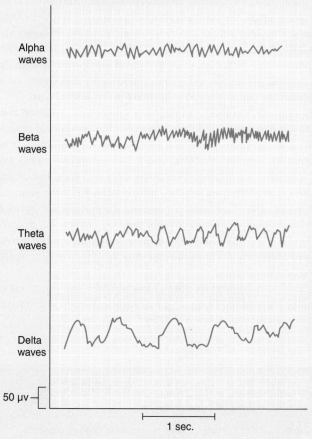

FIGURE 11F Brain waves record fluctuating electrical changes in the brain.

---

dura mater descends to the upper surface of the coccyx. This cord, called the *filum terminale*, anchors the spinal cord (fig.11.14*a*).

Two grooves, a deep *anterior median fissure* and a shallow *posterior median sulcus*, extend the length of the spinal cord, dividing it into right and left halves. A cross section of the cord (fig. 11.15) reveals that it consists of white matter surrounding a core of gray matter. The pattern the gray matter produces roughly resembles a butterfly with its wings spread. These posterior and anterior "wings" of gray matter are called the *posterior horns* and the *anterior horns,* respectively. Between them on either side in some regions is a protrusion of gray matter called the *lateral horn*. Motor neurons with relatively large cell bodies in the anterior horns (anterior horn cells) give rise to axons that pass out through spinal nerves to various skeletal muscles. However, the majority of neurons in the gray matter are interneurons (see chapter 10, pp. 366 and 368).

A horizontal bar of gray matter in the middle of the spinal cord, the *gray commissure,* connects the wings of the gray matter on the right and left sides. This bar surrounds the **central canal,** which is continuous with the ventricles of the brain and contains CSF. The central canal is prominent during embryonic development, but it becomes almost microscopic in adulthood.

The gray matter divides the white matter of the spinal cord into three regions on each side—the *anterior, lateral,* and *posterior funiculi.* Each funiculus, or column, consists of longitudinal bundles of myelinated nerve fibers that compose the major pathways called **tracts.**

Functions of the Spinal Cord

The spinal cord has two main functions. It is a center for spinal reflexes, and it is a conduit for impulses to and from the brain.

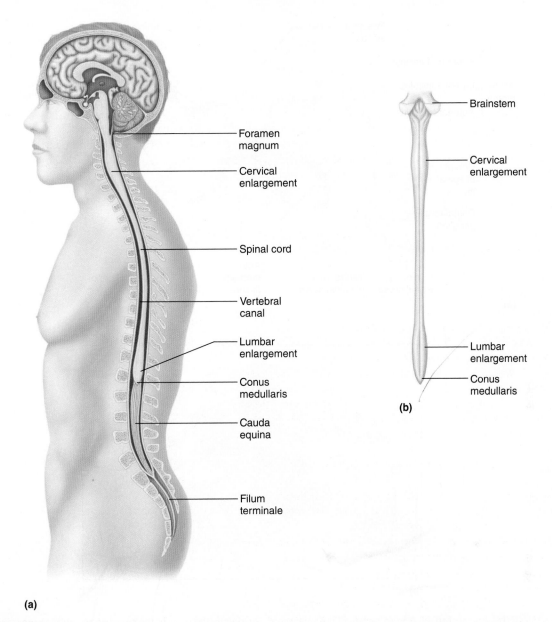

(a)

(b)

FIGURE 11.14 AP|R Spinal cord. (a) The spinal cord begins at the level of the foramen magnum. (b) Posterior view of the spinal cord with the spinal nerves removed.

Reflex Arcs

Communication in the nervous system combines a series of action potentials along the axon of a neuron and synaptic transmission between that neuron and a postsynaptic cell. Two or more neurons involved in such communication constitute a nerve pathway. The simplest of the nerve pathways begins with a sensory receptor and ends with an effector, and includes as few as two neurons. Such a nerve pathway is called a **reflex** (re′fleks).

All reflexes share the same basic components, which together are known as a **reflex arc,** as figure 11.16a shows. A reflex arc begins with a sensory receptor at the dendritic end of a sensory neuron. Impulses on these sensory neurons enter the CNS and constitute a sensory or afferent limb of the reflex. The CNS is a processing center. Afferent neurons may synapse with interneurons, which may in turn connect with other parts of the CNS. Afferent neurons or interneurons ultimately connect with motor neurons,

whose fibers pass outward from the CNS to effectors. (It may help to remember that **eff**erent neurons control **eff**ector organs.)

Reflexes occur throughout the CNS. Those that involve the spinal cord are called spinal reflexes and reflect the simplest level of CNS function. Figure 11.16b shows the general components of a spinal reflex.

Reflex Behavior

Reflexes are automatic responses to changes (stimuli) inside or outside the body. They help maintain homeostasis by controlling many involuntary processes such as heart rate, breathing rate, blood pressure, and digestion. Reflexes also carry out the automatic actions involved in swallowing, sneezing, coughing, and vomiting.

The *patellar reflex* (knee-jerk reflex) is an example of a simple monosynaptic reflex, so-called because it uses only two neurons—a

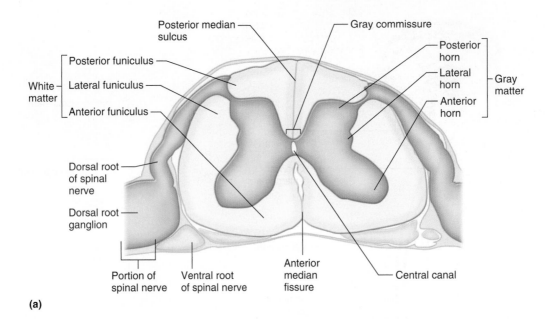

(a)

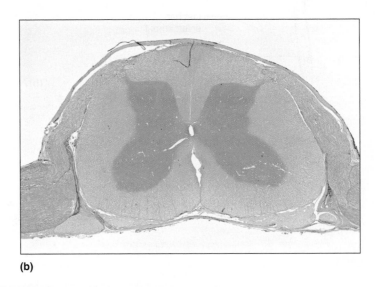

(b)

FIGURE 11.15 **AP|R** Spinal cord. (*a*) A cross section of the spinal cord. (*b*) Identify the parts of the spinal cord in this micrograph (7.5×).

Where would you expect to find the cell bodies of neurons in the above figure?

Answer can be found in Appendix G.

sensory neuron communicating directly to a motor neuron. Striking the patellar ligament just below the patella initiates this reflex. The quadriceps femoris muscle group, attached to the patella by a tendon, is pulled slightly, stimulating stretch receptors in the muscle group. These receptors, in turn, trigger impulses that pass along the peripheral process (see fig. 10.6*c*, p. 366) of the axon of a unipolar sensory neuron, continuing along the central process of the axon into the lumbar region of the spinal cord. In the spinal cord, the sensory neuron axon synapses with a motor neuron. An impulse is then triggered on the motor neuron and is conducted along its axon to the neuromuscular junctions in that motor unit of the quadriceps femoris. The muscle fibers involved respond by contracting, and the reflex is completed as the leg extends (fig. 11.17).

The patellar reflex helps maintain an upright posture. For example, if a person is standing still and the knee begins to bend

in response to gravity, the quadriceps femoris is stretched, the reflex is triggered, and the leg straightens again. Adjustments within the stretch receptors keep the reflex responsive at different muscle lengths.

Another type of reflex, called a *withdrawal reflex* (fig. 11.18), happens when a person touches something painful (and potentially damaging), as in stepping on a tack. Activated skin receptors send impulses to the spinal cord along the axons of sensory neurons. There the sensory neurons synapse with interneurons, which in turn synapse with motor neurons. The motor neurons activate fibers in the flexor muscles of the leg and thigh, which contract in response, pulling the foot away from the painful stimulus.

At the same time, some of the incoming sensory impulses stimulate interneurons that inhibit the action of the antagonistic extensor muscles (reciprocal innervation). This inhibition of

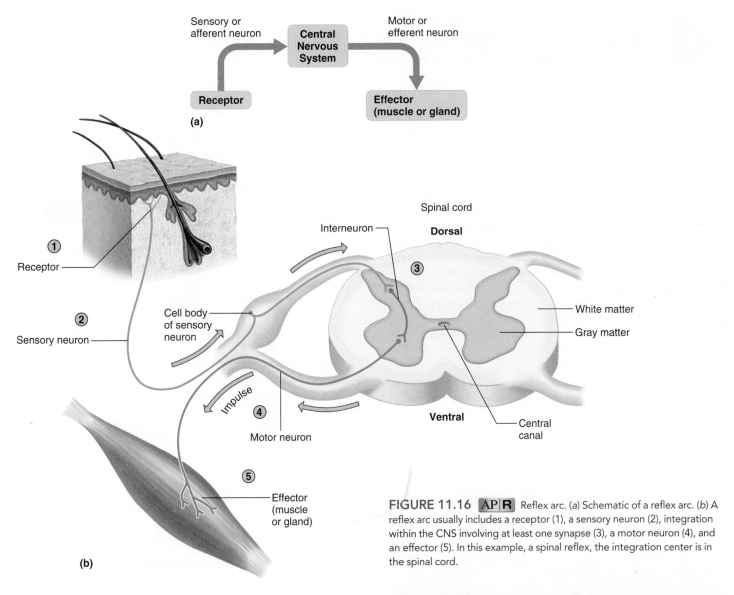

(a)

(b)

FIGURE 11.16 AP|R Reflex arc. (*a*) Schematic of a reflex arc. (*b*) A reflex arc usually includes a receptor (1), a sensory neuron (2), integration within the CNS involving at least one synapse (3), a motor neuron (4), and an effector (5). In this example, a spinal reflex, the integration center is in the spinal cord.

antagonists allows the flexor muscles to effectively withdraw the affected part (fig. 11.19).

While flexor muscles on the affected side (ipsilateral side) contract, the flexor muscles of the limb on the other side (contralateral side) are inhibited. Furthermore, the extensor muscles on the contralateral side contract, helping to support the body weight shifted to that side. This phenomenon, called a *crossed extensor reflex,* is due to interneuron pathways in the spinal cord that allow sensory impulses arriving on one side of the cord to pass across to the other side and produce an opposite effect (fig. 11.19). Reflexes like these can be found at different levels of the spinal cord and in the brain, depending on which body parts are involved.

Concurrent with the withdrawal reflex, other interneurons in the spinal cord carry sensory impulses upward to the brain. The person becomes aware of the experience and may feel pain.

A withdrawal reflex protects because it prevents or limits tissue damage when a body part touches something potentially harmful. **Table 11.6** summarizes the components of a reflex arc. Clinical Application 11.5 discusses some familiar reflexes.

TABLE 11.6 Parts of a Reflex Arc

Part	Description	Function
Receptor	The receptor end of a dendrite or a specialized receptor cell in a sensory organ	Sensitive to a specific type of internal or external change
Sensory neuron	Dendrite, cell body, and axon of a sensory neuron	Conducts an impulse from the receptor into the brain or spinal cord
Interneuron	Dendrite, cell body, and axon of a neuron within the brain or spinal cord	Serves as processing center; conducts an impulse from the sensory neuron to its synapse with a motor neuron
Motor neuron	Dendrite, cell body, and axon of a motor neuron	Conducts an impulse from the brain or spinal cord out to the synapse with an effector
Effector	A muscle or gland	Responds to stimulation by the motor neuron and produces the reflex or behavioral action

Normal reflexes require and reflect normal neuron functions. Therefore, reflexes are commonly used to assess the condition of the nervous system. An anesthesiologist, for instance, initiates a reflex in a patient being anesthetized to determine how the anesthetic drug is affecting nerve functions. A physician assessing a patient with injury to the nervous system observes reflexes to judge the location and extent of damage.

Injury to any component of a reflex arc alters its function. For example, stroking the sole of the foot normally initiates a *plantar reflex*, which flexes the foot and toes. Damage to certain nerve pathways (corticospinal tract) may trigger an abnormal response called the *Babinski reflex*, which is a dorsiflexion that extends the great toe upward and fans apart the smaller toes. If the injury is minor, the response may consist of plantar flexion with failure of the great toe to flex, or plantar flexion followed by dorsiflexion. The Babinski reflex is normally present in infants up to the age of twelve months and may reflect immaturity in their corticospinal tracts.

Other reflexes that may be tested during a neurological examination include the following:

1. The *biceps-jerk reflex* is elicited by extending a person's forearm at the elbow. The examiner places a finger on the inside of the extended elbow over the tendon of the biceps muscle and taps the finger. The biceps contracts in response, flexing at the elbow.
2. The *triceps-jerk reflex* is elicited by flexing a person's forearm at the elbow and tapping the short tendon of the triceps muscle close to its insertion near the tip of the elbow. The muscle contracts in response, extending the elbow.
3. The *abdominal reflexes* are a response to stroking the skin of the abdomen. For example, a dull pin drawn from the sides of the abdomen upward toward the midline and above the umbilicus contracts the abdominal muscles underlying the skin, and the umbilicus moves toward the stimulated region.
4. The *ankle-jerk reflex* is elicited by tapping the calcaneal tendon just above its insertion on the calcaneus. Contraction of the gastrocnemius and soleus muscles causes plantar flexion.
5. The *cremasteric reflex* is elicited in males by stroking the upper inside of the thigh. In response, contracting muscles elevate the testis on the same side.

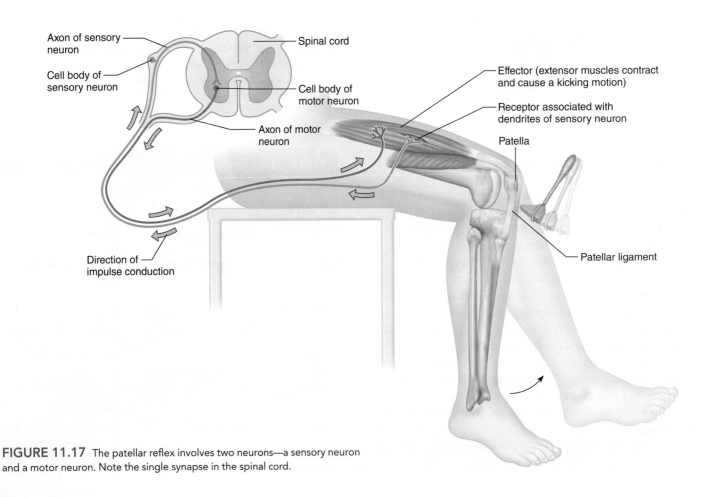

FIGURE 11.17 The patellar reflex involves two neurons—a sensory neuron and a motor neuron. Note the single synapse in the spinal cord.

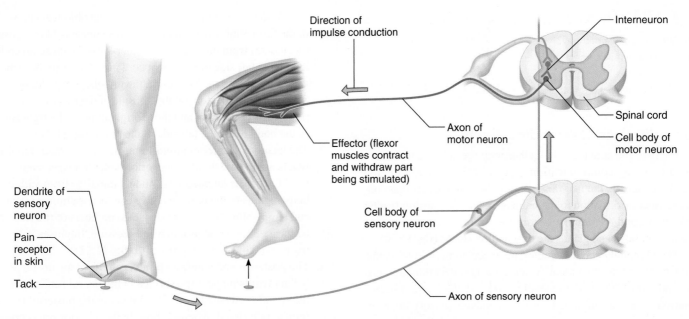

FIGURE 11.18 A withdrawal reflex involves a sensory neuron, an interneuron, and a motor neuron.

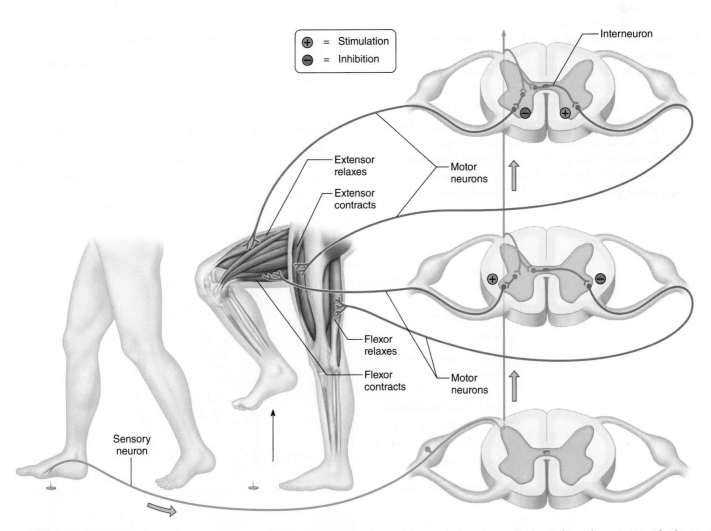

FIGURE 11.19 When the flexor muscle on one side is stimulated to contract in a withdrawal reflex, the extensor muscle on the opposite side also contracts. This helps to maintain balance. For simplicity, only the thigh muscles are shown.

Ascending and Descending Tracts

The tracts of the spinal cord together with the spinal nerves provide a two-way communication system between the brain and body parts outside the nervous system. The spinal tracts that conduct sensory information to the brain are called **ascending tracts;** those that conduct impulses from the brain to motor neurons reaching muscles and glands are **descending tracts** (fig. 11.20).

The ascending and descending tracts are comprised of axons. Many of the names that identify nerve tracts reflect common origins and terminations. For example, a *spinothalamic tract* begins at various levels of the spinal cord and conducts sensory impulses associated with the sensations of pain and touch to the thalamus. A *corticospinal tract* originates in the cerebral cortex and conducts motor impulses on so-called *upper motor neurons* downward through the spinal cord. These impulses control *lower motor neurons* at various levels of the spinal cord whose cell bodies are in the anterior horn and whose axons lead to skeletal muscles.

Ascending pathways typically involve three neurons between a sensory receptor and the final destination of impulses within the brain—the peripheral sensory neuron and two interneurons. Among the major ascending tracts of the spinal cord are the following:

1. **Fasciculus gracilis** (fah-sik′u-lus gras′il-is) and **fasciculus cuneatus** (ku′ne-at-us) are tracts in the posterior funiculi of the spinal cord (fig. 11.20). Their fibers conduct sensory impulses from the skin, muscles, tendons, and joints to the brain, where they are interpreted as sensations of touch, pressure, and body movement.

At the base of the brain in the medulla oblongata most of the fasciculus gracilis and fasciculus cuneatus fibers cross (decussate) from one side to the other—that is, those ascending on the left side of the spinal cord pass across to the right side, and vice versa. As a result, the impulses originating from sensory receptors on the left side of the body reach the right side of the brain, and those originating on the right side of the body reach the left side of the brain (fig. 11.21).

2. The lateral and anterior **spinothalamic** (spi″no-thah-lam′ik) **tracts** are in the lateral and anterior funiculi, respectively (see fig. 11.20). The lateral tracts conduct impulses from various body regions to the brain and give rise to sensations of pain and temperature. Impulses conducted on fibers of the anterior tracts are interpreted as touch and pressure. Impulses in these tracts cross over in the spinal cord (fig. 11.21).

3. The posterior and anterior **spinocerebellar** (spi″no-ser″ĕ-bel′ar) **tracts** lie near the surface in the lateral funiculi of the spinal cord (see fig. 11.20). Fibers in the posterior tracts remain uncrossed, whereas those in the anterior tracts cross over in the medulla. Impulses conducted on their fibers originate in the muscles of the lower limbs and trunk and then travel to the cerebellum. These impulses coordinate muscular movements.

The major descending tracts of the spinal cord include the following:

1. The lateral and anterior **corticospinal** (kor″tĭ-ko-spi′nal) **tracts** occupy the lateral and anterior funiculi, respectively (see fig. 11.20). Most of the fibers of the lateral tracts cross over in the lower medulla oblongata. Some fibers of the anterior tracts cross over at various levels of the spinal cord (fig. 11.22). Axons in the corticospinal tracts conduct motor impulses from the brain and synapse either directly or through interneurons with lower motor neurons, whose axons continue through spinal nerves to various skeletal muscles (fig. 11.22). Thus, they carry instructions that control voluntary movements.

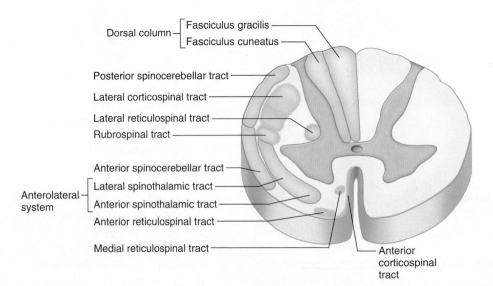

Dorsal column ⎡ Fasciculus gracilis
⎣ Fasciculus cuneatus

Posterior spinocerebellar tract

Lateral corticospinal tract

Lateral reticulospinal tract

Rubrospinal tract

Anterior spinocerebellar tract

Anterolateral system ⎡ Lateral spinothalamic tract
⎣ Anterior spinothalamic tract

Anterior reticulospinal tract

Medial reticulospinal tract

Anterior corticospinal tract

FIGURE 11.20 Major ascending and descending tracts in a cross section of the spinal cord. Ascending tracts are in pink, descending tracts in light brown. (Tracts are shown only on one side.) The pattern varies with the level of the spinal cord. This pattern is representative of the midcervical region.

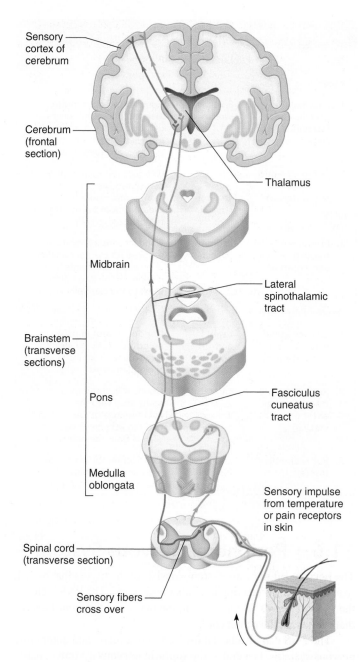

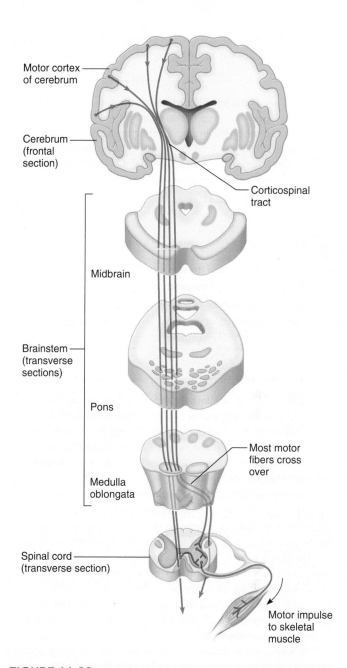

FIGURE 11.21 Sensory impulses originating in skin touch receptors ascend in the fasciculus cuneatus tract and cross over in the medulla of the brain. Pain and temperature information ascends in the lateral spinothalamic tract, which crosses over in the spinal cord.

FIGURE 11.22 Most fibers of the corticospinal tract originate in the cerebral cortex, cross over in the medulla, and descend in the spinal cord, where they synapse with neurons whose fibers lead to spinal nerves supplying skeletal muscles. Some fibers cross over in the spinal cord.

The corticospinal tracts are also called *pyramidal tracts* after the pyramid-shaped regions in the medulla oblongata through which they pass. Other descending tracts are called *extrapyramidal tracts,* and they include the reticulospinal and rubrospinal tracts.

2. The lateral **reticulospinal** (rĕ-tik″u-lo-spi′nal) **tracts** are in the lateral funiculi, whereas the anterior and medial reticulospinal tracts are in the anterior funiculi (see fig. 11.20). Some fibers in the lateral tracts cross over, whereas others remain uncrossed. Those of the anterior and medial tracts remain uncrossed. Motor impulses conducted on the reticulospinal tracts originate in the brain and control muscular tone and activity of sweat glands.

3. The fibers of the **rubrospinal** (roo″bro-spi′nal) **tracts** cross over in the brain and pass through the lateral funiculi (see fig. 11.20). They conduct impulses from the brain to synapses with lower motor neurons, and help to coordinate muscle actions.

Table 11.7 summarizes the nerve tracts of the spinal cord. Clinical Application 11.6 describes injuries to the spinal cord.

A hemi-lesion of the spinal cord (severed on one side) affecting the corticospinal and spinothalamic tracts can cause Brown-Séquard syndrome. Ascending tracts cross over at different levels, so the injured side of the body becomes paralyzed and loses touch sensation. The other side of the body retains movement but loses sensations of pain and temperature.

 PRACTICE

29 Describe the structure of the spinal cord.

30 What are ascending and descending tracts?

31 What is the consequence of fibers crossing over?

32 Name the major tracts of the spinal cord, and list the types of impulses each conducts.

In amyotrophic lateral sclerosis (ALS, Lou Gehrig's disease, or motor neuron disease), motor neurons in the spinal cord, brainstem, and cerebral cortex degenerate. The disease may be due to an inability of the motor neurons or associated astrocytes to counter buildup of oxygen free radicals, or to overreactive microglia that kill motor neurons.

The first symptoms of ALS may be vague and affect only a small part of the body: difficulty speaking, a dragging foot, clumsiness, fatigue, or problems with fine coordination such as turning a key or pulling a zipper. Muscle twitches (fasciculations) may prompt the person to seek medical attention. Diagnosis may take more than a year as a neurologist observes spreading weakness and rules out other conditions, such as multiple sclerosis or a spinal cord tumor. The average age at diagnosis is 55, but adolescents as well as the very elderly have developed ALS.

ALS affects the upper and lower parts of the body, and progresses faster if symptoms begin in the face or neck (bulbar onset) compared to the arms and legs (limb onset). Usually the battle is lost two to five years after diagnosis, typically from respiratory failure, but about 10% of patients live more than a decade with the disease.

ALS has no cure, but a drug (Rilutek [riluzole]) may extend time until respiratory difficulty. Assisted breathing devices may be used, including a ventilator to sustain life. Although about 50% of people with ALS experience some cognitive decline, the mind remains sharp in many people. One patient wrote a novel during his last months, and another remained a brilliant songwriter.

About 10% of ALS cases are inherited, due to mutations in any of several genes. Because prevalence has been increasing over the past few years at a rate faster than the aging of the population can explain, an environmental trigger may combine with an inherited susceptibility to cause the disease.

TABLE 11.7	Nerve Tracts of the Spinal Cord	
Tract	**Location**	**Function**
Ascending Tracts		
1. Fasciculus gracilis and fasciculus cuneatus	Posterior funiculi	Conduct sensory impulses associated with the senses of touch, pressure, and body movement from skin, muscles, tendons, and joints to the brain
2. Spinothalamic tracts (lateral and anterior)	Lateral and anterior funiculi	Conduct sensory impulses associated with the senses of pain, temperature, touch, and pressure from various body regions to the brain
3. Spinocerebellar tracts (posterior and anterior)	Lateral funiculi	Conduct sensory impulses required for the coordination of muscle movements from muscles of the lower limbs and trunk to the cerebellum
Descending Tracts		
1. Corticospinal tracts (lateral and anterior)	Lateral and anterior funiculi	Conduct motor impulses associated with voluntary movements from the brain to skeletal muscles
2. Reticulospinal tracts (lateral, anterior, and medial)	Lateral and anterior funiculi	Conduct motor impulses associated with the maintenance of muscle tone and the activity of sweat glands from the brain
3. Rubrospinal tracts	Lateral funiculi	Conduct motor impulses associated with muscular coordination from the brain

11.6 | Peripheral Nervous System

The **peripheral nervous system** consists of the nerves that branch from the CNS, connecting it to other body parts. The PNS includes the *cranial nerves* that arise from the brain and the *spinal nerves* that arise from the spinal cord.

The PNS can also be subdivided into somatic and autonomic nervous systems. Generally, the **somatic nervous system** consists of the cranial and spinal nerve fibers that connect the CNS to the skin and skeletal muscles, so it plays a role in conscious activities. The **autonomic nervous system** (aw″to nom′ik ner′vus sis′tem) includes fibers that connect the CNS to viscera such as the heart, stomach, intestines, and various glands. The autonomic nervous system controls subconscious actions. Table 11.8 outlines the subdivisions of the nervous system.

Structure of Peripheral Nerves

Nerves are essentially bundles of axons, but they have specific levels of organization. A small amount of loose connective tissue called *endoneurium* surrounds individual axons. Axons are organized in bundles called fascicles. Each fascicle is enclosed in a sleeve of loose connective tissue called the *perineurium*. A group of bundled fascicles, surrounded by an outermost layer of dense connective tissue called the *epineurium*, constitutes a nerve

Thousands of people sustain spinal cord injuries each year. Treatment for spinal cord injury begins as soon as help arrives at the accident scene. Emergency health-care workers establish and maintain the person's ability to breathe, then use a rigid neck collar and carrying board to immobilize the person for transport. In the emergency department, a steroid drug, methylprednisolone, is given within the first 8 hours to minimize inflammation. Surgery may be done to remove bone fragments. Continuing immobilization is crucial because damage continues over days. During this time, the vertebrae are compressed and may break, killing many neurons. Dying neurons release calcium ions, which activate tissue-degrading enzymes. Then white blood cells arrive and produce inflammation that can destroy healthy as well as damaged neurons. Axons tear, myelin coatings are stripped off, and vital connections between neurons and muscle fibers are lost. The tissue cannot regenerate. By the third day, a complete neurological exam and MRIs are done.

The severity of a spinal cord injury depends on the extent and location of damage. Normal spinal reflexes require two-way communication between the spinal cord and the brain. A complete transection (damage through a cross section of the cord) injures nerve pathways, depressing the cord's reflex activities in sites below the injury. At the same time, sensations and muscle tone diminish in the parts that the affected fibers innervate. This condition, called spinal shock, may last for days or weeks, although normal reflex activity may eventually return. If axons are severed, some of the cord's functions may be permanently lost.

Less severe injuries to the spinal cord, such as from a blow to the head, whiplash, or rupture of an intervertebral disc, can compress or distort the cord (fig. 11G). Pain, weakness, and muscular atrophy may develop in the regions the damaged nerve fibers supply. A spinal cord injury increases risk of secondary problems. These include difficulty breathing (if the injury is above the fifth cervical vertebra), development of pneumonia, formation of blood clots, low blood pressure, an irregular heartbeat, pressure ulcers, spasticity, and impaired bowel, bladder, and sexual function.

The two most common causes of spinal cord injury are accidents in the workplace and motor vehicle accidents. Third most common are sports injuries. A spinal cord injury may result from a sudden and unexpected movement. For example, one man suffered a severe spinal cord injury after a powerful wave knocked him down while he was standing in just a foot of water at a shoreline. Regardless of the cause, if nerve fibers in ascending tracts are cut, sensations arising from receptors below the level of the injury are lost. Damage to descending tracts results in loss of motor functions below the level of the injury. For example, if the right lateral corticospinal tract is severed in the neck near the first cervical vertebra, control of the voluntary muscles in the left upper and lower limbs is lost, paralyzing them (hemiplegia). Problems of this type in fibers of the descending tracts produce *upper motor neuron syndrome,* characterized by *spastic paralysis* in which muscle tone increases, with little atrophy of the muscles.

Injury to motor neurons in the anterior horns of the spinal cord results in *lower motor neuron syndrome.* It produces *flaccid paralysis,* a total loss of muscle tone and reflex activity, and the muscles atrophy.

Basic research is exploring ways to limit damage after a spinal cord injury and stimulate regeneration. Studies are identifying molecules in the extracellular matrix (see Clinical Application 5.1, p. 161) and in myelin that inhibit axon regeneration in the CNS and might serve as drug targets, and finding growth factors and other neurotrophins that might promote regeneration. Several clinical trials are investigating the use of stem cells to treat spinal cord injury. These include stem cells taken from a patient's own bone marrow or fat, from cord blood stem cells, or from neural stem cells from a donor.

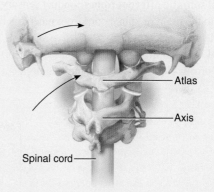

FIGURE 11G A dislocation of the atlas may cause a compression injury to the spinal cord.

(figs. 11.23 and 11.24). Blood vessels in the epineurium and perineurium give rise to a network of capillaries in the endoneurium that provides oxygen and nutrients to the neurons.

The term "muscle fiber" refers to an entire muscle cell, whereas the term "nerve fiber" refers to an axon. The terminology for the connective tissue holding them together, however, is similar. In both cases, for example, fibers are bundled into fascicles, epineurium in nerves corresponds to epimysium in muscles, and so forth (see figs. 11.23, 11.24, 9.2 and 9.3 on p. 294).

Nerve and Nerve Fiber Classification

Recall that nerves are bundles of nerve fibers, or axons. Nerves that have only fibers of sensory neurons, conducting impulses into the brain or spinal cord, are called **sensory nerves.** Nerves that have only fibers involved in motor control are **motor nerves.** Most nerves include both sensory and motor fibers and are called **mixed nerves.**

Nerves originating from the brain that communicate with other body parts are called **cranial nerves,** whereas nerves originating

TABLE 11.8	Subdivisions of the Nervous System
1. Central nervous system (CNS)	
a. Brain	
b. Spinal cord	
2. Peripheral nervous system (PNS)	
a. Cranial nerves arising from the brain	
(1) Somatic fibers connecting to the skin and skeletal muscles	
(2) Autonomic fibers connecting to viscera	
b. Spinal nerves arising from the spinal cord	
(1) Somatic fibers connecting to the skin and skeletal muscles	
(2) Autonomic fibers connecting to viscera	

from the spinal cord that communicate with other body parts are called **spinal nerves.** The nerve fibers in the cranial and spinal nerves can be subdivided further into four groups as follows:

1. **General somatic efferent fibers** conduct motor impulses outward from the brain or spinal cord to skeletal muscles and stimulate them to contract.
2. **General visceral efferent fibers** conduct motor impulses outward from the brain or spinal cord to smooth muscle and glands associated with internal organs.
3. **General somatic afferent fibers** conduct sensory impulses inward to the brain or spinal cord from receptors in the skin and skeletal muscles.
4. **General visceral afferent fibers** conduct sensory impulses to the CNS from blood vessels and internal organs.

The term *general* in each of these categories indicates that the fibers are associated with general structures such as the skin, skeletal muscles, glands, and viscera. Three other groups of fibers, found only in cranial nerves, are associated with more specialized, or *special,* structures:

1. **Special somatic efferent fibers** conduct motor impulses outward from the brain to the muscles used in chewing, swallowing, speaking, and forming facial expressions.
2. **Special visceral afferent fibers** conduct sensory impulses inward to the brain from the olfactory and taste receptors.
3. **Special somatic afferent fibers** conduct sensory impulses inward to the brain from the receptors of sight, hearing, and equilibrium.

Cranial Nerves

Twelve pairs of **cranial nerves** are located on the underside of the brain. Most of the cranial nerves are mixed nerves, but some of those associated with special senses, such as smell and vision, have only sensory fibers. Other cranial nerves that innervate muscles

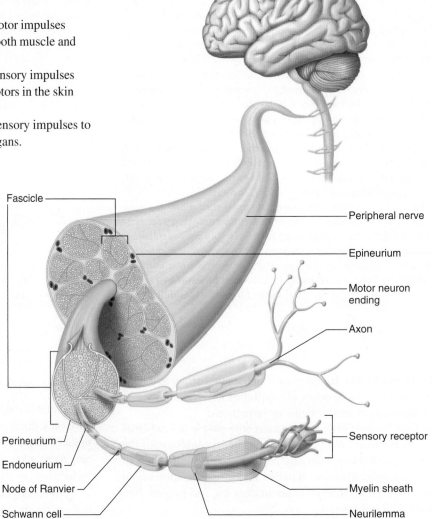

FIGURE 11.23 The structure of a peripheral mixed nerve.

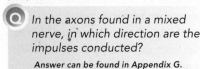 *In the axons found in a mixed nerve, in which direction are the impulses conducted?*

Answer can be found in Appendix G.

and glands are primarily composed of motor fibers. The first pair, which is sensory, has fibers that begin in the nasal cavity and synapse in the frontal lobe of the cerebrum. The second pair, also sensory, originates in the eyes, and their fibers synapse in the thalamus. The remaining cranial nerves attach to the the brainstem, and the direction of their fibers depends on whether they are sensory, motor, or mixed. All of the cranial nerves pass from their sites of attachment through foramina of the skull and lead to areas of the head, neck, and trunk. The cranial nerves that are described as primarily motor do have limited sensory functions because they contain neurons associated with certain receptors (*proprioceptors*) that respond to the changes in length and force of contraction of skeletal muscles. However, because these proprioceptive fibers contribute directly to motor control, cranial nerves whose only sensory component is from such proprioceptors are usually considered motor nerves. This pertains to cranial nerves III, IV, VI, XI, and XII.

Neuron cell bodies from which the sensory fibers in the cranial nerves arise are outside the brain, and most are in groups called *ganglia* (sing., *ganglion*). However, most motor neuron cell bodies are in the gray matter of the brain.

Numbers and names designate cranial nerves. The numbers indicate the order in which the nerves arise from the brain, from anterior to posterior. The names describe primary functions or the general distribution of cranial nerve fibers (fig. 11.25).

The first pair of cranial nerves, the **olfactory** (ol-fak′to-re) **nerves** (I), are associated with the sense of smell and include only sensory neurons. These bipolar neurons, located in the lining of the upper nasal cavity, serve as *olfactory receptor cells*. Axons from these receptor cells pass upward through the cribriform plates of the ethmoid bone, conducting impulses to the olfactory neurons in the *olfactory bulbs*, which are extensions of the cerebral cortex just beneath the frontal lobes. Sensory impulses move from the olfactory bulbs along *olfactory tracts* to cerebral centers where they produce the sensation of smell.

The second pair of cranial nerves, the **optic** (op′tik) **nerves** (II), are sensory and lead from the eyes to the brain. They are associated with vision. The cell bodies of these neurons form ganglion cell layers in the eyes, and their axons pass through the *optic foramina* of the orbits and continue into the visual nerve pathways of the brain (see chapter 12, p. 481).

The third pair of cranial nerves, the **oculomotor** (ok″u-lo-mo′tor) **nerves** (III), arise from the midbrain and pass into the orbits of the eyes. One component of each nerve connects to a number of voluntary muscles, including those that raise the eyelids and four of the six muscles that move the eye.

A second portion of each oculomotor nerve is part of the autonomic nervous system, supplying involuntary muscles inside the eyes. These muscles help adjust the amount of light that enters the eyes and help focus the lenses. This nerve is considered motor, with some proprioceptive fibers.

The fourth pair of cranial nerves, the **trochlear** (trok′le-ar) **nerves** (IV), are the smallest. They arise from the midbrain and conduct motor impulses to a fifth pair of external eye muscles, the *superior oblique muscles,* which are not supplied by the oculomotor nerves. The trochlear nerve is considered motor, with some proprioceptive fibers.

The fifth pair of cranial nerves, the **trigeminal** (tri-jem′i-nal) **nerves** (V), are the largest and arise from the pons. They are mixed nerves, with more extensive sensory portions. Each sensory component includes three large branches, called the ophthalmic, maxillary, and mandibular divisions (fig. 11.26).

The *ophthalmic division* of each the trigeminal nerve consists of sensory fibers that conduct impulses to the brain from the surface of the eye; the tear gland; and the skin of the anterior scalp, forehead, and upper eyelid. The fibers of the *maxillary division* conduct sensory impulses from the upper teeth, upper gum, and upper lip, as well as from the mucous lining of the palate and facial skin. The *mandibular division* includes both motor and sensory fibers. The

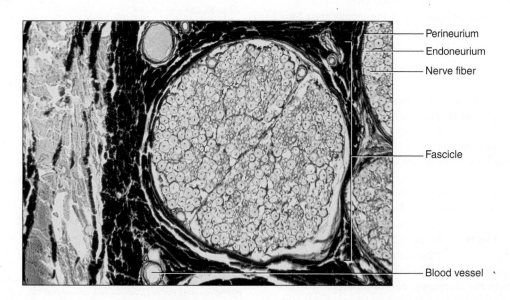

FIGURE 11.24 Scanning electron micrograph of a peripheral nerve in cross section (20×). Note the bundles or fascicles of nerve fibers. Fibers include axons of motor neurons as well as peripheral processes of sensory neurons.

— Perineurium
— Endoneurium
— Nerve fiber

— Fascicle

— Blood vessel

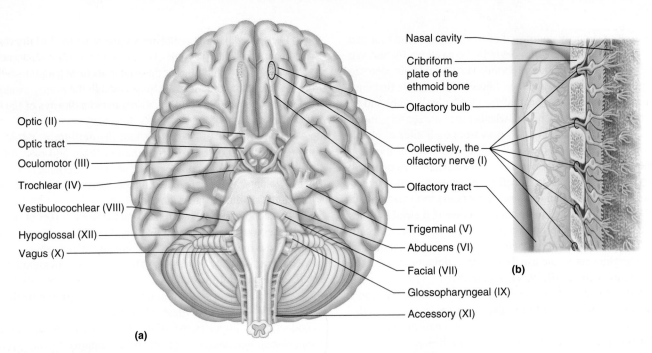

Optic (II)
Optic tract
Oculomotor (III)
Trochlear (IV)
Vestibulocochlear (VIII)
Hypoglossal (XII)
Vagus (X)

Nasal cavity
Cribriform plate of the ethmoid bone
Olfactory bulb
Collectively, the olfactory nerve (I)
Olfactory tract

Trigeminal (V)
Abducens (VI)
Facial (VII)
Glossopharyngeal (IX)
Accessory (XI)

(a)

(b)

FIGURE 11.25 **AP|R** The cranial nerves, except for the first two pairs, arise from the brainstem. (a) They are identified either by numbers indicating their order, their function, or the general distribution of their fibers. (b) An enlargement showing the detail of the olfactory nerve fibers passing through the cribriform plate of the ethmoid bone.

sensory branches conduct impulses from the scalp behind the ears, the skin of the jaw, the lower teeth, the lower gum, and the lower lip. The motor branches supply the muscles of mastication and certain muscles in the floor of the mouth.

A disorder of the trigeminal nerve called *trigeminal neuralgia* (tic douloureux) causes severe recurring pain in the face and forehead on one side. If drugs cannot help, surgery may be used to sever the sensory part of the nerve. However, with surgery the patient loses sensations in other body regions that the sensory branch supplies. After surgery, care must be taken when eating or drinking hot foods or liquids, and the mouth must be inspected daily for food particles or damage to the cheeks from biting. The cause of trigeminal neuralgia is unknown.

The sixth pair of cranial nerves, the **abducens** (ab-du'senz) **nerves** (VI), are small and originate from the pons near the medulla oblongata. They enter the orbits of the eyes and supply motor impulses to the remaining pair of external eye muscles, the *lateral rectus muscles.* This nerve is considered motor, with some proprioceptive fibers.

The seventh pair of cranial nerves, the **facial** (fa'shal) **nerves** (VII), are mixed nerves that arise from the lower part of the pons and emerge on the sides of the face. Their sensory branches are associated with taste receptors on the anterior two-thirds of the tongue, and some of their motor fibers conduct impulses to muscles of facial expression (fig. 11.27). Still other motor fibers of these nerves function in the autonomic nervous system by stimulating secretions from tear glands and certain salivary glands (submandibular and sublingual glands).

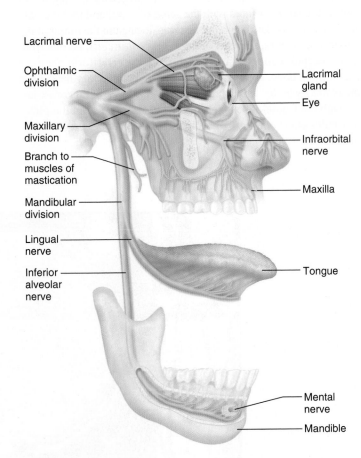

Lacrimal nerve
Ophthalmic division
Maxillary division
Branch to muscles of mastication
Mandibular division
Lingual nerve
Inferior alveolar nerve

Lacrimal gland
Eye
Infraorbital nerve
Maxilla
Tongue
Mental nerve
Mandible

FIGURE 11.26 **AP|R** Each trigeminal nerve has three large branches that supply various regions of the head and face: the ophthalmic division (sensory), the maxillary division (sensory), and the mandibular division (sensory and motor to muscles of mastication).

A common cause of facial paralysis, called Bell's palsy, is usually limited to one side. It occurs from damage to the facial nerve from unknown causes, although it is thought to be linked to viral infections. This condition usually resolves within a few months, but may be permanent.

The eighth pair of cranial nerves, the **vestibulocochlear** (ves-tib″u-lo-kok′le-ar) **nerves** (VIII, acoustic, or auditory, nerves), are sensory nerves that arise from the medulla oblongata. Each of these nerves has two distinct parts—a vestibular branch and a cochlear branch.

The neuron cell bodies of the *vestibular branch* fibers are located in ganglia near the vestibule and semicircular canals of the inner ear. These structures contain receptors that sense changes in the position of the head and, in response, initiate and send impulses to the cerebellum, where they are used in reflexes that maintain equilibrium.

The neuron cell bodies of the *cochlear branch* fibers are located in a ganglion of the cochlea, a part of the inner ear that houses the hearing receptors. Impulses from this branch pass through the medulla oblongata and midbrain on their way to the temporal lobe, where they are interpreted.

The ninth pair of cranial nerves, the **glossopharyngeal** (glos″o-fah-rin′je-al) **nerves** (IX), are associated with the tongue and pharynx. These nerves arise from the medulla oblongata. They are mixed nerves, with predominant sensory fibers that conduct impulses from the lining of the pharynx, tonsils, and posterior third of the tongue to the brain. Fibers in the motor component of the glossopharyngeal nerves innervate certain salivary glands and a constrictor muscle in the wall of the pharynx that functions in swallowing.

The tenth pair of cranial nerves, the **vagus** (va′gus) **nerves** (X), originate in the medulla oblongata and extend downward through the neck into the chest and abdomen. These nerves are mixed, including both somatic and autonomic branches, with the autonomic fibers predominant.

Among the somatic components of the vagus nerves are motor fibers that conduct impulses to muscles of the larynx and pharynx. These fibers are associated with speech and swallowing reflexes that use muscles in the soft palate and pharynx. Vagal sensory fibers conduct impulses from the linings of the pharynx, larynx, and esophagus and from the viscera of the thorax and abdomen to the brain. Autonomic motor fibers of the vagus nerves supply the heart and many smooth muscles and glands in the viscera of the thorax and abdomen (fig. 11.28).

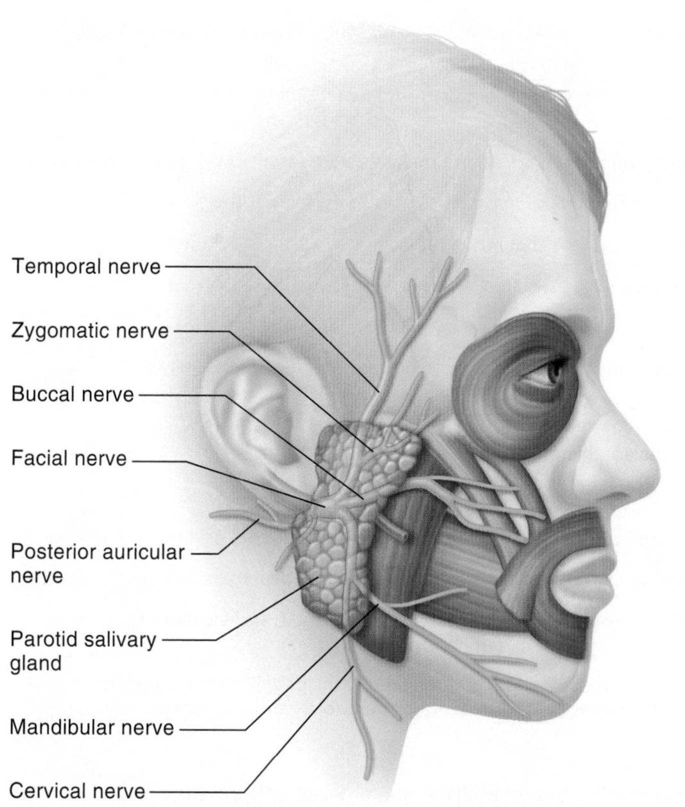

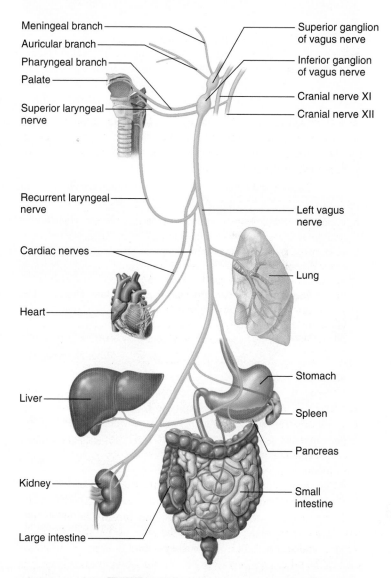

FIGURE 11.27 APR The facial nerves are associated with taste receptors on the tongue and with muscles of facial expression.

FIGURE 11.28 APR The vagus nerves (only the left vagus is shown) extend from the medulla oblongata downward into the chest and abdomen to supply many organs.

The eleventh pair of cranial nerves, the **accessory** (ak-ses′o-re) **nerves** (XI, spinal accessory), originate in the medulla oblongata and the spinal cord. Therefore, these nerves have both cranial and spinal branches. Each *cranial branch* of an accessory nerve joins a vagus nerve and conducts impulses to muscles of the soft palate, pharynx, and larynx. The *spinal branch* descends into the neck and supplies motor fibers to the trapezius and sternocleidomastoid muscles. This nerve is considered motor, with some proprioceptive fibers.

The twelfth pair of cranial nerves, the **hypoglossal** (hi″po-glos′al) **nerves** (XII), arise from the medulla oblongata and pass into the tongue. They primarily consist of fibers that conduct impulses to muscles that move the tongue in speaking, chewing, and swallowing. This nerve is considered motor, with some proprioceptive fibers. Table 11.9 summarizes the functions of the cranial nerves.

 **PRACTICE**

33 Define *peripheral nervous system*.

34 Distinguish between somatic and autonomic nerve fibers.

35 Describe the structure of a peripheral nerve.

36 Distinguish among sensory, motor, and mixed nerves.

37 Name the cranial nerves, and list the major functions of each.

Spinal Nerves

Thirty-one pairs of spinal nerves originate from the spinal cord. All but the first pair are mixed nerves, and they provide two-way communication between the spinal cord and parts of the upper and lower limbs, neck, and trunk.

TABLE 11.9 Functions of Cranial Nerves AP|R

Nerve		Type	Function
I	Olfactory	Sensory	Sensory fibers conduct impulses associated with the sense of smell.
II	Optic	Sensory	Sensory fibers conduct impulses associated with the sense of vision.
III	Oculomotor	Primarily motor	Motor fibers conduct impulses to muscles that raise the eyelids, move the eyes, adjust the amount of light entering the eyes, and focus the lenses.
			Some sensory fibers conduct impulses associated with proprioceptors.
IV	Trochlear	Primarily motor	Motor fibers conduct impulses to muscles that move the eyes.
			Some sensory fibers conduct impulses associated with proprioceptors.
V	Trigeminal	Mixed	
	Ophthalmic division		Sensory fibers conduct impulses from the surface of the eyes, tear glands, scalp, forehead, and upper eyelids.
	Maxillary division		Sensory fibers conduct impulses from the upper teeth, upper gum, upper lip, lining of the palate, and skin of the face.
	Mandibular division		Sensory fibers conduct impulses from the scalp, skin of the jaw, lower teeth, lower gum, and lower lip.
			Motor fibers conduct impulses to muscles of mastication and to muscles in the floor of the mouth.
VI	Abducens	Primarily motor	Motor fibers conduct impulses to muscles that move the eyes.
			Some sensory fibers conduct impulses associated with proprioceptors.
VII	Facial	Mixed	Sensory fibers conduct impulses associated with taste receptors of the anterior tongue.
			Motor fibers conduct impulses to muscles of facial expression, tear glands, and salivary glands.
VIII	Vestibulocochlear	Sensory	
	Vestibular branch		Sensory fibers conduct impulses associated with the sense of equilibrium.
	Cochlear branch		Sensory fibers conduct impulses associated with the sense of hearing.
IX	Glossopharyngeal	Mixed	Sensory fibers conduct impulses from the pharynx, tonsils, posterior tongue, and carotid arteries.
			Motor fibers conduct impulses to salivary glands and to muscles of the pharynx used in swallowing.
X	Vagus	Mixed	Sensory fibers conduct impulses from the pharynx, larynx, esophagus, and viscera of the thorax and abdomen.
			Somatic motor fibers conduct impulses to muscles associated with speech and swallowing; autonomic motor fibers conduct impulses to the viscera of the thorax and abdomen.
XI	Accessory	Primarily motor	
	Cranial branch		Motor fibers conduct impulses to muscles of the soft palate, pharynx, and larynx.
	Spinal branch		Motor fibers conduct impulses to muscles of the neck and back; some proprioceptor input.
XII	Hypoglossal	Primarily motor	Motor fibers conduct impulses to muscles that move the tongue; some proprioceptor input.

Spinal nerves are not named individually but are grouped by the level from which they arise. Each nerve is numbered in sequence (fig. 11.29). On each vertebra the vertebral notches, which are the major parts of the intervertebral foramina, are associated with the inferior portion of their respective vertebrae. For this reason, each spinal nerve, as it passes through the intervertebral foramen, is associated with the vertebra above it. The cervical spinal nerves are an exception because spinal nerve C1 passes superior to the vertebra C1. Therefore, although there are seven cervical vertebrae, there are eight pairs of *cervical nerves* (numbered C1 to C8). There are twelve pairs of *thoracic nerves* (numbered T1 to T12), five pairs of *lumbar nerves* (numbered L1 to L5), five pairs of *sacral nerves* (numbered S1 to S5), and one pair of *coccygeal nerves* (Co).

Each spinal nerve, except for the first pair, emerges from the cord by two short branches, or roots, which lie within the vertebral column. The roots arising from the superior part of the spinal cord

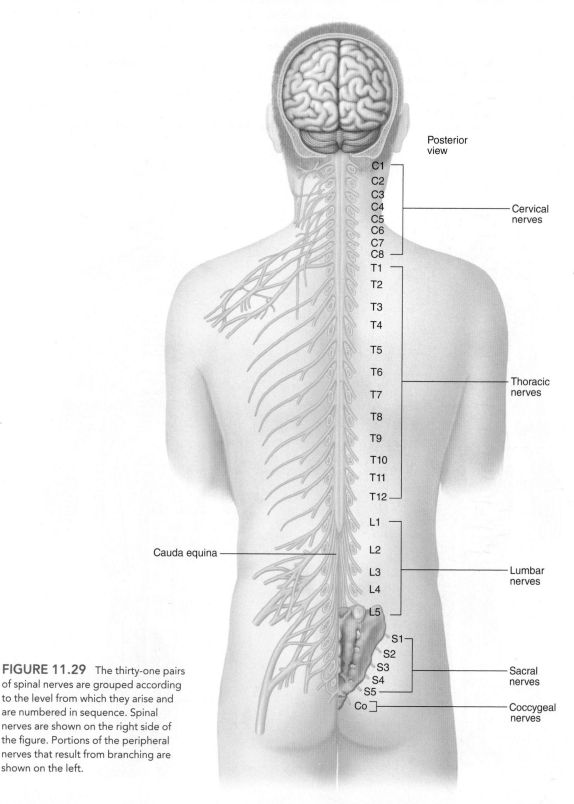

Posterior view

Cervical nerves

Thoracic nerves

Cauda equina

Lumbar nerves

Sacral nerves

Coccygeal nerves

FIGURE 11.29 The thirty-one pairs of spinal nerves are grouped according to the level from which they arise and are numbered in sequence. Spinal nerves are shown on the right side of the figure. Portions of the peripheral nerves that result from branching are shown on the left.

pass outward almost horizontally, whereas those from the inferior portions of the spinal cord descend at sharp angles. This anatomical feature is a consequence of growth. In early life, the spinal cord extends the entire length of the vertebral column, but with age, the column grows more rapidly than the cord. Thus, the adult spinal cord ends at the level between the first and second lumbar vertebrae, so the roots associated with the lumbar, sacral, and coccygeal nerves descend to their exits beyond the end of the cord, still within the vertebral canal. These descending roots form a structure called the *cauda equina* (horse's tail) (fig. 11.29).

The **ventral root** (anterior, or motor, root) of each spinal nerve consists of axons from the motor neurons whose cell bodies lie within the gray matter of the cord. The **dorsal root** (posterior, or sensory, root) can be identified by an enlargement called the *dorsal root ganglion.* This ganglion contains the cell bodies of the sensory neurons whose axons (peripheral processes) conduct impulses inward from peripheral body parts. The axons (central

processes) of these neurons continue through the dorsal root and into the spinal cord, where they form synapses with other neurons or ascend to the brain (see fig. 11.21). A ventral root and a dorsal root unite to form a spinal nerve, which extends outward from the vertebral canal through an *intervertebral foramen* (the dorsal root is usually absent from the first pair of spinal nerves).

Each spinal nerve below C1 contains sensory fibers that reach the skin, and the region innervated is called a *dermatome.* (Similarly, the collection of muscles innervated by motor nerve fibers of a particular spinal nerve is called a *myotome.*) Dermatomes are highly organized, but they vary considerably in size and shape, as figure 11.30 indicates. A map of the dermatomes is useful in localizing the sites of injuries to dorsal roots or to the spinal cord.

Just beyond its foramen, each spinal nerve branches. One of these parts, the small *meningeal branch,* reenters the vertebral canal through the intervertebral foramen and supplies (innervates)

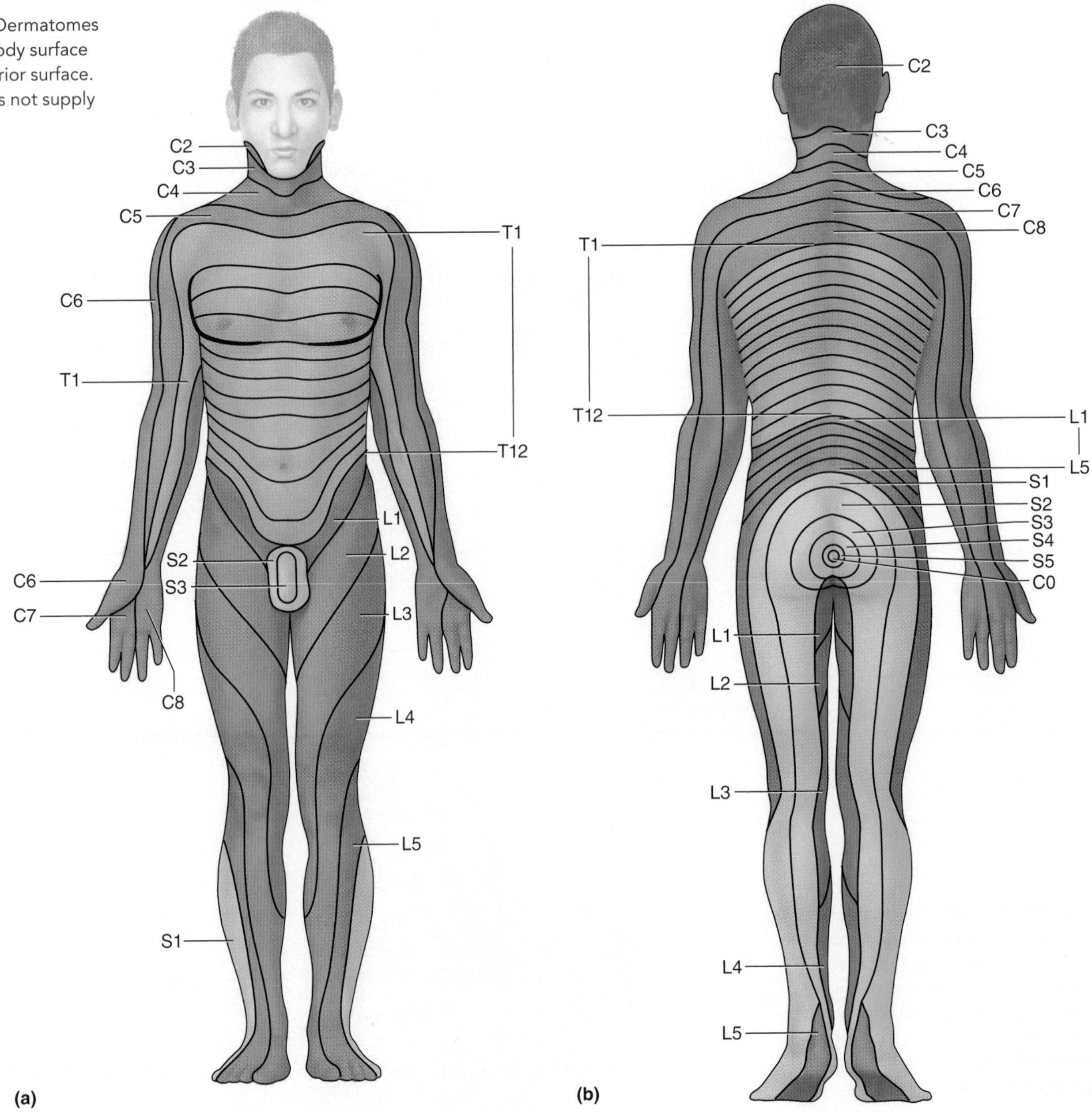

FIGURE 11.30 Dermatomes (a) on the anterior body surface and (b) on the posterior surface. Spinal nerve C1 does not supply any skin area.

(a)

(b)

the meninges and blood vessels of the cord, as well as the intervertebral ligaments and the vertebrae. A *dorsal branch* (dorsal ramus) of each spinal nerve turns posteriorly and innervates the muscles and skin of the back, as figure 11.31 shows. The main portion of the nerve, the *ventral branch* (ventral ramus), continues forward to supply muscles and skin on the front and sides of the trunk and limbs. The spinal nerves in the thoracic and lumbar regions have a fourth branch, or *visceral branch,* which is part of the autonomic nervous system. Motor fibers associated with this branch form synapses in ganglia (paravertebral ganglia) adjacent to the vertebral column.

Ventral branches of the spinal nerves form complex networks called **plexuses** instead of continuing directly to the peripheral body parts, except in the thoracic regions T2 through T12. In a plexus, the fibers of various spinal nerves are sorted and recombined in a way that enables fibers associated with a particular peripheral body part to reach it in the same peripheral nerve, even though the fibers originate from different spinal nerves (fig. 11.32).

> The specific spinal segment or segments that supply a body part is called "segmental innervation." For example, the biceps brachii is innervated by the musculocutaneous nerve with axons originating from C5, C6, and C7. Similarly, the gastrocnemius is innervated by the tibial nerve with axons originating from S1 and S2.

Cervical Plexuses

The ventral branches of the first four cervical nerves form the **cervical plexuses**, which lie deep in the neck on either side. Fibers from these plexuses supply the muscles and skin of the neck. In addition, fibers from the third, fourth, and fifth cervical nerves pass into the right and left **phrenic** (fren′ik) **nerves,** which conduct motor impulses to the muscle fibers of the diaphragm.

Brachial Plexuses

The ventral branches of the lower four cervical nerves and the first thoracic nerve give rise to **brachial plexuses.** These networks of nerve fibers are deep in the shoulders between the neck and the axillae (armpits). The major branches emerging from the brachial plexuses include the following (fig. 11.33):

1. *Musculocutaneous nerves* supply muscles of the arms on the anterior sides and the skin of the forearms.
2. *Ulnar nerves* supply muscles of the forearms and hands and the skin of the hands.
3. *Median nerves* supply muscles of the forearms and muscles and skin of the hands.
4. *Radial nerves* supply muscles of the arms on the posterior sides and the skin of the forearms and hands.
5. *Axillary nerves* supply muscles and skin of the anterior, lateral, and posterior regions of the arm.

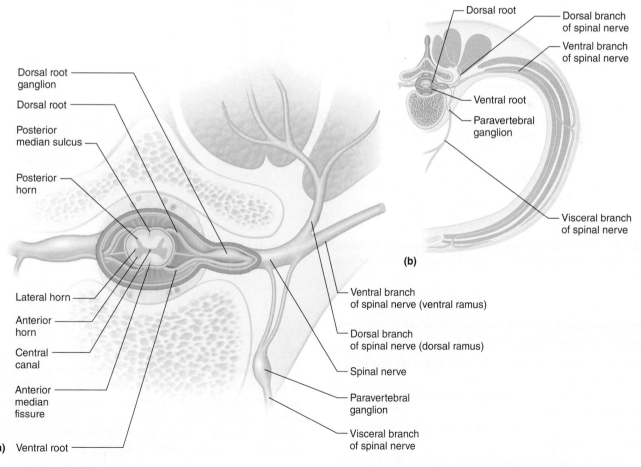

FIGURE 11.31 **AP|R** Spinal nerve. (*a*) Each spinal nerve has a dorsal and ventral branch. (*b*) The thoracic and lumbar spinal nerves also have a visceral branch.

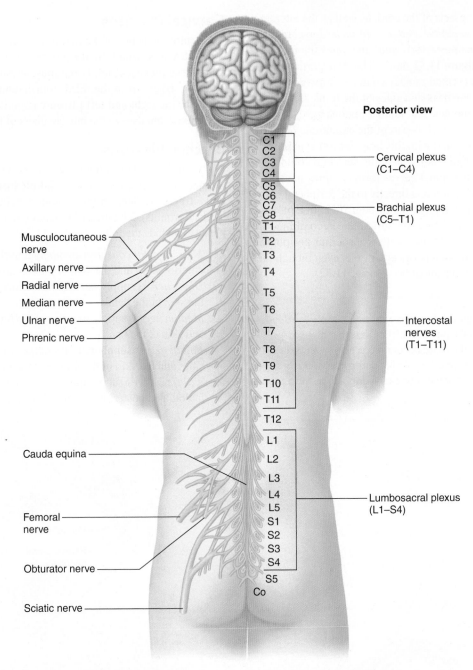

Posterior view

Cervical plexus
(C1–C4)

Brachial plexus
(C5–T1)

Musculocutaneous
nerve

Axillary nerve

Radial nerve

Median nerve

Ulnar nerve

Phrenic nerve

Intercostal
nerves
(T1–T11)

Cauda equina

Femoral
nerve

Obturator nerve

Sciatic nerve

Lumbosacral plexus
(L1–S4)

C1 C2 C3 C4 C5 C6 C7 C8 T1 T2 T3 T4 T5 T6 T7 T8 T9 T10 T11 T12 L1 L2 L3 L4 L5 S1 S2 S3 S4 S5 Co

FIGURE 11.32 The ventral branches of the spinal nerves in the thoracic region give rise to intercostal nerves. Those in other regions combine to form complex networks called plexuses.

Other nerves associated with the brachial plexus that innervate various skeletal muscles include the following:

1. The *lateral* and *medial pectoral nerves* supply the pectoralis major and pectoralis minor muscles.
2. The *dorsal scapular nerve* supplies the rhomboid major and levator scapulae muscles.
3. The *lower subscapular nerve* supplies the subscapularis and teres major muscles.
4. The *thoracodorsal nerve* supplies the latissimus dorsi muscle.
5. The *suprascapular nerve* supplies the supraspinatus and infraspinatus muscles.

Lumbosacral Plexuses

The ventral branches of the lumbar and first four sacral nerves form the **lumbosacral** (lum″bo-sa′kral) **plexuses**. These networks of nerve fibers extend from the lumbar and sacral regions of the back into the pelvic cavity, giving rise to a number of motor and sensory fibers associated with the lower abdominal wall, external genitalia, buttocks, thighs, legs, and feet. The major branches of these plexuses include the following (fig. 11.34):

1. The *obturator nerves* supply the adductor muscles of the thighs.

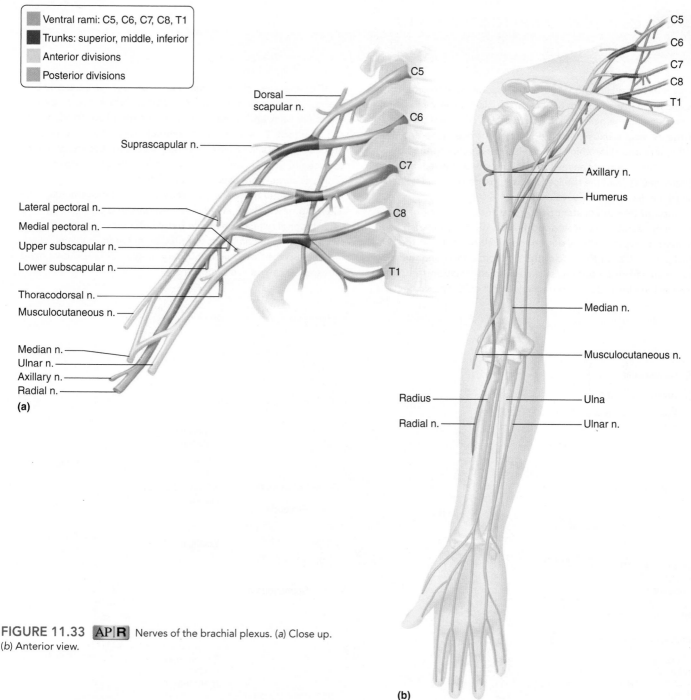

Ventral rami: C5, C6, C7, C8, T1
Trunks: superior, middle, inferior
Anterior divisions
Posterior divisions

Dorsal scapular n.

Suprascapular n.

Lateral pectoral n.
Medial pectoral n.
Upper subscapular n.
Lower subscapular n.

Thoracodorsal n.
Musculocutaneous n.

Median n.
Ulnar n.
Axillary n.
Radial n.

C5
C6
C7
C8
T1

(a)

C5
C6
C7
C8
T1

Axillary n.
Humerus

Median n.

Musculocutaneous n.

Radius
Radial n.

Ulna
Ulnar n.

(b)

FIGURE 11.33 **AP|R** Nerves of the brachial plexus. (*a*) Close up. (*b*) Anterior view.

2. The *femoral nerves* divide into many branches, supplying motor impulses to muscles of the anterior thighs and receiving sensory impulses from the skin of the thighs and legs.

3. The *sciatic nerves* are the largest and longest nerves in the body, formed by the common fibular nerve and the tibial nerve wrapped in a connective tissue sheath. They pass downward into the buttocks and descend into the thighs, where they re-emerge as the *tibial* and *common fibular nerves*. The many branches of these nerves supply muscles and skin in the thighs, legs, and feet.

Other nerves associated with the lumbosacral plexus that innervate various skeletal muscles include the following:

1. The *pudendal nerve* supplies the muscles of the perineum.

2. The *inferior* and *superior gluteal nerves* supply the gluteal muscles and the tensor fasciae latae muscle.

Except for T1, the ventral branches of the thoracic spinal nerves do not enter a plexus. Instead, they extend into spaces between the ribs as **intercostal** (in″ter-kos′tal) **nerves** (T2–T11) or under the ribs as subcostal nerves (T12). These nerves supply motor impulses to the intercostal muscles and the upper abdominal wall muscles. They also receive sensory impulses from the skin of the thorax and abdomen. Clinical Application 11.7 discusses injuries to the spinal nerves.

Spinal Nerve Injuries

Birth injuries, dislocations, vertebral fractures, stabs, gunshot wounds, and pressure from tumors can all injure spinal nerves. Suddenly bending the neck, called whiplash, can compress the nerves of the cervical plexuses, causing persistent headache and pain in the neck and skin, which the cervical nerves supply. If a broken or dislocated vertebra severs or damages the axons reaching the phrenic nerves from the cervical plexuses, partial or complete paralysis of the diaphragm may result.

Intermittent or constant pain in the neck, shoulder, or upper limb may result from prolonged flexion of the arm, such as in painting or typing. This is due to too much pressure on the brachial plexus. Called *thoracic outlet syndrome,* this condition may also result from a congenital skeletal malformation that compresses the plexus during upper limb and shoulder movements.

Degenerative changes may compress an intervertebral disc in the lumbar region, producing *sciatica,* which causes pain in the lower back and gluteal region that can radiate to the thigh, calf, and foot. Sciatica is most common in middle-aged people, particularly distance runners. It usually compresses spinal nerves between L2 and S1, which contain fibers of the sciatic nerve. Rest, drugs, or surgery are used to treat sciatica.

In *carpal tunnel syndrome,* repeated hand movements, such as typing or weeding, inflame the tendons that pass through the carpal tunnel, which is a space between bones in the wrist. The swollen tendons compress the median nerve in the wrist, sending pain up the upper limb. Surgery or avoiding repetitive hand movements can relieve symptoms.

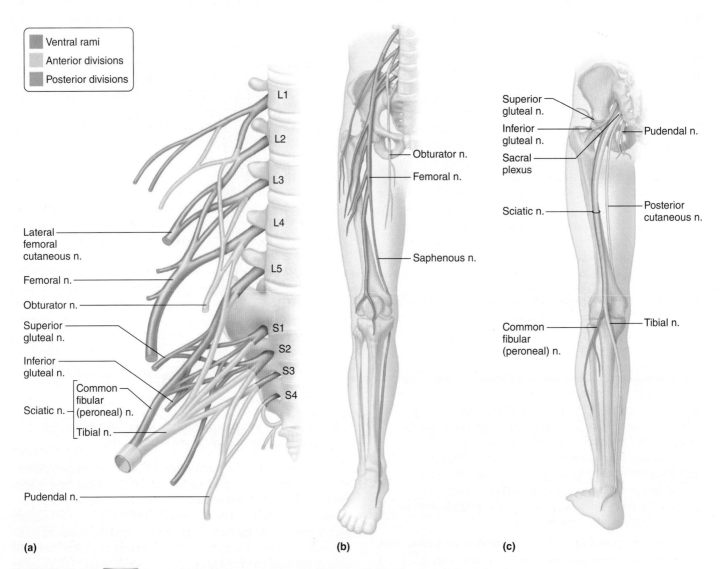

FIGURE 11.34 AP|R Nerves of the lumbosacral plexus. (*a*) Close up. (*b*) Anterior view. (*c*) Posterior view.

38 How are spinal nerves grouped?

39 Describe how a spinal nerve emerges from the spinal cord.

40 Name and locate the major nerve plexuses.

11.7 | Autonomic Nervous System

The autonomic nervous system is the part of the PNS that functions independently (autonomously) and continuously, without conscious effort. This system controls visceral activities by regulating the actions of smooth muscles, cardiac muscles, and various glands. It oversees heart rate, blood pressure, breathing rate, body temperature, and other visceral activities that aid in maintaining homeostasis. Portions of the autonomic nervous system also respond during times of emotional stress and prepare the body to meet the demands of strenuous physical activity.

General Characteristics

Reflexes in which sensory signals originate from receptors in the viscera and in the skin regulate autonomic activities. Afferent nerve fibers conduct these signals to neural centers in the brain or spinal cord. In response, motor impulses leave these centers on efferent nerve fibers in cranial and spinal nerves. Typically, these efferent fibers lead outside the CNS to ganglia. The impulses they conduct are integrated in the ganglia and are relayed to organs (muscles or glands) that respond by contracting, secreting, or being inhibited. The integrative function of the ganglia provides the autonomic nervous system with some degree of independence from the brain and spinal cord, and the visceral efferent nerve fibers associated with these ganglia comprise the autonomic nervous system.

The autonomic nervous system includes two divisions, the **sympathetic** (sim″pah-thet′ik) and **parasympathetic** (par″ah-sim″pah-thet′ik) **divisions.** Many organs receive input from both divisions. Impulses from one division may activate an organ, and impulses from the other division inhibit it. Thus, the divisions may function antagonistically, regulating the actions of some organs by alternately activating or inhibiting them. One example of this, discussed in chapter 15 (p. 587), concerns the baroreflex control of heart rate. In that reflex, parasympathetic activity decreases heart rate and sympathetic activity increases it. Moment to moment changes between the two keep heart rate at a level appropriate to maintaining homeostasis.

Neither division is entirely excitatory or inhibitory. Each activates some organs and inhibits others. The sympathetic division speeds the heart but slows the digestive system. The parasympathetic division slows the heart but activates digestion.

The sympathetic division is always operating, usually without our noticing, but is most active during energy-expending, stressful, or emergency situations (so-called "fight or flight"). Conversely, the parasympathetic division is most active under ordinary, restful conditions (often described as "rest and digest"). It also counters the effects of the sympathetic division and restores the body to a resting state following a stressful experience. For example, during an emergency the sympathetic division increases heart and breathing rates. Following the emergency, the parasympathetic division decreases these activities.

Autonomic Nerve Fibers

The motor pathways of the autonomic nervous system have a distinct pattern. In the motor pathways of the somatic nervous system, a single neuron typically links the CNS and a skeletal muscle. In the autonomic system, motor pathways include two neurons, as figure 11.35 shows. The cell body of one neuron is in the brain or spinal cord. Its axon, the **preganglionic** (pre″gang-gle-on′ik) **fiber,** leaves the CNS and synapses with one or more neurons whose cell bodies are within an autonomic ganglion. The axon of such a second neuron is called a **postganglionic** (pōst″gang-gle-on′ik) **fiber,** and it extends to a visceral effector.

Sympathetic Division

In the sympathetic division (thoracolumbar division), the preganglionic fibers originate from neurons in the lateral horn of the spinal cord. These neurons are found in all of the thoracic segments and in the upper two lumbar segments of the cord (T1–L2). Their axons exit through the ventral roots of spinal nerves along with various somatic motor fibers.

Preganglionic fibers extend a short distance, then leave the spinal nerves through branches called *white rami* (sing., *ramus*) and enter sympathetic ganglia. Two groups of such ganglia, called **sympathetic chain ganglia** (paravertebral ganglia), form chains along the sides of the vertebral column. These ganglia, with the fibers that connect them, compose the **sympathetic trunks** (fig. 11.36).

The paravertebral ganglia lie just beneath the parietal pleura in the thorax and beneath the parietal peritoneum in the abdomen (see chapter 1, p. 20). Although these ganglia are some distance from the viscera they help control, other sympathetic ganglia are nearer to the viscera. The *collateral ganglia,* for example, are in the abdomen, closely associated with certain large blood vessels (fig. 11.37).

Some of the preganglionic fibers that enter paravertebral ganglia synapse with neurons in these ganglia. Other fibers extend through the ganglia and pass up or down the sympathetic trunk and synapse with neurons in ganglia at higher or lower levels in the chain. Still other fibers pass through to collateral ganglia before they synapse. Typically, a preganglionic axon will synapse with several other neurons in a sympathetic ganglion. This is an example of divergence (see chapter 10, pp. 383 and 385).

The axons of the second neurons in sympathetic pathways, the postganglionic fibers, extend from the sympathetic ganglia to visceral effectors. Most axons leaving paravertebral ganglia pass through branches called **gray rami** and return to a spinal nerve before extending to an effector (fig. 11.37). These branches appear gray because the postganglionic axons generally are unmyelinated, whereas nearly all of the preganglionic axons in the white rami are myelinated.

An important exception to the usual arrangement of sympathetic fibers is in a set of preganglionic fibers that pass through the sympathetic ganglia and extend to the medulla of each adrenal gland. These fibers terminate in the glands on special hormone-secreting cells that release **norepinephrine** (20%) and **epinephrine** (80%)

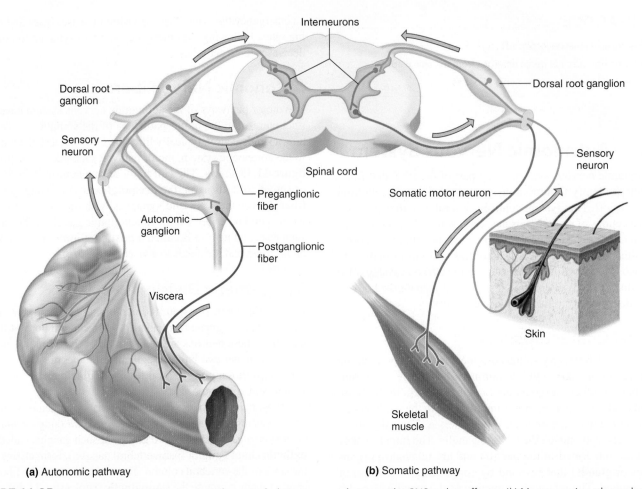

FIGURE 11.35 Motor pathways. (*a*) Autonomic pathways include two neurons between the CNS and an effector. (*b*) Most somatic pathways have a single neuron between the CNS and an effector. In both cases the motor fibers pass through the ventral root of the spinal cord.

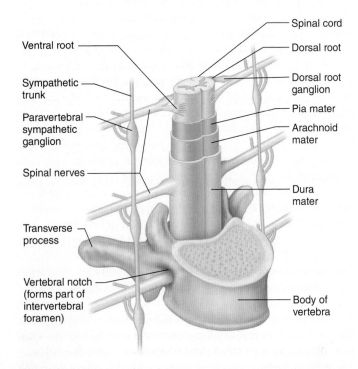

FIGURE 11.36 A chain of paravertebral ganglia extends along each side of the vertebral column.

when they are stimulated. Chapter 13 (pp. 508–510) discusses the functions of the adrenal medulla and its hormones. Figure 11.38 shows the sympathetic division.

Parasympathetic Division

The preganglionic fibers of the parasympathetic division (craniosacral division) arise from neurons in the midbrain, pons, and medulla oblongata of the brainstem and from part of the sacral region (S2–4) of the spinal cord (fig. 11.39). From there, the preganglionic fibers lead outward on cranial or sacral nerves to ganglia near or in various organs *(terminal ganglia)*. The short postganglionic fibers continue from the ganglia to specific muscles or glands in these organs (fig. 11.40). Most parasympathetic preganglionic axons are myelinated, and the parasympathetic postganglionic fibers are unmyelinated.

The parasympathetic preganglionic fibers associated with parts of the head are included in the oculomotor, facial, and glossopharyngeal nerves. Those fibers that innervate organs of the thorax and upper abdomen are parts of the vagus nerves. (The vagus nerves carry about 75% of all parasympathetic fibers.) Preganglionic fibers arising from the sacral region of the spinal cord lie in the branches of the second through the fourth sacral spinal nerves, and they conduct impulses to the viscera in the pelvic cavity (see fig. 11.39).

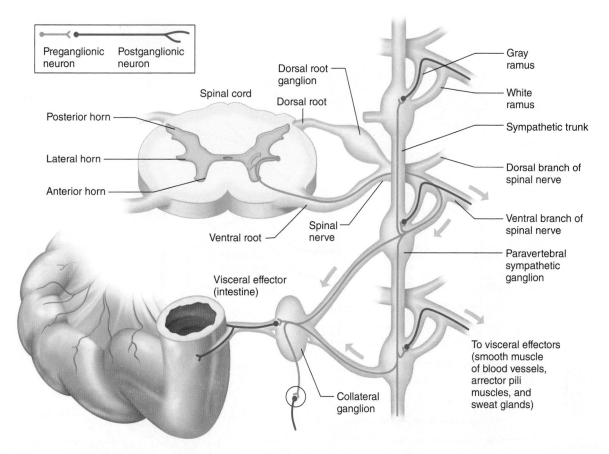

Preganglionic neuron — Postganglionic neuron

Spinal cord

Posterior horn
Lateral horn
Anterior horn

Dorsal root ganglion
Dorsal root
Ventral root
Spinal nerve

Gray ramus
White ramus
Sympathetic trunk
Dorsal branch of spinal nerve
Ventral branch of spinal nerve
Paravertebral sympathetic ganglion

Visceral effector (intestine)

To visceral effectors (smooth muscle of blood vessels, arrector pili muscles, and sweat glands)

Collateral ganglion

FIGURE 11.37 Sympathetic fibers leave the spinal cord in the ventral roots of spinal nerves, enter paravertebral ganglia, and synapse with other neurons that extend to visceral effectors.

 PRACTICE

41 What is the general function of the autonomic nervous system?

42 How are the divisions of the autonomic system distinguished?

43 Describe a sympathetic nerve pathway and a parasympathetic nerve pathway.

Autonomic Neurotransmitters

The different postganglionic neurotransmitters (mediators) are responsible for the different effects that the sympathetic and parasympathetic divisions have on organs. The preganglionic neurons of the sympathetic and parasympathetic divisions all secrete acetylcholine, and for this reason they are called **cholinergic** (ko″lin-er′jik). The parasympathetic postganglionic fibers are also cholinergic (one exception, parasympathetic neurons that secrete nitric oxide, is described in chapter 22, p. 835). Most sympathetic postganglionic neurons, however, secrete norepinephrine (noradrenalin) and are called **adrenergic** (ad″ren-er′jik) (fig. 11.40). Exceptions to this include the sympathetic postganglionic neurons that stimulate sweat glands and a few sympathetic neurons to blood vessels in skin (which cause vasodilation); these neurons secrete acetylcholine and therefore are cholinergic (adrenergic sympathetic fibers to blood vessels cause vasoconstriction).

Most organs receive innervation from both sympathetic and parasympathetic divisions, usually with opposing actions. For example, the sympathetic nervous system increases heart rate and dilates pupils, whereas parasympathetic stimulation decreases heart rate and constricts pupils. However, this is not always the case. For example, the diameters of most blood vessels lack parasympathetic innervation and are thus regulated by the sympathetic division. Smooth muscle in the walls of these vessels is continuously stimulated by sympathetic impulses; thereby being maintained in a state of partial contraction called *sympathetic tone*. Decreasing sympathetic stimulation allows the muscular walls of such blood vessels to relax, increasing their diameters (vasodilation). Conversely, increasing sympathetic stimulation vasoconstricts vessels. Table 11.10 summarizes the effects of autonomic stimulation on various visceral effectors.

Actions of Autonomic Neurotransmitters

The actions of autonomic neurotransmitters result from their binding to protein receptors in the membranes of effector cells, such as in the case of stimulation at neuromuscular junctions (see chapter 9, p. 297) and synapses (see chapter 10, p. 378). Receptor binding alters the membrane. For example, the membrane's permeability to certain ions may increase, and in smooth muscle cells, an action potential followed by muscular contraction may result. Similarly, a gland cell may respond to a change in its membrane by secreting a product.

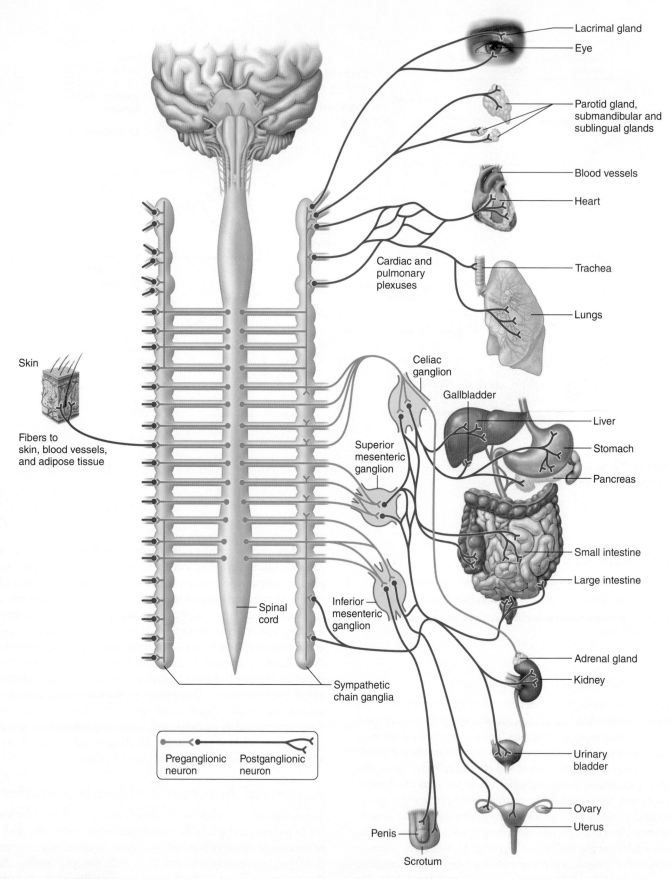

FIGURE 11.38 **AP|R** The preganglionic fibers of the sympathetic division of the autonomic nervous system arise from the thoracic and lumbar regions of the spinal cord. A preganglionic fiber directly innervates the adrenal medulla.

Labels in figure: Lacrimal gland; Eye; Parotid gland, submandibular and sublingual glands; Blood vessels; Heart; Cardiac and pulmonary plexuses; Trachea; Lungs; Celiac ganglion; Gallbladder; Liver; Stomach; Pancreas; Superior mesenteric ganglion; Small intestine; Large intestine; Skin; Fibers to skin, blood vessels, and adipose tissue; Inferior mesenteric ganglion; Adrenal gland; Kidney; Spinal cord; Sympathetic chain ganglia; Urinary bladder; Ovary; Uterus; Penis; Scrotum; Preganglionic neuron; Postganglionic neuron

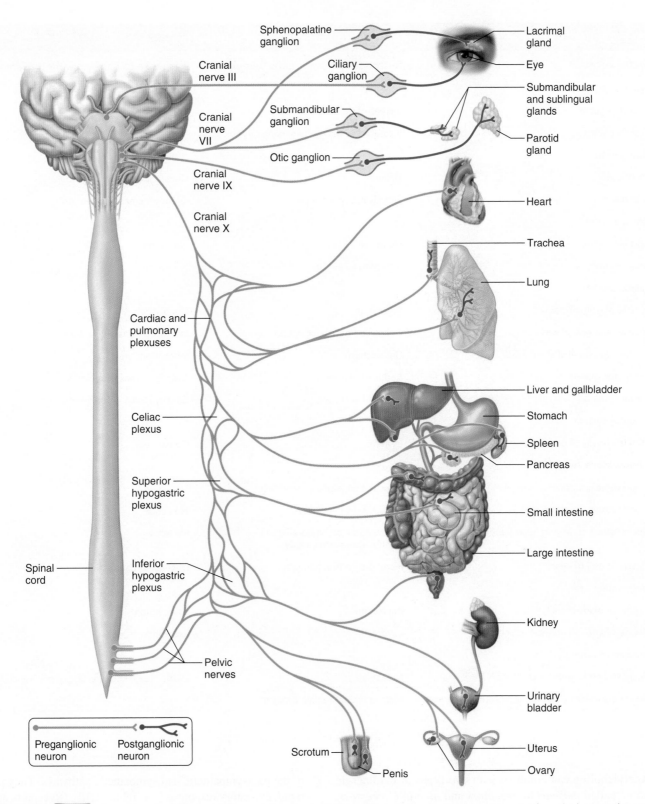

FIGURE 11.39 **AP|R** The preganglionic fibers of the parasympathetic division of the autonomic nervous system arise from the brain and sacral region of the spinal cord.

TABLE 11.10 Effects of Autonomic Stimulation on Various Visceral Effectors

Effector Location	Response to Sympathetic Stimulation	Response to Parasympathetic Stimulation
Integumentary system		
Apocrine glands	Increased secretion	No action
Eccrine glands	Increased secretion (cholinergic effect)	No action
Special senses		
Iris of eye	Dilation	Constriction
Tear gland	Slightly increased secretion	Greatly increased secretion
Endocrine system		
Adrenal cortex	No action	No action
Adrenal medulla	Increased secretion	No action
Digestive system		
Muscle of gallbladder wall	Relaxation	Contraction
Muscle of intestinal wall	Decreased peristaltic action	Increased peristaltic action
Muscle of internal anal sphincter	Contraction	Relaxation
Pancreatic glands	Reduced secretion	Greatly increased secretion
Salivary glands	Reduced secretion	Greatly increased secretion
Respiratory system		
Muscle in walls of bronchioles	Dilation	Constriction
Cardiovascular system		
Blood vessels supplying skeletal muscles	Constriction (alpha adrenergic)	No action
Blood vessels supplying skin	Constriction	No action
Blood vessels supplying heart (coronary arteries)	Constriction (alpha adrenergic) Dilation (beta adrenergic)	No action
Muscle in wall of heart	Increased contraction rate	Decreased contraction rate
Urinary system		
Muscle of bladder wall	Relaxation	Contraction
Muscle of internal urethral sphincter	Contraction	Relaxation
Reproductive systems		
Blood vessels to penis and clitoris	No action	Dilation leading to erection of penis and clitoris
Muscle associated with internal reproductive organs	Male emission, female orgasm	

Acetylcholine can combine with two types of cholinergic receptors, called *muscarinic receptors* and *nicotinic receptors.* These receptor names come from *muscarine,* a toxin from a fungus that can activate muscarinic receptors, and *nicotine,* a toxin in tobacco that can activate nicotinic receptors. The muscarinic receptors are in the membranes of effector cells at the ends of all postganglionic parasympathetic nerve fibers and at the ends of the cholinergic sympathetic fibers. Responses from these receptors are excitatory and relatively slow. The nicotinic receptors are in the synapses between the preganglionic and postganglionic neurons of the parasympathetic and sympathetic pathways. They produce rapid, excitatory responses (see Table 11.10). (Receptors at neuromuscular junctions of skeletal muscles are also nicotinic.)

Epinephrine and norepinephrine are the two chemical mediators of the sympathetic nervous system. The adrenal gland releases both as hormones, but only norepinephrine is released as a neurotransmitter by the sympathetic nervous system. These biochemicals can then bind adrenergic receptors of effector cells.

The two major types of adrenergic receptors are *alpha* and *beta* (both types have subgroups). Exciting them elicits different

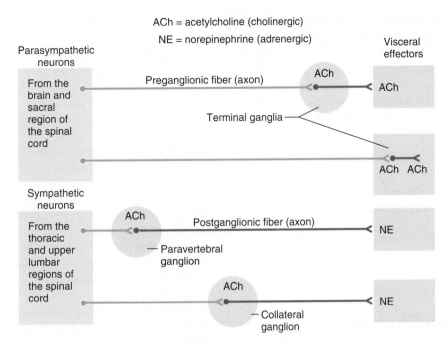

ACh = acetylcholine (cholinergic)

NE = norepinephrine (adrenergic)

FIGURE 11.40 Most sympathetic fibers are adrenergic and secrete norepinephrine at the ends of the postganglionic fiber; parasympathetic fibers are cholinergic and secrete acetylcholine at the ends of the postganglionic fibers. Two arrangements of parasympathetic postganglionic fibers are seen in both cranial and sacral portions. Similarly, sympathetic paravertebral and collateral ganglia are seen in both the thoracic and lumbar portions of the autonomic nervous system.

responses in the effector organs. For example, stimulation of the alpha receptors in vascular smooth muscle causes vasoconstriction, whereas stimulation of the beta receptors in bronchial smooth muscle causes relaxation leading to bronchodilation. Furthermore, although norepinephrine has a somewhat stronger effect on alpha receptors, both of these mediators can stimulate both types of receptors. Consequently, the way each of these adrenergic substances influences effector cells depends on the relative numbers of alpha and beta receptors in the cell membranes.

Many drugs influence autonomic functions. Some, like ephedrine, enhance sympathetic effects by stimulating release of norepinephrine from postganglionic sympathetic nerve endings. Others, like reserpine, inhibit sympathetic activity by preventing norepinephrine synthesis. Another group of drugs, which includes pilocarpine, produces parasympathetic effects, and some, like atropine, block the action of acetylcholine on visceral effectors.

Terminating Autonomic Neurotransmitter Actions

The enzyme acetylcholinesterase rapidly decomposes the acetylcholine that cholinergic fibers release. (Recall that this decomposition also occurs at the neuromuscular junctions of skeletal muscle, chapter 9, p. 299.) Thus, acetylcholine usually affects the postsynaptic membrane for only a fraction of a second.

Much of the norepinephrine released from adrenergic fibers is removed from the synapse by active transport back into the nerve endings. The enzyme monoamine oxidase, in mitochondria, then inactivates norepinephrine. This may take a few seconds, during which some molecules may diffuse into nearby tissues or the bloodstream, where other enzymes decompose them. However, some norepinephrine molecules may escape decomposition and remain active for awhile. For these reasons, norepinephrine is likely to produce a more prolonged effect than acetylcholine. In fact, the adrenal medulla releases norepinephrine and epinephrine into the blood in response to sympathetic stimulation that may trigger sympathetic responses in organs throughout the body lasting up to 20 minutes.

Control of Autonomic Activity

Control of the autonomic nervous system rests mainly in the hypothalamus, although impulse integration in the autonomic nervous system's ganglia provides some independence. In contrast to control of skeletal muscle in the somatic nervous system, which is voluntary, control of the autonomic nervous system is involuntary. Through the autonomic nervous system, the hypothalamus helps regulate body functions such as temperature, hunger, thirst, and water and electrolyte balance.

In addition to the hypothalamic activity, important autonomic reflex centers involve the medulla oblongata and the spinal cord. For example, reflex centers in the medulla oblongata for cardiac, vasomotor, and respiratory activities receive sensory impulses from viscera on sensory fibers in the vagus nerve, and use autonomic nerve pathways to stimulate motor responses in smooth muscle, cardiac muscle, and glands.

Still higher levels in the brain can have important impacts on autonomic function. Connections from the cerebral cortex, including the limbic system, can influence autonomic centers and increase both sympathetic and parasympathetic activities. A familiar example involving the parasympathetic division is salivating at the anticipation of eating, an action that originates in the cortex. An example involving the sympathetic division is the increase in heart rate and blood pressure when a person becomes agitated or upset, which is again a result of input from the cerebral cortex. Subsequent chapters discuss autonomic regulation of particular organs.

 PRACTICE

44 Distinguish between cholinergic and adrenergic fibers.

45 Explain how the fibers of one autonomic division can control the actions of a particular organ.

46 Which neurotransmitters does the autonomic nervous system use?

47 Describe two types of cholinergic receptors and two types of adrenergic receptors.

11.8 | Life-Span Changes

The redundancies and overlap of function in our nervous systems ensure that we can perceive and interact with the environment for many decades. In a sense, aging of this organ system begins before birth, as apoptosis, a form of programmed cell death, carves out the structures that will remain in the brain. This normal dying off of neurons continues throughout life. When brain apoptosis fails, disease results. For example, the brains of individuals who die of schizophrenia as young adults contain the same numbers of neurons as do newborns. These extra neurons may produce the extra dopamine that can lead to hallucinations, the hallmark of this illness.

By age thirty, the die-off of neurons accelerates somewhat, although pockets of neural stem cells lining the ventricles retain the capacity to give rise to cells that differentiate as neurons and neuroglia. Over an average lifetime, the brain shrinks by about 10%, with more loss in gray matter than white. Neuron loss is uneven. Many cells die in the temporal lobe, for example, but very few die in the brainstem. By age ninety the frontal cortex

has lost about half its neurons, but this deficit doesn't necessarily hamper function.

The nervous system changes over time in several ways. The number of dendritic branches in the cerebral cortex falls. Signs of slowing neurotransmission include decreasing levels of neurotransmitters, the enzymes necessary to synthesize them, and the numbers of postsynaptic receptors. The rate of action potential propagation may decrease by 5% to 10%. Nervous system disorders that may begin to cause symptoms in older adulthood include stroke, depression, Alzheimer disease, Parkinson disease, and multi-infarct dementia.

Noticeable signs of a normally aging nervous system include fading memory and slowed responses and reflexes. Decline in function of the sympathetic nervous system may cause transient drops in blood pressure, which, in turn, may cause fainting. By the seventh decade, waning ability of nerves in the ankles to respond to vibrations from walking may affect balance, raising the risk of falling. Poor eyesight, anemia, inner ear malfunction, and effects of drugs also contribute to poor balance in the later years. Because of these factors, nearly a third of individuals over age sixty-five have at least one serious fall a year.

Changes in sleep patterns accompany aging, reflecting the functioning of the reticular activating system. Older individuals generally sleep fewer hours per night than they once did, experiencing transient difficulty in getting to sleep and staying asleep, with more frequent movements when they are sleeping. Many have bouts with insomnia, sometimes not sleeping more than an hour or two a night. Changing electroencephalogram patterns indicate that stage IV slow-wave sleep as well as REM sleep diminish. All of these changes may result in daytime sleepiness.

 PRACTICE

48 How does aging of the nervous system begin even before birth?

49 What are some diseases that affect the aging nervous system?

50 What are some of the physical and functional signs of an aging nervous system? ∎

Chapter Summary

11.1 Overview of Divisions of the Nervous System (page 390)

The central nervous system (CNS) consists of the brain and spinal cord.

1. The brain oversees sensation and perception, movement, and thinking.
2. The brainstem connects the brain and spinal cord, allowing communication between the two.
3. The spinal cord provides communication between the CNS and the PNS.

11.2 Meninges (page 390)

Bone and protective membranes called **meninges** surround the brain and spinal cord.

1. The meninges consist of a **dura mater, arachnoid mater, and pia mater.**
2. **Cerebrospinal fluid** (CSF) occupies the space between the arachnoid and pia maters.

11.3 Ventricles and Cerebrospinal Fluid (page 391)

Ventricles, filled with cerebrospinal fluid, are connected cavities in the cerebral hemispheres and brainstem.

1. **Choroid plexuses** in the walls of the ventricles secrete CSF.

2. Ependymal cells of the choroid plexus regulate the composition of CSF.
3. CSF circulates through the ventricles and is reabsorbed into the blood of the dural sinuses.
4. CSF helps maintain a stable ion concentration in the CNS and provides a pathway to the blood for waste.

11.4 Brain (page 395)

The **brain** is the largest and most complex part of the nervous system. It contains nerve centers associated with sensations. The brain issues motor commands and carries on higher mental functions.

1. Brain development
   a. Brain structure reflects the way it forms.
   b. The brain develops from a neural tube with three cavities—the forebrain, midbrain, and hindbrain.
   c. The cavities persist as ventricles, and the walls give rise to structural and functional regions.
2. Structure of the cerebrum
   a. The **cerebrum** consists of two **cerebral hemispheres** connected by the **corpus callosum.**
   b. Its surface is marked by ridges and grooves; sulci divide each hemisphere into lobes.
   c. The **cerebral cortex** is a thin layer of gray matter near the surface.
   d. White matter consists of myelinated nerve fibers that connect neurons and communicate with other body parts.
3. Functions of the cerebrum
   a. The cerebrum provides higher brain functions, such as thought, reasoning, interpretation of sensory impulses, control of voluntary muscles, and memory storage.
   b. The cerebral cortex has sensory, association, and motor areas.
   c. Areas that interpret sensory impulses from the skin are in the **parietal lobes** near the central sulcus; other specialized sensory areas are in the **temporal lobes** and **occipital lobes.**
   d. Association areas analyze and interpret sensory impulses and provide memory, reasoning, verbalizing, judgment, and emotions.
   e. The primary motor regions lie in the **frontal lobes** near the central sulcus. Other areas of the frontal lobes control special motor functions.
   f. One cerebral hemisphere usually dominates for certain intellectual functions.
   g. Short-term memory is probably bioelectrical. Long-term memory is thought to be encoded in patterns of synaptic connections.
4. Basal nuclei
   a. **Basal nuclei** are masses of gray matter deep within the cerebral hemispheres.
   b. The neurons of the basal nuclei interact with other brain areas to facilitate voluntary movement.
5. Diencephalon
   a. The **diencephalon** includes the **thalamus** and **hypothalamus.**
   b. The thalamus selects incoming sensory impulses and relays them to the cerebral cortex.
   c. The hypothalamus is important in maintaining homeostasis.

   d. The **limbic system** produces emotional feelings and modifies behavior.
6. Brainstem
   a. The **brainstem** extends from the base of the brain to the spinal cord.
   b. The brainstem consists of the **midbrain, pons,** and **medulla oblongata.**
   c. The midbrain contains reflex centers associated with eye and head movements.
   d. The pons conducts impulses between the cerebrum and other parts of the nervous system.
   e. The medulla oblongata conducts all ascending and descending impulses and contains several vital reflex centers including those involved with the cardiovascular and respiratory systems.
   f. The **reticular formation** filters incoming sensory impulses, arousing the cerebral cortex into wakefulness in response to meaningful impulses.
   g. Normal sleep results from decreasing activity of the reticular formation, and paradoxical sleep occurs when some parts of the brain, but not others, receive activating impulses.
7. Cerebellum
   a. The **cerebellum** consists of two hemispheres connected by the vermis.
   b. A thin cortex of gray matter surrounds the white matter of the cerebellum.
   c. The cerebellum functions primarily as a reflex center, coordinating skeletal muscle movements and maintaining equilibrium.

11.5 Spinal Cord (page 409)

The **spinal cord** is a nerve column that extends from the brain into the vertebral canal. It terminates at the level between the first and second lumbar vertebrae.

1. Structure of the spinal cord
   a. The spinal cord is composed of thirty-one segments, each of which gives rise to a pair of **spinal nerves.**
   b. It is characterized by a cervical enlargement, a lumbar enlargement, and two deep longitudinal grooves that divide it into right and left halves.
   c. White matter surrounds a central core of gray matter.
   d. The white matter is composed of bundles of myelinated nerve fibers.
2. Functions of the spinal cord
   a. The spinal cord is the center for spinal reflexes.
      (1) **Reflex**es are automatic, subconscious responses to changes.
      (2) They help maintain homeostasis.
      (3) The knee-jerk reflex employs only two neurons. Other reflexes involve more neurons.
      (4) Withdrawal reflexes are protective.
   b. The spinal cord is a two-way communication system between the brain and structures outside the nervous system.
      (1) **Ascending tracts** conduct sensory impulses to the brain; **descending tracts** conduct motor impulses to muscles and glands.
      (2) Many of the fibers in the ascending and descending tracts cross over in the spinal cord or brain.

11.6 Peripheral Nervous System (page 418)

The peripheral nervous system consists of cranial and spinal nerves that branch out from the brain and spinal cord to all body parts. It is subdivided into the **somatic nervous system** and the **autonomic nervous system.**

1. Structure of peripheral nerves
   a. A nerve consists of a bundle of nerve fibers surrounded by connective tissues.
   b. The connective tissues form an outer epineurium, a perineurium enclosing bundles of nerve fibers, and an endoneurium surrounding each fiber.
2. Nerve and nerve fiber classification
   a. Nerves are cordlike bundles of nerve fibers. Nerves are classified as **sensory nerves, motor nerves,** or **mixed nerves,** depending on fiber type.
   b. Nerve fibers in the CNS are subdivided into groups with general and special functions.
3. Cranial nerves
   a. Twelve pairs of **cranial nerves** connect the brain to parts in the head, neck, and trunk.
   b. Most cranial nerves are mixed, but some are pure sensory and others are primarily motor.
   c. The names of cranial nerves indicate their primary functions or the general distributions of their fibers.
   d. Some cranial nerve fibers are somatic and others are autonomic.
4. Spinal nerves
   a. Thirty-one pairs of **spinal nerves** originate from the spinal cord. All but the first pair are mixed nerves.
   b. These mixed nerves provide a two-way communication system between the spinal cord and the upper limbs, lower limbs, neck, and trunk.
   c. Spinal nerves are grouped according to the levels from which they arise, and they are numbered sequentially.
   d. Each nerve emerges by a **dorsal root** and a **ventral root.**
      (1) A dorsal root contains sensory fibers and has a dorsal root ganglion.
      (2) A ventral root contains motor fibers.
   e. Just beyond its foramen, each spinal nerve divides into several branches.
   f. Most spinal nerves combine to form **plexuses** that direct nerve fibers to a particular body part.

11.7 Autonomic Nervous System (page 431)

The autonomic nervous system functions without conscious effort. It regulates visceral activities that maintain homeostasis.

1. General characteristics
   a. Autonomic functions are reflexes controlled from centers in the hypothalamus, brainstem, and spinal cord.
   b. Autonomic nerve fibers are associated with ganglia where impulses are integrated before distribution to effectors.
   c. The integrative function of the ganglia provides a degree of independence from the CNS.
   d. The autonomic nervous system consists of the visceral efferent fibers associated with these ganglia.
   e. The autonomic nervous system is subdivided into two divisions—sympathetic and parasympathetic.
   f. The **sympathetic division** prepares the body for stressful and emergency conditions.
   g. The **parasympathetic division** is most active under ordinary conditions.

2. Autonomic nerve fibers are efferent, or motor.
3. Sympathetic division
   a. Sympathetic fibers leave the spinal cord and synapse in ganglia.
   b. **Preganglionic fibers** pass through white rami to reach paravertebral ganglia.
   c. Paravertebral ganglia and interconnecting fibers comprise the **sympathetic trunks.**
   d. Preganglionic fibers synapse with paravertebral or collateral ganglia.
   e. Most **postganglionic fibers** pass through **gray rami** to reach spinal nerves before passing to effectors.
   f. A special set of sympathetic preganglionic fibers passes through ganglia and extends to the adrenal medulla.
4. The parasympathetic division includes the parasympathetic fibers that begin in the brainstem and sacral region of the spinal cord and synapse in ganglia near various organs or in the organs themselves.
5. Autonomic neurotransmitters
   a. Sympathetic and parasympathetic preganglionic fibers secrete acetylcholine.
   b. Most sympathetic postganglionic fibers secrete norepinephrine and are **adrenergic;** postganglionic parasympathetic fibers secrete acetylcholine and are **cholinergic.**
   c. The different effects of the autonomic divisions are due to the different neurotransmitters the postganglionic fibers release.
6. Actions of autonomic neurotransmitters
   a. Neurotransmitters combine with receptors and alter cell membranes.
   b. There are two types of cholinergic receptors and two types of adrenergic receptors.
   c. How cells respond to neurotransmitters depends upon the number and type of receptors in their membranes.
   d. Acetylcholine acts very briefly; norepinephrine and epinephrine may have more prolonged effects.
7. Terminating autonomic neurotransmitters actions
   a. Acetylcholinesterase breaks down ACh.
   b. Norepinephrine is transported back into presynaptic neurons.
8. Control of autonomic activity
   a. The central nervous system largely controls the autonomic nervous system.
   b. The medulla oblongata uses autonomic fibers to regulate cardiac, vasomotor, and respiratory activities.
   c. The hypothalamus uses autonomic fibers in regulating visceral functions.
   d. The limbic system and cerebral cortex control emotional responses through the autonomic nervous system.

11.8 Life-Span Changes (page 438)

Aging of the nervous system is a gradual elimination of cells and, eventually, slowed functioning.

1. Apoptosis of brain neurons begins before birth.
2. Neuron loss among brain regions is uneven.
3. In adulthood, numbers of dendrites in the cerebral cortex fall, as more generally neurotransmission slows.
4. Nervous system changes in older persons increase the risk of falling.
5. Sleep problems are common in the later years.

CHAPTER ASSESSMENTS

11.1 Overview of Divisions of the Nervous System

1 Explain the general functions of the brain and spinal cord, and their interrelationship. (p. 390)

11.2 Meninges

2 Name the layers of the meninges, and explain their functions. (pp. 390–391)

11.3 Ventricles and Cerebrospinal Fluid

3 Describe the relationship among the cerebrospinal fluid, the ventricles, the choroid plexuses, and arachnoid granulations. (pp. 391–395)

4 List the functions of cerebrospinal fluid. (p. 395)

11.4 Brain

5 Describe the events of brain development. (p. 395)

6 Which choice lists the parts of the brainstem? (p. 395)
   a. midbrain, pons, and medulla oblongata
   b. forebrain, midbrain, and hindbrain
   c. sulci and fissures
   d. frontal, parietal, and temporal lobes

7 Describe the structure of the cerebrum. (pp. 395 and 397)

8 Define *cerebral cortex.* (p. 397)

9 Describe the location and function of the sensory areas of the cortex. (p. 399)

10 Explain the function of the association areas of the lobes of the cerebrum. (pp. 399–401)

11 Describe the location and function of the motor areas of the cortex. (pp. 401–402)

12 Broca's area controls _____. (p. 402)
   a. memory
   b. defecation
   c. understanding grammar
   d. movements used in speaking

13 Explain *hemisphere dominance.* (p. 402)

14 Distinguish between short-term and long-term memory. (pp. 402–403)

15 Explain the conversion of short-term to long-term memory. (pp. 402–403)

16 The _____ conducts sensory information from other parts of the nervous system to the cerebral cortex. (p. 403)
   a. pineal gland
   b. hypothalamus
   c. thalamus
   d. basal nuclei

17 List the parts of the limbic system, and explain its functions. (pp. 403 and 405)

18 Name the functions of the midbrain, pons, and medulla oblongata. (pp. 405–407)

19 Describe the location and function of the reticular formation. (p. 407)

20 Distinguish between normal and paradoxical sleep. (pp. 407–408)

21 The cerebellum _____. (p. 409)
   a. communicates with the rest of the CNS
   b. creates awareness of the body's location in space
   c. coordinates skeletal muscle activity
   d. all of the above

11.5 Spinal Cord

22 Describe the structure of the spinal cord. (pp. 409–410)

23 List the two main functions of the spinal cord. (p. 410)

24 Distinguish between a reflex arc and a reflex. (p. 411)

25 Which of the choices is the correct sequence of events in a reflex arc? (p. 411)
   a. effectors to motor neurons to interneurons to CNS to sensory receptor
   b. sensory receptor to CNS to interneurons to motor neurons to effectors
   c. effectors to CNS to interneurons to motor neurons to sensory receptor
   d. sensory receptor to motor neurons to CNS to interneurons to effectors

26 Describe a withdrawal reflex. (pp. 412–413)

27 Indicate whether each nerve tract is ascending or descending: (pp. 416–417)
   a. rubrospinal   d. fasciculus gracilis
   b. corticospinal   e. reticulospinal
   c. spinothalamic   f. spinocerebellar

28 Explain the consequences of nerve fibers crossing over. (p. 416)

11.6 Peripheral Nervous System

29 Distinguish between the somatic and autonomic nervous systems. (p. 418)

30 Describe the connective tissue and nervous tissue making up a peripheral nerve. (pp. 418–419)

31 Which of the following conduct sensory impulses to the CNS from receptors in muscle or skin? (p. 420)
   a. general somatic efferent fibers
   b. general somatic afferent fibers
   c. general visceral afferent fibers
   d. general visceral efferent fibers

32 Draw the underside of a brain and label the cranial nerves. (pp. 421–424)

33 Match the cranial nerve with its function(s). Functions may be used more than once. (pp. 421–424)

   (1) olfactory nerve
   (2) optic nerve
   (3) oculomotor nerve
   (4) trochlear nerve
   (5) trigeminal nerve
   (6) abducens nerve
   (7) facial nerve
   (8) vestibulocochlear nerve
   (9) glossopharyngeal nerve
   (10) vagus nerve
   (11) accessory nerve
   (12) hypoglossal nerve

   A. conducts impulses to muscles used in swallowing
   B. conducts impulses to muscles that move the tongue
   C. conducts impulses to muscles that move the eyes
   D. conducts impulses to viscera
   E. conducts impulses to muscles of facial expression
   F. conducts impulses to muscles of neck
   G. conducts impulses associated with hearing
   H. conducts impulses to muscles that raise eyelids
   I. conducts impulses associated with sense of smell
   J. conducts impulses from upper and lower teeth
   K. conducts impulses associated with vision

34 Explain how the spinal nerves are grouped and numbered. (p. 425)

35 Define *cauda equina.* (p. 426)

36 Describe the parts of a spinal nerve and their functions. (pp. 426–427)

37 Define *plexus,* and locate the major plexuses of the spinal nerves. (pp. 427–428)

11.7 Autonomic Nervous System

38 The autonomic portion of the PNS functions _____.
(p. 431)
   a. consciously
   b. voluntarily
   c. without conscious effort
   d. dependently
39 Contrast the sympathetic and parasympathetic divisions of the autonomic nervous system. (p. 431)
40 Distinguish between a preganglionic fiber and a postganglionic fiber. (p. 431)
41 Define *paravertebral ganglion*. (p. 431)
42 Trace a sympathetic nerve pathway through a ganglion to an effector. (p. 431)
43 Trace a parasympathetic nerve pathway. (p. 432)

44 Distinguish between cholinergic and adrenergic nerve fibers. (p. 433)
45 Define *sympathetic tone*. (p. 433)
46 Explain how autonomic neurotransmitters influence the actions of effector cells. (pp. 433 and 436–437)
47 Distinguish between alpha adrenergic and beta adrenergic receptors. (pp. 436–437)

11.8 Life-Span Changes

48 Explain the effects of apoptosis on the developing brain. (p. 438)
49 List three ways that the nervous system changes as we age. (p. 438)
50 Describe sleep problems that may accompany aging. (p. 438)

 INTEGRATIVE ASSESSMENTS/CRITICAL THINKING

Outcomes 4.4, 11.4

1. In planning treatment for a patient who has had a cerebrovascular accident (CVA), why would it be important to know whether the CVA was caused by a ruptured or obstructed blood vessel?

Outcomes 7.6, 7.7, 11.2, 11.3

2. If a physician plans to obtain a sample of spinal fluid from a patient, what anatomical site can be safely used, and how should the patient be positioned to facilitate this procedure?

Outcomes 11.4, 11.5

3. What functional losses would you expect to observe in a patient who has suffered injury to the right occipital lobe of the cerebral cortex? To the right temporal lobe?

4. Brown-Seguard syndrome is due to an injury on one side of the spinal cord. It is characterized by paralysis below the injury and on the same side as the injury, and by loss of sensations of temperature and pain on the opposite side. How would you explain these symptoms?

Outcomes 11.5, 11.6

5. The biceps-jerk reflex employs motor neurons that exit from the spinal cord in the fifth spinal nerve (C5), that is, fifth from the top of the cord. The triceps-jerk reflex involves motor neurons in the seventh spinal nerve (C7). How might these reflexes be used to help locate the site of damage in a patient with a neck injury?

Outcome 11.7

6. What symptoms might the sympathetic division of the autonomic nervous system produce in a patient experiencing stress?

 ONLINE STUDY TOOLS

Connect Interactive Questions Reinforce your knowledge using assigned interactive questions covering the parts of the nervous system.

Connect Integrated Activity Could you help a neurologist determine the location of a lesion in the central or peripheral nervous system?

LearnSmart Discover which chapter concepts you have mastered and which require more attention. This adaptive learning tool is personalized, proven, and preferred.

Anatomy & Physiology Revealed Go more in depth into the human body by exploring cadaver dissections of the brain and spinal cord, as well as viewing animations on physiological processes.

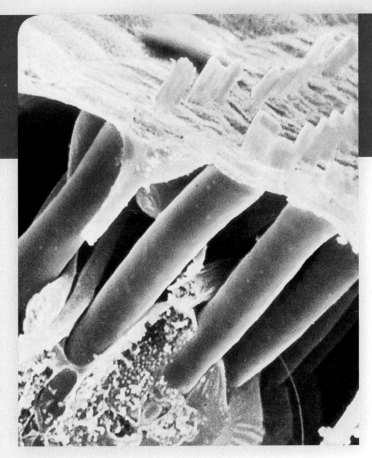

The spiral organ, in the inner ear, has rows of hair cells. Each row bears up to 100 hairs, which translate sound into neural messages that travel to the brain. In this falsely colored micrograph, the cells appear red and the hairs yellow (2,700×).

THE WHOLE PICTURE

The senses provide information about what is happening in the world outside and inside the body, whether it is the feel of clothing against your skin as you are getting dressed, an unusual smell, or the awareness that your heart is pounding in your chest when you are excited.

Despite the fact that all action potentials are the same, we are able to distinguish among a wide variety of stimuli. What we hear, smell, taste, and see, and how those senses combine, reflect both the parts of the brain that process the incoming information, and the rich collection of receptors tuned in to the world. At the same time, the nervous system also "tunes out" unimportant information. If your brain seems overloaded now, imagine how it would be if you were aware of all of the sounds, smells, and other sensory stimuli present in your environment all of the time!

aprevealed.com

Module 7: Nervous System

Nervous System III
Senses

 LEARNING OUTCOMES

After you have studied this chapter, you should be able to:

12.1 Introduction to Sensory Function
1 Differentiate between general senses and special senses. (p. 444)

12.2 Receptors, Sensation, and Perception
2 Name the five types of receptors and state the function of each. (p. 444)
3 Explain how receptors trigger sensory impulses. (p. 444)
4 Explain sensation and sensory adaptation. (p. 445)

12.3 General Senses
5 Describe the differences among receptors associated with the senses of touch, pressure, temperature, and pain. (pp. 446–447)
6 Describe how the sensation of pain is produced. (p. 449)
7 Explain the importance of stretch receptors in muscles and tendons. (pp. 449–451)

12.4 Special Senses
8 Explain the relationship between the senses of smell and taste. (p. 452)
9 Describe how the sensations of smell and taste are produced and interpreted. (pp. 452–456)
10 Name the parts of the ear and explain the function of each part. (pp. 456–461)
11 Distinguish between static and dynamic equilibrium. (pp. 464 and 466)
12 Describe the roles of the accessory organs to the eye. (pp. 468–470)
13 Name the parts of the eye and explain the function of each part. (pp. 470–475)
14 Explain how the eye refracts light. (pp. 475 and 478)
15 Distinguish between rods and cones, and discuss their respective visual pigments. (pp. 478–481)
16 Explain how the brain perceives depth and distance. (p. 481)
17 Describe the visual nerve pathways. (p. 481)

12.5 Life-Span Changes
18 Describe aging-associated changes that diminish the senses. (p. 482)

UNDERSTANDING WORDS

aud-, to hear: *aud*itory—pertaining to hearing.
choroid, skinlike: *choroid* coat—middle, vascular layer of the eye.
cochlea, snail: *cochlea*—coiled tube in the inner ear.
corn-, horn: *cornea*—transparent outer layer in the anterior portion of the eye.
iris, rainbow: *iris*—colored, muscular part of the eye.
labyrinth, maze: *labyrinth*—complex system of connecting chambers and tubes of the inner ear.

lacri-, tears: *lacri*mal gland—tear gland.
lut-, yellow: macula *lut*ea—yellowish spot on the retina.
macula, spot: *macula* lutea—yellowish spot on the retina.
malle-, hammer: *malle*us—one of the three bones in the middle ear.
ocul-, eye: orbicularis *ocul*i—muscle associated with the eyelid.
olfact-, to smell: *olfact*ory—pertaining to the sense of smell.

palpebra, eyelid: levator *palpebra*e superioris—muscle associated with the eyelid.
photo-, light: *photo*receptors—specialized structures in the eye responsive to light.
scler-, hard: *scler*a—tough, outer protective layer of the eye.
therm-, heat: *therm*oreceptor—receptor sensitive to changes in temperature.
tympan-, drum: *tympan*ic membrane—eardrum.
vitre-, glass: *vitre*ous humor—clear, jellylike substance in the eye.

12.1 | Introduction to Sensory Function

Our senses not only make our lives meaningful, connecting us to the sights, sounds, smells, tastes, and textures of the outside world, but also help our bodies maintain homeostasis by providing information about what is happening on the inside. Sensory receptors are the portals that link our nervous systems to all of these events. The **general senses** are those with receptors widely distributed throughout the body, including the skin, various organs, and joints. The **special senses** have more specialized receptors and are confined to structures in the head, such as the eyes and ears.

All senses work in basically the same way. Sensory receptors are specialized cells (receptor cells) or multicellular structures that collect information from the environment. In some senses, such as smell and taste, the receptor cells have specific molecules on the cell membrane (membrane receptors). Stimulated receptor cells in turn stimulate neurons to conduct impulses along sensory fibers to the brain. There the cerebral cortex interprets the information.

Chapter 11 (p. 419) uses the terms *axon* and *nerve fiber* synonymously. Also recall from chapter 10 (p. 366) that unipolar neurons, which include most sensory neurons, are unusual in that the portion of the neuron associated with the dendrites, called a peripheral process, functions like an axon (see fig. 10.6, p. 366). Because of this, and for simplicity, we call the neuron processes that bring sensory information into the CNS "sensory fibers" or "afferent fibers," no matter the type of neuron.

12.2 | Receptors, Sensation, and Perception

The body's ability to interpret different sensory events depends partly on receptors that respond to specific stimuli. Sensory receptors are diverse but share certain features. Each type of receptor is particularly sensitive to a distinct type of environmental change and is much less sensitive to other forms of stimulation. The raw form in which these receptors send information to the brain is called **sensation.** The way the brain interprets this information is called **perception.**

Receptor Types

Five types of sensory receptors are recognized, based on their sensitivities to specific stimuli:

1. **Chemoreceptors** (ke″mo-re-sep′torz) respond to changes in the concentration of chemicals. Receptors associated with the senses of smell and taste are of this type. Chemoreceptors in internal organs detect changes in the blood concentrations of oxygen, hydrogen ions, glucose, and other chemicals.
2. **Pain receptors,** also called nociceptors (no″se-sep′torz), respond to tissue damage. Triggering stimuli include exposure to excess mechanical, electrical, thermal, or chemical energy.
3. **Thermoreceptors** (ther″mo-re-sep′torz) sense temperature change.
4. **Mechanoreceptors** (mek″ah-no re-sep′torz) are of several types and sense mechanical forces by detecting changes that deform the receptors. They include a number of receptors in the skin that respond to physical contact, and several receptors in the ear that provide information about balance and vibrations from sound. **Proprioceptors** (pro″pre-o-sep′torz) sense changes in the tensions of muscles and tendons; **baroreceptors** (bar″o-re-sep′torz), also called pressoreceptors, in certain blood vessels detect changes in blood pressure; and **stretch receptors** in the lungs sense degree of inflation.
5. **Photoreceptors** (fo″to-re-sep′torz) in the eyes respond to light energy of sufficient intensity.

Sensory Impulses

Sensory receptors can be ends of neurons or other types of cells that are near neuron extensions. In either case, stimulation locally changes the cells' membrane potentials (receptor potentials), generating a graded electric current that reflects the intensity of stimulation (see chapter 10, p. 377).

If a receptor is a neuron and the change in membrane potential reaches threshold, an action potential is generated and is propagated along the afferent fiber. However, if the receptor is another type of cell, its receptor potential must be transferred to a neuron to trigger an action potential. Peripheral nerves conduct the resulting impulses to the central nervous system (CNS), where they are analyzed and interpreted in the brain.

Sensation and Perception

A sensation occurs when sensory neurons reach threshold and the resulting action potentials cause the brain to become aware of that sensory event. A perception occurs when the brain interprets those sensory impulses. Thus, pain is a sensation, but realizing that you have just stepped on a tack is a perception. Table 12.1 outlines the pathways from sensing to perceiving an apple using the special senses: smell, taste, sight, and hearing.

At the same time that a sensation forms, the cerebral cortex interprets it as coming from the receptors being stimulated. This process, which is closely related to perception, is called **projection,** because the brain projects the sensation back to its apparent source. Projection allows a person to pinpoint the region of stimulation. In this way, we perceive that the eyes see an apple, the nose smells it, and the ears hear the teeth crunch into it.

Because all of the impulses conducted on sensory fibers into the CNS are alike, the resulting sensation depends on which region of the cerebral cortex receives the impulse. For example, impulses reaching one region are always interpreted as sounds, and those reaching another are always sensed as touch. (Some receptors, such as those that measure oxygen levels in the blood, do not trigger sensations.)

> Imagine two cell phones in front of you, one on the left and one on the right. The ring tones are identical and equally loud. In fact, there is no way to tell them apart other than by their location. Yet from experience you know that the one on the left always handles calls from your best friend, and the one on the right is always someone else. Similarly, impulses that transmit sensations are just as alike as the two cell phone ring tones, but they are interpreted differently because they go to different parts of the brain.

RECONNECT

To Chapter 11, Functions of the Cerebrum, pages 399–401.

Sensory receptors are specialized to respond to certain types of stimuli, but they may respond to other stimuli that are strong enough. In either case the sensations are the same. Pain receptors, for example, can be stimulated by heat, cold, or pressure, but the sensation is always the same because, in each case, the same part of the brain interprets the resulting impulses as pain. Similarly, a stimulus other than light, such as a sharp blow to the head, may trigger impulses in visual receptors. When this happens, the person may "see stars," even though no light is entering the eye, because any impulses reaching the visual cortex are interpreted as light. Receptors respond only to specific stimuli, and the brain typically creates the correct sensation for that particular stimulus.

Sensory Adaptation

The brain must prioritize the sensory input it receives, or incoming unimportant information would be overwhelming. For example, until this sentence prompts you to think about it, you are probably unaware of the pressure of your clothing against your skin, or the background noise in the room. This ability to ignore unimportant stimuli is called **sensory adaptation** (sen'so-re ad"ap-ta'shun). It may reflect a decreased response to a particular stimulus from the receptors (peripheral adaptation) or along the CNS pathways leading to the sensory regions of the cerebral cortex (central adaptation). Once adaptation to a particular stimulus occurs, a sensation will occur only if the strength of the stimulus changes.

PRACTICE

1　Distinguish between general and special senses.
2　List the five general types of sensory receptors.
3　What do all types of receptors have in common?
4　Explain how a sensation is different from a perception.
5　What is sensory adaptation?

CAREER CORNER
Audiologist

The young man awoke in a military hospital just as a medical professional approached. She pointed to her pin, which said "audiology" above her name, and started to speak, but he couldn't hear a word she was saying. When the man pointed to his ears, she pulled out a pad and pen and wrote "blast injury—IED" and showed it to him. The man realized he'd been exposed to an improvised explosive device, and the pressure of the bomb had damaged his ears.

The audiologist gave her new patient printed information on blast injuries that explained how she would assess the severity of his hearing loss: visual inspection for eardrum injury, and computed tomographic scans to examine bones of the middle ears and the cochlea in the inner ears. Hearing would return if the damage was confined to the outer or middle ears, but a cochlear implant might be necessary if the inner ear was damaged.

Audiology grew out of the need for specialists to treat hearing loss among soldiers returning from World War II. Today audiologists examine, diagnose, and treat individuals who have problems with hearing or balance. They fit hearing aids, remove wax, and provide protective devices for people who are exposed to loud sounds, such as construction workers and musicians. Patients range from young children with hearing loss from repeated ear infections, to the very elderly.

An audiologist uses a device called an audiometer to test how loud a sound must be for a patient to hear it, and whether the person can distinguish sounds. For patients with dizziness, an audiologist can determine whether the vestibular system in the ear is implicated.

Audiologists may work in private practice or in hospitals, clinics, schools, the military, long-term care facilities or forensics laboratories. A doctoral degree in audiology (Au.D.), typically earned after college, takes three years plus a year of externship.

TABLE 12.1 Information Flow from the Environment Through the Nervous System

Information Flow	Smell	Taste	Sight	Hearing
Sensory receptors	Olfactory receptor cells	Taste bud receptor cells	Rods and cones in retina	Hair cells in cochlea
↓	↓	↓	↓	↓
Impulse in sensory fibers	Olfactory nerve fibers	Sensory fibers in various cranial nerves	Optic nerve fibers	Auditory nerve fibers
↓	↓	↓	↓	↓
Impulse reaches CNS	Cerebral cortex	Cerebral cortex	Midbrain and cerebral cortex	Midbrain and cerebral cortex
↓	↓	↓	↓	↓
Sensation (new experience, recalled memory)	A pleasant smell	A sweet taste	A small, round, red object	A crunching sound
↓	↓	↓	↓	↓
Perception	The smell of an apple	The taste of an apple	The sight of an apple	The sound of biting into an apple

12.3 | General Senses

General senses are those whose sensory receptors are widespread, associated with the skin, muscles, joints, and viscera. These senses can be divided into three groups:

1. **Exteroreceptive senses** are associated with changes at the body surface. They include the senses of touch, pressure, temperature, and pain.
2. **Interoceptive** (visceroceptive) **senses** are associated with changes in viscera (blood pressure stretching blood vessels, an ingested meal stimulating pH receptors in the small intestine, and so on).
3. **Proprioceptive senses** are associated with changes in muscles and tendons and in body position.

Touch and Pressure Senses

The senses of touch and pressure derive from three types of receptors (fig. 12.1). As a group, these receptors sense mechanical forces that deform or displace tissues. The touch and pressure receptors include the following:

1. **Free nerve endings,** the simplest receptors, are common in epithelial tissues, where they lie between epithelial cells. They are responsible for the sensation of itching (fig. 12.1a).
2. **Tactile (Meissner's) corpuscles** are small, oval masses of flattened connective tissue cells in connective tissue sheaths. Two or more sensory fibers branch into each corpuscle and end within it as tiny knobs.

   Tactile corpuscles are abundant in hairless areas of skin, such as the lips, fingertips, palms, soles, nipples, and external genital organs. They provide fine touch, such as distinguishing two points on the skin where an object touches, to judge its texture (fig. 12.1b).
3. **Lamellated (Pacinian) corpuscles** are relatively large, ellipsoidal structures composed of connective tissue fibers and cells. They are common in the deeper dermal tissues of the hands, feet, penis, clitoris, urethra, and breasts and also in

the connective tissue capsules of synovial joints (fig. 12.1c). Heavier pressure and stretch stimulate lamellated corpuscles. They also detect vibrations in tissues.

Temperature Senses

Temperature receptors (thermoreceptors) include two groups of free nerve endings in the skin. Those that respond to warmer temperatures are *warm receptors,* and those that respond to colder temperatures are *cold receptors.*

The warm receptors are most sensitive to temperatures above 25°C (77°F) and become unresponsive at temperatures above 45°C (113°F). Temperatures near and above 45°C also trigger pain receptors, producing a burning sensation.

Cold receptors are most sensitive to temperatures between 10°C (50°F) and 20°C (68°F). Temperature dropping below 10°C also stimulates pain receptors, producing a freezing sensation.

At intermediate temperatures, the brain interprets sensory input from different combinations of warm and cold receptors as an intermediate temperature sensation. Both types of receptors rapidly adapt, so within about a minute of continuous stimulation, the sensation of warm or cold begins to fade. This is why we quickly become comfortable after jumping into a cold swimming pool or submerging into a steaming hot tub.

Sense of Pain

Pain receptors (nociceptors) consist of free nerve endings. These receptors are widely distributed throughout the skin and internal tissues, except in the nervous tissue of the brain, which lacks pain receptors. Pain receptors protect in that they are stimulated when tissues are damaged and may prevent further damage. Most pain sensations are perceived as unpleasant, signaling the individual to remove the source of the stimulation.

More than one type of change stimulates most pain receptors. However, some pain receptors are most sensitive to mechanical damage, while others are particularly sensitive to temperature extremes. Some pain receptors are most responsive to chemicals,

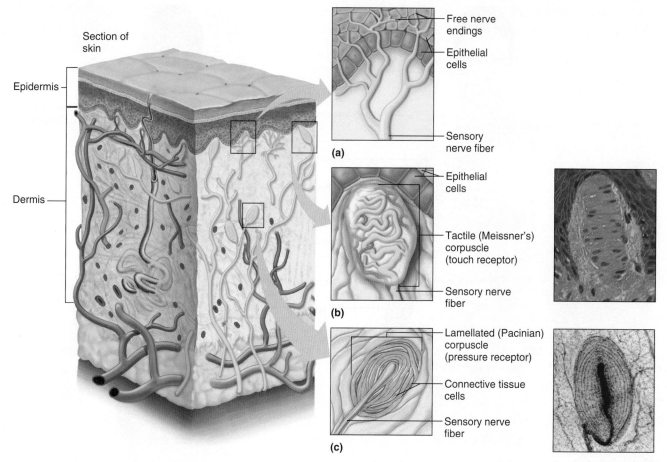

FIGURE 12.1 Touch and pressure receptors include (a) free ends of sensory nerve fibers, (b) a tactile corpuscle (with 225× micrograph), and (c) a lamellated corpuscle (with 50× micrograph).

such as hydrogen ions, potassium ions, or specific breakdown products of proteins, histamine, and acetylcholine. A deficiency of blood flow (ischemia) and the resulting deficiency of oxygen (hypoxia) in a tissue also trigger pain sensation. For example, the sharp pain of a muscle cramp results from interruption of blood flow as the sustained contraction squeezes capillaries, as well as from the stimulation of mechanoreceptors. Additionally, when blood flow is interrupted, pain-stimulating chemicals accumulate. Increasing blood flow through the sore tissue may relieve the resulting pain, and this is why heat is sometimes applied to reduce muscle soreness. The heat dilates blood vessels and thus promotes blood flow, which helps reduce the concentration of the pain-stimulating substances. In some conditions, accumulating chemicals lower the thresholds of pain receptors, making inflamed tissues more sensitive to heat or pressure.

Pain receptors adapt very little, if at all. Once a pain receptor is activated, even by a single stimulus, it may continue to send impulses into the CNS for some time. Clinical Application 3.1 (p. 89) describes the genetic basis of how people vary in their sensitivity to pain.

Visceral Pain

As a rule, pain receptors are the only receptors in viscera whose stimulation produces sensations. Pain receptors in these organs respond differently to stimulation than those associated with surface tissues. For example, localized damage to intestinal tissue during surgical procedures may not elicit any pain sensations, even in a conscious person. However, when visceral tissues are subjected to more widespread stimulation, such as when intestinal tissues are stretched or when the smooth muscle in the intestinal walls undergoes spasms, a strong pain sensation may follow. Once again, the resulting pain comes from stimulation of mechanoreceptors and from decreased blood flow accompanied by lower tissue oxygen levels and accumulation of pain-causing chemicals stimulating chemoreceptors.

Visceral pain may feel as if it is coming from some part of the body other than the part being stimulated, in a phenomenon called **referred pain.** For example, pain originating in the heart may be referred to the left shoulder or the medial surface of the left upper limb. Pain from the lower esophagus, stomach, or small intestine may seem to be coming from the upper central (epigastric) region of the abdomen. Pain from the urogenital tract may be referred to the lower central (hypogastric) region of the abdomen or to the sides between the ribs and the hip (fig. 12.2).

Referred pain may derive from *common nerve pathways* that sensory impulses coming both from skin areas and from internal organs use. Pain impulses from the heart seem to be conducted over the same nerve pathways as those from the skin of the left shoulder and the inside of the left upper limb, as shown in **figure 12.3.**

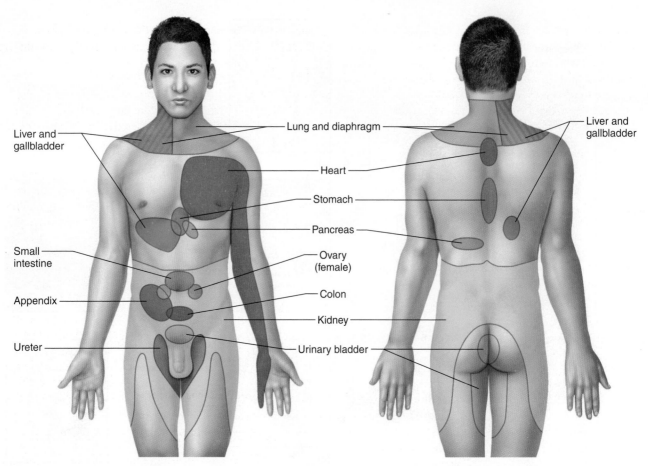

FIGURE 12.2 Surface regions to which visceral pain may be referred.

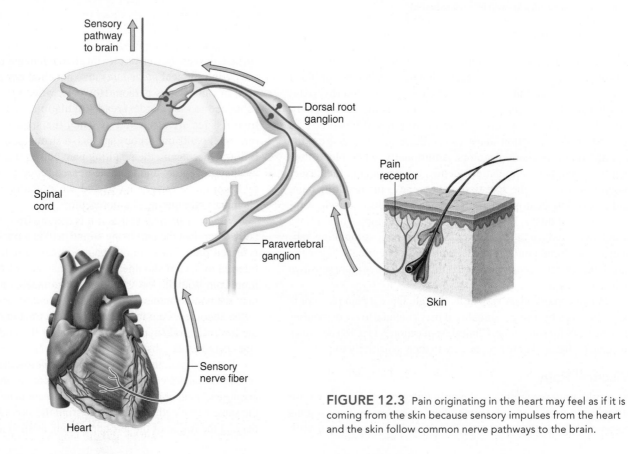

FIGURE 12.3 Pain originating in the heart may feel as if it is coming from the skin because sensory impulses from the heart and the skin follow common nerve pathways to the brain.

During a heart attack, the cerebral cortex may incorrectly interpret the source of the impulses as the shoulder and the medial surface of the left upper limb, rather than the heart.

Pain originating in the parietal layers of the thoracic and abdominal membranes—parietal pleura, parietal pericardium, or parietal peritoneum—is usually not referred. Instead, such pain is felt directly over the area being stimulated.

> Neuropathic pain is an overreaction to a stimulus that would ordinarily cause pain, or a pain response to a normally innocuous stimulus. Reflex sympathetic dystrophy is a form of neuropathic pain that causes an intense burning sensation in a hand or foot, even if the extremity is paralyzed or has been amputated. During the Civil War, it was called "causalgia." Union Army Surgeon S. Weir Mitchell described causalgia as "the most terrible of all tortures."

Pain Pathways

The axons (fibers) that conduct impulses away from pain receptors are of two main types: fast pain fibers and slow pain fibers.

The *fast pain fibers* (also known as A-delta fibers) are myelinated. They conduct impulses rapidly, at velocities up to 30 meters per second. These impulses are associated with the immediate sensation of sharp pain, which typically seems to originate in a local area of skin. This type of pain seldom continues after the pain-producing stimulus stops.

The *slow pain fibers* (C fibers) are unmyelinated. They conduct impulses more slowly than fast pain fibers, at velocities up to 2 meters per second. These impulses cause a delayed, dull, aching pain sensation that may be widespread and difficult to pinpoint. Such pain may continue after the original stimulus ceases. Although immediate pain is usually sensed as coming from the surface, delayed pain is felt in deeper tissues as well as in the skin. Visceral pain impulses are typically carried on C fibers.

Usually, an event that stimulates pain receptors triggers impulses on both fast and slow pain fibers. The result is a dual sensation—a sharp, pricking pain, then a dull, aching pain that is more intense and may worsen over time. Chronic pain that resists relief and control can be debilitating.

Pain impulses that originate from tissues of the head reach the brain on sensory fibers of the fifth, seventh, ninth, and tenth cranial nerves (see chapter 11, pp. 421–423). All other pain impulses are conducted on sensory fibers of spinal nerves, and they pass into the spinal cord by way of the dorsal roots of these spinal nerves.

Upon reaching the spinal cord, these sensory fibers enter the gray matter of the posterior horn. There they synapse with neurons whose axons cross over to the opposite side of the spinal cord at that level and then ascend in the anterior and lateral spinothalamic tracts (anterolateral system). These pathways also involve one or more interneurons before reaching the long fibers that cross over and ascend to the brain.

In the brain, most of the pain fibers synapse in the reticular formation (see chapter 11, p. 407), and from there impulses are conducted on fibers of still other neurons to the thalamus, hypothalamus, and cerebral cortex. Fibers of the spinothalamic tracts bring pain and temperature information directly to the thalamus.

Regulation of Pain Pathways

Sensation of pain begins when pain impulses reach the level of the thalamus—that is, even before they reach the cerebral cortex. However, the cerebral cortex must judge the intensity of pain and locate its source, and it is also responsible for emotional and motor responses to pain.

Still other parts of the brain, including areas of gray matter in the midbrain, pons, and medulla oblongata, regulate the flow of pain impulses from the spinal cord (see chapter 11, pp. 405–407). Impulses from special neurons in these areas descend in the lateral funiculus to various levels within the spinal cord. The impulses stimulate the ends of certain nerve fibers to release biochemicals that can block pain signals by inhibiting presynaptic nerve fibers in the posterior horn of the spinal cord.

Among the inhibiting substances released in the posterior horn are neuropeptides called *enkephalins* and the amine *serotonin* (see chapter 10, pp. 380–382). Enkephalins can suppress both acute and chronic pain impulses. Therefore, they can relieve strong pain sensations, much as morphine and other opiate drugs do. In fact, enkephalins were discovered because they bind to the same receptors on neuron membranes as does morphine. Serotonin stimulates other neurons to release enkephalins.

> Cannabinoids are chemicals in the plant *Cannibus sativa*, the source of marijuana, that may relieve pain. Neurons in areas of the brain, brainstem, and peripheral nervous system have receptors for cannabinoids that bind a type of neurotransmitter made in the body called anandamide. A synthetic version of the compound in marijuana responsible for most of marijuana's effects (delta-9-tetrahydrocannabinol), as well as the plant itself, are used to alleviate nausea and vomiting in people receiving cancer chemotherapy and to boost appetite in people who have AIDS.

The *endorphins* are another group of neuropeptides with pain-suppressing, morphinelike actions. They are found in the pituitary gland and in regions of the nervous system, such as the hypothalamus, that relay pain information. Enkephalins and endorphins are released in response to extreme pain, providing natural pain control. Clinical Application 12.1 discusses treatments for severe pain.

 PRACTICE

6   Describe three types of touch and pressure receptors.

7   Describe thermoreceptors.

8   What types of stimuli excite pain receptors?

9   What is referred pain?

10 Explain how neuropeptides control pain.

Proprioception

Proprioceptors are mechanoreceptors that send information to the CNS about body position and the length and tension of skeletal

About 25% of all people have moderate to severe pain, and a quarter of them may be undertreated. Pain remedies include nonsteroidal anti-inflammatory drugs (NSAIDs) such as aspirin (fig. 12A), ibuprofen, and COX-2 inhibitors; opiates (fig. 12B); and drugs such as acetaminophen, which is the top-selling painkiller. Acetaminophen does not have the side effects of gastrointestinal irritation of the NSAIDs nor is it addicting like the opiates, but in large doses it can damage the liver.

Several painkillers are being reformulated as particles that are up to 20 times smaller than older preparations of the drugs. The new versions of old standards, such as indomethacin, dissolve faster and relieve pain faster, so that patients can take lower doses for shorter periods, making the drugs safer.

In 2004 a new painkiller became available. Ziconotide is a synthetic version of a peptide that the marine cone snail *Conus magus* releases to paralyze its fish prey (fig. 12C). When researchers noticed that the natural peptide binds to a type of calcium channel protein on spinal cord neurons that receive pain impulses, the effort began to turn the snail's weapon into a pain reliever. Ziconotide is delivered by catheter into the cerebrospinal fluid and is prescribed to relieve intractable chronic pain.

The types of patients in greatest need of pain relief are people with cancer or chronic pain syndromes. Cancer patients take NSAIDs, weak narcotics such as hydrocodone, strong narcotics such as morphine, and opiates delivered directly to the spinal cord via an implanted reservoir. Narcotics are much more likely to be addicting when they are abused to induce euphoria than when they are taken to relieve severe pain. Patients may use devices to control the delivery of pain medications. Anti-anxiety medications can ease the perception of pain.

Some types of chronic pain are lower back pain, migraine, and myofascial syndrome (inflammation of muscles and their fascia). Treatment approaches include NSAIDs, stretching exercises, injection of local anesthetic drugs into cramping muscles, and antidepressants to raise serotonin levels in the CNS. Chronic pain may also be treated with electrodes implanted near the spinal cord; transcutaneous electrical nerve stimulation (TENS), which also places electrodes on pain-conducting nerves; and an invasive nerve block, which interrupts a pain signal by freezing or by introducing an anesthetic drug.

FIGURE 12A Painkillers come from nature. Aspirin derives from bark of the willow tree.

FIGURE 12B Poppies are the source of opiate drugs.

FIGURE 12C A newer analgesic for extreme chronic pain is based on a peptide from the marine cone snail *Conus magus*.

---

muscles. Recall that lamellated corpuscles function as pressure receptors in joints. The other main proprioceptors are stretch receptors: muscle spindles and Golgi tendon organs.

Muscle spindles are located throughout skeletal muscles. Each spindle consists of several small, modified skeletal muscle fibers (intrafusal fibers) enclosed in a connective tissue sheath. Each intrafusal fiber has near its center a specialized nonstriated region with the end of a sensory nerve fiber wrapped around it (fig. 12.4a).

The striated portions of the intrafusal fiber contract to keep the spindle taut at different muscle lengths. Thus, if the whole muscle is stretched, the muscle spindle is also stretched, triggering sensory impulses on its nerve fiber. These sensory fibers synapse in the spinal cord with lower motor neurons leading back to the same muscle. In this way, stretch of the muscle spindle triggers impulses that contract the skeletal muscle of which it is a part. This action, called a **stretch reflex,** opposes the lengthening of the muscle and helps maintain the desired position of a limb despite gravitational or other forces tending to move it (see chapter 11, pp. 411–412).

A leg cramp in the middle of the night may be so painful that it makes you leap out of bed. The contraction of a cramp may result from local changes in electrolyte levels in a part of a muscle that inappropriately trigger impulses from the muscle spindle.

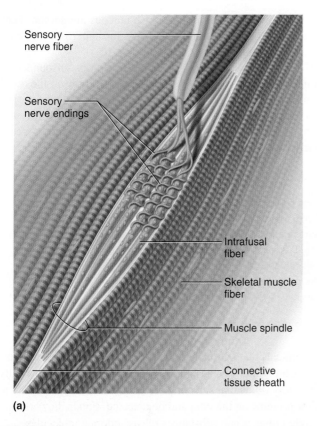

(a)

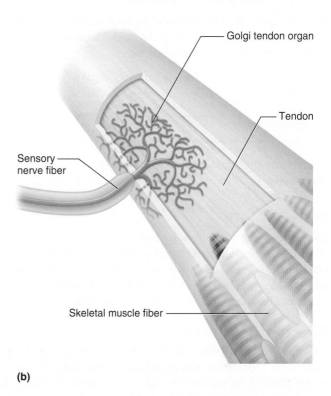

(b)

FIGURE 12.4 Stretch receptors maintain posture. (a) Increased muscle length stimulates muscle spindles, which stimulate muscle contraction. (b) Golgi tendon organs occupy tendons, where they inhibit muscle contraction.

Golgi tendon organs are in tendons close to their attachments to muscles. Each Golgi tendon organ connects to a set of skeletal muscle fibers and is innervated by a sensory neuron (fig. 12.4b). Golgi tendon organs have high thresholds, and increased tension stimulates them. Sensory impulses from them produce a reflex that inhibits contraction of the muscle whose tendon they occupy. Therefore, the Golgi tendon organs stimulate a reflex with an effect that is the opposite of a stretch reflex. The Golgi tendon reflex also helps maintain posture, and it protects muscle attachments from being pulled away from their insertions by excessive tension. Table 12.2 summarizes some of the receptors of the general senses and their functions.

Visceral Senses

Receptors in internal organs include lamellated corpuscles and free nerve endings. The information these receptors convey includes the sense of fullness after eating a meal, the discomfort of intestinal gas, and the pain that signals a heart attack.

 PRACTICE

11 Describe a muscle spindle.

12 Explain how muscle spindles help maintain posture.

13 Where are Golgi tendon organs located?

14 What is the function of Golgi tendon organs?

TABLE 12.2	Receptors Associated with General Senses	
Type	**Function**	**Sensation**
Free nerve endings (mechanoreceptors)	Detect changes in pressure	Touch, pressure
Tactile corpuscles (mechanoreceptors)	Detect objects moving over the skin	Touch, texture
Lamellated corpuscles (mechanoreceptors)	Detect changes in pressure	Deep pressure, vibrations, fullness in viscera
Free nerve endings (thermoreceptors)	Detect changes in temperature	Heat, cold
Free nerve endings (pain receptors)	Detect tissue damage	Pain
Free nerve endings (mechanoreceptors)	Detect stretching of tissues, tissue spasms	Visceral pain
Muscle spindles (mechanoreceptors)	Detect changes in muscle length	None
Golgi tendon organs (mechanoreceptors)	Detect changes in muscle tension	None

12.4 | Special Senses

Special senses are those whose sensory receptors are part of large, complex sensory organs in the head. These senses and their respective organs include the following:

- smell ⟶ olfactory organs
- taste ⟶ taste buds
- hearing ⟶ ears
- equilibrium ⟶
- sight ⟶ eyes

Clinical Application 12.2 discusses an unusual type of sensory abnormality.

Sense of Smell

The ability to detect the strong scent of a fish market, the antiseptic odor of a hospital, the aroma of a ripe melon—and thousands of other smells—is possible thanks to a yellowish patch of tissue the size of a quarter high up in the nasal cavity. This fabric of sensation is a layer of as many as 6 million specialized cells.

Olfactory Receptors

Olfactory receptor cells, and their membrane receptor molecules, sense odors. The receptor cells are similar to those for taste in that they are chemoreceptors sensitive to chemicals dissolved in liquids. The two chemical senses function closely together and aid in food selection, because we smell food at the same time we taste it. It is often difficult to tell what part of a food sensation is due to smell and what part is due to taste. For this reason, an onion tastes different when sampled with the nostrils closed, because much of the usual onion sensation is due to odor. Similarly, if copious mucous secretions from an upper respiratory infection cover the olfactory receptors, food may seem tasteless. About 75% to 80% of flavor derives from the sense of smell.

Olfactory Organs

The olfactory organs, which contain the olfactory receptor cells, also include epithelial supporting cells. These organs appear as yellowish brown masses within pinkish mucous membrane. They cover the upper parts of the nasal cavity, the superior nasal conchae, and a portion of the nasal septum (fig. 12.5).

The olfactory receptor cells are bipolar neurons surrounded by columnar epithelial cells. These neurons have knobs at the distal ends of their dendrites covered with hairlike cilia. The cilia project into the nasal cavity and are the sensitive portions of the receptor cells (fig. 12.6). Each of a person's 12 million olfactory receptor cells has ten to twenty cilia.

> Dogs can sense subtle odors that people emit when becoming ill with certain conditions. Service dogs are used to sense imminent seizures, drops in blood glucose, and accelerated heart rate. Dogs can detect a chemical in urine that prostate cancer cells release. In one study a dog identified a man whose prostate biopsy was normal. The biopsy was in error—not the dog. Cats have excellent olfaction, too.

Chemicals that stimulate olfactory receptor cells, called odorant molecules, enter the nasal cavity and dissolve at least partially in the watery fluids that surround the cilia before they can bond to receptor proteins on the cilia and be detected. Foods, flowers, and many other objects and substances release odorant molecules.

An odorant molecule may bind to several of the almost 400 types of olfactory membrane receptors that are part of the olfactory receptor cells, depolarizing these cells and thereby generating action potentials if the depolarization reaches threshold. Note the distinction between sensory receptors and membrane receptors. Sensory receptors may be as small as individual cells or as large as complex organs such as the eye or ear. Sensory receptors respond to sensory stimuli. In contrast, membrane receptors are molecules such as proteins and glycoproteins on the cell membranes. They enable cells, such as olfactory receptor cells, to respond to specific molecules.

FIGURE 12.5 AP|R Olfactory receptors. (*a*) Columnar epithelial cells support olfactory receptor cells, which have cilia at their distal ends. (*b*) The olfactory area is associated with the superior nasal concha.

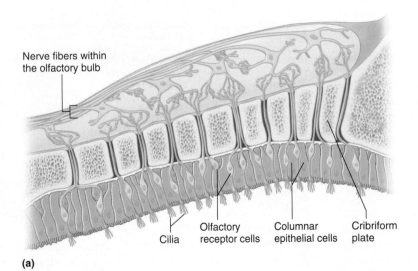

(a)

Nerve fibers within the olfactory bulb

Cilia — Olfactory receptor cells — Columnar epithelial cells — Cribriform plate

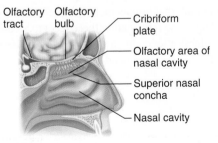

(b)

Olfactory tract — Olfactory bulb — Cribriform plate — Olfactory area of nasal cavity — Superior nasal concha — Nasal cavity

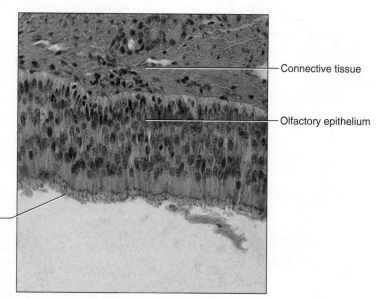

FIGURE 12.6 **AP|R** Light micrograph of the olfactory epithelium (250×).

Connective tissue

Olfactory epithelium

Olfactory receptor cell cilia

Olfactory Pathways

Once olfactory receptor cells are stimulated, impulses are conducted along their axons through tiny openings in the cribriform plates of the ethmoid bone. These fibers (which collectively form the first cranial nerves) synapse with neurons located in the enlargements of the **olfactory bulbs,** which are structures that lie on either side of the crista galli of the ethmoid bone (see fig. 7.24, p. 220, and fig. 12.5).

In the olfactory bulbs, the sensory impulses are analyzed, and as a result, additional impulses travel along the **olfactory tracts** to portions of the limbic system (see chapter 11, pp. 403 and 405), a brain center for memory and emotions. This is why we may become nostalgic over a scent from the past. A whiff of the perfume that grandma used to wear may bring back a flood of memories. The input to the limbic system also explains why odors can alter mood so easily. For example, the scent of new-mown hay or rain on a summer's morning generally makes us feel good. The main interpreting areas for the olfactory impulses (olfactory cortex) are deep within the temporal lobes and at the bases of the frontal lobes, anterior to the hypothalamus.

Olfactory Stimulation

Biologists are not certain how stimulated receptors encode specific smells, but a leading hypothesis is that each olfactory receptor cell has many copies of only one type of olfactory receptor membrane protein, but that a receptor protein can bind several related types of odorant molecules. In addition, any one odorant molecule can bind several types of olfactory receptors. The brain interprets this binding information as an olfactory code. In a simplified example, banana might stimulate receptors 2, 4, and 7; garlic, receptors 1, 5, and 9.

The olfactory organs are high in the nasal cavity above the usual pathway of inhaled air, so in order to smell a faint odor, a person may have to sniff and force air up to the receptor areas. Olfaction undergoes sensory adaptation rather rapidly, so the intensity of an odor drops about 50% within a second following the stimulation. Within a minute, the receptors may become almost insensitive to a given odor. This is why the odor of a fish market becomes tolerable quickly. However, olfactory receptors that have adapted to one scent remain sensitive to others.

The olfactory receptor neurons are the only nerve cells in direct contact with the outside environment, and as such are prone to damage. Fortunately, basal cells along the basement membrane of the olfactory epithelium regularly divide and yield cells that differentiate to replace lost olfactory receptor neurons. These neurons are unusual in that they are regularly replaced when damaged.

 PRACTICE

15 Where are the olfactory receptors located?

16 Trace the pathway of an olfactory impulse from a receptor to the cerebrum.

Sense of Taste

Taste buds are the special organs of taste. They resemble orange sections and associate on the surface of the tongue with tiny elevations called **papillae** (figs. 12.7 and 12.8). Taste buds are also scattered in the roof of the mouth, the linings of the cheeks, and the walls of the pharynx.

Taste Receptors

Each taste bud includes a group of modified epithelial cells, the **taste cells** (gustatory cells), that function as sensory receptors.

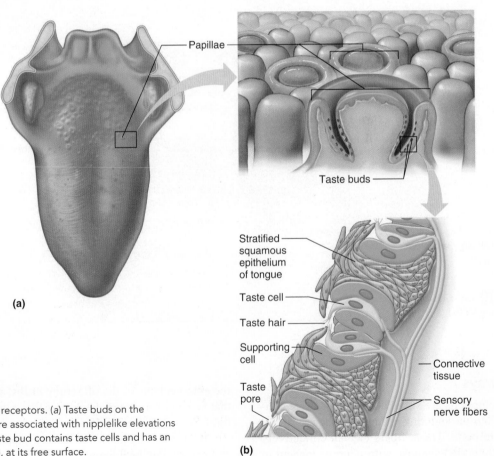

(a)

Papillae

Taste buds

Stratified squamous epithelium of tongue

Taste cell

Taste hair

Supporting cell

Taste pore

Connective tissue

Sensory nerve fibers

(b)

FIGURE 12.7 Taste receptors. (a) Taste buds on the surface of the tongue are associated with nipplelike elevations called papillae. (b) A taste bud contains taste cells and has an opening, the taste pore, at its free surface.

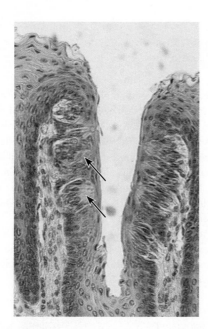

FIGURE 12.8 AP|R A light micrograph of some taste buds (arrows) (225×).

Each of our 10,000 taste buds houses 50 to 150 taste cells. The taste bud also includes epithelial supporting cells. The entire structure of a taste bud is somewhat spherical, with an opening, the **taste pore,** on its free surface. Tiny projections (microvilli), called **taste hairs,** protrude from the outer ends of the taste cells and jut through the taste pore. These taste hairs are the sensitive parts of the receptor cells.

Interwoven among and wrapped around the taste cells is a network of sensory fibers whose ends closely contact the receptor cell membranes. A stimulated receptor cell triggers an impulse, which is conducted along a nearby fiber into the brain.

A chemical to be tasted must dissolve in saliva, which is the watery fluid surrounding the taste buds. The salivary glands supply this fluid. To demonstrate the importance of saliva, blot your tongue and try to taste some dry food; then repeat the test after moistening your tongue with saliva.

The sense of taste derives from combinations of chemicals binding specific receptor cells. This takes place on taste hair surfaces. The binding of the chemical alters membrane polarization, generating impulses on nearby sensory nerve fibers. The degree of change is directly proportional to the concentration of the stimulating substance.

Taste Sensations

The five primary taste sensations are sweet, sour, salty, bitter, and umami (oo-mom′ee). Each of the many flavors we experience results from one of the primary sensations or from a combination of them. The way we experience flavors may also reflect the concentration of chemicals as well as the sensations of smell, texture (touch), and temperature. Furthermore, chemicals in some foods—such as capsaicin in chili peppers—may stimulate pain receptors that cause a burning sensation.

Experiments indicate that each taste receptor cell is most sensitive to one taste stimulus. Salts and acids act through specific ion channels. The sensations of bitter, sweet, and umami involve specific membrane receptors.

Taste cells for each of the five taste sensations are in all areas of the tongue, but are distributed such that each sensation seems to arise most strongly from a particular region. Sensitivity to a sweet stimulus peaks at the tip of the tongue, whereas responsiveness to sour is greatest at the margins of the tongue, and to bitter at the back. Receptors particularly responsive to salt are widely distributed.

Sweet receptors are usually stimulated by carbohydrates, but a few inorganic substances, including some salts of lead and beryllium, also elicit sweet sensations. Acids stimulate *sour receptors.* The intensity of a sour sensation is roughly proportional to the concentration of the hydrogen ions in the substance being tasted. Ionized inorganic salts mainly stimulate *salt receptors.* The quality of the sensation that each salt produces depends upon the type of positively charged ion that it releases into solution, such as Na^+ from table salt. A variety of chemicals stimulates *bitter receptors,* including many organic compounds. Inorganic salts of magnesium and calcium produce bitter sensations, too. Extreme sensitivity to bitter tastes is inherited, which is why diet colas taste sweet to some people but are bitter to others. Twenty-five types of bitter receptors have been identified. Quite a few of them detect flavors unique to fermented foods.

One group of bitter compounds of medical interest are the *alkaloids,* which include a number of poisons such as strychnine, nicotine, and morphine. Spitting out bitter substances may be a protective mechanism to avoid ingesting poisonous alkaloids in foods.

The taste sensation called *umami* has long been recognized in Japan (the word means "delicious" in Japanese) but has only recently come to the attention of Western taste researchers. Umami arises from the binding of certain amino acids, including glutamic acid and aspartic acid, to specific receptors. The flavor enhancer monosodium glutamate (MSG), used in many prepared foods, is formed from glutamic acid and also stimulates umami receptors.

Taste receptors, like olfactory receptors, rapidly undergo sensory adaptation. The resulting loss of taste can be avoided by moving bits of food over the surface of the tongue to stimulate different receptors at different moments.

> We experience a series of flavors as we savor some foods because of a sequence of chemical reactions that take place in foods as we chew them. Imagine biting into a juicy ripe peach. The fruit's tissues tear, first releasing aromatic hydrocarbons. Thirty seconds later, products of fatty acid breakdown appear and, finally, several alcohols are released. This gradual flood of stimulating molecules is one reason why we enjoy eating so much.

Although taste cells are close to the surface of the tongue and are therefore exposed to environmental wear and tear, taste cells are modified epithelial cells that divide continually and differentiate into new taste cells. A taste cell functions for only about three days before it is replaced.

CLINICAL APPLICATION 12.3
Smell and Taste Disorders

Imagine a spicy slice of pizza or freshly brewed coffee, and your mouth waters in anticipation. But for millions of people, the senses of smell and taste are dulled, distorted, or gone. Many more of us get some idea of their plight when a cold temporarily stifles these senses.

Compared to the loss of hearing or sight, being unable to taste or smell may seem more an oddity than an illness. People with such ailments would probably disagree. In some situations, a poor or absent sense of smell can be dangerous, such as in a house on fire.

The direct connection between the outside environment and the brain makes the sense of smell vulnerable to damage. Smell and taste disorders can be triggered by colds and flu, allergies, nasal polyps, swollen mucous membranes inside the nose, a head injury, chemical exposure, a nutritional or metabolic problem, or a disease. In many cases, a cause cannot be identified.

Drugs can alter taste and smell in many ways, affecting cell turnover, the neural conduction system, the status of receptors, and changes in nutritional status. Drugs containing sulfur atoms, for example, squelch taste. They include the anti-inflammatory drug penicillamine, the antihypertensive drug captopril (Capoten), and transdermal (patch) nitroglycerin to treat chest pain. The anti-biotic tetracycline and the antiprotozoan metronidazole (Flagyl) impart a metallic taste. Cancer chemotherapy and radiation treatment often alter taste and smell.

Exposure to toxic chemicals can affect taste and smell, too. For example, excess exposure to organic solvents in paint thinner can cause *cacosmia*, the association of an odor of decay with normally inoffensive stimuli.

Taste Pathways

Sensory impulses from taste receptor cells in the anterior two-thirds of the tongue travel on fibers of the facial nerve (VII); impulses from receptors in the posterior one-third of the tongue and the back of the mouth pass along the glossopharyngeal nerve (IX); and impulses from receptors at the base of the tongue and the pharynx travel on the vagus nerve (X). These cranial nerves conduct the impulses into the medulla oblongata. From there, impulses ascend to the thalamus and are directed to the gustatory cortex of the cerebrum, located in the insula. Clinical Application 12.3 and table 12.3 discuss disorders of smell and taste.

PRACTICE

17 Why is saliva necessary to taste?

18 Name the five primary taste sensations.

19 What characteristic of taste receptors helps maintain a sense of taste with age?

20 Trace a sensory impulse from a taste receptor to the cerebral cortex.

RECONNECT
To Chapter 11, Cranial Nerves, pages 421–424.

Sense of Hearing

The organ of hearing, the *ear*, has outer (external), middle, and inner (internal) sections. In addition to making hearing possible, the ear provides the sense of equilibrium.

Outer (External) Ear

The outer ear consists of all of the structures that face the outside. These include an outer, funnel-like structure called the **auricle** (pinna), an S-shaped tube, the **external acoustic** (ah-kōōs′tik) **meatus** (external auditory canal) that leads inward for about 2.5 cen-

TABLE 12.3	Types of Smell and Taste Disorders	
	Smell	**Taste**
Loss of sensation	Anosmia	Ageusia
Diminished sensation	Hyposmia	Hypogeusia
Heightened sensation	Hyperosmia	Hypergeusia
Distorted sensation	Dysosmia	Dysgeusia

timeters, and the **tympanic membrane** (eardrum) (fig. 12.9). The meatus terminates with the tympanic membrane. AP|R

The external acoustic meatus passes into the temporal bone. Near this opening, hairs guard the tube. The opening and tube are lined with skin that has many modified sweat glands called *ceruminous glands,* which secrete wax (cerumen). The hairs and wax help keep large foreign objects, such as insects, out of the ear.

Ear wax may be "wet" or "dry." The wet type has a higher lipid content and is brown. It is found among people of European or African ancestry. The dry form is grey and is found among Asians and American Indians. A single DNA base difference determines whether ear wax is wet or dry.

The transfer of vibrations through matter produces sound. These vibrations travel in waves, much like ripples on the surface of a pond. The higher the wave, the louder the sound. The more waves per second, the higher the frequency, or pitch, of the sound. Just as vibrating strings on a guitar or reeds on an oboe provide the sounds of these musical instruments, vibrating vocal folds (vocal cords) in the larynx provide the sounds of the human voice. The

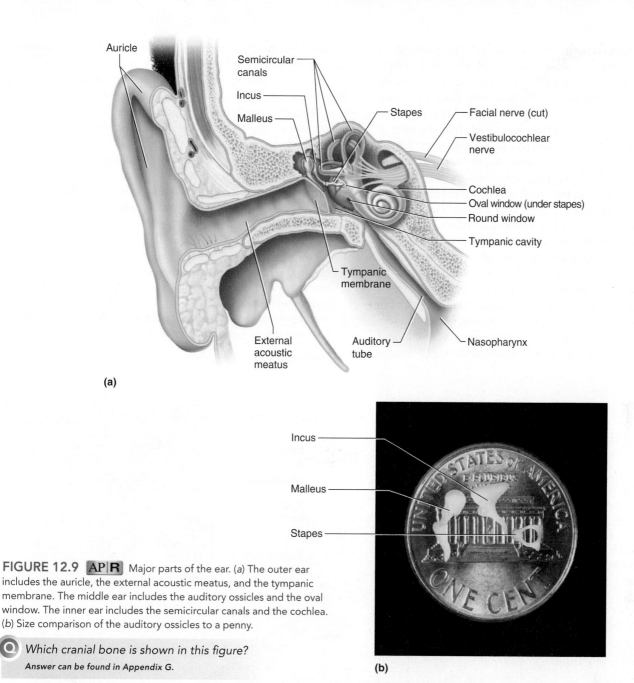

Auricle

Semicircular canals

Incus

Malleus

Stapes

Facial nerve (cut)

Vestibulocochlear nerve

Cochlea

Oval window (under stapes)

Round window

Tympanic cavity

Tympanic membrane

External acoustic meatus

Auditory tube

Nasopharynx

(a)

Incus

Malleus

Stapes

(b)

FIGURE 12.9 AP|R Major parts of the ear. (*a*) The outer ear includes the auricle, the external acoustic meatus, and the tympanic membrane. The middle ear includes the auditory ossicles and the oval window. The inner ear includes the semicircular canals and the cochlea. (*b*) Size comparison of the auditory ossicles to a penny.

Q *Which cranial bone is shown in this figure?*
Answer can be found in Appendix G.

auricle of the ear helps collect sound waves traveling through air and directs them into the external acoustic meatus.

After entering the meatus, the sound waves pass to the end of the tube and alter the pressure on the tympanic membrane. The tympanic membrane is a semitransparent membrane covered by a thin layer of skin on its outer surface and by mucous membrane on the inside. It has an oval margin and is cone-shaped, with the apex of the cone directed inward. The tympanic membrane vibrates back and forth in response to sound waves, reproducing the vibrations of the sound-wave source.

Middle Ear

The **middle ear,** or the **tympanic cavity,** is an air-filled space in the temporal bone that separates the outer and inner ears. It is bounded by the tympanic membrane laterally and the inner ear

medially and houses three small bones called **auditory ossicles** (aw'di-to"re os'i-klz).

The three auditory ossicles, called the *malleus,* the *incus,* and the *stapes,* are attached to the wall of the tympanic cavity by tiny ligaments and are covered by mucous membrane. These bones bridge the tympanic membrane and the inner ear, transferring vibrations between these parts. Specifically, the malleus is attached to the tympanic membrane, helping to maintain its conical shape. When the tympanic membrane vibrates, the malleus vibrates in unison with it. The malleus vibrates the incus, and the incus passes the movement on to the stapes. Ligaments hold the stapes to an opening in the wall of the tympanic cavity called the **oval window** (fig. 12.9). Vibration of the stapes, which acts like a piston at the oval window, transfers the vibrations to a fluid within the inner ear. These vibrations of the fluid stimulate the hearing receptors.

In addition to transferring vibrations, the auditory ossicles form a lever system that helps increase (amplify) the force of the vibrations as they pass from the tympanic membrane to the oval window. Also, because the ossicles transfer vibrations from the large surface of the tympanic membrane to a much smaller area at the oval window, the vibrational force strengthens as it travels from the outer to the inner ear. As a result, the pressure (per square millimeter) that the stapes applies at the oval window is about twenty-two times greater than that which sound waves exert on the tympanic membrane.

The middle ear also has two small skeletal muscles attached to the auditory ossicles that are controlled by a reflex. One of them, the *tensor tympani*, is inserted on the medial surface of the malleus and is anchored to the cartilaginous wall of the auditory tube. When it contracts, it pulls the malleus inward. The other muscle, the *stapedius*, is attached to the posterior side of the stapes and the inner wall of the tympanic cavity. It pulls the stapes outward when it contracts (fig. 12.10). These muscles are the effectors in the **tympanic reflex,** which is elicited in about one-tenth of a second after a loud, external sound. The reflex contracts the muscles, and the malleus and stapes move. As a result, the bridge of ossicles

in the middle ear becomes more rigid, reducing its effectiveness in transferring vibrations to the inner ear.

The tympanic reflex reduces pressure from loud sounds that might otherwise damage the hearing receptors. Ordinary vocal sounds also elicit the tympanic reflex, such as when a person speaks or sings. This action muffles the lower frequencies of such sounds, improving the hearing of higher frequencies, which are common in human vocal sounds. In addition, the tensor tympani muscle maintains tension on the tympanic membrane. This is important because a loose tympanic membrane would not be able to effectively transmit vibrations to the auditory ossicles.

The muscles of the middle ear take 100 to 200 milliseconds to contract. For this reason, the tympanic reflex cannot protect the hearing receptors from the effects of very sudden loud sounds, such as from an explosion or a gunshot. On the other hand, this protective mechanism can reduce the effects of intense sounds that arise slowly, such as the roar of thunder.

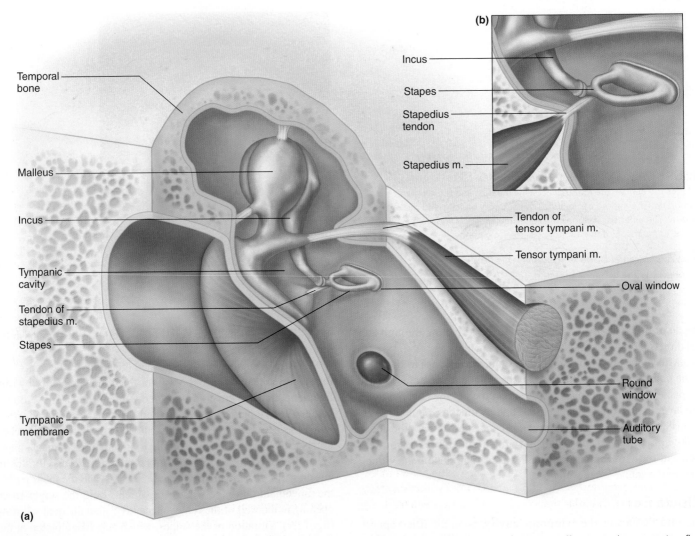

FIGURE 12.10 Two small muscles attached to the (a) malleus and (b) stapes, the tensor tympani and the stapedius, are effectors in the tympanic reflex. Figure 12.9 does not show these muscles. (*m.* stands for muscle.)

Auditory Tube

An **auditory** (aw'di-to"re) **tube** (eustachian tube) connects each middle ear to the throat. This tube allows air to pass between the tympanic cavity and the outside of the body by way of the throat (nasopharynx) and mouth. It helps maintain equal air pressure on both sides of the tympanic membrane. This is necessary for normal hearing (fig. 12.10).

The function of the auditory tube becomes noticeable during rapid change in altitude. As a person descends from a high altitude, the air pressure on the outside of the tympanic membrane steadily increases. This may push the tympanic membrane inward, out of its normal position, impairing hearing.

When the air pressure difference on the sides of the tympanic membrane is great enough, some air may force its way up through the auditory tube into the middle ear. This equalizes the pressure on both sides of the tympanic membrane, which moves back into its regular position, causing a popping sound as normal hearing returns. A reverse movement of air ordinarily occurs when a person ascends from a low altitude.

The auditory tube is usually closed by valvelike flaps in the throat, which may inhibit air movements into the middle ear. Swallowing, yawning, or chewing aid in opening the flaps and can hasten equalization of air pressure.

Signs of a middle ear infection (otitis media) in a toddler include irritability, fever, and tugging on the painful ear. Using an instrument called an otoscope reveals a red and bulging tympanic membrane.

Ear infections occur because the mucous membranes that line the auditory tubes are continuous with the linings of the middle ears, enabling bacteria infecting the throat or nasal passages to reach the ear. The bacteria that most commonly cause middle ear infection are *Streptococcus pneumoniae*, *Haemophilus influenzae*, and *Moraxella catarrhalis*. This route to infection is greater in young children because their auditory tubes are shorter than those in adults.

Acute otitis media is treated with antibiotics, but these might not be prescribed at first for a child older than two years who is not in severe pain, because such infections tend to clear up on their own in a few days. A child with recurrent otitis media might have brief surgery to fit the affected ear with a tympanostomy tube, to lower the risk of hearing loss. The tube drains the ear through a small hole in the tympanic membrane. **AP|R**

Inner (Internal) Ear

The inner ear is a complex system of intercommunicating chambers and tubes called a **labyrinth** (lab'i-rinth). Each ear has two such regions—the bony labyrinth and the membranous labyrinth.

The *bony (osseous) labyrinth* is a cavity within the temporal bone; the *membranous labyrinth* is a tube that lies within the bony labyrinth and has a similar shape (fig. 12.11a). Between the bony and membranous labyrinths is a fluid called *perilymph,* secreted by cells in the wall of the bony labyrinth. In the membranous labyrinth is a slightly different fluid called *endolymph.*

The parts of the labyrinths include a **cochlea** (kok'le-ah) that functions in hearing and three **semicircular canals** that provide a sense of equilibrium. A bony chamber called the **vestibule,** between the cochlea and the semicircular canals, houses membranous structures that serve both hearing and equilibrium.

The cochlea is a tube shaped a bit like a snail shell, coiled around a bony core, the *modiolus.* A thin, bony shelf *(spiral lamina)* extends out from the core and coils around it within the tube (fig. 12.11b). A portion of the membranous labyrinth in the cochlea, called the *cochlear duct* (scala media), runs inside the tube opposite the spiral lamina, and together these structures divide the tube into upper and lower compartments.

The upper compartment, called the *scala vestibuli,* leads from the oval window to the apex of the spiral. The lower compartment, the *scala tympani,* extends from the apex of the spiral to a membrane-covered opening in the wall of the inner ear facing the tympanic cavity, called the **round window.** These compartments form the bony labyrinth of the cochlea, and they are filled with perilymph. The cochlear duct between them is filled with endolymph. At the apex of the cochlea, beyond the tip of the cochlear duct, a small opening, the helicotrema, connects the perilymph in the upper and lower compartments and allows the fluid pressures in them to equalize (fig. 12.12).

The cochlear duct ends as a closed sac at the apex of the cochlea. The duct is separated from the scala vestibuli by a *vestibular membrane* (Reissner's membrane) and from the scala tympani by a *basilar membrane* (fig. 12.12). Clinical Application 12.4 describes an effective treatment for hearing loss called a cochlear implant.

The basilar membrane extends from the bony shelf of the cochlea and forms the floor of the cochlear duct. It has many thousands of elastic fibers and becomes thinner from the base of the cochlea to its apex. Vibrations entering the perilymph at the oval window travel along the scala vestibuli and pass through the vestibular membrane to enter the endolymph of the cochlear duct, where they move the basilar membrane. After passing through the basilar membrane, the vibrations enter the perilymph of the scala tympani, and movement of the membrane covering the round window dissipates their force into the air in the tympanic cavity (fig. 12.12).

The **spiral organ** (organ of Corti), which contains about 16,000 hearing receptor cells, is on the superior surface of the basilar membrane and stretches from the apex to the base of the cochlea. The receptor cells, called **hair cells,** are in four parallel rows, with many hairlike processes known as stereovilli (also called stereocilia) that extend into the endolymph of the cochlear duct. Above these hair cells is a *tectorial membrane,* attached to the bony shelf of the cochlea. It passes like a roof over the receptor cells, contacting the tips of their hairs (figs. 12.13 and 12.14).

Different frequencies of vibration move different regions along the length of the basilar membrane. A particular sound frequency bends the hairs of a specific group of receptor cells, activating them. Other frequencies activate receptor cells elsewhere along the cochlea (fig. 12.15, and see fig. 12.12). If sound activates receptors at different places along the basilar membrane simultaneously, we hear multiple tones at the same time.

The greater the deflection of the basilar membrane pushing the hair cells upward against the tectorial membrane, the louder the sound. However, recall from chapter 10 (p. 377) that action

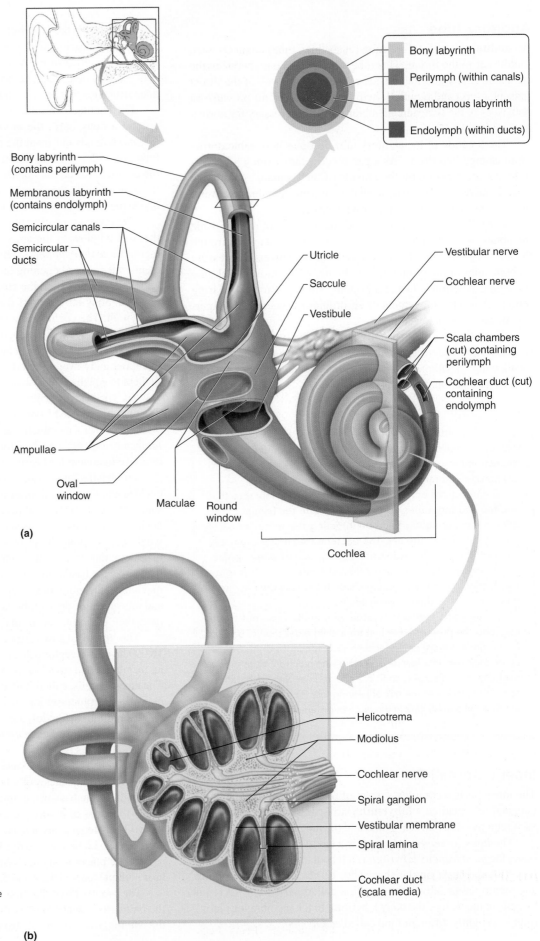

Bony labyrinth
Perilymph (within canals)
Membranous labyrinth
Endolymph (within ducts)

Bony labyrinth
(contains perilymph)

Membranous labyrinth
(contains endolymph)

Semicircular canals

Semicircular
ducts

Utricle

Saccule

Vestibule

Vestibular nerve

Cochlear nerve

Scala chambers
(cut) containing
perilymph

Cochlear duct (cut)
containing
endolymph

Ampullae

Oval
window

Maculae

Round
window

Cochlea

(a)

Helicotrema

Modiolus

Cochlear nerve

Spiral ganglion

Vestibular membrane

Spiral lamina

Cochlear duct
(scala media)

FIGURE 12.11 AP|R In the inner ear (a) perilymph separates the bony labyrinth from the membranous labyrinth, which contains endolymph. Note that areas of bony labyrinth have been removed to reveal underlying structures. (b) The spiral lamina coils around a bony core, the modiolus.

(b)

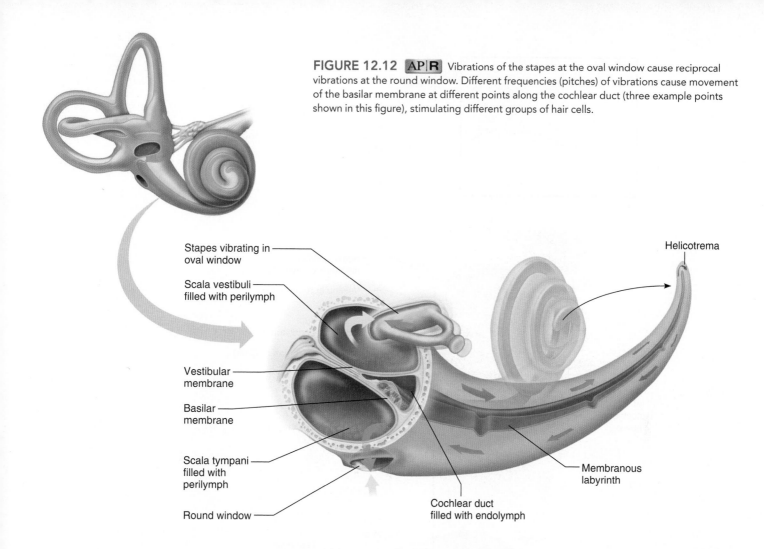

FIGURE 12.12 **AP|R** Vibrations of the stapes at the oval window cause reciprocal vibrations at the round window. Different frequencies (pitches) of vibrations cause movement of the basilar membrane at different points along the cochlear duct (three example points shown in this figure), stimulating different groups of hair cells.

Stapes vibrating in oval window

Scala vestibuli filled with perilymph

Vestibular membrane

Basilar membrane

Scala tympani filled with perilymph

Round window

Cochlear duct filled with endolymph

Membranous labyrinth

Helicotrema

potentials are all-or-none. More intense stimulation of the hair cells causes more action potentials per second to reach the brain, and we perceive a louder sound.

Hearing receptor cells are epithelial cells, but they respond to stimuli somewhat like neurons (see chapter 10, pp. 371–372). For example, when a receptor cell is at rest, its membrane is polarized. When its hairs bend, selective ion channels open and its cell membrane depolarizes. The membrane then becomes more permeable, specifically to calcium ions. The receptor cell has no axon or dendrites, but it does have neurotransmitter-containing vesicles in the cytoplasm near its base. As calcium ions diffuse into the cell, some of these vesicles fuse with the cell membrane and release neurotransmitter to the outside. The neurotransmitter stimulates the dendritic ends of nearby sensory neurons, and in response their axons conduct impulses along the cochlear branch of the vestibulocochlear nerve (cranial nerve VIII) to the brain.

The ear of a young person with normal hearing can detect sound waves with frequencies varying from about 20 to 20,000 or more vibrations per second. The range of greatest sensitivity is between 2,000 and 3,000 vibrations per second (fig. 12.15).

Auditory Pathways

The cochlear branches of the vestibulocochlear nerves enter the auditory nerve pathways that extend into the medulla oblongata and proceed through the midbrain to the thalamus. From there

they pass into the auditory cortices of the temporal lobes of the cerebrum, where they are interpreted. On the way, some of these fibers cross over, so that impulses arising from each ear are interpreted on both sides of the brain. Consequently, damage to a temporal lobe on one side of the brain is not necessarily accompanied by complete hearing loss in the ear on that side (fig. 12.16).

Table 12.4 summarizes the pathway of vibrations through the parts of the middle and inner ears. Clinical Application 12.5 examines types of hearing loss.

Units called *decibels* (dB) measure sound intensity as a logarithmic scale. The decibel scale begins at 0 dB, which is the intensity of the sound least perceptible by a normal human ear. A sound of 10 dB is 10 times as intense as the least perceptible sound; a sound of 20 dB is 100 times as intense; and a sound of 30 dB is 1,000 times as intense. A whisper has an intensity of about 40 dB, normal conversation measures 60–70 dB, and heavy traffic produces about 80 dB. Listening to a music device through earbuds exposes the ears to 110 dB. A sound of 120 dB, such as a rock concert, produces discomfort; and a sound of 140 dB, such as a jet plane at take-off, causes pain. Frequent or prolonged exposure to sounds with intensities above 90 dB can cause permanent hearing loss.

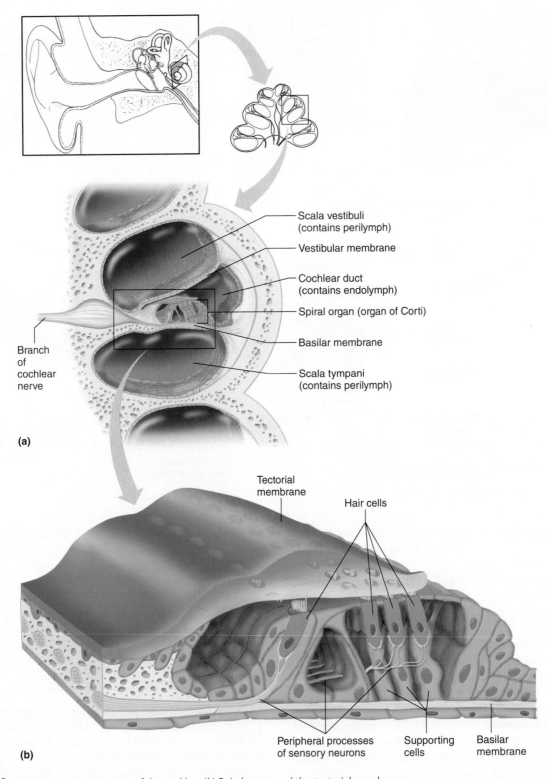

(a)

Scala vestibuli (contains perilymph)

Vestibular membrane

Cochlear duct (contains endolymph)

Spiral organ (organ of Corti)

Basilar membrane

Scala tympani (contains perilymph)

Branch of cochlear nerve

Tectorial membrane

Hair cells

Peripheral processes of sensory neurons

Supporting cells

Basilar membrane

(b)

FIGURE 12.13 Cochlea. (*a*) Cross section of the cochlea. (*b*) Spiral organ and the tectorial membrane.

CLINICAL APPLICATION 12.4

Getting a Cochlear Implant

The little boy probably lost his hearing when he suffered a high fever at eight weeks of age. When he was nine months old and still didn't babble like his age-mates, his parents suspected he might be deaf. With hearing aids he did well at a preschool for the deaf. Then his parents read about the cochlear implant, a device that does not magically restore hearing, but enables a person to hear certain sounds. Unlike a hearing aid that amplifies sound, a cochlear implant directly stimulates the auditory nerve.

The boy received his cochlear implant when he was three years old. Before three is the best time because the brain is rapidly processing speech and hearing as the person masters language. Some children who receive the devices are only one year old. However, even people who lose their hearing as adults can benefit from cochlear implants, because they link the sounds they hear through the device to memories of what sounds were like, perhaps using clues from other senses.

The cochlear implant consists of a part inserted under the skin above the ear that leads to two dozen electrodes placed near the auditory nerve in the cochlea, the snail-shaped part of the inner ear. A headset includes a microphone lodged at the back of the ear to pick up incoming sounds and a fanny pack containing a speech processor that digitizes the sounds into coded signals. A transmitter on the headset sends the coded signals, as FM radio waves, to the implant, which changes them to electrical signals and delivers them to the cochlea. Here, the auditory nerve is stimulated, and it conducts neural messages to the brain's cerebral cortex, which interprets the input as sound.

The boy's audiologist (see the Career Corner, p. 445) turned on the speech processor a month after the surgery. At first, the youngster heard low sounds and sometimes responded with a low hum. He grabbed at the processor, realizing it was the source of the sound. Gradually, the child learned from context what certain sounds meant. About 325,000 people worldwide have received cochlear implants since the devices became available in 1984.

FIGURE 12D A cochlear implant enables this child to detect enough sounds to effectively communicate.

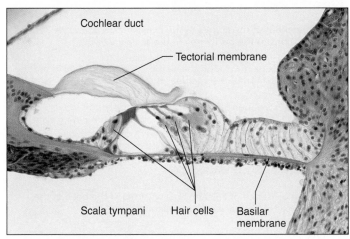

(a)

FIGURE 12.14 AP|R Spiral organ. (a) A micrograph of the spiral organ and the tectorial membrane (400×). (b) A scanning electron micrograph of hair cells in the spiral organ, looking down on the "hairs" (bright yellow) (6,700×).

(b)

FIGURE 12.15

How the cochlea might look if it could be straightened out. Receptors in regions of the cochlear duct sense different frequencies of vibration, expressed in cycles per second (cps).

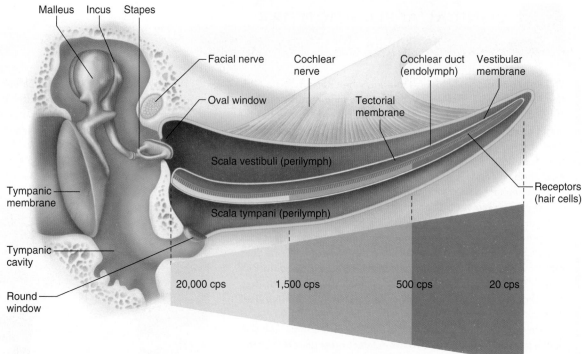

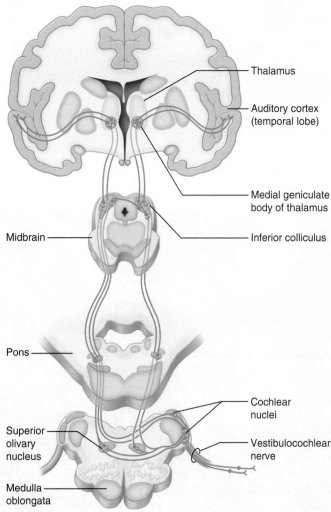

FIGURE 12.16 The auditory nerve pathway extends into the medulla oblongata, proceeds through the midbrain to the thalamus, and passes into the auditory cortex of the cerebrum.

TABLE 12.4	Steps in the Generation of Sensory Impulses from the Ear AP|R

1. Sound waves enter the external acoustic meatus.

2. Waves of changing pressures cause the tympanic membrane to reproduce the vibrations coming from the sound-wave source.

3. Auditory ossicles amplify and conduct vibrations to the end of the stapes.

4. Movement of the stapes at the oval window conducts vibrations to the perilymph in the scala vestibuli.

5. Vibrations pass through the vestibular membrane and enter the endolymph of the cochlear duct.

6. Different frequencies of vibration in endolymph move specific regions of the basilar membrane, stimulating specific sets of receptor cells.

7. A receptor cell depolarizes; its membrane becomes more permeable to calcium ions.

8. In the presence of calcium ions, vesicles at the base of the receptor cell release neurotransmitter.

9. Neurotransmitter stimulates nearby sensory neurons.

10. Sensory impulses are conducted along fibers of the cochlear branch of the vestibulocochlear nerve.

11. The auditory cortex of the temporal lobe interprets the sensory impulses.

About 8% of people have some degree of hearing loss. Several factors can impair hearing, including interference with conduction of vibrations to the inner ear (*conductive deafness*) or damage to the cochlea or the auditory nerve and its pathways (*sensorineural deafness*). Disease, injury, and heredity all can impair hearing. There are more than 100 forms of inherited deafness, and many are part of syndromes. About 95% of cases of hearing loss are conductive. One cause is accumulated dry wax or a foreign object in the ear, which plugs the acoustic meatus. Changes in the tympanic membrane or auditory ossicles can also block hearing. The tympanic membrane may harden as a result of disease, becoming less responsive to sound waves, or an injury may tear or perforate it. Sudden pressure changes, very loud sounds, infection, or sticking an object into the ear may rupture the tympanic membrane.

A common disorder of the auditory ossicles is *otosclerosis*, in which new bone deposited abnormally around the base of the stapes interferes with the movement of the ossicles that is necessary to conduct vibrations to the inner ear. Surgery often can restore some hearing by chipping away the bone that holds the stapes in position or replacing the stapes with a wire or plastic substitute.

Two tests used to diagnose conductive deafness are the Weber test and the Rinne test. In the Weber test, the handle of a vibrating tuning fork is pressed against the forehead. A person with normal hearing perceives the sound coming from directly in front, whereas a person with sound conduction blockage in one middle ear hears the sound coming from the impaired side.

In the Rinne test, a vibrating tuning fork is held against the bone behind the ear. After the sound is no longer heard by conduction through the bones of the skull, the fork is moved to just in front of the external acoustic meatus. In middle ear conductive deafness, the vibrating fork can no longer be heard, but a normal ear will continue to hear its tone.

Very loud sounds can cause sensorineural deafness. If exposure is brief, hearing loss may be temporary, but when exposure is repeated and prolonged, such as occurs in foundries, near jackhammers, or on a firing range, impairment may be permanent. Frequent and prolonged listening to very loud music through earbuds can cause desensitization—becoming able to tolerate louder and louder sounds—and damage inner ear hair cells.

Sensorineural hearing loss begins as the hair cells develop blisterlike bulges that eventually pop. The tissue beneath the hair cells swells and softens until the hair cells, and sometimes the neurons, leaving the cochlea become blanketed with scar tissue and degenerate. Other causes of sensorineural deafness include tumors in the CNS, brain damage as a result of vascular accidents, and the use of certain drugs. Exposure to blasts, such as in combat, can damage all parts of the ear.

Hearing loss and other ear problems can begin gradually. Signs include:

- difficulty hearing people talking softly
- inability to understand speech when there is background noise
- ringing or a sensation of fullness in the ears
- dizziness
- loss of balance
- difficulty distinguishing high-pitched sounds from each other
- the need to turn up the volume on a television or phone

New parents should notice whether their infant responds to sounds in a way that indicates normal hearing. Hearing exams are part of a well-baby visit to a doctor. If the baby's responses indicate a possible problem, the next step is to see an audiologist, who identifies and measures hearing loss.

Often a hearing aid can help people with conductive hearing loss. A hearing aid has a tiny microphone that picks up sound waves and converts them to electrical signals, which are then amplified. An ear mold holds the device in place, either behind the outer ear, in the outer ear, or in the ear canal.

PRACTICE

21 Describe the outer, middle, and inner ears.

22 Explain how sound waves are transmitted through the parts of the ear.

23 Describe the tympanic reflex.

24 Distinguish between the osseous and membranous labyrinths.

25 Explain the function of the spiral organ.

Sense of Equilibrium

The feeling of equilibrium (balance) derives from two senses—**static equilibrium** (stat′ik e′kwĭ-lib′re-um) and **dynamic equilibrium** (di-nam′ik e′kwĭ-lib′re-um). Different sensory organs provide these two components of equilibrium. The organs associated with static equilibrium sense the position of the head, maintaining stability and posture when the head and body are still. When the head and body suddenly move or rotate, the organs of dynamic equilibrium detect the motion and aid in maintaining balance.

Static Equilibrium

The organs of static equilibrium are in the **vestibule,** a bony chamber between the semicircular canals and the cochlea. The membranous labyrinth inside the vestibule consists of two expanded chambers—a **utricle** (u′trĭ-kl) and a **saccule** (sak′ūl). The larger utricle communicates with (is continuous with) the saccule and the membranous portions of the semicircular canals; the saccule, in turn, communicates with the cochlear duct (fig. 12.17).

The utricle and saccule each has a small patch of hair cells and supporting cells called a **macula** (mak′u-lah) on its wall. When the

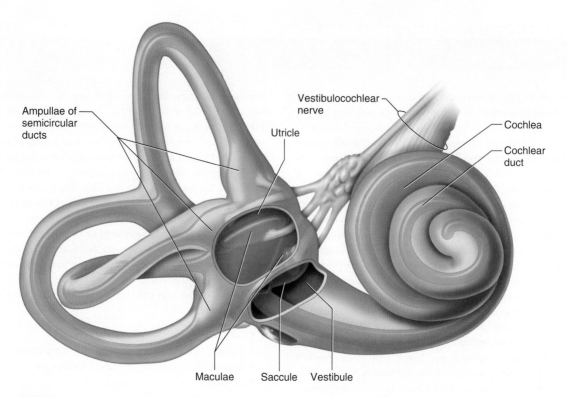

FIGURE 12.17 AP|R The saccule and utricle, expanded portions of the membranous labyrinth, are in the bony chamber of the vestibule. (Compare with figure 12.11. The oval window is not shown in this figure.)

head is upright, the hairs of the macula in the utricle project vertically, while those in the saccule project horizontally. In both the utricle and saccule, the hairs contact a sheet of gelatinous material (otolithic membrane) that has crystals of calcium carbonate (otoliths) embedded on its surface. These particles add weight to the gelatinous sheet, making it more responsive to changes in position. The hair cells, which are the sensory receptors, have dendrites of sensory neurons wrapped around their bases. These neurons are associated with the vestibular portion of the vestibulocochlear nerve.

Gravity stimulates hair cells to respond. This usually happens when the head bends forward, backward, or to one side. Such movements tilt the gelatinous mass of one or more maculae, and as the material sags in response to gravity, the hairs projecting into it bend. This action stimulates the hair cells, and they signal their associated neurons (figs. 12.18 and 12.19). The resulting impulses are conducted into the CNS by means of the vestibular branch of the vestibulocochlear nerve, informing the brain of the head's position. The brain responds by sending motor impulses to skeletal muscles, which may contract or relax appropriately to maintain balance.

The maculae also participate in the sense of dynamic equilibrium. For example, if the head or body is thrust forward or backward abruptly, the gelatinous mass of the maculae lags slightly behind, and the hair cells are stimulated. In this way, the maculae aid the brain in detecting movements such as falling and in maintaining posture while walking.

Dynamic Equilibrium

Each semicircular canal follows half of a roughly circular path about 6 millimeters in diameter. The three bony semicircular canals lie at right angles to each other and occupy three different planes in space.

Two of them, the *anterior canal* and the *posterior canal,* are oriented vertically, whereas the third, the *lateral canal,* is horizontal.

Suspended in the perilymph of each bony canal is a membranous semicircular duct that ends in a swelling called an **ampulla** (am-pul'ah). The ampullae communicate with the utricle of the vestibule (see fig. 12.17).

An ampulla contains a ridge that crosses the tube and houses a sensory organ. Each of these organs, called a **crista ampullaris,** has a number of sensory hair cells and supporting cells. As in the maculae, the hairs of the hair cells extend upward into a dome-shaped gelatinous mass called the *cupula.* Also, the hair cells are connected at their bases to dendrites of neurons that make up part of the vestibular branch of the vestibulocochlear nerve (fig. 12.20).

When the head or torso moves, the semicircular canals move as well, but initially the fluid inside the membranous ducts tends to remain stationary because of inertia. (Imagine turning rapidly while holding a full glass of water.) This bends the cupula in one or more of the semicircular canals in a direction opposite that of the head and torso movement, and the hairs embedded in it also bend. The moving of the hairs stimulates the hair cells to signal their associated neurons, and as a result, impulses are conducted to the brain (fig. 12.21). Movements in different directions affect different combinations of semicircular canals. The brain interprets impulses originating from these different combinations as different movements.

Parts of the cerebellum are particularly important in interpreting impulses from the semicircular canals. Analysis of such information allows the brain to predict the consequences of rapid body movements, and by modifying signals to appropriate skeletal muscles, the cerebellum can maintain balance.

FIGURE 12.18 The maculae respond to changes in head position. (*a*) Macula of the utricle with the head in an upright position. (*b*) Macula of the utricle with the head bent forward.

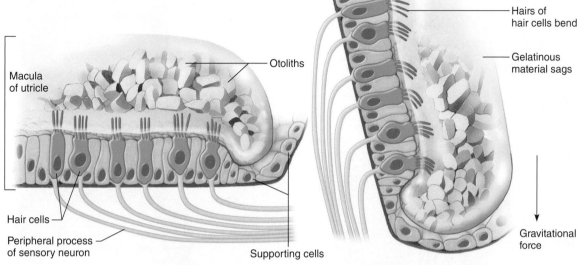

Macula of utricle

Otoliths

Hairs of hair cells bend

Gelatinous material sags

Hair cells

Peripheral process of sensory neuron

Supporting cells

Gravitational force

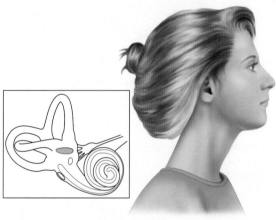

(a) Head upright

(b) Head bent forward

Other sensory structures aid in maintaining equilibrium. Various proprioceptors, particularly those associated with the joints of the neck, inform the brain about the position of body parts. The eyes detect changes in posture that result from body movements. Such visual information is so important that even if the organs of equilibrium are damaged, keeping the eyes open and moving slowly may be sufficient to maintain normal balance.

Motion sickness is a disturbance of the inner ear's sensation of balance. Nine out of ten people have experienced this nausea and vomiting, usually when riding in a car or on a boat. Astronauts suffer a form of motion sickness called space adaptation syndrome.

Motion sickness is thought to result when visual information contradicts the inner ear's sensation that a person is motionless. Consider a woman riding in a car. Her inner ears tell her that she is not moving, but the passing scenery tells her eyes that she is moving. The problem is compounded if she tries to read. The brain reacts to these seemingly contradictory sensations by causing a feeling of queasiness, which may lead to vomiting.

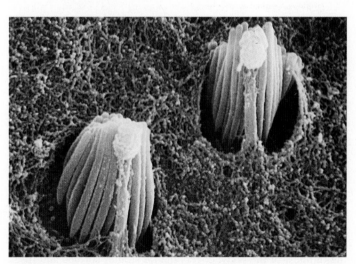

FIGURE 12.19 Scanning electron micrograph of hairs of hair cells, such as those in the utricle and saccule (8,000×).

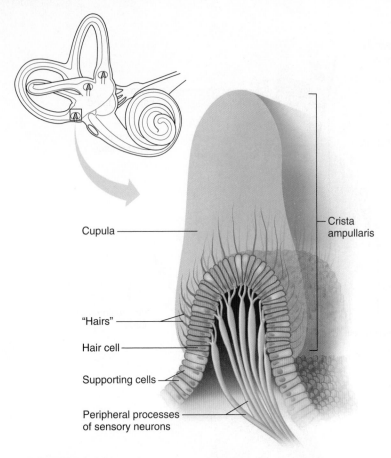

FIGURE 12.20 A crista ampullaris is in the ampulla of each semicircular canal.

(a) Head in still position

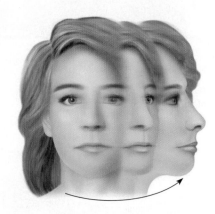

(b) Head rotating

 PRACTICE

26 Distinguish between the senses of static and dynamic equilibrium.

27 Which structures provide the sense of static equilibrium? of dynamic equilibrium?

28 How does sensory information from other receptors help maintain equilibrium?

Sense of Sight

A number of accessory organs assist the visual receptors in the eyes. These include the eyelids and lacrimal apparatus that help protect the eyes and a set of extrinsic muscles that move them.

Visual Accessory Organs

Each eye, lacrimal gland, and associated extrinsic muscles are housed in the orbital cavity of the skull. The orbit, lined with the periosteum of various bones, also contains fat, blood vessels, nerves, and connective tissues.

Each **eyelid** (palpebra) is composed of four layers—skin, muscle, connective tissue, and conjunctiva. The skin of the eyelid, the thinnest of the body, covers the lid's outer surface and fuses with its inner lining near the margin of the lid (fig. 12.22).

The muscles that move the eyelids include the *orbicularis oculi* and the *levator palpebrae superioris*. Fibers of the orbicularis oculi encircle the opening between the lids and spread out

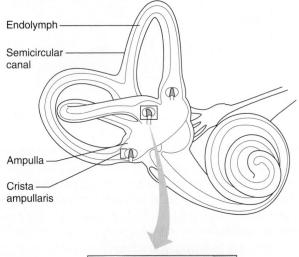

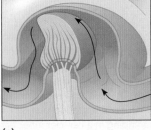

(c)

FIGURE 12.21 Equilibrium. (a) When the head is stationary, the cupula of the crista ampullaris remains upright. (b) When the head is moving rapidly, (c) the cupula bends opposite the motion of the head, stimulating sensory receptors.

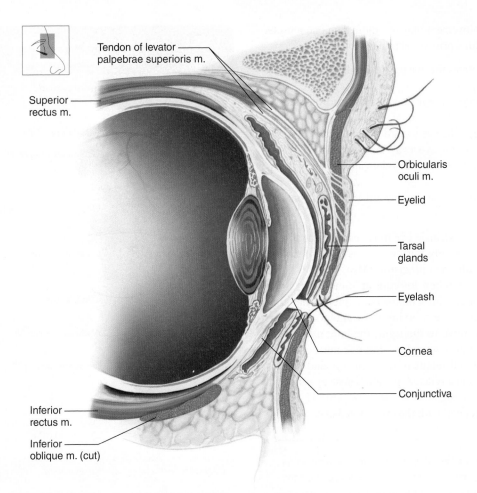

Superior
rectus m.

Tendon of levator
palpebrae superioris m.

Orbicularis
oculi m.

Eyelid

Tarsal
glands

Eyelash

Cornea

Conjunctiva

Inferior
rectus m.

Inferior
oblique m. (cut)

FIGURE 12.22 Sagittal section of the closed eyelids and the anterior portion of the eye.

onto the cheek and forehead. This muscle acts as a sphincter that closes the lids when it contracts. Fibers of the levator palpebrae superioris muscle arise from the roof of the orbit and are inserted in the connective tissue of the upper lid. When these fibers contract, the upper lids are raised, and the eye opens.

The connective tissue layer of the eyelid, which helps give it form, contains many modified sebaceous glands (tarsal glands). Ducts carry the oily secretions of these glands to openings along the borders of the lids. This secretion helps keep the lids from sticking together.

The **conjunctiva** is a mucous membrane that lines the inner surfaces of the eyelids and folds back to cover the anterior surface of the eyeball, except for its central portion (cornea). Although the tissue that lines the eyelids is relatively thick, the conjunctiva that covers the eyeball is very thin. It is also freely movable and transparent, so that blood vessels are clearly visible beneath it.

Most children who arrive at school with "pinkeye" are sent home. Pinkeye is a form of *conjunctivitis,* or inflammation of the conjunctiva. When caused by bacteria, it is highly contagious. It is not usually contagious when caused by a virus.

The **lacrimal apparatus** consists of the *lacrimal gland,* which secretes tears, and a series of *ducts,* which carry the tears into the nasal cavity (fig. 12.23). The gland is in the orbit, superior and lateral to the eye. It secretes tears continuously, which pass through tiny tubules and flow downward and medially across the eye.

Two small ducts (superior and inferior canaliculi) collect tears, and their openings (puncta) can be seen on the medial borders of the eyelids. From these ducts, the fluid moves into the *lacrimal sac,* which lies in a deep groove of the lacrimal bone, and then into the *nasolacrimal duct,* which empties into the nasal cavity.

Glandular cells of the conjunctiva also secrete a tearlike liquid that, together with the secretion of the lacrimal gland, moistens and lubricates the surface of the eye and the lining of the lids. Tears contain an enzyme, *lysozyme,* that has antibacterial properties, reducing the risk of eye infections.

Tear glands secrete excessively when a person is upset or when the conjunctiva is irritated. Tears spill over the edges of the eyelids, and the nose fills with fluid. When a person cries, parasympathetic nerve fibers conduct motor impulses to the lacrimal glands.

The **extrinsic muscles** of the eye arise from the bones of the orbit and are inserted by broad tendons on the eye's tough outer surface. Six such muscles move the eye in different directions (fig. 12.24).

Although any given eye movement may use more than one muscle, each one is associated with a primary action, as follows:

1. **Superior rectus**—rotates the eye upward and toward the midline.
2. **Inferior rectus**—rotates the eye downward and toward the midline.
3. **Medial rectus**—rotates the eye toward the midline.
4. **Lateral rectus**—rotates the eye away from the midline.
5. **Superior oblique**—rotates the eye downward and away from the midline.
6. **Inferior oblique**—rotates the eye upward and away from the midline.

The motor units of the extrinsic eye muscles have the fewest muscle fibers (five to ten) of any muscles in the body, enabling them to move the eyes with great precision. Also, the eyes move together so that they align when looking at something. Such alignment is the result of complex motor adjustments that contract certain eye muscles while relaxing their antagonists. For example, when the eyes move to the right, the lateral rectus of the right eye and the medial rectus of the left eye must contract. At the same time, the medial rectus of the right eye and the lateral rectus of the left eye must relax. A person whose eyes are not coordinated well enough to align has *strabismus*. Table 12.5 summarizes the muscles associated with the eyelids and eye.

> One eye deviating from the line of vision produces double vision (diplopia). If it persists, the brain may eventually suppress the image from the deviated eye, and the turning eye may experience partial vision loss (suppression amblyopia). Treating the eye deviation early in life with exercises, eyeglasses, and surgery can prevent such monocular vision loss. Vision screening programs for children can detect this problem.

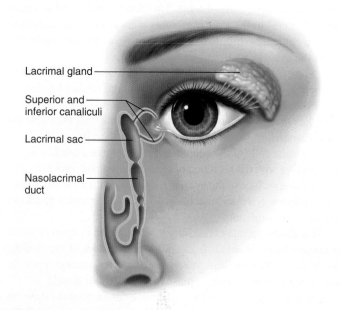

Lacrimal gland

Superior and inferior canaliculi

Lacrimal sac

Nasolacrimal duct

FIGURE 12.23 The lacrimal apparatus consists of a tear secreting gland and a series of ducts.

TABLE 12.5 Muscles Associated with the Eyelids and Eyes

Skeletal Muscles	Innervation	Function
Muscles of the eyelids		
Orbicularis oculi	Facial nerve (VII)	Closes eye
Levator palpebrae superioris	Oculomotor nerve (III)	Opens eye
Extrinsic muscles of the eyes		
Superior rectus	Oculomotor nerve (III)	Rotates eye upward and toward midline
Inferior rectus	Oculomotor nerve (III)	Rotates eye downward and toward midline
Medial rectus	Oculomotor nerve (III)	Rotates eye toward midline
Lateral rectus	Abducens nerve (VI)	Rotates eye away from midline
Superior oblique	Trochlear nerve (IV)	Rotates eye downward and away from midline
Inferior oblique	Oculomotor nerve (III)	Rotates eye upward and away from midline

Smooth Muscle	Innervation	Function
Ciliary muscles	Oculomotor nerve (III) parasympathetic fibers	Relax suspensory ligaments
Iris, constrictor muscle	Oculomotor nerve (III) parasympathetic fibers	Constrict pupils
Iris, dilator muscle	Sympathetic fibers	Dilate pupils

 PRACTICE

29 Explain how the eyelid is moved.
30 Describe the conjunctiva.
31 What is the function of the lacrimal apparatus?
32 Describe the function of each extrinsic eye muscle.

Structure of the Eye

The eye is a hollow, spherical structure about 2.5 centimeters in diameter. Its wall has three distinct layers—an outer *fibrous tunic,* a middle *vascular tunic,* and an inner *nervous tunic.* The spaces in the eye are filled with fluids that support its wall and internal structures that help maintain its shape. **Figure 12.25** shows the major parts of the eye.

The Outer Tunic

Light is often described as moving in straight lines called *rays* of light. The anterior sixth of the outer tunic bulges forward as the transparent **cornea** (kor´ne-ah), the window of the eye. The cornea helps focus entering light rays. It is largely composed of connective

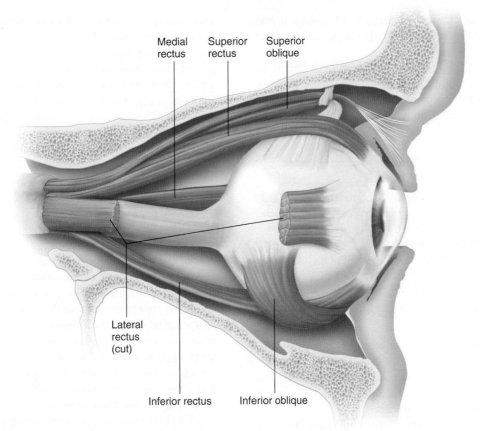

FIGURE 12.24 **AP|R** Extrinsic muscles of the right eye (lateral view).

Q *Are the extrinsic muscles of the eye under voluntary control?*
Answer can be found in Appendix G.

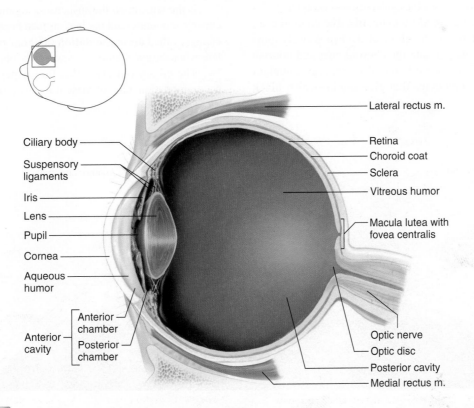

FIGURE 12.25 **AP|R** Transverse section of the right eye (superior view).

tissue with a thin surface layer of epithelium. The cornea is transparent because it contains no blood vessels and the collagen fibers form unusually regular patterns.

The cornea is well supplied with sensory nerve fibers that enter its margin and radiate toward its center. These fibers are associated with many pain receptors that have very low thresholds. Cold receptors are also abundant in the cornea, but heat and touch receptors are not.

The most common cause of blindness worldwide is loss of transparency of the cornea. Each year, 40,000 corneal transplants are performed in the United States. Corneal transplants do not evoke an immune response, but they are effective only if the transplanted tissue includes stem cells that are normally found in a cell layer, called the limbus, which separates the cornea from the conjunctiva. The cornea itself does not contain stem cells—their nuclei are so large that they would block light rays, impairing vision.

The cornea is continuous with the **sclera** (skle′rah), the white portion of the eye. The sclera makes up the posterior five-sixths of the outer tunic and is opaque due to many large, seemingly disorganized collagen and elastic fibers. The sclera protects the eye and is an attachment for the extrinsic muscles.

In the back of the eye, the **optic** (op′tik) **nerve** and blood vessels pierce the sclera. The dura mater that encloses these structures is continuous with the sclera.

The Middle Tunic

The middle, or vascular, tunic of the eyeball (uveal layer) includes the **choroid coat,** the ciliary body, and the iris. The choroid coat, in the posterior five-sixths of the globe of the eye, loosely joins the sclera. Blood vessels pervade the choroid coat and nourish surrounding tissues. The choroid coat also contains abundant pigment-producing melanocytes that give it a brownish-black appearance. The melanin of these cells absorbs excess light and helps keep the inside of the eye dark.

The **ciliary body,** which is the thickest part of the middle tunic, extends anteriorly and inward from the choroid coat and forms a ring within the front of the eye. In the ciliary body are radiating folds called *ciliary processes* and groups of smooth muscle cells that constitute the *ciliary muscles.* Figure 12.26 shows these structures.

Many strong but delicate fibers, called *suspensory ligaments* (zonular fibers), extend inward from the ciliary processes and hold the transparent **lens** in position. The distal ends of these fibers are attached along the margin of a thin capsule that surrounds the lens. The body of the lens, which lacks blood vessels, lies directly behind the iris and pupil and is composed of specialized epithelial cells.

The cells of the lens originate from a single layer of epithelium beneath the anterior portion of the lens capsule. The cells divide, and the new cells on the surface of the lens capsule differentiate into specialized columnar epithelial cells called *lens fibers,* which constitute the substance of the lens. Lens fiber production continues slowly throughout life, thickening the lens from front to back. Simultaneously, the deeper lens fibers are compressed toward the center of the structure (fig. 12.27). More than 90% of the proteins in a lens cell are lens crystallins, which aggregate into the fibers. These proteins, along with the absence of organelles that scatter light (mitochondria, endoplasmic reticula, and nuclei), provide the transparency of the lens.

The lens capsule is a clear, membranelike structure largely composed of intercellular material (fig. 12.28). It is quite elastic, a quality that keeps it under constant tension. As a result, the lens can assume a globular shape. However, the suspensory ligaments attached to the margin of the capsule are also under tension, and they pull outward, flattening the capsule and the lens.

If the tension on the suspensory ligaments relaxes, the elastic capsule rebounds, and the lens surface becomes more convex. This change, called **accommodation** (ah-kom″o-da′shun), occurs in the lens when the eye focuses to view a close object.

The ciliary muscles relax the suspensory ligaments during accommodation. One of these muscles forms a circular sphincter-

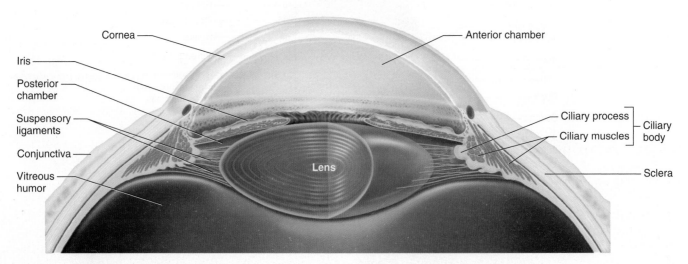

FIGURE 12.26 **AP|R** Anterior portion of the eye.

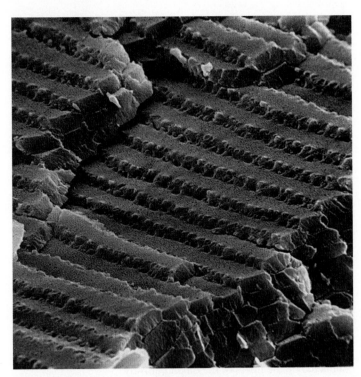

FIGURE 12.27 A scanning electron micrograph of the long, flattened lens fibers (2,650×). Note the fingerlike junctions where one fiber joins another.

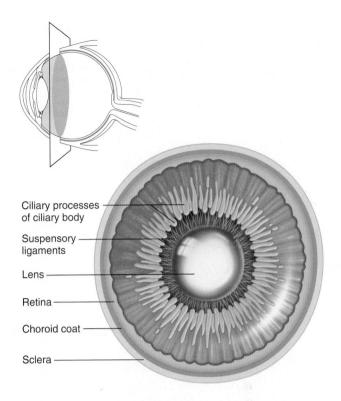

FIGURE 12.28 Lens and ciliary body viewed from behind.

- Ciliary processes of ciliary body
- Suspensory ligaments
- Lens
- Retina
- Choroid coat
- Sclera

like structure around the ciliary processes. The other muscle extends back from fixed points in the sclera to the choroid coat. When the circular muscle contracts, the diameter of the ring formed by the ciliary processes decreases; when the other muscle contracts, the choroid coat is pulled forward, and the ciliary body shortens. Both of these actions relax the suspensory ligaments, thickening the lens. In this thickened state, the lens is focused for viewing objects closer than before (fig. 12.29a).

To focus on a distant object, the ciliary muscles relax, increasing tension on the suspensory ligaments. The lens thins again (fig. 12.29b).

 PRACTICE

33 Describe the outer and middle tunics of the eye.

34 What factors contribute to the transparency of the cornea?

35 How does the shape of the lens change during accommodation?

36 Why would reading for a long time lead to "eye fatigue," whereas looking at something distant is restful?

The **iris** is a thin diaphragm mostly composed of connective tissue and smooth muscle. Seen from the outside, it is the colored portion of the eye. The iris extends forward from the periphery of the ciliary body and lies between the cornea and the lens. It divides the space separating these parts, called the *anterior cavity,* into an *anterior chamber* (between the cornea and the iris) and a *posterior chamber* (between the iris and the vitreous humor, occupied by the lens).

The epithelium on the inner surface of the ciliary body continuously secretes a watery fluid called **aqueous humor** into the posterior chamber. The fluid circulates from this chamber through

the **pupil,** which is a circular opening in the center of the iris, and into the anterior chamber (fig. 12.30). Aqueous humor fills the space between the cornea and the lens, providing nutrients and maintaining the shape of the front of the eye. It leaves the anterior chamber through veins and a special drainage canal, called the scleral venous sinus (canal of Schlemm), in the wall of the anterior chamber at the junction of the cornea and the sclera (fig. 12.30).

The smooth muscle cells of the iris form two groups, a *circular set* and a *radial set*. These control the size of the pupil, through which light passes. The circular set (pupillary constrictor) acts as a sphincter, and when it contracts, the pupil gets smaller and the intensity of the light entering decreases. When the radial set (pupillary dilator) contracts, the diameter of the pupil increases and the intensity of the light entering increases.

The sizes of the pupils change constantly in response to pupillary reflexes triggered by such factors as light intensity, gaze, accommodation, and variations in emotional state. For example, bright light elicits a reflex, and impulses are conducted along parasympathetic nerve fibers to the pupillary constrictors of the irises. The pupils constrict in response. Conversely, in dim light, impulses are conducted on sympathetic nerve fibers to the pupillary dilators of the irises, and the pupils dilate (fig. 12.31).

 RECONNECT

To Chapter 11, Autonomic Nervous System, pages 431–436.

The amount and distribution of melanin in the irises and the density of the tissue in the body of the iris determine eye color. If melanin is present only in the epithelial cells on the iris's posterior surface,

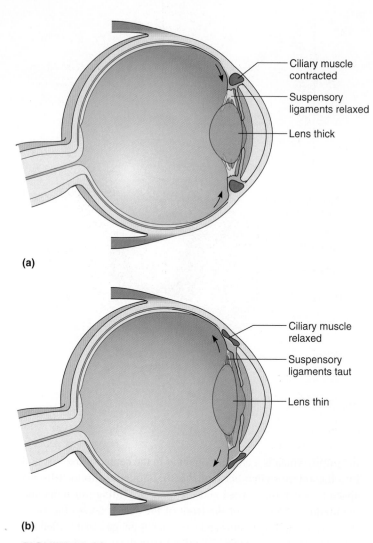

(a)

- Ciliary muscle contracted
- Suspensory ligaments relaxed
- Lens thick

(b)

- Ciliary muscle relaxed
- Suspensory ligaments taut
- Lens thin

FIGURE 12.29 In accommodation, (a) the lens thickens as the ciliary muscles contract. (b) The lens thins as the ciliary muscles relax.

the iris reflects more colors of light, and appears blue or green. When the same distribution of melanin is denser in the body of the iris, eye color is gray. When melanin is within the body of the iris as well as in the posterior epithelial covering, the iris appears brown.

The Inner Tunic

The inner tunic of the eye consists of the **retina** (ret′ĭ-nah), which contains the visual receptor cells (photoreceptors). This nearly transparent sheet of tissue is continuous with the optic nerve in the back of the eye and extends forward as the inner lining of the eyeball. It ends just behind the margin of the ciliary body.

The retina is thin and delicate, but its structure is complex. It has distinct layers, including retinal pigment epithelium, neurons, nerve fibers, and limiting membranes (figs. 12.32 and 12.33).

There are five major groups of retinal neurons. The nerve fibers of three of these groups—the *photoreceptors, bipolar neurons,* and *ganglion cells*—provide a direct pathway for impulses triggered in the photoreceptors to the optic nerve and brain. The nerve fibers of the other two groups of retinal cells, called *horizontal cells* and *amacrine cells,* pass laterally between retinal cells (see fig. 12.32). The horizontal and amacrine cells modify the pattern of impulses conducted on the fibers of the direct pathway.

In the central region of the retina is a yellowish spot called the **macula lutea** that occupies about 1 square millimeter. A depression in its center, called the **fovea centralis,** is in the region of the retina that produces the sharpest vision.

Just medial to the fovea centralis is an area called the **optic disc** (fig. 12.34). Here the nerve fibers from the retina leave the eye and become parts of the optic nerve. A central artery and vein also pass through at the optic disc. These vessels are continuous with capillary networks of the retina, and together with vessels in the underlying choroid coat, they supply blood to the cells of the inner tunic. The optic disc lacks receptor cells, so it is commonly referred to as the *blind spot* of the eye.

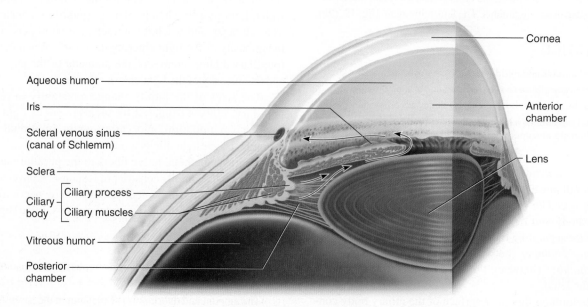

- Cornea
- Aqueous humor
- Iris
- Scleral venous sinus (canal of Schlemm)
- Sclera
- Ciliary body — Ciliary process — Ciliary muscles
- Vitreous humor
- Posterior chamber
- Anterior chamber
- Lens

FIGURE 12.30 Aqueous humor (arrows), secreted into the posterior chamber, circulates into the anterior chamber and leaves it through the scleral venous sinus (canal of Schlemm).

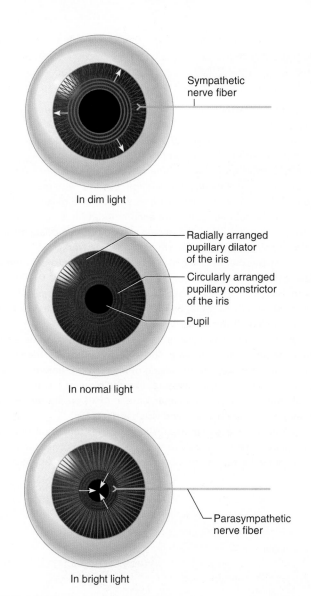

FIGURE 12.31 Dim light stimulates the radial muscle of the iris to contract, and the pupil dilates. Bright light stimulates the circular muscle of the iris to contract, and the pupil constricts.

In dim light

In normal light

Sympathetic nerve fiber

Radially arranged pupillary dilator of the iris

Circularly arranged pupillary constrictor of the iris

Pupil

In bright light

Parasympathetic nerve fiber

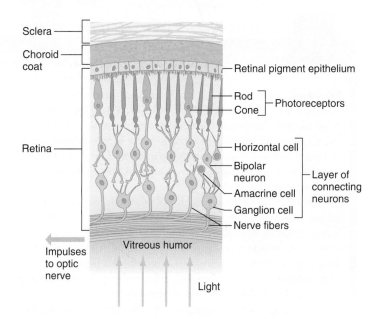

FIGURE 12.32 The retina consists of several cell layers.

Sclera

Choroid coat

Retina

Retinal pigment epithelium

Rod
Cone — Photoreceptors

Horizontal cell

Bipolar neuron

Amacrine cell — Layer of connecting neurons

Ganglion cell

Nerve fibers

Vitreous humor

Impulses to optic nerve

Light

The space enclosed by the lens, ciliary body, and retina is the largest compartment of the eye and is called the *posterior cavity*. It is filled with a transparent, jellylike fluid called **vitreous humor,** which with some collagen fibers comprise the **vitreous body.** The vitreous body supports the internal structures of the eye and helps maintain its shape.

In summary, light entering the eye must pass through the cornea, aqueous humor, lens, vitreous humor, and several layers of the retina before it reaches the photoreceptors (see fig. 12.32). Table 12.6 summarizes the layers of the eye.

 PRACTICE

37 Explain the origin of aqueous humor and trace its path through the eye.

38 How is the size of the pupil regulated?

39 Describe the structure of the retina.

Light Refraction

When a person sees an object, either the object is giving off light, or it is reflecting light from another source. Rays of light enter the eye, and an image of what is seen focuses upon the retina. The light rays must bend to be focused, a phenomenon called **refraction** (re-frak′shun).

Refraction occurs when light rays pass at an angle from a medium of one density into a medium of a different density. For example, as **figure 12.35** shows, when light passes from a less-dense medium such as air into a denser medium such as glass, or from air into the cornea of the eye, the light bends toward a line perpendicular to the surface between these substances. The light bends again when it reenters the air. When the surface between such refracting media is curved, a lens is formed. A lens with a *convex* surface causes light waves to converge, and a lens with a *concave* surface causes light waves to diverge (fig. 12.36). Note how the first example is similar to the shape of both the cornea and the lens of the eye. Clinical Application 12.6 discusses familiar visual problems resulting from abnormal refraction.

In the previous example, light rays were bent going from air to glass and from glass to air. Inside the eye, however, the lens and various fluids are more alike in density than air is compared to glass. About 75 percent of the refraction occurs when light passes from the air through the convex surface of the cornea. The light is refracted again by the convex surface of the lens and to a lesser extent at the surfaces of the fluids in the eye chambers. The lens is the only one of these structures that can change its amount of refraction to achieve a focused image (accommodation).

If the shape of the eye is normal, light rays are focused sharply upon the retina, much as a motion-picture image is focused on a screen for viewing. Unlike the motion-picture image, however, the one formed on the retina is upside down and reversed from left to right (fig. 12.37). When the visual cortex of the cerebrum interprets such an image, it corrects the reversals and objects are seen in their real positions.

For many people, after the age of forty-five, it seems as if the print in magazines and on medicine bottles suddenly becomes too small to read. The problem is not in the print, but in a lessening of the elastic quality of the lens capsule. In the condition presbyopia, or farsightedness of age, eyes remain focused for distant vision. Eyeglasses or contact lenses can usually make up for the eyes' loss of refracting power.

Other visual problems result from eyeballs that are too long or too short for sharp focusing. If an eyeball is too long, light waves focus in front of the retina, blurring the image. In other words, the refracting power of the eye, even when the lens is flattened, is too great. Although a person with this problem may be able to focus on close objects by accommodation, distance vision is invariably poor. For this reason, the person is said to be *nearsighted*. Eyeglasses or contact lenses with concave surfaces that focus images farther from the front of the eye treat nearsightedness (myopia).

If an eye is too short, light waves are not focused sharply on the retina because their point of focus lies behind it. A person with this condition may be able to bring the image of distant objects into focus by accommodation, but this requires contraction of the ciliary muscles at times when these muscles are at rest in a normal eye. Still more accommodation is necessary to view closer objects, and the person may suffer from ciliary muscle fatigue, pain, and headache when doing close work.

Most people with short eyeballs are unable to accommodate enough to focus on very close objects. They are *farsighted*. Eyeglasses or contact lenses with *convex* surfaces can remedy this condition (hyperopia) by focusing images closer to the front of the eye (figs. 12E and 12F).

Another refraction problem, *astigmatism*, reflects a defect in the curvature of the cornea or the lens. The normal cornea has a spherical curvature, like the inside of a ball; most astigmatic corneas have an elliptical curvature, like the bowl of a spoon. As a result, some portions of an image are in focus on the retina, but other portions are blurred, and vision is distorted.

Without corrective lenses, astigmatic eyes tend to accommodate back and forth reflexly to sharpen focus. The consequence of this continual action is often ciliary muscle fatigue and headache.

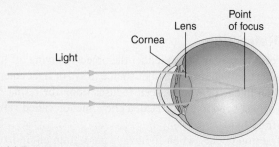

(a) Eye too long (myopia)

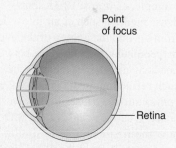

(b) Normal eye

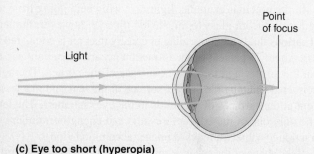

(c) Eye too short (hyperopia)

FIGURE 12E Point of focus. (*a*) Myopia (nearsightedness)—If an eye is too long, the focus point of images lies in front of the retina. (*b*) In a normal eye, the focus point is on the retina. (*c*) Hyperopia (farsightedness)—If an eye is too short, the focus point lies behind the retina.

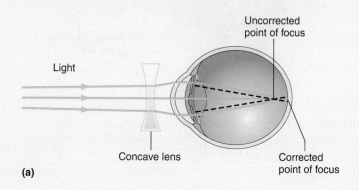

(a)

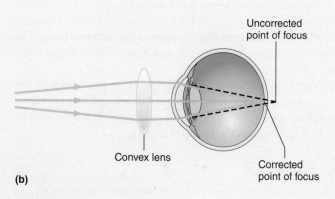

(b)

FIGURE 12F Corrective lenses. (*a*) A concave lens corrects nearsightedness. (*b*) A convex lens corrects farsightedness.

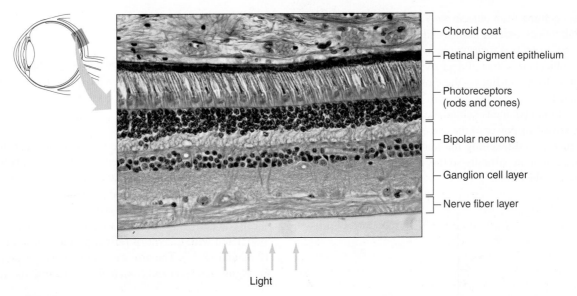

FIGURE 12.33 [AP|R] Note the layers of cells and nerve fibers in this light micrograph of the retina (75×).

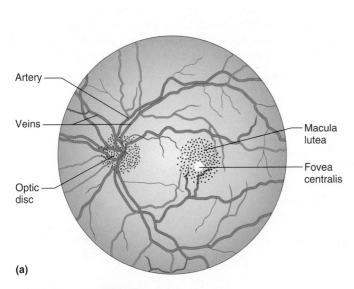

(a)

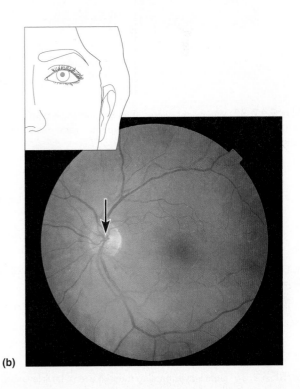

(b)

FIGURE 12.34 [AP|R] The retina. (a) Major features of the retina. (b) Nerve fibers leave the retina in the area of the optic disc (arrow) to form the optic nerve in this magnified view (53×).

Layer/Tunic	Posterior Portion	Function	Anterior Portion	Function
Outer layer	Sclera	Protection	Cornea	Light transmission and refraction
Middle layer	Choroid coat	Blood supply; pigment prevents reflection	Ciliary body, iris	Accommodation; controls light intensity
Inner layer	Retina	Photoreception; impulse conduction	None	

TABLE 12.6 Layers of the Eye

Light rays coming from objects more than 20 feet away travel in nearly parallel lines, and the cornea and the lens in its more flattened or "at-rest" condition focus the light rays on the retina. Light arriving from objects less than 20 feet away, however, reaches the eye along more divergent lines—the closer the object, the more divergent the lines.

Divergent light rays focus behind the retina unless something increases the refracting power of the eye. Accommodation accomplishes this increase, thickening the lens. As the lens thickens, light rays converge more strongly so that light coming from close objects focuses on the retina.

PRACTICE

40 What is refraction?

41 What parts of the eye provide refracting surfaces?

42 Why is it necessary to accommodate for viewing close objects?

Photoreceptors

The photoreceptors of the eye are modified neurons of two distinct types. One group of receptor cells, called **rods,** have long, thin projections at their terminal ends. The cells of the other group, called **cones,** have short, blunt projections. The retina has about 100 million rods and 3 million cones.

Rods and cones occupy a deep layer of the retina, closely associated with a layer of retinal pigment epithelium (see figs. 12.32 and 12.33). The projections from the receptors extend into the pigmented layer and contain light-sensitive visual pigments.

> *Albinism* is an inherited condition in which an enzyme required to produce pigment is missing, causing very pale, highly sun-sensitive skin. More severe forms of albinism also affect the eyes, making vision blurry and intolerant to light. A person may squint even in very faint light. This separate extra sensitivity arises because light reflects inside the lenses, overstimulating visual receptors. The eyes of many people with albinism also dart about uncontrollably, in a condition called *nystagmus.*

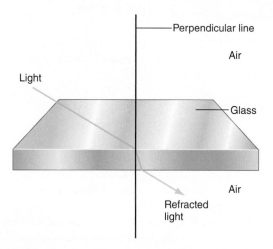

FIGURE 12.35 When light passes at an oblique angle from air into glass, the light waves bend toward a line perpendicular to the surface of the glass.

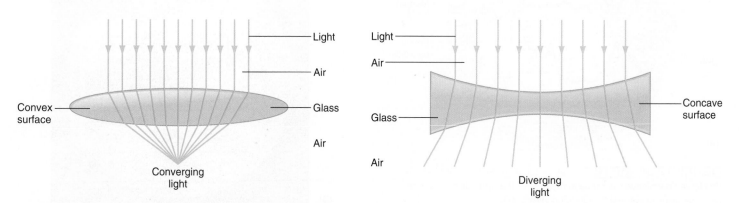

FIGURE 12.36 Light waves passing through a lens. (*a*) A lens with a convex surface causes light waves to converge. (*b*) A lens with a concave surface causes them to diverge.

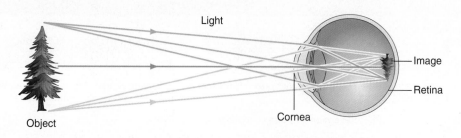

FIGURE 12.37 The image of an object forms upside down on the retina.

The retinal pigment epithelium absorbs light that the receptor cells do not absorb, and with the pigment of the choroid coat, keeps light from reflecting off the surfaces inside the eye. The retinal pigment epithelium also stores vitamin A, which the receptor cells use to synthesize visual pigments.

The visual receptors are stimulated only when light reaches them. A light image focused on an area of the retina stimulates some receptors, which conduct impulses to the brain. However, the impulse leaving each activated receptor provides only a small portion of the information required for the brain to interpret a complete scene.

Rods and cones function differently. Rods are hundreds of times more sensitive to light than are cones, and as a result, rods provide vision in dim light. In addition, rods produce colorless (black and white) vision, whereas cones can detect colors.

Cones provide sharp images, whereas rods produce more general outlines of objects. This is because nerve fibers associated with many rods may converge, so their impulses may be conducted to the brain on a single nerve fiber (see chapter 10, pp. 382–383). Thus, if light stimulates a rod, the brain cannot tell which one of many receptors has been stimulated. Such convergence is much less common among cones, so when a cone is stimulated, the brain is able to pinpoint the stimulation more accurately (fig. 12.38).

The area of sharpest vision, the fovea centralis in the macula lutea, lacks rods but has densely packed cones with few or no converging fibers. Also, the overlying layers of the retina, as well as the retinal blood vessels, are displaced to the sides in the fovea, which more fully exposes the receptors to incoming light. Consequently, to view something in detail, a person moves the eyes so that the important part of an image falls upon the fovea centralis.

The concentration of cones decreases in areas farther away from the macula lutea, whereas the concentration of rods increases in these areas. Also, the degree of convergence among the rods and cones increases toward the periphery of the retina. As a result, the visual sensations from images focused on the sides of the retina are blurred compared with those focused on the central portion of the retina.

A forceful blow to the bony orbit can displace structures in and around the eye. The suspensory ligaments may tear, and the lens may become dislocated into the posterior cavity, or the retina may pull away from the underlying vascular choroid coat. Once the retina is detached, photoreceptor cells may die from lack of oxygen and nutrients. Unless such a *detached retina* is repaired surgically, visual loss or blindness may result.

Visual Pigments

Rods and cones contain light-sensitive pigments that decompose when they absorb light energy. The light-sensitive pigment in rods is **rhodopsin** (ro-dop′sin), or visual purple, and it is embedded in membranous discs stacked in these receptor cells (fig. 12.39). A single rod cell may have 2,000 interconnected discs, derived from the cell membrane. In the presence of light, rhodopsin molecules break down into molecules of a colorless protein called *opsin* and

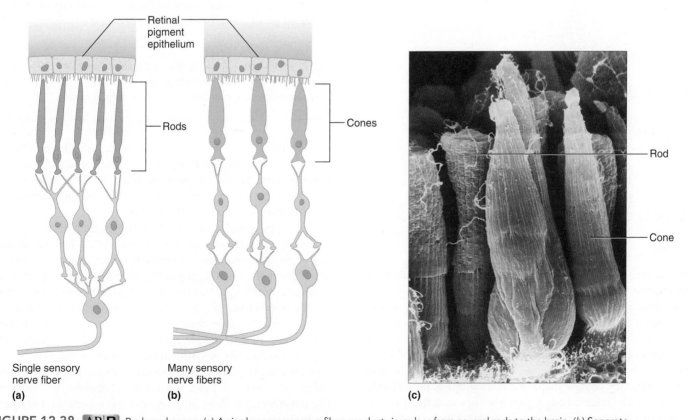

FIGURE 12.38 AP|R Rods and cones. (*a*) A single sensory nerve fiber conducts impulses from several rods to the brain. (*b*) Separate sensory nerve fibers conduct impulses from cones to the brain. (*c*) Scanning electron micrograph of rods and cones (1,350×).

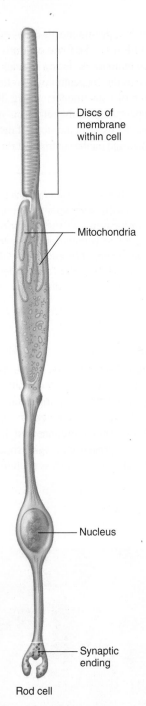

Rod cell

FIGURE 12.39 Rhodopsin is embedded in discs of membrane stacked in the rod cells.

Discs of
membrane
within cell

Mitochondria

Nucleus

Synaptic
ending

this series breaks down cGMP, and as the concentration of cGMP decreases, sodium and calcium channels close, and the receptor cell membrane hyperpolarizes (see chapter 10, p. 375). The degree of hyperpolarization is directly proportional to the intensity of the light stimulating the receptor cells.

The hyperpolarization reaches the synaptic end of the cell, inhibiting release of neurotransmitter. Decreased release of neurotransmitter by photoreceptor cells triggers action potentials in nearby retinal neurons. Consequently, impulses are conducted away from the retina, through the optic nerve, and into the brain, where they are interpreted as vision. This process is essentially the same in both rods and cones.

In bright light, nearly all of the rhodopsin in the rods decomposes, sharply reducing the sensitivity of these receptors (the rhodopsin loses its purplish color as a result, and is said to have "bleached"). The cones continue to function, however, and in bright light, we therefore see in color. In dim light, rhodopsin can be regenerated from opsin and retinal faster than it is broken down. This regeneration requires cellular energy, which ATP provides (see chapter 4, p. 126). Under these conditions, the rods continue to function and the cones remain unstimulated. Hence, we see only shades of gray in dim light.

The light sensitivity of an eye whose rods have converted the available opsin and retinal to rhodopsin increases about 100,000 times, and the eye is said to be *dark adapted*. A person needs dark-adapted eyes to see in dim light. For example, when going from daylight into a darkened theater, it may be difficult to see well enough to locate a seat, but soon the eyes adapt to the dim light, and vision improves. Later, leaving the theater and entering the sunlight may cause discomfort or even pain. This occurs at the moment that most of the rhodopsin decomposes in response to the bright light. Consequently, the light sensitivity of the eyes decreases greatly, and they become *light adapted*.

Too little vitamin A in the diet reduces the amount of retinal, impairing rhodopsin production and sensitivity of the rods. The result is poor vision in dim light, called nightblindness.

The light-sensitive pigments of cones, called **iodopsins,** are similar to rhodopsin in that they are composed of retinal combined with a protein; the protein, however, differs from the protein in the rods. The three sets of cones in the retina all contain an abundance of one of three different visual pigments.

Daylight, as well as so-called white light from indoor lighting, is actually a mixture of light of different colors. A familiar illustration of this phenomenon is a rainbow, which appears seemingly out of nowhere when light passing through water vapor in the air after a storm separates into its component colors. The colors are described either by a physical property, called a *wavelength,* or simply by the name, such as blue, red, or green. The shortest wavelengths of visible light are perceived as violet, whereas the longest wavelengths of visible light are seen as red. One type of cone pigment (erythrolabe) is most sensitive to red light, another (chlorolabe) to green light, and a third (cyanolabe) to blue light. The sensitivities of these pigments overlap somewhat. For example, both red and green light pigments are sensitive to orange light. Red pigment absorbs orange light most effectively.

a yellowish organic molecule called *retinal* (retinene) synthesized from vitamin A.

In darkness, a nucleotide called *cyclic guanosine monophosphate* (cGMP) keeps sodium and calcium channels open in portions of the receptor cell membranes. The resultant release of neurotransmitter by the rods inhibits the generation of action potentials along the optic nerve. When rhodopsin molecules absorb light, they change shape and release opsin, in trillionths of a second. The released opsin then becomes an active enzyme, which activates a second enzyme (transducin), which, in turn, activates another enzyme (phosphodiesterase). The third enzyme of

The color perceived depends upon which cones the light in a given image stimulates. If all three types of cones are stimulated with equal intensity, the light is perceived as white, and if none are stimulated, it is seen as black.

Examination of the retinas of different people reveals that individuals have unique patterns of these three cone types, all apparently able to provide color vision. Some parts of the retina are even normally devoid of one particular type, yet the brain integrates information from all over to "fill in the gaps," creating a continuous overall image. People who lack a cone type due to a mutation are colorblind.

Stereoscopic Vision

Stereoscopic vision (stereopsis) simultaneously perceives distance, depth, height, and width of objects. Such vision is possible because the pupils are 6–7 centimeters apart. Consequently, close objects (less than 20 feet away) produce slightly different retinal images. That is, the right eye sees a little more of one side of an object, while the left eye sees a little more of the other side. The visual cortex superimposes and interprets the two images. The result is the perception of a single object in three dimensions (fig. 12.40).

Stereoscopic vision requires vision with two eyes (binocular vision). Therefore, a person with only one functional eye is less able to accurately judge distance and depth. To compensate, a person with one eye can use the relative sizes and positions of familiar objects as visual clues.

Visual Pathways

The axons of the ganglion cells in the retina leave the eyes, forming the *optic nerves* (see chapter 11, p. 421). Just anterior to the pituitary gland, these nerves give rise to the X-shaped *optic chiasma,* and in the chiasma, some of the fibers cross over. Specifically, the fibers from the nasal (medial) half of each retina cross over, whereas those from the temporal (lateral) halves do not. In this way, fibers from the nasal half of the left eye and the temporal half of the right eye form the right *optic tract;* fibers from the nasal half of the right eye and the temporal half of the left eye form the left optic tract.

The nerve fibers continue in the optic tracts, and before they reach the thalamus, a few of them enter nuclei that function in various visual reflexes. Most of the fibers, however, enter the thalamus and synapse in a part of its posterior portion (lateral geniculate body). From this region, the visual sensory fibers enter pathways called *optic radiations* that lead to the visual cortex of the occipital lobe (fig. 12.41).

> Each visual cortex receives impulses from each eye, so a person may develop partial blindness in both eyes if either visual cortex is injured. For example, if the right visual cortex (or the right optic tract) is injured, sight may be lost in the temporal side of the right eye and the nasal side of the left eye. Similarly, damage to the central portion of the optic chiasma, where fibers from the nasal sides of the eyes cross over, blinds the nasal sides of both eyes.

Sensory fibers not leading to the thalamus conduct visual impulses downward into the brainstem. These impulses are important for controlling head and eye movements associated with visually tracking an object; for controlling the simultaneous movements of both eyes; and for controlling certain visual reflexes, such as those that move the muscles of the iris.

 PRACTICE

43 Distinguish between the rods and the cones of the retina.

44 Explain the roles of visual pigments.

45 What factors make stereoscopic vision possible?

46 Trace the pathway of visual impulses from the retina to the visual cortex of the occipital lobe.

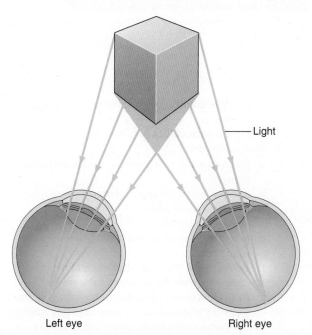

FIGURE 12.40 Stereoscopic vision results from formation of two slightly different retinal images.

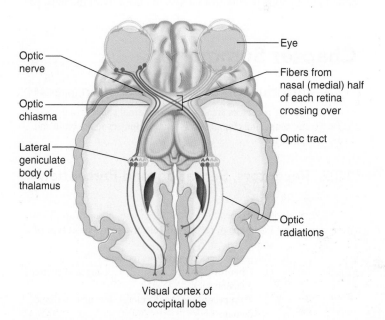

FIGURE 12.41 AP|R The visual pathway includes the optic nerve, optic chiasma, optic tract, and optic radiations.

Optic nerve
Optic chiasma
Lateral geniculate body of thalamus
Eye
Fibers from nasal (medial) half of each retina crossing over
Optic tract
Optic radiations
Visual cortex of occipital lobe

12.5 | Life-Span Changes

We often first become aware of aging-associated changes through diminished senses. By age forty, a book may need to be held farther away from the eyes. By the fifties, the senses of smell and taste may begin to diminish, which usually reflects anosmia, a loss of olfactory receptors.

By age sixty, a quarter of the population experiences noticeable hearing loss, and from ages sixty-five to seventy-four, the percentage reaches a third. Half of all people over age eighty-five cannot hear adequately. Age-related hearing loss may be the result of decades of cumulative damage to the sensitive hair cells of the spiral organ in the inner ear. It becomes more difficult to hear high pitches, as well as particular sounds, such as *f, g, s, t, sh, th, z,* and *ch.* Hearing loss may also be due to a degeneration or failure of nerve pathways to the brain. This condition, called presbycusis, may affect the ability to understand speech. It gradually worsens. Tinnitus, a ringing or roaring in the ears, is also more common among older adults. Hearing aids can often restore some hearing. A person to whom the ordinary sounds of life are hopelessly garbled may show symptoms of paranoia, depression, or social withdrawal.

Vision may decline with age for several reasons. "Dry eyes" are common. Too few tears, or poor quality tears, lead to itching and burning eyes, and diminished vision. In some cases, too many tears result from oversensitivity to environmental effects, such as wind, intense light, or a change in temperature.

With age, tiny dense clumps of gel or crystal-like deposits form in the vitreous humor. When these clumps cast shadows on the retina, the person sees small moving specks in the field of vision. These specks, or *floaters,* are most apparent when looking at a plain background, such as the sky or a blank wall. Also with age, the vitreous humor may shrink and pull away from the retina. This may mechanically stimulate receptor cells of the retina, and the person may see flashes of light.

The inability to read small print up close as one gets older, called presbyopia, results from a loss of elasticity in the lens, preventing it from changing shape easily. After age seventy, the pupil cannot dilate as well as it once did, halving the amount of light that can enter the eye. Brighter lights can counter this effect.

Glaucoma develops in the eyes as a person ages when the rate of aqueous humor formation exceeds the rate of its removal. Fluid accumulates in the anterior chamber of the eye, raising the fluid pressure. As this pressure is transmitted to all parts of the eye, in time, the blood vessels that supply the receptor cells of the retina may squeeze shut, cutting off the nutrient and oxygen supply. The result may eventually be permanent blindness.

Drugs, or traditional or laser surgery to promote the outflow of aqueous humor, can treat glaucoma if it is diagnosed early. However, because glaucoma in its early stages typically produces no symptoms, discovery of the condition usually depends on measuring the intraocular pressure using an instrument called a *tonometer.*

A common eye disorder particularly in older people is *cataract.* The lens or its capsule slowly becomes cloudy, opaque, and discolored, adding a yellowish tinge to a person's view of the world. Clear images cannot focus on the retina, and in time, the person may become blind. Removing the lens with a laser and replacing it with an artificial implant can treat cataract. Afterward, patients report that their surroundings are no longer yellow.

Several conditions affect the retinas of an older person. In a form of age-related macular degeneration, and also in people with diabetes who develop diabetic retinopathy, tiny blood vessels extend into the macula, making images in the center of the visual field appear wavy. Retinal detachment becomes more common. Despite these various problems, many older individuals continue to enjoy sharp, functional senses well into the upper decades of life.

 PRACTICE

47 Why do smell and taste diminish with age?

48 What are some causes of age-related hearing loss?

49 Describe visual problems likely to arise with age. ∎

Chapter Summary

12.1 Introduction to Sensory Function (page 444)

Sensory receptors are sensitive to internal and external environmental changes and initiate impulses to the brain and spinal cord.

12.2 Receptors, Sensation, and Perception
(page 444)

1. Receptor types
   a. Each type of receptor is sensitive to a distinct type of stimulus.
   b. The major types of receptors include:
      (1) **Chemoreceptors,** sensitive to changes in chemical concentration.
      (2) **Pain receptors** (nociceptors), sensitive to tissue damage.
      (3) **Thermoreceptors,** sensitive to temperature changes.
      (4) **Mechanoreceptors,** sensitive to mechanical forces.
      (5) **Photoreceptors,** sensitive to light.
2. Sensory impulses
   a. When receptors are stimulated, membrane potentials change.
   b. Receptor potentials are transferred to nerve fibers, triggering action potentials.
3. Sensation and perception
   a. **Sensation** is an awareness resulting from sensory stimulation.
   b. **Perception** is when a particular part of the sensory cortex interprets the sensory stimulation.
   c. **Projection** occurs when the cerebral cortex interprets a sensation to come from the receptors being stimulated.
4. **Sensory adaptation** is the adjustment of sensory receptors to continuous stimulation. Impulses are triggered at slower rates.

12.3 General Senses (page 446)

General senses receive information from receptors in skin, muscles, joints, and viscera. They can be grouped as **exteroceptive, interoceptive,** and **proprioceptive senses.**

1. Touch and pressure senses
   a. **Free nerve endings,** common in epithelial tissues, are responsible for the sensation of itching.
   b. **Tactile corpuscles** are the receptors for the sensations of light touch.
   c. **Lamellated corpuscles** are the receptors for the sensations of heavy pressure and vibrations.
2. Temperature senses
   a. Thermoreceptors include two sets of free nerve endings that are heat and cold receptors.
   b. Combinations of input from both receptor types are interpreted as intermediate temperatures.
3. Sense of pain
   a. Pain receptors
      (1) Pain receptors are free nerve endings that tissue damage stimulates.
      (2) Pain receptors provide protection; do not adapt rapidly; and can be stimulated by changes in temperature, mechanical force, and chemical concentration.
   b. The only receptors in viscera that provide sensations are pain receptors.
      (1) These receptors are most sensitive to certain chemicals and lack of blood flow.
      (2) The sensations they produce may feel as if they come from some other part of the body (**referred pain**).
   c. Pain pathways
      (1) The two main types of pain fibers are acute pain fibers and chronic pain fibers.
      (2) Acute pain fibers are fast conducting; chronic pain fibers are slower conducting.
      (3) Pain impulses are processed in the dorsal horn of the spinal cord, and they ascend in the spinothalamic tracts.
      (4) Within the brain, pain information passes through the thalamus or the reticular formation before being relayed to the cerebral cortex.
   d. Regulation of pain impulses
      (1) Awareness of pain occurs when impulses reach the thalamus.
      (2) The cerebral cortex judges the intensity of pain and locates its source.
      (3) Certain neuropeptides synthesized in the CNS inhibit pain impulses.
      (4) Impulses descending from the brain cause neurons to release pain-relieving substances, such as enkephalins and serotonin.
      (5) Endorphin is a pain-relieving biochemical produced in the brain.
4. Proprioception
   a. **Stretch receptors** provide information about the condition of muscles and tendons.
   b. **Muscle spindles** are stimulated when a muscle is relaxed, and they initiate a reflex that contracts the muscle.
   c. **Golgi tendon organs** are stimulated when muscle tension increases, and they initiate a reflex that relaxes the muscle.
5. Visceral receptors include lamellated corpuscles and free nerve endings.

12.4 Special Senses (page 452)

Special senses have receptors in complex sensory organs of the head.

1. Sense of smell
   a. Olfactory receptors
      (1) Olfactory receptors are chemoreceptors that chemicals dissolved in nasal secretions stimulate.
      (2) Olfactory receptors function with taste receptors and aid in food selection.
   b. Olfactory organs
      (1) The olfactory organs consist of receptor cells and supporting cells in the nasal cavity.
      (2) **Olfactory receptor cells** are neurons with cilia containing membrane receptor proteins.
   c. Olfactory pathways.
      (1) Smell information travels from the olfactory receptor cells through the **olfactory bulbs** and **olfactory tracts.**
      (2) This information goes to interpreting centers in the limbic system.
   d. Olfactory stimulation
      (1) Olfactory impulses may result when an odorant molecule stimulates a distinct set of receptor cells.
      (2) Olfaction undergoes rapid sensory adaptation.
      (3) Olfactory receptors are often damaged by environmental factors and are replaced from a pool of stem cells.
2. Sense of taste
   a. Taste receptors
      (1) **Taste buds** consist of receptor cells and supporting cells.
      (2) **Taste cells** have taste hairs that sense particular chemicals dissolved in water.
      (3) Taste hairs are the chemosensitive structures that trigger impulses to the brain.
   b. Taste sensations
      (1) The five primary taste sensations are sweet, sour, salty, bitter, and umami.
      (2) Various taste sensations result from the stimulation of one or more sets of taste cells.
      (3) Each of the five primary types of taste cells is particularly sensitive to a certain group of chemicals.
   c. Taste pathways
      (1) Sensory information from taste receptors travels on fibers of the facial, glossopharyngeal, and vagus nerves.
      (2) This information is carried to the medulla oblongata and ascends to the thalamus and then to the gustatory cortex in the parietal lobes.
3. Sense of hearing
   a. The outer ear includes the **auricle,** the **external acoustic meatus,** and the **tympanic membrane.** It collects sound waves created by vibrating objects and transfers the vibrations to the middle ear.
   b. Middle ear
      (1) **Auditory ossicles** of the middle ear transfer sound waves from the tympanic membrane to the **oval window** of the inner ear. They also increase the force of these waves.
      (2) Skeletal muscles attached to the auditory ossicles provide the **tympanic reflex,** which protects the inner ear from the effects of loud sounds.

c. **Auditory tubes** connect the middle ears to the throat and help maintain equal air pressure on both sides of the tympanic membranes.

d. Inner ear
   (1) The inner ear consists of a complex system of connected tubes and chambers—the **bony labyrinth** and the **membranous labyrinth.** It includes the **cochlea,** which houses the **spiral organ.**
   (2) The **spiral organ** includes the hearing receptors that are stimulated by vibrations in the fluids of the inner ear.
   (3) Different frequencies of vibrations stimulate different sets of receptor cells; the human ear can detect sound frequencies from about 20 to 20,000 vibrations per second.

e. Auditory pathways
   (1) The axons from hearing receptors are found in the cochlear branch of the vestibulocochlear nerves.
   (2) Auditory information travels into the medulla oblongata, midbrain, and thalamus and is interpreted in the temporal lobes of the cerebrum.

4. Sense of equilibrium
   a. **Static equilibrium** maintains the stability of the head and body when they are motionless. The organs of static equilibrium are in the **vestibule.**
   b. **Dynamic equilibrium** balances the head and body when they are suddenly moved or rotated. The organs of this sense are in the **ampullae** of the semicircular canals.
   c. Other structures that help maintain equilibrium include the proprioceptors associated with certain joints and the eyes.

5. Sense of sight
   a. Visual accessory organs include the **eyelids** and **lacrimal apparatus** that protect the eye and the **extrinsic muscles** that move the eye.
   b. Structure of the eye
      (1) The wall of the eye has an outer, a middle, and an inner tunic that function as follows:
         (a) The outer layer (**sclera**) is protective, and its transparent anterior portion (**cornea**) refracts light entering the eye.
         (b) The middle layer (**choroid coat**) is vascular and has pigments that help keep the inside of the eye dark.
         (c) The inner layer (**retina**) includes visual receptor cells.
      (2) The **lens** is a transparent, elastic structure. The ciliary muscles control its shape.
      (3) The **iris** is a muscular diaphragm that controls the amount of light entering the eye; the **pupil** is an opening in the iris.
      (4) Spaces in the eye are filled with fluids (**aqueous** and **vitreous humors**) that help maintain its shape.

c. Light refraction
   (1) Light rays are primarily refracted by the cornea and lens to focus an image on the retina.
   (2) The lens must thicken to focus on close objects.

d. Visual receptors
   (1) The photoreceptors are **rods** and **cones.**
   (2) Rods are responsible for colorless vision in dim light, and cones provide color vision.

e. Visual pigments
   (1) A light-sensitive pigment in rods (**rhodopsin**) decomposes in the presence of light and triggers a complex series of reactions that initiate action potentials on the optic nerve.
   (2) Three types of cones provide color vision. Each type has a different light-sensitive pigment, and each type is sensitive to a different wavelength of light; the color perceived depends on which cones are stimulated.

f. Stereoscopic vision
   (1) **Stereoscopic vision** provides perception of distance and depth.
   (2) Stereoscopic vision occurs because of the formation of two slightly different retinal images that the brain superimposes and interprets as one image in three dimensions.
   (3) A one-eyed person uses relative sizes and positions of familiar objects to judge distance and depth.

g. Visual pathways
   (1) Nerve fibers from the retina form the optic nerves.
   (2) Some fibers cross over in the optic chiasma.
   (3) Most of the fibers enter the thalamus and synapse with others that continue to the visual cortex of the occipital lobe.
   (4) Other impulses pass into the brainstem and function in various visual reflexes.

12.5 Life-Span Changes (page 482)

Diminished senses are often one of the first noticeable signs of aging.

1. Age-related hearing loss may reflect damage to hair cells of the spiral organ, degeneration of nerve pathways to the brain, or tinnitus.

2. Age-related visual problems include dry eyes, floaters and light flashes, presbyopia, glaucoma, cataracts, macular degeneration, and retinal detachment.

CHAPTER ASSESSMENTS

12.1 Introduction to Sensory Function

1 Explain the difference between a general sense and a special sense. (p. 444)

12.2 Receptors, Sensation, and Perception

2 Match each sensory receptor to the type of stimulus to which it is likely to respond: (p. 444)

(1) chemoreceptor   A. approaching headlights
(2) pain receptor   B. a change in blood pressure
(3) thermoreceptor   C. the smell of roses
(4) mechanoreceptor   D. an infected tooth
(5) photoreceptor   E. a cool breeze

3 Explain the difference between a sensation and a perception. (p. 444)

4 Explain how sensory receptors stimulate sensory impulses. (p. 444)

5 Explain the projection of a sensation. (p. 445)

6 Define *sensory adaptation*. (p. 445)

7 You fill up the tub to take a hot bath, but the water is too hot. You test it a second and third time within a few seconds, and it feels okay. Which of the following is the most likely explanation? (p. 446)

   a. The water has cooled down unusually quickly.
   b. Your ability to sense heat has adapted.
   c. Your nervous system is suddenly not functioning properly.
   d. Someone added ice cubes to your bath.

12.3 General Senses

8 Explain how general senses can be grouped. (p. 446)

9 Describe the functions of free nerve endings, tactile corpuscles, and lamellated corpuscles. (p. 446)

10 Describe the functions of the two classes of thermoreceptors. (p. 446)

11 Compare pain receptors with the other types of somatic receptors. (p. 446)

12 List the conditions likely to stimulate visceral pain receptors. (p. 447)

13 Define *referred pain*, and provide an example. (pp. 447 and 449)

14 Contrast the pathways involved in the production of acute and chronic pain. (p. 449)

15 Explain how neuropeptides relieve pain. (p. 449)

16 Distinguish between muscle spindles and Golgi tendon organs. (pp. 450–451)

12.4 Special Senses

17 Explain how the senses of smell and taste function together to create the perception of the flavors of foods. (p. 452)

18 Which two of the following are part of the olfactory organs? (p. 452)

   a. olfactory receptors
   b. columnar epithelial cells in the nasal mucosa
   c. the nose
   d. the brain

19 Trace each step in the pathway from an olfactory receptor to the interpreting center of the cerebrum. (pp. 453–454)

20 Salivary glands are important in taste because _____. (p. 455)

   a. they provide the fluid in which food molecules dissolve
   b. the taste receptors are located in salivary glands
   c. salivary glands are part of the brain
   d. lamellar corpuscles are activated

21 Name the five primary taste sensations and indicate a specific stimulus for each. (p. 455)

22 Explain why taste sensation is less likely to diminish with age than olfactory sensation. (p. 455)

23 Trace each step in the pathway from a taste receptor to the interpreting center of the cerebrum. (p. 456)

24 Match the ear area with the associated structure: (pp. 456–459)

(1) outer ear   A. cochlea
(2) middle ear   B. tympanic membrane
(3) inner ear   C. auditory ossicles

25 Trace each step in the pathway from the external acoustic meatus to hearing receptors. (pp. 456–459)

26 Describe the functions of the auditory ossicles. (p. 457)

27 Identify the parts of the tympanic reflex, explain how they work, and explain the importance of this reflex. (p. 458)

28 The function of the auditory tube is to _____. (p. 459)

   a. equalize air pressure on both sides of the tympanic membrane
   b. conduct sound vibrations to the tympanic membrane
   c. contain the hearing receptors
   d. contain the auditory ossicles

29 Distinguish between the bony and membranous labyrinths. (p. 459)

30 Describe the cochlea and its function. (p. 459)

31 Which of the following best describes hearing receptor "hair cells"? (p. 459)

   a. They are neurons.
   b. They lack ion channels.
   c. They are epithelial, but function like neurons.
   d. They are made of keratin.

32 Explain how a hearing receptor stimulates a sensory neuron. (p. 461)

33 Trace each step in the pathway from the spiral organ to the interpreting centers of the cerebrum. (pp. 459–461)

34 Describe the organs of static and dynamic equilibrium and their functions. (pp. 465–466)

35 Explain how the sense of vision helps maintain equilibrium. (p. 467)

36 Match the visual accessory organ with its function: (p. 468)

(1) eyelid   A. moves the eye
(2) conjunctiva   B. covers the eye
(3) lacrimal gland   C. lines the eyelids
(4) extrinsic muscle D. produces tears

37 Name the three layers of the eye wall and describe the functions of each layer. (pp. 472–474)

38 Explain the mechanisms of pupil constriction and pupil dilation. (p. 473)

39 Distinguish between the fovea centralis and the optic disc. (p. 474)

40 The following are compartments in the eye. In which one is vitreous humor found? (p. 475)

   a. anterior chamber
   b. posterior chamber
   c. anterior cavity
   d. posterior cavity

41 Explain how light is focused on the retina. (pp. 475 and 478)

42 Explain why looking at a close object causes fatigue in terms of how accommodation is accomplished. (pp. 475 and 478)

43 Distinguish between rods and cones. (pp. 478–479)

44 Explain why cone vision is generally more acute than rod vision. (p. 479)

45 Describe the function of rhodopsin. (pp. 479–480)

46 Explain why rod vision may be more important in dim light than in bright light. (p. 480)

47 Describe the relationship between light wavelength and color vision. (pp. 480–481)
48 Define *stereoscopic vision*. (p. 481)
49 Explain why a person with normal binocular vision is able to judge distance and depth of close objects more accurately than a person who has lost one eye. (p. 481)
50 Trace each step in the pathway from the retina to the visual cortex. (p. 481)

12.5 Life-Span Changes
51 Explain the basis of fading senses of smell and taste with aging. (p. 482)
52 List three causes of hearing loss associated with aging. (p. 482)
53 Explain five problems that can interfere with vision as a person ages. (p. 482)

 # INTEGRATIVE ASSESSMENTS/CRITICAL THINKING

Outcomes 2.2, 11.4, 12.2, 12.3, 12.4

1. Positron emission tomography (PET) scans of the brains of people who have been blind since birth reveal high neural activity in the visual centers of the cerebral cortex when these people read Braille. When sighted individuals run their fingers over the raised letters of Braille, their visual centers do not show increased activity. Explain these findings.

Outcomes 6.4, 11.5, 12.2

2. Why are some serious injuries, like a bullet entering the abdomen, relatively painless, but others, such as a burn, considerably more painful?

Outcomes 11.4, 12.2, 12.4

3. Loss of the sense of smell often precedes the major symptoms of Alzheimer disease and Parkinson disease. What additional information is needed to use this association to prevent or treat these diseases?

Outcomes 12.2, 12.3

4. A patient with heart disease experiences pain at the base of the neck and in the left shoulder and upper limb during exercise. How would you explain the likely origin of this pain to the patient?

Outcomes 12.2, 12.4

5. People who are deaf due to cochlear damage do not suffer motion sickness. Why not?

6. Labyrinthitis is an inflammation of the inner ear. What symptoms would you expect in a patient with this disorder?

 # ONLINE STUDY TOOLS

 |

Connect Interactive Questions Reinforce your knowledge using assigned interactive questions covering the general senses (touch, pressure, temperature, and pain) and special senses (smell, taste, hearing, balance, and vision).

Connect Integrated Activity Can you predict the effects on vision of injuries at various locations along the visual pathways?

LearnSmart Discover which chapter concepts you have mastered and which require more attention. This adaptive learning tool is personalized, proven, and preferred.

Anatomy & Physiology Revealed Go more in depth into the human body and explore the structures associated with your senses of hearing and vision.

In this electron micrograph of a hormone-secreting cell from the adrenal cortex, the upper portion of the cell has been removed to reveal the nucleus (brown) and the edge of the cell membrane (grey) (14,400×).

13

Endocrine System

 LEARNING OUTCOMES

After you have studied this chapter, you should be able to:

13.1 General Characteristics of the Endocrine System
1. Distinguish between endocrine and exocrine glands. (p. 488)
2. Explain what makes a cell a target cell for a hormone. (p. 488)
3. List some important functions of hormones. (pp. 488–489)

13.2 Hormone Action
4. Describe how hormones can be classified according to their chemical composition. (pp. 489–490)
5. Explain how steroid and nonsteroid hormones affect their target cells. (pp. 490–496)

13.3 Control of Hormonal Secretions
6. Discuss how negative feedback mechanisms regulate hormone secretion. (p. 496)
7. Explain how the nervous system controls hormone secretion. (p. 496)

13.4–13.9 Pituitary Gland–Other Endocrine Glands
8. Name and describe the locations of the major endocrine glands, and list the hormones that they secrete. (pp. 498–515)
9. Describe the actions of the various hormones and their contributions to homeostasis. (pp. 500–515)
10. Explain how the secretion of each hormone is regulated. (pp. 500–515)

13.10 Stress and Its Effects
11. Distinguish between physical and psychological stress. (p. 517)
12. Describe the general stress response. (pp. 517–518)

13.11 Life-Span Changes
13. Describe some of the changes associated with aging of the endocrine system. (pp. 519–521)

THE WHOLE PICTURE

The hormones that the endocrine system produces have many and diverse effects on the body. A new mother hears her baby cry and her breasts release milk. A busy executive attends meetings all day without a break to eat but still maintains enough blood glucose to function. The early morning smell of smoke and the clang of fire alarms in a dormitory elevate heart rates, dilate airways, and send students streaming from the building.

The endocrine system, like the nervous system, is all about communication as cells secrete chemicals that act on other cells. Whereas the nervous system releases neurotransmitter molecules at synapses, the endocrine system releases its characteristic substances, hormones, into the bloodstream.

Some hormones act briefly, while others have quite long-lasting effects. Hormone actions may be subtle, their levels requiring constant adjustment to maintain homeostasis. Other hormones have dramatic effects, from overseeing the gradual growth and development of an infant into a toddler, an adolescent, and then an adult, to the immediate excitement of participating in an athletic competition.

aprevealed.com

Module 8: Endocrine System

cort-, bark, rind: adrenal *cortex*—outer portion of an adrenal gland.

-crin, to secrete: endo*crine*—internal secretion.

diuret-, to pass urine: *diuret*ic—substance that promotes urine production.

endo-, inside: *endo*crine gland—gland that internally secretes into a body fluid.

exo-, outside: *exo*crine gland—gland that secretes to the outside through a duct.

horm-, impetus, impulse: *horm*one—substance that a cell secretes that affects another cell.

hyper-, above: *hyper*thyroidism—condition resulting from an above-normal secretion of thyroid hormone.

hypo-, below: *hypo*thyroidism—condition resulting from a below-normal secretion of thyroid hormone.

lact-, milk: pro*lact*in—hormone that promotes milk production.

med-, middle: adrenal *med*ulla—middle section of an adrenal gland.

para-, beside: *para*thyroid glands—set of glands near the surface of the thyroid gland.

toc-, birth: oxy*toc*in—hormone that stimulates the uterine muscles to contract during childbirth.

-tropic, influencing: adrenocortico*tropic* hormone—a hormone secreted by the anterior pituitary gland that stimulates the adrenal cortex.

vas-, vessel: *vas*opressin—substance that contracts blood vessel walls.

13.1 | General Characteristics of the Endocrine System

Endocrine means "internal secretion." The **endocrine system** is so named because the cells, tissues, and organs that compose it, collectively called endocrine glands, secrete substances into the internal environment. The secreted substances, called **hormones,** diffuse from the interstitial fluid into the bloodstream and eventually act on cells, called **target cells,** some distance away (fig. 13.1*a*). In contrast to endocrine secretions, exocrine secretions are released externally. Secretions from **exocrine glands** enter tubes or ducts that lead to body surfaces. Two examples of exocrine secretions are stomach acid reaching the lumen of the digestive tract and sweat released at the skin's surface (fig. 13.1*b*).

Other glands secrete substances into the internal environment that are not hormones by the traditional definition, but they function similarly as messenger molecules and are sometimes termed "local hormones." These include **paracrine** secretions, which enter the interstitial fluid but affect only nearby cells, and **autocrine** secretions, which affect only the cell secreting the substance.

Cells of the endocrine system and the nervous system both communicate using chemical signals that bind to receptor molecules. Table 13.1 summarizes some similarities and differences between the two systems. In contrast to the nervous system, which releases neurotransmitter molecules into synapses, the endocrine system releases hormones into the bloodstream, which carries these messenger molecules everywhere. However, the endocrine system is also precise, because only target cells can respond to a particular type of hormone (fig. 13.2). A hormone's target cells have specific receptors that are not on other cells. These receptors are proteins or glycoproteins with binding sites for a specific hormone. The non-endocrine chemical messengers, paracrine and autocrine substances, also bind to specific receptors. This chapter includes examples of them.

Endocrine glands and their hormones help regulate metabolic processes. They control the rates of certain chemical reactions; aid in transporting substances through membranes; and help regulate water balance, electrolyte balance, and blood pressure. Endocrine hormones also play vital roles in reproduction, development, and growth.

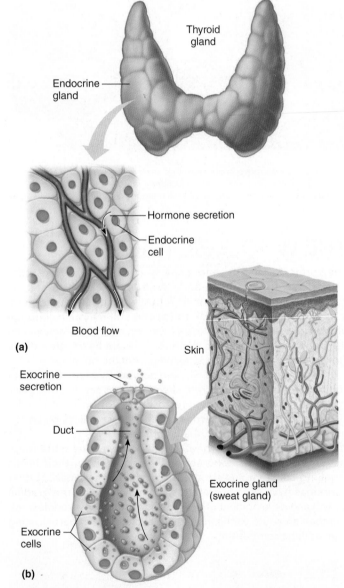

FIGURE 13.1 Types of glands. (*a*) Endocrine glands release hormones into the internal environment (body fluids). (*b*) Exocrine glands secrete to the outside environment through ducts that lead to body surfaces.

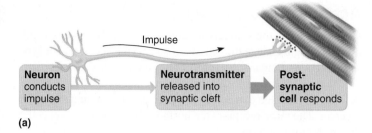

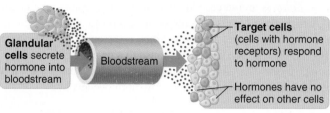

(a)

(b)

FIGURE 13.2 Chemical communication. (*a*) Neurons release neurotransmitters into synapses, affecting postsynaptic cells. (*b*) Glands release hormones into the bloodstream. Blood carries hormone molecules throughout the body, but only target cells respond.

 Which is more specific in terms of which cells are affected, a neurotransmitter or a hormone?

Answer can be found in Appendix G.

Small groups of specialized cells produce some hormones. However, the larger endocrine glands—the pituitary gland, thyroid gland, parathyroid glands, adrenal glands, and pancreas—are the subject of this chapter (fig. 13.3). Subsequent chapters discuss several other hormone-secreting glands and tissues.

13.2 | Hormone Action

Hormones are released into the extracellular spaces surrounding endocrine cells. From there, they diffuse into the bloodstream and are carried to all parts of the body.

TABLE 13.1	A Comparison Between the Nervous System and the Endocrine System	
	Nervous System	**Endocrine System**
Cells	Neurons	Glandular epithelium
Chemical signal	Neurotransmitter	Hormone
Specificity of action	Receptors on postsynaptic cell	Receptors on target cell
Speed of onset	Seconds	Seconds to hours
Duration of action	Very brief unless neuronal activity continues	May be brief or may last for days even if secretion ceases

Chemistry of Hormones

Hormones are organic compounds. They are of two major types: steroids, or steroidlike substances; and nonsteroids, which include amines, peptides, proteins, and glycoproteins. Hormones can stimulate changes in target cells even in extremely low concentrations.

Steroid Hormones

Steroids (ste′roidz) are lipids that include complex rings of carbon and hydrogen atoms (fig. 13.4*a*). Steroids differ by the types and numbers of atoms attached to these rings and the ways they are joined (see fig. 2.16, p. 72). All steroid hormones are derived from cholesterol (see chapter 2, p. 72). They include sex hormones such as testosterone and the estrogens, and secretions of the adrenal cortex (the outer portion of the adrenal gland), including aldosterone and cortisol. Vitamin D is a modified steroid and when converted to the active form in the kidneys and liver becomes a hormone (see "Parathyroid Hormone" on pp. 507–508, and chapter 18, p. 706).

Nonsteroid Hormones

Hormones called *amines,* including norepinephrine and epinephrine, are derived from the amino acid tyrosine. These hormones are also synthesized in the adrenal medulla (the inner portion of the adrenal gland) (fig. 13.4*b*).

CAREER CORNER

Diabetes Educator

The young woman had been controlling her type 2 diabetes mellitus for several years with a strict low-carbohydrate diet, exercise whenever she could manage it, and oral medications. But her blood glucose levels over the past three months had been too elevated, and so her physician advised that she begin injecting insulin once a day. The woman was very upset about having to inject herself, and wondered how she would travel for business and maintain her daily treatment. Fortunately her physician prescribed regular sessions with a diabetes educator, who helped the patient adjust to her change in treatment plan.

A diabetes educator is a health-care professional with a degree in nursing, dietetics, pharmacy, or another related field. The educator is knowledgeable about the many aspects of maintaining stable blood glucose levels and treating diabetes, and is also expert in counseling and communicating complex medical information.

Living with diabetes requires more self-care, on a daily basis, than living with many other diseases. The diabetes educator helps the patient understand how to maintain healthful diet and exercise habits, how to time her monitoring of her blood glucose level and take various medications, and how to recognize and reduce risks of diabetes-related complications. A diabetes educator can tailor a care plan to the individual, for although diabetes is a common condition, no two patients have exactly the same experience.

Diabetes educators work in doctor's offices, clinics, and hospitals. They may provide one-on-one care, or lead groups in discussing how to live with diabetes.

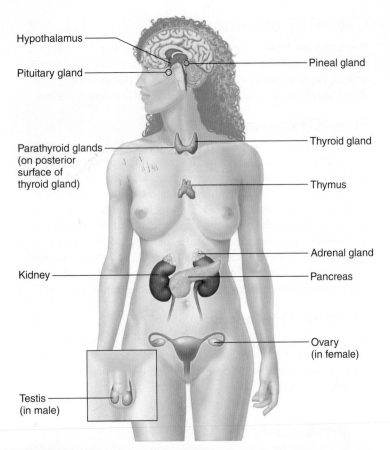

Hypothalamus

Pituitary gland

Pineal gland

Parathyroid glands (on posterior surface of thyroid gland)

Thyroid gland

Thymus

Adrenal gland

Kidney

Pancreas

Ovary (in female)

Testis (in male)

FIGURE 13.3 Locations of major endocrine glands.

Protein hormones, like all proteins, are composed of long chains of amino acids that are linked and folded into specific molecular structures (see chapter 2, pp. 72–75, and fig. 13.4c). They include the hormone secreted by the parathyroid gland and some of those secreted by the anterior pituitary gland. Certain other hormones secreted from the anterior pituitary gland are *glycoproteins*, which consist of carbohydrates joined to proteins.

The *peptide* hormones are short chains of amino acids (fig. 13.4d). This group includes hormones associated with the posterior pituitary gland and some produced in the hypothalamus.

Another group of compounds, called *prostaglandins* (pros″tah-glan′dinz), are paracrine substances. They regulate neighboring cells. Prostaglandins are lipids (20-carbon fatty acids that include 5-carbon rings) and are synthesized from a type of fatty acid (arachidonic acid) in cell membranes (fig. 13.4e). Prostaglandins are produced in a wide variety of cells, including those of the liver, kidneys, heart, lungs, thymus, pancreas, brain, and reproductive organs.

Table 13.2 lists the names and abbreviations of some of the hormones discussed in this chapter. Table 13.3 and figure 13.4 summarize the chemical composition of hormones.

 PRACTICE

1 What is a hormone?

2 How do endocrine glands and exocrine glands differ?

3 How are hormones chemically classified?

Actions of Hormones

Hormones exert their effects by altering metabolic processes. A hormone might change the activity of an enzyme necessary for synthesizing a particular substance or alter the rate at which particular chemicals are transported through cell membranes. A hormone delivers its message to a cell by uniting with the binding site of its receptor. The more receptors the hormone binds on its target cells, the greater the response.

The number of receptors on target cells may change. *Upregulation* is an increase in the number of receptors on a target cell, which often occurs as a response to a prolonged decrease in the level of a hormone. *Downregulation* is the opposite, a decrease in the number of receptors in response to a prolonged increase in hormone levels. Therefore, the number of receptors changes in ways that maintain an appropriate response to hormone level.

Steroid Hormones and Thyroid Hormones

Steroid hormones are poorly soluble in water. They are carried in the bloodstream bound to plasma proteins in a way that they are released in sufficient quantity to affect their target cells. Unlike amine, peptide, and protein hormones, steroid hormones are soluble in the lipids that make up the bulk of cell membranes. For this reason, these hormones can diffuse into cells relatively easily and are able to enter any cell in the body, although only target cells will respond.

 RECONNECT

To Chapter 3, Membrane Transport, pages 98–100.

Thyroid hormones are also poorly soluble in water and are transported in the blood bound to plasma proteins. They have often been described as entering cells the same way that steroids enter, but recent evidence suggests that specific membrane transport mechanisms may move them.

Once inside a target cell, steroid and thyroid hormones combine (usually in the nucleus) with specific protein receptors. The resulting *hormone-receptor complex* binds to particular DNA sequences, either activating or repressing specific genes. Activated genes are transcribed into messenger RNA (mRNA) molecules. The mRNAs enter the cytoplasm, where they direct the synthesis of specific proteins, which may be enzymes, transport proteins, or even hormone receptors. The activities of these hormones produce the cellular changes associated with the particular hormone (fig. 13.5, table 13.4, and Clinical Application 13.1). An example of a steroid hormone is **aldosterone** (al′do-ster-ōn″, al-dos′ter-ōn), which is secreted from the adrenal glands and stimulates the kidneys to retain sodium. In response to aldosterone, cells that form tubules in the kidney begin to synthesize more Na$^+$/K$^+$ pumps, which are the proteins that actively transport these ions across the cell membrane, in this case returning sodium to the bloodstream.

 RECONNECT

To Chapter 4, Nucleic Acids and Protein Synthesis, pages 138–142.

In some cases, steroid hormones may repress a particular gene, so it is not transcribed. The cellular response results from decreased levels of the encoded protein.

TABLE 13.2 Hormone Names and Abbreviations

Source	Name	Abbreviation	Synonym
Hypothalamus	Corticotropin-releasing hormone	CRH	
	Gonadotropin-releasing hormone	GnRH	Luteinizing hormone-releasing hormone (LHRH)
	Somatostatin	SS	Growth hormone release-inhibiting hormone (GHRIH)
	Growth hormone-releasing hormone	GHRH	
	Prolactin release-inhibiting hormone	PIH	Dopamine
	Prolactin-releasing factor*	PRF*	
	Thyrotropin-releasing hormone	TRH	
Anterior pituitary gland	Adrenocorticotropic hormone	ACTH	Corticotropin
	Follicle-stimulating hormone	FSH	Follitropin
	Growth hormone	GH	Somatotropin (STH)
	Luteinizing hormone	LH	Lutropin, interstitial cell-stimulating hormone (ICSH)
	Prolactin	PRL	
	Thyroid-stimulating hormone	TSH	Thyrotropin
Posterior pituitary gland	Antidiuretic hormone	ADH	Vasopressin
	Oxytocin	OT	
Thyroid gland	Calcitonin		
	Thyroxine	T_4	Tetraiodothyronine
	Triiodothyronine	T_3	
Parathyroid gland	Parathyroid hormone	PTH	Parathormone
Adrenal medulla	Epinephrine	EPI	Adrenalin
	Norepinephrine	NE	Noradrenalin
Adrenal cortex	Aldosterone		
	Cortisol		Hydrocortisone
Pancreas	Glucagon		
	Insulin		
	Somatostatin	SS	

*"Factor" is used because specific prolactin-releasing hormones have not yet been identified.

Nonsteroid Hormones

A nonsteroid hormone, such as an amine, peptide, or protein, combines with specific receptor molecules on the target cell membrane. Each receptor molecule is a protein that has a *binding site* and an *activity site.* The hormone combines with the binding site, which causes the receptor's activity site to interact with other membrane proteins. The hormone that triggers this first step in what becomes a cascade of biochemical activity is considered a *first messenger.* The biochemicals in the cell that induce the changes leading to the hormone effect are called *second messengers.*

Many hormones use **cyclic adenosine monophosphate** (cyclic AMP, or cAMP) as a second messenger. In this mechanism, a hormone binds to its receptor, and the resulting hormone-receptor complex activates a protein called a **G protein,** which then activates

TABLE 13.3 Types of Hormones

Type of Compound	Formed from	Examples
Amines	Amino acids	Norepinephrine, epinephrine
Peptides	Amino acids	ADH, OT, TRH, SS, GnRH
Proteins	Amino acids	PTH, GH, PRL
Glycoproteins	Protein and carbohydrate	FSH, LH, TSH
Steroids	Cholesterol	Estrogens, testosterone, aldosterone, cortisol

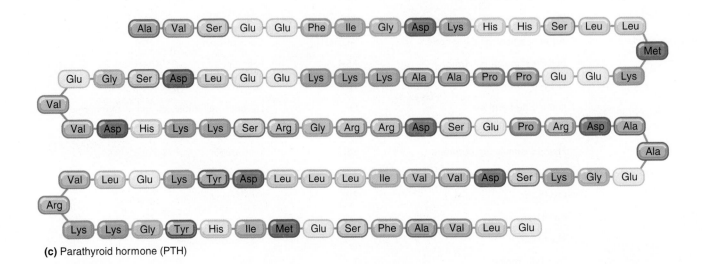

FIGURE 13.4 Structural formulas of (*a*) a steroid hormone (cortisol) and (*b*) an amine hormone (norepinephrine). Amino acid sequences of (*c*) a protein hormone (PTH) and (*d*) a peptide hormone (oxytocin). Structural formula of (*e*) a prostaglandin (PGE₂).

(a) Cortisol

(b) Norepinephrine

(c) Parathyroid hormone (PTH)

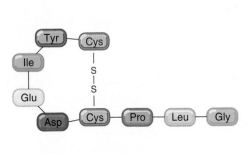

(d) Oxytocin

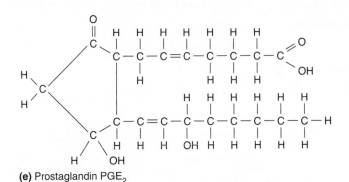

(e) Prostaglandin PGE₂

an enzyme called **adenylate cyclase** (ah-den′ĭ-lāt si′klās), which is an integral membrane protein with its active site facing the inside of the cell. Activated adenylate cyclase removes two phosphates from ATP and reconnects the exposed oxygen, forming *cyclic AMP* (fig. 13.6). Cyclic AMP activates another set of enzymes called **protein kinases** (ki′nās-ez) that transfer phosphate groups from ATP molecules to protein substrate molecules, a process called *phosphorylation*. Phosphorylation activates some proteins, and renders others inactive.

Next, the phosphorylated proteins alter various cellular processes, bringing about the effect of that particular hormone (fig. 13.7). The response of a cell to a hormone is determined by the type of membrane receptors present and by the types of protein substrate molecules in the cell. Table 13.5 summarizes these events.

TABLE 13.4	Sequence of Steroid Hormone Action

1. Endocrine gland secretes steroid hormone.

2. Blood carries hormone molecules throughout the body.

3. Steroid hormone diffuses through target cell membrane and enters cytoplasm or nucleus.

4. Hormone combines with a receptor molecule in the cytoplasm or nucleus.

5. Steroid hormone-receptor complex binds to DNA and promotes transcription of messenger RNA.

6. Messenger RNA enters the cytoplasm and directs protein synthesis.

7. Newly synthesized proteins produce the steroid hormone's specific effects.

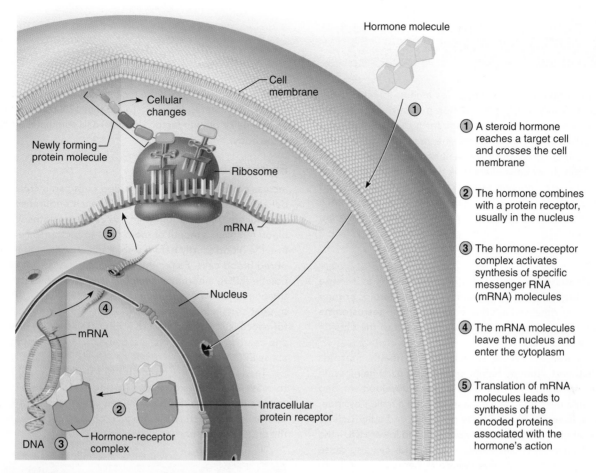

FIGURE 13.5 **AP|R** Steroid hormones. Thyroid hormones, which are non-steroids, act in much the same way, but may enter the cell by a different mechanism.

1. A steroid hormone reaches a target cell and crosses the cell membrane

2. The hormone combines with a protein receptor, usually in the nucleus

3. The hormone-receptor complex activates synthesis of specific messenger RNA (mRNA) molecules

4. The mRNA molecules leave the nucleus and enter the cytoplasm

5. Translation of mRNA molecules leads to synthesis of the encoded proteins associated with the hormone's action

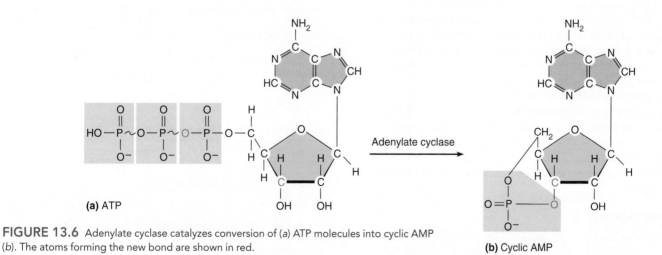

FIGURE 13.6 Adenylate cyclase catalyzes conversion of (a) ATP molecules into cyclic AMP (b). The atoms forming the new bond are shown in red.

Cellular responses to second messenger activation include altering membrane permeabilities, activating enzymes, promoting synthesis of certain proteins, stimulating or inhibiting specific metabolic pathways, promoting cellular movements, and initiating secretion of hormones and other substances. A specific example is the action of epinephrine to raise blood sugar during periods of physical stress. Epinephrine acts through the second messenger cAMP to increase the activity of the enzyme that breaks down liver glycogen, increasing the number of glucose molecules that can diffuse out of liver cells and enter the bloodstream.

Another enzyme, phosphodiesterase, quickly and continuously inactivates cAMP, so its action is short-lived. For this reason, a continuing response in a target cell requires a continuing signal from hormone molecules binding receptors in the target cell membrane.

Abusing Hormones to Improve Athletic Performance

Abuse of performance-enhancing drugs among athletes, both amateur and professional, is a common problem and not new (table 13A). Athletes have used drugs for a competitive edge since the earliest Olympics, when cocaine, heroin, morphine, and strychnine were the drugs of choice. Amphetamines joined the list after World War II, when soldiers took these stimulants to mask the fatigue that accompanies great exertion. The drugs boosted energy in sports, too. Today, the general focus of performance enhancement is misuse of certain hormones. Three examples are described here.

Steroids

Athletes who abuse steroids seek the hormone's ability to increase muscular strength. Abusers are caught when the steroids or their metabolites are detected in urine or when natural testosterone levels plummet in a negative feedback response to the outside supply of the hormone.

Abusing steroids carries serious risks to health. Steroids hasten adulthood, stunting height and causing early hair loss. In males, excess steroid hormones lead to breast development, and in females to a deepened voice, hairiness, and a male physique. The kidneys, liver, and heart may be damaged, and atherosclerosis may develop because steroids raise LDL and lower HDL—the opposite of a healthy cholesterol profile. Steroids can also cause psychiatric symptoms, including delusions, depression, and violent behavior.

For some athletes, illicit steroid use cannot be detected because of a natural mutation that deletes part of a gene. The result is a block in the conversion of testosterone from fat soluble to water soluble, and the hormone is not excreted into the urine as it normally is. This mutation is responsible for variations seen in different populations in the amount of testosterone excreted in the urine. In one telling experiment, researchers in Sweden injected fifty-five male volunteers with a high dose of testosterone, and seventeen of the men showed no traces of the steroid in their urine!

The urine ratio of testosterone to an inactive form of the hormone called epitestosterone can reveal the taking of exogenous (outside) testosterone. The body normally produces testosterone and epitestosterone in about equal amounts. The presence of much more testosterone than epitestosterone therefore indicates "doping."

Growth Hormone

Some athletes take human growth hormone (HGH) preparations instead of, or to supplement, the effects of steroids. HGH enlarges muscles; steroids strengthen them. HGH has been available as a prescription drug since 1985, and it is used as such to treat children with certain forms of inherited dwarfism. However, HGH is available from other nations without prescription and can be obtained illegally to enhance athletic performance. Unlike steroids, HGH has a half-life of only seventeen to forty-five minutes, which means that it becomes so scant that it is undetectable in body fluids within an hour. At first only elite athletes abused HGH

to enhance performance. One recent study of young weight lifters in the United States, however, found that 12% of them took HGH or insulin-like growth factor-1 (which HGH stimulates synthesis of in the body).

Erythropoietin

Increasing the number of red blood cells can increase oxygen delivery to muscles and thereby enhance endurance. Athletes introduced "blood doping" in 1972. The athletes would have blood removed a month or more prior to performance, then reinfuse the blood shortly before a competition, boosting the number of red blood cells. Easier than blood doping is to take erythropoietin (EPO), a hormone secreted from the kidneys that signals the bone marrow to produce more red blood cells. EPO is used to treat certain forms of anemia. Using it to improve athletic performance is ill advised. In 1987, EPO abuse led to heart attacks and death in twenty-six cyclists. Some runners and swimmers also abuse EPO.

Testing for drugs or for mutations that enable athletes to mask abuse is expensive. Testing for biomarkers that are the breakdown products of abused drugs may be a less costly approach to detecting the practice of cheating by manipulating the endocrine system.

TABLE 13A	List of Prohibited Drugs
Anabolic agents	
Anabolic androgenic steroids	
Other anabolic agents	
Hormones and related substances	
EPO	
Growth factors	
Gonadotropins (LH, hCG in males)	
Insulin	
Corticotropins	
Beta-2 agonists	
Hormone antagonists and modulators	
Aromatase inhibitors	
Selective estrogen receptor modulators	
Other antiestrogens	
Myostatin inhibitors	
Diuretics	

Source: World Anti-Doping Agency

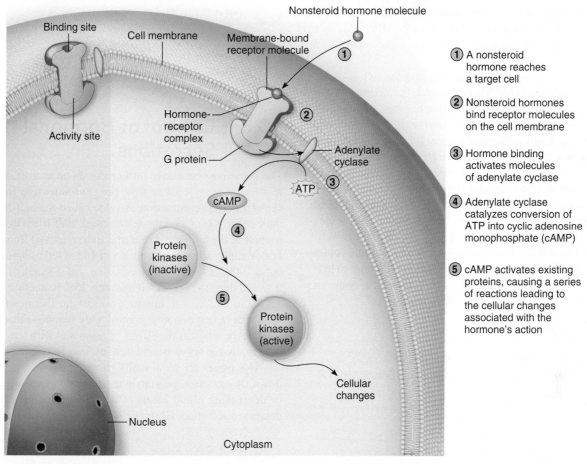

FIGURE 13.7 AP|R Nonsteroid hormone action.

Labels in figure:
- Binding site
- Cell membrane
- Membrane-bound receptor molecule
- Nonsteroid hormone molecule
- Activity site
- Hormone-receptor complex
- G protein
- Adenylate cyclase
- ATP
- cAMP
- Protein kinases (inactive)
- Protein kinases (active)
- Cellular changes
- Nucleus
- Cytoplasm

1. A nonsteroid hormone reaches a target cell

2. Nonsteroid hormones bind receptor molecules on the cell membrane

3. Hormone binding activates molecules of adenylate cyclase

4. Adenylate cyclase catalyzes conversion of ATP into cyclic adenosine monophosphate (cAMP)

5. cAMP activates existing proteins, causing a series of reactions leading to the cellular changes associated with the hormone's action

TABLE 13.5	Sequence of Actions of Nonsteroid Hormone Using Cyclic AMP

1. Endocrine gland secretes nonsteroid hormone.

2. Blood carries hormone molecules throughout the body.

3. Hormone combines with receptor site on membrane of its target cell, activating G protein.

4. Adenylate cyclase molecules are activated in target cell's membrane.

5. Adenylate cyclase converts ATP into cyclic AMP.

6. Cyclic AMP activates protein kinases.

7. Protein kinases activate protein substrates in the cell that change metabolic processes.

8. Cellular changes produce the hormone's effects.

An abnormality in cAMP-mediated signaling can lead to symptoms from many endocrine glands. In McCune-Albright syndrome, for example, a defect in the G protein that activates adenylate cyclase results in conversion of ATP to cAMP even without hormonal stimulation. As a result, cells in the pituitary, thyroid, gonads, and adrenal glands secrete hormones in excess. One symptom is precocious puberty. Infant girls menstruate, and boys as young as six years produce mature sperm. Abnormal or deficient G proteins also cause colorblindness and retinitis pigmentosa.

Hormones whose actions require cyclic AMP include releasing hormones from the hypothalamus; thyroid-stimulating hormone (TSH), adrenocorticotropic hormone (ACTH), follicle-stimulating hormone (FSH), and luteinizing hormone (LH) from the anterior pituitary gland; antidiuretic hormone (ADH) from the posterior pituitary gland; parathyroid hormone (PTH) from the parathyroid glands; norepinephrine and epinephrine from the adrenal glands; calcitonin from the thyroid gland; and glucagon from the pancreas.

Certain nonsteroid hormones use second messengers other than cAMP. For example, a second messenger called diacylglycerol (DAG), like cAMP, activates a protein kinase, leading to a cellular response.

In another mechanism, a hormone binding its receptor increases calcium ion concentration in the cell. Such a hormone may stimulate transport of calcium ions inward through the cell membrane or induce release of calcium ions from cellular storage sites via a second messenger called inositol triphosphate (IP3). The calcium ions combine with the protein *calmodulin* (see chapter 9, p. 307), altering its molecular structure in a way that activates the molecule. Activated calmodulin can then interact with enzymes, altering their activities and thus eliciting diverse responses.

Still another hormonal mechanism uses *cyclic guanosine monophosphate* (cyclic GMP, or cGMP). Like cAMP, cGMP is a nucleotide derivative and functions in much the same manner as a second messenger.

The cellular response to a hormone operating through a second messenger is greatly amplified. This is possible because many second messenger molecules can be activated in response to just a few hormone-receptor complexes, and the enzymes that are activated as a result can repeatedly catalyze reactions. Because existing proteins are activated, the response is fast. Cells are highly sensitive to changes in the concentrations of nonsteroid hormones because of such rapid amplification. Cellular response to a steroid hormone (and thyroid hormones) is directly proportional to the number of hormone-receptor complexes that form. Some amplification occurs because more than one mRNA may be transcribed from an activated gene and each mRNA may be translated into multiple copies of a protein. The response is much slower than the response to hormones acting through second messengers because protein synthesis takes some time, although the response lasts longer.

 PRACTICE

4 How does a steroid hormone act on its target cells?

5 How does a nonsteroid hormone act on its target cells?

6 What is a second messenger?

Prostaglandins

Prostaglandins are paracrine substances, acting locally, that are in small amounts, but potent. They are not stored in cells but are synthesized just before they are released. They are rapidly inactivated.

Some prostaglandins regulate cellular responses to hormones. For example, different prostaglandins can either activate or inactivate adenylate cyclase in cell membranes, thereby controlling production of cAMP and altering the cell's response to a hormone.

Prostaglandins produce a variety of effects. Some prostaglandins relax smooth muscle in the airways of the lungs and in the blood vessels, dilating these passageways. Yet other prostaglandins can contract smooth muscle in the walls of the uterus, causing menstrual cramps and labor contractions. They stimulate secretion of hormones from the adrenal cortex and inhibit secretion of hydrochloric acid from the wall of the stomach. Prostaglandins also influence movements of sodium ions and water in the kidneys, help regulate blood pressure, and have powerful effects on both male and female reproductive physiology. When tissues are injured, prostaglandins promote inflammation.

Understanding prostaglandin function has medical applications. Drugs such as aspirin and certain steroids that relieve the joint pain of rheumatoid arthritis inhibit production of prostaglandins in the synovial fluid of affected joints. Daily low doses of aspirin may reduce the risk of heart attack by altering prostaglandin activity.

 PRACTICE

7 What are prostaglandins?

8 Describe one function of prostaglandins.

9 List effects of prostaglandins.

13.3 | Control of Hormonal Secretions

The body must be able to turn processes on and off. For example, removal of acetylcholine from the neuromuscular junction stops skeletal muscle contraction. A value called the half-life is used to indicate the rate of a hormone's removal. Half-life is the time it takes for half of the hormone molecules to be removed from the plasma. For example, a hormone with a half-life of ten minutes would start out at 100% of its blood concentration, and if secretion were to stop, it would drop to 50% in ten minutes, 25% in another ten minutes, 12.5% in another ten minutes, and so on. Hormones with short half-lives (a few minutes) control body functions that turn on and off quickly, whereas the effects of hormones with longer half-lives, such as thyroid hormone and steroids, may last for days.

Hormones are continually excreted in the urine and broken down by enzymes, primarily in the liver. Therefore, increasing or decreasing blood levels of a hormone requires increased or decreased secretion. Hormone secretion is precisely regulated.

Control Sources

Control of hormone secretion is essential to maintaining the internal environment. In a few cases, primarily in the reproductive systems, positive feedback affects this control.

Generally, hormone secretion is controlled in three ways, all of which employ **negative feedback** (see chapter 1, p. 16). In each case, an endocrine gland or the system controlling it senses the concentration of the hormone the gland secretes, a process the hormone controls, or an action the hormone has on the internal environment (fig. 13.8).

1. The hypothalamus, which constantly receives information about the internal environment, controls the anterior pituitary gland's release of hormones. Many anterior pituitary hormones affect the activity of other glands. Hormones that act on other glands are called **tropic hormones** (fig. 13.8a) (fig. 13.9).

2. The nervous system directly stimulates some glands. The adrenal medulla, for example, secretes its hormones (epinephrine and norepinephrine) in response to impulses from preganglionic sympathetic neurons. The secretory cells replace the postganglionic sympathetic neurons, which would normally secrete norepinephrine alone as a neurotransmitter (see fig. 13.8b).

3. Another group of glands responds directly to changes in the composition of the internal environment. For example, when the blood glucose level rises, the pancreas secretes insulin, and when the blood glucose level falls, it secretes glucagon (see "Hormones of the Pancreatic Islets" on pp. 514–515 and fig. 13.8c).

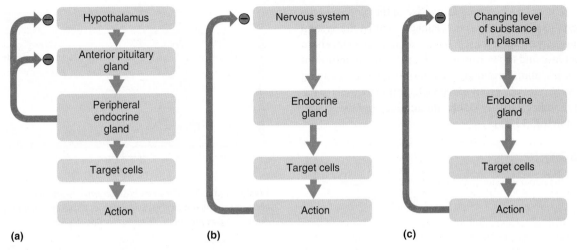

FIGURE 13.8 Examples of endocrine system control: (*a*) one way the hypothalamus controls the anterior pituitary, (*b*) the nervous system controls some glands directly, and (*c*) some glands respond directly to changes in the internal environment. ⊖ indicates negative feedback inhibition.

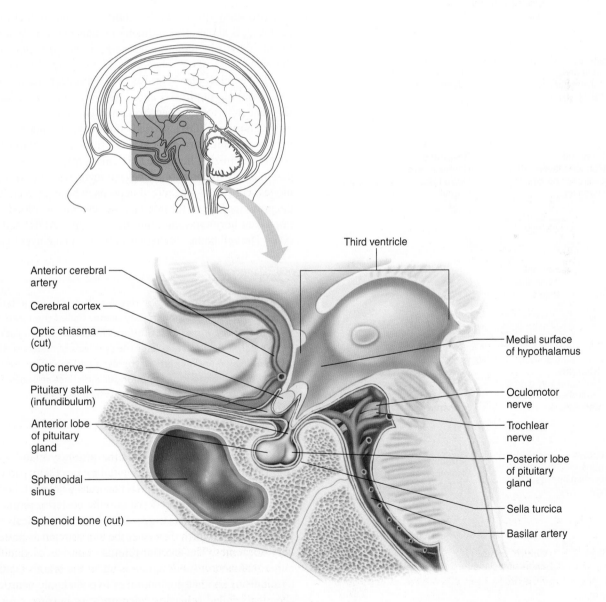

FIGURE 13.9 **AP|R** The pituitary gland is attached to the hypothalamus and lies in the sella turcica of the sphenoid bone.

In each of these cases, as hormone levels rise in the blood and the hormone exerts its effects, negative feedback inhibits the system and hormone secretion decreases. Then, as hormone levels in the blood decrease and the hormone's effects wane, inhibition of the system ceases, and secretion of that hormone increases again (fig. 13.10). As a result of negative feedback, hormone levels in the bloodstream remain relatively stable, fluctuating slightly around an average value (fig. 13.11).

 PRACTICE

10 How does the nervous system help regulate hormonal secretions?

11 How does a negative feedback system control hormonal secretion?

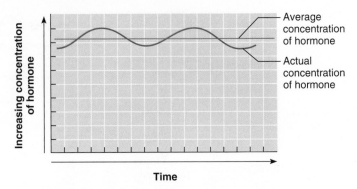

FIGURE 13.11 As a result of negative feedback, hormone concentrations remain relatively stable, although they may fluctuate slightly above and below average concentrations.

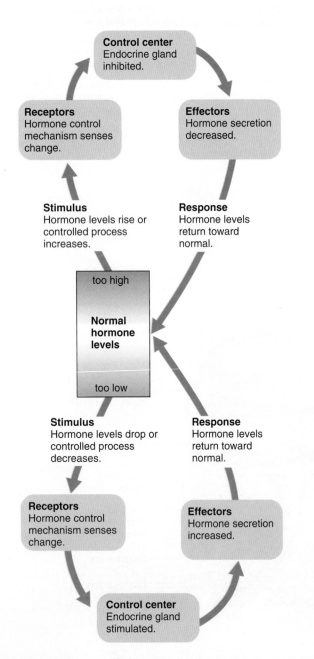

FIGURE 13.10 Hormone secretion is under negative feedback control.

13.4 | Pituitary Gland

The **pituitary** (pǐ-tu′ǐ-tār″e) **gland** (hypophysis), at the base of the brain, is about one centimeter in diameter. It is attached to the hypothalamus by the pituitary stalk, or *infundibulum,* and lies in the sella turcica of the sphenoid bone, as figure 13.9 shows.

The pituitary gland consists of two distinct portions: an *anterior lobe* (adenohypophysis) and a *posterior lobe* (neurohypophysis). The anterior lobe secretes a number of hormones, including growth hormone (GH), thyroid-stimulating hormone (TSH), adrenocorticotropic hormone (ACTH), follicle-stimulating hormone (FSH), luteinizing hormone (LH), and prolactin (PRL). The cells that make up the posterior lobe (pituicytes) do not synthesize hormones. However, specialized neurons, whose axon endings enter the posterior lobe of the pituitary, secrete into the bloodstream two important hormones: antidiuretic hormone (ADH) and oxytocin (OT). The cell bodies of these neurons are in the hypothalamus.

In the fetus, a narrow region develops between the anterior and posterior lobes of the pituitary gland. Called the *intermediate lobe* (pars intermedia), it produces melanocyte-stimulating hormone (MSH), which regulates the synthesis of melanin—the pigment in skin and in parts of the eyes and brain. In most adults this intermediate lobe is no longer a distinct structure, but its secretory cells persist in the two remaining lobes.

The brain controls most of the pituitary gland's activities (fig. 13.12). In fact, the posterior pituitary is actually part of the nervous system. The axons of certain neurons whose cell bodies are in the hypothalamus extend down into the posterior pituitary gland. Impulses on these axons trigger the release of chemicals from their axon terminals, which then enter the bloodstream as posterior pituitary hormones. The anterior pituitary consists of glandular cells rather than neurons, but it, too, is under the brain's control. Axon terminals of another population of hypothalamic neurons release chemicals called **releasing hormones** (or in some cases release-inhibiting hormones). These releasing hormones are carried in the

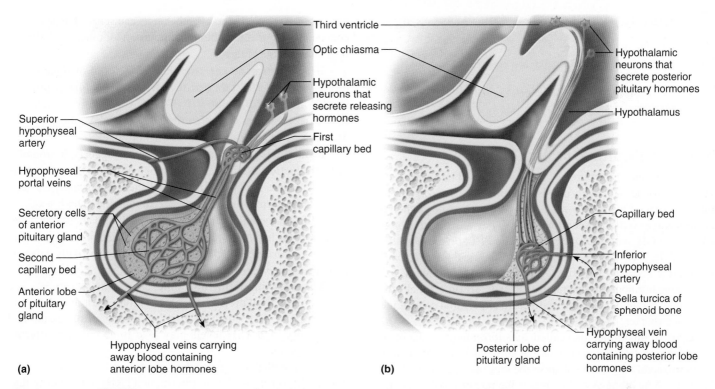

(a)

Third ventricle

Optic chiasma

Hypothalamic neurons that secrete releasing hormones

First capillary bed

Superior hypophyseal artery

Hypophyseal portal veins

Secretory cells of anterior pituitary gland

Second capillary bed

Anterior lobe of pituitary gland

Hypophyseal veins carrying away blood containing anterior lobe hormones

(b)

Hypothalamic neurons that secrete posterior pituitary hormones

Hypothalamus

Capillary bed

Inferior hypophyseal artery

Sella turcica of sphenoid bone

Hypophyseal vein carrying away blood containing posterior lobe hormones

Posterior lobe of pituitary gland

FIGURE 13.12 **AP|R** Pituitary gland. (*a*) Hypothalamic releasing hormones stimulate cells of the anterior lobe to secrete hormones. (*b*) Impulses originating in the hypothalamus stimulate nerve endings in the posterior lobe of the pituitary gland to release hormones.

blood via a capillary bed associated with the hypothalamus. The vessels merge to form the **hypophyseal** (hi″po-fiz′e-al) **portal veins** that pass downward along the pituitary stalk and give rise to a second capillary bed in the anterior lobe. In this way, substances released into the blood from the hypothalamus are carried directly to the anterior lobe. The hypothalamus, therefore, functions as an endocrine gland, yet it also controls other endocrine glands. This dual activity is also true of the anterior pituitary.

> The arrangement of two capillaries in series is unusual and is called a *portal system*. It exists in three places in the body: the hepatic portal vein connects intestinal capillaries to special liver capillaries called sinusoids, the efferent arteriole of kidney nephrons connects two sets of capillaries, and the hypophyseal portal vein gives rise to a capillary net in the anterior lobe of the pituitary gland.

Each of the hypothalamic releasing hormones acts on a specific population of cells upon reaching the anterior lobe of the pituitary gland. Some of the resulting actions are inhibitory (prolactin release-inhibiting hormone and somatostatin), but most stimulate the anterior pituitary to release hormones that stimulate the secretions of peripheral endocrine glands. In many of these cases, important negative feedback relationships regulate hormone levels in the bloodstream. **Figure 13.13** shows this general relationship.

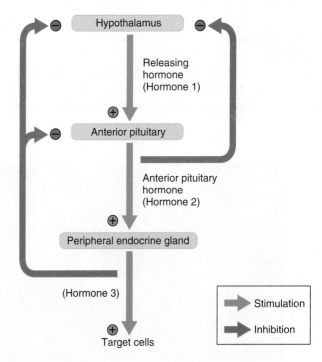

Hypothalamus

Releasing hormone (Hormone 1)

Anterior pituitary

Anterior pituitary hormone (Hormone 2)

Peripheral endocrine gland

(Hormone 3)

Target cells

Stimulation

Inhibition

FIGURE 13.13 Hypothalamic control of the peripheral endocrine glands may use as many as three types of hormones, with multiple negative feedback controls. (⊕ = stimulation; ⊖ = inhibition)

 **PRACTICE**

12 Where is the pituitary gland?

13 List the hormones that the anterior and posterior lobes of the pituitary gland secrete.

14 Explain how the hypothalamus controls the actions of the pituitary gland.

Anterior Pituitary Hormones

The anterior lobe of the pituitary gland is enclosed in a capsule of connective tissue and largely consists of epithelial tissue organized in blocks around many thin-walled blood vessels. The epithelial tissue has five types of secretory cells. They are *somatotropes* that secrete GH, *mammatropes* that secrete PRL, *thyrotropes* that secrete TSH, *corticotropes* that secrete ACTH, and *gonadotropes* that secrete FSH and LH (figs. 13.14 and 13.15). In males, LH (luteinizing hormone) is also referred to as ICSH (interstitial cell-stimulating hormone) because it affects the interstitial cells of the testes (see chapter 22, p. 838).

Growth hormone, also called *somatotropin* (or somatotrophic hormone, STH), is a protein that stimulates cells to enlarge and more rapidly divide. It enhances the movement of amino acids through cell membranes and increases the rate of protein synthesis. GH also decreases the rate at which cells use carbohydrates and increases the rate at which they use fats.

Growth hormone secretion varies during the day, peaking during sleep. Two biochemicals from the hypothalamus control its secretion. They are released alternately, exerting opposite effects. *Growth hormone-releasing hormone* (GHRH) stimulates secretion of GH, and *somatostatin* (SS) inhibits secretion.

Nutritional state can affect control of GH. More GH is released during periods of protein deficiency and abnormally low blood glucose concentration. Conversely, when blood protein and glucose concentrations increase, growth hormone secretion decreases. Apparently the hypothalamus can sense changes in the concentrations of certain blood nutrients and it releases GHRH in response to some of them.

Growth hormone can stimulate elongation of bone tissue directly, but its effect on cartilage requires a mediator protein, insulin-like growth factor-1 (IGF-1). Growth hormone releases IGF-1 from the liver and other tissues. Clinical Application 13.2 discusses some clinical uses of growth hormone.

Prolactin is a protein, and as its name suggests, it promotes milk production. No normal physiological role for this hormone in human males has been firmly established. Abnormally elevated levels of the hormone can disrupt sexual function in both sexes.

Prolactin secretion is mostly under inhibitory control by dopamine from the hypothalamus, which is also called *prolactin release-inhibiting hormone* (PRIH). The hypothalamus likely releases more than one *prolactin-releasing factor* (PRF).

Thyroid-stimulating hormone, also called *thyrotropin,* is a glycoprotein. It controls secretion of certain hormones from the thyroid gland. TSH can also stimulate growth of the gland, and abnormally high TSH levels may lead to an enlarged thyroid gland, or *goiter.*

The hypothalamus partially regulates TSH secretion by producing thyrotropin-releasing hormone (TRH). Circulating thyroid hormones help regulate TSH secretion by inhibiting release of TRH and TSH. Therefore, as the blood concentration of thyroid hormones increases, secretion of TRH and TSH declines (fig. 13.16).

External factors influence release of TRH and TSH. Exposure to extreme cold, for example, increases hormonal secretion. Emotional stress can either increase or decrease TRH and TSH secretion, depending upon circumstances.

> After the thyroid gland is removed to treat cancer, an endocrinologist monitors a patient's TSH level to determine the appropriate daily dose of synthetic thyroid hormone. Enough thyroid hormone must be taken to suppress secretion of TSH. TSH is a useful marker for thyroid hormone levels because of negative feedback, even though TSH itself would be ineffectual because the thyroid gland is no longer present.

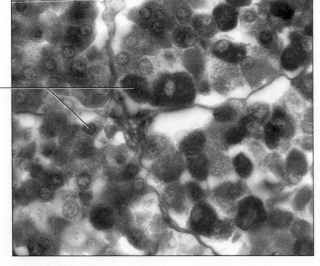

Secretory cells

FIGURE 13.14 **AP|R** Light micrograph of the anterior pituitary gland (240×).

 **PRACTICE**

15 How does growth hormone affect the cellular metabolism of carbohydrates, fats, and proteins?

16 What are the functions of prolactin?

17 How is TSH secretion regulated?

Adrenocorticotropic (ah-dre″no-kor″te-ko-trōp′ik) **hormone** also called "*corticotropin,*" is a peptide that controls the manufacture and secretion of certain hormones from the outer layer (cortex) of the adrenal gland. The secretion of ACTH is stimulated by *corticotropin-releasing hormone* (CRH), which the hypothalamus

Growth Hormone Ups and Downs

Insufficient secretion of human growth hormone (HGH) during childhood produces *hypopituitary dwarfism.* Body proportions and mental development are normal, but because secretion of other anterior pituitary hormones is also below normal, additional hormone deficiency symptoms may appear. For example, a child with growth hormone deficiency might not develop adult sexual features without hormone therapy.

HGH used as a drug can treat hypopituitary dwarfism if administration begins before the bones completely ossify. In children it is also used to treat conditions that include very short stature but in which HGH is not deficient, such as chronic renal failure, Turner syndrome, intrauterine growth retardation, and Prader-Willi syndrome. In adults HGH is used to treat short bowel syndrome, muscle wasting in people who have AIDS, and rare pituitary tumors. The Food and Drug Administration has not approved HGH to slow aging or to build muscle in healthy individuals.

The reputation of HGH as an anti-aging agent stems from a 1990 study in which a dozen men over age sixty who received the hormone showed slight improvements in muscle mass and bone mineral density. HGH given to older individuals can increase muscle mass and decrease fat but, it does not improve strength and likely just replaces some fat with water. Excess HGH can cause joint pain and swelling and increased risk of diabetes mellitus. HGH used for approved indications is injected. Oral preparations are digested in the stomach before they can exert an effect.

Oversecretion of growth hormone in childhood may result in *gigantism,* in which height may eventually exceed 8 feet. Gigantism is usually caused by a tumor of the pituitary gland, which secretes excess pituitary hormones, including HGH. As a result, a person with gigantism may have other metabolic disturbances. An inherited form of gigantism is seen in families in Ireland.

Growth hormone oversecretion in an adult after the epiphyses of the long bones have ossified causes a condition called acromegaly. The person does not grow taller, but soft tissues continue to enlarge and bones thicken, producing a large tongue, nose, hands, and feet, and a protruding jaw. The heart and thyroid enlarge. Early symptoms include headache, joint pain, fatigue, and depression. A pituitary tumor or abuse of growth hormone (as a drug) can cause acromegaly.

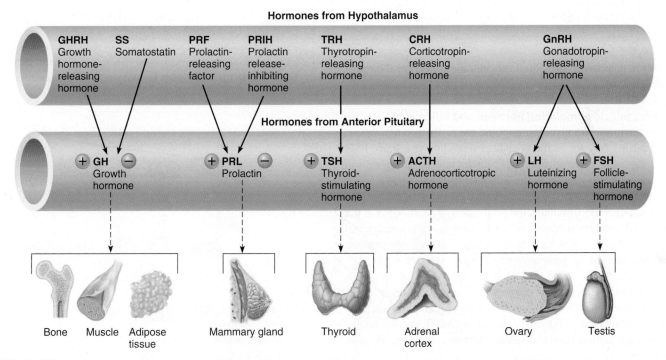

FIGURE 13.15 Hormones released from the hypothalamus, the corresponding hormones released from the anterior lobe of the pituitary gland, and their target organs.

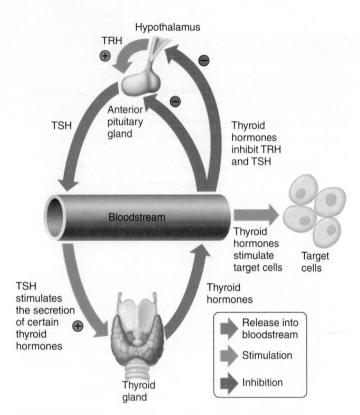

FIGURE 13.16 Thyrotropin-releasing hormone (TRH) from the hypothalamus stimulates the anterior pituitary gland to release thyroid-stimulating hormone (TSH), which stimulates the thyroid gland to release hormones. These thyroid hormones reduce the secretion of TSH and TRH by negative feedback. (⊕ = stimulation; ⊖ = inhibition)

releases in response to decreased concentrations of adrenal cortical hormones. Stress can increase secretion of ACTH by stimulating release of CRH.

Both **follicle-stimulating hormone** and **luteinizing** (lu′te′in-īz″ing) **hormone** are glycoproteins and are called *gonadotropins,* which means they act on the gonads or reproductive organs. FSH controls growth and development of follicles that house egg cells in the ovaries. It also stimulates the follicular cells to secrete a group of female sex hormones, collectively called *estrogen* (or estrogens).

In males, FSH stimulates the production of sperm cells in the testes. LH promotes secretion of sex hormones in both males and females and is essential for release of egg cells from the ovaries. Other functions of the gonadotropins and their interactions are discussed in chapter 22 (pp. 838, 849–850).

Gonadotropin secretion is under the control of *gonadotropin-releasing hormone* (GnRH). Gonadotropins are absent in the body fluids of infants and children. Secretion of GnRH, and the resulting increase in blood levels of gonadotropins, mark the the onset of puberty.

 PRACTICE

18 What is the function of ACTH?

19 What is a gonadotropin?

20 Describe the functions of FSH and LH in a female and in a male.

Posterior Pituitary Hormones

The posterior lobe of the pituitary largely consists of neuroglia *(pituicytes)* and the axons of hypothalamic neurons. This is different from the anterior lobe, which is primarily glandular epithelium. The neuroglia support the axons. The secretions of these neurons function not as neurotransmitters but as hormones (see fig. 13.12).

Specialized neurons in the hypothalamus produce the two hormones associated with the posterior pituitary—**antidiuretic** (an″tĭ-di″u-ret′ik) **hormone** (also known as *vasopressin*) and **oxytocin** (ok″sĭ-to′sin). These hormones are transported down axons through the pituitary stalk to the posterior pituitary and are stored in vesicles (secretory granules) near the ends of the axons. The hormones are released into the blood in response to action potentials conducted on the axons of the neurosecretory cells. Therefore, posterior pituitary hormones are synthesized in the hypothalamus, and their secretion is controlled by the hypothalamus, but they are named for where they enter the bloodstream.

Antidiuretic hormone and oxytocin are short polypeptides with similar sequences (fig. 13.17). A *diuretic* is a chemical that increases urine production. An *antidiuretic,* then, is a chemical that decreases urine formation. ADH produces its antidiuretic effect by reducing the volume of water that the kidneys excrete. In this way, ADH plays an important role in regulating the concentration of body fluids (see chapter 20, p. 787).

Frequent and copious urination often follows drinking alcoholic beverages, because ethyl alcohol inhibits ADH secretion. A person must replace the lost body fluid to maintain normal water balance. Drinking too much beer can actually lead to dehydration because the body loses more water than it takes in.

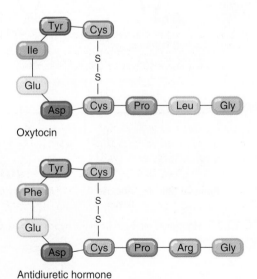

Oxytocin

Antidiuretic hormone

FIGURE 13.17 AP|R The structure of oxytocin differs from that of ADH by only two amino acids, yet they function differently.

A sufficient concentration of ADH contracts certain smooth muscle, including the smooth muscle in the walls of blood vessels. As a result, vascular resistance and blood pressure may increase. (This is why ADH is also called vasopressin.) Although ADH is seldom at high enough levels to cause high blood pressure, its secretion increases following severe blood loss. In this situation, ADH's vasoconstrictor effect may help to minimize the drop in blood pressure that results from profuse bleeding and help return blood pressure toward normal.

ADH's two effects—vasoconstriction and water retention—are possible because the hormone binds two different receptors on target cells. The binding of ADH to V1 receptors increases the concentration of the second messenger inositol triphosphate, which increases the intracellular calcium ion concentration in the smooth muscle of blood vessel walls, leading to vasoconstriction. The second receptor, V2, is on parts of the kidneys' microscopic tubules called collecting ducts. ADH binding there activates the cAMP second messenger system, which ultimately causes collecting duct cells to reabsorb water that would otherwise be excreted in the urine.

The hypothalamus regulates ADH secretion. Certain neurons in this part of the brain, called *osmoreceptors,* sense changes in the concentration of body fluids. For example, if a person is dehydrating due to a lack of water intake, the solutes in blood become more concentrated. The osmoreceptors, sensing the resulting increase in osmotic pressure, signal the posterior pituitary to release ADH, which causes the kidneys to retain water. On the other hand, if a person drinks a large volume of water, body fluids become more dilute, which inhibits the release of ADH. In response, the kidneys excrete a more dilute urine until the concentration of body fluids returns to normal.

Blood volume also affects ADH secretion. Increased blood volume stretches the walls of certain blood vessels, stimulating volume receptors that signal the hypothalamus to inhibit release of ADH. However, if hemorrhage decreases blood volume, these receptors are stretched less and therefore send fewer inhibiting impulses. As a result, ADH secretion increases and the kidneys conserve water, countering further volume loss.

Diabetes insipidus impairs ADH regulation of water balance. A baby with the condition first displayed symptoms at five months of age—he drank huge volumes of water. By thirteen months, he had become severely dehydrated, despite nearly continuous drinking. The boy was drinking sufficient fluids, but his kidneys could not retain the water because ADH V2 receptors on the kidney collecting ducts were defective. The hormone could bind, but the receptor failed to trigger cAMP formation. The boy's ADH was still able to constrict blood vessels because the V1 receptors were unaffected. A high-calorie diet and providing lots of water preserved the boy's mental abilities, but he remained small for his age. Tumors and injury affecting the hypothalamus and posterior pituitary can also cause diabetes insipidus.

Oxytocin has an antidiuretic action, but less so than ADH. In addition, oxytocin can contract smooth muscle in the uterine wall, playing a role in the later stages of childbirth. The uterus becomes more sensitive to oxytocin's effects during pregnancy. Stretching of uterine and vaginal tissues late in pregnancy, as the fetus grows, initiates sensory impulses to the hypothalamus, which then signals the posterior pituitary to release oxytocin, which, in turn, stimulates the uterine contractions of labor.

In the breasts, oxytocin contracts certain cells near the milk-producing glands and their ducts. In lactating breasts, this action forces liquid from the milk glands into the milk ducts and ejects the milk.

The mechanical stimulation of suckling initiates sensory impulses that travel to the mother's hypothalamus, which signals the posterior pituitary to release oxytocin, which, in turn, stimulates milk release. Milk is normally not ejected from the milk glands and ducts until the baby suckles. The fact that milk is ejected from both breasts in response to suckling is a reminder that all target cells respond to a hormone.

Oxytocin may be given intravenously to stimulate uterine contractions, inducing labor, if the uterus is not sufficiently contracting to expel a fully developed fetus. Oxytocin may also be administered to the mother following childbirth to ensure that the uterine smooth muscle contracts enough to squeeze broken blood vessels closed, minimizing bleeding.

Oxytocin has no established function in males, although it is present in the male posterior pituitary and it may stimulate the movement of sperm and certain fluids in the male reproductive tract during sexual activity. Oxytocin is called the "cuddle hormone" because studies on pregnant women show that higher levels of the hormone during pregnancy correlate to more intense maternal bonding behavior with the infant, such as more eye contact, touching, and singing. Table 13.6 reviews the hormones of the pituitary gland.

 PRACTICE

21 What is the function of ADH?

22 How is the secretion of ADH controlled?

23 What effects does oxytocin produce in females?

13.5 | Thyroid Gland

The **thyroid gland** (thi'roid gland), as figure 13.18 shows, is a vascular structure that consists of two large lateral lobes connected by a broad **isthmus** (is'mus). The thyroid lies just below the larynx (voicebox) on either side and anterior to the trachea (windpipe). The gland is specialized to remove iodine from the blood.

Structure of the Gland

A capsule of connective tissue covers the thyroid gland, which is made up of many secretory parts called *follicles.* Cavities in the follicles are lined with a single layer of cuboidal epithelial cells, called follicular cells. A clear viscous substance called *colloid* fills

TABLE 13.6 | Hormones of the Pituitary Gland

Anterior Lobe	Action	Source of Control
Growth hormone (GH)	Stimulates increase in size and rate of division of body cells; enhances movement of amino acids through membranes; promotes growth of long bones	Secretion inhibited by somatostatin (SS) and stimulated by growth hormone-releasing hormone (GHRH) from the hypothalamus
Prolactin (PRL)	Sustains milk production after birth; amplifies the effect of LH in males	Secretion inhibited by prolactin release-inhibiting hormone (PIH) and may be stimulated by yet to be identified prolactin-releasing factor (PRF) from the hypothalamus
Thyroid-stimulating hormone (TSH)	Controls secretion of hormones from the thyroid gland	Thyrotropin-releasing hormone (TRH) from the hypothalamus
Adrenocorticotropic hormone (ACTH)	Controls secretion of certain hormones from the adrenal cortex	Corticotropin-releasing hormone (CRH) from the hypothalamus
Follicle-stimulating hormone (FSH)	Development of egg-containing follicles in ovaries; stimulates follicular cells to secrete estrogen; in males, stimulates production of sperm cells	Gonadotropin-releasing hormone (GnRH) from the hypothalamus
Luteinizing hormone (LH)	Promotes secretion of male and female sex hormones; releases egg cell in females	Gonadotropin-releasing hormone (GnRH) from the hypothalamus
Posterior Lobe	**Action**	**Source of Control**
Antidiuretic hormone (ADH)	Causes kidneys to reduce water excretion; in high concentration, raises blood pressure	Hypothalamus in response to changes in body fluid concentration and blood volume
Oxytocin (OT)	Contracts smooth muscle in the uterine wall; forces liquid from the milk glands into the milk ducts, ejects milk	Hypothalamus in response to stretching uterine and vaginal walls and stimulation of breasts

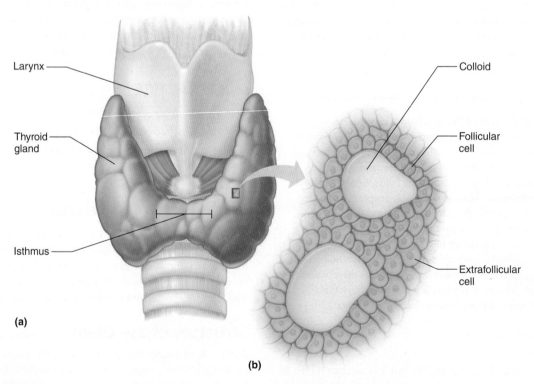

(a)

(b)

FIGURE 13.18 AP|R Thyroid gland. (a) The thyroid gland consists of two lobes connected anteriorly by an isthmus. (b) Follicular cells secrete thyroxine and triiodothyronine. Extrafollicular cells secrete calcitonin.

the cavities. It consists primarily of a glycoprotein called *thyro-globulin*. The follicular cells produce and secrete hormones that are either stored in the colloid or released into nearby capillaries (fig. 13.19). Other hormone-secreting cells, called *extrafollicular cells* (C cells), lie outside the follicles.

Thyroid Hormones

The thyroid gland produces three important hormones. The follicular cells synthesize two of these, which have marked effects on the metabolic rates of body cells. The extrafollicular cells produce the third type of hormone, which influences blood concentrations of calcium and phosphate ions.

The two thyroid hormones that affect cellular metabolic rates are **thyroxine** (thi-rok′sin) and **triiodothyronine** (tri″i-o″do-thi′ro-nēn). Thyroxine, or tetraiodothyronine, is also called T_4 because it includes four atoms of iodine. Triiodothyronine is also called T_3 because it includes three atoms of iodine (fig. 13.20). These hormones help regulate the metabolism of carbohydrates, lipids, and proteins. Specifically, thyroxine and triiodothyronine increase the rate at which cells release energy from carbohydrates, enhance the rate of protein synthesis, and stimulate breakdown and mobilization of lipids. These hormones are the major factors determining how many calories the body must consume at rest to maintain life, which is measured as the *basal metabolic rate* (BMR). The two hormones are essential for normal growth and development and for maturation of the nervous system. TSH from the anterior pituitary gland controls the levels of thyroid hormones.

Follicular cells require iodine salts (iodides) to produce thyroxine and triiodothyronine. Such salts are normally obtained from foods, and after they have been absorbed from the intestine, the blood carries some of them in the form of iodide (I^-) to the thyroid gland. An efficient active transport protein called the *iodide pump* moves the iodides into the follicular cells, where they are converted to iodine and concentrated. The iodine, with the amino acid tyrosine, is used to synthesize these thyroid hormones.

Follicular cells synthesize thyroglobulin, whose protein portion includes molecules of tyrosine, many of which have already had iodine attached by an enzymatic reaction. As the protein part of thyroglobulin folds into its tertiary structure, bonds form between some of the tyrosine molecules, forming and storing potential thyroid hormones. The follicular cells take up molecules of thyroglobulin by endocytosis, break down the protein, and release the individual thyroid hormones into the bloodstream. When the thyroid hormone levels in the bloodstream drop below a certain level, accessing thyroglobulin accelerates, returning thyroid hormone levels to normal.

Thyroxine (T_4) accounts for at least 95% of circulating thyroid hormones, but once in the blood, most of the T_3 and T_4 combine with blood proteins (alpha globulins). It is the small fraction of hormone molecules that are not protein-bound (so called "free" hormone) that act on target cells. Thus T_3, which has a 50-fold higher free concentration in the plasma, is physiologically more important. Additionally, T_3 is nearly five times more potent than T_4, and about a third of T_4 is converted to T_3 in peripheral tissues.

The thyroid gland produces **calcitonin,** which is usually not referred to as a "thyroid hormone" because it is synthesized by the C cells, distinct from the gland's follicles. Calcitonin plays a role in the control of blood calcium and phosphate ion concentrations. It helps lower concentrations of calcium and phosphate ions by decreasing the rate at which they leave the bones and enter extracellular fluids by inhibiting the bone-destroying activity of osteoclasts. At the same time, calcitonin increases the rate at which calcium and phosphate ions are deposited in bone matrix by stimulating activity of osteoblasts (see chapter 7, p. 202). Calcitonin also increases the excretion of calcium ions and phosphate ions by the kidneys.

A high blood calcium ion concentration stimulates calcitonin secretion. This may occur following absorption of calcium ions from a recent meal. Certain hormones also prompt calcitonin secretion, such as gastrin, released from active digestive organs.

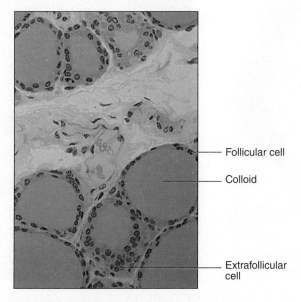

FIGURE 13.19 AP|R A light micrograph of thyroid gland tissue (240×). The open spaces that follicular cells surround are filled with colloid.

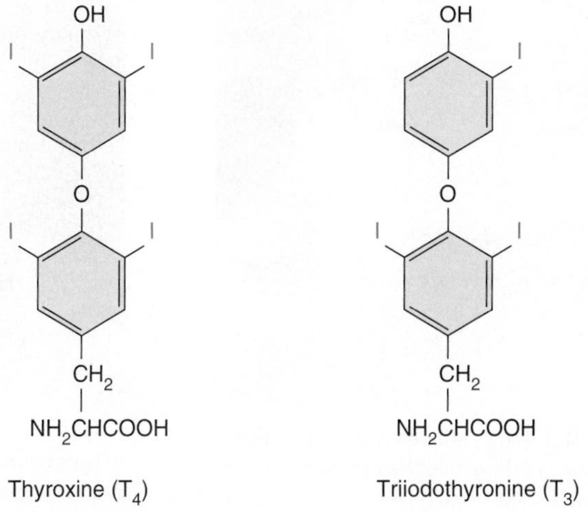

- Follicular cell
- Colloid
- Extrafollicular cell

FIGURE 13.20 The hormones thyroxine and triiodothyronine have very similar molecular structures.

Thyroxine (T_4)

Triiodothyronine (T_3)

Calcitonin helps prevent prolonged elevation of the blood calcium ion concentration after eating.

Research suggests that calcitonin may be most important during early growth and physiological stress. In the young, calcitonin stimulates the increase in bone deposition associated with growth. In females, it helps protect bones from resorption during pregnancy and lactation, when calcium is required for growth of the fetus and synthesis of breast milk.

Table 13.7 summarizes the actions and sources of control of the thyroid hormones. From Science to Technology 2.1 (chapter 2, p. 61), table 13.8, and figures 13.21, 13.22, and 13.23 discuss disorders of the thyroid gland.

PRACTICE

24 Where is the thyroid gland located?

25 Which hormones of the thyroid gland affect carbohydrate metabolism, the mobilization of lipids, and protein synthesis?

26 What substance is essential for the production of thyroxine and triiodothyronine?

27 How does calcitonin influence the concentrations of blood calcium and phosphate ions?

TABLE 13.7	Hormones of the Thyroid Gland AP|R	
Hormone	Action	Source of Control
Thyroxine (T_4)	Increases rate of energy release from carbohydrates; increases rate of protein synthesis; accelerates growth; stimulates activity in the nervous system	TSH from the anterior pituitary gland
Triiodothyronine (T_3)	Same as above, but five times more potent than thyroxine	Same as above
Calcitonin	Lowers blood calcium and phosphate ion concentrations by inhibiting release of calcium and phosphate ions from bones and by increasing the rate at which calcium and phosphate ions are deposited in bones; increases excretion of calcium by the kidneys	Elevated blood calcium ion concentration, digestive hormones

13.6 | Parathyroid Glands

The **parathyroid glands** (par"ah-thi'roid glandz) are on the posterior surface of the thyroid gland, as figure 13.24 shows. Most individuals have four of them—a superior and an inferior gland associated with each of the thyroid's bilateral lobes. The parathyroid glands secrete a hormone that regulates the concentrations of calcium and phosphate ions in the blood.

Structure of the Glands

Each parathyroid gland is a small, yellowish brown structure covered by a thin capsule of connective tissue. The body of the gland consists of many tightly packed secretory cells closely associated with capillary networks (fig. 13.25).

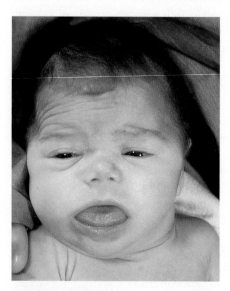

FIGURE 13.21 Infantile hypothyroidism is due to an underactive thyroid gland during infancy and childhood.

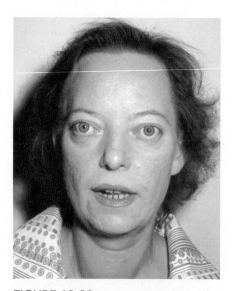

FIGURE 13.22 Graves disease (a form of hyperthyroidism) may cause the eyes to protrude (exopthalmia).

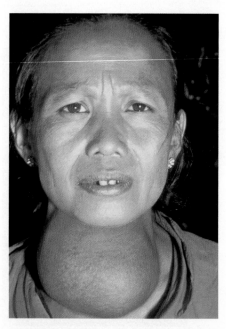

FIGURE 13.23 An iodine deficiency causes simple (endemic) goiter and results in high levels of TSH.

TABLE 13.8 Disorders of the Thyroid Gland

Condition	Mechanism/Symptoms
Hyperthyroid	
Hyperthyroidism	High metabolic rate, sensitivity to heat, restlessness, hyperactivity, weight loss, protruding eyes, goiter
Graves disease	Autoantibodies (against self) bind TSH receptors on thyroid cell membranes, mimicking action of TSH, overstimulating gland (hyperthyroidism); exopthalmia (protrusion of the eyes) and goiter
Hypothyroid	
Hashimoto disease	Autoantibodies (against self) destroy thyroid cells, resulting in hypothyroidism
Hypothyroidism (infantile)	Stunted growth, abnormal bone formation, intellectual disability, sluggishness
Hypothyroidism (adult)	Low metabolic rate, sensitivity to cold, sluggishness, poor appetite, swollen tissues, mental dullness
Simple goiter	Deficiency of thyroid hormones due to iodine deficiency; because no thyroid hormones inhibit pituitary release of TSH, thyroid is overstimulated and enlarges but functions below normal (hypothyroidism)

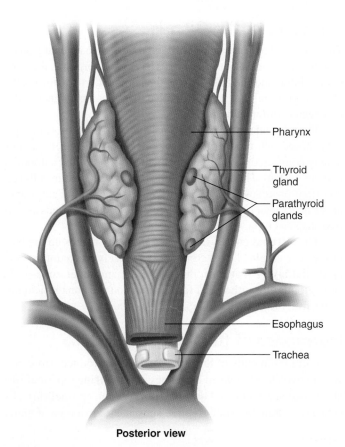

FIGURE 13.24 The parathyroid glands are embedded in the posterior surface of the thyroid gland.

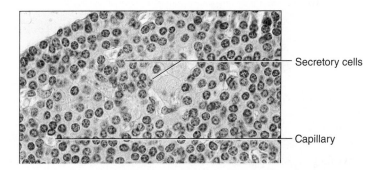

FIGURE 13.25 AP|R Light micrograph of the parathyroid gland (205×).

Parathyroid Hormone

The parathyroid glands secrete a protein, **parathyroid hormone** (PTH), or *parathormone* (see fig. 13.4c). This hormone increases blood calcium ion concentration and decreases blood phosphate ion concentration through actions in the bones, kidneys, and intestines.

The extracellular matrix of bone tissue contains a considerable amount of calcium phosphate and calcium carbonate. PTH stimulates bone resorption by osteoclasts and inhibits the activity of osteoblasts (see chapter 7, pp. 202, 205, and 207). As bone resorption increases, calcium and phosphate ions are released into the blood. At the same time, PTH causes the kidneys to conserve blood calcium ions and to excrete more phosphate ions in the urine. PTH also indirectly stimulates absorption of calcium ions from food in the intestine by influencing metabolism of vitamin D.

Vitamin D (cholecalciferol) synthesis begins when intestinal enzymes convert dietary cholesterol into the inactive form, provitamin D (dehydrocholesterol). This provitamin is largely stored in the skin, and exposure to the ultraviolet wavelengths of sunlight changes it to vitamin D. Some vitamin D also comes from foods.

The liver changes vitamin D to hydroxycholecalciferol, which is carried in the bloodstream or stored in tissues. When PTH is present, hydroxycholecalciferol can be changed in the kidneys into the active form of vitamin D (dihydroxycholecalciferol), which controls absorption of calcium ions from the intestine (fig. 13.26).

A negative feedback mechanism operating between the parathyroid glands and the blood calcium ion concentration regulates

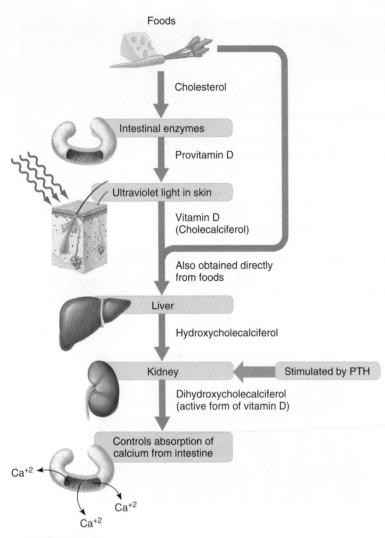

Foods

Cholesterol

Intestinal enzymes

Provitamin D

Ultraviolet light in skin

Vitamin D
(Cholecalciferol)

Also obtained directly
from foods

Liver

Hydroxycholecalciferol

Kidney ← Stimulated by PTH

Dihydroxycholecalciferol
(active form of vitamin D)

Controls absorption of
calcium from intestine

Ca⁺²   Ca⁺²   Ca⁺²

FIGURE 13.26 Mechanism by which PTH promotes calcium absorption in the intestine.

secretion of PTH (fig. 13.27). As the concentration of blood calcium ions rises, less PTH is secreted; as the concentration of blood calcium ions drops, more PTH is released.

The opposite effects of calcitonin and PTH maintain calcium ion homeostasis (see fig. 7.14, p. 211). This is important in a number of physiological processes. For example, if the blood calcium ion concentration drops below the normal range (hypocalcemia), the nervous system becomes abnormally excitable, and impulses may be triggered spontaneously. As a result, muscles, including

the respiratory muscles, may undergo tetanic contractions, and the person may suffocate. In contrast, an abnormally high concentration of blood calcium ions (hypercalcemia) depresses the nervous system. Consequently, muscle contractions are weak and reflexes are sluggish. Table 13.9 lists disorders of the parathyroid glands.

 PRACTICE

28 Where are the parathyroid glands located?

29 How does parathyroid hormone help regulate the concentrations of blood calcium and phosphate ions?

30 How does the negative feedback system of the parathyroid glands differ from that of the thyroid gland?

13.7 | Adrenal Glands AP|R

The **adrenal glands** (suprarenal glands) are closely associated with the kidneys. A gland sits atop each kidney like a cap and is embedded in the mass of adipose tissue that encloses the kidney.

Structure of the Glands

The adrenal glands are shaped like pyramids. Each adrenal gland is vascular and consists of two parts. The central portion is the adrenal medulla, and the outer part is the adrenal cortex (fig. 13.28). These regions are not sharply divided, but they are distinct in that they secrete different hormones.

The **adrenal medulla** (ah-dre′nal me-dul′ah) consists of irregularly shaped cells grouped around blood vessels. These cells are intimately connected with the sympathetic division of the autonomic nervous system. The adrenal medullary cells (chromaffin cells) are modified postganglionic neurons, and preganglionic autonomic nerve fibers lead to them directly from the central nervous system (see chapter 11, pp. 431–432).

The **adrenal cortex** (ah-dre′nal kor′teks) makes up the bulk of the adrenal gland. It is composed of closely packed masses of epithelial layers that form outer, middle, and inner zones—the zona glomerulosa, the zona fasciculata, and the zona reticularis, respectively (fig.13.28 and fig. 13.29).

Hormones of the Adrenal Medulla

The chromaffin cells of the adrenal medulla produce, store, and secrete two closely related hormones, **epinephrine** (ep″ĭ-nef′rin), also called adrenalin, and **norepinephrine** (nor″ep-ĭ-nef′rin), also called noradrenalin. Both of these substances are a type of amine

TABLE 13.9	Disorders of the Parathyroid Glands		
Condition	Symptoms/Mechanism	Cause	Treatment
Hyperparathyroidism	Fatigue, muscular weakness, painful joints, altered mental functions, depression, weight loss, bone weakening. Increased PTH secretion overstimulates osteoclasts.	Tumor	Remove tumor, correct bone deformities
Hypoparathyroidism	Muscle cramps and seizures. Decreased PTH secretion reduces osteoclast activity, diminishing blood calcium ion concentration.	Inadvertent surgical removal; injury	Calcium salt injections, massive doses of vitamin D

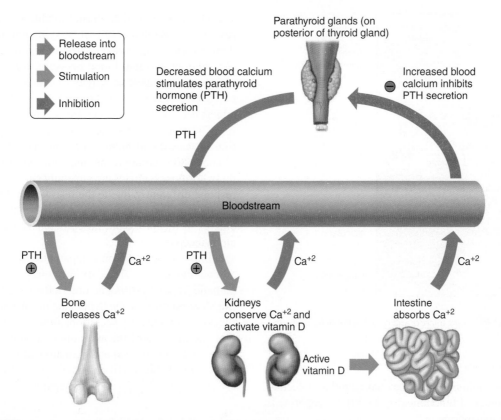

FIGURE 13.27 **AP|R** Parathyroid hormone (PTH) stimulates bone to release calcium (Ca^{+2}) and the kidneys to conserve calcium. It indirectly stimulates the intestine to absorb calcium. The resulting increase in blood calcium concentration inhibits secretion of PTH by negative feedback. (⊕ = stimulation; ⊖ = inhibition)

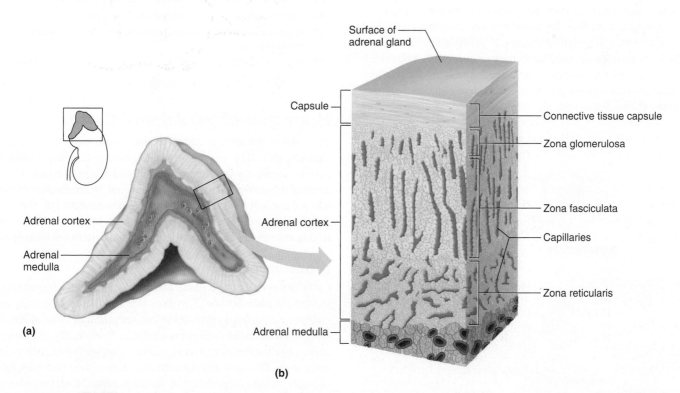

FIGURE 13.28 **AP|R** Adrenal glands. (*a*) An adrenal gland consists of an outer cortex and an inner medulla. (*b*) The cortex consists of three layers, or zones, of cells.

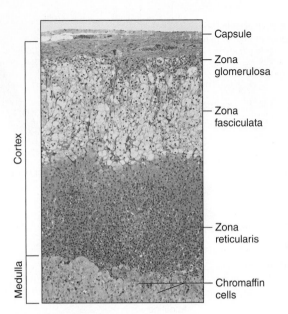

- Capsule
- Zona glomerulosa
- Zona fasciculata
- Zona reticularis
- Chromaffin cells

Cortex

Medulla

FIGURE 13.29 AP|R Light micrograph of the adrenal medulla and the adrenal cortex (75×).

hormone called a *catecholamine,* and they have similar molecular structures and physiological functions (fig. 13.30). Epinephrine is synthesized from norepinephrine.

The synthesis of catecholamines begins with the amino acid tyrosine. In the first step of the pathway, an enzyme (tyrosine hydroxylase) in the secretory cells catalyzes a reaction that converts tyrosine into a substance called *dopa.* A second enzyme (dopa decarboxylase) catalyzes a reaction that modifies dopa into dopamine, and a third enzyme (dopamine betahydroxylase) catalyzes a reaction that alters dopamine to form norepinephrine. Still another enzyme (phenylethanolamine N-methyltransferase) then catalyzes conversion of norepinephrine to epinephrine. About 15% of the norepinephrine is stored unchanged. The hormones occupy tiny vesicles (chromaffin granules), much like neurotransmitters are stored in vesicles in neurons.

Norepinephrine

Epinephrine

FIGURE 13.30 Epinephrine and norepinephrine have similar molecular structures and similar functions.

The effects of the adrenal medullary hormones generally resemble those that result when sympathetic neurons stimulate their effectors: increased heart rate and force of cardiac muscle contraction, elevated blood pressure, increased breathing rate, and decreased digestive activity (see table 11.10, p. 436). The hormonal effects last up to ten times longer than the neurotransmitter effects because the hormones are slowly removed from the tissues.

The ratio of the two hormones in the adrenal medullary secretion varies with different physiological conditions, but usually it is about 80% epinephrine and 20% norepinephrine. Although these hormones' effects are generally similar, certain effector cells respond differently, due to the relative numbers of alpha and beta receptors in their membranes. Both hormones can stimulate both classes of receptors, but norepinephrine has a greater effect on alpha receptors.

Impulses arriving on sympathetic nerve fibers stimulate the adrenal medulla to release its hormones at the same time as sympathetic impulses stimulate other effectors. As a rule, all of these impulses originate in the hypothalamus in response to stress. Thus, the adrenal medullary secretions function together with the sympathetic division of the autonomic nervous system in preparing the body for energy-expending action, also called "fight or flight" responses. Table 13.10 compares some of the differences in the effects of epinephrine and norepinephrine.

RECONNECT

To Chapter 11, Sympathetic Division, pages 431–432.

PRACTICE

31 Describe the location and structure of the adrenal glands.

32 Name the hormones the adrenal medulla secretes.

33 What general effects do hormones secreted by the adrenal medulla produce?

34 What usually stimulates release of hormones from the adrenal medulla?

Hormones of the Adrenal Cortex

The cells of the adrenal cortex produce more than thirty different steroids, including several hormones (corticosteroids). Unlike the adrenal medullary hormones, without which a person can survive, some of those released by the cortex are vital. In the absence of these adrenal cortical secretions, without extensive electrolyte therapy a person usually dies within a week. The most important adrenal cortical hormones are aldosterone, cortisol, and certain sex hormones.

Aldosterone

Cells in the outer zone (zona glomerulosa) of the adrenal cortex synthesize **aldosterone.** This hormone is called a *mineralocorticoid* because it helps regulate the concentration of mineral electrolytes, such as sodium and potassium ions. Specifically, aldosterone causes the kidney to conserve sodium ions and to excrete potassium ions. The cells that secrete aldosterone respond directly to changes in the composition of blood plasma. However, whereas an increase in plasma potassium strongly stimulates these cells, a decrease in plasma sodium only slightly stimulates them. Control

TABLE 13.10 Comparative Effects of Epinephrine and Norepinephrine

Structure or Function Affected	Epinephrine	Norepinephrine
Heart	Heart rate increases Force of contraction increases	Heart rate increases Force of contraction increases
Blood vessels	Vasodilation, especially important in skeletal muscle at onset of fight or flight	Vasoconstriction in skin and viscera shifts blood flow to other areas, such as exercising skeletal muscle
Systemic blood pressure	Some increase due to increased cardiac output	Some increase due to increased cardiac output and vasoconstriction (offset in some areas, such as exercising skeletal muscle, by local vasodilation due to other factors)
Airways	Dilation	Some dilation
Reticular formation of brainstem	Activated	Little effect
Liver	Promotes breakdown of glycogen to glucose, increasing blood sugar level	Little effect on blood glucose level
Metabolic rate	Increases	Increases

of aldosterone secretion is indirectly linked to plasma sodium level by the **renin-angiotensin system.**

Groups of specialized kidney cells (juxtaglomerular cells) are able to respond to changes in blood pressure and the plasma sodium ion concentration. If the level of either of these factors decreases, the cells release an enzyme called **renin** (re'nin). Renin reacts with a blood protein called **angiotensinogen** (an″je-o-ten-sin'o-jen) to catalyze the partial breakdown of angiotensinogen into a peptide called **angiotensin I.** Another enzyme (angiotensin-converting enzyme, or ACE), found primarily in lung blood vessels,

catalyzes a reaction that converts angiotensin I into another form, **angiotensin II,** which is carried in the bloodstream (fig. 13.31). When angiotensin II reaches the adrenal cortex, it stimulates the release of aldosterone. ACTH is necessary for the adrenal gland to respond to this and other stimuli.

Aldosterone, in conserving sodium ions, indirectly retains water by osmosis. This action helps maintain both the blood sodium ion concentration and blood volume (fig. 13.31). Angiotensin II is also a powerful vasoconstrictor and thereby helps maintain systemic blood pressure.

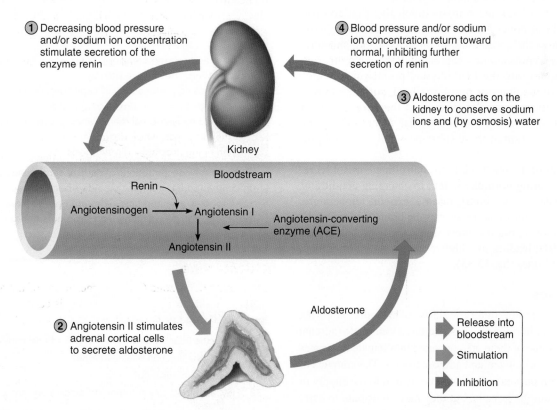

1 Decreasing blood pressure and/or sodium ion concentration stimulate secretion of the enzyme renin

4 Blood pressure and/or sodium ion concentration return toward normal, inhibiting further secretion of renin

3 Aldosterone acts on the kidney to conserve sodium ions and (by osmosis) water

Kidney

Bloodstream

Renin

Angiotensinogen ⟶ Angiotensin I

Angiotensin-converting enzyme (ACE)

Angiotensin II

Aldosterone

2 Angiotensin II stimulates adrenal cortical cells to secrete aldosterone

Release into bloodstream

Stimulation

Inhibition

FIGURE 13.31 Aldosterone increases blood volume and pressure by promoting conservation of sodium ions and water (steps *1–4*).

ACE inhibitors are a class of drugs used to treat some forms of high blood pressure (hypertension). They work by binding to the active site on angiotensin-converting enzyme, blocking formation of angiotensin II and preventing inactivation of bradykinin, a vasodilator. Both effects dilate blood vessels, lowering blood pressure.

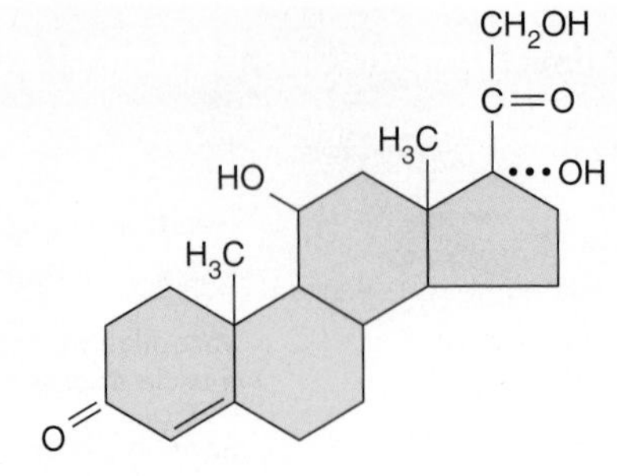

Cortisol

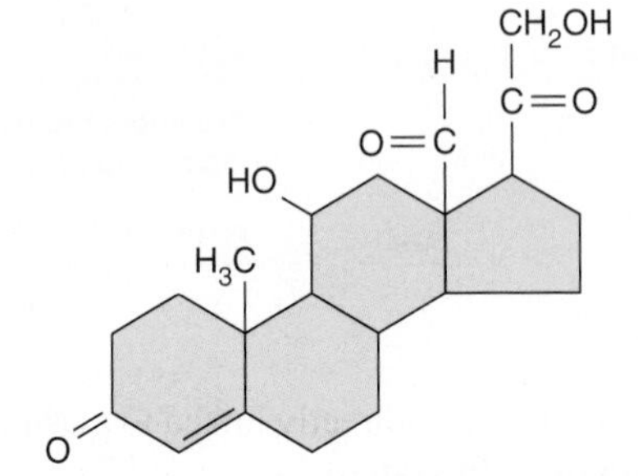

Aldosterone

FIGURE 13.32 Cortisol and aldosterone are steroids with similar molecular structures.

Cortisol

Cortisol (hydrocortisone) is a *glucocorticoid,* which means it affects glucose metabolism. It is produced primarily in the middle zone (zona fasciculata) of the adrenal cortex and has a molecular structure similar to aldosterone (fig. 13.32). In addition to affecting glucose, cortisol influences protein and fat metabolism. Among the more important actions of cortisol are the following:

1. It inhibits the synthesis of protein in various tissues, increasing the blood concentration of amino acids.
2. It stimulates liver cells to synthesize glucose from noncarbohydrates (gluconeogenesis), such as circulating amino acids and glycerol, thus increasing the blood glucose concentration.
3. It promotes the release of fatty acids from adipose tissue, increasing the use of fatty acids and decreasing the use of glucose as energy sources, thus preserving glucose availability for the brain.

Cortisol's actions help keep the blood glucose concentration within the normal range between meals. These actions are important because just a few hours without food can exhaust liver glycogen, which is another major source of glucose.

A negative feedback mechanism much like that controlling the thyroid hormones T_3 and T_4 regulates cortisol release. It involves the hypothalamus, anterior pituitary gland, and adrenal cortex. The hypothalamus secretes CRH (corticotropin-releasing hormone) into the hypophyseal portal veins, which carry the CRH to the anterior pituitary gland, stimulating it to secrete ACTH. In turn, ACTH stimulates the adrenal cortex to release cortisol. Cortisol inhibits release of both CRH and ACTH. As concentration of these substances falls, cortisol production drops.

The set point of the feedback loop controlling cortisol secretion changes, adapting hormone output to changing conditions. For example, under stress—injury, disease, extreme temperature, or emotional upset—information reaches the brain concerning the situation. In response, brain centers signal the hypothalamus to release more CRH, leading to a higher concentration of cortisol until the stress subsides (fig. 13.33).

Sex Hormones

Cells in the inner zone (zona reticularis) of the adrenal cortex mostly produce sex hormones. These hormones are male (adrenal androgens), but some of them are converted into female hormones (estrogens) by the skin, liver, and adipose tissues. The amounts of these adrenal hormones are very small compared to the supply of sex hormones from the gonads, but they may contribute to early development of the reproductive organs. Adrenal androgens may also play a role in the female sex drive. Table 13.11 summarizes the actions of the cortical hormones. Clinical Application 13.3 discusses some of the effects of a malfunctioning adrenal gland on health.

Cortisol and related compounds are used as drugs to reduce inflammation. They relieve pain by:

- decreasing permeability of capillaries, preventing leakage of fluids that swell surrounding tissues
- stabilizing lysosomal membranes, preventing release of their enzymes, which destroy tissue
- inhibiting prostaglandin synthesis

Because the concentrations of cortisol compounds used to stifle inflammation have significant side effects, these drugs can be used for only a limited time. They are used to treat autoimmune disorders, allergies, asthma, and recipients of organ transplants or tissue grafts.

PRACTICE

35 Name the important hormones of the adrenal cortex.

36 What is the function of aldosterone?

37 What does cortisol do?

38 How are blood concentrations of aldosterone and cortisol regulated?

President John F. Kennedy's bronze complexion may have resulted not from sunbathing, but from a disorder of the adrenal glands. When he ran for president in 1960, Kennedy knew he had *Addison disease*, but his staff kept his secret, for fear it would affect his career. Kennedy had almost no adrenal tissue, but he functioned by receiving mineralocorticoids and glucocorticoids, the standard treatment.

In Addison disease, the adrenal cortex does not secrete hormones sufficiently. This may be due to immune system attack (autoimmunity) or an infection such as tuberculosis. Signs and symptoms include decreased blood sodium, increased blood potassium, low blood glucose level (hypoglycemia), dehydration, low blood pressure, frequent infections, fatigue, nausea and vomiting, loss of appetite, and increased skin pigmentation. Some sufferers experience salt cravings—one woman reported eating many bowls of salty chicken noodle soup, with pickles and briny pickle juice added! Without treatment, death comes within days from severe disturbances in electrolyte balance.

Cushing syndrome is hypersecretion of cortisol from any cause, such as an adrenal tumor or oversecretion of ACTH by the anterior pituitary. The condition may also result from taking corticosteroid drugs for many years, such as to treat asthma or rheumatoid arthritis. Tissue protein level plummets, due to muscle wasting and loss of bone tissue. Blood glucose level remains elevated, and excess sodium is retained. As a result, tissue fluid increases, blood pressure rises, and the skin appears puffy. The skin may appear thin due to inhibition of collagen synthesis by the excess cortisol. Adipose tissue deposited in the face and back produce a characteristic "moon face" and "buffalo hump." Increase in adrenal sex hormone secretion may masculinize a female, causing growth of facial hair and a deepening voice. Other symptoms include extreme fatigue, sleep disturbances, skin rashes, headache, and leg muscle cramps.

Treatment of Cushing syndrome attempts to reduce ACTH secretion. This may entail removing a tumor in the pituitary gland or partially or completely removing the adrenal glands.

Both Addison disease and Cushing syndrome are rare, and for this reason, they are often misdiagnosed, or, in early stages, the patient's report of symptoms is not taken seriously. Addison disease affects thirty-nine to sixty people of every million, and Cushing syndrome affects five to twenty-five people per million.

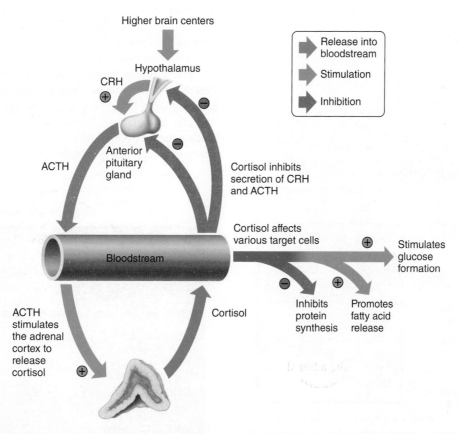

FIGURE 13.33 Negative feedback regulates cortisol secretion, similar to the regulation of thyroid hormone secretion (see fig. 13.16). (⊕ = stimulation; ⊖ = inhibition)

TABLE 13.11 Hormones of the Adrenal Cortex AP|R

Hormone	Action	Factors Regulating Secretion
Aldosterone	Helps regulate the concentration of extracellular electrolytes by conserving sodium ions and excreting potassium ions	Electrolyte concentrations in body fluids and renin-angiotensin system
Cortisol	Decreases protein synthesis, increases fatty acid release, and stimulates glucose synthesis from noncarbohydrates	CRH from the hypothalamus and ACTH from the anterior pituitary gland
Adrenal androgens	Supplement sex hormones from the gonads; may be converted into estrogens	

13.8 | Pancreas

The **pancreas** (pan′kre-as) consists of two major types of secretory tissues. This organization reflects the dual function of the pancreas as an exocrine gland that secretes digestive juice through a duct, and an endocrine gland that releases hormones into body fluids.

Structure of the Gland

The pancreas is an elongated, somewhat flattened organ located posterior to the stomach and partly between the parietal peritoneum and the posterior abdominal wall (retroperitoneal) (**fig. 13.34**). A duct that attaches the pancreas to the first section of the small intestine (duodenum) transports its digestive juice into the intestine. Chapter 17 (pp. 668–669) discusses the digestive functions of the pancreas.

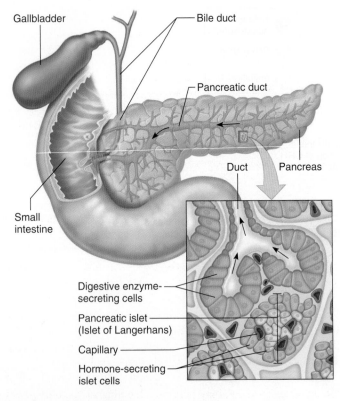

FIGURE 13.34 AP|R The hormone-secreting cells of the pancreas are grouped in clusters, or islets, closely associated with blood vessels. Other pancreatic cells secrete digestive enzymes into ducts.

The endocrine portion of the pancreas consists of cells grouped around blood vessels. These groups, called *pancreatic islets* (islets of Langerhans), include three distinct types of hormone-secreting cells—*alpha cells,* which secrete glucagon; *beta cells,* which secrete insulin; and *delta cells,* which secrete somatostatin (fig.13.34 and fig. 13.35).

Hormones of the Pancreatic Islets

Glucagon is a protein that stimulates the liver to break down glycogen into glucose (glycogenolysis) and to convert noncarbohydrates, such as amino acids, into glucose (gluconeogenesis). Glucagon also stimulates breakdown of fats into fatty acids and glycerol.

In a negative feedback system, a low concentration of blood glucose stimulates release of glucagon from the alpha cells. When the blood glucose concentration returns toward normal, glucagon secretion decreases (**fig. 13.36**). This mechanism prevents hypoglycemia when the blood glucose concentration is relatively low, such as between meals, or when glucose is being used rapidly, such as during exercise.

The hormone **insulin** is also a protein, and its main effect is exactly opposite that of glucagon. Insulin stimulates the liver to form glycogen from glucose and inhibits conversion of noncarbohydrates into glucose. Insulin also has the special effect of promoting the facilitated diffusion (see chapter 3, pp. 100–101) of glucose through the membranes of cells bearing insulin receptors.

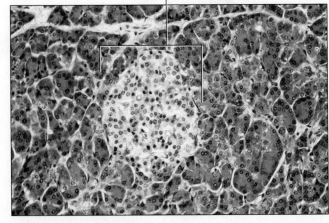

FIGURE 13.35 AP|R Light micrograph of a pancreatic islet (80×).

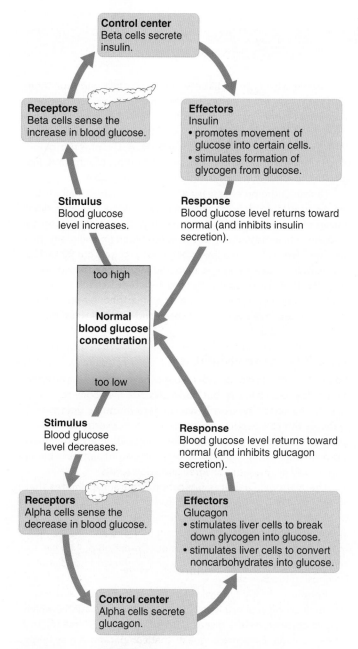

Control center
Beta cells secrete insulin.

Receptors
Beta cells sense the increase in blood glucose.

Effectors
Insulin
• promotes movement of glucose into certain cells.
• stimulates formation of glycogen from glucose.

Stimulus
Blood glucose level increases.

Response
Blood glucose level returns toward normal (and inhibits insulin secretion).

too high

Normal blood glucose concentration

too low

Stimulus
Blood glucose level decreases.

Response
Blood glucose level returns toward normal (and inhibits glucagon secretion).

Receptors
Alpha cells sense the decrease in blood glucose.

Effectors
Glucagon
• stimulates liver cells to break down glycogen into glucose.
• stimulates liver cells to convert noncarbohydrates into glucose.

Control center
Alpha cells secrete glucagon.

FIGURE 13.36 AP|R Insulin and glucagon function together to stabilize blood glucose concentration. Negative feedback responding to blood glucose concentration controls the levels of both hormones.

Q *Four hours after a meal, would you expect blood levels of insulin to be high or low? What about blood levels of glucagon?*

Answers can be found in Appendix G.

These cells include those of adipose tissue, liver, cardiac muscle, and skeletal muscle (glucose uptake by active muscle is not dependent on insulin). Insulin action decreases the concentration of blood glucose, promotes transport of glucose and amino acids into cells, and increases glycogen and protein synthesis. It also stimulates adipose cells to synthesize and store fat.

An enzyme called glucokinase enables pancreatic cells to "sense" the blood glucose level, which is important in regulating synthesis of glucagon and insulin. In one form of a rare type of diabetes mellitus, maturity-onset diabetes of the young (MODY), a mutation in a gene encoding glucokinase impairs the ability of beta cells to assess when to produce insulin. Other mutations that cause MODY alter insulin's structure, secretion, or cell surface receptors or the ability of liver cells to form glycogen in response to insulin. MODY is treated with drugs or dietary modification.

A negative feedback system sensitive to the concentration of blood glucose regulates insulin secretion. When glucose concentration is relatively high, such as after a meal, the beta cells release insulin. By promoting formation of glycogen in the liver and entrance of glucose into adipose and muscle cells, insulin helps prevent excessive rise in blood glucose concentration (hyperglycemia). Then, when the glucose concentration falls, between meals or during the night, insulin secretion decreases.

If the blood glucose concentration falls, insulin secretion decreases. As blood insulin levels drop, less glucose enters adipose and muscle cells, and the glucose remaining in the blood is available for cells, such as nerve cells, that do not require insulin to take up glucose. Neurons readily tap the energy in a continuous supply of glucose to produce ATP.

Hypoglycemia, or low blood glucose level due to excess insulin in the bloodstream, causes episodes of shakiness, weakness, and anxiety. Following a diet of frequent, small meals low in carbohydrates and high in protein can often control symptoms by preventing the surges of insulin that lower the blood glucose level. Hypoglycemia is most often seen when a person with diabetes mellitus injects too much insulin, but it can also reflect a tumor of the insulin-producing cells of the pancreas, or it may occur transiently following strenuous exercise.

Neurons, including those of the brain, obtain glucose by a facilitated diffusion mechanism not dependent on insulin, but rather only on the glucose concentration gradient. A consequence of this independence is that neurons are particularly sensitive to changes in blood glucose concentration. Conditions that cause such changes are likely to affect brain functions.

Insulin and glucagon secretion respond in opposite directions to changes in blood glucose level. When one increases, the other decreases (fig. 13.36). These hormones function together to maintain a relatively constant blood glucose concentration, despite great variations in the amounts of ingested carbohydrates.

Somatostatin (similar to the hypothalamic hormone), which the delta cells release, may help regulate glucose metabolism by inhibiting secretion of glucagon and insulin, but its inhibitory effect in the pancreas is not strong. **Table 13.12** summarizes the hormones of the pancreatic islets, and Clinical Application 13.4

Diabetes mellitus is a metabolic disease that arises from a lack of insulin or an inability of cells to recognize insulin. The persistent elevated blood glucose level can damage the eyes, heart, kidneys, and peripheral nerves. For a person with diabetes, controlling blood glucose requires daily monitoring and action. According to data from the National Health and Nutrition Examination Survey, in the United States about 21 million people have diabetes, and the prevalence is on the rise. From 1988 to 1994, 6.2% of the population had diabetes, from 1999 to 2004 the percentage was 8.8%, and from 2005 to 2010 it was 9.9%. Many people have the condition and do not yet know that they do.

Both insulin deficiency and impaired insulin response disturb carbohydrate, protein, and fat metabolism. Because insulin helps glucose cross some cell membranes, diabetes impairs movement of glucose into adipose and resting skeletal muscle cells. At the same time, the formation of glycogen, which is a long chain of glucose molecules, declines. As a result, blood glucose concentration rises (hyperglycemia). When it reaches a certain level, the kidneys begin to excrete the excess. Glucose in the urine (glycosuria) raises the urine's osmotic pressure, and too much water is excreted. Excess urine output causes dehydration and extreme thirst (polydipsia).

Diabetes mellitus also hampers protein and fat synthesis. Glucose-starved cells increasingly use proteins for energy, and as a result, tissues waste away as weight drops, hunger increases, exhaustion becomes overwhelming, children stop growing, and wounds do not heal. Changes in fat metabolism cause fatty acids and ketone bodies to accumulate in the blood, which lowers pH (acidosis). Dehydration and acidosis may harm brain cells, causing disorientation, coma, and eventually death.

The two most common forms of diabetes mellitus are type 1 (insulin-dependent or juvenile diabetes) and type 2 (non-insulin-dependent or maturity-onset diabetes).

Type 1 Diabetes Mellitus

Type 1 diabetes mellitus usually appears before age twenty. It is an autoimmune disease: the immune system destroys the beta cells of the pancreas (see chapter 16, p. 639).

People with type 1 diabetes must carefully monitor their blood glucose levels. They do this in two ways. Every three months, a laboratory test checks the levels of hemoglobin molecules in the blood that bind glucose. This measurement is called "A1C" and should be between 6% and 7%. A1C represents blood glucose level over the preceding three months, which is the life span of the red blood cells that transport hemoglobin.

The second type of test for a person with type 1 diabetes is self-monitoring of blood glucose. A person uses a test kit to draw a drop of blood, applies it to a test strip, then uses a meter to read the concentration of glucose in the blood (in milligrams per deciliter). Normal plasma levels of glucose range from 90 to 130 mg/dL before meals and less than 180 mg/dL one to two hours after meals. Most people with type 1 diabetes check their glucose this way two to four times a day. Smartphone apps record blood glucose levels with foods eaten and exercise done during the testing period.

Type 2 Diabetes Mellitus

About 90–95% of people with diabetes have type 2, in which the beta cells produce insulin but body cells lose the ability to recognize it. The condition usually has milder symptoms than type 1 diabetes but can become much more severe over time if blood glucose is not rigorously controlled. Most affected individuals are overweight when symptoms begin. Type 2 diabetes is being increasingly diagnosed in children and adolescents.

The National Diabetes Education Program estimates that about 79 million people in the United States have prediabetes, which is blood glucose levels above the normal range but not yet indicative of type 2 diabetes, corresponding to an A1C from 5.7 to 6.4. A diabetes educator (see Career Corner on p. 489) can be very helpful in guiding an individual new to the world of diabetes in choosing foods wisely.

People with any type of diabetes must monitor and regulate their blood glucose levels to forestall complications, which include coronary artery disease, peripheral nerve damage, and retinal damage. Evidence suggests that complications may begin even before blood glucose levels indicate disease.

TABLE 13.12	Hormones of the Pancreatic Islets	
Hormone	**Action**	**Source of Control**
Glucagon	Stimulates the liver to break down glycogen and convert noncarbohydrates into glucose; stimulates breakdown of fats	Blood glucose concentration
Insulin	Promotes formation of glycogen from glucose, inhibits conversion of noncarbohydrates into glucose, and enhances movement of glucose through adipose and muscle cell membranes, decreasing blood glucose concentration; promotes transport of amino acids into cells; enhances synthesis of proteins and fats	Blood glucose concentration
Somatostatin	Helps regulate carbohydrates	Not determined

and From Science to Technology 13.1 discuss diabetes mellitus, which is a disruption of the control of glucose metabolism that affects millions of people.

 PRACTICE

39 Name the endocrine portion of the pancreas.

40 What is the function of glucagon?

41 What is the function of insulin?

42 How are the secretions of glucagon and insulin controlled?

43 Why are nerve cells particularly sensitive to changes in blood glucose concentration?

13.9 | Other Endocrine Glands

Other organs that produce hormones are part of the endocrine system, too. They include the pineal gland; the thymus; reproductive organs; and certain cells of the digestive tract, the heart, and the kidneys.

The **pineal gland** (pin′e-al gland) is a small, oval structure deep between the cerebral hemispheres, where it attaches to the upper portion of the thalamus near the roof of the third ventricle. It largely consists of specialized *pineal cells* and supportive neuroglia (see fig. 11.11*b*, p. 406).

The pineal gland secretes a hormone, **melatonin,** that is synthesized from serotonin. Varying patterns of light and dark outside the body control the gland's activities. In the presence of light, action potentials from the retina travel to the hypothalamus, then to the reticular formation, and then downward into the spinal cord. From here, the impulses travel along sympathetic nerve fibers back into the brain, and finally they reach the pineal gland, where they decrease melatonin secretion. In the absence of light, impulses from the retina decrease and secretion of melatonin increases.

Melatonin secretion is part of the regulation of **circadian rhythms,** which are patterns of repeated activity associated with cycles of night and day, such as sleep/wake rhythms. Melatonin binds to two types of receptors on brain neurons, one that is abundant and one that is scarce. The major receptors are on cells of the suprachiasmatic nucleus, a region of the hypothalamus that regulates the circadian clock. Binding to the second, less abundant type of receptor induces sleepiness.

The **thymus** (thi′mus), which lies in the mediastinum posterior to the sternum and between the lungs, is large in young children

The fact that melatonin secretion responds to day length explains why traveling across several time zones produces the temporary insomnia of jet lag. Some clinical trials have found that taking melatonin 30 minutes before bedtime in the new location, for several days, hastens return to a normal sleep schedule, decreases daytime fatigue, and increases alertness. Studies have found that melatonin helps about half of people who take it for jet lag.

but shrinks with age. This gland secretes a group of hormones, called **thymosins,** that affect production and differentiation of certain white blood cells (T lymphocytes). The thymus plays an important role in immunity and is discussed in chapter 16 (p. 623).

The reproductive organs that secrete important hormones include the **testes,** which produce testosterone; the **ovaries,** which produce estrogens and progesterone; and the **placenta,** which produces estrogens, progesterone, and a gonadotropin. Chapters 22 and 23 discuss these glands and their secretions (pp. 838, 849, and 876).

The digestive glands that secrete hormones are generally associated with the linings of the stomach and small intestine. The small intestine alone produces dozens of hormones, many of which have not been well studied. Chapter 17 (pp. 665, 669, and 674) describes these structures and their secretions.

Other organs that produce hormones include the heart, which secretes two *natriuretic peptides* (chapter 15, p. 586), and the kidneys, which secrete *erythropoietin* that stimulates red blood cell production (chapter 14, p. 533). Clinical Application 13.1 (p. 494) discusses abuse of EPO to improve athletic performance.

 PRACTICE

44 Where is the pineal gland located?

45 What is the function of the pineal gland?

46 Where is the thymus gland located?

13.10 | Stress and Its Effects

Factors that change the body's internal or external environment are potentially life threatening. Sensory receptors detecting such changes trigger impulses that reach the hypothalamus, initiating physiological responses that resist a loss of homeostasis. These responses include increased activity in the sympathetic division of the autonomic nervous system and increased secretion of adrenal hormones. A factor capable of stimulating such a response is called a **stressor,** and the condition it produces in the body is called **stress.**

Types of Stress

Stressors may be physical or psychological, or a combination. *Physical stress* threatens tissues. Extreme heat or cold, decreased oxygen concentration, infections, injuries, prolonged heavy exercise, and loud sounds inflict physical stress. Unpleasant or painful sensations often accompany physical stress.

Psychological stress results from thoughts about real or imagined dangers, personal losses, unpleasant social interactions (or lack of social interactions), or any threatening factors. Feelings of anger, fear, grief, anxiety, depression, and guilt cause psychological stress. Psychological stress may also stem from pleasant stimuli, such as friendly social contact, feelings of joy or happiness, or sexual arousal. The factors that produce psychological stress vary greatly from person to person. A situation that is stressful to one person may not affect another, and what is stressful at one time may not be at another time.

The sweet-smelling urine that is the hallmark of type 1 diabetes mellitus was noted as far back as an Egyptian papyrus from 1500 B.C. In A.D. 96 in Greece, Aretaeus of Cappadocia described the condition as a "melting down of limbs and flesh into urine." One of the first people to receive insulin as a drug was a three-year-old boy whose body could not produce the hormone (fig. 13A). In December 1922, before treatment, he weighed only fifteen pounds. The boy rapidly improved after beginning insulin treatment, doubling his weight in just two months.

In 1921, Canadian physiologists Sir Frederick Grant Banting and Charles Herbert Best discovered the link between lack of insulin and diabetes. They induced diabetes symptoms in dogs by removing their pancreases, then cured them by administering insulin from other dogs' healthy pancreases. A year later, people with diabetes began to receive insulin extracted from pigs or cattle. The medication allowed them to control their disease.

In 1982, pure human insulin became available by genetically altering bacteria to produce the human protein (recombinant DNA technology). Human insulin helps people with type 1 diabetes who are allergic to the product from pigs or cows. Today, people receive insulin in several daily injections, from an implanted insulin pump, and/or in aerosol form; a transdermal insulin delivery system (skin patch) is in development.

Providing new pancreatic islets is a longer-lasting treatment for type 1 diabetes. An initial technical challenge was separating islets from cadaver pancreases, and then collecting enough beta cells, which account for only 2% of pancreas cells. Many patients' immune systems rejected transplants. By the 1990s, automated islet isolation and new anti-rejection drugs helped. In 1996 in Germany, and then in 1999 in Edmonton, Canada, islet transplantation to treat type 1 diabetes began.

FIGURE 13A Before and after insulin treatment. The boy in his mother's arms is three years old but weighs only 15 pounds because he has untreated type 1 diabetes mellitus. The inset shows the same child after two months of receiving insulin. His weight had doubled!

Responses to Stress

The hypothalamus controls the response to stress, termed the *general adaptation* (or *general stress) syndrome*. This response proceeds through two stages: the immediate "alarm" stage and the long-term "resistance" stage, and works to maintain homeostasis.

Recall that the hypothalamus receives information from nearly all body parts, including visceral receptors, the cerebral cortex, the reticular formation, and the limbic system. In the immediate stage, the hypothalamus activates the "fight or flight" response. Specifically, sympathetic impulses from the hypothalamus raise the blood glucose concentration, the level of blood glycerol and fatty acids, heart rate, blood pressure and breathing rate, and dilate the air passages. The response also shunts blood from the skin and digestive organs into the skeletal muscles and increases secretion of epinephrine from the adrenal medulla. The epinephrine, in turn, intensifies these sympathetic responses and prolongs their effects (fig. 13.37).

In the resistance stage, the hypothalamus secretes corticotropin-releasing hormone (CRH). This stimulates the anterior pituitary gland to secrete ACTH, which increases the adrenal cortex's secretion of cortisol. Cortisol supplies cells with amino acids and extra energy sources and allows glucose to be spared for brain tissue (fig. 13.37). Stress can also stimulate release of glucagon from the pancreas, growth hormone (GH) from the anterior pituitary gland, antidiuretic hormone (ADH) from the posterior pituitary gland, and renin from the kidneys.

Glucagon and growth hormone help mobilize energy sources, such as glucose, glycerol, and fatty acids, and stimulate cells to take up amino acids, facilitating repair of injured tissues. ADH stimulates the kidneys to retain water. This action decreases urine output and helps to maintain blood volume, which is important if a person is bleeding or sweating heavily. Renin, by increasing angiotensin II levels, helps stimulate the kidneys to retain

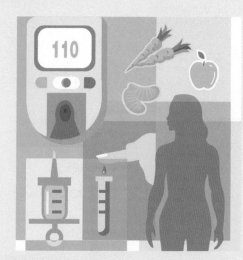

FIGURE 13B A person with either common form of diabetes mellitus must monitor his or her blood glucose level and be very diligent about proper diet and exercise.

Since 2000, several hundred people have received islet transplants in a procedure called the Edmonton protocol, which introduces islets into a vein in the liver. By a year after transplant, from 50% to 68% of patients do not need to receive additional insulin, but by five years after the procedure, fewer than 10% of total patients are free of daily insulin supplementation. The procedure is risky—12% of patients hemorrhage, and 4% develop blood clots in the liver vein. These risks, plus the apparent short-term improvement, have prompted physicians to carefully evaluate which patients are most likely to benefit from the few years of insulin independence that the procedure may offer. Researchers

are also investigating implants of stem cells and progenitor cells from a patient's body.

Treatment for type 2 diabetes includes following a low-carbohydrate, high-protein diet; regular aerobic and weight-bearing exercise; and maintaining a healthy body weight (fig. 13B). Several oral drugs can help control blood glucose levels, which can delay the onset of complications. A new drug that increases insulin production is injected once a week. Some people with type 2 diabetes may need to inject insulin. Another treatment for type 2 diabetes is gastric bypass surgery, which removes parts of the stomach and small intestine to help people lose weight. Isolated reports since the 1950s noted cases of morbidly obese people with diabetes having gastric bypass surgery, and then, within days and with lasting effect, not needing to inject insulin. By the 1980s, doctors noticed that some patients who had normal regulation of blood glucose before the surgery, a few months after began to experience confusion, altered behavior, seizures, and unconsciousness—signs of low blood glucose.

Today the effect of gastric bypass surgery on blood glucose levels is well recognized, and in some surgical centers more than 90% of patients with diabetes have levels of blood glucose within the normal range without medication just days or a few weeks after surgery. One proposed mechanism is that food reaching the small intestine has larger pieces than normal because of the smaller stomach capacity, and when the small intestine requires more energy to digest, it comes from blood glucose. Another hypothesis is that the shortened digestive tract alters the microbiome (gut bacteria) in a way that lowers blood glucose levels. Surgeons are now performing the surgery on people with severe diabetes who are only moderately overweight, and someday gastric bypass, or a way to mimic its effects or mechanism, may be a frontline treatment for type 2 diabetes.

sodium (through aldosterone), and through the vasoconstrictor action of angiotensin II contributes to maintaining blood pressure. Table 13.13 summarizes the body's reactions to stress.

 PRACTICE

47 What is stress?

48 Distinguish between physical stress and psychological stress.

49 Describe the general adaptation syndrome.

13.11 | Life-Span Changes

With age, the glands of the endocrine system generally decrease in size and increase in the proportion of each gland that is fibrous in nature. At the cellular level, lipofuscin pigment accumulates

as glands age. Hormone levels change with advancing years. Treatments for endocrine disorders associated with aging supplement deficient hormones, remove part of an overactive gland, or use drugs to block the action of an overabundant hormone.

Aging affects different hormones in characteristic ways. For growth hormone, the surge in secretion that typically occurs at night lessens with age. Lower levels of GH are associated with declining strength in the skeleton and muscles with advancing age. Levels of antidiuretic hormone increase with age due to slowed elimination by the liver and kidneys, rather than increased synthesis, stimulating the kidneys to reabsorb more water.

The thyroid gland shrinks with age, as individual follicles shrink and more abundant fibrous connective tissue separates them. Thyroid nodules, which may be benign or cancerous, become more common with age, and are often first detected upon autopsy. Although blood levels of T_3 and T_4 may diminish with

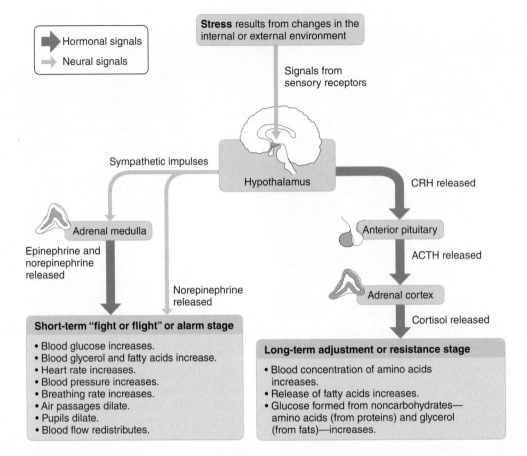

FIGURE 13.37 During the alarm stage of stress, the hypothalamus helps prepare the body for "fight or flight" by triggering sympathetic impulses to various organs. It also stimulates epinephrine release, intensifying the sympathetic responses. In the resistance stage, the hypothalamus stimulates the adrenal cortex to release cortisol, which promotes longer-term responses that resist the effects of stress.

Legend:
- ➡ Hormonal signals
- ➡ Neural signals

Stress results from changes in the internal or external environment

Signals from sensory receptors

Hypothalamus

Sympathetic impulses

CRH released

Adrenal medulla

Epinephrine and norepinephrine released

Norepinephrine released

Anterior pituitary

ACTH released

Adrenal cortex

Cortisol released

Short-term "fight or flight" or alarm stage
- Blood glucose increases.
- Blood glycerol and fatty acids increase.
- Heart rate increases.
- Blood pressure increases.
- Breathing rate increases.
- Air passages dilate.
- Pupils dilate.
- Blood flow redistributes.

Long-term adjustment or resistance stage
- Blood concentration of amino acids increases.
- Release of fatty acids increases.
- Glucose formed from noncarbohydrates— amino acids (from proteins) and glycerol (from fats)—increases.

TABLE 13.13	Major Events in the General Stress Syndrome

1. In response to stress, impulses are conducted to the hypothalamus.

2. Sympathetic impulses originating from the hypothalamus increase blood glucose concentration, blood glycerol concentration, blood fatty acid concentration, heart rate, blood pressure, and breathing rate. They dilate air passages, shunt blood into skeletal muscles, and increase secretion of epinephrine from the adrenal medulla.

3. Epinephrine intensifies and prolongs sympathetic actions.

4. The hypothalamus secretes CRH, which stimulates secretion of ACTH by the anterior pituitary gland.

5. ACTH stimulates release of cortisol by the adrenal cortex.

6. Cortisol increases the concentration of blood amino acids, releases fatty acids, and stimulates formation of glucose from noncarbohydrate sources.

7. Secretion of glucagon from the pancreas and growth hormone from the anterior pituitary increase.

8. Glucagon and growth hormone aid mobilization of energy sources and stimulate uptake of amino acids by cells.

9. Secretion of ADH from the posterior pituitary increases.

10. ADH promotes the retention of water by the kidneys, which increases blood volume.

11. Renin increases blood levels of angiotensin II, which acts as a vasoconstrictor and also stimulates the adrenal cortex to secrete aldosterone.

12. Aldosterone stimulates sodium retention by the kidneys.

age, in general, the thyroid gland's control over the metabolism of various cell types is maintained throughout life. Calcitonin levels decline with age, which raises the risk of osteoporosis.

Parathyroid function differs between the sexes with age. Secretion peaks in males at about age fifty, whereas in women, the level of parathyroid hormone decreases until about age forty, after which it rises and contributes to osteoporosis risk. Fat accumulates between the cells of the parathyroid glands.

The adrenal glands illustrate the common theme of aging-related physical changes, yet continued function. Fibrous connective tissue, lipofuscin pigment, and increased numbers of abnormal cells characterize the aging adrenal glands. However, thanks to the fine-tuning of negative feedback systems, blood levels of glucocorticoids and mineralocorticoids usually remain within the normal range, although the ability to maintain homeostasis of osmotic pressure, blood pressure, acid-base balance, and sodium and potassium ion distributions may falter with age.

The most obvious changes in the aging endocrine system involve blood glucose regulation. The pancreas may be able to maintain secretion of insulin and glucagon, but lifestyle changes, such as increase in fat intake and less exercise, may increase the blood insulin level. The development of insulin resistance—the decreased ability of muscle, liver, and fat cells to take in glucose even in the presence of insulin—reflects impaired ability of these target cells to respond to the hormone, rather than compromised pancreatic function. Blood glucose buildup may signal the pancreas to secrete more insulin, setting the stage for type 2 diabetes mellitus.

The daily fall and rise of melatonin may level out with age, which can affect the sleep/wake cycle. People usually require less sleep as they age. Changes to the tempo of the body clock may, in turn, affect secretion of other hormones.

The thymus begins to noticeably shrink before age twenty, with accompanying declining levels of thymosins. By age sixty, thymosin secretion is nil. The result is a slowing of the maturation of B and T cells, which increases susceptibility to infections as a person ages.

 PRACTICE

50 What general types of changes occur in the glands of the endocrine system with aging?

51 How do the structures and functions of particular endocrine glands change over a lifetime? ∎

Chapter Summary

13.1 General Characteristics of the Endocrine System (page 488)

The nervous system and the **endocrine system** work together to control body functions.

1. Endocrine glands secrete their products (**hormones**) into body fluids (the internal environment); **exocrine glands** secrete their products into ducts that lead to the outside of the body.
2. A hormone's **target cells** have specific receptors.
3. Hormones from endocrine glands regulate metabolic processes.

13.2 Hormone Action (page 489)

Endocrine glands secrete hormones into the bloodstream, which carries them to all parts of the body.

1. Chemistry of hormones
   a. Steroid hormones are lipids that include complex rings of carbon and hydrogen atoms.
   b. Nonsteroid hormones are amines, peptides, and proteins.
2. Actions of hormones
   a. Steroid hormones and thyroid hormones
      (1) Steroid hormones and thyroid hormones enter target cells and combine with receptors to form complexes.
      (2) These complexes activate specific genes in the nucleus, which direct synthesis of specific proteins.
      (3) The degree of cellular response is proportional to the number of hormone-receptor complexes formed.
   b. Nonsteroid hormones
      (1) Nonsteroid hormones combine with receptors in the target cell membrane.
      (2) A hormone-receptor complex stimulates membrane proteins, such as adenylate cyclase, to induce the formation of second messenger molecules.
      (3) A second messenger, such as **cAMP,** activates **protein kinases.**
      (4) Protein kinases activate certain protein substrate molecules, which, in turn, change cellular processes.
      (5) The cellular response to a nonsteroid hormone is amplified because the enzymes induced by a small number of hormone-receptor complexes can catalyze formation of a large number of second messenger molecules.
3. Prostaglandins
   a. Prostaglandins are paracrine substances that have powerful hormonelike effects, even in small amounts.
   b. Prostaglandins modulate hormones that regulate formation of cyclic AMP.

13.3 Control of Hormonal Secretions (page 496)

The concentration of each hormone in the body fluids is precisely regulated.

1. Some endocrine glands secrete hormones in response to releasing hormones that the hypothalamus secretes.
2. Some endocrine glands secrete in response to nervous stimulation.
3. Some endocrine glands secrete in response to changes in the plasma concentration of a substance.
4. In a negative feedback system, a gland is sensitive to the physiological effect that its hormone brings about.
5. When the physiological effect reaches a certain level, it inhibits the gland.
6. As the gland secretes less hormone, the physiological effect is lessened.

13.4 Pituitary Gland (page 498)

The **pituitary gland,** attached to the base of the brain, has an anterior lobe and a posterior lobe. **Releasing hormones** from the hypothalamus control most pituitary secretions.

Endocrine System

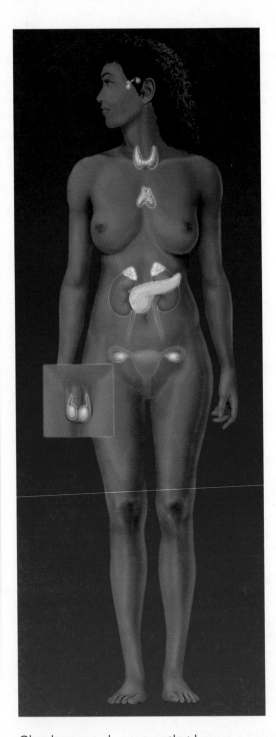

Integumentary System

Melanocytes produce skin pigment in response to hormonal stimulation.

Skeletal System

Hormones act on bones to control calcium balance.

Muscular System

Hormones help increase blood flow to exercising muscles.

Nervous System

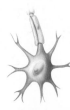

Neurons control the secretions of the anterior and posterior pituitary glands and the adrenal medulla.

Cardiovascular System

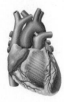

Hormones are carried in the bloodstream; some have direct actions on the heart and blood vessels.

Lymphatic System

Hormones stimulate lymphocyte production.

Digestive System

Hormones help control digestive system activity.

Respiratory System

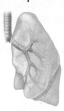

Decreased oxygen causes hormonal stimulation of red blood cell production; red blood cells transport oxygen and carbon dioxide.

Urinary System

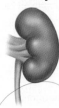

Hormones act on the kidneys to help control water and electrolyte balance.

Reproductive System

Sex hormones play a major role in development of secondary sex characteristics, egg, and sperm.

Glands secrete hormones that have a variety of effects on cells, tissues, organs, and organ systems.

1. Anterior pituitary hormones
   a. The anterior pituitary consists largely of epithelial cells, and it secretes GH, PRL, TSH, ACTH, FSH, and LH.
   b. **Growth hormone** (GH)
      (1) Growth hormone stimulates body cells to grow and divide.
      (2) Growth hormone-releasing hormone and somatostatin from the hypothalamus control GH secretion.
   c. **Prolactin** (PRL)
      (1) PRL promotes breast development and stimulates milk production.
      (2) A normal function of prolactin in males has not been established.
      (3) Prolactin release-inhibiting hormone from the hypothalamus restrains secretion of prolactin, whereas prolactin-releasing factors are thought to promote its secretion.
   d. **Thyroid-stimulating hormone** (TSH)
      (1) TSH controls secretion of hormones from the thyroid gland.
      (2) The hypothalamus, by secreting thyrotropin-releasing hormone, regulates TSH secretion.
   e. **Adrenocorticotropic hormone** (ACTH)
      (1) ACTH controls the secretion of certain hormones from the adrenal cortex.
      (2) The hypothalamus, by secreting corticotropin-releasing hormone, regulates ACTH secretion.
   f. **Follicle-stimulating hormone** (FSH) and **luteinizing hormone** (LH) are gonadotropins that affect the reproductive organs.
2. Posterior pituitary hormones
   a. The posterior lobe of the pituitary gland largely consists of neuroglia and nerve fibers that originate in the hypothalamus.
   b. The two hormones of the posterior pituitary are produced in the hypothalamus.
   c. **Antidiuretic hormone** (ADH)
      (1) ADH causes the kidneys to excrete less water.
      (2) In high concentration, ADH constricts blood vessel walls, raising blood pressure.
      (3) The hypothalamus regulates ADH secretion.
   d. **Oxytocin** (OT)
      (1) OT has an antidiuretic effect and can contract smooth muscle in the uterine wall.
      (2) OT also contracts certain cells associated with production and ejection of milk from the milk glands of the breasts.

13.5 Thyroid Gland (page 503)

The **thyroid gland** is located in the neck and consists of two lateral lobes.
1. Structure of the gland
   a. The thyroid gland consists of many hollow secretory parts called follicles.
   b. The follicles are fluid filled and store the **thyroxine** and **triiodothyronine** the follicular cells secrete.
   c. Extrafollicular cells secrete **calcitonin.**
2. Thyroid hormones
   a. Thyroxine and triiodothyronine
      (1) These hormones increase the rate of metabolism, enhance protein synthesis, and stimulate lipid breakdown.

      (2) These hormones are needed for normal growth and development and for maturation of the nervous system.
   b. Calcitonin
      (1) Calcitonin lowers blood calcium and phosphate ion concentrations.
      (2) This hormone prevents prolonged elevation of calcium after a meal.

13.6 Parathyroid Glands (page 506)

The **parathyroid glands** are on the posterior surface of the thyroid gland.
1. Structure of the glands
   a. Each gland is small and yellow-brown, within a thin connective tissue capsule.
   b. Each gland consists of secretory cells well supplied with capillaries.
2. **Parathyroid hormone** (PTH)
   a. PTH increases blood calcium ion concentration and decreases blood phosphate ion concentration.
   b. PTH stimulates resorption of bone tissue, causes the kidneys to conserve calcium ions and excrete phosphate ions, and indirectly stimulates absorption of calcium ions from the intestine.
   c. A negative feedback mechanism operating between the parathyroid glands and the blood regulates these glands.

13.7 Adrenal Glands (page 508)

The **adrenal glands** are located atop the kidneys.
1. Structure of the glands
   a. Each adrenal gland consists of a medulla and a cortex.
   b. The **adrenal medulla** and **adrenal cortex** are distinct in that they secrete different hormones.
2. Hormones of the adrenal medulla
   a. The adrenal medulla secretes **epinephrine** and **norepinephrine.**
   b. These hormones are synthesized from tyrosine and are chemically similar.
   c. These hormones produce effects similar to those of the sympathetic nervous system.
   d. Sympathetic impulses originating from the hypothalamus stimulate secretion of these hormones.
3. Hormones of the adrenal cortex
   a. The cortex produces several types of steroids that include hormones.
   b. **Aldosterone**
      (1) It causes the kidneys to conserve sodium ions and water and to excrete potassium ions.
      (2) It is secreted in response to increased potassium ion concentration or the presence of **angiotensin II.**
      (3) By conserving sodium ions and water, it helps maintain blood volume and pressure.
   c. **Cortisol**
      (1) It inhibits protein synthesis, releases fatty acids, and stimulates glucose formation from noncarbohydrates.
      (2) A negative feedback mechanism involving secretion of CRH from the hypothalamus and ACTH from the anterior pituitary gland controls its level.

d. Sex hormones
   (1) These hormones are of the male type although some can be converted into female hormones.
   (2) They supplement the sex hormones produced by the gonads.

13.8 Pancreas (page 514)

The **pancreas** secretes digestive juices as well as hormones.

1. Structure of the gland
   a. The pancreas is posterior to the stomach and is attached to the small intestine.
   b. The endocrine portion, called the pancreatic islets (islets of Langerhans), secretes **glucagon, insulin,** and **somatostatin.**
2. Hormones of the pancreatic islets
   a. Glucagon stimulates the liver to produce glucose, increasing concentration of blood glucose. It also breaks down fat.
   b. Insulin activates facilitated diffusion of glucose through cell membranes, stimulates its storage, promotes protein synthesis, and stimulates fat storage.
   c. Facilitated diffusion of glucose into nerve cells does not depend on insulin.
   d. Somatostatin inhibits insulin and glucagon release.

13.9 Other Endocrine Glands (page 517)

1. Pineal gland
   a. The **pineal gland** is attached to the thalamus near the roof of the third ventricle.
   b. Postganglionic sympathetic nerve fibers innervate it.
   c. It secretes **melatonin,** which regulates some **circadian rhythms.**
2. Thymus
   a. The **thymus** lies posterior to the sternum and between the lungs.
   b. It shrinks with age.
   c. It secretes **thymosins,** which affect the production of certain lymphocytes that, in turn, provide immunity.
3. Reproductive organs
   a. The **testes** secrete testosterone.

b. The **ovaries** secrete estrogens and progesterone.
c. The **placenta** secretes estrogens, progesterone, and a gonadotropin.
4. The digestive glands include certain glands of the stomach and small intestine that secrete hormones.
5. Other hormone-producing organs include the heart and kidneys.

13.10 Stress and Its Effects (page 517)

Stress occurs when the body responds to **stressors** that threaten the maintenance of homeostasis. Stress responses include increased activity of the sympathetic nervous system and increased secretion of adrenal hormones.

1. Types of stress
   a. Physical stress results from environmental factors that are harmful or potentially harmful to tissues.
   b. Psychological stress results from thoughts about real or imagined dangers. Factors that produce psychological stress vary with the individual and the situation.
2. Responses to stress
   a. Responses to stress maintain homeostasis.
   b. The hypothalamus controls a general adaptation (or general stress) syndrome.

13.11 Life-Span Changes (page 519)

With age, endocrine glands shrink and accumulate fibrous connective tissue, fat, and lipofuscin, but hormonal activities usually remain within the normal range.

1. GH levels even out, as muscular strength declines.
2. ADH levels increase due to slowed breakdown.
3. The thyroid shrinks but control of metabolism continues.
4. Decreasing levels of calcitonin and increasing levels of parathyroid hormone increase osteoporosis risk.
5. The adrenal glands show aging-related changes, but negative feedback maintains functions.
6. Muscle, liver, and fat cells may develop insulin resistance.
7. Changes in melatonin secretion affect the body clock.
8. Thymosin production declines, hampering resistance to disease.

CHAPTER ASSESSMENTS

13.1 General Characteristics of the Endocrine System

1 Contrast the definitions of *endocrine gland* and *exocrine gland*. (p. 488)
2 Explain the specificity of a hormone for its target cell. (p. 488)
3 List six general functions of hormones. (p. 488)

13.2 Hormone Action

4 Explain how hormones can be grouped on the basis of their chemical composition. (pp. 489–490)
5 List the steps of steroid hormone action. (p. 490)
6 List the steps of the action of most nonsteroid hormones. (pp. 491–496)
7 Explain how prostaglandins are similar to hormones and how they are different. (p. 496)

13.3 Control of Hormonal Secretions

8 Diagram the three mechanisms that control hormone secretion, including negative feedback in each mechanism. (p. 496)

13.4 Pituitary Gland

9 Describe the location and structure of the pituitary gland. (p. 498)
10 List the hormones that the anterior pituitary secretes. (p. 498)
11 Explain two ways that the brain controls pituitary gland activity. (pp. 498–499)
12 Releasing hormones come from which one of the following? (p. 498)
   a. thyroid gland
   b. anterior pituitary gland
   c. posterior pituitary gland
   d. hypothalamus
13 Match the following hormones with their actions (the choices from the list on the right may be used more than once): (pp. 500–503)

(1) growth hormone	A.	milk synthesis
(2) thyroid-stimulating hormone	B.	cell division
(3) prolactin	C.	metabolic rate
(4) adrenocorticotropic hormone	D.	acts on gonads
(5) follicle-stimulating hormone	E.	controls secretion
(6) luteinizing hormone		of adrenal cortical hormones

14 Explain how growth hormone produces its effects. (p. 500)
15 Describe the control of growth hormone secretion. (p. 500)
16 Describe the anatomical differences between the anterior and posterior lobes of the pituitary gland. (pp. 500 and 502)
17 Name and describe the functions of the posterior pituitary hormones. (pp. 502–503)
18 Under which of the following conditions would you expect an increase in antidiuretic hormone secretion? (pp. 502–503)
   a. An individual ingests excess water.
   b. The posterior pituitary is removed because it has a tumor.
   c. An individual is rescued after three days in the desert without food or water.
   d. An individual receives an injection of synthetic antidiuretic hormone.

13.5 Thyroid Gland

19 Describe the location and structure of the thyroid gland. (pp. 503–505)
20 Match the hormones from the thyroid gland with their descriptions. (pp. 505–506)

(1) thyroxine	A.	most potent at controlling metabolism
(2) triiodothyronine	B.	regulates blood calcium
(3) calcitonin	C.	has four iodine atoms

21 Define *iodide pump*. (p. 505)
22 Diagram the control of thyroid hormone secretion. (pp. 505–506)

13.6 Parathyroid Glands

23 Describe the location and structure of the parathyroid glands. (p. 506)
24 Explain the general function of parathyroid hormone. (pp. 507–508)
25 Diagram the regulation of parathyroid hormone secretion. (pp. 507–508)

13.7 Adrenal Glands

26 Distinguish between the adrenal medulla and the adrenal cortex. (p. 508)
27 Match the adrenal hormones with their source and actions: (pp. 508–512)

(1) cortisol	A.	cortex; sodium retention
(2) aldosterone	B.	cortex; male sex hormones
(3) epinephrine	C.	medulla; fight-or-flight response
(4) androgens	D.	cortex; gluconeogenesis

28 Diagram control of aldosterone secretion. (pp. 510–511)
29 Diagram control of cortisol secretion. (p. 512)

13.8 Pancreas

30 Describe the location and structure of the pancreas. (p. 514)
31 List the hormones the pancreatic islets secrete and their general functions. (pp. 514–515)
32 Diagram the control of pancreatic hormone secretion. (pp. 514–515)

13.9 Other Endocrine Glands

33 Describe the location and general function of the pineal gland. (p. 517)
34 Describe the location and general function of the thymus. (p. 517)
35 Name five additional hormone-secreting organs. (p. 517)

13.10 Stress and Its Effects

36 Distinguish between a stressor and stress. (p. 517)
37 List several factors that cause physical and/or psychological stress. (p. 517)
38 Describe hormonal and nervous responses to stress. (pp. 518–519)

13.11 Life-Span Changes

39 Levels of which hormones decrease with age? Which increase? (pp. 519–521)

INTEGRATIVE ASSESSMENTS/CRITICAL THINKING

Outcomes 2.2, 13.2, 13.3, 13.4, 13.5

1. When a nuclear reactor explodes, a great plume of radioactive isotopes erupts into the air and may spread for thousands of miles. Most of the isotopes emitted immediately following the blast are of the element iodine. Which of the glands of the endocrine system would be most seriously—and immediately—affected by exposure to the isotopes, and how do you think this will become evident in the nearby population?

Outcomes 4.5, 13.3, 13.8

2. Why might oversecretion of insulin reduce glucose uptake by nerve cells?

Outcomes 13.2, 13.4, 13.10, 13.11

3. A young mother feels shaky, distracted, and generally ill. She lives with her mother, who is dying. A friend tells the young woman, "It's just stress, it's all in your head." Is it?

Outcomes 13.3, 13.4

4. Growth hormone is administered to people who have pituitary dwarfism. Parents wanting their normal children to be taller have requested the treatment for them. Do you think that this is a wise request? Why or why not?

Outcomes 13.3, 13.4, 13.5, 13.7, 13.9

5. An adult has had her anterior pituitary removed. Which hormone supplements will she require?

Outcomes 13.4, 13.7

6. The adrenal cortex of a patient who has lost a large volume of blood will increase secretion of aldosterone. What effect will this increased secretion have on the patient's blood concentrations of sodium and potassium ions?

Outcomes 13.4, 13.7, 13.10

7. What problems might result from the prolonged administration of cortisol to a person with severe inflammatory disease?

 # ONLINE STUDY TOOLS

Connect Interactive Questions Reinforce your knowledge using assigned interactive questions covering the location and structure of endocrine glands, actions of hormones, and the body's response to stress.

Connect Integrated Activity Can you apply your new knowledge to discussing various hormonal imbalances?

LearnSmart Discover which chapter concepts you have mastered and which require more attention. This adaptive learning tool is personalized, proven, and preferred.

Anatomy & Physiology Revealed Go more in depth into the human body by exploring cadaver dissections, viewing histological slides, and watching animations related to endocrine glands and hormonal secretions.

14

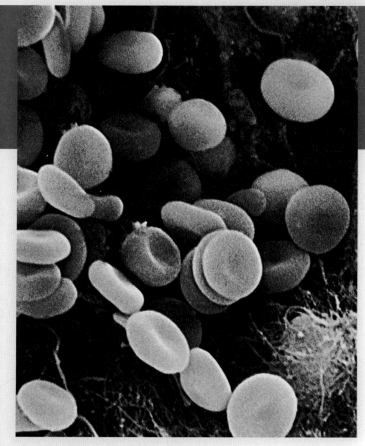

A micrograph of blood—a complex mixture of cells, cell fragments, and many types of dissolved biochemicals that provides nutrients, oxygen, and other vital substances to our cells (1,800×).

Blood

 LEARNING OUTCOMES

After you have studied this chapter, you should be able to:

14.1 Characteristics of Blood

1 Distinguish among the formed elements of blood and the liquid portion of blood. (pp. 528–529)

14.2 Blood Cells

2 Describe the origin of blood cells. (p. 529)

3 Explain the significance of red blood cell counts and how they are used to diagnose disease. (p. 531)

4 Discuss the life cycle of a red blood cell. (pp. 531, 533, and 535)

5 Summarize the control of red blood cell production. (p. 533)

6 Distinguish among the five types of white blood cells, and give the function(s) of each type. (pp. 536–537)

7 Describe a blood platelet, and explain its functions. (pp. 539–540)

14.3 Plasma

8 Describe the functions of each of the major components of plasma. (pp. 541–542)

14.4 Hemostasis

9 Define *hemostasis,* and explain the mechanisms that help to achieve it. (pp. 542–543)

10 Review the major steps in blood coagulation. (pp. 543–544)

11 Explain how to prevent blood coagulation. (p. 547)

14.5 Blood Groups and Transfusions

12 Explain blood typing and how it is used to avoid adverse reactions following blood transfusions. (pp. 548–551)

13 Describe how blood reactions may occur between fetal and maternal tissues. (p. 551)

THE WHOLE PICTURE

Blood signifies life, and for good reason—it has many vital functions. This complex mixture of cells, cell fragments, and dissolved biochemicals carries nutrients, oxygen, wastes, and hormones; helps maintain the stability of the intestinal fluid; and distributes heat. The blood, heart, and blood vessels form the cardiovascular system and link the body's internal and external environments.

Module 9: Cardiovascular System

UNDERSTANDING WORDS

agglutin-, to glue together: *agglutin*ation—clumping of red blood cells.

bil-, bile: *bil*irubin—pigment excreted in the bile.

-crit, to separate: hemato*crit*—percentage by volume of red blood cells in a blood sample, determined by separating the red blood cells from the plasma.

embol-, stopper: *embol*ism—a mass lodging in and obstructing a blood vessel.

erythr-, red: *erythr*ocyte—red blood cell.

hema-, blood: *hema*tocrit—percentage of red blood cells in a given volume of blood.

hemo-, blood: *hemo*globin—red pigment responsible for the color of blood.

hepa-, liver: *hepa*rin—anticoagulant secreted by liver cells.

leuko-, white: *leuko*cyte—white blood cell.

-lys, to break up: fibrino*lys*in—protein-splitting enzyme that can digest fibrin.

macro-, large: *macro*phage—large phagocytic cell.

-osis, abnormal condition: leukocyt*osis*—condition in which white blood cells are overproduced.

-poie, make, produce: erythro*poie*tin—hormone that stimulates the production of red blood cells.

poly-, many: *poly*cythemia—overproduction of red blood cells.

-stasis, halt, make stand: hemo*stasis*—process that stops bleeding from damaged blood vessels.

thromb-, clot: *thromb*ocyte—blood platelet involved in the formation of a blood clot.

14.1 | Characteristics of Blood

Blood is a type of connective tissue whose cells are suspended in a liquid extracellular matrix. Blood is vital in carrying substances between body cells and the external environment, thereby promoting homeostasis.

Blood volume varies with body size, changes in fluid and electrolyte concentrations, and the amount of adipose tissue.

Blood volume is typically about 8% of body weight. An average-size adult has about 5 liters of blood.

Whole blood is slightly heavier and three to four times more viscous than water. Its cells include red blood cells and white blood cells. Blood also contains cellular fragments called platelets (fig. 14.1). The cells and platelets are termed "formed elements" of the blood, in contrast to the liquid portion.

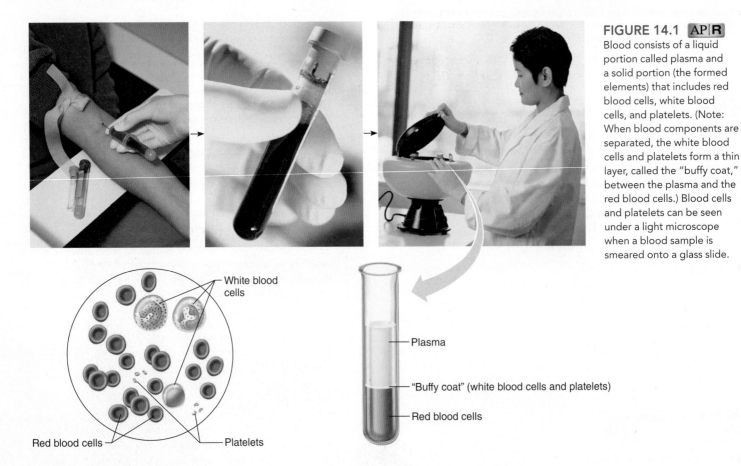

FIGURE 14.1 AP|R

Blood consists of a liquid portion called plasma and a solid portion (the formed elements) that includes red blood cells, white blood cells, and platelets. (Note: When blood components are separated, the white blood cells and platelets form a thin layer, called the "buffy coat," between the plasma and the red blood cells.) Blood cells and platelets can be seen under a light microscope when a blood sample is smeared onto a glass slide.

White blood cells

Red blood cells

Platelets

Peripheral Blood Smear

Plasma

"Buffy coat" (white blood cells and platelets)

Red blood cells

Centrifuged Blood Sample

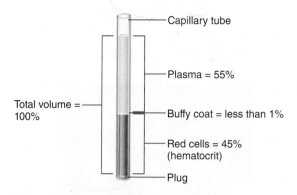

Capillary tube

Plasma = 55%

Buffy coat = less than 1%

Red cells = 45%
(hematocrit)

Plug

Total volume =
100%

FIGURE 14.2 If a blood-filled capillary tube is centrifuged, the red cells pack in the lower portion and the percentage of red cells (hematocrit) can be determined. Values shown are within the normal range for healthy humans.

If a blood sample stands in a tube for a while and is prevented from clotting, the cells separate from the liquid portion and settle to the bottom. Centrifuging the sample quickly packs the cells into the lower part of the centrifuge tube, as figure 14.2 shows. The percentages of cells and liquid in the blood sample can then be calculated.

Most blood samples are about 45% red blood cells by volume. This percentage is called the **hematocrit** (HCT), or **packed cell volume** (PCV). The white blood cells and platelets account for less than 1%. The remaining blood sample, about 55%, is the clear, straw-colored **plasma** (plaz′mah). Plasma is a complex solution that includes water, amino acids, proteins, carbohydrates, lipids, vitamins, hormones, electrolytes, and cellular wastes (fig. 14.3). Appendix C, Laboratory Tests of Clinical Importance, lists values for the hematocrit and other blood tests commonly performed on healthy individuals. Clinical Application 14.1 discusses the use of universal precautions for handling blood and other body fluids.

 PRACTICE

1 What factors affect blood volume?

2 What are the major components of blood?

3 How is hematocrit determined?

14.2 | Blood Cells

The Origin of Blood Cells

Hematopoiesis (he″mat-o-poi-e′sis) is the formation of blood cells. Blood cells originate in red bone marrow from **hematopoietic stem cells,** also known as hemocytoblasts (he″mo-si′to-blastz) (fig. 14.4). A stem cell can divide to give rise to specialized (more differentiated) cells as well as more stem cells. As hematopoietic stem cells divide, the new cells, myeloid and lymphoid stem cells, respond to different secreted growth factors, called **hematopoietic growth factors,** that turn on some genes and turn off others. This exposure to growth factors ultimately sculpts the distinctive formed elements of blood, including the cellular components of the immune system.

 RECONNECT
To Chapter 3, Stem and Progenitor Cells, pages 112–113.

Characteristics of Red Blood Cells

Red blood cells, also called **erythrocytes** (ĕ-rith′ro-sītz), are tiny, approximately 7.5 μm in diameter. They are biconcave discs, thin near their centers and thicker around their rims (fig. 14.5). This distinctive shape is an adaptation for the red blood cell's function of transporting gases—it increases the surface area through which gases can diffuse. The shape also places the cell membrane closer to oxygen-carrying **hemoglobin** (he″mo-glo′bin) molecules in the cell. A red blood cell's shape and flexibility enable it to readily squeeze through the narrow capillaries.

Each red blood cell is about one-third hemoglobin by volume. This protein imparts the color of blood. The rest of the cell mainly consists of membrane, water, electrolytes, and enzymes. When hemoglobin combines with oxygen, the resulting *oxyhemoglobin* is bright red; when the oxygen is released, the resulting *deoxyhemoglobin* is darker. Blood rich in deoxyhemoglobin may appear bluish when it is viewed through blood vessel walls.

 CAREER CORNER
Blood Bank Technologist

When a large truck jackknifed on an icy highway, the resulting ten-car pile-up sent many injured people to the emergency department of the closest hospital. Alerts requesting blood donations went out on social and traditional media right away, and soon people began to arrive to donate blood.

At the hospital, one blood bank technologist helped to collect blood from the donors, while another working in a laboratory processed and typed the blood so that physicians could match blood to patients, on the lookout for rare blood types. The technologist in the laboratory also tested the blood for viral contamination and presence of antibodies, and when too many donors showed up, prepared the extra blood for storage, separating its components.

In addition to expertise in blood banking, blood bank technologists must have coursework that covers molecular testing, immunology, genetics, regenerative medicine, cell therapies, transfusion practices, and transplantation, and be aware of complications that may arise from giving one person's blood to another. Computer skills, manual dexterity, and normal color vision are important for the job.

Blood bank technologists work in blood donor centers, transfusion services, research institutions, reference laboratories, transplant centers, and suppliers of blood bank equipment. They must have a degree and certification in medical technology and complete a 12- or 24-month training program, or a four-year degree in a science with work experience at a blood bank plus the postgraduate program. The American Society for Clinical Pathology (ASCP) certifies blood bank technologists.

The career has room for growth. Blood bank technologists may become regulatory experts involved with clinical trials, administrators of laboratories, technical/procedural advisors, quality assessors and managers, or educators to train other health-care professionals in blood transfusion medicine.

CLINICAL APPLICATION 14.1

Universal Precautions

Blood can contain more than cells, nutrients, proteins, and water—a single drop from an infected individual can harbor billions of viruses. In the wake of the AIDS epidemic, in 1988 the U.S. Centers for Disease Control and Prevention (CDC) devised "universal precautions," which are specific measures that health-care workers should take to prevent transmission of bloodborne infectious agents in the workplace (fig. 14A). The CDC singled out HIV and the hepatitis B virus. The guidelines grew out of earlier suggestions for handling patients suspected to have been exposed to viruses. The term *universal* refers to the assumption that *any* patient may have been exposed to a pathogen that can be transmitted in a body fluid such as blood.

Attention to safety in the health-care setting can prevent transmission of infectious diseases. The World Health Organization estimates that 4% to 7% of new infections worldwide are transmitted via unsafe injections. Specific recommendations include:

- use of personal protective equipment, such as gloves, goggles, and masks
- engineering controls, such as fume hoods and sharps containers
- work-practice controls, such as handwashing before and after performing procedures

Universal precautions were designed for, and work well in, preventing transmission of viral illnesses in settings already relatively safe, such as clinics. This isn't necessarily the case for outbreaks, natural disasters, and combat zones. For example, in 2005 several pediatric nurses who aided neighbors infected with the Marburg virus in the isolated town of Uige in Angola, in southwestern Africa, died from this hemorrhagic fever along with hundreds of other people.

Headache, fever, vomiting, and diarrhea begin three to nine days after exposure to the Marburg virus. Then the person bleeds from all body openings, internally and under the skin. Plummeting blood pressure kills most infected individuals within a week, and anyone contacting their blood is in danger of infection. Infected indi-

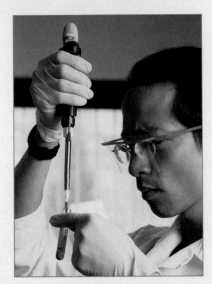

FIGURE 14A Health-care workers wear personal protective equipment to shield themselves from body fluids containing disease-causing viruses.

 Which protective gear is worn by this health-care worker?
Answer can be found in Appendix G.

viduals must be isolated and not touched, but the scourge spreads because many family members become infected while tending to their loved ones.

In the 2005 outbreak, contaminated medical equipment contributed to the rapid and deadly spread of the infection. Untrained clinic workers reused needles, and some people reused needles and intravenous equipment in their homes. However, even universal precautions might not have contained this outbreak, because the body fluids were so copious and infectious.

Prolonged oxygen deficiency (hypoxia) causes *cyanosis*, in which the skin and mucous membranes appear bluish due to an abnormally high blood concentration of deoxyhemoglobin. Exposure to low temperature may also result in cyanosis by constricting superficial blood vessels. This slows blood flow, allowing removal of more oxygen than usual from blood flowing through the vessels, thereby increasing the amount of deoxyhemoglobin.

Red blood cells have nuclei during their early stages of development but extrude the nuclei as they mature, providing more space for hemoglobin. Because mature red blood cells lack nuclei, they cannot synthesize proteins or divide. Mature red blood cells pro-

duce ATP through glycolysis only and use none of the oxygen they carry because they also lack mitochondria. As long as cytoplasmic enzymes function, these cells can carry on vital energy-releasing processes. With time, however, red blood cells become less active, more rigid, and more likely to be damaged or worn. Eventually the spleen and liver remove older red blood cells from the circulation.

PRACTICE

4 How do blood cells form?

5 Describe a red blood cell.

6 How does the biconcave shape of a red blood cell make possible its function?

7 What is the function of hemoglobin?

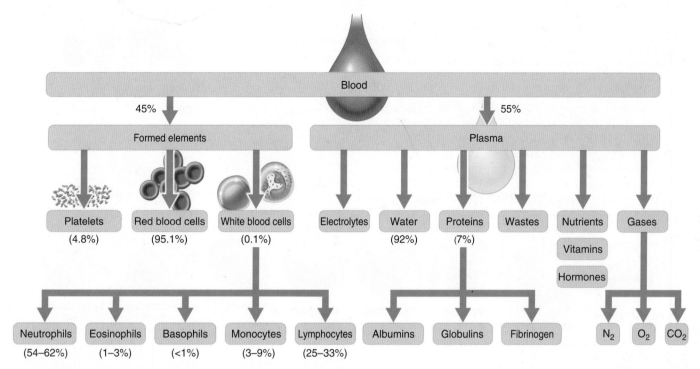

FIGURE 14.3 Blood composition. Blood is a complex mixture of formed elements in a liquid extracellular matrix, plasma. Note that water and proteins account for 99% of the plasma.

In sickle cell disease, a single DNA base mutation changes one amino acid in the protein part of hemoglobin, causing the molecule to attach within itself at many more points. The abnormal hemoglobin makes the red blood cells sticky and likely to deform into sickle shapes as they move through the bloodstream (see figure 14.7b). The fragile cells lodge in narrow blood vessels, cutting off local blood supplies. Once a blockage occurs, the falling oxygen level hastens sickling and the shape change spreads, blocking circulation further. The result is great pain in the blocked body parts, particularly in the hands, feet, and bones. As red blood cells die too quickly, their depletion causes the profound fatigue of anemia while the spleen becomes overworked in breaking down the cells, increasing the risk of infection.

Children with sickle cell disease are typically diagnosed at birth. The spleen is usually damaged early on, so they receive antibiotics daily for years to prevent infection. Hospitalization for blood transfusions may be necessary for painful sickling "crises" of blocked circulation.

A drug, hydroxyurea, is used to activate production of a form of hemoglobin that more effectively binds oxygen, normally produced only in the fetus. Fetal hemoglobin slows sickling, which enables red blood cells to reach the lungs—where fresh oxygen restores the cells' normal shapes. A bone marrow transplant or an umbilical cord stem cell transplant from a donor can completely cure sickle cell disease, but has a 5% risk of fatality.

Red Blood Cell Counts

The number of red blood cells in a microliter (μL or mcL or 1 mm³) of blood is called the *red blood cell count* (RBCC or RCC). The typical range for adult males is 4,700,000–6,100,000 cells per microliter, and that for adult females is 4,200,000–5,400,000 cells per microliter. For children, the average range is 4,500,000–5,100,000 cells per microliter. These values may vary slightly with the hospital, physician, and type of equipment used to make blood cell counts, and even in a healthy individual from time to time. The number of red blood cells generally increases for several days following strenuous exercise or an increase in altitude.

Red blood cell counts are routinely consulted to help diagnose and evaluate the courses of certain diseases. Changes in red blood cell count may affect health by altering the blood's *oxygen-carrying capacity*.

Red Blood Cell Production and Its Control

Red blood cell formation (erythropoiesis) initially occurs in the yolk sac, liver, and spleen. After birth, these cells are produced almost exclusively by tissue lining the spaces in bones, which are filled with red bone marrow.

 RECONNECT
To Chapter 7, Blood Cell Formation, pages 210–211.

In the red bone marrow, hematopoietic stem cells divide and give rise to **erythroblasts** (ĕ-rith′ro-blastz). The erythroblasts also divide and give rise to many new cells. The nuclei of these newly formed cells soon shrink and are pinched off from the cell along

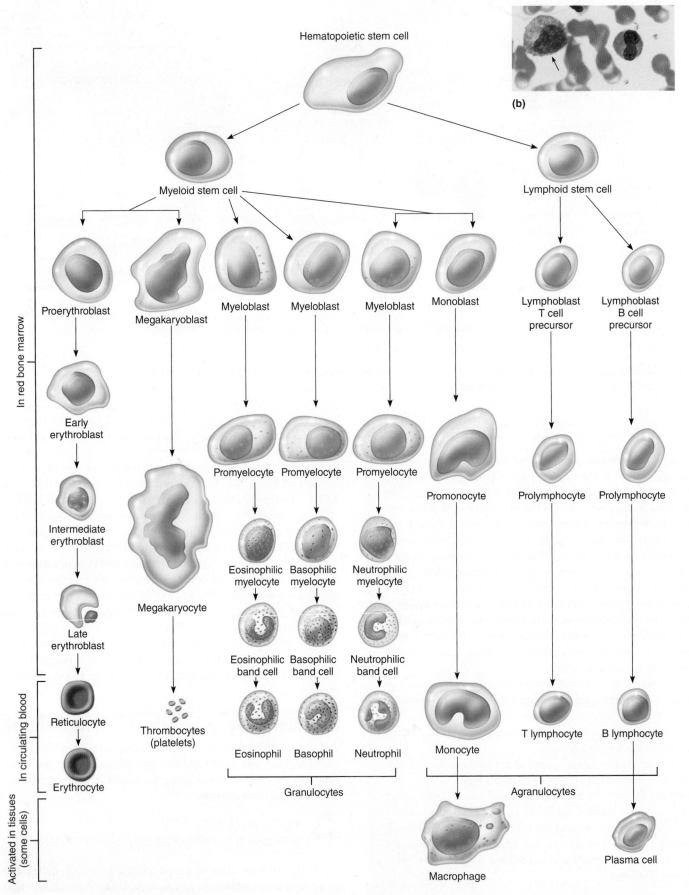

Hematopoietic stem cell

Myeloid stem cell

Lymphoid stem cell

In red bone marrow

Proerythroblast

Megakaryoblast

Myeloblast

Myeloblast

Myeloblast

Monoblast

Lymphoblast
T cell
precursor

Lymphoblast
B cell
precursor

Early
erythroblast

Intermediate
erythroblast

Late
erythroblast

Promyelocyte

Promyelocyte

Promyelocyte

Promonocyte

Prolymphocyte

Prolymphocyte

Eosinophilic
myelocyte

Basophilic
myelocyte

Neutrophilic
myelocyte

Megakaryocyte

Eosinophilic
band cell

Basophilic
band cell

Neutrophilic
band cell

In circulating blood

Reticulocyte

Thrombocytes
(platelets)

Eosinophil

Basophil

Neutrophil

Monocyte

T lymphocyte

B lymphocyte

Erythrocyte

Granulocytes

Agranulocytes

Activated in tissues (some cells)

Macrophage

Plasma cell

(b)

(a)

FIGURE 14.4 **AP|R** Blood cells. (*a*) Development of red blood cells, white blood cells, and platelets from hematopoietic stem cells in bone marrow. (*b*) Light micrograph of a hematopoietic stem cell (arrow) in red bone marrow (500×).

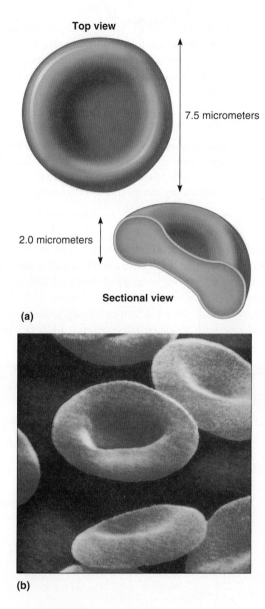

Top view

7.5 micrometers

2.0 micrometers

Sectional view

(a)

(b)

FIGURE 14.5 AP|R   Red blood cells. (*a*) The biconcave shape of a red blood cell makes possible its function of transporting oxygen. (*b*) Falsely colored scanning electron micrograph of human red blood cells (5,000×).

with a thin covering of cytoplasm and cell membrane, in a process called nuclear extrusion. Macrophages (see chapter 5, p. 160) engulf and degrade the extruded materials. For a day or two, some of these young red cells may contain a netlike structure (reticulum) that appears when the cells are stained. These cells are called **reticulocytes** (rĕ-tik′u-lo-sitz). This is the stage in which cells exit the bone marrow to enter the blood. When the remaining organelles degenerate the cells are fully mature. The resulting cells are erythrocytes.

The average life span of a red blood cell is 120 days. During that time, it travels through the body about 75,000 times. Many red blood cells are removed from the circulation each day, yet the number in the circulating blood remains relatively stable. These numbers suggest a homeostatic control of the rate of red blood cell production.

A *negative feedback mechanism* using the hormone **erythropoietin** (e-rith″ro-poi′ĕ-tin) (EPO) controls the rate of red blood cell formation. In response to prolonged oxygen deficiency, EPO is released from the kidneys and to a lesser extent from the liver. (In a fetus, the liver is the main site of EPO production.) At high altitudes, for example, although the percentage of oxygen in the air remains the same, the atmospheric pressure decreases, reducing availability of oxygen. The amount of oxygen delivered to the tissues initially decreases. As **figure 14.6** shows, this drop in oxygen triggers release of EPO, which travels via the blood to the red bone marrow and increases erythrocyte production.

After a few days in the high-altitude environment, many newly formed red blood cells appear in the circulating blood. The increased rate of production continues until the number of erythrocytes is sufficient to supply tissues with oxygen. When the availability of oxygen returns to normal, EPO release decreases and the rate of red blood cell production returns to normal.

RECONNECT

To Chapter 13, Abusing Hormones to Improve Athletic Performance, page 494.

Other conditions can lower oxygen levels and stimulate EPO release. These include loss of blood, which decreases the oxygen-carrying capacity of the cardiovascular system, and chronic lung diseases, which decrease the respiratory surface area available for gas exchange. An excessive increase in red blood cells is *polycythemia*. This increases blood viscosity, slowing blood flow and impairing circulation.

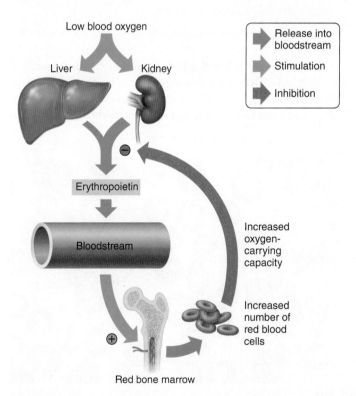

Low blood oxygen

Liver   Kidney

Release into bloodstream

Stimulation

Inhibition

Erythropoietin

Bloodstream

Increased oxygen-carrying capacity

Increased number of red blood cells

Red bone marrow

FIGURE 14.6 Low blood oxygen causes the kidneys and, to a lesser degree, the liver to release erythropoietin. Erythropoietin stimulates target cells in the red bone marrow to increase the production of red blood cells, which carry oxygen to tissues.

8 What is the typical red blood cell count for an adult male? For an adult female?

9 Where are red blood cells produced?

10 How does a red blood cell change as it matures?

11 How is red blood cell production controlled?

Dietary Factors Affecting Red Blood Cell Production

Availability of B-complex vitamins B$_{12}$ and folic acid significantly influences red blood cell production. These vitamins, required for DNA synthesis, are necessary for the growth and division of all cells. Because hematopoietic stem cells frequently divide, they are especially vulnerable to a deficiency of either of these vitamins. Lack of vitamin B$_{12}$ is usually due to a disorder in the stomach lining rather than to a dietary deficiency, because parietal cells in the stomach secrete a substance called *intrinsic factor* required to absorb vitamin B$_{12}$.

In the absence of intrinsic factor, vitamin B$_{12}$ absorption decreases, causing the red bone marrow to form abnormally large, irregularly shaped, thin-membraned fragile red blood cells, resulting in a condition called *pernicious anemia*. B$_{12}$ deficiency can also cause permanent brain damage if not treated promptly with vitamin B$_{12}$ injections. Taking excess folic acid can mask a vitamin B$_{12}$ deficiency by correcting the anemia, but will not prevent the neurological damage.

Iron is required for hemoglobin synthesis. Although much of the iron released during the decomposition of hemoglobin is available for reuse, some iron is lost each day and must be replaced. Only a small fraction of ingested iron is absorbed. Iron absorp-tion is slow, but the rate varies with the total amount of iron in the body. When iron stores are low, absorption rate increases, and when the tissues are becoming saturated with iron, the rate greatly decreases. Table 14.1 summarizes the dietary factors that affect red blood cell production.

Vitamin C increases absorption of iron in the digestive tract. Drinking orange juice with a meal is a good way to boost iron absorption. Drinking tea with a meal reduces absorption of iron because tannic acid in tea binds dietary iron and prevents its absorption.

A deficiency of red blood cells or a reduction in the amount of hemoglobin they contain results in a condition called **anemia.** This reduces the oxygen-carrying capacity of the blood, and the affected person may appear pale and lack energy. Figure 14.7 shows red blood cells of someone who has iron-deficiency anemia (fig. 14.7a) and of someone with sickle cell disease (fig. 14.7b). A pregnant

TABLE 14.1	Dietary Factors Affecting Red Blood Cell Production	
Substance	**Source**	**Function**
Vitamin B$_{12}$ (requires intrinsic factor for absorption via small intestine)	Absorbed from small intestine	DNA synthesis
Iron	Absorbed from small intestine; conserved during red blood cell destruction and made available for reuse	Hemoglobin synthesis
Folic acid	Absorbed from small intestine	DNA synthesis

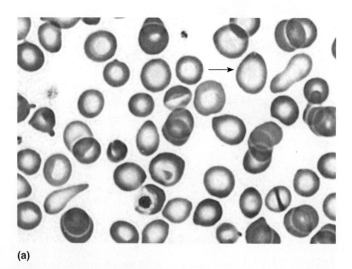

(a)

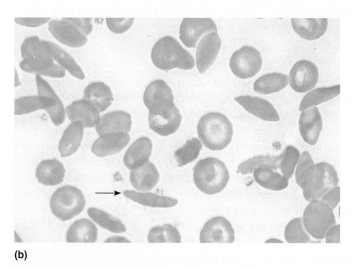

(b)

FIGURE 14.7 Abnormal red blood cells. (a) Light micrograph of erythrocytes with central pallor or paleness (arrow) as seen in iron-deficiency anemia (1,000×). The purple color is due to the stain used in the preparation. (b) Light micrograph of sickled erythrocytes (arrow) from a person with sickle cell disease (1,000×). The pink color results from using a different staining procedure.

woman may become anemic if she doesn't eat enough iron-rich foods, because her blood volume increases due to fluid retention. This increased blood volume decreases the hematocrit. Table 14.2 describes some types of anemia.

 PRACTICE

12 Which vitamins are necessary for red blood cell production?

13 Why is iron required for the formation of red blood cells?

TABLE 14.2	Some Types of Anemia	
Type	**Cause**	**Defect**
Aplastic anemia	Toxic chemicals, radiation	Damaged bone marrow
Hemolytic anemia	Toxic chemicals	Red blood cells destroyed
Iron-deficiency anemia	Dietary lack of iron	Hemoglobin deficient
Pernicious anemia	Inability to absorb vitamin B_{12}	Excess of large, fragile cells
Sickle cell disease	Defective gene	Red blood cells abnormally shaped
Thalassemia	Defective gene	Hemoglobin deficient; red blood cells short-lived

Destruction of Red Blood Cells

Red blood cells are elastic and flexible, and they readily bend as they pass through small blood vessels. With age, however, these cells become more fragile, and may be damaged by passing through capillaries, particularly those in active muscles that must withstand strong forces.

Damaged or worn red blood cells rupture as they pass through the spleen or liver. In these organs, macrophages phagocytize and destroy damaged red blood cells and their contents. Hemoglobin molecules liberated from the red blood cells break down into their four component polypeptide "globin" chains, each surrounding a *heme* group (fig. 14.8a,b).

The heme further decomposes into iron and a greenish pigment called **biliverdin.** The iron, combined with a protein called *transferrin,* may be carried by the blood to the hematopoietic tissue in the red bone marrow and reused in synthesizing new hemoglobin. About 80% of the iron is stored in the liver cells in the form of an iron-protein complex called *ferritin.* In time, the biliverdin is converted to an orange pigment called **bilirubin.** Biliverdin and bilirubin are secreted in the bile as bile pigments (fig. 14.8c,d and fig. 14.9).

The polypeptide globin chains break down into amino acids. The individual amino acids are metabolized by the macrophages or released into the blood. Table 14.3 summarizes the process of red blood cell destruction. Figure 14.9 summarizes the life cycle of a red blood cell.

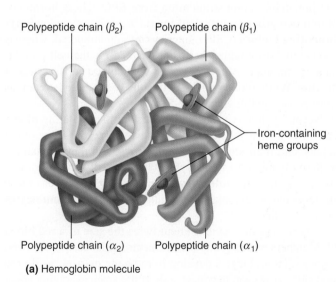

(a) Hemoglobin molecule

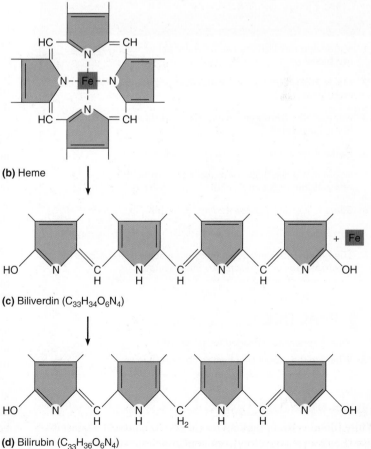

(b) Heme

(c) Biliverdin ($C_{33}H_{34}O_6N_4$)

(d) Bilirubin ($C_{33}H_{36}O_6N_4$)

FIGURE 14.8 Structural formulas. (*a*) When a hemoglobin molecule decomposes, (*b*) the heme groups break down into (*c*) iron (Fe) and biliverdin. (*d*) Most of the biliverdin is then converted to bilirubin.

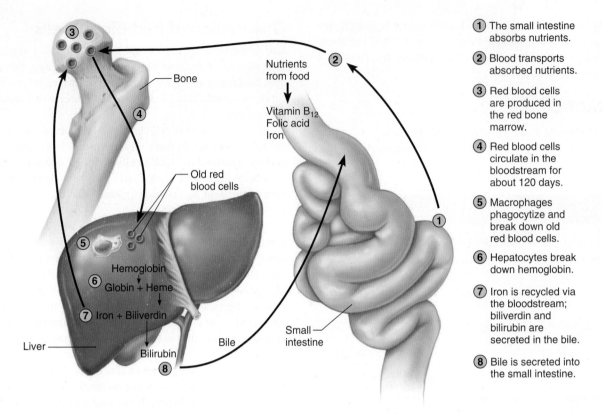

1 The small intestine absorbs nutrients.

2 Blood transports absorbed nutrients.

3 Red blood cells are produced in the red bone marrow.

4 Red blood cells circulate in the bloodstream for about 120 days.

5 Macrophages phagocytize and break down old red blood cells.

6 Hepatocytes break down hemoglobin.

7 Iron is recycled via the bloodstream; biliverdin and bilirubin are secreted in the bile.

8 Bile is secreted into the small intestine.

FIGURE 14.9 **AP|R** Life cycle of a red blood cell.

TABLE 14.3	Major Events in Red Blood Cell Destruction

1. Squeezing through the capillaries of active tissues damages red blood cells.

2. Macrophages in the spleen and liver phagocytize damaged red blood cells.

3. Hemoglobin from the red blood cells is decomposed into heme and globin.

4. Heme is decomposed into iron and biliverdin.

5. Iron is made available for reuse in the synthesis of new hemoglobin or is stored in the liver as ferritin.

6. Some biliverdin is converted into bilirubin.

7. Biliverdin and bilirubin are secreted in bile as bile pigments.

8. The globin is broken down into amino acids metabolized by macrophages or released into the plasma.

 PRACTICE

14 What happens to damaged red blood cells?

15 What are the products of hemoglobin breakdown?

Types of White Blood Cells

White blood cells, or **leukocytes** (lu′ko-sītz), protect against disease. Leukocytes develop from hematopoietic stem cells in the red bone marrow in response to hormones, much as red cells form from precursors upon stimulation from EPO. These hormones fall into two groups—**interleukins** (in″ter-lu-kinz) and **colony-stimulating factors** (CSFs). Interleukins are numbered, whereas most colony-stimulating factors are named for the cell population they stimulate. Blood transports white blood cells to sites of infection. White blood cells may then leave the bloodstream, as described later in this chapter (p. 538).

Normally, five types of white cells are in circulating blood. They differ in size, the nature of their cytoplasm, the shape of the nucleus, and their staining characteristics, and they are named for these distinctions. Leukocytes with markedly granular cytoplasm are called **granulocytes** (gran′u-lo-sītz″), whereas those with less obvious cytoplasmic granules are called **agranulocytes** (a-gran′u-lo-sītz).

A typical granulocyte is about twice the size of a red blood cell. Members of this group include neutrophils, eosinophils, and basophils. Granulocytes develop in red bone marrow, as do red blood cells. However, they have a short life span, averaging about twelve hours.

Neutrophils (nu′tro-filz) have fine cytoplasmic granules that appear light purple with a combination of acid and base stains. The nucleus of a more mature neutrophil is lobed and consists of two to five sections (or segments, so these cells are sometimes called *segs*) connected by thin strands of chromatin (fig. 14.10). They are also called *polymorphonuclear leukocytes* (PMNs) due to the variation of nucleus shape from cell to cell. Younger neutrophils may be called *bands* because their nuclei are curved bands. Neutrophils are the first white blood cells to arrive at an infection site. These

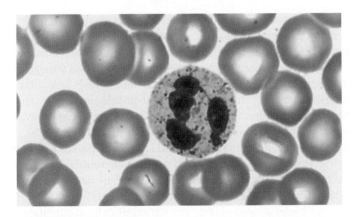

FIGURE 14.10 [AP|R] A neutrophil has a lobed nucleus with two to five segments (2,000×).

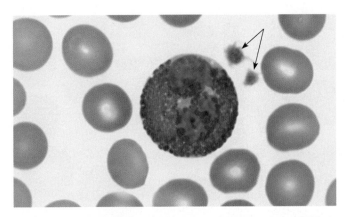

FIGURE 14.11 [AP|R] An eosinophil has red-staining cytoplasmic granules (2,000×). Note the platelets indicated by the arrows.

cells phagocytize bacteria, fungi, and some viruses. Neutrophils account for 54% to 62% of the leukocytes in a typical blood sample from an adult.

Eosinophils (e″o-sin′o-filz) contain coarse, uniformly sized cytoplasmic granules that stain deep red in acid stain (fig. 14.11). The nucleus usually has two lobes (bilobed). Eosinophils moderate allergic reactions and defend against parasitic worm infestation. These cells make up 1% to 3% of the total number of circulating leukocytes.

Basophils (ba′so-filz) are similar to eosinophils in size and in the shape of their nuclei. However, they have fewer, more irregularly shaped cytoplasmic granules than eosinophils, and these granules appear deep blue in basic stain (fig. 14.12). A basophil's granules can obscure a view of the nucleus. Basophils migrate to damaged tissues where they release *histamine*, which promotes inflammation (discussed in the next section, p. 538), and *heparin*, which inhibits blood clotting, actions that increase blood flow to injured tissues. Basophils usually account for less than 1% of the leukocytes.

The leukocytes of the agranulocyte group include monocytes and lymphocytes. Monocytes generally arise from red bone marrow. Lymphocytes form in the organs of the lymphatic system as well as in the red bone marrow.

Monocytes (mon′o-sītz), the largest of the white blood cells, are two to three times greater in diameter than red blood cells. Their nuclei are spherical, kidney-shaped, oval, or lobed (fig. 14.13). Monocytes leave the bloodstream and become *macrophages* that phagocytize bacteria, dead cells, and other debris in the tissues. They usually make up 3% to 9% of the leukocytes in a blood sample and live for several weeks or even months.

Lymphocytes (lim′fo-sītz), the smallest of the white blood cells, are only slightly larger than erythrocytes. A typical lymphocyte has a large, spherical nucleus surrounded by a thin layer of cytoplasm (fig. 14.14). The major types of lymphocytes are *T cells* and *B cells,* which are both important in *immunity.* T cells directly attack microorganisms, tumor cells, and transplanted cells (see chapter 16, pp. 629–630). B cells produce *antibodies* (see chapter 16, p. 630), which are proteins that attack foreign molecules. Lymphocytes account for 25% to 33% of the circulating leukocytes. They may live for years.

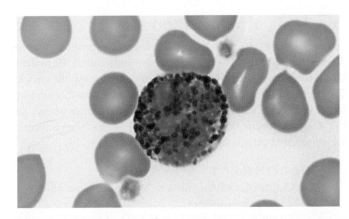

FIGURE 14.12 [AP|R] A basophil has cytoplasmic granules that stain deep blue (2,000×).

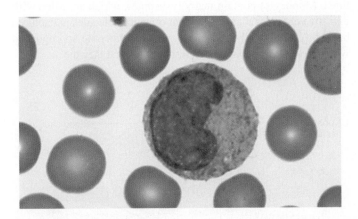

FIGURE 14.13 [AP|R] A monocyte is the largest of the blood cells (2,000×).

 PRACTICE

16 Which hormones are necessary for the differentiation of white blood cells from hematopoietic stem cells in red bone marrow?

17 Distinguish between granulocytes and agranulocytes.

18 List five types of white blood cells, and explain how they differ from one another.

19 Describe the function of each type of white blood cell.

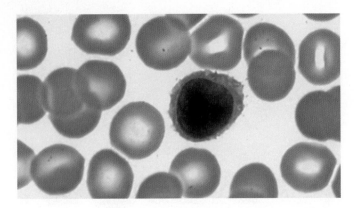

FIGURE 14.14 AP|R The lymphocyte, the smallest of the white blood cells, has a large, round nucleus (2,000×).

Functions of White Blood Cells

Leukocytes can squeeze between the cells that form the walls of the smallest blood vessels. This movement, called **diapedesis** (di″ah-pĕ-de′sis), allows the white blood cells to leave the circulation (fig. 14.15). A series of proteins called cellular adhesion molecules help guide leukocytes to the site of injury. Once outside the blood, leukocytes move through interstitial spaces using a form of self-propulsion called *ameboid motion.*

 RECONNECT
To Chapter 3, Cellular Adhesion Molecules, pages 88–89.

The most mobile and active phagocytic leukocytes are neutrophils and monocytes. Neutrophils cannot ingest particles much larger than bacterial cells, but monocytes can engulf larger structures. Monocytes contain many lysosomes, which are filled with digestive enzymes that break down organic molecules in captured bacteria. Neutrophils and monocytes can become so engorged with digestive products and bacterial toxins that they die.

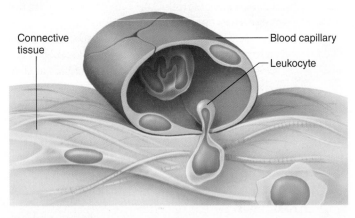

FIGURE 14.15 In a type of movement called diapedesis, leukocytes squeeze between the endothelial cells of a capillary wall and enter the tissue space outside the blood vessel.

Q *What is a monocyte called once it has left the bloodstream and entered the tissues?*

Answer can be found in Appendix G.

When microorganisms invade human tissues, basophils respond by releasing biochemicals that dilate local blood vessels. For example, histamine dilates smaller blood vessels and makes the smallest vessels more permeable. As more blood flows through the smallest vessels, the tissues redden and copious fluids leak into the interstitial spaces. This response, called an inflammatory reaction (inflammation), produces swelling that delays the spread of invading microorganisms into other regions (see chapter 16, p. 626). At the same time, damaged cells release chemicals that attract leukocytes. This phenomenon is called **positive chemotaxis** (poz′ĭ-tiv ke″mo-tak′sis) and, when combined with diapedesis, it quickly brings many white blood cells into inflamed areas (fig. 14.16).

As bacteria, leukocytes, and damaged cells accumulate in inflamed tissue, a thick fluid called *pus* forms and remains while the invading microorganisms are active. If the pus is not moved to the outside of the body or into a body cavity, it may remain trapped in the tissues for some time. Eventually, surrounding cells absorb it.

 A GLIMPSE AHEAD | **To Chapter 16**

White blood cells protect against infection in various ways. Neutrophils and monocytes remove any foreign particles from the body as part of the innate (nonspecific) defense against disease. As part of the adaptive (specific) defense, lymphocytes participate in the formation of specific antibody proteins in the immune response.

PRACTICE

20 How do white blood cells reach microorganisms outside blood vessels?

21 Which white blood cells are the most active phagocytes?

22 How do white blood cells fight infection?

White Blood Cell Counts

The procedure used to count white blood cells is similar to that used for counting red blood cells. However, before a *white blood cell count* (WBCC or WCC) is made, the red blood cells in the blood sample are destroyed so they will not be mistaken for white blood cells. Normally, a microliter of blood includes 3,500 to 10,500 white blood cells.

The total number and percentages of different white blood cell types are of clinical interest. A rise in the number of circulating white blood cells may indicate infection. A total number of white blood cells exceeding 10,500 per microliter of blood constitutes **leukocytosis** (lu″ko-si-to′sis), indicating acute infection, such as appendicitis. Leukocytosis may also follow vigorous exercise, emotional disturbances, or great loss of body fluids.

A total white blood cell count below 3,500 per microliter of blood is called **leukopenia** (lu″ko-pe′ne-ah). Such a deficiency may accompany typhoid fever, influenza, measles, mumps, chickenpox, AIDS, or poliomyelitis. Leukopenia may also result from anemia or from lead, arsenic, or mercury poisoning.

A *differential white blood cell count* lists percentages of the various types of leukocytes in a blood sample. This test is useful because the relative proportions of white blood cells may change in particular diseases. The number of neutrophils, for instance, usually

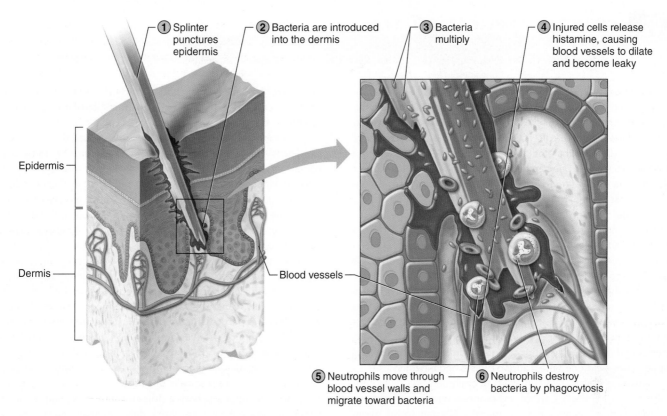

① Splinter punctures epidermis

② Bacteria are introduced into the dermis

③ Bacteria multiply

④ Injured cells release histamine, causing blood vessels to dilate and become leaky

Epidermis

Dermis

Blood vessels

⑤ Neutrophils move through blood vessel walls and migrate toward bacteria

⑥ Neutrophils destroy bacteria by phagocytosis

FIGURE 14.16 When bacteria invade the tissues, leukocytes migrate into the region and destroy the microbes by phagocytosis.

increases during bacterial infections, and eosinophils may become more abundant during certain parasitic infections and allergic reactions. In HIV infection and AIDS, the number of helper T cells, a type of lymphocyte, drops sharply.

Table 14.4 lists some disorders that alter the numbers of particular types of white blood cells. Clinical Application 14.2 examines leukemia, which is cancer of the white blood cells.

 PRACTICE

23 What is the normal human white blood cell count?

24 Distinguish between leukocytosis and leukopenia.

25 What is a differential white blood cell count?

Platelets AP|R

Platelets (plăt′letz), or **thrombocytes** (throm′bo-sītz), are not complete cells (see fig.14.11). They arise from very large cells in the red bone marrow called **megakaryocytes** (meg″ah-kar′o-sītz). The megakaryocytes develop long cellular extensions that break off in small sections in the bone marrow. These small pieces fragment, when they reach the circulation, to form the platelets. Megakaryocytes, and therefore platelets, develop from hematopoietic stem cells (see fig. 14.4) in response to the hormone **thrombopoietin** (throm″bo-poi′ĕ-tin).

Each platelet is surrounded by membrane, but lacks a nucleus and is less than half the size of a red blood cell. It is capable of ameboid movement and may live for about ten days. In normal blood, the *platelet count* varies from 150,000 to 350,000 platelets per microliter.

TABLE 14.4	Abnormal White Blood Cell Numbers
White Blood Cell Population Change	**Illness**
Elevated lymphocytes	Hairy cell leukemia, whooping cough, mononucleosis
Elevated eosinophils	Tapeworm infestation, hookworm infestation, allergic reactions
Elevated monocytes	Typhoid fever, malaria, tuberculosis
Elevated neutrophils	Bacterial infections
Too few helper T cells (lymphocytes)	AIDS

Thrombocytopenia (throm″bo-si″to-pe′ne-ah) occurs when the platelet count drops below 150,000 platelets per microliter of blood. Symptoms include bleeding easily; capillary hemorrhages throughout the body; and small, bruiselike spots on the skin called petechiae. Thrombocytopenia is a common side effect of cancer chemotherapy and radiation treatments and can be a complication of pregnancy, leukemia, bone marrow transplantation, infectious disease, cardiac surgery, or anemia. Conventional treatment is transfusion of platelets. Treatment with thrombopoietin (TPO) stimulates formation and maturation of megakaryocytes and thereby boosts platelet levels.

When the twenty-three-year-old had a routine physical examination, she expected reassurance that her healthy lifestyle had indeed been keeping her healthy. What she got, a few days later, was a shock. Instead of having 3,500 to 10,500 white blood cells per microliter of blood, she had more than ten times that number—and many of the cells were cancerous. She had chronic myeloid leukemia (CML). Her red bone marrow was flooding her circulation with too many granulocytes, most of them poorly differentiated (fig. 14B).

Another type of leukemia is lymphoid, in which the cancer cells are lymphocytes, produced in lymph nodes. Both myeloid and lymphoid leukemia can cause fatigue, headaches, nosebleeds and other bleeding, frequent respiratory infections, fever, bone pain, bruising, and other signs of slow blood clotting.

The symptoms arise from the disrupted proportions of the blood's formed elements and their malfunction. Immature white blood cells increase the risk of infection. Leukemic cells crowd out red blood cells and their precursors in the red marrow, causing anemia and the resulting fatigue. Platelet deficiency (thrombocytopenia) slows clotting time, causing bruises and bleeding.

Leukemia is also classified as acute or chronic. An acute condition appears suddenly, symptoms progress rapidly, and without treatment, death occurs in a few months. Chronic forms begin more slowly and may remain undetected for months or even years or, in rare cases, decades. Without treatment, life expectancy after symptoms develop is about three years.

Traditional cancer treatments destroy any cell that is actively dividing. A newer drug, Gleevec, specifically targets cancer cells by nestling into ATP-binding sites on a version of an enzyme called tyrosine kinase that is found only on cancer cells. If cancer cells become resistant to Gleevec, even newer drugs are available that target the cancer cells in different ways. People with leukemia have other options. Bone marrow and stem cell transplants may cure the condition.

Another way that leukemia treatment is improving is refining diagnosis, based on identifying the proteins that leukemia cells produce. This information is used to predict which drugs are most likely to be effective, and which will cause intolerable side effects, in particular individuals. For example, some people with acute lymphoblastic leukemia (ALL), diagnosed on the basis of the appearance of the cancer cells in a blood smear, do not respond to standard chemotherapy. Their cells produce different proteins than the cancer cells of patients who do respond. The nonresponders actually have a different form of leukemia, called mixed-lineage leukemia. For them, different drugs may work.

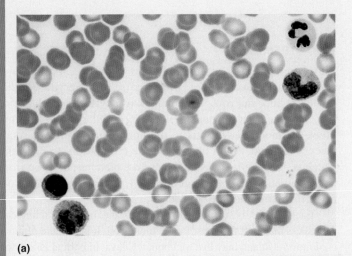

(a)

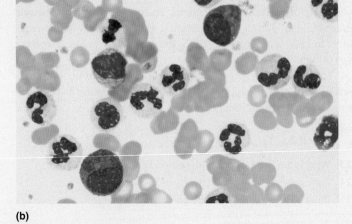

(b)

FIGURE 14B Leukemia and blood cells. (a) Normal blood cells (700×). (b) Blood cells from a person with granulocytic leukemia, a type of myeloid leukemia (700×). Note the increased number of leukocytes.

Platelets help repair damaged blood vessels by sticking to broken surfaces. They release **serotonin,** which contracts smooth muscle in the vessel walls, reducing blood flow. Table 14.5 summarizes the characteristics of blood cells and platelets.

PRACTICE

26 What is the normal human blood platelet count?

27 What is the function of blood platelets?

The absolute numbers for red blood cell, white blood cell, and platelet counts can vary depending on how they are measured and the instruments used to measure them. For this reason, different sources may present different, but very similar, ranges of normal values.

Component	Description	Number Present	Function
TABLE 14.5	**Cellular Components of Blood**		
Red blood cell (erythrocyte)	Biconcave disc without a nucleus, about one-third hemoglobin	4,700,000–6,100,000 per microliter, male 4,200,000 to 5,400,000 per microliter, female	Transports oxygen and carbon dioxide
White blood cell (leukocyte)		3,500 to 10,500 per microliter	Destroys pathogenic microorganisms and parasites and removes worn cells
Granulocytes	About twice the size of red blood cells; cytoplasmic granules are present		
Neutrophil	Nucleus with two to five lobes; cytoplasmic granules stain light purple in combined acid and base stains	54%–62% of white blood cells present	Phagocytizes small particles
Eosinophil	Nucleus bilobed; cytoplasmic granules stain red in acid stain	1%–3% of white blood cells present	Kills parasites and moderates allergic reactions
Basophil	Nucleus lobed; cytoplasmic granules stain blue in basic stain	Less than 1% of white blood cells present	Releases heparin and histamine
Agranulocytes	Cytoplasmic granules are absent		
Monocyte	Two to three times larger than a red blood cell; nucleus shape varies from spherical to lobed	3%–9% of white blood cells present	Phagocytizes large particles
Lymphocyte	Only slightly larger than a red blood cell; its nucleus nearly fills cell	25%–33% of white blood cells present	Provides immunity
Platelet (thrombocyte)	Cellular fragment	150,000 to 350,000 per microliter	Helps control blood loss from broken vessels

14.3 | Plasma

Plasma is the clear, straw-colored, liquid part of the blood in which the cells and platelets are suspended. It is approximately 92% water and contains a complex mixture of organic and inorganic biochemicals, including amino acids, nucleic acids, carbohydrates, and a great variety of lipids. Functions of plasma include carrying nutrients, gases, hormones, and vitamins; helping to regulate fluid and electrolyte balance; and maintaining a favorable pH.

Plasma Proteins

By weight, **plasma proteins** are the most abundant dissolved substances (solutes) in plasma. These proteins remain in the blood and interstitial fluids and ordinarily are not used as energy sources. The three main types of plasma proteins are albumins, globulins, and fibrinogen. The groups differ in composition and function.

Albumins (al-bu′minz) are the smallest of the plasma proteins, yet account for 60% of these proteins by weight. They are synthesized in the liver, and because they are so plentiful, albumins are an important determinant of the *osmotic pressure* of the plasma.

Recall from chapter 3 (p. 101) that the presence of an impermeant solute on one side of a selectively permeable membrane creates an osmotic pressure and that water always moves toward a greater osmotic pressure. Plasma proteins are too large to pass through the capillary walls, so they are impermeant, and they create an osmotic pressure that holds water in the capillaries despite blood pressure forcing water out of capillaries by filtration (see chapter 3, pp. 102–103). The term *colloid osmotic pressure* is used to describe this osmotic effect due to the plasma proteins.

If the concentration of plasma proteins falls, tissues swell, a condition called *edema*. Low plasma protein may result from starvation or a protein-deficient diet or from an impaired liver that cannot synthesize plasma proteins. As the concentration of plasma proteins drops, so does the colloid osmotic pressure, allowing water to leave the blood vessels and to accumulate in the interstitial spaces.

Albumins and other plasma proteins help regulate water movement between the blood and the tissues by maintaining the colloid osmotic pressure of plasma. In doing so, they help control blood volume, which, in turn, directly affects blood pressure (see chapter 15, p. 586). For this reason, it is important that the concentration of plasma proteins remains relatively stable. Albumins also bind and transport certain molecules, such as bilirubin, free fatty acids, many hormones, and certain drugs.

Globulins (glob′u-linz), which make up about 36% of the plasma proteins, can be subdivided into *alpha, beta,* and *gamma globulins.* The liver synthesizes alpha and beta globulins, which have a variety of functions, including transport of lipids and fat-soluble vitamins. Lymphatic tissues produce the gamma globulins, which are the antibodies in the immune response (see chapter 16, p. 632).

Fibrinogen (fi-brin′o-jen), which constitutes about 4% of the plasma proteins, plays a primary role in blood coagulation. Synthesized in the liver, it is the largest of the plasma proteins. The function of fibrinogen is discussed later in this chapter under the section "Blood Coagulation" on page 543. Table 14.6 summarizes the characteristics of the plasma proteins.

TABLE 14.6 Plasma Proteins

Protein	Percentage of Total	Origin	Function
Albumins	60%	Liver	Help maintain colloid osmotic pressure
Globulin	36%		
Alpha globulins		Liver	Transport lipids and fat-soluble vitamins
Beta globulins		Liver	Transport lipids and fat-soluble vitamins
Gamma globulins		Lymphatic tissues	Constitute the antibodies of immunity
Fibrinogen	4%	Liver	Plays a key role in blood coagulation

 PRACTICE

28 List three types of plasma proteins.

29 How do albumins help maintain water balance between the blood and the tissues?

30 Which of the globulins functions in immunity?

31 What is the role of fibrinogen?

Gases and Nutrients

The most important *blood gases* are oxygen and carbon dioxide. Plasma also contains a considerable amount of dissolved nitrogen, which ordinarily has no physiological function. Chapter 19 (pp. 758–760) discusses blood gases and their transport.

As a rule, blood gases are evaluated using a fresh sample of whole blood obtained from an artery. This blood is cooled to decrease the rates of metabolic reactions, and an anticoagulant is added to prevent clotting. Laboratory tests determine the levels of oxygen and carbon dioxide, measure blood pH, and calculate the plasma bicarbonate concentration. Such information is used to diagnose and treat disorders of circulation, respiration, and electrolyte balance. Appendix C lists average values for these laboratory tests.

The *plasma nutrients* include amino acids, simple sugars, nucleotides, and lipids absorbed from the digestive tract. For example, plasma carries glucose from the small intestine to the liver, where it may be stored as glycogen or converted to fat. If the blood glucose concentration drops below the normal range, glycogen may be broken down into glucose, as chapter 13 describes (p. 514).

Like glucose, recently absorbed amino acids are carried to the liver. Here they may be used to manufacture proteins or deaminated and used as an energy source (see chapter 18, p. 698).

Plasma lipids include the familiar fats (triglycerides), phospholipids, and cholesterol, but researchers are identifying hundreds and possibly thousands of additional lipids. Free fatty acids in the plasma are associated with albumins, and more complex lipids join with other proteins to form lipoproteins. Abnormal levels of certain plasma lipids serve as biomarkers of specific metabolic, cardiovascular, or neurological diseases.

 PRACTICE

32 Which gases are in plasma?

33 Which nutrients are in plasma?

Nonprotein Nitrogenous Substances

Molecules that contain nitrogen atoms but are not proteins comprise a group called **nonprotein nitrogenous substances** (NPNs). In plasma, this group includes amino acids, urea, uric acid, creatine (kre′ah-tin), and creatinine (kre-at′ĭ-nin). Amino acids come from protein digestion and amino acid absorption. Urea and uric acid are products of protein and nucleic acid catabolism, respectively, and creatinine results from the metabolism of creatine. Creatine is present as **creatine phosphate** in muscle and brain tissues as well as in the blood, where it stores energy in phosphate bonds, much like those of ATP molecules. Chapter 9 (p. 300) discussed creatine phosphate.

Normally, the concentration of nonprotein nitrogenous substances in plasma remains relatively stable because protein intake and use are balanced with excretion of nitrogenous wastes. Because about half of the NPN substances is urea, which the kidneys ordinarily excrete, a rise in the blood urea nitrogen (BUN) may suggest a kidney disorder. Excess protein catabolism or infection may also elevate BUN.

Plasma Electrolytes

Plasma contains a variety of *electrolytes* that are absorbed from the intestine or released as by-products of cellular metabolism. They include sodium, potassium, calcium, magnesium, chloride, bicarbonate, phosphate, and sulfate ions. Of these, sodium and chloride ions are the most abundant. Bicarbonate ions are important in maintaining the osmotic pressure and the pH of plasma, and like other plasma constituents, they are regulated so that their blood concentrations remain relatively stable. These electrolytes are discussed in chapter 21 (pp. 809–812) in connection with water and electrolyte balance.

 PRACTICE

34 What is a nonprotein nitrogenous substance?

35 Why does kidney disease increase the blood concentration of these substances?

36 What are the sources of plasma electrolytes?

14.4 | Hemostasis

Hemostasis (he″mo-sta′sis) is the stoppage of bleeding, which is vitally important when blood vessels are damaged. Following an injury to the blood vessels, several actions may help to limit or prevent blood loss, including blood vessel spasm, platelet plug formation, and blood coagulation. These mechanisms are most effective in minimizing blood losses from small vessels. Injury to a larger vessel may result in a severe hemorrhage that requires special treatment.

Vascular Spasm

Cutting or breaking a small blood vessel stimulates *vascular spasm,* the rapid contraction of smooth muscle in the blood vessel walls. Blood loss lessens almost immediately, and the ends of the severed vessel may close completely. This effect results from direct stimulation of the vessel wall as well as from reflexes elicited by pain receptors in the injured tissues.

Although the reflex response may last only a few minutes, the effect of the direct stimulation usually continues for about thirty minutes. By then, a blockage called a *platelet plug* has formed, and blood is coagulating. Also, platelets release serotonin, which contracts smooth muscle in the blood vessel walls. This vasoconstriction further helps to reduce blood loss.

Platelet Plug Formation

Platelets adhere to exposed ends of injured blood vessels. They stick to any rough surface, particularly to the collagen in connective tissue underlying the endothelial lining of blood vessels.

When platelets contact collagen, their shapes change as many spiny processes begin to extend from their membranes. At the same time, platelets adhere to each other, forming a platelet plug in the vascular break. A plug may control blood loss from a small break, but a larger break may require a blood clot to halt bleeding. Figure 14.17 shows the steps in platelet plug formation.

 PRACTICE

37 What is hemostasis?

38 How does vascular spasm help control bleeding?

39 Describe the formation of a platelet plug.

Blood Coagulation

Coagulation (ko-ag″u-la′shun), the most effective hemostatic mechanism, forms a *blood clot* in a series of reactions, each one activating the next in a *cascade.* Coagulation may occur in two ways: extrinsically or intrinsically. Release of biochemicals from broken blood vessels or damaged tissues triggers the **extrinsic clotting mechanism** (tissue factor pathway). Blood contact with foreign surfaces in the absence of tissue damage stimulates the **intrinsic clotting mechanism** (contact activation pathway). The following sections describe these responses.

Blood coagulation is complex and uses many biochemicals called *clotting factors.* They are designated by Roman numerals indicating the order of their discovery. Vitamin K is necessary for some clotting factors to function. Whether the blood coagulates depends on the balance between factors that promote coagulation (procoagulants) and others that inhibit it (anticoagulants). Normally, the anticoagulants prevail, and the blood does not clot. However, as a result of injury (trauma), biochemicals that favor coagulation may increase in concentration, and the blood may coagulate.

The major event in blood clot formation is conversion of the soluble plasma protein *fibrinogen* (factor I) into insoluble threads of the protein **fibrin** (fig. 14.18). Activation of certain plasma proteins by still other protein factors triggers conversion of fibrinogen to fibrin. Table 14.7 summarizes the three primary hemostatic mechanisms: blood vessel spasm, platelet plug formation, and blood coagulation.

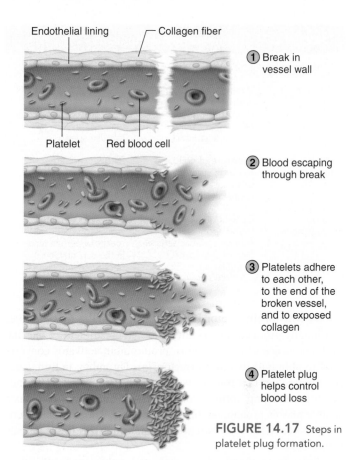

① Break in vessel wall

② Blood escaping through break

③ Platelets adhere to each other, to the end of the broken vessel, and to exposed collagen

④ Platelet plug helps control blood loss

FIGURE 14.17 Steps in platelet plug formation.

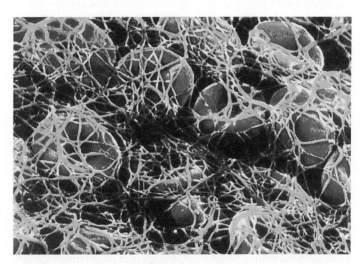

FIGURE 14.18 A scanning electron micrograph of red blood cells caught in a meshwork of fibrin threads (2,800×).

Extrinsic Clotting Mechanism

The extrinsic clotting mechanism is triggered when blood contacts damaged blood vessel walls or tissues outside blood vessels. Such damaged tissues release *tissue thromboplastin* (factor III), that is associated with disrupted cell membranes. Tissue thromboplastin activates factor VII, which combines with and activates factor X. Further, factor X combines with and activates factor V. These reactions, which also require calcium ions (factor IV), lead the platelets to produce and release *prothrombin activator.*

TABLE 14.7	Hemostatic Mechanisms	
Mechanism	Stimulus	Effect
Vascular spasm	Direct stimulus to vessel walls or to pain receptors; platelets release serotonin, a vasoconstrictor	Smooth muscle in vessel walls contracts reflexly; vasoconstriction helps maintain prolonged vascular spasm
Platelet plug formation	Exposure of platelets to rough surfaces or to collagen of connective tissue	Platelets adhere to rough surfaces and to each other, forming a plug
Blood coagulation	Cellular damage and blood contact with foreign surfaces activate factors that favor coagulation	Blood clot forms as a result of a series of reactions, terminating in the conversion of fibrinogen into fibrin

Prothrombin (factor II) is an alpha globulin that the liver continually produces and is thus a normal constituent of plasma. In the presence of calcium ions, prothrombin activator converts prothrombin into **thrombin** (factor IIa). Thrombin, in turn, catalyzes a reaction that fragments fibrinogen (factor I). The fibrinogen fragments join, forming long threads of fibrin. Fibrinogen is a soluble plasma protein, but fibrin is insoluble. Thrombin also activates factor XIII, which strengthens and stabilizes fibrin threads (fig. 14.19).

Once fibrin threads form, they stick to exposed surfaces of damaged blood vessels, creating a meshwork that entraps blood cells and platelets. The resulting mass is a blood clot, which may block a vascular break and prevent further blood loss.

The amount of prothrombin activator in the blood is directly proportional to the degree of tissue damage. Once a blood clot begins to form, it promotes additional clotting, because thrombin also acts directly on prothrombin activator, causing prothrombin to form still more thrombin. This type of self-initiating action is an example of a **positive feedback mechanism,** in which the original action stimulates more of the same type of action. Such a mechanism produces unstable conditions and can operate for only a short time in a living system, because life requires the maintenance of a stable internal environment (see chapter 1, p. 15).

Normally, blood flow throughout the body prevents formation of a massive clot in the cardiovascular system by rapidly carrying excess thrombin away and keeping its concentration too low to promote further clotting. In addition, a substance called *antithrombin*, in the blood and on the surfaces of endothelial cells that line blood

In *disseminated intravascular coagulation (DIC)*, clotting is abnormally activated in several regions of the cardiovascular system. This condition is usually associated with bacterial infection or bacterial toxins in the blood or with a disorder causing widespread tissue damage. Many small clots form and obstruct blood flow into various tissues and organs, particularly the kidneys. As plasma clotting factors and platelets are depleted, severe bleeding occurs.

vessels, limits thrombin formation. Consequently, blood coagulation usually occurs only in blood that is standing still or moving slowly, and clotting ceases where a clot contacts circulating blood.

Intrinsic Clotting Mechanism

Unlike the extrinsic clotting mechanism, the intrinsic clotting mechanism can begin without tissue damage. Activation of a substance called the *Hageman factor* (factor XII) initiates intrinsic clotting. This happens when blood is exposed to a foreign surface such as collagen in connective tissue instead of the smooth endothelial lining of intact blood vessels or when blood is stored in a glass container. Activated factor XII activates factor XI, which activates factor IX. Factor IX then joins with factor VIII and platelet phospholipids to activate factor X. These reactions, which also require calcium ions, lead to the production of prothrombin activator. The subsequent steps of blood clot formation are the same as those described for the extrinsic mechanism (fig. 14.19). Table 14.8 compares extrinsic and intrinsic clotting mechanisms. Table 14.9 lists the clotting factors, their sources, and clotting mechanisms.

Laboratory tests commonly used to evaluate blood coagulation mechanisms include *prothrombin time* (PT) and *partial thromboplastin time* (PTT). These tests measure the time it takes for fibrin threads to form in a sample of blood plasma. The prothrombin time test checks the extrinsic clotting mechanism, whereas the partial thromboplastin test evaluates intrinsic clotting.

Fate of Blood Clots

A newly formed blood clot soon begins to retract as the tiny extensions from the platelet membranes adhere to strands of fibrin within the clot and contract. The blood clot shrinks, pulling the edges of the broken vessel closer together and squeezing a fluid called **serum** from the clot. Serum is essentially plasma minus all

TABLE 14.8	Blood Coagulation	
Steps	Extrinsic Clotting Mechanism	Intrinsic Clotting Mechanism
Trigger	Damage to vessel or tissue	Blood contacts foreign surface
Initiation	Tissue thromboplastin	Hageman factor
Series of reactions involving several clotting factors and calcium ions (Ca^{+2}) lead to the production of:	Prothrombin activator	Prothrombin activator
Prothrombin activator and calcium ions cause the conversion of:	Prothrombin to thrombin	Prothrombin to thrombin
Thrombin causes fragmentation, then joining of:	Fibrinogen to fibrin	Fibrinogen to fibrin

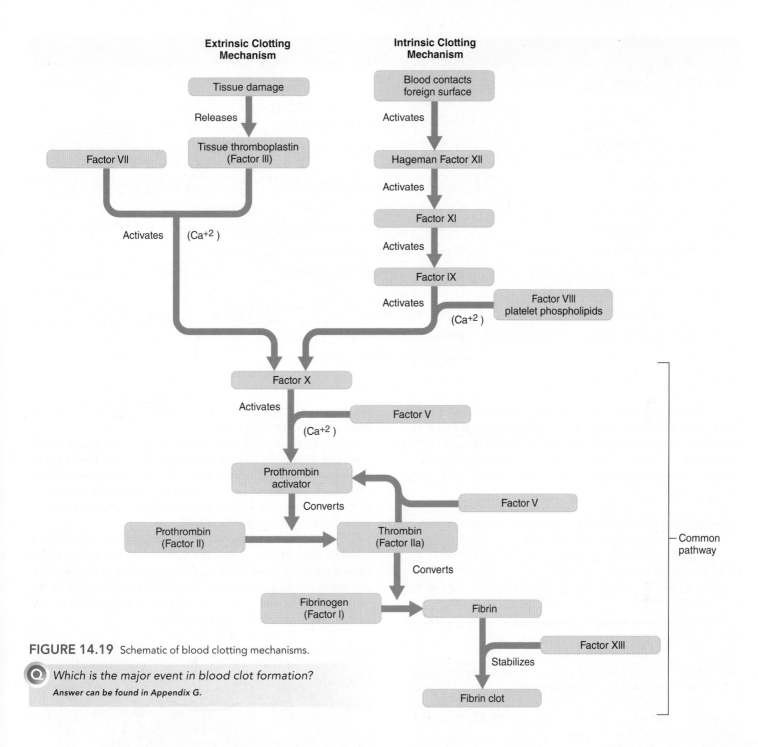

Extrinsic Clotting Mechanism

Tissue damage

Releases

Tissue thromboplastin (Factor III)

Factor VII

Activates   (Ca^{+2})

Intrinsic Clotting Mechanism

Blood contacts foreign surface

Activates

Hageman Factor XII

Activates

Factor XI

Activates

Factor IX

Activates

Factor VIII platelet phospholipids

(Ca^{+2})

Factor X

Activates

Factor V

(Ca^{+2})

Prothrombin activator

Converts

Prothrombin (Factor II)

Factor V

Thrombin (Factor IIa)

Converts

Common pathway

Fibrinogen (Factor I)

Fibrin

Factor XIII

Stabilizes

Fibrin clot

FIGURE 14.19 Schematic of blood clotting mechanisms.

Which is the major event in blood clot formation?
Answer can be found in Appendix G.

of its fibrinogen and most other clotting factors. Platelets associated with a blood clot also release *platelet-derived growth factor* (PDGF), which stimulates smooth muscle cells and fibroblasts to repair damaged blood vessel walls.

Fibroblasts (see chapter 5, p. 160) invade blood clots that form in ruptured vessels, producing fibrous connective tissue throughout the clots, which helps strengthen and seal vascular breaks. Many clots, including those that form in tissues as a result of blood leakage (hematomas), disappear in time. In clot dissolution, fibrin threads absorb a plasma protein called *plasminogen* (profibrinolysin). Then a substance called plasminogen activator released from the lysosomes of damaged tissue cells converts plas-

minogen to *plasmin*. Plasmin is a protein-splitting enzyme that can digest fibrin threads and other proteins associated with blood clots. Plasmin formation may dissolve a whole clot. However, clots that fill large blood vessels are seldom removed naturally.

A blood clot abnormally forming in a vessel is a **thrombus** (throm'bus). A clot that dislodges, or a fragment of a clot that breaks loose and is carried away by the blood flow, is called an **embolus** (em'bo-lus). Generally, emboli continue to move until they reach narrow places in vessels where they may lodge and block blood flow, causing an **embolism.** Other than a blood clot, an embolus could be an air bubble, a fat globule, or plaque—any obstructive object in the bloodstream.

TABLE 14.9	Clotting Factors		
Component		**Source**	**Mechanism(s)**
I	(fibrinogen)	Synthesized in liver	Extrinsic and intrinsic
II	(prothrombin)	Synthesized in liver, requires vitamin K	Extrinsic and intrinsic
III	(tissue thromboplastin)	Damaged tissue	Extrinsic
IV	(calcium ions)	Plasma electrolyte	Extrinsic and intrinsic
V	(proaccelerin)	Synthesized in liver, released by platelets	Extrinsic and intrinsic
VII	(serum prothrombin conversion accelerator)	Synthesized in liver, requires vitamin K	Extrinsic
VIII	(antihemophilic factor)	Released by platelets and endothelial cells	Intrinsic
IX	(plasma thromboplastin component)	Synthesized in liver, requires vitamin K	Intrinsic
X	(Stuart-Prower factor)	Synthesized in liver, requires vitamin K	Extrinsic and intrinsic
XI	(plasma thromboplastin antecedent)	Synthesized in liver	Intrinsic
XII	(Hageman factor)	Synthesized in liver	Intrinsic
XIII	(fibrin-stabilizing factor)	Synthesized in liver, released by platelets	Extrinsic and intrinsic

There is no clotting factor VI. The chemical once thought to be factor VI is apparently a combination of activated factors V and X.

A blood clot forming in a vessel that supplies a vital organ, such as the heart (coronary thrombosis) or the brain (cerebral thrombosis), blocks blood flow and kills tissues the vessel serves (*infarction*) and may be fatal. A blood clot that travels and then blocks a vessel that supplies a vital organ, such as the lungs (pulmonary embolism), affects the portion of the organ that the blocked blood vessel supplies.

Abnormal clot formations are often associated with conditions that change the endothelial linings of vessels. For example, in *atherosclerosis* (ath"er-o"skle-ro'sis), accumulations of fatty deposits change arterial linings, sometimes initiating inappropriate clotting. Atherosclerosis is the most common cause of thrombosis in medium-sized arteries (fig. 14.20).

Coagulation may also occur in blood flowing too slowly. The concentration of clot-promoting substances may increase to a critical level instead of being carried away by more rapidly moving blood, and a clot may form. This imbalance is the usual cause of thrombosis in veins. Clinical Application 14.3 discusses deep vein thrombosis.

> Drugs based on "clot-busting" biochemicals can be lifesavers. *Tissue plasminogen activator* (tPA) may restore cerebral circulation blocked by a clot if given within 3 to 4 1/2 hours of a stroke. A drug derived from bacteria, called *streptokinase*, may also be successful. Another plasminogen activator used as a drug is *urokinase*, an enzyme produced by certain kidney cells.

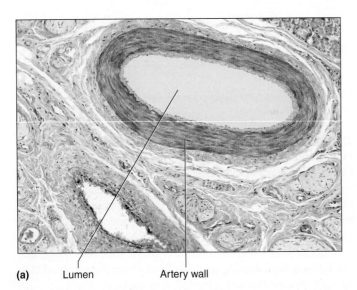

(a) Lumen   Artery wall

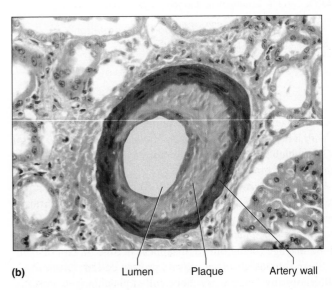

(b) Lumen   Plaque   Artery wall

FIGURE 14.20 Artery cross sections. (*a*) Light micrograph of a normal artery (50×). (*b*) The inner wall of an artery changed as a result of atherosclerosis (100×).

After a transcontinental flight, a man complained of cramps behind his knees. He mentioned the unfamiliar sensations to his traveling companion, who urged him to call his doctor. The nurse who took the call told the man to seek immediate medical attention, but because he was already starting to feel better, he didn't take her advice. Instead, he continued his trip, reaching his destination after a short bus ride. Three days later, the man suddenly collapsed and died from a pulmonary embolism, a complication of the condition called *deep vein thrombosis* (DVT). He had several risk factors for DVT: prolonged periods of immobility on flights; dehydration; and an inherited clotting disorder that he had not known about called factor V Leiden. Other risk factors for DVT are prolonged immobility due to surgery; oral contraceptive use; hormone replacement therapy (estrogen); surgery (of the abdomen, pelvis, or limbs); and cancer (of the ovaries, pancreas, colon, liver, or stomach, or lymphoma).

In DVT, stagnant blood pools, leading to clot formation, typically in the femoral or popliteal veins or in the deep veins of the pelvis. Symptoms occur in half of all affected individuals. These include deep muscle pain, redness, swelling, and possibly discoloration and dilation of surface veins (phlebitis). Part of the clot may break off hours or days after it forms and follow the path

of circulation, lodging in the pulmonary arteries. This is a pulmonary embolism, and it is life threatening. Symptoms include chest pain, anxiety, racing pulse, sweating, cough with bloody sputum, and loss of consciousness. In the United States, approximately 2 million people a year develop DVT, and 200,000 die from pulmonary embolism.

Guidelines from the American College of Chest Physicians recommend preventive measures against DVT for patients at higher risk. These actions include taking anticoagulants if immobilization is expected and wearing compression stockings that help keep blood flowing in the legs. Doing exercises while immobilized during travel is a good idea for everyone, and some airlines advise passengers on how to exercise on cramped flights, such as by curling the toes up and down and moving the ankles (fig. 14C).

FIGURE 14C Exercising the toes and ankles on a long flight can lower the risk of deep vein thrombosis.

PRACTICE

40 Distinguish between extrinsic and intrinsic clotting mechanisms.

41 Which clotting factor is converted to fibrin?

42 What prevents the formation of massive clots throughout the cardiovascular system?

43 Distinguish between a thrombus and an embolus.

44 How might atherosclerosis promote the formation of blood clots?

Prevention of Coagulation

In a healthy cardiovascular system, the endothelium of the blood vessels partly prevents spontaneous blood clot formation. This smooth lining lowers the risk of platelets and clotting factors accumulating. Endothelial cells also produce a prostaglandin (see chapter 13, p. 496) called *prostacyclin* (PGI_2), which inhibits the adherence of platelets to the inner surface of healthy blood vessel walls.

Several mechanisms confine clotting to the area where it is needed. When a clot is forming, fibrin threads latch onto or *adsorb* thrombin, thus helping prevent the spread of the clotting reaction. A plasma alpha globulin, *antithrombin,* inactivates additional thrombin by binding to it and blocking its action on fibrinogen. In addition, basophils and mast cells in the connective tissue surrounding capillaries secrete the anticoagulant *heparin.* This substance interferes with formation of prothrombin activator, prevents the action of thrombin on fibrinogen, and promotes removal of thrombin by antithrombin and fibrin adsorption. Heparin and another compound, Coumadin (warfarin), are used as drugs to prevent abnormal clotting.

Heparin-secreting cells are particularly abundant in the liver and lungs, where capillaries trap small blood clots that commonly form in the slow-moving blood of veins. These cells continually secrete heparin, preventing additional clotting in the cardiovascular system. Table 14.10 summarizes clot-inhibiting factors.

PRACTICE

45 How does the lining of a blood vessel help prevent blood clot formation?

46 What is the function of antithrombin?

47 How does heparin help prevent blood clot formation?

TABLE 14.10 Factors That Inhibit Blood Clot Formation

Factor	Action	Factor	Action
Smooth lining of blood vessel	Prevents activation of intrinsic blood clotting mechanism	Antithrombin in plasma	Interferes with the action of thrombin
Prostacyclin	Inhibits adherence of platelets to blood vessel wall	Heparin from mast cells and basophils	Interferes with the formation of prothrombin activator
Fibrin threads	Adsorbs thrombin		

14.5 | Blood Groups and Transfusions

Early blood transfusion experiments, which date from the late 1600s, used lamb blood. By the 1800s, human blood was being used with unpredictable results—some recipients were cured, but some were killed when their kidneys failed under the strain of handling clumping red blood cells when blood types were incompatible. So poor was the success rate that, by the late 1800s, many nations banned transfusions.

Around this time, Austrian physician Karl Landsteiner began investigating why transfusions sometimes worked and sometimes did not. In 1900, he determined that blood was of differing types and that only certain combinations of them were compatible. In 1910, identification of the ABO blood antigen gene explained the observed blood type incompatibilities. Today, thirty-three different genes are known to contribute to the surface features of red blood cells, which determine compatibility between blood types.

Antigens and Antibodies

An **antigen** (an'ti-jen) is any molecule that triggers an immune response. When the immune system encounters an antigen not found on the body's own cells, it will attack that antigen by producing protein **antibodies** (an'ti-bod"ez) against that antigen. Antibodies are carried in the plasma. In a transfusion reaction, antigens (*agglutinogens*) on the surface of the donated red blood cells react with antibodies (*agglutinins*) in the plasma of the recipient, resulting in **agglutination** (clumping) of the donated red blood cells.

Only a few of the 33 known antigens on red blood cell membranes can produce serious transfusion reactions. These include the antigens of the ABO group and those of the Rh group. Avoiding the mixture of certain types of antigens and antibodies prevents adverse transfusion reactions. From Science to Technology 14.1 discusses genetic testing to prevent transfusion mismatches.

A mismatched blood transfusion quickly produces telltale signs of agglutination—anxiety; breathing difficulty; facial flushing; headache; and severe pain in the neck, chest, and lumbar area. Red blood cells burst, releasing free hemoglobin. Liver cells and macrophages phagocytize the hemoglobin, breaking it down into heme and globin. Some of the heme is recycled. The rest of the heme is converted to bilirubin, which may sufficiently accumulate to cause the yellow skin of jaundice. Free hemoglobin carried in the plasma and reaching the kidneys may ultimately cause them to fail.

ABO Blood Group

The *ABO blood group* is based on the presence (or absence) of two major antigens on red blood cell membranes—*antigen A* and *antigen B*. A and B antigens are carbohydrates attached to glycolipids projecting from the red blood cell surface. The carbohydrate portions of red blood cell antigens are best known for being the basis of blood types, but the fact that they are not found exclusively on red blood cells indicates that they have other functions in addition to marking the surfaces of red blood cells as coming from "self" or not. In epithelial cells, for example, these same A and B antigens provide points of cellular adhesion and cell membrane stability and function as receptors or parts of transport systems across the membrane for various molecules. A person's erythrocytes have one of four antigen combinations: only A, only B, both A and B, or neither A nor B.

An individual with only antigen A has *type A blood;* a person with only antigen B has *type B blood;* one with both antigen A and antigen B has *type AB blood;* and one with neither antigen A nor antigen B has *type O blood*. The resulting ABO blood type is inherited because a gene specifies an enzyme that catalyzes the final step in the synthesis of the A or B antigen. Table 14.11 indicates some of the frequencies of ABO blood types in the diverse population of the United States.

Antibodies that affect the ABO blood groups are synthesized about two to eight months following birth as a result of exposure to foods or microorganisms containing the antigen(s) absent on the individual's red blood cells. The stimulus for their synthesis has not clearly been established. However, whenever antigen A is absent on the red blood cells, an antibody called *anti-A* is produced, and whenever antigen B is absent, an antibody called *anti-B* is produced.

TABLE 14.11 Some ABO Blood Type Frequencies (%) in the United States

Population	Type O	Type A	Type B	Type AB
Caucasian	45	40	11	4
African American	49	27	20	4
American Indian	79	16	4	1
Hispanic	63	14	20	3
Chinese American	42	27	25	6
Japanese American	31	38	21	10
Korean American	32	28	30	10

Typing and matching blood is essential to minimize the chance that the immune response of a blood transfusion recipient will reject the blood. Familiar blood types are A, B, AB, and O. Rh blood types become important in pregnancy, when woman and offspring may be incompatible (see figure 14.23). However, human blood can be classified into at least thirty-three major types based on protein and carbohydrate molecules (antigens) on the surfaces of red blood cells. Each type includes many subtypes, generating hundreds of ways that the topographies of our red blood cells differ from individual to individual.

For more than a century, an approach called serology has been used to type blood into major groups by identifying red blood cell antigens. Different reagents are required to detect different antigens, and so typing by serology beyond a few blood groups is costly and time consuming. It can miss variants of antigens that are small or hidden among the embedded proteins and emanating carbohydrate chains of the cell surface. A more informative way to type blood is to identify the *instructions* for the cell-surface antigens—the genes that encode these proteins.

A consortium of European blood banks and universities, called Bloodgen, has pioneered this more discriminating approach, called genotyping. A tiny device called a BLOODchip detects, with one test, 128 distinct DNA "signatures" that represent all known gene variant combinations that underlie blood

types traditionally defined by serology (fig. 14D). BLOODchip is able to accurately type 99.8% of samples; serology can accurately type 97%. Another device, called BeadChip, tests for a different group of antigens. Other variations on the DNA chip theme are in development.

Accurate blood matching for many blood groups may be life-saving for a person who has a chronic disorder that requires multiple transfusions, such as leukemia or sickle cell disease. After years of receiving blood transfusions, these individuals produce so many antibodies (immune system proteins) against so many types of donor blood that serology is insufficient to identify blood types that they can tolerate. DNA blood typing may eventually become routine.

FIGURE 14D Genetic testing adds precision to blood typing.

Therefore, persons with type A blood have anti-B antibodies in their plasma; those with type B blood have anti-A antibodies; those with type AB blood have neither of these antibodies; and those with type O blood have both anti-A and anti-B antibodies (**fig. 14.21** and **table 14.12**). The anti-A and anti-B antibodies are too large to cross the placenta. If a pregnant woman and her fetus are of different ABO blood types, agglutination in the fetus will not occur.

The major concern in blood transfusion procedures is that the cells in donated blood do not clump due to antibodies in the recipient's plasma. For example, a person with type A blood makes anti-B antibodies and therefore must not receive blood of type B or AB, either of which would clump (**fig. 14.22**). Likewise, a person with type B blood must not be given type A or AB blood, and a person with type O blood must not be given type A, B, or AB blood.

Type AB blood lacks both anti-A and anti-B antibodies, so an AB person can receive a transfusion of blood of any other type. For this reason, type AB people are sometimes called *universal recipients*. However, type A blood, type B blood, and type O blood still contain antibodies (either anti-A and/or anti-B) that could agglutinate type AB cells. Consequently, even for AB individuals, it is always best to use donor blood of the same type as the recipient blood. If the matching type is not available and type A, B, or O is used, it should be transfused slowly and in limited amounts so that the recipient's larger blood volume dilutes the donor blood. This precaution usually avoids serious reactions between the donor's antibodies and the recipient's antigens.

Type O blood does not have antigen A or B. Therefore, theoretically this type could be transfused into people with blood of any other type. Individuals with type O blood are also called *universal donors*. Type O blood, however, does contain both anti-A and anti-B antibodies, and if it is given to a person with blood type A, B, or AB, it too should be transfused slowly and in limited amounts to minimize the chance of an adverse reaction. When type O blood is given to blood types A, B, or AB, it is generally transfused as "packed cells," meaning the plasma has been removed. This minimizes adverse reactions due to the anti-A and anti-B antibodies in the plasma of type O blood. **Table 14.13** summarizes preferred blood types for normal transfusions and permissible blood types for emergency transfusions.

TABLE 14.12	Antigens and Antibodies of the ABO Blood Group	
Blood Type	**Antigen**	**Antibody**
A	A	anti-B
B	B	anti-A
AB	A and B	Neither anti-A nor anti-B
O	Neither A nor B	Both anti-A and anti-B

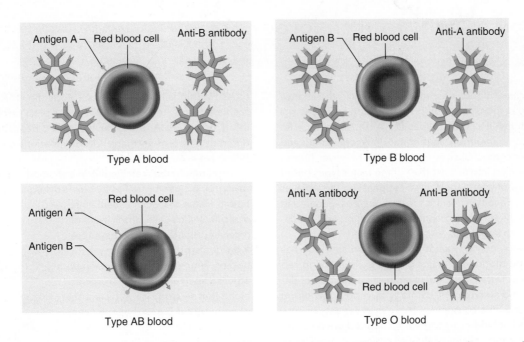

FIGURE 14.21 Different combinations of antigens and antibodies distinguish blood types. (Cells and antibodies not drawn to scale.)

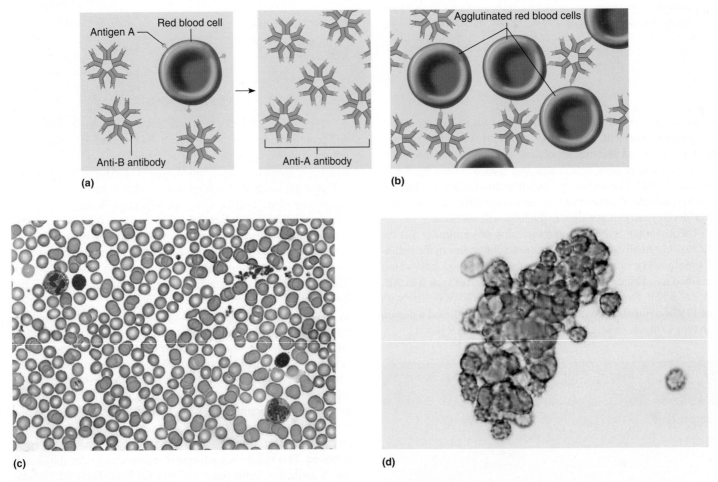

(a)

(b)

(c)

(d)

FIGURE 14.22 Agglutination. (*a*) If red blood cells with antigen A are added to blood containing anti-A antibody, (*b*) the antigens react with the antibodies, causing clumping (agglutination). (*c*) Nonagglutinated blood. (*d*) Agglutinated blood. (Cells and antibodies in *a* and *b* not drawn to scale.)

TABLE 14.13	Preferred and Permissible Blood Types for Transfusions		
Blood Type of Recipient	Preferred Blood Type of Donor	If Preferred Blood Type is Unavailable, Permissible Blood Type of Donor (In an Extreme Emergency)	
A	A	O	
B	B	O	
AB	AB	A, B, O	
O	O	No alternate types	

A person with a rare genetic condition called the Bombay phenotype lacks an enzyme that inserts a particular molecule onto red blood cell surfaces. Without that molecule, the A and B antigens cannot bind. The result is blood that tests as O (because it lacks A and B antigens) but can genetically be of any ABO type—A, B, AB, or O. Although the Bombay phenotype does not affect health, it can sometimes explain a child's ABO type that cannot otherwise be predicted genetically from those of the parents. Taking a cue from the genetic condition, researchers are using enzymes to cleave the A and B antigens from red blood cells, creating more type O blood for transfusions in emergency situations.

 PRACTICE

48 Distinguish between antigens and antibodies.

49 What is the main concern when blood is transfused from one individual to another?

50 Why is a type AB person called a universal recipient?

51 Why is a type O person called a universal donor?

Rh Blood Group

The *Rh blood group* was named after the rhesus monkey in which it was first studied. In humans, this group includes several Rh antigens (factors). The most prevalent of these is *antigen D,* a transmembrane protein. If antigen D or other Rh antigens are present in red blood cell membranes, the blood is said to be *Rh-positive.* Conversely, if red blood cells lack Rh antigens, the blood is called *Rh-negative.* About 15% of Caucasians and 5% of African Americans in the U.S. population are Rh-negative. The remaining ethnic groups are all Rh-positive. The presence (or absence) of Rh antigens is an inherited trait.

Anti-Rh antibodies (anti-Rh) form only in Rh-negative individuals in response to the presence of red blood cells with Rh antigens. This happens, for example, if an Rh-negative person receives a transfusion of Rh-positive blood. The Rh antigens stimulate the recipient's antibody-producing cells to make anti-Rh antibodies. Generally, no serious consequences result from the first transfusion, but if the Rh-negative person—now sensitized to Rh-positive blood—receives another transfusion of Rh-positive blood some months later, the donated red blood cells are likely to agglutinate.

A similar situation of Rh incompatibility arises when an Rh-negative woman is pregnant with an Rh-positive fetus. Her first pregnancy with an Rh-positive fetus would probably be uneventful. However, if at the time of this infant's birth the placental membranes that separated the maternal blood from the fetal blood tear, and some of the infant's Rh-positive blood cells enter the maternal circulation, these cells may stimulate the maternal tissues to produce anti-Rh antibodies (fig. 14.23). This situation may also arise if the pregnancy ends before the birth or fetal cells enter the maternal circulation during an invasive prenatal test such as amniocentesis. If a woman who has already developed anti-Rh antibodies becomes pregnant with a second Rh-positive fetus, these anti-Rh antibodies, called hemolysins, can cross the placental membrane and destroy the fetal red cells. The fetus then develops a condition called *erythroblastosis fetalis* (ĕrith″ro-blas-to′sis fe′tal-iz), also known as hemolytic disease of the newborn (fig. 14.23).

Erythroblastosis fetalis is extremely rare today because obstetricians carefully track Rh status. An Rh-negative woman who might carry an Rh-positive fetus receives an injection of a drug called RhoGAM at week 28 of her pregnancy and after delivery of an Rh-positive baby. RhoGam is a preparation of anti-Rh antibodies, which bind to and shield any Rh-positive fetal cells that might enter the woman's bloodstream and sensitize her immune system. RhoGAM must be given within 72 hours of possible contact with Rh-positive cells—including giving birth, terminating a pregnancy, miscarrying, or undergoing amniocentesis (a prenatal test in which a needle is inserted into the uterus).

 PRACTICE

52 What is the Rh blood group?

53 What are two ways that Rh incompatibility can arise?

Table 14.14 describes a few of the many and diverse inherited disorders that affect the blood. ∎

FIGURE 14.23

Rh incompatibility. If a man who is Rh-positive and a woman who is Rh-negative conceive a child who is Rh-positive, the woman's body may manufacture antibodies that attack future Rh-positive offspring.

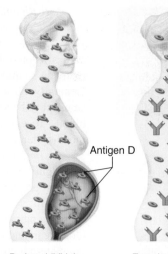

Antigen D

| Rh-negative female with Rh-positive fetus | During childbirth, cells from the Rh-positive fetus may enter the female's bloodstream | Female becomes sensitized— antibodies ( ) form against Rh-positive blood cells between pregnancies | In the next Rh-positive pregnancy, maternal antibodies attack fetal red blood cells |

TABLE 14.14	Some Inherited Disorders of Blood
Disorder	**Abnormality**
Chronic granulomatous disease	Granulocytes cannot produce superoxide, which kills pathogenic bacteria
Erythrocytosis	Reticulocytes have extra EPO receptors, enhancing stamina
Factor V Leiden	Increases risk of abnormal clotting; elevates risk of deep vein thrombosis
Hemophilia (several types)	Lack of specific clotting factor causes bleeding
Hereditary hemochromatosis	Excess absorption of dietary iron into bloodstream deposits iron in various organs
Porphyria variegata	Enzyme deficiency excretes porphyrin ring of hemoglobin into urine; metabolic blockage causes sequence of varied symptoms
Sickle cell disease	Abnormal hemoglobin crystallizes under low-oxygen conditions, sickling red blood cells, which block circulation causing anemia, pain, and other symptoms
Von Willebrand disease	Lack of clotting factor (von Willebrand factor), which stabilizes factor VIII, causes bleeding; less severe than hemophilia

CHAPTER SUMMARY

14.1 Characteristics of Blood (page 528)

Blood is a type of connective tissue in which cells are suspended in a liquid extracellular matrix. Blood volume varies with body size, fluid and electrolyte balance, and adipose tissue content. Blood can be separated into formed elements (mostly red blood cells) and a liquid, **plasma,** portion.

14.2 Blood Cells (page 529)

1. The origin of blood cells
   a. Blood cells develop from **hematopoietic stem cells,** or hemocytoblasts, in red bone marrow.
   b. Blood cells descended from stem cells respond to **hematopoietic growth factors** to specialize.

2. Characteristics of red blood cells
   a. **Red blood cells** are biconcave discs with shapes that provide increased surface area and place their cell membranes close to oxygen-carrying **hemoglobin.**
   b. Red blood cells contain hemoglobin, which combines loosely with oxygen.
   c. The mature form lacks nuclei and mitochondria, but contains enzymes needed for glycolysis.

3. Red blood cell counts
   a. The red blood cell count equals the number of cells per microliter of blood.
   b. The average count may range from approximately 4,000,000 to 6,000,000 cells per microliter.
   c. Red blood cell count determines the oxygen-carrying capacity of the blood and is used in diagnosing and evaluating the courses of certain diseases.

4. Red blood cell production and its control
   a. During fetal development, red blood cells form in the yolk sac, liver, and spleen; after birth, red blood cells are produced by the red bone marrow.
   b. In a person in good health, the number of red blood cells remains relatively stable.
   c. A negative feedback mechanism involving **erythropoietin** from the kidneys and liver controls rate of red blood cell production.
      (1) Erythropoietin is released in response to low oxygen levels.
      (2) High altitude, loss of blood, or chronic lung disease can lower oxygen concentration in the blood.
5. Dietary factors affecting red blood cell production
   a. The availability of vitamin B_{12}, iron, and folic acid affects red blood cell production.
   b. The rate of iron absorption, needed for hemoglobin synthesis, varies with the amount of iron in the body.
6. Destruction of red blood cells
   a. Red blood cells are fragile and are damaged while moving through capillaries.
   b. Macrophages in the spleen and liver phagocytize damaged red blood cells.
   c. Hemoglobin molecules are decomposed, and nearly all of the iron from the heme portion is recycled.
   d. **Biliverdin** and **bilirubin** are pigments, released from the heme portion, excreted in bile.
   e. The globin portion is broken down into amino acids metabolized by macrophages or released into the blood.
7. Types of white blood cells
   a. **Granulocytes** include **neutrophils, eosinophils,** and **basophils.**
   b. **Agranulocytes** include **monocytes** and **lymphocytes.**
8. Functions of white blood cells
   a. **Diapedesis** allows white blood cells to leave the circulation.
   b. Neutrophils and monocytes phagocytize foreign particles.
   c. Basophils release biochemicals that dilate local blood vessels.
   d. Chemicals released by damaged cells attract and stimulate leukocytes.
9. White blood cell counts
   a. Normal total white blood cell counts vary from 3,500 to 10,500 cells per microliter of blood.
   b. The number of white blood cells may change in abnormal conditions such as infections, emotional disturbances, or excessive loss of body fluids.
   c. Because relative proportions of white blood cells may change in particular diseases, the differential white blood cell count, which indicates the percentages of various types of leukocytes, is a useful test.
10. Platelets
   a. **Platelets** are fragments of **megakaryocytes** that enter the circulation.
   b. The normal count varies from 150,000 to 350,000 platelets per microliter.
   c. Platelets help close breaks in blood vessels.

14.3 Plasma (page 541)

Plasma is the liquid part of the blood that is composed of water and a mixture of organic and inorganic substances. It carries nutrients and gases, helps regulate fluid and electrolyte balance, and helps maintain stable pH.

1. Plasma proteins
   a. **Plasma proteins** remain in the blood and interstitial fluids and are not normally used as energy sources.
   b. Three major types exist.
      (1) **Albumins** help maintain the colloid osmotic pressure.
      (2) **Globulins** transport lipids and fat-soluble vitamins and include antibodies that provide immunity.
      (3) **Fibrinogen** functions in blood clotting.
2. Gases and nutrients
   a. Gases in plasma include oxygen, carbon dioxide, and nitrogen.
   b. Plasma nutrients include simple sugars, amino acids, and lipids.
      (1) Glucose is stored in the liver as glycogen and is released whenever the blood glucose concentration falls.
      (2) Amino acids are used to synthesize proteins and are deaminated for use as energy sources.
      (3) Lipoproteins transport lipids.
3. Nonprotein nitrogenous substances
   a. **Nonprotein nitrogenous substances** are composed of molecules that contain nitrogen atoms but are not proteins.
   b. They include amino acids, urea, uric acid, creatine, and creatinine.
      (1) Urea and uric acid are products of catabolism.
      (2) Creatinine results from the metabolism of creatine.
   c. Levels of these substances usually remain stable; an increase may indicate a kidney disorder.
4. Plasma electrolytes
   a. Plasma electrolytes are absorbed from the intestines and are released as by-products of cellular metabolism.
   b. They include ions of sodium, potassium, calcium, magnesium, chloride, bicarbonate, phosphate, and sulfate.
   c. They are important in maintaining the osmotic pressure and pH of plasma.

14.4 Hemostasis (page 542)

Hemostasis is the stoppage of bleeding. Hemostatic mechanisms are most effective in controlling blood loss from small vessels.

1. Vascular spasm (blood vessel spasm)
   a. Smooth muscle in the walls of smaller blood vessels reflexly contracts following injury.
   b. Platelets release serotonin that stimulates vasoconstriction and helps maintain blood vessel spasm.
2. Platelet plug formation
   a. Platelets adhere to rough surfaces and exposed collagen.
   b. Platelets adhere together at the sites of injuries and form platelet plugs in broken vessels.
3. Blood coagulation
   a. Blood **coagulation** or clotting, the most effective means of hemostasis, is a series of reactions wherein each reaction stimulates the next (cascade), which may be initiated by extrinsic or intrinsic mechanisms.
   b. The major event of coagulation is the conversion of soluble fibrinogen into insoluble **fibrin.**
   c. The **extrinsic clotting mechanism** (tissue factor pathway) is triggered when blood contacts damaged tissue.
   d. The **intrinsic clotting mechanism** (contact activation pathway) is triggered when blood contacts a foreign surface.
   e. Clot formation reflects balance between clotting factors that promote or inhibit clotting.

f. A formed clot retracts and pulls the edges of a broken blood vessel closer together.

g. A **thrombus** is an abnormal blood clot in a blood vessel; an **embolus** is a clot or fragment of a clot that moves in a blood vessel.

h. Fibroblasts invade a clot, forming connective tissue throughout.

i. Protein-splitting enzymes may eventually destroy a clot.

4. Prevention of coagulation

a. The smooth inner lining of blood vessels discourages the accumulation of platelets.

b. As a clot forms, fibrin adsorbs thrombin and prevents the reaction from spreading.

c. Antithrombin interferes with the action of excess thrombin.

d. Some cells secrete heparin, an anticoagulant.

14.5 Blood Groups and Transfusions (page 548)

Blood can be typed on the basis of the surface structures of its cells.

1. Antigens and antibodies

a. An **antigen** is any molecule that triggers an immune response.

b. **Antibodies** are protein molecules produced in an immune response to an encounter with an antigen that is not found on the body's own cells.

2. ABO blood group

a. Blood can be grouped according to the presence or absence of antigens A and B.

b. Wherever antigen A is absent, anti-A antibody is present; wherever antigen B is absent, anti-B antibody is present.

c. Preventing the mixing of red blood cells that have an antigen with plasma that contains the corresponding antibody avoids a transfusion reaction.

d. Adverse reactions are due to **agglutination** (clumping) of the red blood cells.

3. Rh blood group

a. Rh antigens are present on the red blood cell membranes of Rh-positive blood; they are absent in Rh-negative blood.

b. An Rh-negative person exposed to Rh-positive blood produces anti-Rh antibodies in response.

c. Mixing Rh-positive red cells with plasma that contains anti-Rh antibodies agglutinates the positive cells.

d. If an Rh-negative female is pregnant with an Rh-positive fetus, some of the positive cells may enter the maternal blood at the time of birth and stimulate the maternal tissues to produce anti-Rh antibodies.

e. Anti-Rh antibodies in maternal blood can cross the placental membrane and react with the red blood cells of an Rh-positive fetus.

CHAPTER ASSESSMENTS

14.1 Characteristics of Blood

1 Blood volume varies with _____. (p. 528)
   a. the amount of adipose tissue
   b. body size
   c. changes in electrolyte concentrations
   d. all of the above

2 Formed elements in blood are _____, _____, and _____. (p. 528)

3 Define *hematocrit*, and explain how it is determined. (p. 529)

4 The liquid portion of the blood is called _____. (p. 529)

14.2 Blood Cells

5 Indicate where blood cells differentiate, and explain the process. (p. 529)

6 Describe a red blood cell. (p. 529)

7 Contrast oxyhemoglobin and deoxyhemoglobin. (p. 529)

8 Explain the significance of red blood cell counts. (p. 531)

9 Describe the life cycle of a red blood cell from production through destruction. (pp. 531, 533, and 535)

10 Define *erythropoietin*, and explain its function. (p. 533)

11 Explain how vitamin B_{12} and folic acid deficiencies affect red blood cell production. (p. 534)

12 List two sources of iron that can be used for the synthesis of hemoglobin. (p. 534)

13 Distinguish between biliverdin and bilirubin. (p. 535)

14 Distinguish between granulocytes and agranulocytes. (p.536)

15 Name five types of leukocytes, and list the major functions of each type. (pp. 536–537)

16 Explain the significance of white blood cell counts as aids to diagnosing disease. (pp. 538–539)

17 _____ are fragments of megakaryocytes that function in _____. (p. 539)

14.3 Plasma

18 The most abundant component of plasma is _____. (p. 541)
   a. waste
   b. oxygen
   c. proteins
   d. water

19 Name three types of plasma proteins, and indicate the function of each type. (p. 541)

20 Name the gases and nutrients in plasma. (p. 542)

21 Define *nonprotein nitrogenous substances*, and name those commonly present in plasma. (p. 542)

22 The most abundant plasma electrolytes are _____ and _____. (p. 542)

23 Name several plasma electrolytes. (p. 542)

14.4 Hemostasis

24 _____ is the term for stoppage of bleeding. (p. 542)

25 Explain how vascular spasm is stimulated following an injury. (pp. 542–543)

26 Platelets adhering to form a plug may control blood loss from a _____ break, but a larger break may require a _____ to halt bleeding. (p. 543)

27 Name a vitamin required for blood clotting. (p. 543)

28 Distinguish between fibrinogen and fibrin. (p. 543)

29 Indicate the trigger and outline the steps for extrinsic clotting and for intrinsic clotting. (pp. 543–544)

30 Describe the major steps leading to the formation of a blood clot. (p. 544)

31 Describe a positive feedback mechanism that operates during blood clotting. (p. 544)

32 Define *serum*. (p. 544)

33 Explain how a blood clot may be removed naturally from a blood vessel. (p. 545)

34 Distinguish between a thrombus and an embolus. (p. 545)
35 Describe how blood coagulation may be prevented. (p. 547)

14.5 Blood Groups and Transfusions

36 Distinguish between an antigen and an antibody. (p. 548)
37 Explain the basis of ABO blood types. (p. 548)
38 Explain why a person with blood type AB is sometimes called a universal recipient. (p. 549)

39 Explain why a person with blood type O is sometimes called a universal donor. (p. 549)
40 Distinguish between Rh-positive and Rh-negative blood. (p. 551)
41 Describe how a person may become sensitized to Rh-positive blood. (p. 551)
42 Describe *erythroblastosis fetalis,* and explain how this condition may develop. (p. 551)

 # INTEGRATIVE ASSESSMENTS/CRITICAL THINKING

Outcomes 3.4, 3.6, 14.2

1. If a patient with inoperable cancer is treated using a drug that reduces the rate of cell division, how might the patient's white blood cell count change? How might the patient's environment be modified to compensate for the effects of these changes?

Outcomes 4.5, 9.3, 13.2, 14.2

2. Erythropoietin is available as a drug, as are drugs that mimic the effects of erythropoietin. Why would athletes abuse these drugs? Why is it difficult to detect erythropoietin that has been taken as a drug?

3. Some athletes have been accused of performing "blood doping" to improve their athletic performance. Why would removing blood a month or so prior to performance, then reinfusing the blood shortly before a competition, boost performance?

Outcome 14.2

4. Hypochromic (iron-deficiency) anemia is common among aging persons admitted to hospitals for other conditions. What environmental and sociological factors might promote this form of anemia?

5. How would you explain to a patient with leukemia, who has a greatly elevated white blood cell count, the importance of avoiding bacterial infections?

Outcomes 14.2, 14.4

6. In the United States between 1977 and 1985, more than 10,000 men contracted the human immunodeficiency virus (HIV) from contaminated factor VIII that they received to treat hemophilia. What are two abnormalities in the blood of these men?

Outcomes 14.3, 14.4

7. Why do patients with liver diseases commonly develop blood clotting disorders?

Outcome 14.5

8. Commercially available antiserum samples containing antibodies for antigens A, B, and D are used to determine the type of a particular patient's blood. The antiserum is mixed with the sample of blood, and if agglutination (clumping) occurs, that means that the antigen with the same name as the antiserum is present on those RBCs. Indicate the blood type (both ABO and Rh) of this individual: What blood type(s) could safely receive blood from this individual? What blood type(s) could this individual safely receive? (Note: The control indicates the absence of agglutination, or clumping.)

| Anti-A antiserum | Anti-B antiserum | Anti-D antiserum | Control |

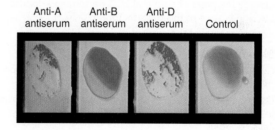

 # ONLINE STUDY TOOLS

 | ■ LEARNSMART® |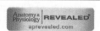

Connect Interactive Questions Reinforce your knowledge using assigned interactive questions covering the components of blood and their functions, as well as the processes of blood cell formation and hemostasis.

Connect Integrated Activity Explore the process of blood cell formation.

LearnSmart Discover which chapter concepts you have mastered and which require more attention. This adaptive learning tool is personalized, proven, and preferred.

Anatomy & Physiology Revealed Go more in depth into the human body by exploring the histology of the formed elements in blood, as well as viewing animations that illustrate blood cell production and hemoglobin recycling.

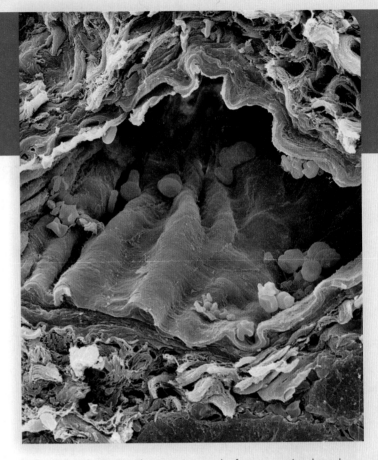

Falsely colored scanning electron micrograph of a cross section through an artery containing red blood cells (400×). The typical thick wall (pink/brown) is adapted to withstand the high pressure of blood that the beating heart pumps throughout the body.

15

Cardiovascular System

 LEARNING OUTCOMES

After you have studied this chapter, you should be able to:

15.1 Overview of the Cardiovascular System

1 Explain the roles of the heart and blood vessels in circulating the blood. (p. 557)

15.2 The Heart

2 Describe the location of the heart within the body. (p. 557)

3 Distinguish between the coverings of the heart and the layers that compose the wall of the heart. (pp. 557–560)

4 Identify and locate the major parts of the heart and discuss the function of each part. (pp. 560–563)

5 Trace the pathway of the blood through the heart and the vessels of coronary circulation. (pp. 561–566)

6 Describe the cardiac cycle and explain how heart sounds are produced. (pp. 566–568)

7 Identify the parts of a normal ECG pattern and discuss the significance of this pattern. (pp. 570–572)

8 Explain control of the cardiac cycle. (p. 572)

15.3 Blood Vessels

9 Compare the structures and functions of the major types of blood vessels. (pp. 577–582)

10 Describe how substances are exchanged between blood in the capillaries and the tissue fluid surrounding body cells. (p. 581)

15.4 Blood Pressure

11 Explain how blood pressure is produced and controlled. (pp. 582–588)

12 Describe the mechanisms that aid in returning venous blood to the heart. (p. 588)

15.5 Paths of Circulation

13 Compare the pulmonary and systemic circuits of the cardiovascular system. (pp. 590–591)

15.6–15.7 Arterial System–Venous System

14 Identify and locate the major arteries and veins. (pp. 591–606)

15.8 Life-Span Changes

15 Describe life-span changes in the cardiovascular system. (p. 608)

THE WHOLE PICTURE

At rest, the heart pumps about 7,000 liters of blood through the body each day, contracting about 2.5 billion times in an average lifetime. This muscular pump forces blood through arteries, which connect to smaller-diameter vessels called arterioles. Arterioles branch into the tiniest tubes, the capillaries, which are sites of nutrient, electrolyte, gas, and waste exchange. Capillaries converge into venules, which in turn converge into veins that return blood to the heart, completing the closed system of blood circulation. These structures—the pump and its vessels—form the cardiovascular system.

Module 9: Cardiovascular System

UNDERSTANDING WORDS

angio-, vessel: *angio*tensin—substance that constricts blood vessels.

ather-, porridge: *ather*osclerosis—deposits of plaque in arteries.

brady-, slow: *brady*cardia—abnormally slow heartbeat.

diastol-, dilation: *diastol*ic pressure—blood pressure when the ventricle of the heart is relaxed.

edem-, swelling: *edem*a—accumulation of fluids in the tissues that causes them to swell.

-gram, something written: electrocardio*gram*—recording of the electrical changes in the myocardium during a cardiac cycle.

lun-, moon: semi*lun*ar valve—valve with crescent-shaped flaps.

myo-, muscle: *myo*cardium—muscle tissue within the wall of the heart.

papill-, nipple: *papill*ary muscle—small mound of muscle projecting into a ventricle of the heart.

phleb-, vein: *phleb*itis—inflammation of a vein.

scler-, hard: arterio*scler*osis—loss of elasticity and hardening of a blood vessel wall.

syn-, together: *syn*cytium—mass of merging cells that act together.

systol-, contraction: *systol*ic pressure—blood pressure resulting from a single ventricular contraction.

tachy-, rapid: *tachy*cardia—abnormally fast heartbeat.

15.1 | Overview of the Cardiovascular System

The term "cardiovascular" refers to both the heart and the blood vessels. The pumping action of the heart moves blood through the body's blood vessels. The blood vessels form two circuits. The **pulmonary** (pul′mo-ner″e) **circuit** sends oxygen-poor blood to the lungs to pick up oxygen and unload carbon dioxide. The **systemic** (sistem′ ik) **circuit** sends oxygen-rich blood and nutrients to all body cells and removes wastes. Without circulation, tissues would lack a supply of oxygen and nutrients, and wastes would accumulate. Such deprived cells soon begin irreversible change, which quickly leads to their death. The heart is really two pumps in one. The right side of the heart pumps blood to the pulmonary circuit and the blood returns to the left side of the heart. At the same time, the left side of the heart pumps blood through the systemic circuit, and that blood returns to the right side of the heart (fig.15.1).

15.2 | The Heart

The **heart** is a hollow, cone-shaped, muscular pump. It is in the mediastinum of the thorax and rests on the diaphragm.

Size and Location of the Heart

Heart size varies with body size. An average adult's heart is generally about 14 centimeters long and 9 centimeters wide (fig. 15.2).

The heart is bordered laterally by the lungs, posteriorly by the vertebral column, and anteriorly by the sternum (fig. 15.3 and reference plates 10, 16, 21, and 22, pp. 47, 50, and 53). The *base* of the heart, which attaches to several large blood vessels, lies beneath the second rib. The heart's inferior end extends downward and to the left, terminating as a bluntly pointed *apex* at the level of the fifth intercostal space. Because of the heart's location, it is possible to detect the *apical heartbeat* by feeling or listening to the chest wall between the fifth and sixth ribs, about 7.5 centimeters to the left of the midline.

Coverings of the Heart

The **pericardium** (per″i-kar′de-um), or pericardial sac, is a covering that encloses the heart and the proximal ends of the large blood vessels to which it attaches. The pericardium consists of an outer fibrous bag, the *fibrous pericardium*, that surrounds a more delicate, double-layered serous membrane. The innermost layer of this serous mem-

CAREER CORNER

Perfusionist

The fifty-six-year-old man has survived three heart attacks, and his heart has continued to weaken. He is in heart failure and is put on a transplant list. When a heart becomes available, the surgical team assembles quickly. One key member is the perfusionist.

When the patient's heart is temporarily stopped so that the surgeon can remove it and replace it with the donor heart, the perfusionist sets up and operates the heart-lung machine that keeps the circulation going, in some cases for many hours. The perfusionist makes sure that the rate of blood flow, the patient's temperature, and blood composition are within normal range. If the surgeon orders that the patient receive drugs or blood products, the perfusionist delivers them through the heart-lung machine.

Training for a career as a perfusionist entails earning a certificate beyond a four-year degree in a field such as biology, nursing, or medical technology. The job requires familiarity with several types of devices and the reasons why they are used. The perfusionist must learn to use several types of heart-lung machines as well as extracorporeal membrane oxygenation (ECMO) machines to assist breathing, perform at least 125 perfusion procedures, and pass a certification exam from the American Board of Cardiovascular Perfusion.

The job requires stamina, concentration, attention to detail, and excellent communication skills. The perfusionist may also order supplies and equipment and advise support staff. These medical professionals work at hospitals and surgical centers. They may go into teaching or work at companies that manufacture equipment used in procedures involving the heart and/or lungs.

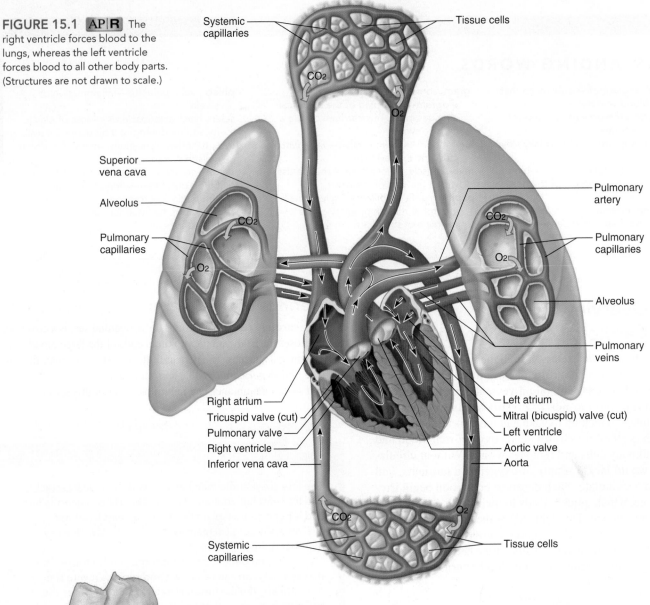

FIGURE 15.1 AP|R The right ventricle forces blood to the lungs, whereas the left ventricle forces blood to all other body parts. (Structures are not drawn to scale.)

Systemic capillaries

Tissue cells

CO_2

O_2

Superior vena cava

Alveolus

CO_2

Pulmonary capillaries

O_2

Pulmonary artery

CO_2

Pulmonary capillaries

O_2

Alveolus

Pulmonary veins

Right atrium

Tricuspid valve (cut)

Pulmonary valve

Right ventricle

Inferior vena cava

Left atrium

Mitral (bicuspid) valve (cut)

Left ventricle

Aortic valve

Aorta

Systemic capillaries

CO_2

O_2

Tissue cells

FIGURE 15.2

Anterior view of a human heart. This photo is not life-size, so a proportionately reduced ruler has been included as a size reference to help the student grasp the true size of the organ.

0 1 2 3 4 5 cm

brane, the *visceral pericardium* (epicardium), covers the heart. At the base of the heart, the visceral pericardium turns back upon itself to become the *parietal pericardium,* which covers the inner surface of the fibrous pericardium (fig. 15.4).

The fibrous pericardium is a tough, protective sac composed of dense connective tissue. It is attached to the central portion of the diaphragm, the posterior of the sternum, the vertebral column, and the large blood vessels emerging from the heart. Between the parietal and visceral serous layers of the pericardium is a space, the *pericardial cavity,* that contains a small volume of serous fluid that the pericardial membranes secrete (see fig. 1.10*b*, p. 19 and reference plates 16 and 17, pp. 50–51 and fig. 15.4). This fluid reduces friction between the pericardial membranes as the heart moves within them.

In *pericarditis,* inflammation of the pericardium is often due to viral or bacterial infection. This painful condition interferes with heart movements.

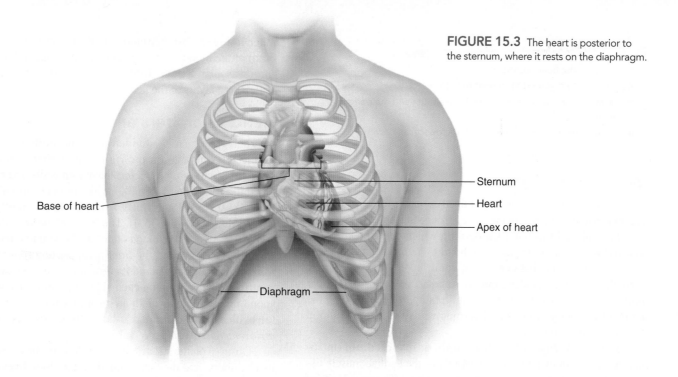

FIGURE 15.3 The heart is posterior to the sternum, where it rests on the diaphragm.

Base of heart

Sternum

Heart

Apex of heart

Diaphragm

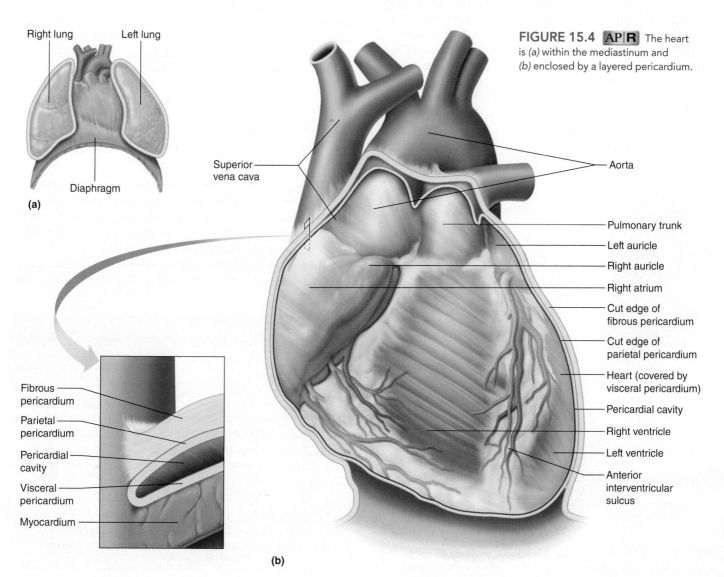

Right lung

Left lung

Diaphragm

(a)

FIGURE 15.4 AP|R The heart is (a) within the mediastinum and (b) enclosed by a layered pericardium.

Superior vena cava

Aorta

Pulmonary trunk

Left auricle

Right auricle

Right atrium

Cut edge of fibrous pericardium

Cut edge of parietal pericardium

Heart (covered by visceral pericardium)

Pericardial cavity

Right ventricle

Left ventricle

Anterior interventricular sulcus

Fibrous pericardium

Parietal pericardium

Pericardial cavity

Visceral pericardium

Myocardium

(b)

 **PRACTICE**

1 Where is the heart located?

2 Where would you listen to hear the apical heartbeat?

3 Distinguish between the visceral pericardium and the parietal pericardium.

4 What is the function of the fluid in the pericardial cavity?

Wall of the Heart

The wall of the heart is composed of three distinct layers: an outer epicardium, a middle myocardium, and an inner endocardium (fig. 15.5).

The **epicardium** (ep"ĭ-kar'de-um), which corresponds to the visceral pericardium, protects the heart by reducing friction. It is a serous membrane that consists of connective tissue covered by epithelium, and it includes capillaries and nerve fibers. The deeper portion of the epicardium may contain fat, particularly along the paths of coronary arteries and cardiac veins that provide blood flow through the myocardium.

The middle layer of the heart wall, or **myocardium** (mi"o-kar'de-um), is thick and consists largely of the cardiac muscle tissue that pumps blood out of the heart chambers. The muscle fibers lie in planes that are separated by connective tissues richly supplied with blood capillaries, lymph capillaries, and nerve fibers.

The inner layer of the heart wall, or **endocardium** (en"do-kar'de-um), consists of epithelium and underlying connective tissue that contains many elastic and collagen fibers. The endocardium also contains blood vessels and covers some specialized cardiac cells called *Purkinje fibers,* described in the section "Cardiac Conduction System" (p. 569).

The endocardium lines all of the heart chambers and covers the structures, such as the heart valves, that project into them. This inner lining is also continuous with the inner linings (endothelium) of the blood vessels attached to the heart and throughout the cardiovascular system. Table 15.1 summarizes the characteristics of the three layers of the heart wall.

Heart Chambers and Valves

Internally the heart is divided into four hollow chambers, two on the left and two on the right. The upper chambers, called **atria** (a'tre-ah) (sing., *atrium*), have thin walls and receive blood returning to the heart. Small, earlike projections called **auricles** (aw'ri-klz) extend anteriorly from the atria, slightly increasing atrial volume (see fig. 15.4). The lower chambers, the **ventricles** (ven'tri-klz), force the blood out of the heart into arteries.

A structure called the *interatrial septum* separates the right from the left atrium. An *interventricular septum* separates the two ventricles. The atrium on each side communicates with its corresponding ventricle through an opening called the **atrioventricular orifice** (a"tre-o-ven-trik'u-lar ori-fis), which is guarded by an *atrioventricular valve* (AV valve).

Grooves on the surface of the heart mark the divisions between its chambers, and they also contain major blood vessels that supply the heart tissues. The deepest of these grooves is the **atrioventricular** (coronary) **sulcus** (a"tre-o-ven-trik'u-lar sul'kus), which encircles the heart between the atria and ventricles. Two **interventricular** (anterior and posterior) **sulci** mark the septum that separates the right and left ventricles (see fig. 15.4).

 PRACTICE

5 Describe the layers of the heart wall.

6 Name and locate the four chambers of the heart.

The right atrium receives blood from two large veins: the *superior vena cava* and the *inferior vena cava.* These veins return blood, which is low in oxygen, from tissues. A smaller vein, the *coronary sinus,* also drains venous blood into the right atrium from the myocardium of the heart.

A large **tricuspid valve** (right atrioventricular valve) guards the atrioventricular orifice between the right atrium and the right

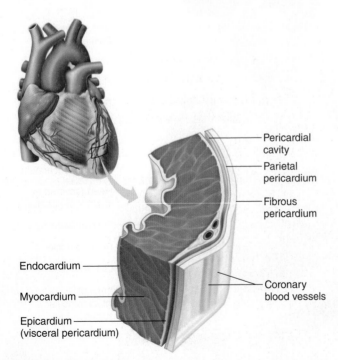

FIGURE 15.5 The heart wall has three layers: an endocardium, a myocardium, and an epicardium.

TABLE 15.1	Wall of the Heart	
Layer	**Composition**	**Function**
Epicardium (visceral pericardium)	Serous membrane of connective tissue covered with epithelium and including blood capillaries, lymph capillaries, and nerve fibers	Forms a protective outer covering; secretes serous fluid
Myocardium	Cardiac muscle tissue separated by connective tissue and including blood capillaries, lymph capillaries, and nerve fibers	Contracts to pump blood from the heart chambers
Endocardium	Membrane of epithelium and underlying connective tissue, including blood vessels and specialized muscle fibers	Forms a protective inner lining of the chambers and valves

ventricle. It is composed of three leaflets, or cusps, as its name implies. This valve permits the blood to move from the right atrium into the right ventricle and prevents it from moving in the opposite direction. The cusps fold passively out of the way against the ventricular wall when the blood pressure is greater on the atrial side, and they close passively when the pressure is greater on the ventricular side (figs. 15.6, 15.7, and 15.8).

Strong, fibrous strings, called *chordae tendineae* (kor'de ten'di-ne), attach to the cusps of the tricuspid valve on the ventricular side. These strings originate from small mounds of cardiac muscle tissue, the **papillary muscles** (pap'ĭ-ler"e mus'elz), that project inward from the walls of the ventricle (see fig. 15.7). The papillary muscles contract when the right ventricle contracts. As the tricuspid valve closes, these muscles pull on the chordae tendineae and prevent the cusps from swinging back (prolapsing) into the right atrium.

The right ventricle has a thinner muscular wall than the left ventricle (see fig. 15.6). This right chamber pumps the blood a fairly short distance to the lungs against a relatively low resistance to blood flow. In contrast, the left ventricle must force the blood to all the other parts of the body against a much greater resistance to flow.

When the muscular wall of the right ventricle contracts, the blood inside its chamber is put under increasing pressure, and the tricuspid valve closes passively. As a result, the only exit for the blood is through the *pulmonary trunk*, which divides to form the left and right *pulmonary arteries* that lead to the lungs. At the base of this trunk is a **pulmonary valve** (pulmonary semilunar valve), which consists of three cusps (fig. 15.8). This valve opens as the right ventricle contracts. However, when the ventricular muscles relax, the blood begins to back up in the pulmonary trunk. This closes the pulmonary valve, preventing backflow into the right ventricle. Unlike the tricuspid valve, the pulmonary valve does not have chordae tendineae or papillary muscles attached to its cusps.

The left atrium receives the blood from the lungs through four *pulmonary veins*—two from the right lung and two from the left lung. The blood passes from the left atrium into the left ventricle through the atrioventricular orifice, which a valve guards. This valve consists of two leaflets and is named the **mitral valve** (shaped like a mitre, a type of headpiece) or bicuspid valve or left atrioventricular valve. It prevents the blood from flowing back into the left atrium from the left ventricle when the ventricle contracts. As with the tricuspid valve, the papillary muscles and the chordae tendineae prevent the cusps of the mitral valve from swinging back (prolapsing) into the left atrium.

When the left ventricle contracts, the mitral valve closes passively, and the only exit is through a large artery called the *aorta*. Its branches distribute blood to all parts of the body.

At the base of the aorta is an **aortic valve** (aortic semilunar valve) which consists of three cusps (fig. 15.8). It opens and allows blood to leave the left ventricle as it contracts. When the ventricular muscles relax, this valve closes and prevents blood from backing up into the left ventricle.

The mitral and tricuspid valves are also called atrioventricular valves because they are between atria and ventricles. The pulmonary and aortic valves are also called semilunar because of the half-moon shapes of their cusps. Table 15.2 summarizes the locations and functions of the heart valves.

TABLE 15.2	Valves of the Heart	
Valve	Location	Function
Tricuspid valve	Right atrioventricular orifice	Prevents blood from moving from the right ventricle into the right atrium during ventricular contraction
Pulmonary valve	Entrance to pulmonary trunk	Prevents blood from moving from the pulmonary trunk into the right ventricle during ventricular relaxation
Mitral valve	Left atrioventricular orifice	Prevents blood from moving from the left ventricle into the left atrium during ventricular contraction
Aortic valve	Entrance to aorta	Prevents blood from moving from the aorta into the left ventricle during ventricular relaxation

Mitral valve prolapse (MVP) affects up to 6% of the U.S. population. In this condition, one (or both) of the cusps of the mitral valve stretches and bulges into the left atrium during ventricular contraction. The valve usually continues to function adequately, but sometimes blood regurgitates (flows back) into the left atrium. Symptoms of MVP include chest pain, palpitations, fatigue, and anxiety.

People with MVP are particularly susceptible to infective endocarditis. This inflammation of the endocardium due to an infection appears as a plantlike growth on the valve. Some people with MVP may be advised by their physicians to take antibiotics before undergoing dental work, to prevent *Streptococcus* bacteria in the mouth from migrating through the bloodstream to the heart and spreading this infection.

 PRACTICE

7 Which blood vessels carry blood into the right atrium?

8 Where does blood go after it leaves the right ventricle?

9 Which blood vessels carry blood into the left atrium?

10 What prevents blood from flowing back into the ventricles when they relax?

Skeleton of the Heart

Rings of dense connective tissue surround the pulmonary trunk and aorta at their proximal ends. These rings are continuous with others that encircle the atrioventricular orifices. They provide firm attachments for the heart valves and for muscle fibers and prevent the outlets of the atria and ventricles from dilating during contraction. The fibrous rings, together with other masses of dense connective tissue in the part of the septum between the ventricles (interventricular septum), constitute the *skeleton of the heart* (fig. 15.8).

Blood Flow Through the Heart

Blood low in oxygen (oxygen-poor blood) and high in carbon dioxide enters the right atrium through the venae cavae and the

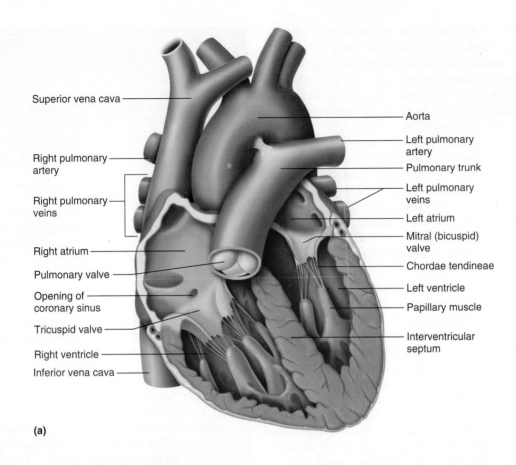

(a)

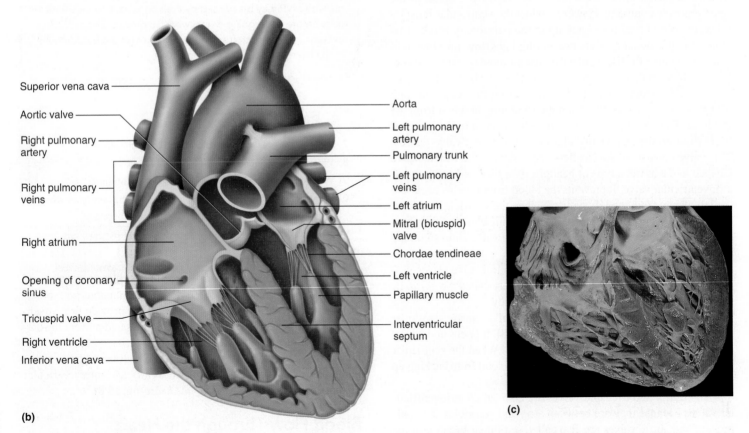

(b)

(c)

FIGURE 15.6 **AP|R** Frontal sections of the heart. (*a*) Drawings show the connection between the right ventricle and the pulmonary trunk and (*b*) the connection between the left ventricle and the aorta, as well as the four hollow chambers. (*c*) A section through a cadaver heart.

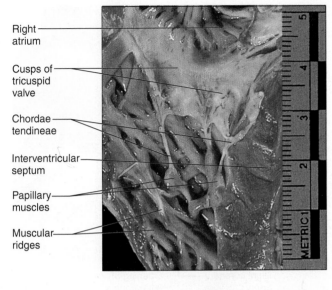

Right atrium

Cusps of tricuspid valve

Chordae tendineae

Interventricular septum

Papillary muscles

Muscular ridges

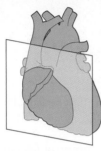

FIGURE 15.7 Photograph of a human tricuspid valve.

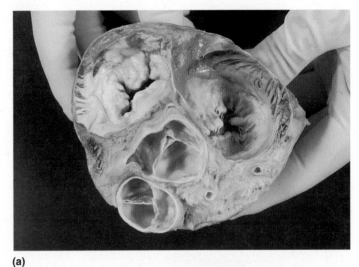

(a)

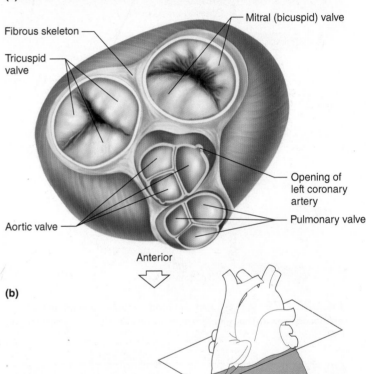

Fibrous skeleton

Tricuspid valve

Mitral (bicuspid) valve

Opening of left coronary artery

Pulmonary valve

Aortic valve

Anterior

(b)

FIGURE 15.8 Heart valves. (*a*) Photograph of a transverse section through the heart, showing the four valves (superior view). (*b*) The skeleton of the heart consists of fibrous rings to which the heart valves are attached (superior view). Note that (*b*) is in anatomically correct orientation.

coronary sinus. As the right atrial wall contracts, the blood passes through the right atrioventricular orifice and enters the chamber of the right ventricle (see fig. 15.1).

When the right ventricular wall contracts, the tricuspid valve closes the right atrioventricular orifice, and the blood passes through the open pulmonary valve into the pulmonary trunk and its branches (pulmonary arteries). From these vessels, blood enters the capillaries associated with the alveoli (microscopic air sacs) of the lungs. Gas exchange occurs between the blood in the capillaries and the air in the alveoli. The oxygen-rich blood, now relatively low in carbon dioxide, returns to the heart through the pulmonary veins that lead to the left atrium.

The left atrial wall contracts, and the blood moves through the left atrioventricular orifice and into the chamber of the left ventricle. When the left ventricular wall contracts, the mitral valve closes the left atrioventricular orifice, and the blood passes through the open aortic valve into the aorta and its branches. Figure 15.9 summarizes the path the blood takes as it passes through the heart to the alveolar capillaries and systemic capillaries, then back to the heart.

Blood Supply to the Heart

The first two branches of the aorta, called the right and left **coronary arteries,** supply blood to the tissues of the heart. Their openings lie just superior to the aortic valve (fig. 15.10).

The right coronary artery passes along the atrioventricular sulcus between the right atrium and the right ventricle. It gives off two major branches—a *posterior interventricular artery,* which travels along the posterior interventricular sulcus and supplies the posterior walls of both ventricles, and a *marginal artery,* which passes along the lower border of the heart. Branches of the marginal artery supply the walls of the right atrium and the right ventricle (figs. 15.11 and 15.12).

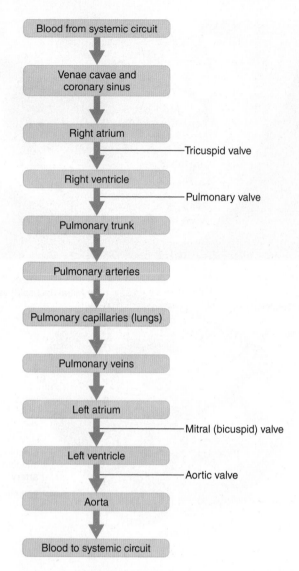

FIGURE 15.9 **AP|R** Path of blood through the heart and pulmonary circuit. The systemic circuit includes the coronary circulation.

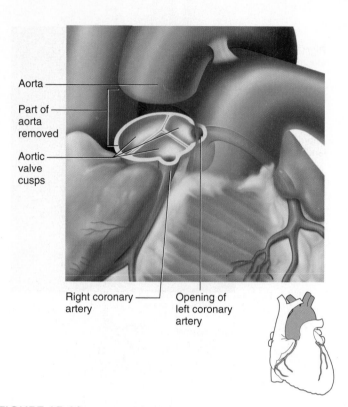

FIGURE 15.10 The openings of the coronary arteries lie just superior to the aortic valve.

interventricular artery (or *left anterior descending artery*), lies in the anterior interventricular sulcus. Its branches supply the walls of both ventricles (figs. 15.11 and 15.12).

The heart must beat continually to supply blood to the tissues. To do this, myocardial cells require a constant supply of oxygen-rich blood. The myocardium contains many capillaries fed by branches of the coronary arteries. The smaller branches of these arteries usually have connections (anastomoses) between vessels that provide alternate pathways for blood, called collateral circulation. These detours in circulation may supply oxygen and nutrients to the myocardium when a coronary artery is blocked.

Magnetic resonance imaging (MRI) can image coronary arteries. Blood flow appears as a bright signal, and areas of diminished or absent blood flow, or blood turbulence, appear as blank areas. This approach is less invasive than *coronary angiography*, in which a catheter is inserted into a blood vessel in the arm, groin, or neck and threaded through various blood vessels until it enters the coronary arteries. Dye is then forced through the catheter, and an X-ray machine takes images (angiograms) showing the condition of the blood vessels as the dye flows through them.

A thrombus or embolus that partially blocks or narrows a coronary artery branch causes a decrease in blood flow called *ischemia*. This deprives myocardial cells of oxygen, producing a painful condition called *angina pectoris*. The pain usually happens during physical activity, when oxygen demand exceeds supply. Pain lessens with rest. Emotional disturbance may also trigger angina pectoris.

Angina pectoris feels like heavy pressure, tightening, or squeezing in the chest, usually behind the sternum or in the anterior upper thorax. The pain may radiate to the neck, jaw, throat, left shoulder, left upper limb, back, or upper abdomen. Profuse perspiration (diaphoresis), difficulty breathing (dyspnea), nausea, or vomiting may occur.

A blood clot may completely obstruct a coronary artery (coronary thrombosis), killing tissue in that part of the heart. This is a *myocardial infarction* (MI) or heart attack.

One branch of the left coronary artery, the *circumflex artery*, following the atrioventricular sulcus between the left atrium and the left ventricle encircles the heart and travels posteriorly. Its branches supply blood to the walls of the left atrium and the left ventricle. Another branch of the left coronary artery, the *anterior*

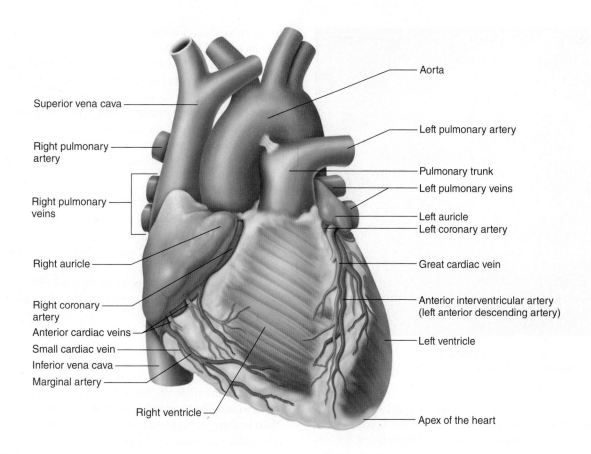

Superior vena cava

Right pulmonary artery

Right pulmonary veins

Right auricle

Right coronary artery

Anterior cardiac veins

Small cardiac vein

Inferior vena cava

Marginal artery

Right ventricle

Aorta

Left pulmonary artery

Pulmonary trunk

Left pulmonary veins

Left auricle

Left coronary artery

Great cardiac vein

Anterior interventricular artery (left anterior descending artery)

Left ventricle

Apex of the heart

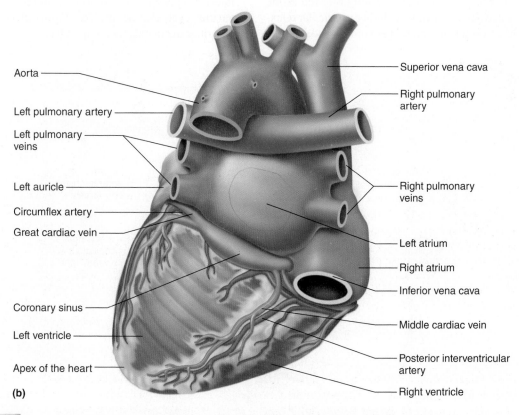

Aorta

Left pulmonary artery

Left pulmonary veins

Left auricle

Circumflex artery

Great cardiac vein

Coronary sinus

Left ventricle

Apex of the heart

Superior vena cava

Right pulmonary artery

Right pulmonary veins

Left atrium

Right atrium

Inferior vena cava

Middle cardiac vein

Posterior interventricular artery

Right ventricle

(b)

FIGURE 15.11 **AP|R** Blood vessels associated with the surface of the heart. (a) Anterior view. (b) Posterior view.

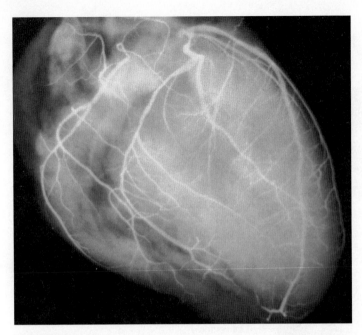

FIGURE 15.12 An angiogram (radiograph) of the coronary arteries is a diagnostic procedure used to examine specific blood vessels.

In most body parts, blood flow in arteries peaks during ventricular contraction. However, blood flow in the vessels of the myocardium is poorest during ventricular contraction. This is because the muscle fibers of the myocardium compress nearby vessels as they contract, interfering with blood flow. Also, the openings into the coronary arteries are partially blocked as the flaps of the aortic valve open. Conversely, during ventricular relaxation, the myocardial vessels are no longer compressed, and the aortic valve does not block the orifices of the coronary arteries. This increases blood flow into the myocardium.

Branches of the **cardiac veins** drain blood that has passed through the capillaries of the myocardium. Their paths roughly parallel those of the coronary arteries. As figure 15.11*b* shows, these veins join the **coronary sinus,** which is an enlarged vein on the posterior surface of the heart in the atrioventricular sulcus. The coronary sinus empties into the right atrium. Figure 15.13 summarizes the path of blood that supplies the tissues of the heart.

 PRACTICE

11 Which structures make up the skeleton of the heart?

12 Review the path of blood through the heart.

13 How does blood composition differ in the right and left ventricle?

14 Which vessels supply blood to the myocardium?

15 How does blood return from the cardiac tissues to the right atrium?

Cardiac Cycle

Contraction of the heart chambers is called **systole** (sis′to-le), whereas relaxation of the heart chambers is called **diastole** (di-as′to-le). The heart chambers function in coordinated fashion. Their actions are regulated so that atria contract, called atrial systole, while ventricles relax, called ventricular diastole; then ventricles contract (ventricular systole) while atria relax (atrial diastole) (fig. 15.14). Then the atria and ventricles both relax for a brief interval. This series of events constitutes a complete heartbeat, or **cardiac cycle** (kar′de-ak si′kl). From Science to Technology 15.1 describes treatments for a failing heart.

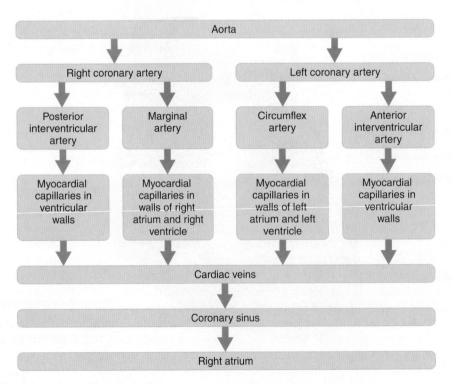

FIGURE 15.13 Path of blood through the coronary circulation.

Medical science offers several ways to aid or even replace a failing heart. In a *heart transplant,* the recipient's failing heart is removed, except for the posterior walls of the right and left atria and their connections to the venae cavae and pulmonary veins. The donor heart is similarly prepared and is attached to the atrial cuffs remaining in the recipient's thorax. Finally, the recipient's aorta and pulmonary arteries are connected to those of the donor heart (fig. 15A).

Donor hearts are scarce. A mechanical half-heart, called a *left ventricular assist device* (LVAD), can in some cases maintain cardiac function and counter deterioration long enough for a heart to become available. An LVAD allows a patient to resume some activities and to exercise, which can increase the chance of success of an eventual heart transplant. Some patients too ill to receive transplants are surviving with permanently implanted LVADs. Developed as a way to keep a patient alive for a short time until a donor heart became available, the LVAD is now considered a long-term therapy.

An implantable replacement heart is available for people who are not candidates for heart transplantation and have less than a month to live. The two-pound, titanium and plastic cardiac stand-in consists of an internal motor-driven hydraulic pump, a battery and electronics package, and an external battery pack. The electronics component manages the rate and force of the pump's actions, tailoring them to the patient's condition. Newer implantable replacement hearts are smaller and can provide up to five years of life.

Stem cell technology may allow researchers to patch failing hearts with new cardiac muscle. Human cardiac muscle tissue can be cultured from reprogrammed somatic cells (see From Science to Technology 3.1, p. 114) or from stem cells. In laboratory dishes, the tissue contracts, as cardiac muscle would in the body. These cells are combined with a synthetic, compatible biomaterial elastic enough to stretch as the cells divide and that will degrade in a con-

trolled way. The idea is that "stem cell heart patches" would consist of scaffolding made of such a biomaterial that would support the cells as they nestle into a damaged heart. As the cells contract, the synthetic portion degrades, leaving the pulsating patch. If the cells originate from the patient, the patch would not be rejected.

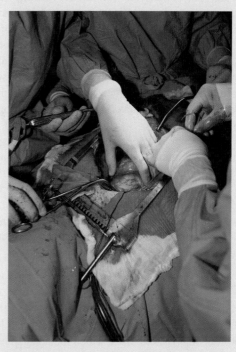

FIGURE 15A A heart transplant can save a life. A heart that might have died with its donor may provide a new lease on life for a recipient, thanks to our understanding of the cardiovascular and immune systems—and a well-trained medical team.

During a cardiac cycle, the pressure in the heart chambers rises and falls. These pressure changes open and close the valves, much like a door opened and closed by the wind. Pressure in the ventricles is low early in ventricular diastole, and the pressure difference between atria and ventricles opens the AV valves. The ventricles fill. About 70% of the returning blood enters the ventricles prior to contraction, and ventricular pressure gradually increases. During atrial systole, the remaining 30% of returning blood is pushed into the ventricles, and ventricular pressure increases. Then, as the ventricles contract, ventricular pressure rises sharply. As soon as the ventricular pressure exceeds atrial pressure, the AV valves close. At the same time, the papillary muscles contract. By pulling on the chordae tendineae, they prevent the cusps of the AV valves from bulging too far into the atria.

During ventricular systole, the AV valves remain closed. The atria are now relaxed, and pressure in the atria is low, even lower than venous pressure. As a result, blood flows into the atria from the large, attached veins. That is, as the ventricles are contracting, the atria are filling, already preparing for the next cardiac cycle (fig. 15.14).

As ventricular systole progresses, ventricular pressure continues to increase until it exceeds the pressure in the pulmonary trunk (right side) and aorta (left side). At this point, the pressure differences across the semilunar valves open the pulmonary and aortic valves, and blood is ejected from each valve's ventricle into these arteries.

As blood flows out of the ventricles, ventricular pressure begins to drop, and it falls even farther as the ventricles relax. When ventricular pressure is lower than the blood pressure in the aorta and pulmonary trunk, the semilunar valves close. The ventricles continue to relax, and as soon as ventricular pressure is less than atrial pressure, the AV valves open, and the ventricles begin to fill once more. Atria and ventricles relax for a brief interval.

Heart Sounds

A heartbeat heard through a stethoscope sounds like "lubb-dupp." These sounds are due to vibrations in the heart tissues produced as the blood flow is suddenly slowed with the contraction and relaxation of the heart chambers and with the closing of the valves.

The first part of a heart sound (*lubb*) occurs during ventricular systole, when the AV valves are closing. The second part (*dupp*) occurs during ventricular diastole, when the pulmonary and aortic valves are closing.

Heart sounds are of particular interest because they can indicate the condition of the heart valves. For example, inflammation of the endocardium (endocarditis) may erode the edges of the valvular cusps. As a result, the cusps may not close completely, and some blood may leak back through the valve, producing an abnormal sound called a *murmur*. The seriousness of a murmur depends on the degree of valvular damage. Many heart murmurs are harmless. Open-heart surgery may be needed to repair or replace severely damaged valves.

Using a stethoscope, it is possible to hear sounds associated with the aortic and pulmonary valves by listening from the second intercostal space on either side of the sternum. The *aortic sound* comes from the right, and the *pulmonic sound* from the left. The sound associated with the mitral valve can be heard from the fifth intercostal space at the nipple line on the left. The sound of the tricuspid valve can be heard at the fifth intercostal space just to the left of the sternum (fig. 15.15).

Cardiac Muscle

Recall that cardiac muscle cells function like those of skeletal muscles, but the cells connect in branching networks (chapter 9, pp. 308–309). The intercalated discs, which include gap junctions, join cardiac muscle cells, allowing action potentials to spread throughout a network of cells that contracts as a unit (chapter 5, p. 170).

A mass of merging cells that act as a unit is called a **functional syncytium** (funk'shun-al sin-sish'e-um). Two such structures are in the heart—in the atrial walls and in the ventricular walls. Portions of the heart's fibrous skeleton separate these masses of cardiac muscle, except for a small area in the right atrial floor. In this region, specialized conduction fibers connect the *atrial syncytium* and the *ventricular syncytium*.

 PRACTICE

16 Describe the pressure changes in the atria and ventricles during a cardiac cycle.

17 What causes heart sounds?

18 What is a functional syncytium?

19 Where are the functional syncytia of the heart?

Cardiac Conduction System

The heart is *autorhythmic,* able to initiate contraction itself without external nervous stimulation. Throughout the heart are clumps and strands of specialized cardiac muscle tissue whose cells contain only a few myofibrils. Instead of contracting, these areas initiate and distribute impulses (cardiac impulses) throughout the myocardium. They comprise the **cardiac conduction system,** which coordinates the events of the cardiac cycle.

A key portion of the conduction system is the **SA node (sinoatrial node** or **sinuatrial node)**, which is a small, elongated mass of specialized cardiac muscle tissue just beneath the epicardium. It is in the right atrium near the opening of the superior vena cava, and its cells are continuous with those of the atrial syncytium.

The cells of the SA node reach threshold spontaneously. Recall from chapter 10 (p. 375) that an action potential in a neuron is triggered by a depolarizing input from presynaptic neurons. In contrast, an SA node reaches threshold and triggers an action potential on its own. A number of changing conditions appear to play a role in stimulating the SA node to reach threshold. These include a progressive increase in permeability to calcium ions and sodium ions and decreasing permeability to potassium ions.

SA node activity is rhythmic. The SA node initiates one impulse after another, more than 80 times a minute in an adult (resting heart rate is usually closer to 70 beats per minute due to inhibition by the parasympathetic nervous system). The SA node is also called the **pacemaker** because it initiates the heart's

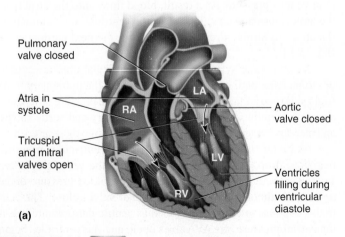

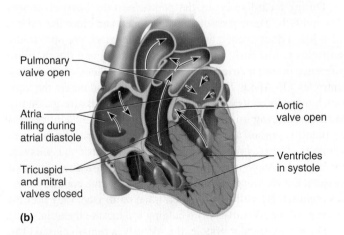

FIGURE 15.14 AP|R A cardiac cycle. The atria (a) empty during atrial systole and (b) fill with blood during atrial diastole.

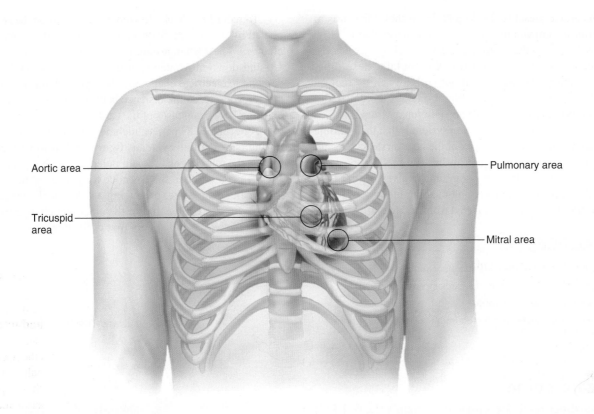

FIGURE 15.15 Thoracic regions where the sounds of each heart valve are most easily heard. Note that where the sound is heard may differ slightly from the anatomical location of the structure.

Aortic area

Tricuspid area

Pulmonary area

Mitral area

rhythmic contractions. From the SA node, bundles of atrial muscle, called *internodal atrial muscle,* preferentially conduct impulses to more distant regions of the atria. Then, because gap junctions connect cardiac muscle cells, the resulting impulse spreads into the surrounding atrial myocardium and stimulates the muscle cells to contract.

The right and left atria contract almost simultaneously. The cardiac impulse does not pass directly into the ventricular syncytium, which is separated from the atrial syncytium by the fibrous skeleton of the heart. Instead the cardiac impulse passes along specialized muscle cells, called junctional fibers, of the cardiac conduction system that lead to a mass of specialized cardiac muscle tissue called the **AV node** (atrioventricular node). This node is in the inferior part of the interatrial septum and just beneath the endocardium. It provides the only normal conduction pathway between the atrial and ventricular syncytia, because the fibrous skeleton does not conduct the impulse.

The junctional fibers that conduct the cardiac impulse into the AV node have small diameters. Small fibers conduct impulses slowly, and thus they delay conduction of the impulse. Therefore, the impulse is delayed further as it moves through the AV node. This allows time for the atria to contract completely so they empty all their blood into the ventricles prior to ventricular systole. Once the cardiac impulse reaches the distal side of the AV node, it passes into a group of large conduction fibers that make up the **AV bundle** (atrioventricular bundle or bundle of His), and the impulse moves rapidly through them. The AV bundle enters the

upper part of the interventricular septum and divides into right and left bundle branches that lie just beneath the endocardium. About halfway down the septum, the branches give rise to enlarged **Purkinje fibers** (pur-kin′je fi′berz). These larger fibers conduct the impulse to distant regions of the ventricular myocardium much faster than cell-to-cell conduction could. As a result, the massive ventricular myocardium contracts as a functioning unit.

The base of the aorta, which includes the aortic valve, protrudes somewhat into the interatrial septum close to the AV bundle. Consequently, inflammatory conditions, such as bacterial endocarditis affecting the aortic valve (aortic valvulitis), may also affect the AV bundle.

If a portion of the bundle is damaged, it may no longer conduct impulses normally. As a result, cardiac impulses may reach the two ventricles at different times so that they fail to contract together. This condition is called a *bundle branch block.*

The Purkinje fibers spread from the interventricular septum into the papillary muscles, which project inward from the ventricular walls, and then continue downward to the apex of the heart. There they curve around the tips of the ventricles and pass upward over the lateral walls of these chambers. Along the way, the Purkinje fibers give off many small branches, which become continuous with cardiac muscle cells. **Figure 15.16** shows the parts of the conduction system and figure 15.17 summarizes them.

The muscle cells in the ventricular walls form irregular whorls. When impulses on the Purkinje fibers stimulate these muscle cells, the ventricular walls contract with a twisting motion (fig. 15.18). This action squeezes blood out of the ventricular chambers and forces it into the aorta and pulmonary trunk.

Another property of the conduction system is that the Purkinje fibers conduct the impulse to the apex of the heart first. As a result, contraction begins at the apex and pushes the blood superiorly toward the aortic and pulmonary semilunar valves, rather than originating the impulse superiorly and pushing blood toward the apex, as it would if the impulse traveled from cell to cell.

 PRACTICE

20 What is the function of the cardiac conduction system?

21 What types of tissues make up the cardiac conduction system?

22 How is a cardiac impulse initiated?

23 How is a cardiac impulse conducted from the right atrium to the other heart chambers?

Electrocardiogram

An **electrocardiogram** (e-lek″tro-kar′de-o-gram″) (ECG, EKG) is a recording of the electrical changes in the myocardium during a cardiac cycle. (This pattern occurs as action potentials stimulate cardiac muscle cells to contract, but it is not the same as individual action potentials.) Because body fluids can conduct electrical currents, such changes can be detected on the surface of the body.

To record an ECG, electrodes are placed on the skin and connected by wires to an instrument that responds to small electrical changes. These changes are recorded on a moving strip of paper. Up-and-down movements, or deflections from the baseline, correspond to electrical changes in the myocardium. The paper moves at a known rate, so the distance between deflections indicates time elapsing between phases of the cardiac cycle.

A normal ECG pattern includes several deflections, or *waves*, during each cardiac cycle, as figure 15.19*a* illustrates. Between cycles, the muscle cells remain polarized, with no detectable electrical changes. When the SA node triggers a cardiac impulse, the atrial cells depolarize, producing an electrical change. A deflection occurs, and at the end of the electrical change the recording returns to the baseline. This first deflection produces a *P wave*,

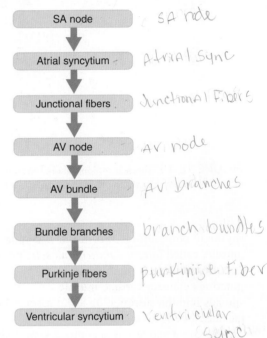

FIGURE 15.17 Path of a cardiac impulse.

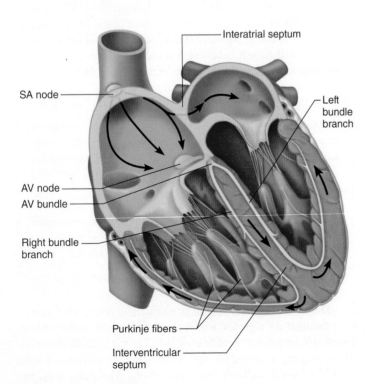

FIGURE 15.16 AP|R Components of the cardiac conduction system.

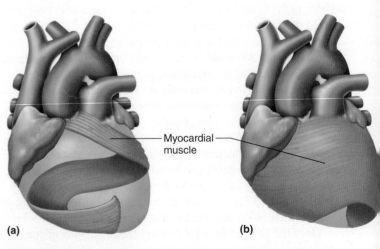

FIGURE 15.18 The muscle cells within the ventricular walls interconnect to form whorled networks. The networks of groups (*a*) and (*b*) surround both ventricles in these anterior views of the heart.

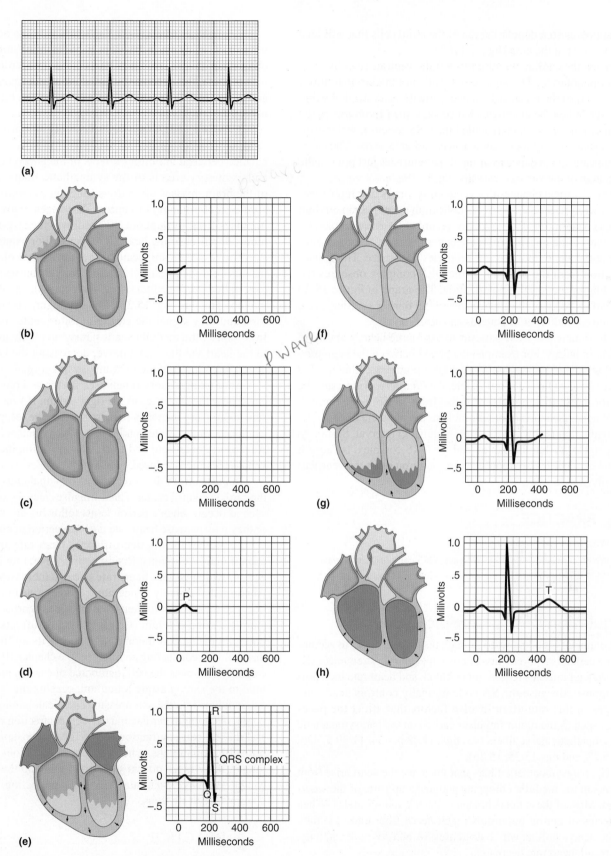

FIGURE 15.19 ECG pattern. (*a*) A normal ECG. In this set of drawings (*b–h*), the yellow areas of the hearts indicate where depolarization is occurring, and the green areas indicate where tissues are repolarizing; the portion of the ECG pattern produced at each step is shown by the continuation of the line on the graph paper.

Q *Which two electrical events occur during the QRS complex?*

Answer can be found in Appendix G.

corresponding to a depolarization of the atrial cells that will lead to contraction of the atria (fig. 15.19*b–d*).

When the cardiac impulse reaches the ventricular cells, they rapidly depolarize. The ventricular walls are thicker than those of the atria, so the electrical change is greater, and the deflection is greater. When the electrical change ends, the pen returns to the baseline. This leaves a mark called the *QRS complex,* which usually consists of a *Q wave,* an *R wave,* and an *S wave.* The complex appears due to depolarization of the ventricles just prior to the contraction of the ventricular walls (fig. 15.19*e,f*).

The electrical changes that accompany ventricular repolarization slowly produce a *T wave* as deflection occurs again, ending the ECG pattern (fig. 15.19*g,h*). The record of the atrial repolarization seems to be missing from the pattern because the atria repolarize at the same time that the ventricles depolarize. The greater deflection from the baseline of the QRS complex obscures the recording of the atrial repolarization. The graph in figure 15.20 summarizes some of the changes that occur during a cardiac cycle with corresponding ECG patterns and heart sounds.

Physicians use ECG patterns to assess the heart's ability to conduct impulses. For example, the period between the beginning of a P wave and the beginning of a QRS complex called the *PQ interval* (or if a Q wave is not visible, the *PR interval*) indicates the time for the cardiac impulse to travel from the SA node through the AV node. Ischemia or other problems affecting the fibers of the AV conduction pathways can prolong this PQ interval. Similarly, injury to the AV bundle can extend the QRS complex, because it may take longer for an impulse to spread throughout the ventricular walls (fig. 15.21).

 PRACTICE

24 What is an electrocardiogram?

25 Which cardiac events do the P wave, QRS complex, and T wave represent?

Regulation of the Cardiac Cycle

The volume of blood pumped through the body changes to accommodate cellular requirements. For example, during strenuous exercise, skeletal muscles require more blood, and heart rate increases in response. Because the SA node normally controls heart rate, changes in this rate often involve factors that affect the pacemaker, such as the motor impulses carried on the parasympathetic and sympathetic nerve fibers (see figs. 11.38 p. 434, 11.39 p. 435, fig. 15.22, and figs.15.35, 15.36).

The parasympathetic fibers that innervate the heart arise from neurons in the medulla oblongata and make up parts of the *vagus nerves.* Most of these fibers branch to the SA and AV nodes. When the series of action potentials reach nerve fiber endings, they secrete acetylcholine, which decreases SA and AV nodal activity. As a result, heart rate decreases.

The vagus nerves continually conduct impulses to the SA and AV nodes, "braking" heart action. Consequently, parasympathetic activity can change heart rate in either direction. An increase in the impulses slows the heart rate, and a decrease in the impulses releases the parasympathetic "brake" and increases heart rate.

Sympathetic fibers reach the heart by means of the *accelerator nerves,* whose branches join the SA and AV nodes as well as other areas of the atrial and ventricular myocardium. The endings of these fibers secrete norepinephrine in response to series of action potentials. Norepinephrine increases the rate and force of myocardial contractions.

Reflexes called *baroreceptor reflexes* arising from the *cardiac control center* of the medulla oblongata maintain balance between inhibitory effects of the parasympathetic fibers and excitatory effects of the sympathetic fibers. In this region of the brain, masses of neurons function as *cardioinhibitor* and *cardioaccelerator reflex centers.* These centers receive sensory impulses from throughout the cardiovascular system and relay motor impulses to the heart in response. For example, receptors sensitive to stretch are located in certain regions of the aorta (aortic arch) and in the carotid arteries (carotid sinuses). These receptors, called *baroreceptors* (pressoreceptors), can detect changes in blood pressure (fig. 15.22). Rising pressure stretches the receptors, and they signal the cardioinhibitor center in the medulla. In response, the medulla sends parasympathetic motor impulses to the heart via the vagus nerve, decreasing the heart rate. This action helps lower blood pressure toward normal.

Another regulatory reflex uses stretch receptors in the venae cavae near the entrances to the right atrium. If venous blood pressure abnormally increases in these vessels, the receptors signal the cardioaccelerator center, and sympathetic impulses reach the heart. As a result, heart rate and force of contraction increase, and the venous pressure is reduced.

Impulses from the cerebrum or hypothalamus also influence the cardiac control center. These impulses may decrease heart rate, such as occurs when a person faints following an emotional upset, or they may increase heart rate during a period of anxiety.

Two other factors that influence heart rate are temperature change and certain ions. Rising body temperature increases heart action, which is why heart rate usually increases during fever. Abnormally low body temperature decreases heart action.

The most important ions that influence heart action are potassium (K^+) and calcium (Ca^{+2}). Potassium affects the electrical potential of the cell membrane, altering its ability to reach the threshold for conducting an impulse (see chapter 10, p. 375). Some calcium ions cross the cell membrane of cardiac muscle cells and bind to the sarcoplasmic reticulum, causing the release of many calcium ions into the sarcoplasm. These calcium ions bind to troponin, resulting in cardiac muscle cell contraction much like skeletal muscle cell contraction. Although homeostatic mechanisms normally maintain the concentrations of these ions within narrow ranges, these mechanisms sometimes fail, and the consequences can be serious or even fatal. Clinical Application 15.1 examines abnormal heart rhythms.

 PRACTICE

26 Which nerves supply parasympathetic fibers to the heart? Which nerves supply sympathetic fibers?

27 How do parasympathetic and sympathetic impulses help control heart rate?

28 How do changes in body temperature affect heart rate?

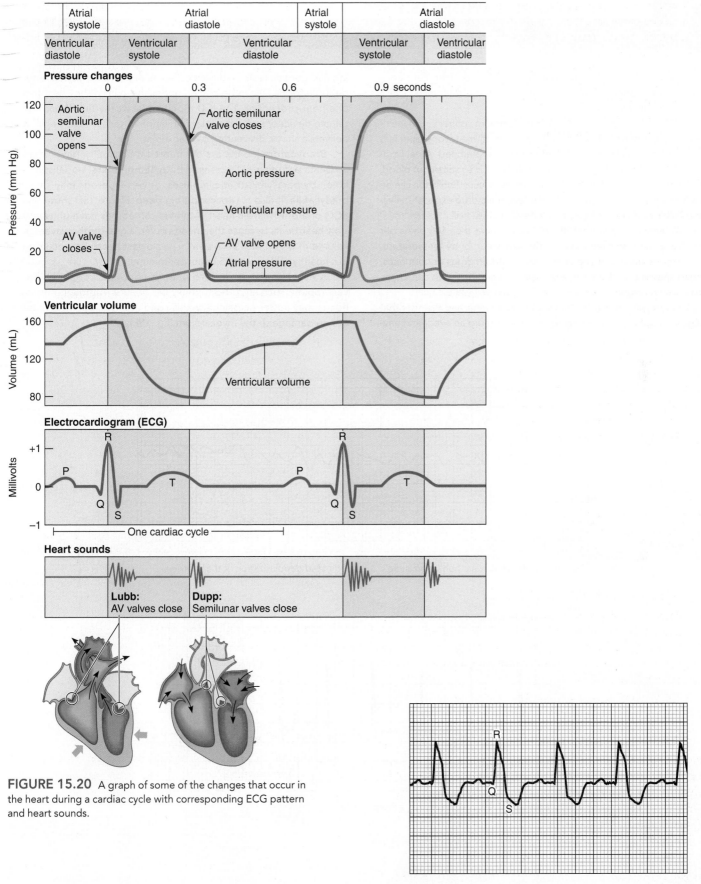

| Atrial systole | Atrial diastole | | Atrial systole | Atrial diastole | |
| Ventricular diastole | Ventricular systole | Ventricular diastole | | Ventricular systole | Ventricular diastole |

Pressure changes

0   0.3   0.6   0.9 seconds

Aortic semilunar valve opens

Aortic semilunar valve closes

Aortic pressure

Ventricular pressure

AV valve closes

AV valve opens

Atrial pressure

Ventricular volume

Ventricular volume

Electrocardiogram (ECG)

R   R

P   T   P   T

Q   Q
S   S

⊢——— One cardiac cycle ———⊣

Heart sounds

Lubb: AV valves close   **Dupp:** Semilunar valves close

FIGURE 15.20 A graph of some of the changes that occur in the heart during a cardiac cycle with corresponding ECG pattern and heart sounds.

R

Q
S

FIGURE 15.21 A prolonged QRS complex may result from damage to the AV bundle fibers.

Each year, thousands of people die from fast or irregular heartbeats. These altered heart rhythms are called *arrhythmias*. There are several types.

In *fibrillation*, small areas of the myocardium contract in an uncoordinated, chaotic fashion (fig. 15B). As a result, the myocardium fails to contract as a whole, and blood is no longer pumped. Atrial fibrillation is not life threatening, because the ventricles still pump blood, but ventricular fibrillation is often deadly. Ventricular fibrillation can be caused by an obstructed coronary artery, toxic drug exposure, electric shock, or traumatic injury to the heart or chest wall.

An abnormally fast heartbeat, usually more than 100 beats per minute at rest, is called *tachycardia*. Increase in body temperature, nodal stimulation by sympathetic fibers, certain drugs or hormones, heart disease, excitement, exercise, anemia, or shock can cause tachycardia. Figure 15C shows the ECG of a tachycardic heart.

For people who know that they have an inherited disorder that causes sudden cardiac arrest (by having suffered an event and having had genetic tests), a device called an implantable cardioverter defibrillator (ICD) can be placed under the skin of the chest in a one-hour procedure. The ICD monitors heart rhythm. When the telltale deviations of ventricular fibrillation or tachycardia begin, it delivers a shock, preventing cardiac arrest.

Bradycardia means a slow heart rate, usually fewer than 60 beats per minute. Decreased body temperature, nodal stimulation by parasympathetic impulses, or certain drugs may cause bradycardia. It also may occur during sleep. Figure 15D shows the ECG of a bradycardic heart. Athletes sometimes have unusually slow heartbeats because their hearts pump a greater-than-average volume of blood with each beat. The slowest heartbeat recorded in a healthy athlete was 25 beats per minute!

A heart chamber *flutters* when it contracts regularly, but very rapidly, such as 250–350 times per minute. Although normal hearts may flutter occasionally, this condition is more likely to be due to damage to the myocardium (fig. 15E).

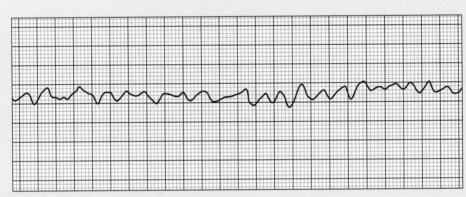

FIGURE 15B Ventricular fibrillation is rapid, uncoordinated depolarization of the ventricles.

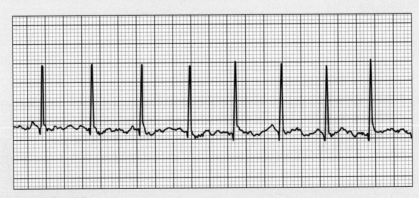

FIGURE 15C Tachycardia is a rapid heartbeat.

A *premature beat* occurs before it is expected in a normal series of cardiac cycles. Cardiac impulses originating from unusual (ectopic) regions of the heart probably cause a premature beat. That is, the impulse originates from a site other than the SA node. Cardiac impulses may arise from ischemic tissues or from muscle cells irritated by disease or drugs.

Any interference or block in cardiac impulse conduction may cause arrhythmia, the type varying with the location and extent of the block. Such arrhythmias may arise from ectopic pacemakers (pacemakers outside the SA node).

The SA node usually initiates 70–80 heartbeats per minute, called a sinus rhythm. If the SA node is damaged, impulses originating in the AV node may take over and travel upward into the atrial myocardium and downward into the ventricular walls, stimulating them to contract. Under the influence of the AV node acting as a *secondary pacemaker*, the heart may continue to pump blood, but at a rate of 40–60 beats per minute, called a nodal rhythm. Similarly, the Purkinje fibers can initiate cardiac impulses, contracting the heart 15–40 times per minute.

An *artificial pacemaker* can treat a disorder of the cardiac conduction system. This device includes an electrical pulse generator and a lead wire that communicates with a portion of the myocardium. The pulse generator contains a permanent battery that provides energy and a microprocessor that can sense the cardiac rhythm and signal the heart to alter its contraction rate.

An artificial pacemaker is surgically implanted beneath the patient's skin in the shoulder. A programmer adjusts its functions from the outside. The first pacemakers, made in 1958, were crude. Today, thanks to telecommunications advances, a physician can check a patient's pacemaker over the phone. A device called a pacemaker-cardioverter-defibrillator attempts to correct ventricular fibrillation should it occur.

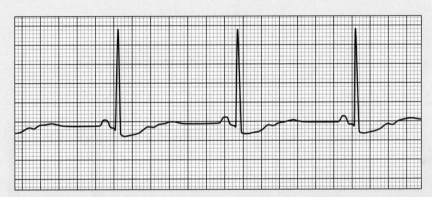

FIGURE 15D Bradycardia is a slow heartbeat.

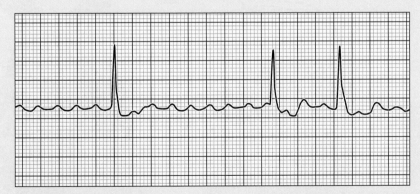

FIGURE 15E Atrial flutter is an abnormally rapid rate of atrial depolarization.

15.3 | Blood Vessels

The blood vessels are organs of the cardiovascular system. They form a closed circuit of tubes that carries blood from the heart to the body cells and back again. These vessels include arteries, arterioles, capillaries, venules, and veins. The arteries and arterioles conduct blood away from the ventricles of the heart and lead to the capillaries, where substances are exchanged between blood and the body cells. Venules and veins return blood from the capillaries to the atria. From Science to Technology 15.2 describes angiogenesis, the formation of new blood vessels in the body.

Researchers create replacement blood vessels from cells and their products plus synthetic materials. One such blood vessel consists of smooth muscle cells seeded onto tubes of a biodegradable polymer. The cells secrete collagen and extracellular matrix, which replace the polymer, and then detergent is applied to remove the cells. The tubes can be stored for long periods of time. Another approach uses rubber tubing and a nutrient solution that prompts the cells to produce extra elastin, which makes the replacement vessels more flexible. Because construction of tissues and organs in the laboratory (part of regenerative medicine) requires a blood supply, tissue engineers are creating cell-lined tubules using 3D printing to function as part of a semi-synthetic, *ex vivo* ("outside the body") cardiovascular system.

FIGURE 15.22 **AP|R** Baroreceptor reflex. (*a*) Schematic of a general reflex arc. Note the similarity to figure 1.6 on page 16. (*b*) Autonomic impulses alter the activities of the SA and AV nodes.

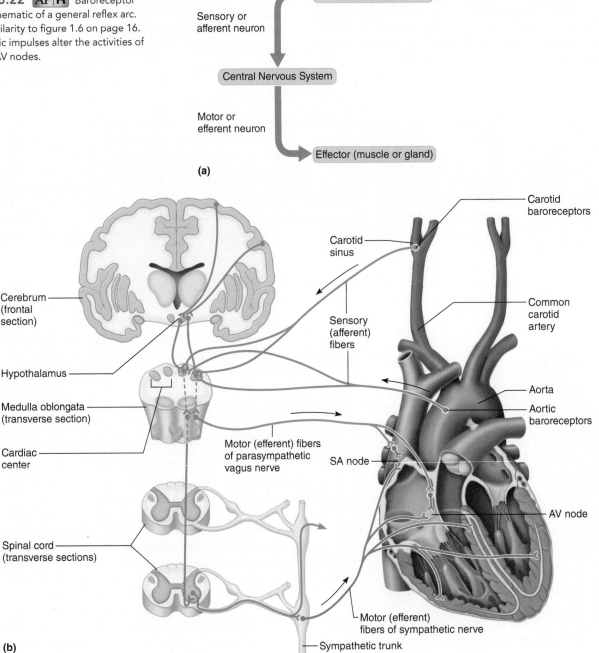

Angiogenesis is the formation of new blood vessels. Under the influence of vascular endothelial growth factor (VEGF), endothelial cells divide and assemble into the tubules that form capillaries as well as the innermost linings of larger blood vessels. In normal development, angiogenesis is crucial to build a blood supply to serve a growing body. New blood vessels deliver nutrients, hormones, and growth factors to tissues and remove wastes. Angiogenesis is also essential for healing. After a heart attack, for example, new vessels form in the remaining healthy cardiac muscle.

As is the case for most biological processes, angiogenesis must be highly controlled. Drugs that increase or decrease the activity of VEGF are used to target several common diseases caused by excess, deficient, or inappropriate angiogenesis. Two specific applications are healing hearts and removing extra capillaries in tumors and in eyes.

Promoting Angiogenesis

A clot blocks a coronary artery. Within seconds, the localized lack of oxygen stimulates muscle cells to activate hypoxia-inducible factor (HIF-1). This protein turns on several genes whose products restore homeostasis by stimulating glycolysis (anaerobic respiration) and by signaling the kidneys to produce erythropoietin, which boosts the red blood cell supply and turns on production of VEGF. Capillaries extend and restore some blood flow to the blocked cardiac muscle. Fibroblast growth factor also assists in angiogenesis.

When natural angiogenesis isn't sufficient, part of the heart dies. Coronary bypass surgery and angioplasty are treatments that restore blood flow by circumventing a blockage or opening up an artery, respectively. However, for patients who cannot undergo these procedures or whose blockages are in vessels too narrow or difficult to reach, targeting angiogenesis may help to save starved heart parts. One approach is to package VEGF in time-release capsules implanted near small vessels while large blood vessels are being surgically bypassed. Another strategy is gene therapy, which delivers the genes that encode VEGF to oxygen-starved areas of the heart.

Preventing Angiogenesis

Once a tumor reaches the size of a pinhead, it secretes VEGF, which stimulates nearby capillaries to branch and extend toward it. At the same time, endothelial cells that are part of the tumor assemble into sheets, roll into tubules, and snake out of the tumor as new capillaries. Other cancer cells wrap around the capillaries, spreading out on this scaffolding into nearby tissues. Some cancer cells enter blood vessels and travel to other parts of the body. For a time, maybe even years, these secondary tumors stay small, adhering to the outsides of the blood vessels that delivered them.

From the observation by many surgeons that when a primary tumor is removed, secondary tumors grow, Harvard researcher Judah Folkman hypothesized that the primary tumor secretes antiangiogenesis factors that keep the secondary tumors small. Once factors that promote or block angiogenesis were discovered in the 1980s, researchers began to study the antiangiogenesis factors that keep secondary tumors small, to develop them as cancer treatments. The first antiangiogenesis drug to treat cancer became available in 2004, for colorectal cancer that has spread to other organs. Today antiangiogenesis drugs are also used to treat age-related macular degeneration, in which extra capillaries extend into the retina and block central vision.

Arteries and Arterioles

Arteries (ar′-te-rēz) are strong, elastic vessels adapted for transporting blood away from the heart under relatively high pressure. These vessels subdivide into progressively thinner tubes and eventually give rise to the finer, branched **arterioles** (ar-te′re-olz).

The wall of an artery consists of three distinct layers, or *tunics,* shown in figure 15.23*a.* The innermost tunic, tunica interna (intima), is composed of a layer of simple squamous epithelium, called *endothelium,* that rests on a connective tissue membrane rich in elastic and collagen fibers.

The endothelial lining of an artery provides a smooth surface that allows blood cells and platelets to flow through without being damaged. Additionally, endothelium helps prevent blood clotting by secreting biochemicals that inhibit platelet aggregation (see chapter 14, p. 547). Endothelium also may help regulate local blood flow by secreting substances that dilate or constrict blood vessels. For example, endothelium releases the gas nitric oxide, which relaxes the smooth muscle of the vessel.

The middle layer, tunica media, makes up the bulk of the arterial wall. It includes smooth muscle cells, which encircle the tube, and a thick layer of elastic connective tissue. The connective tissue gives the vessel a tough elasticity that enables it to withstand the force of blood pressure and, at the same time, to stretch and accommodate the sudden increase in blood volume that accompanies ventricular contraction.

The outer layer, tunica externa (adventitia), is relatively thin and chiefly consists of connective tissue with irregular elastic and collagen fibers. This layer attaches the artery to the surrounding tissues. It also contains tiny vessels (vasa vasorum) that give rise to capillaries and provide blood to the more external cells of the artery wall.

The sympathetic branches of the autonomic nervous system innervate smooth muscle in artery and arteriole walls. *Vasomotor fibers* stimulate the smooth muscle cells to contract, reducing the diameter of the vessel. This is called **vasoconstriction** (vas″o-kon-strik′-shun). If vasomotor impulses are inhibited, the muscle cells relax, and the diameter of the vessel increases. This is called **vasodilation** (vas″o-di-la′shun). Changes in the diameters of arteries and arterioles greatly influence blood flow and blood pressure.

The walls of the larger arterioles have three layers similar to those of arteries (fig. 15.23*c*), but the middle and outer layers thin as

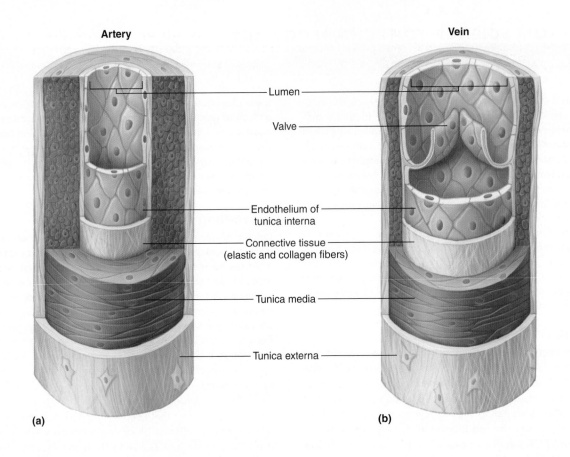

Artery

Vein

Lumen

Valve

Endothelium of
tunica interna

Connective tissue
(elastic and collagen fibers)

Tunica media

Tunica externa

(a)

(b)

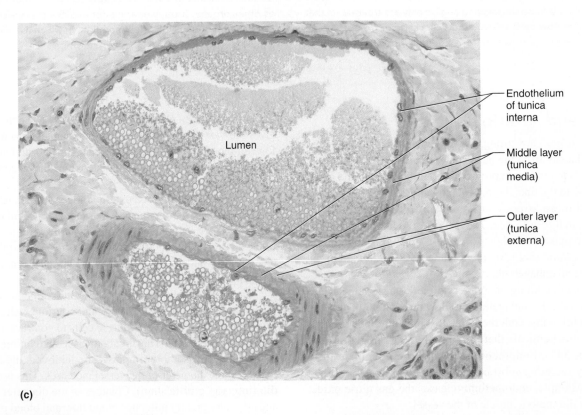

Lumen

Endothelium
of tunica
interna

Middle layer
(tunica
media)

Outer layer
(tunica
externa)

(c)

FIGURE 15.23 AP|R Blood vessels. (a) The wall of an artery. (b) The wall of a vein. (c) Note the structural differences in these cross sections of an arteriole (bottom) and a venule (top) (200×).

the arterioles approach the capillaries. The wall of a very small arteriole consists only of an endothelial lining and some smooth muscle cells, surrounded by a small amount of connective tissue (fig. 15.24). Arterioles, which are microscopic continuations of arteries, give off branches called *metarterioles* that, in turn, join capillaries.

The arteriole and metarteriole walls are adapted for vasoconstriction and vasodilation in that their smooth muscle cells respond to impulses from the autonomic nervous system by contracting or relaxing. In this way, these vessels help control the flow of blood into the capillaries.

 A GLIMPSE AHEAD | To Chapter 22

Local vasodilation, increasing blood flow to the arteries associated with the genitalia, stiffens erectile tissue in both sexes.

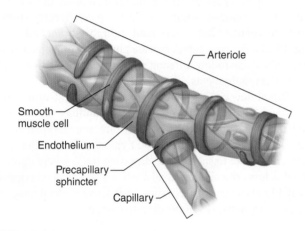

FIGURE 15.24 The smallest arterioles have only a few smooth muscle fibers in their walls. Capillaries lack these fibers.

In some places metarterioles connect directly to venules, and blood entering them can bypass the capillaries. These connections between arteriole and venous pathways, shown in **figure 15.25**, are called *arteriovenous shunts*.

 PRACTICE

29 Describe the wall of an artery.
30 What is the function of the smooth muscle in the arterial wall?
31 How is the structure of an arteriole different from that of an artery?

Capillaries

Capillaries (kap′ĭ-ler″ēz), the smallest-diameter blood vessels, connect the smallest arterioles and the smallest venules. Capillaries are extensions of the inner linings of arterioles in that their walls are endothelium—a single layer of squamous epithelial cells (fig. 15.26a). These thin walls form the semipermeable layer through which substances in the blood are exchanged for substances in the tissue fluid surrounding body cells.

Capillary Permeability

The openings or intercellular passageways in the capillary walls are thin slits where endothelial cells overlap (fig. 15.26b,c). The sizes of these openings, and consequently the permeability of the capillary wall, vary from tissue to tissue. For example, the openings are relatively smaller in the capillaries of smooth, skeletal, and cardiac muscle than they are in capillaries associated with endocrine glands, the kidneys, and the lining of the small intestine.

Capillaries with the largest openings include those of the liver, spleen, and red bone marrow. These capillaries are discontinuous, and the spaces between the individual cells of their walls appear as

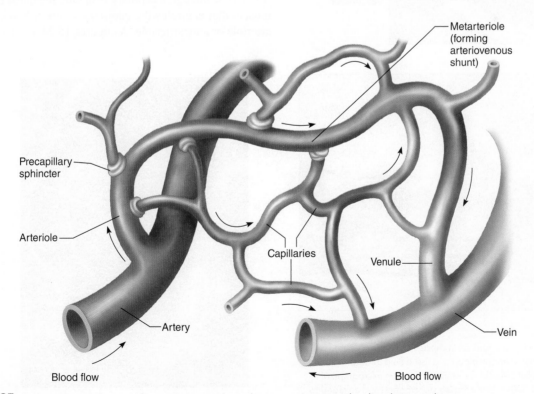

FIGURE 15.25 Some metarterioles provide arteriovenous shunts by connecting arterioles directly to venules.

small cavities (sinusoids) in the organ. Discontinuous capillaries allow large proteins and even intact cells to pass through as they enter or leave the circulation. From Science to Technology 5.1 (p. 150) discusses the blood-brain barrier, which is formed by the protective tight capillaries in the brain. The blood-brain barrier is not present in the pituitary and pineal glands and parts of the hypothalamus.

Capillary Arrangement

The higher a tissue's rate of metabolism, the denser its capillary networks. Muscle and nerve tissues, which use abundant oxygen and nutrients, are richly supplied with capillaries. Tissues with slower metabolic rates have fewer capillaries, such as cartilage, or lack them entirely, such as in the cornea.

> If the capillaries of an adult were unwound and lined up end to end, they would be between 25,000 and 60,000 miles long.

The spatial patterns of capillaries also differ in various body parts. For example, some capillaries pass directly from arterioles to venules, but others lead to highly branched networks (fig. 15.27). Because capillaries are physically organized in diverse ways, blood can follow different pathways through a tissue, attuned to cellular requirements.

Blood flow can vary among tissues as well. During exercise, for example, blood is directed into the capillary networks of the skeletal muscles, where the cells require more oxygen and nutrients. At the same time, the blood bypasses some of the capillary networks in the tissues of the digestive tract, where demand for blood is less critical. Conversely, when a person is relaxing after a meal, blood can be shunted from the inactive skeletal muscles into the capillary networks of the digestive organs.

Regulation of Capillary Blood Flow

Blood flow through a capillary is mainly regulated by the smooth muscle that encircles the capillary where it branches off of an arteriole or a metarteriole. As figures 15.24 and 15.25 show, this

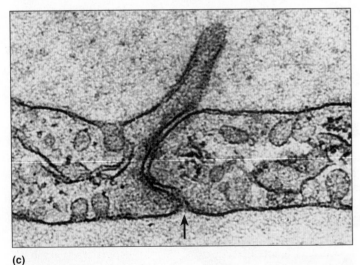

FIGURE 15.26 Capillary structure. (*a*) In capillaries, substances are exchanged by diffusion between the blood and tissue fluid through openings (slits) separating endothelial cells. (*b*) Transmission electron micrograph of a capillary cross section (11,500×). (*c*) Micrograph (*b*) cropped and enlarged to show the narrow slitlike opening at the junction of the cells (62,500×).

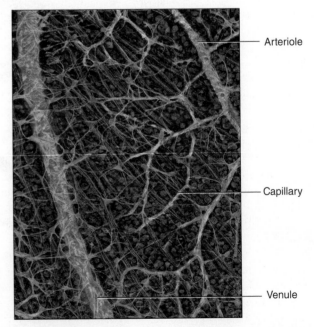

FIGURE 15.27 Falsely colored light micrograph of a capillary network.

band of smooth muscle forms a *precapillary sphincter,* which may close a capillary by contracting or open it by relaxing. A precapillary sphincter responds to the demands of the cells the capillary supplies. When these cells have low concentrations of oxygen and nutrients, the precapillary sphincter relaxes, and blood flow increases; when cellular requirements have been met, the precapillary sphincter may contract again.

 PRACTICE

32 Describe a capillary wall.

33 What is the function of a capillary?

34 What controls blood flow into capillaries?

Exchanges in the Capillaries

The vital function of exchanging gases, nutrients, and metabolic by-products between the blood and the tissue fluid surrounding the cells takes place in the capillaries. The biochemicals exchanged move through the capillary walls by diffusion, filtration, and osmosis.

 RECONNECT

To Chapter 3, Movements Into and Out of the Cell, pages 98–102.

Diffusion is the most important means of transfer between the blood and tissue fluid. Because blood entering systemic capillaries carries high concentrations of oxygen and nutrients, these substances diffuse through the capillary walls and enter the tissue fluid. Conversely, the concentrations of carbon dioxide and other wastes are generally greater in the tissues, and such wastes tend to diffuse into the capillary blood.

The paths these substances follow depend primarily on their solubilities in lipids. Substances that are soluble in lipid, such as oxygen, carbon dioxide, and fatty acids, can diffuse through most areas of the cell membranes that make up the capillary wall because the membranes are largely lipid. Lipid-insoluble substances, such as water, sodium ions, and chloride ions, diffuse through channels in the cell membranes and through the slitlike openings between the endothelial cells that form the capillary wall (fig. 15.28). Plasma proteins generally remain in the blood because they are not soluble in the lipid of the endothelial cell membranes, and they are too large to diffuse through the membrane channels or openings between endothelial cells of most capillaries.

In *filtration, hydrostatic pressure* forces molecules through a membrane. In the capillaries, the blood pressure generated when ventricle walls contract provides the force for filtration.

Blood pressure also moves blood through the arteries and arterioles. This pressure decreases as the distance from the heart increases because of friction (peripheral resistance) between the blood and the vessel walls. For this reason, blood pressure is greater in the arteries than in the arterioles and greater in the arterioles than in the capillaries. It is similarly greater at the arteriolar end of a capillary than at the venular end. The flow is always from high pressure to low pressure.

The walls of arteries and arterioles are too thick to allow blood components to pass through. However, the hydrostatic pressure of the blood pushes water and other small molecules through capillary walls by filtration primarily at the arteriolar ends of capillaries, whereas diffusion takes place along their entire lengths.

The presence of an impermeant solute on one side of a cell membrane creates an osmotic pressure. Plasma proteins trapped in the capillaries create an osmotic pressure that draws water into the capillaries. The term *colloid osmotic pressure* describes this osmotic effect due to the plasma proteins, which include the albumins.

The effect of the plasma colloid osmotic pressure, which favors reabsorption, opposes the action of capillary blood pressure, which favors filtration. Because at the arteriolar end of capillaries the blood pressure is higher (35 mm Hg outward) than the colloid osmotic pressure (24 mm Hg inward), filtration predominates here. At the venular end, the colloid osmotic pressure is essentially unchanged (24 mm Hg inward), but the blood pressure has decreased due to resistance through the capillary (16 mm Hg outward). Thus, at the venular end, fluid reabsorption predominates (fig. 15.28). (The interstitial fluid also has hydrostatic pressure and osmotic pressure, but the values are low and cancel each other; as such, they are omitted from this discussion.)

Normally, more fluid leaves the capillaries than returns to them because the net inward pressure at the venular ends of the capillaries is less than the net outward pressure at the arteriolar ends of the capillaries. Closed-ended vessels called lymphatic capillaries collect the excess fluid and return it through lymphatic vessels to the venous circulation. Chapter 16 (pp. 617–618) discusses this mechanism.

Unusual events may increase blood flow to capillaries, causing excess fluid to enter the spaces between tissue cells (interstitial spaces). This may occur in response to certain chemicals, such as *histamine,* that vasodilate the metarterioles and increase capillary permeability. Enough fluid may leak out of the capillaries to overwhelm lymphatic drainage. Affected tissues become swollen (edematous) and painful.

 PRACTICE

35 Which forces affect the exchange of substances between blood and tissue fluid?

36 Why is the fluid movement out of a capillary greater at its arteriolar end than at its venular end?

37 More fluid leaves the capillary than returns to it, so how is the remainder returned to the vascular system?

If the right ventricle of the heart is unable to pump blood out as rapidly as it enters, other parts of the body may develop edema because the blood backs up into the veins, venules, and capillaries, increasing blood pressure in these vessels. As a result of this increased *back pressure,* osmotic pressure of the blood in the venular ends of the capillaries is less effective at balancing filtration, and the tissues swell. This is true particularly in the lower extremities if the person is upright, or in the back if the person is supine. In the terminal stages of heart failure, edema is widespread, and fluid accumulates in the peritoneal cavity of the abdomen. This painful condition is called *ascites.*

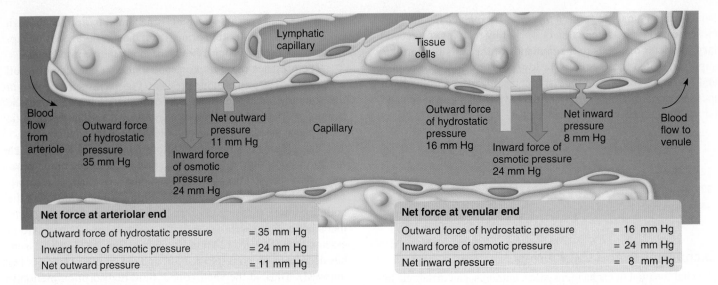

FIGURE 15.28 **AP|R** Water and other substances leave capillaries because of a net outward pressure at the capillaries' arteriolar ends. Water enters at the capillaries' venular ends because of a net inward pressure. Substances move in and out along the length of the capillaries according to their respective concentration gradients.

Q *Which substances do not leave the blood at the arteriolar end of the capillary and whose presence at the venular end of the capillary draws water by osmosis back into the capillary?*

Answer can be found in Appendix G.

Venules and Veins

Venules (ven'ūlz) are the microscopic vessels that continue from the capillaries and merge to form **veins** (vānz). The veins, which transport blood back to the atria, follow pathways that roughly parallel those of the arteries.

The walls of veins are similar to those of arteries in that they are composed of three distinct layers. However, the tunica media of the venous wall is less developed compared to that of the arterial wall. Consequently, veins have thinner walls that have less smooth muscle and less elastic connective tissue than those of comparable arteries, but their lumens have a greater diameter (see fig. 15.23*b*).

Many veins, particularly those in the upper and lower limbs, have flaplike *valves* (called semilunar valves), which project inward from their linings. Valves, shown in **figure 15.29**, are usually composed of two leaflets that close if blood begins to back up

in a vein. These valves aid in returning blood to the heart because they are open as long as the blood flow is toward the heart but prevent flow in the opposite direction.

Veins also function as *blood reservoirs,* which are useful in times of blood loss. For example, in hemorrhage accompanied by a drop in arterial blood pressure, sympathetic impulses reflexly stimulate the muscular walls of the veins to constrict, which helps maintain blood pressure by returning more blood to be pumped by the heart. This mechanism helps ensure a nearly normal blood pressure even when as much as 25% of blood volume is lost. **Figure 15.30** illustrates the relative volumes of blood in the veins and other blood vessels.

Table 15.3 summarizes the characteristics of blood vessels. Clinical Application 15.2 examines disorders of blood vessels.

PRACTICE

38 How does the structure of a vein differ from that of an artery?

39 What are the functions of veins and venules?

40 How does venous circulation help to maintain blood pressure when hemorrhaging causes blood loss?

15.4 | Blood Pressure

Blood pressure is the force the blood exerts against the inner walls of the blood vessels and that circulates the blood. Although this force is present throughout the vascular system, the term *blood pressure* most commonly refers to pressure in arteries supplied by branches of the aorta (systemic arteries). Pressure differences throughout the vascular system keep the blood moving from higher pressure to lower pressure.

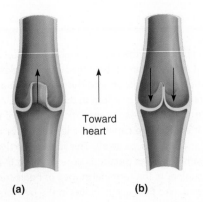

Toward heart

(a)   (b)

FIGURE 15.29 Venous valves. (a) allow blood to move toward the heart, but (b) prevent blood from moving backward away from the heart.

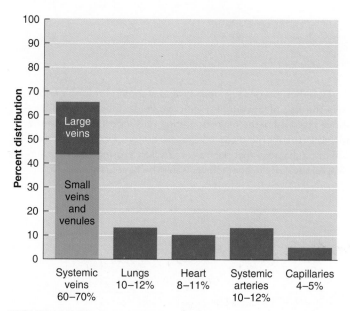

FIGURE 15.30 Most blood is in the veins and venules.

TABLE 15.3	Characteristics of Blood Vessels	
Vessel	**Type of Wall**	**Function**
Artery	Thick, strong wall with three layers—an endothelial lining, a middle layer of smooth muscle and elastic connective tissue, and an outer layer of connective tissue	Carries blood under relatively high pressure from the heart to arterioles
Arteriole	Thinner wall than an artery but with three layers; smaller arterioles have an endothelial lining, some smooth muscle tissue, and a small amount of connective tissue	Connects an artery to a capillary, helps control the blood flow into a capillary by vasoconstricting or vasodilating
Capillary	Single layer of squamous epithelium	Allows nutrients, gases, and wastes to be exchanged between the blood and tissue fluid; connects an arteriole to a venule
Venule	Thinner wall than an arteriole, less smooth muscle and elastic connective tissue	Connects a capillary to a vein
Vein	Thinner wall than an artery but with similar layers; the middle layer is more poorly developed; some have flaplike valves	Carries blood under relatively low pressure from a venule to the heart; valves prevent a backflow of blood; serves as a blood reservoir

Arterial Blood Pressure

Arterial blood pressure rises and falls in a pattern corresponding to the phases of the cardiac cycle. That is, when the ventricles contract (ventricular systole), their walls squeeze the blood inside their chambers and force it into the pulmonary trunk and aorta. As a result, the pressures in these arteries sharply increase. The maximum pressure

achieved during ventricular contraction is called the **systolic pressure.** When the ventricles relax (ventricular diastole), the arterial pressure drops, and the lowest pressure that remains in the arteries before the next ventricular contraction is termed the **diastolic pressure.**

Systemic arterial blood pressure usually is measured using an instrument called a sphygmomanometer (sfig″mo-mah-nom′ĕ-ter) (fig. 15.31). This device consists of an inflatable cuff connected by tubing to a compressible bulb and a pressure gauge. The bulb is used to pump air into the cuff, and a rise in pressure is indicated on the pressure gauge. The pressure in the cuff is expressed in millimeters of mercury (mm Hg) based on older equipment that used a glass tube containing a column of mercury in place of a pressure gauge.

To measure arterial blood pressure, the cuff of the sphygmomanometer is wrapped around the arm so that it surrounds the brachial artery. Air is pumped into the cuff until the cuff pressure exceeds the pressure in that artery, squeezing the vessel closed and stopping its blood flow. At this moment, if the diaphragm of a stethoscope is placed over the brachial artery at the distal border of the cuff, no sounds can be heard from the vessel because the blood flow is interrupted. As air is slowly released from the cuff, the air pressure inside it decreases. When the cuff pressure is just slightly lower than the systolic blood pressure in the brachial artery, the artery opens enough for a small volume of blood to spurt through, producing a sharp sound (Korotkoff's sound) heard through the stethoscope. Turbulence in the narrowed artery causes the sound. The pressure indicated on the pressure gauge when this first tapping sound is heard represents a good estimate of the *arterial systolic pressure* (SP).

As the cuff pressure continues to drop, increasingly louder sounds are heard. Then, when the cuff pressure is just slightly lower than that within the fully opened artery, the sounds become abruptly muffled and disappear. The pressure indicated on the pressure gauge when this happens represents a good estimate of the *arterial diastolic pressure* (DP).

The results of a blood pressure measurement are reported as a fraction, such as 127/74 (with normal being less than 120/less

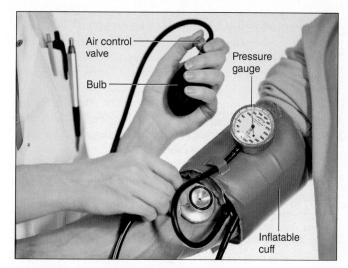

FIGURE 15.31 A sphygmomanometer is used to measure arterial blood pressure. The use of the column of mercury is the most accurate measurement, but due to environmental concerns, it has been replaced by alternative gauges and digital readouts.

In the arterial disease *atherosclerosis* (ath"er-o-sklĕ-ro'sis), deposits of fatty materials, particularly cholesterol, form within and on the inner lining of the arterial walls. Such deposits, called *plaque*, protrude into the lumens of the vessels and interfere with blood flow (fig. 15F). Furthermore, plaque often forms a surface texture that can initiate formation of a blood clot, increasing the risk of developing thrombi or emboli that cause decreased blood flow *(ischemia)*, causing tissue death *(necrosis)* downstream from the obstruction. In addition, the walls of diseased arteries may degenerate, losing their elasticity and becoming hardened or *sclerotic*. In this stage of the disease, a sclerotic vessel may rupture under the force of blood pressure.

Risk factors for developing atherosclerosis include a fatty diet, elevated blood pressure, tobacco smoking, obesity, and lack of physical exercise (see chapter 18, pp. 702–703). Genetic factors may increase the risk of developing atherosclerosis.

If blood pressure dilates a weakened area of an artery wall, a bulge called an *aneurysm* may form and enlarge. If the lining of the vessel tears and allows blood to enter the middle layer of the arterial wall, it is called a *dissecting aneurysm*. An aneurysm may cause symptoms by pressing on nearby organs, or it may rupture and cause great blood loss, which is life-threatening.

Aneurysms may also result from trauma, high blood pressure, certain infections, inherited disorders such as Marfan syndrome, or congenital defects in blood vessels. Common sites of aneurysms include the thoracic and abdominal aorta and an arterial circle at the base of the brain (circle of Willis).

Phlebitis, or inflammation of a vein, is relatively common. It may occur in association with an injury or infection or after surgery, or it may develop for no apparent reason. If inflammation is restricted to a superficial vein, such as the greater or lesser saphenous veins, blood flow may be rechanneled through other vessels. But if it occurs in a deep vein, such as the tibial, peroneal, popliteal, or femoral veins, the consequences can be serious, particularly if the blood in the affected vessel clots and blocks normal circulation (see Clinical Application 14.3, p. 547). This condition, called *thrombophlebitis*, introduces a risk that a blood clot in a vein will detach, move with the venous blood, pass through the heart, and lodge in the pulmonary arterial system in a lung (pulmonary embolism*)*.

Varicose veins are abnormal and irregular dilations in superficial veins, particularly in the legs. This condition is usually associated with prolonged, increased back pressure in the affected vessels due to gravity, such as when a person stands. Crossing the legs or sitting in a chair so that its edge presses against the area behind the knee can obstruct venous blood flow and aggravate varicose veins.

Increased venous back pressure stretches and widens the veins. The valves in these vessels lose their ability to block the backflow of blood, and blood accumulates in the veins upstream from the valves. The resulting increased venous pressure is accompanied by rising pressure in the venules and capillaries that supply the veins. Consequently, tissues in affected regions typically become edematous and painful.

Genetics, pregnancy, obesity, and standing for long periods raise the risk of developing varicose veins. Elevating the legs above the level of the heart or putting on support hosiery before arising in the morning can relieve discomfort. Intravenous injection of a substance that destroys veins (a sclerosing agent) or surgical removal of the affected veins may be necessary.

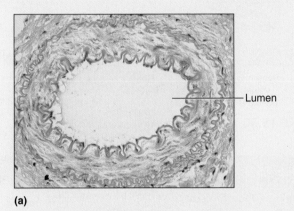

(a)

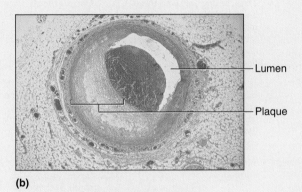

(b)

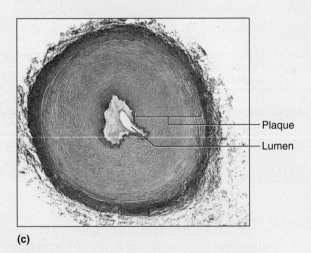

(c)

FIGURE 15F Development of atherosclerosis. (*a*) Normal arteriole (100×). (*b, c*) Accumulation of plaque on the inner wall of the arteriole (*b* and *c* 100×).

than 80). The upper number indicates the systolic pressure in mm Hg, and the lower number indicates the diastolic pressure in mm Hg. Figure 15.32 shows how these pressures decrease as distance from the left ventricle increases.

The surge of blood entering the arterial system during ventricular systole distends the elastic arterial walls, but the pressure begins to drop almost immediately as the contraction ends, and the arterial walls recoil. This alternate expanding and recoiling of the arterial wall can be felt as a *pulse* in an artery that runs close to the body surface. Figure 15.33 shows several sites where a pulse can be detected. The radial artery, for example, courses near the surface at the wrist and is commonly used to sense a person's radial pulse.

The radial pulse rate is equal to the rate at which the left ventricle contracts, and for this reason, it can be used to determine heart rate. A pulse can also reflect blood pressure, because an elevated pressure produces a pulse that feels strong and full, whereas a low pressure produces a pulse that is weak and easily compressed.

The difference between the systolic and diastolic pressures (SP − DP), called the *pulse pressure* (PP), is normally about 40 mm Hg and is what one feels when taking a pulse. The average pressure in the arterial system is also of interest because it represents the average force throughout the cardiac cycle driving blood to the tissues. To approximate this force, called the *mean arterial pressure,* add the diastolic pressure to one-third of the pulse pressure (DP + 1/3PP).

PRACTICE

41 Distinguish between systolic and diastolic blood pressure.

42 In which blood vessel is arterial blood pressure commonly measured?

43 What causes a pulse in an artery?

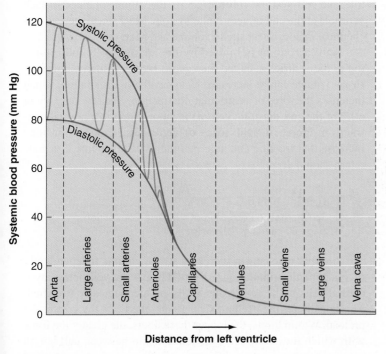

FIGURE 15.32 Blood pressure decreases as the distance from the left ventricle increases. Systolic pressure occurs during maximal ventricular contraction. Diastolic pressure occurs when the ventricles relax.

Factors That Influence Arterial Blood Pressure

Arterial pressure depends on a variety of factors. These include cardiac output, blood volume, peripheral resistance, and blood viscosity (fig. 15.34).

Cardiac Output AP|R

Each ventricular contraction determines the volume of blood that enters the arterial system, which is called the **stroke volume.** In an average-weight male at rest, the stroke volume equals about 70 milliliters. The volume discharged from the ventricle per minute is called the **cardiac output.** It is calculated by multiplying the stroke volume by the heart rate, expressed in beats per minute. [Cardiac output (CO) = stroke volume (SV) × heart rate (HR).] For example, if the stroke volume is 70 milliliters per beat and the heart rate is 72 beats per minute, the cardiac output is 5,040 milliliters per minute.

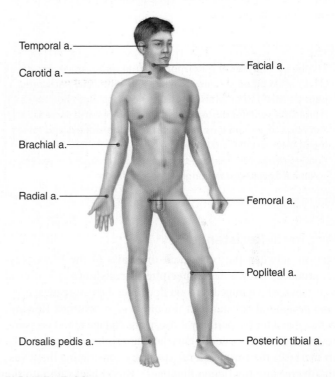

FIGURE 15.33 Sites where an arterial pulse is most easily detected. (*a.* stands for artery.)

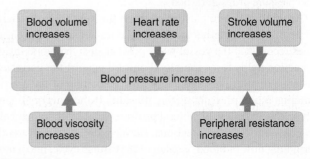

FIGURE 15.34 Some of the factors that influence arterial blood pressure.

Blood pressure varies with the cardiac output. If either the stroke volume or the heart rate increases, so does the cardiac output, and, blood pressure would increase. Conversely, if the stroke volume or the heart rate decreases, the cardiac output decreases, and blood pressure would also decrease.

Blood Volume

Blood volume equals the sum of the formed elements and plasma volumes in the vascular system. Although the blood volume varies somewhat with age, body size, and sex, it is usually about 5 liters for adults or 8% of body weight in kilograms (1 kilogram of water equals 1 liter).

Normally blood pressure is directly proportional to the volume of the blood in the cardiovascular system. Thus, any changes in the blood volume can initially alter the blood pressure. For example, if a hemorrhage reduces blood volume, blood pressure at first drops. If a transfusion restores normal blood volume, normal pressure may be reestablished. Blood volume can also fall if the fluid balance is upset, as happens in dehydration. Fluid replacement can reestablish normal blood volume and pressure.

> The heart secretes the hormones atrial natriuretic peptide (ANP) when blood volume increases and ventricular natriuretic peptide (BNP) when blood pressure increases. Both hormones inhibit the renin-angiotensin system. The result is increased excretion of sodium ions and water from the kidneys and lowered blood volume and blood pressure. Researchers are investigating drugs that increase these peptides to treat the excess volume associated with congestive heart failure.

Peripheral Resistance

Friction between the blood and the walls of the blood vessels produces a force called **peripheral resistance** (pĕ-rif′er-al re-zis′tans), which impedes blood flow. Blood pressure must overcome peripheral resistance if the blood is to continue flowing. Factors that alter the peripheral resistance change blood pressure. For example, when smooth muscle in arteriolar walls contracts, this increases the peripheral resistance by constricting these vessels (decreasing the lumen diameter). Blood backs up into the arteries supplying the arterioles, and the arterial pressure rises. Dilation of the arterioles (increasing the lumen diameter) has the opposite effect—peripheral resistance decreases, and the arterial blood pressure drops in response.

Arterial walls are elastic, so when the ventricles discharge a surge of blood, arteries swell. Almost immediately, the elastic tissues recoil, and the vessel walls press against the blood inside. This action helps force the blood onward against the peripheral resistance in arterioles and capillaries. Recoiling of the arteries maintains blood pressure during diastole. If there were no elasticity in the arterial walls, blood pressure would fall dramatically between ventricular contractions. Elastic recoil also converts the intermittent flow of blood, a result of the cardiac cycle, into a more continuous movement through the blood vessels.

Viscosity

Viscosity (vis-kos′ĭ-te) is the difficulty with which the molecules of a fluid flow past one another. The greater the viscosity, the greater the resistance to flow.

Blood cells and plasma proteins increase blood viscosity. The greater the blood's resistance to flowing, the greater the force needed to move it through the vascular system, so blood pressure rises as blood viscosity increases and drops as blood viscosity decreases.

The viscosity of blood normally remains stable. However, any condition that alters the concentrations of blood cells or plasma proteins may alter blood viscosity. For example, anemia may decrease viscosity and consequently lower blood pressure. Excess red blood cells increase viscosity and blood pressure.

 PRACTICE

44 How is cardiac output calculated?

45 How are cardiac output and blood pressure related?

46 How does blood volume affect blood pressure?

47 What is the relationship between peripheral resistance and blood pressure? between blood viscosity and blood pressure?

Control of Blood Pressure

Blood pressure (BP) is determined by cardiac output (CO) and peripheral resistance (PR) according to this relationship: BP = CO × PR. Maintenance of normal blood pressure therefore requires regulation of these two factors (fig. 15.35).

Cardiac output depends on the stroke volume and heart rate. Stroke volume, the amount of blood pumped in a single beat, is reflected by the difference between **end-diastolic volume** (EDV), the volume of blood in each ventricle at the end of ventricular diastole, and **end-systolic volume** (ESV), the volume of blood in each ventricle at the end of ventricular systole. Mechanical, neural, and chemical factors affect stroke volume and heart rate.

Cardiac output is limited by the amount of blood returning to the ventricles, called the *venous return*. Usually, however, stroke volume can be increased by sympathetic stimulation, which increases the force of ventricular contraction. Because only about 60% of the end-diastolic volume is pumped out in a normal contraction, increasing the force of ventricular contraction may increase that fraction and help maintain stroke volume if venous return should decrease.

 RECONNECT

To Chapter 9, Recording of a Muscle Contraction, page 304 and fig. 9.15 on page 305.

Another mechanism increases stroke volume independently of sympathetic stimulation. As blood enters the ventricles, myocardial cells in the ventricular walls are mechanically stretched. This constitutes the **preload.** The greater the EDV, the greater the preload. Within limits, the longer these fibers, the greater the force with which they contract. This relationship between cell length (due to stretching of the cardiac muscle cell just before contraction) and force of contraction is called the **Frank-Starling law of the heart,** or Starling's law of the heart. The law becomes important, for example, during exercise, when venous blood returns more rapidly to the heart. The more blood that enters the heart from the veins, the greater the ventricular distension, the stronger the contraction, the greater the stroke volume, and the greater the cardiac output.

Conversely, the less blood that returns from the veins, the less the ventricle distends, the weaker the ventricular contraction, and the lesser the stroke volume and cardiac output. This mechanism helps to ensure that the volume of blood discharged from the heart is equal to the volume entering its chambers.

Some blood remains in the ventricles after contraction and stroke volume ejection. This ESV is influenced by preload, contractility of the ventricle, and afterload. **Contractility,** the strength of a contraction at a given preload (EDV), is influenced by autonomic innervation and hormones (epinephrine, norepinephrine, thyroid hormones). Sympathetic stimulation contracts the ventricles more forcefully, increasing the volume ejected and decreasing the ESV. Decreased sympathetic stimulation produces the opposite effect. The force that the ventricles must produce to open the semilunar valves to eject blood is the **afterload.** Increased arterial pressure (hypertension), especially the diastolic pressure, increases afterload and the heart must work harder to eject blood.

Cardiac output and peripheral resistance are controlled in part by baroreceptor reflexes. Recall from chapter 12 (p. 444) that baroreceptors in the walls of the aortic arch and carotid sinuses sense changes in blood pressure. If arterial pressure increases, impulses travel from the receptors to the *cardiac center* of the medulla oblongata. This center relays parasympathetic impulses to the SA node in the heart, and heart rate decreases in response. As a result of this *cardioinhibitor reflex,* cardiac output falls and blood pressure decreases toward the normal level. Figure 15.36 summarizes this mechanism.

Conversely, decreasing arterial blood pressure initiates the *cardioaccelerator reflex,* which sends sympathetic impulses to the SA node. As a result of these cardiostimulatory effects, the heart beats faster, increasing cardiac output and arterial pressure.

Recall that epinephrine increases heart rate (chapter 13, p. 510) and consequently alters cardiac output and blood pressure.

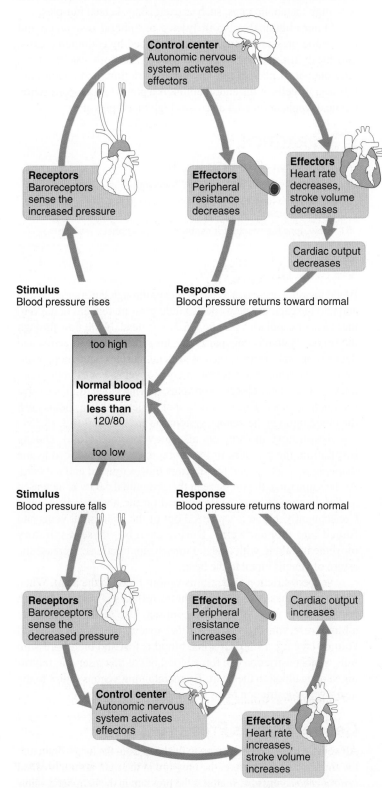

FIGURE 15.35 Controlling cardiac output and peripheral resistance regulates blood pressure. Normal values for blood pressure increase with age.

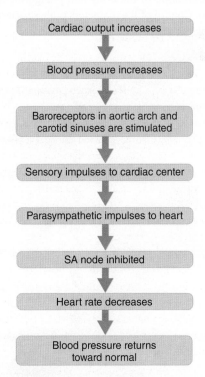

FIGURE 15.36 If blood pressure rises, baroreceptors initiate the cardioinhibitor reflex, which lowers the blood pressure.

Other factors that increase heart rate and blood pressure include emotional responses, such as fear and anger; physical exercise; and a rise in body temperature.

Changes in arteriole diameters regulate peripheral resistance. Because blood vessels with smaller diameters offer a greater resistance to blood flow, factors that vasoconstrict arterioles increase peripheral resistance, raising blood pressure; factors that vasodilate arterioles decrease peripheral resistance, lowering blood pressure (see fig. 15.35).

The *vasomotor center* of the medulla oblongata continually sends sympathetic impulses to the smooth muscle in the arteriole walls, keeping them in a state of tonic contraction, which helps maintain the peripheral resistance associated with normal blood pressure. The vasomotor center responds to changes in blood pressure, so it can increase peripheral resistance by increasing its outflow of sympathetic impulses, or it can decrease such resistance by decreasing its sympathetic outflow. In the latter case, the vessels vasodilate as sympathetic stimulation decreases.

Whenever arterial blood pressure suddenly increases, baroreceptors in the aortic arch and carotid sinuses signal the vasomotor center, and the sympathetic outflow to the arteriole walls falls (fig. 15.37). The resulting vasodilation decreases peripheral resistance, and blood pressure decreases toward the normal level. Similarly, if blood pressure drops, as it would following a hemorrhage, the vasomotor center increases sympathetic outflow. The resulting release of epinephrine and norepinephrine vasoconstricts most systemic vessels, increasing peripheral resistance. This helps return blood pressure toward normal.

The vasomotor center's control of vasoconstriction and vasodilation is especially important in the arterioles of the *abdominal viscera* (splanchnic region). These vessels, if fully dilated, could accept nearly all the blood of the body and plunge the arterial pressure toward zero. Control of the diameters of these arterioles is essential in regulating normal peripheral resistance.

Certain chemicals, including carbon dioxide, oxygen, and hydrogen ions, influence peripheral resistance by affecting precapillary sphincters and smooth muscle in arteriole and metarteriole walls. For example, increasing blood carbon dioxide, decreasing blood oxygen, and lowering blood pH relaxes smooth muscle in the systemic circulation. This increases local blood flow to tissues with high metabolic rates, such as exercising skeletal muscles.

Other chemicals also influence peripheral resistance and thus blood pressure. Nitric oxide, produced by endothelial cells, and bradykinin, formed in the blood, are both vasodilators. Angiotensin II plays a role in vasoconstriction; and endothelin, released by cells of the endothelium, is a powerful vasoconstrictor. Clinical Application 15.3 discusses high blood pressure.

PRACTICE

48 What factors affect cardiac output?

49 Explain the Frank-Starling law of the heart.

50 What is the function of the baroreceptors in the aortic arch and carotid sinuses?

51 How does the vasomotor center control peripheral resistance?

Venous Blood Flow

Blood pressure decreases as blood moves through the arterial system and into the capillary networks, so little pressure remains at the venular ends of capillaries (see fig. 15.32). Instead, blood flow through the venous system is only partly the direct result of heart action and depends on other factors, such as skeletal muscle contraction, respiratory movements, and vasoconstriction of veins. For example, contracting skeletal muscles press on veins, moving blood from one valve section to another. This contraction of skeletal muscles helps push the blood through the venous system toward the heart (fig. 15.38).

Respiratory movements also move venous blood. During inspiration, the pressure in the thoracic cavity is reduced as the diaphragm contracts and the rib cage moves upward and outward. At the same time, the pressure in the abdominal cavity is increased as the diaphragm presses downward on the abdominal viscera. Consequently, blood is squeezed out of the abdominal veins and forced into thoracic veins. During exercise, these respiratory movements, along with skeletal muscle contractions, increase the return of venous blood to the heart.

Venoconstriction also returns venous blood to the heart. When venous pressure is low, sympathetic reflexes stimulate smooth muscle in the walls of veins to contract. The veins also provide a blood reservoir that can adapt its capacity to changes in blood volume (see fig. 15.30). If some blood is lost and blood pressure falls, venoconstriction can force blood out of this reservoir, returning venous blood to the heart. By maintaining venous return to the heart, venoconstriction helps to maintain blood pressure.

Central Venous Pressure

All veins, except those returning to the heart from the lungs, drain into the right atrium. Therefore, the pressure in the right atrium is called *central venous pressure.* It affects the pressure in the peripheral veins. For example, if the heart is failing and beating weakly, the central venous pressure increases and blood backs up in the peripheral veins,

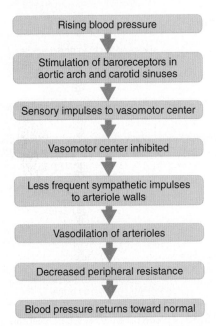

Rising blood pressure

Stimulation of baroreceptors in aortic arch and carotid sinuses

Sensory impulses to vasomotor center

Vasomotor center inhibited

Less frequent sympathetic impulses to arteriole walls

Vasodilation of arterioles

Decreased peripheral resistance

Blood pressure returns toward normal

FIGURE 15.37 Dilating arterioles helps regulate blood pressure.

Hypertension, or high blood pressure, is persistently elevated arterial pressure. It is a prevalent disease of the cardiovascular system in industrialized nations.

High blood pressure with unknown cause is called *essential* (also primary or idiopathic) *hypertension.* Elevated blood pressure that is a consequence of another problem, such as kidney disease, is called *secondary hypertension.*

Kidney disease often produces changes that interfere with blood flow to kidney cells. In response, certain kidney cells may release an enzyme called *renin* that leads to the production of *angiotensin II,* a powerful vasoconstrictor that increases peripheral resistance in the arterial system, raising arterial pressure (fig. 15G). Angiotensin II also stimulates the adrenal cortex to release *aldosterone,* which stimulates the kidneys to retain sodium ions and water. The resulting increase in blood volume contributes to increased blood pressure. Normally, this mechanism ensures that a decrease in blood flow to the kidneys is followed by an increase in arterial pressure, which, in turn, restores blood flow to the kidneys.

In some individuals, high sodium intake leads to increased blood pressure. Obesity also is a risk factor for hypertension because it increases peripheral resistance. Psychological stress, which activates sympathetic nerve impulses that cause generalized vasoconstriction, may also lead to hypertension. Yet another cause of hypertension may be an inability of endothelium to respond to a relaxing factor, leading to vasoconstriction.

Hypertension is called a "silent killer" because it may not have direct symptoms, yet can raise the risk for serious cardiovascular complications. For example, as the left ventricle works harder to pump blood at a higher pressure, the myocardium thickens, enlarging the heart. If coronary blood flow cannot support this overgrowth, parts of the heart muscle die and fibrous tissue replaces them. Eventually, the enlarged and weakened heart fails to maintain adequate output for survival.

Hypertension also contributes to the development of atherosclerosis. As arteries accumulate plaque, a *coronary thrombosis* or a *coronary embolism* may occur. Similar changes in the arteries of the brain increase the chances of a *cerebral vascular accident* (CVA), or stroke, due to a cerebral thrombosis, embolism, or hemorrhage. When an embolus or hemorrhage causes a stroke, paralysis and other functional losses suddenly appear. A thrombus-caused stroke is slower because the clot takes time to develop. It may begin with clumsiness, progress to partial visual loss, then affect speech. One arm may become paralyzed, then a day later, perhaps an entire side of the body is affected. Table 15A lists risk factors for a stroke.

A *transient ischemic attack* (TIA, or "ministroke") is a temporary block in a small artery. Symptoms include difficulty in speaking or understanding speech; numbness or weakness in the face, upper limb, lower limb, or one side; dizziness; falling; an unsteady gait; blurred vision; or blindness. These symptoms typically resolve within twenty-four hours with no lasting effects, but may be a warning of an impending, more serious stroke.

Exercising regularly, maintaining a healthy weight, reducing stress, and limiting dietary sodium can help prevent hypertension. If necessary, medications may include diuretics and/or inhibitors of sympathetic nerve action. Diuretics increase urinary excretion of sodium and water, reducing the volume of body fluids. Sympathetic inhibitors block the synthesis of neurotransmitters, such as norepinephrine, or block receptors on effector cells. Table 15B describes how drugs that treat hypertension work.

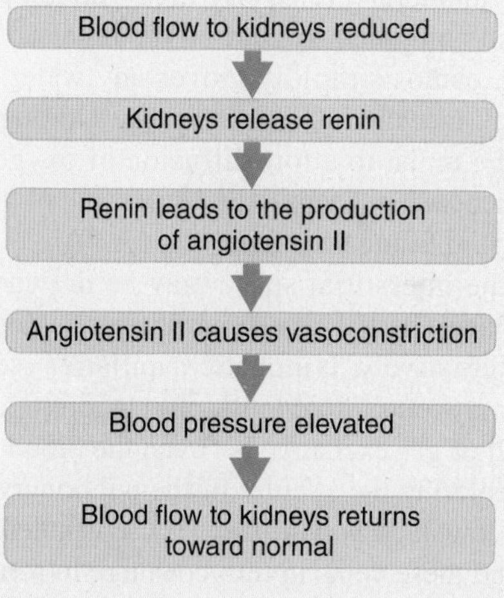

TABLE 15A Risk Factors for Stroke

Alcohol consumption
Diabetes
Elevated serum cholesterol
Family history of cardiovascular disease
Hypertension
Smoking
Transient ischemic attacks

TABLE 15B Drugs to Treat Hypertension

Type of Drug	Mechanism of Action
Angiotensin-converting enzyme (ACE) inhibitors	Block formation of angiotensin II, preventing vasoconstriction and sodium retention
Beta blockers	Lower heart rate, reduce contractility by blocking beta adrenergic receptors on the heart
Calcium channel blockers	Dilate blood vessels by keeping calcium ions out of smooth muscle cells in vessel walls
Diuretics	Increase urine output, lowering blood volume

Blood flow to kidneys reduced

↓

Kidneys release renin

↓

Renin leads to the production of angiotensin II

↓

Angiotensin II causes vasoconstriction

↓

Blood pressure elevated

↓

Blood flow to kidneys returns toward normal

FIGURE 15G Renin stimulates production of angiotensin II, which elevates blood pressure.

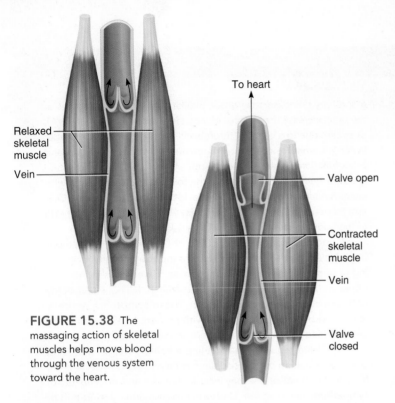

Relaxed skeletal muscle

Vein

To heart

Valve open

Contracted skeletal muscle

Vein

Valve closed

FIGURE 15.38 The massaging action of skeletal muscles helps move blood through the venous system toward the heart.

raising their pressure, too. This can lead to peripheral edema because the resulting higher capillary hydrostatic pressure favors movement of fluid into the tissues. However, if the heart is beating normally, the central venous pressure and the pressure in the venous network remain low.

An increase in blood volume or widespread venoconstriction also increases blood flow into the right atrium, elevating the central venous pressure. Clinical Application 15.4 discusses the effects of exercise on the heart and blood vessels.

Blood or tissue fluid accumulating in the pericardial cavity (pericardial effusion) increases pressure, causing a potentially deadly condition called *acute cardiac tamponade*. Increasing pressure around the heart may compress it, interfering with the flow of blood into its chambers and preventing pumping action. An early symptom of acute cardiac tamponade is increased central venous pressure, with visible engorgement of the veins in the neck. Other symptoms include anxiety, rapid or difficulty breathing, light-headedness, palpitations, pallor, and chest pain. Acute cardiac tamponade has several causes, including bacterial or viral infection, injury, acute myocardial infarction, advanced lung cancer, and dissecting aortic aneurysm.

 PRACTICE

52 What is the function of the venous valves?

53 How do skeletal muscles affect venous blood flow?

54 How do respiratory movements affect venous blood flow?

55 What factors stimulate venoconstriction?

15.5 Paths of Circulation

The blood vessels form two circuits (fig. 15.39). The pulmonary circuit sends oxygen-poor blood to the lungs to pick up oxygen and unload carbon dioxide. The systemic circuit sends oxygen-rich blood and nutrients to all body cells and removes wastes.

The pathways described in the following sections are those of an adult. Chapter 23 (pp. 883–887) describes the somewhat different fetal pathways.

Pulmonary Circuit

Blood enters the pulmonary circuit as it leaves the right ventricle through the pulmonary trunk. The pulmonary trunk extends upward and posteriorly from the heart. About 5 centimeters above its origin, the pulmonary trunk divides into the right and left pulmonary arteries, which penetrate the right and left lungs, respectively. In the lungs, they diverge into *lobar branches* (three on the right side and two on the left) that accompany the main divisions of the bronchi (airways) into the lobes of the lungs. After repeated divisions, the lobar branches give rise to arterioles that continue into the capillary networks associated with the walls of the alveoli (air sacs).

The blood in the arteries and arterioles of the pulmonary circuit is low in oxygen and high in carbon dioxide. Gases are exchanged between the blood and the air as the blood moves through the *alveolar capillaries,* discussed in chapter 19 (pp. 755–756).

The right ventricle contracts with less force than the left ventricle. Therefore, the arterial pressure in the pulmonary circuit is less than that in the systemic circuit. As a result, the alveolar capillary pressure is low.

The force that moves fluid out of an alveolar capillary is 23 mm Hg; the force pulling fluid into it is 22 mm Hg. Thus, such a capillary has a net filtration pressure of 1 mm Hg. This pressure propels a slight, continuous flow of fluid into the narrow interstitial space between the alveolar capillary and the alveolus.

The epithelial cells of the alveoli are so tightly joined that sodium, chloride, and potassium ions, as well as glucose and urea, enter the interstitial space but usually do not enter the alveoli. This helps maintain a relatively high osmotic pressure in the interstitial fluid compared to fluid on the inner alveolar surface. Consequently, osmosis rapidly moves any water that gets into the alveoli back into the interstitial space. Although the alveolar surface must be moist to allow diffusion of oxygen and carbon dioxide, this mechanism keeps excess water out of the alveoli, preventing them from filling with fluid (fig. 15.40).

Fluid in the interstitial space may be drawn back into the alveolar capillaries by the somewhat higher osmotic pressure of the blood. Alternatively, lymphatic capillaries (see chapter 16, pp. 617–618) may return fluid to the circulation (fig. 15.40).

As a result of gas exchange between the blood and the alveolar air, blood entering the venules of the pulmonary circuit is rich in oxygen and low in carbon dioxide. These venules merge to form small veins, and these veins in turn converge to form larger veins. Four *pulmonary veins,* two from each lung, return blood to the left atrium. This completes the vascular loop of the pulmonary circuit.

We know that exercise is good for the heart. Yet each year, a few individuals die of sudden cardiac arrest while shoveling snow, running, or engaging in some other strenuous activity. The explanation for this apparent paradox is that exercise is good for the heart—but only if the exercise is a regular part of life.

The cardiovascular system adapts to aerobic exercise. The aerobically conditioned athlete experiences increases in heart pumping efficiency, blood volume, blood hemoglobin concentration, and the number of mitochondria in muscle fibers. These adaptations improve oxygen delivery to, and use by, muscle tissue.

An athlete's heart typically changes in response to these increased demands and may enlarge 40% or more. Myocardial mass increases, the ventricular cavities expand, and the ventricle walls thicken. At rest, stroke volume increases, and heart rate and blood pressure decrease. To a physician unfamiliar with a conditioned cardiovascular system, a trained athlete may appear to be abnormal.

The cardiovascular system responds beautifully to a slow, steady buildup in exercise frequency and intensity. It does not react well to sudden demands, such as when a person who never exercises suddenly shovels snow or runs 3 miles. Sedentary people have two- to sixfold increased risk of cardiac arrest while exercising than when not; people in shape have little or no excess risk while exercising.

For exercise to benefit the cardiovascular system, the heart rate must be elevated to 70% to 85% of its "theoretical maximum" for 30 to 60 minutes at least three to four times a week, according to the American Heart Association. To calculate your theoretical maximum, subtract your age from 220. If you are eighteen years old, your theoretical maximum is 202 beats per minute. A rate of 141–172 beats per minute is 70% to 85% of this value. Examples of activities that raise heart rate are tennis, skating, skiing, handball, vigorous dancing, hockey, basketball, biking, and fast walking.

It is wise to consult a physician before starting an exercise program. People over the age of thirty are advised to have a stress test, which is an electrocardiogram taken while exercising. (The standard electrocardiogram is taken at rest.) An arrhythmia that appears only during exercise may indicate heart disease that has not yet produced symptoms. The American Heart Association suggests that after a physical exam, a sedentary person wishing to start an exercise program begin with 30 minutes of activity (perhaps broken into two 15-minute sessions at first) at least five times per week.

Pulmonary edema, in which lungs fill with fluid, can accompany a failing left ventricle or a damaged mitral valve. A weak left ventricle may be unable to move the normal volume of blood into the systemic circuit. Blood backing up into the pulmonary circuit increases pressure in the alveolar capillaries, flooding the interstitial spaces with fluid. Increasing pressure in the interstitial fluid may rupture the alveolar membranes, and fluid may enter the alveoli more rapidly than it can be removed. This reduces the alveolar surface available for gas exchange, and the person may suffocate.

Systemic Circuit

Oxygen-rich blood moves from the left atrium into the left ventricle. Contraction of the left ventricle forces this blood into the systemic circuit, which includes the aorta and its branches that lead to all of the body tissues, as well as the companion system of veins that returns blood to the right atrium.

PRACTICE

56 Distinguish between the pulmonary and systemic circuits of the cardiovascular system.

57 Trace the path of blood through the pulmonary circuit from the right ventricle.

58 Explain why the alveoli normally do not fill with fluid.

15.6 | Arterial System

The **aorta** is the largest diameter artery in the body. It extends upward from the left ventricle, arches over the heart to the left, and descends just anterior and to the left of the vertebral column.

Principal Branches of the Aorta

The part of the aorta attached to the heart is called the aortic root. From there, the first part of the aorta is called the *ascending aorta*. At the root are the three cusps of the aortic valve, and opposite each cusp is a swelling in the aortic wall called an **aortic sinus.** The right and left *coronary arteries* arise from two of these aortic sinuses. The elastic recoil of the aortic wall following contraction of the left ventricle helps drive blood flow into these arteries.

Three major arteries originate from the *arch of the aorta* (aortic arch). They are the brachiocephalic artery, the left common carotid artery, and the left subclavian artery. The aortic arch has baroreceptors that detect changes in blood pressure. Several small structures called **aortic bodies** lie in the epithelial lining of the aortic arch. The aortic bodies house chemoreceptors that sense blood pH and blood levels of oxygen and carbon dioxide.

The **brachiocephalic** (brak"e-o-sĕ-fal'ik) **artery** supplies blood to the tissues of the upper limb and head, as its name suggests. It is the first branch from the aortic arch and rises through the mediastinum to a point near the junction of the sternum and the right clavicle. There it divides, giving rise to the right **common carotid** (kah-rot'id) **artery,** which transports blood to the right

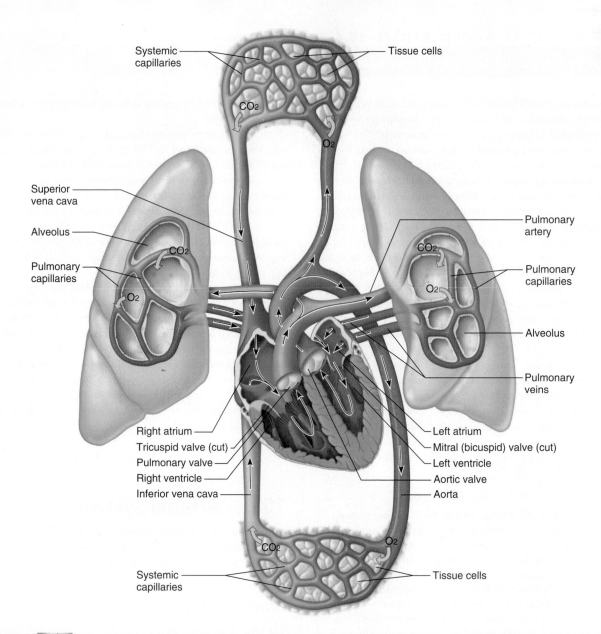

FIGURE 15.39 **AP|R** The cardiovascular system transports blood between the body cells and organs such as the lungs, intestines, and kidneys that communicate with the external environment. Vessels in the pulmonary circuit carry blood from the heart to the lungs and back to the heart, replenishing oxygen (O_2) and releasing the metabolic waste carbon dioxide (CO_2). Vessels of the systemic circuit supply all of the other cells. (Structures are not drawn to scale.)

side of the neck and head, and the right **subclavian** (sub-kla′ve-an) **artery,** which leads into the right arm. Branches of the subclavian artery also supply blood to parts of the shoulder, neck, and head.

The left *common carotid artery* and the left *subclavian artery* are respectively the second and third branches of the aortic arch. They supply blood to regions on the left side of the body corresponding to those supplied by their counterparts on the right (fig. **15.41** and reference plates 21, 22, and 23, pp. 53–54).

The upper part of the *descending aorta* is left of the midline, but it gradually extends medially and lies directly anterior to the vertebral column at the level of the twelfth thoracic vertebra. The part of the descending aorta above the diaphragm is the **thoracic aorta** (tho-ras′ik a-or′tah). It gives off many small branches to the thoracic wall and the thoracic viscera. These branches, the *bronchial, pericardial,* and *esophageal arteries,* supply blood to

the structures for which they were named. Other branches become *mediastinal arteries,* supplying various tissues in the mediastinum, and *posterior intercostal arteries,* which pass into the thoracic wall.

Below the diaphragm, the descending aorta becomes the **abdominal aorta,** and it branches to the abdominal wall and several abdominal organs. These branches include the following:

1. The **celiac** (se′le-ak) **artery** gives rise to the left *gastric, splenic,* and *hepatic arteries,* which supply upper portions of the digestive tract, the spleen, and the liver, respectively. (*Note:* The hepatic artery supplies the liver with about one-third of its blood flow, and this blood is oxygen-rich. The remaining two-thirds of the liver's blood flow arrives by means of the hepatic portal vein and is oxygen-poor.)

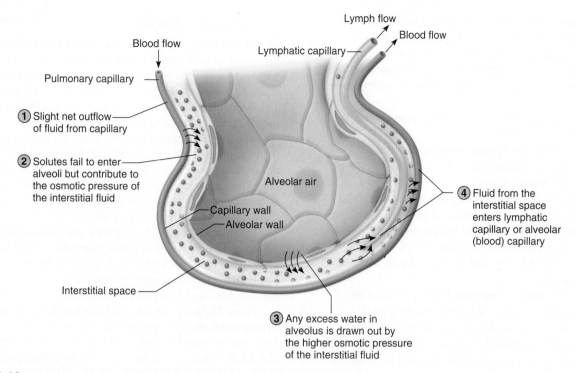

FIGURE 15.40 Cells of the alveolar wall are tightly joined. The relatively high osmotic pressure of the interstitial fluid draws water out of the alveoli.

In the figure, the following labels appear:

Lymph flow

Blood flow

Lymphatic capillary

Blood flow

Pulmonary capillary

Pulmonary capillary

① Slight net outflow of fluid from capillary

② Solutes fail to enter alveoli but contribute to the osmotic pressure of the interstitial fluid

Alveolar air

Capillary wall

Alveolar wall

④ Fluid from the interstitial space enters lymphatic capillary or alveolar (blood) capillary

Interstitial space

③ Any excess water in alveolus is drawn out by the higher osmotic pressure of the interstitial fluid

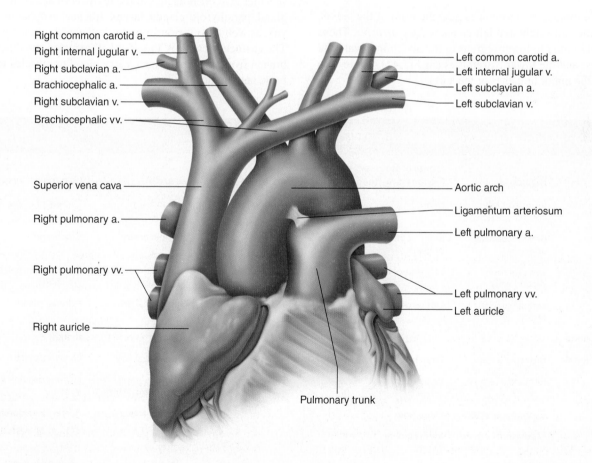

FIGURE 15.41 **AP|R** The major blood vessels associated with the heart. (*a.* stands for artery, *v.* stands for vein, *vv.* stands for veins.)

In the figure, the following labels appear:

Right common carotid a.
Right internal jugular v.
Right subclavian a.
Brachiocephalic a.
Right subclavian v.
Brachiocephalic vv.

Left common carotid a.
Left internal jugular v.
Left subclavian a.
Left subclavian v.

Superior vena cava

Aortic arch

Ligamentum arteriosum

Right pulmonary a.

Left pulmonary a.

Right pulmonary vv.

Left pulmonary vv.

Left auricle

Right auricle

Pulmonary trunk

2. The **phrenic** (fren′ik) **arteries** are paired arteries that supply blood to the diaphragm.
3. The **superior mesenteric** (mes″en-ter′ik) **artery** is a large, unpaired vessel that branches to many parts of the intestinal tract, including the jejunum, ileum, cecum, ascending colon, and transverse colon.
4. The pair of **suprarenal** (soo″prah-re′nal) **arteries** supplies blood to the adrenal glands.
5. The **renal** (re′nal) **arteries** pass laterally from the aorta into the kidneys. Each artery then divides into several lobar branches in the kidney tissues.
6. **Gonadal** (go′nad-al) **arteries** are in the female and male. In a female, paired *ovarian arteries* arise from the aorta and pass into the pelvis to supply the ovaries. In a male, *spermatic arteries* originate in similar locations. They course downward and pass through the body wall by way of the *inguinal canal* to supply the testes.
7. Branches of the **inferior mesenteric artery** lead to the descending colon, the sigmoid colon, and the rectum.
8. Three or four pairs of *lumbar arteries* arise from the posterior surface of the aorta in the region of the lumbar vertebrae. These arteries supply muscles of the skin and the posterior abdominal wall.
9. The **middle sacral artery,** a small vessel, descends medially from the aorta along the anterior surfaces of the lower lumbar vertebrae. It transports blood to the sacrum and coccyx.

The abdominal aorta terminates near the brim of the pelvis, where it divides into right and left *common iliac arteries*. These vessels supply blood to lower regions of the abdominal wall, the pelvic organs, and the lower extremities (fig. 15.42). Table 15.4 summarizes the major branches of the aorta.

Arteries to the Brain, Head, and Neck

Branches of the subclavian and common carotid arteries supply blood to structures in the brain, head, and neck (figs. 15.43 and 15.44). The main divisions of the subclavian artery to these regions are the vertebral, thyrocervical, and costocervical arteries. The common carotid artery communicates with these regions by means of the internal and external carotid arteries.

The **vertebral arteries** arise from the subclavian arteries in the base of the neck near the tips of the lungs. They pass upward through the foramina of the transverse processes of the cervical vertebrae and enter the skull by way of the foramen magnum. Along their paths, these vessels supply blood to vertebrae and to their associated ligaments and muscles.

In the cranial cavity, the vertebral arteries unite to form a single *basilar artery*. This vessel passes along the ventral brainstem and gives rise to branches leading to the pons, midbrain, and cerebellum. The basilar artery terminates by dividing into two *posterior cerebral arteries* that supply parts of the occipital and temporal lobes of the cerebrum. The posterior cerebral arteries also help form the **cerebral arterial circle** *(circle of Willis)* at the base of the brain, which connects the vertebral artery and internal carotid artery systems (fig. 15.45). The union of these systems provides alternate pathways for blood to circumvent blockages and reach brain tissues. It also equalizes blood pressure in the brain's blood supply. The circle is complete in only 20% to 30% of the population.

The **thyrocervical** (thi″ro-ser′vĭ-kal) **arteries** are short vessels that give off branches at the thyrocervical axis to the thyroid gland, parathyroid glands, larynx, trachea, esophagus, and pharynx, as well as to various muscles in the neck, shoulder, and back. The **costocervical** (kos″to-ser′vĭ-kal) **arteries,** the third vessels to branch from the subclavians, carry blood to muscles in the neck, back, and thoracic wall.

TABLE 15.4	**Major Branches of the Aorta** AP|R				
Portion of Aorta	Branch	General Regions or Organs Supplied	Portion of Aorta	Branch	General Regions or Organs Supplied
Ascending aorta	Right and left coronary arteries	Heart	Abdominal aorta	Celiac artery	Organs of upper digestive tract
Arch of aorta	Brachiocephalic artery	Right upper limb, right side of head		Phrenic artery	Diaphragm
	Left common carotid artery	Left side of head		Superior mesenteric artery	Portions of small and large intestines
	Left subclavian artery	Left upper limb		Suprarenal artery	Adrenal gland
Descending aorta				Renal artery	Kidney
Thoracic aorta	Bronchial artery	Bronchi		Gonadal artery	Ovary or testis
	Pericardial artery	Pericardium		Inferior mesenteric artery	Lower portions of large intestine
	Esophageal artery	Esophagus		Lumbar artery	Posterior abdominal wall
	Mediastinal artery	Mediastinum		Middle sacral artery	Sacrum and coccyx
	Posterior intercostal artery	Thoracic wall		Common iliac artery	Lower abdominal wall, pelvic organs, and lower limb

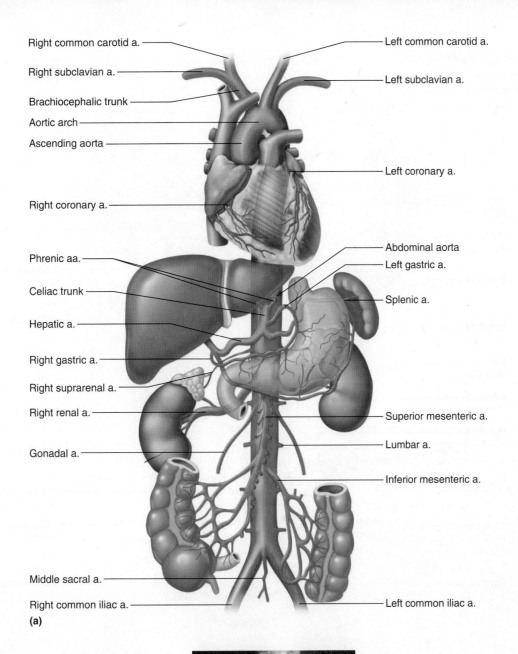

Right common carotid a.
Right subclavian a.
Brachiocephalic trunk
Aortic arch
Ascending aorta

Right coronary a.

Phrenic aa.
Celiac trunk
Hepatic a.
Right gastric a.
Right suprarenal a.
Right renal a.
Gonadal a.

Middle sacral a.
Right common iliac a.
(a)

Left common carotid a.
Left subclavian a.

Left coronary a.

Abdominal aorta
Left gastric a.
Splenic a.

Superior mesenteric a.
Lumbar a.
Inferior mesenteric a.

Left common iliac a.

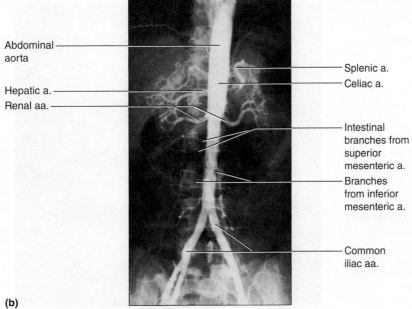

Abdominal aorta

Hepatic a.
Renal aa.

Splenic a.
Celiac a.

Intestinal branches from superior mesenteric a.

Branches from inferior mesenteric a.

Common iliac aa.

(b)

FIGURE 15.42 AP|R

Abdominal aorta.
(a) Its major branches.
(b) Angiogram (radiograph).
(a. stands for artery, aa. stands for arteries.)

FIGURE 15.43 AP|R The major arteries of the head and neck. The clavicle has been removed. (a. stands for artery.)

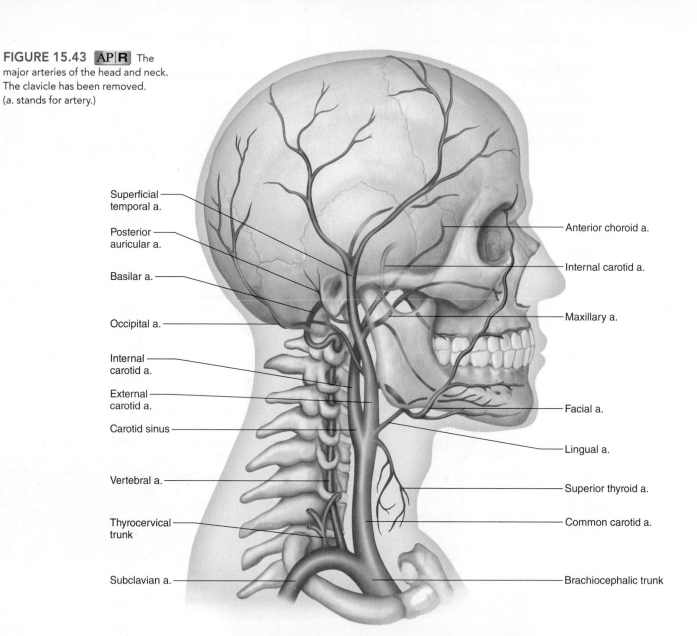

Superficial temporal a.

Posterior auricular a.

Basilar a.

Occipital a.

Internal carotid a.

External carotid a.

Carotid sinus

Vertebral a.

Thyrocervical trunk

Subclavian a.

Anterior choroid a.

Internal carotid a.

Maxillary a.

Facial a.

Lingual a.

Superior thyroid a.

Common carotid a.

Brachiocephalic trunk

FIGURE 15.44 AP|R An angiogram of the arteries associated with the head. (a. stands for artery.)

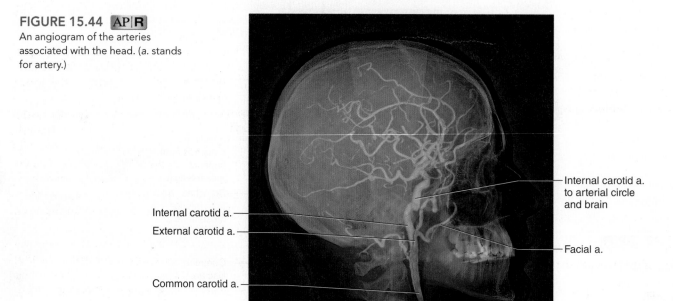

Internal carotid a.

External carotid a.

Common carotid a.

Internal carotid a. to arterial circle and brain

Facial a.

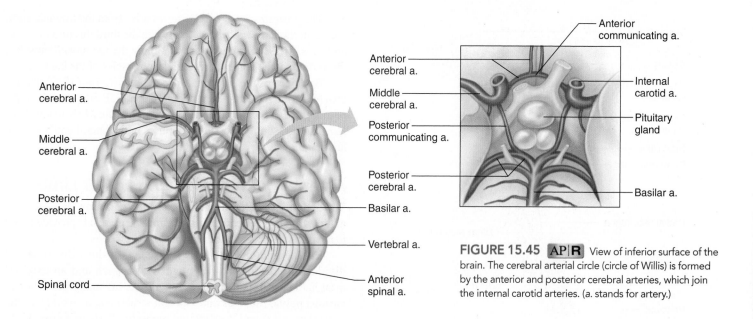

Labels on left illustration:
Anterior cerebral a.
Middle cerebral a.
Posterior cerebral a.
Spinal cord
Basilar a.
Vertebral a.
Anterior spinal a.

Labels on right illustration:
Anterior cerebral a.
Middle cerebral a.
Posterior communicating a.
Posterior cerebral a.
Anterior communicating a.
Internal carotid a.
Pituitary gland
Basilar a.

FIGURE 15.45 **AP|R** View of inferior surface of the brain. The cerebral arterial circle (circle of Willis) is formed by the anterior and posterior cerebral arteries, which join the internal carotid arteries. (*a.* stands for artery.)

The left and right *common carotid arteries* ascend deeply in the neck on either side. At the level of the upper laryngeal border, they divide to form the internal and external carotid arteries.

The **external carotid artery** courses upward on the side of the head, giving off branches to structures in the neck, face, jaw, scalp, and base of the skull. The main vessels that originate from this artery include the following:

1. *superior thyroid artery* to the hyoid bone, larynx, and thyroid gland
2. *lingual artery* to the tongue, muscles of the tongue, and salivary glands beneath the tongue
3. *facial artery* to the pharynx, palate, chin, lips, and nose
4. *occipital artery* to the scalp on the back of the skull, the meninges, the mastoid process, and various muscles in the neck
5. *posterior auricular artery* to the ear and the scalp over the ear

The external carotid artery terminates by dividing into the *maxillary* and *superficial temporal arteries*. The maxillary artery supplies blood to the teeth, gums, jaws, cheek, nasal cavity, eyelids, and meninges. The superficial temporal artery extends to the parotid salivary gland and to various surface regions of the face and scalp.

The **internal carotid artery** begins lateral to the external carotid artery, then follows a deep course upward along the pharynx to the base of the skull. Entering the cranial cavity, it provides the major blood supply to the brain. The major branches of the internal carotid artery include the following:

1. *ophthalmic artery* to the eyeball and to various muscles and accessory organs within the orbit
2. *posterior communicating artery* that forms part of the cerebral arterial circle
3. *anterior choroid artery* to the choroid plexus within the lateral ventricle of the brain and to nerve structures in the brain

The internal carotid artery terminates by dividing into the *anterior* and *middle cerebral arteries*. The middle cerebral artery passes through the lateral tissue and supplies the lateral surface of the cerebrum, including the primary motor and sensory areas of the face and upper limbs, the optic radiations, and the speech area (see chapter 11, pp. 401–402). The anterior cerebral artery extends anteriorly between the cerebral hemispheres and supplies the medial surface of the brain.

Near the base of each internal carotid artery is an enlargement called a **carotid sinus.** Like the aortic arch, these structures contain baroreceptors that control blood pressure. A number of small epithelial masses, called **carotid bodies,** are also in the wall of the carotid sinus. The carotid bodies have chemoreceptors that act with those of the aortic bodies in monitoring blood chemistry, to regulate circulation and respiration.

Arteries to the Shoulder and Upper Limb

The subclavian artery, after giving off branches to the neck, continues into the arm (fig. 15.46). It passes between the clavicle and the first rib and becomes the axillary artery.

The **axillary artery** supplies branches to structures in the axilla and the chest wall, including the skin of the shoulder; part of the mammary gland; the upper end of the humerus; the shoulder joint; and muscles in the back, shoulder, and chest. As this vessel leaves the axilla, it becomes the brachial artery.

The **brachial artery** courses along the humerus to the elbow. It gives rise to a *deep brachial artery* that curves posteriorly around the humerus and supplies the triceps brachii muscle. Shorter branches pass into the muscles on the anterior side of the arm, whereas others descend on each side to the elbow and connect with arteries in the forearm. The resulting arterial network allows blood to reach the forearm even if a portion of the distal brachial artery becomes obstructed.

In the elbow, the brachial artery divides into an ulnar artery and a radial artery. The **ulnar artery** leads downward on the ulnar side of the forearm to the wrist. Some of its branches join the anastomosis around the elbow joint, whereas others supply blood to flexor and extensor muscles in the forearm.

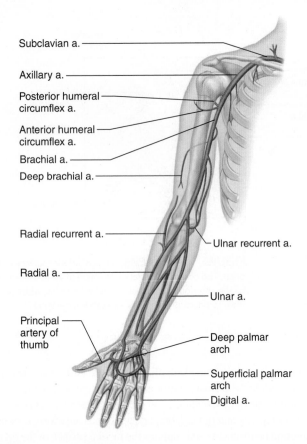

Subclavian a.
Axillary a.
Posterior humeral circumflex a.
Anterior humeral circumflex a.
Brachial a.
Deep brachial a.
Radial recurrent a.
Ulnar recurrent a.
Radial a.
Ulnar a.
Principal artery of thumb
Deep palmar arch
Superficial palmar arch
Digital a.

FIGURE 15.46 **AP|R** The major arteries to the shoulder and upper limb. (a. stands for artery.)

Blood from the brachial artery flows into which artery (arteries) in the forearm?
Answer can be found in Appendix G.

The **radial artery,** which is a continuation of the brachial artery, extends along the radial side of the forearm to the wrist. As it nears the wrist, it comes close to the surface and provides a convenient vessel for taking the pulse (radial pulse). Branches of the radial artery join the anastomosis of the elbow and supply the lateral muscles of the forearm.

At the wrist, the branches of the ulnar and radial arteries join to form a network of vessels. Arteries arising from this network supply blood to structures in the hand.

Arteries to the Thoracic and Abdominal Walls

Blood reaches the thoracic wall through several vessels. These include branches from the subclavian artery and the thoracic aorta (fig. 15.47).

The subclavian artery contributes to this supply through a branch called the **internal thoracic artery.** This vessel originates in the base of the neck and passes downward on the pleura and behind the cartilages of the upper six ribs. It gives off two *anterior intercostal arteries* to each of the upper six intercostal spaces; these two arteries supply the intercostal muscles, other intercostal tissues, and the mammary glands.

The *posterior intercostal arteries* arise from the thoracic aorta and enter the intercostal spaces between the third through the eleventh ribs. These arteries branch to supply the intercostal muscles, the vertebrae, the spinal cord, and deep muscles of the back.

Branches of the *internal thoracic* and *external iliac arteries* provide blood to the anterior abdominal wall. Paired vessels originating from the abdominal aorta, including the *phrenic* and *lumbar arteries,* supply blood to structures in the lateral and posterior abdominal wall.

Arteries to the Pelvis and Lower Limb

The abdominal aorta divides to form the **common iliac** (il′e-ak) **arteries** at the level of the pelvic brim. These vessels provide blood to the pelvic organs, gluteal region, and lower limbs.

Each common iliac artery descends a short distance and divides into an internal (hypogastric) branch and an external branch. The **internal iliac artery** gives off many branches to various pelvic muscles and visceral structures, as well as to the gluteal muscles and the external genitalia. Parts of figure 15.48 show important branches of this vessel, including the following:

1. *iliolumbar artery* to the ilium and muscles of the back
2. *superior and inferior gluteal arteries* to the gluteal muscles, pelvic muscles, and skin of the buttocks
3. *internal pudendal artery* to muscles in the distal portion of the alimentary canal, the external genitalia, and the hip joint
4. *superior* and *inferior vesical arteries* to the urinary bladder; in males, these vessels also supply the seminal vesicles and the prostate gland
5. *middle rectal artery* to the rectum
6. *uterine artery* to the uterus and vagina

The **external iliac artery** provides the main blood supply to the lower limbs (fig. 15.49). It passes downward along the brim of the pelvis and gives off two large branches—an *inferior epigastric artery* and a *deep circumflex iliac artery*. These vessels supply the muscles and skin in the lower abdominal wall. Midway between the pubic symphysis and the anterior superior iliac spine, the external iliac artery becomes the femoral artery.

The **femoral** (fem′or-al) **artery,** which passes fairly close to the anterior surface of the upper thigh, gives off many branches to muscles and superficial tissues of the thigh. These branches also supply the skin of the groin and the lower abdominal wall. Important subdivisions of the femoral artery include the following:

1. *superficial circumflex iliac artery* to the lymph nodes and skin of the groin
2. *superficial epigastric artery* to the skin of the lower abdominal wall
3. *superficial* and *deep external pudendal arteries* to the skin of the lower abdomen and external genitalia
4. *deep femoral artery* (the largest branch of the femoral artery) to the hip joint and muscles of the thigh
5. *deep genicular artery* to distal ends of thigh muscles and to an anastomosis around the knee joint

As the femoral artery reaches the proximal border of the space behind the knee (popliteal fossa), it becomes the **popliteal**

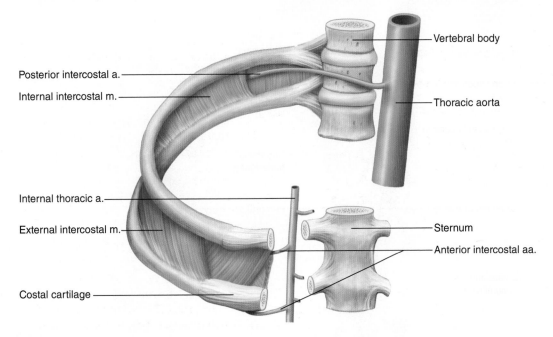

FIGURE 15.47 **AP|R** Arteries that supply the thoracic wall. (*a.* stands for artery, *aa.* stands for arteries , *m.* stands for muscle.)

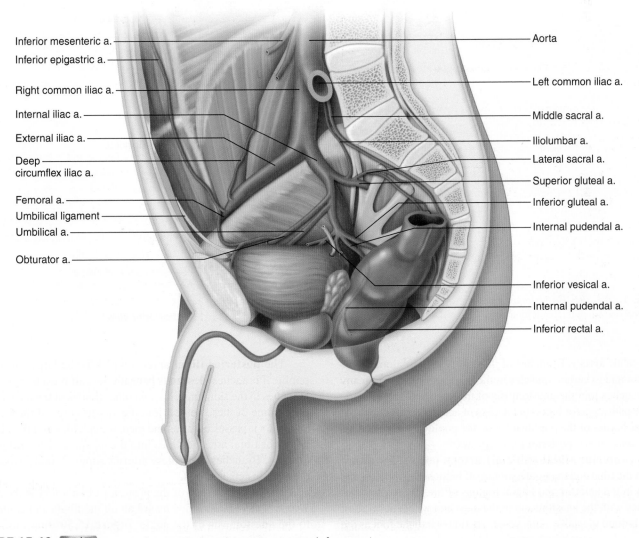

FIGURE 15.48 **AP|R** Arteries that supply the pelvic region. (*a.* stands for artery.)

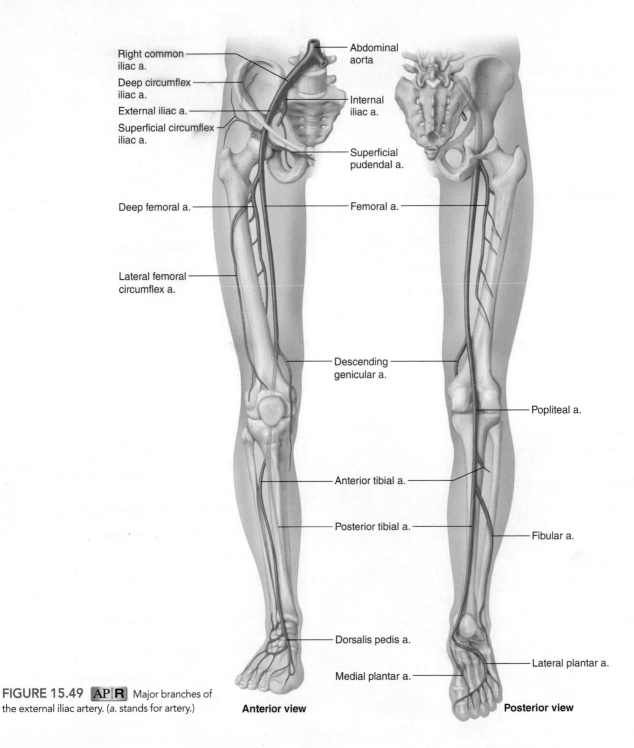

Right common iliac a.

Deep circumflex iliac a.

External iliac a.

Superficial circumflex iliac a.

Abdominal aorta

Internal iliac a.

Superficial pudendal a.

Deep femoral a.

Femoral a.

Lateral femoral circumflex a.

Descending genicular a.

Popliteal a.

Anterior tibial a.

Posterior tibial a.

Fibular a.

Dorsalis pedis a.

Lateral plantar a.

Medial plantar a.

Anterior view

Posterior view

FIGURE 15.49 AP|R Major branches of the external iliac artery. (a. stands for artery.)

(pop″lĭ-te′al) **artery.** Branches of this artery supply blood to the knee joint and to certain muscles in the thigh and calf. Also, many of its branches join the anastomosis of the knee and help provide alternate pathways for blood in the case of arterial obstructions. At the lower border of the popliteal fossa, the popliteal artery divides into the anterior and posterior tibial arteries.

The **anterior tibial** (tib′e-al) **artery** passes downward between the tibia and the fibula, giving off branches to the skin and muscles in the anterior and lateral regions of the leg. It also communicates with the anastomosis of the knee and with a network of arteries around the ankle. This vessel continues into the foot as the *dorsalis pedis artery,* which supplies blood to the instep and toes.

The **posterior tibial artery,** which is the larger of the two popliteal branches, descends beneath the calf muscles, giving off branches to the skin, muscles, and other tissues of the leg along the way. Some of these vessels join the anastomoses of the knee and ankle. As it passes between the medial malleolus and the heel, the posterior tibial artery divides into the *medial* and *lateral plantar arteries.* Branches from these arteries supply blood to tissues of the heel, instep, and toes.

The largest branch of the posterior tibial artery is the *fibular artery,* which extends downward along the fibula and contributes to the anastomosis of the ankle. **Figure 15.50** shows the major vessels of the arterial system.

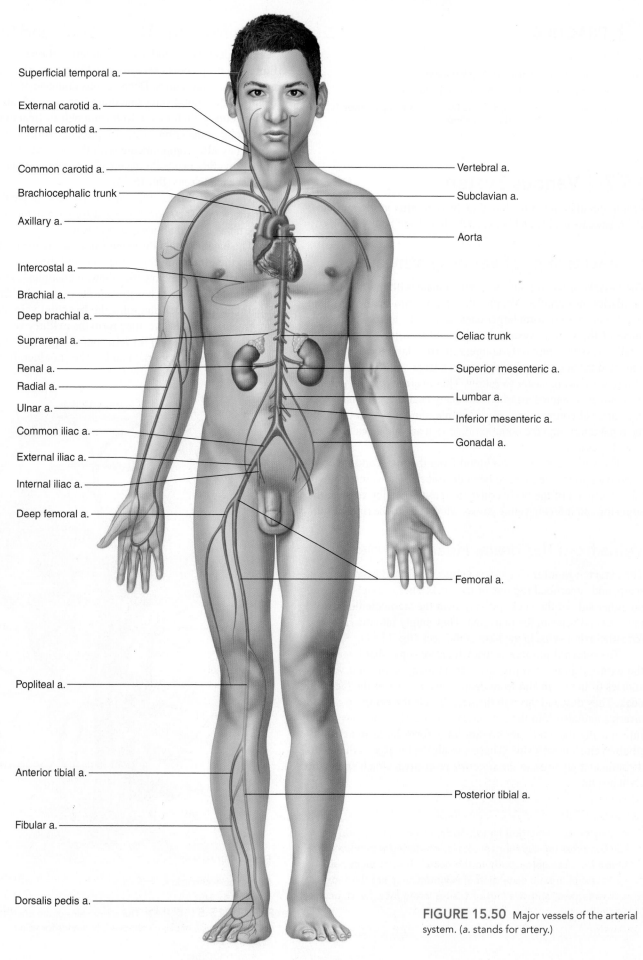

Superficial temporal a.

External carotid a.

Internal carotid a.

Common carotid a.

Brachiocephalic trunk

Axillary a.

Intercostal a.

Brachial a.

Deep brachial a.

Suprarenal a.

Renal a.

Radial a.

Ulnar a.

Common iliac a.

External iliac a.

Internal iliac a.

Deep femoral a.

Popliteal a.

Anterior tibial a.

Fibular a.

Dorsalis pedis a.

Vertebral a.

Subclavian a.

Aorta

Celiac trunk

Superior mesenteric a.

Lumbar a.

Inferior mesenteric a.

Gonadal a.

Femoral a.

Posterior tibial a.

FIGURE 15.50 Major vessels of the arterial system. (*a.* stands for artery.)

59 Name the parts of the aorta.

60 Name the vessels that arise from the aortic arch.

61 Name the branches of the thoracic and abdominal aorta.

62 Which vessels supply blood to the head? To the upper limb? To the abdominal wall? To the lower limb?

15.7 | Venous System

Venous circulation returns blood to the heart after gases, nutrients, and wastes are exchanged between the blood and body cells.

Characteristics of Venous Pathways

The vessels of the venous system originate with the merging of capillaries into venules. Venules then merge into small veins, and small veins meet to form larger ones. Unlike the arterial pathways, those of the venous system are difficult to follow because the smaller vessels commonly connect in irregular networks. Many unnamed tributaries may join to form a large vein. The pathways of larger veins are easier to follow. These veins typically parallel the courses of named arteries, and many bear the same names as their arterial counterparts. For example, the renal vein parallels the renal artery, and the common iliac vein accompanies the common iliac artery.

The veins that carry the blood from the lungs and myocardium back to the heart have been described. The veins from all the other parts of the body converge into two major vessels, the **superior** and **inferior venae cavae,** which lead to the right atrium.

Veins from the Brain, Head, and Neck

The **external jugular** (jug'u-lar) **veins** drain blood from the face, scalp, and superficial regions of the neck. These vessels descend on either side of the neck, passing over the sternocleidomastoid muscles and beneath the platysma. They empty into the *right* and *left* **subclavian veins** in the base of the neck (fig. 15.51).

The **internal jugular veins,** which are somewhat larger than the external jugular veins, arise from many veins and venous sinuses of the brain and from deep veins in parts of the face and neck. They descend through the neck beside the common carotid arteries and also join the subclavian veins. These unions of the internal jugular and subclavian veins form large **brachiocephalic veins** on each side. These vessels then merge in the mediastinum and give rise to the *superior vena cava,* which enters the right atrium.

> A lung cancer, enlarged lymph node, or an aortic aneurysm can compress the superior vena cava, interfering with return of blood from the upper body to the heart. This produces pain; shortness of breath; distension of veins draining into the superior vena cava; and swelling of tissues in the face, head, and upper limbs.

Veins from the Upper Limb and Shoulder

A set of deep veins and a set of superficial ones drain the upper limb. The deep veins generally parallel the arteries in each region and have similar names. Deep venous drainage of the upper limbs begins in the digital veins that drain into pairs of **radial veins** and **ulnar veins,** which merge to form a pair of **brachial veins.** The superficial veins connect in complex networks just beneath the skin. They also communicate with the deep vessels of the upper limb, providing many alternate pathways through which the blood can leave the tissues (fig. 15.52).

The major vessels of the superficial network are the basilic and cephalic veins. They arise from anastomoses in the palm and wrist on the ulnar and radial sides, respectively.

The **basilic** (bah-sil'ik) **vein** passes along the back of the forearm on the ulnar side for a distance and then curves forward to the anterior surface below the elbow. It continues ascending on the medial side until it reaches the middle of the arm. There it deeply penetrates the tissues and joins the **brachial vein.** As the basilic and brachial veins merge, they form the **axillary vein.**

The **cephalic** (sĕ-fal'ik) **vein** courses upward on the lateral side of the upper limb from the hand to the shoulder. In the shoulder,

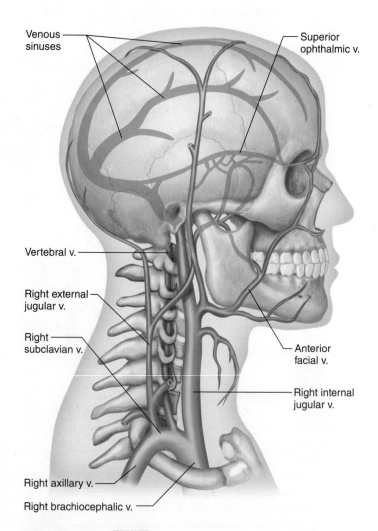

FIGURE 15.51 AP|R The major veins of the brain, head, and neck. The clavicle has been removed. (*v.* stands for vein.)

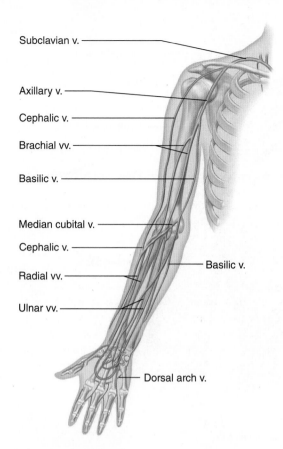

Subclavian v.

Axillary v.

Cephalic v.

Brachial vv.

Basilic v.

Median cubital v.

Cephalic v.

Radial vv.

Ulnar vv.

Basilic v.

Dorsal arch v.

FIGURE 15.52 [AP|R] The major veins of the upper limb and shoulder. (*v.* stands for vein, *vv.* stands for veins.)

Blood from the brachial and basilic veins drains into which vein(s)?

Answer can be found in Appendix G.

it pierces the tissues and joins the axillary vein, which beyond the axilla becomes the **subclavian vein.**

In the bend of the elbow, a *median cubital vein* ascends from the cephalic vein on the lateral side of the forearm to the basilic vein on the medial side. This large vein is usually visible beneath the skin. It is often used as a site for *venipuncture,* when it is necessary to remove a sample of blood for examination or to add fluids to the blood.

Veins from the Abdominal and Thoracic Walls

Tributaries of the brachiocephalic and azygos veins drain the abdominal and thoracic walls. For example, the *brachiocephalic vein* receives blood from the *internal thoracic vein,* which generally drains the tissues the internal thoracic artery supplies. Some *intercostal veins* also empty into the brachiocephalic vein (fig. 15.53).

The **azygos** (az′ĭ-gos) **vein** originates in the dorsal abdominal wall and ascends through the mediastinum on the right side of the vertebral column to join the superior vena cava. It drains most of the muscular tissue in the abdominal and thoracic walls.

Tributaries of the azygos vein include the *posterior intercostal veins* on the right side, which drain the intercostal spaces, and

the *superior* and *inferior hemiazygos veins,* which receive blood from the posterior intercostal veins on the left. The right and left *ascending lumbar veins,* with tributaries that include vessels from the lumbar and sacral regions, also connect to the azygos system.

Veins from the Abdominal Viscera

Veins transport blood directly to the atria of the heart, except for portal veins, such as those that drain the abdominal viscera (fig. 15.54). They originate in the capillary networks of the stomach, intestines, pancreas, and spleen and transport blood from these organs through a **hepatic portal** (por′tal) **vein** to the liver (fig. 15.55). There the blood enters capillary-like **hepatic sinusoids** (hĕ-pat′ik si′nŭ-soidz). This unique venous pathway is called the **hepatic portal system,** which allows blood to flow from the gastrointestinal organs to the liver before returning to the heart.

The tributaries of the hepatic portal vein include the following vessels:

1. right and left *gastric veins* from the stomach
2. *superior mesenteric vein* from the small intestine, ascending colon, and transverse colon
3. *splenic vein* from a convergence of several veins draining the spleen, the pancreas, and a portion of the stomach; as well as its largest tributary, the *inferior mesenteric vein,* from the descending colon, sigmoid colon, and rectum

About 80% of the blood flowing to the liver in the hepatic portal system comes from the capillaries in the stomach and intestines and is oxygen-poor, but nutrient-rich (see chapter 17, pp. 670–671). The liver handles these nutrients in a variety of ways. It regulates blood glucose concentration by polymerizing excess glucose into glycogen for storage or by breaking down glycogen into glucose when blood glucose concentration drops below normal.

The liver helps regulate blood concentrations of recently absorbed amino acids and lipids by modifying them into forms cells can use, by oxidizing them, or by changing them into storage forms. The liver also stores certain vitamins and detoxifies harmful substances.

Blood in the hepatic portal vein nearly always contains bacteria that have entered through intestinal capillaries. Large *Kupffer cells* lining the hepatic sinusoids phagocytize these microorganisms, removing them from the portal blood before it leaves the liver.

After passing through the hepatic sinusoids of the liver, the blood in the hepatic portal system travels through a series of merging vessels into **hepatic veins.** These veins empty into the *inferior vena cava,* returning the blood to the general circulation.

Other veins empty into the inferior vena cava as it ascends through the abdomen. They include the *lumbar, gonadal, renal, suprarenal,* and *phrenic veins.* These vessels drain regions that arteries with corresponding names supply.

Veins from the Lower Limb and Pelvis

Veins that drain the blood from the lower limb can be divided into deep and superficial groups, as in the upper limb (fig. 15.56). The deep veins of the leg, such as the paired *anterior* and *posterior* **tibial veins,** have names that correspond to the arteries they accompany. At the level of the knee, these vessels form a single trunk, the

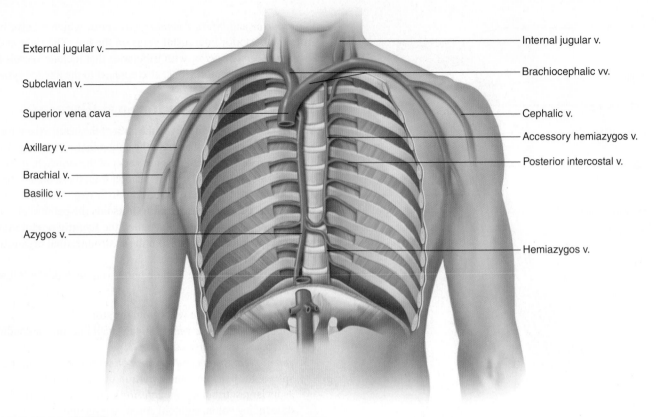

FIGURE 15.53 AP|R Veins that drain the thoracic wall. (*v.* stands for vein, *vv.* stands for veins.)

Labels (top figure):
- External jugular v.
- Subclavian v.
- Superior vena cava
- Axillary v.
- Brachial v.
- Basilic v.
- Azygos v.
- Internal jugular v.
- Brachiocephalic vv.
- Cephalic v.
- Accessory hemiazygos v.
- Posterior intercostal v.
- Hemiazygos v.

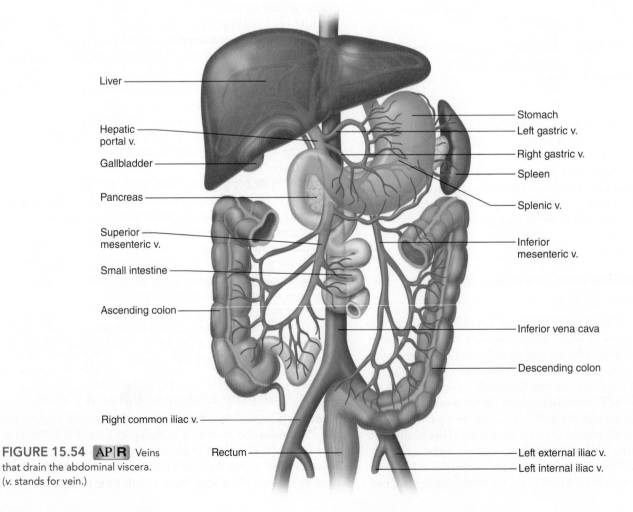

FIGURE 15.54 AP|R Veins that drain the abdominal viscera. (*v.* stands for vein.)

Labels (bottom figure):
- Liver
- Hepatic portal v.
- Gallbladder
- Pancreas
- Superior mesenteric v.
- Small intestine
- Ascending colon
- Right common iliac v.
- Rectum
- Stomach
- Left gastric v.
- Right gastric v.
- Spleen
- Splenic v.
- Inferior mesenteric v.
- Inferior vena cava
- Descending colon
- Left external iliac v.
- Left internal iliac v.

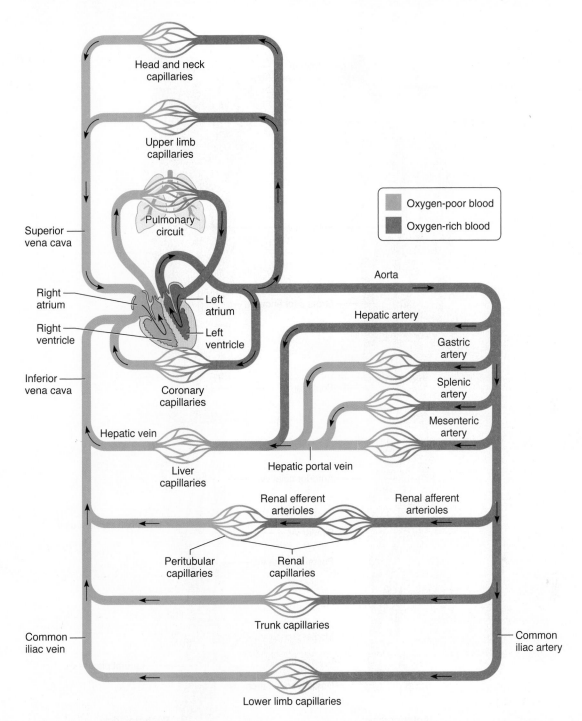

FIGURE 15.55 In this schematic drawing of the cardiovascular system, note how the hepatic portal vein drains one set of capillaries and leads to another set. A similar relationship exists in the kidneys.

popliteal vein. This vein continues upward through the thigh as the **femoral vein,** which, in turn, becomes the **external iliac vein.**

The superficial veins of the foot, leg, and thigh connect to form a complex network beneath the skin. These vessels drain into two major trunks: the small and great saphenous veins.

The **small saphenous** (sah-fe′nus) **vein** begins in the lateral portion of the foot and passes upward behind the lateral malleolus. It ascends along the back of the calf, enters the popliteal fossa, and joins the popliteal vein.

The **great saphenous vein,** the longest vein in the body, originates on the medial side of the foot. It ascends in front of the medial

malleolus and extends upward along the medial side of the leg and thigh. In the thigh just below the inguinal ligament, it deeply penetrates and joins the femoral vein. Near its termination, the great saphenous vein receives tributaries from a number of vessels that drain the upper thigh, groin, and lower abdominal wall.

In addition to communicating freely with each other, the saphenous veins communicate extensively with the deep veins of the leg and thigh. Blood can thus return to the heart from the lower extremities by several routes.

In the pelvic region, vessels leading to the **internal iliac veins** transport blood away from organs of the reproductive, urinary,

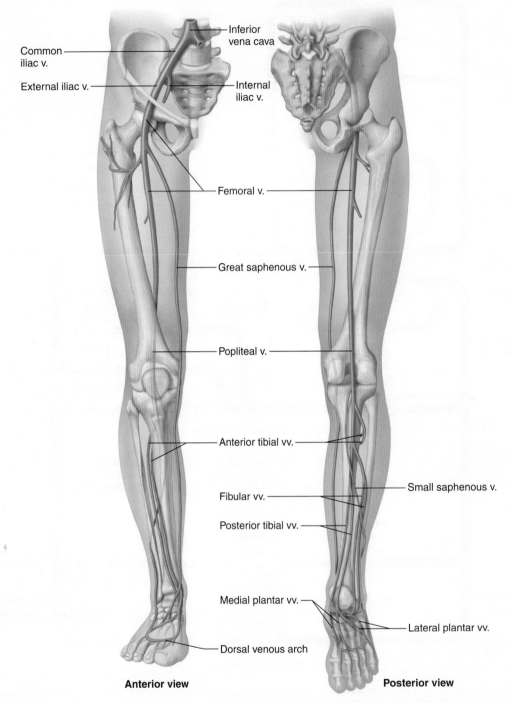

Inferior
vena cava

Common
iliac v.

External iliac v.

Internal
iliac v.

Femoral v.

Great saphenous v.

Popliteal v.

Anterior tibial vv.

Small saphenous v.

Fibular vv.

Posterior tibial vv.

Medial plantar vv.

Lateral plantar vv.

Dorsal venous arch

Anterior view

Posterior view

FIGURE 15.56 The major veins of the lower limb and pelvis. (*v.* stands for vein, *vv.* stands for veins.)

and digestive systems. These veins are formed by tributaries corresponding to the branches of the internal iliac artery, such as the *gluteal, pudendal, vesical, rectal, uterine,* and *vaginal veins*. Typically, these veins have many connections and form complex networks (plexuses) in the regions of the rectum, urinary bladder, and prostate gland (in the male) or uterus and vagina (in the female).

The internal iliac veins originate deep within the pelvis and ascend to the pelvic brim. There they unite with the right and left external iliac veins to form the **common iliac veins.** These vessels,

in turn, merge to produce the *inferior vena cava* at the level of the fifth lumbar vertebra. **Figure 15.57** shows the major vessels of the venous system.

PRACTICE

63 Name the veins that return blood to the right atrium.

64 Which major veins drain blood from the head? From the upper limbs? From the abdominal viscera? From the lower limbs?

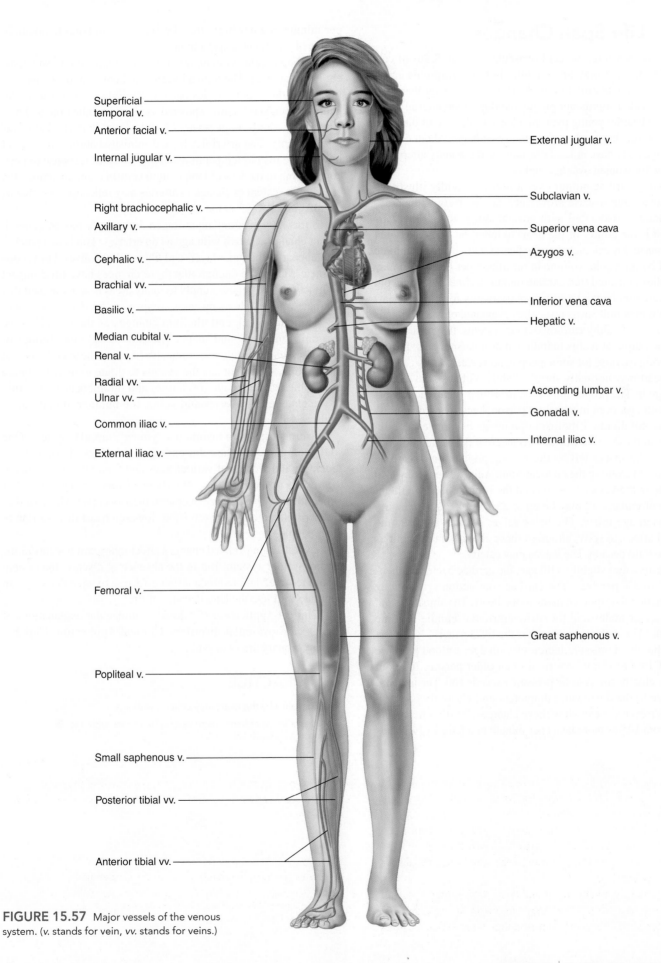

Superficial temporal v.

Anterior facial v.

Internal jugular v.

Right brachiocephalic v.

Axillary v.

Cephalic v.

Brachial vv.

Basilic v.

Median cubital v.

Renal v.

Radial vv.

Ulnar vv.

Common iliac v.

External iliac v.

Femoral v.

Popliteal v.

Small saphenous v.

Posterior tibial vv.

Anterior tibial vv.

External jugular v.

Subclavian v.

Superior vena cava

Azygos v.

Inferior vena cava

Hepatic v.

Ascending lumbar v.

Gonadal v.

Internal iliac v.

Great saphenous v.

FIGURE 15.57 Major vessels of the venous system. (*v.* stands for vein, *vv.* stands for veins.)

15.8 | Life-Span Changes

The years take a toll on the cardiovascular system. Signs of cardiovascular disease may appear long before symptoms arise. Autopsies of soldiers killed in the Korean and Vietnam Wars, for example, revealed significant plaque buildup in the arterial walls of otherwise healthy young men. Incidence of disease of the heart and blood vessels increases exponentially with age. About 60% of men over age sixty have at least one narrowed coronary artery; the same is true for women over age eighty.

Assessing cardiac output over a lifetime vividly illustrates how cardiovascular disease prevalence can interfere with studying the changes associated with normal aging. Recall that cardiac output is the ability of the heart to meet the body's oxygen requirements and is calculated as the heart rate in beats per minute multiplied by the stroke volume in milliliters per beat. For many years, studies indicated that cardiac output declines with age, but when researchers began to screen participants for hidden heart disease with treadmill stress tests, then evaluated only individuals with completely healthy cardiovascular systems, they discovered that cardiac output at rest is maintained as a person ages. It does decline during exercise for some people, however.

The heart may normally shrink slightly with age, but disease may enlarge it. The proportion of the heart that is cardiac muscle declines with age, even in a healthy person, because cardiac muscle cells do not divide. Lipofuscin pigments become especially prominent in these cells. Fibrous connective tissue and adipose tissue fill in the spaces left by the waning population of cardiac muscle cells, thickening the endocardium. Adipose cells may also accumulate in the ventricle walls and the septum between them. The left ventricular wall may be up to 25% thicker at age eighty than it was at age thirty. The heart valves thicken and become more rigid after age sixty, although these changes may begin as early as the third decade. The valves may calcify.

The heart slows slightly with age, the cardiac cycle lengthening by 2% to 5% per year. The cardiac conduction system may remain functional despite changes to the heart. The sinoatrial and atrioventricular nodes and the atrioventricular bundle become more elastic. These changes may alter the ECG pattern.

Systolic blood pressure increases with age; a blood pressure reading of 140/90 is not abnormal in an older person. In about 40% of the elderly, the systolic pressure exceeds 160. The increase may be due to the decreasing diameters and elasticity of arteries. Regular exercise can slow these changes. Resting heart rate declines from 145 or more beats per minute in a fetus to 140 beats per minute in a newborn, then levels out in an adult to about 70 (range of 60–99) beats per minute.

In the vascular system, age-related changes are most apparent in the arteries. The tunica interna thickens and, over time, the lumens of the larger arteries narrow. Rigidity increases as collagen, calcium, and fat are deposited as elastin production declines. Arterial elasticity at age seventy is only about half of what it was at age twenty. The arterioles have diminished ability to contract in response to cold temperatures and to dilate in response to heat, contributing to the loss of temperature control common among the elderly. The extent of change in arteries may reflect stress—that is, not all arteries "age" at the same rate.

Veins may accumulate collagen and calcify but, in general, do not change as much with age as do arteries. Thickened patches may appear in the inner layer, and fibers in the valves, but venous diameters are large enough that these changes have little impact on function. The venous supply to many areas is so redundant that alternate veins can in many cases compensate for damaged ones.

The once-sleek endothelium changes as the cells become less uniform in size and shape. The endothelial inner linings of blood vessels are important to health because these cells release nitric oxide, which signals the vessels to dilate to increase blood flow, which counters atherosclerosis and thrombosis. In addition to the changes in arteries and veins, the number of capillaries declines with age.

Exercise can help maintain a "young" vascular system. One study compared the vascular endothelial linings of athletic and sedentary individuals of various ages and found that the status of the vessels of the exercising elderly were very similar to those of either athletic or sedentary people in their twenties. Many studies have correlated regular exercise to lowered heart disease risk in older people.

Overall, aging-related changes affect many components of the cardiovascular system. But in the absence of disease, the system is so fine-tuned and redundant that effective oxygen delivery can continue well into the later decades of life.

Clinical Application 15.5 looks at molecular explanations of certain cardiovascular disorders. Clinical Application 15.6 discusses coronary artery disease.

 PRACTICE

65 Explain why the heart may enlarge with age.

66 Describe what happens to resting heart rate with age. ■

Several environmental and inherited factors contribute to cardiovascular disease, including poor diet, sedentary lifestyle, and genetic predisposition. Disorders of the heart and blood vessels caused by single genes are rare, but understanding how they arise can provide insights useful in developing treatments for more prevalent forms of disease. For example, widely used cholesterol-lowering drugs called statins were developed based on understanding an inherited condition, familial hypercholesterolemia, which produces faulty LDL (low-density lipoprotein) receptors and that affects one in a million children.

A Connective Tissue Defect

Just after midnight, the college basketball player lay in her bed watching TV. At about 2 A.M., her roommate heard sounds of disturbed breathing and could not rouse the young player. The nineteen-year-old athlete died within an hour of reaching the hospital. Her aorta had burst. She had *Marfan syndrome,* an inherited condition that also caused the characteristics that led her to excel in her sport—her great height and long fingers.

In Marfan syndrome, an abnormal form of a connective tissue protein called fibrillin weakens the aorta wall, dilating the aortic root (fig. 15H). Identifying the mutation can make it possible to locate and surgically repair a weakened aorta, or to take a drug that can slow the effect of the disease on the aorta. This can prevent a tragic first symptom—sudden death.

A Myosin Defect

Each year, one or two seemingly healthy young people die suddenly during a sports event, usually basketball. The cause of death is often *familial hypertrophic cardiomyopathy,* an inherited overgrowth of the myocardium. The defect in this disorder is different from that behind Marfan syndrome. It is an abnormality in one of the myosin chains that comprise cardiac muscle. Again, detecting the responsible gene can alert affected individuals to their increased risk of sudden death. They can adjust the type of exercise they do to avoid stressing the cardiovascular system.

A Metabolic Block

Inherited heart disease can strike early in life. Jim D. died at four days of age, two days after suffering cardiac arrest. Two years later, his parents had another son. Like Jim, Kerry seemed normal at birth, but when he was thirty-six hours old, his heart rate plummeted, he had a seizure, and he stopped breathing. He was resuscitated. A blood test revealed excess long-chain fatty acids, indicating inability to use fatty acids. Hunger triggered the symptoms because the boys could not use fatty acids for energy, as healthy people do. Kerry survived for three years by following a diet low in fatty acids and eating frequently. Once he became comatose because he missed a meal. Eventually, he died of respiratory failure.

Jim and Kerry had inherited a deficiency of a mitochondrial enzyme that processes long-chain fatty acids. Because long-chain fatty acids are a primary energy source for cardiac muscle, their hearts failed.

Controlling Cholesterol

Low-density lipoprotein (LDL) receptors on liver cells admit cholesterol into the cells, keeping the lipid from building up in the bloodstream and occluding arteries. When LDL receptors bind cholesterol, they activate a negative feedback mechanism that temporarily halts the cell's production of cholesterol. In the severe form of *familial hypercholesterolemia,* a person inherits two defective copies of the gene encoding the LDL receptors. Yellowish lumps of cholesterol are deposited behind the knees and elbows, and heart failure usually causes death in childhood. People who inherit one defective gene have a milder form of the illness. They develop coronary artery disease in young adulthood, but can delay symptoms by following a heart-healthy diet and regularly exercising. These people have half the normal number of LDL receptors.

In Niemann-Pick type C disease, a defective protein disturbs the fate of cholesterol inside cells. Normally, the protein escorts cholesterol out of a cell's lysosomes, which triggers the negative feedback mechanism that shuts off cholesterol synthesis. When the protein is absent or malfunctions, the cell keeps producing cholesterol and LDL receptors. Coronary artery disease develops, and is typically fatal in childhood.

FIGURE 15H The symptoms of Marfan syndrome, including a progressive dilation of the aortic root, arise from an abnormal form of the connective tissue protein fibrillin that in turn increases signaling by a protein called transforming growth factor B (TGF-beta). A drug originally developed to treat hypertension, losartan, lowers levels of TGF-beta and slows the life-threatening ballooning out of the ascending aorta. Surgery can replace the aortic root with a synthetic graft. Losartan is helpful even after surgery.

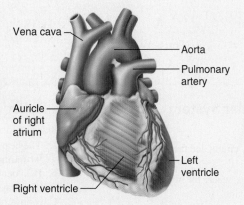
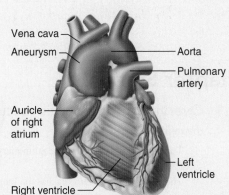

Normal heart   Heart in Marfan syndrome

Dave R., a fifty-two-year-old overweight accountant, had been having occasional chest pains for several months. The mild pain occurred during his usual weekend tennis match, and he attributed it to indigestion. The discomfort almost always diminished after the game, but recently, the pain seemed more severe and prolonged. When Dave asked his physician about the problem, she explained that he was probably experiencing *angina pectoris*, a symptom of *coronary artery disease* (CAD), and suggested an *exercise stress test*. Dave walked on a treadmill, increasing speed and incline while he exercised. An ECG was recorded and his blood pressure monitored. Near the end of the test, when Dave's heart reached the desired rate, a small amount of radioactive thallium-201 was injected into a vein. A *scintillation counter* scanned Dave's heart to determine if branches of his coronary arteries carried the blood marked with the thallium uniformly throughout the myocardium.

The test revealed that Dave was developing CAD. He also had hypertension and high serum cholesterol. The physician advised Dave to lose weight; avoid stress; stop smoking; reduce his intake of foods high in saturated fats, cholesterol, refined carbohydrates, and sodium; and exercise regularly. Dave also started taking medications to lower his blood pressure and relieve the pain of angina.

Six months later, like 1.2 million other people in the United States each year, despite following medical advice, Dave suffered a heart attack (myocardial infarction). Blood flow to part of his myocardium was obstructed, producing oxygen deficiency. The attack began as severe, crushing chest pain, shortness of breath, and sweating. Paramedics stabilized Dave's condition and transported him to a hospital. There, a cardiologist concluded from an ECG that Dave's heart attack was caused by a blood clot blocking a coronary artery (occlusive coronary thrombosis). The cardiologist intravenously administered a thrombolytic ("clot-busting") drug.

A repeat ECG a few hours later showed that the blood vessel remained partially obstructed, so the cardiologist ordered a *coronary angiogram* (see fig. 15.12). In this X-ray procedure, conducted in a cardiac catheterization laboratory, a thin plastic catheter was passed through a guiding sheath inserted into the femoral artery of Dave's right inguinal area. From there, the catheter was pushed into the aorta until it reached the region of the openings to the coronary arteries. Then, dye was injected to track coronary blood flow in sequential X rays.

X-ray fluoroscopy monitored the progress of the catheter. Each time the catheter was in proper position, a radiopaque dye (contrast medium) was released from its distal end into the blood. X-ray images that revealed the path of the dye as it entered a coronary artery and its branches were digitally recorded and later analyzed frame by frame, revealing a single severe narrowing near the origin of Dave's left anterior descending artery. The cardiologist decided to perform *percutaneous transluminal coronary angioplasty* (PTCA) to enlarge the opening (lumen) of that vessel.

The PTCA passed another plastic catheter through the guiding sheath used for the angiogram. This second tube had a tiny deflated balloon at its tip that was placed in the region of the arterial narrowing and inflated briefly under high pressure. The inflated balloon compressed the atherosclerotic plaque (atheroma) obstructing the arterial wall and stretched the blood vessel wall, widening its lumen (recanalization). Blood flow to the myocardial tissue downstream from the obstruction immediately improved.

About 50% of the time, a vessel opened with PTCA becomes occluded again, because the underlying disease persists. To prevent this restenosis, the doctor inserted a *coronary stent,* which is an expandable tube or coil that holds the vessel wall open. The cardiologist had two other options that have a slightly higher risk of causing damage. She might have vaporized the plaque obstructing the vessel with an excimer laser pulse delivered along optical fibers threaded through the catheter. Or, she could have performed atherectomy, in which a cutting device attached to the balloon inserted into the catheter spins, removing plaque by withdrawing it on the catheter tip.

Should the coronary stent fail, or an obstruction block another heart vessel, Dave might benefit from *coronary bypass surgery.* A portion of his internal mammary artery inside his chest wall would be removed and sutured between the aorta and the blocked coronary artery at a point beyond the obstruction, restoring circulation through the myocardium.

Chapter Summary

15.1 Overview of the Cardiovascular System
(page 557)

1. The heart pumps blood to the pulmonary circuit and the systemic circuit.
2. The **pulmonary circuit** reaches the lungs, where the blood picks up oxygen and unloads carbon dioxide.
3. The **systemic circuit** delivers oxygen to all body cells and picks up carbon dioxide to be excreted.

15.2 The Heart (page 557)

1. Size and location of the **heart**
   a. The heart is about 14 centimeters long and 9 centimeters wide.
   b. It is located in the mediastinum and rests on the diaphragm.
2. Coverings of the heart
   a. A layered **pericardium** encloses the heart.
   b. The pericardial cavity is a space between the visceral and parietal layers of the pericardium.

Cardiovascular System

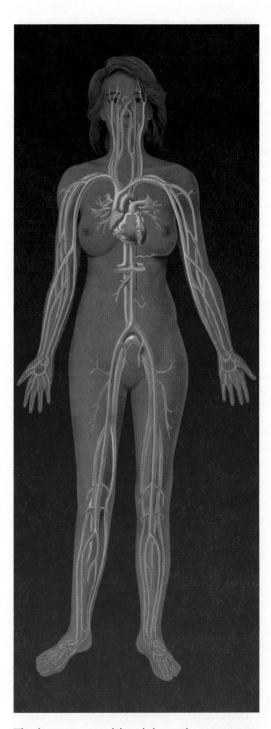

Integumentary System

Changes in skin blood flow are important in temperature control.

Skeletal System

Bones help control plasma levels of calcium ions, which influence heart action.

Muscular System

Blood flow increases to exercising skeletal muscle, delivering oxygen and nutrients and removing wastes. Muscle actions help the blood circulate.

Nervous System

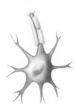

The brain is especially dependent on blood flow for survival. The nervous system helps control blood flow and blood pressure.

Endocrine System

Hormones are carried in the bloodstream. Some hormones directly affect the heart and blood vessels.

Lymphatic System

The lymphatic system returns tissue fluids to the bloodstream.

Digestive System

The digestive system breaks down nutrients into forms readily absorbed by the bloodstream.

Respiratory System

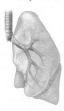

The respiratory system oxygenates the blood and removes carbon dioxide. Respiratory movements help the blood circulate.

Urinary System

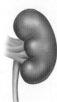

The kidneys clear the blood of wastes. The kidneys help control blood pressure and blood volume.

Reproductive System

Blood pressure is important in normal function of the sex organs.

The heart pumps blood through as many as 60,000 miles of blood vessels, delivering nutrients to, and removing wastes from, all body cells.

3. Wall of the heart
   a. The wall of the heart has three layers.
   b. These layers include an **epicardium**, a **myocardium**, and an **endocardium.**
4. Heart chambers and valves
   a. The heart is divided into four chambers—two **atria** and two **ventricles**—that communicate through **atrioventricular orifices** on each side.
   b. Right chambers and valves
      (1) The right atrium receives blood from the venae cavae and coronary sinus.
      (2) The **tricuspid valve** guards the right atrioventricular orifice.
      (3) The right ventricle pumps blood into the pulmonary trunk.
      (4) A **pulmonary valve** guards the base of the pulmonary trunk.
   c. Left chambers and valves
      (1) The left atrium receives blood from the pulmonary veins.
      (2) The **mitral valve** guards the left atrioventricular orifice.
      (3) The left ventricle pumps blood into the aorta.
      (4) An **aortic valve** guards the base of the aorta.
5. Skeleton of the heart
   a. The skeleton of the heart consists of fibrous rings that enclose the bases of the pulmonary trunk, aorta, and atrioventricular orifices.
   b. The fibrous rings provide attachments for valves and muscle cells and prevent the orifices from excessively dilating during ventricular contractions.
6. Blood flow through the heart
   a. Blood low in oxygen (oxygen-poor blood) and high in carbon dioxide enters the right side of the heart from the venae cavae and coronary sinus and then is pumped into the pulmonary circulation.
   b. After the blood is oxygenated in the lungs and some of its carbon dioxide is removed, the oxygen-rich blood returns to the left side of the heart through the pulmonary veins.
   c. From the left ventricle, the blood moves into the aorta.
7. Blood supply to the heart
   a. The **coronary arteries** supply blood to the myocardium.
   b. It is returned to the right atrium through the **cardiac veins** and **coronary sinus.**
8. Cardiac cycle
   a. The atria contract (atrial systole) while the ventricles relax (ventricular diastole); the ventricles contract (ventricular systole) while the atria relax (atrial diastole).
   b. Pressure in the chambers rises and falls in cycles.
9. Heart sounds
   a. Heart sounds can be described as *lubb-dupp.*
   b. Heart sounds are due to the vibrations that the valve movements produce.
   c. The first part of the sound occurs as AV valves close, and the second part is associated with the closing of pulmonary and aortic valves.
10. Cardiac muscle
   a. Cardiac muscle cells connect to form a **functional syncytium.**
   b. If any part of the syncytium is stimulated, the whole structure contracts as a unit.
   c. Except for a small region in the floor of the right atrium, the fibrous skeleton separates the atrial syncytium from the ventricular syncytium.

11. **Cardiac conduction system**
   a. This system, composed of specialized cardiac muscle tissue, initiates and conducts depolarization waves through the myocardium.
   b. Impulses from the **SA node** pass slowly to the **AV node;** impulses are conducted rapidly along the **AV bundle** and **Purkinje fibers.**
   c. Muscle cells in the ventricular walls form whorls that squeeze blood out of the contracting ventricles.
12. Electrocardiogram
   a. An **electrocardiogram** (ECG, EKG) records electrical changes in the myocardium during a cardiac cycle.
   b. The pattern contains several waves.
      (1) The P wave represents atrial depolarization.
      (2) The QRS complex represents ventricular depolarization.
      (3) The T wave represents ventricular repolarization.
13. Regulation of the cardiac cycle
   a. Physical exercise, body temperature, and concentration of various ions affect heart rate.
   b. Branches of parasympathetic and sympathetic nerve fibers innervate the SA and AV nodes.
      (1) Parasympathetic impulses decrease heart action; sympathetic impulses increase heart action.
      (2) The cardiac center in the medulla oblongata regulates autonomic impulses to the heart.

15.3 Blood Vessels (page 576)

The blood vessels form a closed circuit of tubes that transports blood between the heart and body cells. The tubes include arteries, arterioles, capillaries, venules, and veins.

1. Arteries and arterioles
   a. The **arteries** are adapted to transport blood under relatively high pressure away from the heart.
   b. The **arterioles** are branches of arteries.
   c. The walls of arteries and arterioles consist of layers of endothelium, smooth muscle, and connective tissue.
   d. Autonomic fibers innervate smooth muscle in vessel walls, causing **vasoconstriction** or **vasodilation.**
2. Capillaries
   Capillaries connect arterioles and venules. The capillary wall is a single layer of squamous epithelial cells that forms a semipermeable membrane.
   a. Capillary permeability
      (1) Openings in the capillary walls are thin slits between endothelial cells.
      (2) The sizes of the openings vary from tissue to tissue.
      (3) Endothelial cells of brain capillaries are tightly fused, forming a blood-brain barrier.
   b. Capillary arrangement
      Capillary density varies directly with tissue metabolic rates.
   c. Regulation of capillary blood flow
      (1) Precapillary sphincters regulate capillary blood flow.
      (2) Precapillary sphincters open when cells are low in oxygen and nutrients and close when cellular needs are met.
   d. Exchanges in the capillaries
      (1) Gases, nutrients, and metabolic by-products are exchanged between the capillary blood and the tissue fluid.
      (2) Diffusion provides the most important means of transport.
      (3) Diffusion pathways depend on lipid solubilities.

(4) Plasma proteins generally remain in the blood.

(5) Filtration, due to the hydrostatic pressure of blood, causes a net outward movement of fluid at the arteriolar end of a capillary.

(6) Colloid osmotic pressure causes a net inward movement of fluid at the venular end of a capillary.

(7) Some factors cause fluids to accumulate in the tissues.

3. Venules and veins

a. **Venules** continue from capillaries and merge to form veins.

b. **Veins** transport blood to the heart.

c. Venous walls are similar to arterial walls but are thinner and contain less muscle and elastic tissue.

d. Many veins contain flaplike valves that open, allowing blood to flow to the heart, but close to prevent flow in the opposite direction.

15.4 Blood Pressure (page 582)

Blood pressure is the force blood exerts against the inner walls of blood vessels.

1. Arterial blood pressure

a. The arterial blood pressure is produced primarily by heart action; it rises and falls with phases of the cardiac cycle.

b. **Systolic pressure** occurs when the ventricles contract; **diastolic pressure** occurs when the ventricles relax.

2. Factors that influence arterial blood pressure

a. Cardiac output, blood volume, peripheral resistance, and blood viscosity influence arterial blood pressure.

b. Arterial pressure increases as **cardiac output, blood volume, peripheral resistance,** or blood **viscosity** increases.

3. Control of blood pressure

a. Blood pressure is controlled in part by the mechanisms that regulate cardiac output and peripheral resistance.

b. Cardiac output depends on the volume of blood discharged from the ventricle with each beat **(stroke volume)** and on the heart rate.

(1) The more blood that enters the heart, the stronger the ventricular contraction, the greater the stroke volume, and the greater the cardiac output.

(2) The baroreceptor reflexes involving the cardiac center of the medulla oblongata regulate heart rate.

c. Changes in the diameter of arterioles, controlled by the vasomotor center of the medulla oblongata, regulate peripheral resistance.

4. Venous blood flow

a. Venous blood flow is not a direct result of heart action; it depends on skeletal muscle contraction, breathing movements, and venoconstriction.

b. Venoconstriction can increase venous pressure and blood flow.

5. Central venous pressure

a. Central venous pressure is the pressure in the right atrium.

b. Factors that influence it alter the flow of blood into the right atrium.

c. It affects pressure in the peripheral veins.

15.5 Paths of Circulation (page 590)

1. Pulmonary circuit

a. The pulmonary circuit consists of vessels that transport oxygen-poor blood from the right ventricle to the alveolar capillaries in the lungs, and vessels that transport oxygen-rich blood back to the left atrium.

b. Alveolar capillaries exert less pressure than those of the systemic circuit.

c. Tightly joined epithelial cells of alveolar walls prevent most substances from entering the alveoli.

d. Osmotic pressure rapidly draws water out of alveoli into the interstitial fluid, so alveoli do not fill with fluid.

2. Systemic circuit

a. The systemic circuit is composed of vessels that lead from the left ventricle to all body parts (including vessels supplying the heart itself) and back to the heart.

b. It includes the aorta and its branches as well as the system of veins that return blood to the right atrium.

15.6 Arterial System (page 591)

1. Principal branches of the **aorta**

a. The branches of the ascending aorta include the right and left coronary arteries.

b. The branches of the aortic arch include the **brachiocephalic,** left **common carotid,** and left **subclavian arteries.**

c. The branches of the descending aorta include the thoracic and abdominal groups.

d. The **abdominal aorta** terminates by dividing into right and left common iliac arteries.

2. Arteries to the brain, head, and neck include branches of the subclavian and common carotid arteries.

3. Arteries to the shoulder and upper limb

a. The subclavian artery passes into the arm, and in various regions, it is called the **axillary** and **brachial arteries.**

b. Branches of the brachial artery include the **ulnar** and **radial arteries.**

4. Arteries to the thoracic and abdominal walls

a. Branches of the subclavian artery and thoracic aorta supply the thoracic wall.

b. Branches of the abdominal aorta and other arteries supply the abdominal wall.

5. Arteries to the pelvis and lower limb

a. The **common iliac artery** supplies the pelvic organs, gluteal region, and lower limb.

b. The **femoral artery** of the lower limb becomes the **popliteal artery** that branches into the **anterior** and **posterior tibial arteries.**

15.7 Venous System (page 602)

1. Characteristics of venous pathways

a. The veins return blood to the heart.

b. Larger veins usually parallel the paths of major arteries.

2. Veins from the brain, head, and neck

a. The **jugular veins** drain these regions.

b. Jugular veins unite with **subclavian veins** to form the **brachiocephalic veins.**

3. Veins from the upper limb and shoulder

a. Sets of superficial and deep veins drain the upper limb.

b. Digital veins drain into pairs of **radial veins** and **ulnar veins,** which merge to form a pair of **brachial veins.**

c. The major superficial veins are the **basilic** and **cephalic veins.**

d. Basilic and brachial veins merge to form the **axillary vein.**

e. The median cubital vein in the bend of the elbow is often used as a site for venipuncture.

4. Tributaries of the brachiocephalic and **azygos veins** drain the abdominal and thoracic walls.
5. Veins from the abdominal viscera
   a. The blood from the abdominal viscera generally enters the **hepatic portal system** and is transported to the liver.
   b. The blood in the hepatic portal system is rich in nutrients.
   c. The liver helps regulate the blood concentrations of glucose, amino acids, and lipids.
   d. Phagocytic cells in the liver remove bacteria from the portal blood.
   e. From the liver, hepatic veins transport blood to the inferior vena cava.
6. Veins from the lower limb and pelvis
   a. Sets of deep and superficial veins drain these regions.

b. The deep veins include the **tibial veins,** and the superficial veins include the **saphenous veins.**

15.8 Life-Span Changes (page 608)

1. Plaque buildup may begin early.
2. Fibrous connective tissue and adipose tissue enlarge the heart by filling in when the number and size of cardiac muscle cells fall.
3. Heart rate and output decline slightly with age.
4. Blood pressure increases with age, while resting heart rate decreases with age.
5. Moderate exercise correlates to lowered risk of heart disease in older people.

CHAPTER ASSESSMENTS

15.1 Overview of the Cardiovascular System

1 The _____ circuit picks up oxygen from the lungs and the _____circuit delivers oxygen to the body's cells. (p. 557)

15.2 The Heart

2 The heart is located _____ to the lungs, _____ to the vertebral column, and _____ to the sternum. (p. 557)
   a. medial, posterior, anterior
   b. lateral, posterior, anterior
   c. medial, anterior, posterior
   d. lateral, anterior, posterior
3 Compare the layers of the pericardium and the heart wall. (pp. 557–560)
4 Draw a heart and label the chambers and valves. (pp. 560–561)
5 Blood flows through the vena cavae and coronary sinus into the right atrium, through the _____ to the right ventricle, through the pulmonary valve to the pulmonary trunk into the right and left _____ to the lungs, then leaves the lungs through the pulmonary veins and flows into the _____, through the mitral valve to the _____, and through the _____ to the aorta. (pp. 561–563)
6 List the vessels through which blood flows from the aorta to the myocardium and back to the right atrium. (pp. 563–566)
7 Describe the pressure changes in the atria and ventricles during a cardiac cycle. (pp. 566–567)
8 Explain the origins of heart sounds. (p. 568)
9 Describe the arrangement of cardiac muscle fibers. (p. 568)
10 Distinguish between the roles of the SA node and AV node. (pp. 568–569)
11 Explain how the cardiac conduction system controls the cardiac cycle. (pp. 568–570)
12 Describe and explain the normal ECG pattern. (pp. 570–572)
13 Discuss how the nervous system regulates the cardiac cycle. (p. 572)
14 Describe two factors other than the nervous system that affect the cardiac cycle. (p. 572)

15.3 Blood Vessels

15 Distinguish between an artery and an arteriole. (p. 577)
16 Explain control of vasoconstriction and vasodilation. (p. 577)
17 Describe the structure and function of a capillary. (p. 579)

18 Describe the function of the blood-brain barrier. (p. 580)
19 Explain control of blood flow through a capillary. (pp. 580–581)
20 Relate how diffusion functions in the exchange of substances between blood plasma and tissue fluid. (p. 581)
21 Explain why water and dissolved substances leave the arteriolar end of a capillary and enter the venular end. (p. 581)
22 Describe the effect of histamine on a capillary. (p. 581)
23 Distinguish between a venule and a vein. (p. 582)
24 Explain how veins function as blood reservoirs. (p. 582)

15.4 Blood Pressure

25 Arterial blood pressure peaks when the ventricles contract. This maximum pressure achieved is called the _____. (p. 583)
26 Name several factors that influence blood pressure, and explain how each produces its effect. (pp. 585–586)
27 Describe the control of blood pressure. (pp. 586–588)
28 Which of the following promote the flow of venous blood? (p. 588)
   a. skeletal muscle contraction
   b. breathing
   c. venoconstriction
   d. all of the above
29 Define central venous pressure. (p. 588)

15.5 Paths of Circulation

30 Distinguish between the pulmonary and systemic circuits of the cardiovascular system. (pp. 590–591)
31 Trace the path of blood through the pulmonary circuit. (pp. 590–591)
32 Explain why the alveoli normally do not fill with fluid. (p. 590)

15.6–15.7 Arterial System–Venous System

33 Describe the aorta, and name its principal branches. (pp. 591–594)
34 Discuss the relationship between the major venous pathways and the major arterial pathways to the head, upper limbs, abdominal viscera, and lower limbs. (pp. 594–606)

15.8 Life-Span Changes

35 List and discuss changes in the aging cardiovascular system. (p. 608)

INTEGRATIVE ASSESSMENTS/CRITICAL THINKING

Outcomes 5.5, 9.5, 15.2

1. What structures and properties should an artificial heart have?

Outcome 15.2

2. Why is ventricular fibrillation more likely to be life-threatening than atrial fibrillation?

Outcomes 15.2, 15.3, 15.4, 15.5

3. How might the results of a cardiovascular exam differ for an athlete in top condition and a sedentary, overweight individual?

4. Cigarette smoke contains thousands of chemicals, including nicotine and carbon monoxide. Nicotine constricts blood vessels. Carbon monoxide prevents oxygen binding to hemoglobin. How do these two components of smoke affect the cardiovascular system?

Outcomes 15.2, 15.6

5. If a cardiologist inserts a catheter into a patient's right femoral artery, which arteries will the tube have to pass through to reach the entrance of the left coronary artery?

Outcome 15.3

6. Given the way capillary blood flow is regulated, do you think it is wiser to rest or to exercise following a heavy meal? Cite a reason for your answer.

Outcomes 15.3, 15.4, 15.5, 15.7

7. Cirrhosis of the liver, a disease commonly associated with alcoholism, obstructs blood flow through the hepatic blood vessels. As a result the blood backs up, and the capillary pressure greatly increases in the organs drained by the hepatic portal system. What effects might this increasing capillary pressure produce, and which organs would it affect?

Outcomes 15.5, 15.6, 15.7

8. If a patient develops a blood clot in the femoral vein of the left lower limb and a portion of the clot breaks loose, where is the blood flow likely to carry the embolus? What symptoms are likely?

 # ONLINE STUDY TOOLS

Connect Interactive Questions Reinforce your knowledge using assigned interactive questions covering structures of the heart, locations of the blood vessels, parts of an ECG, factors that affect blood pressure, and more.

Connect Integrated Activity Can you predict the effect(s) of different factors on a patient's blood pressure?

LearnSmart Discover which chapter concepts you have mastered and which require more attention. This adaptive learning tool is personalized, proven, and preferred.

Anatomy & Physiology Revealed Go more in depth into the human body by exploring cadaver dissections of the heart and blood vessels, or by viewing animations demonstrating blood flow or the cardiac cycle.

16

Lymphatic System and Immunity

LEARNING OUTCOMES

After you have studied this chapter, you should be able to:

16.1 Lymphatic Pathways
1 Describe the functions of the lymphatic vessels. (p. 617)
2 Identify and describe the parts of the major lymphatic pathways. (pp. 617–618)

16.2 Tissue Fluid and Lymph
3 Describe how tissue fluid and lymph form, and explain the function of lymph. (p. 620)
4 Explain how lymphatic circulation is maintained, and describe the consequence of lymphatic obstruction. (p. 621)

16.3 Lymphatic Tissues and Lymphatic Organs
5 Describe a lymph node and its major functions. (pp. 621–623)
6 Identify the locations of the major chains of lymph nodes. (p. 622)
7 Discuss the locations and functions of the thymus and spleen. (p. 623)

16.4 Body Defenses Against Infection
8 Distinguish between innate (nonspecific) and adaptive (specific) defenses. (pp. 625–626)
9 List seven innate body defense mechanisms, and describe the action of each mechanism. (pp. 626–627)
10 Explain how two major types of lymphocytes are formed and activated and how they function in immune mechanisms. (pp. 628–630)
11 Identify the parts of an antibody molecule. (p. 632)
12 Discuss the actions of the five types of antibodies. (pp. 634–635)
13 Distinguish between primary and secondary immune responses. (pp. 635–636)
14 Distinguish between active and passive immunity. (p. 636)
15 Explain how hypersensitivity reactions, tissue rejection reactions, and autoimmunity arise from immune mechanisms. (pp. 636–640)

16.5 Life-Span Changes
16 Describe life-span changes in immunity. (p. 640)

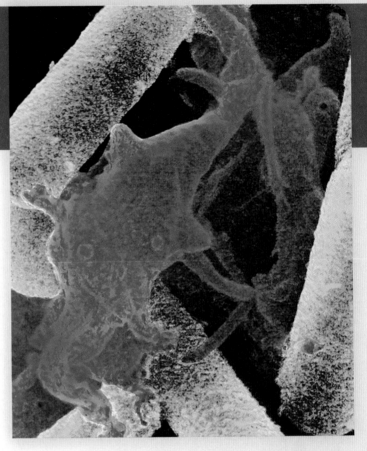

A falsely colored leukocyte (blue) engulfs rod-shaped *Bacillus cereus* bacteria (orange), and will use enzymes to break them down (20,000×).

THE WHOLE PICTURE

The lymphatic system is a vast collection of cells and biochemicals that travel in lymphatic vessels, and the organs and glands that produce them. The lymphatic system includes a network of vessels that assist in circulating body fluids, so it is closely associated with the cardiovascular system. The system performs three important functions: return of interstitial fluid to the bloodstream, lipid absorption, and defense against disease. The term *immune system* is linked with the lymphatic system because many of the cells that are part of the lymphatic system provide not only a defense against disease, but also immunity against future infection.

Anatomy & Physiology | REVEALED
aprevealed.com

Module 10: Lymphatic System

auto-, self: *auto*immune disease—the immune system attacking the body's own tissues.

-gen, become, be produced: aller*gen*—substance that evokes an allergic response.

humor-, moisture, fluid: *humoral* immunity—immunity resulting from antibodies in body fluids.

immun-, free, exempt: *immun*ity—resistance to (freedom from) a specific disease.

inflamm-, to set on fire: *inflamm*ation—localized redness, heat, swelling, and pain in the tissues.

nod-, knot: lymph *nod*ule—small mass of lymphocytes surrounded by connective tissue.

patho-, disease, sickness: *patho*gen—disease-causing agent.

16.1 | Lymphatic Pathways

The **lymphatic** (lim-fat′ik) **pathways** begin as lymphatic capillaries that merge to form lymphatic vessels. These, in turn, lead to larger vessels, trunks, and ducts that unite with the veins in the thorax.

Lymphatic vessels transport excess fluid away from the interstitial spaces in most tissues and return it to the bloodstream (fig. 16.1). Without the lymphatic system, this fluid would accumulate in tissue spaces. Special lymphatic capillaries *(lacteals)* in the lining of the small intestine absorb digested fats, then transport the fats to the venous circulation.

Lymphatic Capillaries

Lymphatic capillaries are microscopic, closed-ended tubes. They extend into the interstitial spaces, forming complex networks that parallel the blood capillaries (fig. 16.2). The walls of lymphatic capillaries, like those of blood capillaries, are formed from a single layer of squamous epithelial cells called *endothelium.* These thin walls allow tissue fluid (interstitial fluid) from the interstitial space to enter the lymphatic capillaries. Once inside lymphatic capillaries, the fluid is called **lymph** (limf). The lymphatic capillaries merge into lymph vessels.

Lymphatic Vessels

The walls of **lymphatic vessels** are similar to those of veins, but thinner. Each lymphatic vessel is composed of three layers: an endothelial lining, a middle layer of smooth muscle and elastic fibers, and an outer layer of connective tissue. The lymphatic vessels are also like some peripheral veins in that they have semilunar valves, which help prevent backflow of lymph (fig. 16.3).

The larger lymphatic vessels lead to specialized organs called **lymph nodes** (limf nōdz). These vessels merge into larger lymphatic trunks after leaving the lymph nodes.

Lymphatic Trunks and Collecting Ducts

The **lymphatic trunks,** which drain lymph from the lymphatic vessels, are named for the regions they serve. For example, the *lumbar trunk* drains lymph from the lower limbs, lower abdominal wall, and pelvic organs; the *intestinal trunk* drains the abdominal viscera; the *intercostal* and *bronchomediastinal trunks* drain lymph from portions of the thorax; the *subclavian trunk* drains the upper limb; and the *jugular trunk* drains portions of the neck and head. These lymphatic trunks then join one of two **collecting ducts**—the thoracic duct or the right lymphatic duct. **Figure 16.4** shows the location of the major lymphatic trunks and collecting ducts, and **figure 16.5** shows a lymphangiogram, which is a radiograph of lymphatic vessels and lymph nodes.

The **thoracic duct** is the wider and longer of the two collecting ducts. It originates as an enlarged sac, the **cisterna chlyi** (sis-ter′nah ki′li) in the abdomen, and passes upward through the diaphragm beside the aorta (fig. 16.6*a*). The thoracic duct ascends anterior to the vertebral column through the mediastinum, and empties into the left subclavian vein near the junction of the left jugular vein. This duct drains lymph from the intestinal, lumbar, and intercostal trunks, as well as from the left subclavian, left jugular, and left bronchomediastinal trunks.

CAREER CORNER

Public Health Nurse

When a hurricane approaching New England veered inland unexpectedly, the historic Stockade district of Schenectady, New York, suddenly was under water as the Mohawk River overflowed its banks. Residents were evacuated, and when they returned weeks later to assess the damage, public health nurses went door to door, warning people about the dangers of exposure to respiratory irritants such as mold and assisting them with getting help.

A public health nurse is a registered nurse with specialized training in community-based care. Training includes courses in public policy, health administration, and public health, and being bilingual (such as being fluent in English and Spanish) is very helpful. Volunteer work at the community level is a good way to prepare for this career.

Public health nurses work in a variety of facilities. They may vaccinate children in schools, talk to senior citizens about coping with chronic illness or physical limitations, educate parents of preschoolers about avoiding or treating scalp lice, or conduct a blood pressure screening at a shopping mall. These health-care professionals are particularly valuable in underserved communities, where they educate the public and help them to access health-care services.

In contrast to a nurse in a hospital setting, a public health nurse must be able to work well with groups. Employment requires passing a national licensing exam.

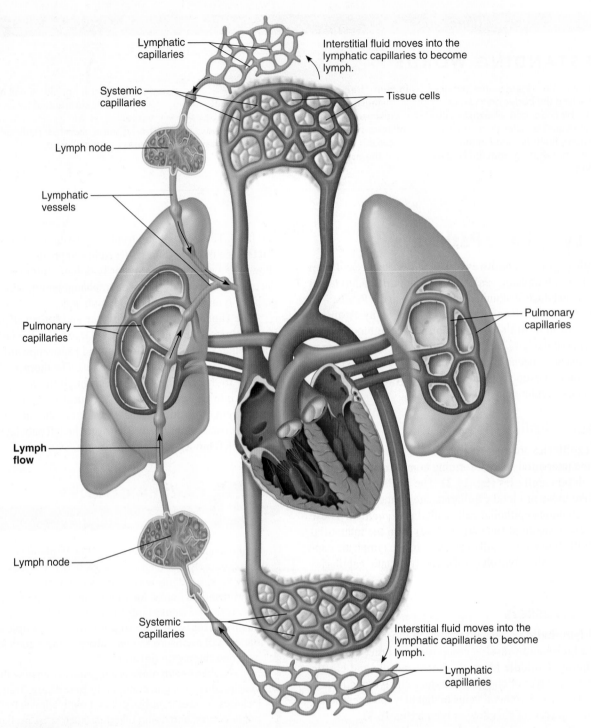

FIGURE 16.1 APR Schematic representation of lymphatic vessels transporting fluid from interstitial spaces to the bloodstream.

The **right lymphatic duct** originates in the right thorax at the union of the right jugular, right subclavian, and right bronchomediastinal trunks. It empties into the right subclavian vein near the junction of the right jugular vein.

Lymph from the lower body regions, the left upper limb, and the left side of the head and neck enters the thoracic duct; lymph from the right side of the head and neck, the right upper limb, and the right thorax enters the right lymphatic duct (fig. 16.6). After lymph leaves the two collecting ducts, it enters the venous system

and becomes part of the plasma just before blood returns to the right atrium. **Figure 16.7** summarizes the lymphatic pathway.

 PRACTICE

1 What is the function of the lymphatic vessels?

2 Distinguish between the thoracic duct and the right lymphatic duct.

3 Through which lymphatic structures would lymph pass in traveling from a lower limb back to the bloodstream?

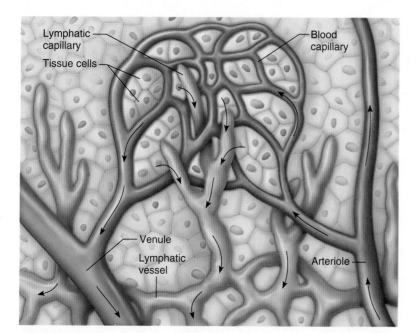

FIGURE 16.2 Lymphatic capillaries are microscopic, closed-ended tubes that originate in the interstitial spaces of most tissues.

Labels in Figure 16.2:
- Lymphatic capillary
- Tissue cells
- Blood capillary
- Venule
- Lymphatic vessel
- Arteriole

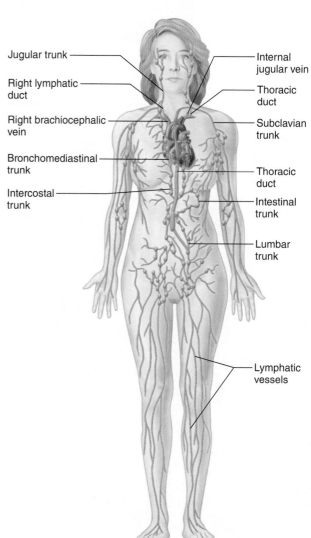

FIGURE 16.4 Lymphatic vessels merge into larger lymphatic trunks, which, in turn, drain into collecting ducts.

Labels in Figure 16.4:
- Jugular trunk
- Right lymphatic duct
- Right brachiocephalic vein
- Bronchomediastinal trunk
- Intercostal trunk
- Internal jugular vein
- Thoracic duct
- Subclavian trunk
- Thoracic duct
- Intestinal trunk
- Lumbar trunk
- Lymphatic vessels

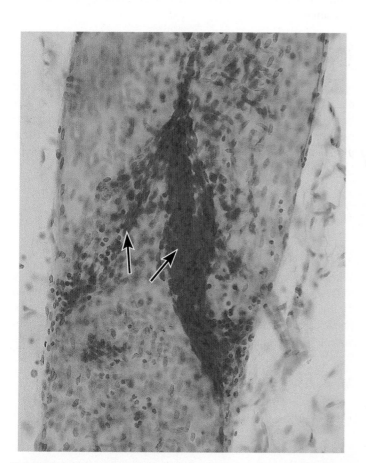

FIGURE 16.3 Light micrograph of the flaplike valve (arrows) within a lymphatic vessel (60×).

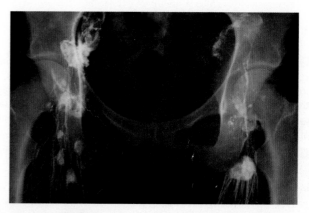

FIGURE 16.5 **AP|R** A lymphangiogram (radiograph) of the lymphatic vessels and lymph nodes of the lumbar and upper pelvic regions.

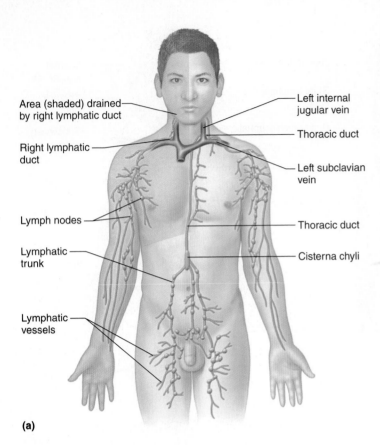

Area (shaded) drained by right lymphatic duct

Right lymphatic duct

Lymph nodes

Lymphatic trunk

Lymphatic vessels

Left internal jugular vein

Thoracic duct

Left subclavian vein

Thoracic duct

Cisterna chyli

(a)

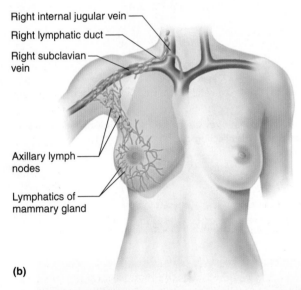

Right internal jugular vein

Right lymphatic duct

Right subclavian vein

Axillary lymph nodes

Lymphatics of mammary gland

(b)

FIGURE 16.6 **AP|R** Lymphatic pathways. (a) The right lymphatic duct drains lymph from the upper right side of the body, whereas the thoracic duct drains lymph from the rest of the body. (b) Lymph drainage of the right breast illustrates a localized function of the lymphatic system. Surgery to treat breast cancer can disrupt this drainage, causing painful swelling (edema) in the arm.

 Which lymphatic duct drains lymph from the right lower limb?

Answer can be found in Appendix G.

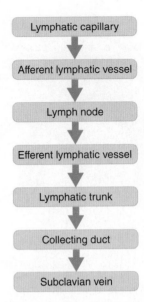

Lymphatic capillary

↓

Afferent lymphatic vessel

↓

Lymph node

↓

Efferent lymphatic vessel

↓

Lymphatic trunk

↓

Collecting duct

↓

Subclavian vein

FIGURE 16.7 The lymphatic pathway. This pathway is true for both right and left sides.

16.2 | Tissue Fluid and Lymph

Lymph is essentially tissue fluid that has entered a lymphatic capillary. Therefore, lymph formation depends upon tissue fluid formation.

Tissue Fluid Formation

Capillary blood pressure filters water and small molecules from the plasma. The resulting fluid has much the same composition as the plasma (including nutrients, gases, and hormones), with the important exception of the plasma proteins, most of which are too large to pass through the blood capillary walls. The osmotic effect of these plasma proteins (called the *plasma colloid osmotic pressure*) helps to draw fluid back into the blood capillaries by osmosis.

RECONNECT
To Chapter 15, Exchanges in the Capillaries, page 581.

Lymph Formation

Filtration from the plasma normally exceeds reabsorption, leading to the net formation of tissue fluid. This increases the tissue fluid hydrostatic pressure, the force which moves tissue fluid into lymphatic capillaries, forming lymph. In this way, lymph formation prevents the accumulation of excess tissue fluid, or *edema* (ě′de-mah).

PRACTICE

4 What is the relationship between tissue fluid and lymph?

5 How do plasma proteins in blood capillaries affect lymph formation?

Lymph Flow

Lymph within lymphatic vessels, like venous blood, is under relatively low hydrostatic pressure. It may not flow readily through the lymphatic vessels without help from contracting skeletal muscles in the limbs, contraction of smooth muscle in the walls of the larger lymphatic trunks, and pressure changes from the action of skeletal muscles used in breathing. Lymph flow peaks during physical exercise, due to the actions of skeletal muscles and pressure changes associated with breathing.

Contracting skeletal muscles compress lymphatic vessels. This squeezing action moves the lymph inside lymphatic vessels. Valves in these vessels prevent backflow, so lymph can only move toward a collecting duct. Additionally, the smooth muscle in the walls of the larger lymphatic trunks contracts rhythmically and compresses the lymph inside, forcing the fluid onward.

Breathing aids lymph circulation by creating a relatively low pressure in the thorax during inhalation. At the same time, the contracting diaphragm increases the pressure in the abdominal cavity. Consequently, lymph is squeezed out of the abdominal vessels and forced into the thoracic vessels. Once again, the valves of the lymphatic vessels prevent lymph backflow.

The continuous movement of fluid from interstitial spaces into blood capillaries and lymphatic capillaries stabilizes the volume of fluid in these spaces. Conditions that interfere with lymph movement cause tissue fluid to accumulate in interstitial spaces, producing edema. This can happen when the extent of a breast tumor requires that the surgeon remove nearby axillary lymph nodes, which can obstruct drainage from the upper limb, causing edema (see fig. 16.6b).

Lymph Function

Lymphatic capillaries in the small intestine play a major role in the absorption of dietary fats (chapter 17, pp. 678–679). Lymph also returns to the bloodstream most of the small proteins that the blood capillaries filtered. At the same time, lymph transports foreign particles, such as bacteria and viruses, to lymph nodes.

Lymphatic capillaries are adapted to receive proteins and foreign particles in a way that blood capillaries are not. The epithelial cells that form the walls of lymphatic capillaries overlap but are not attached to each other. This configuration, shown in figure 16.8, creates flaplike valves in the lymphatic capillary wall. The valves are pushed inward when the hydrostatic pressure is greater on the outside of the lymphatic capillary but close when the hydrostatic pressure is greater on the inside.

The epithelial cells of the lymphatic capillary wall are also attached to surrounding connective tissue cells by thin protein filaments. As a result, the lumen of a lymphatic capillary remains open even when the outside hydrostatic pressure is greater than the hydrostatic pressure inside the lymph capillary.

 PRACTICE

6 What factors promote lymph flow?

7 What is the consequence of lymphatic obstruction?

8 What are the major functions of lymph?

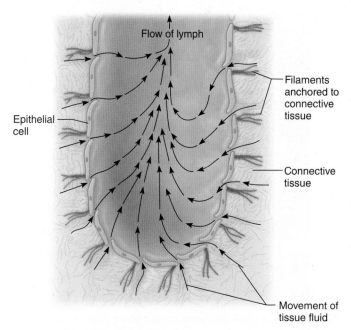

FIGURE 16.8 Tissue fluid enters lymphatic capillaries through flaplike valves between epithelial cells.

16.3 | Lymphatic Tissues and Lymphatic Organs

Lymphatic tissue contains lymphocytes, macrophages, and other cells. The unencapsulated diffuse lymphatic tissue associated with the digestive, respiratory, urinary, and reproductive tracts is called the **mucosa-associated lymphoid tissue (MALT).** Included in the MALT are compact masses of lymphatic tissue, called **lymphatic nodules,** which comprise the *tonsils* and *appendix*. These are described in Chapter 17 (pp. 653–654 and 683–684, respectively). MALT aggregates of lymphatic nodules, called *Peyer's patches,* are scattered throughout the mucosal lining of the distal portion of the small intestine.

The lymphatic organs, including the lymph nodes, thymus, and spleen, are encapsulated lymphatic tissue. A *capsule* of connective tissue with many fibers encloses each organ.

Lymph Nodes

Lymph nodes vary in size and shape but are usually less than 2.5 centimeters long and are somewhat bean-shaped (fig. 16.9). Figure 16.10 illustrates a section of a typical lymph node.

Blood vessels and nerves join a lymph node through the indented region of the node, called the **hilum.** The lymphatic vessels leading to a node (afferent vessels) enter separately at various points on its convex surface, but the lymphatic vessels leaving the node (efferent vessels) exit from the hilum.

The capsule surrounding the lymph node extends into the node and partially subdivides it into compartments. Masses of lymphocytes (B cells) and macrophages in the cortex are contained within these **lymphatic nodules,** also called lymphatic follicles, the functional units of the lymph node. Within the lymph nodules are *germinal centers* where B lymphocytes proliferate.

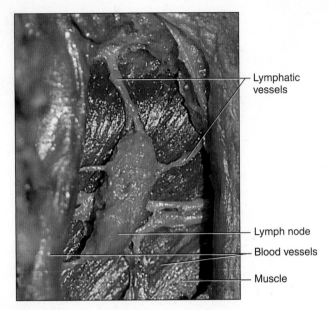

FIGURE 16.9 Lymph enters and leaves a lymph node through lymphatic vessels.

Spaces in a lymph node, called **lymphatic sinuses,** provide a complex network of chambers and channels through which lymph circulates. Lymph enters a lymph node through *afferent lymphatic vessels,* moves slowly through the lymph sinuses, and leaves through *efferent lymphatic vessels* (fig. 16.10a).

> Superficial lymphatic vessels inflamed in response to bacterial infection (*lymphangitis*) appear as red streaks beneath the skin. Inflammation of the lymph nodes, called *lymphadenitis,* often follows. Affected nodes enlarge and may be painful.

Locations of Lymph Nodes

Lymph nodes are found in groups or chains along the paths of the larger lymphatic vessels throughout the body, but they are not in the central nervous system. The major locations of lymph nodes, shown in figure 16.11, are as follows:

1. The lymph nodes in the **cervical region** follow the lower border of the mandible, anterior to and posterior to the ears, and lie deep in the neck along the paths of the larger blood vessels. These nodes are associated with the lymphatic vessels that drain the skin of the scalp and face, as well as the tissues of the nasal cavity and pharynx.
2. Lymph nodes in the **axillary region** receive lymph from vessels that drain the upper limbs, the wall of the thorax, the mammary glands (breasts), and the upper wall of the abdomen.
3. The lymph nodes in the **supratrochlear region** are located superficially on the medial side of the elbow. They often enlarge in children in response to infections acquired through cuts and scrapes on the hands.
4. Lymph nodes in the **inguinal region** receive lymph from the lower limbs, the external genitalia, and the lower abdominal wall.

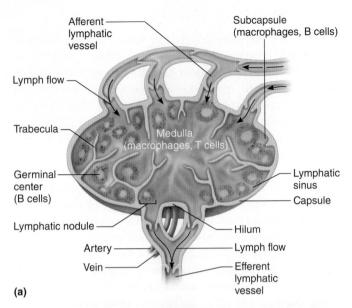

(a)

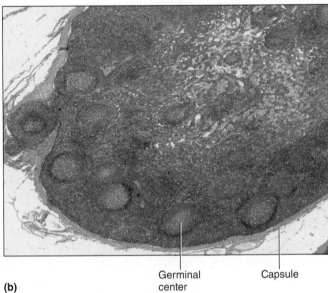

(b)

FIGURE 16.10 AP|R Lymph node. (a) A section through a lymph node. (b) Light micrograph of a lymph node (20×).

5. In the **pelvic cavity** lymph nodes primarily follow the iliac blood vessels. They receive lymph from the lymphatic vessels of the pelvic viscera.
6. The lymph nodes in the **abdominal cavity** form chains along the main branches of the mesenteric arteries and the abdominal aorta and receive lymph from the abdominal viscera.
7. The lymph nodes in the **thoracic cavity** are in the mediastinum and along the trachea and bronchi. They receive lymph from the thoracic viscera and from the internal wall of the thorax.

> The illness described as "swollen glands" refers to enlarged cervical lymph nodes associated with throat or respiratory infection.

Functions of Lymph Nodes

Lymph nodes have two primary functions: filtering potentially harmful particles from lymph before returning it to the bloodstream and monitoring body fluids (immune surveillance) through the actions of lymphocytes and macrophages. The lymph nodes and the red bone marrow are centers for lymphocyte production. Lymphocytes attack viruses, bacteria, and parasitic cells that are brought to the lymph nodes by lymphatic vessels. Macrophages in the lymph nodes engulf and destroy foreign substances, damaged cells, and cellular debris.

 PRACTICE

9 What is the MALT?

10 Distinguish between a lymph node and a lymph nodule.

11 In which body regions are lymph nodes most abundant?

12 What are the major functions of lymph nodes?

Thymus

The **thymus** (thi′mus) is a soft, bilobed gland enclosed in a connective tissue capsule. It is in the mediastinum, anterior to the aortic arch and posterior to the upper part of the body of the sternum, and extends from the root of the neck to the pericardium (fig. 16.12*a*). The thymus varies in size and is usually proportionately larger during infancy and early childhood. After puberty, the thymus shrinks, and in an adult, it may be small (fig. 16.13). In elderly persons, adipose and other connective tissues replace lymphatic tissue in the thymus.

Connective tissues extend inward from the surface of the thymus, subdividing it into lobules (see fig. 16.12*b*). The lobules house many lymphocytes that developed from progenitor cells in the bone marrow. Most of these cells (thymocytes) are inactive; however, some mature into **T lymphocytes,** also called T cells, which leave the thymus and provide immunity. Epithelial cells in the thymus secrete protein hormones called *thymosins,* which stimulate maturation of T lymphocytes.

Spleen

The **spleen** (splēn) is the largest lymphatic organ. It is in the upper left portion of the abdominal cavity, just inferior to the diaphragm, posterior and lateral to the stomach (see fig. 16.12*a* and reference plates 4, 5, and 6, pp. 41–43).

The spleen resembles a large lymph node. It is enclosed in connective tissue that extends inward from the surface and partially subdivides the organ into lobules. The organ also has a hilum on one surface through which blood vessels and nerves enter. However, unlike the lymphatic sinuses of a lymph node, the spaces (venous sinuses) in the spleen are filled with blood instead of lymph.

The tissues in the lobules of the spleen are of two types. *White pulp* is distributed throughout the spleen in tiny islands. This tissue is composed of splenic nodules, which are similar to the lymphatic nodules in lymph nodes and are packed with lymphocytes. The *red pulp,* which fills the remaining spaces of the lobules, includes the venous sinuses and the space around the venous sinuses. This pulp contains abundant red blood cells, which impart its color, plus many lymphocytes and macrophages (fig. 16.14).

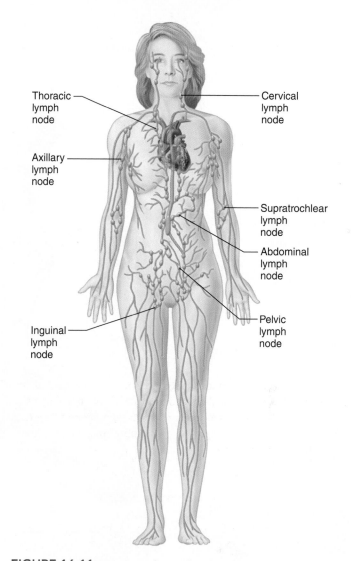

FIGURE 16.11 Locations of major lymph nodes.

During fetal development, pulp cells of the spleen produce blood cells, much as red bone marrow cells do after birth. As the time of birth approaches, the spleen stops producing red blood cells. However, in certain diseases, such as *erythroblastosis fetalis* (see chapter 14, page 551), in which many red blood cells are destroyed, the splenic pulp cells may resume their hematopoietic activity.

The venous sinuses in the red pulp are quite permeable. Red blood cells can squeeze through pores in the sinus walls and enter the surrounding spaces. The older, more fragile red blood cells may rupture during this passage. Macrophages in the splenic sinuses remove the resulting cellular debris.

Macrophages engulf and destroy foreign particles, such as bacteria, that may be carried in the blood as it flows through the splenic sinuses. Lymphocytes of the spleen, like those of the thymus and lymph nodes, also help defend the body against infections. Thus, the spleen filters blood much as the lymph nodes filter lymph. Table 16.1 summarizes the characteristics of the major organs of the lymphatic system.

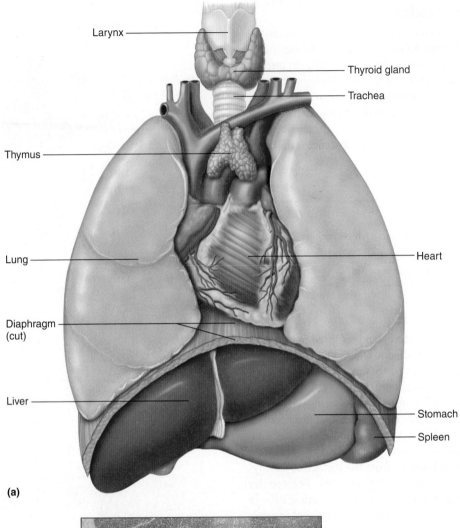

Larynx

Thyroid gland

Trachea

Thymus

Lung

Heart

Diaphragm (cut)

Liver

Stomach

Spleen

(a)

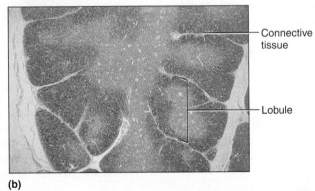

Connective tissue

Lobule

(b)

FIGURE 16.12 AP|R Thymus and spleen. (a) The thymus is bilobed, located between the lungs and superior to the heart. The spleen is located inferior to the diaphragm and posterior and lateral to the stomach. (b) A section through the thymus (15×). Note how the thymus is subdivided into lobules.

TABLE 16.1	Major Organs of the Lymphatic System	
Organ	**Location**	**Function**
Lymph nodes	In groups or chains along the paths of larger lymphatic vessels	Filter foreign particles and debris from lymph; produce and house lymphocytes that destroy foreign particles in lymph; house macrophages that engulf and destroy foreign particles and cellular debris carried in lymph
Thymus	In the mediastinum posterior to the upper portion of the body of the sternum	Houses lymphocytes; differentiates thymocytes into T lymphocytes
Spleen	In the upper left portion of the abdominal cavity, inferior to the diaphragm and posterior and lateral to the stomach	Houses macrophages that remove foreign particles, damaged red blood cells, and cellular debris from the blood; contains lymphocytes

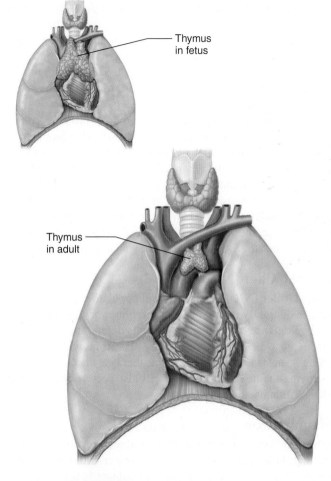

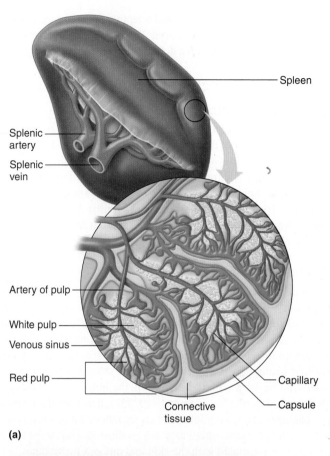

FIGURE 16.13 **AP|R** Compared to other thoracic organs, the thymus in the fetus is large, but in the adult is small. Figure is not to scale.

 PRACTICE

13 Why are the thymus and spleen considered organs of the lymphatic system?

14 What are the major functions of the thymus and the spleen?

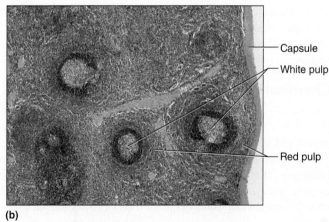

FIGURE 16.14 **AP|R** Spleen. (*a*) The spleen resembles a large lymph node. (*b*) Light micrograph of the spleen (40×).

16.4 | Body Defenses Against Infection

The organs of the lymphatic system also help defend the body against infection by disease-causing agents, or **pathogens** (path′o-jenz). The presence and multiplication of a pathogen in the body may cause an infection. Pathogens include simple microorganisms such as bacteria, complex microorganisms such as protozoa, and spores of multicellular organisms such as fungi. Viruses are pathogens, but they are not considered organisms, because their structure is much simpler than that of a living cell and they must infect a living cell to reproduce. An infection may be present even if an individual feels well.

The human body can prevent entry of pathogens or destroy them if they enter. Some mechanisms are quite general and protect against many types of pathogens, providing **innate** *(nonspecific)* **defense.** These general responses function the same way regardless of the type of pathogen or the number of exposures. Such mechanisms include species resistance, mechanical barriers, chemical barriers (enzyme action, interferon, and complement), natural killer cells, inflammation, phagocytosis, and fever. Other protective mechanisms are very precise, targeting specific pathogens with an **adaptive** *(specific)* **defense,** also called immunity. These more directed responses are carried out by specialized lymphocytes that recognize foreign molecules (nonself antigens) in the body and act against them. Together, innate and adaptive defense mechanisms protect the body against infection. The innate defenses respond quite rapidly, whereas adaptive defenses develop more slowly.

Innate (Nonspecific) Defenses

Species Resistance

Species resistance refers to the fact that a species may be resistant to diseases that affect other species because its cells do not have receptors for the pathogen or its tissues do not provide the temperature or chemical environment that a particular pathogen requires. For example, humans are infected by the infectious agents that cause measles, mumps, gonorrhea, and syphilis, but other animal species are not. Similarly, humans are resistant to canine distemper that affects dogs.

Mechanical Barriers

The skin and mucous membranes lining the passageways of the respiratory, digestive, urinary, and reproductive systems create **mechanical barriers** that prevent the entrance of some infectious agents and provide a *first line of defense.* As long as these barriers remain intact, many pathogens are unable to penetrate them. Another protection is that the epidermis sloughs off, removing superficial bacteria with it. In addition, the mucus coated ciliated epithelium, described in chapter 19 (p. 737), that lines the respiratory passages entraps particles and sweeps them out of the airways and into the pharynx, where they are swallowed. Hair traps infectious agents associated with the skin and mucous membranes, and sweat and mucus rinse away microorganisms. Tears, saliva, and urine also wash away microorganisms before they become firmly attached. The rest of the nonspecific defenses discussed in this section are part of the *second line of defense.*

Chemical Barriers

Enzymes in body fluids provide a **chemical barrier** to pathogens. Gastric juice, for example, contains the protein-splitting enzyme pepsin and has a low pH due to hydrochloric acid in the stomach. The combined effect of pepsin and hydrochloric acid kills many pathogens that enter the stomach. Similarly, tears contain the enzyme lysozyme, which destroys certain bacteria on the eyes. The accumulation of salt from perspiration also kills certain bacteria on the skin.

Interferons (in″ter-fēr′onz) are proteins that lymphocytes and fibroblasts produce in response to viruses or tumor cells. Once released from a virus-infected cell, interferon binds to receptors on uninfected cells, stimulating them to synthesize proteins that block replication of a variety of viruses. Thus, interferon's effect is nonspecific. Interferons also stimulate phagocytosis and enhance the activity of other cells that help to resist infections and the growth of tumors.

Other antimicrobial biochemicals are defensins and collectins. **Defensins** are peptides produced by neutrophils and other types of granular white blood cells in the intestinal epithelium, the urogenital tract, the kidneys, and the skin. Recognition of a nonself cell surface or viral particle triggers the expression of genes that encode defensins. Some defensins make holes in bacterial cell walls and membranes, crippling the microbes.

Collectins are proteins that provide broad protection against bacteria, yeasts, and some viruses. These proteins home in on slight differences in the structures and arrangements of sugars that protrude from the surfaces of pathogens. Collectins detect not only the sugar molecules, but the pattern in which they are clustered, binding much like velcro clings to fabric, thus making the pathogen more easily phagocytized.

Complement (kom′plĕ-ment) is a group of proteins (complement system), in plasma and other body fluids, that interact in an expanding series of reactions or cascade. Complement activation can rapidly occur by the *classical pathway* when a complement protein binds to an antibody attached to its specific antigen (discussed later in this chapter, pp. 634–635), or more slowly by the *alternative pathway* triggered by exposure to foreign antigens, in the absence of antibodies. Activation of complement stimulates inflammation, attracts phagocytes, and enhances phagocytosis.

Natural Killer (NK) Cells

Natural killer (NK) cells are a small population of lymphocytes that are distinctly different from the lymphocytes that provide adaptive defense mechanisms. NK cells defend the body against various viruses and cancer cells by secreting cytolytic ("cell-cutting") substances called **perforins** that lyse the cell membrane, destroying the infected cell. NK cells also secrete chemicals that enhance inflammation.

Inflammation

Inflammation is a reaction that produces localized redness, swelling, heat, and pain. The redness is a result of blood vessel dilation that increases blood flow and volume in affected tissues (hyperemia). This effect, coupled with an increase in permeability of nearby capillaries and subsequent leakage of protein-rich fluid into tissue spaces, swells tissues (edema). The heat comes as blood enters from deeper body parts, which are warmer than the surface. Pain results from stimulation of nearby pain receptors. Most inflammation is a tissue response to pathogen invasion, but physical factors (heat, ultraviolet light) or chemical factors (acids, bases) can also cause it.

White blood cells accumulate at the sites of inflammation, where some of them help control pathogens by phagocytosis. Neutrophils are the first to arrive at the site, followed by monocytes. Monocytes pass through capillary walls (diapedesis), becoming macrophages that remove pathogens from surrounding tissues. In bacterial infections, the resulting mass of white blood cells, bacterial cells, and damaged tissue may form a thick fluid called *pus.*

Fluids called *exudates* also collect in inflamed tissues. These fluids contain fibrinogen and other clotting factors that may stimulate formation of a network of fibrin threads in the affected region. Later, fibroblasts arrive and secrete fibers around the area, enclosing it in a sac of connective tissue. This walling off of the infected area helps inhibit the spread of pathogens and toxins to adjacent tissues.

Once an infection is controlled, phagocytic cells remove dead cells and other debris from the site of inflammation. Cell division replaces lost cells. **Table 16.2** summarizes the process of inflammation.

TABLE 16.2	Major Actions of an Inflammation Response	
Action	**Result**	
Blood vessels dilate. Capillary permeability increases and fluid leaks into tissue spaces.	Tissues become red, swollen, warm, and painful.	
White blood cells invade the region.	Pus may form as white blood cells, bacterial cells, and cellular debris accumulate.	
Tissue fluids containing clotting factors seep into the area.	A clot containing threads of fibrin may form.	
Fibroblasts arrive.	A connective tissue sac may form around the injured tissues.	
Phagocytes are active.	Bacteria, dead cells, and other debris are removed.	
Cells divide.	Newly formed cells replace injured ones.	

TABLE 16.3	Types of Innate (Nonspecific) Defenses	
Type	**Description**	
Species resistance	A species is resistant to certain diseases to which other species are susceptible.	
Mechanical barriers	Unbroken skin and mucous membranes prevent the entrance of some infectious agents. Fluids wash away microorganisms before they can firmly attach to tissues.	
Chemical barriers	Enzymes in various body fluids kill pathogens. pH extremes and high salt concentration also harm pathogens. Interferons induce production of other proteins that block reproduction of viruses, stimulate phagocytosis, and enhance the activity of cells such that they resist infection and the growth of tumors. Defensins damage bacterial cell walls and membranes. Collectins bind to microbes. Complement stimulates inflammation, attracts phagocytes, and enhances phagocytosis.	
Natural killer cells	Distinct type of lymphocyte that secretes perforins that lyse virus-infected cells and cancer cells.	
Inflammation	A tissue response to injury that helps prevent the spread of infectious agents into nearby tissues.	
Phagocytosis	Neutrophils, monocytes, and macrophages engulf and destroy foreign particles and cells.	
Fever	Elevated body temperature indirectly inhibits microbial growth and increases phagocytic activity.	

Phagocytosis

Phagocytosis (fag″o-si-to′sis) removes foreign particles from the lymph as it moves from the interstitial spaces to the bloodstream. Phagocytes in the blood vessels and in the tissues of the spleen, liver, or bone marrow usually remove particles that reach the blood. Recall from chapter 14 (p. 538) that the most active phagocytic cells of the blood are *neutrophils* and *monocytes*. Chemicals released from injured tissues attract these cells (chemotaxis). Neutrophils engulf and digest smaller particles; monocytes phagocytize larger ones.

Monocytes that leave the blood differentiate to become macrophages, which may be *free* or *fixed* in various tissues, including the lymph nodes, spleen, liver, and lungs, or attached to the inner walls of blood and lymphatic vessels. A macrophage can engulf up to 100 bacteria, compared to the twenty or so bacteria that a neutrophil can engulf. Monocytes and macrophages constitute the **mononuclear phagocytic system** (reticuloendothelial system).

Fever

A **fever** is a nonspecific defense that offers powerful protection. A fever begins as a viral or bacterial infection that stimulates lymphocytes to proliferate, producing cells that secrete a substance called *interleukin-1* (IL-1), more colorfully known as *endogenous pyrogen* ("fire maker from within"). IL-1 raises the thermoregulatory set point in the brain's hypothalamus to maintain a higher body temperature.

Fever indirectly counters microbial growth because higher body temperature causes the liver and spleen to sequester iron, which reduces the level of iron in the blood. Because bacteria and fungi require iron for normal metabolism, their growth and reproduction in a fever-ridden body slow and may cease. Also, phagocytic cells attack more vigorously when the temperature rises. For these reasons, low-grade fever of short duration may be a desired natural response, not a symptom to be treated aggressively with medications.

Chemical barriers, natural killer cells, inflammation, phagocytosis, and fever all provide a second line of defense against pathogens. Table 16.3 summarizes the types of innate (nonspecific) defenses.

RECONNECT

To Chapter 6, Clinical Application 6.4, page 191.

PRACTICE

15 What may cause an infection?

16 Explain seven innate (nonspecific) defense mechanisms.

Adaptive (Specific) Defenses or Immunity

The *third line of defense*, **immunity** (ĭ-mu′nĭ-te), is resistance to specific pathogens or to their toxins or metabolic by-products. An immune response is based upon the ability to distinguish molecules that are part of the body ("self") from those that are not ("nonself," or foreign). Such nonself molecules that can elicit an immune response are called **antigens** (an′tĭ-jenz). Lymphocytes and macrophages that recognize specific nonself antigens carry out adaptive immune responses, which include the *cellular immune response* and *humoral immune response*. Figure 16.15 shows a schematic representation of body defenses against pathogens.

A GLIMPSE AHEAD | To Chapter 22

Eggs and sperm have the potential of being recognized as nonself and eliciting an immune response. Fortunately, their development is hidden from the immune system. Eggs develop within follicles in the ovary, where they are protected by layers of cells. Sperm cells developing in the testes are protected by the blood–testis barrier.

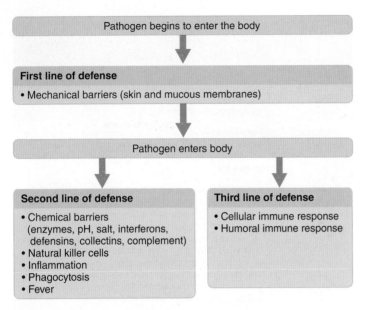

Pathogen begins to enter the body

↓

First line of defense
- Mechanical barriers (skin and mucous membranes)

↓

Pathogen enters body

↓   ↓

Second line of defense
- Chemical barriers
  (enzymes, pH, salt, interferons,
  defensins, collectins, complement)
- Natural killer cells
- Inflammation
- Phagocytosis
- Fever

Third line of defense
- Cellular immune response
- Humoral immune response

FIGURE 16.15 Schematic representation of body defenses against pathogens.

Antigens

Before birth, cells inventory the proteins and other large molecules in the body, learning to identify these as self. The lymphatic system responds to nonself antigens, but not normally to self antigens. Receptors on lymphocyte surfaces enable these cells to recognize nonself antigens.

Antigens may be proteins, polysaccharides, glycoproteins, or glycolipids. The antigens most effective in eliciting an immune response are large and complex, with few repeating parts. Sometimes, a smaller molecule that cannot by itself stimulate an immune response combines with a larger one, which makes it able to do so (become antigenic). Such a small molecule is called a **hapten** (hap′ten). Stimulated lymphocytes react either to the hapten or to the larger molecule of the combination. Hapten molecules are in drugs, such as penicillin; in household and industrial chemicals; in dust particles; and in animal dander.

Lymphocyte Origins

Lymphocyte production begins during fetal development (before birth) and continues throughout life, with red bone marrow releasing unspecialized precursors of lymphocytes into the circulation. About half of these cells reach the thymus, where they remain for a time. Here, these thymocytes specialize into **T lymphocytes,** or **T cells.** (*T* refers to *thymus-derived* lymphocytes.) Some of these T cells constitute 70% to 80% of the circulating lymphocytes in blood. Other T cells reside in lymphatic organs and are particularly abundant in the lymph nodes, the thoracic duct, and the white pulp of the spleen.

Other lymphocytes remain in the red bone marrow until they differentiate fully into **B lymphocytes,** or **B cells.** (Historically, the "B" stands for *bursa of Fabricius,* an organ in the chicken where these cells were discovered.) The blood distributes B cells, which constitute 20% to 30% of circulating lymphocytes (fig. 16.16). B cells settle in lymphatic organs along with T cells and are abundant in the lymph nodes, spleen, bone marrow, and intestinal lining (fig. 16.17).

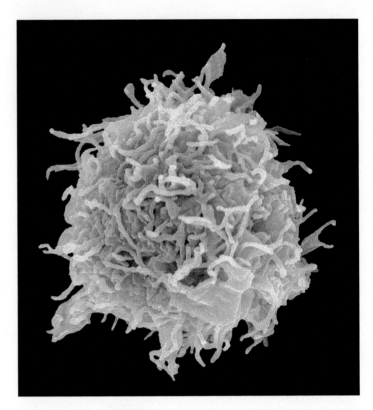

FIGURE 16.16 **AP|R** Falsely colored scanning electron micrograph of a B cell (6,000×).

Each person has millions of varieties of T and B cells. The members of each variety originate from a single early cell, so they are all alike, forming a **clone** (klōn) of cells (genetically identical cells originating from division of a single cell). The members of each variety have a particular type of antigen receptor on their cell membranes that can respond only to a specific antigen. Table 16.4 compares the characteristics of T cells and B cells.

TABLE 16.4	A Comparison of T Cells and B Cells	
Characteristic	**T Cells**	**B Cells**
Origin of undifferentiated cell	Red bone marrow	Red bone marrow
Site of differentiation	Thymus	Red bone marrow
Primary locations	Lymphatic tissues, 70% to 80% of the circulating lymphocytes in blood	Lymphatic tissues, 20% to 30% of the circulating lymphocytes in blood
Primary functions	Provide cellular immune response in which T cells interact directly with the antigens or antigen-bearing agents, to destroy them	Provide humoral immune response in which B cells interact indirectly, producing antibodies that destroy the antigens or antigen-bearing agents

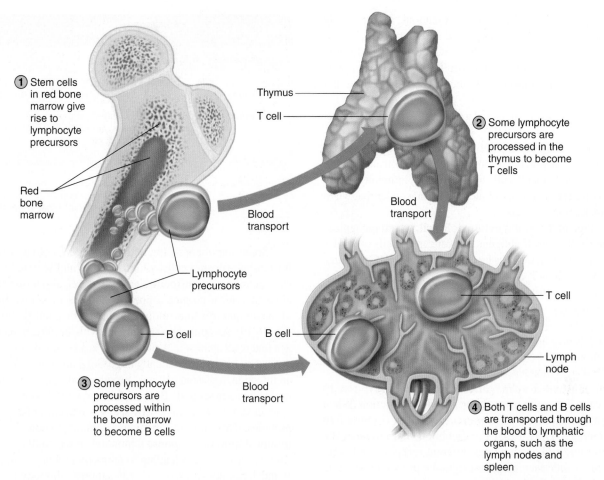

FIGURE 16.17 During fetal development, red bone marrow releases unspecialized lymphocyte precursors, which after processing specialize as T cells (T lymphocytes) or B cells (B lymphocytes). Lymphocyte production continues throughout life, with most T cells arising before puberty.

Image labels (clockwise):
1 Stem cells in red bone marrow give rise to lymphocyte precursors
Thymus
T cell
2 Some lymphocyte precursors are processed in the thymus to become T cells
Blood transport
Red bone marrow
Blood transport
Lymphocyte precursors
T cell
B cell
B cell
Lymph node
3 Some lymphocyte precursors are processed within the bone marrow to become B cells
Blood transport
4 Both T cells and B cells are transported through the blood to lymphatic organs, such as the lymph nodes and spleen

 PRACTICE

17 What is immunity?

18 What is the difference between an antigen and a hapten?

19 How do T cells and B cells originate?

T Cells and the Cellular Immune Response

A lymphocyte must be activated before it can respond to an antigen. T cell activation requires that processed fragments of the antigen be attached to the surface of another type of cell, called an **antigen-presenting cell** (accessory cell). *Dendritic cells* located in the epidermis, mucous membrane, and lymphatic organs may engulf the antigen by receptor-mediated endocytosis, process the antigen, and display antigen fragments on their cell surfaces. Macrophages and B cells can also be antigen-presenting cells.

T cell activation may occur when a macrophage phagocytizes a bacterium, and digests it within a *phagolysosome* formed by the fusion of the vesicle containing the bacterium (phagosome) and a lysosome. Some of the resulting bacterial antigens are displayed on the macrophage's cell membrane near certain protein molecules that are part of a group of proteins called the **major histocompatibility complex** (MHC) or *human leukocyte*

antigens (HLA) because they were first identified on white blood cells. MHC antigens help T cells recognize that a newly displayed antigen is foreign, not self. Class I MHC antigens are in cell membranes of all body cells except red blood cells. Class II MHC antigens are on the surfaces of antigen-presenting cells, thymus cells, and activated T cells.

Activated T cells interact directly with the antigen-presenting cell. Such cell-to-cell contact is called the **cellular immune response,** or cell-mediated immunity.

T cells (and some macrophages) also synthesize and secrete polypeptides called **cytokines** that enhance certain cellular responses to antigens. For example, *interleukin-1* and *interleukin-2* stimulate synthesis of several other cytokines from other T cells. In addition, interleukin-1 helps activate T cells, whereas interleukin-2 causes T cells to proliferate. Other cytokines called *colony-stimulating factors* (CSFs) stimulate production of leukocytes in the red bone marrow, cause B cells to grow and mature, and activate macrophages. Certain cytokine combinations shut off the immune response. Table 16.5 summarizes several cytokine types.

In addition to cytokines, T cells may secrete other substances. These include toxins that kill their antigen-bearing target cells, growth-inhibiting factors that prevent target cell growth, and interferons that inhibit the proliferation of viruses and tumor cells.

Several types of T cells have specific functions. A particularly important specialized type of T cell, called a **helper T cell,** is activated when its antigen receptor combines with a displayed foreign antigen (fig. 16.18). The activated helper T cell stimulates a type of B cell (plasma cell) to produce antibodies specific for the displayed antigen.

A subtype of helper T cell called a CD4 cell is the prime target of HIV, the virus that causes AIDS. (*CD4* stands for the "cluster-of-differentiation" antigen the T cell bears that enables it to recognize a macrophage displaying a foreign antigen.) Considering the central role of CD4 helper T cells in establishing immunity—they stimulate B cells and secrete cytokines—it is no wonder that when HIV harms them, immunity declines.

Another type of T cell is a **cytotoxic T cell,** which recognizes and combines with nonself antigens that cancerous cells or virally infected cells display on their surfaces near certain MHC proteins. Cytokines from helper T cells activate the cytotoxic T cell (fig. 16.18a). Next the cytotoxic T cell proliferates, enlarging its clone of cells. Cytotoxic T cells then bind to the surfaces of antigen-bearing cells, where they release *perforin* protein that cuts pore-like openings, destroying these cells. In this way, cytotoxic T cells continually monitor the body's cells, recognizing and eliminating tumor cells and cells infected with viruses.

Certain cytotoxic T cells, called CD8 T cells, give rise to **memory T cells** that provide for future immune protection. When a CD8 T cell contacts an antigen-presenting cell, it contorts into a dumbbell shape. The side of the dumbbell that contacts the antigen-presenting cell accumulates different receptors and other proteins from the side facing farthest from the provoking antigen. When the CD8 T cell divides, the daughter cell that was the part of the original cell closest to the antigen becomes an active cytotoxic T cell. The daughter cell farther from the antigen becomes a memory T cell. As its name implies, a memory T cell does not respond to an initial exposure to an antigen, but upon subsequent exposure immediately divides and differentiates into a cytotoxic T cell. This memory cell response usually destroys the pathogen before the body responds to it with the signs and symptoms of disease.

 PRACTICE

20 How do T cells become activated?

21 What are some functions of cytokines?

22 Name three types of T cells.

23 How do cytotoxic T cells destroy antigen-bearing cells?

B Cells and the Humoral Immune Response

When a B cell encounters an antigen whose molecular shape fits the shape of the B cell's antigen receptors, it becomes activated (fig. 16.18a). In response to the receptor-antigen combination, the B cell divides repeatedly. Such B cell activation typically requires T cell "help."

When an activated helper T cell encounters a B cell already combined with an identical foreign antigen, the helper T cell releases certain cytokines that stimulate the B cell to proliferate, enlarging its clone of cells (fig. 16.19). The cytokines also attract macrophages and leukocytes into inflamed tissues and help keep them there.

TABLE 16.5	Types of Cytokines
Cytokine	**Function**
Colony-stimulating factors	Stimulate bone marrow to produce lymphocytes
Interferons	Block viral replication, stimulate macrophages to engulf viruses, stimulate B cells to produce antibodies, attack cancer cells
Interleukins	Control lymphocyte differentiation and proliferation
Tumor necrosis factor	Stops tumor growth, releases growth factors, causes fever that accompanies bacterial infection, stimulates lymphocyte differentiation

Some members of the activated B cell's clone differentiate further into **plasma cells,** which produce and secrete large globular proteins called **antibodies** (an'tĭ-bod″ēz), also termed **immunoglobulins** (im″u-no-glob′u-linz). Antibodies are similar in structure to the antigen-receptor molecules on the original B cell's surface (fig. 16.19). An antibody can combine with the antigen on the pathogen and react against it. A plasma cell is an antibody factory, as evidenced by its characteristically huge Golgi apparatus. At the peak of an infection, a plasma cell may produce and secrete 2,000 antibody molecules per second! Body fluids carry antibodies, which then react in various ways to destroy specific antigens or antigen-bearing particles. This antibody-mediated immune response is called the **humoral immune response** ("humoral" refers to fluid). Table 16.6 summarizes the steps leading to antibody production as a result of B and T cell activities. T cells can suppress antibody formation by releasing cytokines that inhibit B cell function.

A single type of B cell carries information to produce a single type of antibody. However, different B cells respond to different antigens on a pathogen's surface. Therefore, an immune response may include several types of antibodies manufactured against a single microbe or virus. This involvement of more than one antibody type is called a **polyclonal response.** From Science to Technology 16.1 discusses how researchers use clones of single B cells to produce single, or monoclonal, antibodies.

The human body can manufacture more than 1,000,000,000 different antibodies, each reacting against a specific antigen. The enormity and diversity of the antibody response defends against many pathogens.

There are a limited number of antibody genes. During the early development of B cells, sections of their antibody genes move to other chromosomal locations, creating new genetic instructions for antibodies. The great diversity of antibody types is increased further because different antibody protein subunits combine. Antibody diversity is like using the limited number of words in a language to compose an almost unlimited variety of stories.

Other members of the activated B cell's clone differentiate further into **memory cells** (fig. 16.19). Like memory T cells, these memory B cells respond rapidly to subsequent exposure to a specific antigen.

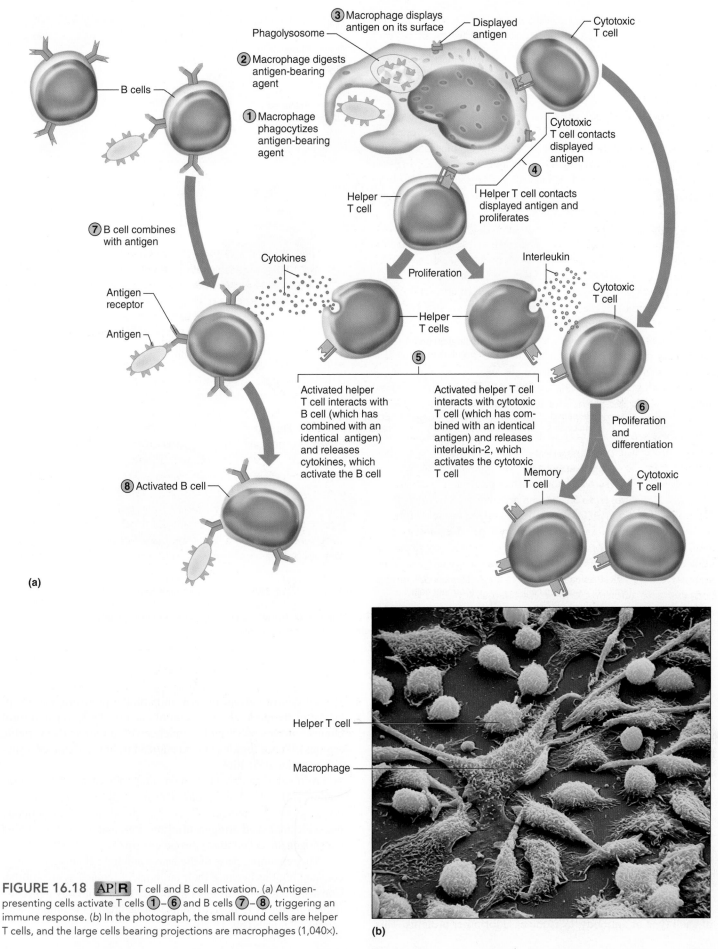

(a)

③ Macrophage displays antigen on its surface

Displayed antigen

Cytotoxic T cell

Phagolysosome

② Macrophage digests antigen-bearing agent

① Macrophage phagocytizes antigen-bearing agent

B cells

Cytotoxic T cell contacts displayed antigen ④

Helper T cell contacts displayed antigen and proliferates

Helper T cell

⑦ B cell combines with antigen

Cytokines

Interleukin

Proliferation

Cytotoxic T cell

Antigen receptor

Antigen

Helper T cells

⑤

Activated helper T cell interacts with B cell (which has combined with an identical antigen) and releases cytokines, which activate the B cell

Activated helper T cell interacts with cytotoxic T cell (which has combined with an identical antigen) and releases interleukin-2, which activates the cytotoxic T cell

⑥ Proliferation and differentiation

⑧ Activated B cell

Memory T cell

Cytotoxic T cell

Helper T cell

Macrophage

(b)

FIGURE 16.18 **AP|R** T cell and B cell activation. (a) Antigen-presenting cells activate T cells ①–⑥ and B cells ⑦–⑧, triggering an immune response. (b) In the photograph, the small round cells are helper T cells, and the large cells bearing projections are macrophages (1,040×).

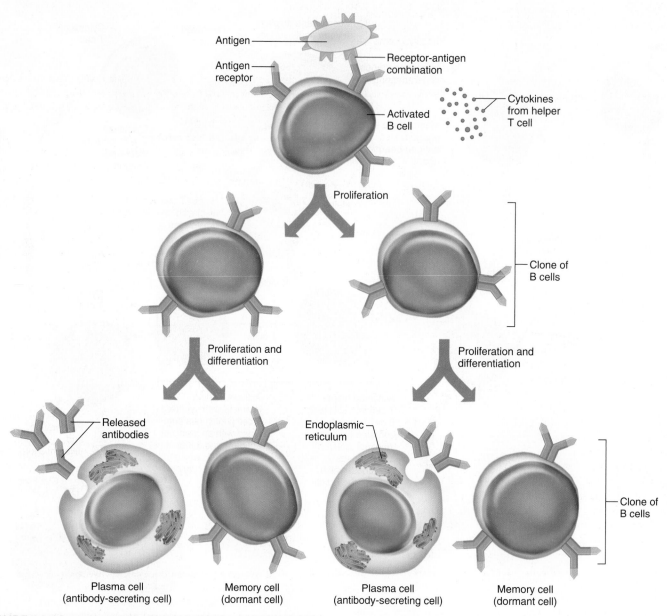

FIGURE 16.19 An activated B cell proliferates after stimulation by cytokines released by helper T cells. The B cell's clone enlarges. Some cells of the clone give rise to antibody-secreting plasma cells and others to dormant memory cells.

Antibody Molecules

Antibodies are soluble, globular proteins that constitute the *gamma globulin* fraction of plasma proteins (see chapter 14, p. 541). Each antibody molecule consists of four chains of amino acids linked by pairs of sulfur atoms that attract by disulfide bonds. The four chains form a Y-shaped structure (fig. 16.20). Two of these amino acid chains are identical **light chains** (L-chains), and two are identical **heavy chains** (H-chains). The heavy chains have about twice as many amino acids as the light chains. The five major types of antibody molecules are distinguished by a particular type of heavy chain. Most of the types of antibody molecules consist of a single Y-shaped structure, but some have as many as five (see fig. 14.21, p. 550).

The sequences of amino acids of the heavy and light chains confer the unique, three-dimensional structure (conformation) of each type of antibody, as is the case for other proteins. This

special conformation, in turn, imparts the physiological properties of the molecule. One end of each of the heavy and light chains consists of variable sequences of amino acids (variable regions). These regions are specialized to fit the shape of a specific antigen molecule.

Antibodies can bind to certain antigens because of the conformation of the variable regions. The antibody contorts, forming a pocket around the antigen. These specialized ends of the antibody molecule are called **antigen-binding sites,** and the parts that bind the antigen are called **idiotypes** (id′e-o-tīpz′).

The remaining parts of the heavy and light chains are termed constant regions because their amino acid sequences are similar. Constant regions impart other properties of the antibody molecule, such as its ability to bond to cellular structures or to combine with certain chemicals (fig. 16.20).

Immunotherapy uses immune system components to fight disease—both the humoral immune response (antibodies) and the cellular immune response (cytokines).

Monoclonal Antibodies

Tapping the specificity of a single B cell and using its single type, or *monoclonal,* antibody to target a specific antigen (such as on a cancer or bacterial cell) awaited finding a way to entice the normally short-lived mature B cells into persisting in culture. In 1975, British researchers Cesar Milstein and Georges Köhler devised *monoclonal antibody* (MAb) technology to capture the antibody-making capacity of a single B cell.

Milstein and Köhler injected a mouse with antigen-laden red blood cells from a sheep. They then isolated a single B cell from the mouse's spleen and fused it with a cancerous white blood cell from a mouse. The result was a fused cell, or *hybridoma,* with a valuable pair of talents: Like the B cell, it produced large amounts of a single antibody type; like the cancer cell, it divided continuously (fig. 16A). Human versions of MAbs are now used.

MAbs are important in basic research, veterinary and human health care, and agriculture. Cell biologists use MAbs to localize and isolate proteins. Diagnostic MAb "kits" detect tiny amounts of a single molecule. Most kits consist of a paper strip containing a MAb, to which the user adds a body fluid. For example, a woman who suspects she is pregnant places drops of her urine onto the paper. A color change ensues if the MAb binds to the hormone human chorionic gonadotropin (see chapter 23, p. 872), indicating pregnancy.

MAbs can highlight a new cancer or detect recurrence. The MAb is attached to a radioactive chemical, which emits a signal when the MAb binds an antigen unique to the cancer cell surface.

MAbs can ferry conventional cancer treatments or radioactive chemicals to tumors and spare healthy tissue. MAbs can also combat cancer directly by blocking receptors for growth factors on cancer cells, so the cells do not receive signals to divide.

Cytokines

Immunotherapy experiments were difficult to do before the 1970s because cytokines and antibodies could be obtained only in small amounts from cadavers. Then recombinant DNA and monoclonal antibody technologies yielded unlimited amounts of pure proteins—just when the AIDS epidemic was making cadavers an unsafe source of biochemicals.

Interferon was the first cytokine tested on a large scale. It is used to treat a few conditions, including a type of leukemia, multiple sclerosis, hepatitis, and genital warts. Interleukin-2 is used to treat metastatic melanoma and metastatic kidney cancer, and colony-stimulating factors boost the white blood cell supply in people receiving drugs to treat cancer or AIDS, or transplant recipients.

Increasingly, cancer treatment consists of combinations of immune system cells or biochemicals, plus standard therapies. Immunotherapy can enable a patient to withstand higher doses of a conventional drug or may be used to directly destroy cancer cells remaining after standard treatment.

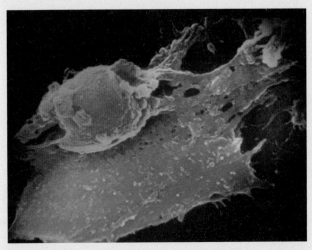

FIGURE 16A Monoclonal antibodies are produced by a type of artificial cell combination called a hybridoma. It consists of a cancer cell (the flat blue cell) fused with a B cell (the round green cell). The cancer cell contributes rapid and continuous division; the B cell secretes a single antibody type (7,000×).

Types of Immunoglobulins

Three of the five major types of immunoglobulins include most of the circulating antibodies. These types are immunoglobulin G, which accounts for about 80% of the antibodies; immunoglobulin A, which makes up about 13%; and immunoglobulin M, responsible for about 6%. The remainder of the antibodies are immunoglobulin D or immunoglobulin E.

Immunoglobulin G (IgG) is in plasma and tissue fluids and is effective against bacteria, viruses, and toxins. IgG also activates complement proteins (see p. 626; also see the next section "Antibody Actions"). Anti-Rh antibodies are examples of IgG and, as described in chapter 14 (p. 551), can cross the placenta.

Immunoglobulin A (IgA) is in exocrine gland secretions. It is in breast milk, tears, nasal fluid, gastric juice, intestinal juice, bile, and urine.

A newborn does not yet have its own antibodies but has IgG that passed through the placenta from the mother that protect the infant against some illnesses to which the mother is immune. The maternally obtained antibodies begin to disappear just about when the infant begins to make its own antibodies. The newborn also receives IgA from colostrum, a yellowish fluid that the mother's breasts secrete for the first few days after giving birth, and then from breast milk. Antibodies in colostrum and breast milk protect against certain digestive and respiratory infections.

TABLE 16.6 Steps in Antibody Production

B Cell Activities

1. Antigen-bearing agents enter tissues.

2. B cell encounters an antigen that fits its antigen receptors.

3. Either alone or more often in conjunction with helper T cells, the B cell is activated. The B cell proliferates, enlarging its clone.

4. Some of the newly formed B cells differentiate further to become plasma cells.

5. Plasma cells synthesize and secrete antibodies whose molecular structure is similar to the activated B cell's antigen receptors.

T Cell Activities

1. Antigen-bearing agents enter tissues.

2. An accessory cell, such as a macrophage, phagocytizes the antigen-bearing agent, and the macrophage's lysosomes digest the agent.

3. Antigens from the digested antigen-bearing agents are displayed on the membrane of the accessory cell.

4. Helper T cell becomes activated when it encounters a displayed antigen that fits its antigen receptors.

5. Activated helper T cell releases cytokines when it encounters a B cell that has previously combined with an identical antigen-bearing agent.

6. Cytokines stimulate the B cell to proliferate.

7. Some of the newly formed B cells give rise to cells that differentiate into antibody-secreting plasma cells.

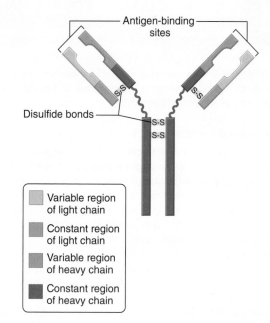

FIGURE 16.20 An antibody (immunoglobulin) molecule consists basically of two identical light chains of amino acids and two identical heavy chains of amino acids.

Immunoglobulin M (IgM) is a type of antibody produced in plasma in response to contact with certain antigens in foods or bacteria. Examples of IgM are the anti-A and anti-B antibodies, described in chapter 14 (pp. 548–549). IgM also activates complement.

Immunoglobulin D (IgD) is on the surfaces of most B cells, especially those of infants. IgD acts as an antigen receptor and is important in activating B cells (see fig. 16.18).

Immunoglobulin E (IgE) appears in exocrine secretions with IgA. It is associated with allergic responses (see "Hypersensitivity Reactions" on pages 636–638). Table 16.7 summarizes the major immunoglobulins and their functions.

 PRACTICE

24 How are B cells activated?

25 How does the antibody response protect against many different infections?

26 What is an immunoglobulin?

27 Describe the structure of an immunoglobulin molecule, and name the five major types of immunoglobulins.

Antibody Actions

In general, antibodies react to antigens in three ways. Antibodies directly attack antigens, activate complement, or stimulate localized changes (inflammation) that help prevent spread of the pathogen.

TABLE 16.7 Characteristics of Major Immunoglobulins

Type	Occurrence	Major Functions
IgG	Plasma and tissue fluid	Defends against bacteria, viruses, and toxins; activates complement
IgA	Exocrine gland secretions	Defends against bacteria and viruses
IgM	Plasma	Reacts with antigens on some red blood cell membranes following mismatched blood transfusions; activates complement
IgD	Surface of most B lymphocytes	B cell activation
IgE	Exocrine gland secretions	Promotes inflammation and allergic responses

In a direct attack, antibodies combine with antigens and cause them to clump (agglutination) or to form insoluble substances (precipitation). Such actions make it easier for phagocytic cells to engulf the antigen-bearing pathogens and eliminate them. In other instances, antibodies bind to and cover the toxic parts of antigen molecules. This neutralizes the effects (neutralization) of the antigen molecules. However, under normal conditions, complement activation is more important in protecting against infection than is direct antibody attack.

When certain IgG or IgM antibodies combine with antigens, they expose reactive sites on the antibody constant regions. This triggers a series of reactions leading to activation of complement proteins. Activated complement proteins coat the antigen-antibody complexes (opsonization), making the complexes more susceptible to phagocytosis; attract macrophages and neutrophils into the region (chemotaxis); agglutinate antigen-bearing cells; rupture cell membranes (lysis) of foreign cells; and neutralize viruses by altering their molecular structure, making them harmless. Other proteins promote inflammation, which helps prevent the spread of infectious agents.

Immunoglobulin E promotes inflammation that may be so intense that it damages tissues. This antibody is usually attached to the membranes of widely distributed *mast cells* (see chapter 5, p. 160). When antigens combine with the antibodies, the resulting antigen-antibody complexes stimulate mast cells to release biochemicals, such as histamine, that cause the changes associated with inflammation, such as vasodilation and edema. Table 16.8 summarizes the actions of antibodies.

 PRACTICE

28 In what general ways do antibodies function?

29 How is complement activated?

30 What are the effects of complement activation?

Immune Responses

The immune response occurs in two parts. A **primary immune response** occurs when B cells and T cells become activated after first encountering the antigens for which they are specialized to react. During this response, plasma cells release antibodies (IgM, followed by IgG). The antibodies are carried by the blood throughout the body, where they help destroy antigen-bearing agents. Production and release of antibodies continues for several weeks, maintaining higher numbers of antibodies in the plasma. The *antibody titer* is the number of circulating antibodies in the bloodstream.

After a primary immune response, some of the B cells produced during proliferation of the clone will serve as *memory cells* (see fig. 16.19). If the same type of antigen is encountered in the future, the clones of these memory cells enlarge, and they can respond rapidly with IgG. These memory B cells along with memory T cells produce a **secondary immune response** (anamnestic response). In lymph nodes, *follicular dendritic cells* may help memory by harboring and slowly releasing viral antigens after an initial infection. This

TABLE 16.8	Actions of Antibodies	
General Action	Type of Effect	Description
Direct Attack		
	Agglutination	Antigens clump
	Precipitation	Antigens become insoluble
	Neutralization	Antigens lose toxic properties
Activation of Complement (Antibodies combined with antigens)	Opsonization	Alters antigen cell membranes so cells are more susceptible to phagocytosis
	Chemotaxis	Attracts macrophages and neutrophils into the region
	Agglutination	Clumps antigen-bearing cells
	Lysis	Allows rapid movement of water and ions into the foreign cell causing osmotic rupture of the foreign cell
	Neutralization	Alters the molecular structure of viruses, making them harmless
Localized Changes		
	Inflammation	Helps prevent the spread of antigens

exposure constantly stimulates memory B cells, which present the antigens to memory T cells, maintaining immunity.

A primary immune response leads to detectable concentrations of antibodies in the plasma within five to ten days after exposure to antigens. If the same type of antigen is encountered again later, a secondary immune response may produce additional antibodies within a day or two (fig. 16.21). Although newly formed

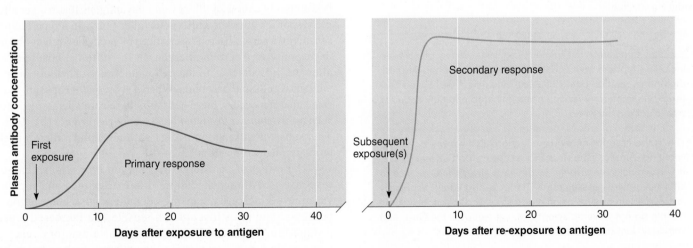

FIGURE 16.21 A primary immune response occurs after the first exposure to an antigen. A secondary immune response occurs after subsequent exposure(s) to the same antigen. The break in the time line represents a lapse between initial and subsequent exposure(s) to the antigen.

 Which immune response produces antibodies to a specific antigen more rapidly, primary or secondary?

Answer can be found in Appendix G.

antibodies may persist in the body for only a few months or years, memory cells live much longer. A secondary immune response may be very long-lasting.

Practical Classification of Immunity

Before vaccines against "childhood diseases" began to be developed in the 1960s, suffering through measles, mumps, rubella, and chickenpox was common. However, most children had each illness only once, thanks to *naturally acquired active immunity*. This form of immunity develops after a primary immune response and is a response to exposure to a live pathogen.

Today, most children in developed countries do not contract measles, mumps, rubella, or chickenpox because they develop another type of active immunity, produced in response to receiving a **vaccine** (vak′sēn). A vaccine is a preparation that includes an antigen that can stimulate a primary immune response against a particular pathogen but does not produce symptoms of the associated infectious disease.

A vaccine might include bacteria or viruses that have been killed or attenuated (weakened) so that they cannot cause a serious infection, or a toxoid, which is a toxin from an infectious organism that has been chemically altered to destroy its dangerous effects. A "subunit" vaccine consists of a single glycoprotein or similar large molecule from the pathogen's surface, which provides enough of a foreign antigen to alert the immune system. A vaccine causes a person to develop *artificially acquired active immunity*.

Individuals receive vaccines, but the ultimate effect is at the population level. If a critical number of people are vaccinated, becoming immune, the infectious agent can no longer easily pass from person to person. This population protection that results from widespread vaccination is called *herd immunity*.

Viruses whose genetic material rapidly mutates present a great challenge to vaccine development because their surfaces, which serve as antigens, change. It is a little like fighting an enemy who is constantly changing disguises. For this reason, pharmaceutical companies must develop a new vaccine against influenza each year. HIV is particularly changeable, which has severely hampered efforts to produce a vaccine.

> Vaccines stimulate active immunity against a variety of diseases, including typhoid fever, cholera, whooping cough (pertussis), diphtheria, tetanus, polio, influenza, hepatitis A and B, and bacterial pneumonia. A vaccine has eliminated naturally acquired smallpox from the world.
>
> For some infections, developing a vaccine is challenging. This is the case for norovirus infection, also known as "winter vomiting disease." The virus and receptors for it on small intestine lining cells are diverse, so the virus always finds a place to infect. Norovirus passes readily in feces and vomit, especially in crowded areas such as on cruise ships and in schools. Some people do not develop symptoms, yet can pass the virus, and immunity does not last until the next winter.

In some situations a person who has been exposed to infection needs protection against a pathogen but lacks time to develop active immunity from a vaccine. This happens with hepatitis A, which is a viral infection of the liver. In such a case, it may be possible to inject the person with antiserum, which has ready-made antibodies from gamma globulin separated from the blood plasma of people who have already developed immunity against the disease.

An injection of antiserum (antibody-rich serum) or antitoxin (antibodies against a toxin) provides *artificially acquired passive immunity*. It is called passive because the recipient's cells do not produce the antibodies. Such immunity is short-term, seldom lasting more than a few weeks. Furthermore, because the recipient's lymphocytes might not have time to react to the pathogens for which protection was needed, susceptibility to infection may persist.

During pregnancy, certain antibodies (IgG) pass from the maternal blood into the fetal bloodstream. Receptor-mediated endocytosis (see chapter 3, p. 104) using receptor sites on cells of the fetal yolk sac accomplishes the transfer. These receptor sites bind to a region common to the structure of IgG molecules. After entering the fetal cells, the antibodies are secreted into the fetal blood. The fetus acquires limited immunity against the pathogens for which the pregnant woman has developed active immunities. Thus, the fetus has *naturally acquired passive immunity*, which may persist for six months to a year after birth. The newborn may naturally acquire passive immunity through breast milk as well. Table 16.9 summarizes the types of immunity.

 PRACTICE

31 Distinguish between a primary and a secondary immune response.

32 Distinguish between active and passive immunities.

Hypersensitivity Reactions

A hypersensitivity reaction is an exaggerated immune response to a non-harmful antigen. In all hypersensitivities, the individual is pre-sensitized to a particular antigen. Some hypersensitivities can affect almost anyone, but others happen only to people with an inherited tendency toward an exaggerated immune response.

A type I hypersensitivity (*immediate-reaction hypersensitivity*) is commonly called an **allergy** and the antigens that trigger allergic responses are called **allergens** (al′er-jenz). In a type I hypersensitivity a person becomes sensitized by producing IgE antibodies in response to a certain allergen. In the initial exposure to the allergen, IgE attaches to the cell membranes of widely distributed mast cells and basophils. When a subsequent exposure to the same allergen occurs, these cells rapidly release allergy mediators such as *histamine, prostaglandin D2*, and *leukotrienes* (fig. 16.22). This subsequent reaction occurs within seconds, therefore the term "immediate-reaction".

The effect of an allergy depends on how widespread the response is. Allergy mediators dilate arterioles and increase vascular permeability, causing edema and also contraction of bronchial and intestinal smooth muscle and increased mucus production. The overall result is an inflammatory reaction responsible for the symptoms of the allergy, such as hives, hay fever, asthma, eczema, or gastric disturbances. Anaphylaxis is a severe form of immediate-reaction hypersensitivity and may be life threatening. Large amounts of histamine and other allergy mediators spread throughout the body. The person may at first

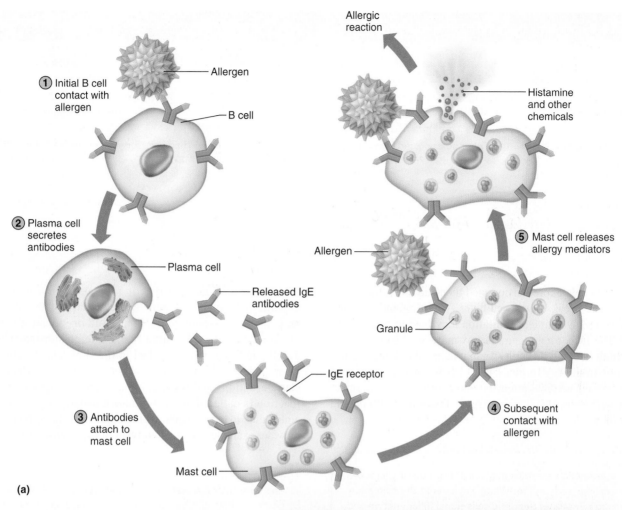

① Initial B cell contact with allergen

Allergen

B cell

② Plasma cell secretes antibodies

Plasma cell

Released IgE antibodies

③ Antibodies attach to mast cell

IgE receptor

Mast cell

(a)

Allergic reaction

Histamine and other chemicals

⑤ Mast cell releases allergy mediators

Allergen

Granule

④ Subsequent contact with allergen

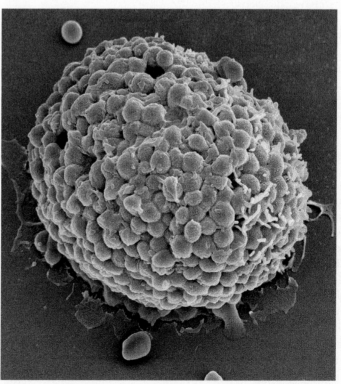

(b)

FIGURE 16.22 AP|R Immediate-reaction allergy.
(a) B cells contact the allergen, and subsequent contact with the allergen leads to mast cells releasing allergy mediators, which cause the symptoms of the allergy attack. (b) A mast cell releases histamine granules (1,750×).

TABLE 16.9   Practical Classification of Immunity

Type	Mechanism	Result
Naturally acquired active immunity	Exposure to live pathogens	Stimulation of an immune response with symptoms of a disease
Artificially acquired active immunity	Exposure to a vaccine containing weakened or dead pathogens or their components	Stimulation of an immune response without symptoms of a disease
Naturally acquired passive immunity	Antibodies passed to fetus from pregnant woman with active immunity or to newborn through colostrum or breast milk from a woman with active immunity	Short-term immunity for newborn without stimulating an immune response
Artificially acquired passive immunity	Injection of antiserum containing specific antibodies or antitoxin	Short-term immunity without stimulating an immune response

feel an inexplicable apprehension, and then suddenly, the entire body itches and breaks out in red hives. Vomiting and diarrhea may follow, and systemic vasodilation may lead to anaphylactic shock. The face, tongue, and larynx may swell from edema, and breathing become difficult. Anaphylactic shock most often results from an allergy to penicillin or from insect stings, but thanks to prompt medical attention and avoidance of allergens by people who know they have allergies, fewer than 100 people a year die of it.

Immediate-reaction allergies may result from the dry, cracked skin of eczema early in life. When tiny cuts in dry skin allow allergens to penetrate, dendritic cells signal inflammation and eczema results. At the same time, the dendritic cells activate immune memory. Years later, inhaling the same allergens that once crossed the broken skin triggers hay fever and asthma.

Peanut allergy is also linked to early breaks in the skin. One in ten people inherit a mutation in a gene that encodes a skin protein called filaggrin, causing the early cracked skin and setting the stage for later allergy. Researchers are investigating whether treatment of dry skin in childhood can prevent development of allergies.

Hypersensitivities that involve IgG (sometimes IgM) antibodies and take one to three hours to develop include type II (*antibody-dependent cytotoxic reactions*) and type III (*immune complex reactions*). In a type II reaction, an antigen binds to a specific cell, stimulating phagocytosis and complement-mediated lysis of the antigen. A transfusion reaction to mismatched blood is a type II hypersensitivity reaction.

In a type III reaction, soluble antibody-antigen complexes form and deposit in certain tissues. There they can trigger tissue damage through phagocytosis or complement binding. Rheumatoid arthritis is an example of a Type III hypersensitivity reaction, and because the antibody-antigen complexes that deposit in the joints trigger an immune response, it is also an example of autoimmunity. Autoimmunity is discussed further on the next page.

A type IV (*delayed-reaction hypersensitivity*) may affect anyone. It results from repeated exposure of the skin to certain chemicals—commonly, household or industrial chemicals or some cosmetics. Eventually the foreign substance activates T cells, many of which collect in the skin. The T cells and the macrophages they attract release chemical factors, which, in turn, cause eruptions and inflammation of the skin (dermatitis). This reaction is called *delayed* because it usually takes about forty-eight hours to begin. Table 16.10 summarizes the responses of, and provides examples for, the four types of hypersensitivity reactions.

The *tuberculin skin test* is used to detect individuals who are suspected of having tuberculosis (TB), who have had it (or a closely related infection), or have been exposed to it. The test (Mantoux test) introduces a tuberculin preparation called *purified protein derivative* (PPD) into the superficial layers of the skin. If the person's T cells have been sensitized to the antigens of the mycobacteria that cause tuberculosis, a Type IV reaction ensues within forty-eight to seventy-two hours, and a localized region of the skin and subcutaneous tissue hardens (indurates). The absence of this reaction (negative result) signifies that the person's T cells have not previously encountered these mycobacterial antigens.

Transplantation and Tissue Rejection

When a car breaks down, replacing the damaged or malfunctioning part often fixes the trouble. The same is sometimes true for the human body. Transplanted tissues and organs include bone, corneas, kidneys, lungs, pancreases, bone marrow, pieces of skin, livers, and hearts. A transplant is risky. The recipient's cells may recognize the donor's tissues as foreign and attempt to destroy the transplanted tissue in a **tissue rejection reaction.** If the transplanted tissue is immunologically active, as is bone marrow, it may produce molecules that harm the recipient's tissue, in a response called graft-versus-host disease (GVHD).

Tissue rejection resembles the cellular immune response against a foreign antigen. The greater the antigenic difference between the cell surface molecules (MHC antigens, discussed

TABLE 16.10	Hypersensitivity Reactions	
Type of Hypersensitivity Reaction	**Response**	**Example**
Type I (immediate-reaction)	Overproduction of IgE antibodies	Hay fever
Type II (antibody-dependent cytotoxic reaction)	Phagocytosis and complement-mediated lysis of antigen	Mismatched blood transfusion
Type III (immune complex reaction)	Phagocytosis and lysis cannot clear antigen-antibody complexes	Autoimmunity
Type IV (delayed-reaction)	T cells and macrophages release chemical factors into the skin	Dermatitis

earlier in this chapter on page 629) of the recipient tissues and the donor tissues, the more rapid and severe the rejection reaction. Matching the cell surface molecules of donor and recipient tissues as closely as possible can minimize the rejection reaction. This means locating a donor whose tissues are antigenically similar to those of the person needing a transplant.

The four major types of grafts (transplant tissue) include:

- *isograft,* tissue from an identical twin
- *autograft,* tissue from elsewhere in a person's body (Technically, this is not a transplant because it is within an individual.)
- *allograft,* tissue from another person who is not an identical twin
- *xenograft,* tissue from a different species, such as pigs and baboons

Table 16.11 presents examples of transplants.

Immunosuppressive drugs are used to reduce the recipient immune system's rejection of transplanted tissue. These drugs suppress T cell or antibody production, thereby dampening the cellular and humoral immune responses, which may also make the recipient more susceptible to infection. Severe side effects, including kidney damage and cancer, may result from use of immunosuppressive drugs. When they first came into use, in the 1980s, immunosuppressive drugs were typically given for the remainder of the recipient's life. That approach is now less common, as cases accumulate of patients who received well-matched transplants and are surviving *without* continued immunosuppressive therapy. These patients received, along with standard transplants such as kidneys, bone marrow stem cells from the donors. Apparently the influx of donor cells establishes a "stable coexistence" with the recipient's cells—they accept each other. When this balance is disturbed, tissue rejection or graft-versus-host disease can result. Researchers hypothesize that bombarding the recipient's body with immunosuppressive drugs immediately after the transplant can disturb this process. The emerging new view is to give the drugs *before* the transplant, and in some cases minimize their use after.

Autoimmunity

The immune system can fail to distinguish self from nonself, producing antibodies, called **autoantibodies,** and cytotoxic T cells that attack and damage the body's tissues and organs. This attack against self is called **autoimmunity.** The signs and symptoms of autoimmune disorders reflect the affected cell types. In autoimmune hemolytic anemia, autoantibodies destroy red blood cells. Autoimmune ulcerative colitis harms colon cells and severe abdominal pain results. Table 16.12 lists some autoimmune disorders.

Why might the immune system attack body tissues? Perhaps a virus, while replicating in a human cell, takes proteins from the host cell's surface and incorporates them onto its own surface. When the immune system "learns" the surface of the virus in order to destroy it, it also learns to attack the human cells that normally bear those particular proteins. Another explanation of autoimmunity is that somehow T cells never learn to distinguish self from nonself, like a soldier unable to recognize his country's uniform.

A third possible route of autoimmunity is when a nonself antigen coincidentally resembles a self antigen. For example, damage to heart valve cells in acute rheumatic fever is due to attack by antibodies present from a recent throat infection with group A streptococcus bacteria. The surfaces of the cells that make up the heart valve resemble those of the bacteria.

Some disorders thought to be autoimmune may have a stranger cause—fetal cells persisting in a woman's circulation for decades. In response to an as yet unknown trigger, the fetal cells, perhaps "hiding" in a tissue such as skin, emerge, stimulating antibody production. The resulting antibodies and symptoms appear to be an autoimmune disorder. The presence of more than one genetically distinct cell population in an individual is called microchimerism ("small mosaic"). Microchimerism that reflects the retention of cells from a fetus may explain the higher prevalence of autoimmune disorders among women. It is seen in a disorder called scleroderma, which means "hard skin."

Scleroderma, which is four times more common in women, typically begins between ages forty-five and fifty-five. It is described as "the body turning to stone." Symptoms include fatigue, swollen joints, stiff fingers, and a masklike face. The hardening may affect blood vessels, the lungs, and the esophagus.

TABLE 16.11	Transplant Types	
Type	**Donor**	**Example**
Isograft	Identical twin	Bone marrow transplant from a healthy twin to a twin who has leukemia
Autograft	Self	Skin graft from one part of the body to replace burned skin
Allograft	Same species	Kidney transplant from a relative or closely matched donor
Xenograft	Different species	Heart valves from a pig

TABLE 16.12 Autoimmune Disorders

Disorder	Symptoms	Antibodies Against
Glomerulonephritis	Lower back pain	Kidney cell antigens that resemble streptococcal bacteria antigens
Graves disease	Restlessness, weight loss, irritability, increased heart rate and blood pressure	Thyroid gland antigens near thyroid-stimulating hormone receptor, causing overactivity
Type I diabetes mellitus	Thirst, hunger, weakness, emaciation	Pancreatic beta cells
Hemolytic anemia	Fatigue and weakness	Red blood cells
Multiple sclerosis	Weakness, incoordination, speech disturbances, visual complaints	Myelin in the white matter of the central nervous system
Myasthenia gravis	Muscle weakness	Receptors for neurotransmitters on skeletal muscle
Pernicious anemia	Fatigue and weakness	Binding site for vitamin B on cells lining stomach
Rheumatic fever	Weakness, shortness of breath	Heart valve cell antigens that resemble streptococcal bacteria antigens
Rheumatoid arthritis	Joint pain and deformity	Cells lining joints
Systemic lupus erythematosus	Red rash on face, prolonged fever, weakness, kidney damage, joint pain	Connective tissue
Ulcerative colitis	Lower abdominal pain	Colon cells

Clues that scleroderma is a delayed response to persisting fetal cells include the following observations:

- It is much more common among women past their reproductive years.
- Symptoms resemble those of graft-versus-host disease (GVHD), in which transplanted tissue produces chemicals that destroy the recipient's tissues.
- Mothers who have scleroderma and their sons have cell surfaces more similar than those of unaffected mothers and their sons. Perhaps the similarity of cell surfaces enables the fetal cells to escape destruction by the woman's immune system. Female fetal cells probably have the same effect, but they are not observable because the method used to detect fetal cells identifies a DNA sequence unique to the Y chromosome (which is only in males.)

Perhaps other disorders considered autoimmune reflect an immune system response to lingering fetal cells.

 PRACTICE

33 How are allergic reactions and immune reactions similar yet different?

34 How does a tissue rejection reaction involve an immune response?

35 How is autoimmunity an abnormal functioning of the immune response?

16.5 | Life-Span Changes

The immune system begins to decline early in life. The thymus reaches its maximal size in adolescence and then slowly shrinks. By age seventy, the thymus is one-tenth the size it was at the age of ten, and the immune system is only 25% as powerful.

The declining strength of the immune response is why elderly people have a higher risk of developing cancer and succumb more easily to infections that they easily fought off at an earlier age, such as influenza, tuberculosis, and pneumonia. HIV infection progresses to AIDS faster in people older than forty. AIDS is more difficult to diagnose in older people, sometimes because physicians do not initially suspect the condition, instead attributing the fatigue, confusion, loss of appetite, and swollen glands to other causes. However, 11% of new cases of AIDS occur in those over age fifty.

Interestingly, numbers of T cells diminish only slightly with increasing age, and numbers of B cells not at all. However, activity levels change for both types of lymphocytes. Because T cell function controls production of B cells, effects on B cells are secondary. The antibody response to antigens slows, and as a result, vaccines may require an extra dose. The proportions of the different antibody classes shift, with IgA and IgG increasing, and IgM and IgE decreasing. A person may produce more autoantibodies than at a younger age, increasing the risk of developing an autoimmune disorder.

Elderly people may not be candidates for certain medical treatments that suppress immunity, such as cancer chemotherapy and steroids to treat inflammatory disorders, because of the declining function of the immune system. Overall, the immune system enables us to survive in a world that is home to many microorganisms. Clinical Application 16.1 looks at the devastation of immunity that is AIDS.

 PRACTICE

36 At about what age is the thymus at its largest?

37 Explain how the strength of the immune response declines in elderly people. ∎

Natural History of a Modern Plague

In the early 1980s, physicians from large cities began reporting cases of rare infections in otherwise healthy young men who also had severe cases of more common infections, such as herpes simplex virus and cytomegalovirus. Some of the men contracted infections known only in nonhuman animals, and conditions known only in individuals with suppressed immune systems, particularly pneumonia caused by the fungus *Pneumocystis jiroveci* and a cancer, Kaposi sarcoma (fig. 16B). The infections were *opportunistic*, which means that they seemed to take advantage of a weakened immune system.

These young men were among the first to have been infected with the human immunodeficiency virus (HIV). Table 16A lists how HIV is, and isn't, spread. The infection typically remains silent for several years and then may progress to *acquired immune deficiency syndrome*, or AIDS, which starts with recurrent fever, weakness, and weight loss. Then, usually after a healthy period, opportunistic infections begin.

How HIV Ravages the Immune System

First HIV crosses a mucosal barrier, such as in the anus or vagina (see figure 3.34, p. 106). Then the virus enters macrophages. In these cells and later in helper T cells, the virus adheres with a surface protein, called gp120, to two receptors on the host cell surface, called CD4 and CCR5. Once the virus enters the cell, a viral enzyme (reverse transcriptase) catalyzes construction of a DNA strand complementary to the viral RNA sequence (the virus has

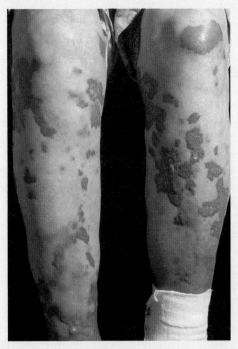

FIGURE 16B Prior to the appearance of AIDS, Kaposi sarcoma was a rare cancer seen only in elderly Jewish and Italian men and in people with suppressed immune systems. In these groups, it produces purplish patches on the lower limbs, but in AIDS patients, Kaposi sarcoma patches appear all over the body and sometimes internally, too. These lower limbs display characteristic lesions.

TABLE 16A   HIV Transmission
How HIV Is Transmitted
Sexual contact, particularly anal intercourse, but also vaginal intercourse and oral sex
Contaminated needles (intravenous drug use, injection of anabolic steroids, accidental needle stick in medical setting)
During birth from infected mother
Breast milk from an infected mother
Receiving infected blood or other tissue (precautions usually prevent this)
How HIV Is Not Transmitted
Casual contact (social kissing, hugging, handshakes)
Objects (toilet seats, deodorant sticks, doorknobs)
Mosquitoes
Sneezing and coughing
Sharing food
Swimming in the same water
Donating blood

—Continued next page

RNA as its genetic material). Synthesis of a second DNA strand complementary to the first results in a viral DNA double helix, which is transported into the cell's nucleus and inserted into a chromosome. The cell then uses the viral DNA sequences to mass-produce pieces of HIV, which are then assembled into new viral particles that burst from the cell.

Once infected helper T cells start to rapidly die, bacterial infections begin because B cells aren't activated to produce antibodies. Much later in infection, HIV variants arise that bind to receptors called CXCR4 on cytotoxic T cells, killing them, too. Loss of these cells renders the body vulnerable to other infections and to cancers. More than 200 drugs are used to treat AIDS-associated infections and cancers.

HIV replicates quickly and its surface features are altered by mutation in ways that evade recognition and attack by the immune system. The virus is especially prone to mutation, because it cannot repair DNA replication errors. The immune system cannot keep up; antibodies against one viral variant are useless against another.

Treatment Approaches

Within days of initial infection, viral variants arise that resist drugs used to treat HIV infection and AIDS. Combining drugs that act at different steps in infection minimizes the number of viruses (viral load) and delays symptom onset and progression. This approach is called "combined antiretroviral therapy" (cART), formerly called "highly active antiretroviral therapy," or HAART. These drugs block HIV from binding to cells, fusing with the cell membrane, entering the cell, replicating viral genetic material, or processing viral proteins to functional sizes. Many people with HIV infection who use antiretroviral drugs can keep their viral levels low enough to stay healthy. The World Health Organization now recommends that cART start within hours of birth for newborns infected with HIV, after case reports demonstrated the ability of the drugs to lower viral RNA to undetectable levels. An ongoing clinical trial is evaluating the long-term efficacy of treating newborns with cART.

A different approach to treating HIV infection mimics the few lucky people who inherit a double dose of a mutation that prevents their cells from making the CCR5 receptor that HIV requires to enter cells. A drug blocks the CCR5 receptor, and bone marrow stem cell transplants replace a patient's cells with cells lacking CCR5. A variation on this strategy alters a patient's own cells, permanently. The idea arose in 2008 when a forty-year-old man with HIV infection and leukemia, known first as "the Berlin patient" but then identifying himself as Timothy, received a stem cell transplant from a donor who had the *CCR5* mutation. Timothy was apparently cured of both the leukemia and the HIV infection as his T cells became incapable of manufacturing the receptor that the virus requires to enter. An ongoing clinical trial in the United States is giving patients their own blood stem cells after the cells have been removed and their *CCR5* genes replaced with nonfunctional ones. So far, this gene therapy has delayed the course of AIDS in a dozen patients.

Prevention

Developing a vaccine for HIV/AIDS has been challenging, both because of the great variability of HIV and the difficulty of testing a vaccine. A live vaccine is too dangerous; a killed vaccine does not generate a sufficient immune response; and insertion of viral genetic material into human chromosomes happens too fast for antibodies or memory cells to form. Some clinical trials did not go on for long enough to see an effect. Many people dropped out of clinical trials or failed to comply with complex drug regimens. HIV infection may be prevented in other ways, including education about reducing risk factors, and giving infected pregnant women antiretroviral drugs to prevent transmission of the virus to offspring.

Lymphatic System

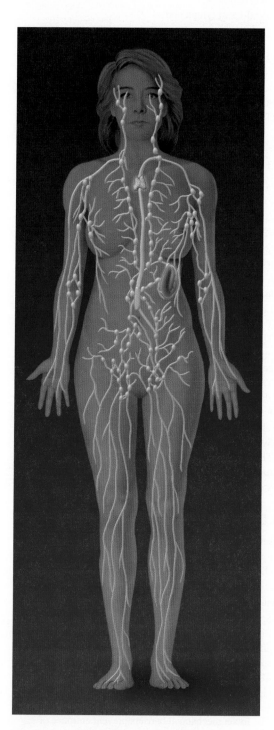

The lymphatic system is an important link between tissue fluid and the plasma; it also plays a major role in the body's response to infection.

Integumentary System

The skin is a first line of defense against infection.

Skeletal System

Cells of the immune system originate in the bone marrow.

Muscular System

Muscle action helps pump lymph through the lymphatic vessels.

Nervous System

Stress may impair the immune response.

Endocrine System

Hormones stimulate lymphocyte production.

Cardiovascular System

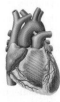

The lymphatic system returns tissue fluid to the bloodstream. Lymph originates as tissue fluid, formed by the action of blood pressure.

Digestive System

Lymph plays a major role in the absorption of fats.

Respiratory System

Cells of the immune system patrol the respiratory system to defend against infection.

Urinary System

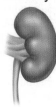

The kidneys control the volume of extracellular fluid, including lymph.

Reproductive System

Special mechanisms inhibit the female immune system in its attack of sperm as foreign invaders.

Chapter Summary

16.1 Lymphatic Pathways (page 617)

1. Lymphatic capillaries
   a. **Lymphatic capillaries** are microscopic, closed-ended tubes that extend into interstitial spaces.
   b. They receive tissue fluid through their thin walls.
2. Lymphatic vessels
   a. **Lymphatic vessels** are formed by the merging of lymphatic capillaries.
   b. They have walls similar to veins, only thinner, and possess valves that prevent backflow of lymph.
   c. The larger lymphatic vessels lead to **lymph nodes** and then merge into lymphatic trunks.
3. Lymphatic trunks and collecting ducts
   a. **Lymphatic trunks** drain lymph from large body regions.
   b. Trunks lead to two **collecting ducts**—the **thoracic duct** and the **right lymphatic duct.**
   c. Collecting ducts empty into the subclavian veins.

16.2 Tissue Fluid and Lymph (page 620)

1. Tissue fluid formation
   a. Tissue fluid originates from plasma and includes water and dissolved substances that have passed through the blood capillary wall.
   b. Tissue fluid generally lacks large proteins, but some smaller proteins are filtered out of blood capillaries into interstitial spaces.
   c. As the protein concentration of tissue fluid increases, colloid osmotic pressure increases.
2. Lymph formation
   a. Increasing hydrostatic pressure in interstitial spaces forces some tissue fluid into lymphatic capillaries. This fluid becomes lymph.
   b. Lymph formation prevents accumulation of excess tissue fluid (edema).
3. Lymph flow
   a. Lymph is under relatively low hydrostatic pressure and may not flow readily without external aid.
   b. Contraction of skeletal muscles, contraction of smooth muscle in the walls of the large lymphatic trunks, and low pressure in the thorax created by breathing movements move lymph.
   c. Any condition that interferes with the flow of lymph results in edema.
4. Lymph function
   a. Lymphatic capillaries in the small intestine absorb and transport lipids.
   b. Lymph eventually returns small protein molecules and fluid to the bloodstream.
   c. It transports foreign particles to the lymph nodes.

16.3 Lymphatic Tissues and Lymphatic Organs (page 621)

Lymphatic tissues include the **mucosa-associated lymphoid tissue (MALT)** which is associated with the digestive, respiratory, urinary, and reproductive tracts. The lymph nodes, thymus, and spleen are lymphatic organs.

1. Lymph nodes
   Lymph nodes are typically bean-shaped, enclosed in connective tissue that extends into the nodes and subdivides them into nodules which contain masses of lymphocytes and macrophages and spaces through which lymph flows.
   a. Locations of lymph nodes
      (1) Lymph nodes aggregate in groups or chains along the paths of larger lymphatic vessels.
      (2) They are in the **cervical, axillary, supratrochlear,** and **inguinal regions** and in the **pelvic, abdominal,** and **thoracic cavities.**
   b. Functions of lymph nodes
      (1) Lymph nodes filter potentially harmful foreign particles from the lymph before it is returned to the bloodstream.
      (2) Lymph nodes are centers for the production of lymphocytes that act against foreign particles.
      (3) They contain macrophages that remove foreign particles from lymph.
2. Thymus
   a. The **thymus** is a soft, bilobed organ within the mediastinum.
   b. It slowly shrinks after puberty.
   c. It is composed of lymphatic tissue subdivided into lobules.
   d. Lobules contain lymphocytes.
   e. **T lymphocytes** leave the thymus and provide immunity.
   f. The thymus secretes thymosins, which stimulate maturation of T lymphocytes.
3. Spleen
   a. The **spleen** is in the upper left portion of the abdominal cavity.
   b. It resembles a large lymph node encapsulated and subdivided into lobules by connective tissue.
   c. Spaces in splenic lobules are filled with blood.
   d. The spleen, which filters foreign particles and damaged red blood cells from the blood, contains many macrophages and lymphocytes.

16.4 Body Defenses Against Infection (page 625)

The presence and reproduction of **pathogens** may cause an infection. Pathogens include bacteria, protozoa, fungi, and viruses. The body has **innate** (nonspecific) and **adaptive** (specific) **defenses** against infection.

1. Innate (nonspecific) defenses
   a. Species resistance
      Each species is resistant to certain diseases that may affect other species but is susceptible to diseases other species may resist.
   b. Mechanical barriers
      (1) **Mechanical barriers** include the skin and mucous membranes.
      (2) Intact mechanical barriers prevent entrance of some pathogens.
      (3) Hair traps infectious agents. Fluids such as tears, sweat, saliva, mucus, and urine wash away micro-organisms before they can firmly attach.

c. **Chemical barriers**
   (1) Enzymes in gastric juice and tears kill some pathogens.
   (2) Low pH in the stomach prevents growth of some bacteria.
   (3) High salt concentration in perspiration kills some bacteria.
   (4) **Interferons** stimulate uninfected cells to synthesize antiviral proteins that block proliferation of viruses, stimulate phagocytosis, and enhance activity of cells that help resist infections and stifle tumor growth.
   (5) **Defensins** make holes in bacterial cell walls and membranes.
   (6) **Collectins** broadly protect against a wide variety of microbes by binding to them, facilitating phagocytosis.
   (7) Activation of **complement** proteins in plasma stimulates inflammation, attracts phagocytes, and enhances phagocytosis.

d. Natural killer (NK) cells
Natural killer cells secrete **perforins,** which destroy cancer cells and cells infected with viruses.

e. Inflammation
   (1) **Inflammation** is a tissue response to damage, injury, or infection.
   (2) The response includes localized redness, swelling, heat, and pain.
   (3) Chemicals released by damaged tissues attract white blood cells to the site.
   (4) Clotting may occur in body fluids that accumulate in affected tissues.
   (5) Connective tissue may form a sac around the injured tissue and thus aid in preventing the spread of pathogens.

f. **Phagocytosis**
   (1) The most active phagocytes in blood are neutrophils and monocytes. Monocytes give rise to macrophages, which may be free or fixed in various tissues.
   (2) Phagocytic cells associated with the linings of blood vessels, in the liver, spleen, and lymph nodes constitute the **mononuclear phagocytic system.**
   (3) Phagocytes remove foreign particles from tissues and body fluids.

g. **Fever**
   (1) Viral or bacterial infection stimulates certain lymphocytes to secrete IL-1, which temporarily raises body temperature.
   (2) Elevated body temperature and the resulting decrease in blood iron level and increased phagocytic activity hamper infection.

2. Adaptive (specific) defenses or immunity
a. **Antigens**
   (1) Before birth, body cells inventory "self" proteins and other large molecules.
   (2) After inventory, lymphocytes develop receptors that allow them to differentiate between nonself (foreign) and self antigens.
   (3) **Haptens** are small molecules that can combine with larger ones, becoming antigenic.

b. Lymphocyte origins
   (1) Lymphocytes originate in red bone marrow and are released into the blood.
   (2) Some reach the thymus where they mature into **T cells.**
   (3) Others, the **B cells,** mature in the red bone marrow.
   (4) Both T cells and B cells reside in lymphatic tissues and organs.
   (5) Varieties of T cells and B cells number in the millions.
   (6) The members of each variety respond only to a specific antigen.
   (7) Originating from a single cell, the members of each variety form a **clone.**

c. T cells and the cellular immune response
   (1) T cells are activated when an **antigen-presenting cell** displays a foreign antigen.
   (2) When a macrophage acts as an accessory cell, it phagocytizes an antigen-bearing agent, digests the agent, and displays the resulting antigens on its cell membrane in association with certain MHC proteins.
   (3) T cells respond to antigens by cell-to-cell contact **(cellular immune response).**
   (4) T cells secrete **cytokines,** such as interleukins, that enhance cellular responses to antigens.
   (5) T cells may also secrete substances that are toxic to their target cells.
   (6) A **helper T cell** becomes activated when it encounters displayed antigens for which it is specialized to react.
   (7) Once activated, a helper T cell stimulates a type of B cell (plasma cell) to produce antibodies for the specific antigen.
   (8) **Cytotoxic T cells** recognize foreign antigens on tumor cells and cells whose surfaces indicate that they are infected by viruses. Stimulated cytotoxic T cells secrete perforin to destroy these cells.
   (9) **Memory T cells** allow for immediate response to second and subsequent exposure to the same antigen.

d. B cells and the humoral immune response
   (1) Sometimes a B cell is activated when it encounters an antigen that fits its antigen receptors, or more often a B cell is activated when stimulated by a helper T cell.
   (2) An activated B cell proliferates (especially when stimulated by a T cell), enlarging its clone.
   (3) Some activated B cells differentiate into antibody-producing **plasma cells.**
   (4) **Antibodies** react against the antigen-bearing agent that stimulated their production **(humoral immune response).**
   (5) An individual's many different B cells defend against many pathogens.
   (6) Other activated B cells differentiate further into B memory cells.
   (7) Antibodies are soluble proteins called **immunoglobulins.**
   (8) They constitute the gamma globulin fraction of plasma.

(9) Each immunoglobulin molecule consists of four linked chains of amino acids.

(10) Variable regions at the ends of these chains are specialized **antigen-binding sites** that react with different antigens.

(11) The five major types of immunoglobulins are IgG, IgA, IgM, IgD, and IgE.

(12) IgG, IgA, and IgM make up most of the circulating antibodies.

(13) Antibodies directly attach to antigens, activate complement, or stimulate local tissue changes that are unfavorable to antigen-bearing agents.

(14) Direct attachment results in agglutination, precipitation, or neutralization.

(15) Activated complement proteins alter infected cells so they become more susceptible to phagocytosis, attract phagocytes, and lyse foreign cell membranes.

e. Immune responses

(1) B cells and T cells first encountering an antigen for which they are specialized to react constitutes a **primary immune response.**

(a) During this response, antibodies are produced for several weeks.

(b) Some T cells and B cells remain dormant as memory cells.

(2) A **secondary immune response** occurs as memory cells rapidly respond to subsequent exposure to an antigen.

f. Practical classification of immunity

(1) A person who encounters a pathogen and has a primary immune response develops naturally acquired active immunity.

(2) A person who receives a **vaccine** containing a dead or weakened pathogen, or part of it, develops artificially acquired active immunity. Herd immunity protects populations.

(3) A person who receives an injection of antiserum or antitoxin has artificially acquired passive immunity.

(4) When antibodies pass through a placental membrane from a pregnant woman to her fetus, the fetus naturally acquires passive immunity.

(5) Active immunity lasts much longer than passive immunity.

g. Hypersensitivity reactions

(1) Hypersensitivity reactions are excessive misdirected immune responses that may damage tissues.

(2) Type I hypersensitivity (immediate-reaction hypersensitivity) is an inborn ability to overproduce IgE in response to an **allergen.**

(a) Allergic reactions result from mast cells bursting and releasing allergy mediators such as histamine.

(b) The released chemicals cause allergy symptoms such as hives, hay fever, asthma, eczema, or gastric disturbances.

(c) In anaphylactic shock, allergy mediators flood the body, causing severe symptoms, including decreased blood pressure and difficulty breathing.

(3) Type II hypersentivity (antibody-dependent cytotoxic hypersensitivity) reactions occur when blood transfusions are mismatched.

(4) Type III hypersensitivity (immune complex hypersensitivity) reactions involve autoimmunity, an immune reaction against self antigens.

(5) Type IV hypersensitivity (delayed-reaction hypersensitivity), which can occur in anyone and inflame the skin, results from repeated exposure to household or industrial chemicals or some cosmetics.

h. Transplantation and tissue rejection

(1) A transplant recipient's immune system may react against the donated tissue in a **tissue rejection reaction.**

(2) Transplants may take place between genetically identical twins, from one body part to another, between unrelated individuals of the same species, or between individuals of different species.

(3) Matching cell surface molecules (MHC antigens) of donor and recipient tissues, transplanting stem cells from the donor, and the use of immunosuppressive drugs can help the body accept the foreign tissue.

i. Autoimmunity

(1) In autoimmune disorders, **autoantibodies** attack the body's tissues.

(2) Autoimmune disorders may result from a previous viral infection, faulty T cell development, or reaction to a nonself antigen that resembles a self antigen.

(3) Retained fetal cells can cause a condition that resembles an autoimmune disorder.

16.5 Life-Span Changes (page 640)

1. The immune system begins to decline early in life, in part due to the shrinking thymus.

2. Numbers of T cells and B cells do not significantly change, but activity levels do.

3. Proportions of the different antibody classes shift.

CHAPTER ASSESSMENTS

16.1 Lymphatic Pathways

1   Describe two functions of the lymphatic vessels. (p. 617)
2   Trace the general pathway of lymph from the interstitial spaces to the bloodstream. (pp. 617–618)

16.2 Tissue Fluid and Lymph

3   Tissue fluid forms as a result of filtration from blood capillaries exceeding _____ , whereas lymph forms due to increasing _____ _____ in the tissue fluid. (p. 620)
4   Explain why physical exercise promotes lymphatic circulation. (p. 621)
5   Explain how a lymphatic obstruction leads to edema. (p. 621)
6   Describe the primary functions of lymph. (p. 621)

16.3 Lymphatic Tissues and Lymphatic Organs

7   Draw a lymph node, and label its parts. (pp. 621–622)
8   On a drawing of the body locate the major body regions containing lymph nodes. (p. 622)
9   Explain the functions of a lymph node. (p. 623)
10 Indicate the locations of the thymus and spleen. (p. 623)
11 Compare and contrast the functions of the thymus and spleen. (p. 623)

16.4 Body Defenses Against Infection

12 Defense mechanisms that prevent the entry of many types of pathogens and destroy them if they enter, provide _____ (nonspecific) defense. Precise mechanisms targeting specific pathogens provide _____ (specific) defense. (p. 625)
13 Define *species resistance*. (p. 626)
14 Identify the barriers that provide the body's first line of defense against infectious agents. (p. 626)
15 Describe how enzymatic actions function as defense mechanisms against pathogens. (p. 626)
16 Distinguish among the chemical barriers (interferons, defensins, collectins, and complement proteins), and give examples of their different actions. (p. 626)
17 _____ _____ _____ are specialized lymphocytes that secrete perforins to lyse cell membranes of virus-infected cells. (p. 626)
18 List the major effects of inflammation and explain why each occurs. (p. 626)
19 Identify the major phagocytic cells in the blood and other tissues. (p. 627)
20 List possible causes of fever, and explain the benefits of fever. (p. 627)
21 Distinguish between an antigen and a hapten. (pp. 627–628)
22 Review the origin of T cells and B cells. (p. 628)
23 Define *clone of lymphocytes*. (p. 628)
24 Explain the cellular immune response including the activation of T cells. (pp. 629–630)

25 Define *cytokine*. (p. 629)
26 List three types of T cells, and describe the function of each in the immune response. (p. 630)
27 Explain the humoral immune response, including the activation of B cells. (p. 630)
28 Explain the function of plasma cells. (p. 630)
29 Draw and label the parts of an antibody molecule. (p. 632)
30 Distinguish between the variable region and the constant region of an antibody molecule. (p. 632)
31 Match the types of antibodies with their function and/or where each is found. (pp. 633–634)

(1) associated with allergic reactions   A. IgA
(2) important in B cell activation, on surfaces   B. IgM
    of most B cells   C. IgG
(3) activates complement, anti-A and   D. IgD
    anti-B in blood   E. IgE
(4) effective against bacteria, viruses, and toxins
    in plasma and tissue fluids
(5) found in exocrine secretions, including breast milk

32 Describe three ways in which an antibody's direct attack on an antigen helps remove that antigen. (p. 634)
33 List the various effects of complement activation. (pp. 634–635)
34 Contrast a primary and a secondary immune response. (p. 635)
35 Contrast active and passive immunity. (p. 636)
36 Define *vaccine*. (p. 636)
37 Explain how a vaccine produces its effect. (p. 636)
38 Describe how a fetus may obtain antibodies from maternal blood. (p. 636)
39 Explain the relationship between a hypersensitivity reaction and an immune response. (p. 636)
40 Distinguish between an antigen and an allergen. (p. 636)
41 Describe how an immediate-reaction hypersensitivity response may occur. (p. 636)
42 List the major events leading to a delayed-reaction hypersentivity response. (p. 637)
43 Explain the relationship between tissue rejection and an immune response. (pp. 637–638)
44 Describe two methods used to reduce the severity of a tissue rejection reaction. (p. 639)
45 Explain the goal of using immunosuppressive drugs before a transplant. (p. 639)
46 Explain the relationship between autoimmunity and an immune response. (p. 639)

16.5 Life-Span Changes

47 Explain the causes for a decline in the strength of the immune response in the elderly. (p. 640)

INTEGRATIVE ASSESSMENTS/CRITICAL THINKING

Outcomes 6.1, 16.1, 16.2, 16.3

1. Why is injecting a substance into the skin like injecting it into the lymphatic system?

Outcomes 16.1, 16.2, 16.3

2. How can removal of enlarged lymph nodes for microscopic examination aid in diagnosing certain diseases?

Outcomes 16.3, 16.4

3. What functions of the lymphatic system would be affected in a person born without a thymus?

Outcome 16.4

4. The immune response is specific, diverse, and has memory. Give examples of each of these characteristics.

5. Some parents keep their preschoolers away from other children to prevent them from catching illnesses. How might these well-meaning parents be harming their children?

6. Why does vaccination provide long-lasting protection against disease, whereas gamma globulin (IgG) provides only short-term protection?

7. Why is a transplant consisting of fetal tissue less likely to provoke an immune rejection response than tissue from an adult?

8. An eighteen-year-old female athlete received a kidney transplant from her brother, who was an HLA match, along with stem cells from his blood. The surgery was so successful that she was able to return to competitive swimming within a year and made the Olympic team. However, she was disqualified due to blood test results that found a male Y chromosome in her blood. Explain one way that this could have happened.

ONLINE STUDY TOOLS

Connect Interactive Questions Reinforce your knowledge using assigned interactive questions covering lymphatic system structures, lymph flow, and innate and adaptive immunity.

Connect Integrated Activity Can you differentiate between an allergic reaction, an autoimmune response, and an immune deficiency?

LearnSmart Discover which chapter concepts you have mastered and which require more attention. This adaptive learning tool is personalized, proven, and preferred.

Anatomy & Physiology Revealed Go more in depth into the human body by exploring the lymphatic organs. Also view animations on the immune response.

Digestive System

LEARNING OUTCOMES

After you have studied this chapter, you should be able to:

17.1 Overview of the Digestive System

1 Which processes are carried out by the digestive system. (p. 650)

2 Name the organs of the digestive system. (p. 650)

3 Describe the structure of the wall of the alimentary canal. (p. 650)

4 Explain how the contents of the alimentary canal are mixed and moved. (pp. 650–651)

5 Discuss the general effects of innervation of the alimentary canal by the sympathetic and parasympathetic divisions of the autonomic nervous system. (pp. 651–652)

17.2 Mouth

6 Describe the functions of the structures associated with the mouth. (pp. 653–656)

7 Describe how different types of teeth are adapted for different functions, and list the parts of a tooth. (p. 656)

17.3–17.9 Salivary Glands—Large Intestine

8 Locate each of these organs and glands of the digestive system; then describe the general structure and function of each. (pp. 658–685)

9 Identify the function of each enzyme secreted by these digestive organs and glands. (pp. 658–680)

10 Describe how digestive secretions are regulated. (pp. 659–680)

11 Explain control of movement of material through the alimentary canal. (pp. 653, 660–667, 682–686)

12 Describe the mechanisms of swallowing, vomiting, and defecating. (pp. 660–661, 667, 686)

13 Explain how the products of digestion are absorbed. (pp. 666, 681–686)

17.10 Life-Span Changes

14 Describe aging-related changes in the digestive system. (p. 688)

Falsely colored transmission electron micrograph (TEM) of microvilli on intestinal cells (20,000×).

THE WHOLE PICTURE

The main part of the digestive system is a long tube that extends through the body. Just as a tunnel through a mountain is filled with air, not the material of the mountain, the inside of the digestive tube is really part of the outside world, not part of the internal environment. Whereas the pH, electrolyte concentrations, and water volume of the internal environment are tightly regulated, they change from the beginning of the tube to its end, and vary with the chemical nature of what we eat.

Material that enters the tube is first broken down into its chemical building blocks, such as simple sugars, amino acids, nucleotides, fatty acids, and glycerol, and then absorbed across the wall of the tube, often by specific transport processes. Your body then recombines those building blocks into the molecules it needs. If you are trying to build muscles, your cells need to synthesize muscle protein, but that doesn't mean you need to eat actin and myosin. As long as all of the necessary amino acids are present in your diet in sufficient amounts, the digestive system will bring them into the internal environment, and your muscles will be able to use them to build the proteins that they need.

Anatomy & Physiology REVEALED®
aprevealed.com

Module 12: Digestive System

aliment-, food: *alimentary* canal—tubelike part of the digestive system.

cari-, decay: dental *caries*—tooth decay.

cec-, blindness: *cecum*—blind-ended sac at the origin of the large intestine.

chym-, juice: *chyme*—semifluid paste of food particles and gastric juice formed in the stomach.

decidu-, falling off: *deciduous* teeth—teeth shed during childhood.

frenul-, bridle, restraint: *frenulum*—membranous fold that anchors the tongue to the floor of the mouth.

gastr-, stomach: *gastric* gland—part of the stomach that secretes gastric juice.

hepat-, liver: *hepatic* duct—duct that carries bile from the liver to the bile duct.

hiat-, opening: esophageal *hiatus*—opening through which the esophagus penetrates the diaphragm.

lingu-, tongue: *lingual* tonsil—mass of lymphatic tissue at the root of the tongue.

peri-, around: *peristalsis*—wavelike ring of contraction that moves material along the alimentary canal.

pyl-, gatekeeper, door: *pyloric* sphincter—muscle that serves as a valve between the stomach and small intestine.

rect-, straight: *rectum*—distal part of the large intestine.

sorpt-, to soak up: ab*sorption*—uptake of substances.

vill-, hairy: *villi*—tiny projections of mucous membrane in the small intestine.

17.1 | Overview of the Digestive System

Digestion (di-jest'yun) is the mechanical and chemical breakdown of foods into forms that cell membranes can absorb. *Mechanical digestion* breaks large pieces into smaller ones without altering their chemical composition. *Chemical digestion* breaks food into simpler chemicals. The organs of the **digestive system** carry out these processes, as well as ingestion, propulsion, absorption, and defecation.

The digestive system consists of the **alimentary canal** (al"i-men'tar-e kah-nal'), extending from the mouth to the anus, and several accessory organs, which release secretions into the canal. The alimentary canal includes the mouth, pharynx, esophagus, stomach, small intestine, large intestine, and anal canal. The accessory organs include the salivary glands, liver, gallbladder, and pancreas. **Figure 17.1** and reference plates 4, 5, and 6 (pp. 41–43) show the major organs of the digestive system.

The digestive system originates from the inner layer (endoderm) of the embryo, which folds to form the tube of the alimentary canal. The accessory organs develop as buds from the tube.

PRACTICE

1 What are the general functions of the digestive system?

2 Which organs constitute the digestive system?

General Characteristics of the Alimentary Canal

The alimentary canal is a muscular tube about 8 meters long that passes through the body's thoracic and abdominopelvic cavities (fig. 17.2). The structure of its wall, how it moves food, and its innervation are similar throughout its length.

Structure of the Wall

The wall of the alimentary canal consists of four distinct layers that vary somewhat from region to region. Although the four-layered structure persists throughout the alimentary canal, certain

regions are specialized for particular functions. Beginning with the innermost tissues, these layers, shown in **figure 17.3**, include the following:

1. The **mucosa** (mu-ko'sah), or mucous membrane has a surface of epithelium, underlying connective tissue (lamina propria), and a small amount of smooth muscle (muscularis mucosae). In some regions, the mucosa is folded, with tiny projections that extend into the passageway, or **lumen,** of the digestive tube. These folds increase the absorptive surface area. The mucosa also has glands into which the lining cells secrete mucus and digestive enzymes. The mucosa protects the tissues beneath it, secretes into the lumen, and absorbs substances from the diet.

2. The **submucosa** (sub"mu-ko'sah) contains considerable loose connective tissue as well as glands, blood vessels, lymphatic vessels, and nerves. Its vessels nourish the surrounding tissues and carry away absorbed materials.

3. The **muscularis externa,** or **muscularis,** which provides movements of the tube, consists of two layers of smooth muscle tissue. The cells of the inner layer encircles the tube. When the circular layer contracts, the diameter of the tube decreases. The cells of the outer muscular layer run lengthwise. When the longitudinal layer contracts, the tube shortens. Coordinated contractions of both muscle layers cause movements associated with digestion and absorption of food.

4. The **serosa** (ser-o'sah), or serous layer, is the outer covering of the tube. It is composed of the *visceral peritoneum,* which is formed of epithelium on the outside and connective tissue beneath it. The cells of the serosa protect underlying tissues and secrete serous fluid, which moistens and lubricates the tube's outer surface. Serous fluid enables the organs in the abdominal cavity to slide freely against one another.

Table 17.1 summarizes the characteristics of the layers of the alimentary canal wall.

Movements of the Tube

The motor functions of the alimentary canal are of two basic types—*mixing movements* and *propelling movements*. Mixing occurs when

TABLE 17.1	Layers of the Wall of the Alimentary Canal	
Layer	Composition	Function
Mucosa	Epithelium, connective tissue, smooth muscle	Protection, secretion, absorption
Submucosa	Loose connective tissue, blood vessels, lymphatic vessels, nerves	Nourishes surrounding tissues, transports absorbed materials
Muscularis	Smooth muscle cells in circular and longitudinal groups	Movements of the tube and its contents
Serosa	Epithelium, connective tissue	Protection, lubrication

smooth muscle in small segments of the tube contracts rhythmically. For example, when the stomach is full, waves of muscular contractions move along its wall from one end to the other (fig. 17.4a). These waves pass every twenty seconds or so and mix foods with the digestive juices secreted by the mucosa. In the small intestine, **segmentation** is a type of movement that aids mixing by alternately contracting and relaxing the smooth muscle in segments of the organ (fig. 17.4b). Because segmentation follows a back-and-forth pattern, materials are not moved along the tract in one direction.

Propelling movements include a wavelike motion called **peristalsis** (per″ĭ-stal′sis), in which a ring of contraction occurs in the wall of the tube and moves progressively along its length (fig. 17.4c). At the same time, the muscular wall just ahead of the ring relaxes—a phenomenon called *receptive relaxation*. As the peristaltic wave of contraction moves along the tube, it pushes the contents of the tube ahead of it. Peristalsis begins when food expands the tube. It causes the sounds that can be heard through a stethoscope applied to the abdominal wall.

A "GI camera" the size of a large vitamin pill can image the alimentary canal, revealing blockages and sites of bleeding and helping to diagnose celiac disease, Crohn's disease, and cancer. The patient swallows the capsule, which contains a camera, a light source, radio transmitter, and batteries. Peristalsis moves it along, and about six hours after swallowing, it transmits images from the small intestine to a device worn on the physician's belt. The information goes to a computer, which downloads still or video images. The disposable device leaves the body in the feces. Millions of people have had the procedure, which has an advantage over endoscopy of not requiring sedation.

Innervation of the Tube

Branches of the sympathetic and parasympathetic divisions of the autonomic nervous system extensively innervate the alimentary canal. Some of these postganglionic fibers, which are associated with the tube's muscular layer, maintain muscle tone and regulate the strength, rate, and velocity of muscular contractions. Many of the postganglionic fibers are organized into a network or nerve plexus within the wall of the canal (see fig. 17.3). The *submucosal plexus* is important in controlling secretions by the gastrointestinal tract. The *myenteric plexus* of the muscular layer is more extensive and controls gastrointestinal motility. The plexuses also include sensory neurons.

Parasympathetic impulses generally increase the activities of the digestive system such as motility and enzyme secretion. Some

of these impulses originate in the brain and are conducted through branches of the vagus nerves to the esophagus, stomach, pancreas, gallbladder, small intestine, and proximal half of the large intestine. Other parasympathetic impulses arise in the sacral region of the spinal cord and supply the distal half of the large intestine.

The effects of sympathetic impulses on digestive actions usually oppose those of the parasympathetic division, meaning

CAREER CORNER

Endoscopy Technician

The young woman had undergone gastric bypass surgery five weeks earlier, and instead of being able to eat more foods, as she had anticipated, she was now unable to keep anything down. She recognized the danger signs and went to the emergency department at the hospital where the bypass had been performed, and was admitted for an endoscopy to find the blockage.

An endoscope is a flexible fiberoptic instrument that is passed through an opening in the body or inserted through an incision. It provides a direct view of internal anatomy. Endoscopy of the gastrointestinal tract can detect inflammation, ulcerations, and other superficial lesions and abnormalities. The endoscopy procedure can also obtain tissue samples for biopsies.

The endoscopy technician works as part of a team, providing support for physicians and nurses throughout procedures. The technician prepares materials, obtains specimens, maintains a sterile field and instruments during procedures, and may administer medications, monitor vital signs, and change dressings. Setting up and cleaning equipment is also part of the job.

Training to become an endoscopy technician begins with a high school diploma. This is followed by an associate of science degree in Surgical Technology, which includes courses in anatomy and physiology and medical terminology. Additional clinical training may take place on the job or in an internship program at a health-care facility. Because the job requires informing and reassuring patients, excellent communication skills are a must. In addition, solid critical thinking skills enable the technician to function as part of a team. The job also requires physical strength and stamina.

Most endoscopy technicians work in hospitals or clinics. Jobs may also be available in long-term care facilities and nursing homes. Hours may be long and unusual, and the pace of performing procedures rapid.

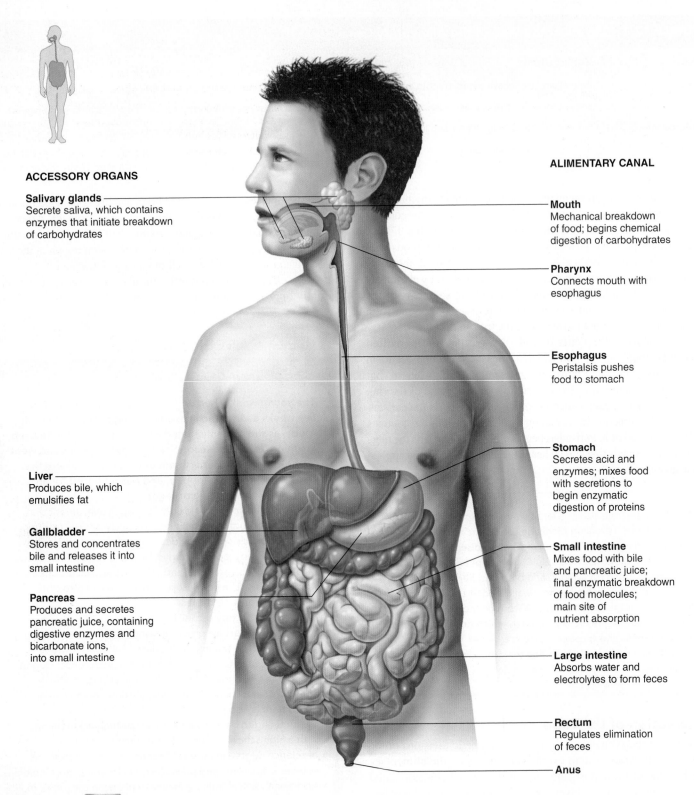

ACCESSORY ORGANS

Salivary glands
Secrete saliva, which contains enzymes that initiate breakdown of carbohydrates

Liver
Produces bile, which emulsifies fat

Gallbladder
Stores and concentrates bile and releases it into small intestine

Pancreas
Produces and secretes pancreatic juice, containing digestive enzymes and bicarbonate ions, into small intestine

ALIMENTARY CANAL

Mouth
Mechanical breakdown of food; begins chemical digestion of carbohydrates

Pharynx
Connects mouth with esophagus

Esophagus
Peristalsis pushes food to stomach

Stomach
Secretes acid and enzymes; mixes food with secretions to begin enzymatic digestion of proteins

Small intestine
Mixes food with bile and pancreatic juice; final enzymatic breakdown of food molecules; main site of nutrient absorption

Large intestine
Absorbs water and electrolytes to form feces

Rectum
Regulates elimination of feces

Anus

FIGURE 17.1 AP|R Organs of the digestive system.

that certain digestive actions are inhibited. For example, sympathetic impulses inhibit mixing and propelling movements, but stimulate contraction of the sphincter muscles in the wall of the alimentary canal, blocking movement of materials through the tube.

 PRACTICE

3 Describe the wall of the alimentary canal.

4 Name the types of movements in the alimentary canal.

5 How do parasympathetic nerve impulses affect digestive actions? What effect do sympathetic nerve impulses have?

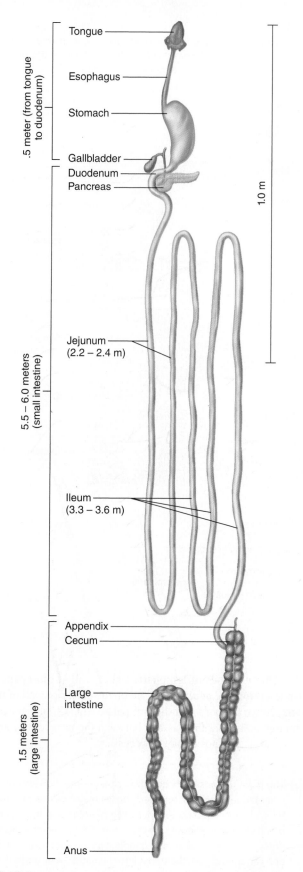

Tongue

.5 meter (from tongue to duodenum)

Esophagus

Stomach

Gallbladder

Duodenum

Pancreas

1.0 m

Jejunum
(2.2 – 2.4 m)

5.5 – 6.0 meters
(small intestine)

Ileum
(3.3 – 3.6 m)

Appendix

Cecum

Large
intestine

1.5 meters
(large intestine)

Anus

FIGURE 17.2 The alimentary canal is a muscular tube about 8 meters long.

What is the distance from the tongue to the duodenum, in English units?

Answer can be found in Appendix G.

17.2 | Mouth

The **mouth,** the first portion of the alimentary canal, receives food and begins digestion by mechanically breaking solid particles into smaller pieces and mixing them with saliva. This action is called *mastication* (mas″tĭ-ka′shun). The mouth also functions as an organ of speech and sensory reception. It is surrounded by the lips, cheeks, tongue, and palate and includes a chamber between the palate and tongue called the *oral cavity,* as well as a narrow space between the teeth, cheeks, and lips called the *oral vestibule* (fig. 17.5 and reference plate 9, p. 46).

Cheeks and Lips

The **cheeks** form the lateral walls of the mouth. They consist of outer layers of skin, pads of subcutaneous fat, muscles associated with expression and chewing, and inner linings of moist, stratified squamous epithelium.

Cells lining the cheek are sampled to provide DNA for genetic testing. A person rubs the inside of the cheek with a cotton swab or swishes with mouthwash and expectorates into a small tube. The sample is sent to a lab, the DNA extracted, and informative DNA sequences identified. DNA tests performed on cheek cells are used in diagnostic medicine, ancestry testing, and forensics.

The **lips** are highly mobile structures that surround the mouth opening. They contain skeletal muscles and sensory receptors useful in judging the temperature and texture of foods. Their normal reddish color is due to the many blood vessels near their surfaces. The external borders of the lips mark the boundaries between the skin of the face and the mucous membrane that lines the alimentary canal.

Tongue

The **tongue** (tung) is a thick, muscular organ that occupies the floor of the mouth and nearly fills the oral cavity when the mouth is closed. Mucous membrane covers the tongue, and a membranous fold called the **lingual frenulum** (ling′gwahl fren′u-lum) connects the midline of the tongue to the floor of the mouth.

The *body* of the tongue is largely composed of skeletal muscle fibers that run in several directions. Muscular action mixes food particles with saliva during chewing and moves food toward the pharynx during swallowing. The tongue also helps move food underneath the teeth for chewing. The surface of the tongue has rough projections, called **papillae** (pah-pil′a) (fig. 17.6). Some of these provide friction, which helps move food. Other papillae contain taste buds, which are used in detecting sweet, salty, bitter, sour, and umami (see chapter 12, pp. 454–455). Although most of the taste buds are located on papillae, some are scattered elsewhere in the mouth, particularly in children.

The posterior region, or *root,* of the tongue is anchored to the hyoid bone. It is covered with rounded masses of lymphatic tissue called **lingual tonsils** (ton′silz) (fig. 17.7).

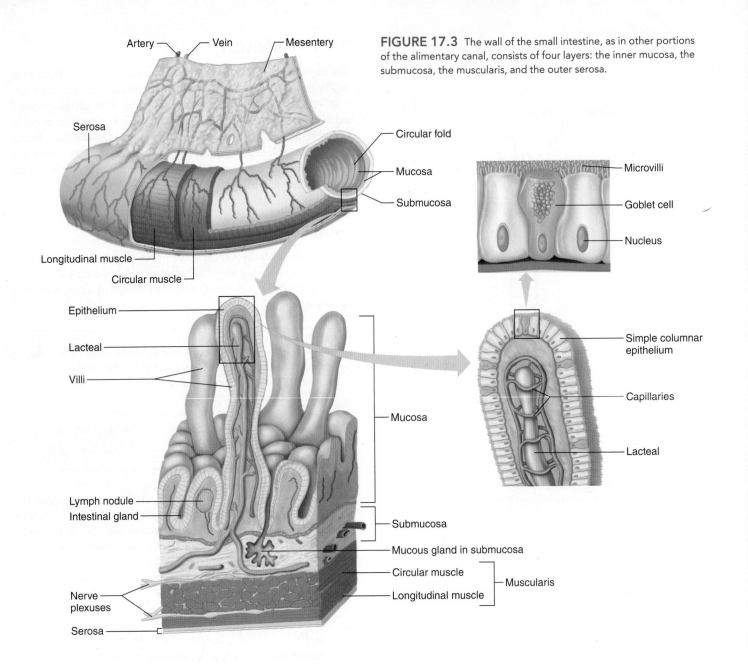

FIGURE 17.3 The wall of the small intestine, as in other portions of the alimentary canal, consists of four layers: the inner mucosa, the submucosa, the muscularis, and the outer serosa.

Labels in figure:
- Artery
- Vein
- Mesentery
- Serosa
- Circular fold
- Mucosa
- Submucosa
- Longitudinal muscle
- Circular muscle
- Epithelium
- Lacteal
- Villi
- Lymph nodule
- Intestinal gland
- Nerve plexuses
- Serosa
- Mucosa
- Submucosa
- Mucous gland in submucosa
- Circular muscle
- Longitudinal muscle
- Muscularis
- Microvilli
- Goblet cell
- Nucleus
- Simple columnar epithelium
- Capillaries
- Lacteal

Palate

The **palate** (pal′at) forms the roof of the oral cavity and consists of a hard anterior part and a soft posterior part. The *hard palate* is formed by the palatine processes of the maxillae in front and the horizontal portions of the palatine bones in back. The *soft palate* forms a muscular arch, which extends posteriorly and downward as a cone-shaped projection called the **uvula** (u′vu-lah).

During swallowing, muscles draw the soft palate and the uvula upward. This action closes the opening between the nasal cavity and the pharynx, preventing food from entering the nasal cavity.

In the back of the mouth, on either side of the tongue and closely associated with the palate, are masses of lymphatic tissue called **palatine** (pal′ah-tīn) **tonsils** (fig. 17.7). These structures lie beneath the epithelial lining of the mouth and, like other lymphatic tissues, help protect the body against infections (see chapter 16, p. 621).

Other masses of lymphatic tissue, called **pharyngeal** (fah-rin′je-al) **tonsils,** or *adenoids,* are on the posterior wall of the pharynx, above the border of the soft palate (fig. 17.7). If the adenoids enlarge and block the passage between the nasal cavity and pharynx, surgical removal may be necessary.

The palatine tonsils are common sites of infection and become inflamed in *tonsillitis.* Infected tonsils may swell so greatly that they block the passageways of the pharynx and interfere with breathing and swallowing. Because the mucous membranes of the pharynx, auditory tubes, and middle ears are continuous, an infection can spread from the throat into the middle ears (otitis media).

When tonsillitis recurs and does not respond to antibiotic treatment, the tonsils may be surgically removed. Such tonsillectomies are done less often today than they were a generation ago because the tonsils' role in immunity is now recognized.

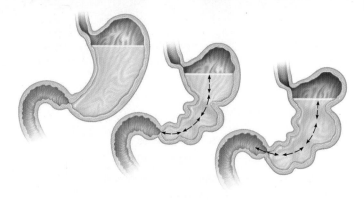

(a)

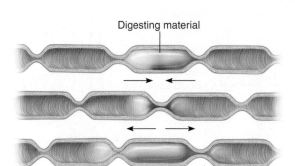

Digesting material

(b)

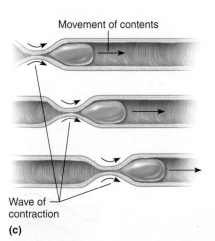

Movement of contents

Wave of contraction

(c)

FIGURE 17.4 Movements through the alimentary canal. (*a*) Mixing movements occur when small segments of the muscular wall of the stomach rhythmically contract. (*b*) Segmentation mixes the contents of the small intestine. (*c*) Peristaltic waves move the contents along the canal.

 PRACTICE

6 What are the functions of the mouth?
7 How does the tongue function as part of the digestive system?
8 Where are the tonsils located?

Teeth

The **teeth** are the hardest structures in the body. They are not considered part of the skeletal system because they have at least two types of proteins that are not found in bone, and their structure is different from that of bone.

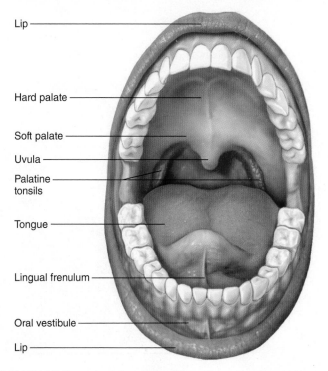

Lip
Hard palate
Soft palate
Uvula
Palatine tonsils
Tongue
Lingual frenulum
Oral vestibule
Lip

FIGURE 17.5 The mouth is adapted for ingesting food and beginning digestion, both mechanically and chemically.

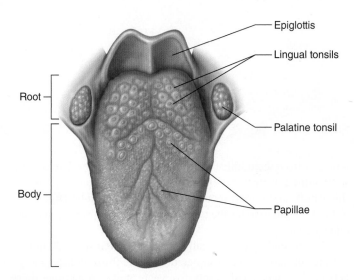

Epiglottis
Lingual tonsils
Palatine tonsil
Root
Body
Papillae

FIGURE 17.6 The surface of the tongue, superior view.

Teeth develop within sockets in the alveolar processes of the mandible and maxillae. They are unique in that two sets form during development (fig. 17.8). The first set, the *primary teeth* (deciduous teeth), usually erupt through the gums (gingiva) at regular intervals between the ages of six months and two to four years. The ten primary teeth are anchored in each jaw from the midline toward the sides in the following sequence: central incisor, lateral incisor, canine (cuspid), first molar, and second molar.

The primary teeth are usually shed in the same order as they erupted. After their roots are resorbed, the *secondary* (permanent) *teeth* push the primary teeth out of their sockets. This secondary set consists of thirty-two teeth—sixteen in each jaw—and they are

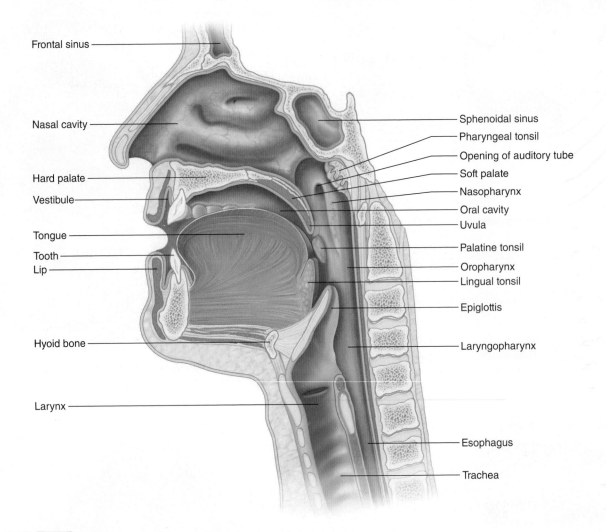

Frontal sinus

Nasal cavity

Hard palate

Vestibule

Tongue

Tooth

Lip

Hyoid bone

Larynx

Sphenoidal sinus

Pharyngeal tonsil

Opening of auditory tube

Soft palate

Nasopharynx

Oral cavity

Uvula

Palatine tonsil

Oropharynx

Lingual tonsil

Epiglottis

Laryngopharynx

Esophagus

Trachea

FIGURE 17.7 AP|R Sagittal section of the mouth, nasal cavity, and pharynx.

arranged from the midline as follows: central incisor, lateral incisor, canine (cuspid), first premolar (bicuspid), second premolar (bicuspid), first molar, second molar, and third molar (fig. 17.9). Table 17.2 summarizes the types and numbers of primary and secondary teeth.

The secondary teeth usually begin to erupt at six years of age, but the set may not be completed until the third molars emerge between seventeen and twenty-five years of age. Sometimes these third molars, also called wisdom teeth, become wedged in abnormal positions in the jaws and fail to erupt. Such *impacted* wisdom teeth may be removed to alleviate pain.

The teeth break food into smaller pieces, which begins mechanical digestion. Chewing increases the surface area of the food particles, enabling digestive enzymes to interact more effectively with nutrient molecules.

Different teeth are adapted to handle food in different ways. The sharp edges of the chisel-shaped *incisors* bite off large pieces of food. The cone-shaped *canines* grasp and tear food. The flattened surfaces of the *premolars* and *molars* grind food particles.

Each tooth consists of two main parts—the *crown,* which projects beyond the gum, and the *root,* which is anchored to the alveolar process of the jaw. These structures meet at the *neck* of the tooth. Glossy, white *enamel* covers the crown. Enamel mainly

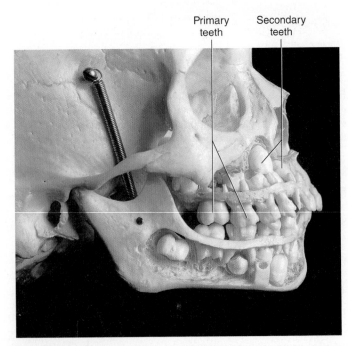

Primary teeth   Secondary teeth

FIGURE 17.8 This partially dissected child's skull reveals primary and developing secondary teeth in the maxilla and mandible.

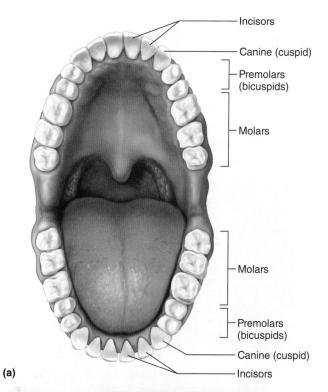

(a)

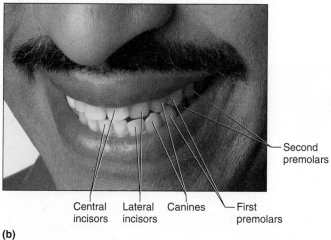

(b)

FIGURE 17.9 AP|R Permanent teeth. (*a*) The secondary teeth of the upper and lower jaws. (*b*) Anterior view of the secondary teeth.

Primary Teeth (Deciduous)		Secondary Teeth (Permanent)	
TABLE 17.2		**Primary and Secondary Teeth**	
Type	**Number**	**Type**	**Number**
Incisor		Incisor	
Central	4	Central	4
Lateral	4	Lateral	4
Canine (cuspid)	4	Canine (cuspid)	4
		Premolar (bicuspid)	
		First	4
		Second	4
Molar		Molar	
First	4	First	4
Second	4	Second	4
		Third	4
Total	20	**Total**	32

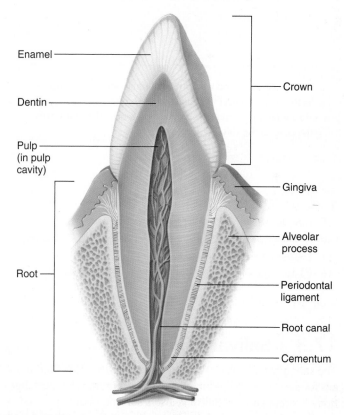

FIGURE 17.10 A section of a tooth.

consists of calcium salts and is the hardest substance in the body. If abrasive action or injury damages enamel, it is not replaced. Enamel may wear away with age.

The bulk of a tooth beneath the enamel is composed of a living cellular tissue called *dentin,* which is similar to bone but harder. Dentin, in turn, surrounds a central cavity (pulp cavity) that is filled with pulp. Pulp is a mass of tissue including blood vessels, nerves, and connective tissue. Blood vessels and nerves reach the central cavity through tubular *root canals,* which extend into the root. Tooth loss is most often associated with diseases of the gums (gingivitis) and the dental pulp (endodontitis).

A thin layer of bonelike material called *cementum,* surrounded by a *periodontal ligament* (periodontal membrane), encloses the root. This ligament, composed of collagen, passes between the cementum and the bone of the alveolar process, firmly attaching the tooth to the jaw. The ligament also contains blood vessels and nerves near the surface of the cementum-covered root (fig. 17.10). Clinical Application 17.1 describes the effect of bacteria on teeth. Table 17.3 summarizes the mouth parts and their functions.

Sticky foods, such as caramel, lodge between the teeth and in the crevices of molars, feeding bacteria such as *Actinomyces, Strepto-coccus mutans*, and *Lactobacillus*. These microbes metabolize carbohydrates in the food, producing acid by-products that destroy tooth enamel and dentin (fig. 17A). The bacteria also produce sticky substances that hold them in place.

If a person eats candy, for example, but does not brush the teeth soon afterward, the acid-forming bacteria may cause tooth decay, creating cavities within the enamel called *dental caries*. Unless a dentist cleans and fills the cavity that forms where enamel is destroyed, the damage will spread to the underlying dentin.

The following may help prevent dental caries:

1. Brush and floss teeth regularly.
2. Have regular dental exams and cleanings.
3. Talk with your dentist about receiving a fluoride treatment. Fluoride is added to the water supply in many communities. Fluoride is incorporated into the enamel's chemical structure, strengthening it.
4. The dentist may apply a sealant to children's and adolescents' teeth where crevices might hold onto decay-

causing bacteria. The sealant is a coating that keeps acids from eating away at tooth enamel.

For information on how to care for teeth over a lifetime, see the American Dental Association's website.

FIGURE 17A *Actinomyces* bacteria (falsely colored) clinging to teeth release acids that decay tooth enamel (1,250×).

Extracted primary and wisdom teeth may one day provide stem cells that can be used to regenerate tooth roots and supporting periodontal ligaments or even replace teeth. The stem cells are in the pulp of the developing tooth.

PRACTICE

9 How do primary teeth differ from secondary teeth?

10 How are types of teeth adapted to provide specialized functions?

11 Describe the structure of a tooth.

12 Explain how a tooth is attached to the bone of the jaw.

TABLE 17.3	Mouth Parts and Their Functions in Digestion	
Part	**Location**	**Function**
Cheeks	Form lateral walls of mouth	Hold food in mouth; muscles chew food
Lips	Surround mouth opening	Contain sensory receptors used to judge characteristics of foods
Tongue	Floor of mouth	Mixes food with saliva; moves food toward pharynx; contains taste receptors
Palate	Forms roof of mouth	Holds food in mouth; directs food to pharynx
Teeth	In sockets of mandible and maxillae	Break food particles into smaller pieces; help mix food with saliva during chewing

17.3 | Salivary Glands

The **salivary** (sal′ĭ-ver-e) **glands** secrete *saliva*. This fluid moistens food particles, helps bind them, and begins the chemical digestion of carbohydrates. Saliva is also a solvent, dissolving foods so that they can be tasted. Saliva helps cleanse the mouth and teeth. Bicarbonate ions (HCO_3^-) in saliva help buffer acid in the mouth, keeping the pH near neutral, between 6.5 and 7.5. This is a favorable range for the action of the salivary enzyme and protects the teeth from exposure to acids in foods.

Many minor salivary glands are scattered throughout the mucosa of the tongue, palate, and cheeks. They continuously secrete fluid, moistening the lining of the mouth. The three pairs of major salivary glands are the parotid glands, the submandibular glands, and the sublingual glands.

Salivary Secretions

The different salivary glands have varying proportions of two types of secretory cells, *serous cells* and *mucous cells*. Serous cells produce a watery fluid that contains a digestive enzyme, **salivary amylase** (am′ĭ-lās). This enzyme splits starch and glycogen molecules into disaccharides, starting the chemical digestion of carbohydrates. Mucous cells secrete a thick liquid called **mucus,** which binds food particles and acts as a lubricant during swallowing.

Branches of both sympathetic and parasympathetic nerves innervate the salivary glands, as they do other digestive structures. Impulses arriving on sympathetic fibers stimulate the gland cells to secrete a small volume of viscous saliva. Parasympathetic impulses, on the other hand, elicit the secretion of a large volume of watery saliva. Such parasympathetic impulses are activated reflexly when a person sees, smells, tastes, or even thinks about pleasant foods. Conversely, if food looks, smells, or tastes unpleasant, parasympathetic activity is inhibited. Less saliva is produced, and swallowing may become difficult.

Major Salivary Glands

The **parotid** (pah-rot′id) **glands** are the largest of the major salivary glands. Each gland lies anterior to and somewhat inferior to each ear, between the skin of the cheek and the masseter muscle. A *parotid duct* (Stensen's duct) passes from the gland inward through the buccinator muscle, entering the mouth just opposite the upper second molar on either side of the jaw. The parotid glands secrete a clear, watery fluid rich in salivary amylase because their secretory cells are primarily serous cells (figs. 17.11 and 17.12*a*).

The **submandibular** (sub″man-dib′u-lar) **glands** are in the floor of the mouth on the inside surface of the mandible. The secretory cells of these glands are about equally serous and mucous. Consequently, the submandibular glands secrete a more viscous fluid than the parotid glands (see figs. 17.11 and 17.12*b*). The ducts of the submandibular glands (Wharton's ducts) open inferior to the tongue, near the lingual frenulum.

The **sublingual** (sub-ling′gwal) **glands** are the smallest of the major salivary glands. They are in the floor of the mouth inferior to the tongue. Because the secretory cells of the sublingual glands are primarily the mucous type, their secretions, which enter the mouth through many separate ducts (Rivinus's ducts), are thick and stringy (see figs. 17.11 and 17.12*c*). Table 17.4 summarizes the characteristics of the major salivary glands.

TABLE 17.4	The Major Salivary Glands		
Gland	Location	Duct	Type of Secretion
Parotid glands	Anterior to and somewhat inferior to the ears between the skin of the cheeks and the masseter muscles	Parotid ducts pass through the buccinator muscles and enter the mouth opposite the upper second molars	Clear, watery serous fluid, rich in salivary amylase
Submandibular glands	In the floor of the mouth on the inside surface of the mandible	Ducts open inferior to the tongue near the frenulum	Some serous fluid with some mucus; more viscous than parotid secretion
Sublingual glands	In the floor of the mouth inferior to the tongue	Many separate ducts	Primarily thick, stringy mucus

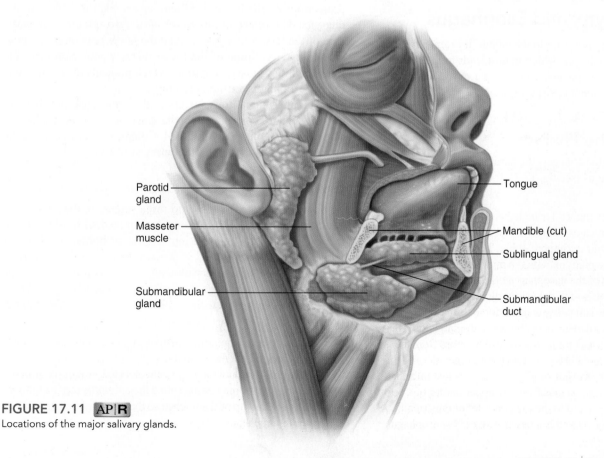

Parotid gland

Masseter muscle

Submandibular gland

Tongue

Mandible (cut)

Sublingual gland

Submandibular duct

FIGURE 17.11 AP|R
Locations of the major salivary glands.

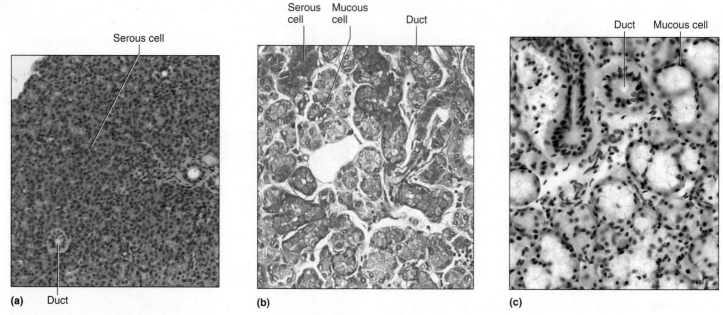

FIGURE 17.12 Light micrographs of (*a*) the parotid salivary gland (100×), (*b*) the submandibular salivary gland (180×), and (*c*) the sublingual salivary gland (200×).

 PRACTICE

13 What is the function of saliva?

14 What stimulates the salivary glands to secrete saliva?

15 Where are the major salivary glands located?

17.4 | Pharynx and Esophagus

The pharynx is a cavity posterior to the mouth. It extends from the nasal cavity to the esophagus, which in turn leads to the stomach. The pharynx and the esophagus do not assist in food digestion. However, both are important passageways, and their muscular walls function in swallowing.

Structure of the Pharynx

The **pharynx** (far′ingks) connects the nasal and oral cavities with the larynx and esophagus (see fig. 17.7). It can be divided into the following parts:

1. The **nasopharynx** (na″zo-far′ingks) is superior to the soft palate. It communicates with the nasal cavity and provides a passageway for air during breathing. The auditory tubes, which connect the pharynx with the middle ears, open through the walls of the nasopharynx (see chapter 12, p. 459).
2. The **oropharynx** (o″ro-far′ingks) is posterior to the oral cavity. It is posterior to the soft palate and inferior to the nasopharynx, projecting downward to the upper border of the epiglottis. This portion is a passageway for food moving downward from the mouth and for air moving to and from the nasal cavity.
3. The **laryngopharynx** (lah-ring″go-far′inks) is just inferior to the oropharynx. It extends from the upper border of the epiglottis downward to the lower border of the cricoid cartilage of the larynx and is a passageway to the esophagus.

The muscles in the walls of the pharynx form inner circular and outer longitudinal groups (fig. 17.13). The circular muscles, called *constrictor muscles*, pull the walls inward during swallowing. The *superior constrictor muscles*, attached to bony processes of the skull and mandible, curve around the upper part of the pharynx. The *middle constrictor muscles* arise from projections on the hyoid bone and fan around the middle of the pharynx. The *inferior constrictor muscles* originate from cartilage of the larynx and pass around the lower portion of the pharyngeal cavity. Some of the lower inferior constrictor muscle fibers contract most of the time, which prevents air from entering the esophagus during breathing.

The pharyngeal muscles are skeletal muscles, but they are under voluntary control only in the sense that swallowing (deglutition) can be voluntarily initiated. Complex reflexes control the precise actions of these muscles during swallowing.

Swallowing Mechanism

Swallowing can be divided into three stages. In the first stage, which is voluntary, food is chewed and mixed with saliva. Then, the tongue rolls this mixture into a mass, or **bolus,** and forces it into the oropharynx.

The second stage of swallowing begins as food reaches the oropharynx and stimulates sensory receptors around the pharyngeal opening. This triggers the swallowing reflex, illustrated in figure 17.14, which includes the following actions:

1. The soft palate (including the uvula) raises, preventing food from entering the nasal cavity.
2. The hyoid bone and the larynx are elevated. A flaplike structure attached to the larynx, called the *epiglottis* (ep″ĭ-glot′is), closes off the top of the trachea so that food is less likely to enter the trachea.

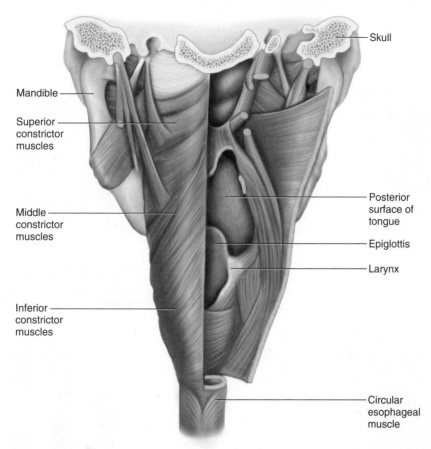

Mandible

Superior constrictor muscles

Middle constrictor muscles

Inferior constrictor muscles

Skull

Posterior surface of tongue

Epiglottis

Larynx

Circular esophageal muscle

FIGURE 17.13 Muscles of the pharyngeal wall, posterior view.

3. The tongue is pressed against the soft palate and uvula, sealing off the oral cavity from the nasal cavity.
4. The longitudinal muscles in the pharyngeal wall contract, pulling the pharynx upward toward the food.
5. The lower portion of the inferior constrictor muscles relaxes, opening the esophagus.
6. The superior constrictor muscles contract, stimulating a peristaltic wave to begin in other pharyngeal muscles. This wave forces the food into the esophagus.

The swallowing reflex momentarily inhibits breathing. Then, during the third stage of swallowing, peristalsis transports the food in the esophagus to the stomach.

Esophagus

The **esophagus** (ĕ-sof'ah-gus) is a straight, collapsible tube about 25 centimeters long. It provides a passageway for food, and its muscular wall propels food from the pharynx to the stomach. The esophagus begins at the base of the laryngopharynx and descends through the thorax posterior to the trachea, passing through the mediastinum. It penetrates the diaphragm through an opening, the *esophageal hiatus* (ĕ-sof″ah-je′al hi-a′tus), and is continuous with the stomach on the abdominal side of the diaphragm (fig. 17.15 and reference plates 17 and 23, pp. 51 and 54).

Mucous glands are scattered throughout the submucosa of the esophagus (fig. 17.16). Their secretions moisten and lubricate the inner lining of the tube.

Just superior to the point where the esophagus joins the stomach, some of the cells of the circular smooth muscle layer have increased sympathetic muscle tone and form the **lower esophageal sphincter** (loh′er ĕ-sof″ah-je′al sfingk′ter), or cardiac sphincter (fig. 17.17). These cells usually remain contracted and close the entrance to the stomach. In this way, they help prevent regurgitation of the stomach contents into the esophagus. When peristaltic waves in the esophagus reach the stomach, the smooth muscle cells of the sphincter temporarily relax and allow the swallowed food to enter.

 PRACTICE

16 Describe the regions of the pharynx.

17 List the major events of swallowing.

18 What is the function of the esophagus?

17.5 | Stomach APR

The **stomach** (stum′ak) is a J-shaped, pouchlike organ, about 25–30 centimeters long, which hangs inferior to the diaphragm in the upper-left portion of the abdominal cavity (see figs. 17.1 and 17.15; reference plates 4 and 5, pp. 41–42). The stomach has a capacity of about one liter or more. Its inner lining has thick gastric folds (rugae) of the mucosal and submucosal layers that unfold when the wall is distended. The stomach receives food from the esophagus, mixes it with gastric juice, initiates the digestion of

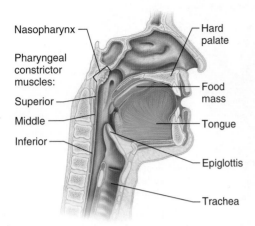

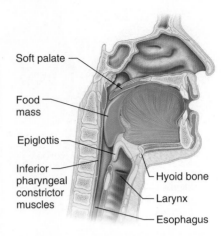

(a) The tongue forces food into the pharynx.

(b) The soft palate, hyoid bone, and larynx are raised, the tongue is pressed against the palate, the epiglottis closes, and the inferior constrictor muscles relax so that the esophagus opens.

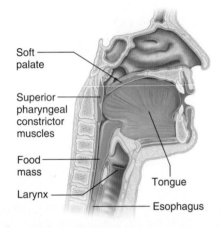

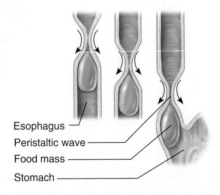

(c) Superior constrictor muscles contract and initiate a peristaltic wave that forces food into the esophagus.

(d) A peristaltic wave moves food through the esophagus to the stomach.

FIGURE 17.14 Steps in the swallowing reflex. (The mucosa in (*a*), (*b*), and (*c*) has been removed to reveal the underlying muscles.)

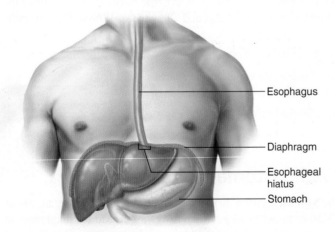

FIGURE 17.15 AP|R The esophagus is a passageway between the laryngopharynx and the stomach.

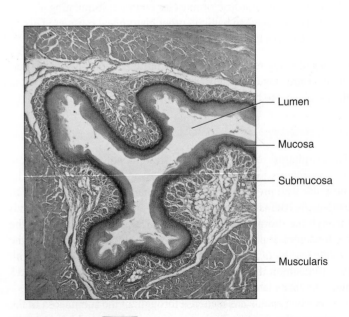

FIGURE 17.16 AP|R This cross section of the esophagus shows its muscular wall (10×).

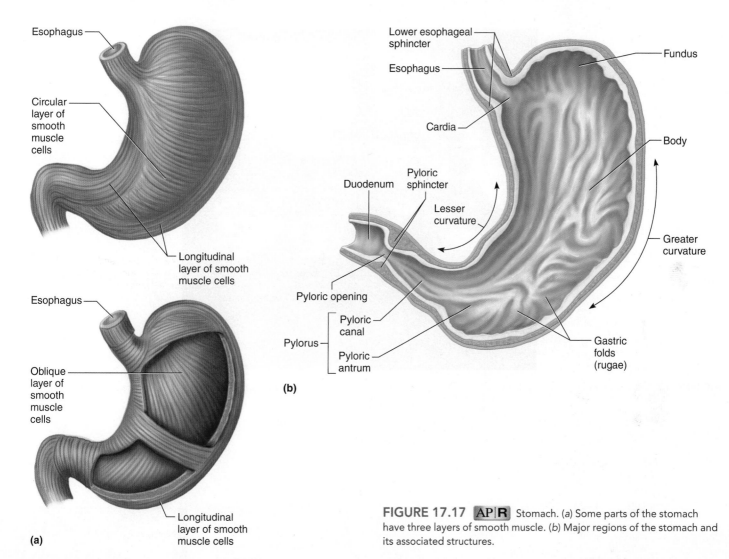

FIGURE 17.17 AP|R Stomach. (*a*) Some parts of the stomach have three layers of smooth muscle. (*b*) Major regions of the stomach and its associated structures.

proteins, carries on limited absorption, and moves food into the small intestine.

In addition to the two layers of smooth muscle—an inner circular layer and an outer longitudinal layer—which are also in other regions of the alimentary canal, some parts of the stomach have an additional inner oblique layer, which strengthens the stomach wall and help the mixing and churning. This third innermost muscular layer is most highly developed near the opening of the esophagus and in the body of the stomach (fig. 17.17).

Parts of the Stomach

The stomach, shown in figure 17.17 and **figure 17.18**, can be divided into the cardia, fundus, body, and pylorus. The *cardia* is a small area near the esophageal opening. The *fundus,* which balloons superior to the cardia, is a temporary storage area and sometimes fills with swallowed air. This produces a gastric air bubble, which may be used as a landmark on a radiograph of the abdomen. The dilated *body,* which is the main part of the stomach, lies between the fundus and pylorus. The *pyloric antrum* is a funnel-shaped portion that narrows and becomes the *pyloric canal* as it approaches the small intestine.

At the end of the pyloric canal, a thickening of the circular layer of smooth muscle forms the powerful **pyloric sphincter.** This sphincter acts as a valve that controls gastric emptying into the small intestine.

Hypertrophic pyloric stenosis is a birth defect in which muscle overgrowth blocks the pyloric canal and causes a newborn to vomit with increased force. To diagnose the condition, a radiograph is taken of the area after the infant drinks formula containing a radiopaque barium compound. Surgical splitting of the muscle blocking the passageway from stomach to small intestine is necessary to enable the infant to eat normally. Pyloric stenosis can also occur later in life as a result of ulcers or cancer.

Gastric Secretions

The mucous membrane that forms the inner lining of the stomach is thick. Its surface is studded with many small openings, called *gastric pits,* that are located at the ends of tubular **gastric glands** (oxyntic glands) (**fig. 17.19**). Although their structure and

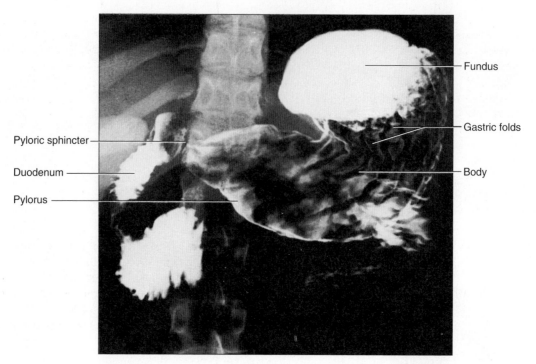

FIGURE 17.18 Radiograph of a stomach. (*Note:* A radiopaque compound the patient swallowed appears white in the radiograph.)

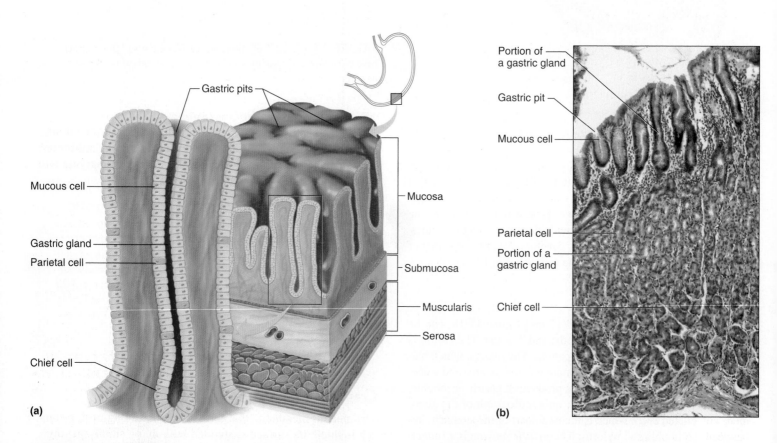

(a)

Gastric pits

Mucous cell

Gastric gland

Parietal cell

Chief cell

Mucosa

Submucosa

Muscularis

Serosa

(b)

Portion of a gastric gland

Gastric pit

Mucous cell

Parietal cell

Portion of a gastric gland

Chief cell

FIGURE 17.19 **AP|R** Lining of the stomach. (*a*) Gastric glands include mucous cells, parietal cells, and chief cells. The mucosa of the stomach is studded with gastric pits that are the openings of the gastric glands. (*b*) A light micrograph of cells associated with the gastric glands (60×).

the composition of their secretions vary in different parts of the stomach, gastric glands generally contain three types of secretory cells. One type, the *mucous cell*, found in the necks of the glands near the openings of the gastric pits, secretes mucus. The other types, *chief cells* (peptic cells) and *parietal cells* (oxyntic cells), reside in the deeper parts of the glands (fig. 17.19). The chief cells secrete digestive enzymes, and the parietal cells release a solution containing hydrochloric acid. The secretions from the mucous cells, chief cells, and parietal cells together form **gastric juice** (gas'trik jōōs).

Pepsin is by far the most important of the digestive enzymes in gastric juice. The chief cells secrete pepsin in the form of an inactive, nonerosive enzyme precursor called **pepsinogen.** When pepsinogen contacts the hydrochloric acid from the parietal cells, it breaks down rapidly and forms the active enzyme called pepsin. Pepsin, in turn, can also break down pepsinogen to release more pepsin.

Pepsin begins the digestion of nearly all types of dietary protein into polypeptides. This enzyme is most active in an acidic environment, which is provided by the hydrochloric acid in gastric juice.

Gastric juice contains small quantities of a fat-splitting enzyme, *gastric lipase.* Its action is inhibited somewhat by the low pH of gastric juice.

Much of what we know about the stomach's functioning comes from a French-Canadian explorer, Alexis St. Martin, who in 1822 accidentally shot himself in the abdomen. His extensive injuries eventually healed, but a hole, called a fistula, was left, allowing observers to look at his stomach in action. A U.S. Army surgeon, William Beaumont, spent eight years watching food digesting in the stomach, and noted how the stomach lining changed in the process.

In 1984, our knowledge of digestive function expanded when medical resident Barry Marshall at Royal Perth Hospital in western Australia performed a daring experiment. His mentor, J. Robin Warren, had hypothesized that a bacterial infection causes gastritis (inflammation of the stomach lining) and peptic ulcers (sores in the lining of the esophagus, stomach, or small intestine). At the time, these conditions were attributed to poor diet and stress. Marshall drank "swamp water"—billions of bacteria. He developed gastritis, which, fortunately, cleared up. A colleague who repeated the experiment developed an ulcer and required antibiotics. After a decade of debate, the medical community finally concurred that the bacterium *Helicobacter pylori*, which thrives under acidic conditions, causes many cases of gastritis and peptic ulcers. A short course of antibiotics and acid-lowering drugs has replaced lifelong treatments. Marshall and Warren were awarded a Nobel Prize in 2005 for their discovery.

The mucous cells of the gastric glands *(mucous neck cells)* and the mucous cells, associated with the stomach's inner surface, release a viscous, alkaline secretion that coats the inside of the stomach wall. This coating is especially important because pepsin can digest the proteins of stomach tissues, as well as those in foods. The coating normally prevents the stomach from digesting itself.

Another component of gastric juice is **intrinsic factor** (in-trin'sik fak'tor). The parietal cells of the gastric glands secrete intrinsic factor, which is required for vitamin B_{12} absorption from the small intestine. Table 17.5 summarizes the components of gastric juice.

PRACTICE

19 Where is the stomach located?

20 What are the secretions of the chief cells and parietal cells?

21 Why doesn't the stomach digest itself?

Regulation of Gastric Secretions

Gastric juice is produced continuously. However, the rate varies considerably and is controlled both neurally and hormonally. In the gastric glands, specialized cells closely associated with the parietal cells secrete the hormone *somatostatin,* which inhibits acid secretion. However, acetylcholine (ACh) released from synaptic knobs in response to parasympathetic impulses arriving on the vagus nerves suppresses the secretion of somatostatin and stimulates the gastric glands to secrete abundant gastric juice, which is rich in hydrochloric acid and pepsinogen. These parasympathetic impulses also stimulate certain stomach cells, mainly in the pyloric region, to release a peptide hormone called **gastrin,** which increases the secretory activity of gastric glands (fig.17.20). Furthermore, parasympathetic impulses and gastrin promote release of *histamine* from gastric mucosal cells, which, in turn, stimulates additional gastric secretion.

Gastric secretion occurs in three stages—the cephalic, gastric, and intestinal phases. The *cephalic phase* begins before food reaches the stomach and possibly even before eating. In this stage, parasympathetic reflexes operating through the vagus nerves stimulate gastric secretion at the taste, smell, sight, or thought of food. The greater the hunger, the greater the gastric secretion.

The *gastric phase* of gastric secretion, which accounts for most of the secretory activity, starts when food enters the stomach. The presence of food and the distension of the stomach wall trigger the stomach to release gastrin, which stimulates production of more gastric juice.

As food enters the stomach and mixes with gastric juice, the pH of the contents rises, which enhances gastrin secretion. Consequently, the pH of the stomach contents drops. As the pH approaches 3.0, secretion of gastrin is inhibited. When the pH reaches 1.5, gastrin secretion ceases.

For the stomach to secrete hydrochloric acid, hydrogen ions are actively transported into the stomach. Negatively charged chloride ions, attracted by the positively charged hydrogen ions, move from the blood into the stomach. An equivalent number of alkaline bicarbonate ions are released into the blood. Following a meal, the blood concentration of bicarbonate ions increases, and the urine excretes excess bicarbonate ions. This phenomenon is called the *alkaline tide.*

The *intestinal phase* of gastric secretion begins when food leaves the stomach and enters the small intestine. When food first contacts the intestinal wall, it stimulates intestinal cells to release a hormone, *intestinal gastrin,* that briefly enhances gastric gland secretion.

TABLE 17.5	Major Components of Gastric Juice AP|R	
Component	Source	Function
Pepsinogen	Chief cells of the gastric glands	Inactive form of pepsin
Pepsin	Formed from pepsinogen in the presence of hydrochloric acid	A protein-splitting enzyme that digests nearly all types of dietary protein into polypeptides
Hydrochloric acid	Parietal cells of the gastric glands	Provides the acid environment needed for production and action of pepsin
Mucus	Mucous cells	Provides a viscous, alkaline protective layer on the stomach's inner surface
Intrinsic factor	Parietal cells of the gastric glands	Aids in vitamin B_{12} absorption in the small intestine

The intestinal phase primarily inhibits gastric juice secretion. As more food moves into the small intestine, a sympathetic reflex triggered by acid in the upper part of the small intestine inhibits secretion of gastric juice from the stomach wall. At the same time, proteins and fats in this region of the intestine stimulate release of the peptide hormone **cholecystokinin** (ko"le-sis"to-ki'nin) **(CCK)** from the intestinal wall, which decreases gastric motility. Similarly, fats in the small intestine stimulate intestinal cells to release *intestinal somatostatin,* which inhibits release of gastric juice. Overall, these actions decrease gastric secretion and motility as the small intestine fills with food. Table 17.6 summarizes the phases of gastric secretion.

PRACTICE

22 What controls gastric juice secretion?

23 Distinguish among the cephalic, gastric, and intestinal phases of gastric secretion.

24 What is the function of cholecystokinin?

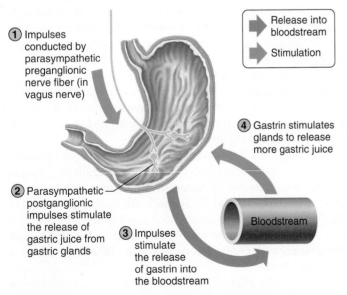

FIGURE 17.20 The secretion of gastric juice is regulated in part by parasympathetic impulses that stimulate the release of gastric juice and gastrin.

① Impulses conducted by parasympathetic preganglionic nerve fiber (in vagus nerve)

Release into bloodstream

Stimulation

④ Gastrin stimulates glands to release more gastric juice

② Parasympathetic postganglionic impulses stimulate the release of gastric juice from gastric glands

③ Impulses stimulate the release of gastrin into the bloodstream

Bloodstream

Gastric Absorption

Gastric enzymes begin breaking down proteins. However, the stomach wall is not well-adapted to absorb digestive products. The stomach absorbs only small volumes of water and certain salts, as well as certain lipid-soluble drugs. Most nutrients are absorbed in the small intestine. Alcohol, which is not a nutrient, is absorbed both in the small intestine and stomach.

Mixing and Emptying Actions

Food stretches the smooth muscle layers of the stomach wall. The stomach may enlarge, but its muscle layers maintain their tone, and internal pressure of the stomach normally is unchanged. When a person eats more than the stomach can comfortably hold, the internal pressure may rise enough to stimulate pain receptors. The result is a stomachache. Clinical Application 17.2 discusses this common problem along with its associated indigestion.

Following a meal, the mixing movements of the stomach wall aid in producing a semifluid paste of food particles and gastric juice called **chyme** (kīm). Peristaltic waves push the chyme toward the pylorus of the stomach, and as chyme accumulates near the pyloric sphincter, it begins to relax. Stomach contractions push chyme a little (5–15 milliliters) at a time into the small intestine. These stomach contraction waves push most of the chyme backward into the stomach, mixing it further. The lower esophageal sphincter prevents reflux of stomach contents into the esophagus. Figure 17.21 illustrates this process.

TABLE 17.6	Phases of Gastric Secretion AP|R
Phase	Action
Cephalic phase	The sight, taste, smell, or thought of food triggers parasympathetic reflexes. Gastric juice is secreted in response.
Gastric phase	Food in stomach chemically and mechanically stimulates release of gastrin, which, in turn, stimulates secretion of gastric juice; reflex responses also stimulate gastric juice secretion.
Intestinal phase	As food enters the small intestine, it stimulates intestinal cells to release intestinal gastrin, which, in turn, briefly promotes the secretion of gastric juice from the stomach wall. This phase primarily inhibits gastric juice secretion.

If a person eats large amounts of food too quickly, perhaps because it can take up to 20 minutes for the brain's hypothalamus to sense that the stomach is full, a stomachache can result. A feeling of fullness becomes abdominal pain and then heartburn, as stomach contents back up into the esophagus (gastric reflux). The material from the stomach causes inflammation of the esophagus (esophagitis), but the perception that the pain is from the heart has given rise to the term "heartburn."

An antacid product may provide temporary relief from gastric reflux by quickly raising the pH of the stomach contents. Most antacids include a compound containing sodium, calcium, magnesium, or aluminum. Some products also contain simethicone, which breaks up gas bubbles in the digestive tract.

Avoiding gastric reflux and heartburn is a more healthful approach than gorging and then reaching for medication. Some tips:

- Avoid large meals. The more food, the more acid the stomach produces.
- Eat slowly so that stomach acid secretion is more gradual.
- Do not lie down immediately after eating. Being upright enables gravity to help food move along the alimentary canal.
- If prone to indigestion or heartburn, avoid caffeine, which increases stomach acid secretion.
- Nicotine and alcohol irritate the stomach lining and relax the lower esophageal sphincter. This makes it easier for gastric reflux to occur.

The rate at which the stomach empties depends on the fluidity of the chyme and the type of food. Liquids usually pass through the stomach rapidly, but solids remain until they are well mixed with gastric juice. Fatty foods may remain in the stomach three to six hours; foods high in proteins move through more quickly; carbohydrates usually pass through more rapidly than either fats or proteins.

As chyme fills the duodenum (the proximal portion of the small intestine), internal pressure on the organ increases, stretching the intestinal wall. These actions stimulate sensory receptors in the wall, triggering an **enterogastric reflex** (en-ter-o-gas′trik re′fleks). The name of this reflex, like those of other digestive reflexes, describes the origin and termination of reflex impulses. That is, the enterogastric reflex begins in the small intestine *(entero)* and ends in the stomach *(gastro)*. As a result of the enterogastric reflex, fewer parasympathetic impulses arrive at the stomach, decreasing peristalsis, and

intestinal filling slows (fig. 17.22). If chyme entering the intestine is fatty, the intestinal wall releases the hormone cholecystokinin, which further inhibits peristalsis.

Vomiting results from a complex reflex that empties the stomach in the reverse of the normal direction. Irritation or distension in the stomach or intestines can trigger vomiting. Sensory impulses travel from the site of stimulation to the *vomiting center* in the medulla oblongata. The responses initiated by the vomiting center include taking a deep breath, raising the soft palate and thus closing the nasal cavity, closing the opening to the trachea (glottis), relaxing the lower esophageal sphincter, contracting the diaphragm so it presses downward over the stomach, and contracting the abdominal wall muscles to increase pressure inside the abdominal cavity. These motor responses squeeze the stomach from all sides, forcing its contents upward and out via the open pathway through the esophagus, pharynx, and mouth.

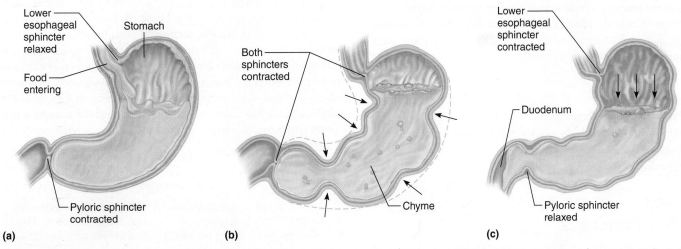

(a)   (b)   (c)

FIGURE 17.21 Stomach movements. (*a*) As the stomach fills, its muscular wall stretches, but the pyloric sphincter remains closed. (*b*) Mixing movements combine food and gastric juice, creating chyme. (*c*) Peristaltic waves move the chyme toward the pyloric sphincter, which relaxes and admits some chyme into the duodenum.

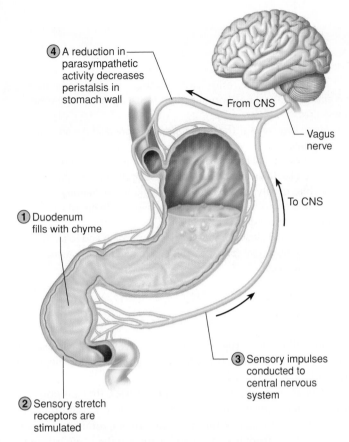

④ A reduction in parasympathetic activity decreases peristalsis in stomach wall

From CNS

Vagus nerve

To CNS

① Duodenum fills with chyme

③ Sensory impulses conducted to central nervous system

② Sensory stretch receptors are stimulated

FIGURE 17.22 The enterogastric reflex partially regulates the rate at which chyme leaves the stomach.

Several types of stimuli can promote activity in the vomiting center. These include certain drugs (emetics); toxins in contaminated foods; rapid changes in body motion; and sensory impulses from the labyrinths of the inner ears, which can produce motion sickness. Stimulation of higher brain centers through sights, sounds, odors, tastes, emotions, or mechanical stimulation of the back of the pharynx can also trigger vomiting.

Nausea emanates from activity in the vomiting center or in nerve centers near it. During nausea, stomach movements usually are diminished or absent, and duodenal contents may move back into the stomach.

 PRACTICE

25 How is chyme produced?

26 What factors influence how quickly chyme leaves the stomach?

27 Describe the enterogastric reflex.

28 Describe the vomiting reflex.

17.6 | Pancreas

The **pancreas** was discussed as an endocrine gland in chapter 13 (pp. 514–515). It also has an exocrine function—secretion of a digestive fluid called **pancreatic juice** (pan″kre-at′ik jōōs).

Structure of the Pancreas

The pancreas is closely associated with the small intestine and is posterior to the parietal peritoneum, or retroperitoneal. It extends horizontally across the posterior abdominal wall, with its head in the C-shaped curve of the duodenum (portion of the small intestine) and its tail against the spleen (fig. 17.23 and reference plate 19, p. 52).

Pancreatic acinar cells produce pancreatic juice and make up the bulk of the pancreas. These cells form clusters called *acini* (*acinus,* sing.) around tiny tubes into which they release their secretions. The smaller tubes unite to form larger ones, which join a *pancreatic duct* that extends the length of the pancreas and transports pancreatic juice to the small intestine. The pancreatic duct usually connects with the duodenum at the same place where the bile duct from the liver and gallbladder joins the duodenum, although divisions and other connections may be present (see figs. 13.34 on p. 514 and 17.23).

The main pancreatic duct and bile duct join at a short, dilated tube called the *hepatopancreatic ampulla* (ampulla of Vater). A band of smooth muscle, called the *hepatopancreatic sphincter* (sphincter of Oddi), surrounds this ampulla. This sphincter controls the movement of pancreatic juice and bile into the small intestine.

Pancreatic Juice

Pancreatic juice contains enzymes that digest carbohydrates, fats, proteins, and nucleic acids. The carbohydrate-digesting enzyme, **pancreatic amylase,** splits molecules of starch or glycogen into disaccharides. The fat-digesting enzyme, **pancreatic lipase,** breaks triglyceride molecules into fatty acids and monoglycerides. (A monoglyceride molecule consists of one fatty acid bound to glycerol.)

The protein-splitting (proteolytic) enzymes are **trypsin, chymotrypsin,** and **carboxypeptidase.** Each of these enzymes splits the bonds between particular combinations of amino acids in proteins. No single enzyme can split all possible amino acid combinations. Therefore, several enzymes are necessary to completely digest protein molecules.

The proteolytic enzymes are stored in tiny cellular structures called *zymogen granules.* These enzymes, like gastric pepsin, are secreted in inactive forms and must be activated by other enzymes after they reach the small intestine. For example, the pancreatic cells release inactive **trypsinogen,** which is activated to trypsin when it contacts the enzyme *enterokinase,* which the mucosa of the small intestine secretes. Trypsin, in turn, activates chymotrypsin and carboxypeptidase. The need for activation prevents enzymatic digestion of proteins in the secreting cells and the pancreatic ducts.

Painful *acute pancreatitis* results when pancreatic enzymes become active before they are secreted, and digest parts of the pancreas. Blockage of the release of pancreatic juice can cause pancreatitis, as can alcoholism, gallstones, certain infections, traumatic injuries, and the side effects of some drugs.

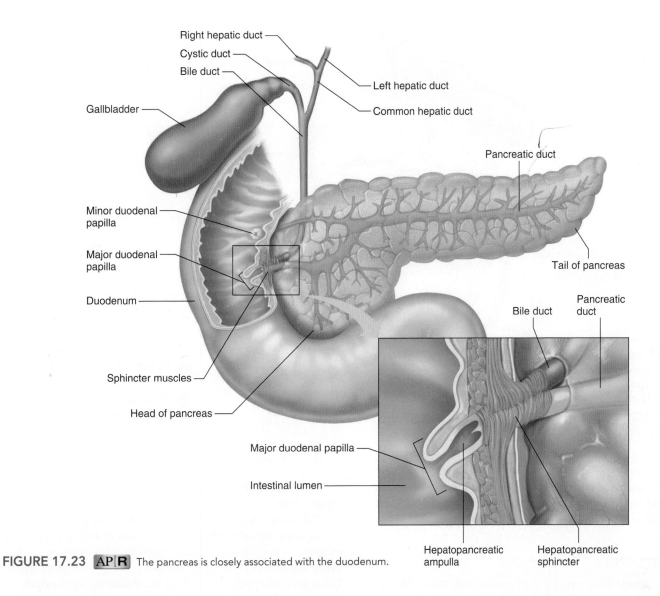

Right hepatic duct

Cystic duct

Bile duct

Left hepatic duct

Common hepatic duct

Gallbladder

Pancreatic duct

Minor duodenal papilla

Major duodenal papilla

Tail of pancreas

Duodenum

Bile duct

Pancreatic duct

Sphincter muscles

Head of pancreas

Major duodenal papilla

Intestinal lumen

Hepatopancreatic ampulla

Hepatopancreatic sphincter

FIGURE 17.23 AP|R The pancreas is closely associated with the duodenum.

Pancreatic juice contains two types of **nucleases,** which are enzymes that break down nucleic acid molecules into nucleotides. A high concentration of bicarbonate ions makes the juice alkaline, which provides a favorable environment for the actions of the digestive enzymes and helps neutralize acidic chyme as it arrives from the stomach. The alkaline environment in the small intestine also blocks the action of pepsin, which might otherwise damage the duodenal wall.

Regulation of Pancreatic Secretion

The nervous and endocrine systems regulate release of pancreatic juice, much as they regulate gastric and small intestinal secretions. For example, during the cephalic and gastric phases of gastric secretion, parasympathetic impulses stimulate the pancreas to release digestive enzymes. A peptide hormone, **secretin,** stimulates the pancreas to secrete abundant fluid when acidic chyme enters the duodenum. Secretin is released into the blood from the duodenal mucous membrane in response to the acid in chyme. The pancreatic juice secreted at this time contains few, if any, digestive enzymes but has a high concentration of bicarbonate ions. These ions neutralize the acid in chyme (fig. 17.24).

Proteins and fats in chyme in the duodenum also stimulate the release of *cholecystokinin* from the intestinal wall. As in the case of secretin, cholecystokinin reaches the pancreas via the bloodstream. Pancreatic juice secreted in response to cholecystokinin has a high concentration of digestive enzymes.

 PRACTICE

29 Where is the pancreas located?

30 List the enzymes in pancreatic juice.

31 What are the functions of the enzymes in pancreatic juice?

32 What regulates secretion of pancreatic juice?

17.7 | Liver

The **liver,** the largest internal organ, is located in the right upper quadrant of the abdominal cavity, just inferior to the diaphragm. A fold of visceral peritoneum called the *coronary ligament* attaches the liver to the diaphragm on its superior surface. The liver is partially surrounded by the ribs and extends from the level of the fifth

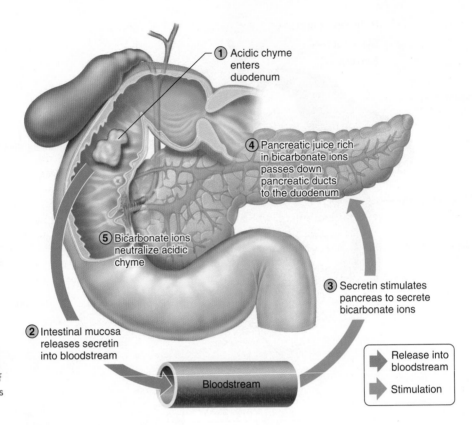

FIGURE 17.24 Acidic chyme entering the duodenum from the stomach stimulates the release of secretin, which, in turn, stimulates the release of pancreatic juice.

In figure labels:

1 Acidic chyme enters duodenum

4 Pancreatic juice rich in bicarbonate ions passes down pancreatic ducts to the duodenum

5 Bicarbonate ions neutralize acidic chyme

3 Secretin stimulates pancreas to secrete bicarbonate ions

2 Intestinal mucosa releases secretin into bloodstream

Bloodstream

Release into bloodstream

Stimulation

intercostal space to the lower margin of the ribs. It is reddish brown in color and well supplied with blood vessels (figs. 17.25 and 17.26 and reference plates 10, 17, and 24, pp. 47, 51, and 55).

Liver Structure

A fibrous capsule encloses the liver, and ligaments divide the organ into a large *right lobe* and a smaller *left lobe*. The *falciform ligament* is a fold of visceral peritoneum that separates the lobes and fastens the liver to the abdominal wall anteriorly. The liver also has two minor lobes, the *quadrate lobe,* near the gallbladder, and the *caudate lobe,* close to the vena cava (fig. 17.26). The area where the four lobes meet and blood vessels and ducts enter or exit the liver is the *porta hepatis.*

Each lobe is separated into many tiny **hepatic lobules,** which are the liver's functional units (fig. 17.27). A lobule consists of many *hepatic cells* (hepatocytes) radiating outward from a *central vein.* Blood-filled channels called **hepatic sinusoids** separate plate-like groups of these cells from each other. Blood from the digestive tract, carried in the *hepatic portal vein* (see chapter 15, p. 603), brings newly absorbed nutrients into the sinusoids (fig. 17.28). The oxygen-rich blood from the hepatic artery mixes freely with the blood containing nutrients and flows through the liver sinusoids, which provides oxygen and nourishment to the hepatic cells.

Often blood in the hepatic portal vein contains some bacteria that have entered through the intestinal wall. However, large **Kupffer cells,** fixed to the inner lining (endothelium) of the hepatic sinusoids, remove most of the bacteria from the blood by phagocytosis. Then the blood passes from the sinusoids into

the *central veins* of the hepatic lobules and exits the liver via the hepatic veins.

Within the hepatic lobules are many fine *bile canaliculi,* which carry bile secreted by hepatic cells to *bile ductules.* The ductules of neighboring lobules unite to form intrahepatic bile ducts, which then converge to ultimately form the **hepatic ducts.** These ducts merge, in turn, to form the **common hepatic duct.**

Liver Functions

The liver carries on many important metabolic activities. From Science to Technology 17.1 discusses a bioengineered liver. Recall from chapter 13 (p. 514) that the liver plays a key role in carbohydrate metabolism by helping maintain concentration of blood glucose within the normal range. Liver cells responding to hormones such as insulin and glucagon lower the blood glucose level by polymerizing glucose to glycogen and raise the blood glucose level by breaking down glycogen to glucose or by converting noncarbohydrates into glucose.

The liver's effects on lipid metabolism include oxidizing fatty acids at an especially high rate (see chapter 18, pp. 697–698); synthesizing lipoproteins, phospholipids, and cholesterol; and converting excess portions of carbohydrate molecules into fat molecules. The blood transports fats synthesized in the liver to adipose tissue for storage.

The most vital liver functions are probably those related to protein metabolism. They include deaminating amino acids; forming urea (see chapter 18, p. 698); synthesizing plasma proteins such as clotting factors (see chapter 14, pp. 541 and 543); and converting certain amino acids into other amino acids.

FROM SCIENCE TO TECHNOLOGY 17.1

Replacing the Liver

Life without a liver is not possible. Although the liver is capable of regeneration if even a small part of it (25 - 30%) remains healthy, a person can survive only a few days once the liver stops functioning. For example survival is generally only weeks or a few months after cancer, spreads to the liver.

Livers are in great demand for transplant, but they are scarce. Each year in the United States, only about 6,000 of the 16,000 or so individuals requiring livers survive long enough to undergo a transplant. A person can receive part of a liver donated by a living relative or other close match. The donor lives because only a portion of the liver is needed for survival, and enough is left by the procedure to allow regeneration to occur.

A promising solution to the liver shortage is an "extracorporeal liver assist device" (ELAD). It can take over the liver's blood-

cleansing function until a cadaver organ becomes available, or enable remaining functional liver tissue to stimulate enough regeneration to restore health.

An ELAD is a "bioartificial" liver because it has synthetic as well as biological components. The device consists of four cartridges filled with hollow fibers that house millions of human liver cells (hepatocytes). A patient's plasma is separated from the blood and passed through the device, where the liver cells remove toxins and add liver-secreted products, such as clotting factors. The plasma is then reunited with the formed elements of the blood, and the blood reinfused into the patient. It is clinical trials in the United States. An ELAD is similar to an artificial kidney used in hemodialysis (see Clinical Application 20.3 on page 792).

FIGURE 17.25 AP|R This transverse section of the abdomen reveals the liver and other organs in the upper part of the abdominal cavity.

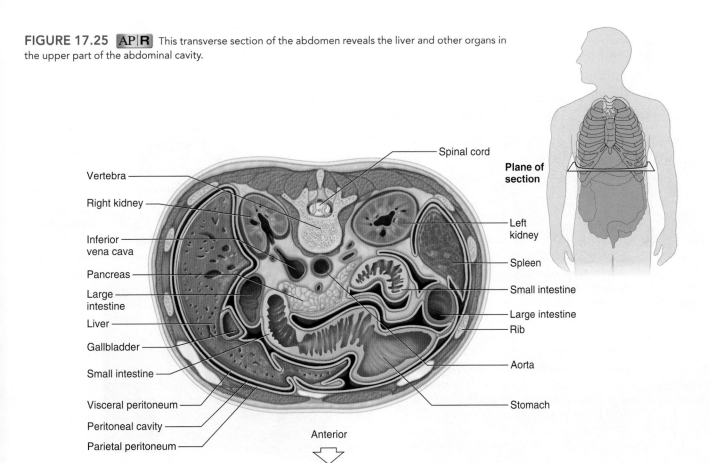

Spinal cord
Vertebra
Right kidney
Inferior vena cava
Pancreas
Large intestine
Liver
Gallbladder
Small intestine
Visceral peritoneum
Peritoneal cavity
Parietal peritoneum
Anterior

Plane of section
Left kidney
Spleen
Small intestine
Large intestine
Rib
Aorta
Stomach

Bacteria in the intestine produce ammonia, which is carried in the blood to the liver, where it reacts to yield urea. When this liver function fails, concentration of blood ammonia sharply rises, causing *hepatic coma*, a condition that can lead to death.

The liver also stores many substances, including glycogen, iron, and vitamins A, D, and B_{12}. Extra iron from the blood combines with a protein (apoferritin) in liver cells, forming *ferritin*. The iron is stored in this form until blood iron concentration falls, when some of the iron is released. Thus, the liver is important in iron homeostasis.

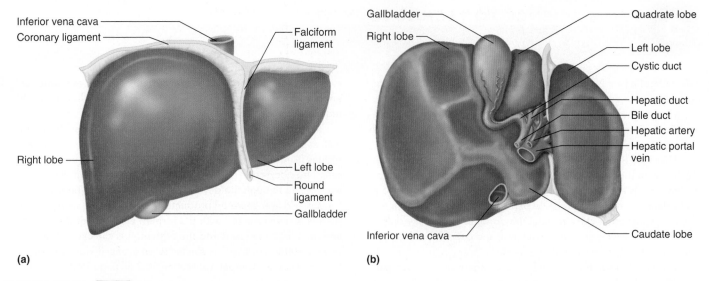

(a)

(b)

FIGURE 17.26 AP|R Lobes of the liver, viewed (a) anteriorly and (b) inferiorly.

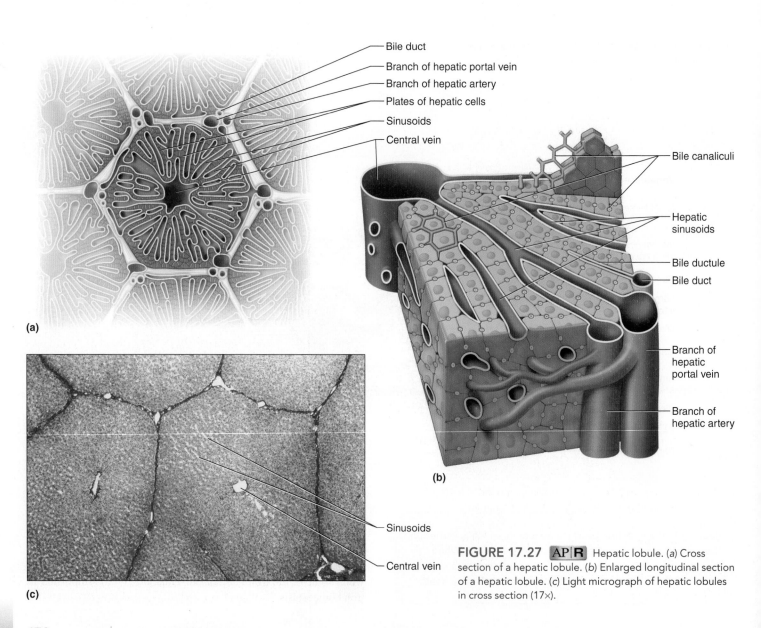

(a)

(b)

(c)

FIGURE 17.27 AP|R Hepatic lobule. (a) Cross section of a hepatic lobule. (b) Enlarged longitudinal section of a hepatic lobule. (c) Light micrograph of hepatic lobules in cross section (17×).

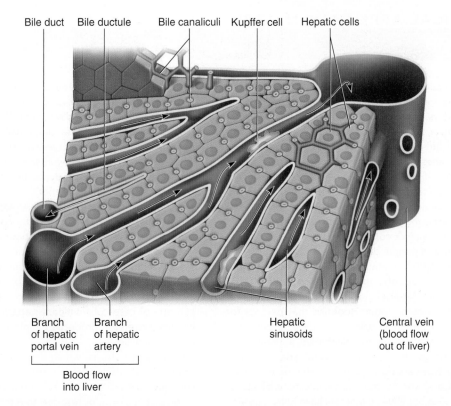

Bile duct   Bile ductule   Bile canaliculi   Kupffer cell   Hepatic cells

Branch of hepatic portal vein   Branch of hepatic artery   Hepatic sinusoids   Central vein (blood flow out of liver)

Blood flow into liver

FIGURE 17.28 AP|R The paths of blood and bile in a hepatic lobule.

Liver cells help destroy damaged red blood cells and phagocytize foreign antigens. The liver removes toxic substances such as alcohol and certain drugs from the blood (detoxification). The liver can also serve as a blood reservoir, storing 200 to 400 milliliters of blood. The liver's role in digestion is to secrete bile. Table 17.7 summarizes the major functions of the liver. Clinical Application 17.3 discusses hepatitis, an inflammation of the liver.

 PRACTICE

33   Locate the liver.

34   Review liver functions.

35   How does the liver aid in digestion?

Composition of Bile

Bile (bīl) is a yellowish-green liquid that is continuously secreted from hepatic cells. In addition to water, bile contains *bile salts, bile pigments, cholesterol,* and *electrolytes.* Bile salts are the most abundant and are the only bile components that have a digestive function.

Hepatic cells use cholesterol to produce bile salts, and in secreting these salts, they release some cholesterol into the bile. Cholesterol by itself has no special function in bile or in the alimentary canal.

Bile pigments (bilirubin and biliverdin) are breakdown products of hemoglobin from red blood cells (see chapter 14, p. 535). These pigments are normally secreted in the bile. The yellowish skin, sclerae, and mucous membranes of jaundice result from excess deposition of bile pigments.

TABLE 17.7	Major Functions of the Liver
General Function	**Specific Function**
Carbohydrate metabolism	Polymerizes glucose to glycogen; breaks down glycogen to glucose; converts noncarbohydrates to glucose
Lipid metabolism	Oxidizes fatty acids; synthesizes lipoproteins, phospholipids, and cholesterol; converts excess portions of carbohydrate molecules into fat molecules
Protein metabolism	Deaminates amino acids; forms urea; synthesizes plasma proteins; converts certain amino acids into other amino acids
Storage	Stores glycogen, vitamins A, D, and B_{12}, iron, and blood
Blood filtering	Removes damaged red blood cells and foreign substances by phagocytosis
Detoxification	Removes toxins from the blood
Secretion	Produces and secretes bile

Jaundice, a yellowing of the skin, sclerae of the eyes, and mucous membranes due to accumulation of bile pigments, has several causes. In *obstructive jaundice,* bile ducts are blocked, perhaps by gallstones or tumors. In *hepatocellular jaundice,* the liver is diseased, as in cirrhosis or hepatitis. In *hemolytic jaundice,* red blood cells are destroyed too rapidly, as happens with an incompatible blood transfusion or a blood infection.

Hepatitis is an inflammation of the liver. It has several causes, but the various types have similar symptoms.

For the first few days, hepatitis may resemble the flu, producing mild headache, low fever, fatigue, lack of appetite, nausea and vomiting, and sometimes stiff joints. By the end of the first week, more distinctive symptoms arise: a rash, pain in the right upper quadrant of the abdomen, dark and foamy urine, and pale feces. The skin and sclera of the eyes begin to turn yellow (jaundice) due to accumulating bile pigments. Great fatigue may continue for two or three weeks, and then gradually the person begins to feel better.

This is hepatitis in its most common, least dangerous, acute form. About half a million people develop hepatitis in the United States each year, and 6,000 die. In a rare form called *fulminant hepatitis*, symptoms are sudden and severe, along with altered behavior and personality. Medical attention is necessary to prevent kidney or liver failure or coma. Hepatitis that persists for more than six months is termed chronic.

Only rarely does hepatitis result from alcoholism, autoimmunity, or the use of certain drugs. Usually, one of several types of viruses causes hepatitis. Viral types are distinguished by the route of infection, surface features, and whether the viral genetic material is DNA or RNA. Hepatitis B virus has DNA; the others have RNA. The viral types are classified as follows:

Hepatitis A spreads by contact with food or objects contaminated with virus-containing feces, including diapers. The course of hepatitis A is short and mild.

Hepatitis B spreads by contact with virus-containing body fluids, such as blood, saliva, or semen. It may be transmitted by blood transfusions, hypodermic needles, or sexual activity.

Hepatitis C accounts for about half of all known cases of hepatitis. This virus is primarily transmitted in blood—by sharing razors or needles, from pregnant woman to fetus, or through blood transfusions or use of blood products. Ten to twenty percent of people known to be infected with hepatitis C virus develop symptoms.

Hepatitis D occurs in people already infected with the hepatitis B virus. It is blood-borne and associated with blood transfusions and intravenous drug use. About 20% of individuals infected with this virus die from the infection.

Hepatitis E virus is usually transmitted in water contaminated with feces. It most often affects visitors to developing nations.

Hepatitis G is rare but seems to account for a significant percentage of cases of fulminant hepatitis. In people with healthy immune systems, it produces symptoms so mild that they may not even be noticed.

Antibiotic drugs, which are effective against bacteria, do not treat hepatitis because most cases are viral. Usually, the person must just wait out the symptoms. Hepatitis C infection may resolve on it's own, or respond to 6 to 12 months treatment with a form of interferon or any of several new drugs. As many as 300 million people worldwide are hepatitis carriers. They do not have symptoms but can infect others. Five percent of carriers eventually develop liver cancer.

Gallbladder

The **gallbladder** is a pear-shaped sac in a depression on the inferior surface of the liver. It is connected to the **cystic duct,** which, in turn, joins the common hepatic duct (see fig. 17.23 and reference plate 19, p. 52). The gallbladder has a capacity of 30–50 milliliters, is lined with columnar epithelial cells, and has a strong muscular layer in its wall. It stores bile between meals, concentrates bile by reabsorbing water, and contracts to release bile into the small intestine.

The **bile duct** (common bile duct) is formed by the union of the common hepatic and cystic ducts. It leads to the duodenum, where the hepatopancreatic sphincter muscle guards its exit (see fig. 17.23). This sphincter normally remains contracted, so as bile collects in the bile duct it backs up into the cystic duct. When this happens, the bile flows into the gallbladder, where it is stored.

Bile salts, bile pigments, and cholesterol become increasingly concentrated as the gallbladder lining reabsorbs water and electrolytes. The cholesterol normally remains in solution, but under certain conditions it may precipitate and form solid crystals. If cholesterol continues to come out of solution, the crystals enlarge, forming *gallstones* (**fig. 17.29**). This may happen if the bile is too concentrated, hepatic cells secrete too much cholesterol, or the gallbladder is inflamed (cholecystitis). Gallstones in the bile duct may block the flow of bile, causing obstructive jaundice and considerable pain. Clinical Application 17.4 discusses disorders of the gallbladder.

Regulation of Bile Release

Normally, bile does not enter the duodenum until *cholecystokinin* stimulates the gallbladder to contract. The intestinal mucosa releases this hormone in response to proteins and fats in the small intestine. (Recall from earlier in this chapter, p. 669, that cholecystokinin also stimulates pancreatic enzyme secretion.) The hepatopancreatic sphincter usually remains contracted until a peristaltic wave in the

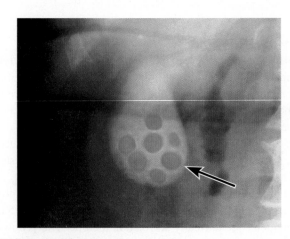

FIGURE 17.29 Radiograph of a gallbladder that contains gallstones (arrow).

CLINICAL APPLICATION 17.4

Gallbladder Disease

Molly G., an overweight, forty-seven-year-old college administrator and mother of four, had pain in the right upper quadrant of her abdomen (see fig. 1.25b, p. 31). Sometimes the discomfort seemed to radiate around to her back and move upward into her right shoulder (see fig. 12.2, p. 448). She usually felt this pain after her evening meal, but it occasionally happened at night, awakening her. After an episode of severe pain accompanied by sweating (diaphoresis) and nausea, Molly visited her physician, who discovered tenderness in the epigastric region (see fig. 1.25a, p. 31). She decided that Molly's symptoms could indicate acute cholecystitis—an inflammation of the gallbladder. Ultrasonography of the gallbladder or an X ray of the gallbladder called a *cholecystogram* could provide the evidence required to diagnose the painful problem.

Molly took tablets containing a contrast medium the night before the X-ray procedure, which allowed time for the small intestine to absorb the substance and for it to reach the liver and be excreted into the bile. Later, contrast medium would be concentrated in the bile and stored in the gallbladder and would make the contents of the gallbladder opaque to X rays.

Molly's cholecystogram revealed several stones (calculi) in her gallbladder, a condition called *cholelithiasis* (see fig. 17.29). Because Molly's symptoms were worsening, her physician recommended *cholecystectomy*—surgical removal of the gallbladder.

An incision was made in Molly's right subcostal region and her gallbladder excised from the liver. Fortunately, the cystic duct and hepatic ducts did not have stones (see fig. 17.26b). Unfortunately, Molly's symptoms persisted following her recovery from surgery. So her surgeon ordered a *cholangiogram*—an X-ray series of the bile ducts. This study showed a residual stone at the distal end of Molly's bile duct (see fig. 17.23).

The surgeon extracted the residual stone using a *fiberoptic endoscope*, a long, flexible tube passed through the esophagus and stomach and into the duodenum. This instrument enables a surgeon to observe features of the gastrointestinal tract directly through the eyepiece of the endoscope or on a monitor. A surgeon can also perform manipulations using specialized tools passed through the endoscope to its distal end.

In Molly's case, the surgeon performed an *endoscopic papillotomy*—an incision of the hepatopancreatic sphincter made by applying an electric current to a wire extending from the end of the endoscope (see fig. 17.23). She then removed the exposed stone by manipulating a tiny basket at the tip of the endoscope. Many patients undergo only the endoscopic procedure to remove the gallbladder, performed on an outpatient basis.

---

duodenal wall approaches. Just before the wave hits, the sphincter relaxes, and bile squirts into the duodenum (fig. 17.30). Table 17.8 summarizes the hormones that control digestion.

Functions of Bile Salts

Bile salts aid digestive enzymes. Molecules of fats are lipophilic and tend to clump when in a water solution. Bile salts have both a water soluble end and a lipid soluble end, and break up fat clumps by allowing the fat and water to interact, much like the action of soap or detergent. The result is smaller fat droplets. This process is called emulsification. The smaller fat droplets greatly increase the total surface area of the fatty substance, giving lipase molecules better access to the fat molecules and resulting in more effective fat digestion.

Bile salts enhance absorption of fatty acids and cholesterol by forming complexes (micelles) that are very soluble in chyme and that epithelial cells can more easily absorb. The fat-soluble vitamins A, D, E, and K are also absorbed in the presence of bile salts. Lack of bile salts results in poor lipid absorption and vitamin deficiencies.

The mucous membrane of the small intestine reabsorbs nearly all of the bile salts, along with fatty acids. The blood carries bile

TABLE 17.8	Hormones of the Digestive Tract	
Hormone	Source	Function
Gastrin	Gastric cells, in response to food	Increases secretory activity of gastric glands
Intestinal gastrin	Cells of small intestine, in response to chyme	Increases secretory activity of gastric glands
Somatostatin	Gastric cells	Inhibits secretion of acid by parietal cells
Intestinal somatostatin	Intestinal wall cells, in response to fats	Inhibits secretion of acid by parietal cells
Cholecystokinin	Intestinal wall cells, in response to proteins and fats in the small intestine	Decreases secretory activity of gastric glands and inhibits gastric motility; stimulates pancreas to secrete fluid with a high digestive enzyme concentration; stimulates gallbladder to contract and release bile
Secretin	Cells in the duodenal wall, in response to acidic chyme entering the small intestine	Stimulates pancreas to secrete fluid with a high bicarbonate ion concentration

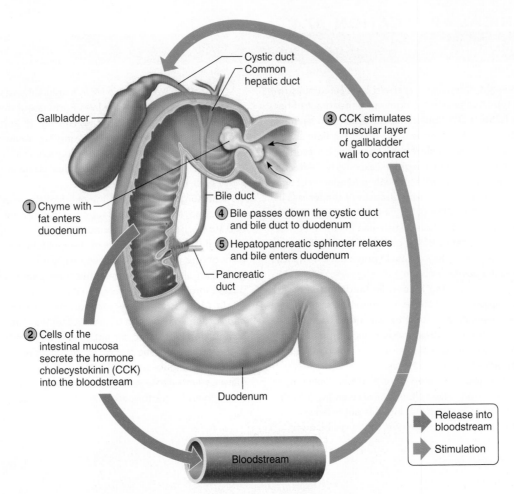

FIGURE 17.30 Fatty chyme entering the duodenum stimulates the gallbladder to release bile.

Which other organ, besides the gallbladder, responds to cholecystokinin stimulation, and what is the response of that organ to cholecystokinin stimulation?

Answer can be found in Appendix G.

salts to the liver, where hepatic cells resecrete them into the bile ducts. Liver cells synthesize bile salts, which replace the small amounts lost in the feces.

 PRACTICE

36 Explain how bile forms.

37 Describe the function of the gallbladder.

38 How is secretion of bile regulated?

39 How do bile salts function in digestion?

17.8 | Small Intestine

The **small intestine** is a tubular organ that extends from the pyloric sphincter to the beginning of the large intestine. With its many loops and coils, the small intestine fills much of the abdominal cavity (see fig. 17.1 and reference plates 4 and 5, pp. 41–42). The small intestine is 5.5–6.0 meters (18–20 feet) long in a cadaver when the muscular wall lacks tone, but it may be only half this long in a living person.

The small intestine receives chyme from the stomach and secretions from the pancreas and liver. It completes digestion of the nutrients in chyme, absorbs the products of digestion, and transports the remaining material to the large intestine.

Parts of the Small Intestine

The small intestine, shown in figures 17.31 and 17.32 and in reference plates 12 and 18 (pp. 48 and 51), consists of three portions: the duodenum, the jejunum, and the ileum.

The **duodenum** (du″o-de′num), which is about 25 centimeters long and 5 centimeters in diameter, lies posterior to the parietal peritoneum (retroperitoneal). It is the shortest and most fixed portion of the small intestine. The duodenum follows a C-shaped path as it passes anterior to the right kidney and the upper three lumbar vertebrae.

The remainder of the small intestine is mobile and lies free in the peritoneal cavity. The proximal two-fifths of this portion of the

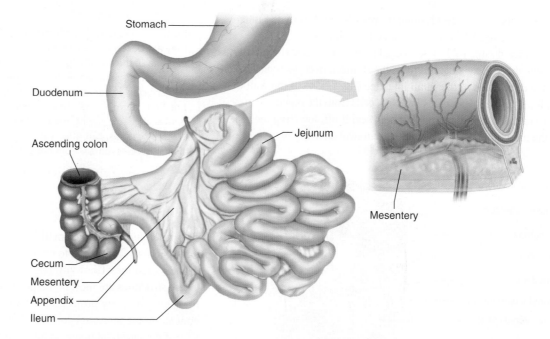

FIGURE 17.31 AP|R
The three parts of the small intestine are the duodenum, the jejunum, and the ileum. Veins are not visible from this view of the small intestine.

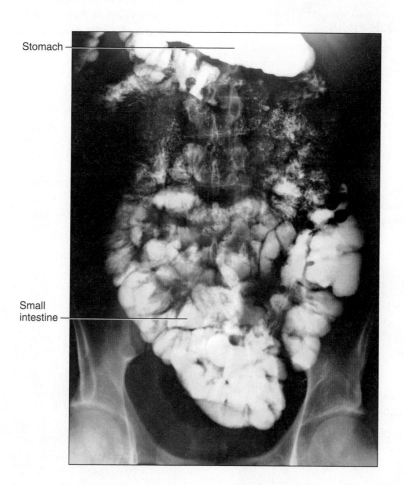

FIGURE 17.32 AP|R
Radiograph showing a normal small intestine containing a radiopaque substance that the patient ingested.

small intestine is the **jejunum** (jĕ-joo′num), and the remainder is the **ileum** (il′e-um). There is no distinct separation between the jejunum and ileum, but the diameter of the jejunum is typically greater, and its wall is thicker, more vascular, and more active than that of the ileum. The ileum has lymph nodules (Peyer's patches, p. 621) and

a more diverse and extensive "microbiome"—the bacterial species that live in the human body.

The jejunum and ileum are suspended from the posterior abdominal wall by a double-layered fold of peritoneum called **mesentery** (mes′en-ter″e) (fig. 17.33). The mesentery supports the blood vessels, nerves, and lymphatic vessels that supply the intestinal wall (see fig. 17.31 and reference plate 5, p. 42).

A filmy, double fold of peritoneal membrane called the *greater omentum* drapes like an apron from the stomach over the transverse colon and the folds of the small intestine (**fig. 17.34**). If the wall of the alimentary canal becomes infected, cells from the omentum may adhere to the inflamed region and help seal it off, lowering the risk that the infection will spread to the peritoneal cavity.

Most of the cells within a human body are not actually human—they are bacterial, constituting our microbiome. The ten trillion bacteria of the human "gut microbiome" have long been known to help us digest certain foods. The mouth houses more than 600 species of bacteria, and the large intestine 6,800. One of the first microbiome studies examined soiled diapers from babies regularly during their first year. Newborns start out with clean intestines, but after various bacteria come and go, very similar species remain from baby to baby by the first birthday.

Structure of the Small Intestinal Wall

The inner surface of the small intestine has a velvety appearance throughout its length, due to many tiny projections of mucous membrane called **intestinal villi** (figs. 17.35 and 17.36; see fig. 17.3). These structures are most numerous in the duodenum and the proximal jejunum. They project into the lumen of the alimentary canal, contacting the intestinal contents. Villi greatly increase the surface area of the intestinal lining, aiding absorption of digestive products.

Each villus consists of a layer of simple columnar epithelium and a core of connective tissue containing blood capillaries, a lymphatic capillary called a **lacteal,** and nerve fibers (see fig. 17.35).

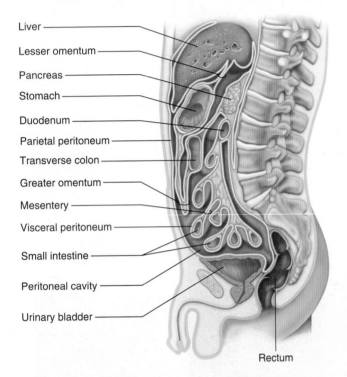

FIGURE 17.33 AP|R Mesentery formed by folds of the peritoneal membrane suspends portions of the small intestine from the posterior abdominal wall.

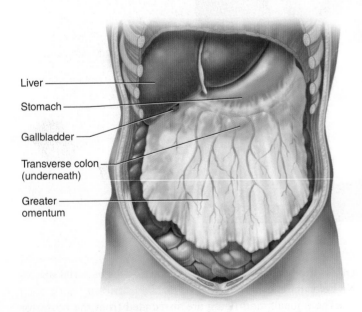

FIGURE 17.34 AP|R The greater omentum hangs like an apron over the abdominal organs.

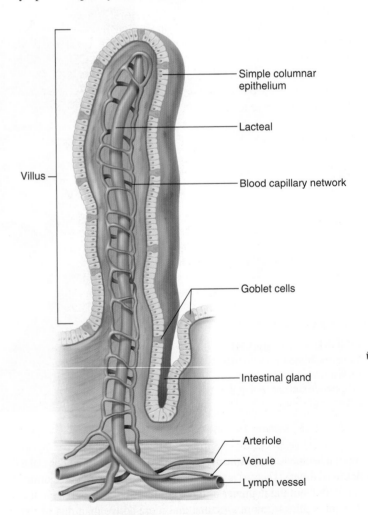

FIGURE 17.35 AP|R Structure of a single intestinal villus.

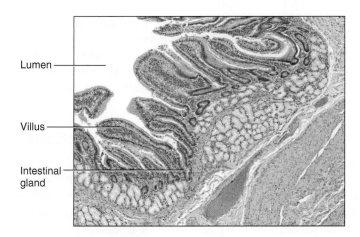

FIGURE 17.36 AP|R Light micrograph of intestinal villi from the wall of the duodenum (50×).

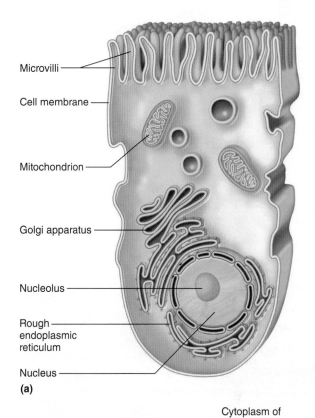

Microvilli

Cell membrane

Mitochondrion

Golgi apparatus

Nucleolus

Rough endoplasmic reticulum

Nucleus

(a)

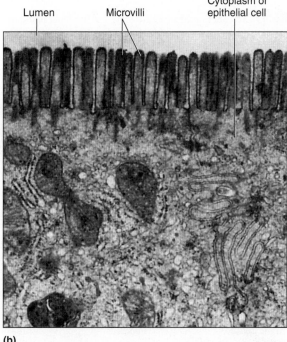

Lumen   Microvilli   Cytoplasm of epithelial cell

(b)

FIGURE 17.37 AP|R Intestinal epithelium. (a) Microvilli increase the surface area of intestinal epithelial cells. (b) Transmission electron micrograph of microvilli (16,000×).

At their free surfaces, the epithelial cells have many fine extensions called *microvilli* that form a brushlike border and greatly increase the surface area of the intestinal cells, further enhancing absorption (fig. 17.37; see fig. 17.3). The blood capillaries and lacteals carry away absorbed nutrients, and nerve fibers conduct impulses to stimulate or inhibit villus activities.

Between the bases of adjacent villi are tubular **intestinal glands** (crypts of Lieberkühn) that dip into the mucous membrane. The deeper layers of the small intestinal wall are much like those of other parts of the alimentary canal in that they include the submucosa, the muscularis, and the serosa layers.

The lining of the small intestine has many circular folds of mucosa, called *plicae circulares,* that are especially well developed in the lower duodenum and upper jejunum (fig. 17.38). With the villi and microvilli, these folds help increase the surface area of the intestinal lining.

The epithelial cells that form the lining of the small intestine are continually replaced. New cells form in the intestinal glands by cell division and migrate outward onto the villus surface. When the migrating cells reach the tip of the villus, they are shed. As a result, nearly one-quarter of the bulk of feces consists of dead epithelial cells from the small intestine. This *cellular turnover* renews the small intestine's epithelial lining every three to six days.

Secretions of the Small Intestine

Mucus-secreting goblet cells are abundant throughout the mucosa of the small intestine. In addition, many specialized *mucus-secreting glands* (Brunner's glands) in the submucosa in the proximal portion of the duodenum secrete a thick, alkaline mucus in response to certain stimuli.

The intestinal glands at the bases of the villi secrete large volumes of a watery fluid (see fig. 17.35). The villi rapidly reabsorb this fluid, which carries digestive products into the villi.

The fluid the intestinal glands secrete has a nearly neutral pH (6.5–7.5), and it lacks digestive enzymes. However, the epithelial cells of the intestinal mucosa have digestive enzymes embedded in the membranes of the microvilli on their luminal surfaces. These enzymes break down food molecules just before absorption takes place. The enzymes include **peptidases,** which split

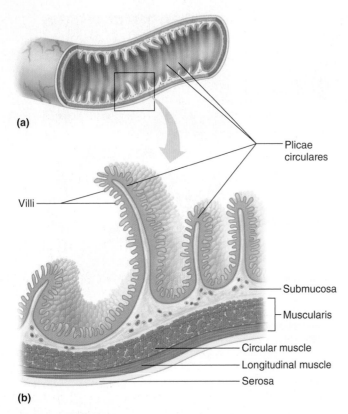

(a)

Villi

Plicae circulares

Submucosa

Muscularis

Circular muscle

Longitudinal muscle

Serosa

(b)

FIGURE 17.38 Section of small intestine. (*a*) The inner lining of the small intestine contains many circular folds, the plicae circulares. (*b*) A longitudinal section through some of these folds.

peptides into their constituent amino acids; **sucrase, maltase,** and **lactase,** which split the disaccharides sucrose, maltose, and lactose into the monosaccharides glucose, fructose, and galactose; and **intestinal lipase,** which splits fats into fatty acids and glycerol. Table 17.9 summarizes the sources and actions of the major digestive enzymes.

Regulation of Small Intestinal Secretions

Mucus protects the intestinal wall in the same way it protects the stomach lining. Therefore, mucus secretion increases in response to mechanical stimulation and the presence of irritants, such as gastric juice. Stomach contents entering the small intestine stimulate the duodenal mucous glands to release mucus.

Direct contact with chyme chemically and mechanically stimulates goblet cells and intestinal glands to secrete their products. Distension of the intestinal wall activates the nerve plexuses therein and stimulates parasympathetic reflexes that also trigger release of small intestine secretions.

 PRACTICE

40 Describe the parts of the small intestine.

41 What is the function of an intestinal villus?

42 Distinguish between intestinal villi and microvilli.

43 What is the function of the intestinal glands?

44 List intestinal digestive enzymes.

TABLE 17.9	Summary of the Major Digestive Enzymes AP|R	
Enzyme	**Source**	**Digestive Action**
Salivary Enzyme		
Salivary amylase	Salivary glands	Begins carbohydrate digestion by breaking down starch and glycogen to disaccharides
Gastric Enzymes		
Pepsin	Gastric glands	Begins protein digestion
Gastric lipase	Gastric glands	Begins butterfat digestion
Pancreatic Enzymes		
Pancreatic amylase	Pancreas	Breaks down starch and glycogen into disaccharides
Pancreatic lipase	Pancreas	Breaks down fats into fatty acids and glycerol
Trypsin, chymotrypsin	Pancreas	Breaks down proteins or partially digested proteins into peptides
Carboxypeptidase	Pancreas	Breaks down peptides into amino acids
Nucleases	Pancreas	Breaks down nucleic acids into nucleotides
Intestinal Enzymes		
Peptidase	Mucosal cells	Breaks down peptides into amino acids
Sucrase, maltase, lactase	Mucosal cells	Breaks down disaccharides into monosaccharides
Intestinal lipase	Mucosal cells	Breaks down fats into fatty acids and glycerol
Enterokinase	Mucosal cells	Converts trypsinogen into trypsin

Adults who have *lactose intolerance* do not produce sufficient lactase to adequately digest lactose, or milk sugar. Undigested lactose increases osmotic pressure of the intestinal contents and draws water into the intestines. At the same time, intestinal bacteria metabolize undigested sugar, producing organic acids and gases. The overall result is bloating, intestinal cramps, and diarrhea.

Digestive difficulties following milk drinking have been noted for thousands of years, but were associated with lactose only a century ago. Genetic evidence suggests that lactose intolerance may be the "normal" condition, with the ability to digest lactose the result of a mutation that occurred recently in our past and became advantageous when agriculture brought dairy foods to human populations. The ability to digest lactose has increased with the increased use of dairy foods.

Absorption in the Small Intestine

Villi greatly increase the surface area of the intestinal mucosa, making the small intestine the most important absorbing organ of the alimentary canal. The small intestine is so effective in absorbing digestive products, water, and electrolytes, that very little absorbable material reaches the organ's distal end.

RECONNECT
To Chapter 2, Organic Substances, pages 70–76.

Carbohydrate digestion begins in the mouth with the activity of salivary amylase and is completed in the small intestine by enzymes from the intestinal mucosa and pancreas (fig. 17.39). The resulting monosaccharides are absorbed by facilitated diffusion

or active transport into the villi and enter blood capillaries (see chapter 3, pp. 100–101 and 103).

Protein digestion begins in the stomach as a result of pepsin activity and is completed in the small intestine by enzymes from the intestinal mucosa and the pancreas (fig. 17.40). Protein molecules are ultimately broken down into amino acids, which are then absorbed into the villi by active transport and enter the circulation.

Fat molecules are digested almost entirely by enzymes from the pancreas and intestinal mucosa (fig. 17.41). The resulting fatty acid molecules are absorbed in the following steps: (1) The fatty acid molecules dissolve in the epithelial cell membranes of the villi and diffuse through them. (2) The endoplasmic reticula of the cells use the fatty acids to resynthesize fat molecules similar to those previously digested. (3) These fats collect in clusters that become encased in protein. (4) The resulting large molecules of lipoprotein are called *chylomicrons,* and they make their way to the lacteals of the villi. (5) Periodic contractions of smooth muscle in the villi help empty the lacteals into the cysterna chyli (see fig. 16.6a, p. 620), which is an expansion of the thoracic duct. The lymph carries the chylomicrons to the bloodstream (fig. 17.42).

Chylomicrons in the blood transport dietary fats to muscle and adipose cells. Similarly, very-low-density lipoprotein (VLDL) molecules, produced in the liver, transport triglycerides synthesized from excess dietary carbohydrates. As VLDL molecules reach adipose cells, an enzyme, *lipoprotein lipase,* catalyzes reactions that unload their triglycerides, converting VLDL to low-density lipoprotein (LDL) molecules. Because most of the triglycerides have been removed, LDL molecules have a higher cholesterol content than the original VLDL molecules. Cells in the peripheral tissues obtain cholesterol by using receptor-mediated endocytosis to remove LDL particles from plasma (see chapter 3, pp. 104–105).

Unlike LDL, which delivers cholesterol to tissues, high-density lipoprotein (HDL) removes cholesterol from tissues and

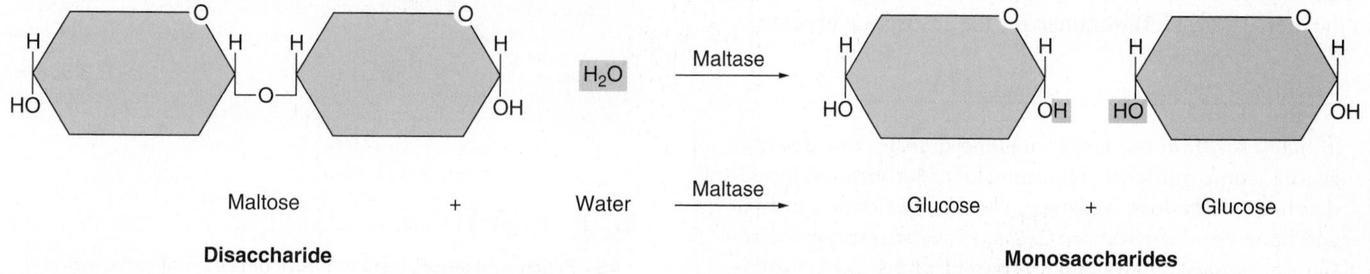

FIGURE 17.39 Digestion breaks down complex carbohydrates into disaccharides, which are then broken down into monosaccharides, which are small enough for intestinal villi to absorb. The monosaccharides then enter the bloodstream.

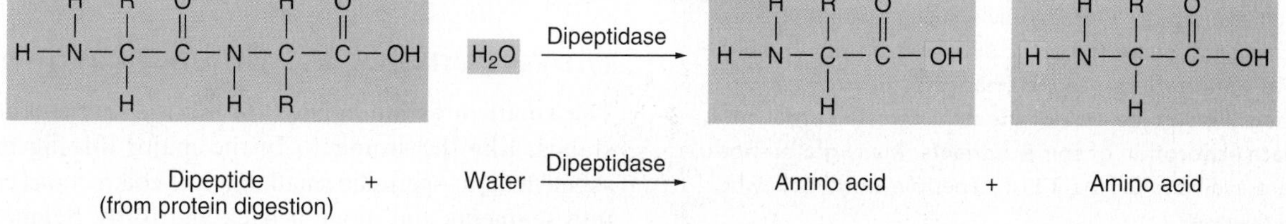

FIGURE 17.40 The amino acids that result from dipeptide digestion are absorbed by intestinal villi and enter the blood.

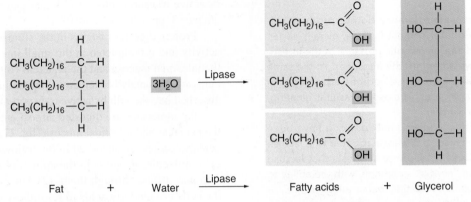

FIGURE 17.41 Fatty acids and glycerol result from fat digestion. Intestinal villi absorb them, and most are resynthesized into fat molecules before they enter the lacteals.

delivers it to the liver. The liver produces the basic HDL framework and secretes HDL molecules into the bloodstream. Circulating HDL picks up cholesterol from peripheral tissues and returns it to the liver, where it enters cells by receptor-mediated endocytosis. The liver disposes of the cholesterol it obtains in this manner by secreting it into bile or by using it to synthesize bile salts.

The intestine reabsorbs much of the cholesterol and bile salts in bile, which are then transported back to the liver, and the secretion-reabsorption cycle repeats. During each cycle, some of the cholesterol and bile salts escape reabsorption, reach the large intestine, and are excreted as part of the feces.

The intestinal villi also absorb electrolytes and water. Most electrolyte absorption is by facilitated diffusion and active transport, and water by osmosis. Thus, the intestinal contents may be hypertonic at first, but as nutrients and electrolytes are absorbed, the contents become slightly hypotonic to the cells. Then, water follows the nutrients and electrolytes into the villi by osmosis. In most cases the electrolytes are absorbed at a steady rate, but in some cases, as with iron and calcium, absorption will increase when body stores are low, as long as these substances are present in the diet. Table 17.10 summarizes the absorption process.

In *malabsorption* the small intestine digests, but does not absorb, some nutrients. Symptoms of malabsorption include diarrhea, weight loss, weakness, vitamin deficiencies, anemia, and bone demineralization. Causes of malabsorption include surgical removal or rerouting of a portion of the small intestine, obstruction of lymphatic vessels due to a tumor, or interference with the production and release of bile as a result of liver disease and enzyme deficiency. Another cause of malabsorption is a reaction to *gluten*, which is a composite of two types of proteins that are found in certain grains, such as wheat, barley, and rye. This condition, called *celiac disease*, is an autoimmune response to eating gluten that damages or destroys microvilli, reducing the surface area of the small intestine, preventing adequate absorption of some nutrients. Many gluten-free products are available for the 1 in 100 people worldwide who have celiac disease.

TABLE 17.10	Intestinal Absorption of Nutrients	
Nutrient	**Absorption Mechanism**	**Means of Circulation**
Monosaccharides	Facilitated diffusion and active transport	Blood in capillaries
Amino acids	Active transport	Blood in capillaries
Fatty acids and glycerol	Facilitated diffusion of glycerol; diffusion of fatty acids into cells or blood capillaries	
	(a) Most fatty acids are resynthesized into fats and incorporated in chylomicrons	Lymph in lacteals
	(b) Some fatty acids with relatively short carbon chains are absorbed without being changed back into fats	Blood in capillaries
Electrolytes	Diffusion and active transport	Blood in capillaries
Water	Osmosis	Blood in capillaries

 PRACTICE

45 Which substances resulting from digestion of carbohydrate, protein, and fat molecules does the small intestine absorb?

46 Which ions does the small intestine absorb?

47 Describe how fatty acids are absorbed and transported.

Movements of the Small Intestine

The small intestine carries on mixing movements and peristalsis, like the stomach. In the major mixing movement—segmentation—periodic small, ringlike contractions cut the chyme into segments and move it back and forth. Segmentation also slows the movement of chyme through the small intestine.

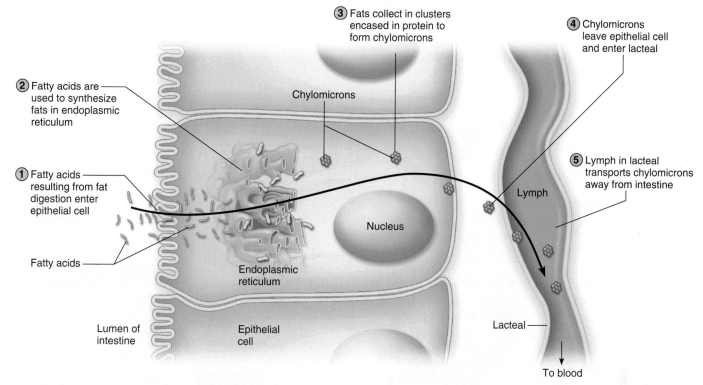

③ Fats collect in clusters
encased in protein to
form chylomicrons

④ Chylomicrons
leave epithelial cell
and enter lacteal

Chylomicrons

② Fatty acids are
used to synthesize
fats in endoplasmic
reticulum

① Fatty acids
resulting from fat
digestion enter
epithelial cell

⑤ Lymph in lacteal
transports chylomicrons
away from intestine

Lymph

Nucleus

Fatty acids

Endoplasmic
reticulum

Lumen of
intestine

Epithelial
cell

Lacteal

To blood

FIGURE 17.42 Fat absorption involves several steps.

Peristaltic waves propel chyme through the small intestine. These waves are usually weak, and they stop after pushing the chyme a short distance. Consequently, chyme moves slowly through the small intestine, taking from three to ten hours to travel its length.

As might be expected, parasympathetic impulses enhance both mixing and peristaltic movements, and sympathetic impulses inhibit them. Reflexes involving parasympathetic impulses to the small intestine sometimes originate in the stomach. For example, food in the stomach distends its wall, triggering the gastroenteric reflex, which greatly increases peristaltic activity in the small intestine. Another reflex begins when the duodenum fills with chyme, stretching its wall. This reflex speeds movement through the small intestine. These reflexes make room for new chyme from the stomach.

If the small intestine wall becomes overdistended or irritated, a strong *peristaltic rush* may pass along the organ's entire length. This movement sweeps the contents into the large intestine so quickly that water, nutrients, and electrolytes that would normally be absorbed are not. The result is *diarrhea,* characterized by frequent defecation and watery stools. Prolonged diarrhea causes dehydration and imbalances in electrolyte concentrations.

At the distal end of the small intestine, the **ileocecal sphincter** joins the small intestine's ileum to the large intestine's cecum. Normally, this sphincter remains constricted, preventing the contents of the small intestine from entering the large intestine, and at the same time keeping the contents of the large intestine from backing up into the ileum. However, after a meal, a gastroileal reflex increases peristalsis in the ileum and relaxes the sphincter, forcing some of the contents of the small intestine into the cecum.

 PRACTICE

48 Describe the movements of the small intestine.

49 What is a peristaltic rush?

50 What stimulus relaxes the ileocecal sphincter?

17.9 | Large Intestine

The **large intestine** is so named because its diameter is greater than that of the small intestine. This part of the alimentary canal is about 1.5 meters long, and it begins in the lower right side of the abdominal cavity where the ileum joins the cecum. From there, the large intestine ascends on the right side, crosses obliquely to the left, and descends into the pelvis. At its distal end, it opens to the outside of the body as the anus.

The large intestine absorbs ingested water and electrolytes remaining in the alimentary canal. Additionally it reabsorbs and recycles water and remnants of digestive secretions. The large intestine also forms and stores feces.

Parts of the Large Intestine

The large intestine consists of the cecum, the colon, the rectum, and the anal canal. Figures 17.43 and 17.44 and reference plates 11, 12, 18, and 25 (pp. 47, 48, 51, and 56) depict the large intestine.

The **cecum,** at the beginning of the large intestine, is a dilated, pouchlike structure that hangs slightly inferior to the ileocecal opening. Projecting downward from it is a narrow tube with a closed end called the **appendix.** The human appendix has no known digestive function, but it contains lymphatic tissue. Removal of the

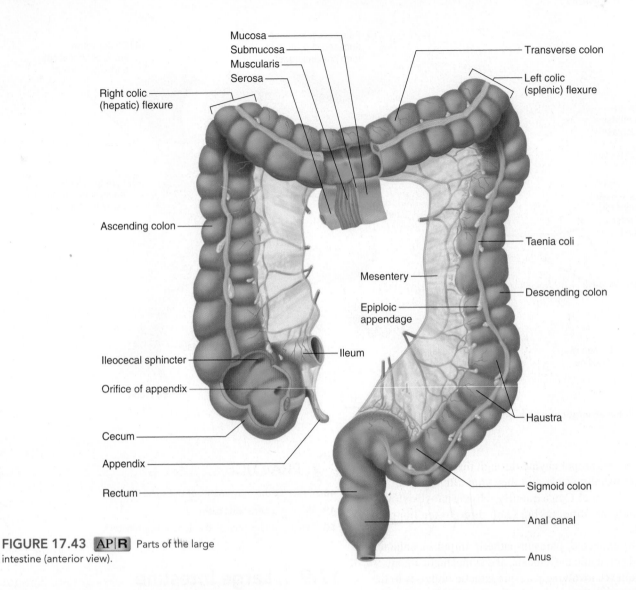

FIGURE 17.43 AP|R Parts of the large intestine (anterior view).

Labels on figure:
Mucosa
Submucosa
Muscularis
Serosa
Right colic (hepatic) flexure
Transverse colon
Left colic (splenic) flexure
Ascending colon
Taenia coli
Mesentery
Descending colon
Epiploic appendage
Ileum
Ileocecal sphincter
Orifice of appendix
Haustra
Cecum
Appendix
Rectum
Sigmoid colon
Anal canal
Anus

appendix raises the risk of recurrence of certain infections, such as *Clostridium difficile*. This evidence suggests that the human appendix may be a storage place for the healthy "gut" microbiome.

> In *appendicitis,* the appendix becomes inflamed. Surgery is often required to remove the appendix before it ruptures. If it does rupture, this may allow contents of the large intestine including any infectious organisms to enter the abdominal cavity and cause a serious inflammation of the peritoneum called *peritonitis.*

The **colon** is divided into four parts—the ascending, transverse, descending, and sigmoid colons. The **ascending colon** begins at the cecum and extends upward against the posterior abdominal wall to a point just inferior to the liver. There it turns sharply to the left (as the right colic, or hepatic, flexure) and becomes the **transverse colon.** The transverse colon is the longest and most movable part of the large intestine. It is suspended by a fold of peritoneum and sags in the middle below the stomach. As the transverse colon approaches the spleen, it turns abruptly downward (as the left colic, or splenic, flexure) and becomes the **descending colon.** At the brim of the pel-

vis, the colon makes an S-shaped curve, called the **sigmoid colon,** and then becomes the rectum.

The **rectum** lies next to the sacrum and generally follows its curvature. The peritoneum firmly attaches it to the sacrum, and it ends about 5 centimeters inferior to the tip of the coccyx, where it becomes the anal canal (fig. 17.45).

The **anal canal** is formed by the last 2.5 to 4.0 centimeters of the large intestine. The mucous membrane in the canal is folded into a series of six to eight longitudinal *anal columns.* At its distal end, the canal opens to the outside as the **anus.** Two sphincter muscles guard the anus—an *internal anal sphincter muscle,* composed of smooth muscle under involuntary control, and an *external anal sphincter muscle,* composed of skeletal muscle under voluntary control.

> *Hemorrhoids* are enlarged and inflamed branches of the rectal vein in the anal columns that cause intense itching, sharp pain, and sometimes bright red bleeding. The hemorrhoids may be internal or bulge out of the anus. Causes of hemorrhoids include anything that puts prolonged pressure on the delicate rectal tissue, including obesity, pregnancy, constipation, diarrhea, and liver disease.

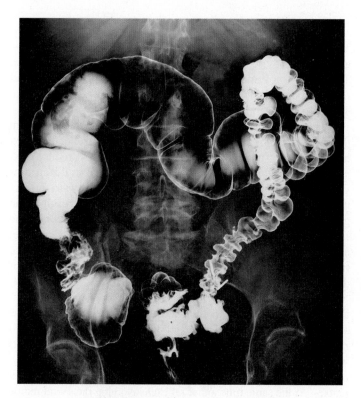

FIGURE 17.44 APR Radiograph of the large intestine containing a radiopaque substance that the patient ingested.

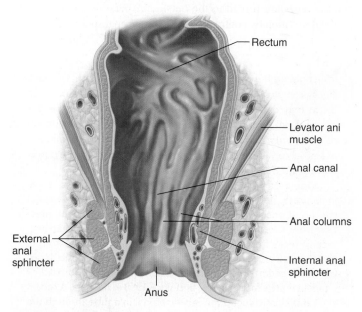

FIGURE 17.45 The rectum and the anal canal are at the distal end of the alimentary canal.

 PRACTICE

51 What is the general function of the large intestine?

52 Describe the parts of the large intestine.

53 Distinguish between the internal sphincter muscle and the external sphincter muscle of the anus.

Structure of the Large Intestinal Wall

The wall of the large intestine includes the same types of tissues found in other parts of the alimentary canal but also has some unique features (fig. 17.46). The large intestinal wall lacks the villi and plicae circularis characteristic of the small intestine. The longitudinal layer of smooth muscle cells is not uniformly distributed throughout the large intestinal wall. Instead the smooth muscle cells are mostly in three distinct bands (taeniae coli) that extend the entire length of the colon. These bands exert tension lengthwise on the wall, creating a series of pouches (haustra). The large intestinal wall also has small collections of fat (epiploic appendages) in the serosa on its outer surface (see fig. 17.43).

Functions of the Large Intestine

The large intestine has little or no digestive function. This is in contrast to the small intestine, which secretes digestive enzymes and absorbs the products of digestion. However, the mucous membrane that forms the inner lining of the large intestine includes many tubular glands. Structurally, these glands are similar to those of the small intestine, but they are composed almost entirely of goblet cells (fig. 17.47). Consequently, mucus is the large intestine's only significant secretion.

Mechanical stimulation from chyme and parasympathetic impulses control the rate of mucus secretion. In both cases, the goblet cells respond by increasing mucus production, which, in turn, protects the intestinal wall against the abrasive action of materials passing through it. Mucus also holds particles of fecal matter together, and, because it is alkaline, mucus helps control the pH of the large intestinal contents. This is important because acids are sometimes released from the feces as a result of bacterial activity.

Chyme entering the large intestine usually has few nutrients remaining in it and mostly consists of materials not digested or absorbed in the small intestine. It also contains water, electrolytes, mucus, and bacteria.

Absorption in the large intestine is normally limited to water and electrolytes, and this usually occurs in the proximal half of the tube. Electrolytes such as sodium ions can be absorbed by active transport, while water follows passively, crossing the

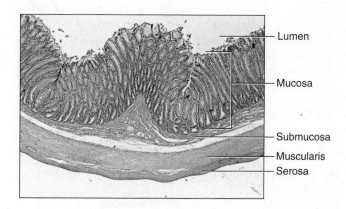

FIGURE 17.46 APR Light micrograph of the large intestinal wall (64×).

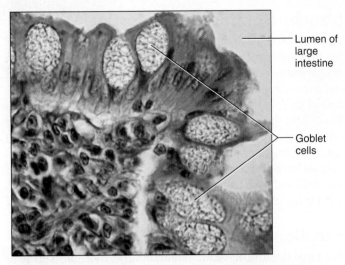

FIGURE 17.47 Light micrograph of the large intestinal mucosa (560×).

Q *Note the many goblet cells in the mucosa of the large intestine. Why are there so many more of these cells in the large intestinal wall than in the small intestinal wall?*

Answer can be found in Appendix G.

mucosal layer by osmosis. About 90% of the water that enters the large intestine is absorbed, and little sodium or water is lost in the feces.

The many bacteria that normally inhabit the large intestine, called *intestinal flora,* break down some of the molecules that escape the actions of human digestive enzymes. For instance, cellulose, a complex carbohydrate in food of plant origin, passes through the alimentary canal almost unchanged, but colon bacteria can break down cellulose and use it as an energy source. These bacteria, in turn, synthesize vitamins, such as K, B_{12}, thiamine, and riboflavin, which the intestinal mucosa absorbs. Bacterial actions in the large intestine may produce intestinal gas (flatus).

Intestinal gas contains nitrogen and oxygen taken in while breathing and eating, plus variable amounts of methane (CH_4), carbon dioxide (CO_2), and hydrogen contributed from the bacterial fermentation of undigested food. The characteristic odor comes from bacterial action on the nitrogen and sulfur in proteins, which yields pungent-smelling ammonia (NH_3) and foul hydrogen sulfide (H_2S). Most people release a half liter of intestinal gas a day. Foods rich in sulfur-containing amino acids make intestinal gas more foul. These foods include beans, broccoli, bran, brussels sprouts, cabbage, cauliflower, and onions.

 PRACTICE

54 How does the structure of the large intestine differ from that of the small intestine?

55 Which substances does the large intestine absorb?

56 Which useful substances do bacteria inhabiting the large intestine produce?

Movements of the Large Intestine

The movements of the large intestine—mixing and peristalsis—are similar to those of the small intestine, although usually slower. The mixing movements break the fecal matter into segments and turn it so that all portions are exposed to the intestinal mucosa. This helps absorb water and electrolytes.

The peristaltic waves of the large intestine are different from those of the small intestine. Instead of occurring frequently, they happen only two or three times each day. These waves produce *mass movements* in which a large section of the intestinal wall constricts vigorously, forcing the intestinal contents to move toward the rectum. Typically, mass movements follow a meal, as a result of the gastrocolic reflex initiated in the small intestine. Irritation of the intestinal mucosa can also trigger such movements. For instance, a person suffering from an inflamed colon (colitis) may experience frequent mass movements. Clinical Application 17.5 examines conditions affecting the large intestine.

When it is appropriate to defecate, a person usually can initiate a *defecation reflex* by holding a deep breath and contracting the abdominal wall muscles. This action increases the internal abdominal pressure and forces feces into the rectum. As the rectum fills, its wall distends, triggering the defecation reflex, which stimulates peristaltic waves in the descending colon. The internal anal sphincter relaxes. At the same time, other reflexes involving the sacral region of the spinal cord strengthen the peristaltic waves, lower the diaphragm, close the glottis, and contract the abdominal wall muscles. These actions further increase the internal abdominal pressure and squeeze the rectum. The external anal sphincter is signaled to relax, and the feces are forced to the outside. A person can voluntarily inhibit defecation by contracting the external anal sphincter.

Investigating the gut microbiome can improve health, because illness can alter the bacterial populations within us. Antibiotic drugs may cause diarrhea by temporarily changing the gut microbiome. Probiotics are live microorganisms, such as bacteria and yeasts, that, when ingested, confer a health benefit. For example, certain *Lactobacillus* strains added to yogurt help protect against *Salmonella* foodborne infection.

Probiotics typically consist of one or a few microbial species. In contrast, "fecal transplantation" alters hundreds of bacterial species at once. In the procedure, people with recurrent infection from *Clostridium difficile*, which causes severe diarrhea, receive feces from a healthy donor. The bacteria transferred in the donated feces reconstitute a healthy gut microbiome, treating the infection. Fecal transplantation has been performed by enema in cattle for a century, and since 1958 in humans. A recent clinical trial introduced the fecal material in a tube through the nose to the small intestine, but soon the material may be delivered in a capsule.

Feces

Feces (fe'sēz) are composed of materials not digested or absorbed, along with water, electrolytes, mucus, and bacteria. Usually the feces are about 75% water, and their color derives from bile pigments altered by bacterial action.

Disorders of the Large Intestine

The large intestine (colon) is the source of familiar digestive discomforts as well as more serious disorders.

Diverticulosis and Inflammatory Bowel Disease

In diverticulosis, parts of the intestinal wall weaken, and the inner mucous membrane protrudes through, forming small sacs on the outer intestinal surface. If chyme accumulates in the outpouching and becomes infected and inflamed (diverticulitis), antibiotics or surgical removal of the area may become necessary. Lack of dietary fiber may increase the risk of developing diverticulosis.

Inflammatory bowel disease is a group of disorders that includes ulcerative colitis and Crohn's disease. The disorders differ by the site and extent of inflammation and ulceration of the intestines. In the United States, about 100,000 people suffer the abdominal cramps and diarrhea of ulcerative colitis, and 500,000 individuals have similar symptoms of Crohn's disease.

Ulcerative colitis affects the mucosa and submucosa of the distal large intestine and the rectum. In about 25% of cases, the disease extends no farther than the rectum. Bloody diarrhea and cramps may last for days or weeks and may recur frequently or only rarely. The severe diarrhea leads to weight loss and electrolyte imbalances and may develop into colon cancer or affect other organs, including the skin, eyes, or liver. The inflamed and ulcerous tissue is continuous.

Crohn's disease is more extensive than ulcerative colitis, infiltrating the small and large intestines and penetrating all tissue layers. In contrast to the uniformity of ulcerative colitis, affected portions of intestine in Crohn's disease are interspersed with unaffected areas, producing a "cobblestone" effect after many years. The ileum and cecum are affected in about 40% of individuals, only the small intestine is involved in 30% of cases, and only the large intestine is involved in 25% of cases. Rarely, the disease affects more proximal structures of the gastrointestinal tract. The diarrhea is often not bloody, and complications such as cancer are rare.

Overall, about 20% of people with inflammatory bowel disease have symptoms that fall between the descriptions of ulcerative colitis and Crohn's disease, and are considered to have "indeterminate colitis." Autoimmunity, infection, or a genetic predisposition may contribute to causing inflammatory bowel disease. Surgery and drugs that suppress the immune response are used to treat some cases of inflammatory bowel disease.

Colorectal Cancer

Cancer of the large intestine or rectum, known as *colorectal cancer*, is the fourth most prevalent cancer in the United States and the second most common cause of cancer death. More than 30,000 new cases are diagnosed each year, and more than 56,000 people die of the condition. It tends to run in families.

Symptoms of colorectal cancer include

- a change in frequency or consistency of bowel movements
- blood in the feces
- a narrowing of feces
- abdominal discomfort or pain
- weight loss
- fatigue
- unexplained vomiting

Diagnostic tests, described in table 17A, may detect colorectal cancer. These tests are of two general types—the fecal occult blood test performed on a stool sample and imaging the large intestinal wall. An experimental test screens the DNA from cells in feces for mutations associated with colorectal cancer.

Fiberoptic colonoscopy is a test commonly performed on people over age 50, when the risk of colorectal cancer increases. Under sedation, a flexible lit tube is inserted into the rectum, and polyps and tumors are identified and removed. People with a family history of colon cancer should be screened at an earlier age. Fiberoptic colonoscopy takes less than an hour. A newer procedure, computed tomographic colonography (virtual colonoscopy), requires the same preparatory bowel cleansing, but does not require sedation and is faster. However, if a lesion is detected, the more invasive approach must be used to remove the suspicious tissue.

Treatment for colorectal cancer removes the affected tissue. If a large portion of the intestine is removed, surgery is used to construct a new opening to release feces. The free end of the intestine is attached to an opening created through the skin of the abdomen, and a bag is attached to the opening to collect the fecal matter. This procedure is called a colostomy.

For people who have a certain inherited form of colon cancer (familial adenomatous polyposis), nonsteroidal anti-inflammatory drugs (NSAIDs) called Cox-2 inhibitors are used to treat the disease and may even help to prevent it in those identified by a genetic test to be at high risk. However, these drugs increase the risk of heart disease in certain individuals.

TABLE 17A	Diagnostic Tests for Colorectal Cancer
Diagnostic Test	**Description**
Digital rectal exam	Physician palpates large intestine and rectum
Double-contrast barium enema	X-ray exam following ingestion of contrast agent highlights blockages in large intestine
Fecal occult blood test	Blood detected in feces sample
Colorectal cancer gene test (experimental)	Mutations associated with colorectal cancer detected in DNA of cells shed with feces
Sigmoidoscopy	Endoscope views rectum and lower colon
Colonoscopy	Endoscope views rectum and entire colon

The pungent odor of the feces results from a variety of compounds that bacteria produce. These compounds include phenol, hydrogen sulfide, indole, skatole, and ammonia.

 PRACTICE

57 How does peristalsis in the large intestine differ from peristalsis in the small intestine?

58 List the major events of defecation.

59 Describe the composition of feces.

17.10 | Life-Span Changes

Changes to the digestive system associated with the passing years are slow and slight, so most people can enjoy eating a variety of foods as they grow older. Maintaining healthy teeth is vital to obtaining adequate nutrition. This requires frequent dental check-ups, cleanings, and plaque removal, plus care of the gums. Tooth loss due to periodontal disease becomes more likely after age thirty-five.

Despite regular dental care, some signs of aging may affect the teeth. The enamel often thins from years of brushing, teeth grinding, and eating acidic foods. Thinning enamel may make the teeth more sensitive to hot and cold foods. At the same time, the cementum may thicken. The dentin heals more slowly and enlarges as the pulp shrinks. Loss of neurons in the pulp may make it more difficult to notice tooth decay. The gums recede, creating more pockets to harbor the bacteria whose activity contributes to periodontal disease. The teeth may loosen as the bones of the jaw weaken. On a functional level, older people sometimes do not chew their food thoroughly and swallow larger chunks of food that may present a choking hazard.

A common complaint of older individuals is "dry mouth," or xerostomia. This condition is not a normal part of aging—studies have shown that the oldest healthy people make just as much saliva as healthy younger people. Dry mouth is common, however, because it is a side effect of more than 400 medications, many of which are more likely to be taken by the elderly. These medications include antidepressants, antihistamines, and drugs that treat cancer or hypertension. In addition, radiation and chemotherapy used to treat cancer can cause mouth sores and tooth decay.

The gastrointestinal tract gradually becomes less efficient with age. Slowing peristalsis may cause frequent heartburn as food backs up into the esophagus. The stomach lining thins with age, and secretion of hydrochloric acid, pepsin, and intrinsic factor decline. Exit of chyme from the stomach slows. Overall, these changes may affect the rate at which certain medications are absorbed.

The small intestine is the site of absorption of nutrients, so it is here that noticeable signs of aging on digestion arise. Subtle shifts in the microbial species that inhabit the small intestine alter the rates of absorption of particular nutrients. With age, the small intestine becomes less efficient at absorbing vitamins A, D, and K and the mineral zinc. This raises the risk of deficiency symptoms—effects on skin and vision due to a lack of vitamin A; weakened bones from inadequate vitamin D; impaired blood clotting seen in vitamin K deficiency; and slowed healing, decreased immunity, and altered taste evidenced in zinc deficiency.

Many people who have inherited lactose intolerance begin to notice the telltale cramping after eating dairy foods in the middle years. They must be careful that by avoiding dairy products, they do not also lower their calcium intake. Less hydrochloric acid also adversely affects the absorption of calcium, as well as iron. Too little intrinsic factor may lead to vitamin B_{12} deficiency anemia.

The lining of the large intestine changes too, thinning and containing less smooth muscle and mucus. A dampening of the responsiveness of the smooth muscle to neural stimulation slows peristalsis, ultimately causing constipation. Compounding this common problem is a loss of elasticity in the walls of the rectum and declining strength and responsiveness of the internal and external sphincters.

The accessory organs to digestion age too, but not necessarily in ways that affect health. Both the pancreas and the liver are large organs with cells to spare, so a decline in their secretion abilities does not usually hamper digestion. Only 10% of the pancreas and 20% of the liver are required to digest foods. However, the liver may not be able to detoxify certain medications as quickly as it once did. The gallbladder becomes less sensitive to cholecystokinin, but in a classic feedback response, cells of the intestinal mucosa secrete more of it into the bloodstream. The gallbladder also retains its ability to contract. The bile ducts widen in some areas, but the end of the bile duct narrows as it approaches the small intestine. As long as gallstones do not become entrapped in the ducts, the gallbladder generally functions well into the later years.

 PRACTICE

60 Describe the effects of aging on the teeth.

61 Which conditions might be caused by the slowing of peristalsis in the digestive tract that occurs with aging? ∎

Digestive System

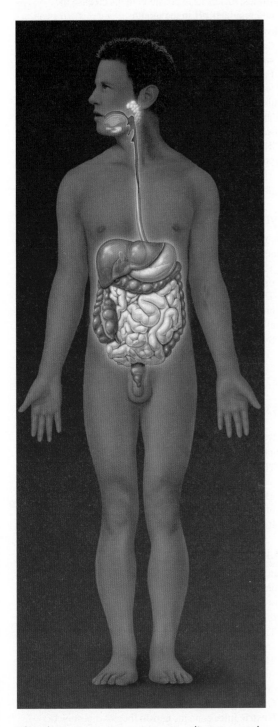

The digestive system ingests, digests, and absorbs nutrients for use by all body cells.

Integumentary System

Vitamin D activated in the skin plays a role in absorption of calcium from the digestive tract.

Skeletal System

Bones are important in mastication. Calcium absorption is necessary to maintain bone matrix.

Muscular System

Muscles are important in mastication, swallowing, and the mixing and moving of digestion products through the gastrointestinal tract.

Nervous System

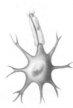

The nervous system can influence digestive system activity.

Endocrine System

Hormones can influence digestive system activity.

Cardiovascular System

The bloodstream carries absorbed nutrients to all body cells.

Lymphatic System

The lymphatic system plays a major role in the absorption of fats.

Respiratory System

The digestive system and the respiratory system share anatomical structures.

Urinary System

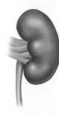

The kidneys and liver work together to activate vitamin D.

Reproductive System

In a woman, nutrition is essential for conception and normal development of an embryo and fetus.

Chapter Summary

17.1 Overview of the Digestive System
(page 650)

Digestion is the process of mechanically and chemically breaking down foods so that they can be absorbed. The **digestive system** consists of an **alimentary canal** and several accessory organs that carry out the processes of ingestion, propulsion, digestion, absorption, and defecation.

1. General characteristics of the alimentary canal
   a. Structure of the wall
      (1) The wall consists of four layers.
      (2) These layers include the **mucosa, submucosa, muscularis,** and **serosa.**
   b. Movements of the tube
      (1) Motor functions include mixing and propelling movements.
      (2) **Peristalsis** is responsible for propelling movements.
      (3) The wall of the tube undergoes receptive relaxation just ahead of a peristaltic wave.
   c. Innervation of the tube
      (1) The tube is innervated by branches of the sympathetic and parasympathetic divisions of the autonomic nervous system.
      (2) Parasympathetic impulses generally increase digestive activities; sympathetic impulses generally inhibit digestive activities.
      (3) Sympathetic impulses contract certain sphincter muscles, controlling movement of digesting food through the alimentary canal.

17.2 Mouth (page 653)

The **mouth** is adapted to receive food and begin digestion by mechanically breaking up solid particles (mastication). It also serves as an organ of speech and sensory perception.

1. Cheeks and lips
   a. **Cheeks** form the lateral walls of the mouth.
   b. **Lips** are highly mobile and have a variety of sensory receptors useful in judging the characteristics of food.
2. Tongue
   a. The **tongue** is a thick, muscular organ that mixes food with saliva and moves it toward the pharynx.
   b. The rough surface of the tongue handles food and has taste buds.
   c. **Lingual tonsils** are located on the root of the tongue.
3. Palate
   a. The **palate** comprises the roof of the mouth and includes hard and soft portions.
   b. The soft palate, including the uvula, closes the opening to the nasal cavity during swallowing.
   c. **Palatine tonsils** are located on either side of the tongue in the back of the mouth.
   d. Tonsils consist of lymphatic tissues.
4. Teeth
   a. Two sets of **teeth** develop in sockets of the mandibular and maxillary bones.
   b. There are twenty primary and thirty-two secondary teeth.
   c. Teeth mechanically break food into smaller pieces, increasing the surface area exposed to digestive actions.
   d. Different types of teeth are adapted to handle foods in different ways, such as biting, grasping, or grinding.
   e. Each tooth consists of a crown and root and is composed of enamel, dentin, pulp, nerves, and blood vessels.
   f. A tooth is attached to the alveolar process by the periodontal ligament.

17.3 Salivary Glands (page 658)

Salivary glands secrete saliva, which moistens food, helps bind food particles, begins chemical digestion of carbohydrates, makes taste possible, helps cleanse the mouth, and regulates pH in the mouth.

1. Salivary secretions
   a. Salivary glands include serous cells that secrete **salivary amylase** and mucous cells that secrete **mucus.**
   b. Parasympathetic impulses stimulate the secretion of a large volume of watery saliva.
2. Major salivary glands
   a. The **parotid glands** are the largest, and they secrete saliva rich in amylase.
   b. The **submandibular glands** in the floor of the mouth produce viscous saliva containing amylase.
   c. The **sublingual glands** in the floor of the mouth primarily secrete mucus.

17.4 Pharynx and Esophagus (page 660)

The pharynx and esophagus serve as passageways.

1. Structure of the pharynx
   a. The **pharynx** is divided into a **nasopharynx, oropharynx,** and **laryngopharynx.**
   b. The muscular walls of the pharynx contain fibers in circular and longitudinal groups.
2. Swallowing mechanism
   a. Swallowing (deglutition) occurs in three stages.
      (1) Food is mixed with saliva and forced into the pharynx.
      (2) Involuntary reflex actions move the food into the esophagus.
      (3) Peristalsis transports food in the esophagus to the stomach.
   b. Swallowing reflexes momentarily inhibit breathing.
3. Esophagus
   a. The **esophagus** passes through the mediastinum and penetrates the diaphragm.
   b. The **lower esophageal sphincter,** at the distal end of the esophagus, help prevent regurgitation of food from the stomach.

17.5 Stomach (page 661)

The **stomach** receives food, mixes it with gastric juice, carries on a limited amount of absorption, and moves food into the small intestine.

1. Parts of the stomach
   a. The stomach is divided into the cardia, fundus, body, and pylorus.
   b. The **pyloric sphincter** serves as a valve between the stomach and the small intestine.
2. Gastric secretions
   a. **Gastric glands** secrete gastric juice.
   b. **Gastric juice** contains **pepsin** (begins digestion of proteins), hydrochloric acid, lipase, and **intrinsic factor.**
3. Regulation of gastric secretions
   a. Parasympathetic impulses and the hormone **gastrin** enhance gastric secretion.

b. The three stages of gastric secretion are the cephalic, gastric, and intestinal phases.

c. The presence of food in the small intestine reflexly inhibits gastric secretions.

4. Gastric absorption

   a. The stomach is not well adapted for absorption.

   b. A few substances such as water and other small molecules are absorbed through the stomach wall.

5. Mixing and emptying actions

   a. As the stomach fills, its wall stretches, but its internal pressure remains unchanged.

   b. Mixing movements aid in producing chyme; peristaltic waves move **chyme** into the pylorus.

   c. The muscular wall of the pylorus regulates chyme movement into the small intestine.

   d. The rate of emptying depends on the fluidity of the chyme and the type of food present.

   e. The upper part of the small intestine fills, and an **enterogastric reflex** inhibits peristalsis in the stomach.

   f. Vomiting results from a complex reflex that has many stimuli.

17.6 Pancreas (page 668)

The **pancreas** is closely associated with the duodenum.

1. Structure of the pancreas

   a. It produces pancreatic juice secreted into a pancreatic duct.

   b. The pancreatic duct leads to the duodenum.

2. Pancreatic juice

   a. Pancreatic juice contains enzymes that can split carbohydrates, proteins, fats, and nucleic acids.

   b. Pancreatic juice has a high bicarbonate ion concentration that helps neutralize chyme and causes the intestinal contents to be alkaline.

3. Regulation of pancreatic secretion

   a. **Secretin** from the duodenum stimulates the release of pancreatic juice that contains few digestive enzymes but has a high bicarbonate ion concentration.

   b. Cholecystokinin from the intestinal wall stimulates the release of pancreatic juice that has a high concentration of digestive enzymes.

17.7 Liver (page 669)

The **liver** is located in the upper-right quadrant of the abdominal cavity.

1. Liver structure

   a. The liver is a highly vascular organ, enclosed in a fibrous capsule, and divided into lobes.

   b. Each lobe consists of **hepatic lobules,** the functional units of the liver.

   c. Bile from the lobules is carried by bile ductules to **hepatic ducts** that unite to form the **common hepatic duct.**

2. Liver functions

   a. The liver has many functions. It metabolizes carbohydrates, lipids, and proteins; stores some substances; removes toxic substances from the blood (detoxifies); and secretes bile.

   b. Bile is the only liver secretion that directly affects digestion.

3. Composition of bile

   a. **Bile** contains bile salts, bile pigments, cholesterol, and electrolytes.

   b. Only the bile salts have digestive functions.

   c. Bile pigments are products of red blood cell breakdown.

4. Gallbladder

   a. The **gallbladder** stores bile between meals.

   b. A sphincter muscle controls release of bile from the **bile duct.**

   c. Gallstones may form within the gallbladder.

5. Regulation of bile release

   a. Cholecystokinin from the small intestine stimulates bile release.

   b. The sphincter muscle at the base of the bile duct relaxes as a peristaltic wave in the duodenal wall approaches.

6. Functions of bile salts

   a. Bile salts emulsify fats and aid in the absorption of fatty acids, cholesterol, and certain vitamins.

   b. Bile salts are reabsorbed in the small intestine.

17.8 Small Intestine (page 676)

The **small intestine** extends from the pyloric sphincter to the large intestine. It receives secretions from the pancreas and liver, completes digestion of nutrients, absorbs the products of digestion, and transports the residues to the large intestine.

1. Parts of the small intestine

   a. The small intestine consists of the **duodenum, jejunum,** and **ileum.**

   b. The small intestine is suspended from the posterior abdominal wall by mesentery.

2. Structure of the small intestinal wall

   a. The wall is lined with villi that greatly increase the surface area and aid in mixing and absorption.

   b. Microvilli on the free ends of epithelial cells increase the surface area even more.

   c. Intestinal glands are located between the villi.

   d. Circular folds in the lining of the intestinal wall also increase its surface area.

3. Secretions of the small intestine

   a. **Intestinal glands** secrete a watery fluid that lacks digestive enzymes but provides a vehicle for moving chyme to the villi.

   b. Digestive enzymes embedded in the surfaces of microvilli split molecules of sugars, proteins, and fats.

4. Regulation of small intestinal secretions

   Secretion is stimulated by chyme and parasympathetic reflexes stimulated by distension of the small intestinal wall.

5. Absorption in the small intestine

   a. Blood capillaries in the villi also absorb water and electrolytes.

   b. Blood capillaries in the villi absorb monosaccharides and amino acids.

   c. Fatty acids diffuse into small intestinal epithelial cells where they are processed to form chylomicrons, which enter the lacteals of the villi.

6. Movements of the small intestine

   a. Movements include mixing by segmentation and peristalsis.

   b. Overdistension or irritation may stimulate a peristaltic rush and result in diarrhea.

   c. The **ileocecal sphincter** controls movement of the intestinal contents from the small intestine into the large intestine.

17.9 Large Intestine (page 683)

The **large intestine** absorbs water and electrolytes and forms and stores feces.

1. Parts of the large intestine
   a. The large intestine consists of the **cecum,** colon, rectum, and anal canal.
   b. The **colon** is divided into **ascending, transverse, descending,** and **sigmoid** portions.
2. Structure of the large intestinal wall
   a. The large intestinal wall resembles the wall in other parts of the alimentary canal.
   b. The large intestinal wall has a unique layer of longitudinal muscle fibers, arranged in distinct bands, that extend the entire length of the colon.
3. Functions of the large intestine
   a. The large intestine has little or no digestive function, although it secretes mucus.
   b. Mechanical stimulation and parasympathetic impulses control the rate of mucus secretion.
   c. The large intestine absorbs water and electrolytes.
   d. Many bacteria inhabit the large intestine, where they break down some undigestible substances, such as cellulose, and synthesize vitamins K, B_{12}, thiamine, and riboflavin.

4. Movements of the large intestine
   a. Movements are similar to those in the small intestine.
   b. Mass movements occur two to three times each day.
   c. A reflex stimulates defecation.
5. Feces
   a. The large intestine forms and stores **feces.**
   b. Feces consist of water, undigested material, mucus, and bacteria.
   c. The color of feces is due to bile pigments that have been altered by bacterial actions.

17.10 Life-Span Changes (page 688)

1. Older people sometimes do not chew food thoroughly because thinning enamel makes teeth more sensitive to hot and cold foods, gums recede, and teeth may loosen.
2. Slowing peristalsis in the digestive tract may cause heartburn and constipation.
3. Aging affects nutrient absorption in the small intestine.
4. Accessory organs to digestion also age, but not necessarily in ways that affect health.

CHAPTER ASSESSMENTS

17.1 Overview of the Digestive System

1 Functions of the digestive system include _____. (p. 650)
  a. ingestion
  b. mechanical and chemical digestion
  c. absorption
  d. all of the above
2 List the major parts of the alimentary canal, then separately list the accessory organs of the digestive system. (p. 650)
3 Contrast the composition of the four layers in the wall of the alimentary canal. (p. 650)
4 Distinguish between mixing movements and propelling movements. (pp. 650–651)
5 Define *peristalsis*. (p. 651)
6 Explain the relationship between peristalsis and receptive relaxation. (p. 651)
7 Describe the effects of parasympathetic and sympathetic impulses on the alimentary canal. (p. 651)

17.2 Mouth

8 Discuss the functions of the mouth and its parts. (pp. 653–656)
9 Distinguish among lingual, palatine, and pharyngeal tonsils. (pp. 653–654)
10 Compare the primary and secondary teeth. (pp. 655–656)
11 The teeth best adapted for grasping and tearing food are the _____. (p. 656)
  a. incisors
  b. canines
  c. premolars
  d. molars
12 Describe the structure of a tooth. (pp. 656–657)
13 Explain how a tooth is anchored in its socket. (p. 657)

17.3–17.9 Salivary Glands–Large Intestine

14 Discuss the digestive functions of saliva. (p. 659)
15 Describe the locations of the major salivary glands. (p. 659)
16 Explain how secretions of the salivary glands differ. (p. 659)
17 Name and locate the three major regions of the pharynx. (p. 660)
18 Discuss the mechanism of swallowing. (pp. 660–661)

19 Explain the function of the esophagus. (p. 661)
20 Describe the structure of the stomach. (pp. 661 and 663)
21 List the components of gastric juice, and describe the function of each component. (pp. 663 and 665)
22 Explain how gastric secretions are regulated. (pp. 665–661)
23 Describe the mechanism that controls the emptying of the stomach. (p. 667)
24 Describe the enterogastric reflex. (p. 667)
25 Explain the mechanism of vomiting. (p. 667)
26 Describe the location of the pancreas and the pancreatic duct. (p. 668)
27 Explain how pancreatic secretions are regulated. (p. 669)
28 Describe the structure of the liver. (pp. 669–670)
29 Trace the path of bile from a bile canaliculus to the small intestine by filling in the blanks: bile canaliculi carry secretions from hepatic cells to _____, which unite to form intrahepatic bile ducts, which converge to become the _____ that merge to form the common hepatic duct, which joins the cystic duct to form the _____, which empties into the small intestine (duodenum). (pp. 670 and 674)
30 List the major functions of the liver. (pp. 670–673)
31 Describe the composition of bile. (p. 673)
32 Explain how gallstones form. (p. 674)
33 Explain the regulation of bile release. (pp. 674–675)
34 Discuss the functions of bile salts. (p. 675)
35 Describe the locations of the parts of the small intestine. (p. 676)
36 Describe the functions of the intestinal villi. (pp. 678–679)
37 Match the organ or gland with the enzyme(s) it secretes. Enzymes may be used more than once. An organ or gland may secrete more than one enzyme. (pp. 658–680)

(1) salivary glands (serous cells)	A. peptidase
(2) stomach (chief cells)	B. amylase
(3) pancreas (acinar cells)	C. nuclease
(4) small intestine (mucosal cells)	D. lipase
	E. pepsin
	F. trypsin, chymotrypsin, carboxypeptidase
	G. sucrase, maltase, lactase

38 Match the enzyme(s) with its (their) function(s). Functions may be used more than once. (pp. 658–680)

(1) peptidase
(2) amylase
(3) nuclease
(4) lipase
(5) pepsin
(6) trypsin, chymotrypsin
(7) carboxypeptidase
(8) sucrase, maltase, lactase

A. begins protein digestion
B. breaks fats into fatty acids and glycerol
C. breaks down proteins into peptides
D. breaks down starch and glycogen into disaccharides
E. breaks down peptides into amino acids
F. breaks down nucleic acids into nucleotides
G. breaks down disaccharides into monosaccharides

39 Explain the regulation of the secretions of the small intestine. (p. 680)
40 Discuss absorption of amino acids, monosaccharides, glycerol, fatty acids, electrolytes, and water from substances in the small intestine. (pp. 681–682)
41 Explain control of the movement of the intestinal contents. (p. 683)
42 Describe the locations of the parts of the large intestine. (pp. 683–684)
43 Explain the general functions of the large intestine. (pp. 685–686)
44 Explain the defecation reflex. (p. 686)

17.10 Life-Span Changes

45 How does digestive function change with age? (p. 688)
46 What are the effects of altered rates of absorption, due to aging, in the small intestine? (p. 688)

 # INTEGRATIVE ASSESSMENTS/CRITICAL THINKING

Outcomes 17.3, 17.6, 17.8, 17.9

1. Several years ago, an extract from kidney beans was sold in health-food stores as a "starch blocker." Advertisements claimed that one could eat a plate of spaghetti, yet absorb none of it, because starch-digesting enzyme function would be blocked. The kidney bean product indeed kept salivary amylase from functioning. However, people who took the starch blocker developed abdominal pain, bloating, and gas. Suggest a reason for these ill effects.

Outcomes 17.5, 17.8

2. How would removal of 95% of the stomach (subtotal gastrectomy) to treat severe ulcers or cancer affect digestion and absorption? How would the patients have to alter eating habits? Why? Do you think that people should have this type of surgery to treat life-threatening obesity?

3. What effect is a before-dinner alcoholic cocktail likely to have on digestion? Why are such beverages inadvisable for people with ulcers?

4. What type of acid-base imbalance is likely to develop if the stomach contents are repeatedly lost by vomiting over a prolonged period? Which acid-base imbalance may develop as a result of prolonged diarrhea?

Outcomes 17.6, 17.7

5. Why might a person with inflammation of the gallbladder (cholecystitis) also develop inflammation of the pancreas (pancreatitis)?

 # ONLINE STUDY TOOLS

Connect Interactive Questions Reinforce your knowledge using assigned interactive questions covering the anatomy of the digestive organs, and the functions of the digestive enzymes.

Connect Integrated Activity Could you help a radiologist identify organs of the digestive system on a CT scan?

LearnSmart Discover which chapter concepts you have mastered and which require more attention. This adaptive learning tool is personalized, proven, and preferred.

Anatomy & Physiology Revealed Go more in depth into the human body by exploring the cadaver dissection of the digestive system, identifying organs on a CT scan, or viewing animations covering the stomach or liver.

18 Nutrition and Metabolism

People can satisfy nutritional requirements in a wide variety of ways.

LEARNING OUTCOMES

After you have studied this chapter, you should be able to:

18.1–18.3 Carbohydrates–Proteins

1   List the major sources of carbohydrates, lipids, and proteins. (pp. 695–699)

2   Describe how cells use carbohydrates, lipids, and proteins. (pp. 695–699)

3   Identify examples of positive and negative nitrogen balance. (p. 699)

18.4 Energy Expenditures

4   Explain how energy values of foods are determined. (p. 701)

5   Explain the factors that affect an individual's energy requirements. (pp. 701–702)

6   Contrast the physiological impact of positive and negative energy balance. (p. 702)

7   Explain what *desirable weight* means. (p. 702)

18.5 Appetite Control

8   Explain how hormones control appetite. (p. 704)

18.6 Vitamins

9   List the fat-soluble and water-soluble vitamins and summarize the general functions of each vitamin. (pp. 705–711)

18.7 Minerals

10 Distinguish between a vitamin and a mineral. (p. 711)

11 List the major minerals and trace elements and summarize the general functions of each. (pp. 713–717)

18.8 Healthy Eating

12 Describe an adequate diet. (p. 718)

13 Distinguish between primary and secondary malnutrition. (p. 720)

18.9 Life-Span Changes

14 List the factors that may lead to inadequate nutrition later in life. (pp. 723–724)

THE WHOLE PICTURE

The human body requires a constant supply of energy, and body structures must develop, grow, and occasionally be repaired. To accomplish all of this the body requires fuel as well as materials to serve as building blocks and to drive the body's metabolic processes, anabolism and catabolism. Nutrients from a balanced diet fulfill these requirements.

Nutrients are chemicals obtained from the environment that an organism requires for survival. There are two major classes of nutrients. The macronutrients, needed in larger quantities, are the carbohydrates, lipids, and proteins. Micronutrients are required in small daily amounts and include vitamins and minerals. Nutrients that human cells cannot synthesize and that must be obtained in the diet, such as certain amino acids, are called essential nutrients. Water, which the body cannot produce in sufficient amounts to survive, as well as minerals and oxygen, are additional examples of essential nutrients.

Module 12: Digestive System

bas-, base: *bas*al metabolic rate—metabolic rate of body under resting (basal) conditions.

calor-, heat: *calor*ie—unit used to measure heat or energy content of foods.

carot-, carrot: *carot*ene—yellowish plant pigment that imparts the color of carrots and other yellowish plant tissues.

lip-, fat: *lip*ids—fat or fatlike substance insoluble in water.

mal-, bad, abnormal: *mal*nutrition—poor nutrition resulting from lack of food or failure to adequately use available foods.

-meter, measure: calori*meter*—instrument used to measure the caloric content of food.

nutri-, nourish: *nutri*ent—substance needed to nourish cells.

obes-, fat: *obes*ity—condition in which the body has excess fat.

pell-, skin: *pell*agra—vitamin deficiency condition characterized by inflammation of the skin and other symptoms.

18.1 | Carbohydrates

Carbohydrates are organic compounds and include the sugars and starches. The energy held in their chemical bonds is used to power cellular processes.

Carbohydrate Sources

Carbohydrates are ingested in a variety of forms. Complex carbohydrates include the *polysaccharides* ("many sugars"), such as starch from plant foods and glycogen from meats. Most foods containing starch and glycogen have many other nutrients, including valuable vitamins and minerals. The simple carbohydrates include *disaccharides* ("double sugars") from milk sugar, cane sugar, beet sugar, and molasses and *monosaccharides* ("single sugars") from honey and fruits. Digestion ultimately breaks complex carbohydrates down to monosaccharides, which are small enough to be absorbed into the bloodstream.

> Sugar substitutes provide concentrated sweetness, so fewer calories are needed to sweeten a food compared to table sugar (sucrose). Stevia is extracted from leaves of an herb, and is 30 times as sweet as table sugar. Aspartame, a dipeptide, is 200 times as sweet; the artificial sweetener saccharin is 300 times as sweet; and sucralose is 600 times as sweet as sucrose. Sucralose is derived from sucrose, and includes chloride.

Cellulose is a complex carbohydrate abundant in our food—it provides the crunch to celery and the crispness to lettuce. We cannot digest cellulose, and most of it passes through the alimentary canal largely unchanged. However, cellulose provides bulk (also called fiber or roughage) against which the muscular wall of the digestive system can push, facilitating the movement of intestinal contents. *Hemicellulose, pectin,* and *lignin* are other plant carbohydrates that provide fiber.

Carbohydrate Use

The monosaccharides absorbed from the digestive tract include *fructose, galactose,* and *glucose.* Liver enzymes catalyze reactions that convert fructose and galactose into glucose (fig. 18.1). Recall that glucose is the carbohydrate most commonly oxidized in glycolysis for cellular fuel.

RECONNECT

To Chapter 4, Cellular Respiration, pages 127–128.

Some excess glucose is polymerized to form *glycogen* (glycogenesis), which the liver and muscles store as a glucose reserve. Glycogen can be rapidly broken down to yield glucose (glycogenolysis) when it is required to supply energy. However, the body can store only a certain amount of glycogen, and excess glucose reacts to form fat (lipogenesis), which is stored in adipose tissue (fig. 18.2).

Many cells can also oxidize fatty acids to obtain energy. However, some cells, such as neurons, normally require a continuous supply of glucose for survival. (Under some conditions, such as prolonged starvation, other fuel sources may become available for neurons.) Even a temporary decrease in the glucose supply

CAREER CORNER
Registered Dietitian

The young woman has decided to have weight loss surgery, and one of the first steps to prepare for the procedure is to lose 10% of her body weight. A registered dietitian (RD) helps the patient do this by discussing the types of foods that a person with type 2 diabetes mellitus should and should not eat. The RD uses plastic models of foods to demonstrate the makeup of appropriate meals, and then helps the patient develop a meal plan that will start her on the road to success, and also be enjoyable.

Dietitians work in diverse settings. These include businesses, hotels, food-service corporations, community agencies, schools, senior centers, health-care facilities, prisons, and restaurants. Registered dietitians can also have careers in research or teaching other health-care professionals, or open their own businesses. Educating the public about healthful food choices can be a major part of the job.

Training to become a registered dietitian includes earning a bachelor's degree in an accredited program, completing a supervised practice program typically lasting six to twelve months, and passing a national exam. Continuing education courses are necessary throughout the career of a registered dietitian to stay up to date with new nutritional findings and medical practices related to diet, and changing government recommendations.

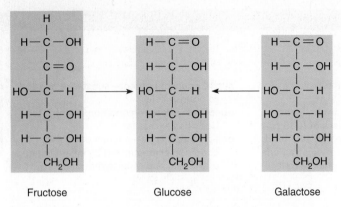

FIGURE 18.1 Liver enzymes catalyze reactions that convert the monosaccharides fructose and galactose into glucose.

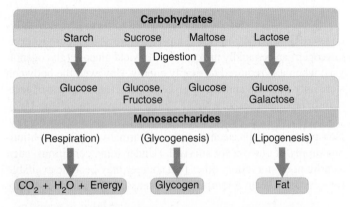

FIGURE 18.2 AP|R Monosaccharides from foods are used for energy, stored as glycogen, or reacted to produce fat.

may seriously impair nervous system function. Consequently, the body requires a minimum amount of carbohydrate. If foods do not provide an adequate supply of carbohydrates, the liver may convert some noncarbohydrates, such as amino acids from proteins or glycerol from fats, into glucose—a process called *gluconeogenesis.* The requirement for glucose has physiological priority over the need to synthesize certain other substances, such as proteins, from available amino acids.

Cells also use carbohydrates as starting materials for synthesizing such vital biochemicals as the 5-carbon sugars *ribose* and *deoxyribose,* which are the sugars required for the production of the nucleic acids RNA and DNA. The disaccharide *lactose* (milk sugar) is synthesized when the breasts are actively secreting milk.

Carbohydrate Requirements

Carbohydrates provide the primary source of fuel for cellular processes. The need for carbohydrates varies with individual energy requirements. Physically active individuals require more carbohydrates than those who are sedentary. The minimal requirement for carbohydrates in the human diet is unknown, but getting at least 125 to 175 grams daily is probably necessary to avoid protein breakdown and to avoid metabolic disorders resulting from excess fat use. An average diet includes 200 to 300 grams of carbohydrates daily.

An adult's liver stores about 100 grams of glycogen, and muscle tissue stores another 200 grams, providing enough reserve to meet energy demands for about twelve hours when the person is resting. Whether these stores are filled depends on diet. People consume widely varying amounts of carbohydrates, often reflecting economic conditions. In the United States, a typical adult's diet supplies about 50% of total body energy from carbohydrates. In Asian countries where rice is a staple, carbohydrates contribute even more to the diet.

 PRACTICE

1 List several common sources of carbohydrates.
2 Explain what happens to excess glucose in the body.
3 Name two uses of carbohydrates other than supplying energy.
4 Why do daily requirements for carbohydrates vary from person to person?

18.2 | Lipids

Lipids are organic compounds that include fats, oils, and fat-like substances such as phospholipids and cholesterol (see chapter 2, pp. 71–72). They supply energy for cellular processes and help build structures, such as cell membranes. The most common dietary lipids are the fats called *triglycerides* (tri-glis′er-īdz), consisting of three fatty acid molecules and a glycerol molecule (see fig. 2.14, p. 71).

Lipid Sources

Triglycerides are found in plant- and animal-based foods. Saturated fats contain no double bonds between the carbons of their fatty acid molecules and are mainly found in foods of animal origin, such as meat, eggs, milk, and lard, as well as in palm and coconut oils. Unsaturated fats contain fatty acid molecules with double bonds in their carbon chains and are in seeds, nuts, and plant oils. Monounsaturated fats (in which fatty acids contain one double bond), such as those in olive, peanut, and canola oils, are the healthiest. Saturated fats in excess are a risk factor for cardiovascular disease.

Cholesterol is abundant in liver and egg yolk and, to a lesser extent, in whole milk, butter, cheese, and meats. Foods of plant origin do not contain cholesterol. A label on a plant-based food claiming that it is "cholesterol-free" states the obvious.

Be wary of claims that a food product is "99% fat-free." This usually refers to percentage by weight—not calories, which is what counts. A 99% fat-free creamy concoction may be largely air and water, and therefore in that form, fat comprises very little of it. But when the air and the water are removed, as happens in the body, the fat percentage may skyrocket.

Lipid Use

The lipids in foods are phospholipids, cholesterol, and, most commonly, fats (triglycerides). Lipids provide a variety of physiological functions; however, fats mainly supply energy. Gram for gram, fats contain more than twice as much chemical energy as carbohydrates or proteins.

Before a triglyceride molecule can release energy, it must undergo hydrolysis. This happens when digestion breaks triglycerides down into fatty acids and glycerol. After being absorbed, these products are carried by the lymph to the blood, then on to tissues. As figure 18.3 shows, some of the resulting fatty acid portions can then form molecules of acetyl coenzyme A (acetyl CoA) by a series of reactions called **beta oxidation,** which occurs in the mitochondria.

In the first phase of beta oxidation, fatty acids are activated. This change requires energy from ATP and a special group of enzymes called thiokinases. Each of these enzymes can act upon a fatty acid that has a particular carbon chain length.

Once fatty acid molecules have been activated, other enzymes called **fatty acid oxidases** in mitochondria break them down. This phase of the reactions removes successive two-carbon segments of fatty acid chains. In the liver, some of these segments react to produce acetyl coenzyme A molecules. Excess acetyl CoA molecules can be converted into compounds called **ketone bodies,** such as acetone, which later may be changed back to acetyl coenzyme A. In either case, the resulting acetyl coenzyme A can be oxidized in the citric acid cycle. The glycerol parts of the triglyceride molecules can also enter metabolic pathways leading to the citric acid cycle, or they can be used to synthesize glucose.

When ketone bodies form faster than they can be decomposed, some of them are eliminated through the lungs and kidneys. When this happens, the ketone acetone may impart a fruity odor to the breath and urine. Ketones are released when a person fasts, forcing body cells to metabolize a large amount of fat. People with diabetes mellitus may develop a serious imbalance in pH called ketoacidosis when acetone and other acidic ketones accumulate.

Glycerol and fatty acid molecules resulting from the hydrolysis of fats can also combine to form fat molecules in anabolic reactions and be stored in fat tissue. Additional fat molecules can be synthesized from excess glucose or amino acids.

The liver can convert fatty acids from one form to another. However, the liver cannot synthesize certain fatty acids, called **essential fatty acids,** which must be obtained in the diet. *Linoleic acid,* for example, is an essential fatty acid required to synthesize phospholipids, which, in turn, are necessary for constructing cell membranes and myelin sheaths, and for transporting circulating lipids. Good sources of linoleic acid include corn oil, cottonseed oil, and soy oil. *Linolenic acid* is another essential fatty acid.

The liver uses free fatty acids to synthesize triglycerides, phospholipids, and lipoproteins that may then be released into the bloodstream (fig. 18.4). These lipoproteins are large and consist of a surface layer of phospholipid, cholesterol, and protein surrounding a triglyceride core. The protein constituents of lipoproteins in the outer layer, called *apoproteins* or apolipoproteins, can combine with receptors on the membranes of specific target cells. Lipoprotein molecules vary in the proportions of the lipids they contain.

Lipids are less dense than proteins. As the proportion of lipids in a lipoprotein increases, the density of the particle decreases. Conversely, as the proportion of lipids decreases, the density increases. Lipoproteins are classified on the basis of their densities, which reflect their composition. *Very-low-density lipoproteins* (VLDL) have a high concentration of triglycerides. *Low-density lipoproteins* (LDL) have a high concentration of cholesterol and are the major cholesterol-carrying lipoproteins. *High-density lipoproteins* (HDL) have a relatively high concentration of protein and a lower concentration of lipids.

In addition to regulating circulating lipids, the liver controls the total amount of cholesterol in the body by synthesizing and releasing it into the blood or by removing cholesterol from the blood and excreting it into the bile. The liver uses cholesterol to produce bile salts. Cholesterol is not used as an energy source, but it does provide structural material for cell and organelle membranes, and it furnishes starting materials for the synthesis of certain sex hormones and hormones produced by the adrenal cortex.

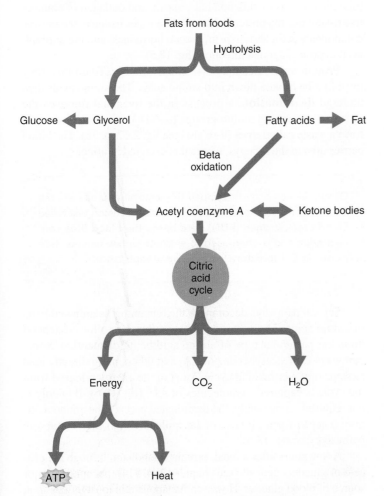

FIGURE 18.3 The body digests fat from foods into glycerol and fatty acids, which may enter catabolic pathways and provide energy.

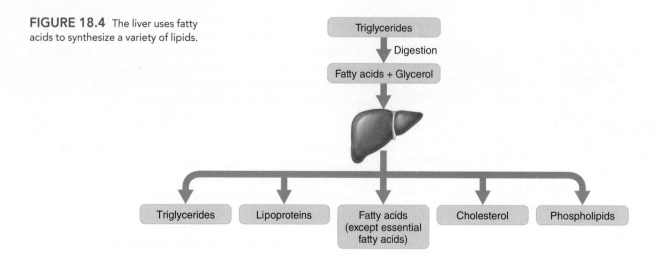

FIGURE 18.4 The liver uses fatty acids to synthesize a variety of lipids.

Adipose tissue stores excess triglycerides. If the blood lipid concentration drops (in response to fasting, for example), some of these stored triglycerides are hydrolyzed into free fatty acids and glycerol and are released into the bloodstream.

Lipid Requirements

The lipid content of human diets varies widely. A person who eats mostly burgers, fries, and shakes may consume 50% or more of total daily calories from fat. For a vegetarian, the percentage may be far lower. The USDA and the American Heart Association recommend that lipid intake not exceed 30% of the total daily calories.

> The types and locations of the chemical bonds between carbon atoms of fatty acid molecules affect how healthful the fat is. Monounsaturated fats (such as from avocado and olives) promote cardiovascular health, whereas saturated fats (such as those in butter or lard) raise the risk of heart disease. The site of the double bond that contributes to a fat's degree of unsaturation is also important. Omega-3 fatty acids, which have double bonds between the third and fourth carbons, are more healthful than omega-6 fatty acids, with double bonds between the sixth and seventh carbons. Omega-3 fatty acids are found in fish; omega-6 fatty acids are in red meat.

The amounts and types of fats required for health vary with individuals' habits, goals, and stage of life. For example, infant formula contains two long-chain polyunsaturated fatty acids (docosahexaenoic acid and arachidonic acid) that are in breast milk and are vital for development of the infant's nervous system, particularly the eyes. Fat intake must be sufficient throughout life to support absorption and transport of fat-soluble vitamins. Most adults who eat a variety of foods obtain adequate fats.

 PRACTICE

5 Which foods commonly supply lipids?
6 Which fatty acids are essential nutrients?
7 What is the role of the liver in the use of lipids?
8 What are the functions of cholesterol?

18.3 | Proteins

Proteins are composed of amino acids. They have a wide variety of functions. When dietary proteins are digested, the resulting amino acids are absorbed and carrried by the blood to cells. Many of these amino acids are used to form new protein molecules, as specified by DNA base sequences. These new proteins include enzymes that control the rates of metabolic reactions; clotting factors; the keratins of skin and hair; elastin and collagen of connective tissue; plasma proteins that regulate water balance; the muscle components actin and myosin; certain hormones; and the antibodies that protect against infection (fig. 18.5).

Protein molecules may also supply energy. To do this, they must first be broken down into amino acids. The amino acids then undergo **deamination,** a process in the liver that removes the nitrogen-containing amino groups ($—NH_2$), which then react to form a waste called **urea** (u-re′ah) (see fig. 2.17, p. 73). The blood carries urea to the kidneys, where it is excreted in urine.

> Certain kidney disorders impair the removal of urea from the blood, raising the blood urea concentration. A blood test called blood urea nitrogen (BUN) determines the blood urea concentration and is often used to evaluate kidney function (see Appendix C, Laboratory Tests of Clinical Importance).

Several pathways decompose the remaining deaminated parts of amino acids. The specific pathways that are followed depend upon the particular type of amino acid being dismantled. Some pathways form acetyl coenzyme A, and others more directly lead to steps of the citric acid cycle. Most of the energy released from the cycle is captured in molecules of ATP (fig. 18.6). If energy is not required immediately, the deaminated parts of the amino acids may react to form glucose or fat molecules in other metabolic pathways (see fig. 18.5).

A few hours after a meal, protein catabolism, through the process of gluconeogenesis (see chapter 13, p. 514), becomes a major source of blood glucose. However, metabolism in most tissues soon shifts away from glucose and toward fat catabolism as a source of ATP. Thus, energy needs are met in a way that spares proteins for

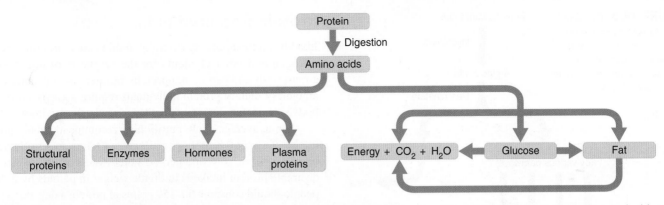

FIGURE 18.5 Proteins are digested to their constituent amino acids. These amino acids are then linked, following genetic instructions, to build new proteins. Free amino acids are also used to supply energy under certain conditions.

tissue building and repair, rather than being broken down and reassembled into carbohydrates to supply energy. Using structural proteins to generate energy causes the tissue-wasting of starvation.

Protein Sources

Foods rich in proteins include meats, seafood, poultry, cheese, nuts, milk, eggs, and cereals. Legumes, including beans and peas, contain less protein.

The human body can synthesize many amino acids (nonessential amino acids). However, eight amino acids the adult body needs (ten required for growing children) cannot be synthesized sufficiently or at all, and they are called **essential amino acids.** This term refers only to dietary intake, because all amino acids are required for normal protein synthesis. Table 18.1 lists the amino acids in foods and indicates those that are essential.

All twenty types of amino acids must be in the body at the same time for growth and tissue repair to occur. In other words, if the diet lacks one essential amino acid, the cells cannot synthesize protein. Essential amino acids are not stored. Those not used to make proteins are oxidized as energy sources or are converted into carbohydrates or fats.

Proteins are classified as complete or incomplete based on the amino acid types they provide. **Complete proteins** have adequate amounts of the essential amino acids to maintain human body tissues and promote normal growth and development. Certain proteins in milk, meat, and eggs are complete. **Incomplete proteins**

lack one or more of the essential amino acids, and cannot by themselves maintain human tissues or support normal growth and development. Zein in corn, for example, has too little of the essential amino acids tryptophan and lysine to be complete. A *partially complete protein* does not have enough amino acid variety to promote growth, but it has enough to maintain life. A protein in wheat called gliadin is a partially complete protein because it has very little of the amino acid lysine.

> Many plant proteins have too little of one or more essential amino acids to provide adequate nutrition for a person. However, combining appropriate plant foods can supply an adequate diversity of dietary amino acids. For example, beans are low in methionine but have enough lysine. Rice lacks lysine but has enough methionine. A meal of beans and rice offers enough of both types of amino acids.

 PRACTICE

9 How do cells use proteins?

10 Which foods are rich sources of protein?

11 Why are some amino acids called essential?

12 Distinguish between a complete protein and an incomplete protein.

Nitrogen Balance

A healthy adult continuously builds up and breaks down proteins. Proteins are metabolized at different rates in different tissues, but the overall gain of body proteins equals the loss, producing a state of **dynamic equilibrium** (di-nam′ik e″kwĭ-lib′re-um). Because proteins have a high percentage of nitrogen, dynamic equilibrium also brings **nitrogen balance** (ni′tro-jen bal′ans), in which the amount of nitrogen taken in equals the amount excreted.

A person who is starving has a *negative nitrogen balance* because the amount of nitrogen excreted as a result of amino acid oxidation exceeds the amount the diet replaces. Conversely, a growing child, a pregnant woman, or an athlete in training is likely to have a *positive nitrogen balance* because more protein is being built into new tissue and less is being used for energy or excreted.

TABLE 18.1	Amino Acids in Foods	
Alanine	Glycine	Proline
Arginine (ch)	Histidine (ch)	Serine
Asparagine	Isoleucine (e)	Threonine (e)
Aspartic acid	Leucine (e)	Tryptophan (e)
Cysteine	Lysine (e)	Tyrosine
Glutamic acid	Methionine (e)	Valine (e)
Glutamine	Phenylalanine (e)	

Eight essential amino acids (e) cannot be synthesized by human cells and must be provided in the diet. Two additional amino acids (ch) are essential in growing children.

FIGURE 18.6 The body digests proteins from foods into amino acids, but must deaminate these smaller molecules before they can be used as energy sources.

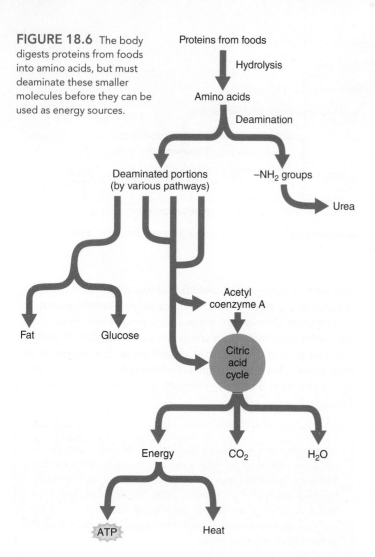

Proteins from foods
↓ Hydrolysis
Amino acids
↓ Deamination

Deaminated portions (by various pathways)

−NH_2 groups
→ Urea

Fat   Glucose

Acetyl coenzyme A
↓
Citric acid cycle
↓
Energy CO_2 H_2O
↓
ATP   Heat

Protein Requirements

In addition to supplying essential amino acids, proteins provide nitrogen and other elements for the synthesis of nonessential amino acids and certain nonprotein nitrogenous substances. The amount of dietary protein individuals require varies according to body size, metabolic rate, and nitrogen balance condition.

For an average adult, nutritionists recommend a daily protein intake of about 0.8 gram per kilogram (0.4 gram per pound) of body weight or 10% of a person's diet. Another way to estimate desirable protein intake is to divide weight in pounds by 2. Most people should consume 60–150 grams of protein a day. For a pregnant woman, who needs to maintain a positive nitrogen balance, the recommendation adds 30 grams of protein per day. Similarly, a nursing mother requires an additional 20 grams of protein per day to maintain milk production.

Protein deficiency causes tissue wasting and also decreases the level of plasma proteins, which decreases the colloid osmotic pressure of the plasma. As a result, fluids collect in the tissues, producing *nutritional edema*. Table 18.2 summarizes the sources, requirements, and uses for carbohydrate, lipid, and protein nutrients.

RECONNECT
To Chapter 15, Exchanges in the Capillaries, page 581.

PRACTICE

13 What is a negative nitrogen balance? A positive nitrogen balance?

14 How can inadequate nutrition cause edema?

TABLE 18.2	Carbohydrate, Lipid, and Protein Nutrients				
Nutrient	**Sources and RDA* for Adults**	**Calories per Gram**	**Use**	**Conditions Associated with**	
				Excesses	**Deficiencies**
Carbohydrate	Primarily from starch and sugars in foods of plant origin and from glycogen in meats 225–325 g	4.1	Oxidized for energy; used in production of ribose, deoxyribose, and lactose; stored in liver and muscles as glycogen; converted to fats and stored in adipose tissue	Obesity, dental caries, nutritional deficits	Metabolic acidosis
Lipid	Meats, eggs, milk, lard, plant oils 44–78 g	9.5	Oxidized for energy; production of triglycerides, phospholipids, lipoproteins, and cholesterol, stored in adipose tissue; glycerol portions of fat molecules may be used to synthesize glucose	Obesity, increased serum cholesterol, increased risk of heart disease	Weight loss, skin lesions
Protein	Meats, cheese, nuts, milk, eggs, cereals, legumes 50–175 g	4.1	Production of protein molecules used to build cell structure and to function as enzymes or hormones; used in the transport of oxygen, regulation of water balance, control of pH, formation of antibodies; amino acids may be broken down and oxidized for energy or converted to carbohydrates or fats for storage	Obesity	Extreme weight loss, wasting, anemia, growth retardation

*RDA = recommended dietary allowance (based on 2,000 calories per day).

18.4 | Energy Expenditures

Carbohydrates, fats, and proteins supply energy, which is required for all metabolic processes and therefore important to cell survival. If the diet is deficient in energy-supplying nutrients, structural molecules may gradually be consumed, leading to death. On the other hand, excess intake of energy-supplying nutrients may lead to obesity, which also threatens health.

Energy Values of Foods

The amount of potential energy a food contains is expressed as **calories** (kal′o-rēz), which are units of heat. Although a calorie is defined as the amount of heat required to raise the temperature of a gram of water by 1 degree Celsius (°C), the calorie used to measure food energy is 1,000 times greater. This *large calorie* (Cal.) equals the amount of heat required to raise the temperature of a kilogram (1,000 grams) of water by 1°C (from 14.5°C to 15.5°C) and is also equal to 4.184 joules. A joule is the international unit of heat and energy. A large calorie is also called a *kilocalorie,* but it is customary in nutritional studies to refer to it as a calorie.

Figure 18.7 shows a bomb calorimeter, which is a device used to measure the caloric contents of foods. It consists of a metal chamber submerged in a known volume of water. A food sample is dried, weighed, and placed in a nonreactive dish inside the chamber. The chamber is filled with oxygen and submerged in the water. Then, the food is ignited and allowed to completely oxidize. Heat released from the food raises the temperature of the surrounding water, and the change in temperature is measured. Because the volume of the water is known, the amount of heat released from the food can be calculated in calories.

Caloric values determined in a bomb calorimeter are somewhat higher than the amount of energy that metabolic oxidation releases, because nutrients generally are not completely absorbed from the digestive tract. Also, the body does not completely oxidize amino acids, but excretes parts of them in urea or uses them to synthesize other nitrogenous substances. When such losses are considered, cellular oxidation yields on the average about 4.1 calories from 1 gram of carbohydrate, 4.1 calories from 1 gram of protein, and 9.5 calories from 1 gram of fat. More than twice as much energy is derived from equal amounts by weight of fats as from either proteins or carbohydrates. This is one reason why avoiding fatty foods helps weight loss, if intake of other nutrients does not substantially increase. Fats encourage weight gain because they add flavor to food, which can cause overeating. However, fatty foods satisfy hunger longer than carbohydrate-rich foods.

 PRACTICE

15 What term designates the potential energy in a food?

16 What is the energy value of a gram of carbohydrate? a gram of protein? a gram of fat?

Energy Requirements

The amount of energy required to support metabolic activities for twenty-four hours varies from person to person. The factors that influence individual energy needs include a measurement called the basal metabolic rate, the degree of muscular activity, body temperature, and rate of growth.

The **basal metabolic rate** (ba′sal met″ah-bol′ik rāt), or BMR, measures the rate at which the body expends energy under *basal conditions*—when a person is awake and at rest; after an overnight fast; and in a comfortable, controlled environment. Tests of thyroid function can be used to estimate a person's BMR.

The amount of oxygen the body consumes is directly proportional to the amount of energy released by cellular respiration. The BMR indicates the total amount of energy expended in a given time to support the activities of such organs as the brain, heart, lungs, liver, and kidneys.

The BMR for an average adult indicates a requirement for approximately 1 calorie of energy per hour for each kilogram of body weight. However, this requirement varies with sex, body size, body temperature, and level of endocrine gland activity. For example, because heat loss is directly proportional to the body surface area, and a smaller person has a greater surface area relative to body mass, he or she will have a higher BMR. Males typically have higher metabolic rates than females. As body temperature, blood level of thyroxine, or blood level of epinephrine increase, so does the BMR. The BMR can also increase when the level of physical activity increases during the day.

Maintaining the BMR usually requires the body's greatest expenditure of energy. The energy required to support voluntary muscular activity comes next, though this amount varies greatly with the type of activity (table 18.3). For example, the energy to maintain posture while sitting at a desk might require 100 calories per hour above the basal need, whereas running or swimming might require 500–600 calories per hour.

Maintenance of body temperature may require additional energy expenditure, particularly in cold weather. In this case, extra

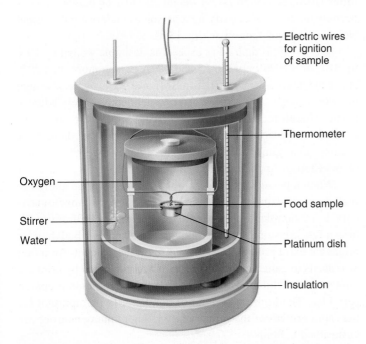

FIGURE 18.7 A bomb calorimeter measures the caloric content of a food sample.

Electric wires for ignition of sample

Thermometer

Oxygen

Food sample

Stirrer

Water

Platinum dish

Insulation

TABLE 18.3	Calories Used During Various Activities
Activity	**Calories (per Hour)**
Walking up stairs	1,100
Running (jogging)	570
Swimming	500
Vigorous exercise	450
Slow walking	200
Dressing and undressing	118
Sitting at rest	100

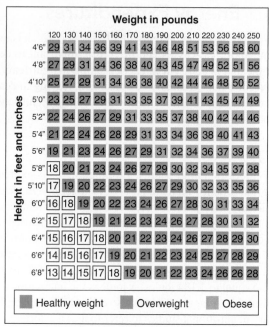

FIGURE 18.8 Body mass index (BMI). BMI can be calculated. In this chart the calculations have been done. The uncolored squares indicate lower than healthy weight according to this index.

What is your body mass index? Based on whether you are lower than healthy weight, healthy weight, overweight, or obese, what might you do with regards to your diet and activity level to achieve/maintain healthy weight?

Answer can be found in Appendix G.

energy is expended in involuntary muscular contractions, such as shivering, or through voluntary muscular actions, such as walking. Growing children and pregnant women, because their bodies are actively producing new tissues, also require more calories.

Energy Balance

A state of **energy balance** exists when caloric intake in the form of foods equals caloric expenditure from the basal metabolic rate and muscular activities. Under these conditions, body weight remains constant, except perhaps for slight variations due to changes in water content.

If, however, caloric intake exceeds expenditure, a *positive energy balance* occurs, and tissues store excess nutrients. This increases body weight because 3,500 excess calories are stored as a pound of fat. Conversely, if caloric expenditure exceeds input, the energy balance is negative, and stored materials are mobilized from the tissues for oxidation, causing weight loss. To maintain weight, calories in must equal calories expended.

Desirable Weight

The most obvious and common nutritional disorders reflect calorie imbalances, which may result from societal and geographic factors. Obesity is prevalent in nations where food is plentiful and diverse. The tendency to become obese may be a holdover from thousands of years ago, when the ability to store energy in the form of fat was a survival advantage when food supplies were scarce or erratic. Today in many African nations, natural famines combined with political unrest cause mass starvation. Starvation is considered later in the chapter.

It is difficult to determine a desirable body weight. In the past, weight standards were based on average weights and heights in a certain population, and the degrees of underweight and overweight were expressed as percentage deviations from these averages. These standards reflected the gradual gain in weight as people age. Then medical researchers recognized that such an increase in weight after the age of twenty-five to thirty years is not necessary and may not be healthy. This lead to developing standards of *desirable weights*. Today a measurement termed **body mass index** (BMI) is used to assess weight considering height, and has become the basis of classifying a person as underweight, normal weight, overweight, or obese (fig. 18.8).

Overweight is defined as exceeding desirable weight by 10% to 20%, or a BMI between 25 and 30. A person more than 20% above the desired weight or with a BMI over 30, is *obese,* although **obesity** (o-bēs′ĭ-te) is more correctly defined as excess adipose tissue. Therefore, overweight and obesity are not the same. For example, as figure 18.9 shows, an athlete or a person whose work requires heavy muscular activity may be overweight, but not obese. Clinical Application 18.1 discusses obesity.

When a person needs to gain weight, diet can be altered to include more calories and to emphasize particular macronutrients. For example, a person recovering from a debilitating illness might consume more carbohydrates, whereas a bodybuilder might eat extra protein to hasten muscle development. An infant also needs to gain weight rapidly, best accomplished by drinking human milk, which has more total carbohydrate than prepared formulas. The high fat content of human milk is important for the rapid growth of the infant's brain, where many neurons are ensheathed in lipids.

In the United States, obesity is common. Nearly a third of all adults are obese, defined as 20% above "ideal" weight based on population statistics considering age, sex, and build, or a body mass index above 30. Obesity raises risks for type 2 diabetes, digestive disorders, heart disease, kidney failure, hypertension, stroke, and cancers of the gallbladder and female reproductive organs. The body has to support the extra weight—miles of extra blood vessels are needed to nourish the additional pounds. Obesity is the second leading cause of preventable death, following cigarette smoking.

Obesity refers to extra pounds of fat. The proportion of fat in a human body ranges from 5% to more than 50%, with "normal" for males falling between 12% and 23% and for females between 16% and 28%. An elite athlete may have a body fat level as low as 4%. Fat distribution also affects health. Excess poundage above the waist is linked to increased risk of heart disease, type 2 diabetes, hypertension, and lipid disorders. The body mass index (BMI) accounts somewhat for a person's build (see fig. 18.8). A person who weighs 170 pounds and is 6 feet tall is slim, whereas a person of the same weight who is 5 feet tall is obese. The tall person's BMI is 23; the short person's is 33.5.

Both heredity and the environment contribute to obesity. Dozens of genes interact to control energy balance and therefore body weight. The observation that identical twins reared in different households can grow into adults of vastly different weights indicates that environment influences weight too. Even the environment before birth can affect body weight later. Individuals born at full term, but undernourished as fetuses, are at high risk of obesity. Physiological changes that countered starvation in the uterus cause obesity when they persist.

Certain genes encode proteins that connect sensations in the gastrointestinal tract with centers in the hypothalamus that control hunger and satiety. It is how we satisfy those signals—what we eat—that provides the environmental component to body weight. A certain set of gene variants may have led to a trim figure in a human many thousands of years ago, when food had to be hunted or gathered—and meat was leaner. Today those same gene variants do not foster slimness in a person who takes in many more calories than he or she expends.

Treatments for Obesity

Diet and Exercise

A pound of fat contains 3,500 calories of energy, so that pound can be shed by an appropriate combination of calorie cutting and exercise. This might mean eating 500 fewer calories per day or exercising to expend an extra 500 calories each day.

The National Heart, Lung, and Blood Institute recommends that overweight and obese individuals aim to lose 10% of their body weight over a six-month period. For people with BMIs from 27 to 35, that means a decrease of 300 to 500 calories per day, to lose one-half to one pound of body weight per week. For people with BMIs exceeding 35, a decrease of up to 500 to 1,000 calories per day will translate to a loss of one to two pounds of body weight per week.

Drug Therapy

Some physicians recommend drug therapy if the BMI exceeds 30 or if it exceeds 27 and the person also has hypertension, type 2 diabetes mellitus, or hyperlipidemia. Several types of "diet drugs" are no longer in use because they are dangerous. Amphetamines, for example, carried the risk of addiction, and the combination of fenfluramine and phentermine damaged heart valves.

Newer antiobesity drugs target fat in diverse ways. Tetrahydrolipostatin, marketed as Orlistat and Xenical, inhibits the function of pancreatic lipase, preventing the digestion and absorption of about a third of dietary fat, which is eliminated in loose feces. This effect is not disruptive as long as the person follows a low-fat diet. Future weight control drugs may manipulate appetite-control hormones, such as ghrelin and leptin.

Surgery

For people with BMIs above 40, or above 35 in addition to an obesity-related disorder, bariatric (weight loss) surgery can lead to weight loss. Three types of procedures are done. In laparoscopic adjustable gastric banding, a silicone band ties off part of the stomach, limiting its capacity to hold food. The band can be inflated or deflated in a doctor's office by adding or removing saline. The band may need to be removed if it slips out of place, it erodes the stomach lining, the body rejects it, or the port to add or remove fluid becomes displaced. In the second type of bariatric surgery, sleeve gastrectomy, a surgeon removes about 75% of the stomach, leaving a banana-shaped "sleeve." This sleeve procedure is irreversible. The third type of bariatric surgery is gastric bypass, in which part of the stomach is stapled shut, forming a pouch surgically connected to the jejunum, bypassing the duodenum.

All three bariatric procedures lead to decreased hunger, greatly reduced food intake, and some decrease in the absorption of nutrients. A special diet, liquid at first, must be followed. Many patients who have had bariatric surgery report improvement in or disappearance of type 2 diabetes, back pain, arthritis, varicose veins, sleep apnea, and hypertension. However, people can regain the weight and lose the benefits if they fail to adhere to the dietary restrictions. From Science to Technology 13.1 (pp. 518–519) discusses the effect of gastric bypass surgery on blood glucose control.

(a)

(b)

FIGURE 18.9 Weight. (*a*) An obese person is overweight and has excess adipose tissue. (*b*) An athlete may be overweight due to muscle overgrowth but is not considered obese. Many athletes have very low percentages of body fat.

 PRACTICE

17 What is the basal metabolic rate?

18 What factors influence the BMR?

19 What is energy balance?

20 Distinguish between being overweight and being obese.

18.5 | Appetite Control

We eat to obtain the nutrients that power the activities of life. Eating is a complex, finely tuned homeostatic mechanism that balances nutrient intake with nutrient use. Too few nutrients, and disorders associated with malnutrition result. Too many nutrients, and obesity is the consequence. Several factors influence food intake, including smell, taste, and texture of food; neural signals triggered by stretch receptors in the stomach; stress; and hormones.

Appetite is the drive that compels us to seek food. Several types of interacting hormones control appetite by affecting part of the hypothalamus called the *arcuate nucleus* (table 18.4).

Insulin, secreted from the pancreas, regulates fat stores by stimulating cells called adipocytes to take up glucose and store fat, and by stimulating certain other cells to take up glucose and link it to form glycogen, a storage carbohydrate. Eating stimulates adipocytes to secrete the hormone **leptin,** which acts on target cells in the hypothalamus. Leptin secretion suppresses appetite by inhibiting release of the hypothalamic neurotransmitter **neuropeptide Y,** which stimulates eating. The interaction between leptin and neuropeptide Y is a negative feedback response to ingesting calories. Leptin also stimulates metabolic rate (fig. 18.10). Low leptin levels indicate depleted fat stores, a condition in which metabolism slows to conserve energy and appetite increases. Inherited leptin deficiency is very rare, but the resultant loss of this appetite "brake" results in obesity. In this special case, leptin injections enable these individuals to reduce their weight.

Conversely, **ghrelin,** a hormone that the stomach secretes, enhances appetite by stimulating the release of neuropeptide Y. Therefore, a compound that blocks ghrelin production or activity might help people lose weight. Fasting and a low-calorie diet stimulate ghrelin production. The success of gastric bypass surgery may be due in part to decreased ghrelin secretion resulting from loss of stomach tissue.

Adiponectin is a protein hormone synthesized in adipose cells and secreted into the bloodstream that mediates response to insulin and regulates fatty acid catabolism. Adiponectin also has an anti-inflammatory effect. The hormone increases basal metabolic rate (calories burned at rest) without affecting appetite. Because levels of adiponectin are higher in people who do not have diabetes and have healthy body weights, some studies suggest boosting its levels may help in weight loss. Although companies sell adiponectin in pill form, it is a protein and is therefore broken down in the digestive tract before it can exert an effect. The hormone is marketed as a dietary supplement, not as a drug.

 PRACTICE

21 Describe how hormones control appetite.

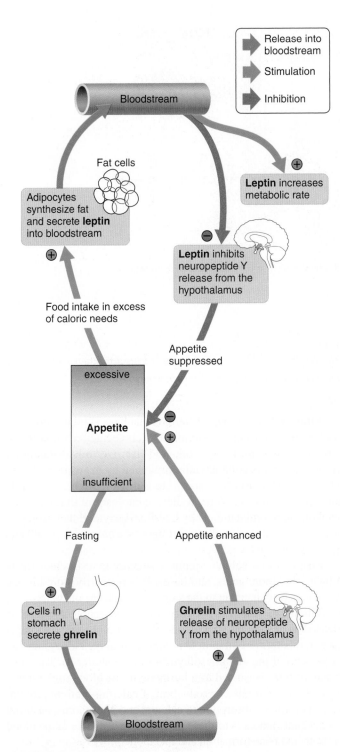

	Release into bloodstream
	Stimulation
	Inhibition

Bloodstream

Fat cells

Adipocytes synthesize fat and secrete **leptin** into bloodstream

⊕

Leptin increases metabolic rate

⊕

⊖

Leptin inhibits neuropeptide Y release from the hypothalamus

Food intake in excess of caloric needs

Appetite suppressed

excessive

Appetite

⊖
⊕

insufficient

Fasting

Appetite enhanced

⊕

Cells in stomach secrete **ghrelin**

Ghrelin stimulates release of neuropeptide Y from the hypothalamus

⊕

Bloodstream

FIGURE 18.10 Appetite control is complex. Illustrated here are the effects of leptin and ghrelin on appetite.

18.6 | Vitamins

Vitamins (vi′tah-minz) are organic compounds (other than carbohydrates, lipids, and proteins) required in small amounts for normal metabolism that body cells cannot synthesize in adequate amounts. Vitamins are essential nutrients that must come directly from foods or indirectly from **provitamins,** which are precursor substances.

TABLE 18.4	Substances That Control Appetite	
Substance	**Site of Secretion**	**Function**
Insulin	Pancreas	Stimulates adipocytes to admit glucose and store fat; glycogen synthesis
Leptin	Adipocytes	Suppresses appetite and increases metabolic rate after eating
Neuropeptide Y	Hypothalamus	Enhances appetite
Ghrelin	Stomach	Enhances appetite

Vitamins are classified on the basis of whether they are soluble in fats (or fat solvents) or in water. *Fat-soluble* vitamins are A, D, E, and K; the *water-soluble* group includes the B vitamins and vitamin C. Table 18.5 lists, and corrects, some common misconceptions about vitamins.

Fat-Soluble Vitamins

Fat-soluble vitamins dissolve in fats. Therefore they associate with lipids and are influenced by the same factors that affect lipid absorption. For example, bile salts in the intestine promote absorption of fat-soluble vitamins. The fat-soluble vitamins are stored in moderate quantities in various tissues, which is why excess intake can lead to overdose. These vitamins resist the effects of heat, so cooking and food processing do not usually destroy them.

 **PRACTICE**

22 What are vitamins?

23 How are vitamins classified?

24 How do bile salts affect the absorption of fat-soluble vitamins?

Vitamin A exists in several forms, including retinol and retinal (retinene). Body cells synthesize this vitamin from a group of yellowish plant pigments, which are provitamins called *carotenes* (fig. 18.11). Excess vitamin A or its precursors are mainly stored in the liver, which regulates their concentration in the body. An

TABLE 18.5	Vitamin Fallacies and Facts
Fallacy	**Fact**
The more vitamins, the better	Too much of a water-soluble vitamin results in excretion of the vitamin through urination; too much of a fat-soluble vitamin can harm health
A varied diet provides all needed vitamins	Many people do benefit from vitamin supplements, particularly pregnant and breast-feeding women
Vitamins provide energy	Vitamins do not directly supply energy; they aid in the release of energy from carbohydrates, fats, and proteins

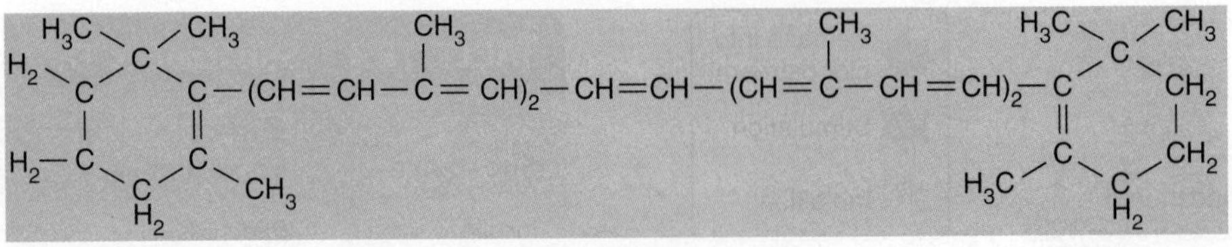

Beta carotene

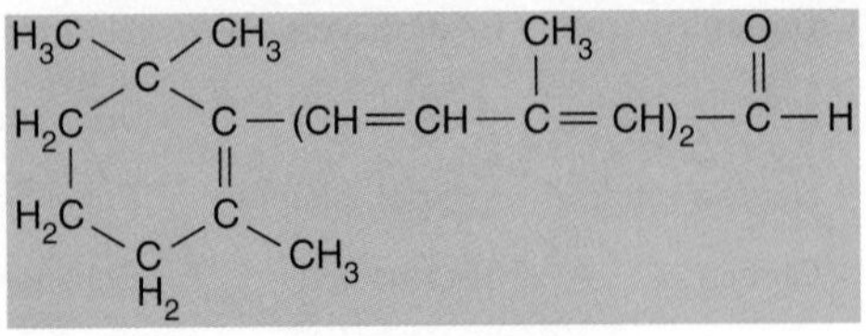

Retinal (retinene)

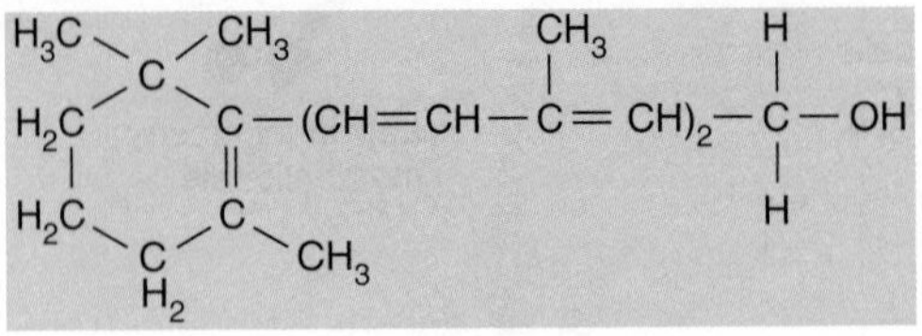

Retinol

FIGURE 18.11 A molecule of beta carotene can react to form two molecules of retinal, which, in turn, can react to form retinol.

adult's liver stores enough vitamin A to supply body requirements for a year. Infants and children usually lack such reserves and are therefore more likely to develop vitamin A deficiencies if their diets are inadequate.

Vitamin A is relatively stable to the effects of heat, acids, and bases. However, it is readily destroyed by oxidation and is unstable in light.

Vitamin A is important in vision. Retinal is used to synthesize *rhodopsin* (visual purple) in the rods of the retina and light-sensitive pigments in the cones (pp. 479–480). The vitamin also functions in the synthesis of mucoproteins and mucopolysaccharides, in development of normal bones and teeth, and in maintenance of epithelial cells in skin and mucous membranes. Vitamin A and beta carotenes also act as **antioxidants** (an'tĭ-ok'sĭ-dant) by readily combining with oxygen and certain oxygen-containing molecules that have unshared electrons, which makes them highly reactive and damaging to cellular structures. These unstable molecules are called oxygen free radicals, and they accumulate in certain diseases and with age.

RECONNECT
To Chapter 2, Bonding of Atoms, page 64.

Only foods of animal origin such as liver, fish, whole milk, butter, and eggs are sources of vitamin A. However, the vitamin's precursor, carotene, is widespread in leafy green vegetables and in yellow or orange vegetables and fruits.

Excess vitamin A produces the peeling skin, hair loss, nausea, headache, and dizziness of *hypervitaminosis A.* Chronic overdoses of the vitamin may inhibit growth and break down bones and joints. "Megadosing" on fat-soluble vitamins is particularly dangerous during pregnancy. Some forms of vitamin A, in excess, can cause birth defects.

A deficiency of vitamin A causes *night blindness,* in which a person cannot see normally in dim light. Xeropthalmia, a dryness of the conjunctiva and cornea, is due to vitamin A deficiency. Vitamin A deficiency also causes degenerative changes in certain epithelial tissues, and the body becomes more susceptible to infection.

PRACTICE

25 Which chemical in the body is the precursor to vitamin A?

26 Which foods are good sources of vitamin A?

Vitamin D is a group of steroids that have similar properties. One of these substances, vitamin D_3 (cholecalciferol), is found in foods such as milk, egg yolk, and fish liver oils. Vitamin D_2 (ergocalciferol) is commercially produced by exposing a steroid obtained from yeasts (ergosterol) to ultraviolet light. Vitamin D can also be synthesized from dietary cholesterol that has been metabolized to provitamin D by intestinal enzymes, then stored in the skin and exposed to ultraviolet light (see chapter 6, p. 189 and fig. 13.26 on p. 508).

Like other fat-soluble vitamins, vitamin D resists the effects of heat, oxidation, acids, and bases. It is primarily stored in the liver and is less abundant in the skin, brain, spleen, and bones.

Vitamin D stored in the form of hydroxycholecalciferol is released as needed into the blood. When parathyroid hormone is present, this form of vitamin D is converted in the kidneys into an active form of the vitamin (dihydroxycholecalciferol). This substance, in turn, is carried as a hormone in the blood to the intestines where it stimulates production of calcium-binding protein. Here, it promotes absorption of calcium and phosphorus, ensuring that adequate amounts of these minerals are available in the blood for tooth and bone formation and metabolic processes.

Natural foods are poor sources of vitamin D, so the vitamin is often added to them during processing. For example, homogenized, nonfat, and evaporated milk are typically fortified with vitamin D. *Fortified* means essential nutrients have been added to a food where they originally were absent or scarce. *Enriched* means essential nutrients have been partially replaced in a food that has lost nutrients during processing.

Excess vitamin D, or *hypervitaminosis D,* produces diarrhea, nausea, and weight loss. Over time too much of this vitamin may calcify certain soft tissues and irreversibly damage the kidneys.

In children, vitamin D deficiency results in *rickets,* in which the bones, teeth, and abdominal muscles do not develop normally (fig. 18.12). In adults or in the elderly who have little exposure to sunlight, vitamin D deficiency may lead to *osteomalacia,* in which the bones decalcify and weaken due to disturbances in calcium and phosphorus metabolism. Just five minutes of sun exposure two to three times a week can maintain skeletal health without elevating skin cancer risk. Because many older people stay indoors, the Institute of Medicine suggests that daily vitamin D intake increase with age (table 18.6).

 PRACTICE

27 What are the functions of vitamin D?

28 Which foods are good sources of vitamin D?

Vitamin E includes a group of compounds, the most active of which is *alpha-tocopherol.* This vitamin is resistant to the effects of heat, acids, and visible light but is unstable in bases and in the presence of ultraviolet light or oxygen. Vitamin E is a strong antioxidant.

Vitamin E is found in all tissues but is primarily stored in the muscles and adipose tissue. It is also highly concentrated in the pituitary and adrenal glands.

The precise functions of vitamin E are unknown, but it is thought to act as an antioxidant by preventing oxidation of vitamin A and polyunsaturated fatty acids in the tissues. It may also help maintain the stability of cell membranes.

Vitamin E is widely distributed among foods. Its richest sources are oils from cereal seeds such as wheat germ. Other good sources are salad oils, margarine, shortenings, fruits, nuts, and vegetables. Excess vitamin E may cause nausea, headache, fatigue, easy bruising and bleeding, and muscle weakness. This vitamin is so easily obtained that deficiency conditions are rare.

Vitamin K, like the other fat-soluble vitamins, is in several chemical forms. One of these, vitamin K_1 (phylloquinone), is found in foods, whereas another, vitamin K_2, is produced by bacteria *(Escherichia coli)* that normally inhabit the human intestinal tract. These vitamins resist the effects of heat but are destroyed by oxidation or by exposure to acids, bases, or light. The liver stores them to a limited degree.

Vitamin K primarily functions in the liver, where it is necessary for the formation of several proteins needed for blood clotting, including *prothrombin* (see chapter 14, pp. 543–544). Consequently, deficiency of vitamin K prolongs blood clotting time and may increase risk of hemorrhage. Excess vitamin K may occur in formula-fed infants, causing jaundice, hemolytic anemia, and hyperbilirubinemia.

The richest sources of vitamin K are leafy green vegetables. Other good sources are egg yolk, pork liver, soy oil, tomatoes, and cauliflower. Table 18.7 summarizes the fat-soluble vitamins and their properties.

About 1 in every 200 to 400 newborns develops vitamin K deficiency because of an immature liver, poor transfer of vitamin K through the placenta, or lack of intestinal bacteria that can synthesize this vitamin. The deficiency causes "hemorrhagic disease of the newborn," in which abnormal bleeding occurs two to five days after birth. Injections of vitamin K shortly after birth prevent this condition. Adults may develop vitamin K deficiency if they take antibiotic drugs that kill the intestinal bacteria that manufacture the vitamin. People with cystic fibrosis may develop vitamin K deficiency, and/or deficiency of other fat-soluble vitamins, because they cannot digest fats well.

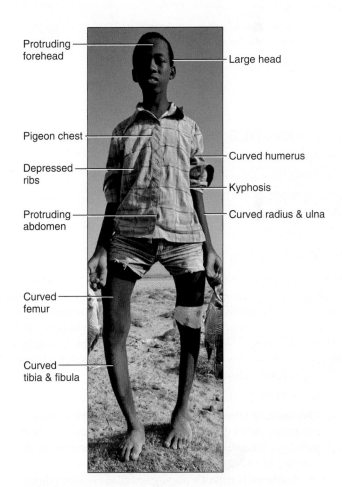

FIGURE 18.12 Vitamin D deficiency in childhood causes bone deformities.

 Why would vitamin D deficiency cause bone deformities?
Answer can be found in Appendix G.

TABLE 18.6	Vitamin D Requirements Increase with Age
Age Range (Years)	**International Units of Vitamin D**
0–1	400
1–70	300
70+	800

Source: National Institutes of Health (NIH)

TABLE 18.7 Fat-Soluble Vitamins

Vitamin	Characteristics	Functions	Sources and RDA* for Adults	Conditions Associated with	
				Excesses	Deficiencies
Vitamin A	Exists in several forms; synthesized from carotenes; stored in liver; stable in heat, acids, and bases; unstable in light	An antioxidant necessary for synthesis of visual pigments, mucoproteins, and mucopolysaccharides; for normal development of bones and teeth; and for maintenance of epithelial cells	Liver, fish, whole milk, butter, eggs, leafy green vegetables, yellow and orange vegetables and fruits 600 IU**	Nausea, headache, dizziness, hair loss, birth defects	Night blindness, degeneration of epithelial tissues
Vitamin D	A group of steroids; resistant to heat, oxidation, acids, and bases; stored in liver, skin, brain, spleen, and bones	Promotes absorption of calcium and phosphorus; promotes development of teeth and bones	Produced in skin exposed to ultraviolet light; in milk, egg yolk, fish liver oils, fortified foods 400 IU	Diarrhea, calcification of soft tissues, renal damage	Rickets, bone decalcification and weakening
Vitamin E	A group of compounds; resistant to heat and visible light; unstable in presence of oxygen and ultraviolet light; stored in muscles and adipose tissue	An antioxidant; prevents oxidation of vitamin A and polyunsaturated fatty acids; may help maintain stability of cell membranes	Oils from cereal seeds, salad oils, margarine, shortenings, fruits, nuts, and vegetables 30 IU	Nausea, headache, fatigue, easy bruising and bleeding	Rare, uncertain effects
Vitamin K	Exists in several forms; resistant to heat but destroyed by acids, bases, and light; stored in liver	Required for synthesis of prothrombin, which functions in blood clotting	Leafy green vegetables, egg yolk, pork liver, soy oil, tomatoes, cauliflower 55–70 µg	Jaundice in formula-fed newborns	Prolonged clotting time

*RDA = recommended daily allowance.

**IU = international unit.

 PRACTICE

29 Where is vitamin E stored?

30 Which foods are good sources of vitamin E?

31 Name two sources of vitamin K.

32 What is the function of vitamin K?

Water-Soluble Vitamins

The water-soluble vitamins include the B vitamins and vitamin C. Cooking and food processing destroy some of them. The **B vitamins** are several compounds essential for normal cellular metabolism. They help oxidize (remove electrons from) carbohydrates, lipids, and proteins during cellular respiration. The B vitamins are usually in the same foods, so they are called the *vitamin B complex*. Members of this group differ chemically and functionally.

The B-complex vitamins include the following:

1. **Thiamine,** or **vitamin B₁,** in its pure form is a crystalline compound called thiamine hydrochloride. Exposure to heat and oxygen destroys it, especially in alkaline environments. (See fig. 18.19 for its molecular structure.)

   Thiamine is part of a coenzyme called *cocarboxylase,* which oxidizes carbohydrates. Specifically, thiamine is required for pyruvic acid to enter the citric acid cycle (see chapter 4, p. 129). Without this vitamin, pyruvic acid accumulates in the blood. Thiamine also functions as a coenzyme in the synthesis of the sugar ribose, which is part of the nucleic acid RNA.

Thiamine is primarily absorbed through the wall of the duodenum and is transported by the blood to body cells. Only small amounts are stored in the tissues, and excess is excreted in the urine.

Vitamin B₁ oxidizes carbohydrates, and therefore cellular requirements vary with caloric intake. The recommended daily allowance (RDA) for adults is 1.2 milligrams (mg). Good sources of thiamine are lean meats, liver, eggs, whole-grain cereals, leafy green vegetables, and legumes.

Excess thiamine is not as common as excesses of fat-soluble vitamins, due to the excretion of thiamine in urine. Toxicity effects include vasodilation, cardiac dysrhythmias, headache, weakness, and convulsions.

A mild deficiency of thiamine produces loss of appetite, fatigue, and nausea. Prolonged deficiency leads to a disease called *beriberi,* which causes gastrointestinal disturbances, mental confusion, muscular weakness and paralysis, and heart enlargement. In severe cases, the heart may fail.

In developed nations, beriberi affects mostly people with chronic alcoholism who have substituted alcohol for foods. Because thiamine is required for the metabolic oxidation of alcohol, people with alcoholism are particularly likely to develop a thiamine deficiency.

2. **Riboflavin,** or **vitamin B₂,** is a yellowish brown crystalline substance that is relatively stable to the effects of heat, acids, and oxidation but is destroyed by exposure to bases and ultraviolet light. This vitamin is part of several enzymes and coenzymes known as *flavoproteins.* One such coenzyme, FAD, is an electron carrier in the citric acid cycle and electron transport chain of aerobic respiration. Flavoproteins are essential for the oxidation of glucose and fatty acids and for cellular growth. An active transport system controls the amount of riboflavin entering the intestinal mucosa. Riboflavin is carried in the blood combined with proteins called *albumins.* Excess riboflavin in the blood is excreted in the urine, turning it yellow-orange, and any that remains unabsorbed in the intestine is lost in the feces.

   The amount of riboflavin the body requires varies with caloric intake. About 1.3 mg is sufficient to meet daily cellular requirements.

   Riboflavin is widely distributed in foods, and rich sources include meats and dairy products. Leafy green vegetables, whole-grain cereals, and enriched cereals provide lesser amounts. Vitamin B₂ deficiency produces dermatitis and blurred vision.

3. **Niacin** or **vitamin B₃,** also known as *nicotinic acid,* is in plant tissues and is stable in the presence of heat, acids, and bases. After ingestion, it is converted to a physiologically active form called *niacinamide* (fig. 18.13). Niacinamide is the form of niacin in foods of animal origin.

   Niacin functions as part of two coenzymes (coenzyme I, also called NAD [fig. 18.14], and coenzyme II, called NADP) essential in glucose oxidation. These coenzymes are electron carriers in glycolysis, the citric acid cycle, and the electron transport chain, as well as in the synthesis of proteins and fats. They are also required for the synthesis of the sugars (ribose and deoxyribose) that are part of nucleic acids.

   Niacin is readily absorbed from foods, and human cells synthesize it from the essential amino acid *tryptophan.* Consequently, the daily requirement for niacin varies with tryptophan intake. The RDA for niacin is 14 mg for women and 16 mg for men.

   Rich sources of niacin (and tryptophan) include liver, lean meats, peanut butter, and legumes. Milk is a poor source of niacin but a good source of tryptophan.

   Excess niacin can cause acute toxicity with effects such as flushing, wheezing, vasodilation, headache, diarrhea, and vomiting. Chronic toxicity effects include liver problems.

   Historically, niacin deficiencies have been associated with diets based on corn and corn products, which are very low in niacin and lack tryptophan. Niacin deficiency causes *pellagra,* which produces dermatitis, inflammation of the digestive tract, diarrhea, and mental disorders.

   Pellagra is rare today, but it was a serious problem in the rural South of the United States in the early 1900s. Pellagra is less common in cultures that extensively treat corn with lime (CaCO₃) to release niacin bound to protein. Like beriberi, pellagra affects people who consume alcoholic beverages instead of food.

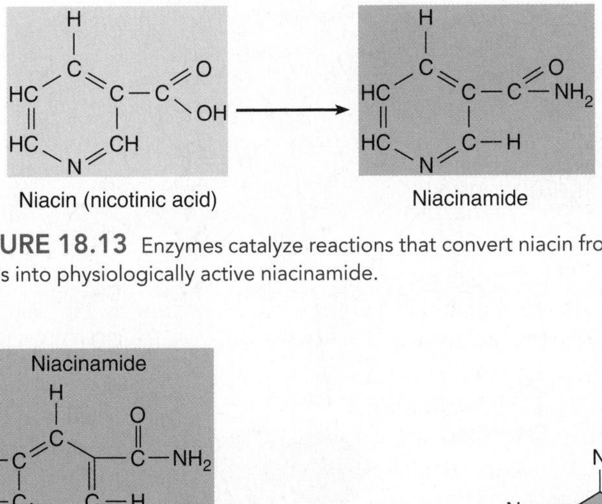

4. **Pantothenic acid,** or **vitamin B₅,** is a yellowish oil destroyed by heat, acids, and bases. It functions as part of a complex molecule called *coenzyme A,* which, in turn, reacts with intermediate products of carbohydrate and fat metabolism to yield *acetyl coenzyme A,* which enters the citric acid cycle. Pantothenic acid is therefore essential to cellular energy release.

   A daily adult intake of 5 mg of pantothenic acid is adequate. Most diets provide sufficient amounts, and therefore deficiencies are rare. Good sources of pantothenic acid include meats, whole-grain cereals, legumes, milk, fruits, and vegetables.

5. **Vitamin B₆** is a group of three compounds, *pyridoxine, pyridoxal,* and *pyridoxamine,* which are chemically similar (fig. 18.15). These compounds have similar actions and are fairly stable in the presence of heat and acids. Oxidation or exposure to bases or ultraviolet light destroys them. The vitamin B₆ compounds function as coenzymes essential in several metabolic pathways, including those that synthesize proteins, amino acids, antibodies, and nucleic acids, as well as the reaction of tryptophan to produce niacin.

   Vitamin B₆ functions in the metabolism of nitrogen-containing substances. Thus, the requirement for this vitamin varies with the protein content of the diet rather than with caloric intake. The RDA for vitamin B₆ ranges from 1.3 to 1.7 mg depending upon age, but because it is so abundant in foods, deficiency is rare. Good sources of vitamin B₆ include liver, meats, bananas, avocados, beans, peanuts, whole-grain cereals, and egg yolk. Excess vitamin B₆ produces burning pains, numbness, clumsiness, diminished reflexes, and paralysis.

6. **Biotin,** or **vitamin B₇,** is a simple compound that is unaffected by heat, acids, and light but may be destroyed by oxidation or bases. (See fig. 18.19 for the molecular structure of biotin.) It

FIGURE 18.13 Enzymes catalyze reactions that convert niacin from foods into physiologically active niacinamide.

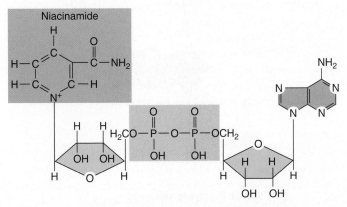

NAD or nicotinamide adenine dinucleotide

FIGURE 18.14 Niacinamide is incorporated into molecules of NAD.

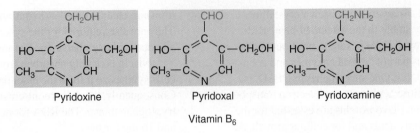

FIGURE 18.15 Vitamin B₆ includes three similar chemical compounds.

is a coenzyme in metabolic pathways for amino acids and fatty acids. It also plays a role in the synthesis of the purine nitrogenous bases of nucleic acids (adenine and guanine).

Metabolically active organs such as the brain, liver, and kidneys store some biotin. Bacteria that inhabit the intestinal tract synthesize biotin. The vitamin is widely distributed in foods, and dietary deficiencies are rare. Good sources include liver, egg yolk, nuts, legumes, and mushrooms. Excess biotin does not produce toxic effects.

7. **Folacin or vitamin B₉,** also known as *folic acid,* is a yellow crystalline compound that exists in several forms. It is easily oxidized in an acid environment and is destroyed by heat in alkaline solutions; consequently, this vitamin may be lost in stored or cooked foods.

Folacin is readily absorbed from the digestive tract and is stored in the liver, where it is converted to a physiologically active substance called *folinic acid.* Folinic acid functions as a coenzyme necessary for the metabolism of certain amino acids and for the synthesis of DNA. It also acts with cyanocobalamin in producing normal red blood cells.

Good sources of folacin include liver, leafy green vegetables, whole-grain cereals, and legumes. Because excess folacin is excreted in the urine, toxicity is rare. Folacin deficiency leads to *megaloblastic anemia,* in which the number of normal red blood cells is reduced and large, nucleated red cells appear. Folacin deficiency has been linked to neural tube defects, in which the tube that becomes the central nervous system in a fetus does not completely close. Neural tube defects include spina bifida and anencephaly. Taking synthetic folic acid supplements just before and during pregnancy can greatly reduce the risk of a neural tube defect.

Naturally occurring folate is a mixture of compounds that collectively have the same activity as synthetic folic acid, also called pteroylmonoglutamic acid. However, synthetic folic acid is much more stable and enters the bloodstream much more readily. This difference has led to confusion. For example, 200 micrograms of synthetic folic acid is prescribed to treat anemia, but the same effect requires 400 micrograms of folate from foods. Synthetic folic acid used to enrich grain foods has a greater effect on health than folate.

8. **Cyanocobalamin, or vitamin B₁₂,** has a complex molecular structure, including a single atom of the element *cobalt* (fig. 18.16). In its pure form, this vitamin is red. It is stable to

the effects of heat but is inactivated by light or strong acids or strong bases.

Secretion of *intrinsic factor* from the parietal cells of the gastric glands regulates cyanocobalamin absorption. Intrinsic factor combines with cyanocobalamin and facilitates its transfer through the epithelial lining of the small intestine and into the blood. Calcium ions must be present for the process to take place.

Inability of the gastric glands to secrete adequate amounts of intrinsic factor impairs vitamin B₁₂ absorption. This leads to *pernicious anemia,* which produces abnormally large red blood cells, called macrocytes, when bone marrow cells do not divide properly because of defective DNA synthesis.

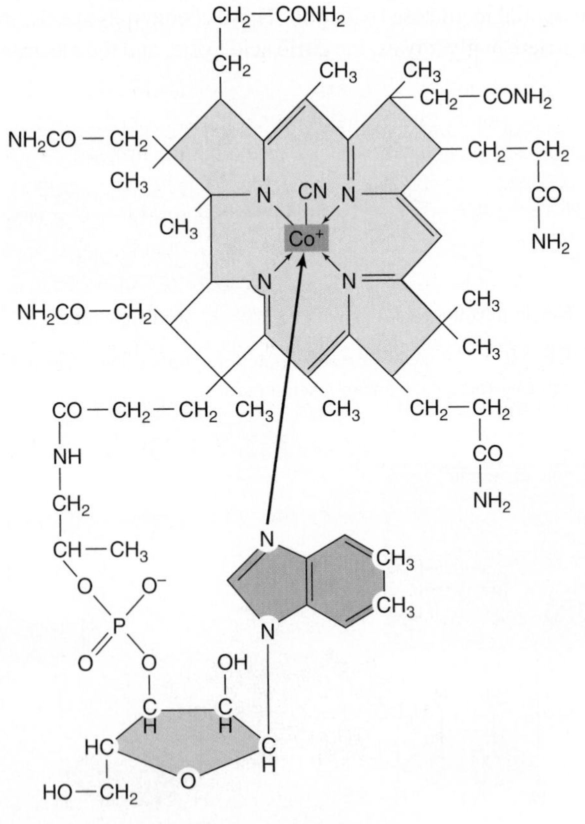

Vitamin B₁₂ (cyanocobalamin)

FIGURE 18.16 Vitamin B₁₂, which has the most complex molecular structure of the vitamins, contains cobalt (Co).

Various tissues store cyanocobalamin, particularly those of the liver. An average adult has a reserve sufficient to supply cells for three to five years. This vitamin is essential for the functions of all cells. It is part of coenzymes required for the synthesis of nucleic acids and the metabolism of carbohydrates and fats. Vitamin B$_{12}$ is important to erythrocyte production. Cyanocobalamin also helps form myelin in the central nervous system.

Only foods of animal origin contain cyanocobalamin. Good sources include liver, meats, milk, cheese, and eggs. Excessive intake does not appear to be toxic. In most countries, dietary lack of this vitamin is rare, although strict vegetarians may develop a deficiency.

 PRACTICE

33 Which biochemicals comprise the vitamin B complex?

34 Which foods are good sources of vitamin B complex?

35 What is the general function of each member of the vitamin B complex?

Vitamin C, or **ascorbic acid,** is a crystalline compound that has six carbon atoms. Chemically, it is similar to the monosaccharides (fig. 18.17). Vitamin C is one of the least stable vitamins in that oxidation, heat, light, or bases destroy it. However, vitamin C is fairly stable in acids.

Ascorbic acid is necessary for the production of the connective tissue protein *collagen,* for conversion of folacin to folinic acid, and in the metabolism of certain amino acids. It also promotes iron absorption and synthesis of certain hormones from cholesterol.

Overall, vitamin C is not stored in any great amount, but the adrenal cortex, pituitary gland, and intestinal glands contain high concentrations of it. Excess vitamin C is excreted in the urine or oxidized.

Individual requirements for ascorbic acid may vary. Ten mg per day is sufficient to prevent deficiency symptoms, and 80 mg per day saturates the tissues within a few weeks. The RDA for vitamin C is 75 mg for women and 90 mg for men.

Ascorbic acid is fairly widespread in plant foods, with high concentrations in citrus fruits and tomatoes. Leafy green vegetables are also good sources.

Prolonged deficiency of ascorbic acid leads to *scurvy,* which is more likely to affect infants and children. Scurvy impairs bone

development and causes swollen, painful joints. The gums may swell and bleed easily, resistance to infection is lowered, and wounds heal slowly (fig. 18.18). If a woman takes large doses of ascorbic acid during pregnancy, the newborn may develop symptoms of scurvy when the daily dose of the vitamin drops after birth because it is no longer delivered through the placenta. Table 18.8 summarizes the water-soluble vitamins and their characteristics.

 PRACTICE

36 What are functions of vitamin C?

37 Which foods are good sources of vitamin C?

Millions of Americans regularly take vitamin supplements. Consumer spending on vitamins and minerals is well into the billions of dollars annually. This practice has led to clinical signs of excess vitamin and mineral toxicity. Iron-containing vitamins are the most toxic, especially in acute pediatric ingestions.

18.7 | Minerals

Carbohydrates, lipids, proteins, and vitamins are organic compounds. In contrast, dietary **minerals** are inorganic elements essential in human metabolism. Most plants extract minerals from soil, and humans obtain them from plant foods or from animals that have eaten plants.

Characteristics of Minerals

Minerals contribute about 4% of body weight and are most concentrated in the bones and teeth. The minerals *calcium* and *phosphorus* are very abundant in these tissues.

Minerals are usually incorporated into organic molecules. Examples include phosphorus in phospholipids, iron in hemoglobin, and iodine in thyroxine. However, some minerals are part of inorganic compounds, such as the calcium phosphate of bone. Other minerals are free ions, such as sodium, chloride, and calcium ions in the blood.

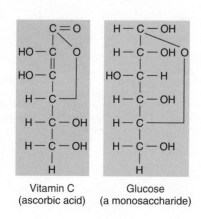

FIGURE 18.17 Vitamin C is chemically similar to some 6-carbon monosaccharides.

Vitamin C
(ascorbic acid)

Glucose
(a monosaccharide)

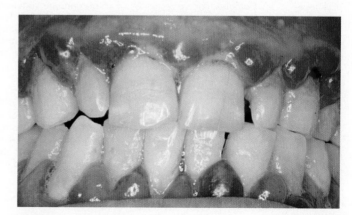

FIGURE 18.18 Vitamin C deficiency causes bleeding gums and other symptoms of scurvy.

TABLE 18.8 | Water-Soluble Vitamins

Vitamin	Characteristics	Functions	Sources and RDA* for Adults	Conditions Associated with	
				Excesses	Deficiencies
Thiamine (Vitamin B$_1$)	Destroyed by heat and oxygen, especially in an alkaline environment	Part of coenzyme required for oxidation of carbohydrates; coenzyme required for ribose synthesis	Lean meats, liver, eggs, whole-grain cereals, leafy green vegetables, legumes 1.2 mg	Uncommon, vasodilation, cardiac dysrhythmias	Beriberi, muscular weakness, enlarged heart
Riboflavin (Vitamin B$_2$)	Stable to heat, acids, and oxidation; destroyed by bases and ultraviolet light	Part of enzymes and coenzymes such as FAD, required for oxidation of glucose and fatty acids and for cellular growth	Meats, dairy products, leafy green vegetables, whole-grain cereals 1.3 mg	None known	Dermatitis, blurred vision
Niacin (Vitamin B$_3$, Nicotinic acid)	Stable to heat, acids, and bases; converted to niacinamide by cells; synthesized from tryptophan	Part of coenzymes NAD and NADP required for oxidation of glucose and synthesis of proteins, fats, and nucleic acids	Liver, lean meats, peanut butter, legumes 14 mg for females 16 mg for males	Flushing, vasodilation, wheezing, liver problems	Pellagra, dermatitis, diarrhea, mental disorders
Pantothenic acid (Vitamin B$_5$)	Destroyed by heat, acids, and bases	Part of coenzyme A required for oxidation of carbohydrates and fats	Meats, whole-grain cereals, legumes, milk, fruits, vegetables 5 mg	None known	Rare, loss of appetite, mental depression, muscle spasms
Vitamin B$_6$	Group of three compounds; stable to heat and acids; destroyed by oxidation, bases, and ultraviolet light	Coenzyme required for synthesis of proteins and various amino acids, for conversion of tryptophan to niacin, for production of antibodies, and for nucleic acid synthesis	Liver, meats, bananas, avocados, beans, peanuts, whole-grain cereals, egg yolk 1.3 mg <50 yrs 1.7 mg >50 yrs	Numbness, clumsiness, paralysis	Rare, convulsions, vomiting, seborrhea lesions
Biotin (Vitamin B$_7$)	Stable to heat, acids, and light; destroyed by oxidation and bases	Coenzyme required for metabolism of amino acids and fatty acids and for nucleic acid synthesis	Liver, egg yolk, nuts, legumes, mushrooms 0.3 mg	None known	Rare, elevated blood cholesterol, nausea, fatigue, anorexia
Folacin (Vitamin B$_9$, Folic acid)	Occurs in several forms; destroyed by oxidation in an acid environment or by heat in an alkaline environment; stored in liver where it is converted into folinic acid	Coenzyme required for metabolism of certain amino acids and for DNA synthesis; promotes production of normal red blood cells	Liver, leafy green vegetables, whole-grain cereals, legumes 0.4 mg	None known	Megaloblastic anemia
Cyanoco-balamin (Vitamin B$_{12}$)	Complex, cobalt-containing compound; stable to heat; inactivated by light, strong acids, and strong bases; absorption regulated by intrinsic factor from gastric glands; stored in liver	Part of coenzyme required for synthesis of nucleic acids and for metabolism of carbohydrates; plays role in myelin synthesis; required for normal red blood cell production	Liver, meats, milk, cheese, eggs 2.4 µg	None known	Pernicious anemia
Ascorbic acid (Vitamin C)	Chemically similar to monosaccharides; stable in acids but destroyed by oxidation, heat, light, and bases	Required for collagen production, conversion of folacin to folinic acid, and metabolism of certain amino acids; promotes absorption of iron and synthesis of hormones from cholesterol	Citrus fruits, tomatoes, leafy green vegetables 75 mg for females 90 mg for males	Exacerbates gout and kidney stone formation	Scurvy, lowered resistance to infection, wounds heal slowly

*RDA = recommended daily allowance.

Source: nutrition.gov.

Minerals compose parts of the structural materials of all body cells. They also constitute portions of enzyme molecules, contribute to the osmotic pressure of body fluids, and play vital roles in impulse conduction in neurons, muscle fiber contraction, blood coagulation, and maintenance of the pH of body fluids. The physiologically active form of minerals is the ionized form, such as Ca^{+2}.

Homeostatic mechanisms regulate the concentrations of minerals in body fluids so that excretion of minerals matches intake. Mineral toxicity may result not only from consumption of too much of a mineral, but also from overexposure to industrial pollutants, household chemicals, or certain drugs. Certain diseases lead to mineral toxicity. Hemochromatosis, for example, is an inherited form of "iron overload" resulting from excess iron absorption. Injuries may also lead to mineral toxicity, such as severe trauma causing hyperkalemia (high plasma potassium).

 PRACTICE

38 How do minerals differ from other nutrients?

39 What are the major functions of minerals?

Major Minerals

Calcium and phosphorus account for nearly 75% by weight of the mineral elements in the body. Therefore, they are considered **major minerals** (macrominerals). Other major minerals, each of which accounts for 0.05% or more of body weight, include potassium, sulfur, sodium, chlorine, and magnesium. Descriptions of the major minerals follow:

1. **Calcium (Ca)** is widely distributed in cells and body fluids, even though 99% of the body's supply is in the inorganic salts of the bones and teeth. It is essential for neurotransmitter release, muscle fiber contraction, the cardiac action potential, and blood coagulation. Calcium also activates certain enzymes.

    The amount of calcium absorbed varies with a number of factors. For example, the proportion of calcium absorbed increases as the body's need for calcium increases. Vitamin D and high protein intake promote calcium absorption. Increased motility of the digestive tract or an excess intake of fats decreases absorption. Daily intake of 1,000 to 1,200 mg of calcium covers adult requirements even with variations in absorption.

    Only a few foods contain significant amounts of calcium. Milk and milk products and fish with bones, such as salmon or sardines, are the richest sources. Leafy green vegetables, such as mustard greens, turnip greens, and kale, are good sources, but because very large servings of these vegetables are necessary to obtain sufficient minerals, most people must regularly consume milk or milk products to get enough calcium.

    Calcium toxicity is rare, but overconsumption of calcium supplements can deposit calcium phosphate in soft tissues. Calcium deficiency in children causes stunted growth, misshapen bones, and enlarged wrists and ankles. In adults, such a deficiency may remove calcium from the bones, thinning them and raising risk of fracture. Calcium is required for normal closing of the sodium channels in nerve cell membranes.

Because of this, too little calcium (hypocalcemia) can cause tetany. Extra calcium demands in pregnancy can cause cramps.

2. **Phosphorus (P)** accounts for about 1% of total body weight, most of it in the calcium phosphate of bones and teeth. The remainder serves as structural components and plays important roles in nearly all metabolic reactions. Phosphorus is a constituent of nucleic acids, many proteins, some enzymes, and some vitamins. It is also in the phospholipids of cell membranes, in the energy-carrying molecule ATP, and in the phosphates of body fluids that regulate pH. (Review the molecular structure of ATP in fig. 4.7, p. 126.)

    The recommended daily adult intake of phosphorus is 700 mg, and because this mineral is abundant in protein foods, diets adequate in proteins are also adequate in phosphorus. Phosphorus-rich foods include meats, cheese, nuts, whole-grain cereals, milk, and legumes.

 PRACTICE

40 Which are the most abundant minerals in the body?

41 What are the functions of calcium?

42 What are the functions of phosphorus?

43 Which foods are good sources of calcium and phosphorus?

3. **Potassium (K)** is widely distributed throughout the body and is concentrated inside cells rather than in extracellular fluids. On the other hand, sodium, which has similar chemical properties, is concentrated outside cells. The ratio of potassium to sodium in a cell is 10:1, whereas the ratio outside the cell is 1:28.

    Potassium helps maintain intracellular osmotic pressure and pH. It is a cofactor for enzymes that catalyze reactions of carbohydrate and protein metabolism. Potassium is vital in establishing the membrane potential and in impulse conduction in neurons.

    Nutritionists recommend a daily adult intake of 4.7 grams (4,700 mg) of potassium. This mineral is in many foods, so a typical adult diet provides between 2 and 6 grams each day. Excess potassium in the blood is uncommon because of the uptake of potassium by body cells and the excretion of potassium in urine. Potassium deficiency due to diet is rare, but it may occur for other reasons. For example, when a person has diarrhea, the intestinal contents may pass through the digestive tract so rapidly that potassium absorption is greatly reduced. Vomiting or using diuretic drugs may also deplete potassium. Such losses may cause muscular weakness, cardiac abnormalities, and edema.

    Foods rich in potassium are avocados, dried apricots, meats, milk, peanut butter, potatoes, and bananas. Citrus fruits, apples, carrots, and tomatoes provide lesser amounts.

 PRACTICE

44 How is potassium distributed in the body?

45 What is the function of potassium?

46 Which foods are good sources of potassium?

4. **Sulfur (S)** is responsible for about 0.25% of body weight and is widely distributed throughout tissues. It is abundant in skin, hair, and nails. Most sulfur is part of the amino acids *methionine* and *cysteine.* Other sulfur-containing compounds include thiamine, insulin, and biotin (fig. 18.19). In addition, sulfur is a constituent of mucopolysaccharides in cartilage, tendons, and bones and of sulfolipids in the liver, kidneys, salivary glands, and brain.

No daily requirement for sulfur has been established. However, a diet providing adequate amounts of protein will also likely meet the body's sulfur requirement. Good food sources of this mineral include meats, milk, eggs, and legumes.

5. **Sodium (Na)** makes up about 0.15% of adult body weight, and is widely distributed throughout the body. Only about 10% of this mineral is inside cells, and about 40% is in the extracellular fluids. The remainder is bound to the inorganic salts of bone.

Active transport readily absorbs sodium from foods. The kidneys regulate the blood concentration of sodium under the influence of the adrenal cortical hormone *aldosterone,* which causes the kidneys to reabsorb sodium while expelling potassium.

Sodium makes a major contribution to the solute concentration of extracellular fluids and thus helps regulate water movement between cells and their surroundings. It is necessary for impulse conduction in neurons and helps to move substances, such as chloride ions, through cell membranes (see chapter 21, p. 809).

The usual human diet probably provides more than enough sodium to meet the body's requirements. Sodium

toxicity, which shrinks cells, including those of the brain, requires unusual ingestion of additional sodium, such as drinking ocean water. Sodium may be lost as a result of diarrhea, vomiting, kidney disorders, sweating, or using diuretics. Sodium loss may cause a variety of symptoms, including nausea, cramps, and convulsions.

The amount of sodium naturally present in foods varies greatly, and it is commonly added to foods in the form of table salt. In some geographic regions, drinking water contains significant concentrations of sodium. Foods high in sodium include cured ham, sauerkraut, and cheese.

 PRACTICE

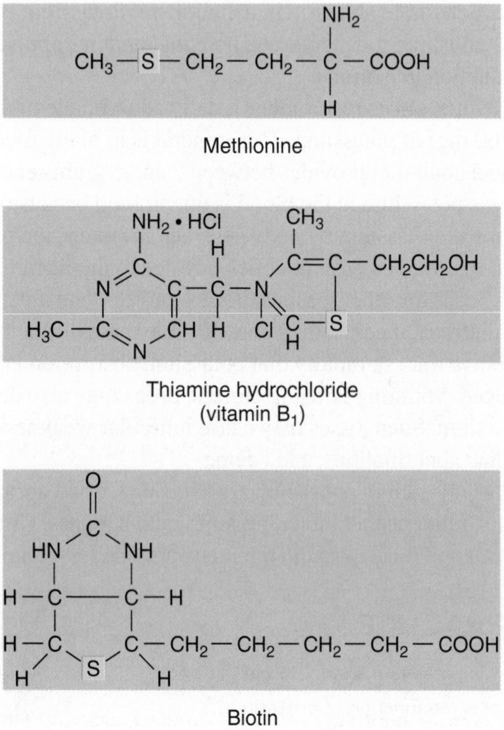

47 In which compounds and tissues of the body is sulfur found?

48 What are the functions of sodium?

6. **Chlorine (Cl)** in the form of chloride ions is found throughout the body and is most highly concentrated in cerebrospinal fluid and in gastric juice. With sodium, chlorine helps to regulate pH and maintain electrolyte balance and the solute concentration of extracellular fluids. Chlorine is also essential for the formation of hydrochloric acid in gastric juice and in the transport of carbon dioxide by red blood cells.

Chlorine and sodium are usually ingested in table salt, and as in the case for sodium, an ordinary diet usually provides considerably more chlorine than the body requires. Vomiting, diarrhea, kidney disorders, sweating, or using diuretics can deplete chlorine in the body.

7. **Magnesium (Mg)** is responsible for about 0.05% of body weight and is found in all cells. It is particularly abundant in bones in the form of phosphates and carbonates.

Magnesium is important in ATP-forming reactions in mitochondria, as well as in breaking down ATP to ADP. Therefore, it is important in providing energy for cellular processes.

Magnesium absorption in the intestinal tract adapts to dietary intake of the mineral. When the intake of magnesium is high, a smaller percentage is absorbed from the intestinal tract, and when the intake is low, a larger percentage is absorbed. Absorption increases as protein intake increases, and decreases as calcium and vitamin D intake increase. Bone tissue stores a reserve supply of magnesium, and excess is excreted in the urine.

The recommended daily allowance of magnesium is 310 to 420 mg, with higher amounts for males. A typical diet usually provides only about 120 mg of magnesium for every 1,000 calories, barely meeting the body's needs. Good sources of magnesium include milk and dairy products (except butter), legumes, nuts, and leafy green vegetables. Table 18.9 summarizes the major minerals.

 PRACTICE

49 Where are chloride ions most highly concentrated in the body?

50 What are the functions of magnesium?

$CH_3-S-CH_2-CH_2-\overset{\overset{\displaystyle NH_2}{|}}{\underset{\underset{\displaystyle H}{|}}{C}}-COOH$

Methionine

Thiamine hydrochloride
(vitamin B₁)

Biotin

FIGURE 18.19 Three essential sulfur-containing nutrients.

TABLE 18.9 Major Minerals

Mineral	Distribution	Functions	Sources and RDA* for Adults	Conditions Associated with	
				Excesses	Deficiencies
Calcium (Ca)	Mostly in the inorganic salts of bones and teeth	Structure of bones and teeth; essential for neurotransmitter release, muscle fiber contraction, the cardiac action potential, and blood coagulation; activates certain enzymes	Milk, milk products, leafy green vegetables 1,000 mg <50 yrs 1,200 mg >50 yrs	Kidney stones, deposition of calcium phosphate in soft tissues	Stunted growth, misshapen bones, fragile bones, tetany
Phosphorus (P)	Mostly in the inorganic salts of bones and teeth	Structure of bones and teeth; component of nearly all metabolic reactions; in nucleic acids, many proteins, some enzymes, and some vitamins; in cell membrane, ATP, and phosphates of body fluids	Meats, cheese, nuts, whole-grain cereals, milk, legumes 700 mg	None known	Stunted growth
Potassium (K)	Widely distributed; tends to be concentrated inside cells	Helps maintain intracellular osmotic pressure and regulate pH; required for impulse conduction in neurons	Avocados, dried apricots, meats, peanut butter, potatoes, bananas 4,700 mg	Uncommon	Muscular weakness, cardiac abnormalities, edema
Sulfur (S)	Widely distributed; abundant in skin, hair, and nails	Essential part of certain amino acids, thiamine, insulin, biotin, and mucopolysaccharides	Meats, milk, eggs, legumes No RDA established	None known	None known
Sodium (Na)	Widely distributed; mostly in extracellular fluids and bound to inorganic salts of bone	Helps maintain osmotic pressure of extracellular fluids; regulates water movement; plays a role in impulse conduction in neurons; regulates pH and transport of substances across cell membranes	Table salt, cured ham, sauerkraut, cheese 2,300 mg	Hypertension, edema, body cells shrink	Nausea, cramps, convulsions
Chlorine (Cl)	Closely associated with sodium; most highly concentrated in cerebrospinal fluid and gastric juice	Helps maintain osmotic pressure of extracellular fluids, regulates pH; maintains electrolyte balance; forms hydrochloric acid; aids transport of carbon dioxide by red blood cells	Same as for sodium No RDA established	Vomiting	Cramps
Magnesium (Mg)	Abundant in bones	Required in metabolic reactions in mitochondria that produce ATP; plays a role in the breakdown of ATP to ADP	Milk, dairy products, legumes, nuts, leafy green vegetables 310–420 mg	Diarrhea	Neuromuscular disturbances

*RDA = recommended daily allowance.
Source: nutrition.gov.

Trace Elements

Trace elements (microminerals) are essential minerals found in minute amounts, each making up less than 0.005% of adult body weight. They include iron, manganese, copper, iodine, cobalt, zinc, fluorine, selenium, and chromium.

Iron (Fe) is most abundant in the blood, but is stored in the liver, spleen, and bone marrow and is found to some extent in all cells. Iron enables *hemoglobin* molecules in red blood cells to carry oxygen (fig. 18.20). Iron is also part of *myoglobin*, which stores oxygen in muscle cells. In addition, iron assists in vitamin A synthesis, is incorporated into a number of enzymes, and is included in the cytochrome molecules that participate in ATP-generating reactions.

An adult male requires from 0.7 to 1 mg of iron daily, and a female needs 1.2 to 2 mg. A typical diet supplies about 10 to 18 mg of iron each day, but only 2% to 10% of the iron is absorbed. For some people, this may not be enough iron. Eating foods rich in vitamin C along with iron-containing foods can increase absorption of this important mineral.

Liver is the only rich source of dietary iron, and because liver is not a popular food, iron is one of the more difficult nutrients to obtain from natural sources in adequate amounts. Foods that contain some iron include lean meats, dried apricots, raisins, and prunes, enriched whole-grain cereals, legumes, and molasses.

> Pregnant women require extra iron to support the formation of a placenta and the growth and development of a fetus. Iron is required for the synthesis of hemoglobin in a fetus as well as in a pregnant woman, whose blood volume increases by a third.

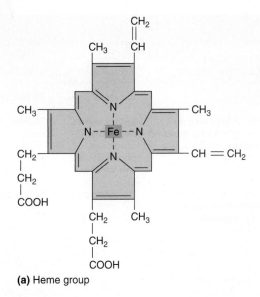

(a) Heme group

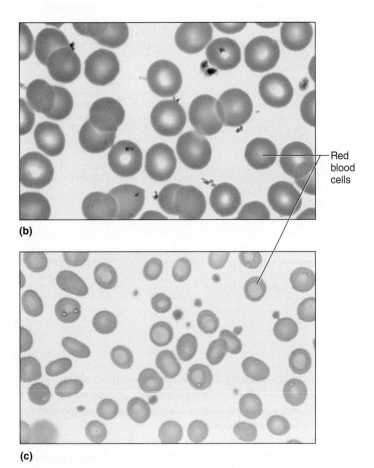

Red blood cells

(b)

(c)

FIGURE 18.20 Iron in hemoglobin. (*a*) A hemoglobin molecule contains four heme groups, each of which houses a single iron atom (Fe) that can combine with oxygen. The red blood cells in (*b*) are normal (1,250×), but many of those in (*c*) are small and pale (1,250×). They contain too little hemoglobin, which may be the result of a diet lacking in iron.

Manganese (Mn) is most concentrated in the liver, kidneys, and pancreas. It is necessary for normal growth and development of skeletal structures and other connective tissues. Manganese is part of enzymes essential for the synthesis of fatty acids and cholesterol, for urea formation, and for the normal functions of the nervous system.

The daily requirement for manganese is 2.5 to 5 mg. The richest sources include nuts, legumes, and whole-grain cereals. Leafy green vegetables and fruits are good sources of manganese too.

PRACTICE

51 What is the primary function of iron?

52 How is manganese used?

53 Which foods are good sources of manganese?

Copper (Cu) is in all body tissues but is most highly concentrated in the liver, heart, and brain. It is essential for hemoglobin synthesis, bone development, melanin production, and formation of myelin in the nervous system.

A daily intake of 0.9 to 3 mg of copper is sufficient to supply cells. A typical adult diet has about 2 to 5 mg of this mineral, so adults seldom develop copper deficiencies. Foods rich in copper include liver, oysters, crabmeat, nuts, whole-grain cereals, and legumes.

Iodine (I) is highly concentrated in the thyroid gland, but is found in very small quantities in other tissues. Its only known function is as an essential component of thyroid hormones. (Fig. 13.20 on p. 505 shows the molecular structures of these hormones, thyroxine and triiodothyronine.)

A daily intake of 0.15 mg of iodine is adequate for most adults. The iodine content of foods varies with the iodine content of soils in different geographic regions. In many places, people use *iodized* table salt to season foods, which prevents iodine deficiency.

Cobalt (Co) is widely distributed in the body because it is an essential part of cyanocobalamin (vitamin B_{12}). It is also necessary for the synthesis of several important enzymes.

The amount of cobalt required in the daily diet is unknown. This mineral is found in a great variety of foods, and the quantity in the average diet is apparently sufficient. Good sources of cobalt include liver, lean meats, and milk.

Zinc (Zn) is most concentrated in the liver, kidneys, and brain. It is part of many enzymes involved in digestion, respiration, and bone and liver metabolism. It is also necessary for normal wound healing and for maintaining the integrity of the skin.

The daily requirement for zinc is about 11 to 15 mg, and most diets provide enough. Only some may be absorbed, so zinc deficiencies may occur. Meat provides the most zinc. Cereals, legumes, nuts, and vegetables provide lesser amounts.

Fluorine (F), as part of the compound fluoroapatite, replaces hydroxyapatite in teeth, strengthening the enamel and preventing dental caries. **Selenium (Se)** is stored in the liver and kidneys. It is a constituent of certain enzymes and participates in heart function. This mineral is found in lean meats, whole-grain cereals,

TABLE 18.10 Trace Elements

Trace Element	Distribution	Functions	Sources and RDA* for Adults	Conditions Associated with Excesses	Conditions Associated with Deficiencies
Iron (Fe)	Primarily in blood; stored in liver, spleen, and bone marrow	Part of hemoglobin molecule; assists in vitamin A synthesis; incorporated into a number of enzymes	Liver, lean meats, dried apricots, raisins, enriched whole-grain cereals, legumes, molasses 10–18 mg	Liver damage	Anemia
Manganese (Mn)	Most concentrated in liver, kidneys, and pancreas	Part of enzymes required for fatty acids and cholesterol synthesis, urea formation, and normal functioning of the nervous system	Nuts, legumes, whole-grain cereals, leafy green vegetables, fruits 2.5–5 mg	None known	None known
Copper (Cu)	Most highly concentrated in liver, heart, and brain	Essential for hemoglobin synthesis, bone development, melanin production, and myelin formation	Liver, oysters, crabmeat, nuts, whole-grain cereals, legumes 0.9–3 mg	Rare	Rare
Iodine (I)	Concentrated in thyroid gland	Essential component for synthesis of thyroid hormones	Food content varies with soil content in different geographic regions; iodized table salt 0.15 mg	Autoimmune thyroid disease	Decreased synthesis of thyroid hormones
Cobalt (Co)	Widely distributed	Component of cyanocobalamin; required for synthesis of several enzymes	Liver, lean meats, milk No RDA established	Heart disease	Pernicious anemia
Zinc (Zn)	Most concentrated in liver, kidneys, and brain	Component of enzymes involved in digestion, respiration, bone metabolism, liver metabolism; necessary for normal wound healing and maintaining integrity of the skin	Meats, cereals, legumes, nuts, vegetables 11–15 mg	Slurred speech, problems walking	Depressed immunity, loss of taste and smell, learning difficulties
Fluorine (F)	Primarily in bones and teeth	Component of tooth enamel	Fluoridated water 1.5–4 mg	Mottled teeth	None known
Selenium (Se)	Concentrated in liver and kidneys	Component of certain enzymes	Lean meats, cereals, onions 0.05–0.07 mg	Vomiting, fatigue	None known
Chromium (Cr)	Widely distributed	Essential for use of carbohydrates	Liver, lean meats, yeast 0.05–2 mg	None known	None known

*RDA = recommended daily allowance.

and onions. **Chromium (Cr)** is widely distributed throughout the body and regulates glucose use. It is found in liver, lean meats, yeast, and pork kidneys. Table 18.10 summarizes the characteristics of trace elements.

The term "dietary supplement" traditionally refers to minerals, vitamins, carbohydrates, proteins, and fats—the micronutrients and macronutrients. Clinical Application 18.2 discusses the more commercial meaning of "dietary supplement."

 PRACTICE

54 How is copper used?

55 What is the function of iodine?

56 Why might zinc deficiencies be common?

A compulsive disorder that may result from mineral deficiency is *pica*, in which people consume large amounts of nondietary substances such as ice chips, soil, sand, laundry starch, clay and plaster, and even hair, toilet paper, matchheads, inner tubes, mothballs, and charcoal. The condition is named for the magpie bird, *Pica pica*, which eats a range of odd things.

Pica affects people of all cultures and was noted as early as 40 B.C. The connection to dietary deficiency stems from the observation that slaves suffering from pica in the colonial United States recovered when their diets improved, particularly when given iron supplements.

Displayed prominently among the standard vitamin and mineral preparations in the pharmacy or health food store is a dizzying collection of products (fig. 18A). Some obviously come from organisms (bee pollen and shark cartilage), some have chemical names (glucosamine with chondroitin), and some names are historical or cultural (St. John's wort). These "dietary supplements" are classified as neither food nor drug, but despite regulatory language, they may contain active compounds that function as pharmaceuticals in the human body.

By United States law dietary supplements include: "a product (other than tobacco) that is intended to supplement the diet that bears or contains one or more of the following dietary ingredients: a vitamin, a mineral, an herb or other botanical, an amino acid, a dietary substance for use by man to supplement the diet by increasing the total daily intake, or a concentrate, metabolite, constituent, extract, or combinations of these ingredients."

Labels cannot claim that a dietary supplement diagnoses, prevents, mitigates, treats, or cures any specific disease. Instead, the language is positive and guarded, as in claims that valerian root "promotes restful sleep" and that echinacea and goldenseal "may help support the immune system." The U.S. Food and Drug Administration (FDA)

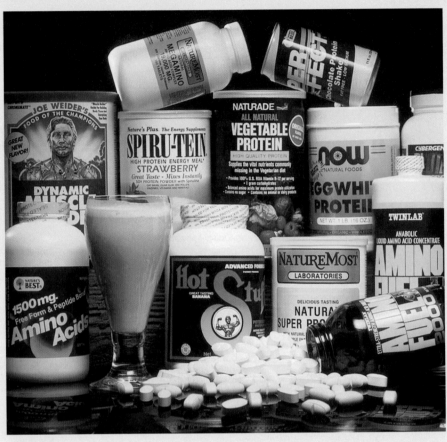

FIGURE 18A Some dietary supplements are natural substances that function as drugs in the human body.

18.8 | Healthy Eating

An *adequate diet* provides sufficient energy (calories), essential fatty acids, essential amino acids, vitamins, and minerals to support optimal growth and to maintain and repair body tissues. Because individual nutrient requirements vary greatly with age, sex, growth rate, level of physical activity, and stress, as well as with genetic and environmental factors, it is not possible to design a diet adequate for everyone. However, nutrients are so widely found in foods that consuming satisfactory amounts and combinations is usually possible despite individual food preferences, assuming that foods are available.

In countries with adequate food supplies, most healthy individuals can obtain nourishment by eating a variety of foods and limiting fat intake. People who do not eat meat products can also be well-nourished, but they must pay more attention to food choices to avoid developing nutrient deficiencies. For example, eliminating red meat also means eliminating an excellent source of iron, copper, zinc, and vitamin B$_{12}$. The fiber that often makes up much of a vegetarian's diet, although very healthful in many ways, also decreases absorption of iron. Therefore, a vegetarian must obtain sufficient iron from nonmeat sources. It is easy for a vegetarian to receive proper **nutrition** (adequate nutrients) by being aware of the sources, actions, and interactions of nutrients. Fortified foods, green leafy vegetables, and whole grains provide many of the nutrients that are also in meat. Table 18.11 lists types of vegetarian diets.

It is difficult to track the different nutrients in a diet and be certain that an adequate amount of each is consumed daily. Nutritionists have devised several ways to help consumers make healthy food choices, recognizing that people can meet dietary requirements in many and diverse ways. Most familiar is the RDA guideline that has appeared on several tables in the chapter. *RDA* allows only certain specific food and health claims that are supported by evidence. These include:

- Dietary calcium decreases the risk of osteoporosis.
- A low-fat diet lowers the risk of some cancers.
- A diet low in saturated fat and cholesterol lowers the risk of coronary heart disease.
- Fruits and vegetables reduce the risk of cancers and heart disease.
- Lowering sodium intake lowers blood pressure.
- Folic acid lowers the risk of neural tube defects.

Some dietary supplements are of uncertain value. Pyruvic acid and ATP, which are marketed to boost energy levels, for example, are abundant in the cellular respiration pathways—a person doesn't need to take more. DNA sold in a form that is merely expensive brewer's yeast is unnecessary because any food consisting of cells is packed with DNA.

Taking dietary supplements may be dangerous if their active ingredients interact with certain drugs. An example is the active ingredient in St. John's wort, hypericin, which affects levels of nearly half of all prescription drugs by either decreasing or increasing their effectiveness, or making side effects more likely.

The FDA does not require clinical trials to demonstrate that dietary supplements are safe and effective, as the agency does for drugs. The FDA can act only after a dietary supplement has been linked to illness, injury, or death. This was the case for a supplement marketed to boost energy in athletes that initially caused 97 cases of liver failure, including one death, in Hawaii, but the effects of that product may be much more widespread. The agency has investigated more than 500 dietary supplements and found that many contain drugs, including stimulants and steroids, that have never been tested in humans!

In the United States, people use more than 85,000 combinations of ingredients marketed as dietary supplements, and spend more than $32 billion a year on them. Be certain to tell your healthcare provider about the dosage and frequency of use of any and all dietary supplements you take. These are powerful chemicals that affect physiology. Table 18A lists some medical conditions that may result from use of specific dietary supplements.

TABLE 18A	Health Conditions Associated with Dietary Supplements
Health Condition	**Supplements**
Anxiety, panic attacks	Ginseng, hawthorn, sibutramine, yohimbe
Arrhythmia	Ephedra, horny goat weed, oleander
Cancers	Anabolic steroids (liver), beta-carotene (lung), vitamin E (prostate)
Gastrointestinal upset	Echinacea, fenugreek, guggul, hawthorn, horny goat weed, oleander, saw palmetto
Intracranial hemorrhage	Ginkgo biloba
Liver toxicity	Aegeline, chaparral, comfrey, germander, kava
Mood change	Belladonna, ginseng, kratom, St. John's wort, yohimbe
Myocardial infarction	Ephedra, sibutramine
Rash and allergic reactions	Bishop's weed, chrysanthemum, echinacea, fenugreek, guggul, St. John's wort, willow bark
Stroke	Ephedra, ginkgo, sibutramine

stands for United States Recommended Daily Allowance. An RDA is the upper limit of another measurement, the Recommended Dietary Allowance, which lists optimal calorie intake for each sex at various ages, and the amounts of vitamins and minerals needed to avoid deficiency or excess conditions. The RDA values on food packages are set high, ensuring that most people who follow them receive sufficient amounts of each nutrient. Government panels meet every five years to evaluate the RDAs in light of new data.

Placing foods into groups is a simpler way to follow a healthy diet. For years, diagrams called *food pyramids* organized foods according to suggested proportions in the diet, often in serving sizes. One food pyramid developed by the United States Department of Agriculture (USDA) dominated, but other pyramids offered more specific suggestions geared to age, health, ethnicity, food preferences such as vegetarianism, or weight loss goals. Past pyramids can seem strange in light of today's **MyPlate**

(developed by the USDA) replacement, depicted in figure 18.21. Clinical Application 18.3 discusses some ways that understanding nutrition can help athletic performance.

TABLE 18.11	Types of Vegetarian Diets
Type	**Food Restrictions**
Vegan	No animal foods
Ovo-vegetarian	Eggs allowed; no dairy or meat
Lacto-vegetarian	Dairy allowed; no eggs or meat
Lacto-ovo-vegetarian	Dairy and eggs allowed; no meat
Pesco-vegetarian	Dairy, eggs, and fish allowed; no other meat
Semivegetarian	Dairy, eggs, chicken, and fish allowed; no other meat

An endurance athlete and a sedentary individual have different nutritional requirements. A diet that is 60% or more carbohydrate, 18% protein, and 22% fat supports a lifestyle that includes frequent, strenuous activity.

Macronutrients

Athletes should get the bulk of their carbohydrates from vegetables and grains to avoid cholesterol. They should eat frequently, because the muscles can store only 1,800 calories worth of glycogen.

Athletes need to consume only 25% more protein than less-active individuals. The American Dietetic Association suggests that athletes eat 1 gram of protein per kilogram of weight per day, compared to 0.8 gram for nonathletes. Athletes should not rely solely on meat for protein, because these foods can be high in fat. Protein supplements may be necessary for young athletes at the start of training, under a doctor's supervision. Too little protein in an athlete is linked to "sports anemia," in which hemoglobin levels decline and blood may appear in the urine.

Water

A sedentary person loses a quart of water a day as sweat; an athlete may lose 2 to 4 quarts of water an hour! To stay hydrated, athletes should drink 3 cups of cold water two hours before an event, then 2 more cups fifteen minutes before the event, and small amounts every fifteen minutes during the event. They should drink afterward too. Another way to determine water needs is to weigh in before and after training. For each pound lost, athletes should drink a pint of water. They should avoid alcohol, which increases fluid loss. However, athletes should also avoid drinking too much water during competition, which can cause hyponatremia (too little sodium in the bloodstream, see Clinical Applications 21.1 and 21.2, on pp. 810–811 and 813, respectively).

Vitamins and Minerals

If an athlete eats an adequate, balanced diet, vitamin supplements are not needed. Supplements of sodium and potassium are usu-ally not needed either, because the active body naturally conserves these nutrients. To be certain of enough sodium, athletes may want to salt their food; to get enough potassium, they can eat bananas, dates, apricots, oranges, or raisins.

A healthy pregame meal should be eaten two to five hours before the game, provide 500 to 1,500 calories, and include 4 or 5 cups of fluid. It should also be high in carbohydrates, which taste good, provide energy, and are easy to digest.

Creatine

Creatine is advertised to increase energy stores and provide a safe alternative to steroids for bulking up muscles. This is deceptive.

Creatine may be obtained from foods, through supplements, or by synthesis from the amino acids arginine, glycine, and methionine. Creatine, in the form of creatine phosphate, provides energy to muscle cells by phosphorylating ADP to generate ATP. Creatine is converted to its metabolite, creatinine, at such a constant rate that the excretion of creatinine in the urine is used as a marker for normal kidney function.

Do creatine supplements enhance performance? The emerging picture suggests that during peak exertion, especially repetitive peak exertion (such as multiple sprints), conditions in which creatine levels may become depleted, supplemental creatine may be advantageous very briefly. However, muscle mass may appear to increase in athletes taking creatine supplements because creatine draws water into muscle cells by osmosis, not because new muscle tissue forms. Creatine use may be dangerous. The disturbance in water distribution may create problems if the athlete encounters extreme heat—sweating becomes inadequate to effectively cool the body. Swelled muscle cells may burst, causing a potentially fatal condition called rhabdomyolysis. The Food and Drug Administration has received many adverse event reports of muscle cramps, seizures, diarrhea, loss of appetite, muscle strains, dehydration and even deaths, associated with creatine use among athletes.

PRACTICE

57 What is an adequate diet?

58 What factors influence individual needs for nutrients?

Malnutrition

Malnutrition (mal″nu-trish′un) is poor nutrition that results from a lack of essential nutrients or an inability to utilize them. It may result from *undernutrition* and produce the symptoms of deficiency diseases, or it may be due to *overnutrition* arising from excess nutrient intake.

A variety of factors can lead to malnutrition. For example, a deficiency condition may stem from lack of availability or poor quality of food. On the other hand, malnutrition may result from overeating or taking too many vitamin supplements. Malnutrition from diet alone is called *primary malnutrition.*

Secondary malnutrition occurs when an individual's characteristics make a normally adequate diet insufficient. For example, a person who does not secrete enough bile salts is likely to develop a deficiency of fat-soluble vitamins, because bile salts promote absorption of fats. Severe and prolonged emotional stress may lead to secondary malnutrition, because stress can change hormonal concentrations in ways that break down amino acids or excrete nutrients, and affect appetite.

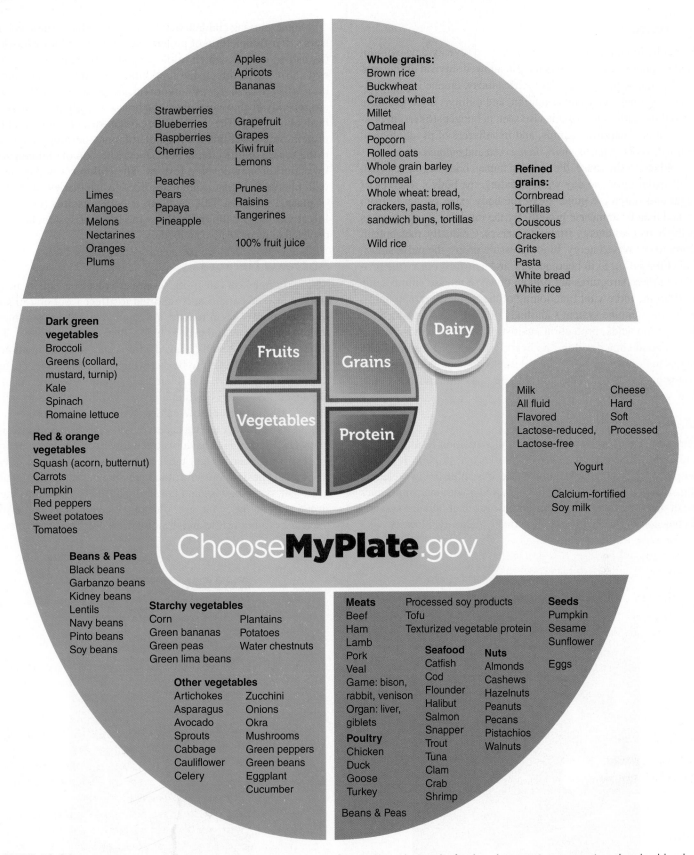

Fruits

Apples
Apricots
Bananas

Strawberries
Blueberries
Raspberries
Cherries

Grapefruit
Grapes
Kiwi fruit
Lemons

Limes
Mangoes
Melons
Nectarines
Oranges
Plums

Peaches
Pears
Papaya
Pineapple

Prunes
Raisins
Tangerines

100% fruit juice

Grains

Whole grains:
Brown rice
Buckwheat
Cracked wheat
Millet
Oatmeal
Popcorn
Rolled oats
Whole grain barley
Cornmeal
Whole wheat: bread, crackers, pasta, rolls, sandwich buns, tortillas

Wild rice

Refined grains:
Cornbread
Tortillas
Couscous
Crackers
Grits
Pasta
White bread
White rice

Dairy

Milk
All fluid
Flavored
Lactose-reduced,
Lactose-free

Cheese
Hard
Soft
Processed

Yogurt

Calcium-fortified
Soy milk

Vegetables

Dark green vegetables
Broccoli
Greens (collard, mustard, turnip)
Kale
Spinach
Romaine lettuce

Red & orange vegetables
Squash (acorn, butternut)
Carrots
Pumpkin
Red peppers
Sweet potatoes
Tomatoes

Beans & Peas
Black beans
Garbanzo beans
Kidney beans
Lentils
Navy beans
Pinto beans
Soy beans

Starchy vegetables
Corn
Green bananas
Green peas
Green lima beans
Plantains
Potatoes
Water chestnuts

Other vegetables
Artichokes
Asparagus
Avocado
Sprouts
Cabbage
Cauliflower
Celery
Zucchini
Onions
Okra
Mushrooms
Green peppers
Green beans
Eggplant
Cucumber

Protein

Meats
Beef
Ham
Lamb
Pork
Veal
Game: bison, rabbit, venison
Organ: liver, giblets

Poultry
Chicken
Duck
Goose
Turkey

Beans & Peas

Processed soy products
Tofu
Texturized vegetable protein

Seafood
Catfish
Cod
Flounder
Halibut
Salmon
Snapper
Trout
Tuna
Clam
Crab
Shrimp

Nuts
Almonds
Cashews
Hazelnuts
Peanuts
Pecans
Pistachios
Walnuts

Seeds
Pumpkin
Sesame
Sunflower

Eggs

ChooseMyPlate.gov

FIGURE 18.21 MyPlate, developed by the United States Department of Agriculture, depicts the foods and appropriate proportions that should make up a healthy diet.

Starvation

A healthy human can stay alive for fifty to seventy days without food. In prehistoric times, this margin allowed survival during seasonal famines. In some areas of Africa today, famine is not a seasonal event but a constant condition, and millions of people have starved to death. Starvation is also seen in hunger strikers, in prisoners of concentration camps, and in sufferers of psychological eating disorders such as *anorexia nervosa* and *bulimia*.

Whatever the cause, the starving human body begins to digest itself. After only one day without eating, the body's reserves of sugar and starch are gone. Next, the body extracts energy from fat and then from muscle protein. By the third day, hunger ceases as the body uses energy from fat reserves. Gradually, metabolism slows to conserve energy, blood pressure drops, the pulse slows, and chills set in. Skin becomes dry and hair falls out as the proteins in these structures are broken down to release amino acids used for the more vital functioning of the brain, heart, and lungs. When the immune system's antibody proteins are dismantled for their amino acids, protection against infection declines. Mouth sores and anemia develop, the heart beats irregularly, and bone begins to degenerate. After several weeks without food, coordination is gradually lost. Near the end, the starving human is blind, deaf, and emaciated.

Marasmus and Kwashiorkor

Marasmus (mah-raz'mus) is a form of starvation resulting from undernourishment with a lack of calories and protein. It causes people to lose so much weight that they resemble living skeletons (fig. 18.22*a*). Children under the age of two with marasmus may die of measles or other infectious diseases as their immune systems become too weakened to fight off normally mild viral illnesses.

Some starving children have protruding bellies. These youngsters suffer from a form of protein starvation called **kwashiorkor** (kwash"-e-or'kor), which in the language of Ghana means "the evil spirit which infects the first child when the second child is born" (fig. 18.22*b*). Kwashiorkor typically appears in a child who has recently been weaned from the breast, usually because of the birth of a sibling. The switch from protein-rich breast milk to the protein-poor gruel that is the staple of many developing nations is the source of this protein deficiency. The skin of children with kwashiorkor may develop lesions. The children's bellies swell with filtered fluid that is not reabsorbed by capillaries due to lack of plasma proteins. This swelling is called **ascites** (ah-si'tēz). Infections overwhelm the body as the immune system becomes depleted of its protective antibodies.

Anorexia Nervosa

Anorexia nervosa (an"o-rek'se-ah ner'vo-sah) is self-imposed starvation. The condition is reported to affect 1 out of 250 adolescents, most of them female, although the true number among males is not known and may be higher than has been thought. The sufferer, typically a well-behaved adolescent girl from an affluent family, perceives herself to be overweight and eats barely enough to survive (fig. 18.23). She is terrified of gaining weight and usually loses 25% of her original body weight. In addition to eating only small amounts of low-calorie foods, she may further lose weight by vomiting, taking laxatives and diuretics, or exercising intensely. Her eating behavior is often ritualized. She may meticulously arrange her meager meal on her plate or consume only a few foods. She develops low blood pressure, a slowed or irregular heartbeat, constipation, and constant chilliness. She stops menstruating as her body fat level plunges. Like any starving person, the hair becomes brittle and the skin dries out. To conserve body

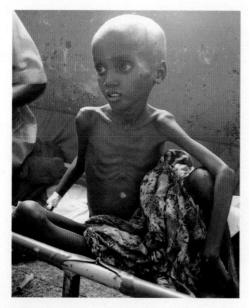
(a)

(b)

FIGURE 18.22 Two types of starvation in the young. (*a*) This child, suffering from marasmus, did not have adequate nutrition as an infant. (*b*) This child suffers from kwashiorkor. Although he may have received adequate nourishment from breast milk early in life, he became malnourished when his diet switched to a watery, white extract from cassava that looks like milk but has very little protein. The lack of protein in the diet causes edema and the ascites that swells his belly.

FIGURE 18.23 Perceiving herself overweight, this young woman is tying the measuring tape extraordinarily tight around her waist to have a waist measurement as small as possible.

heat, she may develop soft, pale, fine body hair called *lanugo*, normally seen only on a fetus.

When the person with anorexia reaches an obviously emaciated state, her parents usually have her hospitalized, where she is fed intravenously so that she does not starve to death or die suddenly of heart failure due to an electrolyte imbalance. She also receives psychotherapy and nutritional counseling. Despite these efforts, 15% to 21% of people with anorexia die.

What causes anorexia nervosa is unknown. One hypothesis is that the person is rebelling against approaching womanhood. Indeed, her body is astonishingly childlike, and she has often ceased to menstruate. She typically has low self esteem and believes that others, particularly her parents, are controlling her life. Her weight is something that she can control. Genetics and chemical imbalances in the brain may also elevate the risk of developing anorexia. Cases of anorexia nervosa are increasing in the elderly, possibly as a result of diminished smell and taste sensations, as well as difficulties chewing and swallowing. This eating disorder can be a onetime, short-term experience or a lifelong obsession.

Bulimia

A person suffering from **bulimia** (bu-lim'e-ah) is often of normal weight. She eats whatever she wants, often in huge amounts, but she then rids her body of the thousands of extra calories by vomiting, taking laxatives, or exercising frantically. For an estimated one in five college students, the majority of them female, "bingeing and purging" appears to be a way of coping with stress.

A dentist, who can observe the tooth decay caused by frequent vomiting, may be the first to spot bulimia. The backs of the hands of a person with bulimia may bear telltale scratches from efforts to induce vomiting. In addition, the throat is raw and the

esophageal lining ulcerated from the stomach acid forced forward by vomiting. The binge and purge cycle is hard to break, even with psychotherapy and nutritional counseling.

 PRACTICE

59 What is primary malnutrition? Secondary malnutrition?

60 What happens to the body during starvation?

61 How do marasmus and kwashiorkor differ?

62 How do anorexia nervosa and bulimia differ?

18.9 | Life-Span Changes

Dietary requirements remain generally the same throughout life, but the ability to acquire nutrients may change. The basal metabolic rate (BMR) changes with age. It rises from birth to about age five and then declines until adolescence, when it peaks again. During adulthood, the BMR drops in parallel to decreasing activity levels and shrinking muscle mass. In women, it may spike during pregnancy and breastfeeding, when caloric requirements likewise increase. Table 18.12 shows changes in energy requirements for adults who are healthy and engage in regular, light exercise.

People of all ages gain weight when energy intake exceeds energy output, and lose weight when energy output exceeds energy intake. Age fifty seems to be a key point in energy balance. Before this age, most adults have a positive energy balance and maintain a constant weight, but after fifty their weight may creep up. However, being aware of a decrease in activity and curbing food consumption enables many people over the age of fifty to maintain their weight.

Changing nutrition with age often reflects effects of medical conditions, many of which are more common among older people, and social and economic circumstances. Medications can dampen appetite directly through side effects such as nausea or altered taste perception or affect a person's mood in a way that prevents eating. Poverty may take a greater nutritional toll on older people who either cannot get out to obtain food or who give whatever is available to younger people.

Medical conditions that affect the ability to obtain adequate nutrition include depression, tooth decay and periodontal disease, diabetes mellitus, lactose intolerance, and alcoholism. These conditions may lead to deficiencies that are not immediately obvious. Vitamin A deficiency, for example, may take months or years to

TABLE 18.12	Energy Requirements Decline with Age	
	CAL/DAY	
Age	**Female**	**Male**
23–50	2,000	2,700
51–74	1,800	2,400
75+	1,600	2,050

become noticeable because the liver stores this fat-soluble vitamin. Calcium depletion may not produce symptoms, even as the mineral is taken from bones. The earliest symptom of malnutrition, fatigue, may easily be attributed to other conditions or ignored.

Evidence for vitamin D deficiency related to sun avoidance has a long history. The link between lack of sunlight and development of rickets was noted in 1822, and a century later, researchers realized that sun exposure helps reverse the disease in children.

Other evidence comes from diverse sources, such as women who wear veils and naval personnel serving three-month tours of duty on submarines.

 **PRACTICE**

63 List factors that affect nutrient acquisition as people age. ■

Chapter Summary

18.1 Carbohydrates (page 695)

Carbohydrates are organic compounds primarily used to supply cellular energy.

1. Carbohydrate sources
   a. Carbohydrates are ingested in a variety of forms.
   b. Polysaccharides, disaccharides, and monosaccharides are carbohydrates.
   c. **Cellulose** is a polysaccharide that human enzymes cannot digest, but it provides bulk that facilitates movement of intestinal contents.
2. Carbohydrate use
   a. Carbohydrates are absorbed as monosaccharides.
   b. Enzymes in the liver catalyze reactions that convert fructose and galactose into glucose.
   c. Oxidation releases energy from glucose.
   d. Excess glucose is stored as glycogen or combined to produce fat.
   e. Some cells, such as neurons, require a continuous supply of glucose to survive.
   f. If glucose is scarce, amino acids may react to produce glucose.
   g. Most carbohydrates supply energy; some are used to produce sugars (ribose, deoxyribose, lactose).
3. Carbohydrate requirements
   a. Humans survive with a wide range of carbohydrate intakes.
   b. Poor nutritional status is usually related to low intake of nutrients other than carbohydrates.

18.2 Lipids (page 696)

Lipids are organic compounds that supply energy and are used to build cell structures. They include fats, phospholipids, and cholesterol.

1. Lipid sources
   a. Triglycerides are obtained from foods of plant and animal origins.
   b. Cholesterol is mostly obtained in foods of animal origin.
2. Lipid use
   a. Before fats can be used as an energy source, they must be broken down into glycerol and fatty acids.
   b. **Beta oxidation** decomposes fatty acids.
      (1) Beta oxidation activates fatty acids and breaks them down into segments of two carbon atoms each.
      (2) Fatty acid segments are converted into acetyl coenzyme A, which can then be oxidized in the citric acid cycle.
   c. The liver and adipose tissue control triglyceride metabolism.
   d. Liver enzymes can alter the molecular structures of fatty acids.
   e. Linoleic acid and linolenic acid are **essential fatty acids** which must be obtained through the diet.
   f. The liver regulates cholesterol level by synthesizing or excreting it.
3. Lipid requirements
   a. Humans survive with a wide range of lipid intakes.
   b. The amounts and types of lipids needed for health are unknown.
   c. Fat intake must be sufficient to support absorption and transport of fat-soluble vitamins.

18.3 Proteins (page 698)

Proteins are broken down in digestion. The resulting amino acids can be used to form new protein molecules such as enzymes, clotting factors, keratin, elastin, collagen, actin, myosin, hormones, and antibodies, or can be used as energy sources. Before amino acids can be used as energy sources, they must be deaminated, forming the waste urea. During starvation, tissue proteins may be used as energy sources, causing the tissues to waste away.

1. Protein sources
   a. Proteins are mainly obtained from meats, dairy products, cereals, and legumes.
   b. Eight amino acids are essential for adults, whereas ten are essential for growing children.
   c. All **essential amino acids** must be present at the same time for growth and repair of tissues to take place.
   d. **Complete proteins** contain adequate amounts of all the essential amino acids needed to maintain the tissues and promote growth.
   e. **Incomplete proteins** lack adequate amounts of one or more essential amino acids.
2. Nitrogen balance
   a. In healthy adults, the gain of protein equals the loss of protein, and a **nitrogen balance** exists.
   b. A starving person has a negative nitrogen balance; a growing child, a pregnant woman, or an athlete in training usually has a positive nitrogen balance.
3. Protein requirements
   a. Proteins and amino acids are needed to supply essential amino acids and nitrogen for the synthesis of nitrogen-containing molecules.
   b. The consequences of protein deficiencies are particularly severe among growing children.

18.4 Energy Expenditures (page 701)

Energy is of prime importance to survival and may be obtained from carbohydrates, fats, or proteins.

1. Energy values of foods
   a. The potential energy values of foods are expressed in **calories.**
   b. When energy losses due to incomplete absorption and incomplete oxidation are taken into account, 1 gram of carbohydrate or 1 gram of protein yields about 4 calories, whereas 1 gram of fat yields about 9 calories.
2. Energy requirements
   a. The amount of energy required varies from person to person.
   b. Factors that influence energy requirements include **basal metabolic rate,** muscular activity, body temperature, and nitrogen balance.
3. Energy balance
   a. **Energy balance** exists when caloric intake equals caloric output.
   b. If energy balance is positive, body weight increases; if energy balance is negative, body weight decreases.
4. Desirable weight
   a. The most common nutritional disorders involve caloric imbalances.
   b. Average weights of persons 25–30 years of age are desirable for older persons as well.
   c. **Body mass index** assesses weight taking height into account.
   d. A BMI between 25 and 30 indicates overweight, and above 30, **obesity.**

18.5 Appetite Control (page 704)

1. **Appetite** is the drive that compels us to eat. Food powers the activities of life.
2. Hormones control appetite by affecting the arcuate nucleus, a part of the hypothalamus.
3. **Leptin** and **ghrelin** are hormones that affect appetite.

18.6 Vitamins (page 705)

Vitamins are organic compounds (other than carbohydrates, lipids, and proteins), that cannot be synthesized by body cells in adequate amounts and are essential for normal metabolic processes.

1. Fat-soluble vitamins
   a. General characteristics
      (1) Fat-soluble vitamins are carried in lipids and are influenced by the same factors that affect lipid absorption.
      (2) They resist the effects of heat; thus, they are not destroyed by cooking or food processing.
   b. Vitamin A
      (1) **Vitamin A** exists in several forms, is synthesized from carotenes, and is stored in the liver.
      (2) It is an **antioxidant** required for production of visual pigments.
   c. Vitamin D
      (1) **Vitamin D** is a group of related steroids.
      (2) It is found in certain foods and is produced commercially; it can also be synthesized in the skin.
      (3) When needed, vitamin D is converted by the kidneys to an active form that functions as a hormone and promotes the intestine's absorption of calcium and phosphorus.
   d. Vitamin E
      (1) **Vitamin E** is an antioxidant.
      (2) It is stored in muscles and adipose tissue.
      (3) It prevents breakdown of polyunsaturated fatty acids and stabilizes cell membranes.
   e. Vitamin K
      (1) **Vitamin K** is in foods and is produced by intestinal bacteria.
      (2) Some vitamin K is stored in the liver.
      (3) It is used to produce prothrombin, required for blood clotting.
2. Water-soluble vitamins
   a. General characteristics
      (1) Water-soluble vitamins include the B vitamins and vitamin C.
      (2) Cooking or processing food destroys some water-soluble vitamins.
      (3) **B vitamins** make up a group called the vitamin B complex and oxidize carbohydrates, lipids, and proteins.
   b. Vitamin B complex
      (1) **Thiamine (vitamin B_1)**
         (a) Thiamine functions as part of coenzymes that oxidize carbohydrates and synthesize ribose.
         (b) Small amounts are stored in the tissues; excess is excreted in the urine.
         (c) Quantities needed vary with caloric intake.
      (2) **Riboflavin (vitamin B_2)**
         (a) Riboflavin functions as part of several enzymes and coenzymes essential to the oxidation of glucose and fatty acids.
         (b) Its absorption is regulated by an active transport system; excess is excreted in the urine.
         (c) Quantities required vary with caloric intake.
      (3) **Niacin (vitamin B_3 or nicotinic acid)**
         (a) Niacin functions as part of coenzymes required for the oxidation of glucose and for the synthesis of proteins and fats.
         (b) It can be synthesized from tryptophan; daily requirement varies with the tryptophan intake.
      (4) **Pantothenic acid (vitamin B_5)**
         (a) Pantothenic acid functions as part of coenzyme A; thus, it is essential for energy-releasing mechanisms.
         (b) Most diets provide sufficient amounts; deficiencies are rare.
      (5) Vitamin B_6
         (a) **Vitamin B_6** is a group of compounds that function as coenzymes in metabolic pathways that synthesize proteins, certain amino acids, antibodies, and nucleic acids.
         (b) Its requirement varies with protein intake.
      (6) **Biotin (vitamin B_7)**
         (a) Biotin is a coenzyme required for the metabolism of amino acids and fatty acids, and for nucleic acid synthesis.

(b) It is stored in metabolically active organs, including the brain, liver, and kidneys.

(7) **Folacin (vitamin B₉ or folic acid)**

(a) Liver enzymes catalyze reactions that convert folacin to physiologically active folinic acid.

(b) It is a coenzyme needed for the metabolism of certain amino acids, DNA synthesis, and the production of normal red blood cells.

(8) **Cyanocobalamin (vitamin B₁₂)**

(a) The cyanocobalamin molecule contains cobalt.

(b) Its absorption is regulated by the secretion of intrinsic factor from the gastric glands.

(c) It functions as part of coenzymes needed for nucleic acid synthesis and for the metabolism of carbohydrates and fats.

(d) It is important to erythrocyte production and myelin formation in the central nervous system.

c. **Ascorbic acid (vitamin C)**

(1) Vitamin C is similar chemically to monosaccharides.

(2) It is required for collagen production, the metabolism of certain amino acids, and iron absorption.

(3) It is not stored in large amounts; excess is excreted in the urine.

18.7 Minerals (page 711)

1. Characteristics of minerals

    a. **Minerals** account for about 4% of body weight.

    b. Minerals are usually incorporated into organic molecules, although some are in inorganic compounds or are free ions.

    c. They compose structural materials, function in enzymes, and play vital roles in various metabolic processes.

    d. Homeostatic mechanisms regulate mineral concentrations.

    e. The physiologically active form of minerals is the ionized form.

2. **Major minerals**

    a. Calcium

    (1) **Calcium** is essential for forming bones and teeth, neurotransmitter release, contracting muscle fibers, the cardiac action potential, clotting blood, and activating various enzymes.

    (2) Existing calcium concentration, vitamin D, protein intake, and motility of the digestive tract affect calcium absorption.

    b. Phosphorus

    (1) **Phosphorus** is incorporated into the salts of bones and teeth.

    (2) It participates in nearly all metabolic reactions as a constituent of nucleic acids, proteins, and some vitamins.

    (3) It also is in the phospholipids of cell membranes, in ATP, and in phosphates of body fluids.

    c. Potassium

    (1) Potassium is concentrated inside cells.

    (2) It maintains osmotic pressure, regulates pH, and plays a role in impulse conduction in neurons.

    d. Sulfur

    (1) **Sulfur** is incorporated into two of the twenty amino acids.

    (2) It is also in thiamine, insulin, biotin, and mucopolysaccharides.

    e. Sodium

    (1) Most **sodium** is in extracellular fluids or is bound to the inorganic salts of bone.

    (2) The kidneys, under the influence of aldosterone, regulate the blood concentration of sodium.

    (3) Sodium helps maintain solute concentration and regulates water balance.

    (4) It is essential for impulse conduction in neurons and moving substances through cell membranes.

    f. Chlorine

    (1) **Chlorine** is closely associated with sodium as chloride ions.

    (2) It acts with sodium to help maintain osmotic pressure, regulate pH, and maintain electrolyte balance.

    (3) Chlorine is essential for hydrochloric acid formation and for carbon dioxide transport by red blood cells.

    g. Magnesium

    (1) **Magnesium** is abundant in the bones as phosphates and carbonates.

    (2) It functions in ATP production and in the breakdown of ATP to ADP.

    (3) A reserve supply of magnesium is stored in the bones; excesses are excreted in the urine.

3. **Trace elements**

    a. Iron

    (1) **Iron** is part of hemoglobin in red blood cells and myoglobin in muscles.

    (2) A reserve supply of iron is stored in the liver, spleen, and bone marrow.

    (3) It is required to catalyze vitamin A formation; it is also incorporated into various enzymes and the cytochrome molecules.

    b. Manganese

    (1) Most **manganese** is concentrated in the liver, kidneys, and pancreas.

    (2) It is necessary for normal growth and development of skeletal structures and other connective tissues; it is essential for the synthesis of fatty acids, cholesterol, and urea.

    c. Copper

    (1) Most **copper** is concentrated in the liver, heart, and brain.

    (2) It is required for hemoglobin synthesis, bone development, melanin production, and myelin formation.

    d. Iodine

    (1) **Iodine** is most highly concentrated in the thyroid gland.

    (2) It is an essential component of thyroid hormones.

    (3) It is often added to foods as iodized table salt.

    e. Cobalt

    (1) **Cobalt** is widely distributed throughout the body.

    (2) It is an essential part of cyanocobalamin and is required for the synthesis of several enzymes.

f. Zinc
  (1) **Zinc** is most concentrated in the liver, kidneys, and brain.
  (2) It is a component of several enzymes that take part in digestion, respiration, and metabolism.
  (3) It is necessary for normal wound healing.
g. Fluorine
  (1) The teeth concentrate **fluorine.**
  (2) It is incorporated into enamel and prevents dental caries.
h. Selenium
  (1) The liver and kidneys store **selenium.**
  (2) It is a component of certain enzymes.
i. Chromium
  (1) **Chromium** is widely distributed throughout the body.
  (2) It regulates glucose use.

18.8 Healthy Eating (page 718)

1. An adequate diet provides sufficient energy and essential nutrients to support optimal growth, as well as maintenance and repair, of tissues.
2. Individual needs vary so greatly that it is not possible to design a diet adequate for everyone.
3. Devices to help consumers make healthy food choices include Recommended Daily Allowances, Recommended Dietary Allowances, food group plans such as **"MyPlate,"** and food labels.

4. **Malnutrition**
  a. Poor nutrition is due to lack of foods or failure to wisely use available foods.
  b. Primary malnutrition is due to poor diet.
  c. Secondary malnutrition is due to an individual characteristic that makes a normal diet inadequate.
5. Starvation
  a. A person can survive fifty to seventy days without food.
  b. A starving body digests itself, starting with carbohydrates, then fats, then proteins.
  c. Symptoms include low blood pressure, slow pulse, chills, dry skin, hair loss, and poor immunity. Finally, vital organs cease to function.
  d. **Marasmus** is undernutrition involving a lack of calories and protein.
  e. **Kwashiorkor** is protein starvation.
  f. **Anorexia nervosa** is a self-starvation eating disorder.
  g. **Bulimia** is an eating disorder characterized by bingeing and purging.

18.9 Life-Span Changes (page 723)

1. Basal metabolic rate rises in early childhood, declines, then peaks again in adolescence, with decreasing activity during adulthood.
2. Weight gain, at any age, occurs when energy in exceeds energy out, and weight loss occurs when energy out exceeds energy in.
3. Changing nutrition with age reflects medical conditions and social and economic circumstances.

 CHAPTER ASSESSMENTS

18.1–18.3 Carbohydrates–Proteins

1. Identify dietary sources of carbohydrates. (p. 695)
2. Summarize the importance of cellulose in the diet. (p. 695)
3. Explain what happens to excess glucose in the body. (p. 695)
4. Explain why a temporary drop in blood glucose concentration may impair nervous system functioning. (pp. 695–696)
5. List some factors that affect an individual's need for carbohydrates. (p. 696)
6. Identify dietary sources of lipids. (p. 696)
7. Define *beta oxidation*. (p. 697)
8. Explain how fats may provide energy. (p. 697)
9. Describe the liver's role in fat metabolism. (p. 697)
10. Review the major functions of cholesterol. (p. 697)
11. Define *deamination*, and explain its importance. (p. 698)
12. Identify dietary sources of proteins. (p. 699)
13. Distinguish between essential and nonessential amino acids. (p. 699)
14. Explain why all of the essential amino acids must be present for growth. (p. 699)
15. Distinguish between complete and incomplete proteins. (p. 699)

16. _____ is when the amount of nitrogen taken in is equal to the amount excreted. (p. 699)
17. Explain why a protein deficiency may accompany edema. (p. 700)

18.4 Energy Expenditures

18. Define *calorie*. (p. 701)
19. Explain how the caloric values of foods are determined. (p. 701)
20. Define *basal metabolic rate*. (p. 701)
21. List some of the factors that affect the BMR. (p. 701)
22. _____ exists when caloric intake in the form of foods equals caloric output from basal metabolic rate and muscular activities. (p. 702)
23. Distinguish among underweight, desirable weight, overweight, and obesity. (p. 702)

18.5 Appetite Control

24. Define *appetite*. (p. 704)
25. Identify the part of the brain where hormones act, controlling appetite. (p. 704)
26. Explain how leptin and ghrelin influence appetite. (p. 704)

18.6 Vitamins

27 Match the vitamins with their general functions, and indicate if the vitamin is fat-soluble or water-soluble. Functions may be used more than once, and more than one function may be applied to a vitamin. (pp. 706–711)

(1) vitamin A
(2) vitamin B$_1$ (thiamine)
(3) vitamin B$_2$ (riboflavin)
(4) vitamin B$_3$ (niacin)
(5) vitamin B$_5$ (pantothenic acid)
(6) vitamin B$_6$
(7) vitamin B$_7$ (biotin)
(8) vitamin B$_9$ (folacin)
(9) vitamin B$_{12}$ (cyanocobalamin)
(10) vitamin C (ascorbic acid)
(11) vitamin D
(12) vitamin E
(13) vitamin K

A. part of coenzyme A in oxidation of carbohydrates
B. coenzyme in ribose synthesis
C. necessary for synthesis of visual pigments
D. required for synthesis of prothrombin
E. required to produce collagen
F. required to synthesize nucleic acids
G. promotes normal red blood cell production
H. plays a role in myelin synthesis
I. antioxidant, helps stabilize cell membranes
J. promotes development of teeth and bones
K. required to produce antibodies
L. required for oxidation of glucose
M. part of coenzymes to synthesize proteins and fats

18.7 Minerals

28 Match the minerals/elements with their functions, and indicate whether each is a major mineral or a trace element required for nutrition. Functions may be used more than once, and more than one function may be applied to a mineral or trace element. (pp. 713–717)

(1) calcium
(2) chlorine
(3) chromium
(4) cobalt
(5) copper
(6) fluorine
(7) iodine
(8) iron
(9) magnesium
(10) manganese
(11) phosphorus
(12) potassium
(13) selenium
(14) sodium
(15) sulfur
(16) zinc

A. essential for the use of glucose
B. component of certain enzymes
C. component of tooth enamel
D. component of teeth and bones
E. helps maintain intracellular osmotic pressure
F. essential part of certain amino acids
G. helps maintain extracellular fluid osmotic pressure
H. necessary for normal wound healing
I. component of cyanocobalamin
J. essential for synthesis of thyroid hormones
K. required in metabolic reactions associated with ATP production
L. component of hemoglobin
M. essential for hemoglobin synthesis and melanin production
N. required for cholesterol synthesis and urea formation

18.8 Healthy Eating

29 Define *adequate diet*. (p. 718)
30 Explain various methods to eat an adequate diet. (pp. 718–719)
31 Define *malnutrition*. (p. 720)
32 Contrast primary and secondary malnutrition. (p. 720)
33 Discuss bodily changes during starvation. (p. 722)
34 Distinguish among marasmus, kwashiorkor, anorexia nervosa, and bulimia. (pp. 722–723)

18.9 Life-Span Changes

35 Factors that may lead to inadequate nutrition later in life include _____. (p. 723)
a. medical conditions
b. social circumstances
c. economic circumstances
d. all of the above
36 Name some medical conditions that affect the ability to obtain adequate nutrition as a person ages. (pp. 723–724)

INTEGRATIVE ASSESSMENTS/CRITICAL THINKING

Outcomes 4.4, 9.3, 18.1, 18.2, 18.3, 18.4, 18.8

1. Which of the diets described in the following chart would be most appropriate for an athlete training for a triathlon (biking, swimming, running event)? What is a problem with all of these diets for such a person?

Diet	Total Calories/Day	% Fat	% Carbohydrate	% Protein
I	1,450	10–20	70	17
II	1,450	25	60	15
III	1,400	60	10	30

Outcomes 4.4, 18.1, 18.2, 18.3, 18.4, 18.6

2. A young man takes several vitamin supplements each day, claiming that they give him energy. Is he correct? Why or why not?

Outcomes 13.8, 18.1, 18.2, 18.3, 18.4, 18.5

3. Why does the blood sugar concentration of a person whose diet is low in carbohydrates remain stable?

Outcomes 18.1, 18.2, 18.3, 18.4

4. Using nutrient tables, calculate the number of grams of carbohydrate, lipid, and protein that you eat in a typical day, and the total calories in these foods. Suggest ways to improve your diet.

Outcomes 18.1, 18.2, 18.3, 18.4, 18.5, 18.6, 18.7, 18.8, 18.9

5. How do you think the nutritional requirements of a healthy twelve-year-old boy, a twenty-four-year-old pregnant woman, and a healthy sixty-year-old man differ?

Outcomes 18.1, 18.2, 18.3, 18.6, 18.7

6. Examine the label information on the packages of a variety of breakfast cereals. Which types of cereals provide the best sources of carbohydrates, lipids, proteins, vitamins, and minerals? Which major nutrients are lacking in these cereals?

Outcome 18.3

7. If a person decided to avoid eating meat and other animal products, such as milk, cheese, and eggs, what foods might be included in the diet to provide essential amino acids?

 # ONLINE STUDY TOOLS

Connect Interactive Questions Reinforce your knowledge using assigned interactive questions covering the sources and metabolism of nutrients.

Connect Integrated Activity Using your understanding of the functions of various nutrients, can you relate nutritional disorders to the correct nutrient?

LearnSmart Discover which chapter concepts you have mastered and which require more attention. This adaptive learning tool is personalized, proven, and preferred.

19

Respiratory System

LEARNING OUTCOMES

After you have studied this chapter, you should be able to:

19.1 Overview of the Respiratory System

1 Identify the general functions of the respiratory system. (p. 731)

2 Explain why respiration is necessary for cellular survival. (p. 731)

19.2 Organs of the Respiratory System

3 Name and describe the locations of the organs of the respiratory system. (pp. 732–741)

4 Describe the functions of each organ of the respiratory system. (pp. 732–741)

19.3 Breathing Mechanism

5 Explain how inspiration and expiration are accomplished. (pp. 741, 744–746)

6 Describe each of the respiratory air volumes and capacities. (pp. 747–748)

7 Show how alveolar ventilation rate is calculated. (p. 749)

8 List several nonrespiratory air movements, and explain how each occurs. (p. 749)

19.4 Control of Breathing

9 Locate the respiratory areas, and explain control of normal breathing. (pp. 750–751)

10 Discuss how various factors affect breathing. (pp. 752–753)

19.5 Alveolar Gas Exchanges

11 Describe the structure and function of the respiratory membrane. (pp. 754–755)

12 Explain the importance of partial pressure in diffusion of gases. (pp. 755–756)

19.6 Gas Transport

13 Explain how the blood transports oxygen and carbon dioxide. (pp. 758–760)

14 Describe gas exchange in the pulmonary and systemic circuits. (pp. 758–760)

19.7 Life-Span Changes

15 Describe the effects of aging on the respiratory system. (p. 762)

Falsely colored scanning electron micrograph of the trachea lining consisting of mucous-secreting goblet cells (brown) amid ciliated epithelium (red) (3,700×).

THE WHOLE PICTURE

The respiratory system uses skeletal muscles, which are under voluntary control, but unless you are blowing up party balloons or playing a wind instrument, you may not be aware of the air moving in and out of your lungs. Parts of the brainstem control breathing automatically, constantly bringing in oxygen to support aerobic production of ATP and constantly eliminating just the right amount of carbon dioxide to maintain homeostasis.

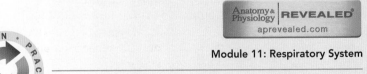

Anatomy & Physiology **REVEALED**®
aprevealed.com

Module 11: Respiratory System

alveol-, small cavity: *alveolus*—microscopic air sac in a lung.

bronch-, windpipe: *bronchus*—primary branch of the trachea.

carcin-, spreading sore: *carcinoma*—type of cancer.

carin-, keel-like: *carina*—ridge of cartilage between the right and left bronchi.

cric-, ring: *cricoid* cartilage—ring-shaped mass of cartilage at the base of the larynx.

epi-, upon: *epiglottis*—flaplike structure that partially covers the opening into the larynx during swallowing.

exhal-, to breathe out: *exhalation*—to expel air from the lungs.

hem-, blood: *hemoglobin*—pigment in red blood cells.

inhal-, to breathe in: *inhalation*—to take air into the lungs.

phren-, diaphragm: *phrenic* nerve—nerve associated with the cervical plexuses that stimulates the muscle fibers of the diaphragm to contract.

tuber-, swelling: *tuberculosis*—infectious disease in which fibrous masses form in the lungs.

19.1 | Overview of the Respiratory System

The respiratory system consists of passages that filter incoming air and transport it into the body, into the lungs, and to the many microscopic air sacs where gases are exchanged. The entire process of exchanging gases between the atmosphere and body cells is called **respiration** (res″pĭ-ra′shun). It consists of several events:

- movement of air in and out of the lungs, commonly called breathing, or *ventilation*
- exchange of gases between the air in the lungs and the blood, called *external respiration*
- transport of gases by the blood between the lungs and body cells
- exchange of gases between the blood and the body cells, also called *internal respiration*
- oxygen (O_2) use and production of carbon dioxide (CO_2) by body cells as part of the process of *cellular respiration*

Respiration occurs on a macroscopic level—it is a function of an organ system. However, the reason that body cells must exchange gases—take up oxygen and release carbon dioxide—is apparent at the cellular and molecular levels.

 RECONNECT

To Chapter 4, Aerobic Reactions, pages 129–131.

Cellular respiration enables cells to harness energy held in the chemical bonds of nutrient molecules. In aerobic reactions, cells liberate energy from these molecules by removing electrons and channeling them through a series of carriers called the electron transport chain, yielding ATP. At the end of this chain, electrons bind oxygen atoms and hydrogen ions to produce water molecules. Without oxygen, these reactions cease.

The aerobic reactions also produce CO_2. CO_2 is a metabolic waste that combines with water to form carbonic acid, helping to maintain blood pH. Too much CO_2, however, lowers blood pH, compromising homeostasis. The respiratory system both provides oxygen for aerobic reactions and eliminates CO_2 at the appropriate rate to maintain the pH of the internal environment.

 PRACTICE

1   What is respiration?

19.2 | Organs of the Respiratory System

The organs of the respiratory system can be divided into two groups, or tracts. The *upper respiratory tract* includes the nose, nasal cavity, sinuses, and pharynx. The *lower respiratory tract* includes the larynx, trachea, bronchial tree, and lungs (fig. 19.1).

 CAREER CORNER

Respiratory Therapist

The teachers welcome an education session from the local hospital's respiratory therapist. Several children in the elementary school have asthma, and the teachers have requested this visit so that they can better understand the condition. After the respiratory therapist's presentation, the teachers feel more confident that they can help the students, and also explain to their classmates what is happening.

Respiratory therapists assess patients who are experiencing breathing problems and treat them, under supervision of a physician. They work in a variety of situations. A respiratory therapist might help a patient breathe on his or her own following anesthesia, instruct a patient newly diagnosed with emphysema on how to use supplemental oxygen, check on how patients on a hospital floor are breathing, or teach parents of a young child with cystic fibrosis how to apply pressure to the child's chest to shake free the thick mucus that builds up in the child's lungs. The overall goal is to help patients on breathing assistance be as independent as possible.

A respiratory therapist is an expert in how the heart and lungs function. Educational requirements include an associates degree and certification, plus continuing education.

The respiratory structures that air passes through face the outside environment, and except for the parts where exchange of gases takes place, they are lined with mucous membrane.

Nose

The nose is covered with skin and is supported internally by muscle, bone, and cartilage. Its two *nostrils* (external nares) provide openings through which air can enter and leave the nasal cavity. Many internal hairs in these openings prevent entry of large particles carried in the air.

Nasal Cavity

The **nasal cavity,** a hollow space behind the nose, is divided medially into right and left portions by the **nasal septum.** This cavity is separated from the cranial cavity by the cribriform plate of the ethmoid bone and from the oral cavity by the hard palate.

The nasal septum may bend during birth or shortly before adolescence. Such a *deviated septum* may obstruct the nasal cavity, making breathing difficult.

As figure 19.2 shows, **nasal conchae** (turbinate bones) curl out from the lateral walls of the nasal cavity on each side, forming passageways called the *superior, middle,* and *inferior meatuses* (see chapter 7, pp. 220 and 222). The nasal chonchae support the mucous membrane that lines the nasal cavity and help increase its surface area.

The upper posterior portion of the nasal cavity, below the cribriform plate, is slitlike, and its lining contains the olfactory receptors that provide the sense of smell. The remainder of the cavity conducts air to and from the nasopharynx.

The mucous membrane lining the nasal cavity has pseudostratified ciliated epithelium rich in mucous-secreting goblet cells (see chapter 5, pp. 153–154). It also includes an extensive network of blood vessels and normally appears pinkish. As air passes over the membrane, heat radiates from the blood and warms the air, adjusting its temperature to that of the body. At the same time, evaporation of water from the mucous lining moistens the

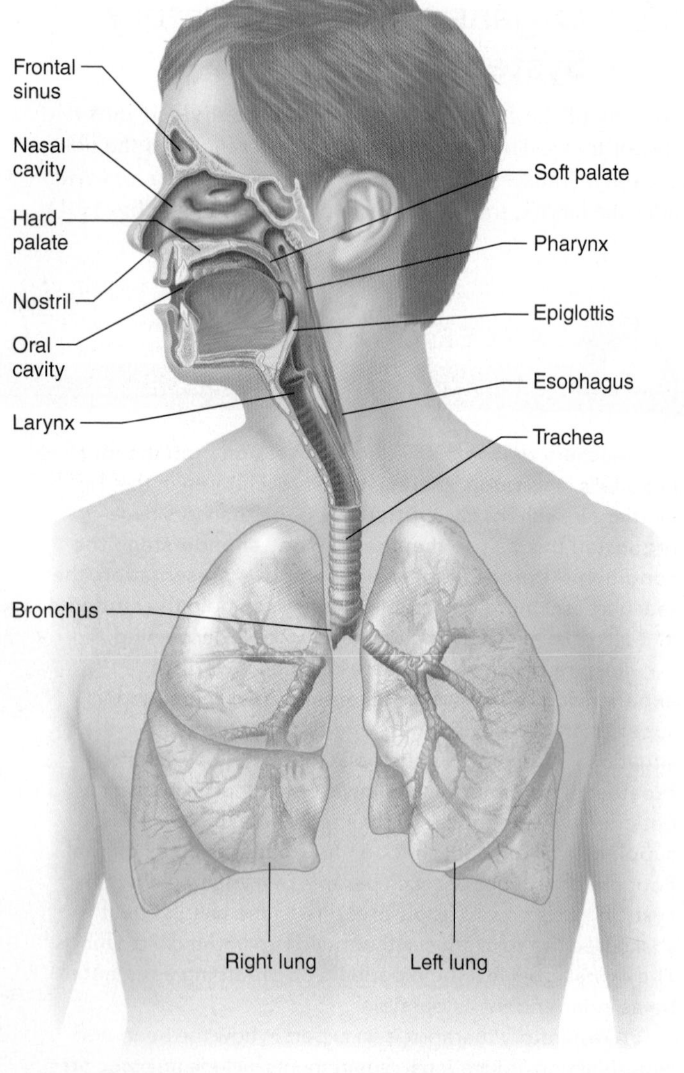

FIGURE 19.1 Parts of the respiratory system.

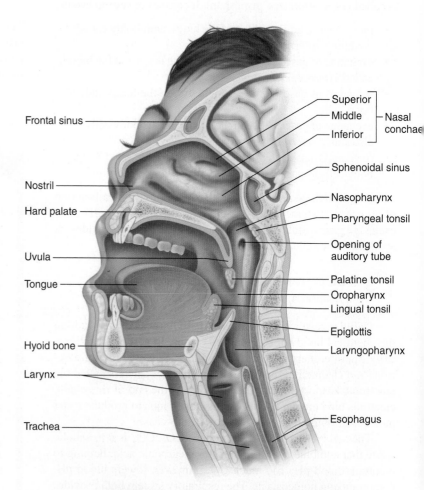

FIGURE 19.2 AP|R Major structures associated with the respiratory tract in the head and neck.

air. The sticky mucus secreted by the mucous membrane entraps dust and other small particles entering with the air.

As the cilia of the epithelial cells move, they push a thin layer of mucus toward the pharynx (fig. 19.3), where the mucus and any entrapped particles are swallowed. In the stomach, gastric juice destroys microorganisms in the mucus, including pathogens. In this way, the mucous membrane keeps particles from reaching the lower air passages, preventing respiratory infections. Clinical Application 19.1 discusses how cigarette smoking impairs the function of the respiratory system, beginning with the mucus and cilia.

 PRACTICE

2 Which organs constitute the respiratory system?
3 What is the function of the mucous membrane that lines the nasal cavity?
4 What is the function of the cilia on the cells that line the nasal cavity?

Sinuses

Recall from chapter 7 (pp. 218, 220, and 221) that the *sinuses* (paranasal sinuses) are air-filled spaces in the *frontal, sphenoid, ethmoid,* and *maxillary bones* of the skull (fig. 19.4). The sinuses reduce the weight of the skull. They also serve as resonant chambers that affect the quality of the voice. The sinuses open into the nasal cavity and are lined with mucous membranes that are continuous with the lining of the nasal cavity, allowing mucus secretions to drain from the sinuses into the nasal cavity. Membranes that are inflamed and swollen because of nasal infections or allergic reactions (sinusitis) may block this drainage, increasing pressure in a sinus and causing headache.

The sinuses reduce the weight of the skull. They also serve as resonant chambers that affect the quality of the voice.

> It is possible to illuminate a person's frontal sinus in a darkened room by holding a small flashlight just beneath the eyebrow. Similarly, holding the flashlight in the mouth illuminates the maxillary sinuses.

 PRACTICE

5 Where are the sinuses located?
6 What are the functions of the sinuses?

Pharynx

The **pharynx** (far'inks) is posterior to the nasal cavity, oral cavity, and larynx. It is a passageway for food moving from the oral cavity to the esophagus and for air passing between the nasal cavity and the larynx. The pharynx also aids in producing the sounds of speech. It can be divided into the nasopharynx, the oropharynx, and the laryngopharynx (see fig. 19.2):

1. The **nasopharynx** (na"zo-far'inks) is superior to the soft palate. It communicates with the nasal cavity and provides a passageway for air during breathing. The auditory tubes, which connect the pharynx with the middle ears, open through the walls of the nasopharynx (see chapter 12, p. 459).
2. The **oropharynx** (o"ro-far'inks) is posterior to the mouth and soft palate and inferior to the nasopharynx, projecting downward to the upper border of the epiglottis. It is a passageway both for food moving downward from the mouth and for air moving to and from the nasal cavity.
3. The **laryngopharynx** (lah-ring"go-far'inks) is just inferior to the oropharynx. It extends from the upper border of the epiglottis downward to the lower border of the cricoid cartilage of the larynx. The laryngopharynx is continuous with both the esophagus and the larynx.

Larynx

The **larynx** (lar'inks) is an enlargement in the airway superior to the trachea (see reference plates 9 and 21, pp. 46 and 53). It is a passageway for air moving in and out of the trachea and prevents foreign objects from entering the trachea. The larynx also houses the *vocal cords*.

The larynx is composed of a framework of muscles and cartilages bound by elastic tissue. The largest of the cartilages are the thyroid, cricoid, and epiglottic cartilages (fig. 19.5). These structures are

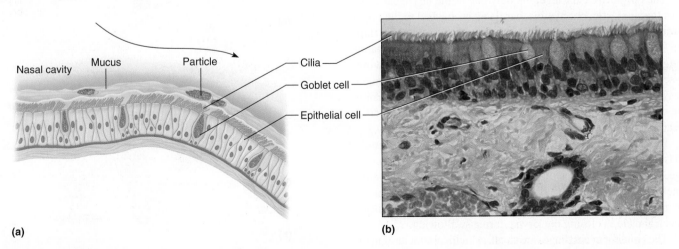

(a)   (b)

FIGURE 19.3 **AP|R** Mucus movement in the respiratory tract. (a) Cilia move mucus and trapped particles from the nasal cavity to the pharynx. (b) Micrograph of ciliated epithelium in the respiratory tract (275×).

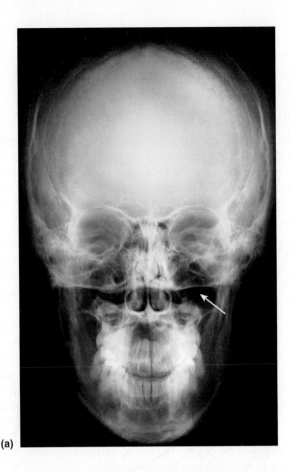

(a)

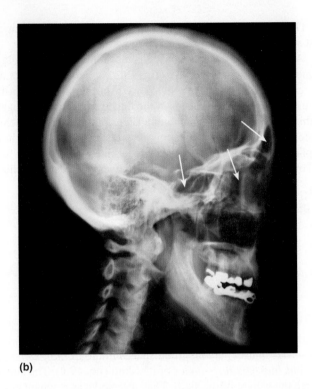

(b)

FIGURE 19.4 Radiograph of a skull (*a*) anterior view and (*b*) lateral view, showing air-filled sinuses (arrows) within the bones. Sinuses include: maxillary sinuses in (a); sphenoidal, ethmoidal, and frontal sinuses in (b).

single. The other laryngeal cartilages—the arytenoid, corniculate, and cuneiform cartilages—are paired.

The **thyroid cartilage** was named for the thyroid gland that covers its lower area. This cartilage is the shieldlike structure that protrudes in the front of the neck and is also called the "Adam's apple." The protrusion typically is more prominent in males than in females because of an effect of male sex hormones on the development of the larynx.

The **cricoid cartilage** lies inferior to the thyroid cartilage. It marks the lowermost portion of the larynx.

The **epiglottic cartilage,** the only one of the laryngeal cartilages that is elastic, not hyaline, cartilage, is attached to the upper border of the thyroid cartilage and is the central part of a flaplike structure called the **epiglottis.** The epiglottis usually stands upright and allows air to enter the larynx. During swallowing, muscular contractions raise the larynx, and the base of the tongue presses the epiglottis downward, partially covering the opening into the larynx, helping to prevent foods and liquids from entering the air passages.

Posteriorly on the larynx, the pyramid-shaped **arytenoid cartilages** are superior to and on either side of the cricoid cartilage. Attached to the tips of the arytenoid cartilages are the tiny, conelike **corniculate cartilages.** These cartilages are attachments for muscles that help regulate tension on the vocal cords during speech and aid in closing the larynx during swallowing.

The **cuneiform cartilages** are small, cylindrical structures in the mucous membrane between the epiglottic and the arytenoid cartilages. They stiffen the soft tissues in this region.

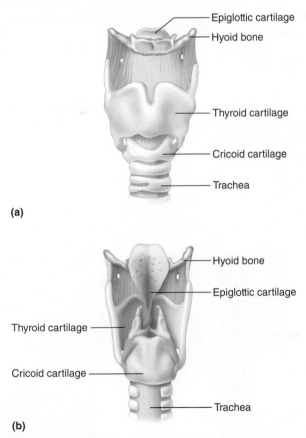

(a)

- Epiglottic cartilage
- Hyoid bone
- Thyroid cartilage
- Cricoid cartilage
- Trachea

(b)

- Hyoid bone
- Epiglottic cartilage
- Thyroid cartilage
- Cricoid cartilage
- Trachea

FIGURE 19.5 AP|R Larynx. (*a*) Anterior and (*b*) posterior views of the larynx.

The Effects of Cigarette Smoking on the Respiratory System

Damage to the respiratory system from cigarette smoking is slow, progressive, and potentially deadly. A healthy respiratory system is continuously cleansed. The mucus produced by the respiratory tubes traps dirt and pathogens, which cilia sweep toward the mouth, where they can be eliminated. Smoking greatly impairs this action. With the first inhalation of smoke, the beating of cilia slows. With time, the cilia become paralyzed and, eventually, degenerate, leading to "smoker's cough" when the cilia no longer effectively remove mucus, which must be coughed up. Coughing is usually worse in the morning because mucus has accumulated during sleep.

Smokers produce excess mucus and it accumulates, clogging the air passageways. Pathogens normally removed now have easier access to the respiratory surfaces, and the resulting lung congestion favors their growth. This is why smokers have respiratory infections more often than nonsmokers. A lethal chain reaction begins. Cough leads to chronic bronchitis, and increased mucus production and bronchial lining thickening compromises breathing, especially expiration (exhalation). As expiration becomes more difficult, air gets trapped in the alveoli (microscopic air sacs in the lungs). Alveolar walls may be destroyed. This is the beginning of smoking-induced *emphysema,* which is fifteen times more common among individu-

als who smoke a pack of cigarettes a day than among nonsmokers.

Cellular changes induced by smoking set the stage for the development of lung cancer. First, cells in the outer border of the bronchial lining begin to divide more rapidly than usual. Eventually, these cells displace the ciliated cells. Their nuclei begin to resemble those of cancerous cells—large and distorted with abnormal numbers of chromosomes. Up to this point, the damage can be repaired if smoking ceases. If smoking continues, these cells may eventually break through the basement membrane and begin dividing within the lung tissue, forming a tumor with the potential of spreading throughout lung tissue (figs. 19A and 19B) and beyond, such as to the brain or bones. While more than 80% of lung cancer cases are due to cigarette smoking, only 20% of smokers develop the cancer. Genes may influence which smokers develop cancer.

Smoking endangers more than the smoker. Exposure to environmental tobacco smoke (ETS)—also called secondhand smoke—may be as dangerous as smoking. ETS has two sources: sidestream smoke comes from lit cigarettes, cigars, or pipes, and mainstream smoke is exhaled by smokers. The smoke contains more than 4,000 chemical compounds, including irritants, carcinogens, mutagens, and many toxins.

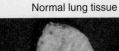
Normal lung tissue

Cancerous lung tissue

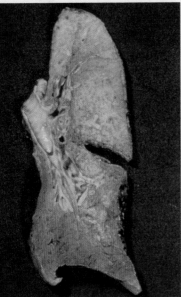

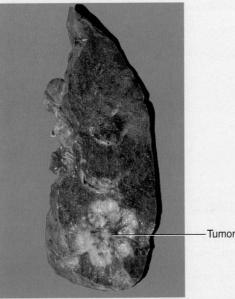

Tumor

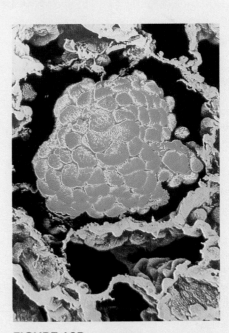

FIGURE 19A The lung on the left is normal. A cancerous tumor has invaded the lung on the right, taking up much of the lung space.

FIGURE 19B Lung cancer may begin as a tiny tumor (orange) growing in an alveolus, a microscopic air sac (125×). This image is falsely colored.

Inside the larynx, two pairs of horizontal folds composed of muscle tissue and connective tissue with a covering of mucous membrane extend inward from the lateral walls. The upper folds (vestibular folds) are called *false vocal cords* because they do not

produce sounds. Muscle fibers within these folds help close the larynx during swallowing.

The lower folds of the larynx are the *true vocal cords.* They have elastic fibers and are responsible for vocal sounds, which are created when air is forced between these folds, vibrating them. This action generates sound waves, which are formed into words by changing the shapes of the pharynx and oral cavity and by using the tongue and lips. Figure 19.6 shows both pairs of cords.

Changing tension on the vocal cords, by contracting or relaxing laryngeal muscles, controls *pitch* (musical tone) of the voice. Increasing the tension produces a higher pitch, and decreasing the tension creates a lower pitch.

The *intensity* (loudness) of a vocal sound depends upon the force of the air passing over the vocal cords. Stronger blasts of air result in greater vibration of the vocal cords and louder sound.

The true vocal cords and the opening between them form the **glottis** (glot'is). During normal breathing the vocal cords remain relaxed, and the glottis is an open triangular slit. When a person swallows food or liquid, muscles in the false vocal folds close the glottis. Along with the epiglottis pressing downward to partially cover the glottis, this action helps prevent food or liquid from entering the trachea (fig. 19.7). The mucous membrane that lines the larynx continues to filter incoming air by entrapping particles as waving cilia move them toward the pharynx by ciliary action.

 PRACTICE

7 What part of the respiratory tract is shared with the alimentary canal?

8 Describe the structure of the larynx.

9 How do the vocal cords produce sounds?

10 What is the function of the epiglottis? Of the glottis?

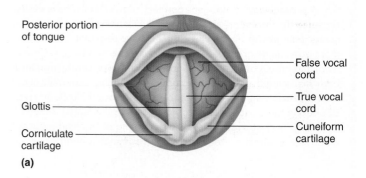
(a)

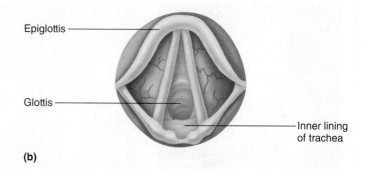
(b)

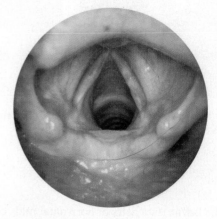
(c)

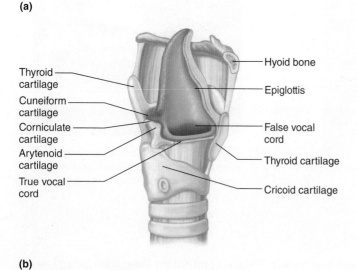
(a)

(b)

FIGURE 19.6 AP|R Larynx. (a) Frontal section and (b) sagittal section of the larynx.

FIGURE 19.7 The vocal cords as viewed from above with the glottis (a) closed and (b) open. (c) Photograph of the glottis (open) and vocal cords.

Trachea

The **trachea** (windpipe) is a flexible cylindrical tube about 2.5 centimeters in diameter and 12.5 centimeters in length. It extends downward anterior to the esophagus and into the thoracic cavity, where it splits into right and left main bronchi (fig. 19.8 and reference plate 9, p. 46).

The inner wall of the trachea is lined with a ciliated mucous membrane that has many goblet cells. This membrane continues to clean the incoming air and to move entrapped particles upward into the pharynx where the mucus can be swallowed.

Within the tracheal wall are about twenty C-shaped pieces of hyaline cartilage, one above the other. The open ends of these incomplete rings are directed posteriorly, and the gaps between their ends are filled with smooth muscle and connective tissues (figs. 19.9 and 19.10). These cartilaginous rings prevent the trachea from collapsing and blocking the airway. At the same time, the soft tissues that complete the rings in the back allow the nearby esophagus to expand as food moves through it on the way to the stomach.

A blocked trachea can cause asphyxiation in minutes. If swollen tissues, excess secretions, or a foreign object obstruct the trachea, making a temporary, external opening in the trachea, so that a tube can be inserted to bypass the obstruction, can be lifesaving. This procedure, shown in figure 19.11, is called a *tracheostomy*.

Bronchial Tree

The **bronchial tree** (brong′ke-al trē) consists of branched airways leading from the trachea to the microscopic air sacs in the lungs. Its branches begin with the right and left **main (primary) bronchi,** which arise from the trachea at the level of the fifth thoracic vertebrae. A ridge of cartilage called the *carina* separates the openings of the main bronchi (see fig. 19.8). Each bronchus, accompanied by large blood vessels, enters its respective lung.

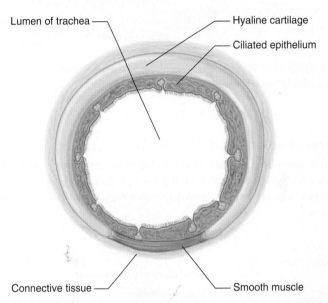

FIGURE 19.9 Cross section of the trachea. Note the C-shaped ring of hyaline cartilage in the wall. The cartilage is absent in the posterior part of the tracheal wall.

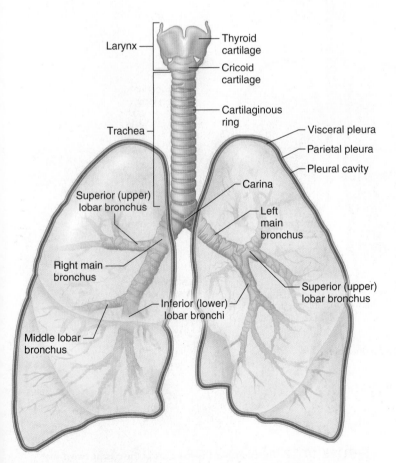

FIGURE 19.8 AP|R The trachea transports air between the larynx and the bronchi.

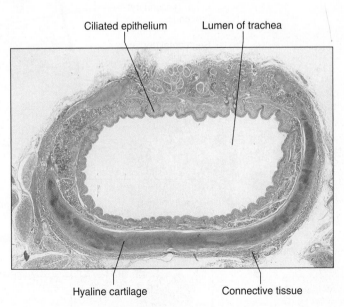

FIGURE 19.10 AP|R Light micrograph of a section of the tracheal wall (100×).

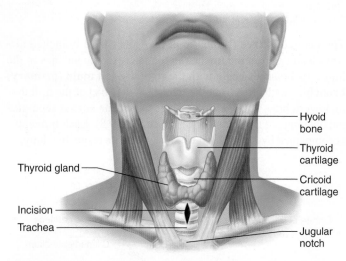

FIGURE 19.11 A tracheostomy may be performed to allow insertion of a tube to bypass an obstruction.

Branches of the Bronchial Tree

Each main bronchus divides into **lobar,** or **secondary, bronchi** (two on the left and three on the right) a short distance from its origin. The secondary bronchi branch repeatedly. The airways appear as an upside-down tree when stripped of their associated blood vessels and tissues (figs. 19.12 and 19.13). The successive divisions of these branches from the trachea to the microscopic air sacs are as follows:

1. **Right** and **left main (primary) bronchi** branch directly off the trachea.
2. Three **lobar (secondary) bronchi** branch from the right main bronchus, and two branch from the left.
3. **Segmental (tertiary) bronchi** supply portions of the lungs called *bronchopulmonary segments.* In most individuals ten such segments are in the right lung and eight are in the left lung.
4. **Intralobular bronchioles** are small branches of the segmental bronchi that enter the basic units of the lung—the *lobules.*
5. **Terminal bronchioles** branch from an intralobular bronchiole. Fifty to eighty terminal bronchioles occupy a lobule of the lung.
6. Two or more **respiratory bronchioles** branch from each terminal bronchiole. These structures, which are short and about 0.5 millimeter in diameter, are called "respiratory" because a few air sacs bud from their sides, enabling them to take part in gas exchange.

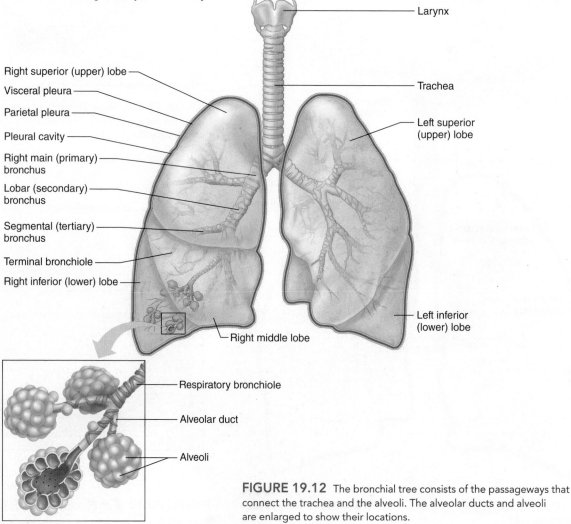

FIGURE 19.12 The bronchial tree consists of the passageways that connect the trachea and the alveoli. The alveolar ducts and alveoli are enlarged to show their locations.

7. **Alveolar ducts** branch from each respiratory bronchiole (fig. 19.14).
8. **Alveolar sacs** are thin-walled, closely packed outpouchings of the alveolar ducts (fig. 19.14).
9. **Alveoli** (al-ve'o-li) are thin-walled, microscopic air sacs that open to an alveolar sac. Air can diffuse freely from the alveolar ducts, through the alveolar sacs, and into the alveoli. The alveoli are the site of gas exchange between the inhaled air and the bloodstream (fig. 19.15).

Dust particles, asbestos fibers, and other pollutants travel at speeds of 200 centimeters per second in the trachea but slow to 1 centimeter per second when deep in the lungs. Gravity deposits such particles, particularly at branchpoints in the respiratory tree. It is a little like traffic backing up at an exit from a highway.

In severe cases of the inherited illness *cystic fibrosis*, airways become clogged with thick, sticky mucus, which attracts bacteria. As damaged white blood cells accumulate at the infection site, their DNA may leak out and further clog the area. A treatment that moderately eases breathing is deoxyribonuclease (DNase), an enzyme that degrades accumulating extracellular DNA.

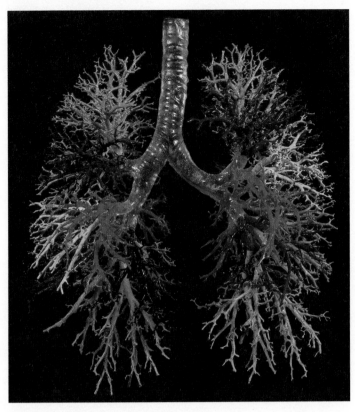

FIGURE 19.13 Falsely colored plastic cast of the bronchial tree.

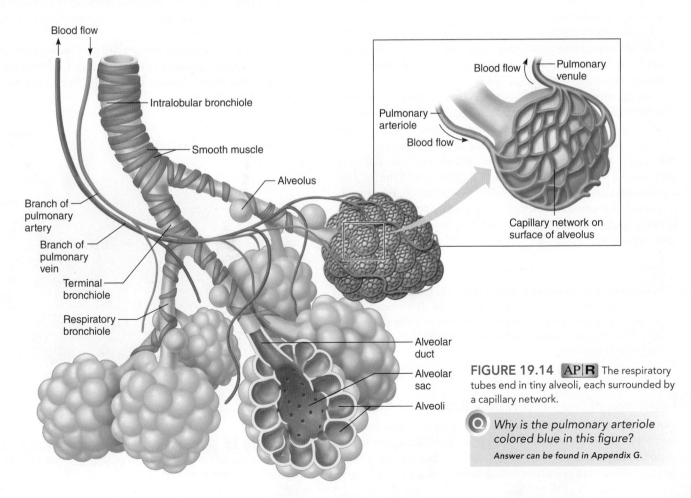

FIGURE 19.14 APR The respiratory tubes end in tiny alveoli, each surrounded by a capillary network.

Q *Why is the pulmonary arteriole colored blue in this figure?*
Answer can be found in Appendix G.

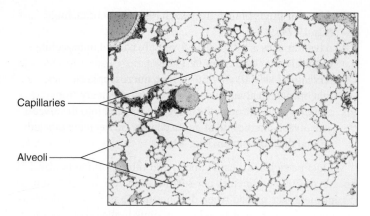

FIGURE 19.15 AP|R Light micrograph of alveoli (250×).

Structure of the Respiratory Tubes

The structure of a main bronchus is similar to that of the trachea, but the C-shaped cartilaginous rings are replaced with cartilaginous plates where the main bronchus branches. These plates are irregularly shaped and completely surround the tube. However, as the branching tubes become thinner, the amount of cartilage decreases. It finally disappears in the bronchioles, which have diameters of about 1 millimeter. Also as the tubes thin, layers of surrounding smooth muscle just beneath the mucosa become more prominent. This muscular layer persists in the walls to the ends of the respiratory bronchioles, and only a few muscle fibers are in the walls of the alveolar ducts.

Elastic fibers are scattered among the smooth muscle cells and are abundant in the connective tissue that surrounds the respiratory tubes. These fibers play an important role in breathing, as is explained later in this chapter (p. 746).

As the tubes become smaller in diameter, the type of cells that line them changes. The lining of the larger tubes consists of pseudostratified, ciliated columnar epithelium and mucus-secreting goblet cells. However, along the way, the number of goblet cells and the height of the other epithelial cells decline, and cilia become scarcer. In the finer tubes, beginning with the respiratory bronchioles, the lining is cuboidal epithelium; in the alveoli, it is simple squamous epithelium closely associated with a dense network of capillaries. The mucous lining gradually thins, until none appears in the alveoli.

> A flexible optical instrument called a *fiberoptic bronchoscope* is used to examine the trachea and bronchial tree. This procedure (bronchoscopy) is used in diagnosing tumors or other pulmonary diseases and to locate and remove aspirated foreign bodies in the air passages.

Functions of the Respiratory Tubes and Alveoli

The branches of the bronchial tree are air passages, whose mucous membranes continue to filter incoming air and distribute it to the alveoli in all parts of the lungs. The alveoli, in turn, provide a large surface area of thin epithelial cells through which gas exchanges

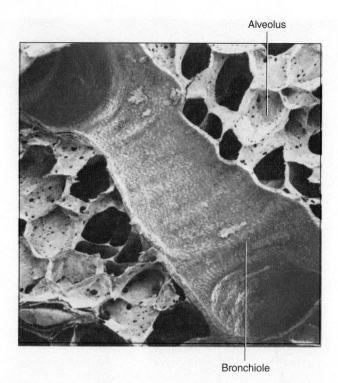

FIGURE 19.16 AP|R Falsely colored scanning electron micrograph of lung alveoli and a bronchiole (70×).

can occur (fig. 19.16). The 300 million alveoli in two human lungs have a combined surface area of 70–80 square meters—nearly half the area of a tennis court.

During gas exchange, oxygen diffuses through the alveolar walls and capillary walls to enter the blood. Carbon dioxide diffuses from the blood through these walls and enters the alveoli (figs. 19.17 and 19.18).

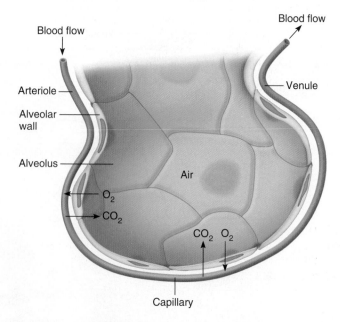

FIGURE 19.17 AP|R Oxygen (O_2) diffuses from the air in the alveolus into the capillary, while carbon dioxide (CO_2) diffuses from blood in the capillary into the alveolus.

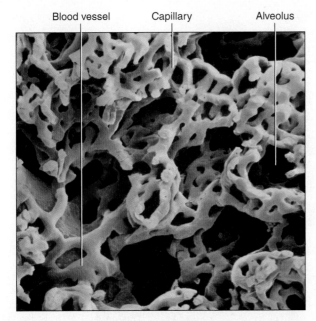

FIGURE 19.18 Falsely colored scanning electron micrograph of casts of alveoli and associated capillary networks. These casts were prepared by filling the alveoli and blood vessels with resin and later removing the soft tissues by digestion, leaving only the resin casts (420×). *Tissues and Organs: A Text-Atlas of Scanning Electron Microscopy,* by Richard G. Kessel and Randy Kardon. © 1979 W. H. Freeman and Company.

Several techniques enable a person who has stopped breathing to survive. In *artificial respiration,* a person blows into the mouth of a person who has stopped breathing. The oxygen in the rescuer's exhaled breath can keep the victim alive.

In *extracorporeal membrane oxygenation,* blood is pumped out of the body and across a gas-permeable membrane that adds oxygen and removes CO_2, simulating lung function. Such a device can keep a person alive until he or she recovers from other problems, but is too costly and cumbersome to maintain life indefinitely.

A lung assist device, called an *intravascular oxygenator,* consists of hundreds of tiny porous hair-thin fibers surgically implanted in the inferior vena cava. Here, deoxygenated blood returning to the heart receives oxygen and is rid of CO_2—but only at about 30% the capacity of a healthy respiratory system.

 PRACTICE

11 What is the function of the cartilaginous rings in the tracheal wall?

12 How do the right and left bronchi differ in structure?

13 List the branches of the bronchial tree.

14 Describe structural changes in the respiratory tubes as their diameters decrease.

15 How are gases exchanged in the alveoli?

Lungs

The lungs are soft, spongy, cone-shaped organs in the thoracic cavity. The right and left lungs are separated medially by the heart

and the mediastinum, and they are enclosed by the diaphragm and the thoracic cage (see fig. 1.10 on p. 19, fig. 19.19, and reference plates 16, 17, and 21 on pp. 50, 51, and 53).

Each lung occupies most of the thoracic space on its side and is suspended in the cavity by a bronchus and some large blood vessels. These tubular structures enter the lung on its medial surface through a region called the **hilum.** A layer of serous membrane, the *visceral pleura,* is firmly attached to the surface of each lung, and this membrane folds back at the hilum to become the *parietal pleura.* The parietal pleura, in turn, forms part of the mediastinum and lines the inner wall of the thoracic cavity (fig. 19.20).

No significant space exists between the visceral and parietal pleurae, because they are essentially in contact with each other. The potential (possible) space between them, called the **pleural cavity,** contains only a thin film of serous fluid that lubricates the adjacent pleural surfaces, reducing friction as they move against one another during breathing. This fluid also helps hold the pleural membranes together.

The right lung is larger than the left lung. Fissures divide the right lung into three parts, the superior, middle, and inferior lobes. The left lung is divided into two parts, a superior lobe and an inferior lobe.

A lobar bronchus of the bronchial tree supplies each lobe. A lobe also has connections to blood and lymphatic vessels and is enclosed by connective tissues. Connective tissue further subdivides a lobe into **lobules,** each of which contains terminal bronchioles together with their alveolar ducts, alveolar sacs, alveoli, nerves, and associated blood and lymphatic vessels.

Table 19.1 summarizes the characteristics of the major parts of the respiratory system. Clinical Application 19.2 considers substances that irritate the lungs.

 PRACTICE

16 Where are the lungs located?

17 What is the function of the serous fluid in the pleural cavity?

18 How does the structure of the right lung differ from that of the left lung?

19 What types of structures make up a lung?

19.3 | Breathing Mechanism

Breathing, also called ventilation, is the movement of air from outside the body into the bronchial tree and alveoli, followed by a reversal of this air movement. The actions responsible for these air movements are termed **inspiration** (in″spĭ-ra′shun), or inhalation, and **expiration** (ek″spi-ra′shun), or exhalation. One inspiration plus the following expiration is called a **respiratory cycle.**

Inspiration

Atmospheric pressure, the pressure of the air around us, is the force that moves air into the lungs. At sea level, this pressure is sufficient to support a column of mercury about 760 millimeters high in a tube. Therefore, normal air pressure equals 760 millimeters (mm) of mercury (Hg). (Units in common usage include: 760 mm Hg = 760 Torr = 1 atmosphere.)

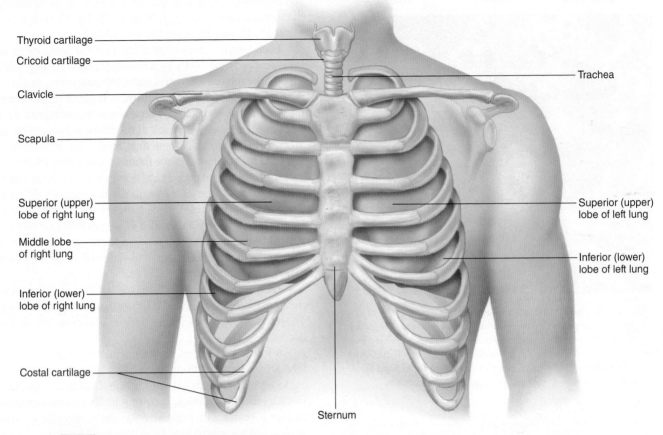

FIGURE 19.19 AP|R Location of the lungs in the thoracic cavity.

Thyroid cartilage
Cricoid cartilage
Clavicle
Scapula
Superior (upper) lobe of right lung
Middle lobe of right lung
Inferior (lower) lobe of right lung
Costal cartilage
Sternum
Trachea
Superior (upper) lobe of left lung
Inferior (lower) lobe of left lung

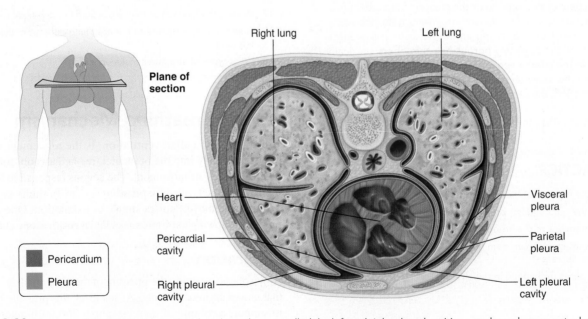

Plane of section

Right lung
Left lung
Heart
Pericardial cavity
Right pleural cavity
Visceral pleura
Parietal pleura
Left pleural cavity

Pericardium
Pleura

FIGURE 19.20 The potential spaces between the pleural membranes, called the left and right pleural cavities, are shown here as actual spaces.

The lungs are sensitive to inhaled particles. Such exposures can cause a variety of symptoms, both acute and chronic, that range in severity from a persistent cough to cancer.

Asbestos

Asbestos, a naturally occurring mineral, was once widely used in buildings and in various products because it resists burning and chemical damage. Asbestos easily crumbles into fibers, which, when airborne, can enter human respiratory passages. Asbestos-related problems include asbestosis (shortness of breath resulting from scars in the lungs), lung cancer, and mesothelioma (a cancer of the pleural membrane).

Asbestos fibers longer than 5 micrometers (0.0002 inch) and thinner than 2 micrometers (0.00008 inch) can cause illness when inhaled. Because asbestos causes illness only if the fibers are airborne, in some situations it is safer to encapsulate asbestos in a building and leave it in place than to remove it. The risk of becoming ill rises with duration of exposure to asbestos.

Berylliosis

Beryllium is an element used in fluorescent powders, metal alloys, and in the nuclear power industry. A small percentage of workers exposed to beryllium dust or vapor develop an immune response that damages the lungs. Symptoms include cough, shortness of breath, fatigue, loss of appetite, fevers and night sweats, and weight loss. Radiographs show granuloma scars in the lungs, and pulmonary function tests and listening to breath sounds with a stethoscope reveal impaired breathing.

Symptoms of berylliosis typically begin about a decade after the first exposure. A blood test that detects antibodies to beryllium distinguishes the condition from other lung ailments. Workers who do not have symptoms but know that they were exposed to beryllium can have periodic blood tests and chest radiographs to detect the condition early. The steroid drug prednisone is used to control symptoms.

A Disorder with Many Names

Repeatedly inhaling dust of organic origin can cause a lung irritation called extrinsic allergic alveolitis. An acute form of this reaction impairs breathing and causes a fever a few hours after encountering dust. In the chronic form, lung changes occur gradually over several years. The condition is associated with several occupations and has a variety of colorful names, including bathtub refinisher's lung, maple bark stripper disease, popcorn worker's lung, and wheat weevil disease.

Air Pollution

Air pollution threatens health when the particles are small enough to penetrate deeply into the lower respiratory system. Through sneezes and coughs the human respiratory tract easily ejects par-ticles larger than 10 micrometers in diameter. But particles 2.5 to 10 micrometers in diameter can enter the airways, and the finest particulates, with diameters 0.09 to 0.24 micrometers, pass the respiratory system's initial barriers and may reach the alveoli.

The World Health Organization (WHO) recommends a safe upper limit for 2.5-micrometer particulates (called microfine airborne pollutants) of 25 micrograms per cubic meter. One of the most heavily polluted cities is Beijing, China, where the level regularly exceeds the WHO air quality index's maximum of 500. In China the source of the microfine airborne pollutants is largely soot from diesel engines and industrial operations, but these particulates also are in cigarette smoke. Indoor exposure is mostly from wood-burning stoves. Cities in Iran, India, and Pakistan top the WHO's list of most-polluted cities. In the United States the top three most polluted cities are Bakersfield and Los Angeles in California, and Phoenix, Arizona.

Exposure to microfine airborne pollutants worsens asthma and irritates the eyes and lungs. It is also associated with increased risk of heart disease and lung cancer. WHO estimates that breathing polluted air, either outdoors or indoors, kills more than seven million people a year. The U.S. Environmental Protection Agency advises that for microfine airborne pollutant levels above 200 on the air quality index, "people with heart or lung disease, older adults, and children should avoid all physical activity outdoors. Everyone else should avoid prolonged or heavy exertion."

FIGURE 19C Urban aerosols may compromise lung function, at least temporarily. This image highlights the sources of some of the debris that covered lower Manhattan after the September 11, 2001 attacks. Red and yellow represent concrete dust, and purple indicates a mineral, gypsum, found in wallboard.

TABLE 19.1 Parts of the Respiratory System

Part	Description	Function
Nose	Part of the face centered above the mouth and inferior to the space between the eyes	Nostrils provide entrance to nasal cavity; internal hairs begin to filter incoming air
Nasal cavity	Hollow space behind nose	Conducts air to pharynx; mucous lining filters, warms, and moistens incoming air
Sinuses	Hollow spaces in various bones of the skull	Reduce weight of the skull; serve as resonant chambers
Pharynx	Chamber posterior to the nasal cavity, oral cavity, and larynx	Passageway for air moving from nasal cavity to larynx and for food moving from oral cavity to esophagus
Larynx	Enlargement at the top of the trachea	Passageway for air; prevents foreign objects from entering trachea; houses vocal cords
Trachea	Flexible tube that connects larynx with bronchial tree	Passageway for air; mucous lining continues to filter air
Bronchial tree	Branched tubes that lead from the trachea to the alveoli	Conducts air to the alveoli; mucous lining continues to filter incoming air
Lungs	Soft, cone-shaped organs that occupy a large portion of the thoracic cavity	Contain the air passages, alveoli, blood vessels, connective tissues, lymphatic vessels, and nerves of the lower respiratory tract

Air pressure is exerted on all surfaces in contact with the air, and because people breathe air, the inside surfaces of their lungs are also subjected to pressure. In other words, when the respiratory muscles are at rest, the pressures on the inside of the lungs and alveoli and on the outside of the thoracic wall are about the same (fig. 19.21).

Pressure and volume are related in an opposite, or inverse, way (this is known as Boyle's law). For example, pulling back on the plunger of a syringe increases the volume inside the barrel, lowering the air pressure inside. Atmospheric pressure then pushes outside air into the syringe (fig. 19.22*a*). In contrast, pushing on the plunger of a syringe reduces the volume inside the syringe, increasing the pressure inside and forcing air out of the syringe into the atmosphere (fig. 19.22*b*). Air moves into and out of the lungs in much the same way.

If the pressure inside the lungs and alveoli (intra-alveolar pressure) decreases, atmospheric pressure pushes outside air into the airways. This happens during resting inspiration, and it uses muscle fibers in the dome-shaped *diaphragm.*

The diaphragm is just inferior to the lungs. It consists of an anterior group of skeletal muscle fibers (costal fibers) that originate from the ribs and sternum, and a posterior group of skeletal muscle fibers (crural fibers) that originate from the vertebrae. Both groups of muscle fibers are inserted on a tendinous central portion of the diaphragm (reference plate 21, p. 53).

The muscle fibers of the diaphragm are stimulated to contract by impulses conducted on the *phrenic nerves,* which are associated with the cervical plexuses. When this happens, the diaphragm moves downward, the thoracic cavity enlarges, and the intra-alveolar pres-

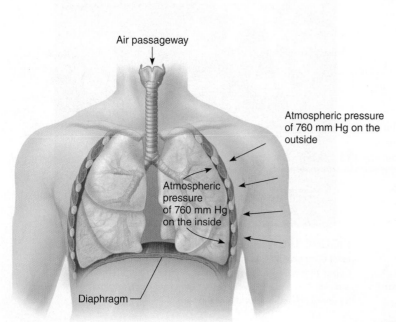

FIGURE 19.21 When the lungs are at rest, the pressure on the inside of the lungs is equal to the pressure on the outside of the thorax.

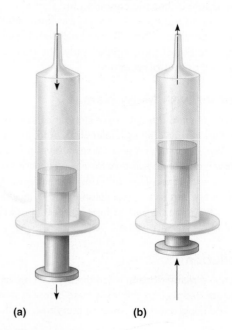

FIGURE 19.22 Moving the plunger of a syringe causes air to move (*a*) in or (*b*) out of the syringe. Air movements in and out of the lungs occur in much the same way.

sure falls about 2 mm Hg below atmospheric pressure. In response to this decreased intra-alveolar pressure, atmospheric pressure forces air into the alveoli (fig. 19.23).

While the diaphragm is contracting and moving downward, the *external (inspiratory) intercostal muscles* and certain thoracic muscles may be stimulated to contract. This action elevates the ribs and the sternum, increasing the size of the thoracic cavity even more. The intra-alveolar pressure falls further, and atmospheric pressure forces more air into the alveoli.

Lung expansion in response to movements of the diaphragm and chest wall depends on movements of the pleural membranes. Any separation of the pleural membranes decreases pressure in the intrapleural space, resisting further separation and holding these membranes together. In addition, only a thin film of serous fluid separates the parietal pleura on the inner wall of the thoracic cavity from the visceral pleura attached to the surface of the lungs. The water molecules in this fluid greatly attract the pleural membranes and each other, helping to hold the moist surfaces of the pleural membranes tightly together, much as a wet coverslip sticks to a microscope slide. As a result of these factors, when the intercostal muscles move the thoracic wall upward and outward, the parietal pleura moves too, and the visceral pleura follows it. These movements help expand the lung in all directions.

Although the moist pleural membranes help expand the lungs, the moist inner surfaces of the alveoli have the opposite effect. In the alveoli, the attraction of water molecules to each other creates a force called **surface tension** that makes it difficult to inflate the alveoli and may collapse them. Certain alveolar cells, however, synthesize a mixture of lipoproteins called **surfactant,** which is secreted continuously into alveolar air spaces. Surfactant reduces the alveoli's tendency to collapse, especially when lung volumes are low, and eases inspiratory efforts to expand the alveoli. Table 19.2 summarizes the steps of inspiration.

TABLE 19.2	Major Events in Inspiration
1.	Impulses are conducted on phrenic nerves to muscle fibers in the diaphragm, contracting them.
2.	As the dome-shaped diaphragm moves downward, the thoracic cavity expands.
3.	At the same time, the external intercostal muscles may contract, raising the ribs and expanding the thoracic cavity further.
4.	The intra-alveolar pressure decreases.
5.	Atmospheric pressure, greater than intra-alveolar pressure, forces air into the respiratory tract through the air passages.
6.	The lungs fill with air.

Surfactant is particularly important in the minutes after birth, when the newborn's lungs inflate for the first time. Premature infants often suffer respiratory distress syndrome because they do not produce sufficient surfactant. Physicians inject synthetic surfactant into the tiny lungs through an endotracheal tube. A ventilator especially geared to an infant's size assists breathing.

If a person needs to take a deeper than normal breath, the diaphragm and external intercostal muscles contract more forcefully. Additional muscles, such as the pectoralis minors, sternocleidomastoids, and the scalenes, can also pull the thoracic cage farther upward and outward, enlarging the thoracic cavity and decreasing intra-alveolar pressure even more (fig. 19.24).

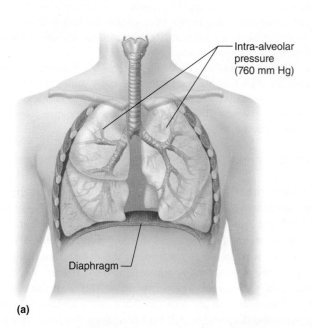

(a)

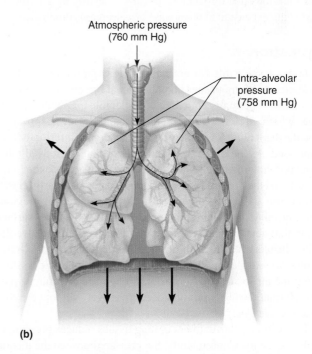

(b)

FIGURE 19.23 AP|R Normal inspiration. (*a*) Prior to inspiration, the intra-alveolar pressure is 760 mm Hg. (*b*) The intra-alveolar pressure decreases to about 758 mm Hg as the thoracic cavity enlarges, and atmospheric pressure forces air into the airways.

FIGURE 19.24 AP|R

Maximal inspiration. (a) Shape of the thorax at the end of normal inspiration. (b) Shape of the thorax at the end of maximal inspiration, aided by contraction of the sternocleidomastoid and pectoralis minor muscles.

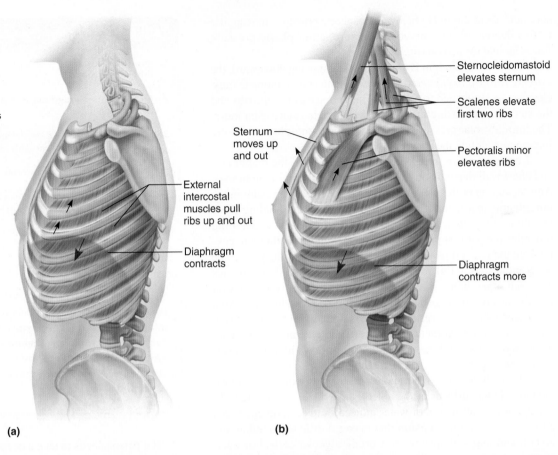

External intercostal muscles pull ribs up and out

Diaphragm contracts

Sternocleidomastoid elevates sternum

Scalenes elevate first two ribs

Sternum moves up and out

Pectoralis minor elevates ribs

Diaphragm contracts more

(a)

(b)

The ease with which the lungs can expand as a result of pressure changes during breathing is called *compliance* (distensibility). In a normal lung, compliance decreases as lung volume increases, because an inflated lung is more difficult to expand than a lung at rest. Conditions that obstruct air passages, destroy lung tissue, or impede lung expansion in other ways also decrease compliance.

Expiration

The forces responsible for normal resting expiration come from *elastic recoil* of lung tissues and abdominal organs and from surface tension. The lungs contain a considerable amount of elastic tissue, which stretches as the lungs expand during inspiration. When the diaphragm lowers, the abdominal organs inferior to it are compressed. As the diaphragm and the external intercostal muscles relax following inspiration, the elastic tissues cause the lungs to recoil, and they return to their original shapes. Similarly, elastic tissues cause abdominal organs to spring back into their previous shapes, pushing the diaphragm upward. At the same time, surface tension that develops between the moist surfaces of the alveolar linings shrinks alveoli. Each of these factors increases the intra-alveolar pressure about 1 mm Hg above atmospheric pressure, forcing the air inside the lungs out through the respiratory passages. Normal resting expiration occurs passively without the contraction of muscles.

The recoil of elastic fibers in lung tissues reduces pressure in the pleural cavity. Consequently, the pressure between the pleural membranes (intrapleural pressure) is typically about 4 mm Hg less than atmospheric pressure.

The visceral and parietal pleural membranes are held closely together because of the low intrapleural pressure, and no significant space normally separates them in the pleural cavity. However, if the thoracic wall is punctured, atmospheric air may enter the pleural cavity and intrapleural pressure becomes equal to atmospheric pressure. This condition, called *pneumothorax*, can collapse the lung on the affected side because of the lung's elasticity, creating a substantial space between the parietal and visceral pleural membranes.

Pneumothorax may be treated by covering the chest wound with an impermeable bandage, passing a tube (chest tube) through the thoracic wall into the pleural cavity, and applying suction to the tube. The suction reestablishes negative pressure in the cavity and the collapsed lung expands.

A person can exhale more air than normal by contracting the posterior *internal (expiratory) intercostal muscles*. These muscles pull the ribs and sternum downward and inward, increasing the air pressure in the lungs, forcing more air out. Also, the *abdominal wall muscles,* including the external and internal obliques, the transversus abdominis, and the rectus abdominis, squeeze the abdominal organs inward. In this way, the abdominal wall muscles can increase pressure in the abdominal cavity and force the diaphragm still higher against the lungs, pushing additional air out of the lungs (fig. 19.25). Table 19.3 summarizes the steps in expiration.

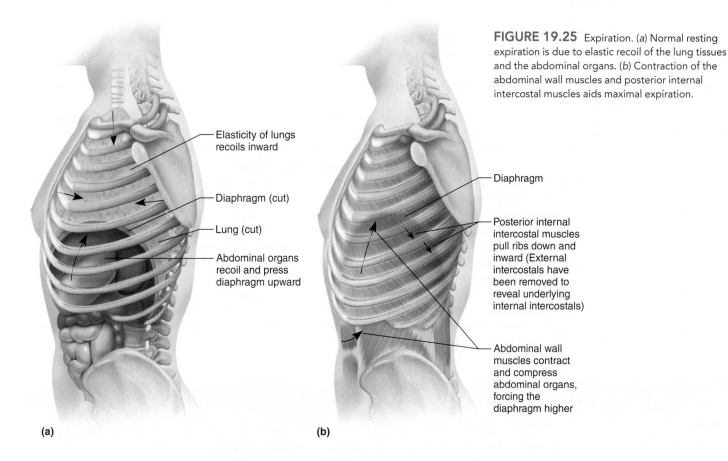

FIGURE 19.25 Expiration. (*a*) Normal resting expiration is due to elastic recoil of the lung tissues and the abdominal organs. (*b*) Contraction of the abdominal wall muscles and posterior internal intercostal muscles aids maximal expiration.

Elasticity of lungs recoils inward

Diaphragm (cut)

Lung (cut)

Abdominal organs recoil and press diaphragm upward

(a)

Diaphragm

Posterior internal intercostal muscles pull ribs down and inward (External intercostals have been removed to reveal underlying internal intercostals)

Abdominal wall muscles contract and compress abdominal organs, forcing the diaphragm higher

(b)

TABLE 19.3	Major Events in Expiration

1. The diaphragm and external respiratory muscles relax.

2. Elastic tissues of the lungs, stretched during inspiration, suddenly recoil, and surface tension pulls in on alveolar walls.

3. Tissues recoiling around the lungs increase the intra-alveolar pressure.

4. Air is forced out of the lungs.

PRACTICE

20 Describe the events in inspiration.

21 How does surface tension aid in expanding the lungs during inspiration?

22 What forces are responsible for normal expiration?

Respiratory Air Volumes and Capacities

Different volumes of air, called **respiratory volumes,** can be moved in or out of the lungs. The measurement of such air volumes is called *spirometry* and uses a device called a spirometer (fig. 19.26). Three distinct **respiratory volumes** can be measured using spirometry, and a fourth (residual volume) cannot.

The **tidal volume** is the volume of air that enters or leaves the airways and alveoli during a respiratory cycle. About 500 millili-

ters (mL) of air enter during a normal, resting inspiration. On the average, the same volume leaves during a normal, resting expiration. Thus, the **resting tidal volume** is about 500 mL (fig. 19.27).

During forced maximal inspiration, a volume of air in addition to the resting tidal volume, called the **inspiratory reserve volume** (complemental air), enters the lungs. It equals about 3,000 mL.

During a maximal forced expiration, about 1,100 mL of air in addition to the resting tidal volume can be expelled from the lungs. This volume is called the **expiratory reserve volume** (supplemental

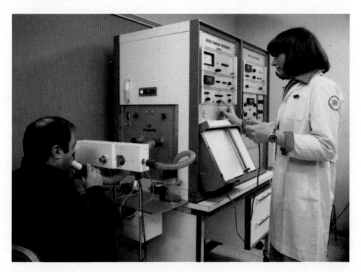

FIGURE 19.26 A spirometer measures respiratory air volumes.

air). However, even after the most forceful expiration, about 1,200 mL of air remains in the lungs. This is the **residual volume** and can only be measured using special gas dilution techniques.

As figure 19.27 depicts, each resting inspiration adds about 500 mL of air to about 2,300 mL of air already in the lungs. Normally, this newly inhaled air mixes completely with air already in the lungs. This mixing prevents the oxygen and carbon dioxide concentrations in the alveoli from fluctuating greatly with each breath.

Four **respiratory capacities** can be calculated by combining two or more of the respiratory volumes. If the inspiratory reserve volume (3,000 mL) is combined with the tidal volume (500 mL) and the expiratory reserve volume (1,100 mL), the total is termed the **vital capacity** (4,600 mL). This capacity is the maximum volume of air a person can exhale after taking the deepest breath possible.

The tidal volume (500 mL) plus the inspiratory reserve volume (3,000 mL) gives the **inspiratory capacity** (3,500 mL). This is the maximum volume of air a person can inhale following a resting expiration. Similarly, the expiratory reserve volume (1,100 mL) plus the residual volume (1,200 mL) equals the **functional residual capacity** (2,300 mL), which is the volume of air that remains in the lungs following a resting expiration.

The vital capacity plus the residual volume equals the **total lung capacity** (about 5,800 mL) (fig. 19.27). This total varies with age, sex, and body size.

A spirometer can be used to evaluate the course of respiratory illnesses, such as emphysema, pneumonia, lung cancer, and bronchial asthma (see fig. 19.26). Table 19.4 summarizes respiratory air volumes and capacities.

Some of the air that enters the respiratory tract during breathing does not reach the alveoli. This volume (about 150 mL) remains in the passageways of the trachea, bronchi, and bronchioles. Gases are not exchanged through the walls of these passages, so this air is said to occupy *anatomic dead space.*

Occasionally, alveoli in some regions of the lungs do not function, due to poor blood flow in the adjacent capillaries. This creates *alveolar dead space.* The anatomic and alveolar dead space volumes combined equal *physiologic dead space.* In normal lungs, the anatomic and physiologic dead spaces are essentially the same (about 150 mL).

TABLE 19.4	Respiratory Air Volumes and Capacities	
Name	Volume*	Description
Tidal volume (TV)	500 mL	Volume of air moved in or out of the lungs during a respiratory cycle
Inspiratory reserve volume (IRV)	3,000 mL	Maximum volume of air that can be inhaled in addition to resting tidal volume
Expiratory reserve volume (ERV)	1,100 mL	Maximum volume of air that can be exhaled in addition to resting tidal volume
Residual volume (RV)	1,200 mL	Volume of air that remains in the lungs even after a maximal expiratory effort
Inspiratory capacity (IC)	3,500 mL	Maximum volume of air that can be inhaled following exhalation of resting tidal volume: IC = TV + IRV
Functional residual capacity (FRC)	2,300 mL	Volume of air that remains in the lungs following exhalation of resting tidal volume: FRC = ERV + RV
Vital capacity (VC)	4,600 mL	Maximum volume of air that can be exhaled after taking the deepest breath possible: VC = TV + IRV + ERV
Total lung capacity (TLC)	5,800 mL	Total volume of air that the lungs can hold: TLC = VC + RV

*Values are typical for a tall, young adult.

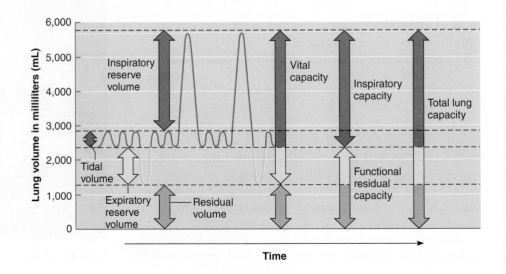

FIGURE 19.27 Respiratory volumes and capacities.

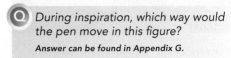

During inspiration, which way would the pen move in this figure?

Answer can be found in Appendix G.

PRACTICE

Alveolar Ventilation

The volume of air moved into the respiratory passages each minute is called the *minute ventilation*. It equals the tidal volume multiplied by the breathing rate. For example, if the tidal volume is 500 mL and the breathing rate is 12 breaths per minute, the minute ventilation is 500 mL × 12, or 6,000 mL per minute. However, for each breath much of the new air remains in the physiologic dead space.

For each respiratory cycle, the volume of new air that does reach the alveoli and is available for gas exchange (in this example) is calculated by subtracting the physiologic dead space (150 mL) from the tidal volume (500 mL). The resulting volume (350 mL) multiplied by the breathing rate (12 breaths per minute) is the *alveolar ventilation rate* (4,200 mL per minute). This is the more important value physiologically because it affects the concentrations of oxygen and carbon dioxide in the alveoli and thus available for gas exchange with the blood.

Increasing respiratory rate by itself will increase alveolar ventilation, but each respiratory cycle must fill the dead space before new air reaches the alveoli. The expiratory reserve volume and especially the large inspiratory reserve volume allow for an increase in tidal volume, and therefore an increase in alveolar ventilation independently of dead space. This becomes important at times when increased gas exchange is needed, as during exertion. Clinical Application 19.3 discusses respiratory problems that affect ventilation.

Nonrespiratory Air Movements

Air movements other than breathing are called *nonrespiratory movements*. They clear air passages (coughing, sneezing) or express emotions (laughing, crying).

Nonrespiratory movements usually result from *reflexes*, although sometimes they are initiated voluntarily. A cough, for example, can be produced through conscious effort or may be triggered by a foreign object in an air passage.

Coughing involves taking a deep breath, closing the glottis, and forcing air upward from the lungs against the closure. Then the glottis suddenly opens, and a blast of air is forced upward from the lower respiratory tract. Usually this rapid rush of air is of sufficient force to dislodge the object that triggered the reflex.

> The most sensitive areas of the air passages are in the larynx, the carina, and in regions near the branches of the major bronchi. The distal portions of the bronchioles (respiratory bronchioles), alveolar ducts, and alveoli do not have a nerve supply. Consequently, before any material in these parts can trigger a cough reflex, it must be moved into the larger passages of the respiratory tract.

A *sneeze* is much like a cough, but it clears the upper respiratory passages rather than the lower ones. Usually a mild irritation in the lining of the nasal cavity forces a blast of air up through the glottis. The air is directed into the nasal passages by depressing the uvula, closing the opening between the pharynx and the oral cavity.

In *laughing*, a person takes a breath and releases it in a series of short expirations. *Crying* consists of similar movements, and sometimes it is necessary to note a person's facial expression to distinguish laughing from crying.

A *hiccup* is caused by sudden inspiration due to a spasmodic contraction of the diaphragm while the glottis is closed. Air striking the vocal folds causes the sound of the hiccup. We do not know the function, if any, of hiccups.

Yawning is familiar to everyone, yet its significance and contagiousness remain poorly understood. Evidence points away from a role in increasing oxygen intake. Yawning, and its effect of getting others yawning, may be rooted in brainstem mechanisms that maintain alertness. Table 19.5 summarizes the characteristics of nonrespiratory air movements.

PRACTICE

TABLE 19.5	Nonrespiratory Air Movements	
Air Movement	**Mechanism**	**Function**
Coughing	Deep breath is taken, glottis is closed, and air is forced against the closure; suddenly the glottis is opened, and a blast of air passes upward	Clears lower respiratory passages
Sneezing	Same as coughing, except air moving upward is directed into the nasal cavity by depressing the uvula	Clears upper respiratory passages
Laughing	Deep breath is released in a series of short expirations	Expresses happiness
Crying	Same as laughing	Expresses sadness
Hiccuping	Diaphragm contracts spasmodically while glottis is closed	No useful function known
Yawning	Deep breath is taken	Some hypotheses, but no established function
Speech	Air is forced through the larynx, causing vocal cords to vibrate; actions of lips, tongue, and soft palate form words	Vocal communication

Injuries to the respiratory center or to spinal tracts that conduct motor impulses may paralyze breathing muscles. Paralysis may also be due to a disease that affects the central nervous system and injures motor neurons, such as *poliomyelitis*. Sometimes, other muscles, by increasing their responses, can compensate for functional losses of a paralyzed muscle. Otherwise, mechanical ventilation is necessary. More common disorders that decrease ventilation are bronchial asthma and emphysema.

Bronchial asthma is usually an allergic reaction to foreign antigens in the airways, such as from inhaled pollen or material on dust mites. Normally, cells of the larger airways secrete abundant mucus, which traps allergens. Ciliated columnar epithelial cells move the mucus up and out of the bronchi, then up and out of the trachea. However, in the smaller airways, mucus and edematous secretions resulting from the allergic response accumulate because fewer cells are ciliated. The allergens and secretions irritate airway smooth muscle, causing bronchoconstriction. Breathing becomes increasingly difficult, and inhalation produces a characteristic wheezing sound as air moves through narrowed and partially clogged passages.

A person with asthma usually finds it harder to force air out of the lungs than to bring it in. This is because inspiration expands the lungs opening the air passages. Expiration, on the other hand, is due to elastic recoil of stretched tissues which compresses the airways, further impairing air movement through the narrowed air passages.

Increase in the prevalence of asthma in the United States may be due to a too-clean environment, especially for children. Many studies have shown that children who are with others and contract minor respiratory infections, as well as children raised with cats or dogs, are less likely to develop asthma than are children who do not have these exposures. This association of a primed immune system with lower risk of developing asthma is called the "hygiene hypothesis."

Emphysema is a progressive, degenerative disease that destroys alveolar walls. As a result, clusters of small air sacs merge into larger chambers, which greatly decreases the total surface area of the alveoli, thereby reducing the volume of gases that can be exchanged across their walls. At the same time, the alveolar walls lose their elasticity, and the capillary networks associated with the alveoli diminish (fig. 19D).

A person with emphysema finds it increasingly difficult to exhale because of the loss of tissue elasticity. Abnormal muscular efforts are required to compensate for the lack of elastic recoil that normally contributes to expiration.

Only 3% of the 2 million people in the United States who have emphysema inherit the condition; most of that minority has a deficiency of an enzyme, alpha-1 antitrypsin. The majority of the other cases are due to smoking or exposure to other respiratory irritants. Emphysema is a type of chronic obstructive pulmonary disease (COPD). The other major form of COPD is chronic bronchitis.

Drug combinations treat emphysema by relaxing the muscles associated with the airways. An experimental treatment for severe emphysema is lung volume reduction surgery. As its name suggests, the procedure reduces lung volume, by removing 20 to 35% of the most damaged areas of each lung. The treatment opens collapsed airways in the remaining parts of the lung and eases breathing. So far, it seems to noticeably improve lung function and quality of life for many patients.

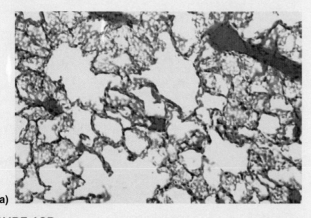
(a)

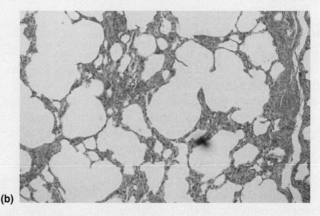
(b)

FIGURE 19D Comparison of lung tissues. (*a*) Normal lung tissue (100×). (*b*) As emphysema develops, alveoli coalesce, forming larger chambers (100×).

19.4 | Control of Breathing

Normal breathing is a rhythmic, involuntary act that continues when a person is unconscious. However, the respiratory muscles can be voluntarily controlled too (Take a deep breath and consider this!).

Respiratory Areas

Groups of neurons in the brainstem comprise the **respiratory areas**, which control breathing. These areas periodically initiate impulses that travel on cranial and spinal nerves to breathing muscles, causing inspiration and expiration. The respiratory areas also adjust the rate and depth of breathing to meet cellular requirements for supply of oxygen and removal of CO_2, even during strenuous physical exercise.

The components of the respiratory areas are widely scattered throughout the pons and medulla oblongata. However, two parts of the respiratory areas are of special interest. They are the respiratory center of the medulla and the respiratory group of the pons (fig. 19.28).

The **medullary respiratory center** includes two bilateral groups of neurons that extend throughout the length of the medulla oblongata. They are called the ventral respiratory group and the dorsal respiratory group.

Current evidence suggests that the basic rhythm of breathing arises from the *ventral respiratory group*. Two different populations of neurons in this area have been implicated in maintaining inspiration and expiration, depending on the experimental conditions. Indeed, it has been suggested that such an important physiological process might occur in more than one way.

The *dorsal respiratory group* stimulates the inspiratory muscles, primarily the diaphragm. The dorsal respiratory group also helps process sensory information related to the respiratory system and may play a role in certain cardiopulmonary reflexes that affect respiratory rhythm.

Neurons in another part of the brainstem, the pons, form the *pontine respiratory group* (formerly called the *pneumotaxic center*). They may contribute to the rhythm of breathing by limiting inspiration (fig. 19.29).

A condition called *sleep apnea* causes one to momentarily stop breathing while sleeping. In infants, apnea is usually *central*, due to a problem with respiratory control centers and is responsible for some cases of sudden infant death. In adults, apnea is usually *obstructive*, caused by airway blockage, and is often associated with snoring.

Babies who have difficulty breathing just after birth are sent home with monitoring devices, which sound an alarm when the child stops breathing, alerting parents to resuscitate the infant. The position in which the baby sleeps seems to affect the risk of sleep apnea—sleeping on the back or side is safest during the first year of life.

Adults with sleep apnea may cease breathing for ten to twenty seconds, hundreds of times a night. The frequent cessation in breathing is associated with snoring. Sleep apnea in adults causes fatigue, headache, depression, and drowsiness during waking hours.

Sleep apnea is diagnosed in a sleep lab, where breathing during slumber is monitored for several consecutive hours. One treatment for obstructive sleep apnea, *nasal continuous positive airway pressure*, straps a device onto the nose at night that maintains air flow into the respiratory system.

PRACTICE

31 Where are the respiratory areas?

32 Describe how the respiratory areas maintain a normal breathing rhythm.

33 Explain how the breathing rhythm may be changed.

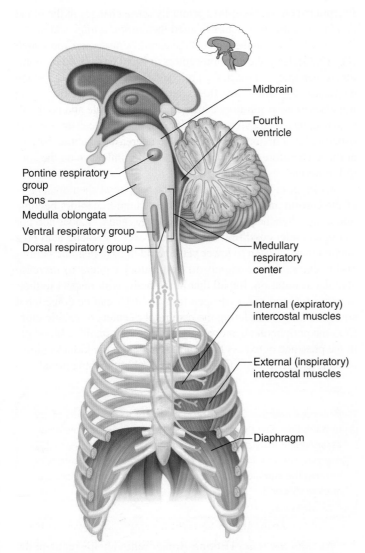

FIGURE 19.28 The respiratory areas are located in the pons and the medulla oblongata.

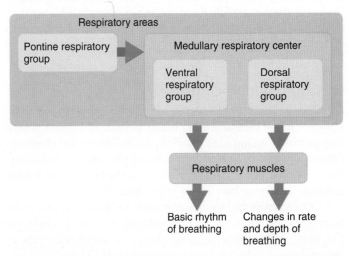

FIGURE 19.29 The medullary respiratory center and the pontine respiratory group control breathing.

Partial Pressure

In a mixture of gases such as air, each gas accounts for a portion of the total pressure the mixture produces. The amount of pressure each gas contributes is called the **partial pressure** of that gas and is proportional to its concentration. For example, because air is 21% oxygen, oxygen accounts for 21% of the atmospheric pressure (21% of 760 Hg), or 160 mm Hg (.21 × 760 = 160). Thus, the partial pressure of oxygen, symbolized P_{O_2}, in atmospheric air is 160 mm Hg. Similarly, the partial pressure of carbon dioxide (P_{CO_2}) in air is 0.3 mm Hg.

Gas molecules from the air may enter, or dissolve, in liquid. This is what happens when CO_2 is added to a carbonated beverage or when inspired gases dissolve in the blood in the alveolar capillaries. Using partial pressures simplifies the calculation of the concentration of a dissolved gas. The partial pressure of a gas dissolved in a liquid equals the partial pressure of that gas in the air with which the liquid has equilibrated. For example, the P_{O_2} in a glass of water that has been on a desk for a while must be 160 mm Hg, the same as in the air around it. Instead of referring to concentrations of oxygen and carbon dioxide in the body fluids, we will refer to P_{O_2} and P_{CO_2}.

Factors Affecting Breathing

A number of factors influence breathing rate and depth. These include P_{O_2} and P_{CO_2} in body fluids, the degree to which lung tissues are stretched, emotional state, and level of physical activity. The receptors involved include mechanoreceptors that sense stretch as well as central and peripheral chemoreceptors.

Under normal circumstances the main control of respiratory activity is not the level of oxygen in the blood. The limited role of P_{O_2} may be surprising, considering the importance of oxygen for sustaining life. However, most blood oxygen is carried bound to the hemoglobin molecules in red blood cells, and so-called oxygen-poor systemic venous blood still has, on average, 75% of the oxygen it had when it was fully oxygenated.

Usually this large excess of oxygen "frees up" respiratory control from responding to blood oxygen levels. Thus, the respiratory system responds predominantly to blood P_{CO_2} and hydrogen ion concentration, which are important in maintaining the pH of the internal environment, especially in the brain. The receptors that respond to P_{O_2} and P_{CO_2} are chemoreceptors, and they fall into two categories, central and peripheral.

Central Chemoreceptors

Low blood P_{O_2} has little effect on the **central chemoreceptors.** They are in the ventral portion of the medulla oblongata, where they primarily monitor the pH of the brain. The central chemoreceptors also respond to changes in blood pH, but only indirectly, because hydrogen ions do not easily cross the blood-brain barrier and therefore do not have access to this area.

CO_2 can cross the blood-brain barrier, and as a result CO_2 in the brain is in equilibrium with CO_2 in the blood. If plasma P_{CO_2} rises, the CO_2 easily diffuses into the brain, where it combines with water in the cerebrospinal fluid to form carbonic acid (H_2CO_3):

$$CO_2 + H_2O \rightarrow H_2CO_3$$

The carbonic acid quickly ionizes, releasing hydrogen ions (H^+) and bicarbonate ions (HCO_3^-):

$$H_2CO_3 \rightarrow H^+ + HCO_3^-$$

The central chemoreceptors respond to the released hydrogen ions, not to the CO_2. However, because CO_2 and H^+ are so closely linked, it may help to think of them as equivalent in terms of acidity in the body. Breathing rate and tidal volume increase when a person inhales air rich in CO_2 or when body cells produce excess CO_2. These changes increase alveolar ventilation, exhaling more CO_2, and the blood P_{CO_2} and hydrogen ion concentration return toward normal.

> Adding CO_2 to air can stimulate the rate and depth of breathing. Ordinary air is about 0.04% CO_2. If a person inhales air containing 4% CO_2, breathing rate usually doubles.

Peripheral Chemoreceptors

Peripheral chemoreceptors primarily sense changes in the blood P_{O_2}, in specialized structures called the *carotid bodies* and *aortic bodies,* which are in the walls of the carotid sinuses and aortic arch (fig. 19.30). Here the chemoreceptors are well positioned to monitor oxygen that will reach the brain and the rest of the body. When decreased P_{O_2} stimulates these peripheral receptors, they send impulses to the respiratory center, and breathing rate and tidal volume rise, increasing alveolar ventilation. This mechanism does not usually play a major role until the P_{O_2} decreases to about 50% of normal. Therefore, oxygen has only a minor influence on the control of normal respiration.

Changes in blood pH stimulate the peripheral chemoreceptors of the carotid and aortic bodies. These chemoreceptors become important when the pH of the blood changes and CO_2 does not. For example, under conditions of strenuous exercise, lactic acid production may threaten to lower pH. Should this happen, the peripheral chemoreceptors signal the respiratory centers to increase alveolar ventilation. Recall that in the body, with respect to their impact on blood pH, hydrogen ions and CO_2 can be considered equivalent. By stimulating the respiratory centers to exhale more CO_2, the peripheral chemoreceptors effectively stabilize blood pH in the presence of excess lactic acid. Gradually the kidneys eliminate the extra hydrogen ions, and blood CO_2 returns to normal.

> Patients who have COPD gradually adapt to high levels of carbon dioxide. For them, low oxygen levels may serve as a necessary respiratory stimulus. If such a patient is placed on 100% oxygen, the low arterial P_{O_2} may be corrected sufficiently to remove the stimulus. As a result, breathing may be suppressed or even stopped.

An *inflation reflex* (Hering-Breuer reflex) helps regulate the depth of breathing. This reflex occurs when stretch receptors in the visceral pleura, bronchioles, and alveoli are stimulated as

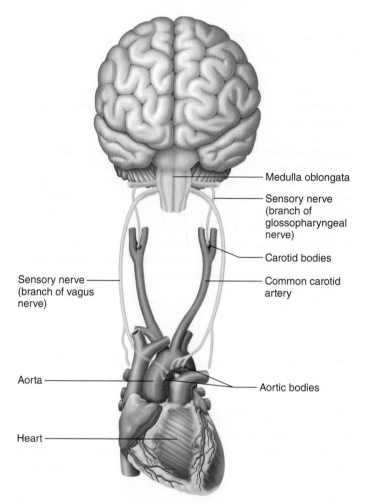

FIGURE 19.30 Decreased P_{O_2} stimulates peripheral chemoreceptors in the carotid and aortic bodies.

During childbirth, for example, concentration on controlling her breathing can distract the mother from her pain.

Holding one's breath makes the blood concentrations of carbon dioxide and hydrogen ions rise and the concentration of oxygen fall. These changes (primarily the increased CO_2) stimulate the chemoreceptors, and soon the need to inhale overpowers the desire to hold the breath—much to the relief of parents when young children threaten to hold their breath until they turn blue! However, a person can increase the breath-holding time by breathing rapidly and deeply in advance. (This could be dangerous, see box that follows.) This action, termed **hyperventilation** (hi″per-ven″tĭ-la′shun), is specifically breathing that lowers the blood CO_2 concentration below normal. Following hyperventilation, it takes longer than usual for the carbon dioxide concentration to reach the level needed to override the conscious effort of breath holding.

Table 19.6 discusses factors affecting breathing. Clinical Application 19.4 focuses on one influence on breathing—exercise.

A person who is emotionally upset may hyperventilate, become dizzy, and lose consciousness. This is due to a lowered CO_2 concentration followed by a rise in pH (respiratory alkalosis) and a localized vasoconstriction of cerebral arterioles, decreasing blood flow to nearby brain cells. Hampered oxygen supply to the brain causes fainting. A person should never hyperventilate to help hold the breath while swimming, because the person may lose consciousness under water and drown.

lung tissues are stretched. The sensory impulses of the reflex travel via the vagus nerves to the respiratory areas in the brainstem and shorten the duration of inspiratory movements. This action prevents overinflation of the lungs during forceful breathing (fig. 19.31).

Emotional upset or strong sensory stimulation may alter the normal breathing pattern. Gasping and rapid breathing are familiar responses to fear, anger, shock, excitement, horror, surprise, sexual stimulation, or even the chill of stepping into a cold shower. Because control of the respiratory muscles is voluntary, we can alter our breathing pattern consciously or stop it for a short time.

 PRACTICE

34 What does the term partial pressure refer to?
35 Which chemical factors affect breathing?
36 Describe the inflation reflex.
37 How does hyperventilation decrease respiratory rate?

19.5 | Alveolar Gas Exchanges

The tubelike parts of the respiratory system move air in and out of the air passages. The alveoli are the sites of the vital process of gas exchange between the air and the blood.

TABLE 19.6	Factors Affecting Breathing		
Factors	**Receptors Stimulated**	**Response**	**Effect**
Stretch of tissues	Stretch receptors in visceral pleura, bronchioles, and alveoli	Inhibits inspiration	Prevents overinflation of lungs during forceful breathing
Low plasma P_{O_2}	Chemoreceptors in carotid and aortic bodies	Increases alveolar ventilation	Increases plasma P_{O_2}
High plasma P_{CO_2}	Chemosensitive areas of the respiratory center	Increases alveolar ventilation	Decreases plasma P_{CO_2}
High cerebrospinal fluid hydrogen ion concentration	Chemosensitive areas of the respiratory center	Increases alveolar ventilation	Decreases plasma P_{CO_2}

Moderate to heavy exercise greatly increases the amount of oxygen skeletal muscles use. A young man at rest uses about 250 milliliters of oxygen per minute but may require 3,600 milliliters per minute during maximal exercise. While oxygen use is increasing, CO_2 production increases also. Because decreased blood oxygen and increased blood CO_2 concentrations stimulate the respiratory centers, exercise would be expected to increase breathing rate. Studies reveal, however, that blood oxygen and carbon dioxide levels do not change significantly during exercise, but breathing rate does.

The cerebral cortex and the proprioceptors associated with muscles and joints cause much of the increased breathing rate associated with exercise (see chapter 12, pp. 449–451). The cortex sends stimulating impulses to the respiratory center whenever it signals the skeletal muscles to contract. At the same time, muscular movements stimulate the proprioceptors, triggering a *joint reflex*. In this reflex, sensory impulses are conducted from the proprioceptors to the respiratory center, and the respiratory rate increases.

Exercise increases not only the breathing rate, but also the demand of exercising muscles for blood flow. In this way, exercise taxes both the respiratory and the cardiovascular systems. If either of these systems does not keep pace, the person will begin to feel short of breath. This sensation, however, is usually due to the inability of the cardiovascular system to move enough blood between the lungs and the cells, rather than to the inability of the respiratory system to provide enough air.

Alveoli

Alveoli are microscopic air sacs clustered at the distal ends of the finest respiratory tubes—the alveolar ducts. Each alveolus is a tiny space within a thin wall that separates it from adjacent alveoli. Tiny openings, called **alveolar pores,** in the walls of some alveoli may permit air to pass from one alveolus to another (fig. 19.32). The pores provide alternate air pathways if the passages in some parts of the lung become obstructed.

Phagocytic cells called *alveolar macrophages* are in alveoli and in the pores. These macrophages phagocytize airborne agents, including bacteria, thereby cleaning the alveoli (fig. 19.33).

Respiratory Membrane

Part of the wall of an alveolus is made up of cells (type II cells) that secrete pulmonary surfactant, described earlier. However, the bulk of the wall of an alveolus consists of a layer of simple squamous

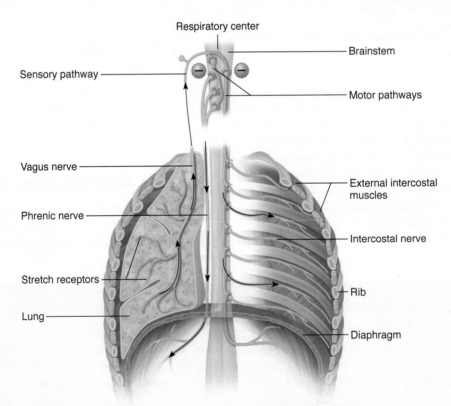

FIGURE 19.31 In the process of inspiration, motor impulses travel from the respiratory center to the diaphragm and external intercostal muscles, which contract and cause the chest and lungs to expand. This expansion stimulates stretch receptors in the lungs to send inhibiting impulses to the respiratory center, preventing overinflation.

epithelium (type I cells). Each alveolus is associated with a dense network of capillaries with walls of simple squamous epithelial cells (fig. 19.33). Thin basement membranes separate the walls of the alveoli from the walls of the capillaries, and in the spaces between them elastic and collagen fibers help support the alveolar walls. Thus, two thicknesses of epithelial cells and basement membranes separate the air in an alveolus and the blood in a capillary. These layers make up the **respiratory membrane** (alveolar-capillary membrane), through which gas exchange occurs between the alveolar air and the blood (figs. 19.34 and 19.35).

Diffusion Through the Respiratory Membrane

Solutes diffuse from regions where they are in higher concentration toward regions where they are in lower concentration. Therefore, to determine the direction of diffusion of a solute, we must know the concentration gradient. For dissolved gases, it is necessary to think in terms of partial pressure gradients. A gas will diffuse from an area of higher partial pressure to an area of lower partial pressure.

 RECONNECT
To Chapter 3, Diffusion, pages 98–100.

When a mixture of gases dissolves in blood, the resulting concentration of each dissolved gas is proportional to its partial pressure. Each gas diffuses between blood and its surroundings from areas of higher partial pressure to areas of lower partial pressure until the partial pressures in the two regions reach equilibrium. For example, the P_{CO_2} of blood entering the pulmonary capillaries is 45 mm Hg, but the P_{CO_2} in alveolar air is 40 mm Hg. Because of the difference in these partial pressures, carbon dioxide diffuses from blood, where its partial pressure is higher, across the respiratory membrane and into alveolar air. When blood leaves the lungs, its P_{CO_2} is 40 mm Hg, the same as the P_{CO_2} of alveolar air. Similarly, the P_{O_2} of blood entering the pulmonary capillaries is 40 mm Hg, but reaches 104 mm Hg as oxygen diffuses from alveolar air into

the blood. Because of the large volume of air always in the lungs, and the oxygen added as long as breathing continues, alveolar P_{O_2} stays relatively constant at 104 mm Hg. Reaching equilibrium, blood leaves the alveolar capillaries with a P_{O_2} of 104 mm Hg. (In an exception to the usual pattern, some systemic venous blood draining the bronchi and bronchioles mixes with this blood before returning to the heart. This contributes to lowering the P_{O_2} of left atrial, left ventricular, and systemic arterial blood to 95 mm Hg.) Clinical Application 19.5 looks at a respiratory effect that occurs under specific conditions—high altitude.

The respiratory membrane is normally thin (about 1 micrometer thick), and gas exchange is rapid. However, a number of factors may affect diffusion across the respiratory membrane. More

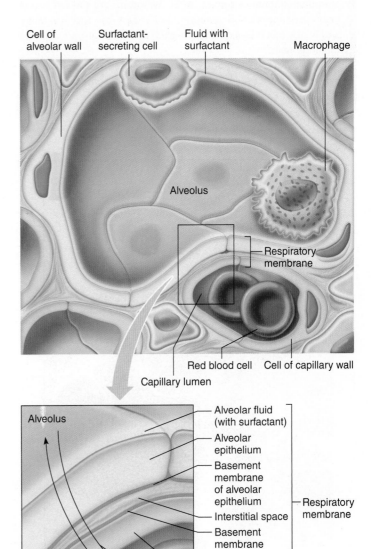

FIGURE 19.33 The respiratory membrane consists of the walls of the alveolus and the capillary.

Cell of alveolar wall · Surfactant-secreting cell · Fluid with surfactant · Macrophage · Alveolus · Respiratory membrane · Red blood cell · Cell of capillary wall · Capillary lumen

Alveolus · Alveolar fluid (with surfactant) · Alveolar epithelium · Basement membrane of alveolar epithelium · Interstitial space · Basement membrane of capillary endothelium · Capillary endothelium · Diffusion of O_2 · Diffusion of CO_2 · Red blood cell · Respiratory membrane · Capillary

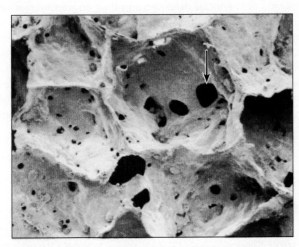

FIGURE 19.32 Alveolar pores (arrow) allow air to pass from one alveolus to another (300×).

Every year, about 100,000 mountain climbers experience varying degrees of altitude sickness, because at high elevations, the proportion of oxygen in air remains the same (about 21%), but the P_{O_2} decreases. When a person ascends rapidly, oxygen diffuses more slowly from the alveoli into the blood, and the hemoglobin becomes less saturated with oxygen. In some individuals, the body's efforts to get more oxygen—increased breathing and heart rate and enhanced red blood cell and hemoglobin production—cannot keep pace with the plummeting oxygen supply.

Severe altitude sickness includes a condition called high-altitude pulmonary edema (HAPE). Symptoms are sudden severe headache, nausea and vomiting, rapid heart rate and breathing, and a cyanotic (blue) cast to the skin, often first apparent under the fingernails.

The hypoxia (low blood oxygen) associated with high altitude can vasoconstrict pulmonary blood vessels. In some people, this shunts blood under high pressure through less constricted vessels in the pulmonary circuit, raising capillary pressure and filtering fluid from the blood vessels into the alveoli, causing the edema. People with severe HAPE also commonly develop high-altitude cerebral edema (HACE).

HAPE is treated by giving oxygen and coming down from the mountain. Delay may prove fatal. Exertion may worsen the symptoms, and victims often need to be carried. Some prescription vasodilators, such as nifedipine, may help reduce the pulmonary hypertension, but they can be dangerous without proper medical attention.

Mountain climbing is an extreme activity that endangers the respiratory system. Regularly exercising at moderately high altitude, however, can strengthen the system. Distance runners, too, often train at high altitudes because of beneficial effects on fitness and performance.

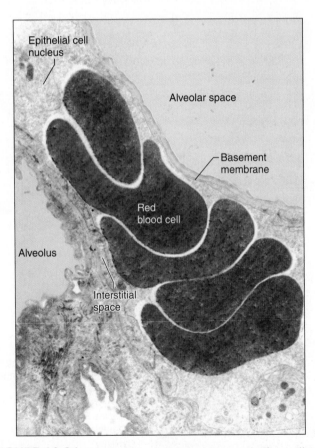

FIGURE 19.34 Falsely colored electron micrograph of a capillary located between alveoli (7,000×).

surface area, shorter distance, greater solubility of gases, and a steeper partial pressure gradient all favor increased diffusion. Diseases that harm the respiratory membrane, such as pneumonia, or reduce the surface area for diffusion, such as emphysema, impair gas exchange. These conditions may require increased P_{O_2} for treatment. Clinical Application 19.6 examines illnesses that result from impaired gas exchange.

The respiratory membrane is normally so thin that certain soluble chemicals other than CO_2 may diffuse into alveolar air and be exhaled. This is why breath analysis can reveal alcohol in the blood or acetone can be smelled on the breath of a person who has untreated diabetes mellitus. Breath analysis may also detect substances associated with kidney failure, certain digestive disturbances, and liver disease.

An optical tool called "frequency combs" can detect trace amounts of many different compounds in an exhaled breath. The sets of chemicals detected and their concentrations can provide clues to health. A special laser sends short pulses of light into a device holding an exhaled breath. An algorithm uses the pattern of light wavelengths absorbed to identify specific chemical compounds that may serve as biomarkers. For example, a signature of nitric oxide, carbon monoxide, nitrites, nitrates, pentane, ethane, and hydrogen peroxide indicates asthma, but a different set of compounds in exhaled breath indicates cystic fibrosis.

 PRACTICE

38 Describe the structure of the respiratory membrane.

39 What causes oxygen and carbon dioxide to move across the respiratory membrane?

Five-year-old Carly became ill very quickly. After twelve hours of flulike symptoms, her temperature soared, her chest began to hurt, and her breathing became rapid and shallow. A chest radiograph indicated that she had *pneumonia*. The bacteria that had caused a mild upper respiratory infection in her sisters had infected Carly's lower respiratory structures.

Antibiotics treated Carly's pneumonia. Many of the 50 million people who died in the 1918 influenza pandemic were not so lucky, in that era before antibiotics. Although viruses can cause pneumonia, most of the flu deaths were actually caused by secondary bacterial infections. Bacteria that normally inhabit upper respiratory structures easily moved downward to the lungs made vulnerable by the viral infection. Fungi can also cause pneumonia, such as the fungus *Pneumocystis jirovecii* that affects people with AIDS.

For all types of pneumonia, events in the infected lung are similar: alveolar linings swell with edema and become abnormally permeable, allowing fluids and white blood cells to accumulate in the air sacs. As the alveoli fill, the surface area available for gas exchange diminishes. Breathing becomes difficult. Untreated, pneumonia can kill.

Tuberculosis is a different type of lung infection, caused by the bacterium *Mycobacterium tuberculosis*. Fibrous connective tissue develops around the sites of infection, forming structures called *tubercles* (fig. 19E). By walling off the bacteria, the tubercles help stop their spread. Sometimes this protective mechanism fails, and the bacteria flourish throughout the lungs and may spread to other organs. In the later stages of infection, other types of bacteria may cause secondary infections. As lung tissue is destroyed, the surface area for gas exchange decreases. In addition, the widespread fibrous tissue thickens the respiratory membrane, further restricting gas exchange. A variety of drugs are used to treat tuberculosis, but in recent years, "multidrug resistant" and "extensively drug resistant" strains have emerged that can kill quickly.

Another type of condition that impairs gas exchange is *atelectasis*. This is the collapse of a lung, or some part of it, with the collapse of the blood vessels that supply the affected region. Obstruction of a respiratory tube, such as by an inhaled foreign object or excess mucus secretion, may cause atelectasis. The air in the alveoli beyond the obstruction is absorbed, and as the air pressure in the alveoli decreases, their elastic walls collapse, and they can no longer function. Fortunately, after a portion of a lung collapses, the functional regions that remain are often able to carry on enough gas exchange to sustain the body cells.

In *acute respiratory distress syndrome* (ARDS), which is a special form of atelectasis, alveoli collapse. Causes include pneumonia and other infections, near drowning, shock, sepsis, aspiration of stomach acid into the respiratory system, or physical trauma to the lungs from an injury or surgical procedure. Anesthetic drugs can cause ARDS by suppressing surfactant production. In response the tiny air sacs collapse. Blood vessels and airways narrow. Delivery of oxygen to tissues is seriously impaired. ARDS is fatal about 60% of the time. (It was until recently called adult respiratory distress syndrome.)

New virally caused acute respiratory disorders have appeared. Severe acute respiratory syndrome (SARS) is a disease caused by a virus (SARS-coronavirus) that infects lower respiratory structures. The disease swept through two dozen nations from winter 2002 until spring 2003, infecting more than 8,000 people and killing about 800 of them. China, Hong Kong, and Taiwan were the hardest hit. Since then the disease has mysteriously vanished.

In 2012 another acute respiratory condition caused by a coronavirus appeared in six nations near the Arabian peninsula. MERS-coronavirus causes the fever, cough, and shortness of breath of Middle Eastern respiratory syndrome (MERS).

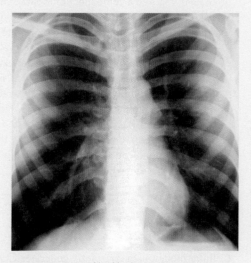

Healthy lungs

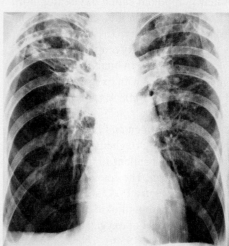

Tuberculosis

FIGURE 19E Healthy lungs appear dark and clear on a radiograph (above on the left). Lungs with tuberculosis have cloudy areas where fibrous tissue grows, walling off infected areas (radiograph above on the right).

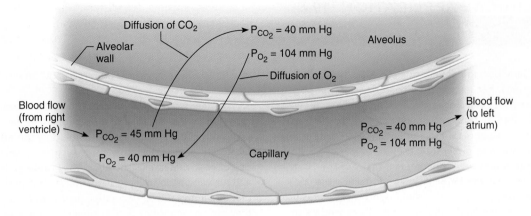

Diffusion of CO₂

Alveolar wall

$P_{CO_2} = 40$ mm Hg

$P_{O_2} = 104$ mm Hg

Diffusion of O₂

Alveolus

Blood flow (from right ventricle)

$P_{CO_2} = 45$ mm Hg

$P_{O_2} = 40$ mm Hg

Capillary

$P_{CO_2} = 40$ mm Hg

$P_{O_2} = 104$ mm Hg

Blood flow (to left atrium)

FIGURE 19.35 **AP|R** Gases are exchanged between alveolar air and capillary blood because of differences in partial pressures.

19.6 | Gas Transport

The blood carries oxygen and carbon dioxide between the lungs and the body cells. As these gases enter the blood, they dissolve in the liquid portion, the plasma, or combine chemically with other atoms or molecules.

Oxygen Transport

Almost all the oxygen (over 98%) is carried in the blood bound to the protein *hemoglobin* in red blood cells. The oxygen bound to iron in hemoglobin provides the color of these blood cells. The remainder of the oxygen is dissolved in the blood plasma.

Hemoglobin consists of two types of components called *heme* and *globin* (see chapter 18, pp. 715–716). Globin is a protein of 574 amino acids in four polypeptide chains. Each chain is associated with a heme group, and each heme group surrounds an atom of iron. Each iron atom can loosely bind an oxygen molecule. As oxygen dissolves in blood, it rapidly combines with hemoglobin, forming a new compound called **oxyhemoglobin** (ok″sĭ-he″mo-glo′bin). Each hemoglobin molecule can bind up to four oxygen molecules.

The P_{O_2} determines the amount of oxygen that hemoglobin binds. The greater the P_{O_2}, the more oxygen binds until the hemoglobin molecules are completely loaded with oxygen, or saturated (fig. 19.36). At normal arterial P_{O_2} (95 mm Hg), hemoglobin is essentially completely saturated.

The chemical bonds between oxygen and hemoglobin molecules can break. As the P_{O_2} decreases, oxyhemoglobin releases oxygen molecules (fig. 19.36). This happens in tissues in which cells have used oxygen in respiration. The free oxygen diffuses from the blood into nearby cells, as figure 19.37 shows.

Increasing the blood level of carbon dioxide (P_{CO_2}), acidity, and temperature all increase the amount of oxygen that oxyhemoglobin releases (figs. 19.38, 19.39, and 19.40). These influences help release more oxygen from the blood to the skeletal muscles during exercise. The increased muscular activity accompanied by increased oxygen use increases the P_{CO_2}, decreases the pH, and raises the local temperature. At the same time, less-active cells receive the usual amount of oxygen.

As described earlier, respiratory control under most circumstances is responding to plasma P_{CO_2} and pH, not P_{O_2}, despite the central role of oxygen in cellular metabolism. Notice, however, in figures 19.36 and 19.37, that the "oxygen-poor" systemic venous blood retains 75% of the oxygen it had when it was "oxygen-rich." This safety margin for O_2 makes it possible for the respiratory system to safely adjust CO_2 levels, and thereby the pH, of the internal environment.

 PRACTICE

40 How is oxygen transported from the lungs to body cells?

41 What factors affect the release of oxygen from oxyhemoglobin?

Carbon monoxide (CO) is a toxic gas produced in gasoline engines and some stoves as a result of incomplete combustion of fuels. It is also a component of tobacco smoke. Carbon monoxide is toxic because it binds hemoglobin in a way that prevents hemoglobin from delivering oxygen to tissues. It may also inhibit aerobic respiration at the cellular level. Treatment for carbon monoxide poisoning is to administer oxygen at high partial pressure to increase the amount of oxygen available for delivery to the tissues. Carbon dioxide (CO_2) is usually given simultaneously to stimulate the respiratory center, which, in turn, increases breathing rate. Rapid breathing, once the carbon monoxide source has been removed, helps reduce the partial pressure of carbon monoxide in the alveoli, favoring diffusion of carbon monoxide out of the blood.

Carbon Dioxide Transport

Cellular metabolism constantly generates CO_2, which diffuses into the systemic capillaries. This CO_2 is carried to the lungs in one of three forms: as CO_2 dissolved in plasma, as part of a compound formed by bonding to hemoglobin, or as part of bicarbonate ions (fig. 19.41).

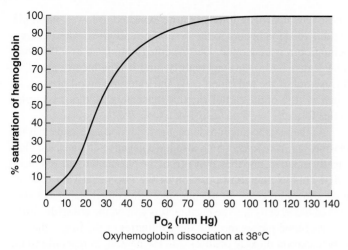

FIGURE 19.36 Hemoglobin is completely saturated at normal systemic arterial P_{O_2} but readily releases oxygen at the P_{O_2} of the body tissues.

The amount of carbon dioxide that dissolves in plasma is determined by its partial pressure. The higher the P_{CO_2} of the tissues, the more carbon dioxide dissolves. However, only about 7% of the carbon dioxide that enters the blood is carried in this form.

Unlike oxygen, which binds the iron atoms of hemoglobin molecules, carbon dioxide bonds with the amino groups (—NH₂) of these molecules. Consequently, oxygen and carbon dioxide do not directly compete for binding sites—a hemoglobin molecule can transport both gases at the same time.

Carbon dioxide binding hemoglobin forms a loosely bound compound called **carbaminohemoglobin** (kar-bam″ĭ-no-he″mo-globin). This molecule readily decomposes in regions where the P_{CO_2} is low, releasing its carbon dioxide. Although this method of transporting carbon dioxide is theoretically quite effective, carbaminohemoglobin forms relatively slowly. About 15% to 25% of the CO_2 that enters the blood is carried this way.

In the most important CO_2 transport mechanism **bicarbonate ions** (HCO_3^-) form. Recall that carbon dioxide reacts with water to form carbonic acid (H_2CO_3). This reaction occurs slowly in the blood plasma, but much of the CO_2 diffuses into the red blood cells. These cells contain an enzyme, **carbonic anhydrase** (kar-bon′ik an-hi′drās), which speeds the reaction between CO_2 and water.

The resulting carbonic acid dissociates almost immediately, releasing hydrogen ions (H^+) and bicarbonate ions (HCO_3^-):

$$CO_2 + H_2O \rightarrow H_2CO_3 \rightarrow H^+ + HCO_3^-$$

FIGURE 19.37 Blood carries oxygen. Oxygen molecules, entering the blood from the alveolus, bond to hemoglobin, forming oxyhemoglobin. In the regions of the body cells, oxyhemoglobin releases oxygen. Much oxygen is still bound to hemoglobin at the P_{O_2} of systemic venous blood.

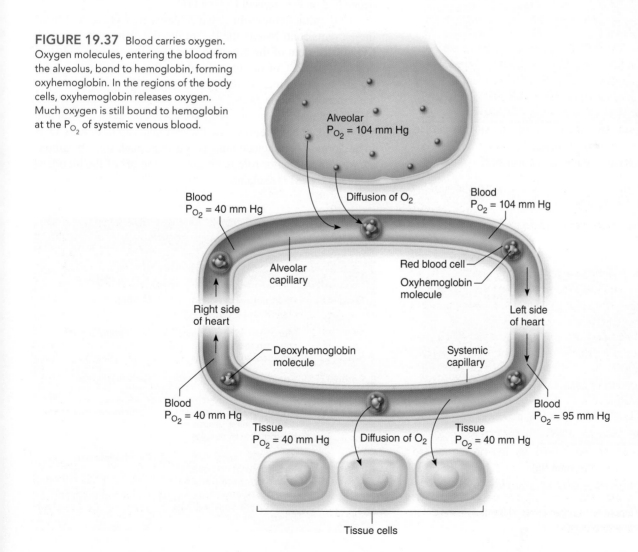

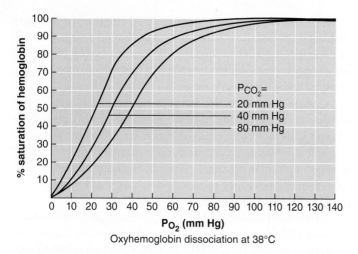

FIGURE 19.38 The amount of oxygen released from oxyhemoglobin increases as the P_{CO_2} increases.

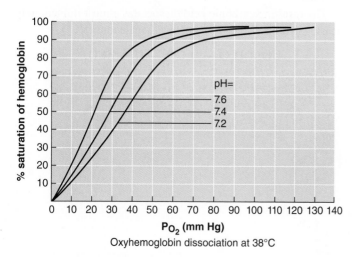

FIGURE 19.39 The amount of oxygen released from oxyhemoglobin increases as the blood pH decreases.

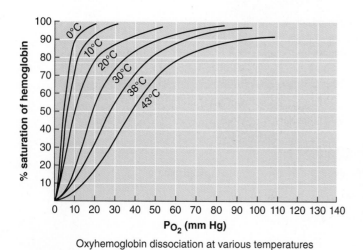

FIGURE 19.40 The amount of oxygen released from oxyhemoglobin increases as the blood temperature increases.

These new hydrogen ions might be expected to lower blood pH, but this reaction occurs in the systemic capillaries, where deoxyhemoglobin is generated. Deoxyhemoglobin is an excellent buffer because hydrogen ions readily bind it. The bicarbonate ions diffuse out of the red blood cells and enter the blood plasma. As much as 70% of the carbon dioxide that enters the blood is carried in this form.

As the bicarbonate ions leave the red blood cells and enter the plasma, *chloride ions,* which also have negative charges, are electrically repelled, and they move from the plasma into the red blood cells. This exchange in position of the two negatively charged ions, shown in figure 19.42, maintains the ionic balance between the red blood cells and the plasma. It is termed the **chloride shift.**

As blood passes through the capillaries of the lungs, the dissolved carbon dioxide diffuses into the alveoli, in response to the relatively low P_{CO_2} of the alveolar air. As the plasma P_{CO_2} drops, hydrogen ions and bicarbonate ions in the red blood cells recombine to form carbonic acid, and under the influence of carbonic anhydrase, the carbonic acid quickly yields new molecules of CO_2 and water:

$$H^+ + HCO_3^- \rightarrow H_2CO_3 \rightarrow CO_2 + H_2O$$

Note that this is the reverse of the reaction described above. Carbonic anhydrase catalyzes this reaction in either direction, depending on the levels of CO_2 and H^+.

Carbaminohemoglobin also releases its CO_2, and both of these events contribute to the P_{CO_2} of the alveolar capillary blood. CO_2 diffuses out of the blood until an equilibrium is established between the P_{CO_2} of the blood and the P_{CO_2} of the alveolar air. Figure 19.43 summarizes this process, and table 19.7 summarizes the transport of blood gases.

The actual percentages vary somewhat between arterial and venous blood, but in either case the vast majority of blood CO_2 is in the form of bicarbonate ion. This is important, since bicarbonate ion plays a major role in controlling the pH of the blood, as chapter 21 (p. 814) explains.

TABLE 19.7	Gases Entering the Blood	
Gas	**Reaction**	**Substance Transported**
Oxygen	Less than 2% dissolves in plasma	Oxygen
	More than 98% combines with iron atoms of hemoglobin molecules	Oxyhemoglobin
Carbon dioxide	About 7% dissolves in plasma	Carbon dioxide
	About 15–25% combines with the amino groups of hemoglobin molecules	Carbaminohemoglobin
	About 70% reacts with water to form carbonic acid; the carbonic acid then dissociates to release hydrogen ions and bicarbonate ions	Bicarbonate ions

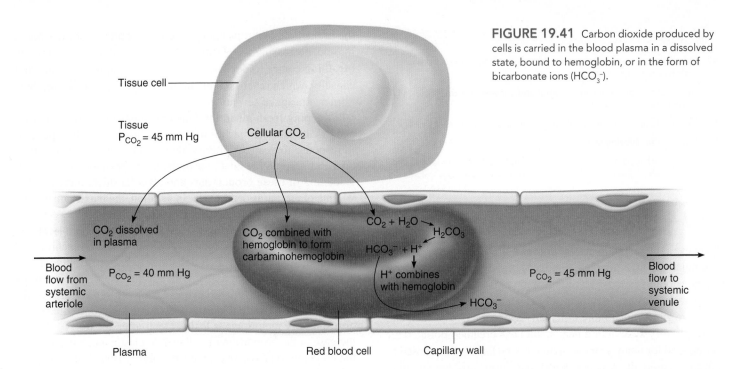

FIGURE 19.41 Carbon dioxide produced by cells is carried in the blood plasma in a dissolved state, bound to hemoglobin, or in the form of bicarbonate ions (HCO_3^-).

Tissue cell

Tissue $P_{CO_2} = 45$ mm Hg

Cellular CO_2

CO_2 dissolved in plasma

CO_2 combined with hemoglobin to form carbaminohemoglobin

$CO_2 + H_2O \rightarrow H_2CO_3$

$HCO_3^- + H^+$

H^+ combines with hemoglobin

HCO_3^-

Blood flow from systemic arteriole

$P_{CO_2} = 40$ mm Hg

$P_{CO_2} = 45$ mm Hg

Blood flow to systemic venule

Plasma

Red blood cell

Capillary wall

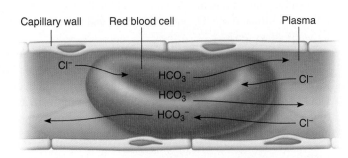

Capillary wall   Red blood cell   Plasma

Cl^-

HCO_3^-

HCO_3^-

HCO_3^-

Cl^-

Cl^-

FIGURE 19.42 As bicarbonate ions (HCO_3^-) diffuse out of the red blood cell, chloride ions (Cl^-) from the plasma diffuse into the cell, maintaining the electrical balance between ions. This exchange of ions is called the chloride shift.

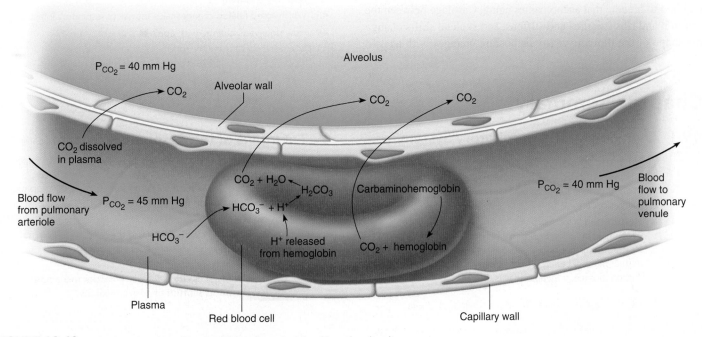

Alveolus

$P_{CO_2} = 40$ mm Hg

CO_2

Alveolar wall

CO_2

CO_2

CO_2 dissolved in plasma

$CO_2 + H_2O$

H_2CO_3

Carbaminohemoglobin

$P_{CO_2} = 40$ mm Hg

Blood flow to pulmonary venule

Blood flow from pulmonary arteriole

$P_{CO_2} = 45$ mm Hg

$HCO_3^- + H^+$

HCO_3^-

H^+ released from hemoglobin

$CO_2 + $ hemoglobin

Plasma

Red blood cell

Capillary wall

FIGURE 19.43 In the lungs, carbon dioxide diffuses from the blood into the alveoli.

42 Describe three ways carbon dioxide can be transported from cells to the lungs.

43 How can hemoglobin carry oxygen and carbon dioxide at the same time?

44 How do bicarbonate ions help buffer the blood (maintain its pH)?

45 What is the chloride shift?

46 How is carbon dioxide released from the blood into the lungs?

19.7 | Life-Span Changes

Changes in the respiratory system over a lifetime reflect both the accumulation of environmental influences and the effects of aging in other organ systems. The lungs and respiratory passageways of a person who has breathed only clean air are pinker and can exchange gases much more efficiently as the years pass than can the respiratory system of a person who has breathed polluted air and smoked for many years. Individuals who have been exposed to foul air are more likely to develop chronic bronchitis, emphysema, and/or lung cancer than people fortunate enough to breathe fresh air. Long-term exposure to particulates in the workplace can also raise the risk of developing respiratory illnesses. Still, many age-associated changes in the respiratory system are unavoidable.

With age, protection of the lungs and airways falters, as ciliated epithelial cells become fewer, and their cilia less active or gone. At the same time, mucus thickens; the swallowing, gagging, and coughing reflexes slow; and macrophages lose their efficiency in phagocytizing bacteria. These changes combine to slow the clearance of pathogens from the lungs and respiratory passages, which increases susceptibility to and the severity of respiratory infections.

Several changes contribute to an overall increase in the effort required to breathe that accompanies aging. Cartilage between the sternum and ribs calcifies and stiffens, and skeletal shifts change the shape of the thoracic cavity as posture changes with age. In the bronchioles, fibrous connective tissue replaces some smooth muscle, decreasing the ability to dilate. As muscles lose strength, breathing comes to depend more upon the diaphragm. The vital capacity, which reaches a maximum by age forty, may drop by a third by the age of seventy years.

Keeping fresh air in the lungs becomes more difficult with age. As the farthest reaches of the bronchiole walls thin, perhaps in response to years of gravity, they do not stay as open as they once did, trapping residual air in the lower portions of the lungs. Widening of these bronchioles and alveolar ducts increases the volume of air that is dead space. In addition, the maximum minute ventilation begins to decrease at around age thirty.

Aging-associated changes occur at the microscopic level too. The number of alveoli is about 24 million at birth, reaching 480 million by adulthood, which translates to about 170 alveoli per cubic millimeter of lung tissue. With advancing age, alveolar walls thin and alveoli may coalesce, decreasing the surface area available for gas exchange. One study estimated that one square foot of respiratory membrane is lost per year after age thirty. Also, an increase in the proportion of collagen to elastin and a tendency of the collagen to cross-link impair the ability of alveoli to expand fully. Oxygen transport from the alveoli to the blood, as well as oxygen loading onto hemoglobin in red blood cells, becomes less efficient. Diffusion of CO_2 out of the blood and through the alveolar walls slows too.

As with other organ systems, the respiratory system undergoes specific changes, but these may be unnoticeable at the whole-body level. A person who is sedentary or engages only in light activity would probably not be aware of the slowing of air flow in and out of the respiratory system. Unaccustomed exercise, however, would quickly reveal how difficult breathing has become with age.

 PRACTICE

47 How does the environment influence the effects of aging on the respiratory system?

48 Which aging-related changes raise the risk of respiratory infection?

49 How do alveoli change with age? ■

Chapter Summary

19.1 Overview of the Respiratory System
(page 731)

The respiratory system includes the passages that transport air to and from the lungs and the air sacs in which gas exchanges occur.

a. **Respiration** is the entire process by which gases are exchanged between the atmosphere and the body cells.

b. Respiration is necessary because cells require oxygen to extract maximal energy from nutrient molecules and release carbon dioxide, a metabolic waste.

19.2 Organs of the Respiratory System (page 731)

The respiratory system is divided into two tracts. The upper respiratory tract includes the nose, nasal cavity, sinuses, and pharynx; the lower respiratory tract includes the larynx, trachea, bronchial tree, and lungs.

1. Nose
   a. Bone and cartilage support the nose.
   b. Nostrils provide entrances for air.
2. Nasal cavity
   a. The **nasal cavity** is a space posterior to the nose.
   b. The **nasal septum** divides it medially.

Respiratory System

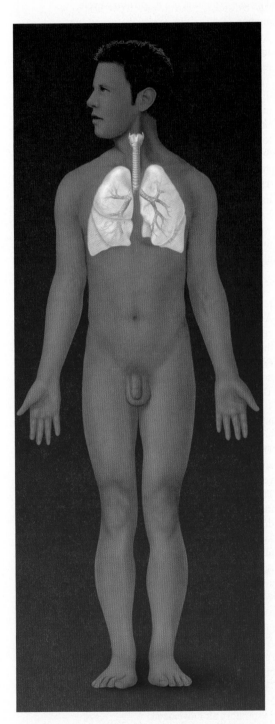

The respiratory system provides oxygen for the internal environment and excretes carbon dioxide.

Integumentary System

Stimulation of skin receptors may alter respiratory rate.

Skeletal System

Bones provide attachments for muscles involved in breathing.

Muscular System

The respiratory system eliminates carbon dioxide produced by exercising muscles.

Nervous System

The brain controls the respiratory system.

Endocrine System

Hormonelike substances control the production of red blood cells that transport oxygen and carbon dioxide.

Cardiovascular System

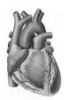

As the heart pumps blood through the lungs, the lungs oxygenate the blood and excrete carbon dioxide.

Lymphatic System

Cells of the immune system patrol the lungs and defend against infection.

Digestive System

The digestive system and respiratory system share openings to the outside.

Urinary System

The kidneys and the respiratory system work together to maintain blood pH. The kidneys compensate for water lost through breathing.

Reproductive System

Respiration increases during sexual activity. Fetal gas exchange begins before birth.

c. **Nasal conchae** divide the cavity into passageways and help increase the surface area of the mucous membrane.

d. Mucous membrane cleans, warms, and moistens incoming air.

e. Particles trapped in the mucus are carried to the pharynx by ciliary action and are swallowed.

3. Sinuses

a. Sinuses are spaces in the bones of the skull that open into the nasal cavity.

b. They are lined with mucous membrane that is continuous with the lining of the nasal cavity.

4. Pharynx

a. The **pharynx** is posterior to the mouth, between the nasal cavity and the larynx.

b. It provides a common passage for air and food.

c. It aids in creating vocal sounds.

5. Larynx

a. The **larynx** is an enlargement at the top of the trachea.

b. It is a passageway for air and helps prevent foreign objects from entering the trachea.

c. It is composed of muscles and cartilages; some of these cartilages are single, whereas others are paired.

d. It contains the vocal cords, which produce sounds by vibrating as air passes over them.

   (1) The pitch of a sound is related to the tension on the cords.

   (2) The intensity of a sound is related to the force of the air passing over the cords.

e. The **epiglottis** helps prevent food and liquid from entering the trachea.

6. Trachea

a. The **trachea** extends into the thoracic cavity anterior to the esophagus.

b. It divides into the right and left main bronchi.

c. The mucous lining continues to clean incoming air.

d. Incomplete cartilaginous rings support the wall.

7. Bronchial tree

a. The **bronchial tree** consists of branched air passages that connect the trachea to the air sacs.

b. The branches of the bronchial tree include **main bronchi, lobar bronchi, segmental bronchi, intralobular bronchioles, terminal bronchioles, respiratory bronchioles, alveolar ducts, alveolar sacs,** and **alveoli.**

c. Structure of the respiratory tubes

   (1) As tubes branch, the amount of cartilage in the walls decreases, and the muscular layer persists to the ends of the respiratory bronchioles.

   (2) Elastic fibers in the walls aid breathing.

   (3) The epithelial lining changes from pseudostratified and ciliated to cuboidal and then to simple squamous as the tubes become progressively smaller.

d. Functions of the respiratory tubes and alveoli include distribution of air and exchange of gases between the alveolar air and the blood.

8. Lungs

a. The left and right lungs are separated by the mediastinum and are enclosed by the diaphragm and the thoracic cage.

b. The visceral pleura is attached to the surface of the lungs; parietal pleura lines the thoracic cavity.

c. The right lung has three lobes, and the left lung has two.

d. Each lobe is composed of lobules that contain alveolar ducts, alveolar sacs, alveoli, nerves, blood vessels, lymphatic vessels, and connective tissues.

19.3 Breathing Mechanism (page 741)

Inspiration and **expiration** movements are accompanied by changes in the size of the thoracic cavity. One inspiration followed by one expiration is a **respiratory cycle.**

1. Inspiration

a. Atmospheric pressure forces air into the lungs.

b. Inspiration occurs when the intra-alveolar pressure is reduced.

c. The intra-alveolar pressure is reduced when the diaphragm moves downward and the thoracic cage moves upward and outward.

d. **Surface tension** holding the pleural membranes together aids lung expansion.

e. **Surfactant** reduces surface tension in the alveoli.

2. Expiration

a. The forces of expiration come from the elastic recoil of tissues and from surface tension in the alveoli.

b. Expiration can be aided by thoracic and abdominal wall muscles that pull the thoracic cage downward and inward and compress the abdominal organs inward and upward.

3. Respiratory air volumes and capacities

a. The amount of air that moves in or out during a respiratory cycle is the **tidal volume.**

b. Additional air that can be inhaled is the **inspiratory reserve volume;** additional air that can be exhaled is the **expiratory reserve volume.**

c. Residual air remains in the lungs and is mixed with newly inhaled air.

d. The **residual volume** is difficult to measure.

e. The **vital capacity** is the maximum amount of air a person can exhale after taking the deepest breath possible.

f. The **inspiratory capacity** is the maximum volume of air a person can inhale following exhalation of the tidal volume.

g. The **functional residual capacity** is the volume of air that remains in the lungs following the exhalation of the tidal volume.

h. The **total lung capacity** is equal to the vital capacity plus the residual volume.

i. Air in the anatomic and alveolar dead spaces is not available for gas exchange.

4. Alveolar ventilation

a. Minute ventilation is tidal volume multiplied by breathing rate.

b. Alveolar ventilation rate is the physiologic dead space subtracted from the tidal volume multiplied by breathing rate.

c. The alveolar ventilation rate is a major factor affecting gas exchange between the alveolar air and the blood.

5. Nonrespiratory air movements

a. Nonrespiratory air movements are air movements other than breathing.

b. They include coughing, sneezing, laughing, crying, hiccupping, and yawning.

19.4 Control of Breathing (page 750)

Normal breathing is rhythmic and involuntary, although the respiratory muscles can be controlled voluntarily.

1. Respiratory areas
   a. The **respiratory areas** are in the brainstem and include parts of the medulla oblongata and pons.
   b. The **medullary respiratory center** includes two groups of neurons.
      (1) The ventral respiratory group provides the basic rhythm of breathing.
      (2) The dorsal respiratory group stimulates inspiratory muscles and processes sensory information regarding the respiratory system. It may play a role in certain cardiopulmonary reflexes.
   c. The pontine respiratory group may contribute to the rhythm of breathing by limiting inspiration.
2. Partial pressure
   a. The **partial pressure** of a gas is determined by the concentration of that gas in a mixture of gases or the concentration of gas dissolved in a liquid.
   b. The partial pressure of a gas dissolved in a liquid equals the partial pressure of that gas in the air with which the liquid has equilibrated.
3. Factors affecting breathing.
   a. Chemicals, lung tissue stretching, and emotional state affect breathing.
   b. Chemosensitive areas (**central chemoreceptors**) are associated with the respiratory center.
      (1) CO_2 combines with water to form carbonic acid, which, in turn, releases hydrogen ions in the CSF.
      (2) Stimulation of these areas increases alveolar ventilation.
   c. **Peripheral chemoreceptors** are in the carotid bodies and aortic bodies of certain arteries.
      (1) These chemoreceptors sense low oxygen levels as well as excess hydrogen ions.
      (2) When oxygen levels are low or blood pH drops, alveolar ventilation increases.
   d. Stretching the lung tissues triggers an inflation reflex.
      (1) This reflex reduces the duration of inspiratory movements.
      (2) This prevents overinflation of the lungs during forceful breathing.
   e. **Hyperventilation** decreases CO_2 levels below normal, but *this is dangerous when associated with breath holding during underwater swimming.*

19.5 Alveolar Gas Exchanges (page 753)

Gas exchange between the air and the blood occurs in the alveoli.

1. Alveoli
   a. The alveoli are tiny sacs clustered at the distal ends of the alveolar ducts.
   b. Some alveoli open into adjacent air sacs through **alveolar pores** that provide alternate pathways for air when passages are obstructed.

2. Respiratory membrane
   a. The **respiratory membrane** consists of the alveolar and capillary walls.
   b. Gas exchange takes place through these walls.
3. Diffusion through the respiratory membrane
   a. Gases diffuse from regions of higher partial pressure toward regions of lower partial pressure.
   b. Oxygen diffuses from the alveolar air into the blood; CO_2 diffuses from the blood into the alveolar air.

19.6 Gas Transport (page 758)

Blood carries gases between the lungs and the body cells.

1. Oxygen transport
   a. Oxygen is mainly transported bound to hemoglobin molecules.
   b. The resulting **oxyhemoglobin** is unstable and releases its oxygen in regions where the P_{O_2} is low.
   c. More oxygen is released as the plasma P_{CO_2} increases, as the blood becomes more acidic, and as the blood temperature increases.
2. Carbon dioxide transport
   a. CO_2 may be carried either as dissolved CO_2 in solution, CO_2 bound to hemoglobin, or as a **bicarbonate ion.**
   b. Most CO_2 is transported as part of bicarbonate ions.
   c. **Carbonic anhydrase** speeds the reaction between CO_2 and water to form carbonic acid as well as the reverse reaction to breakdown carbonic acid into CO_2 and water.
   d. Carbonic acid dissociates to release hydrogen ions and bicarbonate ions.

19.7 Life-Span Changes (page 762)

The lungs, respiratory passageways, and alveoli undergo aging-associated changes exacerbated by exposure to polluted air. However, the increased work required to breathe with age is typically not noticeable unless one engages in vigorous exercise.

1. Exposure to pollutants, smoke, and other particulates raises the risk of developing diseases of the respiratory system.
2. Loss of cilia, thickening of mucus, and impaired macrophages raise the risk of infection.
3. Calcified cartilage, skeletal changes, altered posture, and replacement of smooth muscle with fibrous connective tissue in bronchioles make breathing more difficult. Vital capacity diminishes.
4. Physiologic dead space increases.
5. Alveoli coalesce, decreasing the surface area available for gas exchange.

CHAPTER ASSESSMENTS

19.1 Overview of the Respiratory System

1 List the general functions of the respiratory system. (p. 731)
2 Explain why oxygen is required at the cellular level. (p. 731)

19.2 Organs of the Respiratory System

3 Distinguish between the upper and lower respiratory tracts. (p. 731)
4 Explain how the nose and nasal cavity filter incoming air. (p. 732)
5 Name and describe the locations of the major sinuses. (p. 733)
6 Explain how a sinus headache may occur. (p. 733)
7 The nasopharynx is posterior to the _____. (p. 733)
   a. nasal cavity
   b. oral cavity
   c. larynx
   d. esophagus
8 Name and describe the functions of the cartilages of the larynx. (p. 734)
9 Match the following structures with their descriptions: (pp. 735–741)

(1) true vocal cords	A. serous membrane on lungs
(2) false vocal cords	B. contains the vocal cords
(3) larynx	C. vibrate to make sound
(4) visceral pleura	D. air sacs
(5) alveoli	E. muscular folds that close the glottis

10 Name the successive branches of the bronchial tree, from the main bronchi to the alveoli, and identify their functions. (pp. 738–739)
11 Describe how the structure of the respiratory tubes changes as the branches become finer. (p. 740)
12 Distinguish between the visceral pleura and the parietal pleura. (p. 741)
13 Name the lobes of the lungs and identify their locations. (p. 741)

19.3 Breathing Mechanism

14 Compare the muscles used in a resting inspiration with those used in a forced inspiration. (p. 745)
15 Define *surface tension*, and explain how it works against the breathing mechanism. (p. 745)
16 Define *surfactant*, and explain its function. (p. 745)
17 Define *compliance*. (p. 746)
18 Compare the muscles used (if any) in a resting expiration with those used in a forced expiration. (p. 746)
19 Match the air volumes with their descriptions: (pp. 747–748)

(1) tidal volume	A. volume of air that remains after most forceful expiration
(2) inspiratory reserve volume	B. volume of air, in addition to resting tidal volume, that can enter lungs
(3) expiratory reserve volume	C. volume of air that enters or leaves lungs during a respiratory cycle
(4) residual volume	D. volume of air, in addition to resting tidal volume, that can be expelled from the lungs

20 Distinguish between vital capacity and total lung capacity. (p. 748)
21 Physiologic dead space is equal to _____. (pp. 748–749)
   a. anatomic dead space
   b. anatomic dead space plus alveolar dead space
   c. alveolar dead space
22 Calculate both minute ventilation and alveolar ventilation given the following: (p. 749)
respiratory rate = 12 breaths per minute
tidal volume = 500 mL per breath
physiologic dead space = 150 mL per breath
23 Explain the mechanisms of coughing and sneezing, and give the functions of each. (p. 749)
24 Describe a possible function of yawning. (p. 749)

19.4 Control of Breathing

25 Locate the respiratory areas and name their major components. (pp. 750–751)
26 Explain control of the basic rhythm of breathing. (p. 751)
27 What is the partial pressure of pure oxygen at sea level? (p. 752)
   a. 40 mm Hg
   b. 95 mm Hg
   c. 104 mm Hg
   d. 160 mm Hg
28 Explain the effect increasing CO_2 levels have on the central chemoreceptors. (p. 752)
29 Describe the response of the peripheral chemoreceptors in the carotid and aortic bodies to both CO_2 and hydrogen ions. (p. 752)
30 Describe the inflation reflex. (pp. 752–753)
31 Describe the effects of emotions on breathing. (p. 753)
32 Hyperventilation is which one of the following? (p. 753)
   a. any increase in breathing
   b. an increase in breathing that brings in oxygen too quickly
   c. an increase in breathing that eliminates CO_2 too quickly
   d. an increase in breathing that has no effect on blood gases

19.5 Alveolar Gas Exchanges

33 Describe the respiratory membrane. (pp. 754–755)
34 Explain the relationship between the partial pressure of a gas and its rate of diffusion. (p. 755)
35 Summarize the exchange of oxygen and CO_2 across the respiratory membrane. (p. 755)

19.6 Gas Transport

36 Describe how the blood carries oxygen. (p. 758)
37 List three factors that increase the release of oxygen from hemoglobin. (p. 758)
38 Explain why carbon monoxide is toxic. (p. 758)
39 Give the percentages of the three ways CO_2 is transported in blood. (pp. 758–760)
40 Explain the function of carbonic anhydrase. (p. 760)
41 Define *chloride shift*. (p. 760)

19.7 Life-Span Changes

42 Describe the changes that make it harder to breathe with advancing years. (p. 762)

Outcomes 3.2, 5.2, 5.3, 5.5, 19.2

1. Describe the following structures that are part of the respiratory tubes and state their locations.
   a. pseudostratified epithelium
   b. cuboidal epithelium
   c. simple squamous epithelium
   d. goblet cells
   e. cartilage
   f. smooth muscle
   g. elastic fibers
   h. cilia

Outcomes 19.2, 19.3

2. Patients experiencing asthma attacks are often advised to breathe through pursed (puckered) lips. How might this help reduce the symptoms of asthma?

Outcomes 19.2, 19.3, 19.4, 19.5, 19.6

3. What changes would you expect to occur in the levels of blood oxygen and carbon dioxide in a patient who breathes rapidly and deeply for a prolonged time?

Outcomes 19.2, 19.3, 19.5

4. If a tracheostomy bypasses the upper respiratory passages, how might the air entering the trachea differ from air normally passing through this tube? What problems might this cause for the patient?

5. Certain respiratory disorders, such as emphysema, reduce the capacity of the lungs to recoil elastically. Which respiratory volumes will this condition affect? Explain the impact on gas exchange.

Outcomes 19.4, 19.5, 19.6

6. If a person is receiving supplemental oxygen to restore blood oxygen levels, why might it be better to administer a combination of oxygen and carbon dioxide rather than pure oxygen?

Outcomes 19.4, 19.6

7. Why is it impossible, under normal circumstances, for a person to hold the breath long enough to pass out?

Outcomes 19.5, 19.6

8. What problem might a person with a serious respiratory disorder encounter flying in the passenger compartment of a commercial aircraft that has an air pressure equivalent to an altitude of 8,000 feet?

 # ONLINE STUDY TOOLS

  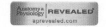

Connect Interactive Questions Reinforce your knowledge using assigned interactive questions covering respiratory structures, the mechanics of breathing, and the process of gas transport to and from tissues.

Connect Integrated Activity Can you predict changes in respiration that will occur in various circumstances?

LearnSmart Discover which chapter concepts you have mastered and which require more attention. This adaptive learning tool is personalized, proven, and preferred.

Anatomy & Physiology Revealed Go more in depth into the human body by exploring the upper respiratory tract and lower respiratory tract structures. Also view animations on the process of breathing.

20

Urinary System

LEARNING OUTCOMES

After you have studied this chapter, you should be able to:

20.1 Overview of the Urinary System
1 Name the organs of the urinary system and list their general functions. (p. 769)

20.2 Kidneys
2 Describe the locations of the kidneys and the structure of the kidney. (pp. 769–771)
3 List the functions of the kidneys. (p. 769)
4 Trace the pathway of blood flow through the major vessels within a kidney. (p. 772)
5 Describe a nephron and explain the functions of its major parts. (pp. 775–776)

20.3 Urine Formation
6 Explain how glomerular filtrate is produced and describe its composition. (p. 779)
7 Explain how various factors affect the rate of glomerular filtration and identify ways that this rate is regulated. (pp. 782–783)
8 Explain tubular reabsorption, and its role in urine formation. (pp. 783–786)
9 Identify the changes in the osmotic concentration of the glomerular filtrate as it passes through the renal tubule. (pp. 783–788)
10 Explain tubular secretion, and its role in urine formation. (pp. 787–788)
11 Identify the characteristics of a countercurrent mechanism, and explain its role in concentrating the urine. (p. 789)
12 Explain how the final composition of urine contributes to homeostasis. (pp. 791–792)

20.4 Storage and Elimination of Urine
13 Describe the structures of the ureters, urinary bladder, and urethra. (pp. 791–794)
14 Explain how micturition occurs, and how it is controlled. (pp. 795–796)

20.5 Life-Span Changes
15 Describe how the components of the urinary system change with age. (pp. 797–798)

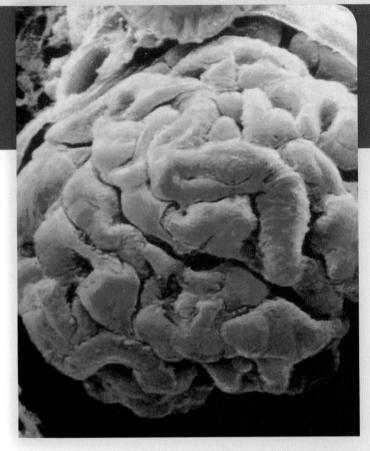

A falsely colored, scanning electron micrograph of part of the kidney shows one of the million or so glomeruli, which is a tangle of capillaries (red), where filtration of plasma begins urine formation (1,200×).

THE WHOLE PICTURE

The main theme in anatomy and physiology is the maintenance of a stable internal environment. Consider how many substances we add to our internal environments every day: the complex of chemicals called "food," medications, or simply water. Despite all of these challenges to maintaining a constant internal environment, the composition of our body fluids remains remarkably steady, thanks largely to the urinary system.

The urinary system is amazingly reliable. The kidneys excrete exactly the right amounts of whatever we take in to ensure that the composition and even the total volume of body fluids stay in a range compatible with life. Each substance is handled separately. If we eat heavily salted French fries, the urine later that day contains more salt than usual. If we don't drink enough water, the microscopic tubules that compose the kidneys retain water, decreasing urine volume. The kidneys are working constantly, yet except for an occasional trip to empty the bladder, we are unaware from moment to moment of all that they do.

Module 13: Urinary System

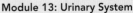

UNDERSTANDING WORDS

af-, to: *afferent* arteriole—arteriole that leads to a nephron.

calyc-, small cup: major *calyces*—cuplike subdivisions of the renal pelvis.

cort-, covering: renal *cortex*—shell of tissue surrounding the inner region of a kidney.

cyst-, bladder: *cystitis*—inflammation of the urinary bladder.

detrus-, to force away: *detrusor* muscle— muscle in the urinary bladder wall that contracts to expel urine.

glom-, little ball: *glomerulus*—cluster of capillaries in a renal corpuscle.

juxta-, near to: *juxtamedullary* nephron— nephron near the renal medulla.

mict-, to pass urine: *micturition*—expelling urine from the urinary bladder.

nephr-, pertaining to the kidney: *nephron*— functional unit of a kidney.

papill-, nipple: renal *papillae*—small elevations that project into a minor calyx.

prox-, nearest: *proximal* tubule—coiled portion of the renal tubule leading from the glomerular capsule.

ren-, kidney: *renal* cortex—outer region of a kidney.

trigon-, triangular shape: *trigone*—triangular area on the internal floor of the urinary bladder.

20.1 | Overview of the Urinary System

A major part of homeostasis is maintaining the composition, pH, and volume of body fluids within normal ranges. The urinary system accomplishes this task. It removes metabolic wastes and excess biochemicals, yet does not deplete the body of essential substances. The urinary system also excretes foreign substances, such as drugs and their metabolites that can be toxic if they remain in the body fluids after they have exerted their therapeutic effects.

The urinary system consists of a pair of kidneys, which remove substances from the blood, form urine, and help regulate certain metabolic processes by secreting hormones; a pair of tubular ureters, which transport urine from the kidneys; a saclike urinary bladder, which collects urine from the ureters and serves as a urine reservoir; and a tubular urethra, which conveys urine to the outside of the body. Figures 20.1 and 20.2 show these organs.

20.2 | Kidneys

A **kidney** is a reddish brown, bean-shaped organ with a smooth surface. It is about 12 centimeters long, 6 centimeters wide, and 3 centimeters thick in an adult, and it is enclosed in a tough, fibrous capsule (tunic fibrosa).

Location of the Kidneys

The kidneys lie on either side of the vertebral column in a depression high on the posterior wall of the abdominal cavity. The upper and lower borders of the kidneys are generally at the levels of the twelfth thoracic and third lumbar vertebrae, respectively, although the positions of the kidneys may vary slightly with changes in posture and with breathing movements (see fig. 20.1). The left kidney is typically about 1.5 to 2 centimeters higher than the right one.

The kidneys are positioned *retroperitoneally* (re"tro-per"ĭ-to-ne'al-le); they are posterior to the parietal peritoneum and against the deep muscles of the back. Connective tissue (renal fascia) and masses of adipose tissue (renal fat) surrounding the kidneys hold them in place (fig. 20.3 and reference plates 18, 19 on pp. 51 and 52).

Functions of the Kidneys

The main function of the kidneys is to regulate the volume and composition of body fluids. In the process, the kidneys remove metabolic wastes as well as excess water and electrolytes from the blood and, along with the ureters, urinary bladder, and urethra, excrete them to the outside as urine. These wastes include nitrogenous and sulfur-containing products of protein metabolism, such as certain metabolic acids.

The kidneys also help control the rate of red blood cell formation by secreting the hormone *erythropoietin* (see chapter 14, p. 533), regulate blood pressure by secreting the enzyme *renin* (see chapter 13, p. 511), and regulate absorption of calcium ions by activating *vitamin D* (see chapter 18, p. 706).

CAREER CORNER

Dialysis Technician

The 83-year-old woman's kidneys have been slowly failing for several years. Now, with her kidneys working at less than 10% of normal, she requires hemodialysis to cleanse her blood of toxins and wastes and remove excess fluid. A nephrologist has prescribed dialysis three times a week, for three to four hours per session, to treat her chronic kidney disease.

A dialysis technician performs the tasks of hemodialysis, including:

- Setting up, cleaning, sterilizing, and operating the dialysis machine.
- Preparing dialysis solutions.
- Monitoring and documenting vital signs.
- Preparing and administering medication.

Dialysis technicians work in hospitals and dialysis centers, as well as assist patients performing peritoneal dialysis in their homes. Training programs typically take two semesters, leading to a written exam for certification and national licensure. Students learn anatomy and physiology, nutrition, transplantation, pharmacology, health information management, medical terminology, and clinical skills related to performing dialysis.

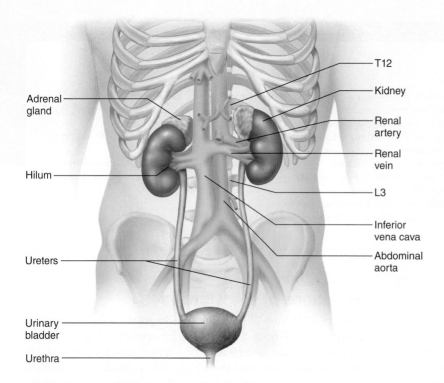

Adrenal gland
Hilum
Ureters
Urinary bladder
Urethra

T12
Kidney
Renal artery
Renal vein
L3
Inferior vena cava
Abdominal aorta

FIGURE 20.1 AP|R The urinary system includes the kidneys, ureters, urinary bladder, and urethra. Notice the relationship of these structures to the major blood vessels.

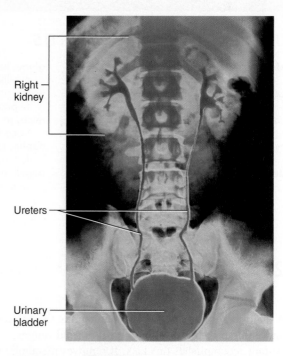

Right kidney
Ureters
Urinary bladder

FIGURE 20.2 AP|R Structures of the urinary system are visible in this falsely colored radiograph (anterior view).

Medical technology can take over some of the roles of a kidney. In *hemodialysis,* a person's blood is rerouted across an artificial membrane that "cleanses" it, removing substances that would normally be excreted in the urine, and adding some substances that the kidney normally produces.

Kidney Structure

The lateral surface of each kidney is convex, but its medial side is deeply concave. The resulting medial depression leads into a hollow chamber called the **renal sinus** (fig. 20.4). Through the entrance to this sinus, termed the *hilum,* pass blood vessels, nerves, lymphatic vessels, and the ureter (fig. 20.5, see fig. 20.1).

The superior end of the ureter expands to form a funnel-shaped sac called the **renal pelvis,** which is located mostly inside the renal sinus and directs the urine formed by the kidney toward the ureter. The pelvis is formed by the convergence of two or three tubes, called *major calyces* (sing., *calyx*), and they, in turn, are formed by the convergence of seven to twenty *minor calyces* (see fig. 20.4). At least one small projection called a *renal papilla* extends into each minor calyx.

The kidney includes two distinct regions: an inner medulla and an outer cortex. The **renal medulla** (re′nal mĕ-dul′ah) is composed of conical masses of tissue called *renal pyramids.* Their bases orient toward the convex surface of the kidney, and their apexes form the renal papillae. The tissue of the medulla appears striated because it consists of microscopic tubules that lead from the cortex to the renal papillae.

The **renal cortex** (re′nal kor′teks), which appears somewhat granular, forms a shell around the medulla. It dips into the medulla

between the renal pyramids, forming *renal columns.* The **renal capsule** is a fibrous membrane that surrounds the cortex and helps maintain the shape of the kidney. It also is protective (see fig. 20.4).

Certain organs, such as the kidneys, can be broken down into subunits, each of which performs the functions of the organ as a whole. In the kidneys, these functional units are called **nephrons** (nef′ronz), and each kidney contains about 1 million of them. Each nephron in turn consists of a **renal corpuscle** (re′nal kor′pusl) and a **renal tubule** (re′nal tu′būl) (see fig. 20.4). Nephrons are responsible for urine formation, and will be discussed in detail later in this chapter.

 PRACTICE

1 Where are the kidneys located?
2 What are the general functions of the kidneys?
3 Describe the structure of a kidney.

About two in every ten patients with renal failure can use a procedure that can be done at home called *continuous ambulatory peritoneal dialysis* instead of hemodialysis. The patient infuses a solution into the abdominal cavity through a permanently implanted tube. The solution stays in for four to eight hours, while it takes up substances that would normally be excreted into urine. Then the patient drains the waste-laden solution out of the tube, replacing it with clean fluid. Infection is a risk associated with this procedure.

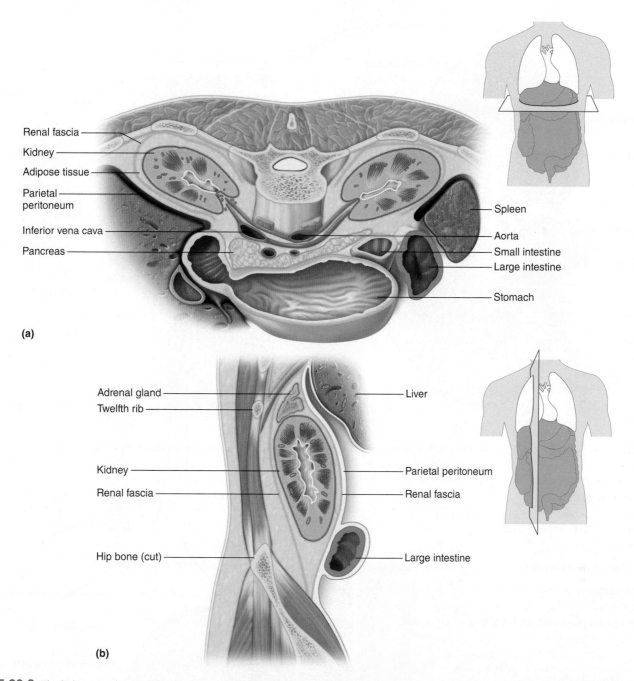

(a)

Renal fascia
Kidney
Adipose tissue
Parietal peritoneum
Inferior vena cava
Pancreas

Spleen
Aorta
Small intestine
Large intestine
Stomach

(b)

Adrenal gland
Twelfth rib

Kidney

Renal fascia

Hip bone (cut)

Liver

Parietal peritoneum

Renal fascia

Large intestine

FIGURE 20.3 The kidneys are located retroperitoneally. (*a*) Transverse section through the posterior abdominal cavity including the kidneys, behind the parietal peritoneum. Adipose and other connective tissues surround and support the kidneys. (*b*) Sagittal section through the posterior abdominal cavity showing the kidney.

Renal Blood Vessels

The **renal arteries,** which arise from the abdominal aorta, transport a large volume of blood to the kidneys (fig. 20.5). When a person is at rest, the renal arteries usually carry 15% to 30% of the total cardiac output into the kidneys, although the kidneys account for less than 1% of body weight. This disproportionately large blood flow reflects the kidney's role in maintaining the volume and composition of the body fluids.

A renal artery enters a kidney through the hilum and gives off several branches, called the *interlobar arteries,* which pass between the renal pyramids. At the junction between the medulla

and the cortex, the interlobar arteries branch to form a series of incomplete arches, the *arcuate arteries* (arciform arteries), which in turn give rise to *cortical radiate arteries* (interlobular arteries). The final branches of the cortical radiate arteries, called **afferent arterioles** (af′er-ent ar-te′re-ōlz), lead to the nephrons, the functional units of the kidneys. Each afferent arteriole gives rise to a tuft of capillaries within the renal corpuscle called the glomerulus (glomerular capillaries).

Blood passes through the glomerular capillaries, and then (minus any filtered fluid) enters an efferent arteriole. (This is different from entering a venule, which is the usual circulatory

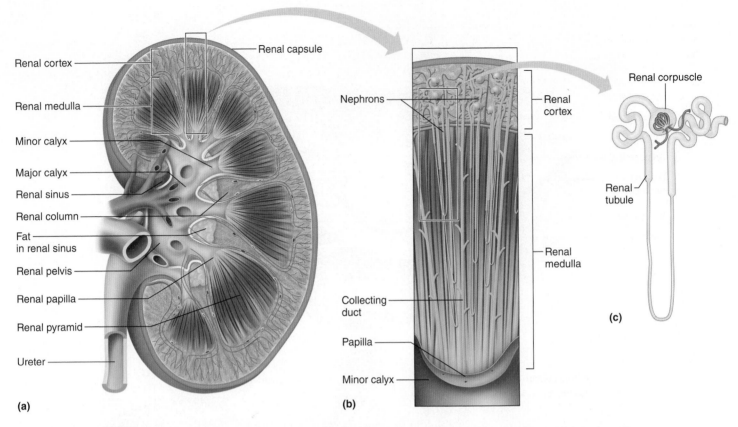

Renal cortex

Renal medulla

Minor calyx

Major calyx

Renal sinus

Renal column

Fat
in renal sinus

Renal pelvis

Renal papilla

Renal pyramid

Ureter

Renal capsule

Nephrons

Renal
cortex

Renal
medulla

Collecting
duct

Papilla

Minor calyx

Renal corpuscle

Renal
tubule

(a)

(b)

(c)

FIGURE 20.4 AP|R The kidney. (a) Longitudinal section of a kidney. (b) A renal pyramid containing nephrons. (c) A single nephron.

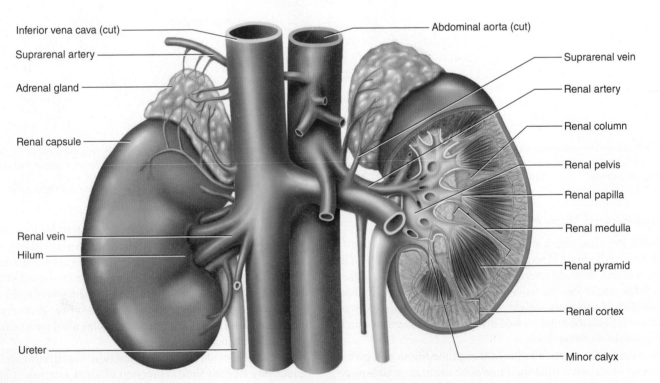

Inferior vena cava (cut)

Suprarenal artery

Adrenal gland

Renal capsule

Renal vein

Hilum

Ureter

Abdominal aorta (cut)

Suprarenal vein

Renal artery

Renal column

Renal pelvis

Renal papilla

Renal medulla

Renal pyramid

Renal cortex

Minor calyx

FIGURE 20.5 Blood vessels are associated with the kidneys and adrenal glands. Note their relationship with the renal pelvis and ureters.

FIGURE 20.6 AP|R

Renal blood vessels. (*a*) a simplified diagram showing the relationship between the nephron and the renal blood vessels. (*b*) Corrosion cast of renal blood vessels. Not all blood vessels associated with the nephron are shown.

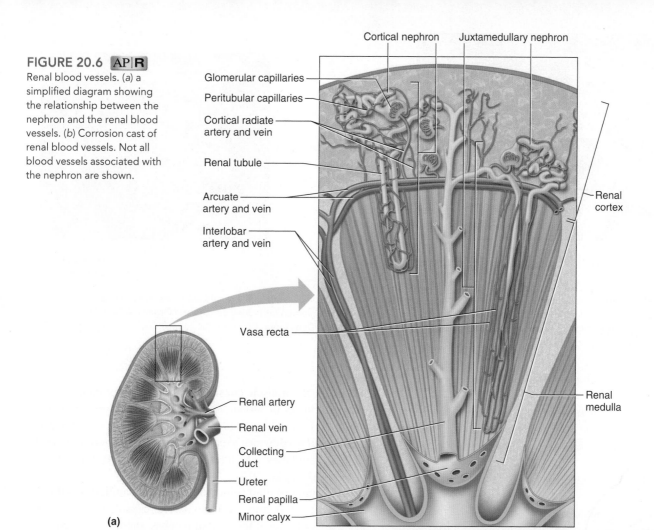

Cortical nephron

Juxtamedullary nephron

Glomerular capillaries

Peritubular capillaries

Cortical radiate artery and vein

Renal tubule

Arcuate artery and vein

Interlobar artery and vein

Vasa recta

Renal cortex

Renal medulla

Renal artery

Renal vein

Collecting duct

Ureter

Renal papilla

Minor calyx

(a)

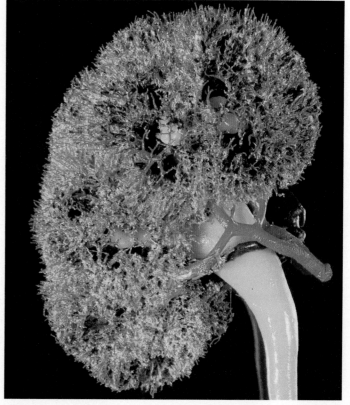

(b)

route.) The efferent arterioles typically lead to complex networks of capillaries in the renal cortex that surround the renal tubules, called the **peritubular capillaries** (per″ĭ-tu′bu-lar kap′ĭ-ler″es) that empty into the venous system of the kidney.

Only 10% to 15% of the renal blood flow reaches the medulla. A small fraction of efferent arterioles form relatively straight, parallel bundles of blood vessels that descend into the renal medulla, form capillary loops, and then ascend back to the cortex. These bundles are called the *vasa recta*. After blood flows through the vasa recta it returns to the renal cortex, where it enters the venous system of the kidney.

Venous blood returns through a series of vessels that generally correspond to the arterial pathways. For example, the venous blood passes through cortical radiate, arcuate, interlobar, and renal veins. The **renal vein** then joins the inferior vena cava as it courses through the abdominal cavity. **Figure 20.6** shows branches of the renal arteries and veins. **Figure 20.7** shows detail of the blood vessels and other structures associated with the nephrons.

Figure 20.8 summarizes the pathway that blood follows as it passes through the blood vessels of the kidney. Blood circulating through two capillaries as it moves through the kidney is another example of a portal system (see fig. 15.55, p. 605).

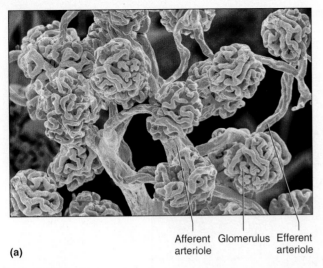

(a)

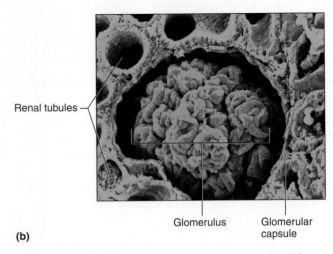

Renal tubules

Glomerulus   Glomerular
                capsule

(b)

FIGURE 20.7 Blood vessels associated with nephrons. (*a*) A scanning electron micrograph of a cast of the renal blood vessels associated with the glomeruli (200×). (*b*) A scanning electron micrograph of a glomerular capsule surrounding a glomerulus (480×). *Tissues and Organs: A Text-Atlas of Scanning Electron Microscopy,* by R. G. Kessel and R. H. Kardon. ©1979 W. H. Freeman and Company.

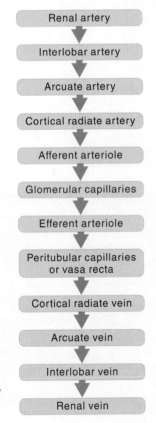

Renal artery
↓
Interlobar artery
↓
Arcuate artery
↓
Cortical radiate artery
↓
Afferent arteriole
↓
Glomerular capillaries
↓
Efferent arteriole
↓
Peritubular capillaries or vasa recta
↓
Cortical radiate vein
↓
Arcuate vein
↓
Interlobar vein
↓
Renal vein

FIGURE 20.8
Pathway of blood through the blood vessels of the kidney and nephron.

RECONNECT
To Chapter 13, Portal system, pages 498–499.
To Chapter 17, Portal system, page 670.

Nephrons

Structure of a Nephron

As described earlier, each kidney contains about 1 million **nephrons.** Each nephron in turn consists of a **renal corpuscle** and a **renal tubule** (see fig. 20.4).

A renal corpuscle consists of a filtering unit composed of a tangled cluster of blood capillaries called a **glomerulus** (glo-mer′u-lus) and a surrounding thin-walled, saclike structure called a **glomerular** (Bowman's) **capsule.** An afferent arteriole gives rise to these capillaries, which lead to an **efferent arteriole** (ef′er-ent ar-te′re-ōl) (see fig. 20.7).

Filtration of fluid from the glomerular capillaries is the first step in urine formation. The glomerular capsule is an expansion at the end of a renal tubule that receives the fluid filtered at the glomerulus. The capsule is composed of two layers of epithelial cells: a visceral layer that closely covers the glomerulus and an outer parietal layer that is continuous with the visceral layer and with the wall of the renal tubule (**fig. 20.9*a***).

The cells of the parietal layer are typical squamous epithelial cells, but those of the visceral layer are highly modified epithelial cells called *podocytes.* Each podocyte has several primary processes extending from its cell body, and these processes, in turn, bear numerous secondary processes, called *pedicels* (foot processes). The pedicels of each cell interdigitate with those of adjacent podocytes, and the clefts between them form a complicated system of *slit pores* (fig. 20.9*b*).

A cascade of cell-to-cell communication forms glomeruli in the embryo. Podocyte precursor cells give rise to podocytes, but also secrete vascular endothelial growth factor (VEGF), which attracts the squamous epithelium that forms the parietal layer. These epithelial cells then produce platelet-derived growth factor (PDGF) and extracellular matrix proteins, which in turn signal certain cells to specialize as mesangial cells. These cells are closely associated with the capillary tuft, and can contract, which decreases the rate at which the glomerulus filters plasma. Mesangial cells also provide structural support to the glomerulus and phagocytize debris.

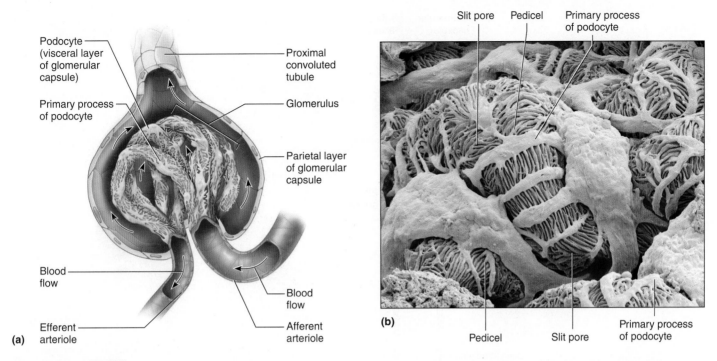

Slit pore   Pedicel   Primary process
of podocyte

Podocyte
(visceral layer
of glomerular
capsule)

Primary process
of podocyte

Blood
flow

Efferent
arteriole

(a)

Proximal
convoluted
tubule

Glomerulus

Parietal layer
of glomerular
capsule

Blood
flow

Afferent
arteriole

(b)

Pedicel   Slit pore   Primary process
of podocyte

FIGURE 20.9 AP|R   (a) The glomerular capsule has a visceral layer and a parietal layer. (b) Scanning electron micrograph of a portion of a glomerulus (5,000×). Note the slit pores between the pedicels.

Filtered fluid enters the renal tubule, which leads away from the glomerular capsule and becomes highly coiled. This coiled portion is the *proximal convoluted tubule.* Following it is the **nephron loop** (loop of Henle). The proximal convoluted tubule dips toward the renal pelvis to become the *descending limb* of the nephron loop. The tubule then curves back toward its renal corpuscle and forms the *ascending limb* of the nephron loop. The ascending limb returns toward its renal corpuscle of origin, where it becomes the *distal convoluted tubule* (fig. 20.10a). This distal portion is shorter and somewhat less coiled than the proximal tubule.

Several distal convoluted tubules merge in the renal cortex to form a *collecting duct* (collecting tubule), which is technically not part of any one nephron but instead collects fluid from multiple nephrons. The collecting duct passes into the renal medulla, widening as it joins other collecting ducts. The resulting tube empties into a minor calyx through an opening in a renal papilla (see figure 20.6). Figures 20.10 and **20.11** show the parts of a nephron. **Figure 20.12** lists the structures involved in urine formation and excretion. Clinical Application 20.1 examines glomerulonephritis, an inflammation of the glomeruli.

Cortical and Juxtamedullary Nephrons

Most nephrons have renal corpuscles in the renal cortex near the surface of the kidney. These are called *cortical nephrons* and generally have short loops that extend only part way into the renal medulla. Another group of nephrons, called *juxtamedullary nephrons,* have corpuscles located deep in the cortex, close to the renal medulla. Their loops extend well into the medulla (see fig. 20.6). Although juxtamedullary nephrons represent only a small fraction of the total, they are important in regulating water balance and are discussed in more detail later in this chapter.

Juxtaglomerular Apparatus

At the end of the nephron loop, before it becomes the distal convoluted tubule, the ascending limb passes between the afferent and efferent arterioles of its own glomerulus and contacts them. Here, the epithelial cells of the ascending limb are tall and densely packed, composing a structure called the *macula densa.*

Close by, in the wall of the afferent arteriole near its attachment to the glomerulus, are large, vascular smooth muscle cells called *juxtaglomerular cells.* Together with the cells of the macula densa, they constitute the **juxtaglomerular apparatus** (juks″tah-glo-mer′u-lar ap″ah-ra′tus). This structure plays a role in regulating the secretion of renin (fig. 20.13; see chapter 13, p. 511).

 PRACTICE

4   Describe the system of vessels that supplies blood to the kidney.

5   Name the parts of a nephron.

6   Distinguish a cortical nephron from a juxtamedullary nephron.

7   Which structures comprise the juxtaglomerular apparatus?

20.3 | Urine Formation

The main function of the nephrons and collecting ducts is to control the composition of body fluids and remove wastes from the blood. The product is **urine,** which is excreted from the body.

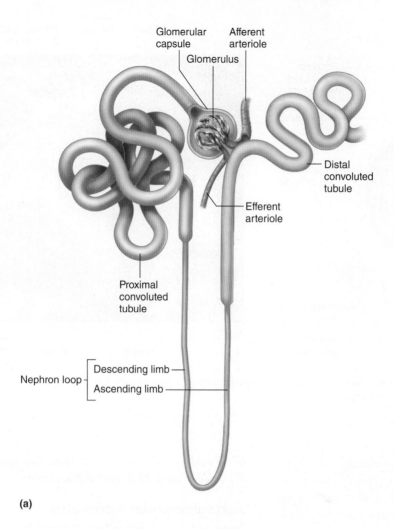

(a)

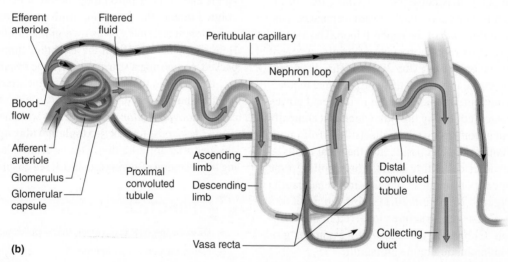

(b)

FIGURE 20.10 AP|R A nephron. (*a*) Structure of a nephron plus the afferent and efferent arterioles. (*b*) Schematic representation of a nephron and the blood vessels associated with it.

CLINICAL APPLICATION 20.1

Glomerulonephritis

Nephritis is an inflammation of the kidney. *Glomerulonephritis* is an inflammation of the glomeruli, and it may be acute or chronic and can lead to renal failure.

Acute glomerulonephritis (AGN) can result from an abnormal immune reaction that develops one to three weeks following bacterial infection by beta-hemolytic *Streptococcus*. The infection is not in the kidneys, but bacterial antigens trigger production of antibodies that form insoluble immune complexes (see chapter 16, p. 638) that travel in the bloodstream and lodge in the kidneys. The antigen-antibody complexes are deposited in and block the glomerular capillaries, which become further obstructed as the inflammatory response sends white blood cells to the region. Capillaries remaining open may become abnormally permeable, allowing plasma proteins and red blood cells to enter the urine.

Most acute glomerulonephritis patients eventually regain normal kidney function. However, in severe cases, renal functions may fail completely. Without treatment, the disease is usually fatal within a week or so.

Chronic glomerulonephritis is a progressive disease in which more and more nephrons are damaged until the kidneys are unable to function. The cause of this condition is not fully understood, but it may also involve formation of antigen-antibody complexes that precipitate and accumulate in the glomeruli. The resulting inflammation is prolonged, and fibrous tissue replaces glomerular membranes, permanently disabling the nephrons. Eventually the kidneys fail.

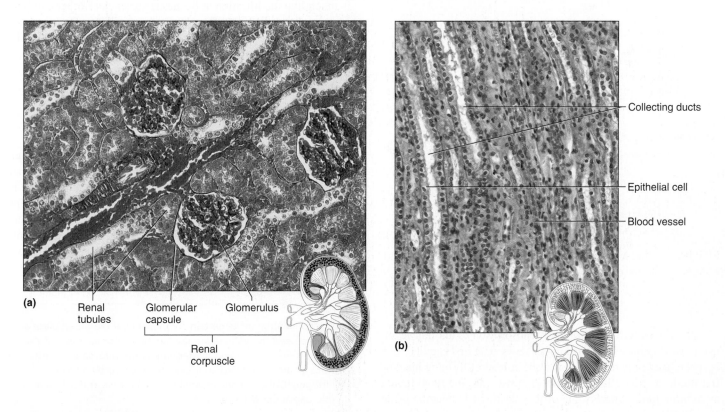

FIGURE 20.11 AP|R Microscopic view of the kidney. (*a*) Light micrograph of a section of the human renal cortex (220×). (*b*) Light micrograph of the renal medulla (80×).

Urine contains wastes, along with excess water and electrolytes. Urine formation involves three steps, which are discussed in the following paragraphs:

1. Glomerular filtration
2. Tubular reabsorption
3. Tubular secretion

Urine formation begins when the glomerular capillaries filter plasma, a process called **glomerular filtration** (glo-mer′u-lar fil-tra′shun). Recall from chapter 15 (p. 581) that the force of blood pressure drives filtration at capillaries throughout the body. Most of this fluid is reabsorbed into the bloodstream by the colloid osmotic pressure of the plasma, leaving only a small volume of interstitial fluid (fig. 20.14*a*).

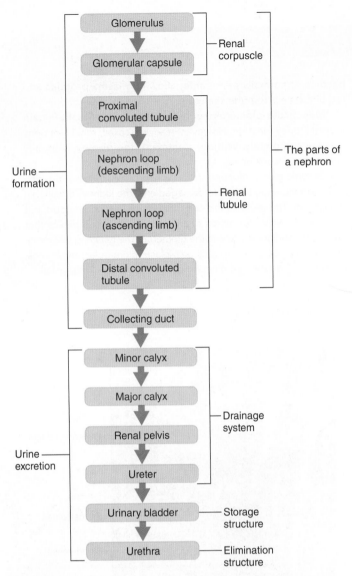

FIGURE 20.12 A summary of the sequence of the structures involved in the formation and elimination of urine.

within the peritubular capillary system into the renal tubule. Tubular secretion removes some substances that the body must eliminate, such as hydrogen ions and certain toxins, faster than filtration alone (fig. 20.14*b*).

Overall, the following relationship determines the volume of substances excreted in the urine:

$$
\begin{array}{r}
\text{glomerular filtration} \\
+ \text{ tubular secretion} \\
- \text{ tubular reabsorption} \\
\hline
\text{urinary excretion}
\end{array}
$$

The final product of these processes is *urine.*

Glomerular Filtration

Urine formation begins when *glomerular filtration* filters water and other small dissolved molecules and ions out of the glomerular capillary plasma and into the glomerular capsules. Large molecules, such as proteins, are restricted primarily because of their size. The filtration of these materials through the capillary walls is much like the filtration at the arteriolar ends of other capillaries throughout the body. The glomerular capillaries, however, are many times more permeable to small molecules than are the capillaries in other tissues, due to the many tiny openings (fenestrae) in their walls (fig. 20.15).

 RECONNECT
To Chapter 15, Exchanges in the Capillaries, page 581.

The glomerular capsule receives the resulting **glomerular filtrate,** which has about the same composition as the filtrate that becomes tissue fluid elsewhere in the body. Glomerular filtrate is mostly water and the same solutes as in blood plasma, except for the larger protein molecules. More specifically, glomerular filtrate includes water, glucose, amino acids, urea, uric acid, creatine, creatinine, sodium, chloride, potassium, calcium, bicarbonate, phosphate, and sulfate ions.

The concentrations of certain components of the blood plasma can be used to evaluate kidney functions. For example, if the kidneys are functioning inadequately, the plasma concentrations of urea (a nitrogenous waste) indicated by a blood urea nitrogen (BUN) test and creatinine may increase as much as tenfold above normal (see Appendix C, Common Tests Performed on Urine).

Nephrons take filtration and reabsorption to another level, using two capillary beds working in series. The first capillary bed is specialized only to filter. However, instead of forming interstitial fluid, the filtered fluid (filtrate) moves into the renal tubule as *tubular fluid,* where some of it is destined to become urine (fig. 20.14*b*).

Glomerular filtration produces 180 liters of fluid, more than four times the total body water, every 24 hours. Obviously this output could not continue for long unless most of this filtered fluid were returned to the internal environment. The kidney accomplishes this by the process of **tubular reabsorption** (too′bu-lar re-ab-sorp′shun), selectively reclaiming just the right amounts of substances, such as water, electrolytes, and glucose, that the body requires (fig. 20.14*b*). Waste products and substances in excess are allowed out of the body.

The reverse of tubular reabsorption is **tubular secretion** (too′bu-lar se-kre′shun), which moves substances from the blood

Filtration Pressure

The main force that moves substances by filtration through the glomerular capillary wall, as in other capillaries, is the hydrostatic pressure of the blood inside. The afferent arterioles have diameters larger than the efferent arterioles. The greater resistance to blood flow of the efferent arteriole causes blood to back up into the glomerular capillaries. This raises the blood pressure in the glomerular capillaries, favoring filtration.

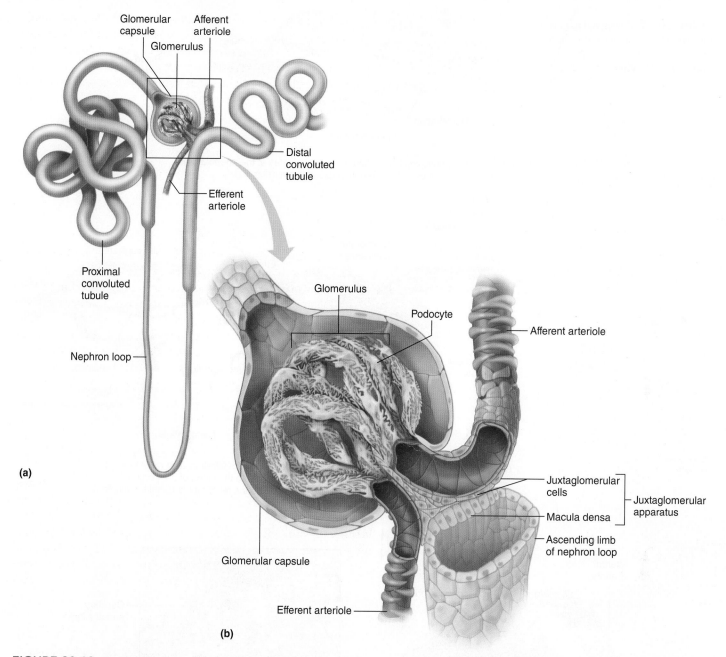

(a)

(b)

FIGURE 20.13 Juxtaglomerular apparatus. (*a*) Location of and (*b*) enlargement of a section of the juxtaglomerular apparatus, which consists of the macula densa and the juxtaglomerular cells.

The colloid osmotic pressure of the plasma caused by plasma proteins and the hydrostatic pressure inside the glomerular capsule also influence glomerular filtration. The plasma colloid osmotic pressure is always higher than that of the glomerular filtrate (except in some types of kidney disease). This draws water back into the glomerular capillaries by osmosis, opposing filtration. Any increase in glomerular capsular hydrostatic pressure also opposes filtration (fig. 20.16).

The result of these forces is called **net filtration pressure,** and it is normally positive, favoring filtration at the glomerulus. Net filtration pressure is calculated as follows:

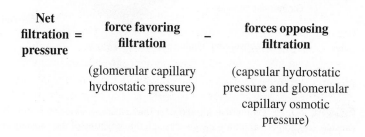

Net filtration pressure = force favoring filtration − forces opposing filtration

(glomerular capillary hydrostatic pressure)

(capsular hydrostatic pressure and glomerular capillary osmotic pressure)

FIGURE 20.14 Compared to most capillaries in the (a) systemic circulation, those in the (b) kidneys are highly specialized for filtration.

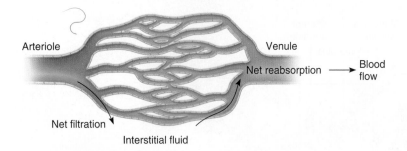

(a) In most systemic capillaries, filtration predominates at the arteriolar end and osmotic reabsorption predominates at the venular end.

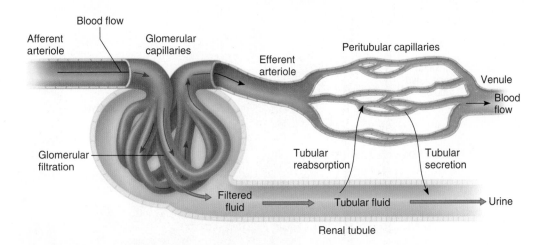

(b) In the kidneys, the glomerular capillaries are specialized for filtration. The renal tubule is specialized to control movements of substances back into the blood of the peritubular capillaries (tubular reabsorption) or from the blood into the renal tubule (tubular secretion).

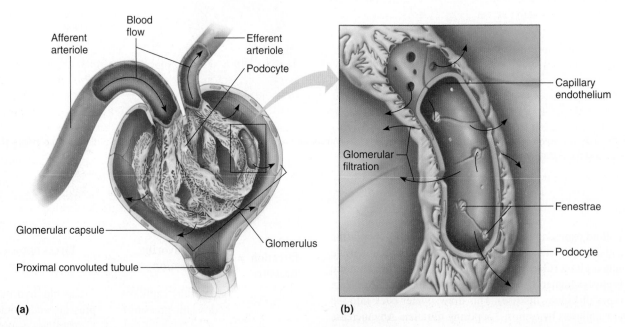

FIGURE 20.15 Glomerular filtration. (a) The first step in urine formation is filtration of substances out of glomerular capillaries and into the glomerular capsule. (b) Glomerular filtrate passes through the fenestrae of the capillary endothelium.

Vasoconstriction of which blood vessel in this figure would decrease glomerular filtration?
Answer can be found in Appendix G.

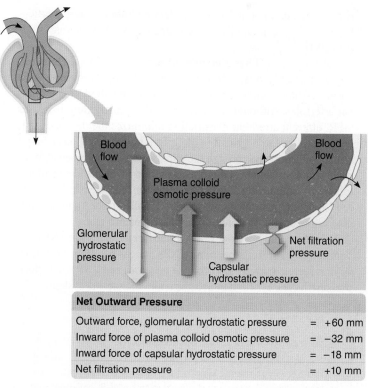

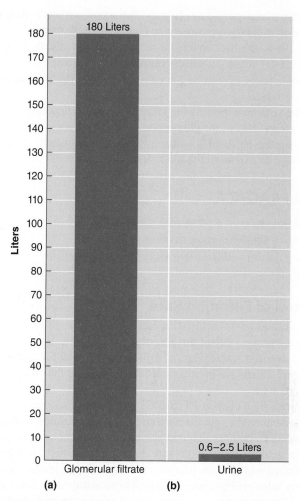

Blood flow

Plasma colloid osmotic pressure

Blood flow

Glomerular hydrostatic pressure

Net filtration pressure

Capsular hydrostatic pressure

Net Outward Pressure

Outward force, glomerular hydrostatic pressure	= +60 mm
Inward force of plasma colloid osmotic pressure	= −32 mm
Inward force of capsular hydrostatic pressure	= −18 mm
Net filtration pressure	= +10 mm

FIGURE 20.16 Normally the glomerular net filtration pressure is positive, causing filtration. The responsible forces include the hydrostatic and osmotic pressure of the plasma and the hydrostatic pressure of the fluid in the glomerular capsule.

FIGURE 20.17 Relative volumes of (a) glomerular filtrate and (b) urine formed in twenty-four hours.

Filtration Rate

At rest, the kidneys receive approximately 25% of the cardiac output, and about 20% of the blood plasma is filtered as it flows through the glomerular capillaries. This means that in an average adult, the glomerular filtration rate for the nephrons of both kidneys is about 125 milliliters per minute, or 180,000 milliliters (180 liters) in twenty-four hours. Estimating the blood plasma volume at about 3 liters, the production of 180 liters of filtrate in twenty-four hours means that an amount equivalent to the entire plasma volume must be filtered through the glomeruli about sixty times each day. Clearly not all of it is excreted as urine (fig. 20.17). Instead, most of the fluid that passes through the renal tubules is reabsorbed and reenters the plasma.

The glomerular filtration rate (GFR) is directly proportional to the net filtration pressure. If the net filtration pressure increases, GFR goes up, and if net filtration pressure decreases, GFR goes down. Consequently, the factors that affect the glomerular hydrostatic pressure, glomerular plasma colloid osmotic pressure, or hydrostatic pressure in the glomerular capsule also affect the rate of filtration (see fig. 20.16). GFR is the most commonly measured index of kidney function.

Normally, glomerular hydrostatic pressure is the most important factor determining net filtration pressure and GFR. Because each glomerular capillary lies between two arterioles—the afferent and efferent arterioles—any change in the diameters of these vessels is likely to change glomerular hydrostatic pressure, affecting glomerular filtration rate. If the afferent arteriole (through which blood enters the glomerulus) constricts, filtration pressure will decrease. In contrast,

if the efferent arteriole (through which blood leaves the glomerulus) constricts, blood backs up into the glomerulus and filtration pressure rises. Vasodilation of these vessels produces opposite effects.

The colloid osmotic pressure of the glomerular plasma also influences net filtration pressure and the rate of filtration. In other systemic capillaries, filtration occurs at the beginning of the capillary, but the osmotic effect of the plasma proteins predominates at the end of the capillary, and most filtered fluid is thus reabsorbed. The small excess remaining eventually becomes lymph.

Much more fluid is filtered by glomerular capillaries than by capillaries elsewhere because of the relatively high hydrostatic pressure in the glomerular capillaries. In fact, as filtration occurs through the capillary wall, proteins remaining in the plasma raise the colloid osmotic pressure within the glomerular capillaries. Despite this, the glomerular capillary hydrostatic pressure is sufficiently great that the net filtration pressure is normally positive. That is, the forces favoring filtration in the glomerular capillaries always predominate. Of course, conditions that lower plasma colloid osmotic pressure, such as a decrease in plasma protein concentration, would increase filtration rate.

The hydrostatic pressure in the glomerular capsule is another factor that may affect net filtration pressure and GFR. This capsular pressure can change as a result of an obstruction, such as a stone in a ureter or an enlarged prostate gland pressing on the urethra. If this occurs, fluids may back up into the renal tubules and raise the hydrostatic pressure in the glomerular capsules. Any increase in capsular pressure opposes glomerular filtration, so filtration rate may significantly decrease.

The volume of plasma the kidneys filter also depends on the large *surface area* of the glomerular capillaries. This surface area is estimated to be about 2 square meters—approximately equal to the surface area of an adult's skin.

 PRACTICE

8 Which processes occur as urine forms?

9 How is filtration pressure calculated?

10 Which factors influence the rate of glomerular filtration?

Control of Filtration Rate

In general, glomerular filtration rate remains relatively constant through a process called **autoregulation.** However, certain conditions override autoregulation. GFR may increase, for example, when body fluids are in excess and decrease when the body must conserve fluid.

Recall from chapter 15 (p. 577) that the sympathetic nervous system controls the vascular smooth muscle of arterioles. Reflexes responding to changes in blood pressure and volume control sympathetic activity. If blood pressure drops, vasoconstriction of both the afferent and efferent arterioles results. Afferent arteriole constriction helps maintain peripheral resistance and systemic pressure, but would lower filtration pressure. Simultaneous constriction of the efferent arteriole counteracts this effect on filtration pressure. Thus, systemic pressure is protected, but GFR remains relatively constant.

> If arterial blood pressure drops greatly, such as during shock, the resulting increase in sympathetic stimulation may cause so much afferent arteriolar constriction that glomerular hydrostatic pressure falls below the level required for filtration, leading to acute renal failure. At the same time, the epithelial cells of the renal tubules may not receive sufficient nutrients to maintain their high rates of metabolism. Cells may die (tubular necrosis), and renal functions may be lost permanently, resulting in chronic kidney disease.

A second control of GFR is the hormonelike **renin-angiotensin system.** The juxtaglomerular cells of the afferent arterioles secrete an enzyme, **renin,** in response to stimulation from sympathetic nerves and pressure-sensitive cells, called **renal baroreceptors,** in the afferent arteriole. These factors stimulate renin secretion if blood pressure drops. The macula densa also controls renin secretion. Cells of the macula densa sense the numbers of sodium, potassium, and chloride ions in the ascending limb of the nephron loop. Decreasing levels of these ions stimulate renin secretion.

Once in the bloodstream, *renin* reacts with the plasma protein *angiotensinogen* to form *angiotensin I.* An enzyme, *angiotensin-*

converting enzyme (ACE), on capillary endothelial cells (particularly in the lungs), rapidly converts angiotensin I to *angiotensin II* (fig. 20.18).

Angiotensin II has a number of renal effects that help maintain sodium balance, water balance, and blood pressure. As a vasoconstrictor, angiotensin II affects both the afferent and efferent arterioles. Although afferent arteriolar constriction decreases GFR, efferent arteriolar constriction minimizes the decrease, thus contributing to autoregulation of GFR. Angiotensin II has a major effect on the kidneys through the adrenal cortical hormone aldosterone, which stimulates sodium reabsorption in the distal convoluted tubule. By stimulating aldosterone secretion, angiotensin II helps to reduce the amount of sodium excreted in the urine. Angiotensin II also stimulates ADH secretion, helping to retain water.

Two cardiac hormones, atrial natriuretic peptide (ANP) and ventricular natriuretic peptide (BNP) increase sodium excretion. ANP is secreted when blood volume increases and BNP when blood pressure increases. Both inhibit the renin-angiotensin system.

> Eating raw spinach or undercooked hamburger, drinking unpasteurized apple cider, and petting animals at country fairs have all led to hemolytic uremic syndrome (HUS). The direct cause is a poison called shigatoxin that a certain strain of *E. coli* (bacteria) produces. Humans encounter the toxin through excrement—in the water used to grow spinach, in beef contaminated during processing, from apples that fell into droppings, and from tiny hands unwashed after petting animals. Food poisoning from toxin-producing *E. coli* begins with sharp abdominal pain and bloody diarrhea, and about 16% of the time HUS develops. About 5% of HUS cases are fatal.
>
> Shigatoxin causes fibrin to form thrombi that obstruct the narrow glomerular capillaries. Platelets join the clumping, depleting the circulation of these cell fragments (thrombocytopenia). Red blood cells in the blocked glomerular capillaries break apart, causing hemolytic anemia. As GFR plummets, the person, typically a child, goes into acute renal failure.

Tubular Reabsorption

If the composition of the glomerular filtrate entering the renal tubule is compared with that of urine, it is apparent that the fluid changes as it passes through the tubule. Table 20.1 compares the concentrations of some of the substances in the blood plasma, glomerular filtrate, and urine. For example, glucose is present in the filtrate but absent in the urine. Such a change in fluid composition is largely the result of *tubular reabsorption,* the process by which substances are transported out of the tubular fluid, through the epithelium of the renal tubule, and into the interstitial fluid. These substances then diffuse into the peritubular capillaries or the vasa recta (fig. 20.19).

In contrast, other substances are more concentrated in urine than they are in the glomerular filtrate (see table 20.1). This difference reflects tubular secretion of certain substances as well as the reabsorption of water from the renal collecting ducts, described in detail later.

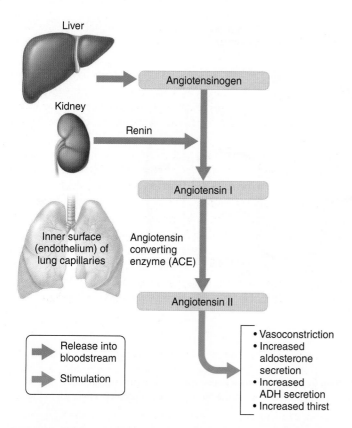

Liver

Kidney

Angiotensinogen

Renin

Angiotensin I

Inner surface (endothelium) of lung capillaries

Angiotensin converting enzyme (ACE)

Angiotensin II

• Vasoconstriction
• Increased aldosterone secretion
• Increased ADH secretion
• Increased thirst

→ Release into bloodstream

→ Stimulation

FIGURE 20.18 The formation of angiotensin II in the bloodstream involves several organs and results in multiple actions that conserve sodium and water.

Tubular reabsorption returns substances to the internal environment. The term *tubular* is used because this process is carried out by the epithelial cells that make up the renal tubules and collecting ducts. In tubular reabsorption, substances must first cross the cell membrane facing the inside of the tubule (mucosal surface) and then the cell membrane facing the interstitial fluid (serosal surface).

The basic rules for movements across cell membranes apply to tubular reabsorption. Substances moving down a concentration gradient must either be lipid soluble or there must be a carrier or channel for that substance. Active transport, requiring ATP, may move substances uphill against a concentration gradient. If active transport is involved at any step of the way, the overall process is considered active. In all other cases, the process is considered passive.

RECONNECT

To Chapter 3, Movements Into and Out of the Cell, pages 98–104.

Peritubular capillary blood is under relatively low pressure because it has already passed through two arterioles and a capillary bed. Also, the wall of the peritubular capillary is more permeable than that of other capillaries. Finally, the relatively high rate of glomerular filtration has increased the protein concentration and, thus, the colloid osmotic pressure of the peritubular capillary plasma. These factors enhance the rate of fluid reabsorption from the renal tubule.

TABLE 20.1	Relative Concentrations of Plasma, Glomerular Filtrate, and Urine Components		
	CONCENTRATIONS (mEq/L)		
Substance	Plasma	Glomerular Filtrate	Urine
Sodium (Na$^+$)	142	142	128
Potassium (K$^+$)	5	5	60
Calcium (Ca^{+2})	4	4	5
Magnesium (Mg^{+2})	3	3	15
Chloride (Cl$^-$)	103	103	134
Bicarbonate (HCO$_3^-$)	27	27	14
Sulfate (SO$_4^{-2}$)	1	1	33
Phosphate (PO$_4^{-3}$)	2	2	40

(mEq/L [milliequivalents per liter] is a commonly used measure of concentration based on how many charges an ion carries. For a substance with a charge of 1, such as Cl$^-$, a milliequivalent is equal to a millimole.)

	CONCENTRATIONS (mg%)		
Substance	Plasma	Glomerular Filtrate	Urine
Glucose	100	100	0
Urea	26	26	1,820
Uric acid	4	4	53
Creatinine	1	1	196

(mg% [mg of a substance per 100 mL of solution] is a commonly used measure of concentration of solute. It is equivalent to mg/deciliter [mg/dL].)

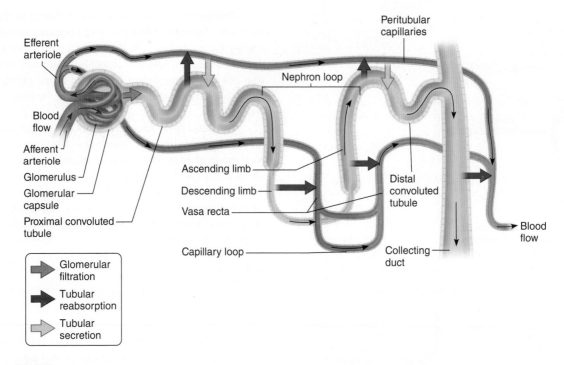

Efferent arteriole

Blood flow

Afferent arteriole

Glomerulus

Glomerular capsule

Proximal convoluted tubule

Peritubular capillaries

Nephron loop

Ascending limb

Descending limb

Vasa recta

Capillary loop

Distal convoluted tubule

Collecting duct

Blood flow

Glomerular filtration

Tubular reabsorption

Tubular secretion

FIGURE 20.19 **AP|R** Two processes in addition to glomerular filtration help to form urine. Tubular reabsorption transports substances from the glomerular filtrate into the blood in the peritubular capillary. Tubular secretion transports substances from the blood in the peritubular capillary into the renal tubule.

Which of the three processes (glomerular filtration, tubular reabsorption, tubular secretion), if increased for a substance, would reduce urinary excretion of that substance?
Answer can be found in Appendix G.

Tubular reabsorption occurs throughout the renal tubule. However, most of it occurs in the proximal convoluted portion. The epithelial cells in this area have many *microvilli* that form a "brush border" on their free surfaces facing the tubular lumen. These tiny extensions greatly increase the surface area exposed to the glomerular filtrate and enhance reabsorption.

Segments of the renal tubule are adapted to reabsorb specific substances, using particular modes of transport. Glucose reabsorption, for example, takes place through the walls of the proximal convoluted tubule by active transport. Water also is rapidly reabsorbed through the epithelium of the proximal convoluted tubule by osmosis; however, portions of the distal convoluted tubule and collecting duct may be almost impermeable to water. This characteristic of the distal convoluted tubule and collecting duct is important in the regulation of urine concentration and volume, as described in a subsequent section.

Recall that active transport requires carrier proteins in a cell membrane. The molecule to be transported binds to the carrier. The carrier changes shape, releases the transported molecule on the other side of the cell membrane, and then returns to its original shape and repeats the process. Such a mechanism has a *limited transport capacity*. That is, it can transport only a certain number of molecules in a given time because the number of carriers is limited.

Usually all of the glucose in the glomerular filtrate is reabsorbed because there are enough carrier molecules to transport it. When the plasma glucose concentration increases to a critical

level, called the **renal plasma threshold,** more glucose molecules are in the filtrate than the active transport mechanism can handle. As a result, some glucose remains in the filtrate and is excreted in the urine. This overwhelming of the active transport mechanism for glucose reabsorption explains why the elevated blood glucose of diabetes mellitus results in glucose in the urine.

An increase in urine volume is called *diuresis*. Nonreabsorbed glucose in the tubular fluid increases the osmotic concentration of the tubular fluid, which reduces the volume of water reabsorbed by osmosis from the proximal tubule and collecting duct. The resultant increase in urine volume is called an *osmotic diuresis*.

Amino acids enter the glomerular filtrate and are reabsorbed in the proximal convoluted tubule. Three active transport mechanisms reabsorb different types of amino acids with similar structures. Normally only a trace of amino acids remains in the urine.

The glomerular filtrate is nearly free of protein, but a number of smaller protein molecules, such as albumin, may squeeze through the glomerular capillaries. These proteins are transported by *endocytosis* through the brush border of epithelial cells lining the proximal convoluted tubule. Once inside an epithelial cell, the proteins are degraded to amino acids, which are moved into the blood of the peritubular capillary.

The epithelium of the proximal convoluted tubule also reabsorbs lactic, citric, uric, and ascorbic (vitamin C) acids; and phosphate, sulfate, calcium, potassium, and sodium ions. Active transport mechanisms with limited transport capacities reabsorb

The Nephrotic Syndrome

The *nephrotic syndrome* is a set of symptoms that often appears in patients with renal diseases. Considerable loss of plasma proteins into the urine (proteinuria) results in widespread edema and increased susceptibility to infection.

Plasma proteins enter the urine because of increased permeability of the glomerular membranes, which accompanies renal disorders such as glomerulonephritis. As a consequence of a decreasing plasma protein concentration (hypoproteinemia), the plasma colloid osmotic pressure falls, increasing net filtration pressure in capillaries throughout the body. This may lead to widespread accumulation of fluid in interstitial spaces in tissues (edema) and in body spaces such as the abdominal cavity (ascites), pleural cavity, pericardial cavity, and joint cavities.

As edema and other fluid accumulation develops, blood volume decreases and blood pressure drops. These changes may activate the renin-angiotensin system, stimulating the adrenal cortex to release aldosterone (see chapter 13, p. 511) and the posterior pituitary to release antidiuretic hormone (see chapter 13, p. 503), which in turn stimulates the kidneys to conserve sodium ions and water. This action helps to restore blood volume and may aggravate the edema.

The nephrotic syndrome sometimes appears in young children who have *lipoid nephrosis*. The cause of this condition is unknown, but it alters the epithelial cells of the glomeruli so that the glomerular membranes enlarge and distort, allowing proteins to be filtered.

---

all of these chemicals. The amounts of substances in the urine will increase if their concentrations in the glomerular filtrate exceed their respective renal plasma thresholds. Clinical Application 20.2 discusses how plasma proteins enter the urine in nephrotic syndrome.

Glucose in the urine is called *glucosuria*. It may follow intravenous administration of glucose, or eating candy, or it may occur in a person with diabetes mellitus. In type 1 diabetes, blood glucose concentration rises because of insufficient insulin secretion from the pancreas (see Clinical Application 13.4, page 516).

One in three people who have diabetes mellitus sustains kidney damage (nephropathy). In the past, large amounts of the protein albumin in the urine indicated high risk of developing kidney damage. Recent studies show that small amounts of albumin (microalbuminuria) predict kidney damage in people with type 1 diabetes. Following a low-protein diet can slow the loss of kidney function.

Sodium and Water Reabsorption

Water reabsorption occurs passively by osmosis, primarily in the proximal convoluted tubule, and is closely associated with the active reabsorption of sodium ions. In the proximal convoluted tubule, if sodium reabsorption increases, water reabsorption increases; if sodium reabsorption decreases, water reabsorption decreases also.

In addition to active mechanisms, reabsorption of some solutes is passively linked to active reabsorption of sodium ions. When the positively charged sodium ions (Na^+) are moved through the tubular wall, negatively charged ions, including chloride ions (Cl^-) and phosphate ions (PO_4^{-3}), two abundant plasma anions, accompany them. This movement of negatively charged ions is due to the electrochemical attraction between particles of opposite electrical charge. Although this movement of negatively charged ions depends on active transport of sodium, it is considered a passive process because it does not require a direct expenditure of cellular energy. Active transport also reabsorbs some negative ions directly, such as bicarbonate (HCO_3^-) and phosphate.

As more sodium ions are reabsorbed into the peritubular capillary along with negatively charged ions, the concentration of solutes in the peritubular blood might be expected to increase. However, because water moves by osmosis through cell membranes from regions of lesser solute concentration (hypotonic) toward regions of greater solute concentration (hypertonic), water is also reabsorbed, following the ions from the renal tubule into the peritubular capillary.

The proximal convoluted tubule reabsorbs about 70% of the filtered sodium, other ions, and water. By the end of the proximal convoluted tubule, osmotic equilibrium is reached, and the remaining tubular fluid is isotonic (fig. 20.20).

Active transport continues to reabsorb sodium ions as the tubular fluid moves through the nephron loop, the distal convoluted tubule, and the collecting duct. Consequently, almost all of the sodium and water (97% to 99%) that enters the renal tubules as part of the glomerular filtrate may be reabsorbed before the urine is excreted. However, aldosterone controls sodium reabsorption, and antidiuretic hormone controls water reabsorption. Under the influence of these hormones, reabsorption of sodium and water can change to keep conditions in the body fluids constant. Chapter 21 (pp. 808 and 809) discusses the specific effects of antidiuretic hormone and aldosterone, respectively.

Recall that the kidneys filter an extremely large volume of fluid (180 liters) each day. If 99% of the glomerular filtrate is reabsorbed, the remaining 1% excreted includes a relatively large amount of sodium and water (table 20.2). On the other hand, if sodium and water reabsorption decrease to 97% of the amount filtered, the amount excreted triples! Therefore, small changes in the tubular reabsorption of sodium and water result in large changes in urinary excretion of these substances.

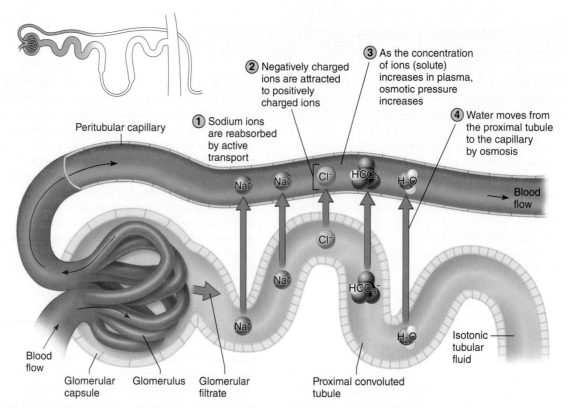

FIGURE 20.20 In the proximal part of the renal tubule, osmosis reabsorbs water in response to reabsorption of sodium and other solutes.

TABLE 20.2	Average Values for Sodium and Water Filtration, Reabsorption, and Excretion		
	Amount Filtered per Day	Amount Reabsorbed per Day (%)	Amount Excreted per Day
Water (L)	180	178.2 (99%)	1.8 (1%)
Na$^+$ (g)	630	626.8 (99.5%)	3.2 (0.5%)

 PRACTICE

11 How is the peritubular capillary adapted for reabsorption?

12 Which substances in glomerular filtrate are not normally present in urine?

13 Which mechanisms reabsorb solutes from glomerular filtrate?

14 Define *renal plasma threshold*.

15 Describe the role of passive transport in urine formation.

Tubular Secretion

In *tubular secretion*, certain substances move from the plasma of the peritubular capillary into the fluid of the renal tubule (see 20.19). Because of tubular secretion, the amount of a particular chemical excreted in the urine may exceed the amount filtered from the plasma in the glomerulus (see table 20.1). This process can be so thorough that in the case of some substances virtually

none can be detected in the blood leaving through the renal vein. The term *tubular* is used because the epithelial cells of the renal tubules carry out the process, as they do in tubular reabsorption.

Active transport mechanisms similar to those that function in reabsorption secrete some substances. However, the secretory mechanisms transport substances in the opposite direction. For example, the epithelium of the proximal convoluted tubules actively secretes certain organic compounds, including penicillin and histamine, into the tubular fluid.

Hydrogen ions are actively secreted throughout the entire renal tubule. Urine is usually acidic by the time it is excreted, although the urinary pH can vary considerably. The secretion of hydrogen ions is important in regulating the pH of body fluids, as chapter 21 (p. 816) explains.

The kidneys respond to certain chemicals, including antibiotics, by eliminating them from the body as rapidly as possible. In some cases, such as elimination of toxins, this is beneficial. In the case of helpful drugs, however, it is undesirable. The drug probenecid is given with certain antibiotics to block their tubular secretion, increasing their levels in the plasma. For example, by blocking tubular secretion of penicillin, probenecid increases plasma concentration of penicillin two to four fold. It also acts on ampicillin, methicillin, and certain other drugs in the penicillin family, as well as certain cephalosporin antibiotics.

Most of the potassium ions in the glomerular filtrate are actively reabsorbed in the proximal convoluted tubule, but control of potassium excretion depends on potassium secretion in the distal convoluted tubule and collecting duct. During this process, the active reabsorption of sodium ions out of the tubular fluid under the influence of aldosterone produces a negative electrical charge in the tubule. Positively charged potassium ions (K^+) move through channels activated by the negative charge in the tubule and enter the tubular fluid (fig. 20.21).

To summarize, urine forms as a result of the following:

- glomerular filtration of materials from blood plasma
- tubular reabsorption of substances, including glucose; water; urea; proteins; amino, lactic, citric, and uric acids; and phosphate, sulfate, calcium, potassium, and sodium ions
- tubular secretion of substances, including penicillin, histamine, phenobarbital, hydrogen ions, ammonia, and potassium ions

 PRACTICE

16 Define *tubular secretion*.

17 Which substances are actively secreted? Passively secreted?

18 How does the reabsorption of sodium affect the secretion of potassium?

Regulation of Urine Concentration and Volume

Hormones such as aldosterone and cardiac natriuretic polypeptides affect the solute concentration of urine, particularly sodium.

However, the ability of the kidneys to maintain the internal environment is largely due to their reabsorbing large volumes of water, which concentrates the urine.

In contrast to conditions in the proximal convoluted tubule, the tubular fluid reaching the distal convoluted tubule is hypotonic because of changes that occur through the loop segment of each nephron. The cells lining the distal convoluted tubule and the collecting duct that follows continue to reabsorb sodium ions (chloride ions follow passively) under the influence of aldosterone, which the adrenal cortex secretes (see chapter 13, p. 511). In addition, the interstitial fluid surrounding the collecting ducts is hypertonic, particularly in the medulla. These might seem to be ideal conditions for water reabsorption as well. However, the cells lining the later portion of the distal convoluted tubule and the collecting duct are impermeable to water unless antidiuretic hormone (ADH) is present. Thus, water inside the tubule may be excreted, forming dilute urine.

Neurons in the hypothalamus produce ADH, as chapter 13 discussed (p. 503). The posterior lobe of the pituitary gland releases ADH in response to decreasing concentration of water in the body fluids or to decreasing blood volume and blood pressure. When ADH reaches the kidney, it stimulates cells in the distal convoluted tubules and collecting ducts to insert proteins called aquaporins into their cell membranes, which form water channels. These channels greatly increase permeability to water. Consequently, water rapidly moves out of these structures by osmosis, especially where the distal tubules and collecting ducts pass through the extremely hypertonic medulla. The urine becomes more concentrated, and water is retained in the internal environment (fig. 20.22).

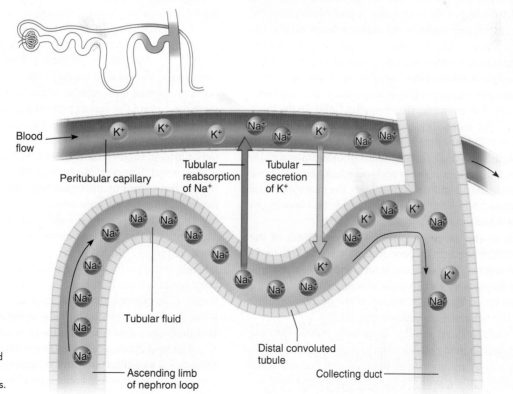

FIGURE 20.21 In the distal convoluted tubule, potassium ions are secreted in response to the active reabsorption of sodium ions.

Blood flow

Peritubular capillary

Tubular reabsorption of Na⁺

Tubular secretion of K⁺

Tubular fluid

Ascending limb of nephron loop

Distal convoluted tubule

Collecting duct

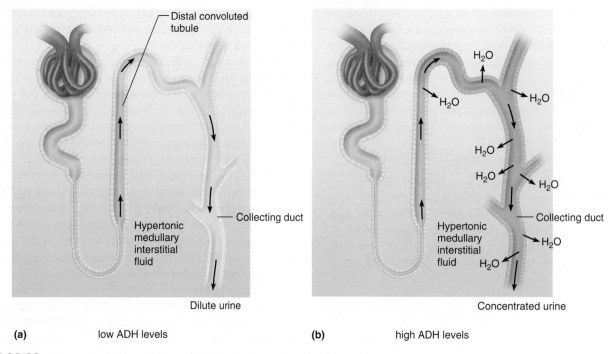

FIGURE 20.22 Urine concentrating mechanism. (*a*) The distal convoluted tubule and collecting duct are impermeable to water, so water may be excreted as dilute urine. (*b*) If ADH is present, however, these segments become permeable, and water is reabsorbed by osmosis into the hypertonic medullary interstitial fluid.

Most nephrons in humans are cortical nephrons, but it is the juxtamedullary nephrons that are most important in regulating water reabsorption. These nephrons set up an osmotic gradient between the collecting duct and the interstitial fluid in the renal medulla that enables water to be reabsorbed.

A **countercurrent mechanism** in the nephron loops of the juxtamedullary nephrons ensures that the medullary interstitial fluid becomes hypertonic. This mechanism is possible because the descending and ascending limbs of the nephron loops lie parallel and close to one another. The mechanism is named partly for the fact that fluid moving down the descending limb results in a current that is counter to that of the fluid moving up in the ascending limb.

The different parts of the nephron loop have important functional distinctions. For example, the epithelial lining in the thick upper portion of the ascending limb (thick ascending limb, or thick segment) is relatively impermeable to water. However, this part of the nephron loop does actively reabsorb sodium and chloride ions (some potassium is actively reabsorbed as well). As these solutes accumulate outside the ascending limb, the interstitial fluid becomes hypertonic, while the tubular fluid inside becomes hypotonic because it is losing solute.

In contrast to the ascending limb, the epithelium of the descending limb (thin segment) is permeable to water, but relatively impermeable to solutes. This segment is surrounded by hypertonic fluid created by the thick ascending limb, so water tends to leave the descending limb by osmosis. The contents of the descending limb become more concentrated, or hypertonic (fig. 20.23).

The very concentrated tubular fluid now moves into the ascending limb, and sodium chloride (NaCl) is again actively reabsorbed into the medullary interstitial fluid, raising the interstitial NaCl concentration further. With the increased interstitial fluid sol-ute concentration, even more water diffuses out of the descending limb, further increasing the salt concentration of the tubular fluid. Each time this circuit is completed, the concentration of NaCl in the ascending limb increases, or multiplies. For this reason, the mechanism is called a **countercurrent multiplier.** The countercurrent multiplier is dependent not only on the countercurrent flow in the nephron loops, but also on the active reabsorption of solute in the thick ascending limbs of the loops.

The descending limb of the loop is permeable to water, so the interstitial fluid at any level of the loop is essentially in equilibrium with the fluid in the tubule. Thus, the concentration gradient in the loop is also found in the interstitial fluid. The solute concentration of both tubular and interstitial fluids progressively increases toward the deeper parts of the renal medulla (fig. 20.23).

In addition to sodium chloride, urea contributes to the high solute concentration of the renal medulla. Urea is a waste product of protein catabolism and is filtered at the glomerulus. Before it is excreted, however, as much as 80% of the urea in the tubular fluid is reabsorbed by the collecting duct and enters the medullary interstitial fluid. This urea eventually is secreted into the ascending limb of the nephron loop and again reaches the collecting duct. This urea "cycle" excretes about 20% of the filtered urea in the urine, but newly filtered urea constantly replaces what is lost. As a result, there is always a significant concentration of urea in the renal medulla contributing to the osmotic gradient for water reabsorption. In humans, these mechanisms create a tubular and interstitial fluid solute concentration near the tip of the loop more than four times the solute concentration of plasma (fig. 20.23).

The vasa recta, another example of a countercurrent mechanism called a **countercurrent exchanger,** prevent blood flow from carrying away the solute concentrated in the renal medulla by the

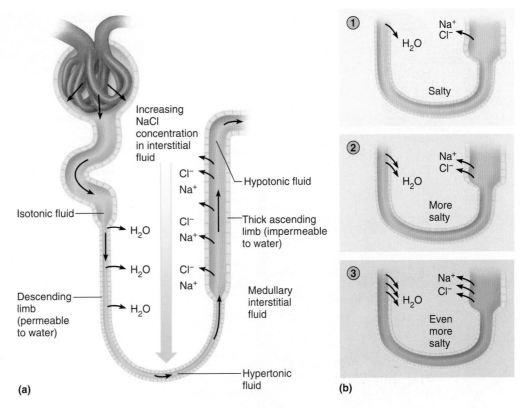

FIGURE 20.23 The countercurrent multiplier. (a) Active solute reabsorption from the ascending limb increases solute concentration in the interstitial fluid of the medulla. The solute concentration of interstitial fluid in the medulla equilibrates with tubular fluid, which loses water in the descending limb, and thus becomes hypertonic by the tip of the nephron loop. The concentrated tubular fluid flows into the ascending limb and again active solute reabsorption occurs causing even more water loss from the descending limb as tubular fluid continues to flow. (b) The countercurrent multiplier progressively increases the solute concentration of the interstitial fluid, up to a maximum near the tip of the loop more than four times that of plasma.

countercurrent multiplier. Recall that the vasa recta are bundles of parallel blood vessels connected by capillary loops and extending deep into the medulla (see figure 20.6). The vasa recta do not actively transport NaCl, so they do not contribute to establishing the high solute concentration in the renal medulla. Rather, they preserve the NaCl gradient established by the countercurrent multiplier despite blood flowing through the renal medulla. As blood flows down the descending portion of the vasa recta, NaCl enters. Then, as the blood flows back up toward the renal cortex, NaCl leaves to reenter the medullary interstitial fluid. Consequently, the bloodstream carries away little of the NaCl, thus preserving the osmotic gradient for water reabsorption. At the same time the vasa recta carry away water reabsorbed from the collecting ducts in the medulla (fig. 20.24).

To summarize, the countercurrent multiplier creates a large osmotic gradient for water reabsorption in the interstitial fluid surrounding the distal convoluted tubules and the collecting ducts of the nephron. The epithelial lining of these structures is impermeable to water, unless ADH is present. The higher the blood levels of ADH, the more permeable the epithelial lining becomes, increasing water reabsorption and concentrating the urine. In this way, soluble wastes and other substances are excreted in minimal water, preserving body water when dehydration is a threat. If the body fluids contain excess water, ADH secretion decreases and the epithelial linings of the distal convoluted tubule and the collecting duct become less permeable to water. Less water is reabsorbed, and the urine becomes more dilute. Table 20.3 summarizes the role of ADH in urine production. Table 20.4 summarizes the functions of parts of the nephron.

Urea and Uric Acid Excretion

Urea is a by-product of amino acid catabolism in the liver. Therefore, the amount of urea that must be eliminated in the urine reflects the amount of protein in the diet. As described earlier, urea enters the renal tubule by filtration and is both reabsorbed and secreted by different portions of the renal tubule. The net effect of these processes is that up to 80% of the filtered urea is reabsorbed, and the remaining 20% is excreted in the urine.

Uric acid is a product of the metabolism of certain nucleic acid bases (the purines adenine and guanine). Active transport completely reabsorbs the filtered uric acid. The approximately 10% of the filtered uric acid excreted in the urine reflects secretion into the renal tubule.

TABLE 20.3	Role of ADH in Regulating Urine Concentration and Volume
1.	Concentration of water in the blood decreases.
2.	Increase in the osmotic pressure of body fluids stimulates osmoreceptors in the hypothalamus.
3.	Hypothalamus signals the posterior pituitary gland to release ADH.
4.	Blood carries ADH to the kidneys.
5.	ADH causes the distal convoluted tubules and collecting ducts to increase water reabsorption by osmosis.
6.	Urine becomes more concentrated, and urine volume decreases.

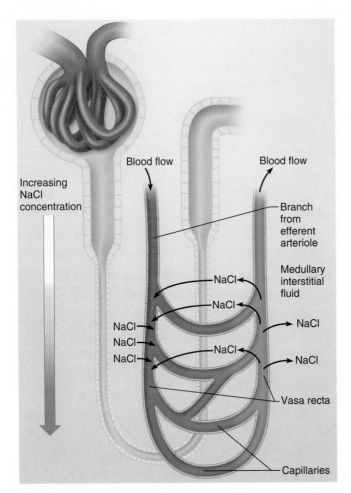

Blood flow

Increasing
NaCl
concentration

Blood flow

Branch
from
efferent
arteriole

Medullary
interstitial
fluid

NaCl
NaCl
NaCl
NaCl
NaCl
NaCl
NaCl
NaCl

Vasa recta

Capillaries

FIGURE 20.24 A countercurrent mechanism in the vasa recta helps maintain the NaCl concentration gradient in the medullary interstitial fluid.

 PRACTICE

19 Describe a countercurrent mechanism.

20 How does the hypothalamus regulate urine concentration and volume?

21 Explain how urea and uric acid are excreted.

Excess uric acid may precipitate in the plasma and be deposited as crystals in joints, causing the inflammation and extreme pain of gout, particularly in the digits and especially in the great toe. Gout has had an interesting history. Hippocrates mentioned it. King Charles I of Spain gave up his vast empire in 1556 due to the painful condition. In 2006, Spanish researchers confirmed the diagnosis by detecting uric acid deposits in the terminal joint of a finger that, for reasons unknown, had been preserved in a small box apart from the rest of the king. Today, gout is treated with drugs that inhibit uric acid reabsorption or block an enzyme in the biosynthetic pathway for uric acid. Limiting foods rich in uric acid, such as meats and seafood, and drinking more water to dilute the urine can help. Gout is inherited, but an attack may not occur unless the person eats large amounts of the offending foods.

TABLE 20.4	Functions of Nephron Components and the Collecting Duct
Part	**Function**
Nephron	
Renal Corpuscle	
Glomerulus	Filtration of water and dissolved substances from the plasma
Glomerular capsule	Receives the glomerular filtrate
Renal Tubule	
Proximal convoluted tubule	Reabsorption of glucose; amino acids; creatine; lactic, citric, uric, and ascorbic acids; phosphate, sulfate, calcium, potassium, and sodium ions by active transport
	Reabsorption of proteins by endocytosis
	Reabsorption of water by osmosis
	Reabsorption of chloride ions and other negatively charged ions by electrochemical attraction
	Active secretion of substances such as penicillin, histamine, creatinine, and hydrogen ions
Descending limb of nephron loop	Reabsorption of water by osmosis
Ascending limb of nephron loop	Reabsorption of sodium, chloride, and potassium ions by active transport
Distal convoluted tubule	Reabsorption of sodium ions by active transport
	Reabsorption of water by osmosis
	Active secretion of hydrogen ions
	Secretion of potassium ions both actively and by electrochemical attraction
Collecting Duct*	Reabsorption of water by osmosis

*The collecting duct is not anatomically part of the nephron, but it is functionally linked in the process of urine formation.

Urine Composition

Urine composition varies depending on the volumes of water and solutes that the kidneys must eliminate from the body or retain in the internal environment to maintain homeostasis. Urinary excretion of any solute may increase at one time and decrease at another. Water excretion is independent of solute excretion. Urine composition varies considerably from time to time because of differences in dietary intake and physical activity. Urine is typically about 95% water and usually also contains urea and uric acid from the catabolism of amino acids and nucleic acids, and creatinine from metabolism of creatine. Urine also contains various ions whose concentrations vary with the diet and may contain a trace of amino acids (see table 20.1). **Appendix C, Common Tests Performed on Urine** lists the normal concentrations of urine components.

Not all abnormal constituents of urine indicate illness. Glucose may enter urine after a sugary meal or toward the end of pregnancy; protein may appear in the urine following vigorous physical exercise; ketones are in urine during a prolonged fast or when a person follows a very low-calorie or low-carbohydrate diet.

The volume of urine produced usually varies between 0.6 and 2.5 liters per day. Such factors as fluid intake; environmental temperature; relative humidity of the surrounding air; and a person's emotional condition, respiratory rate, and body temperature influence the exact urine volume. An output of 50–60 milliliters of urine per hour is considered normal, and an output of less than 30 milliliters per hour may indicate kidney failure.

The kidneys of infants and young children are unable to concentrate urine and conserve water as effectively as those of adults. As a result, they can lose water rapidly, which may lead to dehydration. A 20-pound infant can lose a pound in just a day of an acute viral illness, and this is a sufficiently significant proportion of body weight to warrant hospitalization. Intravenous fluids may be required to restore water and electrolyte balance (see chapter 21, p. 804).

Renal Clearance

The rate at which a chemical is removed from the plasma is called the *renal clearance* of that substance. Tests of renal clearance detect glomerular damage or monitor the progression of renal disease. One such test, the *inulin clearance test,* uses *inulin* (not to be confused with insulin), which is a complex polysaccharide found in certain plant roots. In the test, a known amount of inulin is infused into the blood at a constant rate. The inulin passes freely through the glomerular membranes, so its concentration in the glomerular filtrate equals that in the plasma. In the renal tubule, inulin is not reabsorbed to any significant degree, nor is it secreted. Consequently, the rate at which inulin appears in the urine can be used to calculate the rate of glomerular filtration.

The kidneys also remove creatinine from the blood. Creatinine is produced at a constant rate during muscle metabolism. Like inulin, creatinine is filtered but neither reabsorbed nor secreted by the kidneys. The *creatinine clearance test,* which compares a patient's blood and urine creatinine concentrations, can also be used to calculate the GFR. A significant advantage of this test over the inulin clearance test is that the bloodstream normally has a constant level of creatinine. Therefore, a single measurement of plasma creatinine levels provides a rough index of kidney function. For example, significantly elevated plasma creatinine levels suggest that GFR is greatly reduced. Because nearly all of the creatinine that the kidneys filter normally appears in the urine, a decrease in the rate of creatinine excretion may indicate renal failure.

Another plasma clearance test uses *para-aminohippuric acid* (PAH), which filters freely through the glomerular membranes.

However, unlike inulin, any PAH remaining in the peritubular capillary plasma after filtration is secreted into the proximal convoluted tubules. Therefore, essentially all PAH passing through the kidneys appears in the urine. For this reason, the rate of PAH clearance can be used to calculate the rate of plasma flow through the kidneys. Then, if the hematocrit is known (see chapter 14, p. 529), the rate of total blood flow through the kidneys can also be calculated. Clinical Application 20.3 describes some of the causes and treatments of chronic kidney disease.

 PRACTICE

22 List the normal constituents of urine.

23 What is the normal hourly output of urine? The minimal hourly output?

20.4 | Storage and Elimination of Urine

After urine forms along the nephrons and collecting ducts, it passes through openings in the renal papillae and enters the minor and major calyces of the kidney. From there it passes through the renal pelvis, into a ureter, and into the urinary bladder. The urethra delivers urine to the outside.

Ureters

Each **ureter** is a tubular organ about 25 centimeters long, which begins as the funnel-shaped renal pelvis. It extends downward posterior to the parietal peritoneum and parallel to the vertebral column. In the pelvic cavity, each ureter courses forward and medially to join the urinary bladder from underneath.

The wall of a ureter is composed of three layers. The inner layer, or *mucous coat,* includes several thicknesses of transitional epithelial cells and is continuous with the linings of the renal tubules and the urinary bladder. The middle layer, or *muscular coat,* consists largely of smooth muscle fibers in circular and longitudinal bundles. The outer layer, or *fibrous coat,* is composed of connective tissue (fig. 20.25).

Muscular peristaltic waves, originating in the renal pelvis, help move the urine along the length of the ureter. The presence of urine in the renal pelvis initiates these waves, whose frequency keeps pace

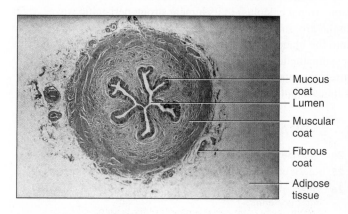

— Mucous coat
— Lumen
— Muscular coat
— Fibrous coat
— Adipose tissue

FIGURE 20.25 **AP|R** Cross section of a ureter (75×).

Charles B., a forty-three-year-old construction worker, had been feeling unusually tired for several weeks, with occasional dizziness and difficulty sleeping. More recently he had noticed a burning pain in his lower back, just below his rib cage, and his urine had darkened. In addition, his feet, ankles, and face were swollen. His wife suggested that he consult their family physician.

Charles had elevated blood pressure (hypertension) and the regions of his kidneys were sensitive to pressure. His urine had excess protein (proteinuria) and blood (hematuria), and his blood had elevated blood urea nitrogen (BUN), elevated serum creatinine, and decreased serum protein (hypoproteinemia) concentrations.

Charles had *chronic glomerulonephritis,* an inflammation of the capillaries in the glomeruli of the nephrons. It is a progressive degeneration with no direct treatment, although drugs can help to control blood pressure. Microscopic examination of a small sample of kidney tissue (biopsy) confirmed the diagnosis.

Despite antihypertensive drugs and careful attention to his diet, Charles's condition deteriorated rapidly. When it appeared that most of his kidney function had been lost (end-stage renal disease, or ESRD), he was offered artificial kidney treatments (hemodialysis).

To prepare Charles for hemodialysis, a vascular surgeon created a fistula in his patient's left forearm by surgically connecting an artery to a vein. The greater pressure of the blood in the artery that now flowed directly into the vein swelled the vein, making it more accessible.

During hemodialysis, a hollow needle was inserted into the vein of the fistula near its arterial connection. This allowed the blood to flow, with the aid of a blood pump, through a tube leading to the blood compartment of a dialysis machine, where the blood passed over a selectively permeable membrane. On the opposite side of the membrane was a dialysate solution with a controlled composition. Negative pressure on the dialysate side of the membrane, created by a vacuum pump, increased the movement of fluid through the membrane. At the same time, waste and excess electrolytes diffused from the blood through the membrane and entered the dialysate solution. The blood was then returned through a tube to the vein of the fistula.

To maintain favorable blood concentrations of waste, electrolytes, and water, Charles had to undergo hemodialysis three times per week, with each treatment lasting three to four hours. Each time he took an anticoagulant to prevent blood clotting, an antibiotic drug to control infections, and an antihypertensive drug.

Charles had to carefully control his intake of water, sodium, potassium, proteins, and total calories between treatments. He was a candidate for a kidney transplant, which could free him from dependence on hemodialysis.

In a transplant, a kidney from a living donor or a cadaver, whose tissues are antigenically similar (histocompatible) to those of the recipient, is placed in the depression on the medial surface of the right or left ilium (iliac fossa). The renal artery and vein of the donor kidney are connected to the recipient's iliac artery and vein, respectively, and the kidney's ureter is attached to the dome of the recipient's urinary bladder.

---

with the rate of urine formation. If urine forms quickly, a peristaltic wave may pass every few seconds; if it forms more slowly, a wave may pass every few minutes.

When a peristaltic wave reaches the urinary bladder, a jet of urine spurts through a fold of mucous membrane that surrounds the opening, and the urine enters. Because of the angle at which the ureters enter the bladder, the bladder wall acts as a valve, allowing urine to enter the bladder from the ureters but preventing it from flowing backward as the bladder fills.

If a ureter becomes obstructed, such as by a small kidney stone (renal calculus) in its lumen, strong peristaltic waves begin in the proximal part of the tube, which may help move the stone into the bladder. The presence of a stone usually also stimulates a sympathetic reflex (ureterorenal reflex) that constricts the renal arterioles and reduces urine production in the affected kidney. Clinical Application 20.4 describes kidney stones.

 PRACTICE

24 Describe the structure of a ureter.

25 How is urine moved from the renal pelvis to the urinary bladder?

26 What prevents urine from backing up from the urinary bladder into the ureters?

27 How does an obstruction in a ureter affect urine production?

Urinary Bladder

The **urinary bladder** is a hollow, distensible, muscular organ in the pelvic cavity, posterior to the pubic symphysis and inferior to the parietal peritoneum (fig. 20.26 and reference plate 8, p. 45). In a female, the bladder contacts the anterior walls of the uterus and vagina, and in a male, the bladder lies posteriorly against the rectum.

The pressure of surrounding organs alters the spherical shape of the bladder. When the bladder is empty, its inner wall forms many folds, but as it fills with urine, the wall becomes smoother. At the same time, the superior surface of the bladder expands upward into a dome.

When greatly distended, the bladder pushes above the pubic crest and into the region between the abdominal wall and the parietal peritoneum. The dome can reach the level of the umbilicus and press against the coils of the small intestine.

The internal floor of the bladder includes a triangular area called the *trigone,* which has an opening at each of its three angles (fig. 20.27). Posteriorly, at the base of the trigone, the openings are those of the ureters. Anteriorly, at the apex of the trigone, is a short, funnel-shaped extension called the *neck* of the bladder, which contains the opening into the urethra. The trigone generally remains in a fixed position, even though the rest of the bladder distends and contracts.

The wall of the urinary bladder consists of four layers. The inner layer, or *mucous coat,* includes several thicknesses of tran-

Kidney stones, which are usually composed of uric acid, calcium oxalate, calcium phosphate, or magnesium phosphate, can form in the collecting ducts and renal pelvis (fig. 20A). Such a stone passing into a ureter causes sudden, severe pain that begins in the region of the kidney and radiates into the abdomen, pelvis, and lower limbs. It may also cause nausea and vomiting, and blood in the urine.

About 60% of kidney stones pass from the body on their own. Other stones were at one time removed surgically but are now shattered with intense sound waves. In this procedure, called *extracorporeal shock-wave lithotripsy (ESWL)*, the patient is placed in a stainless steel tub filled with water. A spark-gap electrode produces underwater shock waves, and a reflector concentrates and focuses the shock-wave energy on the stones. The resulting sandlike fragments then leave in urine.

The tendency to form kidney stones is inherited, particularly the stones that contain calcium, which account for more than half of all cases. Eating calcium-rich foods does not increase the risk, but taking calcium supplements can. People who have calcium oxalate stones can reduce the risk of recurrence by avoiding specific foods: chocolate, coffee, wheat bran, cola, strawberries, spinach, nuts, and tea. Other causes of kidney stones include excess vitamin D, blockage of the urinary tract, or a complication of a urinary tract infection.

It is very helpful for a physician to analyze the composition of the stones, because certain drugs can prevent recurrence. Stones can be collected during surgery or by the person, using a special collection device.

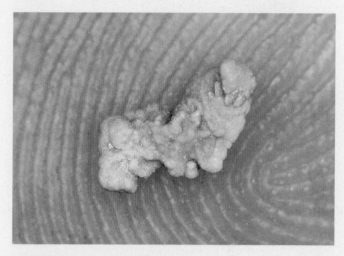

FIGURE 20A This kidney stone is small, held against this fingertip, but it is large enough to cause severe pain.

sitional epithelial cells similar to those lining the ureters and the upper portion of the urethra. The thickness of this tissue changes as the bladder expands and contracts. During distension, the tissue appears to be only two or three cells thick, but during contraction, it appears to be five or six cells thick (see fig. 5.11, p. 157).

The second layer of the bladder wall is the *submucous coat*. It consists of connective tissue and has many elastic fibers.

The third layer of the bladder wall, the *muscular coat*, is primarily composed of coarse bundles of smooth muscle. These bundles are interlaced in all directions and at all depths, and together they comprise the **detrusor muscle** (de-truz′or mus′l). The portion of the detrusor muscle that surrounds the neck of the bladder forms an *internal urethral sphincter*. Sustained contraction of this sphincter muscle prevents the bladder from emptying

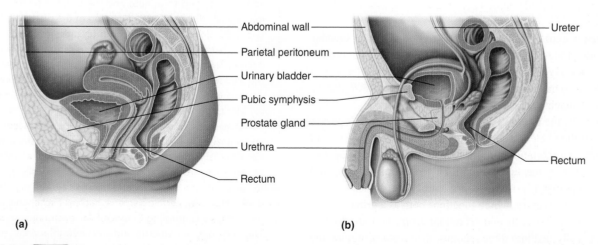

(a) (b)

FIGURE 20.26 APR The urinary bladder is in the pelvic cavity and behind the pubic symphysis. (a) In a female, it contacts the uterus and vagina. (b) In a male, it lies against the rectum.

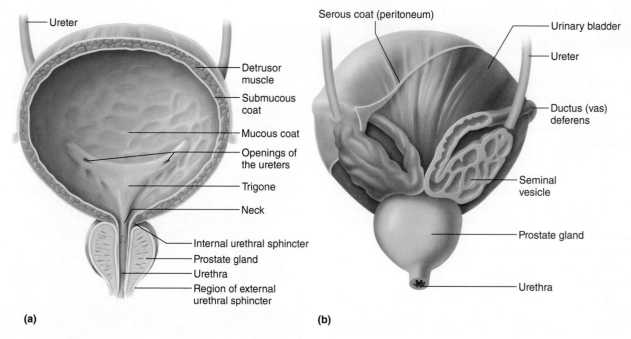

Ureter
Detrusor muscle
Submucous coat
Mucous coat
Openings of the ureters
Trigone
Neck
Internal urethral sphincter
Prostate gland
Urethra
Region of external urethral sphincter

(a)

Serous coat (peritoneum)
Urinary bladder
Ureter
Ductus (vas) deferens
Seminal vesicle
Prostate gland
Urethra

(b)

FIGURE 20.27 AP|R A male urinary bladder. (*a*) Longitudinal section. (*b*) Posterior view.

until the pressure within it increases to a certain level. The detrusor muscle has parasympathetic nerve fibers that function in the reflex that passes urine.

The outer layer of the wall, the *serous coat,* consists of the parietal peritoneum. It is found only on the upper surface of the bladder. Elsewhere, the outer coat is composed of fibrous connective tissue (fig. 20.28). From Science to Technology 5.2 on page 172 discusses a tissue-engineered replacement bladder.

 PRACTICE

28 Describe the trigone of the urinary bladder.

29 Describe the structure of the bladder wall.

30 What type of nerve fiber supplies the detrusor muscle?

Urethra

The **urethra** is a tube that conveys urine from the urinary bladder to the outside of the body. Its wall is lined with mucous membrane and has a thick layer of longitudinal smooth muscle fibers. The urethral wall also has many mucous glands, called *urethral glands,* which secrete mucus into the urethral canal (fig. 20.29).

In a female, the urethra is about 4 centimeters long. It passes forward from the bladder, courses below the pubic symphysis, and empties between the labia minora. Its opening, the *external urethral orifice* (urinary meatus), is anterior to the vaginal opening and is about 2.5 centimeters posterior to the clitoris (fig. 20.30*a*).

In a male, the urethra, which functions both as a urinary canal and a passageway for cells and secretions from the reproductive organs, can be divided into three sections: the prostatic urethra, the intermediate part of urethra, and the spongy urethra (see fig. 20.30*b* and reference plate 20, p. 52).

The **prostatic urethra** is about 2.5 centimeters long and passes from the urinary bladder through the *prostate gland,* located just below the bladder. Ducts from reproductive structures join the urethra in this region.

The **intermediate part of urethra,** or membranous urethra, is about 2 centimeters long. It originates just distal to the prostate gland, passes through the urogenital diaphragm, and is surrounded by the fibers of the external urethral sphincter muscle.

The **spongy urethra** is about 15 centimeters long and passes through the corpus spongiosum of the penis, within erectile tissue. This portion of the urethra terminates with the *external urethral orifice* at the tip of the penis.

The urgent need to urinate and the burning pain of urination may signal inflammation or infection of the bladder and/or ureters. Inflammation of the bladder (cystitis) is more common in women because the urethral pathway is shorter than in men, and bacteria from the external environment may more readily enter the bladder. Inflammation of the ureters is called ureteritis.

In a urinary tract infection (UTI), bacteria ascend from the bladder to the ureters on the continuous linings. Urination is frequent, painful, scant, and may be bloody, with accompanying abdominal pain. Usually pathogenic bacteria in the urinary tract remain outside the cells and are easily killed with antibiotic drugs or prevented from attaching to ureter lining cells by exposure to compounds in cranberry and blueberry juices that the person drinks. However, certain bacteria, such as *Escherichia coli,* enter the lining cells, forming "intracellular bacterial communities." The microbes can emerge and start new infections, explaining why some people experience frequent UTIs.

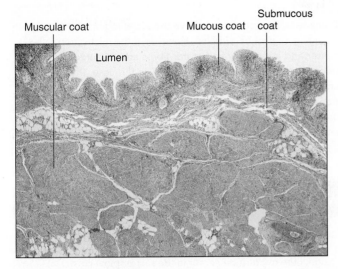

FIGURE 20.28 **AP|R** Light micrograph of the human urinary bladder wall (6×).

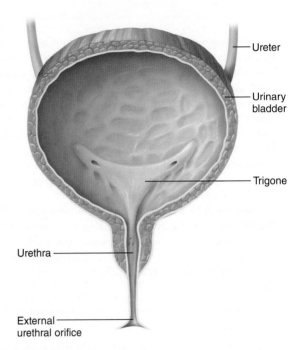

(a)

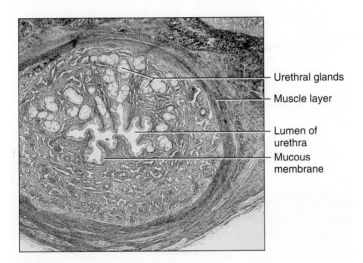

FIGURE 20.29 Cross section through the urethra (10×).

 PRACTICE

31 Describe the structure of the urethra.

32 How does the urethra of a male differ from that of a female?

Micturition

Urine leaves the urinary bladder by the **micturition** (mik″tu-rish′un) or urination **reflex.** The detrusor muscle contracts, and contractions of muscles in the abdominal wall and pelvic floor may help, as well as fixation of the thoracic wall and diaphragm. In micturition, the *external urethral sphincter* also relaxes. This muscle, part of the urogenital diaphragm (see chapter 9, pp. 325 and 327–328), surrounds the urethra about 3 centimeters from the bladder and is composed of voluntary skeletal muscle tissue.

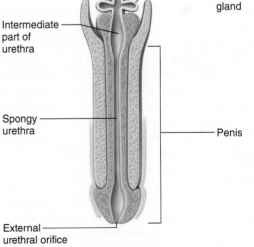

(b)

FIGURE 20.30 Urinary bladder and urethra (a) of the female (longitudinal section) and (b) of the male (longitudinal section).

Incontinence is the loss of control of micturition. Stress incontinence, caused by pressure on the bladder, is particularly common among women who have had children, especially if they have gained weight. An effective treatment is at least two months of doing Kegel exercises, in which a woman contracts the muscles that support the bladder, several times daily. Treatments for severe cases include a tamponlike cone inserted into the vagina to raise the pelvic floor; a small foam pad placed over the urethra to catch small amounts of urine; collagen injections around the urethra to tighten it; and surgery. Many people use absorbent pads.

Nighttime bedwetting was noted as long ago as 1500 B.C. Treatments have ranged from blocking the urethra at night to punishment and ridicule. In many cases, this *nocturnal enuresis* is inherited. Drug treatment and pads to absorb urine help to manage the problem in children, who usually outgrow the condition.

Distension of the bladder wall as it fills with urine stimulates the urge to urinate. The wall expands, stimulating stretch receptors, which eventually triggers the micturition reflex. The effectors of the reflex include the detrusor muscle and the internal urethral sphincter. Conscious control of the external sphincter also plays a role.

The *micturition reflex center* is in the sacral portion of the spinal cord. When sensory impulses from the stretch receptors signal the reflex center, parasympathetic motor impulses travel out to the detrusor muscle, which contracts rhythmically in response. A sensation of urgency accompanies this action.

The urinary bladder may hold up to 600 milliliters of urine. The desire to urinate usually begins when it contains about 150 milliliters. Then, as urine volume increases to 300 milliliters or more, the sensation of fullness becomes increasingly uncomfortable.

As the bladder fills with urine and its internal pressure increases, contractions of its wall intensify. When these contractions become strong enough to force the internal urethral sphincter to open slightly and admit some urine into the urethra, another reflex signals the external urethral sphincter to relax, and the bladder may empty. However, because the external urethral sphincter is composed of skeletal muscle, it is under conscious control, and therefore usually remains contracted until a person decides to urinate. Nerve centers in the brainstem and cerebral cortex that inhibit the micturition reflex aid this control.

When a person decides to urinate, the external urethral sphincter relaxes and inhibition of the micturition reflex lifts. Neural centers in the pons and the hypothalamus heighten the micturition reflex. The detrusor muscle contracts, the internal urethral sphincter relaxes, and urine is excreted through the urethra. Subsequently, the micturition reflex subsides, the detrusor muscle relaxes, the internal urethral sphincter contracts, and the bladder begins to fill with urine again. Table 20.5 outlines the micturition process, and Clinical Application 20.5 discusses urinalysis and health. Table 20.6 lists conditions that result from abnormal development of part of the urinary system. About 1 in 500 newborns has a birth defect affecting the urinary system.

Damage to the spinal cord above the sacral region may abolish voluntary control of urination. However, if the micturition reflex center and its sensory and motor fibers are uninjured, micturition may continue to occur reflexively. In this case, the bladder collects urine until its walls stretch enough to trigger a micturition reflex, and the detrusor muscle contracts and the internal urethral sphincter relaxes in response. This condition is called an *automatic bladder*.

TABLE 20.5 Major Events of Micturition AP|R

1. Urinary bladder distends as it fills with urine.

2. Stretch receptors in the bladder wall are stimulated, and they signal the micturition center in the sacral spinal cord.

3. Parasympathetic nerve impulses travel to the detrusor muscle, which responds by contracting rhythmically.

4. The need to urinate is perceived.

5. Voluntary contraction of the external urethral sphincter and inhibition of the micturition reflex by impulses from the brainstem and the cerebral cortex prevent urination.

6. Following the decision to urinate, the external urethral sphincter is relaxed, and impulses from the pons and the hypothalamus facilitate the micturition reflex.

7. The detrusor muscle contracts, the internal urethral sphincter relaxes, and urine is expelled through the urethra.

8. The micturition reflex subsides, the detrusor muscle relaxes, the internal urethral sphincter contracts, and the bladder begins to fill with urine again.

TABLE 20.6 Developmental Abnormalities of the Urinary System

Condition	Description
Crossed fused ectopia	Fused kidneys that lie on one side of the midline
Horseshoe kidney	Fusion of kidneys at one pole, usually lower, with most of each kidney on opposing side of midline
Nephrotic syndrome	Proteinuria (protein in urine) due to abnormal glomeruli
Oligomeganephronia	Reduced number of nephrons that are abnormally large
Polycystic kidney disease	Cysts form in renal tubules and/or collecting ducts
Renal agenesis	Absence of a kidney
Renal dysplasia	Abnormal kidney structure
Renal hypoplasia	Small kidney with fewer nephrons, but development normal
Tubular dysgenesis	Abnormal formation of proximal tubules
Vesicoureteral reflux	Urine backs up from bladder to ureter or kidney

Urine has long fascinated medical minds. As a folk remedy, urine has been used as a mouthwash, toothache treatment, and a cure for sore eyes. Hippocrates (460–377 BC) was the first to observe that the condition of the urine can provide clues to health, noting that frothy urine denoted kidney disease. During the Middle Ages, health practitioners consulted charts that matched urine colors to diseases. In the seventeenth century, British physicians diagnosed diabetes by having their medical students taste sugar in patients' urine. Today, urine composition is still used as a window on health and also to check for illicit drug use.

Certain inherited disorders can noticeably alter urine. The name *maple syrup urine disease* vividly describes what this inborn error of metabolism does to the urine. This condition, which causes intellectual disability, results from a block in the breakdown pathways for certain amino acids. In *alkaptonuria*, urine turns black when it is left to stand. This condition also produces painful arthritis and blackened ear tips. People with *Wilson disease* have an inherited inability to excrete copper. If they are properly diagnosed and given the drug penicillamine, their urine becomes the color of copper.

Other genetic conditions alter urine without causing health problems. People with *beeturia* excrete dark pink urine after they eat beets. The problem for people with *urinary excretion of odoriferous component of asparagus* is obvious. Parents of newborns who have inherited *blue diaper syndrome* are shocked when they change their child's first diaper. Due to a defect in transport of the amino acid tryptophan in the small intestine, bacteria degrade the unabsorbed tryptophan, producing a compound that turns blue on contact with oxygen.

PRACTICE

33 Describe micturition.

34 How is it possible to consciously inhibit the micturition reflex?

20.5 | Life-Span Changes

The urinary system has sufficient reserve, in both structure and function, to mask aging-related changes. However, the kidneys become slower to remove nitrogenous wastes and toxins and to compensate for changes that might alter homeostasis.

From the outside, the kidneys change with age, appearing scarred and grainy as arterioles serving the cortex constrict and fibrous connective tissue accumulates around the capsules. On the inside, kidney cells begin to die as early as age twenty years, but the gradual shrinkage is not generally noticeable until after age forty. By eighty years, the kidneys have lost about a third of their mass.

Kidney shrinkage is mostly due to the gradual loss of glomeruli. They may atrophy, cease functioning, become blocked with fibrous connective tissue, or untwist. About 5% of glomeruli are abnormal by age forty; 37% are abnormal by age ninety. The progressive shutdown of glomeruli decreases the surface area for filtration, and as a result, glomerular filtration rate (GFR) begins to drop in the fourth decade. By age seventy-five, GFR is about half that in a young adult, falling from about 125 milliliters/minute to about 60. With this decline, proteins are more likely to enter the urine. About a third of the elderly have proteinuria.

Further along the nephron, renal tubules thicken, accumulating fatty coats. They may shorten, forming small outpouches as cell death disrupts their sleek symmetry. Urine may become more dilute as reabsorption of sodium and glucose and other molecules becomes less efficient. The renal tubules also slow in processing certain drugs, which remain in the circulation longer. It becomes harder to clear nonsteroidal anti-inflammatory drugs such as aspirin, as well as opiates, antibiotics, urea, uric acid, creatinine, and various toxins. Slowing kidney function is one reason why a person's age should be considered when prescribing drugs.

Cardiovascular changes slow the journey of blood through the kidneys. A college student's kidneys may process about a fourth of the cardiac output, or about 1,200 milliliters, per minute. Her eighty-year-old grandfather's kidneys can handle about half that volume. Starting at about age twenty, renal blood flow rate diminishes by about 1% per year. The blood vessels that serve the kidneys become slower to dilate or constrict in response to body conditions. The kidneys' release of renin declines, hampering control of blood pressure and sodium and potassium ion concentrations in the blood. The kidneys also become less able to activate vitamin D, which may contribute to the higher prevalence of osteoporosis among the elderly.

The urinary bladder, ureters, and urethra lose elasticity and recoil with age. In the later years, the bladder holds less than half of what it did in young adulthood, and may retain more urine after urination. In the elderly, the urge to urinate may become delayed, so when it does happen, it is sudden. Older individuals have to urinate at night more than younger people.

Controlling bladder function is a challenge at the beginning of life and much later too. A child usually learns to control urination by about age two or three years. Incontinence, the uncontrolled leakage of urine, becomes more common in advanced years, although it is not considered a normal part of aging. It results from loss of muscle tone in the bladder, leading to increased bladder volume, and in the sphincters, which fail to close normally. Incontinence affects 15% to 20% of women over sixty-five and half of all men. In women, incontinence may develop from the stresses of childbirth and the effects of less estrogen during menopause. In males, incontinence usually is a response to an enlarged prostate gland pressing on the urethra and bladder.

PRACTICE

35 How do the kidneys change in appearance with advancing years?

36 What happens to glomeruli as a person ages?

37 How does kidney function change with age?

38 How do aging-related changes in the cardiovascular system affect the kidneys?

39 How do the urinary bladder, ureters, and urethra change with age? ■

Chapter Summary

20.1 Overview of the Urinary System (page 769)

The urinary system maintains homeostasis by regulating the composition, pH, and volume of body fluids. It does this by producing urine. The urinary system includes the kidneys, ureters, urinary bladder, and urethra.

20.2 Kidneys (page 769)

1. Location of the kidneys
   a. The **kidneys** are bean-shaped organs on either side of the vertebral column, high on the posterior wall of the abdominal cavity.
   b. They are posterior to the parietal peritoneum and anchored by adipose and connective tissues.
2. Functions of the kidneys
   a. The kidneys remove metabolic wastes from the blood and excrete them.
   b. They also help regulate red blood cell production, blood pressure, calcium ion absorption, and the volume, composition, and pH of the blood.
3. Kidney structure
   a. A kidney has a hollow **renal sinus.**
   b. The ureter expands into the **renal pelvis,** which, in turn, leads to the major and minor calyces.
   c. Renal papillae project into the minor calyces.
   d. Kidney tissue is divided into a medulla and a cortex.
4. Renal blood vessels
   a. Arterial blood flows through the **renal artery,** interlobar arteries, arcuate arteries, cortical radiate arteries, **afferent arterioles,** glomerular capillaries, **efferent arterioles,** and **peritubular capillaries.**
   b. Venous blood returns through a series of vessels that correspond to those of the arterial pathways.
5. Nephrons
   a. Structure of a nephron
      (1) A **nephron** is the functional unit of the kidney.
      (2) It consists of a **renal corpuscle** and a **renal tubule.**
         (a) The corpuscle consists of a **glomerulus** and a **glomerular capsule.**
         (b) Parts of the renal tubule include the proximal convoluted tubule, the **nephron loop** (ascending and descending limbs), and the distal convoluted tubule.
      (3) The nephron joins a collecting duct, which empties into a minor calyx.
   b. Cortical and juxtamedullary nephrons
      (1) Cortical nephrons are the most numerous and have corpuscles near the surface of the kidney.
      (2) Juxtamedullary nephrons have corpuscles near the medulla.
   c. Juxtaglomerular apparatus
      (1) The **juxtaglomerular apparatus** is between the ascending limb of the nephron loop and the afferent and efferent arterioles.
      (2) It consists of the macula densa and the juxtaglomerular cells.

20.3 Urine Formation (page 775)

Nephrons remove wastes from the blood and regulate water and electrolyte concentrations. **Urine** is the product of **glomerular filtration, tubular reabsorption,** and **tubular secretion.**

1. Glomerular filtration
   a. Urine formation begins when water and dissolved materials are filtered out of the glomerular capillaries.
   b. The glomerular capillaries are much more permeable than the capillaries in other tissues.
   c. Filtration is mainly due to hydrostatic pressure inside the glomerular capillaries.
   d. The osmotic pressure of the blood plasma and hydrostatic pressure in the glomerular capsule also affect filtration.
   e. Filtration pressure is the net force moving material out of the glomerulus and into the glomerular capsule.
   f. The composition of the filtrate is similar to that of tissue fluid.
   g. The rate of filtration varies with the filtration pressure.
   h. Filtration pressure changes with the diameters of the afferent and efferent arterioles.
   i. As the osmotic pressure in the glomerulus increases, filtration decreases.
   j. As the hydrostatic pressure in a glomerular capsule increases, the filtration rate decreases.
   k. The kidneys produce about 125 milliliters of glomerular fluid per minute; most is reabsorbed.
   l. The volume of filtrate varies with the surface area of the glomerular capillaries.
   m. Glomerular filtration rate (GFR) remains relatively constant but may increase or decrease with need.
   n. Autoregulation is the ability of an organ or tissue to maintain a constant blood flow under certain conditions when the arterial blood pressure is changing.
   o. When tubular fluid NaCl concentration decreases, the macula densa causes the juxtaglomerular cells to release renin. This triggers a series of changes leading to vasoconstriction, which may affect GFR, and secretion of aldosterone, which stimulates tubular sodium reabsorption.

Urinary System

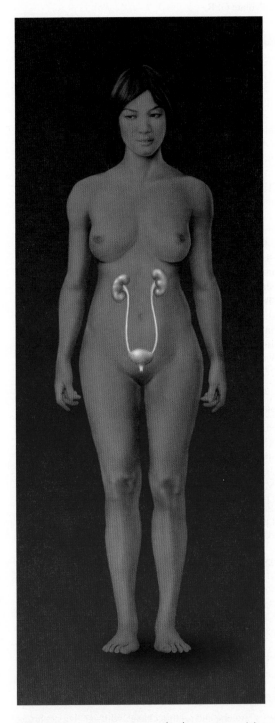

The urinary system controls the composition of the internal environment.

Integumentary System

The urinary system compensates for water loss due to sweating. The kidneys and skin both play a role in vitamin D production.

Skeletal System

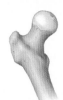

The kidneys and bone tissue work together to control plasma calcium levels.

Muscular System

Muscle tissue controls urine elimination from the bladder.

Nervous System

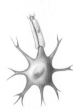

The nervous system influences urine production and elimination.

Endocrine System

The endocrine system influences urine production.

Cardiovascular System

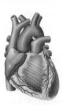

The urinary system controls blood volume. Blood volume and blood pressure play a role in determining water and solute excretion.

Lymphatic System

The kidneys control extracellular fluid (including lymph) volume and composition.

Digestive System

The kidneys compensate for fluids lost by the digestive system.

Respiratory System

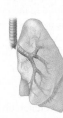

The kidneys and the lungs work together to control the pH of the internal environment.

Reproductive System

The urinary system in males shares organs with the reproductive system. The kidneys compensate for fluids lost from the male and female reproductive systems.

2. Tubular reabsorption
   a. Substances are selectively reabsorbed from the glomerular filtrate.
   b. The peritubular capillary is adapted for reabsorption.
      (1) It carries low-pressure blood.
      (2) It is permeable.
   c. Most reabsorption is in the proximal tubule, where epithelial cells have microvilli.
   d. Different modes of transport reabsorb various substances in particular segments of the renal tubule.
      (1) Glucose and amino acids are reabsorbed by active transport.
      (2) Water is reabsorbed by osmosis.
      (3) Proteins are reabsorbed by endocytosis.
   e. Active transport mechanisms have limited capacities.
   f. If the concentration of a substance in the filtrate exceeds its renal plasma threshold, the excess is excreted in the urine.
   g. Substances that remain in the filtrate are concentrated as water is reabsorbed.
   h. Sodium ions are reabsorbed by active transport.
      (1) Negatively charged ions accompany positively charged sodium ions out of the filtrate.
      (2) Water is passively reabsorbed by osmosis as sodium ions are actively reabsorbed.
3. Tubular secretion
   a. Tubular secretion transports certain substances from the plasma to the tubular fluid.
   b. Some substances are actively secreted.
      (1) These include various organic compounds and hydrogen ions.
      (2) The proximal and distal convoluted tubules secrete hydrogen ions.
   c. Potassium ions are secreted in the distal convoluted tubule and collecting duct.
4. Regulation of urine concentration and volume
   a. ADH from the posterior pituitary gland increases the permeability of the distal convoluted tubule and collecting duct, promoting water reabsorption.
   b. The **countercurrent mechanism** concentrates sodium ions in the renal medulla.
      (1) Sodium and chloride ions are actively reabsorbed in the ascending limb.
      (2) Tubular fluid in the ascending limb becomes hypotonic as it loses solutes.
      (3) Water leaves the descending limb by osmosis.
      (4) Tubular fluid in the descending limb becomes hypertonic as it loses water.
      (5) As this process repeats, the NaCl concentration in the medulla increases.
   c. The vasa recta countercurrent mechanism helps maintain NaCl concentration in the medulla.
   d. The distal convoluted tubule and collecting duct are impermeable to water, which therefore is excreted in urine.
5. Urea and uric acid excretion
   a. **Urea** is a by-product of amino acid metabolism.
      (1) It is reabsorbed and secreted by different portions of the renal tubule.
      (2) Up to 80% of the urea is reabsorbed.
   b. Uric acid results from the metabolism of nucleic acids.
      (1) Most is reabsorbed by active transport.
      (2) Some is secreted into the renal tubule.
6. Urine composition
   a. Urine is about 95% water, and it usually contains urea, uric acid, and creatinine.
   b. It may contain a trace of amino acids and varying amounts of electrolytes, depending upon diet.
   c. The volume of urine varies with the fluid intake, with certain environmental factors, and a person's emotional condition, respiratory rate, and body temperature.
7. Renal clearance
   a. Renal clearance is the rate at which a chemical is removed from the plasma.
   b. The inulin clearance test, creatinine clearance test, and para-aminohippuric acid test can be used to calculate GFR.

20.4 Storage and Elimination of Urine (page 791)

1. Ureters
   a. The **ureter** is a tubular organ that extends from each kidney to the urinary bladder.
   b. Its wall has mucous, muscular, and fibrous layers.
   c. Peristaltic waves in the ureter force urine to the urinary bladder.
   d. Obstruction in the ureter stimulates strong peristaltic waves and a reflex that decreases urine production.
2. Urinary bladder
   a. The **urinary bladder** is a distensible organ that stores urine and forces it into the urethra.
   b. The ureters and urethra open at the three angles of the trigone in the floor of the urinary bladder.
   c. Muscle fibers in the wall form the **detrusor muscle.**
   d. A portion of the detrusor muscle forms an internal urethral sphincter.
3. Urethra
   a. The **urethra** conveys urine from the urinary bladder to the outside.
   b. In females, it empties between the labia minora.
   c. In males, it conveys products of reproductive organs as well as urine.
      (1) Three portions of the male urethra are the **prostatic urethra,** the **intermediate part of the urethra,** and the **spongy urethra.**
      (2) The urethra empties at the tip of the penis.
4. Micturition
   a. **Micturition** is the process of expelling urine.
   b. In micturition, the detrusor muscle contracts and the external urethral sphincter relaxes.
   c. Micturition reflex
      (1) Distension stimulates stretch receptors in the urinary bladder wall.
      (2) The micturition reflex center in the sacral portion of the spinal cord sends parasympathetic motor impulses to the detrusor muscle.
      (3) As the urinary bladder fills, its internal pressure increases, forcing the internal urethral sphincter open.
      (4) A second reflex relaxes the external urethral sphincter, unless its contraction is voluntarily controlled.
      (5) Nerve centers in the brainstem and cerebral cortex aid control of urination.

20.5 Life-Span Changes (page 797)

The kidneys, ureters, and urethra change with age, but nephrons are so numerous that a healthy person is usually unaware of kidney shrinkage and slowed cleansing of the blood.

1. With age, the kidneys appear grainy and scarred.

2. GFR drops significantly with age as glomeruli atrophy, fill with connective tissue, or unwind.
3. Renal tubules accumulate fat on their outsides and become asymmetric. Reabsorption and secretion may slow or become impaired. Drugs remain longer in the circulation as a person ages.

4. Changes in the cardiovascular system slow the rate of processing through the urinary system. The kidneys slow in their response to changes, and are less efficient at activating vitamin D.
5. The urinary bladder, ureters, and urethra lose elasticity, with effects on the urge and timing of urination.

 # CHAPTER ASSESSMENTS

20.1 Overview of the Urinary System
1 Explain why the urinary system is necessary for survival. (p. 769)
2 Identify the organs of the urinary system and list their general functions. (p. 769)

20.2 Kidneys
3 List the functions of the kidneys. (p. 769)
4 Describe the external and internal structure of a kidney. (pp. 770–771)
5 List in correct order the vessels through which blood passes as it travels from the renal artery to the renal vein. (p. 772)
6 Distinguish between a renal corpuscle and a renal tubule. (pp. 775–776)
7 Name in correct order the structures through which fluid passes from the glomerulus to the collecting duct. (pp. 775–776)
8 Distinguish between cortical and juxtamedullary nephrons. (p. 776)
9 Describe the location and structure of the juxtaglomerular apparatus. (p. 776)

20.3 Urine Formation
10 Distinguish among filtration, tubular reabsorption, and tubular secretion as they relate to urine formation. (pp. 777–778)
11 Define *net filtration pressure*. (p. 779)
12 Explain how the diameters of the afferent and efferent arterioles affect the rate of glomerular filtration. (p. 781)
13 Explain how changes in the osmotic pressure of blood plasma affect the glomerular filtration rate. (p. 781)
14 Explain how the hydrostatic pressure of a glomerular capsule affects the rate of glomerular filtration. (pp. 781–782)
15 Define *autoregulation*. (p. 782)
16 Describe the two mechanisms by which the body regulates glomerular filtration rate. (p. 782)
17 Which one of the following is abundant in blood plasma, but present only in small amounts in glomerular filtrate? (p. 782)
  a. sodium ions
  b. water
  c. glucose
  d. protein
18 Explain how the peritubular capillary is adapted for tubular reabsorption. (p. 783)
19 Explain how epithelial cells of the proximal convoluted tubule are adapted for tubular reabsorption. (p. 784)
20 Discuss how tubular reabsorption is selective. (p. 784)
21 Explain why active transport mechanisms have limited transport capacities. (p. 784)
22 Define *renal plasma threshold,* and explain its significance in tubular reabsorption. (p. 784)

23 Explain how amino acids and proteins are reabsorbed. (p. 784)
24 Describe the effect of sodium reabsorption on the reabsorption of negatively charged ions. (p. 785)
25 Explain how sodium reabsorption affects water reabsorption. (p. 785)
26 Explain how the renal tubule is adapted to secrete hydrogen ions. (p. 786)
27 Explain how potassium ions may be passively secreted. (p. 787)
28 The major action of ADH on the kidneys is to increase _____. (p. 787)
  a. water reabsorption by the proximal convoluted tubule
  b. glomerular filtration rate
  c. water reabsorption by the collecting duct
  d. potassium excretion
29 Explain how hypotonic fluid is produced in the ascending limb of the nephron loop. (p. 788)
30 Explain why fluid in the descending limb of the nephron loop is hypertonic. (p. 788)
31 Explain how urine may become concentrated as it moves through the collecting duct. (p. 789)
32 Compare the processes by which urea and uric acid are excreted. (p. 789)
33 List the common constituents of urine and their sources. (p. 790)
34 List some of the factors that affect the daily urine volume. (p. 791)

20.4 Storage and Elimination of Urine
35 Describe the structure and function of a ureter. (p. 791)
36 Explain how the muscular wall of the ureter helps move urine. (pp. 791–792)
37 Describe what happens if a ureter becomes obstructed. (p. 792)
38 Describe the location and structure of the urinary bladder. (pp. 792–793)
39 Define *detrusor muscle*. (p. 793)
40 Distinguish between the internal and external urethral sphincters. (pp. 793 and 795)
41 Describe the micturition reflex. (pp. 795–796)
42 Which movement involves skeletal muscle? (p. 796)
  a. contraction of the internal urethral sphincter
  b. contraction of the external urethral sphincter
  c. ureteral peristalsis
  d. detrusor muscle contraction

20.5 Life-Span Changes
43 Describe changes in the urinary system with age. (pp. 797–798)

INTEGRATIVE ASSESSMENTS/CRITICAL THINKING

Outcomes 14.3, 16.4, 20.2, 20.3

1. Why are people with the nephrotic syndrome, in which plasma proteins are lost into the urine, more susceptible to infection?

Outcomes 15.4, 20.2, 20.3

2. If the blood pressure of a patient in shock as a result of a severe injury decreases greatly, how would you expect the volume of urine to change? Why?

Outcomes 15.4, 20.2, 20.3, 20.4

3. If a patient who has had major abdominal surgery receives intravenous fluids equal to the volume of blood lost during surgery, would you expect the volume of urine produced to be greater than or less than normal? Why?

Outcomes 15.6, 15.7, 17.2, 17.4, 17.5, 17.7, 17.8, 20.2, 20.3, 20.4

4. A physician prescribes oral penicillin therapy for a patient with an infection of the urinary bladder. How would you describe for the patient the route the drug follows to reach the bladder?

Outcomes 16.4, 20.4

5. Inflammation of the urinary bladder is more common in women than in men. What anatomical differences between the female and male urethra explain this observation?

Outcomes 20.2, 20.3

6. What effect would being born with narrowed renal arteries have on the volume and composition of urine?

 # ONLINE STUDY TOOLS

Connect Interactive Questions Reinforce your knowledge using assigned interactive questions covering the anatomy of the kidney and the process of urine formation.

Connect Integrated Activity Can you predict the differences in urine composition associated with different factors such as diet and health status?

LearnSmart Discover which chapter concepts you have mastered and which require more attention. This adaptive learning tool is personalized, proven, and preferred.

Anatomy & Physiology Revealed Go more in depth into the human body by exploring cadaver dissections of the organs of the urinary system, as well as viewing an animation about urine formation.

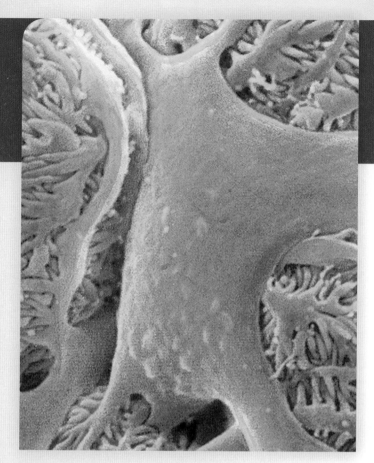

Falsely colored scanning electron micrograph of glomerular podocytes involved in the filtration of water and electrolytes (5,700×).

21

Water, Electrolyte, and Acid-Base Balance

 **LEARNING OUTCOMES**

After you have studied this chapter, you should be able to:

21.1 The Balance Concept
1 Explain the balance concept. (p. 804)
2 Explain the importance of water and electrolyte balance. (p. 804)

21.2 Distribution of Body Fluids
3 Describe how body fluids are distributed in compartments. (p. 804)
4 Explain how fluid composition varies among compartments and how fluids move from one compartment to another. (pp. 805–806)

21.3 Water Balance
5 List the routes by which water enters and leaves the body. (pp. 806–807)
6 Explain the regulation of water input and water output. (pp. 806–808)

21.4 Electrolyte Balance
7 List the routes by which electrolytes enter and leave the body. (p. 809)
8 Explain the regulation of the input and output of electrolytes. (pp. 809–812)

21.5 Acid-Base Balance
9 Explain acid-base balance. (p. 812)
10 Identify how pH number describes the acidity and alkalinity of a body fluid. (p. 812)
11 List the major sources of hydrogen ions in the body. (pp. 812–813)
12 Distinguish between strong acids and weak acids. (p. 813)
13 Explain how chemical buffer systems, the respiratory center, and the kidneys keep the pH of body fluids relatively constant. (pp. 814–816)

21.6 Acid-Base Imbalances
14 Describe the causes and consequences of increase or decrease in body fluid pH. (pp. 817–818)

21.7 Compensation
15 For each of the four main types of acid-base imbalance, explain which system(s)—renal, respiratory, or both—would help to return blood pH toward normal. (p. 819)

THE WHOLE PICTURE

Each of the cells that make up the body has water inside it. In fact, approximately two-thirds of the water in the body is inside cells. Outside the cells is the internal environment of the body, including the interstitial fluid and the plasma of the blood. The synovial fluid of the joints, fluids in the eyes, serous fluids, and others are also on the list of body fluids.

If the internal environment is to remain constant, what happens to the salt, water, and solutes that are ingested at every meal? How does the body compensate for water lost when we sweat, urinate, defecate, and even just exhale? The answer is the concept of *balance,* and it is a simple equation: What goes in must equal what goes out.

aprevealed.com

Module 13: Urinary System

UNDERSTANDING WORDS

de-, separation from: *dehydration*—removal of water from the cells or body fluids.

edem-, swelling: *edema*—swelling due to abnormal accumulation of extracellular fluid.

-emia, a blood condition: *hypoproteinemia*—abnormally low concentration of blood plasma proteins.

extra-, outside: *extracellular fluid*—fluid outside of the body cells.

im- (or **in-**), not: *imbalance*—condition in which factors are not in equilibrium.

intra-, within: *intracellular fluid*—fluid in body cells.

neutr-, neither one nor the other: *neutral*—solution that is neither acidic nor basic.

-osis, a state of: *acidosis*—abnormally high hydrogen ion concentration.

-uria, a urine condition: *ketouria*—ketone bodies in the urine.

21.1 | The Balance Concept

The term *balance* suggests a state of constancy. For water and electrolytes, balance means that the quantities entering the body equal the quantities leaving. Mechanisms that replace lost water and electrolytes and excrete excesses maintain this balance. As a result, the levels of water and electrolytes in the body remain relatively stable at all times, helping to maintain homeostasis.

RECONNECT
To Chapter 1, Homeostasis, page 15.

Water balance and electrolyte balance are interdependent, because electrolytes are dissolved in the water of body fluids. Consequently, anything that alters the concentrations of the electrolytes will alter the concentration of the water by adding solutes to it or by removing solutes from it. Likewise, anything that changes the concentration of the water will change the concentrations of the electrolytes by concentrating or diluting them.

PRACTICE

1 How are fluid balance and electrolyte balance interdependent?

21.2 | Distribution of Body Fluids

Body fluids are not uniformly distributed. Instead, they occupy regions, or *compartments*, of different volumes that contain fluids of varying compositions. The movement of water and electrolytes between these compartments is regulated to stabilize the distribution and the composition of body fluids.

Fluid Compartments

The body of an average adult female is about 52% water by weight and that of an average male is about 63% water by weight. This difference between the sexes is because females generally have more adipose tissue, which has little water. Males generally have more muscle tissue, which contains a great deal of water. Water in the body (about 40 liters), with its dissolved electrolytes, is distributed into two major compartments: an intracellular fluid compartment and an extracellular fluid compartment (fig. 21.1).

The **intracellular** (in"trah-sel'u-lar) **fluid compartment** includes all the water and electrolytes that cell membranes enclose.

In other words, intracellular fluid is the fluid inside cells. In an adult, it accounts for about 63% by volume of total body water.

The **extracellular** (ek"strah-sel'u-lar) **fluid compartment** includes all the fluid outside cells—in tissue spaces (interstitial fluid), blood vessels (plasma), and lymphatic vessels (lymph). Epithelial layers separate a specialized fraction of the extracellular fluid from other extracellular fluids. This *transcellular* (trans-sel'ular) *fluid* includes cerebrospinal fluid of the central nervous system, aqueous and vitreous humors of the eyes, synovial fluid of the joints, and serous fluid in the body cavities. The fluids of the extracellular compartment constitute about 37% by volume of the total body water (fig. 21.2).

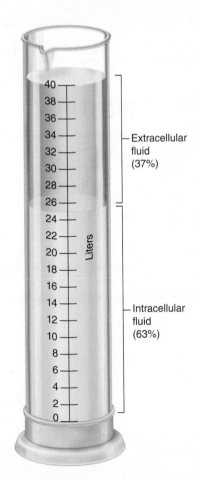

FIGURE 21.1 Of the 40 liters of water in the body of an average adult, about two-thirds is intracellular and about one-third is extracellular.

Components of total body water

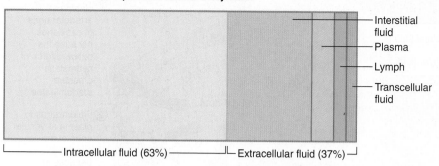

FIGURE 21.2 Approximately two-thirds of the water in the body is inside cells.

- Interstitial fluid
- Plasma
- Lymph
- Transcellular fluid

Intracellular fluid (63%) — Extracellular fluid (37%)

Body Fluid Composition

All of the body fluids are solutions of electrolytes in water. They also include dissolved gases, proteins, and other substances to varying degrees (see chapter 14, p. 529) *Extracellular fluids* generally are similar in composition, including high concentrations of sodium, chloride, calcium, and bicarbonate ions and lesser concentrations of potassium, magnesium, phosphate, and sulfate ions. The blood plasma portion of extracellular fluid has considerably more protein than do either interstitial fluid or lymph.

Intracellular fluid has high concentrations of potassium, phosphate, and magnesium ions. It includes a greater concentration of sulfate ions and lesser concentrations of sodium, chloride, and bicarbonate ions than does extracellular fluid. Intracellular fluid also has a greater concentration of protein than plasma. Figure 21.3 shows these relative concentrations.

PRACTICE

2 Describe the normal distribution of water in the body.

3 Which electrolytes are in higher concentrations in extracellular fluids? In intracellular fluid?

4 How does the concentration of protein vary in body fluids?

Movement of Fluid Between Compartments

Two major factors regulate the movement of water and electrolytes from one fluid compartment to another: hydrostatic pressure and osmotic pressure. For example, as explained in chapter 15 (p. 581), fluid leaves the plasma at the arteriolar ends of capillaries and enters the interstitial spaces because of the net outward force of *hydrostatic pressure* (blood pressure). Fluid returns to the plasma from the interstitial spaces at the venular ends of capillaries because of the net inward force of *colloid osmotic pressure* due to the plasma proteins. Likewise, as mentioned in chapter 16 (p. 620), fluid leaves the interstitial spaces and enters the lymph capillaries due to the hydrostatic pressure of the interstitial fluid. The circulation of lymph returns interstitial fluid to the plasma.

Hydrostatic pressure in the cells and surrounding interstitial fluid is ordinarily equal and remains stable. Therefore, any net fluid movement is likely to result from changes in osmotic pressure (fig. 21.4). Recall that osmotic pressure is due to impermeant solutes on one side of a cell membrane. Sodium (extracellular) and potassium (intra-

cellular) ions function as impermeant solutes and create an osmotic pressure because of the Na^+/K^+ pump. For example, because most cell membranes in the body are freely permeable to water, a decrease in extracellular sodium ion concentration causes a net movement of water from the extracellular compartment into the intracellular compartment by osmosis. The cell swells. Conversely, if the extracellular sodium ion concentration increases, cells shrink as they lose water by osmosis. Although the solute composition of body fluids

CAREER CORNER
Medical Technologist

A medical technologist is the behind-the-scenes professional who tests cells and tissues from patients, enabling physicians to diagnose disease. The tests fall into five areas: immunology, microbiology, chemistry, blood banking, and hematology.

Microbiology skills are used to culture bacteria, fungi, and parasites in body fluids. The polymerase chain reaction amplifies the genetic material of viruses. Flow cytometry can reveal the abnormal proportions of blood cells that underlie an anemia, clotting disorder, mononucleosis, or cancer. Other tests can confirm a pregnancy, detect illicit drug use, or measure levels of substances in body fluids such as cholesterol, glucose, or hormones.

In addition to deep knowledge of microscopic and macroscopic anatomy and physiology, a medical technologist must be comfortable and adept at working with a great variety of instruments and devices, be able to focus on details, be able to stand for long periods, and perhaps most important, be a problem solver. Excellent communication skills and the ability to recognize errors and work efficiently under stress are also important skills.

Medical technologists work in many places, including clinics, hospitals, commercial and government testing laboratories, public health facilities, biotechnology and pharmaceutical companies, *in vitro* fertilization facilities, and research laboratories.

The educational requirement is a bachelor's degree in medical technology approved by the National Accrediting Agency for Clinical Laboratory Sciences or in a related field such as biology or clinical laboratory science plus medical technology training. The Board of Registry of the American Society for Clinical Pathology offers national certification after passing an exam. Licensing requirements vary by state.

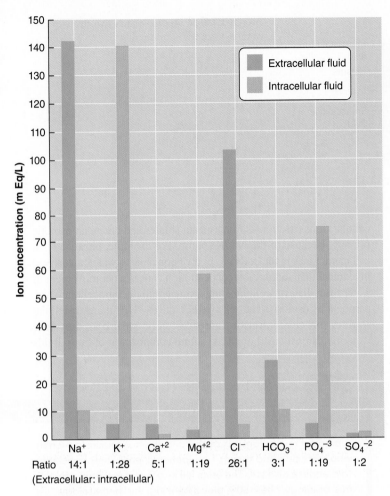

Ratio   14:1   1:28   5:1   1:19   26:1   3:1   1:19   1:2
(Extracellular: intracellular)

FIGURE 21.3 Extracellular fluids have relatively high concentrations of sodium (Na^+), calcium (Ca^{+2}), chloride (Cl^-), and bicarbonate (HCO_3^-) ions. Intracellular fluid has relatively high concentrations of potassium (K^+), magnesium (Mg^{+2}), phosphate (PO_4^{-3}), and sulfate (SO_4^{-2}) ions.

 According to this graph, which cation is most abundant in the extracellular fluid?
Answer can be found in Appendix G.

varies between intracellular and extracellular compartments, water will "follow salt" and distribute by osmosis such that the water concentration (and total solute concentration) is essentially equal inside and outside cells.

Different substances distribute to different compartments. Sodium is largely kept out of cells by sodium pumps in cell membranes. Because water follows trapped solute by osmosis, a liter of isotonic saline (9 grams of NaCl per 1,000 mL of solution) distributes throughout the extracellular space. Glucose, on the other hand, is metabolized by cells. A 5% glucose solution (5 grams of glucose per 100 mL of solution) can be infused slowly intravenously without hemolyzing red blood cells, but as the glucose is metabolized, eventually only the water remains, and water distributes throughout both the intracellular and extracellular compartments.

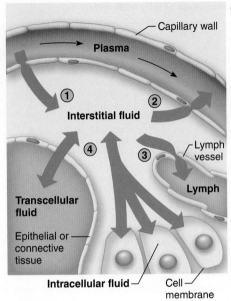

FIGURE 21.4 Net movements of fluid between compartments result from differences in hydrostatic and osmotic pressures.

① Fluid leaves plasma at arteriolar ends of capillaries because the outward force of hydrostatic pressure predominates

② Fluid returns to plasma at venular ends of the capillaries because the inward force of colloid osmotic pressure predominates

③ Hydrostatic pressure within interstitial spaces forces fluid into lymph capillaries

④ Interstitial fluid is in equilibrium with transcellular and intracellular fluids

PRACTICE

5 Which factors control the movement of water and electrolytes from one fluid compartment to another?

6 How does the sodium ion concentration in body fluids affect the net movement of water between the compartments?

21.3 | Water Balance

Water balance exists when water intake (and production by metabolism) equals water output. Homeostasis requires control of both water intake and water output. Ultimately, maintenance of the internal environment depends on thirst centers in the brain to vary water intake and on the kidneys' ability to vary water output.

Water Intake

The volume of water gained each day varies among individuals. An average adult living in a moderate environment takes in about 2,500 milliliters daily. Probably 60% is obtained from drinking water or beverages, and another 30% comes from moist foods. The remaining 10% is a by-product of the oxidative metabolism of nutrients, called **water of metabolism** (fig. 21.5*a*).

Regulation of Water Intake

The primary regulator of water intake is thirst. The intense feeling of thirst derives from a change in either the volume or the osmotic pressure of extracellular fluids and both neural and hormonal input to the brain.

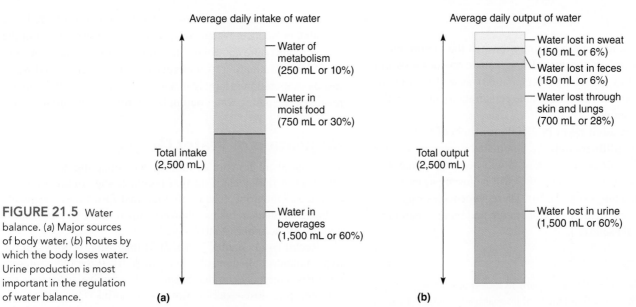

Average daily intake of water

- Water of metabolism (250 mL or 10%)
- Water in moist food (750 mL or 30%)

Total intake (2,500 mL)

- Water in beverages (1,500 mL or 60%)

(a)

Average daily output of water

- Water lost in sweat (150 mL or 6%)
- Water lost in feces (150 mL or 6%)
- Water lost through skin and lungs (700 mL or 28%)

Total output (2,500 mL)

- Water lost in urine (1,500 mL or 60%)

(b)

FIGURE 21.5 Water balance. (a) Major sources of body water. (b) Routes by which the body loses water. Urine production is most important in the regulation of water balance.

Recall from chapter 2 (pp. 64–65) that water molecules are polar. Molecules that have polar regions, such as carbohydrates and proteins, dissolve in water but remain intact. In contrast, electrolytes dissociate in water to release ions. Intact or dissociated, all of these dissolved particles are solutes.

The total solute concentration of a body fluid determines its *osmolarity*. One molecule of glucose yields one dissolved particle, and one molecule of sodium chloride yields two, a sodium ion and a chloride ion. The total number of dissolved particles determines the osmolarity of body solutions, irrespective of the source, so the term *osmoles* is used. Thus, one mole of glucose yields one osmole of dissolved particles, and one mole of sodium chloride yields two osmoles. The total number of osmoles per liter gives the osmolarity of the solution.

> The units for osmolarity are milliosmoles per liter of solution (mOsm/L). Osmolality is a closely related term that is measured clinically, with units of milliosmoles per kilogram of water (mOsm/ kg H_2O). In the body fluids the two are approximately equivalent.

Osmolarity and osmotic pressure sound alike but are significantly different. Osmolarity (and osmolality) are measures of the total solute concentration of a solution. Osmotic pressure is due only to the solutes in a solution that are impermeant, unable to cross the cell membrane, which cause osmosis. The osmoreceptors are sensitive to the osmotic pressure of the body fluids.

As the body loses water, the osmotic pressure of the extracellular fluids increases. Such a change stimulates *osmoreceptors* (oz″mo-re-sep′torz) in the hypothalamus, and in response, the person feels thirsty and seeks water. A thirsty person usually has a dry mouth, caused by loss of extracellular water and the resulting decreased flow of saliva.

The thirst mechanism is normally triggered whenever the total body water decreases by as little as 1%. The act of drinking

and the resulting distension of the stomach wall trigger impulses that inhibit the thirst mechanism. In this way, drinking stops long before the swallowed water is absorbed. This inhibition helps prevent the person from drinking more than is required to replace the volume lost, avoiding development of an imbalance. Table 21.1 summarizes this mechanism.

Another mechanism of stimulating thirst responds to volume loss that occurs as a result of hemorrhage. Stretch receptors in certain parts of the cardiovascular system (volume receptors) stimulate the thirst center directly. In addition, if blood pressure drops sufficiently, the renin-angiotensin system may be activated, and one action of angiotensin II is to stimulate thirst. These mechanisms are not as sensitive as the response to changing osmolality, and they do not act until blood volume and blood pressure decrease by 10% or more.

 PRACTICE

7 What is water balance?

8 Where is the thirst center?

9 What stimulates fluid intake? What inhibits it?

TABLE 21.1	Regulation of Water Intake
1.	The body loses as little as 1% of its water.
2.	An increase in the osmotic pressure of extracellular fluid due to water loss stimulates osmoreceptors in the hypothalamus.
3.	Activity in the hypothalamus causes the person to feel thirsty and to seek water.
4.	Drinking and the resulting distension of the stomach by water stimulate impulses that inhibit the thirst center.
5.	Water is absorbed through the walls of the stomach and small intestine.
6.	The osmotic pressure of extracellular fluid returns to normal.

Water Output

Water normally enters the body only through the mouth, but it can be lost by a variety of routes. These include obvious losses in urine, feces, and sweat (sensible perspiration), as well as evaporation of water from the skin (insensible perspiration) and from the lungs during breathing.

If an average adult takes in 2,500 milliliters of water each day, then 2,500 milliliters must be eliminated to maintain water balance. Of this volume, perhaps 60% will be lost in urine, 6% in feces, and 6% in sweat. About 28% will be lost by evaporation from the skin and lungs (fig. 21.5b). These percentages vary with such environmental factors as temperature and relative humidity and with physical exercise.

Heatstroke is a result of exposure to extreme environmental heat combined with a failure of normal body temperature control that can be quickly fatal. It occurs when the body is exposed to a heat index (heat considering humidity) of more than 105°F and body temperature reaches 104°F. Under these conditions, evaporation of sweat becomes less effective at cooling the body, and organs begin to fail.

The symptoms of heatstroke happen in a sequence. First come headache, dizziness, and exhaustion. Sweating is profuse, then stops, as the skin becomes dry, hot, and red. Respiratory rate rises and the pulse may race up to 180 beats per minute. If the person isn't cooled, neurological symptoms may begin, including disorientation, hallucinations, and odd behavior. Kidney failure and/or heart arrhythmia can prove fatal.

During heat waves, the very young and the very old are more susceptible to heatstroke because their body temperature control mechanisms may be poor. However, heatstroke also affects two groups of young, otherwise healthy individuals—athletes who work out in extreme heat and soldiers deployed to hot climates. For this reason, soldiers must carry drinking water with them at all times and drink throughout the day, whether they feel thirsty or not.

If water intake is insufficient, water output must be reduced to maintain balance. Water lost in sweat is a necessary part of the body's temperature control mechanism; water lost in feces accompanies the elimination of undigested food materials; and water lost by evaporation is largely unavoidable. Therefore, the primary means of regulating water output is control of urine production.

Regulation of Water Output

The renal distal convoluted tubules and collecting ducts are the effectors of the mechanism that regulates the volume of water excreted in the urine. The epithelium that forms these segments of the renal tubule remains relatively impermeable to water unless antidiuretic hormone (ADH) is present.

Recall from chapter 13 (p. 503) that osmoreceptors in the hypothalamus help control release of ADH. If the blood plasma becomes more concentrated because of excessive water loss, the osmoreceptors lose water by osmosis and shrink. This change triggers impulses that signal the posterior pituitary gland to release ADH into the bloodstream. When ADH reaches the kidneys, it increases the permeability of the distal convoluted tubules and collecting ducts to water, increasing water reabsorption, which conserves water. This action resists further osmotic change in the plasma. The *osmoreceptor-ADH mechanism* can reduce a normal urine production of 1,500 milliliters per day to about 500 milliliters per day when the body is dehydrated.

If a person drinks too much water, the plasma becomes less concentrated, and the osmoreceptors swell as they gain extra water by osmosis. In this instance, ADH release is inhibited, and the distal convoluted tubules and collecting ducts remain impermeable to water. Consequently, less water is reabsorbed and more urine is produced. Table 21.2 summarizes this mechanism. Clinical Application 21.1 discusses disorders resulting from water imbalance.

Just as in the case of stimulating thirst, volume receptors also control ADH. Too much volume will inhibit and too little volume will stimulate ADH secretion. In addition, if blood pressure drops sufficiently due to the loss of volume, the renin-angiotensin system may be activated, and one action of angiotensin II is to stimulate ADH secretion.

TABLE 21.2   Events in Regulation of Water Output AP|R

Dehydration	Excess Water Intake
1. Extracellular fluid becomes osmotically more concentrated.	1. Extracellular fluid becomes osmotically less concentrated.
2. Osmoreceptors in the hypothalamus are stimulated by the increase in the osmotic pressure of body fluids.	2. This change stimulates osmoreceptors in the hypothalamus.
3. The hypothalamus signals the posterior pituitary gland to release ADH into the blood.	3. The posterior pituitary gland decreases ADH release.
4. Blood carries ADH to the kidneys.	4. Renal tubules decrease water reabsorption.
5. ADH causes the distal convoluted tubules and collecting ducts to increase water reabsorption.	5. Urine output increases, and excess water is excreted.
6. Urine output decreases, and further water loss is minimized.	

PRACTICE

10 By what routes does the body lose water?

11 What is the primary regulator of water loss?

12 What types of water loss are unavoidable?

13 How does the hypothalamus regulate water balance?

21.4 | Electrolyte Balance

An **electrolyte balance** (e-lek′tro-līt bal′ans) exists when the quantities of electrolytes (molecules that release ions in water) the body gains equal those lost (fig. 21.6).

Electrolyte Intake

The electrolytes of greatest importance to cellular functions release sodium, potassium, calcium, magnesium, chloride, sulfate, phosphate, bicarbonate, and hydrogen ions. These electrolytes are primarily obtained from foods, but they may also be found in drinking water and other beverages. In addition, some electrolytes are byproducts of metabolic reactions.

Regulation of Electrolyte Intake

Ordinarily, a person obtains sufficient electrolytes by responding to hunger and thirst. However, a severe electrolyte deficiency may cause *salt craving,* which is a strong desire to eat salty foods.

Electrolyte Output

The body loses some electrolytes by perspiring (sweat has about half the solute concentration of plasma). The quantities of electrolytes leaving vary with the amount of perspiration. More electrolytes are lost in sweat on warmer days and during strenuous exercise. Varying amounts of electrolytes are lost in the feces. The greatest electrolyte output occurs as a result of kidney function and urine production. The kidneys alter renal electrolyte losses to maintain the proper composition of body fluids, thereby promoting homeostasis.

 PRACTICE

14 Which electrolytes are most important to cellular functions?

15 Which mechanisms ordinarily regulate electrolyte intake?

16 By what routes does the body lose electrolytes?

Regulation of Electrolyte Output

The concentrations of positively charged ions (cations), such as sodium (Na^+), potassium (K^+), and calcium (Ca^{+2}), are particularly important. Certain concentrations of these ions are vital for impulse conduction along an axon, muscle fiber contraction, and maintenance of cell membrane permeability. Potassium is especially important in maintaining the resting potential of nerve and cardiac muscle cells, and abnormal potassium levels may disrupt the functioning of these cells.

Sodium ions account for nearly 90% of the positively charged ions in extracellular fluids. The kidneys and the hormone aldosterone primarily regulate these ions. Aldosterone, which the adrenal cortex secretes, increases sodium ion reabsorption in the distal convoluted tubules and collecting ducts of the nephrons. A decrease in sodium ion concentration in the extracellular fluid stimulates aldosterone secretion via the renin-angiotensin system, as described in chapter 20 on page 782 (see fig. 20.18, p. 783).

Aldosterone also regulates *potassium ions.* An important stimulus for aldosterone secretion is a rising potassium ion concentration, which directly stimulates cells of the adrenal cortex. This hormone enhances the renal tubular secretion of potassium ions at the same time that it stimulates renal tubular reabsorption of sodium ions (fig. 21.7).

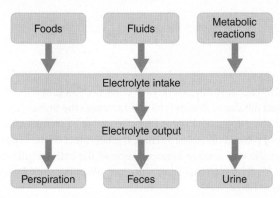

FIGURE 21.6 Electrolyte balance exists when the intake of electrolytes from all sources equals the output of electrolytes.

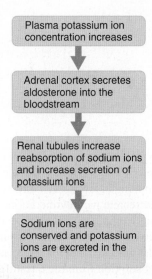

FIGURE 21.7 If the potassium ion concentration increases, the kidneys conserve sodium ions and excrete potassium ions.

Among the more common disorders involving an imbalance in the water of body fluids are dehydration, water intoxication, and edema.

Dehydration

Dehydration is a deficiency condition that occurs when output of water exceeds intake. It is a great problem for athletes, military personnel, and certain industrial workers. Dehydration may develop following excessive sweating or as a result of prolonged water deprivation accompanied by continued water output. In either case, as water is lost, the extracellular fluid becomes more concentrated, and water leaves cells by osmosis (fig. 21A). Dehydration may also accompany illnesses in which prolonged vomiting or diarrhea depletes body fluids, such as the bacterial infection cholera.

During dehydration, the skin and mucous membranes of the mouth feel dry, and body weight drops. Severe hyperthermia may develop as the body temperature regulating mechanism falters due to lack of water for sweat. In severe dehydration, as waste products accumulate in the extracellular fluid, symptoms of cerebral disturbances, including mental confusion, delirium, and coma, may develop. Dehydration is responsible for some of the symptoms of heatstroke.

Infants are more likely to become dehydrated because their kidneys are less efficient at conserving water than those of adults. Elderly people are also especially susceptible to developing water imbalances because the sensitivity of their thirst mechanisms decreases with age, and physical disabilities may make it difficult to obtain adequate fluids.

The treatment for dehydration is to replace the lost water and electrolytes. If only water is replaced, the extracellular fluid may become more dilute than normal, causing cells to swell (fig. 21B). This may produce a condition called water intoxication.

Water Intoxication

Until recently, runners were advised to drink as much fluid as they could while running, particularly in long events. The death of a young woman in a marathon from low blood sodium (*hyponatremia*, "water intoxication") due to excessive fluid intake inspired further study and a reevaluation of this advice. Researchers from Harvard Medical School studied 488 runners from the race, and found that 13% of them developed hyponatremia. The tendency to develop the condition was associated with longer race time, high or low body mass index, and significant weight gain during the race. Drinking sports drinks instead of water does not make a difference because these beverages are mostly water.

In recognition of the possibility of hyponatremia, USA Track and Field, the national governing body for the sport, offers on their website instructions for runners to determine exactly how much to consume during a one-hour training run. The goal is to replace exactly what is lost.

Edema

Edema is an abnormal accumulation of extracellular fluid in the interstitial spaces (fig. 21B). Causes include a decrease in the plasma protein concentration (*hypoproteinemia*), obstructions in lymphatic vessels, increased venous pressure, and increased capillary permeability.

Hypoproteinemia may result from failure of the liver to synthesize plasma proteins; kidney disease (glomerulonephritis) that damages glomerular capillaries, allowing proteins to enter the

FIGURE 21A AP|R If excess extracellular fluids are lost, cells dehydrate by osmosis.

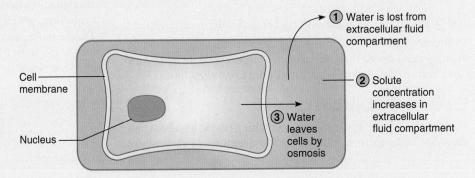

Cell membrane

Nucleus

① Water is lost from extracellular fluid compartment

② Solute concentration increases in extracellular fluid compartment

③ Water leaves cells by osmosis

Recall from chapter 13 (pp. 507–508) that the plasma calcium ion concentration dropping below normal directly stimulates the parathyroid glands to secrete parathyroid hormone. Parathyroid hormone increases activity in bone-resorbing cells (osteoclasts), which increases the concentrations of both calcium and phosphate ions in the extracellular fluids. Parathyroid hormone also indirectly stimulates calcium absorption from the intestine. Concurrently, this hormone causes the kidneys to conserve calcium ions (through increased tubular reabsorption) and increases the urinary excretion of phosphate ions. The increased phosphate excretion offsets the increased plasma phosphate. Thus, the net effect of the hormone is to return the *calcium ion* concentration of the extracellular fluid to normal levels but to maintain a normal *phosphate ion* concentration (fig. 21.8).

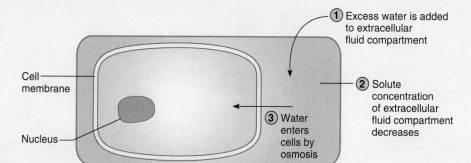

① Excess water is added to extracellular fluid compartment

② Solute concentration of extracellular fluid compartment decreases

③ Water enters cells by osmosis

Cell membrane

Nucleus

FIGURE 21B If excess water is added to the extracellular fluid compartment, cells gain water by osmosis.

urine; or starvation, in which amino acid intake is insufficient to support synthesis of plasma proteins. In each of these instances, the plasma protein concentration is decreased, which decreases plasma colloid osmotic pressure, reducing the normal return of tissue fluid into the venular ends of capillaries. Consequently, tissue fluid accumulates in the interstitial spaces.

Lymphatic obstructions may result from surgery or from parasitic infections of lymphatic vessels, as chapter 16 (p. 621) discussed. Back pressure develops in the lymphatic vessels, which interferes with the normal movement of tissue fluid into them. At the same time, proteins that the lymphatic circulation ordinarily removes accumulate in the interstitial spaces, raising osmotic pressure of the interstitial fluid. This effect attracts still more fluid into the interstitial spaces.

If the outflow of blood from the liver into the inferior vena cava is blocked, the venous pressure in the liver and portal vein greatly increases. This, in turn, raises pressure in liver sinusoids and intestinal capillaries, increasing filtration. Fluid is filtered into the peritoneal cavity. This condition, called *ascites*, distends the abdomen. It is painful.

Edema may also result from increased capillary permeability accompanying *inflammation*. Recall that inflammation is a response to tissue damage and usually releases chemicals such as histamine from damaged cells. Histamine causes vasodilation and increased capillary permeability, so fluid leaks out of the capillary and enters the interstitial spaces. Table 21A summarizes the factors that result in edema.

TABLE 21A	Factors Associated with Edema	
Factor	**Cause**	**Effect**
Low plasma protein concentration	Liver disease and failure to synthesize proteins; kidney disease and loss of proteins in urine; lack of proteins in diet due to starvation	Plasma osmotic pressure decreases; less fluid reabsorbed at venular ends of capillaries by osmosis
Obstruction of lymph vessels	Surgical removal of portions of lymphatic pathways; certain parasitic infections	Back pressure in lymph vessels interferes with movement of fluid from interstitial spaces into lymph capillaries
Increased venous pressure	Venous obstructions or faulty venous valves	Back pressure in veins increases capillary filtration and interferes with return of fluid from interstitial spaces into venular ends of capillaries
Inflammation	Tissue damage	Vasodilation and increased capillary permeability lead to increased filtration

Abnormal increases in blood calcium (hypercalcemia) can result from hyperparathyroidism, in which excess secretion of PTH increases bone resorption. Hypercalcemia may also be caused by cancers, particularly those originating in the bone marrow, breasts, lungs, or prostate gland. Usually the increase in calcium occurs when cancer causes bone tissue to release calcium ions. Symptoms of cancer-induced hypercalcemia include weakness and fatigue, impaired mental function, headache, nausea, increased urine volume (polyuria), and increased thirst (polydipsia).

Abnormal decreases in blood calcium (hypocalcemia) may result from reduced availability of PTH following removal of the parathyroid glands, or from vitamin D deficiency. It may also result from decreased calcium absorption following gastrointestinal surgery or excess calcium excretion due to kidney disease. Hypocalcemia may cause muscle spasms in the airways or cardiac arrhythmias. Administering calcium salts and high doses of vitamin D to promote calcium absorption may correct this condition.

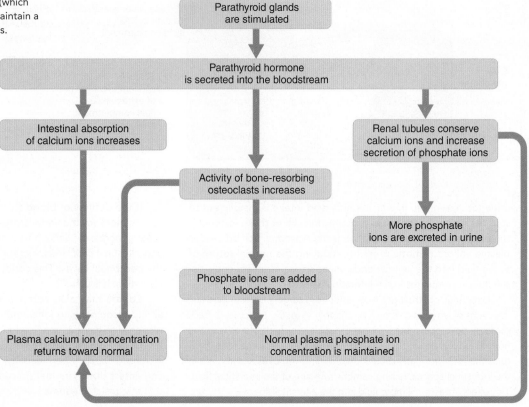

FIGURE 21.8 If the plasma concentration of calcium ions decreases, parathyroid hormone works to increase the calcium ion concentration back to normal. Increased urinary phosphate excretion offsets bone resorption of phosphates (which increases blood phosphate levels) to maintain a normal concentration of phosphate ions.

Generally, the regulatory mechanisms that control positively charged ions secondarily control the concentrations of negatively charged ions (anions). For example, chloride ions (Cl⁻), the most abundant negatively charged ions in the extracellular fluids, are passively reabsorbed from the renal tubules in response to the active reabsorption of sodium ions. That is, the negatively charged chloride ions are electrically attracted to the positively charged sodium ions and accompany them as they are reabsorbed.

Some negatively charged ions, such as phosphate ions (PO_4^{-3}) and sulfate ions (SO_4^{-2}), also are partially regulated by active transport mechanisms that have limited transport capacities. For example, if the extracellular phosphate ion concentration is low, the phosphate ions in the renal tubules are conserved. On the other hand, if the renal plasma threshold is exceeded, excess phosphate will be excreted in the urine. Clinical Application 21.2 discusses symptoms associated with sodium and potassium imbalances.

 PRACTICE

17 How does aldosterone regulate sodium and potassium ion concentration?

18 How is calcium regulated?

19 Which mechanism regulates the concentrations of most negatively charged ions?

21.5 | Acid-Base Balance

Electrolytes that release hydrogen ions are **acids,** as chapter 2 discussed (p. 66). Substances that release ions that combine with hydrogen ions are **bases.** Acid-base balance entails regulation of the hydrogen ion concentration of body fluids. This is important because slight changes in hydrogen ion concentrations can alter the rates of enzyme-controlled metabolic reactions, shift the distribution of other ions, or modify hormone actions. Recall from chapter 2 (pp. 67–68) that pH number indicates the degree to which a solution is acidic or basic (alkaline). The more acid the solution, the lower its pH, and vice versa. The internal environment is normally maintained between pH 7.35 and 7.45.

Sources of Hydrogen Ions

Most of the hydrogen ions in body fluids originate as by-products of metabolic processes, although the digestive tract may directly absorb some hydrogen ions. The major metabolic sources of hydrogen ions include the following. (These are reversible reactions but, for clarity, are presented as the net reaction only. Remember, it is the concentration of H⁺ at equilibrium that determines the pH.)

1. **Aerobic respiration of glucose.** This process produces carbon dioxide and water. Carbon dioxide diffuses out of the

Sodium is the most abundant cation in extracellular fluid, and potassium is the most abundant in intracellular fluid. The renal regulation of sodium is closely related to that of potassium because active reabsorption of sodium (under the influence of aldosterone) is accompanied by secretion (and excretion) of potassium. Thus, it is not surprising that conditions that alter sodium ion balance also affect potassium ion balance. Such disorders can be summarized as follows:

1. *Low blood sodium concentration (hyponatremia)* Possible causes of hyponatremia include prolonged sweating, vomiting, or diarrhea; renal disease in which sodium is inadequately reabsorbed; adrenal cortex disorders in which aldosterone secretion is insufficient to promote the reabsorption of sodium (Addison disease); and drinking too much water.

    Possible effects of hyponatremia include the development of extracellular fluid that is hypotonic and promotes the movement of water into the cells by osmosis. This is accompanied by the symptoms of water intoxication described in Clinical Application 21.1 on page 810.

2. *High blood sodium concentration (hypernatremia)* Possible causes of hypernatremia include excessive water loss by evaporation and diffusion, as may occur during high fever, or increased water loss accompanying diabetes insipidus (not to be confused with diabetes mellitus), in one form of which ADH secretion is insufficient to maintain water conservation by the renal tubules and collecting ducts. Possible effects of hypernatremia include disturbances of the central nervous system, such as confusion, stupor, and coma.

3. *Low blood potassium concentration (hypokalemia)* Possible causes of hypokalemia include use of diuretic drugs that promote potassium excretion; kidney disease; and a drop in extracellular hydrogen ion concentration. Possible effects of hypokalemia include muscular weakness or paralysis, respiratory difficulty, and severe cardiac disturbances, such as atrial or ventricular arrhythmias.

4. *High blood potassium concentration (hyperkalemia)* Possible causes of hyperkalemia include renal disease, which decreases potassium excretion; use of drugs that promote renal conservation of potassium; insufficient secretion of aldosterone by the adrenal cortex (associated with Addison disease); or a shift of potassium from the intracellular fluid to the extracellular fluid, a change that accompanies an increase in plasma hydrogen ion concentration (acidosis) as hydrogen ions begin to move into cells. Possible effects of hyperkalemia include paralysis of the skeletal muscles and severe cardiac disturbances, such as cardiac arrest.

cells and reacts with water in the extracellular fluids to form *carbonic acid:*

$$CO_2 + H_2O \rightarrow H_2CO_3$$

The resulting carbonic acid then ionizes to release hydrogen ions and bicarbonate ions:

$$H_2CO_3 \rightarrow H^+ + HCO_3^-$$

2. **Anaerobic respiration of glucose.** Glucose metabolized anaerobically produces *lactic acid,* which adds hydrogen ions to body fluids.

3. **Incomplete oxidation of fatty acids.** The incomplete oxidation of fatty acids produces *acidic ketone bodies,* which increase hydrogen ion concentration.

4. **Oxidation of amino acids containing sulfur.** The oxidation of sulfur-containing amino acids yields *sulfuric acid* (H_2SO_4), which ionizes to release hydrogen ions.

5. **Hydrolysis of phosphoproteins and nucleic acids.** Phosphoproteins and nucleic acids contain phosphorus. Their oxidation produces *phosphoric acid* (H_3PO_4), which ionizes to release hydrogen ions.

The acids resulting from metabolism vary in strength (fig. 21.9). Thus, their effects on the hydrogen ion concentration of body fluids vary.

Strengths of Acids and Bases

Acids that ionize more completely (release more H⁺) are strong acids, and those that ionize less completely are weak acids. For example, the hydrochloric acid (HCl) of gastric juice is a strong acid and dissociates completely to release a lot of H⁺, but the carbonic acid (H_2CO_3) produced when carbon dioxide reacts with water is weak and dissociates less completely to release less H⁺.

Bases release ions, such as hydroxide ions (OH⁻), which can combine with hydrogen ions, thereby lowering their concentration. Sodium hydroxide (NaOH), which releases hydroxide ions, and sodium bicarbonate ($NaHCO_3$), which releases bicarbonate ions (HCO_3^-), are bases. Strong bases dissociate to release more OH⁻ or its equivalent than do weak bases. Often, the negative ions themselves are called bases. For example, HCO_3^- acting as a base combines with H⁺ from the strong acid HCl to form the weak acid carbonic acid (H_2CO_3).

Regulation of Hydrogen Ion Concentration

Either an acid shift or an alkaline (basic) shift in the body fluids could threaten the internal environment. However, normal metabolic reactions generally produce more acid than base. These reactions include cellular metabolism of glucose, fatty acids, and amino acids. Consequently, the maintenance of acid-base balance

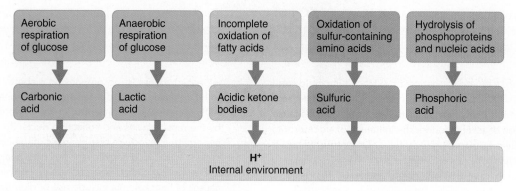

FIGURE 21.9 Some of the metabolic processes that provide hydrogen ions.

usually eliminates acid. This is accomplished in three ways: acid-base buffer systems; respiratory excretion of carbon dioxide; and renal excretion of hydrogen ions.

 PRACTICE

20 Explain why the regulation of hydrogen ion concentration is so important.

21 What are the major sources of hydrogen ions in the body?

Chemical Buffer Systems

Chemical buffer systems are in all body fluids and are based on chemicals that combine with excess acids or bases. Buffers are substances that stabilize the pH of a solution, despite the addition of an acid or a base. More specifically, the chemical components of a buffer system can combine with strong acids to convert them into weak acids. Likewise, these buffers can combine with strong bases to convert them into weak bases. Such reactions help minimize pH changes in the body fluids. The three most important buffer systems in body fluids are the bicarbonate buffer system, the phosphate buffer system, and the protein buffer system.

In the following discussion, associated anions and cations have been omitted for clarity. For example, the weak base sodium bicarbonate ($NaHCO_3$) is represented by bicarbonate (HCO_3^-). Sodium is also the cation associated with the phosphate ions.

1. **Bicarbonate buffer system.** In the bicarbonate buffer system, present in both intracellular and extracellular fluids, the bicarbonate ion (HCO_3^-) acts as a weak base, and carbonic acid (H_2CO_3) acts as a weak acid. In the presence of excess hydrogen ions, bicarbonate ions combine with hydrogen ions to form carbonic acid, minimizing any increase in the hydrogen ion concentration of body fluids:

$$H^+ + HCO_3^- \rightarrow H_2CO_3$$

On the other hand, if conditions are basic or alkaline, carbonic acid dissociates to release bicarbonate ion and hydrogen ion:

$$H_2CO_3 \rightarrow H^+ + HCO_3^-$$

Although this reaction releases bicarbonate ion, it is the increase of free hydrogen ions at equilibrium that is important in minimizing the shift toward a more alkaline pH.

2. **Phosphate buffer system.** The phosphate buffer system is also present in both intracellular and extracellular fluids. However, it is particularly important in the control of hydrogen ion concentration in the intracellular fluid and in renal tubular fluid and urine. This buffer system consists of two phosphate ions, dihydrogen phosphate ($H_2PO_4^-$) and monohydrogen phosphate (HPO_4^{-2}).

In the presence of excess hydrogen ions, monohydrogen phosphate ions act as a weak base, combining with hydrogen ions to form dihydrogen phosphate, minimizing increase in the hydrogen ion concentration of body fluids.

$$H^+ + HPO_4^{-2} \rightarrow H_2PO_4^-$$

On the other hand, if conditions are basic or alkaline, dihydrogen phosphate, acting as a weak acid, dissociates to release hydrogen ion:

$$H_2PO_4^- \rightarrow H^+ + HPO_4^{-2}$$

3. **Protein buffer system.** The protein acid-base buffer system consists of the plasma proteins, such as albumins, and certain proteins in cells, including hemoglobin in red blood cells.

Recall from chapter 2 (p. 73) that proteins are chains of amino acids. Some of these amino acids have freely exposed groups of atoms, called carboxyl groups. If the H^+ concentration drops, a carboxyl group (—COOH) can become ionized, releasing a hydrogen ion, thus resisting the pH change:

$$—COOH \leftrightarrow —COO^- + H^+$$

This is a reversible reaction. (However, the degree to which it is reversible depends on the particular amino acids.) In the presence of excess hydrogen ions, the —COO⁻ parts of the protein molecules accept hydrogen ions and become —COOH groups again. This action decreases the number of free hydrogen ions in the body fluids and again minimizes the pH change.

Some of the amino acids of a protein molecule also have freely exposed amino groups (—NH₂). If the H^+ concentration rises, these amino groups can accept hydrogen ions in another reversible reaction (Once again, the degree to which it is reversible depends on the particular amino acids.):

$$—NH_2 + H^+ \leftrightarrow —NH_3^+$$

In the presence of excess hydroxyl ions (OH⁻), the —NH₃⁺ groups of protein molecules give up hydrogen ions and become —NH₂ groups again. These hydrogen ions then combine with hydroxyl ions to form water molecules. Once again, pH change is minimized. The "R"-groups of certain amino acids (histidine and cysteine) can also function as buffers. Thus, protein molecules can function as acids by releasing hydrogen ions under alkaline conditions or as bases by accepting hydrogen ions under acid conditions. This special property allows protein molecules to operate as an acid-base buffer system.

Hemoglobin is an especially important protein that buffers hydrogen ions. Recall from chapter 19 (p. 759) that carbon dioxide, produced by cellular oxidation of glucose, diffuses through the capillary wall and enters the plasma and then the red blood cells. The red blood cells contain an enzyme, *carbonic anhydrase,* that speeds the reaction between carbon dioxide and water, producing carbonic acid:

$$CO_2 + H_2O \rightarrow H_2CO_3$$

The carbonic acid quickly dissociates, releasing hydrogen ions and bicarbonate ions:

$$H_2CO_3 \rightarrow H^+ + HCO_3^-$$

In the peripheral tissues, where CO_2 is generated, oxygen is used in the metabolism of glucose. Hemoglobin gives up much of its oxygen and is in the form of deoxyhemoglobin. In this form, hemoglobin can bind the hydrogen ions generated in red blood cells, acting as a buffer to minimize the pH change that would otherwise occur.

The above two reactions can be written as a single reversible reaction:

$$CO_2 + H_2O \leftrightarrow H_2CO_3 \leftrightarrow H^+ + HCO_3^-$$

In the peripheral tissues, where CO_2 levels are high, the reaction equilibrium shifts to the right. This generates H^+, which is buffered by hemoglobin, and HCO_3^-, which becomes a plasma electrolyte. In the lungs, where oxygen levels are high, hemoglobin is no longer a good buffer, and it releases its H^+. However, the released H^+ combines with plasma HCO_3, shifting the reaction equilibrium to the left, generating carbonic acid, which quickly dissociates to form CO_2 and water. The water is added to the body fluids, and the CO_2 is exhaled. These shifts occur rapidly because carbonic anhydrase catalyzes the reactions in both directions. Carbonic acid is sometimes called a *volatile acid* because of this relationship to CO_2 (see figs. 19.41 and 19.43, p. 761).

Individual amino acids in body fluids can also function as acid-base buffers by accepting or releasing hydrogen ions. This is possible because every amino acid has an amino group (—NH₂) and a carboxyl group (—COOH).

To summarize, acid-base buffer systems take up hydrogen ions when body fluids are becoming more acidic and give up hydrogen ions when the fluids are becoming more basic (alkaline). Buffer systems convert stronger acids into weaker acids or convert stronger bases into weaker bases, as **table 21.3** summarizes.

In addition to minimizing pH fluctuations, acid-base buffer systems in body fluids buffer each other. Consequently, whenever the hydrogen ion concentration begins to change, the chemical balances in all of the buffer systems change too, resisting the drift in pH.

 PRACTICE

22 What is the difference between a strong acid or base and a weak acid or base?

23 How does a chemical buffer system help regulate pH of body fluids?

24 List the major buffer systems of the body.

Chemical buffer systems only temporarily solve the problem of acid-base balance. Ultimately, the body must eliminate excess acid or base. The lungs (controlled by the respiratory center) and the kidneys accomplish this task.

Respiratory Excretion of Carbon Dioxide

The **respiratory center** in the brainstem helps regulate hydrogen ion concentrations in the body fluids by controlling the rate and depth of breathing. If body cells increase their production of carbon dioxide, carbonic acid production increases. As the carbonic acid dissociates, the concentration of hydrogen ions increases, and the pH of the internal environment begins to drop (see chapter 19, p. 752). Such an increasing concentration of carbon dioxide in the central nervous system and the subsequent increase in hydrogen ion concentration in the cerebrospinal fluid stimulate chemosensitive areas in the respiratory center.

In response to stimulation, the respiratory center increases the depth and rate of breathing so that the lungs excrete more carbon dioxide. Hydrogen ion concentration in body fluids returns toward normal, because the released carbon dioxide is in equilibrium with carbonic acid (fig. 21.10):

$$CO_2 + H_2O \leftrightarrow H_2CO_3 \leftrightarrow H^+ + HCO_3^-$$

TABLE 21.3	Chemical Acid-Base Buffer Systems	
Buffer System	**Constituents**	**Actions**
Bicarbonate system	Bicarbonate ion (HCO_3^-)	Converts a strong acid into a weak acid
	Carbonic acid (H_2CO_3)	Converts a strong base into a weak base
Phosphate system	Monohydrogen phosphate ion (HPO_4^{-2})	Converts a strong acid into a weak acid
	Dihydrogen phosphate ($H_2PO_4^-$)	Converts a strong base into a weak base
Protein system (and amino acids)	—NH₃⁺ group of an amino acid or protein	Releases a hydrogen ion in the presence of excess base
	—COO⁻ group of an amino acid or protein	Accepts a hydrogen ion in the presence of excess acid

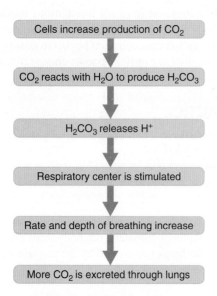

FIGURE 21.10 **AP|R** An increase in carbon dioxide elimination follows an increase in carbon dioxide production.

Conversely, if body cells are less active, concentrations of carbon dioxide and hydrogen ions in body fluids may begin to decrease. In response to the change, the respiratory center decreases breathing rate and depth. This increases the carbon dioxide level in the body fluids, returning the pH toward normal. If the pH drops below normal, the respiratory center is stimulated to increase the rate and depth of breathing.

Activity of the respiratory center, therefore, changes in response to shifts in the pH of the body fluids, reducing these shifts to a minimum. Because most of the hydrogen ions in the body fluids originate from carbonic acid produced when carbon dioxide reacts with water, the respiratory regulation of hydrogen ion concentration is important.

Renal Excretion of Hydrogen Ions

Nephrons help regulate the hydrogen ion concentration of body fluids by excreting hydrogen ions in the urine. Recall from chapter 20 (p. 786) that the epithelial cells of the proximal and distal convoluted tubules and the collecting ducts secrete these ions into the tubular fluid. The tubular secretion of hydrogen ions is linked to tubular reabsorption of bicarbonate ions. In this way, the kidneys also regulate the concentration of bicarbonate ions in body fluids. These mechanisms also help balance the sulfuric acid, phosphoric acid, and organic acids that appear in body fluids as by-products of metabolic processes.

The metabolism of certain amino acids, for example, produces sulfuric and phosphoric acids. Consequently, a diet high in proteins may form excess acid. The kidneys compensate by altering the tubular secretion of hydrogen ions, thus resisting a shift in the pH of body fluids (fig. 21.11). Once hydrogen ions are secreted, phosphates filtered into the fluid of the renal tubule buffer them, aided by ammonia (NH_3).

Through deamination of certain amino acids, the cells of the renal tubules produce ammonia, which diffuses readily through cell membranes and enters the renal tubules. When increase in the hydrogen ion concentration of body fluids is prolonged, the renal tubules increase ammonia production. Ammonia is a weak base, so it can accept hydrogen ions to form *ammonium ions* (NH_4^+):

$$H^+ + NH_3 \rightarrow NH_4^+$$

Cell membranes are impermeable to ammonium ions, which are trapped in the renal tubules as they form and are excreted with the urine. This mechanism helps to transport excess hydrogen ions to the outside and helps prevent the urine from becoming too acidic.

Time Course of Hydrogen Ion Regulation

The various regulators of hydrogen ion concentration operate at different rates. Acid-base buffers function rapidly and can convert strong acids or bases into weak acids or bases almost immediately. For this reason, these chemical buffer systems are called the body's *first line of defense* against shifts in pH.

Physiological buffer systems, such as the respiratory and renal mechanisms, function more slowly and constitute the *second line of defense*. The respiratory mechanism may require several minutes to begin resisting a change in pH, and the renal mechanism may require one to three days to regulate a changing hydrogen ion concentration (fig. 21.12).

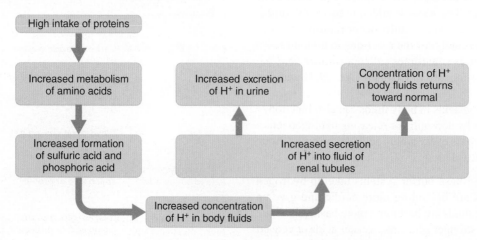

FIGURE 21.11 If the concentration of hydrogen ions in body fluids increases, the renal tubules increase their secretion of hydrogen ions into the urine.

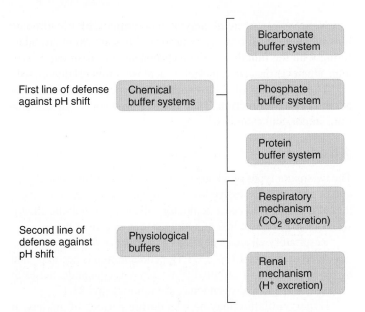

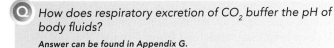

FIGURE 21.12 Chemical buffers act rapidly, while physiological buffers may require several minutes to several days to begin resisting a change in pH.

> **How does respiratory excretion of CO_2 buffer the pH of body fluids?**
> Answer can be found in Appendix G.

PRACTICE

25 How does the respiratory system help regulate acid-base balance?

26 How do the kidneys respond to excess hydrogen ions?

27 How do the rates at which chemical and physiological buffer systems act differ?

21.6 | Acid-Base Imbalances

Chemical and physiological buffer systems ordinarily maintain the hydrogen ion concentration of body fluids within very narrow pH ranges. Abnormal conditions may disturb the acid-base balance. For example, the pH of arterial blood is normally 7.35–7.45. A value below 7.35 produces *acidosis*. A pH above 7.45 produces *alkalosis* (fig. 21.13).

Acidosis results from accumulation of acids or loss of bases, both of which cause abnormal increases in the hydrogen ion con-

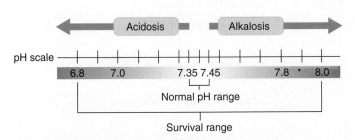

FIGURE 21.13 If the pH of arterial blood drops to 6.8 or rises to 8.0 for more than a few hours, the person usually cannot survive.

centrations of body fluids. Conversely, alkalosis results from a loss of acids or an accumulation of bases accompanied by a decrease in hydrogen ion concentrations (fig. 21.14). If the imbalance is due to a change in respiratory function, it is termed a respiratory disturbance. Any other disturbance in acid-base balance is termed metabolic. Such shifts in the pH of body fluids may be life threatening. A person usually cannot survive if the pH drops to 6.8 or rises to 8.0 for more than a few hours.

Acidosis

The two major types of acidosis are *respiratory acidosis* and *metabolic acidosis*. Factors that increase carbon dioxide levels, which increases the concentration of carbonic acid (the respiratory acid), cause respiratory acidosis. Metabolic acidosis is due to an abnormal accumulation of any other acids in the body fluids or to a loss of bases, including bicarbonate ions. Similarly, the two major types of alkalosis are *respiratory alkalosis* and *metabolic alkalosis*. Excessive loss of carbon dioxide and consequent loss of carbonic acid cause respiratory alkalosis. Metabolic alkalosis is due to excessive loss of hydrogen ions or gain of bases.

In respiratory acidosis, carbon dioxide accumulates. Evidence of respiratory insufficiency, such as labored breathing and cyanosis, may be evident. Respiratory insufficiency can result from factors that hinder alveolar ventilation (fig. 21.15). These factors include the following:

1. Injury to the respiratory center of the brainstem that decreases rate and depth of breathing.
2. Obstructions in air passages that interfere with air movement into and out of the alveoli.
3. Diseases that decrease gas exchange, such as pneumonia or emphysema.

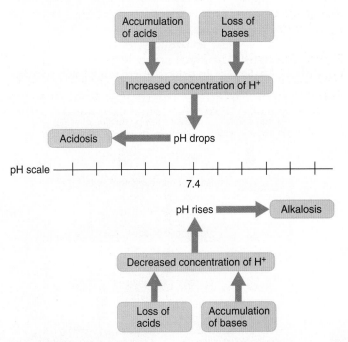

FIGURE 21.14 Acidosis results from accumulation of acids or loss of bases. Alkalosis results from loss of acids or accumulation of bases.

Any of these conditions can increase the level of carbonic acid and hydrogen ions in body fluids, lowering pH.

Metabolic acidosis is due to either accumulation of nonrespiratory acids or loss of bases (fig. 21.16). Factors that may lead to this condition include:

1. Kidney disease reduces glomerular filtration and fails to excrete the acids produced in metabolism (uremic acidosis).
2. Prolonged vomiting loses the alkaline contents of the upper intestine and acidic stomach contents. (Losing only the stomach contents produces metabolic alkalosis.)
3. Prolonged diarrhea causes loss of alkaline intestinal secretions (especially in infants).
4. In diabetes mellitus some fatty acids react to produce ketone bodies, such as *acetoacetic acid, beta-hydroxybutyric acid,* and *acetone*. Normally, these molecules are scarce, and cells oxidize them as energy sources. However, if fats are used at an abnormally high rate, as may occur in diabetes mellitus, ketone bodies may accumulate faster than they can be oxidized, and as a result spill over into the urine (ketonuria). In addition, the lungs may release acetone, which is volatile and imparts a fruity odor to the breath. More seriously, the accumulation of acetoacetic acid and beta-hydroxybutyric acid may lower pH (ketonemic acidosis).

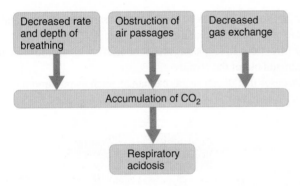

FIGURE 21.15 Some of the factors that lead to respiratory acidosis.

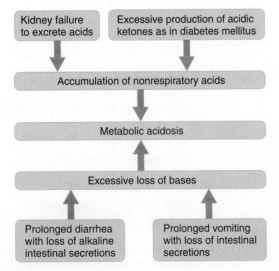

FIGURE 21.16 Some of the factors that lead to metabolic acidosis.

Nonrespiratory acids may also combine with bicarbonate ions in the urine. Too many bicarbonate ions are excreted, interfering with the function of the bicarbonate acid-base buffer system. Whatever the cause, metabolic acidosis shifts pH downward.

The symptoms of acidosis result from depression of central nervous system function. They include drowsiness, disorientation, stupor, and cyanosis.

Alkalosis

The two major types of alkalosis are *respiratory alkalosis* and *metabolic alkalosis*. Excessive loss of carbon dioxide and consequent loss of carbonic acid cause respiratory alkalosis. Metabolic alkalosis is due to excessive loss of hydrogen ions or gain of bases.

Respiratory alkalosis develops as a result of *hyperventilation*, described in chapter 19 (p. 753). Hyperventilation is accompanied by too great a loss of carbon dioxide and consequent decreases in carbonic acid and hydrogen ion concentrations (fig. 21.17).

Hyperventilation may happen during periods of anxiety. It may also accompany fever or poisoning from salicylates, such as aspirin. At high altitudes, hyperventilation may be a response to low oxygen partial pressure. Also, musicians, such as bass tuba players, who must provide a large volume of air when playing sustained passages, sometimes hyperventilate. In each case, rapid, deep breathing depletes carbon dioxide, and the pH of body fluids increases.

Metabolic alkalosis results from a loss of hydrogen ions or from a gain in bases, both accompanied by a rise in the pH of the blood (alkalemia) (fig. 21.18). This condition may occur following gastric drainage (lavage), prolonged vomiting in which only the stomach contents are lost, or the use of certain diuretic drugs. Gastric juice is acidic, so its loss leaves the body fluids with a net increase of basic substances and a pH shift toward alkaline values. Metabolic alkalosis may also develop as a result of ingesting too much antacid, such as sodium bicarbonate, to relieve the symptoms of indigestion.

The symptoms of alkalosis include light-headedness, agitation, dizziness, and tingling sensations. In severe cases, impulses may be triggered spontaneously on motor neurons, and muscles may respond with tetanic contractions (see chapter 9, p. 304).

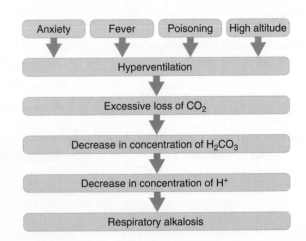

FIGURE 21.17 Some of the factors that lead to respiratory alkalosis.

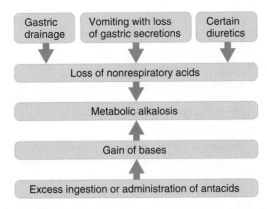

FIGURE 21.18 Some of the factors that lead to metabolic alkalosis.

 PRACTICE

28 What is the difference between a respiratory acid-base disturbance and a metabolic disturbance?

29 How do the symptoms of alkalosis compare with those of acidosis?

21.7 | Compensation

During an acid-base imbalance, the same mechanisms described in this chapter, including chemical buffers, respiratory excretion of CO_2, and renal excretion of hydrogen ions, may resist the shift in pH. This resistance is called **compensation**. For example, in the case of metabolic alkalosis due to excess administration of antacids to a patient, all three mechanisms play a role in compensating. Chemical buffers release hydrogen ions to lower the pH. Also, respiratory control mechanisms inhibit ventilation, thereby retaining CO_2 to help lower the pH toward normal. Finally, the kidneys decrease tubular secretion of hydrogen ions, causing the urine to become more alkaline, reflecting excretion of the excess base.

Another example of compensation is metabolic acidosis caused by uncontrolled diabetes mellitus and the resultant accumulation of acidic ketones in the blood. As blood pH drops, respiratory control mechanisms increase breathing rate and depth, thereby lowering carbon dioxide levels, raising blood pH toward normal. Also, the kidneys begin to excrete the excess hydrogen ions in the urine.

If the disturbance is due to a problem with one of the organ systems that normally compensates, however, the response is limited to the remaining mechanisms. If the problem is respiratory acidosis due to pulmonary disease, the respiratory system is not able to take part in the compensation. In this case, compensatory response depends on chemical buffering and on increased excretion of hydrogen ions by the kidneys. Similarly, in the case of metabolic acidosis caused by kidney disease, compensation depends on chemical buffering and on increased CO_2 excretion by the respiratory system.

PRACTICE

30 What is compensation and how is it achieved? ■

Chapter Summary

21.1 The Balance Concept (page 804)

The maintenance of water and electrolyte balance requires that the quantities of these substances entering the body equal the quantities leaving it. Altering the water balance affects electrolyte balance.

21.2 Distribution of Body Fluids (page 804)

1. Fluid compartments
   a. The **intracellular fluid compartment** includes the fluids and electrolytes cell membranes enclose.
   b. The **extracellular fluid compartment** includes all fluids and electrolytes outside cell membranes.
      (1) Interstitial fluid is in tissue spaces.
      (2) Plasma is part of the blood.
      (3) Lymph is within lymphatic vessels.
      (4) Transcellular fluid is in body cavities.
2. Body fluid composition
   a. Extracellular fluids have high concentrations of sodium, chloride, calcium, and bicarbonate ions, with less potassium, calcium, magnesium, phosphate, and sulfate ions. Plasma contains more protein than does either interstitial fluid or lymph.
   b. Intracellular fluid contains relatively high concentrations of potassium, magnesium, and phosphate ions; it also contains a greater concentration of sulfate ions and lesser concentrations of sodium, chloride, calcium, and bicarbonate ions than does extracellular fluid.
3. Movement of fluid between compartments
   a. Hydrostatic and osmotic pressure regulate fluid movements.
      (1) Fluid leaves plasma because of hydrostatic pressure and returns to plasma because of colloid osmotic pressure.
      (2) Hydrostatic pressure drives fluid into lymph vessels.
      (3) Osmotic pressure regulates fluid movement in and out of cells.
   b. Sodium ion concentrations are especially important in fluid movement regulation.

21.3 Water Balance (page 806)

1. Water intake
   a. The volume of water taken in varies from person to person.
   b. Most water comes from consuming liquid or moist foods.
   c. Oxidative metabolism produces some water.

2. Regulation of water intake
  a. The thirst mechanism is the primary regulator of water intake.
  b. Drinking and the resulting stomach distension inhibit the thirst mechanism.
3. Water output
  a. Water is lost in a variety of ways.
    (1) It is excreted in the urine, feces, and sweat.
    (2) Insensible loss occurs through evaporation from the skin and lungs.
  b. Urine is the primary means of water output.
4. Regulation of water output
  a. The renal distal convoluted tubules and collecting ducts regulate water output.
    (1) ADH from the hypothalamus and posterior pituitary gland stimulates water reabsorption in these segments.
    (2) ADH can reduce normal output of 1,500 milliliters to 500 milliliters per day.
  b. Taking in excess water inhibits the ADH mechanism.

21.4 Electrolyte Balance (page 809)

1. Electrolyte intake
  a. The most important electrolytes in the body fluids are those that release ions of sodium, potassium, calcium, magnesium, chloride, sulfate, phosphate, and bicarbonate.
  b. These ions are obtained in foods and beverages or as by-products of metabolic processes.
2. Regulation of electrolyte intake
  a. Hunger and thirst usually drive consumption of sufficient electrolytes.
  b. Severe electrolyte deficiency may induce a salt craving.
3. Electrolyte output
  a. Electrolytes are lost through perspiration, feces, and urine.
  b. Quantities lost vary with temperature and physical exercise.
  c. The greatest electrolyte loss occurs as a result of kidney functions.
4. Regulation of electrolyte output
  a. Concentrations of sodium, potassium, and calcium ions in the body fluids are particularly important.
  b. The regulation of sodium ions involves the secretion of aldosterone from the adrenal glands.
  c. The regulation of potassium ions also involves aldosterone.
  d. Parathyroid hormone regulates calcium ion concentration in body fluids.
  e. The mechanisms that control positively charged ions secondarily regulate negatively charged ions.
    (1) Chloride ions are passively reabsorbed in renal tubules as sodium ions are actively reabsorbed.
    (2) Some negatively charged ions, such as phosphate ions, are reabsorbed partially by limited-capacity active transport mechanisms.

21.5 Acid-Base Balance (page 812)

Acids are electrolytes that release hydrogen ions. **Bases** combine with hydrogen ions.

1. Sources of hydrogen ions.
  a. **Aerobic respiration of glucose** produces carbon dioxide, which reacts with water to form carbonic acid. Carbonic acid dissociates to release hydrogen and bicarbonate ions.
  b. **Anaerobic respiration of glucose** produces lactic acid.
  c. **Incomplete oxidation of fatty acids** releases acidic ketone bodies.
  d. **Oxidation of amino acids containing sulfur** produces sulfuric acid.
  e. **Hydrolysis of phosphoproteins and nucleic acids** gives rise to phosphoric acid.
2. Strengths of acids and bases
  a. Acids vary in the extent to which they ionize.
    (1) Strong acids, such as hydrochloric acid, ionize more completely.
    (2) Weak acids, such as carbonic acid, ionize less completely.
  b. Bases vary in strength also.
    (1) Strong bases, such as hydroxide ions, combine readily with hydrogen ions.
    (2) Weak bases, such as bicarbonate ions, combine with hydrogen ions less readily.
3. Regulation of hydrogen ion concentration
  a. Acid-base buffer systems minimize pH changes.
    (1) **Chemical buffer systems** are composed of sets of two or more chemicals.
    (2) They convert strong acids into weaker acids or strong bases into weaker bases.
    (3) They include the **bicarbonate buffer system, phosphate buffer system,** and **protein buffer system.**
  b. The respiratory system excretes carbon dioxide.
    (1) The **respiratory center** is in the brainstem.
    (2) It helps regulate pH by controlling the rate and depth of breathing.
    (3) Increasing carbon dioxide and hydrogen ion concentrations stimulates chemoreceptors associated with the respiratory center; breathing rate and depth increase, and carbon dioxide concentration decreases.
    (4) If the carbon dioxide and hydrogen ion concentrations are low, the respiratory center inhibits breathing.
  c. The kidneys excrete hydrogen ions.
    (1) Nephrons secrete hydrogen ions to regulate pH.
    (2) Ammonia produced by renal cells helps transport hydrogen ions to outside the body.
  d. Chemical buffers act rapidly; physiological buffers act more slowly.

21.6 Acid-Base Imbalances (page 817)

1. Acidosis
   a. Respiratory acidosis results from increased levels of carbon dioxide and carbonic acid.
   b. Metabolic acidosis results from accumulation of other acids or loss of bases.
2. Alkalosis
   a. Respiratory alkalosis results from loss of carbon dioxide and carbonic acid.
   b. Metabolic alkalosis results from loss of other acids or gain of bases.

21.7 Compensation (page 819)

1. If an acid-base imbalance occurs, chemical buffering, respiratory mechanisms, and renal mechanisms return pH to normal. This resistance to pH change is called **compensation.**
2. If an acid-base imbalance is due to dysfunction of one of the compensatory mechanisms, another takes over.
   a. In a respiratory disturbance, compensation depends on chemical buffering and renal mechanisms.
   b. In a renal disturbance, compensation depends on chemical buffering and the respiratory mechanism.

CHAPTER ASSESSMENTS

21.1 The Balance Concept

1 Explain how water balance and electrolyte balance are interdependent. (p. 804)

21.2 Distribution of Body Fluids

2 The water and electrolytes enclosed by cell membranes constitute the _____. (p. 804)
   a. transcellular fluid
   b. intracellular fluid
   c. extracellular fluid
   d. lymph
3 Explain how the fluids in the compartments differ in composition. (p. 805)
4 Describe how fluid movements between the compartments are controlled. (p. 805)

21.3 Water Balance

5 Prepare a list of sources of normal water gain and loss to illustrate how the input of water equals the output of water. (pp. 806–807)
6 Define *water of metabolism*. (p. 806)
7 Explain how water intake is regulated. (pp. 806–807)
8 Explain how the kidneys regulate water output. (p. 808)

21.4 Electrolyte Balance

9 Electrolytes in body fluids of importance to cellular functions include _____. (p. 809)
   a. sodium
   b. potassium
   c. calcium
   d. all of the above
10 Explain how electrolyte intake is regulated. (p. 809)
11 List the routes by which electrolytes leave the body. (p. 809)
12 Explain how the adrenal cortex functions to regulate electrolyte balance. (p. 809)

13 Describe the role of the parathyroid glands in regulating electrolyte balance. (p. 810)
14 Describe the mechanisms by which the renal tubules regulate electrolyte balance. (p. 812)

21.5 Acid-Base Balance

15 Define *acid* and *base*. (p. 812)
16 List five sources of hydrogen ions in the body fluids, and name an acid that originates from each source. (pp. 812–813)
17 _____ dissociate to release hydrogen ions more completely. An example is hydrochloric acid. (p. 813)
18 _____ dissociate to release fewer hydroxide ions. (p. 813)
19 Explain how an acid-base buffer system functions. (p. 814)
20 Explain how the bicarbonate buffer system resists changes in pH. (p. 814)
21 Explain why a protein has both acidic and basic properties. (pp. 814–815)
22 Describe how a protein functions as a buffer system. (pp. 814–815)
23 Describe the role of hemoglobin as a buffer. (p. 815)
24 Explain how the respiratory system functions in the regulation of acid-base balance. (pp. 815–816)
25 Explain how the kidneys function in the regulation of acid-base balance. (p. 816)
26 Describe the role of ammonia in the transport of hydrogen ions to the outside of the body. (p. 816)
27 Distinguish between a chemical buffer system and a physiological buffer system. (p. 816)

21.6 Acid-Base Imbalances

28 Distinguish between respiratory and metabolic acid-base imbalances. (pp. 817–818)

21.7 Compensation

29 Explain how the body compensates for acid-base imbalances. (p. 819)

INTEGRATIVE ASSESSMENTS/CRITICAL THINKING

Outcomes 14.3, 19.5, 19.6, 21.5, 21.6

1. An elderly, semiconscious patient is tentatively diagnosed as having acidosis. What components of the arterial blood will be most valuable in determining if the acidosis is of respiratory origin?

Outcomes 15.2, 15.4, 16.2, 21.2

2. If the right ventricle of a patient's heart is failing, increasing the systemic venous pressure, what changes might occur in the patient's extracellular fluid compartments?

Outcomes 17.2, 17.5, 17.8, 17.9, 21.4, 21.5, 21.6

3. Radiation therapy may damage the mucosa of the stomach and intestines. What effect might this have on the patient's electrolyte balance?

Outcomes 17.5, 17.8, 17.9, 19.4, 21.5, 21.6

4. Describe what might happen to the plasma pH of a patient, before compensation occurs, as a result of
   a. prolonged diarrhea.
   b. suction of the gastric contents.
   c. hyperventilation.
   d. hypoventilation.

Outcomes 19.4, 21.5, 21.6, 21.7

5. A student hyperventilates and is disoriented just before an exam. Is this student likely to be experiencing acidosis or alkalosis? How will the body compensate in an effort to maintain homeostasis?

ONLINE STUDY TOOLS

  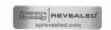

Connect Interactive Questions Reinforce your knowledge using assigned interactive questions covering fluid compartments and the regulation of water, electrolyte, and acid-base balance.

Connect Integrated Activity Can you predict the effects of different types of fluid and electrolyte imbalances?

LearnSmart Discover which chapter concepts you have mastered and which require more attention. This adaptive learning tool is personalized, proven, and preferred.

22

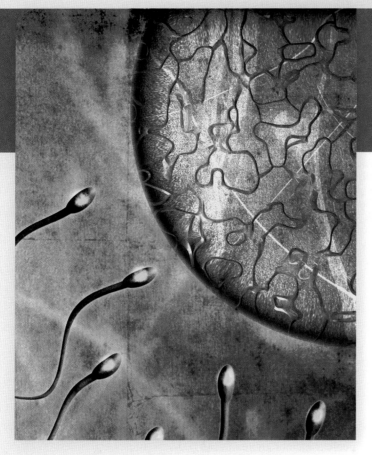

Several sperm approach an egg, the winners of a race that several hundred million sperm began (1,500×). Only one sperm cell can fertilize the egg.

Reproductive Systems

LEARNING OUTCOMES

After you have studied this chapter, you should be able to:

22.1 Meiosis and Sex Cell Production
1 Outline the process of meiosis, and explain how it mixes up parental genes. (p. 824)

22.2 Organs of the Male Reproductive System
2 Describe the structure and function(s) of each part of the male reproductive system. (pp. 826–835)
3 Outline the process of spermatogenesis. (p. 828)
4 Describe semen production and exit from the body. (pp. 828–836)
5 Explain how the tissues of the penis produce an erection. (p. 835)

22.3 Hormonal Control of Male Reproductive Functions
6 Explain how hormones control the activities of the male reproductive organs and the development of male secondary sex characteristics. (pp. 837–839)

22.4 Organs of the Female Reproductive System
7 Describe the the structure and function(s) of each part of the female reproductive system. (pp. 840–848)
8 Outline the process of oogenesis. (p. 841)

22.5 Hormonal Control of Female Reproductive Functions
9 Explain how hormones control the activities of the female reproductive organs and the development of female secondary sex characteristics. (p. 849)
10 Describe the major events during a female reproductive cycle. (pp. 850–852)

22.6 Mammary Glands
11 Review the structure of the mammary glands. (pp. 852–853)

22.7 Birth Control
12 Describe several methods of birth control, including the relative effectiveness of each method. (pp. 854–858)

22.8 Sexually Transmitted Infections
13 List the general symptoms of diseases associated with sexually transmitted infections. (p. 859)

THE WHOLE PICTURE

The male and female reproductive systems are each connected sets of organs and glands. Some of the reproductive organs and glands secrete hormones vital to the development and maintenance of secondary sex characteristics and the regulation of reproductive functions. Reproductive organs produce and nurture sex cells (gametes) and transport them to sites of fertilization.

Module 14: Reproductive System

UNDERSTANDING WORDS

andr-, man: *andr*ogens—male sex hormones.

contra-, against, counter: *contra*ception—prevention of fertilization.

crur-, lower part: *crur*a—diverging parts at the base of the penis by which it attaches to the pelvic arch.

ejacul-, to shoot forth: *ejacul*ation—expulsion of semen from the male reproductive tract.

fimb-, fringe: *fimb*riae—irregular extensions on the margin of the infundibulum of the uterine tube.

follic-, small bag: *follic*le—ovarian structure that contains an egg.

-genesis, origin: spermato*genesis*—formation of sperm cells.

gubern-, to steer, to guide: *gubern*aculum—fibrous cord that guides the descent of a testis.

labi-, lip: *labi*a minora—flattened, longitudinal folds that extend along the margins of the female vestibule.

mamm-, breast: *mamm*ary gland—female accessory gland that secretes milk.

mast-, breast: *mast*itis—inflammation of the mammary gland.

mens-, month: *mens*es—monthly flow of blood from the female reproductive tract.

mons-, an eminence: *mons* pubis—rounded elevation of fatty tissue overlying the pubic symphysis in a female.

oo-, egg: *oo*genesis—formation of an egg.

prim-, first: *prim*ordial follicle—ovarian follicle composed of an oocyte surrounded by a single layer of cells.

puber-, adult: *puber*ty—the time when a person becomes able to reproduce.

zon-, belt: *zon*a pellucida—transparent layer surrounding an oocyte.

22.1 | Meiosis and Sex Cell Production

Male sex cells are **sperm.** Female sex cells are eggs, or **oocytes** (o'o-sītz), which in Latin means "egg cells." Sex cells have one set of genetic instructions carried on 23 chromosomes, compared to two sets on 46 chromosomes in other body cells. When sex cells join at fertilization, the amount of genetic information held in 46 chromosomes is restored. Sperm and oocytes are produced by a special type of cell division called *meiosis.*

A GLIMPSE AHEAD | To Chapter 24

A normal fertilized human egg contains two copies of each gene, one from the sperm and one from the egg. The different possible combinations of variants of these genes explain why siblings are not exactly alike.

Meiosis (mi-o'sis) includes two successive divisions, called the *first* and *second meiotic divisions.* At the beginning of the process each **diploid** cell has two sets of chromosomes (46 chromosomes) that are in the form of 23 homologous chromosome pairs. In the first meiotic division (*meiosis I*) members of each homologous pair are separated from each other. Homologous pairs are the same, gene for gene. They may not be identical, however, because a gene may have variants, and the chromosome that comes from the person's mother may carry a different variant for the corresponding gene from the father's homologous chromosome. Before meiosis I, each chromosome is replicated, so it consists of two DNA molecules called *chromatids.* Each chromatid has the complete genetic information associated with that chromosome. The chromatids of a replicated chromosome attach at regions called *centromeres.*

Each of the cells emerges from meiosis I with one member of each homologous pair, a condition termed **haploid.** That is, a haploid cell has one set of chromosomes (23 chromosomes, one from each homologous pair). The second meiotic division (*meiosis II*) separates the chromatids, producing cells that are still haploid, but whose chromosomes are no longer in the replicated form. After meiosis II, each of the chromatids is an independent chromosome.

The steps of meiosis are clearer when considered in a time sequence (fig. 22.1). However, keep in mind that meiosis, like mitosis, is a continuous process. Considering it in steps simply makes it easier to follow.

First Meiotic Division

Prophase I. Individual chromosomes appear as thin threads in the nucleus that then shorten and thicken. Nucleoli disappear, the nuclear membrane temporarily disassembles, and microtubules begin to build the spindle that will separate the chromosomes. The DNA of the chromosomes has already been replicated.

As prophase I continues, homologous chromosomes pair up side by side and tightly intertwine. During this pairing, called *synapsis,* the chromatids of the homologous chromosomes touch at various points along their lengths. Often, the chromatids break in one or more places and exchange parts, forming chromatids with new combinations of genetic information (fig. 22.2). One chromosome of a homologous pair is maternal and the other is paternal. Therefore, an exchange, or **crossover,** between homologous chromosomes produces chromatids that contain genetic information from both parents.

Metaphase I. During the first metaphase, chromosome pairs line up about midway between the poles of the developing spindle, and they are held under great tension, like two groups of people playing tug-of-war. Each chromosome pair consists of two chromosomes, which equals four chromatids. Each chromosome attaches to spindle fibers from one pole. The chromosome alignment is random with respect to maternal and paternal origin of the chromosomes. Each of the 23 chromosomes contributed from the mother may be on the left or the right, and the same is true for the paternal chromosomes—it is similar to the number of ways that 23 pairs of children could line up, while maintaining the pairs. Chromosomes can line up with respect to each other in many combinations.

Anaphase I. Homologous chromosome pairs separate, and each replicated member moves to one end of the spindle. Each new, or daughter, cell receives only one replicated member of a homologous pair of chromosomes, overall halving the chromosome number.

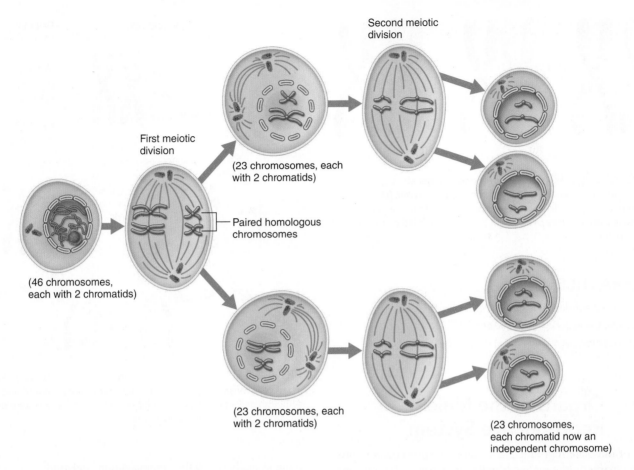

Second meiotic
division

First meiotic
division

(23 chromosomes, each
with 2 chromatids)

Paired homologous
chromosomes

(46 chromosomes,
each with 2 chromatids)

(23 chromosomes, each
with 2 chromatids)

(23 chromosomes,
each chromatid now an
independent chromosome)

FIGURE 22.1 AP|R Sex cells are formed by a special type of cell division, meiosis. This illustration follows two representative pairs of homologous chromosomes.

Telophase I. The original cell divides in two. Nuclear membranes form around the chromosomes, nucleoli reappear, and the spindle fibers disassemble into their constituent microtubules.

Second Meiotic Division

After telophase I, the second meiotic division begins. Meiosis II is similar to a mitotic division (see fig. 22.1). During *prophase II,* chromosomes condense and reappear, still replicated. They move into positions midway between the poles of the developing spindle. In *metaphase II,* the replicated chromosomes attach to spindle fibers. In *anaphase II,* centromeres separate, freeing the chromatids to move to opposite poles of the spindles. The former chromatids are now considered to be chromosomes. In *telophase II,* each of the two cells resulting from meiosis I divides to form two cells. Therefore, each cell undergoing meiosis has the potential to produce four gametes. In males, the gametes mature into four sperm cells. In females, three of the products of meiosis are "cast aside" as polar bodies, and one cell becomes the egg.

Meiosis generates astounding genetic variety. Any one of a person's more than 8 million possible combinations of 23 chromosomes can combine with any one of the more than 8 million combinations of his or her mate, raising the potential variability to more than 70 trillion genetically unique individuals! Crossing over contributes even more genetic variability. **Figure 22.3** illustrates in a simplified manner how maternal and paternal traits reassort during meiosis.

CAREER CORNER
Nurse-Midwife

A couple chooses a certified nurse-midwife (CNM) to deliver their third child because they prefer a home birth. Their first two children had been delivered in a hospital. The children are healthy and the births were uncomplicated.

The nurse-midwife meets with the couple throughout the pregnancy, for at least an hour each time, and is present for the entire labor and birth. She uses soothing words and gentle massage to ease the baby into the world. The nurse-midwife places the newborn on the mother's chest while she completes the initial examination.

A CNM provides pregnancy care, but can also provide primary care, gynecologic care, family planning, treatment for sexually transmitted infections, and newborn care for the first month of life. She or he can do physical exams, prescribe medications, and order and interpret medical tests to provide a diagnosis.

A nurse-midwife is a registered nurse who has a graduate degree in midwifery from an accredited program and has passed a certification or licensing exam. Many nurse-midwives deliver babies in patients' homes, but some are affiliated with hospitals. Nurse-midwives work with physicians who can help in case of emergency.

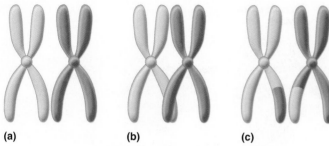

FIGURE 22.2 AP|R Crossing over mixes up genetic traits. (*a*) Homologous chromosome pair, (*b*) chromatids cross over, (*c*) crossing over recombines genes. The different colors indicate that one of the homologous chromosomes comes from the individual's father and one from the mother.

PRACTICE

1 What are the male and female sex cells called?
2 Describe the major events of meiosis.
3 How does meiosis provide genetic variability?

22.2 | Organs of the Male Reproductive System

Organs of the male reproductive system are specialized to produce and maintain the male sex cells, or *sperm cells;* transport these cells and supporting fluids to the outside; and secrete male sex hormones.

The *primary sex organs* (gonads) of this system are the two testes in which the sperm cells (spermatozoa) and the male sex hormones are formed. The other structures of the male reproductive system are termed *accessory sex organs* (secondary sex organs). They include the internal reproductive organs and the external reproductive organs (fig. 22.4; reference plates 3 and 4, pp. 40–41).

Testes

The **testes** (tes′tēz; sing., *testis*) are ovoid structures about 5 centimeters in length and 3 centimeters in diameter. Both testes, each suspended by a spermatic cord, are within the cavity of the saclike *scrotum* (fig. 22.4 and reference plate 12, p. 48).

Descent of the Testes

In a male fetus, the testes originate from masses of tissue posterior to the parietal peritoneum, near the developing kidneys. Usually a month or two before birth, the testes descend to the lower abdominal cavity and pass through the abdominal wall into the scrotum.

The male sex hormone *testosterone,* which the developing testes secrete, stimulates the testes to descend. A fibrous cord called the **gubernaculum** (goo″ber′-nak′u-lum) is attached to each developing testis and extends into the inguinal region of the abdominal cavity. The gubernaculum passes through the abdominal wall and is fastened to the skin of the scrotum. As the body grows the testis

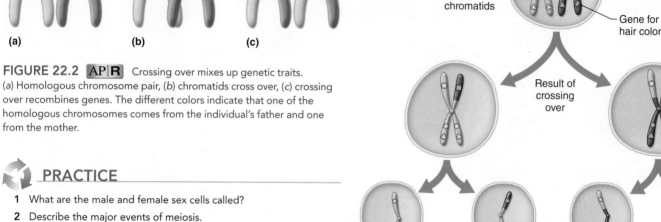

FIGURE 22.3 As a result of crossing over, the genetic information in sex cells varies from cell to cell. Colors represent parent of origin. Although only one eye color gene pair is illustrated, eye color is polygenic (involving more than one gene pair).

remains anchored to the gubernaculum, and guided by the gubernaculum, the testis descends through the **inguinal canal** (ing′gwĭ-nal kah-nal′) of the abdominal wall and enters the scrotum, where it remains. Each testis carries a developing *ductus (vas) deferens,* blood vessels, and nerves. These structures later in development form parts of the **spermatic cord** that suspends the testis in the scrotum (fig. 22.5).

If the testes do not descend into the scrotum, they cannot produce sperm cells because the temperature in the abdominal cavity is too high. If this condition, called *cryptorchidism,* is left untreated, the cells that normally produce sperm cells degenerate, causing infertility.

PRACTICE

4 What are the primary sex organs of the male reproductive system?
5 Describe the descent of the testes.

As a testis descends, a pouch of peritoneum, called the *vaginal process,* moves through the inguinal canal and into the scrotum. In about 25% of males, this pouch stays open, providing a potential passageway through which a loop of intestine may be forced by great abdominal pressure, producing an *indirect inguinal hernia.* If the protruding intestinal loop is so tightly constricted within the inguinal canal that its blood supply stops, the condition is called a *strangulated hernia.* Without prompt treatment, the blood-deprived tissues may die.

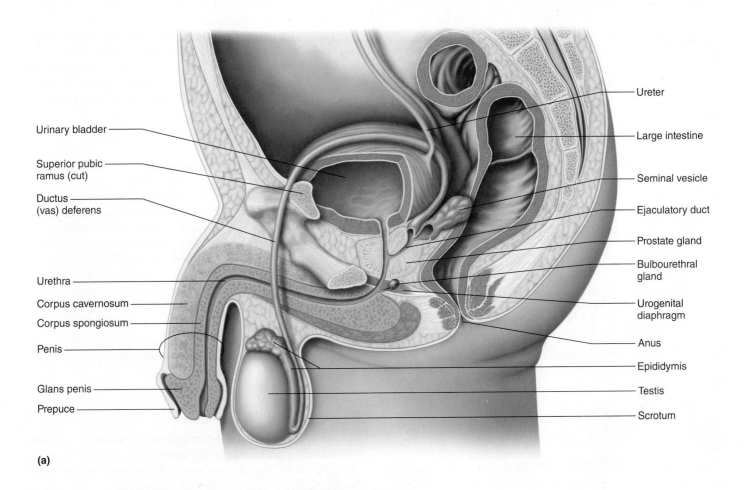

Urinary bladder

Superior pubic ramus (cut)

Ductus (vas) deferens

Urethra

Corpus cavernosum

Corpus spongiosum

Penis

Glans penis

Prepuce

Ureter

Large intestine

Seminal vesicle

Ejaculatory duct

Prostate gland

Bulbourethral gland

Urogenital diaphragm

Anus

Epididymis

Testis

Scrotum

(a)

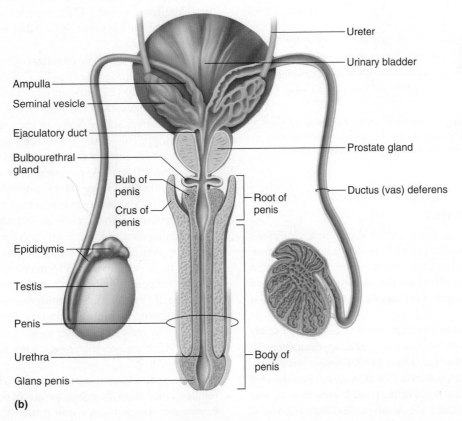

Ampulla

Seminal vesicle

Ejaculatory duct

Bulbourethral gland

Bulb of penis

Crus of penis

Epididymis

Testis

Penis

Urethra

Glans penis

Ureter

Urinary bladder

Prostate gland

Ductus (vas) deferens

Root of penis

Body of penis

(b)

FIGURE 22.4 AP|R Male reproductive organs. (*a*) Sagittal view and (*b*) posterior view. The paired testes are the primary sex organs, and the other reproductive structures, both internal and external, are accessory sex organs.

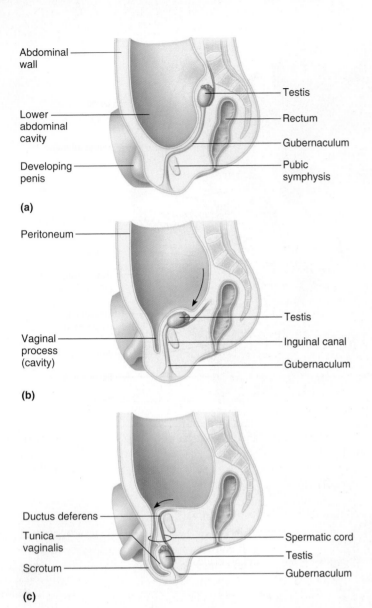

Abdominal wall

Lower abdominal cavity

Developing penis

Testis

Rectum

Gubernaculum

Pubic symphysis

(a)

Peritoneum

Testis

Vaginal process (cavity)

Inguinal canal

Gubernaculum

(b)

Ductus deferens

Tunica vaginalis

Scrotum

Spermatic cord

Testis

Gubernaculum

(c)

FIGURE 22.5 During fetal development, each testis develops near a kidney and then descends through an inguinal canal and enters the scrotum, completing the journey by the eighth gestational month (a–c).

Structure of the Testes

A tough, white, fibrous capsule called the *tunica albuginea* encloses each testis. Along the capsule's posterior border, the connective tissue thickens and extends into the organ, forming a mass called the *mediastinum testis.* From this structure, thin layers of connective tissue, called *septa,* pass into the testis and subdivide it into about 250 *lobules.*

A lobule contains one to four highly coiled, convoluted **semi-niferous tubules** (sem″ĭ-nif′er-us tu′būlz), each approximately 70 centimeters long when uncoiled. These tubules course posteriorly and unite to form a complex network of channels called the *rete testis* (re′te tes′tis). The rete testis is in the mediastinum testis and gives rise to several ducts that join a tube called the *epididymis.* The epididymis, in turn, is coiled on the outer surface of the testis and continues to become the *ductus deferens* (figs. 22.6).

The seminiferous tubules are lined with a specialized stratified epithelium, which includes the **spermatogenic cells** that give rise to the sperm cells. Other specialized cells, called **interstitial** (in″ter-stish′al) **cells** (cells of Leydig), lie between the seminiferous tubules (fig. 22.6 and fig. 22.7). Interstitial cells produce and secrete male sex hormones.

The epithelial cells of the seminiferous tubules can give rise to *testicular cancer,* a common cancer in young men. In most cases, the first sign is a painless testis enlargement or a small lump or area of hardness on the testis. If a biopsy (tissue sample) reveals cancer cells, surgery is performed to remove the affected testis (orchiectomy). Radiation and/or chemotherapy in many cases prevents the cancer from recurring.

PRACTICE

6 Where in the testes are sperm cells produced?

7 Which cells produce male sex hormones?

Formation of Sperm Cells

The epithelium of the seminiferous tubules consists of supporting cells called *sustentacular cells* (Sertoli cells) and spermatogenic cells (fig. 22.7). The sustentacular cells are columnar and extend the full thickness of the epithelium, from its base to the lumen of the seminiferous tubule. The sustentacular cells support, nourish, and regulate the spermatogenic cells, which give rise to sperm cells (spermatozoa).

In the male embryo, undifferentiated spermatogenic cells are called *spermatogonia* (fig. 22.7). Each spermatogonium has 46 chromosomes (23 pairs) in its nucleus, the number for most human body (somatic) cells. Hormones stimulate the spermatogonia to become active. Some of the cells undergo mitosis (see chapter 3, pp. 108–110). Mitotic cell division gives rise to two new cells, with each of these new cells containing 46 chromosomes. One cell (type A) maintains the supply of undifferentiated cells, the other cell (type B) differentiates to become a *primary spermatocyte.* Sperm production arrests at this stage.

At puberty, mitosis resumes, and new spermatogonia form. Testosterone secretion increases, and the primary spermatocytes divide by meiosis, each forming two *secondary spermatocytes.* Each of these cells divides to form two *spermatids,* which mature into sperm cells. Meiosis halves the number of chromosomes in each cell (fig. 22.8). **Spermiogenesis** is the further development of spermatids into sperm (see fig. 22.10). The combined processes of meiosis and spermiogenesis constitute **spermatogenesis** (sper″mah-to-jen′ĕ-sis), taking about 65 to 75 days from start to finish.

The spermatogonia are located within the seminiferous tubules, adjacent to the inside surface of the surrounding basement membrane. As spermatogenesis proceeds, cells in more advanced stages are pushed along the sides of sustentacular cells toward the lumen of the seminiferous tubule (fig. 22.9).

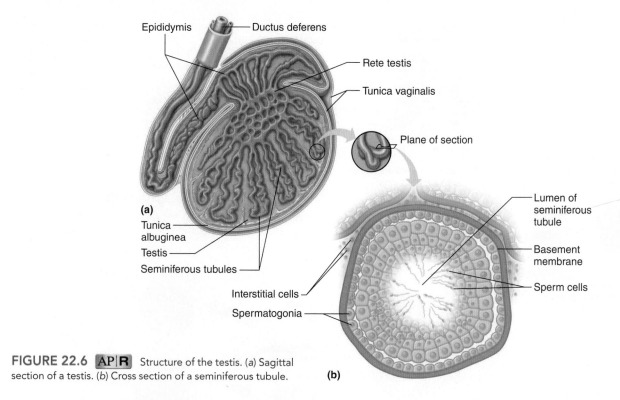

FIGURE 22.6 **AP|R** Structure of the testis. (a) Sagittal section of a testis. (b) Cross section of a seminiferous tubule.

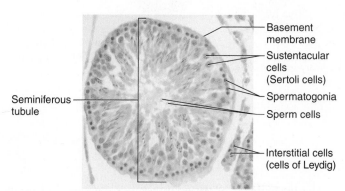

FIGURE 22.7 **AP|R** Light micrograph of a seminiferous tubule (250×).

Near the base of the epithelium, tight junctions fuse the membranous processes of adjacent sustentacular cells (fig. 22.9). These tightly packed cells and their attachments form the *blood-testis barrier*, which prevents some substances from reaching the developing sperm. The blood-testis barrier helps maintain a favorable environment by isolating developing sperm from the male's immune system, which might otherwise react to the sperm as abnormal, nonself cells.

Anton van Leeuwenhoek was the first to view human sperm under a microscope in 1678, concluding that they were parasites in semen. By 1685, he had modified his view, writing that sperm contain a preformed human being and are seeds requiring nurturing in a female to start a new life. Although his interpretation was incorrect, he did identify sperm as playing a role in human reproduction.

Spermatogenesis happens continually in a male, starting at puberty. The resulting sperm cells collect in the lumen of each seminiferous tubule, then pass through the rete testis to the epididymis, where they accumulate and mature.

Structure of a Sperm Cell

A mature sperm cell is a tiny, tadpole-shaped structure about 0.06 millimeter long. It consists of a flattened head, a cylindrical midpiece (body), and an elongated tail (flagellum).

The oval *head* of a sperm cell is primarily composed of a nucleus and contains highly compacted chromatin consisting of 23 chromosomes. A small caplike covering over the head, called the *acrosome*, contains enzymes that aid the sperm cell in penetrating the layers surrounding the oocyte during fertilization (fig. 22.10). An enzyme on the sperm cell membrane also contributes to this process.

The *midpiece* of a sperm has a central, filamentous core and many mitochondria organized in a spiral. The *tail* consists of several microtubules enclosed in an extension of the cell membrane. The mitochondria provide ATP for the lashing movement of the tail that propels the sperm cell. The micrograph in **figure 22.11** shows a few mature sperm cells.

Many toxic chemicals that affect sperm hamper the tail's ability to propel them, preventing the sperm from reaching an egg and exposing it to the chemical. An exception is cocaine, which attaches to thousands of binding sites on a human sperm cell, without apparently harming the cell or impeding its ability to ferry cocaine to an egg. However, it is not known what harm, if any, the drug causes. We do know that fetuses exposed to cocaine in the uterus may suffer a stroke, or, as infants, be unable to react normally to their surroundings.

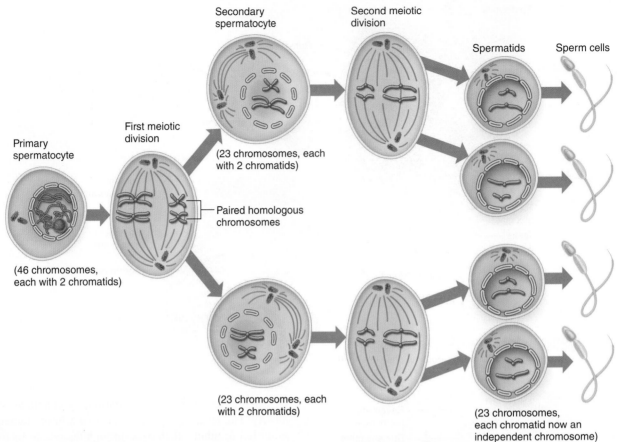

Secondary spermatocyte

First meiotic division

Primary spermatocyte

Second meiotic division

Spermatids

Sperm cells

(46 chromosomes, each with 2 chromatids)

Paired homologous chromosomes

(23 chromosomes, each with 2 chromatids)

(23 chromosomes, each with 2 chromatids)

(23 chromosomes, each chromatid now an independent chromosome)

FIGURE 22.8 During spermatogenesis four sperm cells result from meiosis of a primary spermatocyte. Two representative homologous chromosome pairs are shown.

Why is it important that a sperm possess only 23 chromosomes?
Answer can be found in Appendix G.

PRACTICE

8 Review the events of spermatogenesis.

9 Explain the function of the sustentacular cells in the seminiferous tubules.

10 Describe the structure of a sperm cell.

Male Internal Accessory Reproductive Organs

The internal accessory organs of the male reproductive system are specialized to nurture and transport sperm cells. These structures include the two epididymides, two ductus deferentia, two ejaculatory ducts, the urethra, as well as the two seminal vesicles, prostate gland, and two bulbourethral glands.

Epididymides

The **epididymides** (ep″ĭ-dĭ-dy′mĭdes; sing., *epididymis*) are tightly coiled, threadlike tubes about 6 meters long (see fig. 22.4 and reference plate 12, p. 48). Each epididymis is connected to ducts in a testis. It emerges from the top of the testis, descends along its posterior surface, and then courses upward to become the ductus deferens.

The inner lining of the epididymis is composed of pseudostratified columnar cells that bear nonmotile cilia (fig. 22.12). These cells secrete glycogen and other substances that support stored sperm cells and promote their maturation.

When immature sperm cells reach the epididymis, they are nonmotile. However, as they travel through the epididymis as a result of rhythmic peristaltic contractions, they mature. Following this aging process, the sperm cells can move independently and fertilize egg cells. However, they usually do not move independently until after ejaculation.

Ductus Deferentia

The **ductus deferentia** (duk′tus def′er-en′sha; sing., *ductus deferens*), also called *vasa deferentia,* are muscular tubes about 45 centimeters long lined with pseudostratified columnar epithelium (fig. 22.13). Each ductus deferens originates at the lower end of the epididymis and passes upward along the medial side of a testis to become part of the spermatic cord. The ductus deferens passes through the inguinal canal, enters the abdominal cavity outside the parietal peritoneum, and courses over the pelvic brim. From there, it extends backward and medially into the pelvic cavity, where it ends behind the urinary bladder.

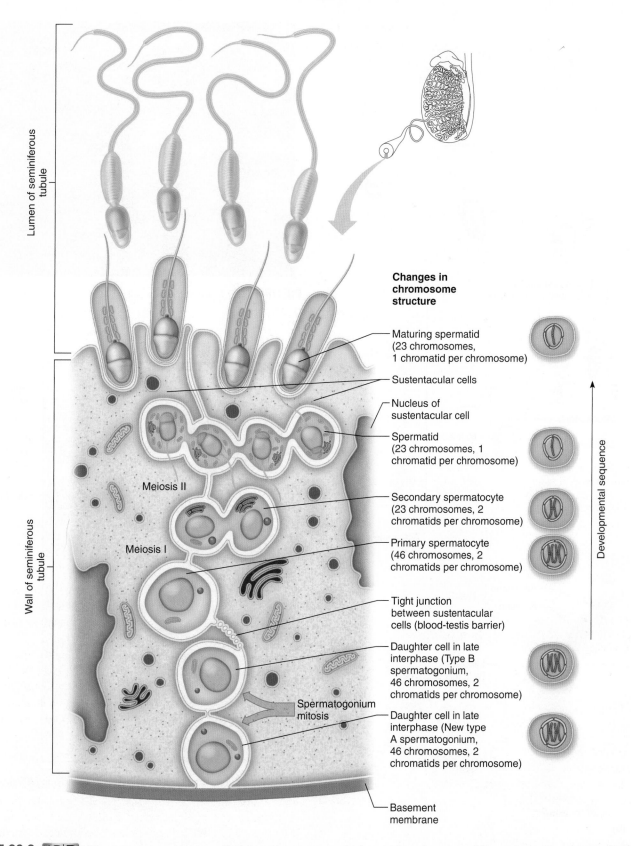

Lumen of seminiferous tubule

Wall of seminiferous tubule

Meiosis II

Meiosis I

Spermatogonium mitosis

Changes in chromosome structure

Maturing spermatid (23 chromosomes, 1 chromatid per chromosome)

Sustentacular cells

Nucleus of sustentacular cell

Spermatid (23 chromosomes, 1 chromatid per chromosome)

Secondary spermatocyte (23 chromosomes, 2 chromatids per chromosome)

Primary spermatocyte (46 chromosomes, 2 chromatids per chromosome)

Tight junction between sustentacular cells (blood-testis barrier)

Daughter cell in late interphase (Type B spermatogonium, 46 chromosomes, 2 chromatids per chromosome)

Daughter cell in late interphase (New type A spermatogonium, 46 chromosomes, 2 chromatids per chromosome)

Basement membrane

Developmental sequence

FIGURE 22.9 AP|R Mitosis in spermatogonia results in type A spermatogonia that continue the germ cell line and type B spermatogonia that give rise to primary spermatocytes. The primary spermatocytes, in turn, give rise to sperm cells by meiosis. Changes in chromosome number and structure are represented by a single pair of chromosomes. Note that as the cells approach the lumen they mature.

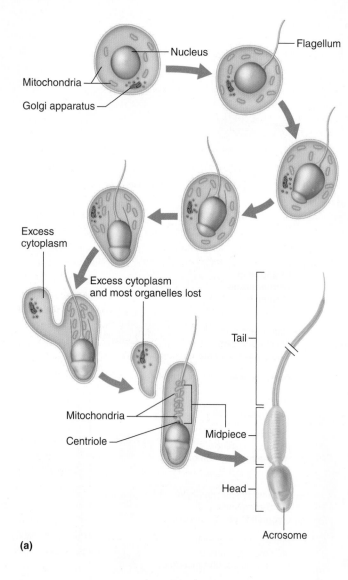

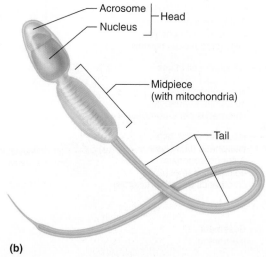

(a)

(b)

FIGURE 22.10 Spermiogenesis—sperm cell maturation. (*a*) The head of the sperm develops largely from the nucleus of the formative cell. (*b*) Parts of a mature sperm cell.

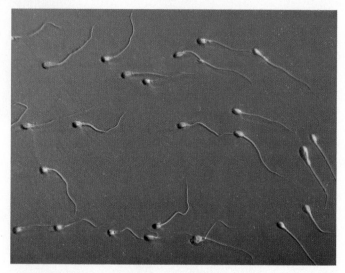

FIGURE 22.11 Human sperm cells (1,400×).

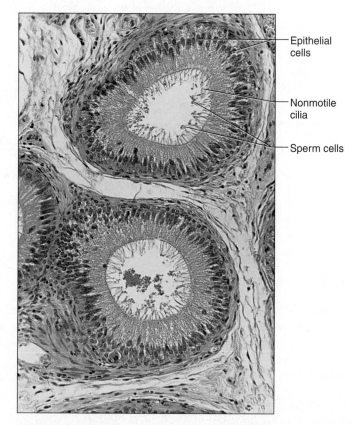

FIGURE 22.12 Cross section of a human epididymis (200×).

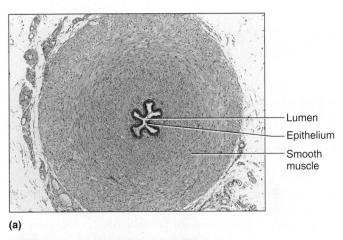

(a)

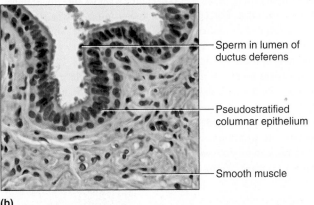

Lumen
Epithelium
Smooth muscle

Sperm in lumen of ductus deferens

Pseudostratified columnar epithelium

Smooth muscle

(b)

FIGURE 22.13 AP|R Ductus (vas) deferens. (*a*) Micrograph of a cross section of the ductus deferens (40×). (*b*) Light micrograph of the wall of the ductus deferens (400×).

Near its termination, the ductus deferens dilates into a portion called the *ampulla.* Just outside the prostate gland, the tube becomes slender again and unites with the duct of a seminal vesicle. The fusion of these two ducts forms an **ejaculatory duct,** which passes through the prostate gland and empties into the urethra through a slit-like opening (see fig. 22.4).

Seminal Vesicles

The **seminal vesicles** (or **seminal glands**) are convoluted, saclike structures about 5 centimeters long, each attached to the ductus deferens on the posterior surface and near the base of the urinary bladder (see fig. 22.4). The glandular tissue lining the inner wall of the seminal vesicle secretes a slightly alkaline fluid. This fluid helps regulate the pH of the tubular contents as sperm cells travel to the outside. Additionally, seminal vesicle fluid neutralizes the acidic secretions of the vagina, helping to sustain the sperm cells that enter the female reproductive tract. Seminal vesicle secretions also include *fructose,* a monosaccharide that provides energy to the sperm cells, and *prostaglandins,* which stimulate muscular contractions of the female reproductive organs, aiding the movement of sperm cells toward the egg cell.

As sperm move through the ductus deferens into the ejaculatory duct, the contents of the seminal vesicles empty into the ejaculatory ducts. This greatly increases the volume of the fluid discharged from the ductus deferens.

PRACTICE

11 Describe the structure of the epididymis.

12 Trace the path of the ductus deferens.

13 What is the function of a seminal vesicle?

Prostate Gland

The **prostate** (pros′tāt) **gland** (see **figs.** 22.4 and **22.14**) is a chestnut-shaped structure about 4 centimeters across and 3 centimeters thick that surrounds the proximal portion of the urethra, just inferior to the urinary bladder. It is composed of many branched tubular glands enclosed in connective tissue. Septa of connective tissue and smooth muscle extend inward from the capsule, separating the tubular glands. The ducts of these glands open into the urethra. The prostate gland releases its secretions into the urethra as smooth muscle contracts in its capsular wall. As this release occurs, the contents of the ductus deferens and the seminal vesicles enter the urethra, which increases the volume of the fluid.

The prostate gland secretes a thin, milky fluid. This slightly acidic secretion also contains citrate, a nutrient for sperm, and prostate-specific antigen (PSA), an enzyme which helps liquefy semen following ejaculation. Clinical Application 22.1 discusses prostate cancer.

Bulbourethral Glands

The **bulbourethral** (bul″bo-u-re′thral) **glands** (Cowper's glands) are two small structures, each about a centimeter in diameter. They are inferior to the prostate gland lateral to the intermediate part of the urethra and are enclosed by muscle fibers of the urogenital diaphragm (see fig. 22.4).

The bulbourethral glands are composed of many tubes whose epithelial linings secrete a mucuslike fluid. This fluid is released in response to sexual stimulation and lubricates the end of the penis in preparation for sexual intercourse (coitus). However, females secrete most of the lubricating fluid for intercourse.

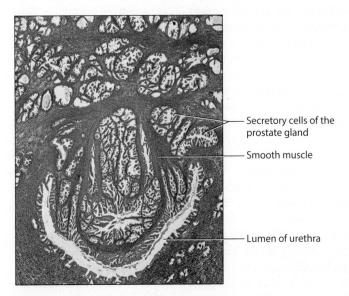

Secretory cells of the prostate gland

Smooth muscle

Lumen of urethra

FIGURE 22.14 Light micrograph of a cross section through the prostate gland (10×).

Each year in the United States about 240,000 men are diagnosed with prostate cancer, and about 30,000 die of the disease. However, many prostate tumors grow so slowly that they are not likely to affect survival and do not require treatment.

The diagnostic process typically begins with a digital rectal exam, in which a physician inserts a finger into the patient's rectum to feel for an enlargement of the prostate gland. This exam is typically coupled with a blood test to detect elevated levels of a biomarker called prostate-specific antigen (PSA). Normally, secretory epithelium in the prostate gland releases PSA, which liquefies the ejaculate. When cancer cells accumulate, more PSA is produced, and it enters capillaries in the prostate.

Health-care organizations' recommendations for PSA screening change often, and range from screening all men to screening none, although any man with symptoms of an enlarged prostate (frequent and slowed urination) should be tested. The controversy is that the value of saving lives following screening must be balanced against the risk that high levels of PSA in the absence of cancer can lead to unnecessary biopsies.

If the physician feels an enlargement of the patient's prostate, or if the PSA level remains high or rises on tests repeated a few months later, the next step is a biopsy procedure to sample cells from several sites in the gland. A cancer detected with a biopsy is assigned a two-digit number, called a Gleason score, which indicates how specialized the cancer cells are. The less specialized the cells, the more aggressive the disease. Imaging technologies can then assess whether the cancer has spread beyond the prostate capsule.

Treatment of prostate cancer may be necessary if the tumor has a high Gleason score or fits the genetic profile of being likely to spread. Treatments include surgery to remove the prostate gland, radiation, and hormones. Adverse effects of treatment include urinary incontinence and erectile dysfunction, which may improve with time. Fortunately, for many men "active surveillance" to regularly monitor the disease is sufficient. This means PSA tests twice a year and a biopsy every one to two years, with treatment only if the condition worsens.

New tests measure the expression of specific genes that change activity dramatically when the disease spreads. In one large study that evaluated a gene expression test, doctors switched 37% of men being treated to active surveillance and 23% on active surveillance to treatment based on test results. It will take a few years to determine how well gene expression profiling improves selection of patients for treatment.

Semen

The fluid the urethra conveys to the outside during ejaculation is called **semen** (se′men). It consists of sperm cells from the testes and secretions of the seminal vesicles, prostate gland, and bulbourethral glands. Semen is slightly alkaline (pH about 7.5), and it includes prostaglandins and nutrients.

The volume of semen released at one time varies from 2 to 5 milliliters. The average number of sperm cells in the fluid is about 120 million per milliliter.

Sperm cells remain nonmotile while they are in the ducts of the testis and epididymis, but begin to swim as they mix with the secretions of accessory glands. However, sperm cells cannot fertilize an egg cell until they enter the female reproductive tract, because here they undergo *capacitation,* which weakens the acrosomal membranes. When sperm cells are placed with egg cells in a laboratory dish to achieve fertilization—a technique called *in vitro* fertilization, discussed in From Science to Technology 23.1 (pp. 870–871)—chemicals are added to simulate capacitation.

Although sperm cells can live for many weeks in the ducts of the male reproductive tract, they usually survive only up to six days after being expelled to the outside, even when they are maintained at body temperature. The ability of a sperm cell to fertilize an oocyte generally lasts only twenty-four to forty-eight hours after the sperm enter the female reproductive tract. On the other hand, sperm cells can be stored and kept viable for years if they are frozen at a temperature below –100°C. Clinical Application 22.2 describes some causes of male infertility.

PRACTICE

14 Where is the prostate gland located?

15 What is the function of the bulbourethral glands?

16 What are the components of semen?

Male External Reproductive Organs

The male external reproductive organs are the scrotum, which encloses two testes, and the penis. The urethra passes through the penis.

Scrotum

The **scrotum** is a pouch of skin and subcutaneous tissue that hangs from the lower abdominal region posterior to the penis. The subcutaneous tissue of the scrotal wall lacks fat but contains a layer of smooth muscle fibers that constitute the *dartos muscle.* Exposure to cold stimulates these muscle fibers to contract, the scrotal skin to wrinkle, and the testes to move closer to the body, where they can absorb heat. In warmer temperatures, the dartos muscle relaxes and the scrotum hangs loosely. The testes move away from the body into an environment 3°C (about 5°F) below body temperature, more conducive to sperm production and survival.

A medial septum divides the scrotum into two chambers, each of which encloses a testis. Each chamber also contains a serous membrane, which covers the front and sides of the testis and the epididymis, helping to ensure that the testis and epididymis move smoothly within the scrotum (see fig. 22.4).

Penis

The **penis** is a cylindrical organ that conveys urine and semen through the urethra to the outside. It is also specialized to enlarge and stiffen, which enables it to enter the vagina during sexual intercourse.

The *body,* or shaft, of the penis is composed of three columns of erectile tissue, which include a pair of dorsally located *corpora cavernosa* and a single, ventral *corpus spongiosum.* A tough capsule of white dense connective tissue called a *tunica albuginea* (too'nĭ-kah al''bu-jin'e-ah) surrounds each column. Skin, a thin layer of subcutaneous tissue, and a layer of connective tissue enclose the penis (fig. 22.15).

The corpus spongiosum, surrounding the urethra, enlarges at its distal end to form a sensitive, cone-shaped **glans penis.** The glans covers the ends of the corpora cavernosa and bears the urethral opening—the *external urethral orifice.* The skin of the glans is very thin, hairless, and contains sensory receptors for sexual stimulation. A loose fold of skin called the *prepuce* (foreskin) originates just posterior to the glans and extends anteriorly to cover the glans as a sheath. The prepuce can be removed by a surgical procedure called *circumcision.*

At the *root* of the penis, the columns of erectile tissue separate. The corpora cavernosa diverge laterally in the perineum and are firmly attached to the inferior surface of the pubic arch by connective tissue. These diverging parts form the *crura* (sing., *crus*) of the penis. The single corpus spongiosum is enlarged between the crura as the *bulb* of the penis, which is attached to membranes of the perineum (see fig. 22.4*b*).

 PRACTICE

17 Describe the structure of the penis.

18 What is circumcision?

Erection, Orgasm, and Ejaculation

During sexual stimulation, parasympathetic impulses from the sacral portion of the spinal cord release the vasodilator nitric oxide, which dilates the arteries leading into the penis, increasing blood flow into erectile tissues. At the same time, the increasing pressure of arterial blood entering the vascular spaces of the erectile tissue compresses the veins of the penis, reducing flow of venous blood away from the penis. Consequently, blood accumulates in the erectile tissues, and the penis swells and elongates, producing an **erection** (fig. 22.16).

In erectile dysfunction (impotence), the penis cannot become erect or sustain an erection. Causes of erectile dysfunction include underlying disease such as diabetes mellitus; paralysis; treatments such as prostate surgery or certain drugs; smoking cigarettes; and drinking alcohol. Development of drugs to treat erectile dysfunction grew out of understanding the physiology of erection. The first drug, Viagra (sildenafil), blocks the enzyme that breaks down cyclic guanosine monophosphate in the erectile tissues, which is necessary for an erection to persist.

The culmination of sexual stimulation is **orgasm** (or'gazm), a pleasurable feeling of physiological and psychological release. Orgasm in the male is accompanied by emission and ejaculation.

Emission (e-mish'un) is the movement of sperm cells from the testes and secretions from the prostate gland and seminal vesicles into the urethra, where they mix to form semen. Emission is a response to sympathetic impulses from the spinal cord, which stimulate peristaltic contractions in smooth muscle in the walls of the testicular ducts, epididymides, ductus deferentia, and ejaculatory ducts. Other sympathetic impulses stimulate rhythmic contractions of the seminal vesicles and prostate gland.

As the urethra fills with semen, sensory impulses are stimulated and pass into the sacral part of the spinal cord. In response, motor impulses are conducted from the spinal cord to certain skeletal muscles at the base of the erectile columns of the penis, rhythmically contracting them. This movement increases the pressure in the erectile tissues and aids in forcing the semen through

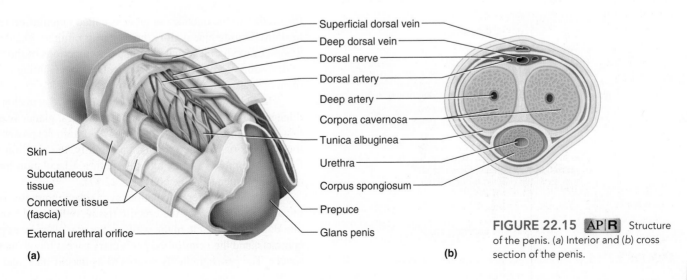

Skin

Subcutaneous tissue

Connective tissue (fascia)

External urethral orifice

(a)

Superficial dorsal vein

Deep dorsal vein

Dorsal nerve

Dorsal artery

Deep artery

Corpora cavernosa

Tunica albuginea

Urethra

Corpus spongiosum

Prepuce

Glans penis

(b)

FIGURE 22.15 AP|R Structure of the penis. (*a*) Interior and (*b*) cross section of the penis.

Male infertility—the inability of sperm cells to fertilize an egg cell—has several causes. If, during fetal development, the testes do not descend into the scrotum, the higher temperature of the abdominal cavity or inguinal canal impedes development of sperm cells in the seminiferous tubules. Certain diseases, such as mumps, may inflame the testes (orchitis), impairing fertility.

The quality and quantity of sperm cells are essential factors in the ability of a man to father a child. If a sperm head is misshapen, if the acrosome is too tough to burst and release enzymes, or if too few sperm cells reach the well-protected egg, fertilization may not happen. The structure of the sperm tail is particularly important. Male infertility can result from sperm tails that are irregularly shaped, coiled, bent, shortened, or absent.

Computer-aided sperm analysis (CASA) is a technique used to evaluate a man's fertility. It can analyze the pathways of up to 200 moving sperm in a few seconds. CASA assesses the number of cells per milliliter of seminal fluid (density), sperm motility, and the size and shape of sperm cell parts (morphology).

In a sperm analysis, a man abstains from intercourse for two to three days, then provides a sperm sample. This may be done either in a clinical setting or at home using a kit ordered from the Internet, with the sample mailed to a lab. The man provides information about his reproductive history and possible exposure to toxins. The CASA system captures images with a digital camera and analyzes and integrates information on sperm density, motility, and morphology (fig. 22A). A computer may also be used to track sperm progress, in a woman's body, toward fertilizing an egg cell (fig. 22B).

Devices are being developed that will enable a man to estimate his sperm count at home. They indicate whether a man's sperm count is above or below the World Health Organization's designation of 20 million sperm per milliliter of ejaculate as the lower criterion for normal fertility. If several tests performed days apart fall below this level, the man should consult a fertility specialist for further testing. Table 22A indicates characteristics of a man's sperm output that are the lowest for which he is still likely to be fertile. It is clear that the male body manufactures many more sperm than are necessary to fertilize an egg.

FIGURE 22A Computer analysis improves the consistency and accuracy of describing sperm density, motility, and morphology, each important in diagnosing male infertility.

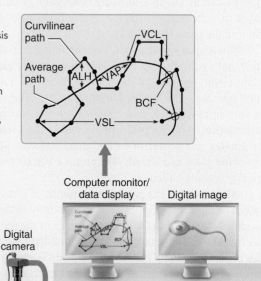

the urethra to the outside—a process called **ejaculation** (e-jak″u-la′shun). During ejaculation, the posterior pituitary gland releases a burst of oxytocin, which stimulates contractions of the epididymides, seminiferous tubules, and prostate gland, aiding the movement of sperm.

The sequence of events during emission and ejaculation is coordinated so that the fluid from the bulbourethral glands is expelled first. This is followed by the release of fluid from the prostate gland, the passage of the sperm cells, and finally, the ejection of fluid from the seminal vesicles into the urethra. Ejaculation forcefully expels the semen from the body (fig. 22.17).

Immediately after ejaculation, sympathetic impulses constrict the arteries that supply the erectile tissue, reducing the inflow of blood. Smooth muscle in the walls of the vascular spaces partially contracts, and the veins of the penis carry excess blood out of these spaces. The penis gradually returns to its flaccid state, and usually

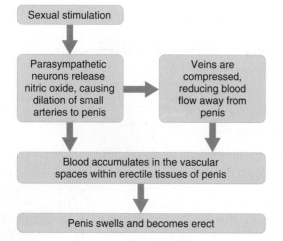

FIGURE 22.16 Mechanism of penile erection in the male.

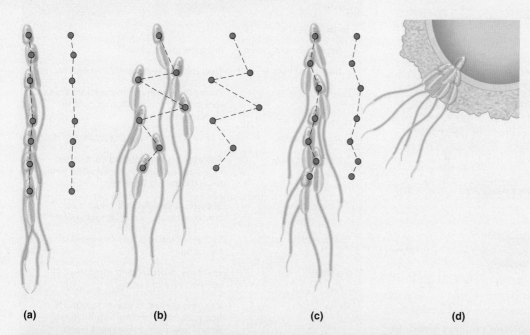

A computer tracks sperm cell movements. In semen, sperm cells swim in a straight line (*a*), but as they are activated by biochemicals in the woman's body, their trajectories widen (*b*). The sperm cells in (*c*) are in the mucus of a woman's cervix, and the sperm cells in (*d*) are attempting to digest through the structures surrounding an egg cell.

(a)　　　　　(b)　　　　　(c)　　　　　(d)

TABLE 22A	Ranges of Semen and Sperm Characteristics in Healthy Men
Characteristic	**Range**
Volume/ejaculate	1.5–5.0 mL
Number of sperm	20–150 million/mL
Concentration of sperm	12–16 million/mL
Sperm vitality	55–63%
Sperm motility	38–42%
Morphologically normal	3–4%

mL = milliliter

another erection and ejaculation cannot be triggered for ten to thirty minutes or longer. Table 22.1 summarizes the functions of the male reproductive organs.

> Spontaneous emission and ejaculation are common in sleeping adolescent males. Changes in hormonal concentrations that accompany adolescent development and sexual maturation cause these events.

 PRACTICE

19 What controls blood flow into penile erectile tissues?

20 Distinguish among orgasm, emission, and ejaculation.

21 Review the events associated with emission and ejaculation.

22.3 | Hormonal Control of Male Reproductive Functions

Hormones secreted by the hypothalamus, the anterior pituitary gland, and the testes control male reproductive functions. These hormones initiate and maintain sperm cell production and oversee the development and maintenance of male sex characteristics.

Hypothalamic and Pituitary Hormones

The male body before ten years of age is reproductively immature with undifferentiated spermatogenic cells in the testes. Then a series of changes leads to development of a reproductively functional adult. The hypothalamus controls many of these changes.

Recall from chapter 13 (pp. 498–499) that the hypothalamus secretes gonadotropin-releasing hormone (GnRH), which

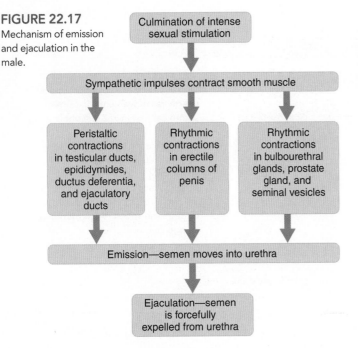

FIGURE 22.17 Mechanism of emission and ejaculation in the male.

Culmination of intense sexual stimulation

↓

Sympathetic impulses contract smooth muscle

↓

Peristaltic contractions in testicular ducts, epididymides, ductus deferentia, and ejaculatory ducts

Rhythmic contractions in erectile columns of penis

Rhythmic contractions in bulbourethral glands, prostate gland, and seminal vesicles

↓

Emission—semen moves into urethra

↓

Ejaculation—semen is forcefully expelled from urethra

TABLE 22.1	Functions of the Male Reproductive Organs AP|R
Organ	**Function**
Testis	
Seminiferous tubules	Produce sperm cells
Interstitial cells	Produce and secrete male sex hormones
Epididymis	Promotes sperm cell maturation; stores sperm cells; conveys sperm cells to ductus deferens
Ductus deferens	Conveys sperm cells to ejaculatory duct
Seminal vesicle	Secretes an alkaline fluid containing nutrients and prostaglandins that helps regulate the pH of semen
Prostate gland	Secretes a slightly acidic fluid that contains citrate, a nutrient for sperm
Bulbourethral gland	Secretes fluid that lubricates end of the penis
Scrotum	Encloses, protects, and regulates temperature of testes
Penis	Conveys urine and semen to outside of body; inserted into the vagina during sexual intercourse; the glans penis is richly supplied with sensory nerve endings associated with feelings of pleasure during sexual stimulation

enters the blood vessels leading to the anterior pituitary gland. In response, the anterior pituitary gland secretes the **gonadotropins** (go-nad″o-trōp′inz) called *luteinizing hormone* (LH) and *follicle-stimulating hormone* (FSH). LH, which in males has been referred to as interstitial cell-stimulating hormone (ICSH), promotes development of the interstitial cells of the testes. The interstitial cells secrete male sex hormones, mainly testosterone. FSH stimulates the sustentacular cells of the seminiferous tubules to proliferate, grow, mature, and respond to the effects of the male sex hormone testosterone. Then, in the presence of FSH and testosterone, sustentacular cells stimulate the spermatogenic cells to undergo spermatogenesis, giving rise to sperm cells (fig. 22.18). The sustentacular cells also secrete a hormone called *inhibin*, which inhibits the anterior pituitary gland by negative feedback. This action prevents oversecretion of FSH.

Male Sex Hormones

Male sex hormones are termed **androgens** (an′dro-jenz). Interstitial cells of the testes produce most male sex hormones, but the adrenal cortex also synthesizes small amounts (see chapter 13, p. 512).

The hormone **testosterone** (tes-tos′tĕ-rōn) is the most important androgen. It is secreted by the testes and transported in the blood, loosely attached to plasma proteins. Like other steroid hormones, testosterone binds receptor molecules, which are usually in the nuclei of its target cells (see chapter 13, p. 490). However, in many target cells, such as those in the prostate gland, seminal vesicles, and male external accessory organs, testosterone is first converted to another androgen called **dihydrotestosterone** (di-hi″dro-tes-tos′tĕ-rōn), which stimulates the cells of these organs. Most androgen molecules that do not reach receptors in target cells are changed by the liver into forms that can be excreted in bile or urine.

Testosterone secretion begins during fetal development and continues for several weeks following birth; then it nearly ceases

during childhood. Between the ages of thirteen and fifteen, a young man's androgen production usually increases rapidly. This phase in development, when an individual becomes reproductively functional, is **puberty** (pu′ber-te). After puberty, testosterone secretion continues throughout the life of a male.

In a group of disorders called male pseudohermaphroditism, testes are present but a block in testosterone synthesis prevents the genetically male fetus from developing male structures. The child appears to be a female, but at puberty, the adrenal glands begin to produce testosterone (as they normally do in both sexes) and masculinization ensues. The voice deepens as muscles build up into a masculine physique. Breasts do not develop, nor does menstruation occur. The clitoris may enlarge so greatly under the adrenal testosterone surge that it looks like a penis.

Actions of Testosterone

Cells of the embryonic testes first produce testosterone after about eight weeks of development. This hormone stimulates the formation of the male reproductive organs, including the penis, scrotum, prostate gland, seminal vesicles, and ducts. Later in development, testosterone causes the testes to descend into the scrotum.

During puberty, testosterone stimulates enlargement of the testes (the primary male sex characteristic) and accessory organs of the

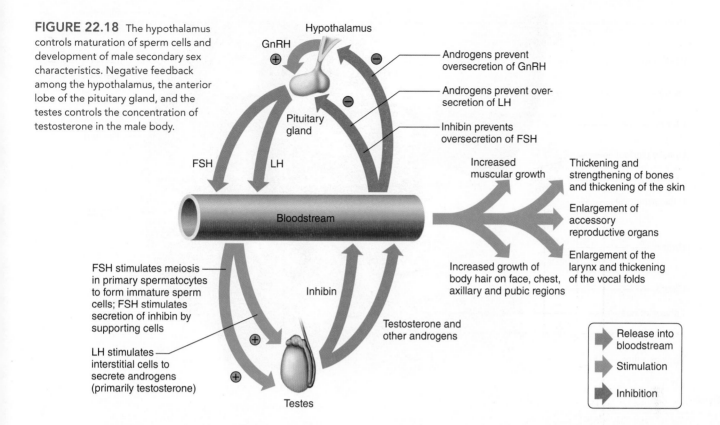

FIGURE 22.18 The hypothalamus controls maturation of sperm cells and development of male secondary sex characteristics. Negative feedback among the hypothalamus, the anterior lobe of the pituitary gland, and the testes controls the concentration of testosterone in the male body.

Hypothalamus

GnRH

Androgens prevent oversecretion of GnRH

Androgens prevent over-secretion of LH

Inhibin prevents oversecretion of FSH

Pituitary gland

FSH   LH

Bloodstream

Increased muscular growth

Thickening and strengthening of bones and thickening of the skin

Enlargement of accessory reproductive organs

Enlargement of the larynx and thickening of the vocal folds

Increased growth of body hair on face, chest, axillary and pubic regions

FSH stimulates meiosis in primary spermatocytes to form immature sperm cells; FSH stimulates secretion of inhibin by supporting cells

Inhibin

Testosterone and other androgens

LH stimulates interstitial cells to secrete androgens (primarily testosterone)

Testes

Release into bloodstream

Stimulation

Inhibition

reproductive system, as well as development of male *secondary sex characteristics,* which are special features associated with the adult male body. Secondary sex characteristics in the male include:

1. increased growth of body hair, particularly on the face, chest, axillary region, and pubic region; growth of hair on the scalp may slow
2. enlargement of the larynx and thickening of the vocal folds, with lowering of the pitch of the voice
3. thickening of the skin
4. increased muscular growth, broadening of the shoulders, and narrowing of the waist
5. thickening and strengthening of bones

Testosterone also increases the rate of cellular metabolism and red blood cell production. For this reason, the average number of red blood cells in a microliter of blood is usually greater in males than in females. Testosterone stimulates sexual activity by affecting certain parts of the brain.

Regulation of Male Sex Hormones

The extent to which male secondary sex characteristics develop is directly related to the amount of testosterone that the interstitial cells secrete. The hypothalamus regulates testosterone output through negative feedback (fig. 22.18).

As the concentration of testosterone in the blood increases, the hypothalamus becomes inhibited, decreasing its stimulation of the anterior pituitary gland by GnRH. As the pituitary's secretion of LH falls in response, the amount of testosterone the interstitial cells release decreases.

As the blood testosterone concentration drops, the hypothalamus becomes less inhibited, and it once again stimulates the anterior pituitary gland to release LH. The increasing secretion of LH causes the interstitial cells to release more testosterone, and blood testosterone concentration increases. Testosterone level decreases somewhat during and after the *male climacteric,* which is a decline in sexual function that accompanies aging. At any given age, the testosterone concentration in the male body is regulated to remain relatively constant.

 PRACTICE

22 Which hormone initiates the changes associated with male sexual maturity?

23 Describe several male secondary sex characteristics.

24 Explain how the secretion of male sex hormones is regulated.

22.4 | Organs of the Female Reproductive System

The organs of the female reproductive system are specialized to produce and maintain the female sex cells, the *egg cells* (or oocytes); transport these cells to the site of fertilization; provide a favorable environment for a developing offspring; move the offspring to the outside; and produce female sex hormones.

The *primary sex organs* (gonads) of this system are the two ovaries, which produce the female sex cells and sex hormones. The *accessory sex organs* (secondary sex organs) of the female reproductive system are the internal and external reproductive organs (fig. 22.19; reference plates 5 and 6, pp. 42–43).

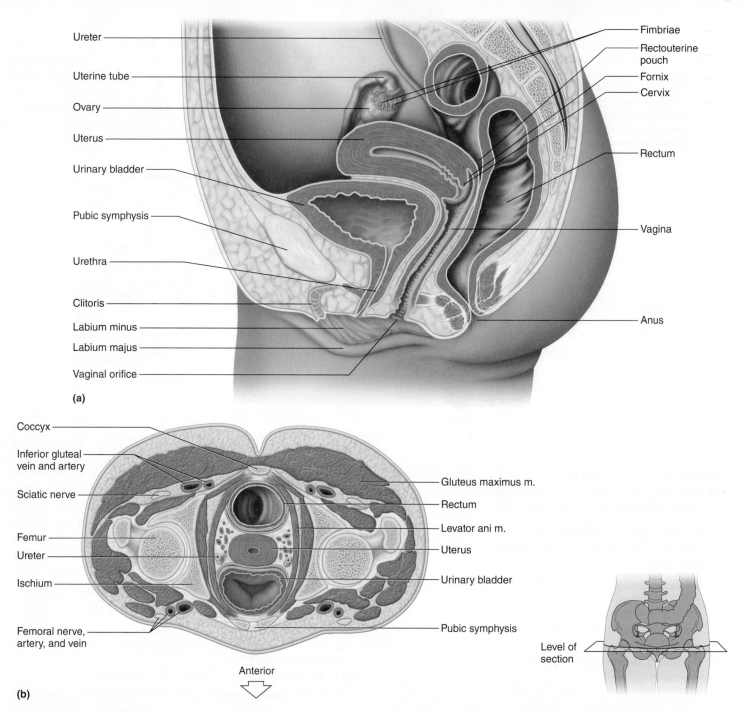

FIGURE 22.19 AP|R The paired ovaries are the primary female sex organs, and the other structures, both internal and external, are accessory sex organs. (*a*) Sagittal view. (*b*) Transverse section of the female pelvic cavity. (*m*. stands for muscle.)

Ovaries

The two **ovaries** are solid, ovoid structures measuring about 3.5 centimeters in length, 2 centimeters in width, and 1 centimeter in thickness. The ovaries lie in shallow depressions (ovarian fossae) on each side in the lateral wall of the pelvic cavity (fig. 22.20).

Ovary Attachments

Several structures help hold each ovary in position. A fold of peritoneum containing ovarian blood vessels and nerves attaches each ovary to the largest of these, called the *broad ligament*.

The broad ligament is also attached to the uterine tubes and the uterus.

A small fold of peritoneum, called the *suspensory ligament*, holds the ovary at its upper end. This ligament also surrounds ovarian blood vessels and nerves. At its lower end, the ovary is attached to the uterus by a rounded, cordlike thickening of the broad ligament called the *ovarian ligament* (fig. 22.20).

Ovary Descent

The ovaries in a female fetus, like the testes in a male fetus, originate from masses of tissue posterior to the parietal peritoneum, near the developing kidneys. During development, these structures descend to locations just inferior to the pelvic brim, where they remain attached to the lateral pelvic wall.

Ovary Structure

The tissues of an ovary can be subdivided into two rather indistinct regions, an inner *medulla* and an outer *cortex*. The ovarian medulla is mostly composed of loose connective tissue and contains relatively larger blood vessels, lymphatic vessels, and nerve fibers. The ovarian cortex consists of more compact tissue and has a granular appearance due to tiny masses of cells called *ovarian follicles*, within which egg cells develop.

A layer of cuboidal epithelial cells covers the free surface of the ovary. Just beneath this epithelium is a layer of dense connective tissue called the *tunica albuginea*.

PRACTICE

25 What are the primary sex organs of the female?

26 Describe the attachments of the ovary.

27 Describe the structure of an ovary.

Primordial Follicles

During prenatal (before birth) development of a female, oogonia divide by mitosis, producing more oogonia. The oogonia develop into *primary oocytes*. Each primary oocyte is closely surrounded by a layer of flattened epithelial cells called *follicular cells,* forming a **primordial follicle.**

Early in development, the primary oocytes begin to undergo meiosis, but the process soon halts and does not continue until the individual reaches puberty. Once the primordial follicles appear,

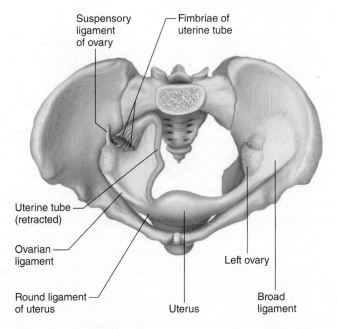

FIGURE 22.20 **AP|R** The ovaries are located on each side against the lateral walls of the pelvic cavity. The right uterine tube is retracted to reveal the ovarian ligament.

no new ones form. Of the 250,000 to 500,000 primordial follicles present at birth, 90% will be lost to apoptosis during infancy and early adulthood. Fewer than 400 or 500 will be released from the ovary during the female's reproductive life, with less than ten going on to form a new individual.

> Children of older mothers are at risk of having abnormal numbers of chromosomes because when meiosis resumes in primary oocytes after decades, spindle fibers can misalign, leading to oocytes with extra or missing chromosomes. Exposure to radiation, viruses, and toxins may contribute to the risk.

Oogenesis

Oogenesis (o'o-jen'ĕ-sis) is the process of egg cell formation. Beginning at puberty, some primary oocytes are stimulated to continue meiosis. The resulting cells, as in the case of sperm cells, have one-half as many chromosomes (23) in their nuclei as their parent cells.

Unlike a primary spermatocyte, when a primary oocyte divides, the cytoplasm is distributed unequally. One of the resulting cells, called a *secondary oocyte,* is large, and the other, called the *first polar body,* is small (fig. 22.21).

The large secondary oocyte represents a future *egg cell* (ovum) that can unite with a sperm cell, becoming fertilized. If this happens, the oocyte divides unequally, producing a tiny *second polar body* and a large fertilized egg cell, or **zygote** (zi'gōt). The polar bodies have no further function, and they begin to degenerate fifteen hours post fertilization.

Formation of polar bodies may appear wasteful, but it has an important biological function. It allows for production of an egg cell that has the massive amounts of cytoplasm and abundant organelles required to carry a zygote through the first few cell divisions, yet the haploid number of chromosomes.

PRACTICE

28 Describe the major events of oogenesis.

29 What is the function of polar body formation?

Follicle Maturation

At puberty, the anterior pituitary gland secretes increased amounts of FSH, and the ovaries enlarge in response. Primordial follicles begin to mature into **primary follicles** (pri'ma-re fol'ĭ-klz). Figure 22.22 traces the maturation of a primordial follicle in an ovary. During early follicle maturation, the primary oocyte within the follicle enlarges, and surrounding follicular cells proliferate by mitosis, giving rise to stratified epithelium composed of *granulosa cells*. A layer of glycoprotein, called the **zona pellucida** (zo'nah pel-u'cĭ-dah), gradually separates the primary oocyte from the granulosa cells.

Meanwhile, the ovarian cells outside the follicle organize into layers. The *inner vascular layer* (theca interna) is largely composed of steroid-secreting cells, plus some loose connective

FIGURE 22.21 During oogenesis, (*a*) a single egg cell (secondary oocyte) results from meiosis of a primary oocyte. If the egg cell is fertilized, it generates a second polar body and becomes a zygote. (Note: The second meiotic division does not occur in the egg cell if it is not fertilized.) (*b*) Light micrograph of a secondary oocyte and a polar body (arrow) (700×).

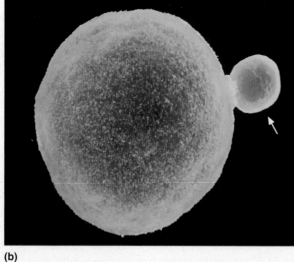

Primary oocyte

(46 chromosomes, each with 2 chromatids)

First meiotic division

Secondary oocyte

(23 chromosomes, each with 2 chromatids)

Sperm cell (23 chromosomes)

Fertilization

Sperm nucleus

Second meiotic division

Zygote (46 chromosomes, 23 from sperm cell and 23 from egg cell; each chromatid now an independent chromosome)

Second polar body degenerating

First polar body (23 chromosomes, each with 2 chromatids)

Second meiotic division

or

First polar body degenerating

Polar bodies degenerating

(a)

(b)

tissue and blood vessels. The *outer fibrous layer* (theca externa) consists of tightly packed connective tissue cells (fig. 22.23).

The follicular cells continue to proliferate, and when they form six to twelve layers of cells, irregular, fluid-filled spaces appear among them. It may take 150 days or more for the primordial follicle to reach the large pre-antral stage of development. The spaces within the follicle soon join to form a single fluid-filled cavity (antrum), and the primary oocyte is pressed to one side of the follicle. At this stage, which may take another sixty-five or more days to complete, the follicle is called an *antral follicle*.

After about fifteen days, the *mature antral follicle* (preovulatory, or Graafian, follicle) is about 20 millimeters in diameter, and its fluid-filled cavity bulges outward on the surface of the ovary, like a blister. The secondary oocyte within the mature follicle is a large, spherical cell, surrounded by a thick zona pellucida, attached to a mantle of follicular cells called the **corona radiata** (ko-ro′nah ra-di-ă′ta). Processes from these follicular cells extend through the zona pellucida and supply nutrients to the oocyte (fig. 22.23).

Although many primary follicles may begin maturing at any given time, one follicle usually outgrows the others. Typically, only the dominant follicle fully develops, and the other follicles

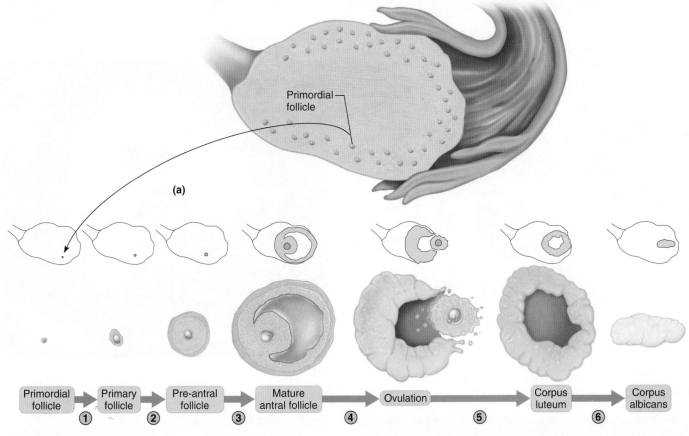

(a)

Primordial follicle

| Primordial follicle | → | Primary follicle | → | Pre-antral follicle | → | Mature antral follicle | → | Ovulation | → | Corpus luteum | → | Corpus albicans |
| ① | | ② | | ③ | | ④ | | ⑤ | | ⑥ | | |

① At puberty, in response to hormonal stimulation, primordial follicles (0.02 mm) begin maturing into primary follicles (0.1 mm).

② It takes two to three months for the primary follicles to reach the pre-antral stage. It takes another 70 additional days for a pre-antral follicle to reach a size of 0.2 mm.

③ In another 65 to 70 days the pre-antral follicle will have become an antral follicle.

④ The antral follicle will grow quite rapidly and mature in about 15 days prior to ovulation (20 mm).

⑤ Following ovulation, the remants of the antral follicle become the corpus luteum.

⑥ If fertilization of the ovulated secondary oocyte does not occur, the corpus luteum degenerates to become the corpus albicans.

(b)

FIGURE 22.22 Follicle maturation in an ovary. (*a*) Ovary, containing primordial follicles awaiting development, with it's associated portion of the uterine tube. (*b*) Stages in development of a follicle. (*c*) Light micrograph of an ovary showing various stages of follicular development occuring simultaneously throughout the cortex (250×).

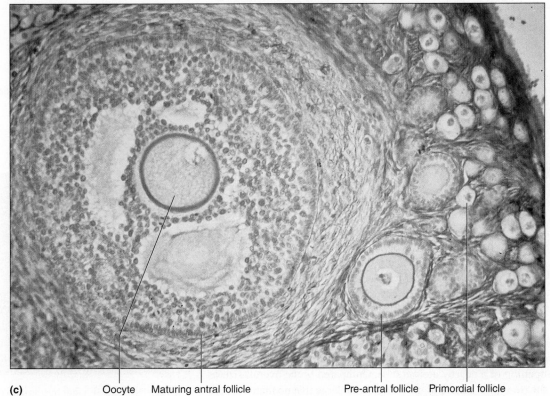

(c)   Oocyte   Maturing antral follicle   Pre-antral follicle   Primordial follicle

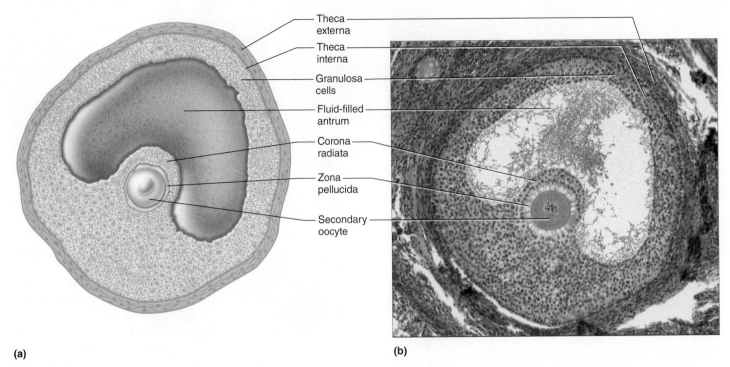

Theca externa
Theca interna
Granulosa cells
Fluid-filled antrum
Corona radiata
Zona pellucida
Secondary oocyte

(a)   **(b)**

FIGURE 22.23 Ovarian follicle. (*a*) Structure of a mature antral (Graafian) follicle. (*b*) Light micrograph of a mature antral follicle (250×).

 Which structure is formed by follicular cells attached to and surrounding the secondary oocyte?

Answer can be found in Appendix G.

degenerate. This entire process from a primordial follicle to a dominant, fully developed mature antral follicle takes almost 300 days. The process is ongoing in the ovaries, such that a new dominant mature antral follicle becomes ready for ovulation approximately every 28 days.

> Certain drugs used to treat female infertility may cause a woman to "superovulate." If more than one follicle fully develops, more than one secondary oocyte is ovulated. If all of the ovulated secondary oocytes are fertilized, multiple births may result.

Ovulation

As a follicle matures, its primary oocyte undergoes meiosis I, giving rise to a secondary oocyte and a first polar body. A process called **ovulation** (o″vu-la′shun) releases the secondary oocyte and first polar body from the mature antral follicle (**fig. 22.24**).

Release of LH from the anterior pituitary gland plays a role in triggering ovulation, during which the mature antral follicle swells and its wall weakens. Eventually the wall ruptures, and the follicular fluid, accompanied by the secondary oocyte, is released from the ovary's surface (fig. 22.24).

After ovulation, the secondary oocyte and one or two layers of follicular cells surrounding it are normally propelled to the opening of a nearby uterine tube. Compared to the overall size of the ovary, the mature antral follicle is so large that no matter where

the secondary oocyte is released from the ovary it will contact the branched extensions (fimbriae) of the associated uterine tube. If the secondary oocyte is not fertilized within hours, it degenerates.

PRACTICE

30 Describe changes that occur in a follicle and its oocyte during maturation.

31 What causes ovulation?

32 What happens to an oocyte following ovulation?

Uterine tube

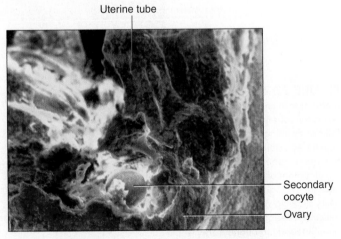

Secondary oocyte
Ovary

FIGURE 22.24 Light micrograph of a follicle during ovulation (72×).

Female Internal Accessory Reproductive Organs

The *internal accessory organs* of the female reproductive system include a pair of uterine tubes, a uterus, and a vagina.

Uterine Tubes

The **uterine tubes** (Fallopian tubes, or oviducts) are suspended by portions of the broad ligament and open near the ovaries. Each tube, about 10 centimeters long and 0.7 centimeters in diameter, passes medially to the uterus, penetrates its wall, and opens into the uterine cavity. This allows for the movement of the egg from the ovary to the uterus.

Near each ovary, a uterine tube expands to form a funnel-shaped **infundibulum** (in"fun-dib'u-lum), which partially encircles the ovary. On its margin, the infundibulum bears a number of irregular, branched extensions called **fimbriae** (fim'bre) (see figs. 22.19*a* and 22.20, fig. 22.25). Although the infundibulum generally does not touch the ovary, one of the larger extensions (ovarian fimbria) connects directly to the ovary. This is why some cases of ovarian cancer actually begin in cells of the fimbriae.

The wall of a uterine tube consists of an inner mucosal layer, a middle muscular layer of smooth muscle, and an outer covering of peritoneum. The mucosal layer is drawn into many longitudinal folds and is lined with simple columnar epithelial cells, some *ciliated* (fig. 22.26). The epithelium secretes mucus, and the cilia beat toward the uterus. These actions help draw the secondary oocyte and expelled follicular fluid into the infundibulum following ovulation. Ciliary action and peristaltic contractions of the tube's muscular layer aid transport of the egg down the uterine tube. Fertilization occurs in the uterine tube.

Uterus

The **uterus** receives the embryo that develops from an egg cell fertilized in the uterine tube and sustains development of the fetus. The uterus is hollow, muscular, and shaped like an inverted pear.

The *broad ligament,* which also attaches to the ovaries and uterine tubes, extends from the lateral walls of the uterus to the pelvic walls and floor, creating a drape across the top of the pelvic cavity (see fig. 22.25). A flattened band of tissue within the broad ligament, called the *round ligament,* connects the upper end of the uterus to the anterior pelvic wall (see figs. 22.20 and 22.25). The *cardinal ligaments* attach the lower portion of the uterus to the lateral walls of the pelvis, and the *uterosacral ligaments* attach the uterus to the sacrum posteriorly.

The size of the uterus changes greatly during pregnancy. In its nonpregnant, adult state, it is about 7 centimeters long, 5 centimeters wide (at its broadest point), and 2.5 centimeters in diameter. The uterus is located medially in the anterior part of the pelvic cavity, superior to the vagina, and usually bends forward over the urinary bladder.

The upper two-thirds, or *body,* of the uterus has a dome-shaped top, called the *fundus,* and is joined by the uterine tubes, which enter its wall at its broadest part. The lower one-third, or neck, of the uterus is called the **cervix.** This tubular part extends downward into the upper part of the vagina. The cervix surrounds the opening called the *cervical orifice* (ostium uteri), through which the uterus opens to the vagina.

The uterine wall is thick and composed of three layers (fig. 22.27). The **endometrium** (en"do-me'tre-um), the inner mucosal layer, is covered with columnar epithelium and contains abundant tubular glands. The **myometrium** (mi"o-me'tre-um), a thick, middle, muscular layer, consists largely of bundles of smooth

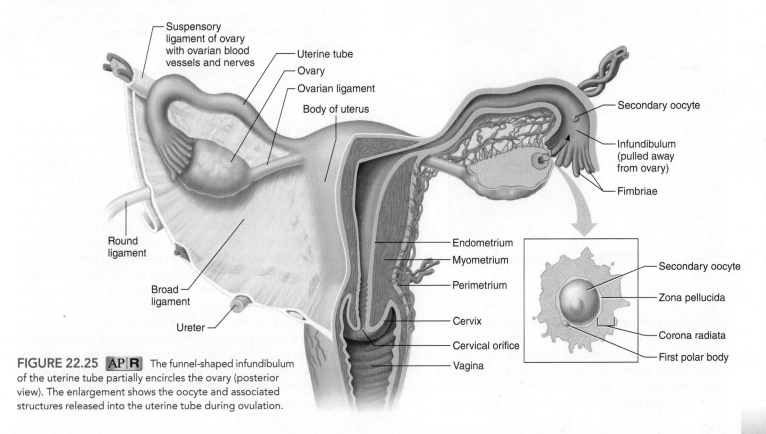

FIGURE 22.25 AP|R The funnel-shaped infundibulum of the uterine tube partially encircles the ovary (posterior view). The enlargement shows the oocyte and associated structures released into the uterine tube during ovulation.

Labels on figure: Suspensory ligament of ovary with ovarian blood vessels and nerves; Uterine tube; Ovary; Ovarian ligament; Body of uterus; Round ligament; Broad ligament; Ureter; Secondary oocyte; Infundibulum (pulled away from ovary); Fimbriae; Endometrium; Myometrium; Perimetrium; Cervix; Cervical orifice; Vagina; Secondary oocyte; Zona pellucida; Corona radiata; First polar body

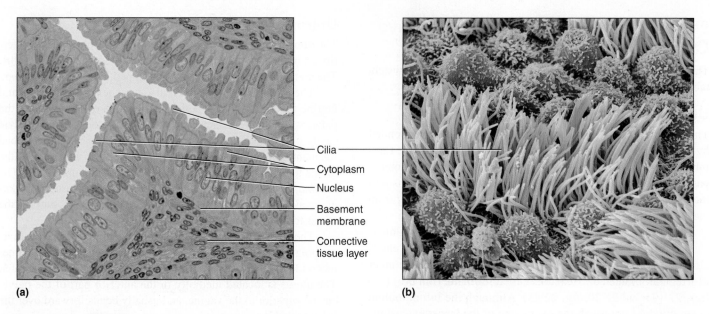

FIGURE 22.26 AP|R Uterine tube. (*a*) Light micrograph of a uterine tube (800×). (*b*) Falsely colored scanning electron micrograph showing nonciliated (pink) cells and ciliated (green) cells that line the uterine tube (1,200×).

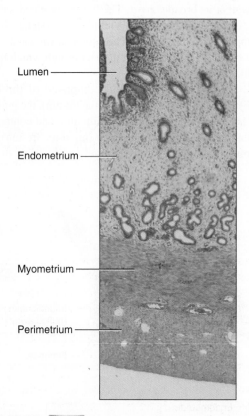

Lumen

Endometrium

Myometrium

Perimetrium

FIGURE 22.27 AP|R Light micrograph of the uterine wall (10×).

muscle fibers in longitudinal, circular, and spiral patterns and is interlaced with connective tissues. During the monthly female reproductive cycles and during pregnancy, the endometrium and myometrium extensively change. The **perimetrium** (per″-ĭ-me′tre-um) consists of an outer serosal layer, which covers the body of the uterus and part of the cervix.

A "pap (Papanicolaou) smear test" identifies changes in cells sampled from the cervix that indicate a localized precancerous condition called carcinoma *in situ*. The U.S. Preventive Services Task Force advises that women aged 21 to 30 years have a Pap smear every three years, no matter when they became sexually active, and after age 30 every five years. A second test performed with a Pap smear detects variants of human papilloma virus (HPV) that cause cervical cancer. Coupling HPV testing to Pap smears avoids performing more invasive procedures on women who receive false positive results on the Pap smear. Many young people are being vaccinated against the cancer-causing variants of HPV.

Vagina

The **vagina** is a fibromuscular tube, about 9 centimeters long, extending from the uterus to the outside of the body (see fig. 22.19*a*). It conveys uterine secretions, receives the erect penis during sexual intercourse, and provides an open channel for the offspring during birth.

The vagina extends upward and back into the pelvic cavity. It is posterior to the urinary bladder and urethra, anterior to the rectum, and attached to these structures by connective tissues. The upper one-fourth of the vagina is separated from the rectum by a pouch (rectouterine pouch). The tubular vagina also surrounds the end of the cervix, and the recesses between the vaginal wall and the cervix are termed *fornices* (sing., *fornix*). The fornices are clinically important because they are thin-walled and allow the physician to palpate the internal abdominopelvic organs during a physical examination. Also, the posterior fornix, which is somewhat longer than the others, provides surgical access to the peritoneal cavity through the vagina.

The *vaginal orifice* is partially closed by a thin membrane of connective tissue and stratified squamous epithelium called the

hymen. A central opening of varying size allows uterine and vaginal secretions to pass to the outside of the body.

The vaginal wall has three layers. The inner *mucosal layer* is stratified squamous epithelium and is drawn into many longitudinal and transverse ridges (vaginal rugae). This layer lacks mucous glands; the mucus in the lumen of the vagina comes from the glands of the cervix and the vestibular glands at the mouth of the vagina.

The middle *muscular layer* of the vagina mainly consists of smooth muscle fibers in longitudinal and circular patterns. At the lower end of the vagina is a thin band of striated muscle. This band helps close the vaginal opening; however, another voluntary muscle (bulbospongiosus) is primarily responsible for closing this orifice.

The outer *fibrous layer* consists of dense connective tissue interlaced with elastic fibers. It attaches the vagina to surrounding organs.

> Four young women born with a rare condition that causes underdevelopment of the uterus and vagina who received tissue-engineered vaginas are doing well with their new organs a decade later. The anatomy and physiology of the replacement parts appear to be normal, and the patients report normal sexual functioning.
>
> To build the vaginas, the tissue engineers began with epithelium and muscle from each patient's external genitalia. The cells were expanded in a culture system and then seeded onto a biodegradable scaffold in the shape of a vagina, designed to fit each patient. Surgeons then fashioned a canal in the correct anatomical location and sutured the replacement vagina into place. The cells divided as blood vessels and nerves infiltrated the implant. The body eventually absorbed the synthetic scaffolding material as the three layers of the new vagina took shape.
>
> Years later, the new vaginas fit perfectly into the young women's reproductive tracts. Tissue-engineered vaginas can also be used in patients who have cancer or injuries of the vagina.

 PRACTICE

33 How does a secondary oocyte move into the infundibulum following ovulation?

34 How is a secondary oocyte moved along a uterine tube?

35 Describe the structure of the uterus.

36 Describe the structure of the vagina.

Female External Reproductive Organs

The *external accessory organs* of the female reproductive system include the labia majora, the labia minora, the clitoris, and the vestibular glands (fig. 22.28). These structures that surround the openings of the urethra and vagina compose the **vulva.**

Labia Majora

The **labia majora** (sing., *labium majus*) enclose and protect the other external reproductive organs. They are homologous structures, similar in origin and function, to the scrotum of the male and are composed of rounded folds of adipose tissue and a thin layer of

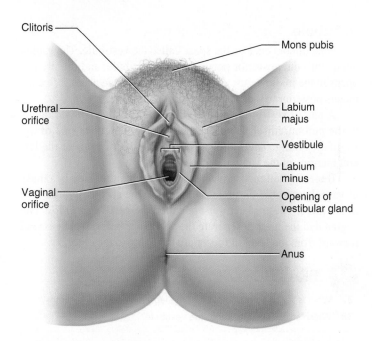

FIGURE 22.28 AP|R Female external reproductive organs and associated structures. The perineum includes the diamond-shaped area of the combined urogenital and anal regions extending from the pubic symphysis to the sacrum.

smooth muscle, covered by skin. On the outside, this skin includes hairs, sweat glands, and sebaceous glands, whereas on the inside, it is thinner and hairless.

The labia majora lie close together and are separated longitudinally by a cleft (pudendal cleft), which includes the urethral and vaginal openings. The labia merge at their anterior ends to form a medial, rounded elevation of adipose tissue called the *mons pubis,* which overlies the pubic symphysis. The labia taper at their posterior ends and merge near the anus.

Labia Minora

The **labia minora** (sing., *labium minus*) are flattened longitudinal folds between the labia majora. They are composed of connective tissue richly supplied with blood vessels, which impart a pinkish appearance. Stratified squamous epithelium covers this tissue. Posteriorly, the labia minora merge with the labia majora. Anteriorly, they converge to form a hoodlike covering around the clitoris.

Clitoris

The **clitoris** (kli'to-ris) is visible as a small projection at the anterior end of the vulva between the labia minora. In most women it is about 2 centimeters long and 0.5 centimeter in diameter, most of which is embedded in surrounding tissues. The clitoris is homologous to the penis and has a similar structure. It is composed of two columns of erectile tissue called *corpora cavernosa.* A septum separates these columns, which are covered with dense connective tissue.

At the root of the clitoris, the corpora cavernosa diverge to form *crura,* which, in turn, attach to the sides of the pubic arch. At the anterior end of the clitoris, a small mass of erectile tissue forms a glans, which is richly supplied with sensory nerve fibers.

Vestibule

The labia minora enclose the space called the **vestibule.** The vagina opens into the posterior portion of the vestibule, and the urethra opens in the midline, just anterior to the vagina and about 2.5 centimeters posterior to the glans of the clitoris.

A pair of **vestibular glands** (Bartholin's glands), homologous to the bulbourethral glands in the male, lie on either side of the vaginal opening. Their ducts open into the vestibule near the lateral margins of the vaginal orifice.

Beneath the mucosa of the vestibule on either side is a mass of vascular erectile tissue homologous to the corpus spongiosum in the male. These structures are called the *vestibular bulbs.* The vagina and the urethra separate the vestibular bulbs, which extend forward from the level of the vaginal opening to the clitoris.

 PRACTICE

37 What is the male counterpart of the labia majora? Of the clitoris?

38 Which structures are within the vestibule?

Erection, Lubrication, and Orgasm

Erectile tissues in the clitoris and around the vaginal entrance respond to sexual stimulation. Following such stimulation, parasympathetic impulses from the sacral portion of the spinal cord release the vasodilator nitric oxide, dilating the arteries associated with the erectile tissues. As a result, blood inflow increases, tissues swell, and the vagina expands and elongates.

If sexual stimulation is sufficiently intense, parasympathetic impulses stimulate the vestibular glands to secrete mucus into the vestibule. This secretion moistens and lubricates the tissues surrounding the vestibule and the lower end of the vagina, facilitating insertion of the penis into the vagina. Mucus secretion continuing during sexual intercourse helps prevent irritation of tissues that might occur if the vagina remained dry.

The clitoris has abundant sensory nerve fibers, which are especially sensitive to local stimulation. The culmination of such stimulation is orgasm, the pleasurable sensation of physiological and psychological release.

Just prior to orgasm, the tissues of the outer third of the vagina engorge with blood and swell. This increases the friction on the penis during intercourse. Orgasm initiates a series of reflexes involving the sacral and lumbar portions of the spinal cord. In response to these reflexes, the muscles of the perineum and the walls of the uterus and uterine tubes contract rhythmically (fig. 22.29). These contractions help transport sperm cells through the female reproductive tract toward the upper ends of the uterine tubes.

Following orgasm, the flow of blood into the erectile tissues decreases, and the muscles of the perineum and reproductive tract relax. Consequently, the organs return to a state similar to that prior to sexual stimulation. Table 22.2 summarizes the functions of the female reproductive organs.

PRACTICE

39 What events result from parasympathetic stimulation of the female reproductive organs?

40 What changes occur in the vagina just prior to and during female orgasm?

41 How do the uterus and the uterine tubes respond to orgasm?

TABLE 22.2	Functions of the Female Reproductive Organs
Organ	**Function**
Ovary	Produces oocytes and female sex hormones
Uterine tube	Conveys secondary oocyte toward uterus; site of fertilization; conveys developing embryo to uterus
Uterus	Protects and sustains embryo during pregnancy
Vagina	Conveys uterine secretions to outside of body; receives erect penis during sexual intercourse; provides open channel for offspring during birth process
Labium majus	Encloses and protects other external reproductive organs
Labium minus	Forms margin of vestibule; protects openings of vagina and urethra
Clitoris	Produces feelings of pleasure during sexual stimulation due to abundant sensory nerve endings in glans
Vestibule	Space between labia minora that contains vaginal and urethral openings
Vestibular gland	Secretes fluid that moistens and lubricates the vestibule

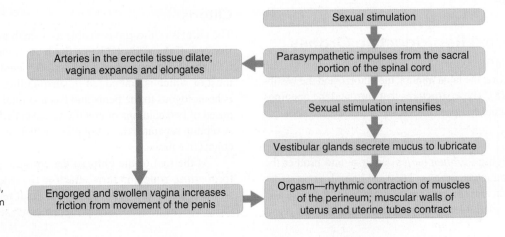

FIGURE 22.29
Mechanism of erection, lubrication, and orgasm in the female.

22.5 | Hormonal Control of Female Reproductive Functions

The hypothalamus, the anterior pituitary gland, and the ovaries secrete hormones that control maturation of female sex cells, development and maintenance of female secondary sex characteristics, and changes that occur during the monthly reproductive cycle.

Female Sex Hormones

A female body is reproductively immature until about ten years of age. Then, the hypothalamus begins to secrete increasing amounts of GnRH, which, in turn, stimulate the anterior pituitary gland to release the gonadotropins FSH and LH. These hormones play primary roles in controlling female sex cell maturation and in producing female sex hormones.

Several tissues, including the ovaries, the adrenal cortices, and the placenta (during pregnancy), secrete female sex hormones. These hormones include the group of **estrogens** (es'tro-jenz) and **progesterone** (pro-jes'te-rōn). *Estradiol* is the most abundant of the estrogens, which also include *estrone* and *estriol*.

The primary source of estrogens in a nonpregnant female is the ovaries, although adipose tissue also synthesizes some estrogens from adrenal androgens. At puberty, under the influence of the anterior pituitary gland, the ovaries secrete increasing amounts of estrogens. Estrogens stimulate enlargement of reproductive organs, including the vagina, uterus, uterine tubes, and ovaries, as well as the external structures; stimulate the endometrium to thicken; and are also responsible for the development and maintenance of female *secondary sex characteristics*. These are listed in **figure 22.30** and include the following:

1. development of the breasts and the ductile system of the mammary glands in the breasts
2. increased deposition of adipose tissue in the subcutaneous layer generally and in the breasts, thighs, and buttocks particularly
3. increased vascularization of the skin

The ovaries are also the primary source of progesterone in a nonpregnant female. This hormone promotes changes in the uterus during the female reproductive cycle, affects the mammary glands, and helps regulate secretion of gonadotropins from the anterior pituitary gland.

Certain other changes in females at puberty are related to *androgen* (male sex hormone) concentrations. For example, increased growth of hair in the pubic and axillary regions is due to androgen secreted by the adrenal cortices. Conversely, development of the female skeletal configuration, which includes narrow shoulders and broad hips, is a response to a low concentration of androgen.

> Female athletes who train for endurance events, such as a marathon, typically maintain about 6% body fat. Male endurance athletes usually have about 4% body fat. This difference in proportion of body fat reflects the actions of sex hormones in males and females. Testosterone promotes deposition of protein throughout the body and especially in skeletal muscles, whereas estrogens promote deposition of adipose tissue in the breasts, thighs, buttocks, and the subcutaneous layer of the skin.

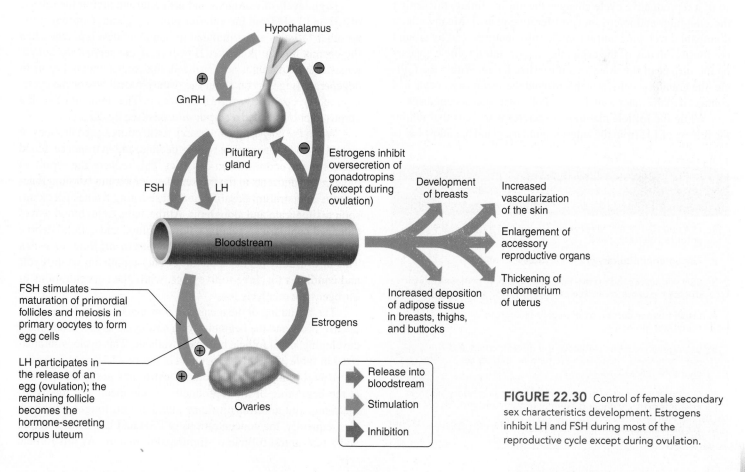

FIGURE 22.30 Control of female secondary sex characteristics development. Estrogens inhibit LH and FSH during most of the reproductive cycle except during ovulation.

PRACTICE

42   What stimulates sexual maturation in a female?

43   Name the major female sex hormones.

44   What are the functions of estrogens?

45   What is the function of androgen in a female?

Female Reproductive Cycle

The female reproductive cycle is characterized by regular, recurring changes in the endometrium, which culminate in menstrual bleeding (menses). Such cycles usually begin near the thirteenth year of life and continue into the early fifties, then cease.

A female's first reproductive cycle, called **menarche** (mě-nar'ke), occurs after the ovaries and other organs of the female reproductive system mature and begin responding to certain hormones. Then, the hypothalamic secretion of GnRH stimulates the anterior pituitary gland to release threshold levels of FSH and LH. As its name implies, FSH stimulates maturation of ovarian follicles. The granulosa cells of the maturing follicles produce increasing amounts of estrogens and some progesterone. LH stimulates certain ovarian cells (theca interna) to secrete precursor molecules (such as testosterone), that are also used to produce estrogens.

In a young female, estrogens stimulate development of various secondary sex characteristics. Estrogens secreted during subsequent reproductive cycles continue development of these traits and maintain them. Table 22.3 summarizes the hormonal control of female secondary sex characteristics.

Increasing concentration of estrogens during the first week or so of a reproductive cycle changes the uterine lining, thickening the glandular endometrium (proliferative phase). Meanwhile, a dominant developing antral follicle fully matures, and by around the fourteenth day of the cycle, the mature antral follicle appears on the surface of the ovary as a blisterlike bulge. Within the follicle, the granulosa cells, which surround the secondary oocyte and connect it to the inner wall, loosen. Follicular fluid accumulates.

While the follicle matures, it secretes estrogens that inhibit the release of LH from the anterior pituitary gland but allow LH to

TABLE 22.3	Hormonal Control of Female Secondary Sex Characteristics
1.	The hypothalamus releases GnRH, which stimulates the anterior pituitary gland.
2.	The anterior pituitary gland secretes FSH and LH.
3.	FSH stimulates the maturation of a follicle; granulosa cells of the follicle produce and secrete estrogens.
4.	LH stimulates certain ovarian cells to secrete estrogen precursor molecules.
5.	Estrogens are responsible for the development and maintenance of most of the female secondary sex characteristics.
6.	Concentrations of androgen affect other secondary sex characteristics, including skeletal growth and growth of hair.
7.	Progesterone, secreted by the ovaries, affects cyclical changes in the uterus and mammary glands.

be stored in the gland. Estrogens also make the anterior pituitary cells more sensitive to the action of GnRH, which is released from the hypothalamus in rhythmic pulses about ninety minutes apart.

Near the fourteenth day of follicular development, the anterior pituitary cells finally respond to the pulses of GnRH and release stored LH. The resulting surge in LH concentration lasts about thirty-six hours. In response to the LH, the primary oocyte within the antral follicle completes meiosis I, giving rise to a secondary oocyte and a polar body. The LH also interacts synergistically with FSH for the induction of complex interactions with prostaglandins, progesterone, plasmin, and proteolytic enzymes, leading to the weakening and rupturing of the bulging follicular wall. This event sends the secondary oocyte and follicular fluid out of the ovary (ovulation).

Following ovulation, the space that contained the follicular fluid fills with blood, which soon clots. Under the influence of LH, the remnants of the follicle within the ovary form a temporary glandular structure in the ovary called a **corpus luteum** (kor'pus loot'e-um) ("yellow body") (see fig. 22.22).

Follicular cells secrete some progesterone during the first part of the reproductive cycle. However, corpus luteum cells secrete abundant progesterone and estrogens during the second half of the cycle. Consequently, as a corpus luteum is established, the blood concentration of progesterone sharply increases.

Progesterone makes the endometrium more vascular and glandular. It also stimulates the uterine glands to secrete more glycogen and lipids (secretory phase). The endometrial tissues fill with fluids containing nutrients and electrolytes, which provide a favorable environment for embryo development.

High levels of estrogens and progesterone inhibit the release of LH and FSH from the anterior pituitary gland. Consequently, no other follicles are stimulated to complete development when the corpus luteum is active. However, if the secondary oocyte released at ovulation is not fertilized, the corpus luteum begins to degenerate (regress) on about the twenty-fourth day of the cycle. Eventually, connective tissue replaces it. The remnant of such a corpus luteum is called a *corpus albicans* (see fig. 22.22).

When the corpus luteum ceases to function, concentrations of estrogens and progesterone rapidly decline, and in response, blood vessels in the endometrium constrict. This reduces the supply of oxygen and nutrients to the thickened endometrium (stratum functionalis and stratum basalis), and these lining tissues (decidua) soon disintegrate and slough off. At the same time, blood leaves damaged capillaries, creating a flow of blood and cellular debris, that passes through the vagina as the *menstrual flow* (menses). This flow usually begins about the twenty-eighth day of the cycle and continues for three to five days, while the concentrations of estrogens are relatively low.

The beginning of the menstrual flow marks the end of a reproductive cycle and the beginning of a new cycle as a new dominant developing antral follicle becomes available. This cycle is summarized in table 22.4 and diagrammed in figure 22.31.

Low blood concentrations of estrogens and progesterone at the beginning of the reproductive cycle mean that the hypothalamus and anterior pituitary gland are no longer inhibited. Consequently, the concentrations of FSH and LH soon increase, and a new antral follicle is stimulated to mature. As this follicle

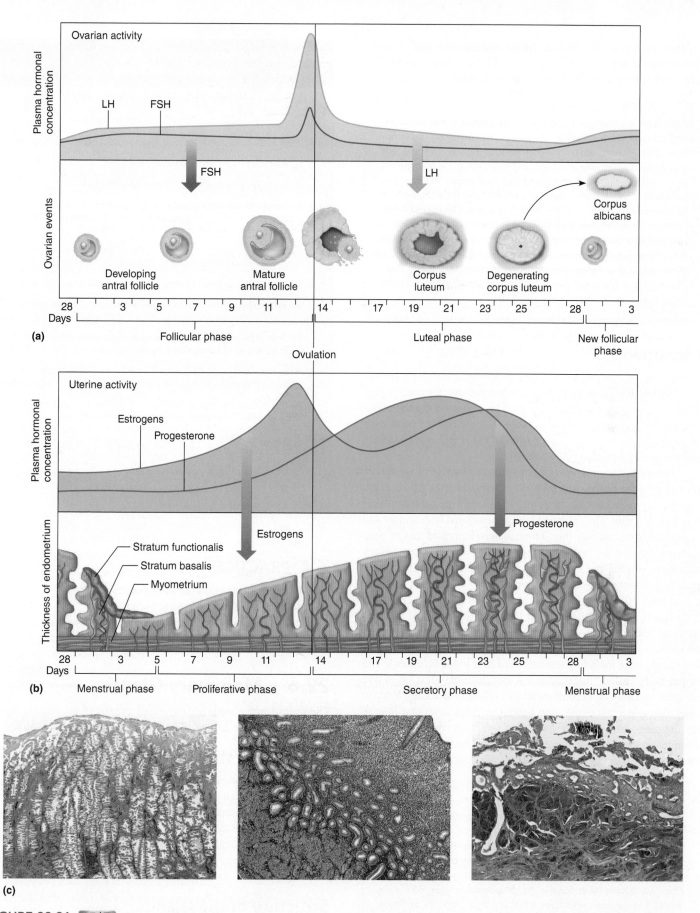

FIGURE 22.31 AP|R Major events in the female reproductive cycle. (*a*) Plasma hormonal concentrations of FSH and LH affect follicle maturation in and ovulation from the ovaries. (*b*) Plasma hormonal concentrations of estrogen and progesterone influence changes in the endometrium. (*c*) Micrographs of the endometrium during (*left panel*) the menstrual phase (3×), (*center panel*) the proliferative phase (40×), and (*right panel*) the secretory phase (14×).

secretes estrogens, the uterine lining undergoes repair, and the endometrium begins to thicken again. Clinical Application 22.3 addresses some causes of infertility in the female.

Elite female athletes and professional dancers may have disturbed reproductive cycles, ranging from diminished menstrual flow (oligomenorrhea) to complete stoppage (amenorrhea). The more active an athlete or dancer, the more likely it is that she will have menstrual irregularities, and this may impair her ability to conceive. Trim elite athletes and dancers have little fat, leading to decreased secretion of the hormone leptin. Decreased leptin secretion is associated with lowered secretion of gonadotropin-releasing hormone from the hypothalamus, which in turn lowers the blood estrogen levels. Adipose tissue itself also produces some estrogen. Adequate estrogen is necessary for fertility. These girls and women are at high risk of developing low bone density and cardiovascular impairments.

Menopause

After puberty, reproductive cycles continue at regular intervals into the late forties or early fifties, when the ovaries start to produce less estrogen and progesterone. This results in the reproductive cycles becoming less predictable. Then within a few months or years, the cycles cease. This period in life is called **menopause** (men'o-pawz), or female climacteric.

Reduced concentrations of estrogens and lack of progesterone may change the female secondary sex characteristics. The breasts, vagina, uterus, and uterine tubes may shrink, and the pubic and axillary hair may thin. The epithelial linings associated with urinary and reproductive organs may thin. There may be increased loss of bone matrix (osteoporosis) and thinning of the skin. Because the pituitary secretions of FSH and LH are no longer inhibited, these hormones may be continuously released for some time.

About 50% of women reach menopause by age fifty, and 85% reach it by age fifty-two. Of these 85%, about 20% have no unusual health effects—they simply stop menstruating. However, about 50% of women experience unpleasant vasomotor signs during menopause, including sensations of heat in the face, neck, and upper body called "hot flashes." A hot flash may last up to five minutes and may be accompanied by chills and sweating. Women may also experience migraine headache, backache, and fatigue during menopause. These vasomotor symptoms may result from changes in the rhythmic secretion of GnRH by the hypothalamus in response to declining concentrations of sex hormones.

To minimize menopause symptoms, some women take hormone replacement therapy (HRT) which consists of estrogen plus progestin. A woman whose uterus has been removed may take estrogen alone, called estrogen replacement therapy (ERT). A doctor prescribes the therapy in any of several forms, including rings, patches, pills, creams, and gels. The lowest effective dose is taken for the shortest possible time. HRT is not advised for women who have a history of or high risk of abnormal blood clotting, heart disease, stroke, breast cancer, or gallbladder disease.

TABLE 22.4	Major Events in a Reproductive Cycle

1. The anterior pituitary gland secretes FSH and LH.

2. FSH stimulates maturation of a dominant follicle.

3. Granulosa cells of the follicle produce and secrete estrogens.
   a. Estrogens maintain secondary sex traits.
   b. Estrogens cause the endometrium to thicken.

4. The anterior pituitary gland releases a surge of LH, which interacts with FSH, progesterone, prostaglandins, plasmin, and proteolytic enzymes leading to ovulation.

5. Following ovulation, follicular and thecal cells become corpus luteum cells, which secrete estrogens and progesterone.
   a. Estrogens continue to stimulate uterine wall development.
   b. Progesterone stimulates the endometrium to become more glandular and vascular.
   c. Estrogens and progesterone inhibit secretion of FSH and LH from the anterior pituitary gland.

6. If the secondary oocyte is not fertilized, the corpus luteum degenerates and no longer secretes estrogens and progesterone.

7. As the concentrations of hormones from the corpus luteum decline, blood vessels in the endometrium constrict.

8. The uterine lining disintegrates and sloughs off, producing a menstrual flow.

9. The anterior pituitary gland is no longer inhibited and again secretes FSH and LH.

10. The reproductive cycle repeats.

 PRACTICE

46 Trace the events of the female reproductive cycle.

47 What causes menstrual flow?

48 What are some changes that may occur at menopause?

22.6 | Mammary Glands

The **mammary glands** are accessory organs of the female reproductive system specialized to secrete milk following pregnancy.

Location of the Glands

The mammary glands are located in the subcutaneous tissue of the anterior thorax within the hemispherical elevations called *breasts*. The breasts overlie the *pectoralis major* muscles and extend from the second to the sixth ribs and from the sternum to the axillae (fig. 22.32a).

A *nipple* is located near the tip of each breast at about the level of the fourth intercostal space. It is surrounded by a circular area of pigmented skin called the *areola* (fig. 22.32b).

Structure of the Glands

A mammary gland is composed of fifteen to twenty irregularly shaped lobes. Each lobe contains glands (alveolar glands), drained

For one out of six couples, trying for parenthood is a time of increasing concern. Infertility is the inability to conceive after a year of trying. A physical cause is found in 90% of cases, and 60% of the time, the abnormality lies in the female's reproductive system. Some medical specialists (reproductive endocrinologists) use the term *subfertility* to distinguish individuals and couples who can conceive unaided, but for whom this may take longer than usual.

One of the more common causes of female infertility is hyposecretion of gonadotropic hormones from the anterior pituitary gland, followed by inability to ovulate (anovulation). Testing the woman's urine for *pregnanediol*, a product of progesterone metabolism, can detect an anovulatory cycle, because the concentration of progesterone normally rises following ovulation. If pregnanediol level in the urine during the latter part of the reproductive cycle does not rise, the woman may not have ovulated.

Fertility specialists can treat absence of ovulation due to too little secretion of gonadotropic hormones by administering hCG (obtained from human placentas) or another ovulation-stimulating biochemical, human menopausal gonadotropin (hMG), which contains LH and FSH and is obtained from the urine of women past menopause. However, either hCG or hMG may overstimulate the ovaries and cause many follicles to release egg cells simultaneously, resulting in multiple births if fertilization occurs.

Another cause of female infertility is *endometriosis*, in which tissue resembling the inner lining of the uterus (endometrium) grows in the abdominal cavity. This may happen if small pieces of the endometrium move up through the uterine tubes during menses and implant in the abdominal cavity. Here the tissue changes as it would in the uterine lining during the reproductive cycle. However, when the tissue begins to break down at the end of the cycle, it cannot be expelled to the outside. Instead, material remains in the abdominal cavity where it may irritate the peritoneum and cause considerable pain. These breakdown products also stimulate formation of fibrous tissue (fibrosis), which may encase the ovary and prevent ovulation or obstruct the uterine tubes. Conception becomes impossible.

Some women become infertile as a result of infections, such as gonorrhea. Infections can inflame and obstruct the uterine tubes or stimulate production of viscous mucus that can plug the cervix and prevent entry of sperm.

The first step in finding the right treatment for a particular patient is to determine the cause of the infertility. Table 22B describes diagnostic tests that a woman having difficulty conceiving may undergo.

Women become infertile if their ovaries are removed, possibly as treatment for persistent masses on the ovaries, or are damaged by cancer treatment. To make future pregnancies possible, these women can have strips of ovarian tissue removed before their cancer treatment begins. The strips are frozen and stored, then thawed and implanted under the skin of the forearm or abdomen or in the pelvic cavity near the ovaries. Secondary oocytes are collected and fertilized *in vitro*.

TABLE 22B	Tests to Assess Female Infertility
Test	**What It Checks**
Hormone levels	If ovulation occurs
Ultrasound	Placement and appearance of reproductive organs and structures
Postcoital test	Cervix examined soon after unprotected intercourse to see if mucus is thin enough to allow sperm through
Endometrial biopsy	Small piece of uterine lining sampled and viewed under microscope to see if it can support an embryo
Hysterosalpingogram	Dye injected into uterine tube and followed with scanner shows if tube is clear or blocked
Laparoscopy	Small, lit optical device inserted near navel to detect scar tissue blocking tubes, which ultrasound may miss

---

by alveolar ducts, which drain into a lactiferous duct that leads to the nipple and opens to the outside. Adipose and irregular dense connective tissues separate the lobes and support the glands, attaching them to the fascia of the underlying pectoral muscles. Other connective tissue, which forms dense strands called *suspensory ligaments,* extends inward from the dermis of the breast to the fascia, helping support the breast. Clinical Application 22.4 discusses diagnosis, treatment, and prevention of breast cancer.

Development of the Breasts

The mammary glands of males and females are similar. As children reach *puberty,* the mammary glands in males do not develop, whereas estrogens stimulate development of the mammary glands in females. The alveolar glands and ducts enlarge, and fat is deposited so that each breast contains adipose tissue, except for in the region of the areola. Chapter 23 (pp. 891–893) describes the hormonal mechanism that stimulates mammary glands to produce and secrete milk.

 PRACTICE

49 Describe the structure of a mammary gland.

50 How does ovarian hormone secretion change the mammary glands?

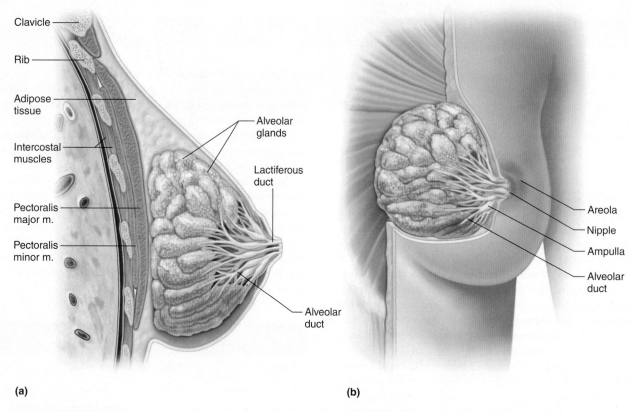

Clavicle

Rib

Adipose
tissue

Intercostal
muscles

Pectoralis
major m.

Pectoralis
minor m.

Alveolar
glands

Lactiferous
duct

Alveolar
duct

Areola

Nipple

Ampulla

Alveolar
duct

(a)

(b)

FIGURE 22.32 **AP|R** Structure of the female breast and mammary glands. (*a*) Sagittal section. (*b*) Anterior view. (*m.* stands for *muscle.*)

22.7 | Birth Control

Birth control is the voluntary regulation of the number of offspring produced and the time they are conceived. This control requires a method of **contraception** (kon″trah-sep′shun) designed to avoid fertilization of an egg cell following sexual intercourse (coitus) or to prevent implantation of a blastocyst. The several methods of contraception have varying degrees of effectiveness.

Coitus Interruptus

Coitus interruptus (also called the withdrawal method) is the practice of withdrawing the penis from the vagina before ejaculation, preventing entry of sperm cells into the female reproductive tract. This method can still result in pregnancy because a male may find it difficult to withdraw just prior to ejaculation. Also, some semen containing sperm cells may reach the vagina before ejaculation occurs.

Rhythm Method

The *rhythm method* (also called timed coitus or natural family planning) requires abstinence from sexual intercourse two days before and one day after ovulation. The rhythm method results in a relatively high rate of pregnancy because accurately identifying infertile ("safe") times to have intercourse is difficult. Another disadvantage of the rhythm method is that it requires adherence to a particular pattern of behavior and restricts spontaneity in sexual activity.

A woman can increase the effectiveness of the rhythm method by measuring and recording her body temperature when she awakens each morning for several months. Body temperature typically rises only about 0.6°F immediately following ovulation, so she has to use a special, extra-sensitive thermometer. However, this technique does not work for all women. More precise may be an "ovulation predictor kit" that detects the surge in LH preceding ovulation.

 PRACTICE

51 Why is coitus interruptus unreliable?

52 Describe the idea behind the rhythm method of contraception.

Mechanical Barriers

Mechanical barrier contraceptives prevent sperm cells from entering the female reproductive tract during sexual intercourse. The *male condom* is a thin latex or natural membrane sheath placed over the erect penis before intercourse to prevent semen from entering the vagina upon ejaculation. A *female condom* resembles a small plastic bag (fig. 22.33*a*). A woman inserts it into her vagina prior to intercourse. The device blocks sperm from reaching the cervix. Some men feel that a condom decreases the sensitivity of the penis during intercourse, and condom use may interrupt spontaneity.

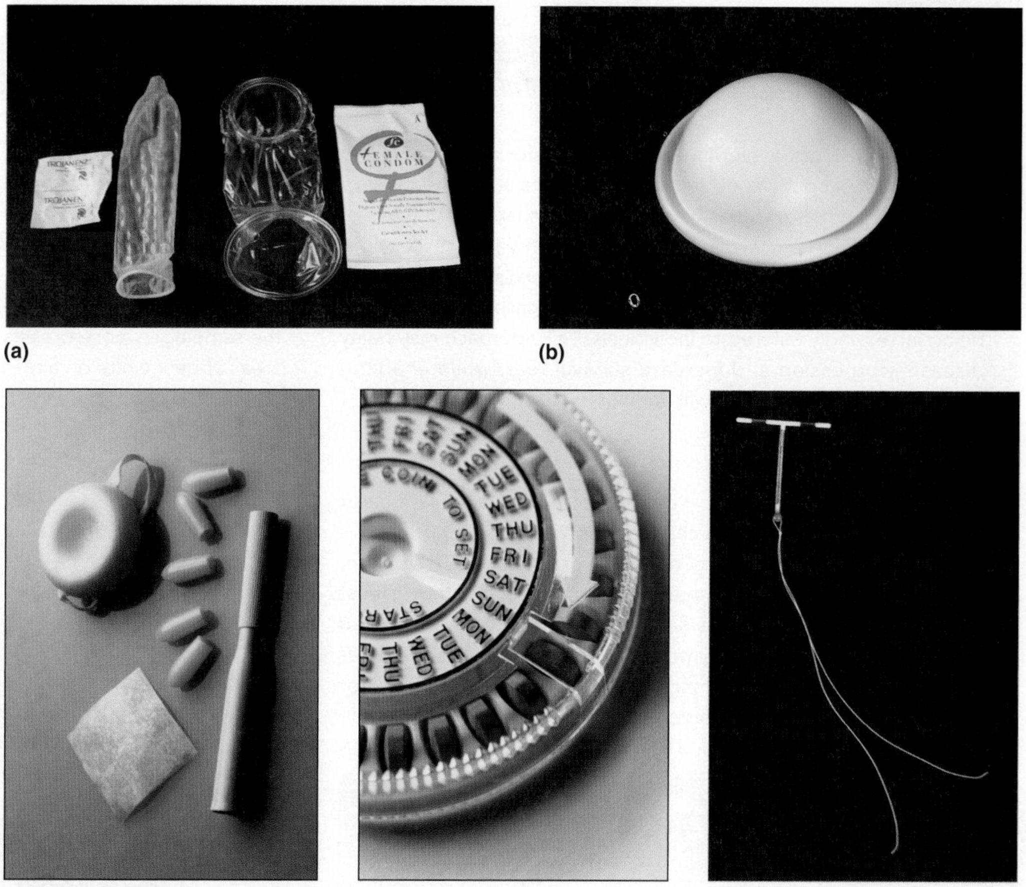

(a)

(b)

FIGURE 22.33 Devices and substances used for birth control include (*a*) male and female condoms, (*b*) a diaphragm, (*c*) spermicide in film, sponge, suppositories, gel (*d*) oral contraceptive, and (*e*) an IUD.

Q Which of these methods of birth control are statistically most effective?

Answer can be found in Appendix G.

(c)

(d)

(e)

However, condoms are inexpensive and may also protect the user from contracting or spreading sexually transmitted infections.

Another mechanical barrier is the *diaphragm.* It is a cup-shaped structure with a flexible ring forming the rim. The diaphragm is inserted into the vagina so that it covers the cervix, preventing sperm cells from entering the uterus (fig. 22.33*b*). To be effective, a diaphragm must be fitted for size by a physician, inserted properly, and used in conjunction with a spermicide applied to the diaphragm surface adjacent to the cervix and to the rim of the diaphragm. The device must be left in position for several hours following sexual intercourse, and can be inserted into the vagina up to six hours before sexual contact.

Similar to but smaller than the diaphragm is the *cervical cap,* which must also be fitted by a physician. A woman inserts it with her fingers before intercourse. Cervical caps have been used for centuries in different cultures and have been made of such varied substances as beeswax, lemon halves, paper, and opium poppy fibers. The cervical shield, which does not need fitting, has a one-way valve that creates suction against the cervix.

Chemical Barriers

Chemical barrier contraceptives include creams, foams, and jellies that have spermicidal properties. These chemicals create an environment in the vagina that is unfavorable for sperm cells (fig. 22.33*c*).

Chemical barrier contraceptives are easy to use but have a high failure rate when used alone. They are more effective when used with a condom or diaphragm. The contraceptive sponge contains a spermicide and blocks the entry of sperm into the uterus.

Combined Hormone Contraceptives

Combined hormone contraceptives deliver estrogen and progestin. Various methods are used to administer the hormones, but all of them work on the same principle with about the same efficacy, although the amounts of the hormones may vary. One device is a small flexible plastic ring that is impregnated with the hormones and is inserted deep into the vagina once a month, remaining in place three out of four weeks. A plastic patch containing the hormones may be applied to the skin on the buttocks, stomach, arm, or upper torso once a week for three out of four weeks. The most commonly used method to deliver the hormones is orally, in pill form (fig. 22.33*d*).

Combined hormone contraceptives contain synthetic estrogen-like and progesterone-like chemicals. These drugs disrupt the normal pattern of gonadotropin (FSH and LH) secretion, preventing follicle maturation and the LH surge that contributes to ovulation. They also thicken cervical mucus to prevent the sperm from joining an egg. Most combined hormone contraceptives cause light monthly bleeding. A newer type of pill causes only four bleeding periods a year.

The lifetime risk of developing breast cancer, for a woman, is 1 in 8 (table 22C). About 1% of breast cancer cases are in men. There are several varieties of breast cancer, distinguished by the types of cells affected and by certain mutations that increase susceptibility to the condition. As research reveals the cellular and molecular characteristics that distinguish types of the disease, treatments are being increasingly tailored to individuals. This approach may delay disease progression and increase survival rate, while enabling patients to avoid drugs that will not work.

Warning Signs

Changes that could signal breast cancer include a small area of thickened tissue; a dimple; a change in contour; or a nipple that is flattened, points in an unusual direction, or produces a discharge. A woman can note these changes by performing a monthly "breast self-exam," in which she lies flat on her back with an arm raised behind her head and feels all parts of each breast. But sometimes breast cancer gives no warning—fatigue and feeling ill may not occur until the disease has spread beyond the breast.

TABLE 22C	Breast Cancer Risk		
By Age	Odds	By Age	Odds
25	1 in 19,608	60	1 in 24
30	1 in 2,525	65	1 in 17
35	1 in 622	70	1 in 14
40	1 in 217	75	1 in 11
45	1 in 93	80	1 in 10
50	1 in 50	85	1 in 9
55	1 in 33	95 or older	1 in 8

After finding a lump, the next step is a physical exam. A health-care provider palpates the breast and does a mammogram, which is an X-ray scan that can pinpoint the location and approximate extent of abnormal tissue (fig. 22C). Women whose breasts have a high percentage of connective tissue and glandular tissue ("dense breasts") are at higher risk of developing cancer than women of the same age whose breasts have more adipose tissue. The dense areas appear white on a mammogram. An ultrasound scan can dis-

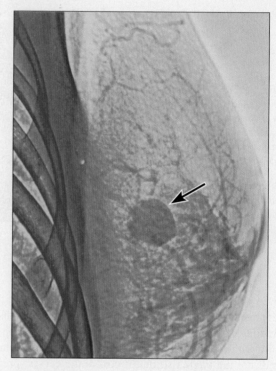

FIGURE 22C AP|R Mammogram of a breast with a tumor (arrow).

If used correctly, combined hormone contraceptives are very effective at preventing pregnancy. However, they may cause nausea, retention of body fluids, increased skin pigmentation, and breast tenderness. Some women, particularly those over thirty-five years of age who smoke, may develop intravascular blood clots, liver disorders, or high blood pressure when using certain types of these contraceptives. The patch, for example, has a higher amount of estrogen than other products and bears a warning label about the association with blood clots.

Similar to, but different from the combined hormone contraceptives is the "minipill," which contains only progestin. The progestin thickens the cervical mucus so the sperm have difficulty reaching the egg. The minipill must be taken every day at approximately the same time for maximum effectiveness. It is slightly less effective than combined hormone contraceptives.

Other Hormone Contraceptives

An intramuscular injection of medroxyprogesterone acetate protects against pregnancy for three months by preventing follicle maturation and release of a secondary oocyte. It also makes the uterine lining less hospitable for a developing embryo. This

A large dose of estrogen in combination with progestin (a synthetic progesterone) can prevent ovulation and may also prevent implantation of a developing embryo in the uterus. Such a "morning-after pill," taken shortly after unprotected intercourse, may prevent pregnancy. However, if the embryo has already implanted, this treatment will not disrupt the pregnancy.

tinguish between a cyst (a fluid-filled sac of glandular tissue) and a tumor (a solid mass). If an area is suspicious, a thin needle is used to take a biopsy (sample) of the tissue, whose cells are scrutinized under a microscope for the telltale characteristics of cancer.

Further tests can identify estrogen and progesterone receptors on the cancer cells, providing information used to guide treatment choices. A very aggressive form of breast cancer is called triple negative, because affected cells do not have the receptors (for estrogen, progesterone, and epidermal growth factor) that many chemotherapies target. Triple negative breast cancer tends to start earlier, spread faster, and recur more often than other types.

Surgery, Radiation, and Chemotherapies

If the breast cells sampled in a biopsy are cancerous, treatment usually begins with surgery. A lumpectomy removes a small tumor and some surrounding tissue; a simple mastectomy removes a breast; and a modified radical mastectomy removes the breast and usually one or two lymph nodes but preserves the pectoral muscles. If cancer cells are detected in lymph nodes, further surgery is performed.

After surgery, many breast cancers are treated with radiation and combinations of chemotherapeutic drugs, plus sometimes newer drugs that are targeted to certain types of breast cancer. Standard chemotherapies kill all rapidly dividing cells, and those used for breast cancer include fluorouracil, doxorubicin, cyclophosphamide, methotrexate, and paclitaxol. Protocols that provide more frequent, lower doses can temper some of the side effects of these powerful drugs.

Newer treatments are easier to tolerate and can be extremely effective. Three types of drugs keep signals (estrogen and growth factors) from stimulating cancer cells to divide:

1. Selective estrogen receptor modulators (SERMs), such as tamoxifen and raloxifene, block estrogen receptors. About half of people with breast cancer have receptors for estrogen on their cancer cells and can benefit from these drugs.

2. Aromatase inhibitors block an enzyme required for tissues other than those of the ovaries to synthesize estrogens. These drugs are used in women who are past menopause, whose ovaries no longer synthesize estrogen. They are prescribed after a five-year course of a SERM.

3. Herceptin (trastuzumab) can help people whose cancer cells bear too many receptors that bind epidermal growth factor. It is a monoclonal antibody, based on an immune system protein. Herceptin blocks the growth factor from signaling cell division. This drug treats a particularly aggressive form of the disease that strikes younger women.

Prevention Strategies

Public health agencies and physicians are debating the exact age at which women should begin having mammograms, and how frequently they should have them thereafter. For this reason, it is important for women to discuss personal mammogram schedules with health-care providers who are familiar with their family histories. Although a mammogram can detect a tumor up to two years before it can be felt, it can also miss some tumors. Thus, breast self-exam is also important in early detection.

Genetic tests can identify women who have inherited certain variants of genes—such as *BRCA1*, *BRCA2*, *p53*, and *HER-2/neu*—that place them at high risk for developing breast cancer. Some of these women elect to have their breasts removed to prevent the disease.

Gene expression profiling is beginning to be used to identify which drugs are most likely to help particular patients with other cancers, such as leukemia, as described in Clinical Application 14.2 (p. 540). Gene expression profiling of breast cancer cells may help identify drugs for breast cancer patients, as well as predict risk of recurrence after surgery. This type of information may help in deciding on follow-up treatment

Only 5% to 10% of all breast cancers arise from an inherited tendency. Much research seeks to identify the environmental triggers that contribute to causing the majority of cases.

---

drug is long-acting; it takes ten to eighteen months after the last injection for the effects to wear off. Use of medroxyprogesterone acetate requires a doctor's care, because of potential side effects and risks.

An implantable rod containing progestin may be inserted under the skin of the arm. The hormone insert prevents follicle maturation and ovulation for up to three years.

 PRACTICE

53 Describe two methods of contraception that use mechanical barriers.

54 What action can increase the effectiveness of chemical contraceptives?

55 How do combined hormone contraceptives, including oral contraceptives, and injectable contraceptives prevent pregnancy?

Intrauterine Devices

An *intrauterine device*, or *IUD*, is a small, solid object that a physician places in the uterine cavity. A copper-containing IUD interferes with fertilization and implantation of a blastocyst, perhaps by inflaming the uterine tissues, for up to ten years (fig. 22.33e). A levonorgestrel-releasing IUD, combining hormone contraception with an IUD, is effective in preventing pregnancy for up to five years.

An IUD may be spontaneously expelled from the uterus or produce abdominal pain or excessive menstrual bleeding. It may also harm the uterus or produce other serious health problems.

Sterilization Methods

Sterilization methods of contraception include surgical procedures that may be performed on the male or female. In the male, a physician removes a small section of each ductus (vas) deferens near the epididymis and ties the cut ends of the ducts (fig. 22.34*a*). This is a *vasectomy*. It has few side effects, but may cause some pain for a week or two. After a vasectomy, sperm cells cannot leave the epididymis, and therefore are not in the semen. However, sperm cells may already be present in the ducts distal to the cuts. For this reason, the sperm count may not reach zero for several weeks.

The corresponding procedure in the female is called *tubal ligation* (fig. 22.34*b*). The uterine tubes are cut and tied so that sperm cells cannot reach an egg cell. In a nonsurgical method of sterilization in the female, a physician inserts a tube through the vagina and uterus into the uterine tubes that delivers a tiny spring-like device. Within three months, scar tissue blocks the uterine tubes so sperm cannot reach the egg to fertilize it.

Sterilization methods do not change hormonal concentrations or sex drives. These procedures provide the most reliable forms of contraception. Reversing them requires microsurgery. Table 22.5 describes several contraceptive approaches and devices and indicates their effectiveness.

 PRACTICE

56 How does an IUD prevent pregnancy?

57 Describe the surgical methods of contraception for a male and for a female.

TABLE 22.5	Birth Control Methods				
	Method	Mechanism	Advantages	Disadvantages	Pregnancies per Year per 100 Women*
	None				85
Barrier and Spermicidal	Condom	Worn over penis or within vagina, keeps sperm out of vagina or from entering cervix	Protection against sexually transmitted infections (latex only)	Disrupts spontaneity, can break, reduces sensation in male	2–12
	Condom and spermicide	Worn over penis or within vagina, keeps sperm out of vagina, and kills sperm that escape	Protection against sexually transmitted infections (latex only)	Disrupts spontaneity, can break, reduces sensation in male	2–5
	Diaphragm and spermicide	Kills sperm and blocks uterus	Inexpensive	Disrupts spontaneity, messy, needs to be fitted by doctor	6–18
	Cervical cap and spermicide	Kills sperm and blocks uterus	Inexpensive, can be left in 24 hours	May slip out of place, messy, needs to be fitted by doctor	6–18
	Spermicidal film, sponge, suppository, foam, or gel	Kills sperm and blocks vagina	Inexpensive, easy to use and carry	Messy, irritates 25% of users (both male and female)	3–21
Hormonal	Combination estrogen and progestin (pill, patch, ring, or injection)	Prevents follicle maturation, ovulation, and implantation	Does not interrupt spontaneity, lowers risk of some cancers, decreases menstrual flow (one type of pill eliminates menstruation)	Raises risk of cardiovascular disease in some women, causes weight gain and breast tenderness	3
	Minipill	Thickens cervical mucus	Does not interrupt spontaneity	Menstrual changes	5
	Medroxyprogesterone acetate	Prevents ovulation, alters uterine lining	Easy to use	Menstrual changes, weight gain	0.3
Behavioral	Rhythm method	No intercourse during fertile times	No cost	Difficult to do, hard to predict timing	20
	Withdrawal (coitus interruptus)	Removal of penis from vagina before ejaculation	No cost	Difficult to do	4–18
Surgical	Vasectomy	Sperm cells never reach penis	Permanent, does not interrupt spontaneity	Requires surgery	0.15
	Tubal ligation	Egg cells never reach uterus	Permanent, does not interrupt spontaneity	Requires surgery, entails some risk of infection	0.4
Other	Intrauterine device	Prevents implantation	Does not interrupt spontaneity	Severe menstrual cramps, increases risk of infection	3

*The lower figures apply when the contraceptive device is used correctly. The higher figures reflect human error in using birth control.

22.8 | Sexually Transmitted Infections

Sexually transmitted infection (STI) is replacing the term *sexually transmitted disease (STD)* because a person can be infected with a pathogen and transmit the pathogen to others, but not develop symptoms of the disease. By the time symptoms appear, it is often too late to prevent complications or the spread of the infection to sexual partners. Many diseases associated with STIs have similar symptoms, and some of the symptoms are also seen in diseases or allergies not sexually related, so it is wise to consult a physician if one or a combination of these symptoms appears:

1. burning sensation during urination
2. pain in the lower abdomen
3. fever or swollen glands in the neck
4. discharge from the vagina or penis
5. pain, itching, or inflammation in the genital or anal area
6. pain during intercourse
7. sores, blisters, bumps, or a rash anywhere on the body, particularly the mouth or genitals
8. itchy, runny eyes

Table 22.6 describes some prevalent diseases associated with sexually transmitted infections. One possible complication of the associated diseases gonorrhea or chlamydia is **pelvic inflammatory disease,** in which bacteria enter the vagina and spread throughout the reproductive organs. The disease begins with intermittent cramps, followed by sudden fever, chills, weakness, and severe cramps. Intravenous antibiotics can stop the infection. The uterus and uterine tubes are often scarred, resulting in infertility and increased risk of ectopic pregnancy, in which the embryo develops in a uterine tube.

Acquired immune deficiency syndrome (AIDS) is a steady deterioration of the body's immune defenses in which the body becomes overrun by infection and possibly cancer. The human immunodeficiency virus (HIV) that causes AIDS is transmitted in body fluids such as semen, blood, and milk. It is most frequently passed during unprotected intercourse or by using a needle containing contaminated blood. Clinical Application 16.1 (pp. 641–642) explores HIV infection.

 PRACTICE

58 Why is the term *sexually transmitted infection* replacing the term *sexually transmitted disease*?

59 Why are sexually transmitted infections sometimes difficult to diagnose?

60 What are some common symptoms of diseases associated with sexually transmitted infections? ■

TABLE 22.6	Some Diseases Associated with Sexually Transmitted Infections						
Associated Disease	Cause	Symptoms	Number of New Cases Reported*	Effects on Fetus	Treatment	Complications
Acquired immune deficiency syndrome (AIDS)	Human immunodeficiency virus (HIV)	Fever, weakness, infections, cancer	32,052	Exposure to HIV and other infections	Drugs to treat or delay symptoms	Body overrun by infection and cancer
Chlamydia infection	*Chlamydia trachomatis* bacteria	Painful urination and intercourse, mucopurulent discharge from penis or vagina	1,422,976 (most cases go undiagnosed)	Premature birth, blindness, pneumonia	Antibiotics	Pelvic inflammatory disease, infertility, arthritis, ectopic pregnancy
Genital herpes	Herpes simplex 2 virus (HSV2)	Genital sores, fever	228,000	Brain damage, stillbirth	Antiviral drug (acyclovir)	Increased risk of cervical cancer
Genital warts	Human papilloma virus (HPV)	Warts on genitals	353,000	None known	Chemical or surgical removal	Increased risk of cervical cancer
Gonorrhea	*Neisseria gonorrhoeae* bacteria	In women, usually none; in men, painful urination	334,826	Blindness, stillbirth	Antibiotics	Arthritis, rash, infertility, pelvic inflammatory disease
Syphilis	*Treponema pallidum* bacteria	Initial chancre usually on genitals or mouth; rash 6 months later; several years with no symptoms as infection spreads; finally damage to heart, liver, nerves, brain	15,667 (primary and secondary)	Miscarriage, premature birth, birth defects, stillbirth	Antibiotics	Dementia

*2012 CDC statistics, Centers for Disease Control and Prevention (U.S.), for all associated diseases except AIDS (2011)

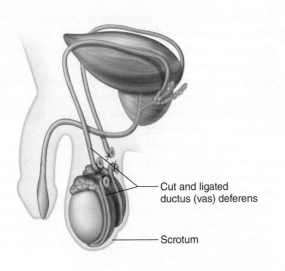

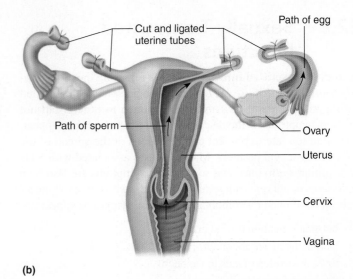

(a)

(b)

FIGURE 22.34 Surgical methods of birth control. (*a*) In a vasectomy, each ductus (vas) deferens is cut and ligated (tied). (*b*) In a tubal ligation, each uterine tube is cut and ligated.

Chapter Summary

22.1 Meiosis and Sex Cell Production (page 824)

1. Sex cells are produced by **meiosis** consisting of two divisions, each progressing through prophase, metaphase, anaphase, and telophase.
   a. In the first meiotic division, homologous, replicated chromosomes (each consisting of two chromatids held together by a centromere) separate, and their number is halved.
   b. In the second meiotic division, the chromatids part, producing four **haploid** cells.
   c. Meiosis leads to genetic variability because of the random alignment of maternally and paternally derived chromosomes in metaphase I and crossing over.

22.2 Organs of the Male Reproductive System (page 826)

The primary male sex organs are the two **testes,** which produce sperm cells and male sex hormones. Accessory sex organs are internal and external.

1. Testes
   a. Descent of the testes
      (1) Testes originate posterior to the parietal peritoneum near the level of the developing kidneys.
      (2) The **gubernaculum** guides the descent of the testes into the lower abdominal cavity and through the **inguinal canal.**
      (3) Undescended testes cannot produce sperm cells because of the high abdominal temperature.
   b. Structure of the testes
      (1) The testes are separated by connective tissue and filled with **seminiferous tubules.**
      (2) The seminiferous tubules unite to form the rete testis that joins the epididymis.

   (3) The seminiferous tubules contain **spermatogenic cells** that give rise to sperm cells.
   (4) The **interstitial cells** that produce male sex hormones are between the seminiferous tubules.
   c. Formation of sperm cells
      (1) The epithelium lining the seminiferous tubules includes sustentacular cells and spermatogenic cells.
         (a) The sustentacular cells support and nourish the spermatogenic cells.
         (b) The spermatogenic cells give rise to spermatogonia.
      (2) **Spermatogenesis** produces sperm cells from spermatogonia that have differentiated to become primary spermatocytes.
         (a) Spermatogenesis produces four sperm cells from each primary spermatocyte.
         (b) Meiosis halves the number of chromosomes in sperm cells (46 to 23).
      (3) Membranous processes of adjacent sustentacular cells form a barrier in the epithelium.
         (a) The barrier separates early and late stages of spermatogenesis.
         (b) It helps provide a favorable environment for differentiating cells.
   d. Structure of a sperm cell
      (1) The sperm head contains a nucleus with 23 chromosomes.
      (2) The sperm body has many mitochondria.
      (3) The sperm tail propels the cell.
2. Male internal accessory reproductive organs
   a. **Epididymides**
      (1) Each epididymis is a tightly coiled tube on the outside of the testis that leads into the ductus deferens.
      (2) They store and nourish immature sperm cells and promote their maturation.

Reproductive Systems

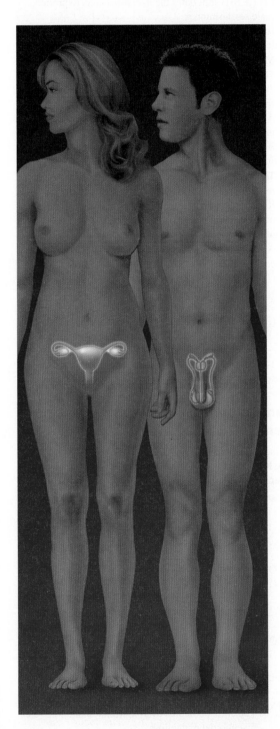

Integumentary System

Skin sensory receptors play a role in sexual pleasure.

Skeletal System

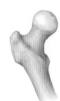

Bones can be a temporary source of calcium during lactation.

Muscular System

Skeletal, cardiac, and smooth muscles all play a role in reproductive processes and sexual activity.

Nervous System

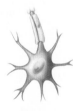

The nervous system plays a major role in sexual activity and sexual pleasure.

Endocrine System

Hormones control the production of eggs in the female and sperm in the male.

Cardiovascular System

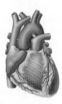

Blood pressure is necessary for the normal function of erectile tissue in the male and female.

Lymphatic System

Special mechanisms inhibit the female immune system from attacking sperm as foreign invaders.

Digestive System

Proper nutrition is essential for the formation of normal gametes.

Respiratory System

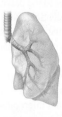

Breathing provides oxygen that assists in the production of ATP needed for egg and sperm development.

Urinary System

Male urinary and reproductive systems share structures. Kidneys help compensate for fluid loss from the reproductive systems.

Gamete production, fertilization, embryonic and fetal development, and childbirth are essential for survival of the species.

b. **Ductus deferentia**
  (1) Each ductus deferens is a muscular tube that forms part of the spermatic cord.
  (2) They pass through the inguinal canal, enter the abdominal cavity, course medially into the pelvic cavity, and end behind the urinary bladder.
  (3) They fuse with the ducts from the seminal vesicles to form the ejaculatory ducts.

c. **Seminal vesicles**
  (1) Each seminal vesicle is a saclike structure attached to the ductus deferens.
  (2) They secrete an alkaline fluid that contains nutrients, such as fructose, and prostaglandins.
  (3) This secretion is added to sperm cells entering the ejaculatory ducts.

d. **Prostate gland**
  (1) The prostate gland surrounds the urethra just inferior to the urinary bladder.
  (2) It secretes a thin, milky fluid containing citrate and prostate-specific antigen.

e. **Bulbourethral glands**
  (1) These glands are two small structures inferior to the prostate gland.
  (2) They secrete a fluid that lubricates the penis in preparation for sexual intercourse.

f. **Semen**
  (1) Semen consists of sperm cells and secretions of the seminal vesicles, prostate gland, and bulbourethral glands.
  (2) This fluid is slightly alkaline and contains nutrients and prostaglandins.
  (3) Sperm cells in semen swim, but cannot fertilize egg cells until they enter the female reproductive tract.

3. Male external reproductive organs

a. **Scrotum**
  (1) The scrotum is a pouch of skin and subcutaneous tissue that encloses the testes.
  (2) Contraction of the dartos muscle in the scrotal wall causes the scrotal skin to wrinkle and the testes to move closer to the pelvic cavity or with relaxation of the muscle fibers the scrotum hangs loosely, regulating the environmental temperature of the testes, important for sperm production and survival.

b. **Penis**
  (1) The penis conveys urine and semen.
  (2) It becomes erect for insertion into the vagina during sexual intercourse.
  (3) Its body is composed of three columns of erectile tissue surrounded by connective tissue.
  (4) The root of the penis is attached to the pelvic arch and membranes of the perineum.

c. Erection, orgasm, and ejaculation
  (1) During erection, vascular spaces in the erectile tissue become engorged with blood as arteries dilate and veins are compressed.
  (2) **Orgasm** is the culmination of sexual stimulation and is accompanied by **emission** and **ejaculation.**
  (3) Semen moves along the reproductive tract as smooth muscle in the walls of the tubular structures contracts by reflex.
  (4) Following ejaculation, the penis becomes flaccid.

22.3 Hormonal Control of Male Reproductive Functions (page 837)

1. Hypothalamic and pituitary hormones
   The male body remains reproductively immature until the hypothalamus releases GnRH, which stimulates the anterior pituitary gland to release **gonadotropins.**
   a. FSH stimulates spermatogenesis.
   b. LH (ICSH) stimulates the interstitial cells to produce male sex hormones.
   c. Inhibin prevents oversecretion of FSH.
2. Male sex hormones
   a. Male sex hormones are called **androgens.**
   b. **Testosterone** is the most important androgen.
   c. Testosterone is converted into **dihydrotestosterone** in some organs.
   d. Androgens that fail to become fixed in tissues are metabolized in the liver and excreted.
   e. Androgen production increases rapidly at **puberty.**
3. Actions of testosterone
   a. Testosterone stimulates the development of the male reproductive organs and causes the testes to descend.
   b. It is responsible for the development and maintenance of male secondary sex characteristics.
4. Regulation of male sex hormones
   a. A negative feedback mechanism regulates testosterone concentration.
      (1) As the concentration of testosterone rises, the hypothalamus is inhibited, and the anterior pituitary gland secretion of gonadotropins is reduced.
      (2) As the concentration of testosterone falls, the hypothalamus signals the anterior pituitary gland to secrete gonadotropins.
   b. The concentration of testosterone remains relatively stable from day to day.

22.4 Organs of the Female Reproductive System (page 839)

The primary female sex organs are the two ovaries, which produce female sex cells and sex hormones. Accessory sex organs are internal and external.

1. **Ovaries**
   a. Ovary attachments
      (1) Several ligaments hold the ovaries in position.
      (2) These ligaments include the broad, suspensory, and ovarian ligaments.
   b. Ovary descent
      (1) The ovaries descend from posterior to the parietal peritoneum near the developing kidneys.
      (2) They attach to the pelvic wall just inferior to the pelvic brim.
   c. Ovary structure
      (1) The ovaries are subdivided into a medulla and a cortex.
      (2) The medulla is composed of connective tissue, blood vessels, lymphatic vessels, and nerves.
      (3) The cortex contains ovarian follicles and is covered by cuboidal epithelium.
   d. Primordial follicles
      (1) During prenatal development, oogonia divide by mitosis to produce more oogonia that develop into primary oocytes.

(2) Each **primordial follicle** contains a primary oocyte and a layer of flattened epithelial cells.

(3) The primary oocyte begins to undergo meiosis, but the process soon halts and does not resume until puberty.

(4) The number of oocytes steadily declines throughout the life of a female.

e. Oogenesis

(1) Beginning at puberty, some primary oocytes are stimulated to continue meiosis.

(2) When a primary oocyte undergoes oogenesis, it gives rise to a secondary oocyte in which the original chromosome number is halved (from 46 to 23).

(3) A secondary oocyte may be fertilized to produce a **zygote.**

f. Follicle maturation

(1) At puberty, FSH initiates follicle maturation.

(2) During maturation of the follicle, the primary oocyte enlarges, the follicular cells proliferate, and a fluid-filled cavity (atrum) appears.

(3) Ovarian cells surrounding the follicle form two layers.

(4) A mature antral follicle contains a secondary oocyte surrounded by a **zona pellucida** and **corona radiata.**

(5) Usually only one follicle fully develops per reproductive cycle.

g. Ovulation

(1) **Ovulation** is the release of a secondary oocyte and first polar body from an ovary.

(2) The secondary oocyte is released when its follicle ruptures.

(3) After ovulation, the secondary oocyte is drawn into the opening of the uterine tube.

2. Female internal accessory reproductive organs

a. **Uterine tubes**

(1) These tubes convey egg cells toward the uterus.

(2) The end of each uterine tube is expanded, and its margin bears irregular extensions.

(3) Ciliated cells that line the tube and peristaltic contractions in the wall of the tube move an egg cell into the tube's opening.

(4) Fertilization of the egg occurs here.

b. Uterus

(1) The **uterus** receives the embryo and sustains it during development.

(2) The uterine wall includes the **endometrium, myometrium,** and **perimetrium.**

c. Vagina

(1) The **vagina** connects the uterus to the vestibule.

(2) It receives the erect penis, conveys uterine secretions to the outside, and provides an open channel for the fetus during birth.

(3) A thin membrane, the **hymen,** partially closes the vaginal orifice.

(4) The vaginal wall consists of a mucosal layer, muscular layer, and outer fibrous layer.

3. Female external reproductive organs

a. Labia majora

(1) The **labia majora** are rounded folds of adipose tissue and skin that enclose and protect the other external reproductive parts.

(2) The anterior ends form a rounded elevation over the pubic symphysis.

b. Labia minora

(1) The **labia minora** are flattened, longitudinal folds between the labia majora.

(2) They are well supplied with blood vessels.

c. Clitoris

(1) The **clitoris** is a small projection at the anterior end of the vulva; it is homologous to the male penis.

(2) It is composed of two columns of erectile tissue.

(3) Its root is attached to the sides of the pubic arch.

d. Vestibule

(1) The **vestibule** is the space between the labia minora that encloses the vaginal and urethral openings.

(2) The **vestibular glands** secrete mucus into the vestibule during sexual stimulation.

e. Erection, lubrication, and orgasm

(1) During periods of sexual stimulation, the erectile tissues of the clitoris and vestibular bulbs become engorged with blood and swollen.

(2) The vestibular glands secrete mucus into the vestibule and vagina.

(3) During orgasm, the muscles of the perineum, uterine wall, and uterine tubes contract rhythmically.

22.5 Hormonal Control of Female Reproductive Functions (page 849)

Hormones from the hypothalamus, anterior pituitary gland, and ovaries play important roles in the control of sex cell maturation, the development and maintenance of female secondary sex characteristics, and changes that occur during the monthly reproductive cycle.

1. Female sex hormones

a. A female body remains reproductively immature until about ten years of age, when gonadotropin secretion increases.

b. The most important female sex hormones are **estrogens** and **progesterone.**

(1) Estrogens are responsible for the development and maintenance of most female secondary sex characteristics.

(2) Progesterone prepares the uterus for pregnancy.

2. Female reproductive cycle

a. The reproductive cycle is characterized by regularly recurring changes in the uterine lining culminating in menstrual flow.

b. **Menarche** is the female's first reproductive cycle.

c. FSH stimulates maturation of ovarian follicles.

d. Granulosa cells of maturing follicles secrete estrogens, responsible for maintaining the secondary sex traits and thickening the uterine lining.

e. In response to the LH surge, just prior to ovulation, the primary oocyte completes meiosis I to give rise to a secondary oocyte and first polar body.

f. Ovulation occurs when LH along with FSH induce complex interactions which weaken and rupture the antral follicle, releasing the secondary oocyte.

g. After ovulation, follicle remnants form the **corpus luteum.**

(1) The corpus luteum secretes estrogens and progesterone, which cause the uterine lining to become more vascular and glandular.

(2) If a secondary oocyte is not fertilized, the corpus luteum begins to degenerate.

(3) As the concentrations of estrogens and progesterone decline, the uterine lining disintegrates, causing menstrual flow.

h. During this cycle, estrogens and progesterone inhibit the release of LH and FSH (except for the large increase at mid-cycle). As the concentrations of estrogens and progesterone decline, the anterior pituitary gland secretes FSH and LH again, stimulating a new reproductive cycle.

3. Menopause

a. **Menopause** is cessation of the female reproductive cycles. Less estrogens and progesterone are produced.

b. The female reproductive organs decrease in size.

22.6 Mammary Glands (page 852)

1. Location of the glands

a. The **mammary glands** are located in the subcutaneous tissue of the anterior thorax within the breasts.

b. The breasts extend between the second and sixth ribs and from sternum to axillae.

2. Structure of the glands

a. The mammary glands are composed of lobes that contain tubular glands.

b. Dense connective and adipose tissues separate the lobes.

c. Ducts connect the mammary glands to the nipple.

3. Development of the breasts

a. Breasts of males do not develop.

b. Estrogens stimulate breast development in females.

(1) Alveolar glands and ducts enlarge.

(2) Fat is deposited in the breasts.

22.7 Birth Control (page 854)

Birth control is the voluntary regulation of the number of children produced and the time they are conceived. This usually involves some method of **contraception.**

1. Coitus interruptus

a. Coitus interruptus is withdrawal of the penis from the vagina before ejaculation.

b. Some semen may be expelled from the penis before ejaculation.

2. Rhythm method

a. Abstinence from sexual intercourse two days before and one day after ovulation is the rhythm method.

b. It is difficult to predict the time of ovulation.

3. Mechanical barriers

a. Males and females can use condoms.

b. Females use diaphragms, cervical caps and cervical shields.

4. Chemical barriers

a. Spermicidal film, sponges, suppositories, foams, and gels are chemical barriers to conception.

b. These provide an unfavorable environment in the vagina for sperm survival.

5. Combined hormone contraceptives

a. A flexible ring inserted deep into the vagina, a plastic patch, or a pill can deliver estrogen and progestin to prevent pregnancy.

b. They disrupt the normal pattern of gonadotropin secretion which prevents follicle maturation and ovulation, and they thicken cervical mucus to prevent the sperm from joining the egg.

c. When used correctly, combined hormone contraceptives are almost 100% effective.

d. Some women develop undesirable side effects.

e. A minipill contains only progestin and must be taken at the same time daily.

6. Other hormone contraceptives

a. Intramuscular injection with medroxyprogesterone acetate every three months prevents pregnancy.

b. An implantable rod containing progestin may prevent pregnancy for up to three years.

7. Intrauterine devices

a. An IUD is a solid object inserted in the uterine cavity.

b. It prevents pregnancy by interfering with fertilization and implantation of a blastocyst.

c. A hormone-releasing IUD can effectively prevent pregnancy for up to five years.

d. It may be expelled spontaneously or produce undesirable side effects.

8. Sterilization methods

a. These are surgical procedures.

(1) Vasectomy is performed in males.

(2) Tubal ligation is performed in females.

b. A nonsurgical method inserts a tiny coil into each uterine tube, causing scarring that blocks fertilization.

c. Sterilization methods are the most reliable contraception.

22.8 Sexually Transmitted Infections (page 859)

1. **Sexually transmitted infections** are passed during sexual contact and may go undetected for years.

2. Many of the diseases associated with sexually transmitted infections share symptoms.

CHAPTER ASSESSMENTS

22.1 Meiosis and Sex Cell Production

1 Male sex cells are called _____, and female sex cells are called _____. (p. 824)

2 Construct a table of the steps in meiosis including the major event(s). (pp. 824–825)

3 Define *crossover,* and explain how it produces genetic variation. (pp. 824–825)

22.2 Organs of the Male Reproductive System

4 Distinguish between the primary and accessory male reproductive organs, then explain how each organ's structure affects the organ's function. (pp. 826–836)

5 Describe the descent of the testes. (p. 826)

6 Define *cryptorchidism.* (p. 826)

7 Explain the function of the sustentacular cells in the testis. (p. 828)

8 List the major steps in spermatogenesis. (p. 828)

9 Describe a sperm cell. (p. 829)

10 Trace the path of the ductus deferens from the epididymis to the ejaculatory duct. (pp. 830 and 833)

11 On a diagram, locate the seminal vesicles, prostate gland, and bulbourethral glands, and describe the composition of their secretions. (p. 833)

12 Describe the composition of semen. (p. 834)

13 Define *capacitation.* (p. 834)

14 Explain the mechanism that produces an erection of the penis. (p. 835)

15 Define *orgasm.* (p. 835)

16 Distinguish between emission and ejaculation. (pp. 835–836)

17 Explain the mechanism of ejaculation. (pp. 835–836)

22.3 Hormonal Control of Male Reproductive Functions

18 Describe the role of GnRH in the control of male reproductive functions. (p. 838)

19 Distinguish between androgen and testosterone. (p. 838)

20 Define *puberty.* (p. 838)

21 Discuss the actions of testosterone. (pp. 838–839)

22 List several male secondary sex characteristics. (p. 839)

23 Explain the regulation of testosterone secretion. (p. 839)

22.4 Organs of the Female Reproductive System

24 Distinguish between the primary and accessory female reproductive organs, then explain how each organ's structure affects the organ's function. (pp. 840–848)

25 Describe how the ovaries are held in position. (p. 840)

26 Describe the descent of the ovaries. (pp. 840–841)

27 Define *primordial follicle.* (p. 841)

28 List the steps in oogenesis. (p. 841)

29 Describe how a follicle matures. (pp. 841–844)

30 Distinguish between a pre-antral and an antral follicle. (pp. 841–842)

31 Define *ovulation.* (p. 844)

32 Define *vulva.* (p. 847)

33 Describe the process of an erection in the female reproductive organs. (p. 848)

22.5 Hormonal Control of Female Reproductive Functions

34 Describe the role of GnRH in regulating female reproductive functions. (p. 849)

35 List several female secondary sex characteristics. (p. 849)

36 Explain how a reproductive cycle is initiated. (p. 850)

37 Summarize the major events in a reproductive cycle. (pp. 850–852)

38 A female's first reproductive cycle is _____, whereas _____ is when the female's reproductive cycles cease. (pp. 850–852)

22.6 Mammary Glands

39 Describe the structure of a mammary gland. (pp. 852–853)

22.7 Birth Control

40 Match the birth control method with its description. (pp. 854–858)

(1) withdrawal	A. kills sperm
(2) rhythm method	B. keeps sperm out of vagina or from entering cervix
(3) condom	
(4) spermicide	C. prevents implantation of a blastocyst
(5) estrogen/ progesterone	D. no intercourse during fertile times
(6) IUD	E. penis removed from vagina before ejaculation
(7) vasectomy	
(8) tubal ligation	F. sperm cells never reach penis
	G. prevents follicle maturation and ovulation
	H. oocytes never reach uterus

22.8 Sexually Transmitted Infections

41 Common symptoms of diseases associated with sexually transmitted infections include _____. (p. 859)

a. a burning sensation during urination

b. discharge from penis or vagina

c. sores, blisters, or rash on genitals

d. all of the above

42 If left untreated, a complication of the diseases gonorrhea and chlamydia is _____. (p. 859)

INTEGRATIVE ASSESSMENTS/CRITICAL THINKING

Outcomes 13.4, 13.7, 22.2, 22.3

1. What changes, if any, might occur in the secondary sex characteristics of an adult male following removal of one testis? Following removal of both testes? Following removal of the prostate gland?

Outcomes 13.4, 13.7, 22.2, 22.3, 22.4, 22.5, 22.7

2. Understanding the causes of infertility can be valuable in developing new birth control methods. Cite a type of contraceptive based on each of the following causes of infertility: (a) failure to ovulate due to a hormonal imbalance; (b) a large fibroid tumor that disturbs the uterine lining; (c) endometrial tissue blocking uterine tubes; (d) low sperm count (too few sperm per ejaculate).

Outcomes 13.4, 13.7, 22.4, 22.5

3. What effect would removal of an ovary have on a woman's reproductive cycles? What effect would removal of both ovaries have?

Outcomes 13.4, 13.7, 22.4, 22.5, 22.7

4. Does a tubal ligation cause a woman to enter menopause prematurely? Why or why not?

Outcomes 22.2, 22.3

5. As a male reaches adulthood, what will be the consequences if his testes have remained undescended since birth? Why?

Outcomes 22.2, 22.4

6. Some men are unable to become fathers because their spermatids do not mature into sperm. Injection of their spermatids into their partner's secondary oocytes sometimes results in conception. Men have fathered healthy babies this way. Why would this procedure work with spermatids but not with primary spermatocytes?

Outcomes 22.4, 22.5

7. Sometimes a sperm cell fertilizes a polar body rather than a secondary oocyte. An embryo does not develop, and the fertilized polar body degenerates. Why is the polar body unable to support development of an embryo?

ONLINE STUDY TOOLS

Connect Interactive Questions Reinforce your knowledge using assigned interactive questions covering the anatomy of the reproductive systems and the hormonal control of male and female reproductive functions.

Connect Integrated Activity Can you differentiate the physiological events that occur in spermatogenesis from those that occur in oogenesis?

LearnSmart Discover which chapter concepts you have mastered and which require more attention. This adaptive learning tool is personalized, proven, and preferred.

Anatomy & Physiology Revealed Go more in depth into the human body exploring the structures of the reproductive organs in a cadaver or viewing animations on spermatogenesis and the female reproductive system.

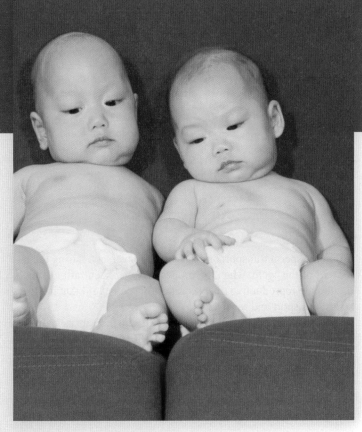

"Higher multiples" such as quadruplets face health risks, but twins usually do very well.

THE WHOLE PICTURE

A sperm cell and an egg cell (secondary oocyte) unite, forming a zygote, and the journey of pregnancy begins. Following thirty-eight weeks of cell division, growth, and specialization into distinctive tissues and organs, a new human being enters the world.

Humans grow, develop, and age. Growth is an increase in size. As a human grows, cell numbers increase as a result of mitosis, followed by enlargement of the newly formed cells and of the body. The diverse parts of our bodies—the many specialized cells that form tissues, organs, and organ systems—originated from just one cell. That initial cell began with a unique combination of genes from the gametes of both parents.

Development, which includes growth, is the continuous process by which an individual changes from one life phase to another. These life phases include a prenatal period, which begins with the fertilization of an egg cell and ends at birth, and a postnatal period, which begins at birth and ends with death.

Module 14: Reproductive System

23

Pregnancy, Growth, and Development

LEARNING OUTCOMES

After you have studied this chapter, you should be able to:

23.1 Fertilization
1 Trace the movement of sperm toward an egg. (p. 868)
2 Describe, in detail, fertilization. (p. 869)

23.2 Pregnancy and the Prenatal Period
3 List and provide details of the major events of cleavage. (pp. 870–871)
4 Describe implantation. (pp. 871–872)
5 Describe the extraembryonic membranes. (pp. 873–876)
6 Describe the formation and function of the placenta. (pp. 873–876)
7 Explain how the primary germ layers originate, and list the structures each layer produces. (pp. 876–878)
8 Define *fetus*, and describe the major events of the fetal stage of development. (pp. 881–882)
9 Trace the path of blood through the fetal cardiovascular system. (pp. 883–887)
10 Discuss the hormonal and other changes in the maternal body during pregnancy. ((pp. 887–889)
11 Explain the role of hormones in the birth process and milk production. (pp. 890–892)

23.3 Postnatal Period
12 Name the postnatal stages of development of a human, and indicate the general characteristics of each stage. (pp. 894–897)
13 Describe the major cardiovascular and physiological adjustments in the newborn. (pp. 894–895)

23.4 Aging
14 Distinguish between passive and active aging. (p. 899)
15 Contrast life span and life expectancy. (p. 899)

UNDERSTANDING WORDS

allant-, sausage: *allantois*—tubelike structure extending from the yolk sac into the connecting stalk of an embryo.

chorio-, skin: *chorion*—outermost extraembryonic membrane.

cleav-, to divide: *cleavage*—period of development when a zygote divides, producing increasingly smaller cells.

ect-, outside: *ectoderm*—outermost germ layer of the embryo.

lacun-, pool: *lacuna*—space between the chorionic villi that fills with maternal blood.

lanug-, down: *lanugo*—fine hair covering the fetus.

mes-, middle: *mesoderm*—middle germ layer of the embryo.

morul-, mulberry: *morula*—embryonic structure consisting of a solid ball of about sixteen cells that resembles a mulberry.

nat-, to be born: pre*nat*al—before birth.

ne-, new, young: *ne*onatal—the first four weeks after birth.

post-, after: *post*natal—after birth.

pre-, before: *pre*natal—before birth.

sen-, old: *sen*escence—becoming old.

troph-, well fed: *troph*oblast—cellular layer that surrounds the inner cell mass and helps nourish it.

umbil-, navel: *umbil*ical cord—structure attached to the fetal navel (umbilicus) that connects the fetus to the placenta.

23.1 | Fertilization

The union of a secondary oocyte and a sperm cell is called **fertilization** (fer″tĭ-lĭ-za′shun), or conception. This usually takes place in the infundibulum of a uterine tube.

Transport of Sex Cells

Before fertilization can occur, a secondary oocyte must be ovulated and enter a uterine tube. During sexual intercourse, the male deposits semen containing sperm cells in the vagina near the cervix. To reach the secondary oocyte, the sperm cells must then move upward through the uterus and uterine tube (fig. 23.1). Prostaglandins in the semen stimulate lashing of sperm tails, and muscular contractions within the walls of the uterus and uterine tube aid the sperm cells' journey. Also, under the influence of high concentrations of estrogens during the first part of the reproductive cycle, the uterus and cervix contain a thin, watery secretion that promotes sperm transport and survival. Conversely, during the latter part of the cycle, when the progesterone concentration is high, the female reproductive tract secretes a viscous fluid that hampers sperm transport and survival.

Sperm movement is inefficient. Even though as many as 200 million to 600 million sperm cells may be deposited in the vagina by a single ejaculation, only a few hundred ever reach a secondary oocyte. The journey to and through a uterine tube takes less than an hour following sexual intercourse. Many sperm cells may reach a secondary oocyte, but usually only one sperm cell fertilizes it (fig. 23.2). If a second sperm were to enter, the fertilized ovum would have three sets of chromosomes and be very unlikely to develop.

A secondary oocyte may survive for only twelve to twenty-four hours following ovulation. In contrast, sperm cells may survive for up to six days after ejaculation into the woman's body, but they are best able to fertilize an oocyte within twenty-four to forty-eight hours after entering the female reproductive tract. Consequently,

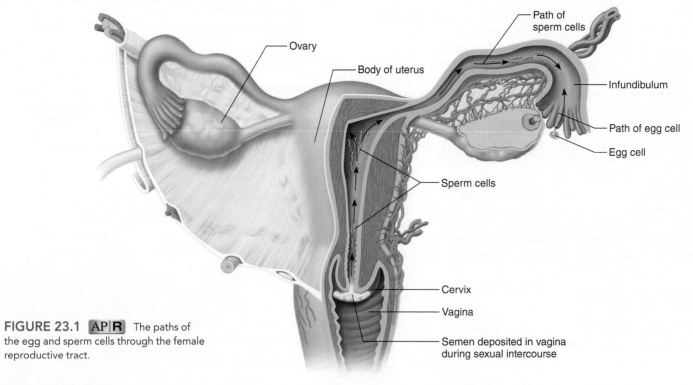

FIGURE 23.1 **AP|R** The paths of the egg and sperm cells through the female reproductive tract.

Labels: Ovary · Body of uterus · Path of sperm cells · Infundibulum · Path of egg cell · Egg cell · Sperm cells · Cervix · Vagina · Semen deposited in vagina during sexual intercourse

sexual intercourse that takes place between forty-eight hours before ovulation and twenty-four hours after ovulation is more likely to result in fertilization.

Sperm Cell Joins Secondary Oocyte

Sperm cells first invade the *corona radiata,* which is a layer that consists of follicular cells that adhere to the secondary oocyte's surface. An enzyme associated with the sperm cell membrane (hyaluronidase) aids this initial penetration (fig. 23.3). The sperm next bind a specific class of proteins on the *zona pellucida,* which is a membrane rich in glycoproteins that is the layer closest to the secondary oocyte's cell membrane. With this contact, the acrosome releases enzymes by exocytosis that digest the material of the zona pellucida. The sperm now have direct access to the oocyte.

Hundreds of sperm take part in removing the layers that surround the oocyte, but only one sperm will actually fertilize the oocyte. That sperm cell binds to and then fuses with the oocyte membrane. Then the sperm head enters the secondary oocyte, leaving the mitochondria-rich middle section and tail outside. Sperm entry triggers lysosome-like vesicles just beneath the oocyte cell membrane to release enzymes that harden the zona pellucida. This reduces the chance that other sperm cells will penetrate.

> The exact moment of sperm-oocyte contact is an interaction between a protein on the sperm head and a receptor that extends from the oocyte's cell membrane. Researchers identified the sperm protein, naming it Izumo1 after a Japanese marriage shrine, in 2005, and the receptor on the oocyte, naming it Juno for the Roman goddess of fertility, in 2014. So far the two proteins that orchestrate conception have been found in several mammalian species. The genes that encode the proteins are in the human genome, so the search is on to observe Izumo1 and Juno as human gametes meet.

The sperm nucleus enters the secondary oocyte's cytoplasm and swells. The secondary oocyte then divides unequally to form a large cell, whose nucleus contains one copy of the female's genetic contribution, and a tiny second polar body, which is later expelled (see fig. 22.21, p. 842). Meiosis is completed. The approaching nuclei from the two sex cells are called **pronuclei,** until they join. When the pronuclei unite, their nuclear membranes disassemble, and their chromosomes mingle, finishing the process of fertilization.

Each sex cell has 23 chromosomes, so the product of fertilization is a cell with 46 chromosomes—the usual number in a human body cell (somatic cell). This cell, called a **zygote** (zi′gōt), is the first cell of the future offspring. From Science to Technology 23.1 describes assisted routes to conception.

 ## PRACTICE

1 Where in the female reproductive system does fertilization take place?

2 What factors aid the movements of the secondary oocyte and sperm cells through the female reproductive tract?

3 List the events of fertilization.

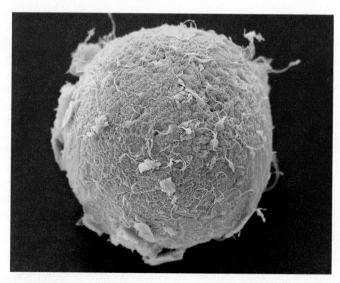

FIGURE 23.2 Scanning electron micrograph of sperm cells on the surface of a secondary oocyte (650×). Only one sperm cell actually fertilizes a secondary oocyte.

 ### CAREER CORNER
Physician's Assistant

The 52-year-old man had spent many summers working as a landscaper, and now he is at his internist's office, where a physician's assistant (PA) is examining several patches of scaly light-colored skin on the man's arms. One area looks as if it could be cancerous, so the PA removes the skin and prepares a biopsy slide to send to a laboratory for analysis. Then he teaches the patient how to care for the wound.

Three days a week the PA examines patients at the office, ordering tests, diagnosing conditions, prescribing medications, and providing counseling and treatments. On the other two days the PA visits patients at assisted living facilities, doing routine checkups and diagnosing and treating uncomplicated problems such as infections and minor injuries.

A PA practices medicine under the supervision of a physician, and has national certification and state licensure. Training takes three years. Applicants must have completed the same prerequisites as medical school requirements, including courses in anatomy and physiology, chemistry, microbiology, and biology. Three years of health-care experience is required by most programs.

Training includes the prerequisite subjects in greater depth, and adds pathophysiology, biochemistry, psychology, pharmacology, physical diagnosis, clinical laboratory science, and bioethics. A minimum of 2,000 hours of clinical rotations provides experience in pediatrics, family medicine, internal medicine, obstetrics and gynecology, emergency medicine, general surgery, and psychiatry. Continuing medical education is required to maintain skills and licensure.

Medical practices that employ PAs can see more patients. In underserved areas, a PA may communicate with the supervising physician online or by phone. Several factors—including an aging population, a physician shortage, and an increase in prevalence of chronic diseases—make the job market for PAs very promising.

Conception requires the meeting and merging of sperm cell and egg cell, which naturally occurs in the woman's uterine tube. Abnormal production of gametes or blockages that impede this meeting of cells can result in infertility (see Clinical Applications 22.2 and 22.3, pp. 836–837, 853). Assisted reproductive technologies (ART) can help couples conceive. The procedures usually involve a laboratory technique and sometimes participation of a third person. Many of these techniques are costly and may take several attempts to work. In the United States about 1% of the 4 million or so births each year result from an ART. Here is a look at some of the procedures.

Donated Sperm—Intrauterine Insemination

In intrauterine ("artificial") insemination (IUI), a doctor places donated sperm in a woman's reproductive tract. A woman might seek IUI if her partner is infertile or carries a gene for an inherited illness that could affect a child's health, or if she desires to be a single parent or raise a child with a female partner.

Millions of babies have been born worldwide as a result of IUI. The first human IUIs by donor were done in the 1890s. For many years, donating sperm was a way for male medical students to earn a few extra dollars. By 1953, sperm could be frozen and stored. Today, sperm banks provide the cells to physicians who perform the procedure. Since 1983 sperm banks have asked donors if they wish to be contacted by their children years later. The first such meetings occurred in 2002, and thousands of others have followed. Today several websites offer DNA tests that help people find their sperm donor fathers.

A woman or couple choosing IUI can select sperm from a catalog that lists characteristics of the donors, including blood type;

hair, skin, and eye color; build; and even educational level and interests. Not all of these traits are inherited. On rare occasion IUI has led to dilemmas (table 23A).

In Vitro Fertilization

In *in vitro* fertilization (IVF), which means "fertilization in glass," a sperm cell fertilizes an egg cell outside of the woman's body. The fertilized egg cell divides two or three times, producing a four- or eight-celled embryo, and is then introduced into the egg donor's

TABLE 23A	Assisted Reproductive Dilemmas
1.	A physician in California used his own sperm for intrauterine insemination of fifteen patients and told them that he had used sperm from anonymous donors.
2.	Cases of sperm donors fathering more than 150 children each led to sperm donation limits.
3.	A man who donated sperm when he was healthy later developed a genetic disease. Each of the eighteen children conceived using his sperm faces a 1 in 2 chance of having inherited spinocerebellar ataxia.
4.	A plane crash killed the wealthy parents of two early embryos frozen in a hospital in Melbourne, Australia. Adult children of the couple were asked to share their estate with the embryos.
5.	Two Rhode Island couples sued a fertility clinic for misplacing embryos.
6.	A man sued his ex-wife for possession of their frozen embryos as part of the divorce settlement.

A couple expecting a child can estimate the approximate time of fertilization (conception) by adding 14 days to the date of the onset of the last menstrual period. They can predict the time of birth by adding 266 days (38 weeks) to the fertilization date. Most babies are born within 10 to 15 days of this calculated time.

Tracking a pregnancy's progress can be confusing, because some health-care providers measure 40 weeks from the last menstrual period, rather than the more accurate 38 weeks from fertilization. Obstetricians can, however, estimate the date of conception by scanning the embryo with ultrasound and comparing the crown-to-rump length to known values that are the average for each day of gestation. This approach is inaccurate if an embryo is smaller or larger than usual.

23.2 | Pregnancy and the Prenatal Period

Pregnancy (preg'nan-se) is the presence of a developing offspring in the uterus. Pregnancy consists of three periods called trimesters, each about three months long.

The **prenatal period** (prena'tal pe're-od) of development of the offspring usually lasts for thirty-eight weeks from conception. It can be divided into an embryonic stage and a fetal stage.

Embryonic Stage

The **embryonic stage** extends from fertilization through the eighth week of prenatal development. During this time the placenta forms, the main internal organs develop, and the external body structures appear.

Period of Cleavage

The fertilization sequence of events is completed when the pronuclei of a sperm cell and a secondary oocyte join, forming a zygote. Thirty hours later, the zygote undergoes mitosis, giving rise to two new cells. These cells in turn divide to form four cells, which then divide into eight cells, and so forth. The divisions are rapid, with little time for the cells to grow, so the resulting cells of these initial divisions are increasingly smaller (fig. 23.4). This rapid cell division and distribution of the zygote's cytoplasm into progressively smaller cells is called **cleavage** (klēv'ij), and the resulting cells are called *blastomeres*. The ball of cells

(or another woman's, surrogate) uterus. If all goes well, a pregnancy begins.

A woman might undergo IVF if her ovaries and uterus function but her uterine tubes are blocked. To begin, she takes a hormone that hastens maturity of several secondary oocytes. Using a laparoscope to view the ovaries and uterine tubes, a physician removes a few of the largest egg cells and transfers them to a dish, then adds chemicals similar to those in the female reproductive tract, and sperm cells.

If a sperm cell cannot penetrate the secondary oocyte *in vitro*, it may be sucked up into a tiny syringe and injected into the secondary oocyte using a tiny needle (fig. 23A). This variant of IVF, called intracytoplasmic sperm injection (ICSI), has a 68% fertilization rate. It can help men with very low sperm counts, high numbers of abnormal sperm, or injuries or illnesses that prevent ejaculation. However, if ICSI is done to help a man become a father who has an inherited form of infertility, he may be passing the problem on to a son. In ICSI, minor surgery removes testicular tissue, from which viable sperm are isolated and injected into eggs. A day or so later, a physician transfers some of the resulting balls of eight or sixteen cells to the woman's uterus. The birth rate following IVF for a woman implanted with her own egg is about 26%, compared with 31% for natural conceptions.

Gamete Intrafallopian Transfer

One reason that IVF rarely works is the artificial fertilization environment. A procedure called GIFT (gamete intrafallopian transfer) moves fertilization to the woman's body. A woman takes a superovulation drug for a week and then has several of her largest eggs removed.

A man donates a sperm sample, and a physician separates the most active cells. The collected eggs and sperm are deposited together in the woman's uterine tube, at a site past any obstruction so that implantation can occur. GIFT is 22% successful. In another variation called zygote intrafallopian transfer (ZIFT), a physician places an *in vitro* fertilized secondary oocyte in a woman's uterine tube.

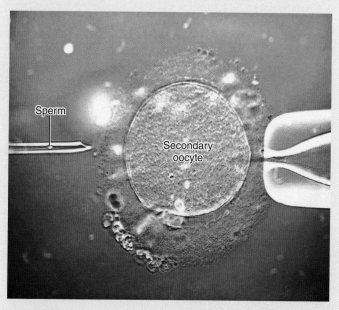

FIGURE 23A Intracytoplasmic sperm injection (ICSI) is injection of a single sperm cell into the cytoplasm of an egg.

---

that results from these initial cell divisions is called a *cleavage embryo*. From Science to Technology 23.2 describes genetic tests of blastomeres.

The tiny mass of cells moves through the uterine tube to the uterine cavity, aided by the beating of cilia on the surface of the epithelium lining the tube and by weak peristaltic contractions of smooth muscles in the tubular wall. Secretions from the epithelial lining bring nutrients to the developing organism.

The trip to the uterus takes about three days. By then the structure consists of a solid ball, called a **morula** (mor'u-lah), of about sixteen cells (fig. 23.4). The morula remains unattached within the uterine cavity for about three days. Cell division continues, and the solid ball of cells gradually hollows out. During this stage, the zona pellucida of the original secondary oocyte degenerates. The now-hollow structure, called a **blastocyst** (blas'to-sist), adheres to the endometrium. **Figure 23.5** diagrams this process.

By the end of the first week of development, the blastocyst superficially implants in the endometrium (**fig. 23.6a**). Up until this point, the cells that will become developing offspring are pluripotent stem cells, which means they can give rise to several specialized types of cells, as well as yield additional stem cells.

Within the blastocyst, cells in one region group to form an *inner cell mass* (or embryoblast) that eventually gives rise to the **embryo proper** (em'bre-o prop'er)—the body of the developing offspring. The cells that form the wall of the blastocyst make up the *trophoblast*, which develops into structures that assist the development of the embryo proper.

> If two ovarian follicles release secondary oocytes simultaneously, and both are fertilized, the resulting zygotes develop into fraternal (dizygotic) twins, which are no more alike genetically than any nontwin siblings. Twins may develop from a single fertilized oocyte (monozygotic twins) if two inner cell masses form within a blastocyst and each produces an embryo. Twins of this type usually share a single placenta, and they are genetically identical. They are always of the same sex and similar in appearance.

Implantation

The blastocyst attaches to the uterine lining, aided by its secretion of proteolytic enzymes that digest part of the endometrium

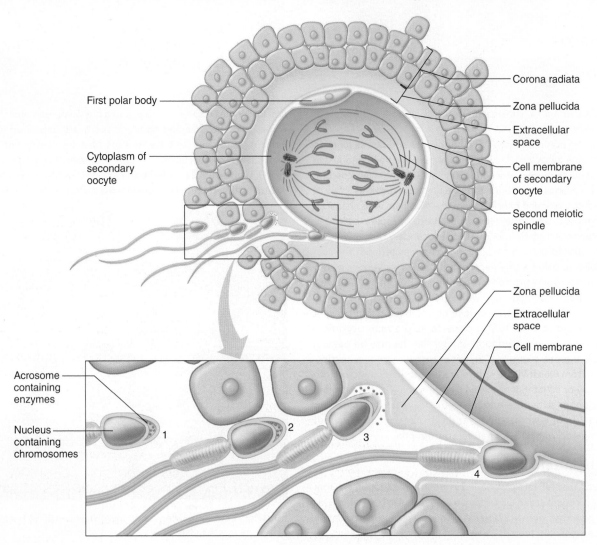

First polar body

Corona radiata

Zona pellucida

Extracellular space

Cytoplasm of secondary oocyte

Cell membrane of secondary oocyte

Second meiotic spindle

Zona pellucida

Extracellular space

Cell membrane

Acrosome containing enzymes

Nucleus containing chromosomes

1

2

3

4

FIGURE 23.3 Steps in fertilization: (*1*) The sperm cell reaches the corona radiata surrounding the secondary oocyte. (*2*) An enzyme on the sperm surface digests a path through the corona radiata. (*3*) The sperm binds to proteins in the zona pellucida, triggering the release of digestive enzymes from the acrosome by exocytosis. (*4*) Once the sperm is through the zona pellucida, the sperm head fuses with the oocyte membrane, and the sperm nucleus enters the oocyte. Although only a single sperm can actually fertilize the egg, hundreds of sperm are necessary to break down these barriers encountered along the way.

How many chromosomes are contained in the secondary oocyte prior to fertilization?

Answer can be found in Appendix G.

(fig. 23.6*b,c*). Cells of the trophoblast begin to produce tiny, fingerlike extensions (microvilli) that grow into the endometrium. At the same time, growth of the endometrium envelopes the blastocyst until it is completely embedded in the uterine wall. This nestling of the blastocyst into the uterine lining is called **implantation** (im″-plan-ta′shun). It begins toward the end of the first week and completes during the second week of development (fig. 23.7).

The trophoblast secretes the hormone **human chorionic gonadotropin** (hCG), which maintains the corpus luteum during the early stages of pregnancy and keeps the immune system from rejecting the blastocyst. This hormone also stimulates synthesis of other hormones from the developing placenta. The **placenta** (plah-sen′tah) is a vascular structure, formed by the cells surrounding the embryo and cells of the endometrium, that anchors the embryo to the uterine wall and exchanges nutrients, gases, and wastes between the maternal blood and the embryo's blood.

If an embryo implants in tissues outside the uterus, such as those of a uterine tube, an ovary, the cervix, or an organ in the abdominal cavity, the result is an *ectopic pregnancy*. A fertilized egg implanted in the uterine tube is a *tubal pregnancy*. The tube usually ruptures as the embryo enlarges, resulting in severe pain and heavy vaginal bleeding. Treatment is prompt surgical removal of the embryo and repair or removal of the damaged uterine tube.

PRACTICE

4 What is cleavage?

5 How does a blastocyst attach to the endometrium?

6 How does the endometrium respond to the activities of the blastocyst?

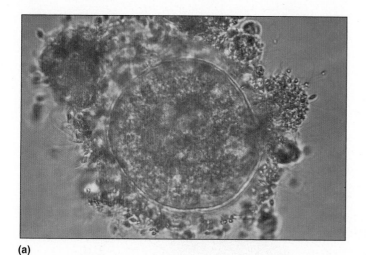

(a)

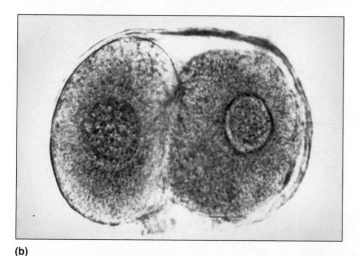

(b)

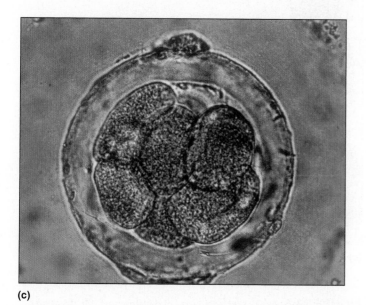

(c)

FIGURE 23.4 Light micrographs of (a) a human secondary oocyte surrounded by follicular cells and sperm cells (250×), (b) the two-cell stage (600×), and (c) a morula (500×).

Extraembryonic Membrane Formation and Placentation

As the embryo implants in the uterus, proteolytic enzymes from the trophoblast break down endometrial tissue, providing nutrients for the developing embryo. A second layer of cells begins to line the trophoblast, and together these two layers form a structure called the **chorion** (ko′re-on), the outermost extraembryonic membrane. Soon slender projections grow out from the trophoblast, including the new cell layer, eroding their way into the surrounding endometrium by continuing to secrete proteolytic enzymes. These projections become increasingly complex and form the highly branched **chorionic villi,** which are well established by the end of the fourth week (fig. 23.8). Figure 24.13 (p. 919) depicts prenatal genetic tests performed on chorionic villi.

Continued secretion of proteolytic enzymes forms irregular spaces called **lacunae** in the endometrium around and between the chorionic villi. These spaces fill with maternal blood from endometrial blood vessels eroded by enzyme action. At the same time, embryonic blood vessels carrying blood to and from the embryo extend through the connecting stalk, which attaches the embryo to the developing placenta. Capillary networks from the embryo form in the developing chorionic villi. These embryonic vessels allow nutrient exchange with blood in the lacunae, meeting the increased nutrient demands of the growing embryo.

While the placenta is forming from the chorion, a second membrane, called the **amnion** (am′ne-on), develops around the embryo proper. It appears during the second week. Its margin is attached around the edge of the flattened inner cell mass or **embryonic disc.** Fluid, called **amniotic fluid,** fills the space (*amniotic cavity*) between the amnion and the embryonic disc. The amniotic fluid provides a watery environment in which the embryo can grow freely without being compressed by surrounding tissues. The amniotic fluid also protects the embryo from being jarred by the movements of the woman's body, and maintains a stable temperature for embryonic and fetal development. Figure 24.13 (p. 919) illustrates amniocentesis, which examines chromosomes in fetal cells shed into the amniotic fluid.

If a pregnant woman repeatedly ingests an addictive substance that crosses the placenta, her newborn may suffer from withdrawal symptoms when amounts of the chemical the fetus was accustomed to receiving suddenly plummet after birth. Newborn addiction occurs with certain drugs of abuse, such as heroin, and with certain prescription drugs used to treat anxiety.

The margins of the amnion fold, enclosing the embryo in the amnion and amniotic fluid. The amnion envelopes the tissues on the underside of the embryo, particularly the connecting stalk, by which it is attached to the chorion and the developing placenta. In this manner, the **umbilical cord** (um-bil′ĭ-kal kord) forms (figs. 23.9 and 23.10). The fully developed umbilical cord is about 1 centimeter in diameter and about 55 centimeters in length. It originates at the

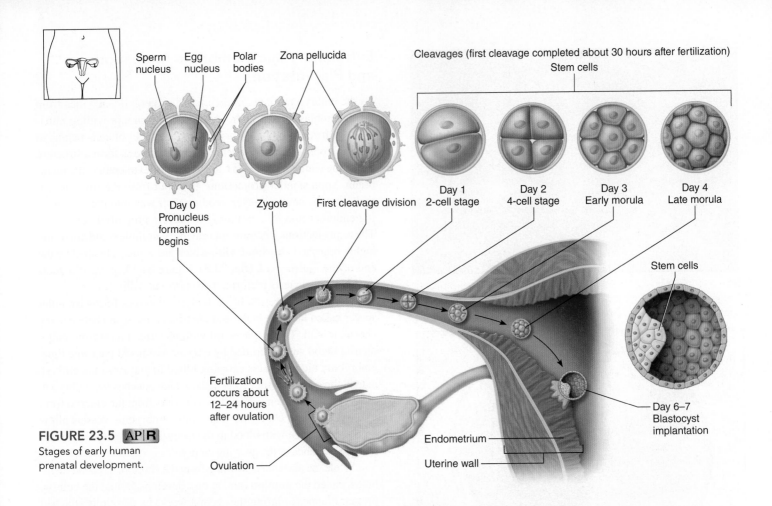

Sperm nucleus

Egg nucleus

Polar bodies

Zona pellucida

Cleavages (first cleavage completed about 30 hours after fertilization)

Stem cells

Day 0
Pronucleus formation begins

Zygote

First cleavage division

Day 1
2-cell stage

Day 2
4-cell stage

Day 3
Early morula

Day 4
Late morula

Stem cells

Fertilization occurs about 12–24 hours after ovulation

Day 6–7
Blastocyst implantation

Endometrium

Uterine wall

Ovulation

FIGURE 23.5 AP|R
Stages of early human prenatal development.

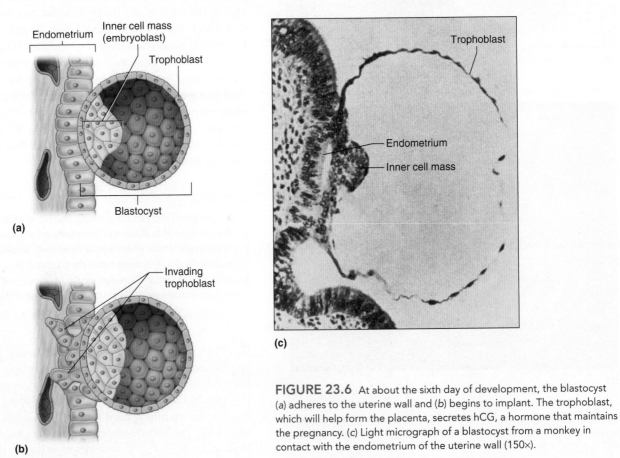

Endometrium

Inner cell mass (embryoblast)

Trophoblast

Blastocyst

(a)

Invading trophoblast

(b)

Trophoblast

Endometrium

Inner cell mass

(c)

FIGURE 23.6 At about the sixth day of development, the blastocyst (a) adheres to the uterine wall and (b) begins to implant. The trophoblast, which will help form the placenta, secretes hCG, a hormone that maintains the pregnancy. (c) Light micrograph of a blastocyst from a monkey in contact with the endometrium of the uterine wall (150×).

Six-year-old Molly would probably have died within a year or two of Fanconi anemia, had she not received a very special gift from her baby brother Adam—his umbilical cord stem cells. Not only was Adam free of the mutation that causes the anemia, but his cell surfaces matched those of his sister, making a transplant likely to succeed. But the parents didn't have to wait until Adam's birth in August 2000 to know that his cells could save Molly—they knew when he was a mere eight-celled cleavage embryo (fig. 23B).

When a compatible bone marrow donor for Molly could not be found, her parents turned to preimplantation genetic diagnosis (PGD). Following *in vitro* fertilization, described in From Science to Technology 23.1, a single cell was removed from each of several eight-celled cleavage embryos. The cells were probed to detect those free of the disease-causing mutation. They were also scrutinized for the HLA genes, which control rejection of a transplanted organ, and the ball of cells that would become Adam was chosen. The cleavage embryo divided in the laboratory until it was about 120 cells, and then it was implanted into Lisa's uterus. A month after Adam was born, physicians infused the umbilical cord stem cells into his sister. Today Molly is healthy.

PGD works because up until a certain point in early development of an embryo, a cell or two can be removed, yet the remainder of the embryo can continue to develop normally in a uterus. PGD, implemented in 1989, at first helped a few families avoid devastating inherited illnesses in their sons. Then in 1992, a baby was born free of the cystic fibrosis that made her brother very ill, thanks to PGD. In 1994 came another milestone, when a girl was conceived and selected to provide umbilical cord stem cells that cured her teenage sister's leukemia. This case became the basis of a popular novel.

Thousands of children have been born following PGD, free of the disorders that run in their families. PGD also enables couples who repeatedly lose early embryos due to chromosome abnormalities to select chromosomally normal embryos. PGD is becoming routine with IVF, because it ensures that only embryos with normal-appearing chromosomes or lacking a family's mutation are implanted. PGD is often confused with genetic modification (engineering), which it is not. It is selection of an embryo.

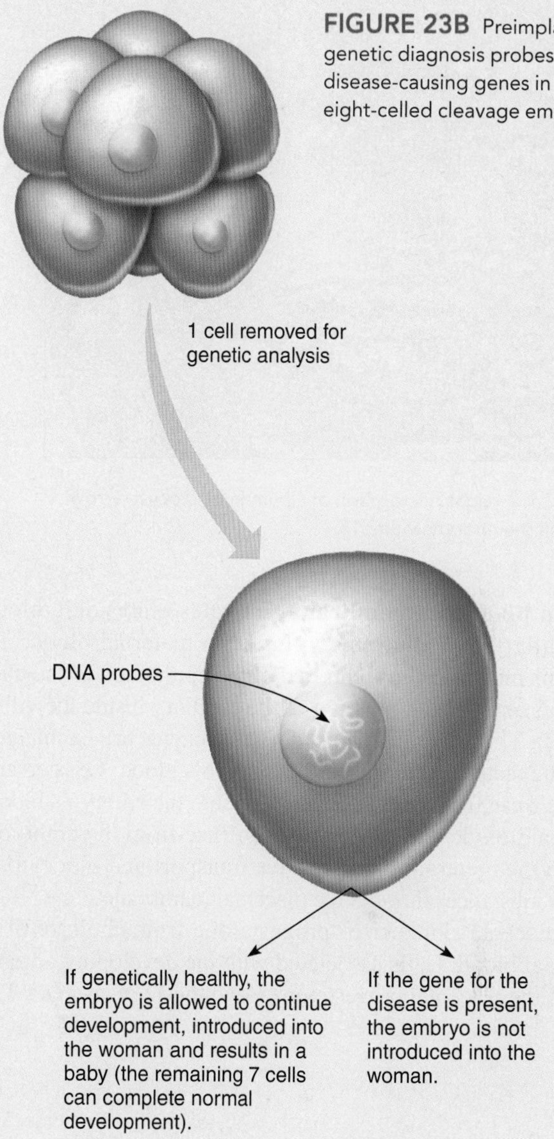

FIGURE 23B Preimplantation genetic diagnosis probes disease-causing genes in an eight-celled cleavage embryo.

1 cell removed for genetic analysis

DNA probes

If genetically healthy, the embryo is allowed to continue development, introduced into the woman and results in a baby (the remaining 7 cells can complete normal development).

If the gene for the disease is present, the embryo is not introduced into the woman.

---

umbilicus of the embryo and inserts into the center of the placenta. The umbilical cord suspends the embryo in the amniotic cavity. The cord contains three blood vessels—two umbilical arteries and one umbilical vein—that transport blood between the embryo and the placenta (fig. 23.11).

Eventually the amniotic cavity becomes so enlarged that the membrane of the amnion contacts the thicker chorion around it. The two membranes fuse into an amniochorionic membrane. In addition to the chorion and amnion, two other extraembryonic membranes form during development. They are the yolk sac and the allantois.

The **yolk sac** forms during the second week, and it is attached to the underside of the embryonic disc (see figs. 23.9 and 23.10). This structure forms blood cells in the early stages of development and gives rise to the cells that later become sex cells. The yolk sac also produces stem cells of the bone marrow, which are precur-

sors to many cell types, but predominantly to blood cells. Parts of the yolk sac form the embryonic digestive tube as well. Part of the membrane derived from the yolk sac becomes incorporated into the umbilical cord, and the remainder lies in the cavity between the chorion and the amnion near the placenta. The **allantois** (ah-lan to-is) forms during the third week as a tube extending from the early yolk sac into the connecting stalk (see figs. 23.9 and 23.10). It, too, forms blood cells and gives rise to the umbilical blood vessels.

The disc-shaped area where the chorion still contacts the uterine wall develops into the placenta. The embryonic portion of the placenta is composed of parts of the chorion and its villi; the maternal portion is composed of the area of the uterine wall (decidua basalis) where the villi attach (fig. 23.11). The fully formed placenta is a reddish-brown disc about 20 centimeters long and 2.5 centimeters thick, weighing about 0.5 kilogram.

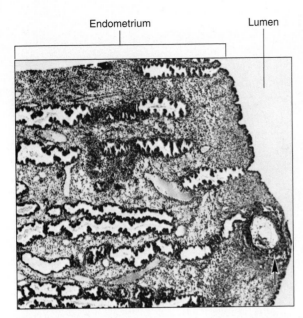

Endometrium   Lumen

FIGURE 23.7 Light micrograph of a human blastocyst (arrow) implanting in the endometrium (18×).

A thin **placental membrane** separates embryonic blood in the capillary of a chorionic villus from maternal blood in a lacuna. This membrane is composed of the epithelium of the chorionic villus and the endothelium of the capillary inside the villus (fig. 23.12). Through this membrane, substances are exchanged between the maternal blood and the embryo's blood. Oxygen and nutrients diffuse from the maternal blood into the embryo's blood, and carbon dioxide and other wastes diffuse from the embryo's blood into the maternal blood. Active transport and pinocytosis also move substances through the placental membrane.

The placenta synthesizes progesterone from cholesterol in the maternal blood. Cells associated with the developing adrenal glands use the placental progesterone to synthesize estrogens. The estrogens, in turn, promote changes in the maternal uterus and breasts and influence maternal metabolism and the development of organs in the embryo.

 PRACTICE

7 Describe the structure of a chorionic villus.
8 What is the function of amniotic fluid?
9 What types of cells and other structures are derived from the yolk sac?
10 What is the function of the placental membrane?

Gastrulation and Organogenesis

Gastrulation is the movement of cells within the embryonic disc to form multiple layers. By the end of the second week the embryonic disc consists of two distinct layers: an outer *ectoderm* and an inner *endoderm*. A short time later, through a process called gastrulation, a third layer of cells, the *mesoderm,* forms between the ectoderm and endoderm. These three layers of cells are called the **primary germ layers** (pri′ma-re jerm la′erz) of the primordial embryo (see fig. 23.8). They are the primitive tissues from which all organs form in a process called **organogenesis.** At this point the embryo is termed a **gastrula** (gas′troo-lah). Table 23.1 summarizes the stages of early human prenatal development.

Gastrulation is an important process in prenatal development because a cell's fate is determined by which layer it is in. The cells of the ectoderm and endoderm are epithelia. The mesoderm is loosely organized connective tissue. *Ectodermal cells* give rise to the nervous system, parts of special sensory organs, the epidermis, hair, nails, glands of the skin, and linings of the mouth and anal canal. *Mesodermal cells* form all types of muscle tissue, bone tissue, bone marrow, blood, blood vessels, lymphatic vessels, internal reproductive organs, kidneys, and the epithelial linings of the body cavities. *Endodermal cells* produce the epithelial linings of the digestive tract, respiratory tract, urinary bladder, and urethra

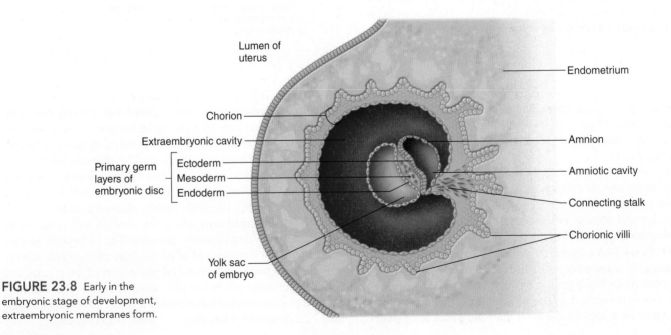

FIGURE 23.8 Early in the embryonic stage of development, extraembryonic membranes form.

Lumen of uterus

Endometrium

Chorion

Extraembryonic cavity

Primary germ layers of embryonic disc
- Ectoderm
- Mesoderm
- Endoderm

Amnion

Amniotic cavity

Connecting stalk

Chorionic villi

Yolk sac of embryo

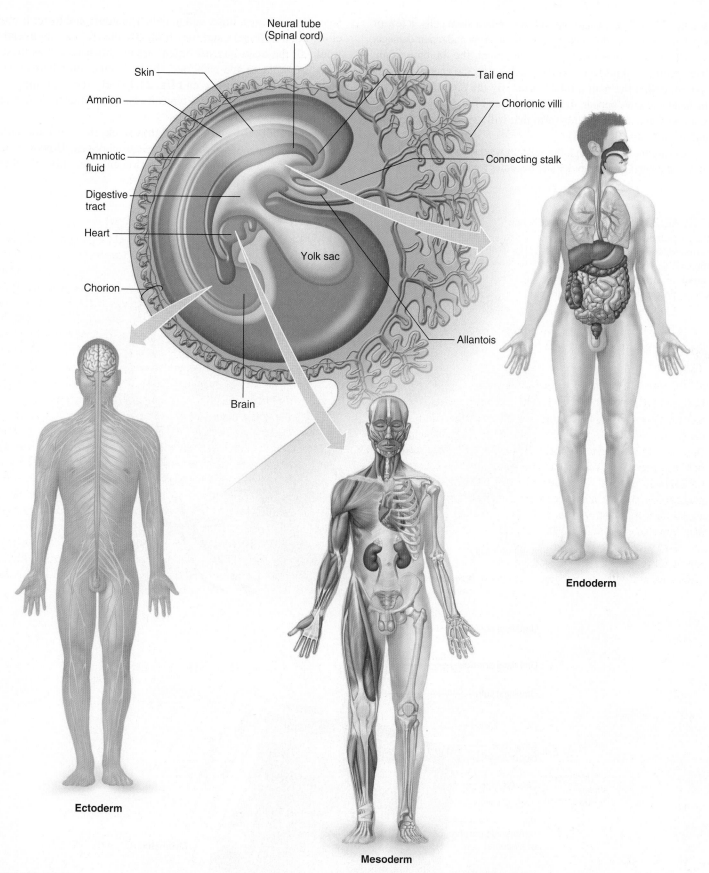

Neural tube
(Spinal cord)

Skin

Amnion

Amniotic
fluid

Digestive
tract

Heart

Chorion

Brain

Tail end

Chorionic villi

Connecting stalk

Yolk sac

Allantois

Ectoderm

Mesoderm

Endoderm

FIGURE 23.9 The primary germ layers collectively serve as primitive tissues from which all organs form.

(see fig. 23.9). The primary germ layers retain stem cells, a few of which persist in the adult, enabling tissues to grow and repair damage.

During the fourth week of development, the flat embryonic disc becomes cylindrical, and the precursor of the central nervous system, called the neural tube, forms. By the end of week four, the head and jaws appear, the heart beats and forces blood through blood vessels, and tiny buds form that will give rise to the upper and lower limbs (fig. 23.13).

During the fifth through the seventh weeks, as figure 23.14a–e shows, the head grows rapidly and becomes rounded and erect. The face develops eyes, nose, and mouth. The upper and lower limbs elongate, and fingers and toes form. By the end of the seventh week, all the main internal organs are established, and as these structures enlarge and elaborate, the body takes on a humanlike appearance (fig. 23.14f, g and fig. 23.15). By the beginning of the eighth week, the embryo typically is 30 millimeters long and weighs less than 5 grams.

Until about the end of the eighth week, the chorionic villi cover the entire surface of the former trophoblast. However, as the embryo and the surrounding chorion enlarge, only villi that

FIGURE 23.10 As the amnion develops, it surrounds the embryo, and the umbilical cord begins to form from structures in the connecting stalk.

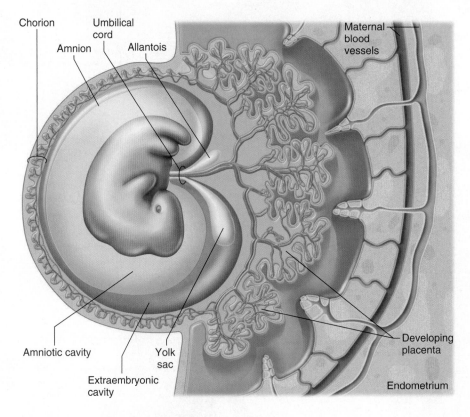

Chorion
Umbilical cord
Amnion
Allantois
Maternal blood vessels
Amniotic cavity
Yolk sac
Extraembryonic cavity
Developing placenta
Endometrium

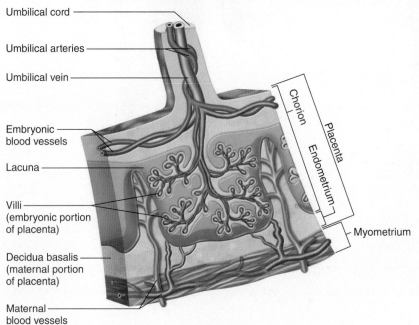

Umbilical cord
Umbilical arteries
Umbilical vein
Embryonic blood vessels
Lacuna
Villi (embryonic portion of placenta)
Decidua basalis (maternal portion of placenta)
Maternal blood vessels
Chorion
Endometrium
Placenta
Myometrium

FIGURE 23.11 The placenta consists of an embryonic portion and a maternal portion.

TABLE 23.1 Stages and Events of Early Human Prenatal Development

Stage	Time Period	Principal Events
Zygote	12–24 hours following ovulation	Secondary oocyte fertilized, meiosis is completed; zygote has 46 chromosomes and is genetically distinct
Cleavage	30 hours to third day	Mitosis increases cell number
Morula	Third to fourth day	Solid ball of cells
Blastocyst	Fifth day through second week	Hollow ball consisting of the trophoblast (outside), which implants and helps form the placenta, and inner cell mass, which flattens to form the embryonic disc
Gastrula	End of second week	Primary germ layers form

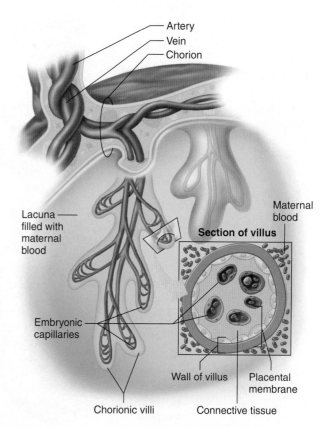

FIGURE 23.12 The placental membrane consists of the endothelium of an embryonic capillary and the epithelial wall of a chorionic villus, as illustrated in the section of the villus (lower part of the figure).

contact the endometrium endure. The others degenerate, and the areas of the chorion where they were attached become smooth. The region of the chorion still in contact with the uterine wall is restricted to the placenta.

The embryonic stage concludes at the end of the eighth week. It is the most critical period of development, because during it, the embryo implants in the uterine wall and all the essential external and internal body parts form. Disturbances to development during the embryonic stage can cause major malformations or malfunctions. This is why early prenatal care is important.

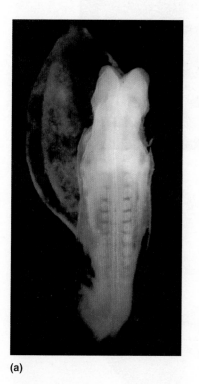

(a)

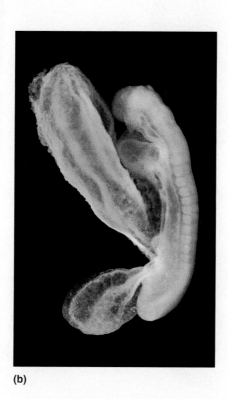

(b)

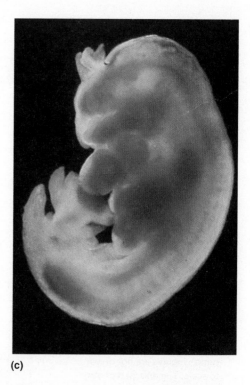

(c)

FIGURE 23.13 Embryo. (a) A human embryo at three weeks, posterior view; (b) at three and one-half weeks, lateral view; (c) at about four weeks, lateral view. (Figures are not to scale.)

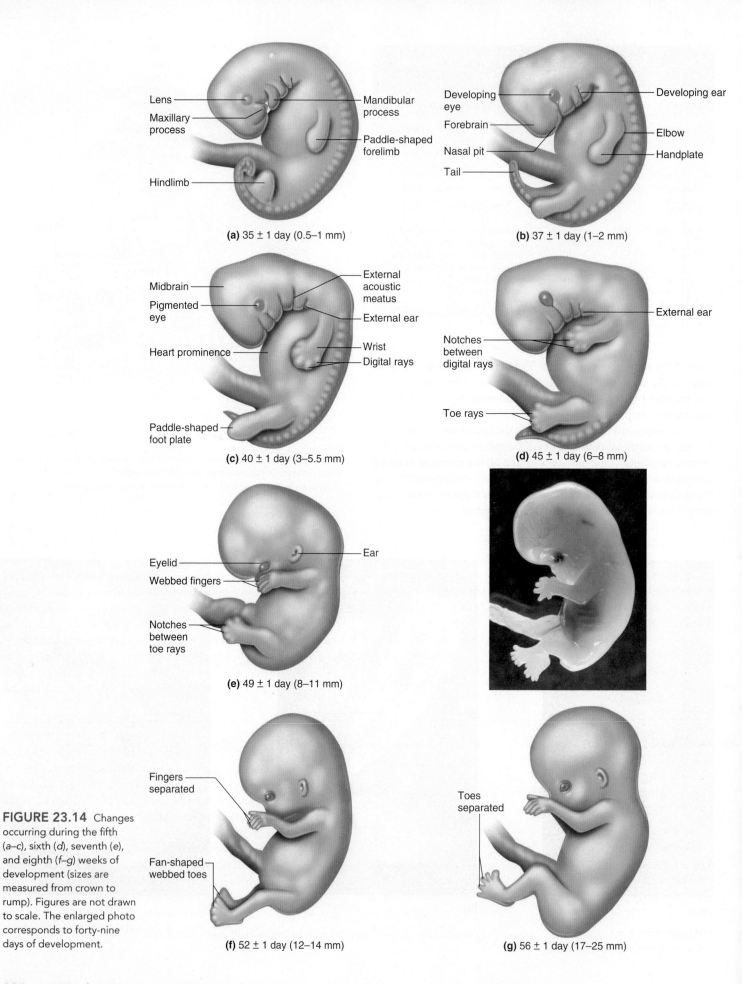

(a) 35 ± 1 day (0.5–1 mm)

Lens
Maxillary process
Hindlimb
Mandibular process
Paddle-shaped forelimb

(b) 37 ± 1 day (1–2 mm)

Developing eye
Forebrain
Nasal pit
Tail
Developing ear
Elbow
Handplate

(c) 40 ± 1 day (3–5.5 mm)

Midbrain
Pigmented eye
Heart prominence
Paddle-shaped foot plate
External acoustic meatus
External ear
Wrist
Digital rays

(d) 45 ± 1 day (6–8 mm)

Notches between digital rays
Toe rays
External ear

(e) 49 ± 1 day (8–11 mm)

Eyelid
Webbed fingers
Notches between toe rays
Ear

(f) 52 ± 1 day (12–14 mm)

Fingers separated
Fan-shaped webbed toes

(g) 56 ± 1 day (17–25 mm)

Toes separated

FIGURE 23.14 Changes occurring during the fifth (a–c), sixth (d), seventh (e), and eighth (f–g) weeks of development (sizes are measured from crown to rump). Figures are not drawn to scale. The enlarged photo corresponds to forty-nine days of development.

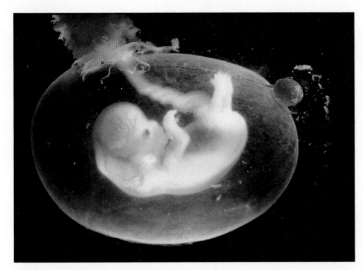

FIGURE 23.15 By the beginning of the eighth week of development, the embryonic body is recognizable as a human.

Factors that cause congenital malformations by affecting an embryo during its period of rapid growth and development or the specialization of structures during fetal development are called **teratogens.** Such agents include drugs, viruses, radiation, and even large amounts of otherwise healthful substances, such as fat-soluble vitamins. Each prenatal structure has a time in development, called its *critical period,* when it is sensitive to teratogens (fig. 23.16).

A critical period may extend over many months or just a day or two. Neural tube defects, for example, are traced to day twenty-eight in development, when a sheet of ectoderm folds to form the neural tube. When this process is disrupted, an opening may remain in the spine (spina bifida). In contrast, the critical period for the developing brain begins when the anterior neural tube begins to swell into a brain, and continues throughout gestation. This is why so many teratogens affect the brain. Clinical Application 23.1 discusses some teratogens and their effects.

 RECONNECT

To Chapter 11, Brain Development, page 395.

 PRACTICE

11 What is gastrulation?

12 Which tissues and structures develop from ectoderm? from mesoderm? from endoderm?

13 How do teratogens cause birth defects?

Fetal Stage

The **fetal stage** begins at the end of the eighth week of prenatal development and lasts until birth. During this period, growth is rapid, and body proportions change considerably. At the beginning of the fetal stage, the head is disproportionately large, and the lower limbs are relatively short. Gradually, proportions come to more closely resemble those of a child (fig. 23.17).

(a) When physical structures develop

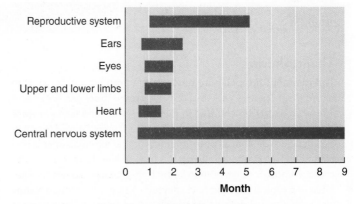

(b) When different teratogens disrupt development

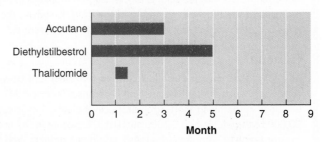

FIGURE 23.16 Critical periods. (a) Structures in the developing embryo and fetus (b) are sensitive to specific teratogens at different times in gestation.

During the third month, body lengthening accelerates, but head growth slows. The upper limbs of the **fetus** (fe'tus) achieve the relative length they will maintain throughout development, and ossification centers appear in most bones. By the twelfth week, the external reproductive organs are distinguishable as male or female. Figure 23.18 illustrates how these external reproductive organs of the male and female differentiate from precursor structures.

In the fourth month, the body grows rapidly and reaches a length of up to 20 centimeters and weighs about 170 grams. The lower limbs lengthen considerably, and the skeleton continues to ossify. The fetus has hair, nipples, and nails, and may even scratch itself. A four-month-old fetus will startle and turn away from a bright light flashed on a pregnant woman's belly, and may also react to sudden loud noises.

In the fifth month, growth slows. The lower limbs achieve their final relative proportions. Skeletal muscles contract, and the pregnant woman may feel fetal movements. Some hair grows on the fetal head, and fine, downy hair called lanugo covers the skin. A cheesy mixture of sebum from the sebaceous glands and dead epidermal cells (vernix caseosa) also coats the skin. The fetus, about 30 centimeters long and weighing about 450 grams, curls into the fetal position.

During the sixth month, the fetus gains substantial weight. Eyebrows and eyelashes appear. The skin is wrinkled and translucent. Blood vessels in the skin cause a reddish appearance.

In the seventh month, the skin becomes smoother as fat is deposited in the subcutaneous tissues. The eyelids, which fused during the third month, reopen. At the end of this month, the fetus is about 40 centimeters long.

Thalidomide

The idea that the placenta protects the embryo and fetus from harmful substances was tragically disproven between 1957 and 1961, when 10,000 children in Europe, Africa, and Canada were born with malformed limbs. Doctors soon identified a mild tranquilizer, thalidomide, that all of the mothers of the deformed infants had taken early in pregnancy during the time of limb formation, as the cause. Although some women in the United States did use thalidomide in a clinical trial and had affected children, the United States was spared a thalidomide disaster because an astute government physician noted adverse effects of the drug on monkeys in experiments, and she blocked marketing of the drug. However, thalidomide is used today to treat leprosy and certain blood disorders.

Rubella

The virus that causes rubella (German measles) is a powerful teratogen. Australian physicians first noted its effects in 1941, and a rubella epidemic in the United States in the early 1960s caused 20,000 birth defects and 30,000 stillbirths. Exposure in the first trimester leads to cataracts, deafness, and heart defects, and later exposure causes learning disabilities, speech and hearing problems, and type 1 diabetes mellitus. Successful vaccination programs that provide maternal immunity to the virus have greatly lowered the incidence of "congenital rubella syndrome" in many countries.

Alcohol

A pregnant woman who has just one or two alcoholic drinks a day, or perhaps many drinks at a crucial time in prenatal development, risks *fetal alcohol syndrome (FAS)* or the more prevalent *fetal alcohol effects* in her unborn child. The effects of small amounts of alcohol at different stages of pregnancy are not yet well understood, and because each woman metabolizes alcohol slightly differently, women are advised to avoid drinking alcohol entirely when pregnant or when trying to become pregnant.

A child with FAS has a small head, misshapen eyes, and a flat face and nose (fig. 23C). Growth is slow before and after birth. Intellect is impaired, ranging from minor learning disabilities to severe intellectual disability. Teens and young adults with FAS are short and have small heads. Many individuals remain at the early grade-school level of intellectual development. They often lack social and communication skills, such as understanding the consequences of actions, forming friendships, taking initiative, and interpreting social cues. People with FAS are more likely to have seizures.

Problems in children of alcoholic mothers were noted by Aristotle more than twenty-three centuries ago. Today in the United States, fetal alcohol syndrome is the third most common cause of intellectual disability in newborns. Each year in the United States about 5,000 children are born with FAS, and many more are born with milder "alcohol-related effects."

Cigarettes

Chemicals in cigarette smoke stress a fetus. Carbon monoxide crosses the placenta and binds to fetal hemoglobin in a way that prevents the hemoglobin from delivering oxygen to the fetal tissues. Other chemicals in smoke prevent nutrients

---

In the final trimester, fetal brain cells rapidly form networks, as organs specialize and grow. A layer of fat is laid down beneath the skin. The testes of males descend from regions near the developing kidneys, through the inguinal canal, and into the scrotum (see chapter 22, p. 826). The digestive and respiratory systems mature last, which is why premature infants may have difficulty digesting milk and breathing.

> Each year in the U.S., 1 in 8 babies are born preterm, defined as delivery before 37 weeks of gestation. Such premature infants' survival chances increase directly with age and weight, and parallel increasing maturity of the lungs. A baby born at 23 weeks has an almost 30% chance of survival; at 24 weeks a 50% to 60% chance; at 25 weeks a 75% chance; and at 27 to 28 weeks, a greater than 90% chance.

Approximately 266 days after a single sperm fertilized a secondary oocyte, a baby is ready to be born. It is *full-term*. It is about 50 centimeters long and weighs 2.7 to 3.6 kilograms. The skin has lost its downy hair but is still coated with sebum and dead epidermal cells. The scalp is usually covered with hair; the fingers and toes have well-developed nails; and the skull bones are largely ossified. The fetus is usually positioned upside down (fig. 23.19) with its head toward the cervix *(vertex position)*.

The birth of a live, healthy baby is against the odds, considering human development from the beginning. Of every 100 secondary oocytes exposed to sperm, 84 are fertilized. Of these, 69 implant in the uterus, 42 survive one week or longer, 37 survive six weeks or longer, and only 31 are born alive. Of those that do not survive to birth, about half have chromosomal abnormalities too severe to maintain life. Table 23.2 summarizes the stages of prenatal development.

 PRACTICE

14 What major changes happen during the fetal stage of development?

15 When is the sex of a fetus visible externally?

16 How is a fetus usually positioned in the uterus as birth nears?

from reaching the fetus. Smoke-exposed placentas lack important growth factors. The result of these assaults is poor growth before and after birth. Cigarette smoking during pregnancy is linked to spontaneous abortion, stillbirth, prematurity, and low birth weight.

Nutrients and Malnutrition

Certain nutrients in large amounts, particularly vitamins, act in the body as drugs. The acne medication *isotretinoin* (Accutane) is a derivative of vitamin A that causes spontaneous abortions and defects of the fetal heart, nervous system, and face. A psoriasis drug based on vitamin A, as well as excesses of vitamin A itself, also cause birth defects because some forms of the vitamin are stored in body fat for up to three years after ingestion.

Malnutrition during pregnancy causes intrauterine growth retardation (IUGR). Malnutrition before birth also causes shifts in metabolism to make the most of calories from food. This protective action, however, sets the stage for developing obesity and associated disorders, such as type 2 diabetes mellitus and cardiovascular disease, in adulthood.

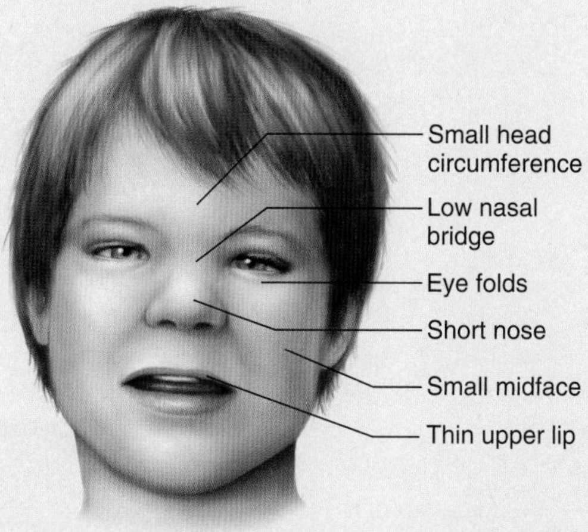

Occupational Hazards

The workplace can be a source of teratogens. Women who work with textile dyes, lead, certain photographic chemicals, semiconductor materials, mercury, and cadmium have increased rates of spontaneous abortion and delivering children with birth defects. Men whose jobs expose them to sustained heat, such as smelter

FIGURE 23C Fetal alcohol syndrome. Some children whose mothers drank alcohol during pregnancy have characteristic flat faces. Women who drink while pregnant have a chance of having a child affected to some degree by prenatal exposure to alcohol.

- Small head circumference
- Low nasal bridge
- Eye folds
- Short nose
- Small midface
- Thin upper lip

workers, glass manufacturers, and bakers, may produce sperm that can fertilize a secondary oocyte, but possibly lead to spontaneous abortion or a birth defect. A virus or a toxic chemical carried in semen may also cause a birth defect.

A fetus can receive medical treatments. Tubes can be inserted to remove abnormal fluid accumulations, such as in the bladder or in the brain, and tumors removed. Correcting the neural tube defect spina bifida in a fetus is more successful in enabling the child to eventually walk than correcting the defect after birth.

Fetal Blood and Circulation

Throughout fetal development, the maternal blood supplies oxygen and nutrients and carries away wastes. These substances diffuse between the maternal and fetal blood through the placental membrane, and the umbilical blood vessels carry them to and from the fetus (fig. 23.20). The fetal blood and cardiovascular system are adapted to this intrauterine environment. For example, the concentration of oxygen-carrying hemoglobin in the fetal blood is about 50% greater than in the maternal blood, and fetal hemoglobin has a greater attraction for oxygen than does adult hemoglobin. As a result of these adaptations, at the oxygen

partial pressure of the placental capillaries, fetal hemoglobin can carry 20% to 30% more oxygen than adult hemoglobin. Different genes encode the protein subunits of hemoglobin in embryos, fetuses, and individuals after birth. The different subunits have different affinities for oxygen.

In the fetal cardiovascular system, the *umbilical vein* transports blood rich in oxygen and nutrients from the placenta to the fetal body. This vein enters the body through the umbilical ring and continues along the anterior abdominal wall to the liver. About half the blood it carries passes into the liver, and the rest enters a vessel called the **ductus venosus** (duk'tus ven-o'sus), which bypasses the liver.

The ductus venosus extends a short distance and joins the inferior vena cava. There, oxygen-rich blood from the placenta mixes with oxygen-poor blood from the lower parts of the fetal body. This mixture continues through the inferior vena cava to the right atrium.

RECONNECT

To Chapter 15, Blood Flow Through the Heart, pages 562–563.

FIGURE 23.17 During development, body proportions change considerably.

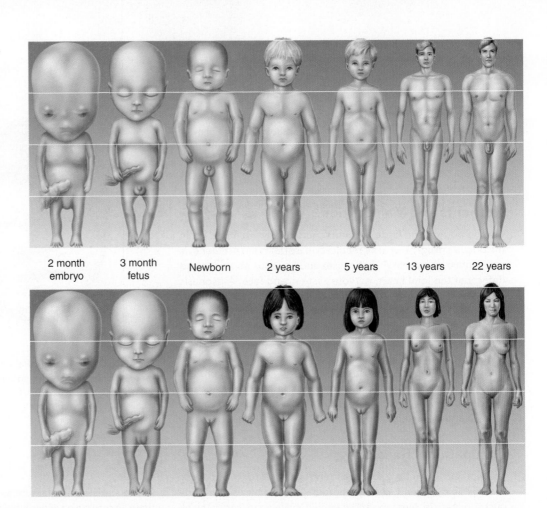

2 month embryo 3 month fetus Newborn 2 years 5 years 13 years 22 years

TABLE 23.2	Stages of Prenatal Development	
Stage	**Time Period**	**Major Events**
Embryonic stage	First week	Cells undergo mitosis, blastocyst forms; inner cell mass appears; blastocyst implants in uterine wall *Size: 0.25 inch (0.63 centimeter), weight: 1/120 ounce (0.21 gram)
	Second through eighth week	Inner cell mass becomes embryonic disc; primary germ layers form, embryo proper becomes cylindrical; main internal organs and external body structures appear; placenta and umbilical cord form, embryo proper is suspended in amniotic fluid *Size: 0.75 inch (1.9 centimeters) as measured from crown to rump, weight: 1/30 ounce (0.8 gram)
Fetal stage	Ninth through twelfth week	Ossification centers appear in bones; sex organs differentiate; nerves and muscles coordinate so that the fetus can move its limbs *Size: 2.13 inches (5.41 centimeters) as measured from crown to rump, weight: 1/2 ounce (14 grams)
	Thirteenth through sixteenth week	Body grows rapidly; ossification continues *Size: 4.57 inches (11.6 centimeters) as measured crown to rump, weight: 3.5 ounces (100 grams)
	Seventeenth through twentieth week	Muscle movements are stronger, and woman may be aware of slight flutterings; skin is covered with fine downy hair (lanugo) and coated with sebum mixed with dead epidermal cells (vernix caseosa) *Size: 6.46 inches (16.4 centimeters) as measured from crown to rump, weight: 10.5 ounces (300 grams)
	Twenty-first through thirty-eighth week	Body gains weight, subcutaneous fat deposited; eyebrows and lashes appear; eyelids reopen; testes descend *Size: 19.6 inches (49.8 centimeters) as measured from crown to heel, weight: 6 to 10 pounds (2.7 to 4.5 kilograms)

* Measurements pertain to the end of the designated time period.

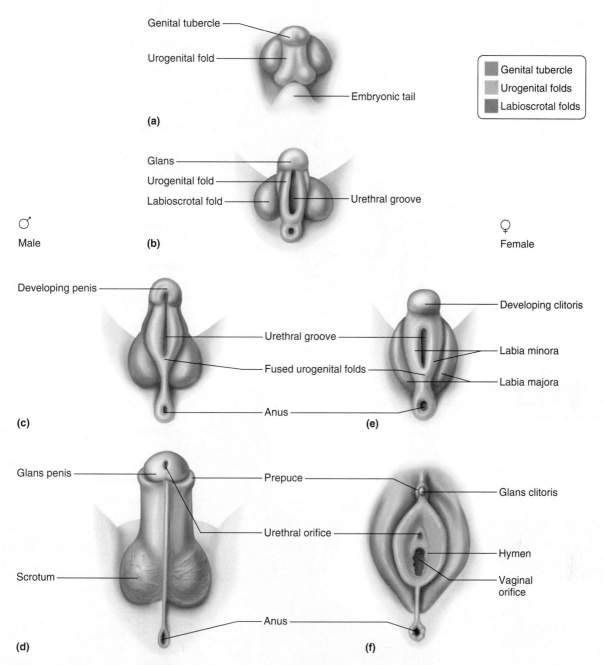

FIGURE 23.18 Formation of external reproductive organs. (*a* and *b*) The genital tubercle, urogenital fold, and labioscrotal folds, which appear during the fourth week of development, may differentiate into (*c* and *d*) male external reproductive organs or (*e* and *f*) female external reproductive organs.

Postnatally, the blood from the right atrium enters the right ventricle and is pumped through the pulmonary trunk and pulmonary arteries to the lungs. In the fetus, however, the lungs are not functioning in gas exchange, and the blood largely bypasses them. As blood from the inferior vena cava enters the fetal right atrium, much of it is shunted directly into the left atrium through an opening in the atrial septum. This opening is called the **foramen ovale** (fo-ra′men o-val′e), and the blood passes through it because the blood pressure in the right atrium is greater than that in the left atrium. Furthermore, a small valvelike structure (septum primum) on the left side of the atrial septum overlaps the foramen ovale and helps prevent blood from moving in the reverse direction.

The rest of the fetal blood entering the right atrium, including a large proportion of the oxygen-poor blood entering from the superior vena cava, passes into the right ventricle and out through the pulmonary trunk. Only a small volume of blood enters the pulmonary circuit because the lungs are collapsed and their blood vessels have a high resistance to blood flow. However, enough blood reaches the lung tissues to sustain them.

Most of the blood in the pulmonary trunk bypasses the lungs by entering a fetal vessel called the **ductus arteriosus** (duk′tus ar-te″re-o′sus), which connects the pulmonary trunk to the descending portion of the aortic arch. As a result of this connection, the blood with a relatively low oxygen concentration, return-

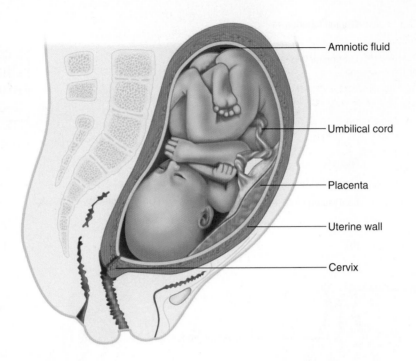

FIGURE 23.19 A full-term fetus is usually positioned with its head near the cervix.

Amniotic fluid

Umbilical cord

Placenta

Uterine wall

Cervix

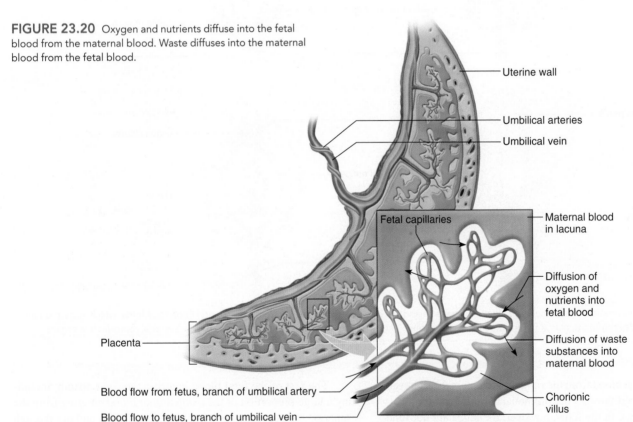

FIGURE 23.20 Oxygen and nutrients diffuse into the fetal blood from the maternal blood. Waste diffuses into the maternal blood from the fetal blood.

Uterine wall

Umbilical arteries

Umbilical vein

Fetal capillaries

Maternal blood in lacuna

Diffusion of oxygen and nutrients into fetal blood

Diffusion of waste substances into maternal blood

Placenta

Chorionic villus

Blood flow from fetus, branch of umbilical artery

Blood flow to fetus, branch of umbilical vein

ing to the heart through the superior vena cava, bypasses the lungs and does not enter the part of the aorta that branches to the heart and brain. (This might seem odd, but remember that fetal blood becomes oxygenated in the placenta, not in the fetal lungs.)

The more highly oxygenated blood that enters the left atrium through the foramen ovale mixes with a small amount of oxygen-poor blood returning from the pulmonary veins. This mixture

moves into the left ventricle and is pumped into the aorta. Some of it reaches the myocardium through the coronary arteries, and some reaches the brain tissues through the carotid arteries.

Blood carried by the descending aorta includes the less-oxygenated blood from the ductus arteriosus. Some of the blood is carried into the branches of the aorta that lead to the lower regions of the body. The rest passes into the *umbilical arteries,* which branch from the internal iliac arteries and lead to the placenta. There the blood is reoxygenated (figs. 23.21 and 23.22).

In most newborns the umbilical cord contains two arteries and one vein. On rare occasion a newborn has only one umbilical artery. This condition is often associated with other cardiovascular, urogenital, or gastrointestinal disorders. The vessels in the severed cord are routinely counted following a birth to rule out these conditions.

Table 23.3 summarizes the major features of fetal circulation. At the time of birth, important adjustments must occur in the cardiovascular system when the placenta ceases to function and the newborn begins to breathe.

 **PRACTICE**

17 Which umbilical vessel carries oxygen-rich blood to the fetus?

18 What is the function of the ductus venosus?

19 How does fetal circulation allow blood to bypass the lungs?

Maternal Changes During Pregnancy

During a typical reproductive cycle, the corpus luteum degenerates about two weeks after ovulation. Consequently, concentrations of estrogens and progesterone decline rapidly, the uterine

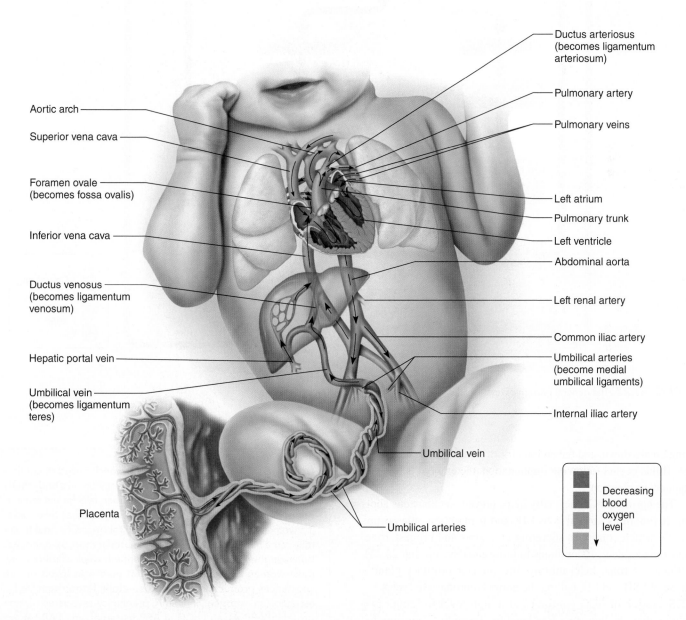

FIGURE 23.21 The general pattern of fetal circulation is shown anatomically. (Position of the right atrium has been shifted slightly to reveal the foramen ovale.)

Which structures are unique to fetal circulation?

Answer can be found in Appendix G.

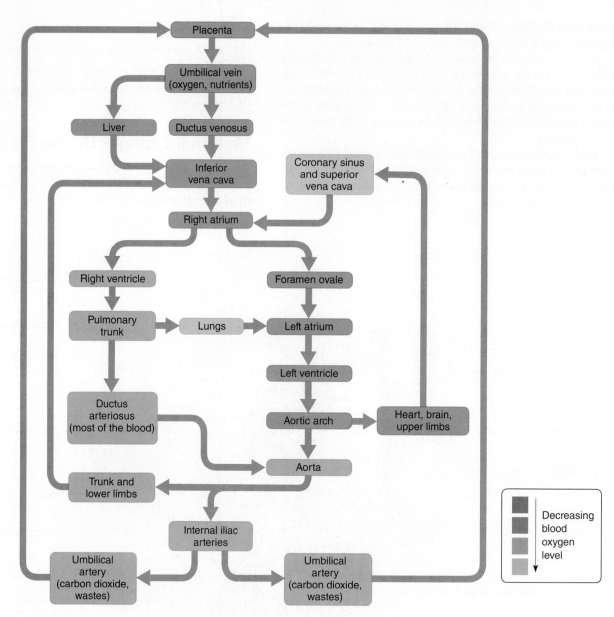

FIGURE 23.22 The general pattern of fetal circulation is shown schematically.

lining breaks down, and the endometrium sloughs off as menstrual flow. If this occurs following implantation, the embryo is lost in a spontaneous abortion.

The hormone hCG normally helps prevent spontaneous abortion. It functions similarly to LH, and it maintains the corpus luteum, which continues secreting estrogens and progesterone. Thus, the uterine wall continues to grow and develop (fig. 23.23). At the same time, hCG inhibits the anterior pituitary gland's release of FSH and LH, halting the normal reproductive cycles.

Secretion of hCG continues at a high level for about two months, then declines to a low level by the end of four months. Although the corpus luteum persists throughout pregnancy, its function as a source of hormones becomes less important after the first three months (first trimester), when the placenta secretes sufficient estrogens and progesterone (fig. 23.24).

Detecting hCG in a woman's urine or blood is used to confirm pregnancy. The level of hCG in a pregnant woman's body fluids peaks at fifty to sixty days of gestation, then falls for the remainder of her pregnancy. Later on, measuring hCG has other uses. If a woman miscarries but her blood still shows hCG, fetal tissue may remain in her uterus, and this material must be removed. Between the fifteenth week and twentieth week of pregnancy, most women have a blood test that measures levels of five substances produced by the fetus—alpha fetoprotein (AFP), estradiol (an estrogen), inhibin A, pregnancy-associated plasma protein (PAPP-A), and hCG. If AFP, estradiol, and PAPP-A are low and inhibin A and hCG are elevated, the fetus is at risk of having an extra chromosome 21 (Down syndrome). Further tests, discussed in Chapter 24 (pp. 919–920), are necessary for a definitive diagnosis.

TABLE 23.3	Fetal Cardiovascular Adaptations
Adaptation	**Function**
Fetal blood	Hemoglobin has greater oxygen-carrying capacity than adult hemoglobin
Umbilical vein	Carries nutrient-rich, oxygen-rich blood from the placenta to the fetus
Ductus venosus	Conducts about half the blood from the umbilical vein directly to the inferior vena cava, bypassing the liver
Foramen ovale	Conveys a large proportion of the blood entering the right atrium from the inferior vena cava, through the atrial septum, and into the left atrium, bypassing the lungs
Ductus arteriosus	Conducts some blood from the pulmonary trunk to the aorta, bypassing the lungs
Umbilical arteries	Carry the blood containing carbon dioxide and wastes from the internal iliac arteries to the placenta

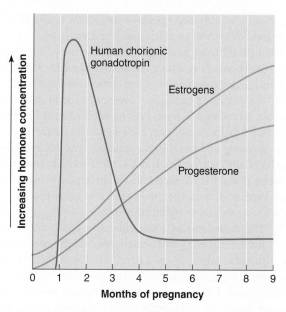

FIGURE 23.24 Relative concentrations of three hormones in maternal blood during pregnancy.

For the remainder of the pregnancy, *placental estrogens* and *placental progesterone* maintain the uterine wall. The placenta also secretes a hormone called **placental lactogen** that, with the aid of placental estrogens and progesterone, stimulates breast development and prepares the mammary glands to secrete milk. Placental progesterone and a polypeptide hormone called *relaxin* from the corpus luteum and the placenta inhibit the smooth muscle in the myometrium, suppressing uterine contractions until the birth process begins.

The high concentration of placental estrogens during pregnancy enlarges the vagina and the external reproductive organs. Also, relaxin relaxes the connective tissue of the pubic symphysis and sacroiliac joints. Relaxin action allows for greater movement at these joints, aiding passage of the fetus through the birth canal.

Other hormonal changes during pregnancy include increased secretion of aldosterone from the adrenal cortex and of parathyroid hormone from the parathyroid glands. Aldosterone promotes renal reabsorption of sodium, leading to fluid retention. Parathyroid hormone helps to maintain a high concentration of maternal blood calcium, because fetal demand for calcium can cause hypocalcemia, which promotes cramps. Table 23.4 summarizes the hormonal changes of pregnancy.

 PRACTICE

20 What are the sources of the hormones that sustain the uterine wall during pregnancy?

21 What other hormonal changes occur during pregnancy?

 RECONNECT
To Chapter 13, Parathyroid Glands, pages 506–508.

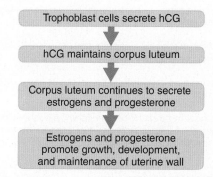

FIGURE 23.23 Mechanism that preserves the uterine lining during early pregnancy.

TABLE 23.4	Hormonal Changes During Pregnancy

1. Following implantation, cells of the trophoblast begin to secrete hCG.

2. hCG maintains the corpus luteum, which continues to secrete estrogens and progesterone.

3. As the placenta develops, it secretes abundant estrogens and progesterone.

4. Placental estrogens and progesterone
   a. stimulate the uterine lining to continue development.
   b. maintain the uterine lining.
   c. inhibit secretion of FSH and LH from the anterior pituitary gland.
   d. stimulate development of the mammary glands.
   e. inhibit uterine contractions (progesterone).
   f. enlarge the reproductive organs (estrogens).

5. Relaxin from the corpus luteum also inhibits uterine contractions and relaxes the pelvic ligaments.

6. The placenta secretes placental lactogen that stimulates breast development.

7. Aldosterone from the adrenal cortex promotes reabsorption of sodium.

8. Parathyroid hormone from the parathyroid glands helps maintain a high concentration of maternal blood calcium.

A woman's body changes as it responds to the increased requirements of a growing offspring. The uterus enlarges greatly, extending upward from its normal location in the pelvic cavity, in some cases eventually reaching the level of the ribs. The abdominal organs are displaced upward and compressed against the diaphragm. The enlarging uterus also presses on the urinary bladder. As a result, a pregnant woman may be unable to eat large meals, may develop heartburn, and may have to urinate frequently.

The growing and developing placenta requires more blood, and as the offspring enlarges, it needs more oxygen and produces more waste that must be excreted. The pregnant woman's blood volume, cardiac output, breathing rate, and urine production all increase to accommodate offspring growth.

The pregnant woman must eat more to obtain adequate nutrition for the offspring and herself, specifically supplying sufficient vitamins, minerals, and proteins. The offspring tissues have a greater capacity to capture available nutrients than do the maternal tissues. Consequently, if the pregnant woman's diet is inadequate, her body will usually show symptoms of a deficiency condition before offspring growth is adversely affected.

Maternal nausea and vomiting in pregnancy may shield a fetus from foods that might contain toxins or pathogens. The condition affects two in three pregnancies and coincides with the time in gestation when a woman's immune system is at its weakest. An analysis of more than 80,000 pregnant women found that they avoid foods that spoil easily, such as eggs and meats, as well as coffee and alcohol. Yet many pregnant women eat more fruits and vegetables than usual. In societies where the diet is mostly grains with little if any meat, incidence of morning sickness is much lower than in groups with more varied, and possibly dangerous, diets. Rates of morning sickness are highest in Japan, where raw fish is a dietary staple, and European countries, where undercooked meat is often eaten.

Birth Process

Pregnancy terminates with the *birth process* (parturition). It is complex. Progesterone plays a major role in its start. During pregnancy, this hormone suppresses uterine contractions. As the placenta ages, progesterone concentration in the uterus declines, which stimulates synthesis of a prostaglandin that promotes uterine contractions. At the same time, the cervix begins to thin and then open. Changes in the cervix may begin a week or two before other signs of labor.

Stretching of the uterine and vaginal tissues late in pregnancy also stimulates the birth process. This action initiates impulses to the hypothalamus, which in turn signals the posterior pituitary gland to release the hormone **oxytocin** (see chapter 13, p. 503), which stimulates powerful uterine contractions. Combined with the greater excitability of the myometrium due to the decline in progesterone secretion, stimulation by oxytocin aids *labor* in its later stages.

During labor, muscular contractions force the fetus through the birth canal. Waves of rhythmic contractions that begin at the top of the uterus and continue down its length force the contents of the uterus toward the cervix.

The fetus is usually positioned head downward, so labor contractions force the head against the cervix. This action stretches the cervix, which elicits a reflex that stimulates still stronger labor contractions. Thus, a *positive feedback system* operates in which uterine contractions produce more intense uterine contractions until effort is maximal (fig. 23.25). At the same time, dilation of the cervix reflexly stimulates an increased release of oxytocin from the posterior pituitary gland.

As labor continues, positive feedback stimulates abdominal wall muscles to contract, helping to propel the fetus through the birth canal (cervix, vagina, and vulva) to the outside. Table 23.5 summarizes some of the factors promoting labor. Figure 23.26 illustrates the steps of the birth process.

Following birth of the fetus, the placenta, which initially remains inside the uterus, separates from the uterine wall and is pushed by uterine contractions through the birth canal. This

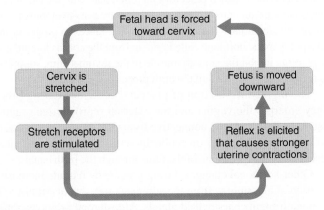

FIGURE 23.25 A positive feedback mechanism operates in the birth process.

TABLE 23.5	Factors Contributing to the Labor Process

1. As the time of birth approaches, secretion of progesterone declines, and its inhibiting effect on uterine contractions lessens.

2. Decreasing progesterone concentration stimulates synthesis of prostaglandins, which initiate labor.

3. Stretching uterine tissues stimulates release of oxytocin from the posterior pituitary gland.

4. Oxytocin stimulates uterine contractions and aids labor in its later stages.

5. As the fetal head stretches the cervix, a positive feedback mechanism results in stronger and stronger uterine contractions and a greater release of oxytocin.

6. Positive feedback stimulates abdominal wall muscles to contract with greater and greater force.

7. The fetus is forced through the birth canal to the outside.

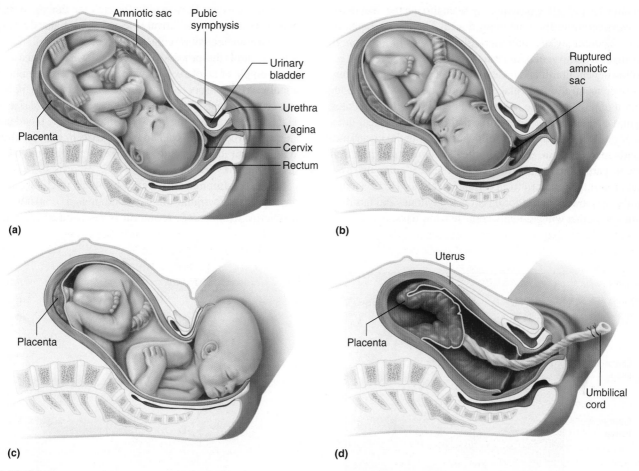

FIGURE 23.26 Stages in birth. (*a*) Fetal position before labor, (*b*) dilation of the cervix, (*c*) expulsion of the fetus, (*d*) expulsion of the placenta.

expelled placenta, called the *afterbirth,* is accompanied by bleeding, because vascular tissues are damaged in the process. However, continued contraction of the uterus compresses the bleeding vessels, minimizing blood loss. Oxytocin stimulates this contraction. Breastfeeding also contributes to returning the uterus to its original, prepregnancy size, because suckling by the newborn stimulates the mother's posterior pituitary gland to release oxytocin.

> An infant passing through the birth canal can stretch and tear perineal tissues between the vulva and anus. Before the birth is complete, a physician may make an incision along the midline (clinical perineum) from the vestibule to within 1.5 centimeters of the anus. This procedure, called an *episiotomy,* ensures that the perineal tissues are cut cleanly rather than torn, which aids healing.

For several weeks following childbirth, a process called *involution* shrinks the uterus. The endometrium sloughs off and is discharged through the vagina as a bloody and then yellowish discharge over the course of a few weeks. An epithelial lining characteristic of a nonpregnant female then returns.

 PRACTICE

22 Describe the role of progesterone in initiating labor.

23 Explain how dilation of the cervix affects labor.

24 Explain how bleeding is naturally controlled after the placenta is expelled.

Milk Production and Secretion

During pregnancy, placental estrogens and progesterone stimulate further development of the mammary glands. Estrogens cause the ductile systems to grow and branch, and deposit abundant fat around them. Progesterone stimulates the development of the alveolar glands at the ends of the ducts. Placental lactogen also promotes these changes.

The breasts may double in size during pregnancy because of hormonal activity. At the same time, glandular tissue replaces the adipose tissue of the breasts. Beginning about the fifth week of pregnancy, the anterior pituitary gland releases increasing amounts of *prolactin*. Prolactin is synthesized from early pregnancy throughout gestation, peaking at the time of birth. However, milk secretion does not begin until after birth. This is because during pregnancy, placental progesterone inhibits milk production, and placental lactogen blocks the action of prolactin (see chapter 13, p. 500). Consequently, even though the mammary glands can secrete milk, none is produced. The micrographs in figure 23.27 compare the mammary gland tissues of a nonpregnant woman with those of a lactating woman.

Following childbirth and the expulsion of the placenta, maternal blood concentrations of placental hormones rapidly decline. The action of prolactin is no longer inhibited. Prolactin stimulates the mammary glands to secrete milk. This hormonal effect does not occur until two or three days following birth. In the meantime, the glands secrete a thin, watery fluid called *colostrum*. It is rich in proteins, particularly antibodies from the mother's immune system that protect the newborn from certain infections. However, this "first milk" has lower concentrations of carbohydrates and fats than the mature milk that will gradually come in a few days.

Milk does not flow readily through the ductile system of the mammary gland, but is actively ejected as specialized *myoepithelial cells* surrounding the alveolar glands contract (fig. 23.28). A reflex action controls this process and is elicited when the breast is suckled or the nipple or areola is otherwise mechanically stimulated. Then, impulses from sensory receptors in the breasts travel to the hypothalamus, which signals the posterior pituitary gland to release oxytocin. The oxytocin reaches the breasts by means of the blood and stimulates the myoepithelial cells in both breasts to contract (fig. 23.29). Within about thirty seconds, milk is ejected from the breast into a suckling infant's mouth.

Sensory impulses triggered by mechanical stimulation of the nipples also signal the hypothalamus to continue secreting prolactin. In this way, prolactin is released as long as breast-

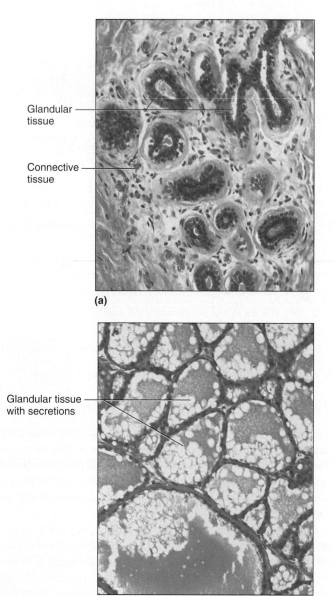

Glandular tissue

Connective tissue

(a)

Glandular tissue with secretions

(b)

FIGURE 23.27 Mammary glands. (*a*) Light micrograph of a mammary gland in a nonpregnant woman (160×). (*b*) Light micrograph of an active (lactating) mammary gland (160×).

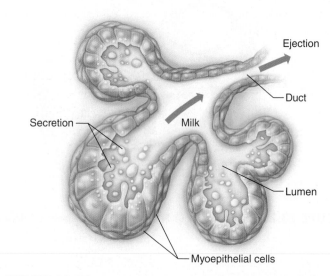

Ejection

Duct

Secretion

Milk

Lumen

Myoepithelial cells

FIGURE 23.28 Mechanism that releases milk from the breasts.

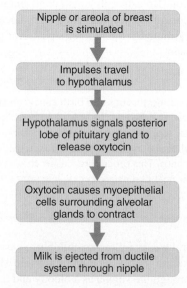

Nipple or areola of breast is stimulated

Impulses travel to hypothalamus

Hypothalamus signals posterior lobe of pituitary gland to release oxytocin

Oxytocin causes myoepithelial cells surrounding alveolar glands to contract

Milk is ejected from ductile system through nipple

FIGURE 23.29 Myoepithelial cells contract to release milk from an alveolar gland.

feeding continues. However, if stimulation of the nipple does not occur regularly, the hypothalamus inhibits secretion of prolactin, and within about one week the mammary glands stop producing milk.

> A woman who is breastfeeding feels her milk "let down," or flood her breasts, when her infant suckles. If the baby nurses on a regular schedule, the mother may feel the letdown shortly before the baby is due to nurse. The connection between mind and hormonal control of lactation is so strong that if a nursing mother hears a baby cry, her milk may flow.

To wean a nursing child, it is best to stop breastfeeding gradually, by eliminating one feeding per day each week, for example. If a woman stops nursing abruptly, her breasts will become painfully engorged for several days.

A woman who is breastfeeding usually does not ovulate for several months, because prolactin suppresses release of gonadotropins from the anterior pituitary gland. When a woman stops breastfeeding, the anterior pituitary gland stops secreting prolactin. Then FSH is released, and the reproductive cycle is activated. A new mother will be fertile approximately two weeks prior to the return of her menstrual period.

Table 23.6 summarizes the hormonal control of milk production, and table 23.7 lists some agents that adversely affect lactation or harm the child. Clinical Application 23.2 explains the benefits of breastfeeding.

 PRACTICE

25 How does pregnancy affect the mammary glands?
26 What stimulates the mammary glands to produce milk?
27 What causes milk ejection?
28 What happens to milk production if milk is not regularly removed from the breast?

TABLE 23.6 Hormonal Control of the Mammary Glands

Before Pregnancy (Beginning of Puberty)	Following Childbirth
Ovarian hormones secreted during reproductive cycles stimulate alveolar glands and ducts of mammary glands to develop.	1. Placental hormonal concentrations decline, so the action of prolactin is no longer inhibited.
During Pregnancy	2. The breasts begin producing milk.
1. Estrogens cause the ductile system to grow and branch.	3. Mechanical stimulation of the breasts releases oxytocin from the posterior pituitary gland.
2. Progesterone stimulates development of alveolar glands.	4. Oxytocin stimulates release of milk from ducts.
3. Placental lactogen promotes development of the breasts.	5. As long as breastfeeding continues, more prolactin is released; if the nipple is not stimulated regularly, milk production ceases.
4. Prolactin is secreted throughout pregnancy, but placental progesterone inhibits milk production and placental lactogen blocks the action of prolactin.	

TABLE 23.7 Agents Contraindicated During Breastfeeding

Agent	Use	Effect on Lactation or Baby
Doxorubicin, methotrexate	Cancer chemotherapy, psoriasis, rheumatoid arthritis	Immune suppression
Cyclosporine	Immune suppression in transplant patients	Immune suppression
Radioactive isotopes	Cancer diagnosis and therapy	Radioactivity in milk
Phenobarbitol	Anticonvulsant	Sedation, spasms on weaning
Oral contraceptives	Birth control	Decreased milk production
Caffeine (large amounts)	Food additive	Irritability, poor sleeping
Cocaine	Drug of abuse	Intoxication, seizures, vomiting, diarrhea
Ethanol (alcohol) (large amounts)	Drug of abuse	Weak, drowsy; infant decreases in length but gains weight; decreased milk ejection reflex
Heroin	Drug of abuse	Tremors, restlessness, vomiting, poor feeding
Nicotine	Drug of abuse	Diarrhea, shock, increased heart rate; lowered milk production
Phencyclidine	Drug of abuse	Hallucinations

23.3 | Postnatal Period

Following birth, both mother and newborn experience physiological and anatomical changes. The **postnatal period** of development lasts from birth until death. It can be divided into the neonatal period, infancy, childhood, adolescence, adulthood, and senescence. Dying is also part of the life cycle.

Neonatal Period

The **neonatal** (ne"o-na'tal) **period,** which extends from birth to the end of the first four weeks, begins abruptly at birth. At that moment, physiological adjustments must occur quickly because the newborn (neonate) must suddenly do for itself what the mother's body had been doing for it. The newborn must respire, obtain and digest nutrients, excrete wastes, and regulate body temperature. However, a newborn's most immediate need is to obtain oxygen and excrete carbon dioxide, so the first breath is critical.

The first breath must be particularly forceful, because the newborn's lungs are collapsed and the airways are small, offering considerable resistance to air movement. Also, surface tension adheres the moist membranes of the lungs. However, the lungs of a full-term fetus continuously secrete *surfactant* (see chapter 19, p. 745), which reduces surface tension. After the first powerful breath begins to expand the lungs, breathing eases.

Several factors stimulate a newborn's first breath: increasing concentration of carbon dioxide, decreasing pH, low oxygen concentration, drop in body temperature, and mechanical stimulation

during and after birth. Also, in response to the stress to the fetus during birth, blood concentrations of epinephrine and norepinephrine rise significantly (see chapter 13, pp. 509–510). These hormones promote normal breathing by increasing the secretion of surfactant and dilating the airways.

For energy, the fetus primarily uses glucose and fatty acids in the pregnant woman's blood. The newborn, on the other hand, is suddenly without an external source of nutrients. The mother will not produce mature milk for two to three days, by which time the infant's gastrointestinal tract will be able to digest it. The early milk, *colostrum,* is an adaptation to the state of the newborn's digestive physiology. The newborn has a high metabolic rate, and its liver, not fully mature, may be unable to supply enough glucose to support metabolism. Instead, the newborn uses stored fat for energy.

A newborn's kidneys are usually unable to concentrate urine, so the baby excretes dilute urine. This may cause dehydration and a water and electrolyte imbalance. Also, certain homeostatic control mechanisms may not function adequately. For example, during the first few days of life, body temperature may respond to slight stimuli by fluctuating above or below the normal level.

When the placenta ceases to function and breathing begins, the newborn's cardiovascular system changes. Following birth, the umbilical vessels constrict. The umbilical arteries close first, and if the umbilical cord is not clamped or severed for a minute or so, blood continues to flow from the placenta to the newborn through the umbilical vein, adding to the newborn's blood volume.

The proximal portions of the umbilical arteries persist in the adult as the *superior vesical arteries* that supply blood to the urinary bladder. The more distal portions become solid cords (lateral umbilical ligaments). The umbilical vein becomes the cordlike *ligamentum teres* that extends from the umbilicus to the liver in an adult. The ductus venosus constricts shortly after birth and appears in the adult as a fibrous cord (ligamentum venosum) superficially embedded in the wall of the liver.

The foramen ovale closes as a result of changes in blood pressure in the right and left atria. As blood ceases to flow from the umbilical vein into the inferior vena cava, the blood pressure in the right atrium falls. Also, as the lungs expand with the first breathing movements, resistance to blood flow through the pulmonary circuit decreases, more blood enters the left atrium through the pulmonary veins, and blood pressure in the left atrium increases.

As the blood pressure in the left atrium rises and that in the right atrium falls, the valvelike septum primum on the left side of the atrial septum closes the foramen ovale. In most individuals this valve gradually fuses with the tissues along the margin of the foramen. In an adult, a depression called the *fossa ovalis* marks the site of the past opening.

The ductus arteriosus, like other fetal vessels, constricts after birth. Once this happens, blood can no longer bypass the lungs by moving from the pulmonary trunk directly into the aorta. In an adult, a cord called the *ligamentum arteriosum* represents the ductus arteriosus.

In *patent ductus arteriosus* (PDA), the ductus arteriosus does not close completely. After birth, the metabolic rate and oxygen consumption in neonatal tissues increase, in large part to maintain body temperature. If the ductus arteriosus remains open, bypassing the pulmonary circuit, the newborn's blood oxygen concentration may be too low to adequately supply body tissues, including the myocardium. If PDA is not corrected surgically, the heart may fail, even though the myocardium is normal.

Changes in the newborn's cardiovascular system are gradual. Although constriction of the ductus arteriosus may be functionally

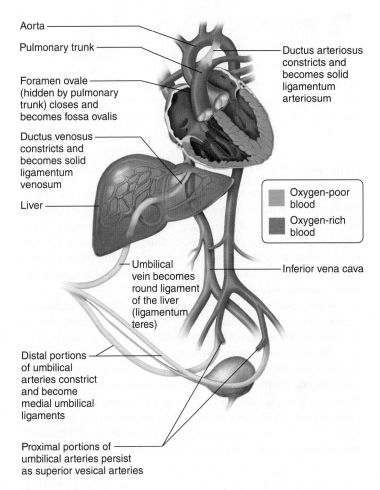

FIGURE 23.30 Major changes occur in the newborn's cardiovascular system.

complete within fifteen minutes, the permanent closure of the foramen ovale may take up to a year. These cardiovascular changes are illustrated in figure 23.30 and summarized in table 23.8.

Fetal hemoglobin production falls after birth. By the time an infant is four months old, most of the circulating hemoglobin is the adult type.

TABLE 23.8	Cardiovascular Adjustments in the Newborn	
Structure	**Adjustment**	**In the Adult**
Umbilical vein	Constricts	Becomes ligamentum teres that extends from the umbilicus to the liver
Ductus venosus	Constricts	Becomes ligamentum venosum superficially embedded in the wall of the liver
Foramen ovale	Closes by valvelike septum primum as blood pressure in right atrium decreases and blood pressure in left atrium increases	Valve fuses along margin of foramen ovale and is marked by a depression called the fossa ovalis
Ductus arteriosus	Constricts	Becomes ligamentum arteriosum that extends from the pulmonary trunk to the aorta
Umbilical arteries	Distal portions constrict	Proximal portions function as superior vesical arteries; distal portions become lateral umbilical ligaments

29 Define *neonatal period*.

30 What factors stimulate the first breath?

31 What does a newborn use for energy during its first few days?

32 How do the kidneys of a newborn differ from those of an adult?

33 What is the fate of the foramen ovale? of the ductus arteriosus?

Infancy

The period of continual development extending from the end of the first four weeks to one year is called **infancy.** During this time, the infant grows rapidly and may triple its birth weight. Teeth begin to erupt through the gums, and the muscular and nervous systems mature so that movements become increasingly coordinated. The infant is soon able to visually follow objects, reach for and grasp objects, and sit, crawl, and stand.

Infancy also brings the beginning of the ability to communicate. The infant learns to smile, laugh, and respond to some sounds. By the end of the first year, the infant may be able to say two or three words.

The infant and young child have particular nutritional requirements to fuel rapid growth. In addition to an energy source, the body requires proteins to provide the amino acids necessary to form new tissues, calcium and vitamin D to promote the development and ossification of skeletal structures (see chapter 7, pp. 207–209), iron to support blood cell formation, and vitamin C for production of structural tissues such as cartilage and bone.

Childhood

Childhood begins at the end of the first year and ends at puberty. During this period, growth continues at a rapid rate. The primary teeth erupt, and then secondary teeth replace them (see chapter 17, pp. 655–656). The child develops voluntary muscular control and learns to walk, run, and climb. Bladder and bowel controls are established. The child learns to communicate effectively by speaking, and later, learns to read, write, and think. At the same time, the child is maturing emotionally.

 PRACTICE

34 Define *infancy*.

35 What developmental changes characterize infancy?

36 Define *childhood*.

37 What developmental changes characterize childhood?

Adolescence

Adolescence is the period of development between puberty and adulthood when anatomical and physiological changes result in reproductively functional individuals (see chapter 22, pp. 837–839 and 849). Most of these changes are hormonally controlled, and they include the appearance of secondary sex characteristics as well as growth spurts in the muscular and skeletal systems.

Females usually experience the changes of adolescence at younger ages than males. Early in adolescence, females may be taller and stronger than their male peers. However, females attain full growth at earlier ages, and in late adolescence the average male is taller and stronger than the average female.

The periods of rapid growth in adolescence, which usually begin between the ages of eleven and thirteen in females and between thirteen and fifteen in males, increase demands for certain nutrients. It is not uncommon for a teenager to consume a huge plate of food, go back for more—and still remain thin. In addition to energy sources, foods must provide ample amounts of proteins, vitamins, and minerals to support growth of new tissues. Adolescence also brings increasingly refined motor skills, intellectual ability, and emotional maturity.

Adulthood

Adulthood (maturity) extends from adolescence to old age. As we age, we become gradually aware of certain declining functions—yet other abilities remain adequate. The "Life-Span Changes" sections in previous chapters have chronicled the effects of aging on particular organ systems. It is interesting to note the peaks of particular structures or functions throughout an average human life.

By age eighteen, the human male is producing the most testosterone he will ever have, and as a result his sex drive is strong. In the twenties, muscle strength peaks in both sexes. Hair is at its fullest, each hair its thickest. By the end of the third decade of life, signs of aging may appear as a loss in the elasticity of facial skin, producing small wrinkles around the mouth and eyes. Height is already starting to decrease, but not yet at a detectable level.

The age of thirty is a developmental turning point. After this, hearing often becomes less acute. Heart muscle begins to thicken. The elasticity of the ligaments between the small bones in the back lessens, setting the stage for the slumping posture that becomes apparent in later years. Researchers estimate that beginning roughly at age thirty, the human body becomes functionally less efficient by about 0.8% every year.

During their forties, many people weigh 10 to 20 pounds (4.5 to 9 kilograms) more than they did at the age of twenty, thanks to a slowing of metabolism and decrease in activity level. They may be 1/8 inch (0.3 centimeter) shorter. Hair may be graying as melanin production wanes, and some hair may fall out. Vision may become farsighted. The immune system is less efficient, making the body more prone to infection and cancer. Skeletal muscles lose strength as connective tissue appears within them; the cardiovascular system is strained as the lumens of arterioles and arteries narrow with fatty deposits; skin loosens and wrinkles as elastic fibers in the dermis break down.

The early fifties bring further declines. Nail growth slows, taste buds die, and the skin continues to lose elasticity. For most people, the ability to see close objects becomes impaired, but for the nearsighted, vision improves. Women stop menstruating, although interest in sex continues (see chapter 22, p. 852). Delayed or reduced insulin release by the pancreas, in response to a glucose load, may lead to diabetes. By the decade's end, muscle mass and weight begin to decrease. A male produces less semen but is still sexually active. His voice may become higher as his vocal cords degenerate. A man has half the strength in his upper limb muscles and half the lung function as he did at age twenty-five. He is about 3/4 inch (2 centimeters) shorter.

The sixty-year-old may experience minor memory losses. A few million of the person's billions of brain cells have been lost over his or her lifetime, but for the most part, intellect remains sharp. By age seventy, height decreases a full inch (2.5 centimeters). Sagging skin and loss of connective tissue, combined with continued growth of cartilage, make the nose, ears, and eyes more prominent.

Senescence

Senescence (se-nes'ens) is the process of growing old. It is a continuation of the degenerative changes that begin during adulthood. The body becomes less able to cope with the demands placed on it by the individual and by the environment.

Senescence is a result of the normal wear and tear of body parts over many years. For example, the cartilage covering the ends of bones at joints may wear away, leaving the joints stiff and painful. Other degenerative changes are caused by disease processes that interfere with vital functions, such as gas exchanges or blood circulation. Metabolic rate and distribution of body fluids may change. The rate of division of certain cell types declines, and immune responses weaken. The person becomes less able to repair damaged tissue and more susceptible to disease.

Decreasing efficiency of the central nervous system accompanies senescence. The person may lose some intellectual functions. Also, the physiological coordinating capacity of the nervous system may decrease, and homeostatic mechanisms may fail to operate effectively. Sensory functions decline with age also.

Death usually results, not from these degenerative changes, but from mechanical disturbances in the cardiovascular system, failure of the immune system, or disease processes that affect vital organs. **Table 23.9** summarizes the major phases of postnatal life and their characteristics, and **table 23.10** lists some aging-related changes.

From 65% to 80% of all deaths in the United States take place in hospitals, often with painful and sometimes unwanted interventions to prolong life. The medical community is trying to remedy shortcomings in the treatment of the dying. Medical training is increasing emphasis on providing palliative care for the terminally ill. Such care seeks to make a patient comfortable, even if the treatment does not cure the disease or extend life.

The End of Life

Nearing the end of life is a personal process, influenced by belief as well as circumstance. However, if the person has been chronically ill and is receiving comfort (palliative) care, certain signs of impending death may appear, in many cases in a sequence. A person may exhibit some or all of these signs.

Health-care professionals view the dying process in two stages—preactive dying and active dying. Preactive dying may take up to three months. During this time, some people are aware of what is happening and begin the psychological process of coming to terms with their mortality. A month or more before death, the person starts to withdraw, losing interest in news from the outside world and possibly

TABLE 23.9	Stages in Postnatal Development	
Stage	**Time Period**	**Major Events**
Neonatal period	Birth to end of fourth week	Newborn begins to carry on respiration, obtain nutrients, digest nutrients, excrete wastes, regulate body temperature, and make cardiovascular adjustments
Infancy	End of fourth week to one year	Growth rate is high; teeth begin to erupt; muscular and nervous systems mature so that coordinated activities are possible; communication begins
Childhood	One year to puberty	Growth rate is high; primary teeth erupt and are replaced by secondary teeth; high degree of muscular control is achieved; bladder and bowel controls are established; intellectual abilities mature
Adolescence	Puberty to adulthood	Person becomes reproductively functional and emotionally more mature; growth spurts occur in skeletal and muscular systems; high levels of motor skills are developed; intellectual abilities increase
Adulthood	Adolescence to old age	Person remains relatively unchanged anatomically and physiologically; degenerative changes begin to occur
Senescence	Old age to death	Degenerative changes continue; body becomes less able to cope with demands; death usually results from mechanical disturbances in the cardiovascular system or from diseases that affect vital organs

requesting that visits from friends and relatives cease or shorten. He or she sleeps more and might not even get out of bed on some days. Conversation lags. Gradually the loss of interest in everyday activities extends to eating. This parallels physical changes, such as difficulty swallowing, that make eating increasingly difficult. The person might first give up eating meats, then fibrous vegetables, until it is clear that softer foods are preferred. The person might eat and drink astonishingly little, and the family might feel the need to try to force eating—which could cause the dying person discomfort. Dry mouth is common. The caregiver can provide ice chips or popsicles, or frequently wet the mouth with a swab.

Active dying presents a distinct set of signs, which might appear only on the day before death, or might begin up to two weeks earlier. During this phase the person sleeps often but can easily be awakened. Even if sleep is deep, the person can hear, because this is the last sense to fade. He or she may confuse time, place, and identities. A nurse might be mistaken for a relative, or an adult child might not be recognized. An actively dying person may go back in time, talking to a deceased spouse, for example. Signs of agitation appear, such as picking lint on the blanket or thrashing the arms about. Appetite may be nil.

In active dying, the organ systems slowly shut down. Cardiovascular signs include falling blood pressure (systolic

TABLE 23.10	Aging-Related Changes
Organ System	Aging-Related Changes
Integumentary system	Degenerative loss of collagen and elastin fibers in dermis; decreased production of pigment in hair follicles, hair eventually turns white; reduced activity of sweat and sebaceous glands; skin thins, wrinkles, and becomes drier
Skeletal system	Degenerative loss of bone matrix; bones become thinner, less dense, and more likely to fracture; stature may shorten due to compression of intervertebral discs and vertebrae
Muscular system	Loss of skeletal muscle fibers; degenerative changes in neuromuscular junctions; loss of muscular strength
Nervous system	Degenerative changes in neurons; loss of dendrites and synaptic connections; accumulation of lipofuscin in neurons; decreases in sensation; decreasing efficiency in processing and recalling information; decreasing ability to communicate; diminished senses of smell and taste; loss of elasticity of lenses and consequent loss of ability to accommodate for close vision
Endocrine system	Reduced hormonal secretions; decreased metabolic rate; reduced ability to cope with stress; reduced ability to maintain homeostasis
Cardiovascular system	Degenerative changes in cardiac muscle; decrease in lumen diameters of arteries and arterioles; decreased cardiac output; increased resistance to blood flow; increased blood pressure
Lymphatic system	Decrease in efficiency of immune system; increased incidence of infections and neoplastic diseases; increased incidence of autoimmune diseases
Digestive system	Decreased motility in gastrointestinal tract; reduced secretion of digestive juices; reduced efficiency of digestion
Respiratory system	Degenerative loss of elastic tissue in lungs; fewer alveoli; reduced vital capacity; increase in dead air space; reduced ability to clear airways by coughing
Urinary system	Degenerative changes in kidneys; fewer functional nephrons; reductions in filtration rate, tubular secretion, and tubular reabsorption
Reproductive systems	
Male	Reduced secretion of sex hormones; enlargement of prostate gland; decrease in sexual energy
Female	Degenerative changes in ovaries; decrease in secretion of sex hormones; menopause; regression of secondary sex characteristics

below 70, diastolic below 50). The pulse may race or slow, or alternate. Poor circulation, which redirects the blood supply to the body's core, ushers in peripheral changes. The limbs feel cool to the touch, and the person may note numbness. The extremities become pale, then take on a bluish tinge. Skin areas under pressure, such as the undersides of the limbs, become mottled (discolored). Sensitivity to touch and pain declines.

The slowing circulation affects muscles. Poor ability to cough and swallow causes secretions to build up in the lungs. Secretions may not be suctioned, because this increases their rate of accumulation. The person can be repositioned to provide some relief. Congestion is intermittent. One day it may be so severe that eating is impossible; the next day breathing may ease. Cheyne-Stokes breathing—shallow mouth-breathing interspersed with increasingly long periods of apnea—is common. The normal rate of 16 to 20 breaths per minute may speed to more than 50, slow precipitously, perhaps pausing for 10 to 30 seconds, and then the person gasps and breathes rapidly again. As the throat muscles relax, exhalation over the vocal cords causes a passive moaning sound—this does not indicate that the person is in pain. A day or two before death, breathing may become loud and labored.

Control of body temperature is altered, and the person may have elevated body temperature or feel cold. The skin may be alternately flushed, then blue with an internal chill. Sweating is common, and as death nears, the skin takes on a yellowish pallor.

In the day or two before death, the signs intensify, although a last burst of energy may occur. A bedridden person may suddenly wish to be propped up in the living room and see people, or, after weeks of barely eating, suddenly request an ice cream sundae. Meanwhile, respiration and circulation slow, and decreased oxygen delivery may cause restlessness and agitation. The pulse becomes thready. Right before death the person may lose control of the bladder or bowels. Breathing becomes more irregular, with longer periods between breaths, and the lung rattling intensifies. Consciousness fades in and out. The eyes may be unfocused and appear glassy, or may be only partially open, and may tear frequently. Finally the person can no longer respond. After one, two, or three long, last breaths, the eyes become fixed and open, the pupils dilate, the jaw relaxes, and the mouth may slightly open. The journey of life has ended.

 PRACTICE

38 How does the body change during adolescence?

39 What changes occur during adulthood?

40 What changes accompany senescence?

41 What are the signs of preactive and active dying?

23.4 | Aging

The aging process is difficult to analyze because of the intricate interactions of the body's organ systems. Breakdown of one structure ultimately affects the functioning of others. The medical field of gerontology examines the biological changes of aging at the molecular, cellular, organismal, and population levels. Aging is both passive and active.

Passive Aging

Aging as a passive process is a breakdown of structures and slowing of functions. At the molecular level, passive aging is seen in the degeneration of the elastin and collagen proteins of connective tissues, causing skin to sag and muscle to lose its firmness.

During a long lifetime, biochemical abnormalities accumulate. Mistakes occur throughout life when DNA replicates in dividing cells. Usually repair enzymes correct this damage immediately. But over many years, exposure to chemicals, viruses, and radiation disrupts DNA repair mechanisms so that the error burden becomes too great to be fixed. The cell may die as a result of faulty genetic instructions.

DNA continues to mutate, even in the oldest old. A study of a woman who died at age 115 in 2005 revealed that at that age, her white blood cells had 450 mutations, whereas her brain neurons had no mutations. This finding indicated that mutations were continuing in rapidly dividing cell types—and with no apparent ill effects.

Another sign of passive aging at the biochemical level is the breakdown of lipids. As aging membranes leak during lipid degeneration, a fatty, brown pigment called lipofuscin accumulates. Mitochondria also begin to break down in older cells, decreasing the supply of chemical energy to power cell functions.

The cellular degradation associated with aging may be set into action by highly reactive chemicals called **free radicals.** The most abundant type of free radical forms from oxygen atoms. A molecule that is a free radical has an unpaired electron in its outermost valence shell. This causes the molecule to grab electrons from other molecules, destabilizing them, and a chain reaction of chemical instability begins that could kill the cell. Free radicals are a by-product of normal metabolism and also form by exposure to radiation or toxic chemicals. The bile pigment bilirubin protects against free radicals. Enzymes that usually inactivate free radicals diminish in number and activity in the later years. One such enzyme is *superoxide dismutase* (SOD).

Active Aging

Aging also entails new activities or the appearance of new substances. Lipofuscin granules, for example, may be considered an active sign of aging, but they result from the passive breakdown of lipids. Another example of active aging is autoimmunity, in which the immune system turns against the body, attacking its cells as if they were invading organisms.

Active aging begins before birth, as certain cells die as part of the developmental program encoded in the genes. This process of programmed cell death, called **apoptosis** (ap″o-tō′sis), occurs regularly in the embryo, degrading certain structures to pave the way for new ones. The number of neurons in the fetal brain, for example, is halved as those that make certain synaptic connections are spared from death. In the fetal thymus, T cells that do not recognize "self" cell surfaces die, thereby building the immune system. Throughout life, apoptosis enables organs to maintain their characteristic shapes.

Mitosis and apoptosis are opposite, but complementary, processes. As organs grow, the number of cells in some regions increases, but in others the number decreases. Cell death is not a phenomenon only of the aged. It is a normal part of life. Clinical Application 23.3 discusses characteristics shared by people who live past 100.

The Human Life Span

The human *life span*—the length of time that a human can theoretically live—is 120 years. Although most people succumb to disease or injury long before that point, in many countries the fastest-growing age group is those over age eighty. Many of these "oldest old," having passed the age when cancer and cardiovascular disease typically strike, are quite healthy.

Life expectancy is a realistic projection of how long an individual will live, based on epidemiological information. In the United States, life expectancy currently is 76.3 years for men and 81.1 years for women. Yet in some African nations being decimated by infectious diseases, life expectancy is in the thirties. Some life expectancy estimates account for future health-care progress. For a female born today, life expectancy exceeds 87 and for a male, 83.

Medical advances have greatly contributed to improved life expectancy. Antibiotics have tamed some once-lethal infections, drugs enable many people with cancer to survive, and such advances as beta-blocker drugs and coronary bypass surgery have extended the lives of people with heart disease. However, the rise of new or renewed infectious diseases, such as AIDS, polio, and measles, also indicates that we cannot yet conquer all illnesses. Although we can alter our environment more than other species can, some forces of nature remain beyond our control. **Table 23.11** lists the top 10 causes of death in the United States.

 PRACTICE

42 How is aging a passive process?

43 How is aging an active process?

44 Distinguish between life span and life expectancy. ∎

TABLE 23.11	The Ten Leading Causes of Death in the United States, 2013	
Cause		**Percent of total deaths**
1. Heart disease		24.2
2. Cancer		23.3
3. Chronic lower respiratory diseases		5.6
4. Stroke		5.2
5. Unintentional injuries		4.9
6. Alzheimer disease		3.4
7. Diabetes mellitus		2.8
8. Nephritis, nephrotic syndrome, and nephrosis		2.0
9. Influenza and pneumonia		1.9
10. Suicide		1.6

Source: Centers for Disease Control and Prevention, 2013.

People who live past 100 years are called centenarians, and past 110, supercentenarians. About 70 supercentenarians are alive at any given time, most of them female. Most centenarians enjoy excellent health and remain active and engaged in activities, then succumb rapidly to diseases that typically claim people decades earlier. Some never get these disorders at all. Researchers hope that learning which gene variants and environmental factors centenarians share will lead to a better understanding of the common disorders of later adulthood—heart disease, cancers, stroke, type 2 diabetes mellitus, and dementias.

While the environment seems to play an important role in the deaths of people ages 60 to 85, after that age genes predominate. Someone who dies at age 65 of lung cancer can probably blame decades of smoking, but a smoker who dies at age 101 of the same type of cancer probably had protective gene variants. Centenarians have higher levels of HDL cholesterol than other people, which researchers estimate adds 20 years of life. Factors that vary widely among centenarians include religion, education level, socioeconomic status, ethnicity, and diet. They share normal or low weight, not smoking, handling stress well, and not having dementia. Women who gave birth late in life are overrepresented, perhaps because their reproductive systems, and the rest of their bodies, age slowly.

Children and siblings of centenarians also tend to be long-lived, suggesting a large inherited component. The saying of the New England Centenarian Study, one of the longest-running programs, is "The older you get, the healthier you've been," rather than associating extreme old age with sickness.

Centenarians have luckily inherited two general types of gene variants: those that are directly protective, and normal variants of genes that, when mutant, cause disease. Specifically, single genes important in aging affect:

- control of glucose metabolism and insulin secretion
- immune system functioning
- cell cycle control
- lipid (cholesterol) metabolism

FIGURE 23D Researchers are discovering clues to good health by probing the genomes of centenarians.

- response to stress
- production of antioxidant enzymes

Several studies are identifying the gene variants and lifestyle practices that contribute to living long and well. The New England study is amassing the "healthy standard genome." Investigators at the Coriell Institute in New Jersey are probing the genomes of the oldest old in nursing homes. So far, what these people share is never having had heart disease and never having smoked. Several have had cancer, indicating that this disorder is survivable. Researchers at the University of Pittsburgh are pursuing gene variants that preserve cognition. The National Institute on Aging is currently seeking families with more than two long-lived members living near Boston, New York, or Pittsburgh to participate in the Long Life Family Study, which will collect and study genetic and health information for clues to longevity. Perhaps this information will help the majority of us, who have not inherited longevity gene variants, discover how to control our environments in ways that promote long and healthy lives.

Chapter Summary

23.1 Fertilization (page 868)

Fertilization occurs with the union of a secondary oocyte and a sperm cell.

1. Transport of sex cells
   a. The secondary oocyte is ovulated and enters a uterine tube.
   b. A sperm cell moves, by its tail lashing and muscular contraction in the female reproductive tract, into a uterine tube.

2. Sperm cell joins secondary oocyte
   a. With the aid of enzymes, a sperm cell penetrates the corona radiata and zona pellucida.
   b. When a sperm cell fuses with the secondary oocyte's membrane, changes in the oocyte cell membrane and the zona pellucida prevent entry of additional sperm.
   c. Completion of meiosis II forms the second polar body.
   d. Fusion of the **pronuclei** of a sperm and a secondary oocyte completes fertilization.
   e. The product of fertilization is a **zygote** with 46 chromosomes.

23.2 Pregnancy and the Prenatal Period

(page 870)

Pregnancy is the presence of a developing offspring in the uterus. The **prenatal period** of the offspring is divided into the embryonic stage and the fetal stage.

1. **Embryonic stage** (fertilization through the eighth week)
   a. During **cleavage** the zygote undergoes mitosis, and the newly formed cells divide mitotically.
      (1) Each subsequent division produces smaller and smaller cells.
      (2) A solid ball of cells (**morula**) forms, and it becomes a hollow ball called a **blastocyst.**
      (3) The inner cell mass (or embryoblast) that gives rise to the **embryo proper** forms within the blastocyst.
      (4) The blastocyst implants in the uterine wall.
         (a) Enzymes digest the endometrium around the blastocyst.
         (b) Fingerlike processes from the blastocyst penetrate the endometrium.
      (5) Cleavage lasts through the first week of development.
      (6) The trophoblast secretes human chorionic gonadotropin (hCG), which helps maintain the corpus luteum, helps protect the blastocyst against being rejected, and stimulates the developing **placenta** to secrete hormones.
   b. Extraembryonic membrane formation and placentation
      (1) The trophoblast and its lining layer of cells form the **chorion.**
      (2) **Chorionic villi** develop and are surrounded by spaces (**lacunae**) filled with maternal blood.
      (3) A fluid-filled **amnion** develops around the embryo proper.
      (4) The **umbilical cord** is formed as the amnion envelopes the tissues attached to the underside of the embryo.
         (a) It suspends the embryo in the amniotic cavity.
         (b) The umbilical cord includes two arteries and a vein.
      (5) The chorion and amnion fuse.
      (6) The **yolk sac** forms on the underside of the embryonic disc.
         (a) It gives rise to blood cells and cells that later form sex cells.
         (b) It helps form the digestive tube.
      (7) The **allantois** extends from the yolk sac into the connecting stalk.
         (a) It forms blood cells.
         (b) It gives rise to the umbilical vessels.
      (8) The placenta develops in the disc-shaped area where the chorion contacts the uterine wall.
         (a) The embryonic portion consists of the chorion and its villi.
         (b) The maternal portion consists of the endometrium.
      (9) The **placental membrane** consists of the epithelium of the chorionic villi and the endothelium of the capillaries inside the chorionic villi.
         (a) Oxygen and nutrients diffuse from the maternal blood through the placental membrane and into the fetal blood.
         (b) Carbon dioxide and other wastes diffuse from the fetal blood through the placental membrane and into the maternal blood.
   c. Gastrulation and organogenesis
      (1) **Gastrulation** is the movement of cells within the embryonic disc to form layers.
      (2) The cells of the embryonic disc fold inward, forming a **gastrula** that has two and then three primary germ layers.
      (3) The primitive tissues in the germ layers will form organs in the process of **organogenesis.**
         (a) Ectoderm gives rise to the nervous system, portions of the skin, the lining of the mouth, and the lining of the anal canal.
         (b) Mesoderm gives rise to muscles, bones, blood vessels, lymphatic vessels, reproductive organs, kidneys, and linings of body cavities.
         (c) Endoderm gives rise to linings of the digestive tract, respiratory tract, urinary bladder, and urethra.
      (4) The embryo develops a head, face, upper limbs, lower limbs, and mouth, and appears more humanlike.
      (5) By the beginning of the eighth week, the embryo is recognizable as a human.
      (6) **Teratogens** can cause congenital malformations by affecting an embryo or fetus in periods of rapid growth, development, or specialization of body structures.
2. **Fetal stage** (the end of the eighth week through birth)
   a. The body enlarges, upper and lower limbs reach final relative proportions, the skin is covered with sebum and dead epidermal cells, the skeleton continues to ossify, external reproductive organs are distinguishable as male or female, muscles contract, and fat is deposited in subcutaneous tissue.
   b. The **fetus** is full term at approximately 266 days.
      (1) It is about 50 centimeters long and weighs 2.7–4.5 kilograms.
      (2) It is positioned with its head toward the cervix.
   c. Fetal blood and circulation promote reception of oxygen and nutrients from maternal blood and wastes being carried away by maternal blood.
      (1) Umbilical vessels carry blood between the placenta and the fetus.
      (2) Fetal blood carries a greater concentration of oxygen than does maternal blood because the concentration of oxygen-carrying hemoglobin is greater in fetal blood, and fetal hemoglobin has greater affinity for oxygen.
      (3) Blood enters the fetus through the umbilical vein and partially bypasses the liver by means of the **ductus venosus.**
      (4) Blood enters the right atrium and partially bypasses the lungs by means of the **foramen ovale.**
      (5) Blood entering the pulmonary trunk partially bypasses the lungs by means of the **ductus arteriosus.**
      (6) Blood enters the umbilical arteries from the internal iliac arteries.
3. Maternal changes during pregnancy
   a. Embryonic cells produce hCG that maintains the corpus luteum, which continues to secrete estrogens and progesterone.

b. Placental tissue produces high concentrations of estrogens and progesterone.
   (1) Estrogens and progesterone maintain the uterine wall and inhibit secretion of FSH and LH.
   (2) Progesterone and relaxin inhibit contractions of uterine muscles.
   (3) Estrogens enlarge the vagina.
   (4) Relaxin helps relax the connective tissue of the pelvic joints.
c. The placenta secretes **placental lactogen** that stimulates the development of the breasts and mammary glands.
d. During pregnancy, increasing secretion of aldosterone promotes retention of sodium and body fluid, and increasing secretion of parathyroid hormone helps maintain a high concentration of maternal blood calcium.
e. The uterus greatly enlarges.
f. The woman's blood volume, cardiac output, breathing rate, and urine production increase.
g. The woman's dietary needs increase, but if intake is inadequate, fetal tissues have priority for use of available nutrients.

4. Birth process
   a. Several events occur at birth.
      (1) A decreasing concentration of progesterone and the release of prostaglandins may initiate the birth process.
      (2) The posterior pituitary gland releases **oxytocin.**
      (3) Uterine muscles are stimulated to contract, and labor begins.
      (4) Positive feedback causes stronger contractions and greater release of oxytocin.
   b. Following the birth of the infant, placental tissues are expelled.

5. Milk production and secretion
   a. During pregnancy, the breasts change.
      (1) Estrogens cause the ductile system to grow.
      (2) Progesterone causes development of alveolar glands.
      (3) Prolactin is released during pregnancy, but progesterone inhibits milk production.
   b. Following childbirth, the concentrations of placental hormones decline.
      (1) The action of prolactin is no longer blocked.
      (2) The mammary glands begin to secrete milk.
   c. Reflex response to mechanical stimulation of the nipple causes the posterior pituitary gland to release oxytocin, which releases milk from the alveolar ducts.
   d. As long as milk is removed from the breasts, more milk is produced; if milk is not removed, production ceases.
   e. During the period of milk production, the reproductive cycle may be inhibited.

23.3 Postnatal Period (page 894)

The **postnatal period** includes the neonatal period, infancy, childhood, adolescence, adulthood, and senescence.

1. Neonatal period
   a. The **neonatal period** extends from birth to the end of the fourth week.
   b. The newborn must begin to respire, obtain nutrients, excrete wastes, and regulate its body temperature.
   c. The first breath must be powerful to expand the lungs.
      (1) Surfactant reduces surface tension.
      (2) A variety of factors stimulate the first breath.

d. The liver is immature and unable to supply sufficient glucose, so the newborn depends primarily on stored fat for energy.
e. Immature kidneys cannot concentrate urine very well.
   (1) The newborn may become dehydrated.
   (2) Water and electrolyte imbalances may develop.
f. Homeostatic mechanisms may function imperfectly, and body temperature may be unstable.
g. The cardiovascular system changes when placental circulation ceases.
   (1) Umbilical vessels constrict.
   (2) The ductus venosus constricts.
   (3) The foramen ovale is closed by a valve as blood pressure in the right atrium falls and blood pressure in the left atrium rises.
   (4) The ductus arteriosus constricts.

2. Infancy
   a. **Infancy** extends from the end of the fourth week to one year of age.
   b. Infancy is a period of rapid growth.
      (1) The muscular and nervous systems mature, and coordinated activities become possible.
      (2) Communication begins.
   c. Rapid growth depends on an adequate intake of proteins, vitamins, and minerals in addition to energy sources.

3. Childhood
   a. **Childhood** extends from the end of the first year to puberty.
   b. Primary teeth erupt and are replaced by secondary teeth.
   c. It is a period of rapid growth, development of muscular control, and establishment of bladder and bowel control.

4. Adolescence
   a. **Adolescence** extends from puberty to adulthood.
   b. Physiological and anatomical changes result in a reproductively functional individual.
   c. Females may be taller and stronger than males in early adolescence, but the situation reverses in late adolescence.
   d. Adolescents develop high levels of motor skills, their intellectual abilities increase, and they continue to mature emotionally.

5. Adulthood
   a. **Adulthood** extends from adolescence to old age.
   b. The adult remains relatively unchanged physiologically and anatomically for many years.
   c. After age thirty, degenerative changes begin.
      (1) Skeletal muscles lose strength.
      (2) The cardiovascular system becomes less efficient.
      (3) The skin loses elasticity.
      (4) The capacity to produce sex cells declines.

6. Senescence
   a. **Senescence** is growing old.
   b. Degenerative changes continue, and the body becomes less able to cope with demands.
   c. Changes occur because of prolonged use, effects of disease, and cellular alterations.
   d. An aging person usually loses some intellectual functions, sensory functions, and physiological coordinating capacities.
   e. Death usually results from mechanical disturbances in the cardiovascular system or from disease processes that affect vital organs.

7. The end of life
   a. Certain signs may appear in sequence when a person dies of a chronic illness.

b. Preactive dying takes up to three months. The person withdraws socially and appetite wanes.

c. Active dying takes up to two weeks. The person rests, may become confused or agitated, and eats very little. Gradually the organ systems shut down. The skin becomes mottled as circulation slows and congestion and loud breathing occur.

23.4 Aging (page 898)

1. Passive aging
   a. Passive aging entails breakdown of structures and slowing or failure of functions.
   b. Connective tissue breaks down.
   c. DNA errors accumulate.

d. Lipid breakdown in aging membranes releases lipofuscin.
   e. **Free radical** damage escalates.
2. Active aging
   a. In autoimmunity, the immune system attacks the body.
   b. **Apoptosis** is a form of programmed cell death. It occurs throughout life, shaping organs.
3. The human life span
   a. The theoretical maximum life span is 120 years.
   b. Life expectancy, based on real populations, is 76.3 years for men and 81.1 years for women in the United States, and may be lower in poorer nations and those ravaged by infectious diseases.
   c. Medical technology makes life expectancy more closely approach life span.

CHAPTER ASSESSMENTS

23.1 Fertilization

1 Describe how sperm cells move in the female reproductive tract. (p. 868)
2 Summarize the events occurring after the sperm cell head enters the oocyte's cytoplasm. (p. 869)

23.2 Pregnancy and the Prenatal Period

3 Define *pregnancy*. (p. 870)
4 Describe the process of cleavage. (pp. 870–871)
5 Distinguish between a morula and a blastocyst. (p. 871)
6 Describe the formation of the inner cell mass, and explain its significance. (p. 871)
7 Explain what happens when the blastocyst nestles into the endometrium. (p. 872)
8 List the functions of hCG. (p. 872)
9 Distinguish between the chorion and amnion. (p. 873)
10 Explain the function of the amniotic fluid. (p. 873)
11 Describe the formation of the umbilical cord. (pp. 873 and 875)
12 Explain how the yolk sac and the allantois are related, and list the functions of each. (p. 875)
13 Describe the formation of the placenta, and explain its functions. (pp. 875–876)
14 Describe the placental membrane. (p. 876)
15 Explain how gastrulation forms the primary germ layers. (p. 876)
16 List the structures formed from the primitive tissues of the ectoderm, mesoderm, and endoderm. (pp. 876–878)
17 Explain why the embryonic period is so important. (p. 879)
18 Give the time frame for the fetus, listing the major changes that occur during fetal development. (pp. 881–882)
19 Describe a full-term fetus. (p. 882)
20 Explain how the fetal cardiovascular system is adapted to intrauterine existence. (pp. 883–887)
21 Compare the properties of fetal hemoglobin to those of adult hemoglobin. (p. 883)
22 Trace the pathway of blood from the placenta to the fetus and back to the placenta. (pp. 883–887)

23 Explain the hormonal changes in the maternal body during pregnancy. (pp. 887–889)
24 Describe the nonhormonal changes in the maternal body during pregnancy. (p. 890)
25 Describe the role of progesterone in initiating the birth process. (p. 890)
26 Discuss the events of the birth process. (pp. 890–891)
27 Explain positive feedback and the role of hormones in expelling the fetus and the afterbirth. (p. 890)
28 Detail the roles of prolactin and oxytocin in milk production and secretion. (pp. 891–893)

23.3 Postnatal Period

29 Explain why a newborn's first breath must be particularly forceful. (p. 894)
30 List some of the factors that stimulate the first breath. (p. 894)
31 Explain why newborns tend to develop water and electrolyte imbalances. (p. 894)
32 Discuss the cardiovascular changes in the newborn. (pp. 894–895)
33 Distinguish between a newborn and an infant. (pp. 894 and 896)
34 Describe the characteristics of an infant. (p. 896)
35 Distinguish between a child and an adolescent. (p. 896)
36 Define *adulthood*. (p. 896)
37 List some of the degenerative changes that begin during adulthood. (pp. 896–897)
38 Define *senescence*. (p. 897)
39 List some of the factors that promote senescence. (p. 897)
40 Contrast preactive dying and active dying. (pp. 897–898)

23.4 Aging

41 Discuss the signs of passive and active aging and the physiological causes of these signs. (p. 899)
42 _____ is the length of time a human can theoretically live, whereas _____ is the realistic projection of how long an individual will live. (p. 899)

INTEGRATIVE ASSESSMENTS/CRITICAL THINKING

Outcomes 6.5, 7.12, 8.4, 9.8, 11.8, 12.5, 13.11, 15.8, 16.5, 17.10, 18.9, 19.7, 20.5, 23.3, 23.4

1. If an aged relative came to live with you, what special provisions could you make in your household environment and routines that would demonstrate your understanding of the changes brought on by aging?

Outcomes 15.2, 15.5, 23.2, 23.3

2. What symptoms may appear in a newborn if its ductus arteriosus fails to close?

Outcomes 15.4, 23.3, 23.4

3. Why is it important for a middle-aged adult who has neglected physical activity for many years to have a physical examination before beginning an exercise program?

Outcomes 16.4, 23.1

4. Why can twins resulting from a single fertilized secondary oocyte exchange blood or receive organ transplants from each other without rejection, while twins resulting from two fertilized secondary oocytes in some cases cannot?

Outcomes 23.2

5. Why do toxins cause more severe medical problems in a child if exposure was during the first eight weeks of prenatal development rather than during the later weeks?

6. What technology would enable a fetus born in the fourth month to survive in a laboratory setting? (This is not yet possible.)

ONLINE STUDY TOOLS

Connect Interactive Questions Reinforce your knowledge using assigned interactive questions covering fertilization through birth and the postnatal period.

Connect Integrated Activity Learn more about potential risks to fetal development.

LearnSmart Discover which chapter concepts you have mastered and which require more attention. This adaptive learning tool is personalized, proven, and preferred.

Anatomy & Physiology Revealed Go more in depth as you view animations on development from ovulation to implantation.

Note the similar profiles of the four generations of this family. Researchers have discovered the genes that determine the distances between facial features, molding our appearances.

24

Genetics and Genomics

 LEARNING OUTCOMES

After you have studied this chapter, you should be able to:

24.1 Genes and Genomes
1 Distinguish among gene, chromosome, and genome. (p. 906)
2 Identify the two processes that transfer genetic information between generations. (p. 906)
3 Explain how the environment influences how genes are expressed. (p. 907)

24.2 Modes of Inheritance
4 Describe a karyotype, and explain what it represents. (p. 908)
5 Explain the basis of multiple alleles of a gene. (p. 909)
6 Distinguish between heterozygous and homozygous; genotype and phenotype; dominant and recessive. (p. 909)
7 Distinguish between autosomal recessive and autosomal dominant inheritance. (p. 910)

24.3 Factors That Affect Expression of Single Genes
8 Explain how and why the same genotype can have different phenotypes among individuals. (p. 912)

24.4 Multifactorial Traits
9 Describe and give examples of how genes and the environment determine traits. (pp. 912–913)

24.5 Matters of Sex
10 Describe how and when sex is determined. (p. 914)
11 Explain how X-linked inheritance differs from inheritance of autosomal traits. (pp. 914–915)
12 Discuss factors that affect how phenotypes may differ between the sexes. (p. 916)

24.6 Chromosome Disorders
13 Describe three ways that chromosomes can be abnormal. (p. 916)
14 Explain how prenatal tests provide information about chromosomes. (pp. 919–920)

24.7 Genomics and Health Care
15 Contrast the value of whole genome sequencing with that of single gene tests. (p. 920)
16 Explain how understanding gene expression patterns can improve and personalize health care. (pp. 920–921)

THE WHOLE PICTURE

Genetics is the study of inherited traits, their variation, and how they are transmitted between generations. Inherited traits range from obvious physical characteristics, such as red hair and freckles, to many aspects of health, including disease. Within the 3 billion bases of the human genome lies the information that cells use to synthesize proteins, which in turn provide traits. The unit of genetic information is a gene, which encodes a protein. Genes interact with each other and environmental influences to help mold who we are.

Personal genetic information is increasingly guiding diagnosis and treatment choices. Tens of thousands of people have already had their genomes sequenced, although all of the information has yet to be interpreted.

Anatomy & Physiology | REVEALED®
aprevealed.com

chromo-, color: *chromosome*—a "colored body" in a cell's nucleus that includes the genes.

gen-, born: *genetics*—the study of inheritance of characteristics.

hetero-, other, different: *heterozygous*—different members of a gene pair.

hom-, same, common: *homologous* chromosomes—pair of chromosomes that have similar genetic information.

karyo-, nucleus: *karyotype*—a chart that displays chromosomes in size order.

mono-, one: *monosomy*—a chromosome type present in only one copy.

phen-, show, be seen: *phenotype*—physical appearance or health condition that results from the way genes are expressed in an individual.

tri-, three: *trisomy*—three copies of a chromosome.

24.1 | Genes and Genomes

Genetics (jĕ-net′iks) is the study of inheritance of characteristics. **Genes** are sequences of DNA that encode particular proteins. They contribute to our physical diversity, including eye, skin, and hair color; many aspects of health; abilities and talents; and hard-to-define characteristics such as personality traits. We often equate the study of genetics with disease, but it is more accurately described as the study of inherited variation. Our genomes are more than 99.9% alike in DNA sequence; within that less than 0.1% of genetic variation lies our individuality.

The human genome is written in the language of DNA molecules. Recall from figure 4.20 (p. 135) that DNA consists of sequences of the nucleotide building blocks A (adenine), G (guanine), C (cytosine),

and T (thymine). In somatic cells, two copies of the human genome of 3.2 billion building blocks each are dispersed among the 46 chromosomes (fig. 24.1).

The transfer of genetic information from one generation to the next occurs through the processes of meiosis and fertilization, when one copy of the genome from each parent join. The cells that give rise to eggs and sperm, like all somatic cells, are diploid, with two copies of each of the 23 chromosomes. Eggs and sperm are haploid, with one set of chromosomes. The joining of an egg and a sperm reconstitutes the diploid state.

RECONNECT

To Chapter 22, Formation of Sperm Cells, page 828, and Oogenesis, page 841.

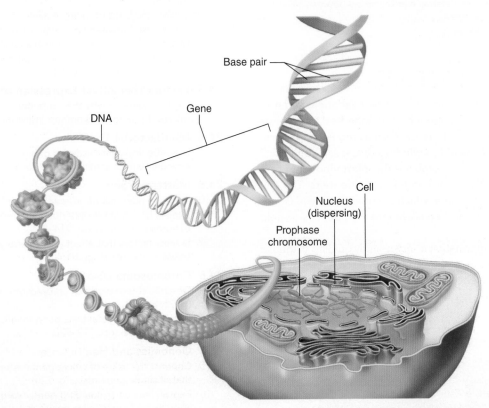

FIGURE 24.1 From DNA to gene to chromosome. Chromosomes in a cell about to divide consist of sister chromatids, each of which is a continuous DNA double helix and associated proteins. Chromosomes condense just prior to cell division, becoming visible under a microscope. The DNA winds around proteins, enabling the long molecule to fit inside the cell.

The human genome includes 20,325 protein-encoding genes, constituting the *exome*. Different cell types access (through transcription and translation) different subsets of the genome to produce particular proteins and in this way take on the distinctive characteristics of the hundreds of types of specialized cells in the body. Messenger RNA molecules can represent parts of different genes, so that the 20,325 genes actually encode 100,000 to 200,000 different proteins. It is a little like creating a great variety of stories from a set of words. The exome accounts for only a little less than 2% of the 3.2 billion DNA bases of the human genome. The remaining sequences control how the exome is used. The first human genomes were sequenced by the year 2000 and took thousands of researchers more than a decade, costing $3 billion. Today a human genome can be sequenced in under a day, for about $1,000. However, researchers are still analyzing what all of our genes do, how genes interact, and how we vary genetically.

Genetic information functions at several levels. It is encoded in DNA and expressed in RNA and protein; affects cells and tissues; affects the individual; and is also passed to the next generation. **Figure 24.2** depicts the inherited disease cystic fibrosis (CF) viewed at several levels.

RECONNECT

To Chapter 4, Nucleic Acids and Protein Synthesis, pages 132–142.

Until recently, the field of medical genetics dealt mostly with rare disorders that could be traced to the malfunction or absence of single genes. However, information from the human genome sequence and about which genes are expressed (transcribed and translated) under specific circumstances is providing a new view of physiology as a complex interplay of gene functions.

Discovering how diseases are related at the genetic level is enabling physicians to "repurpose" existing drugs. This is the case for progeria, an inherited disease that causes a young child to show signs of greatly accelerated aging.

Shortly after birth, in a child with progeria, weight gain slows and hair thins. Teeth never erupt, joints stiffen and bones weaken, and the skin wrinkles and toughens. Blood vessels stiffen with premature atherosclerosis, fat pockets shrink, and connective tissue hardens. The child usually dies during adolescence, from a heart attack or stroke.

Fewer than 250 children worldwide are known to have progeria. When researchers in 2003 identified the mutant gene, they realized, from the nature of the molecular defect, that several existing drugs might alleviate certain symptoms. They tested a failed cancer drug, and several children began to gain weight, and their arteries grew less stiff as the blockages shrunk. Hearing improved and bones became stronger. An unanticipated outcome was that these children, who face a life expectancy of 14 years, lived at least 1.6 years longer.

Understanding how a mutation causes a disease is an important first step in finding a treatment. The drug that helps children with progeria is now being tested in people who develop atherosclerosis as part of aging.

Genes are important, but they do not determine who we are, nor do they act alone. The environment influences how genes are expressed—under which circumstances they are transcribed into mRNA and those mRNA molecules translated into protein. The environment includes the chemical, physical, social and biological factors surrounding an individual, and also the conditions to which a cell is exposed. A common medical condition may result from inheriting risk factors and exposure to environmental influences. For example, development of cardiovascular disease may reflect not only inheritance of specific gene variants that control blood pressure, blood clotting, and cholesterol metabolism, but also lifestyle influences such as stress, smoking, poor diet, and lack of physical exercise that may affect the expression of those genes in negative ways. The inaccurate idea that "we are our genes" is called "genetic determinism."

PRACTICE

1. What is the chemical composition of the genetic material?
2. Define *gene, genome,* and *exome.*
3. Explain how the human genome encodes more proteins than it has genes.

CAREER CORNER

Genetic Counselor

The pregnant patient is anxious after receiving a phone call reporting the test result that she is a carrier for cystic fibrosis. She soon meets with a genetic counselor, who explains how the disease is inherited and describes testing options. The next step is to test the father-to-be, and when he is found not to have a cystic fibrosis mutation, the couple relaxes. Had he also carried a mutation, the counselor explained, each child would have had a 25% chance of inheriting the disease, which could have been detected with further testing.

A genetic counselor is a health-care professional trained in genetics, psychology, statistics, and counseling techniques to provide guidance to patients taking genetic tests or with inherited diseases in their families. A certified genetic counselor has a master's degree in the field, but nurses, social workers, physicians, and PhD geneticists also provide genetic counseling. The counselor explains inheritance patterns, assesses risks for specific individuals in a family, recommends tests, and interprets the results of tests.

Genetic counselors work in medical centers, in physicians' offices, in research laboratories, and at companies that provide genetic tests. They may specialize in prenatal care or work in clinics for patients with specific diseases, such as familial cancers, hereditary forms of blindness, or inherited blood disorders. Genetic counselors are hired to help in the ongoing effort to identify the functions of all of the genes in the human genome, a process called annotation. This entails searching disease and mutation databases and reading the scientific and medical literature to connect reports and information to DNA sequences. Genetic counselors are at the forefront of making practical sense of the human genome sequences.

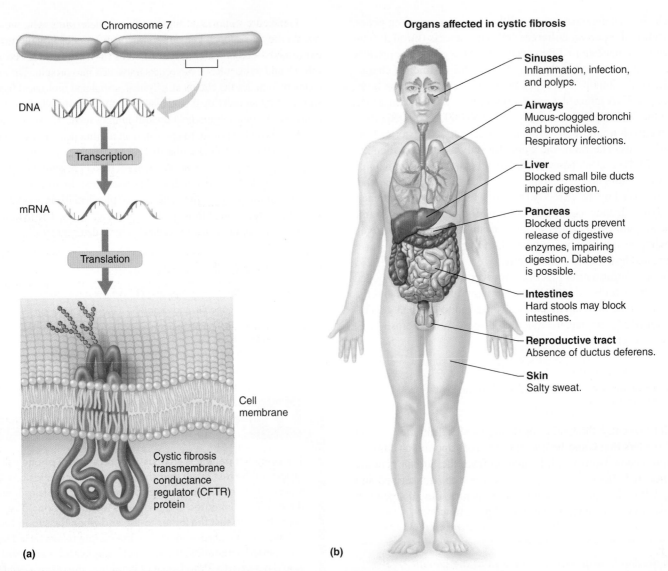

FIGURE 24.2 From gene to protein to person. (*a*) The gene encoding the CFTR protein, and causing cystic fibrosis when mutant, is on the seventh largest chromosome. CFTR protein folds into a channel that regulates the flow of chloride ions into and out of cells lining the respiratory tract, pancreas, intestines, and elsewhere. (*b*) In most cases of cystic fibrosis, the CFTR protein is missing an amino acid, which alters its folding and results in very thick mucus and other secretions in the places highlighted in the illustration. The sticky secretions impair breathing and increase the risk of certain respiratory infections. Source: Data from M. C. Iannuzi and F. S. Collins, "Reverse Genetics and Cystic Fibrosis," in *American Journal of Respiratory Cellular and Molecular Biology*, 2:309–316 (1990).

24.2 | Modes of Inheritance

Genetics has the power of prediction. Knowing how genes are distributed in meiosis and the combinations in which they can join at fertilization makes it possible to calculate the probability that a certain trait will appear in the offspring of two particular individuals. The patterns in which genes are transmitted in families are termed *modes of inheritance*. An educator, monk, and gardener named Gregor Mendel deciphered the modes of inheritance in the 1860s by breeding pea plants and following easily visible traits over several generations. The laws he derived apply to humans because our cells are diploid, like those of peas and most other multicellular organisms.

Chromosomes and Genes Are Paired

The basis of making genetic predictions is the way in which genes are carried as parts of chromosomes. Charts called **karyotypes** display the 23 chromosome pairs in size order (fig. 24.3). Pairs 1 through 22 are **autosomes** (aw'to-sōmz), which do not carry genes that determine sex. The other two chromosomes, the X and the Y, include genes that determine sex and are called **sex chromosomes.** A female is XX and a male is XY. (See the section of this chapter entitled "Matters of Sex.")

Each chromosome includes hundreds or thousands of genes. A human somatic cell has two copies of each chromosome, and therefore two copies of each gene. The two members of a gene pair are

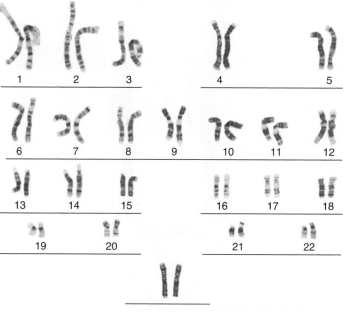

1 2 3 4 5
6 7 8 9 10 11 12
13 14 15 16 17 18
19 20 21 22

Sex chromosomes

FIGURE 24.3 A normal human karyotype has 22 pairs of autosomes aligned in size order, plus the sex chromosomes. This karyotype is from a female; note the two *X* chromosomes.

ⓠ *How would a karyotype for a male look different from this one?*

Answer can be found in Appendix G .

located at the same position on homologous chromosomes. Because a gene consists of hundreds of nucleotide building blocks, it can exist in variant forms, called **alleles** (ah-lēlz′), that differ in DNA sequence. A person who has two identical alleles of a particular gene is **homozygous** (ho″mo-zi′gus) for that gene. A person with two different alleles for a gene is **heterozygous** (het″er-o-zi′gus) for it. A gene may have multiple alleles, but an individual person can normally have a maximum of two alleles for a particular gene.

> The allele that causes most cases of cystic fibrosis was discovered in 1989, and a test was developed to detect it. However, other alleles were soon discovered. Today, more than 1,600 mutations (changes) in the cystic fibrosis gene are known. Different allele combinations produce different combinations and severities of symptoms.

The combination of gene variants (alleles) in a person's genome constitutes the **genotype** (je′no-tīp). That is, all humans have the same types of genes, but differ in the specific variants. The appearance or health condition of the individual that develops as a result of the ways the genes are expressed is the **phenotype** (fe′no-tīp). An allele is **wild type** if its associated phenotype is either normal function or the most common expression in a particular population. Wild type is indicated with a + sign. An allele

that is a change from wild type, perhaps producing an uncommon phenotype, is **mutant.** Disease-causing alleles are mutant, but not all mutations cause disease. Some mutations, such as the one that prevents HIV from entering cells, can be advantageous.

Dominant and Recessive Inheritance

For many genes, in heterozygotes, one allele determines the phenotype. Such an allele whose action masks that of another allele is termed **dominant.** The allele whose expression is masked is **recessive.** For genes with only two alleles, the dominant ones are usually indicated with a capital letter.

An allele that causes a disease can be recessive or dominant. It may also be *autosomal* (carried on a nonsex chromosome) or *X-linked* (carried on the X chromosome) or *Y-linked* (carried on the Y chromosome). The mode of inheritance refers to whether a trait is dominant or recessive, autosomal or carried on a sex chromosome. This designation has important consequences in predicting the chance that offspring will inherit an illness or trait. The following rules emerge:

1. An autosomal condition is equally likely to affect either sex. X-linked characteristics affect many more males than females.
2. A person most likely inherits a recessive condition from two parents who are heterozygotes (carriers). The parents are usually healthy. For this reason, recessive conditions can "skip" generations.
3. A person who inherits a dominant condition has at least one affected parent. Therefore, dominant conditions do not skip generations. If, by chance, a dominant trait does not appear in a generation in a particular family, it does not reappear in subsequent generations, as a recessive trait might. (An exception is if the dominant allele arises as a new mutation in the sperm or egg.)

Cystic fibrosis (CF) is an example of an autosomal recessive disorder. The wild type allele for the *CFTR* gene, which is dominant over the disease-causing allele, specifies formation of chloride channels in the cell membranes of cells lining the structures depicted in figure 24.2. Certain recessive mutant alleles disrupt the structure and/or function of the chloride channels.

An individual who inherits two identical mutant alleles has CF and is homozygous recessive. A person inheriting one recessive mutant allele plus a dominant wild type allele is a carrier and transmits the disease-causing allele in half of the gametes. A person who has two wild type alleles is homozygous dominant and does not have or carry CF. The three possible genotypes are associated with only two phenotypes, because carriers and homozygous dominant individuals do not have the illness. A person who has two different mutant alleles is called a compound heterozygote, and may or may not have symptoms, depending upon the effects of the allele combination on the chloride channels.

Using logic, understanding how chromosomes and genes are distributed among gametes in meiosis, and knowing that mutant alleles that cause CF are autosomal recessive, we can predict genotypes and phenotypes in the next generation. **Figure 24.4** illustrates two carriers of CF. Half of the man's sperm contain a specific mutant allele, as do half of the woman's eggs. Sperm and eggs combine at random, so each offspring has a

- 25% chance of inheriting two wild type alleles (homozygous dominant, healthy, and not a carrier)
- 50% chance of inheriting a mutant allele from either parent (heterozygous and a carrier, but healthy)
- 25% chance of inheriting a mutant allele from each parent (homozygous recessive, has CF)

Genetic counselors use two tools to explain inheritance to families: Punnett squares and pedigrees. A **Punnett square** is a table that symbolizes the logic used to deduce the probabilities of genotypes in offspring. The mother's alleles (for a specific gene) are listed atop the four boxes comprising the square, and the father's alleles are listed along the left side. Each box records an allele combination at fertilization. (When more than one gene is considered, the Punnett square has more boxes.)

A **pedigree** is a diagram that depicts family relationships and known genotypes and phenotypes. Circles are females and squares are males; shaded-in symbols represent people who have a trait or condition; half-shaded symbols denote carriers. Roman numerals indicate generations. Figure 24.4 and figure 24.5 show Punnett squares and pedigrees.

A person who has an autosomal recessive illness has parents who are carriers—they do not have the illness. Or, if the phenotype is mild, a parent might be homozygous recessive and affected. A person who has an autosomal dominant condition typically has an affected parent, and need inherit only one copy of the mutant allele to have the associated phenotype. In contrast, expression of an autosomal recessive condition requires inheriting two mutant alleles.

> Certain recessive alleles that cause illness remain in a population because they protect carriers from another condition, such as an infectious disease. For example, in carriers of sickle cell disease (see fig. 4.26, p. 143), not enough red blood cells are deformed to block circulation under normal atmospheric conditions, but enough are sickle shaped to prevent malaria parasites from infecting many red blood cells. Carriers for sickle cell disease and certain other inherited anemias do not easily contract malaria. If they do contract malaria, the symptoms are mild.

An example of an autosomal dominant condition is Huntington disease (HD). Symptoms usually begin in the late thirties or early forties and include loss of coordination, uncontrollable dancelike movements, behavioral changes, and cognitive decline. Figure 24.5 shows the inheritance pattern for HD. If one parent has the mutant allele, half of his or her gametes will have it. Assuming the other parent does not have a mutant allele, each child conceived has a 1 in 2 chance of inheriting the disease allele and, eventually, developing the condition.

Most of the 3,000 or so known human inherited disorders are autosomal recessive. These conditions tend to produce symptoms in childhood. Autosomal dominant conditions often begin to cause

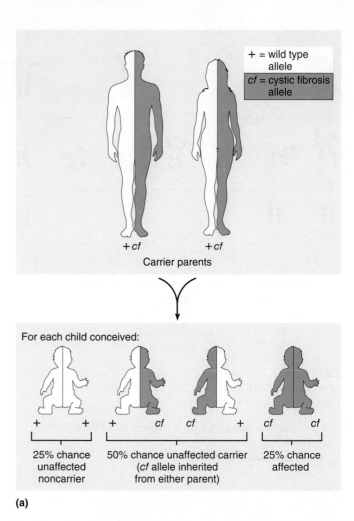

FIGURE 24.4 Cystic fibrosis is autosomal recessive. (a) Each child of carrier parents has a 25% chance of being unaffected and not a carrier, a 50% chance of being an unaffected carrier, and a 25% chance of being affected. Sexes are affected with equal frequency. A Punnett square (b) and a pedigree (c) also display this information. Symbols in the pedigree with both black and white indicate unaffected carriers (heterozygotes). The pedigree illustrates the makeup of one possible family.

Q *If a mother is homozygous wild type for the CFTR gene, what is the probability that her child will have CF?*
Answer can be found in Appendix G.

symptoms in adulthood, and they remain in populations because people have children before knowing they could pass on the illness. For some autosomal dominant disorders, genetic tests can reveal that a disease-causing genotype has been inherited, even before symptoms begin.

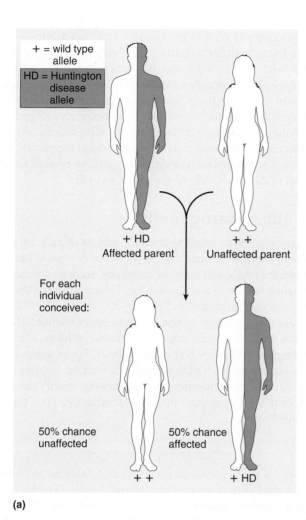

+ = wild type allele

HD = Huntington disease allele

+ HD
Affected parent

+ +
Unaffected parent

For each individual conceived:

50% chance unaffected

+ +

50% chance affected

+ HD

(a)

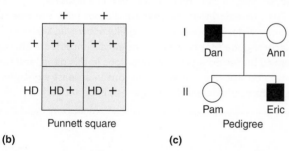

Punnett square

(b)

I
Dan — Ann

II
Pam — Eric

Pedigree

(c)

FIGURE 24.5 Huntington disease is autosomal dominant. (*a*) A person with one HD allele develops the disease. A Punnett square (*b*) and pedigree (*c*) depict the inheritance of HD. The pedigree symbols for HD are completely filled in to indicate that the person is affected. Autosomal dominant conditions affect both sexes.

People of all ages take genetic tests. A couple contemplating parenthood may take blood tests that detect carriers of nearly 500 single-gene diseases. Newborns are routinely screened for 50 or more genetic diseases. Children with unusual sets of symptoms have their exomes sequenced to identify mutations. Genetic tests can indicate which drugs are most likely to be effective in particular patients. Genetic tests are also used to evaluate forensic evidence and to trace ancestry.

Different Dominance Relationships

Most genes exhibit complete dominance or recessiveness. Interesting exceptions are incomplete dominance and codominance.

In **incomplete dominance,** the heterozygous phenotype is intermediate between that of either homozygote. For example, in familial hypercholesterolemia (FH), a person with two disease-causing alleles completely lacks LDL (low-density lipoprotein) receptors on liver cells that take up cholesterol from the bloodstream (fig. 24.6). A person with one disease-causing allele (a heterozygote) has half the normal number of cholesterol receptors. Someone with two wild type alleles has the normal number of receptors. The associated phenotypes parallel the number of receptors—those with two mutant alleles develop severe disease as children, individuals with one mutant allele may become ill in young or middle adulthood, and people with two wild type alleles do not develop this type of hereditary heart disease.

Different alleles that are both expressed in a heterozygote are **codominant.** For example, two of the three alleles of the I gene, which determines ABO blood type, are codominant (see fig. 14.21, p. 550). The I gene encodes the enzymes that place the A and B antigens on red blood cell surfaces. The three alleles are I^A, I^B, and i. People with type A blood may be either genotype $I^A I^A$ or $I^A i$; type B corresponds to $I^B I^B$ or $I^B i$; type AB to $I^A I^B$; and type O to ii. The I^A and I^B alleles are codominant.

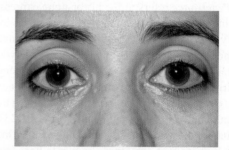

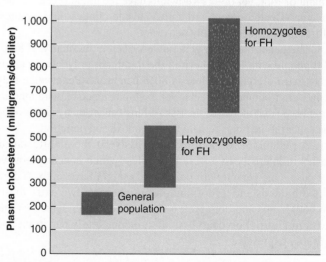

FIGURE 24.6 Incomplete dominance is seen in the plasma cholesterol levels of heterozygotes and homozygotes for familial hypercholesterolemia (FH). This condition increases serum cholesterol level in the blood, raising the risk of developing cardiovascular disease. The photograph shows cholesterol deposits near the eyes of a person who is a homozygote for the disease-causing allele.

RECONNECT

To Chapter 14, Blood Groups and Transfusions, pages 548–551.

PRACTICE

4 Distinguish between autosomes and sex chromosomes.

5 Distinguish between genotype and phenotype.

6 Distinguish between wild type and mutant alleles.

7 How do the modes of transmission of autosomal recessive and autosomal dominant inheritance differ?

8 Distinguish between incomplete dominance and codominance.

24.3 | Factors That Affect Expression of Single Genes

The expression of most genotypes varies from person to person, due to the effects of the environment and other genes. Even identical twins may not exhibit a phenotype in the exact same way. The terms *penetrance, expressivity,* and *pleiotropy* are used to describe some of these distinctions of genotype expression.

Penetrance and Expressivity

Penetrance refers to the all-or-none expression of a genotype in an individual, whereas *expressivity* refers to degrees of a phenotype. Some disease-causing allele combinations are **completely penetrant,** which means that everyone who inherits a particular genotype expresses the associated phenotype. A genotype is **incompletely penetrant** if some individuals do not express the associated phenotype. Polydactyly, having extra fingers or toes, is incompletely penetrant (see fig. 7.47, p. 239). Some people who inherit the autosomal dominant allele have more than five digits on a hand or foot, yet others known to have the allele (because they have an affected parent and child) have ten fingers and ten toes. Most traits are incompletely penetrant.

The penetrance of a gene is described numerically. If 80 of 100 people who have inherited the dominant polydactyly allele have extra digits, the allele is 80% penetrant. Penetrance may reflect environmental effects. For example, a person who inherits a genotype that increases the risk of lung cancer many not develop the cancer if the lungs remain free of smoke and pollution.

A phenotype is **variably expressive** if the manifestation varies in intensity in different people—which is nearly always the case. One person with polydactyly might have an extra digit on both hands and a foot; another might have two extra digits on both hands and both feet; a third person might have just one extra fingertip. Polydactyly is both incompletely penetrant and variably expressive.

Pleiotropy

A single gene that effects several phenotypic traits exhibits **pleiotropy** (ple´ot-ro-pe). Family members who have different sets of symptoms can appear to have different illnesses, when they actually have the same pleiotropic disorder.

Pleiotropy is seen in genetic diseases that affect a single protein found in different parts of the body. This is the case for Marfan syndrome, an autosomal dominant defect in an elastic connective tissue protein called fibrillin. The protein's abundance in the lens of the eye; in the bones of the limbs, fingers, and ribs; and in the aorta explains the symptoms of lens dislocation, long limbs, spindly fingers, and a caved-in chest. The most serious symptom is weakening in the aorta wall, which can burst the vessel. If the weakening is detected early, a synthetic graft can be used to patch that part of the vessel wall, saving the person's life.

Genetic Heterogeneity

The same phenotype resulting from the actions of different genes is called **genetic heterogeneity** (jĕ-net´ik het″er-o-je-ne´ĭ-te). For example, the nearly 200 forms of hereditary deafness are each due to impaired actions of a different gene. Different genes affect different aspects of hearing.

Genetic heterogeneity occurs when genes encode different enzymes that catalyze the same biochemical pathway, or encode different proteins that are part of the pathway. For example, eleven biochemical reactions lead to blood clot formation. Clotting disorders may result from mutations in the genes that specify any of the enzymes that catalyze these reactions, leading to several types of bleeding disorders.

> A couple was arrested on suspicion of child abuse because their toddler suffered repeated broken bones. Tests for the gene known to cause osteogenesis imperfecta, "brittle bone disease," which can break bones even in a fetus (see fig. 7.6b, p. 204), revealed that the parents did not have the disease-causing allele. They were found guilty. However, discovery of a second gene that causes osteogenesis imperfecta enabled them to be tested again, for mutations in this gene, and their conviction was overturned. Mutations in at least eight genes cause the brittle bones of osteogenesis imperfecta.

PRACTICE

9 Distinguish between penetrance and expressivity.

10 What is pleiotropy?

11 What is genetic heterogeneity?

24.4 | Multifactorial Traits

Most, if not all, characteristics and disorders considered "inherited" reflect input from the environment as well as genes. Characteristics molded by one or more genes plus the environment are termed **multifactorial traits.** They are also called "complex traits," but this is a less precise definition.

A trait determined by more than one gene is termed **polygenic.** Most polygenic traits are also influenced by the environment.

Usually, several genes each contribute to differing degrees toward molding an overall phenotype. A polygenic trait, with many degrees of expression because of the input of several genes, is said to be continuously varying. Height, skin color, and eye color are polygenic traits (figs. 24.7, 24.8, and 24.9).

Although the expression of a polygenic trait is continuous, we can categorize individuals into classes and calculate the frequencies of the classes. When we do this and plot the frequency for each phenotype class, a bell-shaped curve results. This curve indicating continuous variation of a polygenic trait is strikingly similar for different characteristics, such as fingerprint patterns, height, eye color, and skin color. Even when different numbers of genes contribute to the phenotype, the curve is the same shape.

Eye color illustrates how interacting genes can mold a single trait. The colored part of the eye, the iris, darkens as melanocytes produce the pigment melanin. Unlike melanin in skin melanocytes, the pigment in the eye tends to stay in the cell that produces it. Blue eyes have just enough melanin to make the color opaque, and dark blue or green, brown or black eyes have increasingly more melanin in the iris.

Because the genes that determine eye color are present in all cells, forensic scientists can determine the eye color of a corpse from tissue elsewhere on the body.

Two genes (*OCA2* and *HERC2*) control melanin synthesis and deposition. The alleles of the two genes interact additively, producing distinct eye colors. Figure 24.9 depicts how this might happen to account for five colors—light blue, deep blue or green, light brown, medium brown, and dark brown/black. If each dominant allele contributes a certain amount of pigment, then the greater the number of such alleles, the darker the eye color. The bell curve arises because there are more ways to inherit light brown eyes, with any two dominant alleles, than there are ways to inherit the other colors.

Other genes modify the expression of the two genes that control melanin synthesis. Two genes add greenish hues. Overlying the colors and tones are specks and flecks, streaks and rings, and regions of dark versus light that arise from the way pigment is laid down onto the distinctive peaks and valleys at the back of the iris.

Height and skin color are multifactorial as well as polygenic, because environmental factors influence them: good nutrition enables a person to reach the height dictated by genes, and sun exposure affects skin color. Most of the more common illnesses, including heart disease, diabetes mellitus, hypertension, and cancers, are multifactorial.

 PRACTICE

12 How does polygenic inheritance make possible many variations of a trait?

13 How can two genes specify five phenotypes?

14 How may the environment influence gene expression?

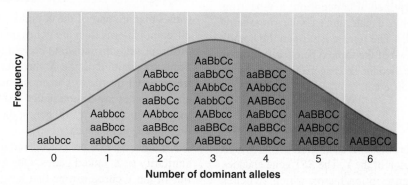

FIGURE 24.7 Height is a polygenic trait. These students in a genetics class lined up by height to illustrate the continuously varying nature of the trait.

FIGURE 24.8 Variations in skin color. A model of three genes, with two alleles each, can explain some of the hues of human skin. In actuality, this trait likely involves many more than three genes. The mid-range colors are more common.

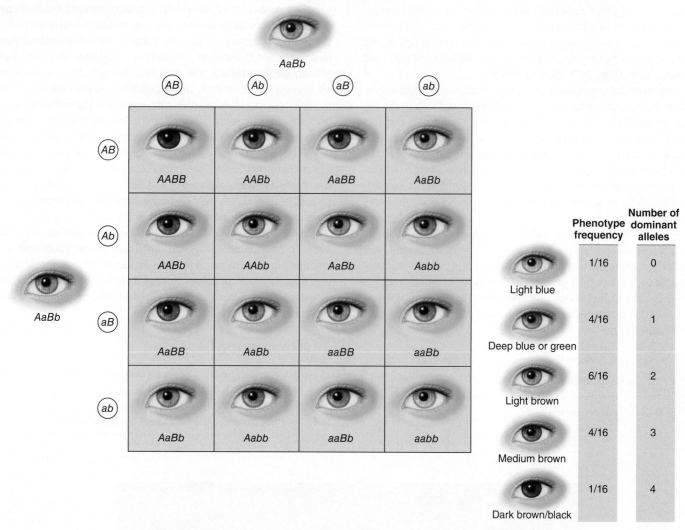

FIGURE 24.9 Variations in eye color. A model of two genes, with two alleles each, can explain five human eye colors. If eye color is controlled by two genes *A* and *B*, each of which comes in two allelic forms *A* and *a* and *B* and *b*, then the lightest color would be genotype *aabb*; the darkest, *AABB*.

24.5 | Matters of Sex

Human somatic (body or nonsex) cells include an X and a Y chromosome in males and two X chromosomes in females. All eggs carry a single X chromosome, and sperm carry either an X or a Y chromosome. Sex is determined at conception: a Y-bearing sperm fertilizing an egg conceives a male, and an X-bearing sperm conceives a female (fig. 24.10). The female is termed the homogametic sex because she has two of the same type of sex chromosome, and the human male is called the heterogametic sex because his two sex chromosomes are different.

Sex Determination

Maleness derives from a Y chromosome gene called *SRY,* which stands for sex-determining region of the Y. The *SRY* gene encodes a type of protein called a transcription factor, which switches on other genes. *SRY* activates transcription of genes that direct development of male structures in the embryo, while suppressing

formation of female structures. **Figure 24.11** shows the sex chromosomes. A female develops in the absence of the *SRY* transcription factor, and if a gene called *Wnt4* is expressed.

Genes on the Sex Chromosomes

Genes that are part of the X and Y chromosomes are inherited in different patterns than are genes on the autosomes because of the different sex chromosome constitutions of males and females. Traits transmitted on the X chromosome are X-linked, and those transmitted on the Y chromosome are Y-linked. The two sex chromosomes are dramatically different in size. The X chromosome has more than 1,500 genes; the Y chromosome has only 231 protein-encoding genes.

Y-linked genes are considered in three groups, based on their similarity to X-linked genes. One group consists of genes at the tips of the Y chromosome that have counterparts on the X chromosome. These genes encode a variety of proteins that function in both sexes, participating in or controlling such activities as

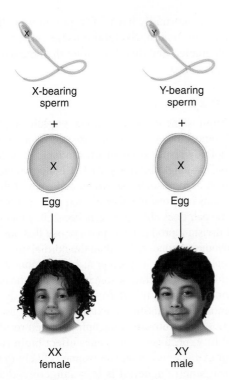

FIGURE 24.10 Sex determination. An egg contributes an X chromosome, and a sperm, either an X or a Y. If an X-bearing sperm fertilizes an egg, the zygote is female (XX). If a Y-bearing sperm fertilizes an egg, the zygote is male (XY).

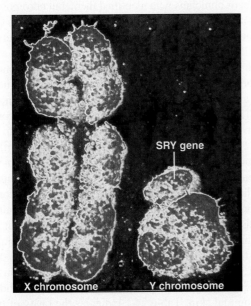

FIGURE 24.11 The X and Y chromosomes. The *SRY* gene, at one end of the short arm of the Y chromosome, starts the cascade of gene activity that directs development of a male (31,000×).

bone growth, the synthesis of hormones and receptors, and energy metabolism. The members of the second functional group of Y chromosome genes are similar in DNA sequence to certain genes on the X chromosome, but they are not identical. These genes are expressed in nearly all tissues, including those found only in males. The third group of genes includes those unique to the Y chromosome. Many of them control male fertility, such as the *SRY*

gene. Some cases of male infertility arise from tiny deletions of these male-specific parts of the Y chromosome. Other genes in this group encode proteins that participate in cell cycle control; proteins that regulate gene expression; enzymes; and protein receptors for immune system biochemicals.

Y-linked genes are transmitted only from fathers to sons, because only males have Y chromosomes. The differences in inheritance patterns of X-linked genes between females and males result from the fact that any gene on the X chromosome of a male is expressed in his phenotype, because he has no second allele on a second X chromosome to mask its expression. The human male is **hemizygous** for X-linked traits because he has only one copy of each X chromosome gene. Red-green colorblindness is a recessive X-linked trait.

A male always inherits his Y chromosome from his father and his X chromosome from his mother. A female inherits one X chromosome from each parent. If a mother is heterozygous for a particular X-linked gene, then her son has a 50% chance of inheriting either allele from her. X-linked genes are therefore passed from mother to son.

The transmission pattern of hemophilia A illustrates X-linked recessive inheritance. This clotting disorder is passed from carrier mother to affected son with a risk of 50%, because he can inherit either her normal allele or the mutant one. A daughter has a 50% chance of inheriting the hemophilia allele and being a carrier like her mother and a 50% chance of not inheriting the allele.

> The contribution of the X chromosome is equal in males and females, even though males have one X and females have two, because one X is silenced in every somatic cell of a female. She is a mosaic, with genes from her father's X chromosome expressed in some cells, and genes from her mother's X chromosome expressed in others.

A woman would not know that she is a carrier of an X-linked recessive trait unless she has an affected son, or if there is a biochemical test that reveals that her cells have half the normal amount of the gene's protein product. For a daughter to inherit an X-linked recessive disorder or trait, her father must be affected and her mother must be a carrier or affected. The daughter would need to inherit one affected X chromosome from each parent.

For X-linked recessive traits that seriously impair health, affected males may not feel well enough to have children. Therefore, X-linked recessive conditions that are nearly as common among females as males tend to have mild phenotypes. Colorblindness is a mild X-linked trait—men who are colorblind are as likely to have children as men with full color vision.

Dominant disease-causing alleles on the X chromosome are rare. Males are usually much more severely affected than females, who have a second X that provides protection. In a condition called *incontinentia pigmenti,* for example, a female has swirls of pigment in the skin where melanin in the epidermis extends into the dermis, and may have abnormal teeth, sparse hair, visual problems, and seizures. Males are so severely affected that they do not survive to be born.

Gender Effects on Phenotype

Certain autosomal traits are expressed differently in males and females, due to differences between the sexes. A **sex-limited trait** affects a structure or function present in only males or only females. Such a gene may be X-linked or autosomal. Beard growth and breast size are sex-limited traits. A woman cannot grow a beard because she does not manufacture sufficient hormones required for facial hair growth, but she can pass to her sons the genes that specify heavy beard growth.

In **sex-influenced inheritance,** an allele is dominant in one sex but recessive in the other. Again, such a gene may be X-linked or autosomal. This difference in expression reflects hormonal differences between the sexes. For example, a gene for hair growth pattern has two alleles, one that produces hair all over the head and another that causes pattern baldness. The baldness allele is dominant in males but recessive in females, which is why more men than women are bald. A heterozygous male is bald, but a heterozygous female is not. A bald woman would have two mutant alleles.

About 1% of human genes exhibit **genomic imprinting,** in which the expression of a disorder differs depending upon which parent transmits the disease-causing gene or chromosome. The phenotype may differ in degree of severity, in age of onset, or even in the nature of the symptoms. The physical basis of genomic imprinting is that methyl (—CH_3) groups cover the gene inherited from one parent, blocking it from being transcribed and translated.

 PRACTICE

15 Which chromosomes and genes determine sex?

16 What are the three functional classes of genes on the Y chromosome?

17 Why do X-linked recessive conditions appear most commonly in males?

18 How can gender affect gene expression?

24.6 | Chromosome Disorders

Deviations from the normal human chromosome number of 46 produce syndromes because of the excess or deficit of genes. Rearrangement of chromosomes, such as an inversion of a section of a chromosome, or two nonhomologous chromosomes exchanging parts, may also cause symptoms. This may happen if the rearrangement disrupts a vital gene or if it results in "unbalanced" gametes that contain too little or too much genetic material. Chromosome number abnormalities may involve single chromosomes or entire sets of chromosomes.

Polyploidy

The most extreme upset in chromosome number is having an entire extra set or sets, which is called **polyploidy** (pol'e-ploi"de). This results from formation of a diploid, rather than a normal haploid, gamete. For example, if a haploid sperm fertilizes a diploid egg, the fertilized egg is *triploid,* with three copies of each chromosome. Most human polyploids cease developing as embryos or fetuses, but occasionally an infant survives for a few days, with many abnormalities.

Some organs normally have a few polyploid cells, with no adverse effects on health. Skeletal muscle cells, for example, have multiple nuclei, and therefore more than the usual number of each chromosome.

Aneuploidy

A normal chromosome number is termed **euploid** (u'ploid). Cells missing a chromosome or having an extra one are **aneuploid** (an'u-ploid). Aneuploidy results from an error in meiosis called **nondisjunction** (non"dis-jungk'shun) (fig. 24.12). In normal meiosis, pairs of homologous chromosomes separate, and each of the resulting gametes contains only one member of each pair. In nondisjunction, a chromosome pair does not separate, either at the first or at the second meiotic division, producing a sperm or egg that has two copies of a chromosome or none, rather than the normal one copy. When such a gamete fuses with its partner at fertilization, the resulting zygote has either 47 or 45 chromosomes, instead of the normal 46.

The symptoms that result from aneuploidy depend upon which chromosome is missing or extra, and which genes are part of the chromosome. Autosomal aneuploidy often results in intellectual disability; because many genes affect brain function the absence of or extra copies of any autosome is likely to affect such a gene. Extra genetic material is less dangerous than missing material. This is why most children born with the wrong number of chromosomes have an extra one, called a **trisomy** (tri'so-me), rather than a missing one, called a **monosomy** (mon'o-so-me).

Aneuploid conditions have historically been named for the researchers or clinicians who identified them, but chromosome designations are more precise and for this reason are more clinically useful. Down syndrome, for example, refers to a distinct set of symptoms usually caused by trisomy 21. It is the most common autosomal aneuploid. However, the syndrome may also arise from one copy of chromosome 21 exchanging parts with a different chromosome in a gamete and the fertilized ovum receiving excess chromosome 21 material. This exchange of genetic material is a type of chromosomal aberration called a *translocation.* Knowing whether an individual has trisomy 21 or translocation Down syndrome is important, because the probability of trisomy 21 recurring in a sibling is about 1 in 100, but the chance of translocation Down syndrome recurring is much greater. Clinical Application 24.1 takes a closer look at trisomy 21.

Trisomies 13 and 18 are the next most common autosomal aneuploids and usually result in miscarriage. An infant with trisomy 13 has an underdeveloped face, extra and fused fingers and toes, heart defects, small adrenal glands, and a cleft lip or palate. An infant with trisomy 18 suffers many of the problems seen in trisomy 13, plus a peculiar positioning of the fingers and flaps of extra abdominal skin. **Table 24.1** indicates the rarity of trisomies

TABLE 24.1	Comparing and Contrasting Trisomies 13, 18, and 21	
Type of Trisomy	Incidence at Birth	Percent of Conceptions That Survive 1 Year After Birth
13 (Patau)	1/12,500–1/21,700	<5%
18 (Edward)	1/6,000–1/10,000	<5%
21 (Down)	1/800–1/826	85%

The most common autosomal aneuploid is *trisomy 21,* which is an extra chromosome 21. The characteristic slanted eyes and flat face prompted Sir John Langdon Haydon Down to coin the term *mongolism* when he described the syndrome in 1886. As the medical superintendent of a facility for the profoundly intellectually disabled, Down noted that about 10% of his patients resembled people of the Mongolian race. The resemblance is coincidental. Males and females of all races can have the syndrome.

A person with Down syndrome (either trisomy or translocation) is short and has straight, sparse hair and a tongue protruding through thick lips. The eyes slant and have upward "epicanthal" skin folds in the inner corners. Ears are abnormally shaped. The hands have an unusual pattern of creases, the joints are loose, and reflexes and muscle tone are poor. Developmental milestones (such as sitting, standing, and walking) are slow, and toilet training may take several years. Intelligence varies greatly, from profound intellectual disability to being able to follow simple directions, read, and use a computer. Some colleges specialize in educating people with Down syndrome (fig. 24A).

Down syndrome is associated with many physical problems, including heart and kidney defects, susceptibility to infections, and blockages in the digestive system. An affected child is fifteen times more likely to develop leukemia than a healthy child, but this is still a low figure. People with Down syndrome have a risk of developing Alzheimer disease of 25%, compared to 6% for the general population.

The likelihood of giving birth to a child with trisomy 21 Down syndrome increases dramatically with the age of the mother (table 24A). However, 80% of children with trisomy 21 are born to women under age thirty-five, because younger women are more likely to become pregnant and until recently less likely to have prenatal testing. About 95% of cases of trisomy 21 can be traced to nondisjunction in the egg.

The age factor in Down syndrome may be because meiosis in the female is completed after conception. The older a woman is, the longer her oocytes have been arrested on the brink of completing meiosis. During this time, the oocytes may have been exposed to chromosome-damaging chemicals or radiation.

In 1910, life expectancy for people with Down syndrome was only to age nine. Many of the medical problems that people with Down syndrome suffer are treatable, raising life expectancy to sixty years. New treatments or drugs repurposed to treat trisomy 21 Down syndrome may become available because researchers now have a way to study the condition in cells growing in laboratory glassware. Researchers use the gene that normally turns off one X chromosome in females (see the blue box on page 915) to turn off the third copy of chromosome 21 in stem cells made from skin fibroblasts of people with Down syndrome. This approach, called genome editing, may reveal how the abnormalities associated with trisomy 21 Down syndrome begin.

FIGURE 24A Many people with Down syndrome can learn, go to school, and hold jobs. This young lady is learning from a chef.

TABLE 24A	Risk of Trisomy 21 Increases with Maternal Age	
Maternal Age	Trisomy 21 Risk	Risk for Any Aneuploidy
20	1/1,667	1/526
24	1/1,250	1/476
28	1/1,053	1/435
30	1/952	1/385
32	1/769	1/322
35	1/378	1/192
36	1/289	1/156
37	1/224	1/127
38	1/173	1/102
40	1/106	1/66
45	1/30	1/21
48	1/14	1/10

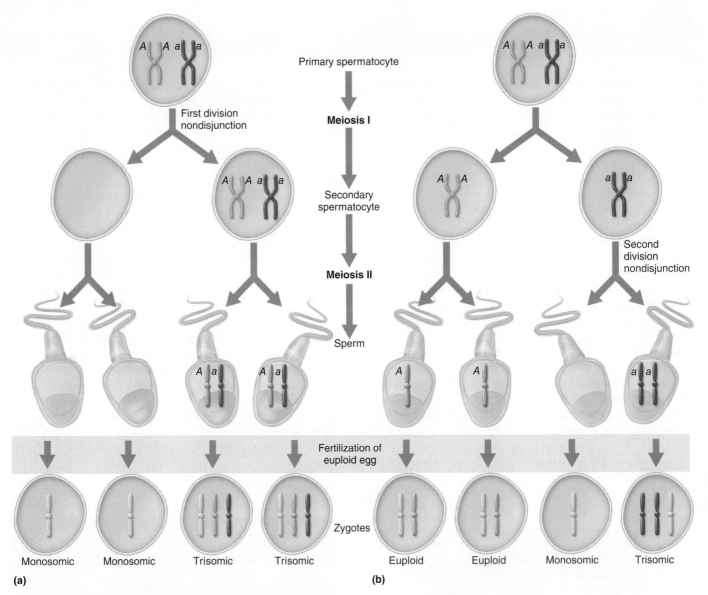

(a)

(b)

FIGURE 24.12 Extra or missing chromosomes constitute aneuploidy. Unequal division of chromosome pairs into sperm and egg cells can occur at either the first or the second meiotic division. (*a*) A single pair of chromosomes is unevenly partitioned into the two cells arising from the first division of meiosis in a male. The result: two sperm cells that have two copies of the chromosome and two sperm cells that have no copies of that chromosome. When a sperm cell with two copies of the chromosome fertilizes a normal egg cell, the zygote produced is trisomic for that chromosome; when a sperm cell lacking the chromosome fertilizes a normal egg cell, the zygote is monosomic for that chromosome. Symptoms depend upon which chromosome is involved. (*b*) This nondisjunction occurs at the second meiotic division, leaving two of the mature sperm normal, and two aneuploid. Egg cells can undergo nondisjunction as well, leading to zygotes with extra or missing chromosomes when they are fertilized by normal sperm cells.

13, 18, and 21 and that it is rarer still for an affected newborn to survive infancy. However, a few individuals with trisomy 18 have survived into their twenties. Trisomies of autosomes other than 13, 18, and 21 do not develop beyond the embryonic period.

Sex chromosome aneuploids are less severely affected than are autosomal aneuploids. XO syndrome (Turner syndrome) affects 1 in 2,000 newborn girls, but these represent only 1% of XO conceptions. The other 99% presumably do not complete prenatal development. Often the only symptom is delayed sexual development. With hormone supplements, life can be fairly normal, except for infertility.

About 1 in every 1,000 to 2,000 females has an extra X chromosome in each cell, which is called triplo-X. Often the only associated characteristics of triplo-X are great height and menstrual irregularities. Males with an extra X chromosome have XXY syndrome (Klinefelter syndrome). Like XO females, XXY males may not realize they have an unusual number of chromosomes until they have fertility problems and their chromosomes are checked. Associated characteristics are sexual underdevelopment (rudimentary testes and prostate glands and no pubic or facial hair), growth of breast tissue, long limbs, and large hands and feet. XXY syndrome affects 1 in every 500 to 2,000 male births.

One male in 1,000 has an extra Y chromosome, called XYY syndrome, or Jacobs syndrome. A fertilized ovum that has one Y chromosome and no X chromosome has never been observed. Presumably, when a zygote lacks an X chromosome, so much genetic material is missing that only a few, if any, cell divisions are possible.

Prenatal Tests Detect Chromosome Abnormalities

Several types of tests performed on pregnant women can identify anatomical or physiological features of fetuses that are often found with a specific chromosomal problem or detect the abnormal chromosomes (fig. 24.13). An ultrasound scan, for example, can reveal the fusion of the eyes, cleft lip and/or palate, malformed nose, and extra fingers and toes that indicate trisomy 13 (fig. 24.14). A blood test performed on the woman during the fifteenth week of pregnancy detects levels of certain biochemicals in serum. Levels of these maternal serum markers, which include alpha fetoprotein, an estrogen, pregnancy-associated plasma protein A, and human chorionic gonadotropin, can indicate an underdeveloped liver, which is associated with trisomies 13, 18, and 21.

Detecting abnormal maternal serum marker levels is indirect and imprecise, and so it is considered to be a screening rather than a diagnostic test. Maternal serum marker screening is gradually being replaced with analysis of cell-free fetal DNA. Pieces of fetal DNA are normally in a pregnant woman's bloodstream, and because these pieces are smaller than maternal DNA, they can be separated. If the ratio of number of pieces from one chromosome type is 50% greater than the number of pieces from the other chromosomes, then a trisomy is indicated. A fetus with trisomy 21, for example, would have 50% more DNA pieces in the maternal bloodstream from chromosome 21 compared to the number of pieces from the other chromosomes, representing the extra chromosome.

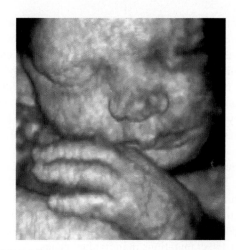

FIGURE 24.14 In an ultrasound exam, sound waves are bounced off the embryo or fetus, and the pattern of deflected sound waves is converted into an image. Ultrasound reveals the face of this healthy 13-week fetus.

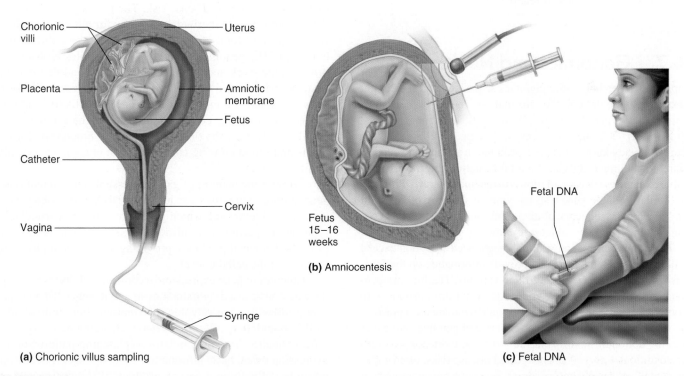

(a) Chorionic villus sampling

(b) Amniocentesis

(c) Fetal DNA

FIGURE 24.13 Three ways to check a fetus's chromosomes. (a) Chorionic villus sampling removes cells that would otherwise develop into the placenta. Because these cells descended from the fertilized ovum, they should have the same chromosomes as the fetus. (b) Amniocentesis draws out amniotic fluid. Fetal cells shed into the fluid are collected and their chromosomes examined. (c) Analysis of small pieces of cell-free fetal DNA in a pregnant woman's circulation allows comparison to detect extra or absent chromosomes. Fetal DNA is tested anytime after the tenth week of gestation, and may replace the more invasive tests.

After maternal serum marker screening and/or ultrasound indicates increased risk, the patient is offered either of two invasive tests, to provide a definitive diagnosis. In **chorionic villus sampling** (CVS), a physician collects chorionic villus cells through the cervix as early as 10 weeks. The basis of the test is that, theoretically, these cells are genetically identical to fetal cells because they too descend from the fertilized ovum. However, a mutation can occur in a villus cell only, or a fetal cell only, creating a false positive or false negative test result, respectively.

In **amniocentesis,** a needle inserted into the amniotic sac withdraws about 5 milliliters of fluid. Fetal fibroblasts in the sample are cultured and a karyotype constructed, which reveals extra, missing, or translocated chromosomes or smaller abnormalities. However, additional tests on amniotic fluid or fibroblast DNA are necessary to detect mutations in individual genes.

Cell-free fetal DNA testing may replace CVS and amniocentesis. It is also called non-invasive prenatal testing, or NIPT, but this designation also includes ultrasound scans. Researchers have sequenced genomes from fetuses using cell-free fetal DNA. Table 24.2 and figure 24.13 summarize the tests used to visualize fetal chromosomes as a window to health status.

 PRACTICE

19 Why do deviations from the normal chromosome number of 46 affect health?

20 Distinguish between polyploidy and aneuploidy.

21 How do extra sets of chromosomes or extra individual chromosomes arise?

22 How are fetal chromosomes examined?

24.7 | Genomics and Health Care

People may have their entire genomes sequenced, but they might not know what to do with the information. The situation is similar to owning a car. A warning light on the dashboard flashes, telling the driver to check the brakes, steering, or oil. Perhaps the motorist already knows there is a problem, from an odd sound or different handling of the car. Would it be helpful for the driver to be able to name every car part, down to the tiniest screw? Probably not. Similarly, single-gene tests or symptoms may be more useful than knowing the sequence of all 3.2 billion DNA bases that constitute a human genome, at least for now.

The number of human genomes sequenced will soon exceed a million, and their analysis will provide information on the functions of our 20,325 genes and their variants. Health-care professionals are learning how and under what circumstances to incorporate genetic and genomic information into medical practice. One impending challenge is that exome and genome sequences provide "actionable" information—that is, evidence of treatable conditions—as well as unexpected disease risks, often called "incidental" or "unsolicited" findings. The potential consequences of clinical exome and genome sequencing are a little like taking a car to a mechanic to fix one problem and finding another. However, exome and genome sequencing are already solving diagnostic mysteries, as Clinical Application 24.2 explains.

TABLE 24.2	Prenatal Tests		
Procedure	Time (Weeks)	Source	Information Provided
Ultrasound	Any time	Applied externally or through vagina	Growth rate, head size, size and location of organs
Maternal serum markers	15–16	Maternal blood	Small liver may indicate increased risk of trisomy
CVS	10–14	Chorionic villi	Karyotype of cell from chorionic villus
Amniocentesis	15 and beyond	Fetal cells in amniotic fluid	Karyotype of cell from fetus
Cell-free fetal DNA	10 weeks on	Maternal blood	Abnormal chromosome number

To test the practicality of whole genome sequencing, a forty-year-old biomedical engineer reported on his personal genome sequence in a medical journal. He sought sequencing because a young cousin had died of sudden cardiac arrest, and although the engineer's heart and blood vessels were healthy, he wanted to know if he shared any gene variants with his unfortunate cousin that might elevate his disease risk. The man's genome revealed that he indeed had inherited a tendency for atherosclerosis and inappropriate blood clotting that could cause a thrombus or embolism. His genome also revealed that cholesterol-lowering drugs would work and not have side effects, and that one antiplatelet drug would be effective but another would not be. This "personalized medicine" approach will likely enable the engineer to avoid his cousin's fate. But like the car analogy, he could have learned the risks from standard blood tests for lipid levels and blood clotting combined with clues from a detailed family history.

Considering patterns of gene expression is a different type of genetic information than a sequence of DNA bases. Gene expression means identifying which genes are active in particular cell types under particular conditions. This approach monitors the proteins that a cell produces, providing snapshots of physiology in action at the cellular level.

Changes in gene expression accompany all diseases, not just the rare ones caused by mutations in single genes. Gene expression profiles provide valuable information about individuals as well as populations. Within a person, changing gene expression reflects health. Among populations, changing patterns of gene expression reveal basic information about diseases. For example, white blood cells in a person produce different types of proteins when the body is fighting an infection than when it is not. Comparing people, a muscle cell from someone with diabetes mellitus expresses different genes than a muscle cell from a person whose blood glucose level is controlled.

The exome is the part of the genome that encodes protein. It is less than 2% of the genome—45 million of the 3.2 billion DNA bases—but includes 85% of the mutations known to affect health.

Knowing the exome sequence of a patient with an unusual set of symptoms can help a physician make a diagnosis by revealing what is abnormal at the molecular level. Consider the following examples.

A Boy with Dissolving Intestines

By the time he was four years old, the boy had already had 100 surgeries, including removal of his colon. His gastrointestinal tract had been riddled with holes, starting at age two with an abscess in his rectum. Feces leaked from his intestines. He was fed by tube, and weighed less than 20 pounds. Doctors thought he was suffering from inflammatory bowel disease, but his symptoms seemed much too severe.

Researchers at the medical center where the boy was being treated had received funding to perform one of the very first exome-sequencing experiments, planned for 2014. When the boy's doctors implored the researchers to help, the date was moved up to 2009, and the boy became the first person to receive a diagnosis based on the results of exome sequencing.

The procedure uncovered more than 16,000 variations in gene sequences, and narrowed those down to 32 "candidate genes." The boy's symptoms were explained by a mutation in only one of those genes—a gene that normally prevents cells of the gastrointestinal tract from dying during infection. The disease corresponding to the mutation was known, but as an immune system disorder, not a digestive one. It was treatable with a cord-blood stem cell transplant, but until the exome sequencing, nobody knew that this condition was what the boy had. He received stem cells from an anonymous donor, rapidly gained weight, and today is much improved.

In the Clinic

Each of the twelve children in a study conducted at a large medical center had developmental delay or intellectual disability, plus birth defects. None had a known inherited disease, or had been exposed to trauma or a toxin that might have caused the problems. Exome sequencing identified causative mutations for seven of the children. The information is practical. For example, one child's symptoms were due to a mutation in a gene that encodes a sodium channel protein, and the child would have reacted severely to certain drugs. The parents now know to avoid these drugs.

A larger study sequenced and analyzed hundreds of exomes from patients whom doctors had been unable to diagnose, and found answers for a fourth of them. The researchers had to narrow down 400 to 700 gene variants per patient! A two-year-old boy who had severe feeding difficulties and poor weight gain had a mutation known to cause Marfan syndrome, although his symptoms were atypical. One symptom of Marfan syndrome is an enlarged aorta, the largest artery. An ultrasound revealed that the boy indeed had a bulge in the wall of the aorta near the heart, which could have caused sudden death. He was successfully treated.

As the cost of exome sequencing falls, it may become a standard part of diagnostic medicine. Studies are under way to sequence exomes to better understand the physiology behind many diseases.

Analyzing gene expression for the whole exome (the protein-encoding part of the genome) can suggest new ways to treat diseases by identifying biochemical changes that traditional tests do not reveal. Such a discovery sometimes leads to a new use for an existing drug. For example, statin drugs used to lower serum cholesterol level are being tested on people with or at high risk of developing Alzheimer disease, because people with heart disease due to high serum cholesterol share gene expression patterns with people who have Alzheimer disease.

Physicians are increasingly using tests of gene expression to select or track responses to treatment. In an approach called *pharmacogenetics*, tests based on gene expression profiling predict which drugs will be most effective and least likely to cause adverse effects in a particular patient. Physicians may use this approach to select anticoagulants, antidepressants, cancer medications, and cholesterol-lowering drugs. This is a more personalized approach to medicine than trying one drug at a time, based on the fact that each drug helps some people.

Overall, examining our genes and what they do provides a new type of window into the functioning, and malfunctioning, of the human body. In the coming years, as more individual genomes are sequenced and compared, detailed portraits of the living chemistry within our cells, and the chemical crosstalk among cells, will enhance our view of anatomy and physiology.

 RECONNECT

To Chapter 4, Nucleic Acids and Protein Synthesis, pages 132–142.

 PRACTICE

23 State the benefits and limitations of knowing one's genome sequence.

24 Explain how gene expression profiling provides information about physiology. ∎

Chapter Summary

24.1 Genes And Genomes (page 906)

1. **Genetics** is the study of inheritance of characteristics and human variation.
2. **Genes** are DNA sequences that encode proteins.
3. Genetic information is passed from generation to generation through meiosis and fertilization, when the haploid genomes of the parents join.
4. The protein-encoding portion of the genome is small and called the exome.
5. The human genome assembles many more proteins than there are genes by combining gene parts.
6. Genetic information functions at the biochemical, cell, tissue, individual, family, and population levels.
7. Genes influence each other and their expression responds to environmental influences.

24.2 Modes of Inheritance (page 908)

1. Chromosomes and genes are paired
   a. Chromosome charts are called **karyotypes.**
   b. Chromosomes 1 through 22, numbered in decreasing size order, are **autosomes.** They do not have genes that determine sex.
   c. The X and Y chromosomes are **sex chromosomes.** They have genes that determine sex.
   d. An **allele** is an alternate form (variant) of a gene. A person can have a maximum of two different alleles for a gene, but the gene can have many alleles, because a gene consists of many building blocks, any of which may be altered.
   e. A person with a pair of identical alleles for a particular gene is **homozygous;** if the alleles are different, the individual is **heterozygous.**
   f. The combination of gene variants in an individual's cells constitutes the **genotype;** the appearance or traits of the individual is the **phenotype.**
   g. A **wild type** allele provides normal function or the most common expression. A **mutant** allele causes disease or an unusual trait; it is a change from the wild type condition.
2. Dominant and recessive inheritance
   a. In a heterozygote, an allele expressed when the other is not expressed is **dominant.** The masked allele is **recessive.**
   b. Recessive and dominant genes may be autosomal or X-linked or Y-linked.
   c. An autosomal recessive condition affects both sexes and may skip generations. The homozygous dominant and heterozygous individuals have normal phenotypes. An affected individual inherits one mutant allele from each parent. The mutant alleles may be different.
   d. An autosomal dominant condition affects both sexes and does not skip generations. A person inherits it from one affected parent.
   e. **Punnett squares** and **pedigrees** depict modes of inheritance.
3. Different dominance relationships
   a. In incomplete dominance, a heterozygote has a phenotype intermediate between those of both homozygotes.
   b. In codominance, each allele in the heterozygote is expressed.

24.3 Factors That Affect Expression of Single Genes (page 912)

1. Penetrance and expressivity
   a. A genotype is **incompletely penetrant** if not all individuals inheriting it express the phenotype.
   b. A genotype is **variably expressive** if it is expressed to different degrees in different individuals.
2. Pleiotropy
   a. In **pleiotropy,** a disorder has several symptoms, different subsets of which may affect individuals.
   b. Pleiotropy results from a gene product that is part of more than one biochemical reaction or is in several organs or structures.
3. Genetic heterogeneity
   a. **Genetic heterogeneity** refers to a phenotype resulting from mutation in more than one gene.
   b. The same symptoms may result from mutations in genes whose products are enzymes in the same biochemical pathway.

24.4 Multifactorial Traits (page 912)

1. A trait caused by the action of one or more genes and the environment is **multifactorial.**
2. A trait caused by the action of a single gene is monogenic, and by the action of more than one gene, **polygenic.**
3. Height, skin color, eye color, and many common illnesses are multifactorial traits.
4. A frequency distribution for a polygenic trait forms a bell curve.

24.5 Matters of Sex (page 914)

A female has two X chromosomes; a male has one X and one Y chromosome. The X chromosome has many more genes than the Y.

1. Sex determination
   a. A male zygote forms when a Y-bearing sperm fertilizes an egg. A female zygote forms when an X-bearing sperm fertilizes an egg.
   b. A gene on the Y chromosome, *SRY,* switches on genes in the embryo that promote development of male characteristics.
   c. Lack of *SRY* and activation of *Wnt4* direct development as female.
2. Genes on the sex chromosomes
   a. Genes on the sex chromosomes follow different inheritance patterns than those on autosomes.
   b. Y-linked genes are in three functional groups: those with counterparts on the X; those similar to genes on the X; and genes unique to the Y, many of which affect male fertility. Y-linked genes pass from fathers to sons.
   c. Males are **hemizygous** for X-linked traits; they can have only one copy of an X-linked gene, because they have only one X chromosome.
   d. Females can be heterozygous or homozygous for genes on the X chromosome, because they have two copies of it.
   e. A male inherits an X-linked trait from a carrier mother. These traits are more common in males than in females.
   f. A female inherits an X-linked mutant gene from her carrier mother and/or from her father if he is well enough to have children.
   g. Dominant X-linked traits are typically lethal to males before birth.

3. Gender effects on phenotype
   a. **Sex-limited** traits affect structures or functions seen in only one sex and may be autosomal.
   b. **Sex-influenced** traits are dominant in one sex and recessive in the other.
   c. In **genomic imprinting,** the severity, age of onset, or nature of symptoms varies according to which parent transmits a mutation.

24.6 Chromosome Disorders (page 916)

Extra, missing, or rearranged chromosomes or parts of them can cause syndromes, because they alter the balance of genetic material or disrupt a vital gene.

1. Polyploidy
   a. **Polyploidy** is an extra chromosome set.
   b. Polyploidy results from fertilization with or by a diploid gamete.
   c. Polyploids do not survive long.
2. Aneuploidy
   a. Cells with the normal chromosome number are **euploid.** Cells with an extra or missing chromosome are **aneuploid.**
   b. Aneuploidy results from **nondisjunction,** in which a chromosome pair does not separate, in meiosis I or meiosis II, producing a gamete with a missing or extra chromosome. At fertilization, a **monosomy** or **trisomy** results.
   c. A cell with an extra chromosome is trisomic. A cell with a missing chromosome is monosomic. Individuals with trisomies are more likely to survive to be born than those with monosomies.
   d. Autosomal aneuploids are more severe than sex chromosome aneuploids.
3. Prenatal tests detect chromosome abnormalities
   a. Ultrasound can detect large-scale structural abnormalities and assess growth.
   b. Maternal serum marker tests indirectly detect a small fetal liver, which can indicate a trisomy.
   c. **Chorionic villus sampling** obtains and examines chorionic villus cells, which descend from the fertilized egg and therefore are presumed to be genetically identical to fetal cells.
   d. **Amniocentesis** samples and examines fetal chromosomes from fetal cells in amniotic fluid.
   e. Cell-free fetal DNA testing detects aneuploids from fetal DNA in the maternal circulation.

24.7 Genomics and Health Care (page 920)

1. People can now have their entire genomes sequenced, but focusing on specific genes may be more useful.
2. Exome and genome sequencing reveal actionable as well as unsolicited findings.
3. Comparing profiles of gene expression reveals aspects of physiology and makes possible more personalized treatments for disease.

 # CHAPTER ASSESSMENTS

24.1 Genes and Genomes

1 Which choice places the structures in order of increasing size? (pp. 906–907)
   a. genome, chromosome, gene, DNA base
   b. DNA base, gene, exome, genome
   c. gene, DNA base, genome, chromosome
   d. population, family, individual, organ, tissue, cell, DNA
2 Discuss the origin of the 23 chromosome pairs in a diploid human cell. (p. 906)
3 Explain how a certain number of genes hold enough information to encode a greater number of proteins. (p. 907)
4 Explain how genes can respond to environmental factors. (p. 907)

24.2 Modes of Inheritance

5 Explain the two factors considered in determining mode of inheritance. (p. 908)
6 Which is a chromosome chart? (p. 908)
   a. karyotype
   b. pedigree
   c. Punnett square
   d. genotype
7 Distinguish between
   a. autosome and sex chromosome
   b. homozygous and heterozygous
   c. phenotype and genotype
   d. mutant and wild type
   e. dominant and recessive
   f. incomplete dominance and codominance. (pp. 908–911)
8 Explain how a gene might have hundreds of alleles. (p. 909)

24.3 Factors That Affect Expression of Single Genes

9 Explain the distinction between penetrance and expressivity. (p. 912)
10 A single gene disorder that produces several symptoms is _____. (p. 912)
11 A single syndrome that has more than one genetic cause exhibits _____. (p. 912)
   a. pleiotropy
   b. incomplete penetrance
   c. genetic heterogeneity
   d. variable expressivity

24.4 Multifactorial Traits

12 Define *multifactorial trait.* (p. 912)
13 List three multifactorial traits. (p. 913)
14 Explain why the frequency distributions of different multifactorial traits give very similar bell curves. (p. 913)
15 Give an example of how the environment can influence a multifactorial trait. (p. 913)

24.5 Matters of Sex

16 Explain how genes and chromosomes determine sex. (pp. 914–915)
17 Explain why Y-linked traits are passed only from fathers to sons. (p. 915)
18 Explain why the inheritance pattern of X-linked traits differs in males and females. (p. 915)
19 Distinguish between a sex-limited and a sex-influenced trait. (p. 916)
20 Define *genomic imprinting.* (p. 916)

24.6 Chromosome Disorders

21 State whether trisomy 21 Down syndrome is euploid, aneuploid, or polyploid. (p. 916)

22 In nondisjunction _____. (p. 916)
   a. multiple sets of chromosomes are inherited
   b. gametes receive only even-numbered chromosomes or only odd-numbered chromosomes
   c. a chromosome pair does not separate during meiosis
   d. two eggs fertilize each other

23 Describe three types of prenatal tests. (pp. 919–920)

24.7 Genomics and Health Care

24 Discuss the benefits and limitations of whole genome sequencing. (p. 920)

25 Distinguish the types of information provided in an exome or genome sequence and in a gene expression profile. (p. 920)

 # INTEGRATIVE ASSESSMENTS/CRITICAL THINKING

Outcomes 4.6, 24.1, 24.2

1. Bob and Joan know from a blood test that they are each heterozygous (carriers) for the autosomal recessive gene that causes sickle cell disease. If their first three children are healthy, what is the probability that their fourth child will have the disease?

Outcomes 6.1, 24.4

2. Why are medium-brown skin colors more common than very fair or very dark skin?

Outcomes 6.2, 24.2, 24.3

3. A balding man undergoes a treatment that transfers some of the hair from the sides of his head, where it is still plentiful, to the top. Is he altering his phenotype or his genotype?

Outcomes 24.1, 24.2

4. A young couple is devastated when their second child is diagnosed with CF. Their older child is healthy, and no one else in the family has CF. How is this possible?

Outcomes 24.1, 24.2, 24.3

5. CF is an autosomal recessive disorder that varies widely in phenotype, from great difficulty in breathing and digesting, to frequent bouts of sinusitis, bronchitis, and/or pneumonia. In the Maxwell family, the mother, Matilda, has a mild case—she often suffers from respiratory infections. Her husband Jake is a heterozygote, and testing reveals that he carries an allele associated with severe disease. Of the four Maxwell children, Katie and Jim have fairly severe disease, although Jim has to be hospitalized more frequently than Katie. The other two children, Emily and Rose, do not have CF.
   a. Explain how Katie and Jim can have severe cases of CF when one parent does not have the disease and the other has a mild case.
   b. Explain how Katie and Jim have different severities of the illness.
   c. What is the probability that Emily or Rose is a carrier, like their father?

Outcomes 24.4, 24.7

6. Describe a genetic test or technology that can predict whether a person who smokes is likely to develop lung cancer.

Outcome 24.5

7. In Hunter syndrome, lack of an enzyme leads to build up of sticky carbohydrates in the liver, spleen, and heart. The individual is also deaf and has unusual facial features. Hunter syndrome is inherited as an X-linked recessive condition. Intellect is usually unimpaired, and life span can be normal. A man who has mild Hunter syndrome has a child with a woman who is a carrier (heterozygote).
   a. What is the probability that a son inherits the syndrome?
   b. What is the chance that a daughter inherits it?
   c. What is the chance that a girl would be a carrier?

Outcome 24.7

8. Would you have your own genome sequenced? Cite a reason for your answer.

 # ONLINE STUDY TOOLS

Connect Interactive Questions Reinforce your knowledge using assigned interactive questions covering inheritance and gene expression.

Connect Integrated Activity Can you assist a genetic counselor in explaining the inheritance of various inherited disorders?

LearnSmart Discover which chapter concepts you have mastered and which require more attention. This adaptive learning tool is personalized, proven, and preferred.

APPENDIX A

Scientific Method

Our knowledge of how the human body works is based on centuries of careful observation and experimentation. A way of thinking and organizing information, called the **scientific method**, guides the acquisition and interpretation of new information about the human body. The scientific method is used to either support or challenge a scientific theory, which is a systematically organized body of knowledge that applies to a variety of situations. It is a key part of a general process known as *scientific inquiry*, which includes other ways of investigating the natural world. The scientific method is a framework in which to consider ideas and evidence that involves use of familiar skills of observing, questioning, reasoning, predicting, testing, interpreting, and concluding.

The scientific method provides a sequence of logical steps that address a specific question. A proposed answer to the question is stated as a *hypothesis*, or "educated guess." More than a hunch, a hypothesis is based on existing knowledge and is often phrased in the form "if this is true, then that must follow." It is specific and testable, and should examine only one changeable factor, or *variable*. Even though the hypothesis may seem attractive and may make sense, it is not accepted until sufficient experimental evidence supports it. Indeed, it may be proven wrong. Experimental findings that go against the hypothesis are just as important as those that support it.

The process of drug development illustrates the use of the scientific method in anatomy and physiology. Consider a hypothesis to test a new cancer treatment:

> *If a drug blocks the receptor proteins to which a growth factor binds on cancer cells, then the cells may stop dividing and halt progression of the cancer.*

This hypothesis is based on past experiments that identified the specific receptors and growth factors that fuel a particular type of cancer and described how they interact, and observations that the cells of certain cancers have excess growth factor receptors compared to healthy cells. Using that knowledge and the hypothesis, researchers design an experiment (clinical trial), which is a test that yields information:

> *One group of patients receives the experimental treatment (new drug) and a second group (the control) receives an existing treatment.*

Experiments are carefully planned to make the results as meaningful as possible. In drug development, the control group provides a point of comparison for the new treatment, because the participants

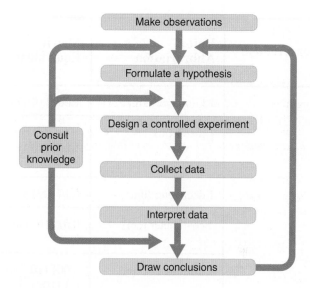

are selected to be as alike as possible. In this way, differences due to the drug can stand out. Studies should also be "double blinded" whenever possible, which means that neither participants nor researchers know which patients receive which treatment until the end of the study. Outcomes are specific. At the end of six months, if significantly more of the patients taking the experimental drug show measurable signs of improvement compared to the control group, then further study is warranted. Repeating scientific findings is essential to confirming experimental results. Before a new drug is prescribed for patients, it is tested on thousands of individuals in multiple clinical trials.

The scientific method is not a linear process, but a cycle, because results and conclusions often suggest new questions to explore. Something about the new cancer drug, for example, may suggest how it could be "repurposed" to treat a different condition. Testing begins anew.

In scientific investigations, so-called "negative" results are just as important as results that support the hypothesis. If the drug being tested doesn't work, researchers can either modify the hypothesis (perhaps the drug only works on certain subtypes of the disease) or develop more effective versions of the drug. Medical journals provide diverse examples of how the scientific method lies behind the drugs and procedures that have vastly improved the lives of many of us.

APPENDIX B

Metric Measurement System and Conversions

Measurement	Unit & Abbreviation	Metric Equivalent	Conversion Factor Metric to English (approximate)	Conversion Factor English to Metric (approximate)
Length	1 kilometer (km)	1,000 (10^3) m	1 km = 0.62 mile	1 mile = 1.61 km
	1 meter (m)	100 (10^2) cm 1,000 (10^3) mm	1 m = 1.1 yards = 3.3 feet = 39.4 inches	1 yard = 0.9 m 1 foot = 0.3 m
	1 decimeter (dm)	0.1 (10^{-1}) m	1 dm = 3.94 inches	1 inch = 0.25 dm
	1 centimeter (cm)	0.01 (10^{-2}) m	1 cm = 0.4 inches	1 foot = 30.5 cm 1 inch = 2.54 cm
	1 millimeter (mm)	0.001 (10^{-3}) m 0.1 (10^{-1}) cm	1 mm = 0.04 inches	
	1 micrometer (μm)	0.000001 (10^{-6}) m 0.001 (10^{-3}) mm		
Mass	1 metric ton (t)	1,000 (10^3) kg	1 t = 1.1 ton	1 ton = 0.91 t
	1 kilogram (kg)	1,000 (10^3) g	1 kg = 2.2 pounds	1 pound = 0.45 kg
	1 gram (g)	1,000 (10^3) mg	1 g = 0.04 ounce	1 pound = 454 g 1 ounce = 28.35 g
	1 milligram (mg)	0.001 (10^{-3}) g		
	1 microgram (μg)	0.000001 (10^{-6}) g		
Volume (liquids and gases)	1 liter (L)	1,000 (10^3) mL	1 L = 1.06 quarts	1 gallon = 3.78 L 1 quart = 0.95 L
	1 milliliter (mL)	0.001 (10^{-3}) L 1 cubic centimeter (cc or cm³)	1 mL = 0.03 fluid ounce 1 mL = 1/5 teaspoon 1 mL = 15–16 drops	1 quart = 946 mL 1 fluid ounce = 29.6 mL 1 teaspoon = 5 mL
Time	1 second (s)	1/60 minute	same	same
	1 millisecond (ms)	0.001 (10^{-3}) s	same	same
Temperature	Degrees Celsius (°C)		°F = 9/5°C + 32	°C = 5/9 (°F − 32)

APPENDIX C

Laboratory Tests of Clinical Importance

Common Tests Performed on Blood

Test	Normal Values* (Adult)	Clinical Significance
Albumin (serum)	3.2–5.5 g/100 mL	Values increase in multiple myeloma and decrease with proteinuria and as a result of severe burns.
Albumin-globulin ratio, or A/G ratio (serum)	1.5:1 to 2.5:1	Ratio of albumin to globulin is lowered in kidney diseases and malnutrition.
Ammonia	80–110 mg/100 mL (12–55 µ mol/L)	Values increase in severe liver disease, pneumonia, shock, and congestive heart failure.
Amylase (serum)	4–25 units/mL	Values increase in acute pancreatitis, intestinal obstructions, and mumps. They decrease in chronic pancreatitis, cirrhosis of the liver, and toxemia of pregnancy.
Bilirubin, total (serum)	0–1.0 mg/100 mL	Values increase in conditions causing red blood cell destruction or biliary obstruction.
Blood urea nitrogen, or BUN (plasma or serum)	8–25 mg/100 mL (2.5–9.3 mmol/L)	Values increase in various kidney disorders and decrease in liver failure and during pregnancy.
Calcium (serum)	8.5–10.5 mg/100 mL	Values increase in hyperparathyroidism, hypervitaminosis D, and respiratory conditions that cause a rise in CO_2 concentration. They decrease in hypoparathyroidism, malnutrition, and severe diarrhea.
Carbon dioxide (serum)	24–30 mEq/L	Values increase in respiratory diseases, intestinal obstruction, and vomiting. They decrease in acidosis, nephritis, and diarrhea.
Chloride (serum)	100–106 mEq/L	Values increase in nephritis, Cushing syndrome, dehydration, and hyperventilation. They decrease in metabolic acidosis, Addison disease, diarrhea, and following severe burns.
Cholesterol, total (serum)	120–220 mg/100 mL (below 200 mg/100 mL recommended by the American Heart Association)	Values increase in diabetes mellitus and hypothyroidism. They decrease in pernicious anemia, hyperthyroidism, and acute infections.
Cholesterol, high-density lipoprotein (HDL)	Men: 30–70 mg/100 mL Women: 30–80 mg/100 mL	Values increase in liver disease. Decreased values are associated with an increased risk of atherosclerosis.
Cholesterol, low-density lipoprotein (LDL)	62–185 mg/100 mL	Increased values are associated with an increased risk of atherosclerosis.
Creatine (serum)	0.2–0.8 mg/100 mL	Values increase in muscular dystrophy, nephritis, severe damage to muscle tissue, and during pregnancy.
Creatinine (serum)	0.6–1.5 mg/100 mL	Values increase in various kidney diseases.
Ferritin (serum)	Men: 10–270 mg/100 mL Women: 5–280 mg/100 mL	Values correlate with total body iron store. They decrease with iron deficiency.
Globulin (serum)	2.3–3.5 g/100 mL	Values increase as a result of chronic infections.
Glucose (plasma)	70–110 mg/100 mL	Values increase in diabetes mellitus, liver diseases, nephritis, hyperthyroidism, and pregnancy. They decrease in hyperinsulinism, hypothyroidism, and Addison disease.
Hematocrit (whole blood)	Men: 40–54% Women: 37–47% Children: 35–49% (varies with age)	Values increase in polycythemia due to dehydration or shock. They decrease in anemia and following severe hemorrhage.
Hemoglobin (whole blood)	Men: 14–18 g/100 mL Women: 12–16 g/100 mL Children: 11.2–16.5 g/100 mL (varies with age)	Values increase in polycythemia, obstructive pulmonary diseases, congestive heart failure, and at high altitudes. They decrease in anemia, pregnancy, and as a result of severe hemorrhage or excessive fluid intake.
Iron (serum)	50–150 µg/100 mL	Values increase in various anemias and liver disease. They decrease in iron-deficiency anemia.

*These values may vary with hospital, physician, and type of equipment used to make measurements.

Test	Normal Values* (Adult)	Clinical Significance
Iron-binding capacity (serum)	250–410 µg/100 mL	Values increase in iron-deficiency anemia and pregnancy. They decrease in pernicious anemia, liver disease, and chronic infections.
Lactic acid (whole blood)	0.6–1.8 mEq/L	Values increase with muscular activity and in congestive heart failure, severe hemorrhage, and shock.
Lactic dehydrogenase, or LDH (serum)	70–200 U/L	Values increase in pernicious anemia, myocardial infarction, liver disease, acute leukemia, and widespread carcinoma.
Lipids, total (serum)	450–850 mg/100 mL	Values increase in hypothyroidism, diabetes mellitus, and nephritis. They decrease in hyperthyroidism.
Magnesium	1.3–2.1 mEq/L	Values increase in renal failure, hypothyroidism, and Addison disease. They decrease in renal disease, liver disease, and pancreatitis.
Mean corpuscular hemoglobin (MCH)	26–32 pg/RBC	Values increase in macrocytic anemia. They decrease in microcytic anemia.
Mean corpuscular volume (MCV)	86–98 µ mm^3/RBC	Values increase in liver disease and pernicious anemia. They decrease in iron-deficiency anemia.
Osmolality	275–295 mOsm/kg	Values increase in dehydration, hypercalcemia, and diabetes mellitus. They decrease in hyponatremia, Addison disease, and water intoxication.
Oxygen saturation (whole blood)	Arterial: 96–100% Venous: 60–85%	Values increase in polycythemia and decrease in anemia and obstructive pulmonary diseases.
P$_{CO_2}$	35–45 mmHg	Values increase in respiratory diseases, intestinal obstruction, and vomiting. They decrease in acidosis, nephritis, and diarrhea.
pH (whole blood)	7.35–7.45	Values increase due to mild vomiting, Cushing syndrome, and hyperventilation. They decrease as a result of hypoventilation, severe diarrhea, Addison disease, and diabetic acidosis.
P$_{O_2}$	75–100 mmHg	Values increase in polycythemia. They decrease in anemia and obstructive pulmonary diseases.
Phosphatase acid (serum)	Men: 0.13–0.63 Sigma U/mL Women: 0.01–0.56 Sigma U/mL	Values increase in cancer of the prostate gland, hyperparathyroidism, certain liver diseases, myocardial infarction, and pulmonary embolism.
Phosphatase, alkaline (serum)	Adults: 13–39 U/L Children: up to 104 U/L	Values increase in hyperparathyroidism (and in other conditions that promote resorption of bone), liver diseases, and pregnancy.
Phosphorus (serum)	3.0–4.5 mg/100 mL	Values increase in kidney diseases, hypoparathyroidism, acromegaly, and hypervitaminosis D. They decrease in hyperparathyroidism.
Platelet count (whole blood)	150,000–350,000/µL	Values increase in polycythemia and certain anemias. They decrease in acute leukemia and aplastic anemia.
Potassium (serum)	3.5–5.0 mEq/L	Values increase in Addison disease, hypoventilation, and conditions that cause severe cellular destruction. They decrease in diarrhea, vomiting, diabetic acidosis, and chronic kidney disease.
Protein, total (serum)	6.0–8.4 g/100 mL	Values increase in severe dehydration and shock. They decrease in severe malnutrition and hemorrhage.
Prothrombin time (serum)	12–14 sec (one stage)	Values increase in certain hemorrhagic diseases, liver disease, vitamin K deficiency, and following the use of various drugs.
Red cell count (whole blood)	Men: 4,700,000–6,100,000/µL Women: 4,200,000–5,400,000/µL Children: 4,500,000–5,100,000/µL (varies with age)	Values increase as a result of severe dehydration or diarrhea, and decrease in anemia, leukemia, and following severe hemorrhage.
Red cell distribution width (RDW)	8.5–11.5 microns	Variation in cell width changes with pernicious anemia.
Sedimentation rate, erythrocyte (whole blood)	Men: 1–13 mm/hr Women: 1–20 mm/hr	Values increase in infectious diseases, menstruation, pregnancy, and as a result of severe tissue damage.
Serum glutamic pyruvic transaminase (SGPT)	Men: 6–24 U/L Women: 4–17 U/L	Values increase in liver disease, pancreatitis, and acute myocardial infarction.

*These values may vary with hospital, physician, and type of equipment used to make measurements.

Common Tests Performed on Blood—*continued*

Test	Normal Values* (Adult)	Clinical Significance
Sodium (serum)	135–145 mEq/L	Values increase in nephritis and severe dehydration. They decrease in Addison disease, myxedema, kidney disease, and diarrhea.
Thromboplastin time, partial (plasma)	35–45 sec	Values increase in deficiencies of blood factors VIII, IX, and X.
Thyroid-stimulating hormone (TSH)	0.5–5.0 µU/mL	Values increase in hypothyroidism and decrease in hyperthyroidism.
Thyroxine, or T_4 (serum)	4–12 µg/100 mL	Values increase in hyperthyroidism and pregnancy. They decrease in hypothyroidism.
Transaminases, or SGOT (serum)	7–27 units/mL	Values increase in myocardial infarction, liver disease, and diseases of skeletal muscles.
Triglycerides	40–150 mg/100 mL	Values increase in liver disease, nephrotic syndrome, hypothyroidism, and pancreatitis. They decrease in malnutrition and hyperthyroidism.
Triiodothyronine, or T_3 (serum)	75–195 ng/100 mL	Values increase in hyperthyroidism and decrease in hypothyroidism.
Uric acid (serum)	Men: 2.5–8.0 mg/100 mL Women: 1.5–6.0 mg/100 mL	Values increase in gout, leukemia, pneumonia, toxemia of pregnancy, and as a result of severe tissue damage.
White blood cell count, differential (whole blood)	Neutrophils 54–62% Eosinophils 1–3% Basophils <1% Lymphocytes 25–33% Monocytes 3–7%	Neutrophils increase in bacterial diseases; lymphocytes and monocytes increase in viral diseases; eosinophils increase in collagen diseases, allergies, and in the presence of intestinal parasites.
White blood cell count, total (whole blood)	3,500–10,500/µL	Values increase in acute infections, acute leukemia, and following menstruation. They decrease in aplastic anemia and as a result of drug toxicity.

*These values may vary with hospital, physician, and type of equipment used to make measurements.

Common Tests Performed on Urine

Test	Normal Values* (Adult)	Clinical Significance
Acetone and acetoacetate	0	Values increase in diabetic acidosis.
Albumin, qualitative	0 to trace	Values increase in kidney disease, hypertension, and heart failure.
Ammonia	20–70 mEq/L	Values increase in diabetes mellitus and liver diseases.
Bacterial count	Under 10,000/mL	Values increase in urinary tract infection.
Bile and bilirubin	0	Values increase in melanoma and biliary tract obstruction.
Calcium	Under 300 mg/24 hr	Values increase in hyperparathyroidism and decrease in hypoparathyroidism.
Creatinine (24 hours)	15–25 mg/kg body weight/day	Values increase in infections, and decrease in muscular atrophy, anemia, leukemia, and kidney diseases.
Creatinine clearance (24 hours)	100–140 mL/min	Values increase in renal diseases.
Glucose	0	Values increase in diabetes mellitus and various pituitary gland disorders.
Hemoglobin	0	Blood may occur in urine as a result of extensive burns, crushing injuries, hemolytic anemia, or blood transfusion reactions.
17-hydroxycorticosteroids	3–8 mg/24 hr	Values increase in Cushing syndrome and decrease in Addison disease.
Osmolality	850 mOsm/kg	Values increase in hepatic cirrhosis, congestive heart failure, and Addison disease. They decrease in hypokalemia, hypercalcemia, and diabetes insipidus.
pH	4.6–8.0	Values increase in urinary tract infections and chronic renal failure. They decrease in diabetes mellitus, emphysema, and starvation.
Phenylpyruvic acid	0	Values increase in phenylketonuria.
Specific gravity (SG)	1.003–1.035	Values increase in diabetes mellitus, nephrosis, and dehydration. They decrease in diabetes insipidus, glomerulonephritis, and severe renal injury.
Urea	25–35 g/24 hr	Values increase as a result of excessive protein breakdown. They decrease as a result of impaired renal function.
Urea clearance	Over 40 mL blood cleared of urea/min	Values increase in renal diseases.
Uric acid	0.6–1.0 g/24 hr as urate	Values increase in gout and decrease in various kidney diseases.
Urobilinogen	0–4 mg/24 hr	Values increase in liver diseases and hemolytic anemia. They decrease in complete biliary obstruction and severe diarrhea.

*These values may vary with hospital, physician, and type of equipment used to make measurements.

APPENDIX D

Periodic Table of the Elements

Legend:
- 9 / F / Fluorine / 19.00 — with labels: Atomic number (9), Atomic mass (19.00)

1 / 1A																	18 / 8A
1 **H** Hydrogen 1.008	2 / 2A											13 / 3A	14 / 4A	15 / 5A	16 / 6A	17 / 7A	2 **He** Helium 4.003
3 **Li** Lithium 6.941	4 **Be** Beryllium 9.012											5 **B** Boron 10.81	6 **C** Carbon 12.01	7 **N** Nitrogen 14.01	8 **O** Oxygen 16.00	9 **F** Fluorine 19.00	10 **Ne** Neon 20.18
11 **Na** Sodium 22.99	12 **Mg** Magnesium 24.31	3 / 3B	4 / 4B	5 / 5B	6 / 6B	7 / 7B	8 / 8B	9 / 8B	10 / 8B	11 / 1B	12 / 2B	13 **Al** Aluminum 26.98	14 **Si** Silicon 28.09	15 **P** Phosphorus 30.97	16 **S** Sulfur 32.07	17 **Cl** Chlorine 35.45	18 **Ar** Argon 39.95
19 **K** Potassium 39.10	20 **Ca** Calcium 40.08	21 **Sc** Scandium 44.96	22 **Ti** Titanium 47.88	23 **V** Vanadium 50.94	24 **Cr** Chromium 52.00	25 **Mn** Manganese 54.94	26 **Fe** Iron 55.85	27 **Co** Cobalt 58.93	28 **Ni** Nickel 58.69	29 **Cu** Copper 63.55	30 **Zn** Zinc 65.39	31 **Ga** Gallium 69.72	32 **Ge** Germanium 72.59	33 **As** Arsenic 74.92	34 **Se** Selenium 78.96	35 **Br** Bromine 79.90	36 **Kr** Krypton 83.80
37 **Rb** Rubidium 85.47	38 **Sr** Strontium 87.62	39 **Y** Yttrium 88.91	40 **Zr** Zirconium 91.22	41 **Nb** Niobium 92.91	42 **Mo** Molybdenum 95.94	43 **Tc** Technetium (98)	44 **Ru** Ruthenium 101.1	45 **Rh** Rhodium 102.9	46 **Pd** Palladium 106.4	47 **Ag** Silver 107.9	48 **Cd** Cadmium 112.4	49 **In** Indium 114.8	50 **Sn** Tin 118.7	51 **Sb** Antimony 121.8	52 **Te** Tellurium 127.6	53 **I** Iodine 126.9	54 **Xe** Xenon 131.3
55 **Cs** Cesium 132.9	56 **Ba** Barium 137.3	57 **La** Lanthanum 138.9	72 **Hf** Hafnium 178.5	73 **Ta** Tantalum 180.9	74 **W** Tungsten 183.9	75 **Re** Rhenium 186.2	76 **Os** Osmium 190.2	77 **Ir** Iridium 192.2	78 **Pt** Platinum 195.1	79 **Au** Gold 197.0	80 **Hg** Mercury 200.6	81 **Tl** Thallium 204.4	82 **Pb** Lead 207.2	83 **Bi** Bismuth 209.0	84 **Po** Polonium (210)	85 **At** Astatine (210)	86 **Rn** Radon (222)
87 **Fr** Francium (223)	88 **Ra** Radium (226)	89 **Ac** Actinium (227)	104 **Rf** Rutherfordium (257)	105 **Db** Dubnium (260)	106 **Sg** Seaborgium (263)	107 **Bh** Bohrium (262)	108 **Hs** Hassium (265)	109 **Mt** Meitnerium (266)	110 **Ds** Darmstadtium (269)	111 **Rg** Roentgenium (272)	112 **Cn** Copernicium	(113)	114 **Fl** Flerovium	(115)	116 **Lv** Livermorium	(117)	(118)

Key:
- Metals
- Metalloids
- Nonmetals

Lanthanides and Actinides:

58 **Ce** Cerium 140.1	59 **Pr** Praseodymium 140.9	60 **Nd** Neodymium 144.2	61 **Pm** Promethium (147)	62 **Sm** Samarium 150.4	63 **Eu** Europium 152.0	64 **Gd** Gadolinium 157.3	65 **Tb** Terbium 158.9	66 **Dy** Dysprosium 162.5	67 **Ho** Holmium 164.9	68 **Er** Erbium 167.3	69 **Tm** Thulium 168.9	70 **Yb** Ytterbium 173.0	71 **Lu** Lutetium 175.0
90 **Th** Thorium 232.0	91 **Pa** Protactinium (231)	92 **U** Uranium 238.0	93 **Np** Neptunium (237)	94 **Pu** Plutonium (242)	95 **Am** Americium (243)	96 **Cm** Curium (247)	97 **Bk** Berkelium (247)	98 **Cf** Californium (249)	99 **Es** Einsteinium (254)	100 **Fm** Fermium (253)	101 **Md** Mendelevium (256)	102 **No** Nobelium (254)	103 **Lr** Lawrencium (257)

The 1–18 group designation has been recommended by the International Union of Pure and Applied Chemistry (IUPAC) but is not yet in wide use.

Elements up to 122 have been reported. Of these, some of the higher-numbered elements have only been observed under laboratory conditions, and have not been confirmed.

APPENDIX E

Cellular Respiration

Glycolysis

Figure E.1 illustrates the chemical reactions of glycolysis. In the early steps of this metabolic pathway, the original glucose molecule is altered by the addition of phosphate groups (*phosphorylation*) and by the rearrangement of its atoms. ATP supplies the phosphate groups and the energy to drive these reactions. The result is a molecule of fructose bound to two phosphate groups (fructose-1,6-bisphosphate). This molecule is split through two separate reactions into two 3-carbon molecules (glyceraldehyde-3-phosphate). Since each of these is converted to pyruvic acid, the following reactions, 1 through 5, must be counted twice to account for breakdown of a single glucose molecule.

1. An inorganic phosphate group is added to glyceraldehyde-3-phosphate to form 1,3-bisphosphoglyceric acid, releasing two hydrogen atoms, to be used in ATP synthesis, described later.

2. 1,3-bisphosphoglyceric acid is changed to 3-phosphoglyceric acid. As this occurs, some energy in the form of a high-energy phosphate is transferred from the 1,3-bisphosphoglyceric acid to an ADP molecule, phosphorylating the ADP to ATP.

3. A slight alteration of 3-phosphoglyceric acid forms 2-phosphoglyceric acid.

4. A change in 2-phosphoglyceric acid converts it into phosphoenolpyruvic acid.

5. Finally, a high-energy phosphate is transferred from the phosphoenolpyruvic acid to an ADP molecule, phosphorylating it to ATP. A molecule of pyruvic acid remains.

Overall, one molecule of glucose is ultimately broken down to two molecules of pyruvic acid. Also, a total of four hydrogen atoms are released (step 1 above), and four ATP molecules form (two in step 2 and two in step 5, above). However, because two molecules of ATP are used early in glycolysis, there is a net gain of only two ATP molecules during this phase of cellular respiration.

In the presence of oxygen, each pyruvic acid molecule is oxidized to an acetyl group, which then combines with a molecule of coenzyme A (obtained from the vitamin pantothenic acid) to form acetyl coenzyme A. As this occurs, two more hydrogen atoms and one carbon dioxide molecule are released for each molecule of acetyl coenzyme A formed. The acetyl coenzyme A is then broken down by means of the citric acid cycle, which figure E.2 illustrates.

Because obtaining energy for cellular metabolism is vital, disruptions in glycolysis or the reactions that follow it can devastate health.

Citric Acid Cycle

An acetyl coenzyme A molecule enters the citric acid cycle by combining with a molecule of oxaloacetic acid to form citric acid. As citric acid is produced, coenzyme A is released and thus can be used again to form acetyl coenzyme A from pyruvic acid. The citric acid is then changed by a series of reactions back into oxaloacetic acid, and the cycle may repeat.

Steps in the citric acid cycle release carbon dioxide and hydrogen atoms. More specifically, for each glucose molecule metabolized in the presence of oxygen, each of the two molecules of acetyl coenzyme A enters a citric acid cycle. The two cycles release a total of four carbon dioxide molecules and sixteen hydrogen atoms. At the same time, two more molecules of ATP form.

The released carbon dioxide dissolves in the cytoplasm and leaves the cell, eventually entering the bloodstream. Most of the hydrogen atoms released from the citric acid cycle, and those released during glycolysis and during the formation of acetyl coenzyme A, supply electrons used to produce ATP.

ATP Synthesis

Note that in figures E.1 and E.2 various metabolic reactions release hydrogen atoms. The electrons of these hydrogen atoms contain much of the energy associated with the chemical bonds of the original glucose molecule. To keep this energy in a usable form, these hydrogen atoms, with their high energy electrons, are passed in pairs to *hydrogen carriers*. One of these carriers is NAD^+ (nicotinamide adenine dinucleotide). When NAD^+ accepts a pair of hydrogen atoms, the two electrons and one hydrogen nucleus bind to NAD^+ to form NADH, and the remaining hydrogen nucleus (a hydrogen ion) is released as follows:

$$NAD^+ + 2H \rightarrow NADH + H^+$$

NAD^+ is a coenzyme obtained from a vitamin (niacin), and when it combines with the energized electrons it is said to be *reduced*. Reduction results from the addition of electrons, often as part of hydrogen atoms. Another electron acceptor, FAD (flavine adenine dinucleotide), acts in a similar manner, combining with two electrons and two hydrogen nuclei to form $FADH_2$ (fig. E.2). In their reduced states, the hydrogen carriers NADH and $FADH_2$ now hold most of the energy once held in the bonds of the original glucose molecule.

Figure 4.13 (p. 131) shows that NADH can release the electrons and hydrogen nucleus. Since this reaction removes electrons, the resulting NAD^+ is said to be *oxidized*. Oxidation results from the removal of electrons, often as part of hydrogen atoms; it is the opposite of reduction. The two electrons this reaction releases pass to a series of electron carriers. The regenerated NAD^+ can once again accept electrons, and is recycled.

The molecules that act as electron carriers comprise an **electron transport chain,** described in chapter 4 (pp. 129–131). As electrons are passed from one carrier to another, the carriers are alternately reduced and oxidized as they accept or release electrons. The transported electrons gradually lose energy as they proceed down the chain.

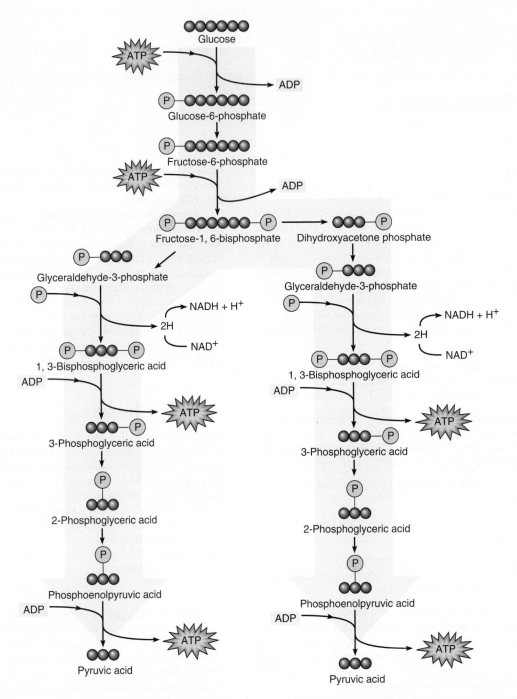

FIGURE E.1 Chemical reactions of glycolysis. There is a net production of 2 ATP molecules from each glucose molecule. The four hydrogen atoms released provide high-energy electrons that may be used to generate ATP in the electron transport chain, described later.

Among the members of the electron transport chain are several proteins, including a set of iron-containing molecules called **cytochromes.** The chain is located in the inner membranes of the mitochondria (fig. E.3; also see chapter 3, p. 92). The folds of the inner mitochondrial membrane provide surface area on which the energy reactions take place. In a muscle cell, the inner mitochondrial membrane, if stretched out, may be as much as forty-five times as long as the cell membrane!

Note in figure E.3, that as electrons pass through the electron transport chain, hydrogen ions are forced into the space between the inner and outer mitochondrial membranes. This sets up a concentration gradient for the hydrogen ions to diffuse back into the mitochondrial matrix via the enzyme complex *ATP synthase.* Much of their energy is used by the synthase to synthesize ATP.

The final cytochrome of the electron transport chain (cytochrome oxidase) gives up a pair of electrons and causes two hydrogen ions (formed at the beginning of the sequence) to combine with an atom of oxygen. This process produces a water molecule:

$$2e^- + 2H^+ + 1/2\ O_2 \rightarrow H_2O$$

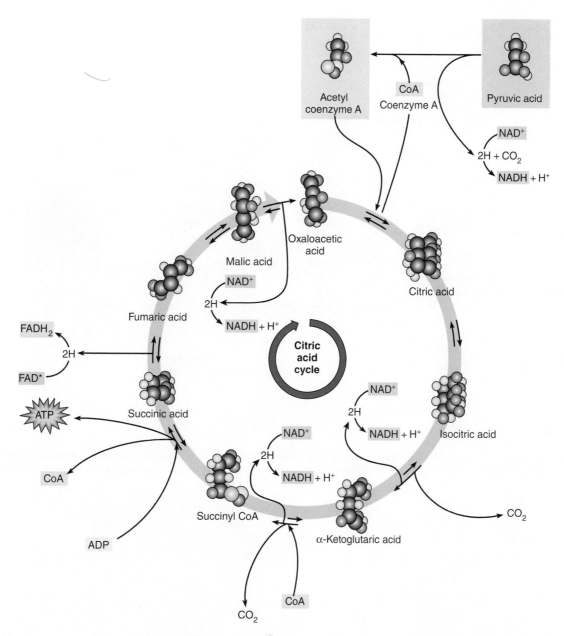

FIGURE E.2 Chemical reactions of the citric acid cycle. NADH and FADH₂ molecules carrying hydrogens are highlighted.

Thus, oxygen is the final electron acceptor. In the absence of oxygen, electrons cannot pass through the electron transport chain, NAD⁺ cannot be regenerated, and aerobic respiration halts.

As electrons pass through the electron transport chain, energy is released. Some of this energy is used by a mechanism involving the ATP synthase to combine phosphate and ADP by a high-energy bond (phosphorylation), forming ATP.

Note in figures E.1 and E.2 that twelve pairs of hydrogen atoms are released during the complete breakdown of one glucose molecule—two pairs from glycolysis, two pairs from the conversion of pyruvic acid to acetyl coenzyme A (one pair from each of two pyruvic acid molecules), and eight pairs from the citric acid cycle (four pairs for each of two acetyl coenzyme A molecules).

High-energy electrons from ten pairs of these hydrogen atoms produce thirty ATP molecules in the electron transport chain. Two pairs enter the chain differently and form four ATP molecules. Because this process of forming ATP involves both the oxidation of hydrogen atoms and the bonding of phosphate to ADP, it is called oxidative phosphorylation. Also, there is a net gain of two ATP molecules during glycolysis, and two ATP molecules form by direct enzyme action in two turns of the citric acid cycle. Thus, a maximum of thirty-eight ATP molecules form for each glucose molecule metabolized.

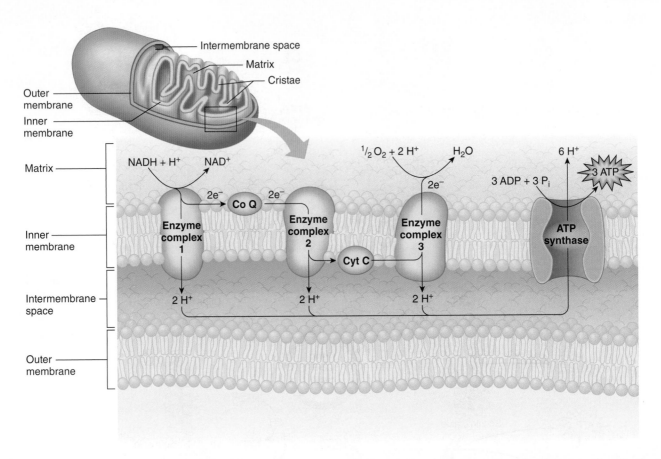

FIGURE E.3 High-energy electrons moving down the electron transport chain force hydrogen ions into the space between the mitochondrial membranes. This sets up a gradient for hydrogen ions to diffuse back into the matrix by way of ATP synthase, which converts much of their energy into ATP.

A Closer Look at DNA and RNA Structures

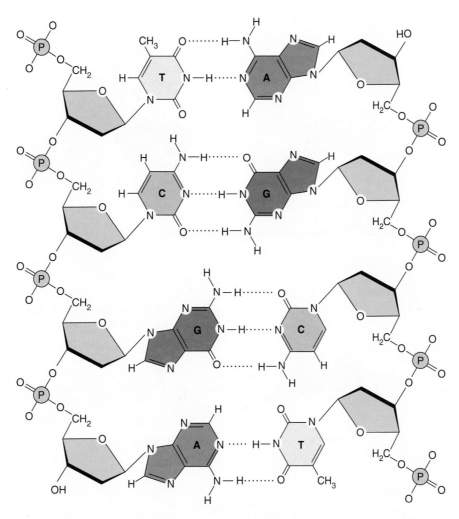

FIGURE F.1 The nucleotides of a double-stranded DNA molecule pair so that an adenine (A) of one strand hydrogen bonds to a thymine (T) of the other strand, and a guanine (G) of one strand hydrogen bonds to a cytosine (C) of the other. The dotted lines represent hydrogen bonds.

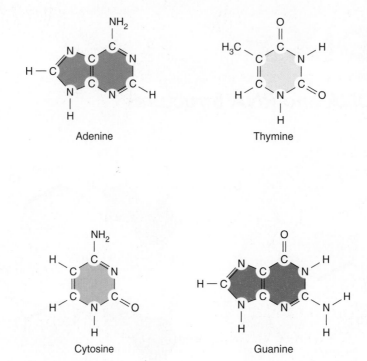

FIGURE F.2 The deoxyribonucleotides contain adenine, thymine, cytosine, or guanine.

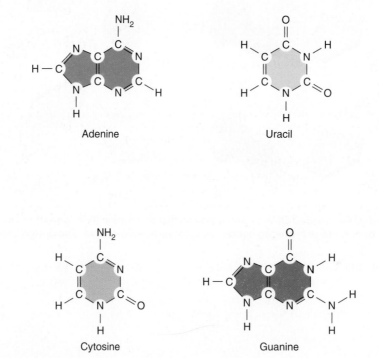

FIGURE F.3 The ribonucleotides contain adenine, uracil, cytosine, or guanine.

APPENDIX G

Figure Question Answers

Chapter 1

Figure 1.7: The heat would come on until the temperature reached the new set point.
Figure 1.21: The hand is more lateral than the hip (remember that anatomical position is the reference).

Chapter 2

Figure 2.8: The number of protons equals the number of electrons.
Figure 2.10: The solution at pH 6.4 has 100 times the hydrogen ion concentration of a solution at pH 8.4.

Chapter 3

Figure 3.3: Nucleus (nuclear envelope) mitochondria, lysosomes, vesicles, endoplasmic reticulum, cell membrane, golgi apparatus
Figure 3.28: ATP

Chapter 4

Figure 4.10: Glycolysis occurs in the cytosol and cellular respiration occurs in the mitochondria.
Figure 4.24: They are all informational molecules, but differ in their structures.

Chapter 5

Figure 5.1: Gap junction
Figure 5.28: Fluid

Chapter 6

Figure 6.1: Blood, muscle tissue, nervous tissue, areolar connective tissue
Figure 6.3: Palms of hands, soles of feet
Figure 6.6: Lunula

Chapter 7

Figure 7.2: Right femur
Figure 7.3: Cross section
Figure 7.14: Receptors, control center (set point), effectors
Figure 7.50: Female

Chapter 8

Figure 8.5: Cartilaginous joint
Figure 8.11: Humerus
Figure 8.13: Frontal section

Chapter 9

Figure 9.8: Diffusion
Figure 9.10: The length does not change, only the degree to which they overlap.

Chapter 10

Figure 10.1: A phospholipid bilayer—the cell membrane
Figure 10.3: Cellular differentiation
Figure 10.14: ATP

Chapter 11

Figure 11.15: In the gray matter and in the dorsal root ganglion
Figure 11.23: In both directions—sensory impulses travel toward the CNS and motor impulses travel away from the CNS.

Chapter 12

Figure 12.9: Temporal bone
Figure 12.24: Yes, voluntary control—look up at the ceiling if you have a question about this!

Chapter 13

Figure 13.2: They are equally specific because only cells with receptors, either for the neurotransmitter or the hormone, will respond.
Figure 13.36: As blood glucose drops several hours after a meal, insulin levels will decrease in response, but glucagon levels will rise in response.

Chapter 14

Figure 14A: Gloves, goggles, labcoat
Figure 14.15: Macrophage
Figure 14.19: The conversion of soluble fibrinogen into insoluble fibrin

Chapter 15

Figure 15.19: Ventricular depolarization and atrial repolarization
Figure 15.28: Plasma proteins including the albumins
Figure 15.46: Radial and ulnar arteries
Figure 15.52: Axillary vein

Chapter 16

Figure 16.6: Thoracic duct
Figure 16.21: Secondary response

Chapter 17

Figure 17.2: 19.7 inches (based on the measurements in the figure)
Figure 17.30: The pancreas is stimulated to secrete digestive enzymes.
Figure 17.47: The major secretion of the large intestinal wall is mucus to lubricate and protect against the abrasiveness of material moving through the tube, to bind fecal matter and control pH of the large intestinal contents. Other cells in the small intestinal wall secrete digestive enzymes and hormones, in addition to those cells secreting mucus.

Chapter 18

Figure 18.8: BMI will vary from individual to individual. Suggestions for adjusting diet to achieve/maintain healthy weight include: lower than healthy weight-increase calorie intake, especially proteins; healthy weight-continue with your current diet unless your activity level changes; overweight-decrease your calorie intake and increase your activity level; obese-see your physician for advice on diet (lowering calorie intake, particularly carbohydrates and lipids) and exercise (you may have medical conditions that limit your physical activity or require you to slowly build up your activity level).
Figure 18.12: Bone deformities result from insufficient calcium deposition in the bones. Vitamin D is important for the absorption of calcium in the small intestine.

Chapter 19

Figure 19.14: Pulmonary arterioles carry oxygen-poor blood from the heart to the lungs.
Figure 19.27: The pen would move upward.

Chapter 20

Figure 20.15: Vasoconstriction of the afferent arteriole
Figure 20.19: Tubular reabsorption

Chapter 21

Figure 21.3: Sodium ion
Figure 21.12: Because of the bicarbonate-buffer reaction ($CO_2 + H_2O \leftrightarrow H_2CO_3 \leftrightarrow HCO_3^- + H^+$), CO_2 and hydrogen ions are in balance with each other. Exhaling more CO_2 will lower the hydrogen ion concentration in the plasma (raising the pH) and exhaling less CO_2 will raise the hydrogen ion concentration (lowering the pH). Respiratory adjustments help to keep the pH of the internal environment in the range 7.35–7.45.

Chapter 22

Figure 22.8: It is important for the sperm and egg to each possess 23 chromosomes, so that upon fertilization of the egg by the sperm, the developing zygote will have the full complement of 46 chromosomes (no more, no less) found in human body cells.
Figure 22.23: Corona radiata
Figure 22.33: Combined hormone contraceptives, intrauterine device

Chapter 23

Figure 23.3: 23 chromosomes each with two chromatids
Figure 23.21: Umbilical vein, ductus venosus, foramen ovale, ductus arteriosus, umbilical arteries

Chapter 24

Figure 24.3: The male karyotype would have one X chromosome and one Y chromosome.
Figure 24.4: Zero chance of having the disease

GLOSSARY

Each word in this glossary is followed by a phonetic guide to pronunciation. In this guide, any unmarked vowel that ends a syllable or stands alone as a syllable has the long sound. Thus, *play* would be spelled *pla*. Any unmarked vowel that is followed by a consonant has the short sound. *Tough,* for instance, is spelled *tuf.* If a long vowel appears in the middle of a syllable (followed by a consonant), it is marked with the macron (ˉ), the sign for a long vowel. Thus, the word *plate* would be phonetically spelled plāt. Similarly, if a vowel stands alone or ends a syllable, but has the short sound, it is marked with a breve (ˇ).

A

abdominal (ab-dom´ĭ-nal) Pertaining to the portion of the body between the diaphragm and the pelvis.

abdominal cavity (ab-dom´ĭ-nal kav´ĭ-te) Space between the diaphragm and the pelvic inlet that contains the abdominal viscera.

abdominopelvic cavity (ab-dom″ĭ-no-pel´vik kav´ĭ-te) Space between the diaphragm and the pelvic outlet that contains the abdominal and pelvic viscera.

abduction (ab-duk´shun) Movement of a body part away from the midline.

absorption (ab-sorp´shun) Passage of substances through membranes and into body fluids.

accessory organ (ak-ses´o-re or´gan) Organ that supplements the functions of other organs.

accommodation (ah-kom″o-da´shun) Adjustment of the lens of the eye for close or distant vision.

acetylcholine (as″ĕ-til-ko´lēn) Type of neurotransmitter, which is a biochemical secreted into the synaptic clefts at the axon ends of neurons; ACh.

acetylcholinesterase (as″ĕ-til-ko″lin-es´ter-ās) Enzyme that catalyzes breakdown of acetylcholine.

acetyl coenzyme A (as´ĕ-til ko-en´zīm) Intermediate compound produced from the oxidation of carbohydrates, proteins, and fats.

acid (as´id) Substance that ionizes in water to release hydrogen ions.

acidosis (as″ĭ-do´sis) Decrease in pH of body fluids below pH 7.35.

acoustic (ah-koōs´tik) Pertaining to sound.

acromial (ah-kro´me-al) Pertaining to the shoulder.

ACTH Adrenocorticotropic hormone.

actin (ak´tin) Protein in a muscle fiber that forms the thin filaments that slide between filaments of the protein myosin, shortening the muscle fibers.

action potential (ak´shun po-ten´shal) Sequence of electrical changes in part of a nerve cell or muscle cell exposed to a stimulus that exceeds threshold.

activation energy (ak″tĭ-va´shun en´er-je) Energy required to initiate a chemical reaction.

active site (ak´tiv sīt) Part of an enzyme molecule that temporarily binds a substrate.

active transport (ak´tiv trans´port) Process that requires energy and a carrier molecule to move a substance across a cell membrane against the concentration gradient.

adaptive defense (a-dap´tiv de-fenc) Specific defenses T and B lymphocytes carry out.

adduction (ah-duk´shun) Movement of a body part toward the midline.

adenoids (ad´ĕ-noids) The pharyngeal tonsils in the nasopharynx.

adenosine diphosphate (ah-den´o-sēn di-fos´fāt) Molecule produced when adenosine triphosphate loses a terminal phosphate; ADP.

adenosine triphosphate (ah-den´o-sēn tri-fos´fāt) Organic molecule that transfers energy, used in cellular processes; ATP.

adenylate cyclase (ah-den´ĭ-lāt si´klās) Enzyme activated when certain hormones bind receptors on cell membranes. It catalyzes the circularization of ATP to cyclic AMP.

ADH Antidiuretic hormone.

adipose tissue (ad´ĭ-pōs tish´u) Fat-storing tissue.

adolescence (ad″o-les´ens) Period of life between puberty and adulthood.

ADP Adenosine diphosphate.

adrenal cortex (ah-dre´nal kor´teks) Outer part of the adrenal gland.

adrenal gland (ah-dre´nal gland) Endocrine gland on the superior portion of each kidney.

adrenalin (ah-dren´ah-lin) Epinephrine. Main hormone of the adrenal medulla.

adrenal medulla (ah-dre´nal me-dul´ah) Inner part of the adrenal gland.

adrenergic fiber (ad″ren-er´jik fi´ber) Axon that secretes the neurotransmitter norepinephrine at its terminal.

adrenocorticotropic hormone (ah-dre″no-kor″te-ko-trōp´ik hor´mōn) Hormone the anterior pituitary secretes that stimulates activity in the adrenal cortex; ACTH.

adulthood (ah-dult´hood) Period of life between adolescence and senescence.

aerobic (a″er-ōb´ik) In the presence of molecular oxygen.

afferent (af´er-ent) Directed toward a central part of a structure.

afterload (af´ter-lōd) The force required to open the semilunar valves to eject blood from the ventricles; determined largely by arterial pressure.

agglutination (ah-gloō″ti-na´shun) Clumping of blood cells in response to a reaction between an antibody and an antigen.

agonist (ag´o-nist) A muscle that causes a particular movement.

agranulocyte (a-gran´u-lo-sīt) Nongranular leukocyte.

albumin (al-bu´min) Plasma protein that contributes to the colloid osmotic pressure of plasma.

aldosterone (al-dos´ter-ōn) Adrenal cortical hormone that regulates sodium and potassium ion concentrations and fluid volume.

alimentary canal (al″i-men´tar-e kah-nal´) Tubular part of the digestive tract from the mouth to the anus.

alkaloid (al´kah-loid) Any of a group of organic compounds that are bitter and may have toxic effects.

alkalosis (al″kah-lo´sis) Increase in the pH of body fluids above 7.45.

allantois (ah-lan´to-is) Structure in the embryo from which the umbilical cord blood vessels develop.

allele (ah-lēl) One of two or more different forms of a gene.

allergen (al´er-jen) Chemical that triggers an allergic reaction.

all-or-none response (al´or-nun´ re-spons´) A response that occurs completely or not at all, such as a neuron generating an action potential in response to a stimulus of threshold strength.

alpha receptor (al´fah re-sep´tor) Receptor on an effector cell membrane that binds epinephrine or norepinephrine.

alveolar duct (al-ve´o-lar dukt) Fine tube that conducts air to and from an alveolar sac of the lungs.

alveolar pore (al-ve´o-lar pōr) Small opening in the walls between alveoli that permits air to pass from one alveolus to another.

alveolar process (al-ve′o-lar pros′es) Projection on the border of the jaw that holds the bony sockets of the teeth.

alveolar sac (al-ve′o-lar sak) A space at the center of a cluster of alveoli.

alveolus (al-ve′o-lus) Any of the 300 million microscopic air sacs of the lungs (pl., *alveoli*).

amine (am′in) Type of nitrogen-containing organic compound, including hormones from the adrenal medulla.

amino acid (ah-me′no as′id) Organic compound that includes an amino group (–NH$_2$) and a carboxyl group (–COOH); the structural unit of a protein molecule.

amniocentesis (am″ne-o-sen-te′sis) Removal of a sample of amniotic fluid through the abdominal wall of a pregnant woman. Fetal cells in it are cultured and checked for chromosome complement.

amnion (am′ne-on) Extraembryonic membrane that encircles a fetus and amniotic fluid.

amniotic cavity (am″ne-ot′ik kav′ĭ-te) Fluid-filled space surrounded by the amnion.

amniotic fluid (am″ne-ot′ik floo′id) Fluid in the amniotic cavity that surrounds the developing fetus.

ampulla (am-pul′ah) Expansion at the end of each semicircular canal that houses a crista ampullaris.

amylase (am′ĭ-lās) Enzyme that hydrolyzes polysaccharides.

anabolism (ah-nab′o-liz″em) Synthesis of larger molecules from smaller ones; anabolic metabolism.

anaerobic (an-a″er-ōb′ik) Occurring in the absence of molecular oxygen.

anal canal (a′nal kah-nal′) The most distal two or three inches of the large intestine that open to the outside at the anus.

anaphase (an′ah-fāz) Stage in mitosis when sister chromatids separate and move as chromosomes to opposite poles of the cell.

anatomical position (an″ah-tom′ĭ-kal po-zish′un) Body posture with the body erect, face forward, arms at the sides with palms facing forward.

anatomy (ah-nat′o-me) Branch of science dealing with the form and organization of body parts.

androgen (an′dro-jen) Male sex hormone such as testosterone.

anemia (ah-ne′me-ah) Deficiency of red blood cells or hemoglobin.

aneuploid (an′u-ploid) Cell with one or more extra or missing chromosomes.

aneurysm (an′u-rizm) Abnormal bulge in a blood vessel.

angiotensin I (an″je-o-ten′sin wun) A peptide released by the action of renin on angiotensinogen.

angiotensin II (an″je-o-ten′sin too) Vasoconstricting biochemical generated in response to reduced blood pressure or blood volume.

angiotensinogen (an″je-o-ten-sin′o-jen) Liver serum globulin that renin converts to angiotensin I.

anion (an′i-on) Atom or molecule carrying a net negative charge due to one or more extra electrons.

antagonist (an-tag′o-nist) A muscle that opposes a particular movement.

antebrachial (an″te-bra′ke-al) Pertaining to the forearm.

antecubital (an″te-ku′bi-tal) Pertaining to the front of the elbow joint.

anterior (an-te′re-or) Pertaining to the front.

anterior pituitary (an-te′re-or pi-tu′i-tār″e) Front lobe of the pituitary gland.

antibody (an′ti-bod″e) Protein that B cells of the immune system produce in response to a nonself antigen that then reacts with the antigen; immunoglobulin.

anticoagulant (an″tĭ-ko-ag′u-lant) Biochemical that inhibits blood clotting.

anticodon (an″ti-ko′don) Three contiguous nucleotides of a transfer RNA molecule that are complementary to a specific mRNA codon.

antidiuretic hormone (an″tĭ-di″u-ret′ik hor′mōn) Hormone of the posterior pituitary gland that reduces water excretion by the kidneys; ADH; vasopressin.

antigen (an′tĭ-jen) Chemical that triggers an immune response.

antigen-binding site (an′tĭ-jen-bīn′ding sīt) Specialized ends of antibodies that bind specific antigens.

antigen-presenting cell (an′tĭ-jen-pre-sen′ting cel) The cell that displays an antigen to the cells of the immune system so they can defend the body against that particular antigen.

antioxidant (an″tĭ-ok′sĭ-dant) Substance that inhibits oxidation of another substance.

antithrombin (an″tĭ-throm′bin) Plasma protein that inhibits the action of thrombin, and thus inhibits blood clotting.

anus (a′nus) Inferior opening of the digestive tube.

aorta (a-or′tah) Systemic artery that receives blood directly from the left ventricle.

aortic body (a-or′tik bod′e) Structure associated with the wall of the aortic arch that contains chemoreceptors.

aortic sinus (a-or′tik si′nus) Swelling in the aortic wall, behind each cusp of the semilunar valve.

aortic valve (a-or′tik valv) Flaplike structures in the wall of the aorta near its origin that prevent aortic blood from returning to the left ventricle of the heart.

apocrine gland (ap′o-krin gland) Type of gland whose secretions have parts of secretory cells.

aponeurosis (ap″o-nu-ro′sis) Sheet of connective tissue that attaches muscles to bone and other muscles.

apoptosis (ap″o-to′-sis) Programmed cell death.

appendicular (ap″en-dik′u-lar) Pertaining to the upper or lower limbs.

appendix (ah-pen′diks) Small, tubular appendage of lymphatic tissue that extends outward from the cecum; vermiform appendix.

aqueous humor (a′kwe-us hu′mor) Watery fluid that fills the anterior cavity of the eye.

arachnoid granulation (ah-rak′noid gran″u-la′shun) Any of several fingerlike structures that projects from the subarachnoid space of the meninges into blood-filled dural sinuses and reabsorbs cerebrospinal fluid.

arachnoid mater (ah-rak′noid ma′ter) Delicate, weblike middle layer of the meninges.

arbor vitae (ar′bor vi′ta) Treelike pattern of white matter in the cerebellum.

areola (ah-re′o-lah) Pigmented region surrounding the nipple of the mammary gland or breast.

areolar tissue (ah-re′o-lar tish′u) Connective tissue composed mainly of fibers.

arrector pili muscle (ah-rek′tor pil′i mus′l) Smooth muscle in the skin associated with a hair follicle.

arrhythmia (ah-rith′me-ah) An irregular rate or rhythm of the heartbeat.

arteriole (ar-te′re-ōl) Small branch of an artery that communicates with a capillary network.

arteriosclerosis (ar-te″re-o-sklĕ-ro′sis) Condition in which the walls of arteries thicken and lose their elasticity, hardening of the arteries.

artery (ar′ter-e) Vessel that transports blood from the heart.

arthritis (ar-thri′tis) Joint inflammation.

articular cartilage (ar-tik′u-lar kar′tĭ-lij) Hyaline cartilage that covers the ends of bones in synovial joints.

articulation (ar-tik″u-la′shun) The union of two or more bones, a joint.

ascending colon (ah-send′ing ko′lon) Portion of the large intestine that passes upward on the right side of the abdomen from the cecum to the lower edge of the liver.

ascending tract (ah-send′ing trakt) Group of axons in the spinal cord that conducts sensory impulses upward to the brain.

ascites (ah-si′tez) Serous fluid accumulation in the abdominal cavity.

ascorbic acid (as-kor′bik as′id) One of the water-soluble vitamins; vitamin C.

assimilation (ah-sim″ĭ-la′shun) Chemically changing absorbed substances in the body.

association area (ah-so″se-a′shun a′re-ah) Region of the cerebral cortex controlling memory, reasoning, judgment, and emotions.

astigmatism (ah-stig′mah-tizm) Visual defect due to errors in refraction caused by abnormal, assymetric curvatures in the surface of the cornea or lens.

astrocyte (as′tro-sīt) Type of neuroglia that connects neurons to blood vessels in the CNS.

atherosclerosis (ath″er-o-sklĕ-ro′sis) Accumulation of fatty substances on the inner linings of arteries.

atmospheric pressure (at″mos-fer′ik presh′ur) Pressure exerted by the weight of the air, about 760 mm of mercury at sea level.

atom (at′om) Smallest particle of an element that has the properties of that element.

atomic number (ah-tom′ik num′ber) Number of protons in an atom of an element.

atomic weight (ah-tom′ik wāt) Number of protons plus the number of neutrons in an atom of an element.

ATP Adenosine triphosphate.

ATPase Enzyme that transfers the energy stored in terminal phosphate bonds of ATP molecules.

atrial natriuretic peptide (a′trē-al na″trē-u-ret′ik pep′tīd) Polypeptide hormone from the atria that increases sodium excretion.

atrioventricular bundle (a″tre-o-ven-trik′ u-lar bun′dl) Group of specialized muscle fibers that conducts impulses from the atrioventricular node to the Purkinje fibers in the ventricular muscle of the heart, AV bundle, bundle of His.

atrioventricular node (a″tre-o-ven-trik′u-lar nōd) Specialized mass of cardiac muscle fibers in the interatrial septum of the heart that conducts cardiac impulses from the sinoatrial node to the AV bundle; AV node.

atrioventricular orifice (a″tre-o-ven-trik′u-lar or′i-fis) Opening between the atrium and the ventricle on each side of the heart.

atrioventricular sulcus (a″tre-o-ven-trik′u-lar sul′kus) Groove on the surface of the heart that marks the division between an atrium and a ventricle.

atrioventricular valve (a″tre-o-ven-trik′u-lar valv) Cardiac valve between an atrium and a ventricle.

atrium (a′tre-um) Either of the two upper chambers of the heart (pl., *atria*).

atrophy (at′ro-fe) Shrinking of an organ or tissue.

auditory (aw′di-to″re) Pertaining to the ear or the sense of hearing.

auditory ossicle (aw′di-to″re os′i-kl) A bone of the middle ear.

auditory tube (aw′di-to″re tūb) Tube that connects the middle ear cavity to the pharynx; eustachian tube.

auricle (aw′ri-kl) Flaplike structure on the anterior wall of the atrium.

autoantibody (aw″to-an′tĭ-bod″e) An antibody produced against oneself.

autocrine (aw′to-krin) Hormone that acts on the same cell that secreted it.

autoimmunity (aw″to-ĭ-mū′ni-tē) An immune response against a person's own tissues.

autonomic nervous system (aw″to-nom′ik ner′vus sis′tem) Part of the nervous system that controls the viscera.

autoregulation (aw″to-reg″u-la′shun) Ability of an organ or tissue to maintain a local process despite changing local conditions, as in autoregulation of glomerular filtration rate in the kidney.

autosome (aw′to-sōm) Any one of the 44 chromosomes that does not include a gene that determines sex.

AV bundle (bun′dl) Atrioventricular bundle.

AV node (nōd) Atrioventricular node.

axial (ak′se-al) Pertaining to the head, neck, and trunk.

axial skeleton (ak′se-al skel′ĕ-ton) The part of the skeleton that supports and protects the organs of the head, neck, and trunk.

axillary (ak′sĭ-ler″e) Pertaining to the armpit.

axon (ak′son) A nerve fiber; conducts an impulse away from a neuron cell body.

axonal transport (ak′so-nal trans′port) Transport of substances from the neuron cell body to an axon terminal.

B

ball-and-socket joint (bawl-and-sok′et joint) A bone with a spherical head on one end joined with a bone that has a cup-shaped cavity; spheroidal joint.

baroreceptor (bar″o-re-sep′tor) Sensory receptor in a blood vessel wall stimulated by changes in blood pressure; pressoreceptor.

basal metabolic rate (ba′sal met″ah-bo′lic rāt) Rate of metabolic reactions when the body is at rest; BMR.

basal nuclei (bas′al nu′kle-i) Masses of gray matter deep within a cerebral hemisphere.

base (bās) Substance that ionizes in water, releasing hydroxide ions (OH^-) or other ions that combine with hydrogen ions.

basement membrane (bās′ment mem′brān) Layer of nonliving material that anchors epithelial tissue to underlying connective tissue.

basophil (ba′so-fil) White blood cell containing cytoplasmic granules that react with basic stain.

beta oxidation (ba′tah ok″sĭ-da′shun) Chemical process that breaks fatty acids down to form acetyl coenzyme A, which can enter the citric acid cycle.

beta receptor (ba′tah re-sep′tor) Receptor on an effector cell membrane that combines mainly with epinephrine and only slightly with norepinephrine.

bicarbonate buffer system (bi-kar′bo-nāt buf′er sis′tem) System in which carbonic acid acts as a weak acid and bicarbonate ion acts as a weak base that weakens a strong base and a strong acid, respectively; resists a change in pH.

bicarbonate ion (bi-kar′bon-āt i′on) HCO_3^-.

bicuspid tooth (bi-kus′pid tooth) Premolar specialized for grinding hard particles of food.

bile (bīl) Fluid secreted by the liver and stored in the gallbladder.

bile duct (bīl dukt) Tube that transports bile from the cystic duct and common hepatic duct to the duodenum.

bilirubin (bil″ĭ-roo′bin) A bile pigment produced from hemoglobin breakdown.

biliverdin (bil″ĭ-ver′din) A bile pigment produced from hemoglobin breakdown.

biochemistry (bi″o-kem′is-tre) Branch of science dealing with the chemistry of living organisms.

biotin (bi′o-tin) A water-soluble vitamin; member of the vitamin B complex.

bipolar neuron (bi-po′lar nu′ron) A nerve cell with two cell processes associated with its cell body.

blastocyst (blas′to-sist) Early stage of prenatal development when the embryo is a hollow ball of cells.

blood (blud) A connective tissue, consisting of cells in a liquid matrix called plasma, that circulates through the heart and vessels carrying substances throughout the body.

B lymphocyte (B lim′fo-sīt) Lymphocyte that produces and secretes antibodies that bind and destroy nonself molecules; B cell.

BMI Body mass index.

BMR Basal metabolic rate.

body mass index (bŏdē mas in′-dex) A measure of relative weight (underweight, normal weight, overweight, obesity) calculated as weight in kilograms divided by the square of height in meters; BMI.

bolus (bo′lus) Mass of food or liquid, such as enters the esophagus during swallowing.

bond (bond)   Connection between atoms in a compound.

bone (bōn)   Any of the 206 individual parts of the skeleton composed of cells and inorganic, mineral matrix; also a connective tissue and a type of cell.

brachial (bra′ke-al)   Pertaining to the arm.

bradycardia (brad″e-kar′de-ah)   An abnormally slow heart rate (or pulse rate).

brainstem (brān′stem)   Portion of the brain that includes the midbrain, pons, and medulla oblongata.

Broca's area (bro′kahz a′re-ah)   Region of the frontal lobe that coordinates complex muscular actions of the mouth, tongue, and larynx, making speech possible.

bronchial tree (brong′ke-al trē)   The bronchi and their branches that carry air between the trachea and the alveoli of the lungs.

bronchiole (brong′ke-ōl)   Small branch of a bronchus in the lung.

bronchus (brong′kus)   Branch of the trachea that leads to a lung (pl., *bronchi*).

buccal (buk′al)   Pertaining to the mouth and inner lining of the cheeks.

buffer (buf′er)   Substance that can react with a strong acid or base to form a weaker acid or base, and thus resist a change in pH.

bulbourethral gland (bul″bo-u-re′thral gland)   Gland that secretes a viscous fluid into the male urethra during sexual excitement; Cowper's gland.

bulimia (bu-lim′e-ah)   Disorder of binge eating followed by purging.

bulk element (bulk el′ĕ-ment)   Basic chemical required in abundance.

bursa (bur′sah)   Saclike, fluid-filled cushioning structure, lined with synovial membrane, near a joint (pl., *bursae*).

bursitis (bur-si′tis)   Inflammation of a bursa.

C

calcitonin (kal″sĭ-to′-nin)   Thyroid hormone that helps regulate blood calcium concentration.

calorie (kal′o-re)   Unit that measures heat energy and the energy contents of foods.

calorimeter (kal″o-rim′ĕ-ter)   Device that measures the heat energy content of foods, bomb calorimeter.

canaliculus (kan″ah-lik′u-lus)   Microscopic canal that connects lacunae of bone tissue (pl., *canaliculi*).

capacitation (kah-pas″ĭ-ta′shun)   Activation of a sperm cell to fertilize an egg cell.

capillary (kap′ĭ-ler″ē)   Small blood vessel that connects an arteriole and a venule and allows diffusional exchange of nutrients and wastes.

carbaminohemoglobin (kar-bam″ĭ-no-he″mo-glo′bin)   Bonded carbon dioxide and hemoglobin.

carbohydrate (kar″bo-hi′drāt)   Organic compound consisting of carbon, hydrogen, and oxygen, in a 1:2:1 ratio.

carbonic anhydrase (kar-bon′ik an-hi′drās)   Enzyme that catalyzes the reaction between carbon dioxide and water to form carbonic acid.

carbon monoxide (kar′bon mon-ok′sīd)   Toxic gas that binds hemoglobin, forming a relatively stable compound that can no longer carry oxygen; CO.

carboxypeptidase (kar-bok″se-pep′ti-dās)   Protein-splitting enzyme in pancreatic juice.

cardiac center (kar′de-ak sen′ter)   Neurons in the medulla oblongata that control heart rate.

cardiac conduction system (kar′de-ak kon-duk′shun sis′tem)   System of specialized cardiac muscle fibers that conduct cardiac impulses from the SA node throughout the myocardium.

cardiac cycle (kar′de-ak si′kl)   Sequence of myocardial contraction and relaxation that constitutes a complete heartbeat.

cardiac muscle tissue (kar′de-ak mus′el tish′u)   Specialized muscle tissue found only in the heart.

cardiac output (kar′de-ak owt′poot)   The volume of blood per minute that the heart pumps (stroke volume in milliliters multiplied by the heart rate in beats per minute).

cardiac vein (kar′de-ak vān)   Any of the blood vessels that returns blood from the venules of the myocardium to the coronary sinus.

cardiovascular (kar″de-o-vas′ku-lar)   Pertaining to the heart and blood vessels.

carina (kah-ri′nah)   Cartilaginous ridge between the openings of the right and left bronchi.

carotene (kar′o-tēn)   Yellow, orange, or reddish pigment in plants and a precursor of vitamin A.

carotid bodies (kah-rot′id bod′ēz)   Masses of chemoreceptors in the wall of the carotid artery near the carotid sinus.

carpal (kar′pal)   Pertaining to the wrist or any of the individual wrist bones.

carpus (kar′pus)   Wrist, the wrist bones as a group.

cartilage (kar′tĭ-lij)   Type of connective tissue in which cells are in lacunae separated by a semisolid extracellular matrix.

cartilaginous joint (kar″tĭ-laj′ĭ-nus joint)   Two or more bones joined by cartilage.

catabolism (ka-tab′o-lizm)   Breakdown of large molecules; catabolic metabolism.

catalyst (kat′ah-list)   Chemical that increases the rate of a chemical reaction, but is not permanently altered by the reaction.

catecholamine (kat″ĕ-kol′am-in)   Type of organic compound that includes epinephrine, norepinephrine, and dopamine.

cation (kat′i-on)   Atom or molecule carrying a net positive charge due to a deficiency of electrons.

cecum (se′kum)   Pouchlike part of the large intestine attached to the small intestine.

celiac (se′le-ak)   Pertaining to the abdomen.

cell (sel)   The structural and functional unit of an organism.

cell body (sel bod′e)   Part of a nerve cell that includes a cytoplasmic mass and a nucleus, and from which processes extend.

cell cycle (sel sī-kl)   Life cycle of a cell consisting of G_1 (growth), S (DNA synthesis), G_2 (growth), mitosis (nuclear division), and cytokinesis (cytoplasmic division).

cell membrane (sel mem′brān)   The selectively permeable outer boundary of a cell consisting of a phospholipid bilayer embedded with proteins.

cellular adhesion molecules (sel′u-lar ad-hee′zhon mol′ĕ-kūlz)   Proteins that guide movement of cells; CAMs.

cellular immune response (sel′u-lar ĭ-mūn re-spons′)   The body's attack by T cells and their secreted products on nonself antigens.

cellular respiration (sel′u-lar res″pĭ- ra′shun)   A biochemical pathway that releases energy from organic compounds.

cellulose (sel′u-lōs)   Polysaccharide abundant in plant tissues that human digestive enzymes cannot break down.

cementum (se-men′tum)   Bonelike material that surrounds the root of a tooth.

central canal (sen′tral kah-nal′)   Tiny channel in bone tissue that contains a blood vessel; Haversian canal. Also a tube in the spinal cord continuous with brain ventricles that contains cerebrospinal fluid.

central nervous system (sen′tral ner′vus sis′tem)   The brain and spinal cord; CNS.

centriole (sen′tre-ōl)   Cellular structure built of microtubules that organizes the mitotic spindle.

centromere (sen′tro-mēr)   Region of a chromosome where spindle fibers attach during mitosis; region where chromatids are attached.

centrosome (sen′tro-sōm)   Cellular organelle consisting of two centrioles.

cephalic (sĕ-fal′ik)   Pertaining to the head.

cerebellar cortex (ser′ĕ-bel′ar kor′teks)   Outer layer of the cerebellum.

cerebellar peduncles (ser″ĕ-bel′ar pe-dung′klz)   Bundles of nerve fibers connecting the cerebellum and the brainstem.

cerebellum (ser″ĕ-bel′um)   Part of the brain that coordinates skeletal muscle movement.

cerebral aqueduct (ser′ĕ-bral ak′wĕ-dukt″)   Canal that connects the third and fourth ventricles of the brain.

cerebral cortex (ser′ĕ-bral kor′teks)   Outer layer of the cerebrum.

cerebral hemisphere (ser′ĕ-bral hem′ĭ-sfēr)   Either of the large, paired structures that constitute the cerebrum.

cerebrospinal fluid (ser″ĕ-bro-spi′nal floo′id)   Fluid in the ventricles of the brain, the subarachnoid space of the meninges, and the central canal of the spinal cord; CSF.

cerebrum (ser′ĕ-brum)   Part of the brain in the upper part of the cranial cavity that provides higher mental functions.

cerumen (sĕ-roo′men)   Waxlike substance produced by cells that line the external ear canal; ear wax.

cervical (ser′vĭ-kal)   Pertaining to the neck; also referring to the neck of the uterus.

cervix (ser′viks)   Narrow, inferior end of the uterus that leads into the vagina.

chemical buffer system (kem-ĭ-kal buf′er sis′tem)   Pair of chemicals, one a weak acid, the other a weak base, that resists pH changes.

chemoreceptor (ke″mo-re-sep′tor)   Receptor stimulated by the binding of certain chemicals.

chemotaxis (ke″mo-tak′sis)   Attraction of leukocytes to chemicals released from damaged cells.

chief cell (chēf sel)   Cell type in a gastric gland that secretes digestive enzymes.

childhood (child′hood)   Period of life between infancy and adolescence.

chloride shift (klo′rīd shift)   Movement of chloride ions from the blood plasma into red blood cells as bicarbonate ions diffuse out of the red blood cells into the plasma.

cholecystokinin (ko″le-sis″to-ki′nin)   Hormone the small intestine secretes that stimulates release of pancreatic juice from the pancreas and bile from the gallbladder.

cholesterol (ko-les′ter-ol)   A lipid produced in the body and acquired from food that cells use to synthesize steroid hormones.

cholinergic fiber (ko″lin-er′jik fi′ber)   Axon that secretes acetylcholine at its terminal.

chondrocyte (kon′dro-sīt)   A cartilage cell.

chordae tendineae (kor′de ten′dĭ-ne)   Fibrous strings attached to the cusps of the tricuspid and mitral valves in the heart.

chorion (ko′re-on)   Extraembryonic membrane that forms the outermost covering around a fetus and contributes to formation of the placenta.

chorionic villus (ko″re-on′ik vil′us)   Projection that extends from the outer surface of the chorion and helps attach an embryo to the uterine wall.

choroid coat (ko′roid kōt)   Vascular, pigmented middle layer of the wall of the eye.

choroid plexus (ko′roid plek′sus)   Mass of specialized capillaries that secretes cerebrospinal fluid into a brain ventricle.

chromatid (kro′mah-tid)   One longitudinal half of a replicated chromosome.

chromatin (kro′mah-tin)   The complex of DNA and protein making up the cell's 46 chromosomes.

chromatophilic substance (kro″mah-to-fil′ik sub′stans)   Membranous sacs in the cytoplasm of nerve cells that have ribosomes attached to their surfaces; Nissl bodies.

chromosome (kro′mo-sōm)   Threadlike structure built of DNA and protein that condenses and becomes visible in a cell's nucleus during mitosis.

chylomicron (kil″o-mi′kron)   Microscopic droplet of fat encased in protein that forms during fat absorption.

chyme (kīm)   Semifluid mass of partially digested food that passes from the stomach to the small intestine.

chymotrypsin (ki″mo-trip′sin)   Protein-splitting enzyme in pancreatic juice.

cilia (sil′e-ah)   Microscopic, hairlike extensions of the exposed surfaces of most cells.

ciliary body (sil′e-er″e bod′e)   Structure associated with the choroid layer of the eye that secretes aqueous humor and contains the ciliary muscle.

circadian rhythm (ser″kah-de′an rithm)   Pattern of repeated activity associated with the cycles of night and day.

circular muscles (ser′ku-lar mus′lz)   Muscles whose fibers are organized in circular patterns, typically around an opening or in the wall of a tube; sphincter muscles.

circumduction (ser″kum-duk′shun)   Movement of a body part, such as a limb, so that the end follows a circular path.

cisterna (sis-ter′nah)   Enlarged portion of the sarcoplasmic reticulum near the actin and myosin filaments of a muscle fiber.

citric acid cycle (sit′rik as′id si′kl)   Series of chemical reactions that oxidizes certain molecules, transferring energy; Krebs cycle.

cleavage (klēv′ij)   Early successive divisions of the zygote into a ball of progressively smaller cells.

clitoris (kli′to-ris)   Small erectile organ in the anterior vulva of the female, corresponding to the penis in the male.

clone (klōn)   Group of cells that descend from a single cell and are therefore genetically identical to it.

CNS Central nervous system.

coagulation (ko-ag″u-la′shun)   Blood clotting.

cochlea (kok′le-ah)   Part of the inner ear that has hearing receptors.

codominant (ko-dom′ĭ-nant)   Alleles that are each expressed in a heterozygote.

codon (ko′don)   Set of three contiguous nucleotides of a messenger RNA molecule that specifies a particular amino acid.

coenzyme (ko-en′zīm)   Small, nonprotein organic molecule required for the activity of a particular enzyme.

coenzyme A (ko-en′zīm)   A molecule that reacts with a 2-carbon acetyl group, forming acetyl coenzyme A, which enters the citric acid cycle.

cofactor (ko′fak-tor)   Small organic molecule or ion that must combine with an enzyme for activity.

collagen (kol′ah-jen)   Protein in the white fibers of connective tissues and in bone matrix.

collateral (ko-lat′er-al)   A branch of an axon or blood vessel.

collectin (ko-lek′tin) Protein that protects against bacteria, yeasts, and some viruses.

collecting duct (ko-lek′ting dukt)   In the lymphatic system, ducts into which lymphatic trunks drain; in the kidney, tubule that receives fluid from several nephrons.

colon (kolon)   Part of the large intestine.

colony-stimulating factor (ko′le-ne stim′yu-lay″ting fak′tor)   Protein that stimulates differentiation and maturation of white blood cells.

colorblindness (kul′erblīnd′nes)   Inherited inability to distinguish certain colors.

colostrum (ko-los′trum)   The first secretion of a woman's mammary glands after she gives birth.

compact bone (kom′pakt bōn)   Dense bone tissue in which cells are organized in osteons without apparent spaces.

complement (kom′plĕ-ment)   Group of proteins activated when an antibody binds an antigen; enhances reaction against nonself substances.

complementary base pair (kom″plĕ-men′ tă-re bās pār)   Hydrogen bonded adenine and thymine (A–T) or guanine and cytosine (G–C) in DNA. Adenine bonds to uracil (A–U) in RNA.

completely penetrant (kom-plēt′le pen′e-trent)   A genotype that affects everyone who inherits it.

complete protein (kom-plēt pro′tēn)   Protein that contains adequate amounts of the essential amino acids to maintain body tissues and to promote normal growth and development.

compound (kom′pownd)   Substance composed of two or more chemically bonded elements.

concentric contraction (kon-sen′-trik kon-trak′-shun)   A contraction that shortens a muscle.

condom (kon′dum)   Latex sheath used to cover the penis or line the vagina, preventing sperm from entering the uterus during sexual intercourse. Also used to minimize the risk of transmitting infection.

conduction (kon-duk′shun)   Movement of body heat into the molecules of cooler objects in contact with the body surface.

condyle (kon′dīl)   Rounded process of a bone, usually part of a joint.

condylar joint (kon′dǐ-lar joint)   Ovoid projection on one bone fits into a complementary elliptical cavity on another; ellipsoidal joint.

cone (kōn)   Photoreceptor in the retina of the eye that can detect color.

conformation (kon-for-ma′shun)   Three-dimensional structure of a protein, determined by its amino acid sequence and attractions and repulsions between amino acids in different parts of the molecule.

conjunctiva (kon″junk-ti′vah)   Mucous membrane covering the inside of the eyelid and much of the anterior surface of the eye.

connective tissue (kŏ-nek′tiv tish′u)   Basic tissue type that consists of cells within an extracellular matrix, including bone, cartilage, blood, loose and dense connective tissues.

contraception (kon″trah-sep′shun)   Behavior or device that prevents fertilization or implantation.

contractility (kon″trak-til′ǐ-te)   A measure of the force generated by contraction of the heart muscle with a given volume of blood in the ventricles.

contralateral (kon″trah-lat′er-al)   On the opposite side of the body.

convection (kon-vek′shun)   Transmission of heat from one substance to another through the circulation of heated air particles.

convergence (kon-ver′jens)   Two or more neurons forming synapses with the same neuron.

cornea (kor′ne-ah)   Transparent anterior portion of the outer layer of the eye wall.

corona radiata (ko-ro′nah ra-di-ă′ta)   Follicular cells surrounding the zona pellucida of an ovum.

coronary artery (kor′o-na″re ar′ter-e)   An artery that supplies blood to the wall of the heart.

coronary sinus (kor′o-na″re si′nus)   Large vessel on the posterior surface of the heart into which the cardiac veins drain.

corpus callosum (kor′pus kah-lo′sum)   Mass of white matter in the brain composed of axons connecting the right and left cerebral hemispheres.

corpus luteum (kor′pus lu′te-um)   Structure that forms from the tissues of a ruptured ovarian follicle and secretes progesterone and estrogens.

cortex (kor′teks)   Outer portion of an organ such as the adrenal gland, cerebrum, or kidney.

cortical nephron (kor′tǐ-kl nef′ron)   Nephron with a corpuscle in the renal cortex near the surface.

cortisol (kor′ti-sol)   Glucocorticoid hormone secreted by the adrenal cortex.

costal (kos′tal)   Pertaining to the ribs.

countercurrent mechanism (kown′ter-kar″ent me′kĕ-nǐ″zm)   Maintains a hypertonic interstitial fluid in the renal medulla. Part of the process by which the kidneys concentrate urine.

covalent bond (ko′va-lent bond)   Chemical bond formed by electron sharing between atoms.

coxal (kok′sel)   Pertaining to the hip.

cranial (kra′ne-al)   Pertaining to the cranium. The part of the skull that does not include the face.

cranial cavity (kran′e-al kav′i-te)   Space in the cranium containing the brain.

cranial nerve (kra′ne-al nerv)   Any of the nerves that arise from the brain or brainstem.

cranium (kra′ne-um)   Structure that encloses and protects the brain, formed by eight of the skull bones; brain case.

creatine phosphate (kre′ah-tin fos′fāt)   Molecule in muscle that stores energy.

crest (krest)   Ridgelike projection of a bone.

cricoid cartilage (kri′koid kar′tǐ-lij)   A ringlike cartilage that forms the lower end of the larynx.

crista ampullaris (kris′tah am-pul-lah′ris)   Sensory organ in a semicircular canal that functions in the sense of dynamic equilibrium.

crossover (kroso′ver)   Exchange of genetic material between homologous chromosomes during meiosis.

crural (krur′al)   Pertaining to the leg.

cubital (ku′bi-tal)   Pertaining to the elbow.

cuspid (kus′pid)   A canine tooth.

cutaneous (ku-ta′ne-us)   Pertaining to the skin.

cyanosis (si″ah-no′sis)   Bluish skin coloration due to decreased blood oxygen level.

cyclic adenosine monophosphate (sik′lik ah-den′o-sēn mon″o-fos′fāt)   A second messenger molecule in a signal transduction pathway; cyclic AMP or cAMP.

cystic duct (sis′tik dukt)   Tube that connects the gallbladder to the bile duct.

cytokine (si′to-kīn)   Type of polypeptide secreted by a T lymphocyte that enhances cellular responses to antigens.

cytokinesis (si-to-kin-e′sis)   Division of the cytoplasm during the cell cycle.

cytoplasm (si′to-plazm)   The contents of a cell including the gel-like cytosol and organelles, excluding the nucleus, enclosed by the cell membrane.

cytoskeleton (si′to-skel″e-ton)   System of protein tubules and filaments that reinforces a cell's three-dimensional form and provides scaffolding and transport pathways for organelles.

cytosol (si′to-sol)   Fluid portion of the cytoplasm.

D

deamination (de-am″ǐ-na′shun)   Removing amino groups ($-NH_2$) from amino acids.

decomposition (de-kom″po-zish′un)   The breakdown of molecules.

deep (dēp)   More internal, not near the surface.

defecation (def″ĕ-ka′shun)   Discharge of feces from the rectum through the anus.

defensin (di-fen′sin)   Antimicrobial peptide.

dehydration (de″hi-dra′shun)   Excess water loss from the internal environment.

dehydration synthesis (de″hi-dra′shun sin′thĕ-sis)   Anabolic process that joins small molecules by releasing the equivalent of a water molecule; synthesis.

dendrite (den′drīt)   Process of a neuron that receives input from other neurons.

dental caries (den′tal kar′ēz)   Destruction of the enamel and dentin of teeth; tooth decay.

dentin (den′tin)   Bonelike substance that forms the bulk of a tooth beneath the enamel.

deoxyhemoglobin (de-ok″sǐ-he″mo-glo′bin)   Hemoglobin that is not saturated with oxygen.

deoxyribonucleic acid (dē-ok′si-rī″bō-nu-klē′ik as′id)   The genetic material; a double-stranded polymer of nucleotides, each containing a phosphate group, a nitrogenous base (adenine, thymine, guanine, or cytosine), and the sugar deoxyribose; DNA.

depolarization (de-po″lar-ǐ-za′shun)   The membrane of a neuron becoming less negative (more positive) than the resting potential.

depression (de-presh′un)   Downward displacement.

dermatome (der′mah-tōm)   Area of the body supplied by sensory nerve fibers associated with a particular dorsal root of a spinal nerve.

dermis (der′mis)   The thick layer of the skin beneath the epidermis.

descending colon (de-send′ing ko′lon) Part of the large intestine that passes downward along the left side of the abdominal cavity to the brim of the pelvis.

descending tract (de-send′ing trakt) Group of axons that conducts impulses downward from the brain through the spinal cord to motor neurons controlling muscles and glands.

desmosome (des′mo-sōm) Specialized junction between cells, which serves as a "spot weld."

detrusor muscle (de-trūz′or mus′l) Muscular layer of the wall of the urinary bladder.

dextrose (dek′strōs) Glucose.

diabetes insipidus (di″ah-be′tĕz in-sip′ĭ-dus) Extremely copious urine produced due to a deficiency of antidiuretic hormone or lack of ADH response.

diabetes mellitus (di″ah-be′tĕz mel-li′tus) Elevated glucose in the urine and blood due to a deficiency of insulin or a poor response to it.

diapedesis (di″ah-pĕ-de′sis) Movement of leukocytes between the cells of blood vessel walls.

diaphragm (di′ah-fram) A sheetlike structure largely composed of skeletal muscle and connective tissue that separates the thoracic and abdominal cavities; also a contraceptive device inserted in the vagina.

diaphysis (di-af′ĭ-sis) Shaft of a long bone.

diastole (di-as′to-le) Phase of the cardiac cycle when a heart chamber wall relaxes.

diastolic pressure (di-a-stol′ik presh′ur) Lowest arterial blood pressure during the cardiac cycle; occurs during diastole.

diencephalon (di″en-sef′ah-lon) Part of the brain in the region of the third ventricle that includes the thalamus and hypothalamus.

differentiation (dif′er-en″she-a′shun) Cell specialization.

diffusion (dĭ-fu′zhun) Random movement of molecules from a region of higher concentration toward one of lower concentration.

digestion (di-jest′yun) Breaking down of large nutrient molecules into molecules small enough to be absorbed; hydrolysis.

digital (di′ji-tal) Pertaining to the finger or toe.

dihydrotestosterone (di-hi″dro-tes-tos′ter-ōn) Hormone produced from testosterone that stimulates certain cells of the male reproductive system.

dipeptide (di-pep′tīd) Molecule composed of two joined amino acids.

diploid (dip′loid) Body cell with a full, paired, set of chromosomes (half from each parent); in humans 46 chromosomes (2 × 23).

disaccharide (di-sak′ah-rīd) Sugar produced by the union of two monosaccharides.

distal (dis′tal) Farther from the trunk or point of attachment; opposite of proximal.

diuretic (di″u-ret′ik) Substance that increases urine production.

divergence (di-ver′jens) A single neuron having synapses with two or more postsynaptic cells.

DNA Deoxyribonucleic acid; the genetic material.

DNA repair Reactions that enable a cell to correct certain types of mutations that occur during replication.

dominant allele (dom′ĭ-nant ah-lēl′) An allele that masks another allele.

dorsal root (dor′sal rōot) Structure containing axons of sensory neurons that emerges from the spinal cord to form part of each spinal nerve.

dorsal root ganglion (dor′sal rōot gang′gle-on) Mass of sensory neuron cell bodies forming an enlargement in the dorsal root associated with a spinal nerve.

dorsiflexion (dor″si-flek′shun) Ankle movement that brings foot closer to shin.

dorsum (dor′sum) Pertaining to the back surface of the body or a body part.

ductus arteriosus (duk′tus ar-te″re-o′sus) Blood vessel that connects the pulmonary artery and the aorta in a fetus.

ductus deferens (duk′tus def′er-ens) Tube that leads from the epididymis to the urethra of the male reproductive tract (pl., *ductus deferentia*); vas deferens.

ductus venosus (duk′tus ven-o′sus) Blood vessel that connects the umbilical vein and the inferior vena cava in a fetus.

duodenum (du″o-de′num) First portion of the small intestine that leads from the stomach to the jejunum.

dural sinus (du′ral si′nus) Blood-filled channel formed by the division of the dura mater into two layers.

dura mater (du′rah ma′ter) Tough outer layer of the meninges.

dynamic equilibrium (di-nam′ik e″kwĭ-lib′re-um) Maintenance of balance when the head and body are suddenly moved or rotated.

E

eccentric contraction (ek-sen′trik kon-trak′shun) A lengthening contraction, in which the force in a muscle is less than an opposing force.

ECG Electrocardiogram; EKG.

ectoderm (ek′to-derm) Outermost primary germ layer in the embryo.

edema (ĕ-de′mah) Fluid accumulation in tissue spaces.

effector (ĕ-fek′tor) A muscle or gland that effects change in the body.

efferent (ef′er-ent) Directed away from the central part of a structure.

ejaculation (e-jak″u-la′shun) Discharge of semen from the male urethra.

ejaculatory duct (e-jak′u-lah-to″re dukt) Tube, formed by the joining of the ductus deferens and the tube from the seminal vesicle, that carries sperm to the urethra.

elastic cartilage (e-las′tik kar′tĭ-lij) Opaque, flexible connective tissue with many elastic fibers.

elastic fiber (e-las′tik fi′ber) Stretchy, yellow connective tissue fiber consisting of the protein elastin.

electrocardiogram (e-lek″tro-kar′de-o-gram″) Recording of the electrical activity associated with the cardiac cycle; ECG or EKG.

electrolyte (e-lek′tro-līt) Substance that ionizes in a water solution.

electrolyte balance (e-lek′tro-līt bal′ans) Condition when the quantities of electrolytes entering the body equal those leaving it.

electron (e-lek′tron) A small, negatively charged particle that encircles the nucleus of an atom.

electron shell (e-lek′tron shel) The space occupied by an electron or several electrons encircling the nucleus of an atom at a particular energy level.

electron transport chain (e-lek′tron trans′pohrt) Series of oxidation-reduction reactions that takes high-energy electrons from glycolysis and the citric acid cycle to form water and ATP.

element (el′ĕ-ment) Any of the fundamental chemical substances, each characterized by a distinct type of atom.

elevation (el-e-va-shun) Upward movement of a body part.

embolus (em′bo-lus) Blood clot, gas bubble, or other object carried in circulation that may obstruct a blood vessel.

embryo (em′bre-o) A prenatal stage of development from fertilization through the eighth week of development, when the rudiments of all organs are present.

embryonic disc (em″brē-on′ik disk) Flattened area in the developing embryo from which the germ layers arise.

emission (e-mish′un) Movement of sperm cells and prostate and seminal vesicle secretions into the urethra.

emulsification (e-mul″sĭ-fĭ′ka′shun) Breaking up of fat globules into smaller droplets by the action of bile salts.

enamel (e-nam′el) Hard covering on the exposed surface of a tooth.

end-diastolic volume (end-di-a-stol′ik vol′ūm) Blood volume remaining in the ventricles at the end of ventricular diastole.

endocardium (en″do-kar′de-um) Inner lining of the heart chambers.

endochondral bone (en′do-kon′dral bōn) Bone that begins as hyaline cartilage that is subsequently replaced by bone tissue.

endocrine gland (en′do-krĭn gland) Gland that secretes hormones directly into the blood; hormone secreting gland.

endocytosis (en″do-si-to′sis) Process by which a cell membrane envelopes a particle and draws it into the cell in a vesicle.

endoderm (en′do-derm) Innermost primary germ layer in the embryo.

endolymph (en′do-limf) Fluid in the membranous labyrinth of the inner ear.

endometrium (en″do-me′tre-um) Inner lining of the uterus.

endomysium (en″do-mis′e-um) Layer of connective tissue surrounding each skeletal muscle fiber.

endoneurium (en″do-nu′re-um) Layer of loose connective tissue that surrounds individual nerve fibers.

endoplasmic reticulum (en-do-plaz′mic rē-tik′u-lum) Organelle composed of a network of connected membranous tubules and vesicles.

endorphin (en-dor′fin) Any group of neuropeptides synthesized in the pituitary gland and hypothalamus that suppresses pain.

endosteum (en-dos′tē-um) Tissue lining the medullary cavity in a bone.

endothelium (en″do-the′le-um) Layer of epithelial cells that forms the inner lining of blood vessels and heart chambers.

end-systolic volume (end-sis-tol′ik vol′ūm) Blood volume remaining in the ventricles at the end of ventricular systole.

energy (en′er-je) An ability to move something, and thus do work.

energy balance (en′er-je bal′ans) When the caloric intake of the body equals its caloric expenditure.

enkephalin (en-kef′ah-lin) Any group of neuropeptides in the brain and spinal cord that inhibits pain impulses.

enterogastric reflex (en-ter-o-gas′trik re′fleks) Inhibition of gastric peristalsis and gastric secretions when food enters the small intestine.

enzyme (en′zīm) Protein that catalyzes a specific biochemical reaction.

eosinophil (e″o-sin′o-fil) White blood cell containing cytoplasmic granules that turn deep red with acidic stain.

ependyma (ĕ-pen′dĭ-mah) Neuroglia that line the ventricles of the brain.

epicardium (ep″ĭ-kar′de-um) Visceral part of the pericardium on the surface of the heart.

epicondyle (ep″ĭ-kon′dīl) Projection of a bone above a condyle.

epidermis (ep″ĭ-der′mis) Outer, epithelial layer of the skin.

epididymis (ep″ĭ-did′ĭ-mis) Highly coiled tubule that leads from the seminiferous tubules of the testis to the ductus deferens (pl., *epididymides*).

epidural space (ep″ĭ-du′ral spās) Space between the dural sheath of the spinal cord and the bone of the vertebral canal.

epigastric region (ep″ĭ-gas′trik re′jun) Upper middle part of the abdomen.

epiglottis (ep″ĭ-glot′is) Flaplike, cartilaginous structure attached at the back of the tongue near the entrance to the trachea.

epimysium (ep″ĭ-mis′e-um) Outer layer of connective tissue surrounding a skeletal muscle.

epinephrine (ep″ĭ-nef′rin) Hormone the adrenal medulla secretes during times of stress; adrenalin.

epineurium (ep″ĭ-nu′re-um) Outermost layer of connective tissue surrounding a nerve.

epiphyseal plates (ep″ĭ-fiz′e-al plātz) Cartilaginous layers between the epiphyses and diaphysis of a long bone that grow, lengthening the bone.

epiphysis (e-pif′ĭ-sis) Either end of a long bone.

epithelial membrane (ep″ĭ-the′le-al mem′brān) Thin layer of epithelial tissue and connective tissue lining a cavity or covering a surface.

epithelial tissue (ep″ĭ-the′le-al tish′u) One of the basic types of tissue; it covers all free body surfaces.

equilibrium (e″kwĭ-lib′re-um) State of balance between two opposing forces or processes.

erection (ĕ-rek′shun) The filling of penile or clitoral tissues with blood, stiffening and elevating the structure.

erythroblast (ĕ-rith′ro-blast) A red blood cell at an immature stage.

erythroblastosis fetalis (ĕ-rith″ro-blas-to′sis fe-tal′is) Life-threatening massive agglutination in the fetus or neonate due to the mother's anti-Rh antibodies reacting with the baby's Rh-positive red blood cells.

erythrocyte (ĕ-rith′ro-sīt) Red blood cell.

erythropoietin (ĕ-rith″ro-poi′ĕ-tin) Kidney and liver hormone that promotes red blood cell formation; EPO.

esophageal hiatus (ĕ-sof″ah-je′al hi-a′tus) Opening in the diaphragm through which the esophagus passes.

esophagus (ĕ-sof′ah-gus) Tubular part of the digestive tract connecting the pharynx to the stomach.

essential amino acid (ĕ-sen′shal ah-me′no as′id) Amino acid required for health that body cells cannot synthesize in adequate amounts; must be obtained in diet.

essential fatty acid (ĕ-sen′shal fat′e as′id) Fatty acid required for health that body cells cannot synthesize in adequate amounts; must be obtained in diet.

essential nutrient (ĕ-sen′shal nu′trē-ent) Nutrient necessary for growth, normal functioning, and maintaining life that the diet must supply because the body cannot synthesize it.

estrogens (es′tro-jenz) Group of hormones (including estradiol, estrone, and estriol) that stimulates the development of female secondary sex characteristics and produces an environment suitable for fertilization, implantation, and growth of an embryo.

euploid (u′ploid) Having a balanced set of chromosomes.

eumelanin (u-mel′ah-nin) Brownish-black pigment.

evaporation (e″vap′o-ra-shun) Changing a liquid into a gas.

eversion (e-ver′zhun) Turning the plantar surface of the foot outward, away from midline.

exchange reaction (eks-chānj re-ak′shun) Chemical reaction in which parts of two types of molecules trade positions.

excretion (ek-skre′shun) Elimination of metabolic wastes from the internal environment.

exocrine gland (ek′so-krin gland) Gland that secretes its products into a duct or onto an outside body surface.

exocytosis (eks″o-si-to′sis) Transport of substances out of a cell in a membrane-bounded vesicle.

expiration (ek″spĭ-ra′shun) Breathing out; exhalation.

expiratory reserve volume (eks-pi′rah-to″re re-zerv′ vol′ūm) Volume of air that can be exhaled in addition to the resting tidal volume.

extension (ek-sten′shun) Movement increasing the angle between parts at a joint.

exteroreceptive sense (eks″ter-o-re-sep′tiv sens) Associated with changes at the body surface.

extracellular fluid (eks″trah-selu′-lar floo′id) Body fluids outside cells.

extracellular matrix (eks″trah-selu′-lar ma′trix) Fibers and ground substance in the spaces between cells, especially between connective tissue cells.

F

facilitated diffusion (fah-sil′ĭ-tāt′id dĭ-fu′zhun) Diffusion in which a protein channel or a carrier molecule transports a substance across a cell membrane from a region of higher concentration to a region of lower concentration.

facilitation (fah-sil″ĭ-tā′shun) Subthreshold stimulation of a neuron that increases the likelihood that it will reach threshold with further stimulation.

fascia (fash′e-ah) Sheet of dense connective tissue that encloses a muscle.

fascicle (fas′ĭ-k′l) Small bundle of muscle fibers.

fat (fat) An organic molecule that includes glycerol and fatty acids; adipose tissue.

fatty acid (fat′e as′id) One of the building blocks of a fat molecule.

fatty acid oxidase (fat′e as′id ok′si-days″) An enzyme that catalyzes the removal of hydrogen or electrons from a fatty acid molecule.

feces (fe′sēz) Material expelled from the digestive tract during defecation.

femoral (fem′or-al) Pertaining to the thigh.

fertilization (fer″tĭ-lĭ-za′shun) Union of an egg cell and a sperm cell.

fetus (fe′tus) Prenatal human after eight weeks of development.

fever (fe′ver) Elevation of body temperature due to an elevated set point.

fibrillation (fi″brī-la′shun) Uncoordinated contraction of cardiac muscle fibers.

fibrin (fi′brin) Insoluble, fibrous protein formed from fibrinogen during blood coagulation.

fibrinogen (fi-brin′o-jen) Plasma protein converted into fibrin during blood coagulation.

fibroblast (fi′bro-blast) Cell that produces fibers in connective tissues.

fibrocartilage (fi″bro-kar′ti-lij) Strongest and most durable cartilage; made up of cartilage cells and many collagen fibers.

fibrous joint (fi′brus joint) Two or more bones joined by dense connective tissue.

filtration (fil-tra′shun) Movement of molecules through a membrane as a result of hydrostatic pressure.

fimbria (fim′bre-ah) Fringelike process at the distal end of the infundibulum of a uterine tube.

fissure (fish′ūr) Deep groove separating parts, such as a fissure separating lobes of the cerebrum.

flagellum (flah-jel′um) Relatively long, motile process that extends from the surface of a sperm cell.

flexion (flek′shun) Bending at a joint to decrease the angle between bones.

folacin (fōl′ah-sin) B complex vitamin necessary for biosynthesis; folic acid.

follicle (fol′ĭ-kl) Pouchlike depression or cavity.

follicle-stimulating hormone (fol′ĭ-kl stim′u-la″ting hor′mōn) Hormone secreted by the anterior pituitary gland to stimulate development of an ovarian follicle in a female or production of sperm cells in a male; FSH.

follicular cells (fŏ-lik′u-lar selz) Ovarian cells that surround a developing egg cell and secrete female sex hormones.

fontanel (fon″tah-nel′) Membranous region between certain developing cranial bones in the skull of a fetus or infant.

foramen (fo-ra′men) An opening, usually in a bone or membrane (pl., *foramina*).

foramen magnum (fo-ra′men mag′num) Larger opening in the occipital bone of the skull through which the spinal cord passes.

foramen ovale (fo-ra′men o-val′e) Opening in the interatrial septum of the fetal heart.

forebrain (fōr′brān) Anteriormost part of the developing brain; gives rise to the cerebrum and basal nuclei.

fossa (fos′ah) Depression in a bone or other part.

fovea centralis (fo′ve-ah sen-tral′is) Depressed region of the retina, consisting of densely packed cones, that provides the greatest visual acuity.

fracture (frak′chur) Break in a bone.

Frank-Starling law of the heart (frank star′ling law of thĕ hart) The relatively longer the cardiac muscle fibers are stretched by filling, the greater the force of contraction.

free radical (frē rad′eh-kel) Highly reactive by-product of metabolism that can damage tissue.

frontal (frun′tal) Pertaining to the forehead.

FSH Follicle-stimulating hormone.

functional residual capacity (funk′shun-al re-zid′u-al kah-pas′i-te) Volume of air remaining in the lungs after a normal quiet respiration.

functional syncytium (funk′shun-al sin-sish′e-um) A mass of cells performing as a unit; those of the heart are joined electrically.

G

galactose (gah-lak′tōs) Monosaccharide component of the disaccharide lactose.

gallbladder (gawl′blad-er) Saclike organ associated with the liver that stores and concentrates bile.

ganglion (gang′gle-on) Mass of neuron cell bodies located outside the central nervous system (pl., *ganglia*).

gastric gland (gas′trik gland) Any of the glands in the stomach lining that secretes gastric juice.

gastric juice (gas′trik joo-s) Secretion of the gastric glands in the stomach containing mucus, digestive enzymes, and hydrochloric acid.

gastrin (gas′trin) Hormone secreted by the stomach that stimulates gastric juice secretion.

gastrula (gas′troo-lah) Embryonic stage after the blastula; cells differentiate and aggregate into endoderm, mesoderm, and ectoderm.

gene (jēn) Part of a DNA molecule that encodes the information to synthesize a protein; the unit of inheritance.

general sense (jen′er-al sens) A sense detected through receptors widely distributed throughout the body.

genetic code (jĕ-net′ik kōd) Information for synthesizing proteins, encoded in the nucleotide sequence of DNA molecules.

genetic heterogeneity (jĕ-net′ik het″er-o-je-ne′ĭ-te) A trait or condition that is inherited from more than one gene.

genetics (jĕ-net′iks) The study of the transmission of inherited traits.

genital (jen′i-tal) Pertaining to the genitalia (external organs of reproduction).

genome (jeh′nōm) Complete set of genetic instructions for an organism.

genomics (je-nom′iks) Study of all of the genetic information in an individual organism.

genotype (je′no-tīp) The alleles of a particular gene in an individual.

glans penis (glanz pe′nis) Enlarged mass of corpus spongiosum at the end of the penis; may be covered by the foreskin.

gliding joint (glīd′eng joint) Two bones with nearly flat surfaces joined.

globin (glo′bin) Protein part of a hemoglobin molecule.

globulin (glob′u-lin) Type of protein in blood plasma.

glomerular capsule (glo-mer′u-lar kap′sŭl) Double-walled enclosure of the glomerulus of a nephron; Bowman's capsule.

glomerular filtrate (glo-mer′u-lar fil′trāt) Water and solutes filtered out of the glomerular capillaries in the kidney.

glomerular filtration (glo-mer′u-lar fil-tra′shun) Filtration of plasma by the glomerular capillaries of the nephron.

glomerulus (glo-mer′u-lus) Capillary tuft partially enclosed by the glomerular capsule of a nephron.

glottis (glot′is) The true vocal cords and the slitlike opening between them.

glucagon (gloo′kah-gon) Hormone secreted by the pancreatic islets that releases glucose from glycogen.

glucocorticoid (gloo″ko-kor′tĭ-koid) Any of several hormones that the adrenal cortex secretes that affects carbohydrate, fat, and protein metabolism.

gluconeogenesis (gloo″ko-ne″o-jen′ĕ-sis) Synthesis of glucose from noncarbohydrates such as amino acids.

glucose (gloo′kōs) Monosaccharide in blood that is the primary source of cellular energy.

gluteal (gloo′te-al) Pertaining to the buttocks.

glycerol (glis′er-ol) Organic compound that is a building block for fat molecules.

glycogen (gli′ko-jen) Polysaccharide that stores glucose in the liver and muscles.

glycolysis (gli-kol′ĭ-sis) The energy-releasing breakdown of glucose to produce 2 pyruvic acid molecules and a net of 2 ATP.

glycoprotein (gli″ko-pro′te-in) Compound composed of a carbohydrate and a protein.

goblet cell (gob′let sel) Epithelial cell specialized to secrete mucus.

goiter (goi′ter) Enlarged thyroid gland.

Golgi apparatus (gol′jē ap″ah-ra′tus) Organelle that prepares and modifies cellular products for secretion.

Golgi tendon organs (gol′jē ten′dun or′ganz) Sensory receptors that sense tension and are involved in reflexes that help maintain posture.

gomphosis (gom-fo′sis) Type of joint in which a cone-shaped process is fastened in a bony socket.

gonad (go′nad) A sex cell-producing organ; an ovary or testis.

gonadotropin (go-nad″o-trōp′in) Hormone that stimulates activity in the gonads.

G protein (g pro′tēn) Organic compound which activates an enzyme bound to the inner surface of the cell membrane, eliciting a signal.

granulocyte (gran′u-lo-sīt) Leukocyte with granules in its cytoplasm.

gray matter (grā mat′er) Regions of the central nervous system that contain cell bodies and unmyelinated axons, and thus appears gray.

gray ramus (grā ra′mus) Short nerve branch containing postganglionic axons returning to a spinal nerve.

growth (grōth) Process by which a structure enlarges.

growth hormone (grōth hor′mōn) Hormone released by the anterior lobe of the pituitary gland that promotes growth of the organism; GH.

gubernaculum (goo″ber′nak′u-lum) A cord that connects two structures; guides the descent of the testes during development in the male.

gyrus (ji′rus) Elevation on the brain's surface caused by infolding; convolution.

H

hair cell (hār sel) Mechanoreceptor in the inner ear that triggers action potentials in fibers of the auditory nerve in response to sound vibrations.

hair follicle (hār fol′ĭ-kl) Tubelike depression in the skin where a hair develops.

haploid (hap′loid) Cell with half the normal number of chromosomes, in humans 23.

hapten (hap′ten) Small molecule that combines with a larger one, forming an antigen.

haustra (haws′trah) Pouches in the wall of the large intestine.

hematocrit (he-mat′o-krit) The percentage by volume of red blood cells in a sample of whole blood; packed cell volume (PCV).

hematopoiesis (hem″ah-to-poi-e′sis) Production of blood cells from dividing stem and progenitor cells; hemopoiesis.

hematopoietic stem cell (hem″ah-to-poi-e′tik stem sel) Stem cell that gives rise to blood cells; hemocytoblast.

heme (hēm) Iron-containing part of a hemoglobin molecule.

hemizygous (hem″ĭ-zi′gus) The genotype of having only one copy of a particular gene, such as certain genes on the X chromosome in males.

hemoglobin (he″mo-glo′bin) Oxygen-carrying pigment in red blood cells.

hemostasis (he″mo-sta′sis) Stoppage of bleeding.

hepatic (hĕ-pat′ik) Pertaining to the liver.

hepatic lobule (hĕ-pat′ik lob′ul) Functional unit of the liver.

hepatic sinusoid (hĕ-pat′ik si′nŭ-soid) Vascular channel in the hepatic lobules.

heterozygous (het″er-o-zi′gus) Different alleles in a gene pair.

hilum (hi′lum) Depression where vessels, nerves, and other structures (bronchus, ureter, etc.) enter an organ.

hindbrain (hīnd′brān) Posteriormost part of the developing brain that gives rise to the cerebellum, pons, and medulla oblongata.

hinge joint (hinj joint) Two bones joined where the convex end of one bone fits into the complementary concave end of another.

hippocampus (hip″o-kam′pus) Part of the cerebral cortex where memories form.

histamine (his′tah-min) Substance released from mast cells and basophils that promotes inflammation.

holocrine gland (ho′lo-krin gland) Gland whose secretion contains entire secretory cells.

homeostasis (ho″me-o-sta′sis) Dynamic state in which the body's internal environment is maintained within the normal range.

homeostatic mechanism (ho″me-o-stat′ik mek′ah-nizm) Any of the control systems that help maintain a normal internal environment in the body.

homozygous (ho″mo-zi′gus) Identical alleles in a gene pair.

hormone (hor′mōn) Substance secreted by an endocrine gland and transported in the blood, which has actions on its target cells.

human chorionic gonadotropin (hu′man ko″re-on′ik gon″ah-do-tro′pin) Hormone, secreted by an embryo, that helps support pregnancy; hCG.

humoral immune response (hu′mor-al i-mūn′ ri-spons′) Circulating antibodies' destruction of pathogens bearing nonself antigens.

hyaline cartilage (hi′ah-līn kar′tĭ-lij) Semitransparent, flexible connective tissue with ultrafine collagen fibers.

hydrogen bond (hi′ dro-jen bond) Weak bond between a hydrogen atom and an atom of oxygen or nitrogen.

hydrolysis (hi-drol′ĭ-sis) Enzymatically adding parts of a water molecule to split a bond.

hydrostatic pressure (hi″dro-stat′ik presh′ur) Pressure exerted by fluids, such as blood pressure.

hymen (hi′men) Membranous fold of tissue that partially covers the vaginal opening.

hyperextension (hi″per-ek-sten′shun) Extension beyond the anatomical position; also, injurious extension beyond the normal range of motion.

hyperglycemia (hi″per-gli-se′me-ah) Elevated blood glucose.

hyperkalemia (hi″per-kah-le′me-ah) Elevated blood potassium.

hypernatremia (hi″per-nah-tre′me-ah) Elevated blood sodium.

hyperparathyroidism (hi″per-par″ah-thi′roi-dizm) Oversecretion of parathyroid hormone.

hyperpolarization (hi″per-po″lar-i-za′shun) Increase in the negativity of the resting potential of a cell membrane; further from threshold.

hypertension (hi″per-ten′shun) Persistently elevated blood pressure.

hyperthyroidism (hi″per-thi′roi-dizm) Oversecretion of thyroid hormones.

hypertonic (hi″per-ton′ik) Solution with a greater osmotic pressure than the solution (usually body fluids) to which it is compared.

hypertrophy (hi-per′tro-fe) Enlargement of an organ or tissue.

hyperventilation (hi″per-ven″tĭ-la′shun) Deep and rapid breathing that lowers the blood CO_2 level.

hypochondriac region (hi″po-kon′dre-ak re′jun) Part of the abdomen on either side of the epigastric region; just inferior to the ribs.

hypoglycemia (hi″po-gli-se′me-ah) Low blood glucose.

hypokalemia (hi″po-kah-le′me-ah) Low blood potassium.

hyponatremia (hi″po-nah-tre′me-ah) Low blood sodium.

hypoparathyroidism (hi″po-par″ah-thi′roi-dizm) Undersecretion of parathyroid hormone.

hypophysis (hi-pof′i-sis) Pituitary gland.

hypoproteinemia (hi″po-pro″te-ĭ-ne′me-ah) Low blood proteins.

hypothalamus (hi″po-thal′ah-mus)   Part of the brain located below the thalamus and forming the floor of the third ventricle.

hypothyroidism (hi″po-thi′roi-dizm)   Undersecretion of thyroid hormones.

hypotonic (hi″po-ton′ik)   Solution with a lower osmotic pressure than the solution (usually body fluids) to which it is compared.

hypoxia (hi-pok′se-ah)   Deficiency of oxygen.

I

idiotype (id′e-o-tīp′)   Parts of an antibody's antigen binding site that are complementary in conformation to a particular antigen.

ileocecal sphincter (il″e-o-se′kal sfingk′ ter)   Ring of smooth muscle fibers at the distal end of the ileum where it joins the cecum.

ileum (il′e-um)   Part of the small intestine between the jejunum and the cecum.

ilium (il′e-um)   One of the fused bones of a hip bone.

immunity (ĭ-mu′nĭ-te)   Resistance to the effects of specific disease-causing agents.

immunoglobulin (im″u-no-glob′u-lin)   Globular plasma protein that functions as an antibody.

immunosuppressive drugs (im″u-no-sŭ-pres′iv drugz)   Substances that suppress the immune response against transplanted tissue.

impermeant solute (im-per′me-ant sol′ ut)   A solute unable to cross a cell membrane.

implantation (im″plan-ta′shun)   Embedding of a blastocyst in the lining of the uterus.

impulse (im′puls)   Bioelectric signal conducted along a neuron axon or muscle fiber.

incisor (in-si′zor)   One of the front teeth adapted for cutting food.

inclusion (in-kloo′zhun)   Inert chemicals in the cytoplasm.

incomplete dominance (in″kom-plēt′ do′meh-nents)   Heterozygote whose phenotype is intermediate between the phenotypes of the two homozygotes.

incompletely penetrant (in″kom-plēt′le pen′e-trent)   When the frequency of genotype expression is less than 100%.

inert (in-ert′)   Elements that do not react with other elements.

infancy (in′fan-se)   Period of life from the fifth week after birth through the end of the first year.

infection (in-fek′shun)   Invasion and multiplication of microorganisms or infectious agents in body tissues.

inferior (in-fēr′e-or)   Situated below something else; pertaining to the lower surface of a part.

inflammation (in″flah-ma′shun)   Tissue response to stress that includes pain, warmth, redness, and swelling.

infundibulum (in″fun-dib′u-lum)   Stalk attaching the pituitary gland to the base of the brain; also, the funnel-shaped end of a uterine tube.

inguinal (ing′gwĭ-nal)   Pertaining to the groin.

inguinal canal (ing′gwĭ-nal kah-nal′)   Passage in the lower abdominal wall through which a testis descends into the scrotum.

inguinal region (ing′gwĭ-nal re′jun)   Part of the abdomen on either side of the pubic region; iliac region.

inhibin (in′hib′in)   Hormone secreted by cells of the testes and ovaries that inhibits the secretion of FSH from the anterior pituitary gland.

innate defense (in-nāt′ de-fens′)   Inborn, nonspecific defense that blocks entry of or destroys pathogens.

inorganic (in″or-gan′ik)   Chemical substances that do not include both carbon and hydrogen atoms.

insertion (in-ser′shun)   End of a muscle attached to a more movable part.

inspiration (in″spĭ-ra′shun)   Breathing in; inhalation.

inspiratory capacity (in-spi′rah-to″re kah-pas′i-te)   Volume of air that can be inhaled after a resting expiration.

inspiratory reserve volume (in-spi′rah-to″re re-zerv′ vol′ūm)   Amount of air that can be inhaled in addition to the resting tidal volume.

insula (in′su-lah)   Cerebral lobe deep within the lateral sulcus.

insulin (in′su-lin)   Hormone the pancreatic islets secrete that stimulates cells to take up glucose.

integumentary (in-teg-u-men′tar-e)   Pertaining to the skin and its accessory organs.

intercalated disc (in-ter″kah-lāt′ed disk)   Connection between cardiac muscle cells that includes gap junctions.

intercellular junction (in″ter-sel′u-lar junk′shun)   Site of union between cells.

interferon (in″ter-fēr′on)   Subgroup of cytokines (immune system chemicals) that inhibit viral multiplication and tumor growth.

interleukin (in″ter-lu′kin)   Subgroup of cytokines (immune system chemicals) with varied effects.

internal environment (in-ter′nal en-vi′ron-ment)   Conditions inside the body, surrounding the cells.

interneuron (in″ter-nu′ron)   Neuron between a sensory neuron and a motor neuron; internuncial or association neuron.

interoceptive sense (in″ter-o-sep′tiv sens)   Detecting changes in the viscera; visceroreceptive sense.

interphase (in′ter-fāz)   Period between cell divisions when a cell metabolizes and prepares to divide.

interstitial cell (in″ter-stish′al sel)   Hormone-secreting cell between the seminiferous tubules of the testis; cell of Leydig.

intervertebral disc (in″ter-ver′tĕ-bral disk)   Fibrocartilage structure between the bodies of adjacent vertebrae.

intestinal gland (in-tes′tĭ-nal gland)   Tubular gland at the base of a villus in the intestinal wall.

intracellular fluid (in″trah-sel′u-lar floo′id)   Fluid inside cells.

intramembranous bone (in″trah-mem′brah-nus bōn)   Bone that forms from membranelike layers of primitive connective tissue.

intrauterine device (in″trah-u′ter-in de-vīs)   Object placed in the uterine cavity to prevent implantation of a fertilized ovum; IUD.

intrinsic factor (in-trin′sik fak′tor)   Substance that gastric glands produce that promotes intestinal absorption of vitamin B_{12}.

inversion (in-ver′zhun)   Turning the plantar surface of the foot inward, toward midline.

involuntary (in-vol′un-tār″e)   Not consciously controlled; functions automatically.

ion (i′on)   Electrically charged atom or molecule.

ionic bond (i-on′ik bond)   Chemical bond that results between two ions formed by transfer of electrons; electrovalent bond.

ipsilateral (ip″sī-lat′er-al)   On the same side of the body.

iris (i′ris)   Colored, muscular part of the eye around the pupil that regulates its size.

ischemia (is-ke′me-ah)   Deficiency of blood flow in a body part.

isometric contraction (i″so-met′rik kon-trak′shun)   Muscular contraction that does not change the muscle length.

isotonic (i″so-ton′ik)   Solution with the same osmotic pressure as the solution (usually body fluids) to which it is compared.

isotonic contraction (i″so-ton′ik kon-trak′shun)   Muscular contraction that changes muscle length.

isotope (i′so-tōp)   Atom that has the same number of protons in its nucleus as other atoms of that element but has a different number of neutrons.

isthmus (is′mus)   Narrow connection between two parts.

J

jejunum (jĕ-joo′num)   Part of the small intestine between the duodenum and the ileum.

joint (joy′nt)   Union of two or more bones; articulation.

joint capsule (joint kap′sul)   An envelope, attached to the end of each bone at the joint, enclosing the cavity of a synovial joint.

juxtaglomerular apparatus (juks″tah-glo-mer′u-lār ap″ah-ra′tus)   A structure in the nephron including cells in the wall of the afferent arteriole (juxtaglomerular cells) and cells in the ascending limb of the nephron loop (macula densa) that helps control renin secretion.

juxtamedullary nephron (juks″tah-med′u-lār-e nef′ron)   A nephron with its corpuscle in the cortex, but near the renal medulla.

K

karyotype (kar′ē-o-tīp)   A chart of the chromosomes arranged in size-ordered homologous pairs. The human karyotype has 23 chromosome pairs.

keratin (ker′ah-tin)   Intracellular protein in epidermis, hair, and nails.

keratinization (ker″ah-tin″ĭ-za′shun)   Process by which cells form fibrils of keratin and harden.

ketone body (ke′tōn bod′e)   Compound produced during fat catabolism, including acetone, acetoacetic acid, and betahydroxybutyric acid.

Kupffer cell (koop′fer sel)   Large, fixed phagocyte in the liver that removes bacterial cells from the blood.

kwashiorkor (kwash″e-or′kor)   Starvation caused by switching from breast milk to food low in protein.

L

labor (la′bor)   Process of childbirth.

labyrinth (lab′ĭ-rinth)   System of connecting tubes in the inner ear, including the cochlea, vestibule, and semicircular canals.

lacrimal gland (lak′rĭ-mal gland)   Tear-secreting gland.

lactase (lak′tās)   Enzyme that catalyzes breakdown of lactose into glucose and galactose.

lacteal (lak′te-al)   Lymphatic capillary associated with a villus of the small intestine.

lactic acid (lak′tik as′id)   Organic compound formed from pyruvic acid in the anaerobic pathway of cellular respiration.

lactose (lak′tōs)   A disaccharide in milk; milk sugar.

lacuna (lah-ku′nah)   Small chamber or cavity.

lamella (lah-mel′ah)   Layer of matrix surrounding the central canal of an osteon.

lamellated corpuscle (lah-mel′a-ted kor′pusl)   Sensory receptor deep in the dermis providing perception of pressure; Pacinian corpuscle.

large intestine (lahrj in-tes′tin)   Part of the gastrointestinal tract from the ileum to the anus, divided into the cecum, colon, rectum, and anal canal.

laryngopharynx (lah-ring″go-far′ingks)   Lower part of the pharynx, posterior to the larynx, that leads to the esophagus.

larynx (lar′ingks)   Structure between the pharynx and trachea that houses the vocal cords.

latent period (la′tent pe′re-od)   Time between the application of a stimulus and the beginning of a response in a muscle fiber.

lateral (lat′er-al)   Pertaining to the side, away from midline.

lateral region (lat′er-al re′jun)   Part of the abdomen on either side of the umbilical region; lumbar region.

leptin (lep′tin)   Hormone, produced by fat cells, that binds receptors in the hypothalamus, suppressing hunger.

leukocyte (lu′ko-sīt)   White blood cell.

leukocytosis (lu″ko-si-to′sis)   Too many white blood cells in the blood.

leukopenia (lu″ko-pe′ne-ah)   Too few white blood cells in the blood.

lever (lev′er)   Simple mechanical device consisting of a rod, fulcrum, weight, and a force that is applied to some point on the rod.

ligament (lig′ah-ment)   Cord or sheet of connective tissue binding two or more bones at a joint.

limbic system (lim′bik sis′tem)   Connected structures in the brain that produce emotional feelings.

linea alba (lin′e-ah al′bah)   Narrow band of tendinous connective tissue in the midline of the anterior abdominal wall.

lingual (ling′gwal)   Pertaining to the tongue.

lingual frenulum (ling′gwal fren′ u-lum)   Fold of tissue that anchors the tongue to the floor of the mouth.

lipase (lī′pās)   Fat-digesting enzyme.

lipid (lip′id)   Group of organic compounds that includes fats (triglycerides), phospholipids, and steroids.

lipoprotein (lip″o-pro′te-in)   A complex of lipid and protein.

liver (liv′er)   Large, dark red organ in the upper right quadrant of the abdomen that detoxifies blood, stores glycogen and fat-soluble vitamins and synthesizes proteins, including clotting factors and enzymes.

lobule (lob′ul)   Small, well-defined part of an organ.

long-term synaptic potentiation (long-term sĭ-nap′tik po-ten′she-a-shun)   Theory that frequent, repeated stimulation of the same neurons in the hippocampus strengthens their synaptic connections.

lower esophageal sphincter (loh′er ĕ-sof′ah-je′al sfingk′ter)   Ring of smooth muscle, at the distal end of the esophagus where it joins the stomach, that prevents food from re-entering the esophagus when the stomach contracts; cardiac sphincter.

lower limb (loh′er lim)   Either inferior appendage consisting of the thigh, leg, and foot.

lumbar (lum′bar)   Pertaining to the lower back.

lumen (lu′men)   Hollow part of a tubular structure such as a blood vessel or intestine.

luteinizing hormone (lu′te-in-īz″ing hor′mōn)   A hormone that the anterior pituitary gland secretes that controls formation of the corpus luteum in females and stimulates sex hormone secretion in both sexes.

lymph (limf)   Fluid, derived from interstitial fluid, that the lymphatic vessels carry.

lymph node (limf nōd)   Mass of lymphoid tissue located along the course of a lymphatic vessel.

lymphocyte (lim′fo-sīt)   Type of white blood cell that provides immunity; B cell or T cell.

lysosome (li′so-sōm)   Organelle that contains digestive enzymes.

M

macromolecule (mak′ro-mol′ĕ-kūl)   Very large molecule, such as a protein, starch, or nucleic acid.

macronutrient (mak′ro-nu′tre-ent)   Nutrient (carbohydrate, lipid, and protein) required in a large amount.

macrophage (mak′ro-fāj)   Large phagocytic cell.

macula (mak′u-lah)   Hair cells and supporting cells associated with an organ of static equilibrium.

macula lutea (mak′u-lah lu′te-ah)   Yellowish depression in the retina of the eye associated with acute vision.

major histocompatibility complex (ma′jŏr his″to-kom-pat″ĭ-bil′ĭ-te kom′pleks)   Cluster of genes that code for cell surface proteins; MHC.

major mineral (ma′jor min″er-al)   Inorganic substance necessary for metabolism that is part of a group that accounts for 75% of the mineral elements in the body; macromineral.

malignant (mah-lig′nant)   The power to threaten life; cancerous.

malnutrition (mal″nu-trish′un)   Symptoms resulting from lack of specific nutrients.

maltase (mawl′tās)   Enzyme that catalyzes breakdown of maltose into glucose.

maltose (mawl′tōs)   Disaccharide composed of two glucose molecules.

mammary (mam′ar-e)   Pertaining to the breast.

mammillary body (mam′ĭ-lar″e bod′e)   One of two small, rounded bodies posterior to the hypothalamus involved with reflexes associated with the sense of smell.

marasmus (mah-raz′mus)   Starvation due to profound nutrient deficiency.

marrow (mar′o) Connective tissue within spaces in bones that includes hematopoietic stem cells.

mast cell (mast sel) Cell to which antibodies formed in reponse to allergens attach, causing the cell to release allergy mediators which cause symptoms.

mastication (mas′tĭ-ka′shun) Chewing actions.

matter (mat′er) Anything that has weight and occupies space.

meatus (me-a′tus) Passageway.

mechanoreceptor (mek′ah-no-re-sep′tor) Sensory receptor sensitive to mechanical stimulation, such as changes in pressure or tension.

medial (me′de-al) Toward or near the midline.

mediastinum (me″de-ah-sti′num) The region in the thoracic cavity between the lungs.

medulla (mĕ-dul′ah) Inner part of an organ.

medulla oblongata (mĕ-dul′ah ob″long-gah′tah) Part of the brainstem between the pons and the spinal cord.

medullary cavity (med′u-lār″e kav′ĭ-te) Cavity containing red or yellow marrow within the diaphysis of a long bone.

medullary respiratory center (med′u-lār″e re-spi′rah-to″re sen′ter) Area of the brainstem that controls the rate and depth of breathing.

megakaryocyte (meg″ah-kar′e-o-sīt) Large bone marrow cell that breaks apart to yield blood platelets.

meiosis (mi-o′sis) Cell division that halves the chromosome number, producing egg and sperm cells (gametes).

melanin (mel′ah-nin) Dark pigment generally found in skin and hair.

melanocyte (mel′ah-no-sīt) Melanin-producing cell.

melatonin (mel″ah-to′nin) Hormone that the pineal gland secretes.

membrane potential (mem′brān po-ten′shal) Stored electrical energy across a cell membrane due to the unequal distribution of ions on the two sides of the membrane.

memory cell (mem′o-re sel) B lympho-cyte or T lymphocyte produced in a primary immune response that can be activated rapidly if the same antigen is encountered again.

memory consolidation (mem′o-re kon-sol″ĭ-da′shun) Conversion of short-term memories to long-term memories.

menarche (mĕ-nar′ke) A female's first reproductive cycle.

meninx (me′-ninks) Any of the three layers of membrane that covers the brain and spinal cord (pl., *meninges*).

meniscus (men-is′kus) Fibrocartilage that separates the articulating surfaces of bones in the knee (pl., *menisci*).

menopause (men′o-pawz) Cessation of the female reproductive cycles.

menses (men′sēz) Shedding of blood and tissue from the uterine lining at the end of a female reproductive cycle.

mental (men′tal) Pertaining to the chin body region.

merocrine gland (mer′o-krĭn gland) A gland whose cells remain intact while secreting; a type of sweat gland.

mesentery (mes′en-ter″e) Fold of peritoneal membrane that attaches abdominal organs to the posterior abdominal wall.

mesoderm (mez′o-derm) Middle primary germ layer of the embryo.

messenger RNA (mes′in-jer RNA) RNA that carries information for a protein's amino acid sequence from the nucleus of a cell to the cytoplasm; mRNA.

metabolic pathway (met″ah-bol′ik path′wa) Series of linked, enzymatically controlled chemical reactions.

metabolism (mĕ-tab′o-lizm) The combined chemical reactions in cells that use or release energy.

metacarpal (met″ah-kar′pal) Any of the five bones of the hand between the wrist and finger bones.

metaphase (met′ah-fāz) Stage in mitosis when chromosomes align in the middle of the cell

metatarsal (met″ah-tar′sal) Any of the five foot bones between the ankle and toe bones.

microfilament (mi″kro-fil′ah-ment) Rod of the actin protein that provides structural support or movement; part of the cytoskeleton.

microglia (mi-krog′le-a) Neuroglia of the CNS that support neurons and phagocytize bacteria and cellular debris.

micronutrient (mi-kro-nu′tre-ent) Nutrient (vitamin or mineral) required in small amount.

microtubule (mi′kro-tu′būl) Hollow rod of the protein tubulin; part of the cytoskeleton.

microvillus (mi″kro-vil′us) Any of the many cylindrical processes that extends from some epithelial cell membranes, increasing membrane surface area (pl., *microvilli*).

micturition (mik″tu-rish′un) Urination.

midbrain (mid′brān) Small region of the brainstem between the diencephalon and the pons.

mineral (min′er-al) Inorganic element essential in human metabolism.

mineralocorticoid (min″er-al-o-kor′tĭ-koid) Type of hormone the adrenal cortex secretes that affects electrolyte concentrations in body fluids.

mitochondrion (mi″to-kon′dre-on) Organelle housing enzymes that catalyze aerobic reactions of cellular respiration (pl., *mitochondria*).

mitosis (mi-to′sis) Division of a somatic cell forming two genetically identical somatic cells.

mitral valve (mi′tral valv) Heart valve between the left atrium and the left ventricle; bicuspid valve.

mixed nerve (mikst nerv) Nerve that includes both sensory and motor nerve fibers.

molar (mo′lar) Rear tooth with a flattened surface adapted for grinding food.

molecular formula (mo-lek′u-lar for′mu-lah) Abbreviation for the number of atoms of each element in a compound.

molecule (mol′ĕ-kūl) Particle composed of two or more joined atoms.

monoamine oxidase (mon″o-am′ēn ok′sĭ-dās) Enzyme that inactivates monoamine neurotransmitters (such as norepinephrine).

monocyte (mon′o-sīt) Type of white blood cell that can leave the bloodstream and become a macrophage.

monosaccharide (mon″o-sak′ah-rīd) Single sugar, such as glucose or fructose.

monosomy (mon′o-so″me) Cell missing one chromosome.

morula (mor′u-lah) Early stage in prenatal development; solid ball of cells.

motor area (mo′tor a′re-ah) Region of the frontal lobe of the brain that controls voluntary movement.

motor end plate (mo′tor end plāt) Specialized part of a muscle fiber membrane at a neuromuscular junction.

motor nerve (mo′tor nerv) Nerve that consists of axons of motor neurons.

motor neuron (mo′tor nu′ron) Neuron that conducts impulses from the central nervous system to an effector.

mucosa (mu-ko′sah) Innermost layer of the alimentary canal.

mucous membrane (mu′kus mem′brān) Type of membrane that lines tubes and body cavities that open to the outside of the body.

mucus (mu′kus) Fluid secretion of mucous cells.

multiple motor unit summation (mul′tĭ-pl mo′tor u′nit sum-mā′shun) An increase in the number of activated motor units as a muscle contracts.

multipolar neuron (mul″tĭ-po′lar nu′ron) Nerve cell that has many processes.

muscle fiber (mus′el fi′ber) Skeletal muscle cell.

muscle spindle (mus′el spin′dul) Mechanoreceptor that senses changes in muscle length.

muscle tissue (mus′el tish′u) Contractile tissue of filaments of actin and myosin, which slide past each other, shortening cells.

muscle tone (mus′el tōn) Ongoing contraction of some fibers in otherwise resting skeletal muscle.

mutagen (mu′tah-jen) Agent that can cause mutations.

mutant (mu′tant)   Allele for a certain gene that has been altered from the "normal" condition.

mutation (mu-ta′shun)   Change in a gene.

myelin (mi′ĕ-lin)   Lipid material that forms a sheathlike covering around some axons.

myocardium (mi″o-kar′de-um)   Muscle layer of the heart.

myofibril (mi″o-fi′bril)   Any of the thread-like bundles of filaments in striated muscle cells.

myoglobin (mi″o-glo′bin)   Oxygen-storing pigment in muscle tissue.

myometrium (mi″o-me′tre-um)   Layer of smooth muscle tissue in the uterine wall.

myopia (mi-o′pe-ah)   Nearsightedness.

myosin (mi′o-sin)   Protein that, with actin, forms the filaments that interact to contract muscle fibers.

N

nail (nāl)   Protective plate at the distal end of a finger or toe.

nasal cavity (na′zal kav′ĭ-te)   Space within the nose.

nasal concha (na′zal kong′kah)   Any of the six shelflike bones or bony processes that extend from the wall of the nasal cavity; a turbinate bone.

nasal septum (na′zal sep′tum)   Wall of bone and cartilage that separates the nasal cavity into two parts.

nasopharynx (na″zo-far″ingks)   Part of the pharynx posterior to the nasal cavity.

natural killer cell (nat′u-ral kil′er sel)   Lymphocyte that bursts an infected or cancerous cell.

negative feedback (neg′ah-tiv fēd′bak)   A mechanism that restores the level of a biochemical or other condition in the internal environment.

neonatal (ne″o-na′tal)   The first four weeks after birth.

nephron (nef′ron)   Any of the one million functional units of a kidney, consisting of a renal corpuscle and a renal tubule.

nerve (nerv)   Bundle of axons in the peripheral nervous system.

nerve cell (nerv sel)   Neuron.

nervous tissue (ner′vus tish′u)   Neurons and neuroglia composing the brain, spinal cord, and nerves.

net filtration pressure (fil-tra′shun presh′ur)   The driving force for glomerular filtration in the kidneys.

neurilemma (nur″ĭ-lem′ah)   Living cellular layer formed from Schwann cells on the exterior of some axons.

neurofibril (nu″ro-fi′bril)   Fine cytoplasmic thread that extends from the cell body into the axon of a neuron.

neuroglia (nu-ro′gle-ah)   Specialized cells of the nervous system that produce myelin, maintain the ionic environment, provide growth factors that support neurons, provide structural support, and play a role in cell-to-cell communication.

neuromodulator (nu″ro-mod′u-lā-tor)   Substance that alters a neuron's response to a neurotransmitter.

neuromuscular junction (nu″ro-mus′ku-lar jungk′shun)   Synapse between a motor neuron and a skeletal muscle fiber; myoneural junction.

neuron (nu′ron)   Nerve cell.

neuronal pool (nu′ro-nal pool)   A group of nerve cells in the central nervous system that synapse with each other.

neuropeptide (nu″ro-pep′tīd)   Peptide in the brain that functions as a neurotransmitter or neuromodulator.

neurosecretory cell (nu″ro-se-kre′to-re sel)   Cell in the hypothalamus that conducts impulses like a neuron but secretes into the bloodstream.

neurotransmitter (nu″ro-trans-mit′er)   Chemical that an axon secretes, into a synapse, that stimulates or inhibits an effector (muscle or gland) or other neuron.

neutral (nu′tral)   Neither acidic nor alkaline; pH 7.

neutron (nu′tron)   Electrically neutral subatomic particle.

neutrophil (nu′tro-fil)   Type of phagocytic white blood cell containing cytoplasmic granules that react with neutral pH stain.

niacin (ni′ah-sin)   Vitamin of the B-complex group; nicotinic acid.

nitrogen balance (ni′tro-jen bal′ans)   Condition in which the amount of nitrogen ingested equals the amount excreted.

node of Ranvier (nōd of Ron′vee-ay)   Any of the many gaps in the myelin sheath along axons of neurons of the peripheral nervous system.

nondisjunction (non″dis-jungk′shun)   A pair of chromosomes that incorrectly remains together rather than separating during meiosis.

nonprotein nitrogenous substance (non-pro′tēn ni-troj′ĕ-nus sub′stans)   A nitrogen-containing molecule that is not a protein.

norepinephrine (nor″ep-ĭ-nef′rin)   Neurotransmitter released from the axons of some nerve fibers; a hormone from the adrenal medulla; noradrenaline.

normal range (nor′mal rānj)   Range of values for a particular measurement obtained from a sample of the healthy population.

nuclear envelope (nu′kle-ar en′vĕ-lōp)   Double bilayer membrane surrounding the cell nucleus and separating it from the cytoplasm.

nuclear pore (nu′kle-ar pōr)   Protein-lined channel in the nuclear envelope.

nuclease (nu′kle-ās)   Enzyme that catalyzes decomposition of nucleic acids.

nucleic acid (nu-kle′ik as′id)   A molecule that is composed of bonded nucleotides; RNA or DNA.

nucleolus (nu-kle′o-lus)   Small structure in the cell nucleus that contains RNA and proteins and is the site of ribosome production (pl., *nucleoli*).

nucleoplasm (nu′kle-o-plazm″)   Contents of the cell nucleus.

nucleotide (nu′kle-o-tīd″)   Building block of a nucleic acid molecule, consisting of a sugar, nitrogenous base, and phosphate group.

nucleus (nu′kle-us)   The dense core of an atom that is composed of protons and neutrons; cellular structure enclosed by a double bilayer nuclear envelope and containing DNA; masses of interneuron cell bodies in the CNS (pl., *nuclei*).

nutrient (nu′tre-ent)   Chemical that the body requires from the environment.

nutrition (nu-trish′un)   Study of the sources, actions, and interactions of nutrients.

O

obesity (o-bēs′ĭ-te)   Excess adipose tissue; a body mass index greater than 30.

occipital (ok-sip′i-tal)   Pertaining to the lower, back portion of the head.

olfactory (ol-fak′to-re)   Pertaining to the sense of smell.

oligodendrocyte (ol″ĭ-go-den′dro-sīt)   Type of neuroglia in the CNS that produces myelin.

oncogene (ong′ko-jēn)   Gene that normally controls cell division but when overexpressed leads to cancer.

oocyte (o′o-sīt)   Cell formed by oogenesis; egg cell.

oogenesis (o′o-jen′ĕ-sis)   Formation of an egg cell.

optic chiasma (op′tik ki-az′mah)   X-shaped structure on the underside of the brain formed by optic nerve fibers that cross over.

optic disc (op′tik disk)   Region in the retina of the eye where sensory fibers exit, becoming part of the optic nerve.

oral (o′ral)   Pertaining to the mouth.

orbital (or′bĭ-tal)   Pertaining to the cavities containing the eyes; region in the atom containing electrons.

organ (or′gan)   Structure consisting of two or more tissues with a specialized function.

organelle (or″gah-nel′)   Any of the structures in cells that has a specialized function.

organic (or-gan′ik)   A molecule that contains both carbon and hydrogen.

organism (or′gah-nizm)   An individual living thing.

organ system (or′gan sis′tem)   Group of organs coordinated to carry on a specialized function.

orgasm (or′gaz-em)   An intense sensation that is the culmination of sexual stimulation.

origin (or′ĭ-jin)   End of a muscle that attaches to a relatively immovable part.

oropharynx (o″ro-far′ingks)   Part of the pharynx posterior to the oral cavity.

osmoreceptor (oz″mo-re-sep′tor)   Receptor that senses changes in the osmotic pressure of body fluids.

osmosis (oz-mo′sis)   Movement of water through a selectively permeable membrane toward a greater concentration of impermeant solute.

osmotic pressure (oz-mot′ik presh′ur)   Hydrostatic pressure needed to stop osmosis.

ossification (os″ĭ-fĭ-ka′shun)   Formation of bone tissue.

osteoblast (os′te-o-blast″)   Bone-forming cell.

osteoclast (os′te-o-klast″)   Cell that breaks down bone matrix.

osteocyte (os′te-o-sīt)   Mature bone cell.

osteon (os′te-on)   Cylinder-shaped unit containing bone cells and matrix lamellae that surround a central canal; Haversian system.

osteoporosis (os″te-o-po-ro′sis)   Condition in which bones break easily because bone matrix is lost faster than it is replaced.

otic (o′tik)   Pertaining to the ear.

otolith (o′to-lith)   A small particle of calcium carbonate associated with the receptors of static equilibrium.

oval window (o′val win′do)   Opening to the inner ear, covered by the stapes.

ovarian (o-va′re-an)   Pertaining to the ovary.

ovary (o′var-e)   Primary female reproductive organ; egg cell-producing organ.

ovulation (o″vu-la′shun)   Release of an egg cell from a mature ovarian follicle.

oxidation (ok″sĭ-da′shun)   Removal of electrons; opposite of reduction.

oxygen debt (ok′sĭ-jen det)   Amount of oxygen muscle cells require after physical exercise to restore levels of glucose, ATP, and creatine phosphate.

oxyhemoglobin (ok″sĭ-he″mo-glo′bin)   Compound formed when oxygen binds hemoglobin.

oxytocin (ok″sĭ-to′sin)   Posterior pituitary hormone that contracts smooth muscles in the uterus and mammary gland myoepithelial cells.

P

pacemaker (pās′māk-er)   Mass of specialized cardiac muscle tissue that controls the rhythm of the heartbeat; the sinoatrial node.

packed cell volume (pakt sel vol′ūm)   Percent volume of red blood cells in a sample of centrifuged blood; hematocrit.

pain receptor (pān re″sep′tor)   Sensory nerve ending that conducts impulses interpreted as pain.

palate (pal′at)   Roof of the mouth.

palatine (pal′ah-tīn)   Pertaining to the palate.

palmar (pahl′mar)   Pertaining to the palm of the hand.

pancreas (pan′kre-as)   Glandular organ in the abdominal cavity that secretes hormones and digestive enzymes.

pantothenic acid (pan″to-the′nik as′id)   Vitamin of the B-complex group; vitamin B$_5$.

papilla (pah-pil′ah)   Tiny, nipplelike projection.

papillary muscles (pap′ĭ-ler″e mus′elz)   Muscles that extend inward from the ventricular walls of the heart and to which the chordae tendineae attach.

paracrine (par′ah-krin)   Type of endocrine secretion in which the hormone affects nearby cells.

paradoxical sleep (par″ah-dok′se-kal slēp)   Sleep in which some areas of the brain are active, producing dreams and rapid eye movements.

paranasal sinus (par″ah-na′zal si-nus)   Air-filled cavity in a cranial or facial bone, lined with mucous membrane and connected to the nasal cavity.

parasympathetic division (par″ah-sim″pah-thet′ik dĭ-vizh′un)   Part of the autonomic nervous system that arises from the brain and sacral region of the spinal cord.

parathyroid gland (par″ah-thi′roid gland)   One of four small endocrine glands embedded in the posterior part of the thyroid gland.

parathyroid hormone (par″ah-thi′roid hor′mōn)   Hormone secreted by the parathyroid glands that helps regulate the level of blood calcium and phosphate ions; PTH.

parietal (pah-ri′ĕ-tal)   Pertaining to the wall of a cavity.

parietal cell (pah-ri′ĕ-tal sel)   Cell of a gastric gland that secretes hydrochloric acid and intrinsic factor.

parietal peritoneum (pah-ri′ĕ-tal per″ĭ-to-ne′um)   Membrane that lines the inner surface of the abdominal cavity.

parietal pleura (pah-ri′ĕ-tal ploo′rah)   Membrane that lines the inner wall of the thoracic cavity.

parotid glands (pah-rot′id glandz)   Large salivary glands deep to the sides of the face just in front and below the ears.

partial pressure (par′shal presh′ur)   Pressure one gas contributes in a mixture of gases.

patellar (pah-tel′ar)   Pertaining to the kneecap.

pathogen (path′o-jen)   Disease-causing agent.

pectoral (pek′tor-al)   Pertaining to the chest.

pectoral girdle (pek′tor-al ger′dl)   Part of the skeleton that supports and attaches the upper limbs.

pedal (ped′al)   Pertaining to the foot.

pedigree (ped′ĭ-gre)   Chart that displays relationships among family members and their inherited traits and disorders.

pelvic (pel′vik)   Pertaining to the pelvis.

pelvic cavity (pel′vik kav′ĭ-te)   Hollow place within the ring formed by the sacrum and hip bones.

pelvic girdle (pel′vik ger′dl)   Part of the skeleton to which the lower limbs attach.

pelvic inflammatory disease (pel′vik in-flam′ah-tore dĭ-zēz′)   Ascending infection of the upper female genital tract.

pelvis (pel′vis)   Basin-shaped structure formed by the sacrum and hip bones.

penis (pe′nis)   Male external reproductive organ through which the urethra passes.

pepsin (pep′sin)   Protein-splitting enzyme that the gastric glands secrete.

pepsinogen (pep-sin′o-jen)   Inactive form of pepsin.

peptidase (pep′tĭ-dās)   Enzyme that catalyzes the breakdown of polypeptides.

peptide (pep′tīd)   Compound composed of two or more amino acids.

peptide bond (pep′tīd bond)   Bond between the carboxyl group of one amino acid and the amino group of another.

perception (per-sep′shun)   Mental interpretation of sensory stimulation.

perforating canal (per′fo-rāt″eng kah-nal′)   Transverse channel that connects central canals in compact bone; Volkmann's canal.

perforin (per′fo-rin)   Protein that natural killer cells release that forms pores in the cell membrane of infected cell.

pericardial (per″ĭ-kar′de-al)   Pertaining to the pericardium.

pericardium (per″ĭ-kar′de-um)   Serous membrane that surrounds the heart.

perichondrium (per″ĭ-kon′dre-um)   Layer of dense connective tissue that encloses cartilaginous structures.

perilymph (per-ĭ-limf)   Fluid in the space between the membranous and osseous labyrinths of the inner ear.

perimetrium (per-ĭ-me′tre-um)   Outer serosal layer of the uterine wall.

perimysium (per″ĭ-mis′e-um)   Sheath of connective tissue that encloses a bundle of skeletal muscle fibers or a fascicle.

perineal (per″ĭ-ne′al)   Pertaining to the perineum, the diamond-shaped area extending from the coccyx to the pubic symphysis.

perineurium (per″ĭ-nu-re-um) Layer of connective tissue that encloses a bundle of nerve fibers in a nerve.

periodontal ligament (per″e-o-don′tal lig′ah-ment) Dense connective tissue that surrounds a tooth and attaches it to the jawbone.

periosteum (per″e-os′te-um) Dense connective tissue covering the surface of a bone.

peripheral nervous system (pě-rif′er-al ner′vus sis′tem) Parts of the nervous system outside the brain and spinal cord; PNS.

peripheral protein (pě-rif′er-al pro′tēn) Globular protein associated with the outer surface of the cell membrane.

peripheral resistance (pě-rif′er-al re-zis′tans) Resistance to blood flow due to friction between blood and blood vessel walls.

peristalsis (per″ĭ-stal′sis) Rhythmic waves of muscular contraction in the walls of certain tubular organs.

peritoneal cavity (per″ĭ-to-ne′al kav′ĭ-te) The potential space between the parietal and visceral peritoneal membranes.

peritoneum (per″ĭ-to-ne′um) Serous membrane that lines the abdominal cavity and encloses the abdominal viscera.

peritubular capillary (per″ĭ-tu′bu-lar kap′ĭ-ler″e) Capillary that surrounds a renal tubule and functions in tubular reabsorption and tubular secretion during urine formation.

peroxisome (pě-roks′ĭ-sōm) Membranous cytoplasmic vesicle that contains enzymes that catalyze reactions that produce and decompose hydrogen peroxide.

phagocyte (fag′o-sīt) Cell that ingests particulate matter.

phagocytosis (fag″o-si-to′sis) Process by which a cell engulfs solids from its surroundings.

phalanx (fa′langks) Bone of a finger or toe (pl., *phalanges*).

pharynx (far′ingks) Portion of the digestive tube posterior to the nasal and oral cavities, as well as the larynx.

phenotype (fe′no-tīp) A trait or health condition caused by the expression of a gene or genes.

pheomelanin (fe″o-mel′ah-nin) A reddish-yellow pigment.

phosphate buffer system (fos′fāt buf′er sis′tem) A mixture in which sodium monohydrogen phosphate acts as a weak base and sodium dihydrogen phosphate acts as a weak acid; resists changes in pH.

phospholipid (fos″fo-lip′id) Molecule consisting of two fatty acids and a phosphate group bound to a glycerol molecule.

phosphorylation (fos″for-ĭ-la′shun) Metabolic process that adds a phosphate to an organic molecule.

photoreceptor (fo″to-re-sep′tor) Sensory receptor sensitive to light; a rod or cone of the eye.

pH scale (pH skāl) Shorthand notation for the hydrogen ion concentration used to indicate the acidic or alkaline condition of a solution; values range from 0 to 14.

physiology (fiz″e-ol′o-je) The study of body functions.

pia mater (pi′ah ma′ter) Inner layer of meninges that encloses the brain and spinal cord.

pineal gland (pin′e-al gland) Small structure in the central part of the brain that secretes the hormone melatonin, which affects certain biological rhythms.

pinocytosis (pin″o-si-to′sis) Process by which a cell engulfs droplets of fluid from its surroundings.

pituitary gland (pĭ-tu′ĭ-tār″e gland) Endocrine gland attached to the base of the brain that consists of anterior and posterior lobes; the hypophysis.

pivot joint (piv′ut joint) End of a bone moving within a ring formed by another bone and connective tissue.

placenta (plah-sen′tah) Structure that attaches the fetus to the uterine wall, delivering nutrients to and removing wastes from the fetus.

placental lactogen (plah-sen′tahl lak′to-jen) Hormone secreted by the placenta that inhibits maternal insulin activity during pregnancy.

plantar (plan′tar) Pertaining to the sole of the foot.

plantar flexion (plan′tar flek′shun) Ankle movement that brings the foot farther from the shin.

plasma (plaz′mah) Fluid portion of the blood.

plasma cell (plaz′mah sel) Type of antibody-producing cell that forms when activated B cells proliferate.

plasma protein (plaz′mah pro′tēn) Protein dissolved in blood plasma.

plasmin (plaz′min) Protein-splitting enzyme that can digest fibrin in a blood clot.

platelet (plāt′let) Cellular fragment found in the blood that helps blood clot.

pleiotropy (plē′o-tro-pē) Gene that has several expressions (phenotypes).

pleural (ploo′ral) Pertaining to the pleura or membranes surrounding the lungs.

pleural cavity (ploo′ral kav′ĭ-te) Potential space between pleural membranes.

pleural membrane (ploo′ral mem′brān) Serous membrane that encloses the lungs and lines the chest wall.

plexus (plek′sus) Network of interlaced nerves or blood vessels.

pluripotent (ploo-rip′o-tent) Cell able to differentiate to yield several specialized cell types.

PNS Peripheral nervous system.

polar body (po′lar bod′e) Small, nonfunctional cell that is a product of meiosis in the female.

polarization (po″lar-ĭ-za′shun) Electrical charge difference across a cell membrane surface due to an unequal distribution of positive and negative ions on either side.

polar molecule (po′lar mol′ĕ-kūl) Molecule in which the electrons and protons are not distributed symmetrically.

polygenic (pol″ĕ-jēn′ik) An inherited trait that results from the actions of more than one gene.

polymorphonuclear leukocyte (pol″e-mor″fo-nu′kle-ar lu′ko-sīt) White blood cell with an irregularly lobed nucleus; neutrophil.

polynucleotide (pol″e-noo′-kle-o-tīd) Compound formed by the union of many nucleotides; a nucleic acid.

polypeptide (pol″e-pep′tīd) Compound formed by the union of many amino acid molecules.

polyploidy (pol′e-ploi″de) A cell with one or more extra sets of chromosomes.

polysaccharide (pol″e-sak′ah-rīd) Carbohydrate composed of many joined monosaccharides.

pons (ponz) Part of the brainstem above the medulla oblongata and below the midbrain.

popliteal (pop″lĭ-te′al) Pertaining to the region behind the knee.

positive chemotaxis (poz′ĭ-tiv ke″mo-tak′sis) Movement of a cell toward the greater concentration of a substance.

positive feedback (poz′ĭ-tiv fēd′bak) Process by which changes cause additional similar changes, producing unstable conditions.

posterior (pos-tēr′e-or) Toward the back; opposite of anterior.

postganglionic fiber (pōst″gang-gle-on′ik fi′ber) Autonomic nerve fiber on the distal side of a ganglion.

postnatal (pōst-na′tal) After birth.

postsynaptic neuron (pōst″sĭ-nap′tik nu′ron) Neuron on the "receiving" side of a synapse.

postsynaptic potential (pōst″sĭ-nap′tik po-ten′shal) Membrane polarization is decreased (excitatory) or increased (inhibitory) in the postsynaptic neuron with repeated stimulation over an excitatory or inhibitory pathway so that the neuron will either fire or not respond.

preganglionic fiber (pre″gang-gle-on′ik fi′ber) Autonomic nerve fiber bringing input into a ganglion.

pregnancy (preg′nan-se) Condition in which a female has a developing offspring in her uterus.

preload (pre′lōd) The volume of blood filling the relaxed ventricles prior to their contraction.

prenatal (pre-na′tal)   Before birth.

presbyopia (pres″be-o′pe-ah)   Loss of the eye's ability to accommodate due to declining elasticity in the lens; farsightedness of age.

presynaptic neuron (pre″sĭ-nap′tik nu′ron)   The neuron releasing neurotransmitter at a synapse, on the "sending" side of a synapse.

primary germ layers (pri′ma-re jerm lā′erz)   Three layers of cells in the embryo that divide and differentiate into specific tissues and organs; endoderm, mesoderm, and ectoderm.

primary immune response (pri′ma-re ĭ-mūn′ re-spons′)   Immune system's response to an initial encounter with a nonself antigen.

primary sex organs (pri′ma-re seks or′ganz)   Organs that produce sex cells; testes in males and ovaries in females.

prime mover (prīm moo-v′er)   Muscle that provides most of a particular body movement; agonist.

primordial follicle (pri-mor′de-al fol′lĭ-kl)   Egg enclosed by a single layer of cells in the ovary.

product (prod′ukt)   The result of a chemical reaction.

progenitor cell (pro-jen′ĭ-tor sel)   Daughter cell of a stem cell whose own daughter cells are restricted to follow specific lineages.

progesterone (pro-jes′tĕ-rōn)   Female hormone secreted by the corpus luteum of the ovary and by the placenta.

projection (pro-jek′shun)   Process by which the brain causes a sensation to seem to come from the region of the body being stimulated.

prolactin (pro-lak′tin)   Hormone secreted by the anterior pituitary gland that stimulates the production of milk in the mammary glands; PRL.

pronation (pro-na′shun)   Downward or backward rotation of the palm.

prophase (pro′fāz)   Stage of mitosis when chromosomes become distinguishable when stained and viewed under a microscope.

proprioceptor (pro″pre-o-sep′tor)   Sensory receptor that detects change in muscle, tendon, and body position.

prostaglandins (pros′tah-glan′dins)   Group of compounds that have powerful, hormonelike effects.

prostate gland (pros′tāt gland)   Gland surrounding the male urethra below the urinary bladder that secretes fluid into semen prior to ejaculation.

protein (pro′tēn)   Nitrogen-containing organic compound composed of many bonded amino acid molecules.

protein buffer system (pro′tēn buf′er sis′tem)   Amino acids of a protein in a solution accept or donate hydrogen ions to resist changes in pH.

protein kinase (pro′tēn ki′nās)   Enzyme that catalyzes phosphorylation of a protein.

prothrombin (pro-throm′bin)   Plasma protein that functions in blood clotting.

proton (pro′ton)   Positively charged particle in an atomic nucleus.

protraction (pro-trak′shun)   Forward movement of a body part.

provitamin (pro-vi′tah-min)   Precursor of a vitamin.

proximal (prok′sĭ-mal)   Closer to the trunk or point of attachment; opposite of distal.

pseudostratified columnar epithelium (soo″do-strat′ĭ-fīd co-lum′nar ep″ĭ-the′le-um)   Single layer of cells appearing as more than one layer because the nuclei occupy different positions in the cells.

puberty (pu′ber-te)   Stage of development in which the reproductive organs become functional.

pubic region (pu′bik re′jun)   Lower middle portion of the abdomen; hypogastric region.

pulmonary circuit (pul′mo-ner″e ser′kit)   System of blood vessels that transports blood between the heart and the lungs.

pulmonary valve (pul′mo-ner″e valv)   Valve leading from the right ventricle to the pulmonary trunk; pulmonary semilunar valve.

pulse (puls)   Surge of blood pressure felt through the walls of arteries due to the contraction of the heart ventricles.

Punnett square (pun′it skwair)   A grid diagram that displays possible genotypes in offspring based on parental gametes.

pupil (pu′pil)   Opening in the iris through which light enters the eye.

purine (pu rēn)   Type of nitrogen-containing base that is part of DNA and RNA and has two organic rings; adenine and guanine.

Purkinje fibers (pur-kin′je fi′berz)   Specialized muscle fibers that conduct the cardiac impulse from the AV bundle into the ventricular walls.

pyloric sphincter (pi-lor′ik sfingk′ter)   Ring of smooth muscle in the stomach wall where it joins the duodenum; controls food entry into the duodenum.

pyrimidine (pe-ri′mĭ-dēn)   Type of nitrogen-containing base that is part of DNA and RNA and has two organic rings; thymine, cytosine, and uracil.

pyruvic acid (pi-roo′vik as′id)   Intermediate product of carbohydrate oxidation.

R

radiation (ra′de-a′shun)   Form of energy that includes visible light, ultraviolet light, and X rays; the means by which body heat is lost as infrared rays.

radioactive (ra″de-o-ak′tiv)   Refers to an atom that releases energy at a constant rate.

rate-limiting enzyme (rāt lim′ĭ-ting en′zīm)   Enzyme, usually present in small amounts, that controls the rate of a metabolic pathway by regulating one of its steps.

reactant (re-ak′tant)   A chemical that takes part in a chemical reaction. A starting material.

receptor (re″sep′tor)   Specialized cell or structure that provides information about the environment; also, cell surface molecule that binds other molecules, called ligands, thereby conveying a signal to inside the cell.

receptor-mediated endocytosis (re″sep′tor-me′de-ā-tid en″do-si-to′sis)   Selective cellular uptake of molecules that bind to a specific receptor type.

recessive allele (re-ses′iv ah-lēl)   Form of a gene not expressed if the dominant form is also present.

recruitment (re-krōōt′ment)   Increase in number of motor units taking part in a muscle contraction.

rectum (rek′tum)   Terminal end of the digestive tube between the sigmoid colon and the anus.

red blood cell (red blud sel)   Disc-shaped cell, lacking a nucleus, that is packed with the oxygen-carrying molecule hemoglobin; erythrocyte.

red fiber (red fi′ber)   Slow-contracting postural skeletal muscle fiber that contains abundant myoglobin.

red marrow (red mar′o)   Blood cell-forming tissue in spaces within bones.

referred pain (re-ferd′ pān)   Pain that feels as if it is originating from a part other than the site being stimulated.

reflex (re′fleks)   Automatic response to a stimulus.

reflex arc (re′fleks ark)   Components of a reflex, consisting of a sensory receptor, sensory neuron, interneuron (usually), motor neuron, and effector.

refraction (re-frak′shun)   Bending of light as it passes between media of different densities.

refractory period (re-frak′to-re pe′re-od)   Period following stimulation during which a neuron or muscle fiber will not respond to another stimulus.

relaxin (re-lak′sin)   Hormone from the corpus luteum that causes relaxation of the joints of the pelvic girdle.

renal (re′nal)   Pertaining to the kidney.

renal corpuscle (re′nal kor′pusl)   Part of a nephron that consists of a glomerulus and a glomerular capsule (Bowman's capsule).

renal cortex (re′nal kor′teks)   Outer part of a kidney.

renal medulla (re′nal mě-dul′ah)   Inner part of a kidney.

renal pelvis (re′nal pel′vis)   Cavity in a kidney that channels urine to the ureter.

renal plasma threshold (re′nal plaz′mah thresh′old) Concentration of a substance in plasma at which it begins to be excreted in the urine.

renal tubule (re′nal tu′būl) Tubular part of a nephron that extends from the renal corpuscle to the collecting duct.

renin (re′nin) Enzyme that kidneys release that helps maintain blood pressure, plasma sodium, and blood volume.

renin-angiotensin system (re′nin-an″je-o-ten′sin sis′tem) A series of reactions involving an enzyme, renin, that ultimately forms angiotension II, a vasoconstrictor and stimulator of aldosterone and antidiuretic hormone secretion.

repair enzyme (re-pār′ en′zīm) Protein that removes mismatched nucleotides from a section of DNA and replaces them with complementary nucleotides.

replication (rep″lĭ-ka′shun) Copying of a DNA molecule.

reproduction (re″pro-duk′shun) Offspring formation.

residual volume (re-zid′u-al vol′ūm) Volume of air remaining in the lungs after the most complete expiration.

resorption (re-sorp′shun) Decomposition of a structure and return of the components to the bloodstream.

respiration (res″pĭ-ra′shun) Exchange of gases between the atmosphere and body cells; breathing.

respiratory area (re-spi′rah-to″re ah′re-a) Groups of neurons in the brainstem that control breathing depth and rate.

respiratory capacity (re-spi′rah-to″re kah-pas′ĭ-te) The sum of any two or more respiratory volumes.

respiratory cycle (re-spi′rah-to″re si′kl) An inspiration followed by an expiration.

respiratory membrane (re-spi′rah-to″re mem′brān) Membrane composed of a capillary wall, an alveolar wall, and their respective basement membranes through which blood and inspired air exchange gases.

respiratory volume (re-spi′rah-to″re vol′ūm) Any one of several distinct volumes of air that can be moved in or out of the lungs.

resting potential (res′ting po-ten′shal) Stored electrical energy due to the difference in electrical charge between the inside and outside of an undisturbed nerve cell membrane.

resting tidal volume (res′ting tīd′al vol′ūm) Volume of air entering and leaving the body in a respiratory cycle at rest.

reticular fiber (rĕ-tik′u-lar fi′ber) Thin, collagen fiber.

reticular formation (rĕ-tik′u-lar fōr-ma′shun) Complex network of nerve fibers and islands of gray matter in the brainstem that arouses the cerebrum.

reticulocyte (rĕ-tik′u-lo-sīt) Immature red blood cell that has a network of fibrils in its cytoplasm.

retina (ret′ĭ-nah) Inner layer of the eye wall that contains the photoreceptors.

retinal (ret′ĭ-nal) A form of vitamin A; retinene.

retinene (ret′ĭ-nēn) Chemical precursor of rhodopsin, a visual pigment.

retraction (rĕ-trak′shun) Movement of a part toward the back.

retroperitoneal (ret″ro-per″ĭ-to-ne′al) Behind the peritoneum.

reversible reaction (re-ver′sĭ-b′l re-ak′shun) Chemical reaction in which the products can react, yielding the original reactants.

rhodopsin (ro-dop′sin) Light-sensitive pigment in the rods of the retina; visual purple.

riboflavin (ri″bo-fla′vin) A vitamin of the B-complex group; vitamin B_2.

ribonucleic acid (ri″bo-nu-kle′ik as′id) Single stranded polymer of nucleotides, in which each nucleotide includes a phosphate group, a nitrogen-containing base (adenine, uracil, cytosine, or guanine), and the sugar ribose; RNA.

ribose (ri′bōs) 5-carbon sugar in RNA.

ribosomal RNA (ri-bo-sōm′al) Type of RNA that forms part of the ribosome; rRNA.

ribosome (ri′bo-sōm) Organelle composed of RNA and protein that provides structural support and enzymes for protein synthesis.

RNA Ribonucleic acid.

rod (rod) Type of light receptor that provides colorless (black and white) vision.

rotation (ro-ta′shun) Movement turning a body part on its longitudinal axis.

round window (rownd win′do) Membrane-covered opening between the inner ear and the middle ear.

S

saccule (sak′ūl) Saclike cavity that makes up part of the membranous labyrinth of the inner ear.

sacral (sa′kral) Pertaining to the five fused (pelvic) vertebrae at the inferior end of the spinal column.

saddle joint (sad′l joint) Two bones joined, each with a convex and concave surface that are complementary.

sagittal (saj′ĭ-tal) Plane or section that divides a structure into right and left portions.

salivary gland (sal′ĭ-ver-e gland) Any of the glands, associated with the mouth, that secrete saliva.

salt (sawlt) Compound produced by a reaction between an acid and a base.

saltatory conduction (sal′tah-tor-e kon-duk′shun) Impulse conduction along an axon that seems to jump from one node to the next.

SA node (nōd) Sinoatrial node.

sarcolemma (sar″ko-lem′ah) Cell membrane of a muscle fiber.

sarcomere (sar′ko-mēr) Structural and functional unit of a myofibril.

sarcoplasm (sar′ko-plazm) Cytoplasm in a muscle fiber.

sarcoplasmic reticulum (sar″ko-plaz′mik rĕ-tik′u-lum) Membranous network of channels and tubules in a muscle fiber, corresponding to the endoplasmic reticulum of other cells.

satellite cells (sat′il-līte selz) Glia in the peripheral nervous system that support neurons in ganglia.

saturated fat (sat′u-rāt″ed fat) Fat molecule that contains only fatty acid molecules with as many hydrogen atoms as possible, and therefore no double-bonded carbon atoms.

Schwann cell (shwahn sel) Type of neuroglia that surrounds an axon of a peripheral neuron, forming the neurilemma and myelin.

sclera (skle′rah) White fibrous outer layer of the eyeball.

scrotum (skro′tum) Pouch of skin in males that encloses the testes.

sebaceous gland (sĕ-ba′shus gland) Skin gland that secretes sebum.

sebum (se′bum) Oily secretion of the sebaceous glands.

secondary immune response (sek′un-der″e i-mun′ re-spons′) Immune system's response to subsequent encounters with a nonself antigen.

secretin (se-kre′tin) Hormone from the small intestine that stimulates the pancreas to release pancreatic juice rich in bicarbonate ion.

secretion (se-kre′shun) Substance produced in and released from a gland cell.

segmentation (seg″men-ta′shun) Alternating contraction and relaxation of circular muscle fibers that mix intestinal contents.

selectively permeable (se-lĕk′tiv-le per′me-ah-b′l) Membrane that allows some types of molecules through but not others.

semen (se′men) Fluid containing sperm cells and secretions discharged from the male reproductive tract at ejaculation.

semicircular canal (sem″ĭ-ser′ku-lar kah-nal′) Tubular structure in the inner ear that contains the receptors providing the sense of dynamic equilibrium.

seminal vesicle (sem″ĭ-nal ves′ĭ-kel) One of a pair of pouches that adds fructose and prostaglandins to sperm as semen forms.

seminiferous tubule (sem″ĭ-nif′er-us tu′būl) Tubule within the testes where sperm cells form.

senescence (sĕ-nes′ens) Aging.

sensation (sen-sa'shun)   An awareness that impulses associated with a sensory event have reached the brain.

sensory adaptation (sen'so-re ad"ap-ta'shun)   Repeated stimulation, resulting in sensory receptors becoming unresponsive or inhibition along the CNS pathways leading to sensory regions of the cerebral cortex.

sensory nerve (sen'so-re nerv)   Nerve composed of sensory nerve fibers.

sensory neuron (sen'so-re nu'ron)   Neuron that conducts an impulse from a receptor to the central nervous system.

sensory receptor (sen'so-re re"sep'tor)   Specialized structure associated with the peripheral end of a sensory neuron specific to detecting a particular sensation and triggering an impulse in response.

serosa (sēr-o'sah)   A serous membrane.

serotonin (se"ro-to'nin)   Vasoconstrictor that blood platelets release when blood vessels break, controlling bleeding. Also a neurotransmitter.

serous cell (se'rus sel)   Glandular cell that secretes a watery lubricating fluid (serous fluid).

serous fluid (se'rus floo'id)   Lubricating secretion from a serous cell.

serous membrane (se'rus mem'brān)   Membrane that lines a cavity without an opening to the outside of the body.

serum (se'rum)   Fluid portion of coagulated blood.

sesamoid bone (ses'ah-moid bōn)   Any of the round bones that form within tendons adjacent to joints.

set point (set point)   Target value of a physiological measure maintained automatically in the body.

sex chromosome (seks kro'mo-sōm)   Either of two chromosomes that carries genes responsible for the development of characteristics associated with maleness or femaleness; an X or Y chromosome.

sex-influenced inheritance (seks-in"floo-enst in-her'ĭ-tens)   Transmission of a trait dominant in one sex but recessive in the other.

sex-limited trait (seks'-lim'it-ed tra'-t)   A characteristic seen in one sex only.

sexually transmitted infection (sek'shoo-ah-le trans-mi'ted in-fek'shun)   Infection transmitted from one individual to another by direct contact during sexual activity.

sigmoid colon (sig'moid ko'lon)   S-shaped part of the large intestine between the descending colon and the rectum.

signal transduction (sig'nahl trans-duk'shun)   Series of biochemical reactions that allows cells to receive and respond to messages reaching the cell membrane.

single nucleotide polymorphism (sing'el nu'kle-o-tīd pol"e-mor'fiz-em)   A variant base at a particular place in the genome that occurs in at least 1% of a population.

sinoatrial node (si"no-a'tre-al nōd)   Specialized tissue in the wall of the right atrium that initiates cardiac cycles; the pacemaker; SA node.

sinus (si'nus)   Cavity or space in a bone or other body part.

skeletal muscle (skel'ĕ-tal mus'l)   Voluntary muscle tissue; muscles attached to bones.

sliding filament model (slī'ding fil'eh-ment ma'dul)   Muscles shorten when the thin (actin) and thick (myosin) filaments slide past each other, pulling on the muscle ends.

small intestine (smawl in-tes'tin)   Part of the digestive tract extending from the stomach to the cecum; consisting of the duodenum, jejunum, and ileum.

smooth muscle (smoo-th mus'l)   Type of involuntary muscle tissue; lacks striations.

sodium/potassium pump (so'de-um-po-tas'e-um pump)   Active transport mechanism that concentrates sodium ions outside and potassium ions inside a cell membrane.

solute (sol'ūt)   Chemical dissolved in a solution.

solvent (sol'vent)   Liquid portion of a solution in which a solute is dissolved.

somatic cell (so-mat'ik sel)   Any cell of the body other than the sex cells.

somatic nervous system (so-mat'-ik ner'vus sis'tem)   Motor pathways of the peripheral nervous system that lead to skeletal muscles.

somatostatin (so-mat'o-sta'tin)   Hormone secreted by the pancreatic islets that inhibits the release of glucagon and insulin.

somatotropin (so"mah-to-tro'pin)   Growth hormone.

special sense (spesh'al sens)   Sense that stems from receptors associated with specialized sensory organs, such as the eyes and ears.

species resistance (spe'sēz re-zis'tans)   Natural ability of one type of organism to resist infection by pathogens that cause disease in another type of organism.

spermatic cord (sper-mat'ik kord)   Structure consisting of blood vessels, nerves, the ductus deferens, and other vessels extending from the abdominal inguinal ring to the testis.

spermatid (sper'mah-tid)   Intermediate stage in sperm cell formation.

spermatocyte (sper-mat'o-sīt)   Early stage in sperm cell formation.

spermatogenesis (sper"mah-to-jen'ĕ-sis)   Sperm cell production.

spermatogonium (sper"mah-to-go'ne-um)   Undifferentiated spermatogenic cell in the outer part of a seminiferous tubule.

sphygmomanometer (sfīg'mo-mah-nom'ĕ-ter)   Instrument used for measuring blood pressure.

spinal cord (spi'nal kord)   Part of the central nervous system extending from the brainstem through the vertebral canal.

spinal nerve (spi'nal nerv)   Nerve that arises from the spinal cord.

spiral organ (spī rul or'gan)   Organ in the cochlear duct containing the receptors for hearing. It consists of hair cells and supporting cells.

spleen (splēn)   Large organ in the upper left quadrant of the abdomen that processes old red blood cells.

spongy bone (spunj'e bōn)   Bone that consists of bars and plates separated by irregular spaces; cancellous bone.

squamous (skwa'mus)   Flat or platelike.

starch (starch)   Polysaccharide common in foods of plant origin.

static equilibrium (stat'ik e'kwĭ-lib're-um)   Maintenance of balance when the head and body are not in motion.

stem cell (stem sel)   Undifferentiated cell that can divide to yield two daughter stem cells or a stem cell and a progenitor cell.

stereoscopic vision (ster"e-o-skop'ik vizh'un)   The ability, using two eyes, to perceive objects as three-dimensional; depth perception.

sternal (ster'nal)   Pertaining to the sternum.

steroid (ste'roid)   Type of organic molecule including complex rings of carbon and associated hydrogen and oxygen atoms.

stomach (stum'ak)   Digestive organ between the esophagus and the small intestine.

strabismus (strah-biz'mus)   Eyes not coordinated enough to align.

stratified (strat'ĭ-fīd)   Organized in layers.

stratum basale (stra'tum ba'sal)   Deepest layer of the epidermis, where cells divide; stratum germinativum.

stratum corneum (stra'tum kor'ne-um)   Outer, horny layer of the epidermis.

stress (stres)   Response to factors perceived as life-threatening.

stressor (stres'or)   Factor that can stimulate a stress response.

stretch receptor (strech re-sep'ter)   Sensory receptor that responds to a change in length or volume.

stretch reflex (strech re'fleks)   Muscle contraction in response to stretching the muscle.

stroke volume (strōk vol'ūm)   Volume of blood the ventricle discharges with each heartbeat.

structural formula (struk′cher-al fōr′mu-lah) Representation of the way atoms bond in a molecule, using symbols for each element and lines to indicate chemical bonds.

subarachnoid space (sub″ah-rak′noid spās) Space in the meninges between the arachnoid mater and the pia mater.

subatomic particles (sub″ah-tom′ik par′tĕ-kalz) Parts of an atom, including protons, electrons, and neutrons.

subcutaneous layer (sub″ku-ta′ne-us la′yer) Loose connective tissue layer beneath the skin; hypodermis.

sublingual (sub-ling′gwal) Beneath the tongue.

submucosa (sub″mu-ko′sah) Layer of the alimentary canal beneath the mucosa.

substrate (sub′strāt) Molecule on which an enzyme acts.

sucrase (su′krās) Digestive enzyme that catalyzes the breakdown of sucrose.

sucrose (soo′krōs) Disaccharide; table sugar.

sugar (shoog′ar) Sweet-tasting carbohydrate.

sulcus (sul′kus) Shallow groove, such as that between gyri on the brain surface (pl., *sulci*).

summation (sum-ma′shun) Increased force of contraction by a skeletal muscle fiber when a twitch occurs before the previous twitch relaxes.

superficial (soo″per-fish′al) Near the surface of the body or a specified body structure.

superior (su-pe′re-or) Structure above another structure.

supination (soo″pĭ-na′shun) Upward or forward rotation of palm of hand.

surface tension (ser′fas ten′shun) Force that holds moist membranes together due to the attraction of water molecules.

surfactant (ser-fak′tant) Substance produced by the lungs that reduces the surface tension in alveoli.

suture (soo′cher) Immovable joint, such as that between flat bones of the skull.

sweat gland (swet gland) Exocrine gland in skin that secretes a mixture of water, salt, and wastes such as urea.

sympathetic nervous system (sim″pah-thet′ik ner′vus sis′tem) That part of the autonomic nervous system that arises from the thoracic and lumbar regions of the spinal cord.

symphysis (sim′fĭ-sis) Slightly movable joint between bones separated by a pad of fibrocartilage.

synapse (sin′aps) Functional connection between the axon of one neuron and the dendrite or cell body of another neuron or the membrane of another cell type.

synaptic cleft (sĭ-nap′tik kleft) A narrow extracellular space between the cells at a synapse.

synaptic knob (sĭ-nap′tik nob) Tiny enlargement at the end of an axon that secretes a neurotransmitter.

synaptic potential (sĭ-nap′tik po-ten′shal) A change in the membrane potential of the postsynaptic cell.

synaptic transmission (sĭ-nap′tik trans-mish′un) Neurotransmitter from a presynaptic neuron affecting the postsynaptic cell.

synchondrosis (sin″kon-dro′sis) Type of joint in which bands of hyaline cartilage unite bones.

syndesmosis (sin″des-mo′sis) Type of joint in which bundles of connective tissue unite bones.

synergist (sin′er-jist) Muscle that assists the action of an agonist.

synovial fluid (sĭ-no′ve-al floo′id) Lubricating fluid that the synovial membrane secretes.

synovial joint (sĭ-no′ve-al joint) Freely movable joint.

synovial membrane (sĭ-no′ve-al mem′brān) Membrane that forms the inner lining of the capsule of a freely movable joint.

synthesis (sin′thĕ-sis) Building large molecules by bonding smaller ones.

systemic circuit (sis-tem′ik ser′kit) Vessels that transport blood between the heart and all body tissues except the lungs.

systole (sis′to-le) Phase of the cardiac cycle when a heart chamber wall contracts.

systolic pressure (sis-tol′ik presh′ur) Arterial blood pressure reached during the systolic phase of the cardiac cycle.

T

tachycardia (tak″e-kar′de-ah) Abnormally rapid heartbeat.

tactile corpuscle (tak′til kor′pus-l) Sensory receptor close to the surface of the skin that is sensitive to light touch; Meissner's corpuscle.

target cell (tar′get sel) Cell with specific receptors for a hormone on which that hormone exerts its effect.

tarsal (tahr′sal) Pertaining to the ankle or any of the individual ankle bones.

tarsus (tar′sus) The ankle bones as a group; the ankle.

taste bud (tāst bud) Organ containing receptors associated with the sense of taste.

telophase (tel′o-fāz) Stage in mitosis when newly formed nuclei separate.

tendon (ten′don) Cordlike or bandlike mass of dense connective tissue that connects a muscle to a bone.

teratogen (ter′ah-to-jen) Chemical or other environmental agent that causes a birth defect.

testis (tes′tis) Primary male reproductive organ; sperm cell-producing organ (pl., *testes*).

testosterone (tes-tos′tĕ-rōn) Male sex hormone secreted by the interstitial cells of the testes.

tetanic contraction (tĕ-tan′ik kon-trak′shun) Continuous, forceful skeletal muscle contraction with little or no relaxation.

thalamus (thal′ah-mus) Mass of mostly gray matter at the base of the cerebrum, bulging from the wall on either side of the third ventricle.

thermoreceptor (ther″mo-re-sep′tor) Sensory receptor sensitive to temperature changes; warm and cold receptors.

thiamine (thi′ah-min) Vitamin of the B-complex group; vitamin B_1.

thoracic (tho-ras′ik) Pertaining to the chest.

thoracic cavity (tho-ras′ik kav′i-te) Space containing organs in the upper trunk.

threshold potential (thresh′old po-ten′shal) Level of potential at which an action potential is triggered in a neuron.

threshold stimulus (thresh′old stim′u-lus) Stimulation level that must be reached to elicit an action potential in a neuron or a muscle fiber.

thrombin (throm′bin) Blood-clotting enzyme that catalyzes formation of fibrin from fibrinogen.

thrombocyte (throm′bo-sīt) Blood platelet.

thrombopoietin (throm′bo-poi′ĕ-tin) A hormone that stimulates megakaryocytes to proliferate and yield platelets.

thrombus (throm′bus) Blood clot that remains where it forms in a blood vessel.

thymosins (thi′mo-sins) Group of peptides the thymus gland secretes that increases production of certain types of white blood cells.

thymus (thi′mus) Glandular organ in the mediastinum, above the heart.

thyroid gland (thi′roid gland) Endocrine gland just below the larynx and in front of the trachea that secretes thyroid hormones.

thyroid-stimulating hormone (thi′roid-stim″u-lāt′eng hor′mōn) Hormone secreted from the anterior pituitary gland that controls secretion from the thyroid gland; TSH.

thyroxine (thi-rok′sin) Major hormone secreted by the thyroid gland; T4.

tidal volume (tīd′al vol′ūm) Volume of air entering and leaving the lungs in a respiratory cycle.

tissue (tish′u) Assembled group of similar cells that performs a specialized function.

titin (ti′tin) Protein that attaches myosin filaments to z lines in muscle.

T lymphocyte (T lim′fo-sīt) Type of white blood cell that interacts with antigen-bearing cells and particles and secretes cytokines, contributing to the cellular immune response; T cell.

tonsil (ton′sil) Collection of lymphatic tissue in the throat.

total lung capacity (toh′tal lung kah-pas′i-te) The volume of air in completely filled lungs.

totipotent (to-tip′o-tent) Ability of a cell to differentiate into any type of cell. Only a fertilized ovum is totipotent.

trabecula (trah-bek′u-lah) Branching bony plate that separates irregular spaces within spongy bone.

trace element (trās el′ĕ-ment) Basic chemical substance needed in small quantity.

trachea (tra′ke-ah) Tubular organ that leads from the larynx to the bronchi.

transcellular fluid (trans″sel′u-lar floo′-id) Part of the extracellular fluid, including the fluid within special body cavities.

transcription (trans-krip′shun) Manufacturing a complementary RNA from DNA.

transcytosis (trans″si-to′sis) Combination of receptor-mediated endocytosis and exocytosis that moves particles through a cell layer.

transfer RNA (trans′fer RNA) RNA molecule that carries a specific amino acid to a ribosome in protein synthesis; tRNA.

translation (trans-la′ shun) Assembly of an amino acid chain according to the sequence of base triplets in an mRNA molecule.

transverse (trans-vers′) Plane that divides a structure into superior and inferior portions.

transverse colon (trans-vers′ ko′lon) Part of the large intestine that extends across the abdominal cavity from right to left below the stomach.

transverse tubule (trans-vers′ tu′būl) Membranous channel that extends deep into the cell from a muscle fiber membrane.

triad (tri′ad) Group of three structures; in muscle, two cisternae and a transverse tubule.

tricuspid valve (tri-kus′pid valv) Heart valve between the right atrium and the right ventricle.

trigger zone (trig′ger zōn) Sensitive part of an axon where an action potential is generated.

triglyceride (tri-glis′er-īd) Lipid composed of three fatty acids and a glycerol molecule; fat.

triiodothyronine (tri″i-o″do-thi′ro-nēn) One hormone secreted by the thyroid gland; T_3.

trisomy (tri′so-me) Condition in which a cell has three copies of a particular chromosome instead of two.

trochanter (tro-kan′ter) Large process on the femur.

trophoblast (trof′o-blast) Outer cells of a blastocyst that give rise to cells that become part of the placenta and extraembryonic membranes.

tropic hormone (trōp′ik hor′mōn) Hormone that stimulates another endocrine gland to release a hormone.

tropomyosin (tro″po-mi′o-sin) Protein that blocks muscle contraction until its position is modified by troponin when calcium ions are present.

troponin (tro′po-nin) Protein that regulates muscle contraction by altering the position of tropomyosin.

trypsin (trip′sin) Enzyme in pancreatic juice that breaks down protein molecules.

trypsinogen (trip-sin′o-jen) Substance pancreatic cells secrete that is enzymatically cleaved to yield trypsin.

tubercle (tu′ber-kl) Small, rounded process on a bone.

tuberosity (tu″bĕ-ros′i-te) Knoblike process on a bone.

tubular reabsorption (too′bu-lar re″ab-sorp′shun) Transport of substances out of the renal tubule into the interstitial fluid from which the substances diffuse into peritubular capillaries.

tubular secretion (too′bu-lar se-kre′shun) Movement of substances out of the peritubular capillaries into the renal tubule for excretion in the urine.

tumor (too′mor) Tissue mass formed when cells lose division control.

tumor suppressor gene (too′mor sŭ-pres′or jēn) A gene (DNA sequence) that encodes a protein that ordinarily inhibits cell division.

twitch (twich) Single contraction of a muscle fiber followed by relaxation.

tympanic membrane (tim-pan′ik mem′brān) Thin membrane that covers the auditory canal and separates the outer ear from the middle ear; the eardrum.

U

ultratrace element (ul′trah-trās el′ĕ-ment) Basic chemical substance needed in very small quantity.

umbilical (um-bil′ĭ-kal) Pertaining to the navel.

umbilical cord (um-bil′ĭ-kal kord) Cordlike structure that connects the fetus to the placenta.

umbilical region (um-bil′ĭ-kal re′jun) Central portion of the abdomen.

umbilicus (um-bil′ĭ-kus) Region to which the umbilical cord was attached; the navel.

unipolar neuron (un″ĭ-po′lar nu′ron) Neuron that has a single process associated with its cell body.

unsaturated fat (un-sat′u-rāt″ed fat) Fat molecule that includes one or more unsaturated fatty acids, each with one or more double bonds.

upper limb (uh′per lim) Superior appendage consisting of the arm, forearm, and hand.

urea (u-re′ah) Nonprotein nitrogenous substance resulting from protein catabolism.

ureter (u-re′ter) Muscular tube that carries urine from the kidney to the urinary bladder.

urethra (u-re′thrah) Tube leading from the urinary bladder to the outside of the body.

uric acid (u′rik as′id) Product of nucleic acid metabolism in the body.

urine (u′rin) Wastes and excess water removed from the blood and excreted by the kidneys into the ureters to the urinary bladder and out of the body through the urethra.

uterine (u′ter-in) Pertaining to the uterus.

uterine tube (u′ter-in tūb) Tube that extends from the uterus on each side toward an ovary; fallopian tube or oviduct.

uterus (u′ter-us) Hollow muscular organ in the female pelvic cavity where a fetus develops.

utricle (u′trĭ-kl) Chamber in the inner ear associated with static equilibrium.

uvula (u′vu-lah) Fleshy portion of the soft palate that hangs down above the root of the tongue.

V

vaccine (vak′sēn) Preparation that includes antigens that stimulate an immune response to prevent an infectious disease.

vagina (vah-ji′nah) Tubular organ that leads from the uterus to the vestibule of the female reproductive tract.

variably expressive (va′re-ah-ble eks″pres′iv) Phenotype whose symptoms vary in intensity in different people.

varicose veins (var′ĭ-kos vānz) Abnormally swollen and enlarged veins, especially in the legs.

vasa recta (va′sah rek′tah) Branch of the peritubular capillary that parallels the structure of the nephron loop.

vasoconstriction (vas″o-kon-strik′shun) Decrease in the diameter of a blood vessel.

vasodilation (vas″o-di-la′shun) Increase in the diameter of a blood vessel.

vasomotor center (vas″o-mo′tor sen′ter) Neurons in the brainstem that control the diameter of the arteries.

vasopressin (vas″o-pres′in) Antidiuretic hormone.

vein (vān) Vessel that carries blood toward the heart.

vena cava (vēn′ah kāv′ah) One of two large veins that conveys oxygen-poor blood to the right atrium of the heart.

ventral root (ven′tral roo-t) Structure containing axons of motor neurons that emerges from the spinal cord to form part of each spinal nerve.

ventricles (ven'trĭ-klz) Cavities, such as brain ventricles filled with cerebrospinal fluid; also heart ventricles that contain blood.

venule (ven'ūl) Vessel that carries blood from capillaries to a vein.

vertebral (ver'te-bral) Pertaining to the bones of the spinal column.

vertebral canal (ver'te-bral kah-nal') Hollow area in vertebrae containing the spinal cord.

vesicle (ves'ĭ-kal) Membranous, cytoplasmic sac.

vestibule (ves'tĭ-būl) Bony chamber in the inner ear; also, space into which the vagina and urethra open in the female.

villus (vil'us) Tiny, fingerlike projection that extends from the inner lining of the small intestine into the lumen (pl., *villi*).

viscera (vis'er-ah) Organs in a body cavity.

visceral (vis'er-al) Pertaining to the organs in a body cavity.

visceral peritoneum (vis'er-al per″ĭ-to-ne'um) Membrane that covers organ surfaces in the abdominal cavity.

visceral pleura (vis'er-al ploo'rah) Membrane that covers the surfaces of the lungs.

viscosity (vis-kos'ĭ-te) Tendency for a fluid to resist flowing due to the internal friction of its molecules.

vital capacity (vi'tal kah-pas'ĭ-te) The maximum volume of air a person can exhale after taking the deepest breath possible.

vitamin (vi'tah-min) Organic nutrient other than a carbohydrate, lipid, or protein needed for normal metabolism that the body cannot synthesize in adequate amounts and must therefore be obtained in the diet.

vitreous body (vit're-us bod'e) Collagen fibers and fluid in the posterior cavity of the eye.

vitreous humor (vit're-us hu'mor) Fluid between the lens and the retina of the eye.

vocal cords (vo'kal kordz) Folds of tissue in the larynx that produce sounds when they vibrate.

voluntary (vol'un-tār″e) Consciously controlled.

vulva (vul'vah) External female reproductive parts that surround the vaginal opening.

W

water balance (wot'er bal'ans) When the volume of water entering and produced by the body is equal to the volume leaving it.

water of metabolism (wot'er uv mĕ-tab'o-lizm) Water produced as a by-product of aerobic metabolism.

Wernicke's area (ver'ni-kez air'i-ah) Region in the temporal lobe, usually in the left hemisphere, important in understanding and formulating written and spoken language.

white blood cell (whīt blud sel) Cell that helps fight infection; leukocyte.

white fiber (whīt fi'ber) Fast-contracting skeletal muscle fiber type.

wild type (wīld tīp) Phenotype or allele that is the most common for a certain gene in a population.

X

X-linked trait (eks-linkt'trāt) Trait determined by a gene on an X chromosome.

x-ray (eks' ray) Used as a verb: to photograph using X radiation.

X ray (eks ray) Used as a noun: a photograph produced by X radiation; radiogram. May also be used as an adjective: X-ray image.

Y

yellow marrow (yel'o mar'o) Fat storage tissue in the medullary cavities of certain bones.

yolk sac (yōk sak) Membranous sac connected to the embryo that contains stem cell precursors to many cell types.

Z

zona pellucida (zo'nah pel-u'cĭ-dah) Thick, transparent, noncellular layer surrounding a secondary oocyte.

zygote (zi'gōt) Cell produced by the fusion of an egg and sperm; a fertilized egg cell; fertilized ovum.

zymogen granule (zi-mo'jen gran'-ūl) Cellular structure that stores inactive forms of protein-splitting enzymes in a pancreatic cell.

CREDITS

Photo Credits

Front Matter

Page vi: Courtesy of Fran Simon; p. vi: Courtesy of Michael's Photography; p. vi: Courtesy of Dr. Wendy Josephs; p. v (top): Courtesy of Leslie Day; p. v (bottom): Courtesy of Gary Pilcher; p. xvii (left): Courtesy of Leslie Day; p. xvii (right): Courtesy of Gary Pilcher; p. xix (preview): © Comstock Images/Jupiterimages RF; p. xix (1): © Brand X Pictures/PunchStock RF; p. xix (2): © Dr. Tim Evans/SPL/Science Source; p. xix (3): © K.R. Porter/Science Source; p. xix (4): © Biophoto Associates/SPL/Science Source; p. xx (5, 6): © Science Photo Library/Getty Images RF; p. xx (7): © McGraw-Hill Education/Dennis Strete, photographer; p. xx (8): © Keith Brofsky/Getty Images RF; p. xx (9): © CNRI/SPL/Science Source; p. xxi (10): © Nancy Kedersha/UCLA/SPL/Science Source; p. xxi (11): © Secchi-Lecaque/Roussel-UCLAF/CNRA/SPL/Science Source; p. xxi (12): © SPL/Science Source; p. xxi (13): © Prof. P. Motta/Dept. of Anatomy/University of "La Sapienza" Rome/SPL/Science Source; p. xxi (14): © Image Source RF; p. xxii (15): © Science Photo Library/Getty Images RF; p. xxii (16): © Dr. Kari Lounatmaa/SPL/Science Source; p. xxii (17): © Science Photo Library/Getty Images RF; p. xxii (18): © Digital Vision/Getty Images RF; p. xxii (19): © Susumu Nishinaga/SPL/Science Source; p. xxii (20): © Prof. P. Motta/Dept. of Anatomy/University "La Sapienza" Rome/SPL/Science Source; p. xxiii (21): © Prof. P.M. Motta & M. Castellucci/SPL/Science Source; p. xxiii (22): © Brand X Pictures/Corbis RF; p. xxiii (23): © Big Cheese Photo/Jupiterimages RF; p. xxiii (24): © Barros & Barros/Getty Images.

Preview

Page 1: © Comstock Images/Jupiterimages RF; p. 4: Courtesy of John W. Hole, Jr.

Chapter 1

Opener: © Brand X Pictures/PunchStock RF; 1.1: © Classic Image/Alamy; 1A: © Keith Brofsky/Getty Images RF; 1B: © Jim Wehtje/Getty Images RF; 1C: © SPL/Science Source; 1.4: © McGraw-Hill Education/Carol D. Jacobson, Ph.D., Dept. of Veterinary Anatomy, Iowa State University; 1.21 (left, right): © McGraw-Hill Education/Aaron Roeth,

photographer; 1.22 (top left, right): © McGraw-Hill Education/Karl Rubin, 1.22 (bottom left): © Living Art Enterprises/Science Source; 1.22 (middle): © McGraw-Hill Education/Joe DeGrandis, photographer; 1.23a: © Patrick J. Lynch/Science Source; 1.23b: © Biophoto Associates/Science Source; 1.23c: © A. Glauberman/Science Source; 1.25a–b: © I love images/Male Body/Alamy RF; Plates 8–25: © McGraw-Hill Education/Karl Rubin.

Chapter 2

Opener: © Dr. Tim Evans/SPL/Science Source; 2A: © Mark Antman/The Image Works; 2Ba: © Southern Illinois University School of Medicine; 2.7, 2.23a–e: Courtesy of John W. Hole, Jr.; 2.23f: Courtesy Hyland Optical Technologies Corp.; 2Da: © SPL/Science Source; 2Db: © CNRI/SPL/Science Source; 2E: © WDCN/Univ. College London/Science Source.

Chapter 3

Opener: © K.R. Porter/Science Source; 3.4: © Steve Allen/Getty Images RF; 3.5a: © M. I. Walker/Science Source; 3.5b: © Biophoto Associates/Science Source; 3.5c: © Susumu Nishinaga/Science Source; 3.9a: © Don W. Fawcett/Science Source; 3.10a: © Biophoto Associates/Science Source; 3.12: © Bill Longcore/Science Source; 3.13: © Dr. Gopal Murti/Science Source; 3.14a: © Don W. Fawcett/Science Source; 3.15a: © Oliver Meckes/Science Source; 3.16: © Juergen Berger/Science Source; 3.17: © Dr. Alexey Khodjakov/Science Source; 3.18b, 3.19b: © Dr. Gopal Murti/Science Source; 3.25a–c: © David M. Phillips/Science Source; 3.36a–e: © Ed Reschke; 3.37a–c: © Steve Gschmeissner/Science Source; 3.38: © Tony Brain/Science Source; 3.42: © Peter Skinner/Science Source.

Chapter 4

Opener: © Biophoto Associates/SPL/Science Source; 4.16: © Corbis RF.

Chapter 5

Opener: © Science Photo Library/Getty Images RF; 5.2a: © McGraw-Hill Education/Al Telser, photographer; 5.2b: © Victor P. Eroschenko; 5.3b–c: © McGraw-Hill Education/Al Telser, photographer; 5.4b, 5.5b: © Victor P. Eroschenko; 5.6: © Fawcett, Hirokawa, Heuser/Science Source; 5.7b: © McGraw-Hill Education/

Dennis Strete, photographer; 5.8b–5.10b: © McGraw-Hill Education/Al Telser, photographer; 5.11b, d: © Ed Reschke; 5.14: © McGraw-Hill Education/Al Telser, photographer; 5.15: © Juergen Berger/Science Source; 5.16: © Biology Pics/Science Source; 5.17: © Steve Gschmeissner/Science Source; 5.18: © Prof. P. Motta/Univ. "La Sapienza"/Science Source; 5.19b–5.21b: © McGraw-Hill Education/Al Telser, photographer; 5.22b: © McGraw-Hill Education/Dennis Strete, photographer; 5.23b–5.25b: © McGraw-Hill Education/Al Telser, photographer; 5.26b: © Victor P. Eroschenko; 5.27b: © McGraw-Hill Education/Dennis Strete, photographer; 5.27c: © Prof. P. Motta/Univ. "La Sapienza"/Science Source; 5.28b: © McGraw-Hill Education/Dennis Strete, photographer; 5.29b–5.32b & p. 176: © McGraw-Hill Education/Al Telser, photographer.

Chapter 6

Opener: © Science Photo Library/Getty Images RF; 6.1–6.3b: © McGraw-Hill Education/Al Telser, photographer; 6.4a: © Don W. Fawcett/Science Source; 6Aa: ©Science Photo Library/Getty Images RF; 6Ab: © Science Photo Library/Alamy RF; 6Ac: © McGraw-Hill Education; 6.5: © Medical-on-Line/Alamy; 6.7b: © McGraw-Hill Education/Al Telser, photographer; 6B: © Anthony Saint James/Getty Images RF; 6.8: ©CNRI/SPL/Science Source; 6.9: © McGraw-Hill Education/Al Telser, photographer; 6.11: © Astrid & Hanns-Frieder Michler/Science Source; 6C: © McGraw-Hill Education; 6.14: © Mauro Fermariello/Science Source; 6.16: © Ingram Publishing RF.

Chapter 7

Opener: © McGraw-Hill Education/Dennis Strete, photographer; 7.3a: © Ed Reschke/Getty Images; 7.3b–c: Courtesy of John W. Hole, Jr.; 7.5: © Andrew Syred/Science Source; 7.6a: © Biophoto Associates/Science Source; 7.6b: Photo courtesy of T.D. Gelehrter and F.S. Collins; 7.7: © Secchi, Lecaque, Roussel, Uclaf, CNRA/SPL/Science Source; 7.10b: © McGraw-Hill Education/Al Telser, photographer; 7.11: © Biophoto Associates/Science Source; 7.12: © Ed Reschke/Peter Arnold/Getty Images; 7.13: Courtesy of John W. Hole, Jr.; 7C: © Michael Klein/Getty Images; 7.31: © Thinkstock/Jupiterimages RF; 7.36: © Stockbyte/PunchStock RF;

7.39b: © Thinkstock/Jupiterimages RF; 7.41b, 7.43d, 7.46c: Courtesy, Dale Butler; 7.47: © Dr. Charles Eaton/e-hand Electronic Textbook of Hand Surgery (www.e-hand.com); 7.48b: © Image Source/Getty Images RF; 7.51a, 7.54a, 7.55b: Courtesy, Dale Butler; 7.56 (left, right): Courtesy of Pat Hesse; Plates 26–54: Courtesy of John W. Hole, Jr.

Chapter 8

Opener: © Keith Brofsky/Getty Images RF; 8.3a–b: Courtesy of John W. Hole, Jr.; 8.10–8.12 (all): © J & J Photography; 8.13b, 8.15b: © Dr. Ronald Bergman; 8.17a–b: Courtesy of John W. Hole, Jr.; 8.18b, 8.20b: © Dr. Ronald Bergman; 8.22: © Alexander Tsiaras/Science Source; 8.23a–b: © Chris Priest/Science Source.

Chapter 9

Opener: © CNRI/SPL/Science Source; 9.3: © Ed Reschke/Getty Images; 9.5a: © H.E. Huxley; 9.8b: © McGraw-Hill Education/Carol D. Jacobson Ph.D., Dept. Veterinary Anatomy, Iowa State University; 9.10b: © H.E. Huxley; 9.19: © Dr. Paul Heidger, University of Iowa College of Medicine; 9.20a: © Oscar Moncho/agefotostock RF; 9.20b: © Sheila Terry/Science Source; 9.20c: © View Stock/Getty Images RF; p. 345 (left, right) : © Corbis RF; Plates 55–65: © Dr. Kent M. Van De Graaff; Plates 66–75: © McGraw-Hill Education/Karl Rubin.

Chapter 10

Opener: © Nancy Kedersha/UCLA/SPL/Science Source; 10.1: © McGraw-Hill Education/ Al Telser, photographer; 10.4b: © Biophoto Associates/Science Source; 10.5: © Dennis Emery; 10.9: © Thomas Deerinck, NCMIR/Science Source; 10.12 (right): © Don W. Fawcett/Science Source.

Chapter 11

Opener: © Secchi-Lecaque/Roussel-UCLAF/CNRA/SPL/Science Source; 11A: © Paul Jasienski/Getty Images; 11Ca–b: © Dr. Kent M. Van De Graaff; 11.6b: © Martin Rotker/Science Source; 11.8: © McGraw-Hill Education/Rebecca Gray, photographer/Don Kincaid, dissections; 11D: © AFP/Getty Images; 11E: Courtesy National Human Genome Research Institute/Kondi Wong, Armed Forces Institute of Pathology; 11.19b: © McGraw-Hill Education/Photo and dissection by Christine Eckel; 11.15b: © Carolina Biological Supply Company/Phototake; 11.24: © Manfred Kage/Science Source.

Chapter 12

Opener: © SPL/Science Source; 12.1b: © Ed Reschke; 12.1c: © Ed Reschke/Getty Images;

12A: © Imagebroker/Alamy RF; 12B: © Pallava Bagla/Corbis; 12C: © K.S. Matz; 12.6: © Dwight Kuhn; 12.8: © Victor B. Eichler, Ph.D.; 12.9b: © J & J Photography; 12D: © Gene J. Puskar/AP Photo; 12.14a: © Biophoto Associates/Science Source; 12.14b: © Dr. Fred Hossler; 12.19: Courtesy of Dean E. Hillman; 12.27: © SPL/Science Source; 12.33: © Ed Reschke/Peter Arnold/Getty Images; 12.34b: © Mediscan/Corbis; 12.38c: © Frank S. Werblin, Ph.D.

Chapter 13

Opener: © Prof. P. Motta/Dept. of Anatomy/University of "La Sapienza" Rome/SPL/Science Source; 13.14: © Michael Ross/Science Source; 13.19: © Dr. Fred Hossler; 13.21: © Medical-on-Line/Alamy; 13.22: © imagingbody.com; 13.23: © Lester V. Bergman/Corbis; 13.25: © Victor P. Eroschenko; 13.29: © Ed Reschke/Getty Images; 13.35: © Victor P. Eroschenko; 13B: Wellcome Library, London. Creative Commons License http://creativecommons.org/licenses/by/4.0; 13C: © Imagezoo/PunchStock RF.

Chapter 14

Opener: © Image Source RF; 14.1 (left): © Keith Brofsky/Getty Images RF; 14.1 (middle): © Comstock/PunchStock RF; 14.1 (right): © Keith Brofsky/Getty Images RF; 14A: © Photodisc Collection/Getty Images RF; 14.4b: © McGraw-Hill Education/Al Telser, photographer; 14.5b: © Bill Longcore/Science Source; 14.7a: © McGraw-Hill Education; 14.7b: © Cultura Science/Michael J. Klein, M.D./Oxford Scientific/Getty Images; 14.10–14.14: © Ed Reschke; 14Ba: © McGraw-Hill Education/Al Telser, photographer; 14Bb: © Andrew Syred/SPL/Science Source; 14.18: © SPL/Science Source; 14.20a–b: © McGraw-Hill Education/Al Telser, photographer; 14.22c: © Biophoto Associates/Science Source; 14.22d: © Ed Reschke/Getty Images; p. 555: © Image Source/Getty Images RF.

Chapter 15

Opener: © Science Photo Library/Getty Images RF; 15.2: © McGraw-Hill Education/Photo and dissection by Christine Eckel; 15.6c, 15.7: © McGraw-Hill Education; 15.8a: © McGraw-Hill Education/Karl Rubin, photographer; 15.12: © David Bassett/Science Source; 15A: © F. Schussler/PhotoLink/Getty Images RF; 15.23c: © McGraw-Hill Education/Al Telser, photographer; 15.26b–c: © Don W. Fawcett/Science Source; 15.27: © Thomas Deerinck, NCMIR/Science Source; 15.31: © Johner/SuperStock RF; 15Fa: © Ed Reschke/Peter Arnold/Getty Images; 15Fb: © Image Source/Getty Images; 15Fc: © Alfred Pasieka/Getty Images; 15.42b: © Dr. Kent M. Van De Graaff; 15.46: © PDSN/Phototake.

Chapter 16

Opener: © Dr. Kari Lounatmaa/SPL/Science Source; 16.3: © McGraw-Hill Education/Dennis Strete, photographer; 16.5: © Ed Reschke/Getty Images; 16.9: © Dr. Kent M. Van De Graaff; 16.10b: © McGraw-Hill Education/Al Telser, photographer; 16.12b: © McGraw-Hill Education/Dennis Strete, photographer; 16.14b: © McGraw-Hill Education/Al Telser, photographer; 16.16: © Science Photo Library/Getty Images RF; 16.18b: © Manfred Kage/Science Source; 16A: Merck Archives. Merck, Sharp & Dohme Corp., 2013/Photo by Phillip A. Harrington; 16.22b: © Steve Gschmeissner/Science Source; 16B: © Zeva Oelbaum.

Chapter 17

Opener: © Science Photo Library/Getty Images RF; 17.8: © McGraw-Hill Education/Rebecca Gray, photographer; 17.9b: © Nick Koudis/Getty Images RF; 17A: © SPL/Science Source; 17.12a: © McGraw-Hill Education/Al Telser, photographer; 17.12b: © Biophoto Associates/Science Source; 17.12c: © McGraw-Hill Education/Al Telser, photographer; 17.16: © Ed Reschke/Getty Images; 17.18: © Dr. Kent M. Van De Graaff; 17.19b: © McGraw-Hill Education/Al Telser, photographer; 17.27c: © Victor P. Eroschenko; 17.29: © Southern Illinois University/Science Source; 17.32: © tbradord/Getty Images; 17.36, 17.37b: © McGraw-Hill Education/Al Telser, photographer; 17.44: © Jim Wehtje/Getty Images RF; 17.46: © Ed Reschke/Getty Images; 17.47: © Ed Reschke.

Chapter 18

Opener: © Digital Vision/Getty Images RF; 18.9a: © McGraw-Hill Education/Lars A. Niki, photographer; 18.9b: © PhotoLink/Getty Images RF; 18.12: © Jeff Rotman/Alamy; 18.18: © BioPhoto/Science Source; 18.20b: © McGraw-Hill Education/Al Telser, photographer; 18.20c: © Biophoto Associates/Science Source; 18A: © Charles D. Winters/Science Source; 18.21: USDA. USDA does not endorse any products, services, or organizations; 18.22a: © AFP/Getty Images; 18.22b: © Copyright 1997 IMS Communications Ltd./Capstone Design. All Rights Reserved; 18.23: © Brand X Pictures/PunchStock RF.

Chapter 19

Opener: © Susumu Nishinaga/SPL/Science Source; 19.3b: © Biophoto Associates/Science Source; 19.4a–b: © Anthony Bradshaw/Getty Images; 19A (left): © St. Bartholomew's Hospital/Science Source; 19A (right): © Biophoto Associates/Science Source; 19B: © Moredun Animal Health/SPL/Science

Source; 19.7c: © CNRI/Science Source; 19.10: © Biophoto Associates/Science Source; 19.13: © Medical-on-Line/Alamy; 19.15: © McGraw-Hill Education/Al Telser, photographer; 19.16: © Motta & Macchiarelli/Anatomy Dept./Univ. "La Sapienza", Rome/Science Source; 19.18: Courtesy of the American Lung Association; 19C: R.N. Clark, R.O. Green, G.A. Swayze, G. Meeker, S. Sutley, T.M. Hoefen, K.E. Livo, G. Plumlee, B. Pavri, C. Sarture, S. Wilson, P. Hageman, P. Lamothe, J.S. Vance, J. Boardman, I. Brownfield, C. Gent, L.C. Morath, J. Taggart, P.M. Theodorakos, and M. Adams, 2001, Environmental Studies of the World Trade Center area after the September 11, 2001 attack. U.S. Geological Survey, Open File Report OFR-01-0429; 19.26: © Edward Lettau/Science Source; 19Da–b: © Victor B. Eichler, Ph.D.; 19.32: Courtesy of the American Lung Association; 19.34: © imagingbody.com; 19E (left): © AFIP/Science Source; 19E (right): © Biophoto Associates/Science Source.

Chapter 20

Opener: © Prof. P. Motta/Dept. of Anatomy/ University "La Sapienza" Rome/SPL/Science Source; 20.2: © CNRI/SPL/Science Source; 20.6b: © Lester V. Bergman/Corbis; 20.7a: © Susumu Nishinaga/Science Source; 20.7b: Courtesy of R.B. Wilson M.D., Department of Pathology and Microbiology, University of Nebraska Medical Center; 20.9b: © Thomas Deerinck, NCMIR/Science Source; 20.11a:

© Biophoto Associates/Science Source; 20.11b: © McGraw-Hill Education/Al Telser, photographer; 20.25: © Biophoto Associates/ Science Source; 20A: © Stephen J. Krasemann/ Science Source; 20.28: © McGraw-Hill Education/Al Telser, photographer; 20.29: © Ed Reschke/Getty Images.

Chapter 21

Opener: © Prof. P.M. Motta & M. Castellucci/ SPL/Science Source.

Chapter 22

Opener: © Brand X Pictures/Corbis RF; 22.7: © McGraw-Hill Education/Al Telser, photographer; 22.11: © Pascal Goetgheluck/ Science Source; 22.12: © Image Source RF; 22.13a–b: © McGraw-Hill Education/Al Telser, photographer; 22.14: © Kage Mikrofotografie/ Phototake; 22.21b: © Prof. P.M. Motta/Univ. "La Sapienza", Rome/Science Source; 22.22c: © Biophoto Associates/Science Source; 22.23b: © McGraw-Hill Education/Al Telser, photographer; 22.24: © P.M. Motta & J. Van Blerkom/Science Source; 22.26a: © McGraw-Hill Education/Al Telser, photographer; 22.26b: © Science Photo Library/Getty Images RF; 22.27: © McGraw-Hill Education/Carol D. Jacobson, Ph.D., Department of Veterinary Anatomy, Iowa State University; 22.31c (left-right): © Alvin Telser, Ph.D.; 22.33a–b: © McGraw-Hill Education/Jill Braaten,

photographer; 22.33c: © PhotoLink/Getty Images RF; 22.33d: © Don Farrall/Getty Images RF; 22.33e: © McGraw-Hill Education/ Jill Braaten, photographer; 22C: © Southern Illinois University School of Medicine.

Chapter 23

Opener: © Big Cheese Photo/Jupiterimages RF; 23.2: © Eye of Science/Science Source; 23A: © Science Photo Library/Getty Images RF; 23.4a: © A. Tsiara/Science Source; 23.4b: © Omikron/Science Source; 23.4c: © Petit Format/Nestle/Science Source; 23.6c, 23.7: Courtesy of Ronan O'Rahilly; 23.13a: Human Developmental Anatomy Center, National Museum of Health and Medicine, Silver Spring, MD 20910; 23.13b: © Omikron/Science Source; 23.13c: © Petit Format/Nestle/Science Source; 23.14e: © ISM/Phototake; 23.15: © Dr. G. Moscoso/Science Source; 23.27a: © Biophoto Associates/Science Source; 23.27b: © Biophoto Associates/Science Source; 23D: © Mitch Wojnarowicz/The Image Works.

Chapter 24

Opener: © Barros & Barros/Getty Images; 24.3: Courtesy Genzyme Corporation; 24.6: © Medical-on-Line/Alamy; 24.7: © McGraw-Hill Education/ Photo by David Hyde and Wayne Falda; 24.11: © Biophoto Associates/Science Source; 24A: © Michael Greenlar/The Image Works; 24.14: © Bernard Bennot/SPL/Science Source.

INDEX

This text contains a substantial number of supporting graphics. To ease index use, regular page numbers indicate both discussion and figure on the same page; page numbers with f indicate additional figures on a different page; t indicates a table; **bold** refer to illustrations (reference plates).

A

A bands, 295
Abdomen, **51–52**
Abdominal aorta, **47,** 592, 594, 595f
Abdominal cavity, 18, 19f, **47, 50–52,** 622
    female torso, **42, 43, 44**
    male torso, **40, 41**
Abdominal reflexes, 414
Abdominal region, 30, 31f, 32f
Abdominal viscera, **55–56,** 588, 603, 604f
    female torso, **42, 43, 44**
    male torso, **40, 41**
Abdominal wall, 320f, 325, 328t, 329f, 598, 603, 604f, 746
Abdominopelvic cavity, 19f
Abdominopelvic membranes, 20, 21f
Abducens nerves, 422, 424t
Abduction, 274, 275f
Abductor brevis muscle, **43**
Abductor muscles, **43,** 318–19, 327f
Abductor pollicis longus muscle, 328f
ABO blood groups, 548–51, 911
Absolute refractory period, 378
Absorption, 650
    large intestine, 685–86
    organ systems, 23–24
    small intestine, 681–82
Accelerator nerves, 572
Accessory nerves, 424
Accessory structures
    digestive system, accessory organs of, 652f
    external accessory organs, 847
    of joints, 270–71
    sex organs, 839
    of skin, 184–88, 195
    visual accessory organs, 468–70
Accidents, and spinal cord injuries, 419
Accommodation, of eye, 472–73, 474f
ACE inhibitors, 512
Acetabul- (prefix), 200
Acetabulum, 200, 240f, 241t, 281
Acetaminophen, 450
Acetic acid, 129
Acetoacetic acid, 818
Acetone, 697, 818
Acetyl coenzyme A (acetyl CoA), 128f, 129, 709

Acetylcholine (ACh), 297, 380, 381t, 436, 437, 512
Acetylcholinesterase, 300, 381, 437
Achilles tendon. *See* Calcaneal tendon
Acid, 66–68, 812
Acid indigestion, 667
Acid-base balance, 812–19
Acid-base buffer systems, 814–15
Acidic ketone bodies, 813
Acidosis, 67–68, 804, 813, 817–18
Acini, 668
Acne, 186, 188
Acquired immune deficiency syndrome (AIDS), 286, 539t, 641, 757, 859
Acromegaly, 210, 501
Acromial region, 30, 32f
Acromioclavicular joint, 278t
Acromion process, 233, 234f, **348**
Acrosome, 829, 872f
ACTH. *See* Adrenocorticotropic hormone
Actin, 97f, 293
Action potentials, 297, 361, 375–77
Activation energy, 123
Active aging, 899
Active dying, 897
Active sites, 123
Active transport, 99, 102–3, 107t, 787f
Acute cardiac tamponade, 590
Acute cholecystitis, 675
Acute glomerulonephritis (AGN), 778
Acute lymphoblastic leukemia (ALL), 540
Acute pain fibers (A-delta fibers), 449
Acute pancreatitis, 668
Acute respiratory distress syndrome (ARDS), 757
Adaptive (specific) defenses, 625, 627–40
Addiction, drug, 89
Addison disease, 513, 813
Adduction, 275f, 276
Adductor brevis muscle, 331f, 333, 334t
Adductor longus muscle, **41–42,** 331f, 333, 334t, **356**
Adductor magnus muscle, 331f, 333, 333f, 334t
Adductor muscles, of thigh, 331f, 333
A-delta fibers, 449
Adenoids, 654
Adenosine diphosphate (ADP), 126, 303
Adenosine triphosphate (ATP)
    aerobic reactions, 129
    cellular metabolism, 125–27
    cellular respiration, 92, 127
    chemical energy release, 126–27
    dietary supplements, 719
    muscle contraction, 299f, 300–302, 303
Adenylate cyclase, 492, 493f
ADH. *See* Antidiuretic hormone; Antidiuretic hormone

Adip- (prefix), 149
Adiponectin, 704
Adipose (fat) tissue, 149, 163, 169t, 189
Adolescence, 896, 897t. *See also* Children
ADP. *See* Adenosine diphosphate
Adrenal Androgens, 514t
Adrenal cortex, 488, 491t, 508
Adrenal cortex disorders, 513
Adrenal cortex hormones, 510–12, 513, 514t
Adrenal gland, **43,** 431, 436, 508–14, 520, 522, 773
Adrenal medulla, 488, 491t, 496, 508–10
Adrenergic receptors, 433, 436–37
Adrenocorticotropic hormone (ACTH), 488, 491t, 500, 504t, 512, 518
Adrenoleukodystrophy (ALD), 94
Adulthood, 896–97
Aer- (prefix), 121
Aerobic, 302
Aerobic exercise, 308, 340
Aerobic reactions, 127, 129–32
Aerobic respiration, 128f, 812–13
Af- (prefix), 769
Afferent arterioles, 769, 772
Afferent fibers, 420
Afferent lymphatic vessels, 622
Africa, and famines, 722
Afterbirth, 891
Afterload, and blood pressure, 587
Age and aging. *See also* Elderly; Infants; Life-span changes
    allergic reactions, 638
    changes in organ systems, 519–20, 522, 898t
    chromosome defects in children of older mothers, 841, 920
    heatstroke, 808
    prescription of drugs, 798
    process of, 27, 896–97, 898–99
    skin, 192–93, 195
    testosterone concentration in males, 839
"Age spots," 193
Age-related macular degeneration, 482, 577
Ageusia, 456t
Agglutin- (prefix), 528
Agglutination, 528, 548, 550f, 634, 635t
Agglutinins, 548
Agglutinogens, 548
Agonist, 312, 384
Agranulocytes, 536, 541t
Agriculture, "organic" in, 68
AIDS, 286, 539t, 641, 757, 859
Air pollution, 743
Air pressure, 741, 744
Alanine, 141t, 699t
Alb- (prefix), 178
Albinism, 178, 183, 185, 478
Albumin, 75, 541, 542t, 709

Alcohol and pregnancy, 882–83
Alcoholism, 708
"Alcohol-related effects," 882
Aldosterone, 490, 491t, 510–11, 512f, 514t, 589, 714, 809, 889
Ali, Muhammed, 404
Aliment- (prefix), 650
Alimentary canal, 650–52, 653f
Alkaline tide, 665
Alkaloids, 455
Alkalosis, 67, 817, 818
Alkaptonuria, 797
Allant- (prefix), 868
Allantois, 868, 874
Alleles, 909
Allergen, 617, 636, 637f
Allergies and allergic reactions, 160, 539t, 636–38, 639t
Allograft, 192, 639
All-or-none response, 305, 377
Alopecia areata, 186
Alpha cells, 514
Alpha fetoprotein (AFP), 919
Alpha globulin, 541, 542t
Alpha radiation, 60, 62
Alpha receptors, 436–37
Alpha waves, 410
Alpha-synuclein, 404
Alpha-tocopherol, 707
Alternative pathway, of complement, 626
Altitude sickness, 756
Alveol- (prefix), 731
Alveolar arch, 221, **259**
Alveolar capillaries, 590
Alveolar dead space, 748
Alveolar ducts, 738f, 739, 853
Alveolar gas exchanges, 753–57, 758f
Alveolar glands, 156, 852
Alveolar macrophages, 754
Alveolar pores, 754, 755f
Alveolar process, 221, 222
Alveolar sacs, 739
Alveolar ventilation, 749
Alveolar wall, 593f
Alveolus (alveoli), 593f, 731, 738f, 739, 740, 741f, 754, 762
Alzheimer disease, 27, 75, 150, 899t
Amacrine cells, 474
Ameboid motion, 538
Amenorrhea, 852
American Dietetic Association, 720
American Heart Association, 591, 698
Amine hormones, 489, 491t, 492f
Amino acids, 73, 380. *See also* Protein
    acid-base balance, 813–14
    dehydration synthesis, 121–22
    digestion of, 682t
    structure of, 73–74
Amino group, amino acid, 73
Ammonia, 686
Ammonium ions, 816
Amniocentesis, 919f, 920
Amniochorionic membrane, 874
Amnion, 873–74, 878f
Amniotic cavity, 874

Amniotic fluid, 874
Amoeba, 15
Amphetamines, 494, 703
Amphiarthrotic joints, 268
Ampicillin, 787
Ampulla, 466, 833
Amylase, 123
Amyotrophic lateral sclerosis (ALS), 370, 418
An- (prefix), 121
Ana- (prefix), 121
Anabolic steroids, 494
Anabolism, 121–22, 123
Anaerobic, 302
Anaerobic environment, 298
Anaerobic respiration, 121, 813
Anal canal, 684, 685f
Anal columns, 684
Anal sphincter, 330f, 684
Anaphase, 108, 824, 825
Anaphylactic shock, 636
Anatomic dead space, 748
Anatomical neck, 233, 236f
Anatomical position, 27, 28f
Anatomy, 11. *See also specific topics*
    history of, 10, 11, 791, 797, 829
    terminology of, 27–32
    vs. physiology, 11–12
Anatomy and physiology, overview, 9–36
    anatomical terminology, 27–32
    anatomy *vs.* physiology, 11–12
    characteristics of life, 14
    levels of organization, 12, 14t
    life-span changes, 27
    maintenance of life, 14–18
    medical science origins, 10–11
    organization of human body, 18–26
    summary, assessment, and tools, 33–36
Andr- (prefix), 824
Androgenic alopecia, 186
Androgens, 188, 512, 824, 838, 849
Anemia, 534–35, 640t, 688, 710, 720, 873
Anencephaly, 397, 710
Anerobic reactions, 127–28
Anesthesiologist assistant, 59
Anesthetics and anesthesia, 379
Aneuploidy, 916, 918–19
Aneurysm, 584
Angina pectoris, 564, 610
Angio-, 557
Angiogenesis, 577
Angiogram, 566f
Angiotensin, 511, 557, 588, 589, 783, 784f
Angiotensinogen, 511
Angiotension-converting enzyme (ACE), 783
Anions, 64
Ankle, 278t
Ankle-jerk reflex, 414
Anorexia nervosa, 722–23
Anosmia, 456t, 482
Antacids, 667
Antagonist, 312, 384
Antecubital region, 30, 32f
Anterior, and relative position, 27, 28f, 32f
Anterior canal, 466
Anterior cavity, 473

Anterior cerebral artery, 597
Anterior chamber, 473, 474f
Anterior choroid artery, 596f, 597
Anterior crest, 243
Anterior cruciate ligament, 284
Anterior facial vein, 608f
Anterior fontanel, **265**
Anterior funiculus, 410
Anterior group, thigh muscles, 328, 332
Anterior horns, 410
Anterior inferior iliac spine, 240f
Anterior intercostal arteries, 598, 599f
Anterior interventricular artery, 564
Anterior longitudinal ligaments, 227
Anterior median fissure, 410
Anterior pituitary hormones, 491t, 500–502
Anterior sacral foramina, 229
Anterior superior iliac spine, **38, 40,** 239, 240f
Anterior tibial artery, 339f, 600, 600f
Anterior tibial vein, 603, 608f
Anti-angiogenesis drugs, 111
Antibiotics, 787
Antibodies, 537, 548, 630, 635–36
Antibody titer, 635
Antibody-dependent cytotoxic reactions, 638, 639t
Antibrachial region, 30, 32f
Anticodon, 139
Antidepressants, 381t
Antidiuretic hormone (ADH)
    life-span changes, 519
    pituitary gland, 491t, 498, 502, 504t
    regulation of water output, 808
    stress response, 518
    synthesis and actions of, 502–3, 504t
    urinary volume and concentration, 788–90
Antigen-presenting cell, 629, 631f
Antigens, 548, 627, 628
Antihistamines, 150
Antioxidants, 706
Antiparallel strands, of DNA, 132, 134f
Antithrombin, 544, 547
Antral follicle, 842, 843f, 844f
Anul- (prefix), 268
Anular ligament, 268, 280
Anus, 330f, 652f, 684
Aorta, 19f, **45, 47, 50**–55, 163, 561, 562f, 591–92, 594, 599f. *See also* Abdominal aorta; Arch of aorta; Descending aorta; Thoracic aorta
Aortic aneurysm, 163
Aortic bodies, 591, 752
Aortic sinus, 591
Aortic sound, 568
Aortic valve, 561, 563f, 564f
Aortic valvulitis, 569
Apex, of heart, 557
Aphasia, 402
Apical heartbeat, 557
Apical surface, epithelial tissue, 150
Aplastic anemia, 535t
Apo- (prefix), 83
Apocrine glands, 156, 158f, 159t, 188t, 197
Aponeuroses, 292, 320f
Apoproteins, 697

Apoptosis, 83, 111, 113, 438, 899
Append- (prefix), 10
Appendicitis, 684
Appendicular portion of body, 18, 28
Appendicular skeleton, 212, 213–15, 214f
Appendix, 621, 683–84
Appendix, **41–42,** 621, 683–84
Appetite and appetite control, 703, 704, 705f, 705t. *See also* Diet
Apraxia, 402
Aquaporin, 100
Aqueous humor, 473, 474f
Arachidonic acid, 698
Arachnoid granulations, 395
Arachnoid mater, 391
Arbor vitae, 408
Arch of aorta, **41, 43–44, 53,** 591, 593f, 594t, 595f
Arcuate arteries, 772
Arcuate nucleus, 704
Arcuate popliteal ligament, 284
Areola, **38,** 852
Areolar connective tissue, 163, 168–69, 169t
Areolar tissue, 164f
Arginine, 141t, 699t
Arm, muscles of, 320–23, 326f, 327f, 328f, **348, 350, 356**
Aromatase inhibitors, 855
Arrector pili muscle, 185
Arrhythmias, 574–75
Arsenic, 129
Arsenicosis, 129
Arterial blood pressure, 583, 585–86
Arterial system, 591–601
Arteries, 577–79, 583t, 594, 596–601, 596f. *See also* Cardiovascular system; *specific arteries*
Arterioles, 577–79, 583t, 588f
Arteriovenous shunts, 579
Arth- (prefix), 268
Arthology, 268
Arthritis, 271, 282, 286, 287t, 640t
Arthrocentesis, 271, 315
Arthroscopy, 281
Articular cartilage, 200, 269
Articulations, 267
Artificial insemination, 870
Artificial pacemaker, 575
Artificial respiration, 741
Artificially acquired immunity, 636, 638t
Arytenoid cartilages, 734
Asbestos, 743
Ascending aorta, 591, 594t, 595f
Ascending colon, **41–42, 56,** 684
Ascending limb of nephron loop, 776, 790f, 791t
Ascending lumbar veins, 603, 608f
Ascending reticular activating system, 407
Ascending tract, 416–18
Ascites, 581, 722, 811
Ascorbic acid (Vitamin C), 209, 534, 711, 712t
-Ase (suffix), 123
Asparagine, 141t, 699t
Aspartame, 695
Aspartic acid, 141t, 699t

Asphyxiation, 737
Aspirin, 450, 496
Assisted reproductive technologies (ART), 870
Association areas, of brain, 399–401
Asthma, 750
Astigmatism, 476
Astr- (prefix), 360
Astrocytes, 360, 368, 370t
Atelectasis, 757
Ather- (prefix), 557
Atherectomy, 610
Atherosclerosis, 161, 546, 557, 584
Athletes and athletic performance.
        See also Exercise
    female reproductive system, 849, 852
    heatstroke, 808
    hormones, 494
    male reproductive system, 849
    mountain climbing, 756
    nutrition, 704f, 720
    Parkinson disease, 404
    traumatic brain injury, 392
    weight lifting, 308
Atlantoaxial joint, 278t
Atlantooccipital joint, 278t
Atlas, of cervical vertebrae, 228
Atmospheric pressure, 15
Atomic number, 60
Atomic radiation, 60
Atomic weight, 60
Atoms, 12, 14t, 58–59, 58t, 63–65
ATP. *See* Adenosine triphosphate
ATPase, 300
Atrial natriuretic peptide (ANP), 517, 586, 783
Atrial syncytium, 568
Atrioventricular bundle (AV bundle), 569
Atrioventricular node (AV node), 569, 575, 576f
Atrioventricular orifice, 560
Atrioventricular sulcus, 560
Atrioventricular valve (AV valve), 560, 561
Atrium (atria), **42, 47, 53,** 560, 568f
Atrophy, 210, 308
Atropine, 437
Aud- (prefix), 444
Audiologist, 445
Auditory, 444
Auditory ossicles, 457
Auditory pathways, 461, 464f
Auditory tube, 459
Auricles, 456, 457f, 560, 593f
Auricular artery. *See* Posterior auricular artery
Auricular surfaces, 229
Auto- (prefix), 617
Autoantibodies, 639
Autocrine secretions, 488
Autograft, 192, 639
Autoimmune disorders, 298, 617, 640t.
        See also AIDS; HIV
Autoimmune ulcerative colitis, 639
Autoimmunity, 638, 639–40, 899
Automatic bladder, 796
Autonomic nerve fibers, 431
Autonomic nervous system, 431–38.
        See also Nervous system

autonomic neurotransmitters, 433–37
control of, 437–38
digestive system, 651–52
functions of, 361
general characteristics, 418, 431
nerve fibers, 431
parasympathetic division, 432–33
sympathetic division, 431–32
Autonomic neurotransmitters, 433, 436–37
Autonomic pathways, 432f
Autophagy, 93
Autoregulation, of glomerular filtration, 783
Autorhythmic conduction, of heart, 568
Autosomes and autosomal inheritance, 908, 909–10, 911f
Ax- (prefix), 200, 360
Axial portion of body, 18, 28
Axial skeleton, 200, 212, 213t, 214f
Axillary artery, **41,** 597
Axillary border, 233, 234f
Axillary nerves, 427
Axillary region, 30, 32f, 622
Axillary vein, **41,** 608f
Axillary vein, 602
Axis, of cervical vertebrae, 228
Axon hillock, 363
Axon terminal, 363
Axonal transport, 363
Axons, 360
    neurons, 170, 172f, 363
    regeneration, 370
    use of term, 444
Azygos vein, **51,** 603, 608f

B

B cells, 537, 628, 629f, 630, 631f, 632f
Babinski reflex, 414
Back, muscles of, 319f
Back pressure, 581
Bacteria, 286, 539t, 658, 673, 686
Balance, 804. *See also* Body fluid balance
Baldness, 185, 186, 195, 916
Ball-and-socket joint, 272, 273f, 274t
Banting, Frederick Grant, 518
Bariatric surgery, 703
Baroreceptor reflex, 572, 576f
Baroreceptors, 444, 572, 587f
Bartholin's glands, 848
Bas- (prefix), 695
Basal body, 94
Basal cell carcinoma, 182
Basal metabolic rate (BMR), 505, 695, 701, 723
Basal nuclei, 403, 405f, 409t
Base, 66–68, 812–19
Basement membrane, 150, 161, 178
Basilar artery, 594, 596f
Basilar membrane, 459, 461f
Basilic vein, 602, 608f
Basophils, 537, 541t
BeadChip tests, 549
Beans, 699
Beaumont, William, 665
Bee pollen, 718

Beeturia, 797
Bell's palsy, 423
Benign tumor, 111
Beriberi, 708
Berylliosis, 743
Beryllium, 60t
Best, Charles Herbert, 518
Beta amyloid, 27, 75
Beta cells, 514
Beta endorphins, 382
Beta globulins, 541, 542t
Beta interferons, 367
Beta oxidation, 697
Beta radiation, 60, 62
Beta receptors, 436–37
Beta waves, 410
Beta-hydroxybutyric acid, 818
Bi- (prefix), 360
Bicarbonate buffer system, 814, 815t
Bicarbonate ions, 69t, 759, 761f, 784t, 806f, 813, 814, 815
Biceps brachii muscle, 311, 312, 321, 322f, 324t, **348–50, 353, 356**
Biceps femoris muscle, 332f, 333f, 334f, 335, 337f, 338f, **349, 350, 357–58**
Biceps-jerk reflex, 414
Bil- (prefix), 528
Bilateral, as relative position, 27, 28f
Bile, 673–75
Bile canaliculi, 670
Bile duct, 674
Bile ductules, 670
Bile pigments, 673
Bile salts, 673, 675
Bilirubin, 528, 534, 535f, 673
Biliverdin, 535, 673
Bio- (prefix), 58
Biochemistry, 58
Biogenic amines, 380t
Biotin (Vitamin B$_7$), 709–10, 712t, 714
Bipolar neurons, 360, 365, 366f, 368t, 474
Birth control, 824, 854–58, 858t, 860f
Birth defects, 882–83
Birth process, 890–91
Bitter receptors, 455
Bladder. *See* Urinary bladder
-Blast (suffix), 200
Blastocyst, 854, 871–72, 875f, 876f, 879t
Blastomere, 871
Blast-related brain injury, 392
Blasts, 113
Blebs, 113
Blind spot, 474
Blindness, 472, 481
Blood, 527–55. *See also* Cardiovascular system; Plasma; Red blood cells; White blood cells
    blood groups and transfusions, 548–51, 911
    clotting of, 543–47, 548t, 915
    as connective tissue, 167–68, 169t
    defined, 528
    disorders of, 552t
    fetal circulation, 883, 885–87, 888f, 889t
    hemostasis, 542–47
    path of through heart, 561, 563, 564f

summary, assessment, and tools, 552–55
    supply of to heart, 563–64, 566
    terminology, 528
    volume of, 503, 586
Blood bank technologist, 529
Blood cells, 101, 102f, 210, 529–40
Blood doping, 494
Blood filtering, in liver, 673
Blood gases, 542
Blood groups, 548–51, 911
Blood platelets, 539–40
Blood pressure
    ACE inhibitors, 512
    aging, 608
    arterial, 583, 585–86
    capillary blood flow, 580–81, 582f
    kidney failure, 783
    regulation, 586–88
Blood reservoirs, veins as, 582
Blood sugar levels, 16, 210
Blood typing and matching, 549
Blood urea nitrogen (BUN), 542, 698, 779
Blood vessels, 183, 576–82, 583t, 584. *See also* Arteries; Capillaries; Veins
Blood-brain barrier, 150, 394
BLOODchip, 549
Bloodgen, 549
Blood-testis barrier, 829
Blue diaper syndrome, 797
BMI (body mass index), 702, 703
Body cavities, 18–19
Body covering, 22
Body fluid balance, 803–22
    acid-base balance, 812–19
    balance as concept, 804
    body fluid distribution, 804–6
    compensation, 819
    electrolyte balance, 69, 804, 809–12
    summary, assessments, and tools, 819–22
    water balance, 68, 804, 806–8, 810–11
    water compartments and distribution, 805–6
    water intake, 806–7
    water output, 807–8
Body fluids, 804–6
Body mass index (BMI), 702, 703
Body movement, and skeletal muscle actions, 310–11
Body regions, 28, 30
Body sections, 28, 29f
Body temperature
    aging process, 195
    aging-related changes, 898
    cold, 189
    elevated, 191
    energy requirements, 701–2
    heat, 15, 303
    heatstroke, 808
    homeostasis, 16, 17f
    hyperthermia, 189, 191, 810
    hypothermia, 189, 190
    normal human, 189, 190f
    ovulation, 855
    regulation, 16, 17f, 189, 190f, 192f
Body weight. *See* Diet; Obesity
Bolus, 660

Bomb calorimeter, 701
Bombay phenotype, 551
Bonds, 63
Bone, 200–211. *See also* Skeletal system
    age-related changes, 246–47
    classification of, 200
    as connective tissue, 166–67, 168f, 169t
    development and growth, 202, 204–10
    endochondral, 204
    function, 210–11
    number of, 212, 213f, 213t
    osteoporosis and osteopenia, 212
    shapes, 200
    structure, 200–202, 203f
    summary, assessment, and tools, 246, 248–51
Bone density, 212
Bone marrow, 201, 210–11, 239
Bone matrix, 245
Bone morphogenetic proteins, 210
Bone remodeling, 207–8
Boron, 60t
Borrelia burgdorferi, 281
"Botox" (botulinum toxin), 193
Botulism, 298
Boyle's law, 744
Brachial artery, 597
Brachial plexus, **41**
Brachial plexuses, 427–28, 429f
Brachial region, 30, 32f
Brachial veins, 602, 608f
Brachialis muscle, 321, 324t, **350, 356**
Brachiocephalic artery, **42, 53–54,** 591, 593f
Brachiocephalic trunk, 595f, 596f, 601f
Brachiocephalic veins, **41, 43,** 593f, 602, 603, 608f
Brachioradialis muscle, 321, 324t, 326f, 327f, **348, 350, 356**
Brady- (prefix), 557
Bradycardia, 557, 574
Bradykinin, 588
Brain, **49,** 395–409. *See also specific parts of the brain*
    arteries, 594, 596f, 597
    basal nuclei, 403
    brainstem, 405–8
    cerebellum, 408–9
    cerebral cortex, 397–403
    cerebrum, 395, 397
    development of, 395, 397t
    diencephalon, 403, 405
    imaging, 78
    injury to, 392
    life-span changes, 438
    parts of, 409t
    regulation of pain impulses, 449
    traumatic brain injury (TBI), 392
    veins, 602
Brain death, 410
Brain tumors, 370
Brain waves, 410
Brainstem, **46,** 395, 405–8, 409t
Breast, **38,** 852, 854f, 856f. *See also* Mammary glands
Breast cancer, 854–55

Breastfeeding, 891, 892–93, 894
Breathing, 735, 741–54, 894, 898.
    See also Respiration
"Brittle bone disease," 912
Broad ligament, 840, 845
Broca's area, 402
Bronch- (prefix), 731
Bronchial arteries, 592, 594t
Bronchial asthma, 750
Bronchial tree, 738–40, 744t
Bronchioles, 741f
Bronchomediastinal trunk, 617, 619f
Bronchopulmonary segments, 738
Bronchoscopy, 740
Bronchus, **43,** 731. *See also* Left bronchus
Brown-Séquard syndrome, 418
Buccal region, 30, 32f
Buccinator muscle, 313t, 314–15, **352**
Buffers, 67, 68, 814–15
Bulbospongiosus muscle, 327–28, 330, 330t
Bulbourethral gland, 833, 838t
Bulimia, 722, 723
Bulk elements, 59
Bundle branch block, 569
Burn technician, 179
Burns, skin, 192, 193f
Burs- (prefix), 268
Bursa, 272, 277
Bursa of Fabricius, 628
Bursitis, 286
Buttocks, **349**

C

Cacosmia, 456
Cadaver dissection, 10, **45–54,** 347, **352–58**
Caffeine, 893t
Calat- (prefix), 292
Calcaneal tendon, 338, 338f, **351, 358**
Calcaneal tuberosity, 243
Calcaneus, 243, 244f, 338f, **351**
Calcitonin, 491t, 504f, 505, 506t
Calcium
    bone development, 208
    bone loss and supplements, 245
    electrolyte balance, 811
    heart action, 572
    hormones, 505–6
    nutrition, 711, 713, 715t
    osteoporosis, 163, 212, 520
    storage in bone, 211
    urine and urine tests, 779, 784t
Calcium channel blockers, 308
Calcium ion channels, 89
Calcium ions, 69t, 806f, 810, 812f
Calluses, 179
Calmodulin, 307, 495
Calor- (prefix), 695
Calorie, 695, 701
Calorimeter, 695
Calyc- (prefix), 769
CAM. *See* Cellular adhesion molecules
Canal- (prefix), 200
Canaliculus (caniculi), 167, 168f, 200, 201

Cancer. *See also specific types of cancer*
    angiogenesis, 577
    basement membranes, 150
    carcinomas, 156
    causes of death, 899t
    characteristics of cells, 111–12
    chronic pain, 450
    colon, 687
    extracellular matrix, 161
    immunotherapy, 633
    liver, 671
    lung, 735
    radiation, 62
    testicular, 828
Canine teeth, 656, 657t
Cannabinoids, 449
Capacitation, 834
Capillaries, 579–81, 582f, 583t, 617, 781f
Capitulum, 233, 236f
Capsule, of connective tissue, 621
Captopril, 456
Carbaminohemoglobin, 759
Carbohydrates
    digestion of, 681
    liver and metabolism of, 670, 673t
    metabolic pathways and storage, 131–32
    nutrition, 695–96, 700t
    organic chemicals in cells, 70, 77t
Carbon, 60t
Carbon dioxide
    breathing rate and tidal volume, 752
    chemical constituents of cells, 69, 69t
    citric acid cycle, 129
    concentration gradient, 99–100
    respiratory excretion of, 815–16
    size of molecule, 77f
    transport, 758–61
Carbon monoxide, 69, 758
Carbonate ions, 69t
Carbonic acid, 813, 814, 815
Carbonic anhydrase, 759, 815
Carboxyl group, 73, 814
Carboxypepidase, 668, 680t
Carcin- (prefix), 731
Carcinoma, 156, 731
Carcinoma in situ (CIS), 846
Cardi- (prefix), 10
Cardia, 663
Cardiac center, of medulla oblongata, 407, 587
Cardiac conduction system, 568–70, 608
Cardiac control center, 572
Cardiac cycle, 566–67, 568f, 572, 573f
Cardiac muscle fibers, 568
Cardiac muscle tissue, 170, 171f, 172t
Cardiac muscles, 308–9
Cardiac output (CO), 585–86, 587f
Cardiac veins, 566
Cardinal ligaments, of uterus, 845
Cardioaccelerator and cardioinhibitor reflexes, 572, 587
Cardiology, 30
Cardiovascular disease, 609, 907. *See also* Heart disease
Cardiovascular system, 556–615. *See also* Blood
    arterial system, 591–601

    autonomic stimulation, 436t
    blood pressure, 582–90
    blood vessels, 576–82
    death and end of life, 897–98
    exercise, 591, 610, 754
    fetal blood and circulation, 883, 885–87, 888f, 889t
    heart actions, 568–75
    infants and development of, 895
    interconnections with other systems, 194, 247, 341, 383, 521, 611, 643, 689, 763, 799, 861
    life-span changes, 608, 898t
    major functions of, 23, 24f, 25t
    paths of circulation, 590–91, 592
    structure of heart, 557–58, 560–61
    summary, assessments, and tools, 610, 612–15
    venous system, 602–7
Career Corner
    anesthesiologist assistant, 59
    audiologist, 445
    blood bank technologist, 529
    burn technician, 179
    cytotechnologist, 83
    diabetes educator, 489
    dialysis technician, 769
    Emergency Medical Technician (EMT), 11
    endoscopy technician, 651
    genetic counselor, 907
    massage therapist, 293
    medical technologist, 805
    nurse-midwife, 825
    occupational therapist, 391
    perfusionist, 557
    personal trainer, 121
    pharmacist, 361
    physical therapy assistant, 269
    physician's assistant, 869
    public health nurse, 617
    radiologic technologist, 201
    registered dietitian, 695
    respiratory therapist, 731
    tissue recovery technician, 149
Cari- (prefix), 650
Carina, 738
Carot- (prefix), 695
Carotene, 183, 695, 706
Carotid artery, **50.** *See also* External carotid artery; Internal carotid artery; Left common carotid artery
Carotid bodies, 597, 752
Carotid canal, 216f, 218, 219f, 224t, **256–58**
Carotid sinus, 596f, 597
Carp- (prefix), 200
Carpal bones, 200, 213, 235f, 236, 238f, **350**
Carpal region, 30, 32f
Carpal tunnel syndrome, 430
Carpometacarpal joint, 278t
Carpus, 236, 238f
Carrier molecule, 100–101
CART (combined antiretroviral therapy), 642
Cartilage, 165–66, 167f, 169t
Cartilaginous joints, 269, 274t
Cascade, and blood coagulation, 543
Cat, 452

Cata- (prefix), 121
Catabolism, 121, 122–23, 133f
Catalase, 92, 123
Catalysis, 123
Catalyst, 66
Cataract, 482
Catecholamines, 510
Cations, 64
Cauda equina, **47–48,** 409, 426
Caudate lobe, of liver, 670
Caudate nucleus, 403
Causalgia, 449
Cavities (dental), 650, 658
CCK. *See* Cholecystokinin
CD$_4$ cell, 630
Cec- (prefix), 650
Cecum, **41–42,** 650, 683
Celiac artery, **43,** 592, 594t, 595f
Celiac disease, 682
Celiac region, 30
Cell body, 363
Cell cycle, 107–10
Cell membrane, 83, 86–89, 98t
Cell membrane potential, 372–78
Cell membrane proteins, 87–89
Cell surface proteins, 87t
Cells, 82–119. *See also* Stem cells
    as basic units, 83
    cell cycle, 107–10
    cell death, 113, 115f
    chemical constituents of, 68–78
    composite, 83, 84f, 85, 85f
    connective tissue and types of, 159
    control of cell division, 110–12
    division of, 27, 110–11
    levels of organization, 12–13
    life span, 27
    movements into and out of cell, 99–105,
        106f, 107t
    nervous system, 363–71
    nucleus, 83, 97, 98f
    organized into tissues, 149–50
    progenitor cells, 112–13, 115f
    shape and functions of, 84f
    sizes of, 83, 84f
    specialty cell structure and function,
        112–13, 115f
    structure and function, 83–97, 98t
    summary, assessment, and tools, 116–19
Cellular adhesion molecules (CAM), 87–89, 90f
Cellular immune response, 627, 629
Cellular metabolism, 120–47. *See also*
    Metabolism
    cellular respiration, 127–32
    changes in genetic information, 142–43
    control of reactions, 123–25
    energy for, 125–27
    mutations, 142–43
    nucleic acids, 132–38
    processes of, 121–23
    protein synthesis, 139–41
    summary, assessment, and tools, 144–47
Cellular reprogramming, 83
Cellular respiration, 92, 127–32, 731
Cellulose, 695

Cementum, 657
Centenarians, and aging, 27, 900
Centers for Disease Control and Prevention
        (CDC), 530
Central canal of compact bone, 201
Central canal of spinal cord, 369, 410
Central canals, 167
Central chemoreceptors, 752–53
Central nervous system (CNS). *See also*
        Autonomic nervous system; Peripheral
        nervous system
    brain, 395–409
    functions of, 361
    life-span changes, 438
    meninges, 390–91, 393f
    neuroglia of, 368–70
    neurons, 367f
    senescence, 897
    spinal cord, 409–18
    ventricles and cerebrospinal fluid, 391–95
Central tendon, 327, 330t
Central vein, of liver, 670
Central venous pressure, 588, 590
Centrioles, 94, 95f
Centromere, 108, 824
Centrosome, 94, 95f, 98t
Cephal- (prefix), 390
Cephalic phase, of gastric secretion, 665, 666t
Cephalic region, 30, 32f
Cephalic vein, 602, 603, 608f
Cephalosporin, 787
Cerebellar ataxia, 870
Cerebellar cortex, 408
Cerebellar peduncles, 409
Cerebellum, **45–46,** 408–9
Cerebr- (prefix), 10
Cerebral aqueduct, 392, 394, 405
Cerebral arterial circle, 594
Cerebral arteries, 594, 597
Cerebral cortex, 397, 399–403, 449
Cerebral hemispheres, 395, 397, 399f
Cerebral lobes, 402t
Cerebral peduncles, 405
Cerebral vascular accident (CVA), 397, 589
Cerebrospinal fluid (CSF), 391–95, 396
Cerebrum, **45–46,** 395, 397, 409t
Ceroid pigments, 27
Ceruminous glands, 188t, 456
Cervical cancer, 846
Cervical cap, 856f, 857, 858t
Cervical curvature, 226f, 227
Cervical enlargement, 409
Cervical nerves, 425
Cervical orifice, 845
Cervical plexuses, 427
Cervical region, 32f, 622
Cervical vertebrae, **46,** 226, 228, 229f, 231t
Cervix, 845, 891f
CFTR gene, 908f, 909–10
Chambers, of heart, 560–61
Chaperones, and protein synthesis, 141
Charles I (King of Spain), 791
Cheeks, 653, 658t
Chemical barrier birth control, 856f, 857, 858t
Chemical barriers to infection, 626

Chemical bonds, 59, 63–65
Chemical buffer systems, 814–15, 817f
Chemical digestion, 650
Chemical reactions, 65–66
Chemistry, 57–81
    chemical constituents of cells, 68–78
    hormones, 489–90
    structure of matter, 58–68
    summary, assessment, and tools, 79–81
Chemoreceptors, 444, 752–53
Chemotaxis, 627, 634, 635t
Chemotherapy, for breast cancer, 855
Cheyne-Stokes breathing, 898
Chiasm- (prefix), 390
Chiasma, 390
Chief cells, 665
Children. *See also* Adolescence; Development,
        human; Infants
    adipose tissues, 163
    dehydration, 792
    development, 896, 897t
    ear infections, 459
    genetic tests, 911
    lipoid nephrosis, 786
    nocturnal enuresis, 795
    ossification timetable, 207t
    skull, **266**
Chlamydia infection, 859t
Chloride ion channels, 89
Chloride ions, 69t, 760, 761f, 784t, 806f, 812
Chloride shift, 760, 761f
Chlorine, 64, 714, 715t
Cholangiogram, 675
Cholecystectomy, 675
Cholecystitis, 675
Cholecystogram, 675
Cholecystokinin (CCK), 666, 669, 674, 675t
Cholelithiasis, 675
Cholera, 810
Cholesterol, 86, 105, 609, 673, 681–82, 696
Cholinergic receptors, 433, 436, 437f
Chondr- (prefix), 149
Chondrocyte, 149, 166
Chondrodysplasia, 163t
Chondromucoprotein, 166
Chordae tendineae, 561
Chorio- (prefix), 868
Chorion, 868, 873
Chorionic villi, 873, 879f
Chorionic villus sampling (CVS), 919f, 920
Choroid- (prefix), 444
Choroid artery. *See* Anterior choroid artery
Choroid coat, 444, 472, 474, 477t
Choroid plexus, 369, 390, 394
Chromatids, 108, 824
Chromatin, 97, 98t
"Chromatin remodeling," 141
Chromatophilic substance, 363
Chromium, 717
Chromo- (prefix), 905
Chromosome disorders, 916–20. *See also*
    Genetic disorders
Chromosomes, 94, 95f, 97, 108, 109f, 120f,
    135f, 905, 908–9. *See also* DNA;
    Genetics; RNA

Chronic glomerulonephritis, 772, 778
Chronic granulomatous disease, 552t
Chronic kidney failure, 772, 783. *See also*
    Kidney disease
Chronic myeloid leukemia (CML), 540
Chronic obstructive pulmonary diseases
    (COPD), 750, 752, 899t
Chronic pain, 449, 450
Chronic pain fibers (C fibers), 449
Chronic traumatic encephalopathy (CTE), 392
Chylomicrons, 681
Chym- (prefix), 650
Chyme, 650, 666–67, 668f, 670f, 676f, 682–83,
    685
Chymotrypsin, 668, 680t
Cigarette smoke and smoking, 384, 735,
    882–83. *See also* Nicotine
Cilia, 94, 95–96, 98t, 152, 733, 845, 865f
Ciliary body, 472, 473f, 477t
Ciliary muscles, 470t, 472–73, 474f
Ciliary processes, 472–73
Circadian rhythms, 517
Circular fibers, 650
Circular muscles, 473
Circular set, 473
Circular sulcus, 397
Circumcision, 835
Circumduction, of joint, 275f, 276
Circumflex artery, 564
Cirrhosis, 161
Cisterna chlyi, 617
Cisternae, 90, 296
Citric acid cycle, 127, 128f, 129, 130f, 131f,
    302
Classical pathway, of complement, 626
-Clast (suffix), 200
Clav- (prefix), 200
Clavicle, **38–39,** 200, 213, 233, 234f, 237t, **349**
Cleav- (prefix), 868
Cleavage, 868, 870–71, 879t
Cleft palate, 221
Clinical Application
    acne, 188
    adrenal cortex disorders, 513
    arrhythmias, 574–75
    athletes and nutrition, 720
    blood vessel disorders, 584
    bone fragility, 212
    brain waves, 410
    centenarians and aging, 900
    cerebrospinal fluid (CSF) pressure, 396
    chronic kidney failure, 772
    cigarette smoking effects, 735
    clomerulonephritis, 778
    cochlear implant, 463
    collagen abnormalities, 163
    coronary artery disease, 610
    deep vein thrombosis (DVT), 547
    dental caries, 658
    diabetes mellitus, 516
    dietary supplements, 718–19
    disease at organelle level, 94
    Down syndrome, 917
    drug addiction, 384
    elevated body temperature, 191

exercise, 591, 754
exome sequencing, 921
extracellular matrix (ECM), 161
faulty ion channels and disease, 89
female infertility, 853–55
fractures of bone, 208–9
gallbladder disease, 675
gas exchange impairment, 757
hair loss, 186
hearing loss, 463, 465
heartburn, 667
hepatitis, 674
high altitude and breathing, 756
HIV/AIDS, 641–42
hormones, athletic performance and abuse
    of, 494
human growth hormone (HGH), 501
human milk for human babies, 894
hypertension, 589
"inborn errors of metabolism," 126
indoor tanning and skin cancer, 182
joint disorders, 286–87
joint replacement, 282
large intestine disorders, 687
leukemia, 540
lung irritants, 743
male infertility, 836–37
migraine, 362
molecular causes of cardiovascular
    disease, 609
multiple sclerosis (MS), 367
myasthenia gravis, 298
nephrotic syndrome, 786
nerve impulse conduction, 379
obesity, 703
opiates, 382
osteoporosis and osteopenia, 212
pain treatment, 450
Parkinson disease, 404
prostate cancer, 834
reflexes, uses of, 414
refraction disorders of the eye, 476
respiratory disorders, 750
skeletal muscle use and misuse, 308
smell disorders, 456
sodium and potassium imbalance, 813
spinal cord injuries, 419
spinal nerve injuries, 430
synesthesia, 453
taste disorders, 456
TMJ (temporomandibular joint) syndrome,
    315
traumatic brain injury (TBI), 392
ultrasonography and MRI, 13
universal precautions, 530
urinalysis, 797
vertebral column disorders, 230
water balance disorders, 810–11
Clitoris, 330f, 847, 848t
Clone, 628
Closed fracture, 208
Clostridium botulinum, 193, 298
Clotting, of blood, 543–47, 548t, 915
Clubfoot, 245
Co- (prefix), 121

Coagulation, of blood, 543–47. *See also*
    Clotting, of blood
Cobalt, 710, 716, 717t
Cocaine, 381t, 384, 829, 893t
Cocarboxylase, 708
Coccygeal nerves, 425
Coccygeus, 327, 330t
Coccyx, **48,** 212, 229, 230f, 231t, 330f, **349**
Cochlea, 444, 457f, 459, 461f, 462f, 464f
Cochlear branch fibers, 423, 424t
Cochlear duct, 459, 461f, 464f
Cochlear implant, 463
Codominant inheritance, 911
Codons, 138, 141t, 142
Coenzyme A, 709
Coenzymes, 121, 125, 709
Cofactors, 125
Coitus interruptus, 854, 858t
Cold and body temperature, 189
Cold receptors, 446
Cold War ionizing radiation, 62
Collagen, 27, 77f, 162, 163, 201, 209, 711
Collagen fibers, 162, 162t, 165
Collateral axons, 363
Collateral ganglia, 431
Collaterals, 363
Collecting ducts, 617–18, 619f, 776, 789f, 791t
Collectins, 626
Colloid, 503
Colloid osmotic pressure, 541, 581, 805
Colon, **45, 55,** 684. *See also* Ascending colon;
    Descending colon; Sigmoid colon;
    Transverse colon
Colonoscopy, 687
Colony-stimulating factors (CSF), 536, 629,
    630f
Color. *See* Eye color; Skin
Colorblindness, 481, 915
Colorectal cancer, 687
Colostrum, 633, 892, 894
Columnar epithelium
    pseudostratified, 153–54, 159t
    simple, 152–53, 154f, 159t
    stratified, 156, 159t
Coma, 407
Combined hormone contraceptives, 857, 858t
Comminuted fracture, 208
Common carotid artery, **39–40, 42–44, 53,** 591,
    592, 593f, 595f, 596f, 597. *See also* Left
    common carotid artery
Common fibular nerves, 429
Common hepatic duct, 670
Common iliac artery, **42–43,** 594, 598, 599f
Common iliac vein, 606, 608f
Common nerve pathways, 447
Compact bone, 199f, 201–2, 203f, 246
Compartment syndrome, 293
Compartments, 293, 804–6
Compensation (acid-base imbalance), 819
Complement, 626
Complementary base pairs, 132, 133–34
Complete proteins, 699
Complete tetanic contraction, 304, 305f
Completely penetrant genes, 912
Complex carbohydrates, 70

Compliance, and lungs, 746
Composite cells, 83, 84f, 85
Compound (open) fracture, 208
Compound glands, 156, 157f, 158t
Compound heterozygote, 909
Compounds, 59, 62–63
Computer-aided sperm analysis (CASA), 836
Computerized tomography (CT), 78
Concave surface, 475
Concentration gradient, 103
Concentration gradient, diffusion, 99, 100f
Concentric muscle contraction, 306
Conception, 868–70
Concussion, 392
Condom, 855, 856f, 858t
Conduction, 189
Conductive deafness, 465
Condyl- (prefix), 200
Condylar canal, 219f
Condylar joint, 272, 273f, 274t
Condyle, 200, 215t, 233. *See also* Lateral
        condyle; Mandibular condyle; Medial
        condyle; Occipital condyle
Cones, of eye, 478–79
Confocal microscope, 86
Conformation, of amino acid, 73, 74f
"Congenital rubella syndrome," 882
Conjunctiva, 469
Conjunctivitis, 469
Connective tissue fibers, 162
Connective tissue proper, 162
Connective tissues, 149, 159–68
        adipose (fat) tissue, 163, 169t
        areolar connective tissue, 163, 169t
        blood, 167–68, 169t
        bone, 166–67, 168f, 169t
        cartilage, 165–66, 167f, 169t
        categories of, 162
        characteristics, 151t, 159–60, 169t
        connective tissue fibers, 162
        dense irregular connective tissue, 164–65, 169t
        dense regular connective tissue, 164, 165f,
                169t
        elastic connective tissue, 165, 166f, 169t
        extracellular matrix (ECM), 161
        major cell types, 160
        reticular connective tissue, 164, 165f
        skeletal muscle, 292–93
        SUB, 609
Constrictor muscles, 660
Continuous ambulatory peritoneal dialysis, 771
Contra- (prefix), 824
Contraception, 824, 854–58
Contractility, 587
Contraction, of muscles, 297–303, 307–8
Contralateral, and relative position, 28
Control center, homeostasis, 16, 17f, 190f
Conus magus, 450
Conus medullaris, 409
Convection, 189
Convergence, and neuronal pools, 384–85
Convex surface, 475
Coordination, organ systems, 22–23
Copper, 716, 717t
Corac- (prefix), 200

Coracobrachialis muscle, **39–41,** 318, 322t
Coracohumeral ligament, 277, 279f
Coracoid process, 233, 234f
Coriell Institute (New Jersey), 900
Corn, 179
Corn- (prefix), 444
Cornea, 444, 470–72, 477t
Corniculate cartilages, 734
Corona radiata, 843, 869, 872f
Coronal suture, 216f, 217f, 218, **252–53, 262**
Coronary angiography, 564, 610
Coronary arteries, 563, 564f, 591, 594t, 595f
Coronary artery disease (CAD), 610
Coronary bypass surgery, 610
Coronary embolism, 589
Coronary ligament, 669
Coronary sinus, 560, 566
Coronary stent, 610
Coronary thrombosis, 589
Coronoid fossa, 235
Coronoid process, 217f, 222, 236, 237f, **259**
Corpora cavernosa, 835, 847
Corpora quadrigemina, 406
Corpus albicans, 843f, 850
Corpus callosum, **46, 49,** 397, 402
Corpus luteum, 843f, 850
Corpus spongiosum, 835
Cort- (prefix), 488, 769
Cortex, of ovary, 841
Cortical bone, 201, 203f
Cortical nephrons, 776
Cortical radiate arteries, 772
Corticospinal tract, 414, 416–17, 418t
Corticotropes, 500
Corticotropin-releasing hormone (CRH), 491t,
        500, 501f, 512, 518
Cortisol, 491t, 492f, 512, 513, 514t, 518, 520f
Costal cartilage, 21f, **51–52,** 232f, 234f
Costal region, 30
Costocervical arteries, 594
Coughing, 749
Coumadin (warfarin), 547
Countercurrent exchanger, 789
Countercurrent mechanism, 789, 791f
Countercurrent multiplier, 789, 790f
Covalent bond, 64, 65f
Cox-2 inhibitors, 286, 687
Coxal region, 32f
Cran- (prefix), 10
Cranial branch of accessory nerve, 424
Cranial cavity, 19f, 222f, **263–64**
Cranial nerves, 418, 419, 420–24
Cranial sutures, 270f
Cranium, 212, 215–25
Creatine, 720
Creatine phosphate, 301, 542
Creatinine, 779, 784t, 792
Creatinine clearance test, 792
Cremasteric reflex, 414
Crest, 215t
CRH. *See* Corticotropin-releasing hormone
Crib- (prefix), 200
Cribiform plate, 200, 218, 220, 221f, **260, 264**
Cric- (prefix), 731
Cricoid cartilage, 731, 734

Criminology and DNA, 136
-Crin (suffix), 488
Crista ampullaris, 466, 468f
Crista galli, 220, 221f **260, 263–64**
Cristae, 92
Crit- (prefix), 528
Critical period, in development, 881
Crohn's disease, 687
Cross over, of chromosomes, 824, 826f
Cross section, 28, 30, 30f
Cross-bridge cycling, 300
Crossed extensor reflex, 414
Crossed fused ectopia, 797t
Crown, of tooth, 656
Cruciate ligaments, 284
Crur- (prefix), 824
Crura, 824, 835, 847
Crural region, 30, 32f
Crying, 749t, 750
Cryptorchidism, 826
CT. *See* Computerized tomography
Cubital vein. *See* Median cubital vein
Cuboid bone, 244f
Cuboidal epithelium
        simple, 152, 153f, 159t
        stratified, 154, 155f, 159t
Cuneiform cartilages, 244f, 734
Cupula, 466, 468f
Curare, 381t
Cushing syndrome, 513
Cutaneous carcinoma, 182
Cutaneous melanomas, 182
Cutaneous membranes, 168, 169
Cuts, in skin, 190, 191f
Cyanide, 131
Cyanocobalamin (Vitamin B_{12}), 710–11, 712t
Cyanosis, 183, 529
Cyclic adenosine monophosphate (cAMP), 491,
        493, 495
Cyclic guanosine monophosphate (cGMP),
        480, 496
Cyclins, 110
Cyclosporine, 893t
Cyst- (prefix), 769
Cysteine, 141t, 699t, 714
Cystic duct, **42,** 674
Cystic fibrosis, 75, 89, 707, 739, 873, 908f,
        909–10
Cystitis, 769, 795
Cyt- (prefix) or -cyt (suffix), 83, 149
Cytocrine secretion, 179
Cytokines, 629, 630t, 633
Cytokinesis, 107, 108, 109f, 110
Cytoplasm, 83, 85, 90–96, 97f
Cytoplasmic organelles, 90–96, 97f
Cytoskeleton, 90
Cytosol, 83, 90, 128f, 129
Cytotechnologist, 83
Cytotoxic T cell, 630

D

Dairy foods, 681, 686, 706, 891–93, 894
Dark adapted eye, 480

Dartos muscle, 834
DaTscan, 404
Daughter cells, 107, 110
De- (prefix), 121, 804
Deafness, 463, 465, 482
Deamination, 121, 698
Death, 897–98, 899t. *See also* Apoptosis;
　　Brain death
Death, and rigor mortis, 301
Decibels (dB), 461
Decidu- (prefix), 650
Deciduous teeth, 650, 657t
Decomposition (chemical reactions), 66
Decubitus ulcer, 179
Deep, as relative position, 28
Deep brachial artery, 597
Deep circumflex iliac artery, 598, 599f
Deep external pudendal artery, 598
Deep fascia, 293
Deep femoral artery, 598
Deep fibular nerve, 339f
Deep genicular artery, 598
Deep partial-thickness burn, 192
Deep vein thrombosis (DVT), 547
Defecation and defecation reflex, 686
Defensins, 626
Defibrillators, 574
Dehydration, 792, 804, 810
Dehydration synthesis, 121–22
Dehydrocholesterol, 210
Delayed-reaction allergy, 638, 639t
Delta cells, 514
Delta waves, 410
Deltoid muscle, **38–39,** 312, 319f, 320f, 321f,
　　322t, **348–49, 353–55**
Deltoid tuberosity, 233, 236f
Denaturation, of protein, 75
Dendr- (prefix), 360
Dendrites, 170, 172f, 360, 363
Dendritic cells, 629
Dens (odontoid process), 228
Dense connective tissue, 162
Dense irregular connective tissue, 164–65, 169t
Dense regular connective tissue, 164, 165f, 169t
Dental caries, 650, 658
Dentate nucleus, 409
Dentin, 657
Deoxyhemoglobin, 529, 760
Deoxyribonucleic acid. *See* DNA
Deoxyribose, 696
Depolarized membrane, 375
Deposition, and bone remodeling, 207
Depression
　　of joint, 276, 276f
　　psychiatric, and neurotransmitters, 381t
Derm- (prefix), 178
Dermal papillae, 183
Dermatology, 30
Dermatomes, 426f
Dermis, 178, 183–84
Descending aorta, **43–44,** 592, 594t
Descending colon, **41–42,** 684
Descending limb of nephron loop, 776, 789,
　　790f, 791t
Descending tract, 416–18

Desmosome, 149, 150f, 151t
Detached retina, 479
Detoxification, and liver, 673
Detrus- (prefix), 769
Detrusor muscle, 769, 793
Development, human, 870–93. *See also* Age
　　and aging; Elderly; Growth; Infants
　　abnormalities of urinary system, 797t
　　adolescence, 896, 897t
　　adulthood, 896–97
　　aging process, 898–99
　　body proportions, 884f
　　brain, 395, 397t
　　children, 896, 897t
　　defined, 867
　　end of life and dying, 897–98
　　fetus, **265,** 881, 883, 886f
　　postnatal period, 868, 894–98
　　prenatal period, 868
Deviated septum, 732
Di- (prefix), 58
Diabetes educator, 489
Diabetes insipidus, 503, 813
Diabetes mellitus, 516, 518–19, 640t, 697, 786,
　　818, 899t
Diabetic retinopathy, 482
Diacylglycerol (DAG), 495
Dialysis, 772
Dialysis technician, 769
Diapedesis, 538, 626
Diaphragm, **40–44, 47, 51–56**
　　birth control, 855, 856f, 857, 858t
　　body cavities, 18, 19f
　　heart, 559f
　　respiratory system, 744–45
Diaphysis, 200
Diarrhea, 683, 818
Diarthrotic joints, 268, 269
Diastol- (prefix), 557
Diastolic pressure, 557, 583, 585
Diencephalon, 397t, 403, 405, 409t
Diet. *See also* Diet drugs; Dietary supplements;
　　Foods; Nutrition
　　carbohydrates, 696
　　healthy eating, 718–23
　　kidney function, 786
　　lipids, 698
　　obesity, 702, 703
　　pregnancy, 890
　　protein requirements, 700
　　red blood cell production, 534–35
　　sodium, 714
　　urine composition, 791–92
　　vegetarian, 711, 718, 719t
Diet drugs, 703
Dietary supplements, 718–19
Differential white blood cell count (DIFF), 538
Differentiated cells, 83
Differentiation, 112
Diffusion, 99–101, 107t, 581, 755–56
Diffusional equilibrium, 99
Digestion, 650–52
Digestive system, 649–93
　　accessory organs of, 652f
　　alimentary canal, 650–52

autonomic stimulation, 436t
epithelial tissues lining, 153
esophagus, 661
interconnections with other systems, 194,
　　247, 341, 383, 521, 611, 643, 689,
　　763, 799, 861
large intestine, 683–88
life-span changes, 688, 898t
liver, 669–75
major functions of, 23–24, 25t
mouth, 653–58
pancreas, 668–69
pharynx, 660–61
salivary glands, 658–59
small intestine, 676–83
stomach, 661, 663–68
summary, assessments, and tools, 690–93
Digital region, 30, 32f
Dihydrogen phosphate, 814
Dihydrotestosterone, 838
Dipeptide, 122
Diplopia, 470
Disaccharides, 70, 681f, 695
Disease, 713. *See also* Genetic disorders;
　　Infection; Medical science; Viruses;
　　specific diseases
　　faulty ion channels, 89
　　at organelle level, 94
　　stem cells for treatment of, 114
Dislocation, of shoulder, 279
Dissacharide, 122f
Dissecting aneurysm, 584
Disseminated intravascular clotting, 544
Distal, as relative position, 28
Distal convoluted tubule, 776, 788f, 791t
Distal epiphysis, 200
Distal phalanx, 244f, **350, 351**
Distal radioulnar joint, 278t
Diuresis, 785
Diuret- (prefix), 488
Diuretics, 488, 502, 808
Divergence, of nerve impulses, 385
Diverticulosis, 687
DNA (deoxyribonucleic acid), 132.
　　See also Genetics
　　aging process, 899
　　blood typing, 549
　　in cell nucleus, 97
　　cellular metabolism, 132–34
　　dietary supplements, 719
　　genetic code, 134, 136
　　genetic information, 132–34, 135f
　　mutations, 142–43
　　replication, 108, 134, 135, 137f
　　RNA compared to, 136, 138, 138t
　　structure of, 76, 132–34, 135f
DNA damage response, 143
DNA profiling, 136
DNA repair, 143
DNA tests, 653
Docosahexaenoic acid, 698
Dogs, and sense of smell, 452
Dominant follicle, 843–44
Dominant hemisphere, 402
Dominant inheritance, 909–11

Donor stem cells, 114
Dopa, 510
Dopamine, 380t, 381t, 384, 403, 404, 510
Dors- (prefix), 10
Dorsal- (prefix), 10
Dorsal branch, 427
Dorsal flexors, 336, 337
Dorsal respiratory group, 751
Dorsal root, 426
Dorsal root ganglion, 426
Dorsal scapular nerve, 428
Dorsalis pedis artery, 600, 600f
Dorsiflexion, 274, 275f
Dorsum region, 30, 32f
Double helix, of DNA, 135f
Down, John Langdon Haydon, 920
Down syndrome, 916, 917, 920
Downregulation, of hormones, 490
Doxorubicin, 893t
Drug abuse, 384, 708, 874. *See also* Cocaine;
    Heroin
Drugs. *See also specific types of drugs*
    for acne treatment, 188
    addiction and faulty ion channels, 89
    age and prescription of, 798
    autonomic functions, 437
    breastfeeding and contraindications for, 893t
    cortisol compounds, 512
    derivation from plants, 123
    HIV/AIDS, 642
    nanotechnology and the blood-brain barrier,
        150
    neurotransmitter levels altered by, 381t
    obesity and diet drugs, 703
    opiates, 382
    pharmacogenetics, 921
    smell and taste disorders, 456
Dry mouth, 688
Dual reuptake inhibitors, 381t
Duchenne muscular dystrophy, 142, 297
Ducts, of eye, 469
Ductus arteriosus, 885, 889t, 895t
Ductus deferens, **41,** 826, 827f, 828, 830, 833,
    838t, 860f
Ductus venosus, 883, 889t, 895t
Duodenum, 21f, **42–43,** 676, 677f
Dura mater, **49,** 390–91
Dural sinuses, 391, 394f
Dwarfism, 501
Dying, 897–98
Dynamic equilibrium, 465, 466–68, 699
Dysgeusia, 456t
Dyslexia, 401
Dysosmia, 456t
Dystrophic epidermolysis bullosa, 163t
Dystrophin, 297

E

Ear, 456–68
Ear infections, 218, 459, 654
Eccentric muscle contraction, 306
Eccrine sweat glands, 186–87
Echinacea, 718

Ect- (prefix), 868
Ectoderm, 868, 876
Ectodermal cells, 876
Ectopic pregnancy, 872
Edem-, 557, 804
Edema
    definition and causes of, 541, 557, 804
    heart failure, 581
    inflammation, 626
    lymph formation, 620
    nephrotic syndrome, 786
    nutritional, 700
    pulmonary, 591
    water balance disorders, 810–11
Edmonton protocol, for diabetes, 519
EEG (electroencephalogram), 410
Effectors, 361, 413t, 436t
Effectors, homeostasis, 16, 17f, 190f
Efferent arteriole, 775
Efferent fibers, 420
Efferent lymphatic vessels, 621
Egg cells, 839, 842f
Egypt (ancient), 518
Ejacul- (prefix), 824
Ejaculation, 824, 835–36, 838f
Ejaculatory duct, 833
Elastic cartilage, 169t
Elastic cartilage, 166, 167f
Elastic connective tissue, 165, 166f, 169t
Elastic fibers, 162t, 163
Elastic recoil, of lung tissue, 746
Elastin, 27, 162
Elbow joint, 278t, 279–80, 280f
Elderly. *See also* Age and aging; Age and aging
    arthritis, 286
    disorders of vertebral column, 230
    falls and falling, 246, 438
    fragility fractures, 212
    heatstroke, 808
    hip replacement, 283
    medical treatments suppressing immunity,
        640
    skull, **266**
    water imbalances, 810
Electrical synapses, 379
Electrocardiogram, 570–72
Electrocorticogram (EcoG), 410, 557
Electroencephalogram (EEG), 410
Electrolyte balance, 69, 804, 809–12
Electrolytes, 66, 68, 542, 673, 682, 686
Electromyograph, 315
Electron shells, 63
Electron transport chain, 127, 128f, 129–31,
    131f, 302
Electrons, 58t, 59–60, 63–65
Elements, 58–59, 59t
Elevation, of joint, 276, 276f
Ellipsoidal joint, 272
Embol- (prefix), 528
Embolism, 528, 545
Embolus, 545
Embryo, 841, 871–81
Embryonic disc, 874, 878
Embryonic intramembranous bone
    development, 205f

Embryonic stage, of development, 870–81,
    876f, 884t
Emergency Medical Technician (EMT), 11
-Emia (suffix), 804
Emission, of sperm cells, 835, 838f
Emphysema, 735, 750
Emulsification, 675
Enamel, of tooth, 656–57
Encephalitis, 390
End of life, 897–98
End-diastolic volume (EDV), 586
Endo- (prefix), 83, 488
Endocarditis, 561, 568
Endocardium, 560
Endochondral bones, 204, 206f
Endochondral ossification, 204, 205t, 206f
Endocrine glands, 156, 488, 490f, 517
Endocrine system, 487–526. *See also*
    Hormones
    adrenal glands, 508–14
    autonomic stimulation, 436t
    control of hormonal secretions, 496–99
    general characteristics of, 488–89
    hormone action, 489–96
    interconnections with other systems, 194,
        247, 341, 383, 521, 611, 643, 689,
        763, 799, 861
    life-span changes, 519–20, 522, 898t
    major functions of, 25t
    organs of, 23
    pancreas, 514–17
    parathyroid glands, 506–8
    pituitary gland, 498–503
    stress, 517–19
    summary, assessments, and tools, 522–26
    thyroid gland, 503–6
Endocrinology, 30
Endocytosis, 103–4, 107t, 785
Endoderm, 876
Endodermal cells, 876
Endodontitis, 657
Endogenous pyrogen, 627
Endolymph, 459, 460f
Endometrial biopsy, 853t
Endometriosis, 853
Endometrium, 845, 851f, 875f
Endomysium, 293
Endoneurium, 418–19
Endoplasmic reticulum (ER), 83, 90, 91f, 98t
Endorphins, 380t, 382, 449
Endoscopic papillotomy, 675
Endoscopy technician, 651
Endosteum, 201
Endothelium, 577, 617
End-stage renal disease (ESRD), 772
End-systolic volume (ESV), 586
Energy. *See also* Adenosine triphosphate (ATP)
    balance of, 702
    for cellular metabolism, 121, 125–27
    nutrition, 701–2
Energy shells, 63
Enkephalins, 380t, 382, 449
Enriched foods, 706
Enterogastric reflex, 667, 668f
Enterokinase, 668, 680t

Environmental factors. *See also* Body temperature; Toxins
  cancer, 112
  cardiovascular disease, 609, 907
  genetics, 907, 913
  human life span, 27, 900
  migraine, 362
  obesity, 703
Environmental tobacco smoke (ETS), 735
Enzyme-catalyzed reactions, 124
Enzymes, 73, 87t, 121, 123–25, 143, 680t. *See also specific enzymes*
Eosinophils, 537, 539t, 541t
Ependyma, 336, 360, 370t
Ephedrine, 437
Epi- (prefix), 149, 178, 731
Epicardium, 558, 560
Epicondyle, 215t, 233, 235. *See also* Lateral epicondyle; Medial epicondyle
Epicranial aponeurosis, 314, 314f
Epicranius, 312–13
Epidemiology, 30
Epidermal dendritic cells, 189
Epidermal growth factor, 111
Epidermis, 178–83
Epidermis layers, 180t
Epididymis (epididymides), **41, 48,** 828, 830, 832f, 838t
Epidural space, 391, 393f
Epigastric arteries, 598, 599f
Epigastric region, 30, 31f
Epiglottic cartilage, 734
Epiglottis, 660, 731, 734
Epilepsy, 381t, 410
Epimysium, 292
Epinephrine, 431–32, 436, 491t, 493, 508, 510, 511t, 520f, 587–88
Epineurium, 418–19
Epiphyseal plate, 204–5, 206f, 207
Epiphysis, 200
Episiotomy, 891
Epitestosterone, 494
Epithelial cells, 733, 832f, 845, 846f, 876
Epithelial membranes, 168
Epithelial tissues, 149, 150–59
  characteristics, 150–52, 151t, 159t
  glandular epithelium, 156, 157f, 158–59, 159t
  pseudostratified columnar epithelium, 153–54, 159t
  simple columnar epithelium, 152–53, 154f, 159t
  simple cuboidal epithelium, 152, 153f, 159t
  simple squamous epithelium, 152, 159t
  stratified squamous epithelium, 154, 155f, 159t
  stratified columnar epithelium, 156, 159t
  stratified cuboidal epithelium, 154, 155f, 159t
  transitional epithelium, 156, 157f, 159t
Equilibrium, 465–68
Erectile dysfunction, 835
Erection, of penis, 835, 836f
Erector spinae muscle, 316t, 318, **348**
Erg- (prefix), 292

Eryhr- (prefix), 528
Erythrocystosis, 552t
Erythroblastosis fetalis, 551, 623
Erythroblasts, 531
Erythrocyte, 528, 529
Erythropoietin (EPO), 494, 517, 528, 533, 769
Escherichia coli, 783, 795
Esophageal arteries, 592, 594t
Esophageal hiatus, 650, 661
Esophageal sphincter. *See* Lower esophageal sphincter
Esophagus, 21f, **43–44, 46–47, 50–51, 54,** 652f, 661, 662f
Essential amino acids, 699
Essential fatty acids, 697
Essential hypertension, 589
Essential nutrients, 694
Estradiol, 849
Estriol, 849
Estrogen replacement therapy (ERT), 852
Estrogens
  bone development and growth, 210
  female sex hormones, 512, 849, 851f
  follicular cells and secretion of, 502
  hair loss and baldness, 186
  oral contraceptives, 857
  pregnancy, 889, 891, 893t
Estrone, 849
Ethanol, 893t
Ethmoid bone, 217f, 218, 220, 221f, **252–54, 260, 262–64**
Ethmoidal air cells, 217f, 220, **260, 262**
Ethmoidal sinus, **49, 263,** 733
Eukaryotic cells, 86
Eumelanin, 185
Euploid chromosome number, 916
Eustachian tube, 459
Evaporation, 189, 807
Eversion, 276
Eversion, of joint, 276f
Evertor, and leg muscles, 340
Evolution, 681, 890
Excess post-exercise oxygen consumption, 302
Exchange reaction, 66
Excitation-contraction coupling, 298–99
Excitatory postsynaptic potential (EPSP), 379–80
Excretion, 24
Excretory system, 24. *See also* Defecation; Urinary system
Exercise. *See also* Athletes and athletic performance
  blood sugar levels, 210
  body temperature, 189
  breathing, 754
  cardiovascular system, 591, 610, 754
  deep vein thrombosis, 547
  energy expenditure, 702t
  joints, 286
  muscular system, 308, 340
  osteoporosis and bone health, 212, 246
Exhal- (prefix), 731
Exhalation, 731
Exo- (prefix), 488
Exocrine glands, 156, 157f, 488

Exocytosis, 92, 103, 105, 106f, 107t
Exome, 132, 907, 921
Exome sequencing, 921
Exons, 138
Exopthalmia, 506f
Exosomes, 105
Expiration, 741, 746–47
Expiratory reserve volume (ERV), 747–48
Expressivity, of genes, 912
Extension, 274, 275f. *See also* Extensor muscles
Extensor carpi radialis brevis muscle, 324t, 325, 327f, **356**
Extensor carpi radialis longus muscle, 324t, 325, 326f, **348, 356**
Extensor carpi ulnaris muscle, 324t, 325, 327f, **356**
Extensor digitorum longus muscle, 336f, 337, 339t, **351, 357–58**
Extensor digitorum muscle, 312, 324t, 325, 327f, **348, 350, 356**
Extensor hallucis longus muscle, 337, 339f, 339t, 340
Extensor muscles, 318, 323, 325, 335
Extensor retinaculum muscle, 325, 327f, 336f, 337f, 340, **356**
External accessory organs, 847
External acoustic meatus, 200, 216f, 217f, 218, 219f, **253, 255, 260,** 456, 457f
External anal sphincter, 327, 330f, 684
External carotid artery, 596f, 597
External iliac arteries, 598, 599f, 600f
External iliac vein, 605, 608f
External intercostal muscles, **40,** 320f, 599f, 745
External jugular vein, **40,** 602, 608f
External oblique muscle, **38–40,** 312, 320f, 325, 328t, 329f, **349, 353–54**
External respiration, 731
External urethral orifice, 795, 835
External urethral sphincter, 795
Exteroreceptive senses, 446
Extra- (prefix), 804
Extracellular fluid, 15, 804, 805, 806f, 810–11
Extracellular fluid compartment, 804
Extracellular matrix (ECM), 159–60, 161, 162t, 166, 167
Extracorporeal membrane oxygenation, 741
Extracorporeal shock-wave lithotripsy (ESWL), 794
Extracorporealliver assist device (ELAD), 671
Extrafollicular cells (C cells), 505
Extrapyramidal tracts, 417
Extrinsic allergic alveolitis, 743
Extrinsic clotting mechanism, 543–44
Extrinsic muscles, of eye, 470, 471f
Eye, **49,** 215, 216f, 436t, 468–81. *See also* Colorblindness
Eye color, 913, 914f
Eyelid, 468, 469f

F

Facet, 215t, 228
Facial artery, 596f, 597
Facial bones, 212

Facial expression, muscles of, 313–15
Facial nerves, 422, 423f, 424t
Facial skeleton, 220–24, 224t
Facial vein. *See* Anterior facial vein
Facial wrinkles, 193
Facilitated diffusion, 100–101, 107t
Facilitation, and neuronal pools, 383
Factor V Leiden, 552t
Fainting, 753
Falciform ligament, **40, 55,** 670
Fallopian tubes, 845, 846f
Falls, by elderly, 246, 438
False pelvis, 241
False ribs, 231, 232f
False vocal cords, 735
Falx cerebelli, 391t, 408
Falx cerebri, **49,** 391t, 397
Familial hypercholesterolemia (FH), 609, 911f
Familial hypertrophic cardiomyopathy, 609
Famines, 722
Fanconi anemia, 873
Fasc- (prefix), 292
Fascia, 292
Fascicle, 292, 293t, 294f
Fasciculus, 292
Fasciculus cuneatus and gracilis, 416, 417f, 418t
Fast pain fibers (A-delta fibers), 449
Fast-twitch muscle fibers, 306–7
Fatal familial insomnia, 408t
Fat-free foods, 696
Fats
    adipose connective tissue, 163
    anabolism, 121–22
    athletes and body fat, 849
    carbohydrate storage, 131–32
    digestion of, 681–82, 683f, 697f
Fat-soluble vitamins, 705–7, 708t
Fatty acid oxidases, 697
Fatty acids, 71, 77f, 682f, 682t, 813
Faulty ion channels
    disease, 89
FDA. *See* Food and Drug Administration
Fecal occult blood test, 687
Fecal transplantation, 686
Feces, 688, 807, 809
Feedback. *See* Negative feedback; Positive feedback systems
Female condom, 855, 856f, 858t
Females
    differences in male skeletons, 241t
    hormonal control of reproductive functions, 849–52
    infertility, 853
    organs of reproductive system, 839–48, 885f
    pelvis, 240f, 241
    reproductive system, 25
    torso, **38, 42, 43, 44**
    urinary bladder and urethra, 795, 796f
Femoral arteries, **39–40, 42, 52,** 334f, **356,** 598, 599f
Femoral nerves, **39–41,** 429
Femoral region, 30, 32f
Femoral vein, **38–40, 42, 52,** 334f, **356,** 605, 608f

Femur, **52,** 215, 241, 242f, 243f, 245t, 281f, 334f, **350**
Ferritin, 535, 671
Fertilization, 868–70, 872f. *See also* In vitro fertilization; Infertility
Fetal alcohol syndrome (FAS), 882–83
Fetal DNA sorting, 919, 920t
Fetal intramembraneous bone development, 205f
Fetal skeleton, 204f
Fetal stage, of development, 881–87
Fetus, **265,** 881, 883, 886f. *See also* Development, human; Pregnancy
Fever, 191, 627
Fiberoptic bronchoscope, 740
Fiberoptic colonoscopy, 687
Fiberoptic endoscope, 675
Fibrilin, 609
Fibrillation, 574
Fibrin, 543
Fibrinogen, 541, 542t, 543
Fibrinolysin, 528
Fibroblasts, 160, 162t, 171, 545
Fibrocartilage, 166, 167f, 169t, 208–9, 271f
Fibrous coat, 792
Fibrous joints, 268, 269f, 274t
Fibrous layer, of vagina, 847
Fibrous pericardium, 20, 21f, 557–58
Fibrous tunic, 470
Fibula, 215, 242f, 243, 245t, 268f, 337f, 339f, **350**
Fibular artery, 600, 600f
Fibular collateral ligament, 284
Fibularis brevis muscle, 337f, 338f, **358**
Fibularis longus muscle, 337f, 339t, 340, **350, 357–58**
Fibularis retinacula muscle, 337f, 338f, 340
Fibularis tertius muscle, 336f, 337, 337f, 339t, **358**
Fight or flight, 431
"Fight-or-Flight" response, 518, 520f
Filaggrin, 638
Filtration, 101–2, 107t, 581
Filum terminale, 410
Fimb- (prefix), 824
Fimbriae, 824, 845
Fingerprints, 183
Fingers. *See* Phalanx
Fingers, amino acid structure, 75
First line of defense, 816
First-degree burn, 192
Fissure in bone, 215t
Fissure in cerebrum, 397
Fissured fracture, 208
Fixed cells, 160
Flacc- (prefix), 390
Flaccid paralysis, 390, 419
Flagella, 94, 95–96, 98t
Flared ilium, 240f
Flat bones, 200
Flavin adenine dinucleotide (FAD), 709
Flavoproteins, 709
Flexion, 274, 275f. *See also* Flexor muscles
Flexor carpi ranialis muscle, 324, 328t, **350**
Flexor carpi ulnaris muscle, 324, **350**

Flexor digitorum longus muscle, 339–40
Flexor digitorum profundus, 324t, 325
Flexor digitorum superficialis, 324t, 325
Flexor hallucis longus muscle, 339f
Flexor muscles, 318, 321, 324–25, 335, 338f
Flexor pollicis longus muscle, 328f
Flexor retinaculum muscle, 326f, 338f, 340
Floaters, in eye, 482
Floating ribs, 231, 232f
Fluid balance. *See* Body fluid balance
Fluid compartments, 804
Fluoride, 658
Fluorine, 60t, 716, 717t
Folic acid (folacin, Vitamin B$_9$), 534, 710, 712t
Folinic acid, 710
Folkman, Judah, 577
Follic- (prefix), 178, 824
Follicles, 503, 824, 843f, 844f, 851f
Follicle-stimulating hormone (FSH), 491t, 502, 504t, 838, 849, 850, 851f
Follicular cells, 635, 841, 874f
Fontanels, 215t, 225, **265,** 268
Food and Drug Administration (FDA), 718–19, 720, 834
Food poisoning, 298
Food pyramids, 719
Foods. *See also* Diet; Nutrition
    carbohydrate sources, 695
    energy values, 701
    enriched and fortified products, 706
    "fat-free" foods, 696
    hemolytic uremic syndrome, 783
    lipid sources, 695
    pregnancy, 890
    product labels, 718
    protein sources, 699
    as requirement for life, 14, 15t
Foot, 243, 244f, 245, 336–40, **351**
Foot processes, 775
Foramen lacerum, 219f, 221f, 224t, **256–58, 263–64**
Foramen magnum, 216f, 218, 219f, 221f, 224t, **256–58, 260, 263–64**
Foramen ovale, 88, 219f, 221f, 224t, **256–58, 261, 263–64,** 889t, 895t
Foramen rotundum, 221f, 224t, **261, 264**
Foramen spinosum, 219f, 221f, 224t, **256–58, 261, 263–64**
Foramina (foramen), 215t, 222. *See also* Infraorbital foramen; Jugular foramen; Mandibular foramen; Mastoid foramen; Obturator foramen
Forearm, muscles of, 320–23, 326f, 327f, 328f, **350, 356**
Forebrain, 395, 397t, 403
Forensics, and DNA profiles, 136
Formed elements, of blood, 528, 531f
Fornices, 846
Fortified foods, 706
Fossa, 215t. *See also* Coronoid fossa; Iliac fossa; Infraspinous fossa; Mandibular fossa; Olecranon fossa; Supraspinous fossa
Fossa ovalis, 895
Fourth ventricle, 392
Fov- (prefix), 200

Fovea, 215t
Fovea capitis, 200, 241, 242f, 243f
Fovea centralis, 474
Fox, Michael J., 404
Fractures, of bone, 208–9, 212, 246
Fragility fractures, 212
Frank-Starling law of the heart, 586–87
Free nerve endings, 446, 451
Free radicals, 706, 899
Free surface, epithelial tissue, 150
Frenul- (prefix), 650
Frenulum, 650
"Frequency combs," 756
From Science to Technology
    angiogenesis alteration, 577
    blood typing and matching, 549
    computerized tomography (CT), 78
    DNA profiles and the Innocence Project, 136
    heart transplant or replacement, 567
    immunotherapy, 633
    ionizing radiation, 62
    liver transplant or replacement, 671
    metabolomes, 124
    nanotechnology and the blood-brain barrier, 150
    positron emission tomography (PET), 78
    radioactive isotopes reveal physiology, 61
    stem cells for disease, 114
    tissue engineering a urinary bladder, 172
    treatment of diabetes, 518–19
Frontal bone, **46,** 216f, 217f, 218, 219f, 221f, **252–55, 260, 262–64, 347**
Frontal eye field, 402
Frontal lobe, **49,** 397, 400, 402t
Frontal plane, 30
Frontal region, 28, 29f, 30, 32f
Frontal sinus, **46, 49,** 216f, 217f, 218, 220f, **262–64,** 733
Frontal suture, **265**
Frontalis muscle, **352**
Fructose, 695, 696f, 833
FSH. See Follicle-stimulating hormone
Full-thickness burn, 192
Fulminant hepatitis, 674
Function in the study of anatomy, 11
Function of major organs, 25t
Functional electrical stimulation, 308
Functional regions of cerebral cortex, 399, 401f
Functional residual capacity (FRC), 748, 748t
Functional syncytium, 568
Fundus, 663, 845
Fungus, 757
Funi- (prefix), 390
Funiculus, 390. See also Anterior funiculus; Lateral funiculus; Posterior funiculus

G

G protein, 491–92, 495
GABA, 380t, 381t
Galactose, 695, 696f
Gallbladder, **40–42, 52,** 652f, 674, 675, 688
Gallbladder disease, 675
Gallstones, 674

Gamete intrafallopian transfer (GIFT), 871
Gamma globulins, 541, 542t
Gamma radiation, 60, 62
Gangli- (prefix), 390
Ganglion (ganglia), 365, 390, 421
Ganglion cells, 474
Gap junctions, 149, 150f, 151t
Gas exchange, alveolar, 753–57, 758f
Gas transport, and respiratory system, 758–61
Gases, in blood, 542
Gastr- (prefix), 650
Gastric absorption, 666
Gastric arteries, 592, 595f
Gastric bypass surgery, 519
Gastric enzymes, 680t
Gastric glands, 650, 663–65
Gastric juice, 665, 666t
Gastric lipase, 665, 680t
Gastric phase, of gastric secretion, 665, 666t
Gastric pits, 663, 665
Gastric secretions, 663, 665–66
Gastric veins, 603, 604f
Gastrin, 665, 675t
Gastritis, 665
Gastrocnemius muscle, 333f, 336f, 338, 339f, 339t, **350, 351, 358**
Gastroenterology, 31
Gastrula, 876
Gastrulation, 876, 878–79, 881
Gated channels, 372, 373f
Gen- (prefix) or -gen (suffix), 617, 906
Gender. See Female; Male
Gene expression, 132, 912, 920–21
Gene expression profiling, 855
Gene therapy. See Genetic counseling; Genetic disorders; Genetic tests
General senses, 444, 446–51
General somatic afferent fibers, 420
General somatic efferent fibers, 420
General stress syndrome, 518, 520t
General visceral afferent fibers, 420
General visceral efferent fibers, 420
Genes, 75, 132, 906
-Genesis (suffix), 824
Genetic code, 132, 134, 136, 143
Genetic counseling, 910
Genetic counselor, 907
Genetic determinism, 907
Genetic disorders, 552t, 797, 870, 873, 907, 916–20
Genetic heterogeneity, 912
Genetic tests, 653, 855, 911
Genetics, 905–24. See also Chromosomes; DNA; Genetic disorders; Genome; RNA
    aging, 27
    cardiovascular disease, 609
    cellular metabolism, 142–43
    changes in genetic information, 142–43
    chromosome disorders, 916–20
    cigarette smoking and cancer, 735
    defined, 906
    expression of single genes, 912
    human life span, 900
    modes of inheritance, 908–11
    multifactorial traits, 912–13

obesity, 703
personalized medicine, 920
sensitivity to pain, 447
sex chromosomes and sex determination, 914–16
summary, assessments, and tools, 922–24
Genicular arteries, 598
Genital herpes, 859t
Genital region, 30, 32f
Genital warts, 859t
Genome, 132, 142, 906–7
Genomic imprinting, 916
Genomics, 920–21
Genotype, 909
Geophagy, 717
Geriatrics, 31
Germany, and ionizing radiation, 62
Germinal centers, 621
Gerontology, 31
GH. See Growth hormone
Ghrelin, 704, 705f, 705t
GI camera, 651
Gigantism, 210, 501
Gigaxonin, 96
Gingivitis, 657
Gland, 156
Glandular epithelium, 156, 157f, 158–59
Glans penis, 835
Glatiramer, 367
Glaucoma, 482
Gleason score, 834
Gleevec, 540
Glen- (prefix), 200, 268
Glenohumeral joints, 278t
Glenohumeral ligaments, 277, 279f
Glenoid cavity, 200, 233, 234f, 268
Glenoidal labrum, 268, 279
Glia- (prefix), 149
Gliadin, 699
Gliding joints, 272
Globin, 758
Globulins, 541, 542t
Globus pallidus, 403
Glom- (prefix), 769
Glomerular capsule, 775, 776f, 791t
Glomerular filtration, 779–83
Glomerular fitration rate (GFR), 782–83, 797–98
Glomerulonephritis, 640t, 778, 786
Glomerulus, 769, 775, 776f, 791t, 797
Glossopharyngeal nerves, 423, 424t
Glottis, 736
Glucagon, 491t, 514, 515f, 516t
Glucocorticoid, 512
Glucokinase, 515
Gluconeogenesis, 696
Glucosamine with chondroitin, 718
Glucose
    aerobic respiration, 812–13
    anaerobic respiration, 813
    carbohydrate storage, 131–32
    compartments, 806
    endocrine system, 512, 515, 522
    facilitated diffusion, 100–101
    homeostasis, 17

molecular structure, 77f
nutrition, 695, 696f
urine and urine tests, 784t, 792
Glucosuria, 786
Glutamate, 380t
Glutamic acid, 141t, 699t
Glutamine, 141t, 699t
Gluteal arteries, 598, 599f. *See also* Superior
   gluteal artery
Gluteal gait, 333
Gluteal nerves, 429
Gluteal region, 30, 32f
Gluteal tuberosity, 243f
Gluteal vein, 606
Gluten, 682
Gluteus maximus muscle, **52,** 330f, 332, 332f,
   333f, 334t, **349, 354, 357**
Gluteus medius muscle, 332, 332f, 333f, 334t
Gluteus minimus muscle, 332, 332f, 334t
Glyc- (prefix), 58
Glycerol, 682f, 682t, 697f
Glycerol molecule, 71
Glycine, 77f, 141t, 699t
Glycogen, 70, 121, 695
Glycolysis, 127–29, 131f, 302
Glycoprotein hormones, 490, 491t
GnRH. *See* Gonadotropin-releasing hormone
Goblet cells, 153–54, 159
Goiter, 500, 506f, 507t
Goldenseal, 718
Golgi apparatus, 90, 91f, 92, 98t
Golgi tendon organs, 451
Gomphosis, 268, 270f, 274t
Gonadal arteries, 594, 595f
Gonadal veins, 603, 608f
Gonadotropes, 500
Gonadotropin-releasing hormone (GnRH),
   491t, 501f, 502, 838, 839, 849, 850
Gonadotropins, 502, 838
Gonorrhea, 859t
Gout, 271, 287t, 791
Gracilis muscle, **41–42,** 331f, 333, 333f, 334t,
   338f, **357**
Graft-*versus*-host disease (GVHD), 638, 640
-Gram (suffix), 292, 557
Granulations, 190
Granulocyte colony stimulating factor (G-CSF),
   111
Granulocytes, 536, 541t
Granulosa cells, 841
Grapheme-color-type synesthesia, 453
Graves disease, 506f, 507t, 640t
Gravity, 59, 466
Gray commissure, 410
Gray matter, **49**
Gray matter, in brain, 363
Gray rami, 431
Great saphenous vein, **38–39,** 334f, 339f, 605,
   606f, 608f
Greater omentum, **40, 45, 55,** 678
Greater palatine foramen, 219f, 224t, **256, 258**
Greater sciatic notch, 239, 240f
Greater trochanter, 242, 243f, **349**
Greater tubercle, 233, 236f
Greater wing, of sphenoid bone, **261**

Greece (ancient), 518, 791, 797
Greenstick fracture, 208
Ground substance, 159, 162t
Growth, 867. *See also* Development
   bone development, 202, 204–10
   as characteristic of life, 14t
   characteristics of life, 25t
Growth factors, 111, 190
Growth hormone (GH), 210, 491t, 494,
   500, 504t, 519. *See also* Human
   growth hormone
Growth hormone-releasing hormone (GHRH),
   491t, 500, 501f
Gubern- (prefix), 824
Gubernaculum, 824, 826, 827f, 828f
Gut microbiome, 686
Gynecology, 31
Gyrus (gyri), 397

H

H zones, 295
HAART (highly active antiretroviral therapy),
   642
Haemophilus influenza type b, 391
Hageman factor, 544
Hair, 177f, 184–85, 195
Hair bulb, 185
Hair bulge, 185
Hair cells, 459
Hair follicles, 178, 184–86, 195
Hair loss, 186, 916
Hair papilla, 185
Hair root, 185
Hair shaft, 185
Hairpins, amino acid structure, 75
Hairy cell leukemia, 539t
Half-life, of hormone, 496
Hamstring muscles, 335, **349**
Hand, 207f, 235f, 236, 238f, 323–25, **350, 356**
Haploid cell, 824
Hapten, 628
Hard palate, 221, 654
Harvard University, 810
Hashimoto disease, 507t
Haversian canals, 167
Haversian system, 201
Head, **45–46, 49.** *See also* Skull
   arteries, 594, 596f, 597
   femur, 241, 242f
   fibula, 243, 337f
   humerus, 233
   lateral view, **352**
   of muscle, 311
   muscles in, 314f, 316–18
   radius, 235
   sagittal section of, **45, 46**
   surface anatomy, **347**
   transverse section of, **49**
   veins, 602
Headache, migraine, 362
Healing of wounds and burns, 190–92, 193f
Health care, and genomics, 920–21
Hearing, sense of, 456–68, 482. *See also*

Deafness
Hearing aids, 463, 465, 482
Hearing loss, 463, 465
Heart, **41, 45, 50, 53,** 557–75, 581, 608,
   885–87. *See also* Blood pressure;
   Cardiovascular system; Heart disease
Heart attack, 610
Heart disease, 161, 574, 581, 591, 610, 899t,
   920. *See also* Cardiovascular disease
Heart failure, 161, 581
Heartburn, 667
Heat, 15, 303. *See also* Body temperature
Heat loss, 189
Heatstroke, 808
Height, 245, 913
Helicobacter pylori, 665
Helium, 60t
Helper T cells, 630, 631f, 632f
Hem- (prefix), 731
Hema- (prefix), 528
Hematocrit (HCT), 528, 529, 529f
Hematology, 31
Hematoma, 208
Hematopoiesis, 200, 210
Hematopoietic growth factors, 529
Hematopoietic stem cells, 529, 532f
Heme, 758
Hemiazygous veins, 603, 604f
Hemicellulose, 695
Hemiplegia, 419
Hemispheres of cerebral cortex, 402
Hemizygous gene, 915
Hemo- (prefix), 528
Hemochromatosis, 713
Hemocytoblasts, 529
Hemodialysis, 769, 771, 772
Hemoglobin, 528, 731
   bone marrow and blood cell formation, 210
   chemical buffer systems, 814, 815
   composition of red blood cells, 529
   iron, 715, 716f
   molecular structure of, 535f
   oxygen transport, 758–60
   oxyhemoglobin, 529, 758
Hemolysis, 101, 102f
Hemolytic anemia, 535t, 640t
Hemolytic jaundice, 673
Hemolytic uremic syndrome (HUS), 783
Hemophilia, 552t, 915
Hemorrhagic disease of the newborn, 707
Hemorrhoids, 684
Hemostasis, 528, 542–47
Hepa- (prefix), 528
Heparin, 160, 528, 537, 547
Hepat- (prefix), 650
Hepatic arteries, 592, 595f
Hepatic cells, 670
Hepatic coma, 671
Hepatic ducts, 650, 670
Hepatic lobules, 670, 672f, 673f
Hepatic portal system, 603, 604f
Hepatic portal vein, 603, 604f, 605f, 670
Hepatic sinusoids, 603, 604f, 670
Hepatic veins, 603
Hepatitis, 674

Hepatocellular jaundice, 673
Hepatopancreatic ampulla, 668
Hepatopancreatic sphincter, 668, 674
Herd immunity, 636
Hereditary hemochromatosis, 552t
Herniated disc, 230
Heroin, 384, 874, 893t
Hetero- (prefix), 905
Heterogeneity, genetic, 912
Heterozygous chromosomes, 905, 909
Hiat- (prefix), 650
Hiccups, 749
High altitude, and breathing, 756
High-density lipoprotein (HDL), 681–82, 697
Hilum, 621, 741, 770
Hindbrain, 395, 397t
Hinge joint, 272, 273f, 274t
Hip bones, **52,** 239–41, 245t
Hip joint, 278t, 281–83
Hip pointer, 239
Hippocampus, 403
Hippocrates, 791, 797
Hist- (prefix), 149
Histamine, 160, 380t, 537, 581, 636, 665
Histidine, 141t, 699t
Histology, 31, 149
Histones, 134, 135f
History of study of human body, 10, 518, 791, 797
HIV, 105, 106f, 143, 636, 641
HMG-CoA reductase, 105
Holocrine glands, 156, 157f, 158t, 159t
Hom- (prefix), 905
Homeo- (prefix), 10
Homeostasis
    bone tissue, 207–8
    temperature, 15–18, 26f, 189, 190f, 191
Homeostatic mechanism, 16, 191
Homologous chromosomes, 824, 826f, 830f, 905
Homozygous alleles, 909
Hookworm infestation, 539t
Horizontal cells, 474
Horm- (prefix), 488
Hormone replacement therapy (HRT), 852
Hormone-receptor complex, 490
Hormones, 23, 488. *See also* Endocrine system; Estrogens; Steroid hormones; *specific hormones*
    actions of, 489–96
    androgens, 188, 512, 824, 838, 849
    athletic performance and abuse of, 494
    cell division, 111
    cell division controls, 112
    control of secretions, 496–99
    digestive tract, 675t
    female reproductive functions, 849–52
    male reproductive functions, 837–39
    mammary glands, 893t
    nonsteroid, 489–96
    pregnancy, 888–89
    smooth muscle contraction, 307–8
Horseshoe kidney, 797t
"Hot flashes," 852
Human anatomy and physiology. *See* Anatomy and physiology

Human chorionic gonadotropin (hCG), 853, 872, 875f, 888, 919
Human genome sequences, 142
Human growth hormone (HGH), 494, 501
Human immunodeficiency virus (HIV), 105, 106f, 143, 636, 641
Human leukocyte antigens (HLA), 629
Human menopausal gonadotropin (hMG), 853
Human Metabolome Database, 124
Human organism, 11
Human papilloma virus (HPV), 846
Humans
    anatomy and physiology, 11–12
    genome, 132, 906–7
    genome sequences, 142
    life span, 27, 899
    major elements in body, 59t
    number of cells, 83
    organization of body, 18–26
Humerus, **41,** 213, 233, 234f, 235, 236f, 237t, 319, 322f, **348, 350**
Humor- (prefix), 617
Humoral immune response, 627, 630
Humoral immunity, 617
Huntington disease (HD), 370, 381t, 910, 911f
Hyal- (prefix), 149
Hyaline cartilage, 149, 166, 169t, 737
Hybridoma, 633
Hydrocephalus, 396
Hydrochloric acid, 66, 665, 666t, 813
Hydrogen, 60, 60t. *See also* Hydrogen ions
Hydrogen bond, 64, 65, 66f
Hydrogen ions, 67, 69t, 752, 787, 812–16, 815
Hydrogen peroxide, 123
Hydrogen sulfide, 686
Hydrogenation, of unsaturated fat, 71
Hydrolysis, 122–23, 132f, 813
Hydrophilic, 76
Hydrophobic, 76
Hydrostatic pressure, 15, 101, 581, 779–82, 783, 805
Hydroxide ions, 67, 813
Hydroxyapatite, 211
Hydroxycholecalciferol, 507, 706
Hydroxyl ions, 815
Hydroxyurea, 531
Hygiene hypothesis, 750
Hymen, 847
Hyoid bone, 212, 213t, 214f
Hyper- (prefix), 83, 292, 488
Hypercalcemia, 811
Hyperemia, 626
Hyperextension, 274
Hypergeusia, 456t
Hyperglycemia, 515
Hypericin, 718, 719
Hyperkalemia, 713, 813
Hypernatremia, 813
Hyperopia, 476
Hyperosmia, 456t
Hyperparathyroidism, 508t, 811
Hyperpolarized membrane, 375
Hypersomnia, 381t
Hypertension, 587, 589. *See also* Blood pressure

Hyperthermia, 189, 191, 810
Hyperthyroidism, 488, 506f, 507t
Hypertonic solution, 83, 101, 102f
Hypertrophic pyloric stenosis, 663
Hypertrophy, 210, 292, 308
Hyperventilation, 753, 818
Hypervitaminosis, 706, 707
Hypo- (prefix), 83, 488
Hypocalcemia, 811
Hypochondriac region, 30, 31f
Hypodermic injections, 178
Hypogastric region, 30
Hypogeusia, 456t
Hypoglossal branch, 224t
Hypoglossal nerves, 424, 424t
Hypoglycemia, 515
Hypokalemia, 813
Hypomimia, 404
Hyponatremia, 720, 810, 813
Hypoparathyroidism, 508t
Hypophia, 404
Hypophyseal portal veins, 499
Hypopituitary dwarfism, 501
Hypoproteinemia, 786, 804, 810–11
Hyposmia, 456t
Hypothalamus, **46**
    autonomic nervous system, 437
    diencephalon, 403
    homeostasis, 16, 17f, 190f
    hormone names and abbreviations, 491t
    hormone secretion, 496–99, 501f, 502f, 504t
    male reproductive functions, 837–38, 839f
    stress response, 518, 520f
Hypothermia, 189, 190
Hypothyroidism, 488, 506f, 507t
Hypotonic solution, 101, 102f
Hypoxia, 447, 530, 756
Hypoxia-inducible factor (HIF-1), 577
Hysterosalpingogram, 853t

I

I bands, 295
Ileocecal sphincter, 683
Ileum, **42,** 676, 677f
Iliac arteries, 598
Iliac crest, 239, 240f, **349**
Iliac fossa, 239, 240f
Iliac region, 30
Iliac spine. *See* Anterior inferior iliac spine; Anterior superior iliac spine
Iliacus muscle, 331f, 332, 334t, **356**
Iliofemoral ligament, 281
Iliolumbar artery, 598, 599f
Iliocostalis cervicis, 316t, 317f
Iliocostalis lumorum, 316t, 317f
Iliocostalis thoracis, 316t, 317f
Iliotibial tract, 332f, 338f, **350**
Ilium, 239, 240f
Im- (prefix), 804
Imbalance, 804
Immediate-reaction allergy, 636, 637f, 638, 639t
Immun- (prefix), 617
Immune complex reactions, 638, 639t

Immune responses, 635–36
Immune system and immunity, 617, 627, 636, 638t. *See also* Lymphatic system
Immunoglobulin, 630, 633–34, 635t
Immunology, 31
Immunotherapy, 633
Impacted wisdom teeth, 656
Implant resection arthroplasty, 283
Implantable cardioverter defibrillator (ICD), 574
Implantation, of blastocyst, 871–72, 875f
Impluse, nerve, 22
Impulse conduction, 377t, 378, 379
Impulse processing, and nervous system, 382, 384–85
Impulses, 360. *See also* Action potentials
In vitro fertilization (IVF), 834, 870–71
"Inborn errors of metabolism," 94, 126
Incisive foramen, 219f, 224t
Incisive fossa, **256, 258**
Incisors, 656, 657t
Inclusions, 90
Incomplete dominance, 911
Incomplete protein, 699
Incompletely penetrant genes, 912
Incontinence, 795, 798
Incontinentia pigmenti, 915
Incus, 457
Indirect inguinal hernia, 826
Indoor tanning and skin cancer, 182
Induced mutations, 142
Inert atoms, 63
Infant formula, 894
Infants. *See also* Children; Development, human
　adipose tissue, 163
　antibodies, 633
　dehydration, 792, 808
　drug abuse, 829, 874
　fetus, **265,** 881, 883, 886f
　genetic tests, 911
　jaundice, 183
　ossification timetable, 207t
　postnatal period, 894–96, 897t
　premature, 882
　respiratory distress syndrome, 745
　skull, 225, **265, 266**
Infarction, 546
Infection, 626, 899. *See also* Bacteria; Viruses; *specific types of infections and disorders*
Inferior, and relative position, 27, 28f
Inferior articular process, 227
Inferior constrictor muscles, 660
Inferior epigastric artery, 598, 599f
Inferior gluteal artery, 598, 599f
Inferior gluteal nerve, 429
Inferior hemiazygos vein, 603, 604f
Inferior meatuses, 732
Inferior mesenteric artery, **43,** 594, 595f, 599f
Inferior mesenteric vein, 603, 604f
Inferior nasal concha, **46,** 220f, 222, 224t, **254**
Inferior oblique muscle, 470, 470t
Inferior orbital fissure, 224t, **254**
Inferior peduncle, 409
Inferior pudendal artery, 598, 599f
Inferior rectus muscle, 470, 470t

Inferior vena cava, **51–52, 54,** 560, 602, 603, 606, 608f
Inferior vesical artery, 598, 599f
Infertility, 836–37, 853. *See also* In vitro fertilization
Inflamm- (prefix), 617
Inflammation, 190, 617, 626, 627t, 634, 635t
Inflammatory bowel disease, 687
Inflation reflex, 752
Influenza, 757, 899t
Infraorbital foramen, 224t, **254–55.** *See also* Supraorbital foramen
Infrapatellar bursa, 284
Infraspinatus muscle, 320, 321f, 322t, **348, 353, 355**
Infraspinous fossa, 233, 234f
Infundibulum, 403, 498, 845
Inguinal canal, **40,** 594, 826, 827f, 828f
Inguinal region, 30, 31f, 622
Inhal- (prefix), 731
Inhalation, 731
Inheritance, modes of, 908–11. *See also* Genetics
Inhibin, 838
Inhibitory postsynaptic potential (IPSD), 379–80
Initial segment, 375
Injectable contraception, 857
Injury, as cause of death, 392, 419, 899t
Innate (nonspecific) defenses, 625, 626–27
Inner cell mass, 871
Inner ear, 457f, 459–61, 460f
Inner tunic, 474–75
Inner vascular layer, 841–42
Innocence Project, 136
Inorganic chemicals, 68–69
Inorganic salts, 69
Inositol triphosphate (IP3), 495
Insertion, of skeletal muscle, 274, 311
Insomnia, 381t, 408t, 438
Inspiration, 741, 744–46
Inspiratory capacity (IC), 748, 748t
Inspiratory reserve volume (IRV), 747, 748t
Institute of Medicine, 707
Insula, 397, 400–401
Insulin, 516t
　appetite control, 704, 705t
　facilitated diffusion, 101
　homeostasis, 16, 17
　life-span changes, 522
　nanotechnology and delivery of, 150
　pancreas and secretion of, 491t
　sulfur as component of, 714
　synthesis and actions of, 514–15
　treatment of diabetes, 518–19
Insulin-like growth factor-1 (IGF-1), 500
Integral protein, 87t
Integration of organ systems, 22–23
Integrin, 88–89
Integumentary system, 177–98. *See also* Skin
　aging-related changes, 898t
　autonomic stimulation, 436t
　composition of, 177
　interconnections with other systems, 194, 247, 341, 383, 521, 611, 643, 689, 763, 799, 861

　life-span changes, 192–93, 195
　major functions of, 22, 25t
　summary, assessment, and tools, 195–98
Intensity, of vocal sound, 736
Inter- (prefix), 83, 200, 292
Interatrial septum, 560
Intercalated disc, 149, 170, 292, 308–9
Intercarpal joint, 278t
Intercellular junctions, 149, 150f, 151t
Intercondylar eminence, 243f
Intercostal arteries, 594t, 598, 599f
Intercostal muscles. *See* External intercostal muscles; Internal intercostal muscles
Intercostal nerves, 428f, 429
Intercostal trunk, 617, 619f
Intercostal veins, 603
Interferons, 626, 630t
Interleukins, 536, 627, 629
Interlobar arteries, 772
Intermediary metabolism, 123
Intermediate cuneiform, 244f
Intermediate fibers, 307
Intermediate filaments, 96–97
Internal acoustic meatus, 221f, 224t, **262**
Internal anal sphincter, 327, 684
Internal carotid artery, 596f, 597
Internal environment, 15, 82
Internal iliac artery, 598, 599f
Internal iliac vein, 605, 606f, 608f
Internal intercostal muscles, **40,** 320f, 599f, 746
Internal jugular vein, **39–40, 43,** 593f, 602, 608f
Internal oblique muscle, **39–40,** 320f, 325, 328t, 329f
Internal pudendal artery, 598, 599f
Internal respiration, 731
Internal thoracic artery, 598
Internal thoracic vein, 603, 604f
Internal urethral sphincter, 793
Internal vena cava, **43**
Interneurons, 366, 368, 368t, 413t, 415f
Internodal atrial muscle, 569
Interosseous ligament, 268
Interosseous membrane, 268
Interphalangeal joint, 278t
Interphase, 83, 107f, 108, 109f
Interstitial cells, 828, 838t
Interstitial cell-stimulating hormone (ICSH), 500
Interstitial fluid, 15, 16f
Interstitial matrix, 161
Intertarsal joint, 278t
Intertubercular groove, 233, 236f
Interventricular arteries, 563–64
Interventricular foramina, 392
Interventricular septum, 560
Interventricular sulci, 560
Intervertebral discs, **47–48,** 200, 212, 271f
Intervertebral foramen, 227, 426
Intervertebral joint, 278t
Intestinal enzymes, 680t
Intestinal flora. *See* Microflora
Intestinal gas, 686
Intestinal gastrin, 665, 675t
Intestinal glands, 679
Intestinal lipase, 680, 680t

Intestinal phase, of gastric secretion, 665–66
Intestinal somatostatin, 666, 675t
Intestinal trunk, 617, 619f
Intestinal villi, 678–79, 682f
Intra- (prefix), 200, 804
Intracellular fluid, 15, 16f, 804, 805
Intracellular fluid compartment, 804
Intracytoplasmic sperm injection (ICSI), 871
Intradermal injections, 178
Intralobular bronchioles, 738
Intramembranous bone, 202
Intramembranous bones, 200, 204
Intramembranous ossification, 204, 205t
Intramuscular injections, 178
Intrauterine devices (IUDs), 857, 858t
Intrauterine growth retardation (IDGR), 883
Intrauterine insemination, 870
Intravascular oxygenator, 741
Intravenous fluids, and dehydration, 792
Intrinsic clotting mechanism, 543, 544
Intrinsic factor, 534, 665, 666t, 710
Introns, 138
Inulin, 792
Inulin clearance test, 792
Inversion, of joint, 276, 276f
Invertor, and leg muscles, 340
Involuntary muscle, 170
Involuntary muscle actions, 170
Involution, and childbirth, 891
Iodide pump, 505
Iodine, 62, 716, 717t
Iodopsins, 480
Ion channels, 89, 100–101
Ionic bonds, 64
Ionizing radiation, 62
Ions, 58t, 63–64, 89, 372, 375–77. See also
    specific ions
Ipsilateral, and relative position, 28
Iris, of eye, 444, 470t, 473, 475f, 477t
Iron, 534, 715, 716f, 717t
Iron deficiency anemia, 535t
Irregular bones, 200
Ischemia, 447, 564, 584
Ischial spine, 239, 240f
Ischial tuberosity, 239, 240f, **349**
Ischiocavernous muscle, 328, 330t
Ischiofemoral ligament, 281
Ischium, 239
Iso- (prefix), 58, 83, 292
Isograft, 639
Isoleucine, 141t, 699t
Isometric muscle contraction, 306
Isotonic muscle contraction, 292, 306
Isotonic solution, 83, 101, 102f
Isotope, 60
Isotretinoin (Accutane), 883
Isthmus, 503
IVF (In vitro fertilization), 834, 870–71
Izumo1 (protein), 869

J

Jacobs syndrome, 919
Jaundice, 183, 673

Jejunum, **42,** 676, 677f
Jet lag, 517
Johns Hopkins School of Medicine, 382
Joint capsule, 270, 282
Joint reflex, 754
Joints, 267–90. See also Cartilaginous joints;
    Fibrous joints; Skeletal system; Synovial
    joints
  of body, 278t
  disorders of, 286–87
  life-span changes, 285
  replacement of, 282
  summary, assessments, and tools, 288–90
  types of, 268–72, 274t
  types of movements, 274–77
Jugular foramen, 218, 219f, 221f, 224t, **256–58,**
    263–64
Jugular notch, 232f, **349**
Jugular trunk, 617, 619f
Jugular veins, **50, 53–54.** See also External
    jugular vein; Internal jugular vein
Juno (protein), 869
Juvenile rheumatoid arthritis, 287t
Juxta- (prefix), 769
Juxtaglomerular apparatus, 776, 780f
Juxtaglomerular cells, 776, 780f
Juxtamedullary nephron, 769, 776

K

Kaposi sarcoma, 641
Karyo- (prefix), 906
Karyokinesis, 108
Karyotype, 905, 908, 909f
Kawasaki disease, 287t
Kegel exercises, 795
Kennedy, John F., 513
Kerat - (prefix), 178
Keratin, 96, 154, 178
Keratinization, 179
Keratinocytes, 178
Ketoacidosis, 697
Ketones, 697, 792, 818
Ketouria, 804
Kidney, **43, 51–52,** 698, 769–76, 778f, 797–98,
    809f, 894
Kidney disease, 772, 783, 818, 899t
Kidney stones, 794
Kilocalorie, 701
Kinases, 110
Klinefelter syndrome, 918
Knee joint, 278t, 281–84, 285f, **350**
Köhler, Georges, 633
Korotkoff's sound, 583
Krabbe disease, 94
Kupffer cells, 603, 670
Kwashiorkor, 722
Kyphosis, 94, 230

L

Labi- (prefix), 824
Labia majora, 847, 848t
Labia minora, 824, 847, 848t

Laboratory tests, of clinical importance.
    See Genetic tests; Prenatal tests
Labr- (prefix), 268
Labyrinth, 444, 459, 460f
Lacri- (prefix), 444
Lacrimal apparatus, 469, 470f
Lacrimal bone, 217f, 222, 224t, **253–55**
Lacrimal gland, 314, 444, 469
Lacrimal sac, 469
Lact- (prefix), 488
Lactase, 123, 680, 680t
Lacteals, 617, 679
Lactic acid, 128, 302, 303f, 813
Lactic acid threshold, 302
Lactobacillus strains, 686, 894
Lactoferrin, 894
Lactogen, 893t
Lacto-ovo-vegetarian, 719t
Lactose, 681, 696
Lactose intolerance, 681, 688
Lacto-vegetarian, 719t
Lacun- (prefix), 868
Lacuna (lacunae), 166, 168f, 201, 868, 873
Lambdoid suture, 216f, 217f, 218, 219f, **253**
Lamell- (prefix), 200
Lamellae, 167, 168f, 200
Lamellated (Pacinian) corpuscles, 446, 451t
Laminae, 227
Laminectomy, 230
Laminin, 96
Landsteiner, Karl, 548
Langerhans cells, 189
Lanug- (prefix), 868
Lanugo, 868, 881
Laparoscopic adjustment gastric banding, 703
Laparoscopy, 853t
Large intestine, **51,** 652f, 683–88
Large intestine disorders, 687
Laryngopharynx, 660, 733
Larynx, **39, 42, 46, 53,** 733–36, 744t
Laten- (prefix), 292
Latent period, 292, 304
Lateral, and relative position, 27, 28f
Lateral border, 233, 234f
Lateral canal, 466
Lateral condyle, 242, 243f
Lateral cuneiform, 244f
Lateral curvature, 231
Lateral epicondyle, 236f, 242, 243f, **348, 350**
Lateral funiculus, 410
Lateral geniculate nucleus, 403
Lateral horn, 410
Lateral malleolus, 243, **351**
Lateral pectoral nerves, 428
Lateral plantar artery, 600, 600f
Lateral pterygoid muscle, 314f, 315
Lateral rectus muscle, 422, 470, 470t
Lateral region, 30, 31f
Lateral sacral artery, 599f
Lateral sulcus, 397
Lateral ventricles, **46, 49,** 392
Latissimus dorsi muscle, **39–40,** 318, 319f,
    322t, **348, 353–55**
Laughing, 749t, 750
Leeuwenhoek, Anton van, 829

Left anterior descending arteries.
    See Interventricular arteries
Left atrium, **42**
Left brachiocephalic vein, **53**
Left bronchus, **54**
Left common carotid artery, **53**, 594t, 595f
Left hemisphere, 402
Left lobe, of liver, 670
Left main bronchi, 738
Left subclavian artery, **54**, 594t, 595f
Left ventricle, **42, 49, 52, 53**, 558f, 562f
Left ventricular assist device (LVAD), 567
Leg, muscles of, 331f, 332f, 333f, 335–36,
    337–39t, **349, 356–58**
-Lemm (suffix), 360
Lens, of eye, 472, 473f, 474f
Lens fibers, 472, 473f
Leptin, 704, 705f, 705t
Lesser sciatic notch, 240f
Lesser trochanter, 242, 243f
Lesser tubercle, 233, 236f
Lesser wing, of sphenoid bone, **261**
Leucine, 141t, 699t
Leukemia, 540, 873
Leuko- (prefix), 528
Leukocyte, 528, 536, 538f, 539f
Leukocytosis, 528, 538
Leukopenia, 538
Leukotrienes, 636
Levator ani, 325, 327, 330t
Levator palpebrae superioris, 468, 470t
Levator scapulae muscle, 318, 319f, 320t, **352**
Levels of organization, anatomy and
    physiology, 12, 14t
Levers, 310
Levodopa, 404
Lewy bodies, 404
Lexical-gustatory synesthesia, 453
LH. *See* Luteinizing hormone
Life. *See also* Life expectancy; Life-span
    changes
    characteristics of, 14, 25t
    maintenance of, 14–18
Life cycle, of red blood cells, 535f, 536f
Life expectancy, 899, 920
Life span, human, 899
Life-span changes. *See also* Age and aging
    anatomy and physiology, 27
    cardiovascular system, 608
    digestive system, 688
    endocrine system, 519–20, 522
    joints, 285
    lymphatic system, 640
    muscular system, 340
    nervous system, 438
    nutrition, 723–24
    respiratory system, 762
    senses, 482
    skeletal system, 245–46, 898t
    skin, 192–93, 195
    urinary system, 797–98
Ligaments
    collagen fibers, 162, 164
    joint, 270, 279–81
Ligamentum arteriosum, 593f, 895

Ligamentum nuchae, 317f
Ligamentum teres, 895
Ligands, 104
Light adapted eye, 480
Light microscope, 87, 87f
Light refraction, 475–78
Lignin, 695
Limbic system, 403, 405
Limited transport capacity, 785
Linea, 215t
Linea alba, **39**, 320f, 325, 329f, **353**
Linea aspera, 242, 243f
Lingu- (prefix), 650
Lingual artery, 596f, 597
Lingual frenulum, 653
Lingual tonsils, 650, 653
Linoleic acid, 697
Linolenic acid, 697
Lip- (prefix), 58, 695
Lipase, 123
Lipids
    aging process, 899
    groups of, 73t
    liver and metabolism of, 670, 673t
    nutrition, 695, 696–98, 700t
    structure and function of, 71, 77t
Lipofuscin, 27, 899
Lipoid nephrosis, 786
Lipoprotein lipase, 681
Lips, 653, 658t
Lissencephaly, 397
Lithium, 60t
Liver, **41–42, 45, 47, 51–52, 55–56**
    carbohydrates, 696
    digestive system, 652f, 669–75
    life-span changes, 688
    lipids, 697, 698f
Liver cells, 131
Liver fibrosis, 161
"Liver spots," 193
Liver transplants, 671
Lobar branches, 590
Lobar (secondary) bronchi, 738
Lobes. *See* Frontal lobe; Occipital lobe; Parietal
    lobe; Quadrate lobe; Temporal lobe
Lobules of lung, 738, 741
Lobules of testes, 828
-Logy (suffix), 10
Long bones, 200
Long head biceps brachii muscle, **39**
Long Life Family Study, 900
Longissimus capitis muscle, 316t, 317f
Longissimus cervicis muscle, 316t
Longissimus thoracis muscle, 316t, 317f
Longitudinal fibers, 650
Longitudinal fissure, 397
Longitudinal ligaments, 227
Longitudinal section, 28, 30f
Long-QT syndrome, 89
Long-term memory, 402
Long-term potentiation, 402–3
Loops, amino acid structure, 75
Loose connective tissue, 162
Lordosis, 231
Lorenzo's Oil (film 1992), 94

Losartan, 609f
Lou Gehrig's disease. *See* Amyotrophic lateral
    sclerosis
Low-density lipoprotein (LDL), 104, 105, 609,
    681–82, 697, 911, 913
Lower esophageal sphincter, 661
Lower limbs, 213t, 215, 241–45, 245t, 598,
    600, 603, 605–6
Lower motor neuron syndrome, 419
Lower motor neurons, 416
Lower respiratory tract, 731
Lower subscapular nerve, 428
LSD, 384
Lumbar arteries, 594, 595f, 598
Lumbar curvature, 226f, 227, 231
Lumbar enlargement, 409
Lumbar nerves, 425
Lumbar puncture, 396
Lumbar region, 30, 32f
Lumbar trunk, 617, 619f
Lumbar veins, 603
Lumbar vertebrae, 226, 229, 229f, 231t
Lumbar vertebral body, **47–48**
Lumbosacral plexuses, 428–29, 430f
Lumen, 152, 650
Lumpectomy, 855
Lun- (prefix), 557
Lung cancer, 735
Lung irritants, 743
Lung volume reduction surgery, 750
Lungs, **41–42, 50–51, 53,** 735f, 741, 742f,
    744, 757f
Lut- (prefix), 444
Luteinizing hormone (LH), 491t, 502, 504t,
    838, 839, 844, 849, 850, 851f
Lyme disease, 281, 286, 287t
Lymph, 617, 621. *See also* Endolymph
Lymph nodes, 617, 621–23, 624t
Lymphadenitis, 622
Lymphangiogram, 619f
Lymphangitis, 622
Lymphatic capillaries, 617, 619f, 621f
Lymphatic nodules, 621
Lymphatic obstructions, 811
Lymphatic pathways, 617–20
Lymphatic sinuses, 622
Lymphatic system, 616–48. *See also* Immune
    system and immunity; Lymph nodes;
    Lymphocytes
    adaptive (specific) defenses, 627–40
    body defenses against infection, 625–40
    innate (nonspecific) defenses, 626–27
    interconnections with other systems, 194,
        247, 341, 383, 521, 611, 643, 689,
        763, 799, 861
    life-span changes, 640, 898t
    lymphatic pathways, 617–20
    major functions of, 23, 24f, 25t
    organs, 621–25
    summary, assessments, and tools, 644–48
    thymus and spleen, 623
    tissue fluid and lymph, 620–21
    urinary system, 643
Lymphatic trunks, 617–18, 619f, 620f
Lymphatic vessels, 617, 618f, 619f, 620f, 622f

Lymphocytes, 537, 538f, 539t, 541t, 628
Lys- (prefix) and -lys (suffix), 83, 528
Lysine, 141t, 699
Lysis, 634, 635t
Lysosomes, 83, 93, 94, 98t, 104f
Lysozyme, 469
-Lyt (suffix), 58

M

M cells, 105
M line, 295
Macro- (prefix), 149, 528
Macromolecule, 12, 14t
Macronutrients, 694, 720
Macrophages, 149, 160, 162t, 528, 537, 623
Macula, 444, 465–66, 466f
Macula densa, 776, 780f
Macula lutea, 444, 474
"Mad cow disease," 75
Magnesium, 60t, 69t, 714, 715t, 784t
Magnesium ions, 806f
Magnetic resonance imaging (MRI), 13, 564
Main (primary) bronchi, 738
Mainstream smoke, 735
Major calyces, 769, 770
Major histocompatibility complex (MHC), 629
Major minerals, 713–14
Mal- (prefix), 695
Malabsorption, 682
Malaria, 539t, 910
Male climacteric, 839
Male condom, 855, 856f, 858t
Males
    anatomic illustrations, **39, 41**
    differences in female skeleton, 240f, 241, 241t
    hormonal control of reproductive functions, 837–39
    infertility, 836–37
    organs of reproductive system, 826–37, 885f
    pelvic cavity, sagittal section of, **47**
    pelvic cavity, transverse section of, **52**
    reproductive system, 25
    torso, **39, 40, 41**
    urinary bladder and urethra, 794f, 795
Malignant tumor, 111
Malle- (prefix), 444
Malleolus. See Lateral malleolus; Medial malleolus
Malleus, 444, 457, 458f
Malnutrition, 10, 695, 720, 883
Maltase, 123, 680, 680t
Mamm- (prefix), 824
Mammary glands, **38,** 92f, 188t, 824, 852–53, 856f, 892, 893t. See also Breast
Mammary region, 30, 32f
Mammatropes, 500
Mammillary bodies, 403
Mammogram, 854–55
Mandible, **46,** 215, 216f, 217f, 222, 223f, 224t, **259, 262, 347**
Mandibular condyle, 217f, 222, **259**
Mandibular division, 421, 424t
Mandibular foramen, 222, 224t, **259**

Mandibular fossa, 216f, 218, 219f, **255–57**
Mandibular ramus, **259**
Manganese, 716, 717t
Mania, 381t
Mannitol, 808
Manometer, 396
Manubrium, 231, **349**
Maple syrup urine disease, 797
Marasmus, 722
Marburg virus, 530
Marfan syndrome, 163t, 609, 912, 921
Marginal artery, 563
Marijuana, 449
Marrow, 201, 210–11, 239
Marshall, Barry, 665
Mass, 58, 59
Mass movements, 686
Massage therapist, 293
Masseter muscle, 315, **347, 352**
Mast- (prefix), 824
Mast cells, 148f, 160, 162t, 635, 637f
Mastectomy, 855
Mastication, 653
Mastication, muscles of, 315
Mastitis, 824
Mastoid foramen, 219f. See also Stylomastoid foramen
Mastoid process, 216f, 217f, 218, 241t, **253, 257, 260, 347**
Mastoiditis, 218
Maternal serum markers, 919, 920t
Matter, 58
Matter, structure of, 58–68
Mature follicle, 841–44
Maturity-onset diabetes of the young (MODY), 515
Maxillae (maxilla), **46,** 217f, 220, 224t, **252–56, 347**
Maxillary artery, 596f, 597
Maxillary division, 421, 424t
Maxillary region, of skull, **258**
Maxillary sinus, 217f, 221, **262–63,** 733
McCune-Albright syndrome, 495
Mean arterial pressure, 585
Meat- (prefix), 200
Meatus, 215t. See also External acoustic meatus; Internal acoustic meatus
Mechanical barriers
    birth control, 855, 856f, 857, 858t
    infection, 626
Mechanical digestion, 650
Mechanoreceptors, 444, 451t
Med- (prefix), 488
Medial, and relative position, 27, 28f
Medial border, 233, 234f
Medial condyle, 242, 243f
Medial cuneiform, 244f
Medial epicondyle, 236f, 242, 243f, **350**
Medial malleolus, 243, **351**
Medial pectoral nerves, 428
Medial plantar artery, 600, 600f
Medial pterygoid muscle, 314f, 315
Medial rectus muscle, **49,** 470, 470t
Median cubital vein, 603, 608f
Median fissure. See Anterior median fissure

Median nerves, 328f, 427
Median palatine suture, 219f, **256, 258**
Median sacral crest, 229
Mediastinal arteries, 592, 594t
Mediastinum, 18, 19f, 559f
Mediastinum testis, 828
Medical science. See also Disease; Drugs; Surgery
    fetus, 883
    genetics, 907
    human history, 10–11
    immunity in elderly, 640
    nuclear medicine, 62
    origins of, 10–11
    personalized medicine, 920
    space travel, 210, 467
    treatment of dying, 897
Medical technologist, 805
Medroxyprogesterone acetate, 857
Medulla, of ovary, 841
Medulla oblongata, 406–7, 409t, 423f, 751f
Medullary cavity, 201
Medullary respiratory center, 751
Megadosing, with vitamins, 706
Megakaryocytes, 539
Megaloblastic anemia, 710
Meiosis, 108, 824–25, 830f, 831f
Melan- (prefix), 178
Melanin, 178, 180, 183, 473–74, 498, 913
Melanocytes, 180, 181f, 183
Melanosomes, 180
MELAS (mitochondrial encephalomyopathy, lactic acidosis, and strokelike episodes), 94
Melatonin, 517, 520
Membrane potential, 372, 374f
Membranes, 168–69. See also Cell membrane; Mucous membrane; Pericardial membrane; Placental membrane; Pleural membrane; Respiratory membrane; Serous membrane; Synovial membrane; Tectorial membrane
Membranous labyrinth, 459, 460f, 466f
Membranous urethra, 795
Memory, 402–3, 438, 897
Memory B cells, 630, 635
Memory consolidation, 403
Memory T cells, 629
Menarche, 850
Mendel, Gregor, 908
Mening- (prefix), 390
Meningeal branch, 426–27
Meninges, 390–91, 393f
Meningitis, 391
Meniscus (menisci), 272, 272f, 284, 285f
Menopause, 186, 246, 852
Mens- (prefix), 824
Menses and menstruation, 824, 850–52
Mental foramen, 217f, 222, 224t, **259**
Mental region, 30, 32f
Mental retardation (Down syndrome), 916, 917, 920
Merocine glands, 156, 158f, 159t, 188t
Merocrine sweat glands, 186, 187f
Mes- (prefix), 868

Mesangial cells, 775
Mesenchymal cells, 202
Mesenteric arteries. *See* Inferior mesenteric artery; Superior mesenteric artery
Mesenteric veins. *See* Inferior mesenteric vein; Superior mesenteric vein
Mesentery, **42, 56,** 678
Mesoderm, 868, 876
Mesodermal cells, 876
Mesothelium, 168
Messenger RNA (mRNA), 136, 138, 142
Meta- (prefix), 10
Metabolic acidosis, 817–18
Metabolic alkalosis, 818, 819f
Metabolic pathways, 124–25
Metabolism. *See also* Cellular metabolism
    cardiovascular disease, 609
    characteristics of life, 14
    energy requirements, 701–2
    inborn errors of, 94, 126
    liver, 670, 673t
Metabolites, primary and secondary, 123
Metabolome, 124
Metacarpal bones, 213, 235f, 236, 238f, **350**
Metacarpophalangeal joint, 278t
Metacarpus, 236, 238f
Metaphase, 108, 109f, 824, 825
Metaphysis, 200
Metarterioles, 579
Metatarsal bones, 215, 242f, 243, 244f, 245t, **351**
Metatarsophalangeal joint, 278t
Metatarsus, 243
Metencephalon, 397t
Meter- (suffix), 695
Methicillin, 787
Methionine, 699, 714
Methylprednisolone, 419
Microchimerism, 69
Microfilaments, 96, 98t
Microflora, 686
Microglia, 368–69, 370t
Micrographia, 404
Micrometers, cell size, 83
Micronutrients, 694
Microorganisms. *See* Bacteria; Viruses
Microscopes, 86, 87f, 829
Microscopic structure of bone, 201–2
Microtubules, 96, 97f, 98t
Microvilli, 153, 154f, 679, 784
Mict- (prefix), 769
Micturition, 769, 795–96, 797t
Micturition reflex center, 795–96
Midbrain, 395, 397t, 405–6, 409t
Middle cerebral artery, 597
Middle constrictor muscles, 660
Middle ear bones, 213t, 457–58
Middle ear cavities, 18, 20
Middle Eastern respiratory syndrome (MERS), 757
Middle meatus, 732
Middle nasal concha, 220, 220f, **254, 260**
Middle peduncle, 409
Middle phalanx, 244f, **350, 351**
Middle rectal artery, 598

Middle sacral artery, 594, 595f, 599f
Middle tunic, 472–74
Midwife (nurse), 825
Migraine, 362
Military, and heatstroke, 808
Milk, 891–93, 894
Milliosmoles, 809
Milstein, Cesar, 633
Mineralcorticoid, 510
Minerals, 711, 713–17, 720
Minor calyces, 770
Minoxidil, 186
Minute ventilation, 749
Misfolding, protein, 75
Missense mutation, 142
Mit - (prefix), 83
Mitchell, S. Weir, 449
Mitochondria, 92–93, 94, 98t, 128f, 129
Mitosis, 83, 107, 108, 831f, 899
Mitoxantrone, 367
Mitral valve, 559, 561t, 563f
Mitral valve prolapse (MVP), 561
Mixed nerves, 419, 420f
Mixed-lineage leukemia, 540
Mixing movements, in digestive system, 650–51, 655f, 667f, 682–83, 686
Modiolus, 459, 460f
Molars, 656, 657t
Molecular formula, 62
Molecules, 12, 14t, 58t, 62, 609
Mono- (prefix), 58, 906
Monoamine oxidase, 381, 437
Monoamine oxidase inhibitors, 381t
Monoamines, 380
Monoclonal antibodies, 633
Monocytes, 537, 539t, 541t, 627
Monohydrogen phosphate, 814
Mononuclear phagocytic system, 627
Mononucleosis, 539t
Monosaccharides, 70, 121, 122f, 681f, 682t, 695, 696f
Monosomy, 906, 916
Monounsaturated fats, 696, 698
Monounsaturated fatty acids, 71, 72
Mons- (prefix), 824
Mons pubis, **38,** 824, 847
"Morning sickness," 890
"Morning-after pill," 857
Morphine, 382, 384
Morphology, 12
Morul- (prefix), 868
Morula, 868, 871, 874f, 879t
Motifs, amino acid structure, 75
Motion sickness, 467, 668
Moto- (prefix), 360
Motor areas, of cerebral cortex, 400–401
Motor end plate, 297
Motor nerves, 419
Motor neurons, 297, 360, 368, 368t, 413t, 414f, 415f
Motor speech area, 402
Motor units, recruitment of, 304–5
Motosis, 109f
Mountain climbing, 756
Mouth, 653–58

Movement
    as characteristic of life, 14t
    joints and type of, 274–77
    organ systems, 22, 25t
    into and out of cell, 99–105, 106f, 107t
MRI. *See* Magnetic resonance imaging
mRNA, 136, 138, 142
Mucin, 158
Mucosa, of alimentary canal, 650, 651t, 654f, 686f
Mucosa-associated lymphoid tissues (MALT), 621
Mucosal layer, of vagina, 847
Mucous cells, 658, 665
Mucous coat, 792, 793
Mucous membrane, 168, 169, 650, 732–33
Mucous neck cells, 665
Mucous-secreting glands, 652f, 679
Mucus, 153, 154, 158–59, 169, 658, 666t
Multi- (prefix), 360
Multicellular glands, 156, 157f, 158t
Multifactorial genetic traits, 912–13
Multiple motor unit summation, 305
Multiple sclerosis (MS), 367, 640t
Multipolar neuron, 360, 365, 366f, 368t
Multiunit smooth muscle, 307
Murmur, heart, 568
Muscarine, 436
Muscarinic receptors, 436
Muscle cells and glucose storage, 131
Muscle cramps, 303
Muscle fatigue, 303
Muscle fiber, 293–96
Muscle fibers, 170, 183–84, 395, 419, 568
Muscle spindles, 450, 451t
Muscle strain, 296
Muscle tissue, 149, 151t, 170, 172t
Muscle tone, 306
Muscles, 293t. *See also* Muscular system; Skeletal muscles; *specific muscles*
Muscular coat, 792, 793
Muscular dystrophy, 142, 297
Muscular hypertrophy, 292
Muscular layer of alimentary canal, 650, 651t, 654f
Muscular layer of vagina, 847
Muscular system, 291–358. *See also* Muscle fiber; Muscle tissue; Skeletal muscles
    cardiac muscle, 308–9
    interconnections with other systems, 194, 247, 341, 383, 521, 611, 643, 689, 763, 799, 861
    joints and type of movement, 274
    life-span changes, 340, 898t
    major functions of, 22, 25t
    reference plates, **347–58**
    smooth muscles, 307–8
    summary and review, 340, 342–46
    tissue characteristics, 309t
Musculocutaneous nerve, **41,** 427
Mut- (prefix), 121
Mutagen, 142
Mutant phenotype, 909
Mutations, genetic, 111, 121, 142–43, 909
My Plate, 719, 721f

Myasthenia gravis, 298, 640t
Mycobacterium tuberculosis, 757
Myelencephalon, 397t
Myelin, 363, 367
Myelin sheath, 363
Myelinated axons, 363, 365f, 366f, 371f, 379f
Myenteric plexus, 651
Myo- (prefix), 292, 557
Myoblasts, 113
Myocardial infarction, 564, 610
Myocardium, 557, 560
Myoepithelial cells, 892
Myofibril, 292
Myofibrils, 96, 293, 294f
Myoglobin, 302, 715
Myogram, 292, 304, 305f
Myometrium, 845–46
Myopia, 476
Myosin, 293, 296f, 609
Myotome, 426

N

NAD, 127–28, 129, 709
NADH, 127–28, 129
Nail bed, 184
Nail plate, 184
Nails, 184
Nanotechnology, blood-brain barrier and, 150
Narcolepsy, 408t
Narcotics, 450
Nasal bones, 217f, 218, 220, 222, 224t, **252–54, 347**
Nasal cavity, 18, **45,** 220f, 732–33, 744t
Nasal conchae, 220, 224t, 732. *See also* Inferior nasal concha; Middle nasal concha; Superior nasal concha
Nasal continuous positive airway pressure, 751
Nasal region, 30, 32f
Nasal septum, **49,** 732
Nasolacrimal duct, 469
Nasopharynx, 459, 660, 733
Nat- (prefix), 868
Natalizumab, 367
National Heart, Lung, and Blood Institute, 703
National Institute on Aging, 340, 900
National Library of Medicine, 10
Natural killer (NK) cells, 626
Naturally acquired immunity, 636, 638t
Nausea, 668
Navicular bone, 244f
Ne- (prefix), 868
Neck, **45, 46, 50,** 241, 242f, **347,** 594, 596f, 597, 602
Neck, of bladder, 793
Necrosis, 113, 584
Negative feedback, 16, 125, 496–98, 512, 513f, 515, 533
Negative nitrogen balance, 699
Neon, 60t
Neonatal period, 868, 894–95, 897t
Neonatology, 31
Nephr- (prefix), 769

Nephritis, 778
Nephrology, 31
Nephron, 769, 770–71, 773f, 774f, 775–76, 777f, 791t
Nephron loop, 776
Nephropathy, 786
Nephrotic syndrome, 786, 797t
Nerve cell, 360. *See also* Neurons
Nerve fibers, 412–20, 431, 444
Nerve impulses, 360. *See also* Action potentials
Nerves, classification of, 419–20
Nervous system, 359–486. *See also* Central nervous system
    autonomic nervous system, 431–38
    brain, 395–409
    cell membrane potential, 372–78
    cerebrospinal fluid (CSF), 391–95, 396
    classification of cells, 363–71
    description of cells, 363
    divisions of, 389–438
    hormone secretion, 496–99
    impulse processing, 382, 384–85
    information flow from environment, 446t
    interconnections with other systems, 194, 247, 341, 383, 521, 611, 643, 689, 763, 799, 861
    life-span changes, 438, 898t
    major functions of, 22–23, 25t
    meninges, 390–91
    overview, 390
    peripheral nervous system (PNS), 418–30
    senses, 443–86
    somatic nervous system, 361, 418
    spinal cord, 409–18
    structure and function, 361, 362f, 390
    subdivisions of, 420t
    summary, assessment, and tools, 438–42, 482–86
    summary and review, 385–88
    synapse, 371–72
    synaptic transmission, 378–82
    ventricles, 369, 391–95
Nervous tissue, 149, 151t, 170–71, 172f, 172t
Nervous tunic, 470
Net filtration pressure, 780
Neur- (prefix), 149
Neural stem cells, 371
Neural tube defects, 397, 710, 881, 883
Neurilemma, 360, 363
Neurilemmal sheath, 363
Neurofibrils, 363
Neurofilaments, 363
Neuroglia, 149, 171, 360, 368–70
Neurology, 31
Neuroma, 370
Neuromodulators, 381
Neuromuscular junction, 297
Neuronal pools, 382, 384
Neurons, 149, 170, 172f, 360, 363–71. *See also* Bipolar neurons; Interneurons; Motor neurons; Sensory neurons; Unipolar neurons
Neuropathic pain, 449
Neuropeptide Y, 704, 705t
Neuropeptides, 380t, 381–82

Neurotransmitters, 22–23, 297, 361, 371–72, 380–81, 382t, 433, 436–37
Neutr- (prefix), 804
Neutral solution, 804
Neutralization, 634, 635t
Neutron, 58t
Neutrons, 59–60
Neutrophils, 536, 537f, 539t, 541t, 627
New England Centenarian Study, 900
Niacin (Vitamin B$_3$), 709, 712t
Niacinamide, 709
Nicotinamide adenine dinucleotide. *See* NAD
Nicotine, 381t, 384, 436, 893t. *See also* Cigarette smoke and smoking
Nicotine transdermal patch, 178
Nicotinic receptors, 436
Niemann-Pick type C disease, 609
Nifedipine, 756
Nightblindness, 480, 706
Nipple, **38,** 852
Nitric oxide, 69, 380t, 588
Nitrogen, 60t
Nitrogen balance, 699
Nocturnal enuresis, 795
Nod- (prefix), 617
Nodes of Ranvier, 363
Nodule, 617
Nondisjunction, 916
Nonelectrolytes, 68
Nonprotein nitrogenous substances (NPNs), 542
Non-rapid eye movement sleep, 407
Nonrespiratory air movements, 749
Nonsense mutation, 142
Nonsteroid hormones, 489–96
Nonsteroidal anti-inflammatory drugs (NSAIDs), 286, 450
Nonverbal functions, of brain, 402
Norepinephrine
    adrenal medulla and synthesis of, 491t, 508, 510
    as autonomic neurotransmitter, 433, 436
    drug addiction, 384
    drugs that alter, 381t
    effects of, 511t
    representative actions of, 380t
    smooth muscle contraction, 307
    structural formula, 492f
    sympathetic nervous system, 431–32
    termination and inactivation of, 437
Normal range, of homeostasis, 18
Norovirus, 636
Nose, 732, 744t
Nostrils, 732
Nuclear envelope, 83, 97, 98f, 98t
Nuclear medicine, 62
Nuclear pores, 97, 98f
Nucleases, 669, 680t
Nuclei, 368
Nucleic acids, 75–76, 77t, 132–38, 813. *See also* DNA; RNA
Nucleolus, 97, 98f, 98t
Nucleotides, 75, 132, 133f, 134f
Nucleus cuneatus, 407
Nucleus gracilis, 407

Nucleus of atom, 58–59
Nucleus of cell, 85, 97, 98f
Nurse-midwife, 825
Nutri- (prefix), 695
Nutrients, 542, 682t, 694, 695, 883. *See also* Diet; Food
Nutrition, 694–729. *See also* Diet; Food
   adolescence, 896
   appetite control, 704
   carbohydrates, 695–96
   dietary supplements, 718–19
   energy expenditures, 701–2
   energy requirements and metabolism, 701–2
   growth hormone, 500
   healthy eating, 718–23
   infants and young children, 896
   life-span changes, 723–24
   lipids, 696–98
   minerals, 711, 713–17
   proteins, 698–700
   summary, assessments, and tools, 724–29
   vitamins, 705–11
Nutritional edema, 700
Nystagmus, 478

O

Obes- (prefix), 695
Obesity, 695, 702, 703, 704f
Oblique fracture, 208
Oblique muscles. *See* External oblique muscle; Inferior oblique muscle; Internal oblique muscle; Superior oblique muscles
Oblique popliteal ligament, 284
Oblique section, 28, 30f
Obstetrics, 31
Obstructive jaundice, 673
Obstructive sleep apnea syndrome, 408t
Obturator foramen, 240f, 241
Obturator nerves, 428
Occipital artery, 596f, 597
Occipital bone, 216f, 217f, 218, 219f, 221f, **253, 256–58, 260, 262–63, 347**
Occipital condyle, 216f, 218, 219f, **256–58, 262**
Occipital lobe, **49,** 397, 400, 402t
Occipital region, 30, 32f, **257**
Occipitalis muscle, **352**
Occupational hazards, and birth defects, 881, 883
Occupational therapist, 391
Octet rule, 63
Ocul- (prefix), 444
Oculomotor nerves, 421, 424t
Odont- (prefix), 200
Odontoid process, 200
Olecranon fossa, 235, 235f
Olecranon process, 235f, 236, **349**
Olfact- (prefix), 444
Olfactory, 444
Olfactory bulbs, 421, 453
Olfactory foramina, 218, 221f
Olfactory nerves, 421, 424t
Olfactory organs, 452
Olfactory pathways, 453–54

Olfactory receptors, 421, 452
Olfactory stimulation, 454
Olfactory tracts, 421
Oligo- (prefix), 360
Oligodendrocyte, 363
Oligoendrocytes, 360, 368, 370t
Oligomeganephronia, 797t
Oligomenorrhea, 852
Olive, 406
Omega-3 and omega-6 fatty acids, 698
Oncogenes, 111
Oncology, 31
Oo- (prefix), 824
Oocyte, 824, 839, 845f, 868, 872f. *See also* Primary oocyte; Secondary oocyte
Oogenesis, 824, 841, 842f
Open fracture, 208
Ophthalmic artery, 597
Ophthalmic division, 421, 424t
Ophthalmology, 31
Opiates, 382, 450
Opportunistic infections, 641
Oprthopedics, 31
Opsin, 479–80
Opsonization, 634, 635t
Optic canal, 221f, 224t, **264**
Optic chiasma, 403, 481
Optic disc, 474
Optic foramina, 421
Optic nerves, **49,** 421, 424t, 472, 481
Optic radiations, 481
Optic tract, 481
Optic tracts, 403
Oral cavity, 20, 21f, **46,** 653
Oral contraceptives, 856f, 857, 893t
Oral region, 30, 32f
Oral vestibule, 653
Orb- (prefix), 10
Orbicularis oculi muscle, 313t, 314, **352,** 444, 468, 470t
Orbicularis oris muscle, 313t, 314, **352**
Orbit of eye, 215, 216f
Orbital cavities, 18, 20, 21f
Orbital fissures. *See* Inferior orbital fissure; Superior orbital fissure
Orbital surface, **260**
Organ systems
   absorption and excretion, 23–24
   body covering, 22
   cardiovascular system, 556–615
   digestive system, 649–93
   endocrine system, 487–526
   homeostasis, 17–18, 26f
   integration and coordination, 22–23
   integumentary system, 177–98
   levels of organization, 12, 14t
   lymphatic system, 616–48
   major functions of, 25t
   muscular system, 291–358
   nervous system, 359–486
   reproduction, 25
   reproductive system, 823–66
   respiratory system, 730–67
   skeletal system, 199–266
   support and movement, 22

   transport, 23, 24f
   urinary system, 768–802
Organelle, 12, 14t, 83, 94
Organic chemicals, 68, 70–77, 77t
Organism, 12, 14t
Organization
   of human body, 18–26
   levels of, 12, 14t
Organogenesis, 876, 878–79, 881
Orgasm, 835, 848
Origin, of skeletal muscles, 274, 311
Oropharynx, 660, 733
Os- (prefix), 149
-Osis (suffix), 528, 804
Osmolarity, 809
Osmoles, 809
Osmoreceptor, 503, 807
Osmoreceptor-ADH mechanism, 808
Osmosis, 101, 102f, 107t, 805–6
Osmotic diuresis, 785
Osmotic pressure, 101, 541, 581, 780, 805
Osseus labyrinth, 459, 460f
Osseus tissue, 149
Ossification, and bone development, 207t
Osteoarthritis, 286, 287t
Osteoblasts, 167, 200, 202, 204, 245
Osteocalcin, 210
Osteoclasts, 113, 200, 207, 207f, 208–10, 245
Osteocytes, 149, 167, 168f, 200, 204
Osteogenesis imperfecta, 163t, 204f, 912
Osteomalacia, 208, 707
Osteon, 167, 168f, 201
Osteonectin, 209
Osteopenia, 212
Osteopontin, 209
Osteoporosis, 163, 212, 520
Otic region, 30, 32f
Otitis media, 218, 459, 654
Otolaryngology, 31
Otosclerosis, 465
Outer ear, 456–57
Outer fibrous layer, 842
Outer tunic, 470–72
Ov- (prefix), 268
Oval window, 457, 461f
Ovarian arteries, 594
Ovarian follicles, 841, 844f
Ovarian ligament, 840, 841
Ovary, **46,** 517, 840–44, 845f, 848t, 849
Overnutrition, 720
Overweight, 702. *See also* Obesity
Ovo-vegetarian, 719t
Ovulation, 843f, 844, 845f, 851f
"Ovulation predictor kit," 855
Oxidation, 126–27, 813
Oxygen, 60t
   chemical constituents of cells, 69, 69t
   concentration gradient, 99, 100f
   control of breathing, 752, 756
   muscular contraction, 302
   as requirement for life, 15
   transport, 758–60
Oxygen debt, 302–3
Oxygen-carring capacity, of red blood cells, 531
Oxyhemoglobin, 529, 758

Oxytocin (OT), 488, 491t, 492f, 498, 502, 503, 504t, 890, 893t

P

P wave, 572
Pacemaker, 568
Packed cell volume (PCV), 529
Pain
    ion channels, 89
    receptors, 444, 446–47, 451t
    sense of, 446–49
    treatment of, 450
Palate, 654, 658t
Palatine bone, 219f, 220f, 221, 224t, **256, 258**
Palatine processes, 221, **256, 258**
Palatine tonsils, 654
Pallidotomy, 404
Palmar region, 30, 32f
Palmaris longus muscle, 324t, 325, **350**
Palpebrae superioris, 444
Pancreas, 17, **43, 51–52,** 491t, 514–17, 652f, 668–69, 688
Pancreatic acinar cells, 668
Pancreatic amylase, 668, 680t
Pancreatic duct, 668
Pancreatic enzymes, 680t
Pancreatic islets, 514–19
Pancreatic juice, 668–69, 670
Pancreatic lipase, 668, 680t
Pantothenic acid (Vitamin B$_5$), 709, 712t
Pap smear test, 846
Papil- or papill- (prefixes), 557, 769
Papillae, 454, 653
Papillary layer, 183
Papillary muscles, 557, 561
Para- (prefix), 488
Para-amino hippuric acid (PAH), 792
Paracrine secretions, 488
Paranasal sinuses, 215, 216f
Parasomnias, 408t
Parasympathetic division, of autonomic nervous system, 431, 432, 435f
Parasympathetic fibers, 437f
Parasympathetic impulses, and digestive system, 651, 658, 683
Parathyroid gland, 488, 491t, 506–8, 520. *See also* Hyperparathyroidism; Hypoparathyroidism
Parathyroid hormone (PTH), 210, 491t, 492f, 507–8, 509f, 810, 812f
Paravertebral ganglia, 432f
Pariet - (prefix), 10
Parietal bone, 216f, 217f, 218, 221f, **252–53, 262–63, 347**
Parietal cells, 665
Parietal lobe, 397, 400, 402t
Parietal pericardium, 20, 21f, 558
Parietal pleura, 20, 21f, 741
Parkinson disease, 381t, 404
Parotid duct, 659
Parotid gland, **352,** 659, 660f
Partial pressure, 752–53, 755–56, 758f
Partial thromboplastin time (PTT), 544

Partially complete protein, 699
Passive aging, 899
Passive processes, for movements in and out of cells, 107t
Passive transport, cellular, 99
Patella, 215, 242, 245t, 331f, 336f, **350, 357–58**
Patellar dislocation, 242
Patellar ligament, 242, 243f, 284, 331f, 335, 336f, **350, 358**
Patellar reflex, 411–12, 414f
Patellar region, 30, 32f
Patellar tendon, 335
Patent ductus arteriosus (PDA), 895
Patho- (prefix), 617
Pathogens, 617, 625, 628f
Pathologic fracture, 208
Pathology, 31
Pattern baldness, 186
PCP, 384
Peanut allergy, 638
Pectin, 695
Pectineus muscle, 333, 334t
Pectoral girdle, 213, 231, 233, 234f, 237t, 318, 319–21f
Pectoral nerves, 428
Pectoral region, 30, 32f
Pectoralis major muscle, **38–39, 50,** 312, 318, 320t, 322t, **349, 353,** 852
Pectoralis minor muscle, **39,** 318, 320f
Pedal region, 30, 32f
Pediatrics, 31
Pedicels, 775
Pedicles, 227
Pedigrees (genetic), 910, 911f
Peduncles. *See* Inferior peduncle; Middle peduncle; Superior peduncle
Pell- (prefix), 695
Pellagra, 695, 709
Pelv- (prefix), 10
Pelvic brim, 240f, 241
Pelvic cavity, 18, 19f, 20, 21f, **42, 43, 44, 47, 48, 52,** 622
Pelvic diaphragm, 325, 327
Pelvic floor, muscles of, 325, 327–28, 330f
Pelvic girdle, 213t, 215, 238–41, 245t
Pelvic inflammatory disease, 859
Pelvic inlet, 241
Pelvic outlet, 241
Pelvic region, 30, 32f
Pelvis, 215, 269, 598–600, 603, 605–6
Penetrance, and gene expression, 912
Penicillin, 787
Penile urethra, 795
Penis, **41,** 330f, 835, 838t
Pepsin, 665, 666t, 680t
Pepsinogen, 665, 666t
Peptidase, 679–80, 680t
Peptide, 380
Peptide bond, 122
Peptide hormones, 490, 491t, 492f
Perception, 444–45
Percutaneous transluminal coronary angioplasty (PTCA), 610
Perforating canals, 202
Perforins, 626, 629

Perfusionist, 557
Peri- (prefix), 10, 360, 650
Pericardial arteries, 592, 594t
Pericardial cavity, 18, 19f, 20, 21f, **50,** 558
Pericardial membrane, 20, 21f
Pericardial sac, **40**
Pericarditis, 558
Pericardium, **52,** 557–58, 559f
Perichondrium, 166
Perilymph, 459, 460f
Perimetrium, 846
Perimysium, 292
Perineum, 847f
Perineurium, 418–19
Periodontal ligament, 268, 657
Periosteum, 200
Peripheral chemoreceptors, 752–53
Peripheral nervous system (PNS), 360, 361, 367f, 370, 418–30. *See also* Nervous system
    cranial nerves, 420–24
    nerve and nerve fiber classification, 419–20
    nerve structure, 418–19
    spinal nerves, 424–30
Peripheral resistance (PR), 586, 587f
Peristaltic rush, 683
Peristaltic waves, 655f, 667f, 683, 686
Peristalsis, 307, 650, 651
Peritoneal cavity, 20, 21f
Peritoneal membranes, 20, 21f
Peritonitis, 684
Peritubular capillary system, 772, 776, 779, 785f
Permeable membrane, 99, 101
Pernicious anemia, 534, 535t, 640t, 710
Peroxisomes, 92, 94, 98t
Perpendicular plate, 220, **260**
Persistent vegetative state, 407
Personal trainer, 121
Personalized medicine, and genetics, 920
Pesco-vegetarian, 719t
PET. *See* Positron emission tomography
Peyer's patches, 621
PH and pH scale, 67–68, 752, 814–19. *See also* Acid; Base
Phag- (prefix), 83, 149
Phagocytes, 103, 149
Phagocytosis, 83, 103, 104, 107t, 171, 627
Phagolysosome, 103
Phagosome, 103
Phalanx (phalanges), of fingers and toes, 213, 215, 235f, 236, 237t, 238f, 242f, 244f, 245. *See also* Distal phalanx; Middle phalanx; Proximal phalanx
Pharmacist, 361
Pharmacogenetics, 921
Pharmacology, 31. *See also* Drugs
Pharyngeal tonsils, 654
Pharyngeal wall, 661
Pharynx, **50,** 652f, 660–61, 733, 744t
Phen- (prefix), 906
Phencyclidine, 893t
Phenobarbitol, 893t
Phenotype, 906, 909, 916
Phenylalanine, 141t, 699t

Pheomelanin, 185
Phleb- (prefix), 557
Phlebitis, 557, 584
Phosphate buffer system, 814, 815t
Phosphate ions, 69t, 784t, 806f, 810, 812
Phosphodiesterase, 493
Phospholipid bilayer, 86, 87, 88f
Phospholipid bilayer, nucleus, 98f
Phospholipids, 72, 73t. *See also* Lipids
Phosphoproteins, 813
Phosphoric acid, 813
Phosphorus, 211, 711, 713, 715t
Phosphorylation, 126, 492
Photo- (prefix), 444
Photoreceptors, 444, 478–79
Phren- (prefix), 731
Phrenic arteries, 594, 595f, 598
Phrenic nerves, **54**, 430, 731, 744
Phrenic veins, 603
Physical stress, 517
Physical therapy assistant, 269
Physician's assistant (PA), 869
Physiologic dead space, 748
Physiological buffer systems, 817f
Physiological steady state, 100
Physiology, 11–12. *See also* Anatomy and
    physiology, overview
Pia mater, 391
Pica, 717
Pilocarpine, 437
Pineal gland, 403, 517
Pino- (prefix), 83
Pinocytosis, 83, 103, 104f, 107t
Piriformis muscle, 332f, 333, 334t
Pitch, and vocal cords, 736
Pitching injury, 236
Pituicytes, 502
Pituitary dwarfism, 210
Pituitary gigantism, 210
Pituitary gland, 403, 491t, 497f, 498–503, 504t,
    837–38, 839f
Pivot joint, 272, 273f, 274t
PKU, 126
Placenta, 517, 872, 878f, 891
Placental estrogens, 889
Placental lactogen, 889
Placental membrane, 876, 879f
Placental progesterone, 889
Placentation, 873–74, 876
Plane joints, 272, 273f, 274t
Plantar arteries, 600, 600f
Plantar flexion, 274, 275f
Plantar flexors, 338–40
Plantar reflex, 414
Plantaris, 338f, 339
Plants
    carbohydrates, 695
    drug sources, 450
    "organic" in agriculture, 68
    proteins, 699
Plaque, 584
Plasma, 15, 16f, 167, 169f, 529, 541–42
Plasma cells, 630
Plasma colloid osmotic pressure, 620
Plasma electrolytes, 542

Plasma nutrients, 542
Plasma proteins, 541, 542t, 814
Plasmin, 545
Plasminogen, 545
Plasminogen activator (tPA), 545
Platelet, 168, 169f, 541t
Platelet count, 539
Platelet plug, 542, 543f, 544t
Platelet-derived growth factor (PDGF), 545,
    775
Platysma, 313t, 315
Pleiotropy, 912
Pleur- (prefix), 10
Pleural cavity, 19f, 20, 21f, **43**, 741, 742f
Pleural membrane, 20, 21f, 742f
Plex- (prefix), 390
Plexuses, 427
Plicae circulares, 679
Pluripotent stem cells, 113, 871
Plutonium, 62
Pneumocystis jirovecii, 641, 757
Pneumonia, 757, 899t
Pneumotaxic center, 751
Pneumothorax, 746
Podiatry, 32
Podocytes, 775
Poie- (prefix), 200
Polar body, 841, 842f
Polar molecules, 64–65, 66f, 76
-Pole (suffix), 528
Poliomyelitis, 750
Poly- (prefix), 58, 528
Polycystic kidney disease, 797t
Polycythemia, 528, 533
Polydactyly, 236, 239f, 912
Polygenic traits, 912–13
Polymorphonuclear leukocytes (PMN), 536
Polypeptide, 122
Polyploidy, 916
Polyribosome, 141
Polysaccharides, 70, 121, 695
Polyunsaturated fatty acids, 71
Pons, 405, 406, 409t, 751f
Pontine respiratory group, 751
Popliteal artery, 598, 600, 600f
Popliteal ligament. *See* Oblique popliteal
    ligament
Popliteal veins, 605, 608f
Por- (prefix), 178
Pores, of skin, 178, 187
Porphyria variegata, 552t
Porta hepatis, 670
Portal system, of pituitary gland, 499
Positive chemotaxis, 538
Positive energy balance, 702
Positive feedback systems, 16–17, 544, 890.
    See also Negative feedback
Positive nitrogen balance, 699
Positron emission tomography (PET), 78
Post- (prefix), 868
Postcoital test, for infertility, 853t
Posterior, and relative position, 27, 28f
Posterior auricular artery, 596f, 597
Posterior canal, 466
Posterior cavity, 475

Posterior cerebral arteries, 594, 597f
Posterior chamber, 473, 474f
Posterior communicating artery, 597
Posterior curciate ligament, 284
Posterior funiculus, 410
Posterior group, thigh muscles, 328, 332–33
Posterior horns, 410
Posterior inferior iliac spine, 240f
Posterior intercostal arteries, 592, 598, 599f
Posterior intercostal veins, 603
Posterior interventricular artery, 563
Posterior longitudinal ligaments, 227
Posterior median sulcus, 410
Posterior pituitary gland, 403
Posterior pituitary hormones, 491t, 502–3
Posterior region, 32f
Posterior sacral foramina, 229
Posterior superior iliac spine, 239, 240f, **349**
Posterior tibial artery, 339f, 600, 600f
Posterior tibial veins, 603
Postganglionic fiber, 431
Postnatal period, 867, 868, 894–98
Postsynaptic neuron, 371
Posttraumatic stress disorder (PTSD), 392
Potassium, 572, 713, 715t, 784t, 813
    electrolyte balance, 813
    sodium-potassium pump, 103, 372, 374f
Potassium ion channels, 89
Potassium ions, 69t, 372, 379, 787–88, 806f, 809
PQ interval, 572
Pre- (prefix), 868
Preactive dying, 897
Precapillary sphincters, 581
Precipitation, 634, 635t
Prediabetes, 516
Preembryonic stage, of development, 884t
Preganglionic fibers, 431, 432, 434f, 435f
Pregnancy, 868–93. *See also* Fetus; Infertility
    conception, 868–70
    dietary iron, 715
    fertilization, 868–70, 872f
    fetus, **265**, 881, 883, 886f
    maternal changes, 887–90
    prenatal development, 870–93
    Rh blood groups, 551, 552f
    ultrasonography, 13
Pregnanediol, 853
Preimplantation genetic diagnosis (PGD), 873
Preload, and blood pressure, 586
Premature beat, 575
Premature infants, 882
Premolars, 656, 657t
Prenatal period, of development, 867, 868,
    870–93
Prenatal tests, for chromosome abnormalities,
    919–20
Prepatellar bursa, 268, 284
Prepuce, 835
Presbycusis, 482
Presbyopia, 476, 482
Pressure. *See also* Blood pressure
    air, 741, 744
    atmospheric, 15
    hydrostatic, 15, 101, 581, 779–82, 783, 805
    osmotic, 101, 541, 581, 780, 805

as requirement for life, 15
sense of, 446, 447f
Pressure ulcers, 179
Presynaptic neuron, 371
Prevention
of breast cancer, 855
of HIV /AIDS, 642
PRF. *See* Prolactin-releasing factor
PRIH. *See* Prolactin release-inhibiting hormone
Prim- (prefix), 824
Primary follicle, 841
Primary germ layers, 876–78
Primary immune response, 635
Primary malnutrition, 720
Primary metabolites, 123
Primary motor areas of cerebral cortex, 401
Primary oocyte, 841
Primary ossification center, 204
Primary sex organs, 839
Primary spermatocyte, 828
Primary structure, of amino acid, 73, 74f
Primary teeth, 655
Prime mover, 312
Primordial follcle, 824, 841, 843f
Prion, 75
PRL. *See* Prolactin
Pro- (prefix), 83
Probenecid, 787
Process, 200, 215t
Procollagen, 163
Products, of chemical reactions, 66
Progenitor cells, 112–13, 115f
Progeria, 96, 907
Progesterone, 849, 851f, 889, 890, 893t
Progestin, 857
Projection, and sensation, 445
Prokaryotic cells, 86
Prolactin (PRL), 488, 491t, 500, 504t, 891–93
Prolactin release-inhibiting hormone (PRIH), 491t, 500, 501f
Prolactin-releasing factor (PRF), 491t, 500, 501f
Proline, 141t, 699t
Pronation, of joint, 275f, 276
Pronator quadratus muscle, 323, 324t
Pronator teres muscle, 323, 324t, 326f
Propelling movements, in alimentary canal, 650–51
Prophase, 83, 108, 109f, 824, 825
Proprioceptive senses, 446
Proprioceptors and proprioception, 421, 444, 449–51
Prostacyclin, 547
Prostaglandins, 490, 492f, 496, 636, 833
Prostate cancer, 452, 833, 834
Prostate gland, **52,** 795, 833, 838t
Prostate specific antigen (PSA), 833, 834
Prostatic urethra, 795
Protease, 123
Proteasomes, 75
Protein, 122. *See also* Amino acids
anabolism, 122
cell membrane, 87–89
digestion of, 681
liver and metabolism of, 670, 673t

nutrition, 698–700
organic compounds, 73–75, 77t
synthesis of, 139–41, 142t
urine, 792
Protein buffer system, 814–15
Protein fibers, 159
Protein hormones, 490, 491t, 492f
Protein kinases, 492
Protein misfolding, 75
Proteinuria, 786, 798
Prothrombin, 544, 707
Prothrombin activator, 543
Prothrombin time (PT), 544
Protons, 58t, 59–60
Protraction, of joint, 276, 276f
Provitamins, 705
Prox- (prefix), 769
Proximal, and relative position, 28
Proximal convoluted tubule, 776, 791t
Proximal epiphysis, 200
Proximal phalanx, 244f, **350, 351**
Proximal radioulnar joint, 278t
Proximal tubule, 769
PSA (prostate specific antigen), 833, 834
Pseudo- (prefix), 149
Pseudo stratified epithelium, 149
Pseudohermaphroditism, 838
Psoas major muscle, 328, 331f, 334t
Psoas minor muscle, 331f
Psoriasis, 179
Psychiatry, 32, 381t, 392, 438
Psychological stress, 517
Pteroylmonoglutamic acid, 710
Pterygoid muscles, 314f, 315
PTH. *See* Parathyroid hormone
PTSD (posttraumatic stress disorder), 392
Puber- (prefix), 824
Puberty, 824, 838, 853
Pubic arch, 239, 240f, 241t
Pubic crest, 240f
Pubic symphysis, **43–44, 48,** 239, 240f, 269, 278t, 330f
Pubic tubercle, 240f
Pubis, 239
Public health nurse, 617
Public region, 30, 31f
Pubofemoral ligament, 281
Pudendal arteries, 598. *See also* Inferior pudendal artery; Internal pudendal artery
Pudendal nerve, 429
Pudendal vein, 606
Pulmonary arteries, **42, 47,** 561, 593f
Pulmonary circuit, 557, 590–91
Pulmonary edema, 591
Pulmonary trunk, **41–42, 53,** 561, 562f, 593f
Pulmonary valve, 561, 561t, 563f
Pulmonary veins, **42,** 561, 590
Pulmonic sound, 568, 569f
Pulse, and blood pressure, 585
Punnett squares, 910, 911f
Pupil, of eye, 473, 475f
Purified protein derivative (PPD), 638
Purines, 134
Purkinje fibers, 560, 569
Pus, 538, 626

Putamen, 403
Pyl- (prefix), 650
Pyloric antrum, 663
Pyloric canal, 663
Pyloric sphincter, 650, 663, 667f
Pyramidal cells, 401
Pyramidal tracts, 417
Pyridoxal, 709
Pyridoxamine, 709
Pyridoxine, 709
Pyrimidines, 133
Pyruvic acid, 129, 302, 719

Q

Q wave, 572
QRS complex, 572
Quadrants of abdomen, 30, 31f
Quadrate lobe, 670
Quadratus lumborum muscle, 316t, 317f, 318
Quadriceps femoris muscle group, 331f, 335
Quaternary structure, of protein, 74f, 75

R

R group, amino acid, 74
R wave, 572
Radial artery, 328f, **350,** 598
Radial collateral ligament, 280, 280f
Radial muscles, 473
Radial nerves, 328f, 427
Radial notch, 237f
Radial set, 473
Radial tuberosity, 235, 237f
Radial veins, 602, 608f
Radiation, 62, 189
gamma, 60, 62
solar, 182
ultraviolet, 142, 192
Radioactive isotopes, 60, 62, 893t
Radiologic technologist, 201
Radiology, 32
Radioulnar joints, 278t
Radius, 213, 234f, 235, 237f, 237t, 310, **349**
Radon, 62
Ramus, 215t, 222, **259**
Rapid eye movement (REM) sleep, 407–8
Rate-limiting enzyme, 125
Reactants, 66
Receptive relaxation, 651
Receptor cells, 474
Receptor proteins, 87t
Receptor-mediated endocytosis, 104–5, 107t
Receptors, 16, 413t
adrenergic, 433, 436–37
alpha, 436–37
baroreceptors, 444, 572, 587f
beta, 436–37
bitter, 455
cell membrane, 87
chemoreceptors, 444, 752–53
cholinergic, 433, 436, 437f
cold, 446
homeostasis, 16, 17f, 190f

mechanoreceptors, 444, 451t
muscarinic, 436
nicotinic, 436
olfactory, 421, 452
osmoreceptor, 503, 807
pain, 444, 446–47, 451t
photoreceptors, 444, 478–79
sensory, 361, 399, 411, 416
sour, 455
stretch, 444, 451f
sweet, 455
synaptic transmission, 378
taste, 454–55
thermoreceptors, 444, 446, 451t
types of, 444
visual, 478–79
warm, 446
Recessive inheritance, 909–11, 915
Recommended Daily Allowance (RDA),
    718–19
Recruitment, of motor units, 304–5
Rect - (prefix), 650
Rectal artery, 598, 599f
Rectal vein, 606
Rectum, **42, 48, 52,** 330f, 650, 652f, 684, 685f
Rectus abdominis muscle, **38–39, 48, 51–52,**
    320f, 325, 328t, 329f, **353**
Rectus femoris muscle, **39, 41–42, 52,** 331f,
    332f, 334f, 335, **350, 356**
Red blood cell count (RBCC), 531
Red blood cells. *See also* Hemoglobin
    anemia, 534–35, 640t, 688, 710, 720, 873
    in blood tissue, 167–68, 169f, 541t
    characteristics of, 529–35
    destruction of, 535, 536t
    life cycle of, 535f, 536f
    microscopy, 87f
    production of, 531, 533–35
    size of, 84f
Red fibers, 306
Red marrow, 201, 210, 211, 239
Red nucleus, 406
Red pulp, of spleen, 623
Referred pain, 447
Reflex arcs, 411, 413f
Reflex(es). *See also specific reflexes*
    behavior, 411–13, 414–15f
    defecation, 686
    micturition, 795
    nonrespiratory movements, 749
    uses of, 414
Refraction disorders of the eye, 476
Refractory period, of nerve impulse, 378
Registered dietitian, 695
Regulation
    autoregulation, of glomerular filtration, 783
    bile release, 674–75
    body temperature, 189–90
    capillary blood flow, 580–81, 582f
    cardiac cycle, 572
    electrolyte intake and output, 809–12
    gastric secretions, 665
    hydrogen ion concentration, 813–16
    male sex hormones, 839
    metabolic pathways, 124–25

pain impulses, 449
pancreatic secretion, 669
small intestine secretions, 680
urine concentration and volume, 788–90
water intake and output, 806–7, 808
Relative position, terms of, 27–28
Relative refractory period, 378
Relaxation, of muscles, 299f, 300–301
Relaxin, 889
Releasing hormones, 498
REM sleep, 407–8
Ren- (prefix), 769
Renal agenesis, 797t
Renal arteries, 594, 595f, 772, 773f, 774f
Renal baroceptors, 783
Renal blood vessels, 772–75, 777f
Renal capsule, 770
Renal clearance, 792
Renal columns, 770
Renal corpuscle, 771, 775, 791t
Renal cortex, 769, 770, 778f
Renal dysplasia, 797t
Renal excretion, of hydrogen ions, 816
Renal hypoplasia, 797t
Renal medulla, 770, 778f
Renal papillae, 769, 770
Renal pelvis, 770
Renal plasma threshold, 785
Renal pyramids, 770, 773f
Renal sinus, 770
Renal tubules, 770, 775, 785f, 787f, 798
Renal veins, 603, 608f, 772, 773f, 774f
Renin, 511, 518–19, 589, 783
Renin-angiotensin system, 511, 783
Repair enzymes, 143
Replacement, of joints, 282
Replication, of DNA, 108, 135, 137f
Reproduction
    as characteristic of life, 14t
    organ systems, 25
Reproductive system, 823–66
    autonomic stimulation, 436t
    birth control, 854–58, 858t
    hormonal control of female, 849–52
    hormonal control of male, 837–39
    interconnections with other systems, 194,
        247, 341, 383, 521, 611, 643, 689,
        763, 799, 861
    life span changes, 898t
    major functions of 25t, 26
    mammary glands, 852–53
    meiosis, 824–25
    organs of female, 839–48, 885f
    organs of male, 826–37, 885f
    summary, assessments, and tools, 860,
        862–66
Reserpine, 381t, 437
Residual volume (RV), 748, 748t
Resorption, and bone remodeling, 207
Respiration, 731. *See also* Breathing;
        Respiratory system
    aerobic, 128f, 812–13
    anaerobic, 121, 813
    cellular, 92, 127–32, 731
    as characteristic of life, 14t, 25t

Respiratory acidosis, 817–18
Respiratory alkalosis, 818
Respiratory areas, 750–51
Respiratory bronchioles, 739
Respiratory capacities, 748
Respiratory center, 407, 815–16
Respiratory cycle, 741, 747
Respiratory disorders, 745, 750, 757
Respiratory distress syndrome, 746, 757
Respiratory excretion, of carbon dioxide,
    815–16
Respiratory membrane, 755–56
Respiratory system, 730–67. *See also*
        Respiration
    alveolar gas exchanges, 753–57, 758f
    autonomic stimulation, 436t
    breathing mechanism, 741–49
    cardiovascular system, 611, 763
    control of breathing, 750–53
    gas transport, 758–61
    interconnections with other systems, 194,
        247, 341, 383, 521, 643, 689, 763,
        799, 861
    life-span changes, 762, 898t
    major functions of, 24, 25t
    organs of, 731–41
    reasons for breathing, 731
    summary, assessments, and tools, 762,
        764–67
Respiratory therapist, 731
Respiratory tubes, 738–40
Respiratory volumes, 747–49
Responsiveness, as characteristic of life,
    14t, 25t
Resting potential, 374–75
Resting tidal volume, 747, 748t
Restless legs syndrome, 408t
Rete testis, 828
Reticul- (prefix), 292
Reticular connective tissue, 164, 165f
Reticular fibers, 162t, 163
Reticular formation, 407
Reticular layer, 183
Reticulocytes, 533
Reticulospinal tracts, 417, 418t
Retina, 474–79
Retinal, 480, 706
Retinal detachment, 482
Retinal pigment epithelium, 478–79
Retraction, of joint, 276, 276f
Retroperitoneal fat, **52**
Retroperitoneal positioning, of kidneys, 769,
    771f
Rett syndrome, 114
Reuptake, 381
Reversible reaction, 66
Rh blood types, 551
Rheumatic fever, 639
Rheumatoid arthritis, 271, 286, 287t, 640t
Rhodopsin, 479–80, 706
RhoGAM, 551
Rhomboid major and minor muscles, 318, 319f,
    320t, **353**
Rhythm method, of contraception, 854–55,
    858t

Rhythmicity, of smooth muscles, 307
Riboflavin (Vitamin B$_2$), 709, 712t
Ribose, 696
Ribosomal RNA (rRNA), 141
Ribosomes, 83, 90, 98t
Ribozymes, 141
Ribs, **50–53**, 213, 231, 233f, 234f
Rice, 699
Rickets, 208, 707
Right atrium, **42, 53**
Right brachiocephalic vein, 608f
Right lobe, of liver, 670
Right lymphatic duct, 618, 619f, 620f
Right main bronchi, 738
Right ventricle, **42, 52, 53,** 558f, 562f
Rigor mortis, 301
Riluzole, 418
Rinne test, 465
Risk factors
    for atherosclerosis, 584
    for stroke, 589
RNA (ribonucleic acid), 136. *See also*
    Chromosomes; Genetics
    cellular metabolism, 136, 138
    DNA compared to, 138t
    structure of, 76
RNA polymerase, 136
Rods, of eye, 478–79
Root canals, 657
Rotation, of joint, 275f, 276
Rotator cuff, 277
Rotator muscles, 320, 323
Rough endoplasmic reticulum, 90, 91f
Round bones, 200
Round ligament, of uterus, 845
Round window, 459, 461f
Rubella virus, 882
Rubrospinal tracts, 417, 418t
"Rule of nines," for burns, 192, 193f
Ruptured disc, 230

S

S wave, 572
Sacchar- (prefix), 58
Saccharin, 695
Saccule, 465, 466f
Sacral artery. *See* Middle sacral artery
Sacral canal, 229
Sacral curvature, 226f, 227
Sacral foramina. *See* Anterior sacral foramina
Sacral hiatus, 229
Sacral nerves, 425
Sacral promontory, 229
Sacroiliac joints, 229, 239
Sacrum, **48,** 212, 229, 230f, 231t, 241t, **349**
Saddle joint, 272, 273f, 274t
Safety, and universal precautions, 530
Sagittal plane, 28, 29f
Sagittal section, **46–48**
Sagittal suture, 216f, 218, **253**
Saliva, 455, 658
Salivary amylase, 658, 680t
Salivary glands, 111, 652f, 658–59, 660f

Salivary secretions, 658–59
Salt, 66, 67f, 69, 211, 673, 675. *See also*
    Sodium chloride
Salt craving, 809
Salt receptors, 455
Saltator- (prefix), 360
Saltatory conduction, 360, 378
Saphenous veins. *See* Great saphenous vein;
    Small saphenous vein
Sarco- (prefix), 292
Sarcolemma, 293
Sarcomeres, 295, 300f
Sarcoplasm, 292, 293, 296f
Sarcoplasmic reticulum, 292, 295–96f, 296
Sartorius muscle, **38–39, 42–43, 52,** 331f, 332f,
    333f, 334f, 335, 338t, **350, 356**
Satellite cells, 370
Saturated fats, 71–72, 696, 698
Saturated fatty acids, 72
Scab, 190, 191f
Scala tympani, 459
Scala vestibuli, 459
Scalenes, 316t, 318, 320f
Scalp, **46, 49**
Scanning electron microscope (SEM), 86, 87f
Scapula, 213, 233, 234f, 237t, 321f, **349**
Scar tissue, 190, 191f
Schizophrenia, 381t, 438
Schwann cells, 363, 365f, 370
Sciatic nerve, 334f
Sciatic nerves, 429
Sciatica, 430
Scientific method, 10–11
Scintillation counter, 610
Scler- (prefix), 444, 557
Sclera, 444, 472
Scleroderma, 287t, 639–40
Sclerotic arteries, 584
Scoliosis, 230
Scrotum, **41, 48,** 330f, 826, 828f, 834, 838t
Scurvy, 711
Seb- (prefix), 178
Sebaceous glands, 178, 185–88
Sebum, 186
Second line of defense, 816
Secondary hypertension, 589
Secondary immune response, 635
Secondary malnutrition, 720
Secondary metabolites, 123
Secondary oocyte, 841, 842f, 869, 872f, 874f
Secondary ossification centers, 204
Secondary pacemaker, 575
Secondary sex characteristics, 839, 849, 850t
Secondary spermatocytes, 828
Secondary structure, of amino acid, 74f, 75
Secondary teeth, 655–56, 657f, 657t
Second-degree burn, 192
Secondhand smoke, 735
Secretin, 669, 670f, 675t
Segmental (tertiary) bronchi, 738
Segmental innervation, 427
Segmentation, of alimentary canal, 651, 655f
Selectin, 88
Selective estrogen receptor modulators
    (SERM), 855

Selective serotonin reuptake inhibitors (SSRIs),
    381t
Selectively permeable membrane, 87, 101
Selenium, 716–17
Self-renewal, 112
Sella turcica, 218, 220f, 221f, **261, 263–64**
Sellar joint, 272, 273f
Semen, 834, 835t
Semicircular canals, 457f, 459
Semilunar valve, 557
Semimembranosus muscle, **357–58**
Semimembranous muscle, 334f, 335, 336t, 338f
Seminal vesicle, 833, 838t
Seminiferous tubules, 828, 829f, 838t
Semispinalis capitis muscle, 316, 317f
Semitendinosus muscle, 334f, 335, 338f, **349,
    357**
Semivegetarian, 719t
Sen- (prefix), 868
Senescence, 868, 897
Sens- (prefix), 360
Sensation, 444–45
Senses, 443–86
    general types of, 446–51, 451t
    hearing, 456–68, 482
    life-span changes, 482
    pain, 446–49
    receptors, sensation, and perception, 444–45
    sight, 468–81
    smell, 452–54
    special forms of, 452–81
    summary, assessments, and tools, 482–86
    taste, 454–56, 482
    touch, 446, 447f
Sensorineural deafness, 465
Sensory adaptation, 445
Sensory area, of cerebrum, 399
Sensory impulses, 444
Sensory nerves, 419
Sensory neurons, 360, 366, 368t, 413t, 414f,
    415f
Sensory receptors, 361, 399, 411, 416
Sensory speech area, 399
Septa, 828
Serine, 141t, 699t
Serology, 549
Serosa, of alimentary canal, 650, 651t, 654f
Serotonin, 380t, 381t, 449, 540
Serous cells, 658
Serous coat, 793
Serous fluid, 158, 168
Serous layer, 650, 651t, 654f
Serous membranes, 20, 21f, 168
Serratus anterior muscle, **38–39,** 318, 320t,
    322f, **349, 353**
Sertoli cells, 828
Serum, 544
Sesamoid bones, 200, 244f
Set point, 16
Severe acute respiratory syndrome (SARS), 757
Sex cells, transport of, 868–69
Sex chromosomes, 908, 914–16
Sex determination, 914, 915f
Sex hormones, 512
Sex-influenced inheritance, 916

Sex-limited trait, 916
Sexual intercourse, 868
Sexual stimulation, 835–36, 848
Sexually transmitted infection (STIs), 859
Shark cartilage, 718
Shigatoxin, 783
Shivering, 16, 189, 195
Shock, and blood pressure, 783
Short bones, 200
Short head biceps brachii muscle, **39**
Short-term memory, 402
Shoulder, 597–98, 602–3
    muscles of, 319f, 321f, 323f
Shoulder joint, 277, 278t, 279, 279f
Sickle cell disease, 143, 202, 531, 534, 535t,
    552t, 910
Sidestream smoke, 735
Sight, sense of, 468–81
Sigmoid colon, **43, 48, 56,** 684
Sigmoidoscopy, 687t
Signal transduction, 87
Simple cuboidal epithelium, 152, 153f, 159t
Simple epithelium, 152, 153f
Simple glands, 156, 157f, 158t
Simple squamous epithelium, 152, 159t
Sinoatrial (SA) node, 568, 575, 576f
Sinuses, 215t, 733, 734f, 744t
Skeletal muscles. *See also* Muscular system
    actions of, 310–12
    anterior view of superficial, 313f
    contraction, 297–303
    eye, 470t
    major muscles of, 312–40
    muscular responses, 303–7
    posterior view of superficial, 313f
    structure of, 292–96
    tissue, 149, 151t, 170, 172t
    tissue characteristics, 309t
Skeletal system, 199–266. *See also* Bone;
    Joints; Skeletal muscles
    bone, 200–211
    differences in male and female skeletons,
        241t
    divisions of, 212–15
    interconnections with other systems, 247,
        341, 383, 521, 611, 643, 689, 763,
        799, 861
    life-span changes, 245–46, 898t
    lower limbs, 241–45, 245t
    major functions of, 22–23, 25t
    organization of, 212–15
    pectoral girdle, 231, 234f, 237t
    pelvic girdle, 238–41, 245t
    reference plates on the skull, 252–66
    skull, 215–25, **252–66**
    summary, assessment, and tools, 246,
        248–51
    thoracic cage, 231, 233f
    upper limb, 233–38, 234f, 237t
    vertebral column, 226–30, 231t
Skeleton, of heart, 561
Skin, 177–98. *See also* Integumentary system
    accessory structures, 184–88, 195
    dermis, 183–84
    epidermis, 178–83

exposure to sunlight, 162
    functions of, 189–90
    hair follicles, 184–85
    healing of wounds and burns, 190–92, 193f
    life-span changes, 192–93, 195
    lymphatic capillaries, 622
    nails, 184
    skin glands, 185–88
    summary, assessment, and tools, 195–98
    tissue composition, 178–84
Skin cancer, 142, 182
Skin color, 183, 913
Skin glands, 185–88
Skull, **49,** 212
    bones of, 213t, 215–25, **252–66**
    children, **266**
    cranial sutures, 270f
    cranium, 212, 215–25
    divisions of skeleton, 212
    elderly, **266**
    fetus, **265**
    frontal section, 223f
    infant and children, 225
    joints, 278t
    male *versus* female, 241t
    sinuses, 734f
Sleep and sleep disorders, 407–8, 438
Sleep apnea, 751
Sleep paralysis, 408t
Sliding filament model, 299f, 300
Slit pores, 775
Slow pain fibers (C fibers), 449
Slow-twitch muscle fibers, 306–7
Slow-wave sleep, 407
Small intestine, **41, 45, 47–48, 51, 55–56,** 652f,
    676–83, 688
Small saphenous vein, 339f, 605, 608f
Smell, sense of, 452–54, 456, 482
Smoking. *See* Cigarette smoke and smoking
Smooth endoplasmic reticulum, 90, 91f
Smooth muscle, 307–8, 309t, 470t
Smooth muscle tissue, 170, 171f, 172t
Sneezing, 749
Sodium, 60t
    electrolyte balance, 805–6, 813
    nutrition, 714, 715t
    urine, 784t
Sodium bicarbonate, 813, 814
Sodium chloride, 64, 66, 67f, 789, 791f
Sodium hydroxide, 66, 813
Sodium ions, 69t, 372, 379, 786, 806f, 809
Sodium/potassium pump, 103
Sodium-potassium pump, 103, 372, 374f
Soft palate, 654
Solar radiation, 182
Soleus muscle, 336f, 338f, 339, **351, 357–58**
Som- (prefix), 83
Somatic nervous system, 361, 418
Somatic pathways, 432f
Somatostatin (SS), 491t, 500, 501f, 515, 516t,
    665, 675t
Somatotropes, 500
Somatotropin (STH), 500
Sorpt- (prefix), 650
Sound waves, 461

Sour receptors, 455
Space travel and medicine, 210, 467
Spastic paralysis, 419
Special senses, 444, 452–81
Special somatic afferent fibers, 420
Special somatic efferent fibers, 420
Special visceral afferent fibers, 420
Specialized connective tissues, 162–63
Species resistance, 626
SPECT (single photon emission computed
    tomography), 404
Speech, 749t. *See also* Motor speech area
Sperm, 824, 826, 832f, 837t, 868–69, 872f,
    874f. *See also* Spermatogenesis
Sperm banks, 870
Spermatic arteries, 594
Spermatic cord, **48,** 826, 827f, 828f
Spermatids, 828
Spermatogenesis, 824, 828, 830f
Spermatogenic cells, 828
Spermatogonia, 828, 831f
Spermicide, 856f, 857, 858t
Spermiogenesis, 828, 832f
Spermiogram, 836
Sphenoid bone, 217f, 218, 219f, 221f, **252–53,**
    255–58, 261–64
Sphenoidal sinus, **46,** 217f, 218, 220f, **261–63,**
    733
Spheroidal joint, 272
Sphincter. *See* Anal sphincter; External urethral
    sphincter; Hepatopancreatic sphincter;
    Ileocecal sphincter; Lower esophageal
    sphincter; Precapillary sphincters;
    Pyloric sphincter
Sphincter muscle, 314
Sphincter urethrae muscle, 328, 330t
Sphygmomanometer, 583
Spina bifida, 71, 227, 397, 881, 883
Spinal branch of accessory nerve, 424, 424t
Spinal cord, **45, 47, 50–52,** 398f, 400f, 409–18,
    796. *See also* Vertebra; Vertebral column
Spinal cord injuries, 419
Spinal nerve injuries, 430
Spinal nerves, 409, 418, 419–20, 424–30
Spinalis capitis muscle, 317f
Spinalis cervicis muscle, 316t
Spinalis thoracis muscle, 316t, 317f
Spine, 215t. *See also* Scapula; Spinal cord;
    Vertebra
Spinocerebellar tracts, 416, 418t
Spinothalmic tract, 416, 418t
Spinous process, 227
Spiral fracture, 208
Spiral lamina, 459, 460f
Spiral organ, 459, 462f, 463f
Spirometer, 747, 748
Spleen, **41–42, 52,** 623, 624f, 624t, 625f
Splenic artery, 592, 595f
Splenic vein, 603, 604f
Splenius capitis muscle, 316, **352**
Splenius capitus muscle, 317f
Spondylolisthesis, 229
Spongy bone, 201, 202, 245
Spontaneous fracture, 208
Spontaneous mutations, 142

Sprains, 286
Squam- (prefix), 149
Squamous carcinoma, 182
Squamous epithelium, 149, 152
    simple, 152, 159t
    stratified, 154, 155f, 159t
Squamous suture, 216f, 217f, 218, **255**
St. John's Wort, 718, 719
St. Martin, Alexis, 665
Stanford University, 382
Stapedius, 458
Stapes, 457, 458f, 461f
Starvation, 722–23
-Stasis (suffix), 10, 528
Static equilibrium, 465–66
Statins, 105, 609
Steere, Allen, 286
Stem cells
    baldness therapy, 186
    healing of burns, 192
    heart transplant, 567
    neural, 371
    pluripotent, 113, 871
    spinal cord injuries, 419
    structure and function, 112–13, 115f
    tooth extraction as source of, 658
    transplant, 211
    treatment of disease, 114, 211
Stereoscopic vision, 481
Sternoclavicular joint, 278t
Sternocleidomastoid muscle, **38–39**, 312, 314f,
    316, 320f, **347, 349, 352–53**
Sternocostal joint, 278t
Sternum, 21f, **46–47**, 213, 231, 232f, 234f, **349,**
    559f
Steroid hormones, 489, 490, 491t, 492f, 492t,
    493f, 494
Steroid molecules, 72. *See also* Lipids
Steroids, 73t
Stevia, 695
Stickler syndrome, 163t
Stomach, **41–42, 45, 47**, 652f, 661, 663–68,
    688
Stomachache, 667
STOP codon, 141t
Strabismus, 470
Strains, 296
Strangulated hernia, 826
Strat- (prefix) and -strat (suffix), 121, 149
Stratified squamous epithelium, 154, 155f, 159t
Stratified cuboidal epithelium, 154, 155f, 159t
Stratified epithelial tissue, 149
Stratified epithelium, 152, 153f
Stratum basale, 178, 179, 180t
Stratum corneum, 179, 180t
Stratum germinativum, 178
Stratum granulosum, 179, 180t
Stratum lucidum, 179, 180t
Stratum spinosum, 179, 180t
Strength training, 340
Strep A infection, 287t
Streptococcus bacteria, 778
Streptokinase, 546
Stress and stress response, 517–19. *See also*
    Posttraumatic stress disorder

Stretch receptors, 444, 451f
Stretch reflex, 450
Stria- (prefix), 149
Striations, 149, 170
Stroke, 397, 589, 899t. *See also* Cerebral
    vascular accident; Heatstroke
Stroke volume, and cardiac output, 585–86
Structural formulas, 64, 65f
Styloid process, 216f, 217f, 218, 219f, 235,
    236, 237f, **349–50**
Stylomastoid foramen, 224t, **256–58**. *See also*
    Mastoid foramen
Sub- (prefix), 121
Subacromial bursa, 279
Subarachnoid space, 391, 394f
Subatomic particles, 12, 14t
Subclavian artery, **41, 43–44, 56**, 592,
    593f, 594t, 595f, 596f. *See also* Left
    subclavian artery
Subclavian trunk, 617, 619f
Subclavian veins, **40–40, 43,** 593f, 602, 603,
    608f
Subcoracoid bursa, 279
Subcutaneous fascia, 293
Subcutaneous injections, 178
Subcutaneous layer, 178
Subcutaneous tissue, **49**
Subdeltoid bursa, 279
Subdural hematoma, 391
Subfertility, 853
Sublingual glands, 659, 660f
Submandibular glands, 659, 660f
Submucosa, of alimentary canal, 650, 651t, 654f
Submucosal plexus, 651
Submucous coat, 793
Subscapular bursa, 279
Subscapular fossa, 234f
Subscapular nerves, 428
Subscapularis muscle, **40**, 320, 322t
Subserous fascia, 293
Substance P, 380t, 382
Substantia nigra, 404
Substrate, 121, 123–24
Sucralose, 695
Sucrase, 123, 680, 680t
Sudden cardiac arrest, 574, 591, 920
Sugar concentration gradient, 99
Sugars, 70, 695
Sulcus, 397
Sulfate ions, 69t, 784t, 806f, 812
Sulfur, 714, 715t
Sulfuric acid, 813
Summation, of muscle fiber, 304, 305f
Sun exposure, and vitamin D, 707
Sunburn, 192
Sunlight exposure, 162
Sunscreens and sunblock, 182
Superficial, and relative position, 28
Superficial circumflex iliac artery, 598
Superficial digital flexor muscle, **350**
Superficial epigastric artery, 598
Superficial fibular nerve, 339f
Superficial partial-thickness burn, 192
Superficial pudendal arteries, 598
Superficial temporal artery, 597

Superficial tranversus perinei muscle, 327, 330t
Superior, and relative position, 27, 28f
Superior articular process, 227
Superior border, 233, 234f
Superior constrictor muscles, 660
Superior gluteal artery, 598, 599f
Superior gluteal nerve, 429
Superior hemiazygos vein, 603
Superior iliac spine. *See* Anterior superior iliac
    spine; Posterior superior iliac spine
Superior meatus, 732
Superior mesenteric artery, **43, 56**, 594, 595f
Superior mesenteric vein, **43, 56**, 603, 604f
Superior nasal concha, 220f
Superior oblique muscles, 421, 470, 470t
Superior orbital fissure, 221f, 224t, **254, 261,
    264**
Superior peduncle, 409
Superior rectus muscle, 470
Superior thyroid artery, 596f, 597
Superior vena cava, **42–43, 53**, 560, 593f, 602,
    608f
Superior vesical artery, 598, 894–95
Superovulation, 844
Superoxide dismutase (SOD), 899
Supination, of joint, 275f, 276
Supinator muscle, 323, 324t
Support for organ systems, 22
Suppression amblyopia, 470
Supraorbital foramen, 216f, 218, 224t, **252,
    254–55**. *See also* Infraorbital foramen
Supraorbital notch, 216f, 218, **252, 260, 347**
Supraorbital ridge, 241t
Suprapatellar bursa, 284
Suprarenal arteries, 594, 595f
Suprarenal veins, 603
Suprascapular nerve, 428
Suprascapular notch, 234f
Supraspinatus muscle, 318–19, 322f
Supraspinous fossa, 233, 234f
Supratrochlear region, 622
Surface anatomy, **347–51**
Surface area, of glomerular capillaries, 783
Surface tension, 745
Surfactant, 745, 894
Surgery. *See also* Medical science; Transplants
    and transplantation
    birth control, 857–58
    induced hypothermia, 190
    obesity, 703
Surgical neck, 233, 236f
Suspensory ligaments, 472, 840, 853
Sustained muscle contraction, 305–6
Sustentacular cells, 828
Sutur- (prefix), 268
Sutural ligament, 268
Sutures
    of bone, 212, 215t, 218
    of joint, 268, 274t
Suturing, 190
Swallowing mechanism, 660–61, 662f
Sweat, 16, 488, 807, 809
Sweat glands, 186–87
Sweet receptors, 455
Sympathetic chain ganglia, 431

Sympathetic division, of autonomic nervous system, 431–32, 433f, 434f
Sympathetic fibers, 437f
Sympathetic impulses, and digestive system, 651–52, 658, 683
Sympathetic tone, 433
Sympathetic trunks, 431
Symphysis, 269, 271f, 274t
Syn- (prefix), 58, 268, 292, 360, 557
Synapse, 297, 360, 361, 371–72, 373f. *See also* Synaptic transmission
Synapsis, 824
Synaptic cleft, 297, 363, 371, 372f
Synaptic knobs, 363, 380f
Synaptic potentials, 379–80
Synaptic transmission, 371, 378–81
Synarthrotic joints, 268
Synchondrosis, 269, 270f, 274t
Syncytium, 309, 557, 568
Syndesm, (prefix), 268
Syndesmosis, 268, 268f, 274t
Synergist, 292, 312
Synesthesia, 453
Synostosis, 269
Synovial cavity, 271
Synovial fluid, 269, 271
Synovial joints, 269–72, 273f, 274t, 277–85
Synovial membrane, 169, 271, 280
Synthesis, 66
Synthetic folic acid, 710
Syphils, 859t
Systemic circuit, 557, 564f, 591
Systemic lupus erythematosus, 287t, 640t
Systol- (prefix), 557
Systolic pressure, 557, 583, 585

T

T cells, 367, 537, 623, 628, 629–30, 631f
T lymphocytes. *See* T cells
T wave, 572
Tachy- (prefix), 557
Tachycardia, 557, 574
Tactile (Meissner's) corpuscles, 446, 451t
Talus, 243, 244f
Tanning, and skin cancer, 182
Tapeworm infestation, 539t
Tardive dyskinesia, 381t, 404
Target cells, 23, 488
Targeted muscle reinnervation, 308
Tarsals (tarsus), 215, 242f, 243, 244f, 245t, **351**
Tarsometatarsal joint, 278t
Taste, 454–56, 482
Taste buds, 454–55
Taste cells, 454
Taste hairs, 455
Taste pathways, 456
Taste pore, 454f, 455
Taste receptors, 454–55
Taste sensations, 455
Tattoos, 184
Tay-Sachs disease, 363
TBI (traumatic brain injury), 392
Tear gland, 436t, 470f

Tears and tear gland, 469, 470f
Technology. *See* From Science to Technology
Tectorial membrane, 459, 462f, 463f
Teeth, 650, 655–58, 688
Telencephalon, 397t
Telomeres, 110
Telophase, 108, 109f, 825
Temperature, 15. *See also* Body temperature; Heat
    homeostatic mechanism, 16, 191
    regulation, 16, 17f, 189, 190f
    sense of, 446
Temporal artery, 597
Temporal bone, 216f, 217f, 218, 219f, 221f, **252–53, 256–57, 260, 347**
Temporal lobe, **49,** 397, 400, 402t
Temporal process, 217f, 221
Temporalis muscle, **49,** 314f, 315, **347, 352**
Temporomandibular joint, 278t, 315
Temporomandibular joint (TMJ) syndrome, 315
Tendinitis, 292
Tendons, 164, 292
Tenosynovitis, 292
Tensor fasciae latae muscle, **42–43,** 331, 332f, 333, 334t, **356**
Tensor tympani, 458
Tentorium cerebelli, 391t, 397
Teratogens, 881
Teres major muscle, **40,** 318, 321f, 322t, **348, 354–55**
Teres minor muscle, 320, 322t, **354–55**
Terminal bronchioles, 738
Terminal ganglia, 432
Terminology. *See also specific prefixes and suffixes*
    anatomy and anatomical position, 11, 27–32
    blood, 528
    body fluid balance, 820
    for body parts, 312
    cardiovascular system, 557
    cells and cellular metabolism, 83, 121
    chemistry, 58
    digestive system, 650
    endocrine system, 488
    genetics, 906
    hormone names and abbreviations, 491t
    integumentary system, 178
    joints, 268
    lymphatic system and immunity, 617
    nervous system, 360, 390, 444
    nutrition and metabolism, 695
    pregnancy and growth, 868
    reproductive system, 824
    respiratory systems, 731
    senses, 444
    skeletal muscles, 292
    skeletal system, 200
    skin, 178
    tissues, 149
    urinary system, 769
Tertiary structure, of amino acid, 74f, 75
Testicular cancer, 828
Testis (testes), **41, 45, 48,** 517, 826–29, 838t
Testosterone, 210, 494, 826, 838–39, 849
Tetan- (prefix), 292

Tetanic contraction, 292
Texting, and muscular movements, 325
Thalamotomy, 404
Thalamus, **46, 49,** 403
Thalassemia, 535t
Thalidomide, 882
Therm- (prefix), 444
Thermoreceptors, 444, 446, 451t
Thermostat, 16, 17f
Theta waves, 410
Thiamine (Vitamin B$_1$), 708, 712t, 714
Thigh, muscles of, 328, 331–34, **349, 356–57**
Third ventricle, **49,** 392
Third-degree burn, 192
Thirst center, 806–7
Thoracic aorta, **54,** 592, 594t, 599f
Thoracic artery. *See* Internal thoracic artery
Thoracic cage, 213, 231, 232f, 233f
Thoracic cavity, 18, 19f, **47, 50–54,** 622
    female torso, **42, 43, 44**
    male torso, **41**
Thoracic curvature, 226f, 227, 231
Thoracic duct, 617, 619f, 620f
Thoracic membranes, 20, 21f
Thoracic muscles, **354**
Thoracic nerves, 425, 427
Thoracic outlet syndrome, 430
Thoracic vein. *See* Internal thoracic vein
Thoracic vertebrae, 226, 229f, 228, 231t, 232f
Thoracic viscera
    female torso, **42, 43, 44**
    male torso, **41**
Thoracic wall, 598, 603, 604f
Thoracodorsal nerve, 428
Thorax, **50–51, 53–54, 355**
Threonine, 141t, 699t
Threshold potential, 375, 379
Threshold stimulus, and muscular responses, 303–4
Thromb- (prefix), 528
Thrombin, 544
Thrombocyte, 528, 539
Thrombocytopenia, 539
Thrombophlebitis, 584
Thrombopoietin (TPO), 539
Thrombus, 545
Thymosins, 517, 623
Thymus, 517, 522, 623, 624f, 624t, 625f
Thyrocervical arteries, 594
Thyroglobulin, 505
Thyroid artery. *See* Superior thyroid artery
Thyroid cartilage, **40–41, 349,** 734
Thyroid gland, **39–41, 53,** 62, 491t, 503–6, 507t, 519–20. *See also* Hyperthyroidism; Hypothyroidism
Thyroid hormone, 490
Thyroid-stimulating hormone (TSH), 491t, 500, 504t
Thyrotropes, 500
Thyrotropin-releasing hormone (TRH), 491t, 500, 501f, 502f
Thyroxine, 210, 491t, 504f, 505, 506t
Tibia, 215, 242–43, 245t, 268f, 336f, 339f, **351**
Tibial artery. *See* Anterior tibial artery; Posterior tibial artery

Tibial collateral ligament, 284
Tibial nerve, 339f
Tibial nerves, 429
Tibial tuberosity, 242, 243f, **350**
Tibial vein. *See* Anterior tibial vein; Posterior
    tibial vein
Tibialis anterior muscle, 336f, 337, 339t, 340f,
    350, 351, 357–58
Tibialis posterior muscle, 339f, 339t, 340, **351**
Tibiofibular joint, 278t
Tidal volume, 747, 748t
Tight junctions, 149
Tinnitus, 482
Tissue, 12, 14t, 148–76
    cells organized into, 149–50
    engineering, 172, 847
    fluid formation, 620
    muscle tissue, 149, 151t, 170, 172t
    nervous tissue, 149, 151t, 170–71, 172f, 172t
    sections of, 149–50, 151f
    summary, assessment, and tools, 173–76
    transplantation and rejection of, 638–39
    types of membranes, 168–69
Tissue composition, 178–84
Tissue plasminogen activator (tPA), 546
Tissue recovery technician, 149
Tissue thromboplastin, 543
Tissues
    connective, 149, 151t, 159–68, 169t
    epithelial, 149, 150–59
Titin, 295
TMJ (temporomandibular joint) syndrome, 315
Toc- (prefix), 488
Toes, **351**. *See also* Phalanx
-Tomy (suffix), 10
Tongue, **45–46, 50,** 653, 655f, 658t
-Tonic (suffix), 292
Tonometer, 482
Tonsils, 621, 654
Tonsilitis, 654
Torso
    female, **38, 42, 43**
    male, **39, 40, 41, 45**
Total lung capacity (TLC), 748, 748t
Totipotent cell, 112–13
Touch, sense of, 446, 447f
Toxicology, 32
Toxins
    "Botox" (botulinum toxin), 193
    chemicals, 456
    kidneys, 787
    minerals, 713
    sperm, 829
    teratogens, 881
Trabecular bone, 201, 245, 246
Trace elements, 59, 715–17
Trachea, **42–43, 45–47, 53,** 737, 738, 744t
Tracheostomy, 737
Trans fats, 71
Transcellular fluid, 804
Transcranial magnetic stimulation, 362
Transcription of RNA, 136, 142t
Transcutaneous electrical nerve stimulation
    (TENS), 450
Transcytosis, 105, 106f, 107t

Transdermal patch, 178
Transfer RNA (tRNA), 139–41
Transferrin, 535
Transforming growth factor B (TGF-B), 609f
Transfusions, of blood, 548–51
Transient ischemic attack (TIA), 397, 589
Transitional epithelium, 156, 157f, 159t
Translation of RNA, 138, 142t
Translocation, of chromosomes, 916
Transmissible spongiform encephalopathies, 75
Transmission, of HIV /AIDS, 641t
Transmission electron microscope (TEM),
    86, 87f
Transplants and transplantation
    bone marrow, 211
    heart, 567
    kidney, 772
    liver, 671
    pancreas, 518–19
    stem cells, 211
    tissue rejection, 638–39
    types, 639t
Transport, organ systems, 23, 24f. *See also*
        Active transport; Electron transport
        chain; Gas transport; Limited transport
        capacity; Sex cells
Transverse colon, **41–42, 47, 52, 56,** 684
Transverse fissure, 397
Transverse foramina, 228
Transverse fracture, 208
Transverse humeral ligament, 279
Transverse plane, 28, 29f
Transverse process, 227
Transverse section of thoracic cavity, 21f, **50–52**
Transverse tubules, 296, 296f
Transversus abdominis muscle, **39,** 320f,
    325, 328t
Trapezius muscle, **38,** 318, 319f, 320t, **348–49,**
    353–55
Trastuzumab, 855
Traumatic brain injury (TBI), 392
Travel
    deep vein thrombosis, 547
    jet lag, 517
    space travel and medicine, 210, 467
TRH. *See* Thyrotropin-releasing hormone
Tri- (prefix), 905
Triad, 296
Triangle of auscultation, 318
Triceps brachii muscle, 311, 321f, 322f, 324t,
    348, 353–56
Triceps-jerk reflex, 414
Tricuspid valve, 560, 561t, 563f
Tricyclic antidepressants, 381t
Trigeminal nerves, 421–22, 424t
Trigeminal neuralgia, 422
Trigger zone, 375
Triglycerides, 71, 73t, 696. *See also* Lipids
Trigon- (prefix), 769
Trigone, 769, 793
Triiodothyronine, 491t, 504f, 505, 506t
Triple negative breast cancer, 855
Triploid chromosome number, 916
Triplo-X, 918
Triptans, 362

Trisomy, 905, 916, 920
Tristearin, 71
Trochanter, 215t
Trochlea, 233, 236f
Trochlear nerves, 421, 424t
Trochlear notch, 236, 237f
Trochoid joint, 272
Troph- (prefix) or -troph (suffix), 292, 868
Trophoblast, 868, 871, 872, 875f
Tropic- (prefix), 488
Tropic hormones, 496
Tropomyosin, 296
Troponin, 296
True pelvis, 241
True ribs, 231, 232f
True vocal cords, 735–36
Trunk, 28, **354–55**
Trypsin, 668, 680t
Trypsinogen, 668
Tryptophan, 141t, 381t, 699t, 709, 797
TSH. *See* Thyroid-stimulating hormone
Tubal ligation, 858, 860f
Tubal pregnancy, 872
Tuber- (prefix), 731
Tubercle, 215t, 231, 757
Tuberculin skin test, 638
Tuberculosis, 539t, 638, 731, 757
Tuberosity, 215t
Tubular dysgenesis, 797t
Tubular fluid, 779
Tubular glands, 156
Tubular reabsorption, 779, 783–87
Tubular secretion, 779, 785f, 787–88
Tubulin, 96, 97f
Tumor, 111
Tumor necrosis factor, 630t
Tumor suppressor genes, 111
Tunic, of artery, 577
Tunica albuginea, 828, 835, 841
Tunica externa, 577
Tunica media, 577
Turner syndrome, 918
Twins, 871
Twitch, of muscle, 304, 305f
Tympan- (prefix), 444
Tympanic cavity, 457
Tympanic membrane, 444, 456, 457f
Tympanic reflex, 458
Type 1 diabetes mellitus, 516
Type 2 diabetes mellitus, 516
Typhoid fever, 539t
Tyrosine, 141t, 699t

U

Ulcer, 665
Ulcerative colitis, 640t, 687
Ulna, 213, 234f, 235f, 236, 237f, 237t, 328f,
    350
Ulnar artery, 32f, 597
Ulnar collateral ligament, 279, 280f
Ulnar nerves, 328f, 427
Ulnar veins, 602, 608f
Ultrasonography, 13, 853t, 919, 920t

Ultratrace elements, 59
Ultraviolet radiation, 142, 192
Umami, 455
Umbil- (prefix), 868
Umbilical arteries, 874, 886–87, 895t
Umbilical cord, 868, 874, 878f
Umbilical region, 30, 31f
Umbilical vein, 874, 883, 889f, 895t
Umbilicus, **38, 349, 353**
Undernutrition, 720
Uni- (prefix), 360
Unicellular glands, 156, 157f, 158t
United States Department of Agriculture
    (USDA), 698, 719
Universal precautions, 530
Universal recipients and donors, 549
Unmyelinated axons, 363, 365f, 366f
Unsaturated fats, 71, 696
Unsaturated fatty acids, 71
Upper limbs, 213, 233–38, 234f, 237t, 597–98,
    602–3
Upper motor neuron syndrome, 419
Upper motor neurons, 416
Upper respiratory tract, 731
Upregulation, of hormones, 490
Uranium, 62
Urbanization, and human health, 10
Urea, 698, 784t, 790
Ureteritis, 795
Ureters, **42–43,** 792–93
Urethra, **52,** 330f, 795, 796f, 846
Urethral glands, 795
Urethral orifice, 330f
Urethral sphincter. *See* External urethral
    sphincter
-Uria (suffix), 804
Uric acid, 784t, 790, 791
Urinalysis, 797
Urinary bladder, **41, 48,** 172, 793, 794f, 795f,
    796f
Urinary excretion, 779, 791
Urinary system, 768–802
    autonomic stimulation, 436t
    description of, 769, 770f
    developmental abnormalities, 797t
    elimination of urine, 792–96
    excretion of odoriferous component
        (asparagus), 797
    interconnections with other systems, 194,
        247, 341, 383, 521, 611, 643, 689,
        763, 799, 861
    kidneys, 769–76
    life-span changes, 797–98, 898t
    major functions of, 24, 25t
    regulation, 788–90
    summary, assessment, and tools, 798,
        800–802
    urine formation, 778–92
Urinary tract infection (UTI), 795
Urine, 778–92, 807
Urogenital diaphragm, 325, 327–28, 330f
Urokinase, 546
Urology, 32
USA Track and Field, 810

Uterine artery, 598
Uterine tube, **42,** 845, 846f, 848t, 860f
Uterine vein, 606
Uterus, **42–43,** 152–53, 845–46, 848t
Utricle, 465
Uvula, 654

V

Vaccines and vaccination, 636, 882
Vagina, 846–47, 848t
Vaginal orifice, 330f, 846–47
Vaginal process, 826
Vaginal vein, 606
Vagus nerves, **54,** 423, 424t, 572
Valerian root, 718
Valine, 699t
Valium, 381t, 384
Valves of heart, 560–61
Valves of veins, 582
Variably expressive genes, 912
Varicose veins, 584
Vas- (prefix), 488
Vas deferens. *See* Ductus deferens
Vasa recta, 789–90, 791f
Vascular endothelial growth factor (VEGF),
    577, 775
Vascular smooth muscle, 307
Vascular spasm, 542, 544t
Vascular tunic, 470
Vasectomy, 857–58, 860f
Vasoconstriction, 577
Vasodilation, 577
Vasomotor center, of medulla oblongata,
    407, 588
Vasomotor fibers, 577
Vasopressin, 488, 503
Vastus intermedius muscle, **43,** 331f, 334f, 335
Vastus lateralis muscle, **41–42, 52,** 331f, 332f,
    333f, 334f, 335, 338f, **350, 356–57**
Vastus medialis muscle, **41–42,** 331f, 334f, 335,
    350, 356
Vegan, 719t
Vegetarian diets, 711, 718, 719t
Veins, 578f, 582, 583t. *See also* Venous system;
    specific veins
Vena cava. *See* Inferior vena cava; Internal vena
    cava; Superior vena cava
Venipuncture, 603
Venous blood flow, 588
Venous return, 586
Venous system, 602–7. *See also* Veins
Ventilation, 731
Ventral branch, 427, 428f
Ventral respiratory group, 751
Ventral root, 426
Ventricles, **47.** *See also* Lateral ventricles; Left
    ventricle; Right ventricle; Third ventricle
    brain development, 395
    heart, 558f, 560, 562f
    nervous system, 369, 391–95
Ventricular fibrillation, 574
Ventricular syncytium, 568
Venules, 578f, 582, 583t

Vermis, 408
Vertebra (vertebrae), 18, 21f, 30, 212. *See also*
    Cervical vertebrae; Thoracic vertebrae;
    Vertebral column
Vertebra prominens, 228
Vertebral arch, 227
Vertebral arteries, **50,** 594, 596f
Vertebral body, **51–52**
Vertebral border, 233, 234f
Vertebral canal, 18, 19f, 226
Vertebral column, **45,** 212, 213t, 226–30, 231,
    231t, 316–18. *See also* Spinal cord
Vertebral column disorders, 230
Vertebral spine, **348**
Vertebrochondral ribs, 232f
Vertebrocostal joint, 278t
Very-low-density lipoprotein (VLDL), 681, 697
Vesic- (prefix), 83
Vesical arteries, 598, 599f
Vesical vein, 606
Vesicle trafficking, 90, 381
Vesicles, 83, 90, 98t
Vesicoureteral reflux, 797t
Vestibular branch fibers, 423, 424t
Vestibular bulbs, 848
Vestibular glands, 848, 848t
Vestibular membrane, 459
Vestibule, 459, 465, 848, 848t
Vestibulocochlear nerves, 423, 424t
Viagra (sildenafil), 835
Vill- (prefix), 650
Villi, 650. *See also* Microvilli
Vinca alkaloids, 123
Viruses, 86, 674. *See also* Human
    immunodeficiency virus; Human
    papilloma virus; Influenza; Marburg virus
Viruses and multiple sclerosis, 367
Viscera, 18. *See also* Abdominal viscera
Visceral branch, 427
Visceral effectors, effects of autonomic
    stimulation on, 436t
Visceral pain, 447–49
Visceral pericardium, 20, 21f, 558
Visceral peritoneum, 20, 21f, 650
Visceral pleura, 20, 21f, 741
Visceral senses, 451
Visceral smooth muscle, 307
Visceroreceptive senses, 446
Viscosity, and blood pressure, 586
Visible Human Project, 10
Vision. *See* Eye; Sight
Visual accessory organs, 468–70
Visual cortex, 481
Visual pathways, 481
Visual pigments, 479–81
Visual receptors, 478–79
Vital capacity (VC), 748, 748t
Vitamin, 125. *See also specific vitamins*
    deficiencies of, 688
    dietary supplements, 718–19
    nutrition, 705–11
Vitamin A, 209, 480, 688, 706, 708t, 883
Vitamin B, 534, 708–11
    biotin (Vitamin B_7), 709–10, 712t, 714
    cyanocobalamin (Vitamin B_{12}), 710–11, 712t

folic acid (folacin, Vitamin B$_9$), 534, 710, 712t
niacin (Vitamin B$_3$), 709, 712t
pantothenic acid (Vitamin B$_5$), 709, 712t
riboflavin (Vitamin B$_2$), 709, 712t
thiamine (Vitamin B$_1$), 708, 712t, 714
Vitamin B$_6$, 712t
Vitamin C, 209, 534, 711, 712t
Vitamin D
 bone and deficiencies of, 208–9, 246, 688, 707
 characteristics and functions of, 706, 708t
 dietary cholesterol and synthesis of, 507, 706
 excesses of, 707, 708t
 kidneys, 769
 production of, 189, 195
 requirements, 707
 skin cells and sun exposure, 195, 507, 707
 steroid hormones, 489
Vitamin E, 707, 708t
Vitamin K, 543, 688, 707, 708t
Vitre- (prefix), 444
Vitreous body, 475
Vitreous humor, 444, 475, 482
Vocal cords, 735–36
Volatile acid, 815
Volkmann's canals, 202
Voltage-gated potassium channels, 376
Voltage-gated sodium channels, 375–76
Voluntar- (prefix), 292
Voluntary muscle, 170, 292
Vomer bone, 219f, 222, 223f, 224t, **252, 254, 256–58**
Vomiting and vomiting center, 667, 818
Von Willebrand disease, 552t
Vulva, 847

W

Wall. *See also* Abdominal wall; Alveolar wall; Pharyngeal wall; Thoracic wall
 of alimentary canal, 650, 654f
 of heart, 560
 of large intestine, 685
 of small intestine, 654f, 678–79
Wandering cells, 160

Warm receptors, 446
Warren, J. Robin, 665
Water. *See also* Body fluid balance
 chemical constituents of cells, 68, 69t
 compartments and distribution of, 805–6
 intake of, 806–7
 molecular structure, 63, 64–65, 66f, 77f
 as neutral, 67, 69t
 nutrition, 720
 osmosis, 101
 output of, 807–8
 reabsorption in urinary system, 786–87
 as requirement for life, 14, 15t
Water balance, 68, 804, 806–8, 810–11
Water intoxication, 810
Water of metabolism, 806
Water-soluble vitamins, 705, 708–11, 712t
Weber test, 465
Weight, 59, 702, 704f. *See also* Obesity
Weightlifting, 308
Wernicke's area, 399
Whiplash, 430
White blood cell count (WBCC), 538–39
White blood cells, 84f, 167–68, 169f, 536–39, 541t
White fibers, 307
White matter, in brain, **49,** 363
White pulp, of spleen, 623
White rami, 431
Whooping cough, 539t
Wild type, 909
Wilson disease, 797
Wisdom teeth, 656
Withdrawal reflex, 412–13, 415f
World Health Organization (WHO), 530, 642, 743, 836
Wounds, in skin, 190, 191f
Wrinkled skin, 162
Wrinkles in skin, 193
Wrist, 278t

X

X chromosome, 914–15
Xenograft, 639

Xeroma pigmentosum, 143
Xerophthalmia, 706
Xerostomia, 688
Xiphoid process, 231, 232f, **349**
X-linked inheritance, 909
X-linked recessive disorders, 915
XO syndrome, 918
X-ray fluoroscopy, 610
XXY syndrome, 918
XYY syndrome, 919

Y

Y chromosome, 914–15
Yawning, 749t, 750
Yellow marrow, 201, 210–11
Y-linked inheritance, 909
Yogurt, 686
Yolk sac, 874
Yucca Mountain (Nevada), 62

Z

Z lines, 295
Zein, 699
Ziconotide, 450
Zinc, 716, 717t
Zippers, amino acid structure, 75
Zon- (prefix), 824
Zona fasciculata, 512
Zona glomerulosa, 510
Zona pellucida, 824, 841, 869, 872f
Zona reticularis, 512
Zone of calcified cartilage, 207
Zone of hypertrophic cartilage, 205
Zone of proliferating cartilage, 205
Zone of resting cartilage, 205
Zygomatic arch, 218, 219f, **253, 255, 347, 352**
Zygomatic bone, 217f, 221, 224t, **252–53, 254–56**
Zygomatic process, 217f, 218, **260**
Zygomaticus major and minor, 313t, 314f, 315
Zygote, 841, 842f, 869, 879t
-Zym (suffix), 121
Zymogen granules, 668

Aids to Understanding Words

acetabul-, vinegar cup: *acetabul*um
adip-, fat: *adip*ose tissue
aer-, air: *aer*obic respiration
af-, to: *af*ferent arteriole
agglutin-, to glue together: *agglutin*ation
alb-, white: *alb*inism
aliment-, food: *aliment*ary canal
allant-, sausage-shaped: *allant*ois
alveol-, small cavity: *alveol*us
an-, without: *an*aerobic respiration
ana-, up: *ana*bolic
andr-, man: *andr*ogens
angio-, vessel: *angio*tensin
anul-, ring: *anul*ar ligament
append-, to hang something:
    *append*icular
arth-, joint: *arth*rology
astr-, starlike: *astr*ocyte
ather-, porridge: *ather*osclerosis
aud-, to hear: *aud*itory
auto-, self: *auto*immune disease
ax-, axis: *ax*ial skeleton
bas-, base: *bas*al metabolic rate
bi -, two: *bi*polar neuron
bil-, bile: *bil*irubin
bio-, life: *bio*chemistry
-blast, budding: osteo*blast*
brady-, slow: *brady*cardia
bronch-, windpipe: *bronch*us
burs-, bag, purse: prepatellar *burs*a
calat-, something inserted: inter*calat*ed disc
calor-, heat: *calor*ie
calyc-, small cup: *calyc*es
canal-, channel: *canal*iculus
carcin-, spreading sore: *carcin*oma
cardi-, heart: peri*cardi*um
cari-, decay: dental *cari*es
carin-, keel-like: *carin*a
carot-, carrot: *carot*ene
carp-, wrist: *carp*als
cata-, down: *cata*bolic
cec-, blind: *cec*um
cephal-, head: en*cephal*itis
cerebr-, brain: *cerebr*um
chiasm-, cross: optic *chiasm*a
chondr-, cartilage: *chondr*ocyte
chorio-, skin: *chorio*n

choroid, skinlike: *choroid* plexus
chromo-, color: *chromo*some
chym-, juice: *chym*e
-clast, broken: osteo*clast*
clav-, bar: *clav*icle
cleav-, to divide: *cleav*age
co-, with: *co*enzyme
cochlea, snail: *cochlea*
condyl-, knob: *condyl*e
contra-, against, counter: *contra*ception
corac-, beaklike: *corac*oid process
corn-, horn: *corn*ea
cort-, covering: *cort*ex
cran-, helmet: *cran*ial
cribr-, sievelike: *cribr*iform plate
cric-, ring: *cric*oid cartilage
-crin, to secrete: endo*crin*e
crist-, ridge: *crist*a galli
-crit, to separate: hemato*crit*
crur-, lower part: *crur*a
cut-, skin: sub*cut*aneous
cyst-, bladder: *cyst*itis
cyt-, cell: *cyt*oplasm
-cyt, cell: osteo*cyt*e
de-, undoing: *de*amination
decidu-, falling off: *decidu*ous
dendr-, tree: *dendr*ite
derm-, skin: *derm*is
detrus-, to force away: *detrus*or muscle
di-, two: *di*saccharide
diastol-, dilation: *diastol*e
diuret-, to pass urine: *diuret*ic
dors-, back: *dors*al
ect-, outside: *ect*oderm
edem-, swelling: *edem*a
ejacul-, to shoot forth: *ejacul*ation
embol-, stopper: *embol*us
-emia, a blood condition:
    hypoprotein*emia*
endo-, within: *endo*plasmic reticulum
ependym-, tunic: *ependym*a
epi-, upon: *epi*thelial tissue
erg-, work: syn*erg*ist
erythr-, red: *erythr*ocyte
exo-, outside: *exo*crine gland
extra-, outside: *extra*cellular
fasc-, bundle: *fasc*iculus

fimb-, fringe: *fimb*riae
flacc-, flabby: *flacc*id paralysis
follic-, small bag: hair *follic*le
fov-, pit: *fov*ea
frenul-, bridle, restraint: *frenul*um
funi-, small cord or fiber: *funi*culus
gangli-, swelling: *gangli*on
gastr-, stomach: *gastr*ic gland
-gen, be produced: aller*gen*
-genesis, origin: spermato*genesis*
germ-, to bud or sprout: *germ*inal
glen-, joint socket: *glen*oid cavity
-glia, glue: neuro*glia*
glom-, little ball: *glom*erulus
glyc-, sweet: *glyc*ogen
-gram, something written:
    electrocardio*gram*
gubern-, to steer, to guide: *gubern*aculum
hema-, blood: *hema*toma
hemo-, blood: *hemo*globin
hepa-, liver: *hepa*tic duct
hetero-, other, different: *hetero*zygous
hiat-, opening: esophageal *hiat*us
hist-, web, tissue: *hist*ology
hol-, entire, whole: *hol*ocrine gland
hom-, same, common: *hom*ologous
    chromosomes
homeo-, same: *homeo*stasis
horm-, impetus, impulse: *horm*one
humor-, fluid: *humor*al
hyal-, resemblance to glass: *hyal*ine
    cartilage
hyper-, above: *hyper*tonic
hypo-, below: *hypo*tonic
im-, (or in-), not: *im*balance
immun-, free: *immun*ity
inflamm-, to set on fire: *inflamm*ation
inhal-, to breathe in: *inhal*ation
inter-, between: *inter*phase
intra-, inside: *intra*membranous
iris, rainbow: *iris*
iso-, equal: *iso*tonic
juxta-, near to: *juxta*medullary nephron
karyo-, nucleus: *karyo*type
kerat-, horn: *kerat*in
labi-, lip: *labi*a
labr-, lip: glenoidal *labr*um